KU-733-247

COMPREHENSIVE MEDICINAL CHEMISTRY II

HS 615.19 COMP
v.4

COMPREHENSIVE MEDICINAL CHEMISTRY II

Editors-in-Chief

Dr John B Taylor

Former Senior Vice-President for Drug Discovery, Rhône-Poulenc Rorer, Worldwide, UK

Professor David J Triggle

State University of New York, Buffalo, NY, USA

Volume 4

COMPUTER-ASSISTED DRUG DESIGN

Volume Editor

Dr Jonathan S Mason

Lundbeck Research, Valby, Copenhagen, Denmark

ELSEVIER

AMSTERDAM BOSTON HEIDELBERG LONDON NEW YORK OXFORD
PARIS SAN DIEGO SAN FRANCISCO SINGAPORE SYDNEY TOKYO

Elsevier Ltd.
The Boulevard, Langford Lane, Kidlington, Oxford OX5 1GB, UK

First edition 2007

Copyright © 2007 Elsevier Ltd. All rights reserved

1.05 PERSONALIZED MEDICINE © 2007, D Gurwitz
2.12 HOW AND WHY TO APPLY THE LATEST TECHNOLOGY © 2007, A W Czarnik
3.40 CHEMOGENOMICS © 2007, H Kubinyi
4.12 DOCKING AND SCORING © 2007, P F W Stouten

The following articles are US Government works in the public domain and not subject to copyright:
1.08 NATURAL PRODUCT SOURCES OF DRUGS: PLANTS, MICROBES, MARINE ORGANISMS, AND ANIMALS
6.07 ADDICTION

No part of this publication may be reproduced, stored in a retrieval system or transmitted in any form or by any means electronic, mechanical, photocopying, recording or otherwise without the prior written permission of the Publisher

Permissions may be sought directly from Elsevier's Science & Technology Rights Department in Oxford, UK: phone (+44) (0) 1865 843830; fax (+44) (0) 1865 853333; email: permissions@elsevier.com. Alternatively, you can submit your request online by visiting the Elsevier web site at http://elsevier.com/locate/permissions, and selecting *Obtaining permission to use Elsevier material*

Notice
No responsibility is assumed by the publisher for any injury and/or damage to persons or property as a matter of products liability, negligence or otherwise, or from any use or operation of any methods, products, instructions or ideas contained in the material herein. Because of rapid advances in the medical sciences, in particular, independent verification of diagnoses and drug dosages should be made

British Library Cataloguing in Publication Data
A catalogue record for this book is available from the British Library

Library of Congress Catalog Number: 2006936669

ISBN-13: 978-0-08-044513-7
ISBN-10: 0-08-044513-6

For information on all Elsevier publications visit our website at books.elsevier.com

Printed and bound in Spain

06 07 08 09 10 10 9 8 7 6 5 4 3 2 1

Working together to grow libraries in developing countries

www.elsevier.com | www.bookaid.org | www.sabre.org

ELSEVIER BOOK AID International Sabre Foundation

Disclaimers

Both the Publisher and the Editors wish to make it clear that the views and opinions expressed in this book are strictly those of the Authors. To the extent permissible under applicable laws, neither the Publisher nor the Editors assume any responsibility for any loss or injury and/or damage to persons or property as a result of any actual or alleged libellous statements, infringement of intellectual property or privacy rights, whether resulting from negligence or otherwise.

Knowledge and best practice in this field are constantly changing. As new research and experience broaden our knowledge, changes in practice, treatment and drug therapy may become necessary or appropriate. Readers are advised to check the most current information provided (i) on procedures featured or (ii) by the manufacturer of each product to be administered, to verify the recommended dose or formula, the method and duration of administration, and contraindications. It is the responsibility of the practitioner, relying on their own experience and knowledge of the patient, to make diagnoses, to determine dosages and the best treatment for each individual patient, and to take all appropriate safety precautions. To the fullest extent of the law, neither the Publisher, nor Editors, nor Authors assume any liability for any injury and/or damage to persons or property arising out or related to any use of the material contained in this book.

Contents

Core Concepts and Methods – Target Structure-Based

Core Methods and Applications – Ligand and Structure-Based

Applications to Drug Discovery – Lead Discovery

Applications to Drug Discovery – Ligand-Based Lead Optimization

Applications to Drug Discovery – Target Structure-Based

New Directions

Contents of all Volumes

Volume 6 Therapeutic Areas I: Central Nervous System, Pain, Metabolic Syndrome, Urology, Gastrointestinal and Cardiovascular

Volume 7 Therapeutic Areas II: Cancer, Infectious Diseases, Inflammation & Immunology and Dermatology

Volume 8 Case Histories and Cumulative Subject Index

Preface

The first edition of *Comprehensive Medicinal Chemistry* was published in 1990 and was intended to present an integrated and comprehensive overview of the then rapidly developing science of medicinal chemistry from its origins in organic chemistry. In the last two decades, the field has grown to embrace not only all the sophisticated synthetic and technological advances in organic chemistry but also major advances in the biological sciences. The mapping of the human genome has resulted in the provision of a multitude of new biological targets for the medicinal chemist with the prospect of more rational drug design (CADD). In addition, the development of sophisticated in silico technologies for structure–property relationships (ADMET) enables a much better understanding of the fate of potential new drugs in the body with the subsequent development of better new medicines.

It was our ambitious aim for this second edition, published 16 years after the first edition, to provide both scientists and research managers in all relevant fields with a comprehensive treatise covering all aspects of current medicinal chemistry, a science that has been transformed in the twenty-first century. The second edition is a complete reference source, published in eight volumes, encompassing all aspects of modern drug discovery from its mechanistic basis, through the underlying general principles and exemplified with comprehensive therapeutic applications. The broad scope and coverage of *Comprehensive Medicinal Chemistry II* would not have been possible without our panel of authoritative Volume Editors whose international recognition in their respective fields has been of paramount importance in the enlistment of the world-class scientists who have provided their individual 'state of the science' contributions. Their collective contributions have been invaluable.

Volume 1 (edited by Peter D Kennewell) overviews the general socioeconomic and political factors influencing modern R&D in both the developed and developing worlds. Volume 2 (edited by Walter H Moos) addresses the various strategic and organizational aspects of modern R&D. Volume 3 (edited by Hugo Kubinyi) critically reviews the multitude of modern technologies that underpin current discovery and development activities. Volume 4 (edited by Jonathan S Mason) highlights the historical progress, current status, and future potential in the field of computer-assisted drug design (CADD). Volume 5 (edited by Bernard Testa and Han van de Waterbeemd) reviews the fate of drugs in the body (ADMET), including the most recent progress in the application of 'in silico' tools. Volume 6 (edited by Michael Williams) and Volume 7 (edited by Jacob J Plattner and Manoj C Desai) cover the pivotal roles undertaken by the medicinal chemist and pharmacologist in integrating all the preceding scientific input into the design and synthesis of viable new medicines. Volume 8 (edited by John B Taylor and David J Triggle) illustrates the evolution of modern medicinal chemistry with a selection of personal accounts by eminent scientists describing their lifetime experiences in the field, together with some illustrative case histories of successful drug discovery and development.

We believe that this major work will serve as the single most authoritative reference source for all aspects of medicinal chemistry for the next decade and it is intended to maintain its ongoing value by systematic electronic upgrades. We hope that the material provided here will serve to fulfill the words of Antoine de Saint-Exupery (1900–44) and allow future generations of medicinal chemists to discover the future.

'As for the future, your task is not to foresee it but to enable it'
Citadelle (1948)

John B Taylor and David J Triggle

Preface to Volume 4

It has been an honor to bring together many of the key players in the area of computer-assisted drug design (CADD)/ computational chemistry in volume 4 of this second edition of *Comprehensive Medicinal Chemistry*. This includes both the pioneers of this field of research and newer players, all focused on using computational methods to aid the drug discovery process and driving new approaches to tackle the ever-changing and evolving needs of drug discovery. The spirit of this volume is to bring together the historical, current, and future perspectives of many of the experts in this field, from both academia and industry. The very broad and diverse computational methods used to impact the drug discovery process are covered mainly in this volume, together with the in silico chapters in Volume 5 for ADMET (absorption, distribution, metabolism, excretion, and toxicity) and some chapters in Volume 3 that review related core technologies such as protein x-ray crystallography and NMR approaches and chemoinformatics/databases, etc.

The first chapter has several tables that link the methods/approaches to all the appropriate chapters/volumes. The reader will find that there is a purposeful overlap of some concepts and methods between the various chapters, the goal being to leverage the viewpoints of several experienced drug discovery experts/groups to provide a broader perspective of key approaches, with related practical advice on the particular methods used or preferred by that group. Although there are many new methods and approaches since the first edition, all the methods are still of relevance, even if they are now used in different ways.

The drug discovery process is a multiobjective optimization problem, having to balance many different properties, such as potency, selectivity and ADMET, and approaches now use multiple 'activity' models (such as quantitative structure–activity relationship (QSAR) models or statistical modeling methods, e.g., Bayesian, that can better handle large and noisy data sets) in design and optimization. Clearly, structure-based drug design has now become a major CADD approach, enabling many new insights and impacts. Many aspects are discussed in this volume, and many further exciting advances are expected in the future, from both the computational (e.g., routine prediction of binding affinities) and experimental (e.g., structures of human G protein-coupled receptor (GPCR) target proteins) viewpoints.

Many thanks are given to all the authors for their significant work in providing the chapters for this volume, which we hope will help and inspire all readers to understand and investigate further how CADD approaches can enhance the drug discovery process.

Jonathan S Mason

Editors-in-Chief

John B Taylor, DSc, was formerly Senior Vice President for Drug Discovery at Rhône-Poulenc Rorer. He obtained his BSc in chemistry from the University of Nottingham in 1956 and his PhD in organic chemistry at the Imperial College of Science and Technology with Nobel Laureate Professor Sir Derek Barton in 1962. He subsequently undertook postdoctoral research fellowships at the Research Institute for Medicine and Chemistry in Cambridge (US) with Sir Derek and at the University of Liverpool (UK), before entering the pharmaceutical industry.

During his career in the pharmaceutical industry Dr Taylor spent more than 30 years covering all aspects of research and development in an international environment. From 1970 to 1985 he held a number of positions in the Hoechst Roussel organization, ultimately as research director for Roussel Uclaf (France). In 1985 he joined Rhône-Poulenc Rorer holding various management positions in the research groups worldwide before becoming Senior Vice President for Drug Discovery in Rhône-Poulenc Rorer.

Dr Taylor is the co-author of two books on medicinal chemistry and has more than 50 publications and patents in medicinal chemistry. He was joint executive editor for the first edition of Comprehensive Medicinal Chemistry, a visiting professor for medicinal chemistry at the City University (London) from 1974 to 1984 and was awarded a DSc in medicinal chemistry from the University of London in 1991.

David J Triggle, PhD, is the University Professor and a Distinguished Professor in the School of Pharmacy and Pharmaceutical Sciences at the State University of New York at Buffalo. Professor Triggle received his education in the UK with a BSc degree in chemistry at the University of Southampton and a PhD degree in chemistry at the University of Hull working with Professor Norman Chapman. Following postdoctoral fellowships at the University of Ottawa (Canada) with Bernard Belleau and the University of London (UK) with Peter de la Mare he assumed a position in the School of Pharmacy at the University at Buffalo. He served as Chairman of the Department of Biochemical Pharmacology from 1971 to 1985 and as Dean of the School of Pharmacy from 1985 to 1995. From 1996 to 2001 he served as Dean of the Graduate School and from 1999 to 2001 was also the University Provost. He is currently the University Professor, in which capacity he teaches bioethics and science policy, and is President of the Center for Inquiry Institute, a secular think tank located in Amherst, New York.

Professor Triggle is the author of three books dealing with the autonomic nervous system and drug–receptor interactions, the editor of a further dozen books, some 280 papers, some 150 chapters and reviews, and has presented over 1000 invited lectures worldwide. The Institute for Scientific Information lists him as one of the 100 most highly cited scientists in the field of pharmacology. His principal research interests have been in the areas of drug–receptor interactions, the chemical pharmacology of drugs active at ion channels, and issues of graduate education and scientific research policy.

Editor of Volume 4

Jonathan S Mason studied at the University of London, Queen Mary, where he obtained a BSc in 1976 and PhD in 1979 in organic chemistry. He has subsequently worked in the pharmaceutical industry, starting as a medicinal chemist at Rhône-Poulenc Rorer (now Sanofi-Aventis), then in 1984 initiating and growing computational chemistry groups in the United Kingdom, France, and the United States. In 1997, he was appointed as Director of the Bristol-Myers Squibb Computer-Assisted Drug Design teams, based in Princeton, New Jersey, and then in 1999 as Director of a new integrated Structural Biology and Modeling department comprising 33 scientists, including protein crystallography and protein NMR. From 2001 to 2005, he was leading a team of 45 scientists involved in Computational Chemistry, Structural Biology (from clone to x-ray structure), Medicinal Informatics and Knowledge Discovery (including bioinformatics) as Executive Director of the Medicinal Informatics, Structure & Design Department at Pfizer Global Research & Development, Sandwich, United Kingdom. He is currently a Divisional Director at Lundbeck Research Denmark, jointly responsible for Medicinal Chemistry Research, leading the departments of Early Lead Generation (Chemistry) and Computational Chemistry. He has been a pioneer in the development and use of three-dimensional (3D) pharmacophore fingerprint profiling methods, together with other molecular similarity and diversity approaches, for both ligands and protein targets (e.g., for virtual screening, ligand docking, target-class library design, and high-throughout screening (HTS) analysis). Recently, he has driven the use of biological fingerprints to tackle attrition-related problems, including lead selection/differentiation, and enable new chemogenomic approaches. Other activities include structure-based drug design (including de novo approaches), absorption, distribution, metabolism, and excretion (ADME)/safety prediction, large-scale bioactivity data integration and target-class mining, and the development of intuitive desktop tools for discovery scientists.

Contributors to Volume 4

D J Abraham
Virginia Commonwealth University, Richmond, VA, USA

M Afshar
Ariana Pharmaceuticals, Paris, France

I L Alberts
De Novo Pharmaceuticals, Cambridge, UK

A A Alex
Pfizer Global Research and Development, Sandwich, UK

M Baroni
Molecular Discovery, Pinner, UK

N Barton
GlaxoSmithKline Pharmaceuticals plc, Harlow, UK

F E Blaney
GlaxoSmithKline Pharmaceuticals plc, Harlow, UK

D G Brown
Pfizer Global Research and Development, Sandwich, UK

J N Burrows
AstraZeneca R&D, Södertälje, Sweden

Q Cao
Pfizer Research Technology Center, Cambridge, MA, USA

E Carosati
University of Perugia, Italy

C L Cavallaro
Bristol-Myers Squibb, Princeton, NJ, USA

G Cruciani
University of Perugia, Italy

P M Dean
De Novo Pharmaceuticals, Cambridge, UK

A M Doweyko
Bristol-Myers Squibb, Princeton, NJ, USA

J B Dunbar Jr
Pfizer Inc., Michigan Laboratories, Ann Arbor, MI, USA

N Eswar
University of California at San Francisco, San Francisco, CA, USA

M M Flocco
Pfizer Global Research and Development, Sandwich, UK

S Garland
GlaxoSmithKline Pharmaceuticals plc, Harlow, UK

V J Gillet
University of Sheffield, Sheffield, UK

A Good
Bristol-Myers Squibb, Wallingford, CT, USA

E Griffen
AstraZeneca R&D, Macclesfield, UK

S Haider
University of Oxford, Oxford, UK

L H Hall
Eastern Nazarene College Quincy, MA, USA

L M Hall
Hall Associates Consulting, Quincy, MA, USA

M M Hann
GlaxoSmithKline R&D, Stevenage, UK

C Hansch
Pomona College, Claremont, CA, USA

M J Hartshorn
Astex Therapeutics, Cambridge, UK

A L Hopkins
Pfizer Global Research and Development, Sandwich, UK

L B Kier
Virginia Commonwealth University, Richmond, VA, USA

R T Kroemer
Sanofi-Aventis, Centre de Recherche de VA, Vitry-sur-Seine, France

G Lange
Bayer CropScience, Frankfurt, Germany

A Lanoue
Ariana Pharmaceuticals, Paris, France

A R Leach
GlaxoSmithKline Research and Development, Stevenage, UK

G R Marshall
Washington University, St. Louis, MO, USA

Y C Martin
Abbott Laboratories, Abbott Park, IL, USA

J S Mason
Lundbeck Research, Valby, Copenhagen, Denmark

D Moras
Institut de Génétique et de Biologie Moléculaire et Cellulaire, Illkirch, France

D Motiejunas
EML Research, Heidelberg, Germany

C W Murray
Astex Therapeutics, Cambridge, UK

C Oostenbrink
Vrije Universiteit, Amsterdam, The Netherlands

A Pannifer
Pfizer Global Research and Development, Sandwich, UK

G V Paolini
Pfizer Global Research and Development, Sandwich, UK

S D Pickett
GlaxoSmithKline, Stevenage, UK

J-P Renaud
AliX, Illkirch, France

A Sali
University of California at San Francisco, San Francisco, CA, USA

J Sallantin
CNRS, Montpellier, France

M S P Sansom
University of Oxford, Oxford, UK

D M Schnur
Bristol-Myers Squibb, Princeton, NJ, USA

C Selassie
Pomona College, Claremont, CA, USA

R V Stanton
Pfizer Research Technology Center, Cambridge, MA, USA

P F W Stouten
Nerviano Medical Sciences, Nerviano, Italy

C M Taylor
Washington University, St. Louis, MO, USA

A J Tebben
Bristol-Myers Squibb, Princeton, NJ, USA

B Tehan
GlaxoSmithKline Pharmaceuticals plc, Harlow, UK

N P Todorov
De Novo Pharmaceuticals, Cambridge, UK

A Tropsha
University of North Carolina at Chapel Hill, Chapel Hill, NC, USA

W F van Gunsteren
Eidgenössische Technische Hochschule, Zürich, Switzerland

M M H van Lipzig
Vrije Universiteit, Amsterdam, The Netherlands

R C Wade
EML Research, Heidelberg, Germany

I Wall
GlaxoSmithKline Pharmaceuticals plc, Harlow, UK

P Willett
University of Sheffield, Sheffield, UK

J-M Wurtz
Institut de Génétique et de Biologie Moléculaire et Cellulaire, Illkirch, France

4.01 Introduction to the Volume and Overview of Computer-Assisted Drug Design in the Drug Discovery Process

J S Mason, Lundbeck Research, Valby, Copenhagen, Denmark

© 2007 Elsevier Ltd. All Rights Reserved.

4.01.1 Introduction to the Volume

This volume, together with the in silico chapters in Volume 5 for ADMET (absorption, distribution, metabolism, excretion, and toxicity), covers the very broad and diverse computational chemistry methods used to impact the drug discovery process. Together with Volume 3, that reviews related core technologies such as protein x-ray crystallography and nuclear magnetic resonance (NMR) approaches, and has chemoinformatics chapters on databases, the many roles and points of impact of computer-assisted drug design (CADD) in the drug discovery process are reviewed and put in perspective. This area is known by many related names, such as computer-aided drug design, computational chemistry, and molecular modeling, all of which have been the names of departments with which the author has been involved, together with the broader in silico term. Chemoinformatics, depending on the focus and organizational issues, can be part of a CADD group or separate, but is part of the core CADD methodology. The focus of all these departments, and of this volume, is the application of approaches to impact the drug discovery process positively, tackling key issues such as productivity and attrition, through the use of computational and/or data-mining methods. While these methods can be used to explain observations, the goal and driving force are to affect decisions made, such as the synthesis of new compounds or the choice of compounds to screen or on which to do follow-up studies.

The spirit of this volume is to bring together the historical, current, and future perspective of many of the experts in this field, from both academia and industry. As this is a relatively young field in which there is no definitive 'right' way to approach a problem or use the various methods, there is a purposeful overlap of some concepts and methods between the various chapters, where the viewpoints of several experts or groups experienced in drug discovery have been leveraged to provide the reader with a broader perspective of key approaches, with related practical advice on the particular methods used or preferred by that group.

The volume is divided into seven sections:

I. Introduction to Computer-Assisted Drug Design
II. Ligand-Based Core Concepts and Methods
III. Target Structure-Based Core Concepts and Methods
IV. Ligand and Target Structure-Based Core Methods and Applications
V. Lead Discovery and Ligand-Based Drug Discovery Applications
VI. Target Structure-Based Drug Discovery Applications
VII. New Directions

Section I provides perspectives with a historical emphasis from three of the major players in the area, on CADD (Garland Marshall and Christina Taylor) and more specifically on quantitative structure–activity relationships (QSARs; Corwin Hansch and Cynthia Selassie) and structure-based drug design (SBDD; Don Abraham).

The following five sections address the diverse methods and approaches used in CADD and their application to drug discovery. Any categorization is arbitrary and has limitations, as a combination of many computational processes is generally used to address any task. In this volume a separation of ligand-based (Sections II and V) and structure-based (Sections III and VI) core concepts/methods and applications has been made where this is appropriate for CADD applications. Some key approaches are best described addressing both ligand- and target structure-based core methods and applications together; thus library design and the use of quantum mechanical (QM) methods are not subdivided and appear in Section III. Lead discovery involves both ligand- and structure-based approaches; thus key approaches such as virtual screening are grouped in Section V, together with the applications of approaches such as pharmacophore modeling and QSAR, where the boundary between a use for lead discovery or lead optimization is again arbitrary. The core concepts and methods are discussed first (Sections II–IV), as these are the building blocks for the applications to drug discovery (Sections V and VI). Components of both methods and applications will appear in all chapters, thus the reader is encouraged to read chapters in all sections that are potentially relevant to a area of interest, whether the interest be in the method or the drug discovery applications. The tables and figures in this chapter attempt to map and classify the various methods, although this is nonexhaustive, and there are many additional links and cross-fertilizations of methods.

Key impacts from the use of CADD approaches span the whole drug discovery process. This used to be viewed as a serial process, starting with target identification and validation (target equity), but now the importance of building chemical equity in parallel is recognized, i.e., the need for target validation and hit/lead identification to have both a suitable screen and suitable compounds. Suitable compounds need to have properties that enable activity to be measured on the target of interest and, other than for tool compounds, that are suitable for rapid follow-up to an in vivo active lead/candidate. Whether tool or lead, suitable properties and selectivity are essential to enable the desired target to be modulated in a relevant way for in vivo validation. This is illustrated by the way companies now focus investment on building their screening files as well as in target identification and high-throughput screening (HTS). For example, Pfizer recently made a very large investment in building a suitable screening file, in which millions of compounds were synthesized through a major 'file enrichment' initiative, fit for purpose in terms of both their properties and parallel chemistry heritage to enable rapid hit to lead follow-up.

High rates of attrition as compounds move through the drug discovery process contribute significantly to the low overall productivity of the pharmaceutical industry, and the concept of lead equity in addition to lead selectivity is now also important, e.g., by differentiation to existing drugs, failed drugs, or other candidates. **Figure 1** illustrates this modified view of the drug discovery process.

Some of the key impacts from CADD approaches can thus be categorized according to this expanded view of the drug discovery process, as illustrated in **Table 1**. Some key CADD approaches are mapped in **Figure 2** to the extended drug discovery process stages shown in **Figure 1**.

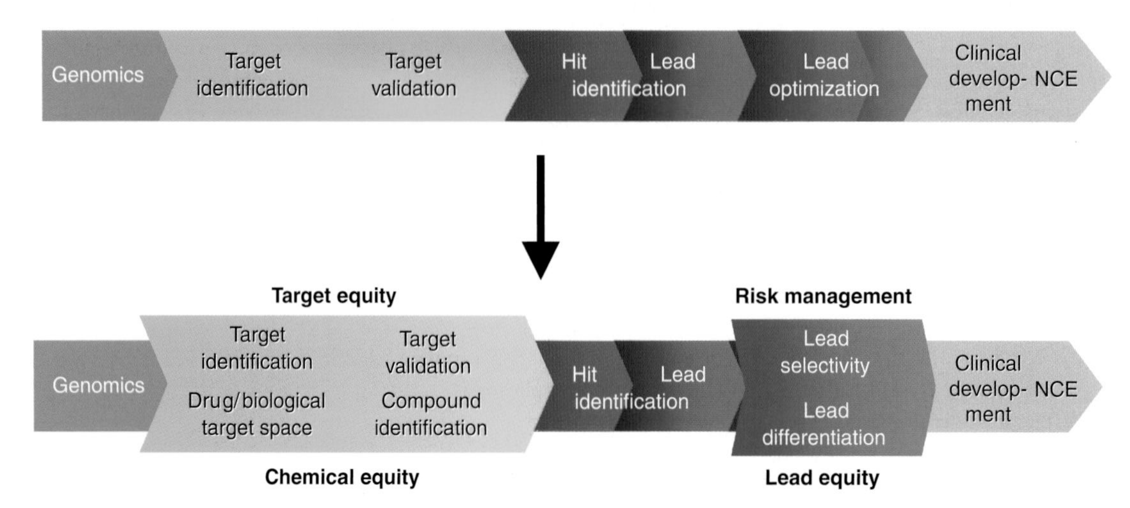

Figure 1 The drug discovery process, showing the more recent evolution where target equity and compound equity are parallel early key components, and lead equity is now considered. Both of these contribute to efforts to tackle productivity and attrition (risk management). NCE, new chemical entity.

Table 1 Some key drug design deliverables that can be impacted by the application of CADD methods, mapped to the extended drug discovery process stages from **Figures 1** and **2**

Target equity	Hit/lead identification	Hit/lead optimization, lead selectivity
The identification of druggable targets. Chemogenomic approaches	The identifiction of hit compounds suitable to become leads/ candidate: • Via in silico/virtual screening approaches • Through the use of multiuse screening subsets (focused or general) • Through the use of triage/ analysis methods on high-throughput screening (HTS) data	The design of compounds and compound libraries for the optimization of hits to leads and leads to candidates, in terms of: • Biological activity on the primary target(s) • Selectivity versus 'off-target' pharmacological activities • Pharmaceutical and ADME properties • Absence of toxicologic/safety issues
Target validation		Lead differentiation
The identification of tool compounds for target validation		The design of compounds with improved chance of success (reduced attrition risks), including: • Prediction of in vivo adverse effects • Differentiation of compounds (relative to other candidates and/or existing drugs) in a biologically relevant way
Compound equity		
The identification/design of suitable compounds for the screening file		

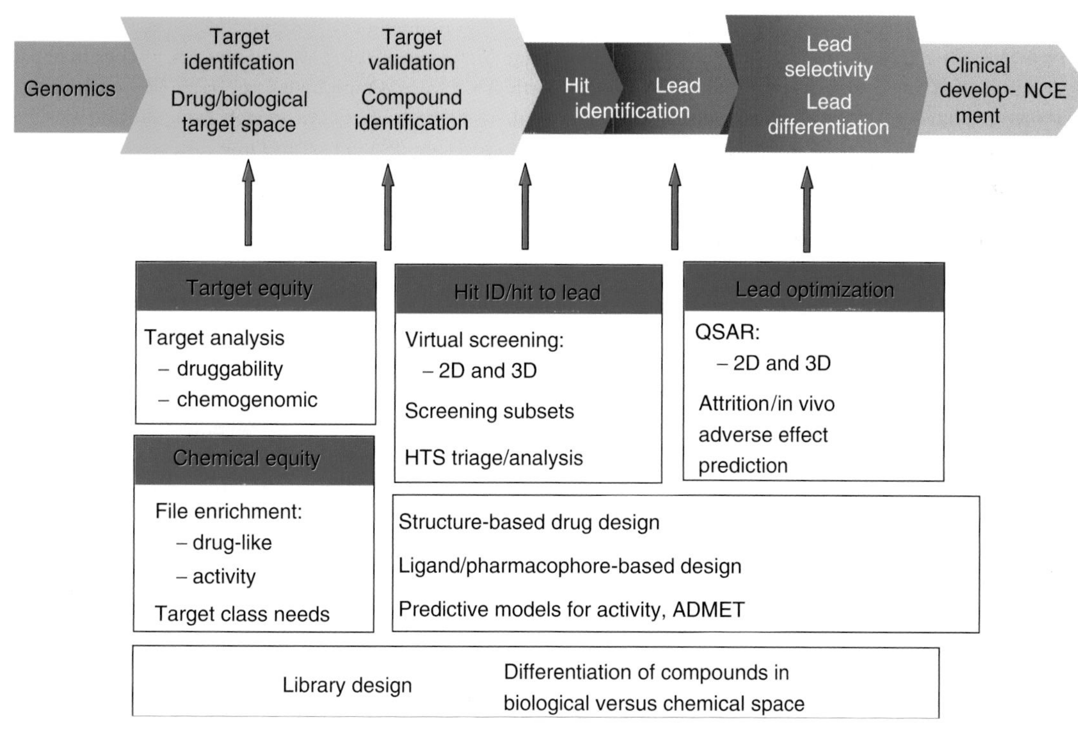

Figure 2 Some CADD applications that impact the drug discovery process.

The use of predictive models and design approaches (e.g., SBDD) thus address many aspects of the properties a compound needs to have to be a drug candidate, and the goal is now often a multidimensional/-parametric/-objective optimization, where multiple properties are optimized more effectively in a parallel rather than serial way. Thus methods such as QSAR can be key to enabling large virtual libraries of compounds that could be synthesized to be systematically analyzed to identify compounds that meet multiple objectives, e.g., biological activity on a desired target, selectivity against other targets, and suitable ADME-related properties such as solubility, metabolic stability, and cytochrome P450 (CYP) inhibition profile. While the structure–activity relationship (SAR) of a single endpoint can be potentially manually learned by the user and applied to the selection of new compounds, an argument against the use of QSAR methods is that the model does not direct ideas outside the structural training range of activity (which is not true of more three-dimensional (3D) methods, in which 'scaffold hopping' is quite common). Once the goal is to find a solution to match or balance between multiple endpoints, and to be able to analyze thousands to millions of virtual compounds (that could potentially be synthesized), then the computational use of multiple models (including QSAR models) becomes key. The reader is thus encouraged to consider the continuing value of QSAR together with the many other predictive models both in the context of what they provide directly (prediction of an activity or property for a new compound (one or many)) and their combined use in the many 'multidimensional' optimization needs in the hit to lead and lead optimization parts of the drug discovery process. The author has experience of how multiple 3D QSAR models (comparative molecular field analysis), one for activity and one for CYP2D6 inhibition, were applied successfully to solve a CYP2D6 inhibition problem through the systematic evaluation of multiple scaffold modifications.

All the main CADD/in silico impact points are covered in this volume except for the key area of pharmaceutical and ADMET properties, that is mainly addressed in Volume 5, together with Chapter 4.31 for the use of protein structures for the metabolizing CYP enzymes. Chemoinformatics and related databases are discussed in Chapters 3.12–3.18 of Volume 3. The mapping of key impacts from the use of CADD approaches to the volumes and chapters of *Comprehensive Medicinal Chemistry* is shown in **Table 2**, and the mapping of CADD approaches to the volume chapters in **Table 3**.

4.01.2 Computer-Assisted Drug Design and the Evolving Drug Discovery Process

Although some CADD methods may appear to some people to be less prominent and relevant today, the new challenges that we face now are able to build on such methods, and use them in ways to enable powerful new approaches to impact drug design. Large virtual libraries and multiple scaffolds (particularly with 3D methods) can be rapidly evaluated, to identify compound(s) that best meet all the criteria, for example using multiple QSAR models for a multidimensional optimization (*see* 4.30 Multiobjective/Multicriteria Optimization and Decision Support in Drug Discovery).

New developments have also made some methods more applicable to, and effective for, drug discovery needs again. Thus de novo design (*see* 4.13 De Novo Design), which was very popular with the arrival of target protein structures, was used less for a period due to synthetic chemistry issues around the designed molecules (e.g., feasibility and time to synthesize a single molecule) and the advent of HTS and combinatorial chemistry that were thought to be able to solve hit/lead generation problems. The need for hit/lead identification still remains for many targets, and the CADD approaches have greatly evolved, such that synthetically feasible structures are generated, and parallel (combinatorial) chemistry methods can be used to synthesize libraries versus singletons. A combined experimental–de novo design approach is also successfully used now, where fragment screening (e.g., by x-ray crystallographic or NMR spectroscopic approaches) seeds the structure-based (de novo) design process. CADD methods are often deployed to prioritize the fragments to screen for a particular target. The fragment screening methods themselves have evolved, using fragments (including templates or reagents from combinatorial/parallel chemistry) that are easier to do follow-up chemistry on, with parallel chemistry protocols enabling the binding hypothesis to be explored rapidly with multiple compounds.

The use of large-scale simulations, such as via molecular dynamics (MD: *see* 4.25 Applications of Molecular Dynamics Simulations in Drug Design; 4.10 Comparative Modeling of Drug Target Proteins), for example in optimizing protein–ligand complexes and in the prediction of ligand-binding energies, also appears relatively less dominant and generally used. This may be due to limited success in the past, where expensive computers and less refined force fields were limiting, and the desire was to process large numbers of compounds (e.g., large databases or large numbers of virtual compounds for the design of combinatorial libraries). Better force fields, docking algorithms and handling of water/solvation, together with combined QM–force field approaches to model ligand–active site interactions, coupled with much more powerful and cheaper computing power than was ever possible before and many more target protein structures, will likely lead to a return of MD simulations as a core method widely used by computational chemists. All QM approaches (*see* 4.16 Quantum Mechanical Calculations in Medicinal Chemistry: Relevant Method or a Quantum Leap Too Far?) also benefit from the much increased computational power, and can already be applied to large systems in reasonable timeframes.

Table 2 The mapping of key impacts from the use of CADD approaches to volume chapters of *Comprehensive Medicinal Chemistry*

Topic	*Example CADD methods*	*Main volume chapter that addresses this topic*
1. The identification of druggable targets	Binding site analysis	4.17 Chemogenomics in Drug Discovery – The Druggable Genome and Target Class Properties
2. The identification/design of suitable compounds for the screening file	Activity and drug-like properties Chemistry–biological space analisis	4.17 Chemogenomics in Drug Discovery – The Druggable Genome and Target Class Properties 4.18 Lead Discovery and the Concepts of Complexity and Lead-Likeness in the Evolution of Drug Candidates
3. The identification of tool compounds for target validation 4. The identification of hit compounds suitable to become leads/candidates: a. Via in silico/virtual screening approaches	Virtual screening Virtual screening De novo design	4.19 Virtual Screening 4.07 Predictive Quantitative Structure–Activity Relationship Modeling 4.08 Compound Selection Using Measures of Similarity and Dissimilarity 4.10 Comparative Modeling of Drug Target Proteins 4.13 De Novo Design
b. Through the use of multiuse screening subsets (focused or general) c. Through the use of triage/analysis methods on high-throughput screening (HTS) data	Design of subsets HTS analysis/triage	4.20 Screening Library Selection and High-Throughput Screening Analysis/Triage
5. The design of compounds and compound libraries for the optimization of hits to leads and leads to candidates, in terms of:	Pharmacophore modeling	4.06 Pharmacophore Modeling: 1 – Methods 4.21 Pharmacophore Modeling: 2 – Applications 4.02 Introduction to Computer-Assisted Drug Design – Overview and Perspective for the Future
a. Biological activity on the primary target(s)	QSAR	4.03 Quantitative Structure–Activity Relationship – A Historical Perspective and the Future 4.07 Predictive Quantitative Structure–Activity Relationship Modeling 4.22 Topological Quantitative Structure–Activity Relationship Applications: Structure Information Representation in Drug Discovery 4.23 Three-Dimensional Quantitative Structure–Activity Relationship: The State of the Art 4.30 Multiobjective/Multicriteria Optimization and Decision Support in Drug Discovery
	Library design	4.14 Library Design: Ligand and Structure-Based Principles for Parallel and Combinatorial Libraries 4.15 Library Design: Reactant and Product-Based Approaches
	Structure-based drug design (SBDD)	4.04 Structure-Based Drug Design – A Historical Perspective and the Future 4.24 Structure-Based Drug Design – The Use of Protein Structure in Drug Discovery 4.31 New Applications for Structure-Based Drug Design

continued

Table 2 Continued

Topic	*Example CADD methods*	*Main volume chapter that addresses this topic*
b. Selectivity versus 'off-target' pharmacological activities	As for 5a. above + biological fingerprints	As for 5a. above + 4.32 Biological Fingerprints
c. Pharmaceutical and absorption, distribution, metabolism, and excretion (ADME) properties	In silico predictive models/QSAR Protein structure-based	Volume 5: 5.22–5.36; 5.41 (see **Table 3** for chapter titles); 4.31 New Applications for Structure-Based Drug Design
d. Absence of toxicologic/safety issues	Fragment/QSAR and modeling approaches	Volume 5: 5.40 In Silico Models to Predict QT Prolongation 5.39 Computational Models to Predict Toxicity
6. The design of compounds with improved chance of success (reduced attrition risks), including:		
a. Prediction of in vivo adverse effects	In silico ADMET	Volume 5: 5.22–5.41
b. Differentiation of compounds (relative to other candidates and/or existing dugs) in a biologically relevant way	Biological fingerprints	4.32 Biological Fingerprints
7. The use of protein structure to provide new insights	4.04 Structure-Based Drug Design – A Historical Perspective and the Future 4.24 Structure-Based Drug Design – The Use of Protein Structure in Drug Discovery 4.26 Seven Transmembrane G Protein-Coupled Receptors: Insights for Drug Design from Structure and Modeling 4.27 Ion Channels: Insights for Drug Design from Structure and Modeling 4.28 Nuclear Hormone Receptors: Insights for Drug Design from Structure and Modeling 4.29 Enzymes: Insights for Drug Design from Structure 4.31 New Applications for Structure-Based Drug Design	

Large amounts of bioactivity data, for thousands to millions of compounds, are now being generated (e.g., by HTS) or electronically assembled/curated (e.g., from journal publications) that can now be rapidly analyzed by probabilistic modeling methods such as Bayesian statistical modeling that can handle 'noisy' data (experimental errors causing significant problems for some methods). Thus predictive models can be generated using both active/inactive data from HTS (e.g., from % inhibition) or from large amounts of IC_{50} data (see Paolini *et al.*[5] for a Laplacian-modified Bayesian classifier approach to analyse 617 694 experimental activities from 238 655 compounds covering 698 targets, using data from an integration of internal screening data and external data from journals and FCFP_6 functional class fingerprints in Scitegic Pipeline Pilot). The models can be used on large (e.g., virtual libraries or HTS sets) or small compound sets to increase the yield of active compounds synthesized or the yiled of interesting compounds from HTS (e.g., to identify potential false negatives and positives).

4.01.3 The Essential Partnership of In Silico and Experimental

4.01.3.1 Partnership with Structural Biology – A Case History

A key partner technology for CADD is structural biology, a process that enables the biophysical determination of ligand binding to a target protein, structural characterization of the target proteins, and the structures of the ligand protein-binding site complexes to be determined by x-ray crystallography or NMR methods.

The x-ray crystallographic determination of the 3D structure of target proteins and bound ligands has provided major impacts of SBDD, enabling structure-based virtual screening to identify novel hits and in understanding ligand–protein interactions, leading to the design of new compounds. These experimental data can often identify or confirm (from

Table 3 The mapping of CADD approaches to volume chapters of *Comprehensive Medicinal Chemistry*

CADD methods	*Related volume chapter*
Binding site analysis	4.09 Structural, Energetic, and Dynamic Aspects of Ligand–Receptor Interactions
	4.11 Characterization of Protein-Binding Sites and Ligands Using Molecular Interaction Fields
	4.17 Chemogenomics in Drug Discovery – The Druggable Genome and Target Class Properties
Structure-based drug design (SBDD)	4.02 Introduction to Computer-Assisted Drug Design – Overview and Perspective for the Future
	4.04 Structure-Based Drug Design – A Historical Perspective and the Future
	4.10 Comparative Modeling of Drug Target Proteins
	4.18 Lead Discovery and the Concepts of Complexity and Lead-Likeness in the Evolution of Drug Candidates
	4.24 Structure-Based Drug Design – The Use of Protein Structure in Drug Discovery
	4.25 Applications of Molecular Dynamics Simulations in Drug Design
	4.26 Seven Transmembrane G Protein-Coupled Receptors: Insights for Drug Design from Structure and Modeling
	4.27 Ion Channels: Insights for Drug Design from Structure and Modeling
	4.28 Nuclear Hormone Receptors: Insights for Drug Design from Structure and Modeling
	4.29 Enzymes: Insights for Drug Design from Structure
	4.31 New Applications for Structure-Based Drug Design
	3.17 The Research Collaboratory for Structural Bioinformatics Protein Data Bank
De novo design	4.13 De Novo Design
Virtual screening	4.19 Virtual Screening
	4.07 Predictive Quantitative Structure–Activity Relationship Modeling
	4.08 Compound Selection Using Measures of Similarity and Dissimilarity
	4.10 Comparative Modeling of Drug Target Proteins
	4.12 Docking and Scoring
Library design	4.14 Library Design: Ligand and Structure-Based Principles for Parallel and Combinatorial Libraries
	4.15 Library Design: Reactant and Product-Based Approaches
Design of subsets	4.20 Screening Library Selection and High-Throughput Screening Analysis/Triage
Chemistry–biological space analysis	4.08 Compound Selection Using Measures of Similarity and Dissimilarity
	4.17 Chemogenomics in Drug Discovery – The Druggable Genome and Target Class Properties
	4.18 Lead Discovery and the Concepts of Complexity and Lead-Likeness in the Evolution of Drug Candidates
Chemoinformatics	3.12 Chemoinformatics
	3.13 Chemical Information Systems and Databases
	3.14 Bioactivity Databases
In silico predictive models	4.02 Introduction to Computer-Assisted Drug Design – Overview and Perspective for the Future
	4.05 Ligand-Based Approaches: Core Molecular Modeling

continued

Table 3 Continued

CADD methods	*Related volume chapter*
Pharmacophore modeling	4.06 Pharmacophore Modeling: 1 – Methods
	4.21 Pharmacophore Modeling: 2 – Applications
QSAR	4.03 Quantitative Structure–Activity Relationship – A Historical Perspective and the Future
	4.07 Predictive Quantitative Structure–Activity Relationship Modeling
	4.22 Topological Quantitative Structure–Activity Relationship Applications: Structure Information Representation in Drug Discovery
	4.23 Three-Dimensional Quantitative Structure–Activity Relationship: The State of the Art
	4.30 Multiobjective/Multicriteria Optimization and Decision Support in Drug Discovery
QM approaches	4.16 Quantum Mechanical Calculations in Medicinal Chemistry: Relevant Method or a Quantum Leap Too Far?
Ligand 3D experimental structures	3.18 The Cambridge Crystallographic Database
Protein 3D experimental structures	3.17 The Research Collaboratory for Structural Bioinformatics Protein Data Bank
Biological fingerprints/lead differentiation	4.32 Biological Fingerprints
In silico ADMET	5.22 Use of Molecular Descriptors for Absorption, Distribution, Metabolism, and Excretion Predictions
	5.23 Electrotopological State Indices to Assess Molecular and Absorption, Distribution, Metabolism, Excretion, and Toxicity Properties
	5.24 Molecular Fields to Assess Recognition Forces and Property Spaces
	5.25 In Silico Prediction of Ionization
	5.26 In Silico Predictions of Solubility
	5.27 Rule-Based Systems to Predict Lipophilicity
	5.28 In Silico Models to Predict Oral Absorption
	5.29 In Silico Prediction of Oral Bioavailability
	5.30 In Silico Models to Predict Passage through the Skin and Other Barriers
	5.31 In Silico Models to Predict Brain Uptake
	5.32 In Silico Models for Interactions with Transporters
	5.33 Comprehensive Expert Systems to Predict Drug Metabolism
	5.34 Molecular Modeling and Quantitative Structure–Activity Relationship of Substrates and Inhibitors of Drug Metabolism Enzymes
	5.36 In Silico Prediction of Plasma and Tissue Protein Binding
	5.39 Computational Models to Predict Toxicity
	5.40 In Silico Models to Predict QT Prolongation
	5.41 The Adaptive In Combo Strategy

computational predictions) many interesting and unexpected binding modes. The iterative use of experimental structure determination and computational analysis and design to leverage the information are powerful examples of the needed partnership of experimental and in silico approaches. Protein NMR methods also allow many experimental studies that do not need the structure of the protein to be determined, for example in detecting just the binding of a ligand (including fragments) to a protein site, providing a screening method that can seed further design (directly and/or with binding structure determination by x-ray or NMR). Chapters 3.19–3.25, 3.38–3.42, 4.24–4.29, and 4.31 discuss the methods, applications, and some target class learnings.

A case history of experiences using multiple approaches in different companies for the identification and optimization of orally bioavailable inhibitors of the factor Xa serine protease enzyme[1] illustrates this partnership and the use of many CADD methods described in this volume. The initial goal was the identification of suitable nonpeptidic small-molecule hits for oral administration. A structure of the enzyme was available (apo, no ligand), and this was used for de novo design. Through early access to the Skelgen de novo structure generation program,[2] multiple structural frameworks were generated to fill and bridge the S1 and S4 pockets. Through significant design chemistry iterations, novel active (K_i 400–900 nM) compounds were designed and synthesized. A recent brief evaluation of the same evolved software program–target combination showed encouragingly that de novo structure generation can now generate directly much more synthetically feasible and drug-like molecules. Combined pharmacophore and shape constrained 3D database-searching methods were found useful to dock compounds into the active site. Recent developments in this area enable an automated and systematic use of site-based pharmacophores for docking, virtual screening of ligands and proteins, and protein clustering,[3] and have shown the power of the use of pharmacophores as descriptors relevant for molecular recognition that enable ligands and proteins to be analyzed in the same frame of reference. Such an unbiased use of all pharmacophore combinations can be important to avoid missing unexpected binding modes. In the factor Xa example, early structure–activity led to a dogma that a basic group was required in the S1 pocket (**Figure 3**), and binding driven by a small hydrophobic area deep in the S1 pocket (**Figure 3**, top right), identified by computational sampling, was discounted until a transformational x-ray structure of a modified compound containing a chlorobenzothiophene showed that a neutral group could bind in the S1 pocket, despite the acidic aspartate residue, as shown in **Figure 3** on the right, and the presence of a basic group in the ligand. Experimental structures are important for identifying or confirming such 'reverse' or nonstandard binding modes; computational analysis can identify such possibilities before structure is possible or on structures that can be synthesized, and their use is important to leverage known structural information and challenge SARs. In the factor Xa example, CADD methods had already provided understanding of the S4 pocket binding; the electrostatics had been investigated using QM methods to rationalize the binding of dibasic compounds, with a favorable π-cation interaction being postulated. QM methods were also found to be very useful in the prediction

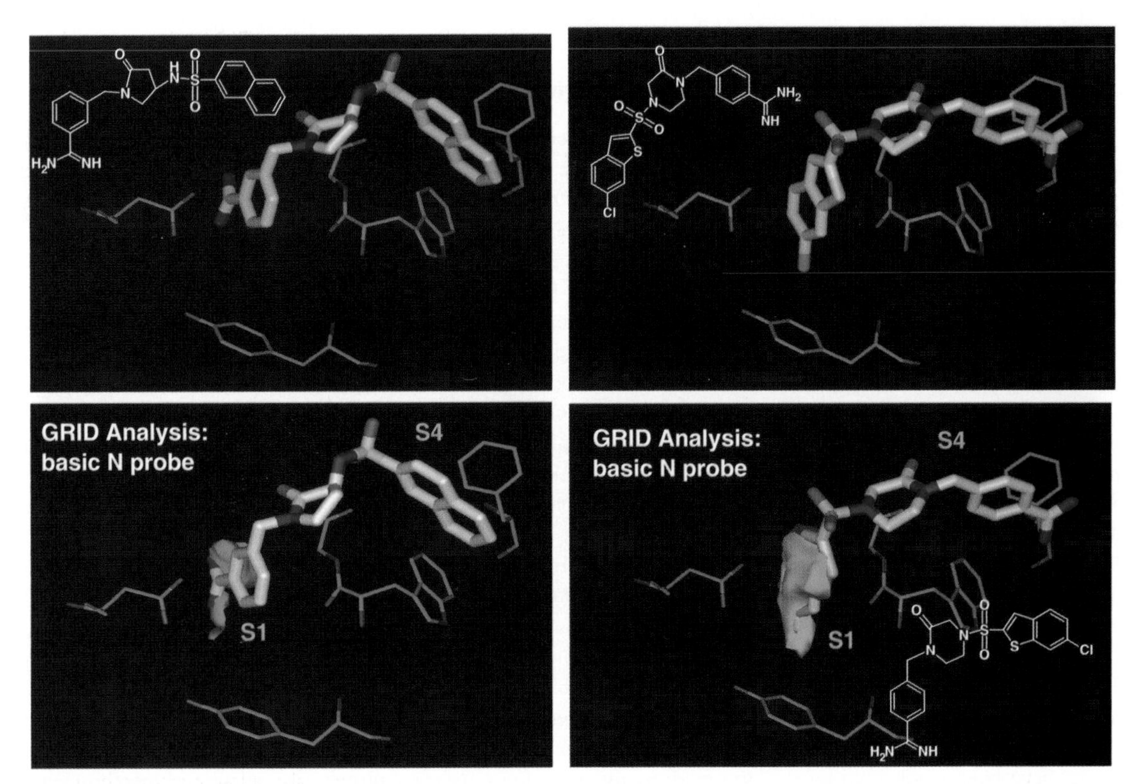

Figure 3 The x-ray structure of the active site of factor Xa serine protease showing the different binding modes of two potent ligands docked into the site (upper), together with the areas for favorable interaction of a basic nitrogen (lower left) and an aromatic hydrogen (lower right) from GRID analyses. The structure on the left has the 'expected' binding of the benzamidine moiety to the aspartate-containing S1 pocket (x-ray structure of the complex), while the structure on the right has the 'reverse' binding mode (from a modeled structure, confirmed by x-ray).

on pK_as of new heterocyclic fragments for the S1 pocket, which predicted poorly using the fragment/substituent-based methods that had no related compounds in the training set.

In a different approach to the identification of factor Xa inhibitors, x-ray structure and in silico approaches were again used together to drive the identification and generation of novel structures with desired properties. Structure-based virtual screening identified potent hits, and a fragment-based approach was used to identify suitable S1 pocket-binding fragments of low basicity. Virtual screening of many possible fragments identified a small number of possible candidates, whose binding was investigated by x-ray crystallography, that confirmed both the binding and orientation. Libraries of compounds were synthesized building off the new S1 pocket fragment, and to investigate other hits; the structure of the enzyme was used to guide the library design, and again the iterative use of crystal structures of ligand complexes was important. Within series, limited success at predicting binding energies was achieved. This is a key area of continued need and methods development.

In the chapters of this volume and related volumes the diverse computational methods that can be and are used in real life to progress drug discovery projects such as this are discussed.

4.01.3.2 Partnership with Pharmacological Profiling

Biological testing is clearly a key step to validate any compound design or proposed active from in silico searching (virtual screening). Pharmacological profiling against multiple targets provides insights into the polypharmacology of many drugs,[4] and provides a powerful and biologically relevant descriptor for compounds that can be used to cluster them and to look for correlations or associations with in vivo effects, such as adverse events (*see* 4.32 Biological Fingerprints). The understanding of the broad biological activity profile of compounds, compared to information on a few assays, provides key data sets to enable the use of the most appropriate descriptors and methods to analyze, search, and cluster structures in 'biological space.' It also raises challenges for concepts such as similarity, as although structurally similar compounds often have similar 'primary' biological activity (the 'pharmacophore' for which is kept relatively core), activities on other targets can be dramatically different. For example, adding a solubilizing group to an otherwise relatively unchanged structure (or the change of an alkyl to ether group) creates many new potential pharmacophic interactions.

4.01.4 Future Challenges and Perspectives

CADD is an exciting and continually evolving area that leverages new data and methods to provide approaches that tackle the ever-changing needs of drug discovery. The scope continues to grow, and applications now span the whole drug discovery process (e.g., with attrition prediction and the modeling of in vitro data to predict in vivo effects, such as the use of biological fingerprints: Chapter 4.32). The availability of experimental data for model building for multiple endpoints or selectivity targets enables CADD to tackle the needed multidimensional optimization challenge, and a combination of models for the different endpoints can be used, together with a variety of methods (*see* 4.30 Multiobjective/Multicriteria Optimization and Decision Support in Drug Discovery). This creates the need for improved methods to achieve such optimization using predictive models, with both a prediction and confidence level in the value being important. Models that alone may be of less perceived value as the user has a good understanding of the structure–activity can be used simultaneously in a multidimensional way to search exhaustively large numbers of possible structures and structural changes to identify more optimal solutions (e.g., potency with selectivity and desired properties). The methods and approaches used have expanded enormously, and now include areas such as data mining; the amount of data that is now electronically available, such as structure–activity data, is ever increasing, with the addition now of records from published journals and patents, as well as commercial databases and in-house data from HTS. Probabilistic modeling approaches like Bayesian statistical models that can be rapidly used to analyze large and noisy data sets, to give predictive models that can be applied to the next iteration of experimental work enable CADD to further impact areas such as HTS analysis, library design, and virtual screening.

Protein structures continue to become more available, and the methods and force fields are also developing for SBDD, with one goal to be able to predict accurately ligand-binding energies/affinities. SBDD will thus continue to be an increasingly applicable enabling approach in drug discovery, considering both the target and off-/antitargets. The existence and importance of polypharmacology for many drugs, affecting potentially the efficacy and the adverse effects, have been highlighted by the systematic analyses of drugs against multiple targets[4] (e.g., from the Cerep BioPrint biological fingerprinting). Such data, that show how very similar compounds, that could be considered the same 'chemotype' as there are only relatively small changes to substituents, can have very different broad biological profiles, offer new challenges to CADD to explain and predict such differences (e.g., by pharmacophoric or shape changes), with the realization that 'selectivity targets' may be in quite different target classes.

This volume contains methods that relate to the design of drug molecules, methods that are both more established and quite new, and the reader is encouraged to consider them all, combined in any ways, to tackle the specific needs of each medicinal chemistry project. The very broad set of approaches that relate to CADD continues to develop, with innovative new methods continually appearing. As CADD evolves to embrace the new methods and new data available, the impacts on drug discovery should also continue to expand, across the whole process from the discovery of suitable targets and hits to differentiation and attrition of clinical candidates.

References

1. Ewing, W. R.; Becker, M. R.; Manetta, V. E.; Davis, R. S.; Pauls, H. W.; Mason, H.; Choi-Sledeski, Y. M.; Green, D.; Cha, D.; Spada, A. P. et al. *J. Med. Chem.* **1999**, *42*, 3557–3571.
2. Todorov, N. P.; Dean, P. M. *J. Comput. Aided Des.* **1997**, *11*, 175–192.
3. Perruccio, F.; Mason, J. S.; Sciabola, S.; Baroni, M. FLAP: 4-Point Pharmacophore Fingerprints from GRID. In *Molecular Interaction Fields: Applications in Drug Discovery and ADME Prediction*; Cruciani, G., Ed.; Wiley: Weinheim, 2005.
4. Hopkins, A. L.; Mason, J. S.; Overington, J. P. *Curr. Opin. Struct. Biol.* **2006**, *16*, 127–136.
5. Paolini, G. V.; Shapland, R. H. B.; van Hoorn, W. P.; Mason, J. S.; Hopkins, A. L. *Nat. Biotechnol.* **2006**, *24*, 805–815.

Biography

Jonathan S Mason, studied at the University of London, Queen Mary, where he obtained a BSc in 1976 and PhD in 1979 in organic chemistry. He has subsequently worked in the pharmaceutical industry, starting as a medicinal chemist at Rhône-Poulenc Rorer (now Sanofi-Aventis), then in 1984 initiating and growing computational chemistry groups in the United Kingdom, France, and the United States. In 1997, he was appointed as Director of the Bristol-Myers Squibb Computer-Assisted Drug Design teams, based in Princeton, New Jersey, and then in 1999 of a new integrated Structural Biology and Modeling department of 33 scientists including protein crystallography and protein NMR. From 2001 to 2005, he was leading a team of 45 scientists involved in Computational Chemistry, Structural Biology (from clone to x-ray structure), Medicinal Informatics and Knowledge Discovery (including bioinformatics) as Executive Director of the Medicinal Informatics, Structure & Design Department at Pfizer Global Research & Development, Sandwich, United Kingdom. He is currently a Divisional Director at Lundbeck Research Denmark, jointly responsible for Medicinal Chemistry Research, leading the departments of Early Lead Generation (Chemistry) and Computational Chemistry. He has been a pioneer in the development and use of three-dimensional (3D) pharmacophore fingerprint profiling methods, together with other molecular similarity and diversity approaches, for both ligands and protein targets (e.g., for virtual screening, ligand docking, target-class library design, and high-throughout screening (HTS) analysis). Recently, he has driven the use of biological fingerprints to tackle attrition-related problems including lead selection/differentiation and enable new chemogenomic approaches. Other activities include structure-based drug design (including de novo approaches), absorption, distribution, metabolism, and excretion (ADME)/safety prediction, large-scale bioactivity data integration and target-class mining, and the development of intuitive desktop tools for discovery scientists.

© 2007 Elsevier Ltd. All Rights Reserved
No part of this publication may be reproduced, stored in any retrieval system or transmitted in any form by any means electronic, electrostatic, magnetic tape, mechanical, photocopying, recording or otherwise, without permission in writing from the publishers

Comprehensive Medicinal Chemistry II
ISBN (set): 0-08-044513-6
ISBN (Volume 4) 0-08-044517-9; pp. 1–11

4.02 Introduction to Computer-Assisted Drug Design – Overview and Perspective for the Future

G R Marshall and C M Taylor, Washington University, St. Louis, MO, USA

© 2007 Elsevier Ltd. All Rights Reserved.

4.02.1 Introduction

What is computer-assisted drug design (CADD), and why is it important? There is no clear definition, although a consensus view has emerged. Simply, CADD is the coalescence of information on chemical structures, their properties, and their interactions with biological macromolecules. Further, these data are transformed into knowledge intended to aid in making better decisions for drug discovery and development.

Assistance from computational chemistry and bioinformatics is necessary to handle the vast and ever-increasing relevant data to be analyzed. One aspect is the sheer magnitude of chemical information that must be processed. For example, Chemical Abstracts Service adds over three-quarters of a million new compounds to its database annually for which large amounts of physical and chemical property data are available. Some groups generate hundreds of

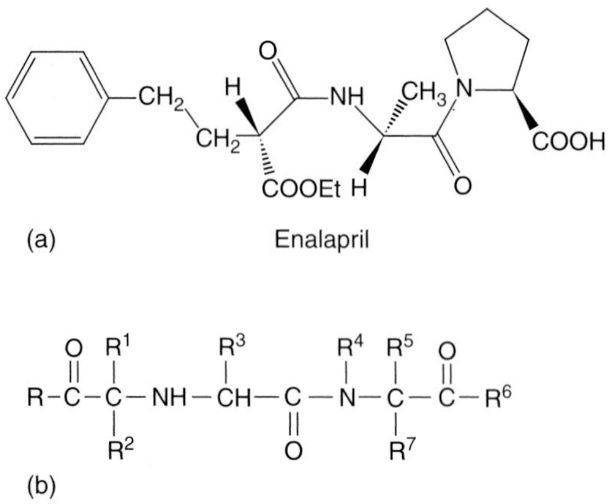

(a) Enalapril

(b)

Figure 1 (a) Chemical structure of enalapril; (b) reference formula described by the patent.

thousands to millions of compounds on a regular basis, through combinatorial chemistry, to be screened in a variety of in vitro and in vivo biological assays. Even more compounds are generated and screened in silico, based on models of molecular recognition, in the search for a 'magic bullet' for a given disease. Both of these processes for generating information about chemistry have their own limitations. Further, computational chemistry has to establish relevant criteria by which compounds of interest are selected for synthesis and testing, as experimental approaches used to test compounds have practical limitations, despite automation. However, the accuracy of affinity prediction with current computational methodology is now approaching sufficient precision to be of utility.

The magnitude of the problem can be illustrated by the following example. Given a patent for a prodrug angiotensin-converting enzyme (ACE) inhibitor, enalapril (**Figure 1a**), it may be useful to estimate the number of compounds covered by its patents. **Figure 1b** shows the reference Markush formula as described by the patent. One can simply enumerate the members of each substituent class and combine them combinatorially. More than 59 trillion compounds are included in the patent.

Of the compounds included in the patent, how many could be predicted to lack druglike properties based on similarity in properties to known orally active drugs? How many would be predicted to be inactive based on the known structure–activity data available on ACE inhibitors, such as captopril, at the time the patent was filed? How many would be predicted to be inactive now that a crystal structure of an ACE–inhibitor complex has been published? Given the structure–activity relationships (SARs) available on the inhibitors, what could one determine regarding the active site of ACE? What novel classes of compounds could be suggested based on the SAR of inhibitors, or based on the new crystal structure of the complex? Do the most potent compounds share a set of properties that can be identified and used to optimize a novel lead structure? Can a predictive equation relating properties and affinity for the isolated enzyme be established? Can a similar equation relating properties and in vivo bioassay effectiveness in lowering blood pressure be established? These are representative of questions facing the current drug-design community and significant applications of CADD.

4.02.2 Historical Evolution

With the advent of computers and the ability to store and retrieve chemical information, serious efforts to compile relevant databases and construct information retrieval systems began. Collecting crystal structure information for small molecules was one of the first efforts that had a substantial long-term inpact. The Cambridge Structural Database (CSD) stores crystal structures of small molecules and provides a fertile resource for geometrical data on molecular fragments for calibration of force fields and validation of results from computational chemistry.[1,2] As protein crystallography gained momentum, the need for a common repository of macromolecular structural data led to the Protein Data Bank (PDB), originally located at Brookhaven National Laboratories[3] and presently at Rutgers University. These efforts focused on the accumulation and organization of experimental results on the three-dimensional structure of molecules, both large and small. Todd Wipke created MDL and MACCS to fulfill the growing need for a chemical information system that could handle the increasing numbers of small molecules generated in industry.

With the advent of computers and the availability of oscilloscopes, displaying a rotatable structure with three-dimensional perspective was an obvious progression. Cyrus Levinthal and colleagues utilized the primitive computer graphics facilities at Massachusetts Institute of Technology (MIT) to generate rotating images of proteins and nucleic acids, which provided insight into the three-dimensional aspects of these structures without having to build physical models. His 1965 paper in *Scientific American* inspired others to explore computer graphics as a means of coping with the three-dimensional nature of chemistry. Physical models (Dreiding stick figures, Corey–Pauling–Koltun (CPK) models, etc.) were useful and widely accepted tools for medicinal chemists. However, physical overlap of two or more compounds was difficult, and exploration of the potential energy surface was hard to correlate with multiple conformations of a physical model.

As more chemical data accumulated with its implicit information content, a multitude of approaches began to extract useful information. Certainly, the shape and variability in geometry of molecular fragments from CSD was mined to provide fragments of functional groups for a variety of purposes. As compounds were tested for biological activity in a given assay, the desire to distill the essence of the chemical requirements for such activity to guide optimization by concepts, such as bioisosteres and pharmacophores, was generated. Initially, efforts focused on congeneric series as the common scaffold presumably eliminating the molecular alignment problem with the assumption that all molecules bound with a common orientation of the scaffold. This was the intellectual basis of the Hansch approach (quantitative structure–activity relationship, QSAR),[4] in which substituent parameters from physical chemistry were used to correlate chemical properties with biological activity for a series of compounds with the same substitution pattern on the congeneric scaffold.[4,5]

Considerable literature developed around the ability of numerical indices, derived from graph-theoretical considerations, to correlate with SAR data.[6] The ability of various indices proved to be useful parameters in QSAR equations.[7–9] Ioan Motoc correlated various numerical indices with more physically relevant variables, such as surface area and molecular volume. Since computational time was at a premium during the early days of QSAR and such numerical indices could be calculated with minimal computations, they played a useful role and continue to be used predictively today. Work using QSAR led to the development of three-dimensional pharmacophores as descriptors of molecular-recognition motifs both for medicinal chemistry, as well as for discerning protein–ligand interactions.

4.02.3 Unknown Receptor

4.02.3.1 Pharmacophores

The success of QSAR led to efforts to extend the domain to noncongeneric series, where the structural similarity between molecules active in the same bioassay was not obvious. The work of Beckett and Casey[10] on opioids to define parts of active molecules (pharmacophoric groups) essential for efficacy was seminal. Kier further developed the concept of pharmacophore and applied it to rationalize the SAR of several systems.[11] Gund and Wipke implemented the first in silico screening methodology with a program to screen a molecular database for pharmacophoric patterns in 1974.[12,13]

An early example of pharmacophore development involved superimposing apomorphine, chlorpromazine, and butaclamol such that the amines are aligned while maintaining the coplanarity of an aromatic ring.[14] This led to a plausible hypothesis of receptor-bound conformations at the dopamine receptor. Least-squares fitting of atomic centers did not allow such an overlap, but the use of the centroid of the aromatic ring with normals to the plane for least-squares fitting accomplished the desired overlap.

There still continues to be method development to generate overlaps of hydrogen-bond donors and acceptors, aromatic rings, etc. to formulate a pharmacophore hypothesis from a set of active compounds for a given receptor/enzyme. One method developed early at Washington University involved minimization of distances between groups in different molecules assigned by the investigator with no intermolecular interactions. In effect, adding springs caused the groups to overlap as the energy of the entire set of molecules was minimized, excluding any interatomic interactions with the exception of those imposed by the springs (**Figure 2**).

The results were dependent on the starting conformations of the set of molecules being minimized, and multiple starting conformations were used to generate multiple pharmacophoric hypotheses. Alternative algorithms utilize multiple distance constraints on the molecular ensemble. They also embedded the matrix of constraints into three dimensions utilizing distance geometry[15] or systematically determined the set of sterically allowed conformers of each molecule and compared their pharmacophoric patterns in three dimensions.[14] A distinct advantage generally exists in using internal coordinates in comparison of molecules, as internal distances are invariant to global rotations and translations.

4.02.3.2 Active-Analog Approach

The early work by medicinal chemists to try and rationalize their SAR with three-dimensional models, as well as the success of Hansch and others in correlating SAR with physical properties, led to exploration of molecular modeling as a

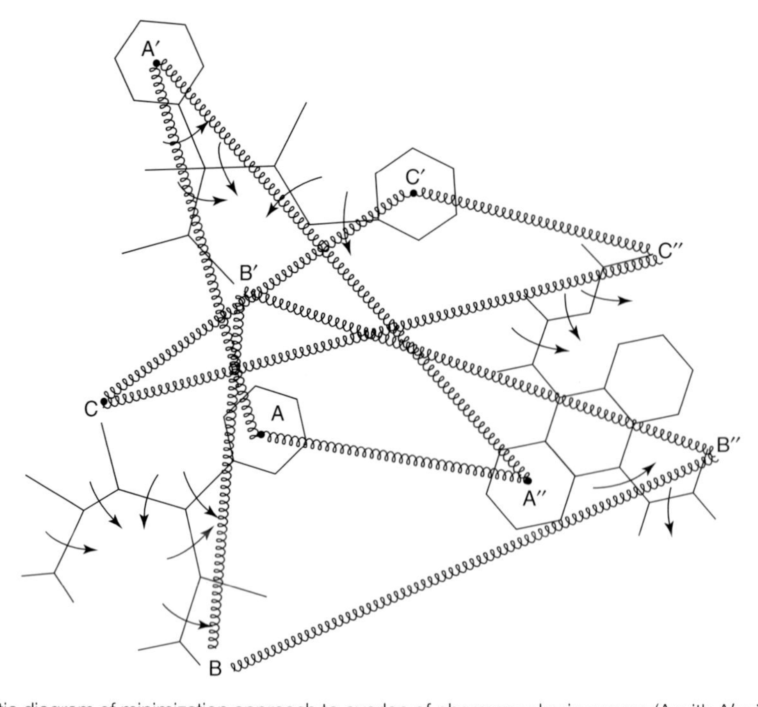

Figure 2 Schematic diagram of minimization approach to overlap of pharmacophoric groups (A with A′ with A″, B with B′ with B″, C with C′ with C″) by introduction of constraints (springs) with intermolecular interactions ignored and only intramolecular interactions considered.

means of combining the two approaches. Clearly, overall physical properties such as hydrophilicity, steric volume, charge, and molar refractivity would be more meaningful in the context of a specific subsite within the receptor, rather than when considered as an overall molecular average. One expected models with greater resolution and with the ability to discriminate between stereoisomers, for example, as a result of the inclusion of geometrically sensitive parameters. By 1979, the group at Washington University had developed a systematic computational approach to the generation of pharmacophore hypotheses, the Active-Analog Approach, which was disclosed at the American Chemical Society (ACS) National Meeting that year.[14] Many, more sophisticated variations[16,17] of this approach have subsequently been developed to generate three-dimensional hypotheses regarding molecular recognition.

The basic premise for the Active-Analog Approach was that each compound tested against a biological assay was a three-dimensional question for the receptor. But each molecule was, in general, flexible and could present a plethora of possible three-dimensional arrays of interactive chemical groups. By computationally analyzing the sets of possible three-dimensional pharmacophoric patterns associated with each active molecule, one could find those three-dimensional pharmacophoric patterns common to a set of actives. In the simplest case, each inactive molecule would be geometrically precluded from presenting the given pharmacophoric pattern common to active molecules by steric or cyclic constraints. In practice, inactives capable of presenting the hypothetical pharmacophoric pattern were found often, so another rationale for their inactivity was necessary. Aligning each active molecule to the candidate pharmacophoric pattern allowed determination of the volume requirements of the set of actives. An inactive compound could present the correct pharmacophoric pattern if it competed for extra volume that was occupied by the receptor. When an inactive was aligned with the pharmacophore as scaffold, subtraction of the active volume space could identify such novel requirements in inactives.

Earlier, a Gaussian representation of molecular volume[18] that readily allowed mathematical manipulation of atomic volumes was developed. A data set with a set of rigid bicyclic amino acids that inhibited *S*-adenosylmethionine, the enzyme that synthesizes the active methyl donor, provides the best example of this rationalization.[19] In this case, the amino acid portion provided a common frame of reference that revealed the compounds that lost ability to inhibit the enzyme shared a small volume not required by active compounds (presumably required by an atom of the enzyme). No other plausible suggestion for the data set has ever been suggested because the physical properties of this series of actives and inactives were effectively identical, and the amino acid portion was clearly essential for enzyme recognition.

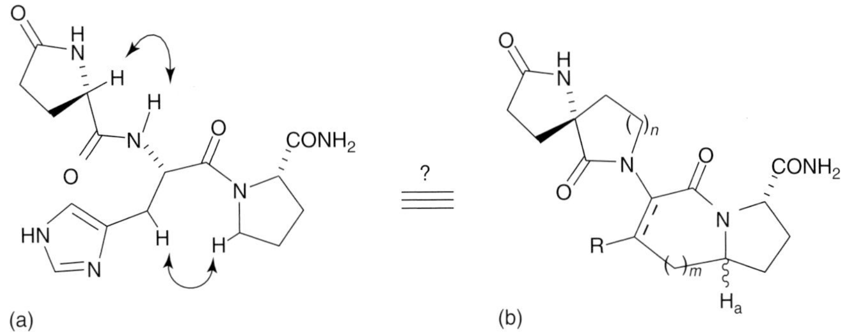

Figure 3 Analysis of TRH (a) analogs by the Active-Analog Approach by Font and Marshall led to a proposal for the receptor-bound conformation compatible with internal cyclization to generate polycyclic analogs (b).

Two other examples of receptor mapping from analysis of SAR data were published on the glucose sensor[20] and on the γ-amino-butyric acid (GABA) receptor.[21]

The thesis work of Font on the tripeptide thyrotropin-releasing hormone (TRH), pyroglutamyl-histidyl-prolineamide, is the example of the earliest determination of the receptor-bound conformation of a biologically active peptide using the Active-Analog Approach (**Figure 3**). Only six torsional angles needed to be specified to determine the backbone conformation and relative position of the imidazole ring of the bioactive conformation. Two alternative conformers were consistent with the conformational constraints required by the set of constrained analogs analyzed. Font designed several polycyclic analogs to test his hypothesis that were intractable for the synthetic procedures available at the time. These compounds served as a catalyst for the design of some novel electrochemical approaches by Moeller of Washington University.[22–25] Once the compounds could be prepared, their activity fully supported the receptor-bound conformation derived a decade before.[23,25]

4.02.3.3 Active-Site Modeling

Crystal structures of protein–ligand complexes (the set of complexes of thermolysin with a variety of inhibitors determined in the Matthews laboratory, for example[26]) clearly show a major limitation of the pharmacophore assumption. In complexes, ligands do not optimize overlap of similar chemical functionality, but find a way to maintain correct hydrogen-bonding geometry, for example, while accommodating other molecular interactions (**Figure 4**).

In 1985, ACE was an object of intense interest in the pharmaceutical industry, as captopril and enalapril, the first two approved drugs inhibiting ACE, were being extensively used to treat hypertension. Thus, each pharmaceutical company was contending to design novel chemical structures that inhibited ACE and minimized side effects to gain a piece of the market. Inhibitors of ACE served as a test bed for the Active-Site Mapping where one tries to deduce the receptor-bound conformation of a series of active analogs based on the assumption of a common binding site. Analysis of the minimum energy conformations of eight ACE inhibitors had revealed a common low-energy conformation of the Ala-Pro segment.[297] By including additional geometrical parameters, the carboxyl group of enalopril could include the zinc atom with optimal geometry from crystal structures of zinc–carboxyl complexes. Similarly, the sulfhydryl group of captopril could be expanded to include the zinc site as well with additional parameters to allow for appropriate geometrical variation. It is much more reasonable to assume that the groups involved in chemical catalysis and substrate recognition in the enzyme have a relatively stable geometrical relationship, in contrast to chemical groups in a set of diverse ligands. Mayer *et al.*[27] analyzed 28 ACE inhibitors of diverse chemical structure available by 1987, as well as two inactive compounds with appropriate chemical functionality. Based on these data, a unique conformation for the core portion of each molecule interacting with a hypothetical ACE active site was deduced; the two inactive compounds were geometrically incapable of appropriate interaction.

After nearly two decades of attempts, the crystal structure of the complex of lisinopril with ACE was finally determined.[28] The common backbone conformation of ACE inhibitors and the location of the zinc atom, hydrogen-bond donor, and cationic site of the enzyme determined by Mayer *et al.*[27] essentially overlaps that seen in the crystal structure of the complex (**Figure 5**), arguing that, at least for this case, the assumptions regarding the relative geometric stability of groups important in catalysis or recognition are valid.[29]

Figure 4 Schematics of (a) pharmacophore modeling with assumed ligand groups A=A′, B=B′, and C=C′; and of (b) active-site modeling with receptor groups X=X′, Y=Y′, and Z=Z′.

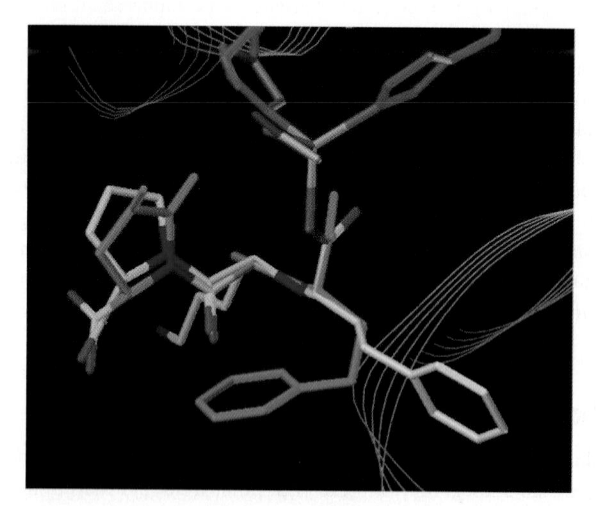

Figure 5 Overlap[29] of crystal structure of complex of the inhibitor lisinopril with ACE and the predicted enzyme-bound conformation of ACE inhibitors by Mayer *et al.*[27] Note overlap between positions of pharmacophoric groups interacting with zinc (orange), C-terminal carboxyl, and carbonyl oxygen of amide, the groups targeted by active-site modeling. The phenyl group common to enalapril analogs such as lisinopril (white ring) was not constrained (green ring) by analogs available at the time of the analysis in 1987. Reproduced with permission from Kuster, D. J.; Marshall, G. R. *J. Comput.-Aided Mol. Des.* **2005**, *19*, 609–615.

4.02.3.4 Statistical Modeling

Because QSAR provides information that relates biological activity to molecular properties, a logical extension to QSAR for drug design is to include three-dimensional data in the correlation as well. The combination of QSAR with three-dimensional structural information is known as 3D-QSAR. The success of comparative molecular field analysis (CoMFA), a type of 3D-QSAR by Tripos,[30] in generating predictive models was entirely due to a new statistical approach, partial least squares of latent variables (PLS),[31] applied to chemistry by Wold of the University of Umeå, Sweden. CoMFA probes various interactions at many different points around the molecule. Different types of probes are used to create a set of values in multiple dimensions, and principal component analysis is preformed on the values to provide a correlation of three-dimensional structure and biological activity. The concept that one could extract useful correlations from situations where more variables than observations were present was revolutionary at the time. Traditional linear regression analysis protects the user from chance correlations when too many variables are used. PLS regression recognizes and corrects for cross-correlation between variables and avoids chance correlations in models by systematically determining the sensitivity of the predictability of a model to omission of training data.[32]

A seminal paper by Cramer[33] examined the principal components derived from examining the physical property data of a large set of chemicals from the *Handbook of Chemistry and Physics*. In effect, only two principal components were responsible for a significant amount of the variance of the data in the model derived. The derivation of chemical principles in terms of those two properties provides a great deal of simplification. An analogous example is the simplification that arises from using internal coordinates, i.e., distances between atoms rather than coordinates, in structural comparisons, enabling the specific global orientation of each molecule to be eliminated from consideration.

4.02.4 Modeling and Docking onto Structural Targets

4.02.4.1 Known versus Unknown Targets

Protein targets are extremely important for current CADD methodologies. Intellectually, the application of molecular modeling of protein targets has dichotomized into two methods. The first deals with biological systems where no atomic-level structural information is known. The second involves systems that have become relatively common, where a three-dimensional structure is known from crystallography or nuclear magnetic resonance (NMR) spectroscopy. The Washington University group has spent most of its efforts over the last three decades focused on the common problem encountered where one has little structural information on the therapeutic target. Others, such as Goodford and Kuntz, have taken the lead in developing approaches to therapeutic targets where the structure of the target was available at atomic resolution through x-ray crystallography. The seminal work of Goodford and colleagues[34] on designing inhibitors of the 2,3-diphosphorylglycerate (DPG) binding site on hemoglobin for the treatment of sickle-cell disease stimulated many others to obtain crystal structures of their therapeutic targets.

Advances in molecular biology have provided the means of cloning and expressing proteins in sufficient quantities to screen a variety of conditions for crystallization. In addition, automated crystallization techniques and the protein structure initiative have led to a large increase in crystal structures. Thus, it is almost expected by management, or study section, that a crystal structure is available for any therapeutic target of interest. Unfortunately, many therapeutic targets such as G protein-coupled receptors (GPCRs) are still significant challenges to structural biology, with only the dark-adapted rhodopsin structure available.[35] Further, very large proteins and many protein–protein complexes still remain elusive.

4.02.4.2 Protein Structure Prediction

Despite the Levinthal paradox,[36,37] which suggests the futility of attempting to predict the structure of a small protein based solely on sequence, efforts to predict the structure of proteins from sequence continue, with increasing evidence that predictions are becoming more reliable.[38,39] The realization that all combinations of dihedral angles are not systematically explored by a protein in solution, rather a funnel-like potential energy landscape guides the process, was a major advance.[40,41] In general, the prediction process can be dichotomized into conformer generation and conformer scoring. Obviously, one must generate a set of candidate folds that contain the correct fold at some level of resolution to be identified and refined. Homology modeling, where one has a crystal structure of a homologous sequence, has proven a powerful approach that can generate useful models. Other approaches assemble models from homologous peptide segments.[42] Baker's group has had considerable success with this approach[38] and recently designed, expressed, and crystallized a small protein with a novel fold.[43] Galaktionov developed a novel ab initio approach to fold prediction based on constraints from the contact matrix of predicted folds (**Figure 6**). The approach restricted possible folds to those with densities similar to those seen in experimental structures.[44,45] By eliminating extended or overly compact folds, the fold space could be efficiently explored to generate sets of C_α atoms for further consideration. To generate a low-resolution structural model, a polyalanine chain was threaded through the C_α atoms, and the orientation of the peptide planes was optimized for hydrogen bonding. A scoring function that utilized amino acid side-chain information from the sequence was needed to evaluate the polyalanine traces and determine folds worthy of full atomic representation and refinement.

A low-resolution scoring function, ProVal,[46] developed with PLS uses a multipole representation of side chains centered on the C_α and C_β atoms. ProVal can distinguish the correct structure from plausible decoy folds in a large percentage of the 28 test cases studied (**Figure 7a**). For 18 of the protein sets (~64%), the crystal structure scored best. In 24 sets (~86% of the cases), including the previous 18, the crystal structure ranked in the top 5%, and the crystal structure was ranked in the top 10% in all 28 cases. A second objective was used to obtain a favorable correlation between root mean square values for the C_α atoms (CRMS value) of decoys, the experimental structure, and the calculated score that was obtained for many of the test sets (**Figure 7b**). In effect, ProVal can eliminate approximately 90% of 'good' fold predictions from further consideration without specifying the coordinate position of

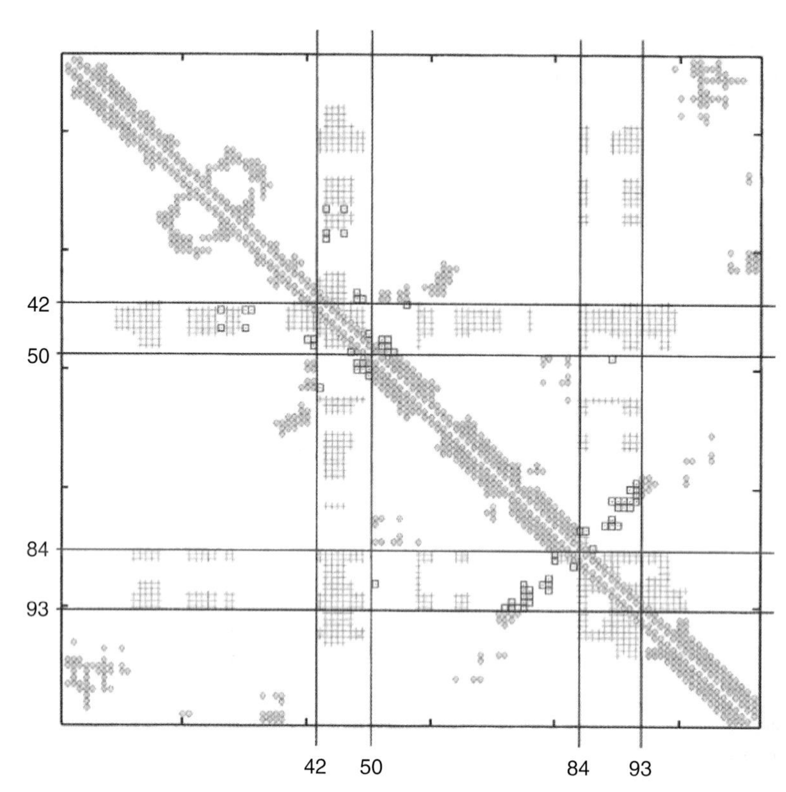

Figure 6 Residue–residue contact matrix for predicted three-dimensional structure of 3c2c (blue and green lines). The constant part, *Ac*, is shown in red, the 'noncontact' matrix *An*, is shown in green, and predicted variable contacts *Ax*, are shown in blue. Numbers correspond to the predicted loops. Reproduced with permission from Galaktinov, S.; Nikiforovich, G. V.; Marshall, G. R. *Biopolymers* **2001**, *60*, 153–168.

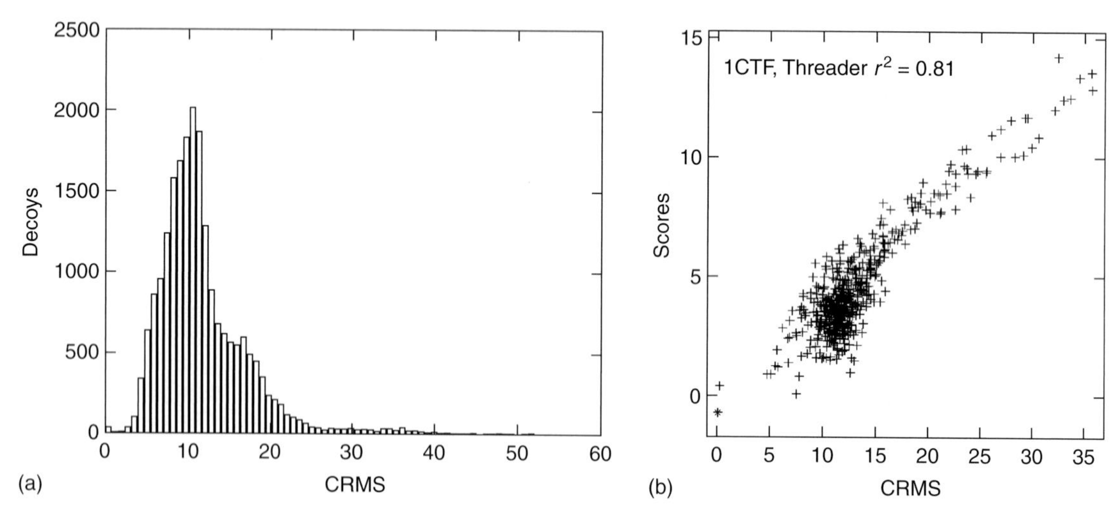

Figure 7 (a) Distribution of C_α RMSD (CRMS) of 28 sets of decoy folds from crystal structures; (b) ProVal score versus CRMS for one decoy set generated by Threader 2.5[47] for 1CTF (Brookhaven PDB code).[46] Reproduced with permission from Berglund, A.; Head, R. D.; Welsh, E.; Marshall, G. R. *Proteins* **2004**, *54*, 289–302.

side chains past the C_β atom. The details of atomic resolution are avoided because of the precision required to pack side chains efficiently without van der Waals overlap. As the quality of fold predictions increases, however, the ability to discriminate between alternative folds becomes more difficult and requires a high-resolution force field including multipole electrostatics and polarization. A recent report from the Baker group suggests high-resolution structures (<1.5 Å RMSD) can be achieved for small protein domains, with conformational sampling remaining the primary bottleneck.[39] This degree of accuracy may be sufficient to refine the model to obtain the precision needed for CADD.

4.02.4.3 Molecular Mechanics and Molecular Simulations

Whether we are discussing intermolecular events such as binding or aggregation, or intramolecular events such as folding or conformational changes, the change in free energy of the system between the two states dominates what is observed:

$$\Delta G = \Delta H - T\Delta S \quad [1]$$

where ΔG is the change in free energy folding or binding, ΔH the change in enthalpy, ΔS the change in entropy, and T the temperature.

With crystal structures of protein complexes becoming increasingly available, attempts to understand interactions that stabilize such complexes and those that determine specificity have been possible by characterizing protein–protein interfaces.[48–53] Force fields, which characterize atomic interactions by classical physics, provide a basis for molecular modeling and simulations.[54] Improvements in this approach, through use of multipole electrostatics and polarizability, have finally provided a tool that can calculate thermodynamic properties of systems to within experimental error.[55,56] Force field potentials have been developed based on statistical arguments and an inverse Boltzmann approximation to derive energies of atomic interactions.[57,58] Protein complex stability and specificity are often characterized by the contact surface area in protein–protein complexes that generally varies from 500 to 5000 $Å^2$, with some even larger.[52,59] Initially, recognition sites on proteins were often claimed to be flat and indistinguishable in shape from other parts of the protein surface.[50,60] As a significant part of the interface of complexes appears to be hydrophobic, it is not surprising that hydrophobic amino acid residues are enriched at the interface in hydrophobic patches of 200–400 $Å^2$.[50,51,61] The relationship between buried surface area and affinity has been reviewed.[62] Thus, it is now feasible to calculate the affinity of one protein for a ligand or another protein with confidence, assuming adequate degrees of freedom within the interacting molecules are explored. Combinations of molecular mechanics with quantum mechanics now allow exploration of chemical reactions in which bond lengths are variables.[63]

4.02.4.4 Docking/Prediction of Affinity

To prioritize synthesis and testing of a compound, an accurate estimate of the binding affinity is of practical utility. As determination of crystal structures has become more commonplace, the efforts to design ligands to compliment a cavity on a molecular surface have become more sophisticated. The pioneering efforts[34] of Goodford and colleagues to design compounds that bind to the diphosphorylglycerol site on hemoglobin were dominated by chemical intuition, physical models, and very primitive modeling systems. The development of DOCK by Kuntz *et al.*[64] was a major innovation, and sons of DOCK (AutoDock,[65,66] DREAM,[67] etc.) are readily available over the web for exploring possible complex formation; these next-generation programs include solvation approximations[68] and flexible ligands.[69,70] Another major innovation was the application of distance geometry as a means of generating three-dimensional coordinates from a set of distance constraints (bond lengths, sum of van der Waals radii, experimental distance constraints from NMR, etc.).[15] Goodford has developed the use of probe atoms and chemical groups in GRID to map the binding site and identify optimal binding subsites.[71] This was a prelude to experimentally determining subsite binding with subsequent assembly of fragments either by crystallography,[72,73] NMR,[74] or computation. There are now available structure-based design tools,[75–79] and a software package, RACHEL, which has been commercialized by Tripos.

Although many limitations exist with current docking methodology, many difficulties relate to complexity in estimating changes in entropy as well as lack of multipole and polarizability terms in the force-field electrostatics employed. Interpolation based on experimental data can be effective as with CoMFA and other predictive models, and simulation techniques that slightly alter a related compound with known activity into the compound of interest often give useful predictions of affinity. Accurate de novo prediction of compound activity is, however, still elusive. Oprea and Marshall have recently reviewed this topic.[80]

The affinity prediction problem consists of both searching and scoring. Adequate sampling of docking poses with a quick, low-resolution scoring function is often used to filter out candidate poses for refinement by a higher-resolution method. This assumes that the local minima in the more refined function correspond to those found at low resolution. Ideally, sampling would be done with the most accurate scoring function available, but this is often computationally intractable due to the computational complexity of more accurate scoring functions.

Head *et al.* developed a PLS-based model VALIDATE[81] to scale the relative contributions of entropy and enthalpy to binding affinity for a variety of complexes whose crystal structures had been determined. Molecular mechanics were used to calculate several parameters most correlated with enthalpy of binding; changes in surface area, number of rotatable bonds fixed, and other parameters more related to the entropy of binding were also included in the model. Although the principal components of the model were dominated by two terms (ΔH and ΔS), several other terms had significant weight for the relative accuracy of the model derived. Doing the statistical mechanics with a next-generation force field that includes polarizability and explicit solvation has potential for yielding significant improvement. A scoring function that quickly discards compounds with low affinity, however, is still desired as witnessed by the effort expended to develop rapid scoring functions for virtual screening.[69,82–85]

The accuracy of scoring functions makes predictions and reproduction of docked flexible ligand conformations nontrivial. Sampling alternative binding modes at the same site, alternative binding sites, and alternative amino acid conformations in the active site increases the degrees of freedom and yields computational complexity. Nevertheless, the number of programs that attempt to predict the correct binding mode to molecules of known structure continues to grow. Out of 11 scoring functions tested,[86] nearly all were not able to accurately predict correct binding from a set of decoys for 100 ligand–protein complexes with a RMSD of $\leqslant 2.0$ Å. Inclusion of ligand polarizability in a quantum mechanical/molecular mechanical (QM/MM) approach generally improves the RMSD of the ligand in test complexes.[87]

4.02.4.5 Attaining Accuracy through Structural Flexibility

An important activity in the field of protein structure prediction and protein–protein docking is the community-wide experiments on critical assessment of structure prediction (CASP)[298] and critical assessment of predicted interactions (CAPRI).[299] These experiments allow a comparison of different computational methods on a set of prediction targets (experimentally determined structures unknown to the predictors). The protein–protein docking category was introduced at CASP2[88,89] and has been successfully continued in CAPRI.[90] The CAPRI paradigm solicits yet unpublished structures of cocrystallized protein–protein complexes from experimentalists (primarily, x-ray crystallographers) and distributes the separately crystallized structures of the components, when available, to the community of predictors. The CAPRI experiment is conducted on a continued basis, updated with the availability of new prediction targets. Currently, approximately 3 years from its inception, five rounds of CAPRI have been completed. CAPRI generated great interest in the protein docking and protein modeling communities in docking methodologies. In the most recent rounds of CAPRI, 24 groups submitted structure that ranged from acceptable to highly accurate.[91]

It has become obvious that static docking is feasible with the current improvement in sampling and force fields that are being introduced. In addition, it has also become obvious that docking of two rigid proteins, in general, cannot provide atomic resolution details for real protein–protein complexes. Dynamics and flexibility are inherent in protein structures, and ignoring or minimizing this intrinsic complication renders most current docking approaches problematic. For example, CheY, a protein involved in chemotaxis, complexes with a variety of modulatory proteins in bacterial two-component systems. The conserved N-terminal CheY domain receives the phosphoryl group and activates target proteins,[92] whereas the C-terminal α_4-β_5-α_5 surface is the CheY interface for protein–protein interactions. CheA/CheY is the best biochemically and structurally characterized two-component system. Convergent binding sites on CheY, with three other proteins, CheA, FliM, and CheZ, have been structurally determined.[93–96] CheY surface loops interact with its protein partners and provide the main difference in the various structures (**Figure 8**). The concept of induced fit is applicable in this case.

Numerous other examples of dramatic changes in conformation upon complex formation can be cited, such as the change in β-hairpin flap position (**Figure 9a**) in human immunodeficiency virus (HIV) protease on inhibitor binding[97] and the dramatic helix distortion (**Figure 9b** and **9c**) of the calmodulin dumbbell[98] on the binding of helical peptides to calmodulin.[99] Until docking algorithms accommodate flexibility in both partners, there is limited chance that the predicted complex will reflect reality at the atomic level, even if the binding surfaces at the interface are predicted more or less correctly.

4.02.5 Scaffold Development

Proteins provide structural integrity and molecular recognition, as well as catalytic processing of most chemical conversions within biological systems. It has become clear with the advent of genomics and proteomics how complex protein interactions are within living systems. The number of expressed proteins in humans has recently been

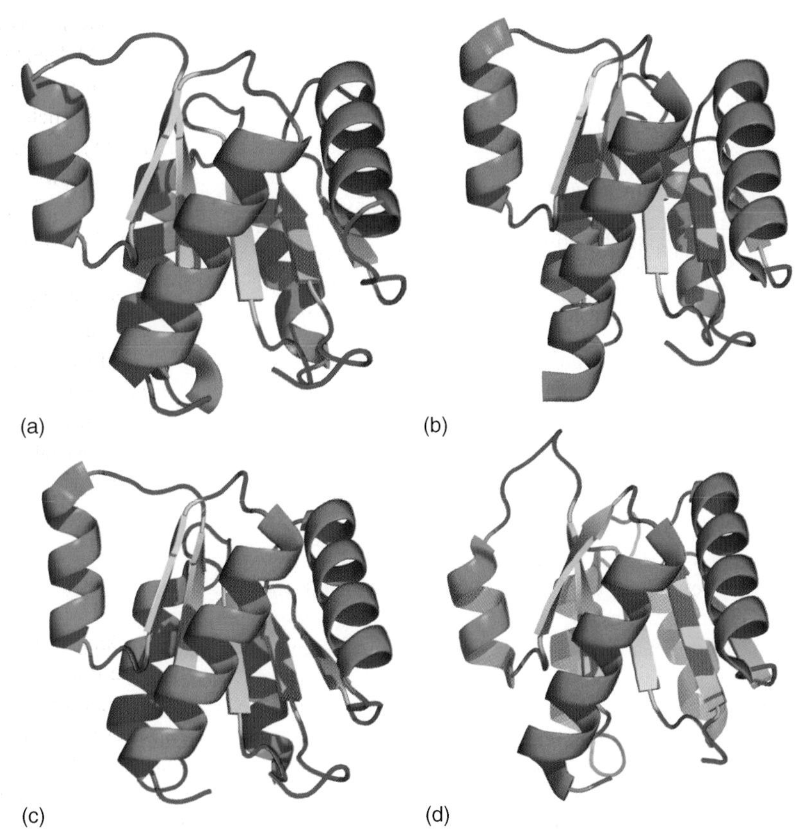

Figure 8 Crystal structures of CheY from four complexes with different proteins to show variation in loop conformation (green ribbon, top left): (a) 1A0O, (b) 1F4V, (c) 1KMI, (d) 1CHN. (Rendered using PyMol.)

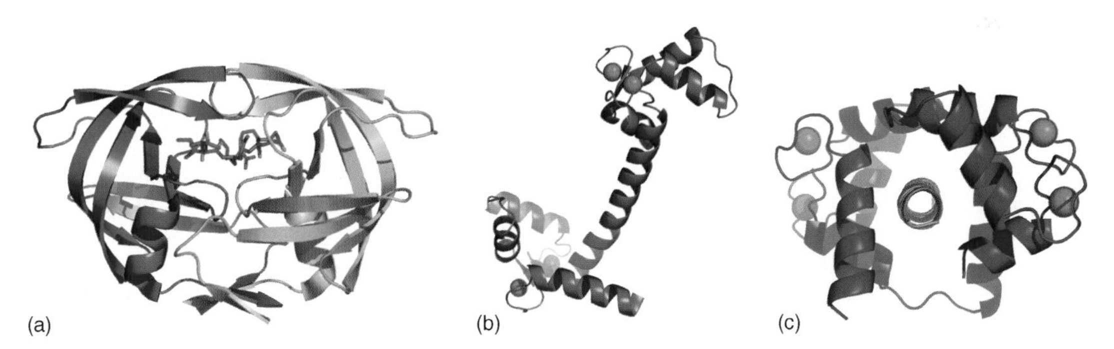

Figure 9 (a) Ribbon diagram of HIV protease with inhibitor (MVT-101, sticks; 4HVP, PDB) bound. The two β-hairpin flaps (on top) have closed down to complete the binding site for the inhibitor, MVT-101; this change is seen with essentially every inhibitor bound to HIV protease. (b) Crystal structure of calmodulin (ribbon diagram, 3CLN, PDB). Note the long blue helix connecting two calcium-binding domains. (c) Calmodulin–peptide complex; notice the dramatic change in the structure of the major helix to allow calcium-binding sites to enfold helical peptide. Calcium ions are shown in red. (Rendered using PyMol.)

estimated to be approximately 25 000.[100] In *Escherichia coli*, fruit fly, and yeast,[101,102] the number of expressed proteins is less, around 6000–15 000, but each protein, on average, has been estimated to interact with up to six other proteins.[103] In addition, protein expression and protein–protein interactions were dynamic when changes were monitored during the cell cycle in yeast.[104] Despite their dynamic nature, proteins and protein–protein interactions have assisted in targeting particular residues for drug design and have provided scaffolds, or the basis for scaffolds, on which drugs can be designed.

4.02.5.1 Recognition 'Hot Spots'

Side-chain recognition is dominant in peptide–receptor complexes. For instance, aromatic residues have a rigid arrangement and large surface area, causing a great deal of potential free energy and a very low entropic cost that results from binding.[105] Some side chains within an interface play a more significant role than others in the energetics of binding and determination of the relative orientation of the two proteins. In the human growth hormone/receptor complex, eight of the 31 side chains involved in the interface accounted for approximately 85% of the binding energy, providing the genesis of the 'hot spot' theory[106] as a basis for inhibitor design and drug discovery. The recognition of the hormone somatostatin by its GPCRs further emphasizes the importance of side chains in peptide recognition, as many of the amide bonds can be reduced,[107] the direction of the peptide backbone can be reversed, and even the whole peptide backbone can be replaced by a saccharide with recognition retained at the receptor.[108,109]

Charged groups, particularly the planar guanidinium of Arg and the carboxyl groups of Glu and Asp, are also often essential recognition 'hot spots' in peptide messages, providing a geometrical array of hydrogen-bond donors and/or acceptors. Asn and Gln also have planar groups with distinctive hydrogen-bond geometries. These observations are also valid within the interfaces of protein–protein interactions,[53] as formation of salt bridges across the intermolecular interface is highly favorable.[110,111] The studies support those of Marshall *et al.* derived from SAR studies on peptide ligands; tryptophan is most highly enriched at almost fourfold at protein interfaces (Trp having the largest planar surface), followed by Arg, Tyr, Ile, Asp, and His, respectively, with 50% enrichment for His.[53]

Experimentally, information on 'hot spots' has been obtained by systematically mutating side chains within the interface to alanine and determining the changes in binding affinity. Bogan and Thorn[112] collected a database of 2325 alanine mutants for which the change in free energy of binding upon mutation to alanine had been measured.[300] Analysis of the database by Bogan and Thorn[112] generated several observations; amino acid side chains in hot spots are located near the center of protein–protein interfaces, are generally solvent-inaccessible, and are self-complementary across protein–protein interfaces, i.e., they align and pack against one another. Out of 31 contact residues involved in the interaction of growth hormone with its receptor, for example, two tryptophan residues of the receptor accounted for over 75% of the free energy of binding, as determined by mutation to alanine.[113] Polar residues have also been localized at hot spots in protein–protein recognition.[53]

4.02.5.2 Protein Engineering

Protein engineering involves designing novel three-dimensional protein scaffolds onto which amino acid side chains can be introduced to obtain functionality for molecular recognition and catalysis. Applications could include therapeutics, biomaterials, biosensors, industrial catalysts, nano materials, etc.[114–118] Despite the exciting applications, protein engineering is difficult, due to the inability to predict the three-dimensional protein structure from amino acid sequence alone. Within the conformational space a protein fold can adopt, there are often small energy differences between several different low-energy protein folds. Many low-energy alternative folds can bury hydrophobic surface in a compact structure and satisfy most internal hydrogen-bonding groups. An accurate scoring potential and adequate sampling of configurational space for each candidate fold is needed to approximate the entropy of the hydrated system and to help distinguish between alternative folds. When there are no constraints on a protein's function and interactions within a given fold are maximized, particular folds have been stabilized by optimization of the amino acid sequences.[119] Nevertheless, optimization utilizing a particular scaffold does not guarantee the sequence will fold to the desired state. In fact, an alternative fold, perhaps a novel fold, may be stabilized to a greater extent than the original fold, causing the alternative fold to form.

Many examples of protein engineering utilize prefolded subunits as scaffolds, simply transferring and/or evolving binding sites or scrambling protein domains to generate new functionality.[120–127] This cut-and-paste approach assumes that the scaffolds and domains are sufficiently stable and that they will retain the same three-dimensional structures, despite any perturbation in amino acid sequences involved. The Hellinga group has focused on bacterial periplasmic binding proteins as functional scaffolds for grafting desired binding sites[128] with considerable success. Metal-binding sites,[129] such as those for zinc,[130] have been appended[131,132] and eliminated[133] from proteins. Others have used covalent modification of existing proteins to generate enzyme activity,[134,135] or modified existing enzymes to cause a predicted change in function.[136,137]

Several recent examples of successful cut-and-paste applications are given below to illustrate the power of this approach. These, and many other examples from the literature, illustrate the creative way in which novel proteins with designed properties can be generated by utilization of pre-existing protein building blocks.

a. *Conversion of protein scaffold to enzyme.* Dwyer *et al.*[138] designed a novel triose phosphate isomerase active site within the bacterial ribose-binding protein. By knowing the geometry of amino acids residues required for catalytic activity in triose phosphate isomerases, an appropriate geometry was found in the ribose-binding protein that could accommodate the active site. Kaplan and DeGrado[139] have reported phenol oxidase activity in a de novo designed four-helix bundle containing two iron-binding sites (**Figure 10a** and **10b**).

b. *Change of specificity of bacterial receptors as biosensors.* Looger *et al.*[140] attached a fluorophore to the ribose-binding protein to signal ribose binding. The binding site was then mutated to selectively recognize TNT and other small ligands instead of similar ligand analogs shown in **Figure 10c**. Similar results were reported for the nerve gas agent soman.[141]

c. *Ligand activation of protein splicing.* Some bacteria have the capability of splicing two protein segments together through excision of a small intervening cysteine protease or intein.[142] By fusing a domain of the receptor responsible for thyroid hormone recognition to an intein, Skretas and Wood[143] generated an expression system that allows an inactive protein to be expressed within a cell, and then spliced together to generate an activate protein by the addition of thyroid hormone.

d. *Flexible loops in zinc fingers.* Another example is the use of the zinc finger domain as a scaffold for protein mimicry. These small proteins associated with DNA regulation also function in protein recognition.[144] Sharpe *et al.*[145] demonstrated that approximately 70% of the residues in a small zinc finger could be mutated to alanine without disruption of the fold. A binding site for the co-repressor CtBP2 was grafted onto a flexible loop of a plant homeodomain motif, Mi2β, and yielded the expected change of function.[146]

e. *Surface residues of cystine knot proteins.* Venoms from snakes and invertebrates often contain small proteins (cystine knots) with multiple disulfide bridges.[147,148] These can be used as scaffolds by grafting critical side chains from one partner in a protein–protein interaction to the appropriate residues on the cystine knot to mimic the recognition surface and inhibit the protein–protein interaction.[149]

Considerable effort has been expended to design proteins from first principles to test the functional level of our understanding of protein motifs, stability, and protein–protein interactions.[150,151] One of the first serious efforts was that of the Richardsons[152] to design a novel β-barrel protein, betabellin. They enlisted the synthetic expertise of Erickson to prepare the designed protein, and iterative cycles of design, synthesis, and disappointment ensued until the 15th iteration in design[153] gave the desired structure. A major difficulty with designing proteins with β-sheet proteins is their tendency to aggregate.[154] Examples of several other successful miniproteins that have been engineered

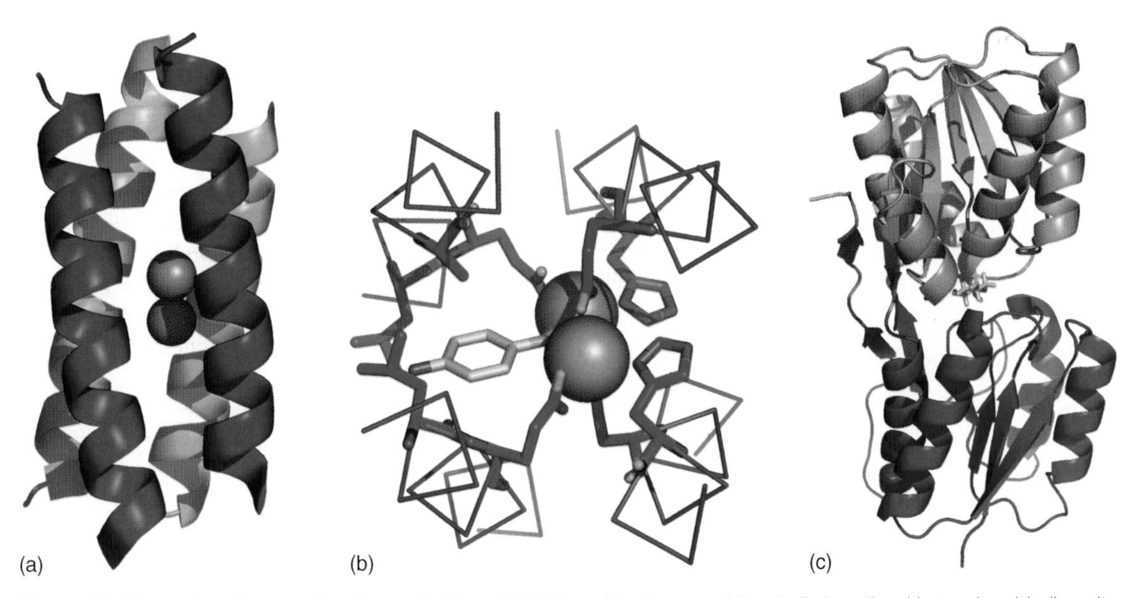

Figure 10 Examples of cut-and-paste applications. (a) Schematic diagram of four-helix bundle with two iron-binding sites (adapted from 1JMO). (b) Model of active site with substrate interaction with diiron catalytic site (adapted from 1JMO). (c) Conversion of ribose-binding protein (C) into a biosensor for TNT that can discriminate TNB, 2, 4-DNT, and 2,6-DNT; a biosensor that can discriminate L-lactate from D-lactate or pyruvate; and a biosensor that can discriminate serotonin from tryptamine and tryptophan (Rendered using Pymol.).

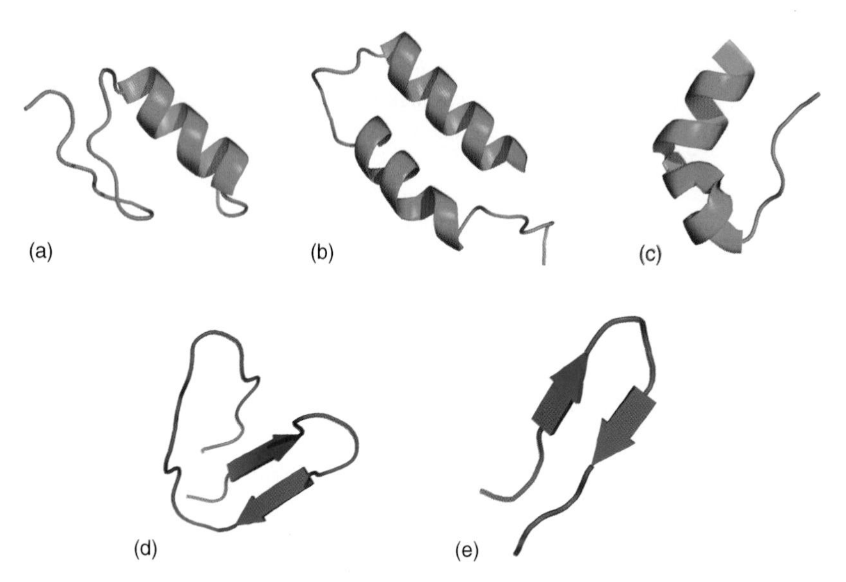

Figure 11 Examples of miniproteins that have been designed (optimized) to have the folds shown. Top – predominately helical: (a) Mayo *et al.*,[158] (b) Wells *et al.*,[164] (c) Andersen *et al.*[163] Bottom – predominately sheet, (d) Imperiali *et al.*,[133,165] (e) Serrano *et al.*[166,167] (Rendered using PyMol.)

are shown in **Figure 11**. Miniproteins are discrete autonomously folding proteins less than 50 amino acids[155] and can be comprised of α-helices, helical bundles, β-sheet, β-hairpins, etc.[156–163] (These references are meant to be illustrative, and many other examples are becoming common in the literature.)

Oligomerization of designed units to explore the basis of protein–protein interactions has also been an active field. Helix bundles have been the goal of many efforts in de novo protein design.[168] Hecht and co-workers have used binary patterning of polar and nonpolar amino acids[169] to guide folding in accord with the lattice model studies of Dill and co-workers.[170] Helix bundles have also been designed by DeGrado and co-workers.[171–177] Recently, the DeGrado group has generated catalytic activity by adding a diiron catalytic center.[139] Hodges,[178–181] Kim,[182–185] and Keating[186–188] have thoroughly explored the physical basis of coiled-coil interactions (**Figure 12**). Further, DeGrado[189,190] and Kim[182,183] have successfully designed coiled-coil systems.

The Imperiali group has focused on transforming a zinc finger structure into a structure that assumes the same ββα fold without the need to bind zinc.[133] Recently, the ββα fold was subjected to selection conditions favoring oligomerization.[159] The recently solved crystal structure revealed a homotetramer with each monomer containing a ββα secondary structure.[191] Computational design on the same system by the Keating group has led to a hetero-tetramer (**Figure 13**).[192] This work provides a practical demonstration of the use of positive and negative strategies to prevent homotetramerization while stabilizing heterotetramerization. Bolon *et al.*[193] also discuss the results of the use of positive strategies to stabilize a particular fold, combined with negative strategies to destabilize alternative folds.

4.02.5.3 Reducing Entropy of the System

One significant difference between peptide–ligand binding and protein–protein interfaces, of course, is that interfaces between proteins are most often composed of amino acid sequences that are not contiguous, while the small size of peptides usually means that the peptide member has an interacting surface composed of adjacent amino acids, such as occur in a reverse turn, or on the surface of an α-helix. One additional difference is that the peptide has little, if any, intrinsic structure in solution, and the entropic cost of formation of the complex is, therefore, much greater. Developing inhibitors of protein–protein complexes where the interacting surfaces arise from discontinuities in the peptide backbone is naturally more problematic. The problems associated with inhibiting protein–protein interactions has been recently reviewed.[59,194,195]

Work on miniproteins has led us to focus on using preorganization to steer a sequence into the designed fold by manipulating entropy. Upon folding into its native state, the free energy associated with folding can be expressed as[196]

$$\Delta G_{\text{folding}} = \Delta H_{\text{chain}} - T\Delta S_{\text{chain}} + \Delta G_{\text{solvent}} \quad [2]$$

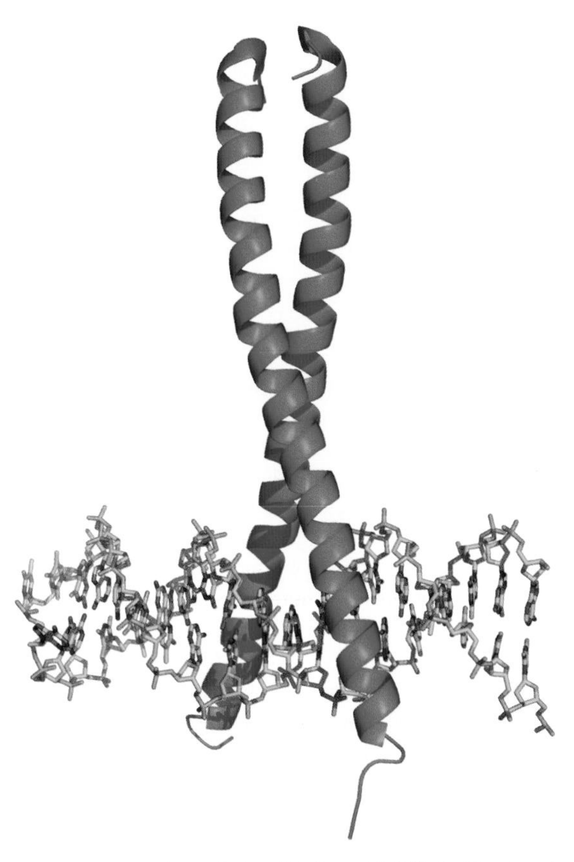

Figure 12 bZIP transcription factor containing the coiled-coil helix motif (1YSA). (Rendered using PyMol.)

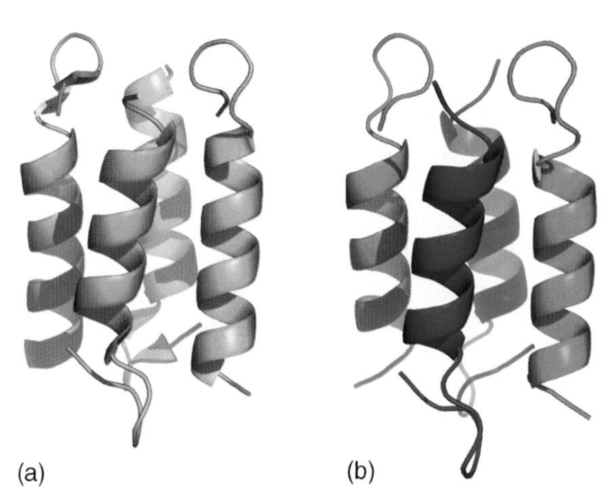

Figure 13 (a) Selected ββα homotetramer (1SN9); (b) x-ray crystal structure of the designed heterotetramer (1XOF). (Rendered using PyMol.)

where ΔH_{chain}, ΔS_{chain}, and $\Delta G_{solvent}$ represent changes in intrachain binding energy, chain entropy, and solvation free energy when the protein folds. Because the change in entropy when a protein folds is unfavorable, conformationally restricting the polypeptide toward the folded structure should reduce the entropic cost of binding and lead to a more stable structure. However, reducing the entropy of the chain was difficult because such semirigid segments of the protein chain could not be incorporated using protein expression and had to be done via chemical synthesis of the entire protein. Recent development of expressed protein ligation[197,198] offers a potential solution to this logistical bottleneck of combining unusual amino acids[197] and organic fragments with expressed protein segments.

Given the fact that side-chain recognition dominates biological interactions, one can utilize peptidomimetics in protein design to preorganize segments of the protein fold without worrying about recognition of the peptide backbone. In addition, incorporation of peptidomimetics will tend to prevent degradation of the designed protein, prolonging its half-life and increasing the probability of oral bioavailability. In essence, incorporation of peptidomimetics in the place of secondary structures will force the resulting hybrid (chimeric) protein to adopt the desired fold by eliminating other potential folds that do not have the same pattern of secondary structural elements aligned in identical locations. The repertoire of secondary structural peptidomimetics to be incorporated into designed proteins is quite comprehensive.

4.02.5.4 Developing β-Sheet and β-Turn Scaffolds

A significant motif of peptide recognition involves interactions with the peptide backbone as commonly seen in proteolytic enzymes. The linear recognition motif in proteolytic enzymes allows alignment of the peptide backbone within the active site to give precise orientation to the enzymatic functional groups responsible for amide bond hydrolysis. In protein–protein recognition, β-strand recognition usually manifests itself as β-sheet assembly in complex formation. Several groups have been prominent in the design/development of reverse-turn mimetics,[199,200] particularly heterochiral chimeric dipeptide analogs of Pro-D-pro and D-pro-Pro as β-strand nucleation sites,[201,202] α,α-dialkyl amino acids as helical nucleation sites,[203] and metal complexes of chiral azacrowns as reverse-turn mimetics.[199] Others have developed semirigid β-strand analogs[204] and, more recently, organic semirigid α-helical segments.[205–207] The basic approach provides structural preorganization by introducing nucleation sites to limit conformational flexibility and thus decrease entropy lost on binding.

As an example, HIV protease is a homodimer with an active-site aspartyl residue contributed from each monomer. A four-stranded β-sheet structurally stabilizes the dimer and is responsible for over 80% of the stabilization energy.[208] Several attempts have been made to inhibit dimer formation with β-sheet mimetics.[209–211] Work on the HIV-1 protease dimer and ribonuclease A illustrates the preorganization approach applied to β-turns (**Figure 9a**). Structural preorganization using an unusual bicyclic dipeptide reverse-turn analog (BTD) (**Figure 14a**) showed the anticipated increase in stability when it was incorporated into HIV protease via chemical synthesis.[196] In the HIV-1 protease homodimer, a BTD was introduced in both monomers, replacing the dipeptide unit Gly-Gly[16,17] that forms a reverse turn in the HIV protease/MVT-101 inhibitor crystal structure.[97] The mimetic forced a reverse turn at the same location in the unfolded state, reducing the entropic penalty for folding and resulting in an approximate 1.5 kcal mol^{-1} increase in protein stability.[196] Similar results were obtained[212] by incorporation of a single β-peptide reverse-turn mimic, *R*-nipecotic-*S*-nipecotic acid dipeptide (**Figure 14b**), at the Asn113-Pro114 type-VI reverse-turn position of RNase by expressed protein ligation. Full enzymatic activity was retained and the thermal stability was increased by 1.2 ± 0.3 °C. All attempts to replace Pro114 in RNase have resulted in less stable proteins; replacement with the unnatural 5,5′-dimethylproline (**Figure 14c**), known to stabilize the *cis*-amide bond, retained enzymatic activity and enhanced the thermal stability by 2.8 ± 0.3 °C.[213]

4.02.5.5 Developing α-Helical Scaffolds

The most common secondary structure element in molecular recognition is the α-helix. Besides the obvious significant role of α-helices play in expression regulation by binding to nucleic acids, α-helices have been shown to play a major

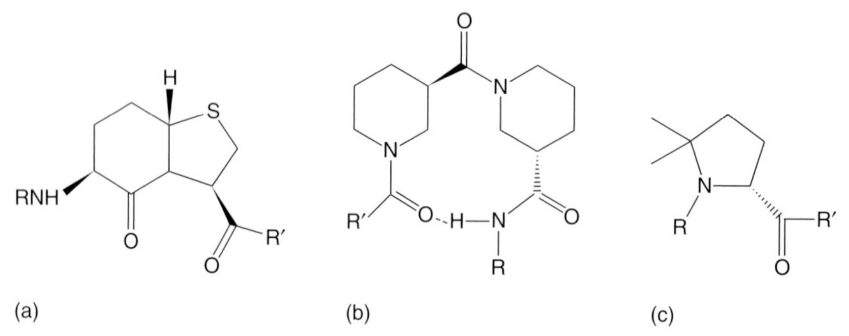

Figure 14 (a) Bicyclic reverse-turn mimetic BTD used by Baca *et al.*[196] to replace Gly-Gly[16,17] in synthetic HIV protease; (b) *R*-nipecotic-*S*-nipecotic reverse-turn mimetic; and (c) 5,5′-dimethylproline.

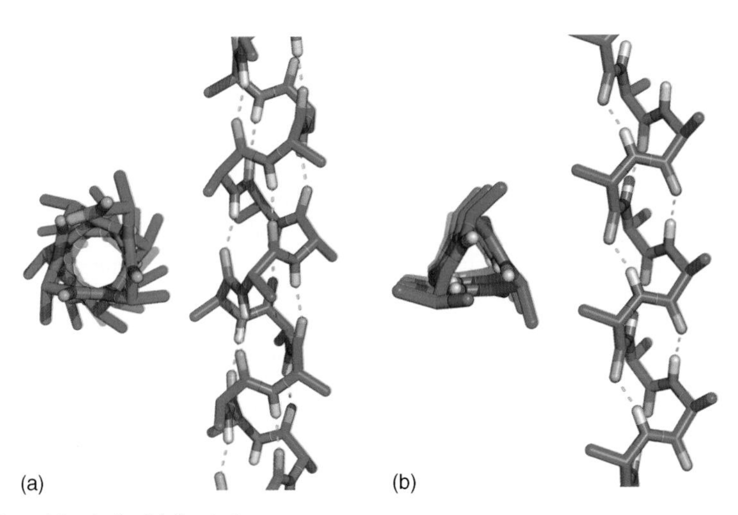

Figure 15 Ideal helices: (a) α-helix, (b) 3_{10}-helix.

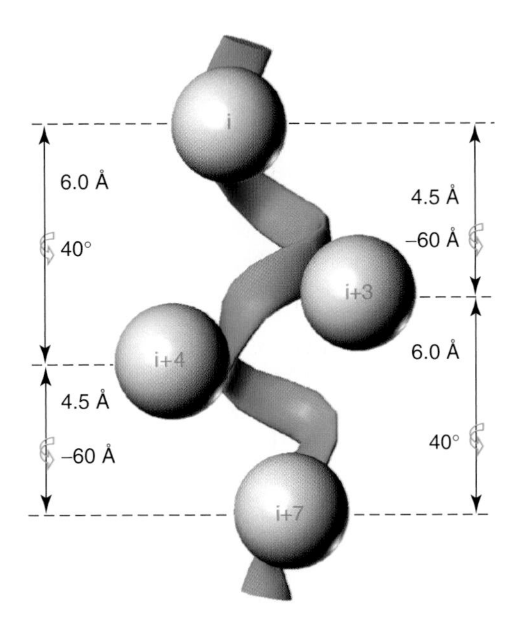

Figure 16 Idealized α-helix geometry with 4.5 or 6 Å rise between residues i, $i+3$, $i+4$, and $i+7$ of an α-helix, all located on same face of the helix.

role in protein–protein recognition. Helices are found in proteins predominately as α-helices (3.6 residues/turn) and occasionally as 3_{10}-helices (3 residues/turn) as shown in **Figure 15**. Transitions between the two in isolated helices are extremely facile; the barrier has been estimated[214] between the two conformations of the helices shown in **Figure 15** to be on the order of 3–4 kcal mol^{-1}. Helix geometry of the peptide backbone allows the single hydrogen bonds of either helical conformer to make two weaker bifurcated hydrogen bonds in the transition state. The small energy barrier suggests that small peptide helices can easily be induced to bind in either conformer or a mixture of the two helical states by their receptor. Helical peptidomimetics that can distinguish between the two helical states by fixing the position and relative orientation of side chains have great potential for drug design.

The problem of organic helix mimetics is simply to design an appropriate organic scaffold with limited conformations that can orient attached R-groups to correctly mimic the orientation seen in α- and 3_{10}-helices (**Figure 16**). There are 3.6 residues per turn of the α-helix, with a rise of 1.5 Å per residue. Since an α-helix often interacts by a common surface of the i, $i+3$, $i+4$, and $i+7$ residues, the characteristic axial rise between these residues is 4.5 or 6.0 Å, respectively. Looking down the helical axis, residues are projected at $-60°$ and $40°$ for $i \rightarrow i+3$ and $i \rightarrow i+4$ interactions, respectively. Just as reverse turns are common recognition motifs in biological systems, helix recognition is common in DNA/RNA–protein[215–217] and protein–protein interactions.

We have chosen to simplify the potential surface of a prospective peptide therapeutic by using preorganization to direct the conformation to that required for three-dimensional complex formation with the target protein of the pathogen. It is estimated that elimination of a single rotational degree of freedom by preorganization stabilizes the receptor-bound conformation and should enhance affinity by approximately 1.2–1.6 kcal mol^{-1}, assuming complete loss of rotational freedom.[218] Thus, preorganization of a seven-residue helical segment with multiple (12–14) rotational degrees of freedom in the peptide backbone, for example, fixed into its bound conformation should enhance the binding affinity by several orders of magnitude. For this to be a viable option, we must demonstrate and develop appropriate methodology for the routine synthesis of helical peptidomimetics. Jacoby suggested 2,6,3′,5′-substituted biphenyls as better than allenes, alkylidene, cycloalkanes, and spiranes as α-helix mimetics of side-chain positions i, $i+1$, $i+3$, and $i+4$.[205] The Hamilton group suggested the terphenyl scaffold as α-helix mimetics of side-chain positions i, $i+1$, $i+3$, and $i+4$.[207] Kelso *et al.* have used a motif of HXXXH complexed with $Pd(en)^{2+}$ to preorganize peptides into α-helices.[219] Taylor has advocated the use of side-chain lactams for a similar purpose.[220]

A good prototype for experimental study of helix recognition is ribonuclease A and/or RNase S in which the amide bond between residues 20 and 21 of ribonuclease A had been cleaved by subtilisin (**Figure 17**).[221] The cleaved enzyme remained enzymatically active and the 20-residue S-peptide can be reversibly dissociated. The energetics of recognition by the S-peptide were studied[222] using calorimetry, and a mutation of Met13 to glycine was found to eliminate half the binding energy. Some of the loss in binding energy must be due to increased entropy of the glycine mutant in the free peptide. The active site of the enzyme consists of two imidazole residues contributed by the S-protein and the S-peptide. Reconstitution of enzymatic activity implies the correct orientation of the two histidines in the binding mode.

The binding of a helix of p53 to a hydrophobic cleft on the surface of Hdm2 provides another example of current biological interest. More recently, the Schepartz group has suggested helical β-peptides as helical peptidomimetics and demonstrated their utility in the p53/hdm2 system.[223] Recently, Vassilev *et al.*[224] have described a low-molecular-weight drug candidate for the treatment of cancer that inhibits the p53/hdm2 complex by mimicking the interaction of three amino acids crucial for binding of the helix.[225] A short helical octapeptide had previously been described[226,227] as a

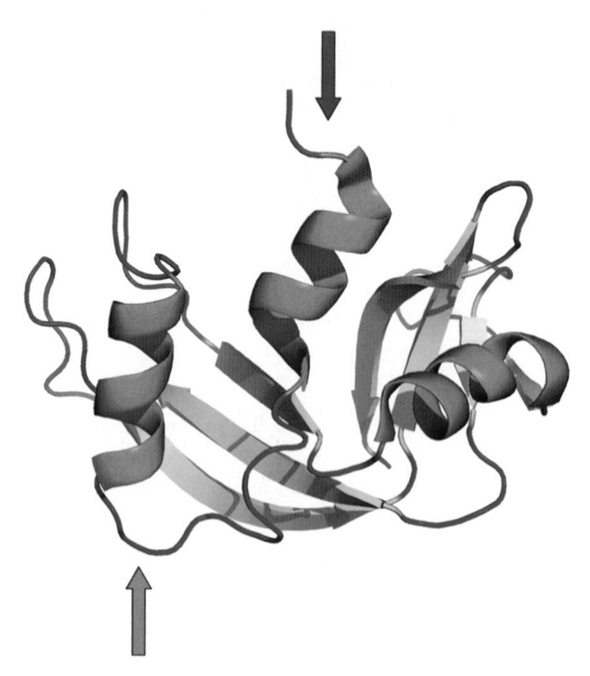

Figure 17 Ribonuclease S derived from ribonuclease A by treatment with subtilisin to cleave the amide bond between residues 20 and 21 to generate a reversible helix-binding system whose readout is enzymatic activity.[221] Structure of ribonuclease A (PDB 1KF5), with secondary structural elements is shown. Helices are red, sheets are yellow, and loop regions are green. The blue arrow indicates the beginning of the α-helix of S-peptide, at Ala4. The red arrow indicates the subtilisin cleavage site. (Rendered using PyMol.)

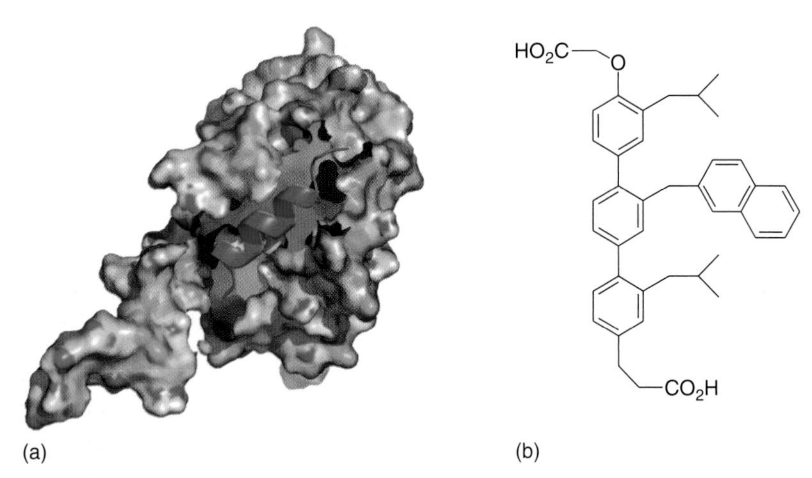

Figure 18 Helical peptidomimetic (b) based on terphenyl scaffold for inhibition of Bcl-xL complex (a) with helical peptide by the Hamilton group.[206]

nanomolar inhibitor that contained two α,α-dialkylamino acids to stabilize ions of the peptide backbone. The whole peptide backbone can be replaced by a saccharide with recognition retained at the receptor.[228] Similar studies on the TRH, Glp-His-Pro-NH_2, led to a central nervous system (CNS)-active analog with the backbone amides replaced by a cyclohexyl scaffold;[229] the analog did not cause a release of thyrotropin (TSH). Our own studies[23,25,230] to determine the bioactive conformation of TRH have led to polycyclic analogs that retain activity at the endocrine receptor responsible for TSH release, but do not show the anticipated increase in binding affinity expected if the receptor-bound conformation were mimicked exactly.

Surface mimetics of the Bak BH3 domain are attractive therapeutic targets in oncology to bind to Bcl-xL and preclude it from complexing with proapoptotic Bak. The NMR structure[231] shows the 16-residue BH3 domain peptide (random coil in solution) from Bak residues 72 to 87, $K_D \cong 300$ nM, binds in a helical conformation to a hydrophobic cleft on the surface of Bcl-xL (**Figure 18**). From alanine scans of the Bak-peptide, four hydrophobic residues (Val74, Leu78, Ile81, Ile85) along one edge of the helix appear to be critically involved in binding. Furthermore, it has been shown that the α-helix propensity of these peptides is decisive for strong binding.[232] The Hamilton group (Yale) has used this system to demonstrate the elegant concept of helical peptidomimetics and its application to therapeutic targets. The terphenyl analog with two carboxylic acids and the isobutyl;1-naphthalenemethylene;isobutyl sequence (**Figure 18b**) shows the strongest binding to Bcl-xL with a K_D value of 114 nM.[206] This work was extended with another class of helical scaffolds, the trispyridylamide (**Figure 19**). In this case, those analogs had maximal binding affinities only in the low μM.[233]

A terphenyl helical peptidomimetic has been shown[234] to mimic the most recently introduced therapeutic Fuzeon (Enfuvirtide), a 36-residue peptide targeting gp41 and preventing HIV/cell-membrane fusion. One mimetic effectively disrupts N36/C34 coiled-coil complexation with an IC_{50} value of 13.18 ± 2.54 mg mL^{-1}. Further, the effects of the antagonist on HIV-1-mediated fusion were studied using a dye-transfer, cell-fusion assay; the compound inhibits HIV-1-mediated cell-to-cell fusion with an IC_{50} value of 15.70 ± 1.30 mg mL^{-1}.

Templates that mimic the surface of 3_{10}-helices have virtually been ignored and represent a new direction for research to explore the 3_{10}-helix as a recognition motif. In 1972, Marshall and Bosshard[235] computationally predicted that the achiral α,α-dialkyl amino acid, α-methylalanine, or aminoisobutyric acid (Aib) would severely restrict the Φ,Ψ torsion angles of that residue to those associated with right- or left-handed α- and 3_{10}-helices. Subsequent experimental validation of that prediction is abundant, from our laboratory[236–238] as well as others.[239–243] For instance, dialkyl amino acids were introduced into p53 analogs to induce a helical peptide in water that enhanced binding to hdm2 (**Figure 20**).[226]

4.02.5.6 Combinatorial Chemistry and High-Throughput Screening

Combinatorial chemistry[244] has taught us that many different chemical structures can interact with a given GPCR and yield therapeutic candidates with nanomolar affinities upon optimization. Frequently, nonpeptide lead compounds have been isolated using high-throughput screening (HTS) against peptide receptors.[245] The concept of privileged organic

Figure 19 Trispyridylamides as potential helical peptidomimetics.[233]

Figure 20 Enhanced binding activity of p53(16-27) analogs to hdm2 resulted when two Aib-like residues were included to induce helicity. NMR conformation of the most potent analog in water. Reprinted with permission from Garcia-Echeverria, C.; Chene, P.; Blommers, M. J.; Furet, P. *J. Med. Chem.* **2000**, *43*, 3205–3208. Copyright (2000) American Chemical Society.)

scaffolds emerged from these ideas. The privileged organic scaffold concept[246,247] utilizes a chemical scaffold proven successful in one receptor system, such as a GPCR, to generate a library which can then be successfully screened against other GPCRs. Combinatorial chemists often use this concept to develop leads that interact with GPCRs, integrins, or other receptors.[248] Certainly, the wide variety of organic scaffolds that has resulted from screening against particular receptors has provided multiple opportunities for lead optimization and circumvention of patent protection.

4.02.5.7 Privileged Organic Scaffolds

Privileged organic scaffolds[249] include, for instance, the benzodiazepine scaffolds (**Figure 21**) that have been utilized by Evans *et al.*[246] Benzodiazepine scaffolds are thought to mimic a reverse turn,[250] and their analogs continue to generate leads against multiple peptide receptors.[251–254] Further, Haskell-Luevano *et al.*[255] screened a library of 951

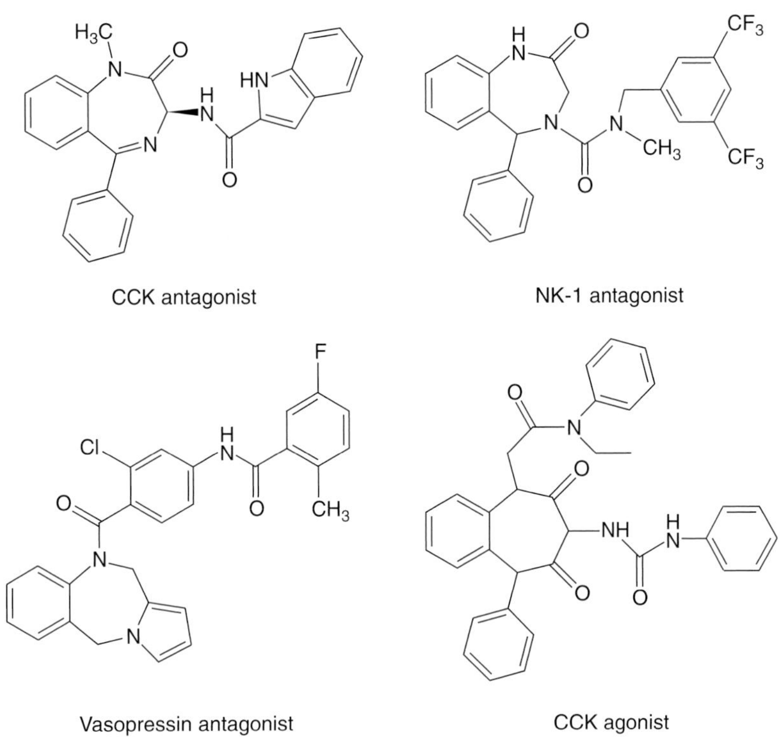

CCK antagonist

NK-1 antagonist

Vasopressin antagonist

CCK agonist

Figure 21 Examples of the benzodiazepine scaffolds suggested as privileged in deriving drugs that interact with GPCRs.

compounds based on the β-turn motif and identified the first two nonpeptidic heterocyclic micromolar agonists associated with the melanocortin-1 receptor. Another example is the rapid identification of selective agonists of the five somatostatin-receptor subtypes through combinatorial chemistry.[256] Rigid arrangement of atoms and the resulting large, fixed surface area of the important 'side chains' (in peptide terms) in the privileged scaffolds maximize the potential free energy of interaction, as the entropic cost of assuming a specific geometry recognizable by the receptor has already been paid in these rigid compounds.

Introduction of conformational constraints, however, is a double-edged sword. The more one rigidifies a ligand, the less likely it will reorient the functionality responsible for complementary interaction with the receptor. If the structure of the receptor–ligand complex to be mimicked is known, one can attempt to design a preorganized ligand that can present a similar binding surface to the receptor. However, many of the known privileged scaffolds, such as those for reverse-turn mimetics,[24,257,258] present synthetic difficulties in orienting the peptidomimetic side chains on the scaffold.[257,259] Cyclic peptide scaffolds, on the other hand, can serve as privileged scaffolds, since data obtained by SAR studies of the parent peptide, and efforts to determine the receptor-bound conformation, are often decisive in library design and lead optimization.

The Bartlett group, while researching α-amylase inhibitors,[260] designed cyclic peptides to mimic a turn motif from an inhibitory protein, tendamistat. Kessler has advocated cyclic heterochiral penta- and hexapeptides as conformational scaffolds for probing receptor recognition. In their system, an amino acid recognition motif, such as RGD or LDT for integrin receptors, is systematically shifted around cyclic peptide backbone structures containing different chiralities to spatially sample various conformations,[261–272] ultimately yielding compounds with nanomolar affinities and high selectivity. Solution conformational studies of the active cyclic peptides were performed to identify the conformational preferences associated with the most active scaffold. The Kessler group has used the results from their 'spatial screening' with constrained cyclic peptides to guide the development of selective nanomolar nonpeptide molecule inhibitors for αvβ6, αvβ5, and αvβ3 integrins.[272] One peptidic αvβ3 inhibitor, cRazaGDfNMe(V), was reported in phase II clinical studies and formed the basis for design of nanomolar nonpeptidic clinical candidates.[273] Porcelli *et al.*[274] utilized this cyclic peptide approach to discover a novel substance P antagonist. While the evidence is clear that such an approach works, expediency has led industry to largely abandon this approach in favor of combinatorial chemistry and HTS. Further, cyclic peptides are not widely used as conformational templates because cyclopentapeptides do not represent a conformationally rigid scaffold,[275] making translation from an active cyclopentapeptide to a nonpeptide drug problematic. Tran *et al.* have recently described[276] a novel method of topological classification of reverse turns that focuses on the orientation of the four α–β vectors of amino acids in the turn. This provides a foundation for reclassification of turns and for generating a database of potential peptidomimetics.

4.02.5.8 Genome-Wide Modeling and Docking of Models

A diverse library of highly constrained peptide mimetics could be used as parallel probes to interrogate the receptor regarding three-dimensional recognition. The diversity of unambiguous three-dimensional orientations of side chains in such a library provides measure of conformational information content present. It is our belief that development of such libraries would provide an efficient way to determine whether side-chain recognition is present in a therapeutic target.

With rapid progress in the experimental structural determination of proteins, currently, about one-half to two-third of individual protein structures can be modeled by relatively accurate template-based techniques.[277–281] This percentage is expected to grow significantly in the near future as more fold templates are determined. At the same time, new experimental and computational techniques yield genome-wide maps of protein–protein interactions with increasingly greater precision.[282–284] Combination of these two factors paves the way for future genome-wide structural modeling of protein–protein interactions. Such modeling will reveal deep insights into the complexity of life at the molecular level. It will also lead to structural characterization of drug targets and facilitate structure-based drug design. To become practical, however, a database of models of protein–protein interactions will require development of advanced high-throughput docking/modeling approaches and much more accurate methods of building genome-wide maps of protein–protein interactions that include the inherent flexibility of proteins.

4.02.5.9 High-Throughput Screens

Development of automation and in vitro high-throughput biological screens has had a dramatic impact on lead discovery. Molecular biology has provided the tools for identification and validation of therapeutic targets, cloning and expression of sufficient protein to accommodate high-throughput screening and determining the impact of elimination of the therapeutic gene by knockout mutations.

Once the ability to screen libraries developed, the pressure on medicinal chemists increased to generate large quantities of compounds for screening. Ironically, combinatorial chemistry developed utilizing the technology of solid phase organic chemistry. Solid phase chemistry[285] was developed by Merrifield of Rockefeller University as an automated method to assemble polypeptides and was later adapted for automated DNA synthesis.[286] Many of the reactions utilized in modern synthetic chemistry have been adapted to solid phase organic chemistry for the synthesis of combinatorial libraries used for HTS in the pharmaceutical industry. More rational approaches based on chemoinformatics were developed to design combinatorial libraries and select candidates for screening based on properties that have proven to be associated with successful therapeutics in the past.

4.02.5.10 Diversity and Similarity

Molecular recognition is essential in biological processes. One assumes that similar molecules are more likely to interact with a given receptor site than molecules that differ dramatically in size, shape, or electronic distribution. This has led to the desire to compare molecules computationally prior to biological testing to prioritize them and test only those molecules most likely to have the desired activity. For example, in a random screening program, a compound may be found which generates the desired biological effect. One would like to examine the company's compound library of 500 000 compounds to select 20 compounds most likely to show the desired activity. Often one wishes to transcend a congeneric series (not choose 20 analogs of the same basic structure) to eliminate untoward side effects or to gain freedom to operate, so comparisons must be done using three-dimensional representation. Alternatively, if one is using combinatorial chemistry in a lead discovery effort, then one may want to explore as diverse a set of potential ligands as possible with a given number of assays. This leads to the concept of chemical diversity space.

One relevant concern has been to prioritize the order of screening, or to decide which compound libraries to purchase for screening. One metric that has been developed is the complementary concepts of diversity and similarity. Given two compounds, how do you quantitate how divergent the two structures are? Some major problems are the choice of a relevant metric, what parameters to consider, how the parameters are scaled, etc. There is no generally relevant set of parameters to explain all observations, and one should expect that a given subset of parameters will be more relevant to one problem than to another. It should be pointed out that one is focused on properties of molecules in the absence of the receptor in contrast to the detailed focus on the complex in drug-design studies. Further, many approaches to similarity comparisons fail to consider chirality, a common discriminator at receptors.

4.02.5.11 Real-World Pharmaceutical Issues

In reality, the approaches discussed above are all focused on the relatively simple part of developing a therapeutic, namely, lead generation. The reality of drug development is that there are many ways to interact with a given active site on a macromolecule. For example, look at the diversity of the structures capable of inhibiting HIV protease or ACE. The real difficulty is in predicting absorption, distribution, metabolism, and excreation (ADME) that determine the pharmacokinetics, dosage regime, and quantity of drug required. Toxicity is very problematic to predict and can become the ultimate filter that eliminates many compounds from clinical studies and the major determinate of therapeutic ratio. A recent article by Stouch *et al.*[287] presents a thoughtful analysis of a validation effort for four such ADME/toxicity models. Oprea *et al.*[288,289] have compared drug leads with compounds in development and in the marketplace and shown that compounds increase in molecular weight and partition coefficient (log P) as they progress to the bedside. In silico approaches certainly have their place in the pharmaceutical industry as one more tool to increase the probability of success.[290]

4.02.5.12 Prediction of Absorption, Distribution, Metabolism, and Excretion

Methods for estimating the molecular properties that correlate with ADME problems have also been a very active arena for chemoinformatics. By studying the isozymes of cytochrome P450 enzymes, for example, certain molecular signatures for metabolic stability can be discerned. In a similar way, properties such as lipophilicity, pK_a, number of hydrogen-bond donors and acceptors, etc. correlate with oral bioavailability. For any drug development effort, oral bioavailability is often a requirement to compete in the marketplace with drugs already available.

4.02.5.13 Failures to Accurately Predict

Why do we still have difficulty in developing useful predictive models? Where are the sources of noise? From the point of view of molecular modeling and computational chemistry, the potential functions in common use have intrinsic significant errors in electrostatics. Estimating the entropy of binding is complex unless one is willing to sample solvent configurations sufficiently to adequately represent the partition function. Solvation models such as generalized born/surface area (GB/SA)[291] are certainly better than ignoring the significant impact that desolvation has on energetics. Multiple binding modes are not uncommon, but often too difficult to consider while modeling. The normal assumption of rigid receptor sites or, at the very least, limited dynamical exploration of the structure seen in the crystal are inherently dangerous. An excellent demonstration of the fallacy of assuming stability of receptor structure comes from work by Abraham. His group designed a compound to bind to an allosteric site on hemoglobin, but it actually displaced core hydrophobic residues to optimize its interactions.[292] Receptors, at least GPCRs, have multiple conformations and probably different modes of activation and coupling with different G proteins. The role of dimerization of GPCRs has recently been shown to be important for a variety of receptors.[293] In summary, we routinely apply Occam's razor for convenience, or as a rough approximation where we hope the results will withstand scrutiny by comparison with experimental results. The reality of biological systems is that Mother Nature never shaved with Occam's razor, and we cannot expect significant signal-to-noise from systems that have not been calibrated with the tools we are applying. If one is not using accurate ab initio methods, then one must remember the old dictum: to extrapolate is human, to interpolate is correct, but only within a relevant data set.

4.02.6 Summary and Perspective for the Future

Computer-assisted drug design is the science of determining important aspects of molecular structures related to optimal and specific interaction with a given therapeutic target. One can contrast the atomic level concerns of drug design where interaction with another molecule is of primary importance with the set of physical attributes related to ADME, for example. In the latter case, interaction with a variety of macromolecules provide a set of molecular filters which can average out specific geometrical details, allowing significant models to be developed by consideration of molecular properties alone.

Sitting at the interface between the revolution in microelectronics where central processing units (CPUs) are now a commodity, and the revolution in molecular biology and genomics that provides a plethora of therapeutic targets and interesting conundrums to consider, has been both exciting and humbling. It is quite clear that access to adequate CPU cycles is becoming less of a problem, and more accurate representation of atomic forces are essential. Nevertheless, calculations of entropy, or folding of proteins, require more accurate approximations than commonly used due to their computational complexity.

GRM's research for the past 40 years at Washington University has been funded through National Institutes of Health grants; CMT was supported by an National Institutes of Heath, Institutional National Research Service Award (5-T32-EY13360-06) from the National Eye Institute. The authors have benefited from the efforts of exceptional mentors, collaborators, and students. This chapter is an extension and update of ones previously published by GRM on chemoinformatics,[294] on protein docking,[295] and on protein and enzyme mimetics.[296]

References

1. Allen, F. H.; Motherwell, W. D. *Acta Crystallogr., Sect. B* **2002**, *58*, 407–422.
2. Allen, F. H.; Davies, J. E.; Galloy, J. J.; Johnson, O.; Kennard, O.; Macrea, C. F.; Mitchell, E. M.; Mitchell, G. F.; Smith, J. M.; Watson, D. G. *J. Chem. Inf. Comput. Sci.* **1991**, *31*, 187–204.
3. Abola, E. E.; Bernstein, F. C.; Koetzle, T. F. The Protein Data Bank. In: *The Role of Data in Scientific Progress*; Glaeser, P. S., Ed.; Elsevier: New York, 1985.
4. Fujita, T.; Iwasa, J.; Hansch, C. *J. Am. Chem. Soc.* **1964**, *86*, 5175–5180.
5. Hansch, C.; Leo, A. *Substituent Constants for Correlation Analysis in Chemistry and Biology*; Wiley and Sons: New York, 1979.
6. Kier, L. B.; Hall, L. H. *Molecular Connectivity in Chemistry and Drug Research*; Academic Press: New York, 1976.
7. Motoc, I.; Marshall, G. R.; Labanowski, J. *Z. Naturforsch., A* **1985**, *40*, 1121–1127.
8. Motoc, I.; Marshall, G. R. *Z. Naturforsch., A* **1985**, *40*, 1114–1120.
9. Motoc, I.; Marshall, G. R.; Dammkoehler, R. A.; Labanowski, J. *Z. Naturforsch., A* **1985**, *40*, 1108–1113.
10. Beckett, A. H.; Casey, A. F. *J. Pharm. Pharmacol.* **1954**, *6*, 986–999.
11. Kier, L. B.; Aldrich, H. S. *J. Theor. Biol.* **1974**, *46*, 529–541.
12. Gund, P.; Wipke, W. T.; Langridge, R. *Comput. Chem. Res. Educ. Technol.* **1974**, *3*, 5–21.
13. Gund, P. *Prog. Mol. Subcell. Biol.* **1977**, *11*, 117–143.
14. Marshall, G. R.; Barry, C. D.; Bosshard, H. E.; Dammkoehler, R. A.; Dunn, D. A. The Conformational Parameter in Drug Design: The Active Analog Approach. In *Computer-Assisted Drug Design*; Olson, E. C., Christoffersen, R. E., Eds.; American Chemical Society: Washington, DC, 1979, pp 205–226.
15. Crippen, G. M.; Havel, T. M. *Distance Geometry and Conformational Calculations*; John Wiley & Sons: Chichester, UK, 1981.
16. Van Drie, J. H.; Weininger, D.; Martin, Y. C. *J. Comput. Aided Mol. Design* **1989**, *3*, 225–251.
17. Martin, Y. C.; Bures, M. G.; Danaher, E. A.; DeLazzer, J.; Lico, I.; Pavlik, P. A. *J. Comput. Aided Mol. Design* **1993**, 7, 83–102.
18. Marshall, G. R.; Barry, C.D. Functional Representation of Molecular Volume for Computer-Aided Drug Design; Abstracts American Crystallographic Association: Honolulu, HI, 1979.
19. Sufrin, J. R.; Dunn, D. A.; Marshall, G. R. *Mol. Pharmacol.* **1981**, *19*, 307–313.
20. Weaver, D. C.; Barry, C. D.; McDaniel, M. L.; Marshall, G. R.; Lacy, P. E. *Mol. Pharmacol.* **1979**, *16*, 361–368.
21. Klunk, W. E.; Kalman, B. L.; Ferrendelli, J. A.; Covey, D. F. *Mol. Pharmacol.* **1982**, *23*, 511–518.
22. Rutledge, L. D.; Perlman, J. H.; Gershengorn, M. C.; Marshall, G. R.; Moeller, K. D. *J. Med. Chem.* **1996**, *39*, 1571–1574.
23. Simpson, J. C.; Ho, C. M. C.; Shands, E. F. B.; Gershengorn, M. C.; Marshall, G. R.; Moeller, K. D. *Bioorg. Med. Chem.* **2002**, *10*, 291–302.
24. Slomczynska, U.; Chalmers, D. K.; Cornille, F.; Smythe, M. L.; Beusen, D. D.; Moeller, K. D.; Marshall, G. R. *J. Org. Chem.* **1996**, *61*, 1198–1204.
25. Tong, Y. S.; Olczak, J.; Zabrocki, J.; Gershengorn, M. C.; Marshall, G. R.; Moeller, K. D. *Tetrahedron* **2000**, *56*, 9791–9800.
26. DePriest, S. A.; Mayer, D.; Naylor, C. B.; Marshall, G. R. *J. Am. Chem. Soc.* **1993**, *115*, 5372–5384; Holden, H. M.; Tronrud, D. E.; Monzingo, A. F.; Weaver, L. H.; Matthews, B. W. *Biochemistry* **1987**, *26*, 8542–8553.
27. Mayer, D.; Naylor, C. B.; Motoc, I.; Marshall, G. R. *J. Comput. Aided Mol. Design* **1987**, *1*, 3–16.
28. Natesh, R.; Schwager, S. L.; Sturrock, E. D.; Acharya, K. R. *Nature* **2003**, *421*, 551–554.
29. Kuster, D. J.; Marshall, G. R. J. *Comput. Aided. Mol. Design* **2005**, *19*, 609–615.
30. Cramer, R. D., III.; Patterson, D. E.; Bunce, J. D. *J. Am. Chem. Soc.* **1988**, *110*, 5959–5967.
31. Wold, S.; Abano, C.; Dunn, W. J., III.; Esbensen, K.; Hellberg, S.; Johansson, E.; Lindberg, W.; Sjostrom, M. *Analysis* **1984**, *12*, 477–485.
32. Cramer, R. D., III.; Bunce, J.; Patterson, D.; Frank, I. *Quant. Struct.-Act. Relat.* **1988**, 7, 18–25.
33. Cramer, I.; Richard, D. *J. Am. Chem. Soc.* **1980**, *10*, 1837–1849.
34. Goodford, P. J. *J. Med. Chem.* **1984**, *27*, 557–564.
35. Palczewski, K.; Takashi, K.; Tetsuya, H.; Behnke, C.; Motoshima, H.; Fox, B.; Le Trong, I.; Teller, D.; Okada, T.; Stenkamp, R. et al. *Science* **2000**, *289*, 739–745.
36. Zwanzig, R.; Szabo, A.; Bagchi, B. *Proc. Natl. Acad. Sci. USA* **1992**, *89*, 20–22.
37. Karplus, M. *Fold. Des.* **1997**, *2*, S69–S75.
38. Simons, K. T.; Bonneau, R.; Ruczinski, I.; Baker, D. *Proteins* **1999**, *37*, 171–176.
39. Bradley, P.; Misura, K. M. S.; Baker, D. *Science* **2005**, *309*, 1868–1871.
40. Shoemaker, B. A.; Portman, J. J.; Wolynes, P. G. *Proc. Natl. Acad. Sci. USA* **2000**, *97*, 8868–8873.
41. Tsai, C.-J.; Kumar, S.; Ma, B.; Nussinov, R. *Protein Sci.* **1999**, *8*, 1181–1190.
42. Bystroff, C.; Baker, D. *J. Mol. Biol.* **1998**, *281*, 565–577.
43. Kuhlman, B.; Dantas, G.; Ireton, G. C.; Varani, G.; Stoddard, B. L.; Baker, D. *Science* **2003**, *302*, 1364–1368.
44. Galaktionov, S.; Nikiforovich, G. V.; Marshall, G. R. *Biopolymers* **2001**, *60*, 153–168.
45. Galaktionov, S. G.; Marshall, G. R. In Proceeding of 27th Hawaii International Conference on System Sciences; IEEE Computer Society Press: Washington, DC, 1994.
46. Berglund, A.; Head, R. D.; Welsh, E.; Marshall, G. R. *Proteins* **2004**, *54*, 289–302.
47. Jones, D. T.; Taylor, W. R.; Thornton, J. M. *Nature* **1992**, *358*, 86–89.
48. Janin, J. *Biochimie* **1995**, 77, 497–505.

49. Hubbard, S. J.; Argos, P. *Protein Sci.* **1994**, *3*, 2194–2206.
50. Jones, S.; Thornton, J. M. *Proc. Natl. Acad. Sci. USA* **1996**, *93*, 13–20.
51. Lijnzaad, P.; Argos, P. *Proteins* **1997**, *28*, 333–343.
52. Lo Conte, L.; Chothia, C.; Janin, J. *J. Mol. Biol.* **1999**, *285*, 2177–2198.
53. Hu, Z.; Ma, B.; Wolfson, H.; Nussinov, R. *Proteins* **2000**, *39*, 331–342.
54. Ponder, J. W.; Case, D. A. *Adv. Protein Chem.* **2003**, *66*, 27–85.
55. Ren, P.; Ponder, J. W. *J. Phys. Chem. B* **2002**, *107*, 5933–5947.
56. Grossfield, A.; Ren, P.; Ponder, J. W. *J. Am. Chem. Soc.* **2003**, *125*, 15671–15682.
57. Miyazawa, S.; Jernigan, R. L. *Proteins* **1999**, *34*, 49–68.
58. Zhang, C.; Liu, S.; Zhou, H.; Zhou, Y. *Protein Sci.* **2004**, *13*, 400–411.
59. Veselovsky, A. V.; Ivanov, Y. D.; Ivanov, A. S.; Archakov, A. I.; Lewi, P.; Janssen, P. *J. Mol. Recognit.* **02002**, *15*, 405–422.
60. Ma, B.; Elkayam, T.; Wolfson, H.; Nussinov, R. *Proc. Natl. Acad. Sci. USA* **2003**, *100*, 5772–5777.
61. Tsai, C.-J.; Lin, S. L.; Wolfson, H. J.; Nussinov, R. *J. Mol. Biol.* **1996**, *260*, 604–620.
62. Houk, K. N.; Leach, A. G.; Kim, S. P.; Zhang, X. *Angew. Chem. Int. Ed. Engl.* **2003**, *42*, 4872–4897.
63. Guallar, V.; Friesner, R. A. *J. Am. Chem. Soc.* **2004**, *126*, 8501–8508.
64. Kuntz, I. D.; Blaney, J. M.; Oatley, S. J.; Langridge, R.; Ferrin, T. E. *J. Mol. Biol.* **1982**, *161*, 269–288.
65. Goodsell, D. S.; Lauble, H.; Stout, C. D.; Olson, A. *J. Proteins* **1993**, *17*, 1–10.
66. Morris, G. M.; Goodsell, D. S.; Halliday, R. S.; Huey, R.; Hart, W. E.; Belew, R. K.; Olson, A. J. *J. Comput. Chem.* **1998**, *19*, 1639–1662.
67. Makino, S.; Ewing, T. J.; Kuntz, I. D. *J. Comput. Aided Mol. Design* **1999**, *13*, 513–532.
68. Shoichet, B. K.; Leach, A. R.; Kuntz, I. D. *Proteins* **1999**, *34*, 4–16.
69. Lamb, M. L.; Burdick, K. W.; Toba, S.; Young, M. M.; Skillman, K. G.; Zou, X. Q.; Arnold, J. R.; Kuntz, I. D. *Proteins* **2001**, *42*, 296–318.
70. Abagyan, R.; Totrov, M. *Curr. Opin. Chem. Biol.* **2001**, *5*, 375–382.
71. Goodford, P. J. *J. Med. Chem.* **1985**, *28*, 849–856.
72. Ringe, D. *J. Nucl. Med.* **1995**, *36*, 28S–30S.
73. Ringe, D.; Mattos, C. *Med. Res. Rev.* **1999**, *19*, 321–331.
74. Shuker, S. B.; Hajduk, P. J.; Meadows, R. P.; Fesik, S. W. *Science* **1996**, *274*, 1531–1534.
75. Ho, C. M.; Marshall, G. R. *J. Comput. Aided Mol. Design* **1990**, *4*, 337–354.
76. Ho, C. M. W.; Marshall, G. R. *J. Comput. Aided Mol. Design* **1993**, *7*, 623–647.
77. Ho, C. M.; Marshall, G. R. *J. Comput. Aided Mol. Design* **1993**, *7*, 3–22.
78. Ho, C. M. W.; Marshall, G. R. In *Proceedings 27th Hawaii International Conference System Sciences*; IEEE Computer Society Press: Washington, DC, 1994.
79. Ho, C. M.; Marshall, G. R. *J. Comput. Aided Mol. Design* **1995**, *9*, 65–86.
80. Marshall, G. R.; Arimoto, R.; Ragno, R.; Head, R. D. *Abstracts American Chemical Society* **2000**, *219*, 056–COMP.
81. Head, R. D.; Smythe, M. L.; Oprea, T. I.; Waller, C. L.; Green, S. M.; Marshall, G. R. *J. Am. Chem. Soc.* **1996**, *118*, 3959–3969.
82. Muegge, I.; Martin, Y. C. *J. Med. Chem.* **1999**, *42*, 791–804.
83. Meng, E. C.; Shoichet, B. K.; Kuntz, I. D. *J. Comput. Chem.* **1992**, *13*, 505–524.
84. Bohm, H. J. *J. Comput. Aided Mol. Design* **1994**, *8*, 243–256.
85. Sun, Y.; Ewing, T. J. A.; Skillman, A. G.; Kuntz, I. D. *J. Comput. Aided Mol. Design* **1998**, *12*, 597–604.
86. Wang, R.; Lu, Y.; Wang, S. *J. Med. Chem.* **2003**, *46*, 2287–2303.
87. Cho, A. E.; Guallar, V.; Berne, B. J.; Friesner, R. *J. Comput. Chem.* **2005**, *26*, 915–931.
88. Dixon, J. S. *Proteins* **1997**, 198–204.
89. Vakser, I. A. *Proteins* **1997**, 226–230.
90. Janin, J.; Henrick, K.; Moult, J.; Ten Eyck, L.; Sternberg, M. J. E.; Vajda, S.; Vakser, I.; Wodak, S. J. *Proteins* **2003**, *52*, 2–9.
91. Mendez, R.; Leplae, R.; Lensink, M. F.; Wodak, S. J. *Proteins* **2005**, *60*, 150–169.
92. Volz, K. *Biochemistry* **1993**, *32*, 11741–11753.
93. Zhu, X.; Volz, K.; Matsumura, P. *J. Biol. Chem.* **1997**, *272*, 23758–23764.
94. Zhao, R.; Collins, E. J.; Bourret, R. B.; Silversmith, R. E. *Nat. Struct. Biol.* **2002**, *9*, 570–575.
95. Welch, M.; Chinardet, N.; Mourey, L.; Birck, C.; Samama, J. P. *Nat. Struct. Biol.* **1998**, *5*, 25–29.
96. Lee, S. Y.; Cho, H. S.; Pelton, J. G.; Yan, D.; Henderson, R. K.; King, D. S.; Huang, L.; Kustu, S.; Berry, E. A.; Wemmer, D. E. *Nat. Struct. Biol.* **2001**, *8*, 52–56.
97. Miller, M.; Schneider, J.; Sathyanarayana, B. K.; Toth, M. V.; Marshall, G. R.; Clawson, L.; Selk, L.; Kent, S. B.; Wlodawer, A. *Science* **1989**, *246*, 1149–1152.
98. Babu, Y. S.; Bugg, C. E.; Cook, W. J. *J. Mol. Biol.* **1988**, *204*, 191–204.
99. Meador, W. E.; Means, A. R.; Quiocho, F. A. *Science* **1992**, *257*, 1251–1255.
100. International Human Genome Sequencing Consortium. *Nature* **2004**, *431*, 931–945.
101. Stuart, J. M.; Segal, E.; Koller, D.; Kim, S. K. *Science* **2003**, *302*, 249–255.
102. Tong, A. H. Y.; Lesage, G.; Bader, G. D.; Ding, H.; Xu, H.; Xin, X.; Young, J.; Berriz, G. F.; Brost, R. L.; Chang, M. et al. *Science* **2004**, *303*, 808–813.
103. Giot, L.; Bader, J. S.; Brouwer, C.; Chaudhuri, A.; Kuang, B.; Li, Y.; Hao, Y. L.; Ooi, C. E.; Godwin, B.; Vitols, E. et al. *Science* **2003**, *302*, 1727–1736.
104. de Lichtenberg, U.; Jensen, L. J.; Brunak, S.; Bork, P. *Science* **2005**, *307*, 724–727.
105. Marshall, G. R. *Biopolymers* **2001**, *60*, 246–277.
106. Clackson, T.; Wells, J. A. *Science* **1995**, *267*, 383–386.
107. Sasaki, Y.; Murphy, W. A.; Heiman, M. L.; Lance, V. A.; Coy, D. H. *J. Med. Chem.* **1987**, *30*, 1162–1166.
108. Boer, J.; Gottschling, D.; Schuster, A.; Semmrich, M.; Holzmann, B.; Kessler, H. *J. Med. Chem.* **2001**, *44*, 2586–2592.
109. Hirschmann, R.; Hynes, J., Jr.; Cichy-Knight, M. A.; van Rijn, R. D.; Sprengeler, P. A.; Spoors, P. G.; Shakespeare, W. C.; Pietranico-Cole, S.; Barbosa, J.; Liu, J. et al. *J. Med. Chem.* **1998**, *41*, 1382–1391.
110. Drozdov-Tikhomirov, L. N.; Linde, D. M.; Poroikov, V. V.; Alexandrov, A. A.; Skurida, G. I. *J. Biomol. Struct. Dyn.* **2001**, *19*, 279–284.
111. Xu, D.; Tsai, C. J.; Nussinov, R. *Protein Eng.* **1997**, *10*, 199–1012.
112. Bogan, A. A.; Thorn, K. S. *J. Mol. Biol.* **1998**, *280*, 1–9.
113. Cunningham, B. C.; Wells, J. A. *J. Mol. Biol.* **1993**, *234*, 554–563.

114. Ghirlanda, G.; Lear, J. D.; Lombardi, A.; DeGrado, W. F. *J. Mol. Biol.* **1998**, *281*, 379–391.
115. Lombardi, A.; Summa, C. M.; Geremia, S.; Randaccio, L.; Pavone, V.; DeGrado, W. F. *Proc. Natl. Acad. Sci. USA* **2000**, *97*, 6298–6305.
116. Moffet, D. A.; Certain, L. K.; Smith, A. J.; Kessel, A. J.; Beckwith, K. A.; Hecht, M. H. *J. Am. Chem. Soc.* **2000**, *122*, 7612–7613.
117. Ryadnov, M. G.; Woolfson, D. N. *Nat. Mater.* **2003**, *2*, 329–332.
118. Petka, W. A.; Harden, J. L.; McGrath, K. P.; Wirtz, D.; Tirrell, D. A. *Science* **1998**, *281*, 389–392.
119. Shifman, J. M.; Mayo, S. L. *J. Mol. Biol.* **2002**, *323*, 417–423.
120. Chin, J. W.; Schepartz, A. *Angew. Chem. Int. Ed. Engl.* **2001**, *40*, 3806–3809.
121. Gemperli, A. C.; Rutledge, S. E.; Maranda, A.; Schepartz, A. *J. Am. Chem. Soc.* **2005**, *127*, 1596–1597.
122. Golemi-Kotra, D.; Mahaffy, R.; Footer, M. J.; Holtzman, J. H.; Pollard, T. D.; Theriot, J. A.; Schepartz, A. *J. Am. Chem. Soc.* **2004**, *126*, 4–5.
123. Montclare, J. K.; Schepartz, A. *J. Am. Chem. Soc.* **2003**, *125*, 3416–3417.
124. Rutledge, S. E.; Volkman, H. M.; Schepartz, A. *J. Am. Chem. Soc.* **2003**, *125*, 14336–14347.
125. Schneider, T. L.; Mathew, R. S.; Rice, K. P.; Tamaki, K.; Wood, J. L.; Schepartz, A. *Org. Lett.* **2005**, 7, 1695–1698.
126. Vogel, C.; Bashton, M.; Kerrison, N. D.; Chothia, C.; Teichmann, S. A. *Curr. Opin. Struct. Biol.* **2004**, *14*, 208–216.
127. Holm, L.; Sanders, C. *Proteins* **1998**, *33*, 88–96.
128. Dwyer, M. A.; Hellinga, H. W. *Curr. Opin. Struct. Biol.* **2004**, *14*, 495–504.
129. Barondeau, D. P.; Getzoff, E. D. *Curr. Opin. Struct. Biol.* **2004**, *14*, 765–774.
130. Alberts, I. L.; Nadassy, K.; Wodak, S. J. *Protein Sci.* **1998**, 7, 1700–1716.
131. Lu, Y.; Berry, S. M.; Pfister, T. D. *Chem. Rev.* **2001**, *101*, 3047–3080.
132. Maglio, O.; Nastri, F.; Pavone, V.; Lombardi, A.; DeGrado, W. F. *Proc. Natl. Acad. Sci. USA* **2003**, *100*, 3772–3777.
133. Struthers, M. D.; Cheng, R. P.; Imperiali, B. *Science* **1996**, *271*, 342–345.
134. Tann, C. M.; Qi, D.; Distefano, M. D. *Curr. Opin. Chem. Biol.* **2001**, *5*, 696–704.
135. Qi, D.; Tann, C. M.; Haring, D.; Distefano, M. D. *Chem. Rev.* **2001**, *101*, 3081–3111.
136. Penning, T. M.; Jez, J. M. *Chem. Rev.* **2001**, *101*, 3027–3046.
137. Baltzer, L.; Nilsson, J. *Curr. Opin. Biotechnol.* **2001**, *12*, 355–360.
138. Dwyer, M. A.; Looger, L. L.; Hellinga, H. W. *Science* **2004**, *304*, 1967–1971.
139. Kaplan, J.; DeGrado, W. F. *Proc. Natl. Acad. Sci. USA* **2004**, *101*, 11566–11570.
140. Looger, L. L.; Dwyer, M. A.; Smith, J. J.; Hellinga, H. W. *Nature* **2003**, *423*, 185–190.
141. Allert, M.; Rizk, S. S.; Looger, L. L.; Hellinga, H. W. *Proc. Natl. Acad. Sci. USA* **2004**, *101*, 7907–7912.
142. Paulus, H. *Annu. Rev. Biochem.* **2000**, *69*, 447–496.
143. Skretas, G.; Wood, D. W. *Protein Sci.* **2005**, *14*, 523–532.
144. Liew, C. K.; Simpson, R. J.; Kwan, A. H.; Crofts, L. A.; Loughlin, F. E.; Matthews, J. M.; Crossley, M.; Mackay, J. P. *Proc. Natl. Acad. Sci. USA* **2005**, *102*, 583–588.
145. Sharpe, B. K.; Liew, C. K.; Kwan, A. H.; Wilce, J. A.; Crossley, M.; Matthews, J. M.; Mackay, J. P. *Structure* **2005**, *13*, 257–266.
146. Kwan, A. H.; Gell, D. A.; Verger, A.; Crossley, M.; Matthews, J. M.; Mackay, J. P. *Structure* **2003**, *11*, 803–813.
147. Craik, D. J.; Daly, N. L.; Bond, T.; Waine, C. *J. Mol. Biol.* **1999**, *294*, 1327–1336.
148. Millard, E. L.; Daly, N. L.; Craik, D. J. *Eur. J. Biochem.* **2004**, *271*, 2320–2326.
149. Fairlie, D. P.; West, M. L.; Wong, A. K. *Curr. Med. Chem.* **1998**, *5*, 29–62.
150. Baltzer, L.; Nilsson, H.; Nilsson, J. *Chem. Rev.* **2001**, *101*, 3153–3163.
151. Moffet, D. A.; Hecht, M. H. *Chem. Rev.* **2001**, *101*, 3191–3203.
152. Richardson, J. S.; Richardson, D. C. *Trends Biochem. Sci.* **1989**, *14*, 304–309.
153. Lim, A.; Makhov, A. M.; Bond, J.; Inouye, H.; Connors, L. H.; Griffith, J. D.; Erickson, B. W.; Kirschner, D. A.; Costello, C. E. *J. Struct. Biol.* **2000**, *130*, 363–370.
154. Richardson, J. S.; Richardson, D. C. *Proc. Natl. Acad. Sci. USA* **2002**, *99*, 2754–2759.
155. Imperiali, B.; Ottesen, J. J. *J. Pept. Res.* **1999**, *54*, 177–184.
156. Hill, R. B.; DeGrado, W. F. *Struct. Fold Des.* **2000**, *8*, 471–479.
157. Struthers, M. D.; Cheng, R. P.; Imperiali, B. *Science* **1996**, *271*, 342–345.
158. Dahiyat, B. I.; Mayo, S. L. *Science* **1997**, *278*, 82–87.
159. Mezo, A. R.; Ottesen, J. J.; Imperiali, B. *J. Am. Chem. Soc.* **2001**, *123*, 1002–1003.
160. Ottesen, J. J.; Imperiali, B. *Nat. Struct. Biol.* **2001**, *8*, 535–539.
161. Kortemme, T.; Ramirez-Alvarado, M.; Serrano, L. *Science* **1998**, *281*, 253–256.
162. Cochran, A. G.; Skelton, N. J.; Starovasnik, M. A. *Proc. Natl. Acad. Sci. USA* **2001**, *98*, 5578–5583.
163. Neidigh, J. W.; Fesinmeyer, R. M.; Andersen, N. H. *Nat. Struct. Biol.* **2002**, *9*, 425–430.
164. Starovasnik, M. A.; Braisted, A. C.; Wells, J. A. *Proc. Natl. Acad. Sci. USA* **1997**, *94*, 10080–10085.
165. Ottesen, J. J.; Imperiali, B. *Nat. Struct. Biol.* **2001**, *8*, 535–539.
166. Pastor, M. T.; Lopez de la Paz, M.; Lacroix, E.; Serrano, L.; Perez-Paya, E. *Proc. Natl. Acad. Sci. USA* **2002**, *99*, 614–619.
167. Lopez de la Paz, M.; Goldie, K.; Zurdo, J.; Lacroix, E.; Dobson, C. M.; Hoenger, A.; Serrano, L. *Proc. Natl. Acad. Sci. USA* **2002**, *99*, 16052–16057.
168. Wei, Y.; Liu, T.; Sazinsky, S. L.; Moffet, D. A.; Pelczer, I.; Hecht, M. H. *Protein Sci.* **2003**, *12*, 92–102.
169. West, M. W.; Hecht, M. H. *Protein Sci.* **1995**, *4*, 2032–2039.
170. Dill, K. A. *Protein Sci.* **1999**, *8*, 1166–1180.
171. DeGrado, W. F.; Regan, L.; Ho, S. P. *Cold Spring Harb. Symp. Quant. Biol.* **1987**, *52*, 521–526.
172. Lovejoy, B.; Choe, S.; Cascio, D.; McRorie, D. K.; DeGrado, W. F.; Eisenberg, D. *Science* **1993**, *259*, 1288–1293.
173. Walsh, S. T.; Lee, A. L.; DeGrado, W. F.; Wand, A. J. *Biochemistry* **2001**, *40*, 9560–9569.
174. Bryson, J. W.; Desjarlais, J. R.; Handel, T. M.; DeGrado, W. F. *Protein Sci.* **1998**, 7, 1404–1414.
175. Ghirlanda, G.; Lear, J. D.; Ogihara, N. L.; Eisenberg, D.; DeGrado, W. F. *J. Mol. Biol.* **2002**, *319*, 243–253.
176. Zhu, Y.; Alonso, D. O.; Maki, K.; Huang, C. Y.; Lahr, S. J.; Daggett, V.; Roder, H.; DeGrado, W. F.; Gai, F. *Proc. Natl. Acad. Sci. USA* **2003**, *100*, 15486–15491.
177. Lahr, S. J.; Engel, D. E.; Stayrook, S. E.; Maglio, O.; North, B.; Geremia, S.; Lombardi, A.; DeGrado, W. F. *J. Mol. Biol.* **2005**, *346*, 1441–1454.
178. Chana, M. S.; Tripet, B. P.; Mant, C. T.; Hodges, R. *J. Pept. Res.* **2005**, *65*, 209–220.
179. Litowski, J. R.; Hodges, R. S. *J. Biol. Chem.* **2002**, *277*, 37272–37279.

180. Adamson, J. G.; Zhou, N. E.; Hodges, R. S. *Curr. Opin. Biotechnol.* **1993**, *4*, 428–437.
181. Monera, O. D.; Zhou, N. E.; Kay, C. M.; Hodges, R. S. *J. Biol. Chem.* **1993**, *268*, 19218–19227.
182. O'Shea, E. K.; Lumb, K. J.; Kim, P. S. *Curr. Biol.* **1993**, *3*, 658–667.
183. Oakley, M. G.; Kim, P. S. *Biochemistry* **1998**, *37*, 12603–12610.
184. Harbury, P. B.; Zhang, T.; Kim, P. S.; Alber, T. *Science* **1993**, *262*, 1401–1407.
185. Lumb, K. J.; Kim, P. S. *Biochemistry* **1995**, *34*, 8642–8648.
186. Fong, J. H.; Keating, A. E.; Singh, M. *Genome Biol.* **2004**, *5*, R11.
187. Keating, A. E.; Malashkevich, V. N.; Tidor, B.; Kim, P. S. *Proc. Natl. Acad. Sci. USA* **2001**, *98*, 14825–14830.
188. Newman, J. R.; Keating, A. E. *Science* **2003**, *300*, 2097–2101.
189. Ogihara, N. L.; Weiss, M. S.; Degrado, W. F.; Eisenberg, D. *Protein Sci.* **1997**, *6*, 80–88.
190. Lombardi, A.; Bryson, J. W.; DeGrado, W. F. *Biopolymers* **1996**, *40*, 495–504.
191. Ali, M. H.; Peisach, E.; Allen, K. N.; Imperiali, B. *Proc. Natl. Acad. Sci. USA* **2004**, *101*, 12183–12188.
192. Ali, M. H.; Taylor, C. M.; Grigoryan, G.; Allen, K. N.; Imperiali, B.; Keating, A. E. *Structure* **2005**, *13*, 225–234.
193. Bolon, D. N.; Grant, R. A.; Baker, T. A.; Sauer, R. T. *Proc. Natl. Acad. Sci. USA* **2005**, *102*, 12724–12729.
194. Chrunyk, B. A.; Rosner, M. H.; Cong, Y.; McColl, A. S.; Otterness, I. G.; Daumy, G. O. *Biochemistry* **2000**, *39*, 7092–7099.
195. Berg, T. *Angew. Chem. Int. Ed. Engl.* **2003**, *42*, 2462–2481.
196. Baca, M.; Alewood, P. F.; Kent, S. B. *Protein Sci.* **1993**, *2*, 1085–1091.
197. Ayers, B.; Blaschke, U. K.; Camarero, J. A.; Cotton, G. J.; Holford, M.; Muir, T. W. *Biopolymers* **1999**, *51*, 343–354.
198. Muir, T. W.; Sondhi, D.; Cole, P. A. *Proc. Natl. Acad. Sci.* **1998**, *95*, 6705–6710.
199. Reaka, A. J.; Ho, C. M.; Marshall, G. R. *J. Comput.-Aided Mol. Des.* **2002**, *16*, 585–600.
200. Lee, H. B.; Zaccaro, M. C.; Pattarawarapan, M.; Roy, S.; Saragovi, H. U.; Burgess, K. *J. Org. Chem.* **2004**, *69*, 701–713.
201. Chalmers, D. K.; Marshall, G. R. *J. Am. Chem. Soc.* **1995**, *117*, 5927–5937.
202. Takeuchi, Y.; Marshall, G. R. *J. Am. Chem. Soc.* **1998**, *120*, 5363–5372.
203. Hodgkin, E. E.; Clark, J. D.; Miller, K. R.; Marshall, G. R. *Biopolymers* **1990**, *30*, 533–546.
204. Smith, A. B., III.; Favor, D. A.; Sprengeler, P. A.; Guzman, M. C.; Carroll, P. J.; Furst, G. T.; Hirschmann, R. *Bioorg. Med. Chem.* **1999**, 7, 9–22.
205. Jacoby, E. *Bioorg. Med. Chem. Lett.* **2002**, *12*, 891–893.
206. Kutzki, O.; Park, H. S.; Ernst, J. T.; Orner, B. P.; Yin, H.; Hamilton, A. D. *J. Am. Chem. Soc.* **2002**, *124*, 11838–11839.
207. Orner, B. P.; Ernst, J. T.; Hamilton, A. D. *J. Am. Chem. Soc.* **2001**, *123*, 5382–5383.
208. Todd, M. J.; Semo, N.; Freire, E. *J. Mol. Biol.* **1998**, *283*, 475–488.
209. Shultz, M. D.; Chmielewski, J. *Bioorg. Med. Chem. Lett.* **1999**, *9*, 2431–2436.
210. Zutshi, R.; Brickner, M.; Chmielewski, J. *Curr. Opin. Chem. Biol.* **1998**, *2*, 62–66.
211. Bowman, M. J.; Chmielewski, J. *Biopolymers* **2002**, *66*, 126–133.
212. Arnold, U.; Hinderaker, M. P.; Nilsson, B. L.; Huck, B. R.; Gellman, S. H.; Raines, R. T. *J. Am. Chem. Soc.* **2002**, *124*, 8522–8523.
213. Arnold, U.; Hinderaker, M. P.; Koditz, J.; Golbik, R.; Ulbrich-Hofmann, R.; Raines, R. T. *J. Am. Chem. Soc.* **2003**, *125*, 7500–7501.
214. Smythe, M. L.; Huston, S.; Marshall, G. R. *J. Am. Chem. Soc.* **1993**, *115*, 11594–11595.
215. Pabo, C. O.; Sauer, R. T. *Annu. Rev. Biochem.* **1984**, *53*, 293–321.
216. Muller, C. W. *Curr. Opin. Struct. Biol.* **2001**, *11*, 26–32.
217. Draper, D. E. *J. Mol. Biol.* **1999**, *293*, 255–270.
218. Marshall, G. R.; Head, R. H.; Ragno, R. Affinity Prediction: The Sina Qua Non. In *Thermodynamics in Biology*; Di Cera, E., Ed.; Oxford University Press: Oxford, UK, 2000, pp 87–111.
219. Kelso, M. J.; Beyer, R. L.; Hoang, H. N.; Lakdawala, A. S.; Snyder, J. P.; Oliver, W. V.; Robertson, T. A.; Appleton, T. G.; Fairlie, D. P. *J. Am. Chem. Soc.* **2004**, *126*, 4828–4842.
220. Taylor, J. W. *Biopolymers* **2002**, *66*, 49–75.
221. Finn, F. M.; Hofmann, K. *Acc. Chem. Res.* **1973**, *6*, 169–176.
222. Varadarajan, R.; Connelly, P. R.; Sturtevant, J. M.; Richards, F. M. *Biochemistry* **1992**, *31*, 1421–1426.
223. Kritzer, J. A.; Lear, J. D.; Hodsdon, M. E.; Schepartz, A. *J. Am. Chem. Soc.* **2004**, *126*, 9468–9469.
224. Vassilev, L. T.; Vu, B. T.; Graves, B.; Carvajal, D.; Podlaski, F.; Filipovic, Z.; Kong, N.; Kammlott, U.; Lukacs, C.; Klein, C. et al. *Science* **2004**, *303*, 844–848.
225. Kussie, P. H.; Gorina, S.; Marechal, V.; Elenbaas, B.; Moreau, J.; Levine, A. J.; Pavletich, N. P. *Science* **1996**, *274*, 948–953.
226. Garcia-Echeverria, C.; Chene, P.; Blommers, M. J.; Furet, P. *J. Med. Chem.* **2000**, *43*, 3205–3208.
227. Chene, P.; Fuchs, J.; Bohn, J.; Garcia-Echeverria, C.; Furet, P.; Fabbro, D. *J. Mol. Biol.* **2000**, *299*, 245–253.
228. Hirschmann, R. *Angew. Chem. Int. Ed. Engl.* **1991**, *30*, 1278–1301.
229. Olson, G. L.; Cheung, H. C.; Chiang, E.; Madison, V. S.; Sepinwall, J.; Vincent, G. P.; Winokur, A.; Gary, K. A. *J. Med. Chem.* **1995**, *38*, 2866–2879.
230. Rutledge, L. D.; Perlman, J. H.; Gershengorn, M. C.; Marshall, G. R.; Moeller, K. D. *J. Med. Chem.* **1996**, *39*, 1571–1574.
231. Sattler, M.; Liang, H.; Nettesheim, D.; Meadows, R. P.; Harlan, J. E.; Eberstadt, M.; Yoon, H. S.; Shuker, S. B.; Chang, B. S.; Minn, A. J. et al. *Science* **1997**, *275*, 983–986.
232. Chin, J. W.; Schepartz, A. *Angew. Chem. Int. Ed. Engl.* **2001**, *40*, 3806–3809, 3922–2925.
233. Ernst, J. T.; Becerril, J.; Park, H. S.; Yin, H.; Hamilton, A. D. *Angew. Chem. Int. Ed. Engl.* **2003**, *42*, 535–539.
234. Ernst, J. T.; Kutzki, O.; Debnath, A. K.; Jiang, S.; Lu, H.; Hamilton, A. D. *Angew. Chem. Int. Ed. Engl.* **2002**, *41*, 278–281.
235. Marshall, G. R.; Bosshard, H. E. *Circul. Res.* **1972**, *31*, 143–150.
236. Van Roey, P.; Smith, G. D.; Balasubramanian, T. M.; Czerwinski, E. W.; Marshall, G. R.; Mathews, F. S. *Int. J. Pept. Protein. Res.* **1983**, *22*, 404–409.
237. Van Roey, P.; Smith, G. D.; Balasubramanian, T. M.; Redlinski, A. S.; Marshall, G. R. *Int. J. Pept. Protein. Res.* **1982**, *19*, 499–505.
238. Marshall, G. R.; Hodgkin, E. E.; Langs, D. A.; Smith, G. D.; Zabrocki, J.; Leplawy, M. T. *Proc. Natl. Acad. Sci. USA* **1990**, *87*, 487–491.
239. Prasad, B. V.; Balaram, P. *CRC Crit. Rev. Biochem.* **1984**, *16*, 307–348.
240. Toniolo, C.; Crisma, M.; Formaggio, F.; Valle, G.; Cavicchioni, G.; Precigoux, G.; Aubry, A.; Kamphuis, J. *Biopolymers* **1993**, *33*, 1061–1072.
241. Andersen, N. H.; Liu, Z.; Prickett, K. S. *FEBS Lett.* **1996**, *399*, 47–52.
242. Higashimoto, Y.; Kodama, H.; Jelokhani-Niaraki, M.; Kato, F.; Kondo, M. *J. Biochem.* **1999**, *125*, 705–712.
243. Inai, Y.; Ousaka, N.; Okabe, T. *J. Am. Chem. Soc.* **2003**, *125*, 8151–8162.

244. *A Practical Guide To Combinatorial Chemistry*, 1st ed.; Czarnik, A. W., DeWitt, S. H., Eds.; American Chemical Society: Washington, DC, 1997.
245. Wiley, R. A.; Rich, D. H. *Med. Res. Rev.* **1993**, *13*, 327–384.
246. Evans, B. E.; Rittle, K. E.; Bock, M. G.; DiPardo, R. M.; Freidinger, R. M.; Whitter, W. L.; Lundell, G. F.; Veber, D. F.; Anderson, P. S.; Chang, R. S. et al. *J. Med. Chem.* **1988**, *31*, 2235–2246.
247. Patchett, A. A.; Nargund, R. P. *Annu. Rep. Med. Chem.* **2000**, *35*, 289–298.
248. Horton, D. A.; Bourne, G. T.; Smythe, M. L. *Chem. Rev.* **2003**, *103*, 893–930.
249. Schneider, J. P.; Kelly, J. W. *Chem. Rev.* **1995**, *95*, 2169–2187.
250. Ripka, W. C.; DeLucca, G. V.; Bach, A. C., II.; Pottorf, R. S.; Blaney, J. M. *Tetrahedron* **1993**, *49*, 3593–3608.
251. Blackburn, B. K.; Lee, A.; Baier, M.; Kohl, B.; Olivero, A. G.; Matamoros, R.; Robarge, K. D.; McDowell, R. S. *J. Med. Chem.* **1997**, *40*, 717–729.
252. Shigeri, Y.; Ishikawa, M.; Ishihara, Y.; Fujimoto, M. *Life Sci.* **1998**, *63*, L151–L160.
253. Dziadulewicz, E. K.; Brown, M. C.; Dunstan, A. R.; Lee, W.; Said, N. B.; Garratt, P. J. *Bioorg. Med. Chem. Lett.* **1999**, *9*, 463–468.
254. Miller, W. H.; Alberts, D. P.; Bhatnagar, P. K.; Bondinell, W. E.; Callahan, J. F.; Calvo, R. R.; Cousins, R. D.; Erhard, K. F.; Heerding, D. A.; Keenan, R. M. et al. *J. Med. Chem.* **2000**, *43*, 22–26.
255. Haskell-Luevano, C.; Rosenquist, A.; Souers, A.; Khong, K. C.; Ellman, J. A.; Cone, R. D. *J. Med. Chem.* **1999**, *42*, 4380–4387.
256. Rohrer, S. P.; Birzin, E. T.; Mosley, R. T.; Berk, S. C.; Hutchins, S. M.; Shen, D. M.; Xiong, Y. S.; Hayes, E. C.; Parmar, R. M.; Foor, F. et al. *Science* **1998**, *282*, 737–740.
257. Hanessian, S.; McNaughton-Smith, G.; Lombart, H.-G.; Lubell, W. D. *Tetrahedron* **1997**, *53*, 12789–12854.
258. Cornille, F.; Slomczynska, U.; Smythe, M. L.; Beusen, D. D.; Moeller, K. D.; Marshall, G. R. *J. Am. Chem. Soc.* **1995**, *117*, 909–917.
259. Halab, L.; Gosselin, F.; Lubell, W. D. *Biopolymers* **2000**, *55*, 101–122.
260. Etzkorn, F. A.; Guo, R.; Lipton, M. A.; Goldberg, S. D.; Bartlett, P. A. *J. Am. Chem. Soc.* **1994**, *116*, 10412–10425.
261. Kessler, H.; Kutscher, B. *Tetrahedron* **1985**, *26*, 177–180.
262. Aumailley, M.; Gurrath, M.; Muller, G.; Calvete, J.; Timpl, R.; Kessler, H. *FEBS Lett.* **1991**, *291*, 50–54.
263. Gurrath, M.; Muller, G.; Kessler, H.; Aumailley, M.; Timpl, R. *Eur. J. Biochem.* **1992**, *210*, 911–921.
264. Muller, G.; Gurrath, M.; Kessler, H. *J. Comput.-Aided Mol. Des.* **1994**, *8*, 709–730.
265. Pfaff, M.; Tangemann, K.; Muller, B.; Gurrath, M.; Muller, G.; Kessler, H.; Timpl, R.; Engel, J. *J. Biol. Chem.* **1994**, *269*, 20233–20238.
266. Haubner, R.; Gratias, R.; Diefenbach, B.; Goodman, S. L.; Jonczyk, A.; Kessler, H. *J. Am. Chem. Soc.* **1996**, *118*, 7461–7472.
267. Haubner, R.; Schmitt, W.; Holzemann, G.; Goodman, S. L.; Jonczyk, A.; Kessler, H. *J. Am. Chem. Soc.* **1996**, *118*, 7881–7891.
268. Haubner, R.; Finsinger, D.; Kessler, H. *Angew. Chem. Int. Ed. Engl.* **1997**, *36*, 1374–1389.
269. Wermuth, J.; Goodman, S. L.; Jonczyk, A.; Kessler, H. *J. Am. Chem. Soc.* **1997**, *119*, 1328–1335.
270. Dechantsreiter, M. A.; Planker, E.; Matha, B.; Lohof, E.; Holzemann, G.; Jonczyk, A.; Goodman, S. L.; Kessler, H. *J. Med. Chem.* **1999**, *42*, 3033–3040.
271. Boer, J.; Gottschling, D.; Schuster, A.; Semmrich, M.; Holzmann, B.; Kessler, H. *J. Med. Chem.* **2001**, *44*, 2586–2592.
272. Goodman, S. L.; Holzemann, G.; Sulyok, G. A.; Kessler, H. *J. Med. Chem.* **2002**, *45*, 1045–1051.
273. Sulyok, G. A.; Gibson, C.; Goodman, S. L.; Holzemann, G.; Wiesner, M.; Kessler, H. *J. Med. Chem.* **2001**, *44*, 1938–1950.
274. Porcelli, M.; Casu, M.; Lai, A.; Saba, G.; Pinori, M.; Cappelletti, S.; Mascagni, P. *Biopolymers* **1999**, *50*, 211–219.
275. Nikiforovich, G. V.; Kover, K. E.; Zhang, W. J.; Marshall, G. R. *J. Am. Chem. Soc.* **2000**, *122*, 3262–3273.
276. Tran, T. T.; McKie, J.; Meutermans, W. D. F.; Bourne, G. T.; Andrews, P. R.; Smythe, M. L. J. *Comput.-Aided Mol. Des.* **2005**, *19*, 551–566.
277. Eswar, N.; John, B.; Mirkovic, N.; Fiser, A.; Ilyin, V. A.; Pieper, U.; Stuart, A. C.; Marti-Renom, M. A.; Madhusudhan, M. S.; Yerkovich, B. et al. *Nucleic Acids Res.* **2003**, *31*, 3375–3980.
278. John, B.; Sali, A. *Nucleic Acids Res.* **2003**, *31*, 3982–3992.
279. Zhang, Y.; Arakaki, A. K.; Skolnick, J. *Proteins* **2005**, *61*, 91–98.
280. Zhang, Y.; Skolnick, J. *Nucleic Acids Res.* **2005**, *33*, 2302–2309.
281. Zhang, Y.; Skolnick, J. *Proc. Natl. Acad. Sci. USA* **2005**, *102*, 1029–1034.
282. Uetz, P.; Giot, L.; Cagney, G.; Mansfield, T. A.; Judson, R. S.; Knight, J. R.; Lockshon, D.; Narayan, V.; Srinivasan, M.; Pochart, P. et al. *Nature* **2000**, *403*, 623–627.
283. Aebersold, R. *J. Infect. Dis.* **2003**, *187*, S315–S320.
284. Salwinski, L.; Eisenberg, D. *Curr. Opin. Struct. Biol.* **2003**, *13*, 377–382.
285. Merrifield, R. B. *J. Am. Chem. Soc.* **1963**, *85*, 2149–2154.
286. Caruthers, M. H. *Science* **1985**, *230*, 281–285.
287. Stouch, T. R.; Kenyon, J. R.; Johnson, S. R.; Chen, X.-Q.; Dowyko, A.; Li, Y. *J. Comput.-Aided Drug Des.* **2003**, *17*, 83–92.
288. Oprea, T. I. *Curr. Opin. Chem. Biol.* **2002**, *6*, 384–389.
289. Oprea, T. I. *J. Comput.-Aided Drug Des.* **2002**, *16*, 325–334.
290. Koppal, T. *Drug Disc. Dev.* **2003**, *6*, 28–32.
291. Still, W. C.; Tempczyk, A.; Hawley, R. C.; Hendrickson, T. *J. Am. Chem. Soc.* **1990**, *112*, 6127–6129.
292. Wireko, F. C.; Kellogg, G. E.; Abraham, D. J. *J. Med. Chem.* **1991**, *34*, 758–767.
293. Park, P. S.; Filipek, S.; Wells, J. W.; Palczewski, K. *Biochemistry* **2004**, *43*, 15643–15656.
294. Marshall, G. R. Introduction to Chemoinformatics in Drug Discovery: A Personal View. In *Cheminformatics in Drug Discovery*; Oprea, T. I., Ed.; Wiley-VCH Verlag: Weinheim, Germany, 2005, pp 1–22.
295. Marshall, G. R.; Vakser, I. A. Protein–Protein Docking Methods. In *Proteomics and Protein–Protein Interactions: Biology, Chemistry, Bioinformatics and Drug Design*; Waksman, G., Ed.; Springer-Verlag: New York, 1999, pp 115–146.
296. Marshall, G. R.; Riley, D. P. Design of Protein and Enzyme Mimetics. In *Handbook of Theoretical and Computational Nanotechnology*; Rieth, M., Sommers, W., Eds.; American Scientific Publishers: in press, 2005.
297. Andrews, P. R.; Carson, J. M.; Caselli, A.; Spark, M. J.; Woods, R. *J. Med. Chem.* **1985**, *28*, 393–399.
298. Critical Assessment of Structure Prediction (CASP). http://predictioncenter.org/ (accessed April 2006).
299. Critical Assessment of Predicted Interactions (CAPRI). http://Capri.ebi.ac.uk/ (accessed April 2006).
300. Alanines Scanning Energetics database (ASEdb). http://thornlab.cgr.harvard.edu/hotspot/index.php (accessed April 2006).

Biographies

Garland R Marshall is Professor of Biochemistry and Molecular Biophysics at Washington University's School of Medicine. Professor Marshall is formerly the Director of the Center for Molecular Design in the Institute for Biomedical Computing. He is a founding member of the new Center for Computational Biology, a joint effort by the Departments of Biochemistry and Molecular Biophysics, of Biomedical Engineering and of Genetics. Professor Marshall also serves on the Steering Committee for Graduate Studies in Computational Biology. His pioneering effort led to a software package SYBYL for computer-aided drug design that was licensed to Tripos, Inc., a software company that he founded in 1979. He terminated his participation with Tripos in 1987 to devote his full attention to his academic research. In 1995, he founded Metaphore Pharmaceuticals, Inc. to focus on metals as therapeutics.

Christina M Taylor is postdoctoral research scholar in Garland Marshall's laboratory at Washington University School of Medicine in St Louis. She received her BSc degree in chemistry from University of Missouri–Rolla in 1999. In September 2005, she received her PhD from the Massachusetts Institute of Technology in Amy Keating's laboratory.

© 2007 Elsevier Ltd. All Rights Reserved
No part of this publication may be reproduced, stored in any retrieval system or transmitted in any form by any means electronic, electrostatic, magnetic tape, mechanical, photocopying, recording or otherwise, without permission in writing from the publishers

Comprehensive Medicinal Chemistry II
ISBN (set): 0-08-044513-6

ISBN (Volume 4) 0-08-044517-9; pp. 13–41

4.03 Quantitative Structure–Activity Relationship – A Historical Perspective and the Future

C Hansch and C Selassie, Pomona College, Claremont, CA, USA

© 2007 Elsevier Ltd. All Rights Reserved.

4.03.1 Introduction

An understanding of the many and diverse interactions of various chemicals with biological macromolecules as determined by their intermolecular forces, i.e., hydrophobic, electrostatic, polar, and steric, was critical to the formulation and development of the quantitative structure–activity relationship (QSAR) paradigm. The availability of rapidly expanding libraries of compounds, and the increased number of biological targets, created a need for a more organized approach to molecular design and development. The use of correlation analysis was useful in helping to mathematically delineate the importance of certain structural attributes of chemicals to their biological activities. In this chapter, we will take a somewhat chronological look at the development of the QSAR paradigm.

Perhaps the earliest attempt at correlating structure with function was during the time of the Renaissance. Paracelsus (1493–1541), the famous alchemist, physician, and astrologer who is sometimes called the 'father of toxicology,' wrote that the 'dose makes the poison.' Thus he established that the dose was as important as the nature of the substance.[1] Many, many years later Crum-Brown and Fraser, in a study of a series of alkaloids, expounded on the relationship between physiological activity and chemical composition and constitution.[2] Following closely on the work of Richet, who associated toxicity inversely with water solubility, Meyer and Overton independently established a strong relationship between olive oil/water partition coefficients and the narcotic action of a group of organic compounds.[3–5]

The period immediately following World War II saw the dawning of a new area in structure–activity relationships (SARs). The focus shifted from similarities in groups or atom types to a more generalized emphasis on the overall properties of molecules. This approach was exemplified by the work of Albert *et al.*, who examined the effects of ionization/electron distribution and steric access on the potencies of a multitude of aminoacridines. They also established that not all drug receptors were on proteins but could also be present on nucleic acids.[6] At the same time, Bell and Roblin undertook a thorough study of the antibacterial activities of a series of sulfanilamides in terms of their ionizations.[7]

4.03.2 Quantitative Structure–Activity Relationship Parameters

Meanwhile, critical contributions of Hammett to developments in physical organic chemistry began to play an important role in not only defining side chain reactions in benzene derivatives, but also delineating reactivities in condensed aromatic and heterocyclic systems.[8,9]

4.03.2.1 Electronic Effects

Hammett's study of the relationships between reaction rates and equilibrium constants was inspired by Bronsted's approach to general acid and base catalysis. Experimental work in Hammett's laboratory and extensive literature analysis of logarithmic reaction relationships led to his critical publication in 1935.[8] At the same time, Burkhardt was independently examining the effects of substituents on organic reactions. He proposed the use of the ionization of substituted benzoic acids as a reference series.[10] In 1937, Hammett published his seminal study on the effects of *meta-* and *para-*substituents on the rate constants or equilibrium constants of side chain reactions of benzene derivatives.[11] The Hammett equation is now a well-established empirical relationship for correlating structures and reactivity (equilibrium and reaction rate constants) for *meta-* and *para-*substituted benzene derivatives. It is generally described in the following form:

$$\log(K_X/K_H) \text{ or } \log(k_X/k_H) = \rho\sigma \qquad [1]$$

K_X and K_H represent the ionization constants for the substituted and unsubstituted benzoic acids, respectively while k_X and k_H are the rate constants for the side chain reactions of substituted and unsubstituted benzene derivatives, respectively. Two parameters that result from Hammett treatments are the substituent constant, σ and the reaction constant, ρ. Over the years, substituents constants and their applications have been determined, calculated, thoroughly analyzed and reviewed in great detail while reaction constants have come into their own, particularly in mechanistic studies of organic reactions as well as biological interactions.[12,13] One might say that the Hammett equation helped lay the foundation for the study of correlation analysis and QSAR.

The operational definition of σ is a measure of the size of the electronic effect of a substituent and thus represents the change in the electronic charge distribution in the benzene nucleus. Electron-withdrawing substitutents are characterized by positive values while electron-releasing substituents have negative values. The proportionality or reaction constant, ρ, is a measure of the susceptibility of any given reaction to substituent effects; a positive value signifies an enhancement of the reaction by electron-withdrawal at the reaction site while a negative value implies that the reaction is assisted by electron-releasing substituents. The Hammett equation holds up well for *meta-* and *para-*substituted benzene derivatives but often fails when it comes to *ortho*-substituted analogs. The latter presents a complex problem and may be resolved by separation into at least two components: the pure steric effect and the field effect transmitted through space. This was elegantly described by Fujita and Nishioka in an integrated approach[14]:

$$\log K_X = \rho\sigma + \delta E_S^{\text{ortho}} + fF_{\text{ortho}} + C \quad [2]$$

Despite the success of the Hammett equation and its extended applications, it does have some limitations which have been described in detail.[15] The most critical one involves the influence of resonance or direct conjugation between the substituent and the reaction site on reaction rates. This led to the development of the modified parameters, σ^+ (for electrophilic reaction centers) and σ^- (for nucleophilic reaction centers) that result in the following modified linear free-energy relationships[16,17]:

$$\log(K_X/K_H) = (\rho^+)(\sigma^+) \text{ or } (\rho^-)(\sigma^-) \quad [3]$$

Concurrently, Taft and co-workers factored the Hammett σ_M and σ_P into their inductive (σ_I) and resonance (σ_R) components:

$$\sigma_P = \sigma_I + \sigma_R \quad [4]$$

This factorization led to the articulation of the dual-substituent approach described by eqn [5].

$$\log K_X/K_H = \rho_I\sigma_I + \rho_R\sigma_R \quad [5]$$

With their rigid structures and absence of multiple bonds, bicyclo[2.2.2]octane carboxylic acids provided an appropriate system for the determination of σ_I values that were measured in 50% ethanol for solubility reasons.[18] Thus accurate σ_P values and properly scaled σ_I values allowed for the calculation of reasonable σ_R values. However, as Exner has pointed out, this method of calculation makes σ_R subject to errors in both σ_P and σ_I.[19]

The success of the Hammett approach to the quantitation and delineation of electronic effects led to further modifications and refinements of these attributes in aromatic and alkyl systems. Most of the Hammett σ constants applied to aromatic systems and the dearth of appropriate constants in aliphatic systems led Taft in 1956 to devise σ^* (polar) constants that were defined as follows[20]:

$$\sigma^* = \frac{1}{2.48}\{\log(k_X/k_H)_{\text{base}} - \log(k_X/k_H)_{\text{acid}}\} \quad [6]$$

log k_X and k_H represent the rate constants of base and acid catalyzed hydrolysis of XCH_2COOR and CH_3COOR respectively. σ^* is closely related to σ_I as seen in eqn [7].[21]

$$\sigma_{I(X)} = 0.45\sigma^*_{(CH_2X)} \quad [7]$$

In 1963, Charton, working with a series of substituted acetic acids and armed with σ_I values initially developed by Taft,[22] proceeded to define σ_I according to eqn [8].[23] The pK_a values of a series of aliphatic compounds including 2- or 3-substituted propanoic acids, substituted methylamines and 2-substituted ethylamines were used to determine secondary σ_I constants.

$$\sigma_{I,X} = b(pK_{a,X}) + d \quad [8]$$

The proliferation of various σ scales led to Swain and Lupton's development of the F (field) and R (resonance) constants in 1968 where f and r are empirical sensitivities for each reaction.[24]

$$\sigma = fF + rR \quad [9]$$

Critical assumptions made in this analysis included the lack of resonance in 4-X-bicyclo[2.2.2]octane-carboxylic acids ($r = 0$), the assignment of $R = 0$, for the $N^+(CH_3)_3$ substituent and the assignment of $r = 1$ for the dissociation of *para*-substituted benzoic acids. Using the R value of zero for the $N^+(CH_3)_3$ substituent, f was defined by eqn [9]. However, they failed to normalize their F and R values with respect to the Hammett constants obtained from the ionization of benzoic acids in water at 25 °C. Hansch, Leo, and Taft corrected this oversight and obtained the following relationships described by eqns [10] and [11].[25]

$$F = \sigma_I = 1.297\ \sigma_M - 0.385\ \sigma_P + 0.033 \quad [10]$$

$$n = 38, \quad r = 0.968, \quad s = 0.046$$

$$R = \sigma_P - 0.921F \quad [11]$$

Using eqn [11], the R values for many substituents could easily be calculated. There has been criticism of Swain and Lupton's approach but, in general, the field/inductive parameter F does not significantly differ from other Hammett type (e.g., σ_I) values. These parameters (F and R) however, have found limited use in physical-organic and biological QSAR.

4.03.2.2 Steric Effects

During the 1950s, the extension and refinement of electronic parameters developed rapidly, but progress to quantify steric effects proceeded much more slowly. The problems of steric hindrance were first recognized in the nineteenth century and various semiquantitative attempts to address these issues were undertaken in the first half of the twentieth century. A major breakthrough was made by Taft, who clearly recognized the need to separate electronic and steric effects. As mentioned earlier, Taft assiduously studied the hydrolysis of esters of substituted acetic acids (XCH_2COOR) under acidic and basic conditions. Thus the first steric parameter to be numerically defined was Taft's E_S parameter.[22] See eqn [12] where k_X and k_H are the same as previously defined in eqn [6]:

$$E_S = \log(k_X/k_H)_{acid} \quad [12]$$

A few years later, Hancock modified Taft's E_S scale by correcting for hyperconjugation according to eqn [13], where n_H represents the number of α-hydrogens and 0.306 is a constant derived from molecular orbital calculations[26]:

$$E_S^O = E_S + 0.306(n_H - 3) \quad [13]$$

Sigma constants and E_S values then found extensive use in quantifying reaction rates in physical organic chemistry.

In 1975, Charton defined a steric constant υ, based on a fundamental attribute of substituents, their van der Waals radii:[27–28]

$$\upsilon_X = \gamma_X - 1.20 \quad [14]$$

1.20 is the radius of hydrogen, while γ_X represents the minimum van der Waals radius of top symmetrical substituents. However with complex or unsymmetrical substituents, υ cannot be easily ascertained merely from the geometry; one has to resort to reactivity data in the form of ester hydrolysis. The relationship between Taft's E_S and Charton's υ was clearly demonstrated in eqn [15].[29]

$$E_S = -2.06(\pm 0.86)\upsilon - 0.19(\pm 0.10) \quad [15]$$

$$n = 104, \quad r = 0.978, \quad s = 0.250$$

At about the same time, Fujita *et al.* refined the Taft–Hancock steric constants to more accurately reflect the steric constraints of a highly substituted nonsymmetrical carbon such as $CR^1R^2R^3$.[30,31] Thus, E_S^C was expressed as a weighted sum of the individual E_S^C values of R_1, R_2 and R_3. The R groups are classified according to size such that $R_1 > R_2 > R_3$ (see eqn [16]):

$$E_S^C(CR_1R_2R_3) = \sum_{i=1}^{3} a_1 E_S^C(R_i) + E \quad [16]$$

The steric parameters developed up to this point focused mainly on describing intramolecular steric hindrance in organic reactions. Increased emphasis on intermolecular interactions led to the development of the STERIMOL parameters L, B_1 to B_4 by Verloop *et al.*, a few years later.[32,33] They based their parameters on standard bond angles and bond distances; L represented the length of the substituent while B_1–B_4 represented various perpendicular widths in four different directions. While this parameterization has been very useful in helping to delineate QSAR in biochemical systems, the need for an adequate number of data points in the dataset under study limited the utility of the approach.

This prompted Verloop to reduce the number of width parameters to two, B_1 and B_5, corresponding to the minimum and maximum widths orthogonal to L, respectively.[34]

Molar refractivity (MR) has often been included as a steric parameter. Tute has aptly dubbed it as the most 'chameleon-like' parameter[35] and, despite 40 years of usage, it still remains an elusive descriptor which defies easy interpretability in terms of QSAR. Pauling and Pressman first suggested that MR could be used to model the dispersion forces affecting hapten–antibody interactions since MR was directly related to molecular polarizability, α.[36]

$$\mathrm{MR} = 4\pi \mathrm{N}\alpha/3 \qquad [17]$$

$$\mathrm{MR} = (n^2 - 1)/(n^2 + 2) \bullet \mathrm{MW}/d \qquad [18]$$

MR is usually defined by the Lorentz/Lorenz eqn [18], where n is the refractive index, d is the density, and MW is the molecular weight of a compound. Since n of organic liquids does not vary much, the molar volume term constitutes from 75% to 80% of MR. Although van de Waterbeemd and Testa have shown a strong correlation between MR and van der Waals volume,[37] MR has been shown to provide superior correlations, particularly in ligand–receptor interactions where it is not adequately replaced by molar volume.[38,39] This suggests that polarizability is important when dealing with interactions in polar space. This has been borne out in extensive molecular graphics studies by Hansch and Blaney with various ligand–receptor systems.[40]

MR does not distinguish shape but, as a crude measure of bulk, it can pinpoint areas of steric constraint. Thus, negative coefficients with scaled MR terms imply steric hindrance at that site while positive coefficients suggest that bulky substitutents, or parts of a pharmacophore, either enhance the anchoring of a ligand in an opportune location or participate in strong dipolar interactions in polar space.[39]

4.03.2.3 Hydrophobicity to the Rescue

One of the earliest attempts to describe biological interactions in a quantitative sense was made by Hansen in 1962.[41] He examined the toxicity of a series of substituted benzoic acids in terms of their σ values and obtained a reasonable but fortuitous correlation as was later realized.[42] Against this backdrop, in the early 1960s we 'plodded' our way through our studies based on the interactions of X-phenoxyacetic acids with oat shoots. We were fortunate to have Robert Muir, a biologist with expertise in plant growth regulators, working in our chemistry building. Since 2,4-D (2,4-dichlorophenoxyacetic acid) was known to cause cell elongation at low concentrations and act as a weed killer at high concentrations, we focused on trying to decipher the molecular features pertinent to growth promoting activity. The two essential pharmacophoric features were determined to be the OCH_2COOH side chain and low electron density at the *ortho* position. Fukui the Japanese Nobelist agreed with our ideas and was instrumental in helping to lure Toshio Fujita of Kyoto University to do a postdoctoral stint at Pomona College. Borrowing on some inspiration from Taft's approach to correlating reaction rates in organic compounds, we used a linear combination of three parameters (the limit of our Clary DE-60 computer) to derive our first QSAR in 1962.[43] Our earlier attempts to utilize Hammett sigma constants were not fruitful although it was apparent that electron withdrawing groups were more effective at promoting auxin activity than were the electron-releasing substituents. It occurred to us that penetration to the site of action was important and so we began our now longstanding 'infatuation' with partition coefficients and our tentative steps to measure the partitioning of this series of X-phenoxyacetic acids between octanol and water at 30 °C. We obtained the following QSAR in this, our first study[43]:

$$\log 1/C = 4.08\pi - 2.14\pi^2 + 2.78\sigma + 3.36 \qquad [19]$$

C is the concentration of auxin that induced 10% growth in the Avena cylinder test. For simplicity in calculations, we used π values that we defined for the first time as the difference between the partition coefficient of the substituted phenoxyacetic acid and the parent structure, i.e., $\pi = \log P_X - \log P_H$. The parabolic dependence on π confirmed our earlier hypothesis that there should be an ideal hydrophilic–lipophilic balance for a given auxin so that it would not remain trapped in the aqueous phase or organic phase of the cell but move rapidly to the site of action. Although this example represents the first time that we derived the parabolic model, we did not have a mathematical justification for its usage; this would be established at a later date. In discussions with other investigators revolving around the correlation of biological activity and chemical structure we surmised that in both plant and animal systems where steric effects and metabolism were minimized, Hammett functions and partition coefficients would be critical to establishing structure activity relationships. Forty-five years later, this dictum still holds!

4.03.2.3.1 Octanol: a fortuitous choice!

The preliminary work in the Hansch laboratory on the measurement of octanol/water partition coefficients was started by Maloney but revved into high gear by Fujita who showed that Veldstra[44] was partially on the right track when he argued that the distribution of chemicals between fatty and aqueous phases of a biological system could totally account for the variation in activities. Veldstra however, discounted the role of electron density in these interactions. Since this important discovery, the partitioning of a bioactive solute between polar and nonpolar phases has been found to be critical to predicting activity and absorption, distribution, metabolism, and excretion (ADME) of drugs, pesticides, environmental toxicants, and various other xenobiotics. The octanol/water system has become the standard for this measurement. In 1964, the need for time intensive and costly partition constraints led Fujita, Iwasa, and Hansch to propose the first calculation methodology for $\log P$.[45] They considered $\log P$ to be an additive-constitutive, free-energy based property that was equal to the $\log P$ of the parent molecule plus a π term for each substituent. Since this defining moment, the measurement, calculation, computational refinement, and utility of $\log P$ has soared to new heights and consequently been the subject of numerous conferences, meetings, and publications. It has been implicated in all types of biological interactions and at least 65% of all the QSAR in the Comparative QSAR (C-QSAR) database have a hydrophobic component.

The choice of octanol as the nonpolar solvent was, in part, a lucky guess, but it was based first on the initial observations of Meyer's son who suggested oleyl alcohol as a potential solvent and second on our need for a practical and economical, easily purified solvent. Thirty thousand measurements later, this choice turned out to be prescient. Its critical attributes include low cost, nontoxicity, chemical unreactivity, excellent mimicry of biomembranes (enhanced by its amphiphilicity and hydrogen-bonding capability) and appreciable water content ($2.3\,\mathrm{M\,L^{-1}}$), which means that polar groups can retain some measure of hydration upon transfer from the aqueous phase. It competes equally well with water for a solute's hydrogen bond donors, but compensating for this shortcoming can be overcome in the rare cases where that property is important.

4.03.3 Development of Partition Coefficient Systems

In 1968 after 15 years in industry, Leo returned to Pomona College to initiate and then expand the Medchem Project. His guidance and expertise in the measurement and calculation of partition coefficients is unparallel in the field. Despite the voluminous number of octanol/water partition coefficients that were being measured in the laboratory,[46] Leo quickly concluded that it would be impossible with any reasonable level of resources to measure all the $\log P_{oct}$ values that were needed to understand chemicobiological interactions as the constraints of time and resources would not be able to meet this demand. There were three solutions to this pressing problem: automate the experimental procedures, collate, curate existing data, and develop a $\log P$ database, and develop software to compute $\log P_{oct}$ from structures. Leo has played pivotal roles in all three arenas.

4.03.3.1 Experimental Methods of Determining Partition Coefficients

The traditional shake-flask method, albeit slow and laborious, has been used to reliably measure many thousands of precise octanol/water partition coefficients.[46,47] An excellent analysis of this method is described by Taylor.[48] However, the time and resources required for this methodology limits its effectiveness, particularly with the advent of combinatorial synthesis and generation of diverse libraries. The widespread application of partition coefficients in ADME, toxicology, environmental chemistry, etc.,[49] mandates the development of $\log P$ methodology that is inclusive (spans the hydrophobicity scale), rapid, reproducible, cost-effective, and portrays an accurate reflection of partition phenomena.

A number of systems have emerged that are rapid, offer a broad dynamic range, and can yield new perspectives on drug partitioning into membranes. The filter probe extractor method of Tomlinson *et al.* has been shown to be rapid, efficient, and effective for water-immiscible systems over a wide range of concentrations and phase volume ratios.[50] Unger *et al.* were the first group to extensively study the use of reverse-phase high-performance liquid chromatography (RP-HPLC) for the determination of $\log P_{oct}$ by using an octanol saturated aqueous mobile phase.[51,52] Valuable contributions were made by Taylor's group, who first demonstrated that octanol could adhere to an inert support such as Hyflo-supercel, and who also emphasized the need for a range of appropriate standards for calibration purposes, alternating strong proton donor and acceptor standards to ensure the integrity of the hydrogen bonding characteristics of the column.[53,54]

Countercurrent chromatography has also been used to accurately and rapidly measure $\log P_{oct}$ coefficients using octanol and water as the stationary and mobile phases.[55] Several modifications such as the use of an octanol mobile phase and concurrent but variable speed movements of the mobile and stationary phases have extended the measurable $\log P_{oct}$ range.[56,57] Refinements in RP-HPLC allow for the rapid and facile determination of $\log P_{oct}$ values of neutral drugs with wide lipophilicity ranges and low sensitivity to impurities.[58] Recent developments in immobilized artificial membrane (IAM) chromatography suggest IAM retentions are akin to $\log P_{oct}$ coefficients albeit on a reduced lipophilicity scale.[59]

In addition to the widespread use and success of the octanol–water solvent system, other immiscible solvent pairs have also been examined. The water–chloroform and water–cyclohexane systems have been used to assess the lipophilicity of nucleic acid bases.[60] Leahy *et al.* have advocated the use of a quartet of model solvents (octanol, chloroform, any alkane, and propylene glycol dipelargonate) to better simulate partitions in vivo.[61]

4.03.3.2 Calculating log *P* from Structures

In 1969, Leo, being acutely aware of the demand and need for log *P* values, slowly began to collect, curate, and collate physicochemical data from in-house and various literature sources. His passion for this outcome led to the creation of the 24-pound bulky ledgers that made their way all over the world. Eventually with the College's acquisition of an IBM 360, collection of this data was streamlined and stored on tapes or microfiche. During this time, Leo's germ of an idea about calculating partition coefficients from computer-generated structural fragments took hold. This led to the development and evolution of the CLOGP program that displaced the error-prone manual procedures.[62–64] With the addition of Weininger to the Medchem team, the WLN system for computerized chemical structure storage was supplanted by a new information system, SMILES (Simplified Molecular Input Line Systems) and a thesaurus-oriented database system, THOR, for data entry and retrieval was added.[65,66] Today, with its ease of usage accuracy and facile structure entry, CLOGP remains one of the most highly utilized and reliable computer programs for partition coefficient calculations.

4.03.3.3 Rekker's Fragmental Approach

In the early 1970s, Rekker and his colleagues developed the first fragmental-based calculation approach.[67,68] Rather than adding substituent hydrophobicity to that of a known parent structure, as in the original Hansch–Fujita scheme, they defined a new constant, *f*, that represented the hydrophobic contribution of that constituent part of a structure to its overall hydrophobicity. Correction factors, *F*, were needed to account for fragment interactions thus:

$$\log P = \sum a_n f_n + \sum b_m F_m \qquad [20]$$

In eqn [20],[69] *a* denotes the number of occurrences of fragment *f* of type 'n' and *b* is the number of occurrences of correction factor *F* of type 'm.' The driving forces for this approach were twofold: simplification of log *P* calculations and inclusion of calculations for aliphatic and aromatic systems under one umbrella. It differed from the Hansch–Leo approach in that it was 'reductionist' in scope as opposed to the prevailing 'constructionist' tack.[69] However, some criticisms were leveled at this approach mainly because of Rekker's attempt to express the factors, *F*, as a multiple of some 'magic constant,' C_M. In order to improve the program's performance, Rekker *et al.* in the mid-1980s, tabulated new, more accurate and consistent *f* values, as well as an adjusted value of C_M.[70]

Since these early developments, numerous calculation procedures based on either substructure approaches or whole-molecule approaches have been developed and assessed for their uses and limitations in terms of their predictive powers. For a more comprehensive analysis of these developments, see the excellent review by Mannhold and van de Waterbeemd.[71] For a more recent comparison of commonly used fragmental-based programs such as Clog P, ACDlogPDB, and KowWin in QSAR analysis one is referred to the chapter by Machatha and Yalkowsky.[72] The validity of five log *P* calculation programs (ACD/logP, CLOGP, DOWWIN, XLOGP, and QLOGP) has also been assessed by Gargadennec *et al.*[73] and it appears that particular solvation behavior affects overall performance.

4.03.3.4 Calculations Based on Whole-Molecule Properties

Newer methodologies utilize a whole-molecule or global approach, which focuses on the molecular attributes of the molecule. They include molecular lipophilicity potentials, topological indices, and molecular descriptors such as surface area, charge densities and molecular electrostatic potentials.[71] Molecular lipophilicity potentials address molecular flexibility issues in polar solvents and lipoidal phases.[74] Recently, Chuman *et al.* have utilized accessible surface area and solvation energy differences to adequately predict $\log P_{oct}$ values.[75] Raevskii and Grigor'ev have developed a two parametric equation based on polarizability and hydrogen bond acceptor ability to calculate $\log P$.[76] Topological descriptors and connectivity indices developed extensively by Kier and Hall have also been used to calculate partition coefficients.[77]

4.03.4 Development of Quantitative Structure–Activity Relationship Models

4.03.4.1 Hansch Approach

In 1964, after analyzing a large number of data sets, Hansch and Dunn formulated a more generalized approach to SARs. The critical breakthrough was the usage of a linear combination of different physicochemical parameters to delineate biological activity.[78]

$$\log 1/C = a \log P + \rho\sigma + \cdots + \text{constant} \quad [21]$$

C is the molar concentration that evokes a certain biological response. The utility of this model which was developed by using multiple linear regression (MLR) analysis has been demonstrated in all types of biological systems that run the gamut from interactions of chemicals with isolated receptors to toxicological interactions in whole animals.[78,79]

During this period, we continued to observe instances where the linear model was inadequate and not adept at explaining the biological data. At this juncture, we analyzed a number of datasets and proposed a model which stressed the nonlinear relationship of hydrophobicity (as expressed by $\log P$) with biological activity.[80]

$$\log 1/C = a \log P + b(\log P)^2 + \cdots + \text{constant} \quad [22]$$

This parabolic model confirmed our earliest observations that drug transport/distribution was indeed a two-step process. We likened the movement of drug molecules in a biological system to a random walk process from the site of administration to the target receptor.[81]

Our first attempt to justify mathematically the parabolic relationship between drug effectiveness and lipophilic character was carried out in 1969 with the help of Bentley, a distinguished statistician in the Mathematics Department at Pomona College. Once again luck, in the form of Bentley, played a critical role in helping us derive the now classical mathematical model of the nonlinear dependence of biological activity on $\log P$ for nonequilibrium conditions using a single lipid barrier between two aqueous compartments.[82,83] In this model, both the more polar compounds and more hydrophobic compounds are limited in their abilities to reach the target receptor. Only a compound with an intermediate hydrophobicity or optimum $\log P$ will arrive at the receptor and wield its biological effects. The concept of an optimum $\log P$ or cutoff was not novel. A few years earlier, Ferguson had determined that the chemical potential of isonarcotic substances, acting via mostly physical mechanisms, were the same up to a critical cutoff point.[84,85] In later years, other models were developed to more adequately reflect the complex interactions involving the partitioning of drugs in the random walk process. These models will be briefly addressed later.

4.03.4.2 Free and Wilson's de novo Contributions

In 1964, while we were laboring on the delineation of the Hansch linear model, Free and Wilson formulated a generalized mathematical model (eqn [23]) that proposed that the biological activity was additive in nature; it was the sum of the activity of the reference compound and the group contributions of substituents, on various positions of this reference molecule:

$$\log 1/C = \sum a_i x_i + \mu \quad [23]$$

log 1/C was the biological activity, a_i was the contribution of the ith substituent, and x_i acquired a value of 1 if it was present and 0 if it was absent. The term μ was the calculated biological activity of the reference/parent molecule.[86,87] This mathematical model incorporated symmetry equations to minimize linear dependence between the variables. Although this approach was easy to apply, it had its drawbacks, mostly centered on the large number of parameters and subsequent loss of the statistical degrees of freedom. It also did not allow for extrapolation to substituents or substructural fragments that were not present in the training set. In 1971, in an attempt to deal with these limitations, Fujita and Ban proposed a simplified approach that solely focused on the additivity of group contributions:[88]

$$\log A/A_O = \sum G_i X_i \quad [24]$$

In Hammett form, A and A_O represented the biological activity of the substituted and unsubstituted compounds respectively, while G_i was the activity contribution of the ith substituent and X_i had a value of 1 or 0 that corresponded to the presence or absence of that substituent. The advantages of this ease-of-use approach have been well addressed by Kubinyi.[89] Despite its user-friendly approach, the Free–Wilson method has not seen extensive usage. Nevertheless, in combination with Hansch analysis (mixed approach), it can offer valuable insights in terms of SAR studies.[90] In a recent QSAR study on sulfamate-based estrone sulfatase inhibitors, Verma utilized a mixed approach to combine a large and varied series of nonsteroidal sulfamate-based compounds.[91] The following QSAR was formulated for all 101 compounds in the study:

$$\log 1/C = -0.39 \text{ Clog P} + 0.71 \text{ CMR} - 2.39\, I + 0.72 \qquad [25]$$

$$n = 101, \quad r^2 = 0.832, \quad q^2 = 0.814, \quad s = 0.650$$

The indicator variable I acquires a value of 1 for the presence of esters of cinnamic acid/or phenyl propionic acid and 0 for all other types of substitution.

4.03.5 Development of Nonlinear Models

After Hansch's development of the parabolic model, a number of other methodolgies were formulated to more clearly reflect biological phenomena in nonlinear situations.

4.03.5.1 Franke's Contributions

Franke made two significant contributions to ligand–receptor interactions. In 1970, he was one of the first scientists to recognize the importance of the coefficients with the hydrophobicity terms in the general expression[92]:

$$\log K_i = a_B \bullet \log P + b_B \qquad [26]$$

He suggested that if $a_B = 0.5 \pm 0.2$, a ligand would bind on the surface of a receptor and if $a_B = 1 \pm 0.2$, it would be transferred into the interior of a hydrophobic pocket.[93] Extensive QSAR analyses from the Hansch laboratory were in agreement with these observations.[94] Ten years later, the advent of molecular graphics and the availability of x-ray crystal structures of adequate resolution allowed us to clearly link the coefficients of the hydrophobic parameter with the extent of desolvation at the receptor site.[95]

Secondly, in 1973, Franke also formulated an empirical model based on a linear ascending portion and a parabolic part. His model was later described by a modified eqn [24][93,96]:

$$\log 1/C = a\{\log(P > P_X)\}^2 + b \bullet \log P + c \qquad [27]$$

In this equation, biological activity was linear up to a limiting value of $\log P_X$ and then assumed nonlinearity beyond that point. The formulation of other theoretical models such as those of Higuchi and Davis, and of Hyde, dealing with biological systems under equilibrium/pseudoequilibrium states, led to further consideration of kinetically based models such as the one proposed by McFarland.[97–99] This latter model was characterized by symmetrical curves with linear ascending and descending slopes flanking a middle parabolic part within the range of optimum hydrophobicity.[99]

4.03.5.2 Kubinyi's Model

The most significant development in the formulation of models that addressed nonlinearity in biological systems was based on computer simulations by Kubinyi of a drug being transported via aqueous and lipoidal compartment systems.[100,101]

$$\log 1/C = a \log P - b \log(B \bullet P + 1) + c \qquad [28]$$

In eqn [28], B represented the ratio of the volumes of the lipoidal phase and the aqueous phase. The tepee-shaped curve had an ascending slope, a and a descending slope, $b - a$. In the last 25 years, the bilinear approach has been applied to not only QSAR with hydrophobicity parameters but also to numerous QSAR delineating other parameters such as molar refractivity, pK_a, and molecular weight with great precision and insight.[48,102,103]

4.03.6 Early Substituent Selection Methods

As QSAR analysis came into widespread use in the 1970s, early practitioners quickly recognized the perils of improper substituent selection. Craig suggested the use of simple two-dimensional 'Craig' plots, e.g., σ versus π, not only to guide further substituent selection once an initial analysis had been carried out, but also to ensure adequate spanning of chemical space.[104] The Topliss operational scheme was developed in 1972 for both aliphatic and aromatic systems with a fixed number of substituents.[105] One began with a 4-chloro analog and then sequential moves were made depending on the magnitude of the biological activity of the 4-chloro analog compared to the parent molecule. Depending on whether activity was enhanced, decreased, or remained the same, one then moved on to the next appropriate substituent in terms of its σ or π value. This type of systematic analysis resulted in the generation of a Topliss decision tree.

The stepwise nature of the approach predicated on its 'choose and test' routine could be slow in terms of biological testing, but Topliss' batchwise scheme with an initial set of five analogs was more useful.[106] When Unger joined the Hansch group, a computer-based batch selection method was developed; cluster analysis, which advocated the testing of one member from each cluster to determine the dominant substituent followed by an emphasis on the members of its particular cluster which were chosen for their wide-ranging physicochemical attributes.[107] In order to ease the substituent selection process, an extensive set of aromatic substituents was composed with their associated physicochemical parameters.[108]

Another methodology that addressed sampling issues in screening was fractional factorial design, first proposed by the Wold group.[109] The utility of this approach was based on the prior calculation via principal component analysis of the principal properties of the aromatic substituents in the study. Since it was not possible to construct a factorial design based on exactly classified high and low levels, the substituents were generally classified according to the magnitude and sign of their principal properties. Using three variables at two levels, (+) for high and (−) for low, involved $2^3 = 8$ experiments. The outcome was then evaluated by partial least squares (PLS).[109] This method allowed for minimization of the compounds utilized in a QSAR study, which subsequently provided good predictability.

4.03.7 Evolution of other Shape Descriptors

Verloop's STERIMOL descriptors focused specifically on substituent shape. Simon *et al.* chose to examine the shape of the whole molecule by employing 'superimposition' techniques to determine the minimal steric difference (MSD) and the minimum topological difference (MTD). They defined 'hypermolecules' which were then used with hydrophobic and electronic parameters in regression analysis.[110,111] Treatment of the hypermolecules as rigid entities, limited the utility of this approach. Hopfinger's molecular shape analysis (MSA) was built on Motoc's Monte Carlo version of MSD by using molecular mechanics to determine accessible conformations and their superimposition on each other.[112,113] Overlap volumes (V_O) were determined and used as de novo descriptors in standard regression analysis.[114]

Distance geometry approach was developed by Ghosh and Crippen in the early 1980s.[115] Low-energy conformations of the three-dimensional structures of the ligands were determined and upper and lower boundaries of the various ligand points (atoms/groups) were characterized. The binding sites (empty/occupied) were defined as well as different binding modes and then the site points were classified according to the nature and intensity of the interactions. The minimum number of site points was then determined and empirical parameters were calculated using least square procedures.[116] Crippen ably demonstrated the utility of this approach in the QSAR of dihydrofolate reductase inhibitors.[117]

4.03.8 Developments in the 1980s

4.03.8.1 Advent of Molecular Graphics

Although high-performance computations were a key component in the development of QSAR algorithms, they were not widely utilized in macromolecular modeling due to the prohibitive cost and the lack of adequate software. Fortunately, the situation began to change in 1980 with the development of high-performance molecular graphics software and dramatic decreases in costs for the appropriate hardware. Blaney was instrumental in helping to demonstrate the coherence between a computer graphics visualization of ligand–receptor interactions and QSAR models.[118] The simultaneous development of real-time interactive color graphics, Connolly's sophisticated molecular surface program, and Bash *et al.*'s van der Waal's dot-surface program dramatically altered the molecular modeling landscape. Using x-ray crystallography structures of various receptors, it became possible to visualize the binding sites and the interactions with appropriate ligands and this both substantiated and augmented the results obtained from the

QSAR analysis. One of the first molecular graphics–QSAR studies focused on the interactions of X-phenyl hippurates with the cysteine protease, papain.[119] The associated QSAR is described by

$$\log 1/K_{\mathrm{M}} = 1.03\pi'_3 + 0.57\sigma + 0.61\mathrm{MR}_4 + 3.80. \qquad [29]$$

$$n = 25, \quad r = 0.907, \quad s = 0.208$$

The coefficient with the hydrophobic term suggested that total desolvation of the substituent occured on binding to the receptor. The π'_3 signified that the more hydrophobic of the *meta*-substituents contacted hydrophobic space while the less hydrophobic substituent gravitated toward the aqueous environment. The MR_4 term implied that binding may have occurred in polar space and that the bulk of the substituent helped to anchor the molecule firmly in its binding pocket. The graphics analysis substantiated these results and revealed that the *para*-substituents were in close proximity to the highly polar amide moiety of Gln142. The ρ value (the coefficient of the σ term) was consistent with values observed in studies with other serine and cysteine proteases.[120] Extensive collaborations on ligand–receptor interactions with members of the Langridge group including Blaney, McClarin, and Klein resulted in formulation of some basic empirical rules: coefficients around 0.5 with the hydrophobic parameter suggested that partial desolvation on a surface was operative while coefficients around 1.0 established that total desolvation in a hydrophobic pocket was occurring.[118]

The ability to actually visualize ligand–receptor interactions was a phenomenal event in the marriage of molecular graphics and QSAR. It undoubtedly changed the outlook and direction of molecular design.

4.03.8.2 Quantitative Structure–Activity Relationship in Cell Culture Systems

Another important development in the 1980s was the successful application of the QSAR paradigm to results obtained in various in vitro systems: cellular systems and bacterial cultures. Previously, there was a general feeling that the complexity of cellular systems would not allow for the formulation of QSAR models with consistency. Results with various mammalian cell lines and bacterial cultures proved otherwise.[121,122] A few representative QSAR examples, for the inhibition of growth of 4,6-diamino-1,2-dihydro-2,2-dimethyl-1-(3-X-phenyl)-s-triazines (I) in bacterial cultures and murine leukemia cells, respectively, are described as follows.

4.03.8.2.1 Inhibition of MTX-sensitive *Lactobacillus casei* cells by I[121]

$$\log 1/C = 0.80\pi'_3 - 1.06\log(\beta \cdot 10^{\pi_{3'}} + 1) - 0.94\,\mathrm{MR_Y} + 0.80\,\mathrm{I} + 4.37 \qquad [30]$$

$$n = 34, \quad r = 0.929, \quad s = 0.371, \quad \log\beta = -2.45, \quad \pi'_{3\mathrm{optimum}} = 2.94$$

4.03.8.2.2 Inhibition of growth of L5178Y (murine leukemia cells) by I[122]

$$\log 1/C = 1.34\pi_3 - 1.69\log(\beta \cdot 10^{\pi_{3'}} + 1) + 0.58\,\mathrm{I} + 0.75\sigma + 7.87 \qquad [31]$$

$$n = 37, \quad r = 0.942, \quad s = 0.254, \quad \log\beta = -2.49, \quad \pi_{3\mathrm{optimum}} = 0.82$$

The usefulness of these studies in in vitro systems was twofold. It the determination of the optimum $\log P$ in cellular systems and also encouraged comparisons with the QSAR for isolated receptors to assess the degree of correspondence between the two systems and gauge if cellular transport was of overriding importance in cellular/organismal toxicity.

4.03.9 New Quantitative Structure–Activity Relationship Approaches

The advent of combinatorial chemistry and high-throughput screening (HTS) has greatly challenged existing and conventional QSAR methodologies and has led to the development of faster and more efficient approaches, particularly for large and diverse data sets. Methods that fall under this umbrella include molecular hologram QSAR (HQSAR), binary QSAR, inverse QSAR, and comparative QSAR (C-QSAR), which will be discussed below.

4.03.9.1 Molecular Hologram Quantitative Structure–Activity Relationship (HQSAR)

This method is strictly based on the two-dimensional connectivity information and requires no explicit three-dimensional information for the ligands.[123] HQSAR involves the identification of all branched, cyclic, and overlapping substructural fragments in sets of molecules that are relevant to biological activity. Each fragment is assigned an integer value by utilizing a cyclic redundancy check algorithm. The structure fragments from each original molecule consist of a minimum and maximum number of atoms which are fragment size parameters. The generated strings are then hashed into a fixed length array. This array constitutes a molecular hologram and the associated bin occupancies form the descriptor variables. The computed molecular holograms for a data set yield a data matrix pertaining to the number of compounds and the length of the molecular hologram. A matrix of biological activities is also created. By using the PLS technique of Wold[124] a relationship is developed between substructural features in the dataset and biological activity. Leave-one-out (LOO) cross-validation is used to determine the number of components that yield the best predictive model. PLS leads to the formulation of a model that correlates biological activity with the appropriate molecular hologram bin value as described by:

$$\text{Biological activity} = \sum_{i=1}^{L} X_{iL} C_L + C_O \qquad [32]$$

where, X_{iL} is the occupancy value of the hologram of compound I at position or bin L, while C_L is the coefficient for that bin. As with other QSAR methods, it is important to choose parameters wisely. Key parameters include hologram length, fragment size, and fragment type (atom versus bonds, etc.). The speed and utility of this approach in terms of predictivity and critical atomic contributions is well described in several recent applications.[125,126]

4.03.9.2 Binary Quantitative Structure–Activity Relationship (Binary QSAR)

Since HTS often produces a large amount of screening data which classifies biological activity as 'active' or 'inactive,' the usage of conventional approaches is ruled out. Binary QSAR presents an approach that can ably handle this type of response. It was first introduced by Labute,[127] who correlated binary measurement data (1 = active, 0 = inactive) with the molecular descriptors of compounds. A probability distribution for actives and inactives is then determined, based on Bayes' Theorem.

An old problem that has to be dealt with in binary QSAR is variable selection, since hundreds of molecular descriptors are now available. Borrowing strategies from conventional QSAR has led to the use of genetic algorithms as the variable selection method of choice in recent studies by Gao *et al.* on inhibitors/ligands binding to the estrogen receptor, carbonic anhydrase II, and monoamine oxidase.[128] In this and other studies, molecular descriptors run the gamut from physicochemical properties ($\log P$) to connectivity and electrotopological indices and molecular diversity (BCUT) metrics.[129] The quality of a binary QSAR is assessed by three measures of performance: accuracy on actives, inactives, and all of the compounds. Gao *et al.* have shown that genetic algorithms-based, derived binary QSAR models have higher predictivity and use far fewer descriptors, which is of critical importance in virtual screening of large libraries.[128]

4.03.9.3 Inverse Quantitative Structure–Activity Relationship (Inverse QSAR)

In 1996, the Tropsha group introduced a new approach to rational design of targeted chemical libraries that was tagged as inverse QSAR. It utilizes a preconstructed QSAR to generate a virtual compound library with high, predicted, biological activity. The optimization of the library was obtained by utilizing genetic algorithms. The key steps in this approach include description, evaluation, and optimization.[130]

Various descriptors such as Kier and Hall's topological descriptors, or the amino acid Z descriptors (Z_1 = hydrophobicity, Z_2 = bulk, Z_3 = electronic), isotropic surface area (ISA), and electronic charge index (ECI) in a string, have been used to model the structures of virtual bradykinin-potentiating pentapeptides.[130] In this study, the effectiveness of the genetic algorithms optimization method was well established by the statistics. There are certain limitations to this approach including the sheer number of structural descriptors that can potentially be sampled, as well as the inability to provide solution molecules that are absent from the database. Faulon's group has addressed these issues by introducing the signature descriptor.[131] The signature of a molecule is defined as the linear combination of its atomic signatures. They recently applied this descriptor in an inverse QSAR study of intercellular adhesion molecules-1 (ICAM) inhibitory peptides.[132] Two of the predicted inhibitors were synthesized, tested, and confirmed as the strongest inhibiting peptides. Some of the newly designed compounds can be combined with the original members of the training set to construct a more focused training set, which bodes well for the generation of higher-quality lead compounds.

4.03.9.4 Comparative Quantitative Structure–Activity Relationship (Comparative QSAR)

Up to the early 1980s, a few hundred QSAR had been slowly accumulated, which grew by leaps and bounds with support from R.J. Reynolds and the National Institutes of Health in the 1990s. Hoekman was instrumental in designing the database which allowed for facile loading, search, and retrieval of QSAR pertaining to biological or physical systems. The intricacies of the C-QSAR program are described elsewhere.[133,134] The availability of such an organized analysis of a large body of various chemical reactions and biological interactions brought the realization that the usage of certain standard, well-developed, transparent parameters could allow for a systematized approach to comparative QSAR.

The development of a science of QSAR is based on the ability to make lateral comparisons between physicochemical-based models and/or biological models in order to validate original hypotheses. Thus the process of 'lateral validation' is critical to the establishment and enhancement of mechanistic interpretation of various QSAR models.[135,136] One of the first inroads in this area was a comparative QSAR study of radical reactions of benzene derivatives in chemistry and biology by Hansch and Gao.[137] Some unexpected discoveries have resulted from the availability of such a massive database. Two examples will be described here: one, involving QSAR pinpointing allosteric effects and another focusing on a simple valence electron parameter (NVE) that is surprisingly effective at delineating polarizability effects.[138–140]

Li *et al.* recently examined the structure-based design, synthesis, and enzymatic evaluation of a series of 22 indazole derivatives and inhibitors of methylthioadenosine (MTA) nucleosidase.[138] Using their data, Mekapati *et al.* obtained the following QSAR for the inhbition of binding of 5-biarylsulfonamides of trisubstituted indazoles to *Escherichia coli* MTA nucleosidase.[139]

$$\log 1/K_i = -1.64\text{Clog P} + 0.15\text{Clog P}^2 + 0.941I + 10.89 \quad [33]$$

$$n = 19, \quad r^2 = 0.900, \quad s = 0.231, \quad q^2 = 0.846, \quad \text{Minimum Clog P}: 5.33$$

An inverted parabola, corresponding to inhibitory potency, is seen as hydrophobicity increases from 4.35 till it reaches 5.33 and then activity starts to increase. I is an indicator variable which acquired a value of 1 if a chloro substituent is present on the indazole ring. The inverse parabolic relationship between activity and either Clog P or CMR has been observed in a substantial number of QSAR. It suggests that MTA nucleosidase, a key regulatory enzyme that mediates the expression of virulence factors, may operate via an allosteric mechanism. The crystal structure of the enzyme reveals a very prominent hydrophobic 'northern' pocket which is a low substrate affinity site and a hydrophobic channel near the dimer interface, which could well be the high substrate affinity site. These results suggest that the most active inhibitor may bind strongly to the enzyme's low substrate affinity site and induce an allosteric effect.

A new, rapid, easily calculable polarizability parameter, NVE, was developed and shown to be effective at delineating various chemicobiological interactions.[140] NVE represents the total number of valence electrons, i.e., $n_\sigma + n_\pi + n_n$. It shows excellent correlations with molecular polarizability, α ($n = 146, r^2 = 0.987$), as well as calculated molar refraction ($n = 146, r^2 = 0.992$).

4.03.10 New Descriptors or a New Look at Old Descriptors

In the same manner that HTS and combinatorial chemistry have affected QSAR methodologies, they have also impacted the descriptors used to formulate QSAR models. There is obviously a great demand for easily accessible descriptors and phenomenal advances in computational methodologies have contributed to the successful deployment of such descriptors. These descriptors may be physicochemical in nature or of a structural bent. The evolution of some of these descriptors, and their uses and limitations, are briefly examined below. A comprehensive compilation of diverse types of QSAR parameters is available.[141,142]

4.03.10.1 Hydrogen Bond Descriptors

The quantitative characterization of the ability of chemicals to act as hydrogen bond donors (HBD) or acceptors (HBA) is critical to ligand–receptor interactions and bioavailability since it also determines the solubility/partitioning of drugs between various phases. It is well established that the earliest quantitative scales of solvent hydrogen bond acidity and basicity were developed by Taft, Kamlet, and Abraham,[143,144] while studies on solute descriptors were undertaken by Kamlet, Taft, Abraham, and co-workers, who established the solvatochromism approach.[145–147] Raevsky has proposed the use of enthalpy factors (E_d and E_a) and free-energy factors to characterize the hydrogen bond binding ability of compounds. This thermodynamic data has been incorporated in the HYBOT program which contains the HYBOT.Database and the HYBOTFactor program.[148] Ghafourian and Dearden have utilized atomic charges and orbital energies to describe hydrogen bond donor capabilities.[149] These parameters were shown to correlate well with Abraham's hydrogen bond acidity parameter (α_Z^H). Recently, Katritzky's group has introduced descriptors that can characterize basicity and acidity features of both solutes and solvents by examining a nonconjugated aliphatic series.[150] Chemical structures are represented by topological distances, geometric distances, electronegativities, and hardness attributes of the sigma orbitals. These descriptors are denoted by B^* (basicity) and A^* (acidity). Excellent correlations were obtained for partition coefficients in different solvents. The correlation between $\log K_{oct}$ and various descriptors is shown below:

$$\log K_{oct} = 0.839 - 1.711B^* - 0.71A^* + 1.283\alpha + 0.169P^* \qquad [34]$$

$$n = 90, \quad r^2 = 0.970, \quad F = 682, \quad s = 0.233$$

where B^* and A^* represent the hydrogen bond basicity and acidity descriptors, respectively, while α represents molecular polarizability, and P^* is a dipolarity term. It remains to be seen whether this approach can also be applied to aromatic species. It should be noted that $\log P_{oct}$ does not depend on solute acidity, i.e., H donor strength, so eqn [34] needs further validation.

4.03.10.2 Quantum Chemical Indices

It is well established that one of the most widely used and successful empirical approaches to quantification of electronic effects in quantitative structure–property/activity (QSPR)/QSAR studies has been embodied in the Hammett equation. The limitations of the Hammett equation coupled with the lack of availability of particular or promising substituents has led to the development of computational-based electronic descriptors derived from semiempirical quantum chemical and molecular topological parameters by Tanji *et al.*[151] The following relationship was obtained for the training set:

$$\sigma = -2.480(\Delta N) - 7.894(\sigma E) - 0.605\{D_X \bullet EA_H / D_H \bullet EA_X\} + 0.009[\varnothing S_X] + 0.028(\Sigma\pi) + 0.279 \qquad [35]$$

$$n = 150, \quad r^2 = 0.958, \quad s = 0.073, \quad F = 657.6$$
$$\text{PRESS} = 0.371, \quad \text{PRESS/SSY} = 0.010$$

where ΔN represents the approximate fractional number of electrons transferred during a general acid–base reaction, ΔE denotes the global energy variations that occur in benzoic acids due to the interaction of frontier orbitals upon substitution, while EA_H, D_H, EA_X, and D_X correspond to the electron affinities and distance terms, respectively of benzoic acid and the substituted benzoic acid, respectively. S_X is the sum of the electrotopological state indices for the atoms of a given substituent group ($\sum_{i=1}^{X}$ Si). Since the electrotopological state is fundamentally insensitive to group position, Ø was introduced to characterize certain structural and electronic properties of the substituent atom directly attached to the phenyl ring and based on its topological distance, principal quantum number (N), and its group number (G). The descriptor $\Sigma\pi$ is a count of all p electrons present in a given substituent group. Internal and external validation procedures, as well as extensive correlations with reaction data, suggest that this computational approach provides a convenient and effective alternative to experimental measurement.

The use of quantum chemical descriptors has also been advocated by Katritzky's group.[152] Thus, semiempirical methods such as extended Huckel theory (EHT), complete, intermediate and modified intermediate neglect of differential overlap (CNDO, INDO, MINDO), PM3, and AM1 have all been used to derive various descriptors. The AM1 method has seen the most extensive usage because of the speed of calculations, the availability of parameterization for a variety of atoms, its reliability and ability to accurately calculate electronic effects.[153] Quantum chemical indices and hydrophobicity have been critical in delineating the QSAR for the mutagenicity of various organic compounds in our laboratory.[154,155] Recent studies from our laboratory have utilized bond dissociation energies (BDE) to describe the radical-based cytotoxicities in mammalian cells of a large number of substituted phenols.[156] The BDE descriptor addresses the ease of homolytic cleavage of the OH bond in phenols with electron-donating substituents.

$$\text{Log } 1/\text{IC}_{50} = -0.21\ \text{BDE} + 0.21 \log P + 3.11 \qquad [36]$$

$$n = 52, \quad r^2 = 0.920, \quad q^2 = 0.909, \quad s = 0.202$$

4.03.10.3 Topological Descriptors

Topological descriptors have garnered considerable interest in the last few years because of the ease of generation and the speed with which these computations can be completed. They have been extensively used in modeling physicochemical properties as well as biological activities.[157] Hall and Hall have examined three ADME processes (aqueous solubility, genotoxicity, human intestinal absorption) and formulated appropriate artificial neural network (ANN) models using *E*-state descriptors and connectivity indices and in some cases calculated log *P* using the CSLogP program. In a fish toxicity study, Rose and Hall utilized multiregression analysis and *E*-state indices to develop a robust model that predicted fish toxicity of a miscellaneous set of 92 organic compounds.[158]

There has been criticism of topological descriptors on two fronts. It concerns the nature and relevancy of the relationship between the descriptor and the dependent variable, which in QSPR and QSAR may be a property or biological activity.[159] Charton has addressed this issue by suggesting that topological descriptors are composite parameters that are counts of numbers of atoms, bond, branching, and electrons but not measures of electron accessibility.[160] Meanwhile, Hall and Hall have criticized the mechanism-based approach to QSAR development, and strongly advocated the use of a structure information-based approach that is not susceptible to erroneous assumptions with regard to conformations, hypotheses approximations, and dynamic crossing of membranes.[157] Despite criticism, there has been renewed interest in not only utilizing the well-developed *E*-state descriptors but also in developing new graph theoretical descriptors to deal with questions of similarity and diversity and 'druglike' and 'non druglike' phenomena.[161]

4.03.11 Recent Applications of Quantitative Structure–Activity Relationship

The two areas that have seen an upsurge in applying QSAR techniques and descriptors to chemical–biological interactions are ADME phenomena and predictive toxicity. One cannot underestimate the critical role played by lipophilicity in governing pharmacokinetic and pharmcodynamic processes. Recent refinements of log *P* predictive tools allow for the lipophilicity determination of each conformer of a molecule. In an elegant study using molecular lipophilicity potentials, Gaillard *et al.* were able to explain differences between calculated and experimental log *P* values in terms of folded and extended conformations.[162] Kraszni *et al.* were the first to report on experimental conformer-specific partition coefficients in octanol/water of amphetamine and clenbuterol.[163] The observed $\log P_{oct}$ value of only one of the rotamers of clenbuterol approached the predicted value, which suggests that the predictive method (KOWWIN) is adept at approximating lipophilicity of the conformer where intramolecular interactions are minimal or absent. Recently, Caron and Ermondi have used VolSurf-generated two-dimensional molecular descriptors to compute neutral $\log P^{N}_{alkane}$ and $\Delta\log P^{N}_{oct-alk}$ as well as $\log D^{pH}_{alkane}$ for the ionized species.[164] In addition to its value in predicting brain permeability and fraction of dose absorbed orally by humans ($\Delta\log P^{N}_{oct-alk}$) as well as artificial membrane permeability ($\log P^{N}_{alk}$), this approach has a high degree of precision and thus reliability. Lipophilicity data from different solvents can easily be compared, compound flexibility is always considered and chemical delineation is based on the GRID force field which is in continuous development by incorporation of newly obtained x-ray crystallography data.

The search for the 'druglikeness' of potential drug molecules has led to an emphasis on in vitro ADME screens to flush out compounds that do not first pass muster in terms of the Lipinski's 'rule of five.'[165] Prediction of intestinal absorption has focused on Caco-2 monolayers as model systems.[165] Size and polar surface area were determined to be significant in influencing permeability. QSAR analyses from our laboratory[166] indicate that hydrophobicity is also of paramount importance in intestinal absorption with an optimum log *P* between 2.35 and 3.80 for five different sets of drugs.

4.03.11.1 Penetration of Drugs through the Blood–Brain Barrier

Delineation of penetration of the blood–brain barrier (BBB) continues to draw the attention of medicinal chemists because of its importance to the design of drugs targeting the central nervous system (CNS), as well as the difficulty and high costs of experimental testing. Rapid scoring is necessary to sweep through extensive databases. Many attempts have been made at predicting log BBB[167], but only a few more recent ones will be addressed. Clark used Clog P and polar surface area (PSA) to accurately predict BBB penetration of 55 compounds,[168] while Feher *et al.* used three descriptors (Clog P, PSA, and the number of hydrogen bond acceptors) to formulate their simple model. Iyer *et al.* used a unique approach to addressing this problem.[169] In addition to utilizing the properties of the solutes, they also incorporated the attributes of the cellular membrane into their model. Thus, a set of membrane–solute intermolecular properties were computed via molecular dynamic simulations and used to develop a relevant QSAR/MI-QSAR model:

$$\log BB = -0.0015 - 0.0235\ PSA + 0.1673\ Clog\ P - 0.0076\ E_{MS} + 0.0388\ E_{SS} + 0.01\ \Delta E_{TT} - 0.0037\ \Delta E_{TT(S/B)} \qquad [37]$$

$$n = 56, \quad r^2 = 0.855, \quad q^2 = 0.792$$

The descriptors are defined as follows: Clog P is the calculated $\log P_{oct}$, PSA is the polar surface area, E_{MS} is the interaction energy between the solute and dimyristoylphosphatidylcholine (DMPC) monolayer, E_{SS} is the torsion energy of the solute, ΔE_{TT} reflects the change in the 1,4 nonbonded interaction energy of the system, and $\Delta E_{TT(S/B)}$ is the change in the sum of the total stretching and bending energy of the complex system. Thus the three critical physicochemical parameters in membrane interaction, blood–brain partition include the relative polarity of the solute, the strength of the binding interaction between solute and membrane, and the conformational flexibility of the solute in the membrane. This approach is not only predictive in scope but mechanistic in nature.

Ma *et al.* utilized this membrane-based approach to develop a predictive model for BBB penetration of organic compounds.[170] They used five descriptors: Clog P, PSA, Balaban index, ΔE_{total}, and $\Delta E_{torsion}$. The energy terms are related to changes in the potential energy and dihedral torsion energy of the solute–membrane–water complex and the DMPC-water model, respectively. Hou and Xu used three descriptors, Slog P, PSA, and excess molecular weight (MW-360), to model BBB penetration.[171] Slog P is an atom-based calculation model for octanol/water partition coefficients developed by the Crippen group.[172] All these models emphasize the importance of hydrophobicity and polar surface area of the solutes in BBB portioning behavior. The Iyer and Ma models take it a step further and consider the relevancy of the interactions of the solutes with phospholipid-rich areas of cellular membranes in BBB distribution.

4.03.11.2 Toxicity Considerations Using Quantitative Structure–Activity Relationship

In the last decade, QSARs have been used aggressively to shed light on a number of toxicological endpoints such as acute toxicity, mutagenicity, carcinogenicity and apoptosis.[173–176] Cronin and Schultz have advocated for the development of QSAR models that are clearly transparent and mechanistically comprehensible.[174] One must keep in mind that these types of data are inherently highly variable and thus, QSAR models based on them should not be expected to explain more than 80% of the data. Descriptors should be simple and accurate and should not defy comprehension/interpretability. Even with large volumes of data, the availability of well-defined and transparent descriptors makes their utility in mechanistically based model development a necessity.

In an excellent recent review on SAR of chemical mutagens and carcinogens, Benigni stressed the importance of careful evaluation and interpretation of QSAR models.[177] The systematic comparison of QSAR models in the form of lateral validation enhances understanding of a particular chemicobiological interaction and opens it up to judicious mechanistic interpretation. Learning about the potential toxicity of a new chemical entity early on in the drug development process is both time and cost effective and useful in terms of risk assessment.

4.03.12 Future Perspectives

It has been 45 years since the formal beginning of QSAR and great strides have been made in that time. The types of QSAR approaches have increased and some have undergone refinement, while the number of parameters and new descriptors have grown astronomically. The applications of QSAR have evolved from 'simple' systems such as narcosis to more complex problems in 'multi-omics' and drug design. QSAR is far from being a finished science; it still retains the ability to predict biological activities or properties as well as the susceptibility to mechanistic interpretation. Enhancing the predictability of a model raises the specter of appropriate validation procedures.[178,179] He and Jurs[179] emphasized the importance of the reliability of a QSAR's model prediction using acute, aquatic, toxicity data from fathead minnows. They demonstrated a strong relationship between the similarity of their query compounds to the appropriate training set compounds of the QSAR model and the predictive ability of that model. What other criteria should be utilized to assess the reliability of a QSAR? It is clear that the goodness of fit (r^2, s, F) and the goodness of prediction are of critical importance. Thus, leave-one-out cross-validation and Y scrambling procedures both adequately fulfill these requirements.[180,181] Finally, lateral validation of a QSAR with well-honed descriptors that have withstood the test of time enhances confidence in its outcomes.[182,183] With superior hardware and software, these methods can easily be utilized in QSAR studies.

Understanding and modeling the action of larger macromolecules such as DNA and RNA in more complex biochemical systems, e.g., signaling pathways, remain a challenge. But improvements in hardware and software all point to closure in this gap of knowledge. In cellular systems or whole organisms, chemicobiological complexity increases as a result of ADME and toxicity concerns. It is expected that these two areas will see the greatest growth in terms of QSAR modeling. Caution must be exercised in the disentangling of the variables that affect these complex processes. The examples of modeling BBB penetration via QSAR is only the tip of the iceberg. QSAR of metabolism of diverse organic compounds by the various cytochrome P450 isozymes[184] underscores the complexity of this process. Extensive studies to delineate the factors that impact acute and chronic toxicity of organic molecules have yet to be carried out in mammalian systems, although some headway has been made in aquatic organisms.[185]

Lastly, as QSAR databases continue to grow, these repositories of information will be invaluable in the early stages of drug design and development. They hold great promise in terms of data mining; input of the structure of a potential lead molecule for a new target could reveal whether similar structures are in the database and the types of chemical, pharmacological, and toxicological activities that are associated with them. Similarity scoring could be used to isolate and focus on related structures in the database. This could result in appreciable savings of time, resources, and money.

QSAR will continue to play a role in molecular design, particularly at the lead optimization stage. It can complement other strategies like data mining/docking, or pharmacophore modeling. Perhaps two-dimensional QSAR will evolve so that it is able to utilize multiple conformers of ligands and their attendant physicochemical descriptors (particularly since conformer-specific $\log P_{OW}$ now appears to be measurable) to disentangle the variables that modulate biological activity.

Acknowledgments

We would like to thank the National Institutes of Health [NIH grant #R01ES7595 (CH/CS)], R. J. Reynolds (RJR-082300), and BioByte Corp. (BB-011005) for their financial assistance over many years.

References

1. Rozman, K. K.; Doull, J. *Toxicology* **2001**, *160*, 191–196.
2. Crum-Brown, A.; Fraser, T. R. *Trans. Roy. Soc. Edinburgh* **1868**, *25*, 151–203.
3. Richet, M. C. *C. R. Soc. Biol. (Paris)* **1893**, *45*, 775–776.
4. Meyer, H. *Arch. Exp. Pathol. Pharmakol.* **1899**, *42*, 109–118.
5. Overton, C. E. *Gustav Fischer*; Jena: Switzerland, 1901; English translation by Lipnick, R. L.; Chapman and Hall: London, 1991.
6. Albert, A.; Goldacre, R. J.; Phillips, J. *J. Chem. Soc.* 1948, 2240–2249.
7. Bell, P. H.; Roblin, R. O., Jr. *J. Am. Chem. Soc.* **1942**, *64*, 2905–2917.
8. Hammett, L. P. *Chem. Rev.* **1935**, *17*, 125–136.
9. Hammett, L. P. *Physical Organic Chemistry*, 2nd ed.; McGraw-Hill: New York, 1970.
10. Burkhardt, G. N. *Nature* **1935**, *136*, 684.
11. Hammett, L. P. *J. Am. Chem. Soc.* **1937**, *59*, 96–103.
12. Jaffe, H. H. *Chem. Rev.* **1953**, *53*, 191–261.
13. Exner, O. *Correlation Analysis of Chemical Data*, 2nd ed.; Plenum Press: New York, 1988, pp 55–138.
14. Fujita, T.; Nishioka, T. *Prog. Phys. Org. Chem.* **1976**, *12*, 49–89.
15. Selassie, C. In *Burger's Medicinal Chemistry and Drug Discovery, Vol. 1, Drug Discovery*, 6th ed.; Abraham, D. J., Ed.; John Wiley: Hoboken, NJ, 2003, pp 13–14.
16. Okamoto, Y.; Brown, H. C. *J. Org. Chem.* **1957**, *22*, 485–494.
17. Johnson, C. D. *The Hammett Equation*; Cambridge University Press: London, 1973, pp 28–31.
18. Roberts, J. D.; Moreland, W. T., Jr. *J. Am. Chem. Soc.* **1953**, *75*, 2167–2173.
19. Exner, O. *Correlation Analysis of Chemical Data*, 2nd ed.; Plenum Press: New York, 1988, pp 147–152.
20. Taft, R. W. In *Steric Effects in Organic Chemistry*; Newman, M. S., Ed.; John Wiley: New York, 1956, pp 556–675.
21. Bowden, K. In *Comprehensive Medicinal Chemistry*, Vol. 4, *Quantitative Drug Design*; Hansch, C., Sammes, P. G., Taylor, J. B., Eds.; Pergamon Press: Oxford, 1990, pp 211–212.
22. Taft, R. W., Jr. *J. Am. Chem. Soc.* **1952**, *74*, 3120–3128.
23. Charton, M. *J. Org. Chem.* **1964**, *29*, 1222–1227.
24. Swain, C. G.; Lupton, E. C., Jr. *J. Am. Chem. Soc.* **1968**, *90*, 4328–4337.
25. Hansch, C.; Leo, A.; Taft, R. W. *Chem. Rev.* **1991**, *91*, 165–195.
26. Hancock, C. K.; Meyers, E. A.; Yager, B. J. *J. Am. Chem. Soc.* **1961**, *83*, 4211–4213.
27. Charton, M. *J. Am. Chem. Soc.* **1975**, *97*, 1552–1556.
28. Charton, M. *J. Org. Chem.* **1976**, *41*, 2217–2220.
29. Hansch, C.; Leo, A. In *Substituent Constants for Correlation Analysis in Chemistry and Biology*; Wiley-Interscience: New York, 1979, p 10.
30. Fujita, T.; Takayama, C.; Nakajama, M. *J. Org. Chem.* **1973**, *38*, 1623–1630.
31. Fujita, T.; Iwamura, H. In *Topics in Current Chemistry*; Charton, M., Motoc, I., Eds.; Springer-Verlag: Berlin, Germany, 1983, pp 119–157.
32. Verloop, A.; Hoogenstraten, W.; Tipker, J. In *Drug Design.*; Ariens, E. J., Ed.; Academic Press: New York, 1976; Vol. III, pp 176–207.
33. Verloop, A. *The STERIMOL Approach to Drug Design*; Marcel Dekker: New York, 1987.

34. Verloop, A. In *QSAR and Strategies in the Design of Bioactive Compounds*, Proceedings of the 5th European Symposium on QSAR; Seydel, J. K., Ed.; VCH: Weinheem, Germany, 1985, pp 98–104.
35. Tute, M. S. In *Comprehensive Medicinal Chemistry, Vol. 4, Quantitative Drug Design*; Hansch, C., Sammes, P. G., Taylor, J. B., Eds.; Pergamon Press: Oxford, 1990, p 18.
36. Pauling, L.; Pressman, D. *J. Am. Chem. Soc.* **1945**, *67*, 1003–1012.
37. Van der Waterbeemd, H.; Testa, B. *Adv. Drug. Res.* **1987**, *16*, 85–225.
38. Hansch, C.; Grieco, C.; Silipo, C.; Vittoria, A. *J. Med. Chem.* **1977**, *20*, 1420–1435.
39. Selassie, C. D.; Klein, T. E. In *Comparative QSAR*; Devillers, J., Ed.; Taylor and Francis: New York, 1998, pp 235–284.
40. Hansch, C.; Blaney, J. M. In *Drug Design: Fact or Fantasy?*; Jolles, G., Wooldridge, K. R. H., Eds.; Academic Press: London, 1984, pp 185–208.
41. Hansen, O. R. *Acta. Chem. Scand.* **1963**, *16*, 1593–1600.
42. Hansch, C.; Fujita, T. *J. Am. Chem. Soc.* **1964**, *86*, 1616–1626.
43. Hansch, C.; Maloney, P. P.; Fujita, T.; Muir, R. M. *Nature* **1962**, *194*, 178–180.
44. Veldstra, H. *Ann. Rev. Plant Physiol. Plant Mol. Biol.* **1953**, *4*, 151–178.
45. Fujita, T.; Iwasa, J.; Hansch, C. *J. Am. Chem. Soc.* **1964**, *86*, 5175–5180.
46. Leo, A.; Hansch, C.; Elkins, D. *Chem. Rev.* **1971**, *71*, 525–616.
47. Hansch, C.; Leo, A. *Exploring QSAR: Fundamentals and Applications in Chemistry and Biology;* American Chemical Society: Washington, DC, 1995, pp 118–121.
48. Taylor, P. In *Comprehensive Medicinal Chemistry, Vol. 4, Quantitative Drug Design*; Hansch, C., Sammes, P. G., Taylor, J. B., Eds.; Pergamon Press: Oxford, 1990, pp 270–271.
49. Leo, A. J.; Hansch, C. *Persp. Drug Disc. Design* **1999**, *17*, 1–25.
50. Tomlinson, E. *J. Pharm. Sci.* **1982**, *71*, 602–604.
51. Unger, S. H.; Cheung, P. S.; Chiang, G. H.; Cook, J. R. In *Partition Coefficient, Determination and Estimation*; Dunn, W. J., III, Block, J. H., Pearlman, R. S., Eds.; Pergamon Press: Elmsford, NY, 1986, pp 69–82.
52. Unger, S. H.; Cook, J. R.; Hollenbery, J. S. *J. Pharm. Sci.* **1978**, *67*, 1364–1367.
53. Mirrlees, M. S.; Moulton, S. J.; Murphy, C. T.; Taylor, P. J. *J. Med. Chem.* **1976**, *19*, 615–619.
54. Lewis, S. J.; Mirrlees, M. S.; Taylor, P. J. *Quant. Struct.-Act. Relat.* **1983**, *2*, 100–111.
55. Berthod, A.; Bully, M. *Anal. Chem.* **1991**, *63*, 2508–2512.
56. Vallat, P.; El-Tayar, N.; Testa, B.; Slacanin, I.; Martson, A.; Hostettmann, K. *J. Chromatogr.* **1990**, *504*, 411–419.
57. Berthod, A.; Carda-Broch, S.; Garcia-Alvarex-Coque, M. C. *Anal. Chem.* **1999**, *71*, 879–888.
58. Lombardo, F.; Shalaeva, M. Y.; Tupper, K. A.; Gao, F.; Abraham, M. H. *J. Med. Chem.* **2000**, *43*, 2922–2928.
59. Vrakas, D.; Hadjipavlou-Litina, D.; Tsantili-Kakoulidou, A. *J. Pharm. Biomed. Anal.* **2005**, *39*, 908–913.
60. Shih, P.; Pederson, L. G.; Gibbs, P. R.; Wolfenden, R. *J. Mol. Biol.* **1998**, *280*, 421–430.
61. Leahy, D. E.; Taylor, P. J.; Wait, A. R. *Quant. Struct.-Act. Relat.* **1989**, *8*, 17–31.
62. Leo, A. In *Methods in Enzymology.*; Langone, J., Ed.; Academic Press: San Diego, CA, 1991; Vol. 202, pp 544–591.
63. Leo, A. *Chem. Rev.* **1993**, *93*, 1281–1306.
64. Leo, A. In *Comprehensive Medicinal Chemistry, Vol. 4, Quantitative Drug Design*; Hansch, C., Sammes, P. G., Taylor, J. B., Eds.; Pergamon Press: Oxford, 1990, pp 302–319.
65. Weininger, D. *J. Chem. Info. Comput. Sci.* **1988**, *28*, 31–36.
66. Weininger, D. *User's Manual V 3.54*; Pomona College Medchem Project: Claremont, CA, 1989.
67. Nys, G. G.; Rekker, R. F. *Chim. Ther.* **1973**, *8*, 521–535.
68. Nys, G. G.; Rekker, R. F. *Eur. J. Med. Chem.* **1974**, *9*, 361–375.
69. Rekker, R. F. *Quant. Struct.-Act. Relat.* **1992**, *11*, 195–199.
70. Rekker, R. G.; Mannhold, R.; Bijloo, G.; deVries, G.; Dross, K. *Quant. Struct.-Act. Relat.* **1998**, *17*, 537–548.
71. Mannhold, R.; van de Waterbeemd, H. *J. Comput.-Aided Mol. Design* **2001**, *15*, 337–354.
72. Machatha, S. G.; Yalkowsky, S. H. *Intl. J. Pharm.* **2005**, *294*, 185–192.
73. Gargadennec, S.; Burgot, G.; Burgot, J. L.; Mannhold, R.; Rekker, R. F. *Pharm. Res.* **2005**, *22*, 875–882.
74. Gaillard, P.; Carrupt, P. A.; Testa, B.; Boudon, A. *J. Comput.-Aided Mol. Design* **1994**, *8*, 83–96.
75. Chuman, H.; Mori, A.; Tanaka, H.; Yamagami, C.; Fujita, T. *J. Pharm. Sci.* **2004**, *93*, 2681–2697.
76. Raevskii, O. A.; Grigor'ev, V. Y. *Pharm. Chem. J.* **1999**, *33*, 274–277.
77. Huunskonen, J. J.; Villa, A. E.; Tetko, I. V. *J. Pharm. Sci.* **1999**, *88*, 229–233.
78. Hansch, C.; Dunn, W. J., III. *J. Pharm. Sci.* **1972**, *61*, 1–19.
79. Hansch, C.; Leo, A. *Exploring QSAR: Fundamentals and Applications in Chemistry and Biology*; American Chemical Society: Washington, DC, 1995.
80. Hansch, C. *Acc. Chem. Res.* **1969**, *2*, 232–239.
81. Hansch, C.; Clayton, J. M. *J. Pharm. Sci.* **1973**, *62*, 1–21.
82. Hansch, C.; Steward, A. R.; Anderson, S. M.; Bentley, D. L. *J. Med. Chem.* **1968**, *11*, 1–11.
83. Penniston, J. T.; Beckett, L.; Bentley, D. L.; Hansch, C. *Mol. Pharmacol.* **1969**, *5*, 333–341.
84. Ferguson, J. *Chem. Ind.* **1964**, *20*, 818–824.
85. Ferguson, J. *Proc. R. Soc. Lond. B* **1939**, *127*, 387–404.
86. Free, S. M., Jr.; Wilson, J. W. *J. Med. Chem.* **1964**, 7, 395–399.
87. Kubinyi, H. In *Comprehensive Medicinal Chemistry, Vol. 4, Quantitative Drug Design*; Hansch, C., Sammes, P. G., Taylor, J. B., Eds.; Pergamon Press: Oxford, 1990, pp 589–643.
88. Fujita, T.; Ban, T. *J. Med. Chem.* **1971**, *14*, 148–152.
89. Kubinyi, H. In *QSAR: Hansch Analysis and Related Approaches*; Mannhold, R., Krogsgaard-Larsen, P., Timmerman, H., Eds.; VCH Publishers: New York, 1993, pp 62–68.
90. Hansch, C.; Calef, D. F. *J. Org. Chem.* **1976**, *41*, 1240–1243.
91. Verma, R. P. *Lett. Drug. Design Disc.* **2005**, *2*, 205–218.
92. Franke, R. *Acta Biol. Med. Germ.* **1970**, *25*, 757–788.
93. Franke, R.; Schmidt, W. *Acta Biol. Med. Germ.* **1973**, *31*, 273–287.
94. Hansch, C. *J. Med. Chem.* **1976**, *19*, 1–6.
95. Hansch, C.; Klein, T. E. *Acc. Chem. Res.* **1986**, *19*, 392–400.

96. Kubinyi, H. *Prog. Drug Res.* **1979**, *23*, 97–198.
97. Higuchi, T.; Davis, S. S. *J. Pharm. Sci.* **1970**, *59*, 1376–1383.
98. Hyde, R. M. *J. Med. Chem.* **1975**, *18*, 231–233.
99. McFarland, J. W. *J. Med. Chem.* **1970**, *13*, 1192–1196.
100. Kubinyi, H. *Arzneim.-Forsch.* **1976**, *26*, 1991–1997.
101. Kubinyi, H. *J. Med. Chem.* **1977**, *20*, 625–629.
102. Selassie, C. In *Burger's Medicinal Chemistry and Drug Discovery, Vol. 1, Drug Discovery*, 6th ed.; Abraham, D. J., Ed.; John Wiley: Hoboken, NJ, 2003, pp 31–33.
103. Selassie, C. D.; Hansch, C.; Khwaja, T. A. *J. Med. Chem.* **1990**, *33*, 1914–1919.
104. Craig, P. N. *J. Med. Chem.* **1971**, *14*, 680–684.
105. Topliss, J. G. *J. Med. Chem.* **1972**, *15*, 1006–1011.
106. Topliss, J. G. *J. Med. Chem.* **1977**, *20*, 463–469.
107. Hansch, C.; Unger, S. H.; Forsythe, A. B. *J. Med. Chem.* **1973**, *16*, 1217–1222.
108. Hansch, C.; Leo, A.; Unger, S. H.; Kim, K. H.; Nikaitaini, D.; Lien, E. J. *J. Med. Chem.* **1973**, *16*, 1207–1216.
109. Hellberg, S.; Sjostrom, M.; Skagerberg, B.; Wold, S. *J. Med. Chem.* **1987**, *30*, 1126–1135.
110. Simon, Z. In *3D QSAR in Drug Design*; Kubinyi, H., Ed.; Escom: Leiden, the Netherlands, 1993, pp 307–319.
111. Badilescu, I. I.; Craescu, T.; Niculescu-Duvaz, I.; Simon, Z. *Neoplasma* **1980**, *27*, 261–269.
112. Motoc, I. *Top. Curr. Chem.* **1983**, *114*, 93–105.
113. Hopfinger, A. J. *J. Am. Chem. Soc.* **1980**, *102*, 7196–7206.
114. Hopfinger, A. J. *J. Med. Chem.* **1983**, *26*, 990–996.
115. Crippen, G. M. *J. Med. Chem.* **1979**, *22*, 988–997.
116. Ghose, A.; Crippen, G. M. In *Computational Medicinal Chemistry, Vol. 4, Quantitative Drug Design*; Hansch, C., Sammes, P. G., Taylor, J. B., Eds.; Pergamon Press: Oxford, 1990, pp 715–733.
117. Ghose, A. K.; Crippen, G. M. *J. Med. Chem.* **1985**, *28*, 333–346.
118. Blaney, J. M.; Hansch, C. In *Computational Medicinal Chemistry, Vol. 4, Quantitative Drug Design*; Hansch, C., Sammes, P. G., Taylor, J. B., Eds.; Pergamon Press: Oxford, 1990, pp 459–496.
119. Smith, R. N.; Hansch, C.; Kim, K. H.; Omiya, B.; Fukumura, G.; Selassie, C. D.; Jow, P. Y. C.; Blaney, J. M.; Langridge, R. *Arch. Biochem. Biophys.* **1982**, *215*, 319–328.
120. Selassie, C. D.; Klein, T. E. In *3D QSAR in Drug Design*; Kubinyi, H., Ed.; Escom: Leiden, the Netherlands, 1993, pp 257–275.
121. Coats, E. A.; Genther, C. S.; Dietrich, S. W.; Guo, Z. R.; Hansch, C. *J. Med. Chem.* **1981**, *24*, 1422–1429.
122. Khwaja, T. A.; Pentecost, S.; Selassie, C. D.; Guo, Z. R.; Hansch, C. *J. Med. Chem.* **1982**, *25*, 153–156.
123. Heritage, T.; Lowis, D. R. In *Rational Drug Design*; Parrill, A. L., Reddy, M. R., Eds.; American Chemical Society: Washington, DC, 1999, pp 212–225.
124. Lindberg, W.; Persson, J.-A.; Wold, S. *Anal. Chem.* **1983**, *55*, 643–648.
125. Zhang, H.; Li, H.; Liu, C. *J. Chem. Inf. Model* **2005**, *45*, 440–448.
126. Honorio, K. M.; Garratt, R. C.; Andricopulo, A. D. *Bioorg. Med. Chem. Lett.* **2005**, *15*, 3119–3125.
127. Labute, P. In *Proceedings Pacific Symposium on Biocomputing '99*; Altman, R. B., Dunker, A. K., Hunter, L., Klein, T. E., Lauderdale, K., Eds.; World Scientific: River Edge, NJ, 1999, pp 444–455.
128. Gao, H.; Lajiness, M. S.; Van Drie, J. *J. Mol. Graphics Model.* **2002**, *20*, 259–268.
129. Gao, H. *J. Chem. Inf. Comput. Sci.* **2001**, *41*, 402–407.
130. Cho, S. J.; Zheng, W.; Tropsha, A. *J. Chem. Inf. Comput. Sci.* **1998**, *38*, 259–268.
131. Visco, D. P., Jr.; Pophale, R. S.; Rintoul, M. D.; Faulon, J. L. *J. Mol. Graphics Model.* **2002**, *20*, 429–438.
132. Churchwell, C. J.; Rintoul, M. D.; Martin, S.; Visco, D. P., Jr.; Kotu, A.; Larson, R. S.; Sillerud, L. O.; Brown, D. C.; Faulon, J. L. *J. Mol. Graphics Model.* **2004**, *22*, 263–273.
133. Hansch, C.; Hoekman, D.; Leo, A.; Weininger, D.; Selassie, C. D. *Chem. Rev.* **2002**, *102*, 783–812.
134. Kurup, A. *J. Comput.-Aided. Mol. Design* **2003**, *17*, 187–196.
135. Selassie, C. D.; Garg, R.; Kapur, S.; Kurup, A.; Verma, R. P.; Mekapati, S. B.; Hansch, C. *Chem. Rev.* **2002**, *102*, 2585–2605.
136. Hadjipavlou-Litina, D.; Garg, R.; Hansch, C. *Chem. Rev.* **2004**, *104*, 3751–3793.
137. Hansch, C.; Gao, H. *Chem. Rev.* **1997**, *97*, 2995–3059.
138. Li, X.; Chu, S.; Feher, V. A.; Khalili, M.; Nie, Z.; Margosiak, S.; Nikulin, V.; Levin, J.; Sprankle, K. G.; Tedder, M. E. et al. *J. Med. Chem.* **2003**, *46*, 5663–5673.
139. Mekapati, S. B.; Kurup, A.; Verma, R. P.; Hansch, C. *Bioorg. Med. Chem.* **2005**, *13*, 3737–3762.
140. Verma, R. P.; Hansch, C. *Bioorg. Med. Chem.* **2005**, *13*, 2355–2372.
141. Todeschini, R.; Consonni, V. In *Handbook of Molecular Descriptors, Vol. 11, Methods and Principles in Medicinal Chemistry*; Mannhold, R., Kubinyi, H., Timmerman, H., Eds.; Wiley-VCH: Weinheim, Germany, 2000.
142. Hansch, C.; Leo, A.; Hoekman, D. In *Exploring QSAR. Hydrophobic, Electronic and Steric Constants*; American Chemical Society: Washington, DC, 2000.
143. Kamlet, M. J.; Taft, R. W. *J. Am. Chem. Soc.* **1976**, *98*, 377–383.
144. Abraham, M. H. *Chem. Soc. Rev.* **1993**, *22*, 73–83.
145. Kamlet, M. J.; Doherty, R. M.; Veith, G. D.; Taft, R. W.; Abraham, M. H. *Environ. Sci. Technol.* **1986**, *20*, 690–695.
146. Abraham, M. H.; Grellier, P. L.; Prior, D. V.; Duce, P. P.; Morris, J. J.; Taylor, P. J. *J. Chem. Soc., Perkin Trans.* **1989**, *2*, 699–711.
147. Harris, J. M.; Sedaghat-Herati, M. R.; McManus, S. P.; Abraham, M. H. *J. Phys. Org. Chem.* **1988**, *1*, 359–362.
148. Raevsky, O. A. *J. Phys. Org. Chem.* **1997**, *10*, 405–413.
149. Ghafourian, T.; Dearden, J. C. *J. Pharm. Pharmacol.* **2000**, *52*, 603–610.
150. Oliferenko, A. A.; Oliferenko, P. V.; Huddleston, J. G.; Rogers, R. D.; Palyulin, V. A.; Zefirov, N. S.; Katritzky, A. R. *J. Chem. Inf. Comput. Sci.* **2004**, *44*, 1042–1055.
151. Sullivan, J. J.; Jones, A. D.; Tanji, K. K. *J. Chem. Inf. Comput. Sci.* **2000**, *40*, 1113–1127.
152. Karelson, M.; Lobanov, V. S.; Katritzky, A. R. *Chem. Rev.* **1996**, *96*, 1027–1043.
153. Sotomatsu, T.; Murata, Y.; Fujita, T. *J. Comput. Chem.* **1989**, *10*, 94–98.
154. Debnath, A. K.; Shusterman, A. J.; Compadre, R. L. L.; Hansch, C. *Mutat. Res.* **1994**, *305*, 63–72.

155. Debnath, A. K.; Compadre, R. L. L.; Debnath, G.; Shusterman, A. J.; Hansch, C. *J. Med. Chem.* **1991**, *34*, 786–797.
156. Selassie, C. D.; Shusterman, A. J.; Kapur, S.; Verma, R. P.; Zhang, L.; Hansch, C. *J. Chem. Soc., Perkin Trans.* **1999**, *2*, 2729–2733.
157. Hall, L. H.; Hall, L. M. *SAR QSAR Environ. Res.* **2005**, *16*, 13–41.
158. Rose, K.; Hall, L. H.; Kier, L. B. *J. Chem. Inf. Comput. Sci.* **2002**, *42*, 651–666.
159. Estrada, E.; Patlewicz, G.; Uriate, E. *Indian J. Chem.* **2003**, *17*, 197–209.
160. Charton, M. *J. Comput.-Aided Mol. Design* **2003**, *17*, 197–209.
161. Roy, K. *Mol. Diversity* **2004**, *8*, 321–323.
162. Gaillard, P.; Carrupt, P. A.; Testa, B. *Bioorg. Med. Chem. Lett.* **1994**, *4*, 737–742.
163. Krasni, M.; Banyai, I.; Noszal, B. *J. Med. Chem.* **2003**, *46*, 2241–2245.
164. Caron, G.; Ermondi, G. *J. Med. Chem.* **2005**, *48*, 3269–3279.
165. Van de Waterbeemd, H. *Quant. Struct.-Act. Relat.* **1996**, *15*, 480–490.
166. Hansch, C.; Leo, A.; Mekapati, S. B.; Kurup, A. *Bioorg. Med. Chem.* **2004**, *12*, 3391–3400.
167. Clark, D. E.; Pickett, S. D. *Drug Disc. Today* **2000**, *5*, 49–58.
168. Clark, D. E. *J. Pharm. Sci.* **1999**, *88*, 815–821.
169. Iyer, M.; Mishra, R.; Han, Y.; Hopfinger, A. J. *Pharm. Res.* **2002**, *19*, 1611–1621.
170. Ma, X.-L.; Chien, C.; Yang, J. *Acta Pharmacol. Sinica* **2005**, *26*, 500–512.
171. Hou, T. J.; Xu, X. J. *J. Chem. Inf. Comput. Sci.* **2003**, *43*, 2137–2152.
172. Wildman, S. A; Crippen, G. M. *J. Chem. Inf. Comput. Sci.* **1999**, *39*, 868–873.
173. Schultz, T. W.; Cronin, M. T. D.; Walker, J. D.; Aptula, A. O. *J. Mol. Struct. (Theochem.)* **2003**, *622*, 1–22.
174. Cronin, M. T. D.; Schultz, T. W. *Chem. Res. Toxicol.* **2001**, *14*, 1284–1295.
175. Smith, C. J.; Hansch, C.; Morton, M. J. *Mutat. Res.* **1997**, *379*, 167–175.
176. Selassie, C.; Kapur, S.; Verma, R. P.; Rosario, M. *J. Med. Chem.* **2005**, *48*, 7234–7242.
177. Benigni, R. *Chem. Rev.* **2005**, *105*, 1767–1800.
178. Netzeva, T. I.; Worth, A. P.; Aldenberg, T.; Benigni, R.; Cronin, M. T. D.; Gramatica, P.; Jaworska, J. S.; Kahn, S.; Klopman, G.; Marchant, C. A. et al. *Altern. Lab. Anim.* **2005**, *33*, 155–174.
179. He, L.; Jurs, P. C. *J. Mol. Graphics Model.* **2005**, *23*, 503–523.
180. Cramer, R. D.; Bunce, J. D.; Patterson, D. E. *Quant. Struct.-Act. Relat.* **1988**, *7*, 18–25.
181. Tropsha, A.; Gramatica, P.; Gomber, V. K. *QSAR Comb. Sci.* **2003**, *22*, 69–77.
182. Hansch, C.; Hoekman, D.; Gao, H. *Chem. Rev.* **1996**, *96*, 1045–1075.
183. Selassie, C. D.; Garg, R.; Mekapati, S. *Pure Appl. Chem.* **2003**, *75*, 2363–2373.
184. Hansch, C.; Mekapati, S. B.; Kurup, A.; Verma, R. P. *Drug. Metab. Rev.* **2004**, *36*, 105–156.
185. Cronin, M. T. D.; Schultz, T. W. *Chem. Res. Toxicol.* **2001**, *14*, 1284–1295.

Biographies

Corwin Hansch received his undergraduate education at the University of Illinois and his PhD in organic chemistry from New York University in 1944. After working with the DuPont Company, first on the Manhattan Project and then in Wilmington DE, he joined the Pomona College faculty in 1946. He has remained at Pomona except for two sabbaticals: one at the Federal Institute of Technology in Zurich with Prof Prelog and the other at the University of Munich with Prof Huisgen. The Pomona group published the first paper on the QSAR approach relating chemical structure with biological activity in 1962. Since then, QSAR has received widespread attention. Dr Hansch is an honorary fellow of the Royal Society of Chemistry and recently received the American Chemical Society Award for Computers in Chemical and Pharmaceutical Research for 1999. He was the first recipient of the Medaglia Pratesi Award from the Società Chimica Italiana in 2003 and he was also recognized as an outstanding scientific contributor by the Pharmaceutical Society of Japan in 2004.

Cynthia Selassie is a professor of chemistry at Pomona College in Claremont. She obtained her MA degree in chemistry from Duke University and her PhD degree in pharmaceutical chemistry from the University of Southern California, under the aegis of Prof Eric Lien. In 1980, she joined Prof Corwin Hansch as a postdoctoral research associate. In 1990, she joined the faculty at Pomona College as an associate professor of chemistry. Her research interests include development of the QSAR paradigm, its coherence with molecular modeling, as well as its applications to computer-assisted molecular design, multidrug resistance, and toxicity of phenols. She recently completed a sabbatical in the Shoichet Laboratory at the University of California, San Francisco.

© 2007 Elsevier Ltd. All Rights Reserved
No part of this publication may be reproduced, stored in any retrieval system or transmitted in any form by any means electronic, electrostatic, magnetic tape, mechanical, photocopying, recording or otherwise, without permission in writing from the publishers

Comprehensive Medicinal Chemistry II
ISBN (set): 0-08-044513-6

ISBN (Volume 4) 0-08-044517-9; pp. 43–63

4.04 Structure-Based Drug Design – A Historical Perspective and the Future

D J Abraham, Virginia Commonwealth University, Richmond, VA, USA

© 2007 Elsevier Ltd. All Rights Reserved.

4.04.1 Introduction

4.04.1.1 Paul Ehrlich and the Nature of the Receptor

Structure-based drug design (SBDD) remains the most logical and aesthetically pleasing approach in drug discovery paradigms. Paul Ehrlich is credited as the first to propose that receptors were similar to atomic locks and drugs to atomic keys. However, in his Nobel Prize address Ehrlich actually attributes the lock and key specificity to Emil Fischer:

> "The relations between toxin and its antitoxin are strictly specific – tetanus antitoxin neutralizes exclusively tetanus toxin, diphtheria serum only diphtherial toxin, snake serum only snake venom, to mention just a few examples out of hundreds. For this reason it must be assumed that the antipodes enter into a chemical bond which, in view of the strict specificity, is most easily explained by the existence of two groups of distinctive configuration: of groups which according to the comparison made by Emil Fischer fit each other "like lock and key." Considering the stability of the bond on the one hand and the fact on the other that neutralization occurs even in very great dilutions and without the help of chemical agents, it must be assumed that this process is to be attributed to a close chemical relationship and probably represents an analog to actual chemical syntheses."[1]

The lock and key concept has captivated drug discovery scientists through much of the twentieth century. Before x-ray crystallographic analysis revealed an atomic resolution of proteins and nucleic acids it was common for those in drug discovery to build hypothetical receptors using Corey, Pauling, and Koltun space-filling models (CPK models) or Drieding models. Only the side chains of Ehrlich were constructed based on some preliminary biochemical or pharmacological data that suggested pharmacophores. Captopril[2–4] and cilazapril[5] were the first two marketed therapeutic agents designed from a hypothetical binding site.

4.04.1.2 Theory and Methods

SBDD involves the integration of a number of independent sciences. X-ray crystallography and, to a lesser extent, nuclear magnetic resonance (NMR) provide the key ingredient, i.e., a detailed atomic-level description of the binding site. Molecular modeling, synthetic organic chemistry, medicinal chemistry, including structure–activity relationships (SARs) and quantitative SARs (QSARs), molecular biology and knowledge about pharmacokinetics/absorption, distribution, metabolism, and excretion (ADME) for any given scaffold of interest are all allies in SBDD.

4.04.1.2.1 X-ray crystallography

The realization of SBDD can be attributed to Max Perutz for solving the phase problem of imaging protein x-ray diffraction data, at atomic resolution.[6] When Perutz started in the mid-1930s, the molecular structure of sugar had not yet been solved. Perutz's mentor was Sir Lawrence Bragg, who, along with his father, won the 1922 Nobel Prize in physics for uncovering the quantitative nature of diffracting crystals with x-ray beams. Bragg gave Perutz zero chance of solving a protein structure but 100% importance if he succeeded. The discovery not only won Perutz and his graduate student John Kendrew, who worked on myoglobin, a Nobel Prize in Chemistry (1962), but gave birth to the field of structural biology. With the discoveries made by two other researchers in Perutz's group, James Watson and Francis Crick (Nobel Prize in Medicine, 1962), this one research group, along with another group headed by Fred Sanger in the same Medical Research Council Laboratory of Molecular Biology (MRC-LMB), Cambridge, UK (two Nobel Prizes in Chemistry 1958 and 1980 for methodology to sequence proteins and nucleic acids), laid the foundation for the biotechnology revolution that would follow.

4.04.1.2.2 X-ray crystal data collection

It took several decades before crystal structures of proteins and nucleic acids or protein nucleic acid complexes could be solved in a reasonable timeframe, permitting their routine use in drug discovery. Data collection using film for the first protein structures might take a year to measure and record as structure factors. There were also human errors in recording reflection intensities from film by hand. The invention of the diffractometer, an automated controlled hardware and software system to measure and record diffraction intensities, greatly hastened the process. The diffractometer would collect data until the crystal lost diffraction power, operating through the night when one went home to sleep. It was therefore coined by some users as the phantom crystallographer. Even with this advance, the measurement of single diffraction intensity required a minute or more and the 20 000 reflections that were needed in taking a 3 Å set of data on a protein the size of hemoglobin (Hb) took $2\frac{1}{2}$–3 weeks or more. In those days crystallographers had to know how to mount their protein crystals, to align the crystal axes in a reasonable manner to identify the space group, and to initiate data collection. This meant beginning and experienced crystallographers needed to understand the theoretical and physical nature of x-ray diffraction, especially reciprocal space. There could also be occasions when one of the four circles of the diffractometer was not working properly and understanding the geometrical relationship between reciprocal and real space was essential to collect a unique set of data. Today's world with image plate detectors, oscillating crystals, and programs that can identify Bragg reflections in any orientation of the crystal were only a dream. Much credit for these advances go to individuals like Ulie Arndt in Max Perutz's MRC-LMB in Cambridge, UK. Ulie was one of the visionaries who envisioned and built a television type of monitor to collect diffraction intensities simultaneously.[7,8] This advance turned out to be another revolutionary event in structural biology.[9] Investigators today can mount a crystal in any orientation in a stream of nitrogen or helium and only have to type 'auto' or some other start code and the whole process initiates and records a 1.5 Å set of data on a protein of around 50 000 Da overnight. The next advance was for crystallographers to take their crystals to a neutron diffraction facility where even the weakest diffracting crystals produced quality data needed for their solution in extremely short time periods.

4.04.1.2.3 Computing advances

Cyrus Levinthal, one of the earliest visionaries in molecular graphics, recognized the importance of computational power required for modern biology.[10] None of what we have today in structural biology would have happened without an equivalent revolution in computer technology. Computers in the early 1960s were not yet sufficiently advanced to calculate an electron density map at atomic resolution of even a large small molecule within a day.

To give a rough idea of the importance of computers in the revolution that occurred in the output of x-ray structures, consider the following test of the difference between humans and computers. In 1974, it took the author 3 days with a hand calculator (and many mistakes) to calculate one axial Fourier peak (one grid point) of a natural product with molecular weight of 222 using only 240 reflections. Axial peaks are the easiest to calculate since many of the trigonometric sine and cosine terms cancel each other. The advanced computer of that day at the Max Planck Institute of Biochemistry in Martinsreid near Munich took about 14 s to do the whole axis row of 50 grid points. If one were to attempt to calculate one virus electron density map at 3 Å resolution, and at the same speed as the natural products calculation (which is not possible, since more complex trigonometric terms are required for nonaxial peaks), it would require 156 million years.[11] Today with modern supercomputers we can get Fourier maps of any of the largest proteins and viruses in minutes or hours, or at best overnight.

4.04.1.2.4 Molecular models to molecular graphics to molecular modeling/computer-aided drug design

4.04.1.2.4.1 Molecular models

Kendrew and colleagues constructed the first three-dimensional (3D) model of a protein, myoglobin, solved by x-ray crystallography.[12] This was the age before molecular graphics would permit workers to model protein structures into the electron density using 3D visualization. Instead, Kendrew models made of brass with screw connectors that could be adjusted to produce the correct bond angles of the protein fold were used to build up the protein structure. Models of this type were 6-ft (2 m) cubes with 2500 vertical rods that obscured the view, making adjustments difficult. Crystallographers would use plumb lines to measure the vertical direction while the atomic positions in the x and y planes were printed out on the surface of the floor or in the beginning electron density was signified by colored chips attached to the rods. Younger crystallographers today might think of such a procedure as archaic, yet this was not long ago (models were in existence as late as the 1970s to mid-1980s). Richards boxes were the first successful attempts at superimposing the electron density on a grid to construct a 3D model.[13] Richards' idea was to use a half-silvered mirror (similar to those he had observed for conducting magic illusions on the stage). The silver mirror enabled the electron density to be directly superimposed for building a metal molecular model. Richards not only built the device but he, Hal Wyckoff, and colleagues used it to solve the structure of ribonuclease-S.[14,15] The size of the models was reduced as larger molecules were solved by decreasing the scale to 2.5 cm $Å^{-1}$, and later to 1.0 cm $Å^{-1}$. The Richards box became a standard method in protein crystallography until computer modeling was successfully developed.

Byron Rubin, while working as a crystallographer in Jane Richardson's group in the early 1970s, constructed a device known as 'Byron's bender' for bending wire to follow the backbone trace of a protein.[16] These small backbone wire models were easy to manipulate and the only portable models at the time. Eric Martz and Eric Francoeur[17] provide the following example illustrating the importance of portable models from Byron's bender. During a scientific meeting in the mid-1970s (when fewer than two dozen protein structures had been solved), "David Davies brought a Bender model of an immunoglobulin Fab fragment, and Jane and David Richardson brought a Bender model of superoxide dismutase. While comparing these physical models at the meeting, they realized both proteins use a similar fold, despite having only about 9% sequence identity. This incident marked the first recognition of the occurrence of the immunoglobulin superfamily domain in proteins unrelated by sequence. The insight was published in a paper entitled, 'Similarity of three-dimensional structure between the immunoglobulin domain and the copper, zinc superoxide dismutase subunit.'[18]

4.04.1.2.4.2 Molecular graphics

It is not possible in this chapter to review in depth the developments in molecular graphics and modeling that underpin SBDD (see other chapters, e.g., 4.05 Ligand-Based Approaches: Core Molecular Modeling). The earliest visionaries to attempt graphically to depict large molecules and electron density contour maps essential for crystallographic analyses are well described in the following two internet histories: *Molecular Surfaces: A Review* by Michael L. Connolly[19] and *History of Visualization of Biological Macromolecules* by Eric Martz and Eric Francoeur.[17] Both are excellent reviews with outstanding photography and extensive references of the early pioneers' work.

The earliest interactive molecular graphics originated at Massachusetts Institute of Technology, USA, with Project MAC (Mathematics and Computation) in 1964. The first computer representations using an oscilloscope that rotated wire representations of macromolecular structures were developed by Cyrus Levinthal and colleagues. The following individuals; Meyer,[20] Barry and North,[21] Katz and Levinthal,[22] Levinthal *et al.*,[23] Langridge,[24] and Collins *et al.* [25] over time migrated to other locales seeding academic communities with molecular modeling centers. The early systems, because of hardware limitations, generally could model only small molecules. The first molecular modeling programs that visualized macromolecules were so unstable the picture tube could freeze in the middle of an analysis and there was no way to recover the lost work. Early hardware was not only limited to what could be visualized, restricting the size of the molecule, but was almost guaranteed to crash if a key demonstration was set up for visiting colleagues, site visitors, administrators with funds, or when a new research insight was about to be visualized.

The development of raster graphics displays not only improved reliability but expanded the capability to move from line drawings to hard-sphere, CPK-type images. For examples see Porter,[26] Feldmann *et al.*,[27] Max,[28] Palmer *et al.*,[29] and Evans.[30]

An early molecular plotting program ORTEP readily took front and center stage for small-molecule crystallographers. It was developed in 1976 at Oak Ridge via the Thermal Ellipsoid Program of Carroll Johnson.[31] Robert Diamond, at the MRC-LMB, Cambridge, UK, wrote a program Bilder (from the German word for picture) to link molecular structure on the graphics.[32–34] This was the first prototype of an interactive graphics program for analyzing protein structure. Lesk and Hardman made significant contributions during the 1980s, developing programs for drawing the secondary structures of proteins, greatly improving the ease of comparison between protein classes as well as simplifying viewing complex binding sites.[35–37] During this time period, one could, for the first time really, count on the hardware not to crash frequently and for the new software to fit protein structure into electron densities without the disruptions and hardships experienced with earlier physical and computer-based systems. The program Frodo, developed by Alvin Jones, was not only easy to learn and use but had several features that enabled quick comparisons with distances, angles, and coloring of different densities, all in three dimensions.[38–40] A newer version, TOM/Frodo,[41] is still considered by some crystallographers as the most user-friendly of the electron density fitting programs. Viewing the fit of protein and nucleic acid structure into electron densities with vivid colors rivals the beauty of art. In some cases, it may even be more appealing as the flow of the helices, sheets, and turns reveals the fundamental principles of hydropathy in nature and the visual beauty of atomic structure.

4.04.1.2.4.3 Molecular modeling/computer-aided drug discovery

Peter Goodford, at Oxford, wrote the software package GRID, the first tailor-made program for medicinal chemists that could probe for pharmacophore or water-binding sites in proteins.[42] Professor Robert Langridge moved from Princeton to University of California at San Francisco, bringing with him imaging programs used to visualize small biological molecule structures and drugs.[43] His efforts would give rise to a molecular modeling environment where Irvin (Tack) Kuntz's extensively used software program DOCK was invented.[44,45] A very important contribution toward using molecular modeling for SBDD was made by Garland Marshall at Washington University, St Louis, MO, who started Tripos Associates in 1979.[47] Garland and associates developed the very successful commercial package Sybyl, used universally for SBDD. Richard Cramer was responsible for the development of CoMFA, another widely used program for SBDD.[48] Arnold Hagler and Donald MacKay, who co-founded Biosym in 1983/1984, developed the macromolecular imaging and energy minimization program package Insight/Discover that gained favor, especially with structural biologists. For a comprehensive treatise on molecular modeling, see Chapter 4.05.

4.04.1.2.5 Nuclear magnetic resonance

The advent of NMR structure determination of proteins has added a second physical technique to image potential drug receptors if they are not too large. The first NMR structure determination was published in 1985.[49] Clearly, NMR has two advantages over x-ray crystallography: the protein does not need to be crystallized and NMR provides real-time dynamic information, while crystal structures at present provide static and thermodynamically stable structures. A major disadvantage of NMR spectroscopy has been the molecular size limit of around 35 000 Da. Researchers in this field report the development of new techniques, such as TROSY,[50] which seem certain to increase the molecular weights available for analysis. Craik and Clark have written an extensive review of the use of NMR in drug discovery in the sixth edition of *Burger's Medicinal Chemistry and Drug Discovery*.[51]

4.04.1.2.6 Electron diffraction

A third physical technique, electron diffraction, is now available to image some receptors not amenable for resolution via crystallography or NMR; for example, the allosteric acetylcholine receptor.[52] Electron diffraction has also been successfully employed to determine the structure of bacteriorhodopsin using high-resolution electron cryomicroscopy that provided the first structure of a seven-helix membrane protein.[53] From this successful endeavor, G-coupled receptors have been modeled[54] and used for SBDD; for example, serotonin inhibitors.[55] In general, the medium-resolution density distributions can often be interpreted in terms of the chemistry of the structure if a high-resolution model of one or more of the component pieces has already been obtained by x-ray, electron microscopy, or NMR methods. As a result, the use of electron microscopy is becoming a powerful technique for which, in some cases, no alternative approach is possible. Useful reviews include those of Dubochet *et al.*,[56] Amos *et al.*,[57] Walz and Grigorieff,[58] and Baker *et al.*,[59] as well as a book written by Frank.[60] This methodology may well be the only way we can determine the 3D structures of membrane receptors that resist crystallization efforts. Henderson and Baker have published an excellent article on electron cryomicroscopy and its future in the determination of the structures of biological macromolecules,[61] suggesting the reviews cited above.

4.04.1.2.7 Structure-based drug design methodology

SBDD studies usually follow one of two methodologies. The first consists of an x-ray crystal structure determination of a biological target with a known drug or inhibitor, a substrate analog, or a library screen hit. The initial cycle is then reiterated with chemical modifications to the scaffold until a more active agent is discovered. The design of the new agents from the last crystal structure determination can include the rational use of additional moieties to increase activity such as changes that regulate solubility/transport, metabolism, and/or toxicity profiles. The second SBDD methodology is to construct de novo a new ligand that will fit the binding pocket and, if possible, have desired transport, metabolic, and toxicity profiles built into the new molecule. The key to both methods is the requirement of an atomic-level structure of the target and a basic understanding of the binding site. Both of these approaches are often applied to SBDD; for example, FK506-binding protein.[62,63]

In some cases with flexible binding pockets such as human immunodeficiency virus (HIV) protease[64] and carboxypeptidase,[65,66] de novo modeling from the native crystal is not possible due to the inhibitor-induced binding pocket having a different structure from the native protein-binding site.

4.04.1.3 History

The desire to design new drugs from the 3D structure of a protein (receptor) did not immediately motivate medicinal chemists to take up crystallography. Besides, there was as no user-friendly molecular modeling (noncrystallographic) programs available for medicinal chemists to visualize and dock proposed ligands. The first paper suggesting SBDD using the 3D coordinates of proteins was published in 1974.[67] Hb was the first enzyme structure determined and the only one for some time with drug discovery implications due to the mutation at β6 Glu→Val, which produces the pathological disease known as sickle-cell anemia.

The 1971 Cold Spring Harbor Symposium on Quantitative Biology was a landmark conference with the first presentation of protein and virus structures by the world leaders in protein crystallography in attendance. There was only one talk involving a drug, a small-molecule complex deoxyguanosine-actinomycin reported by Sobell and co-workers.[68] No one had yet studied any of the known 3D protein structures as potential drug targets.

Three groups initiated SBDD in the middle to late 1970s. Not surprisingly, two of the first three groups, Peter Goodford's at the former Burroughs Wellcome Laboratories in London and mine in the School of Pharmacy, The University of Pittsburgh, chose Hb. The third group was led by crystallographer David Matthews and colleagues, founders of the startup company Agouron whose sole mission was structure-based drug discovery. Matthews *et al.* were the first to determine the structure of a drug bound to its molecular target, methotrexate/dihydrofolate reductase (DHFR).[69]

4.04.2 Structure-Based Drug Design: Examples

4.04.2.1 Hemoglobin: The First Drug Design Target

Perutz and colleagues, starting with the first low-resolution structures in the late 1950s, took 25 years to solve the high-resolution structures of oxy and deoxy horse (1968, 1970) and human Hb (1983, 1984).[70–73] Perutz and co-workers subsequently determined the structures of a number of mutant Hb that caused different medical maladies.[74–86] The solved structures readily explained at the molecular level the reason for the physical impairments observed clinically. However, these mutants were rare. Sickle-cell anemia, on the other hand, represented one of the largest populations of a genetic disease known to the world. By this time, Goodford *et al.* had made significant progress in using Perutz's Hb coordinates to design and synthesize non-naturally occurring allosteric effectors of Hb that might increase oxygenation to hypoxic tissues (*see* Section 4.04.2.1.2). Manipulating the oxygenation capabilities of Hb, if successful, could combat a number of hypoxic diseases and problems, including stroke, angina, trauma, and bypass surgery.

4.04.2.1.1 Sickle-cell anemia

Perutz's solution of the structure of human Hb at near-atomic resolution opened the door to structure-based durg discovery. Sickle-cell anemia, a β6 Glu→Val mutant, provided what was thought at the time to be an ideal target for SBDD. In fact, more has been known about sickle-cell anemia at the molecular and physiological levels than any other disease state, yet only one limited-acting anticancer drug, hydroxyurea, has been approved for treating the disease. The disease arises from the polymerization of sickle Hb, HbS, in red cells. The ideal method to reverse the polymerization would employ binding a stereoselective inhibitor at a Hb polymerization contact site. Unfortunately, massive efforts by a number of groups failed to discover an effective drug. There are three primary reasons for the lack of success: (1) the large amount of Hb in humans; (2) the lack of a deep binding pocket on the surface that would be free energy attractive for selective binding; and (3) the lack of patients in the USA (around 60 000) making the cost-to-profit ratio prohibitive to large pharmaceutical houses (because of the cost-to-profit ratio, sickle cell falls into the orphan drug category).

The amount of drug needed to treat sickle-cell anemia chronically is the largest barrier for a SBDD solution. The concentration of Hb in red cells is 5 mM. A sickle-cell patient with 4 L of blood and a 25% hematocrit has approximately 5 mm of HbS or 322.5 g of receptor target. Discovering an effective stereospecific HbS polymerization inhibitor that only needs to bind to 15–30% of HbS is possible since as low as 15% fetal Hb in homozygous HbS red cells can significantly inhibit polymerization. The amount of drug needed to treat sickle-cell anemia chronically, via inhibiting polymerization over a lifetime, would be in the 1–10 g daily dose range.

Our group was the first to discover a number of structure-based polymerization inhibitors.[87–89] Our PNAS paper in 1983[89] with Perutz and Phillips was cited by Michael Hahn (from GlaxoSmithKline) at the recent Keystone meeting on SBDD as the first example of fragment-based analysis. Goodford and co-workers were the first to develop an antisickling agent (BW12C), based on SBDD, which reached clinical trials.[90–93] While BW12C was shown not to bind as designed,[94] its discovery clearly indicated that ideas using the 3D structure of a protein could produce viable clinical candidates in a very acceptable timeframe. The demise of BW12C as well as many other potential antisickling agents was due to its inability to be used in the quantities needed to treat over 1 lb (0.45 kg) of mutant HbS.

The attempt by Goodford and co-workers to target the N-terminal residue of the alpha chains of Hb to disrupt the sickle-cell polymer from forming has merits. Efforts to find an agent that binds near the sickle mutation at beta 6 or at its receptor cavity in the polymer, beta Phe-85 and beta Leu-88 to inhibit the sickling process have failed miserably.[88] The reason is that the Val receptor cavity is very shallow and only a small amount of free energy of polymerization results from this hydrophobic interaction. There are 20 polymer contacts involved in the polymerization, as Love's group determined from their structure determination of deoxy HbS.[95–97] None of these are located in deep cavities. Those working in structure-based drug discovery quickly became aware from the x-ray crystal structures of a number of enzymes that, with a few exceptions, all binding sites are located in deep pockets. The lower dielectric constant in these cavities greatly increases the binding energies of buried electrostatic interactions, hydrogen bonds, and hydrophobic interactions due to the release of water molecules.[98] This is the reason receptor pockets are receptor pockets. While tight lock-and-key fits are not required, a maximum fit is always a primary goal in SBBD (*see* Section 4.04.3.1). Therefore, Goodford's attempts to bind to a specific surface site such as the N-terminal amino group, where an aldehyde can form a transient covalent Schiff base bond, is one possible solution, if the aldehyde: (1) is specific for Hb; (2) is nontoxic at chronic therapeutic doses; and (3) is not metabolized or excreted quickly.

Using Goodford's strategy we advanced one aldehyde, vanillin, to a phase Ia safety trial for treating sickle-cell anemia. Vanillin had a very remarkable low toxicity profile[99] but failed, due to metabolic first-pass conversion to vanillic acid.[100] Very recently, we have discovered a nontoxic breakdown product in foods and plants, 5-hydroxy-furfural (5-HMF). 5-HMF has shown remarkable activity and specificity for Hb in vitro and in vivo in genetically engineered sickle-cell mice.[101] Crystal-binding studies in our laboratories with 5-HMF bound to Hb showed the reason for its unusually strong affinity and specificity for Hb. 5-HMF has the unique ability to attract a sheath of water molecules that appears to increase its affinity and confer specificity over the N-terminal amino groups of proteins throughout the biological milieu. Other aldehydes we have studied do not show this water sheath[102] and are not as specific for Hb. 5-HMF is in preclinical development for phase Ia and Ib sickle-cell anemia trials. It appears to possess all three of the requirements for an aldehyde antisickling agent stated above. We are currently exploring the role of water binding and its effect on increasing binding constants in other systems.[98,103–105]

We discovered several potent antigelling molecules in vitro over a decade but all had toxicity issues.[87,88,106–110] The most important lesson we learned from these failures was to start SBDD with a nontoxic or very-low-toxicity scaffold.

4.04.2.1.2 Hemoglobin allosteric effectors

It is not widely known that Hb actually carries a reservoir of unused oxygens. Normally humans function during routine daily living by extracting a little above one of the four bound oxygens. The natural Hb allosteric effector, 2,3-diphosphoglycerate (2,3-DPG), is synthesized in red cells from glucose in increasing concentrations during hypoxic conditions, right-shifting the oxygen-binding curve to deliver more oxygen. The binding site of 2,3-DPG determined by Arnone[111] lies on the dyad axis at the mouth of the β-subunit cleft interacting with the N-terminal βVal1, βLys82, and βHis143 of deoxy Hb. A more recent study at a higher resolution by Richard *et al.* found DPG to interact with the residues βHis2 and βLys82.[112]

Athletes who live at high altitudes and compete at lower altitudes or at sea level have an increased oxygen-carrying capability that permits a higher degree of physical output. This idea has been exploited by a number of national Olympic teams. The reason for an increase in ability to compete at lower altitudes is due to an increase in red cell mass and an increase in 2,3-DPG.[113,114]

Goodford and colleagues were the first to recognize the importance of SBDD of an allosteric effector of Hb that might bind to the 2,3-DPG or other sites and increase oxygen delivery to tissues.[115–122] 2,3-DPG cannot be used therapeutically since it has five polar charges and does not penetrate red cells. It is locked into the red cell after metabolic synthesis from glucose. An effective and safe allosteric effector that can enter red cells would have great value in treating hypoxic diseases such as angina, stroke, and trauma from accidents or on the battlefield, or be used for transplant or pulmonary by-pass surgery, or to enhance radiation treatment of hypoxic tumors. Such an agent might also find use to increase the shelf-life of blood; outdated blood results from the decay of 2,3-DPG in red cells. Goodford and colleagues synthesized and studied bis-arylhydroxysulfonic acids designed for the 2,3-DPG-binding site.[118,120] Unfortunately, Goodford's program was terminated and he did not have a chance to pursue this idea effectively.

Our group employing SBDD principles discovered an effective Hb allosteric effector, RSR 13 (efaproxiral)[123–126] that fulfilled Goodford's original goals for an allosteric effector which might be used to increase oxygenation to hypoxic tissues. Efaproxiral has advanced to phase III clinical trials for radiation treatment of breast cancer metastasis to brain.[127] The allosteric mechanism for this allosteric effector and its structure activity was worked out at the molecular level.[128,129] RSR 13 was also found to be promising, in a phase II study, for pulmonary bypass surgery.[130] Not surprisingly, RSR 13 has been a topic of discussion for blood doping for athletic events.[131]

4.04.2.1.3 Blood substitutes and hemoglobin cross-linking agents

SBDD has more recently been applied to designing Hb blood substitutes with ideal oxygen-delivering properties. The earlier cross-linking derivatives were found to have problems in both delivering oxygen and exhibiting toxicity in humans.[132–134]

The first cross-linking agent that possessed potential as an Hb-based blood substitute was described by Walder *et al.*[135] Chatterjee *et al.* identified the binding site to deoxy Hb, and found the two Lys99α side chains were cross-linked.[136] One of the derivatives was proposed as a blood substitute,[137] and has been explored commercially.[138] Another cross-linked Hb engineered by Komiyama and colleagues, at the MRC-LMB in Cambridge, was developed into a blood substitute clinically investigated at Somatogen, now Baxter.[139] Both were designed with SBDD principles and are currently in clinical trials.

4.04.2.1.4 Hemoglobin as a model drug receptor, as a model for structure-based design of allosteric effectors, and as a model for other important allosteric proteins

Working with Max Perutz in the early and mid-1980s, we discovered Hb binds a number of drugs.[140] Most of these bind to the Hb central water cavity in the deoxy, tense state (T-state). They are released during the transition to oxy Hb (the relaxed state or R-state) that has a narrowed and smaller central water cavity. Nonnaturally occurring organic molecules are attracted to the central water cavity and act as allosteric effectors (*see* Section 4.04.2.1.2). Those that increase hydrogen bonding across the central water cavity in the T-state structure greatly increase oxygen delivery to tissues, as described in Section 4.04.2.1.2. There are other allosteric effectors that bind to the R-state, resulting in a high-affinity Hb that left-shifts the oxygen-binding curve. These might be effective in treating sickle-cell anemia, since only deoxy HbS polymerizes.

We were able to solve a puzzle to explain how two allosteric effectors can bind to the same binding site and have opposite effects on oxygen delivery.[141] Those effectors that increased the number of hydrogen bonds across the tetramer symmetry diad increased oxygen delivery while those that disrupt the normal water bridge across the same diad axis decrease oxygen delivery. Hydrophobic drugs and agents are readily accommodated in the central water cavity as well as in hydrophobic binding pockets under the surface of the tetramer[140] that readily bind xenon.[142]

It has been said: "As hydrogen is to quantum mechanics, Hb is to proteins." The fact that more is known about the allosteric states of Hb at the molecular level has permitted it to be a model to understand other allosteric proteins involved in important biological functions such as P53 and the acetyl choline receptor. Mattevi *et al.* have published the similarity of the Hb allosteric structures with other allosteric proteins.[143]

4.04.2.2 Antifolate Targets

Thymidylate and purine nucleotides are essential precursors to DNA and RNA. Therefore, folate-dependent enzymes were among the first useful targets for the development of anticancer and anti-inflammatory drugs (e.g., methotrexate) and antiinfectives (trimethoprim, pyrimethamine). Normally the reduced form of folate (tetrahydrofolate) acts as a one-carbon donor in a wide variety of biosynthetic transformations. DHFR transforms dihydrofolate to tetrahydrofolate which is essential in the synthesis of purine nucleotides and thymidylate required for synthesis of DNA and RNA. Tetrahydrofolate in turn is a reductant used by thymidylate synthase (TS) and is converted stoichiometrically back to dihydrofolate. The regeneration of dihydrofolate and tetrahydrofolate via folate-dependent enzymes provided the Agrouron group with targets for SBDD, starting with determining the bound structures of known and newly designed anticancer and anti-inflammatory drugs (e.g., methotrexate) and antiinfectives (trimethoprim, pyrimethamine).

4.04.2.2.1 Dihydrofolate reductase

DHFR was chosen by crystallographers at Argouron as one of their first SBDD efforts. Matthews *et al.* solved the structure of methotrexate bound to DHFR[69] using the bacterial enzyme. Further studies on DHFR by Matthews and co-workers[144] provided the template for the iterative approach to SBDD that was successful in forwarding new molecules to clinical trials. An excellent review by Kisliuk[145] focuses on deaza antifolates which are: (1) presently under clinical development; and (2) less developed compounds which represent novel approaches. Topics covered relate to all subtopics in this section: DHFR, TS and glycinamide ribonucleotide formyltransferase (GARFT). In addition to inhibition of target enzymes, antifolate membrane transport into cells, and conversion to poly-L-gamma-glutamate forms are shown to have important considerations in drug design along with the reverse processes, cellular hydrolysis of antifolate poly-L-gamma-glutamates to monoglutamates and the extrusion of the monoglutamates through the cell membrane.

4.04.2.2.2 Thymidylate synthase

The studies on DHFR[144] led to the SBDD of agents for TS.[146] TS was shown to undergo substantial conformational change when ligands bind. Using the de novo design process described in Section 4.04.1.2.7, a new and structurally unique inhibitor was discovered that entered clinical trials. The other bootstrap method using a known inhibitor described in Section 4.04.1.2.7 also produced a clinical candidate.

When the design of inhibitors of human TS at Agouron Pharmaceuticals began, the amounts of the human enzyme required for crystallographic study were unavailable. Bacterial TS served as a substitute, with the assumption the conserved natures of both active sites would not affect SBDD.

4.04.2.2.2.1 Structure-guided optimization

A quinazoline ring of a weak inhibitor (compound 1: **Figure 1**) bound to *Escherichia coli* TS was found to bind in a protein crevice surrounded by hydrophobic residues on top of the pyrimidine of the nucleotide.[147,148] The structure of the complex showed a right-angle bend of the inhibitor extending a D-glutamate arm to the surface of the enzyme. Hydrogen bonds were made with several enzyme side chains, the terminal carboxylate, and several tightly bound waters. This weak inhibitor, like folate and most folate analogs, is transported through a system that recognizes a D-glutamate moiety. Compound 1 failed as an anticancer drug because of its insolubility and resulting nephrotoxicity.[149]

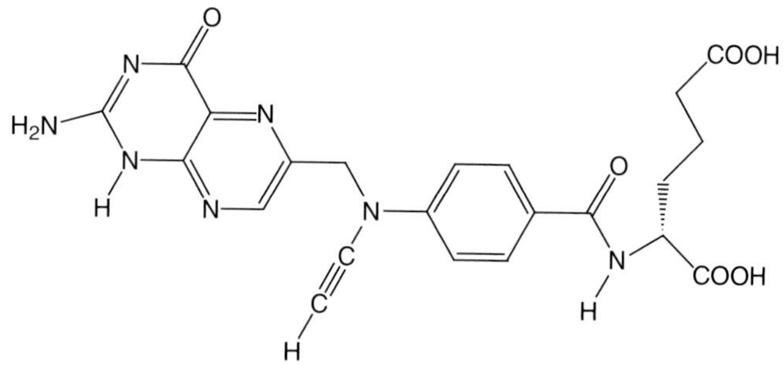

Figure 1 Compound 1.

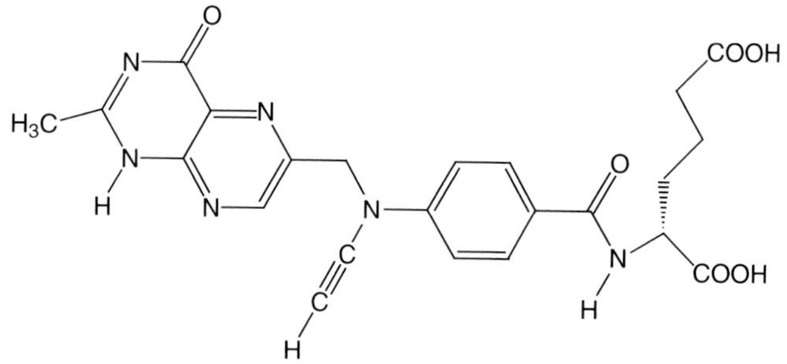

Figure 2 Compound 2.

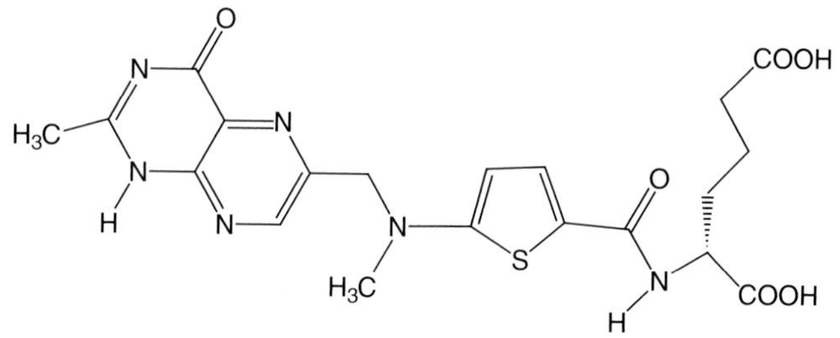

Figure 3 Tomudez, ZD 1694, Astra Zeneca.

The intracellular concentrations of molecules with a D-glutamate moiety are elevated and trapped via the addition of several glutamates by a cellular enzyme. TS inhibitors were designed at Agouron with the goal of obtaining a molecule that could enter cells passively and avoid the need for transport or polyglutamylation. The first were designed by structure-aided modification of known antifolates, and others were designed de novo. Starting with compound 2 (**Figure 2**), the glutamate moiety was deleted from the structure. (Compound 2, the 2-desamino-2-methyl analog of compound 1, and designed from a precursor inhibitor, had been found much more water-soluble than compound 1, and eventually led to AstraZeneca's Tomudex (**Figure 3**),[149,150] which is now approved for the treatment of colorectal cancer in Canadian and European markets.) Removal of the glutamate reduced the potency by 2–3 orders of magnitude. The crystal structure solved using compound 1 indicated potential interactions were exploited by substituents such as the *m*-CF_3 and a phenyl moiety was added to interact with Phe176 and Ile79. Interestingly, when substituents are combined they do not necessarily produce the expected sum of free energy for binding. Also structures of the complexes with several of these compounds revealed that ideal placement of one group does not always accommodate the best interaction for another. This is a general problem for rigid scaffolds. Three analogs had significant activity in vitro, which could be reversed by exogenous thymidine. Thymitaq (AG337 hydrochloride, also known as nolatrexed; **Figure 4**) has recently completed enrollment in a pivotal phase III liver cancer trial.[151] It was the result of numerous SBDD iterations and is remarkably simpler than the starting molecule, compound 1. This example dramatically shows that binding sites can accommodate a variety of scaffolds. For a more complete review of this work and sections below, see Hardy *et al.*[152]

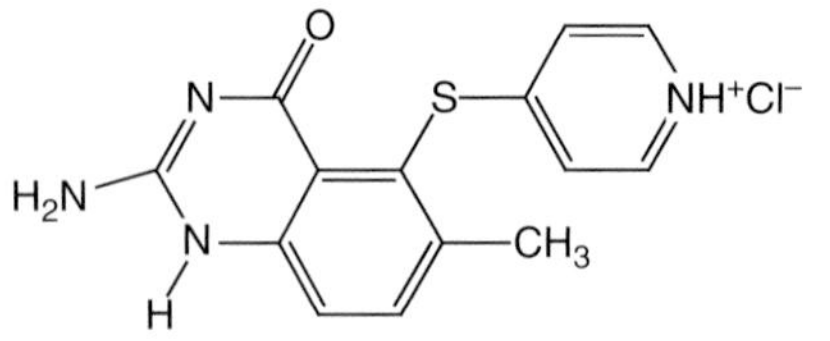

Figure 4 Thymitaq AG337, Agouron.

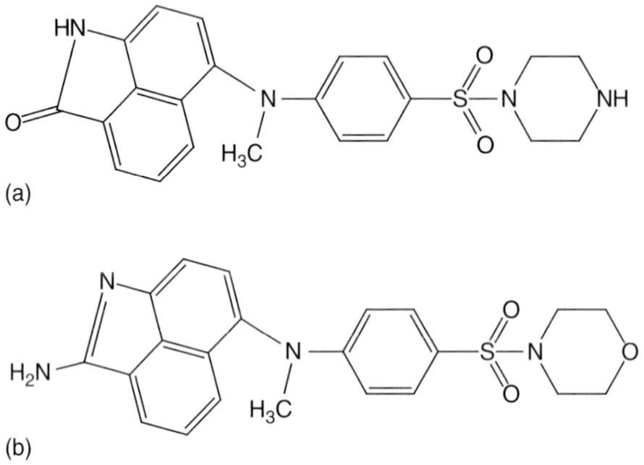

Figure 5 (a) De novo compound 3, Agouron; (b) compound 4, AG331, Agouron.

4.04.2.2.2.2 De novo lead generation

The de novo design effort starts with the design of a group or entire molecule that can be modeled into the active site. In this case, Peter Goodford's program GRID[42] was employed to locate a possible binding site for an aromatic ring system at the TS active site.[153] Naphthalene was inserted into the site and optimized to a favorable binding mode. Additions to the naphthalene scaffold were inserted to provide hydrogen-bonding groups to interact with the enzyme (benz[*cd*]indole), including hydrogen bonding to a tightly bound water. Elaboration from the opposite edge of the naphthalene core to extend into the top of the active site cavity, toward bulk solvent, resulted in a de novo compound that had a K_{is} value of 3 μM for inhibition of human TS with about 10-fold less potency against the bacterial enzyme.

The x-ray structure of the de novo molecule, compound 3 (**Figure 5**) bound to *E. coli* TS, found it was located deeper into the active site and an oxygen in the de novo compound displaced the tightly bound water instead of hydrogen bonding to it. This forced the backbone amide oxygen of Ala263 to move by about 1 Å. Replacement of the oxygen in the de novo compound with nitrogen provided a significant increase in inhibitory potency. Further rational structural changes resulted in recovery of the displaced water, and restoration of the original position of the Ala263 carbonyl oxygen. The process yielded AG331, compound 4 (**Figure 5**), which has a K_{is} value of 12 nM for inhibition of human TS and has entered clinical trials as an antitumor agent.[154]

4.04.2.2.3 **Reiterative design: glycinamide ribonucleotide formyl-transferase**

GARFT catalyzes the N-formylation of glycinamide ribonucleotide. *N*-10-formyltetrahydrofolate serves as the one-carbon donor. GARFT is an essential step in the synthesis of purine nucleotides, and a target for blocking the proliferation of malignant cells. This project fit well into the expansion of the early folate research at Agouron. It nicely illustrates the iterative methodology in SBDD, where not only increasing binding and selectivity are important, but also the incorporation of preferable ADME properties and the understanding of the enzyme mechanism exploited in the design stage. Another interesting option for the proposed work was attempting to unravel the GARFT mechanism from the enzyme structure. If successful, the design of new inhibitors could possibly modulate or exploit any relationship between binding and catalytic events in the design of new inhibitors. Lometrexol (**Figure 6a**) and Pemetrexed (**Figure 6b**), potent GARFT inhibitors, have been shown to be effective antitumor agents in clinical trials.[154,155] These were designed through traditional medicinal chemistry approaches, where folate analogs were synthesized and tested as

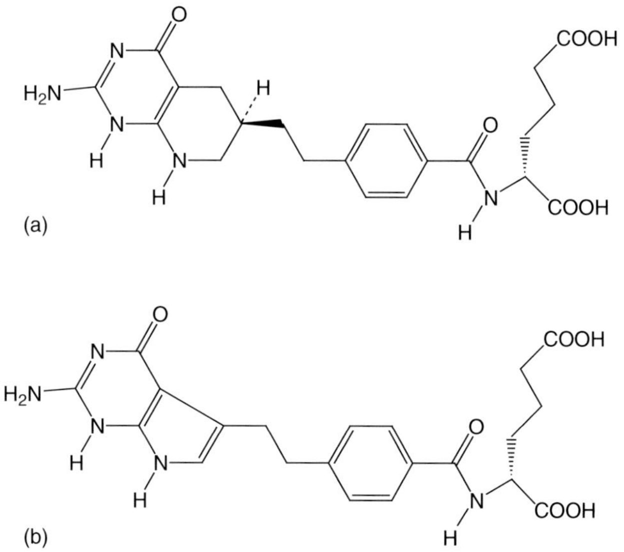

(a)

(b)

Figure 6 (a) Lometrexol, LY309887, AG2034, Agouron; (b) pemetrexed disodium, Eli Lilly.

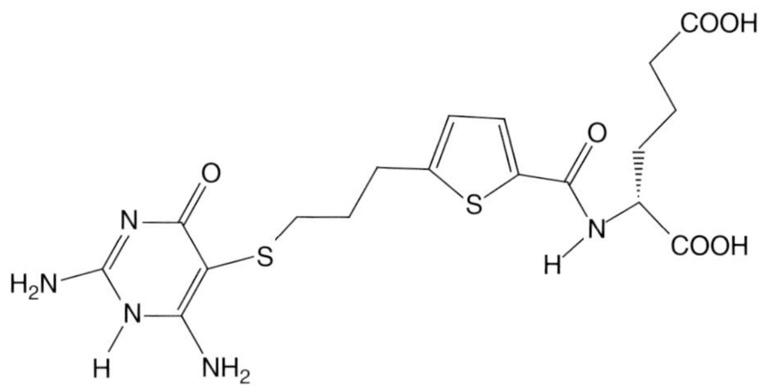

Figure 7 Compound 5.

inhibitors of tumor cell growth or against the activity of different folate-dependent enzymes.[156–158] A potent bisubstrate analog inhibitor of GARFT, from glycinamide ribonucleotide, and a folate analog were apparently catalyzed by the enzyme itself.[159]

The design of new GARFT inhibitors at Agouron commenced with the structure of the complex between the *E. coli* enzyme and 5-deazatetrahydrofolate.[160] Recombinant approaches afforded the active and soluble fragment of a multifunctional human protein that contained the GARFT activity,[161] and its structure was also solved in complex with new novel inhibitors.[162] Comparison studies validated the use of the bacterial enzyme as a model for the human GARFT.

Previous SARs were employed in the design of new and novel GARFT inhibitors for substituents around the core of Lometrexol, including some GARFT inhibitors where the ring containing N5 was opened.[163] The structure of the bacterial GARFT–inhibitor complex revealed several important features. Within the GARFT active site, the pyrimidine portion of the pteridine was fully buried, forming hydrogen bonds with conserved amino acids. As expected, the D-glutamate moiety was largely solvent-exposed with no potential for modeling additional interactions. The importance of the role of the D-glutamate in pharmacodynamic processes required its retention. Clearly the methylene group in 5-deazatetrahydrofolate that replaces the naturally occurring N5 in tetrahydrofolate in the active site might be replaced with a bulkier hydrophobic atom or group. To test this idea, a series of 5-thiapyrimidinones were synthesized. Compound 5 (**Figure 7**) exhibited an activity of 30–40 nM in both a biochemical assay for human GARFT inhibition and a cell-based antiproliferation assay. These analogs were more readily prepared than the corresponding cyclic derivatives. A crystal structure of human GARFT, complexed with the inhibitor and glycinamide ribonucleotide, confirmed the structural homology between *E. coli* and human enzymes.

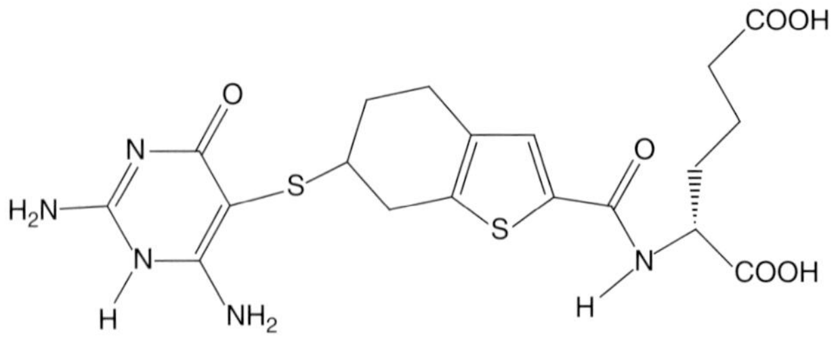

Figure 8 Compound 6.

Deletion of just one methylene in the thio bridge resulted in compounds with much lower activity. Analogs such as compound 6 (**Figure 8**) were obvious attempts to restrict conformational flexibility and fill the active site more fully. The importance of performing reiterative cycles of structure determination with new inhibitors and not relying solely on computational methods was confirmed. Molecular mechanics calculations failed to predict the conformation on the 5-thiamethylene group to GARFT correctly, due to unforeseen conformational flexibility of the enzyme active site revealed by an x-ray structure of this complex. Several functional criteria in addition to GARFT inhibition and cell-based assays were evaluated during cycles of optimization. These included the ability of exogenous purine to rescue cells (which indicates selective GARFT inhibition), and the ability of the inhibitors to function as substrates for enzymes involved in the transport and cellular accumulation of antifolate drugs. Balancing these criteria has resulted in the choice of two inhibitors, AG2034[164] and AG2037 respectively, for clinical development at Pfizer (in 1999 Agouron Pharmaceuticals was acquired by Warner-Lambert, which was subsequently acquired by Pfizer). It is as yet unclear whether the considerable toxicity of these and other GARFT inhibitors will allow these compounds to be approved as new anticancer drugs.

4.04.2.3 Summary of Other More Recent Structure-Based Drug Discovery Targets

Table 1 summarizes the increasing interest in and some successes of SBDD. This increase will undoubtedly continue with the initiation of the National Institutes of Health (NIH) Road Map in Structural Biology that fosters the ability to uncover protein folds extensively (*see* Section 4.04.4.2).

4.04.3 Structure-Based Drug Design Take-Home Messages

4.04.3.1 The First 10 Years at Agouron

Michael Varney[166] provided the following highlights of lessons learned during 10 years of protein SBDD at Agouron. In general these conclusions, presented at the 26th National Medicinal Chemistry Symposium, summarize the fundamentals helpful for all who employ structure-based drug discovery.

1. There is no substitute for experience: whether it be compound design mode, creation of database queries, analysis of output from a database search, or analysis of the results of a computer-based design experiment, the input from a person experienced in structure-based design is essential.
2. Go big early: filling as much of the available space and/or not leaving empty space in an active site will enhance the likelihood of a compound being an inhibitor.
3. There are multiple solutions to the same inhibition problem: structurally speaking, more than one template can satisfy enough of the space and functional group complementarity needs to serve as leads for elaboration to useful drugs.
4. Having the structure of a protein is not equivalent to having a drug: in fact, having a solved structure of a potent inhibitor bound to the target protein is only the very beginning of the game.
5. Solubility matters: the ratio of the solubility to the inhibition constant of a compound is critical to the success of the crystallization experiment. Using structural information, this ratio can be manipulated.
6. Integrated molecular biology: a dedicated molecular biology group is essential to both a constant protein supply and to the development of the right constructs for crystallization.
7. Prior art: structural information can be very useful in deciding ways to avoid competitors prior art.

Table 1 Summary of structure-based drug discovery targets

Protein class	*Target*	*Successful drug/clinical trials*
Peptidase	Angiotensin-converting enzyme	Captopril, therapy for hypertension and congestive heart failure. Cardiovasular and renal diseases[165]
Aspartic protease	HIV protease	The first marketed HIV-P inhibitor (saquinavir)
Aspartic protease	HIV protease	The C2 symmetric compound (A-77003) was synthesized at Abbott and entered clinical trials as an antiviral agent for intravenous treatment of AIDS[27]
Aspartic protease	HIV protease	The discovery of indinavir (L-735,524) was the result of the successful application of SBDD at Merck. During an iterative optimization process, the physicochemical properties of HIV-P inhibitors were modified within constraints that were established structurally. Crixivan (the sulfate) was successfully launched for use as an antiviral drug
Aspartic protease	HIV protease	Viracept was developed in a collaboration between scientists at Lilly and Agouron, and Viracept is marketed by Pfizer as the mesylate salt of nelfinavir. For both amprenavir and Viracept iterative SBDD methods were used to alter the physicochemical properties of the drug molecule while maintaining potency by optimizing interactions with the active site of the enzyme. Fortunately, the bound inhibitors appear to be in low energy conformers, so that minimal conformational energy costs must be paid for binding
Aspartic protease	HIV protease	Amprenavir (Agenerase, also known as VX-478), one of the most recent additions to the HIV-P inhibitors approved for human antiviral treatment, and differs significantly from earlier inhibitors was specifically designed by Vertex scientists to minimize molecular weight in order to increase oral bioavailability
Aspartic protease	β-Amyloid precursor protein, human beta-secretase (BACE-1), gamma-secretase	Alzheimer's, identification of therapeutic strategies to prevent or cure amyloid-related disorders, gamma-secretase inhibitors
Metalloprotease	Matrix metalloproteases	Osteoarthritis, and rheumatoid arthritis
Serine protease	Coagulation factor Xa	Factor Xa inhibitors are reported to have a better safety/efficacy profile compared to other anticoagulative drugs
Serine protease	Coagulation factor VII, VIIa	Structure-based designs of the P3 moiety in the peptide mimetic factor VIIa inhibitor successfully lead to novel inhibitors with selectivity for FVIIa/TF
Serine protease	Thrombin	Thrombin inhibitors are potentially useful in medicine for their anticoagulant and antithrombotic effects. The significant efforts at Merck to use SBDD approaches to develop orally available inhibitors of thrombin, which have yielded compounds that have entered clinical trials, have been reviewed.
Serine protease	Tryptase	Numerous proinflammatory cellular activities in vitro, and in animal models, tryptase provokes broncho-constriction and induces a cellular inflammatory infiltrate characteristic of human asthma. SBDD has resulted in selective, potent and orally bioavailable 4-(3-aminomethyl phenyl)piperidinyl-1-amides
Serine protease precursor	Anticoagulant protein C	Anticoagulation
Thiol protease	Caspase-1	Inhibitors of interleukin-1beta converting enzyme (ICE)
Aminidase	Neuraminidase (int B virus)	Influenza neuraminidase (A/Tokyo/67)

continued

Table 1 Continued

Protein class	*Target*	*Successful drug/clinical trials*
Anhydrase	Carbonic anhydrase 2 (CA 2), 5 (CA 5)	CA 2 target for the treatment of cancer, and it has recently been implicated in the delivery of sulfamate-containing drugs, CA 5, involved in lipogenesis
Cyclooxygenase	COX-1, 2	Comparative computer modeling of the x-ray crystal structures of isoforms COX-1 and COX-2 led to the design of COX-2 selectivity into the nonselective inhibitor flurbiprofen
Deaminase	Adenosine deaminase	Involved in various mammalian regulation processes, as well as in chronic human diseases, they have been proposed to play a role in pathogenicity for *Streptococci*, in vivo efficacy in models of inflammation and lymphoma
Deaminase	Cytosine deaminase	Antimicrobial drug design and gene therapy applications against tumors
Dehydrogenase	Inosine monophosphate dehydrogenase 2	Drug target for the control of parasitic infections
Dehydrogenase	Dihydroorotate dehydrogenase	The enzyme is a promising target for drug design in different biological and clinical applications for cancer and arthritis
Dismutase	Human superoxide dismutase	To improve the therapeutic effectiveness of human Cu, Zn superoxide dismutase (HSOD) by targeting it to cell surfaces and increasing its circulatory half-life
Helicase	DNA helicase pcra [Sa]	Helicases as nucleic acid unwinding machines
Hydrolase	Acetylcholinesterase	Target of nerve agents, insecticides and therapeutic drugs, in particular the first generation of anti-Alzheimer drugs
Hydrolase	Alpha-amylase	Controlling blood glucose levels
Hydrolase	Neuraminidase	Antiviral chemotherapeutic agents against respiratory viruses
Hydrolase	Phospholipase A_2	Anti-inflammatory
Kinase	p38 MAP kinase	Cancer, Alzheimer's disease, diabetes, and rheumatoid arthritis. High-throughput crystallography and lead compounds for p38 MAP kinase starting from fragments
Kinase	Thymidine kinase (HHV)	Inhibitors of herpes simplex virus type 1 (HSV-1) thymidine kinase and *Mycobacterium tuberculosis* thymidylate kinase
Lipase	Phospholipase A_2	Target protein for nonsteroidal anti-inflammatory drugs (NSAIDs)
Oxidoreductase	Aldose reductase	Treatment of diabetic complications
Oxidoreductase	Inosine monophosphate dehydrogenase	IMPDH-dependent hyperproliferative diseases, control of parasitic infections, immunoregulation and organ transplantation
Peptidase	Angiotensin-converting enzyme	Cardiovascular and renal diseases
Phosphatase	Calcineurin A	Intracellular signal transductions, calcineurin-mediated immunosuppressants
Phosphorylase	Purine nucleoside phosphorylase	Schistosomiasis, inhibition of the T-cell response, broad-spectrum antimicrobial activity
Synthetase	Dihydropteroate synthetase [Sa]	Target for the sulfonamide class of antibiotics
Topoisomerase	DNA topoisomerase 1, poly(ADP-ribose) polymerase-1 (PARP-1)	Topoisomerase antitumor efficacy
Transcriptase	HIV reverse transcriptase	Recent SBDD for HIV
Transferase	Guanine phosphoribosyltransferase	Design novel antiparasitic agents and for the treatment of Chagas disease
Nuclear receptor	Retinoic acid receptor	Retinoids have chemotherapeutic roles in dermatology and oncology, but their usefulness is restricted by the high toxicity of retinoic acid and its hydrophobic analogs

Table 1 Continued

Protein class	*Target*	*Successful drug/clinical trials*
Nuclear receptor	Retinoid X receptor	Normalization of differentiation and anti-inflammatory activities
Nuclear receptor	Vitamin D receptor	Inhibit the differentiation of HL-60 cells induced by 1 alpha,25-dihydroxyvitamin D_3, antitumor effects of 1 alpha,25-dihydroxyvitamin D_3 analogs, vitamin D receptor mutant associated with rickets
Nuclear receptor	Androgen	Therapy of breast and prostate cancer
Nuclear receptor	Estrogen receptor	Breast cancer
Nuclear receptor	Progesterone receptor	Prostate cancer
Nuclear receptor	Glucocorticoid receptor	New anti-inflammatory treatments
Receptor	Glutamate receptor 1	Treatment of neurodegenerative diseases such as Alzheimer's
Receptor	Human fibrinogen receptor (GPIIb-IIIa)	Fusion-type fibrinogen receptor antagonist
Receptor	Tumor necrosis factor receptor 1	Anti-inflammatory activities in clinical studies, particularly in rheumatoid arthritis
Hormone	Follicle stimulating hormone	Long acting gonadotropins with enhanced shelf lives
Hormone	Luteinizing hormone	Candidates for long acting gonadotropins with enhanced shelf lives
Hormone	Parathyroid hormone	Antiresorptive activity, parathyroid hormone analogs
Immunophilin	FK506-binding protein	Motor neuron disease, myelofibrosis with myeloid metaplasia
Integrin	β_2-Integrin family	As a target in several different inflammatory diseases, drugs for the treatment of asthma
Integrin	LFA-1 (leukocyte function-associated antigen-1) is a member of the β_2-integrin family	Anti-inflammatory drug target
Interleukin	Interleukins	Suppression of apoptosis in M1 cells, target in the treatment of asthma
Macrophage	Granulocyte-macrophage colony stimulating factor	Proinflammatory effects in autoimmune syndromes
Macrophage	Macrophage colony stimulating factor 1	Therapeutic agents in the area of hematoregulation
Peptide	Neuropeptide Y	Treatment of depression
Picornavirus	Human rhinovirus (HRV)	Human rhinovirus
Water channel	Aquaporin 1	Aquaporin water channels in central nervous system, involvement of water channels in synaptic vesicle swelling. No SBDD yet

AIDS, acquired immune deficiency syndrome; HHV, human herpesvirus; HL, human leukemia; MAP, mitogen-activated protein; TF, tissue factor.

8. Computational tools are inadequate: accurately calculating the absolute binding energy for any compound is essentially impossible. Estimating the relative binding of any two compounds (i.e., scoring designs) is better but there is much room for improvement.
9. Covalent interactions are difficult to model: designing inhibitors that interact with the target in a covalent sense presents special challenges, particularly from the perspective of computational modeling.
10. Physical and biological property manipulation: one of the great strengths of SBDD is that one can use structural information to guide the modification of physical and biological properties without compromising enzyme binding.
11. Designing to disrupt protein–protein interactions is a pipe dream: since these types of interaction involve no chemistry, SBDD would not be the tool of choice to approach this type of problem. Screening combinatorial libraries is probably more suited.

12. Every water molecule is special: the global concept of booting all water molecules is only a concept. The reality is that every water molecule is unique and has a unique environment. In fact, in many situations leaving a water molecule might be the preferred mode of inhibiting a particular enzyme. For example, there is no isosteric replacement for a water molecule that donates two H bonds and accepts one.
13. Not all hydrogen bonds are created equal: often attempts have been made to justify the booting of water molecules from active sites using the entropic advantage of having the water in bulk solvent and assuming that, since the H-bond inventory summed to zero, the only component of interest was the entropy. Although the H-bond inventory may sum to zero, that does not by definition mean the enthalpic component of the H bonds sum to zero, whether induced or discrete, are hard to anticipate and exploit.
14. Iterative cycle is essential: often small structural changes can result in major changes in binding mode. These major changes, if undetected, can lead to unproductive SAR. Put another way, a compound with an alternate binding mode can be considered a new series of compounds. In addition, the protein structure is compound-specific.
15. Dipoles: electrostatics beyond obvious H bonds and ion-pairing are very difficult to model and visualize. That is, dipoles, whether induced or discrete, are hard to anticipate and exploit.
16. Synthetic accessibility is essential in any design exercise: people with synthetic organic chemistry skills tend to be good designers. They don't design compounds that can't be made. The synthesis requirement is even more important if a combinatorial chemistry approach is part of the program.
17. There are very few, if any, single variable design changes: once a structure of a complex has been solved, often the molecule appears to need just one little change to be the 'perfect' compound. This approach often fails because any change to a structure influences all parts of the molecule.
18. Allow for serendipity: care must be taken not to design your way out of or around serendipity. That is, don't reject an idea based on the crystal structure or a calculation if it seems to make intuitive sense.
19. Combine technologies where possible: combinatorial chemistry coupled with SBDD can be a powerful combination. Just as on a compound-by-compound basis, the structural information can help guide the design of libraries. In addition, the numbers game can cover the inadequacies in the computational tools.
20. There is more than one path to a structurally novel inhibitor: structurally novel inhibitors can be designed by totally de novo methods into an empty active site, piece by piece from a known structure or a combination of the two.
21. New small-molecule structures lead to new protein structures: the conformational structure of a protein is dependent upon the structure of the bound small molecule. Alternate protein conformations can be accessed by binding structurally distinct inhibitor.

4.04.3.2 Hemoglobin Drug Discovery

We have published our own take-home messages from our drug discovery efforts in an academic setting.[129] While some of these are similar to those listed above (Section 4.04.3.1), others are unique since they extend beyond the structure-based drug discovery phase.

1. Collaborative efforts across academic units were a vital key for successful translation from the bench to the bedside.
2. Use of a known low-toxicity scaffold (drug) for building molecular specificity was the most important advance permitting us to overcome toxicity issues due to the large doses required to treat almost $1\frac{1}{2}$ lb (0.7 kg) of the in vivo receptor (Hb).
3. When in doubt as to which molecule to forward as a clinical candidate, consider metabolism and toxicity profiles as well as biological activity.
4. Having a dedicated startup company to champion moving a molecule through all the steps to a new drug application made the difference as large pharmaceutical companies are less likely to champion a compound from academia.
5. Serendipity still continues to play an important role in drug discovery. Max Perutz in the preface of his book "I Wish I'd Made You Angrier Earlier"[167] quotes the four German Gs of Paul Ehrlich that scientists need which were all evident in our case study: Geschick (skill), Geduld (patience), Geld (money), and Gluck (luck).

4.04.4 The Future

4.04.4.1 High-Throughput Crystallography

Crystallography continues to undergo dramatic changes.[168] New and improved methodologies are continually evolving in robotics, x-ray sources, computational power, crystal-growing screens, molecular genetics, and preparation of adequate sources of desired targets. More importantly, these improved technologies are integrated in both industrial and academic laboratories. The advent of companies specifically organized and funded to perform high-throughput crystallography has been expanding. Structural genomic programs abound and readily feed into SBDD applications.[168] Will the future for structure-based drug discovery live up to its elegance and early successful efforts in discovering new therapeutic agents? The answer will be 'yes,' and even better than that!

4.04.4.2 Combinatorial Chemistry, High-Throughput Screening (HTS), and Structure-Based Drug Design

The norm, especially in the largest pharmaceutical companies, has been to make sure any new technologies (fads?) for discovering lead molecules found in a competitor's company must be had by all, just in case. Looking back at the mania that arose from the birth of combinatorial chemical libraries coupled with HTS, one must ask, as I have heard said a number of times from colleagues, "Was all the excitement and efforts directed toward combinatorial chemistry/high-throughput screening really worth it?" The answer at this time is no. As of 2002 only a few have entered preclinical and clinical trials.[170]

There have been a number of road blocks with screening of combinatorial libraries. The first is that HTS usually employs a single purified target protein or receptor isolated from cellular contents. When only a single target is assayed, the results are devoid of the robust milieu pharmacologists used with whole-cell or quick looks in vivo in rats. The second is that current HTS methodologies identify large numbers of leads. However, most leads are too insoluble to become drugs. This problem multiplies when chemists are brought in to make analogs with suitable ADME properties. It is now in vogue to use the Lapinski rule of five and SBDD findings for a given system to design the next round of libraries.

Some thought at the beginning – a one-million combinatorial library would contain every known drug. It was a surprise that such a library actually had no known drugs. Campbell has noted a logical reason for the failure of large libraries. "One factor in this perceived lack of success may be that combinatorial chemistry has been widely used to generate thousands of similar compounds, but limited synthetic options have not allowed access to the rich molecular architectures often required for interaction with biological targets. In response, emphasis has now shifted to building compound files with maximum structural richness and diversity, since designed libraries should have higher hit rates."[171]

A recent review by Kubinyi summarizes the problems with combinatorial chemistry with suggestions for success in the future. "Lack of success with early combinatorial chemistry and HTS approaches resulted from inappropriate compound selection. We are now aware that screening compounds should be either 'lead-like' or 'drug-like' and have the potential to be orally available. However, there is a growing tendency to misuse such terms and to overestimate their importance, and to overemphasize ADME problems in clinical failure. Sometimes, this goes hand-in-hand with an uncritical application of high-throughput in silico methods. Structure-based and computer-aided approaches can only be as good as the medicinal chemistry they are based on. The search for new drugs, especially in lead optimization, is an evolutionary process that is only likely to be successful if new methods merge with classical medicinal chemistry knowledge."[172] SBDD and combinatorial chemistry integration are at hand but it is too early to predict the success of this marriage.

4.04.4.3 The National Institutes of Health Structural Biology Road Map, Canonical Structures for Protein Folding, and Structure-Based Drug Design

The NIH Road Map for Structural Biology as provided in the NIH website states it "is a strategic effort to create a 'picture' gallery of the molecular shapes of proteins in the body. This research investment will involve the development of rapid, efficient, and dependable methods to produce protein samples that scientists can use to determine the 3D structure, or shape, of a protein. The new effort will catalyze what is currently a hit-or-miss process into a streamlined routine, helping researchers clarify the role of protein shape in health and disease." Clearly an effort to sample the whole human genome is in progress. It is hoped that there will be a definitive number of folds that proteins form (canonical structures). One will then be able to acquire a desired protein target in three dimensions directly from the human genome. This knowledge of the active site for any protein in the genome has been acknowledged as the primary future economic engine for the future of biotechnology.[173–176] SBDD will therefore have the key focus for the future of drug discovery.

4.04.4.4 Real-Time Crystallography

Protein crystallography is normally a static tool that visualizes end-state structures at stable free energies and cannot be used to follow biological reactions in real time. However, in the 1980s Ringe and Petsko[177] studied protein dynamics by x-ray crystallography and Hajdu and colleagues used Laue diffraction that is capable of recording entire protein crystal data sets in a millisecond.[178] Recently, picosecond time-resolved x-ray crystallography has been able to probe protein function in real time.[179] When combined with low-temperature techniques, such methods could be used to determine the structures of catalytic intermediates. With this development it will be possible for medicinal chemists to use SBDD to discover transition-stated analogs with a probable higher rate of success than with the static receptor design methods. This technique might also be valuable in assaying flexible binding sites during the initial binding of inhibitors or substrates.

4.04.4.5 Single-Molecule Imaging: Determining the Three-Dimensional Structure of Receptors without the Need to Crystallize Them

Electron microscopy is another principal method of macromolecular structure determination that uses scattering techniques, as discussed in Section 4.04.1.2.6. The most important difference between electron microscopy and x-ray crystallography is that electron microscopy specimens are 1–10 nm thick, whereas scattering or absorption of a similar fraction of an illuminating x-ray beam requires crystals that are 100–500 μm thick. The second advantage is that electrons are readily focused while x-rays cannot be directly imaged. As a result, Henderson and Baker[61] point out: "electron lenses are greatly superior to x-ray lenses and can be used to produce a magnified image of an object as easily as a diffraction pattern. This then allows the electron to be switched back and forth instantly between imaging and diffraction modes so the image of a single molecule at any magnification can be obtained as conveniently as the electron diffraction pattern of a thin crystal." The number of inelastic events for electrons scattered by biological structures at all electron energies of interest exceeds the number of elastic events by a factor of 3–4, so each elastically scattered electron deposits 60–80 eV of energy. The amount of information in a single biological image of a macromolecule is therefore limited. Unfortunately, at this time, the 3D atomic structure cannot be determined from a single molecule but theoretically requires the averaging of the information from at least 10 000 molecules: this has not yet been achieved in practice.[180] Obviously, crystals used for x-ray or neutron diffraction contain many magnitudes of diffracting molecules.

Henderson and Baker comment on the current trends in the field[61]: "(1) increased automation, including the recording of micrographs, the use of spotscan procedures in remote microscope operation,[181,182] and in every aspect of image processing; (2) production of better electronic cameras (e.g., CCD or pixel detectors); and (3) increased use of dose-fractionated, tomographic tilt series, to extend EM studies to the domain of larger supramolecular and cellular structures."[183,184] The increased capabilities in the future of electron microscopy may well enable a large number of interesting biological targets to be imaged where crystallography cannot be employed. If these advances are successful we may one day see imaging of a single macromolecule of biological interest for SBDD. This would be the holy grail for structural biology.

References

1. Ehrlich, P. *Int. Arch. Allergy Appl. Immunol.* **1954**, *5*, 67–86.
2. Cushman, D. W.; Cheung, H. S.; Sabo, E. F.; Rubin, B.; Ondetti, M. A. *Fed. Proc.* **1979**, *38*, 2778–2782.
3. Cushman, D. W.; Cheung, H. S.; Sabo, E. F.; Ondetti, M. A. *Prog. Cardiovasc. Dis.* **1978**, *21*, 176–182.
4. Cushman, D. W.; Cheung, H. S.; Sabo, E. F.; Ondetti, M. A. *Biochemistry* **1977**, *16*, 5484–5491.
5. Patchett, A. A.; Harris, E.; Tristram, E. W.; Wyvratt, M. J.; Wu, M. T.; Taub, D.; Peterson, E. R.; Ikeler, T. J.; Ten, B. J.; Payne, L. G. et al. *Nature* **1980**, *288*, 280–283.
6. Perutz, M. F. *Science* **1963**, *140*, 863–869.
7. Arndt, U. W.; Leigh, J. B.; Mallett, J. F.; Twinn, K. E. *J. Sci. Instrum.* **1969**, *2*, 385–387.
8. Arndt, U. W.; Crowther, R. A.; Mallett, J. F. *J. Sci. Instrum.* **1968**, *1*, 510–516.
9. Arndt, U. W. *Methods Enzymol.* **2003**, *368*, 21–42.
10. Levinthal, C. *Ann. NY Acad. Sci.* **1984**, *426*, 171–180.
11. Abraham, D. J. X-Ray Crystallography and Drug Design. In *Computer-Aided Drug Design*; Perun, T. J., Propst, C. L., Eds.; Marcel Dekker: New York, 1989, pp 93–132.
12. Kendrew, J. C.; Bodo, G.; Dintzis, H.; Parrish, R.; Wyckoff, H.; Phillips, D. *Nature* **1958**, *181*, 662–666.
13. Richards, F. M. *Methods Enzymol.* **1985**, *115*, 145–154.
14. Wyckoff, H. W.; Hardman, K. D.; Allewell, N. M.; Inagami, T.; Tsernoglou, D.; Johnson, L. N.; Richards, F. M. *J. Biol. Chem.* **1967**, *242*, 3749–3753.
15. Wyckoff, H. W.; Hardman, K. D.; Allewell, N. M.; Inagami, T.; Johnson, L. N.; Richards, F. M. *J. Biol. Chem.* **1967**, *242*, 3984–3988.
16. Rubin, B.; Richardson, J. S. *Biopolymers* **1972**, *11*, 2381–2385.
17. Martz, E.; Francoeur, E. History of Visualization of Biological Macromolecules 21, 2004. http://www.umass.edu/microbio/rasmol/history.htm (accessed Aug 2006).
18. Richardson, J. S.; Richardson, D. C.; Thomas, K. A.; Silverton, E. W.; Davies, D. R. *J. Mol. Biol.* **1976**, *102*, 221–235.
19. Connolly, M. L. Molecular Surfaces: A Review – Part 4. Molecular Graphics 22, 1996. http://www.netsci.org/ (accessed Aug 2006).
20. Meyer, E. F., Jr. *Nature* **1971**, *232*, 255–257.
21. Barry, C. D.; North, A. C. *Biochem. J.* **1971**, *121*, 3P.

22. Katz, L.; Levinthal, C. *Annu. Rev. Biophys. Bioeng.* **1972**, *1*, 465–504.
23. Levinthal, C.; Wodak, S. J.; Kahn, P.; Dadivanian, A. K. *Proc. Natl. Acad. Sci. USA* **1975**, *72*, 1330–1334.
24. Langridge, R. *Fed. Proc.* **1974**, *33*, 2332–2335.
25. Collins, D. M.; Cotton, F. A.; Hazen, E. E., Jr.; Meyer, E. F., Jr.; Morimoto, C. N. *Science* **1975**, *190*, 1047–1053.
26. Porter, T. K. *Comput. Graphics* **1978**, *12*, 282–285.
27. Feldmann, R. J.; Bing, D. H.; Furie, B. C.; Furie, B. *Proc. Natl. Acad. Sci. USA* **1978**, *75*, 5409–5412.
28. Max, N. L. *Comput. Graphics* **1979**, *13*, 165–173.
29. Palmer, T. C.; Hausheer, F. H.; Saxe, J. D. *J. Mol. Graph.* **1989**, *7*, 160–168.
30. Evans, S. V. *J. Mol. Graph.* **1993**, *11*, 134–138.
31. Johnson, C. K. *ORTEP-II, a Fortran Thermal-Ellipsoid Plot Program for Crystal Structure Illustrations*; Technical Report ORNL-5138; Oak Ridge National Laboratory: Tennessee, 1976.
32. Diamond, R. *Acta Crystallogr.* **1971**, *A27*, 436–452.
33. Diamond, R. *Biomol. Struct. Conform. Funct. Evol.* **1981**, *1*, 567–588.
34. Diamond, R. Bilder: An Interactive Graphics Program for Biopolymers. In *Computational Crystallography*; Sayre, D., Ed.; Clarendon Press: Oxford, 1981, pp 318–325.
35. Lesk, A. M.; Hardman, K. D. *Science* **1982**, *216*, 539–540.
36. Lesk, A. M.; Hardman, K. D. *Methods Enzymol.* **1985**, *115*, 381–390.
37. Lesk, A. M. *Protein Architecture: A Practical Approach*; IRL Press: Oxford, 1991.
38. Jones, T. A. FRODO: A Graphics Fitting Program for Macromolecules. In *Computational Crystallography*; Sayre, D., Ed.; Clarendon Press: Oxford, 1981, pp 303–317.
39. Jones, T. A. *J. Appl. Crystallogr.* **1978**, *11*, 268–272.
40. Jones, T. A. *Methods Enzymol.* **1985**, *115*, 157–171.
41. Roussel, A.; Fontecilla-Camps, J. C.; Cambillau, C. *J. Mol. Graph.* **1990**, *8*, 86–88, 91.
42. Goodford, P. J. *J. Med. Chem.* **1985**, *28*, 849–857.
43. Langridge, R.; Ferrin, T. E.; Kuntz, I. D.; Connolly, M. L. *Science* **1981**, *211*, 661–666.
44. Shoichet, B. K.; Kuntz, I. D. *Protein Eng.* **1993**, *6*, 723–732.
45. Shoichet, B. K.; Stroud, R. M.; Santi, D. V.; Kuntz, I. D.; Perry, K. M. *Science* **1993**, *259*, 1445–1450.
47. Marshall, G. Professor of Biochemistry and Molecular Biophysics at Washington University's School of Medicine, 2005. http://bp.wustl.edu/#program-faculty (accessed Aug 2006).
48. Cramer, R. D., III.; Patterson, D. E.; Bunce, J. D. *Prog. Clin. Biol. Res.* **1989**, *291*, 161–165.
49. Williamson, M. P.; Havel, T. F.; Wuthrich, K. *J. Mol. Biol.* **1985**, *182*, 295–315.
50. Pervushin, K.; Riek, R.; Wider, G.; Wuthrich, K. *Proc. Natl. Acad. Sci. USA* **1997**, *94*, 12366–12371.
51. Craik, D. J.; Clark, R. J. NMR and Drug Discovery. In *Burger's Medicinal Chemistry and Drug Discovery*; 6th ed.; Abraham, D. J., Ed.; John Wiley: Hoboken, NJ, 2003; Vol. 1, pp 507–582.
52. Coluquhoun, D.; Shelley, C.; Hatton, C.; Unwin, N.; Sivilotti, L. Nicotenic Acetyl Choline Receptors. In *Burger's Medicinal Chemistry and Drug Discovery*; 6th ed.; Abraham, D. J., Ed.; John Wiley: Hoboken, NJ, 2003; Vol. 1, pp 357–406.
53. Henderson, R.; Baldwin, J. M.; Ceska, T. A.; Zemlin, F.; Beckmann, E.; Downing, K. H. *J. Mol. Biol.* **1990**, *213*, 899–929.
54. Henderson, R.; Schertler, G. F. *Philos. Trans. R. Soc. Lond. B Biol. Sci.* **1990**, *326*, 379–389.
55. Westkaemper, R. B.; Glennon, R. A. *Pharmacol. Biochem. Behav.* **1991**, *40*, 1019–1031.
56. Dubochet, J.; Adrian, M.; Chang, J. J.; Homo, J. C.; Lepault, J.; McDowall, A. W.; Schultz, P. *Q. Rev. Biophys.* **1988**, *21*, 129–228.
57. Amos, L. A.; Henderson, R.; Unwin, P. N. *Prog. Biophys. Mol. Biol.* **1982**, *39*, 183–231.
58. Walz, T.; Grigorieff, N. *J. Struct. Biol.* **1998**, *121*, 142–161.
59. Baker, T. S.; Olson, N. H.; Fuller, S. D. *Microbiol. Mol. Biol. Rev.* **1999**, *63*, 862–922.
60. Frank, J. *Three-Dimensional Electron Microscopy of MacromolecularAssemblies*; Academic Press: San Diego, CA, 1996.
61. Henderson, R.; Baker, T. S. Electron Cryomicroscopy of Biological Macromolecules. In *Burger's Medicinal Chemistry and Drug Discovery*; 6th ed.; Abraham, D. J., Ed.; John Wiley: Hoboken, NJ, 2003; Vol. 1, pp 611–632.
62. Muegge, I.; Martin, Y. C.; Hajduk, P. J.; Fesik, S. W. *J. Med. Chem.* **1999**, *42*, 2498–2503.
63. Rotstein, S. H.; Murcko, M. A. *J. Med. Chem.* **1993**, *36*, 1700–1710.
64. Wlodawer, A.; Vondrasek, J. *Annu. Rev. Biophys. Biomol. Struct.* **1998**, *27*, 249–284.
65. Lipscomb, W. N.; Reeke, G. N., Jr.; Hartsuck, J. A.; Quiocho, F. A.; Bethge, P. H. *Philos. Trans. R. Soc. Lond B Biol. Sci.* **1970**, *257*, 177–214.
66. Quiocho, F. A.; Bethge, P. H.; Lipscomb, W. N.; Studebaker, J. F.; Brown, R. D.; Koenig, S. H. *Cold Spring Harb. Symp. Quant. Biol.* **1972**, *36*, 561–567.
67. Abraham, D. J. *Intra Sci. Chem. Rep.* **1974**, *8*, 1–9.
68. Sobell, H. M.; Jain, S. C.; Sakore, T. D.; Ponticello, G. *Cold Spring Harb. Symp. Quant. Biol.* **1972**, *36*, 263–270.
69. Matthews, D. A.; Alden, R. A.; Bolin, J. T.; Freer, S. T.; Hamlin, R.; Xuong, N.; Kraut, J.; Poe, M.; Williams, M.; Hoogsteen, K. *Science* **1977**, *197*, 452–455.
70. Bolton, W.; Perutz, M. F. *Nature* **1970**, *228*, 551–552.
71. Fermi, G.; Perutz, M. F.; Shaanan, B.; Fourme, R. *J. Mol. Biol.* **1984**, *175*, 159–174.
72. Perutz, M. F.; Muirhead, H.; Cox, J. M.; Goaman, L. C. *Nature* **1968**, *219*, 131–139.
73. Shaanan, B. *J. Mol. Biol.* **1983**, *171*, 31–59.
74. Huang, Y.; Pagnier, J.; Magne, P.; Baklouti, F.; Kister, J.; Delaunay, J.; Poyart, C.; Fermi, G.; Perutz, M. F. *Biochemistry* **1990**, *29*, 7020–7023.
75. Nagai, K.; Perutz, M. F.; Poyart, C. *Proc. Natl. Acad. Sci. USA* **1985**, *82*, 7252–7255.
76. Perutz, M. F.; Pulsinelli, P. D.; Ranney, H. M. *Adv. Exp. Med. Biol.* **1972**, *28*, 3–18.
77. Pulsinelli, P. D.; Perutz, M. F.; Nagel, R. L. *Proc. Natl. Acad. Sci. USA* **1973**, *70*, 3870–3874.
78. Perutz, M. F.; Pulsinelli, P.; Eyck, L. T.; Kilmartin, J. V.; Shibata, S.; Iuchi, I.; Miyaji, T.; Hamilton, H. B. *Nat. New Biol.* **1971**, *232*, 147–149.
79. Weatherall, D. J.; Clegg, J. B.; Callender, S. T.; Wells, R. M.; Gale, R. E.; Huehns, E. R.; Perutz, M. F.; Viggiano, G.; Ho, C. *Br. J. Haematol.* **1977**, *35*, 177–191.
80. Honig, G. R.; Vida, L. N.; Rosenblum, B. B.; Perutz, M. F.; Fermi, G. *J. Biol. Chem.* **1990**, *265*, 126–132.
81. Perutz, M. F.; Fermi, G.; Fogg, J.; Rahbar, S. *J. Mol. Biol.* **1988**, *201*, 459–461.
82. Tucker, P. W.; Perutz, M. F. *J. Mol. Biol.* **1977**, *114*, 415–420.

83. Anderson, N. L.; Perutz, M. F. *Nat. New Biol.* **1973**, *243*, 274–275.
84. Perutz, M. F. *Nat. New Biol.* **1973**, *243*, 180.
85. Asakura, T.; Adachi, K.; Wiley, J. S.; Fung, L. W.; Ho, C.; Kilmartin, J. V.; Perutz, M. F. *J. Mol. Biol.* **1976**, *104*, 185–195.
86. Phillips, S. E.; Hall, D.; Perutz, M. F. *J. Mol. Biol.* **1981**, *150*, 137–141.
87. Abraham, D. J.; Mehanna, A. S.; Williams, F. L. *J. Med. Chem.* **1982**, *25*, 1015–1017.
88. Abraham, D. J.; Mokotoff, M.; Sheh, L.; Simmons, J. E. *J. Med. Chem.* **1983**, *26*, 549–554.
89. Abraham, D. J.; Perutz, M. F.; Phillips, S. E. *Proc. Natl. Acad. Sci. USA* **1983**, *80*, 324–328.
90. Merrett, M.; Stammers, D. K.; White, R. D.; Wootton, R.; Kneen, G. *Biochem. J.* **1986**, *239*, 387–392.
91. Beddell, C. R.; Goodford, P. J.; Kneen, G.; White, R. D.; Wilkinson, S.; Wootton, R. *Br. J. Pharmacol.* **1984**, *82*, 397–407.
92. Keidan, A. J.; Franklin, I. M.; White, R. D.; Joy, M.; Huehns, E. R.; Stuart, J. *Lancet* **1986**, *1*, 831–834.
93. Fitzharris, P.; McLean, A. E.; Sparks, R. G.; Weatherley, B. C.; White, R. D.; Wootton, R. *Br. J. Clin. Pharmacol.* **1985**, *19*, 471–481.
94. Wireko, F. C.; Abraham, D. J. *Proc. Natl. Acad. Sci. USA* **1991**, *88*, 2209–2211.
95. Padlan, E. A.; Love, W. E. *J. Biol. Chem.* **1985**, *260*, 8272–8279.
96. Padlan, E. A.; Love, W. E. *J. Biol. Chem.* **1985**, *260*, 8280–8291.
97. Wishner, B. C.; Ward, K. B.; Lattman, E. E.; Love, W. E. *J. Mol. Biol.* **1975**, *98*, 179–194.
98. Cozzini, P.; Fornabaio, M.; Marabotti, A.; Abraham, D. J.; Kellogg, G. E.; Mozzarelli, A. *Curr. Med. Chem.* **2004**, *11*, 3093–3118.
99. Abraham, D. J.; Mehanna, A. S.; Wireko, F. C.; Whitney, J.; Thomas, R. P.; Orringer, E. P. *Blood* **1991**, 77, 1334–1341.
100. Farthing, D.; Sica, D.; Abernathy, C.; Fakhry, I.; Roberts, J. D.; Abraham, D. J.; Swerdlow, P. *J. Chromatogr. B Biomed. Sci. Appl.* **1999**, *726*, 303–307.
101. Abdulmalik, O.; Safo, M. K.; Chen, Q.; Yang, J.; Brugnara, C.; Ohene-Frempong, K.; Abraham, D. J.; Asakura, T. *Br. J. Haematol.* **2005**, *128*, 552–561.
102. Safo, M. K.; Abdulmalik, O.; nso-Danquah, R.; Burnett, J. C.; Nokuri, S.; Joshi, G. S.; Musayev, F. N.; Asakura, T.; Abraham, D. J. *J. Med. Chem.* **2004**, *47*, 4665–4676.
103. Burnett, J. C.; Kellogg, G. E.; Abraham, D. J. *Biochemistry* **2000**, *39*, 1622–1633.
104. Cozzini, P.; Fornabaio, M.; Marabotti, A.; Abraham, D. J.; Kellogg, G. E.; Mozzarelli, A. *J. Med. Chem.* **2002**, *45*, 2469–2483.
105. Fornabaio, M.; Spyrakis, F.; Mozzarelli, A.; Cozzini, P.; Abraham, D. J.; Kellogg, G. E. *J. Med. Chem.* **2004**, *47*, 4507–4516.
106. Orringer, E. P.; Blythe, D. S.; Whitney, J. A.; Brockenbrough, S.; Abraham, D. J. *Am. J. Hematol.* **1992**, *39*, 39–44.
107. Abraham, D. J.; Mehanna, A. S.; Williams, F. S.; Cragoe, E. J., Jr.; Woltersdorf, O. W., Jr. *J. Med. Chem.* **1989**, *32*, 2460–2467.
108. Abraham, D. J.; Patwa, D. C. *Clin. Chem.* **1985**, *31*, 649–650.
109. Abraham, D. J.; Kennedy, P. E.; Mehanna, A. S.; Patwa, D. C.; Williams, F. L. *J. Med. Chem.* **1984**, *27*, 967–978.
110. Kennedy, P. E.; Williams, F. L.; Abraham, D. J. *J. Med. Chem.* **1984**, *27*, 103–105.
111. Arnone, A. *Nature* **1972**, *237*, 146–149.
112. Richard, V.; Dodson, G. G.; Mauguen, Y. *J. Mol. Biol.* **1993**, *233*, 270–274.
113. Levine, B. D.; Stray-Gundersen, J. *J. Appl. Physiol.* **1997**, *83*, 102–112.
114. Arnaud, J.; Gutierrez, N. *Am. J. Physiol. Anthropol.* **2005**, *63*, 307–314.
115. Goodford, P. J.; St-Louis, J.; Wootton, R. *Br. J. Pharmacol.* **1980**, *68*, 741–748.
116. Beddell, C. R.; Goodford, P. J.; Stammers, D. K.; Wootton, R. *Br. J. Pharmacol.* **1979**, *65*, 535–543.
117. Goodford, P. J.; St-Louis, J.; Wootton, R. *J. Physiol.* **1978**, *283*, 397–407.
118. Goodford, P. J. *Br. J. Pharmacol.* **1978**, *62*, 428P–429P.
119. Goodford, P. J.; Norrington, F. E.; Paterson, R. A.; Wootton, R. *J. Physiol.* **1977**, *273*, 631–645.
120. Brown, F. F.; Goodford, P. J. *Br. J. Pharmacol.* **1977**, *60*, 337–341.
121. Beddell, C. R.; Goodford, P. J.; Norrington, F. E.; Wilkinson, S.; Wootton, R. *Br. J. Pharmacol.* **1976**, *57*, 201–209.
122. Bedell, C. R.; Goodford, P. J.; Norrington, F. E.; Wilkinson, S. *Br. J. Pharmacol.* **1973**, *48*, 363P–364P.
123. Abraham, D. J.; Wireko, F. C.; Randad, R. S.; Poyart, C.; Kister, J.; Bohn, B.; Liard, J. F.; Kunert, M. P. *Biochemistry* **1992**, *31*, 9141–9149.
124. Randad, R. S.; Mahran, M. A.; Mehanna, A. S.; Abraham, D. J. *J. Med. Chem.* **1991**, *34*, 752–757.
125. Safo, M. K.; Moure, C. M.; Burnett, J. C.; Joshi, G. S.; Abraham, D. J. *Protein Sci.* **2001**, *10*, 951–957.
126. Wireko, F. C.; Kellogg, G. E.; Abraham, D. J. *J. Med. Chem.* **1991**, *34*, 758–767.
127. Suh, J. H.; Stea, B.; Nabid, A.; Kresl, J. J.; Fortin, A.; Mercier, J. P.; Senzer, N.; Chang, E. L.; Boyd, A. P.; Cagnoni, P. J. et al. *J. Clin. Oncol.* **2006**, *24*, 106–114.
128. Abraham, D. J.; Kister, J.; Joshi, G. S.; Marden, M. C.; Poyart, C. *J. Mol. Biol.* **1995**, *248*, 845–855.
129. Abraham, D. J. Drug Discovery in Academia – A Case Study. In *Chemoinformatics in Drug Discovery*; Oprea, T. I., Ed.; Wiley-VCH: Weinheim, 2005, pp 457–484.
130. Kilgore, K. S.; Shwartz, C. F.; Gallagher, M. A.; Steffen, R. P.; Mosca, R. S.; Bolling, S. F. *Circulation* **1999**, *100*, II351–II356.
131. Paranka, N. *Special Workshop Blood Doping in Sports and Detection Strategies: 7. Hemoglobin Modifiers; Is RSR 13 the Next Aerobic Enhancer?* ISLH XIVth International Symposium, 2006. http://www.tallonediachille.it/download/tesiGS.pdf (accessed Aug 2006).
132. Bauman, R. A.; Przybelski, R. J.; Bounds, M. J. *Physiol. Behav.* **1991**, *50*, 205–211.
133. Feola, M.; Simoni, J.; Canizaro, P. C. *Artif. Organs* **1991**, *15*, 243–248.
134. Frantantoni, J. C. *Transfusion* **1991**, *31*, 369–371.
135. Walder, J. A.; Zaugg, R. H.; Walder, R. Y.; Steele, J. M.; Klotz, I. M. *Biochemistry* **1979**, *18*, 4265–4270.
136. Chatterjee, R.; Welty, E. V.; Walder, R. Y.; Pruitt, S. L.; Rogers, P. H.; Arnone, A.; Walder, J. A. *J. Biol. Chem.* **1986**, *261*, 9929–9937.
137. Snyder, S. R.; Welty, E. V.; Walder, R. Y.; Williams, L. A.; Walder, J. A. *Proc. Natl. Acad. Sci. USA* **1987**, *84*, 7280–7284.
138. Campanini, B.; Bruno, S.; Raboni, S.; Mozzarelli, A. Oxygen Delivery by Allosteric Effectors of Hemoglobin, Blood Substitutes, and Plasma Expanders. In *Burger's Medicinal Chemistry and Drug Discovery*; 6th ed.; Abraham, D. J., Ed.; John Wiley: Hoboken, NJ, 2003; Vol. 1, pp 385–442.
139. Komiyama, N.; Tame, J.; Nagai, K. *Biol. Chem.* **1996**, *377*, 543–548.
140. Perutz, M. F.; Fermi, G.; Abraham, D. J.; Poyart, C.; Bursaux, E. *J. Am. Chem. Soc.* **1986**, *108*, 1064–1978.
141. Abraham, D. J.; Safo, M. K.; Boyiri, T.; Danso-Danquah, R. E.; Kister, J.; Poyart, C. *Biochemistry* **1995**, *34*, 15006–15020.
142. Schoenborn, B. P. *Nature* **1965**, *208*, 760–762.
143. Mattevi, A.; Rizzi, M.; Bolognesi, M. *Curr. Opin. Struct. Biol.* **1996**, *6*, 824–829.
144. Matthews, D. A.; Bolin, J. T.; Burridge, J. M.; Filman, D. J.; Volz, K. W.; Kraut, J. *J. Biol. Chem.* **1985**, *260*, 392–399.
145. Kisliuk, R. L. *Curr. Pharm. Des.* **2003**, *9*, 2615–2625.

146. Appelt, K.; Bacquet, R. J.; Bartlett, C. A.; Booth, C. L.; Freer, S. T.; Fuhry, M. A.; Gehring, M. R.; Herrmann, S. M.; Howland, E. F.; Janson, C. A. *J. Med. Chem.* **1991**, *34*, 1925–1934.
147. Matthews, D. A.; Villafranca, J. E.; Janson, C. A.; Smith, W. W.; Welsh, K.; Freer, S. *J. Mol. Biol.* **1990**, *214*, 937–948.
148. Montfort, W. R.; Perry, K. M.; Fauman, E. B.; Finer-Moore, J. S.; Maley, G. F.; Hardy, L.; Maley, F.; Stroud, R. M. *Biochemistry* **1990**, *29*, 6964–6977.
149. Takemura, Y.; Jackman, A. L. *Anticancer Drugs* **1997**, *8*, 3–16.
150. Almog, R.; Waddling, C. A.; Maley, F.; Maley, G. F.; Van, R. P. *Protein Sci.* **2001**, *10*, 988–996.
151. Jackson, S. Thumitaq completes enrollment in pivotal phase III liver cancer study, 4–19–2005. http://www.quakerbio.com/news/pcpr/2005/Thymitaq%204.19.05.pdf (accessed Aug 2006).
152. Hardy, L. W.; Safo, M. K.; Abraham, D. J. Structure-Based Drug Design. In *Burger's Medicinal Chemistry and Drug Discovery*; 6th ed.; Abraham, D. J., Ed.; John Wiley: Hoboken, NJ, 2003; Vol. 1, pp 417–469.
153. Varney, M. D.; Marzoni, G. P.; Palmer, C. L.; Deal, J. G.; Webber, S.; Welsh, K. M.; Bacquet, R. J.; Bartlett, C. A.; Morse, C. A.; Booth, C. L. *J. Med. Chem.* **1992**, *35*, 663–676.
154. Newell, D. R. *Semin. Oncol.* **1999**, *26*, 74–81.
155. Norman, P. *Curr. Opin. Invest. Drugs* **2001**, *2*, 1611–1622.
156. Taylor, E. C. *Adv. Exp. Med. Biol.* **1993**, *338*, 387–408.
157. Beardsley, G. P.; Moroson, B. A.; Taylor, E. C.; Moran, R. G. *J. Biol. Chem.* **1989**, *264*, 328–333.
158. Piper, J. R.; McCaleb, G. S.; Montgomery, J. A.; Kisliuk, R. L.; Gaumont, Y.; Thorndike, J.; Sirotnak, F. M. *J. Med. Chem.* **1988**, *31*, 2164–2169.
159. Greasley, S. E.; Marsilje, T. H.; Cai, H.; Baker, S.; Benkovic, S. J.; Boger, D. L.; Wilson, I. A. *Biochemistry* **2001**, *40*, 13538–13547.
160. Almassy, R. J.; Janson, C. A.; Kan, C. C.; Hostomska, Z. *Proc. Natl. Acad. Sci. USA* **1992**, *89*, 6114–6118.
161. Kan, C. C.; Gehring, M. R.; Nodes, B. R.; Janson, C. A.; Almassy, R. J.; Hostomska, Z. *J. Protein Chem.* **1992**, *11*, 467–473.
162. Varney, M. D.; Palmer, C. L.; Romines, W. H., III; Boritzki, T.; Margosiak, S. A.; Almassy, R.; Janson, C. A.; Bartlett, C.; Howland, E. J.; Ferre, R. *J. Med. Chem.* **1997**, *40*, 2502–2524.
163. Shih, C.; Gossett, L. S.; Worzalla, J. F.; Rinzel, S. M.; Grindey, G. B.; Harrington, P. M.; Taylor, E. C. *J. Med. Chem.* **1992**, *35*, 1109–1116.
164. Bissett, D.; McLeod, H. L.; Sheedy, B.; Collier, M.; Pithavala, Y.; Paradiso, L.; Pitsiladis, M.; Cassidy, J. *Br. J. Cancer* **2001**, *84*, 308–312.
165. Tzakos, A. G.; Gerothanassis, I. P. *ChemBioChem* **2005**, *6*, 1089–1103.
166. Varney, M. D. Copy of the presentation that he made in 1998 to a medicinal chemistry symposium concerning the lessons learned in ten years of using SBDD methods. Abraham, D. J. 2003. Ref Type: Personal Communication: June 13–19, 1998; 26th National Medicinal Chemistry Symposium; Richmond, USA.
167. Perutz, M. F. *I Wish I Would Have Made You Angrier Earlier: Essays on Science, Scientists, and Humanity*; Cold Spring Harbor Laboratory Press: Plainview, NY, 1998, p X.
168. Hartshorn, M. J.; Murray, C. W.; Cleasby, A.; Frederickson, M.; Tickle, I. J.; Jhoti, H. *J. Med. Chem.* **2004**, *48*, 403–413.
169. Overview of Focus Area 1: Enhance Biotechnology through Research and Innovation to Expand Biotechnology-related Businesses and the Kentucky Economy, 2006. http://www.rgs.uky.edu/area1.pdf (accessed Aug 2006).
170. Borman, S. *Chem. Eng. News* **2002**, *80*, 43–57.
171. Campbell, S. F. *Clin. Sci. (Lond.)* **2000**, *99*, 255–260.
172. Kubinyi, H. *Nat. Rev. Drug Disc.* **2003**, *2*, 665–668.
173. The New Jersey Initiative in Structural Genomics and Bioinformatics (NJISGB), Connecting Gene Sequence to Function by 3D Structure Determination – A New Paradigm for Drug Discovery, 1999. http://www-nmr.cabm.rutgers.edu/structuralgenomics/concept.html (accessed Aug 2006).
174. The Australian Government's Vision for Australian Biotechnology, 2002. http://www.biotechnology.gov.au/assets/documents/bainternet/BA%5FBiotech%5Fstrategy20050520161600%2Epdf (accessed Aug 2006).
175. Overview of Focus Area 1: Enhance Biotechnology through Research and Innovation to Expand Biotechnology-Related Businesses and the Kentucky Economy, 2006. http://www.rgs.uky.edu/area1.pdf (accessed Aug 2006).
176. Acharya, T.; Daar, A. S.; Thorsteinsdóttir, H.; Dowdeswell, E.; Singer, P. A. *PLoS Medicine* **2004**, *1*, 40; Genomics and Global Health: A report of the genomics working group of the ience and technology task force of the United Nations Millennium Project, 2004. http://www.medicine.plosjournals.org/perlserv?request = get-document&doi = 10.1371/journal.pmed.0010040 (accessed Aug 2006).
177. Ringe, D.; Petsko, G. A. *Methods Enzymol.* **1986**, *131*, 389–433.
178. Hajdu, J.; Machin, P. A.; Campbell, J. W.; Greenhough, T. J.; Clifton, I. J.; Zurek, S.; Gover, S.; Johnson, L. N.; Elder, M. *Nature* **1987**, *329*, 178–181.
179. Schotte, F.; Soman, J.; Olson, J. S.; Wulff, M.; Anfinrud, P. A. *J. Struct. Biol.* **2004**, *147*, 235–246.
180. Henderson, R. *Q. Rev. Biophys.* **1995**, *28*, 171–193.
181. Kisseberth, N.; Whittaker, M.; Weber, D.; Potter, C. S.; Carragher, B. *J. Struct. Biol.* **1997**, *120*, 309–319.
182. Hadida-Hassan, M.; Young, S. J.; Peltier, S. T.; Wong, M.; Lamont, S.; Ellisman, M. H. *J. Struct. Biol.* **1999**, *125*, 235–245.
183. Baumeister, W.; Grimm, R.; Walz, J. *Trends Cell Biol.* **1999**, *9*, 81–85.
184. McEwen, B. F.; Downing, K. H.; Glaeser, R. M. *Ultramicroscopy* **1995**, *60*, 357–373.

Biography

Donald J Abraham is Professor and Chair of the Department of Medicinal Chemistry at Virginia Commonwealth University and is the Director of the Institute for Structural Biology and Drug Discovery. He holds a BS degree in chemistry from Pennsylvania State University, an MS in organic chemistry from Marshall University, and a PhD in organic chemistry from Purdue University. Dr Abraham's research is interdisciplinary and focuses on structure-based drug design including x-ray crystallography, molecular modeling, synthetic medicinal chemistry, and structural function studies involving allosteric proteins. Targeted therapeutic areas of research include sickle cell anemia, radiation oncology, ischemic cardiovascular diseases, stroke, cancer, and Alzheimer's disease.

© 2007 Elsevier Ltd. All Rights Reserved
No part of this publication may be reproduced, stored in any retrieval system or transmitted in any form by any means electronic, electrostatic, magnetic tape, mechanical, photocopying, recording or otherwise, without permission in writing from the publishers

Comprehensive Medicinal Chemistry II
ISBN (set): 0-08-044513-6

ISBN (Volume 4) 0-08-044517-9; pp. 65–86

4.05 Ligand-Based Approaches: Core Molecular Modeling

A R Leach, GlaxoSmithKline Research and Development, Stevenage, UK

© 2007 Elsevier Ltd. All Rights Reserved.

4.05.1 Introduction

Before the advent of computer-based modeling chemists had long used mechanical models such as those invented by Dreiding or by Corey, Pauling, and Koltun (CPK) to elucidate molecular structures. A particularly celebrated use of models in this era was the elucidation of the helical structure of DNA by Watson and Crick.[1] However, mechanical models do suffer from some obvious drawbacks; one limitation that is particularly pertinent to this chapter is their inability to provide a quantitative value of the energy of a particular structure.

Most molecular modeling studies can be considered to involve three basic components. First, the molecular system under study needs to be described. Many molecular descriptors are used in computational chemistry, according to the problem being addressed. The second component constitutes the algorithm or algorithms that use the molecular description to manipulate the system or to derive a mathematical model that relates the descriptors to some other (measurable) property. An example that is considered later in this chapter is the minimization algorithm, which provides a minimum-energy conformation of the molecule from a starting structure. The third component is the computational infrastructure (hardware, operating systems, software, etc.) that enables the calculations to be performed. The emphasis here will be on the first of these two components. Nevertheless, it is important to recognize that the unremitting pace of improvements in computer power now enable calculations to be performed that would have been impossible only a few years previously, providing the impetus for yet further improvements in the basic methodologies.

This chapter is organized into three sections. First, the use of the empirical force field approach (also known as molecular mechanics) is discussed. This is the most widely used energy model for drug design applications. Second, some of the methods that can be used to manipulate molecular conformations within the computer are considered, with particular emphasis on the analysis of conformational space to identify the most likely structure or structures that a molecule may adopt. Finally, methods for comparing and overlaying the 3D structures of molecules are considered in order to try and rationalize why, for example, molecules of very different chemical structure may have similar biological activity.

The fundamental nature of the techniques to be discussed means that they have been the subject of much research over many years. It will therefore not be possible to provide more than a general overview, with the emphasis being on those methods that are used in contemporary drug discovery applications. More comprehensive descriptions can be found in the literature citations or in molecular modeling textbooks.[2]

4.05.2 Molecular Mechanics and Empirical Force Field Methods

A key requirement in molecular modeling is to be able to calculate the energy of an arrangement of atoms and/or molecules in 3D space. There exist a variety of methods that can be used to perform such calculations. The most 'fundamental' way to tackle this problem is to use quantum mechanics, wherein the Schrödinger equation is solved for the distribution of electrons and atoms in the system in order to derive a wave function from which other properties can be derived. As will be described elsewhere in this volume (*see* 4.16 Quantum Mechanical Calculations in Medicinal Chemistry: Relevant Method or a Quantum Leap Too Far?) a variety of different quantum mechanical methods are applicable to the systems and problems typically encountered in drug design. Quantum mechanical methods have some clear advantages in that they are much less reliant on empirical parameters for the system being studied (see below) and they are also able to provide information on some key properties (e.g., electric multipoles, electrostatic potentials, ionization potentials, etc.) that are dependent upon a knowledge of the electronic distribution and which cannot be calculated using other techniques. However, they do have the significant drawback of being relatively time consuming to perform and so are rarely used for calculations involving large systems and/or for those on large numbers of molecules.

Empirical force field methods[3,4] (also known as molecular mechanics) ignore the electronic motions in the system and calculate the energy solely as a function of the positions of the atoms. The method uses a very simple model of the intra- and intermolecular interactions within a molecular system with the energy being partitioned into contributions from processes such as the stretching of bonds, bending of angles, rotations about single bonds, and steric and electrostatic interactions between pairs of nonbonded atoms (**Figure 1**). The simplest type of force field encountered in drug design applications contains just these four contributions; a common functional form is as follows:

$$E = \sum_{\text{bonds}} \frac{k_i}{2}(l_i - l_{i,0})^2 + \sum_{\text{angles}} \frac{k_i}{2}(\theta_i - \theta_{i,0})^2 + \sum_{\text{torsions}} \frac{V_n}{2}(1 + \cos(n\omega - \gamma)) + \sum_{i=1}^{N}\sum_{j>i}^{N}\left(4\varepsilon_{ij}\left[\left(\frac{\sigma_{ij}}{r_{ij}}\right)^{12} - \left(\frac{\sigma_{ij}}{r_{ij}}\right)^{6}\right] + \frac{q_i q_j}{4\pi\varepsilon_0 r_{ij}}\right) \quad [1]$$

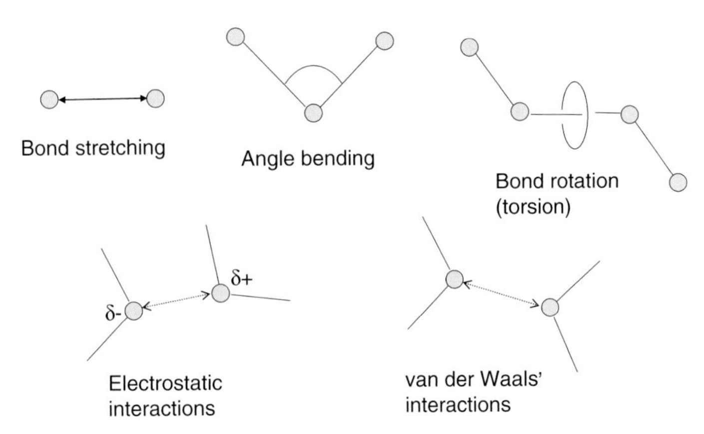

Figure 1 An illustration of the various terms that contribute to a typical force field. (Adapted with permission from Leach, A. R. *Molecular Modelling: Principles and Applications*; Prentice-Hall: Harlow, UK, 2001.)

We shall examine each of these terms in more detail below, but there are two important features about any force field method that need to be emphasized. The first of these is that molecular mechanics is an empirical approach; there is no inherent reason why any one functional form is necessarily better than any other. The method incorporates many empirical parameters (e.g., k_i, V_n, ε_{ij}, q_i in the above equation), which must be derived from some source; typically from a combination of experimental and theoretical data. The second key assumption is that of transferability. This enables a set of parameters developed and tested using a relatively small number of test systems to be applied to a much wider range of molecules. In particular, parameters developed from data on small molecules can be used to study much larger molecules and molecular systems such as protein–drug complexes. However, the applicability of different force fields can vary quite widely. Some force fields are designed for a rather limited range of molecular types (e.g., peptides and proteins) whereas others are applicable to compounds from the entire periodic table. Drug design applications typically require force fields that can deal with a wide range of organic molecules, consistent with the wide variation in molecular structures encountered in medicinal chemistry.

One concept common to most force fields is that of 'atom type.' Quantum mechanical calculations only require the atomic numbers of the nuclei in the system to be specified together with information on the overall charge on the system and the spin multiplicity. In force field applications the atom type usually includes information not only about the nature of the atom (i.e., the atomic number) but also about its hybridization state, and sometimes about its local environment. Thus, most force fields distinguish between sp^3, sp^2, and sp hybridized carbon atoms with their tetrahedral, trigonal, and linear geometries, respectively. Some force fields make yet further distinctions, for example, by characterizing the neighboring environment (e.g., aromatic carbon atoms in 5- and 6-membered rings may be assigned different atom types).

In the next sections we will examine in some detail the terms encountered in force fields commonly used in drug design. First we consider those intramolecular terms involving groups of atoms that are bonded together and then we consider the nonbonded terms.

4.05.2.1 Bond Stretching and Compressing

The energy required to compress or stretch a bond generally varies in the manner shown in **Figure 2**. Of the various functional forms that have been suggested to model this curve that due to Morse is perhaps the best-known:

$$E = D_e\{1 - \exp[-a(l - l_0)]\}^2 \qquad [2]$$

D_e is the depth of the potential energy minimum and a is a constant related to the frequency of the bond vibration and the masses of the atoms. The Morse potential is not usually used in molecular mechanics. In part this is because it requires specification of three parameters for each bond. It is also relatively time consuming to compute. More importantly, in most drug discovery applications we deal with molecular geometries in which the bond lengths (and also the bond angles) stay close to their equilibrium values (i.e., near the bottom of the well in **Figure 2**). Processes involving the breaking of bonds are invariably tackled using quantum mechanics. In the equilibrium region the energy required to deform a bond is well approximated by a much simpler quadratic function (also illustrated in **Figure 2**).

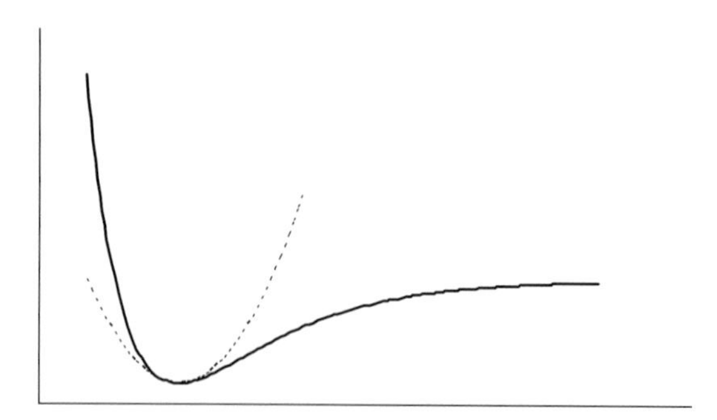

Figure 2 Comparison of the Morse curve (solid line), which well approximates the potential energy curve for a bond over its full range, and the Hooke's law harmonic approximation (dotted line).

This leads to the widely used Hooke's law equation that appears in many molecular mechanics force fields, which use functions of the following form:

$$E = \frac{k}{2}(l - l_0)^2 \qquad [3]$$

where k is a constant for that type of bond, l is the bond length, and l_0 is the reference bond length. A subtle point is that l_0 is not necessarily the equilibrium bond length (i.e., the length that the bond adopts in some minimum-energy conformation). Rather, it is the hypothetical bond length that is adopted when all other terms in the force field are zero. The equilibrium bond length is the value found in a minimum energy structure of the molecule, when the other terms in the force field also contribute.

It usually requires a significant amount of energy to cause a bond to undergo a major deviation from its reference value. This is reflected in the relatively high values for the force constant, k. Moreover, force constants generally follow the trend that deformation of single bonds requires less energy than for double bonds, which in turn require less energy than triple bonds.

More sophisticated force fields use functions that try to incorporate some representation of the asymmetric nature of the experimental function that ultimately leads to dissociation of the bond. This can be achieved via the incorporation of cubic and higher terms.

4.05.2.2 Angle Bending

The angle bending term is applied to all bonded sequences of three atoms in a molecule. As with the bond term, a quadratic Hooke's law expression is frequently used to represent this relationship, giving an expression of the following form:

$$E = \frac{k}{2}(\theta - \theta_0)^2 \qquad [4]$$

As with the corresponding bond term, k is a force constant, θ is the value of the angle subtended between the two bonds at the third common atom, and θ_0 is the reference angle. It requires rather less energy to deform an angle than to stretch a bond and so the force constant values are correspondingly lower. As with the bond stretching term, refinements of the angle bending contribution usually involve the incorporation of higher (cubic etc.) terms.

4.05.2.3 Torsion Potentials

In most drug-like molecules the bond lengths and bond angles deviate little from their reference values. Changes in the intramolecular structure arise rather from rotations about bonds. Classic examples of the variation in energy with rotation about a bond include the three minimum and three maximum energy conformations of ethane or the *cis* and *trans* conformations of butadiene. It is vitally important that a force field can accurately reproduce such energy profiles. Within a typical force field of the form shown in eqn [1] the rotational profiles are governed by a combination of the nonbonded (van der Waals' and electrostatic) terms between the so-called 1,4 atoms at the termini of each torsion angle (defined as the angle between the two planes 1,2,3 and 2,3,4 for four atoms 1,2,3,4 in a bonded sequence) and the torsional potential.

Torsional potentials are invariably expressed as a cosine series expansion. Two equivalent expressions for this are the following:

$$E = \sum_{n=0}^{N} \frac{V_n}{2}[1 + \cos(n\omega - \gamma)] \qquad [5]$$

$$E = \sum_{n=0}^{N} C_n \cos(\omega)^n \qquad [6]$$

In these equations ω is the torsion angle, V_n is a constant, n is the multiplicity, and γ is the phase factor. V_n is sometimes referred to as the barrier height but this is incorrect, obviously so when there is more than one term in the expansion. Moreover, other contributions to the real barrier height arise due to other terms in the force field such as the nonbonded interactions between the 1,4 atoms. The multiplicity corresponds to the number of energy minima as the bond is rotated through 360°. The phase factor determines where the torsion potential passes through its minimum value. To provide a simple illustration of these terms consider rotation about a single bond between two sp^3 carbon atoms. For this bond $n = 3$ and $\gamma = 0°$, giving a threefold potential with minima at torsion angle values of +60°, −60°, and 180°. A double bond between two sp^2 atoms would have $n = 2$ and $\gamma = 180°$ giving minima at 0° and 180°. The greater barrier height for the double bond would also be reflected in a larger value of V_n. More complex torsional profiles can be produced using combinations of terms. For example, the tendency of O–C–C–O bonds to adopt gauche conformations can be reproduced using two terms of the following form such as the following from the AMBER force field:

$$E = 0.25(1 + \cos 3\omega) + 0.25(1 + \cos 2\omega) \qquad [7]$$

This function has deeper minima at torsion angles of ±60° with a shallower minimum at 180° (**Figure 3**).

4.05.2.4 Out-of-Plane Bending Terms and Cross Terms

In addition to the bond, angle, torsion, and nonbonded terms a force field may also contain other contributions designed to ensure that the predicted geometries and energies are consistent with experimental observations. One of the most common of these is the improper torsion, the use of which can be illustrated using cyclobutanone (**Figure 4**). Experimentally, it is found the oxygen atom in this molecule lies in the plane formed by the adjoining carbon atom and

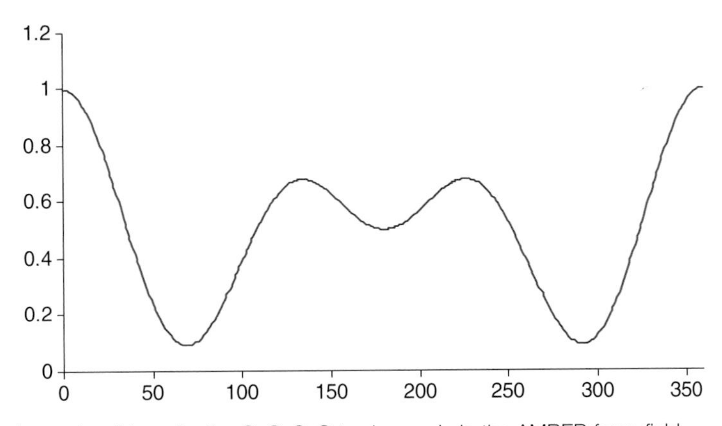

Figure 3 Variation in the torsional term for the O–C–C–O torsion angle in the AMBER force field.

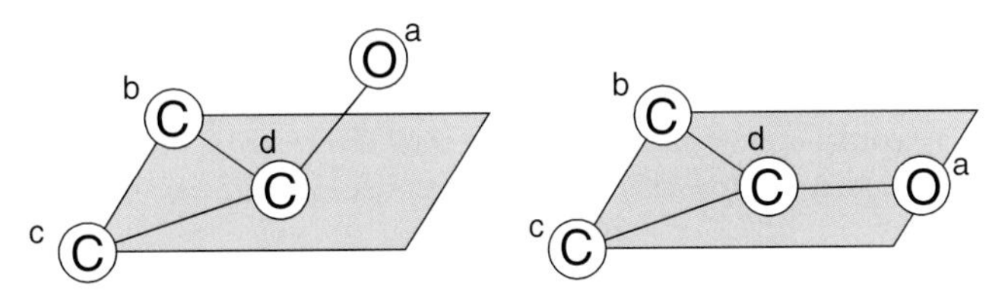

Figure 4 Schematic illustration of the use of an improper torsion term to ensure that the oxygen atom in cyclobutanone remains in the same plane as the attached carbon atom and its neighbors.

the two bonded carbon atoms. If a standard force field containing just bond stretching and angle bending terms were used to model cyclobutanone then the oxygen atom would be predicted to lie out of this plane in the equilibrium geometry, in order to maintain the C–C–O angles close to their reference values of 120° (**Figure 4**). In the experimental structure these angles are closer to 133° giving a conformation that maximizes the π-bonding energy. In order to achieve the desired geometry a so-called out-of-plane bending term is used. The most common way to incorporate such a contribution is via the use of an 'improper' torsion, so named because it involves four atoms that are not bonded in a sequence. Thus, for the cyclobutanone molecule shown in **Figure 4** an improper torsion term of the form $k(1 - 2\cos 2\omega)$ involving the atoms a–b–c–d could be used to keep the torsion angle so defined at 0° or 180°. Other ways to address out-of-plane bending involve calculation of the angle between the bond in question (e.g., bond a–d in **Figure 4**) and the plane (b–c–d) or of the height of the key atom above the plane. However, the improper torsion approach is more common as it can be included more easily alongside the regular torsion angles in the overall energy calculation. A common use of out-of-plane bending terms is to ensure that aromatic rings are maintained in an approximately planar geometry by the use of an improper torsion involving atoms on opposite sides of the ring. Another common use of improper torsion terms is to maintain the required stereochemistry in so-called united atom force fields, in which certain types of hydrogen atoms are not explicitly represented, but are treated as part of the nonhydrogen atom to which they are bonded. Typically, just those hydrogen atoms attached to nonpolar atoms were treated this way. United atom force fields were popular for early simulations of large biomolecules due to the limitations in computer power then available; so-called all-atom force fields are much more common today and are generally preferred.

The earliest molecular mechanics force fields designed to reproduce geometries and relative energies originated from spectroscopic force fields developed to predict vibrational spectra. A common feature of the latter are 'cross terms' that reflect the coupling between the various types of intramolecular motions. For example, as the value of a bond angle is reduced then it is found that the adjacent bonds stretch slightly to compensate for the increased 1,3 interaction. In principle one may require cross terms between all contributions to the force field. However, it is generally found (at least for structural properties) that only a few cross terms are important though if one wishes to use a force field to predict other properties (e.g., vibrational frequencies) then additional cross terms may be required. Most cross terms involve combinations of just two terms, such as stretch–stretch, stretch–bend, or stretch–torsion, but it is also possible to use cross terms involving more than two internal coordinates such as bend–bend–torsion. Some force fields also contain a Urey-Bradley 1,3 term, in which a harmonic potential is applied to the two terminal atoms in a bond angle. Urey-Bradley terms may be used in place of, or to supplement, the more common Hooke's law angle contribution.

The presence of cross terms has been suggested as one way to classify force fields.[5] A class I force field is one restricted to harmonic terms and without any cross terms. A class II force field includes anharmonic terms and explicit cross terms to account for the coupling between coordinates. These additional features enable the force field to predict the properties of more unusual systems (e.g., those that are highly strained) and also predict properties such as vibrational spectra. A class III force field increases the precision yet further and would be able to account for the very small influences on structure due to electronegativity and hyperconjugation.

4.05.2.5 Nonbonded Interactions

Nonbonded forces lie at the heart not only of the interactions between independent molecules and atoms but also play an important role in determining the structure of individual molecules. In contrast to the force field terms we have described so far, nonbonded interactions do not depend upon any particular bonding relationship between the atoms involved. They are conveniently divided into electrostatic and van der Waals' interactions, each of which is typically modeled as a function of an inverse power of the distance.

4.05.2.6 Simple Models for Incorporating Electrostatic Interactions

Most molecules have an uneven charge distribution due to differences in the electronegativities of their constituent atoms. A particularly common way to represent and reproduce the charge distribution in a molecule, especially in drug discovery applications, is through the use of fractional point charges. If these charges are restricted to the atomic centers they are referred to as 'partial atomic charges' and the total electrostatic interaction energy within the system (e.g., within a single molecule or between two or more molecules) is calculated using Coulomb's law:

$$E = \sum_{i=1}^{N} \sum_{j>i}^{N} \frac{q_i q_j}{4\pi\varepsilon_0 r_{ij}} \qquad [8]$$

A wide variety of methods has been used for calculating partial atomic charges. A common approach is to calculate a set of charges that reproduces the molecular electrostatic potential as determined from quantum mechanics calculations.[6–9] This involves a least-squares fitting procedure or a Lagrangian multiplier method with the electrostatic potential being sampled at a series of points around the molecule. The points where the potential is fitted can be chosen in a number of ways; the key requirement is to concentrate on the region where it is most important to model the intermolecular interactions correctly, that is just beyond the van der Waals' radii of the atoms involved. Typically, a series of shells is used at increasing distances from the molecule. An additional constraint that must be incorporated into such a charge fitting procedure is that the sum of the partial atomic charges must equal the overall molecular charge. In some cases it may also be desired to force the charges on symmetrically equivalent atoms to be the same. A problem sometimes observed with the original fitting methods was that artificially high charges were assigned to some of the atoms, especially to carbon atoms buried inside the molecule. The RESP (restrained electrostatic potential fit) procedure was designed to address this limitation through the use of hyperbolic restraints on the atomic charges during the fitting process.[10–12] A further complication is that the charges may vary with the conformation of the molecule.[13] Attempts to address this particular problem include the use of charges weighted according to the relative population of a series of conformations[14] or models where the charges vary with the conformation.

The other major choice when determining a set of electrostatic-potential-fitted charges is the level of quantum mechanical theory used for the electrostatic calculation. In the earliest implementations of such methods the available computer power restricted the level of theory that could be used. This limitation no longer applies and the most widely used approach today is the 6–31G* level of ab initio theory. It is also possible to use different (lower) levels of theory and then to scale the results to the higher levels of theory[15–18] or to use more sophisticated correction methods[19–23] 6–31G* charges generally provide good results for calculations in condensed phases. In part this is due to a fortuitous overestimate of the 'real' charge, which compensates for the lack of explicit polarization effects in the simple Coulomb model. Polarization will be discussed in more detail below; suffice it to say at this stage that one manifestation of this effect is that the molecular dipoles calculated using 6–31G* charges are higher than the real values.

Another approach to deriving charges is to fit these to reproduce the energies and geometries of interaction (derived using ab initio calculations) between small molecular species chosen to be representative of the intermolecular interactions occurring in larger molecules. Some force fields also force certain atom types to have the same charge in all situations in order to promote parameter transferability.

To derive charges for large molecules it is not feasible to perform quantum mechanical calculations on the entire molecule. Moreover, to do so would not be particularly efficient, especially for molecules such as proteins that are composed of a small set of building blocks. In such cases partial atomic charges are derived for relevant fragments, which when combined provide a charge model for the entire system. Often such fragments are chosen such that their overall charge is an integral value.

Although charges derived from quantum mechanical calculations are often the method of choice, this is not necessarily a feasible approach when one wishes to derive charges in an automated fashion for a large number of molecules. In such circumstances rapid charge determination methods such as that devised by Gasteiger and Marsili[24] are frequently employed. The Gastiger–Marsili approach is based on the principle of partial equalization of orbital electronegativity, in which negative charge is gradually moved from less electronegative to more electronegative elements. A related approach is the charge equilibration method[25] employed in the universal force field (UFF), which is based on the principle of equalizing the atomic chemical potentials, again via the movement of charge (electrons) from regions of low electronegativity (high electrochemical potential) to high electronegativity (low electrochemical potential). The charges in UFF are dependent upon the conformation and so are able to automatically vary during calculations such as molecular dynamics simulations.

The partial atomic charge model continues to be the most widely used approach to the calculation of electrostatic interactions in molecular mechanics force fields within drug discovery. However, there are some well-known limitations with this approach, especially with certain types of systems. One such limitation is the interactions between molecules containing aromatic rings and other π systems, which are very common in drug-like molecules and the biological macromolecules such as proteins and DNA with which they interact. A variety of geometries are observed for aromatic dimers yet not all of these are correctly predicted by partial atomic charge force fields. One simple approach to address this problem was suggested by Hunter and colleagues, who placed charges not only at the atomic nuclei but also above and below each atom in an aromatic ring, perpendicular to the ring plane.[26] This was able to correctly predict the stable geometries of various aromatic systems and has subsequently been extended to cover a much wider range of compounds.[27,28] More common is the addition of charges coincident with the location of lone pairs on key functional groups, in order to better reproduce hydrogen-bonding geometries and energies. It is also worth noting that not all force fields use the point charge model; for example, the MM2/MM3/MM4 programs determine electrostatic energies by calculating the interaction between bond dipoles (though these cannot deal with charged molecules or fragments, in which case they are augmented by point charges). Also worthy of note is the distributed multipole approach (DMA) developed by Stone,[29–32] in which the charge distribution is described by a series of multipole moments located on the atoms (and sometimes on other sites). However, distributed multipole models have not been widely incorporated into force fields, not least because of the additional computational effort required and the complications involved in calculating atomic forces (required for applications such as energy minimization or molecular dynamics simulations).

4.05.2.7 Polarization

Polarization is the phenomenon whereby the charge distribution in a molecule is modified due to the presence of an external field. The primary effect of the external field (which in the case of molecular calculations arises from the influence of neighboring atoms and molecules) is to induce a dipole in the molecule. This induced dipole is proportional to the electric field E with the constant of proportionality being the polarizability, α:

$$\boldsymbol{\mu} = \alpha \mathbf{E} \quad [9]$$

The additional energy due to the polarization contribution is given by:

$$E_{\text{pol}} = -0.5\alpha E^2 \quad [10]$$

Polarization is particularly important for calculations performed in condensed phases, especially in water. In a fixed-charge force field the effects of polarization are described only in an averaged way. This is done by having atomic charges that result in dipole moments that are larger (typically by 10–20%) than those actually observed for the isolated molecules in the gas phase as found for the charges derived from the 6–31G* electrostatic potential fitting procedure. This approach has indeed been rather successful in providing values for bulk properties that agree closely with experiment, but it is nevertheless an approximation that does lead to problems, most notably for situations such as the solvation of ions.

Explicit incorporation of polarization can be achieved in a number of ways.[33,34] Perhaps the simplest approach to polarization is to enable point dipoles to be induced at each of the atomic centers. In this case the total electric field at each atom is given by the sum of the fields due to the permanent atomic partial charges together with the field due to the other induced dipoles; the whole system is then iterated to self-consistency. An alternative method that fits naturally into a molecular dynamics simulation is the fluctuating charge model.[36] Here the atomic charges fluctuate according to the principle of electronegativity equalization, similar to the UFF charge equilibration method mentioned above. In the fluctuating charge model the charges are considered as dynamically fluctuating variables (with fictitious masses) that evolve along with the atomic nuclei during the simulation. This method is particularly attractive because the charges naturally evolve; there is no need to determine a new set of charges at each step of the simulation. It is also possible to incorporate fluctuating charges and inducible dipoles into one model. The optimal functional form remains to be determined; one study has suggested that dipoles are required to avoid errors in important cases[37] and another that the additional expense of including fluctuating charges is not justified compared to the additional complexity introduced.[38] A third model uses both permanent and inducible point dipoles together with fluctuating and fixed charges to perform protein simulations in liquid water.[39]

The additional computational burden associated with polarization means that polarization effects have usually been included only when their effect is likely to be particularly significant. For example, there has been much effort spent on the development of polarizable force fields for liquid water[40] (due to its unique properties and its central role in biological processes), and aqueous solutions (for example, solvated ions). More recent publications have described the development of polarizable force fields for peptides and proteins.[41] Indeed, the field is now reaching a sufficient degree of maturity that attention is being paid to the development of polarizable force fields intended for more general use, including (in principle) the whole range of chemical types present in drug-like molecules.[42,43]

4.05.2.8 The Dielectric Constant and Implicit Solvation Models

It will be noticed that the Coulomb law expression for the electrostatic interaction (eqn [8]) includes in its denominator the dielectric constant, ε. For systems in vacuo and for systems that explicitly include all interacting species this is appropriate. However, the number of solvent molecules that need to be included in such calculations is invariably very large, making the calculations time consuming. It is therefore desirable to have methods that can mimic solvent effects without needing to include the solvent explicitly. In the case of biological systems the screening effect of the solvent is even more pronounced due to the high dielectric constant of water. One very simple way in which implicit solvent effects have been included is through the use of a distance-dependent dielectric model, in which the relatively permittivity is assumed to be proportional to the distance between two charges. This has the effect of replacing the inverse distance relationship in Coulomb's law with an inverse square distance. However, the distance-dependent dielectric has no physical basis and does not give results comparable with more exact approaches[44] so is not generally recommended, unless there is no alternative.

More realistic ways to include implicit solvent effects have been developed, the two most common approaches being the generalized Born model and Poisson–Boltzmann calculations, both of which continue to be the subject of significant development.[45]

4.05.2.9 The Generalized Born Model

In 1920 Born[46] derived a formula for the free energy of solvation of placing a charge within a spherical solvent cavity. This free energy equals the difference in electrostatic work to charge the ion in a medium of dielectric constant ε and in vacuo and is given by:

$$\Delta G_{\mathrm{elec}} = -\frac{q^2}{2a}\left(1-\frac{1}{\varepsilon}\right) \qquad [11]$$

where q is the charge and a is the radius of the cavity. Onsager[47] extended the Born model to cover a dipole in a spherical cavity, wherein the solute dipole induces a dipole in the surrounding medium, which in turn induces an electric field (called the reaction field) within the cavity. The reaction field interacts with the solute dipole, providing additional stabilization of the system. A drawback with the basic theory is the restriction to a spherical (or elliptical) cavity; nevertheless, reaction field corrections to gas-phase energies have been implemented successfully in quantum mechanics calculations and do have the advantage of providing analytical expressions for the first and second derivatives of the energy (which makes geometry optimization calculations more efficient).

More generally applicable are models in which the molecule is represented by a more realistic shape, derived either from the van der Waals' radii of the atoms or from the molecular surface. A number of such methods have been described for use in both quantum mechanics and force field models; here we will focus on the most widely used method, the generalised Born model.[48]

The starting point for the generalized Born model is the total electrostatic free energy of a system of particles with radii a_i and charges q_i, which equals the sum of the Coulomb energy and the Born free energy of solvation in a medium of relative permittivity ε:

$$G_{\mathrm{elec}} = \sum_{i=1}^{N}\sum_{j>i}^{N}\frac{q_i q_j}{\varepsilon \rho_{\iota\varphi}} - \frac{1}{2}\left(1-\frac{1}{\varepsilon}\right)\sum_{i=1}^{N}\frac{q_i^2}{a_i} \qquad [12]$$

This equation can be manipulated to provide the following result:

$$G_{\mathrm{elec}} = \sum_{i=1}^{N}\sum_{j>i}^{N}\frac{q_i q_j}{r_{ij}} - \left(1-\frac{1}{\varepsilon}\right)\sum_{i=1}^{N}\sum_{j>i}^{N}\frac{q_i q_j}{r_{ij}} - \frac{1}{2}\left(1-\frac{1}{\varepsilon}\right)\sum_{i=1}^{N}\frac{q_i^2}{a_i} \qquad [13]$$

The second and third terms in the above equation equal the difference between the total electrostatic free energy in solution and the normal Coulomb interaction between the charges in vacuo. The final simplification of the generalized Born model is to combine these two terms into one with the following form:

$$\Delta G_{\mathrm{elec}} = -\frac{1}{2}\left(1-\frac{1}{\varepsilon}\right)\sum_{i=1}^{N}\sum_{j>i}^{N}\frac{q_i q_j}{f(r_{ij}, a_{ij})} \qquad [14]$$

$f(r_{ij}, a_{ij})$ thus depends on the distance r_{ij} between the two atoms and the Born radii a_i. Various forms are possible for the function f; that proposed by Still and colleagues in the original paper was:

$$f(r_{ij}, a_{ij}) = \sqrt{r_{ij}^2 + a_{ij}^2 e^{-D}} \quad a_{ij} = \sqrt{a_i a_j} \quad D = r_{ij}^2/(2a_{ij})^2 \qquad [15]$$

This function has a number of useful properties: it returns the Born expression when $i=j$; for two charges close together (i.e., a dipole) the expression is close to the Onsager result; and for two charges separated by a long distance the result is close to the sum of the Coulomb and Born expressions. A further advantage is that it can be differentiated analytically. It is nevertheless necessary to determine appropriate values of the Born radii. For a spherical solute with a charge at its center this is straightforward; the Born radius equals the van der Waals' radius of the solute. Unfortunately, for real molecules the Born radius of an atom depends on the positions and volumes of the other atoms in the solute and the way in which they displace the solvent dielectric; the Born radius can be considered an average distance from the charge to the boundary of the dielectric medium. The Born radii can also vary as the conformation changes. Several methods have been proposed for determining Born radii or for making other improvements to the method, such as the use of analytical expressions,[49] a surface area-based version,[50] and various modifications to improve the performance for macromolecules.[51–53] It has also been shown[54] that when 'perfect' radii can be assigned the generalized Born model provides an excellent approximation when compared to 'exact' approaches based on Poisson theory (see below).

In the generalized Born/surface area (GB/SA) approach the total solvation free energy is given as the sum of the solvent polarization energy G_{pol}, a solvent–solvent cavity term G_{cav}, and a solute–solvent van der Waals' term G_{vdw}. G_{pol} is given by the generalized Born equation and the sum of G_{cav} and G_{vdw} is determined from the solvent-accessible solvent area:

$$G_{\mathrm{cav}} + G_{\mathrm{vdw}} = \sum \sigma_k SA_k \qquad [16]$$

SA_k is the solvent-accessible surface area of all atoms of type k and σ_k is an empirical parameter. Evidence for the validity of this equation comes from the linear variation of the solvation free energy with solvent-accessible surface area for saturated hydrocarbons (for which G_{pol} is zero). Thus, the GB/SA model comprises the generalized Born equation together with the solvent-accessible surface area term. The GB/SA approach has provided the inspiration for other applications, one example being the widely used SMx series of solvation models based on semiempirical quantum mechanics calculations.[55–60]

4.05.2.10 The Poisson–Boltzmann Equation

The Poisson–Boltzmann equation[61] is derived from two components: the Poisson equation, which relates the variation in electrostatic potential in a medium of constant dielectric to the charge density, and the Boltzmann distribution, which governs the ion distribution in the system. The full Poisson–Boltzmann equation is a nonlinear differential equation that is usually approximated as a series expansion with just the first term being retained, as follows:

$$\nabla \cdot \varepsilon(\mathbf{r})\nabla\phi(\mathbf{r}) - \kappa'\phi(\mathbf{r}) = -4\pi\rho(\mathbf{r}) \qquad [17]$$

where $\varepsilon(\mathbf{r})$ is the variation in dielectric constant with position, $\phi(\mathbf{r})$ is the electric potential, $\rho(\mathbf{r})$ is the charge density, and κ' is a constant related to the ionic strength of the solution.

The difficulty lies in solving such an equation; before computers were available only simple shapes (e.g., spheres, cylinders) could be considered. With the advent of computers it was recognized that numerical methods could be applied, of which the numerical finite difference approach first introduced by Warwicker and Watson[62] has been particularly popular. Here, a cubic lattice is superimposed onto the system and the surrounding solvent. Values of the electrostatic potential, charge density, dielectric constant, and ionic strength are assigned to each lattice point. The derivatives in the Poisson–Boltzmann equation are then determined numerically using a finite difference approach. The potential at each grid point influences the potential at neighboring grid points and so it is necessary to iterate the calculation until convergence is achieved.

Calculations based on the Poisson–Boltzmann equation have proved particularly useful for understanding the electrostatic properties of biological macromolecules.[63,64] The method has also been used to study ligand binding to proteins, either alone or in combination with other force field calculations.[65–68] There continue to be some practical challenges in applying the method, many of which arise from the numerical approximations that are employed. The boundaries between regions of low and high dielectric are especially challenging; one possible solution is to use a nonuniform grid, which contains more points in the boundary regions.[69,70]

4.05.2.11 van der Waals' Interactions

van der Waals' forces constitute the final contribution to the general force field introduced in eqn [1] and are named in honor of van der Waals, who quantified the deviations from ideal gas behavior that are demonstrated by systems such as the noble gases. The conundrum is that in such systems there are no permanent multipole moments and so the electrostatic interactions described above cannot be used to invoke the deviations from ideal-gas behavior. If the

Figure 5 Schematic illustration of the Lennard-Jones potential, showing the steep repulsion as atoms approach each other, the minimum, and the asymptotic approach to zero potential energy as the atoms move further apart.

interaction energy between two rare gas atoms is studied experimentally (using, for example, a molecular beam) then two distinct regions are observed. At infinite distance their interaction energy is zero; as they approach each other the energy decreases (i.e., they attract). The curve passes through a minimum and then rapidly increases (**Figure 5**).

The attractive contribution arises from dispersive forces (sometimes called London forces[71]). These are due to instantaneous dipoles, which then induce dipoles in neighboring atoms, giving rise to an attractive effect. The repulsive contribution has a quantum mechanical origin in the Pauli principle, which forbids any two electrons in a system from having the same set of quantum numbers. This has the effect of reducing the electron density in the internuclear region as two atoms approach closely. This reduced electron density leads to repulsion between the incompletely shielded nuclei.

The mathematical models used to reproduce this behavior usually involve two terms, one corresponding to the attractive and one to the repulsive parts of the potential. The Lennard-Jones 12–6 function is perhaps the best known of these, with the following functional form:

$$E = 4\varepsilon\left[\left(\frac{\sigma}{r}\right)^{12} - \left(\frac{\sigma}{r}\right)^{6}\right] \quad [18]$$

There are two adjustable parameters in this equation: the collision diameter σ (the separation for which the energy is zero) and the well depth ε. An equivalent expression writes the equation in terms of the separation at which the energy passes through a minimum, r_m:

$$E = \varepsilon\left[\left(\frac{r_m}{r}\right)^{12} - 2\left(\frac{r_m}{r}\right)^{6}\right] = \frac{A}{r^{12}} - \frac{C}{r^{6}} \quad [19]$$

The r^{-6} contribution has a reasonable theoretical basis that is demonstrated in some of the simple models of the dispersive interactions. By contrast, the r^{-12} term has no such theoretical basis and is used primarily to facilitate rapid calculation (by squaring the r^{-6} term). Various alternative formulations of the van der Waals' interaction have been suggested. Some of these involve the use of a different power (e.g., 9 or 10) for the repulsive part of the potential in order to give a less steep curve. Halgren[72–74] proposed the use of a 'buffered 14–7' potential based on the following general functional form:

$$E = e_{ij}\left(\frac{1+\delta}{\rho_{ij}+\delta}\right)^{n-m}\left(\frac{1+\gamma}{\rho_{ij}^{m}+\gamma} - 2\right) \quad [20]$$

In this equation $\rho_{ij} = r_{ij}/r_{m,ij}$ and δ and γ are constants. This equation returns the Lennard-Jones expression for $n = 12$, $m = 6$, $\delta = \gamma = 0$. In Halgren's 14–7 potential $n = 14$, $m = 7$, $\delta = 0.07$, and $\gamma = 0.12$. The rationale for using such a seemingly complicated expression includes: the desire for the potential to have a finite value as the interatomic distance approaches zero (to eliminate the problems that can arise when trying to calculate the energy of very strained systems); to give a more accurate representation of the dispersion interaction; and to provide the flexibility to reduce the repulsive component without significantly changing the distance at which the potential crosses zero or the depth of the energy minimum (by modifying the value of the parameter δ).

Two alternative options for modeling the van der Waals' contribution are the Buckingham potential that has three adjustable parameters and in which the r^{-12} term is replaced with a theoretically more reasonable (but computationally more expensive) exponential term:

$$E = \varepsilon\left[\frac{6}{\alpha - 6}\exp[-a(r/r_m - 1)] - \frac{\alpha}{\alpha - 6}\left(\frac{r_m}{r}\right)^6\right] \quad [21]$$

and the Hill potential,[75] which is an exponential-6 potential with just two parameters; the minimum energy radius r_m and the well depth ε:

$$E = -2.25\varepsilon(r_m/r)^6 + 8.28 \times 10^5\,\varepsilon\,\exp(-r/0.0736r_m) \quad [22]$$

Determining the most appropriate and accurate parameters for the van der Waals' forces is often difficult and time consuming. In contrast to the other terms in the force field, it is common for the van der Waals' parameters to be dependent solely upon the atomic number and not upon the atom type (e.g., the same van der Waals' parameters are used for all carbon atoms, irrespective of hybridization). A further common approximation is to assume that the parameters for interactions between unlike atoms can be derived from the parameters for the pure atoms using 'mixing rules.' Commonly used are the Lorentz–Berthelot mixing rules, in which the collision diameter σ_{AB} for the A-B interaction equals the arithmetic mean for the two pure species, and the well depth ε_{AB} is the geometric mean:

$$\sigma_{AB} = 0.5(\sigma_{AA} + \sigma_{BB}); \quad \varepsilon_{AB} = \sqrt{\varepsilon_{AA}\varepsilon_{BB}} \quad [23]$$

The collision diameter is sometimes derived as the geometric mean $\sqrt{(\sigma_{AA}\sigma_{BB})}$ rather than the arithmetic mean, for example, in the OPLS (optimized parameters for liquid simulations) force field.

4.05.2.12 Cutoffs

For large systems the total number of pairwise nonbonded interactions can be extremely large, and to include all of them in the energy calculation would be very time consuming. It is therefore common practice to restrict the calculation of pairwise interactions just to those pairs of atoms that lie within some distance threshold. Interactions between pairs of atoms further apart are ignored. Such an approach introduces only small errors in the case of van der Waals' forces, which act over a relatively short range. However, this can become a significant limitation for the much longer ranged electrostatic interactions and can introduce serious errors and instabilities. As computers have become more powerful so the use of longer cutoff values has become possible. Techniques such as shifted potentials and switching functions are generally recommended to address the discontinuities that sharp cutoffs can introduce. More recently, methods such as the Ewald summation[76] have started to be introduced into simulations of biological systems; this provides a way that effectively enables all of the electrostatic interactions within the system under study to be included. The Ewald calculation is computationally rather expensive but efficient algorithms have been developed that partly address this issue.[77–79] The use of the Ewald method to date has been restricted to detailed molecular dynamics simulations of proteins and DNA rather than being used in practical drug discovery applications; it remains to be seen whether this situation will change in the future.

4.05.2.13 Comparison and Practical Application of Force Fields

The choice of force field is often the first, and frequently the most critical, decision that must be made prior to any modeling calculation. Many of the popular modeling packages provide access to a number of the common published force fields (though one needs to take great care that the implementation is true to the original). Deriving a force field is not a trivial undertaking. It is also important to recognize that the various force fields have been developed for different reasons and with different applications in mind. Allinger's MM2,[80] MMP2,[81,82] MM3,[83–87] and MM4[88–92] force fields represent an evolving series designed to provide accurate structures, energies, and latterly vibrational frequencies for organic molecules, based on fitting to high-quality experimental data. AMBER[93–96] and CHARMM[97,98] are widely used for the simulation of biological macromolecules (proteins and DNA). AMBER has a historical emphasis on parameter transferability and the use of charges derived from electrostatic potential fitting; the CHARMM force fields have focused on more refined parameters, which may limit transferability, though this may be ameliorated to an extent by its use of a modular approach wherein the sum of the charges on functional entities (rings and common chemical functional groups) are required to sum to integer values. GROMOS[99] is also designed for biomolecular simulations. The CFF93[100] and CFF95[101,102] class II force fields are based on fitting[103] to data derived from ab initio calculations; they include a significant number of cross terms. CFF93 provided the basis for a force field of wider scope, particularly in the materials science area (PCFF: polymer consistent force field) that in turn led to COMPASS,[104] in which both ab initio calculated and experimental data were employed in order to improve the performance in condensed phase simulations. CVFF[105,106] also employed the fitting approach using experimental data that included crystal structure parameters, sublimation energies, and dipole moments to provide a force field for the study of proteins and protein–ligand complexes. Two force fields designed to have widespread applicability (including inorganic systems) are the UFF[107] and the Dreiding force field.[108] Both of these use rules for deriving the necessary parameters based on a limited set of mostly atomic parameters. The Tripos force field adopts a relatively straightforward functional form and is designed to provide good geometries and relative energies for a wide variety of organic compounds.[109] Some of the goals of the Merck molecular force field (MMFF[110]) were to achieve high accuracy for small molecules and proteins and to have applicability to a wide variety of systems of interest to drug discovery. Of all the force fields described in this paragraph, this is the only one to be developed (at least initially) in a drug discovery environment; the others were all the product of academic laboratories or software companies. OPLS[111–113] is a continually evolving force field based on the use of statistical mechanics simulations of liquids and aqueous solutions.

As can be deduced from the above brief summary, different philosophies have been proposed for force field and parameter development over the years.[114] All start with the data that will be used to guide the process. The geometries and relative conformational energies of certain key molecules are invariably included in this data set. Some force fields also include vibrational frequencies. One parameterization approach that has been particularly influential is the reproduction of thermodynamic properties using liquid simulations, most notably espoused by Jorgensen in his OPLS force fields. Another trend has been the incorporation of data obtained from high-level quantum mechanics calculations; such calculations provide a wealth of data and can be particularly useful when experimental data is sparse or nonexistent. This may apply, for example, when one wishes to derive the torsional potentials for a new chemical entity or molecular fragment, a common problem in drug discovery.

Having decided on the data to be used in the parameterization and on the basic functional form, the derivation of the actual parameters can commence. Several approaches are possible. The general principle behind the consistent force fields of Lifson and colleagues[115] was to use a mathematical least-squares fitting process to determine the set of parameters that gives the optimal fit to the data, as defined by the sum of squares of differences between the observed and calculated values. Somewhat more common is a stepwise approach, in which the parameters are defined in stages. A common strategy here is to separate the 'hard' degrees of freedom (e.g., bond stretching and angle bending) from the 'soft' degrees of freedom (nonbonded and torsional contributions). Indeed, the bond stretching and angle bending terms are often transferred without change from one force field to another. For the soft degrees of freedom it is common to establish the electrostatic model first and then to define the van der Waals' parameters, either from liquid simulations or by reproducing the interactions between small rigid molecules. Considerable progress continues to be made in the accuracy and applicability of current force fields, yet significant challenges remain.[116–119]

A major problem that often arises when using molecular mechanics is the problem of missing parameters for the molecule(s) that one wishes to study. Ideally, a series of calculations would be performed on the appropriate representative fragments, to determine geometries, energies, and barriers to torsional rotation, in order to calculate values for the missing parameters in a manner consistent with the rest of the force field. In other cases parameters will be estimated based on the most similar values already present in the force field. These parameter estimates can sometimes be quite reasonable, but it is always good practice to check the values assigned and to be vigilant for strange or anomalous results. Some of the main force field packages now provide software tools that help the user determine new parameters for missing fragments in a manner consistent with the existing set. For this reason, force fields where the original parameterization is done using a defined 'recipe' can have an advantage when being used to model new molecules. Nevertheless, a comprehensive comparison[120] of several widely available force fields found that whilst some of them worked well at a general level, all of them encountered problems with some molecular systems, emphasizing the need to carefully test a force field when it is applied to molecules for which it was not originally developed.

4.05.3 Potential Energy Surfaces

Molecular mechanics provides a way to quantify how the energy of a system changes with its configuration (i.e., the arrangement of atoms within it). Such changes in configuration may be due, for example, to changes in a molecule's conformation or to the approach of two molecules to each other. The way in which the energy of the molecular system

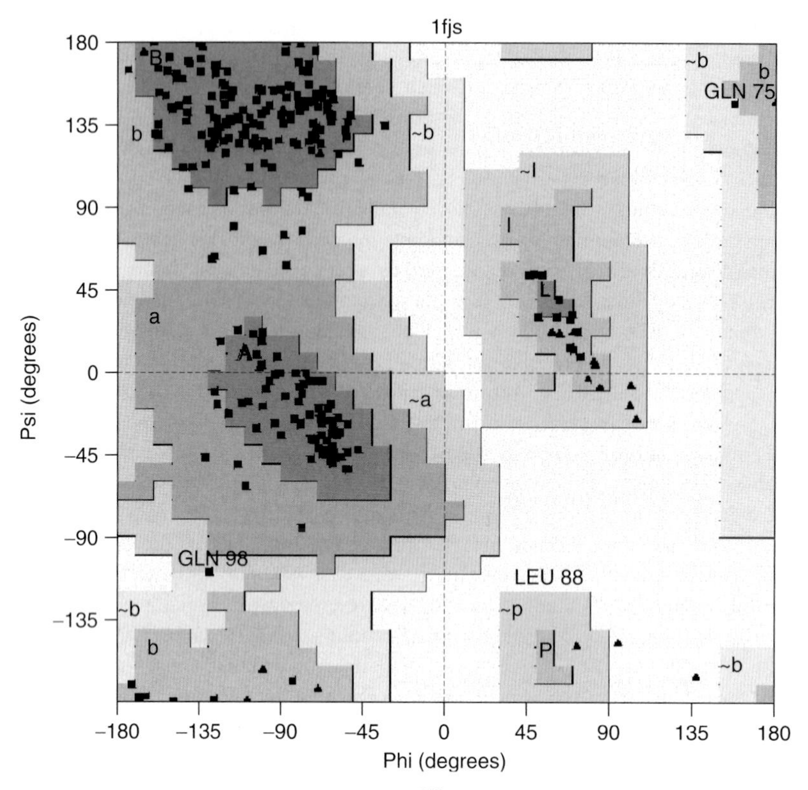

Figure 6 A Ramachandran plot for the serine protease factor Xa[202] showing the distribution of observed phi/psi torsion angles superimposed on the most highly populated regions as observed across many crystal structures (the darker the shade of gray the higher the probability). R indicates the amino acid side chain. Figure generated using Procheck.[203]

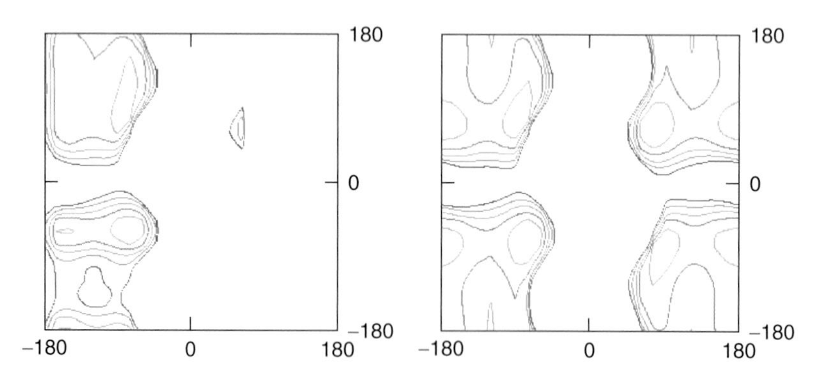

Figure 7 Theoretical Ramachandran plot generated for the amino acids alanine (left) and glycine (right) using the AMBER force field.

changes with the coordinates is referred to as the potential energy surface or the energy 'hypersurface.' A system with N atoms will have $3N$ Cartesian coordinates or $3N$-6 internal coordinates, and so only in the very simplest of systems (e.g., the approach of two argon atoms) will it be possible to visualize how the entire energy surface changes with configuration. An alternative is to visualize just a part of the energy surface in which all but one or two key coordinates are kept fixed. A classic example of this is the Ramachandran plot, which describes how the potential energy of an amino acid changes with the two key backbone torsion angles, ϕ and ω (**Figures 6** and **7**).[121]

There are certain regions on the energy surface that are of particular interest in molecular modeling. Minimum points correspond to stable arrangements of atoms; any movement away from the minimum gives a configuration with a higher energy. The minimum with the very lowest energy is called the global energy minimum. However, it is very important to recognize that many properties require a consideration of more than just the global minimum energy configuration. Saddle points may also be of interest. A saddle point is the highest point on the pathway between two minima. Both minima and saddle points are stationary points on the energy surface, at which the first derivative of the energy is zero with respect to the coordinates.

4.05.4 Energy Minimization

Minimization algorithms can be used to identify minimum energy points on the energy hypersurface. A characteristic of all commonly used minimization methods is that they can only move 'downhill' – they locate the nearest minimum point to the starting location. It is for this reason that when wishing to locate the minimum energy conformations of a molecule in a conformational analysis we need some other algorithm or procedure for generating the starting points for subsequent minimization.

Various types of minimization algorithm are in common use in molecular modeling.[122] Many of these are adapted from the standard minimization methods that are widely used in numerical analysis.[123–125] Minimization algorithms can generally be divided into three categories, depending on the extent to which the method uses information about the derivatives of the energy surface.

The simplex algorithm uses no information about the derivatives of the energy surface. Rather, it generates a mathematical construction called a simplex from which it then uses a set of rules to generate a new point, which will (hopefully) be of lower energy. For a function of N variables the simplex contains $N+1$ vertices. This then defines a new simplex ready for the next iteration. The most common move is reflection (see **Figure 8**), in which the new point is obtained by reflecting the point with highest energy through the plane containing the remaining points. The algorithm gradually explores the energy surface in this fashion, homing in on the region of lowest energy. The simplex method is robust (i.e., it is quite tolerant of high-energy initial structures) but can be very slow in operation.

An example of a method that uses information about the first derivative of the energy surface (the gradient) is the steepest descents algorithm. At each iteration this method calculates the direction of the gradient at the current point. The method then moves in this direction, either using a one-dimensional line search to identify the position of the minimum in the specified direction or by taking a step of arbitrary size in that direction (**Figure 9**). This defines the next starting point at which a new gradient is determined and the process is repeated. The steepest descents method is often found to be robust and can be a particularly good method when the initial configuration is highly strained. However, it does have the drawback that when moving down a long and narrow valley the method will take many small steps towards the minimum. This is because at each iteration the steepest descents method moves in a direction orthogonal to the previous iteration. An alternative method that attempts to correct this deficiency is the conjugate gradients method, in which the gradients in successive iterations are orthogonal but the directions are not. Rather, they take account of previous search directions using the following formula:

$$\mathbf{v}_k = -\mathbf{g}_k + \gamma_k \mathbf{v}_{k-1} \qquad [24]$$

$$\gamma_k = \frac{\mathbf{g}_k \cdot \mathbf{g}_k}{\mathbf{g}_{k-1} \cdot \mathbf{g}_{k-1}} \qquad [25]$$

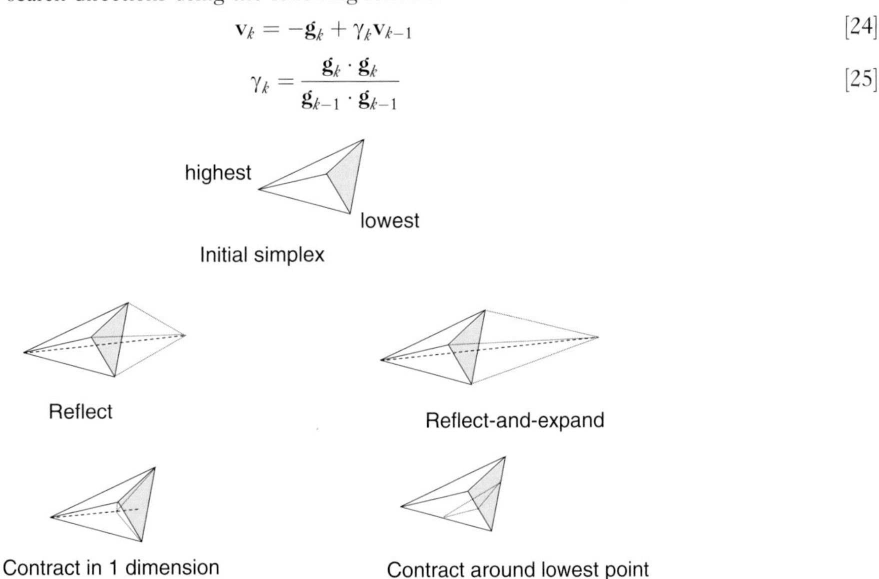

Figure 8 The three basic moves permitted to the simplex algorithm, illustrated for a 4-vertex simplex. (Adapted from Press, W. H.; Flannery, B. P.; Teukolsky, S. A.; Vetterling, W. T. *Numerical Recipes in Fortran*; Cambridge University Press: Cambridge, 1992.)

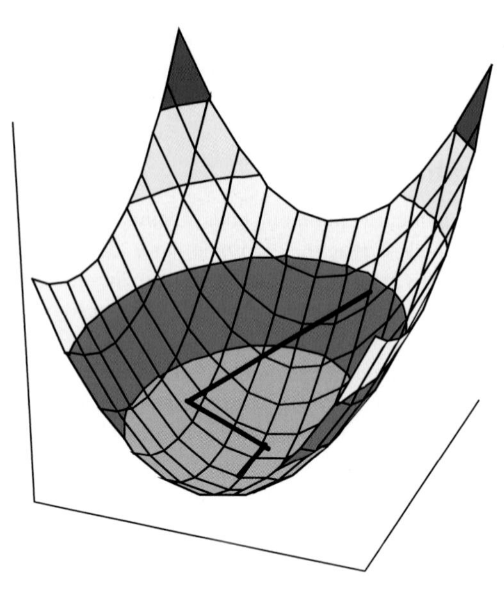

Figure 9 Schematic illustration of the operation of the steepest-descents algorithm, in which each successive move is in the direction of the gradient.

where $\mathbf{v}_k$ is the direction in which the method will move in the current iteration based on the direction for the previous move ($\mathbf{v}_{k-1}$), the current gradient $\mathbf{g}_k$ and the previous gradient $\mathbf{g}_{k-1}$.

A characteristic of the conjugate gradients method is that for a purely quadratic function of M variables the minimum is guaranteed to be reached in no more than M steps. In some cases this can give rise to a much more efficient algorithm than the steepest descents method.

In addition to information about the first derivative (or gradient) of the energy surface some minimization methods can use second derivatives (i.e., the curvature). The simplest of these second-order methods is the Newton–Raphson method, in which the new point is calculated using the following formula:

$$\mathbf{x}' = \mathbf{x} - \mathbf{V}'(\mathbf{x})\mathbf{V}''(\mathbf{x})^{-1} \qquad [26]$$

Here, $\mathbf{x}$ is the vector of current coordinates, $\mathbf{V}'(\mathbf{x})$ is the matrix of first derivatives and $\mathbf{V}''(\mathbf{x})^{-1}$ is the inverse matrix of second derivatives (also known as the inverse Hessian matrix). For a purely quadratic function the Newton–Raphson method will identify the minimum in a single step from the starting point. However, calculation and storage of the Hessian matrix and its inverse can be computationally challenging for systems that contain many atoms and so the pure Newton–Raphson method is generally more suited to relatively small systems. Another requirement for the Newton–Raphson method is that the Hessian matrix of second derivatives must be positive definite (i.e., all of the eigenvalues must be positive). Far from a minimum this requirement may not be satisfied and the Newton–Raphson algorithm may move to points where the energy actually increases (such as saddle points). It is also desirable to be able to calculate the second derivatives using analytical expressions rather than relying on numerical approximations; this is generally the case for molecular mechanics applications but not for all types of quantum mechanical calculation.

The quasi-Newton methods gradually build up the inverse Hessian matrix in successive iterations rather than calculating from scratch at each step. At each iteration the new set of positions $\mathbf{x}'$ are obtained from the current set of coordinates $\mathbf{x}$, the gradient $\mathbf{g}$, and the current approximation to the inverse Hessian matrix $\mathbf{H}$:

$$\mathbf{x}' = \mathbf{x} - \mathbf{H}\mathbf{g} \qquad [27]$$

Having moved to the new positions, the matrix H is updated. Different methods vary according to the formulas they use for this updating step.

Practical considerations to take into account when using energy minimization algorithms include the need to appropriately define the convergence criteria. Thus, whilst methods such as the conjugate gradients, Newton–Raphson, and the quasi-Newton methods are guaranteed to identify the minimum in a specified number of steps for a purely quadratic function, this is only an approximation for the energy surfaces typically encountered in molecular modeling and so more steps than the ideal case will be required. Unless instructed otherwise, a minimization algorithm may 'keep going' forever, getting ever closer to the minimum. It is usual to define some set of criteria that specify when the calculation is 'close enough' to the minimum to stop. Such criteria may be defined in terms of the difference in energy between successive iterations, the difference in coordinates between successive iterations, the root-mean-square gradient or the maximum value of the gradient – or combinations of these.

Routine application of energy minimization using molecular mechanics to small molecules generally proceeds without any problems using the default methods and convergence criteria included in modeling packages. Moreover, a combination of methods may prove more reliable (e.g., using steepest descents or conjugate gradient before Newton–Raphson). More care may be required when minimizing larger systems such as protein–ligand complexes (which may also include many solvent molecules) in order to ensure that all of the strain is removed from the system, for example, prior to a molecular dynamics simulation. Such systems may be minimized in stages, such as the solvent first, then the ligand and the protein side chains, and then the entire system.

The typically smaller size of the systems studied using quantum mechanics means that storing the Hessian matrix is not generally a problem but analytical second derivatives may not always be available, depending on the level of theory employed. The quasi-Newton methods are thus particularly popular here. Another aspect of quantum mechanics is the widespread use of internal coordinates. In such an approach, the geometry of the system is defined not using the usual Cartesian coordinates but rather a construction called a Z-matrix in which the location of each atom is defined in terms of a distance, an angle, and a torsion from three previously defined atoms. There may therefore be several different ways in which the Z-matrix can be defined, some of which are more amenable to efficient operation of the minimization than others.

4.05.4.1 Applications of Energy Minimization: Normal Mode Analysis

A molecule's normal modes of vibration correspond to collective motions of the atoms in a coupled system that can be individually excited. A molecule with N atoms will have $3N$-6 normal modes; water thus has three (illustrated in **Figure 10**). The frequencies and the atomic displacements for each normal mode can be calculated from the Hessian matrix of second derivatives by first converting this to the corresponding force-constant matrix in mass-weighted coordinates:

$$\mathbf{F} = \mathbf{M}^{-1/2}\mathbf{V}''\mathbf{M}^{-1/2} \quad [28]$$

$\mathbf{M}$ is a diagonal matrix of dimension $3N \times 3N$ containing the atomic masses; each element of the matrix $\mathbf{M}^{-1/2}$ contains the inverse square root mass of the relevant atom. The eigenvalues and eigenvectors of the matrix $\mathbf{F}$ provide the normal mode frequencies and relative atomic displacements. If the Hessian matrix is defined in terms of Cartesian coordinates then six of the eigenvalues will be zero (corresponding to translational and rotational motion of the entire system). For a system at a minimum all of the remaining eigenvalues should be positive and at a saddle point there should be just one negative eigenvalue, corresponding to motion over the saddle point from one minimum to another. It is important to calculate the eigenvalues of the Hessian matrix if one wishes to demonstrate that a minimum or saddle point has definitely been located. The determination of saddle points and the pathway from one minimum to another is of particular interest when studying chemical reactions, and methods for their calculation are available in the commonly used quantum chemistry packages.[126] Of course, such techniques are also applicable to the study of transitions between the various minimum energy conformations of flexible molecules and a number of techniques have been described in the literature, including the application to molecules as large as proteins.[127] Normal mode calculations can also be used to estimate the contribution of vibrational terms to the overall free energy via statistical mechanics. However, such calculations are relatively uncommon in drug design, though at least one widely used conformational search algorithm (the low-mode search method) is based on such an approach.

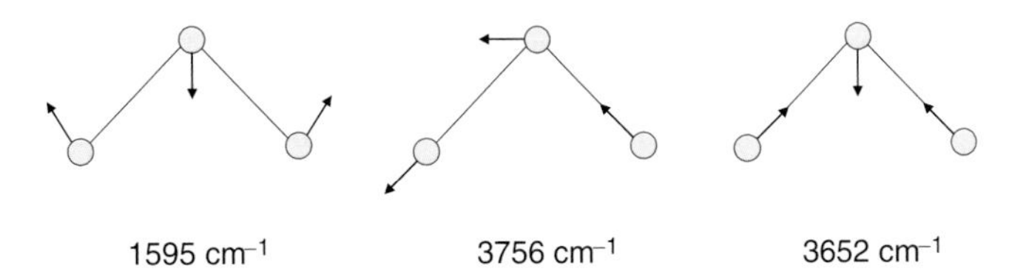

Figure 10 Schematic illustration of the normal modes of water together with their experimental frequencies.

4.05.5 Conformational Analysis and Search

A central requirement in molecular modeling is the identification of a molecule's minimum-energy conformations on the energy surface. This is the goal of a conformational analysis. As stated above, minimization algorithms are only able to locate the energy minimum that lies downhill from the current position. In order to perform a conformational analysis it is therefore necessary to have a method for generating a series of starting points for subsequent energy minimization; this is the objective of a conformational search.

A large number of conformational search algorithms have been developed over the years,[128] but most fall into one of a few general categories. The majority of methods can be considered to adopt either a systematic or a random approach to exploring the energy surface. An additional feature common to many methods is the use of fixed bond lengths and bond angles, with the only degrees of freedom permitted to vary being the torsion angles about single bonds. This has the effect of reducing the scale of the problem.

4.05.5.1 Systematic Search Methods

As the name suggests, a systematic search explores the conformational space of the molecule in a systematic fashion. In the simplest type of systematic search (the 'grid search') each torsion angle is permitted to adopt a defined set of values from which all possible permutations of torsion angle values are generated in a systematic fashion. Each conformation so generated is then subjected to energy minimization to locate the nearest energy minimum. To illustrate with a simple example, suppose a molecule contains three variable torsion angles with the first bond permitted to adopt torsion values of 60°, 180°, and −60°; the second bond values of 0° and 180° and the third bond values of 45°, 135°, −45°, and −135°. The simple grid search will generate a total of 24 conformations with torsion angle values for these three bonds of (60°, 0°, 45°); (60°, 0°, 135°); (60°, 0°, −45°); (60°, 0°, −135°); (60°, 180°, 45°); ... (−60°, 180°, −135°).

As the number of torsion angles in the molecule increases so too does the number of conformations generated by the simple grid search, and the time taken – in an exponential fashion. For this reason the grid search is limited to relatively small molecules, and is an inefficient way to explore conformational space. A major cause of this inefficiency is the fact that the method typically generates many conformations of high energy that are far from an energy minimum. More efficient is to identify such high-energy starting conformations and to eliminate them prior to the time-consuming energy minimization step. Moreover, by carefully choosing the order in which the torsion angles are varied in the molecule, it may be possible to use tree-pruning methods that make the algorithm even more efficient. For example, if the combination (60°, 0°) for the first two torsion angles in our hypothetical example always gives rise to a high-energy interaction then there is no point in generating any more conformations with such a combination; rather, one can proceed directly to the next combination.

With these enhancements the systematic search can become applicable to much larger molecules. A major perceived advantage of the systematic search is its deterministic nature with a defined endpoint. However, it does have some drawbacks. It is necessary to define an appropriate set of values for each torsion angle in the molecule (mindful that the larger the number of torsion increments the longer the search will take). Rings are often difficult for the systematic search to address efficiently; a common approach is to break each ring to give a pseudo-acyclic molecule that can then be treated in the usual way alongside the acyclic bonds. Additional criteria are then used to decide whether the ring can be closed or not. Unfortunately, such ring closure criteria can often be considered only relatively late in the procedure meaning that the search for ring conformations can be rather inefficient.

The Systematic Unbounded Multiple Minimum Method (SUMM)[129] attempts to address another of the drawbacks of the systematic methods, which is that successive conformations are often very similar to each other. This is because a typical systematic search involves just one or two torsion angles being changed between successive conformations. Another limitation is that the torsion increment is usually fixed at the beginning of the search and if it is desired to repeat the search at a higher resolution then one invariably has to rediscover all of the conformations generated previously. SUMM uses a different sequence of changing the torsion angles so that successive conformations correspond to changes to very different parts of the molecule. For example, in our illustration above the second conformation could be (180°, 180°, −45°). The sequence of torsion angles is defined in such a way that if the search is run to completion then all possible combinations will be considered. However, if the search is terminated prematurely then the SUMM protocol means that the conformational space will have been covered to a greater extent when compared to the more traditional systematic search.

4.05.5.2 Random Search Methods

The random search methods[130–133] adopt a very different philosophy to the systematic search. One major difference is that they have no well-defined endpoint. They all follow a similar general approach involving the following steps:

1. Select the next starting point for the current iteration.
2. Apply random change to this starting conformation and minimize.
3. Check to see whether the minimized conformation has been previously generated.
4. Stop, or go to 1.

The simplest random search method involves changes to either the Cartesian coordinates of the atoms or to the torsion angles (or to a subset of atoms or torsions). In the former case a random amount is added to the (x,y,z) coordinates of the atoms in the molecule; in the latter case the torsion angles are set to random values or are changed by a random increment. The resulting structure (which may be highly strained) is then minimized. Several methods have been proposed for deciding how to select the conformation with which to start the next iteration. The simplest approach is to select the conformation generated in the previous iteration. However, this can result in the method becoming fixed in just one region of conformational space. An alternative is to select at random from the structures generated previously, weighting the selection towards those that have been chosen the least. A third method is to bias the selection towards the lowest energy structures. The Metropolis Monte Carlo algorithm is often used in such cases; here, the energy of the new structure is compared to the energy of the structure from the previous iteration. If the current structure has a lower energy then it is used as the new starting point. If it is higher in energy then the Boltzmann factor of the energy difference, $\exp[-(V_{new}-V_{old})/kT]$ is compared with a random number generated between 0 and 1. If the Boltzmann factor is larger than the random number then the latest structure is used as the new starting point; otherwise the previous structure is retained.

One search method that has shown significant promise is the low-mode search.[134] The key feature of this approach is that the changes made to each starting structure are performed in such a way as to encourage movement into the locality of a new energy minimum. In the low-mode search this is done by performing a normal mode analysis on the initial conformation. As described above, such an analysis provides a series of eigenvectors that correspond to low-frequency movements away from the minimum. These low-frequency modes are explored in turn, the atomic coordinates being perturbed in a manner consistent with the corresponding eigenvector. Such a perturbation will usually lead to a structure whose energy exceeds a predefined threshold, in which case the move is terminated. Occasionally, however, it will lead to movement over the saddle point and into the locality of a new energy minimum, which can then be identified using standard energy minimization. If all of the low-frequency modes are exhausted the algorithm switches to a stochastic procedure in which a random mixture of the low-frequency modes is searched. A particular attraction of the low-mode method is that it can be applied to both acyclic and cyclic molecules without requiring any special methods for dealing with ring closures.

As indicated above, there is no in-built mechanism for deciding when to stop a random search. The simplest approach is to stop after a predefined number of iterations. An alternative is to monitor the number of new minimum energy conformations that are generated and to stop when no new conformations are produced after a given number of successive iterations. For these reasons it is inevitable that a random search will generate the same conformation many times.

4.05.5.3 Molecular Dynamics

Molecular dynamics can be used to explore conformational space, and is often the method of choice for large molecules such as proteins. In molecular dynamics the energy surface is explored by solving Newton's laws of motion for the system (*see* 4.25 Applications of Molecular Dynamics Simulations in Drug Design). A common strategy when searching conformational space is to perform the simulation at a very high, physically unrealistic temperature as this enhances the ability of the system to overcome energy barriers. Structures are then selected at regular intervals from the trajectory for subsequent energy minimization.

4.05.5.4 Distance Geometry

A key characteristic of the distance geometry method is its description of a molecule's conformation in terms of interatomic distances, rather than Cartesian or internal coordinates.[135–137] A molecule with N atoms contains $N(N-1)/2$ interatomic distances, most conveniently represented as an $N \times N$ symmetric matrix called a distance matrix. Distance geometry generates a large number of such matrices and then converts these into Cartesian coordinates. An important consideration is that many combinations of interatomic distances are not geometrically possible; the first step in distance geometry is to define upper and lower limits (distance bounds) on each interatomic distance in the molecule. Some of these distance bounds are derived from simple chemical principles. For example, atoms that are bonded together show very little variation in their interatomic distance. Similarly, the distances between atoms that are connected to a common third atom (i.e., are in a 1,3 relationship) are also limited. The distance between atoms that are separated by three bonds (i.e., are in a 1,4 relationship) will vary with the torsion angle of the central bond, in a well-defined fashion. For the remaining pairs of atoms the lower bound is initially assigned to a value corresponding to the sum of the van der Waals' radii of the two atoms and the upper bound to an arbitrarily large value. An iterative process called triangle smoothing is then applied that gradually refines the set of distance bounds until the entire set is self-consistent. The two triangle smoothing criteria are as follows:

$$u_{AC} \leqslant u_{AB} + u_{BC} \tag{29}$$

$$l_{AC} \geqslant l_{AB} - u_{BC} \tag{30}$$

where u_{AB} refers to the upper distance bound between atoms A and B and l_{AB} is the corresponding lower bound.

Having generated a self-consistent bounds matrix the actual process of generating conformations can commence. First, random values are assigned to all of the interatomic distances, between the upper and lower bounds. The resulting distance matrix is then subjected to a process called 'embedding,' in which this distance representation is converted to a set of initial atomic coordinates. The final step involves a refinement of the initial atomic coordinates so that the conformation better satisfies the initial distance bounds. Subsequent conformations are produced by generating a new random distance matrix.

4.05.5.5 Introduction to Evolutionary Algorithms and Simulated Annealing

Two techniques that are widely used throughout computational science and which also play a major role in computational chemistry are simulated annealing and evolutionary algorithms. Both are designed for the optimization of complex problems, but in contrast to the minimization algorithms that were described above they are able to explore the full energy surface and, in principle at least, to identify the globally optimum solution. Both evolutionary algorithms and simulated annealing are at heart random search techniques and cannot be guaranteed to produce the globally optimal solution in each run. Rather, they produce solutions that are often close to the optimal in a reasonable amount of time. Repeated operation using a series of randomly generated and different starting points will identify a series of near-optimal solutions; if the same result is obtained from multiple runs there is a good probability that it does indeed correspond to the true global minimum.

4.05.5.6 Genetic and Evolutionary Algorithms

Evolutionary algorithms are based on concepts of biological evolution. A 'population' of possible solutions to the problem is first created with each solution being scored using a 'fitness function' that indicates how good they are. The population evolves over time and (hopefully) identifies better solutions. Of the various types of evolutionary algorithm[138] the genetic algorithm is the most well known and the one we will briefly describe. The applications of evolutionary algorithms to problems in chemistry continues to grow significantly (see [139–142] and other chapters in this volume).

Each member of the population is encoded by a chromosome, which is often (but not always) a bitstring of 0 s and 1 s. For example, in the application of genetic algorithms to conformational analysis[143–145] the chromosome encodes the values of the torsion angles of the rotatable bonds in the molecule with the fitness function being the energy of the conformation. The initial chromosomes would be generated at random. A new population is then generated from the initial one by the application of genetic operators, the two principal ones being mutation and crossover. In the mutation operator a randomly selected bit is inverted (0 to 1 and vice versa). In the crossover operator two members of the population are randomly selected and a cross position is randomly generated. Two new offspring chromosomes are then created by swapping either side of the cross position (**Figure 11**). Roulette wheel selection may be used to bias the selection of chromosomes for crossover towards the members of the population with the best fitness function values. The process continues for a predefined number of iterations.

A large number of different variants on both the genetic algorithm and other types of evolutionary algorithm are in use. Some of these variants enable a more efficient search; others ensure that the algorithm does not prematurely converge.

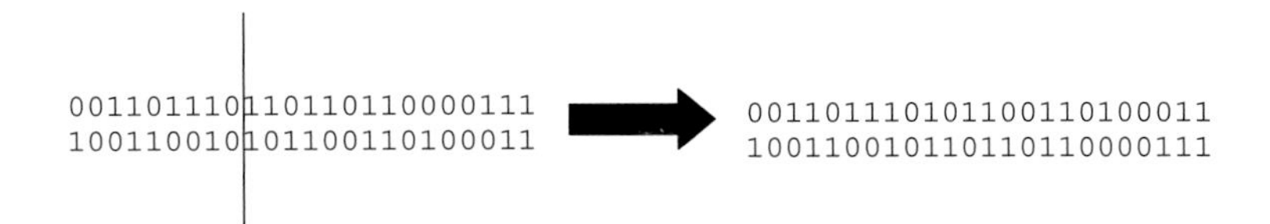

Figure 11 Illustration of the crossover operator in genetic algorithm, in which two new offspring are generated by combining the chromosomes of its two parents at a random crossover point.

4.05.5.7 Simulated Annealing

Simulated annealing[146] is a computational method that mimics the physical process of annealing. In physical annealing the temperature of a molten substance is very slowly reduced so that it crystallizes into a single large crystal, free from any defects (i.e., at the global minimum of free energy). In simulated annealing a cost function is used together with a control parameter that represents the temperature. For conformational analysis the cost function would be the strain energy. The system then explores possible solutions using a Monte Carlo or molecular dynamics simulation; the initial high temperature means that the system is able to occupy high energy regions, to pass over high energy barriers and to explore widely the possible space. The control parameter is then gradually reduced, making the lower energy regions more probable. At absolute zero the system should occupy the lowest energy state. However, this can only be guaranteed in the case of an infinite number of temperature decrements, at each of which the system reaches thermal equilibrium. In practice, therefore, simulated annealing cannot be guaranteed to find the globally optimal solution, but it does usually produce a good solution. As with the genetic algorithm, repeated runs should provide further good solutions.

4.05.5.8 Rule-Based Conformational Search Methods

The desire to search large databases of conformationally flexible molecules has led to a need for methods that can rapidly and automatically explore the conformational space of large numbers of molecules (such databases can contain hundreds of thousands if not millions of structures). Rule-based conformational search methods[147–152] (also called knowledge-based methods or model-building approaches) are designed to address some of the limitations of alternative conformational search procedures, and in particular to reduce the amount of computer time required to perform the analysis. Much of the time required for other conformational search methods is taken exploring regions of conformational space that ultimately prove unfruitful, in the energy minimization of starting conformations that are high in energy, and in the repeated generation of the same set of minimum energy conformations. In a rule-based system the molecule is broken down into fragments, for each of which a defined set of conformations is permitted, typically just those conformations that the fragment is commonly observed to adopt. A conformation of the whole molecule is then assembled by combining together the fragment conformations; the resulting structure may be used directly or may be subjected to a few rounds of energy minimization. Rule-based methods typically explore the conformational space defined by the fragments in a systematic fashion, with various criteria being applied during the search to eliminate higher energy combinations of fragments and to preferentially produce the lower energy conformations.

A critical part of any rule-based system is of course the fragment library or knowledge base used by the program. The Cambridge Structural Database (CSD)[153] has proved to be an invaluable source of information on the conformational preferences of molecular fragments. Fragment libraries can vary greatly in content depending on the size of the fragments and how specifically they are defined. Indeed, there is a subtle balance; the larger and more constrained the fragments the more efficient the search will be, but larger fragments are less likely to have sufficient examples in the database to ensure that the observed distribution is indeed correct and statistically significant. An alternative to the use of experimental data for building the fragment library is to generate such data computationally, for example, by using one of the alternative conformational search techniques. This approach could also be used to fill any gaps where the experimental data was incomplete.

4.05.5.9 Generating a Set of 'Representative' Conformations

Conformational search methods can generate a large number of conformations, even for molecules with relatively few rotatable bonds. Many of these conformations are frequently rather similar to each other, differing only by relatively small rotations in one or a few torsion angles. For some applications it may be undesirable to have to deal with large numbers of conformations; for example, if one wishes to store the structures in a database for some form of 3D virtual screening application. Under such circumstances one would often wish to generate a set of conformations that in some way represent the conformational space. There are two approaches to this problem. The first is to perform the conformational search, generate all of the conformations, and then select a subset. The alternative is to use a conformational search method designed to generate directly a subset of representative conformations.

An obvious requirement when generating a subset of conformations is some means of quantifying how 'different' two conformations are. The most straightforward method is to use a molecular fitting algorithm based on the root mean square distance (RMSD) between pairs of atoms in the two structures:

$$\mathrm{RMSD} = \sqrt{\frac{\sum_{i=1}^{N} d_i^2}{N}} \qquad [31]$$

N is the number of atoms over which the RMSD is measured and d_i is the distance between the (xyz) coordinates of atom i in the two conformations. Molecular fitting is very widely used in molecular modeling and several algorithms to calculate the minimum value of the RMSD for two conformations have been described.[154–156] It is necessary to decide for which set of atoms the fitting procedure will be applied; a common approach is to perform the calculation over all nonhydrogen atoms in the molecule. One potential pitfall arises with molecules containing elements of either local or global symmetry insofar as it may be necessary to consider symmetrically equivalent alternatives when performing the calculation. A simple example would be the three symmetrically equivalent methyl groups in a *t*-butyl group; slightly more complex are 1,4-disubstituted benzene rings, which contain two pairs of symmetrically equivalent carbon atoms (**Figure 12**).

Several alternative methods are available for estimating the difference between any two conformations, some of which will be discussed in more detail later in the section on molecular similarity. An alternative worthy of mention is the use of torsion angles as the criterion. When Cartesian coordinates are used then even a relatively small difference in the torsion angle located in the center of a molecule may give rise to a large difference in the RMSD value. This limitation may be ameliorated by using the differences in torsion angles between the two structures as the basis for the calculation; it is necessary to take account of the cyclic nature of torsion values in this approach.

Having decided upon a method to calculate the differences between any pair of conformations in the set the second stage of the procedure involves the use of a subset selection algorithm to derive a 'representative' subset. Several methods are available for this part of the procedure; clustering algorithms have traditionally been used but other subset selection methods can also be applied (*see* also 4.08 Compound Selection Using Measures of Similarity and Dissimilarity).

This two-stage process for generating a representative set is straightforward to comprehend and implement, but it does suffer from some drawbacks. A practical issue concerns the resources (time and disk storage) required for the initial conformational search and then for the subsequent subset selection algorithm. Methods that directly generate a subset of 'representative' conformations do exist but are less common. One example of such a technique is the 'poling' method, in which a penalty function is introduced into the minimization step to supplement the usual force field energy terms.[157,158] This poling function is designed to penalize a conformation that is too close to any of the conformations already generated. The effect of such a function is illustrated in **Figure 13**. Having generated the first conformation the poling function is introduced to modify the energy hypersurface in the region of this conformation. As each conformation is produced additional 'poles' are added to the penalty function. A natural consequence of the additional terms is that the nature of the energy surface is changed and indeed new minima may be introduced. The minima on the modified energy surface will not necessarily coincide with the 'true' energy minima, but they may lead to a wider range of low-energy conformations being generated that may better represent the accessible conformational space.

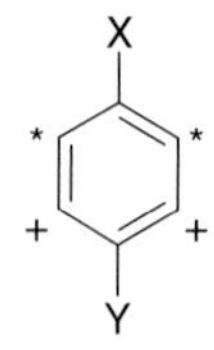

Figure 12 Local symmetry in a 1,4-disubstituted benzene ring needs to be taken into account when performing RMS calculations (i.e., the symmetrically related atoms indicated by an asterisk and a cross).

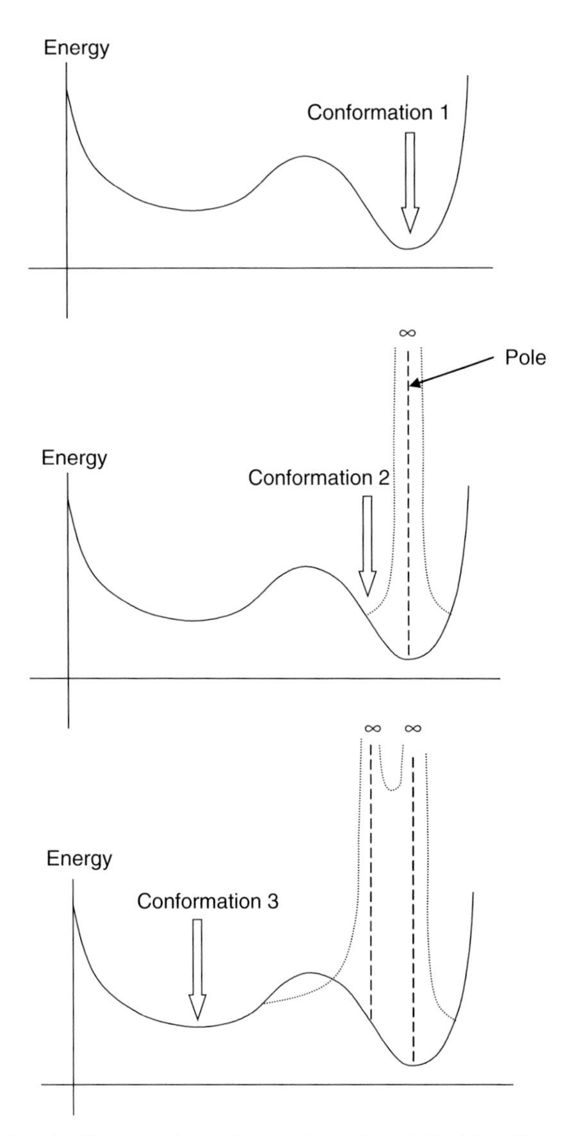

Figure 13 Illustration of the poling algorithm used to produce conformations distinct from those generated previously. (Reprinted from Smellie, A. S.; Teig, W. L.; Towbin, P. *J. Comput. Chem.* **1995**, *16*, 171–187, with permission of John Wiley & Sons.)

4.05.5.10 Comparisons and Evaluations of Conformational Search Methods

Given the many methods available for performing conformational search it is important to understand their applicability, scope, and limitations. Current conformational search applications often fall into one of two categories. The first category comprises those problems where one wishes to perform a very detailed conformational analysis on just one or a small number of complex, biologically active molecules (e.g., cyclic peptides). The goal of such an analysis is to identify the likely bioactive conformation in solution. In the second category are those problems in which a conformational search needs to be performed for a large number of molecules in an automated fashion. There are a number of possible applications in this second category, but the goal is generally to undertake some kind of virtual 3D screening exercise using for example pharmacophore searching, protein-ligand docking or a 3D similarity method. The molecules involved in such an exercise will be typical 'drug-like' molecules but may number as many as several million. It is important in such cases to have available a conformational search method that is automatic (able to operate without human intervention), efficient, and robust.

Published evaluations of conformational search methods often examine a very small number of rather complex molecules.[159] One of the most celebrated of such studies was a comparison between five different search methods[160] (systematic search, Cartesian and torsional random methods, distance geometry, and molecular dynamics). The test molecule was cycloheptadecane ($C_{17}H_{34}$). Although a molecule of minimal pharmaceutical interest, the study did nevertheless provide a rigorous comparison between the different methods and also provided some insights into the nature of the energy hypersurface for this molecule; a total of 262 minimum energy conformations were identified within 3 kcal mol^{-1} of the global energy minimum with the best method (a random torsion angle search) being able to identify 249 of them.

Comparison with experiment is often the preferred measure of success and several studies have used databases of structures solved by x-ray crystallography for this purpose. The CSD contains the structures of over 250 000 organic and inorganic molecules whilst the Protein Data Bank (PDB)[161] contains the structures of proteins and nucleic acids, many with bound ligands. Both databases continue to expand at a rapid rate. The PDB is of particular interest as the ligand conformations it contains correspond to the presumed bioactive conformations of the molecules. However, even though the PDB contains a large number of protein–ligand complexes, not all of these may be considered suitable for a comparative study. It is common to restrict the ligands included to those taken from structures with high resolution (e.g., at least 2.0 Å) and low temperature factors. The ligands should also be reasonably small and drug-like. An example of such a study is that of Boström,[162] who constructed a data set of 32 ligands according to these criteria. Other data sets have also been established, especially for the purpose of evaluating docking programs (*see* 4.12 Docking and Scoring). Structures taken from the public databases may also be supplemented by structures available from in-house drug discovery projects. In the Boström study five different programs were evaluated including methods that generate 'representative' sets of conformations (Catalyst and Confort), two random search methods (flo and the low-mode search) and a rule-based approach (Omega). Each method was applied to the set of molecules and the results analyzed using various criteria. Of particular interest was the ability of each program to generate the bioactive conformation (as measured by the RMSD), but other aspects were also considered, including the number of conformations generated by each method and the speed of calculation. The results suggested that the low-mode method performed best. The rule-based Omega method also gave reasonable results and was of particular interest due to its very rapid search speed.[163]

Other analyses of ligands extracted from the PDB have concentrated on the energetic changes that accompany ligand binding and the extent to which the bound conformation is at the global or a local energy minimum. Such studies[164,165] have shown that ligands rarely bind in the global minimum energy conformation but that the propensity to bind in a local minimum energy conformation is less clear-cut, being dependent on the conformational search method and (especially) on the force field used for the energy calculations and minimization. The strain energy that accompanies binding is found to vary somewhat; the study of Boström and colleagues suggested that for approximately 70% of the systems they studied the bioactive conformation was within 3 kcal mol^{-1} of the global energy minimum whereas Perola and Charifson found that whilst 60% of ligands bound with strain energies less than 5 kcal mol^{-1} at least 10% of ligands bound with strain energies greater than 9 kcal mol^{-1}. This latter study also demonstrated that for more flexible ligands it was important to use higher energy thresholds in order to ensure that the bioactive conformation was included in the conformational ensemble.

4.05.5.11 Three-Dimensional Structure Generation

The need to perform molecular modeling calculations on large numbers of molecules in an automated fashion has resulted in a need for software that can operate automatically and in a high-throughput manner. A good example of such software is in the area of 3D structure generation, where a single low-energy conformation of a molecule is produced directly from its connection table (i.e., solely from information about the atoms present and how they are bonded together). This is a task that historically was performed manually using for example a combination of interactive sketching, the assembly of prestored 3D templates for certain systems (e.g., rings) and rounds of energy minimization. Whilst this is practical for a small number of molecules, it is clearly unfeasible for the many hundreds of thousands (or millions) that it is often required to process today.

Structure generation programs are designed to generate a single, low-energy conformation for a wide range of molecules. The structure should be of sufficient quality that it can be used without the need for further energy minimization. Additional requirements are robustness (i.e., able to run for a long time without crashing), the ability to handle files containing large numbers of molecules, a high conversion rate, and rapid and automatic operation, without requiring any user intervention.

A number of programs can in principle be used for structure generation, including some of those described above for conformational analysis (trivially, by stopping the program after the first conformation has been generated). However, two particular programs have achieved widespread use due to their ability to meet the criteria listed above: CONCORD[166,167] and CORINA.[168,169] Although developed independently, the approach taken by these programs have a number of features in common.[170,171] The first step involves an analysis of the molecule, which is broken into ring systems and acyclic portions. Bond lengths and bond angles are taken from prestored values (e.g., from experimental small-molecule x-ray crystallographic data),[172] which take into account the atom types, hybridization, bond order, and sometimes the neighboring atoms. A conformation for each isolated ring system is then built, often using sets of prestored 3D templates. Fused and bridged ring systems can often pose problems, in which case rules may be applied to identify which combinations of ring templates will lead to an acceptable structure. In some cases suitable ring templates may not be available (e.g., for large rings) in which case a simple conformational analysis may be employed to generate the required structure(s). Acyclic chains are often generated in extended conformations and then the cyclic and acyclic portions are joined together to give a single structure. If this conformation has no problems (e.g., too close interatomic contacts) then it is output and the next molecule considered. If, however, there are problems then an attempt may be made to try and 'resolve' these, perhaps by rotating those single bonds that lie on the path between the offending atoms.

Current structure generation programs have been applied to millions of molecular structures and have therefore been subjected to very extensive testing. For 'typical' drug-like molecules conversion rates over 99% can now be achieved with each structure taking significantly less than one second to be processed.

4.05.6 Three-Dimensional Molecular Alignment

In an increasing number of drug discovery projects the structure of the target protein is available and structure-based techniques such as docking, free energy calculations, and de novo design can be applied. Despite the increasing number of cases where such structures are available, in a significant proportion of projects such structural information about the target is not available but a number of active ligands are known. Ligand-based design methods then become applicable. A crucial component of many of the ligand-based approaches is the need to superimpose, or align, a series of active ligands, ideally to mimic the way in which they would be overlaid in the binding site. Such superimpositions form the basis for techniques such as 3D database searching, 3D quantitative structure–activity relationship (QSAR), and receptor modeling. Note that the 3D similarity between two or more molecules can also be assessed using so-called alignment-free methods, some of which are described in Chapter 4.08.

A large number of techniques have been designed for the 3D alignment of two or more molecules.[173] An obvious prerequisite is the means to quantify the degree of similarity (goodness of fit) between the two conformations when overlaid in 3D space. As the molecules are moved relative to each other the similarity value will change, with the general goal being to find the alignment that gives the optimal similarity value. Many different ways of calculating 3D similarity values have been developed.[174,175] Molecules may be represented by groups of atoms, by their shapes, by projecting properties onto a surface, or by calculating molecular properties over all space or at regular intervals on a grid surrounding the molecule. Further variation comes from the varying degrees of conformational flexibility afforded the ligands during the superimposition process. Many methods deal solely with rigid structures, leaving just the six degrees of translational and rotational degrees of freedom to explore. At the other end of the spectrum are techniques that afford conformational flexibility to both ligands; intermediate between these two extremes are methods in which one ligand is kept rigid (in the presumed bioactive conformation) but the other ligand(s) are allowed to vary both their conformation and their relative orientation to find the best fit. Of course, when using a rigid fitting algorithm, conformational flexibility can be taken into account indirectly by using a number of discrete conformations derived from a previous conformational search procedure.

Of the atom-based methods perhaps the simplest approach to molecular superimposition is to manually define pairs of corresponding atoms in each of two structures, which are then overlaid using RMS fitting. This has the obvious limitation of requiring the atoms for overlay to be defined; as the structures become more dissimilar this becomes increasingly difficult and arbitrary. Extensions to the simple RMSD algorithm include flexible fitting, in which the restriction to rigid conformations is lifted and one of the molecules is permitted to change its conformation (typically by rotating about the free torsion angles in the molecule). If the conformational energy is not taken into account during the torsion rotation procedure, this may lead to unfavorable, high-energy structures. The goal of pharmacophore mapping is to align a series of active molecules in which atoms or functional groups that share a common property (i.e., are bioisosteric) are overlaid in 3D space; this particular variant on the 3D alignment problem is described in detail in Chapter 4.21 and so will not be covered further here.

4.05.6.1 Field-Based Alignment Methods

In one of the earliest proposals for a field-based similarity method Carbó suggested the overlay of electron density maps.[176] The similarity index used in these calculations (also known as the cosine index) is as follows:

$$C_{AB} = \frac{\int P_A P_B}{\left(\int P_A^2\right)^{1/2}\left(\int P_B^2\right)^{1/2}} \qquad [32]$$

P_A and P_B are the electron densities (determined from the square of the wavefunction) and each integral is over all space. A drawback of the Carbó index is that the similarity is dependent only on the degree of overlap; the magnitude of the fields being compared is not taken into account. More fundamentally, the electron density is not a particularly useful measure of similarity as the density peaks near the nuclei and so overlap of the nuclei will dominate the similarity measure. It is also an expensive property to compute, being derived from a quantum mechanics calculation. Much more common is the use of electrostatic fields and/or shape-related properties in molecular alignment calculations. To simplify the problem of integrating over all space the properties are usually mapped onto the vertices of a 3D grid surrounding each molecule. Under such circumstances the Carbó index is calculated as a sum over the n grid points:

$$C_{AB} = \frac{\sum_{i=1}^{n} P_{A,r} P_{B,r}}{\left(\sum_{i=1}^{n} P_{A,r}^2\right)^{1/2}\left(\sum_{i=1}^{n} P_{B,r}^2\right)^{1/2}} \qquad [33]$$

Hodgkin and Richards[177] proposed an alternative similarity measure in an attempt to take account of the magnitude of the fields as well as its shape; this index is also known as the Dice coefficient:

$$S_{AB} = \frac{2\int \phi_A(\mathbf{r})\phi_B(\mathbf{r})}{\int \phi_A^2(\mathbf{r}) + \int \phi_B^2(\mathbf{r})} \qquad [34]$$

A third metric is the Spearman rank-correlation coefficient, in which the Pearson correlation coefficient is calculated for the different grid points:

$$S_{AB} = 1 - \frac{6\sum_{r=1}^{n} d_{\mathrm{r}}^2}{n^3 - n} \qquad [35]$$

d_r is the difference in the property ranking at point r of two structures and n is the total number of points over which the property is measured.[178,179]

The numerical, grid-based similarity calculations can be very time-consuming. Several ways to improve the efficiency of the procedure have been devised. In the 'field-graph method' the minimum-energy regions in the field are identified and used as the basis for deriving possible overlaps using a technique called clique detection. More common is to approximate the field as a series of Gaussian functions of the following form[180–182]:

$$G = \gamma e^{-\alpha|r - R_i|^2} \qquad [36]$$

These Gaussian functions can be substituted for the $1/r$ term in the electrostatic potential term in the relevant similarity calculation. The key advantage of the Gaussian representations is that they provide an analytical solution to the problem, which enables the similarity to be evaluated much more rapidly. Gaussian functions have also been used to provide a more realistic representation of molecular shape when compared to the traditional hard sphere models.[183,184]

Although the grid-based or Gaussian representations are the most popular, alternatives have been proposed. These include gnomonic projection methods, in which the molecule is placed at the center of a sphere and the relevant property or properties projected onto the sphere's surface (**Figure 14**).[185–189] The similarity between two molecules is determined by comparing the spheres, which are usually approximated by tessellated icosahedra or dodecahedra. Two molecules are compared by rotating one sphere (or icosahedron/dodecahedron as appropriate) relative to the other until the differences in the property values over all matched vertices are minimized. A drawback of the gnomic projection method is that the comparisons are only performed in rotational space and so the results are dependent on the centers of projection chosen for each molecule.

Having decided upon a method for calculating the 3D similarity it is necessary to have a mechanism for exploring the orientational (and sometimes conformational) degrees of freedom of the molecules in order to identify the alignment where the similarity is optimal. Minimization algorithms can be used here (e.g., the simplex algorithm[190] or a gradient-based method[191]). As is the case in conformational analysis it is necessary to use a number of starting points in

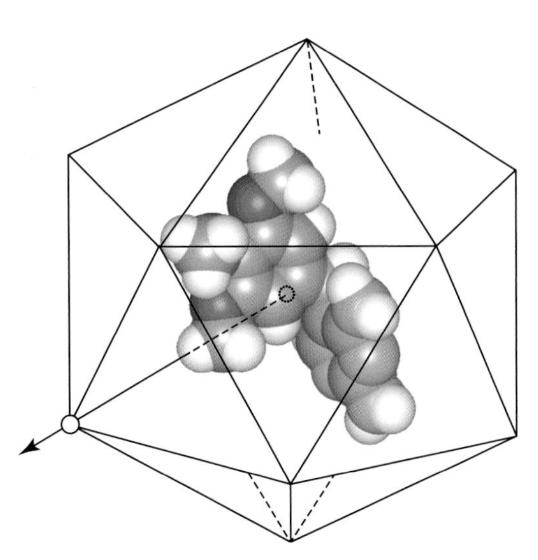

Figure 14 Illustration of the gnomic projection approach to performing 3D molecular superimposition. Redrawn from Leach, A. R; Gillet, V. J. *An Introduction to Chemoinformatics*; Kluwer Academic Publishers: Dordrecht, the Netherlands, 2004, with kind permission of Springer Science and Business Media.

order to try and ensure the entire space is covered and that the globally optimal alignment is identified. Monte Carlo methods are widely used to explore the search space[192] as are genetic algorithms and simulated annealing[193,194] Others use atom-based algorithms to generate the initial alignment[195,196] It is relatively straightforward to incorporate conformational flexibility into these methods by permitting changes to the rotatable bonds as well as the relative molecular orientations.

As indicated previously, it is increasingly desired to use 3D similarity methods to search large databases of molecules as part of 'lead hopping' exercises, library design, or compound acquisition efforts. The large size of such databases means that it is generally not possible to apply the most rigorous methods directly. Rather, approximations are used to speed up parts of the procedure. As indicated above, the problem of conformational flexibility is commonly dealt with via the use of a precalculated, representative set of conformations which are then treated as rigid during the search procedure. One method that is able to perform on-the-fly conformational analysis is the FlexS program,[197,198] in which the query molecule is kept rigid and the test molecule is broken down into a series of molecular fragments in a manner similar to that used in some of the rule-based conformational search procedures. Each fragment is permitted to adopt a small number of defined conformations. First the program superimposes a 'base fragment' onto relevant regions of the query molecule. The remaining fragments are then attached in a stepwise fashion. As each fragment is added the partial molecule is assessed using a multicomponent, Gaussian-based scoring function. The highest-scoring partial solutions are allowed to proceed to the next step, thereby reducing the size of the search space. Another common way to speed up the procedure is to use some form of rapid screening process, or to require the initial overlay to be such that the principal moments of inertia of the two molecules are aligned.

4.05.6.2 Comparison and Evaluation of Similarity Methods

Given the many methods for performing 3D molecular alignment it is of particular interest to have objective ways to compare the different techniques. Two general types of evaluation method have been described. First, one can determine how well the method is able to identify known alignments. Currently, the most definitive way to do this is to identify a series of diverse ligands that are all known to bind to the same protein, the binding modes being determined from x-ray crystal structures. An alternative common form of evaluation is to determine how well the method is able to retrieve known active molecules amongst a large set of inactive compounds. A typical evaluation experiment involves taking an active compound as the search query and ranking a database of known active and inactive compounds in order of decreasing similarity. The perfect similarity method is one in which all of the active molecules are at the top of the list. A method that performs no better than randomly selecting compounds will have the active molecules evenly distributed throughout the ranking list. Most similarity methods lie between the two extremes (though there are sometimes cases where the results are worse than random!). The results from such experiments are often represented as enrichment plots (**Figure 15**).

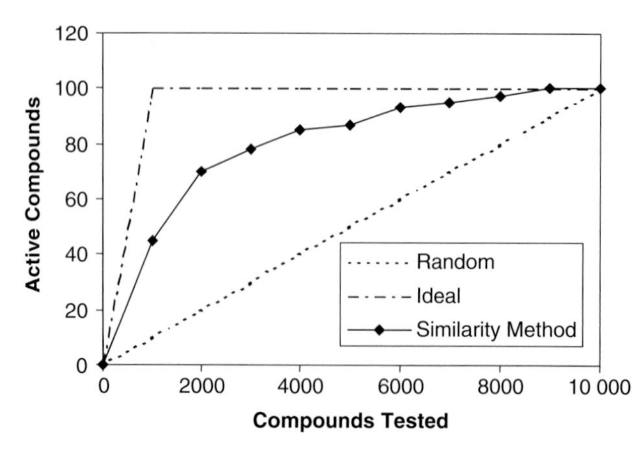

Figure 15 Illustrative example of an enrichment plot used in virtual screening showing the typical performance of a similarity-based selection method compared to random and the ideal. Redrawn from Leach, A. R.; Gillet, V. J. *An Introduction to Chemoinformatics*; Kluwer Academic Publishers: Dordrecht, the Netherlands, 2004, with kind permission of Springer Science and Business Media.

One of the on-going debates within the computational chemistry community concerns the relative performance of 3D similarity methods compared to those based upon molecular graph representations (so-called 2D methods). The latter only consider the atoms in the molecule and the way they are bonded together and none of the conformational properties. As molecular recognition is based on 3D properties one would expect that virtual screening experiments performed using the 3D methods would give superior results to those for 2D methods. However, this is often found not to be the case. A particularly celebrated and much-debated study was performed by Brown and Martin who assessed the ability of a series of 2D and 3D methods to retrieve active and inactive compounds using different clustering techniques.[199] One of the 2D methods was found to be most effective, a similar conclusion being obtained when the methods were assessed in a series of property prediction experiments[200] or in studies performed by other groups.[201] Several reasons have been suggested to try and explain these results. Some of these focus on the databases used for the evaluation and the possibility that they may contain a disproportionate number of close analogs of the query molecules. Such analogs have a high inherent degree of 2D similarity and so introduce a bias in favor of such techniques. The poorer performance of 3D methods may also be due to inadequate representations or incomplete handling of conformational flexibility. One consistent finding, however, is that 3D methods generally find structurally different compounds, which are therefore complementary to the 2D approaches and may in some cases represent a more significant jump in chemical space.

4.05.7 Summary and Conclusions

In this chapter we have surveyed a broad range of techniques that are widely used in computational drug discovery. Some of these methods trace their origins to the earliest days of computational molecular science, and continue to present challenges to the computational chemistry community. As computer power has increased so too has the desire to study larger and more complex systems, or to apply the methods to much larger collections of compounds in virtual screening experiments. New force fields that include polarization and better implicit solvation models will be one area where there continues to be work to do and will hopefully bring new developments in the coming years. Of the other areas considered here, better methods for assessing 3D similarity would be one area where new ideas would have a significant practical impact on drug discovery.

References

1. Watson, J. D.; Crick, F. H. C. *Nature* **1953**, *171*, 737 – 738.
2. Leach, A. R. *Molecular Modelling: Principles and Applications*; Prentice-Hall: Harlow, UK, 2001.
3. Burkert, U.; Allinger, N. L. *Molecular Mechanics*; ACS Monograph 177; American Chemical Society: Washington, DC, 1982.
4. Niketic, S. R.; Rasmussen, K. *The Consistent Force Field: A Documentation*; Springer-Verlag: Berlin, Germany, 1977.
5. Hwang, M. J.; Stockfisch, T. P.; Hagler, A. T. *1994. J. Am. Chem. Soc.* **1994**, *116*, 2515–2525.
6. Cox, S. R.; Williams, D. E. *J. Comput. Chem* **1981**, *2*, 304–323.
7. Chirlian, L. E.; Francl, M. M. *J. Comput. Chem.* **1987**, *8*, 894–905.

8. Singh, U. C.; Kollman, P. A. *J. Comput. Chem.* **1984**, *5*, 129–145.
9. Williams, D. E. Net Atomic Charge and Multipole Models for the Ab Initio Molecular Electric Potential. In *Reviews in Computational Chemistry 2*; Lipkowitz, K. B., Boyd, D. B., Eds.; VCH Publishers: New York, 1991.
10. Bayly, C. I.; Cieplak, P.; Cornell, W. D.; Kollman, P. A. *J. Phys. Chem.* **1993**, *97*, 10269–10280.
11. Cornell, W. D.; Cieplak, P.; Bayly, C. I.; Kollman, P. A. *J. Am. Chem. Soc.* **1993**, *115*, 9620–9631.
12. Wang, J.; Cieplak, P.; Kollman, P. A. *J. Comput. Chem.* **2000**, *21*, 1049–1074.
13. Williams, D. E. *Biopolymers* **1990**, *29*, 1367–1386.
14. Reynolds, C. A.; Essex, J. W.; Richards, W. G. *J. Am. Chem. Soc.* **1992**, *114*, 9075–9079.
15. Luque, F. J.; Ilas, F.; Orozco, M. *J. Comput. Chem.* **1990**, *11*, 416–430.
16. Bezler, B. H.; Merz, K. M., Jr.; Kollman, P. A. *J. Comput. Chem.* **1990**, *11*, 431–439.
17. Aleman, C.; Luque, F.; Orozco, M. *J. Comput. Chem.* **1993**, *14*, 799–808.
18. Ferenczy, G. G.; Reynolds, C. A.; Richards, W. G. *J. Comput. Chem.* **1990**, *11*, 159–169.
19. Storer, J. W.; Giesen, D. J.; Cramer, C. J.; Truhlar, D. G. *J. Comput.-Aided. Mol. Des.* **1995**, *9*, 87–110.
20. Li, J.; Zhu, T.; Cramer, C. J.; Truhlar, D. G. *J. Phys. Chem. A* **1998**, *102*, 1820–1831.
21. Thompson, J. D.; Cramer, C. J.; Truhlar, D. G. *J. Comput. Chem.* **2003**, *24*, 1291–1304.
22. Jakalian, A.; Bush, B. L.; Jack, D. B.; Bayly, C. I. *J. Comput. Chem.* **2000**, *21*, 132–146.
23. Jakalian, A.; Jack, D. B.; Bayly, C. I. *J. Comput. Chem.* **2000**, *23*, 1623–1641.
24. Gasteiger, J.; Marsili, M. *Tetrahedron* **1980**, *36*, 3219–3288.
25. Rappé, A. K.; Goddard, W. A., III. *J. Phys. Chem.* **1991**, *95*, 3358–3363.
26. Hunter, C. A.; Saunders, J. K. M. *J. Am. Chem. Soc.* **1990**, *112*, 5525–5534.
27. Vinter, J. G. *J. Comput.-Aided. Mol. Des.* **1994**, *8*, 653–668.
28. Chessari, G.; Hunter, C. A.; Low, C. M. R.; Packer, M. J.; Vinter, J. G.; Vonta, C. *Chemistry – Eur. J.* **2002**, *8*, 2860–2867.
29. Stone, A. J. *Chem. Phys. Lett.* **1981**, *83*, 233–239.
30. Stone, A. J.; Alderton, M. *Mol. Phys.* **1985**, *56*, 1047–1064.
31. Stone, A. J. *The Theory of Intermolecular Forces*; Oxford University Press: Oxford, 1996.
32. Stone, A. J. *J. Chem. Theory Comput.* **2005**, *1*, 1128–1132.
33. Halgren, T. A.; Damm, W. *Curr. Opin. Struct. Biol.* **2001**, *11*, 236–242.
34. Rick, S. W.; Stuart, S. J. Potentials and Algorithms for Incorporating Polarizability in Computer Simulations. In *Reviews in Computational Chemistry 18*; Lipkowitz, K. B., Boyd, D. B., Eds.; VCH Publishers: New York, 2002, pp 89–146.
36. Rick, S. W.; Stuart, S. J.; Berne, B. J. *J. Chem. Phys.* **1994**, *101*, 6141–6156.
37. Stern, H. A.; Kaminski, G. A.; Banks, J. L.; Zhou, R.; Berne, B. J.; Friesner, R. A. *J. Phys. Chem. B* **1999**, *103*, 4730–4737.
38. Maple, J. R.; Yixiang, C.; Damm, W.; Halgren, T. A.; Kaminski, G. A.; Zhang, L. Y.; Friesner, R. A. *J. Chem. Theory Comput.* **2005**, *1*, 694–705.
39. Harder, E.; Kim, B.; Friesner, R. A.; Berne, B. K. *J. Chem. Theory Comput.* **2005**, *1*, 169–180.
40. Jedlovsky, P.; Richardi, J. *J. Chem. Phys.* **1999**, *110*, 8019–8031.
41. Kaminski, G. A.; Stern, H. A.; Berne, B. J.; Friesner, R. A.; Cao, Y. X.; Murphy, R. B.; Zhou, R.; Halgren, T. A. *J. Comput. Chem.* **2002**, *23*, 1515–1531.
42. Kaminski, G. A.; Stern, H. A.; Berne, B. J.; Friesner, R. A. *J. Phys. Chem.* **2004**, *108*, 621–627.
43. Maple, J. R.; Yixiang, C.; Damm, W.; Halgren, T. A.; Kaminski, G. A.; Zhang, L. Y.; Friesner, R. A. *J. Chem. Theory Comput.* **2005**, *1*, 694–705.
44. Edinger, S. R.; Cortis, C.; Shenkin, P. S.; Friesner, R. A. *J. Phys. Chem. B* **1997**, *7*, 1190–1197.
45. Simonson, T. *Curr. Opin. Struct. Biol.* **2001**, *11*, 243–252.
46. Born, M. *Z. für Physik* **1920**, *1*, 45–48.
47. Onsager, L. *J. Am. Chem. Soc.* **1936**, *58*, 1486–1493.
48. Still, W. C.; Tempczyrk, A.; Hawley, R. C.; Hendrickson, T. *J. Am. Chem. Soc.* **1990**, *112*, 6127–6129.
49. Qiu, D.; Shenkin, P. S.; Hollinger, F. P.; Still, W. C. *J. Phys. Chem.* **1997**, *101*, 3005–3014.
50. Ghosh, A.; Rapp, C. S.; Friesner, R. A. *J. Phys. Chem. B* **1998**, *102*, 10983–10990.
51. Onufriev, A.; Bashford, D.; Case, D. A. *J. Phys. Chem. B* **2000**, *104*, 3712–3720.
52. David, L.; Luo, R.; Gilson, M. K. *J. Comput. Chem.* **2000**, *21*, 295–309.
53. Nymeyer, H.; García, A. E. *Proc. Natl. Acad. Sci. USA* **2003**, *100*, 13934–13939.
54. Onufriev, A.; Case, D. A.; Bashford, D. *J. Comput. Chem.* **2002**, *23*, 1297–1304.
55. Cramer, C. J.; Truhlar, D. G. *J. Am. Chem. Soc.* **1991**, *113*, 8305–8311.
56. Cramer, C. J.; Truhlar, D. G. *Science* **1992**, *256*, 213–217.
57. Hawkins, G. D.; Cramer, C. J.; Truhlar, D. G. *J. Phys. Chem.* **1996**, *100*, 19824–19839.
58. Chambers, C. C.; Hawkins, G. D.; Cramer, C. J.; Truhlar, D. G. *J. Phys. Chem.* **1996**, *100*, 16385–16398.
59. Cramer, C. J.; Truhlar, D. G. Continuum Solvation Models: Classical and Quantum Mechanical Implementations. In *Reviews in Computational Chemistry 6*; Lipkowitz, K. B., Boyd, D. B., Eds.; VCH Publishers: New York, 1995, pp 1–72.
60. Cramer, C. J.; Truhlar, D. G. *Chem. Rev.* **1999**, *99*, 2161–2200.
61. Lamm, G. The Poisson–Boltzmann Equation. In *Reviews in Computational Chemistry 19*; Lipkowitz, K. B., Boyd, D. B., Eds.; VCH Publishers: New York, 2003, pp 147–365.
62. Warwicker, J.; Watson, H. C. *J. Mol. Biol.* **1982**, *157*, 671–679.
63. Honig, B.; Nicholls, A. *Science* **1995**, *268*, 1144–1149.
64. Gilson, M. K.; Honig, B. *Proteins: Struct. Funct. Genet.* **1988**, *4*, 7–18.
65. Srinivasan, J.; Cheatham, T. E.; Cieplak, P.; Kollman, P. A.; Case, D. A. *J. Am. Chem. Soc.* **1998**, *120*, 9401–9409.
66. Massova, I.; Kollman, P. A. *Perspect. Drug Disc. Dev.* **2000**, *18*, 113–135.
67. Kollman, P. A.; Massova, I.; Reyes, C.; Kuhn, B.; Huo, S.; Chong, L.; Lee, M.; Lee, T.; Duan, Y.; Wang, W. et al. *Acc. Chem. Res.* **2000**, *33*, 889–897.
68. Kuhn, B.; Gerber, P.; Schulz-Gasch, T.; Stahl, M. *J. Med. Chem.* **2005**, *48*, 4040–4048.
69. Friedrichs, M.; Zhou, R.; Edinger, S.; Friesner, R. A. *J. Phys. Chem. B* **1999**, *103*, 3057–3061.
70. Holst, M.; Baker, N.; Wang, F. *J. Comput. Chem.* **2000**, *21*, 1319–1342.
71. London, F. *Z. für Physik* **1930**, *63*, 245–279.
72. Halgren, T. A. *J. Am. Chem. Soc.* **1992**, *114*, 7827–7843.

73. Halgren, T. A. *J. Comput. Chem.* **1996**, *17*, 490–519.
74. Halgren, T. A. *J. Comput. Chem.* **1996**, *17*, 520–552.
75. Hill, T. L. *J. Chem. Phys.* **1948**, *16*, 399–404.
76. Darden, T. A.; Perera, L.; Li, L.; Pedersen, L. *Struct. Fold Des.* **1999**, 7, R55–R60.
77. Darden, T. A.; York, D.; Pedersen, L. *J. Chem. Phys.* **1993**, *98*, 10089–10092.
78. Luty, B. A.; David, M. E.; Tironi, I. G.; van Gunsteren, W. F. *Mol. Simul.* **1994**, *14*, 11–20.
79. Luty, B. A.; Tironi, I. G.; van Gunsteren, W. F. *J. Chem. Phys.* **1995**, *103*, 3014–3021.
80. Allinger, N. L. *J. Am. Chem. Soc.* **1977**, *99*, 8127–8134.
81. Allinger, N. L.; Sprague, J. T. *J. Am. Chem. Soc.* **1973**, *95*, 3893–3907.
82. Sprague, J. T.; Tai, J. C.; Yuh, Y.; Allinger, N. L. *J. Comput. Chem.* **1987**, *8*, 581–603.
83. Allinger, N. L.; Yuh, Y. H.; Lii, J. J. *J. Am. Chem. Soc.* **1989**, *111*, 8551–9556.
84. Lii, J.-H.; Allinger, N. L. *J. Am. Chem. Soc.* **1989**, *111*, 8566–8582.
85. Allinger, N. L.; Li, F.; Yan, L. *J. Comput. Chem.* **1990**, *11*, 848–867.
86. Allinger, N. L.; Li, F.; Yan, L.; Tai, J. C. *J. Comput. Chem.* **1990**, *11*, 868–895.
87. Allinger, N. L.; Yan, L. *J. Am. Chem. Soc.* **1993**, *115*, 11918–11925.
88. Allinger, N. L.; Chen, K.; Lii, J.-H. *J. Comput. Chem.* **1996**, *17*, 642–668.
89. Allinger, N. L.; Chen, K.; Katzenelenbogen, J. A.; Wilson, S. R.; Anstead, G. M. *J. Comput. Chem.* **1996**, *17*, 747–755.
90. Nevins, N.; Chen, K.; Allinger, N. L. *J. Comput. Chem.* **1996**, *17*, 669–694.
91. Nevins, N.; Chen, K.; Allinger, N. L. *J. Comput. Chem.* **1996**, *17*, 695–729.
92. Nevins, N.; Chen, K.; Allinger, N. L. *J. Comput. Chem.* **1996**, *17*, 730–746.
93. Weiner, S. J.; Kollman, P. A.; Case, D. A.; Singh, U. C.; Ghio, C.; Alagona, G.; Profeta, S.; Weiner, P. *J. Am. Chem. Soc.* **1984**, *106*, 765–784.
94. Weiner, S. J.; Kollman, P. A.; Nguyen, D. T.; Case, D. A. *J. Comput. Chem.* **1986**, 7, 230–252.
95. Cornell, W. D.; Cieplak, P.; Bayly, C. I.; Gould, I. R.; Merz, K. M., Jr.; Ferguson, D. M.; Spellmeyer, D. C.; Fox, T.; Caldwell, J. W.; Kollman, P. A. *J. Am. Chem. Soc.* **1995**, *117*, 5179–5197.
96. Wang, J.; Wolf, R. M.; Caldwell, J. W.; Kollman, P. A.; Case, D. A. *J. Comput. Chem.* **2004**, *25*, 1157–1174.
97. Brooks, B. R.; Bruccoleri, R. E.; O'Lafson, B. D.; States, D. J.; Swaminathan, D.; Karplus, M. J. *J. Comput. Chem.* **1983**, *4*, 187–217.
98. MacKerell, A. D., Jr.; Bashford, D.; Bellott, M.; Dunbrack, R. L., Jr.; Evanseck, J. D.; Field, M. J.; Fischer, S.; Gao, J.; Guo, H.; Ha, S. *J. Phys. Chem. B* **1998**, *102*, 3586–3616.
99. Daura, X.; Mark, A. E.; van Gunsteren, W. F. *J. Comp. Chem.* **1998**, *19*, 535–547.
100. Hwang, M. J.; Stockfish, T. P.; Hagler, A. T. *J. Am. Chem. Soc.* **1994**, *116*, 2515–2525.
101. Maple, J. R.; Hwang, M.-J.; Stockfisch, T. P.; Dinur, U.; Waldman, M.; Ewig, C. S.; Hagler, A. T. *J. Comput. Chem.* **1994**, *15*, 162–182.
102. Ewig, C. S.; Berry, R.; Dinur, U.; Hill, J.-R.; Hwang, M.-J.; Li, H.; Liang, C.; Maple, J.; Peng, Z.; Stockfish, T. P. et al. *J. Comput. Chem.* **2001**, *22*, 1782–1800.
103. Maple, J. R.; Dinur, U.; Hagler, A. T. *Proc. Natl. Acad. Sci. USA* **1988**, *69*, 5350–5353.
104. Sun, H. *J. Phys. Chem. B* **1998**, *102*, 7338–7364.
105. Lifson, S.; Hagler, A. T.; Dauber, P. *J. Am. Chem. Soc.* **1979**, *101*, 5111–5121.
106. Dauber-Osguthorpe, P.; Roberts, V. A.; Osguthorpe, D. J.; Wolff, J.; Genest, M.; Hagler, A. T. *Proteins: Struct. Funct. Genet.* **1988**, *4*, 31–47.
107. Rappé, A. K.; Casewit, C. J.; Colwell, K. S.; Goddard, W. A., III; Skiff, W. M. *J. Am. Chem. Soc.* **1992**, *114*, 10024–10035.
108. Mayo, S. L.; Olafson, B. D.; Goddard, W. A., Jr. *J. Phys Chem* **1990**, *94*, 8897–8909.
109. Clark, M.; Cramer, R. D., Jr.; Van Opdenbosch, N. *J. Comput. Chem.* **1989**, *10*, 982–1012.
110. Halgren, T. A. *J. Comput. Chem.* **1996**, *17*, 490–519.
111. Jorgensen, W. L.; Tirado-Rives, J. *J. Am. Chem. Soc.* **1988**, *110*, 1657–1666.
112. Jorgensen, W. L.; Maxwell, S. D.; Tirado-Rives, J. *J. Am. Chem. Soc.* **1996**, *118*, 11225–11236.
113. Kaminski, G. A.; Friesner, R. A.; Tirado-Rives, J.; Jorgensen, W. L. *J. Phys. Chem. B* **2001**, *105*, 6474–6487.
114. Bowen, J. P.; Allinger, N. L. Molecular Mechanics: The Art and Science of Parameterisation. In *Reviews in Computational Chemistry 2*; Lipkowitz, K. B., Boyd, D. B., Eds.; VCH Publishers: New York, 1991.
115. Lifson, S.; Warshel, A. *J. Chem. Phys.* **1968**, *49*, 5116–5129.
116. Mackerell, A. D. *J. Comput. Chem.* **2004**, *25*, 1584–1604.
117. Ponder, J. W.; Case, D. A. *Adv. Protein Chem.* **2003**, *66*, 27–85.
118. Jorgensen, W. L.; Tirado-Reeves, J. *Proc. Natl. Acad. Sci. USA* **2005**, *102*, 6665–6670.
119. MacKerell, A. D. Empirical Force Fields for Proteins: Current Status and Future Directions. In *Annual Reports in Computational Chemistry*; Spellmeyer, D. C., Ed.; Elsevier: Amsterdam, 2005, pp 91–102.
120. Halgren, T. A. *J. Comput. Chem.* **1999**, *20*, 730–748.
121. Ramachandran, G. N.; Ramakrishnan, C.; Sasiekharan, V. *J. Mol. Biol.* **1963**, 7, 95–99.
122. Schlick, T. Optimization Methods in Computational Chemistry. In *Reviews in Computational Chemistry 3*; Lipkowitz, K. B., Boyd, D. B., Eds.; VCH Publishers: New York, 1992.
123. Gill, P. E.; Murray, W. *Practical Optimization*; Academic Press: London, UK, 1981.
124. Press, W. H.; Flannery, B. P.; Teukolsky, S. A.; Vetterling, W. T. *Numerical Recipes in Fortran*; Cambridge University Press: Cambridge, UK, 1992.
125. Press, W. H.; Teukolsky, S. A.; Vetterling, W. T.; Flannery, B. P. *Numerical Recipes in C++*; Cambridge University Press: Cambridge, UK, 2002.
126. Schlegel, H. B. *J. Comput. Chem.* **2002**, *24*, 1514–1527.
127. Brooks, B.; Karplus, M. *Proc. Natl. Acad. Sci. USA* **1983**, *80*, 6571–6575.
128. Leach, A. R. A Survey of Methods for Searching the Conformational Space of Small and Medium-Sized Molecules. In *Reviews in Computational Chemistry 2*; Lipkowitz, K. B., Boyd, D. B., Eds.; VCH Publishers: New York, 1991, pp 1–55.
129. Goodman, J. M.; Still, W. C. *J. Comput. Chem.* **1991**, *12*, 1110–1117.
130. Li, Z. Q.; Scheraga, H. A. *Proc. Natl. Acad. Sci. USA* **1987**, *84*, 6611–6615.
131. Saunders, M. *J. Am. Chem. Soc.* **1987**, *109*, 3150–3152.
132. Ferguson, D. M.; Raber, D. J. *J. Am. Chem. Soc.* **1989**, *111*, 4371–4378.
133. Chang, G.; Guida, W. C.; Still, W. C. *J. Am. Chem. Soc.* **1989**, *111*, 4379–4386.
134. Kolossváry, I.; Guida, W. C. *J. Am. Chem. Soc.* **1996**, *118*, 5011–5019.
135. Crippen, G. M. *Distance Geometry and Conformational Calculations*. Chemometrics Research Studies Series 1; Wiley: New York, 1981.

136. Crippen, G. M; Havel, T. F. *Distance Geometry and Molecular Conformation* Chemometrics Research Studies Series 15; Wiley: New York, 1988.
137. Blaney, J. M.; Dixon, J. S. Distance Geometry in Molecular Modeling. In *Reviews in Computational Chemistry 5*; Lipkowitz, K. B., Boyd, D. B., Eds.; VCH Publishers: New York, 1994, pp 299–335.
138. Goldberg, D. E. *Genetic Algorithms in Search, Optimization and Machine Learning*. Addison-Wesley: Reading, MA.
139. Clark, D. E.; Westhead, D. R. *J. Comput.-Aided. Mol. Des.* **1996**, *10*, 337–358.
140. Judson, R. Genetic Algorithms and Their Use in Chemistry. In *Reviews in Computational Chemistry 10*; Lipkowitz, K. B., Boyd, D. B., Eds.; VCH Publishers: New York, 1997, pp 1–73.
141. Jones, G. Genetic and Evolutionary Algorithms. In *The Encyclopedia of Computational Chemistry*; Schleyer, P. V. R., Allinger, N. L., Clark, T., Gasteiger, J., Kollman, P. A., Schaefer, H. F., III., Schreiner, P. R., Eds.; John Wiley & Sons: Chichester, UK, 1998.
142. Clark, D. E., Ed. *Evolutionary Algorithms in Molecular Design*; Wiley-VCH: Weinheim, Germany, 2000.
143. McGarrah, D. B.; Judson, R. S. *J. Comput. Chem.* **1993**, *14*, 1385–1395.
144. Judson, R. S.; Jaeger, W. P.; Treasurywala, A. M.; Peterson, M. L. *J. Comput. Chem.* **1993**, *14*, 1407–1414.
145. Meza, J. C.; Judson, R. S.; Faulkner, T. R.; Treasurywala, A. M. *J. Comput. Chem.* **1996**, *17*, 1142–1151.
146. Kirkpatrick, S.; Gelatt, C. D.; Vecchi, M. P. *Science* **1983**, *220*, 671–680.
147. Dolata, D. P.; Carter, R. E. *J. Chem. Inf. Comput. Sci.* **1987**, *27*, 36–47.
148. Dolata, D. P.; Leach, A. R.; Prout, K. *J. Comput.-Aided. Mol. Des.* **1987**, *1*, 73–85.
149. Leach, A. R.; Prout, K.; Dolata, D. P. *J. Comput.-Aided. Mol. Des.* **1988**, *2*, 107–123.
150. Klebe, G.; Mietzner, T. *J. Comput.-Aided. Mol. Des.* **1994**, *8*, 583–606.
151. Fueston, B. P.; Miller, M. D.; Culberson, J. C.; Nachbar, R. B.; Kearsley, S. K. *J. Chem. Inf. Comput. Sci.* **2001**, *41*, 754–763.
152. OMEGA reference. OpenEye Scientific Software, 3600 Cerrillos Rd. Suite 1107, Santa Fe, NM 87507, USA.
153. Allen, F. H. *Acta Crystallogr, Sect. B* **2002**, *58*, 380–388.
154. Ferro, D. R.; Hermans, J. *Acta Crystallogr., Sect. A* **1977**, *33*, 345–347.
155. Kabsch, W. *Acta Crytallogr., Series A* **1978**, *34*, 827–828.
156. Mackay, A. L. *Acta Crystallogr., Series A* **1984**, *40*, 165–166.
157. Smellie, A. S.; Kahn, S. D.; Teig, S. L. *J. Chem. Inf. Comput. Sci.* **1995**, *35*, 285–294.
158. Smellie, A. S.; Teig, W. L.; Towbin, P. *J. Comput. Chem.* **1995**, *16*, 171–187.
159. Parish, C.; Lombardi, R.; Sinclair, K.; Smith, E.; Goldberg, A.; Rappleye, M.; Dure, M. *J. Mol. Graph. Model.* **2002**, *21*, 129–150.
160. Saunders, M.; Houk, K. N.; Wu, Y.-D.; Still, W. C.; Lipton, M.; Chang, G.; Guida, W. C. *J. Am. Chem. Soc.* **1990**, *112*, 1419–1427.
161. Bernstein, F. C.; Koetzle, T. F.; Williams, G. J. B.; Meyer, E.; Bryce, M. D.; Rogers, J. R.; Kennard, O.; Shikanouchi, T.; Tasumi, M. *J. Mol. Biol.* **1977**, *112*, 535–542.
162. Bostrom, J. *J. Comput.-Aided Mol. Des.* **2001**, *15*, 1137–1152.
163. Bostrom, J.; Greenwood, J. R.; Gottfries, J. *J. Mol. Graph. Model.* **2003**, *21*, 449–462.
164. Boström, J.; Norrby, P.-O.; Liljefors, T. *J. Comput.-Aided Mol. Des.* **1998**, *12*, 383–396.
165. Perola, E.; Charifson, P. S. *J. Med. Chem.* **2004**, *47*, 2499–2510.
166. Rusinko, A., Jr. *Tools for Computer-Assisted Drug Design. Ph.D. Thesis*; University of Texas at Austin: Austin, TX, 1988.
167. Pearlman, R. S. *Chem. Des.Autom. News* **1987**, *2*, 1–2.
168. Gasteiger, J.; Rudolph, C.; Sadowski, J. *Tetrahedron Comput. Methodol.* **1990**, *3*, 537–547.
169. Sadowski, J.; Rudolph, C.; Gasteiger, J. *Anal. Chim. Acta* **1992**, *265*, 233–241.
170. Sadowski, J.; Gasteiger, J. *Chem. Rev.* **1993**, *93*, 2567–2581.
171. Sadowski, J. Three-Dimensional Structure Generation: Automation. In *Encyclopedia of Computational Chemistry*; Schleyer, P. V. R., Allinger, N. L., Clark, T., Gasteiger, J., Kollman, P. A., Schaefer, H. F., III, Schreiner, P. R., Eds.; John Wiley & Sons: Chichester, UK, 1998, pp 2976–2988.
172. Allen, F. H.; Kennard, O.; Watson, D. G.; Brammer, L.; Orpen, A. G.; Taylor, R. *J. Chem. Soc., Perkin Trans. 2* **1987**, S1–S19.
173. Lemmen, C.; Lengauer, T. *J. Comput-Aided Mol. Des.* **2000**, *14*, 215–232.
174. Good, A. C.; Richards, W. G. *Perspect. Drug Disc. Des.* **1998**, *9/10/11*, 321–338.
175. Leach, A. R.; Gillet, V. J. *An Introduction to Chemoinformatics*; Kluwer Academic Publishers: Dordrecht, the Netherlands, 2004.
176. Carbo, R.; Leyda, L.; Arnau, M. *Int. J. Quantum. Chem.* **1980**, *17*, 1185–1189.
177. Hodgkin, E. E.; Richards, W. G. *Int. J. Quantum Chem. Quantum. Biol. Symp.* **1987**, *14*, 105–110.
178. Namasivayam, S.; Dean, P. M. *J. Mol. Graphics* **1986**, *4*, 46–50.
179. Sanz, F.; Manaut, F.; Rodriguez, J.; Loyoza, E.; Ploez-de-Brinao, E. *J. Comput.-Aided Mol. Des.* **1993**, 7, 337–347.
180. Kearsley, S. K.; Smith, G. M. *Tetrahedron Comput. Methodol.* **1990**, *3*, 615–633.
181. Good, A. C.; Hodgkin, E. E.; Richards, W. G. *J. Chem. Inf. Comput. Sci.* **1993**, *32*, 188–192.
182. Good, A. C.; Richards, W. G. *J. Chem. Inf. Comput. Sci.* **1993**, *33*, 112–116.
183. Grant, J. A.; Pickup, B. T. *J. Phys. Chem.* **1995**, *99*, 3503–3510.
184. Grant, J. A.; Gallardo, M. A.; Pickup, B. T. *J. Comput. Chem.* **1996**, *17*, 1653–1666.
185. Chau, P.-L.; Dean, P. M. *J. Mol. Graphics* **1987**, *5*, 97–100.
186. van Geerestein, V. J.; Perry, N. C.; Grootenhuis, P. G.; Haasnoot, C. A. G. *Tetrahedron. Comput. Methodol.* **1990**, *3*, 595–613.
187. Perry, N. C.; van Geerestein, V. J. *J. Chem. Inf. Comput. Sci.* **1992**, *32*, 607–616.
188. Blaney, F. E.; Finn, P.; Phippen, R. W.; Wyatt, M. *J. Mol. Graphics* **1993**, *11*, 98–105.
189. Blaney, F. E.; Edge, C.; Phippen, R. W. *J. Mol. Graphics* **1995**, *13*, 165–174.
190. Good, A. C.; Hodgkin, E. E.; Richards, W. G. *J. Chem. Inf. Comput. Sci.* **1992**, *32*, 188–192.
191. McMahon, A. J.; King, P. M. *J. Comput. Chem.* **1997**, *18*, 151–158.
192. Parretti, M. F.; Kroemer, R. T.; Rothman, J. H.; Richards, W. G. *J. Comput. Chem.* **1997**, *18*, 1344–1353.
193. Wild, D.-J.; Willett, P. *J. Chem. Inf. Comput. Sci.* **1996**, *36*, 159–167.
194. Thorner, D. A.; Wild, D. J.; Willett, P.; Wright, P. M. *J. Chem. Inf. Comput. Sci.* **1996**, *36*, 900–908.
195. Miller, M. D.; Sheridan, R. P.; Kearsley, S. K. *J. Med. Chem.* **1999**, *42*, 1505–1514.
196. Miller, M. D.; Fluder, E. M.; Castonguay, L. A.; Culberson, J. C.; Mosley, R. T.; Prendergast, K.; Kearsley, S. K.; Sheridan, R. P. *Med. Chem. Res.* **1999**, *9*, 513–534.
197. Lemmen, C.; Lengauer, T. *J. Comput.-Aided. Mol. Des.* **1997**, *11*, 357–368.
198. Lemmen, C.; Lengauer, T.; Klebe, G. *J. Med. Chem.* **1998**, *41*, 4502–4520.

199. Brown, R. D.; Martin, Y. C. *J. Chem. Inf. Comput. Sci.* **1996**, *36*, 572–583.
200. Brown, R. D.; Martin, Y. C. *J. Chem. Inf. Comput. Sci.* **1997**, *37*, 1–9.
201. Matter, H.; Pötter, T. *J. Chem. Inf. Comput. Sci.* **1999**, *39*, 1211–1225.
202. Adler, M.; Davey, D. D.; Phillips, G. B.; Kim, S. H.; Jancarik, J.; Rumennik, G.; Light, D. L.; Whitlow, M. *Biochemistry* **2000**, *39*, 12534–12542.
203. Laskowski, R. A.; MacArthur, M. W.; Moss, D. S.; Thornton, J. M. *J. Appl. Crystallogr.* **1993**, *26*, 283–291.

Biography

Andrew R Leach joined Glaxo (now GlaxoSmithKline) in 1994 following postdoctoral research at the University of California, San Francisco and an academic position at the University of Southampton (UK). During his career he has been heavily involved in the development of new computational and cheminformatics methodologies to support drug discovery in areas such as library design, compound acquisition, and structure-based design. He has a BA degree and a PhD in Chemistry from the University of Oxford and is the author of a widely used textbook on molecular modeling. He is currently Director of the GSK Computational Chemistry department in the UK.

© 2007 Elsevier Ltd. All Rights Reserved
No part of this publication may be reproduced, stored in any retrieval system or transmitted in any form by any means electronic, electrostatic, magnetic tape, mechanical, photocopying, recording or otherwise, without permission in writing from the publishers

Comprehensive Medicinal Chemistry II
ISBN (set): 0-08-044513-6

ISBN (Volume 4) 0-08-044517-9; pp. 87–118

4.06 Pharmacophore Modeling: 1 – Methods

Y C Martin, Abbott Laboratories, Abbott Park, IL, USA

© 2007 Elsevier Ltd. All Rights Reserved.

4.06.1 Overview

The word pharmacophore takes on two separate definitions: The first, attributed to Ehrlich, is defined as the set of atoms that a compound must possess for it to be active in a particular biological test.[1] Even today, a century later, this usage of the word is common with medicinal chemists. However, in this chapter three-dimensional (3D) pharmacophores will be explored in detail. A 3D pharmacophore is defined as that set of properties and their arrangement in 3D that a compound must possess for it to be active in a particular biological test.[1,2] As will be seen later in the chapter, some computer methods refine this definition to include a larger set of features of which a molecule needs to match only a subset in order to be active. Implicit in the definition is that all molecules that contain the pharmacophore bind to the target biomolecule in the same manner.

3D database searching, as the name implies, involves searching a database of molecules to discover those that match the 3D properties of the search query. In this chapter, emphasis will be placed on the type of search that relies on the geometric properties of molecules such as distances and angles between points within these molecules. Pure shape-based 3D searching and superposition are discussed in Chapters 4.05 and 4.19.

Pharmacophores form a key strategy in the ligand-based design of biologically active molecules. On the one hand, they typically are a part of 3D quantitative structure–activity relationship (3D QSAR) models derived to forecast potency. The pharmacophore identification can be performed either as a prelude to the actual QSAR analysis or with pharmacophore discovery coupled with the QSAR into one computer program. Additionally, pharmacophores form the query for 3D database searches that have the goal of discovering novel compounds with the desired biological properties. Lastly, one may use a pharmacophore model to superimpose compounds from two quite different series so as to use information from the better-explored series to propose where modifications in the less-explored series will be optimal and where they will not be tolerated.

Table 1 summarizes the ideal information available for pharmacophore detection. It emphasizes that pharmacophore perception and 3D searching represent an interplay between experimental data and computer algorithms, often reflected as collaboration between medicinal and computational chemists. From the experimental side, one must gather a list of all active compounds. If the compounds are conformationally flexible, it is often helpful for conformationally constrained analogs to be synthesized. The design of these analogs may be based on a preliminary pharmacophore model or simple conformational analysis plus an optional 3D database search of the active leads. Because a pharmacophore is assumed to contain the minimum set of features necessary for biological activity, it may be necessary to synthesize analogs to probe this question. Again, a preliminary pharmacophore model might aid the design of such analogs; the model would propose which features could perhaps be eliminated. A pharmacophore model can suggest which enantiomer of a chiral compound matches the other compounds in the model or can be built using the knowledge of which enantiomer is active. All of these points emphasize that the more experimental information is available for the modeling, the more precise the pharmacophore model can become.

However, the experimental and computational work might lead one to discover that all series cannot be included in one model, and to propose that they bind differently to the target and that the structure–activity relationship (SAR) of the two apparent series will not be compatible. At every step along the way, even in the absence of a precise pharmacophore model, the medicinal chemist is gaining insight into the 3D SAR of the compounds of interest. The computational chemist adds information by making the observations more precise, more easily visualized, and more easily transposed into novel compounds either designed or identified in a database.

Table 1 Source of information needed for pharmacophore mapping

Information needed	*Experimental source*	*Computational source*
Set of ligands	In-house and literature data	Literature mining
Points to consider matching	SAR that examines presence or absence of groups	Known types and geometries of intermolecular interactions
Bioactive conformation	Conformationally constrained analogs	Conformational search either within the program or as a prelude to the pharmacophore search
Bioactive stereoisomer	Bioactivity of stereoisomers	Model may predict
Is the pharmacophore correct?	Test activity of compounds predicted to fit the pharmacophore	Does it explain compounds not included? Provide predictions for testing

4.06.2 Pharmacophore Detection[3,4]

4.06.2.1 Overview of the Steps

Figure 1 illustrates the process of determining a pharmacophore. After the candidate molecules are assembled, often a choice is made as to which compounds to examine with the algorithm and which are expected to not add additional information. This choice is made based on the properties of the molecules, both individually and as a set. Once an algorithm identifies one or several pharmacophores, additional molecules are examined for fit and perhaps previously untested molecules are selected for testing.

The algorithms fall into three broad categories. The one of choice for a particular problem depends on the properties of the compounds as well as the software available. Section 4.06.2.6 discusses programs that suggest a bioactive conformation given the set of matching points; Section 4.06.2.7 discusses programs that are devoted to discovering both the matching points and the proposed bioactive conformation; and Section 4.06.2.8 discusses programs that not only determine the matching points and the proposed bioactive conformation but also perform a 3D QSAR analysis on the resulting overlays.

Every pharmacophore hypothesis represents a compromise of different, often conflicting, criteria that reflect the quality of the hypothesis.[3a] For example, the overlap volume of the molecules might be larger if one increases the number of features that should match, or the energy of the conformations included might increase if a tighter match between the features of the compounds is required. Accordingly, there frequently exists a family of related pharmacophore hypotheses even for a well-constrained system. This problem was highlighted in a recent article that described an algorithm that supported their examination of such trade-offs in several literature pharmacophores.[5] The algorithm will be described in more detail in Section 4.06.2.7.2. Additionally, in many cases there is not enough information to produce a unique family of related hypotheses, but rather a program might suggest hypotheses that differ in the features that match or in the bioactive conformations of all of the molecules. The latter situation is not a result of inadequacy of the computer program, but rather that sufficient information was not presented to it.

4.06.2.2 Selection of Compounds to Include

For those methods that do not involve a 3D QSAR, each compound to be included should provide unique information, either based on the conformational possibilities of the compounds or the presence or absence of certain functional groups. Adding redundant information increases the length of the computational analysis and can lead to many

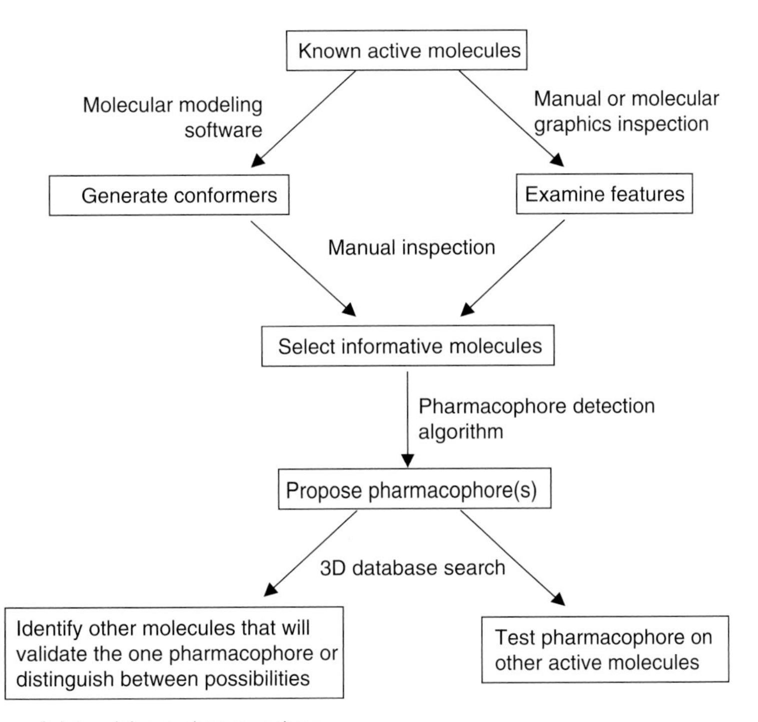

Figure 1 The process of determining a pharmacophore.

proposed bioactive conformations of similar compounds that the user must classify and evaluate. As an example of excluding compounds that provide no new information, there is no need to include dopamine (**1**) in an analysis if **2** and **3** are also included because every model that includes **2** and **3** will also include dopamine.

For such qualitative methods, one can include molecules from diverse sources as long as one is assured or hypothesizes that the biological effects are due to interaction with the same site on the target biomolecule. Thus one can coalesce the knowledge from different series of molecules from different investigators or include a virtual molecule that contains information on both conformational restrictions and SAR. For example, one might delete the exocyclic amino group from **4** because it is known that other compounds of this type that lack this amino group also bind to the D2 receptor.

For such qualitative methods, inactive analogs are used to evaluate a resulting pharmacophore model or to distinguish between possible models. Some algorithms allow one to indicate that specific molecules or a set number of molecules need not fit the model – such a selection might apply to inactive or weakly active compounds.

An interesting example of using structures from many sources, although the binding affinity was remeasured, is the classic work of the noncomputerized pharmacophore definition of D2 dopaminergic agonists[6] following previous key observations on the stereoselectivity of **2** and **3**[7] and knowledge of the bioactive enantiomer of noncatechols such as **4**. The investigators used **2–4** from many sources to define the model shown in **Figure 2**. Because this model was based on a much detailed experimental data, it has stood the test of time and has been able to explain the activity of several new classes of D2 agonists, such as **5**,[8] **6**,[9] and **7**.[10]

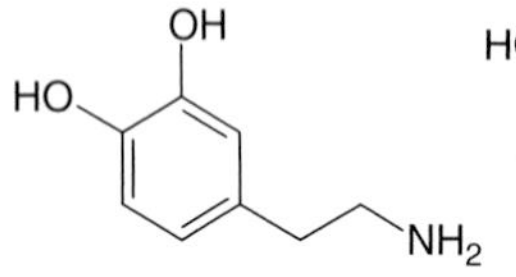

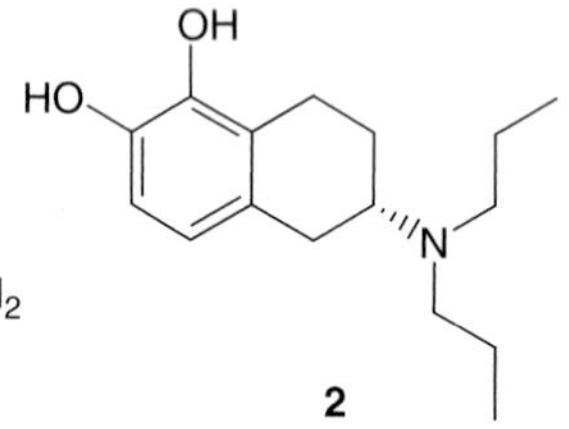

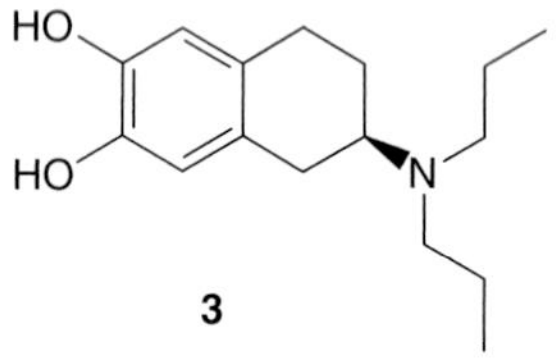

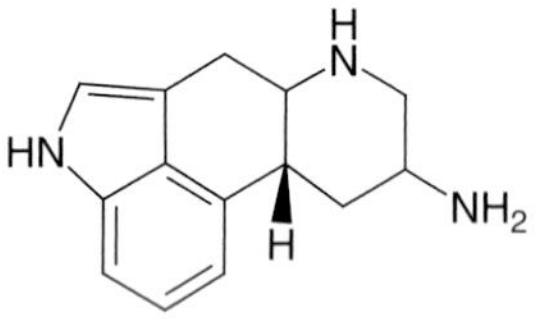

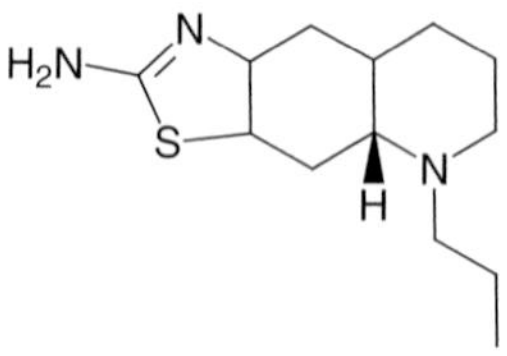

In the case of those methods that also perform a 3D QSAR as well as pharmacophore perception, the analysis can include all molecules for which potency has been measured in the same way, usually in the same laboratory. The algorithm decides when and how to use each compound. For example, Catalyst/Hypogen[11] first identifies candidate pharmacophores from the most potent compounds and then develops hypotheses as to why the less potent and inactive compounds lose potency. It presents the model as a graphical display and also as an underlying equation that can be used to forecast the potency of additional molecules. Because these methods require that biological activity be measured in the same manner for all compounds considered, structurally different compounds that are active in a similar assay cannot be included in the analysis although they can be tested to a proposed model. The usual caveats about performing a QSAR apply in this case.[12,13] Clearly, it is easier to use one program that performs most of the analysis than to move from one program to another.

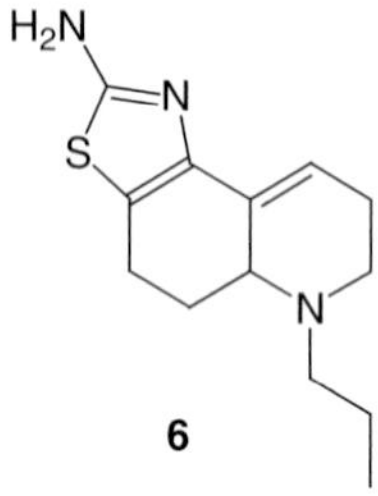

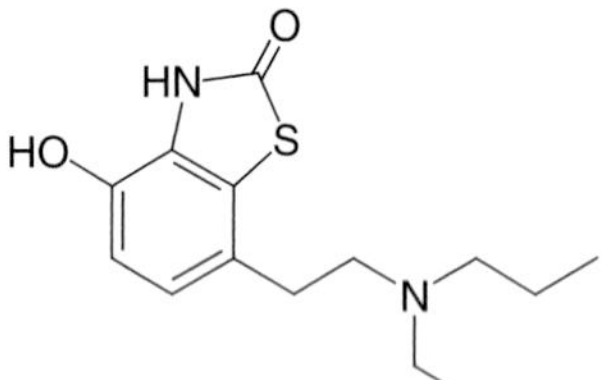

An important distinction between the various methods is whether they consider inactive compounds in the dataset treated by the algorithm. An analog of an active compound may be inactive for a variety of reasons: (1) It may not contain

Figure 2 Our elaboration of the Seeman D2 model.[6]

the appropriate groups in the correct 3D orientation. (2) When superimposed on the 3D pharmacophore, it occupies new regions of space, presumably those occupied by the target biomolecule. (3) Its properties may preclude measurement in the assay of interest. One example would be a compound for which the computational chemist does not realize that it is too insoluble to be tested. (4) Although the compound contains the appropriate groups in the correct 3D orientation, one or more of these atoms may have a markedly different electron density, resulting in much weaker or immeasurable affinity for the target biomolecule. This reason for inactivity is typically revealed by traditional or 3D QSAR.

The classic example of inactive molecules occupying new regions of space is the study of 2-aminonorbornane-2-carboxylic acids as inhibitors of ATP:L-methionine *S*-adenosyltransferase (**8–16**).[14] The amino and carboxyl groups and the α carbon atom form the pharmacophore, the atoms presumed to be recognized by the enzyme. **Figure 3** shows that, when superimposed on the active molecules, each inactive analog occupies new regions of space. The union volume of the active molecules shows the regions of space that have been shown to be acceptable for bioactivity, whereas the new volume occupied by inactive molecules shows the regions to be avoided. The intersection of the regions in space occupied by inactive molecules provides a very conservative estimate of regions required by the protein target.

NH_2 COOH **8** NH_2 COOH **9** COOH NH_2 **10**

NH_2 COOH **11** H_2N COOH **12** H_2N COOH **13**

COOH NH_2 **14** NH_2 COOH **15** NH_2 COOH **16**

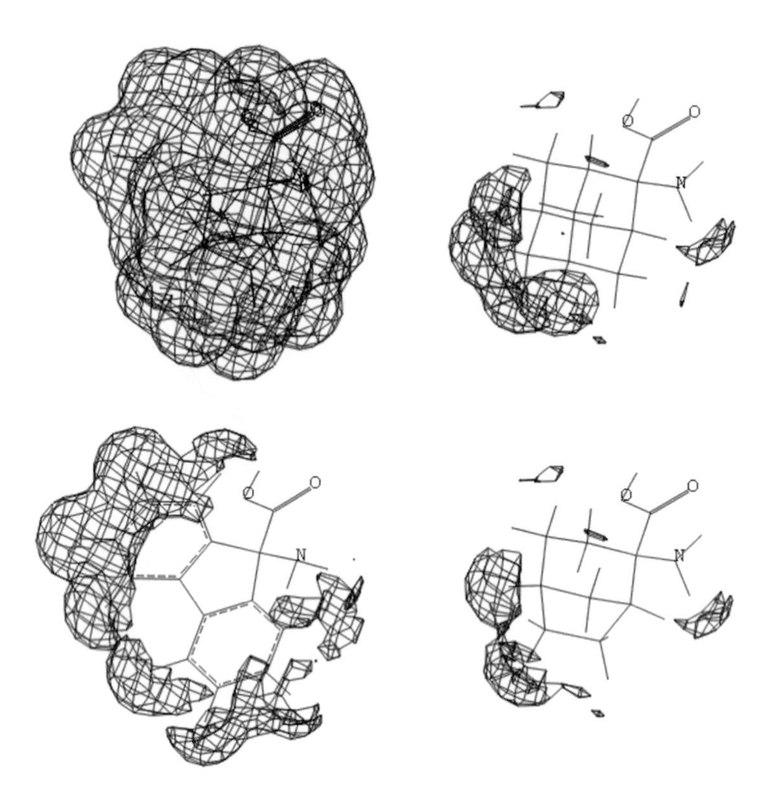

Figure 3 The upper left figure shows the union volume that results from the superposition of the active compounds **8–13**. The remaining figures show the new volume occupied by each inactive analog, **14–16**.

4.06.2.3 Definition of Molecular Features to Include

4.06.2.3.1 Recognition features

Most current pharmacophore detection programs consider for matching the corresponding features of the molecules. In this context features are ligand groups that form some type of intermolecular interaction: namely, hydrogen-bond donors, hydrogen-bond acceptors, positive charge, and negative charge.[15] The programs treat hydrophobic interactions in a number of different ways, usually by considering centers of hydrophobic rings as possible pharmacophore points, but sometimes also with a vector perpendicular to the center of an aromatic ring to represent the possibility of interaction with the π cloud. The various programs differ in the extent to which the user has control over the definition of feature points.

Building on information from small-molecule crystallography, it is recognized that different ligands can bind to a protein in such a way that two atoms that interact with the same protein atom may form this interaction from different directions (**Figure 4**).[16] To accommodate this possibility, pharmacophore-mapping software frequently tries to superimpose not only atoms but also site points that are located at the favored position for an atom in the target biomolecule.[6,17] For example, the site point complementary to an NH hydrogen-bond donor is a hydrogen-bond acceptor 2.5–3.0 Å from the nitrogen atom along the NH-bond vector. **Figure 5** shows typical labels for dopamine as an example.

Although most pharmacophore-mapping programs provide default pharmacophore feature types and distances to corresponding site points, frequently a pharmacophore map must be constructed from molecules that contain an unusual substructure. Searches of the Cambridge Structural Database (CSD)[18] can help one investigate the types of intermolecular contacts favored by the substructure in question.[16,19–25] This database contains the experimentally determined 3D structures and crystal packing of > 150 000 organic compounds. For example, such an analysis revealed that the oxygen of isoxazole is a much weaker hydrogen-bond acceptor than the nitrogen, and that the ether oxygen of an ester rarely forms hydrogen bonds.

4.06.2.3.2 Shape features

Although being able to fit into the macromolecular binding site is a prerequisite for binding, shape is not necessarily considered explicitly in many of the algorithms discussed in this chapter. Common methods to include shape in a

Figure 4 The approach of NH groups to a protein carbonyl group.

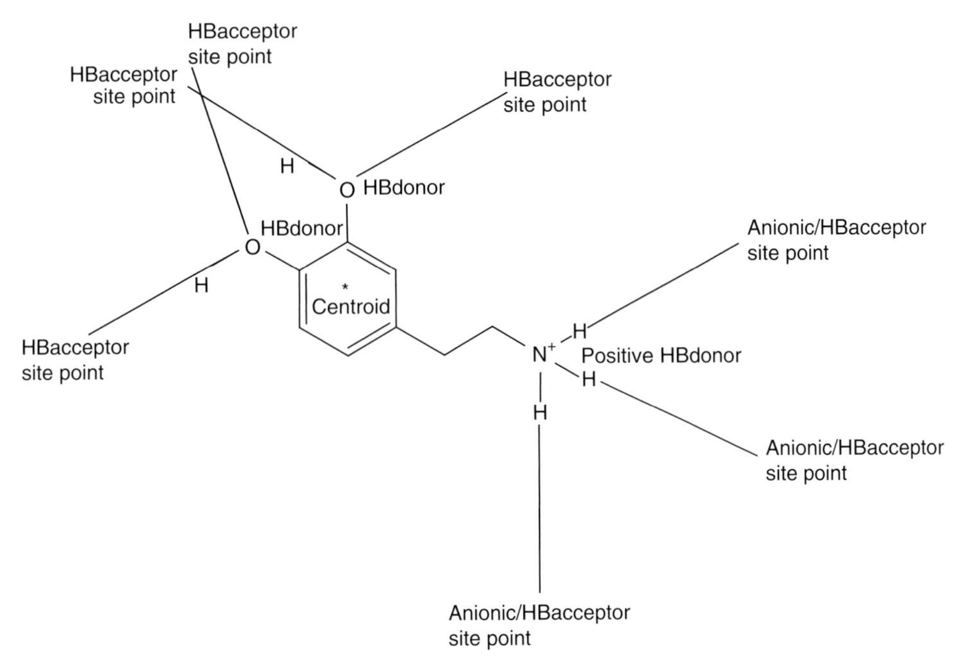

Figure 5 Potential pharmacophore points of dopamine.

pharmacophore are to use centroids of hydrophobic groups or aromatic rings as one of the feature definitions. Alternatively, one may score the proposed pharmacophores based on the union volume of all ligands when superimposed[3a] or score based on some sort of 3D QSAR analysis. More details of shape-based similarity are discussed in Chapter 4.08 and shape-based database searching is discussed in Chapter 4.19. Combined 3D pharmacophore-shape searching is discussed in Section 4.06.3.3.3.

4.06.2.4 Treatment of Conformational Flexibility

A critical decision of every modern pharmacophore-mapping algorithm is the choice of the conformations that will be investigated. For those methods that search for matching conformations given the points to match, the conformational search is the algorithm. However, for other methods, the conformational search can be performed as a preprocessing step or as part of the algorithm.

One advantage of a separate conformational analysis step is that the conformations can be selected by the method most appropriate for that type of structure. For example, in our analysis in the 1980s of dopaminergic compounds, we generated conformations with distance geometry[26] and optimized them MMP2[27] because this gave us the most reliable structure for molecules such as apomorphine that contain the biphenyl substructure. However, for a peptide one would choose some protein or more general force-field,[28–30] and for novel heteroaromatics a quantum-mechanical investigation might be necessary.[31] Further advantages of precalculating conformations are that one can restrict the analysis to only those conformations within some energy window; that one can optimize all degrees of freedom, bond lengths, and bond angles, as necessary; and that the conformational search is done only once.

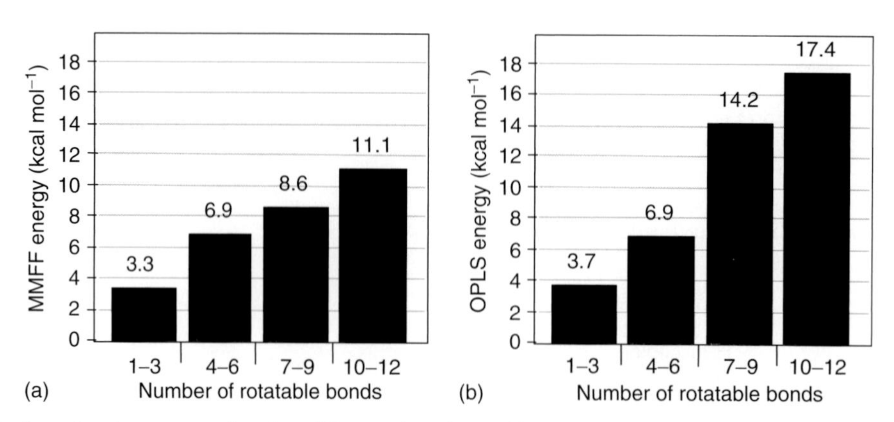

Figure 6 Conformational energy as a function of the number of rotatable bonds in the structure. (a) Energy calculated with the MMFF force-field.[33–37] (b) Energy calculated with the OPLS-AA force-field.[30]

The main advantage of performing the conformational search within the pharmacophore-mapping program is that only the relevant conformations are examined. This can result in a large saving in time and disk space for molecules with many rotatable bonds. If there are concerns about the geometry of certain parts of the molecules, this can be investigated before the pharmacophore mapping is initiated. A disadvantage of performing the conformational search while establishing the pharmacophore is that the conformational search has to be repeated if one decides to do a different analysis such as using a different training set or changing some of the program options which includes different compounds or different options.

The energy cut-off above which conformations will not be considered as a potential bioactive conformation is an important issue. From an investigation of 150 crystal structures of protein–ligand complexes, it was found that over 60% of the ligands do not bind in even a local minimum energy conformation.[32] Although approximately 60% of the ligands bind with strain energies $<5\,\text{kcal mol}^{-1}$, at least 10% of the ligands bind with strain energies $>9\,\text{kcal mol}^{-1}$. **Figure 6** shows how the number of rotatable bonds influences the energy cut-off necessary to retain 90% of the bioactive conformations of the ligands. Comparisons were made with both the MMFF force-field[33–37] and the OPLS-AA force-field.[30] **Figure 6** emphasizes the problem of discovering one unique pharmacophore with a set of conformationally flexible molecules and highlights the role of conformationally constrained analogs. The differences in the cut-offs for the two force-fields illustrate that it is difficult to accurately gauge conformational energy, particularly as conformers move away from the global minimum.

This study also reported that the only molecular feature that correlates with a high-energy bioactive conformation is the degree of hydrophobic collapse in the unbound state, a feature that is more likely as the number of rotatable bonds in a structure increase.[32] The average ratio of the distance between the two most distant atoms in the bound conformation to that in the most extended conformation is 0.85, which indicates that, on average, mostly extended conformations are the bioactive conformations. The energy of the bound conformation does not correlate with the affinity of the molecule for the protein, with the number of hydrogen bonds between the ligand and the protein, nor with the degree of burial of the ligand.

The method by which one generates the conformers for consideration also affects the quality of the pharmacophore result. In the study listed above,[32] it was suggested that one generates conformers with optimized molecular ensemble generation application (Omega),[38] minimizes them with MMFF or OPLS-AA, and applies the cut-offs shown in **Figure 6** to the conformer-ensemble before using them to discover the bioactive conformation.[32] A slightly different study also examined the question of conformational generation and found that again Omega performed well.[39] Chem-X[40] also performed well.

4.06.2.5 Techniques to Describe the Molecule and its Pharmacophore Features to the Computer

Table 2 lists programs that use special enhancements to conformational searching that enable one to propose one or more potential bioactive conformations of molecules that interact with the same macromolecular binding site. These approaches are especially appropriate to use when one has a firm idea or hypothesis of the points that should match and when the molecules are quite flexible.

To view a molecular structure on the computer screen one represents the 3D structure as a set of Cartesian, *XYZ*, coordinates of each atom, its atom type, and the type of bond between the atoms. This type of representation

Table 2 Programs that propose the bioactive conformation given the matching points

Program	*Number of molecules*	*Method of search*	*Ligand features*	*Target features*
Active analog approach[17,57,59]	2–20	Rigid rotation around single bonds, constrained to possible conformations and by distances found in previously searched molecules of the set	Indicated atoms	Site points as supplied by the user
RECEPTOR[59–61]	2–20	Rigid rotation around single bonds as in the 'active analog approach,' but with many computational enhancements to increase the speed 100 times	Indicated atoms	Site points as supplied by the user
Ensemble distance geometry[63]	2–20	Distance geometry generates random sets of conformations in which the corresponding points are superimposed in all molecules	Indicated atoms	Not included in the original paper, but could be included
Boltzmann jump[66]	2–20	The corresponding points in the various molecules are tethered with a strong potential while the Monte Carlo search relaxes other constraints on conformational and translational movement	Indicated atoms	Corresponding site points

corresponds to our experience in the real world. Similarly, the shape of a molecule can be represented as a surface that surrounds it.

A molecule need not be described by its atoms and bonds, but instead one can encode the regions in space that it occupies or with which it would form an attractive interaction. Typically, such a description would involve placing all superimposed 3D structures, one at a time, into a 3D lattice and recording the interaction energy with probes of different character placed at each lattice point. In this way, the molecular structure is transformed from a set of atoms and bonds into interaction fields – it represents the way the structure would be viewed by an interacting molecule. This description is used in the 3D QSAR method of CoMFA.[41]

However, certain computer calculations are more efficient, or indeed possible, if one instead describes the molecule as a distance matrix. Each column and row of the matrix represents an atom, usually indicated by its atomic symbol, and each element of the matrix is the distance between the corresponding atoms. The distance matrix implicitly contains the bonding information because single bonds are longer than double bonds, which in turn are longer than triple bonds. The distances also characterize the conformation.

A different distance matrix is used in the electron conformational method.[42–44] In this approach, the diagonal elements contain the charges or atomic interaction indices of the atom, off-diagonal elements between bonded atoms contain the corresponding bond order, and the remaining off-diagonal elements contain the distance between the corresponding atoms. Pharmacophores are considered to be submatrices that are present in active molecules but absent from inactive analogs.

Distance matrices can be expanded to include distance bounds instead of the distances for an individual conformation. For example, the elements above the diagonal could be the lower bounds of the distance between the corresponding atoms and the elements below the diagonal the upper bounds of the distance between the corresponding atoms. Special algorithms convert one conformation into a distance bounds matrix.[26,45]

For pharmacophore detection one might choose to ignore most of the intramolecular distances and concentrate on only distances between pharmacophoric features. For some applications such distances are not recorded as continuous numbers, but instead binned so that similar distances between the same types of pharmacophoric features are considered identical. If each distance is considered independent of every other distance, a one-dimensional array can be used to record which distance is present in the molecule. Such binning schemes are sometimes called bitmaps and each bin called a key.[46–48] The bitmaps are not limited to distance, but three- and four-point pharmacophores[40,49–53] or angles[54] may be encoded. Because 3D and 4D pharmacophore bitmaps can become very long, special encoding methods to compress the information have also been developed.[53] For some applications all conformations are encoded into one bitmap; for others, conformation is encoded separately with the result that compounds with different numbers of conformations will have different lengths for their bitmaps. If multiple conformations have been encoded into one bitmap, then any particular set of distances may not be present in one conformation. For this reason, pharmacophore detection methods that use such bitmaps must include a step to establish that one conformation of each molecule does include the detected distances.[49,55]

A multidimensional binning scheme can be used to compare the conformational possibilities of molecules. In this case, each dimension represents the distance bins for one particular type of atom pair. There will thus be as many dimensions as there are types of pairs of pharmacophoric features. Each element in this description represents the possibility that the particular molecule has a conformation that has the distance between the particular pair of pharmacophoric features.

A problem with any binning strategy is that the bin boundaries can influence the distinction between 3D structures as would be the case if conformations of two molecules have a very similar distance between two points of interest but these distances set different bins. Consider distances 6.95 and 7.05 Å, which are very similar; however, if the bins were 5–6 and 6–7 Å, then the conformations would set different bins and would be considered to be different. For some applications, the solution to this problem is to repeat the analysis using different bin boundaries to look for alternative solutions. Another approach is to populate the bins using many slightly different conformations. Alternatively, one may make the boundaries of the bins fuzzy by allowing distances that are close to the boundaries to set both bins.[56]

A characteristic of describing a molecule with interpoint distances is that enantiomers have the same interpoint distances. This can be an advantage if one does not know the bioactive enantiomer. In any case, postprocessing is necessary to rule out proposed superpositions that do not recognize the correct enantiomer, or to generate both enantiomers to decide which fits best if one has tested a racemate. However, it is also possible to encode the chirality of the compound as one element of a bitmap.[53]

4.06.2.6 Specialized Methods to Propose a Bioactive Conformation Given the Points to Match

4.06.2.6.1 The active analog approach and receptor

If one has enough SAR information to propose the points to match, the problem becomes simply selecting the proposed bioactive conformation(s) of each compound. One approach to this problem is to perform rigid rotation of all rotatable bonds in each molecule, tabulating which distance bins between the atoms or points of interest are occupied by a conformation of reasonable energy.[57–59] Proposed bioactive conformations thus correspond to conformations that occupy the distance bins occupied by all other molecules of the set. **Figure 7** illustrates the process for a two-distance problem. At the start, all distances are possible. Conformational search of the first molecule eliminates several distances from consideration; the second eliminates more, etc., until only two sets of distances remain for consideration. Postprocessing is necessary to ensure that the distances obey the triangle inequality; any distance AC cannot be longer than the distance of AB plus that of AC. The rigid rotation/distance bin approach can be optimized by a number of heuristics and clever look-ahead strategies that restrict the search of each molecule to relevant conformations.[58–61] An extension of the distance bin approach that not only searches for bioactive conformations but also identifies matching points has been described (**Table 3**).[62]

4.06.2.6.2 Ensemble distance geometry

Distance geometry is a conformational search strategy in which the algorithm starts with a matrix of allowed interpoint (usually all interatomic) distances. It then randomly selects points within the allowed ranges and refines these distance

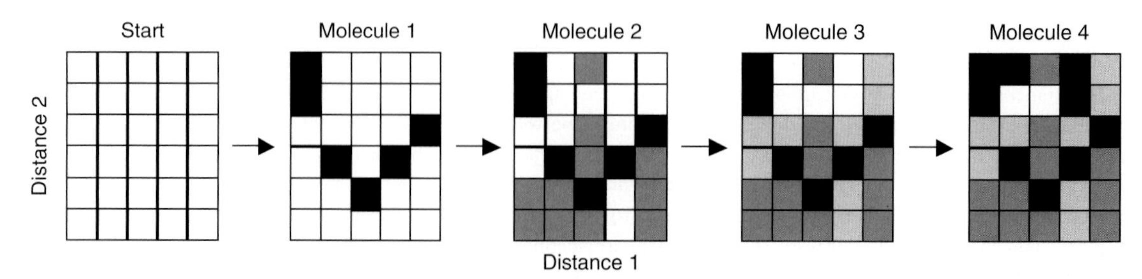

Figure 7 The reduction in pair-wise distances revealed by comparison of different compounds.

Table 3 Programs that propose both the superposition and the bioactive conformation

Program	*Number of molecules*	*Conformations*	*Method of search*	*Scoring*	*Ligand features*	*Target features*
AUTOFIT[69]	2	Precalculated	Exhaustive search	RMS	Functional points, user controlled	Functional points, user controlled
DISCO[74] reimplemented in PRO_LIGAND[162]	2–32	Precalculated	Clique detection of interpoint distances. Certain molecules or number of molecules may be optional	Distance tolerance	Functional points, user controlled	Functional points, user controlled
COMPASS[163]	2–32	Precalculated (example used x-ray structures)	Clique detection of interpoint distances	Distance tolerance	All atoms, properties assigned by the program	Not used
Simulated annealing [164]	<10	Selected by cluster analysis from simulated annealing of torsion angles	Simulated annealing – the sum of the difference in interatomic distances is minimized	The sum of the difference-distance matrix and the RMS for least-squares fit	Atoms	Not used
ChemDBS-3D[49]	2–32	Precalculated 2, 3, and 4 point distance keys	Intersection of 3- or 4-center pharmacophore distance keys	Distance tolerance, possible to generate a conformation for each molecule?	H-bond donors and acceptors, positive atoms, aromatic ring centroids	Not used
Catalyst/ HipHop[94,165,166]	2–32	Precalculated or within the program	Distances and RMS Exhaustive search of superpositions. A feature may be missing in a certain number of compounds	Complex score based on the fraction of ligands that fit the full model, the RMS of fit, and the rarity of the features	Functional points, user controlled	Accessibility of points, H-bond direction and distance
GASP[79,167]	<10	Torsions scanned	Genetic algorithm: chromosome contains torsions and feature matches	Equation that weights feature and volume overlap and van der Waals energy of conformations	Functional points, user controlled	Lone pairs at 1.0 Å plus site points
DANTE[3a]	Not specified	Precalculated	Exhaustive comparison of all 3-element (point, angle, or torsion) pharmacophores	Tightness of corresponding pharmacophores. Also detects when no single pharmacophore explains the data.	Functional points	Not used, but a union shape is also generated with indications of where the SAR has not been explored.

continued

Table 3 Continued

Program	*Number of molecules*	*Conformations*	*Method of search*	*Scoring*	*Ligand features*	*Target features*
MPHIL[168]	<100	Precalculated	Genetic algorithm/clique detection of interpoint distances	Pharmacophore contains K features, with which all the input ligands have m in common. Ideally $K=m$	Atom-based in first implementation	Not used
GAMMA[169]	<10	Torsions scanned followed by directed tweak[154]	Multiobjective genetic algorithm with Pareto optimization to provide multiple solutions with different numbers of matched atoms and fit of the matched atoms	Number of atoms matched and the fit of the matching atoms	Atom-based	Not used
DistComp[170]	5–25	Precalculated	Exhaustive search	Distances common to actives but absent in inactives	User identified features	User identified
Multiobjective genetic algorithm[5]	3–5	Torsions scanned	Multiobjective genetic algorithm with Pareto optimization	Numbers of matched features, average energies of the conformers, and volume of the overlap	Functional points	Site points
GALAHAD[81]	4–10	Precalculated	Multiobjective genetic algorithm with Pareto optimization	Energy of conformations, H-bond overlap, shape overlap	Functional points translated into 2-, 3-, and 4-point keys[53]	Site points translated into 2-, 3-, and 4-point keys[53]
SCAMPI[56]	100–2000	Constrained as distances are discovered[58–61]	Recursive partitioning[87–89]	Sets of distances that separate actives from inactives	Functional points, user controlled	Not used
Modified active analog[62]	2–20	Constrained rigid rotation around single bonds[17]	Examine all pairs of potential features, systematically add one feature to identify the maximum number of matching features present in most of the active compounds	Selectivity of the proposed pharmacophore in identifying actives seeded into a database of presumably inactive similar compounds	Functional points, user controlled	Functional points, user controlled

Chemometric methods[68,171]	>8	Previously generated	Principal components analysis on distances between key groups followed by cluster analysis of three key distances	A cluster that contains mainly active compounds	Functional points selected by the user	Not used
COMPASS [172–175]	~ 100	Previously generated	Neural network	Optimization of fit of observed activity/inactivity to the model	Distance to a corresponding sampling point	100–200 sampling points 2 Å from the union surface of the molecules
PharmID[55]	~ 100	Precalculated	Gibbs sampling as in sequence alignment,[90] modified, followed by clique detection to map back to a conformation	Identity of bits set in the appropriate segment of the bitmap	User defined	User defined
Conformation-mining[176]	3–4	Precalculated	Subshape matching of conformers	Feature maps, how well the chemical properties overlap	Gaussians at features	Not used

bounds so that they obey the triangle inequality.[26,45] Typically one uses the program to generate 100–1000 conformations and then refines the resulting conformations with a program of choice.

Ensemble distance geometry uses distance geometry with three enhancements. (1) All molecules that one wishes to examine for a pharmacophore are included in one distance matrix. (2) The distance matrix has been enhanced such that the lower bounds of all intermolecular distances are set to zero, thus allowing the molecules to superimpose. (3) The upper bound of the distances between the matching points are also set to zero or some small number.[63] The result is that the algorithm produces a set of overlapped conformations ready for optimization.

The problem with distance geometry is that, because the distances for any solution are set at random, many duplicates of some conformations are generated. This requires intervention to get a unique set of conformations. The advantage of ensemble distance geometry over rigid rotation is that ring conformations, bond lengths, and bond angles are not frozen.

4.06.2.6.3 Boltzmann jump

Monte Carlo sampling of conformational space is another popular method to generate conformations of interest.[64,65] It generates a new structure from the previous one by making random changes to one or more torsion angles. If the new structure is lower in energy than the previous one, it is kept. If it is higher in energy than the previous one by ΔE, then it is retained with a probability equal to the Boltzmann factor $e^{\Delta E/RT}$, where R is the gas constant and T the temperature. This procedure allows the search to overcome torsional barriers. In practice, one minimizes the conformations when they differ substantially from the previous one.

By analogy to ensemble distance geometry, ensemble Boltzmann jump sampling tethers the user-defined corresponding points, usually atoms, of the different molecules together or otherwise allows them to interpenetrate.[66] The Monte Carlo search is performed as usual to relax constraints on conformational movement such as rotations about bonds, bond stretching, and angle bending. Additionally, random moves of the molecules are part of the search. Hence, the aim is to minimize the energy of the system calculated from the coincidence of the corresponding points and the approximate conformational energy of each molecule. The advantage of Boltzmann jump over ensemble distance geometry is that the structures are generated in the context of a force-field, hence little if any optimization is needed after the search.

4.06.2.6.4 Chemometric methods

A principal component analysis[67] of the distances between five proposed pharmacophore points in all low-energy conformers of active HMG-CoA reductase inhibitors revealed that only three of these distances was necessary to explain 85.5% of the variance of the 10 distances.[68] Subsequent cluster analysis of the conformers using these three distances produced one cluster of conformations that includes four of the five active compounds and only two of the six inactive compounds. The conformations and distances that correspond to this cluster were selected as the bioactive conformation and pharmacophore distances, respectively. The lack of activity of the two inactive compounds that have the pharmacophore distances are explained by differences in electrostatic potential.

4.06.2.7 Specialized Methods to Detect Corresponding Points

Although one may choose to use conventional minimization techniques[69,70]or compare bitmaps,[49] there are algorithms that are especially appropriate for detecting the conformations and matching points that make up a pharmacophore, **Table 3**. This remains an area of active research.

4.06.2.7.1 Clique detection

The earliest methods to detect the features that can overlap in 3D are an extension of work that showed that clique detection methods efficiently identify the maximum common 3D substructure in a set of molecules.[71,72] Clique detection algorithms use as input a matrix of interpoint distances for each 3D object, a conformation of a molecule in the case of pharmacophore mapping. The algorithms examine this matrix to identify those sets of distances that are common to all 3D objects and for which all distances between these points are included in the set. The early work used each heavy atom as a point in the graph. The resulting solution is the maximum common 3D substructure of the molecules.

The utility of clique detection for pharmacophore mapping required two changes: (1) considering multiple conformations,[73–76]and (2) defining the points to be generalized recognition features (such as hydrogen-bond donors) centered at atoms[75] and site points.[74] By incrementing the tolerance at which distances match, one generates multiple proposed pharmacophores that differ in the points matched, the number of points matched, and/or the conformations included. The disadvantage of clique detection methods is that typically one compound is selected as a reference and other compounds are fit to it; if a poor choice of a reference molecule has been made, then the results can be confusing or useless.

4.06.2.7.2 Genetic algorithms

A genetic algorithm performs its search by analogy to biological evolution.[77] Possible solutions are represented as alleles in a chromosome, one chromosome per molecule. The genetic operators of mutation and crossover operate to optimize some fitness (scoring) function for the whole set of individuals.[78] For example, in the GASP program each molecule is represented by one chromosome that contains alleles to describe each torsion angle and a second set of alleles that identify which atom is matched to a particular atom in a reference molecule. The fitness function of the genetic algorithm is a weighted combination of (1) the number and the similarity of the features that have been overlaid; (2) the volume integral of the overlay; and (3) the van der Waals energy of the molecular conformations defined by the torsion angles encoded in the chromosomes.[79] Other programs use different chromosomes and fitness functions.

All pharmacophore hypotheses are just that, hypotheses. For a scientist, the problem is often to explore how changing one of the constraints would change the hypothesis. For example, if a pharmacophore with three points is found with a (root mean square) (RMS) deviation of 0.5 Å, how does the RMS deviation change if four points are included? Multiobjective optimization techniques offer an efficient method to find such families of solutions.[5,80,81] The technique uses a genetic algorithm for which the fitness function is modified to search for a set of solutions each of which has the optimum value of one fitness criterion, a Pareto optimization. For example, one solution might include the lowest energy conformers but nonideal overlap volume, another might contain the maximum number of matching points but higher energy conformers, and yet another might contain the lowest overlap volume but larger distances between matching points. The investigator then has the option to consider each of these solutions for further work.

4.06.2.7.3 Partial least squares[82,83]

One may use a 3D QSAR study to select the bioactive conformations of the molecules. In multiway PLS,[84] one provides the potency of each molecule and sets of CoMFA descriptors for each potential alignment of the compounds with each conformation of the most potent compound.[85] The multiway PLS identifies the overlay and conformation of the reference compound that best explains the biological potency of the compounds. This overlay was then used to generate a traditional CoMFA, which was shown to have excellent predictive ability on an external test set.

In a related method, 4D QSAR,[86] thousands of conformers of each molecule are first generated by molecular dynamics sampling. For each particular pharmacophore chosen by the investigator, the alignment of each molecule is described by the fraction of time that its conformers occupy each particular type and location of points in a lattice that surrounds the ensemble. Using biological potency as the dependent variable, PLS extracts the lattice occupancies that correlate with biological activity. This analysis is repeated for all potential superpositions. The model with the most predictive power is selected as that which most accurately describes the pharmacophore. The final step is to select as the bioactive conformer of each compound the lowest energy conformer that contributes to the selected PLS features.

4.06.2.7.4 Constrained conformational search

As noted above, one can modify the conformational search used in the active analog approach to also examine possible matching points.[62] This report also included a detailed analysis of how to rank possible pharmacophores using 3D database searching.

4.06.2.7.5 Statistical classification of activities of molecules for pharmacophore identification

Recursive partitioning is a data-mining technique that uses statistical tests to identify descriptors of objects that separate one class from another; in our context it would use molecular descriptors to identify those that separate actives from inactives.[87–89] It works in a tree fashion: one variable is used to make the first split, then all molecules in each subset are further independently divided, one descriptor at a time, until either the compounds are completely classified correctly or further splits will not meet some statistical significance criterion. A recursive partitioning analysis results in a tree in which each branch is the value of some property and each end node is enriched in active or inactive compounds.

In SCAMPI,[56] the molecular descriptors are related to pharmacophore distances. Combined with the recursive partitioning is a constrained conformational search strategy, an enhancement of the ideas in the Active Analog Approach and RECEPTOR.[58–61] Slightly overlapping distance bins are set for pairs of potential pharmacophoric atoms and the most predictive distance–pharmacophore pair is used as the basis for the first split. For all molecules that contain this distance, the conformational search is repeated using this constraint and new bins are set for all three-point pharmacophores. This process continues until no more pharmacophoric points can be added or until a five-point pharmacophore is found. For molecules that do not contain the first distance, the recursive partitioning/conformational search is used to identify a possible second pharmacophore within this subset of compounds.

As noted in **Table 3**, SCAMPI can operate on hundreds to thousands of molecules, but requires a substantial number of compounds for the analysis. This property makes it especially appropriate for investigating possible pharmacophores within a set of actives from high-throughput screening.

4.06.2.7.6 Gibbs sampling of pharmacophore bitmaps

A convenient way to summarize the potential pharmacophores of a molecule is to generate keys in a bitmap or fingerprint that describe the presence or absence of a particular geometric feature such as a distance between two points of a particular character or the angle between three points. In an interesting new approach, each conformer of a molecule populates a set of keys of a predetermined length. The result is that each molecule can be associated with a different length bitmap. By analogy with aligning DNA or protein sequences with Gibbs sampling,[90] one can use a modification of Gibbs sampling to discover which of the keys are common.[55] The modification is to require that the matches are of bits associated with only one conformer. After a bit string alignment is discovered, the corresponding conformations are identified by a clique detection algorithm. An advantage of the Gibbs sampling is that it will detect two binding modes if that is necessary to explain the data.

4.06.2.7.7 EGSITE2 three-dimensional models of the binding site[91]

Behind every pharmacophore model is a, often implicit, model of the binding site with which the compounds interact. Early studies to derive an explicit model used Voronoi polyhedra of different size, shape, and properties to quantitatively explain the observed SARs.[92,93] EGSITE2[91] is an outgrowth of that approach. The algorithm starts by placing all molecules into a uniform macromolecular binding site. It then incrementally partitions the space into regions of different character and fits the ligands into the complementary region by changing both its orientation in space and its conformation. At every stage a regression analysis correlates potency with the total interaction energy of the molecule with the proposed binding site.

4.06.2.8 Methods to Use Pharmacophore Features in Quantitative Structure–Activity Relationship

Although one may choose to evaluate or use a pharmacophore for alignment in more traditional 3D QSAR such as CoMFA,[41] several programs make specific use of pharmacophoric features in a 3D QSAR, using the QSAR to direct the search for the pharmacophore, **Table 4**.

The Hypogen module of Catalyst[11,94] searches first for pharmacophores present in the most potent ligands but not in less potent compounds. It then uses simulated annealing on all compounds to identify one or more equations that fit potency to the presence or absence of pharmacophore features including a weight and a distance dependence for the quality of the fit.

The newer program Phase[95] scores potential pharmacophores by the RMS fit, the deviations from ideal of the cosine of hydrogen-bond angle, the volume overlap, and the energy of the conformations. It includes a penalty if not all molecules fit the hypothesis. Its technique can detect the situation where two binding modes are needed to explain the data. It then uses PLS on subfields generated around the molecule to produce a quantitative model.

In the electron conformational method, the final QSAR is based on the presence or absence of the discovered pharmacophoric groups, corrected for the accessibility of the groups to potential interaction sites on the target biomolecule.[42,43]

4.06.2.9 Strategies to Evaluate a Pharmacophore Hypothesis

Often a pharmacophore-mapping program will provide many possible solutions, some of which are very similar. A summary of the points included and the distances between them can help one identify similar hypotheses.

However, even if only one is suggested, it is important to question the result. Does the proposed model explain other active molecules? Does it select the correct stereoisomer? Does it explain why certain molecules are inactive – perhaps because, although they fit the pharmacophore, they occupy new volume compared with the union volume of the active molecules?

The first step in the evaluation is to superimpose the candidate conformations (and stereoisomers), either to facilitate computer graphics viewing or as a prelude to further calculations. Aside from manual RMS methods several superposition programs can optimize the fit of overall shape between molecules.[96–100]

Table 4 Programs that combine pharmacophore detection and 3D QSAR

Program	*Inactives used*	*Conformations considered*	*Method of search*	*Scoring used*	*Ligand features used*	*Target features used*	*Alignment method*	*3D QSAR method*
Apex-3D[75,76]	Yes	Precalculated	Clique detection	Cliques present in actives but absent in inactives	Functional points, QM charge sites	None	Pharmacophore features and shape	Optional regression analysis on features and physicochemical properties of atoms or whole molecule
Catalyst/ Hypogen[11,94]	Yes	Precalculated, but integrated software supplied[149]	Exhaustive search of most potent ligands; removal of pharmacophores present in many less active compounds; simulated annealing	Correlation of potency with fit to the model	Functional points, user controlled	Excluded volume point/sphere	Pharmacophore features	Fit to the pharmacophore hypothesis
Phase[95]	No	Integrated within the program	Proprietary	RMS fit, hydrogen bond angle, volume overlap, energy, penalty if not all molecules fit	Functional points, user controlled and custom	Extension points along H-bond vector	Pharmacophore features	PLS on fields generated around the molecules
4D QSAR[85,86]	No	Precalculated	Partial least squares (PLS)	PLS selects the superposition and conformation that best explains the data	A set of lattice occupancies or energies for different pharmacophore alignments and conformations	A set of lattice occupancies or energies for different pharmacophore alignments and conformations	User choice, external to the program	PLS on the reduced set of descriptors
Electron-conformational method[42,43]	No	Precalculated with ab initio quantum chemical programs	Examination of matrix for corresponding submatrices	Not specified	A special matrix for each conformation	Charge or atomic interaction indices of an atom, bond orders, and interatomic distances	Implicit	Regression analysis on presence or absence of pharmacophoric substructures and potency detracting groups

continued

Table 4 Continued

Program	*Inactives used*	*Conformations considered*	*Method of search*	*Scoring used*	*Ligand features used*	*Target features used*	*Alignment method*	*3D QSAR method*
EGSITE2[91]	No	Precalculated	Postulate macromolecular binding sites, identify if one conformation of every molecule can bind	Ability to predict the affinity of a test set	Hydrophobic, polarity, and charge of atoms – grouped by the algorithm	Regions of particular property, discovered during the optimization	Atoms 'docked' to complementary sites	Regression analysis of interaction energies with proposed sites
ILP[177]	Include less potent compounds as the 'inactive' class	Precalculated	Inductive logic programming	Correct classification of molecules corrected for complexity of the query	Distances between specific atoms of previously programmed substructures	Not used	Not needed	Presence or absence of a particular distance between 3D substructures

Several criteria can be used to help decide between pharmacophores that match the previous items. If heteroatoms are superimposed, do they project to the same region in space, a site point? What is the overlap volume when all active molecules are superimposed on the pharmacophore – is this more or less than competing models? Does the pharmacophore span most of the smallest molecule? What is the maximum and average overlap RMS when compared with competing models? What is the maximum and average conformational energy of the compounds compared with models? One can also evaluate competing pharmacophores by deriving a 3D QSAR based on the proposed superposition.

If more than one credible hypothesis remains, the user has not supplied enough information to the algorithm. It may be necessary to (synthesize and) test additional compounds. These compounds might constrain the conformation of the molecule or probe the necessity of certain groups in the molecule. 3D searching, described later in this chapter, can be a help in identifying such compounds.

Although a primary use of 3D database searching is to identify molecules that potentially have the biological activity of interest, there are other uses for such databases as well. Such databases provide a simple method to evaluate a pharmacophore hypothesis in that one would expect that searching using the pharmacophore hypothesis should retrieve all of the active molecules. If the pharmacophore-mapping investigation produced more than one potential pharmacophore, 3D searches based on each might highlight the one with the most predictive or discriminatory power, that is one that identifies many actives but few false positives.[62]

4.06.2.10 Limitations of the Assumptions for Pharmacophore Mapping

Because of unavailability of experimental 3D structures of the target biomolecule, one may have no choice but to use pharmacophore-mapping methods to gain insight into the 3D features required for bioactivity. However, protein crystallography studies illustrate the artificiality of the underlying assumptions of pharmacophore mapping. There is abundant evidence that proteins change their structure to accommodate ligands and that this change in structure is not the same for all ligands, even those that bind in essentially the same fashion.[101–109] In addition, even molecules that are very similar frequently bind in different conformations or orientations.[110–112] An interesting example is the binding of **17** and **18** to the 1α subdomain of ricin B.[113] Structural NMR studies showed that the two compounds bind competitively to the same site, but that the *C*-glycoside (**17**) binds in the *anti* conformation whereas the *O*-glycoside binds in the *syn* conformation. A second example is shown in protein crystal structures of histamine *N*-methyltransferase with **19–22**, all potent inhibitors of the enzyme, revealed that all occupy the histamine binding site, but that three aromatic residues of the enzyme become disordered or change conformation to maximize the complementarity between the enzyme and the inhibitors.[107] The foregoing observations reinforce the notion that a pharmacophore map should be considered to be a hypothesis that will suggest further research to support or reject.

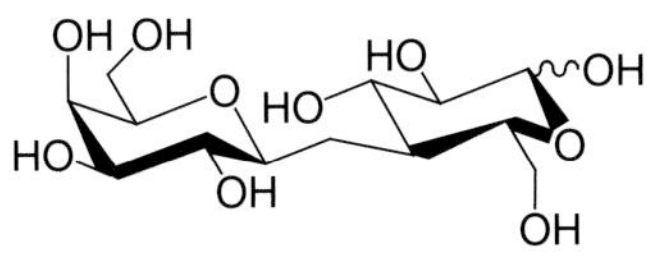

17 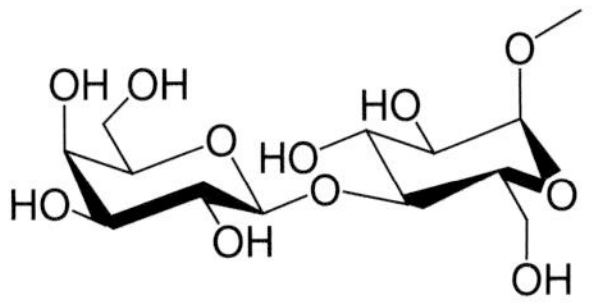 **18**

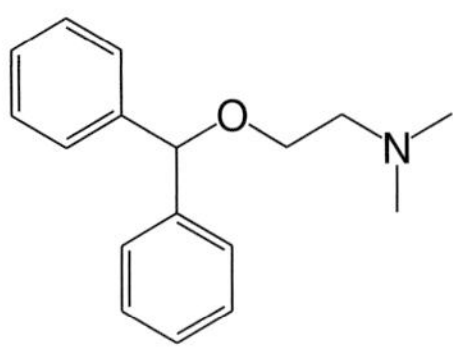

19 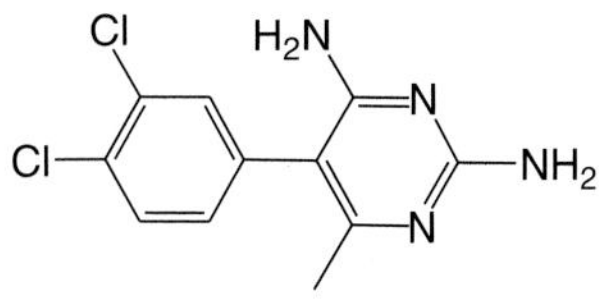 **20**

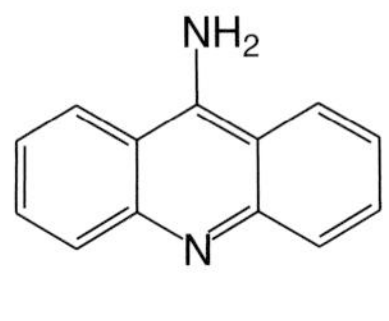

21 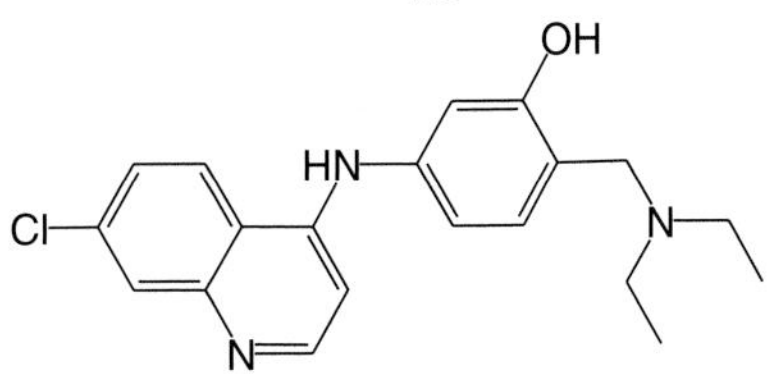 **22**

4.06.3 Three-Dimensional Database Searching

One reason to identify a pharmacophore is to use it to discover other compounds that have the same activity. For this purpose one typically searches a database of 3D structures to identify untested molecules that match the proposed pharmacophore (**Table 5**). The first step in such an investigation is to evaluate the search query on active and inactive molecules – a successful query should recognize all actives and perhaps exclude structurally related inactive molecules.[62] If it does not exclude all inactive molecules, then the user must decide if postprocessing will do so, if the query can be refined by including excluded regions, or if the biological test is inexpensive enough that false positives from the search are easily weeded out experimentally. On a similar note, one might perform 3D searches using competing pharmacophore hypotheses to design a screening library that will distinguish between them. **Table 4** summarizes the capabilities of the various 3D pharmacophore searching programs.

Because of the speed of 3D searching methods, some authors translate either the bound conformation of a ligand or properties of the protein binding site into 3D search queries.[114,115]

Any database of 3D structures can also be used to search for templates that hold functional groups at predetermined distances and angles, thus suggesting conformationally constrained analogs. The program CAVEAT was designed for just this purpose.[116,117] Such a search could be used to design analogs that would distinguish between competing pharmacophore hypotheses, compounds that are selective for one biological activity versus another, or are more potent because conformational restriction decreases the conformational entropy loss on binding. The input structure could also be low energy or the bound conformation of a ligand.

One need not search a 3D database with an explicit pharmacophore, but can search for molecules that are similar in 3D properties to the input molecule.[40,53,118–123] Again, molecules identified by such similarity searching can help one validate or reject a potential pharmacophore.

4.06.3.1 Sources of Compounds

Because the database preparation and screening are done in the computer, one can include any molecular structure in a 3D database. Often one chooses to make a 3D database of the corporate collection. The advantages of such compounds are that they are typically readily available, many of them are proprietary, and their biological profile is known.

One can also make a 3D database of compounds that could be purchased for screening.[124,125] One collection contains references to 23.3 million samples that represent 14.2 unique chemical compounds.[125] The advantage of vendor compounds is that a chemist need not synthesize the compounds; the disadvantages are that they may not be available in a timely manner and that one's competitor can also purchase the compounds thus increasing the chance that a competitor might beat one to a patent on the series.

The National Cancer Institute provides a database of $>200\,000$ chemical compounds that can be requested from them.[126] Most of these compounds are unique to this database.[127] Vendors that provide 3D database searching programs usually provide this database with their software.

Databases of virtual compounds, for example enumerated combinatorial libraries (*see* 4.14 Library Design: Ligand and Structure-Based Principles for Parallel and Combinatorial Libraries; 4.15 Library Design: Reactant and Product-Based Approaches), are also fruitful for searching. A virtual library could be a specific library targeted to the biological target of interest or it could be a virtual library of anything that seems to be synthetically possible.[128–130] If these libraries of virtual compounds are based on proprietary chemistry or proprietary reagents, then one could expect that others would not have the same compounds in their virtual libraries.

To complement the search for conformationally constrained compounds, the CAVEAT system includes a 3D database of 411 000 tricyclic hydrocarbons and another of 110 000 'acyclic' molecules that might be used as linkers. If one of these were to be a hit for a 3D search, then it is expected that the synthetic chemist will substitute heteroatoms at key positions to make the synthesis more facile.

As mentioned above, the CSD contains the experimentally determined 3D structures and crystal packing of $>150\,000$ organic compounds.[18] The CSD contains the 3D structures of many natural products, with the result that the hits from a 3D search of these structures are often quite unique. An analysis of the structures in this database revealed that it contains more distinct ring systems than WDI,[131] ACD,[132] and NCI3D[126] combined.[133] It also contains a large fraction of compounds not found in other databases.[127]

Although one can enter the structures as provided, usually the structures are curated. For example, small counter-ions are typically removed and nitro groups are normalized. Depending on the capabilities of the software that will be used for searching, one may also choose to enumerate all tautomers of the molecules as well. If the compounds are meant to be screened, not as ideas for synthesis, one would usually also eliminate reactive compounds,[134] those predicted not to be permeable into cells,[135] etc.

Table 5 Programs for 3D ligand-based searching

Program	*Source of conformers*	*Query objects*	*Query types*	*Precalculated keys*	*Integrated with pharmacophore perception*	*Shape query*	*Partial matches*	*Integrated with molecular modeling visualization*	*Integrated with 3D QSAR*
MOLPAT[1]	Small molecule crystal structures[18]	Atoms	Distance	No	No	No	No	No	No
SOLON[46,179]	Small molecule crystal structures[18]	Points at atoms of a particular atomic number	Distance	Distances between atoms from input structure	No	No	No	No	No
ALADDIN[180]	Externally generated[18]	User described functional points, lines, torsion angles, planes	Distance, angle, torsion angle, excluded volume	None	No	Excluded spheres	No	No	No
3D Search[136,182]	Externally generated	Points at atoms (based on atomic number, number of bonded atoms and hydrogens, and charge at pH 7 – users can combine types); dummy points at (1) ring centroids, (2) 0.5 Å above and below rings, (3) 1.0 Å along lone-pair/proton vector of heteroatoms including halogens	Distance, angle, torsion angle, excluded volume	Distances from input structure between atoms, dummy points	No	Excluded spheres	No	No	No
CAVEAT[116,117]	Externally generated (original version used crystallographic[18])	Vectors between base atoms and tip replaceable atoms	Distance between base atoms, tip–base–tip angles, and tip–base–base–tip torsions	Distance between base atoms, tip–base–tip angles, and tip–base–base–tip torsions	No	No	No	No	No

continued

Table 5 Continued

Program	*Source of conformers*	*Query objects*	*Query types*	*Precalculated keys*	*Integrated with pharmacophore perception*	*Shape query*	*Partial matches*	*Integrated with molecular modeling visualization*	*Integrated with 3D QSAR*
Conquest[178,183]	Observed crystallographic, public, or proprietary	User described functional points, lines, torsion angles, planes	Distance, angle, torsion angle, excluded volume	Distance between atoms	No	Excluded spheres	No	No	No
ChemDBS-3D[49,156]	Externally generated and stored as keys, conformer generated at search time	Points at atoms based on atomic properties	Distance	Distance bounds of all conformers of input structure	Yes	No	No	Yes	No
UNITY[184]	Externally generated; directed tweak[154] to generate match	User described point, line, torsion angle, plane within 2D substructures	Distance, angle, torsion angle, excluded volume; plus 2D	Distance bounds of all conformers of input structure	Yes	Excluded spheres, can be automatically created	Yes	Yes	Loosely
MACCS 3D[185,186]	Externally generated and stored as keys, conformer generated at search time	User described point, line, torsion angle, plane within 2D substructures	Distance, angle, torsion angle, excluded volume; plus 2D	Unknown type	No	No	Unknown	No	No
Catalyst[160,187,188]	Externally generated[149]	System-defined functional points,[189] exclusion and inclusion points	Distance, RMS	Unknown for geometric queries. Shape keys[190]	Yes	Yes, separate	Yes	Yes	Yes
Phase[95]	Externally generated, can be within the program or at a search time	System-defined functional points			Yes	Yes, automatically created	Yes	Yes	Yes
ROCS[158]	Externally generated	Coordinates of conformation of molecule to be searched	Shape, match of force-field energetics, or pharmacophore types	No	No	Done first	Yes, continuous function	Yes	No

4.06.3.2 Sources of Coordinates

A 3D database searching program obviously needs 3D coordinates to search. The searching mechanism can generate the coordinates, usually a specific conformer from an existing 3D structure, or it can search only registered conformers (*see* 4.19.1.2 Virtual Screening).

4.06.3.2.1 Two- to three-dimensional conversion of structures

Although in principle one could generate the 3D coordinates by any molecular modeling method, 3D database searching of a large variety of molecules was not practical until CONCORD,[136,137] the first practical 3D structure generation program, was invented in 1988. CONCORD converts a structure diagram as stored into a chemical information database such as MACCS/ISIS,[138] or the SMILES,[139,140] or SLN[141] of a molecule to a 3D structure by a combination of templates, rules, and a novel energy minimization algorithm. Shortly thereafter, CORINA, a program with similar capabilities, was produced.[142] These programs have been widely used to convert corporate databases of in-house structures for 3D searching.[143] Currently it takes less than approximately 1 s to produce a 3D structure. A comparison of these programs, and some others that are no longer available, supports their utility, but also underscores the necessity to allow for conformational flexibility.[144,145] More recent enhancements of these programs include the possibility to generate all low-energy conformations of rings, all stereoisomers, and all tautomers.[146]

4.06.3.2.2 Multiple conformations

One method to consider conformational flexibility is to populate the database with all reasonable conformations of each database molecule. Generally, fewer hits were found in searches over databases built with only a CONCORD conformation than with the corresponding database built with the same molecules but including multiple low-energy conformations.[147] Clearly, the method of generating the multiple conformations is an important issue to be considered.

For optimum efficiency, the ensemble of conformations should meet two criteria: no conformation duplicates another and no conformation is high energy. One approach to producing a diverse ensemble of low-energy conformations is to generate conformations by molecular dynamics or Monte Carlo methods, but, once a particular conformation has been identified, it is associated with an artificially high-energy barrier.[148,149] This barrier prevents the algorithm from revisiting this region of conformational space. The conformations for a Catalyst[149] pharmacophore hypothesis and 3D database searches are typically generated in this fashion.

Omega[38] is a newer method that has been found to be not only fast, but also to usually include the bound conformation of a ligand, the bioactive conformation, in its conformational ensemble.[150,151] The method starts by generating a single preliminary 3D conformation from the structure diagram using a distance geometry method, which is then energy minimized. Rotation about single bonds drives the conformational expansion. For this Omega identifies fragments of the molecule in a database of previously examined molecular fragments, builds up the conformations, and discards the higher energy of pairs that are close in RMS fit.

The inventors of CONCORD and CORINA have each developed programs that also claim to efficiently search conformational space.[152,153] As these programs are newer, there has been no reported comparison of them with each other or with Omega.

4.06.3.2.3 Conformations generated during the search

One way to consider conformational flexibility is to generate the required conformation only at search time. The algorithms in UNITY,[154] MACCS-3D,[155] and ChemDBS-3D[156] do just this. At the time that a 3D structure is registered in the database, the algorithm sets keys (*see* Section 4.06.2.5) that correspond to every possible distance between potential pharmacophore atoms in the structure. At search time, a screening step eliminates all molecules except those for which the keys corresponding to the required pharmacophore distances have been set.[47,48,157] For the remaining molecules the database conformation is adjusted to match the search query as closely as possible. Because the algorithms for this can use analytical derivatives to drive the rotatable bonds to the target distance, attempting to generate a matching conformation is very fast.

4.06.3.3 Types of Three-Dimensional Searching

4.06.3.3.1 Geometric searching

By analogy with the discussions of pharmacophore mapping in previous sections of this chapter, one formulates a geometric 3D search in terms of a pharmacophore – the geometric relationships between points, lines, or planes. The points can be described as the locations of atoms of a particular atomic number, or more commonly of a particular interaction type. The points in a search query can also usually be located at some position calculated from the position of more than one atom of the molecule – complementary site points are one example, but points favorable for interaction with a π cloud or metal atoms or exclusions points are also possibilities. The geometric relationships of interest are the distance between two points, the angle between three points, and the torsion angle between four points or two planes.

4.06.3.3.2 Shape searching

There are also computer programs that have been developed to search 3D databases for molecules that match each other in shape. The shape can be the observed or proposed bioactive conformation[98,158] or one that is generated by a specialized algorithm.[129] This chapter will concentrate on pharmacophore 3D searching. Shape searching and docking to a macromolecular active site are discussed in Chapters 4.05 and 4.19.

4.06.3.3.3 Combining shape and geometric searching

A simple method to include shape into a geometric search is to locate exclusion points with reference to the pharmacophore geometric objects. The problem with this approach is that, because there are tolerances on the distances and angles of the pharmacophore, the location of such exclusion points can shift position. For this reason, UNITY[159] and Catalyst[160] have provisions to consider how database molecules fit into the proposed shape of the binding site. To do so in UNITY the user specifies a file that describes the *XYZ* coordinates of union volume of the superimposed active molecules and the corresponding coordinates of the pharmacophore points. Database compounds that match the required geometric queries are superimposed onto the pharmacophore points and attempts are made to modify the conformation so that the molecule fits within the union volume. To consider shape in Catalyst, the user selects one compound to represent the required shape in the coordinate space of the pharmacophore. The program then generates a set of spheres that enclose the conformation of the molecule into which any database molecules must fit. Both programs allow for minor misfit.

The opposite approach is taken in the program ROCS.[158] It first compares all database molecules to the shape query–the shape of a bioactive conformation of a molecule, of a set of superimposed molecules, of the electron density of an active site, or of an active site. The searching is done with a smooth Gaussian function[161] that processes 600–800 comparisons per second. ROCS then scores the hits based on a force-field or feature-type match to the query structure.

4.06.3.4 Sources of Constraints for Three-Dimensional Searching

It is clear from this chapter that one or more pharmacophore models form a useful search query. Sometimes the first results of such a search are surprising – often this happens when all of the molecules that generated the pharmacophore are very similar – and can lead one to add additional constraints to the search. A close integration of molecular modeling and 3D searching helps the researcher envision and describe the additional constraints. Although an initial pharmacophore may be described in terms of distances between points of a certain character, more precision can sometimes be obtained if one recognizes that certain angles are equally appropriate.

At the early phases of pharmacophore identification, one might use a 3D database search to discover compounds that mimic one or another of the low-energy conformations of a compound of interest. If no exact compounds are found, one can search for templates that hold the presumed pharmacophoric groups in the appropriate orientation.

In a similar manner, if one has information on the bound conformation of a ligand one might search for templates that would hold the important functional groups in the correct orientation. One would expect that such analogs would have increased affinity for the target because they lose less entropy upon binding.

4.06.4 Unresolved Challenges

Although pharmacophore modeling is clearly a useful strategy (*see* 4.21 Pharmacophore Modeling: 2 – Applications, for example) to be used within a medicinal chemistry program, there remain problems that, if solved, would increase the efficiency and accuracy of these methods. Although an ideal method would result in one pharmacophore hypothesis and one corresponding bioactive conformation for every molecule, it is useful if a program produces easily testable hypotheses.

The very first problem faced in pharmacophore modeling is the decision as to which compounds should be included. It would be helpful if the computer programs would highlight, before detailed analysis is undertaken, that certain molecules do not add information for the solution of the problem. The user should be prompted which compound(s) to remove from further consideration. A program could also point out that every molecule in the set has certain features and inquire if the user knows if these are essential or if removing them has not been explored – such an inquiry could suggest compounds that should be tested to improve the accuracy of the pharmacophore.

An important computational issue is how high in energy a bioactive conformation can be. Although this problem has been addressed in Section 4.06.2.4, it is not clear that the conformational energies calculated away from low-energy structures are accurate: There is little experimental data on which to calibrate a function that calculates energy away from the minimum because most observations of structures are made on minimum energy conformations. An additional problem is that each conformational program is parameterized differently with the result that the calculated energy of a conformation away from a minimum may differ substantially between programs. Beyond this, the calculated energy differences may depend on the structure of the compound being compared. Further complicating the analysis is the fact that many such programs calculate the energy of a structure in vacuum, with the result that internal hydrogen bonds or hydrophobic interactions may be favored in a way that is not relevant to the relative energy of the protein-bound conformation versus one in water.

Another issue is how one chooses between multiple conformations of a molecule, each of which matches the pharmacophore. Should this be based only on energy? On closeness of fit – if so, to the centroid or to some other compound (which)? Clearly, even if there are only five molecules, each of which has five matching conformations, this still leads to many different overlays of structures.

Another persistent issue with current techniques is the problem of multiple proposed pharmacophores. This is not too serious if they differ only slightly, because one can still formulate a 3D search query from the results. However, one might wish for a computer program that would (1) inform the user that all of the models superimpose a particular atom that can be either a hydrogen-bond donor or acceptor and (2) ask the user if there is any additional information, such as substituting the polar hydrogen with a methyl, to eliminate one or the other. If one does not have such information, the subsequent 3D search may be either too broad, if both possibilities are included, or too narrow, if one is arbitrarily selected.

In some cases a program will suggest several very different overlays. Currently, evaluation of such results is empirical. Should the precision of the overlay of certain features be given more weight than other features? If so, how would such weights be decided? How should one weigh overlapping many hydrogen-bonding and ionic features versus a small union volume of actives? Should the relative energies of the included conformations be considered? Or is it best to conclude that there isn't enough information to select one pharmacophore over the others and that more experimental information is needed?

Pharmacophore-based searching of 3D databases also presents challenges. The first concerns which molecules should be included in such a database. If the only use of such searches is to suggest molecules for further testing, then a database of available molecules is appropriate. However, if the goal of the work is to suggest new molecules, then the database could be skeletons to be searched with geometric constraints, no features, only with the idea that atomic properties will be added when a molecule is actually synthesized, as certain databases of CAVEAT are used.[116,117] Usually, however, it would be desirable to search a database of molecules that could be made using currently known chemistry and available starting materials. This can lead to a database orders of magnitude too large to search with current pharmacophore methods.[191] Could some sort of fragment-based pharmacophore strategy be devised?

Many 3D searching programs present results as pass/fail. If there are too many hits to progress to testing, then the user might do a second search with tighter criteria. However, this cannot be done if it will narrow the limits below those of the original pharmacophore. Post-processing by other considerations remains an art, still unexplored systematically.

References

1. Gund, P. *Prog. Mol. Subcell. Biol.* **1977**, *5*, 117–143.
2. Kier, L. B. Molecular Conformation. *Molecular Orbital Theory in Drug Research*; Academic Press: New York, 1971, pp 164–195.
3. Dror, O.; Shulman, P. A.; Nussinov, R.; Wolfson, H. J. *Curr. Med. Chem.* **2004**, *11*, 71–90.

3a. Van Drie, J. H. *Curr. Pharm. Des.* **2003**, *9*, 1649–1664.

4. Güner, O. F. *Pharmacophore Perception, Development, and Use in Drug Design*; International University Line: La Jolla, CA, 1999, p. 537.
5. Cottrell, S. J.; Gillet, V. J.; Taylor, R.; Wilton, D. J. *J. Comput.-Aided Mol. Des.* **2004**, *18*, 665–682.
6. Seeman, P.; Watanabe, M.; Grigoriadis, D.; Tedesco, J. L.; George, S. R.; Svensson, U.; Lars, J.; Nilsson, G.; Neumeyer, J. L. *Mol. Pharmacol.* **1985**, *28*, 391–399.
7. McDermed, J. D.; Freeman, H. S.; Ferris, R. M. Enantioselective Binding of (+) and (−) 2-Amino-6,7-Dihydroxy-1, 2,3,4-Tetrahydronaphthalenes and Related Agonists to Dopamine Receptors. In *Catecholamines: Basic and Clinical Frontiers*; Usdin, E., Ed.; Pergamon: New York, 1979, p 568.
8. Booher, R. N.; Kornfeld, E. C.; Smalstig, E. B.; Clemens, J. A. *J. Med. Chem.* **1987**, *30*, 580–583.
9. Jaen, J. C.; Caprathe, B. W.; Wise, L. D.; Pugsley, T. A.; Meltzer, L. T.; Heffner, T. G. *Bioorg. Med. Chem. Lett.* **1991**, *1*, 539–544.

10. Weinstock, J.; Gaitanopoulos, D. E.; Stringer, O. D.; Franz, R. G.; Hieble, J. P.; Kinter, L. B.; Mann, W. A.; Flaim, K. E.; Gessner, G. *J. Med. Chem.* **1987**, *30*, 1166–1176.
11. Li, H.; Sutter, J.; Hoffman, R. HypoGen: An Automated System for Generating 3D Predictive Pharmacophore Models. In *Pharmacophore Perception, Development, and Use in Drug Design*; Güner, O. F., Ed.; International University Line: La Jolla, CA, 1999; pp 171–189.
12. Kubinyi, H. *Drug Disc. Today* **1997**, *2*, 457–467.
13. Kubinyi, H. *Drug Disc. Today* **1997**, *2*, 538–546.
14. Sufrin, J. R.; Dunn, D. A.; Marshall, G. R. *Mol. Pharmacol.* **1981**, *19*, 307–313.
15. Pauling, L.; Delbrück, M. *Science* **1940**, *92*, 77–79.
16. Taylor, R.; Kennard, O. *Acc. Chem. Res.* **1984**, *17*, 320–326.
17. Mayer, D.; Naylor, C. B.; Motoc, I.; Marshall, G. R. *J. Comput.-Aided Mol. Des.* **1987**, *1*, 3–16.
18. Allen, F. H. *Acta Crystallogr.* **2002**, *B58*, 380–388.
19. Allen, F. H.; Kennard, O.; Taylor, R. *Acc. Chem. Res.* **1983**, *16*, 146–153.
20. Lommerse, J. P. M.; Stone, A. J.; Taylor, R.; Allen, F. H. *J. Am. Chem. Soc.* **1996**, *118*, 3108–3116.
21. Bruno, I. J.; Cole, J. C.; Lommerse, J.; Rowland, R. S.; Taylor, R.; Verdonk, M. L. *J. Comput.-Aided Mol. Des.* **1997**, *11*, 525–537.
22. Lommerse, J.; Taylor, R. *J. Enzyme Inhib.* **1997**, *11*.
23. Nobeli, I.; Price, S. L.; Lommerse, J.; Taylor, R. *J. Comput. Chem.* **1997**, *18*, 2060–2074.
24. Pitchford, N.; Taylor, R. Crystallographic Databases and Their Use for Studying Intermolecular Interactions. In *Designing Bioactive Molecules: Three-Dimensional Techniques*; Martin, Y. C., Willett, P., Eds.; American Chemical Society: Washington DC, 1997, pp 19–46.
25. Mills, J.; Dean, P. M. *J. Comput.-Aided Mol. Des.* **1996**, *10*, 607–622.
26. Crippen, G. *Distance Geometry and Conformational Calculations*; Research Studies Press: Letchworth, 1981, p 53.
27. Allinger, M. L. *MMP2*; Tripos Inc: St Louis, MO, 1995.
28. DISCOVER; Molecular Simulations: San Diego, CA, 1992.
29. Sybyl Molecular Modeling Software; TRIPOS, Inc.: St. Louis, MO, 1995.
30. Jorgensen, W. L.; Maxwell, D. S.; Tiradorives, J. *J. Am. Chem. Soc.* **1996**, *118*, 11225–11236.
31. Beachy, M. D.; Chasman, D.; Murphy, R. B.; Halgren, T. A.; Friesner, R. A. *J. Am. Chem. Soc.* **1997**, *119*, 5908–5920.
32. Perola, E.; Charifson, P. S. *J. Med. Chem.* **2004**, *47*, 2499–2510.
33. Halgren, T. A. *J. Comput. Chem.* **1996**, *17*, 490–519.
34. Halgren, T. A. *J. Comput. Chem.* **1996**, *17*, 520–552.
35. Halgren, T. A. *J. Comput. Chem.* **1996**, *17*, 616–641.
36. Halgren, T. A. *J. Comput. Chem.* **1996**, *17*, 553–586.
37. Halgren, T. A.; Nachbar, R. B. *J. Comput. Chem.* **1996**, *17*, 587–615.
38. *Omega*, OpenEye Scientific Software, 3600 Cerrillos Rd., Suite 1107: Santa Fe, NM, 87507, USA. www.eyesopen.com/products/applications/omega.html (accessed May 2006).
39. Kristam, R.; Gillet, V. J.; Lewis, R. A.; Thorner, D. *J. Chem. Inf. Model.* **2005**, *45*, 461–476.
40. Mason, J. S.; Morize, I.; Menard, P. R.; Cheney, D. L.; Hulme, C.; Labaudiniere, R. F. *J. Med. Chem.* **1999**, *42*, 3251–3264.
41. Cramer, R. D., III.; Patterson, D. E.; Bunce, J. D. *J. Am. Chem. Soc.* **1988**, *110*, 5959–5967.
42. Bersuker, I. B.; Bahceci, S.; Boggs, J. E.; Pearlman, R. S. *J. Comput.-Aided Mol. Des.* **1999**, *13*, 419–434.
43. Bersuker, I. B.; Bahçeci, S.; Boggs, J. E. *J. Chem. Inf. Comput. Sci.* **2000**, *40*, 1363–1376.
44. Bersuker, I. B. *Curr. Pharm. Des.* **2003**, *9*, 1575–1606.
45. Crippen, G. M. *J. Comp. Phys.* **1977**, *24*, 96–107.
46. Jakes, S. E.; Willett, P. *J. Mol. Graph. Model.* **1986**, *4*, 12–20.
47. Clark, D. E.; Willett, P.; Kenny, P. W. *J. Mol. Graph. Model.* **1992**, *10*, 194–204.
48. Clark, D. E.; Willett, P.; Kenny, P. W. *J. Mol. Graph. Model.* **1993**, *11*, 146–156.
49. Davies, K., Upton, R. *Network Science* **1995**, *1*. http://www.awod.com (accessed May 2006).
50. Pickett, S. D.; Mason, J. S.; Mclay, I. M. *J. Chem. Inf. Comput. Sci.* **1996**, *36*, 1214–1223.
51. Mason, J. S., Cheney, D. L. *Pac. Symp. Biocomput.* **1999**, 456–467.
52. Bradley, E. K.; Beroza, P.; Penzotti, J. E.; Grootenhuis, P. D.; Spellmeyer, D. C.; Miller, J. L. *J. Med. Chem.* **2000**, *43*, 2770–2774.
53. Abrahamian, E.; Fox, P. C.; Naerum, L.; Christensen, I. T.; Thørgersen, H.; Clark, R. D. *J. Chem. Inf. Comput. Sci.* **2003**, *43*, 458–468.
54. Poirrette, A. R.; Willett, P.; Allen, F. H. *J. Mol. Graph. Model.* **1991**, *9*, 203–217.
55. Feng, J.; Sanil, A.; Young, S. S. *J. Chem. Inf. Model.* **2006**, *46*, 1352–1359.
56. Chen, X.; Rusinko, A.; Tropsha, A.; Young, S. S. *J. Chem. Inf. Comput. Sci.* **1999**, *39*, 887–896.
57. Marshall, G. R.; Barry, C. D.; Bosshard, H. E.; Dammkoehler, R. A.; Dunn, D. A. The Conformation Parameter in Drug Design: The Active Analog Approach. In *Computer-Assisted Drug Design*; Olson, E. C., Christoffersen, R. E., Eds.; American Chemical Society: Washington, 1979, pp 205–226.
58. Dammkoehler, R. A.; Karasek, S. F.; Shands, E. F.; Marshall, G. R. *J. Comput.-Aided Mol. Des.* **1989**, *3*, 3–21.
59. Beusen, D. D.; Marshall, G. R. Pharmacophore Definition Using the Active Analog Approach. In *Pharmacophore Perception, Development, and Use in Drug Design*; Güner, O., Ed.; International University Line: La Jolla, CA, 1999, pp 21–45.
60. Dammkoehler, R. A.; Karasek, S. F.; Shands, E.; Marshall, G. R. *J. Comput.-Aided Mol. Des.* **1995**, *9*, 491–499.
61. Beusen, D. D.; Shands, E.; Karasek, S. F.; Marshall, G. R.; Dammkoehler, R. A. *J. Mol. Struct. – Theochem.* **1996**, *370*, 2–3.
62. Van Drie, J. H. *J. Comput.-Aided Mol. Des.* **1997**, *11*, 39–52.
63. Sheridan, R. P.; Nilakantan, R.; Dixon, J. S.; Venkataraghavan, R. *J. Med. Chem.* **1986**, *29*, 899–906.
64. Metropolis, R.; Rosenbluth, A.; Teller, A.; Teller, E. *J. Chem. Phys.* **1953**, *21*, 1087–1092.
65. Barakat, M. T.; Dean, P. M. *J. Comput.-Aided Mol. Des.* **1990**, *4*, 295–316.
66. Hodgkin, E. E.; Miller, A.; Whittaker, M. *J. Comput.-Aided Mol. Des.* **1993**, 7, 515–534.
67. Manly, B. F. J. *Multivariate Statistical Methods. A Primer*, 2nd ed.; Chapman and Hall: London, 1994, p 215.
68. Cosentino, U.; Moro, G.; Pitea, D.; Scolastico, S.; Todeschini, R.; Scolastico, C. *J. Comput.-Aided Mol. Des.* **1992**, *6*, 47–60.
69. Kato, Y.; Inoue, A.; Yamada, M.; Tomioka, N.; Itai, A. *J. Comput.-Aided Mol. Des.* **1992**, *6*, 475–486.
70. Kato, Y.; Itai, A.; Iitaka, Y. *Tetrahedron* **1987**, *43*, 5229–5234.
71. Brint, A. T.; Willett, P. *J. Chem. Inf. Comput. Sci.* **1987**, *27*, 152–158.
72. Takahashi, Y.; Maeda, S.; Sasaki, S.-I. *Anal. Chim. Acta* **1987**, *200*, 363–377.

73. Takahashi, Y.; Akagi, T.; Sasaki, S.-I. *Tetrahedron Comput. Methodol.* **1990**, *3*, 27–35.
74. Martin, Y. C.; Bures, M. G.; Danaher, E. A.; DeLazzer, J.; Lico, I.; Pavlik, P. A. *J. Comput.-Aided Mol. Des.* **1993**, 7, 83–102.
75. Golender, V. E.; Vorpagel, E. R. Computer-Assisted Pharmacophore Identification. In *3D QSAR in Drug Design. Theory Methods and Applications*; Kubinyi, H., Ed.; ESCOM: Leiden, 1993, pp 137–149.
76. Vorpagel, E. R.; Golender, V. E. Apex-3D: Activity Prediction Expert System with 3D QSAR. In *Pharmacophore Perception, Development, and Use in Drug Design*; Güner, O. F., Ed.; International University Line: La Jolla, CA, 1999, pp 128–149.
77. Holland, J. H. *Sci. Am.* **1992**, *267*, 66–72.
78. Payne, A. W. R.; Glen, R. C. *J. Mol. Graph. Model.* **1993**, *11*, 74–91.
79. Jones, G.; Willett, P.; Glen, R. C. GASP: Genetic Algorithm Superimposition Program. In *Pharmacophore Perception, Development, and Use in Drug Design*; Güner, O. F., Ed.; International University Line: La Jolla, CA, 1999, pp 85–107.
80. Handschuh, S.; Gasteiger, J. *J. Mol. Model.* **2000**, *6*, 358–378.
81. Clark, R. D. *Galahad*; Tripos: St. Louis, MO, USA. www.tripos.com (accessed May 2006).
82. Wold, S.; Ruhe, A.; Wold, H.; Dunn, W. J. *SIAM J. Sci. Stat. Comput.* **1984**, *5*, 735–743.
83. Hoskuldsson, A. *J. Chemometr.* **1988**, *2*, 211–228.
84. Bro, R. *J. Chemometr.* **1996**, *10*, 47–61.
85. Hasegawa, K.; Arakawa, M.; Funatsu, K. *Comput. Biol. Chem.* **2003**, *27*, 211–216.
86. Hopfinger, A. J.; Wang, S.; Tokarski, J. S.; Jin, B. Q.; Albuquerque, M.; Madhav, P. J.; Duraiswami, C. *J. Am. Chem. Soc.* **1997**, *119*, 10509–10524.
87. Hawkins, D. M.; Young, S. S.; Rusinko, A. *Quant. Struct.-Act. Relat.* **1997**, *16*, 296–302.
88. Chen, X.; Rusinko, A.; Young, S. S. *J. Chem. Inf. Comput. Sci.* **1998**, *38*, 1054–1062.
89. Young, S. S.; Hawkins, D. M. *SAR QSAR Environ. Res.* **1998**, *8*, 183–193.
90. Lawrence, C. E.; Altschul, S. F.; Boguski, M. S.; Liu, J. S.; Neuwald, A. F.; Wootton, J. C. *Science* **1993**, *262*, 208–214.
91. Crippen, G. M. *J. Med. Chem.* **1997**, *40*, 3161–3172.
92. Boulu, L. G.; Crippen, G. M.; Barton, H. A.; Kwon, H.; Marletta, M. A. *J. Med. Chem.* **1990**, *33*, 771–775.
93. Srivastava, S.; Richardson, W. W.; Bradley, M. P.; Crippen, G. M. Three-Dimensional Receptor Modeling Using Distance Geometry and Voronoi Polyhedra. In *3D QSAR in Drug Design. Theory Methods and Applications*; Kubinyi, H., Ed.; ESCOM: Leiden, 1993, pp 409–430.
94. http://www.accelrys.com/products/catalyst/ (accessed July 2006).
95. *Phase*, Schrödinger, 120 W 45th St.: New York, NY 10036-4041.
96. Kearsley, S. K., Smith, G. M. *Seal. An Alternate Method for the Alignment of Molecular Structures, Qcpe 634*, Quantum Chemistry Program Exchange, Indiana University: Bloomington, IN 47405.
97. Nissink, J. W. M.; Verdonk, M. L.; Kroon, J.; Mietzner, T.; Klebe, G. *J. Comput. Chem.* **1997**, *18*, 638–645.
98. Lemmen, C.; Lengauer, T.; Klebe, G. *J. Med. Chem.* **1998**, *41*, 4502–4520.
99. Lemmen, C.; Lengauer, T. *J. Comput.-Aided Mol. Des.* **1997**, *11*, 357–368.
100. Miller, M. D.; Sheridan, R. P.; Kearsley, S. K. *J. Med. Chem.* **1999**, *42*, 1505–1514.
101. Weber, P. C.; Wendoloski, J. J.; Pantoliano, M. W.; Salemme, F. R. *J. Am. Chem. Soc.* **1992**, *114*, 3197–3200.
102. Weichsel, A.; Montfort, W. R. *Nat. Struct. Biol.* **1995**, *2*, 1095–1101.
103. Verkhivker, G. M.; Bouzida, D.; Gehlhaar, D. K.; Rejto, P. A.; Freer, R. J.; Rose, P. W. *Curr. Opin. Struct. Biol.* **2002**, *12*, 197–203.
104. Birch, L.; Christopher, W. M.; Michael, J. H.; Ian, J. T.; Marcel, L. V. *J. Comput.-Aided Mol. Des.* **2002**, *16*, 855–869.
105. Erickson, J. A.; Jalaie, M.; Robertson, D. H.; Lewis, R. A.; Vieth, M. *J. Med. Chem.* **2004**, *47*, 45–55.
106. Daeyaert, F.; de Jonge, M.; Heeres, J.; Koymans, L.; Lewi, P.; Vinkers, M. H.; Janssen, P. A. J. *Proteins* **2004**, *54*, 526–533.
107. Horton, J. R.; Sawada, K.; Nishibori, M.; Cheng, X. *J. Mol. Biol.* **2005**, *353*, 334–344.
108. Cavasotto, C. N.; Abagyan, R. A. *J. Mol. Biol.* **2004**, *337*, 209–225.
109. Brenk, R.; Meyer, E. A.; Reuter, K.; Stubbs, M. T.; Garcia, G. A.; Diederich, F.; Klebe, G. *J. Mol. Biol.* **2004**, *338*, 55–75.
110. Gao, F.; Bren, N.; Little, A.; Wang, H. L.; Hansen, S. B.; Talley, T. T.; Taylor, P.; Sine, S. M. *J. Biol. Chem.* **2003**, *278*, 23020–23026.
111. Sine, S. M.; Wang, H. L.; Gao, F. *Curr. Med. Chem.* **2004**, *11*, 559–567.
112. Fritz, T. A.; Tondi, D.; Finer-Moore, J. S.; Costi, M. P.; Stroud, R. M. *Chem. Biol.* **2001**, *8*, 981–995.
113. Espinosa, J. F.; Canada, F. J.; Asensio, J. L.; Dietrich, H.; Martinlomas, M.; Schmidt, R. R.; Jimenezbarbero, J. *Angew. Chem. Int. Ed.* **1996**, *35*, 303–306.
114. Gruneberg, S.; Stubbs, M. T.; Klebe, G. *J. Med. Chem.* **2002**, *45*, 3588–3602.
115. Moitessier, N.; Henry, C.; Maigret, B.; Chapleur, Y. *J. Med. Chem.* **2004**, *47*, 4178–4187.
116. Lauri, G.; Bartlett, P. A. *J. Comput.-Aided Mol. Des.* **1994**, *8*, 51–66.
117. Bartlett, P. A.; Shea, G. T.; Telfer, S. J.; Waterman, S. CAVEAT: A Program to Facilitate the Structure-Derived Design of Biologically Active Molecules. In *Chemical and Biological Problems in Molecular Recognition*; Roberts, S. M., Ley, S. V., Campbell, M. M., Eds.; Roy. Soc. Chem.: Cambridge, 1989, pp 182–196.
118. Willett, P. *J. Mol. Recognit.* **1995**, *8*, 290–303.
119. Thorner, D. A.; Willett, P.; Wright, P. M.; Taylor, R. *J. Comput.-Aided Mol. Des.* **1997**, *11*, 163–174.
120. Thorner, D. A.; Wild, D. J.; Willett, P.; Wright, P. M. *J. Chem. Inf. Comput. Sci.* **1996**, *36*, 900–908.
121. Wang, X.; Wang, J. T. L. *J. Chem. Inf. Comput. Sci.* **2000**, *40*, 442–451.
122. Good, A. C. *Internet Journal of Chemistry* **2000**, *3*, NIL_3–NIL_21.
123. Makara, G. M. *J. Med. Chem.* **2001**, *44*, 3563–3571.
124. Irwin, J. J.; Shoichet, B. K. *J. Chem. Inf. Model* **2005**, *45*, 177–182.
125. Iresearch Library, ChemNavigator: San Diego, CA. http://www.chemnavigator.com (accessed May 2006).
126. NCI DTP/2D and 3D Structural Information. http://dtp.nci.nih.gov/docs/3d_database/structural_information/structural_data.html (accessed May 2006).
127. Voigt, J. H.; Bienfait, B.; Wang, S.; Nicklaus, M. C. *J. Chem. Inf. Comput. Sci.* **2001**, *41*, 702–712.
128. Van Drie, J. H.; Nugent, R. A. *SAR QSAR Environ. Res.* **1998**, *9*, 1062–1936X.
129. Andrews, K. M.; Cramer, R. D. *J. Med. Chem.* **2000**, *43*, 1723–1740.
130. Cramer, R. D. *J. Med. Chem.* **2003**, *46*, 374–388.
131. WDI: World Drug Index, Derwent: London.
132. ACD: Available Chemicals Directory; Elsevier MDL, San Ramon, CA.

133. Nilakantan, R.; Bauman, N.; Haraki, K. S. *J. Comput.-Aided Mol. Des.* **1997**, *11*, 447–452.
134. Rishton, G. M. *Drug Disc. Today* **1997**, *2*, 382–384.
135. Lipinski, C. A.; Lombardo, F.; Dominy, B. W.; Feeney, P. J. *Adv. Drug Deliv. Rev.* **1997**, *23*, 3–25.
136. Rusinko, A., III.; Sheridan, R. P.; Nilakantan, R.; Haraki, K. S.; Bauman, N.; Venkataraghavan, R. *J. Chem. Inf. Comput. Sci.* **1989**, *29*, 251–255.
137. Rusinko, A. I.; Skell, J. M.; Balducci, R.; McGarity, C. M.; Pearlman, R. S. *Concord, a Program for the Rapid Generation of High Quality Approximate 3-Dimensional Molecular Structures*, 3.01 ed.; The University of Texas at Austin and Tripos Associates: St. Louis, Missouri.
138. ISIS: Elsevier MDL, San Ramon, CA.
139. Weininger, D.; Weininger, A. *J. Chem. Inf. Comput. Sci.* **1988**, *28*, 31–36.
140. Weininger, D.; Weininger, A.; Weininger, J. L. *J. Chem. Inf. Comput. Sci.* **1989**, *29*, 97–101.
141. Ash, S.; Cline, M. A.; Homer, R. W.; Hurst, T.; Smith, G. B. *J. Chem. Inf. Comput. Sci.* **1997**, *37*, 71–79.
142. Gasteiger, J.; Rudolph, C.; Sadowski, A. *Tetrahedron Comput. Methodol.* **1992**, *3*, 537–547.
143. Sadowski, J.; Gasteiger, J. *Chem. Rev.* **1993**, *93*, 2567–2581.
144. Ricketts, E. M.; Bradshaw, J.; Hann, M.; Hayes, F.; Tanna, N.; Ricketts, D. M. *J. Chem. Inf. Comput. Sci.* **1993**, *33*, 905–925.
145. Sadowski, J.; Gasteiger, J.; Klebe, G. *J. Chem. Inf. Comput. Sci.* **1994**, *34*, 1000–1008.
146. *Stereoplex and Protoplex*, Tripos: St. Louis, MO. www.tripos.com (accessed May 2006).
147. Haraki, K. S.; Sheridan, R. P.; Venkataraghavan, R. *Tetrahedron Comput. Methodol.* **1990**, *3*, 565–573.
148. Huber, T.; Torda, A. E.; van Gunsteren, W. F. *J. Comput.-Aided Mol. Des.* **1994**, *8*, 695–708.
149. Smellie, A.; Teig, S. L.; Towbin, P. *J. Comput. Chem.* **1995**, *16*, 171–187.
150. Boström, J. *J. Comput.-Aided Mol. Des.* **2001**, *15*, 1137–1152.
151. Boström, J.; Greenwood, J. R.; Gottfries, J. *J. Mol. Graph. Model.* **2003**, *21*, 449–462.
152. *Confort*; Tripos, Inc.: St. Louis, MO.
153. *Rotate*; Molecular Networks GmbH: Erlangen, Germany.
154. Hurst, T. *J. Chem. Inf. Comput. Sci.* **1994**, *34*, 190–196.
155. Moock, T. E.; Henry, D. R.; Ozkabak, A. G.; Alamgir, M. *J. Chem. Inf. Comput. Sci.* **1994**, *34*, 184–189.
156. Murrall, N. W.; Davies, E. K. *J. Chem. Inf. Comput. Sci.* **1990**, *30*, 312–316.
157. Clark, D. E., Jones, G., Willett, Kenny, P. W., Glen, R. C., *J. Chem. Inf. Comput. Sci.* **1994**,*34*, 197–206.
158. *ROCS*, OpenEye Scientific Software: Santa Fe, NM. http://www.eyesopen.com/products/applications/rocs.html (accessed May 2006).
159. *Unity Chemical Information Software*, Tripos Associates: St. Louis, MO. www.tripos.com (accessed May 2006).
160. Sprague, P. W. *Perspect. Drug Disc. Des.* **1995**, *3*, 1–20.
161. Grant, J. A.; Gallardo, M. A.; Pickup, B. T. *J. Comput. Chem.* **1996**, *17*, 1653–1666.
162. Waszkowycz, B.; Clark, D. E.; Frenkel, D.; Li, J.; Murray, C. W.; Robson, B.; Westhead, D. R. *J. Med. Chem.* **1994**, *37*, 3994–4002.
163. Takahasi, Y.; Akagi, T.; Sasaki, S.-I. *Tetrahedron Comput. Methodol.* **1990**, *3*, 27–35.
164. Perkins, T. D. J.; Dean, P. M. *J. Comput.-Aided Mol. Des.* **1993**, 7, 155–172.
165. Barnum, D.; Greene, J.; Smellie, A.; Sprague, P. *J. Chem. Inf. Comput. Sci.* **1996**, *36*, 563–571.
166. Clement, O. O.; Mehl, A. T. Hiphop: Pharmacophores Based on Multiple Common Feature Alignments. In *Pharmacophore Perception, Development, and Use in Drug Design*; Güner, O., Ed.; International University Line: La Jolla, CA, 1999, pp 69–84.
167. Jones, G.; Willett, P.; Glen, R. C. *J. Comput.-Aided Mol. Des.* **1995**, *9*, 532–549.
168. Holliday, J. D.; Willett, P. *J. Mol. Graph. Model.* **1997**, *15*, 221–232.
169. Handschuh, S.; Wagener, M.; Gasteiger, J. *J. Chem. Inf. Comput. Sci.* **1998**, *38*, 220–232.
170. Huang, P.; Kim, S.; Loew, G. *J. Comput.-Aided Mol. Des.* **1997**, *11*, 21–28.
171. Cosentino, U.; Moro, G.; Pitea, D.; Todeschini, R.; Brossa, S.; Gualandi, F.; Scolastico, C.; Giannessi, F. *Quant. Stuct.-Act. Relat.* **1990**, *9*, 195–201.
172. Jain, A. N.; Dietterich, T. G.; Lathrop, R. H.; Chapman, D.; Critchlow, R. E., Jr.; Bauer, B. E.; Webster, T. A.; Lozano-Perez, T. *J. Comput.-Aided Mol. Des.* **1994**, *8*, 635–652.
173. Jain, A. N.; Koile, K.; Chapman, D. *J. Med. Chem.* **1994**, *37*, 2315–2327.
174. Jain, A. N.; Harris, N. L.; Park, J. Y. *J. Med. Chem.* **1995**, *38*, 1295–1308.
175. Chapman, D.; Critchlow, R.; Jain, A. N.; Lathrop, R.; Perez, T. L.; Dietterich, T. *Machine-Learning Approach to Modeling Biological Activity for Molecular Design and to Modeling Other Characteristics*; Arris Pharmaceutical Corporation: US, 1996.
176. Putta, S.; Landrum, G. A.; Penzotti, J. E. *J. Med. Chem.* **2005**, *48*, 3313–3318.
177. Marchand, G. N.; Watson, K. A.; Alsberg, B. K.; King, R. D. *J. Med. Chem.* **2002**, *45*, 399–409.
178. Allen, F. H.; Davies, J. E.; Galloy, J. J.; Johnson, O.; Kennard, O.; Macrea, C. F.; Mitchell, E. M.; Mitchell, G. F.; Smith, J. M.; Watson, D. G. *J. Chem. Inf. Comput. Sci.* **1991**, *31*, 187–204.
179. Jakes, S. E.; Watts, N.; Willett, P.; Bawden, D.; Fisher, J. D. *J. Mol. Graph. Model.* **1987**, *5*, 41–48.
180. Van Drie, J. H.; Weininger, D.; Martin, Y. C. *J. Comput.-Aided Mol. Des.* **1989**, *3*, 225–251.
181. Martin, Y. C.; Danaher, E. B.; May, C. S.; Weininger, D. *J. Comput.-Aided Mol. Des.* **1988**, *2*, 15–29.
182. Sheridan, R. P.; Nilakantan, R.; Rusinko, A.; Bauman, N.; Haraki, K.; Ventataraghavan, R. *J. Chem. Inf. Comput. Sci.* **1989**, *29*, 255–260.
183. Bruno, I. J.; Cole, J. C.; Edgington, P. R.; Kessler, M.; Macrae, C. F.; McCabe, P.; Pearson, J.; Taylor, R. *Acta Crystallogr., Sect. B: Struct. Sci.* **2002**, *58*, 389–397.
184. Unity Chemical Information Software, Tripos Associates: St Louis, MO, 1993.
185. Güner, O. F.; Hughes, D. W.; Dumont, L. M. *J. Chem. Inf. Comput. Sci.* **1991**, *31*, 408–414.
186. Christie, B. D.; Henry, D. R.; Wipke, W. T.; Moock, T. E. *Tetrahedron Comput. Methodol.* **1990**, *3*, 653–664.
187. Kuchner, O.; Arnold, F. H. *Trends Biotechnol.* **1997**, *15*, 523–530.
188. Kurogi, Y.; Güner, O. F. *Curr. Med. Chem.* **2001**, *8*, 1035–1055.
189. Greene, J.; Kahn, S.; Savoj, H.; Sprague, P.; Teig, S. *J. Chem. Inf. Comput. Sci.* **1994**, *34*, 1297–1308.
190. Hahn, M. *J. Chem. Inf. Comput. Sci.* **1997**, *37*, 80–86.
191. Andrews, K. M.; Cramer, R. D. *J. Med. Chem.* **2000**, *43*, 1723–1740.

Biography

Yvonne Connolly Martin did her undergraduate work in chemistry and biology at Carleton College in Minnesota and her PhD in chemistry at Northwestern University in Illinois. Her entire career has been at Abbott except for short visits to Pomona College. Since 1970 she has focused on computer-assisted drug design. Her research interests include strategies for increasing the molecular diversity of compound collections; to triage hits from HTS; to predict the potency, binding affinity, or ADME properties of small molecules from their molecular structures; and for the computer design of novel compounds including 3D database searching, de novo design, and pharmacophore mapping.

© 2007 Elsevier Ltd. All Rights Reserved
No part of this publication may be reproduced, stored in any retrieval system or transmitted in any form by any means electronic, electrostatic, magnetic tape, mechanical, photocopying, recording or otherwise, without permission in writing from the publishers

Comprehensive Medicinal Chemistry II
ISBN (set): 0-08-044513-6

ISBN (Volume 4) 0-08-044517-9; pp. 119–147

4.07 Predictive Quantitative Structure–Activity Relationship Modeling

A Tropsha, University of North Carolina at Chapel Hill, Chapel Hill, NC, USA

© 2007 Elsevier Ltd. All Rights Reserved.

4.07.1 Introduction

4.07.1.1 Quantitative Structure–Activity Relationship (QSAR) Modeling in Modern Medicinal Chemistry

At the beginning of its over 40 years of existence as an independent area of research, quantitative structure–activity relationship (QSAR) modeling was viewed strictly as analytical physical chemical approach applicable only to small congeneric series of molecules. The technique was first introduced by Hansch *et al.*[3] on the basis of implications from linear free-energy relationships in general and the Hammett equation in particular.[4] It is based upon the assumption that differences in physicochemical properties account for the differences in biological activities of compounds. According to this approach, the changes in physicochemical properties that affect the biological activities of a set of congeners are of three major types: electronic, steric, and hydrophobic.[5] These structural properties are often described by Hammett electronic constants,[5] Verloop STERIMOL parameters,[6] hydrophobic constants,[7] etc. The quantitative relationships between biological activity (or chemical property) and the structural parameters could be conventionally obtained using multiple linear regression (MLR) analysis. The fundamentals and applications of this method in chemistry and biology have been summarized by Hansch and Leo.[5] This traditional QSAR approach has generated many useful and, in some cases, predictive QSAR equations and led to several documented drug discoveries.[8–10]

Many years of active research in QSAR have dramatically changed the breadth and the depth of this field in all its components including the diversity of target properties, descriptor types, data modeling approaches, and applications. QSAR modeling as an integral part of modern medicinal chemistry is experiencing one of the most exciting periods in its history. The changes have been brought about by an extraordinarily rapid expansion of available biomolecular databases and the growing use of advanced data mining technologies for the analysis of experimental medicinal chemistry data. Today, QSAR researchers are actively expanding the areas of application of QSAR approaches and concepts, with recent examples provided by the use of QSAR approaches as applied to the results of molecular dynamics simulations,[11] or protein sequence or structure classification,[12,13] or scoring functions for protein ligand docking.[14]

These recent developments have dramatically altered our approaches to the analysis of a relationship between chemical structure and biological action. Modern approaches to drug design and discovery are characterized by computational tool integration and a paradigm shift from the analysis of small congeneric series of molecules to the analysis of broad groups of biologically active molecules with applications to the design and discovery of drug-like molecules with optimal absorption, distribution, metabolism, and excretion (ADME), and toxicity properties. The most important changes in QSAR deal with a substantial increase in the size of data sets available for the analysis and an increasing use of QSAR models as virtual screening tools to discover biologically active molecules in chemical databases and virtual chemical libraries.

One of the most characteristic features of the modern age of QSAR as an integral part of drug design and discovery is an unprecedented growth of biomolecular databases, which contain data on chemical structure and in many cases, biological activity (or other relevant drug properties such as toxicity or mutagenicity) of chemicals. Naturally, the growth of molecular databases has been concurrent with the acceleration of the drug discovery process. According to an excellent, historical account of drug discovery,[15] due to high-throughput screening (HTS) technologies, the amount of raw data points obtained by a large pharmaceutical company per year has increased from approximately 200 000 at the beginning of last decade to around 50 million today. The total number of drugs used worldwide is approximately 80 000, which have fewer than 500 characterized molecular targets.[15] Recent estimates suggest that the number of potential targets lies between 5000 and 10 000, approximately 10-fold greater than the number of targets currently pursued (*see* 4.17 Chemogenomics in Drug Discovery – The Druggable Genome and Target Class Properties).[15]

The US National Institutes of Health (NIH) recently initiated an unprecedented public effort (Roadmap[107]) that is poised to boost all aspects of medicinal chemistry. One of the chief stated objectives of the Molecular Libraries Initiative (MLI), a component of Roadmap, is 'to develop and test new algorithms for computational chemistry and virtual screening' to facilitate the discovery of 'chemical probes to study the functions of genes, cells, and biochemical pathways.'[107] The funded Molecular Libraries Screening Center Network (MLSCN) includes 10 centers that are charged 'to screen a minimum of 100 000 compounds in 20 assays that have been adapted for HTS within each center per year' by the end of the pilot 3-year period. A simple estimate suggests that in 3 years, the MLSCN is likely to start depositing no fewer than 12 million screening data points per year! It is hard to evaluate the anticipated dimension of the cumulative library assay data matrix that will result from the MLI. However, given the current plans of the Roadmap to establish an initial collection of 500 000 diverse small molecules that will be eventually screened against hundreds of targets implemented in the MLSCN, it is safe to assume that the initial two-dimensional (2D) data matrix is likely to include at least 500 000 compound rows and at least 120 assay columns. Thus, the Molecular Libraries Roadmap will soon establish an unprecedented, at least in the public sector, collection of biological profiles of chemical compounds. The issues related to storing and accessing this collection will be addressed by the intramural PubChem[16] project. Enormous challenges remain, however, in converting these data into knowledge that will guide future library and compound design efforts and will ultimately have measurable benefits to public health in agreement with the NIH Roadmap goals.

While traditional QSAR modeling has typically been limited to dealing with a maximum of several dozen compounds at a time, rapid generation of large quantities of data requires new methodologies for data analysis. New approaches need to be developed to establish QSAR models for hundreds, if not thousands, of molecules. These new methods should be robust, yet sufficiently computationally efficient to compete with the data generation and the analytical requirements of experimental techniques, such as combinatorial chemistry and HTS.

It is practically impossible to review all, even relatively recent, developments in the field of QSAR in a single chapter. Many reviews discussing different QSAR modeling methodologies have been published,[17,18] and the reader is referred to this collection of general references and publications cited therein for additional in-depth information, and to Chapters 4.03 (Quantitative Structure–Activity Relationship – A Historical Perspective and the Future) and 4.23 (Three-Dimensional Quantitative Structure–Activity Relationship: The State of the Art).

The present chapter concentrates on current trends and developments in QSAR methodology, which are characterized by the growing size of the data sets subjected to the QSAR analysis, use of multiple descriptors of chemical structure, application of both linear and especially nonlinear, optimization algorithms applicable to multidimensional modeling, growing emphasis on rigorous model validation, and application of QSAR models as virtual screening tools in database mining and chemical library design. We begin by establishing general principles of QSAR modeling, emphasizing the common aspects of various QSAR methodologies. We then consider some popular approaches to the derivation of molecular descriptors and optimization algorithms, in the context of three key components of any QSAR investigation: model development, model validation, and model application. We conclude with several remarks on present status and future developments in this exciting research discipline.

4.07.1.2 Key Quantitative Structure–Activity Relationship Concepts

An inexperienced user or sometimes even an avid practitioner of QSAR could be easily confused by the diversity of methodologies and naming conventions used in QSAR studies. 2D or three-dimensional (3D) QSAR, variable selection or artificial neural network (ANN) methods, Comparative molecular field analysis (CoMFA), or binary QSAR present examples of various terms that may appear to describe totally independent approaches, which cannot be generalized or even compared to each other. In fact, any QSAR method can be generally defined as an application of mathematical and statistical methods to the problem of finding empirical relationships (QSAR models) of the form $P_i = \hat{k}(D_1, D_2, \ldots D_n)$, where P_i are biological activities (or other properties of interest) of molecules, $D_1, D_2, \ldots, D_n$ are calculated (or, sometimes, experimentally measured) structural properties (molecular descriptors) of compounds, and $\hat{k}$ is some empirically established mathematical transformation that should be applied to descriptors to calculate the property values for all molecules. The relationship between values of descriptors D and target properties P can be linear (e.g., MLR as in the Hansch QSAR approach), where target property can be predicted directly from the descriptor values, or nonlinear (such as ANNs or classification QSAR methods) where descriptor values are used in characterizing chemical similarity between molecules, which in turn is used to predict compound activity. In general, each compound can be represented by a point in a multidimensional space, in which descriptors $D_1, D_2, \ldots, D_n$ serve as independent coordinates of the compound. The goal of QSAR modeling is to establish a trend in the descriptor values, which parallels the trend in biological activity. All QSAR approaches imply, directly or indirectly, a simple similarity principle, which for a long time has provided a foundation for the experimental medicinal chemistry: compounds with similar structures are expected to have similar biological activities. This implies that points representing compounds with similar activities in multidimensional descriptor space should be geometrically close to each other, and vice versa.

Despite formal differences between various methodologies, any QSAR method is based on a QSAR table, which can be generalized as shown in **Table 1**. To initiate a QSAR study, this table must include some identifiers of chemical structures (e.g., company id numbers, first column of **Table 1**), reliably measured values of biological activity (or any other target property of interest, e.g., solubility, metabolic transformation rate, etc., second column of **Table 1**), and

Table 1 Generalized QSAR table

Structure id	*Target property (EC_{50}, K_i, etc.)*	*Structural properties (descriptors)*			
Compound 1	*P*1	*D*11	*D*12	...	*D*1*n*
Compound 1	*P*2	*D*21	*D*22	...	*D*2*n*
...	...	"	"	"	"
Compound *m*	*Pm*	*Dm*1	*Dm*2	...	*Dmn*

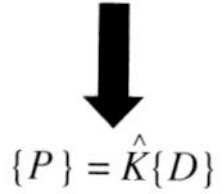

$\{P\} = \hat{K}\{D\}$

calculated values of molecular descriptors in all remaining columns (sometimes, experimentally determined physical properties of compounds could be used as descriptors as well).

The differences in various QSAR methodologies can be understood in terms of the types of target property values, descriptors, and optimization algorithms used to relate descriptors to the target properties and generate statistically significant models. Target properties (regarded as dependent variables in statistical data modeling sense) can generally be of three types: (1) continuous, i.e., real values covering certain range, e.g., IC_{50} values, or binding constants; (2) categorical related, classes of target properties covering certain range of values, e.g., active and inactive compounds, frequently encoded numerically for the purpose of the subsequent analysis as 1 (for active) or 0 (for inactive), or adjacent classes of metabolic stability such as unstable, moderately stable, stable; and (3) categorical unrelated, classes of target properties that do not relate to each other in any continuum, e.g., compounds that belong to different pharmacological classes, or compounds that are classified as drugs versus nondrugs. As simple as it appears, understanding this classification is actually very important since the choice of descriptor types as well as modeling techniques (see below) is often dictated by the type of the target properties. Thus, in general the latter two types require classification modeling approaches whereas the first type of target properties allows the use of linear regression modeling. The corresponding methods of data analysis are referred to as classification or continuous property QSAR.

Chemical descriptors (or independent variables in terms of statistical data modeling) can be typically classified into two types: continuous (i.e., range of real values, e.g., as simple as molecular weight or many molecular connectivity indices); or categorical related (i.e., classes corresponding to adjacent ranges of real values, e.g., counts of functional groups or binary descriptors indicating presence or absence of a chemical functional group or an atom in a molecule). Descriptors can be generated from various representations of molecules, e.g., 2D chemical graphs or 3D molecular geometries, giving rise to the terms of 2D or 3D QSAR, respectively. Understanding these types of descriptors is also important for understanding basic principles of QSAR modeling since as stated above any modeling implies establishing the correlation between chemical similarity between compounds and similarity between their target properties. Chemical similarity is calculated in the descriptors space using various similarity metrics (see excellent reviews by Willett[19]); thus the choice of the metric is dictated in many cases by the descriptor type. For instance, in case of continuous descriptor variables the Euclidean distance in descriptor space is a reasonable choice of the similarity metric whereas in case of binary variables metrics such as the Tanimoto coefficient or Manhattan distance would appear more appropriate.

Finally, correlation methods (which can be used either with or without variable selection) can be classified into two major categories, i.e., linear (e.g., linear regression (LR), or principal component regression (PCR), or partial least squares (PLS)) or nonlinear (e.g., *k* nearest neighbor (*k*NN), recursive partitioning (RP), ANNs, or support vector machines (SVMs). Most of QSAR researchers practice their preferred modeling techniques, and the choice of the technique is frequently coupled with the choice of descriptor types. However, there are recent attempts (discussed in more detail below) to combine various modeling techniques and descriptor types as applied to individual data sets.[20]

In some cases, the types of biological data, the choice of descriptors, and the class of optimization methods are closely related and mutually inclusive. For instance, MLR can only be applied when a relatively small number of molecular descriptors are used (at least five to six times less than the total number of compounds) and the target property is characterized by a continuous range of values. The use of multiple descriptors makes it impossible to use MLR due to a high chance of spurious correlation[21] and requires the use of PLS or nonlinear optimization techniques. However, in general, for any given data set a user can choose between various types of descriptors and various optimization schemes, combining them in a practically mix-and-match mode, to arrive at statistically significant QSAR models in a variety of ways. This situation is in essence analogous to molecular mechanics[22] calculations where different force fields and differently derived parameters are developed in different groups, but the common goal is to compute optimized energies and geometries of molecules from their chemical composition and coordinates of all atoms. Thus in general, all QSAR models can be universally compared in terms of their statistical significance, and, most importantly, their ability to predict accurately biological activities (or other target properties) of molecules not included in the training set (cf. molecular mechanics where different methods are ultimately compared by their ability to reproduce experimental molecular geometries). This concept of the predictive ability as a universal characteristic of QSAR modeling independent of the particulars of individual approaches should be kept in mind as we consider examples of QSAR tools, their applications and pitfalls in the subsequent sections of this chapter.

4.07.2 Molecular Descriptors

It has been said frequently that there are three keys to the success of any QSAR model building exercise: descriptors, descriptors, and descriptors. Many different molecular representations have been proposed, including Hansch-type parameters, topological indices,[23,24] quantum mechanical descriptors,[25] molecular shapes,[26] molecular fields,[27] atomic counts,[28] 2D fragments,[29] 3D fragments,[30] etc. A recent review by Livingstone[31] provides an excellent survey of various 2D and 3D descriptors, along with some associated diversity and similarity functions. Various physicochemical parameters such as the partition coefficient, molar refractivity, and quantum mechanical quantities such as highest occupied molecular orbital (HOMO) and lowest occupied molecular orbital (LOMO) energies have been used to represent molecular identities in early QSAR studies using linear and MLR. However, these descriptors are not suited for the analysis of large numbers of molecules either because of the lack of physicochemical parameters for compounds yet to be synthesized, or because of the computational expenses required by quantum mechanical methods. Recent years have seen the application of various topological descriptors that are usually derived from either 2D or 3D molecular structural information based on the graph theory or molecular topology. These descriptors are generated on the basis of the molecular connectivity, 3D molecular topography, and molecular field properties. We discuss below most popular types of molecular descriptors used in QSAR studies.

4.07.2.1 Topological Descriptors

Two widely applied examples of 2D molecular descriptors are molecular connectivity indices (MCI) and atom pair (AP) descriptors, initially developed by Carhart *et al.*[29] Most 2D QSAR methods have been extensively studied by Randic,[32] and Kier and Hall [33–38] based on graph theoretic indices. Although the physicochemical meaning of these structural indices is unclear, they certainly represent different aspects of molecular structures. These topological indices have been successfully combined with MLR analysis.[39] They have been extensively applied to analytical chemistry, toxicity analysis, and other areas of biological activity prediction.[40–43]

A popular MolConnZ software[44] affords the computation of a wide range of topological indices of molecular structure. These indices include (but are not limited to) the following descriptors: simple and valence path, cluster, path/cluster and chain molecular connectivity indices, kappa molecular shape indices, topological and electro-topological state indices, differential connectivity indices, graph's radius and diameter, Wiener and Platt indices, Shannon and Bonchev-Trinajstić information indices, counts of different vertices, and counts of paths and edges between different kinds of vertices.

Overall, MolConnZ produces over 400 different descriptors. Most of these descriptors characterize chemical structure, but several depend upon the arbitrary numbering of atoms in a molecule and are introduced solely for bookkeeping purposes. In a typical QSAR study, only about a half of all possible MolConnZ descriptors are eventually used after deleting descriptors with zero value or zero variance. **Figure 1** provides a summary of these molecular descriptors and presents some algorithms used in their derivation.

The idea of using atom pairs as molecular features in structure–activity relationship (SAR) studies was first proposed by Carhart *et al.*[29] AP descriptors are defined by their atom types and topological distance bins. An AP is a substructure defined by two atom types and the shortest path separation (or graph distance) between the atoms. The graph distance is defined as the smallest number of atoms along the path connecting two atoms in a molecular structure. The general form of an atom pair descriptor is as follows:

atom type *i* ------ (distance) ------ atom type *j*

where atom chemical types are typically defined by the user. For example, 15 atom types can be defined using SYBYL mol2 format as follows: (1) negative charge center, NCC; (2) positive charge center, PCC; (3) hydrogen bond acceptor, HA; (4) hydrogen bond donor, HD; (5) aromatic ring center, ARC; (6) nitrogen atoms, N; (7) oxygen atoms, O; (8) sulfur atoms, S; (9) phosphorous atoms, P; (10) fluorine atoms, FL; (11) chlorine, bromine, iodine atoms, HAL; (12) carbon atoms, C; (13) all other elements, OE; (14) triple bond center, TBC; (15) double bond center, DBC. Apparently, the total number of pairwise combinations of all 15 atom types is 120. Further, distance bins should be defined to discriminate between identical atom pairs separated by different graph distances and therefore representing different molecular substructures. Thus, 15 distance bins can be introduced in the interval between graph distance zero (i.e., zero atoms separating an atom pair) to 14 and greater. Thus, in this a total of 1800 (120×15) AP descriptors can be generated for any molecular structure. An example of an AP descriptor is shown in **Figure 2**. Frequently, as applied to particular data sets, many of the theoretically possible AP descriptors have zero value (implying that certain atom types or atom pairs are absent in molecular structures).

Dragon descriptors[45] include different groups: constitutional descriptors, topological indices, molecular walk counts, BCUT descriptors, Galvez topological charge indices, 2D autocorrelations, charge indices, aromaticity indices, Randic molecular profiles, geometrical descriptors, radial distribution junction (RDF) descriptors, 3D-MoRSE descriptors,

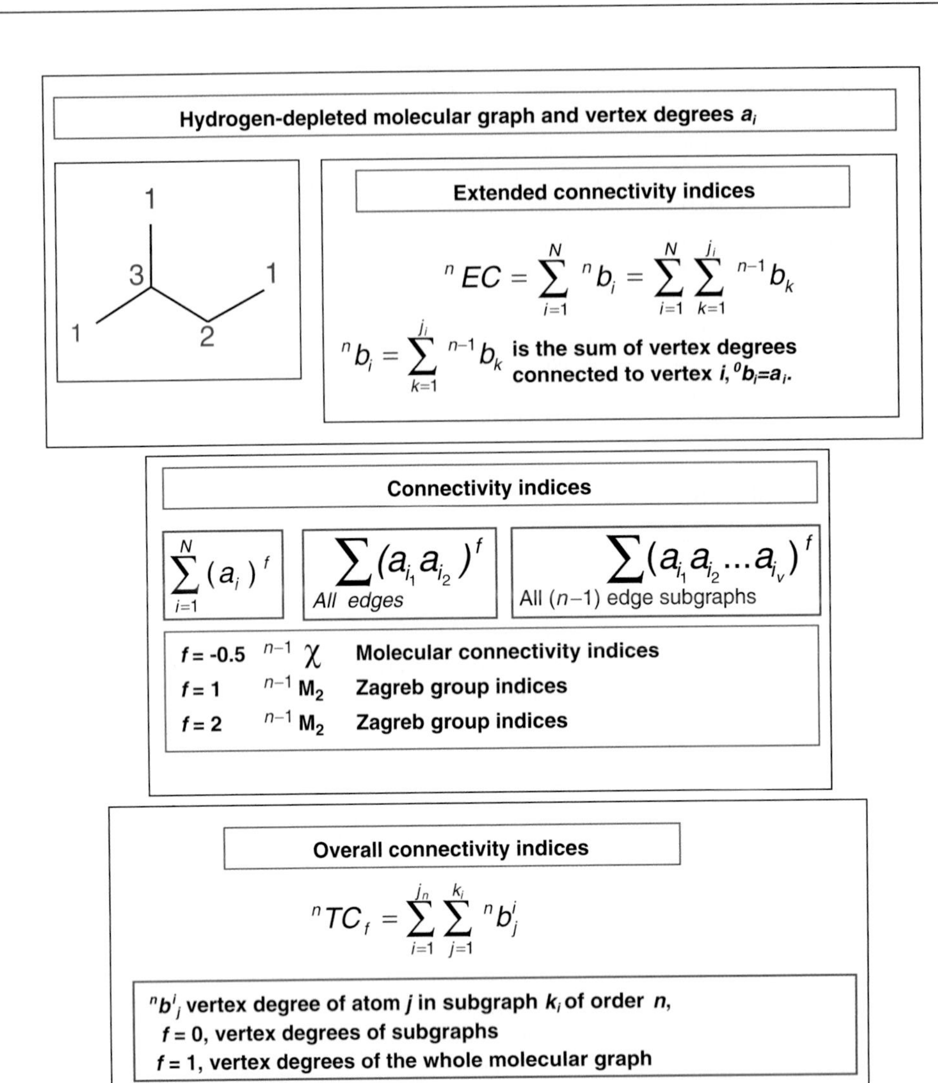

Figure 1 Examples of topological descriptors frequently used in QSAR studies.

N ----- (7) ----- S

Figure 2 An example of an AP descriptor: two atom types, aliphatic nitrogen and aliphatic sulfur, are separated by the shortest chemical graph path of seven.

weighted holistic invariant molecular (WHIM) descriptors, empirical descriptors, GETAWAY descriptors, functional groups, atom-centered fragments, empirical descriptors, and properties.

Interesting recent developments in the area of molecular descriptors have dealt with addressing the inability of topological descriptors to distinguish stereoisomers. To address this deficiency, molecular chirality indices were proposed recently on the basis of 2D molecular graphs.[46] These descriptors afford different values for enantiomers and diastereomers. This work was recently extended to *cis–trans* isomerism.[47] Chirality descriptors have been successfully used in combination with nonchiral descriptors in quantitative structure–property relationship (QSPR) studies of several molecular data sets.[20,46] In all cases, highly predictive QSPR models have been obtained, having better or similar predictive abilities as compared to 3D QSAR methods.

4.07.2.2 Three-Dimensional Descriptors

With the rapid progress of 3D conformational searching of chemical structures, 3D QSAR approaches have been developed to address the problems of 2D QSAR techniques, such as their inability to distinguish stereoisomers.[48,49] These include molecular shape analysis (MSA),[50] distance geometry,[48,49] and Voronoi techniques.[51] The MSA method combined shape descriptors and MLR analysis, while the other two approaches applied atomic refractivity as structural descriptors and the solution of mathematical inequalities to obtain the quantitative relationships. CoMFA[52,53] perhaps is the most popular example of 3D QSAR. It has been widely used in medicinal chemistry and toxicity analysis by elegantly combining the power of molecular graphics and PLS technique.[54,55]

One of the most attractive features of the CoMFA and CoMFA-like methods is that due to the nature of molecular field descriptors these approaches yield models that are relatively easy to interpret in chemical terms. Famous CoMFA contour plots, which are obtained as a result of any successful CoMFA study, tell chemists in rather plain terms how the change in the compound's size or charge distribution as a result of chemical modification correlates with the binding constant or activity. These observations may immediately suggest to a chemist possible ways to modify their molecules in order to increase their potencies. However, as will be demonstrated in the next section, these predictions should be taken with caution only after sufficient work has been done to prove the statistical significance of the models.

VolSurf descriptors are obtained from 3D interaction energy grid maps.[56] They include size and shape descriptors, hydrophilic and hydrophobic regions descriptors, interaction energy moments, and other descriptors. The main advantage of VolSurf descriptors is that they are alignment free.

Molecular Operating Environment (MOE) descriptors[57] include both 2D and 3D molecular descriptors. 2D descriptors include physical properties, subdivided surface areas, atom and bond counts, Kier and Hall connectivity and kappa shape indices, adjacency and distance matrix descriptors, pharmacophore feature descriptors, and partial charge descriptors. 3D molecular descriptors include potential energy descriptors, surface area, volume and shape descriptors, and conformation-dependent charge descriptors.

By analogy with 2D AP descriptors (**Figure 2**), 3D AP descriptors can also be defined using similar atom types and atom pairs and 3D molecular topography; in this case, a physical distance between atom types is used in place of chemical graph distance. The distance between two 'atoms' is measured and then assigned into one or two distance bins. Typically, the width of each distance bin is chosen as 1.0 Å. Since it is also designed to let the adjacent bins have 10% overlap with each other, the actual length of each distance bin is 1.2 Å. Any distance located in the overlap region is assigned to both bins. This 'fuzzy distance' concept is adopted to alleviate the possible unfavorable boundary effects of the distance bins. For example, with strict boundary conditions, a distance of 2.05 Å will be assigned only to bin No. 2, but it can be reasonably argued that it is almost as close to the upper half of bin No. 1 as to bin No. 2. With fuzzy boundary conditions, 2.05 Å belongs to both bin No. 1 and bin No. 2 allowing a possible match to either. All distances larger than 20 Å are assigned into the last bin.

4.07.3 Quantitative Structure–Activity Relationship Modeling Approaches

4.07.3.1 General Classification

Many different approaches to QSAR have been developed since Hansch's seminal work. As briefly discussed above, the major differences between these methods can be analyzed from two viewpoints: (1) the types of structural parameters that are used to characterize molecular identities starting from different representation of molecules, from simple chemical formulas to 3D conformations, and (2) the mathematical procedure that is employed to obtain the quantitative relationship between these structural parameters and biological activity. See Chapter 4.23.2.3.1 for an alignment (although not conformational) independent approach to 3D QSAR.

Based on the origin of molecular descriptors used in calculations, QSAR methods can be divided into three groups. One group is based on a relatively small number (usually many times smaller than the number of compounds in a data set) of physicochemical properties and parameters describing hydrophobic, steric, electrostatic, etc. effects. Usually, these descriptors are used as independent variables in multiple regression approaches. In the literature, these methods are typically referred to as Hansch analysis.

A more recent group of methods is based on quantitative characteristics of molecular graphs (molecular topological descriptors). Since molecular graphs or structural formulas are 'two-dimensional,' these methods are described as 2D QSAR. Most of the 2D QSAR methods are based on graph theoretical indices that are discussed above. Sometimes topological descriptors are also combined with physicochemical properties of molecules. Although these structural indices represent different aspects of molecular structures, and, what is important for QSAR, different structures provide numerically different values of indices, their physicochemical meaning is frequently unclear.

The third group of methods is based on descriptors derived from spatial (3D) representation of molecular structures. Correspondingly, these methods are referred to as 3D QSAR; they have become increasingly popular with the development of fast and accurate computational methods for generating 3D conformations and alignments of chemical structures. Perhaps the most popular example of 3D QSAR is CoMFA, developed by Cramer *et al.*,[58] which has combined the power of molecular graphics and PLS technique and has found wide applications in medicinal chemistry and toxicity analysis.[59] This method is one of the most recent developments in the area of ligand-based receptor modeling. This approach combines traditional QSAR analysis and 3D ligand alignment into a powerful 3D QSAR tool. CoMFA correlates 3D electrostatic and van der Waals fields around sample ligands typically overlapped in their pharmacophoric conformations with their biological activity. This approach has been successfully applied to many classes of ligands.[59]

CoMFA methodology is based on the assumption that, since, in most cases, the drug–receptor interactions are noncovalent, the changes in biological activity or binding constants of sample compounds correlate with changes in the electrostatic and van der Waals fields of these molecules. In order to initiate the CoMFA process, the test molecules should be structurally aligned in their pharmacophoric conformations. This makes the assumption that all bound ligands adopt the exact same conformation, which is unlikely considering accessory side chains may hinder or promote ligand binding depending on van der Waals and electrostatic interactions. After the alignment, steric and electrostatic fields of all molecules are sampled with a probe atom, usually sp3 carbon bearing a + 1 charge, on a rectangular grid that encompasses structurally aligned molecules. The values of both van der Waals and electrostatic interaction between the probe atom and all atoms of each molecule are calculated in every lattice point on the grid using a force field equation and entered into the CoMFA QSAR table. This table thus contains thousands of columns, which makes it difficult to produce a statistically significant model when there are so many possible solutions. A cross-validated r^2 (q^2) that is obtained as a result of this analysis serves as a quantitative measure of the quality and internal predictive ability of the final CoMFA model. The statistical meaning of the q^2 is different from that of the conventional r^2; a q^2 value greater than 0.3 is considered significant.

4.07.3.2 Correlation Approaches

Both 2D and 3D QSAR studies have focused on the development of optimal QSAR models through variable selection. This implies that only a subset of available descriptors of chemical structures, which are the most meaningful and statistically significant in terms of correlation with biological activity, is selected. The optimum selection of variables was first achieved by combining stochastic search methods with correlation methods such as MLR, PLS analysis, or ANNs.[60–65] More specifically, these methods employ either generalized simulated annealing,[60] genetic algorithms,[61] or evolutionary algorithms,[62–65] as the stochastic optimization tool. It has been demonstrated that these algorithms combined with various chemometric tools have effectively improved the QSAR models compared to those without variable selection.

Most of the original QSAR techniques (both 2D and 3D) assumed the existence of a linear relationship between a biological activity and molecular descriptors. However, the assumption of linearity in the SAR may not hold true, especially when a large number of structurally diverse molecules are included in the analysis. Thus, several nonlinear QSAR methods have been proposed in recent years, such as ANNs[67] and *k* nearest neighbors.[68] Such applications, combined with variable selection, represent fast-growing trends in modern QSAR research.

Different QSAR methods have their own strengths and weaknesses. For example, 3D QSAR methods generally result in the diagrams of important molecular fields that can be easily interpreted in terms of specific steric and electrostatic interactions important for the ligand binding to their receptor. However, time-consuming and subjective alignment of molecular structures typically precludes the use of 3D QSAR techniques for the analysis of large data sets. On the other hand, 2D QSAR methods are much faster and more amenable to automation since they require no conformational search and structural alignment. Thus, 2D methods are best suited for the analysis of large numbers of compounds and computational screening of molecular databases; however, the interpretation of the resulting models in familiar chemical terms is frequently difficult if not impossible.

The generality of QSAR modeling approach as drug discovery tool irrespective of descriptor types or optimization algorithms can be best demonstrated in the context of inverse QSAR, which can be defined as designing or discovering molecular structures with a desired property on the basis of QSAR model. In practical terms, inverse QSAR also includes searching for molecules with a desired target property in chemical databases or virtual chemical libraries. These considerations emphasize the universal importance of establishing QSAR model robustness and predictive ability as opposed to concentrating on explanatory power, which has been characteristic feature of many traditional QSAR approaches.

4.07.4 Building Predictive Quantitative Structure–Activity Relationship Models: The Approaches to Model Validation

4.07.4.1 The Importance of Validation

The process of QSAR model development is divided into three key steps: (1) data preparation, (2) data analysis, and (3) model validation. The implementation and relative merit of these steps is generally determined by the researcher's interests and experience, and the availability of software. The resulting models are then frequently employed, at least in theory, to design new molecules based on chemical features or trends found to be statistically significant with respect to underlying biological activity.

The first stage includes the selection of a data set for QSAR studies and the calculation of molecular descriptors. The second stage deals with the selection of a statistical data analysis technique, either linear or nonlinear such as PLS or ANN. A variety of different algorithms and computer software are available for this purpose. In all approaches, descriptors are considered as independent variables, and biological activities as dependent variables.

Typically, the final part of QSAR model development is model validation,[1,69] in which estimates of the predictive power of the model are calculated. This predictive power is one of the most important characteristics of QSAR models. Ideally, it should be defined as the ability of the model to predict accurately the target property (e.g., biological activity) of compounds that were not used in model development. The typical problem of QSAR modeling is that at the time of model building a researcher has only the training set molecules, so predictive ability can be characterized only by statistical characteristics of the training set model, and not by true external validation.

Most QSAR modeling methods implement the leave-one-out (LOO), or leave-some-out, cross-validation procedure. The outcome from this procedure is a cross-validated correlation coefficient q^2, which is calculated according to the following formula:

$$q^2 = 1 - \frac{\sum (y_i - \hat{y}_i)^2}{\sum (y_i - \bar{y})^2} \qquad [1]$$

where y_i, $\hat{y}_i$, and $\bar{y}$ are the actual activities, the estimated activities by LOO cross-validation procedure, and the average activities, respectively. The summations in eqn [1] are performed over all compounds used to build a model (i.e., the training set). Frequently, q^2 is used as the criterion of both robustness and predictive ability of the model. Many authors consider high q^2 (for instance, $q^2 > 0.5$) as an indicator or even as the ultimate proof of the high predictive power of a QSAR model. They do not test the models for their ability to predict the activity of compounds of an external test set (i.e., compounds that have not been used in the QSAR model development). For instance, in several publications[70–73] models were claimed to have high predictive ability in the absence of validation using an external test set. In other examples, models were validated using only one or two compounds that were not used in QSAR model development,[74,75] and the claim was made that these models were highly predictive.

Thus, it is still not common to test QSAR models characterized by a reasonably high q^2 for their ability to accurately predict biological activities of compounds not included in the training set. However, it has been shown[76,77] that various commonly accepted statistical characteristics of QSAR models derived for a training set are insufficient to establish and estimate the predictive power of QSAR models. Contrary to expectations, evidence would seem to indicate that no correlation exists between the LOO cross-validated q^2 and the correlation coefficient R^2 between the predicted and observed activities even when a test set of compounds with known biological activities is available for prediction (**Figure 3**). Furthermore, experience suggests,[1,78] that this phenomenon is characteristic of many data sets and is independent of the descriptor types and optimization techniques used to develop training set models. Several recent publications[69,76,77,79–81] suggest the only way to ensure the high predictive power of a QSAR model is to demonstrate a significant correlation between predicted and observed activities for a validation set of compounds that were not employed in model development.

4.07.4.2 *Y*-Randomization

The *Y*-randomization of response is another important validation approach that is widely used to establish model robustness.[82] This method consists of repeating the QSAR model derivation calculation procedure, but with randomized activities. The subsequent probability assessment of the resultant statistics is then used to gauge the robustness of the model developed with the actual activities. It is often used along with the cross-validation. In many cases, models based on the randomized data have high q^2 values, which can be explained by a chance correlation or structural redundancy.[83] If all QSAR models obtained in the *Y*-randomization test have relatively high R^2 and LOO q^2,

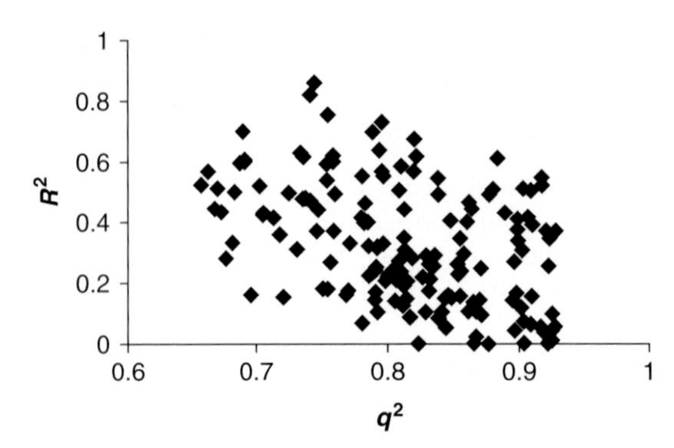

Figure 3 Beware of q^2! External R^2 (for the test set) shows no correlation with the 'predictive' LOO q^2 (for the training set). (Adapted from Golbraikh, A., Tropsha, A. *J. Mol. Graph. Model.* **2002**, *20*, 269–276.)

it implies that an acceptable QSAR model cannot be obtained for the given data set by the current modeling method. A recent publication[20] provides examples of training set models that had high internal q^2 but were still unacceptable based on the *Y*-randomization test criteria.

4.07.4.3 Rational Division of Available Data Sets into Training and Test Sets

We should emphasize that both *Y*-randomization and external validation must be made a mandatory part of model development. This goal can be achieved by a division of an experimental SAR data set into the training and test sets, which are used for model development and validation, respectively. We believe that special approaches should be used to select a training set to ensure the highest significance, robustness, and predictive power of QSAR models.[1,78] Recent reviews and publications describe several algorithms that can be employed for such division.[76–78]

As follows from the above discussion, in order to estimate the true predictive power of a QSAR model, one needs to compare the predicted and observed activities of a sufficiently large external test set of compounds that were not used in the model development. One convenient parameter is an external q^2 defined as follows (similar to eqn [1] for the training set):

$$q^2_{\mathrm{ext}} = 1 - \frac{\sum_{i=1}^{\mathrm{test}} (y_i - \hat{y}_i)^2}{\sum_{i=1}^{\mathrm{test}} (y_i - \bar{y}_{tr})^2} \qquad [2]$$

where y_i and $\hat{y}_i$ are the measured and predicted activities (over the test set), respectively, values of the dependent variable and $\bar{y}_{tr}$ is the averaged value of the dependent variable for the training set; the summations run over all compounds in the test set. Certainly, this formula is only meaningful when $\bar{y}_{tr}$ does not differ significantly from the similar value for the test set.[84] In principle, given the entire collection of compounds with known structure and activity, there is no particular reason to select one particular group of compounds as the training (or test) set; thus, the division of the data set into multiple training and test sets[78] or interchangeable definition of these sets[85] is recommended.

The use of the following statistical characteristics of the test set was also recommended[78]: correlation coefficient R^2 between the predicted and observed activities; coefficients of determination (predicted versus observed activities R_0^2, and observed versus predicted activities $R_0'^2$); slopes k and k' of the regression lines through the origin.

In summary, we consider a QSAR model predictive, if the following conditions are satisfied[78]:

$$q^2 > 0.5 \qquad [3]$$

$$R^2 > 0.6 \qquad [4]$$

$$\frac{(R^2 - R_0^2)}{R^2} < 0.1 \text{ or } \frac{(R^2 - R_0'^2)}{R^2} < 0.1 \qquad [5]$$

$$0.85 \leq k \leq 1.15 \text{ or } 0.85 \leq k' \leq 1.15. \qquad [6]$$

It has been demonstrated[76,78] that all of the above criteria are indeed necessary to adequately assess the predictive ability of a QSAR model.

4.07.4.4 Applicability Domain of Quantitative Structure–Activity Relationship Models

It needs to be emphasized that no matter how robust, significant, and validated a QSAR model may be, it cannot be expected to be applicable to the entire universe of chemicals. Therefore, before any QSAR model is used to predict biological activity of any untested compound, its domain of application must be defined and predictions for only those chemicals that fall into this domain may be considered reliable. Described below are some approaches that aid in defining the applicability domain.

4.07.4.4.1 Extent of extrapolation

For a regression-like QSAR, a simple measure of a chemical being too far from the applicability domain of the model is its leverage, h_i,[86] which is defined as:

$$h_i = x_i^T (X^T X)^{-1} x_i \qquad (i = 1, \ldots, n) \qquad [7]$$

where x_i is the descriptor row-vector of the query compound, and X is the $n \times k - 1$ matrix of k model descriptor values for n training set compounds. The superscript T refers to the transpose of the matrix/vector. The warning leverage h^* is, generally, fixed at $(3 * k)/n$, where n is the number of training compounds, and k is the number of model parameters. A leverage greater than the warning leverage h^* means that the predicted response is the result of substantial extrapolation of the model and, therefore, may not be reliable.[87,88]

4.07.4.4.2 Effective prediction domain

Similarly, for regression-like models, especially when the model descriptors are significantly correlated, Mandel[89] proposed the formulation of effective prediction domain, EPD. It has been demonstrated, with examples, that a regression model is justified inside and on the periphery of the EPD. Clearly, if a compound is determined to be too far from the EPD, its prediction from the model should not be considered reliable.

4.07.4.4.3 Residual standard deviation

Another important approach that can be used to evaluate the applicability domain is the degree-of-fit method developed originally by Lindberg *et al.*[90] and modified subsequently.[91] According to the original method, the predicted y values are considered to be reliable if the following condition is met:

$$s^2 < s_a^2(E_x)F \qquad [8]$$

where s^2 is the residual standard deviation (RSD) of descriptor values generated for a test compound, $s_a^2(E_x)$ is the RSD of the X matrix after dimensions (components) a, and F is the F-statistic at the probability level α and $(p-a)/2$ and $(p-a)(n-a-1)/2$ degrees of freedom. The RSD of descriptor values generated for a test compound is calculated using the following equation:

$$s^2 = ||e||/(p-a) \qquad [9]$$

where p is the number of x-variables, a is the number of components, and $||e||$ is the sum of squared residuals e_i expressed as

$$e_i = x_i - x_i BB' \qquad [10]$$

where x_i is the ith x-variable, and B and B' represent the weight matrix and transposed weight matrix of x variables, respectively. Since the lowest possible value of F is 1.00 at $\alpha = 0.10$ (when both degrees of freedom are equal to infinity), the authors[91] decided to replace F with the degree-of-fit factor f to simplify the above condition. Thus, the modified degree-of-fit condition[91] is as follows: predicted y values are considered to be reliable if

$$s^2 < s_a^2(E_x)f \qquad [11]$$

4.07.4.4.4 Similarity distance

Domain applicability can also be determined based on chemical similarity. Nonlinear methods such as k nearest neighbor (kNN) QSAR[92] employ models based on chemical similarity calculations. As such, a large similarity distance could signal that query compounds are too dissimilar to the training set compounds, and thus are not within the domain of applicability. A proposed[92] cutoff value, D_c (eqn [12]), defines a similarity distance threshold for external compounds:

$$D_c = Z\sigma + y \qquad [12]$$

Here y is the average and σ is the standard deviation of the Euclidean distances of the k nearest neighbors of each compound in the training set in the chemical descriptor space, and Z is an empirical parameter to control the significance level, with the default value of 0.5. If the distance from an external compound to its nearest neighbor in the training set is above D_c, we label its prediction unreliable.

4.07.4.5 Validated Quantitative Structure–Activity Relationship Modeling as an Empirical Data Modeling Approach: Combinatorial Quantitative Structure–Activity Relationship Modeling

We believe QSAR modeling is an empirical, exploratory research area where the models with the best-validated predictive power should be sought by a combinatorial exploration of various groupings of statistical data modeling techniques and different types of chemical descriptors followed by the consensus prediction of activities for external compounds by averaging the predicted activity values resulting from all validated models.[20] This strategy is driven by the concept that if an implicit SAR exists for a given data set, it can be formally manifested via a variety of QSAR models that use different descriptors and optimization protocols. We believe that multiple alternative QSAR models should be developed (as opposed to a single model using some favorite QSAR method) for each data set.

Several popular commercial and noncommercial QSAR software packages provide users with various descriptor types and data modeling capabilities. Practically, every package employs only one (or a few) type of descriptors and, typically, a single or a few molecular modeling techniques. Most commercially available programs provide a relatively easy-to-use interface and allow users to build single models with internal accuracy typically characterized by the q^2. As emphasized in the previous section, training-set-only modeling is insufficient to achieve models with validated predictive power; the QSAR model development process has to be modified to incorporate an independent model validation and applicability domain definition.[77,78] Since the process is relatively fast (and in principle, can be completely automated), these alternative models could be explored simultaneously when making predictions for external data sets. Consensus predictions of biological activity for novel compounds on the basis of several QSAR models, especially when predictions converge, provide more confidence in the activity estimates and better justification for the experimental validation of these compounds. This strategy is outlined in **Figure 4**.

The need to develop and employ the combinatorial QSAR approach is dictated by experience in QSAR modeling, suggesting that QSAR is still an experimental area of statistical data modeling. As such, it is impossible to decide a priori as to which particular QSAR modeling method will prove most successful. Every particular combination of descriptor sets and optimization techniques is likely to capture certain unique aspects of the SAR. Since the ultimate goal is to use the resulting models in database mining to discover diverse biologically active molecules, application of different combinations of modeling techniques and descriptor sets should increase the chances for success as demonstrated in recent publications.[20,93]

4.07.5 Validated Quantitative Structure–Activity Relationship Models as Virtual Screening Tools

Although combinatorial chemistry and HTS have offered medicinal chemists a much broader range of possibilities for lead discovery and optimization, the number of chemical compounds that can be synthesized and tested is still far beyond the capability of today's medicinal chemistry. Therefore, medicinal chemists continue to face the same problem as before: which compounds should be chosen for the next round of synthesis and testing? For chemoinformaticians, the task is to develop and utilize computational approaches to evaluate a very large number of chemical compounds and recommend the most promising ones to bench chemists.

Database mining associated with pharmacophore identification is a common and efficient approach for lead compound discovery. Pharmacophore identification refers to the computational approach to identifying the essential 3D structural features and configurations that are responsible for the biological activity of a series of compounds. Once a pharmacophore model has been developed for a particular set of biologically active molecules, it can be used to search databases of 3D structures with the aim of finding new, structurally different lead molecules with the desired biological activity.[94]

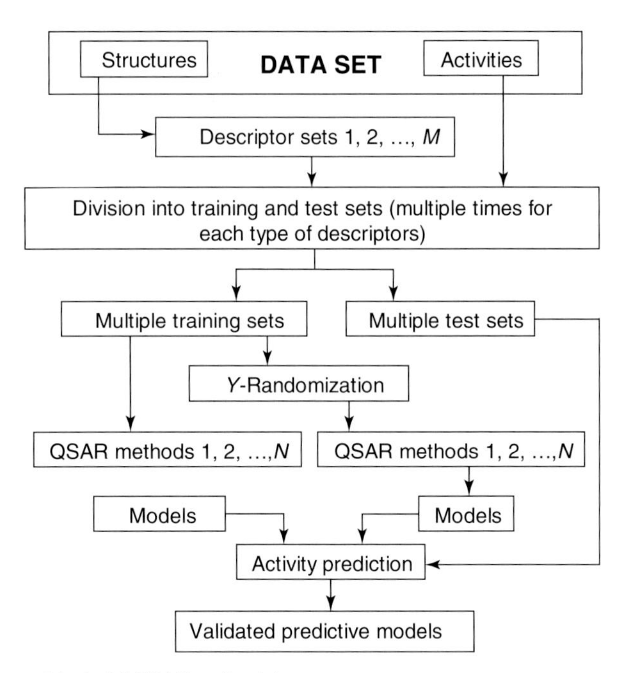

Figure 4 Flowchart of the combinatorial QSAR methodology.

An obvious parallel can be established between the search for pharmacophore elements, which are thought to describe the specificity of drug action, and the identification of a subset of descriptors contributing the most to the correlation with biological activity in a variable-selection QSAR model. Thus, the selection of specific pharmacophore features responsible for biological activity is directly analogous to the selection of specific chemical descriptors that contribute the most to an explanatory QSAR model. It is convenient to establish a concept of the descriptor pharmacophore in the context of variable selection QSAR modeling. Thus, by analogy with the conventional definition of pharmacophores, the descriptor pharmacophore can be defined as a set of descriptor variables implicated in highly statistically significant and predictive QSAR models. It has been demonstrated that QSAR models can be used in database mining, i.e., finding molecular structures that are similar in their activity to the probe molecules or even predicting the activities for the compounds in a database.[95–97] First, a preconstructed QSAR model can be used as a means of screening compounds from existing databases (or virtual libraries) for high-predicted biological activity. Alternatively, variables selected by QSAR optimization can be used for similarity searches to improve the performance of the database mining methods.

It should be noted that despite formal similarity between the common definition of pharmacophores and descriptor pharmacophores, there is also a significant difference in the procedure as well as expected outcome of virtual screening. As mentioned above, traditional approaches to database mining are based on chemical fragment or subfragment-based similarity searches. While this is an efficient approach that has enjoyed certain successes, it limits the chemical diversity of selected compounds to those that are similar to existing ligands. Search methodologies are based on chemical similarity estimated by Euclidean distance (or any other similarity measure) in multidimensional descriptor space (where descriptors are selected from the entire initial space in the process of variable selection model development) combined with quantitative predictions from combinatorial QSPR models. Due to the nature of the descriptors (e.g., whole molecule-based descriptors as opposed to fragments), such searches are more likely to result in accurate prediction of target properties for diverse novel compounds than traditional fragment-based search methodologies. This strategy was successfully tested in recent studies of anticonvulsant agents[2,98] and on the Ames Genotoxicity data set.[99] The approach is outlined in **Figure 5**. It is important to stress that the output of these studies is not models with their statistical characteristics as is typical for most QSAR studies. Rather, the modeling results are the predictions of the target properties for all database or virtual library compounds, which allows for immediate compound prioritization for subsequent experimental verification. Another advantage of using QSAR models for database mining is that this approach affords not only the identification of compounds of interest but also quantitative prediction of compounds' potency. For illustration, we shall discuss recent successes in developing validated predictive models of anticonvulsants[98] and their application to the discovery of novel potent compounds by the means of database mining.[2]

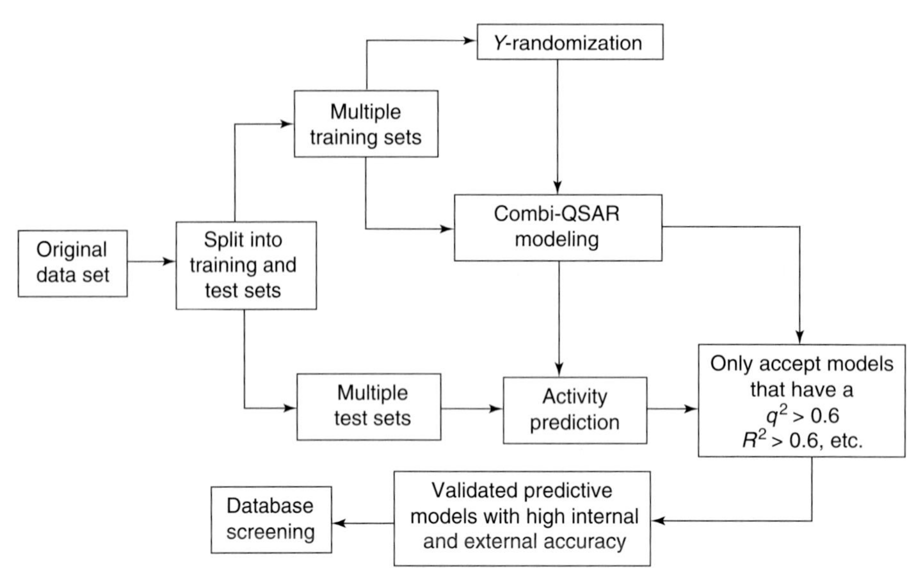

Figure 5 Flowchart of predictive QSAR workflow based on validated combi-QSAR models.

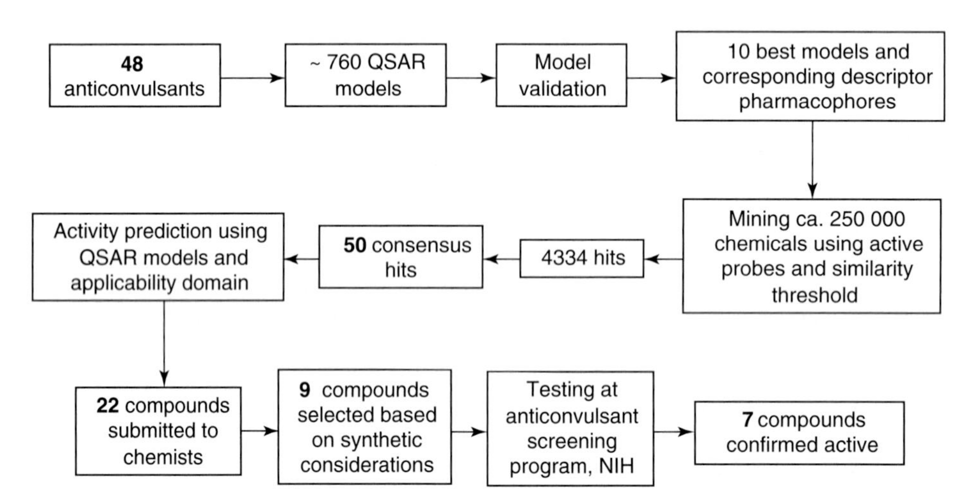

Figure 6 Computer-aided drug discovery workflow based on combination of QSAR modeling and consensus database mining as applied to the discovery of novel anticonvulsants.[2] The workflow emphasizes the importance of model validation and applicability domain in ensuring high hit rates as a result of database mining with predictive QSAR models.

Figure 6 summarizes the approach to using validated QSAR models for virtual screening as applied to the anticonvulsant dataset. Initially, the authors applied variable selection *k*NN and simulated-annealing PLS (SA-PLS) QSAR approaches to a data set of 48 chemically diverse functionalized amino acids (FAA) with anticonvulsant activity to develop validated QSAR models.[98] Both methods used multiple descriptors such as molecular connectivity indices or atom pair descriptors, which are derived from 2D molecular topology. QSAR models with high internal accuracy were generated, with leave-one-out cross-validated R^2 (q^2) values ranging between 0.6 and 0.8. The q^2 values for the actual data set were significantly higher than those obtained for the same data set with randomly shuffled activity values, indicating that models were robust. The original data set was further divided into several training and test sets, and highly predictive models were obtained with q^2 values for the training sets greater than 0.5 and R^2 values for the test sets greater than 0.6.

In the second stage of this process, the validated QSAR models and descriptor pharmacophore concepts were applied[2] to mining of available chemical databases for new lead anticonvulsant agents (**Figure 6**). Two databases have been explored: the National Cancer Institute[100] and the Maybridge[101] databases, including 237 771 and 55 273 chemical structures, respectively. Database mining was performed independently using the 10 best-validated QSAR models with the highest values of both q^2 and R^2. First, chemical similarity searches were performed between the training set compounds and database molecules using descriptor pharmacophores only (i.e., Euclidean similarity was calculated using only descriptors implicated in the 10 best validated *k*NN QSAR models) and over 4300 compounds found within the same similarity threshold in all 10 independent searches were selected as consensus hits (cf. **Figure 2**). Their activities then were predicted using individual QSAR models and the consensus hits with the highest predicted anticonvulsant activity were further explored experimentally.[2]

This study[2,98] presents a practical example of the drug discovery workflow that can be generalized for any data set where sufficient data to develop reliable QSAR models is available. These results certainly appear very promising and reassuring in terms of computational strategies, which emphasize that rigorous validation of QSAR models as well as conservative extrapolation are responsible for a very high hit rate.

4.07.6 Conclusions

A QSAR model describes a mathematical relationship between structural attributes and a property of a set of chemicals. The use of such mathematical relationships to predict the target property of interest for a variety of chemicals prior to, or in lieu of, expensive and labor-intensive experimental measurements has naturally been very enticing. The potential promise of using QSAR models for screening of chemical databases or virtual libraries before their synthesis appears equally attractive to chemical manufacturers, pharmaceutical companies, and government agencies, particularly in times of shrinking resources. Given the growing sizes of chemical databases resulting from combinatorial synthesis and the regulatory and social pressures for timely assessment of health and environmental risks of chemicals, the need for reliable QSAR models is imperative. For instance, environmental agencies in both Europe and the USA require reliable data on the environmental effects and the fate of all industrial chemicals. Traditionally, biological and environmental testing has provided such data; that are available for only a fraction of industrial chemicals, and thousands of industrial chemicals exist that will continue to go untested. Recently, Walker *et al.*[102–106] have addressed this problem by providing a set of guidelines for developing and using QSAR models for environmental risk assessment.

General guiding principles for building robust QSAR models have been described recently as well.[77] Thus, in order to be reliable and predictive, QSAR models should: (1) be statistically significant and robust, (2) be validated by making accurate predictions for external data sets that were not used in the model development, and (3) have their application boundaries defined so they may serve as effective database screening tools. The true power of QSAR results as we have emphasized in this chapter comes from their statistical significance and the ability of the model to predict accurately biological properties of chemical compounds in the training and most importantly test sets. Understanding and practicing these principles in QSAR modeling should help medicinal chemists to prioritize their experimental effort and significantly increase the experimental hit rates.

References

1. Golbraikh, A.; Tropsha, A. *J. Mol. Graph. Model.* **2002**, *20*, 269–276.
2. Shen, M.; Beguin, C.; Golbraikh, A.; Stables, J. P.; Kohn, H.; Tropsha, A. *J. Med. Chem.* **2004**, *47*, 2356–2364.
3. Hansch, C.; Muir, R.; Fujita, T.; Maloney, P.; Geiger, E.; Streich, M. *J. Am. Chem. Soc.* **1963**, *85*, 2817–2824.
4. Hammett, L. P. *Chem. Rev.* **1935**, *17*, 125–136.
5. Hansch, C.; Leo, A. Exploring QSAR. In *Fundamentals and Applications in Chemistry and Biology*; Hellen, S., Ed.; American Chemical Society: Washington, DC, 1995; Vol. 1, 580pp.
6. Verloop, A.; Hoogenstraaten, W.; Tipker, J. In *Drug Design*; Ariens, E. J., Ed.; Academic Press: New York, 1976, pp 165–207.
7. Testa, B.; Seiler, P. *Arzneimittelforschung* **1981**, *31*, 1053–1058.
8. Boyd, D. B. In *Reviews in Computational Chemistry*; Lipkowitz, K. B., Boyd, D. B., Eds.; VCH: New York, 1991, pp 355–371.
9. Norinder, U.; Hogberg, T. *Acta Pharm. Nord.* **1992**, *4*, 73–78.
10. Van de Waterbeemd, W. H.; el Tayar, N.; Testa, B.; Wikstrom, H.; Largent, B. *J. Med. Chem.* **1987**, *30*, 2175–2181.
11. Jorgensen, W. L.; Duffy, E. M. *Bioorg. Med. Chem. Lett.* **2000**, *10*, 1155–1158.
12. Agrafiotis, D. K. *Protein Sci.* **1997**, *6*, 287–293.
13. Giuliani, A.; Benigni, R.; Zbilut, J. P.; Webber, C. L., Jr.; Sirabella, P.; Colosimo, A. *Chem. Rev.* **2002**, *102*, 1471–1492.
14. Deng, W.; Breneman, C.; Embrechts, M. J. *J. Chem. Inf. Comput. Sci.* **2004**, *44*, 699–703.
15. Drews, J. *Science* **2000**, *287*, 1960–1964.
16. PubChem. http://pubchem.ncbi.nlm.nih.gov/ (accessed Aug 2006).
17. Tropsha, A. Recent Trends in Quantitative Structure–Activity Relationships. In *Burger's Medicinal Chemistry and Drug Discovery*; Abraham, D., Ed.; John Wiley: New York, 2003, pp 49–77.
18. T. I. Oprea, 3D-QSAR Modeling in Drug Design. In *Computational Medicinal Chemistry and Drug Discovery*; Tollenaere, J., De Winter, H., Langenaeker, W., Bultinck, P., Eds.; Marcel Dekker: New York, 2004, pp 571–616.
19. Downs, G. M.; Willett, P. Similarity Searching in Databases of Chemical Structures. In *Reviews in Computational Chemistry*; Lipkowitz, K. B., Boyd, D., Eds.; VCH: New York, 1996, pp 1–65.
20. Kovatcheva, A.; Golbraikh, A.; Oloff, S.; Xiao, Y. D.; Zheng, W.; Wolschann, P.; Buchbauer, G.; Tropsha, A. *J. Chem. Inf. Comput. Sci.* **2004**, *44*, 582–595.

21. Topliss, J. G.; Edwards, R. P. *J. Med. Chem.* **1979**, *22*, 1238–1244.
22. Burkert, U.; Allinger, N. L. *Molecular Mechanics*; American Chemical Society: Washington, DC, 1982.
23. Kier, L. B.; Hall, L. H.; Murray, W. J.; Randic, M. *J. Pharm. Sci.* **1975**, *64*, 1971–1974.
24. Kier, L. B.; Murray, W. J.; Randic, M.; Hall, L. H. *J. Pharm. Sci.* **1976**, *65*, 1226–1230.
25. Debnath, A. K.; Lopez de Compadre, R. L.; Debnath, G.; Shusterman, A. J.; Hansch, C. *J. Med. Chem.* **1991**, *34*, 786–797.
26. Jain, A. N.; Koile, K.; Chapman, D. *J. Med. Chem.* **1994**, *37*, 2315–2327.
27. Cramer, R. D., III; Patterson, D. E.; Bunce, J. D. *J. Am. Chem. Soc.* **1988**, *110*, 5959–5967.
28. Burden, F. R.; Winkler, D. A. *J. Med. Chem.* **1999**, *42*, 3183–3187.
29. Carhart, R. E.; Smith, D. H.; Venkataraghavan, R. *J. Chem. Inf. Comput. Sci.* **1985**, *25*, 64–73.
30. Sheridan, R. P.; SanFeliciano, S. G.; Kearsley, S. K. *J. Mol. Graph. Model.* **2000**, *18*, 320–334, 525.
31. Livingstone, D. J. *J. Chem. Inf. Comput. Sci.* **2000**, *40*, 195–209.
32. Randic, M. *J. Am. Chem. Soc.* **1975**, *97*, 6609–6615.
33. Kellogg, G. E.; Kier, L. B.; Gaillard, P.; Hall, L. H. *J. Comput.-Aided Mol. Des.* **1996**, *10*, 513–520.
34. Kier, L. B.; Hall, L. H. *Quant. Struct.–Act. Relat.* **1993**, *12*, 383–388.
35. Hall, L. H.; Mohney, B.; Kier, L. B. *Quant. Struct.–Act. Relat.* **1991**, *10*, 43–51.
36. Hall, L. H.; Kier, L. B. *J. Chem. Inf. Comput. Sci.* **1995**, *35*, 1039–1045.
37. Hall, L. H.; Kier, L. B.; Brown, B. B. *J. Chem. Inf. Comput. Sci.* **1995**, *35*, 1074–1080.
38. Hall, L. H.; Kier, L. B. *J. Mol. Graph. Model.* **2001**, *20*, 4–18.
39. Hall, L. H.; Kier, L. B. The Molecular Connectivity Chi Indexes and Kappa Shape Indexes in Structure–Property Modeling. In *Reviews in Computational Chemistry II*; Lipkowitz, K. B., Boyd, D. B., Eds.; VCH: New York, 1991, pp 367–422.
40. Anker, L. S.; Jurs, P. C.; Edwards, P. A. *Anal. Chem.* **1990**, *62*, 2676–2684.
41. Jurs, P. C.; Ball, J. W.; Anker, L. S.; Friedman, T. L. *J. Chem. Inf. Comput. Sci.* **1992**, *32*, 272–278.
42. Nelson, T. M.; Jurs, P. C. *J. Chem. Inf. Comput. Sci.* **1994**, *34*, 601–609.
43. Stanton, D. T.; Jurs, P. C. *J. Chem. Inf. Comput. Sci.* **1992**, *32*, 109–115.
44. MolConnZ. MolConnZ. [4.05]. 2002. Hall Associates Consulting, Quincy, MA.
45. DRAGON. http://www.disat.unimib.it/chm/Dragon.htm (accessed Aug 2006).
46. Golbraikh, A.; Tropsha, A. *J. Chem. Inf. Comput. Sci.* **2003**, *43*, 144–154.
47. Golbraikh, A.; Bonchev, D.; Tropsha, A. *J. Chem. Inf. Comput. Sci.* **2002**, *42*, 769–787.
48. Crippen, G. M. *J. Med. Chem.* **1980**, *23*, 599–606.
49. Crippen, G. M. *Mol. Pharmacol.* **1982**, *22*, 11–19.
50. Hopfinger, A. J. *J. Am. Chem. Soc.* **1980**, *102*, 7196–7206.
51. Boulu, L. G.; Crippen, G. M. *J. Comb. Chem.* **1989**, *10*, 673–682.
52. Cramer, R. D., III; Patterson, D. E.; Bunce, J. D. *J. Am. Chem. Soc.* **1988**, *110*, 5959–5967.
53. Cramer, R. D., III; Patterson, D. E.; Bunce, J. D. *Prog. Clin. Biol. Res.* **1989**, *291*, 161–165.
54. Kubinyi, H.; Folkers, G.; Martin, Y. C. *Perspect. Drug Disc.* **1998**, *12*, V–VII.
55. Kubinyi, H.; Folkers, G.; Martin, Y. C. *Perspect. Drug Disc.* **1998**, *9–11*, V–VII.
56. Cruciani, G.; Pastor, M.; Guba, W. *Eur. J Pharm. Sci.* **2000**, *11*, S29–S39.
57. MOE. http://www.chemcomp.com/fdept/prodinfo.htm#Cheminformatics. 2005 (accessed Aug 2006).
58. Cramer, R. D.; Patterson, D. E.; Bunce, J. D. *J. Am. Chem. Soc.* **1988**, *110*, 5959–5967.
59. Kubinyi, H.; Folkers, G.; Martin, Y. C. Eds. 3D QSAR in Drug Design. In *Recent Advances*, Kluwer Academic Publishers: Dordrecht, The Netherlands, 1998, Vol. 3, 368pp.
60. Sutter, J. M.; Dixon, S. L.; Jurs, P. C. *J. Chem. Inf. Comput. Sci.* **1995**, *35*, 77–84.
61. Rogers, D.; Hopfinger, A. J. *J. Chem. Inf. Comput. Sci.* **1994**, *34*, 854–866.
62. Kubinyi, H. *Quant. Struct.–Act. Relat.* **1994**, *13*, 285–294.
63. Kubinyi, H. *Quant. Struct.–Act. Relat.* **1994**, *13*, 393–401.
64. Luke, B. T. *J. Chem. Inf. Comput. Sci.* **1994**, *34*, 1279–1287.
65. So, S. S.; Karplus, M. *J. Med. Chem.* **1996**, *39*, 1521–1530.
67. Andrea, T. A.; Kalayeh, H. *J. Med. Chem.* **1991**, *34*, 2824–2836.
68. Zheng, W.; Tropsha, A. *J. Chem. Inf. Comput. Sci.* **2000**, *40*, 185–194.
69. Kubinyi, H.; Hamprecht, F. A.; Mietzner, T. *J. Med. Chem.* **1998**, *41*, 2553–2564.
70. Girones, X.; Gallegos, A.; Carbo-Dorca, R. *J. Chem. Inf. Comput. Sci.* **2000**, *40*, 1400–1407.
71. Bordas, B.; Komives, T.; Szanto, Z.; Lopata, A. *J. Agric. Food Chem.* **2000**, *48*, 926–931.
72. Fan, Y.; Shi, L. M.; Kohn, K. W.; Pommier, Y.; Weinstein, J. N. *J. Med. Chem.* **2001**, *44*, 3254–3263.
73. Suzuki, T.; Ide, K.; Ishida, M.; Shapiro, S. *J. Chem. Inf. Comput. Sci.* **2001**, *41*, 718–726.
74. Recanatini, M.; Cavalli, A.; Belluti, F.; Piazzi, L.; Rampa, A.; Bisi, A.; Gobbi, S.; Valenti, P.; Andrisano, V.; Bartolini, M. et al. *J. Med. Chem.* **2000**, *43*, 2007–2018.
75. Moron, J. A.; Campillo, M.; Perez, V.; Unzeta, M.; Pardo, L. *J. Med. Chem.* **2000**, *43*, 1684–1691.
76. Golbraikh, A.; Tropsha, A. *J. Comput.-Aided Mol. Des.* **2002**, *16*, 357–369.
77. Tropsha, A.; Gramatica, P.; Gombar, V. K. *Quant. Struct.–Act. Relat. Comb. Sci.* **2003**, *22*, 69–77.
78. Golbraikh, A.; Shen, M.; Xiao, Z.; Xiao, Y. D.; Lee, K. H.; Tropsha, A. *J. Comput.-Aided Mol. Des.* **2003**, *17*, 241–253.
79. Novellino, E.; Fattorusso, C.; Greco, G. *Pharm. Acta Helv.* **1995**, *70*, 149–154.
80. Norinder, U. *J. Chemomet.* **1996**, *10*, 95–105.
81. Zefirov, N. S.; Palyulin, V. A. *J. Chem. Inf. Comput. Sci.* **2001**, *41*, 1022–1027.
82. Wold, S.; Eriksson, L. Statistical Validation of QSAR Results. In *Chemometrics Methods in Molecular Design*; van der Waterbeemd, H., Ed.; VCH: New York, 1995, pp 309–318.
83. Clark, R. D.; Sprous, D. G.; Leonard, J. M. Validating Models Based on Large Dataset. In *Rational Approaches to Drug Design*, Proceedings of the 13th European Symposium on Quantitative Structure–Activity Relationship, Aug 27–Sept 1; Höltje, H.-D., Sippl, W., Eds.; Prous Science: Düsseldorf, Germany, 2001, pp 475–485.
84. Oprea, T. I.; Garcia, A. E. *J. Comput.-Aided Mol. Des.* **1996**, *10*, 186–200.
85. Oprea, T. I. *SAR QSAR Environ. Res.* **2001**, *12*, 129–141.

86. Atkinson, A. C. *Plots, Transformations and Regression*; Clarendon Press: Oxford, UK, 1985.
87. Gramatica, P.; Papa, E. *Quant. Struct.–Act. Relat.* **2003**, *22*, 374–385.
88. Gramatica, P.; Pilutti, P.; Papa, E. *Quant. Struct.–Act. Relat.* **2003**, *22*, 364–373.
89. Mandel, J. *J. Res. Nat. Bur. Stand.* **1985**, *90*, 465–476.
90. Lindberg, W.; Persson, J.-A.; Wold, S. *Anal. Chem.* **1983**, *55*, 643–648.
91. Cho, S. J.; Zheng, W.; Tropsha, A. *J. Chem. Inf. Comput. Sci.* **1998**, *38*, 259–268.
92. Zheng, W.; Tropsha, A. *J. Chem. Inf. Comput. Sci.* **2000**, *40*, 185–194.
93. de Cerqueira Lima, P.; Golbraikh, A.; Oloff, S.; Xiao, Y. D.; Tropsha, A. *J. Med. Chem.* **2006**, *46*, 1245–1254.
94. *Pharmacophore Perception, Development, and Use in Drug Design*; IUL: La Jolla, CA, 2000.
95. Tropsha, A.; Cho, S. J.; Zheng, W. "New Tricks for an Old Dog": Development and Application of Novel QSAR Methods for Rational Design of Combinatorial Chemical Libraries and Database Mining. In *Rational Drug Design: Novel Methodology and Practical Applications*; Parrill, A. L., Reddy, M. R., Eds.; American Chemical Society: Washington, DC, 1999, pp 198–211.
96. Tropsha, A.; Zheng, W. *Curr. Pharm. Des.* **2001**, *7*, 599–612.
97. Hoffman, B. T.; Kopajtic, T.; Katz, J. L.; Newman, A. H. *J. Med. Chem.* **2000**, *43*, 4151–4159.
98. Shen, M.; LeTiran, A.; Xiao, Y.; Golbraikh, A.; Kohn, H.; Tropsha, A. *J. Med. Chem.* **2002**, *45*, 2811–2823.
99. Votano, J. R.; Parham, M.; Hall, L. H.; Kier, L. B.; Oloff, S.; Tropsha, A.; Xie, Q.; Tong, W. *Mutagenesis* **2004**, *19*, 365–377.
100. NCI. http://dtp.nci.nih.gov (accessed Aug 2006).
101. Maybridge. http://www.daylight.com (accessed Aug 2006).
102. Walker, J. *Handbook on Quantitative Structure Activity Relationships (QSARs) for Pollution Prevention, Toxicity Screening, Risk Assessment and World Wide Web Applications*; SETAC Press: Pensacola, FL, 2002.
103. Walker, J. *Handbook on Quantitative Structure Activity Relationships (QSARs) for Predicting Physical Properties, Bioconcentration Potential and Environmental Fate of Chemicals*; SETAC Press: Pensacola, FL, 2002.
104. Walker, J. *Handbook on Quantitative Structure Activity Relationships (QSARs) for Predicting Endocrine Disruption Potential of Chemicals*; SETAC Press: Pensacola, FL, 2002.
105. Walker, J. *Handbook on Quantitative Structure Activity Relationships (QSARs) for Predicting Effects of Chemicals on Environmental-Human Health Interactions*; SETAC Press: Pensacola, FL, 2002.
106. Walker, J. *Handbook on Quantitative Structure Activity Relationships (QSARs) for Predicting Ecological Effects of Chemicals*; SETAC Press: Pensacola, FL, 2002.
107. National Institutes of Health. RoadMap. http://nihroadmap.nih.gov/ (accessed Aug 2006).

Biography

Alexander Tropsha is professor and Chair of the Division of Medicinal Chemistry and Natural Products in the School of Pharmacy, UNC-Chapel Hill. He received his MS degree in chemical enzymology in 1982 and his PhD in biochemistry and pharmacology in 1986, both from Moscow State University. Dr Tropsha is a member of several editorial boards, including the *Journal of Chemical Information and Modeling*. He is a permanent member of the National Institutes of Health Biodata Management and Analysis Study Section, and is an elected member of the Board and Vice-Chair of the International QSAR and Cheminformatics Society. His research interests are in the areas of computer-assisted drug design, cheminformatics, and structural bioinformatics. He has authored or co-authored more than 100 peer-reviewed publications and book chapters. His research is supported by grants from the National Institutes of Health, National Science Foundation, Environmental Protection Agency, and industry.

© 2007 Elsevier Ltd. All Rights Reserved
No part of this publication may be reproduced, stored in any retrieval system or transmitted in any form by any means electronic, electrostatic, magnetic tape, mechanical, photocopying, recording or otherwise, without permission in writing from the publishers

Comprehensive Medicinal Chemistry II
ISBN (set): 0-08-044513-6

ISBN (Volume 4) 0-08-044517-9; pp. 149–165

4.08 Compound Selection Using Measures of Similarity and Dissimilarity

V J Gillet and P Willett, University of Sheffield, Sheffield, UK

© 2007 Elsevier Ltd. All Rights Reserved.

4.08.1 Introduction

The concepts of similarity and dissimilarity are used in many disciplines; in medicinal chemistry, they are not just useful but play a key role in the discovery of novel bioactive molecules. This chapter reviews the computational techniques that are used to measure and to exploit (dis)similarity, focusing on those techniques that are sufficiently rapid in execution to enable them to be applied to large-scale databases, containing hundreds of thousands or millions of chemical structures. These structures can be in either two-dimensional (2D) or three-dimensional (3D) form, although, as we shall see, the majority of the techniques that are widely employed are intended for use with databases of 2D structures. In the first part of the chapter, we discuss the concept of molecular similarity, with particular reference to the techniques that are used for similarity searching in chemical databases[1–5]; broader applications of molecular similarity are discussed by Dean[6] and Bender and Glen.[7] After a brief introduction to the concept of similarity searching, and its relationship to other methods for database access, the following sections discuss the types of representation and the types of similarity coefficient that have been reported for similarity searching. We then discuss techniques that can be used to quantify the effectiveness of a similarity searching procedure, and ways in which different types of similarity searching procedure can be combined. In the second part of the chapter, we discuss the related concept of molecular diversity and describe methods that have developed for selecting subsets of compounds.[8–10]

4.08.2 Similarity Searching

Any type of system for retrieving information from a database, whether chemical or not, will offer three distinct access modes: exact match, partial match, and best match. An exact-match search is one in which the database is scanned for identity with a query record, either to establish the presence of the record or to retrieve information that is associated with it, e.g., scanning an electoral roll to see if somebody is eligible to vote or scanning a telephone directory for the telephone number of a specific individual. A partial-match search is one in which the database is scanned for those records that contain the query record (or characteristics that are associated with it), thus identifying a class of records that have some degree of commonality, e.g., scanning an online bibliographic database to find those journal articles that contain some Boolean combination of keywords. Finally, a best-match search is one in which the database is scanned to find those records that are most similar to the query record, based on some quantitative definition of interrecord similarity, e.g., scanning a real-estate database to find houses that best resemble one's dream home. The first two of these access mechanisms have long been a staple of chemoinformatics systems, in which chemical compounds are stored using labeled-graph representations, called connection tables.[11,12] The use of a graph representation facilitates exact-match and partial-match searching in the shape of structure searching and substructure searching.

Structure searching takes as input a specific molecule and reports the presence or absence of that molecule in the database (as is required, for example, for registration and for patent searches) or data associated with that structure, e.g., biological assay results or synthetic details.

Substructure searching retrieves all of those molecules in a database that contain a user-defined query substructure, irrespective of the environment in which the query substructure occurs, e.g., all molecules that contain a particular ring scaffold or a 3D pharmacophoric pattern. Substructure searching is very widely used but does have some inherent characteristics that limit its attractiveness for database access, especially in the lead discovery phase of a drug development program. First, and most importantly, a substructure search system requires that the user who is posing the query must already have acquired a well-defined view of what sorts of structure are expected to be retrieved from the database. For example, a 3D substructure search requires that a pharmacophore is available that describes the geometric constraints associated with binding. Such detailed information is unlikely to be available at the start of a project where perhaps just one or two structurally disparate literature compounds or high-throughput screening (HTS) hits are available. Second, the system will retrieve those molecules, and just those molecules, that satisfy the structural constraints imposed by the query: there may be no molecules or there may be thousands of molecules, but the user cannot control the volume of output. Third, there is no mechanism by which the retrieved molecules can be ranked in order of decreasing probability of activity, something that is of particular importance in the context of a large search output.

The mid-1980s saw the introduction of a form of chemical best-match searching, similarity searching, that alleviates these limitations of substructure searching while still enabling the retrieval of classes of molecules; at the same time, it requires as input just the single molecule that is characteristic of structure searching. The input molecule, which is generally referred to as the target structure or the reference structure, is characterized by one or more descriptors, and these are compared with the corresponding sets of descriptors that represent each of the molecules in the database that is to be searched. Each such comparison results in a quantitative measurement of the degree of similarity between that database structure and the reference structure, and the database is then sorted into decreasing order of these computed similarities. The top-ranked molecules, referred to as the nearest neighbors, are then returned to the user as the output from the search; these are the molecules that are most similar to the reference structure; if the latter was of interest, e.g., it exhibited some particular biological activity, then these nearest neighbors are also probably of interest. This link between chemical and biological similarity has spurred much current interest in similarity searching as a tool for virtual screening,[13,14] where the need for just a single reference structure means that it can be used as a precursor to more sophisticated approaches such as 3D pharmacophore searching, machine learning methods, and flexible ligand docking.

The probabilistic relationship between the reference structure and its nearest neighbors is formalized in the similar property principle, which states that molecules that are structurally similar are likely to have similar properties.[2] This idea, which also underlies research into molecular diversity (as discussed later in this chapter) and chemogenomics[15] under the names of the neighborhood principle[16] and structure–activity relationship homology,[17] respectively, lies at the heart of similarity searching. If the principle holds, then the nearest neighbors of a bioactive reference structure are prime candidates for biological testing, as compared to other molecules that occur further down a ranking of a database and that are less similar to the reference structure.

Whilst there are many exceptions to the principle,[18] a fact that Stahura and Bajorath refer to as the similarity paradox,[19] its appropriateness seems self-evident. After all, if there were not some form of relationship between chemical similarity and biological similarity, then it would be difficult to develop rational approaches for drug discovery, and there have thus been many studies that have reported empirical correlations between chemical similarity and biological activity.[20–24] The approaches that have been developed for the quantitative comparison and evaluation of similarity measures (see below) generally assume that the principle holds, and two recent papers provide conclusive support for the general correctness of the principle. Specifically, both Sheridan *et al.*[25] and He and Jurs[26] have shown that the predictive power of a quantitative structure–activity relationship (QSAR) model is dependent on the extent to which the test-set molecules, for which predictions are required, are similar to the training-set molecules, on which the QSAR model has been based. Their results show clearly that strong similarities between the test-set and training-set molecules enable the making of accurate predictions, as would be expected if the similarity property principle is, indeed, correct.

The extent to which the principle holds for a particular set of compounds will be crucially dependent on the effectiveness of the measure that is used to quantify the degree of similarity between a pair of molecules, and much research into molecular similarity has focused on the key components of a similarity measure that control the effectiveness of searching; the resulting measures are also extensively used in molecular diversity analysis. There are three such components: (1) the representation that is used to characterize the molecules that are being compared; (2) the weighting scheme that is used to assign differing degrees of importance to the various components of these representations; and (3) the similarity coefficient that is used to provide a quantitative measure of the degree of structural relatedness between a pair of structural representations. While there have been a few reports regarding the extent to which the weighting scheme affects the utility of a similarity measure,[20,27–30] much more interest has been shown in the representation and the similarity coefficient, and these are discussed in the next two sections of the chapter.

4.08.2.1 Representations

The many different representations that have been used in similarity search and diversity analysis methods can be categorized according to the type of information they encode. Here they are divided into whole-molecule descriptors, descriptors that can be calculated from 2D representations of molecules, and descriptors that can be calculated from 3D representations. Whole-molecule descriptors consist of single numbers, each of which represents a different property of a molecule, such as its molecular weight or log *P*. A single whole-molecule descriptor is not usually discriminating enough to allow meaningful comparisons of molecules and therefore several different descriptors are usually combined. Descriptors that are derived from the 2D representations of molecules include topological indices and fragment-based descriptors. A topological index is a single number that typically characterizes a structure according to its size and shape. Many different topological indices have been devised and, as for whole-molecule properties, several different indices are usually used in combination. In fragment-based descriptors, a molecule is characterized by the substructural features that it contains. Descriptors that are derived from the 3D representations of molecules include fragment-based descriptors and a variety of more sophisticated representations that encode properties such as shape and electrostatic fields.

Most descriptors result in a global measure of the similarity of two molecules and do not provide a mapping from one molecule to another. However, similarity methods have also been derived that identify local regions of similarity and that involve generating a mapping from the atoms of one molecule onto the atoms of another. For example, graph matching techniques can be applied directly to both 2D and 3D graph representations of molecules to identify the maximum common subgraph that is shared between two molecules. This can then be used to quantify the degree of similarity.

In general, the computational cost of calculating the descriptors increases with their discriminatory power. For example, a simple descriptor like molecular weight is easy to calculate but is not sufficiently discriminating (at least, not when used on its own), whereas descriptors derived using quantum mechanics may describe molecules very accurately; however they are too time-consuming to calculate to be used for database applications. Thus, a suitable compromise has to be made between efficiency and effectiveness, especially when the methods are used to search large databases.

The wide range of different descriptors that have been developed for similarity searching is a reflection of the complex relationship that exists between structure and biological activity. In general, the different representations result in different sets of compounds being identified as near neighbors; however, it is often difficult to predict in advance which descriptors will be most effective in a given situation.[4]

The main types of representations that have been used for similarity searching are described below and include examples of descriptors designed to provide a global measure of similarity as well as descriptors that can be used to identify local similarities.

4.08.2.1.1 Whole-molecule properties

The simplest descriptors are whole-molecule properties and include counts of features such as hydrogen bond donors, hydrogen bond acceptors, ring systems, and rotatable bonds. Many of these features can be defined as substructures and so can be readily calculated from a 2D connection table using substructure search techniques.

Physicochemical property descriptors include some that are easy to calculate, such as molecular weight and molar refractivity, and others that require the use of sophisticated algorithms in order to provide their accurate prediction. For example, several programs have been developed to calculate log *P*, which is widely used to characterize molecules due to the importance of hydrophobicity in drug absorption. The most popular methods are either fragment- or atom-based. Fragment-based methods involve summing contributions from the fragments that exist within a molecule with a correction factor being applied to account for interactions between the fragments. The most widely used program of this type is the ClogP program, developed by Leo.[31] The fragment hydrophobicity values are based on experimental $\log P$ data with the values for missing fragments being estimated.[32] Atom-based methods involve summing contributions from individual atoms rather than fragments.[33–36] The atom contributions are determined from a regression analysis using a training set of compounds for which experimental $\log P$s have been determined. A large number of atom types is needed to describe a range of molecules. For example, in the Ghose and Crippen approach, carbon, hydrogen, oxygen, nitrogen, sulfur, and halogens are classified into 110 atom types depending on their hybridization state and the chemical nature of the neighboring atoms.

4.08.2.1.2 Topological indices

Topological indices have been widely used in various studies. A topological index is a single value that can be calculated from the 2D graph representation of a molecule.[37,38] Many hundreds of different indices have been described in the literature. The simplest topological indices are based on characterizing structures according to their size, degree of branching, and overall shape. More complex indices include consideration of the properties of atoms as well as their connectivities.

The Wiener index[39] involves counting the number of bonds between each pair of atoms in a molecule and summing the distances, D_{ij}, between all such pairs:

$$W = \frac{1}{2}\sum_{i=1}^{N}\sum_{j=1}^{N} D_{ij}$$

The branching index, introduced by Randić,[40] involves calculating the bond connectivity value for each bond as the reciprocal of the square root of the product of the degree of the two atoms in the bond. The branching index then equals the sum of the bond connectivities over all of the bonds in the molecule:

$$\text{Branching Index} = \sum_{\text{bonds}} \frac{1}{\sqrt{\delta_i \, \delta_j}}$$

The related chi indices, developed by Kier and Hall,[41] involve replacing the degree of each atom with terms that encode electronic information such as counts of sigma, pi, and lone pair electrons. The effect is to provide greater discrimination between atoms. A series of indices is developed by summing the values over paths of different lengths. Thus, the zero-order chi index involves a summation over all atoms (i.e., paths of length zero); the first order chi index involves a summation over bonds (as for Randic's branching index); and higher-order chi indices involve summations over sequences of two, three, etc. bonds.

The kappa shape indices[42] characterize aspects of molecular shape by counting the number of paths in a molecule. As for the chi indices, different-order kappa indices exist. For example, the first-order shape index is based on the longest path in a molecule, the second order shape index is based on the number of paths of length two, etc.

As for whole-molecule properties, topological indices are usually combined so that a molecule is represented by a series of different indices. For example, the Molconn-Z program provides access to several hundreds of different descriptors and often these are all used in combination.[43] When descriptors are combined, it is usual to normalize them by the range of values that exist within the dataset prior to using them in order to avoid biases due to different magnitudes of the descriptors. It is also often the case that the different descriptors are correlated. This is particularly true of the topological indices and so a data reduction technique such as principal components analysis is often used to identify a smaller number of principal components that can then be used in similarity calculations.[44]

4.08.2.1.3 Two-dimensional descriptors

The most commonly used descriptors for similarity searching are the 2D fingerprints that were originally developed for substructure searching. In fact, one of the earliest similarity search methods to be reported in the literature is that by Willett *et al.* which is based on the use of 2D fingerprints.[45] These encode the presence or absence of substructural fragments within a molecule as a binary vector. Two different types of fingerprint have been developed: dictionary-based fingerprints and hashed fingerprints.

In dictionary-based fingerprints, each bit in the vector represents a particular substructural fragment contained within a fragment dictionary. For a given molecule, a bit is set to '1' if the fragment it represents is present in the molecule; otherwise the bit is set to '0.' The dictionary is usually composed of a variety of different fragment types, including augmented atoms, atom sequences, bond sequences, and various ring fragments. Augmented atoms consist of a central atom together with its neighboring atoms and bonds. Atom sequences consist of a given number of connected atoms and their intervening bonds. Ring fragments include ring sequences (the sequence of atoms around the ring) and ring fusion sequences (the ring-connectivity counts of the atoms in the ring). Several variants of these fragment types have also been developed in which either the atoms or the bonds are generalized. The fragments that are represented in the dictionary should be chosen following a statistical analysis of the database. For example, fragments that occur very frequently within the database are unlikely to be very discriminating, whereas fragments that occur rarely are unlikely to be useful. Hence, a limitation of dictionary-based fingerprints is that the optimum fragment dictionary is dataset-dependent. Examples of dictionary-based fingerprints that are commonly used in similarity calculations include Barnard Chemical Information (BCI) fingerprints[46] and Molecular Access System (MACCS) structural keys.[47]

Hashed fingerprints do not require the use of a dictionary of fragments and therefore avoid the problem of dataset dependence. They are based on enumerating all paths in a molecule of predefined lengths; for example, the Daylight fingerprint considers linear paths up to a specified length (typically from two to seven atoms).[48] Each path is hashed to a small number of bits (typically four or five) which are then set to '1' in the fingerprint bitstring. Since the paths are mapped into a fixed-length bitstring, collisions can occur whereby a given bit is set by more than one path and, therefore, in contrast to the dictionary-based fingerprints, there is no longer a one-to-one mapping between bit position and fragment.

About the same time as the work of Willett *et al.*, Carhart *et al.*[49] described an approach to similarity searching based on atom pair descriptors. An atom pair comprises two nonhydrogen atoms and their through-bond distance (also known as the topological distance), which is the shortest bond-by-bond path between them. Each atom is described by element type, the number of nonhydrogen atoms to which it is bonded, and its number of π-bonding electrons. Atom pairs can encode more distant information than is the case for the 2D fingerprints described above. Topological torsions are related descriptors which encode sequences of four connected atoms together with their atom types, number of nonhydrogen connections, and π-electrons.[50]

Descriptors that are based on element type have been shown to be effective at identifying close analogs of a query structure; however, it is often more useful to identify compounds that are predicted to be active but that belong to different chemical series than the query compound. Such compounds are especially valuable since they present new patent opportunities and they can also provide alternative lead series should problems occur, such as poor absorption, distribution, metabolism, and excretion (ADME) properties. This is a technique known as scaffold hopping which has become an important goal for similarity searching.[51] It has led to the development of descriptors that are based on the properties of atoms or functional groups rather than exact element type. Thus, the atom pair descriptors were modified to represent the binding properties of atoms rather than their specific element types in what are known as binding property pairs.[52] The atoms are identified as belonging to one of seven classes: (1) cations; (2) anions; (3 and 4) neutral hydrogen bond donors and acceptors; (5) atoms which are both donor and acceptor; (6) hydrophobic atoms; and (7) all others.

More recent descriptors include the Chemically Advanced Template Search (CATS) descriptors, Similog keys, and Scitegic circular fingerprints. The CATS descriptors[53] are based on counts of atom pairs up to 10 bonds distant, with the atoms classified as lipophilic, positive, negative, donor, and acceptor. The 15 possible pairs of atoms and 10 distances give rise to a vector of length 150. The Similog keys[54] represent triplets of atoms. Each atom is represented by the presence or absence of the following four properties: donor, acceptor, bulkiness, and electropositivity (known as the DABE code). A Similog key for a molecule consists of a list of triplets of atoms, with each atom described by its DABE code, the topological distances between the atoms mapped to four intervals, and the frequency of occurrence of the triplet.

The Scitegic circular fingerprints are based on the extended connectivities of the atoms and include the extended connectivity fingerprints (ECFPs) and functional connectivity fingerprints (FCFPs).[55] The extended connectivity of an atom is calculated using a modified version of the Morgan algorithm developed to generate a canonical representation of a connection table.[56] An initial code is assigned to each atom based on its properties (atom type for ECFPs and generalized atom type for FCFPs) and connectivity. Each atom code is then combined with the codes of its immediate neighbors to produce the next order code for the atom. This process is repeated to produce atom codes for different diameters of circular substructures. A molecule is then represented by the atom codes present. Hert *et al.* describe similarity searches based on ECFP_4 descriptors, which are circular fingerprints of diameter four, with the atom codes in the range 2^{-31} to 2^{31} hashed to a fingerprint of 1024 bits.[57]

The whole-molecule and fragment-based descriptors described thus far give rise to vector representations of the molecules, which may consist of binary, integer, or real numbers. When such descriptors are used to compare molecules they result in a global measure of similarity and they do not allow the identification of local regions of similarity. The identification of local similarities requires that a mapping is made from one molecule to another. This can be achieved when molecules are represented as graphs and are compared using graph-matching techniques.[11,12] A 2D chemical structure can be represented as a topological graph where the nodes of the graph represent the atoms and the edges represent the bonds. The similarity between two graphs can be quantified using a maximum common subgraph (MCS) algorithm which identifies the largest substructure in common between the two molecules.[58] As with other graph-matching procedures, identifying the MCS between two structures is too computationally expensive to be used on large databases. However, screening techniques can be used that limit the numbers of compounds that need to undergo the more expensive calculation. Thus, Hagadone developed a method for establishing local similarity known as substructure similarity (or subsimilarity) searching.[59] In subsimilarity searching, a simple, fragment-based similarity search is used to calculate an upper bound to the size of the MCS (in terms of the numbers of constituent atoms or bonds) between the target (sub)structure and each database structure; these upper bounds are then used to prioritize database structures for an MCS search that uses a rapid, but approximate, MCS algorithm. More recently, Raymond *et al.*[60,61] have developed an MCS method called RASCAL that is able to perform thousands of comparisons a second. The method uses a number of sophisticated screening steps to eliminate compounds from the MCS calculation, together with a fast implementation of the MCS algorithm itself that involves first converting the graphs to line graphs.

Initial graph-matching methods involved representing all of the nonhydrogen atoms as nodes with the nodes labeled by element type. These representations allow local similarities to be identified in the form of substructures that are common to the two graphs. Other graph representations have been developed with scaffold hopping in mind. For example, both the feature tree and the reduced graph methods are based on summarizing features within a structure such that several connected atoms are represented as a single node.[62,63] These reduced representations can then be mapped to a fingerprint or they can be compared using graph-matching algorithms.[64–66]

4.08.2.1.4 Three-dimensional descriptors

Molecular recognition is governed by the 3D properties of molecules such as their shape and physicochemical properties and hence it is natural that there should be substantial interest in comparing molecules by their 3D characteristics. 3D methods clearly require the conformational properties of the molecules to be considered and are therefore more computationally demanding than methods based on 2D descriptors. 3D similarity methods can be divided into those that are independent of the relative orientation of the molecules and those that require the molecules to be aligned prior to calculation of their similarity.

As for 2D fingerprints, 3D fingerprints were originally developed for substructure search; however, they have also subsequently been used for similarity searching (and other applications, such as combinatorial library design and diversity analysis, described in Section 4.08.5). The fingerprints encode spatial characteristics of conformations of molecules such as interatomic distances and angles. They may encode the presence or absence of such 3D features or the frequency of occurrence of the features. Distances are usually encoded as ranges, for example, distances in the range of 0–20 Å may be mapped to 10 bins with each bin having a width of 2 Å (0–1.9 Å; 2.0–3.9 Å; 4.0–5.9 Å; etc.). Alternatively, the bins may be nonuniform, with small ranges used to represent the more common distances and larger ranges used to represent less common distances.[67]

The simplest distance-based descriptor is the distance distribution in which each interatomic distance in a molecule increments the count in the appropriate bin. The molecule is represented by the resulting frequency distribution of distances. Distance distributions have also been used to represent triplets of atoms. For example, in the method described by Bemis and Kuntz, a single number is used to represent each triplet of atoms by summing the three squared interatomic distances.[68] A molecule is then represented by the distribution of triplet values. Nilakantan *et al.* developed a similar distance-based descriptor also using triplets of atoms.[69] For each triplet, the distances are sorted into increasing length and are packed into a single 32-bit integer using the formula:

$$n_t = n_1 + 10^3 n_2 + 10^6 n_3$$

Each triplet is hashed to two numbers between 1 and 2048 and the corresponding bits are set 'on' in a bitstring.

The discrimination between molecules can be increased by including atom types so that distances are recorded between specific pairs of atoms, for example, between a carbonyl oxygen and an amine nitrogen. An early review of distance-based similarity measures is provided by Pepperrell and Willett.[70] Other types of spatial features encoded in 3D fingerprints are based on angular information.[71,72] Valence angle descriptors consist of three atoms, *ABC*, together with the angle between them (i.e., the angle between *AB* and *AC*). The atoms may or may not be bonded. Torsion angle descriptors consist of four atoms, *ABCD*, where *BC* is the torsion bond, and the torsion angle is the angle formed between *ABC* and *BCD*. As with valence angles, the atoms may be bonded or nonbonded.

Several of the 2D fingerprints described earlier have 3D counterparts. Thus, the geometric atom pairs and geometric binding property pairs are 3D variants of the atom pair and binding property pair descriptors.[73] In geometric atom pairs the atoms are defined according to element type with the distance between them given as the through-space distance rather than the through-bond distance. In geometric binding property pairs the atoms are described using the same generalized types as for binding property pairs.

Pharmacophore keys have become widely used for both diversity analyses and similarity methods. They are based on the pharmacophoric features present in molecules, that is, the atoms or substructures that are thought to have relevance for receptor binding.[74] These typically include hydrogen bond donors, hydrogen bond acceptors, charged centers, aromatic ring centers and hydrophobic centers. Pharmacophore keys are usually based on combinations of either three or four features, known as three-point pharmacophores and four-point pharmacophores, respectively.[75,76] A binary pharmacophore key represents all possible three- or four-point pharmacophores by enumerating all possible combinations of points together with all possible distances, based on a binning scheme. The pharmacophore key for a molecule is then derived by setting the relevant bits in the key to '1' to indicate the pharmacophores it contains. The binary pharmacophore keys representing two molecules may be compared as for 2D binary fingerprints, but their characteristics are somewhat different.[77] In particular, pharmacophoric keys are typically much longer than 2D fingerprints (which usually consist of around 1000 bits); for example, a key based on six features and six bins results in 5916 possible three-point pharmacophores. Increasing the distance resolution and extending to four-point pharmacophores results in much larger keys. Pharmacophore keys also tend to be relatively sparse compared with 2D fingerprints and small changes in conformation can lead to relatively large changes in the keys. These characteristics have led to the use of modified similarity coefficients compared to 2D fingerprints. Conformational flexibility is usually handled either by generating a separate pharmacophore key for each conformer that is accessible to a molecule or by ORring the keys together to generate a single ensemble pharmacophore key that represents all possible conformers.

Graph-matching approaches can also be applied to 3D graph representations of molecules.[70,78,79] Here the nodes of the graph represent the atoms and the edges represent the interatomic distances between the atoms. Thus, the 3D graph representation of a molecule is a fully connected graph since all of the nodes are connected by edges. The MCS derived from two 3D graphs is the largest set of atoms with matching interatomic distances (within some user-defined tolerances). Calculation of the MCS between two 3D structures is even more time-consuming than for 2D graphs and a more approximate, but much faster, approach has been developed by Pepperrell *et al.*[80] This aims to identify local regions of similarity between two molecules by comparing the 3D environment of each atom in one molecule with the environment of each atom in the other molecule. The distances associated with an atom in one molecule are compared with those of each atom in the second molecule and interatomic similarity scores are calculated. The resulting similarities can be used to identify pairs of atoms which have geometrically similar environments and can be used to calculate a global intermolecular similarity between the two molecules.

Other 3D methods are based on field representations of molecules such as their steric, electrostatic, or hydrophobic fields and require that the molecules are aligned before their similarity can be calculated.[81–83] Field properties are usually represented on a 3D grid which surrounds the molecule with the values of the field being plotted at the grid vertices. For example, the electrostatic charge at a grid vertex can be calculated from atomic charges using the following equation:

$$P_{\mathrm{r}} = \sum_{i=1}^{N} \frac{q_i}{|\mathbf{r} - \mathbf{R}_i|}$$

where q_i is the atomic charge on atom i, at position R_i from the grid point, r. The similarity between two molecules can be calculated by comparing the values at corresponding grid vertices. However, if one molecule is moved relative to the other, the similarity score will typically change. The goal is to find the alignment of the two structures at which the similarity measure reaches a maximum. This can be very time-consuming; for example, the brute-force approach to maximizing the similarity between two molecules would involve rotating and translating one grid relative to the other in all possible ways and recalculating the similarity at each position. The resolution of the grids that is normally used (e.g., $1000 \times 1000 \times 1000$ vertices) makes a direct grid comparison computationally infeasible. Fortunately, methods have been devised for improving the efficiency of this process. For example, Good *et al.* recognized that field representations could be approximated by atom-centered Gaussians allowing the implementation of a fast analytic approach for calculating similarity.[84,85] Since the work of Good *et al.*, several groups have further developed this approach. For example, the field-based similarity searching (FBSS) program uses a genetic algorithm to optimize the alignment of two molecules based on their steric, electrostatic, and hydrophobic fields calculated using Gaussian approximations.[83] Gaussian functions have also been used to provide representations of molecular shape that are more realistic than more traditional 'hard sphere' models.[86] For example, the rapid overlay of chemical structures (ROCS) program uses atom-centered Gaussians to describe molecular shape together with a rapid method for finding the optimum alignment of two molecules. By considering shape only, the method has been shown to be effective at identifying molecules that have similar shape but that belong to different chemical series.[87]

Many different molecular alignment techniques have been developed and a comprehensive review is provided by Lemmen and Lengauer.[88] Despite the efficient implementations that have been devised, alignment-based similarity methods are much more computationally demanding than 3D vector-based methods, even without consideration of conformational flexibility.

4.08.2.2 Similarity Coefficients

Having reviewed many types of structural representation that can be used for similarity searching and diversity applications, we will now consider how the similarity between pairs of such representations can be computed.

There is a need to quantify the degree of similarity (or conversely the distance or dissimilarity) between two records, each characterized by some number of attributes or descriptors, in a wide range of disciplines, including biology, information retrieval, marketing, and psychology, among others. This has led to identical, or near-identical, coefficients being developed for quite different applications, which accounts in part for the variety of different names that are associated with some of the more common coefficients. Comprehensive reviews of similarity coefficients are provided by Hubálek,[89] Gower and Legendre,[90] Ellis *et al.*,[91] and Everitt *et al.*[92] Here, we focus on those that have been widely used for chemical similarity searching; a comparison of more than 20 such coefficients has been reported by Holliday *et al.*[93]

Assume that an object A is described by means of a vector X_A containing n attributes such that

$$X_A = \{x_{1A},\ x_{2A},\ x_{3A}, \ldots\ x_{jA}, \ldots\ x_{nA}\}$$

where x_{jA} is the value of the jth attribute of object A (see **Table 1**). The attribute values may be real numbers that vary over any range (and that may require the application of some standardization factor to ensure that all attributes contribute equally to the final similarity value); for example, a molecule might be represented by a set of experimental and/or computed physicochemical properties as in the study by Downs *et al.*[94] Alternatively, the attributes may take only binary (i.e., present or absent) values; this is very common in similarity searching, where molecules are represented by 2D fingerprints (or fragment bitstrings) (see above) in which bits are switched on or off depending on the presence or absence of particular substructural features. This representation is of particular importance in both similarity and diversity applications as 2D fingerprints have been found to provide a simple but effective way of quantifying the similarity and diversity relationships between sets of molecules; indeed, much of the literature makes the direct assumption that it is this particular type of representation that is being used to characterize the molecules under consideration.

Table 1 Symbols used

i,j	Attributes
A, B	Objects (or molecules)
n	Total number of attributes of an object (e.g., bits in a fingerprint)
X_A	Attribute vector describing object A
x_{jA}	Value of jth attribute in object A
χ_A	Set of 'on' bits in binary vector X_A
a	Number of bits 'on' in molecule A
b	Number of bits 'on' in molecule B
c	Number of bits 'on' in both molecules A and B
d	Number of bits 'off' in both molecules A and B
$S_{A,B}$	Similarity between objects A and B
$D_{A,B}$	Distance between objects A and B

Reprinted in part with permission from *J. Chem. Inf. Comput. Sci.* **1998**, *38*, 983–996. Copyright 1998 American Chemical Society.

Some coefficients quantify the distance, or dissimilarity, between objects, rather than the similarity; such coefficients thus take zero values when comparing objects that have identical representations (the reader should note that identical representations need not necessarily imply identical objects), while others measure similarity directly and have their maximum value for objects that have identical representations. Coefficients often take values in the range from zero to unity, or are normalized so that the values lie in this range. Such normalization is common; it is generally implemented by taking account of the values of the attributes for the two objects that are being compared, and the resulting coefficients (of which there are many) are generally referred to as 'association coefficients'. An association coefficient can be converted into a complementary distance coefficient by subtracting the coefficient's value from unity. The wide range of application domains for similarity have meant that it is possible for a similarity coefficient and its complement to have been developed independently and thus to be known by different names; for example, the Soergel distance coefficient is the complement of the binary Tanimoto (or Jaccard) association coefficient listed in **Table 2**.

There are strong relationships between distance coefficients and metric distances in multidimensional geometric space. For a distance coefficient to be described as a metric it must satisfy several criteria. First, distance values must be nonnegative, and the distance of an object from itself must be zero, i.e.,

$$D_{A,B} \geq 0, \qquad D_{A,A} = D_{B,B} = 0$$

Next, distance values must be symmetric, i.e.,

$$D_{A,B} = D_{B,A}$$

and must also obey the triangular inequality, i.e.,

$$D_{A,B} \leq D_{A,C} + D_{C,B}$$

Finally, the distance between nonidentical objects must be greater than zero, i.e.,

$$A \neq B \quad \Leftrightarrow \quad D_{A,B} > 0$$

A distance coefficient that satisfies the first three criteria above is a pseudometric, and one that does not satisfy the third criterion is nonmetric. It should be noted that satisfaction of all four criteria is not sufficient to imply that the distances involved can be embedded in a Euclidean space of any given dimensionality; the requirements for Euclidean embedding are discussed by Gower.[95]

If only binary attributes are involved then the expressions used for the various similarity and distance measures can be substantially simplified. Given two objects, A and B, which are characterized by vectors X_A and X_B that contain n binary values (e.g., 2D fingerprints in the context of a chemical database), we can write

$$a = \sum_{j=1}^{j=n} x_{jA} \quad \text{(number of bits switched 'on' in } A\text{)}$$

$$b = \sum_{j=1}^{j=n} x_{jB} \quad \text{(number of bits switched 'on' in } B\text{)}$$

$$c = \sum_{j=1}^{j=n} x_{jA}x_{jB} \quad \text{(numbers of bits switched 'on' in both } A \text{ and } B\text{)}$$

$$d = \sum_{j=1}^{j=n} (1 - x_{jA} - x_{jB} + x_{jA}x_{jB}) \quad \text{(number of bits switched 'off' in both } A \text{ and } B\text{)}$$

where

$$n = a + b - c + d$$

This notation is the most commonly encountered in the chemical information literature; different definitions for a and b are given by Gower[95] and by Ellis *et al.*[91]

The above definitions can also be expressed in terms of set theory. Let χ_A denote the set of all elements x_{jA} in vector X_A whose value is 1 (the bits switched 'on'), and let χ_B similarly denote the set of all elements x_{jB} in vector X_B whose

Table 2 Descriptions of some distance metrics and similarity coefficients commonly used in chemical information. Definitions of the symbols used are shown in **Table 1**. Note that the negative lower bound values for the three association coefficients only apply if negative attribute values are possible

Hamming distance	
Other names	• Manhattan distance • City-block distance • Normalized complement for dichotomous data, called simple matching coefficient
Formula for continuous variables	$D_{A,B} = \sum_{j=1}^{j=n} \lvert x_{jA} - x_{jB} \rvert$
Formula for dichotomous variables	$D_{A,B} = a + b - 2c$
Set-theoretic definition	$D_{A,B} = \lvert\chi_A \cup \chi_B\rvert - \lvert\chi_A \cap \chi_B\rvert$
Range	• to 0 (continuous), n to 0 (dichotomous)
Metric properties	• Obeys all four metric properties
Notes	• Equivalent to the squared Euclidean distance for dichotomous variables • Can be normalized to the range 1–0 if the values of all attributes are normalized to this range and the result divided by n
Euclidean distance	
Other names	
Formula for continuous variables	$D_{A,B} = \sqrt{\sum_{j=1}^{j=n} (x_{jA} - x_{jB})^2}$
Formula for dichotomous variables	$D_{A,B} = a + b - 2c$
Set-theoretic definition	$D_{A,B} = \sqrt{\lvert\chi_A \cup \chi_B\rvert - \lvert\chi_A \cap \lvert\chi_B}$
Range	• to 0 (continuous), n to 0 (dichotomous)
Metric properties	• Obeys all four metric properties
Notes	• Frequently used as its square (with which it is, of course, monotonic), which avoids the need to take the square root in the calculation • Monotonic with the Hamming distance in all cases (and its square is equivalent to the Hamming distance for dichotomous variables) • Can be normalized to the range 1–0 if the values of all attributes are normalized to this range and the result divided by n
Soergel distance	
Other names	
Formula for continuous variables	$D_{A,B} = \frac{\sum_{j=1}^{j=n} \lvert x_{jA} - x_{jB} \rvert}{\sum_{j=1}^{j=n} \max(x_{jA}, x_{jB})}$
Formula for dichotomous variables	$D_{A,B} = 1 - \frac{c}{a + b - c} = \frac{a + b - 2c}{a + b - c}$
Set-theoretic definition	$D_{a,B} = \frac{\lvert\chi_A \cup \chi_B\rvert - \chi_A \cap \chi_B}{\chi_A \cup \chi_B}$
Range	• 1–0
Metric properties	• Obeys all four metric properties provided all attributes have nonnegative values
Notes	• For dichotomous variables only, the Soergel distance is identical to the complement of the Tanimoto coefficient
Tanimoto coefficient	
Other names	• Jaccard coefficient
Formula for continuous variables	$S_{A,B} = \frac{\sum_{j=1}^{j=n} x_{jA}x_{jB}}{\sum_{j=1}^{j=n} (x_{jA})^2 + \sum_{j=1}^{j=n} (x_{jB})^2 - \sum_{j=1}^{j=n} x_{jA}x_{jB}}$
Formula for dichotomous variables	$S_{A,B} = \frac{c}{a + b - c}$
Set-theoretic definition	$S_{A,B} = \frac{\lvert\chi_A \cap \chi_B\rvert}{\lvert\chi_A \cup \chi_B\rvert}$
Range	• -0.333 to 1 (continuous), 0 to 1 (dichotomous)
Metric properties	• Complement does not obey the triangular inequality in general, though does obey it if dichotomous variables are used

Table 2 Continued

Notes	• Monotonic with the Dice coefficient • Complement of the dichotomous version is identical to the Soergel distance
Dice coefficient	
Other names	• Czekanowski coefficient • Sørenson coefficient • Essentially equivalent to the Hodgkin index for overlap of electron density functions
Formula for continuous variables	$S_{A,B}\dfrac{2\sum_{j=1}^{j=n} x_{jA}x_{jB}}{\sum_{j=1}^{j=n}(x_{jA})^2 + \sum_{j=1}^{j=n}(x_{jB})^2}$
Formula for dichotomous variables	$S_{A,B} = \dfrac{2c}{a+b}$
Set-theoretic definition	$S_{A,B}\dfrac{2\lvert\chi_A \cap \chi_B\rvert}{\lvert\chi_A\rvert + \lvert\chi_B\rvert}$
Range	• -1 to 1 (continuous), 0 to 1 (dichotomous)
Metric properties	• Complement does not obey the triangular inequality
Notes	• Monotonic with the Tanimoto coefficient
Cosine coefficient	
Other names	• Ochiai coefficient • Essentially equivalent to the carbo index for overlap of electron density functions
Formula for continuous variables	$S_{A,B} = \dfrac{\sum_{j=1}^{j=n} x_{jA}x_{jB}}{\sqrt{\sum_{j=1}^{j=n}(x_{jA})^2 \cdot \sum_{j=1}^{j=n}(x_{jB})^2}}$
Formula for dichotomous variables	$S_{A,B} = \dfrac{c}{\sqrt{a \cdot b}}$
Set-theoretic definition	$S_{A,B} = \dfrac{\lvert\chi_A \cap \chi_B\rvert}{\sqrt{\lvert\chi_A\rvert \cdot \lvert\chi_B\rvert}}$
Range	• -1 to 1 (continuous), 0 to 1 (dichotomous)
Metric properties	• Complement does not obey the triangular inequality
Notes	• Highly correlated with the Tanimoto coefficient, though not strictly monotonic with it

Reprinted in part with permission from *J. Chem. Inf. Comput. Sci.* **1998**, *38*, 983–996. Copyright 1998 American Chemical Society.

value is 1. Then

$$a = \lvert\chi_A\rvert$$

$$b = \lvert\chi_B\rvert$$

$$c = \lvert\chi_A \cap \chi_B\rvert$$

$$d = n - \lvert\chi_A \cup \chi_B\rvert$$

with the number of bits switched 'on' in at least one of the molecules being given by

$$a + b - c = \lvert\chi_A \cup \chi_B\rvert$$

Table 2 uses the definitions above to describe several similarity and distance coefficients that are commonly used in chemoinformatics, with both attribute- and set-based definitions. The coefficients listed in **Table 2** are but a small fraction of those that have been defined in the literature, but many of them are closely related to each other. Indeed, in some cases, a coefficient has been developed independently by more than one author; for example, the well-known Tanimoto coefficient is also known as the Jaccard coefficient. There are also cases where two coefficients are different when applied to the processing of continuous attributes but become equivalent when applied to binary attributes; for example, the Euclidean distance is identical with the Hamming distance for the latter type of data.

Some coefficients are 'monotonic' with each other, that is, they always produce identical similarity rankings of a set of objects, even though the actual similarity values resulting from the coefficients are different. Monotonicity may be representation-dependent, for example, the Tanimoto and Dice coefficients are monotonic when applied to binary attribute vectors but otherwise nonmonotonic; and Whittle *et al.* have shown that a coefficient they had developed for similarity searching, which they called the modulus coefficient, was monotonic with Euclidean distance for binary data but that the two coefficients gave different rankings with continuous data.[96] There are also coefficients that, while not being monotonic with each other, do yield similarity rankings and/or similarity values that are strongly correlated, for example, the cosine and Tanimoto coefficients.[97] Some pairs of coefficients, conversely, exhibit very low correlations, which implies that they are measuring very different types of equivalence when pairs of objects are compared.[91]

Hubálek[89] has reported a detailed study of the monotonicity relationships existing in a set of 43 different coefficients, while Holliday *et al.*[93] have described an analogous study of 22 coefficients that had been applied to fingerprint-based similarity searching. These authors demonstrate that many of the coefficients that have been reported in the literature yield very similar rankings in fingerprint-based similarity searches of the NCI AIDS, ID Alert, and MDL Drug Data Report databases, with many of the pairs of coefficients that were studied showing a very high level of correlation; indeed, their results suggest that there were only three completely distinct classes of coefficient among the 22 coefficients that were studied. Similar results were obtained in a subsequent study that considered not just single coefficients but combinations of similarity coefficients (using the data fusion approach that is discussed further below).[98] Another study by Cheng *et al.*[99] investigated four association coefficients for assessing the degree of relatedness between pairs of different similarity coefficients; their study was used to compare measures in which both the coefficient and the representation were varied, but their approach could also be applied to coefficients based on a common descriptor.

The Hamming distance and the Euclidean distance in **Table 2** are examples of a more general class of distance metrics known as the Minkowski distances. These have the general formula

$$D_{A,B} = \sqrt[t]{\sum_{j=1}^{j=n} \left(|x_{jA} - x_{jB}|\right)^t}$$

where $t=1$ for the Hamming distance and $t=2$ for the Euclidean distance. A fundamental difference between the Hamming and Euclidean distances on the one hand, and association coefficients such as the Tanimoto, Dice, and cosine coefficients on the other, is that the former take the common absence of attributes (or common low values in the case of continuous variables) as evidence of similarity, whereas this is not the case with the association coefficients. The belief that the common absence of an attribute contributes positively toward similarity is a controversial one that has been extensively discussed in the literature (see, for example, the comments of Sokal and Sneath in the context of numerical taxonomy[100]). In the chemical context, James *et al.* have argued that coefficients involving d, the number of bits that are set in the fingerprints of neither A nor B should not be used for similarity searching for two reasons: most molecules have most of the bits in their fingerprints set to zero; and many chemoinformatics systems provide user-control over the length of the fingerprint, thus allowing substantial increases, or decreases, in the value of d.[101]

Comparative studies, as described further below, have demonstrated the general merits of those coefficients that focus just on the bits that have been set to 1, but most of these comparisons have been in the specific context of similarity searching, and slightly different criteria may be required when diversity, rather than similarity, applications are involved. It is for this reason that Fligner *et al.* have suggested a new coefficient that contains two parts: one part is the conventional Tanimoto coefficient, which is based on the fingerprint bits that are set to 1, and the other part is a version of the Tanimoto coefficient that considers only those fingerprint bits that are set to 0, with the overall value being a weighted sum of the two components.[102]

There is a second major difference between association and distance coefficients, which is the former's use of a normalization factor that helps to lower the effect of molecular size in some cases. Thus, in fingerprint-based similarity searching, a large database molecule is much more likely to have bits in common with the reference structure than is a small molecule. It hence seems appropriate for the coefficient to include some size-related factor(s) to avoid the most similar molecules being drawn largely, or even exclusively, from among the largest molecules in the database that is being searched. The inverse problem can arise when dissimilarity-based compound selection procedures are used for molecular diversity analysis. As discussed in detail below, such procedures seek to identify subsets of a database such that the constituent molecules are as dissimilar as possible to each other. Small molecules are likely to contain few substructural fragments and hence few bits switched to 'on' in a fingerprint: since, for example, the Tanimoto coefficient takes no explicit account of features that are absent from both the molecules being compared, and since $c \leq \min(a, b)$ (see the expression for the coefficient in **Table 2**), low similarity (and thus high dissimilarity) values are likely to be obtained with small molecules. There is hence considerable potential for a biasing of the distribution of molecular sizes in the database subset that is selected by a procedure that is based on molecular dissimilarities. An early solution to this problem was to use a composite coefficient involving both the Tanimoto coefficient and the simple matching coefficient[103,104]; more recently, Fligner *et al.*[102] have developed for this purpose the modified form of the Tanimoto coefficient described previously.

Much of the literature on similarity coefficients focusses on their use with 2D fingerprints, but rather different evaluation criteria may be required when continuous, rather than binary, attributes are involved. For example, there has been much discussion as to the appropriateness of the cosine coefficient for the calculation of field-based similarities (or, to be more precise, of the identical coefficient first suggested by Carbo *et al.* for this application[105]), where it has been shown that the coefficient is sensitive to the shape of the molecular fields that are being compared, rather than to their magnitudes.[86,106] Again, in a study of similarity searching in the Dictionary of Natural Products database, Whittle *et al.*[96] found that the Forbes, Russell-Rao, and Kulczynski (2) coefficients performed very poorly when used with continuous-data representations (specifically sets of principal components derived from MOLCONN-Z parameters, and both 2D and 3D autocorrelation vectors), despite their effectiveness for fingerprint-based similarity searching.

Many of the association coefficients in **Table 2** are symmetric in character, but following ideas first proposed by Tversky,[107] there has also been interest in the use of asymmetric similarity coefficients, i.e., those where $S_{A,B} \neq S_{B,A}$. The general form for Tversky similarity for binary data is

$$S_{A,B} = \frac{c}{\alpha(a-c) + \beta(b-c) + c}$$

where α and β are constants whose values are defined by the user. The resulting similarity coefficient is symmetric if α and β are equal, and the specification of certain values can result in one of the well-known association coefficients, e.g., setting $\alpha = \beta = 1/2$ results in the Dice coefficient. The coefficient is asymmetric when α and β have different values. A specific example of this is when $\alpha = 1$ and $\beta = 0$, which corresponds to $S_{A,B} = c/a$ and which represents the fraction of A that is in common with B. The coefficient will have the value of unity when all the features of A are also in B, which corresponds to A being a substructure of B (and one can, of course, have a corresponding measure of the fraction of B that is contained within A). Other such 'subsimilarity' coefficients that can be used for identifying when one structure is contained within another have been described by Hagadone,[59] and Maggiora and co-workers,[82] while the use of such coefficients has been discussed by Wipke and Rodgers,[108] Willett,[109] and Grethe and Hounsell,[110] among others.

4.08.3 Comparison of Similarity Measures

The two previous sections have provided overviews of the many similarity coefficients and structure representations that have been suggested for similarity searching in 2D and, to a lesser extent, 3D databases. The combination of these, together with the various weighting schemes that have been suggested, means that a very large number of different similarity measures could be used for database searching. The range of possibilities means that there is a need to compare different measures, so as to identify those that are most appropriate for some particular application. In this section, we will focus on similarity searching (*see* 4.19 Virtual Screening), but make at least some mention of diversity-based selection (*see* 4.14 Library Design: Ligand and Structure-Based Principles for Parallel and Combinatorial Libraries; 4.15 Library Design: Reactant and Product-Based Approaches), since the same coefficients and representations are generally used for both purposes.

The most important evaluation criterion in the context of similarity searching is the ability to retrieve molecules from a database that prove, on inspection, to be active in the bioassay of interest: only if this is the case will a particular similarity measure be useful for virtual screening. The precise form of the comparison will depend on whether the available biological data are qualitative (e.g., a molecule is categorized as either active or inactive) or quantitative (e.g., an IC_{50} value is available for a molecule), but the general form is based on the similar property principle,[2] which has been introduced previously. If the principle does hold for a particular dataset, i.e., if structurally similar molecules have similar activities, then the nearest-neighbor molecules in a similarity search are expected to have the same activity as a

Table 3 Contingency table describing the output of a search in terms of active molecules and molecules retrieved in a similarity search that returns the n nearest neighbors of the reference structure

		Active		
		Yes	*No*	
Retrieved	Yes	a	$n-a$	n
	No	$A-a$	$N-n-A+a$	$N-n$
		A	$N-A$	N

bioactive reference structure. We can hence evaluate the effectiveness of a similarity measure by the extent to which the similarities resulting from its use mirror similarities in the bioactivity of interest.

Edgar *et al.* provide a detailed review of a range of measures of effectiveness that can be used when qualitative activity data are available, summarizing the various performance measures in terms of the 2×2 contingency table shown in **Table 3**.[111] In this table, it is assumed that a search has been carried out, resulting in the retrieval of the n nearest neighbors at the top of the ranked output. Assume that these n nearest neighbors include a of the A active molecules in the complete database, which contains a total of N molecules. Then the recall, R, is defined to be the fraction of the active molecules that are retrieved, i.e., $R=a/A$, and the precision, P, is defined to be the fraction of the retrieved molecules that are active, i.e., $P=a/n$. A retrieval mechanism should seek to maximize both the recall and the precision of a search so that, in the ideal case, a user would be presented with all of the actives in the database without any additional inactives: needless to say, this ideal is very rarely achieved in practice.

It is inconvenient to have to specify both the recall and precision parameters to quantify the effectiveness of a search, and several single-component performance measures have thus been suggested for comparative experiments. Examples include the enrichment factor, i.e., the number of actives retrieved relative to the number that would have been retrieved if compounds had been picked from the database at random,[52] the numbers of actives that have been retrieved at some fixed position in the ranking, e.g., the top 1% of the ranked database, and the position in the ranking at the point where some specific fraction, e.g., 50%, of the actives have been retrieved.[52,57,112,113] Güner and Henry have described the G-H score: this is a weighted average of recall and precision that has been taken up by several groups since its original introduction for evaluating the effectiveness of 3D pharmacophore searches.[114] Rather than measuring the search performance for some number of nearest neighbors, it may be of interest to study the performance of a measure across the entire ranking resulting from a similarity search. In this case, the most popular approach is the use of a cumulative recall graph, which plots the recall against the number of compounds retrieved (i.e., a/A against n, using the notation of **Table 3**)[52,115] or a receiver operating characteristic curve, which plots the true positives against the false positives for different classifications of the same set of objects (i.e., a against $n-a$, using the notation of **Table 3**).[116,117]

The similar property principle can also be applied to the analysis of datasets for which quantitative bioactivity data are available, most commonly using a 'leave-one-out' approach. Assume that the value of some quantitative property has been measured for each of the molecules in a dataset, and that the value for the reference structure, X, is unknown. A similarity search is carried out to identify X's k nearest neighbors, and the predicted value for X, $P(X)$, is then set equal to the arithmetic mean of the observed property values of its k nearest neighbors. This procedure results in the calculation of a $P(X)$ value for each of the N structures in the dataset (or some subset thereof): an overall figure of merit for the set of searches is then obtained by calculating the product moment correlation coefficient between the sets of N observed and N predicted values. A high, statistically significant value for the correlation coefficient implies that the similarity measure used in the making of the predictions provides an effective way of relating chemical structure to biological activity. This application of the similar property principle was first used by Adamson *et al.*[118] and has since been very extensively applied; a modification of this leave-one-out approach, based on k-nearest neighbor classification, is available for use with qualitative activity data.[119] An extended review of several evaluation methods, using both qualitative and quantitative data, is reported by Dixon and Merz.[120]

Thus far, we have considered only the criteria that can be used for comparative studies, and there is a voluminous literature involving such comparisons. The first quantitative comparison of coefficients was carried out by Adamson and Bush[121] using the similar property principle approach described immediately above. Willett and Winterman used their approach in a detailed comparison of a range of association and distance coefficients.[20] They found that the Tanimoto and cosine coefficients performed rather better in experiments with 16 QSAR datasets than the Hamming and Euclidean distance measures; subsequently, in one of the first operational systems for similarity searching,[45] these authors preferred the Tanimoto coefficient, partly on the basis of a subjective evaluation of the similarity search rankings it produced, and partly because the lack of a square-root calculation made it rather faster. This efficiency factor was a much more important factor in the early 1980s, when the work was done, than would be the case with modern computing equipment. However, the cosine coefficient has the great advantage that it allows the calculation of the average similarity between all pairs of compounds in two disjoint datasets to be carried out extremely rapidly,[97] something that is not possible with most coefficients and something that it is necessary for some similarity and diversity applications.[122–124] As a more recent example of a comparative study using qualitative activity data, Chen and Reynolds[30] report a comparison of both similarity coefficients and fingerprint types in which the Tanimoto coefficient was again found to outperform markedly the Euclidean distance in searches of the NCI AIDS and MDL Drug Data Report databases.

Although the Tanimoto is now the coefficient of choice in most operational systems for fingerprint-based similarity searching, it does have several limitations. Flower was the first to note that the Tanimoto would typically yield low coefficient values when the reference molecule in a similarity search had just a few bits set in its fingerprints (this is typically the case for small molecules but may also include larger, symmetrical molecules for certain types of fingerprint).[125] Subsequent studies by Lajiness[126] and by Dixon and Koehler[104] demonstrated that there was a pronounced sensitivity in the values resulting from the use of the coefficient, especially when small molecules were being considered in a diversity selection procedure. At least in part, this undesirable behavior arises from the fact that the Tanimoto, like most association coefficients, is computed from the ratio of two numbers; when binary fingerprints are being used, these numbers can take only a limited number of integer values, with the result that the Tanimoto coefficient has an inherent bias toward certain values.[127]

A detailed study by Holliday *et al.* of searches of the MDL Drug Data Report database using a range of different coefficients showed that there was a marked preference for certain coefficients to perform well when searching for active molecules of a particular size (as approximated by the number of bits set in a molecule's fingerprint).[128] In particular, the Russell–Rao coefficient performed particularly well when searching for large bioactive molecules, and the Forbes coefficient performed particularly well when searching for small bioactive molecules, with Tanimoto and cosine being the coefficients of choice overall. Detailed analysis showed that these variations could be explained by the way that the mathematical formulas for the various coefficients (such as those shown in **Table 2**) encoded the relative sizes of the reference and database structures, thus permitting a similarity coefficient to be chosen that would maximize performance given some knowledge of the sorts of molecule required in a selection procedure.

Thus far, we have considered the range of similarity coefficients that are available for searching. Studies of the effectiveness of different types of representation are still more common. These studies are normally carried out using qualitative data, in a simulated virtual-screening environment where the true actives are known. A known active is chosen as the reference structure; a similarity search is carried out using, e.g., some particular type of 2D fingerprint, and then the effectiveness of the search measured using the enrichment factor or one of the other performance criteria mentioned above. The search is then repeated several times using different types of fingerprint, and their relative merits determined on the basis of the retrieval performance observed in each case. Given the large number of types of descriptor that are available for similarity searching, as described above, it is perhaps not surprising that there have been very many such comparative studies, with interest being spurred in large part by the detailed studies of Brown and Martin.[21,22] These two much-cited papers compared a number of different structure representations, in both 2D and 3D, when used for the clustering of chemical databases. Clustering, or cluster analysis, can be regarded as repeated similarity searching – in searching, the similarities are computed between a single reference structure and each of the structures in a database, whereas in clustering, each and every database structure is taken in turn as the reference structure – and Brown and Martin's results have hence been very influential in the development of similarity searching.

In particular, much attention has focused on Brown and Martin's finding that simple 2D fingerprints gave a level of performance consistently superior to similarity measures that took account of 3D information. This would appear to be counterintuitive, given the importance of steric effects in determining biological activity, but subsequent experiments have shown that 2D representations are indeed generally superior to 3D representations for similarity-based virtual screening (and similar comments apply to comparisons of techniques for diversity analysis, as discussed further below). There are several reasons why this might be so. For example, conformational effects may best be left unconsidered at the simple similarity level (although they are, of course, of paramount importance in more sophisticated types of processing such as ligand docking); or it may be that the experimental environment, where averaging takes place over large numbers of reference structures, favors approaches that perform at a reasonable level in all (or most) circumstances, which is certainly the case with 2D fingerprints; or it may simply be that the appropriate way of encoding 3D information remains to be identified. Whatever the reason, there is continuing debate as to the relative effectiveness of 2D and 3D measures, with recent contributions involving Schuffenhauer *et al.*,[129] Makara,[130] Cruciani *et al.*,[131] Sheridan and Kearsley,[4] and Cramer *et al.*,[132] among others.

4.08.4 Combination of Similarity Measures

The many evaluations of similarity searching that have been considered thus far have typically involved just a single type of similarity measure: in many cases, indeed, a description of a new type of similarity measure forms the principal focus of the publication. Even where this is not the case, multiple measures have often been employed only as the input to a comparative study that seeks to identify the 'best' measure, using some quantitative performance criterion, as discussed in the previous section. These sorts of studies are very valuable, since they serve to identify measures that are sufficiently effective and efficient to enable the implementation of robust software systems. However, these studies are also inherently limited in that they assume, normally implicitly, that there is some specific type of structural feature, weighting scheme, or whatever that is uniquely well suited to describing the type(s) of biological activity that are being sought in a similarity search. The assumption cannot be expected to be generally valid, given the multifaceted nature of biological activities, and there is thus increasing interest in the use of data fusion methods for combining multiple similarity measures.

In the first book on the subject, Hall defines data fusion as "a process of combining inputs from sensors with information from other sensors, information processing blocks, databases, or knowledge bases."[133] The methods were originally developed for use in defense applications (e.g., establishing the friend-or-foe nature of an incoming missile or aeroplane) but were then rapidly taken up across a wide range of application domains that involve signal processing, including surveillance operations by law enforcement agencies, real-time control of continuous manufacturing processes, and multiimaging systems for the analysis of medical images (see, e.g., Arabnia and Zhu[134]). However, the approach can be used in any context where the availability of information obtained from a number of sensors enables further information to be inferred that might not be obtainable from a single sensor, or where there is a need to obtain an increased level of confidence, since multiple sensors can act together to confirm an event and to reduce any ambiguity surrounding the classification of an event.[135]

The reader may be asking how data fusion could be applied to database searching. This problem was first addressed in the area of information retrieval, which concerns itself with the retrieval of (normally textual) records from large databases – web search engines are the most prominent exemplar of this type of activity. The approach taken was to combine the rankings of a text database produced by different retrieval methods, i.e., the methods were regarded as sensors and the rankings resulting from their use as the sensors' outputs. For example, an early study by Belkin *et al.*[136] combined the results of a series of searches of bibliographic databases, conducted in response to a single query, but employing different indexing and searching strategies. A query was processed using different strategies, each of which was used to produce a ranking of a set of documents in order of decreasing similarity with the query, and the resulting rankings combined using one of several fusion rules (see below). The output of the fusion rule was taken as the document's new similarity score and the fused lists were then re-ranked in descending order of similarity. This basic approach was widely adopted and interest in data fusion for information retrieval continues to the present day, as reviewed recently by Hsu and Taksa.[137]

The work on chemical data fusion reported here is based directly on these previous information retrieval studies, and involves the simple procedure described below, where a user-defined reference structure is searched against a database using several different similarity measures:

1. Execute a similarity search of a chemical database for some particular reference structure using two or more different similarity measures.
2. Note the rank position, r_i, or the similarity score, s_i, of each individual database structure in the ranking resulting from use of the ith similarity measure.
3. Combine the various rank positions or similarity scores using a fusion rule to give a new fused score for each database structure.
4. Rank the fused scores, and then use this ranking to calculate a quantitative measure of the effectiveness of the search for the chosen reference structure.

Typical fusion rules include the maximum, the minimum, and the sum of the rankings, r_i, allocated to each database molecule j by each of the similarity measures. It is possible to use the similarity scores, s_i, rather than the ranks derived from them; use of ranks results in some loss of information but provides a form of standardization that is appropriate given the different magnitudes and the different distributions of the scores that can result from different similarity measures.

The first direct application of this approach to similarity searching was in work by Kearsley *et al.*, who used both similarity-based and rank-based procedures to combine pairs of similarity searches of the Standard Drug File database.[52] They found that substantial improvements in performance could be achieved in simulated property prediction experiments, when compared with the use of a single similarity measure. Similar results were obtained in searches using a range of different types of representation by Ginn *et al.*,[135,138] and the approach is now well established, with recent contributions covering new types of fusion rule[139] and suggestions as to why fusion is able to improve performance over that obtainable with a single reference structure.[140–142] However, the enhancements in search performance that are observed using multiple similarity coefficients are highly variable, with different combinations of coefficients proving to be optimal in searches for different classes of active structures.[98] Thus, while it is normally possible to use data fusion to improve the effectiveness of similarity searching, there are no clear guidelines as to how exactly this can be achieved in the context of a particular reference structure and database, hence emphasizing the need for a firm theoretical basis for the technique. Analogous work is also been undertaken to use data fusion to combine the results of different search algorithms and/or scoring functions for ligand–protein docking, where the approach is referred to as consensus scoring.[142–144]

Thus far, we have considered data fusion to involve the combination of the rankings resulting from searching a database with a single reference structure but with multiple similarity measures. It is, however, increasingly the case that several structurally diverse reference structures may be available, e.g., published competitor compounds or HTS hits, and this has occasioned interest in an alternative formulation of data fusion: the use of a single similarity measure but multiple reference structures. Such an approach seems to have been first suggested by Xue *et al.*[145] and by Schuffenhauer *et al.*,[54] and has been studied in detail more recently by Hert *et al.*[57] and by Whittle *et al.*[96] The latter discuss an approach, which they refer to as group fusion, in contrast to the similarity fusion methods that have been used previously with just a single reference structure. Group fusion uses multiple reference structures, each represented in the same way, with a single similarity measure (a 2D fingerprint and the Tanimoto coefficient in most of their experiments). A suitably modified version of the search procedure shown in the pseudocode above is used to search a database. Let $S(i,j)$ be the value of the coefficient for the similarity between some database molecule j and the ith of the set of known reference structures, then the best results were obtained by giving each database molecule j the fused score $\max\{S(i,j)\}$. Experiments with the MDL Drug Data Report database demonstrate that this approach enables very effective searches to be carried out even when only a few reference structures are available and when the sought actives are structurally diverse. Conventional similarity searching or similarity fusion, conversely, are most effective when the actives are strongly clustered in structural space, suggesting that group fusion provides an effective alternative when this is not the case.

Hert *et al.* have recently applied this finding to the development of an enhanced form of similarity searching, which they refer to as turbo similarity searching (by analogy with the turbocharger of a conventional automobile engine).[146] The similarity property principle implies that the nearest neighbors of a bioactive reference structure are also likely to be active, and the work on group fusion has shown that using multiple active reference structures in a similarity search is more effective than using a single active reference structure. If these two observations are combined, it may be concluded that a similarity search involving both a reference structure and its nearest neighbors is likely to be more effective than a similarity search involving just that reference structure on its own; extended experiments with the MDL Drug Data Report database demonstrate that this is indeed the case, thus providing a simple way of enhancing the performance of conventional similarity searching.

4.08.5 Diversity

The introduction of HTS in the pharmaceutical and agrochemical industries in the 1990s resulted in a dramatic increase in the numbers of compounds that can be tested for activity. Thus there has been enormous interest in the development of computational techniques for the selection of screening sets (*see* 4.20 Screening Library Selection and High-Throughput Screening Analysis/Triage). These methods may be applied to in-house databases to select compounds for screening from the corporate store, they may be applied to sets of compounds available from external compound suppliers in what are known as compound acquisition programs, or they may be used to select virtual compounds for synthesis in combinatorial libraries.

The early trend in HTS experiments was to screen large numbers of compounds in the belief that simply increasing the throughput of biological testing would result in more compounds being taken forward in the drug discovery pipeline. However, the results from these early libraries were disappointing, with either lower hit rates than expected or with the production of hits that had undesirable ADME properties, for example, molecules with high molecular weights or poor solubilities.[147,148] It was soon realized that the numbers of compounds that could potentially exist is so vast that it is not possible to make and test everything. Estimates of the number of drug-like compounds that could exist vary from 10^{40} to 10^{60}. Given the vast size of chemistry space, the low success rates of early HTS and the real costs associated with testing large numbers of samples, the current trend in screening set design has shifted toward smaller, carefully designed sets.

A first step in compound selection is usually to apply computational filters to remove compounds with undesirable properties from further consideration. The simplest computational filters consist of substructural searches that are used to eliminate compounds that contain undesirable functional groups, such as functional groups that are known to be toxic.[149] Other commonly used filters are based on counts of structural features, such as numbers of rotatable bonds, and on physicochemical properties, such as molecular weight and $\log P$. These criteria are in widespread use following the publication of the 'rule of five' by Lipinski *et al.*,[150] which suggests that poor oral absorption is likely for compounds with high molecular weight (>500); $\log P > 5$; number of hydrogen bond donors >5, and number of hydrogen bond acceptors >10. More stringent filters can also be applied to limit the compounds to those that are lead-like.[151,152] The rationale for applying lead-like filters is that the lead optimization stage of drug discovery usually involves adding functionality to the molecules, which typically results in an increase in properties such as molecular weight and hydrogen-bonding groups.

Once the undesirable compounds have been removed, an appropriate compound selection technique can be applied. Diversity analysis is an important concept when selecting compounds for screening against a range of biological targets and for screening against a single target when little is known about it. As for similarity searching, the rationale for diversity analysis is based on the similar property principle: if structurally similar compounds are likely to share the same activity then it should be possible to design a diverse subset of compounds that covers the same biological space as the larger set from which it is derived while minimizing redundancy. This viewpoint forms the basis of the neighborhood concept described by Patterson *et al.*[16] More biased compound sets are appropriate when the compounds are to be screened against a family of targets with related properties, for example, kinases or G protein-coupled receptors. Such compound sets can be designed by placing restrictions on the chemistry space that compounds should occupy; however, it is still important to select a diverse subset of compounds from within the allowed space.

Corporate databases typically contain in the order of 10^6 compounds and therefore represent only a very tiny fraction of the interesting chemical space. Furthermore, they are likely to be biased collections in that they contain families of closely related compounds synthesized using traditional medicinal chemistry techniques that reflect the restricted therapeutic areas explored by a company during its existence. Consequently, most companies are actively engaged in compound acquisition programs with the aim of enriching their in-house collections with compounds that are diverse with respect to the compounds that are already available internally. The last few years have seen the emergence of many companies that supply compounds for purchase with the source of the compounds including both traditional and combinatorial synthesis.

Selecting a diverse subset of compounds from some large collection requires a way of quantifying the structural diversity of a collection of molecules. This can be achieved by defining a chemistry space based on molecular properties and quantifying the amount of chemistry space that is covered by the compounds. Alternatively, the diversity of a subset of compounds can be calculated by considering intermolecular dissimilarities of compounds in the subset. Several diversity indices have been devised that are based on the pairwise dissimilarities or distances of the compounds.

Given a way of quantifying the diversity of a subset of compounds then, theoretically, the most diverse subset of size n compounds in a database containing N compounds could be found by comparing the diversity of all subsets of size n. However, this is usually computationally infeasible since there are $N!/[n!(N-n)!]$ possible subsets to consider. As an example, there are 10^{13} subsets of size 10 contained in 100 compounds. Compound selection methods are usually applied to much larger datasets; for example, they may be used to select a few tens or hundreds of compounds from a large database which could consist of hundreds of thousands or even millions of compounds. Thus, it is clear that identifying the exact solution is usually not possible and that more efficient subset selection methods are required.

The main compound selection methods are discussed below and include clustering, partitioning, or cell-based methods, DBCS, and the use of optimization techniques.

4.08.5.1 Clustering

Cluster analysis, or clustering, is the process of dividing a collection of objects into groups (or clusters) such that the objects within a cluster are highly similar whereas objects in different clusters are dissimilar.[153] When applied to a compound dataset, the resulting clusters provide an overview of the range of structural types within the dataset and a diverse subset of compounds can be selected by choosing one or more compounds from each cluster.[1,154]

The main steps involved in clustering are as follows:

1. Generate descriptors for each compound in the dataset.
2. Calculate the similarity or distance between all pairs of compounds in the dataset.

3. Use a clustering algorithm to group the compounds within the dataset.
4. Select a representative subset by choosing one (or more) compounds from each cluster.

Many different clustering algorithms have been developed and they can be divided into hierarchical and nonhierarchical methods.[154] In hierarchical methods, small clusters of related compounds are grouped together into larger clusters: at one extreme, each compound is in a separate cluster; at the other extreme, all the compounds are in a single cluster. In nonhierarchical methods, compounds are placed in clusters without forming a hierarchical relationship between the clusters. The clustering methods most commonly applied to compound selection include the nonhierarchical Jarvis–Patrick and K-means algorithms and Ward's clustering, which is a hierarchical method.

Jarvis–Patrick clustering involves generating a nearest neighbor list for each compound in the dataset.[155] Compounds are then placed into the same cluster if they have some number of near neighbors in common. Jarvis–Patrick is a relatively fast clustering method; however, the basic algorithm can result in rather skewed clusters, with a small number of very large clusters and a large number of singletons. The effectiveness of the basic method can be improved by taking the position of each compound within the neighbor list into account, instead of simply the number of nearest neighbors, and by restricting a compound's nearest neighbors to those within some threshold distance.

K-means clustering is an example of a relocation clustering method.[156] The first step is to choose a set of K 'seed' compounds, usually selected at random. An initial set of clusters is generated by assigning the remaining compounds to their nearest seed. Next, the centroids of the clusters are calculated and the objects are reassigned (or relocated) to the nearest cluster centroid. The process of calculating cluster centroids and relocating the compounds is repeated until no compounds change clusters, or until a user-defined number of iterations has taken place. The K-means method is dependent upon the initial set of cluster centroids and different results will usually be found for different initial seeds.[157]

Ward's clustering is an agglomerative hierarchical clustering method.[158] The algorithm begins by placing each compound in its own cluster and then proceeds by iteratively merging the most similar clusters together. Thus, the closest two compounds are merged into a single cluster; then the closest two clusters are merged, and so on. The process continues until all compounds are in a single cluster. Once the full hierarchy of clusters has been generated, the next step is to choose a level in the hierarchy that corresponds to a useful set of clusters.[159] Various methods are available for automatically determining an appropriate clustering level. For example, the method described by Kelley *et al.* compares the tightness of the clusters at a particular level with the number of clusters at that level.[160] If the number of clusters at level l is k_l then the Kelley measure is given by:

$$\mathrm{KELLEY}_l = (N-2)\left(\frac{\bar{d}_{wl} - \min(\bar{d}_w)}{\max(\bar{d}_w) - \min(\bar{d}_w)}\right) + 1 + k_l$$

where $\bar{d}_{wl}$ is the mean of the distances between the compounds in the same cluster, with $\min(\bar{d}_{wl})$ and $\max(\bar{d}_{wl})$ the minimum and maximum values of this property across all clusters. N is the total number of points in the dataset. The level that corresponds to the smallest value of the Kelley measure is deemed the optimal level.

Clustering methods can be used to select a diverse subset of compounds from a larger dataset as required in an HTS experiment; however, it is difficult to compare two different datasets as might be required in a compound acquisition program. This is because clustering is based on intermolecular similarities rather than any absolute measure of chemical space. As a consequence, the comparison of two datasets requires that the datasets are first combined and then clustered as a combined unit. The degree of overlap in the two datasets can then be assessed by examining the contents of each cluster. A cluster that is mainly occupied by compounds from one of the datasets indicates a region of space where the datasets differ. Thus a company may decide to purchase supplier compounds from clusters that are sparsely populated by their own in-house compounds.

4.08.5.2 Partitioning

In contrast to clustering, partitioning or cell-based methods provide an absolute measure of the chemistry space covered by a collection of compounds. They require the definition of a low-dimensional chemistry space, for example, one based on a small number of physicochemical properties such as molecular weight, calculated log P, and number of hydrogen bond donors. Each property or descriptor defines an axis of the chemistry space, the range of values for each property is divided into a set of bins, and the combinatorial product of all bins defines the set of cells or partitions that comprise the space. A dataset is mapped on to the space by assigning each molecule to a cell according to its properties.[161,162]

A diverse subset of compounds can be selected by taking one or more compounds from each cell. Partitioning methods are much faster than either clustering or dissimilarity methods. They are also well suited to compound acquisition applications, since a partitioning scheme is defined independently of the compounds that are mapped on to it and so the space occupied by different datasets can be compared easily. However, the methods are limited to low dimensional descriptors, which means that they cannot be used with descriptors such as 2D fingerprints.

One of the first approaches to partitioning was based on six descriptors that represent the hydrophobicity, polarity, shape, hydrogen-bonding properties and aromatic interactions of compounds.[161] Each descriptor was split into two, three, or four partitions to give a total of 576 cells. When a 150 000 compound subset of the corporate database was mapped on to the space, 86% of the cells were occupied and a diverse screening set was selected by taking three compounds from each occupied cell.

BCUT descriptors are a relatively recent class of descriptors that were developed for use in partitioning schemes.[162,163] BCUTs are calculated from matrix representations of a compound's connection table. The diagonal of a matrix represents a property of each of the atoms, such as atomic charge, atomic polarizability, and atomic hydrogen-bonding ability. The off-diagonals are assigned the values 0.1 times the bond type if the atoms are bonded, and 0.001 if the atoms are not bonded. The highest and lowest eigenvalues of each matrix are then extracted for use as descriptors. The three properties described above would thus give rise to six descriptors that describe a six-dimensional space. An initial procedure can be used to identify the most appropriate BCUTs to use for a given dataset in order to tailor the method for datasets with different characteristics.

4.08.5.3 Dissimilarity-Based Compound Selection

DBCS methods involve selecting a subset of compounds directly based on their pairwise dissimilarities. They are iterative procedures whereby one compound is selected in each iteration. The basic algorithm for DBCS is outlined below:

1. Select a compound and place it in the subset.
2. Calculate the dissimilarity between each compound remaining in the dataset and the compounds in the subset.
3. Choose the next compound as that which is most dissimilar to the compounds in the subset.
4. If there are fewer than n compounds in the subset (n being the desired size of the final subset), return to step 2.

Many variants of the basic algorithm have been developed. They differ in the way in which steps 1 and 3 are implemented, with the different methods resulting in different compounds being selected. For example, the first compound in step 1 can be selected by: choosing it at random; choosing that compound that is most dissimilar to the other compounds in dataset (e.g., has the smallest sum of similarities to the other compounds); or choosing that compound that is most representative of the dataset (e.g., has the largest sum of similarities to the other molecules). The selection of the next compound, as indicated in step 3, requires a quantitative definition of the dissimilarity between a single compound in the dataset and the compounds already in the subset. Examples of dissimilarity measures that are used for this task include MaxMin and MaxSum.[164] If there are m compounds in the subset, then the scores for a compound i using these two measures are given by:

$$\text{MaxSum} : \text{score}_i = \sum_{j=1}^{m} D_{i,j}$$

$$\text{MaxMin} : \text{score}_i = \text{minimum}(D_{i,j;j=1,m})$$

where $D_{i,j}$ is the dissimilarity (or distance) between two individual compounds i and j. The compound i that has the largest value of $score_i$ is the one chosen for addition to the subset. Thus MaxSum chooses the compound with the maximum sum of distances to all compounds in subset, whereas MaxMin chooses the compound with the maximum distance to its closest neighbor in the subset.

The basic DBCS algorithm shown above has an expected time complexity of $O(n^2N)$ and, since n is normally some small fraction of N (such as 1% or 5%), this represents a running time that is cubic in N, which makes it extremely demanding of computational resources if the dataset is at all large. Holliday *et al.*[97] have described an efficient implementation of the MaxSum selection algorithm that has time complexity of $O(nN)$; however, various studies have shown that MaxSum can result in the selection of closely related compounds.[165,166] Furthermore, a comparison of different DBCS algorithms carried out by Snarey *et al.*[164] showed MaxMin to be more effective than MaxSum in identifying subsets that exhibit a range of biological activities. Efficient implementations of MaxMin have also been developed with O(nN) time complexity, although they are not as fast as MaxSum. A particularly fast implementation of MaxMin has been developed for use with low-dimensional descriptors that has time complexity of only O($n\log N$).[165]

Sphere exclusion algorithms are closely related to DBCS algorithms; however, they use a dissimilarity threshold to eliminate compounds from consideration that are close to compounds already selected. The threshold defines the radius of an exclusion sphere and each time a compound is selected, all compounds in the dataset that are within the threshold distance of it are excluded. As with DBCS algorithms, variations on the basic method exist. In the method described by Hudson *et al.*, the first compound selected is the one that is most dissimilar to all others in the database, with subsequent compounds chosen based on dissimilarity to the compounds already chosen.[167] Pearlman's method involves choosing both the first and subsequent compounds at random.[162] The use of random numbers results in a different subset being selected each time the algorithm is run. In both cases, the basic algorithm continues until all compounds are either selected or excluded, and hence, in contrast to DBCS, it is not possible to specify the final size of the subset.

The relationship between the DBCS methods and sphere exclusion algorithms is explored in the OptiSim (for Optimizable K-Dissimilarity Selection) program.[168,169] OptiSim makes use of an intermediate pool of molecules, here called Sample. A compound is chosen at random from the dataset. If it has dissimilarity greater than some threshold, t, to a compound in the subset then it is added to the Sample; otherwise it is discarded. When the number of compounds in the Sample set reaches some user-defined number, K, then the 'best' molecule from Sample is added to the subset (the 'best' molecule being that which is most dissimilar to those already in the subset). The remaining $K-1$ compounds from Sample are then set aside but may be reconsidered if the main set runs out before the subset reaches the desired size. The process then repeats. The characteristics of the final subset of selected molecules are determined by the value of K chosen, with values of K equal to 1 corresponding to the sphere exclusion algorithm and K equal to n corresponding to DBCS, respectively.

4.08.5.4 Optimization Techniques

As indicated earlier, the computational intractability of exact methods for selecting the maximally diverse subset of compounds lead to the development of the DBCS and sphere exclusion approximate methods. An alternative approach is to use an optimization algorithm such as a genetic algorithm or simulated annealing. These algorithms can provide effective ways of sampling large search spaces and hence they are well suited to compound selection, provided that they are used with efficient methods of calculating diversity. For example, the Monte Carlo method has been combined with simulated annealing to select a diverse subset of compounds.[170] An initial subset is chosen at random and its diversity is calculated. A new subset is then generated from the first by replacing some of the compounds with others chosen at random. If the new subset is more diverse than the previous subset it is accepted for use in the next iteration; if it is less diverse, then the probability that it is accepted depends on the Boltzmann factor. The process continues for a fixed number of iterations or until no further improvement is observed in the diversity function. An alternative approach is to use a genetic algorithm.[171] Here, each chromosome of the genetic algorithm represents a subset of compounds. Initially the subsets consist of randomly selected compounds. The population of chromosomes then undergoes reproduction, in which the crossover operator is used to mix the information contained in two different chromosomes and the mutation operator randomly changes one compound for another. The fitness of a chromosome is determined using a diversity index and the genetic algorithm iterates until a maximum in diversity is achieved.

A variety of different diversity measures can be used with optimization methods. For example, Gillet *et al.* describe a genetic algorithm in which diversity can be measured as the sum of pairwise dissimilarities or as the number of occupied cells in a partitioning scheme.[172] Mount *et al.* describe a method in which diversity is measured using a spanning tree representation of the subset.[166] A spanning tree is a set of edges that connect a set of nodes without forming any cycles. The nodes are the compounds in the subset and each edge is labeled by the dissimilarity between the two compounds it connects. The minimum spanning tree is the spanning tree that connects all molecules in the subset with the minimum sum of pairwise dissimilarities. The diversity of the subset then equals the sum of the intermolecular similarities along the edges on the minimum spanning tree. The HARPick program of Good and Lewis[173] characterizes the diversity of a subset by the number of three-point pharmacophores it contains.

The principal aim of most of the selection algorithms discussed thus far is to identify the subset of compounds with maximum diversity. Diversity is normally quantified by structural diversity, as embodied in the molecular dissimilarity measures that are used in the various selection algorithms described above. In fact, the rationale for diversity-based compound selection is biological diversity, rather than structural diversity, so as to maximize the amount of SAR information that can be obtained from the chosen subset. Quantitative methods for the measurement of structural diversity are reviewed by Willett.[119] There are, however, many other criteria that it is desirable to consider: whether the algorithms are being used to design screening sets, to select compounds for purchase, or to identify reagents to be used in combinatorial libraries. For example, it is important that compound cost, synthetic feasibility, and drug-like properties are also taken into account. Such criteria can be incorporated into the selection process by applying an initial filtering step[149,174] or they may be included in the fitness function that drives the selection procedure. Multiple design criteria can be incorporated into a fitness function by using a weighted-sum approach.[170,172,175] However, this approach is limited by the difficulty in choosing appropriate weights, especially when the criteria are in competition and the end result is a single somewhat arbitrary compromise solution. These limitations have been addressed through the implementation of a multiobjective genetic algorithm in the program MoSELECT.[176,177] In MoSELECT, multiple objectives are handled independently and a family of equivalent solutions is found, where each solution represents a different compromise in the objectives. This approach allows the relationships between the various objectives to be investigated so that the library designer can make an informed choice on an appropriate compromise solution.

4.08.6 Conclusions

The concept of molecular similarity lies at the heart of medicinal chemistry, but it is only since the mid-1980s that there has been sustained interest in the use of computational methods for the quantification of similarity. The principal focus has been the development of similarity-based tools for compound selection: either the selection of structurally related compounds by similarity searching or, more recently, the selection of structurally unrelated compounds by diversity analysis. In both cases, there is the expectation, or at least the hope, that these structural relationships will correspond to biological relationships, as implied by the similar property principle.

In this chapter, we have summarized the many techniques that have been described in the literature for the quantification of molecular (dis)similarity. Current requirements for improvements in compound acquisition and virtual screening mean that this level of interest will continue for the foreseeable future. In particular, we believe that there is a need for further studies of the ways in which 3D, rather than 2D, structural information can be used in the selection process, and of the use of fusion techniques to ensure that the process uses not just one but several types of information in the measurement of (dis)similarity.

References

1. Willett, P. *Similarity and Clustering in Chemical Information Systems*; Research Studies Press: Letchworth, 1987.
2. Johnson, M. A.; Maggiora, G. M. *Concepts and Applications of Molecular Similarity*; John Wiley: New York, 1990.
3. Downs, G. M.; Willett, P. *Rev. Comput. Chem.* **1995**, 7, 1–66.
4. Sheridan, R. P.; Kearsley, S. K. *Drug Disc. Today* **2002**, 7, 903–911.
5. Nikolova, N.; Jaworska, J. *Quant. Struct.–Act. Relat. Comb. Sci.* **2003**, *22*, 1006–1026.
6. Dean, P. M. *Molecular Similarity in Drug Des.*; Chapman and Hall: Glasgow, 1994.
7. Bender, A.; Glen, R. C. *Org. Biomol. Chem.* **2004**, *2*, 3204–3218.
8. Dean, P. M.; Lewis, R. A. *Molecular Diversity in Drug Design*; Kluwer: Amsterdam, 1999.
9. *J. Mol. Graph. Model.* **2000**, *18*, 317–542.
10. Ghose, A. K.; Viswanadhan, V. N. *Combinatorial Library Design and Evaluation: Principles, Software Tools and Applications in Drug Discovery*; Marcel Dekker: New York, 2001.
11. Leach, A. R.; Gillet, V. J. *An Introduction to Chemoinformatics*; Kluwer: Dordrecht, 2003.
12. Gasteiger, J.; Engel, T. *Chemoinformatics: A Textbook*; Wiley-VCH: Weinheim, 2003.
13. Böhm, H.-J.; Schneider, G. *Virtual Screening for Bioactive Molecules*; Wiley-VCH: Weinheim, 2000.
14. Klebe, G. *Virtual Screening: An Alternative or Complement to High Throughput Screening*; Kluwer: Dordrecht, 2000.
15. Kubinyi, H.; Muller, G. *Chemogenomics in Drug Design*; Wiley-VCH: Weinheim, 2004.
16. Patterson, D. E.; Cramer, R. D.; Ferguson, A. M.; Clark, R. D.; Weinberger, L. E. *J. Med. Chem.* **1996**, *39*, 3049–3059.
17. Frye, S. V. *Chem. Biol.* **1999**, *6*, R3–R7.
18. Kubinyi, H. *Perspect. Drug Disc. Des.* **1998**, *9–11*, 225–232.
19. Stahura, F. L.; Bajorath, J. *Drug Disc. Today* **2002**, 7, S41–S47.
20. Willett, P.; Winterman, V. *Quant. Struct.–Act. Relat.* **1986**, *5*, 18–25.
21. Brown, R. D.; Martin, Y. C. *J. Chem. Inf. Comput. Sci.* **1996**, *36*, 572–584.
22. Brown, R. D.; Martin, Y. C. *J. Chem. Inf. Comput. Sci.* **1997**, *37*, 1–9.
23. Martin, Y. C.; Kofron, J. L.; Traphagen, L. M. *J. Med. Chem.* **2002**, *45*, 4350–4358.
24. Shanmugasundaram, V.; Maggiora, G. M.; Lajiness, M. S. *J. Med. Chem.* **2005**, *48*, 240–248.
25. Sheridan, R. P.; Feuston, B. P.; Maiorov, V. N.; Kearsley, S. K. *J. Chem. Inf. Comput. Sci.* **2004**, *44*, 1912–1928.
26. He, L.; Jurs, P. C. *J. Mol. Graph. Model.* **2005**, *23*, 503–523.
27. Fisanick, W.; Cross, K. P.; Rusinko, A. *J. Chem. Inf. Comput. Sci.* **1992**, *32*, 664–674.
28. Bath, P. A.; Morris, C. A.; Willett, P. *J. Chemometrics* **1993**, 7, 543–550.
29. Turner, D. B.; Willett, P.; Ferguson, A. M.; Heritage, T. W. *SAR QSAR Environ. Res.* **1995**, *3*, 101–130.
30. Chen, X.; Reynolds, C. H. *J. Chem. Inf. Comput. Sci.* **2002**, *42*, 1407–1414.
31. Leo, A. J. *Chem. Rev.* **1993**, *93*, 1281–1306.

32. Leo, A. J.; Hoekman, D. *Perspect. Drug Disc. Des.* **2000**, *18*, 19–38.
33. Ghose, A. K.; Crippen, G. M. *J. Comput. Chem.* **1986**, 7, 565–577.
34. Ghose, A. K.; Viswanadhan, V. N.; Wendoloski, J. J. *J. Phys. Chem. A* **1998**, *102*, 3762–3772.
35. Wildman, S. A.; Crippen, G. M. *J. Chem. Inf. Comput. Sci.* **1999**, *39*, 868–873.
36. Wang, R.; Fu, Y.; Lai, L. *J. Chem. Inf. Comput. Sci.* **1997**, *37*, 615–621.
37. Randić, M. *J. Mol. Graph. Model.* **2001**, *20*, 19–35.
38. Hall, L. H.; Kier, L. B. *J. Mol. Graph. Model.* **2001**, *20*, 4–18.
39. Wiener, H. *J. Am. Chem. Soc.* **1947**, *69*, 17–20.
40. Randić, M. *J. Am. Chem. Soc.* **1975**, *97*, 6609–6615.
41. Kier, L. B.; Hall, H. L. *Molecular Connectivity in Structure–Activity Analysis*; Wiley: New York, 1986.
42. Hall, H. L.; Kier, L. B. The Molecular Connectivity Chi Indexes and Kappa Shape Indexes in Structure–Property Modeling. In *Reviews in Computational Chemistry*; Lipkowitz, K. B., Boyd, D. B., Eds.; VCH Publishers: New York, 1991; Vol. 2, pp 367–422.
43. Molconn-Z. eduSoft, LC, PO Box 1811, Ashland, VA 23005, USA. http://www.eslc.com (accessed Aug 2006).
44. Basak, S. C.; Magnuson, V. R.; Niemi, G. J.; Regal, R. R. *Discrete Appl. Math.* **1988**, *19*, 17–44.
45. Willett, P.; Winterman, V.; Bawden, D. *J. Chem. Inf. Comput. Sci.* **1986**, *26*, 36–41.
46. BCI Barnard Chemical Information Ltd, 46 Uppergate Road, Stannington, Sheffield S6 6BX, UK. http://www.bci.gb.com (accessed Aug 2006).
47. MDL Information Systems, Inc., 14600 Catalina Street, San Leandro, CA 94577, USA. http://www.mdli.com (accessed Aug 2006).
48. Daylight Chemical Information Systems, Inc., 120 Vantis Suite 550, Aliso Viejo, CA 92656, USA. http://www.daylight.com (accessed Aug 2006).
49. Carhart, R. E.; Smith, D. H.; Venkataraghavan, R. *J. Chem. Inf. Comput. Sci.* **1985**, *25*, 64–73.
50. Nilakantan, R.; Bauman, N.; Dixon, J. S.; Venkataraghavan, R. *J. Chem. Inf. Comput. Sci.* **1987**, *27*, 82–85.
51. Böhm, H.-J.; Flohr, A.; Stahl, M. *Drug Disc. Today: Technol.* **2004**, *1*, 217–224.
52. Kearsley, S. K.; Sallamack, S.; Fluder, E. M.; Andose, J. D.; Mosley, R. T.; Sheridan, R. P. *J. Chem. Inf. Comput. Sci.* **1996**, *36*, 118–127.
53. Schneider, G.; Neidhart, W.; Giller, T.; Schmid, G. *Angew. Chem. Int. Ed. Engl.* **1999**, *38*, 2894–2896.
54. Schuffenhauer, A.; Floersheim, P.; Acklin, P.; Jacoby, E. *J. Chem. Inf. Comput. Sci.* **2003**, *43*, 391–405.
55. Scitegic, 9665 Chesapeake Drive, Suite 401, San Diego, CA 92123–1365, USA.
56. Morgan, H. *J. Chem. Doc.* **1965**, *5*, 107–113.
57. Hert, J.; Willett, P.; Wilton, D. J.; Acklin, P.; Azzaoui, K.; Jacoby, E.; Schuffenhauer, A. *J. Chem. Inf. Comput. Sci.* **2004**, *44*, 1177–1185.
58. Raymond, J. W.; Willett, P. *J. Comput.-Aided Mol. Des.* **2002**, *16*, 521–533.
59. Hagadone, T. R. *J. Chem. Inf. Comput. Sci.* **1992**, *32*, 515–521.
60. Raymond, J. W.; Gardiner, E. J.; Willett, P. *Comput. J.* **2002**, *45*, 631–644.
61. Raymond, J. W.; Gardiner, E. J.; Willett, P. *J. Chem. Inf. Comput. Sci.* **2002**, *42*, 305–316.
62. Gillet, V. J.; Willett, P.; Bradshaw, J. *J. Chem. Inf. Comput. Sci.* **2003**, *43*, 338–345.
63. Rarey, M.; Dixon, J. S. *J. Comput.-Aided Mol. Des.* **1998**, *12*, 471–490.
64. Harper, G.; Bravi, G. S.; Pickett, S. D.; Hussain, J.; Green, D. V. S. *J. Chem. Inf. Comput. Sci.* **2004**, *44*, 2145–2156.
65. Barker, E. J.; Buttar, D.; Cosgrove, D. A.; Gardiner, E. J.; Gillet, V. J.; Kitts, P.; Willett, P. *J. Chem. Inf. Model.* **2006**, *46*, 503–511.
66. Takahashi, Y.; Sukekawa, M.; Sasaki, S. *J. Chem. Inf. Comput. Sci.* **1992**, *32*, 639–643.
67. Cringean, J. K.; Pepperrell, C. A.; Poirrette, A. R.; Willett, P. *Tetrahedron Comput. Methodol.* **1990**, *3*, 37–46.
68. Bemis, G. W.; Kuntz, I. D. *J. Comput.-Aided Mol. Des.* **1992**, *6*, 607–628.
69. Nilakantan, R.; Bauman, N.; Venkataraghavan, R. A. *J. Chem. Inf. Comput. Sci.* **1993**, *33*, 79–85.
70. Pepperrell, C. A.; Willett, P. *J. Comput.-Aided Mol. Des.* **1991**, *5*, 455–474.
71. Poirrette, A. R.; Willett, P.; Allen, F. H. *J. Mol. Graph.* **1993**, *11*, 2–14.
72. Poirrette, A. R.; Willett, P.; Allen, F. H. *J. Mol. Graph.* **1991**, *9*, 203–217.
73. Sheridan, R. P.; Miller, M. D.; Underwood, D. J.; Kearsley, S. K. *J. Chem. Inf. Comput. Sci.* **1996**, *36*, 128–136.
74. Pickett, S. D.; Mason, J. S.; McLay, I. M. *J. Chem. Inf. Comput. Sci.* **1996**, *36*, 1214–1223.
75. Davies, K. Using Pharmacophore Diversity to Select Molecules to Test from Commercial Catalogues. In *Molecular Diversity and Combinatorial Chemistry. Libraries and Drug Discovery*; Chaiken, I. M., Janda, K. D., Eds.; American Chemical Society: Washington, DC, 1996, pp 309–316.
76. Mason, J. S.; Morize, I.; Menard, P. R.; Cheney, D. L.; Hulme, C.; Labaudiniere, R. F. *J. Med. Chem.* **1999**, *42*, 3251–3264.
77. Good, A. C.; Cho, S. J.; Mason, J. S. *J. Comput.-Aided Mol. Des.* **2004**, *18*, 523–527.
78. Moon, J. B.; Howe, W. J. *Tetrahedron Comput. Methodol.* **1990**, *3*, 697–711.
79. Raymond, J. W.; Willett, P. *J. Chem. Inf. Comput. Sci.* **2003**, *43*, 908–916.
80. Pepperrell, C. A.; Taylor, R.; Willett, P. *Tetrahedron Comput. Methodol.* **1990**, *3*, 575–593.
81. Good, A. C.; Mason, J. S. *Rev. Comput. Chem.* **1996**, 7, 67–117.
82. Mestres, J.; Rohrer, D. C.; Maggiora, G. M. *J. Comput. Chem.* **1997**, *18*, 934–954.
83. Wild, D. J.; Willett, P. *J. Chem. Inf. Comput. Sci.* **1996**, *36*, 159–167.
84. Good, A. C.; Hodgkin, E. E.; Richards, W. G. *J. Chem. Inf. Comput. Sci.* **1992**, *32*, 188–191.
85. Good, A. C.; Richards, W. G. *J. Chem. Inf. Comput. Sci.* **1993**, *33*, 112–116.
86. Grant, J. A.; Gallardo, M. A.; Pickup, B. T. *J. Comput. Chem.* **1996**, *17*, 1653–1666.
87. Rush, T. S.; Grant, J. A.; Mosyak, L.; Nicholls, A. *J. Med. Chem.* **2005**, *48*, 1489–1495.
88. Lemmen, C.; Lengauer, T. *J. Comput.-Aided Mol. Des.* **2000**, *14*, 215–232.
89. Hubálek, Z. *Biol. Rev. Camb. Philos. Soc.* **1982**, *57*, 669–689.
90. Gower, J. C.; Legendre, P. *J. Classif.* **1986**, *5*, 5–48.
91. Ellis, D.; Furner-Hines, J.; Willett, P. *Perspect. Inf. Manag.* **1994**, *3*, 128–149.
92. Everitt, B. S.; Landau, S.; Leese, M. *Cluster Analysis*, 4th ed.; Edward Arnold: London, 2001.
93. Holliday, J. D.; Hu, C.-Y.; Willett, P. *Combin. Chem. High-Throughput Screen.* **2002**, *5*, 155–166.
94. Downs, G. M.; Willett, P.; Fisanick, W. *J. Chem. Inf. Comput. Sci.* **1994**, *34*, 1094–1102.
95. Gower, J. C. Measures of Similarity, Dissimilarity and Distance. In *Encyclopaedia of Statistical Sciences*; Kotz, S., Johnson, N. L., Read, C. B., Eds.; John Wiley: Chichester, 1982, pp 397–405.
96. Whittle, M.; Gillet, V. J.; Willett, P.; Alex, A.; Loesel, J. *J. Chem. Inf. Comput. Sci.* **2004**, *44*, 1840–1848.
97. Holliday, J. D.; Ranade, S. S.; Willett, P. *Quant. Struct. -Act. Relat.* **1995**, *14*, 501–506.

98. Salim, N.; Holliday, J. D.; Willett, P. *J. Chem. Inf. Comput. Sci.* **2003**, *43*, 435–442.
99. Cheng, C.; Maggiora, G.; Lajiness, M.; Johnson, M. A. *J. Chem. Inf. Comput. Sci.* **1996**, *36*, 909–915.
100. Sokal, R. R.; Sneath, P. H. *Principles of Numerical Taxonomy*; W. H. Freeman: San Francisco, 1963.
101. James, C. A.; Weininger, D.; Delaney, J. Fingerprints – Screening and Similarity. http://www.daylight.com/dayhtml/doc/theory/theory.toc.html (accessed Aug 2006).
102. Fligner, M. A.; Verducci, J. S.; Blower, P. E. *Technometrics* **2002**, *44*, 110–119.
103. Lajiness, M. S. *Perspect. Drug Disc. Des.* **1997**, *7/8*, 65–84.
104. Dixon, S. L.; Koehler, R. T. *J. Med. Chem.* **1999**, *42*, 2887–2900.
105. Carbo, R.; Leyda, L.; Arnau, M. *Int. J. Quant. Chem.* **1980**, *17*, 1185–1189.
106. Reynolds, C. A.; Burt, C.; Richards, W. G. *Quant. Struct. –Act. Relat.* **1992**, *11*, 34–35.
107. Tversky, A. *Psychol. Rev.* **1977**, *84*, 327–352.
108. Wipke, W. T.; Rogers, D. *J. Chem. Inf. Comput. Sci.* **1984**, *24*, 71–81.
109. Willett, P. *J. Chem. Inf. Comput. Sci.* **1985**, *25*, 114–116.
110. Grethe, G.; Hounshell, W. D. Similarity Searching in the Development of New Bioactive Compounds: an Application. In *Chemical Structures 2*; Warr, W. A., Ed.; Springer-Verlag: Berlin, 1993, pp 399–407.
111. Edgar, S. J.; Holliday, J. D.; Willett, P. *J. Mol. Graph. Model.* **2000**, *18*, 343–357.
112. Gillet, V. J.; Willett, P.; Bradshaw, J. *J. Chem. Inf. Comput. Sci.* **1998**, *38*, 165–179.
113. Briem, H.; Lessel, U. F. *Perspect. Drug Disc. Des.* **2000**, *20*, 231–244.
114. Güner, O. F.; Henry, D. R. Metric for Analyzing Hit Lists and Pharmacophores. In *Pharmacophore Perception, Development and Use in Drug Design*; Güner, O. F., Ed.; International University Line: La Jolla, CA, 2000, pp 193–212.
115. Wilton, D.; Willett, P.; Lawson, K.; Mullier, G. *J. Chem. Inf. Comput. Sci.* **2003**, *43*, 469–474.
116. Cuissart, B.; Touffet, F.; Crémilleux, B.; Bureau, R.; Rault, S. *J. Chem. Inf. Comput. Sci.* **2002**, *42*, 1043–1052.
117. Triballeau, N.; Acher, F.; Brabet, I.; Pin, J.-P.; Bertrand, H.-O. *J. Med. Chem.* **2005**, *48*, 2534–2547.
118. Adamson, G. W.; Cowell, J.; McLure, A. H. W.; Town, W. G.; Yapp, A. M.; Lynch, M. F. *J. Chem. Doc.* **1973**, *13*, 153–157.
119. Willett, P. *Methods Mol. Biol.* **2004**, *275*, 51–63.
120. Dixon, S. L.; Merz, K. M. *J. Med. Chem.* **2001**, *44*, 3795–3809.
121. Adamson, G. W.; Bush, J. A. *J. Chem. Inf. Comput. Sci.* **1975**, *15*, 55–58.
122. Pickett, S. D.; Luttman, C.; Guerin, V.; Laoui, A.; James, E. *J. Chem. Inf. Comput. Sci.* **1998**, *38*, 144–150.
123. Trepalin, S. V.; Gerasimenko, V. A.; Kozyukov, A. V.; Savchuk, N. P.; Ivaschenko, A. A. *J. Chem. Inf. Comput. Sci.* **2002**, *42*, 249–258.
124. Sheridan, R. P. *J. Chem. Inf. Comput. Sci.* **2000**, *40*, 1456–1469.
125. Flower, D. R. *J. Chem. Inf. Comput. Sci.* **1988**, *38*, 379–386.
126. Lajiness, M. Molecular Similarity-Based Methods for Selecting Compounds for Screening. In *Computational Chemical Graph Theory*; Rouvray, D., Ed.; Nova Science: New York, 1990, pp 299–316.
127. Godden, J. W.; Xue, L.; Bajorath, J. *J. Chem. Inf. Comput. Sci.* **2000**, *40*, 163–166.
128. Holliday, J. D.; Salim, N.; Whittle, M.; Willett, P. *J. Chem. Inf. Comput. Sci.* **2003**, *43*, 819–828.
129. Schuffenhauer, A.; Gillet, V. J.; Willett, P. *J. Chem. Inf. Comput. Sci.* **2000**, *40*, 295–307.
130. Makara, G. M. *J. Med. Chem.* **2001**, *44*, 3563–3571.
131. Cruciani, G.; Pastor, M.; Mannhold, R. *J. Med. Chem.* **2002**, *45*, 2685–2694.
132. Cramer, R. D.; Jilek, R. J.; Guessregen, S.; Clark, S. J.; Wendt, B.; Clark, R. D. *J. Med. Chem.* **2004**, *47*, 6777–6791.
133. Hall, D. L. *Mathematical Techniques in Multisensor Data Fusion*; Artech House: Northwood, MA, 1992.
134. Arabnia, H. R.; Zhu, D. Proceedings of the International Conference on Multisource-Multisensor Information Fusion, Fusion 98.; CSREA Press: Las Vegas, NY, 1998.
135. Ginn, C. M. R.; Willett, P.; Bradshaw, J. *Perspect. Drug Disc. Des.* **2000**, *20*, 1–16.
136. Belkin, N. J.; Kantor, P.; Fox, E. A.; Shaw, J. B. *Inf. Process. Manag.* **1995**, *31*, 431–448.
137. Hsu, D. F.; Taksa, I. *Inf. Retrieval* **2005**, *8*, 449–480.
138. Ginn, C. M. R.; Turner, D. B.; Willett, P.; Ferguson, A. M.; Heritage, T. W. *J. Chem. Inf. Comput. Sci.* **1997**, *37*, 23–37.
139. Raymond, J. W.; Jalaie, M.; Bradley, P. P. *J. Chem. Inf. Comput. Sci.* **2004**, *44*, 601–609.
140. Wang, R.; Wang, S. *J. Chem. Inf. Comput. Sci.* **2001**, *41*, 1422–1426.
141. Yang, J.-M.; Chen, Y.-F.; Shen, T.-W.; Kristal, B. S.; Hsu, D. F. *J. Chem. Inf. Model.* **2005**, *45*, 1134–1146.
142. Verdonk, M. L.; Berdini, V.; Hartshorn, M. J.; Mooij, W. T. M.; Murray, C. W.; Taylor, R. D.; Watson, P. *J. Chem. Inf. Comput. Sci.* **2004**, *44*, 793–806.
143. Charifsen, P. S.; Corkery, J. J.; Murcko, M. A.; Walters, W. P. *J. Med. Chem.* **1999**, *42*, 5100–5109.
144. Clark, R. D.; Strizhev, A.; Leonard, J. M.; Blake, J. F.; Matthew, J. B. *J. Mol. Graph. Model.* **2002**, *20*, 281–295.
145. Xue, L.; Stahura, F. L.; Godden, J. W.; Bajorath, J. *J. Chem. Inf. Comput. Sci.* **2001**, *41*, 746–753.
146. Hert, J.; Willett, P.; Wilton, D. J.; Acklin, P.; Azzaoui, K.; Jacoby, E.; Schuffenhauer, A. *J. Med. Chem.* **2005**, *48*, 7049–7054.
147. Valler, M. J.; Green, D. *Drug Disc. Today* **2000**, *5*, 286–293.
148. Hann, M. M.; Leach, A. R.; Green, D. V. S. Computational Chemistry, Molecular Complexity and Screening Set Design. In *Chemoinformatics in Drug Discovery*; Oprea, T. I., Ed.; Wiley-VCH: Weinheim, 2004, pp 43–57.
149. Leach, A. R.; Bradshaw, J.; Green, D. V. S.; Hann, M. M.; Delany, J. J. *J. Chem. Inf. Comput. Sci.* **1999**, *39*, 1161–1172.
150. Lipinski, C. A.; Lombardo, F.; Dominy, B. W.; Feeney, P. J. *Adv. Drug Deliv. Rev.* **1997**, *23*, 3–25.
151. Hann, M. M.; Oprea, T. I. *Curr. Opin. Chem. Biol.* **2004**, *8*, 255–263.
152. Oprea, T. I.; Davis, A. M.; Teague, S. J.; Leeson, P. D. *J. Chem. Inf. Comput. Sci.* **2001**, *41*, 1308–1315.
153. Murtagh, F. *Multidimensional Clustering Algorithms*; Physica Verlag: Vienna, 1985.
154. Downs, G. M.; Barnard, J. M. Clustering Methods and their Uses in Computational Chemistry. In *Reviews in Computational Chemistry*; Lipkowitz, K. B., Boyd, D. B., Eds.; VCH: New York, 2002, Vol. 18, pp 1–40.
155. Jarvis, R. A.; Patrick, E. A. *IEEE Trans. Comput.* **1973**, *C-22*, 1025–1034.
156. Forgy, E. *Biometrics* **1965**, *21*, 768–769.
157. Milligan, G. W. *Psychometrika* **1980**, *45*, 325–342.
158. Ward, J. H. *J. Am. Stat. Assoc.* **1963**, *58*, 236–244.
159. Wild, D. J.; Blankley, C. J. *J. Chem. Inf. Comput. Sci.* **2000**, *40*, 155–162.

160. Kelley, L. A.; Gardner, S. P.; Sutcliffe, M. J. *Protein Eng.* **1996**, *9*, 1063–1065.
161. Lewis, R. A.; Mason, J. S.; McLay, I. M. *J. Chem. Inf. Comput. Sci.* **1997**, *37*, 599–614.
162. Pearlman, R. S.; Smith, K. M. *Perspect. Drug Disc. Des.* **1998**, *9–11*, 339–353.
163. Pearlman, R. S.; Smith, K. M. *J. Chem. Inf. Comput. Sci.* **1999**, *39*, 28–35.
164. Snarey, M.; Terrett, N. K.; Willett, P.; Wilton, D. J. *J. Mol. Graph. Model.* **1997**, *15*, 372–385.
165. Agrafiotis, D.; Lobanov, V. S. *J. Chem. Inf. Comput. Sci.* **1999**, *39*, 51–58.
166. Mount, J.; Ruppert, J.; Welch, W.; Jain, A. N. *J. Med. Chem.* **1999**, *42*, 60–66.
167. Hudson, B. D.; Hyde, R. M.; Rahr, E.; Wood, J. *Quant. Struct. –Act. Relat.* **1996**, *15*, 285–289.
168. Clark, R. D. *J. Chem. Inf. Comput. Sci.* **1997**, *37*, 1181–1188.
169. Clark, R. D.; Langton, W. J. *J. Chem. Inf. Comput. Sci.* **1998**, *38*, 1079–1086.
170. Agrafiotis, D. K. *J. Comput.-Aided Mol. Des.* **2002**, *16*, 335–356.
171. Gillet, V. J.; Willett, P.; Bradshaw, J. *J. Chem. Inf. Comput. Sci.* **1997**, *37*, 731–740.
172. Gillet, V. J.; Willett, P.; Bradshaw, J.; Green, D. V. S. *J. Chem. Inf. Comput. Sci.* **1999**, *39*, 169–177.
173. Good, A.; Lewis, R. A. *J. Med. Chem.* **1997**, *40*, 3926–3936.
174. Walters, W. P.; Stahl, M. T.; Murcko, M. A. *Drug Disc. Today* **1998**, *3*, 160–178.
175. Waldman, M.; Li, H.; Hassan, M. *J. Mol. Graph. Model.* **2000**, *18*, 412–426.
176. Gillet, V. J.; Khatib, W.; Willett, P.; Fleming, P. J.; Green, D. V. S. *J. Chem. Inf. Comput. Sci.* **2002**, *42*, 375–385.
177. Gillet, V. J.; Willett, P.; Fleming, P. J.; Green, D. V. S. *J. Mol. Graph. Model.* **2002**, *20*, 491–498.

Biographies

Val J Gillet is Senior Lecturer and Head of the Chemoinformatics Research Group in the Department of Information Studies at the University of Sheffield, UK. She holds MSc and PhD degrees from the University of Sheffield and a first degree in Natural Sciences from Cambridge. Her research interests are in chemoinformatics and computational approaches to drug design, including: combinatorial library design and diversity analysis; the development of novel molecular descriptors; the application of evolutionary algorithms to computational chemistry; the identification of structure–activity relationships; and de novo design. She is organizer of triennial conferences and annual short courses to industry in chemoinformatics and joint author of the first textbook in chemoinformatics.

Peter Willett following a first degree in Chemistry from Oxford, Peter Willett obtained MSc and PhD degrees in Information Science from the Department of Information Studies at the University of Sheffield, UK. He joined the faculty of the Department in 1979, was awarded a Personal Chair in 1991 and a DSc in 1997, and is now the Head of the Department. He was the recipient of the 1993 Skolnik Award of the American Chemical Society Division of Chemical Information, of the 1997 Distinguished Lecturer Award of the New Jersey Chapter of the American Society for Information Science, of the 2001 Kent Award of the Institute of Information Scientists, of the 2002 Lynch Award of the Chemical Structure Association Trust, and of the 2005 American Chemical Society Award for Computers in Chemical and Pharmaceutical Research. He is included in *Who's Who*, is a member of the editorial boards of three international journals, and has written over 430 publications describing novel computational techniques for the processing of chemical, biological, and textual information.

© 2007 Elsevier Ltd. All Rights Reserved
No part of this publication may be reproduced, stored in any retrieval system or transmitted in any form by any means electronic, electrostatic, magnetic tape, mechanical, photocopying, recording or otherwise, without permission in writing from the publishers

Comprehensive Medicinal Chemistry II
ISBN (set): 0-08-044513-6

ISBN (Volume 4) 0-08-044517-9; pp. 167–192

4.09 Structural, Energetic, and Dynamic Aspects of Ligand–Receptor Interactions

D Motiejunas and R C Wade, EML Research, Heidelberg, Germany

© 2007 Elsevier Ltd. All Rights Reserved.

4.09.1 Introduction

Ligand–receptor interactions provide the fundamental basis for the mechanism of action of all drugs. An understanding of the determinants of ligand–receptor recognition, specificity, thermodynamics, and kinetics is essential for rational, structure-based drug design. In this chapter, structural, dynamic, and energetic aspects of ligand–receptor interactions are described and discussed. The binding of low-molecular-weight compounds (ligands) to proteins (receptors) is then addressed. Other receptors, e.g., nucleic acids, will also be considered, as will ligands that are themselves proteins or peptides.

We will describe not only the principles that govern ligand–receptor interactions but also how knowledge of these principles can be translated into models that can be used for computational structure-based drug design. The approximations, applications, and limitations of the computational models will be discussed.

4.09.2 General Principles

Ligand–receptor interactions in vivo are determined by a number of diverse factors. The first issue to consider is whether the ligand and receptor can approach close enough to recognize and bind to each other. Does a ligand have to cross a membrane to reach the same cellular compartment as a receptor? How will an orally administered drug reach the bloodstream and then its target receptor? To answer these questions, the physicochemical properties of the ligand need to be considered as well as its mechanism of transportation to its receptor. This may involve diffusion and active transport, including interaction with transport or membrane pore proteins. These aspects are discussed in detail in Chapter 5.28.

Once the ligand and receptor are sufficiently close, the ligand can diffuse up to and dock into its binding site on the receptor. This requires recognition between the ligand and receptor. This may first be mediated by the long-range electrostatic interactions between the ligand and the receptor and then strengthened by short-range hydrogen bonds and van der Waals' interactions. The relative importance of these terms varies widely from case to case. Water molecules will be displaced upon binding though some may be retained at the interface and mediate binding and influence specificity. Binding is accompanied by conformational changes ranging from modest shifts of a few atoms to movements of whole protein domains.

The binding process described in the preceding paragraph corresponds to the process determining the bimolecular association rate constant, the on-rate, k_{on}. The unbinding process determines the bimolecular dissociation rate constant, the off-rate, k_{off}. For ligand, L, and receptor, R:

$$\mathrm{L} + \mathrm{R} \underset{k_{off}}{\overset{k_{on}}{\leftrightarrow}} \mathrm{LR}$$

The binding affinity is quantified by the equilibrium dissociation constant, $K_d = k_{off}/k_{on} = [\mathrm{L}][\mathrm{R}]/[\mathrm{LR}]$. The selectivity of a receptor for one ligand compared to another is due to different binding affinities and may arise due to differences in on- or off-rates (or both).

The dissociation constant is related to the binding free energy, ΔG by:

$$K_d = C^o \exp(-\Delta G/RT)$$

where R is the universal gas constant and T the temperature in K. K_d is given in units of M (moles per liter) and the free energy is given at the standard state concentration, C^o, of 1 M ($1\,\mathrm{mol\,L^{-1}}$ or 1 solute molecule per $1661\,\mathrm{\AA}^3$ aqueous solvent).

Thermodynamically, ΔG is given by the balance between binding enthalpy, ΔH, and binding entropy, ΔS:

$$\Delta G = \Delta H - T\Delta S$$

Ligand–receptor interactions are often characterized by enthalpy–entropy compensation in which one term favors and the other disfavors binding. While enthalpic contributions include electrostatic, hydrogen bond, and van der Waals' interactions, entropic contributions arise from several sources. Binding is accompanied by an entropic cost due to loss of translational and rotational entropy upon binding. In addition, loss of flexibility upon binding will entropically disfavor binding. On the other hand, the displacement of ordered water molecules upon binding can entropically favor binding. Entropic contributions are discussed in more detail in Section 4.09.3.5.

In the first model of ligand–receptor binding, Emil Fischer proposed that receptor and ligand (specifically, enzyme and substrate) fit together like a lock and key.[1] In this analogy, specificity can be conveniently viewed in terms of distinguishing between locks and keys. However, in the lock and key picture, the receptor and ligand are rigid entities. In reality, binding is accompanied by some degree of conformational change. This can be considered in 'zipper'[2] or 'hand-into-glove' analogies to describe receptor–ligand interactions. The conformational change can be considered to be an induced fit due to binding[3] or the selection of different dominant conformations from the conformational ensemble of the molecule in the bound and unbound states[4] or a combination of both. Some of the binding models are shown in **Figure 1**. Of particular interest for drug design is the observation that drug molecules can bind to parts of the protein that are not revealed at the surface in structures determined for either the unbound form or for a form with a natural ligand bound.[5–7] Conformational changes upon ligand–receptor binding are discussed in more detail in Section 4.09.4.

The solvent surroundings of the ligand and the receptor have a very important influence on their binding. Binding affinities will be very different in a polar solvent like water compared with a more nonpolar environment like methanol or the hydrophobic interior of a lipid membrane. This is because binding always involves the competition between ligand–receptor interactions and ligand–solvent and receptor–solvent interactions. The ionic strength and pH of the surroundings will influence the strength of electrostatic interactions between ligand and receptor. Viscogens and crowding agents can also affect ligand–receptor binding. They may affect the binding kinetics through their viscosity or the binding affinity through alteration of dielectric properties or through local concentration effects.

In the next section, the physical factors affecting ligand–receptor interactions are discussed in more detail, drawing attention to how these can be modeled and how they can be exploited in ligand design.

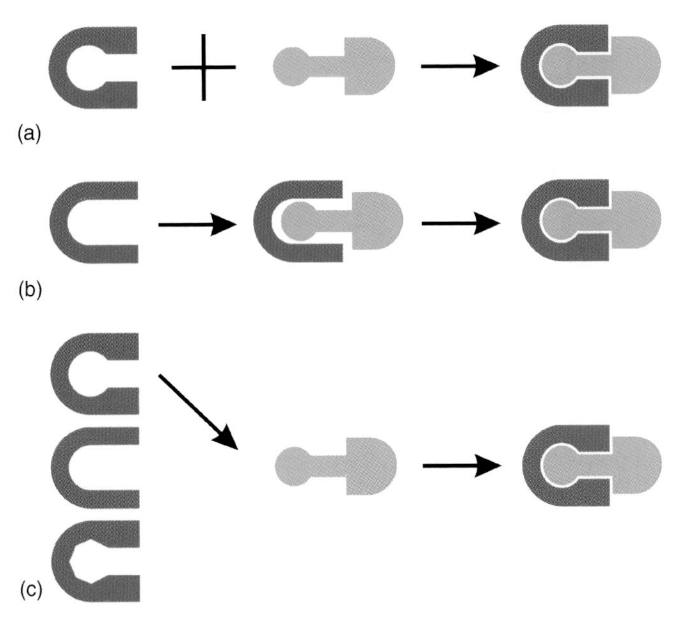

Figure 1 Models of receptor–ligand binding. (a) In the lock-and-key model, the ligand (green) exactly fits into the receptor binding site (violet). (b) The ligand associates weakly with the receptor and induces conformational changes resulting in binding. (c) The ligand binds preferentially to certain conformations in the conformational ensemble of the receptor.

4.09.3 Physical Properties Governing Ligand–Receptor Binding

4.09.3.1 Steric

As the lock-and-key model suggests, shape complementarity is very important for ligand–receptor binding and specificity. The overlap of atoms of the ligand and the receptor is prohibited by electron–electron repulsion. Clashes draw a large energetic penalty. In studies of receptor–ligand interactions, this penalty is usually modeled by the first term in the Lennard-Jones energy function[8,9]:

$$E_{LJ} = \sum \frac{A}{r^{12}} - \frac{C}{r^{6}}$$

The sum is over all ligand–receptor atom pairs and is a function of the distance, r, between the atoms in each pair, which have van der Waals' radii R_l and R_r.

$$A = 0.5C(R_l + R_r)^6$$

and C is given by the Slater-Kirkwood formula[10] and is dependent on atomic polarizability and the number of effective electrons per atom.

The functional form of the repulsive part is purely empirical and was chosen by Lennard-Jones to fit experimental data for rare gases and because of its computational convenience. In some studies of ligand–receptor interactions, it is useful to make the repulsion less abrupt by using a lower power of distance, e.g., r^{-8}. This has the advantage that the repulsive energy is less sensitive to the atomic positions and thus more robust to errors in atomic coordinates or to the effects of atomic motions.

The attractive term in the Lennard-Jones function describes van der Waals' interactions due to dispersive attractions. These induced dipole–induced dipole interactions have an r^{-6} distance dependence and, while less strongly distance dependent than the repulsive interactions, are short range interactions.

Ligand–receptor binding involves the replacement of ligand–water and receptor–water interactions by ligand–receptor and water–water interactions. Whether van der Waals' interactions contribute favorably to ligand–receptor affinity or not depends on the strength of the ligand–receptor van der Waals' interactions compared to the interactions with water. Owing to this balance, any favorable contribution of van der Waals' interactions to binding affinity is typically small.

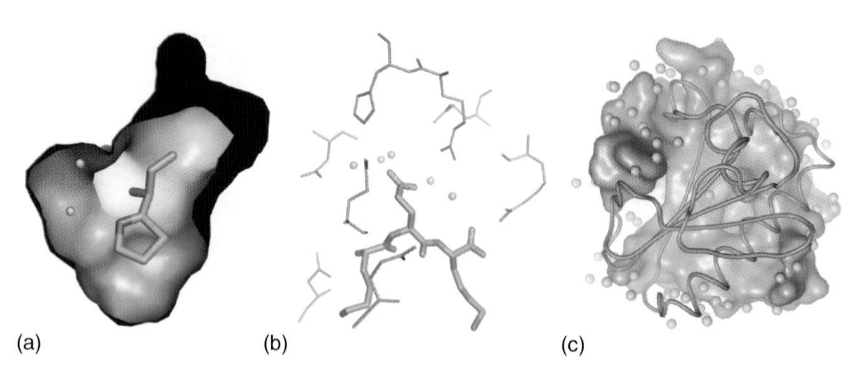

Figure 2 Water in receptor–ligand complexation. (a) Major urinary protein-I (MUP-I) with bound ligand (PDB code 1I06). Two waters (violet spheres) in hydrophobic environment of ligand binding pocket (surface representation) make possible sites for interactions with polar groups present on the lipophilic pheromones (stick representation).[159] (b) The complex of the periplasmic oligopeptide binding protein OppA with peptide (LysGluLys) (PDB code 1JEU). Only active center residues of the protein are shown (stick representation). Water molecules (violet spheres) mediate interaction between OppA (periplasmic oligopeptide binding protein) and its peptide ligands.[160] (c) The protein kinase C interacting protein (PKCI) dimer (PDB code 1KPA) has a dry interface with most crystallographically observed water molecules (violet spheres) forming a ring around the interface.[161]

Nevertheless, consideration of steric complementarity is critical for ligand design. Empty vacuum cavities at the interface between ligand and receptor are energetically unfavorable: Nature abhors a vacuum. Interfacial cavities that are large enough to accommodate one or more water molecules may be hydrated and, if suitably filled, be energetically favorable. This depends not only on the size of the cavity but also on its polarity and ability to make hydrogen bonds with water molecules. Small nonpolar cavities with no hydrogen-bonding groups will be unoccupied whereas nonpolar cavities able to accommodate *c.* four or more water molecules may be hydrated if the water molecules themselves can make a hydrogen bond network.

The packing and the hydration patterns at ligand–receptor interfaces have been studied in crystal structures. Some interfaces are well packed with water molecules totally excluded from them (see, e.g., **Figure 2c**). Ordered interfacial water sites are often observed at positions where water molecules can mediate hydrogen bonds between receptor and ligand (**Figure 2b**). Particularly ordered water sites are sometimes considered to be an intrinsic part of the solute molecule. Interfacial water sites can alter the specificity of a binding site due to their ability to accept and donate up to four hydrogen bonds. As shown in the example in **Figure 2a**, interfacial water molecules may permit a rather hydrophobic receptor site to interact with a ligand and accommodate its hydrogen-bonding ability.

To optimize steric complementarity in ligand design, a means to detect interfacial cavities and crevices is necessary. A solvent accessible surface or a molecular (Connolly) surface can be computed by rolling a spherical probe with a radius of 1.4 Å, corresponding to the size of a water molecule, over the solute molecules.[11] The size of the probe can be varied to represent different molecules, ions, or molecular fragments, and a range of analytical and numerical algorithms can be used to compute surfaces and identify cavities. A useful tool, which employs alpha shape theory to identify pockets and cavities in protein structures, is the CASTp webserver.[12,13,171]

Design of a ligand to fill a crevice or an interfacial space requires not only knowing the size and shape of the space but also its physicochemical properties. Suitable functional groups to add to a molecule can be identified by computing molecular interaction fields for probes with hydrogen-bonding and electrostatic properties as well as steric ones. In the GRID approach,[14,172] an empirical energy function is used to describe the interaction between probe and target, and energetically favorable binding sites for a range of chemical probes, including a water probe, can be mapped out. There are many examples of successful modification of ligands to fill interfacial cavities. The modifications not only need to fill space but also have the appropriate chemical properties. They may also need to displace interfacial water molecules. This was, for example, the approach taken for inhibitors of scytalone dehydratase, which were designed to displace a water molecule from the active site.[15] Further examples are given in the next sections.

4.09.3.2 Electrostatic

Electrostatic interactions play a significant role in ligand–receptor binding. They are long-range, varying with distance as r^{-1}, and therefore can be particularly important for molecular recognition. Each molecule can be considered to have not only a net charge but also a charge distribution over the whole molecule. In modeling studies of ligand–receptor binding, partial charges are usually assigned to atom centers on the basis of ab initio or semiempirical quantum mechanical calculations or simpler approaches such as electronegativity equalization. Like charges repel and unlike charges attract. Consequently, many positive and negative terms contribute to the coulombic electrostatic energy of a ligand–receptor complex, which is given by:

$$E_{el} = \sum \frac{q_l q_r}{4\pi\varepsilon_0\varepsilon\, r}$$

where ε_0 is the permittivity of free space, ε the relative dielectric constant of the surrounding medium, and q_l and q_r are partial atomic point charges in the ligand and receptor, respectively.

Coulomb's law applies for a homogeneous dielectric medium. If all atoms of the system are modeled explicitly, including all water molecules and ions in the solvent, and the system is subjected to molecular dynamics simulation, $\varepsilon = 1$ is usually used. Often, however, the water molecules and ions are treated implicitly and an aqueous solvent is modeled as a continuum having $\varepsilon \approx 80$. The ions are assumed to have a Boltzmann distribution. In this case, the dielectric constant of the solvent medium differs from that of the molecular solutes. The electrostatic potential of this heterogeneous system is described by the Poisson–Boltzmann equation.

The Poisson–Boltzmann equation (or its linearized form) may be solved numerically for biomacromolecules. Finite difference and multigrid methods on grid(s) superimposed on the target molecule are the most commonly used methods to solve the Poisson–Boltzmann equation. Details of solution of the Poisson–Boltzmann equation for biomacromolecules are given in references.[16–19] To solve the Poisson–Boltzmann equation, a suitable value of the dielectric constant for the molecular interior and the definition of the dielectric boundary must be chosen. These are adjustable parameters to be fitted in the context of the complete energy function, the treatment of molecular flexibility, and the properties to be computed.[20,21] The dielectric boundary can be chosen as the van der Waals' surface, the solvent accessible molecular surface (mapped by a solvent probe surface) or the solvent accessible surface (mapped by a solvent probe center). The choice of dielectric boundary definition can significantly impact the magnitude of electrostatic binding free energies.[22,23]

In the Poisson–Boltzmann model, the electrostatic binding free energy consists of ligand charge–receptor charge interaction terms (of similar functional form to coulombic interactions) and ligand and receptor charge desolvation terms, which depend on the square of each partial charge. The charge interaction energy is the first and the charge desolvation energies are the second and third terms on the right-hand side of this equation:

$$E_{el-PB} = \rho_l \Phi_r + \frac{1}{2}\rho_l(\Phi_{\text{bound},\,l} - \Phi_{\text{free},\,l}) + \frac{1}{2}\rho_r(\Phi_{\text{bound},\,r} - \Phi_{\text{free},\,r})$$

ρ_l and ρ_r are the charge distributions over ligand and receptor, respectively, and Φ_l and Φ_r the electrostatic potentials of the ligand and the receptor, respectively, and are computed in the presence (bound) and absence (free) of the other molecule's dielectric cavity. The potential of a charge q_i in homogeneous dielectric ε at distance r is:

$$\Phi_i = \frac{q_i}{4\pi\varepsilon_0\varepsilon\, r}$$

Charge desolvation due to the transfer of a charge from a high to a low dielectric medium disfavors binding. Thus, the overall energetic contribution of electrostatic interactions to the binding affinity of a complex may be small or even unfavorable, while at the same time being very important for specificity. Binding affinity can be improved by systematically optimizing the charge distribution of a ligand considering the balance between charge interaction and charge desolvation terms; an example is the design of improved inhibitors of HIV-1 cell entry.[24]

The computational demands of solution of the Poisson–Boltzmann equation can be prohibitive for some studies of ligand–receptor interactions, so simpler electrostatic models are often used. The effective charge model[25] permits calculation of electrostatic forces between two molecules based on Poisson–Boltzmann calculations for the individual molecules, thus permitting their electrostatic interactions to be computed quickly for many different molecular arrangements. It reproduces the Poisson–Boltzmann forces well at intermolecular separations greater than one water molecule. Other methods to account for the dielectrically discontinuous boundary between molecules and the implicit solvent are the method of images (e.g., assuming an infinite planar boundary[14]) or the generalized Born model.[26,27] A simple and computationally efficient but rather rough approximation is to use Coulomb's law with $\varepsilon = nr$ (n is an integer), i.e., dependent on interatomic distance, r. Most treatments of electrostatics in drug design neglect charge desolvation effects and thus provide an incomplete description of electrostatic interactions.

Specific polarization effects, beyond those modeled by a continuum dielectric model and the movement of atoms, are usually neglected. Many-body effects are also neglected by use of a pair-wise additive energy function. Polarizable forcefields are, however, becoming more common in the molecular mechanics force fields used for molecular dynamics simulations. Furthermore, polarization effects can be considered directly by treating the ligand, and sometimes the surrounding protein, quantum mechanically. The full ab initio quantum mechanical treatment of a protein for computing ligand–receptor binding affinities was recently achieved for a study of ligand binding to the human estrogen receptor alpha.[28]

The strength of electrostatic interactions in a ligand–receptor complex can be investigated experimentally by measuring the ionic strength dependence and pH dependence of binding properties. The strength of electrostatic interactions varies widely. Electrostatic interactions are of clear importance for the recognition and binding properties of highly charged molecules such as nucleic acids. However, all types of bound ligand–receptor complexes can show electrostatic complementarity at the interface. It is not so much the net charge of a molecule as its charge distribution that is important. For example, in many protein–protein complexes, the two proteins have net charges of the same sign but bind with electrostatically complementary interfaces. Importantly, the electrostatic potentials at the interface are complementary, not necessarily the adjacent charges at the interface.[29] This is due to the long-range nature of the electrostatic interactions as well as the effects of the heterogeneous dielectric constant. Complementary electrostatic interactions are shown in **Figure 3a** and **d** for inhibitor–enzyme and peptide-signaling domain complexes, respectively. Solvation effects may make electrostatic interactions quite complex as shown by the examples in **Figure 3b** and **c** for the urokinase-type plasminogen activator with two different inhibitors bound at its active site.[30] In **Figure 3c** the interactions of the inhibitor with the whole active site are such that the carboxylate group of Asp189 and the inhibitor amidine group are too far apart to make a hydrogen-bonded salt link. Their electrostatically favorable interactions are

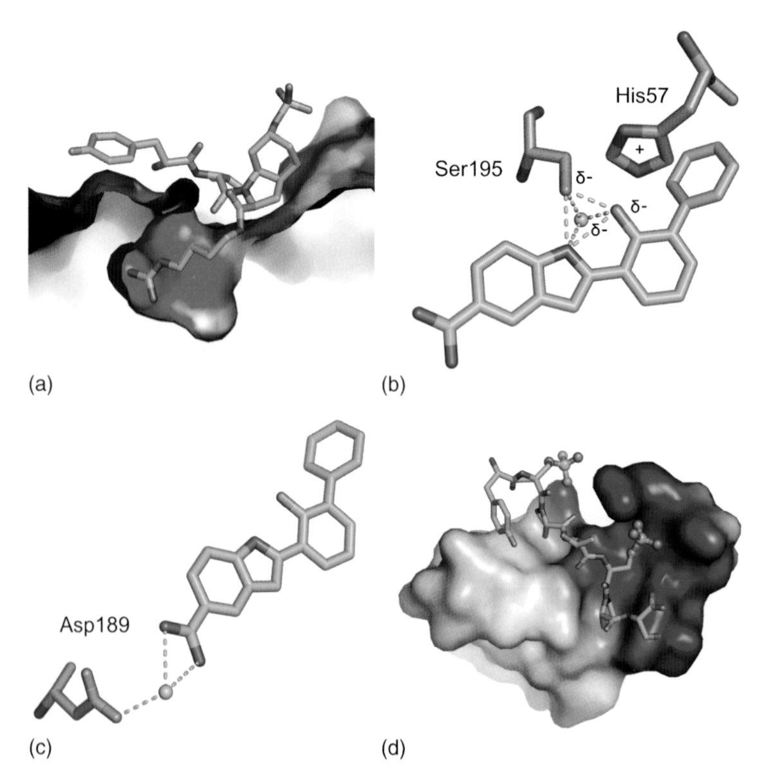

Figure 3 Electrostatic interactions. (a) An inhibitor, Aeruginosin 98-B (stick representation), bound to a negatively charged pocket of bovine trypsin (surface representation) (PDB code 1AQ7).[162] The electrostatic potential of the protein was calculated with UHBD,[173] mapped onto the protein surface and color coded (negative: red; positive: blue). (b) Water (violet sphere) mediated short hydrogen bond network in urokinase-type plasminogen activator (grey sticks) complexed with an inhibitor (green sticks) (PDB code 1GI8). Short hydrogen bonds (dotted lines) are presumed ionic over a large range of pH, with a negative charge distributed among O_{phenol}, O_{oxy}, and O_{Ser195} and a positive charge on His57.[30] (c) Water (violet sphere)-mediated salt bridge between urokinase-type plasminogen activator Asp189 (grey sticks) and inhibitor (PDB code 1GI8).[30] (d) Phosphoserine-proline containing peptide (stick representation, phosphate groups in ball-and-stick representation) bound to a group IV WW domain (surface representation) face with positive electrostatic potential (PDB code 1F8A) (electrostatic potential computed and mapped as in part a).[163]

instead mediated by a water molecule, whose presence reduces the charge desolvation penalty usually associated with salt-link formation. In **Figure 3b** inhibitor–enzyme interactions are again mediated by a water molecule, but this time the water molecule helps to stabilize very short hydrogen bonds.

Electrostatic interactions can be exploited in drug design. An example is provided by the anti-influenza drug Relenza. This compound is an inhibitor of the neuraminidase coat protein of the influenza virus. The design strategy (*see* 4.24.3 Structure-Based Drug Design – The Use of Protein Structure in Drug Discovery and for review, see [31]) was to modify a transition state analog on the basis of the crystal structure of the enzyme so as to optimize binding affinity. By probing the protein with different probes with the GRID program,[14] an interfacial pocket was identified where a positively charged amino or guanidine group could favorably be added as a substituent to the transition state analog.[32] Compounds with these positively charged substituents turned out to be very potent and one of them, Relenza, is used in the clinic. It is delivered to its site of action as a nasal spray: its charged functional groups make it too polar for oral delivery, which requires the compound to be able to pass through membranes.

Subsequent modification to improve inhibitor–protein interactions, the stability of the inhibitor and its bioavailability, resulted in Tamiflu, an anti-influenza drug that is taken orally.[33] It is delivered in an uncharged proform and then converted in the body to the active charged form. This shows how charged moieties in ligands can be optimized for binding to the macromolecular target but also illustrates the need to try to consider simultaneously absorption, distribution, metabolism, excretion, and toxicity (ADMET) properties.

4.09.3.3 Hydrogen Bonds

Hydrogen bonds are specific, short-range, directional nonbonded interactions. They are predominantly electrostatic in character, although charge transfer also contributes to their strength. In molecular mechanics force fields, they are usually treated as resulting from the sum of coulombic terms, and this is possible if polar hydrogen atoms are modeled explicitly.

In some force fields, it is also considered necessary to model the positions and point charges of lone pairs of electrons on hydrogen bond acceptor atoms, e.g., sulfur atoms and carboxylate oxygen atoms, in order to model the appropriate hydrogen-bonding geometry. In other energy functions, there is a hydrogen-bonding term in addition to the electrostatic and Lennard-Jones terms. This hydrogen-bonding term models the distance and angular dependence of hydrogen bonds. The GRID energy function, for example, has been specifically designed to reproduce observed hydrogen-bonding geometries in crystal structures of small molecules and of proteins. It has a hydrogen-bonding energy describing the interaction between a probe (p) (a small ligand or fragment of a ligand) and a target (t) receptor atom that is the product of three terms:

$$E_{hb} = E_r \cdot E_t^{\theta} \cdot E_p^{\phi}$$

E_r is dependent on the separation between target and probe nonhydrogen atoms participating in the hydrogen bond. It has the form:

$$E_r = \frac{M}{r^m} - \frac{N}{r^n}$$

where M and N depend on the chemical nature of the hydrogen-bonding atoms. Possible values of the m and n parameters are $m=6$, $n=4$; $m=8$, $n=6$ (as in GRID); $m=12$, and $n=10$. The angular terms take different functional forms depending on the chemical types of the hydrogen bonding atoms and whether they are in the probe or the target receptor. The angular dependence differs for the same atom type in the probe and in the target. This is because the target includes the interactions of the hydrogen-bonding atom's neighbors whereas these are absent for the probe, which is able to rotate to an orientation that results in an optimal hydrogen bond energy. When multiple hydrogen bonds are possible, the best combination is found by systematic search or analytically.

Hydrogen bonds are critical for the structure and interactions of biological macromolecules. In proteins, hydrogen-bonding patterns are key signatures of secondary structure elements. The same hydrogen-bonding patterns that occur between backbone atoms in adjacent strands in β-sheets in a single protein chain can serve to 'zip' together two proteins that interact via extended strands – see the example in **Figure 4d**. The hydrogen bonds between ligands and receptors can of course also adopt many other arrangements (see examples in **Figure 4**). Hydrogen bonds are important contributors to the specificity of receptor–ligand interactions. The specificity is achieved not only through favorable short-range directionally specific interactions but also because ligand–receptor arrangements that leave hydrogen-bonding capacity unsatisfied are disfavored. It is unfavorable to place the carbonyl oxygen of a ligand so that it is buried pointing into a hydrophobic pocket of a receptor because it cannot make any hydrogen bonds in this location and loses the hydrogen bonds it could make to water molecules in the unbound state.

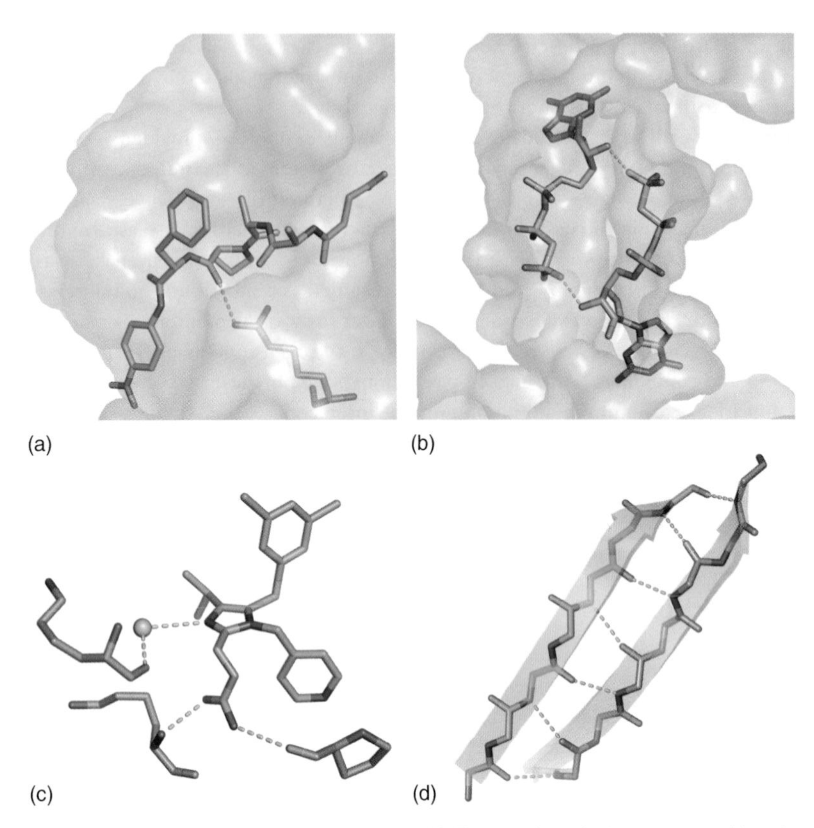

Figure 4 Various intermolecular hydrogen bond interactions. (a) The catalytically important residue Arg55 (grey sticks) of cyclophilin A (CypA) forms a hydrogen bond with its substrate, the succinyl-Ala-Ala-Pro-Phe-*p*-nitroanilide (AAPF) peptide (green sticks) (PDB code 1RMH).[164] (b) Two GTP analogs (stick representation) in the Ffh/FtsY complex interact with each other via hydrogen bonds between the ribose O3′ hydroxyl of one and the γ-phosphate oxygen of the other.[165] (c) Ligand S-1153 (green sticks) forms two direct hydrogen bonds with the HIV-1 reverse transcriptase (PDB code 1EP4) main-chain carbonyl of Pro236 and nitrogen of Lys-103. A water (violet sphere)-mediated hydrogen bond is formed with the carbonyl oxygen of Lys101.[166] (d) Hydrogen bonds between adjacent parallel strands in the kinesin dimer (PDB code 2KIN).

Charged residues in proteins and charged moieties of ligands can engage in hydrogen bonds to other charged groups (salt links or salt bridges) or to polar groups without a net charge (see **Figure 4a** and **b**). While the burying of a charged group on a ligand or protein in a protein has a large charge desolvation penalty, this can be compensated by hydrogen bonds from surrounding polar groups. Indeed, favorable local hydrogen bonds can even overcome an unfavorable electrostatic surface potential at the binding site of a charged ligand. For example, the buried binding sites for phosphate and sulfate ions in bacterial ion active transport receptor proteins have a negative surface potential but nevertheless the anions bind very specifically due to the optimal ligand–receptor hydrogen-bonding arrangement, which leaves no hydrogen bond donor or acceptor unpaired.[34]

Hydrogen bonds between a ligand and its receptor may be mediated by water molecules (see, e.g., **Figure 4c**). The ability of water molecules to donate and accept, as well as their freedom to move, means that the water molecules can serve to weaken the geometric restrictions on hydrogen-bonding interactions and extend their reach. This ability means they can act as a lubricant during binding processes and functional motions, i.e., the water molecules can serve to facilitate conformational changes that maintain or improve hydrogen-bonding network. Such a role has, for example, been proposed for water molecules in the active site gorge of acetylcholinesterase.[35]

Hydrogen bonds are exploited in drug design, particularly to obtain specificity. Docking programs generally model hydrogen bond interactions better than hydrophobic ones (see below). The number of hydrogen bonds in a drug molecule may be limited by requirements on polarity for absorption and permeation. The Lipinski rule-of-five,[36] for example, suggests that compounds with more than five hydrogen bond donors or more than 10 hydrogen bond acceptors are more likely to have poor absorption or permeation characteristics.

4.09.3.4 Hydrophobic Effect

The hydrophobic effect describes the energetic preference of nonpolar molecular surfaces to interact with other nonpolar molecular surfaces and thereby to displace water molecules from the interacting surfaces. The hydrophobic effect is due to both enthalpic and entropic effects. Hydrophobic interactions are short-range attractive interactions that make an important contribution to ligand–receptor binding affinities. They also contribute to specificity but in a less geometrically constrained way than hydrogen-bonding interactions.

In simulations made with an explicit solvent model, there is no need for a special term in the force field to describe the hydrophobic effect. It occurs as a natural consequence of the Lennard-Jones and electrostatic interactions between water molecules and the atoms in the ligand and receptor molecules. However, when the solvent model is an implicit, continuum model or only includes a few particularly ordered water molecules, then hydrophobic interactions should be modeled by an additional term in the energy function. One way to do this is by employing an empirical term dependent on the surface area buried upon binding:

$$E_{SA} = \sum \sigma_i \Delta SA_i$$

The buried surface area may be defined from the molecular (Connolly) surface or the solvent-accessible surface. The coefficients (σ) are dependent on the surface area definition and on the atom type. The coefficients can be assigned according to the polarity of the atoms with polar atoms having opposite sign coefficients to nonpolar atoms.[37]

To identify hydrophobic patches on receptors, the GRID program provides a 'DRY' probe. It can be considered to be like an 'inverse' water probe. It makes a Lennard-Jones interaction in the same way as a water probe. It is also neutral like a water probe and has no electrostatic interaction term. The hydrogen bond energy term is however inverted to reflect the fact that polar parts of the target that are able to make hydrogen bonds will not be energetically favored next to a hydrophobic probe. Hydrophobic patches on proteins detected by the 'DRY' probe and exploited by the hydrophobic sides of ligands are shown in **Figure 5b** and **c** (*see* 4.11 Characterization of Protein-Binding Sites and Ligands Using Molecular Interaction Fields).

4.09.3.5 Entropy

Ligand–receptor binding is accompanied by several types of entropic changes (see [38] for review).

The transformation of two mobile molecules into one mobile complex results in the loss of translational and rotational entropy. Assuming that the ligand is completely immobilized with respect to the receptor, this entropy loss can be estimated from analytical expressions (Sackur–Tetrode equation) for the translational and rotational entropy of molecules in the gas phase. Generally, the entropy loss is less because the ligand still has translational and rotational freedom in the receptor binding site. The contribution of entropy loss to binding affinity is however far from negligible: Murray and Verdonk,[39] estimated that the barrier to binding of small molecules to proteins due to loss of rigid body entropy is

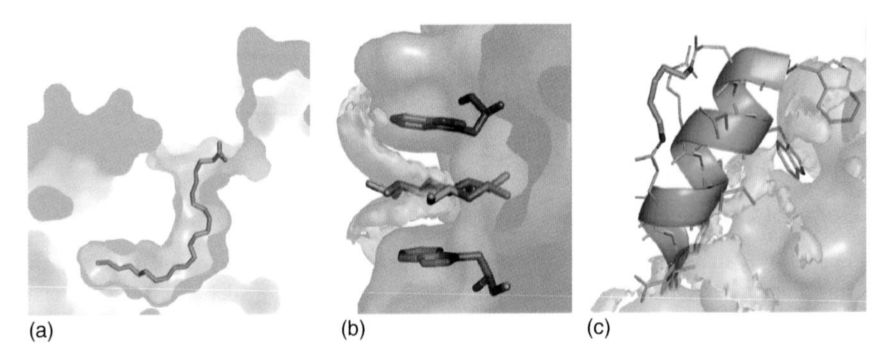

Figure 5 Hydrophobic interactions. (a) Proposed fatty acid ligand $C_{23}H_{48}O_2$ in the active site of cholesterol esterase (PDB code 1LLF). The long alkyl chain of the ligand makes interactions with the hydrophobic environment of the gorge below the reactive site of the enzyme.[167] (b) Two tryptophan residues (grey sticks), in the carbohydrate-binding module of xylanase 10A, provide planar hydrophobic stacking interactions for a glucose disaccharide (green sticks) (PDB code 1I82).[168] Yellow patches indicate favorable regions for the DRY probe as calculated with GRID. (c) Structure of a cross-linked helical peptide, C14linkmid, bound to IQN17, a soluble peptide that contains the HIV-1 gp41 hydrophobic pocket (surface representation) (PDB code 1GZI).[169] Yellow patches indicate favorable regions for the DRY probe as calculated with GRID.

15–20 kJ mol^{-1} i.e., around 3 orders of magnitude in affinity at 298 K. There are several methods to compute rigid body entropy loss from the extent of conformational sampling of the molecules in the bound and unbound states; these usually assume harmonic or quasiharmonic motion in the bound state (see [40] and methods discussed therein).

Binding affects the internal molecular dynamics and this affects the vibrational entropy and the entropy of rotamers. When two molecules bind, it is most often reported that the interface region becomes more ordered upon binding. This results in an entropic penalty. However, it appears that this penalty is quite often neutralized by binding, inducing greater flexibility in the receptor at sites distant to the ligand binding face,[41,42] which may result in allosteric effects at remote sites. Interestingly, regions of proteins that participate in binding are often quite mobile or disordered and become ordered upon binding. Although this is entropically unfavorable, the advantage is that the mobile region of the protein can fish out and adapt to its ligand by a 'fly casting mechanism' that can be kinetically advantageous.[43]

The change in vibrational entropy upon binding can be estimated from normal mode[44] or quasiharmonic[45] analysis of simulated structures. This permits changes in the entropy of harmonic vibrational motions to be computed. Entropic changes in nonharmonic motions, such as rotamer transitions (*see* Section 4.09.4.4.2), can be estimated by computing accessible rotamer states. Upon binding, the number of possible rotamers, n, that an interfacial side chain can adopt usually decreases and the entropic cost can be estimated as:

$$\Delta S = \text{Rn} \ln \Delta n$$

Such an approach has been used to analyze structures of protein–protein complexes by Cole and Waricker[46] who found that, on the basis of computed side chain entropy, the interfaces were less flexible than the rest of the protein surface.

Water molecules are displaced upon protein–ligand binding and their environment is changed; this also has an entropic effect. The displacement of ordered water molecules from an interface into bulk solution makes a favorable entropic contribution. An upper limit for this entropic change is given by the entropy change on transferring a water molecule from ice to water. This was estimated by Dunitz[47] as 28 J K^{-1} mol^{-1} (2 kcal mol^{-1} at 300 K). In general, the contribution is much smaller as the water around protein surfaces, as well as most of the water in protein interiors,[48] has much greater freedom to move than water in ice.

In modeling and simulation, the treatment of solvent entropic contributions depends on the type of model. In an all-atom model with explicit solvent molecules simulated by molecular dynamics methods, the entropic contributions of the solvent arise from the model and no special entropy terms in the force field are necessary. If the solvent is modeled implicitly as a continuum, then a term to account for the change in solvent entropy upon binding should be included in expressions to compute binding affinity. For example, in the GRID method, the entropic cost of displacing a water molecule by a hydrophobic probe is assumed to be 0.848 kcal mol^{-1}.[14] In other approaches, a surface area dependent free energy term includes both entropic and enthalpic contributions resulting from burying molecular surface upon binding.

Entropic effects can be exploited along with enthalpic effects in drug design. To optimize entropic contributions, compounds are usually designed to be relatively rigid with few rotatable bonds (*see* Section 4.09.4.2). They often have hydrophobic patches that interact with the receptor and are entropically favorable. Compounds can be designed to bind to more rigid parts of a protein target. This is entropically advantageous. However, flexibility in parts of small molecules may be useful to allow adaptation to flexible targets and to variants of targets, such as HIV protease.[49]

4.09.4 Molecular Flexibility

4.09.4.1 General Concepts

Conformational flexibility is important in biomolecular binding processes. Even if the binding partners follow the lock-and-key model, minor conformational changes take place. Although in such a case, computational docking methods that neglect flexibility can yield predictions of sufficient accuracy, often some level of flexibility has to be incorporated to model receptor–ligand association successfully. It is common to treat the flexibility of small ligands in molecular modeling studies. However, the flexibility of large receptors and of macromolecular ligands is often neglected or treated only partially. The reason to omit or model only a very limited degree of flexibility is the enormous increase of computational complexity due to the large number of degrees of freedom of macromolecules. Conformational changes in biological molecules occur at different timescales, ranging from picoseconds to hours, with amplitudes extending to hundreds of Angstroms.[50] Naturally, representations of molecular flexibility aim to incorporate only the most biologically relevant motions in computational modeling schemes to achieve acceptable accuracy with minimal computational costs.

In this section, an overview of the types of motions that occur in biomolecules and their relevance to receptor–ligand binding is given. Next, current models of macromolecular association are briefly described. Lastly, the most common computational techniques used to model flexibility in binding studies of small ligands and macromolecular associations are reviewed.

4.09.4.2 Types of Conformational Change upon Binding

4.09.4.2.1 Small-scale motions

Atoms in molecules are constantly moving due to thermal motions. These small movements include bond stretching, bond angle bending, and dihedral angle variations. The timescale of these motions is around 10^{-12} s and the amplitude is less than 1 Å.[50] Although these motions can be simulated in molecular dynamics or Monte Carlo simulations, computational methods often neglect these motions or model them implicitly. First, it is often too computationally demanding to include such small-scale motions in macromolecular docking or virtual library screening. Second, in such cases where only small-scale motions take place, models with rigid bodies or limited flexibility often give reasonable results. Betts and Sternberg[51] identified root mean square deviation (RMSD) changes larger then 0.6 Å for backbone and 1.7 Å for side chain atoms as having a substantial effect on protein–protein docking. About half of the 39 pairs of structures of complexed and unbound proteins they investigated had conformational changes smaller than this threshold and rigid-body docking algorithms could be applied with a certain level of success. Although the docked complexes are too crude for computational ligand design, their level of detail is useful in comparative modeling or predictive docking. For more detailed studies, however, small movements are important. For enzyme catalysis, for example, it was shown that movements of less than 1 Å can alter catalytic rates by several orders of magnitude.[52,53] In studies of the well-characterized enzyme isocitrate dehydrogenase, substitutions in the substrate (replacement of an NH_2 group at the 6 position of the natural substrate NADP by an OH group) or of the ion cofactor (Mg^{2+} substituted by Ca^{2+}), resulted in large changes in reaction velocity (10^{-5} and 10^{-3}, respectively), but only small changes in substrate orientation as revealed by cryo-crystallography.[52]

4.09.4.2.2 Torsion angle rotations

Small ligand molecules often have several rotatable bonds and flexible ring systems. An investigation of features influencing drug-likeness in chemical databases showed that features describing molecular flexibility could distinguish drug-like from non-drug-like molecules with drug-like molecules tending to be more rigid, having $\geqslant 18$ rigid bonds and $\geqslant 3$ rings but also $\geqslant 6$ rotatable bonds.[54] A recent conformational analysis of drug-like ligands binding to proteins shows that many ligands do not bind in a minimum energy conformation.[55] Energetically unfavorable conformational rearrangements can be tolerated in some cases without penalizing the tightness of binding. A correlation is observed between acceptable strain energy and the number of rotatable bonds, while there is no correlation between strain energy and binding affinity. Moreover, analysis of the degree of folding of the bound ligands confirm the general tendency of flexible molecules to bind in an extended conformation.[55] Unsurprisingly, the handling of the conformational flexibility of small molecules is crucial in computational docking techniques and is implemented in most of the current receptor–ligand docking packages.

Motions of protein side chains occur on the pico- to millisecond timescale with displacements of up to 5 Å.[50] These conformational changes may involve rotation of side chains between different rotamers as well as smaller movements necessary to accommodate ligands. Also, movement of side chains can be associated with changes in main-chain conformation. The flexibility scale of amino acid residues as derived from a nonredundant data set of complexed and uncomplexed protein structures has the following order: Lys > Arg, Gln, Met > Glu, Ile, Leu > Asn, Thr, Val, Tyr, Ser, His, Asp > Cys, Trp, Phe.[56] Zavodszky and Kuhn studied side chain flexibility upon protein–ligand binding in 63 complexes that do not undergo major changes in main-chain conformation. On average, 95% of side chain rotations were smaller then 45°, which is not large enough to allow transfer to another rotamer. However, with rigid protein structures, good docking could only be achieved for 46% of these ligands. The resulting ligand orientations were typically less accurate than those obtained when protein flexibility was included.[57] Similarly, in another study of conformational changes in 60 structures of apo and substrate-bound enzymes, mostly small motions were observed.[58] Nevertheless, the results indicated that modeling of side chain flexibility in ligand docking applications is important. Some difference was also observed between catalytic and binding residues as the latter tend to exhibit larger backbone motions, but both catalytic and binding residues show the same amount of side chain flexibility.[58,59]

It is widely accepted that upon association of two macromolecules (protein–protein binding) one or several residues often play the most important role. They are called hot spot or anchor residues.[60–62] Once the anchor residues have docked, an intermediate is formed that allows for additional intermolecular interactions to take place and latch side

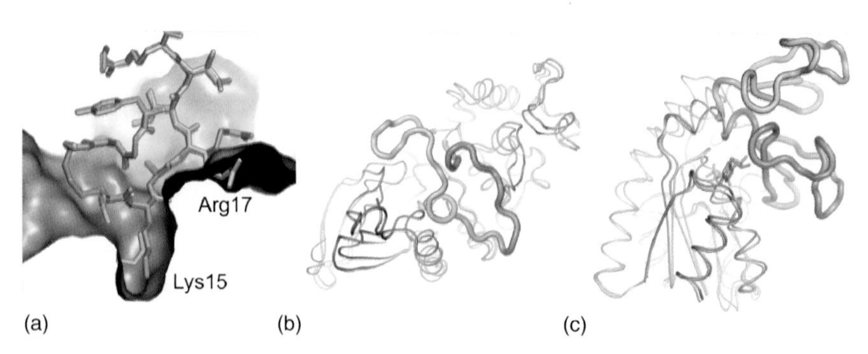

Figure 6 Types of conformational changes upon protein–ligand binding. (a) Side chain flexibility upon formation of the complex of β-trypsin with PBTI. PBTI unbound (PDB code 6PTI) (violet) superimposed onto PBTI bound (PDB code 2PTC) (green), unbound β-trypsin (PDB code 2PTN) (grey) superimposed onto bound complex and shown in surface representation. Residues undergoing significant conformational changes are labeled. (b) Large-scale loop rearrangement upon binding of CDK-2 to cyclin. Unbound CDK-2 (PDB code 1HCK, chain A) (violet) superimposed onto bound CDK-2 (PDB code 1FIN, chain A) (green). Flexible loop region indicated in thick tube representation. (c) ATP-lid domain movement in open versus closed adenylate kinase. Unbound adenylate kinase (PDB code 4AKE, chain A) (violet) superimposed onto bound adenylate kinase (PDB code 1ANK, chain B) (green). ATP-lid domain indicated in thick tube representation, ATP in stick representation.

chains to adopt the most favorable conformation.[62] Clearly, accurate modeling of the flexibility of such residues is important. Camacho and coworkers investigated anchor residues in 39 protein–protein complexes. Their 4 ns MD simulations indicate that the anchor side chains spend a substantial fraction of time (30–70%) in rotameric states that are close to the conformations of the side chain in the complex. Moreover, the preference for the bound state occurs in the absence of the binding partner.[62] A compilation of mutagenesis experiments by Bogan and Thorn[63] indicates that certain hot spot residues are far more important than others for the binding affinity of barstar and barnase. Ehrlich *et al.*,[64] in their docking studies of barstar and barnase, treated five hot spot residues as flexible by performing torsion angle dynamics. Docking with unbound protein structures showed that changing the rotamers of just two hot spot residues to their bound rotamers could have a big effect, resulting in faster and closer association of the proteins. Another example is the well-studied β-trypsin/PBTI complex: the unbound structure of the inhibitor cannot be docked correctly with rigid-body techniques due to the wrong conformations of key side chains such as Lys15 and Arg17 (**Figure 6a**).[65]

4.09.4.2.3 Main-chain and loop motions in proteins

Movement of main-chain and peptide loops is another level of flexibility (**Figure 6b**). Numerous examples of their functional importance are reported. Peptide loops play roles in ligand binding,[66] protein–protein association,[64] signaling cascades,[67] DNA-binding,[68] enzyme catalytic activity,[69] etc. The functional importance of the loops comes from their large flexibility. Unsurprisingly, it is often difficult to resolve these structures with x-ray crystallography or nuclear magnetic resonance (NMR). Nevertheless, to model biological macromolecules accurately, computational methods have to include loops and account for their high flexibility. The ArchDB database contains structurally classified loops extracted from known protein structures and is a good resource for functional and structural studies of protein loops, including comparative modeling, fold recognition, and ab initio prediction.[70,71] Analysis of the conformational changes upon complex formation, by comparing ligand-free and ligand-bound crystal structures, showed that substantial main-chain conformational changes, although less frequent than side chain changes, do occur in certain proteins.[51,56,72] For example, investigation of the interaction of four SH3 domains with the human immunodeficiency virus-1 (HIV-1) Nef protein showed that the RT loop of the SH3 domain has to be highly flexible to form an interaction with the Nef binding site.[66] Another well-studied example is peptidyl-prolyl *cis*/*trans*-isomerase CypA, which catalyzes the interconversion between *cis* and *trans* conformations of X-Pro peptide bonds. The major player in this catalysis is Arg55, for which the observed changes in backbone conformation are likely to be coupled with motions of the catalytically essential side chain.[73,74]

4.09.4.2.4 Large-scale (low-frequency) domain motions in proteins

Large proteins are often built from domains. Large-scale movements occurring between domains are often classified into hinge and shear motions.[75] These motions are controlled by at least one physical hinge. The termini of secondary structure elements (α-helix, β-sheet), as well as short loops, often serve as hinge regions. The center of an interdomain helix was also identified as a connecting region in several cases.[76] Detailed studies of ligand-induced domain closure in five classical enzymes suggested the following scenario: a ligand binds to a dedicated binding domain, and then makes the specific interactions with bending regions and with regions on the closing domain necessary to induce closure.[77] One of the studied examples, adenylate kinase, catalyzes the reaction ATP+AMP→2 ADP. It has a three-domain structure comprising a central core domain connected to two domains on either side: the AMP binding domain, which undergoes a 42° bending, and the ATP-lid domain, which undergoes a 51° bending upon ligand binding (**Figure 6c**). Most of the information on domain movements comes from x-ray crystallography when protein structures are solved in an open conformation without substrate and a closed state with a ligand bound. The DynDom program[78,79] takes two such conformations of the protein and analyzes the conformational change in terms of dynamic domains, interdomain screw axis, and interdomain bending regions. From results of the DynDon program, a database of protein domain motions has been developed to store such movements.[80,81] Another freely accessible database MolMovDB[82] stores data and software relevant to flexibility in proteins and RNA structures. It contains visual representations of different states of molecules as well as textual annotation and information from the literature relating to the motion. It is arranged around a multilevel classification scheme and includes motions of loops, domains, and subunits. A further database is DSMM, which permits location of movies showing biomolecular motions simulated with various computational techniques.[83]

4.09.4.3 Current Models of Flexible Receptor–Ligand Binding

There are several models that describe biomolecular binding mechanisms. The oldest, the lock-and-key (**Figure 1a**) model, assumes that one interacting molecule has a shape that is perfectly matched by another. This model does not account for the flexibility of biological molecules. It also cannot explain how the same protein can bind different ligands with different shapes. Therefore, the induced-fit[3] (**Figure 1b**) model was introduced, which assumes a certain degree of plasticity in binding sites that permits the receptor to accommodate the incoming ligand. In this model, the ligand induces a conformational change at the receptor binding site shifting it toward a bound state.[84] However, the induced fit pathway is kinetically competent only if the ligand and unbound receptor have some degree of pre-existing complementarity, otherwise the initial complex is too short-lived.[85] Another model describes interacting molecules as existing in an ensemble (**Figure 1c**) of conformational states.[4,86–89] The native state of a protein displays an ensemble of conformations, dependent on the topological properties of the protein. Perturbing the protein topology, for instance by mutations, may result in a redistribution of the ensemble populations, even without a change in the flexible regions.[88] In this model, ligands can bind to different conformations and affect the equilibrium of the conformational ensemble. It has been observed that protein ligands with composition, size, and shape that differ from those of the natural ligands may bind with an equal or even higher affinity at the same site.[90–92] Thus, ligands can fall into different categories of different binding affinity and preference for different receptor conformations within an ensemble.[93] The binding process is therefore guided by both the receptor population and by the ligands present in solution, via conformational selection.[88] Actually, a combination of conformational selection and induced fit provides a good description for many interacting molecules. First, preselection between partly fitting structures of a conformational ensemble occurs and this is followed by minor readjustments to the final stable complex.[85]

For proteins that exhibit allosteric behavior, the dynamic population shift model has been proposed.[41] It assumes that proteins exist in a population of conformations. Upon binding, the probability distribution of the ensemble of native states will be redistributed, changing the stability of certain residues throughout the protein molecule and propagating a conformational change at specific residues.[84] For example, for unphosphorylated NtrC, both active and inactive conformations are evident. However, upon phosphorylation, the protein is activated and the equilibrium is shifted toward the active conformation.[94]

Many proteins can bind to structurally diverse sets of ligands. For example, four crystal structures of complexes of HIV-1 reverse transcriptase show a common mode of binding for chemically diverse inhibitors within much the same volume.[95] The BmrR transcriptional regulator of the *Bacillus subtilis* Bmr multidrug transporter, and the Bmr transporter itself, bind multiple structurally dissimilar hydrophobic cations. This binding diversity is thought to arise due to the protein flexibility and because binding affinity is achieved by electrostatic attraction (from a negatively charged glutamic acid residue (134) inside the hydrophobic environment of the protein core), and favorable van der Waals', stacking, and hydrophobic interactions. Notably, and in contrast to the specific anion-binding proteins discussed above,[34] no hydrogen bonds are involved in the binding of inducers by the BmrR.[91]

Different binding modes of structurally similar ligands to one target are also observed. For example, flufenamic acid and diclofenamic acid bind to human transthyretin in opposing orientations.[96]

4.09.4.4 Computational Models

As explained above, flexibility on various scales plays an important role in biomolecular association and function, and therefore has to be accounted for in computational methods. Many computational representations of biomolecules and simulation algorithms have been developed to address the different levels of flexibility with different degrees of detail.

4.09.4.4.1 Implicit models

A simple way to represent a limited degree of flexibility is to use a soft representation of molecules. This can be done by reducing the van der Waals' radii of atoms, or using a softer energy function to allow some degree of atomic clash, or allowing certain atoms to completely penetrate each other.[97–100] The main advantages of these methods are easy implementation and computational efficiency. They can thus be used in the first stages of docking, namely extensive rigid body sampling. A disadvantage is that this approach can address only small conformational changes. Most of the current docking methods have means to account for flexibility in this way.

Ferrari and colleagues[100] evaluated the performance of a soft function in ligand docking versus docking with a hard function and with one or many receptor conformations. Softening was achieved by reducing the steepness of the repulsive term in the Lennard-Jones potential from $1/r^{12}$ to $1/r^{9}$. The parameters in the Lennard-Jones equation were weighted to minimize the effect on the magnitude and position of the energy minimum. The ability of the scoring function to identify known ligands from among 200 000 decoy molecules in the database was tested. The soft potential was always better at identifying known ligands than the hard scoring function when only a single receptor conformation was used. However, the soft function was worse at identifying known leads than the hard function when multiple receptor conformations were used.[100]

Some molecular representations simply exclude particular atoms (e.g., hydrogen), or trim the side chains of flexible amino acid residues.[101–103] The LIGIN[104] program estimates the degree of overlap between the ligand and the side chain of every residue that could move out of the binding site if the residue were to adopt another rotamer. Then, for a user defined number of residues with the highest overlap, the side chains are excluded from the complementarity scoring function. For these residues, only the backbone and Cβ atoms are taken into account. Heifetz and Eisenstein investigated the effect of residue trimming in rigid-body protein–protein docking. The side chains of the most flexible surface accessible residues Lys, Arg, and Gln were trimmed. This lowered the scores of the false-positive solutions, which resulted in better predictions of protein–protein complexes.[101]

Another way to implicitly account for molecular flexibility is to use a reduced representation of the proteins. Li and coworkers[105] used a simplified protein model to predict the structure of antibody–antigen complexes. Each amino acid residue was represented as a sphere of a specific radius,[106] with further modification of the most flexible residues Arg, Lys, Asp, Glu, and Met occurring at the protein surface.[61] Zacharias used a reduced protein model with up to three pseudo atoms per amino acid residue. One pseudo atom represented the protein backbone (located at Cα position) while another one or two represented the side chain. The effective interaction between pseudo atoms is described by a soft distance-dependent Lennard-Jones potential. In contrast to purely residue-based potentials, the description of side chains by two pseudo atoms allows the chemical character of some side chains to be better accounted for. For example, lysine can be considered as an amino acid with a hydrophobic tail and a charged hydrophilic head group.[107,108] This method, implemented into the program ATTRACT, produced several good-quality protein–protein complex predictions in the CAPRI (critical assessment of predicted interactions) 3–5 competitions.[109]

4.09.4.4.2 Selected rotational degrees of freedom

To avoid the complexity of modeling full molecular flexibility, the flexibility of a specific subset of the degrees of freedom can be treated explicitly. It is common to permit rotations around single bonds and consider several conformations to represent the flexibility of ring systems. The initial ligand conformation can be taken from available experimental structures or generated computationally. The selection of the degrees of freedom can be done based on experimentally determined conformations of molecules or can be derived from simulations. In FlexX,[110] the conformational flexibility of the ligand is modeled by a discrete set of preferred torsion angles at acyclic single bonds, and multiple conformations for ring systems. The torsional angles are taken from a database derived from the Cambridge Structure Database (CSD)[111] by Klebe and Mietzner.[112] Multiple conformations for rings are computed with the program CORINA.[113] The GOLD program[114] uses a genetic algorithm to model flexibility in molecular docking. Ligand flexibility is encoded in a chromosome along with potential hydrogen bond interactions. The genetic algorithm then samples the ligand conformation, orientation, and position in the active site and uses a fitness function to select poses in the docking. In the refinement stage of the SPECITOPE program, each single bond of the ligand and receptor side chains is considered as rotatable[115] If overlaps between the ligand and a protein atom occur, rotation around the single bond closest to the bumping atoms is used to resolve the overlap. If it is not possible to resolve an overlap by rotating a ligand side chain, the same approach is applied to a protein side chain involved in the collision.

Probably the most common way to model side chain conformations in a protein is to use rotamer libraries. A rotamer, short for rotational isomer, is a single side chain conformation represented as a set of values, one for each dihedral angle degree of freedom. A rotamer is usually thought to be a local minimum on a potential energy map or an average conformation over some region of dihedral angle space.[116] Although rotamers do not account for the complete flexibility of side chains, they include the orientations with the lowest energies and provide a good estimate of available conformations. Moreover, the use of rotamers is computationally quite efficient.[117,118] Typically, the rotamer libraries are built from an analysis of protein structures from the Protein Data Bank (PDB).[119–121,170] In order to maximize the quality of the rotamer library, structures are carefully filtered. For instance, Lovell and coworkers[121] used only structures with at least 1.7 Å resolution, an R-factor of 20%, and a clash-score <30.[121,122] Rotamer libraries can be backbone-independent or backbone-dependent. The distinctions are made according to whether the dihedral angles of the rotamers and/or their frequencies depend on the local backbone conformation or not.[116]

In one of the early applications of a rotamer library, Leach[123] used single conformations for Gly, Ala, and Pro, 14 side chains were assigned 3–10 rotamers, and Met, Lys, and Arg sampled 21, 51, and 55 rotamers, respectively. More recent methods use larger libraries and more complicated algorithms. Peterson and coworkers[124] built a library that contains 49 042 rotamers and used them to place amino acid residues on an accurate backbone trace with high success. Baker and coworkers used a side chain placement method that employs a simulated annealing algorithm to search through backbone-dependent rotamers.[125] The method also includes the option to expand the standard rotamer library for each residue by including sub-rotamers, i.e., the major rotamer angles + and − 1 standard deviation of those angles. Moreover, the side chain conformations in the unbound proteins are appended to the rotamer library. Off-rotamer side chain conformations are sampled by torsion space minimization.[126] This protocol, implemented in the RosettaDock program, produced promising results in protein–protein docking in CAPRI rounds 3–5.[127]

Alternatively, or together with rotamer libraries, various minimization techniques can be applied to optimize the fit of side chains between receptor and ligand. Often minimization is carried out only for the interface region while keeping the rest of the molecules fixed. Optimization routines can include simulated annealing, steepest-descent minimization, molecular dynamics (MD), Monte Carlo sampling, and other methods. The docking protocol implemented in the HADDOCK program includes a stage of semi-rigid simulated annealing in torsion angle space during which side chains at the interface are allowed to move.[128] Camacho showed that modeling side chains based on the rotamer conformations sampled during MD simulations in explicit solvent of the isolated proteins improves recognition by refining key side chains into native-like conformations amenable for rigid body docking.[129]

4.09.4.4.3 Ensembles of multiple structures

Another way to represent molecular flexibility, including large-scale global motions, is to use multiple structures. An important question is how to combine and represent them all in the docking algorithm. The simplest way is to dock all ligands to all available receptor conformations.[130] However, this may be too expensive computationally and applicable only for small-scale studies. Alternatively, multiple conformers can be represented by combining multiple protein structures or combining probe molecular interaction field maps for multiple protein structures to identify regions conserved in all the structures.[131] FlexE[132] creates an average structure for parts of the proteins, but treats the flexible side chains with a rotamer library. DOCK[133] uses composite grids of two types to represent structural ensembles: geometry-weighted averaged on the structural level and energy-weighted averaged on the ligand–protein interaction energy. Osterberg and coworkers[134] tested four methods to combine multiple target structures within a single interaction energy grid in their docking studies with AutoDock.

An important consideration when combining the protein structures is the dual character of the binding pockets. Binding sites of some proteins are characterized by the presence of regions of high structural stability and regions of low structural stability.[41,86,135] Another important question is how many and which molecular conformers should be included into the ensemble to represent sufficiently molecular flexibility. An obvious source of protein conformations is structures determined experimentally by NMR or x-ray crystallography. However, not many proteins have multiple structures solved by experimental methods; furthermore, using experimental approaches to solve multiple structures is expensive and time-consuming. Thus, computational methods provide an important route to generating multiple structures.

In their study of ligand docking to multiple x-ray conformations of cyclin-dependent kinase 2 (CDK-2) and of heat shock protein 90 (hsp90), Barril and Morley[136] identified two main pitfalls in flexible receptor docking. First, different receptor conformations may not be equally favorable and therefore should not be considered as isoenergetic. Second, the different structures show different binding cavities on the receptor that compete with each other for ligands and are scored relative to each other. Consequently, the results can be sensitive to the adverse effects of an unrealistic or unusual receptor binding site. Both of these issues are likely to be more apparent in docking with computer-generated conformations. Smith and colleagues[137] used MD to investigate the conformational fluctuations of 41 proteins that form binary protein–protein complexes. In each case, 5 ns trajectories were run starting from unbound conformations. The resulting conformations were compared with the protein structures in the complex. It was found that some parts of the proteins do visit conformations close to those observed in bound structures, but at no point does any protein sample its complete bound conformation.

To increase the representation of conformational space by the structural ensemble, conventional MD can be coupled with other methods. Abseher and Nilges[138] first ran 10 independent MD trajectories of 50 ps length and then used principal component restrained simulation (PCR-MD)[138] to generate a more diverse ensemble. The cross docking of ensemble snapshots derived from these simulations of the two partner proteins increased the chances of finding near native solutions. Remarkably, the docking performance of the given combination of receptor and ligand structures was largely uncorrelated with their similarity to the bound conformation. Therefore, the authors proposed that protein–protein binding follows a three-step mechanism of diffusion, selection of free conformers, and refolding.[139] Tatsumi and coworkers[140] used a mixed approach: the global motions of the receptor were represented via harmonic dynamics (HD) motions derived from a preceding MD run, whereas the local flexibility of side chains was modeled by conventional MD.[140]

Using MD to generate an ensemble of structures is computationally demanding. The CONCOORD[141] program can efficiently generate protein conformations around known structures based on geometric restrictions. First, it identifies all interatomic interactions in the starting structure and classifies them depending on their strength. A set of upper and lower geometric bounds is obtained for all interacting pairs of atoms. Second, new structures are generated randomly and corrections are applied until all bounds are satisfied. Mustard and Ritchie used CONCOORD followed by essential dynamics (ED) analysis to generate an ensemble of protein structures sampling the eigenvector directions that were then used in rigid-body protein–protein docking.[142]

Often, the major conformational change upon ligand binding is localized to a specific area of the protein, such as the bending of a loop. In this case, it may be sufficient to generate a conformational ensemble for just the relevant flexible region. Bastard and coworkers[143] generated multiple conformations of a protein loop involved in DNA binding and then used Monte Carlo conformational searches of the relative positions of the protein and DNA and their side chain conformations to iteratively adjust the weight of each loop copy and derive a model of the protein–DNA complex. Cavasotto and colleagues used a normal mode approach to generate multiple receptor structures with the focus on functionally important regions. Interestingly, intermediate-scale loop motions, which occur upon ligand binding to protein kinases, could not be represented by picking the first lowest modes. To overcome this limitation, a measure of the relevance was introduced to select just the most relevant modes for the representation of flexible loops important in ligand binding in protein kinases.[144]

Graph theory algorithms, as incorporated in the program FIRST,[72] provide another approach to identify rigid and flexible regions of the protein structure based on a bond network consisting of covalent bonds, salt bridges, hydrogen bonds, and hydrophobic interactions.[145] This information can then provide geometric constraints for programs that generate conformations to satisfy these constraints.[146–148]

4.09.4.4.4 Water molecules

In addition to the flexible rearrangements of the receptor and ligand that take place upon binding, solvent molecules can also be involved in the process and have a functional role. Water molecules are associated with the native structures of the proteins and in many cases they are implicated as having a direct role in molecular recognition and catalysis.[149] A study of hydrogen bonds and salt bridges in 319 nonredundant high-quality protein–protein interfaces[150] indicated that the quality of the hydrogen bonds in the interfaces is not as good as that generally observed within the chains, and that water molecules mediate noncomplementary donor–donor and acceptor–acceptor pairs, and connect nonoptimally oriented donor–acceptor pairs. Water molecules are often found at DNA–protein interfaces and, from their analysis of the interfacial hydration of the BamHI protein–DNA complex, Fuxreiter and coworkers[151] found evidence that the sequence-specific structure of the water can serve as a hydration fingerprint of a given DNA sequence. Clearly, computational modeling methods should have the means to explicitly incorporate water molecules, when investigating molecular binding processes where water molecules have significant impact on molecular recognition and binding energetics.

Jiang and colleagues[152] constructed a solvated protein rotamer library by adding water molecules to the rotamers of the Dunbrack rotamer library[116] based on the observed distribution of water molecules around backbone and polar side chain atoms. Their free energy function included protein–water interactions and was tested for predicting the positions of water molecules at protein–protein interfaces and the energetic effects on binding. It showed significant improvements when used with crystallographically defined water molecules, and small improvements when used with predicted water sites.[152] The GRID [14]program was also used to predict water sites in ligand binding sites that are used in calculation of free energies of protein–ligand binding[153] and prediction of the binding modes of ligands in crystallographic complexes.[154] An extension to the algorithm of FlexX,[110] called the particle concept,[155] allows inclusion of discrete water molecules during ligand docking. It features the placement, while constructing the ligand–protein complex, of spherical particles (representing water molecules) at favorable positions between the ligand and the protein. The particles make hydrogen bonds with the ligand and the protein, and contribute to the scoring function. Although the overall improvement to docking with this approach was small, it was found to be useful in specific test cases like HIV-1 protease where a single water molecule is known to play a critical role.[155]

4.09.5 Concluding Remarks, Challenges, and Outlook

Ligand–receptor binding is determined by the balance of different energetic contributions: van der Waals', electrostatic, hydrogen-bonding, and the hydrophobic effect. The bulk solvent plays an important part in determining the strength of these interactions. In addition, individual water molecules and ions can be critical in mediating ligand–receptor interactions. Molecular flexibility is crucial for many molecular recognition and binding events, and also has entropic consequences for binding. Biomacromolecular receptors display a wide variety of ligand-binding determinants, meaning that a range of strategies can be adopted to design ligands for receptors.

Although much is known about the factors determining ligand–receptor binding, the prediction of the structures and dynamics of docked complexes and the computation of their binding kinetics and thermodynamics remains difficult. This is primarily due to the large number of degrees of freedom of biological molecules in aqueous solution. Sampling of these degrees of freedom is computationally demanding and this, in particular, hinders the accurate estimation of entropic contributions. Thus, the challenge of computational studies is to build simplified, computationally tractable models of sufficient accuracy to provide useful results (e.g., structures correct to ca. 1 Å RMSD and free energies correct to within 1 kcal mol^{-1}). The choice of suitable simplified models requires careful decisions based on knowledge of the system at hand. Sampling and scoring procedures suitable for studying one ligand–receptor system may not be appropriate or may need to be adapted to study another ligand–receptor system, due to the different relative importance of the physical factors governing binding. For example, target-specific scoring functions for ligand–receptor docking and estimating binding affinities are becoming increasingly common.[156–158] These scoring functions have coefficients/parameters that are adjusted using a training set for the given target before being applied to study the binding of further ligands to the target.

In general, different levels of model are required according to the type of information to be derived. High-throughput virtual screening models must be simple and very computationally efficient but can acceptably yield a high number of false positives. Interesting molecules resulting from such screens can be subjected to more detailed and quantitative simulations. Improvements in all levels of computational models are necessary. Some areas of on-going improvement include the treatment of changes in polarization and protonation upon binding, binding to metal ions, solvation effects, and conformational sampling. These aspects are discussed in other chapters.

Acknowledgments

Financial support from the Klaus Tschira Foundation is gratefully acknowledged.

References

1. Fischer, E. *Ber. Deutsch Chem. Ges.* **1894**, *27*, 2985–2993.
2. Burgen, A. S. V.; Roberts, G. C. K.; Feeny, J. *Nature* **1975**, *253*, 753–755.
3. Koshland, D. E., Jr. *Proc. Natl. Acad. Sci. USA* **1958**, *44*, 98–123.
4. J Tsai, C.; Ma, B.; Nussinov, R. *Proc. Natl. Acad. Sci. USA* **1999**, *96*, 9970–9972.
5. Teague, S. J. *Nat. Rev. Drug Disc.* **2003**, *2*, 527–541.
6. Schames, J. R.; Henchman, R. H.; Siegel, J. S.; Sotriffer, C. A.; Ni, H.; McCammon, J. A. *J. Med. Chem.* **2004**, *47*, 1879–1881.
7. Prade, L.; Jones, A. F.; Boss, C.; Richard-Bildstein, S.; Meyer, S.; Binkert, C.; Bur, D. *J. Biol. Chem.* **2005**, *280*, 23837–23843.
8. Lennard-Jones, J. E. *Cohesion Proc. Phys. Soc.* **1931**, *43*, 461–482.
9. Lennard-Jones, J. E. *Proc. R. Soc. London, Ser. A.* **1924**, *106*, 463–477.
10. Slater, J. C.; Kirkwood, J. G. A. *Phys. Rev.* **1931**, *37*, 682–686.
11. Lee, B.; Richards, F. M. *J. Mol. Biol.* **1971**, *55*, 379–400.
12. Binkowski, T. A.; Naghibzadeh, S.; Liang, J. *Nucleic Acids Res.* **1998**, *31*, 3352–3355.

13. Liang, J.; Edelsbrunner, H.; Woodward, C. *Protein Sci.* **1998**, *7*, 1884–1897.
14. Goodford, P. J. *J. Med. Chem.* **1985**, *28*, 849–857.
15. Chen, J. M.; Xu, S. L.; Wawrzak, Z.; Basarab, G. S.; Jordan, D. B. *Biochemistry* **1998**, *37*, 17735–17744.
16. Fogolari, F.; Brigo, A.; Molinari, H. *J. Mol. Recognit.* **2002**, *15*, 377–392.
17. Baker, N. A. *Methods Enzymol.* **2004**, *383*, 94–118.
18. Bashford, D. *Front Biosci.* **2004**, *9*, 1082–1099.
19. Neves-Petersen, M. *Biotechnol. Annu. Rev.* **2003**, *9*, 315–395.
20. Demchuk, E.; Wade, R. C. *J. Phys. Chem.* **1996**, *100*, 17373–17387.
21. Schutz, C.; Warshel, A. *Proteins* **2001**, *44*, 400–417.
22. Wang, T.; Tomic, S.; Gabdoulline, R. R.; Wade, R. C. *Biophys. J.* **2004**, *12*, 1563–1574.
23. Dong, F.; Vijayakumar, M.; Zhou, H.-Y. *Biophys. J.* **2003**, *85*, 49–60.
24. Green, D. F.; Tidor, B. *Proteins* **2005**, *60*, 644–657.
25. Gabdoulline, R. R.; Wade, R. C. *J. Phys. Chem.* **1996**, *100*, 3868–3878.
26. Still, W. C.; Tempczyk, A.; Hawley, R. C.; Hendrickson, T. *J. Am. Chem. Soc.* **1990**, *112*, 6127–6129.
27. Bashford, D.; Case, D. A. *Annu. Rev. Phys. Chem.* **2000**, *51*, 129–152.
28. Fukuzawa, K.; Kitaura, K.; Uebayasi, M.; Nakata, K.; Kaminuma, T.; Nakano, T. *J. Comput. Chem.* **2005**, *26*, 1–10.
29. McCoy, A. J.; Chandana Epa, V.; Colman, P. M. *J. Mol. Biol.* **1997**, *268*, 570–584.
30. Katz, B. A.; Elrod, K.; Verner, E.; Mackman, R. L.; Luong, C.; Shrader, W. D.; Sendzik, M.; Spencer, J. R.; Sprengeler, P. A.; Kolesnikov, A. *J. Mol. Biol.* **2003**, *329*, 93–120.
31. Wade, R. C. *Structure* **1997**, *5*, 1139–1146.
32. von Itzstein, M.; Wu, W. Y.; Kok, G. B.; Pegg, M. S.; Dyason, J. C.; Jin, B. Q.; Van Phan, T.; Smyth, M. L.; White, H. F.; Oliver, S. W. *Nature* **1993**, *363*, 418–423.
33. Kim, C. U.; Lew, W.; Williams, M. A.; Liu, H.; Zhanĝ, L.; Swaminathan, S.; Bischofberger, N.; Chen, M. S.; Mendel, D. B.; Tai, C. Y. et al. *J. Am. Chem. Soc.* **1997**, *119*, 681–690.
34. Ledvina, P. S.; Tsai, A. L.; Wang, Z.; Koehl, E.; Quiocho, F. A. *Protein Sci.* **1998**, *7*, 2550–2559.
35. Koellner, G.; Kryger, G.; Millard, C. B.; Silman, I.; Sussman, J. L.; Steiner, T. *J. Mol. Biol.* **2000**, *296*, 713–735.
36. Lipinski, C. A.; Lombardo, F.; Dominy, B. W.; Feeney, P. J. *Adv. Drug Deliv. Rev.* **1997**, *23*, 3–25.
37. Eisenberg, D.; McLachlan, A. D. *Nature* **1986**, *319*, 199–203.
38. Gilson, M. K.; Given, J. A.; Bush, B. L.; McCammon, J. A. *Biophys. J.* **1997**, *72*, 1047–1069.
39. Murray, C. W.; Verdonk, M. L. *J. Comput.-Aided Mol. Des.* **2002**, *16*, 741–753.
40. Lu, B.; Wong, C. *Biopolymers* **2005**, *79*, 277–285.
41. Freire, E. *Proc. Natl. Acad. Sci. USA* **1999**, *96*, 10118–10122.
42. Stone, M. J. *Acc. Chem. Res.* **2001**, *34*, 379–388.
43. Shoemaker, B. A.; Portman, J. J.; Wolynes, P. G. *Proc. Natl. Acad. Sci. USA* **2000**, *97*, 8868–8873.
44. Tidor, B.; Karplus, M. *J. Mol. Biol.* **1994**, *238*, 405–414.
45. Luo, H.; Sharp, K. *Proc. Natl. Acad. Sci. USA* **2002**, *99*, 10399–10404.
46. Cole, C.; Warwicker, J. *Protein Sci.* **2002**, *11*, 2860–2870.
47. Dunitz, J. D. *Science* **1997**, *264*, 670.
48. Garcia, A. E.; Hummer, G. *Proteins* **2000**, *38*, 261–272.
49. Ohtaka, H.; Freire, E. *Prog. Biophys. Mol. Biol.* **2005**, *88*, 193–208.
50. Brooks, C. L.; Karplus, M.; Pettitt, B. M. *Proteins: A Theoretical Perspective of Dynamics, Structure, and Thermodynamics*; Wiley: New York, Chichester, UK, 1988.
51. Betts, M. J.; Sternberg, M. J. *Protein Eng.* **1999**, *12*, 271–283.
52. Mesecar, A. D.; Stoddard, B. L.; Koshland, D. E., Jr. *Science* **1997**, *277*, 202–206.
53. Koshland, D. E., Jr. *Nat. Med.* **1998**, *4*, 1112–1114.
54. Oprea, T. I. *J. Comput.-Aided Mol. Des.* **2000**, *14*, 251–264.
55. Perola, E.; Charifson, P. S. *J. Med. Chem.* **2004**, *47*, 2499–2510.
56. Najmanovich, R.; Kuttner, J.; Sobolev, V.; Edelman, M. *Proteins* **2000**, *39*, 261–268.
57. Zavodszky, M. I.; Kuhn, L. A. *Protein Sci.* **2005**, *14*, 1104–1114.
58. Gutteridge, A.; Thornton, J. *J. Mol. Biol.* **2005**, *346*, 21–28.
59. Gutteridge, A.; Thornton, J. *FEBS Lett.* **2004**, *567*, 67–73.
60. Hu, Z.; Ma, B.; Wolfson, H.; Nussinov, R. *Proteins* **2000**, *39*, 331–342.
61. Li, C. H.; Ma, X. H.; Chen, W. Z.; Wang, C. X. *Sheng Wu Hua Xue Yu Sheng Wu Wu Li Xue Bao (Shanghai)* **2003**, *35*, 35–40.
62. Rajamani, D.; Thiel, S.; Vajda, S.; Camacho, C. J. *Proc. Natl. Acad. Sci. USA* **2004**, *101*, 11287–11292.
63. Bogan, A. A.; Thorn, K. *J. Mol. Biol.* **1998**, *280*, 1–9.
64. Ehrlich, L. P.; Nilges, M.; Wade, R. C. *Proteins* **2005**, *58*, 126–133.
65. Lorber, D. M.; Udo, M. K.; Shoichet, B. K. *Protein Sci.* **2002**, *11*, 1393–1408.
66. Arold, S.; O'Brien, R.; Franken, P.; Strub, M. P.; Hoh, F.; Dumas, C.; Ladbury, J. E. *Biochemistry* **1998**, *37*, 14683–14691.
67. Zomot, E.; Kanner, B. I. *J. Biol. Chem.* **2003**, *278*, 42950–42958.
68. Inga, A.; Monti, P.; Fronza, G.; Darden, T.; Resnick, M. A. *Oncogene* **2001**, *20*, 501–513.
69. Lee, M. C.; Deng, J.; Briggs, J. M.; Duan, Y. *Biophys. J.* **2005**, *88*, 3133–3146.
70. Espadaler, J.; Fernandez-Fuentes, N.; Hermoso, A.; Querol, E.; Aviles, F. X.; Sternberg, M. J.; Oliva, B. *Nucleic Acids Res.* **2004**, *32*, D185–D188.
71. Oliva, B.; Bates, P. A.; Querol, E.; Aviles, F. X.; Sternberg, M. J. *J. Mol. Biol.* **1997**, *266*, 814–830.
72. Lei, M.; Zavodszky, M. I.; Kuhn, L. A.; Thorpe, M. F. *J. Comput. Chem.* **2004**, *25*, 1133–1148.
73. Eisenmesser, E. Z.; Bosco, D. A.; Akke, M.; Kern, D. *Science* **2002**, *295*, 1520–1523.
74. Kallen, J.; Mikol, V.; Taylor, P.; Walkinshaw, M. D. *J. Mol. Biol.* **1998**, *283*, 435–449.
75. Gerstein, M.; Lesk, A. M.; Chothia, C. *Biochemistry* **1994**, *33*, 6739–6749.
76. Hayward, S. *Proteins* **1999**, *36*, 425–435.
77. Hayward, S. *J. Mol. Biol.* **2004**, *339*, 1001–1021.
78. Hayward, S.; Berendsen, H. J. *Proteins* **1998**, *30*, 144–154.

79. Hayward, S.; Kitao, A.; Berendsen, H. J. *Proteins* **1997**, *27*, 425–437.
80. Qi, G.; Lee, R.; Hayward, S. *Bioinformatics* **2005**, *21*, 2832–2838.
81. Lee, R. A.; Razaz, M.; Hayward, S. *Bioinformatics* **2003**, *19*, 1290–1291.
82. Echols, N.; Milburn, D.; Gerstein, M. *Nucleic Acids Res.* **2003**, *31*, 478–482.
83. Finnochiaro, G.; Wang, T.; Hoffmann, R.; Gonzalez, A.; Wade, R. C. *Nucleic Acids Res.* **2003**, *31*, 456–457.
84. Goh, C. S.; Milburn, D.; Gerstein, M. *Curr. Opin. Struct. Biol.* **2004**, *14*, 104–109.
85. Bosshard, H. R. *News Physiol. Sci.* **2001**, *16*, 171–173.
86. Luque, I.; Freire, E. *Proteins* **2000**, *41*, 63–71.
87. Ma, B.; Kumar, S.; Tsai, C. J.; Nussinov, R. *Protein Eng.* **1999**, *12*, 713–720.
88. Ma, B.; Shatsky, M.; Wolfson, H. J.; Nussinov, R. *Protein Sci.* **2002**, *11*, 184–197.
89. Pan, H.; Lee, J. C.; Hilser, V. J. *Proc. Natl. Acad. Sci. USA* **2000**, *97*, 12020–12025.
90. DeLano, W. L.; Ultsch, M. H.; de Vos, A. M.; Wells, J. A. *Science* **2000**, *287*, 1279–1283.
91. Vazquez-Laslop, N.; Zheleznova, E. E.; Markham, P. N.; Brennan, R. G.; Neyfakh, A. A. *Biochem. Soc. Trans.* **2000**, *28*, 517–520.
92. Zwahlen, C.; Li, S. C.; Kay, L. E.; Pawson, T.; Forman-Kay, J. D. *EMBO J.* **2000**, *19*, 1505–1515.
93. Carlson, H. A. *Curr. Pharm. Des.* **2002**, *8*, 1571–1578.
94. Volkman, B. F.; Lipson, D.; Wemmer, D. E.; Kern, D. *Science* **2001**, *291*, 2429–2433.
95. Ren, J.; Esnouf, R.; Garman, E.; Somers, D.; Ross, C.; Kirby, I.; Keeling, J.; Darby, G.; Jones, Y.; Stuart, D. et al. *Nat. Struct. Biol.* **1995**, *2*, 293–302.
96. Klabunde, T.; Petrassi, H. M.; Oza, V. B.; Raman, P.; Kelly, J. W.; Sacchettini, J. C. *Nat. Struct. Biol.* **2000**, *7*, 312–321.
97. Ausiello, G.; Cesareni, G.; Helmer-Citterich, M. *Proteins* **1997**, *28*, 556–567.
98. Norel, R.; Petrey, D.; Wolfson, H. J.; Nussinov, R. *Proteins* **1999**, *36*, 307–317.
99. Oostenbrink, C.; van Gunsteren, W. F. *Proc. Natl. Acad. Sci. USA* **2005**, *102*, 6750–6754.
100. Ferrari, A. M.; Wei, B. Q.; Costantino, L.; Shoichet, B. K. *J. Med. Chem.* **2004**, *47*, 5076–5084.
101. Heifetz, A.; Eisenstein, M. *Protein Eng.* **2003**, *16*, 179–185.
102. Palma, P. N.; Krippahl, L.; Wampler, J. E.; Moura, J. J. *Proteins* **2000**, *39*, 372–384.
103. Carter, P.; Lesk, V. I.; Islam, S. A.; Sternberg, M. J. *Proteins* **2005**, *60*, 281–288.
104. Sobolev, V.; Wade, R. C.; Vriend, G.; Edelman, M. *Proteins* **1996**, *25*, 120–129.
105. Li, C. H.; Ma, X. H.; Chen, W. Z.; Wang, C. X. *Proteins* **2003**, *52*, 47–50.
106. Levitt, M. *J. Mol. Biol.* **1976**, *104*, 59–107.
107. Zacharias, M. *Protein Sci.* **2003**, *12*, 1271–1282.
108. Zacharias, M. *Proteins* **2005**, *60*, 252–256.
109. Mendez, R.; Leplae, R.; Lensink, M. F.; Wodak, S. J. *Proteins* **2005**, *60*, 150–169.
110. Rarey, M.; Kramer, B.; Lengauer, T.; Klebe, G. *J. Mol. Biol.* **1996**, *261*, 470–489.
111. Allen, F. H. *Acta Crystallogr., Sect. B* **2002**, *58*, 380–388.
112. Klebe, G.; Mietzner, T. *J. Comput.-Aided Mol. Des.* **1994**, *8*, 583–606.
113. Gasteiger, J.; Rudolph, C.; Sadowski, J. *Tetrah. Comput. Meth.* **1990**, *3*, 537–547.
114. Jones, G.; Willett, P.; Glen, R. C.; Leach, A. R.; Taylor, R. *J. Mol. Biol.* **1997**, *267*, 727–748.
115. Schnecke, V.; Swanson, C. A.; Getzoff, E. D.; Tainer, J. A.; Kuhn, L. A. *Proteins* **1998**, *33*, 74–87.
116. Dunbrack, R. L., Jr. *Curr. Opin. Struct. Biol.* **2002**, *12*, 431–440.
117. Carlson, H. A.; McCammon, J. A. *Mol. Pharmacol.* **2000**, *57*, 213–218.
118. Halperin, I.; Ma, B.; Wolfson, H.; Nussinov, R. *Proteins* **2002**, *47*, 409–443.
119. Dunbrack, R. L., Jr.; Karplus, M. *J. Mol. Biol.* **1993**, *230*, 543–574.
120. Dunbrack, R. L., Jr.; Cohen, F. E. *Protein Sci.* **1997**, *6*, 1661–1681.
121. Lovell, S. C.; Word, J. M.; Richardson, J. S.; Richardson, D. C. *Proteins* **2000**, *40*, 389–408.
122. Word, J. M.; Lovell, S. C.; LaBean, T. H.; Taylor, H. C.; Zalis, M. E.; Presley, B. K.; Richardson, J. S.; Richardson, D. C. *J. Mol. Biol.* **1999**, *285*, 1711–1733.
123. Leach, A. R. *J. Mol. Biol.* **1994**, *235*, 345–356.
124. Peterson, R. W.; Dutton, P. L.; Wand, A. J. *Protein Sci.* **2004**, *13*, 735–751.
125. Kuhlman, B.; Baker, D. *Proc. Natl. Acad. Sci. USA* **2000**, *97*, 10383–10388.
126. Wang, C.; Schueler-Furman, O.; Baker, D. *Protein Sci.* **2005**, *14*, 1328–1339.
127. Schueler-Furman, O.; Wang, C.; Baker, D. *Proteins* **2005**, *60*, 187–194.
128. Dominguez, C.; Boelens, R.; Bonvin, A. M. *J. Am. Chem. Soc.* **2003**, *125*, 1731–1737.
129. Camacho, C. J. *Proteins* **2005**, *60*, 245–251.
130. Bouzida, D.; Arthurs, S.; Colson, A. B.; Freer, S. T.; Gehlhaar, D. K.; Larson, V.; Luty, B. A.; Rejto, P. A.; Rose, P. W.; Verkhivker, G. M. *Pac. Symp. Biocomput.* **1999**, 426–437.
131. Carlson, H. A. *Curr. Opin. Chem. Biol.* **2002**, *6*, 447–452.
132. Claussen, H.; Buning, C.; Rarey, M.; Lengauer, T. *J. Mol. Biol.* **2001**, *308*, 377–395.
133. Knegtel, R. M.; Kuntz, I. D.; Oshiro, C. M. *J. Mol. Biol.* **1997**, *266*, 424–440.
134. Osterberg, F.; Morris, G. M.; Sanner, M. F.; Olson, A. J.; Goodsell, D. S. *Proteins* **2002**, *46*, 34–40.
135. Todd, M. J.; Semo, N.; Freire, E. *J. Mol. Biol.* **1998**, *283*, 475–488.
136. Barril, X.; Morley, S. D. *J. Med. Chem.* **2005**, *48*, 4432–4443.
137. Smith, G. R.; Sternberg, M. J.; Bates, P. A. *J. Mol. Biol.* **2005**, *347*, 1077–1101.
138. Abseher, R.; Nilges, M. *Proteins* **2000**, *39*, 82–88.
139. Grunberg, R.; Leckner, J.; Nilges, M. *Structure (Camb.)* **2004**, *12*, 2125–2136.
140. Tatsumi, R.; Fukunishi, Y.; Nakamura, H. *J. Comput. Chem.* **2004**, *25*, 1995–2005.
141. de Groot, B. L.; van Aalten, D. M.; Scheek, R. M.; Amadei, A.; Vriend, G.; Berendsen, H. J. *Proteins* **1997**, *29*, 240–251.
142. Mustard, D.; Ritchie, D. W. *Proteins* **2005**, *60*, 269–274.
143. Bastard, K.; Thureau, A.; Lavery, R.; Prevost, C. *J. Comput. Chem.* **2003**, *24*, 1910–1920.
144. Cavasotto, C. N.; Kovacs, J. A.; Abagyan, R. A. *J. Am. Chem. Soc.* **2005**, *127*, 9632–9640.
145. Jacobs, D. J.; Rader, A. J.; Kuhn, L. A.; Thorpe, M. F. *Proteins* **2001**, *44*, 150–165.

146. Thorpe, M. F.; Lei, M.; Rader, A. J.; Jacobs, D. J.; Kuhn, L. A. *J. Mol. Graph Model.* **2001**, *19*, 60–69.
147. Zavodszky, M. I.; Lei, M.; Thorpe, M. F.; Day, A. R.; Kuhn, L. A. *Proteins* **2004**, *57*, 243–261.
148. Wells, S.; Menor, S.; Hespenheide, B.; Thorpe, M. F. *Phys. Biol.* **2005**, *2*, 1–10.
149. Quiocho, F. A.; Wilson, D. K.; Vyas, N. K. *Nature* **1989**, *340*, 404–407.
150. Xu, D.; Tsai, C. J.; Nussinov, R. *Protein Eng.* **1997**, *10*, 999–1012.
151. Fuxreiter, M.; Mezei, M.; Simon, I.; Osman, R. *Biophys. J.* **2005**, *89*, 903–911.
152. Jiang, L.; Kuhlman, B.; Kortemme, T.; Baker, D. *Proteins* **2005**, *58*, 893–904.
153. Fornabaio, M.; Spyrakis, F.; Mozzarelli, A.; Cozzini, P.; Abraham, D. J.; Kellogg, G. E. *J. Med. Chem.* **2004**, *47*, 4507–4516.
154. de Graaf, C.; Pospisil, P.; Pos, W.; Folkers, G.; Vermeulen, N. P. *J. Med. Chem.* **2005**, *48*, 2308–2318.
155. Rarey, M.; Kramer, B.; Lengauer, T. *Proteins* **1999**, *34*, 17–28.
156. Ortiz, A. R.; Pisabarro, M. T.; Gago, F.; Wade, R. C. *J. Med. Chem.* **1995**, *38*, 2681–2691.
157. Antes, I.; Merkwirth, C.; Lengauer, T. *J. Chem. Inf. Model.* **2005**, *45*, 1291–1302.
158. Wade, R. C.; Henrich, S.; Wang, T. *Drug Disc. Today: Technologies* **2004**, *1*, 241–246.
159. Timm, D. E.; Baker, L. J.; Mueller, H.; Zidek, L.; Novotny, M. V. *Protein Sci.* **2001**, *10*, 997–1004.
160. Tame, J. R.; Sleigh, S. H.; Wilkinson, A. J.; Ladbury, J. E. *Nat. Struct. Biol.* **1996**, *3*, 998–1001.
161. Lima, C. D.; Klein, M. G.; Weinstein, I. B.; Hendrickson, W. A. *Proc. Natl. Acad. Sci. USA* **1996**, *93*, 5357–5362.
162. Sandler, B.; Murakami, M.; Clardy, J. *J. Am. Chem. Soc.* **1998**, *120*, 595–596.
163. Verdecia, M. A.; Bowman, M. E.; Lu, K. P.; Hunter, T.; Noel, J. P. *Nat. Struct. Biol.* **2000**, 7, 639–643.
164. Zhao, Y.; Ke, H. *Biochemistry* **1996**, *35*, 7356–7361.
165. Focia, P. J.; Shepotinovskaya, I. V.; Seidler, J. A.; Freymann, D. M. *Science* **2004**, *303*, 373–377.
166. Ren, J.; Nichols, C.; Bird, L. E.; Fujiwara, T.; Sugimoto, H.; Stuart, D. I.; Stammers, D. K. *J. Biol. Chem.* **2000**, *275*, 14316–14320.
167. Pletnev, V.; Addlagatta, A.; Wawrzak, Z.; Duax, W. *Acta Crystallogr., Sect. D Biol. Crystallogr.* **2003**, *59*, 50–56.
168. Notenboom, V.; Boraston, A. B.; Kilburn, D. G.; Rose, D. R. *Biochemistry* **2001**, *40*, 6248–6256.
169. Sia, S. K.; Carr, P. A.; Cochran, A. G.; Malashkevich, V. N.; Kim, P. S. *Proc. Natl. Acad. Sci. USA* **2002**, *99*, 14664–14669.
170. Berman, H. M.; Westbrook, J.; Feng, Z.; Gilliland, G.; Bhat, T. N.; Weissig, H.; Shindyalov, I. N.; Bourne, P. E. *Nucleic Acids Res.* **2000**, *28*, 235–242.
171. CASTp webserver. http://cast.engr.uic.edu/cast/ (accessed Aug 2006).
172. GRID. http://www.moldiscovery.com (accessed Aug 2006).
173. Madura, J. D.; Briggs, J. M.; Wade, R. C.; Davis, M. E.; Luty, B. A.; Ilin, A.; Antosiewicz, J.; Gilson, M. K.; Bagheri, B.; Scott, L. R. et al. *Comput. Phys. Commun.* **1995**, *91*, 57–95.

Biographies

Domantas Motiejunas is currently studying for his PhD in the Faculty of Biosciences at the University of Heidelberg. He is carrying out his doctoral work at EML Research, in the Molecular and Cellular Modeling group led by Dr Rebecca Wade. In 2001, he graduated in molecular biology from Vilnius University where he investigated ASR (acid shock RNA) protein function in *Escherichia coli* in the Department of Biochemistry and Biophysics under the leadership of Dr Edita Suziedeliene. In 2002, he entered the International Graduate Program in Molecular and Cellular Biology organized by the University of Heidelberg and the German Cancer Research Center. He carried out his work for a Masters degree at

EML Research on protein–protein docking assisted by sequence conservation and experimental data, and obtained a Masters degree from the University of Heidelberg in 2004. In his PhD project he focuses on the development and application of computational methods to models of macromolecular interactions.

Rebecca C Wade leads the Molecular and Cellular Modeling group at EML Research, a private research institute specializing in applied Information Technology with a strong emphasis on bioinformatics. She studied at the University of Oxford (BA Hons in Physics, 1985; PhD in Molecular Biophysics, 1988), carrying out her doctoral studies in structure-based drug design with Peter Goodford. She did her postdoctoral research in Houston (with Andy McCammon) and Illinois (with Peter Wolynes), primarily in biomolecular simulation, before taking up a position as a group leader in the Structural and Computational Biology Program at the European Molecular Biology Laboratory (EMBL) in Heidelberg in 1992. She set up the Molecular and Cellular Modeling group at EML Research in 2001. The group works on the development and application of computer-aided methods to model and simulate biomolecular interactions. Rebecca Wade is an associate editor of the *Journal of Molecular Recognition*, and a member of the editorial boards of the *Journal of Computer-Aided Molecular Design*, *Journal of Modeling*, and *Molecular Graphics and Biopolymers*. She is a member of the Faculty of 1000: Theory and Simulation Section. She is the recipient of the 2004 Hansch Award of the QSAR and Modeling Society.

© 2007 Elsevier Ltd. All Rights Reserved
No part of this publication may be reproduced, stored in any retrieval system or transmitted in any form by any means electronic, electrostatic, magnetic tape, mechanical, photocopying, recording or otherwise, without permission in writing from the publishers

Comprehensive Medicinal Chemistry II
ISBN (set): 0-08-044513-6

ISBN (Volume 4) 0-08-044517-9; pp. 193–213

4.10 Comparative Modeling of Drug Target Proteins

N Eswar and A Sali, University of California at San Francisco, San Francisco, CA, USA

© 2007 Elsevier Ltd. All Rights Reserved.

4.10.1 Introduction

4.10.1.1 Structure-Based Drug Discovery

Over the past few years, structure-based or rational drug discovery has resulted in a number of drugs on the market and many more in the development pipeline.[1–4] Structure-based methods are now routinely used in almost all stages of drug development, from target identification to lead optimization.[5–8] Central to all structure-based discovery approaches is the knowledge of the three-dimensional (3D) structure of the target protein or complex because the structure and dynamics of the target determine which ligands it binds. The 3D structures of the target proteins are best determined by experimental methods that yield solutions at atomic resolution, such as x-ray crystallography and nuclear magnetic resonance (NMR) spectroscopy.[9] Recent developments in the techniques of experimental structure determination have enhanced the applicability, accuracy, and speed of structural studies.[10,11] Despite these advances, however, structural characterization of sequences remains an expensive and time-consuming task.

4.10.1.2 The Sequence–Structure Gap

The publicly available Protein Data Bank (PDB)[12] currently contains ~33 000 structures and grows at a rate of approximately 40% every 2 years. On the other hand, the various genome-sequencing projects have resulted in ~2.1 million sequences, including the complete genetic blueprints of humans and hundreds of other organisms.[13,14] This achievement has resulted in a vast collection of sequence information about possible target proteins with little or no structural information. Current statistics show that the structures available in the PDB account for only ~1.5% of the sequences in the UniProt database.[13] Moreover, the rate of growth of the sequence information is more than twice that of the structures. Due to this wide sequence–structure gap, reliance on experimentally determined structures limits the number of proteins that can be targeted by structure-based drug discovery.

4.10.1.3 Structure Prediction Addresses the Sequence–Structure Gap

Fortunately, domains in protein sequences are gradually evolving entities that can be clustered into a relatively small number of families with similar sequences and structures.[15,16] For instance, 75–80% of the sequences in the UniProt database have been grouped into fewer than 15 000 domain families.[17,18] Similarly, all the structures in the PDB have been classified into about 1000 distinct folds.[19,20] Computational protein structure prediction methods, such as threading[21] and comparative protein structure modeling,[22,23] strive to bridge the sequence–structure gap by utilizing these evolutionary relationships. The speed, low cost, and relative accuracy of these computational methods have led to the use of predicted 3D structures in the drug discovery process.[24,25] The other class of prediction methods, de novo or ab initio methods, attempts to predict the structure from sequence alone, without reliance on evolutionary relationships. However, despite recent progress in these methods,[26] especially for small proteins with fewer than 100 amino acid residues, comparative modeling remains the most reliable method of predicting the 3D structure of a protein, with an accuracy that can be comparable to a low-resolution, experimentally determined structure.[9]

4.10.1.4 The Basis of Comparative Modeling

The primary requirement for reliable comparative modeling is a detectable similarity between the sequence of interest (target sequence) and a known structure (template). As early as 1986, Chothia and Lesk[27] showed that there is a strong correlation between sequence and structural similarities. This correlation provides the basis of comparative modeling, allows a coarse assessment of model errors, and also highlights one of its major challenges: modeling the structural differences between the template and target structures[28] (**Figure 1**).

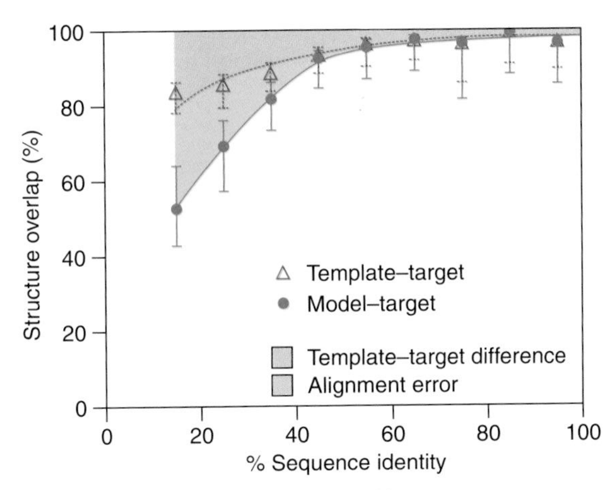

Figure 1 Average model accuracy as a function of sequence identity.[28] As the sequence identity between the target sequence and the template structure decreases, the average structural similarity between the template and the target also decreases (dashed line, triangles).[27] Structural overlap is defined as the fraction of equivalent C^{α} atoms. For the comparison of the model with the actual structure (filled circles), two C^{α} atoms were considered equivalent if they belonged to the same residue and were within 3.5 Å of each other after least-squares superposition. For comparisons between the template structure and the actual target structure (triangles), two C^{α} atoms were considered equivalent if they were within 3.5 Å of each other after alignment and rigid-body superposition. The difference between the model and the actual target structure is a combination of the target–template differences (green area) and the alignment errors (red area). The figure was constructed by calculating 3993 comparative models based on a single template of varying similarity to the targets. All targets had known (experimentally determined) structures.[28]

4.10.1.5 Comparative Modeling Benefits from Structural Genomics

Comparative modeling stands to benefit greatly from the structural genomics initiative.[29] Structural genomics aims to achieve significant structural coverage of the sequence space with an efficient combination of experimental and prediction methods.[30] This goal is pursued by careful selection of target proteins for structure determination by x-ray crystallography and NMR spectroscopy, such that most other sequences are within 'modeling distance' (e.g., >30% sequence identity) of a known structure.[15,16,29,31] The expectation is that the determination of these structures combined with comparative modeling will yield useful structural information for the largest possible fraction of sequences in the shortest possible timeframe. The impact of structural genomics is illustrated by comparative modeling based on the structures determined by the New York Structural Genomics Research Consortium. For each new structure, on average, 100 protein sequences without any prior structural characterization could be modeled at least at the level of the fold.[32] Thus, the structures of most proteins will eventually be predicted by computation, not determined by experiment.

4.10.1.6 Outline

In this review, we begin by describing the various steps involved in comparative modeling. Next, we emphasize two aspects of model refinement, loop modeling and side-chain modeling, due to their relevance in ligand docking and rational drug discovery. We then discuss the errors in comparative models. Finally, we describe the role of comparative modeling in drug discovery, focusing on ligand docking against comparative models. We compare successes of docking against models and x-ray structures, and illustrate the computational docking against models with a number of examples. We conclude with a summary of topics that will impact on the future utility of comparative modeling in drug discovery, including an automation and integration of resources required for comparative modeling and ligand docking.

4.10.2 Steps in Comparative Modeling

Comparative modeling consists of four main steps[23] (**Figure 2a**): (1) fold assignment that identifies similarity between the target sequence of interest and at least one known protein structure (the template); (2) alignment of the target sequence and the template(s); (3) building a model based on the alignment with the chosen template(s); and (4) predicting model errors.

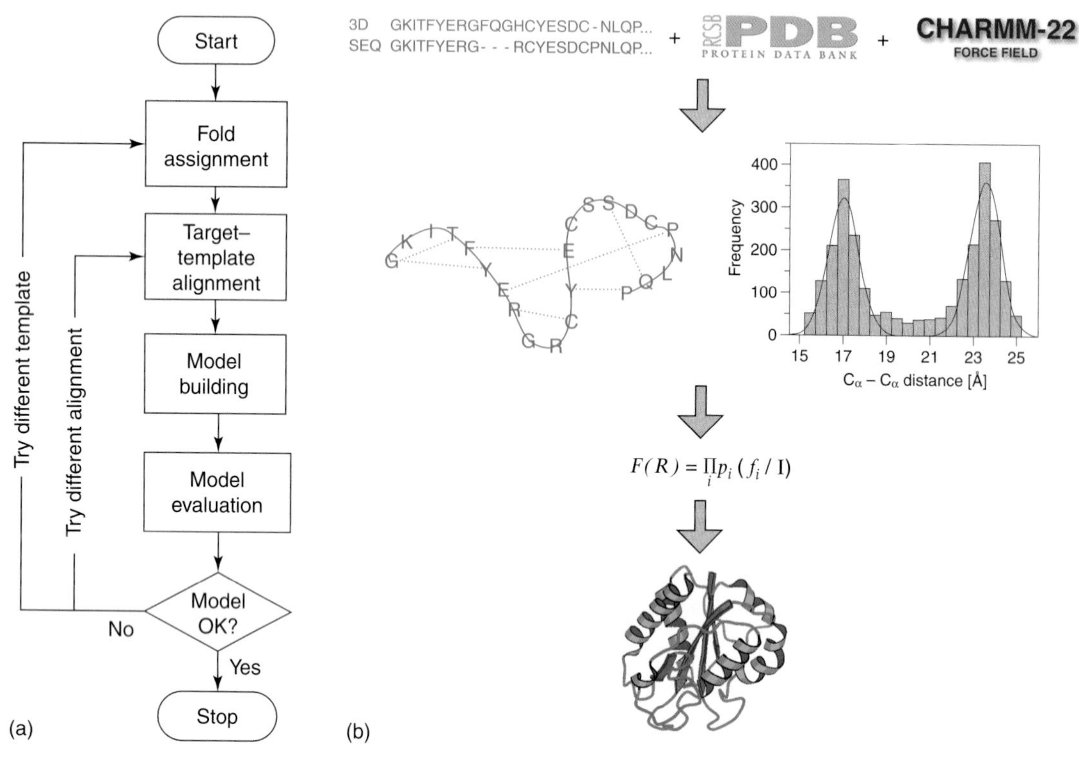

Figure 2 Comparative protein structure modeling. (a) A flowchart illustrating the steps in the construction of a comparative model.[23] (b) Description of comparative modeling by extraction of spatial restraints as implemented in MODELLER.[96] By default, spatial restraints in MODELLER involve: (1) homology-derived restraints from the aligned template structures; (2) statistical restraints derived from all known protein structures; and (3) stereochemical restraints from the CHARMM-22 molecular mechanics force field. These restraints are combined into an objective function that is then optimized to calculate the final 3D structure of the target sequence.

4.10.2.1 Fold Assignment and Sequence–Structure Alignment

Although fold assignment and sequence–structure alignment are logically two distinct steps in the process of comparative modeling, in practice almost all fold assignment methods also provide sequence–structure alignments. In the past, fold assignment methods were optimized for better sensitivity in detecting remotely related homologs, often at the cost of alignment accuracy. However, recent methods simultaneously optimize both the sensitivity and alignment accuracy. Therefore, in the following discussion, we will treat fold assignment and sequence–structure alignment as a single protocol, explaining the differences as needed.

4.10.2.1.1 Fold assignment

As mentioned earlier, the primary requirement for comparative modeling is the identification of one or more known template structures with detectable similarity to the target sequence. The identification of suitable templates is achieved by scanning structure databases, such as PDB,[12] SCOP,[19] DALI,[33] and CATH,[20] with the target sequence as the query. The detected similarity is usually quantified in terms of sequence identity or statistical measures, such as E-value or z-score, depending on the method used.

4.10.2.1.2 Three levels of similarity

Sequence–structure relationships are coarsely classified into three different regimes in the sequence similarity spectrum: (1) the easily detected relationships characterized by >30% sequence identity; (2) the 'twilight zone,'[34] corresponding to relationships with statistically significant sequence similarity in the 10–30% range; and (3) the 'midnight zone,'[34] corresponding to statistically insignificant sequence similarity.

4.10.2.1.3 Sequence–sequence methods

For closely related protein sequences with identities higher than 30–40%, the alignments produced by all methods are almost always largely correct. The quickest way to search for suitable templates in this regime is to use simple pairwise sequence alignment methods such as SSEARCH,[35] BLAST,[36] and FASTA.[35] Brenner *et al.* showed that these methods detect only ~18% of the homologous pairs at less than 40% sequence identity, while they identify more than 90% of the relationships when sequence identity is between 30% and 40%.[37] Another benchmark, based on 200 reference structural alignments with 0–40% sequence identity, indicated that BLAST is able to correctly align only 26% of the residue positions.[46]

4.10.2.1.4 Sequence–profile methods

The sensitivity of the search and accuracy of the alignment become progressively difficult as the relationships move into the twilight zone.[34,38] A significant improvement in this area was the introduction of profile methods by Gribskov and co-workers.[39] The profile of a sequence is derived from a multiple sequence alignment and specifies residue-type occurrences for each alignment position. The information in a multiple sequence alignment is most often encoded as either a position-specific scoring matrix (PSSM)[36,40,41] or as a hidden Markov model (HMM).[42,43] In order to identify suitable templates for comparative modeling, the profile of the target sequence is used to search against a database of template sequences. The profile–sequence methods are more sensitive in detecting related structures in the twilight zone than the pairwise sequence-based methods; they detect approximately twice the number of homologs under 40% sequence identity.[44–46] The resulting profile–sequence alignments correctly align approximately 43–48% of residues in the 0–40% sequence identity range[46,47]; this number is almost twice as large as that of the pairwise sequence methods. Frequently used programs for profile–sequence alignment are PSI-BLAST,[36] SAM,[48] HMMER,[42] and BUILD_PROFILE.[49]

4.10.2.1.5 Profile–profile methods

As a natural extension, the profile–sequence alignment methods have led to profile–profile alignment methods that search for suitable template structures by scanning the profile of the target sequence against a database of template profiles, as opposed to a database of template sequences. These methods have proven to include the most sensitive and accurate fold assignment and alignment protocols to date.[47,50–52] Profile–profile methods detect ~28% more relationships at the superfamily level and improve the alignment accuracy by 15–20% compared to profile–sequence methods.[47,53] There are a number of variants of profile–profile alignment methods that differ in the scoring functions they use.[47,50,53–59] However, several analyses have shown that the overall performances of these methods are comparable.[47,50–52] Some of the programs that can be used to detect suitable templates are FFAS,[60] SP3,[53] SALIGN,[47] and PPSCAN.[49]

4.10.2.1.6 Sequence–structure threading methods

As the sequence identity drops below the threshold of the twilight zone, there is usually insufficient signal in the sequences or their profiles for the sequence-based methods discussed above to detect true relationships.[44] Sequence–structure threading methods are most useful in this regime as they can sometimes recognize common folds, even in the absence of any statistically significant sequence similarity.[21] These methods achieve higher sensitivity by using structural information derived from the templates. The accuracy of a sequence–structure match is assessed by the score of a corresponding coarse model and not by sequence similarity, as in sequence comparison methods.[21] The scoring scheme used to evaluate the accuracy is either based on residue substitution tables dependent on structural features such as solvent exposure, secondary structure type, and hydrogen bonding properties,[53,61–63] or on statistical potentials for residue interactions implied by the alignment.[64–68] The use of structural data does not have to be restricted to the structure side of the aligned sequence–structure pair. For example, SAM-T02 makes use of the predicted local structure for the target sequence to enhance homolog detection and alignment accuracy.[69] Commonly used threading programs are GenTHREADER,[61,70] 3D-PSSM,[71] FUGUE,[63] SP3,[53] and SAM-T02 multitrack HMM.[62,69]

4.10.2.1.7 Iterative sequence–structure alignment

Yet another strategy is to optimize the alignment by iterating over the process of calculating alignments, building models, and evaluating models. Such a protocol can sample alignments that are not statistically significant and identify the alignment that yields the best model. Although this procedure can be time-consuming, it can significantly improve the accuracy of the resulting comparative models in difficult cases.[72]

4.10.2.2 Alignment Errors are Unrecoverable

Regardless of the method used, searching in the twilight and midnight zones of the sequence–structure relationship often results in false negatives, false positives, or alignments that contain an increasingly large number of gaps and alignment errors. Improving the performance and accuracy of methods in this regime remains one of the main tasks of comparative modeling today.[73] It is imperative to calculate an accurate alignment between the target–template pair, as comparative modeling can almost never recover from an alignment error.[74]

4.10.2.3 Template Selection

After a list of all related protein structures and their alignments with the target sequence have been obtained, template structures are prioritized depending on the purpose of the comparative model. Template structures may be chosen purely based on the target–template sequence identity or a combination of several other criteria, such as experimental accuracy of the structures (resolution of x-ray structures, number of restraints per residue for NMR structures), conservation of active-site residues, holo-structures that have bound ligands of interest, and prior biological information that pertains to the solvent, pH, and quaternary contacts. It is not necessary to select only one template. In fact, the use of several templates approximately equidistant from the target sequence generally increases the model accuracy.[75,76]

4.10.3 Model Building

4.10.3.1 Three Approaches to Comparative Model Building

Once an initial target–template alignment is built, a variety of methods can be used to construct a 3D model for the target protein.[23,74,77–80] The original and still widely used method is modeling by rigid-body assembly.[78,79,81] This method constructs the model from a few core regions, and from loops and side chains that are obtained by dissecting related structures. Commonly used programs that implement this method are COMPOSER,[82–85] 3D-JIGSAW,[86] and SWISS-MODEL.[87] Another family of methods, modeling by segment matching, relies on the approximate positions of conserved atoms from the templates to calculate the coordinates of other atoms.[88–92] An instance of this approach is implemented in SegMod.[91] The third group of methods, modeling by satisfaction of spatial restraints, uses either distance geometry or optimization techniques to satisfy spatial restraints obtained from the alignment of the target sequences with the template structures.[93–97] Specifically, MODELLER,[96,98,99] our own program for comparative modeling, belongs to this group of methods.

4.10.3.2 MODELLER: Comparative Modeling by Satisfaction of Spatial Restraints

MODELLER implements comparative protein structure modeling by the satisfaction of spatial restraints that include: (1) homology-derived restraints on the distances and dihedral angles in the target sequence, extracted from its alignment with the template structures[96]; (2) stereochemical restraints such as bond length and bond angle preferences, obtained from the CHARMM-*22* molecular mechanics force field[100]; (3) statistical preferences for dihedral angles and nonbonded interatomic distances, obtained from a representative set of known protein structures[101]; and (4) optional manually curated restraints, such as those from NMR spectroscopy, rules of secondary structure packing, cross-linking experiments, fluorescence spectroscopy, image reconstruction from electron microscopy, site-directed mutagenesis, and intuition (**Figure 2b**). The spatial restraints, expressed as probability density functions, are combined into an objective function that is optimized by a combination of conjugate gradients and molecular dynamics with simulated annealing. This model-building procedure is similar to structure determination by NMR spectroscopy.

4.10.3.3 Relative Accuracy, Flexibility, and Automation

Accuracies of the various model-building methods are relatively similar when used optimally.[102,103] Other factors, such as template selection and alignment accuracy, usually have a larger impact on the model accuracy, especially for models based on less than 30% sequence identity to the templates. However, it is important that a modeling method allows a degree of flexibility and automation to obtain better models more easily and rapidly. For example, a method should allow for an easy recalculation of a model when a change is made in the alignment; it should be straightforward to calculate models based on several templates; and the method should provide tools for incorporation of prior knowledge about the target (e.g., cross-linking restraints and predicted secondary structure).

4.10.4 Refinement of Comparative Models

Protein sequences evolve through a series of amino acid residue substitutions, insertions, and deletions. While substitutions can occur throughout the length of the sequence, insertions and deletions mostly occur on the surface of proteins in segments that connect regular secondary structure segments (i.e., loops). While the template structures are helpful in the modeling of the aligned target backbone segments, they are generally less valuable for the modeling of side chains and irrelevant for the modeling of insertions such as loops. The loops and side chains of comparative models are especially important for ligand docking; thus, we discuss them in the following two sections.

4.10.4.1 Loop Modeling

4.10.4.1.1 Definition of the problem

Loop modeling is an especially important aspect of comparative modeling in the range from 30% to 50% sequence identity. In this range of overall similarity, loops among the homologs vary while the core regions are still relatively conserved and aligned accurately. Loops often play an important role in defining the functional specificity of a given protein, forming the active and binding sites. Loop modeling can be seen as a mini protein folding problem because the correct conformation of a given segment of a polypeptide chain has to be calculated mainly from the sequence of the segment itself. However, loops are generally too short to provide sufficient information about their local fold. Even identical decapeptides in different proteins do not always have the same conformation.[104,105] Some additional restraints are provided by the core anchor regions that span the loop and by the structure of the rest of the protein that cradles the loop. Although many loop-modeling methods have been described, it is still challenging to correctly and confidently model loops longer than approximately 8–10 residues.[98,106]

4.10.4.1.2 Two classes of methods

There are two main classes of loop-modeling methods: (1) database search approaches that scan a database of all known protein structures to find segments fitting the anchor core regions[90,107]; and (2) conformational search approaches that rely on optimizing a scoring function.[108–110] There are also methods that combine these two approaches.[111,112]

4.10.4.1.2.1 Database-based loop modeling

The database search approach to loop modeling is accurate and efficient when a database of specific loops is created to address the modeling of the same class of loops, such as β-hairpins,[113] or loops on a specific fold, such as the hypervariable regions in the immunoglobulin fold.[107,114] There are attempts to classify loop conformations into more general categories, thus extending the applicability of the database search approach.[115–117] However, the database methods are limited because the number of possible conformations increases exponentially with the length of a loop. As a result, only loops up to 4–7 residues long have most of their conceivable conformations present in the database of known protein structures.[118,119] This limitation is made even worse by the requirement for an overlap of at least one residue between the database fragment and the anchor core regions, which means that modeling a 5-residue insertion requires at least a 7-residue fragment from the database.[89] Despite the rapid growth of the database of known structures, it does not seem possible to cover most of the conformations of a 9-residue segment in the foreseeable future. On the other hand, most of the insertions in a family of homologous proteins are shorter than 10–12 residues.[98]

4.10.4.1.2.2 Optimization-based methods

To overcome the limitations of the database search methods, conformational search methods were developed.[108,109] There are many such methods, exploiting different protein representations, objective functions, and optimization or enumeration algorithms. The search algorithms include the minimum perturbation method,[120] molecular dynamics simulations,[111,121] genetic algorithms,[122] Monte Carlo and simulated annealing,[123–125] multiple-copy simultaneous search,[126] self-consistent field optimization,[127] and enumeration based on graph theory.[128] The accuracy of loop predictions can be further improved by clustering the sampled loop conformations and partially accounting for the entropic contribution to the free energy.[129] Another way of improving the accuracy of loop predictions is to consider the solvent effects. Improvements in implicit solvation models, such as the Generalized Born solvation model, motivated their use in loop modeling. The solvent contribution to the free energy can be added to the scoring function for optimization, or it can be used to rank the sampled loop conformations after they are generated with a scoring function that does not include the solvent terms.[98,130–132]

4.10.4.2 Side-Chain Modeling

4.10.4.2.1 Fixed backbone

Two simplifications are frequently applied in the modeling of side-chain conformations.[133] First, amino acid residue replacements often leave the backbone structure almost unchanged,[26] allowing us to fix the backbone during the search for the best side-chain conformations. Second, most side chains in high-resolution crystallographic structures can be represented by a limited number of conformers that comply with stereochemical and energetic constraints.[134] This observation motivated Ponder and Richards[135] to develop the first library of side-chain rotamers for the 17 types of residues with dihedral angle degrees of freedom in their side chains, based on 10 high-resolution protein structures determined by x-ray crystallography. Subsequently, a number of additional libraries have been derived.[136–142]

4.10.4.2.2 Rotamers

Rotamers on a fixed backbone are often used when all the side chains need to be modeled on a given backbone. This approach reduces the combinatorial explosion associated with a full conformational search of all the side chains, and is applied by some comparative modeling[78] and protein design approaches.[143] However, ~15% of the side chains cannot be represented well by these libraries.[144] In addition, it has been shown that the accuracy of side-chain modeling on a fixed backbone decreases rapidly when the backbone errors are larger than 0.5 Å.[145]

4.10.4.2.3 Methods

Earlier methods for side-chain modeling often put less emphasis on the energy or scoring function. The function was usually greatly simplified, and consisted of the empirical rotamer preferences and simple repulsion terms for nonbonded contacts.[138] Nevertheless, these approaches have been justified by their performance. For example, a method based on a rotamer library compared favorably with that based on a molecular mechanics force field,[146] and new methods continue to be based on the rotamer library approach.[147,148] The various optimization approaches include a Monte Carlo simulation,[149] simulated annealing,[150] a combination of Monte Carlo and simulated annealing,[151] the dead-end elimination theorem,[152,153] genetic algorithms,[142] neural network with simulated annealing,[154] mean field optimization,[155] and combinatorial searches.[138,156,157] Several recent papers focused on the testing of more sophisticated potential functions for conformational search[157,158] and development of new scoring functions for side-chain modeling,[159] reporting higher accuracy than earlier studies.

4.10.5 Errors in Comparative Models

The major sources of error in comparative modeling are discussed in the relevant sections above. The following is a summary of these errors, dividing them into five categories (**Figure 3**).

4.10.5.1 Selection of Incorrect Templates

This error is a potential problem when distantly related proteins are used as templates (i.e., less than 30% sequence identity). Distinguishing between a model based on an incorrect template and a model based on an incorrect alignment with a correct template is difficult. In both cases, the evaluation methods (below) will predict an unreliable model. The conservation of the key functional or structural residues in the target sequence increases the confidence in a given fold assignment.

4.10.5.2 Errors due to Misalignments

The single source of errors with the largest impact on comparative modeling is misalignments, especially when the target–template sequence identity decreases below 30%. Alignment errors can be minimized in two ways. Using the profile-based methods discussed above usually results in more accurate alignments than those from pairwise sequence alignment methods. Another way of improving the alignment is iteratively to modify those regions in the alignment that correspond to predicted errors in the model.[75]

4.10.5.3 Errors in Regions without a Template

Segments of the target sequence that have no equivalent region in the template structure (i.e., insertions or loops) are one of the most difficult regions to model. Again, when the target and template are distantly related, errors in the alignment can lead to incorrect positions of the insertions. Using alignment methods that incorporate structural

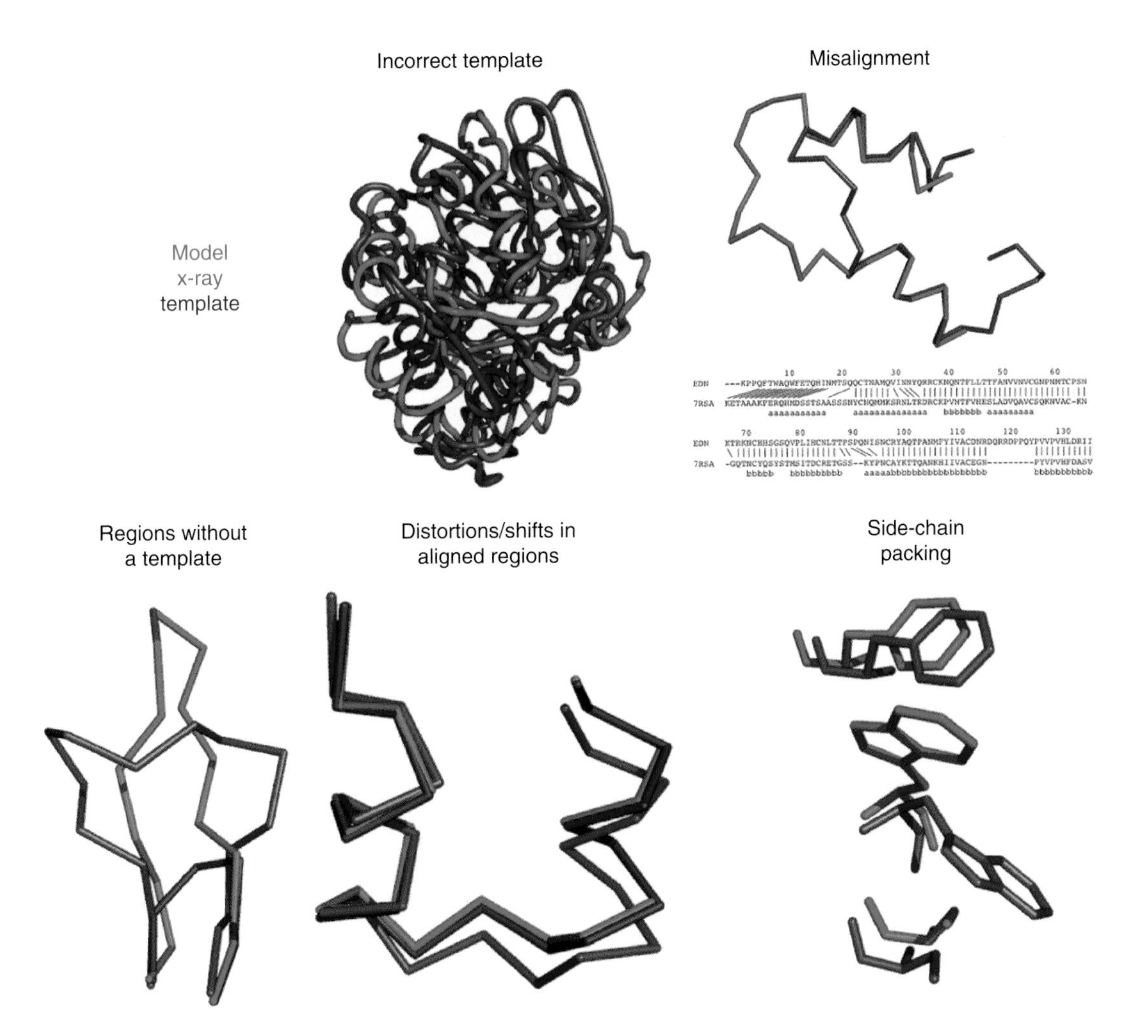

Figure 3 Typical errors in comparative modeling.[23] Shown are the typical sources of errors encountered in comparative models. Two of the major sources of errors in comparative modeling are due to incorrect templates or incorrect alignments with the correct templates. The modeling procedure can rarely recover from such errors. The next significant source of errors arises from regions in the target with no corresponding region in the template, i.e., insertions or loops. Other sources of errors, which occur even with an accurate alignment, are due to rigid-body shifts, distortions in the backbone, and errors in the packing of side chains.

information can often correct such errors. Once a reliable alignment is obtained, various modeling protocols can predict the loop conformation, for insertions of fewer than 8–10 residues.[98,106,111,169]

4.10.5.4 Distortions and Shifts in Correctly Aligned Regions

As a consequence of sequence divergence, the main-chain conformation changes, even if the overall fold remains the same. Therefore, it is possible that in some correctly aligned segments of a model, the template is locally different (<3 Å) from the target, resulting in errors in that region. The structural differences are sometimes not due to differences in sequence, but are a consequence of artifacts in structure determination or structure determination in different environments (e.g., packing of subunits in a crystal). The simultaneous use of several templates can minimize this kind of an error.[75,76]

4.10.5.5 Errors in Side-Chain Packing

As the sequences diverge, the packing of the atoms in the protein core changes. Sometimes even the conformation of identical side chains is not conserved – a pitfall for many comparative modeling methods. Side-chain errors are critical if they occur in regions that are involved in protein function, such as active sites and ligand-binding sites.

4.10.6 Prediction of Model Errors

The accuracy of the predicted model determines the information that can be extracted from it. Thus, estimating the accuracy of a model in the absence of the known structure is essential for interpreting it.

4.10.6.1 Initial Assessment of the Fold

As discussed earlier, a model calculated using a template structure that shares more than 30% sequence identity is indicative of an overall accurate structure. However, when the sequence identity is lower, the first aspect of model evaluation is to confirm whether or not a correct template was used for modeling. It is often the case, when operating in this regime, that the fold assignment step produces only false positives. A further complication is that at such low similarities the alignment generally contains many errors, making it difficult to distinguish between an incorrect template on one hand and an incorrect alignment with a correct template on the other hand. There are several methods that use 3D profiles and statistical potentials,[65,160,161] which assess the compatibility between the sequence and modeled structure by evaluating the environment of each residue in a model with respect to the expected environment, as found in native high-resolution experimental structures. These methods can be used to assess whether or not the correct template was used for the modeling. They include VERIFY3D,[160] PROSAII,[162] HARMONY,[163] ANOLEA,[164] and DFIRE.[165]

Even when the model is based on alignments that have >30% sequence identity, other factors, including the environment, can strongly influence the accuracy of a model. For instance, some calcium-binding proteins undergo large conformational changes when bound to calcium. If a calcium-free template is used to model the calcium-bound state of the target, it is likely that the model will be incorrect, irrespective of the target–template similarity or accuracy of the template structure.[166]

4.10.6.2 Self-Consistency

The model should also be subjected to evaluations of self-consistency to ensure that it satisfies the restraints used to calculate it. Additionally, the stereochemistry of the model (e.g., bond lengths, bond angles, backbone torsion angles, and nonbonded contacts) may be evaluated using programs such as PROCHECK[167] and WHATCHECK.[168] Although errors in stereochemistry are rare and less informative than errors detected by statistical potentials, a cluster of stereochemical errors may indicate that there are larger errors (e.g., alignment errors) in that region.

4.10.7 Evaluation of Comparative Modeling Methods

It is crucial for method developers and users alike to assess the accuracy of their methods. An attempt to address this problem has been made by the Critical Assessment of Techniques for Proteins Structure Prediction (CASP)[170] and the Critical Assessment of Fully Automated Structure Prediction (CAFASP) experiments.[171] However, both CASP and CAFASP assess methods only over a limited number of target protein sequences.[102,172] To overcome this limitation, two additional evaluation experiments have been described, LiveBench[172] and EVA.[173,174] EVA is a large-scale and continuously running web server that automatically assesses protein structure prediction servers in the categories of secondary structure prediction, residue–residue contact prediction, fold assignment, and comparative modeling. The aims of EVA are: (1) to evaluate continuously and automatically blind predictions by prediction servers, based on identical and sufficiently large data sets; (2) to provide weekly updates of the method assessments on the web; and (3) to enable developers, nonexpert users, and reviewers to determine the performance of the tested prediction servers.

4.10.8 Applications of Comparative Models

There is a wide range of applications of protein structure models (**Figure 4**).[1,175–180] For example, high- and medium-accuracy comparative models are frequently helpful in refining functional predictions that have been based on a sequence match alone because ligand binding is more directly determined by the structure of the binding site than by its sequence. It is often possible to predict correctly features of the target protein that do not occur in the template structure.[181,182] For example, the size of a ligand may be predicted from the volume of the binding site cleft and the location of a binding site for a charged ligand can be predicted from a cluster of charged residues on the protein. Fortunately, errors in the functionally important regions in comparative models are many times relatively low because

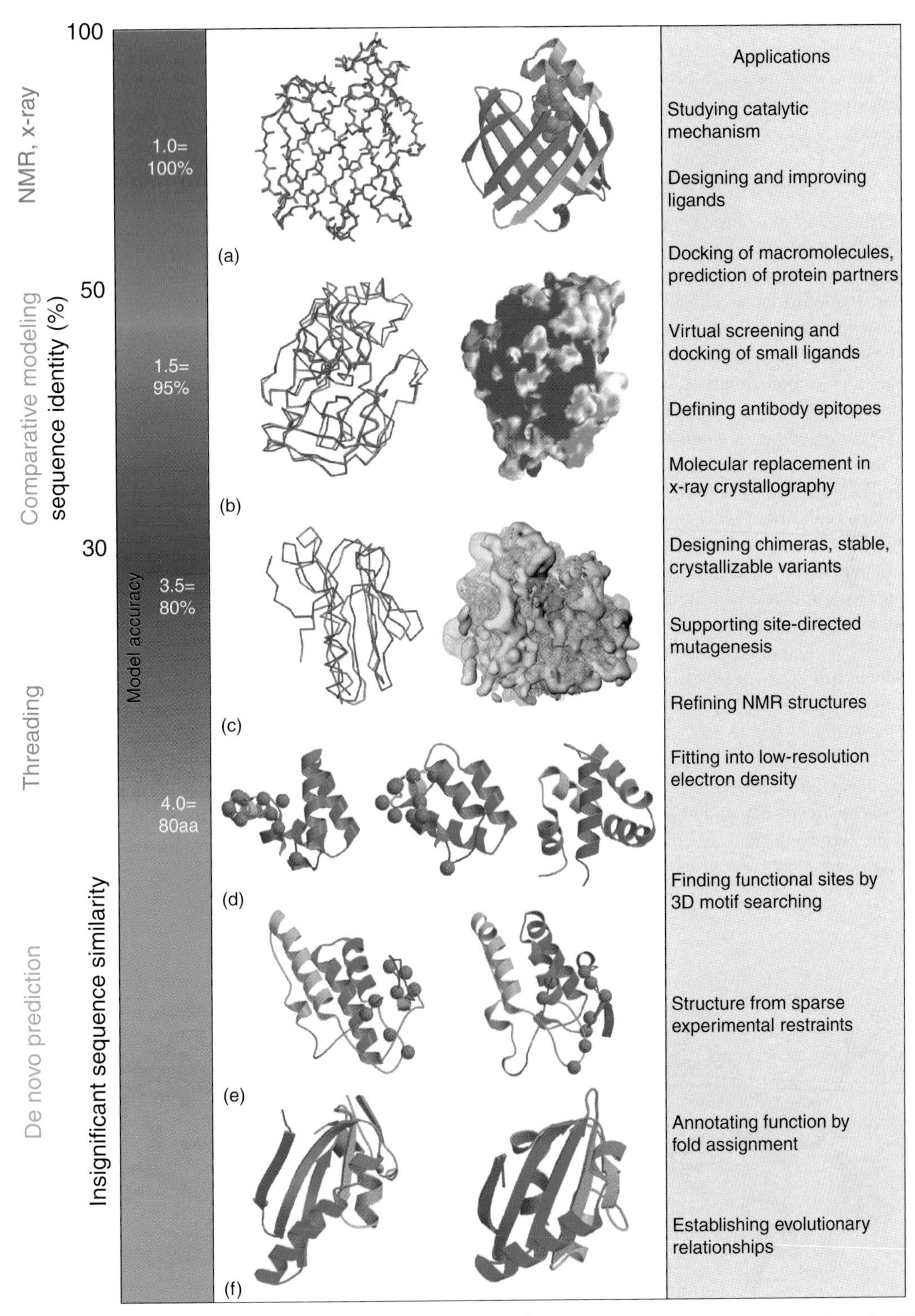

Figure 4 Accuracy and applications of protein structure models.[9] Shown are the different ranges of applicability of comparative protein structure modeling, threading, and de novo structure prediction, their corresponding accuracies, and their sample applications.

the functional regions, such as active sites, tend to be more conserved in evolution than the rest of the fold. Even low-accuracy comparative models may be useful, for example, for assigning the fold of a protein. Fold assignment can be very helpful in drug discovery, because it can shortcut the search for leads by pointing to compounds that have been previously developed for other members of the same family.[183,184]

4.10.8.1 Comparative Models versus Experimental Structures in Virtual Screening

The remainder of this review focuses on the use of comparative models for ligand docking (*see* also Chapter 4.19.2.5).[185–187] It is widely accepted that docking to comparative models is more challenging and less successful than docking to crystallographic structures. However, it seems that surprisingly little work has been done to obtain quantitative information about the accuracy of docking to comparative models, to determine in detail why the results are inferior to those obtained with crystal structures, and to improve methods for docking to comparative models.

We begin our discussion with a study by McGovern and Shoichet[188] that compared the success of docking against three different conformations of 10 enzymes: holo (ligand-bound), apo, and homology modeled. All 10 enzymes had known structures in both the holo and apo form. Comparative models for each of these enzymes were taken from MODBASE, a database of comparative models for all protein sequences that are detectably related to at least one known structure. The models were based on single template structures with sequence identities in the range of 28–87%. Each enzyme had multiple known inhibitors in the MDL Drug Data Report (MDDR) database, a library of drug-like molecules where each molecule has been annotated by the receptor to which it binds. Success of the docking, carried out with the Shoichet group's version of DOCK,[189,190] was assessed by enrichment: the ability to distinguish known inhibitors from a large set of ~100 000 'decoys' relative to random selection. As might be expected, the holo structures were the best at selecting the known ligands from among the MDDR decoys based on the docking score. Unexpectedly, the comparative models often ranked known ligands among the top-scoring database molecules; in four targets, the enrichment was 20 times higher than expected by chance.[188] In one case, purine nucleoside phosphorylase, the modeled structure actually performed better than the holo structure. For the comparative model, 25% of the known ligands were found in the top 1.2% of the ranked database, whereas for the holo conformation, 2.8% of the ranked list had to be searched before 25% of the ligands were found. In another example, the holo structure of thymidylate synthase correctly recognized ligands similar in size to the ligand captured in the x-ray structure, but not ligands that were markedly different from it. In contrast, the binding sites in the modeled conformations were more spacious and could in fact correctly detect and accommodate larger ligands than the holo receptor (**Figure 5**). Thus, it appears that, while x-ray crystallographic structures remain the first choice in docking, many comparative models seem sufficiently accurate to rank highly known ligands from among a very large list of possible alternatives.

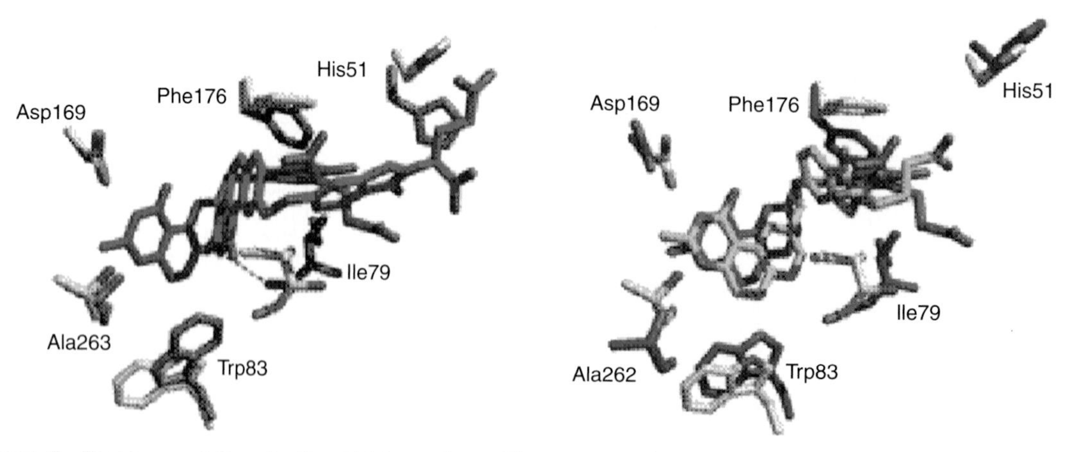

Figure 5 Docking predictions for thymidylate synthase. Shown are the x-ray structure of the holo receptor in gray, the modeled receptor in blue, the docked conformation of the ligand in the holo structure in green, and the docked conformation of the ligand in the modeled structure in yellow. A second holo complex, not used for docking but bound to a larger ligand, is also shown with protein atoms in white and ligand atoms in purple. The ligand in the holo receptor was smaller in size than many of the known ligands in the database. Consequently, while the holo structure yielded better enrichment of ligands that were similar to the native ligand, it was unable to dock larger ligands correctly. The modeled receptors, in contrast, with their more spacious binding sites, showed better competence in such cases. (Courtesy of Brian Shoichet.)

4.10.8.2 Use of Comparative Models to Obtain Novel Drug Leads

Despite problems with comparative modeling and ligand docking, comparative models have been successfully used in practice in conjunction with virtual screening to identify novel inhibitors. We briefly review a few of these success stories' to highlight the potential of the combined comparative modeling and ligand-docking approach to drug discovery (*see* 4.19 Virtual Screening).

Comparative models have been employed to aid rational drug design against parasites for more than 20 years.[122,191–193] As early as 1993, Ring *et al.*[122] used comparative models for computational docking studies that identified low micromolar nonpeptidic inhibitors of proteases in malarial and schistosome parasite lifecycles. Li *et al.*[191] subsequently used similar methods to develop nanomolar inhibitors of falcipain that are active against chloroquine-resistant strains of malaria. In a study by Selzer *et al.*[193] comparative models were used to predict new nonpeptide inhibitors of cathepsin L-like cysteine proteases in *Leishmania major*. Sixty-nine compounds were selected by DOCK 3.5 as strong binders to a comparative model of protein cpB, and of these, 21 had experimental IC_{50} values below 100 mmol L^{-1}. Finally, in a recent study by Que *et al.*[192] comparative models were used to rationalize ligand-binding affinities of cysteine proteases in *Entamoeba histolytica*. Specifically, this work provided an explanation for why proteases ACP1 and ACP2 had substrate specificity similar to that of cathepsin B, although their overall structure is more similar to that of cathepsin D.

Enyedy *et al.*[194] discovered 15 new inhibitors of matriptase by docking against its comparative model. The comparative model employed thrombin as the template, sharing only 34% sequence identity with the target sequence. Moreover, some residues in the binding site are significantly different; a trio of charged Asp residues in matriptase correspond to 1 Tyr and 2 Trp residues in thrombin. Thrombin was chosen as the template, in part because it prefers substrates with positively charged residues at the P1 position, as does matriptase. The comparative model was constructed using MODELLER and refined with MD simulations in CHARMM. The National Cancer Institute database was used for virtual screening that targeted the S1 site with the DOCK program. The 2000 best-scoring compounds were manually inspected to identify positively charged ligands (the S1 site is negatively charged), and 69 compounds were experimentally screened for inhibition, identifying the 15 inhibitors. One of them, hexamidine, was used as a lead to identify additional compounds selective for matriptase relative to thrombin. The Wang group has also used similar methods to discover seven new, low-micromolar inhibitors of Bcl-2, using a comparative model based on the NMR solution structure of $Bcl-X_L$.[195]

Schapira *et al.*[196] discovered a novel inhibitor of a retinoic acid receptor by virtual screening using a comparative model. In this case, the target (RAR-α) and template (RAR-γ) are very closely related; only three residues in the binding site are not conserved. The ICM program was used for virtual screening of ligands from the Available Chemicals Directory (ACD). The 5364 high-scoring compounds identified in the first round were subsequently docked into a full atom representation of the receptor with flexible side chains to obtain a final set of 300 good-scoring hits. These compounds were then manually inspected to choose the final 30 for testing. Two novel agonists were identified, with 50-nanomolar activity.

Zuccotto *et al.*[197] identified novel inhibitors of dihydrofolate reductase (DHFR) in *Trypanosoma cruzi* (the parasite that causes Chagas disease) by docking into a comparative model based on ~50% sequence identity to DHFR in *L. major*, a related parasite. The virtual screening procedure used DOCK for rigid docking of over 50 000 selected compounds from the Cambridge Structural Database (CSD). Visual inspection of the top 100 hits was used to select 36 compounds for experimental testing. This work identified several novel scaffolds with micromolar IC_{50} values. The authors report attempting to use virtual screening results to identify compounds with greater affinity for *T. cruzi* DHFR than human DHFR, but it is not clear how successful they were.

Following the recent outbreak of the severe acute respiratory syndrome (SARS) in 2003, Anand *et al.*[198] used the experimentally determined structures of the main protease from human coronavirus (M^{PRO}) and an inhibitor complex of porcine coronavirus (transmissible gastroenteritis virus, TGEV) M^{PRO} to calculate a comparative model of the SARS coronavirus M^{PRO}. This model then provided a basis for the design of anti-SARS drugs. In particular, a comparison of the active site residues in these and other related structures suggested that the AG7088 inhibitor of the human rhinovirus type 2 3C protease is a good starting point for design of anticoronaviral drugs.[199]

Comparative models of protein kinases combined with virtual screening have also been intensely used for drug discovery.[200–204] The >500 kinases in the human genome, the relatively small number of experimental structures available, and the high level of conservation around the important adenosine triphosphate-binding site make comparative modeling an attractive approach toward structure-based drug discovery.

G protein-coupled receptors are another interesting class of proteins that in principle allow drug discovery through comparative modeling.[205–209] Approximately 40% of current drug targets belong to this class of proteins. However, these proteins have been extremely difficult to crystallize and most comparative modeling has been based on the atomic resolution structure of the bovine rhodopsin.[210] Despite this limitation, a rather extensive test of docking methods with rhodopsin-based comparative models shows encouraging results (*see* 4.26 Seven Transmembrane G Protein-Coupled Receptors: Insights for Drug Design from Structure and Modeling).

4.10.9 Future Directions

Although reports of successful virtual screening against comparative models are encouraging, such efforts are not yet a routine part of rational drug design. Even the successful efforts appear to rely strongly on visual inspection of the docking results. Much work remains to be done to improve the accuracy, efficiency, and robustness of docking against comparative models. Despite assessments of relative successes of docking against comparative models and native x-ray structures,[188,202] surprisingly little has been done to compare the accuracy achievable by different approaches to comparative modeling and to identify the specific structural reasons why comparative models generally produce less accurate virtual screening results than the holo structures. Among the many issues that deserve consideration are the following:

- The inclusion of cofactors and bound water molecules in protein receptors is often critical for success of virtual screening; however, cofactors are not routinely included in comparative models
- Most docking programs currently retain the protein receptor in a completely rigid conformation. While this approach is appropriate for 'lock-and-key' binding modes, it does not work when the ligand induces conformational changes in the receptor upon binding. A flexible receptor approach is necessary to address such induced-fit cases[211,212]
- The accuracy of comparative models is frequently judged by the C^{α} root mean square error or other similar measures of backbone accuracy. For virtual screening, however, the precise positioning of side chains in the binding site is likely to be critical; measures of accuracy for binding sites are needed to help evaluate the suitability of comparative modeling algorithms for constructing models for docking
- Knowledge of known inhibitors, either for the target protein or the template, should help to evaluate and improve virtual screening against comparative models. For example, comparative models constructed from holo' template structures implicitly preserve some information about the ligand-bound receptor conformation
- Improvement in the accuracy of models produced by comparative modeling will require methods that finely sample protein conformational space using a free energy or scoring function that has sufficient accuracy to distinguish the native structure from the nonnative conformations. Despite many years of development of molecular simulation methods, attempts to refine models that are already relatively close to the native structure have met with relatively little success. This failure is likely to be due in part to inaccuracies in the scoring functions used in the simulations, particularly in the treatment of electrostatics and solvation effects. A combination of physics-based energy function with the statistical information extracted from known protein structures may provide a route to the development of improved scoring functions
- Improvements in sampling strategies are also likely to be necessary, for both comparative modeling and flexible docking

4.10.10 Automation and Availability of Resources for Comparative Modeling and Ligand Docking

Given the increasing number of target sequences for which no experimentally determined structures are available, drug discovery stands to gain immensely from comparative modeling and other in silico methods. Despite unsolved problems in virtually every step of comparative modeling and ligand docking, it is highly desirable to automate the whole process, starting with the target sequence and ending with a ranked list of its putative ligands. Automation encourages development of better methods, improves their testing, allows application on a large scale, and makes the technology more accessible to both experts and nonspecialists alike. Through large-scale application, new questions, such as those about ligand-binding specificity, can in principle be addressed. Enabling a wider community to use the methods provides useful feedback and resources toward the development of the next generation of methods.

There are a number of servers for automated comparative modeling (**Table 1**). However, in spite of automation, the process of calculating a model for a given sequence, refining its structure, as well as visualizing and analyzing its family members in the sequence and structure spaces can involve the use of scripts, local programs, and servers scattered across the internet and not necessarily interconnected. In addition, manual intervention is generally still needed to

Table 1 Programs and web servers useful in comparative protein structure modeling

Name	*World Wide Web address*
Databases	
BAliBASE[222]	http://bips.u-strasbg.fr/en/Products/Databases/BAliBASE/
CATH[20]	http://www.biochem.ucl.ac.uk/bsm/cath/
DBALI[215]	http://www.salilab.org/dbali
GENBANK[14]	http://www.ncbi.nlm.nih.gov/Genbank/
GENECENSUS[223]	http://bioinfo.mbb.yale.edu/genome/
MODBASE[32]	http://www.salilab.org/modbase/
PDB[12]	http://www.rcsb.org/pdb/
PFAM[17]	http://www.sanger.ac.uk/Software/Pfam/
SCOP[19]	http://scop.mrc-lmb.cam.ac.uk/scop/
SwissProt[224]	http://www.expasy.org
Uniprot[13]	http://www.uniprot.org
Template search	
123D[225]	http://123d.ncifcrf.gov/
3D pssm[71]	http://www.sbg.bio.ic.ac.uk/~3dpssm
BLAST[36]	http://www.ncbi.nlm.nih.gov/BLAST/
DALI[33]	http://www2.ebi.ac.uk/dali/
FastA[226]	http://www.ebi.ac.uk/fasta33/
FFAS03[60]	http://ffas.ljcrf.edu/
PREDICTPROTEIN[227]	http://cubic.bioc.columbia.edu/predictprotein/
PROSPECTOR[67]	http://www.bioinformatics.buffalo.edu/current_buffalo/skolnick/prospector.html
PSIPRED[228]	http://bioinf.cs.ucl.ac.uk/psipred/
RAPTOR[68]	http://genome.math.uwaterloo.ca/~raptor/
SUPERFAMILY[229]	http://supfam.mrc-lmb.cam.ac.uk/SUPERFAMILY/
SAM-T02[69]	http://www.soe.ucsc.edu/research/compbio/HMM-apps/
SP3[53]	http://phyyz4.med.buffalo.edu/
SPARKS2[230]	http://phyyz4.med.buffalo.edu/
THREADER[231]	http://bioinf.cs.ucl.ac.uk/threader/threader.html
UCLA-DOE FoLD SERVER[232]	http://fold.doe-mbi.ucla.edu
Target–template alignment	
BCM SERVERF[233]	http://searchlauncher.bcm.tmc.edu
BLOCK MAKERF[234]	http://blocks.fhcrc.org/
CLUSTALW[235]	http://www2.ebi.ac.uk/clustalw/
COMPASS[57]	ftp://iole.swmed.edu/pub/compass/
FUGUE[63]	http://www-cryst.bioc.cam.ac.uk/fugue
MULTALIN[236]	http://prodes.toulouse.inra.fr/multalin/
MUSCLE[237]	http://www.drive5.com/muscle
SALIGN[213]	http://www.salilab.org/modeller
SEA[238]	http://ffas.ljcrf.edu/sea/
TCOFFEE[239]	http://www.ch.embnet.org/software/TCoffee.html
USC SEQALN[240]	http://www-hto.usc.edu/software/seqaln
Modeling	
3d-jigsaw[86]	http://www.bmm.icnet.uk/servers/3djigsaw/
COMPOSER[83]	http://www.tripos.com
CONGEN[121]	http://www.congenomics.com/
ICM[123]	http://www.molsoft.com
JACKAL[241]	http://trantor.bioc.columbia.edu/programs/jackal/
DISCOVERY STUDIO	http://www.accelrys.com
MODELLER[96]	http://www.salilab.org/modeller/

continued

Table 1 Continued

Name	*World Wide Web address*
SYBYL	http://www.tripos.com
SCWRL[147]	http://dunbrack.fccc.edu/SCWRL3.php
SNPWEB[213]	http://salilab.org/snpweb
SWISS-MODEL[87]	http://www.expasy.org/swissmod
WHAT IF[242]	http://www.cmbi.kun.nl/whatif/
Prediction of model errors	
ANOLEA[164]	http://protein.bio.puc.cl/cardex/servers/
AQUA[243]	http://urchin.bmrb.wisc.edu/~jurgen/aqua/
BIOTECH[244]	http://biotech.embl-heidelberg.de:8400
ERRAT[245]	http://www.doe-mbi.ucla.edu/Services/ERRAT/
PROCHECK[167]	http://www.biochem.ucl.ac.uk/~roman/procheck/procheck.html
ProsaII[162]	http://www.came.sbg.ac.at
PROVE[246]	http://www.ucmb.ulb.ac.be/UCMB/PROVE
SQUID[247]	http://www.ysbl.york.ac.uk/~oldfield/squid/
VERIFY3D[160]	http://www.doe-mbi.ucla.edu/Services/Verify_3D/
WHATCHECK[168]	http://www.cmbi.kun.nl/gv/whatcheck/
Methods evaluation	
CAFASP[171]	http://cafasp.bioinfo.pl
CASP[248]	http://predictioncenter.llnl.gov
CASA[249]	http://capb.dbi.udel.edu/casa
EVA[174]	http://cubic.bioc.columbia.edu/eva/
LiveBench[172]	http://bioinfo.pl/LiveBench/

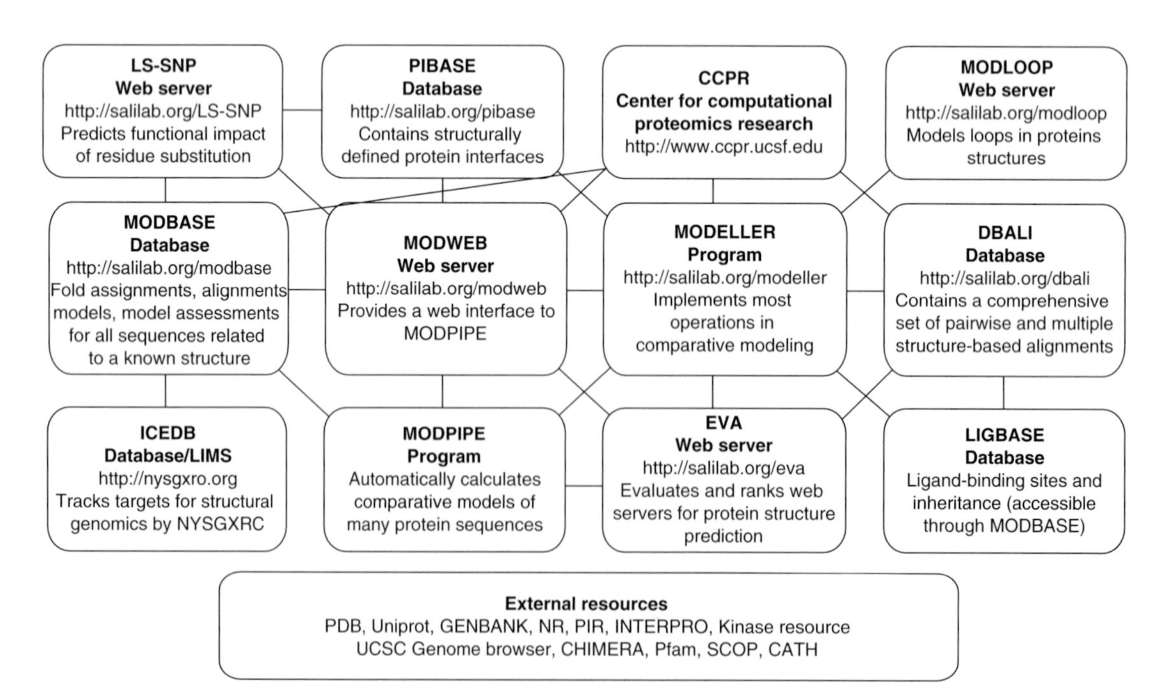

Figure 6 An integrated set of resources for comparative modeling.[32] Various databases and programs required for comparative modeling and docking are usually scattered over the internet, and require manual intervention or a good deal of expertise to be useful. Automation and integration of these resources are efficient ways to put these resources in the hands of experts and nonspecialists alike. We have outlined a comprehensive interconnected set of resources for comparative modeling and hope to integrate it with a similar effort in the area of ligand docking made by the Shoichet group.[220]

maximize the accuracy of the models in the difficult cases. The two main repositories for precomputed comparative models, SWISS-MODEL[87] and MODBASE,[31] begin to address these deficiencies. They provide access to web-based comparative modeling tools, cross-links to other sequence and structure databases, and annotations of sequences and their models.

A schematic of our own attempt at integrating several useful tools for comparative modeling is shown in **Figure 6**.[32,213] MODBASE is a comprehensive database that contains predicted models for domains in approximately one-half of all ~2.1 million known protein sequences. The models were calculated using MODPIPE[28,213] and MODELLER.[96] The web interface to the database allows flexible querying for fold assignments, sequence–structure alignments, models, and model assessments. An integrated sequence–structure viewer, Chimera,[214] allows inspection and analysis of the query results. Models can also be calculated using MODWEB,[213,250] a web interface to MODPIPE and stored in MODBASE to facilitate sharing, presentation, distribution, and annotation. For example, MODBASE contains binding site predictions for small ligands and a set of predicted interactions between pairs of modeled sequences from the same genome. Other resources associated with MODBASE include a comprehensive database of multiple protein structure alignments (DBALI),[215] a server for modeling of loops in protein structures (MODLOOP),[216,251] structurally defined ligand-binding sites,[217] structurally defined binary domain interfaces (PIBASE),[218,252] predictions of ligand-binding sites, interactions between yeast proteins, and functional consequences of human nsSNPs (LS-SNP).[175,219,253]

Compared to protein structure prediction, the attempts at automation and integration of resources in the field of docking for virtual screening are still in their nascent stages. One of the recent successful efforts in this direction is ZINC,[220] a publicly available database of commercially available druglike compounds. ZINC contains more than 3.3 million 'ready-to-dock' compounds organized in several subsets and allows the user to query the compounds by molecular properties and constitution. In the future, ZINC will rely on DOCKBLASTER that will enable end-users to dock the compounds against their target structures using DOCK.[189,190]

In the future, we will no doubt see efforts to improve the accuracy of comparative modeling and ligand docking. But perhaps more importantly, the two techniques will be integrated into a single protocol for more accurate and automated docking of ligands against sequences without known structures. As a result, the number and variety of applications of both comparative modeling and ligand docking will continue to increase.

Acknowledgments

This article is partially based on papers by Jacobson and Sali,[177] Fiser and Sali,[22] and Madhusudhan *et al.*[221] We also acknowledge the funds from Sandler Family Supporting Foundation, NIH R01 GM54762, P01 GM71790, P01 A135707, and U54 GM62529, as well as Sun, IBM, and Intel for hardware gifts.

References

1. Congreve, M.; Murray, C. W.; Blundell, T. L. *Drug Disc. Today* **2005**, *10*, 895–907.
2. Hardy, L., Malikayil, A. *Curr. Drug Disc.* **2003**, 15–20.
3. Lombardino, J. G.; Lowe, J. A., III *Nat. Rev. Drug Disc.* **2004**, *3*, 853–862.
4. van Dongen, M.; Weigelt, J.; Uppenberg, J.; Schultz, J.; Wikstrom, M. *Drug Disc. Today* **2002**, 7, 471–478.
5. Maryanoff, B. E. *J. Med. Chem.* **2004**, *47*, 769–787.
6. Pollack, V. A.; Savage, D. M.; Baker, D. A.; Tsaparikos, K. E.; Sloan, D. E.; Moyer, J. D.; Barbacci, E. G.; Pustilnik, L. R.; Smolarek, T. A.; Davis, J. A. et al. *J. Pharmacol. Exp. Ther.* **1999**, *291*, 739–748.
7. von Itzstein, M.; Wu, W. Y.; Kok, G. B.; Pegg, M. S.; Dyason, J. C.; Jin, B.; Van Phan, T.; Smythe, M. L.; White, H. F.; Oliver, S. W. et al. *Nature* **1993**, *363*, 418–423.
8. Zimmermann, J.; Caravatti, G.; Mett, H.; Meyer, T.; Muller, M.; Lydon, N. B.; Fabbro, D. *Arch. Pharm. (Weinheim)* **1996**, *329*, 371–376.
9. Baker, D.; Sali, A. *Science* **2001**, *294*, 93–96.
10. Arzt, S.; Beteva, A.; Cipriani, F.; Delageniere, S.; Felisaz, F.; Forstner, G.; Gordon, E.; Launer, L.; Lavault, B.; Leonard, G. et al. *Prog. Biophys. Mol. Biol.* **2005**, *89*, 124–152.
11. Pusey, M. L.; Liu, Z. J.; Tempel, W.; Praissman, J.; Lin, D.; Wang, B. C.; Gavira, J. A.; Ng, J. D. *Prog. Biophys. Mol. Biol.* **2005**, *88*, 359–386.
12. Deshpande, N.; Addess, K. J.; Bluhm, W. F.; Merino-Ott, J. C.; Townsend-Merino, W.; Zhang, Q.; Knezevich, C.; Xie, L.; Chen, L.; Feng, Z. et al. *Nucleic Acids Res.* **2005**, *33*, D233–D237.
13. Bairoch, A.; Apweiler, R.; Wu, C. H.; Barker, W. C.; Boeckmann, B.; Ferro, S.; Gasteiger, E.; Huang, H.; Lopez, R.; Magrane, M. et al. *Nucleic Acids Res.* **2005**, *33*, D154–D159.
14. Benson, D. A.; Karsch-Mizrachi, I.; Lipman, D. J.; Ostell, J.; Wheeler, D. L. *Nucleic Acids Res.* **2005**, *33*, D34–D38.
15. Chandonia, J. M.; Brenner, S. E. *Proteins* **2005**, *58*, 166–179.
16. Vitkup, D.; Melamud, E.; Moult, J.; Sander, C. *Nat. Struct. Biol.* **2001**, *8*, 559–566.
17. Bateman, A.; Coin, L.; Durbin, R.; Finn, R. D.; Hollich, V.; Griffiths-Jones, S.; Khanna, A.; Marshall, M.; Moxon, S.; Sonnhammer, E. L. et al. *Nucleic Acids Res.* **2004**, *32*, D138–D141.

18. Mulder, N. J.; Apweiler, R.; Attwood, T. K.; Bairoch, A.; Bateman, A.; Binns, D.; Bradley, P.; Bork, P.; Bucher, P.; Cerutti, L. et al. *Nucleic Acids Res.* **2005**, *33*, D201–D205.
19. Andreeva, A.; Howorth, D.; Brenner, S. E.; Hubbard, T. J.; Chothia, C.; Murzin, A. G. *Nucleic Acids Res.* **2004**, *32*, D226–D229.
20. Pearl, F.; Todd, A.; Sillitoe, I.; Dibley, M.; Redfern, O.; Lewis, T.; Bennett, C.; Marsden, R.; Grant, A.; Lee, D. et al. *Nucleic Acids Res.* **2005**, *33*, D247–D251.
21. Godzik, A. *Methods Biochem. Anal.* **2003**, *44*, 525–546.
22. Fiser, A.; Sali, A. *Methods Enzymol.* **2003**, *374*, 461–491.
23. Marti-Renom, M. A.; Stuart, A. C.; Fiser, A.; Sanchez, R.; Melo, F.; Sali, A. *Annu. Rev. Biophys. Biomol. Struct.* **2000**, *29*, 291–325.
24. Hillisch, A.; Pineda, L. F.; Hilgenfeld, R. *Drug Disc. Today* **2004**, *9*, 659–669.
25. Jorgensen, W. L. *Science* **2004**, *303*, 1813–1818.
26. Bradley, P.; Misura, K. M.; Baker, D. *Science* **2005**, *309*, 1868–1871.
27. Chothia, C.; Lesk, A. M. *EMBO J.* **1986**, *5*, 823–826.
28. Sanchez, R.; Sali, A. *Proc. Natl. Acad. Sci. USA* **1998**, *95*, 13597–13602.
29. Sanchez, R.; Pieper, U.; Melo, F.; Eswar, N.; Marti-Renom, M. A.; Madhusudhan, M. S.; Mirkovic, N.; Sali, A. *Nat. Struct. Biol.* **2000**, *7*, 986–990.
30. Sali, A. *Nat. Struct. Biol.* **1998**, *5*, 1029–1032.
31. Sali, A. *Nat. Struct. Biol.* **2001**, *8*, 482–484.
32. Pieper, U.; Eswar, N.; Braberg, H.; Madhusudhan, M. S.; Davis, F. P.; Stuart, A. C.; Mirkovic, N.; Rossi, A.; Marti-Renom, M. A.; Fiser, A. et al. *Nucleic Acids Res.* **2004**, *32*, D217–D222.
33. Dietmann, S.; Park, J.; Notredame, C.; Heger, A.; Lappe, M.; Holm, L. *Nucleic Acids Res.* **2001**, *29*, 55–57.
34. Rost, B. *Protein Eng.* **1999**, *12*, 85–94.
35. Pearson, W. R. *Methods Mol. Biol.* **1994**, *24*, 307–331.
36. Altschul, S. F.; Madden, T. L.; Schaffer, A. A.; Zhang, J.; Zhang, Z.; Miller, W.; Lipman, D. *J. Nucleic Acids Res.* **1997**, *25*, 3389–3402.
37. Brenner, S. E.; Chothia, C.; Hubbard, T. J. *Proc. Natl. Acad. Sci. USA* **1998**, *95*, 6073–6078.
38. Saqi, M. A.; Russell, R. B.; Sternberg, M. J. *Protein Eng.* **1998**, *11*, 627–630.
39. Gribskov, M.; McLachlan, A. D.; Eisenberg, D. *Proc. Natl. Acad. Sci. USA* **1987**, *84*, 4355–4358.
40. Henikoff, J. G.; Henikoff, S. *Comput. Appl. Biosci.* **1996**, *12*, 135–243.
41. Henikoff, S.; Henikoff, J. G. *J. Mol. Biol.* **1994**, *243*, 574–578.
42. Eddy, S. R. *Bioinformatics* **1998**, *14*, 755–763.
43. Krogh, A.; Brown, M.; Mian, I. S.; Sjolander, K.; Haussler, D. *J. Mol. Biol.* **1994**, *235*, 1501–1531.
44. Lindahl, E.; Elofsson, A. *J. Mol. Biol.* **2000**, *295*, 613–625.
45. Park, J.; Karplus, K.; Barrett, C.; Hughey, R.; Haussler, D.; Hubbard, T.; Chothia, C. *J. Mol. Biol.* **1998**, *284*, 1201–1210.
46. Sauder, J. M.; Arthur, J. W.; Dunbrack, R. L., Jr. *Proteins* **2000**, *40*, 6–22.
47. Marti-Renom, M. A.; Madhusudhan, M. S.; Sali, A. *Protein Sci.* **2004**, *13*, 1071–1087.
48. Karplus, K.; Barrett, C.; Hughey, R. *Bioinformatics* **1998**, *14*, 846–856.
49. Eswar, N., Madhusudhan, M. S., Marti-Renom, M. A., Sali, A.; 8v1 ed. 2005. http://salilab.org/modeller (accessed Aug 2006).
50. Edgar, R. C.; Sjolander, K. *Bioinformatics* **2004**, *20*, 1301–1308.
51. Ohlson, T.; Wallner, B.; Elofsson, A. *Proteins* **2004**, *57*, 188–197.
52. Wang, G.; Dunbrack, R. L., Jr. *Protein Sci.* **2004**, *13*, 1612–1626.
53. Zhou, H.; Zhou, Y. *Proteins* **2005**, *58*, 321–328.
54. Panchenko, A. R. *Nucleic Acids Res.* **2003**, *31*, 683–689.
55. Pietrokovski, S. *Nucleic Acids Res.* **1996**, *24*, 3836–3845.
56. Rychlewski, L.; Zhang, B.; Godzik, A. *Fold Des.* **1998**, *3*, 229–238.
57. Sadreyev, R.; Grishin, N. *J. Mol. Biol.* **2003**, *326*, 317–336.
58. von Ohsen, N.; Sommer, I.; Zimmer, R. *Pac. Symp. Biocomput.* **2003**, 252–263.
59. Yona, G.; Levitt, M. *J. Mol. Biol.* **2002**, *315*, 1257–1275.
60. Jaroszewski, L.; Rychlewski, L.; Li, Z.; Li, W.; Godzik, A. *Nucleic Acids Res.* **2005**, *33*, W284–W288.
61. McGuffin, L. J.; Jones, D. T. *Bioinformatics* **2003**, *19*, 874–981.
62. Karchin, R.; Cline, M.; Mandel-Gutfreund, Y.; Karplus, K. *Proteins* **2003**, *51*, 504–514.
63. Shi, J.; Blundell, T. L.; Mizuguchi, K. *J. Mol. Biol.* **2001**, *310*, 243–257.
64. Bowie, J. U.; Luthy, R.; Eisenberg, D. *Science* **1991**, *253*, 164–170.
65. Sippl, M. J. *J. Mol. Biol.* **1990**, *213*, 859–883.
66. Sippl, M. J. *Curr. Opin. Struct. Biol.* **1995**, *5*, 229–235.
67. Skolnick, J.; Kihara, D. *Proteins* **2001**, *42*, 319–331.
68. Xu, J.; Li, M.; Kim, D.; Xu, Y. *J. Bioinform. Comput. Biol.* **2003**, *1*, 95–117.
69. Karplus, K.; Karchin, R.; Draper, J.; Casper, J.; Mandel-Gutfreund, Y.; Diekhans, M.; Hughey, R. *Proteins* **2003**, *53*, 491–496.
70. Jones, D. T. *J. Mol. Biol.* **1999**, *287*, 797–815.
71. Kelley, L. A.; MacCallum, R. M.; Sternberg, M. J. *J. Mol. Biol.* **2000**, *299*, 499–520.
72. John, B.; Sali, A. *Nucleic Acids Res.* **2003**, *31*, 3982–3992.
73. Moult, J. *Curr. Opin. Struct. Biol.* **2005**, *15*, 285–289.
74. Sanchez, R.; Sali, A. *Curr. Opin. Struct. Biol.* **1997**, 7, 206–214.
75. Sanchez, R.; Sali, A. *Proteins* **1997**, 50–58.
76. Srinivasan, N.; Blundell, T. L. *Protein Eng.* **1993**, *6*, 501–512.
77. Bajorath, J.; Aruffo, A. *Bioconjug. Chem.* **1994**, *5*, 173–181.
78. Blundell, T. L.; Sibanda, B. L.; Sternberg, M. J.; Thornton, J. M. *Nature* **1987**, *326*, 347–352.
79. Browne, W. J.; North, A. C.; Phillips, D. C.; Brew, K.; Vanaman, T. C.; Hill, R. L. *J. Mol. Biol.* **1969**, *42*, 65–86.
80. Johnson, M. S.; Srinivasan, N.; Sowdhamini, R.; Blundell, T. L. *Crit. Rev. Biochem. Mol. Biol.* **1994**, *29*, 1–68.
81. Greer, J. *J. Mol. Biol.* **1981**, *153*, 1027–1042.
82. Nagarajaram, H. A.; Reddy, B. V.; Blundell, T. L. *Protein Eng.* **1999**, *12*, 1055–1062.
83. Sutcliffe, M. J.; Haneef, I.; Carney, D.; Blundell, T. L. *Protein Eng.* **1987**, *1*, 377–384.

84. Sutcliffe, M. J.; Hayes, F. R.; Blundell, T. L. *Protein Eng.* **1987**, *1*, 385–392.
85. Topham, C. M.; McLeod, A.; Eisenmenger, F.; Overington, J. P.; Johnson, M. S.; Blundell, T. L. *J. Mol. Biol.* **1993**, *229*, 194–220.
86. Bates, P. A.; Kelley, L. A.; MacCallum, R. M.; Sternberg, M. J. *Proteins* **2001**, 39–46.
87. Schwede, T.; Kopp, J.; Guex, N.; Peitsch, M. C. *Nucleic Acids Res.* **2003**, *31*, 3381–3385.
88. Bystroff, C.; Baker, D. *J. Mol. Biol.* **1998**, *281*, 565–577.
89. Claessens, M.; Van Cutsem, E.; Lasters, I.; Wodak, S. *Protein Eng.* **1989**, *2*, 335–345.
90. Jones, T. A.; Thirup, S. *EMBO J.* **1986**, *5*, 819–822.
91. Levitt, M. *J. Mol. Biol.* **1992**, *226*, 507–533.
92. Unger, R.; Harel, D.; Wherland, S.; Sussman, J. L. *Proteins* **1989**, *5*, 355–373.
93. Aszodi, A.; Taylor, W. R. *Fold Des.* **1996**, *1*, 325–334.
94. Brocklehurst, S. M.; Perham, R. N. *Protein Sci.* **1993**, *2*, 626–639.
95. Havel, T. F.; Snow, M. E. *J. Mol. Biol.* **1991**, *217*, 1–7.
96. Sali, A.; Blundell, T. L. *J. Mol. Biol.* **1993**, *234*, 779–815.
97. Srinivasan, S.; March, C. J.; Sudarsanam, S. *Protein Sci.* **1993**, *2*, 277–289.
98. Fiser, A.; Do, R. K.; Sali, A. *Protein Sci.* **2000**, *9*, 1753–1773.
99. Fiser, A.; Feig, M.; Brooks, C. L., III; Sali, A. *Acc. Chem. Res.* **2002**, *35*, 413–421.
100. MacKerell, A. D., Jr.; Bashford, D.; Bellott, M.; Dunbrack, R. L., Jr.; Evanseck, J. D.; Field, M. J.; Fischer, S.; Gao, J.; Guo, H.; Ha, S. et al. *J. Phys. Chem. B* **1998**, *102*, 3586–3616.
101. Sali, A.; Overington, J. P. *Protein Sci.* **1994**, *3*, 1582–1596.
102. Marti-Renom, M. A.; Madhusudhan, M. S.; Fiser, A.; Rost, B.; Sali, A. *Struct. (Camb.)* **2002**, *10*, 435–440.
103. Wallner, B.; Elofsson, A. *Protein Sci.* **2005**, *14*, 1315–1327.
104. Kabsch, W.; Sander, C. *Proc. Natl. Acad. Sci. USA* **1984**, *81*, 1075–1078.
105. Mezei, M. *Protein Eng.* **1998**, *11*, 411–414.
106. Jacobson, M. P.; Pincus, D. L.; Rapp, C. S.; Day, T. J.; Honig, B.; Shaw, D. E.; Friesner, R. A. *Proteins* **2004**, *55*, 351–367.
107. Chothia, C.; Lesk, A. M. *J. Mol. Biol.* **1987**, *196*, 901–917.
108. Moult, J.; James, M. N. *Proteins* **1986**, *1*, 146–163.
109. Bruccoleri, R. E.; Karplus, M. *Biopolymers* **1987**, *26*, 137–168.
110. Shenkin, P. S.; Yarmush, D. L.; Fine, R. M.; Wang, H. J.; Levinthal, C. *Biopolymers* **1987**, *26*, 2053–2085.
111. van Vlijmen, H. W.; Karplus, M. *J. Mol. Biol.* **1997**, *267*, 975–1001.
112. Deane, C. M.; Blundell, T. L. *Protein Sci.* **2001**, *10*, 599–612.
113. Sibanda, B. L.; Blundell, T. L.; Thornton, J. M. *J. Mol. Biol.* **1989**, *206*, 759–777.
114. Chothia, C.; Lesk, A. M.; Tramontano, A.; Levitt, M.; Smith-Gill, S. J.; Air, G.; Sheriff, S.; Padlan, E. A.; Davies, D.; Tulip, W. R. et al. *Nature* **1989**, *342*, 877–883.
115. Rufino, S. D.; Donate, L. E.; Canard, L. H.; Blundell, T. L. *J. Mol. Biol.* **1997**, *267*, 352–367.
116. Oliva, B.; Bates, P. A.; Querol, E.; Aviles, F. X.; Sternberg, M. J. *J. Mol. Biol.* **1997**, *266*, 814–830.
117. Ring, C. S.; Kneller, D. G.; Langridge, R.; Cohen, F. E. *J. Mol. Biol.* **1992**, *224*, 685–699.
118. Fidelis, K.; Stern, P. S.; Bacon, D.; Moult, J. *Protein Eng.* **1994**, 7, 953–960.
119. Lessel, U.; Schomburg, D. *Protein Eng.* **1994**, 7, 1175–1187.
120. Fine, R. M.; Wang, H.; Shenkin, P. S.; Yarmush, D. L.; Levinthal, C. *Proteins* **1986**, *1*, 342–362.
121. Bruccoleri, R. E.; Karplus, M. *Biopolymers* **1990**, *29*, 1847–1862.
122. Ring, C. S.; Sun, E.; McKerrow, J. H.; Lee, G. K.; Rosenthal, P. J.; Kuntz, I. D.; Cohen, F. E. *Proc. Natl. Acad. Sci. USA* **1993**, *90*, 3583–3587.
123. Abagyan, R.; Totrov, M. *J. Mol. Biol.* **1994**, *235*, 983–1002.
124. Collura, V.; Higo, J.; Garnier, J. *Protein Sci.* **1993**, *2*, 1502–1510.
125. Higo, J.; Collura, V.; Garnier, J. *Biopolymers* **1992**, *32*, 33–43.
126. Zheng, Q.; Rosenfeld, R.; Vajda, S.; DeLisi, C. *Protein Sci.* **1993**, *2*, 1242–1248.
127. Koehl, P.; Delarue, M. *Nat. Struct. Biol.* **1995**, *2*, 163–170.
128. Samudrala, R.; Moult, J. *J. Mol. Biol.* **1998**, *279*, 287–302.
129. Xiang, Z.; Soto, C. S.; Honig, B. *Proc. Natl. Acad. Sci. USA* **2002**, *99*, 7432–7437.
130. de Bakker, P. I.; DePristo, M. A.; Burke, D. F.; Blundell, T. L. *Proteins* **2003**, *51*, 21–40.
131. DePristo, M. A.; de Bakker, P. I.; Lovell, S. C.; Blundell, T. L. *Proteins* **2003**, *51*, 41–55.
132. Felts, A. K.; Gallicchio, E.; Wallqvist, A.; Levy, R. M. *Proteins-Struct. Funct. Genet.* **2002**, *48*, 404–422.
133. Dunbrack, R. L., Jr. *Curr. Opin. Struct. Biol.* **2002**, *12*, 431–440.
134. Janin, J.; Chothia, C. *Biochemistry* **1978**, *17*, 2943–2948.
135. Ponder, J. W.; Richards, F. M. *J. Mol. Biol.* **1987**, *193*, 775–791.
136. De Maeyer, M.; Desmet, J.; Lasters, I. *Fold Des.* **1997**, *2*, 53–66.
137. Dunbrack, R. L., Jr.; Cohen, F. E. *Protein Sci.* **1997**, *6*, 1661–1681.
138. Dunbrack, R. L., Jr.; Karplus, M. *J. Mol. Biol.* **1993**, *230*, 543–574.
139. Lovell, S. C.; Word, J. M.; Richardson, J. S.; Richardson, D. C. *Proteins* **2000**, *40*, 389–408.
140. McGregor, M. J.; Islam, S. A.; Sternberg, M. J. *J. Mol. Biol.* **1987**, *198*, 295–310.
141. Schrauber, H.; Eisenhaber, F.; Argos, P. *J. Mol. Biol.* **1993**, *230*, 592–612.
142. Tuffery, P.; Etchebest, C.; Hazout, S.; Lavery, R. *J. Biomol. Struct. Dyn.* **1991**, *8*, 1267–1289.
143. Desjarlais, J. R.; Handel, T. M. *J. Mol. Biol.* **1999**, *290*, 305–318.
144. De Filippis, V.; Sander, C.; Vriend, G. *Protein Eng.* **1994**, 7, 1203–1208.
145. Chung, S. Y.; Subbiah, S. *Pac. Symp. Biocomput.* **1996**, 126–141.
146. Cregut, D.; Liautard, J. P.; Chiche, L. *Protein Eng.* **1994**, 7, 1333–1344.
147. Canutescu, A. A.; Shelenkov, A. A.; Dunbrack, R. L., Jr. *Protein Sci.* **2003**, *12*, 2001–2014.
148. Xiang, Z.; Honig, B. *J. Mol. Biol.* **2001**, *311*, 421–430.
149. Eisenmenger, F.; Argos, P.; Abagyan, R. *J. Mol. Biol.* **1993**, *231*, 849–860.
150. Lee, G. M.; Varma, A.; Palsson, B. O. *Biotechnol. Prog.* **1991**, 7, 72–75.
151. Holm, L.; Sander, C. *Proteins* **1992**, *14*, 213–223.

152. Lasters, I.; Desmet, J. *Protein Eng.* **1993**, *6*, 717–722.
153. Looger, L. L.; Hellinga, H. W. *J. Mol. Biol.* **2001**, *307*, 429–445.
154. Hwang, J. K.; Liao, W. F. *Protein Eng.* **1995**, *8*, 363–370.
155. Koehl, P.; Delarue, M. *J. Mol. Biol.* **1994**, *239*, 249–275.
156. Bower, M. J.; Cohen, F. E.; Dunbrack, R. L., Jr. *J. Mol. Biol.* **1997**, *267*, 1268–1282.
157. Petrella, R. J.; Lazaridis, T.; Karplus, M. *Fold Des.* **1998**, *3*, 353–377.
158. Jacobson, M. P.; Kaminski, G. A.; Friesner, R. A.; Rapp, C. S. *J. Phys. Chem. B* **2002**, *106*, 11673–11680.
159. Liang, S.; Grishin, N. V. *Protein Sci.* **2002**, *11*, 322–331.
160. Luthy, R.; Bowie, J. U.; Eisenberg, D. *Nature* **1992**, *356*, 83–85.
161. Melo, F.; Sanchez, R.; Sali, A. *Protein Sci.* **2002**, *11*, 430–448.
162. Sippl, M. J. *Proteins* **1993**, *17*, 355–362.
163. Topham, C. M.; Srinivasan, N.; Thorpe, C. J.; Overington, J. P.; Kalsheker, N. A. *Protein Eng.* **1994**, 7, 869–894.
164. Melo, F.; Feytmans, E. *J. Mol. Biol.* **1998**, *277*, 1141–1152.
165. Zhou, H.; Zhou, Y. *Protein Sci.* **2002**, *11*, 2714–2726.
166. Pawlowski, K.; Bierzynski, A.; Godzik, A. *J. Mol. Biol.* **1996**, *258*, 349–366.
167. Laskowski, R. A.; MacArthur, M. W.; Moss, D. S.; Thornton, J. M. *J. Appl. Crystallogr.* **1993**, *26*, 283–291.
168. Hooft, R. W.; Vriend, G.; Sander, C.; Abola, E. E. *Nature* **1996**, *381*, 272.
169. Coutsias, E. A.; Seok, C.; Jacobson, M. P.; Dill, K. A. *J. Comput. Chem.* **2004**, *25*, 510–528.
170. Zemla, A.; Venclovas, C.; Moult, J.; Fidelis, K. *Proteins* **2001**, 13–21.
171. Fischer, D.; Elofsson, A.; Rychlewski, L.; Pazos, F.; Valencia, A.; Rost, B.; Ortiz, A. R.; Dunbrack, R. L., Jr. *Proteins* **2001**, *45*, 171–183.
172. Bujnicki, J. M.; Elofsson, A.; Fischer, D.; Rychlewski, L. *Protein Sci.* **2001**, *10*, 352–361.
173. Eyrich, V. A.; Marti-Renom, M. A.; Przybylski, D.; Madhusudhan, M. S.; Fiser, A.; Pazos, F.; Valencia, A.; Sali, A.; Rost, B. *Bioinformatics* **2001**, *17*, 1242–1243.
174. Koh, I.-Y. Y.; Eyrich, V. A.; Marti-Renom, M. A.; Przybylski, D.; Madhusudhan, M. S.; Narayanan, E.; Grana, O.; Pazos, F.; Valencia, A.; Sali, A. et al. *Nucleic Acids Res.* **2003**, *31*, 3311–3315.
175. Karchin, R.; Diekhans, M.; Kelly, L.; Thomas, D. J.; Pieper, U.; Eswar, N.; Haussler, D.; Sali, A. *Bioinformatics* **2005**, *21*, 2814–2820.
176. Thiel, K. A. *Nat. Biotechnol.* **2004**, *22*, 513–519.
177. Jacobson, M. P., Sali, A. *Comparative Modeling and Its Applications to Drug Discovery*; Inpharmatica: London, 2004; Vol. 39.
178. Gao, H.; Sengupta, J.; Valle, M.; Korostelev, A.; Eswar, N.; Stagg, S. M.; Van Roey, P.; Agrawal, R. K.; Harvey, S. C.; Sali, A. et al. *Cell* **2003**, *113*, 789–801.
179. Spahn, C. M.; Beckmann, R.; Eswar, N.; Penczek, P. A.; Sali, A.; Blobel, G.; Frank, J. *Cell* **2001**, *107*, 373–386.
180. Blundell, T. L.; Johnson, M. S. *Protein Sci.* **1993**, *2*, 877–883.
181. Chakravarty, S.; Sanchez, R. *Structure (Camb.)* **2004**, *12*, 1461–1470.
182. Chakravarty, S.; Wang, L.; Sanchez, R. *Nucleic Acids Res.* **2005**, *33*, 244–259.
183. von Grotthuss, M.; Wyrwicz, L. S.; Rychlewski, L. *Cell* **2003**, *113*, 701–702.
184. Gordon, R. K.; Ginalski, K.; Rudnicki, W. R.; Rychlewski, L.; Pankaskie, M. C.; Bujnicki, J. M.; Chiang, P. K. *Eur. J. Biochem.* **2003**, *270*, 3507–3517.
185. Evers, A.; Gohlke, H.; Klebe, G. *J. Mol. Biol.* **2003**, *334*, 327–345.
186. Evers, A.; Klebe, G. *Angew. Chem. Int. Ed. Engl.* **2004**, *43*, 248–251.
187. Schafferhans, A.; Klebe, G. *J. Mol. Biol.* **2001**, *307*, 407–427.
188. McGovern, S. L.; Shoichet, B. K. *J. Med. Chem.* **2003**, *46*, 2895–2907.
189. Lorber, D. M.; Shoichet, B. K. *Protein Sci.* **1998**, 7, 938–950.
190. Wei, B. Q.; Baase, W. A.; Weaver, L. H.; Matthews, B. W.; Shoichet, B. K. *J. Mol. Biol.* **2002**, *322*, 339–355.
191. Li, R.; Chen, X.; Gong, B.; Selzer, P. M.; Li, Z.; Davidson, E.; Kurzban, G.; Miller, R. E.; Nuzum, E. O.; McKerrow, J. H. et al. *Bioorg. Med. Chem.* 1996, *4*, 1421–1427.
192. Que, X.; Brinen, L. S.; Perkins, P.; Herdman, S.; Hirata, K.; Torian, B. E.; Rubin, H.; McKerrow, J. H.; Reed, S. L. *Mol. Biochem. Parasitol.* **2002**, *119*, 23–32.
193. Selzer, P. M.; Chen, X.; Chan, V. J.; Cheng, M.; Kenyon, G. L.; Kuntz, I. D.; Sakanari, J. A.; Cohen, F. E.; McKerrow, J. H. *Exp. Parasitol.* **1997**, *87*, 212–221.
194. Enyedy, I. J.; Lee, S. L.; Kuo, A. H.; Dickson, R. B.; Lin, C. Y.; Wang, S. *J. Med. Chem.* **2001**, *44*, 1349–1355.
195. Enyedy, I. J.; Ling, Y.; Nacro, K.; Tomita, Y.; Wu, X.; Cao, Y.; Guo, R.; Li, B.; Zhu, X.; Huang, Y. et al. *J. Med. Chem.* **2001**, *44*, 4313–4324.
196. Schapira, M.; Raaka, B. M.; Samuels, H. H.; Abagyan, R. *BMC Struct Biol* **2001**, *1*, 1.
197. Zuccotto, F.; Zvelebil, M.; Brun, R.; Chowdhury, S. F.; Di Lucrezia, R.; Leal, I.; Maes, L.; Ruiz-Perez, L. M.; Gonzalez Pacanowska, D.; Gilbert, I. H. *Eur. J. Med. Chem.* **2001**, *36*, 395–405.
198. Anand, K.; Ziebuhr, J.; Wadhwani, P.; Mesters, J. R.; Hilgenfeld, R. *Science* **2003**, *300*, 1763–1767.
199. Rajnarayanan, R. V.; Dakshanamurthy, S.; Pattabiraman, N. *Biochem. Biophys. Res. Commun.* **2004**, *321*, 370–378.
200. Diller, D. J.; Li, R. *J. Med. Chem.* **2003**, *46*, 4638–4647.
201. Diller, D. J.; Merz, K. M., Jr. *Proteins* **2001**, *43*, 113–124.
202. Oshiro, C.; Bradley, E. K.; Eksterowicz, J.; Evensen, E.; Lamb, M. L.; Lanctot, J. K.; Putta, S.; Stanton, R.; Grootenhuis, P. D. *J. Med. Chem.* **2004**, *47*, 764–767.
203. Rockey, W. M.; Elcock, A. H. *Proteins* **2002**, *48*, 664–671.
204. Vangrevelinghe, E.; Zimmermann, K.; Schoepfer, J.; Portmann, R.; Fabbro, D.; Furet, P. *J. Med. Chem.* **2003**, *46*, 2656–2662.
205. Becker, O. M.; Shacham, S.; Marantz, Y.; Noiman, S. *Curr. Opin. Drug Disc. Dev.* **2003**, *6*, 353–361.
206. Bissantz, C.; Bernard, P.; Hibert, M.; Rognan, D. *Proteins* **2003**, *50*, 5–25.
207. Bissantz, C.; Logean, A.; Rognan, D. *J. Chem. Inf. Comput. Sci.* **2004**, *44*, 1162–1176.
208. Shacham, S.; Topf, M.; Avisar, N.; Glaser, F.; Marantz, Y.; Bar-Haim, S.; Noiman, S.; Naor, Z.; Becker, O. M. *Med. Res. Rev.* **2001**, *21*, 472–483.
209. Vaidehi, N.; Floriano, W. B.; Trabanino, R.; Hall, S. E.; Freddolino, P.; Choi, E. J.; Zamanakos, G.; Goddard, W. A., III *Proc. Natl. Acad. Sci. USA* **2002**, *99*, 12622–12627.
210. Palczewski, K.; Kumasaka, T.; Hori, T.; Behnke, C. A.; Motoshima, H.; Fox, B. A.; Le Trong, I.; Teller, D. C.; Okada, T.; Stenkamp, R. E. et al. *Science* **2000**, *289*, 739–745.
211. Barril, X.; Morley, S. D. *J. Med. Chem.* **2005**, *48*, 4432–4443.

212. Carlson, H. A.; McCammon, J. A. *Mol. Pharmacol.* **2000**, *57*, 213–218.
213. Eswar, N.; John, B.; Mirkovic, N.; Fiser, A.; Ilyin, V. A.; Pieper, U.; Stuart, A. C.; Marti-Renom, M. A.; Madhusudhan, M. S.; Yerkovich, B. et al. *Nucleic Acids Res.* **2003**, *31*, 3375–3380.
214. Huang, C. C.; Novak, W. R.; Babbitt, P. C.; Jewett, A. I.; Ferrin, T. E.; Klein, T. E. *Pac. Symp. Biocomput.* **2000**, 230–241.
215. Marti-Renom, M. A.; Ilyin, V. A.; Sali, A. *Bioinformatics* **2001**, *17*, 746–747.
216. Fiser, A.; Sali, A. *Bioinformatics* **2003**, *19*, 2500–2501.
217. Stuart, A. C.; Ilyin, V. A.; Sali, A. *Bioinformatics* **2002**, *18*, 200–201.
218. Davis, F. P.; Sali, A. *Bioinformatics* **2005**, *21*, 1901–1907.
219. Mirkovic, N.; Marti-Renom, M. A.; Weber, B. L.; Sali, A.; Monteiro, A. N. *Cancer Res.* **2004**, *64*, 3790–3797.
220. Irwin, J. J.; Shoichet, B. K. *J. Chem. Inf. Model.* **2005**, *45*, 177–182.
221. Madhusudhan, M. S.; Marti-Renom, M. A.; Eswar, N.; John, B.; Pieper, U.; Karchin, R.; Shen, M. Y.; Sali, A. In *The Proteomics Protocols Handbook*; Walker, J. M., Ed.; Humana Press: Totowa, NJ, 2005, pp 831–860.
222. Thompson, J. D.; Plewniak, F.; Poch, O. *Bioinformatics* **1999**, *15*, 87–88.
223. Lin, J.; Qian, J.; Greenbaum, D.; Bertone, P.; Das, R.; Echols, N.; Senes, A.; Stenger, B.; Gerstein, M. *Nucleic Acids Res.* **2002**, *30*, 4574–4582.
224. Boeckmann, B.; Bairoch, A.; Apweiler, R.; Blatter, M. C.; Estreicher, A.; Gasteiger, E.; Martin, M. J.; Michoud, K.; O'Donovan, C.; Phan, I. et al. *Nucleic Acids Res.* **2003**, *31*, 365–370.
225. Alexandrov, N. N.; Nussinov, R.; Zimmer, R. M. *Pac. Symp. Biocomput.* **1996**, 53–72.
226. Pearson, W. R. *Methods Mol. Biol.* **2000**, *132*, 185–219.
227. Rost, B.; Liu, J. *Nucleic Acids Res.* **2003**, *31*, 3300–3304.
228. McGuffin, L. J.; Bryson, K.; Jones, D. T. *Bioinformatics* **2000**, *16*, 404–405.
229. Gough, J.; Karplus, K.; Hughey, R.; Chothia, C. *J. Mol. Biol.* **2001**, *313*, 903–919.
230. Zhou, H.; Zhou, Y. *Proteins* **2004**, *55*, 1005–1013.
231. Jones, D. T.; Taylor, W. R.; Thornton, J. M. *Nature* **1992**, *358*, 86–89.
232. Mallick, P.; Weiss, R.; Eisenberg, D. *Proc. Natl. Acad. Sci. USA* **2002**, *99*, 16041–16046.
233. Worley, K. C.; Culpepper, P.; Wiese, B. A.; Smith, R. F. *Bioinformatics* **1998**, *14*, 890–891.
234. Henikoff, J. G.; Pietrokovski, S.; McCallum, C. M.; Henikoff, S. *Electrophoresis* **2000**, *21*, 1700–1706.
235. Thompson, J. D.; Higgins, D. G.; Gibson, T. J. *Nucleic Acids Res.* **1994**, *22*, 4673–4680.
236. Corpet, F. *Nucleic Acids Res.* **1988**, *16*, 10881–10890.
237. Edgar, R. C. *Nucleic Acids Res.* **2004**, *32*, 1792–1797.
238. Ye, Y.; Jaroszewski, L.; Li, W.; Godzik, A. *Bioinformatics* **2003**, *19*, 742–749.
239. Notredame, C.; Higgins, D. G.; Heringa, J. *J. Mol. Biol.* **2000**, *302*, 205–217.
240. Smith, T. F.; Waterman, M. S. *J. Mol. Biol.* **1981**, *147*, 195–197.
241. Petrey, D.; Xiang, Z.; Tang, C. L.; Xie, L.; Gimpelev, M.; Mitros, T.; Soto, C. S.; Goldsmith-Fischman, S.; Kernytsky, A.; Schlessinger, A. et al. *Proteins* **2003**, *53*, 430–435.
242. Vriend, G. *J. Mol. Graph.* **1990**, *8*, 52–56.
243. Laskowski, R. A.; Rullmannn, J. A.; MacArthur, M. W.; Kaptein, R.; Thornton, J. M. *J. Biomol. NMR* **1996**, *8*, 477–486.
244. Laskowski, R. A.; MacArthur, M. W.; Thornton, J. M. *Curr. Opin. Struct. Biol.* **1998**, *8*, 631–639.
245. Colovos, C.; Yeates, T. O. *Protein Sci.* **1993**, *2*, 1511–1519.
246. Pontius, J.; Richelle, J.; Wodak, S. J. *J. Mol. Biol.* **1996**, *264*, 121–136.
247. Oldfield, T. J. *J. Mol. Graph.* **1992**, *10*, 247–252.
248. Moult, J.; Fidelis, K.; Zemla, A.; Hubbard, T. *Proteins* **2003**, *53*, 334–339.
249. Kahsay, R. Y.; Wang, G.; Dongre, N.; Gao, G.; Dunbrack, R. L., Jr. *Bioinformatics* **2002**, *18*, 496–497.
250. MODWEB. http://salilab.org/modweb (accessed April 2006).
251. MODLOOP. http://salilab.org/modloop (accessed April 2006).
252. PIBASE. http://salilab.org/pibase (accessed April 2006).
253. LS-SNP. http://salilab.org/LS-SNP (accessed April 2006).

Biographies

Eswar Narayanan received his BSc degree in physics from the Loyola College, India, in 1993 for which he was awarded a Gold Medal and Scholarship for academic proficiency. After an MSc degree in Physics from the University of Hyderabad, India, he was awarded a Research Fellowship by the Indian Institute of Science, Bangalore, India, for a PhD under the supervision of Prof C Ramakrishnan at the Molecular Biophysics Unit where he focused on the conformational analysis of protein structures. He then joined the laboratory of Prof Andrej Sali at the Rockefeller University, New York, as a Research Associate where he developed the large-scale protein structure modeling pipeline, MODPIPE. In 2003, he moved along with Prof Sali to the University of California at San Francisco, where, as an Assistant Professional Researcher, he continues to work on the development of methods for protein structure prediction and its application to modeling structures of macromolecular assemblies and modeling drug target proteins for virtual screening.

Andrej Sali received his BSc degree in chemistry from the University of Ljubljana, Slovenia, in 1987. He was awarded the Research Council of Slovenia Scholarship, the Overseas Research Students Award, and the Merck Sharpe and Dohm Academic Scholarship at Birkbeck College, University of London, where he received his PhD in biophysics in 1991, under the supervision of Prof Tom L Blundell. He focused on development of methods for comparative modeling of protein three-dimensional structure and their implementation in the program MODELLER. He then went to the Department of Chemistry at Harvard University as a Jane Coffin Childs Memorial Fund postdoctoral fellow with Prof Martin Karplus, where he continued to develop comparative modeling methods and also studied simple lattice Monte Carlo models of protein folding. From 1995 to 2002, Dr Sali was first an assistant professor and then an associate professor at the Rockefeller University. In 2003, he moved to University of California at San Francisco as a Professor of Computational Biology in the Departments of Biopharmaceutical Sciences and Pharmaceutical Chemistry, and California Institute for Quantitative Biomedical Research. He was a Sinsheimer Scholar, an Alfred P Sloan Research Fellow, and an Irma T Hirschl Trust Career Scientist. Dr Sali is an Editor of Structure and a Founder of Prospect Genomix, now Structural Genomix. He is interested in using computation grounded in the laws of physics and the theory of evolution to study the structure and function of proteins. He is aiming to improve and apply methods for (i) predicting the structures of proteins; (ii) determining the structures of macromolecular assemblies; and (iii) annotating the functions of proteins using their structures.

© 2007 Elsevier Ltd. All Rights Reserved
No part of this publication may be reproduced, stored in any retrieval system or transmitted in any form by any means electronic, electrostatic, magnetic tape, mechanical, photocopying, recording or otherwise, without permission in writing from the publishers

Comprehensive Medicinal Chemistry II
ISBN (set): 0-08-044513-6
ISBN (Volume 4) 0-08-044517-9; pp. 215–236

4.11 Characterization of Protein-Binding Sites and Ligands Using Molecular Interaction Fields

G Cruciani and E Carosati, University of Perugia, Italy
R C Wade, EML Research, Heidelberg, Germany
M Baroni, Molecular Discovery, Pinner, UK

© 2007 Elsevier Ltd. All Rights Reserved.

4.11.1 Introduction

The scientific interest in structure-based ligand design approaches and the number of research applications are rapidly increasing due to the growing amount of three-dimensional (3D) structural information on receptors, channels, enzymes, and transporters becoming available through protein crystallography and nuclear magnetic resonance (NMR).

Structure-based ligand design is based on the observation that most drugs bind to a clearly defined macromolecular target, which is complementary in terms of 3D structure and pharmacophoric chemical groups. The first computational structure-based ligand design methods came into existence in the early 1980s and these and subsequent methods have become powerful tools for modern pharmaceutical research. Indeed, some of the drugs on the market were designed with these approaches. Nowadays, high-throughput protein–ligand structure determination by x-ray crystallography offers another opportunity to speed up and facilitate structure-based ligand/drug design applications.

Structure-based ligand design is based on a firm understanding of the molecular recognition process between a protein's binding site groups and the interacting ligand molecules. When the 3D structure of the binding site is known, structure-based drug design can be performed. The first stage in this process is to determine the characteristics of the protein-binding site. The analysis comprises the quantification of the site's properties, such as the role of water molecules, the location of hydrophobic hot spots, hydrogen bond patterns, and so on. The computational analysis provides data that in turn allow the application of rational strategies to generate de novo virtual structures that are complementary to the binding site. Moreover, visual inspection or statistical analysis of computational outputs enables prioritization of compounds for chemical syntheses and biological assays.

A number of computational techniques have been developed to exploit the relevant information from the x-ray crystallographic structures. One of the pioneering approaches in this field is that in Peter Goodford's GRID program.[1] Goodford was the first to study the protein surface not with a neutral sphere, but rather with a more realistically parameterized probe, with defined hydrogen-bonding geometry. The molecular probes were carefully parameterized using experimental data.[2–4] By computing the energetic interaction of the probes with the protein structure and different positions of the probe on a 3D grid, molecular interaction fields (MIFs) are generated. MIFs can be explored using molecular graphics by contouring the MIFs at various energy levels, with negative-energy-level contours representing the attractive regions for the probe. Although other programs may be used to compute MIFs, in this chapter, MIFs will be described primarily with reference to their calculation with the GRID program, one of the most widely used software programs in the field of structure-based ligand design.

4.11.2 The GRID-Derived Molecular Interaction Fields

A set of MIFs refers to the spatial variation of the interaction energy between a molecular target and a chosen probe. The target may be a molecule or a macromolecule, or even a molecular complex. On the other hand, the probe may be a molecule or a molecular fragment in order to simulate the interaction of any chemical group.

Originally, GRID-derived MIFs were developed to determine energetically favorable binding sites on macromolecules in order to predict where ligands bind to biological macromolecules. Thereby, MIFs can guide structure-based ligand design whenever the target is a protein, a therapeutic agent, or other biologically important macromolecule, gaining a better insight to the factors affecting the binding helps in the design of improved ligands.

However, GRID MIFs are also frequently applied to low-molecular-weight compounds to derive 3D quantitative structure–activity relationships (QSARs) in comparative molecular field analysis (CoMFA)-like[5] and GRID/GOLPE (generating optimal linear PLS estimations) studies.[6,7] Moreover, MIFs are the basis for more complex molecular descriptors well suited to predict pharmacokinetic properties, such as in the VolSurf methodology,[8,9] and metabolic hot spots, such as in the MetaSite procedure.[10] See Cruciani[11] for deeper explanations and applications.

4.11.2.1 Calculation of Molecular Interaction Fields: The Target

Input for a MIF calculation are 3D atomic coordinates of the target, which may consist of single drug-like molecules, molecular arrays such as membranes or crystals, macromolecules such as proteins, nucleic acids, glycoproteins, or polysaccharides, or molecules and ions such as metalloproteins. Biological systems have plenty of water, and very often well-ordered water molecules play a crucial role in important physical and biological functions: the aqueous environment determines on the one hand solvation and hydration, on the other hand ligand recognition, protein or enzyme activation, and so on. Thereby, water molecules may be considered part of the target, and all these biological systems can be computationally handled by means of the program GRID. Aqueous, macromolecular, and mixed environments can be modeled by defining specific dielectric constants, whereas a water-bridged target–probe interaction can be considered by adding water molecules.

It is usually assumed that the aqueous environment surrounding the target has a bulk dielectric of 80, and that the dielectric diminishes toward 1 in the center of a large globular macromolecule. In the GRID energy function, a value of 4 is usually assigned to the relative dielelectric constant of a protein target.

In addition, structural water molecules can be added to the target whenever and wherever it is necessary, and specific orientations of hydrogen bonds can be fixed or not. The addition of water molecules can be derived by experimental knowledge, such as crystal structures; alternatively, coordinates of energy minima, obtained using the water probe, can be used.

Because of the conformational freedom of biological systems, in many computational applications it is very important to treat the target flexibility, at least partially. The simplest way to treat flexibility is to compute MIFs for multiple conformations of the target; these conformations can come from NMR ensembles, conformational searches, or molecular dynamics simulations.

In the GRID program, a more sophisticated way to treat flexibility was developed. According to the types of chemical groups present, the position of a few atoms in the target is adapted to optimize the interaction energy of the probe during calculation of the MIFs. In addition, complete rotational freedom can be given to some chemical bonds. This is routinely done for rotatable hydrogen-bonding donor and acceptor groups and is referred to as the response of the target to the probe. The GRID program also allows for probe-induced switching, such as in the case of histidine tautomers. It is even possible to define specific amino acid side-chains or large chemical groups as movable in response to the probe position.[12] An example of treating flexibility on protein targets with GRID is given in the section on flexible MIFs on protein target, below, while the following section to that is dedicated to the flexibility of small molecules.

4.11.2.2 Calculation of Molecular Interaction Fields: The Probe

GRID probes reflect the specific properties of various chemical groups. Each GRID probe defines one specific atom or chemical group, and the GRID force field is designed to calculate the energies of these particular probes interacting with the target. In the calculation, the probe is located at each point of the grid cage, established throughout and around the target, in order to determine the energy value, E_{xyz}, of the interaction probe target on each point of the cage.

The given array of energy values can be combined into an X matrix in order to apply a statistical approach, especially if GRID is run over a set of targets for 3D QSAR analyses. But it can also be inspected directly: isoenergy contour surfaces (GRID maps) are displayed in 3D on a computer graphics system together with the target structure to identify the regions of attraction and facilitate the interpretation of protein–ligand interaction.

The GRID probe parameters are dependent not just on the chemical element but also on the chemical environment and hybridization. As an example, the character of various oxygen probes is described below, and their interaction with the same target is given in **Figure 1** for comparison purposes.

The carbonyl oxygen probe (O) is one sp^2 oxygen atom carrying a couple of lone pairs, each one accepting a single hydrogen bond. Its size, polarizability, electrostatic charge, and hydrogen-bonding properties were calibrated based on a large amount of crystal structures.[2] The center of the oxygen is placed at each point of the grid cage, a check for unacceptably bad close contacts is made, and nearby hydrogen bond donor atoms on the target are sought: those donors are listed in a sorted way. In the case of hydrogens pointing the wrong way, target atoms are rejected from the list. Conversely, where they are correctly oriented, the probe's oxygen is kept fixed at the grid point, while the probe is rotated until the most favorable interaction occurs, i.e., the best possible hydrogen bonds to the nearby target atoms are formed. Thereafter, the equations reported in the following section are used to compute the energy value for that particular probe at that particular point, and the whole process is repeated systematically until the interaction energy for carbonyl oxygen is known for every grid point on the map. The sp^2 carboxyl oxygen probe (O::) has much greater polarizability and much greater negative charge than the carbonyl probe, which is responsible for stronger interactions with hydrogen-bonding donors from the target. The larger regions of **Figure 1b**, when compared to the regions of **Figure 1a**, reflect this difference. Moreover, the carboxyl probe accepts up to two hydrogen bonds as the carbonyl and all the considerations on the target response are still valid: at each grid point, the probe is rotated to optimize its orientation energetically.

For the aromatic sp^2 hydroxyl probe (OH), the oxygen atom is placed at the grid point as before, but the probe character is different. The probe OH has a higher polarizability, makes hydrogen bonds of a different strength, can accept only one hydrogen bond, and the oxygen is bonded to a hydrogen atom that can donate. This relevant distinction

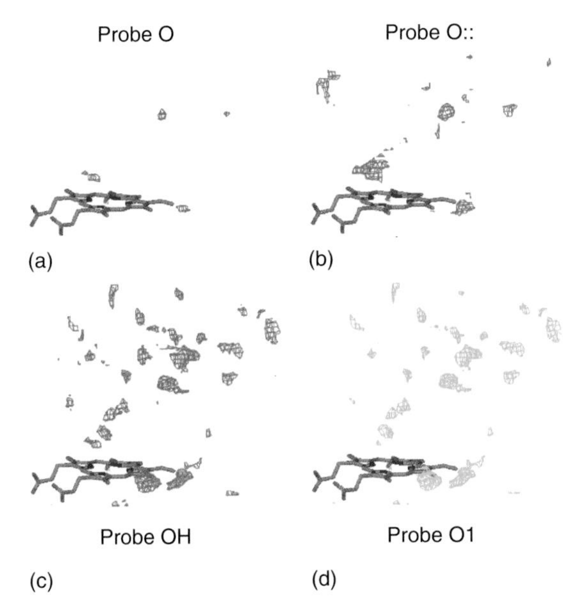

Figure 1 Four probes interact with the CYP2C9 cavity: (a) grey: carbonyl O; (b) red: carboxylate O::; (c) brown: aromatic hydroxyl OH; and (d) yellow: aliphatic hydroxyl O1. Their derived isocontour maps are color-coded, and are all plotted at energy $-8.0\,\text{kcal mol}^{-1}$. The cavity of the CYP2C9 is presented without the residues for clarity. It is worth noting the differences between carbonyl and hydroxyl, and, within the two pairs, O/O:: and OH/O1.

leads to highly differing MIFs (compare the pairs **a–b** and **c–d** in **Figure 1**). In addition, the geometry of the probe–target interaction is constrained toward the sp^2 angle of 120° by simultaneous hydrogen bonds (donor and acceptor). When the probe is being rotated, the probe's donor hydrogen moves round the grid point at a distance of about 1 Å, the length of the O–H bond. Furthermore, the sp^3 aliphatic hydroxyl probe (O1) differs by making its hydrogen bonds at the sp^3 angle of 109° instead of 120°, and by accepting at two lone pairs instead of just one. The marginal difference between the MIFs of **Figure 1c** and **d** corresponds to this small angular variation.

4.11.3 The GRID Force Field

MIFs describe the spatial variation of the probe–target interaction energy and provide information on the favorable and unfavorable sites for a certain probe around a given target. The energy function E_{xyz} at each *xyz* grid position is the sum of many different terms:

$$E_{xyz} = \sum E_{lj} + \sum E_{el} + \sum E_{hb} + S$$

The term E_{lj} is the Lennard-Jones potential, whereas E_{el} is the electrostatic contribution, and E_{hb} is the hydrogen bond potential; each individual term is the sum of all the interactions between the probe and each atom of the target. Finally, S is the entropic term, which is computed when parts of the target are treated as flexible, when the probe interactions are compared to that of water, and for the hydrophobic probe, known as the 'DRY' probe, which is very useful for detecting hydrophobic patches on proteins.

The Lennard-Jones potential describes atomic repulsion and induced dipole-induced dipole dispersive attraction. In the GRID force field it is presented as a 12–6 function, depending on probe target atom pairs (by means of the terms A and B), and on the distance d between the atoms in the pair.

$$E_{lj} = \frac{A}{d^{12}} - \frac{B}{d^6}$$

The positive term A, which refers to the repulsion between the atoms, becomes relevant when they are too close to each other, whereas the negative term B is a measure of attractive forces such as induction and dispersion. The Lennard-Jones contribution is a short-range interaction, so E_{lj} is set to zero whenever the probe and the target atom are more than a certain distance apart, i.e., 8 Å (default). This cut-off is useful for saving calculation time. On the other hand, no cut-off is applied to the electrostatic interaction E_{el}. It does not diminish rapidly with distance, and for each E_{xyz}, the whole target is involved in the electrostatic interaction with the probe. Conversely, the magnitude of E_{el} is critically sensitive to the surrounding environment.

$$E_{el} = \sum_{pt} \frac{q_p q_t}{4\pi\, \varepsilon_0 \varepsilon_d d}$$

In the electrostatic contribution E_{el}, the terms q_p and q_t are partial atomic charges on the probe target pairwise atoms, which are separated by the distance d. The term ε_0 is the permittivity of free space, while ε_d is the relative dielectric constant of the surrounding medium, which is derived from the extent of solvation of the probe target pair. When both probe and target atoms are deep in the protein and close together, they are characterized by high electrostatic interactions because of the small solvent screening effect. On the contrary, when two atoms are well separated and exposed to the water medium, the electrostatic interaction is lowered to account for the solvent dielectric screening.

The hydrogen-bonding term often provides a relevant contribution to the probe–target interaction and defines specificity for such interactions. It is the product of three terms and depends on the chemical nature of the hydrogen-bonding atoms as well as on the length and orientation of the hydrogen bond.

$$E_{hb} = E_d \times E_t \times E_p$$

The term E_d is an 8–6 function depending on the hydrogen-bonding donor–acceptor pair, and on their distance d. The orientational dependence of the hydrogen bond is described by E_t and E_p, which are functions of the angles made by the hydrogen bond at the target and probe atoms, respectively. These angular terms describe the geometry of the hydrogen bond and assume values between 0 and 1. At each grid point, the probe is rotated to the orientation that results in optimal hydrogen bond energy and, in case of multiple hydrogen bonds, the best combination is sought. Whenever the geometry of the interaction is not optimally oriented at the target or at the probe, the value of the corresponding term E_t or E_p is lower than 1, and E_{hb} is consequently lowered.[2]

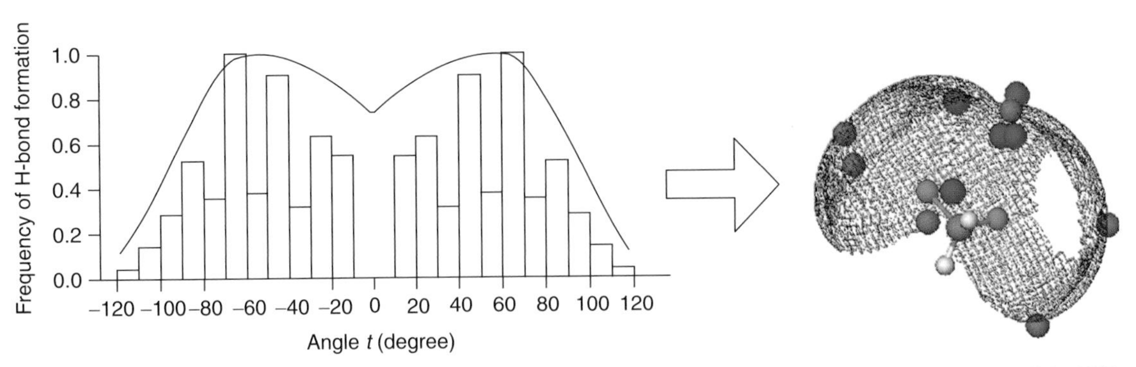

Figure 2 Statistical survey of the hydrogen-bonding interactions involving fluorine atoms from ligand molecules of the PDB. The analysis led to the definition of a proper function of the GRID force field which allowed a correct fitting of the observations. (Reprinted with permission from Carosati, E.; Sciabola, S.; Cruciani, G. *J. Med. Chem.* **2004**, *47*, 5114–5125. Copyright (2004) American Chemical Society.)

The parameterization of the entire GRID force field is empirical and constructed to reproduce experimentally observed structures and experimental data. Initially, the Cambridge Structural Database was the main source for small molecule structures. More recently, the Protein Data Bank (PDB) has become the primary source of structural data for parameterizing hydrogen bond functions[13]: the fluorine hydrogen-bonding pattern was studied in depth through the PDB. A specific angular function was derived from the statistical analysis of the observed hydrogen-bonding interactions, as schematized in **Figure 2**.

A simultaneous effect of probe–target interaction is the displacement of ordered water molecules. The resulting entropic term S represents the contribution to the free energy of solvent water molecules; it can be defined as the entropic cost of reducing the flexibility and solvation of the target when it binds to the probe, and it is calculated starting from the entropic component, called WENT (Water ENTropy), for an ideal flat hydrophobic surface.

When polar groups are present in the target, variation of global energy is due to the interaction between water molecules and target polar groups, especially when hydrogen bonds are formed (EHB), replacing bonds between water molecules. On the other hand, when two hydrophobic surfaces come together there will be a favorable induction–dispersion interaction between the two molecules, and the Lennard-Jones contribution (ELJ) is used to estimate this component. Therefore the entropic term is the result of two opposite effects: the favorable one is due to the well-organized water shell around the target (before the probe–target interaction) and to the dispersion and induction forces (arising from lipophilic contacts), while the unfavorable one is due to the interaction of the water shell with polar groups of the target that disrupts the local ordering of water in the hydration shell. The overall energy of the hydrophobic probe is computed at each grid point with the following equation, in which each term is negative.

$$S = (WENT + ELJ) - EHB$$

The typical Lennard-Jones energy between two touching atoms, such as carbon, oxygen, or nitrogen, is about $-2.0\,\mathrm{kcal\,mol^{-1}}$; the WENT term is fixed at the value $-0.848\,\mathrm{kcal\,mol^{-1}}$. For hydrophobic surfaces the S term may reach the value of $-2.8\,\mathrm{kcal\,mol^{-1}}$. Partly polar surfaces, where the hydrogen-bonding contribution is low, still have negative values for S. However, when the surface is mainly hydrophilic, and the hydrogen-bonding energy is high, the S term has a positive value, and is therefore set to zero.

The hydrophobic regions are well detected by the DRY probe. This can be considered to be like an inverse water probe: it makes a Lennard-Jones interaction in the same way as a water probe; it is also neutral like a water probe; and it has no electrostatic interaction term. The hydrogen bond energy is however inverted to reflect the fact that polar parts of the target that are able to make hydrogen bonds will not be energetically favored next to a hydrophobic probe. In **Figure 3** the cholesterol molecule shows as an example the variation of DRY maps according to the energy values. At weak hydrophobic interactions (**Figure 3a**) the entire molecule is covered by the DRY maps; further images (**Figure 3b–d**) show the progressive diminishing of the hydrophobic regions, and the corresponding interaction energy increasing.

4.11.4 Applications of GRID Molecular Interaction Fields

Finding hot spots for ligand binding is among the difficult challenges of the structure-based ligand design approaches, and several techniques have been introduced worldwide to address this problem. Experimentally, binding sites can be

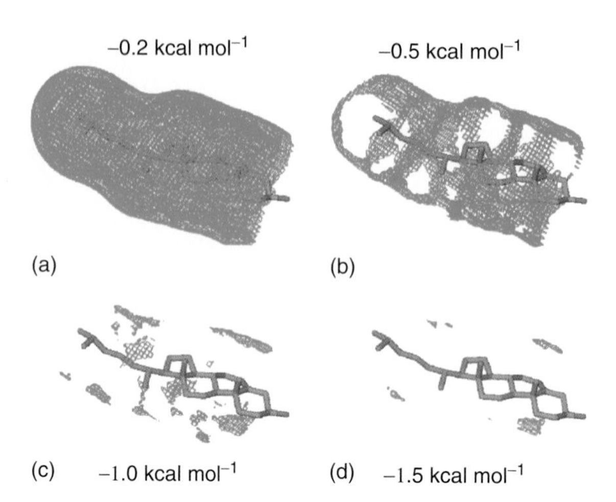

Figure 3 The cholesterol molecule is represented with its GRID DRY maps at four isocontour energy levels (values expressed as kcal mol^{-1}): (a) –0.2; (b) –0.5; (c) –1.0; and (d) –1.5.

explored by using ligand molecules as probes: crystallographically, using solvent or even larger molecules, or by NMR, using small-molecule fragments, that can probe the ability of a particular site to bind to a particular functionality or group of functionalities. Computationally, several programs are used to identify hot spots for ligand binding, and GRID is one of the best known.

The use of GRID-derived MIFs is widespread over several fields of science, as the GRID method was explicitly designed in order to get selective information about binding sites as well as about drug-like molecules. The output can be used in two quite different ways: (1) raw GRID maps, which are intuitively easy to understand, can be visualized in order to interpret interesting features of molecular structures; (2) matrices of numerical data, which can be analyzed statistically, can be easily generated.

In this chapter, we will focus more on the first approach. In structure-based ligand design, the information obtained from the interpretation of the GRID MIFs, i.e., the assessment of hydrophilic and lipophilic regions as well as the mapping of the binding site, offers valuable insights into the structural characteristics of new, potentially interesting compounds, which may be exploited in some of their structural functions before their synthesis. Some applications are briefly presented below.

4.11.4.1 Determination of Water Positions

Proteins, receptors, and enzymes may accommodate ligands and optimize their binding by means of conformational variations as well as by modulating the presence of water molecules at the ligand binding site. Structural water molecules within the protein active sites may affect ligand protein recognition by modifying the active site geometry, therefore contributing to the binding affinity. Some recent interesting GRID applications are described here.

In the framework of a detailed calculation of binding free energy for protein ligand systems,[14,15] Cozzini and co-workers analyzed the interaction of 23 ligands with dimeric human immunodeficiency virus (HIV-1) protease.[16] The primary scope of the authors was to model computationally the free energy of ligand binding to proteins, by means of the program HINT (available from Tripos, Inc., US). In the course of their analyses, they often found the presence of ordered water at the ligand binding site influencing the energetics of binding. Although the resolution was high enough to describe the water network, the analyzed crystallographic structures revealed a highly variable number of water molecules, which ranged from only one to 100 or more water molecules reported in the crystal structures. Thus, the GRID program was used both for checking water locations and for placing water molecules that appear to be missing from the complexes due to the crystallographic uncertainty mentioned above (**Figure 4**). The inclusion of structural water molecules in the analysis gave the most reliable predictions of binding constants.

The well-known GRID accuracy in predicting the location of water molecules within protein-binding sites was used to prepare the targets for several popular docking programs. The scope of de Graaf and co-workers[17] was to evaluate the effect of including water molecules in docking calculations, i.e., to evaluate the structural effects of water molecules on the docking performances. Two sets of therapeutically important protein ligand systems were considered: (1) 19 cytochrome P450 19 thymidine kinases. Docking experiments were carried out: (1) into water-free active sites;

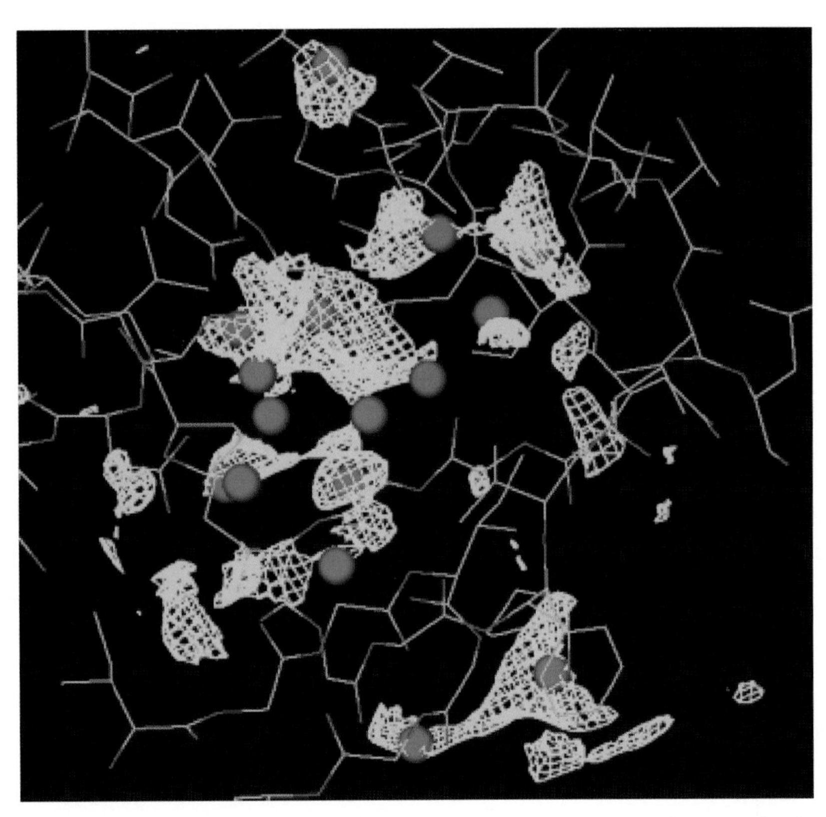

Figure 4 Extract of the binding site of the HIV-1 protease (PDB code 1G6L). The structural water molecules are presented as red balls, whereas the GRID MIF obtained with water probe OH2 and plotted at $-6.5\,\mathrm{kcal\,mol^{-1}}$ is reported as cyan isocontours. For other incomplete crystallographic structures, the GRID OH2 field was used to locate water molecules.

(2) into active sites containing crystallographic water molecules; and (3) into active sites containing water molecules predicted by GRID. The presence of ordered water in the active sites was relevant for the binding, in terms of root mean square deviation (RMSD) accuracy. Results from scenarios considering crystallographically determined and computationally predicted water molecules were better than docking approaches that omitted water molecules. Additionally, the GRID-derived scenario was comparable or better than the scenario featuring x-ray water molecules.

In QSAR or 3D QSAR studies, when some water molecules are assumed to be crucially involved in ligand enzyme binding, it may be more appropriate to include the water molecules as an integral part of the various ligands. In this context, a strategy to deal with such relevant water molecules on 3D QSAR analysis was proposed by Pastor *et al.*,[6] who joined a long-term project of Watson and co-workers, aimed at discovering inhibitors of glycogen phosphorylase (GP) with potential therapeutic activity as antidiabetic drugs.

The ligands of the series mentioned have in common a glucopyranose ring, with different substitutions at the C1 position in α and/or β configurations (**Figure 5**). Their 3D coordinates were in silico superimposed as they appear in the crystal structures, and an extended volume was used to envelop the water molecules to be included in the 3D QSAR analysis. Only water molecules involved in likely close contacts (up to 4 Å from their atomic centers) were retained: with this cut-off criterion, a minimum of five to a maximum of eight water molecules were included for each complex.

From this study, the ligand binding site appears hydrated and containing several water molecules, whose position changes from complex to complex. Most of the water molecules mantain approximately the same positions in every complex, and some serve to bridge hydrogen bonds between the ligand and the enzyme; others have been displaced by the ligands to different positions or completely removed. Together with the water probe OH2, the program GRID was used to calculate energetically favorable positions for water molecules. The inspection of the ligand–water–enzyme complexes revealed mobile water molecules in correspondence of regions where GRID indicated the possibility of several minima. These areas, where a water molecule can establish a large number of hydrogen bonds, are represented by the red clusters on the right of **Figure 5**, mostly due to the different sizes and shapes of the various substituents. Conversely, on the left are more conserved water molecules, which participate in the binding of the glucose ring with GP protein.

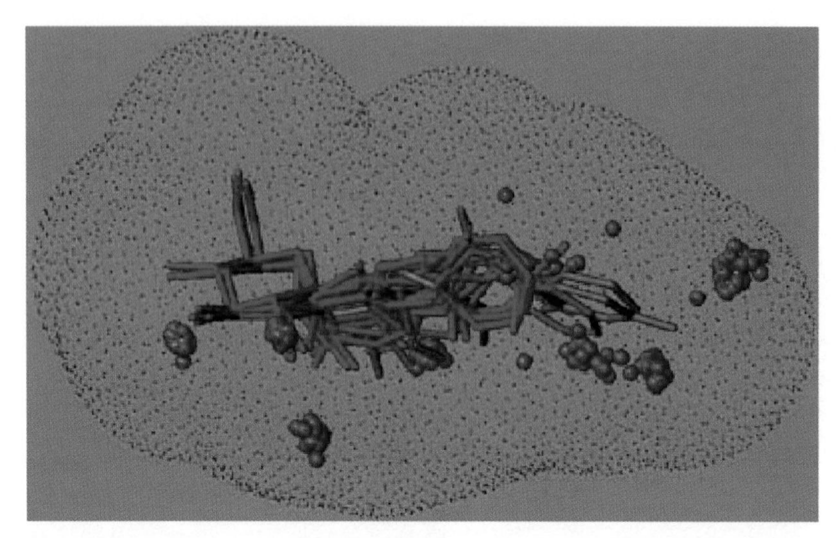

Figure 5 Ligands of the data set (purple) and water molecules (red) included in the QSAR analysis are superimposed. The thin blue points represent the extended volume; hydrogen atoms have been removed for clarity. (Reprinted with permission from Pastor, M.; Cruciani, G.; Watson, K. A. *J. Med. Chem.* **1997**, *40*, 4089–4102. Copyright (1997) American Chemical Society.)

The results of this study using inhibitors of GP clearly indicate that incorporating water molecules into the 3D QSAR analysis improves the quality of the model both with respect to the predictive ability, and with respect to the interpretation. Indeed, incorporating the water molecules gives a more consistent picture of the ligand interaction. However, knowledge of the position of individual water molecules within the ligand binding site is required. Therefore, its application is currently limited to series for which the crystal structures of the ligand–receptor complexes are available.

4.11.4.2 Hydrophobic Patches

Some protein–ligand complexes are often used to evaluate the performance of computational docking and scoring procedures. One of these is the lipid-binding protein, with crystal structure 1LIC in the Brookhaven PDB. The ligand is hexadecane sulfonic acid, a long-chain alkyl sulfonate ligand.

The understanding and correct assessment of the binding mode of this ligand are difficult tasks, and some docking programs fail.[18] Here the ligand protein-binding mode is investigated and the protein target is approached by means of the GRID probes C3 and DRY. The former represents a methyl group and gives an image of the ligand-accessible regions, whereas the latter gives an image of the hydrophobic regions. It is worth recalling the GRID concept of hydrophobicity, which takes into account the entropy of the ligand cavity system by simulating the displacement of water molecules from both ligand and binding site.

It can be clearly seen from **Figure 6** that two interaction modes are likely to occur, according to the paths highlighted by the C3 probe and represented by the yellow color. This contour is due to the Lennard-Jones term. However, the ligand prefers one of the two possibilities (reported as c1), which would be energetically favored (for enthalpy, entropy, or both). The hydrophobic field from the probe DRY, shown by the green color, matches exactly the experimental selection of the ligand.

4.11.4.3 Comparison of Hydrophobic Patches

In the GRID force field, hydrophilicity and hydrophobicity are strictly linked concepts due to the equation adopted for the DRY probe. However, the MIFs obtained with the DRY probe are unique in both their energy scale, which differs from the rest of the probes, and because only the DRY probe accounts for entropy. The importance of hydrophobic forces in recognition processes, together with this special character of the DRY probe, often makes the GRID-derived DRY MIFs the most relevant within the series.

A selectivity analysis conducted by Ridderström and co-workers using the GRID/CPCA (consensus principal component analysis) strategy on four human cytochrome P450 2C enzymes showed that the hydrophobic interactions

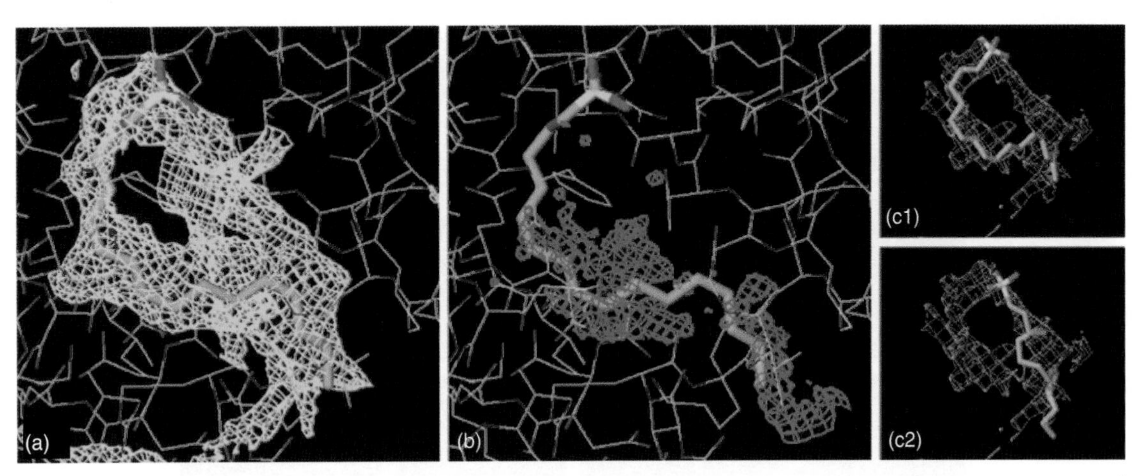

Figure 6 Molecular interaction fields for the protein 1LIC obtained using the C3 probe (a, isocontour energy = $-2.0\,\text{kcal mol}^{-1}$) and the DRY probe (b, isocontour energy = $-0.3\,\text{kcal mol}^{-1}$). All calculations were made without the ligand, which is only added for reference. (c) Schematization of the two possible binding modes (c1 and c2).

and the geometrical features of the active sites are the most important determinants.[19] The four enzymes, CYP2C8, 2C9, 2C18, 2C19, were derived by homology models based on the mammalian CYP2C5 mutant crystal structure from rabbit, which was the first (and the only one when this study was conducted) mammalian CYP structure solved.[20]

The analysis investigating the selectivity of the enzymes and determining the amino acids responsible for the specificity of each CYP2C enzyme was made using GRID/CPCA.[21] In the GRID calculation 10 probes were used: hydrophobic interactions were calculated with the DRY probe; the C3 and NM3 probes described steric interactions; the N1+ and COO– probes were charged; the polar probes consisted of N1, N:, NH=, O, and OH. Two sets of calculations were made: the amino acids in the active sites were considered rigid and flexible. Accordingly, four different types of analyses were performed on the basis of rigid or flexible GRID calculations, and with or without reducing the analysis to the region closest to the heme.

The probes were grouped into four clusters by statistical analysis – five when considering the rigid maps. The DRY probe was clearly distinct from the other nine probes, especially in the rigid analysis with the cut-out pretreatment.

From the analysis of the rigid and flexible MIFs two binding pockets, differing in size and amino acid composition, were revealed for the different enzymes. On the basis of the statistical results, the hydrophobic regions were assessed to be the most important in the selectivity among the CYP2C subfamily of enzymes, together with the geometry of the active sites. In addition, CYP2C8 was the most distinct of the four proteins, due to the large difference in size and to the character, mostly hydrophobic, of the selective amino acids of the active site.

4.11.4.4 Identification of Protein-Binding Sites

In the scientific literature, there are plenty of computational applications where the GRID program has been used to map the binding site of a protein. One of the most popular examples is summarized below. An investigation of the active site of influenza virus sialidase was carried out by means of the GRID program; this procedure has led to the design of potent inhibitors, one of which was subsequently developed by GlaxoSmithKline and marketed as the first neuraminidase-based antiinfluenza drug, under the name of Relenza.

Neuraminidase has been an important target for antiinfluenza therapy for many years. Mark von Itzstein and his group at Monash University (Victoria, Australia) used the program GRID[22] in an attempt to identify binding hot spots in the active site of neuraminidase for guiding compound design. The information extracted by using GRID led to the prediction of how various functional groups, substituted on the basic scaffold, will interact in specific regions within the active site. Summarized below and shown in **Figure 7** is the enzyme–ligand interaction profile, as has been shown by using several probes of the program GRID.

The calculations were originally performed by von Itzstein and co-workers on a cube of 35 Å per side, centered on the active site, with a grid spacing of 0.5 Å.[23] We have repeated the calculations here, starting from the PDB entry 1NNC, for demonstration purposes only. In **Figure 7**, we report some results, which agree with the original calculations from von Itzstein and co-workers.

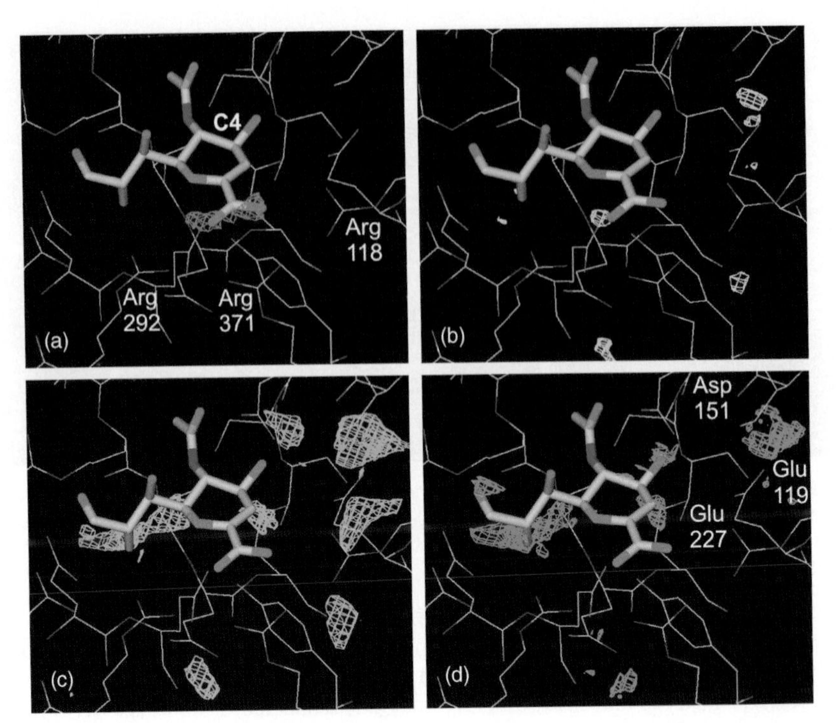

Figure 7 Three different single-atom probes have been used in an effort to determine the functional groups most likely to improve the binding of the starting ligand. (a) Carboxy O:: (red); (b) aliphatic hydroxyl O1 (yellow); and (c) amino N3+ (cyan). All these fields are plotted at $-10.0\,\mathrm{kcal\,mol^{-1}}$. (d) Conversely, for the multiatom aliphatic amidine probe (green), which represents a guanidinium group, the isocontour energy is $-13.0\,\mathrm{kcal\,mol^{-1}}$. All calculations were made without the ligand, which is only added for reference.

4.11.4.4.1 O:: carboxylate probe

There is only one significant region of interaction for the carboxylate probe in the active site (**Figure 7a**). The position of the ligand carboxylate group corresponds very well with this field, plotted at $-10.0\,\mathrm{kcal\,mol^{-1}}$.

4.11.4.4.2 O1 aliphatic hydroxyl probe

Only one region of interaction is present in the active site for the hydroxyl probe. Furthermore, this region corresponds to the position of the ligand carboxylate group: reference to the same energy ($-10.0\,\mathrm{kcal\,mol^{-1}}$) clarifies how the O1 region is significantly smaller than the corresponding one from the carboxyl group, and consequently the assumed interaction with the binding site of the hydroxyl group is weaker (**Figure 7b**). Hence, a hypothetical replacement of carboxylate with a hydroxyl group in the ligand would not provide an increased binding affinity.

4.11.4.4.3 N3+ probe

The positively charged amino probe was chosen according to the supposed optimum pH for an active enzyme (pH 5.5). Three main regions of interaction are in the active site at $-10.0\,\mathrm{kcal\,mol^{-1}}$. One of these is adjacent to the C4 position, another is next to the 4-OH, and the third one is on the floor of the active site, which in **Figure 7c** seems incorrectly to be close to the glycerol chain. The presence of two attractive regions close to position 4 clearly indicated a preference for the binding site of an amino group in such a region, thus suggesting the replacement of the OH at position 4 with a basic nitrogen.

4.11.4.4.4 Aliphatic amidine multiatom probe

This probe showed a similar profile to that obtained with the amino probe N3+. The three areas of interaction in the active site mentioned for the amino probe are still present, though at different isocontour energy values. For this multiatom probe, the MIFs are plotted in **Figure 7d** at $-13.0\,\mathrm{kcal\,mol^{-1}}$: more attractive interaction is due to the energy contributions that arise from the several atoms (mostly three nitrogens) of the probe.

The triarginal cluster (Arg-118, Arg-292, and Arg-371; **Figure 7d**) stabilizes the carboxy substituent of the ligand, whereas Glu-119, Asp-151, and Glu-227 (**Figure 7a**) give rise to a negatively charged region, where a basic group, such as a basic nitrogen or a guanidinium group, is favored. It was clear from this GRID data that replacing the 4-hydroxyl group by an amino group should produce a substantial increase in affinity of this inhibitor for influenza virus sialidase, and further substitution of the 4-hydroxyl group by a more basic 4-guanidinyl group should even further enhance the binding affinity of this analog. Indeed, by means of GRID, the keys of the binding to the active site of the enzyme influenza virus sialidase were identified; not only did some regions of the raw ligand appear to be necessary, but also other regions were found to be very relevant in the design of novel inhibitors for this enzyme.

An interesting investigation conducted by Powers and Shoichet on a series of class C β-lactamase enzymes[24] (AmpC) has been selected as an additional application. These enzymes are studied because they are responsible for the resistance mechanism to β-lactam antibiotics, since they hydrolyze and inactivate penicillins, cephalosporins, and related molecules. The aim was to build a consensus view of where the hot spots for ligand binding are on these enzymes and what sort of ligand groups they recognize.

The authors used nine previously determined x-ray crystal structures, and enriched the set with the structures of four newer inhibitor complexes, also determined by x-ray crystallography. The resulting set of ligands comprised nine boronic acids and four β-lactams. Individually, the structures provided little information about binding hot spots on AmpC. However, when all 13 complexes were superimposed, clear patterns emerged. Several consensus binding sites were identified in this way: an amide site, a hydrophobic site, a carbonyl/hydroxyl site, a carboxylate site, and several conserved water sites. The authors also used the GRID and other computer programs to probe AmpC for functional group binding sites. This allowed them to compare the computational predictions with the experimentally determined complexes and to investigate regions that are not explored in the crystal structures (to learn of new binding sites identified by the programs but not seen crystallographically).

The sites predicted by GRID corresponded best with the experimental structures. Starting with the complexed form of AmpC, GRID successfully predicted most of the experimental binding sites that were examined. The probes used to identify binding sites in the active site region of AmpC were water, hydrophobic, hydroxyl, and the multiatom probe trans amide. In addition, four potential binding site hot spots were identified: three novel amide sites and one new hydrophobic site.

The map that emerged from this study is an assemblage of contributions from different molecules, and no single ligand binds to all of the hot spots identified. Would it be possible to design a molecule that spans all of the consensus sites? This example reports on the power of structure-based ligand design methods.

4.11.4.5 Flexible Molecular Interaction Fields on Protein Target

The protein–ligand complex coded on the Brookhaven PDB as 1ACL was used for demonstration purposes. The long-chain ligand decamethonium is a long and hydrophobic ligand characterized by two ammonium nitrogen atoms on the extremities, and an overall molecular charge of $+2e$. Its site is a narrow and hydrophobic binding cleft, where several hydrophobic amino acids contribute to the stabilization of this long hydrophobic chain. Duplicate calculations were performed by setting the directive 'MOVE' to 0 and to 1, and thereby rigid and flexible MIFs were derived.

Figure 8 shows the MIFs obtained with the water (OH2, cyan) and hydrophobic probes (DRY, green) for both calculations. All calculations were made without the ligand, which is only added for reference. The same isocontour energy was used for both OH2 (**Figure 8a** and **b**, $-7.5\,\mathrm{kcal\,mol^{-1}}$) and DRY plots (**Figure 8c** and **d**, $-0.3\,\mathrm{kcal\,mol^{-1}}$): from the graphical inspection it emerged that the hydrophilic regions are more affected by the 'MOVE' directive with respect to the hydrophobic regions.

Rigid and flexible MIFs from the probe OH2 are respectively shown in **Figure 8a** and **b**, where glutamic acid 199 is highlighted to show its effect on the hydrophilic fields. Since in the flexible case it can adapt its position in response to the probe, a larger region of favorable interaction is the result. Similar behavior occurs in all the hydrophilic patches that surround the binding site, which is conversely highly hydrophobic. The effect of the several amino acids, mainly tyrosines, tryptophans, phenylalanines, and histidines, does not change dramatically when passing from rigid fields (**Figure 8c**) to flexible fields (**Figure 8d**), and the DRY isocontour shapes remain almost unvaried.

4.11.4.6 GRID Molecular Interaction Fields for a Ligand Molecule

Dopamine is a central neurotransmitter that is particularly important in regulating movement. At physiological pH (7.4), it has a protonated nitrogen atom; the cationic nitrogen has three bonded hydrogens, whereas the hydroxyl oxygens have one bonded hydrogen each, both coplanar to the aromatic ring.

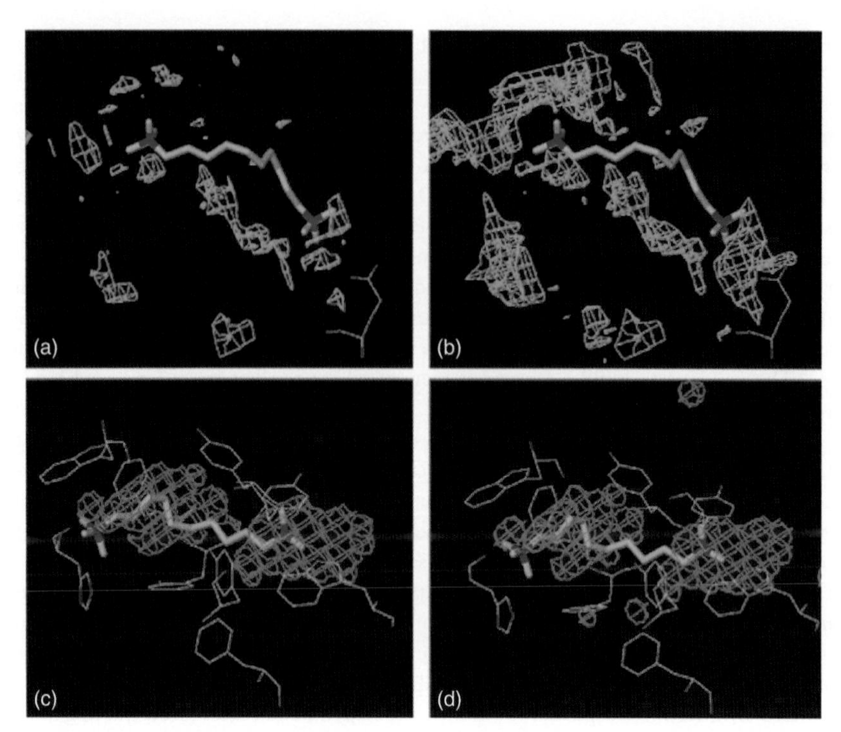

Figure 8 An extract from the binding site of the complex 1ACL. Four MIFs were calculated on the binding cleft with (a) rigid and (b) flexible MIF for the water probe, and (c) rigid and (d) flexible MIF for the hydrophobic probe.

For demonstration purposes, we have chosen this very simple target to illustrate some details of the GRID method, and we will give careful consideration to apparently small details in the GRID maps. The cyan contours in **Figure 9a** were generated by the water probe OH2, which makes good hydrogen bonds to the hydroxyl oxygen atoms as well as to the nitrogen atom of the target, whereas the green contours (**Figure 9b**) were generated using the hydrophobic probe DRY, and show that the central part of the dopamine is nonpolar and very hydrophobic.

The hydrophilic contour around the phenolic oxygen atoms is thinner in the C–O axes, whereas it is enlarged in three regions. Because of the sp^2 character of the oxygens from the target, optimal hydrogen bonds occur along the axes formed at 120° with respect to the C–O bond. Hence, close to the central region it will be a reinforcement of the field due to hydrogen bonds that likely occur at both the oxygen atoms, whereas the two lateral regions are from only one possible hydrogen bond.

The intrinsic flexibility is condensed into the GRID force field and its atom types; however, it is possible to investigate target flexibility in depth. By setting the directive 'MOVE' to 1, all the atoms from the target can be assigned either to the core or to the bead: the core is fixed whereas the bead can move around. The phenolic aromatic ring of the dopamine is the fixed core, while the amino-ethylene group is the bead, which can move in response to the probe. When using the water probe and setting the contour level to $-4.0\,\text{kcal mol}^{-1}$, the resulting map is a very large region, as shown in **Figure 9c**; this region includes the field seen in **Figure 9a**. Further investigation of the map corresponding to $-5.0\,\text{kcal mol}^{-1}$ leads to location of the energy minima where there is a contribution from both the fixed oxygen and the movable nitrogen (**Figure 9d**). This shows how using flexible targets can result in very different MIFs from those obtained with rigid MIFs.

4.11.4.7 Competition between Probe and Water

The decision to remove or to keep crystallographic determined water molecules in the binding site of a ligand is not straightforward, as there are examples of both the successful targeting and displacement of these water molecules by active ligands. When using the GRID program, it sometimes happens that the chosen probe may interact favorably with the target at a certain position, but a water molecule would interact better at the same place. In structure-based ligand design, it is important to know if specifically unfavorable places for certain probes exist on a target, because they can influence the affinity and selectivity of drug–receptor interactions, and the orientation of a ligand at its receptor site.

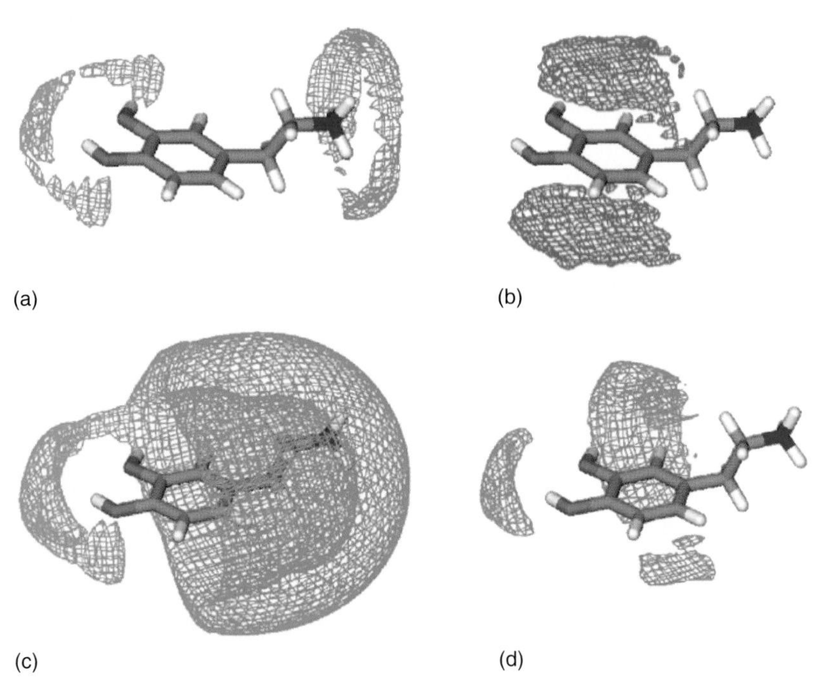

Figure 9 A molecule of dopamine is represented by GRID rigid maps obtained with the probes OH2 (a, isocontour energy $= -4.0\,\text{kcal mol}^{-1}$) and DRY (b, isocontour energy $= -1.0\,\text{kcal mol}^{-1}$). The same molecule was used as target for flexible calculations (directive 'MOVE' = 1) with the water probe; it is plotted with isocontour maps at (c) $-4.0\,\text{kcal mol}^{-1}$ and (d) $-5.0\,\text{kcal mol}^{-1}$.

The net interaction energy of that particular probe at that grid point would then be unfavorable, and GRID can assist with this particular situation.

Hydrogen bond energies in the GRID force field are traditionally computed assuming a straightforward interaction between target (T) and probe (P), as follows:

$$T: +H - P \Rightarrow T: \cdots H - P$$

However, this traditional equation ignores the presence of water, which normally saturates biological systems and can interact with both target and probe. A more comprehensive set of equations is required in order to deal with the water: (i) release of the target group from hydrogen bonding in the aqueous phase; (ii) release of the probe from hydrogen bonding in the aqueous phase; (iii) formation of the hydrogen bond target probe; (iv) formation of hydrogen bonds by the water released in reactions (i) and (ii):

$$\begin{array}{ll} \text{(i)} & T: \cdots H - O - H \Rightarrow T: +HOH \\ \text{(ii)} & P - H \cdots : OH_2 \Rightarrow P - H + OH_2 \\ \text{(iii)} & T: +H - P \Rightarrow T: \cdots H - P \\ \text{(iv)} & H - O - H + : OH_2 \Rightarrow H - O - H \cdots OH_2 \end{array}$$

Care is required in assigning a positive or negative sign to each energy term, are E_{i}, E_{ii}, E_{iii}, and E_{iv}; with these equations the total energy (E_{T}) would be computed as follows:

$$E_{\text{T}} = E_{\text{i}} + E_{\text{ii}} + E_{\text{iii}} + E_{\text{iv}} + N \times \text{WENT}$$

The equation for E_{T} can be easily recast. The energy at each individual grid point, E_{POINT}, is the sum of E_{i} and E_{iii} because it depends on the interaction of the target with the probe, and on the competing interaction of the target with water at the same grid point. In addition, the remaining terms only depend on the properties of water and the properties of the probe. Therefore, the balancing energy, E_{BAL}, is given by summing E_{ii}, E_{iv}, and N times WENT. As mentioned previously, the term WENT is $-0.848\,\text{kcal mol}^{-1}$ and represents the entropic benefit of releasing one bound water molecule into the bulk water phase; as two bound waters are released, $N = 2$. Then, the term E_{BAL} is a constant term that assumes different values according to the probe. Finally, the GRID equation to calculate the overall energy for competition water probe is as follows:

$$E_{T} = E_{POINT} + E_{BAL}$$

This is used when the directive 'LEAU,' determining the way in which water is treated during the computation, is set to 3.

Ligand selectivity may be critically influenced by prebound water molecules, because water is not easily displaced by polar but inappropriate ligand atoms. One misplaced polar atom may give so great a positive energy that it forces the ligand into a completely different orientation, and causes several otherwise apparently acceptable hydrogen bonds to be misaligned.

A well-known example occurs with the enzyme dihydrofolate reductase, which binds the purine ring of its substrate dihydrofolate in one orientation, and the corresponding ring of its inhibitor methotrexate (MTX) after a 180° ring rotation. The significant structural difference between these two ligands is the replacement of a carbonyl oxygen atom in the purine moiety of the substrate by an amino group in MTX, and the inhibitor was originally designed on the assumption that it would bind exactly like the substrate itself.

In **Figure 10**, the MTX-binding cleft is reported with the GRID MIFs obtained in different ways; all the calculations were run without any ligand, and the MTX was added later, in order to facilitate the discussion. Large and strongly favorable interactions are found using the amino probe N2 (**Figure 10a**, green contour, $-9.0\,\mathrm{kcal\,mol^{-1}}$). This MIF confirms the binding mode of MTX, as the amino group from the molecule exactly fits the strong-energy MIF from the corresponding probe N2. Conversely, at the same location no favorable regions are found with the probe O (**Figure 10b**, red contour, $-6.0\,\mathrm{kcal\,mol^{-1}}$), whereas agreement occurs between the carboxylate groups of the MTX and the MIFs from the corresponding probe O.

On the other hand, a water molecule could bind at the place where MTX locates an amino group. Here, large and strongly favorable interactions are found using the probe OH2 (**Figure 10c**, cyan contour, $-9.0\,\mathrm{kcal\,mol^{-1}}$), as with the probe N2. This analysis ended up by simulating the competition between the probe O and the water; to do so,

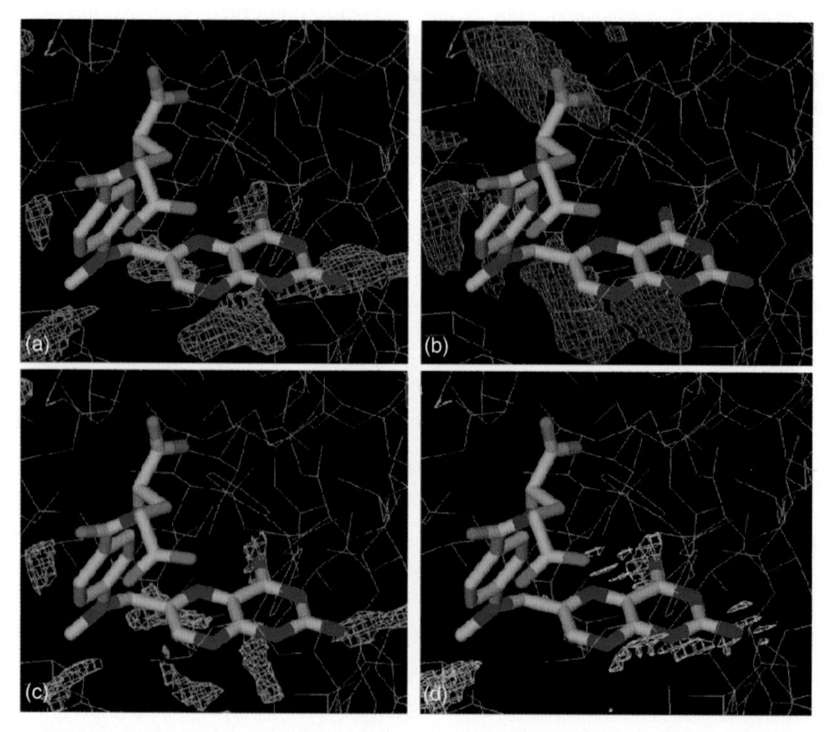

Figure 10 Methotrexate in the binding cleft of dihydrofolatereductase with contours for GRID MIFS computed for several probes and settings. The probes N2 (a, green field) and O (b, red field) are shown at -9.0 and $-6.0\,\mathrm{kcal\,mol^{-1}}$, respectively, whereas the probe OH2 is plotted at $-9.0\,\mathrm{kcal\,mol^{-1}}$ (c, cyan field). Finally, the isocontour obtained for the probe O in the reverse mode (d, yellow field), i.e., setting 'LEAU' to 3, is mapped at $-4.0\,\mathrm{kcal\,mol^{-1}}$.

GRID calculations were carried out with the directive 'LEAU' set to 3. The resulting positive energies indicate that a water molecule at the grid point would interact more favorably with the target than the chosen probe. Thus, the energies and GRID yellow maps of **Figure 10d** show positions on the target where the interactions with the probe O would be disadvantaged relative to water. The contours show that water would compete so well in this part of the binding cleft that any carbonyl oxygen atom of a ligand would be at a disadvantage. Thereby, the carbonyl oxygen of the dihydrofolate would be disadvantaged $-4.0\,\mathrm{kcal\,mol^{-1}}$ if the substrate bound like MTX.

Indeed, it appears as if two backbone carbonyl oxygens of the enzyme are an important structural feature that ensures that the substrate does not bind like MTX, because dihydrofolate reductase can only give a stereochemically correct product if the substrate ring is rotated 180° away from the MTX orientation. Dihydrofolate itself escapes from this predicament by an 180° rotation of the purine ring, and with hindsight one can see that the original design strategy for MTX was unsound. In fact, the rotated orientation of the dihydrofolate ring seems absolutely essential for correct enzyme function, because dihydrofolate reductase could give a hydrogenated product with the wrong stereochemistry if its substrate bound like MTX.

4.11.5 Conclusion

Much has been learned since the first attempts to discover drugs on the basis of available 3D structural data. Nowadays, structure-based ligand design is an established method for creating potential drugs with new structural features, for modifying binding properties, and for elucidating binding modes and structure–activity relationships. In this context, the use of MIFs has been growing continuously, since it aids the understanding of the system under investigation.

This chapter presents the underlying principles of the MIF approach and reports real-life applications. It shows that the use of MIFs in structure-based ligand design has many advantages over other more traditional approaches to designing new drugs. The ability to identify the positions occupied by water molecules or to detect hydrophobic patches is of fundamental importance for mapping protein-binding sites. In addition, since MIFs reflect the individual properties of various chemical groups, they can be used to improve ligand selectivity.

However, the use of MIFs is not an alternative to other methods, but, rather, complements other procedures. For example, MIFs can be used in docking procedures to identify binding sites, to rank the ligand solutions, or to estimate the entropic effect.

Since many methods of generating MIFs exist, it is critically important to choose the right MIF for each project. GRID has been calibrated by studying experimentally observed crystal structures and is particularly well suited for use on wet systems with biologically important (macro)molecules.

There are numerous bottlenecks on the path from gene to protein ligand structure, and from protein ligand structure to drug. We are convinced that MIFs can assist in overcoming many such bottlenecks.

References

1. Goodford, P. J. *J. Med. Chem.* **1985**, *28*, 849–857.
2. Boobbyer, D. N. A.; Goodford, P. J.; McWhinnie, P. M.; Wade, R. C. *J. Med. Chem.* **1989**, *32*, 1083–1094.
3. Wade, R. C.; Clark, K.; Goodford, P. J. *J. Med. Chem.* **1993**, *36*, 140–147.
4. Wade, R. C.; Goodford, P. J. *J. Med. Chem.* **1993**, *36*, 148–156.
5. Cramer, R. D.; Patterson, D. E.; Bunce, J. D. *J. Am. Chem. Soc.* **1988**, *110*, 5959–5967.
6. Pastor, M.; Cruciani, G.; Watson, K. A. *J. Med. Chem.* **1997**, *40*, 4089–4102.
7. Ortiz, A. R.; Pastor, M.; Palomer, A.; Cruciani, G.; Gago, F.; Wade, R. C. *J. Med. Chem.* **1997**, *40*, 1136–1148.
8. Crivori, P.; Cruciani, G.; Carrupt, P. A.; Testa, B. *J. Med. Chem.* **2000**, *43*, 2204–2216.
9. Cruciani, G.; Crivori, P.; Carrupt, P. A.; Testa, B. *J. Mol. Struct.-Teochem.* **2000**, *503*, 17–30.
10. Cruciani, G.; Carosati, E.; De Boeck, B.; Ethirajulu, K.; Mackie, R. J.; Vianello T. *Med. Chem.* **2005**, *48*, 6970–6979.
11. Cruciani G. In *Molecular Interaction Fields. Applications in Drug Discovery and ADME Prediction*; Cruciani, G., Ed.; Wiley-VCH: Weinheim, 2005.
12. Liljefors, T.; Jørgensen, F. S.; Krogsgaard-Larsen, P., Eds. *Rational Molecular Design in Drug Research*, Alfred Benzon Symposium 42, Munksgaard, Copenhagen, 1998.
13. Carosati, E.; Sciabola, S.; Cruciani, G. *J. Med. Chem.* **2004**, *47*, 5114–5125.
14. Cozzini, P.; Fornabaio, M.; Marabotti, A.; Abraham, D. J.; Kellogg, G. E.; Mozzarelli, A. *J. Med. Chem.* **2002**, *45*, 2469–2483.
15. Fornabaio, M.; Cozzini, P.; Mozzarelli, A.; Abraham, D. J.; Kellogg, G. E. *J. Med. Chem.* **2003**, *46*, 4487–4500.
16. Fornabaio, M.; Spyrakis, F.; Mozzarelli, A.; Cozzini, P.; Abraham, D. J.; Kellogg, G. E. *J. Med. Chem.* **2004**, *47*, 4507–4516.
17. de Graaf, C.; Pospisil, P.; Pos, W.; Folkers, G.; Vermeulen, N. P. E. *J. Med. Chem.* **2005**, *48*, 2308–2318.
18. Friesner, R. A.; Banks, J. L.; Murphy, R. B.; Halgren, T. A.; Klicic, J. J.; Mainz, D. T.; Repasky, M. P.; Knoll, E. H.; Shelley, M.; Perry, J. K. et al. *J. Med. Chem.* **2004**, *47*, 1739–1749.
19. Ridderström, M.; Zamora, I.; Fjellström, O.; Andersson, T. B. *J. Med. Chem.* **2001**, *44*, 4072–4081.
20. Williams, P. A.; Cosme, J.; Sridhar, V.; Johnson, E. F.; McRee, D. E. *Mol. Cell.* **2000**, *5*, 121–131.
21. Kastenholz, M. A.; Pastor, M.; Cruciani, G.; Haaksma, E. E. J.; Fox, T. *J. Med. Chem.* **2000**, *43*, 3033–3044.

22. von Itzstein, M.; Wu, W.-Y.; Kok, G. B.; Pegg, M. S.; Dyason, J. C.; Jin, B.; Phan, T. V.; Smythe, M. L.; White, H. F.; Oliver, S. W. et al. *Nature* **1993**, *363*, 418–423.
23. von Itzstein, M.; Dyason, J. C.; Oliver, S. W.; White, H. F.; Wu, W.-Y.; Kok, G. B.; Pegg, M. S. *J. Med. Chem.* **1996**, *39*, 388–391.
24. Powers, R. A.; Shoichet, B. K. *J. Med. Chem.* **2002**, *45*, 3222–3234.

Biographies

Gabriele Cruciani is full professor of Organic Chemistry and Cheminformatrics at the University of Perugia. He currently leads a team comprising 10 students and three PhD students. In 1992, he spent a year as a postdoc with Prof Peter Goodford in the Laboratory of Molecular Biophysics at Oxford University, UK. In 1994–95 he became team laboratory leader of the Biotechnology European Project BIO2-CT94-3025 to design antidiabetic drug inhibitors of the enzyme glycogen phosphorylase *b*, together with Prof Louise Johnson, LMB, Oxford University. In 1999, he spent about 5 months with Bernard Testa and Pierre Alain Carrupt's group (Lausanne) working on metabolism and in silico pharmacokinetic prediction. In 2001, he received the Corvin Hansch Award from the Molecular Modeling Society at the Gordon conference in USA for his work on QSAR and molecular modeling. In 2005, he received the Research Award from Società Chimica Italiana, Organic Chemistry Division. Prof Cruciani is the scientific director of Molecular Discovery company, based in London (UK).

Emanuele Carosati received a degree in chemistry from the University of Perugia in 1998 under the supervision of Prof Sergio Clementi. From 1999 to January 2003, he prepared his PhD, working with Prof Gabriele Cruciani on the study of hydrogen-bonding interactions and GRID parametrization. During 2002, he spent 6 months as a temporary researcher at the Computational and Structural Sciences Group, GlaxoSmithKline, Stevenage, UK, working with Dr Sandeep Modi and Dr Iain McLay. From 2003, he worked as a research fellow at the University of Perugia. His current research interests are focused on 3D-QSAR studies, on the development and application of virtual screening procedures and pharmacophore mapping, and also on the modeling of metabolic issues caused by human cytochromes, such as substrate recognition, metabolic inhibition, and isoform selectivity.

Rebecca C Wade leads the Molecular and Cellular Modeling group at EML Research, a private research institute specializing in applied information technology with a strong emphasis on bioinformatics. She studied at the University of Oxford (BA Hons Physics, 1985; DPhil Molecular Biophysics, 1988), doing her doctoral studies in structure-based drug design with Peter Goodford. She did postdoctoral research in Houston (with Andy McCammon) and Illinois (with Peter Wolynes), primarily in biomolecular simulation, before taking up a position as a group leader in the Structural and Computational Biology Programme at the European Molecular Biology Laboratory (EMBL) in Heidelberg in 1992. She set up the Molecular and Cellular Modeling group at EML Research in 2001. The group works on the development and application of computer-aided methods to model and simulate biomolecular interactions. Rebecca Wade is an associate editor of the *Journal of Molecular Recognition*, and a member of the editorial boards of the *Journal of Computer-Aided Molecular Design*, *Journal of Modeling and Molecular Graphics*, and *Biopolymers*. She is a member of the Faculty of 1000: Theory and Simulation Section. She is the recipient of the 2004 Hansch Award of the QSAR and Modelling Society.

Massimo Baroni is contract professor of Molecular Modeling and Cheminformatrics at the University of Perugia. Since 1990, he has been the scientist responsible for software development at Multivariate Infomatrics Analisys. In 1994, he joined the research center at ISRIM in Terni (Italy) in the laboratory to develop special materials for advanced technologies. In 1990, he received the Thesis Award from the Italian Chemistry Society for his work on QSAR, chemometrics, and modeling. At present, Dr Massimo Baroni is the Chief Scientific Officer at Molecular Discovery, where he is responsible for the company's research scientific products.

© 2007 Elsevier Ltd. All Rights Reserved
No part of this publication may be reproduced, stored in any retrieval system or transmitted in any form by any means electronic, electrostatic, magnetic tape, mechanical, photocopying, recording or otherwise, without permission in writing from the publishers

Comprehensive Medicinal Chemistry II
ISBN (set): 0-08-044513-6

ISBN (Volume 4) 0-08-044517-9; pp. 237–253

4.12 Docking and Scoring

P F W Stouten, Nerviano Medical Sciences, Nerviano, Italy
R T Kroemer, Sanofi-Aventis, Centre de Recherche de VA, Vitry-sur-Seine, France

© 2007 P F W Stouten. Published by Elsevier Ltd. All Rights Reserved.

4.12.1 Introduction

With the increasing number of therapeutic targets, the need for a rapid search for small molecules that may bind to these targets is of crucial importance in the drug discovery process. One way of achieving this is the in silico or virtual screening (VS) of large compound collections to identify a subset of compounds that contains relatively many hits against the target, compared to a randomly selected subset. The compounds that are virtually screened can stem from corporate or commercial compound collections, or from virtual compound libraries. If a three-dimensional (3D) structure or model of the target is available, a commonly used VS technique is high-throughput ligand docking.[1] Here a so-called docking program is used to place many different computer-generated representations of a small molecule in a target structure (or in a user-defined part thereof, e.g., the active site of an enzyme) in a variety of positions and orientations. Each such docking mode is called a 'pose.' In order to identify the energetically most favorable pose (also referred to as 'pose prediction'), each pose is evaluated (scored) based on its complementarity to the target in terms of shape and properties such as electrostatics. A good score indicates that molecule is potentially a good binder. This process is repeated for all molecules in the collection, which are subsequently rank-ordered by their scores (i.e., their predicted affinities). This rank-ordered list is then used to select for purchase, synthesis, or biological investigation only those compounds that are predicted to be most active. Assuming that both the poses and the associated affinity scores have been predicted accurately, this selection will contain a relatively large proportion of active molecules, i.e., it will be 'enriched' with actives compared to a random selection.

High-throughput docking has become increasingly important in the context of drug discovery.[2–4] Despite the technical challenges in reliably predicting the binding mode of a molecule[5] and its binding affinity relative to other compounds,[6] in many cases docking campaigns have yielded significant hit rate improvements compared to random screening.[7–10] One may wonder, however, why in this day and age of 'real' high-throughput screening (HTS) of actual compounds against actual targets in actual assays, one needs VS. There are several reasons for that:

- External compound collections can easily be virtually screened and only compounds that are predicted to inhibit the target will then be acquired.
- Virtual libraries can be screened in silico and the results can be used to select scaffolds and to help design the final library to be synthesized.
- When no HTS is envisioned or has been carried out yet, a discovery effort can be jump-started by VS of corporate and/or public compound collections.
- When no biochemical or other functional assay is available, VS may be the only way of identifying inhibitors of a specific target.
- Compared to HTS, VS is fast and inexpensive.

This chapter details various docking and scoring approaches. It shows how scoring functions can be tuned and combined to optimize affinity predictions. It also highlights problems that can be encountered and possible solutions to those problems. In the penultimate section four success stories are described in detail to give an idea of how docking can have an impact on drug discovery efforts. (Chapter 4.19 details both ligand and structure-based approaches for virtual screening.)

4.12.2 Docking Programs and Algorithms

Determining the correct binding mode of a molecule involves finding the correct orientation and, as most ligand molecules are flexible, the correct conformation of the docked molecule. This implies that the degrees of freedom to be searched include translational and rotational degrees of freedom of the ligand as a whole, as well as its internal degrees of freedom, i.e., predominantly the rotatable bonds. To this end a number of different search algorithms have been developed. In some of these algorithms the ligand is built up incrementally, starting from a docked 'base fragment.' Programs that follow this approach include Hammerhead,[11] DOCK,[12] and FlexX.[13] In other programs, such as AutoDock,[14] GOLD,[15] ICM-Dock,[16] and QXP,[17] the ligand is treated in its entirety. In addition to ligand flexibility, it may be desirable to keep at least part of the receptor flexible in order to allow for conformational changes that are necessary to accommodate the ligand, a phenomenon referred to as 'induced fit.' Because it is computationally expensive, few docking programs allow protein flexibility. Notable exceptions are the latest versions of AutoDock,[18] FlexE,[19] QXP, and Affinity.[20] The way flexibility is handled differs from program to program. For example, FlexE uses multiple receptor conformations; Affinity allows any selection of atoms to be mobile with a user-defined tethered buffer region between the fixed and mobile regions; and QXP allows user-defined parts of the protein, e.g., selected side chains or a particular loop, to move. Searching for the correct binding mode of a molecule is generally carried out by performing a number of trials and keeping those poses that are energetically favorable. The search stops once a certain number of trials have been carried out or a 'sufficient' number of acceptable poses have been found for a molecule. In order to explore a large search space, algorithms have been developed that keep track of previously discovered minima and guide the search into new regions.[21–23] The decision to keep a trial pose is based on the computed ligand-receptor interaction energy (score) of that pose. To identify and rank-order many different poses of a molecule during the search in a reasonable time, several programs calculate a 'dock score' (a crude score based on a simple energy function such as a force field with an electrostatic term and repulsive and attractive Van der Waals terms, which can be evaluated very rapidly) during the docking process, while a more sophisticated function is used to calculate the final 'affinity score' for that molecule. To illustrate some of the radically different approaches that the various docking programs take to address the same issue, we describe the algorithms and scoring functions implemented in three of them (FlexX, GOLD, and QXP) in some detail below. It should be noted that this does not imply that they are better than other docking programs that are available (**Table 1**).

4.12.2.1 Docking Programs: FlexX

FlexX uses an incremental buildup algorithm where ligands are docked starting with a base fragment. Base fragments are generated by severing all noncyclic bonds in a given ligand. All base fragments identified for each ligand serve as starting points for the docking. After placement of a base fragment (in different positions) the complete ligand is constructed by adding the remaining components back on. Each component is added in accordance with a set of predefined, allowed torsion angles, thus allowing for ligand flexibility. At each step the interactions are evaluated and the best solution is selected according to the docking score. The docking score uses the model of molecular interactions developed by Böhm[24,25] and Klebe.[26] For each moiety that can make an interaction, interaction centers and surfaces (usually spheres) are defined. Two moieties interact if the interaction center of one of them is situated at or near the interaction surface of the other one. Different levels of interaction are defined and the program attempts to satisfy high-level interactions (such as hydrogen bonds) first. Subsequently, the docked results are scored using a modified version[27] of the Böhm scoring function.[28] This function takes into account the loss of ligand entropy upon binding (counting the number of rotatable bonds in the ligand), hydrogen bonds, ionic interactions, aromatic interactions, and lipophilic contacts. For more details, see [13,27].

4.12.2.2 Docking Programs: GOLD

GOLD (genetic optimization for ligand docking) utilizes a genetic algorithm (GA),[15,29] that mimics the process of evolution by applying genetic operators to a collection of putative poses for a single ligand (in GA terms, a population of chromosomes). GOLD chromosomes contain four genes. Two of these encode conformational information of the flexible parts of the protein and of the ligand, respectively. Each byte within these genes specifies a rotatable bond. The two remaining genes (feature arrays) encode hydrogen bonds and lipophilic interactions, respectively. Each potential hydrogen-bonding or lipophilic feature of the protein is represented by an array element. Each element either points to a corresponding partner on the ligand or contains an indication that the feature has no partner in the ligand. From the information contained in a chromosome, a 3D pose is generated (referred to as decoding): first a ligand conformation is generated by applying the bond rotations encoded in the chromosome. Thereafter, this conformation is docked into the protein using a least-square (LS) fitting procedure,[30] where the features that should match are defined by the feature arrays. The final decoding step is a second LS fit involving only those feature pairs that are less than a threshold distance of 3 Å apart. The energy of the resulting pose (fitness) consists of three terms: (1) hydrogen-bonding energy; (2) internal energy of the ligand; and (3) steric interaction energy. This steric energy is calculated using a 4–8 potential with a linear cut-off to soften the repulsive term. The linear cut-off is relatively small at the beginning of the GA to allow unhindered exploration of conformational space and reaches its maximum after 75% of the pre-set genetic operations have been performed so as to ensure the absence of steric clashes upon termination of the algorithm.

At the beginning of a docking run the size and location of the binding site are defined. Subsequently, a cavity detection algorithm is employed to calculate concave solvent-accessible surfaces, to which the ligand can bind. All hydrogen bond donor and acceptor atoms and lipophilic points on these surface patches are identified. If a terminal donor or acceptor is bound to the protein via a single bond (and not involved in an intraprotein hydrogen bond), the corresponding bond is defined as rotatable, thus allowing the NH_3^+ and OH groups to move into optimal positions for hydrogen bonding.

The population of chromosomes evolves through sequential application of genetic operations. To improve the efficiency of the algorithm and to avoid premature convergence, the total set of chromosomes is not maintained as a single population, but as a number of populations (each by default consisting of 100 chromosomes) that are arranged as a ring of islands (five by default). Three genetic operators are applied: (1) point mutation of a chromosome; (2) cross-over (i.e., mating of two chromosomes); and (3) migration of a population member from one island to another. According to the operator type, one or two chromosomes are selected with a probability that depends on their position in the fitness-ordered population. This probability is highest for the fittest and decreases linearly. After the genetic operation (other than migration) has been carried out, the resulting new chromosome is decoded and the fitness of the associated pose is evaluated. If the newly generated chromosome is fitter than the least-fit chromosome of the island's

Table 1 Basic features of a selection of popular docking programs cited in this chapter

Program	*Algorithm*	*Ligand BU[a]/E[b]*	*Initial placement & search algorithm*	*Evaluation of poses during search*	*Scoring function(s)*	*Protein flexible*	*Reference*
QXP	MCDOCK+	E	Random placement, Monte Carlo, minimization	Grid E/modified amber FF + contact	As for search	User defined parts move during minimization	Latest algorithms not published yet, older in [17]
	FULLDOCK+	E	Random placement, tree pruning, Monte Carlo, minimization	As above	As above	As above	As above
NWU-Dock		E	Several orientations (via matching features) of pre-calculated conformational ensemble	Energy + desolvation	As for search	No	130,131
ICM-Dock		E	Random placement, Monte Carlo, minimization	Grid energy/force field	Among others solvation and entropic terms	No	16
FlexX		BU	Orient rigid 'base' fragment first, build-up of ligand guided by docking score	Böhm-like molecular interactions	Many	No	13
GOLD		E	Genetic algorithm	Fitness function	As for search	OH, NH_2	15
Glide		E	Random placement of pre-calculated conformers, exhaustive phase space search, followed by Monte Carlo search, minimization	Molecular mechanics grid energy (OPLS-AA)	GlideScore (modified ChemScore)	No	84,113,114
FRED		E	Pre-calculated conformational ensemble, exhaustive translational and rotational search for each conformer	Clash with protein, rank using Chemgauss2 or other scoring function, rigid-body minimization, consensus score for final selection of pose	Same scoring functions as before, in addition Zapbind (area contribution and Poisson–Boltzmann derived electrostatics	No	73

[a] BU – the ligand is built up incrementally during the docking.
[b] E – the ligand is treated in its entirety during the docking.

population, it replaces this least-fit chromosome. After the application of a predefined number (typically 100 000) of genetic operations, the algorithm terminates, saving the poses with the highest scores.

In addition to using the island model, two other measures are taken to avoid convergence to a nonglobal minimum: first, the selection pressure (defined as the relative probability that the fittest chromosome will be selected compared to the average chromosome) is set to the low value of 1.1. Second, the technique of niching is employed: if the root mean square (RMS) distance between all features in any pair of poses on an island is less then 1.0 Å, their encoding chromosomes are considered to share a niche. The number of chromosomes that can occupy a single niche is predefined by the user (default is 2). If, as a result of the addition of a new chromosome, there are more than this predefined number of chromosomes in the same niche, then the least-fit chromosome in the niche is discarded (rather than the least-fit chromosome in the island's entire population).

4.12.2.3 Docking Programs: Quick Explore (QXP)

QXP (Quick Explore[17]) is part of the Flo+ program. It contains two conceptually different docking algorithms: MCDock and ZipDock. The MCDock algorithm is evolutionary in nature and similar to the conformational space annealing method proposed by Lee and Scheraga.[31] For each ligand it applies a user-defined number of repeated cycles of Monte Carlo followed by energy minimization to generate and refine an ensemble of low-energy ligand poses. By adding dissimilar, low-energy poses to the ensemble and by reducing the numbers and sizes of the perturbations as the number of cycles increases, the MCDock procedure is very efficient in finding low-energy solutions. The ensemble is initialized with a single pose and is allowed to grow to 50 poses. For each search cycle a ligand pose is randomly chosen from the ensemble, and subjected to 400 steps of fast Monte Carlo exploration using precalculated potential grids. In each search cycle the best result from the fast exploration is energy-minimized and evaluated for inclusion in the ensemble. Each pose that has an energy less than 50 kJ above the best pose found to date is added to the ensemble. If a similar pose already exists in the ensemble, the pose with the higher energy is removed (analogous to niching; *see* Section 4.12.2.2). If adding a new pose causes the ensemble to grow to 51 poses, the highest-energy pose is discarded. MCDock is evolutionary in that it uses an ensemble of docking pose geometries and that it applies mutations (i.e., random changes of a docking pose) to evolve the ensemble. However, unlike a GA, it does not employ cross-over (i.e., combining different poses to generate new ones).

The ZipDock algorithm was developed to address problems that arise from the fact that random Monte Carlo methods do not provide a systematic search for a globally best solution and that it is difficult to assess how many search cycles are needed. ZipDock attempts to solve this difficulty by carrying out a near-systematic search. First, a representative basis set of (200 by default) ligand conformers is aligned and stored in a conformer tree. With this tree one can in principle generate any conformation of the ligand, by simply combining parts of the various basis set conformers, a process reminiscent of cross-over in a GA. The conformer tree is docked rigidly and as a single entity, using a set of 2000 rotations and translating each of these 2000 instances to the centers of small spheres that fill the binding site. For each combination of rotation and translation, the interaction energy of each atom of each conformer in the tree is evaluated using a potential-energy grid. By combining parts of various conformers from the tree that exhibit favorable energies, one generates new conformers with overall favorable energies. As each low-energy conformer is discovered, it is evaluated using a rapid, approximate scoring method and added to the ensemble of best poses subject to the same energy and dissimilarity criteria that MCDock uses. In terms of pose prediction accuracy, ZipDock and MCDock perform about equally well, but the ligands that fail to dock correctly with MCDock are often different from those that fail with ZipDock.

The FullDock algorithm combines ZipDock and MCDock. First, it systematically scans conformational space by means of up to 10 ZipDock searches per ligand. Thereafter, the best-scoring results are pooled and used as the initial ensemble for a MCDock run to find optimal local minima. The start thus differs significantly from the regular MCDock algorithm described above, which is initialized with a single conformer.

QXP allows the user to define parts of the binding site as flexible. These sections can move during the energy minimization steps of the process and are fully mobile or tethered to their initial positions. When a pose is selected for the next cycle of Monte Carlo perturbation and testing, the program uses the corresponding binding site coordinates (which may be modified as a result of allowing protein flexibility) with a random probability of 90%. Otherwise, it uses the initial input coordinates. This way protein movement carries over and, in doing so, allows for simultaneous conformational searching of both the ligand and the flexible protein sections, while at the same time preventing unrealistic protein motion from propagating indefinitely.

The original MCDock and FullDock programs use a molecular mechanics force field as scoring function. The improved 'plus' incarnations MCDock+ and FullDock+ employ an empirical potency score to score each pose that is generated and to rank different poses of the same ligand. The main terms in this empirical score account for receptor–ligand atom–atom contacts, hydrogen bonds, steric repulsion, desolvation, internal ligand strain, and ligand and receptor entropy. An interesting novelty is that this scoring function has been developed and optimized to predict both the relative potencies of inhibitors in experimental structure–activity relationship series and the crystallographic binding modes of those inhibitors for which complex structures are available.

4.12.3 Scoring Functions

The previous section already alluded to scoring functions. Essentially, they are mathematical constructs used to estimate the binding affinity of a compound for a receptor. For comprehensive overviews of the various scoring schemes used to predict binding affinity, see [32,33].

Many of the scoring functions fall into one of two main groups. One main group comprises knowledge-based scoring functions that are derived using statistics for the observed interatomic contact frequencies and/or distances in a large database of crystal structures of protein–ligand complexes. It can be assumed that only those molecular interactions that are close to the frequency maxima of the interactions in the database favor the binding event and, therefore increase the overall binding affinity, whereas interactions that have been found to occur with low frequency in the database are likely to destabilize binding and decrease the affinity. The observed frequency distributions are converted to what is usually referred to as potentials of mean force or knowledge-based potentials. Several such potentials to predict binding affinity have been developed (e.g., PMF,[34] DrugScore,[35] SmoG,[36] Bleep[37]). All these approaches differ mainly in the size of the training database that was employed and in the types of molecular interaction that were considered.

The other main group contains scoring schemes based on physical interaction terms.[38] These so-called energy component methods are based on the assumption that the change in free energy upon binding of a ligand to its target can be decomposed into a sum of individual contributions:

$$\Delta G_{\text{bind}} = \Delta G_{\text{int}} + \Delta G_{\text{solv}} + \Delta G_{\text{conf}} + \Delta G_{\text{motion}}$$

The individual terms in this equation account for the main energetic contributions to the binding event, as follows: specific ligand–receptor interactions (ΔG_{int}), the interactions of ligand and receptor with solvent (ΔG_{solv}), the conformational changes in the ligand and the receptor (ΔG_{conf}) and the motions in the protein and the ligand during the complex formation (ΔG_{motion}). In principle, a separation into individual terms is only possible if the system of interest is divided into mutually independent variables.[39] However, many of the individual terms are highly correlated with each other and they can affect the binding affinity in more than one way (i.e., positive or negative contribution).[40] Moreover, the free-energy contributions are not calculated as ensemble mean values, but are usually computed from a single structure. Despite these approximations, energy component methods are very appealing as the simplifications result in functions that can be evaluated very rapidly, which is important in a high-throughput docking setting. More importantly, they have also been successfully applied to the prediction of protein–ligand affinity.[41–43]

Two classes of function can be defined within this group of energy-component methods: in first-principle scoring functions the terms are directly derived from physicochemical (statistical mechanics) theory and are not fitted to experimental data.[44,45] The other class comprises empirical or regression-based methods, which assume an often linear statistical relationship between the total free-energy change upon binding (i.e., the binding affinity) and a number of terms that characterize the binding event. Most frequently, these terms include descriptors for hydrogen bonds and ion pairs, the amount of buried and contact surface, and molecular flexibility of the ligands. A training set of crystal structures of ligand–protein complexes and associated binding affinity data is used to optimize the coefficients of the regression equation. Many popular scoring functions have been derived this way (e.g., LUDI,[28] ChemScore,[46] and Validate[47]). Various approaches to derive optimal coefficients for regression-based scoring functions exist. Most of them aim to reproduce experimental binding affinities. However, if the only goal is to classify molecules (i.e., to distinguish binders from nonbinders) and one does not mind sacrificing the correct rank-ordering within the group of binding molecules, optimizing the scoring function by maximizing the score gap between binders and nonbinders is appealing.[48]

4.12.3.1 Developing and Optimizing Scoring Functions

All docking programs come with built-in scoring functions, but there are several reasons why these may not be the best choice for all purposes and why one may develop or optimize a scoring function. Most programs use a single scoring function for initial pose evaluation and acceptance, and also for affinity prediction, but it has been observed that scoring functions optimized for affinity prediction do not necessarily reproduce binding modes best and vice versa. Recently, Giordanetto *et al.* modified the scoring function in the program QXP by adding additional terms and refitting the resulting functions to experimental data. They observed a significant improvement in affinity prediction for their test set.[49] However, when the pose prediction performance of these novel scoring functions was examined, the original (unmodified) QXP scoring function was found to perform much better. As mentioned in Section 4.12.2.3, the original QXP function had been derived by also taking deviations from crystal structure poses into account in the fit procedure. By only using affinity data and descriptors derived from crystallographic complexes, Giordanetto *et al.* improved affinity prediction, but at the expense of pose prediction performance. It may therefore be advisable to generate separate functions for pose and affinity prediction. Also, one needs a fast scoring function for the initial pose evaluation as one quickly needs to accept or reject many generated poses, while a slower scoring function may be acceptable for a more precise assessment of the binding affinity of the surviving poses. For these reasons the ICM-Dock program[50] has two different built-in functions: one for pose evaluation (the docking function) and one for affinity prediction (the scoring function). Alternatively, one can take the poses from a docking program that were scored with its built-in scoring function and rescore them in a separate step outside the docking programs. This process is described in Section 4.12.3.2 below.

The built-in scoring functions of docking programs were developed to work across a large set of target proteins, but that does not necessarily mean that they are the best functions for a particular target. If one has experimental binding affinities or inhibition constants against a specific target for a set of compounds and associated binding mode information (from crystal structures or even from prior docking experiments), one can 'tune' the scoring function to that target. This involves building a regression equation with various scoring function terms and adjusting the regression coefficients to maximize the correlation between observed and calculated affinities. Subsequently this tuned scoring function is applied in docking experiments against that same target. This tuning process is statistical in nature and it is possible that the resulting scoring function contains physically unrealistic terms and coefficients. The risk of straying from a physically meaningful model is that this tuned function may work well only with the same target and very similar compounds as those in the training set. There are several ways of reducing this risk. One can tune scoring functions to a family of related targets (e.g., serine proteases, kinases). Or one can use only scoring function terms that have physical meaning and that are known to work well in validated scoring functions, and make sure that the coefficients remain realistic. Or one can optimize the scoring function by simultaneous minimization of coordinate and affinity deviations, as e.g., the QXP program does. All these approaches would lead to more robust, but possibly somewhat less predictive, scoring functions. Several tuning approaches are described in the two following sections.

4.12.3.1.1 Target-based tuning: three-way partial least-square (PLS)

Scoring functions are most often developed on the basis of energy-minimized ligand-protein co-crystal structures. In doing so, one may encounter the following two problems: first, the resulting scoring functions do not separate correct from supposedly incorrect (decoy) binding modes, and they work well when redocking cognate ligands, but not when docking other well-binding ligands (for a recent example, see [51]). Another unrelated problem is that tuning a scoring function to a particular target or target family could considerably improve docking performance, but especially in the early stages of a project, one may have few crystal structures on which to base such a tuning procedure. Catana *et al.* addressed these problems recently with a novel approach (manuscript submitted for preparation). A total of 129 public domain protein–ligand complexes (from eight protein families) with associated binding affinities (pK_i ranging from 1.5 to 11.5) were selected. QXP[17] was used to dock the 129 ligands to their cognate crystal structure and for each ligand 25 poses were saved. From this saved set a small number (typically 4) of diverse poses were selected on the basis of Cerius2[52] minimum spanning trees, using 15 QXP/Flo + scoring function terms as descriptors. For almost all complexes a pose very similar to the crystallographic one was present among the four selected poses. Subsequently, using the same descriptors as variables, two binding affinity models were developed. One was a straightforward PLS model based on a single energy-minimized crystal structure for each complex. The second was based on the four QXP poses per complex and uses three-way PLS,[53–55] known for its capability to reduce noise, as the statistical technique. The three dimensions of the three-way PLS are the compounds, the descriptors, and the four poses per compound, respectively. The statistics of the two models and a summary of literature models (see [46,47,56–59] and C. McMartin, personal communication) are given in **Table 2**.

The crystallographic model performs better than the best literature model[47] (which was based on 51 complexes and, therefore, relatively easy to fit), but the most important message is that the model, which only uses QXP poses and no crystallographic poses, performs very well, too. In fact, only the model of Head *et al.*[47] has better r^2 and q^2 values. Additional validation of the approach by prediction of an external test set was recently provided by Giordanetto *et al.*[49]

Table 2 Statistics for various three-way PLS pKi models

Model	*Crystallographic*[a]	*QXP poses*[b]	*Literature range*[c]
No. of complexes	129	129	51–200
r^2	0.80	0.76	0.70–0.85
q^2(leave one out)	0.76	0.70	0.65–0.78
RMS pK_i	1.00	1.24	1.02–1.40

[a] Model based on energy-minimized crystal structures.
[b] Model based on four QXP poses per complex.
[c] Ranges for seven literature models.

The approach was developed to work in real-life situations where one has few or no crystal complexes, but one does need at least a crystal structure or homology model of the target. Catana *et al.* used cognate protein structures to dock the ligands to. It remains to be established, however, how well the approach works when ligands are docked to noncognate protein structures (i.e., structures other than those from which they were taken) and how well it works if the training set contains many complexes for which no (nearly) correct pose is identified by the docking procedure. The above statistics show that one can develop adequate generic scoring functions in the absence of complex structures, but the real power of the approach lies in the opportunity to develop scoring functions specifically for a single target or target family, without the need for a large number of crystal structures to calibrate the functions. In addition, the three-way PLS models will provide insight into the most important interaction terms for the targets considered. It should also increase the probability to determine correctly which pose is most similar to the crystallographic pose and to identify alternative poses that may contribute significantly to the observed binding affinity.

4.12.3.1.2 Target-based tuning: other approaches

COMBINE[60] follows a different approach. It carries out a PLS analysis of interaction energies in protein–ligand complexes to identify those interactions that contribute most to the variance in ligand affinity. This information can then be used to predict the affinities of other ligands or to guide structure-based design efforts.[61] Interestingly, in a recent publication it was found that a variation of this approach works better in lead optimization than in hit identification.[62] Another method, the adaptation of fields for molecular comparison (AFMoC) approach, is based on interaction fields for known ligands inside a given receptor-binding site.[63] These fields are correlated to experimental affinities using PLS. In order to prevent the AFMoC scoring function becoming too biased toward the training set, it can be mixed with a more general scoring function. A mixing parameter then determines how much both scoring functions contribute to the final score. In a recent study on inhibitors of 1-deoxyxylulose-5-phosphate (DOXP)-reductoisomerase, this mixing parameter has been found to have a major influence on the results.[64] Setting the parameter to 0.5 (i.e., 50% contribution from the DrugScore scoring function and 50% contribution from the AFMoC fields) allowed for good affinity prediction of related ligands as well as structurally different inhibitors. Yet another, very interesting approach to tuning a scoring function is followed in the latest version of QXP.[17] Using docking results from an initial run with compounds whose experimental binding affinity is known, the user has the option to refit the scoring function to the experimental numbers. This refitted scoring function can subsequently be applied in another docking run, where it serves not only as a scoring function but also as a docking function during the pose generation process.

4.12.3.2 Rescoring

As indicated in Section 4.12.3.1, the scoring functions that come with docking programs do not always yield the best affinity predictions. So, often users resort to rescoring. They take the poses generated by a docking program and apply one or more alternative scoring functions to those poses. Rescoring and tuning are conceptually similar and the distinction is fuzzy. A key difference is that tuning uses scoring function terms as variables to derive an improved scoring function, while rescoring uses the scoring functions themselves. Other, less distinct differences are that one generally refers to tuning when the scoring function is optimized for a target or target family, but is still supposed to be transferable. The tuned function is also often the result of a significant development effort and intended for use both as a docking function (for pose evaluation) and a scoring function (for affinity prediction), while rescoring is a strictly postprocessing effort to improve affinity prediction. As we will see below, rescoring may actually also involve fitting to experimental data, which makes it more similar to tuning. Several approaches have been taken to rescoring. One is built on the physics of the binding process (*see* Section 4.12.3.2.1), while others apply a set of scoring functions and combine the results to generate a final score (*see* Section 4.12.3.2.2). Combining the results can be done simply by giving all scoring functions equal weights (consensus scoring, *see* Section 4.12.3.2.2.1) or, if one has experimental binding or inhibition data, by developing statistical models with optimized weights for the various scoring functions (*see* Section 4.12.3.2.2.2).

4.12.3.2.1 Physics-based scoring: molecular mechanics-Poisson–Boltzmann solvent-accessible surface area (MM-PBSA)

Solvation plays an important role in molecular recognition, but appropriate treatment of solvent effects in scoring functions still remains a major challenge. In many scoring functions these effects are considered only partially, neglected altogether, or included indirectly, as in some knowledge-based scoring schemes.

A more rigorous way of treating solvation effects in the estimation of binding affinities has become known as MM-PBSA or MM-GBSA scoring, where MM stands for molecular mechanics, PB and GB for Poisson–Boltzmann and generalized born, respectively, and SA for solvent-accessible surface area. The MM-PBSA approach has been pioneered by Kollman *et al.*,[65] and the basis is a thermodynamic cycle for complex formation in aqueous solution:

$$\begin{array}{ccccc} \mathrm{P_{aq}} & + & \mathrm{L_{aq}} & \xrightarrow{\Delta G^{\mathrm{b}}_{\mathrm{aq}}} & \mathrm{PL_{aq}} \\ \downarrow -\Delta G^{\mathrm{PL}}_{\mathrm{solv}} & & \downarrow -\Delta G^{\mathrm{PL}}_{\mathrm{solv}} & & \uparrow \Delta G^{\mathrm{PL}}_{\mathrm{solv}} \\ \mathrm{P_{gas}} & + & \mathrm{L_{gas}} & \xrightarrow{\Delta G^{\mathrm{b}}_{\mathrm{gas}}} & \mathrm{PL_{gas}} \end{array}$$

Thus, for $\Delta G^{\mathrm{b}}_{\mathrm{aq}}$, the free energy of binding in solution, one obtains:

$$\Delta G^{\mathrm{b}}_{\mathrm{aq}} = \Delta G^{\mathrm{b}}_{\mathrm{gas}} + \Delta G^{\mathrm{PL}}_{\mathrm{solv}} - \Delta G^{\mathrm{P}}_{\mathrm{solv}} - \Delta G^{\mathrm{L}}_{\mathrm{solv}}$$

Expansion of $\Delta G^{\mathrm{b}}_{\mathrm{gas}}$ yields:

$$\Delta G^{\mathrm{b}}_{\mathrm{aq}} = \Delta E^{\mathrm{intra}}_{\mathrm{gas}} + \Delta E^{\mathrm{elec}}_{\mathrm{gas}} + \Delta E^{\mathrm{VdW}}_{\mathrm{gas}} - T\Delta S + \Delta G^{\mathrm{PL}}_{\mathrm{solv}} - \Delta G^{\mathrm{P}}_{\mathrm{solv}} - \Delta G^{\mathrm{L}}_{\mathrm{solv}}$$

The free energy of solvation for species X, $\Delta G^{\mathrm{X}}_{\mathrm{solv}}$, can be separated in polar and apolar solvation terms $\Delta G^{\mathrm{X}}_{\mathrm{solv}} = \Delta G^{\mathrm{X}}_{\mathrm{solv,p}} + \Delta G^{\mathrm{X}}_{\mathrm{solv,ap}}$, to give the fully expanded equation:

$$\begin{aligned} \Delta G^{\mathrm{b}}_{\mathrm{aq}} = {} & \Delta E^{\mathrm{intra}}_{\mathrm{gas}} + \Delta E^{\mathrm{elec}}_{\mathrm{gas}} + \Delta E^{\mathrm{VdW}}_{\mathrm{gas}} - T\Delta S + \Delta G^{\mathrm{PL}}_{\mathrm{sol,p}} + \Delta G^{\mathrm{PL}}_{\mathrm{solv,ap}} \\ & - \Delta G^{\mathrm{P}}_{\mathrm{solv,p}} - \Delta G^{\mathrm{P}}_{\mathrm{solv,ap}} - \Delta G^{\mathrm{L}}_{\mathrm{solv,p}} - \Delta G^{\mathrm{L}}_{\mathrm{solv,ap}} \end{aligned}$$

$\Delta E^{\mathrm{intra}}_{\mathrm{gas}}$, the change of intramolecular energy for ligand (and protein) upon binding, is sometimes neglected, based on the assumption that the intramolecular energy does not change significantly upon binding. $\Delta E^{\mathrm{elec}}_{\mathrm{gas}}$ and $\Delta E^{\mathrm{VdW}}_{\mathrm{gas}}$, the electrostatic and steric interaction energies between protein and ligand, can be calculated using normal force field expressions, although it would be possible to use continuum electrostatics (PB or GB) for the electrostatic part. Originally, MM-PBSA was intended for, and applied in combination with, molecular dynamics simulations in solution. In this case $T\Delta S$ can be estimated by quasiharmonic analysis of the trajectory,[66] or normal-mode calculation can be used.[67] However, for scoring many ligands of similar type, it should be possible to neglect this term, or to replace it by a simpler calculation, such as an approximation of the buried surface upon binding. $\Delta G^{\mathrm{X}}_{\mathrm{solv,p}}$, the polar free energy of solvation, and $\Delta G^{\mathrm{X}}_{\mathrm{solv,ap}}$, the apolar free energy of solvation of species X, are calculated with the PB or GB approach and using an expression containing a simple surface area term, respectively.[68]

Recently, the first applications of MM-PBSA as a more sophisticated scoring function in the context of docking and VS have become known. In contrast to earlier applications, where it was combined with molecular dynamics (MD) simulations, the recent examples demonstrate its value also for 'snapshot scoring,' i.e., the evaluation of the MM-PB(GB)SA expression for one or a few poses per ligand. These poses had been generated using a conventional docking program and not by means of a lengthy MD simulation. Researchers at Wyeth[69,70] and SGX Pharmaceuticals[71] presented evidence that MM-PBSA scoring can lead to an improvement compared to conventional scoring. It was shown that, given a number of precomputed poses per ligand, re-ranking of the poses with MM-PBSA leads to a better separation between correct and incorrect poses. This improvement was due to a reduction of both false negatives and false positives. Also, it was illustrated that enrichment was significantly higher when MM-PBSA was used to rescore larger databases of docked ligands. Treatment of a substantial number of compounds was computationally feasible, as the compute-intensive part of the MD simulations including explicit water had been replaced by pose generation with a fast docking program.

Kuhn and co-workers recently demonstrated the application of MM-PBSA scoring to several different data sets.[72] Docking was performed using the programs FlexX[13] (ScreenScore function) and FRED[73] (ChemScore function). Prior to rescoring, the complex structures were minimized in the presence of explicit water and counterions. For one data set (neuraminidase) it was shown that the correct inhibitor pose is normally identified by docking, but that the ChemScore function is not able to differentiate between true and false positives. In this case MM-PBSA scoring led to a significant improvement. Also, it was shown that MM-PBSA scoring improved pose prediction and consequently enrichment for p38 MAP kinase inhibitors. Analysis of these cases indicated that a major deficiency of conventional scoring functions is the lack of an energy penalty for the desolvation of mismatched, i.e., polar–apolar protein–ligand interactions, which MM-PBSA can improve upon. The authors concluded that the application of MM-PBSA to a single structure is generally valuable for rescoring after docking and for distinguishing between strong and weak binders.

In another study MM-PBSA scoring was incorporated as the last step in a hierarchical database screening process.[67] In this case 20 poses per ligand were generated using short MD simulations (20 ps after equilibration), starting from the previously docked ligand orientations. Although only a limited number of ligands was considered, an encouraging correlation between experimental and MM-PBSA binding free energies was found, and it was noted that the overall strategy achieved not only high efficiency but also high reliability.

In an earlier study the effects of different solvent models were compared using a data set of 189 protein–ligand complexes.[74] In contrast to the afore-mentioned examples, comparing PB solvation with simpler solvent models – such as GB, constant or distance-dependent dielectric function, in conjunction with the CHARMm force field – did not indicate an advantage of more sophisticated solvation models in terms of the rank correlation between predicted and observed binding energies. Interestingly, the authors also noted that for pose prediction steric complementarity between the ligand and the receptor appears to be the most important factor.

4.12.3.2.2 Composite scoring functions

All scoring functions may exhibit pathological behavior with certain compound classes or functional groups. To minimize the impact of this problem and to reduce statistical noise, composite scoring methods have been introduced.[75–77] Rather than using a single scoring function, several scoring functions are combined such that in order to be classified as a potential binder, a molecule has to be scored well by a number of different scoring functions. Such composite scoring functions come in two flavors. Consensus methods combine the results of various single scoring functions in a predefined, unbiased manner without any training. Statistical composite methods, on the other hand, attempt to optimize the affinity prediction by developing a model that is based on a training set, which is relevant for the target being studied. Both approaches will be described below (for examples of the successful application of composite scoring methods, see [75,78,79]).

4.12.3.2.2.1 Consensus scoring

The premise of consensus scoring is that the more scoring functions agree that a compound is active (or inactive), the more reliable the prediction is. So, a compound that receives a high score from multiple scoring functions is more likely to be a good inhibitor in an actual assay than a compound that receives a high score from only a single function.

Wang and Wang[76] simulated a docking and scoring experiment by taking known binding affinities for a set of compounds to which they added a random error to mimic the behavior of a scoring function. They repeated this process for several scoring functions and subsequently carried out consensus scoring. They found that "consensus scoring outperforms any single scoring for a simple statistical reason: the mean value of repeated samplings tends to be closer to the true value." In other words, consensus scoring will work better than a single scoring function if the scoring functions are largely independent of each other and if the individual scoring functions themselves are equally predictive.

In the rank-by-vote procedure, one lets several (N) scoring functions 'vote' on all compounds. One can use a single or multiple poses, previously generated by one's docking program of choice. For each scoring function, all compounds are rank-ordered by their score and the highest scoring compounds all receive one vote. Subsequently, for each compound the votes from all N scoring functions are added together. All compounds with N votes are predicted to be active. One can also allow one dissenting vote and regard all compounds with at least $N-1$ votes as active, but that tends not to increase enrichments.[79] In traditional rank-by-vote, the number of compounds that receive a vote from each scoring function is user-defined, but fixed. Suppose that one decides that the top 2% of all compounds according to each scoring function receive a vote. It is unlikely that all compounds ranked in the top 2% of one scoring function are the same as the top 2% of another scoring function. So, the number of compounds that will receive the maximum number of N votes will be smaller than 2%. This traditional variant is therefore called reducing. With this variant, the number of compounds that receive N votes is not a priori defined and can vary significantly from one consensus approach to another and from one data set to another. If one wants to compare different rescoring schemes, however, one needs to consider the hit rates for a predefined number of compounds that is the same for each scheme. Also, if one wants to optimize a selection of compounds for synthesis or purchase, predicted affinity is typically only one of the parameters that is considered. Others are, for example, solubility and diversity of the selection. In that case, one needs a rank-ordering of all compounds, not just a subset. For these reasons, a nonreducing variant was recently developed.[79] This variant involves iteratively increasing the number of compounds that receive a vote by descending the rank-ordered list of each scoring function, one compound at a time, until the number of compounds with N votes equals that pre-defined number. This procedure starts by giving a vote only to the single best-scoring compound per scoring function and ends by giving votes to all compounds considered. One can also call this variant a worst-rank consensus scheme because the worst rank a molecule has according to all scoring functions determines its final rank. Another consensus approach uses the mean rank of each compound, i.e., the average value of the ranks of that compound according to each of the scoring functions that are allowed to vote. Unlike rank-by-vote, the mean rank procedure is by definition nonreducing. Neither approach is consistently better than the other. The results depend on the docking program and on the number and the nature of the scoring functions that are chosen to vote. Although the appeal of these methods is that they are unbiased and that their application does not require any experimental binding data, in practice it is important to run some pilot experiments to determine how many and which scoring functions work best (i.e., yield the best enrichments; *see* Section 4.12.4.2.2).

4.12.3.2.2.2 Bayesian statistics

In the preceding paragraph we described how the results of rescoring docking poses with various scoring functions can be combined in an unbiased way. If one has experimental binding affinity data, however, one can optimize the affinity prediction by developing a model that is based on this data (the training set). Recently, Bayesian statistics (BS) has made inroads in drug discovery and developing composite scoring functions is one possible application of BS. In this context, BS considers 'features' of molecules and counts how often each feature occurs in active and in inactive molecules in the training set. Based on this analysis it estimates the probability of each feature to occur in active and in inactive molecules. Subsequently, one predicts the probability to be active for a test set by considering all features of each molecule in the test set and the probability to be active associated with each feature. In practice, one multiplies and scales all these probabilities. In a typical BS application, one uses molecular descriptors that characterize the ligand. If one has a set of scores based on docking results, however, these can be used as descriptors instead. The assumption is that these scores provide a more meaningful characterization than molecular descriptors as they specifically describe the interaction between a ligand and its target rather than only the properties of the ligand itself. An important issue is the exact definition of the features to be used in the BS. For each scoring function one can simply calculate its median value over all compounds. Then one counts the number of actual actives and inactives in the training set with a score above this median and, on the basis of that count, calculates the probability for a compound to be active when it has a score above the median. One does the same for all compounds below the median. This process is then repeated for each scoring function. Rather than using two bins (below the median and above the median), a typical application would use 10 bins for each scoring function (feature). Cotesta *et al.*[79] found that in the majority of cases BS performs better than the individual scoring functions and than the unbiased consensus approaches in terms of enrichment (i.e., hit rate increases; *see* Section 4.12.4.2.2). It also works well in distinguishing between moderately active and very active molecules, making the approach also suitable for lead optimization.

4.12.4 Assessment of Docking Performance

A multitude of approaches and docking programs is available today. This poses the question which docking program one should use, and which docking approach might be most appropriate for a given problem. As outlined above, there are two major challenges faced by docking programs: first, to predict the binding mode of a molecule correctly, and second, to predict binding affinities reliably. The first point is a prerequisite for the second: if the ligand is not docked correctly, it is unlikely that the calculated score is meaningful, apart from fortuitous cases where calculated affinities for different poses are close in energy. As also noted by others, comparison and assessment of docking programs are not easy tasks,[80] and in the following sections we share general considerations as well as some of our personal experiences.

4.12.4.1 Pose Prediction

In many cases it is necessary to evaluate docking programs with respect to their ability to reproduce experimentally known poses in a reliable manner. The traditional way to do this is to calculate the RMS deviation (RMSD) between a pose generated by a docking program and the experimentally observed binding mode. Despite the practical appeal of using RMSDs from a crystal structure to assess pose prediction accuracy, they do not do justice to the complex interactions ligands make with proteins. For that reason, we devised a novel way to evaluate pose prediction accuracy.[81] Here the correctness of a pose was determined by visually comparing the (hydrogen-bonded and other) ligand–protein interactions for that pose with the experimentally observed interactions. In particular this interactions-based accuracy classification (IBAC) scheme was introduced because of the following considerations:

- Differences in the force fields implemented in the docking programs may lead to variations in the predicted poses, resulting in relatively large RMSD values without changing the overall binding modes and interactions.
- If a molecule contains a very flexible moiety that is not involved in key interactions, the differences between docked and experimental binding mode for this part might lead to a high RMSD for the entire molecule, although the overall binding mode of the docked molecule is correct.
- A large, almost symmetric molecule may adopt a nearly correct pose (i.e., correct with the exception of the symmetry-breaking moieties) but may have a very large RMSD.
- The difference between the RMSD and IBAC classification schemes can be significant. This is exemplified by **Figure 1**, which shows the crystallographic pose of an inhibitor bound to CDK-2 and a docking pose generated by the program ICM.[82,83] Despite a low RMSD of 1.62 Å from the experimentally determined binding mode, the pose generated by ICM was deemed incorrect according to the IBAC.

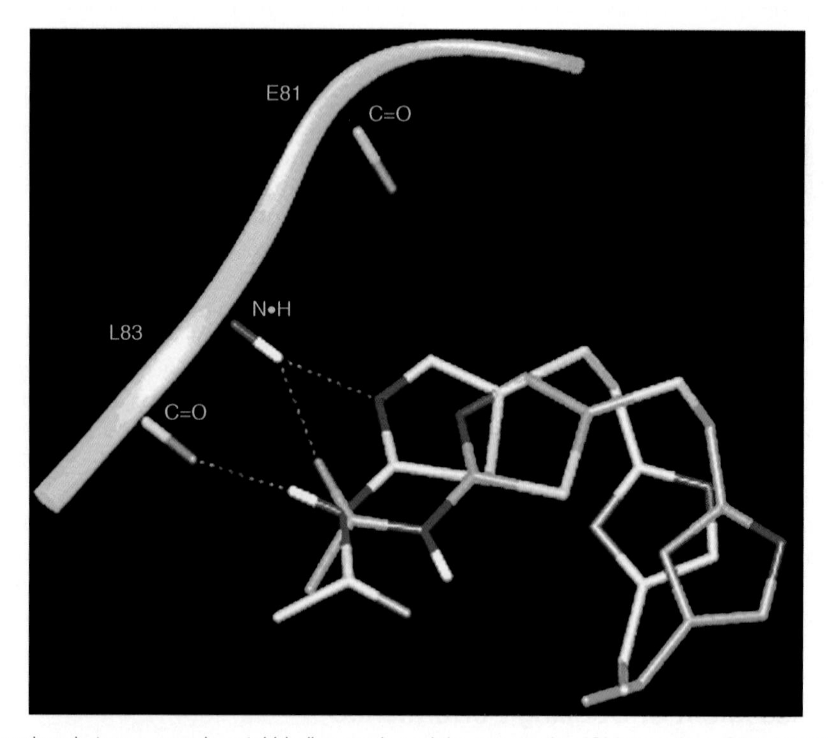

Figure 1 Comparison between experimental binding mode and the top-scoring ICM pose for a CDK-2 inhibitor. The CDK-2 hinge region is displayed as a ribbon with hydrogen bond donor and acceptors. Cyan carbon atoms: experimental binding mode; green carbon atoms: ICM pose classified as incorrect; RMSD = 1.62 Å. (Reprinted with permission from Kroemer, R. T.; Vulpetti, A.; McDonald, J. J.; Rohrer, D. C.; Trosset, J.-Y.; Giordanetto, F.; Cotesta, S.; McMartin, C.; Kihlen, M.; Stouten, P. F. W. *J. Chem. Inf. Comp. Sci.* **2004**, *44*, 871–881. Copyright (2004) American Chemical Society.)

The crystallographic binding mode of the CDK-2 inhibitor involves two hydrogen bonds with L83 (part of the CDK-2 hinge region): the ligand's thiazole nitrogen interacts with L83-NH and its amide-NH with L83-CO (**Figure 1**). In the ICM pose, at first glance it seems that the entire compound has just moved away slightly from the hinge region. However, the hydrogen bond interaction pattern is entirely different. None of the crystallographic hydrogen bonds are retained, while the ligand's amide group has flipped 180° in order to make a single new hydrogen bond interaction between its CO and L83-NH. The thiazole moiety is no longer predicted to interact with the hinge region at all. Also, three of the six hydrophobic interactions originally present in the crystal structure have been lost after docking. Despite the fact that the docked molecule establishes some new interactions with the protein, of the original key interactions, more than 50% are not present anymore. As a consequence, the IBAC regarded the docking pose of this compound as incorrect. Further examples of IBAC–RMSD comparisons can be found in [81].

4.12.4.2 Affinity Prediction

Above we indicated that the assessment of pose prediction performance may not be as straightforward as one might think. To assess properly how well affinities are predicted is also quite complicated. First, there is the problem of which data set to use. It would seem natural to carry out tests with data sets that resemble the data sets that will be used in production mode, but it seems this is not often done. Then there is the issue of building the test set and carrying out the analysis of results such that meaningful conclusions can be drawn that are relevant to the goal of the study. Both these issues are addressed here.

4.12.4.2.1 Data sets

The desired and expected activity range for inhibitors depends on the stage of a drug discovery project. At the hit identification stage, molecules with even weak activity (IC_{50} > 100 nM or > 1 μM) represent a useful source to initiate a medicinal chemistry program, while ligands with nanomolar affinity are searched for during the lead optimization phase. This puts different demands on the computational tools. In practice one rarely finds nanomolar compounds when screening databases of commercially available compounds, so for VS programs to be applicable during hit identification, they should be able to identify micromolar hits among a large number of inactive compounds (hit identification in **Figure 2**). By contrast, if one considers using these programs for lead optimization (e.g., for the design of a combinatorial library built around a potent scaffold), one needs to be able to distinguish potent compounds (< 100 nM) from moderately/weakly active (100 nM–10 μM) and completely inactive (> 10 μM) ones. Consistent with observations made by Charifson *et al.*,[75] one can expect that the ability of docking tools to distinguish active from inactive compounds depends considerably on the activity profile considered. In a recent study this was explicitly considered by defining three activity intervals (A1: IC_{50} < 100 nM; A2: 100 nM < IC_{50} < 1 μM; and A3: 1 μM < IC_{50} < 10 μM) and examining the docking and scoring results for each activity interval separately.[79] A key conclusion of that study was that affinity prediction and enrichment strongly depend on the distribution of activities in the data set.

In that study it was also mentioned that quite often the performance of docking tools is assessed by spiking a large collection of supposedly inactive compounds with several fairly active compounds. Even if these tools are intended for hit identification only, such an evaluation procedure can be considered suboptimal, as commercial or proprietary

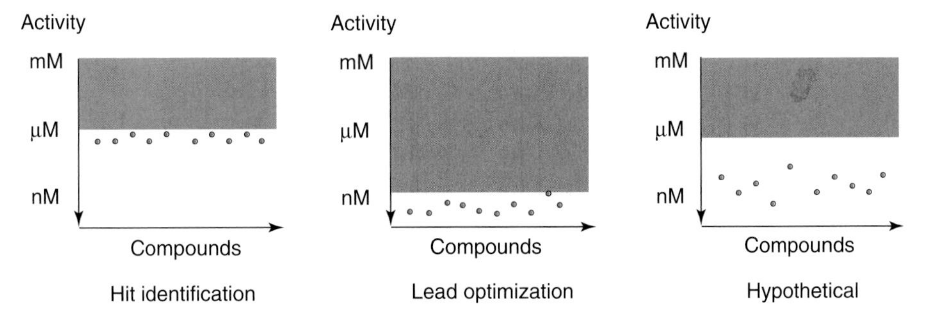

Figure 2 Graphical representation of different activity distributions that can be used in the assessment of docking program performance. The few active compounds are represented by the red dots. The many inactives are represented by the gray areas. Under each distribution the corresponding real-life situation is indicated. The hypothetical situation is rarely encountered in real life, but is often used to assess docking performance nonetheless, as in that situation docking programs/scoring functions exhibit their best performance in terms of separating actives from inactives.

compound collections rarely consist exclusively of nanomolar compounds against a given target on the one hand and fully inactive ones on the other. Similar conclusions were also drawn in another study.[84] A graphical representation of possible activity distributions in datasets for the assessment of docking performance is displayed in **Figure 2**.

Another point that needs to be considered for the assessment of docking program performance is the properties of the compounds, such as molecular weight (MW) or polarity. Some authors have argued that care must be taken that both active and inactive compounds represent an identical or very similar distribution of such properties. This is because charges can introduce an unwanted bias toward certain molecules[85] and because scoring functions in many cases correlate with MW.[86] The counterargument is that there is a physical reason for properties such as MW to increase binding. This can be caused by nonspecific burial of hydrophobic surface of both ligand and receptor. One may not want to pursue high-MW compounds as hits, however, and, rather than selecting compounds with a high calculated affinity, one may select compounds with a high calculated ligand efficiency (i.e., affinity per atom),[87] but that is an issue that is in principle unrelated to affinity prediction performance per se. That aside, it has been observed that with MW as a single descriptor one can preferentially retrieve CDK-2 inhibitors, but the enrichment is on average only 7% (in one case, however, it is 28%) of the enrichment produced by QXP without rescoring (Vulpetti and Stouten, unpublished results).

4.12.4.2.2 Enrichment

Often there is relatively little correlation between experimental and predicted activity (*see* e.g., Section 4.12.6.2). Docking has been successful in many cases nonetheless, because of its ability to preferentially select active compounds, thereby increasing the number of actives in a set of compounds that is selected for experimental testing. This increase is referred to as enrichment and has become a standard measure of quantifying the success of a docking campaign. To determine the enrichment, first the molecules in the data set are ranked by their score. Then one retrieves the top-scoring N_{retr} compounds (generally between 1% and 20% of the data set) and counts the number of active $N_{A,retr}$. The hit rate is defined as the proportion of actives in this retrieved set: $N_{A,retr}/N_{retr}$. This hit rate is compared to the overall random hit rate of the total data set: $N_{A,tot}/N_{tot}$, and the enrichment is defined as the actual hit rate divided by the random hit rate: $(N_{A,retr}/N_{retr})/(N_{A,tot}/N_{tot})$. In many cases enrichment is displayed in the form of enrichment curves, where the number (or percentage) of actives found is plotted against the number of rank-ordered molecules. Recently it was advocated to use receiver operating characteristic curves instead.[88] The advantage of the latter curves is that they are independent of the proportion of actives in the test set. Also, they include information relating to false positives and false negatives in the same plot.

4.12.5 Problems and Improvements

Much work has been invested in the generation of better docking programs and scoring functions over the past years and, although much progress has been made, improvement is still necessary. In this section we highlight some of the fundamental problems in docking and scoring and the ways that researchers have started to address them. Some general approaches to improve the performance of docking and scoring are presented, too.

4.12.5.1 Docking into Flexible Receptors or the Cross-Docking Problem

One of the most challenging problems in docking and scoring is the treatment of flexible receptors. Numerous examples have become known where the same protein adopts different conformations depending on which ligand it binds to.[89,90] As a consequence, docking using a rigid receptor representation corresponding to a single receptor conformation will fail for those ligands that require a different protein conformation in order to bind. An example where relatively small conformational changes can have already a large effect is given by the following cross-docking example: in an in-house evaluation of the pose prediction performance of several docking programs at Pharmacia, a number of publicly known CDK-2 inhibitors was docked into a receptor conformation corresponding to CDK-2 in complex with adenosine triphosphate (ATP) (Protein Data Bank (PDB) code 1QMZ), after the natural ligand had been removed. One of the ligands that was especially difficult to be docked correctly was hymenialdisine.

Examination of the crystal structure of this ligand in complex with CDK-2 (PDB code 1DM2) revealed that 1QMZ has a larger ATP-binding pocket than 1DM2. Moreover, in 1DM2 a hydrogen bond exists between the carbonyl oxygen of the imidazolone moiety of the ligand and one of the two water molecules located behind K33 of the kinase. Also, D145 adopts a different conformation in 1DM2 and forms a hydrogen bond with the ligand. The result of these differences is that, when docked into 1QMZ, hymenialdisine cannot engage in all binding interactions it makes in 1DM2. It can wobble around and adopt several distinctly different, but energetically degenerated poses. This larger,

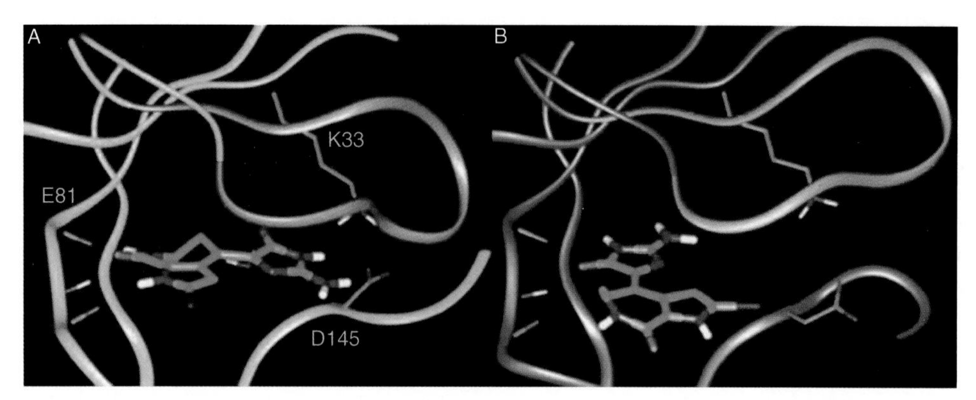

Figure 3 Comparison of hymenialdisine docked into 1DM2 (pane A, correct binding mode) and 1QMZ (pane B, incorrect binding mode). The structures were aligned using the Cα carbons of K33 and ASP145, as well as the main-chain carbonyl carbon of E81 (RMSD = 0.25 Å). Hydrogen bonds with the backbone of the so-called hinge region are displayed as dotted lines. The glycine-rich loop is colored red.

suboptimal binding site may explain the failure of some docking programs to dock hymenialdisine correctly. In order to test the hypothesis that this failure was indeed due to the suboptimal 1QMZ binding pocket, hymenialdisine was docked into the ATP-binding pocket of 1DM2 using QXP. Two runs were carried out: one with the water molecule present and one without the water molecule. In both cases QXP was able to identify the correct binding mode of hymenialdisine and to score the correct pose highest. **Figure 3** displays the correct binding mode found by QXP with the water molecule present. A much more radical example of different conformations adopted by kinases is the transition between the so-called DFG-in or DFG-out conformations, where a stretch of three residues (DFG) is either rotated into or away from the ATP-binding pocket.[91]

In order to deal with the problem of flexible receptors in docking, several approaches have been proposed, which can be grouped roughly into the following categories: (1) letting the receptor or parts thereof move during docking; (2) docking the compounds into several different conformations of the same receptor and aggregating the results; and (3) docking into averaged receptor representations. The borders between these three approaches are sometimes fuzzy and some of the practical implementations known contain elements of more than one of these methods. Examples for each of the three categories are provided here.

Regarding the first category, one well-known program that allows receptor flexibility during docking is QXP.[17] As described in Section 4.12.2.3, the user can specify certain parts of the protein to move during the minimization step at the end of each Monte Carlo cycle during docking. In some cases this can alleviate clashes between the ligand and receptor that would otherwise occur and this can therefore lead to better pose prediction results. Another example has been provided by researchers at Schrödinger, who recently reported on an approach where rigid receptor docking with Glide is iteratively combined with protein structure prediction with Prime.[92] The authors reported that using only Glide the average RMSD to the crystal structure for 21 different complexes was 5.5 Å, and application of their new induced-fit docking (IFD) procedure reduced the average RMSD to 1.4 Å. The tradeoff here is that the IFD procedure requires a relatively large amount of time per ligand and is therefore not applicable to high-throughput docking. Nevertheless, this methodology could prove useful for building and refining homology models, detailed binding studies during lead optimization, and the generation of different conformational hypotheses prior to a larger docking campaign.

A number of approaches have been published recently where ligands are docked into several different conformations of the same receptor, in order to address the problem of receptor flexibility. Cavasotto and Abagyan have presented an algorithm to generate a discrete set of receptor conformations, and each of these structures is then used for rigid receptor docking.[93] Subsequently the results of the multiple docking runs are combined in order to improve enrichment. Combining the results is achieved by merging the hit lists for each of the docking runs and keeping the best rank for each compound. For several protein kinases this procedure led to a significant increase in hit rates compared to the individual results. In another study scoring functions that are more (soft) and less (hard) tolerant to bad geometries were compared in docking runs against one or more conformations of the same receptor.[94] The soft scoring function proved to be superior to the hard potential when a single receptor conformation was used. Conversely, when docking was performed into multiple receptor conformations the hard potential exhibited better performance. In this case multiple flexible regions of the binding site were treated independently, recombining them to generate different discrete conformations.[95] It was also noted that softer scoring functions can increase the likelihood of false positives. Using FlexX as docking program, a comparison of single versus multiple conformer docking was performed for

PTP-1B.[96] Different receptor structures had been created by considering different combinations of side-chain rotamers within the active site. The inhibitors were then docked against all active-site models and for each inhibitor the model with the best interaction energy was identified. This allowed for successful discrimination between correct and incorrect binding modes as well as for an improvement in the ranking of the inhibitors. The FlexE program[19] is based on a united protein description originating from different superimposed conformations of a protein. During the incremental construction of a ligand discrete protein conformations are sampled in a combinatorial fashion. The program was evaluated for 10 proteins represented by 105 crystal structures from the PDB and one modeled structure. For 83% of the ligands the correct pose was found. The results were of a quality comparable to the one obtained by sequentially docking into all conformations separately, but the run times for FlexE docking were much shorter than for the sequential docking.

Receptor averaging is another way of approaching the problem of receptor flexibility. Using AutoDock it was investigated how the interaction energy grids for different receptor conformations can be combined.[18] The study was carried out using complexes of 21 peptidomimetic inhibitors with human immunodeficiency virus-1 (HIV-1) protease. Four different schemes of combining the grids were tested. It turned out that the mean grids performed worst, whereas the energy-weighted methods gave much better results. That using simply an average structure representing several different conformations leads to inferior results has been noted by researchers at Eli Lilly.[97] Their results indicated that docking accuracy decreases significantly when an average structure is used.

McGovern and Shoichet also carried out a very interesting study.[98] The focus of this study was the information loss that occurs when the active-site conformation becomes less defined. To this end, 10 different enzyme-binding sites, represented by their holo, apo, and model structures, were investigated. The MDL Drug Data Report (MDDR) database of 95 000 small molecules containing at least 35 ligands for each of the 10 systems was docked against all 30 structures using the DOCK program.[12] The ability of each structure to preferentially retrieve the known ligands for that enzyme was evaluated. In seven cases, there was clear superiority of the holo structures over the apo and model structures. However, the apo and model structures proved superior for two and one enzymes, respectively. For the latter cases it was postulated that the holo structure may in some cases be overspecialized by induced fit to a particular ligand, and therefore the apo or model structure may be a better choice for the docking experiment.

Summarizing, one may say that to date, protein flexibility remains one of the most challenging problems in docking and scoring. Progress has been made and interesting approaches have been proposed, but it is still an open question whether these techniques have advanced sufficiently to be of substantial help.

4.12.5.2 Water

Water molecules often play a key role in protein–ligand recognition. If one ignores water-mediated interactions during docking then the calculated interaction energy of a given ligand conformation may be too low. If, on the other hand, one retains crystallographically observed water molecules then the binding pose and affinity of a ligand that in reality replaces that water molecule will not be correct. It is notoriously difficult to treat water adequately, as first one needs to identify possible positions for water molecules, where they could interact with the protein and ligand, and subsequently one must be able to predict whether a water molecule is indeed present at that position. Researchers at Astex and the Cambridge Crystallographic Data Centre recently implemented an elegant procedure in the latest version of GOLD to address both these issues.[99] The water positions they consider for a given target are taken from a set of complex structures of that target, but one could also use programs to predict potential water-binding sites.[100,101] Each water molecule can then be present (on) or absent (off). If a water molecule is on, it can make favorable interactions with the ligand and protein, but it pays an entropic penalty for loss of translational and rotational degrees of freedom.[102] The value of this penalty was optimized using a training set of 58 protein–ligand complexes. Considering both the training and test sets, on and off status are correctly predicted for 93% of the water molecules. This increases correct pose prediction rates of water-mediated complexes by 10–12 percentage points, but it decreases correct pose prediction rates for nonwater-mediated complexes by 6–7 percentage points. This latter decrease is readily explained when one assumes that prediction of a water molecule where there should not be one leads to an incorrect binding mode. The expectation is that the correlation of calculated and measured affinities will improve with the inclusion of water molecules in the docking runs, which in turn should improve the enrichments obtained in VS experiments, but this remains to be investigated.

Another approach to dealing with water molecules involved in protein–ligand interactions has been incorporated in the FlexX docking program. This method, referred to as the particle concept, includes the calculation of favorable positions of water molecules inside the active site prior to docking. During the incremental construction phase these water molecules are allowed to occupy the precomputed positions if they can form additional hydrogen bonds with the ligand. The method was tested using a data set of 200 protein–ligand complexes and with pose prediction quality as an evaluation criterion. Similar to the observations made for GOLD, it was found that on average the improvement was minor. Nevertheless, in a number of cases the predicted waters corresponded to the crystallographically observed ones, which led to an improvement in the predictions.[103]

4.12.5.3 Multiple Active-Site Corrections

A possible way of improving docking results is the application of so-called multiple active-site corrections (MASC).[104] Here the underlying idea is that scoring functions are biased toward certain ligand types or characteristics, such as large or hydrophobic ligands. This implies that some ligands are generally predicted to be good binders regardless of whether these ligands will bind to certain active sites or not. Therefore, a simple statistical correction has been introduced, which can be interpreted either as a statistical measure of ligand specificity or as a correction for ligand-related bias in the scoring function (but see also the discussion about MW in Section 4.12.4.2.1). In order to calculate the MASC scores, each ligand is docked into a number of unrelated binding sites of different binding site characteristics. The corrected score (or MASC score) $S^{\bullet}_{ij}$ for molecule i in binding site j is calculated as follows:

$$S^{\bullet}_{ij} = \frac{(S_{ij} - \mu_i)}{\sigma_i}$$

where S_{ij} is the uncorrected score for this molecule. μ_i and σ_i represent mean and standard deviation of the scores for molecule i across the different binding sites. Thus, the MASC score $S^{\bullet}_{ij}$ represents a measure of specificity of molecule i for binding site j, compared to the other binding sites. The MASC scores were tested using FlexX and GOLD and a data set of 15 protein–ligand complexes. Without corrections only in three (FlexX) or four (GOLD) cases the endogenous ligand was identified correctly. After application of the MASC the success rates rose to 11 for both docking programs. In another test, the MASC scores were applied to a database of 600 drug-like molecules mixed with 30 inhibitors of p38α MAP kinase and 30 inhibitors of protein tyrosine phosphatase (PTP-1B). Interestingly, when docking the 660 compounds into the p38α active site, the uncorrected GOLD scores led to an enrichment of PTP-1B inhibitors over the p38α inhibitors. Application of the MASC scores led to a reversal of this trend and the enrichment of true p38 inhibitors was significantly improved. The authors concluded that MASC improves the detection of true positives and reduces the number of false positives in database enrichment studies.

However, in a recent study scientists at OpenEye found that the application of MASC can also lead to a deterioration of the results.[105] When the docking program FRED[73] 2.1 was used in enrichment studies comparing different scoring functions and different targets, it was found that the value of MASC scoring depends heavily on the type of target and the scoring function used. Regarding the change of enrichment for five scoring functions plus consensus scoring, with each scoring scheme averaged across five different targets, only for consensus scoring a significant improvement using MASC scores was observed. Considering each target separately, the picture was also ambiguous. Only for reverse transcriptase did MASC lead to an improvement for all scoring schemes investigated. Other targets exhibited a very mixed behavior, most notably HIV protease, where for three scoring schemes the results after MASC deteriorated significantly, while for two scoring schemes a significant improvement was observed. For another target – the estrogen receptor – MASC scoring either reduced the enrichment or had no effect.

4.12.5.4 Docking with Constraints

By introducing a bias during docking it is possible to influence the way poses are generated and which ones are preferentially kept. For example, in the DockIt program,[106] one can apply distance constraints between ligand and protein atoms that are subsequently used during pose generation via a distance geometry approach. The GA of the GOLD program[107–109] makes it easy to include different types of constraints in the fitness function, thus enabling the generation of biased poses. In the PhDock approach,[110] as implemented in DOCK 4.0,[111] one can perform pharmacophore-based docking by overlaying precomputed conformers of molecules based on to their largest 3D pharmacophore. The pharmacophore is then matched to predefined site points representing putative receptor interactions. Subsequently, all conformers are docked corresponding to the pharmacophore match and the fit of each individual conformer is scored. The advantage of this approach is twofold: speed through a rapid pre-orientation of the molecules to be docked as well as the introduction of bias toward good solutions by defining pharmacophore points that represent favorable interactions with the target binding site. The use of another docking program where one can apply pharmacophoric constraints during docking (FlexX-Pharm[114]) is illustrated in Section 4.12.6.3.

4.12.5.5 PreDock Processing

Another challenge in docking is accounting for the various tautomeric and protomeric states the molecules can adopt. In many databases molecules such as acids or amines are stored in their neutral forms. Considering that they are ionized under physiological conditions it is necessary to ionize them prior to docking. However, while standard ionization is easy to achieve, the problem of tautomer generation is already much more challenging: which tautomer should one use? Or should one use more than one (or all possible) tautomers for a given molecule? Not only for tautomers, but also for different ionization states balanced equilibria between the different forms provide real challenges in docking. One (radical) approach to this would be to generate all possible forms, subsequently to dock all of them, and to choose the relevant form based on the scores. However, it remains to be seen whether such an approach would be beneficial or just generate a large number of false positives.

4.12.5.6 PostDock Processing

One can in principle distinguish between two approaches for introducing bias after the docking: applying postdock filters and using tailor-made (re)scoring functions. The latter have already been described in Section 4.12.3. In many cases postdock filters are conceptually simple and may correspond to certain geometric criteria, such as the presence of certain interactions (e.g., a hydrogen bond with a selected residue or a polar interaction) or the filling of a specified pocket in the active site. Many researchers in the pharmaceutical industry have written their own filters, but nowadays they are an integral part of several commercially available docking programs, such as Glide[84,113,114] and FRED.[73] Implementation and automation of these filters can also reduce the need for the frequently quoted visual inspection of poses after docking and scoring.

The idea of checking for certain receptor–ligand interactions can be extended to entire interaction patterns. These interaction patterns in turn can be compared to a target interaction pattern, as observed in the co-crystal structure of a highly active ligand and its receptor. An interesting method has recently been developed where a structural interaction fingerprint (SIFt) is calculated to provide a unique representation of a binding mode.[115] In this approach each binding-site residue is encoded by a string of 7 bits that represent the different possible interactions with that residue. If such an interaction occurs the corresponding bit is set. All bits together constitute the SIFt of a binding mode. SIFts can then be used to evaluate similarities between different poses of the same or different molecules, to assess how close a pose is to a predefined (e.g., crystallographic) pose and, of course, to filter docking results to find compounds with desirable interactions. Conceptually, SIFt is similar to IBAC (*see* Section 4.12.4.1), but it has taken the idea two steps further in that it is an automated procedure that can also objectively compare dissimilar ligands.

Another way of postprocessing is to use the docking results as input to develop a Bayesian model (*see* Section 4.12.3.2.2.2, where BS is used for rescoring) with the aim of reducing the numbers of false positives and false negatives. Klon *et al.*[116] did this as follows: they rank-ordered the list of compounds at the end of a docking run and designated the top-scoring ones as 'good' and the remaining ones as 'bad.' In the next step fingerprints were calculated for all the compounds and a naive Bayesian classifier was trained. Subsequently all compounds were reranked according to the Bayesian model. Applying this procedure to the results of high-throughput docking to an HIV-1 protease model using Glide, FlexX, and GOLD improved the hit rates in all cases significantly. Despite the impressive results, two caveats are appropriate here. First, the data set investigated consisted of a large number of inactives (the Available Chemicals Directory (ACD)) and a small set of HIV-1 inhibitors from in-house chemistry efforts, thus representing the hypothetical activity distribution in **Figure 2**, which, as mentioned in Section 4.12.4.2.1, is the easiest scenario to achieve enrichment. Second, given that the docking programs have already enriched the 'good' compounds with HIV-1 inhibitor chemistry, it is conceivable that the Bayesian model just separates very similar HIV-1 inhibitor-like compounds from other, chemically very diverse ACD molecules. The authors indicate, however, that the Bayesian model requires a large number of features for this approach to work and that consequently just the similarity between the core structures of the inhibitors cannot explain its success. At any rate, we would suspect the results to be very data set-dependent and further validation of this approach (which is in principle very interesting) with a different data set composition (e.g., with compounds that span a wide range of activities but have similar core structures) would be important.

4.12.6 Docking Applications and Successes

The main application of structure-based virtual screening lies in hit identification, but there are many variations on this theme. In the following sections several examples are given of the successful application of docking and scoring to a variety of different problems. Each of these examples illustrates that docking and scoring is not a stand-alone technique, but that it is normally embedded in a workflow of different in silico as well as experimental techniques, and that careful evaluation before application is a prerequisite for success. See Chapter 4.19 (**Table 3**) for other structure-based virtual screening success and for an extensive review of docking success stories, see [117].

4.12.6.1 Inhibiting β-Catenin–Tcf Protein–Protein Interactions[8]

β-catenin is an intracellular mediator of the Wnt signaling pathway.[118,119] When Wnt signaling is activated, β-catenin together with Tcf proteins functions as a transcriptional activator of a large number of genes. The activation of some of these genes is essential for creating and maintaining the malignant phenotype of colorectal cancer cells.[120,121] Consequently, the β-catenin–Tcf complex has emerged as an attractive anticancer drug target. The interaction between β-catenin and Tcf family members (Tcf3 and Tcf4) extends over a very large surface area of 4800 Å^2. Popular lore has it that trying to disrupt protein–protein interactions with small molecules is a futile exercise as the large interacting surfaces would bind very tightly. However, Clackson and Wells[122] have shown that most of the protein–protein-binding energy is due to interactions with a small number of so-called hot spots, well-characterized patches on the surface of the proteins. By making tight interactions with a hot spot, a small molecule can compete with a protein. Since no biochemical (functional) β-catenin assay was in place at the time, docking experiments were carried out to identify putative small-molecule inhibitors, followed by a medium-throughput biophysical screen to confirm the in silico hits. β-catenin hot spots were identified on the basis of the β-catenin–Tcf3 complex structure[123] (which is more ordered than the β-catenin–Tcf4 structure), mutagenesis data,[124] and a PASS analysis.[125] Some 17 700 Pharmacia & Upjohn compounds were docked to the primary β-catenin hot spot near K435-R469 with QXP,[17] allowing several side chains to move, and the best 10 poses per compound were saved. Poses that exceeded empirical limits for the total association energy (–30 kJ mol^{-1}) and contact energy (–24 kJ mol^{-1}) were eliminated. Past experience had shown that compounds with very unfavorable values for two QXP energy terms (internal ligand strain >15 kJ mol^{-1} and Van der Waals repulsion between ligand and receptor >17 kJ mol^{-1}) are unlikely to be active, irrespective of their calculated overall binding affinity, so these were discarded, too. The surviving poses were rank-ordered and the 3000 top-ranking ones were visually inspected. One would think that the various scoring function terms provide all the information necessary to select compounds. Several research groups have found, however, that visual inspection is essential for the selection of active compounds,[8,126] as it apparently captures aspects that these scoring function terms do not adequately account for. Of the 42 compounds that survived visual inspection, 22 were available in sufficient quantity to be submitted to a biophysical screening funnel (using nuclear magnetic resonance (NMR) WaterLOGSY[127] and isothermal titration calorimetry (ITC)) to measure specific binding to β-catenin, competition with Tcf4 and, finally, binding constants. This process led to the discovery of three Tcf4-competitive compounds, corresponding to a hit rate of 14%. The tightest binder (PNU-74654) has a K_D of 450 nM, while Tcf4 has a K_D of 6 nM. Its structure and proposed binding mode are detailed in **Figure 4**.

Further docking studies with PNU-74654 and binding data on two close analogs supported the binding mode proposed: replacing the methyl group on the furan by a proton led to a clear decrease in binding, while replacing the distal phenyl group by a piperidine moiety abolished all binding. Subsequently, PNU-74654 was used to discover a second set of hits: similarity searches were carried out on the entire Pharmacia collection. The 425 hits were docked to β-catenin.

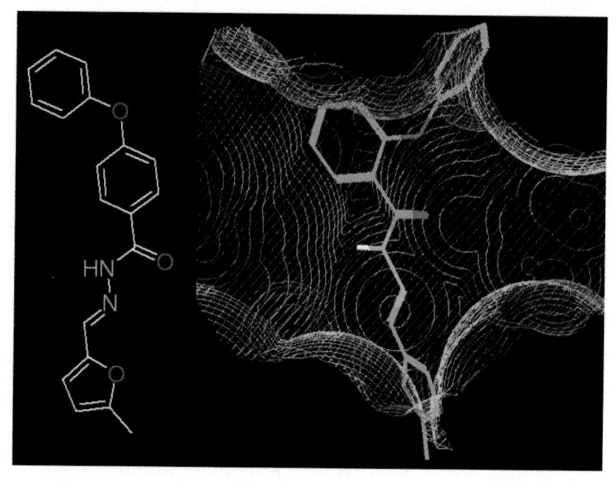

Figure 4 PNU-74654 structure and proposed binding mode to β-catenin.

A total of 44 had good docking energies, of which 22 were available and had adequate solubility. Four of those 22 compete with Tcf4 – an 18% hit rate. Two conclusions can be drawn from this work: (1) by targeting hot spots, small molecules can efficiently interfere with the association of two proteins in vitro; and (2) such molecules can rapidly be identified by combining VS (docking) and biophysical interaction studies (NMR, ITC). It may well be feasible to develop a generally applicable strategy for the identification of inhibitors of protein–protein interactions along the lines of the β-catenin example, and, in doing so, a large number of therapeutic targets may be made accessible for intervention.

4.12.6.2 Virtual and Experimental Screening of Protein Tyrosine Phosphatase-1B[7]

PTP-1B hydrolyzes phosphotyrosines on the insulin receptor, thereby deactivating it. Overproduction of this enzyme has been implicated in the onset of type 2 diabetes, and it is therefore a target for drug discovery.[128,129] Doman *et al.*[7] carried out high-throughput virtual and experimental screens simultaneously, but independently. Their goal was to identify more and more diverse compounds by using two complementary techniques. They also wanted to compare the results of both screens in terms of diversity, drug-likeness, and hit rates. To this end, a corporate collection of approximately 400 000 compounds was screened using high-throughput experimental techniques. Eighty-five of them (0.02%) had IC_{50} values < 100 μM (the most active one having an IC_{50} of 4.2 μM). Concurrently, the Northwestern University version[130,131] of DOCK3.5[132] was used to dock approximately 235 000 commercially available compounds against the crystallographic structure of the 'closed' form of PTP-1B (PDB code 1PTY[133]), after first computationally mutating Ser215 back to wild-type Cys, and assigning it a negative charge.

On the basis of the docking experiments, 365 molecules were selected for experimental verification of activity against the enzyme. A total of 127 (35%) had IC_{50} values < 100 μM (the most active one having an IC_{50} of 1.7 μM). See **Table 3** for a selection of experimentally very active molecules that were found by docking.

Interestingly, the two hit lists contained different and quite dissimilar molecules, reaffirming the notion that the two techniques complement each other. Of the docking hits, 57% did not violate any Lipinski rule,[134] compared to only 23% of the HTS hits. Some 72% of the docking hits passed Pharmacia's exclusion rules for undesirable properties and substructures, compared to only 22% of the HTS hits. Lipinski violations and Pharmacia exclusions had neither been determined for the corporate library used in the HTS, nor for the commercial collection used in the docking experiment. So, it is possible that the underlying physicochemical profile of these collections caused the discrepancy. It cannot be ruled out, however, that the docking experiment preferentially selected compounds with favorable physicochemical characteristics. Hit rates of 35% are impressive, especially when compared to hit rates of 0.02% for the experimental screen. There are several caveats that make the comparison less than straightforward, however:

- First and foremost, the two techniques were applied to two very different compound collections, which, as mentioned above, may have quite dissimilar physicochemical property profiles.
- The HT inhibition assay contained albumin. In the absence of albumin, many inhibitors had significantly higher apparent affinity. It is likely that the HTS hit rate would have been higher, had no albumin been present in the HTS assays.
- One could argue that both the virtual screen and the experimental HTS assay are initial filters prior to hit confirmation and IC_{50} determination. In order to compare VS and HTS, one should consider the hit rates of the confirmation step of both approaches, i.e., they should be calculated with respect to the number of compounds that pass the initial filter. On the other hand, one may be of the opinion that hit rates, expressed as the number of confirmed hits divided by the number of physical samples that must be tested, is the property of interest. We believe that neither type of hit rate is actually a good yardstick to measure efficiency. Important is the absolute number of confirmed active compounds that a process (be it virtual or experimental) ultimately yields, seen in the context of the total costs of carrying out the experiments.

In an absolute sense, a 35% hit rate resulting from a docking experiment is very high, but one sobering piece of information is that docking against the 'open' form of PTP-1B (PDB code 1BZH[135]) did not produce any experimentally confirmed hits, although admittedly only 15–20 of the docking hits were tested. One might think that the open form is not biologically relevant and that binding to it will not inhibit PTP-1B, but that is not the case, as 1BZH is actually a complex of the protein with an inhibitor. In fact, Groves *et al.*,[135] who solved the structure, state: "These results provide a framework for the rational design of higher affinity and more specific phosphotyrosine mimetic inhibitors." A variant of Galileo Galilei's dialogue about the heliocentricity of the solar system is appropriate here:

- Simplicio: You get good results in only one of your two docking experiments. That's random. You might as well flip a coin.

Table 3 Selected PTP-1B docking hits with high experimental inhibition

Structure	*Experimental IC_{50} (μM)*	*Calculated affinity ($kJ\,mol^{-1}$)*	*IC_{50} rank*	*Docking rank*
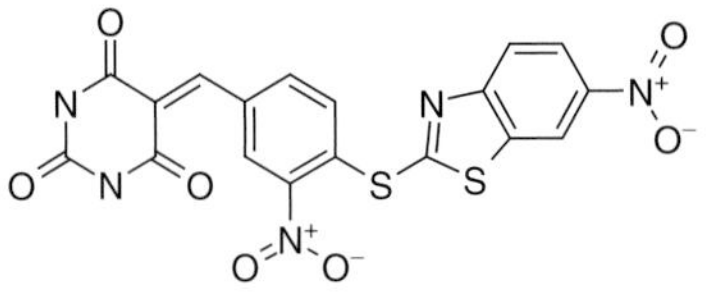	4.1	– 140	1	406
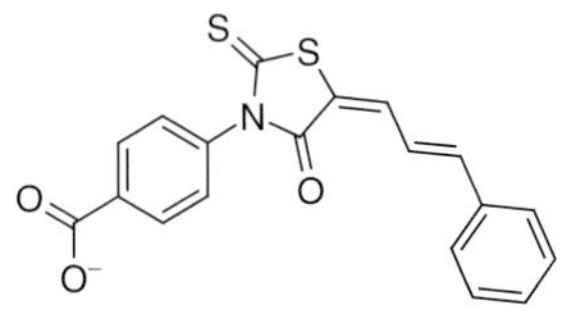	4.4	– 146	2	170
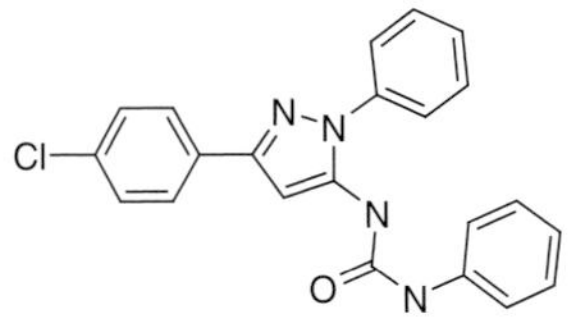	12.0	– 139	5	415
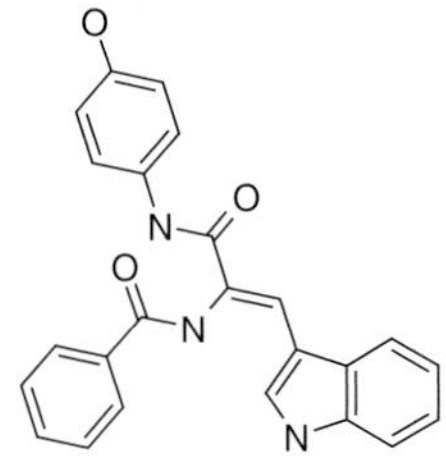	21.5	– 139	7	440
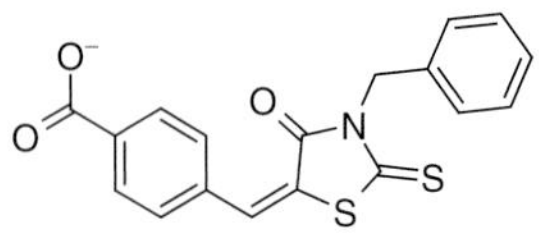	21.6	– 176	8	11

- Salviati: Although one docking experiment failed completely, combining the two still gives a 17% hit rate.
- Sagredo: But you never know beforehand if you are considering a relevant conformation at all. And the success or failure of the docking experiment depends on it.

Despite the very high hit rate, the correlation between docking scores and IC_{50} values is not very good. The red circle in **Table 3** reveals that the compound with the best experimental IC_{50} is only found in place 406 of the list of docking hits, rank-ordered by predicted affinity. The blue circles show that two compounds with nearly identical IC_{50} values can have very different docking ranks. Taken together, the high overall hit rate and the low correlation with experiment indicate that the docking program and scoring function are good at eliminating compounds that do not fit the active site well electrostatically or sterically, but that they are not able to differentiate reliably between two compounds that both fit (and presumably both exhibit measurable inhibition). When taking random samples of the corporate collection and of the docking database, some structural similarity exists. Surprisingly, no similarity existed between the docking hits and the HTS hits, while one would expect the similarity to increase as a result of the bias toward PTP-1B inhibitors.

In summary:

- Structure-based molecular docking was successfully applied to discover novel PTP-1B inhibitors, but the choice of protein conformation to dock against is critical.
- Although the correlation between calculated and experimental values was low, 35% of the 365 top-scoring docking hits were confirmed active.
- Side-by-side application of docking and HTS revealed they are complementary techniques.

The structure-based design of PTP-1B inhibitors for diabetes is also discussed in Chapter 4.25.5.

4.12.6.3 Layered Virtual Screening: Application to Carbonic Anhydrase II and Checkpoint Kinase-1

In some cases structure-based VS can be part of a cascade of different computational techniques in the quest for binders of a given target. We give here two such examples.

The first example of the combination of different VS techniques is given by a study that led to the successful identification of subnanomolar inhibitors of carbonic anhydrase II (CAII).[9,136] A layered strategy including pharmacophore and ligand-based modeling was applied, where the last step consisted of docking and scoring of a number of compounds. The starting point of the study was the high-resolution crystal structure of CAII, and a set of 90 000 molecules (including 35 known CAII inhibitors) from the Maybridge and LeadQuest databases. As a first step, a binding-site analysis was performed, in order to identify key interactions between a putative compound and the receptor. These key interactions were transformed into a 3D pharmacophore model, which was used to search the database of 90 000 molecules. The database scan led to the identification of 3314 molecules. In the second step, these 3314 molecules were rank-ordered using the program FlexS that evaluates their potential binding affinities by comparison with a reference compound. In the third and final step the 100 best-ranking molecules were docked into the binding pocket of CAII using the docking program FlexX. FlexX score and DrugScore were used to predict the binding affinity of the docked molecules and to rank-order them accordingly. The 13 top-scoring compounds were chosen for testing, and three inhibitors with IC_{50} values of <1 nM were identified. Two of the subnanomolar hits were co-crystallized with CAII, and the binding mode generated by FlexX and ranked highest by DrugScore was found to be in good agreement with the experimental structures. The study yielded two important insights. First, it turned out that water molecules played an important role during the docking process. Four conserved water molecules had been identified by superposition of all complex and apo structures of CAII. Inclusion of these solvent molecules in the docking process added to the steric restriction of the binding pocket and led to better solutions. Secondly, despite the successful identification of high-affinity binders, it was observed that overall correlation between the IC_{50} values and the binding affinities predicted by FlexX score or DrugScore was rather poor.

The second example for the integration of structure-based VS into an in silico workflow is provided by a study on checkpoint kinase-1 (Chk1).[137] In this case, in the first stage the in-house compound collection was filtered by general physicochemical properties, such as MW and number of rotatable bonds, followed by removal of compounds with undesired chemical functionality. In the next step the remaining compounds were evaluated by their fit to a pharmacophore representing a minimal kinase binding motif, consisting of two hydrogen bonds (one acceptor, one donor) with the hinge region of the kinase, where the adenine moiety of ATP binds. Approximately 200 000 compounds passed this pharmacophore filter and were subsequently submitted to docking with FlexX-Pharm[112] with the same pharmacophore as constraint. Up to 100 poses were saved for each successfully docked compound. All saved poses were then rescored with a consensus scoring scheme that had been derived in a study for another kinase (CDK-2), which included a combination of the FlexX and PMF scoring functions. Using this scheme and prior knowledge, 250 compounds were retained for visual inspection. After application of this final human filter, 103 compounds were assayed, which yielded 36 active compounds from four different chemical classes with activities ranging from 110 nM to 68 μM.

Several conclusions can be drawn from this study. Integration of structure-based VS in a general in silico workflow leads to a reduction of the number of molecules to be docked, thereby saving a significant amount of computing resources. A tailor-made scoring scheme that has been derived for a related protein can be successfully transferred to the target of interest. Despite all efforts to ensure that docking programs generate realistic poses, we notice a pervasive, industry-wide need to include visual inspection as the final filters in order to remove compounds whose poses are unrealistic.

4.12.7 Conclusions and Outlook

VS for the rapid identification of small-molecule ligands of macromolecular targets is an established technology in drug discovery. High-throughput ligand docking or structure-based VS is a powerful technique to perform such a screen if a 3D structure of the target is available, and docking success stories are abundant. Many docking programs exist, but no single program has emerged yet that outperforms all others in all cases. Generally, programs do an adequate job searching conformational space and generating correct ligand poses (binding modes), but the scoring functions need improvement. This is evidenced by the fact that the correlation between calculated and observed binding affinities is often low and that separate scoring functions are frequently needed for pose evaluation and affinity prediction. Two other problem areas are target flexibility and water molecules. Fortunately, active research is taking place to address all three issues and appreciable progress is being made. Tunable scoring functions, MM-PBSA and other rescoring approaches, explicit protein flexibility (as in Glide/Prime and QXP) and explicit incorporation of water (as in FlexX and GOLD) are just a few examples. We expect that docking tools will continue to improve significantly in the near future.

Despite the docking successes highlighted in this chapter, achieving success is not trivial. A docking campaign cannot be regarded as a black box that one feeds a general compound collection and that automatically produces a collection of high-affinity ligands. Preparing the protein and the ligands, selecting the docking programs and scoring functions, setting and tuning the parameters, and carrying out the postprocessing (including the often necessary visual inspection) require profound expertise.

Docking is especially useful in reducing a collection of virtual compounds down to a manageable number to be synthesized and in selecting compounds from an external collection. Even when experimental HTS is envisioned, however, VS is important, as active compounds are often identified by one technique and not by the other. An added bonus is that VS is fast and inexpensive by any standard. We would recommend, if at all possible, to use docking in parallel with other techniques (experimental HTS, pharmacophore modeling, etc.), to dock to multiple conformations of the target, and to use the docking results to select as many compounds as possible for experimental confirmation.

Nowadays, a large proportion of discovery research is carried out by smaller pharmaceutical companies. Since these smaller companies often do not have the resources to run HTS or to acquire large numbers of compounds, we believe the importance and impact of VS will continue to increase significantly.

References

1. Böhm, H.-J.; Schneider, G., Eds. *Virtual Screening for Bioactive Molecules*; Wiley-VCH: Weinheim, 2000; Vol. 10.
2. Schneider, G.; Böhm, H.-J. *Drug Disc. Today* **2002**, *7*, 64–70.
3. Waszkowycz, B. *Curr. Opin. Drug Disc. Dev.* **2002**, *5*, 407–413.
4. Toledo-Sherman, L. M.; Chen, D. *Curr. Opin. Drug Disc. Dev.* **2002**, *5*, 414–421.
5. Verkhivker, G. M.; Bouzida, D.; Gehlhaar, D. K.; Rejto, P. A.; Arthurs, S.; Colson, A. B.; Freer, S. T.; Larson, V.; Luty, B. A.; Marrone, T. et al. *J. Comput-Aided Mol. Design* **2000**, *14*, 731–751.
6. Stahl, M.; Rarey, M. *J. Med. Chem.* **2001**, *44*, 1035–1042.
7. Doman, T. N.; McGovern, S. L.; Witherbee, B. J.; Kasten, T. P.; Kurumbail, R.; Stallings, W. C.; Connolly, D. T.; Shoichet, B. K. *J. Med. Chem.* **2002**, *45*, 2213–2221.
8. Trosset, J.-Y.; Dalvit, C.; Knapp, S.; Fasolini, M.; Veronesi, M.; Mantegani, S.; Gianellini, L. M.; Catana, C.; Sundström, M.; Stouten, P. F. W. et al. *Proteins* **2006**, *64*, 60–67.
9. Grüneberg, S.; Stubbs, M. T.; Klebe, G. *J. Med. Chem.* **2002**, *45*, 3588–3602.
10. Enyedy, I. J.; Ling, Y.; Nacro, K.; Tomita, Y.; Wu, X.; Cao, Y.; Guo, R.; Li, B.; Zhu, X.; Huang, Y. et al. *J. Med. Chem.* **2001**, *44*, 4313–4324.
11. Welch, W.; Ruppert, J.; Jain, A. N. *Chem. Biol.* **1996**, *3*, 449–462.
12. Ewing, T. J. A.; Kuntz, I. D. *J. Comput. Chem.* **1997**, *18*, 1176–1189.
13. Rarey, M.; Kramer, B.; Lengauer, T.; Klebe, G. *J. Mol. Biol.* **1996**, *261*, 470–489.
14. Morris, G. M.; Goodsell, D. S.; Halliday, R. S.; Huey, R.; Hart, W. E.; Belew, R. K.; Olson, A. J. *J. Comput. Chem.* **1998**, *19*, 1639–1662.
15. Jones, G.; Willett, P.; Glen, R. C.; Leach, A. R.; Taylor, R. *J. Mol. Biol.* **1997**, *267*, 727–748.
16. Totrov, M.; Abagyan, R. *Proteins* **1997**, *S1*, 215–220.
17. McMartin, C.; Bohacek, R. S. *J. Comput.-Aided Mol. Design* **1997**, *11*, 333–344.
18. Österberg, F.; Morris, G. M.; Sanner, M. F.; Olson, A. *J. Proteins* **2002**, *46*, 34–40.
19. Claußen, H.; Buning, C.; Rarey, M.; Lengauer, T. *J. Mol. Biol.* **2001**, *308*, 377–395.
20. Luty, B. A.; Wasserman, Z. R.; Stouten, P. F. W.; Hodge, C. N.; Zacharias, M.; McCammon, J. A. *J. Comput. Chem.* **1995**, *16*, 454–464.
21. Abagyan, R.; Argos, P. *J. Mol. Biol.* **1992**, *225*, 519–532.
22. Smellie, A.; Teig, S. L.; Towbin, P. *J. Comput. Chem.* **1995**, *16*, 171–187.
23. Glover, F. W.; Laguna, M. *Tabu Search*; Kluwer Academic Publishers: Boston, 1998.

24. Böhm, H.-J. *J. Comput.-Aided Mol. Design* **1992**, *6*, 61–78.
25. Böhm, H.-J. *J. Comput.-Aided Mol. Design* **1992**, *6*, 593–606.
26. Klebe, G. *J. Mol. Biol.* **1994**, *237*, 221–235.
27. Kramer, B.; Rarey, M.; Lengauer, T. *Proteins* **1999**, *37*, 228–241.
28. Böhm, H.-J. *J. Comput.-Aided Mol. Design* **1994**, *8*, 243–256.
29. Jones, G.; Willett, P.; Glen, R. C. *J. Mol. Biol.* **1995**, *245*, 43–53.
30. Jones, G.; Willett, P.; Glen, R. C. *J. Comput.-Aided Mol. Design* **1995**, *9*, 532–549.
31. Lee, J.; Scheraga, H. A.; Rackovsky, S. *J. Comput. Chem.* **1997**, *18*, 1222–1232.
32. Gohlke, H.; Klebe, G. *Angew. Chem. Int. Ed.* **2002**, *41*, 2644–2676.
33. Ajay; Murcko, M. A.; Stouten, P. F. W. Recent Advances in the Prediction of Binding Free Energy. In *Practical Application of Computer-Aided Drug Design*, Charifson, P. S., Ed.; Marcel Dekker: New York, 1997, pp 355–410.
34. Muegge, I.; Martin, Y. C. *J. Med. Chem.* **1999**, *42*, 791–804.
35. Gohlke, H.; Hendlich, M.; Klebe, G. *J. Mol. Biol.* **2000**, *295*, 337–356.
36. DeWitte, R. S.; Shakhnovich, E. I. *J. Am. Chem. Soc.* **1996**, *118*, 11733–11744.
37. Mitchell, J. B. O.; Laskowski, R. A.; Alex, A.; Forster, M. J.; Thornton, J. M. *J. Comput. Chem.* **1999**, *20*, 1177–1185.
38. Gohlke, H.; Klebe, G. *Curr. Opin. Struct. Biol.* **2001**, *11*, 231–235.
39. Mark, A. E.; van Gunsteren, W. F. *J. Mol. Biol.* **1994**, *240*, 167–176.
40. Williams, D. H.; Maguire, A. J.; Tsuzuki, W.; Westwell, M. S. *Science* **1998**, *280*, 711–714.
41. Reyes, C. M.; Kollman, P. A. *J. Mol. Biol.* **2000**, *297*, 1145–1158.
42. Kuhn, B.; Kollman, P. A. *J. Am. Chem. Soc.* **2000**, *122*, 3909–3916.
43. Wang, J. M.; Morin, P.; Wang, W.; Kollman, P. A. *J. Am. Chem. Soc.* **2001**, *123*, 5221–5230.
44. Böhm, H.-J.; Stahl, M. *Med. Chem. Res.* **1999**, *9*, 445–462.
45. Stahl, M. *Perspect. Drug Disc. Des.* **2000**, *20*, 83–98.
46. Eldridge, M. D.; Murray, C. W.; Auton, T. R.; Paolini, G. V.; Mee, R. P. *J. Comput.-Aided Mol. Des.* **1997**, *11*, 425–445.
47. Head, R. D.; Smythe, M. L.; Oprea, T. I.; Waller, C. L.; Green, S. M.; Marshall, G. R. *J. Am. Chem. Soc.* **1996**, *118*, 3959–3969.
48. Totrov, M.; Abagyan, R. *Proceedings of the Third Annual International Conference on Computational Molecular Biology*; Istrail, S., Pevzner, P., Waterman, M., Eds.; ACM Press: New York, 1999, pp 312–320.
49. Giordanetto, F.; Cotesta, S.; Catana, C.; Trosset, J.-Y.; Vulpetti, A.; Stouten, P. F. W.; Kroemer, R. T. *J. Chem. Inf. Comput. Sci.* **2004**, *44*, 882–893.
50. ICM-Dock: MolSoft, La Jolla, CA. http://www.molsoft.com/ (accessed Aug 2006).
51. Filikov A. Presentation at 226th ACS National Meeting (New York, Fall 2003).
52. Cerius2: Accelrys, San Diego, CA. http://www.accelrys.com (accessed Aug 2006).
53. Bro, R. *J. Chemometrics* **1996**, *10*, 47–61.
54. Bro, R.; Smilde, A. K.; de Jong, S. *Chemometr. Intell. Lab. Syst.* **2001**, *58*, 3–13.
55. PLS-Toolbox 3.0 for use with MATLAB (2003): Eigenvector Research, Manson, WA. http://www.eigenvector.com (accessed June 2006).
56. Böhm, H.-J. *J. Comput.-Aided Mol. Design* **1998**, *12*, 309–323.
57. Smith, R.; Hubbard, R. E.; Gschwend, D. A.; Leach, A. R.; Good, A. C. *J. Mol. Graph. Model.* **2003**, *22*, 41–53.
58. Wang, R.; Gao, Y.; Lai, L. *J. Mol. Model.* **1998**, *4*, 379–394.
59. Krammer, A.; Kirchhoff, P. D.; Jiang, X.; Venkatachalam, C. M.; Waldman, M. *J. Mol. Graph. Model.* **2005**, *23*, 395–407.
60. Ortiz, A. R.; Pisabarro, M. T.; Gago, F.; Wade, R. C. *J. Med. Chem.* **1995**, *38*, 2681–2691.
61. Wang, T.; Wade, R. C. *J. Med. Chem.* **2001**, *44*, 961–971.
62. Murcia, M.; Ortiz, A. R. *J. Med. Chem.* **2004**, *47*, 805–820.
63. Gohlke, H.; Klebe, G. *J. Med. Chem.* **2002**, *45*, 4153–4170.
64. Silber, K.; Heidler, P.; Kurz, T.; Klebe, G. *J. Med. Chem.* **2005**, *48*, 3547–3563.
65. Kollman, P. A.; Massova, I.; Reyes, C.; Kuhn, B.; Huo, S.; Chong, L.; Lee, M.; Lee, T.; Duan, Y.; Wang, W. et al. *Acc. Chem. Res.* **2000**, *33*, 889–897.
66. Srinivasan, J.; Cheatham, T. E., III; Cieplak, P.; Kollman, P. A.; Case, D. A. *J. Am. Chem. Soc.* **1998**, *120*, 9401–9409.
67. Wang, J.; Kang, X.; Kuntz, I. D.; Kollman, P. A. *J. Med. Chem.* **2005**, *48*, 2432–2444.
68. Sitkoff, D.; Sharp, K. A.; Honig, B. *J. Phys. Chem.* **1998**, *98*, 1978–1983.
69. Alvarez, J. C. Presentation at SMi workshop *High-Throughput Molecular Docking*, London, February 2004.
70. Rush, T. Presentation at SMi conference *From Hit to Lead to Candidate*, London, September 2004.
71. Blaney, J. M. Presentation at SMi workshop *High-Throughput Molecular Docking*, London, February 2004.
72. Kuhn, B.; Gerber, P.; Schulz-Gasch, T.; Stahl, M. *J. Med. Chem.* **2005**, *48*, 4040–4048.
73. FRED: OpenEye Scientific Software, Santa Fe, NM. http://www.eyesopen.com/products/applications/fred.html (accessed June 2006).
74. Ferrara, P.; Gohlke, H.; Price, D. J.; Klebe, G.; Brooks, C. L., III. *J. Med. Chem.* **2004**, *47*, 3032–3047.
75. Charifson, P. S.; Corkery, J. J.; Murcko, M. A.; Walters, W. P. *J. Med. Chem.* **1999**, *42*, 5100–5109.
76. Wang, R.; Wang, S. *J. Chem. Inf. Comput. Sci.* **2001**, *41*, 1422–1426.
77. Terp, G. E.; Johansen, B. N.; Christensen, I. T.; Jørgensen, F. S. *J. Med. Chem.* **2001**, *44*, 2333–2343.
78. Clark, R. D.; Strizhev, A.; Leonard, J. M.; Blake, J. F.; Matthew, J. B. *J. Mol. Graph. Model.* **2002**, *20*, 281–295.
79. Cotesta, S.; Giordanetto, F.; Trosset, J.-Y.; Crivori, P.; Kroemer, R. T.; Stouten, P. F. W.; Vulpetti, A. *Proteins* **2005**, *60*, 629–643.
80. Cole, J. C.; Murray, C. W.; Nissink, J. W. M.; Taylor, R. D.; Taylor, R. *Proteins* **2005**, *60*, 325–332.
81. Kroemer, R. T.; Vulpetti, A.; McDonald, J. J.; Rohrer, D. C.; Trosset, J.-Y.; Giordanetto, F.; Cotesta, S.; McMartin, C.; Kihlen, M.; Stouten, P. F. W. *J. Chem. Inf. Comp. Sci.* **2004**, *44*, 871–881.
82. Abagyan, R. A.; Totrov, M. M.; Kuznetsov, D. *J. Comp. Chem.* **1994**, *15*, 488–506.
83. Totrov, M.; Abagyan, R. Protein-Ligand Docking as an Energy Optimization Problem. In *Drug-Receptor Thermodynamics: Introduction and Applications*; Raffa, R. B., Ed.; John Wiley: New York, 2001, pp 603–624.
84. Halgren, T. A.; Murphy, R. B.; Friesner, R. A.; Beard, H. S.; Frye, L. L.; Pollard, W. T.; Banks, J. L. *J. Med. Chem.* **2004**, *47*, 1750–1759.
85. Morley, S. D.; Afshar, M. *J. Comput.-Aided Mol. Design* **2004**, *18*, 189–208.
86. Verdonk, M. L.; Berdini, V.; Harshorn, M. J.; Mooij, W. T. M.; Murray, C. W.; Taylor, R. D.; Watson, P. J. *Chem. Inf. Comput. Sci.* **2004**, *44*, 793–806.
87. Hopkins, A.; Groom, C.; Alex, A. *Drug Disc. Today* **2004**, *9*, 430–431.

88. Triballeau, N.; Acher, F.; Brabet, I.; Pin, J. P.; Bertrand, H.-O. *J. Med. Chem.* **2005**, *48*, 2534–2547.
89. Teague, S. J. *Nat. Rev. Drug Disc.* **2003**, *2*, 527–541.
90. Murray, C. W.; Baxter, C. A.; Frenkel, A. D. *J. Comput.-Aided Mol. Design* **1999**, *13*, 547–562.
91. Huse, M.; Kuriyan, J. *Cell* **2002**, *109*, 275–282.
92. Sherman, W.; Day, T.; Jacobson, M. P.; Friesner, R. A.; Farid, R. *J. Med. Chem.* **2006**, *49*, 534–553.
93. Cavasotto, C. N.; Abagyan, R. A. *J. Mol. Biol.* **2004**, *337*, 209–225.
94. Ferrari, A. M.; Wei, B. Q.; Costantino, L.; Shoichet, B. K. *J. Med. Chem.* **2004**, *47*, 5076–5084.
95. Wei, B. Q.; Weaver, L. H.; Ferrari, A. M.; Matthews, B. W.; Shoichet, B. K. *J. Mol. Biol.* **2004**, *337*, 1161–1182.
96. Frimurer, T. M.; Peters, G. H.; Iversen, L. F.; Andersen, H. S.; Møller, N. P. H.; Olsen, O. H. *Biophys. J.* **2003**, *84*, 2273–2281.
97. Erickson, J. A.; Jalaie, M.; Robertson, D. H.; Lewis, R. A.; Vieth, M. *J. Med. Chem.* **2004**, *47*, 45–55.
98. McGovern, S. L.; Shoichet, B. K. *J. Med. Chem.* **2003**, *46*, 2895–2907.
99. Verdonk, M. L.; Chessari, G.; Cole, J. C.; Hartshorn, M. J.; Murray, C. W.; Nissink, J. W. M.; Taylor, R. D.; Taylor, R. *J. Med. Chem.* **2005**, *48*, 6504–6515.
100. García-Sosa, A. T.; Mancera, R. L.; Dean, P. M. *J. Mol. Model.* **2003**, *9*, 172–182.
101. Goodford, P. J. A. *J. Med. Chem.* **1985**, *28*, 849–857.
102. Clarke, C.; Woods, R. J.; Gluska, J.; Cooper, A.; Nutley, M. A.; Boons, G.-J. *J. Am. Chem. Soc.* **2001**, *123*, 12238–12247.
103. Rarey, M.; Kramer, B.; Lengauer, T. *Proteins* **1999**, *34*, 17–28.
104. Vigers, G. P. A.; Rizzi, J. P. *J. Med. Chem.* **2004**, *47*, 80–89.
105. McGann, M. Presentation at SMi meeting. *Drug Design*, London, February 2005.
106. DockIt: Metaphorics, Aliso Viejo, CA. http://www.metaphorics.com/ (accessed June 2006).
107. GOLD: Cambridge Crystallographic Data Centre, Cambridge, UK. http://www.ccdc.cam.ac.uk/products/life_sciences/gold/ (accessed June 2006).
108. Verdonk, M. L.; Cole, J. C.; Hartshorn, M. J.; Murray, C. W.; Taylor, R. D. *Proteins* **2003**, *52*, 609–623.
109. Jones, G.; Willett, P.; Glen, R. C. *J. Mol. Biol.* **1995**, *245*, 43–53.
110. Joseph-McCarthy, D.; Thomas, B. E., IV.; Belmarsh, M.; Moustakas, D.; Alvarez, J. C. *Proteins* **2003**, *51*, 172–188.
111. Ewing, T. J. A.; Makino, S.; Skillman, A. G.; Kuntz, I. D. *J. Comput.-Aided Mol. Design* **2001**, *15*, 411–428.
112. Hindle, S. A.; Rarey, M.; Buning, C.; Lengauer, T. *J. Comput.-Aided Mol. Design* **2002**, *16*, 129–149.
113. Glide: Schroedinger, Portland, OR. http://www.schroedinger.com (accessed June 2006).
114. Friesner, R. A.; Banks, J. L.; Murphy, R. B.; Halgren, T. A.; Klicic, J. J.; Mainz, D. T.; Repasky, M. P.; Knoll, E. H.; Shelley, M.; Perry, J. K. et al. *J. Med. Chem.* **2004**, *47*, 1739–1749.
115. Deng, Z.; Chuaqui, C.; Singh, J. *J. Med. Chem.* **2004**, *47*, 337–344.
116. Klon, A. E.; Glick, M.; Davies, J. W. *J. Chem. Inf. Comput. Sci.* **2004**, *44*, 2216–2224.
117. Kubinyi, H. Success Stories of Computer-Aided Design. In *Computer Applications in Pharmaceutical Research and Development*; Ekins, S., Ed., John Wiley: New York, 2006 (in press).
118. Cavallo, R. A.; Cox, R. T.; Moline, M. M.; Roose, J.; Polevoy, G.; Clevers, H.; Peifer, M.; Bejsovec, A. *Nature* **1998**, *395*, 604–608.
119. Roose, J.; Molenaar, M.; Peterson, J.; Hurenkamp, J.; Brantjes, H.; Moerer, P.; Van de Wetering, M.; Destrée, O.; Clevers, H. *Nature* **1998**, *395*, 608–612.
120. Korinek, V.; Barker, N.; Morin, P. J.; Van Wichen, D.; De Weger, R.; Kinzler, K. W.; Vogelstein, B.; Clevers, H. *Science* **1997**, *275*, 1784–1787.
121. Shih, I. M.; Yu, J.; He, T. C.; Vogelstein, B.; Kinzler, K. W. *Cancer Res.* **2000**, *60*, 1671–1676.
122. Clackson, T.; Wells, J. A. *Science* **1995**, *267*, 383–386.
123. Graham, T. A.; Weaver, C.; Mao, F.; Kimelman, D.; Xu, W. *Cell* **2000**, *103*, 885–896.
124. Fasolini, M.; Wu, X.; Flocco, M.; Trosset, J.-Y.; Oppermann, U.; Knapp, S. *J. Biol. Chem.* **2003**, *278*, 21092–21098.
125. Brady, G. P., Jr.; Stouten, P. F. W. *J. Comput.-Aided Mol. Design* **2000**, *14*, 383–401.
126. Wang, J. L.; Liu, D.; Zhang, Z. J.; Shan, S.; Han, X.; Srinivasula, S. M.; Croce, C. M.; Alnemri, E. S.; Huang, Z. *Proc. Natl. Acad. Sci. USA* **2000**, *97*, 7124–7129.
127. Dalvit, C.; Fogliatto, G. P.; Stewart, A.; Veronesi, M.; Stockman, B. *J. Biomol. NMR* **2001**, *21*, 349–359.
128. Elchebly, M.; Payette, P.; Michaliszyn, E.; Cromlish, W.; Collins, S.; Loy, A. L.; Normandin, D.; Cheng, A.; Himms-Hagen, J.; Chan, C. C. et al. *Science* **1999**, *283*, 1544–1548.
129. Møller, N. P.; Iversen, L. F.; Andersen, H. S.; McCormack, J. G. *Curr. Opin. Drug Disc. Dev.* **2000**, *3*, 527–540.
130. Lorber, D. M.; Shoichet, B. K. *Protein Sci.* **1998**, 7, 938–950.
131. Shoichet, B. K.; Leach, A. R.; Kuntz, I. D. *Proteins* **1999**, *34*, 4–16.
132. Gschwend, D. A.; Kuntz, I. D. *J. Comput.-Aided Mol. Design* **1996**, *10*, 123–132.
133. Puius, Y. A.; Zhao, Y.; Sullivan, M.; Lawrence, D. S.; Almo, S. C.; Zhang, Z. Y. *Proc. Natl. Acad. Sci. USA* **1997**, *94*, 13420–13425.
134. Lipinski, C. A.; Lombardo, F.; Dominy, B. W.; Feeney, P. *J. Adv. Drug Del. Rev.* **2001**, *46*, 3–26.
135. Groves, M. R.; Yao, Z. J.; Roller, P. P.; R Burke, T., Jr.; Barford, D. *Biochemistry* **1998**, *37*, 17773–17783.
136. Grüneberg, S.; Wendt, B.; Klebe, G. *Angew. Chem. Int. Ed.* **2001**, *40*, 389–393.
137. Lyne, P. D.; Kenny, P. W.; Cosgrove, D. A.; Deng, C.; Zabludoff, S.; Wendoloski, J. J.; Ashwell, S. *J. Med. Chem.* **2004**, *47*, 1962–1968.

Biographies

Pieter F W Stouten, PhD, Pieter's primary research goal is to speed up the drug discovery and development process by judicious deployment and, if necessary, development of computational techniques. Areas of expertise include structure-based drug design, binding affinity prediction, protein homology modeling, computer programming, crystallography, ADMET property predictions, and virtual screening. He is co-author of 45 peer-reviewed publications and book chapters, and co-inventor on four patents. He has been a member of the American Chemical Society since 1993.

Pieter earned his PhD degree in chemistry from Utrecht University, The Netherlands, in 1989. Until 1993 he was postdoc in Chris Sander's biocomputing group at the European Molecular Biology Laboratory (EMBL), Heidelberg, Germany. There he modeled a wide variety of proteins (ras-p21, BPTI, uteroglobin, adenovirus 5 fiber, crotoxin, etc.) to elucidate some of their key biological properties. He also developed a novel continuum solvent model for protein simulations (now implemented in the AutoDock and Affinity programs). While at EMBL, he enjoyed short-term fellowships at the Institute of Chemistry, University of Wroclaw, Poland and the Protein Engineering Research Institute (PERI), Osaka, Japan.

From 1993 to 1998 Pieter was with DuPont Merck/DuPont Pharma, Wilmington, DE, US, where he focused on structure-based design of inhibitors of CDK-2 and CDK-4 (oncology), and thrombin and factor Xa (cardiovascular area). He was also involved in technology efforts such as the development of PASS (a fast, accurate active site-finder), BForce (a grid/MD-based docking program, marketed by Accelrys as Affinity), and EGALS (a fast, genetic algorithm-based docking program).

In 1998, Pieter joined Pharmacia & Upjohn in Nerviano, Italy, as head of the Molecular Modeling and Design group. In addition to coordinating the modeling activities, he provides project support and is involved in two important technology development areas: biopharmaceutical property predictions (such as solubility) and structure-based virtual screening. After a merger with Monsanto, an acquisition by Pfizer and a spin-off, he now works for Nerviano Medical Sciences.

Romano T Kroemer, PhD, Romano received his PhD in Chemistry in 1993 from the University of Innsbruck, Austria, while he was working at the Sandoz (now Novartis) Research Institute in Vienna. The topic of his PhD work was the investigation of a rearrangement reaction in heterocycles and included both wet lab chemistry and theoretical chemistry.

Romano then remained with Sandoz as a computational chemist, and worked on several in-house drug discovery programs, most notably HIV-1 protease inhibitors and cytochrome P450 inhibitors. During this time he was also actively involved in the development and improvement of 3D-QSAR methodology, in particular new field types – such as indicator fields – in CoMFA and new alignment procedures.

In 1995, Romano took up a position as postdoctoral research assistant at the Physical and Theoretical Chemistry Laboratory at Oxford University, UK, where he worked with W Graham Richards. On this occasion he entered the field of signal transduction, focusing on the prediction of the 3D structure of cytokines and cytokine receptors as well as their interactions. At the same time he got involved in molecular spectroscopy through a collaboration with John P Simons at Oxford and David W Pratt at Pittsburgh, and developed a model for the conformational dependence of the electronic transition moment in flexible benzenes.

In 1998, Romano became a lecturer in physical and computational chemistry at the University of London, UK, where he was leading his own research group. Here his main research activities in the areas of protein modeling, 3D-QSAR and quantum chemistry continued. During his time at Oxford and in London Romano also worked as a consultant for Oxford Molecular and Alizyme. In 1998 Romano completed his habilitation in computational chemistry at the University of Innsbruck, Austria.

In 2001, Romano joined Pharmacia in Milan, Italy. In addition to working on drug discovery projects in the Oncology Department, his research efforts focused on the evaluation and development of docking and scoring algorithms.

In his current position at Sanofi-Aventis near Paris, Romano is in charge of a drug design group. Besides coordinating drug design activities he gets actively involved in the development of novel methods as well as project support. Romano is the (co-)author of several book chapters, review articles and around 60 peer-reviewed publications, predominantly in the field of computational chemistry.

© 2007 Elsevier Ltd. All Rights Reserved
No part of this publication may be reproduced, stored in any retrieval system or transmitted in any form by any means electronic, electrostatic, magnetic tape, mechanical, photocopying, recording or otherwise, without permission in writing from the publishers

Comprehensive Medicinal Chemistry II
ISBN (set): 0-08-044513-6

ISBN (Volume 4) 0-08-044517-9; pp. 255–281

4.13 De Novo Design

N P Todorov, I L Alberts, and P M Dean, De Novo Pharmaceuticals, Cambridge, UK

© 2007 Elsevier Ltd. All Rights Reserved.

4.13.1 Introduction

Automated de novo drug design is a computational process whereby structural coordinate data of the site, together with a design strategy, are used as input to a de novo design algorithm. The algorithm then designs molecular structures to fit the site optimally without further human intervention. Potential candidate ligand structures are delivered as output from the algorithm. Depending on the type of de novo design algorithm, numerous different ligand structures can be generated from the same input and can be postprocessed for synthetic tractability and druglike properties. This approach offers the designer a set of alternative chemotypes from which to choose a number of different ligand series to explore. De novo design methods share a lot of characteristics with methods for virtual screening and library design, discussed in other chapters of this book, and could be viewed as extensions to novel areas in chemical space.

4.13.1.1 Why De Novo Design?

The completion of the sequencing of the human genome has provided drug designers with a plethora of primary information on therapeutic targets. However, for that information to become utilizable for de novo design, it has to be processed in two ways: (1) the gene product (enzyme, receptor, or protein) has to be validated as a therapeutic target, usually by complex biological experiments; and (2) the target protein structure has to be determined to reveal the molecular nature of the target. Once three-dimensional (3D) atomic coordinates are known for the target protein, it is amenable to rational methods for drug design. Protein structure initiatives seek to improve the speed with which protein structures can be solved. If a fast throughput of protein structural data is achieved, the drug design process will need to be accelerated to cope with the expected avalanche of structural data. Automated de novo design methods have the capability of handling this mass of data to generate structures to fit crystallographically determined binding sites.

The huge advantage of de novo design methods is that they can be used to assess in silico large numbers of potential structures for their fit to the site before any synthesis is embarked upon. In that way de novo design offers an overwhelming advantage to pharmaceutical companies seeking to seize a patent estate for the most promising lead compounds.

In a structure-based design paradigm, de novo methods can be used to:

- generate structures to virgin targets (those for which there are no known active compounds)
- generate structures from a proprietary molecular fragment bound into the site
- generate linkers between fragments where the linkers also interact with the site
- generate alternative chemotypes to known active compounds, or to moieties of a compound and thereby perform an automated scaffold-hopping process

In the absence of coordinate data about the site, a situation in which structure-based design is not possible, de novo methods can still be used if active ligands have been discovered, for example by high-throughput screening. In this case, an image of the site has to be built from 3D extended pharmacophore information together with a supersurface taken from the actives. De novo design can then be performed within the constraints of the supersurface and the pharmacophoric points.[1]

The generation of active compounds for a particular target is an initial goal; however, chemical genomics has a more extensive potential value for drug discovery and could lead to personalized medicines. In this case, not only is selectivity for the primary therapeutic target of crucial importance, it is also clear that the processing of the drug by different metabolic pathways could be crucial for absorption, distribution, metabolism, excretion (ADME) properties. Thus, in the future, de novo design methods will have to take into account potential metabolism of the compound as well as its affinity for the target.

4.13.1.2 Overview of De Novo Design

De novo ligand design has matured in the last 15–20 years and has been subject to several outstanding review articles.[2–7] **Figure 1** illustrates the conceptual steps in de novo design. Several problems make de novo ligand design difficult.

First, estimation of binding energy with the target receptor is one problem that is common to methods of virtual screening as well. The rigorous physical chemical methods dealing with this question are often time-consuming and limited by the accuracy of the approximations involved.

Second, in addition there is the combinatorics of the number of ways novel molecules can be assembled from basic building blocks; there is an astronomical number of possibilities, but only a tiny fraction of the possible molecules will

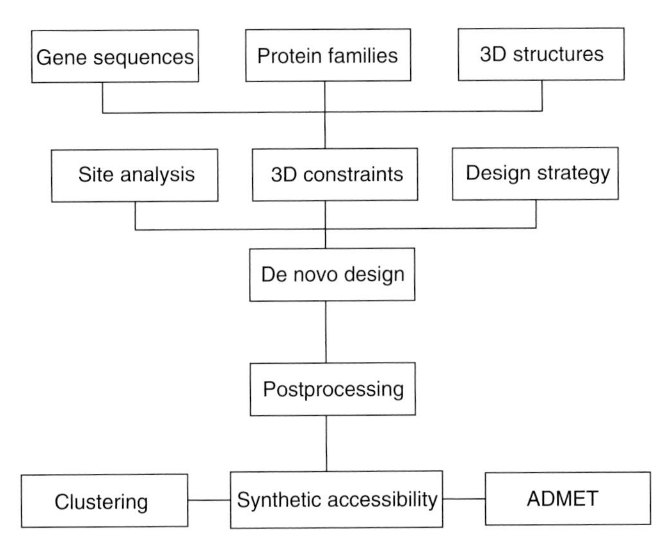

Figure 1 Conceptual steps in de novo drug design.

actually bind to the target of interest. Search, optimization, and heuristic methods that preferentially find the active molecules are required in order to make the process practical.

Third, another problem that is unique to de novo ligand design is how to restrict the possible design products to those that are accessible by known synthetic chemistry and to rank them according to ease of synthesis.

Fourth, other important factors to consider include storage, search, and presentation to the user of the massive data generated by de novo design methods.

The difficulty of these problems has necessitated the development of various strategies to make the process feasible. The procedure is normally divided into stages that are easier to manage.

Sites are first identified on the surface of the target protein. In many cases there are known ligands and crystal structures of their complexes and consequently the active sites are easy to detect. Information from different ligands could be useful for identifying important features of the site that are frequently utilized. This is also done by using rule-based strategies, experimental information about how molecules interact, or force-field and docking strategies of small fragments.

Novel ligands are normally built by linking atoms or fragments derived either theoretically or from crystal data. The diversity of the ligands depends on what building blocks are used and how they are connected together. Different strategies have evolved for this purpose; they attempt to make the procedures more intelligent and closer to the process that a synthetic chemist would use for synthesis of compounds.

The actual strategies for creating a ligand and enforcing its interaction with the active site differ among de novo design methods, but all try to solve the problem of how to limit the search resources efficiently for assessing how well different ligands interact with the site. The consensus appears to be that gradual focusing of the search on to fewer and fewer ligands as the search progresses should work best. It would be wasteful in the beginning of the search to dedicate a large effort to evaluate binding affinity of a random ligand, while this would be highly recommended for a few selected candidates before synthesis. Resources spent on evaluating the binding energy should also be balanced against assessing those needed for synthetic feasibility of a compound.

In the following sections the steps of a ligand design project are followed and described in more detail. References are provided for the reader to examine specific implementations of the general principles involved.

4.13.2 Targets

Targets amenable to de novo design are those where there is coordinate data for the target, or in the absence of target structural data, known active compounds may be used to model an image of the target binding site. Common potential targets are: receptors, enzymes, protein–protein interactions, and nucleic acids. Coordinate data can be derived from x-ray crystal structures where a single conformation is determined from the crystal. Care has to be taken with how that data are used: structure factor information may give clues to regions of flexibility in the protein. Nuclear magnetic resonance (NMR) determinations of structure usually provide numerous conformational models. In the absence of a physical determination of structure, structural models can be provided by homology modeling of closely related proteins.

The evolution of protein structures often reveals that binding sites for common substrates or cofactors are conserved. A clear example is found for adenosine triphosphate (ATP)- or guanosine triphosphate (GTP)-binding sites where the P-loop shows a conserved motif [AG]-x(4)-G-K-[ST].[8] The similarity between binding sites becomes a problem if drugs are to be designed for members of a family of proteins, for example, kinases where the ATP-binding sites are often very similar. In the case of kinases, the issue is how to design selective compounds (*see* Section 4.13.9.3).

The structures of the protein targets could be experimentally obtained via experimental techniques like x-ray and NMR or theoretically derived by homology modeling.[9] In the absence of 3D protein information, pseudoreceptors could be derived from the superposition of ligand binding to the protein of interest. Other biological macromolecules, such as nucleic acids, could also be used as targets.

4.13.3 Site Analysis

The complementarity of tightly binding ligands with the receptor, in terms of shape and physicochemical properties, leads to the concept of receptor site analysis, in which the binding site is examined to identify potential interaction sites or pharmacophore points that define noncovalent interactions, as well as ligand shape and size restraints. Pharmacophore constraints represent regions in space, where ligand hydrogen bond donor, acceptor, or hydrophobic atoms are constrained to be positioned to fulfill specific interactions with the receptor. The utilization of these features reduces the vast ligand search space to specific regions of the binding site involving particular interactions that are expected to provide key requirements for designing strongly binding ligands. Interaction sites are usually defined as hydrogen bond donor, acceptor, or hydrophobic sites. Several approaches have been utilized in de novo drug design to analyze the receptor-binding site and derive these key interaction sites.

4.13.3.1 Site Analysis Using Experimental Data

X-ray crystallographic information of protein–ligand complexes can be used to construct a knowledge base of interaction data, from which interaction sites within a protein-binding site can be derived. For example, SuperStar, developed by the Cambridge Crystallographic Data Centre, uses appropriate intermolecular interaction data from the IsoStar database. This was compiled from analysis of protein–ligand complexes in the Protein Data Bank, small molecule crystal structures in the Cambridge Structural Database (CSD), and theoretical interaction energies. The data can be used to identify hot-spot regions in the receptor-binding site in the form of a 3D map, where favorable interactions with particular functional groups are likely to occur. Regions of high propensity can be designated as pharmacophore points.

4.13.3.2 Derivation of Interaction Sites Using Rules

Grid-based approaches have been utilized to identify interaction sites within protein-binding sites. For example, in the GRID method,[10] a grid is superposed on the binding site and interaction energies of functional groups or probes placed at vertices of the grid are calculated using an empirical force field (*see* 4.11.3 Characterization of Protein-Binding Sites and Ligands Using Molecular Interaction Fields). Suitable probes can be employed to find favorable locations within the binding site cavity for hydrogen bonding and hydrophobic interactions. The accuracy of the GRID approach depends on the resolution of the grid; the finer the grid, the more accurate the sitepoint determination. However, this also leads to an increase in computational cost and so a balance between accuracy and efficiency is usually achieved.

HSITE is a rule-based approach that identifies potential hydrogen-bonding sites with the protein-binding cavity.[11] This scheme uses ideal hydrogen-bonding geometries derived from the analysis of small-molecule crystal structures in the CSD and allows (user-defined) tolerances in hydrogen bond length and angle, to yield a map of hydrogen-bonding interaction regions with the receptor site. A strength-weighted accessible probability score (SWAPS) has been defined that scores the hydrogen-bonding propensity of a sitepoint in terms of an intrinsic hydrogen-bonding ability and hydrogen-bonding accessibility.[12] This has been accomplished again by a survey of the CSD, via IsoStar as well as by using the algorithm HBMAP,[13] which determines the positions of potential hydrogen-bonding sitepoints. The same authors also developed a genetic algorithm-based method for identifying hydrophobic interaction sites in the protein-binding cavity: the method is dependent on the size and shape of the cavity as well as atom type.

In a related approach, LUDI[14] determines hydrogen bonding and hydrophobic protein–ligand interaction sites, which are defined as centers and surfaces, from a set of geometric rules obtained from a survey of nonbonded contact distributions in the CSD. Several other methods have been developed for defining covalent interaction sites as well as bonds to metals.[15]

Water molecules are usually found in crystallographic determinations of protein and protein–ligand complex structures. However, they are typically removed prior to initiation of de novo drug design protocols. Nevertheless, some of these water molecules may exert a crucial effect, in terms of modifying the shape of the binding site as well as forming hydrogen bonds, that may either bridge the interaction between protein and ligand or form part of a hydrogen bond network of water molecules that stabilizes the binding site. Retention of a small subset of bound crystallographic water molecules is known to have a significant effect on the diversity of the chemotypes formed in de novo drug design.[16] In this respect, methods are emerging to identify tightly binding water molecules observed in x-ray crystal structures and in deciding which of these should potentially be retained in drug discovery projects. For example, WaterScore uses a multivariate regression analysis to discriminate between displaceable and bound crystallographic water molecules from a survey of a set of proteins in their apo- and holo-forms.[17]

4.13.3.3 Site Analysis Using Docking and Minimization

Alternative approaches to site analysis involve bathing the binding cavity with functional groups or fragments and using minimization of docking techniques to find the most energetically favorable configurations within the site (**Figure 2**).

The Locus de novo design method uses a set of about 500 diverse fragments derived from a database of about 40 000 structures. Fragments with the best binding energies to the protein under consideration are assembled to generate a large number of potential ligand candidates. The key to the approach is the determination of the protein-binding site and fragment-binding energies. This uses a grand canonical scheme[18] in which the protein is effectively solvated in each of the fragments and an ensemble of ligand configurations is obtained that is within a specified binding energy threshold. At lower binding energy thresholds, the number of fragment orientations in the ensemble decreases and the poses that are consistent with the lowest binding energy represent solutions with the most favorable energetic interactions with the protein. The process explores the protein surface and potential protein-binding sites as well as key interaction sites; these sites are identified as regions where a diverse set of fragments are located with low binding energy thresholds.

The multiple copy simultaneous search (MCSS) method has been employed in de novo design[19] to analyze protein-binding sites. This procedure involves randomly placing multiple copies of functional groups inside the protein-binding site and minimizing these groups to yield a set of local potential minima for each group. In the MCSS, each functional

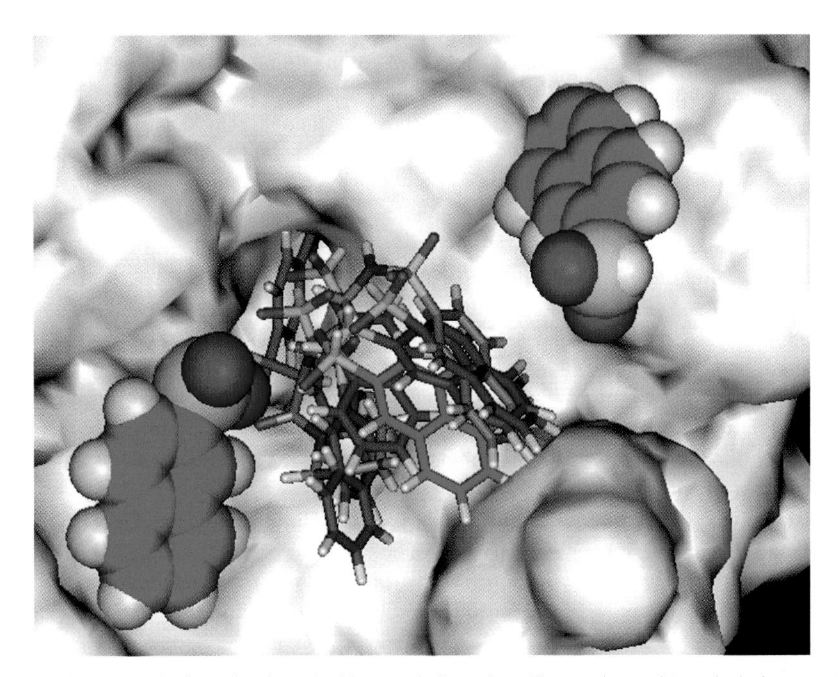

Figure 2 Site analysis of trypsin (1ppc) using docking and clustering. The surface of trypsin is in gray; ligand *N*-alpha-(2-naphthyl-sulphonyl-glycyl)-DL-*p*-amidinophenylalanyl-piperidine (NAPAP) is in gray sticks too. Fragment of NAPAP is docked. The most distant cluster representatives are shown as green creatine phosphokinase models. The cluster representative closest to the position of the fragment in NAPAP is in light-green sticks; other members of the same cluster are in a darker shade of green.

group experiences the average field of the fixed protein; interactions between the replicas are not included. Functional group positions are ranked according to their binding energies, which are estimated using a combination of van der Waals and electrostatic terms. Minima with the lowest interaction energies are selected.

Ligand docking has also been deployed to identify local energy minima and consensus sites within protein-binding cavities. In a recent method,[20] ligand molecules are evenly distributed over the protein surface and the binding energy of each copy is minimized using simplex and adopted basis Newton–Raphson methods. The binding energy is estimated by an empirical function that includes solvation electrostatic terms calculated by the finite-difference Poisson–Boltzmann (PB) approach. The minimized ligands are analyzed and clustered to yield potential consensus interaction sites.

The SEED approach[21] also involves docking of multiple copies of molecular fragments, in which polar groups are required to form at least one hydrogen bond and nonpolar fragments are located in hydrophobic pockets; the binding energy is estimated by a sum of van der Waals and screened electrostatic interactions. The combinatorial connection of the docked or minimized fragments/functional groups provides the basis for several de novo design methods.

Besides minimization and docking, binding sites have also been analyzed by conducting dynamics simulations with both polar and apolar fragments/functional groups and free energy grids[22] (OWFEG method) to identify energetically favorable intraction sites.

4.13.4 Building Blocks for De Novo Design

The most commonly adapted procedure for the construction of ligands is to connect small molecular fragments. There are two aspects in the building process: (1) generation of molecular composition/connectivity; and (2) generation of molecular geometry. The building blocks for de novo ligand design could be either single atoms or molecular fragments. The particular choice of building blocks determines the coverage of chemical space and the chemical complexity of the generated ligands. Rules used to assemble ligands also form part of the procedure.

4.13.4.1 Atom-Based Methods

The main advantage of building ligands from single atoms is that in principle any molecule could be created given certain connection rules. Another advantage is that very little information about building blocks is required before actually initiating the process. The atomic fragments would normally contain information about atom type, hybridization state, and bond order. Ideal bond lengths, bond angles, and torsion angles (with some randomization) are used to attach atoms to the growing ligand. Simple checks are also performed to ensure consistency of the structure. Quick force-field minimization could be used to relieve strain.

The major emphasis in atom-based algorithms is on the construction algorithm. Usually, molecules are built sequentially, adding one atom at a time, and both the connectivity of the molecule and its geometry are calculated during construction. The implementation of the ligand-building process could be complicated due to look-ahead factors such as ring closures, strained structures, and prediction of aromaticity. Very few of the modern methods rely solely on single atoms to build ligands.

It is also possible to decide first on the connectivity of the ligand and then use an external method to generate coordinates and fit the structure into the site. In this case, the information from the ligand interaction with the receptor, directing the ligand growth, is lost and the process is likely to be less efficient unless information from previous structures is retained and reused. Alternatively, only the atom type and hybridization information of the atomic template could be used; the optimized receptor interaction determines the best connectivity and coordinates (conformation) of the ligand[23–25] (MCDNLG, DycoBlock). Such a procedure has to establish reasonable bonded geometry for the ligand using force-field terms in the objective function. This effectively increases the dimension of the problem and makes the convergence slower.

Some of the first methods reported were initially atom-based[26–30] (Optimus, Legend, GenStar, GrowMol) but were later extended to include bigger fragments[31] (GroupBuild). One of the major disadvantages of the atom-based methods is the difficulty of controlling the synthetic tractability of the generated compounds. Due to its nonlocal nature, this is more of a problem for atom-based methods than for fragment-based procedures.

4.13.4.2 Ring and Acyclic Fragments

The use of larger molecular fragments relieves some of the problems associated with using single atoms as building blocks. Normally, fragment sets are derived by analysis of molecules from an established compound collection. This could be either a generally available database such as the National Cancer Institute (NCI) database, or commercial databases such as Available Chemicals Directory (ACD) or World Drug Index (WDI) or the private collection of a pharmaceutical company. Different databases will have a different emphasis on the chemical classes they contain. To some extent this will be reflected in the fragments that are derived, with a consequential effect on the ligands that are generated from those fragments.

The database is processed to identify features of interest in each molecule and using special definitions to derive fragments. Commonly defined fragments are rings and acyclic fragments, molecular frameworks and side chains,[32,33] common ring systems[34] (SHAPES library), ring and acyclic fragments,[35] active rings, and common substituents.[36] The frequency of occurrence in the database is determined as well as property filters. Similar screens are used for the de novo-designed structures and the procedures are described in more detail in Section 4.13.7.

Using fragments for the constuction of ligands makes it somewhat easier to control the diversity and the synthetic feasibility of the putative ligands. This is at the expense of the more restricted coverage of chemical space. The construction method could also take into account the frequency of each fragment and preferentially sample more likely combinations. One potential problem is that the numbers of ring and acyclic or linker fragments are quite different and this could lead to a higher occurrence of ring fragments in the ligands unless careful control is exercised.

The recombination rules for these fragments could be quite simple, like connecting terminal single bonds to single bonds, and double bonds to double bonds, but a more general labeling scheme could also be developed by defining several different types of single bond and only allowing certain combinations to link.

One specialized variation of the fragmentation/recombination method is to derive generalized fragments without atom identities[37,38] that are an average representation of several fragments and only later dress them with specific atom types.[39–41] This procedure aims at managing and reducing the combinatorial problem associated with atom typing. The geometric aspect of this is made possible due to the fact that bond lengths and angles are quite similar in atoms with the same hybridization state. This has previously been used successfully in algorithms for conformational analysis.[42]

4.13.4.3 Reaction-Based Fragments

An interesting application of the fragment derivation procedure is to create an algorithm that splits the input molecules according to known synthetic transforms.[43,44] This provides a major advantage in terms of synthetic feasibility because recombination between fragments occurs according to rules similar to those used in the splitting step and limits the possibility of undesirable products. This type of procedure has become established and favored in the field despite the fact that it could limit drastically the types of molecules the method is able to create. It appears that this could be quite a powerful tool in tackling the synthetic feasibility problem if the right selection of reaction transforms is used.

4.13.4.4 Target-Specific Fragments

It has already been mentioned that the content of the database affects the content of derived fragment set; that in turn affects the chemistry of the generated ligand structures. One way researchers have found that exploits this connection is to derive fragments from a database of known ligands for the target of interest. Such a database will be much smaller than a general database, but the chemical classes of the structures it contains will be highly relevant. Either reaction transforms or fragmentation rules based on rings and linkers could be explored to process such a database. Experiments with this idea have been performed for kinase and G protein-coupled receptor (GPCR) targets,[45–47] with encouraging results.

4.13.4.5 Fragments for Combinatorial Library Design

Combinatorial library design has become an essential part of projects aiming at the discovery of novel ligands for a given target protein. It is possible to utilize de novo design technology in a way that is suitable for library enumeration and selection. In fact, the process of deriving fragments is very similar to that of using reaction transforms, as outlined in the above sections.[48] To present a trivial example, one could consider a tripeptide combinatorial library. If only standard amino acids are used as building blocks, the library will contain 8000 peptides. The amino acids are generated by algorithmically 'digesting' peptide and protein substrates. The N and C ends of the derived amino acids should be labeled so that only the correct recombinations occur. There could be either a single class of amino acids, or a method for capping the N- and the C-terminus provided by the peptide generator of three classes of amino acids, N-capped, C-capped, and central. One of the first examples of de novo ligand design used peptides[49] before extending the method to deal with more general organic molecules. More recent methods[50,51] also revisit this problem, since it provides a well-defined testing ground for any de novo design algorithm.

4.13.5 Ligand Assembly and Design Strategies

Ligand assembly from small building blocks is at the heart of the methods for de novo ligand design. The huge number of ways it is possible to combine molecular fragments makes the search for the compounds active toward the target of interest a great challenge. **Figure 3** illustrates the relationships within de novo design schemes that emerge from examination of published papers.

4.13.5.1 Combinatorial Problems in De Novo Design

There are many combinatorial problems for which no efficient exact solution is known and the search space increases exponentially with the number of problem variables. Combinatorial problems are classified according to their computational complexity as P and NP. Problems in P are solvable by polynomial time-deterministic algorithms while problems in NP are solvable by polynomial time-nondeterministic algorithms. Deterministic algorithms can do only one operation at a time, while nondeterministic algorithms can perform operations simultaneously when faced with several alternatives. Perhaps the most frequently used example of an NP problem is the traveling salesman problem, where a salesman has to find the shortest route to visit a given number of cities, eventually returning to the city where his trip began.

Combinatorial problems abound in the field of de novo ligand design[52] and they can be roughly split into two classes: those related to the ligand and those related to the receptor.

4.13.5.1.1 Ligand-related combinatorial problems

To the best of our knowledge exact estimates and formulas of the number of small ligands binding to a particular receptor are not available; rough estimates are rare. What is established and agreed in the community is that numbers are big.

Some relevant questions to ask relate to the ligand structure in isolation. How many small molecules composed of C, N, O, S, H, Cl, F, I with molecular weight (MW) <600 are there? How many of these are free of steric overcrowding and exist at physiological conditions? How many do not contain certain predefined toxic or reactive substructures? There are certainly more questions like that, but these three are indicative of the area of interest. The history of enumeration problems of this kind started quite early[53] and involves the use of generating functions and, more recently, computer-assisted structure enumeration.[54] Questions of this type are sometimes even easy to answer if restricted to certain types of compounds, e.g., a simple product relation for the members of a combinatorial library.

A second type of question that is more difficult to answer, but also more interesting, is about ligand enumeration, involving its interaction with a receptor. For example, how many ligands bind to a specified receptor at 1 μmol l^{-1}? Is it possible to plot a curve relating the number of ligands versus activity? Computer-aided methods should be able to assist in answering such questions. In particular, several studies[30,55] use computational experiments to estimate the answers

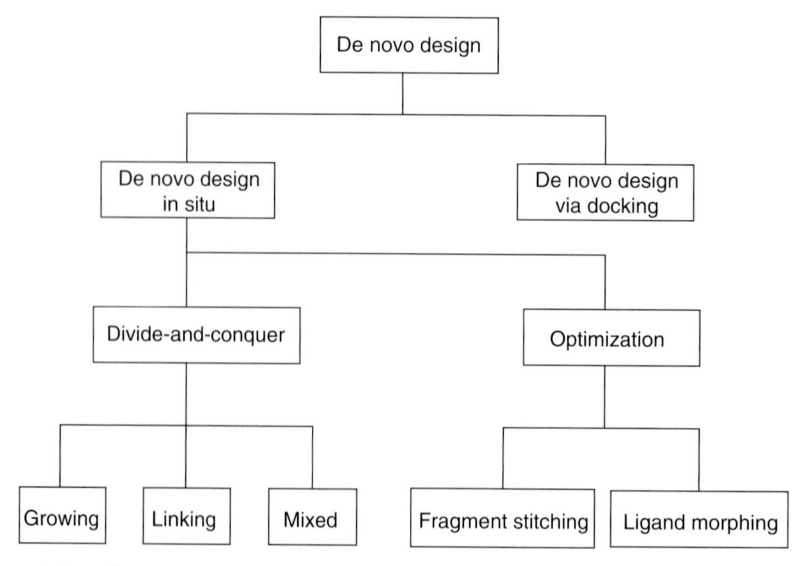

Figure 3 De novo design schemes.

of this type of questions and derive a function for the number of unique structures of thermolysin and human immunodeficiency virus (HIV) protease in an exponential form, $\alpha^{N_{atom}}$, where α is a constant close to 2 and N_{atom} is the number of atoms in the ligand.[30]

4.13.5.1.2 Receptor-related combinatorial problems

Only very recently, methods for de novo ligand design have started to address some of the difficult combinatorial problems associated with the receptor structure. The major ones involve conformational flexibility of the active site, allowing slightly different shape and chemisty to be presented to the ligand at the interface.

A related issue is the presence of strongly bound water molecules which could be displaced by appropriate functional groups in the ligand. The presence, or absence, of such water molecules also modifies the shape of the site; ideally all possible in/out combinations should be considered by the de novo design algorithms.

Another source of combinatorial choices is associated with the strategies medicinal chemists would like to use in order to target specific regions of functional groups in the active site. It would be a very useful addition to a de novo design method to be able to implement strategies targeting specific combinations of regions and groups.

4.13.5.2 Algorithms for Combinatorial Search and Optimization

Combinatorial search and optimization methods are well developed and suited to the incorporation of problem-related knowledge in order to find solutions quickly. There are two types of general methods for solving combinatorial problems depending on the strategy used: (1) divide-and-conquer methods; and (2) relaxation methods. In divide-and-conquer methods, problems are solved by answering subproblems and building up the solution. For example, in the traveling salesman problem, a route for the complete journey could be constructed by adding cities one at a time and taking into account the lengths of these intermediate routes. Well-known search techniques in this class, used in de novo design methods, include the depth-first, branch-and-bound, and A* searches.[56]

In contrast, in relaxation methods, an initial solution would be proposed and used as a baseline for improvements. In the traveling salesman problem, all cities could be initially visited in a random order; to decrease the length of the route, the reverse order in which pairs of cities are visited could be tried. Examples of relaxation methods frequently used for de novo design are simulated annealing[57] and genetic algorithms.[58,59] Systematic examination of all molecular structures up to a certain size in a reasonable amount of time is impossible, except in trivial cases. Furthermore, the configurational degrees of freedom involved in the protein–ligand interaction (translation, rotation, conformation) add extra complexity to the search performed in chemical space. In order to find suitable solutions, different optimization approaches have been proposed for the combined chemical/configurational space.

There are two aspects of molecular representation that are important and differ in different ligand assembly strategies. The first aspect is the molecular composition and connectivity, i.e., what types of atoms are present in the molecule and how the atoms are connected. The second aspect is molecular geometry and conformation. There are two general types of method emerging, depending how these two aspects are handled.

4.13.5.3 De Novo Design via Docking

This type of method for de novo design creates the molecular composition and connectivity and uses an external method to generate the actual geometry and conformation; a docking method then positions the structure in the site. These methods separate the exploration of the different chemical structures from the search for the binding mode of the particular compound under investigation. These methods step through an individual compound, and perform conformational and docking analysis and binding energy evaluation before moving to another compound to investigate. Such methods rely heavily on molecular docking and virtual screening methods for their functionality. They are however not dependent on a preexisting database of compounds: they have an engine capable of generating novel chemical structures which is coupled to a molecular coordinate generator like CONCORD,[60] a docking and a scoring method.

The advantages of this approach are that: (1) it is built on top of other established technologies; and (2) it is thermodynamically correct since it examines and evaluates each receptor–ligand system individually. Several examples of implementations of this strategy have been reported[44,61,62] (Dream + +, Synopsis).

A disadvantage of this type of method is the larger amount of central processing unit time required, since each compound is examined with the same level of detail. While this is recommended for compounds likely to be active, the compounds examined at the start of the procedure could be completely random and resources spent on them could be wasteful. One way to alleviate this problem is to reduce the sampling of the docking method and accuracy of the scoring in the beginning of the search and then gradually bring it back to desired accuracy. Such a strategy has been successfully tried previously in the area of Monte Carlo simulation of liquids.[63] Also, if the molecules in the sequence are similar and have a large common substructure, the binding mode of one of them could be used to guide the position of the other and thus reduce the cost of searching.

It could be argued that the amount of time required to find a good starting compound by methods described next is roughly the same as the time required to dock and evaluate a small number of compounds. Therefore it could be advantageous to find promising compounds first and only then evaluate them more extensively.

4.13.5.4 De Novo Design In Situ

The methods in this class perform the search for the ligand structure and its binding mode at the same time. The composition, connectivity, and geometry of the ligand are created at the same time by linking building blocks in situ in the active site. Although less rigorous from a thermodynamic point of view, these methods tend to arrive at promising solutions faster. If the suggestions generated by these methods are viewed as heuristics and are consequently evaluated by established binding free-energy methods (*see* Section 4.13.6), this could provide a promising approach to de novo ligand design.

De novo design in situ methods could also be subdivided into two classes on the basis of the search method they use. The first class of methods use the divide-and-conquer approach and the second class of methods uses an optimization/relaxation approach.

4.13.5.4.1 Divide-and-conquer methods

The divide-and-conquer methods build ligands in a stepwise manner. The design problem is solved by completing subproblems and combining their solutions. Two basic strategies have emerged: (1) fragment-linking methods; and (2) ligand-growing methods.

4.13.5.4.1.1 Fragment-linking methods

In the fragment-linking methods, fragments are first placed in the site so that they interact favorably with the receptor and later some of them, possibly all, are connected into larger ligands either directly or through link fragments (**Figure 4**).

The advantages of these methods are well known. First, it is ensured by the initial fragment placement that, once constructed, the ligand will hopefully maintain favorable interactions with the receptor site. Second, since the ligand structure is deduced piece by piece, the testing of a large number of inappropriate structures is avoided. Third, since the initial fragments are small and relatively rigid, their placement could be relatively fast.

However it should be remembered that there are several difficulties associated with this procedure as well. First, displacements from the ideal positions of the fragments do not result in crucial differences in the interaction score. This could allow alternative fragment arrangements, in which, despite individual fragments interacting more weakly,

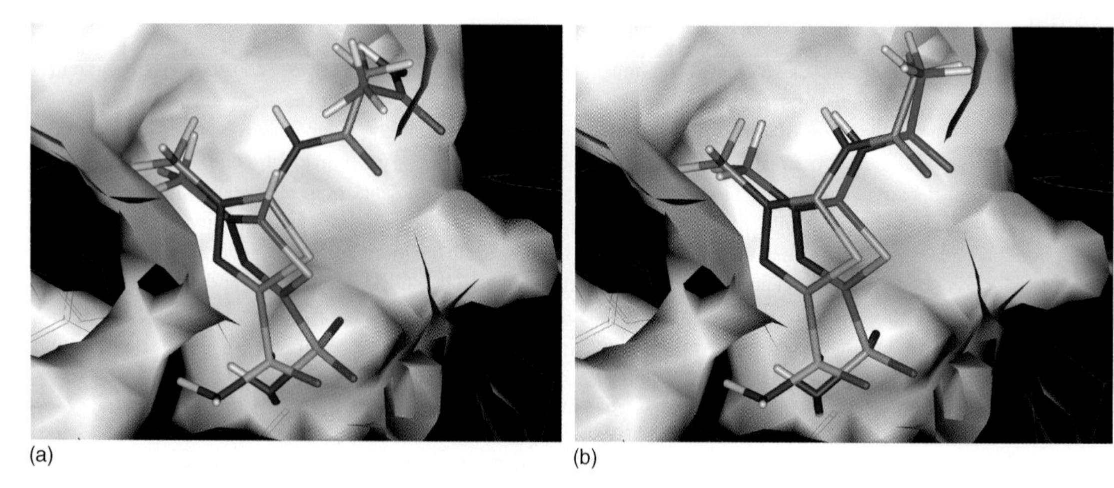

Figure 4 Illustration of fragment-linking methodology for the carbonic anhydrase 1 bzd. (a) Placement of aldehyde and amine fragments (green); native ligand (gray). (b) Fragment linking: designed ligand in green.

the overall interaction of the ligand is stronger; such displacements however lead to a large number of binding modes that are difficult to handle. The number of binding modes is manageable when fragments are sufficiently large, but that inevitably leads to a combinatorial increase in the number of fragments to consider. A heuristic compromise to this problem would be to use generalized fragments[37,38,64] (*see* Section 4.13.4).

Second, the final ligand structure is unknown during the construction stage and therefore it is unknown beforehand whether it would be close to the lowest energy conformation of the ligand; it may turn out, when considering ideal fragment placements, to be impossible to build a low-energy, or even properly linked, ligand.

Third, linking fragments together is another difficult issue. If the distance between two fragments to be connected is too large, linking cannot be achieved directly or by a single fragment, since fragment size is limited. A solution would be to use either several long-chain bridges, stored and folded every time within the constraints of the site and the positions of the two fragments,[65,66] or the link may be grown starting from one of the fragments.[27,67] Later, the conformational flexibility of the linkers could be reduced using a ring-bracing approach.[68] If, on the other hand, the distance is within the limits of a single linking fragment, the prepositioned fragment could be too close to be connected, directly or bridged, within acceptably small geometrical deviations at the junction. If the tolerances are increased, too many linkers become possible with quite distorted and strained geometry that molecular mechanics (MM) minimization might fail to correct. The situation is more ambiguous when more than two fragments have to be connected, or the initially placed fragments are allowed displacements. This could be an advantage because more freedom is allowed, but on the other hand the problem becomes more complex. The simplest strategy to connect more than two fragments seems to be to do it stepwise. First, certain pairs of fragments are connected, then these could be considered as composite fragments that could be linked, either to a fragment that has not yet been used, or to another composite fragment.[69] Careful bookkeeping is essential.

There are several interesting examples in the literature exemplifying this approach reported by Böhm (the Ludi method),[14] Eisen *et al.* (Hook),[70] Tcshinke and Cohen (Newlead),[71] and Leach and Kilvington.[59]

4.13.5.4.1.2 Ligand-growing methods

In this strategy the ligand structure is kept connected from the beginning to the end of construction and thus avoids the issue of fragment linking. This strategy also allows the application of empirical rules during the building process; to some extent it ensures that the final structure is in a low-energy state.

A difficulty in the growing approach is to achieve a strong favorable interaction between the structure and the site. Much depends on where the building is started and the choice of seed atoms for the placement of successive fragments. It is not necessary that at each step the new fragment is added at the best possible position and orientation, since that could prevent an overall stronger interaction after the addition of the next fragments.[29,49] Schemes have been proposed in which a number of alternatives with scores higher than that of the best possible placement are retained at each step and one of the possibilities is then selected.[55,29,30,55,72,73]

There are various possibilities still to be explored in this area by borrowing ideas from simulation methods such as Rosenbluth search[74] or pruned enriched Rosenbluth method.[75]

4.13.5.4.1.3 Mixed methods

It is obviously possible to combine ligand growing with fragment-linking strategies in an attempt to take advantage of the strengths of both strategies,[58,59,76,77] suggesting that they are just the extremes of a spectrum of methods. This could be achieved by starting growth from different fragments that have been positioned in the site until the different components meet; then the components could be connected based on their overlapping fragments,[77] the possibility of forming direct bonds or using linking fragments but with a reduced distance to span. Alternatively, acyclic chains of atoms of different lengths could be fitted between fragments and then the resulting ligand could be grown further by the ring-bracing approach.[58] Yet another way is to dock multiple molecules in the site and recombine moieties from different ligands.[27,47,78]

4.13.5.4.2 **Optimization and evolutionary methods**

This class of methods allows a lot of flexibility in its implementation, and with the appropriate specification of constraints, could be made to resemble both ligand-growing and fragment-linking methods. Essentially, the optimization could start from a random ligand structure. Small changes either in the chemistry of the structure or in its position are introduced in order to obtain a new structure. Repetition of these transitions results in a gradual change of the ligand structure. The suitability of this new structure as a solution of the ligand design problem is evaluated and its score is compared to the previously examined structure(s); this makes it possible to use feedback control to direct the transitions. Once different types of transition and a scoring function have been implemented, standard global optimization methods could be used. In contrast to growing and linking methods that are implementations of the divide-and-conquer strategy, here each structure is viewed as a proposal for a complete solution of the design problem and its suitability is evaluated. Two basic variations of this theme have been reported. In the first one, both the connectivity of the ligand and the positions of the atoms could be changed during transitions. In the second one, only the positions of the atoms change and the optimal connectivity of the ligand is deduced from the atomic position.

4.13.5.4.2.1 Ligand morphing

This method normally starts by linking random fragments together to form the initial ligand. Of course, more sophisticated choices to initialize the structure generation are also implemented and these could reduce search time. Afterwards, transitions include additions, removals, or replacement of fragments in the current ligand by fragments selected from the fragment library. Ligands are always single connected entities. Several implementations of this idea have been investigated by Glen and Payne,[79] Todorov and Dean,[35] and Pegg *et al.*[80] These algorithms derive their efficiency from the ability to remove or replace fragments already included in the ligand. If a growing ligand comes to a dead end in the active site, certain fragments could be removed and replaced while keeping the best-positioned ones still as part of the structure.[81]

4.13.5.4.2.2 Fragment stitching

In another class of methods the building blocks are allowed to move in the site and to connect in ways that favor good interactions with the receptor. The connections between fragments are formed and broken dynamically to reflect the shape and the chemical nature of the receptor structure. The connection between the fragments is not specified beforehand but depends on how close the fragments are to each other and ultimately to the way they interact with the receptor. Such methods have been developed by Pearlman and Murcko (Concepts),[82] Gehlhaar *et al.* (MCDNLG),[23] Stutz and Karplus,[25] and Zhu *et al.*[24] This part of the method resembles how the bond network of a molecule is established by ab initio quantum chemistry methods. Unfortunately, these de novo design methods tend to be slower and the control over the chemistry of the final structure is weaker, especially when small fragments are used as building blocks. They certainly have their attraction due to the simplicity and similarity to the evolution of useful compounds.

4.13.5.5 De Novo Design via Docking versus In Situ

The two classes of methods that have been highlighted here, namely design via docking and design in situ, could be combined or used in sequence. At each stage of an in situ method, docking could be used to study in detail a particular ligand. In fact, such a strategy would have the added benefit of having a very good starting position for the ligand, but a completely unbiased docking run could also be performed and the results from both docking runs considered together. In situ methods could be considered as a limiting case of design via docking with extremely short searches. The results of in situ strategies could be input for more rigorous design via docking.

4.13.5.6 Fragment-Based Design

Problems of scoring and binding energy estimation are well known in both the area of de novo design and virtual compound screening. This has prompted computational and medicinal chemists to seek experimental verification[83,84] of modeling studies as early as possible and has shifted the de novo design strategy from the design of larger high-activity ligands in a single step toward the design of small low-affinity compounds (known under various names: seeds, needles, fragments) which are easier to design.[85] These small weak-binding leads are validated by experimental means, but due to the limited sensitivity of the assays also a minimum size of the compound is required.[86] The concept of ligand efficiency[87] has been found useful for quantifying the potential of a fragment for lead development. Smaller well-fitting fragments are preferable to larger compounds with the same activity. Once active fragments have been found and validated, they could be extended further, either with the help of in silico reagent screening[88,89] or by direct synthesis and testing.

In connection with fragment-based design work, additivity principles in biochemistry in general have been critically examined.[90] The question has been asked: what is the connection between the binding energies of two isolated fragments in relation to the binding energy of a compound containing them? The importance of entropic contributions has been stressed[91] and the implications for the design strategy have also been discussed.[92]

4.13.6 Scoring and Binding Energy Estimation

Scoring functions used in ligand design, as well as in ligand docking and virtual screening, are required to guide the sampling processes employed, as well as to estimate the fitness of the protein–ligand complex, usually in the form of a binding energy.[93] As a consequence of the huge number of iterations conducted in each simulation, and often the processing of a large number of ligands, such scoring functions must be fast, which invariably means that accuracy is compromised, since several terms involved in the full thermodynamic cycle will be ignored. Nevertheless, scoring functions need to be robust in that they rank solutions highly with favorable steric and electrostatic interactions and rank ligands lower that are not complementary to a particular target, or are in an unfavorable configuration. The balance between hydrophobic interactions and hydrogen bonding, which represent the two main proponents of complex formation, is a key requirement for a successful scoring function (*see* 4.12 Docking and Scoring; 4.19.2 Virtual Screening).

4.13.6.1 Knowledge-Based and Empirical Scoring Methods

A large number of scoring functions have been developed over the past 10 years. These can essentially be classified as force-field-based methods, empirical and knowledge-based scoring functions. Force-field-based methods involve potential energy terms corresponding to stretching or compressing bonds, bond angle bending, torsional angle deformations, and nonbonded terms for Coulomb interactions, hydrogen bonding, and van der Waals interactions with binding energies estimated from appropriate energy differences.[94,95] The potential energy terms are usually taken from well-known force fields.

Empirical scoring functions estimate binding energies by a weighted sum of terms, such as hydrogen bonding, hydrophobic contact, and entropic contributions, where the weighting factors are determined by linear regression to appropriate experimental data. Some of the commonly used empirical scoring functions include partial linear potential (PLP),[96] LUDI,[97] ChemScore,[98] and ScreenScore.[99]

Knowledge-based scoring functions involve pair potentials for protein–ligand atoms derived from statistical analyses of protein–ligand crystal structures. Functions of this type have been derived only recently as the number of available x-ray structures has increased, and include potential of mean force (PMF),[100] DrugScore,[101] and SMog2001.[102]

In conjunction with the development of new scoring functions, during the last few years there has been an increasing number of studies that attempt to compare the abilities of several different scoring functions as well as combining multiple scoring functions to yield a consensus score[103] that provides improved results in molecular docking and virtual screening.[99,104]

4.13.6.2 Physical Chemistry Methods

As discussed above, due to the large number of molecules that are typically processed during a structure-based drug design project or in silico virtual screening exercise, fast energy functions are used to score the protein–ligand complex. Virtual screening studies have demonstrated that such fast scoring functions are capable of providing reasonable enrichment values in a set of diverse ligands, but are less successful at ranking compounds of similar chemotype. Several more rigorous methods have been developed that incorporate more of the physics of ligand binding at the expense of increased computational cost.

The main methods of interest are the MM/PBSA[105] and linear interaction energy (LIE) method.[106] MM (Poisson–Boltzmann (PB), generalized born model (GB)) surface area (SA) methods developed by Kollman and colleagues[105] combine molecular mechanics energies with PB or generalized born (GB) for polar solvation energies with solvent accessible surface area for nonpolar solvation energy estimates. Within those methods, variations have been proposed allowing the use of either explicit or implicit water simulations. For example, the MM/GBSA approach described by Rizzo *et al.*[107] uses the GB implicit hydration model, while the implementation of LIE by Huang and Caflisch[108] uses minimization with distance-dependent dielectric instead of molecular dynamics (MD) and finite-difference PB calculation of electrostatic solvation. A solute entropy term is sometimes included to yield a total free energy; binding free energies can be calculated from the difference between the free energy of the complex and that of the protein and ligand components. The explicit approaches adopted are often distinguished by whether the solvated complex is minimized and the binding energy of the single structure is determined, or whether an ensemble of structures is acquired by conducting dynamics simulations and the binding energy is averaged over the ensemble. Kuhn *et al.*[109] find that MM/PBSA applied to a single minimized complex is as successful as the averaged free-energy approach applied to an ensemble in terms of enrichments in virtual screening and in ranking the output from de novo drug design studies.

4.13.6.3 Energy Landscapes

It has been suggested that the energy landscape that is characteristic of ligand–protein binding is funnel-shaped and, in that respect, resembles the energy landscape of protein folding. Native ligands exhibit minimally frustrated pathways to the global minimum, leading to a stable binding mode, whilst nonbinding compounds will have a frustrated energy landscape, leading to multiple modes.[110,111] Native ligands fulfill both thermodynamic stability and kinetic accessibility criteria due to the funnel-shaped binding energy landscape.

4.13.6.4 Scoring in Ligand-Based Design

In the absence of 3D protein structure, ligand-based design could be employed.[54,112] Estimation of binding free energy is less accurate and the method relies on pharmacophore points and pseudoreceptor modeling.[113,114]

4.13.7 Selection of Fragments and Ligands

Ideally, the de novo design methodology should incorporate all requirements that the ligand should satisfy in order to be acceptable from a medicinal chemistry perspective. Often, however, the requirements could be too numerous or too difficult and time-consuming to include during the design stage and are left to be checked at a later stage. Sometimes the requirements are included, but the design objectives change, or an analysis like clustering could only be performed when multiple solutions are known. Filtering, clustering, and other types of analysis are considered as a postdesign stage. Such methods are also useful in the predesign stage for fragment selection as well as virtual screening and library design methods.

4.13.7.1 Filtering

The selection of molecules could be based on properties and functional groups.[115] Commonly used properties include log P, molecular weight, number of hydrogen-bonding groups, number of rotatable bonds, polar surface area, and measures of molecular complexity,[116] and there may be many different compilations of problematic functional groups. These screens try to remove molecules that have questionable absorption, distribution, metabolism, excretion, and toxicity (ADMET) properties, contain reactive groups, and would be problematic in assays.

4.13.7.2 Clustering

Methods for clustering of the designed structures[117,118] and comparison with known ligands[119] also play an important part in the postprocess filtering in order to be able to judge the thoroughness of the coverage of known chemical classes of ligands and to classify interesting novel scaffolds. The designed structures could be compared to binding modes of known inhibitors, both visually and by means of similarity measures considering the size and spatial proximity of the common substructures. It has been shown that algorithms could generate representatives of many inhibitor classes within a very short time and that new similarity measures could be useful for comparing and clustering designed structures.[118,119] Filtering and clustering of the ligands could also be performed based on the interaction pattern with key groups on the receptor.

4.13.7.3 Synthetic Analysis

Another major priority in recent years has been the investigation of ligand construction methods using retrosynthetic analysis and a database of known reactions in order to enhance the synthetic accessibility of the ligands. Due to its complexity, extensive automated synthetic analysis is not possible at the early stages of the design, but it is highly advisable when structures meeting certain minimal design criteria have been obtained and before presenting a list of candidates to a medicinal chemist for inspection. Tools in this area[120] have been developed independently of de novo design methods and could be used off-the-shelf to improve the chemical intelligence of automated de novo design methods.[15,44]

4.13.7.4 Presentation, Storage, and Search

Finally, the issues of presentation via well-designed and intuitive graphical interfaces, visualization, structure storage, search, and retrieval are also of crucial importance for the acceptance and everyday use of de novo design technology.[48,121]

4.13.8 Exemplification of De Novo Design

De novo design methods have been tested for their ability to create structures similar to known inhibitors[24,28,77,119] and suggest novel scaffolds that are consequently synthesized and tested for activity.[44,73,122–124]

Carbonic anhydrase was one of the first examples used to demonstrate the potential of de novo design methodology. Starting from a sulfonamide moiety taken from a known ligand, new structures were synthesized based on an iterative de novo design method.[124] One of the suggestions turned out to be the most potent ligand ever found for that enzyme.

Synopsis is a relatively recent addition to the repertoire of methods for de novo design with special emphasis on synthetic feasibility issues.[44] The application of this method has yielded a number of novel HIV reverse transcriptase ligands with activity in the $\mu mol\,L^{-1}$ range.

Firth-Clark *et al.*[125] report the use of the de novo design algorithm Skelgen[35,119] to design active compounds for estrogen receptor-alpha. Seven crystal structures of the protein were used, showing different conformations of the active site, and thus providing an opportunity to generate many different ligand structures. A total of 14 000 output structures were asked for using two strategies per crystal structure. In all, 13 311 were successfully generated, of which 5492 ligands were unique. From these unique structures, the top 25 scoring structures from each strategy produced 350 structures to be considered for synthesis. These structures were clustered on grounds of similarity using Daylight SMARTS to obtain a diverse set of potential compounds. A retrosynthetic analysis algorithm was used to score the structures for synthetic tractability. High scoring compounds were found in 16 clusters and 35 were chosen for synthesis. Seventeen were synthesized and assayed.

Of the 17 compounds synthesized, five were found to be actives. Prior art searches using MARPAT, MARPATPREV, Chemical Abstract Services (CAS) registry, and Beilstein databases suggested that four of the five actives are novel structures: the fifth was a known compound designed de novo. Thus, with the estrogen receptor using the Skelgen algorithm, de novo design can achieve synthetically tractable compounds with activity, giving a 30% hit rate. Design, synthesis, and biological testing took 6 months in all. These studies provide encouragement for further extensive application of de novo design to disparate targets and open up the scope for de novo methods being widely used to provide novel, patentable compounds in structural genomics.

4.13.9 Emerging New Technologies

4.13.9.1 Flexibility

One of the major problems in computational drug design is incorporating the intrinsic flexibility of protein-binding sites. This is particularly crucial in ligand-binding events, when induced fit can lead to protein structure rearrangements. As a consequence of the huge conformational space available to protein structures, receptor flexibility is rarely considered in ligand design procedures. However, the importance of protein structural flexibility for drug discovery has been discussed in an excellent review by Teague.[126] Several methods have been used to deal with receptor flexibility, primarily in the area of site analysis[127] and ligand docking.[128]

As a result of the rapid increase in the availability of co-crystal structures of proteins with small molecules, the utilization of this information by researchers in projects involving different targets becomes increasingly more important. For instance, the concept of drug discovery through gene families is becoming a focus of efforts, particularly in the pharmaceutical industry, since the knowledge garnered from drug design in a particular member of a family may also be applicable to other members of the same family. In this respect, an approach termed virtual target screening (VTS) developed by Eidogen-Sertanty[129] can be adopted. VTS is the target equivalent of traditional virtual high-throughput screening, which involves screening of a target protein-binding site with a database of small druglike molecules. The difference is that, in VTS, a target protein-binding site is screened against a database of protein-binding sites, which are known to co-crystallize with small molecules and information regarding the similarities in binding sites can be deployed to suggest possible chemotypes or molecular scaffolds that may bind in the new target. In this way, the database of small-molecule binding sites provides information about the binding characteristics of a new target, for which no ligands may currently be available.

4.13.9.1.1 Ensembles

An ensemble of preselected receptor models can be used to describe protein flexibility (**Figure 5**). Proteins exist in an equilibrated ensemble of structures and ligand binding involves an equilibrium shift toward the protein–ligand complex structure. The models can be taken from multiple crystal or NMR structures, snapshots from a molecular dynamics simulation, or homology modeling. During the docking simulation,[130] the binding energy of the ligand is assessed

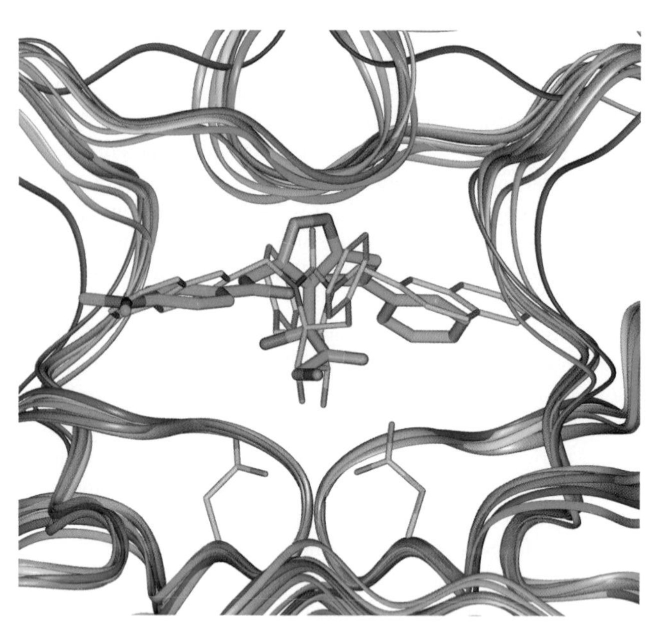

Figure 5 Design of an ensemble of proteins (HIV protease). Ligand xk263 from crystal structure 1hvr and the corresponding protein structure are shown in gray. The ligand designed to bind to the 1hvr structure in the ensemble is shown in green; the ribbons of other HIV protease structures superimposed on 1hvr are shown in color.

against all models in a cross-docking protocol and the lowest energy or a weighted sum of the energies is used.[131–133] For example, Wei *et al.*[134] and Claussen *et al.*[135] identify alternative conformations from overlaying crystal structures and generate protein models, using a combinatoric approach. In another approach,[136,137] an ensemble of conformationally distinct protein–ligand complex conformations is generated by allowing relaxation of the complex through energy minimizations, starting from a number of initial positions for the ligand. In a related approach, the multiple receptor models are averaged, typically by combining interaction grids, and the single averaged structure is used for docking.[138,139] Carlson and co-workers developed a 'dynamic' pharmacophore approach.[140] This involves flooding MD snapshots of a receptor target with probes and minimizing the probes in the site in the spirit of MCSS. The MD snapshots are aligned and clusters of probe molecules from each snapshot are examined to determine 'clusters of clusters' or consensus-binding interaction sites (pharmacophores). This scheme was applied to HIV, starting from the open apo-form and used in virtual screening to identify successfully known HIV inhibitors in a database of druglike molecules.[141] Multiple active site corrections have also been deployed in virtual screening and docking to improve the ranking of ligands to a specific target.[132]

4.13.9.1.2 Soft scoring

Soft docking approaches have been utilized to incorporate protein flexibility implicitly by relaxing the criterion for steric clashes between receptor and ligand atoms.[134] This effectively increases the size of the binding site pocket by allowing closer approach of the ligand. The soft potential can account for relatively small conformational changes; however, it is less successful in docking and virtual screening when larger conformational changes are involved.

4.13.9.1.3 Dynamic

Receptor conformations can also be changed dynamically on the fly during the docking simulation (**Figure 6**). This has been limited to specific degrees of freedom of the protein, such as hydrogen atoms or hydrogen bond donor/acceptor groups. In more recent algorithms this has been extended to complete side chains, while keeping the protein backbone fixed, by identifiying the optimal side-chain torsional angles during the docking procedure[142,143] or using a rotamer library to represent the most likely side-chain orientations.[89,144–146] The latter approach utilizes optimization schemes such as the dead-end elimination theorem[147] and its refinements[148,149] and branch and bound schemes,[150] minimization[151] approaches used in protein modeling to eliminate unfavorable rotameric combinations from the search space as well as clustering methods to handle the large number of possible rotameric combinations. Treatment of only side-chain flexibility may be inadequate in some cases as potentially key backbone movements are neglected, which in

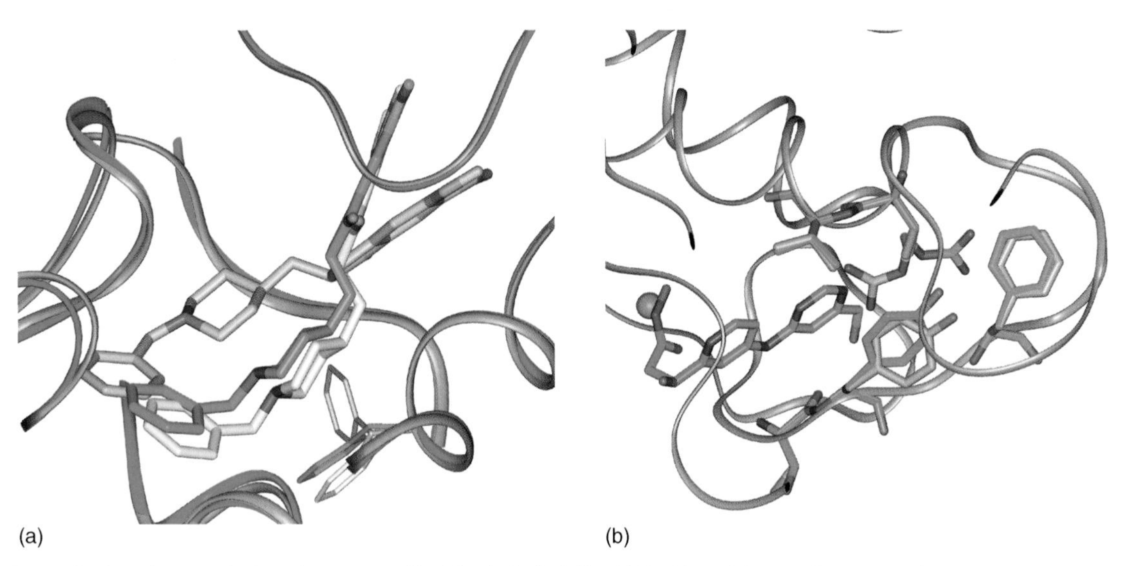

Figure 6 Docking and design of receptors with side-chain flexibility. (a) Docking of donepezil to non-native acetylcholinesterase protein structure 1vot. The crystal structure of the ligand (green) and the top-ranked docking solution are shown in the static 1vot binding site (orange) and the flexible 1vot site (gray). Ribbons and the side-chain conformation of the mobile residue Phe-330 are also shown for the native receptor 1eve (green) and 1vot (orange). (b) Ligand (green) designed in the S1' site of human collagenase (2tcl). The ribbons and the side chains of 2tcl are shown in gray; the final conformations of flexible side chains are shown in green.

turn may influence side-chain conformations,[152] although, in contrast, Najmanovich *et al.*[153] found no correlation between side-chain flexibility and backbone motion upon ligand binding. Extensions for incorporating backbone movements are emerging, including loop closure schemes[154,155] and scaled collective variables approaches that follow low-frequency normal modes of the macromolecular system.[156]

Despite the strong evidence for the impact of receptor flexibility in drug discovery,[157–159] most computational de novo drug design approaches treat the protein as a static system due to the large combinatorial space available to the flexible system. Very few examples of incorporating protein-binding site flexibility in the ligand design process are currently present in the literature. Stultz and Karplus[160] report an extension of the MCSS method in which MCSS minima involving fragments sets are located as described above, and the protein-binding site is subsequently locally optimized in the region of interest. Consensus protein models are derived that can be used in ligand design. Zhu *et al.*[161] discuss an extension of their previous procedure[24] by allowing protein flexibility via MD simulations. The flexible procedure however only allows very limited ligand fragment sets. An in silico reagent screening approach that considers dynamic side-chain flexibility has been reported[89] and this has recently been extended to the ligand design process.[162]

4.13.9.2 Hydration

The size and shape of a protein-binding site can also be modified by tightly binding water molecules. Water molecules exert a local electrostatic effect as a result of their polarity and dipole and this can influence ligand binding. Furthermore, the displacement of tightly bound water molecules by a ligand can lead to significant free-energy penalties. Most in silico ligand design studies employ an unsolvated dry site after the removal of crystallographic water molecules; however, from the above arguments, it is clear that binding-site hydration may exert a significant effect on ligand binding.

A key aspect is the identification of conserved tightly binding water molecules. Crystallographic waters have been analyzed in pairs of ligand-bound and unbound forms of the same protein in order to identify conserved waters that may subsequently be retained in de novo ligand design experiments. The inclusion of explicit crystallographic water molecules in the site has been explored in molecular docking using the phantom particle concept[163] as well as in de novo design.[164] However, the location of crystal waters by resolution of the electron density is a rather uncertain procedure. Typically, isolated regions of electron density are assigned to be waters after all electron density assignments to protein atoms have been conducted. Simulation techniques have been employed to predict the solvated structure of biomolecular systems, usually MD- or MC-based techniques. These approaches require enormous simulation times and, as shown by Guarnieri and Mezei, they may not be able to reach configurations close to the 'real' solvated system, as a consequence of not including multibody interactions. Guanierei and Mezei proposed a grand canonical scheme involving annealing of the chemical potential[165] to overcome these problems. This approach involves removal of water molecules at all possible sites as the chemical potential is changed and includes the effects of multibody configurations. Temperature annealing leads to the formation of additional water clusters and more closely matches known x-ray data. These solvation shells can significantly alter the electrostatic fields around protein surfaces and thus influence ligand-binding events.

4.13.9.3 Specificity

Drug side effects arise from the interaction of a ligand with a number of proteins sharing common motifs. To avoid side effects, ligands have to be specific. Where there is a set of similar proteins, it is important to develop computational methods for finding molecules that bind selectively to individual or to all members of the family.[166] The methods could be applicable to the design of sequence-specific DNA-binding ligands allowing direct targeting of genes.

Electrostatic specificity has been carefully examined[39,167] in order to determine the factors making a ligand specific to particular proteins. One of the interesting conclusions is that the strongest binders are not necessarily the most specific ones. Analysis of interaction maps using principle component analysis could also identify regions in the active site contributing to specificity.[168] Receptor affinity fingerprints[169] could also be helpful in that respect. Selectivity fields have been defined to quantify pairwise selectivity as the difference in the biological activity between two proteins.[170] Recently, de novo design methodology has been applied to the problem of specificity[171] in a three-step procedure. First, putative ligands are created to each of the superimposed members of the protein family. Second, all ligands are docked to all proteins to allow for adjustments due to the different environments provided by the different proteins. Third, the scores from the docking runs are analyzed and used to derive a measure of specificity. Application to cyclin-dependent kinases reveals several candidates with some resemblance to known ligands binding to all members of the family.

4.13.10 Future Perspectives

In the short to medium term, up to the year 2010, there are a number of obvious developments that are needed for the technique of de novo design:

1. The most important is to develop an accurate prediction of affinity of the ligand for its site. Statistically based empirical methods of scoring functions are unlikely to provide a great improvement due to the amounts of uncertainty in deconvoluting an experimental measurement into all its constituent parts. The stumbling blocks are hydration/solvation issues in the protein–ligand complex together with entropy changes. It is more likely that methods for postprocessing the affinity calculations will be developed, such as MM/PBSA, MM/GBSA, or LIE; early indications of the application of such methods is encouraging
2. Static models of the site will be replaced by flexible residues in the binding site region. This should overcome many inadequacies of design within rigid binding sites and perhaps lead to methods for induced fit in the structure-based drug design process. Large-scale protein motions will be harder to incorporate into de novo design due to our weak understanding of protein motion
3. Selectivity in the design of ligands is dependent on the similarities between target proteins. It should be possible to take a set of related proteins, identify similarities and differences, and automate the classification of sitepoints into similar and dissimilar sets. With this information, design strategies could be picked to optimize the best strategies leading to selectivity

In the long-term, de novo design methods will have to be developed to incorporate predictive gene expression and other pharmacogenomic effects into the choice of ligand structures to be taken forward for development. The link between chemical genomics and drug design is an obvious one to make, but it will involve a large amount of rationalization of genomics data and systems biology before the data can be usefully categorized. Perhaps an appropriate starting point would be to link the metabolome to structure generation so that de novo ligands are docked into all drug-metabolizing enzymes for predictive metabolism.[1] In this way, undesired metabolic products could, in principle, be avoided. Ultimately, de novo drug design methods will shift toward utility in the goal of personalized medicine.

References

1. Dean, P. M. Computer-Aided Design of Small Molecules for Chemical Genomics. In *Methods in Molecular Biology, Vol. 310, Chemical Genomics: Reviews and Protocols*; Zanders, E. D., Ed.; Humana Press: Totowa, NJ, 2005, pp 25–39.
2. Lewis, R. A.; Leach, A. R. *J. Comput.-Aided Mol. Des.* **1994**, *8*, 467–475.
3. Murcko, M. A.; Caron, P. R.; Charifson, P. S. *Annu. Rep. Med. Chem.* **1999**, *34*, 297–306.
4. Schneider, G.; Clement-Chomienne, O.; Hilfiger, L.; Schneider, P.; Kirsch, S.; Böhm, H.-J.; Neidhart, W. *Angew. Chem. Int. Ed. Engl.* **2000**, *39*, 4130–4133.
5. Schneider, G.; Fechner, U. *Nat. Drug Disc.* **2005**, *4*, 649–663.
6. Dean, P. M.; Lloyd, D. G.; Todorov, N. P. *Curr. Opin. Drug Disc. Dev.* **2004**, 7, 347–353.
7. Congreve, M.; Murray, C. W.; Blundell, T. L. *Drug Disc. Today* **2005**, *10*, 895–907.
8. Koonin, E. V. (1999) www.expasy.org/prosite/PDOC00017 last updated 1999.
9. Oshiro, C.; Bradley, E. K.; Eksteowicz, J.; Evensen, E.; Lamb, M. L.; Lanctot, J. K.; Putta, S.; Stanton, R.; Grootenhius. *J. Med. Chem.* **2004**, *47*, 768–771.
10. Goodford, P. D. J. *J. Med. Chem.* **1985**, *28*, 764–767.
11. Danziger, D. J.; Dean, P. M. *Proc. R. Soc. Lond.* **1989**, *B236*, 101–113.
12. Kelly, M. D.; Mancera, R. L. *J. Comput.-Aided. Mol. Des.* **2003**, *17*, 401–414.
13. Mills, J. E. J.; Dean, P. M. *J. Comput.-Aided Mol. Des.* **1996**, *10*, 607–622.
14. Böhm, H. J. *J. Comput.-Aided Mol. Des.* **1992**, *6*, 593–606.
15. Gillet, V. J.; Myatt, G.; Zsoldos, Z.; Johnson, A. P. *Perspect. Drug Disc. Des.* **1995**, *3*, 34–50.
16. Mancera, R. *J. Comput.-Aided Mol. Des.* **2002**, *16*, 479–499.
17. Garcia-Sosa, A. T.; Mancera, R. L.; Dean, P. M. *J. Mol. Model.* **2003**, *9*, 172–182.
18. Guarnieri, F.; Mezei, M. *J. Am. Chem. Soc.* **1996**, *118*, 8493–8494.
19. Miranker, A.; Karplus, M. *Proteins* **1991**, *11*, 29–34.
20. Dennis, S.; Kortvelyesi, T.; Vajda, S. *Proc. Natl. Acad. Sci. USA* **2002**, *99*, 4290–4295.
21. Majeux, N.; Scarsi, M.; Caflisch, A. *Proteins* **2001**, *42*, 256–268.
22. Pearlman, D. A. *J. Med. Chem.* **1999**, *42*, 4313–4324.
23. Gehlhaar, D. K.; Moerder, K. E.; Zichi, D.; Sherman, C. J.; Ogden, R. C.; Freer, S. T. *J. Med. Chem.* **1995**, *38*, 466–472.
24. Zhu, J.; Yu, H. B.; Fan, H.; Liu, H. Y.; Shi, Y. Y. *J. Comput.-Aided Mol. Des.* **2001**, *15*, 447–463.
25. Stutz, C.; Karplus, M. *Proteins* **2000**, *40*, 258–289.
26. Lewis, R. A. *J. Mol. Graph.* **1992**, *10*, 131–143.
27. Lewis, R. A.; Poe, D. C.; Huang, C.; Ferrin, T. E.; Lagridge, R.; Kuntz, I. D. *J. Mol. Graph.* **1992**, *10*, 66–78.
28. Nishibata, Y.; Itai, A. *Tetrahedron* **1991**, *47*, 8985–8990.
29. Rotstein, S. H.; Murcko, M. A. *J. Comput.-Aided Mol. Des.* **1993**, 7, 23–43.
30. Bohacek, R. S.; McMartin, C. *J. Am. Chem. Soc.* **1994**, *116*, 5560–5571.
31. Rotstein, S. H.; Murcko, M. A. *J. Med. Chem.* **1993**, *36*, 1700–1710.
32. Bemis, G. W.; Murcko, M. A. *J. Med. Chem.* **1996**, *39*, 2887–2893.
33. Bemis, G. W.; Murcko, M. A. *J. Med. Chem.* **1999**, *42*, 5095–5099.
34. Fejzo, J.; Lepre, C. A.; Peng, J. W.; Bemis, G. W.; Ajay; Murcko, M. A.; Moore, J. M. *Chem. Biol.* **1999**, *6*, 755–769.
35. Todorov, N. P.; Dean, P. M. *J. Comput.-Aided Mol. Des.* **1997**, *11*, 175–192.
36. Ertl, P. *J. Chem. Inf. Comput. Sci.* **2003**, *43*, 374–380.
37. Lewis, R. A.; Dean, P. M. *Proc. R. Soc. Lond.* **1989**, *B236*, 125–140.
38. Gillet, V.; Johnson, A. P.; Mata, P.; Sike, S.; Williams, P. *J. Comput.-Aided. Mol. Des.* **1993**, 7, 127–153.
39. Barakat, M. T.; Dean, P. M. *J. Comput.-Aided Mol. Des.* **1995**, *9*, 448–456.
40. Barakat, M. T.; Dean, P. M. *J. Comput.-Aided Mol. Des.* **1995**, *9*, 457–462.
41. Todorov, N. P.; Dean, P. M. *J. Comput.-Aided Mol. Des.* **1998**, *12*, 335–349.
42. Leach, A. R.; Prout, K.; Dolata, D. P. *J. Comput.-Aided Mol. Des.* **1988**, *2*, 107–123.
43. Lewell, X. Q.; Judd, D. B.; Watson, S. P.; Hann, M. M. *J. Chem. Inf. Comput. Sci.* **1998**, *38*, 511–522.
44. Vinkers, H. M.; de Jonge, M. R.; Daeyaert, F. F. D.; Heeres, J.; Koymans, L. M. H.; van Lenthe, J. H.; Lewi, P. J.; Timmerman, H.; Aken, K. V.; Janssen, P. A. *J. Med. Chem.* **2003**, *46*, 2765–2773.
45. Patchett, A. A.; Nargund, R. P. *Annu. Rep. Med. Chem.* **2000**, *35*, 289–298.
46. Cross, K. P.; Myatt, G.; Yang, C.; Fligner, M. A.; Verducci, J. S.; Blower, P. E. *J. Med. Chem.* **2003**, *46*, 4770–4775.
47. Pierce, A. C.; Rao, G.; Bemis, G. B. *J. Med. Chem.* **2004**, *47*, 2768–2775.
48. Leach, A. R.; Bryce, R. A.; Robinson, A. J. *J. Mol. Graph. Model.* **2000**, *18*, 358–367.
49. Moon, J. B.; Howe, W. J. *Proteins* **1991**, *11*, 314–328.
50. Budin, N.; Ahmed, S.; Majeux, N.; Caflisch, A. *Comb. Chem. High Throughput Screen.* **2001**, *4*, 661–673.
51. Frenkel, D.; Clark, D. E.; Li, J.; Murray, C. W.; Robson, B.; Waszkowycz, B.; Westhead, D. R. *J. Comput.-Aided Mol. Des.* **1995**, *9*, 213–225.
52. Dean, P. M.; Barakat, M. T.; Todorov, N. P. Optimization of Combinatoric Problems in Structure Generation for Drug Design. In *New Perspectives in Drug Design*; Dean, P. M., Jolles, G., Newton, C. G., Eds.; Academic Press: London, 1995.
53. Calay, A. *Philos. Magazine* **1874**, *47*, 444–446.
54. Trinajstic, N.; Nikolic, S.; Knop, J. V.; Muller, W. R.; Szymansky, K. *Computational Graph Theory: Characterization, Enumeration and Generation of Chemical Structures by Computer Methods*; Ellis Horwood: New York, 1991.
55. DeWitte, R. S.; Shakhnovich, E. I. *J. Am. Chem. Soc.* **1996**, *118*, 11733–11744.
56. Reingold, E. M.; Nievergelt, J.; Deo, N. *Combinatorial Algorithms, Theory and Practice*; Prentice-Hall: Englewood Cliffs, NJ, 1977.
57. Kirkpatrick, S.; Gellatt, C. D., Jr.; Vecchi, M. P. *Science* **1983**, *220*, 671–680.
58. Holland, J. H. *Adaptation in Natural and Artificial Systems*; University of Michigan Press: Ann Arbor, MI, 1975.
59. Goldberg, D. E. *Genetic Algorithms in Search, Optimization and Machine Learning*; Addison-Wesley: Reading, MA, 1989.
60. Pearlman, R. S. 3D Molecular Structures: Generation and Use in 3D Searching. In *3D QSAR*; Kubinyi, H., Ed.; ESCOM: Leiden, 1993, pp 41–77.
61. Makiko, S.; Ewing, T. J. A.; Kuntz, I. D. *J. Comput.-Aided Mol. Des.* **1999**, *13*, 513–532.
62. Schneider, G.; Lee, M. L.; Stahl, M.; Schneider, P. *J. Comput.-Aided Mol. Des.* **2000**, *14*, 487–494.

63. Ceperley, D. M.; Dewing, M. *J. Chem. Phys.* **1999**, *110*, 9812–9820.
64. Gillet, V.; Newel, W.; Mata, P.; Myatt, G.; Sike, S.; Zsoldos, Z.; Johnson, A. P. *J. Chem. Inf. Comput. Sci.* **1994**, *34*, 207–217.
65. Coutsias, E. A.; Seok, C.; Jacobson, M. P.; Dill, K. A. *J. Comput. Chem.* **2004**, *25*, 510–528.
66. Shenkin, P. S.; Yarmush, D. L.; Fine, R. M.; Wang, H.; Levinthal, C. *Biopolymers* **1987**, *26*, 2053–2085.
67. Lewis, R. A. *J. Comput.-Aided Mol. Des.* **1990**, *4*, 205–210.
68. Leach, A. R.; Lewis, R. A. *J. Comput. Chem.* **1994**, *15*, 233–240.
69. Leach, A. R.; Kilvington, S. R. *J. Comput.-Aided Mol. Des.* **1994**, *8*, 283–298.
70. Eisen, M. B.; Wiley, D. C.; Karplus, M.; Hubbard, R. E. *Proteins* **1994**, *19*, 199–221.
71. Tschinke, V.; Cohen, N. C. *J. Med. Chem.* **1993**, *36*, 3863–3870.
72. Nishibata, Y.; Itai, A. *J. Med. Chem.* **1993**, *36*, 2921–2928.
73. Honma, T.; Hayashi, K.; Aoyama, T.; Hashimoto, N.; Machida, T.; Fukasawa, K.; Iwama, T.; Ikeura, C.; Ikuta, M.; Suzuki-Takahashi, I. et al. *J. Med. Chem.* **2001**, *44*, 4615–4627.
74. Rosenbluth, M. N.; Rosenbluth, A. W. *J. Chem. Phys.* **1955**, *23*, 356–359.
75. Grassberger, P. *Phys. Rev. E* **1997**, *56*, 3682–3693.
76. Clark, D. E.; Frenkel, D.; Levy, S. A.; Li, J.; Murray, C. W.; Robson, B.; Waszkowycz, B.; Westhead, D. R. *J. Comput. Aided Mol. Des.* **1995**, *9*, 13–32.
77. Law, J. M. S.; Fung, D. Y. K.; Zsoldos, Z.; Simon, A.; Szabo, Z.; Csizmadia, I. G.; Johnson, A. P. *J. Mol. Struct.-Theochem.* **2003**, *666*, 651–657.
78. Ho, C. M. W.; Marshall, G. R. *J. Comput.-Aided Mol. Des.* **1993**, 7, 623–647.
79. Glen, R. C.; Payne, A. W. R. *J. Comput.-Aided Mol. Des.* **1995**, *9*, 181–202.
80. Pegg, S. C. H.; Haresco, J. J.; Kuntz, I. *J. Comput.-Aided Mol. Des.* **2001**, *15*, 911–933.
81. Todorov, N. P.; Dean, P. M. Computational Ligand Design by Free Energy Minimization. In *Biological Physics*; Frauenfelder, H., Hummer, G., Garcia, R., Eds.; American Institute of Physics: Melville, NY, 1999.
82. Pearlman, D. A.; Murcko, M. A. *J. Comput. Chem.* **1993**, *14*, 1184–1193.
83. Pellecchia, M.; Sem, D. S.; Wüthrich, K. *Nat. Rev. Drug Disc.* **2002**, *1*, 211–219.
84. Carr, R. A. E.; Congreve, M.; Murray, C. W.; Rees, D. C. *Drug Disc. Today* **2005**, *10*, 987–992.
85. Böhm, H. J.; Boehringer, M.; Bur, D.; Gmuender, H.; Huber, W.; Klaus, W.; Kostrewa, D.; Kuehne, H.; Luebbers, T.; Meunier-Keller, N. et al. *J. Med. Chem.* **2000**, *43*, 2664–2674.
86. Hann, M. M.; Leach, A. R.; Harper, G. *J. Chem. Inf. Comput. Sci.* **2001**, *41*, 856–864.
87. Hopkins, A. L.; Groom, C. R.; Alex, A. *Drug Disc. Today* **2004**, *9*, 430–431.
88. Sun, Y.; Ewing, T. J.; Skillman, A. G.; Kuntz, I. D. *J. Comput.-Aided Mol. Des.* **1998**, *12*, 597–604.
89. Källblad, P.; Todorov, N. P.; Willems, H.; Alberts, I. L. *J. Med. Chem.* **2004**, *47*, 2761–2767.
90. Dill, K. A. *J. Biol. Chem.* **1997**, *272*, 701–704.
91. Villa, J.; Strajbl, M.; Glennon, T. M.; Sham, Y. Y.; Chu, Z. T.; Warshel, A. *Proc. Natl. Acad. USA* **2000**, *97*, 11899–11904.
92. Murray, C. W.; Verdonk, M. L. *J. Comput.-Aided Mol. Des.* **2002**, *16*, 741–753.
93. Lazaridis, T. *Curr. Org. Chem.* **2002**, *6*, 1319–1332.
94. Head, R. D.; Smythe, M. L.; Oprea, T. I.; Waller, C. L.; Green, S. M.; Marshall, G. R. *J. Am. Chem. Soc.* **1996**, *118*, 3959–3969.
95. Morris, G. M.; Goodsell, D. S.; Halliday, R. S.; Huey, R.; Hart, W. E.; Belew, R. K.; Olson, A. *J. Comput. Chem.* **1998**, *19*, 1639–1662.
96. Gehlhaar, D. K.; Larson, V.; Luty, B. A.; Rose, P. W.; Verkhivker, G. M. *Int. J. Quant. Chem.* **1999**, *72*, 73–84.
97. Böhm, H. J. *J. Comput.-Aided Mol. Des.* **1998**, *12*, 309–323.
98. Eldridge, M. D.; Murray, C. W.; Auton, T. R.; Paolini, G. V.; Mee, R. P. *J. Comput.-Aided Mol. Des.* **1997**, *11*, 425–445.
99. Stahl, M.; Rarey, M. *J. Med. Chem.* **2001**, *44*, 1035–1042.
100. Muegge, I.; Martin, Y. C. *J. Med. Chem.* **1999**, *42*, 791–804.
101. Gohlke, H.; Hendlich, M.; Klebe, G. *Mol. Biol.* **2000**, *295*, 337–356.
102. Ishchenko, A. V.; Shakhnovich, E. I. *J. Med. Chem.* **2002**, *45*, 2770–2780.
103. Charifson, P. S.; Walters, W. P. *J. Comput.-Aided Mol. Des.* **2002**, *16*, 311–323.
104. Wang, R.; Lu, Y.; Wang, S. *J. Med. Chem.* **2003**, *46*, 2287–2303.
105. Srinivasan, J.; Cheatham, T. E., III.; Cieplak, P.; Kollman, P. A.; Case, D. A. *J. Am. Chem. Soc.* **1998**, *120*, 9401–9409.
106. Aqvist, J.; Medina, C.; Samuelson, J.-E. *Protein. Eng.* **1994**, 7, 385–391.
107. Rizzo, R. C.; Toba, S.; Kuntz, I. D. *J. Med. Chem.* **2004**, *47*, 3065–3074.
108. Huang, D.; Caflisch, A. *J. Med. Chem.* **2004**, *47*, 5791–5797.
109. Kuhn, B.; Gerber, P.; Schulz-Gasch, T.; Stahl, M. *J. Med. Chem.* **2005**, *48*, 4040–4048.
110. Verkhivker, G. M.; Bouzida, D.; Gehlaar, D. K.; Rejto, P. A.; Freer, S. T.; Pose, P. W. *Curr. Opin. Struct. Biol.* **2002**, *12*, 197–203.
111. Wang, J.; Verkhivker, G. M. *Phys. Rev. Lett.* **2003**, *90*, 188101.
112. Lloyd, D. G.; Buñemann, C. L.; Todorov, N. P.; Manallack, D. T.; Dean, P. M. *J. Med. Chem.* **2004**, *47*, 493–496.
113. Snyder, J. P.; Rao, S. N.; Koehler, K. F. A. Vedani, Minireceptors and Pseudoreceptors. In *3D QSAR in Drug Design: Theory, Methods and Applications*; Kubinyi, H., Ed.; Escom: Leiden, 1993, pp 336–354.
114. Hahn, M. *J. Med. Chem.* **1995**, *38*, 2080–2090.
115. Charifson, P. S.; Corkery, J. J.; Murcko, M. A.; Walters, W. P. *J. Med. Chem.* **1999**, *42*, 5100–5109.
116. Bertz, S. H. *J. Am. Chem. Soc.* **1981**, *103*, 3599–3601.
117. Deng, Z.; Chuaqui, C.; Singh, J. *J. Med. Chem.* **2004**, *47*, 337–344.
118. Stahl, M.; Mauser, H.; Tsui, M.; Taylor, N. R. *J. Med. Chem.* **2005**, *48*, 4358–4366.
119. Stahl, M.; Todorov, N. P.; James, T.; Mauser, H.; Boehm, H. J.; Dean, P. M. *J. Comput.-Aided Mol. Des.* **2002**, *16*, 459–478.
120. Correy, E. J.; Wipke, W. T. *Science* **1969**, *166*, 178–192.
121. Ertl, P.; Jacob, O. *J. Mol. Struct.* **1997**, *419*, 113–120.
122. McCarthy, J. D.; Hogle, J. M.; Karplus, M. *Proteins* **1997**, *29*, 32–58.
123. Ripka, A. S.; Satyshur, K. A.; Bohacek, R. S.; Rich, D. H. *Org. Lett.* **2001**, *3*, 2309–2312.
124. Grzybowski, B. A.; Ishchenko, A. V.; Kim, C. Y.; Topalov, G.; Chapman, R.; Christianson, D. W.; Whitesides, G. M.; Shakhnovich, E. I. *Proc. Natl. Acad. Sci. USA* **2002**, *99*, 1270–1273.
125. Firth-Clark, S.; Williams, A.; Willems, H. M. G.; Harris, W. J. *Chem. Inf. Model.* **2005**, *46*, 642–647.
126. Teague, S. J. *Nat. Rev. Drug Disc.* **2003**, 527–541.
127. Goodford, P. J. *Rational Mol. Des. Drug Res.* **1998**, *42*, 215–230.

128. Hindle, S. A.; Rarey, M.; Buning, C.; Lengauer, T. *J. Comput.-Aided Mol. Des.* **2002**, *16*, 129–149.
129. Debe, D. A.; Hambly, K. *Curr. Drug Disc.* **2004**, *3*, 15–18.
130. Bouzida, D.; Rejto, P. A.; Arthurs, S.; Colson, A. B.; Freer, S. T.; Bursavich, M. G.; Rich, E. H. *J. Med. Chem.* **2002**, *45*, 541–558.
131. Ragno, R.; Frasca, S.; Manetti, F.; Brizzi, A.; Massa, S. J. *Med. Chem.* **2004**, *48*, 200–212.
132. Vigers, G. P.; Rizzi, J. *J. Med. Chem.* **2004**, *47*, 80–89.
133. Fernandez, M. X.; Kairys, V.; Gilson, M. K. *J. Chem. Inf. Comput. Sci.* **2004**, *44*, 1961–1970.
134. Wei, B. Q.; Weaver, L. H.; Ferrari, A. M.; Matthews, B. W.; Shoichet, B. K. *J. Mol. Biol.* **2004**, *337*, 1161–1182.
135. Claussen, H.; Buning, C.; Rarey, M.; Lengauer, T. *J. Mol. Biol.* **2001**, *308*, 377–395.
136. Apostolakis, J.; Pluckthun, A.; Caflisch, A. *J. Comput. Chem.* **1998**, *19*, 21–37.
137. Cavasotto, C. N.; Abagyan, R. A. *J. Mol. Biol.* **2004**, *337*, 209–225.
138. Knegtel, R. M. A.; Kuntz, I. D.; Oshiro, C. M. *J. Mol. Biol.* **1997**, *266*, 424–440.
139. Österberg, F.; Morris, G. M.; Sanner, M. F.; Olsen, A. J.; Goodsell, D. S. *Proteins: Struct. Funct. Genet.* **2002**, *46*, 34–40.
140. Carlson, H. A.; McCammon, J. A. *Mol. Pharm.* **2000**, *57*, 213–218.
141. Meager, K. L.; Carlson, H. A. *J. Am. Chem. Soc.* **2004**, *126*, 13276–13281.
142. Totrov, M.; Abagyan, R. *Proteins: Struct. Funct. Genet.* **1997**, 215–220.
143. Schnecke, V.; Kuhn, L. A. *Perpect. Drug Disc.* **2000**, *20*, 171–190.
144. Leach, A. R. *J. Mol. Biol.* **1994**, *235*, 345–356.
145. Frimurer, T. M.; Peters, G. H.; Iversen, L. F.; Andersen, H. S.; Moller, N. P.; Olsen, O. H. *Biophys. J.* **2003**, *84*, 2273–2281.
146. Taylor, R. D.; Jewsbury, P. J.; Essex, J. W. *J. Comput. Chem.* **2003**, *24*, 1637–1656.
147. Desmet, J.; De Maeyer, M.; Hazes, B.; Lasters, I. *Nature* **1992**, *356*, 539–542.
148. Lasters, I.; Desmet, J. *Protein Eng.* **1993**, *6*, 717–722.
149. Keller, D.; Shibata, M.; Markus, E.; Ornstein, R.; Rein, R. *Protein Eng.* **1995**, *8*, 893–904.
150. Wernish, L.; Hery, S.; Wodak, S. L. *J. Mol. Biol.* **2000**, *301*, 713–736.
151. Fernandez-Recio, J.; Totrov, M.; Abagyan, R. *Proteins* **2003**, *52*, 113–117.
152. Murray, C. W.; Baxter, C. A.; Frenkel, A. D. *J. Comput.-Aided Mol. Des.* **1999**, *13*, 547–562.
153. Najmanovich, R.; Kuttner, J.; Sobolev, V.; Edelman, E. *Proteins* **2000**, *39*, 261–268.
154. Dunbrack, R. L.; Karplus, M. *J. Mol. Biol.* **1993**, *230*, 543–574.
155. Coutsias, E. A.; Seok, C.; Jacobson, M. P.; Dill, K. A. *J. Comput. Chem.* **2004**, *25*, 510–528.
156. Hassan, S. A.; Mehler, E. L.; Weinstein, H. Structure Calculation of Protein Segments Connecting Domains with Defined Secondary Structure: A Simulated Annealing Monte Carlo Combined with Biased Scaled Collective Variables Technique. In *Lecture Notes in Computational Science and Engineering*; Hark, K., Schlick, T., Eds.; Springer-Verlag: New York, 2002; Vol. 24, pp 197–231.
157. Carson, H. A. *Curr. Opin. Chem. Biol.* **2002**, *6*, 1–6.
158. Bursavich, M. G.; Rich, D. H. *J. Med. Chem.* **2002**, *45*, 542–558.
159. Wong, C. F.; McCammon, J. A. *Annu. Rev. Pharmacol. Toxicol.* **2003**, *43*, 31–45.
160. Stultz, C.; Karplus, M. *Proteins* **1999**, *37*, 512–529.
161. Zhu, J.; Fan, H.; Liu, H.; Shi, Y. *J. Comput.-Aided Mol. Des.* **2001**, *15*, 979–996.
162. Alberts, I. L.; Todorov, N. P.; Dean, P. M. *J. Med. Chem.* **2005**, *48*, 6585–6596.
163. Rarey, M.; Kramer, B.; Lengauer, T. *Proteins* **1999**, *34*, 17–28.
164. Mancera, R. *J. Comput.-Aided Mol Des.* **2002**, *16*, 479–499.
165. Guarnieri, F.; Mezei, M. *J. Am. Chem. Soc.* **1996**, *118*, 8493–8494.
166. Janin, J. *Proteins* **1996**, *25*, 438–445.
167. Kangas, E.; Tidir, B. *J. Chem. Phys.* **2000**, *112*, 9120–9131.
168. Kastenholz, M. A.; Pastor, M.; Cruciani, G.; Haaksma, E. E. J.; Fox, T. *J. Med. Chem.* **2000**, *43*, 3033–3044.
169. Greenbaum, D. C.; Arnold, W. D.; Lu, F.; Hayrapetian, L.; Baruch, A.; Krumrine, J.; Toba, S.; Chehade, K.; Brömme, D.; Kuntz, I. D. *Chem. Biol.* **2002**, *9*, 1085–1094.
170. Baskin, I. I.; Tikhonova, I. G.; Palyulin, V. A.; Zefirov, N. S. *J. Med. Chem.* **2003**, *46*, 4063–4069.
171. Todorov, N. P.; Buñemann, C. L.; Alberts, I. L. *J. Chem. Inf. Model.* **2005**, *45*, 314–320.

Biographies

Nikolay P Todorov, Principal Research Scientist, De Novo Pharmaceuticals, received an MSc degree in Biotechnology and Bio-organic Chemistry from the University of Sofia, Bulgaria, in 1991 for his studies of the thermodynamics of enzymatically catalyzed peptide synthesis. During the course of the degree he also joined the

Shemyakin Institute of Bioorganic Chemistry in Moscow. In 1995 he obtained a PhD degree in Pharmacology from the University of Cambridge, UK, for his work was on the development of computational technology for de novo ligand design for receptors with known crystal structure. Later he joined the TeknoMed project, a collaborative venture between Imperial College, London, Rhône-Poulenc Rorer and the University of Cambridge for the application of emerging technologies to the discovery of new medicines from genomic data. Several patents have been granted for novel ligands designed with the application of the de novo design technology. In April 2000 Dr Todorov joined as a founding member De Novo Pharmaceuticals, a spin-off company from the University of Cambridge supported by a number of venture capital trusts. The research interests of Dr Todorov lie in the application of mathematical and computational methods to biological problems.

Ian L Alberts, Principal Scienticist, De Novo Pharmaceuticals, was awarded his PhD in Theoretical Chemistry from Cambridge in 1988. Since then, he has worked as a postdoctoral research associate in the group of Prof Henry F Schaefer at the University of Georgia, USA, and then had lectureship positions at the University of Edinburgh and Stirling University, Scotland, where he lead the chemistry group. From 1996 to 1999, he also worked as a visiting researcher at the European Bioinformatics Institute, Hinxton, Cambridge. In these positions, Dr Alberts conducted research in several diverse areas, ranging from quantum mechanical investigations of small-molecular systems to empirical studies of biomolecular structures, including protein–ligand complexes. In 2002, Dr Alberts moved back to Cambridge and took up his position at De Novo Pharmaceuticals, where his work has focused on developing novel methodologies for computational drug discovery. He has successfully incorporated protein receptor flexibility into the in silico drug design process and recently developed an algorithm for generating ligands that are selective for specific protein families.

Philip M Dean, Chief Scientific Officer, De Novo Pharmaceuticals, arrived in Cambridge in 1967 to work for a PhD at the Department of Pharmacology. He has spent 33 years of his working life, principally as a research scientist, within the University funded by a series of research awards and fellowships, including a Beit Memorial Fellowship, a Fellowship from King's College, Cambridge, and research awards from the Wellcome Trust and Rhône-Poulenc Rorer. The latter provided $6 million for a three-way research collaboration between Imperial College, Cambridge University, and the company to work on the TeknoMed Project, whose aim was to develop novel computational technologies for de novo drug design to handle the avalanche of genomic material. The Drug Design Group in Cambridge was responsible for the development of novel algorithms in biomolecular structural problems. In April 1999, when the commercial value of this approach began to be clear, Dr Dean co-founded De Novo Pharmaceuticals, an in silico drug design company, with Dr David Bailey. De Novo spun out of the University in April 2000 with £2 million backing from a local consortium of venture capital companies (Avlar BioVentures, Prelude Ventures, and CRIL) and raised a further £16.75 million in 2001. The company has several research collaborations with major pharmaceutical, biotech companies, and research institutions, pursues an active publication policy for research and development, and has research links with a number of University of Cambridge departments.

© 2007 Elsevier Ltd. All Rights Reserved
No part of this publication may be reproduced, stored in any retrieval system or transmitted in any form by any means electronic, electrostatic, magnetic tape, mechanical, photocopying, recording or otherwise, without permission in writing from the publishers

Comprehensive Medicinal Chemistry II
ISBN (set): 0-08-044513-6

ISBN (Volume 4) 0-08-044517-9; pp. 283–305

4.14 Library Design: Ligand and Structure-Based Principles for Parallel and Combinatorial Libraries

D M Schnur, A J Tebben, and C L Cavallaro, Bristol-Myers Squibb, Princeton, NJ, USA

© 2007 Elsevier Ltd. All Rights Reserved.

4.14.1 Introduction

In the early 1990s, the pharmaceutical industry was dazzled by the possibility of combining automation, high-throughput screening (HTS), solid-phase synthesis, and sophisticated methods for identifying compounds in mixtures.[1] This was the era of huge combinatorial endeavors that ranged from tens to hundreds of thousands of compounds per library.[2–4] For peptides, library size even swelled into the millions.[4–6] These libraries, whose sizes dwarfed most typical corporate compound databases, were expected to produce unprecedented numbers of exciting new leads and perhaps even new drugs that would pop directly out of HTS as Aphrodite sprang full-grown from the waves. Obviously, such enthusiastic expectations could not be met,[7] particularly since the design that went into some of these Goliath libraries was minimal at best. As time passed, attempts were made to impose order by the application of various types of experimental design[8,9a–9c] or cell-based approaches,[10a–10c] until a new criterion that reached beyond mere numbers, diversity, became the design fad of the decade. As has often been stated in analogy to beauty, 'diversity is in the eye of the beholder' and, while many methods have been developed for diverse library design,[11] the varying definitions of diversity have remained as diverse as the concept.[12] Naturally, the expectations for diverse large libraries were the same as for the merely large ones. This was another scenario that was bound to disappoint.

In order to create a frame of reference for diversity, the concept of 'chemistry space'[13] emerged. In essence, a chemistry space is a multidimensional matrix defined by the properties of interest. These properties could be based upon various two-dimensional (2D) or three-dimensional (3D) structural descriptors of the reagents or of the products, calculated properties, or upon structure–activity relationship (SAR)/quantitative SAR (QSAR) models.

Even if one imposes a specific definition of diversity on an ultralarge library, it is exceedingly difficult to design such a library within the constraints of a full combinatorial matrix. Large numbers of compounds with similar properties tend to result because of the nature of chemical reactivity. Synthetic reality imposes limitations on the diversity of the reagents that can be used if one expects actually to get products from a given reaction. Typically, a large combinatorial library has a very densely populated core region with significant population at the edges of design space, and large regions that are either very sparsely populated or are void of compounds. The net result is a library that is much less diverse than expected. It might be useful for extracting limited SAR – if the compounds are actually isolated and their HTS results confirmed – but the overall information content of such a library is low relative to the cost of reagents for synthesis and screening, even in HTS formats. The compounds tend to be of the 'methyl, ethyl, butyl, futile' variety – a nonoptimal scenario for general lead-finding. In addition, a significant percentage of the compounds may lie outside drug space because they derive from undesirable combinations of reagents that arise in the full combinatorial synthesis matrix. Making such compounds is a waste of resources, but designing them out of the library further limits its diversity in the property space of choice.

Several solutions to the ultralarge library design issues were found. One was the design of sparse matrices or incomplete combinatorials.[14,15] This requires division of the combinatorial matrix into submatrices so that there can be greater control in the selection of reagents and so that undesirables can be omitted. While this reduces overall synthetic efficiency, it allows for both increased diversity and elimination of at least some of the extremely nondruglike products. Another solution was simply to make smaller libraries and put more effort into property-based design. In fact, the submatrix solution in one form is simply a series of smaller libraries that do not necessarily have to have the same reaction conditions.

It becomes possible with these smaller more 'flexibly designed' libraries (hundreds or a few thousand compounds rather than the tens or hundreds of thousands) to consider the possibility of incorporating absorption, distribution, metabolism, excretion, and toxicology (ADMET) properties, to focus on a receptor target family, or even to design the library for a specific target. As the 1990s progressed into the early twenty-first century, combinatorial chemistry trended away from ultralarge diverse libraries to medium-sized (1–10 K) libraries that were designed to be more druglike, according to Lipinski's 'rule of five'[16] and subsequent analyses.[17a–17c] Using variations upon Evan's original definition of privileged substructure,[18,19] in addition to other methods, designs of target family libraries, particularly for G protein-coupled receptors (GPCRs),[20,21a,21b] but also kinases,[22,23] nuclear hormone receptors,[24] and enzymes such as serine proteases[25] became popular. This trend was greatly aided by the classification of MDL Drug Data Report (MDDR) ligands according to their gene ontology by Schuffenhauer *et al.*[26] and prompted the rise of commercial knowledge databases[27a–27f] of target family ligand information.

While large and medium-sized libraries are suitable for lead-finding and perhaps for the early chemistry space-mapping phase of lead optimization for a particular target or group of targets, small combinatorial libraries of a few hundred compounds and parallel libraries (ranging from dozens to at most a few hundred compounds) have come to play a major role in lead optimization. Because these libraries contain a limited number of compounds, design to maximize information about structure–activity relationships is essential. Experimental design methods that received limited acceptance for large library design are being reexamined.[28] These design methods may also incorporate QSAR[29] and/or ADMET[30] property models.

Another trend which has significantly impacted library design is that of 'small is beautiful' or 'lead-like' libraries.[31,32] Lipinski *et al.*[16] have pointed out that combinatorial chemistry has produced large numbers of increasingly large (molecular weight (MW) greater than 500 and often greater than 700) lipophilic compounds (as demonstrated by high ClogP[33]). These compounds are poor starting points for medicinal chemistry lead optimization because MW and $\log P$ usually continue to increase as potency and selectivity are built into the molecule. As a result, these compounds leave little room for optimization before reaching prohibitively high weights or $\log P$ values. Oprea and others[34,35] have advocated a 'lead-like' approach that starts with smaller compounds that can be optimized through traditional medicinal chemistry methods for increasing potency. Libraries of such leads can be designed for screening and the hits may be used as the basis for subsequent parallel or combinatorial library design as part of the medicinal chemistry optimization process.

4.14.1.1 Design Considerations for Libraries of all Sizes and Types

When designing a combinatorial or parallel library, it is essential to consider the intended use of the library. The library size and makeup of a general screening library are different from that of a target family library, which is, in turn, different from a library intended for a single target or for optimization of a lead. Clearly, the largest library type is the general screening or corporate deck enhancement library. These are most likely to be based on some definition of molecular diversity, but the nature of that diversity will be dependent on whether the library centers on a single scaffold or around a diverse set of scaffolds. The former is mostly likely to be a medium-sized library on the order of 1000–2000 compounds, whereas the latter may range in the low tens of thousands. Similarly, a target family library is likely to be diverse but focused around a set of so-called privileged substructures or around a set of pharmacophores. Typically, these libraries are designed to be in the 10 000–30 000 compound range. Again, if they center on a single scaffold or privileged substructure they will be an order of magnitude smaller. This will depend on whether the library is intended for general target family screening, including de-orphaning of receptors, or if the intent is to enhance a corporate collection. If such libraries are focused on a few receptors the size will obviously be reduced. Libraries for lead optimization generally tend to be on the order of 1000 compounds or less if they are combinatorial. While these libraries may have elements of diversity, they should be designed to yield SAR information if at all possible. More commonly for lead optimization, the library is synthesized in parallel format and size ranges from tens to at most a few hundred compounds. These are extremely SAR-driven libraries and often employ classical SAR approaches such as that of Hansch and Wilson[36] or Topliss.[37] Alternately, they may be designed using statistical experimental design approaches.[9a–9c,10a–10c,28,38]

Whether a library is based on diversity or focused on one or more targets or scaffolds, it is also necessary to consider what constraints will be imposed by ADMET properties. Commonly, cutoffs for MW and log*P* that are based on the application of Lipinski's rules are employed for all library types. Frequently the cutoff of MW < 500 and log*P* < 5 are relaxed to MW < 600 or sometimes 700 and log*P* < 6 or 8 for full combinatorial libraries. There are, of course, two ways of applying these cutoffs: either the cutoffs are applied to the products in a virtual library or the initial reagent lists are culled. The former is preferred since culling reagent results by MW may result in the loss of interesting products that arise from combinations of very small and large reagents.[15] Other ADMET-related properties may also be applied to the library design, particularly if solubility, adsorption, blood–brain barrier (BBB) or other ADMET-based QSAR models[29] are available. In general, such models are not applied to general screening libraries, but their use is becoming increasingly important for lead optimization libraries.[39–41]

Library design methodologies, whether diverse or focused, fall into two basic categories: (1) reagent-based selection methods; and (2) product-based selection methods. While product-based approaches are clearly optimal,[42] synthetic chemists work in reaction space and therefore are concerned with the selection of reagents from commercial or in-house custom sources. Additionally, virtual libraries of products even for single scaffolds may number in the millions or billions of compounds. Methods such as those employed in Pearlman's DiverseSolutions[43] and Schrodinger's CombiGlide[44] provide a compromise of 'reagent biased' product selection, thereby gaining the advantages of both methods.

For small or parallel libraries intended for a single target, combinatorial docking may be a viable focused design option – provided a suitable receptor structure and scoring function exist. The size of the virtual library will also be a determining factor unless the reagent-biased approach (such as that underlying CombiGlide) is employed. When suitable receptor structures are unavailable, ligand-based methods such as pharmacophores or BCUT chemistry spaces with defined active cells typically used for larger libraries may be employed, in addition to QSAR, Hansch, or Topliss-based methods.

All libraries are constrained by available reagents and the synthetic methods employed. An essential part of virtual library generation and, if required, full library enumeration, is preparation of the reagent lists. Known nonreactive and multireactive reagents have to be removed, in addition to any compounds in the list that inadvertently do not contain the appropriate reactive moiety. While many enumeration tools for virtual libraries exist, one of the most popular and most thorough in this regard is OptiveBenchware.[45] The user constructs a reaction with files of reagents in SMILES,[46] SYBYL Mol2,[47] or other MDL SD[48] formats, assigns R-group positions on the 'first' example of each reactant and draws the product with the associated R-group positions (either via ChemDraw[49] or ISISDraw[48]). The user can then edit the reactant lists manually to exclude undesirable reagents in the list or do it automatically through toggle selection from an extensive list of undesirable/desirable fragments that may occur in the reactants. Multiple occurrences of the reactive center can be included or excluded as desired. Sample reactions with the user's choices from the reactant lists can be viewed to verify that the correct products will be enumerated.

4.14.2 Design of Combinatorial and Parallel Libraries

Since there is a great deal of overlap in the methods used to design large and small combinatorial libraries, it is convenient to divide most of the various methods into ligand-based and structure-based methods rather than by library size (*see* 4.15 Library Design: Reactant and Product-Based Approaches for a division based on reactant versus product-based methods; 4.26 Seven Transmembrane G Protein-Coupled Receptors: Insights for Drug Design from Structure and Modeling for library design for GPCR targets).

4.14.2.1 Ligand-Based Methods for Combinatorial and Parallel Libraries

Most of the methods available for ligand-based combinatorial library design have been extensively reviewed.[50,51] Most will be described briefly for purposes of completeness. Emphasis will be placed on newer literature and less reviewed methods.

4.14.2.1.1 Descriptor and property-based diversity methods, principal component analysis, clustering, and design of experiment

The use of property descriptors was among the earliest methods applied to the creation of diverse compound libraries. Since many ligand-derived properties, such as MW, $\log P$, fingerprints, atom pairs,[52] pharmacophores,[53] MolconnZ[54a,54b] topological, and E-state descriptors,[55] are intercorrelated, it is necessary to use methods such as principal component analysis (PCA) to create orthogonal axes. Once these axes have been defined it is possible to use partial least-squares (PLS) regression to define the property space, then distance-based selection of maximally diverse compounds is performed. This type of methodology is exemplified by the tools developed by Hassan *et al.*[56] for Cerius2.[57] Later modifications of the method imposed ADMET constraints such as Lipinski's 'rule of five' to constrain compound selection to 'drug space.'[58]

Another early property-based approach drew upon classical experimental designs, such as full-factorial[10a–10c] and D-optimal[8,9a–9c] designs, to sample diversity space of either the ligands and/or the products. Statistical experimental design is commonly used for process chemistry and other applications where multiple variables have to be optimized simultaneously. The application to analog synthesis was proposed by both Austel[59] and Brannigan *et al.*[10a–10c] Using a two-level, three-parameter full factorial as an example, one chooses a set of properties and assigns a threshold cutoff to each. One can now assign each molecule to a 'bin' according to its properties thus:

	prop1	prop2	prop3
moleculeset1	+	+	+
moleculeset2	+	+	−
moleculeset3	+	−	+
moleculeset4	−	+	+
moleculeset5	−	+	−
moleculeset6	−	−	+
moleculeset7	−	−	−

Selection from the bins may be performed manually by a chemist if a relatively small set of reagents is being sampled (R-group selection) or in an automated fashion by any of a variety of possible methods for products or large reagent sets. In principle, this method allows the user to constrain diversity space by omitting bins that represent nondruglike regions of the property space. It also allows the user to select additional similar compounds or reagents based upon the biological response found for the library or array. In practice, it is necessary to expand the number of 'threshold cutoffs' to allow finer sampling of the properties of interest rather than perform repeated iterative selections for combinatorial libraries. This leads logically to cell-based diversity analysis, particularly for dealing with large virtual libraries, and which will be discussed below.

The statistical design of experiment (DoE) methods, however, are well suited to the design of array libraries for SAR exploration. In this scenario, the chemist designs a monovariate or 'one-by' library. Since only one position is being optimized, the compound selection is easily done in 'reagent space' rather than 'product space.' While the array is monovariate, in the sense that only one R-position is being evaluated, it is in fact multivariate in properties of interest to the chemist: steric bulk or hydrophobicity, hydrogen bonding, polar surface area, etc. Thus, DoE-based approaches provide a reasonable alternative to or complement the traditional medicinal chemistry approaches of Hansch[36] and Topliss.[37] It must be pointed out, however, that the results of such arrays should be used with caution if the chemist plans on a 'best-of-best' approach to optimizing R-groups. It does not necessarily follow that the results of varying different positions should be additive. As shown in **Figure 1**, the optimal combination of R-groups could easily be missed if a combinatorial approach is not applied, either via a combinatorial library or through synthesis of multiple

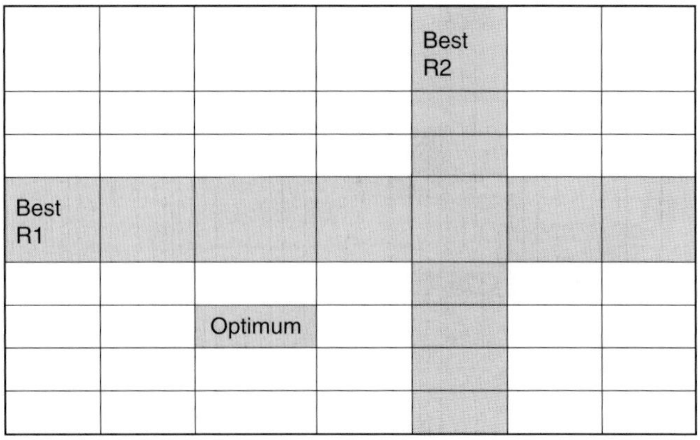

Best R2

Best R1

Optimum

Figure 1 The dangers of monovariate library design using a best-of-best strategy. Properties are nonadditive and the optimum compound is missed.

arrays. Wold and others[8–11] have frequently pointed out that multivariate designs find optimal experimental conditions and compounds missed by traditional one variable (or R-group) at a time (OVAT) approaches historically used for medicinal chemistry optimization. Wold developed tools, including MODDE,[60] for compound design in multivariate scenarios. It has been demonstrated that this approach can uncover nonadditive SAR effects that would normally be missed by the traditional OVAT[61,62] design. Clearly, these methods provide a huge advantage over OVAT library approaches for small, diverse libraries. Their use in the design of parallel and combinatorial libraries has been on the increase.[28] Naturally, scenarios where variation of only a single substituent position may appear additive sometimes arise. For serine proteases, for example, moieties such as hydroxamates in matrix metalloproteinases (MMPs)[63] and the benzamidine for serine proteases[64] determine receptor binding and OVAT optimization of groups extending into the other pockets can be effective. Such an approach is much more risky in scenarios where the binding mode is unknown and no 'warhead' moiety is known to dominate ligand binding to the receptor.

Other property-based methods for diverse selection, such as fingerprints, distance-based clustering, Kohenen maps, and spanning trees, have been reviewed at length.[65] These methods will be not be described here (*see* 4.20 Screening Library Selection and High-Throughput Screening Analysis/Triage).

4.14.2.1.2 Cell-based methods for diverse and focused libraries

While the experimental design strategies described above do, strictly speaking, fall into the category of cell-based methods, the most commonly applied method is exemplified by DiverseSolutions[43] from the Pearlman group at the University of Texas at Austin. Much of the library design functionality, particularly for focused libraries, has subsequently been implemented in OptiveBenchware.[45] Although cell-based methods and the use of DiverseSolutions have been reviewed elsewhere,[66] they are worth discussing in detail with regard to focused design. Cell-based methods divide each axis of a multidimensional property space into bins and thereby divide the space into hypercubes or cells. The known ligands of individual targets or target families can thus be associated with the cells they occupy (**Figure 2**).

One of the basic assumptions regarding descriptor-based methods is that actives possess common properties that cause them to cluster in a 'chemistry space.' Minimally, unknown actives should be found in the same region of chemistry space as the known ligands and, if the descriptors are truly meaningful, the most active compounds should be clustered more tightly than less active compounds. Clearly, the validity of this assumption about 'nearest neighbors' is dependent on the relevance of the chemistry space descriptors to ligand–receptor binding.

One of the difficulties with distance-based methods such as the clustering of Daylight[46] fingerprints is that the result is dependent on pairwise comparisons of all the structures in the virtual library. Addition or deletion of compounds to the original set may change the number, size, and/or membership of the clusters. By contrast, cell-based property spaces, once derived, allow compound addition and deletion without altering the chemistry space. This allows not only comparison of libraries in a chemistry space, as shown in **Figure 2**, but also the definition of hot-spot regions in the chemistry space if in fact actives do cluster.

DiverseSolutions[43] employs a unique set of descriptors called BCUTS[67] that are based upon both connectivity-related and atomic properties such as charge, polarizability, and hydrogen-bonding abilities. Once BCUT descriptors of various 'flavors' are calculated for a fully enumerated virtual library, the optimal chemistry space axes that describe the

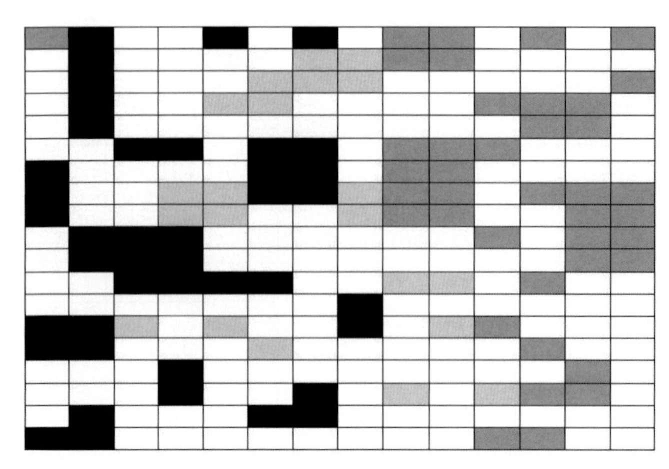

Figure 2 2D representation of two compound sets in a cell-based chemistry space. Cells occupied by set one are shown in black, by set two in dark gray, and by both sets in light gray. Empty cells are white.

library are determined by finding an orthogonal set of four to eight descriptors that distribute the compounds evenly across the space according to a chi-squared algorithm.[68,69] The space is then partitioned into cells or hypercubes by subdividing these axes.

Qualitative observations that BCUTs do appear to correlate with ligand–receptor binding and activity[10a–10c] contributed to the development and implementation of the concept of receptor-relevant subspaces[70] in the DiverseSolutions software package. This concept is so integral to Pearlman's approaches that it needs to be discussed in some detail. Qualitative observations such as those reported by Schnur[10a–10c] and about clustering of actives in 2D or 3D subspaces of five- or six-dimensional chemistry spaces led Pearlman and Smith[70] to develop a novel algorithm for reducing chemistry space dimensionality. (An example of such clustering is shown in **Figure 3**.) Whereas typical methods that reduce dimensionality discard potentially important information, this algorithm identifies which axes (metrics) convey information that may be related to the affinity for a given receptor and also identifies those axes that appear irrelevant and therefore may be safely discarded. This is done by identifying the axes that tightly group active compounds based upon a cluster-breath normalized value of chi-squared, computed either from a simple count of actives per bin along each axis or from an activity-weighted count of those actives. Multiple binding mode possibilities are addressed by allowing more than one cluster per relevant axis. Receptor-relevant subspaces are useful not only for easy graphical visualization of active compounds in the chemistry space, and therefore a visual validation of the chemistry space, but more importantly for the calculation of receptor-relevant distances which are essential for identifying near neighbors of actives in focused library design and for comparing libraries.

Stewart *et al.* have used DiverseSolutions to find a receptor family relevant chemistry space for nuclear hormone receptors that distinguished 907 known nuclear hormone receptor (NHR) ligands from other inactive compounds.[71] Clearly, the next logical step is to use such a space for library design and potentially scaffold-hopping.

In addition to using receptor-relevant subspaces as a design method, Pearlman implemented a list-based nearest-neighbor searching algorithm within DiverseSolutions[14] which can also be used for focused combinatorial library design. Currently available versions of DiverseSolutions and of Pearlman's OptiveBenchware application offer several other novel library design options for focused library design. Ideally suited for target-based library design is a unique cell-based 'fill-in' library design option. A set of known active ligands is used to identify 'promising cells' in chemistry space. The cells sets of promising cells consist of the ligand-occupied cells plus the adjacent surrounding cells up to a user-specified cutoff distance, as shown in **Figure 4**. The chemistry space may have been derived either from a target or target family knowledge database of ligands, the virtual library from which the combinatorial library will be designed, or from a standard corporate chemistry space. A reactant-biased product-based library design algorithm[14] is then used to design a library, of whatever desired size, which best fills these 'promising cells.' The degree of target focus is controlled by the number of bins per axis and the number of cell radii from the known ligand is used to define the size of the 'promising cell.' The focused design approach in DiverseSolutions and OptiveBenchware uses a set of target ligands to score all the compounds in the virtual library based on their distance from the actives and then selects a designed library that optimizes the average virtual activity. An example of using this method to select GPCR compounds for screening to validate the library design approach has been reported by Wang and Saunders.[72] The algorithm also permits the use of externally determined activity scores such as those from docking, QSAR models,

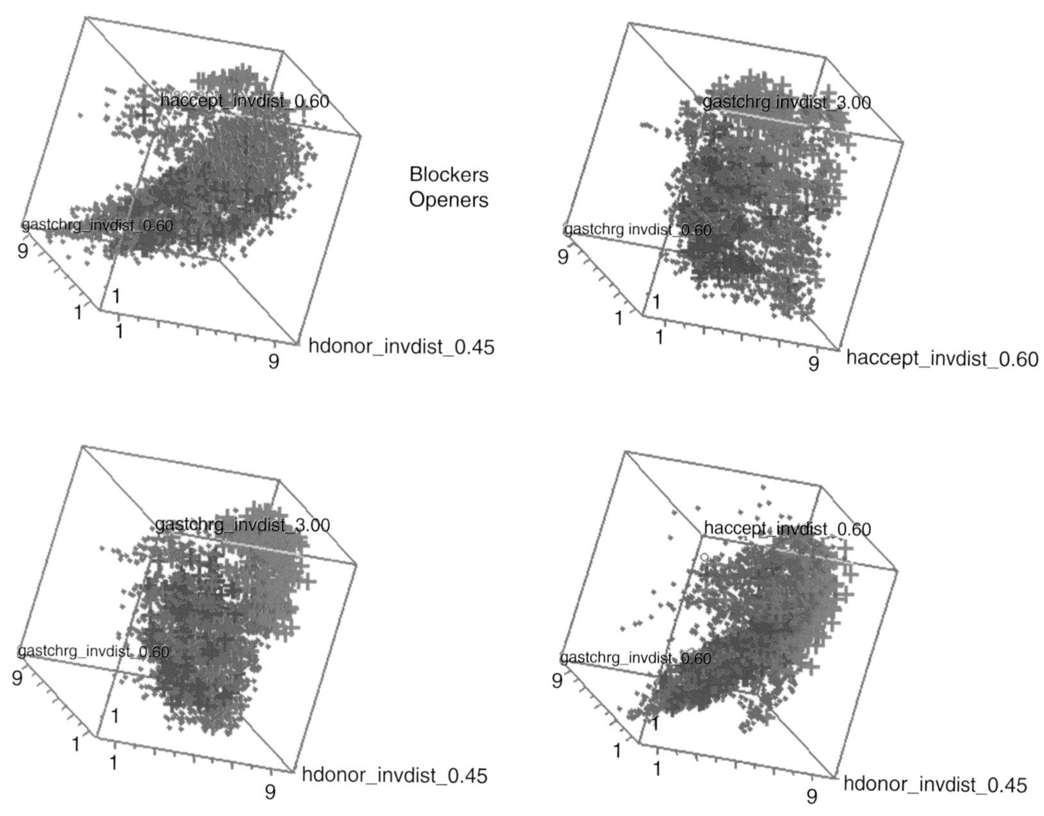

Figure 3 An ion channel-based combinatorial library in 3D subspaces of a 4D DiverseSolutions chemistry space. Channel openers are shown in black, blockers in medium gray, and the rest of the combinatorial library is shown in light gray. Axes are 3D H–hydrogen-suppressed BCUTs based on Gastiger charge, polarizability, and hydrogen bond and acceptor properties. Separation of the openers and blockers is better in some subspaces than in others, suggesting the possibility of a receptor-relevant subspace.

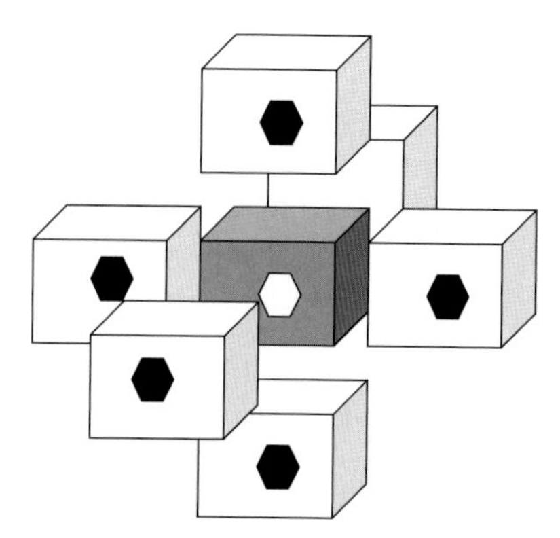

Figure 4 Representation of 'interesting' or promising cells. The interesting cell can be defined as the cell containing the active ligand (gray cell) plus the adjacent cells. One layer is shown, but the number of layers controls the focus or fuzziness of the region defined as interesting.

pharmacophore models, or other sources. Additionally, both DiverseSolutions and OptiveBenchware offer a novel focused/diverse library design option which yields products that are focused with respect to receptor-relevant axes of the chemistry space and are diverse with regard to the receptor-irrelevant axes. Use of this algorithm is best limited to individual targets or at best closely related targets since the ligands for an entire family of receptors are unlikely to have the same receptor-relevant and irrelevant axes.

The receptor relevance of BCUT descriptors has inspired several groups to apply them in conjunction with other methods. Mason and Beno reported the use of simulated annealing to optimize library design using BCUT chemistry space and four-point pharmacophores concurrently[73] and have used chemistry spaces in conjunction with property profiles (Beno and Mason, unpublished results). Pirard and Pickett reported the application of the chemometric method PLS discriminant analysis with BCUT descriptors to classify successfully adenosine triphosphate site-directed kinase inhibitors active against five different protein kinases.[74] Manallack *et al.* used BCUTS as input parameters for neural networks to select compounds that targeted specific gene families.[75] Their training sets were derived from the MDDR[76] and included three classes: (1) protein kinase inhibitors; (2) GPCR class A biogenic amines; and (3) class A peptide-binding type GPCRs.

4.14.2.1.3 Privileged substructures and receptor target family design

As can be noted from the discussion of applications of cell-based methods, a popular type of library design is that of receptor target class or target family design. In fact, creation of combinatorial libraries directed toward families of receptors such as GPCRs, kinases, nuclear hormone receptors, and proteases has largely replaced the generation of libraries based primarily on diversity. Methods for generating receptor target family libraries are basically the same as those used for other focused libraries and have been reviewed previously.[66]

One method that is specific to receptor target family library design is the use of 'privileged substructures.' The original concept was put forth by Evans *et al.*[18] and reviewed by Patchett and Nargund.[19] They described privileged substructures as those found in ligands across a set of diverse receptors. Further elaboration of the privileged substructure was postulated to lead to selectivity toward a specific target receptor. While Evans and Patchett evolved this concept within a relatively narrow class of GPCR ligands, subsequent literature methods for finding privileged substructures and common practice in combinatorial chemistry library design not only expanded on their original analysis, but also modified the definition of 'privileged structure' to that of commonly occurring fragments within ligands associated with a target receptor family.

Clearly, the term 'privileged substructure' has taken on a meaning beyond Evans' original intent. It has become identified with those substructures found to be promiscuous within a given target family and carries the implication that these substructures are specific to that target family. If such substructures exist, off-target affinities might be avoided early in the discovery process and thereby avoid complications as promising compounds are developed into drugs. If these substructures can be identified, they potentially provide cleaner starting points than do the more promiscuous structures.

Various methods have been employed to find so-called target family privileged substructures that are postulated to be selective for a given target family, but promiscuous within that same family of targets. The majority of the commonly used methods are ligand-based and include frameworks analysis,[77] four pointf pharmacophores,[78] and ClassPharmer substructure class generation.[79,80] These fragments have been used not only for target family combinatorial library design,[21a,21b,66,81] but also for virtual screening[82] and for focused screening deck design.[83]

Typically these target family privileged structure analyses have attempted to find minimal ligand substructures that occur frequently within the target family. However, this can very easily lead one away from truly privileged substructures and toward those which are merely 'druglike' and/or promiscuous protein binders. Consider the comparison of the often-cited[66,77,78,80] GPCR privileged substructure, biphenyl, and its analog 2-tetrazolobiphenyl. A substructure search of the 2004 version of the MDDR finds that 2-tetrazolobiphenyl appears in 1046 compounds, all of which fall into the activity classes related to the angiotensin II receptors. The biphenyl substructure is found in 5658 compounds spanning 311 activity classes, which include a significant number of GPCRs, but also a host of other targets. Although biphenyl may be classified as privileged due to its frequent appearance in GPCRs, it is clear that true privilege does not arise until the tetrazole moiety is included. Biphenyl itself is likely only to be a privileged protein binding element.

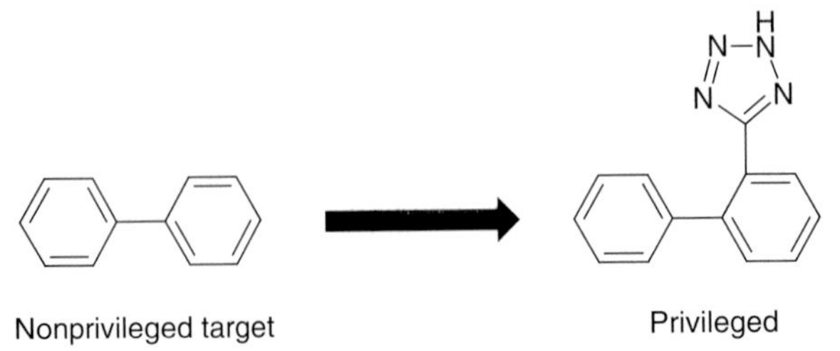

Nonprivileged target

Privileged

While some of the literature studies involving target family privileged substructures compare the fragment occurrence frequency of GPCR privileged substructures with nonGPCRs as a whole among known drugs,[22] little analysis has been done on the selectivity of these substructures with respect to other target families.[66,80] In part, this has been due to the difficulty of collecting or extracting the target family ligand sets from commercial drug databases and corporate collections. Recently however, the publication of an ontology of pharmaceutical ligands by Schuffenhauer *et al.*[26] removed this roadblock by mapping MDDR activities to published target ontologies. Additionally, the need for target family knowledge databases to drive target family-based library design has resulted in a number of commercial target family databases from companies such as Aureus, Jubilant, Sertanty, and Biowisdom.[27a–27f]

Recently, Schnur and co-workers[84] demonstrated that care must be taken in target family library design to ensure that the privileged substructures chosen are truly selective for the target family of interest. In that study, ClassPharmer was used to generate substructures for MDDR-derived ligand sets of class A GPCRs, nuclear hormone receptors, kinases, ion channels, and serine proteases. The compound sets for each receptor family were filtered through the substructure sets for each other receptor family. Both the percentages of occupied substructures for each cross-filtering and the percentages of compounds from each set that occupied other target family substructure sets were examined. Examples of GPCR 'privileged substructures' that were not selective for just the GPCR receptor family were provided.

Nonetheless, privileged substructures, if used as Evans *et al.* originally defined them,[18] can provide useful starting points for library design. Undoubtedly the design literature will continue to be enriched by further examples using not only GPCRs, but other target classes as well. As Evans *et al.* pointed out, it is the elaboration of the privileged substructure that provides target selectivity.[18] Whether a library can be designed that is selective for an entire receptor family is a subject for debate.

4.14.2.1.4 Pharmacophore and atom pair-based methods

3D pharmacophores and their topological equivalents atom pairs[85] and topological torsions (**Figure 5**) are well-established methods of compound selection and library design. Because they are rapidly calculated and easily compared, fingerprints derived from these descriptors are readily applied to the comparison and selection of compound libraries. Several recent reviews have detailed the early work in their use for diversity, focused, and target class library design.[86,87]

The use of 3D pharmacophore descriptors for library design is exemplified in the design of a set of GPCR-targeted libraries based on Ugi chemistry.[78,88] A unique feature of the design was the incorporation of a GPCR privileged substructure in each of the combinatorial products. GPCR privileged substructures were defined as chemical moieties that occur with high frequency in the ligands of multiple GPCRs, like biphenyl tetrazole, indole, and biphenyl-methyl groups.[89] In this example, 502 compounds from the MDDR[90] which were active against a GPCR target and also contained a biphenyl tetrazole moiety were used to generate a 'privileged' four-point 3D pharmacophore fingerprint. A privileged four-point 3D pharmacophore is a four-point 3D pharmacophore in which one of the four molecular features is a privileged moiety, and the other three are members of the standard set of feature types (H-bond donors, acids, etc.) In this case, the privileged feature was represented by the centroid of the biphenyl tetrazole moiety in each compound. The four-point 3D pharmacophore fingerprint for the set of 502 GPCR ligands was the union of the fingerprints of the individual molecules, and represented ~161 000 privileged four-point 3D pharmacophores. Utilizing

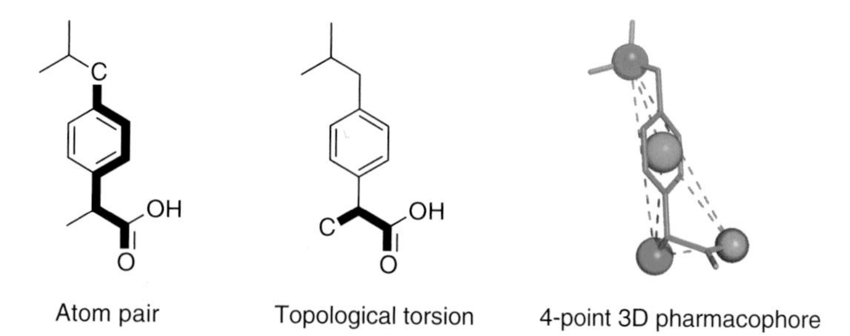

Figure 5 Examples of atom pairs, toplogical torsions, and 3D pharmacophores. The atom pair descriptor is composed of a pair of atoms on a molecule and the number of bonds separating them. Topological torsions are a subset of atom pairs as the atoms can only be separated by three contiguous bonds. Pharmacophores are constructed from pharmacophoric points mapped on to a 3D representation of a molecule and the distances between them. A binary fingerprint can be derived from any of these descriptors by enumerating all possible combinations and setting those bits corresponding to those descriptors present in the molecule.

a simple greedy algorithm, a set of 22 acid reagents (along with 12 aldehydes and eight isonitriles, to yield 2112 products) was selected to maximize the intersection of the privileged four-point 3D pharmacophore fingerprint of the combined combinatorial products with the GPCR privileged 3D pharmacophore fingerprint derived from the MDDR compounds. Approximately 49% of the GPCR privileged 3D pharmacophores found in the GPCR privileged fingerprint were covered by the products of the optimized reagents. Subsequent libraries were designed to cover the four-point 3D pharmacophores present in the GPCR reference fingerprint that were not covered by the original library.

Lamb and co-workers[21a] performed a similar study with typical two-, three-, and four-point pharmacophores. Utilizing a set of 3321 GPCR ligands with reported in vivo activities, a set of 1.8 million two-, three-, and four-point 3D pharmacophores present in at least 10 of the active compounds were identified. An ensemble was constructed and used to score a virtual library, leading to the selection of 7865 products that covered 66% of the 1.8 million GPCR-relevant 3D pharmacophores. The work demonstrated that including pharmacophore frequency count from the analysis of known active ligands improves the odds of identifying 3D pharmacophores that are actually relevant to ligand/receptor binding.

In addition to the analysis of active compounds, incorporating inactive compounds often improves the discriminating power of a pharmacophore ensemble. Utilizing a set of 43 known α_1-adrenergic receptor ligands with K_i values <5 nM (actives), and a set of 62 compounds with K_i values >5 μM against α_1-receptor subtypes (inactives), Bradley and co-workers derived a 3D pharmacophore ensemble model consisting of 500 two-, three-, and four-point pharmacophores with the highest 'information content'. The information content of each pharmacophore is a measure of its ability to distinguish actives from inactives. Those pharmacophores present in several actives, but in few of the inactives will have greater information content than those present in both. The model correctly identified 80% of α_1 actives and falsely identified only 10% of the inactives as active.[91] Bradley and co-workers used the 3D pharmacophore ensemble model to filter a virtual combinatorial library of 3924 N-substituted glycine peptoids[92] containing three known α_1 actives down to a set of 639 products. Using a 'cut-down' technique, a 160-compound combinatorial library was designed in which the number of compounds that passed the ensemble model filter was maximized. This library contained two of the three known actives present in the original 3924-compound virtual library. This represents a substantial enrichment ((2 actives/160 products) $\times$ 100 = 1.25% versus (3 actives/3924 products) $\times$ 100 = 0.076%).

Beno and Mason reported an alternative design approach based on 3D pharmacophore frequency counts that used the same set of 43 known α_1 ligands, and a virtual library of 10 648 N-substituted glycine peptoids.[92,93] The virtual library contained at least three products known to be active at α_1. While Bradley's analysis only accounted for the presence or absence of a pharmacophore, this work also included a frequency of compounds presenting a given pharmacophore. A library of 343 products (7 $R_1 \times 7 R_2 \times 7 R_3$) was selected with a simulated annealing procedure that maximized the similarity of the normalized four-point 3D pharmacophore frequency distributions of the active α_1 ligands and the products comprising the selected virtual library subset. In this case, one of the three known α_1 actives was found in the final library, suggesting that pharmacophore frequency can enhance the discriminating power of the ensemble. This finding was recapitulated in a recent report from Good *et al.*[94] A database was constructed that contained 38 factor Xa inhibitors of 19 chemotypes amongst 9524 randomly selected compounds. Two-, three-, and four-point pharmacophores were generated using Chem-X. From these, a binary fingerprint was generated as well as normalized and filtered fingerprints. The normalized fingerprints were created by translating the pharmacophore bit into a floating point value that reflects the fraction of conformations in the conformational ensemble in which the bit is present. Those that appeared in fewer than half of the conformations were eliminated to create the filtered fingerprint. Virtual screening of the database revealed that the normalized and filtered fingerprints returned approximately twice as many chemotypes as the simple binary fingerprints.

A crystal structure of the receptor can greatly reduce the uncertainties in the derivation of pharmacophores. Several programs, like GRID,[95] SPHGEN,[96] MCSS,[97] and LIGSITE,[98] probe a protein active site and provide a qualitative picture of those regions likely to bind hydrophobic, hydrophilic, and hydrogen-bonding groups. Several groups have extended this analysis toward the derivation of pharmacophores from protein active sites. These pharmacophores can then be utilized in the same manner as those gleaned from ligand-based techniques. Eksterowicz *et al.* reported the use of protein active site-derived pharmacophores as a starting ensemble for informative library design.[99] The site map is constructed from the union of three maps: (1) electrostatic site points (positive, negative, hydrophobic); (2) hydrogen-bonding (donor, acceptor); and (3) aromatic. The electrostatic points are initially placed with the putative active sites with spheres (PASS)[100] algorithm and then colored based on the electrostatic character of the protein. Hydrogen-bonding points are placed by projection of complementary site points from the protein hydrogen-bonding atoms. These points correspond to either the idealized or bifurcated hydrogen-bonding geometries and are removed if a steric clash is found. Aromatic points are placed by repeatedly docking benzene into the binding site. Site points are positioned at the centroids of the top scoring orientations of the benzene rings. From the protein site map, a two-, three-, and four-point pharmacophore fingerprint is generated and used as the starting point for informative library design. The method was validated using a 3800-compound cyclin-dependent kinase 2 (CDK2) data set containing 22 distinct scaffolds. Utilizing a fingerprint derived from 23 CDK2 structures, 21 of the 22 scaffolds were found, 12 of which contained an active structure. An enrichment of 1.7 was reported after four rounds of informative design as compared to 1.1 for DockIt/PLP[101] and 2D.

The SitePrint[102] methodology also derives a 3D pharmacophore descriptor from a protein-binding site, but does so via docking of molecular fragments. The initial set of potential pharmacophore site points is generated with the SPHGEN program within 10 of a user-specified active site residue. The active site is then probed with a set of 18 molecular fragments and those with favorable DOCK force-field scores retained. From those that survive, the best scoring fragment is kept and those within 0.5 deleted. This reduced set of fragments is clustered to produce 10 pharmacophore descriptors consisting of between five and 15 site points. The authors demonstrated that protein pharmacophore descriptors were consistent with those derived from ligands in human immunodeficiency virus (HIV) protease and thrombin. The method was also used to classify a set of 29 proteins which included kinases, nuclear receptors, aspartyl proteases, cysteine proteases, serine proteases, and metalloproteases. The rate of correct classification from the set of 29 proteins ranged from 25% for cysteine proteases to 80% for aspartyl proteases. Although no direct application to ligand design is described, the pharmacophores deduced from SitePrint could easily be incorporated into the methods described above.

The LigandScout[103] method derives pharmacophores from ligands bound into protein active sites and provides a direct link into database searching as it produces pharmacophores that are compatible with the Catalyst[104] program. Because the accuracy of these pharmacophores is dependent on correct ligand structure, the LigandScout procedure begins with chemical perception. The perception routine first detects the appropriate hybridization state of the ligand atoms, followed by assignment of Kekule patterns to functional groups and positioning of double bonds in sp^2 chains. Pharmacophores are then generated from the corrected ligand structures that contain lipophilic, hydrogen bond donors, hydrogen bond acceptors, positive ionizables, and negative ionizables. Since the pharmacophores are in the context of a protein active site, exclusion regions can also be included. Pharmacophores derived from multiple binding sites can be aligned and merged into a single consensus probe. The authors include two examples of the application of LigandScout, human rhinovirus (HRV) serotype 16 and BCR-ABL tyrosine kinase. In the first example, a common pharmacophore model was derived from three complexes consisting of three lipophilic points, two aromatic lipophilic points, one H-bond acceptor and 22 excluded volumes. Three databases were screened with the pharmacophore: a single and a multiple conformer database of ligands extracted from the Protein Data Bank (PDB) and a collection of small molecules sourced from Maybridge. In the single conformer PDB database, eight hits were found, four of which are known HRV binders. A search of the Maybridge database returned 67 hits, but these were not verified by assay. The BCR-ABL pharmacophore was also produced from three complexes, containing four aromatic points and two acceptors and eight excluded volumes. The seven hits found in the PDB single conformer database were all known BCR-ABL-inhibitors. Nineteen hits were returned from the Maybridge search.

The pharmacophore methods described above attempt to mimic the steric environment of the receptor through the use of exclusion volumes. While this can capture the boundaries of the active site, it is much better represented by a description of its shape. Active site shape is implicit in programs that dock ligands into protein structures, but shape screening is also applicable to small molecule screening. Two approaches have been commonly applied in molecular comparison: (1) quantification of shape overlap; and (2) similarity of shape-derived fingerprints. Historically, the utility of shape overlap in the screening of large databases has been hindered by computational expense. However, representation of the shape by a set of overlapping gaussians[105,106] has allowed for comparison at a speed relevant to database screening and library design. The commercially available program ROCS[107] enhanced the algorithms described by Grant *et al.*[106] and implemented them in an easily accessible manner. ROCS aligns and compares the shape of a query molecule to a multiconformer database of candidate molecules and scores the overlay. The overlay can be scored by shape similarity, either simple overlap or as incorporation of chemical 'color' ascribed by the nature of the atoms which comprise the shape. On a modern computer, ROCS is capable of thousands of comparisons per second. ROCS shapes have also been cast into a shape fingerprint.[108] The bits in the fingerprint correspond to a set of reference shapes created from a subset of the molecules in the Cambridge Structural Database (CSD) or MDDR. Shapes of new compounds are compared to the reference shapes and bits turned on if they meet a threshold similarity. Once the fingerprints have been created, typical binary similarity metrics can be used for rapid database screening. The authors observed robust enrichment when searching across the CSD database and noted a speed similar to 2D topological methods.

Putta *et al.*[109] have described a shape that includes pharmacophoric features within the shape and represents the shape/feature space as a binary fingerprint. The process begins by aligning conformations on to a grid, deriving a shape from the grid occupancy, and building a shape catalog of nonredundant shapes. The shape of the first conformation is always kept, and as new conformations are placed on the grid, they are rotated and jiggled to ensure accurate shape comparison. The orientation can be directed by specification of a key feature, which then becomes the origin of the molecular position. For each grid point within the shape, the presence of a pharmacophoric feature is evaluated and captured. The end result is a combined descriptor set containing both shape and the location of pharmacophoric features within it. Once the shape catalog is completed, each molecule is represented by a binary string that captures both the shapes that it can attain and the presence and location of pharmacophoric features within those shapes. These fingerprints are then used for similarity comparison and can be utilized in the informative design process.[110] The method was evaluated with a thrombin data set consisting of 38 thrombin inhibitors embedded in 2418 chemically diverse inactive compounds.[111] A 327-member shape catalog was created from the active compounds and the shape signatures included a positive charge as the key feature. From these, the most informative 500-shape feature descriptors were selected as the ensemble. A 35 462-compound subset of MDDR was screened with this ensemble returning 181 of 540 thrombin actives, resulting in a 23-fold enrichment over random. The method was also applied to a synthetic combinatorial library of 634 compounds containing 64 actives. A modest 1.4-fold enrichment was obtained, as compared to 0.6 for molecular access system (MACCS) keys. This is indicative of the difficulty discriminating amongst a collection of topologically similar molecules.

4.14.2.1.5 Evolutionary programming in library design

The modular nature of parallel and combinatorial libraries is easily translated into a genetic representation (**Figure 6**). Once genetically encoded, it became apparent that a variety of evolutionary optimization methods would be relevant and efficient mechanisms for library optimization. These methods have shown several advantages when compared to the alternate optimization techniques typically employed in library design (i.e., D-optimal, etc). First, these techniques employ iterative sampling of a given library, operating at any given time only on a small fraction of the total population. This enables the analysis and design of very large libraries as they do not require the CPU and memory footprint of whole-library selection algorithms. Second, the function that scores the quality of the design is separate from the operations that select the subsequent sample (generation) from the previous. This disconnect removes the requirement that the method used to evolve the population sample is married to the scoring function, giving the designer the freedom to evaluate several easily. For example, the EA-Inventor program[112] from the Pearlman group is specifically designed to capitalize on this aspect. EA-Inventor uses an evolutionary algorithm to manipulate the population through chemically reasonable modifications which can then be scored via their internal scores or any external program. Although its primary intended use is as a de novo design program, it can operate in library design mode by fixing the core and allowing only R-group modification. The program can then generate the R-groups de novo or the user can specify a list of fragments.

Two evolutionary programming practices have come into common use for library design: genetic algorithms (GAs)[113] and simulated annealing.[114] The application of both algorithms begins with the encoding of the library into a set of chromosomes. These chromosomes are generally encoded such that the individual codons correspond to a specific position on the final product, leading to a unique chromosome for each compound in the library population. The selection process begins by pulling an initial sample from the population, either randomly or through some biased selection, and assessing their fitness via the scoring function. It is the production of the offspring that distinguishes a GA from simulated annealing. In a GA the next generation is evolved using mechanisms paralleling those utilized in nature: retention, deletion, mutation, and crossover. Retention is applied to the fittest parents, whose genes are copied unchanged from one generation to the next. Evolution of the weaker members can be achieved by chance through

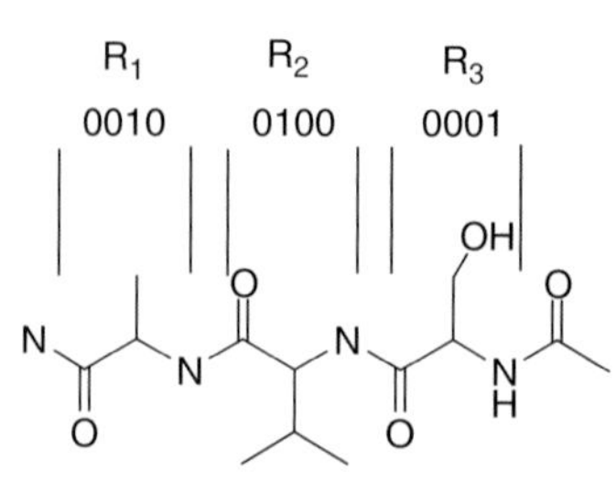

Figure 6 Genetic representation of a three-component tripeptide library. Each four-bit codon represents an individual amino acid in the library. Combination of the three codons results in a unique gene for a given compound.

random mutation, or by combination via crossover. Finally, the weakest can be eliminated though deletion from the population. On the surface, GAs appear to be quite simple. However, much of the complexity arises from their implementation and use. The quality of the solutions and how quickly the algorithm converges is dependent on many parameters, such as the initial sample size, the rate at which members are retained, deleted, or crossed over, and the crossover point. Unfortunately, it is not inherently obvious what the optimal values of these parameters are for a given problem. Nevertheless, parameters have been found enabling their effective use in library design. In contrast to a GA, simulated annealing makes small stochastic changes to the initial set. The resulting 'energy' or score determines whether the members of the new population are accepted or rejected. If a change results in a lower energy, it is accepted. The probability of those with higher energy being accepted is calculated using the Metropolis acceptance criterion:

$$p = e^{(-\mathrm{d}E/kT)}$$

where dE is the change in energy, k the Boltzmann constant, and T the temperature.

Much of the early work in the application of GAs to combinatorial library design was targeted toward optimization of a single objective. In an early proof-of-concept study, Sheridan and Kearsley[115] explored the design of a tripeptoid library scored by similarity to an arbitrary tripeptoid, similarity to a set of tetrapeptide cholecystokin (CCK) antagonists, and against an angiotensin-converting enzyme (ACE) trend vector. The authors employed a best-third method, where only the top third of the population were retained and the rest were killed. The new population was created from three copies of those retained. The first copy was unchanged, a single residue mutation was made to the second, and the third copy was crossed over within itself. The GA was able to identify the target tripeptoid from a possible library of ~20 billion compounds within 23 generations of 300 compounds. Scores of the CCK and ACE runs converged within 21 and 12 generations, respectively. This work was later extended[116] to include additional 2D and 3D descriptors. In this later study, the issue of whether a library-based approach betters a molecule-based approach was also addressed. While the library-based GA did find molecules similar to the molecule-based GA, it took many more generations and was deemed a less efficient search strategy. An interesting twist on the GA process was described by Weber *et al.*,[117] in which the scoring function was the actual biological activities. Selections of 20 were made from a 160 000-compound database of Ugi reaction products, initially a random selection, and then via GA, for testing against thrombin. Although the average EC_{50} of the first generation was $>500\,\mu M$, the GA had identified submicromolar inhibitors within 16 generations, with the most active compound found in generation 18. In a subsequent study,[118] every compound in the 15 360-compound Ugi database was synthesized and tested so as to compare the performance of the GA against a random selection. Some sensitivity to sample size was reported, but at a sample size of 20 the GA was able to identify actives 1.6–2.8-fold more quickly than random. The efficiency of GA's within very large problems has been demonstrated by selection from a potential pool of 64 000 000 natural hexapeptides. Beginning from a population of 24 randomly selected hexapeptides and using activity as the fitness function, the GA was able to advance the average inhibitory activity from 16% to 50% within six generations of 24 peptides and eventually identified a peptide identical to a previously known trypsin inhibitor.[119] A similar strategy was used for the identification of stromelysin hexapeptide substrates.[120] Increasing rates of peptide cleavage were found for the five generations of 60 peptides tested. Although only five cycles were completed, the authors suggested that better substrates could have been found had subsequent generations been produced.

While these early studies provided evidence that the GA could be a powerful tool in library optimization, they were limited to a single objective. Since single-parameter optimization scenarios are rare in the design process, several investigators have pursued multiobjective strategies. Within the GA context, the Gillet group has published the SELECT[121] and MoSELECT[122] programs. Both of these programs optimize on multiple objectives simultaneously, but do so in substantially different ways. The fitness function in the SELECT program consists of a weighted sum of the parameters of interest, with the weightings determined empirically by the user. To prevent convergence to suboptimal solutions, 'niching' is utilized. The GA is allowed to converge to a solution, the center of which is made a niche. An area around the niche is then removed from the search space and the GA continued. The SELECT program was able to find solutions that maximized diversity while optimizing rotatable bond and Andrew's binding energy profiles. The key deficiency in the SELECT program is the weighted combination of multiple objectives into a single score. The assignment of weights is arbitrary and it is often difficult to find weightings that balance the score appropriately. Also, it is possible that one or two objectives can dominate. The MoSELECT program does a true multiobjective optimization using Pareto dominance as the fitness function (**Figure 7**). By mapping the individual objectives on to the Pareto surface, a better sampling of the objective space is achieved because a single objective is far less likely to dominate the score. Niching is applied to ensure even sampling across the Pareto space. The application of this procedure was

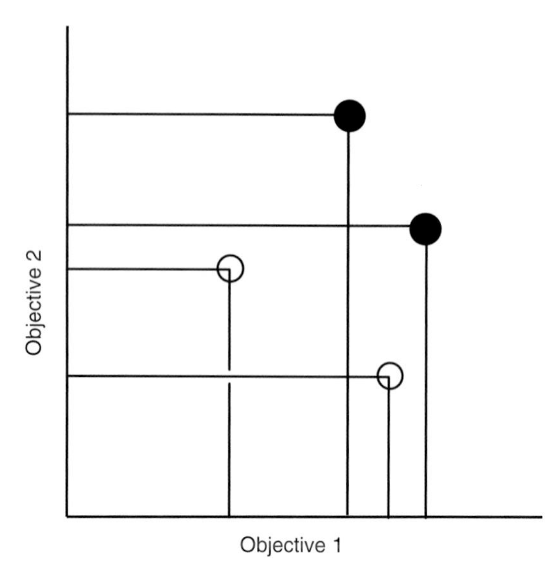

Figure 7 A two-parameter Pareto surface. In this example the desired outcome is minimization of both objectives simultaneously. The less desirable solutions, represented by the dark circles, are 'dominated' by the more desirable solutions, shown as open circles.

exemplified by simultaneously optimizing the diversity, MW, hydrogen bond donor (HBD), hydrogen bond acceptor (HBA), and rotatable bond (RB) of untargeted libraries[122] and similarity, MW, cost, HBD, HBA, and RB for a focused library.[123]

Agrafiotis[124] has shown that Pareto multiobjective optimization does not require a GA, but it works well with simulated annealing. He presented methodology that enabled the selection of multiple compound subsets in sparse array, full array, or plate format optimized against multiple objectives, including diversity and drug-likeness. Simulated annealing was also the optimizer used in the HARPick program reported by Good and Lewis.[125] HARPick puts simulated annealing in front of a very flexible scoring function that allows bench chemists to optimize on pharmacophore profile, diversity, and several molecular properties.

4.14.2.1.6 Quantitative structure–activity relationship-based methods

In general, QSAR-based methods are used to design small focused libraries, parallel arrays and, of course, traditional medicinal chemistry single analogs. While models are usually based on potency for a single target receptor, selectivity or ADME-based models may also be used in library design. The area of QSAR-based library design has been reviewed by Topliss.[37]

One of the more interesting approaches to QSAR-based library design is that employed in the SYBYL[47] comparative molecular field analysis (CoMFA) optimizer. In this approach, the 3D QSAR model derived for a training set of molecules is used as a scoring function to predict activity of new molecules proposed by random addition of fragments to specified R-group positions on a core molecule. The fragment tables used are those provided for the GA module, Leapfrog,[47] and may be edited by the user. Since CoMFA is derived by placing molecules in grid fields, the predictions are very sensitive to the placement of the core scaffold and the conformations of the R-groups. The method, while crude, does provide a complementary ligand-based approach to structure-based docking.

4.14.2.2 Structure-Based Methods for Combinatorial Libraries

In its simplest form, structure-based library design is merely a variation of well-established virtual screening techniques as described in Chapter 4.19 and reviewed by Bohm and Stahl[126] and Beavers and Chen.[127] However, enumeration and docking of an entire library are inefficient and, because of the significant time required, impractical for large libraries. Several methods have been reported recently that address this problem by taking advantage of the modular nature of combinatorial libraries via a 'divide-and-conquer' or a 'dock-and-link' approach. This will be discussed below.

4.14.2.2.1 Docking methods and virtual screening

Two of the ground-breaking papers in the field were those of Kick *et al.*[128] and Haque *et al.*[129] In these studies, inhibitors of the aspartyl proteases cathepsin D and plasmepsin II respectively were designed using CombiBUILD.[130] Instead of fully enumerating the ~1 billion possible compounds, the CombiBUILD program allowed the authors to fix the template in the active site and probe each subsite individually. Combinations of the highest-scoring R-groups were then checked for steric clashes and the 1000 top compounds were selected for synthesis. Upon testing, Kick *et al.* found 67 compounds inhibited cathepsin D > 50% at 1 μM versus 26 from a diverse library. Haque *et al.* screened the directed cathepsin D library against plasmepsin II, leading to the identification of two lead structures. These leads were used as the starting points for a second library selected by varying R_1 while fixing R_2 and R_3 to the substituents found in the initial leads. The second library was screened and 12 R_1 groups were identified and folded into subsequent libraries which included R_2 and R_3 variation. Ultimately several single-digit nanomolar plasmepsin II inhibitors were identified with reasonable MW and ClogP values. In both studies, the position of the core template was well defined as crystal structures were available for both protein targets in complex with pepstatin. The lack of ambiguity in core placement contributed to the success of both campaigns.

Since, in practice, there is rarely a crystal structure available for the core and target of interest, much of the development in structure-based library design has focused on initial template placement followed by stepwise elaboration. FlexX^c [131] was significantly able to enhance the efficiency of FlexX for combinatorial library docking by allowing the user to specify a base fragment. Starting from this base fragment, the algorithm enumerates the library using a tree-based buildup scheme. This requires the base fragment and the subsequent branches of the tree derived from it to be placed only once, eliminating the time-consuming and redundant base placement step. Applying this method to a benzamide library in thrombin and a pyridine and Ugi library for dihydrofolate reductase (DFHR), the FlexX^c combinatorial docking scores were largely equivalent to FlexX sequential docking. The authors did note some dependency on buildup order, particularly in the pyridine library; this was presumably due to steric clashes between R-groups. FlexX^c docked the libraries 25–50 times faster than sequential FlexX. The DREAM++ [132] suite of programs also builds the library starting from a set of placed fragments. The initial fragments are positioned in the binding site with ORIENT++ followed by a conformational search. A set of molecules are selected from these and connected with the next reagent using REACT++ via a predefined set of chemical reactions. Conformations of the newly added piece are generated and evaluated in the context of the binding site. This process is continued until entire molecules have been built. As an example of its use, new inhibitors of HIV protease were created starting from a penicillin core. Although the designed compounds were not tested, they were able to fill the S2′–S2 pockets of the enzyme.

The 'divide-and-conquer' approach has been implemented in several programs, including PRO_SELECT,[133] CombiDOCK,[134] OptiDock,[135] and CombiGlide.[136] PRO_SELECT is a collection of methods that place a user-specified core template, search databases for chemically reasonable reagents, evaluate the fit of R-groups to the receptor at each point of variation, enumerate a combinatorial library based on top-scoring fragments, and score the resulting library. The authors do not focus on a specific program or set of programs that accomplish this, but offer a description of the process and methods they have used in their designs. CombiDOCK[134] takes the CombiBUILD[130] approach but uses the DOCK[137] program for core orientation. Poses of the core scaffold are generated using standard DOCK sphere matching, but using only scaffold atoms. R-groups are then grown from each site individually and scored. In the final step, entire molecules are built from the best-scoring R-groups and those with intramolecular clashes eliminated. The underlying simplifications of CombiDOCK can, however, lead to poor scaffold placement. Docking a small undecorated core into a large binding site can lead to a large number of ambiguous poses and it will often find its way into one of the R-group sites.

Lamb *et al.*[138] enhanced the accuracy of DOCK scaffold placement by incorporating a vector alignment component to the score. Conformations of a methyl-capped scaffold were generated and docked rigidly into the protein active site. Vectors were then grown normal to the concave subsite cavities within the active site and their alignment with the R-group vectors from the template were evaluated. The template orientations which had both a good DOCK and vector score were retained. The template configuration was then fixed and potential R-groups were evaluated at each position sequentially using torsional sampling. Side chains within 10 kcal mol^{-1} of the best were retained and scored with a general born/surface area (GB/SA) solvation correction. Finally, a full library with the five top-scoring side chains was enumerated for each pose, minimized, and free energy scored. The method was validated by designing libraries targeted toward chymotrypsin, elastase, and trypsin. Good alignment of the docked templates with the crystallographic conformation was found and the free-energy scores compared well to experimental data.

OptiDock[135] addresses the template placement issue by selecting a diverse, but representative, set of library members and docking them. Initial core placements are then derived from the whole molecule pose. R-groups are evaluated by individually growing them on a truncated scaffold with its atoms fixed in place and docking using

FlexX.[139] Whole compound scores are calculated by summation of R-group scores for a given core scaffold pose. The authors demonstrated the efficiency of this method by comparing FlexX docking of a library of 10 800 compounds to OptiDock. The FlexX run required 10 days of CPU time versus 81–279 min depending on which 20–80 core placements were sampled. The scores from OptiDOCK were found to correlate well with unconstrained FlexX docking scores for a collagenase, cathepsin D, and thrombin test set. A reasonable correlation was also found between OptiDock energies and pIC_{50} for the thrombin test set.

CombiGlide adds library-docking capabilities to the Glide[140] program. It also follows a core and side chains approach, but allows the user to define a set of representative compounds to define core placement. Once these molecules have been docked, the core positions are extracted and used for side-chain selection. Side chains are grown from the core at each point of substitution sequentially and the top-scoring side chains retained. All combinations are then generated and the resulting combinatorial library docked. Ultimately, the optimal combinatorial library is selected. The program was applied to the selection of a series of p38 mitogen-activated protein (MAP) kinase inhibitors derived from an Ugi library.[136] From a 1.4-million-member virtual library, CombiGlide successfully identified a 384-member optimal library containing several known sub-nM known p38 actives.

A contrasting approach to scaffold placement and R-group growing is identification of optimal fragments for each pocket and then, based on their orientations, linking them together (dock and link). MCSS[141] was used to discover functional groups that bound into three distinct subsites in the binding site of the poliovirus structure.[142] Using the positions of the functional groups as starting points, flexible linkers of appropriate length were selected visually and several 3-monomer libraries were synthesized. Compounds with minimum inhibitory concentrations in the 10 μM range were identified and two were selected for characterization by x-ray crystallography. The crystal structures validated the design, as there was good agreement between the original MCSS minima and the experimental fragment positions. The scaffold–linker–fragment strategy has been used in the design of a phosphodiesterase-4 (PDE4)-focused library starting from the known inhibitor zardaverine.[143] The crystal structure of zardaverine in complex with PDE4 revealed an unoccupied pocket that was accessible from a ligand open valence. Four zardaverine like scaffolds were chosen as anchors and connected to them were 16 functional groups via five alkyl linkers. The virtual library was docked into PDE4 using FlexX. Compounds were selected based on FlexX score and correct placement of the anchor fragment relative to the crystallographic conformation of zardaverine. Of the nine compounds synthesized, five showed inhibition greater than that of zardaverine. The CORGEN[144] program automates this process by utilizing experimental information to place the anchor fragment and demonstrates the advantage of small linker/fragment libraries. The virtual library is generated by elaboration of the anchor with small predefined sets of rings and anchors derived from their frequency of appearance in kinase inhibitors. Validation against three kinase targets and cyclooxygenase-2 showed that the structural motifs of the ligand cores were preserved but variants were also produced.

Structure-based methods have typically only been applicable to those targets for which there is crystallographic information. However, there are a large number of pharmaceutically relevant proteins which have not been crystallized and others that will be quite challenging to crystallize (GPCRs, ion channels, etc.). For many of these problems, homology modeling can provide structural insight but the quality of this data is dependent on how closely related the sequence of interest is to the template structure.[145] GPCRs present a particular challenge as the only relevant template structure available is bovine rhodopsin, which has low sequence homology to other GPCRs. While this could suggest that rhodopsin is not a suitable template for GPCR homology model building, there are now many studies linking models built from rhodopsin to experimental data, lending credibility to the method.[146] Although not applicable across the entire GPCR superfamily, there have been recent reports of successful use of rhodopsin-derived homology models for virtual screening. Bissantz *et al.*[147] constructed models of the dopamine D_3, muscarinic M_1, and vasopressin V_{1A} receptors. A variety of docking and scoring methods were then used to assess whether a set of 10 known ligands could be detected against a background of 990 random molecules from available chemicals directory (ACD). Hit rates of 20–37% were found for the D_3 and V_{1A} receptors by using a combination of either Gold or FlexX for docking and triple scoring with PMF/DOCK/Fresno or FlexX/GOLD/Fresno, respectively. The hit rates for the M_1 receptor were much lower. The models were unable to distinguish known agonists for any of the three receptors. A prospective study with the alpha1A receptor was reported that returned 37 active compounds from 80 tested.[148] The models used in this study were developed via an iterative ligand-guided approach. First, putative contacts between the ligand and receptor were mapped through analysis of mutagenesis and ligand SAR data. An ensemble of crude homology models was then generated and ligands docked into the putative binding site. Those poses consistent with the interaction map were kept and the receptor models refined through minimization with the ligand in place. The models were validated by docking 50 alpha1A antagonists embedded in 990 druglike molecules selected from MDDR. It was found that GOLD docking coupled with peptide mass fingerprinting scoring returned a hit rate 13-fold better than random. A virtual screen of the corporate collection was undertaken, the top 300 scoring compounds clustered, and 80 selected for screening. Thirty-seven of the 80 had potencies of >50% at 10 μM, with the most potent compound being 1.4 nM. Both examples cited here demonstrate the potential utility of homology modeling in virtual screening and, by implication, library design, but also highlight the careful analysis which is an essential component of homology model building.

4.14.2.3 Small is Beautiful

The recent emergence of techniques that provide structural information for weakly bound small fragments has presented another opportunity for anchor/linker/functional group design.[149–151] Because the compounds screened are intended to be starting points, they are simple, small, and low-MW. The hits and their orientations in the active site kickstart the drug discovery process. Although weakly active, these small molecules bind more efficiently[152] and do not have the liabilities typically found in a hit from a whole-molecule HTS. The HTS hits tend to be optimized for a different target, are high in MW, and hydrophobic. Generally they must be pared down before the true 'lead within the lead' is found that can be further optimized. Screening using biophysical techniques also alleviates the problem of false positives as the structural techniques provide direct evidence of compound binding, protein behavior, and aggregation.

NMR as a tool for lead finding was introduced to the SAR by NMR concept.[153] NMR was used to find weakly binding fragments for two subsites within FK506 binding protein (FKBP). Subsequent linking of a 2 and a 100 μM fragment resulted in compounds with K_ds ranging from 19 to 228 nM. This work has also been successfully applied to the discovery of adenosine kinase inhibitors,[154] expanding PTP-1B ligands into a second site[155] and the design of stromelysin inhibitors.[156]

Mass spectroscopy in concert with engineered proteins has been used to detect reactive small fragments 'tethered' to the protein by disulfide linkage.[157] A cysteine is engineered into the protein at a position close to the desired binding site. A set of sulfhydryl-containing ligands are screened against this protein and those with sufficient binding form a covalent bond with the protein and can be detected by the change in mass. The orientation of the fragments is gleaned from x-ray studies of the covalent complex to provide a starting point for drug design. Inhibitors of thymidate synthetase[157] and interleukin-2 receptor antagonists[158] have been elucidated using this protocol.

The use of x-ray crystallography in fragment discovery has been extensively reviewed[150,149,159,160] and the reader is referred to Chapters 4.18 and 4.31 and to these papers for an indepth overview of the state of the art. One recent report that nicely exemplifies the approach is its application to the identification of p38 MAP kinase inhibitors.[161] 2-Amino-3-benzyloxypyridine and 3-(2-(4-pyridyl)ethyl)indole were found in an x-ray screen to be bound within the hinge region and formed a well-conserved hydrogen bond to the backbone amide nitrogen of Met-109. These compounds proved to be convenient starting points for the structure-guided synthesis of more extensively decorated molecules. Probing the lipophilic specificity pocket by replacement of the phenyl moiety from the 2-amino-3-benzyloxypyridine scaffold ultimately resulted in a 65 nM p38 inhibitor which was selective against a panel of kinases. The 3-(2-(4-pyridyl)ethyl)indole was expanded by substitution on the indole 1-, 3-, and 5-positions to yield compounds, of which the most active was 340 nM. In both series, x-ray crystallography verified the binding hypotheses used during compound design. As described in the reviews cited above, there are many other examples of this technique applied to several targets. As a result of the growing body of evidence supporting the use of x-ray, NMR, mass spectrometry, and surface plasmon resonance as lead discovery tools, it has become a viable supplement, and in some cases, an alternative to traditional HTS. Several companies have been formed around this process and most large pharmaceutical companies have an internal effort. Fragment-based lead discovery will almost certainly play a continually expanding role in lead discovery and targeted library design.

4.14.3 Applications of Library Design to Focused Parallel Synthesis

Although many of the same tools and methods described above can also be applied to focused libraries, the design of these smaller libraries operates under a quite different set of parameters. The primary differentiating factor between the larger libraries described previously and smaller, medicinally oriented sets is the amount of information available. The landscape is comparatively rich in the latter case including, at the very least, the precise identity of the target. Each additional piece of information available about the target creates additional restrictions in the library design and these restrictions, in turn, lead to a more focused design. For example, one may have a crystal structure for a target and a series of active compounds. Additionally, it is known that the series has a structure-based metabolic or pharmacokinetic liability, thus placing three restrictions on the design of new entities. Other information that may be available and potentially used in the library design includes: sets of active compounds, the structure of the target, information about binding modes, SAR data, biological screening data, preferred ligand conformations, or even an x-ray crystal structure of the target.

The approach used to design parallel compound sets often depends on the particular information that is available and its inherent quality. When the information consists of sets of known actives and SAR data, ligand-based design methods are often used for design. More information can be generated quickly from this starting point by employing iterative synthesis sets, GAs, or informative design. When more concrete structural information is available, such as a crystal structure of the target or a detailed homology model (such a one based on a highly homologous enzyme), then structure-based design can be used.

A second contrasting factor between larger and smaller libraries is purpose. The purpose of a big library is often to search for something new or different or to serve as a multipurpose tool for a number of individual targets. The questions being asked by focused libraries are often much more specific. The purpose of such a library might be to gain SAR against a known and defined target, to make compounds more potent, to build in target selectivity, to increase solubility, to improve pharmacokinetics, bioavailability, or specificity profiles. In these cases it is imperative to address the question appropriately in the design and also desirable to do it efficiently, with regard to the amount of synthesis required.

With the understanding of these specific requirements, it makes sense that early design attempts that applied pure diversity approaches to parallel medicinal chemistry often met with mixed results. A certain amount of focus is required in order to attack specific challenges in medicinal chemistry programs. This idea is highlighted in an interesting head-to-head comparison where a structure-based design method produced a higher hit rate and led to more potent inhibitors of dihydrofolate reductase (DHFR) than a diversity-based selection method for the same target.[162] This type of result underscores the potential weakness of pure diversity-driven approaches in dealing with libraries that are searching for specific answers.

4.14.3.1 Ligand-Based Methods for Focused Libraries

In the absence of sound structural data, design often defaults to a ligand-based approach utilizing active analogs, SAR data, or computational models as a starting point. There are myriad ways to incorporate this type of information in the computational design of a parallel synthetic library, a selection of which will be detailed below.

Often, however, only vague ligand-based information is available for a target. Although these cases will not be discussed in great detail, it is worth noting some of the design approaches generally used. In these cases information is usually limited to that garnered from an active or small series of actives, such as SAR models, a common structural motif, or a crude pharmacophore model. Design is frequently noncomputational in these scenarios, drawing upon a very limited set of data and incorporating some general element of diversity to explore the unknown features of the target. This type of simplified approach has been used successfully against a number of biological targets[163a–163d] and hints at the advantage that even a basic, systematic design can have over pure chance.

4.14.3.1.1 Fragments

Early computational examples that stepped away from diversity approaches used fragment or fingerprint analyses to filter sizeable virtual libraries down to synthetically reasonable sets of compounds. For example, a fragment-based design employing LUDI[164a,164b] was used to discover a series of micromolar potassium channel inhibitors.[165] Lacking a crystal structure of the desired ion channel, the design started with a published computational model[166a,166b] for the target of interest, the human Kv1.3 potassium channel. LUDI was then used to choose small molecules for synthesis from a library of over 1000 random molecular fragments that were fitted into a model of the active site and evaluated for their ability to form hydrogen bonds or hydrophobic interactions. Fragments suggested by the LUDI analysis were then linked with spacers and modified to reduce conformational rotation. Finally, a library of 400 compounds was synthesized and biological evaluation revealed 12 potassium channel blockers.

4.14.3.1.2 Distance and connectivity

Starting with an active, it is possible to derive a mathematical understanding of its important features by analyzing the distance between key atoms or the way in which they are connected. This information can then be used to design new compounds with similar features. For example, β-turn mimetics have been designed in such a way.[167] Analysis of 11 well-defined β-turns showed that distances between most of the key atoms were highly conserved. This distance information was then used to design a tetrahydro-1,4-benzodiazepine-2-one scaffold which was used as the basis for a 62-compound library. Data for a specific biological target were not disclosed.

Connectivity can be utilized in a similar fashion. Analysis of actives using molecular topology provides graphical information about the connectivity of the molecule. These data can be manipulated to give topological indices. This approach has been applied to the discovery of antihistaminic agents by deriving appropriate topological indices.[168] A virtual library of 9000 compounds was screened against an antihistaminic topological model, leading to the synthesis of seven compounds, six of which showed activity in a rat functional assay.

4.14.3.1.3 Privileged substructures

Related to the geometric analyses above is the concept of privileged substructures. In essence, these are rigorously defined molecular architectures that show biological activity for particular target families. Focused libraries have been designed using privileged substructures as a starting point. Neurokinin-1 receptor antagonists have been discovered from libraries using this method.[169a,169b] A privileged substructure was chosen as the core for the library and coupled with a 'needle' concept, the needle being a molecular fragment or functional group showing specific interactions with one particular biological target. The libraries designed using this strategy provided two chemical series of moderately potent neurokinin-1 antagonists.

4.14.3.1.4 Descriptors

Library design can also be accomplished using descriptor-based methods. This has been applied to the design of osteoclast vacuolar acid pump inhibitors[170] using PCA. A number of descriptors were calculated for a set of amine reagents based on both 2D and 3D structures, and then PCA was used to eliminate redundancy and reduce the property space to a few orthogonal components. Compounds were selected to provide maximum coverage of principal component space or maximum similarity based on closeness in that space. In this way, two sets of reagents were chosen to create a focused, similar subset along with a diverse subset in the same library. In total, 800 compounds were attempted and approximately 400 were actually formed. Unfortunately, biological evaluation failed to uncover any inhibitors more potent than the starting compounds.

Another approach for CC chemokine receptor 3 (CCR3) employed clustering and descriptors in the design.[171] The library was designed taking into account a lead structure, active against both CCR1 and CCR3, and known SAR for the target. Reagent sets were chosen using clustering based on MDL keys, which served as a structure descriptor. The 770-member library led to a novel CCR3 selective antagonist, 2 nM versus the target and showing 950-fold selectivity over CCR1 (**Figure 8**).

Descriptor-based design can be implemented in other ways as well. Pharmacophore descriptors have been used as a basis for library design of potential corticotropin-releasing factor (CRF_1) antagonists.[172] The process was initiated with the creation of a library space biased toward CRF_1 antagonists, created from an initial data set of about 3000 active and 1200 inactive compounds. This space was derived from 3D pharmacophore descriptors. Potential pharmacophores were

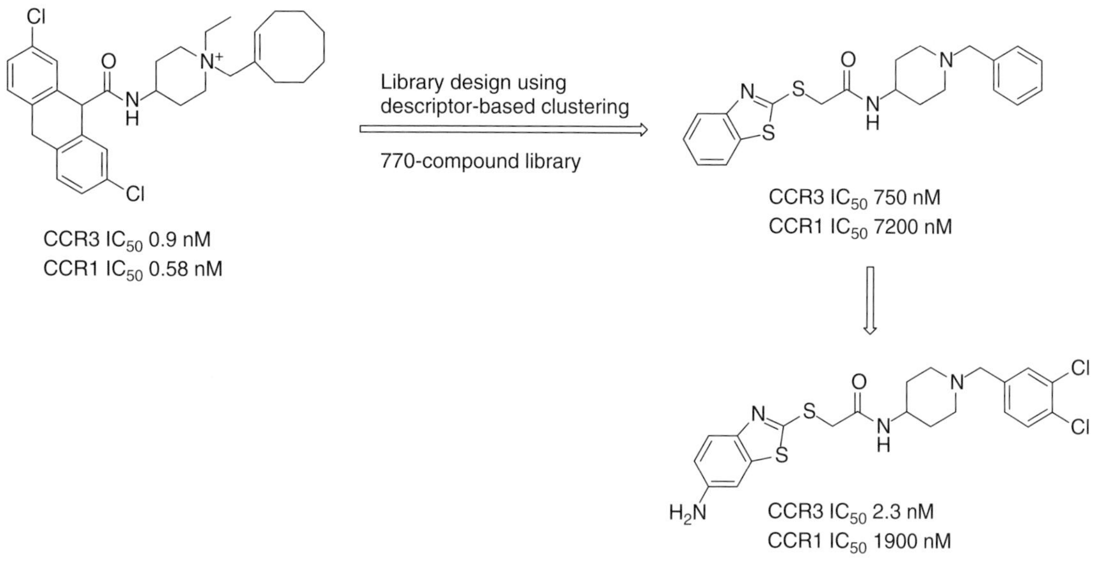

Library design using descriptor-based clustering

770-compound library

CCR3 IC_{50} 0.9 nM
CCR1 IC_{50} 0.58 nM

CCR3 IC_{50} 750 nM
CCR1 IC_{50} 7200 nM

CCR3 IC_{50} 2.3 nM
CCR1 IC_{50} 1900 nM

Figure 8 Implementation of descriptor-based library design in the development of a selective CCR3 antagonist.

then ranked by their ability to differentiate biologically active compounds from inactive compounds. New scaffolds were then checked against the pharmacophores in the CRF_1 design space. Informative design was applied to build the libraries. The computationally chosen cores, along with 12 amines that resulted from the informative design, were then coupled with six anilines and six boronic acids chosen by medicinal chemists. The resulting $12 \times 6 \times 6$ library produced 13 compounds with moderate activity (200–800 nM) against CRF_1.

4.14.3.1.5 Pharmacophore models

Known actives can also serve as a starting point for ligand-based design. In an early foray into pharmacophore-based library design, inhibitors of protein phosphatase PP2A were discovered by using a pharmacophore model to guide the synthesis of an 18-member library.[173] The model was used to direct the choice of the pharmacophore and further exploration was accomplished by random optimization of steric and electronic properties. Although complete biological details were not disclosed, it was revealed that two compounds showed micromolar functional activity against human breast carcinoma cells.

This approach has also been applied to the discovery of growth hormone secretagogues (GHS).[174] A 3D pharmacophore model was constructed for this purpose from a set of peptides and nonpeptides with known GHS activity by using DistComp.[175] The model was used to search against compound databases for novel scaffolds. Synthesis based on the computational results provided GHS actives (precise number undisclosed), one of which was analoged and led to a compound with 1 nM activity (**Figure 9**). The analog effort was guided by fragment QSAR analysis.

Potent agonists and antagonists were found for the nociceptin (N/OFQ) receptor (NOP receptor) using a related approach.[176] Published actives for the receptor were used to build a 2D pharmacophore model for the NOP receptor,

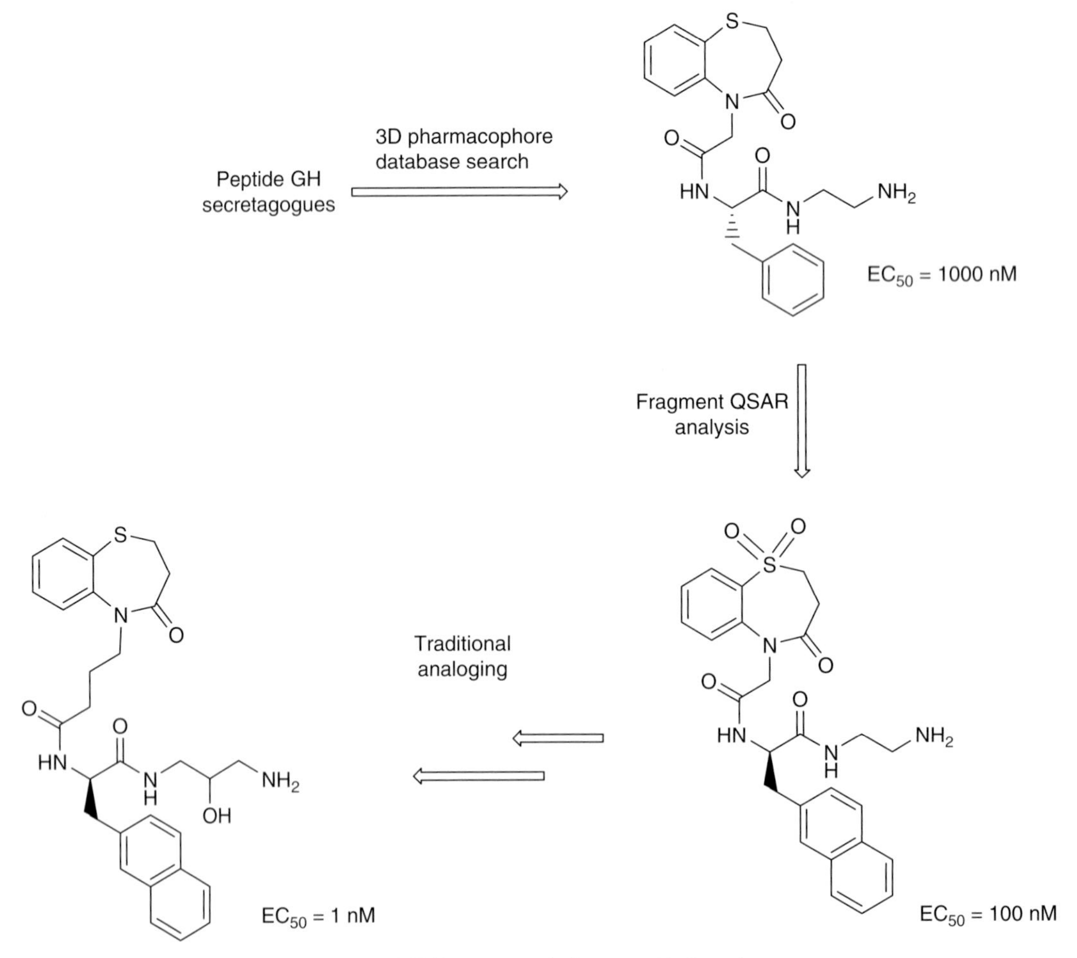

Figure 9 Pharmacophore-based design of GHS compounds from peptide ligands.

which was subsequently used to scan the ACD database. This exercise identified a group of 12 commercially available piperidines which were used as library cores. An alkylation set (192 compounds) and an acylation set (156 compounds) were synthesized with reagent choices guided by the model. Preliminary screening identified 20 actives, which became the basis for further optimization and eventually led to several low-nanomolar agonists of the receptor.

A theoretical model was constructed for $\alpha_V\beta_3$ integrin and successfully used to guide the design of a subsequent chemical library of peptidomimetic carbohydrates.[177] The model was built from three known, highly selective antagonists. They were first subjected to simulated annealing and the resulting low-energy conformers were grouped into families. Next, family representatives were used for molecular dynamics studies to explore the accessible energetic hypersurface. Existing conformers were superimposed to form the model. The pharmacophore model led to the choice of xylose as the core for the peptidomimetic library. The subsequent 126-compound library was screened as sets of mixtures and, after deconvolution, provided a 4 mM inhibitor. The result highlights the effectiveness of this approach in that even large molecules, such as peptides, can be replaced by smaller compounds through an understanding of the key interactions at the receptor.

A pharmacophore model approach has also been applied to the development of DHFR inhibitors.[178] A virtual library of 936 compounds was screened against models for a human and two other species of parasitic DHFR. Compounds were chosen on the basis of fit, via docking, and selectivity for the parasitic enzymes. A set of fragments was then chosen from those used to build the virtual compound set and provided the core for the focused library. Eight of the nine compounds synthesized showed inhibition of *Leishmania major* DHFR, in the range of 1.4–212 nM.

4.14.3.1.6 Combined approaches

In order to incorporate disparate sets of data fully it is often useful to implement library designs that draw on multiple computational methods. In this way, different tools can be used to analyze the various types of data available, with the objective being that the best tool available can be used for each type of data. The multitool approach can also be a serial endeavor, in which data are analyzed and the results are passed on for subsequent processing by another method. This continues in an iterative fashion until the desired goal is achieved. These approaches allow for the incorporation of more data, especially from different sources, into the design but tend to be less general in their applicability than the methods listed above.

One such approach combined two complementary similarity methods, taking advantage of both pharmacophoric patterns and topological similarity in order to find inhibitors of the μ-opiate receptor.[179] Starting with a single lead molecule, the two methods were employed in parallel. The first utilized variations of pharmacophoric elements while retaining similar topology in the molecules. These changes included chain extensions, substitutions, ring openings, and ring closings. The second method took advantage of descriptors based on atom pairs to select library candidates with low dissimilarity scores with respect to the lead compound. The best compounds were then re-ranked using the ComPharm similarity score.[180] Two libraries were then synthesized, screened, and analyzed for SAR. The information was then merged, and a final set of five analogs was prepared using substituents from the first library and the core from the second. The hybrid analog set produced two low-nanomolar inhibitors and seemed to validate the complementary nature of the approaches.

A related methodology utilized an SAR model and topological pharmacophore similarity.[181] The starting point was 153 combinatorial products and their biological performance against two purinergic receptors. The first step was to encode all of the molecules using the chemically advanced template search (CATS)[182] topological pharmacophore descriptor. The 150-dimensional data were then analyzed using a self-organizing map to produce an SAR model that would determine selectivity for the A_{2A} receptor versus the A_1 receptor. This exercise produced two selective substituents, fixing one position of the scaffold. The model was then used to choose the remaining substituents, resulting in a focused library of 17 compounds. The library, on average, showed a threefold increase in binding for the A_{2A} receptor and a 3.5-fold increase in selectivity over the A_1 receptor.

Use of statistical molecular design to vary seven positions of a nonapeptide simultaneously led to the discovery of peptide inhibitors of the FimC/FimH protein–protein interaction.[183] An initial set of building blocks was selected on the basis of molecular properties, using PCA and taking into account known structural information about the target. The set was then refined using the sequences of the native peptide ligands and the application of D-optimal design. These procedures reduced a virtual library of over 57 000 potential peptides to a manageable set of 32. The library was able to deliver highly informative QSAR data and provide the basis for a multivariate QSAR model. Novel peptides were found with an ability to inhibit the protein to the same extent as native peptides.

It has been shown that focused compound sets can also be designed on the basis of known biological data by way of recursive partitioning and the employment of GAs. A retrospective analysis of HTS data for an undisclosed target, along with associated chemical descriptors, was used to build SAR models using recursive partitioning.[184] A GA was then used to predict activity which served as a basis for a virtual library design. This method proved to be efficient, finding the actives from a small number of 'synthesized' compounds while also showing an ability to reach the actives from sets of SAR data that were initially weak.

4.14.3.2 Structure-Based Methods for Focused Libraries

Structural data for a target, in the form of an x-ray crystal structure, NMR data, or an accurate homology model, can be a powerful tool for focused library design. By analyzing the way in which molecules fit into a well-understood site, more accurate computational predictions can be made and potentially libraries with greater biological 'enrichment' can be designed. In addition, this information can be extremely useful for the analysis of SAR and unexpected results, which can be fed back into an iterative design process.

Structural data can be incorporated into the design in a number of ways. In some cases it is used as a map by medicinal chemists to follow in performing noncomputational designs. More sophisticated approaches employ modeling and virtual screening, allowing for the analysis of enormous virtual libraries in a fairly straightforward manner. The power in this approach is synthetic efficiency, as huge numbers of compounds are screened out computationally, and 'smarter' libraries with higher probabilities of success are actually synthesized.

4.14.3.2.1 Homology models

One way in which structural data can be incorporated is in the construction of a homology model. Often this is accomplished by using a related target or different form of the desired target. This has been successfully applied to CDK4.[185] Beginning with an x-ray crystal structure of an activated form of CDK2 and the protein sequences for CDK2 and CDK4, a homology model was created for CDK4. The model was then used for de novo scaffold design through the application of a program called LEGEND. The output of 1000 potential scaffolds was then refined using SEEDS,[186] which chooses synthetically feasible or commercially available scaffolds. By this process 382 commercially available compounds were selected, obtained, and screened. This provided 18 compounds with activity below 500 nM. Using LEGEND and the homology model, a series of focused informer libraries were designed to generate SAR. These efforts eventually culminated in the discovery of a 42 nM inhibitor of CDK4 (**Figure 10**).

The same research group took advantage of their CDK4 homology model in follow-up studies aimed at improving selectivity over other kinases (T Iwama *et al.*, unpublished results). This time the library was constructed from de novo design suggestions provided by LUDI and LeapFrog. The end result was a 1.6 nM inhibitor of CDK4 that showed 110-fold selectivity over CDK2.

A homology model built from the 'wild-type' vitamin D receptor (VDR) has been employed to find inhibitors of a mutant version of the receptor.[187] The library design was based on a known lead and was able to generate analogs with significant activity against the mutant VDR. The most promising compounds had EC_{50} values between 3 and 121 nM; in addition, they had selectivity profiles superior to the original lead.

A similar approach was used to find selective inhibitors of phospholipase A_2.[188] Using the x-ray structural data of related enzymes, a homology model was built for a series of phospholipase enzymes. The models, in conjunction with a

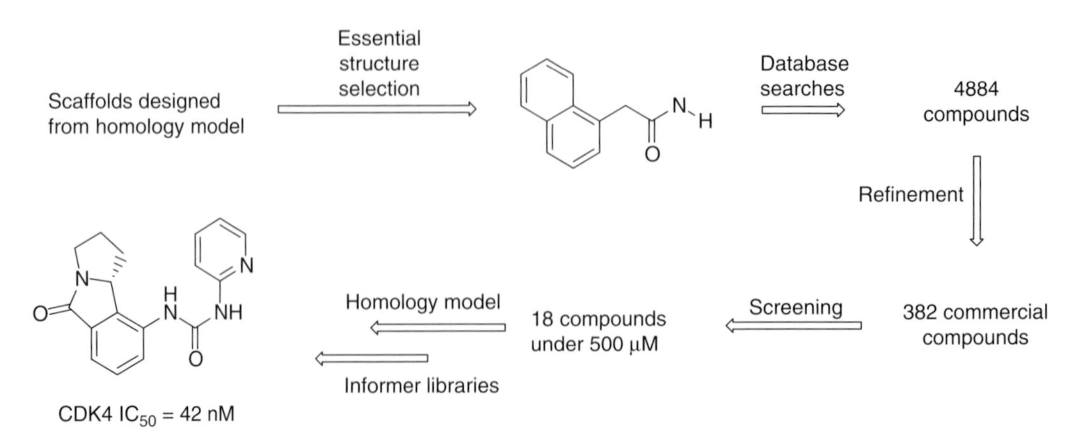

Figure 10 Development of a CDK4 inhibitor employing homology models and de novo design.

known active, were then used to design an 83-member library of indole analogs aimed at improving the selectivity and potency of the series (exact method undisclosed). A number of fairly potent inhibitors were found, though none displayed high selectivity for the desired target.

In an example targeting P-selectin, the homology model was built from the closely related E-selectin,[189] as both are believed to bind to sialyl Lewis in the same manner. Parent scaffolds for use in parallel chemistry were then compared to the P-selectin surface using the Monte Carlo algorithm QXP.[190] Virtual libraries were built from the lowest energy conformations then compared with the model again. The top 200 compounds were screened visually, with the best being used in the library. Using these methods in an iterative fashion, seven libraries were synthesized, leading to a number of compounds that displayed activity in a functional assay.

A homology-based approach has also been applied to bacterial *S*-adenosyl homocysteine/methylthioadenosine nucleosidase.[191] A homology model was used to screen a virtual collection of ~400 000 compounds, providing 1300 of interest. These molecules were then clustered by 2D fingerprints. Compounds were chosen for screening using SYBYL[47] SELECTOR. This route allowed for the development of a 2.8 nM inhibitor of the enzyme from a 1 μM lead structure.

4.14.3.2.2 Pharmacophore models

Structural data can also be used as the foundation for pharmacophore models. X-ray and NMR data were used to refine[192] a previously developed model of protein kinase C.[193] The improved model provided a better understanding of affinity domains located near the active site. The analogs that were produced incorporated side chains capable of interacting with these newly revealed sites, which drove the potency down into the low-nanomolar range. In addition, three compounds showed functional activity against tumor cell lines.

4.14.3.2.3 Docking

Another structure-based approach that can be implemented for library design is docking. Structural data, in the form of a homology model or crystal structure, can be used to model active site of the target. Virtual compounds can then be examined for their ability to interact with the active site. This usually involves the enumeration of large virtual compound sets followed by equally large virtual screening campaigns that eventually lead to small, high-quality synthetic sets. This basic method has been successfully applied in a number of cases. Examples include: using a modified version of DOCK to design cytochrome P450 inhibitors[194]; using PRO_SELECT[134] and iterative library design for selective factor Xa inhibitors[195]; the discovery of selective tissue factor VIIa inhibitors[196a,196b]; and the development of inhibitors of the cysteine protease cruzain.[197]

More sophisticated variations of this method have incorporated other tools to provide a richer understanding of the requirements for inhibition. In the pursuit of a small molecule capable of disrupting the protein binding in the HOX-PBX/DNA (a protein–DNA complex formed between a homeobox protein, pre-B cell homeobox1 and DNA) transcription factor, complex de novo design (LeapFrog) and affinity predictions (FlexX and CScore) were employed in addition to docking.[198] Lastly, MW and the number of rotatable bonds were restricted. A 32-member focused library was synthesized and screened, providing a set of weakly active compounds. Although the inhibitors were weak binders, two were still more potent than the known peptide antagonist. This is notable because the disruption of protein–protein interactions with small molecules is not trivial.

Another multifaceted design utilized a 'scaffold–linker–functional group' approach to generate a virtual library then selected compounds for synthesis by docking.[199] Starting with four fixed scaffolds based on their lead, a virtual library was assembled by attaching variable linkers and a variable functional group. Compounds were then docked into the active site of the enzyme using FlexX. Based on the docking, the library was reduced to a set of 320 compounds which, upon screening, provided a 900-fold increase in activity over the original lead.

4.14.3.2.4 Genetic algorithms

The application of GAs to computational design problems can be quite powerful. By allowing for the dynamic evolution of molecules, they create sets of slowly improving pools of compounds that could potentially reach near-optimum solutions. This can be particularly useful for de novo design by docking. A GA was allowed to drive a refinement process in which LEA3D was used to perform de novo design and FlexX was used to fit molecules into the active site of thymidine monophosphate kinase (TMK).[200] Fragments for the de novo design were derived from databases of known drugs. The 22-member library provided 17 micromolar TMK inhibitors.

4.14.3.2.5 Combined approaches

As was mentioned in the previous section, it is often advantageous to employ multiple design tools when dealing with complex chemical and biological undertakings. This strategy can also be employed when structural information is available. In one particularly interesting example, the structure of the target was known, but the exact binding mode was unclear. In order to explore the binding mode of cholera toxin, the causative agent of cholera, a library was designed.[201] The 20-member library was built upon structural data for the target and produced a number of leads which were, in turn, used to obtain x-ray crystallographic data. In this instance, a focused library and structural data were used to obtain improved structural data about the target.

Statistical molecular design and multivariate design have been paired together in an effort to find thrombin inhibitors.[202] A library of compounds was designed by this route with statistical molecular design providing information about substitution patterns based upon structural data. Multivariate design was then employed to trim the designed library from 108 compounds to 18. The focused library provided important SAR information, as well as showing improved absorption and selectivity over the initial leads.

Multivariate design has also been paired with PCA. An initial lead chemotype was discovered that targeted bacterial pilus biogenesis by use of structure-based design.[203] A virtual library was built upon the scaffold using statistical multivariate design with a goal of generating SAR data.[204] The 84-member set was then reduced to 20 through PCA. The compound set provided the desired SAR information and the binding affinity of seven members was equal to or better than the previous best.

4.14.4 Conclusion

The trend in library design in the twenty-first century has been toward smaller, focused, information-driven libraries in both combinatorial and array formats. Full combinatorial matrices have been displaced by more druglike design formats such as sparse matrices that exclude compounds with undesirable ADME properties. While the libraries and arrays that are actually synthesized have grown smaller, computing grids and CPU clusters have facilitated the development of ligand and receptor structure-based tools that screen virtual libraries of hundreds of thousands of molecules during the design process. Design methods are no longer limited to optimization of just the primary biological activity. Instead, the trend is now multiparametric/multidimensional optimization of properties, including ADME, selectivity among targets, and even simultaneous optimization of several target activities. Library design has developed from the early days of large diverse libraries that were intended to be screened via HTS in a paradigm that was rather analogous to the agricultural industry's 1980s 'spray and pray' herbicide discovery methods to a collection of tools and methods that incorporate available information on receptors, SAR, ADME, etc. The field has now come of age and promises truly to impact drug discovery.

References

1. Burbaum, J. J.; Ohlmeyer, M. H.; Reader, J. C.; Henderson, I.; Dillard, L. W.; Li, G.; Randle, T. L.; Sigal, N. H.; Chelsky, D.; Baldwin, J. J. *Proc. Natl. Acad. Sci. USA* **1995**, *92*, 6027–6031.
2. Blondelle, S. E.; Crooks, E.; Ostresh, J. M.; Houghten, R. A. *Antimicrob. Agents Chemother.* **1999**, *43*, 106–114.
3. Boger, D. L.; Jiang, W.; Goldberg, J. *J. Org. Chem.* **1999**, *64*, 7094–7100.
4. Ferry, G.; Boutin, J. A.; Atassi, G.; Fauchere, J. L.; Tucker, G. C. *Mol. Div.* **1996**, *2*, 135–146.
5. Lutzke, R. A. P.; Eppens, N. A.; Weber, P. A.; Houghten, R. A.; Plasterk, R. H. A. *Proc. Natl. Acad. Sci. USA* **1995**, *92*, 11456–11460.
6. Samson, I.; Kerremans, L.; Rozenski, J.; Samyn, B.; Van Beeumen, J.; Herdewijn, P. *Bioorg. Med. Chem. Lett.* **1995**, *3*, 257–265.
7. Lahana, R. *Drug Disc. Today* **1999**, *4*, 447–448.
8. Martin, E. J.; Blaney, J. M.; Siani, M. A.; Spellmeyer, D. C.; Wong, A. K.; Moos, W. H. *J. Med. Chem.* **1995**, *38*, 1431–1436.
9a. Andersson, P. M.; Sjöström, M.; Wold, S.; Lundstedt, T. *J. Chemometrics* **2001**, *15*, 353–369.
9b. Andersson, P. M.; Linusson, A.; Wold, S.; Sjöström, M.; Lundstedt, T.; Norden, B. Design of Small Libraries for Lead Exploration. In *Molecular Diversity in Drug Design*; Dean, P. M., Lewis, R. A., Eds.; Kluwer Academic: Dordrecht, Netherlands, 1999, pp 197–220.
9c. Linusson, A.; Gottfries, J.; Lindgren, F.; Wold, S. *J. Med. Chem.* **2000**, *43*, 1320–1328.
10a. Schnur, D. M. *J. Chem. Inf. Comput. Sci.* **1999**, *39*, 36–45.
10b. Schnur, D.; Venkatarangan, P. Applications of Cell-Based Diversity Methods to Combinatorial Library Design. In *Combinatorial Library Design and Evaluation*; Ghose, A. K., Viswadhan, V. N., Eds.; Marcel Dekker: New York, NY, 2001, pp 473–501.
10c. Brannigan, L. H.; Grieshaber, M. V.; Schnur, D. M. *ACS Symp. Ser.* **1995**, *606*, 264–281.
11. Lewis, R. A.; Pickett, S. D.; Clark, D. E. *Rev. Comp. Chem.* **2000**, *16*, 1–51.
12. Olsson, T.; Oprea, T. I. *Curr. Opin. Drug. Disc. Dev.* **2001**, *4*, 308–313.
13. Xue, L.; Stahura, F. L.; Bajorath, J. Cell Based Partitioning. In *Chemoinformatics: Methods, and Tools for Discovery, Methods in Molecular Biology*; 275th ed.; Bajorath, J., Ed.; Humana Press: Totowa, NJ, 2004, pp 279–289.
14. Pearlman, R. S. *DiverseSolutions Manual*; Laboratory for Molecular Graphics and Theoretical Modeling College of Pharmacy: University of Texas at Austin, Austin, TX, 1999.

15. Cavallaro, C. L.; Schnur, D.; Pearlman, R. (in preparation).
16. Lipinski, C. A.; Lombardo, F.; Dominy, B. W.; Feeney, P. J. *Adv. Drug Deliv. Rev.* **1997**, *23*, 3–25.
17a. Oprea, T. I. *J. Comput. Aided Mol. Des.* **2002**, *16*, 325–334.
17b. Oprea, T. I. *J. Comput. Aided Mol. Des.* **2000**, *14*, 251–264.
17c. Oprea, T. I.; Gottfries, J. *J. Mol. Graph. Model.* **1999**, *17*, 261–274.
18. Evans, B. E.; Rittle, K. E.; Bock, M. G.; Dipardo, R. M.; Freidinger, R. M.; Whiter, W. L.; Lundell, G. F.; Veber, D. F.; Anderson, P. S.; Chang, R. S. L. et al. *J. Med. Chem.* **1988**, *31*, 2235–2246.
19. Patchett, A.; Nargund, R. P. *Ann. Rep. Med. Chem.* **2000**, *35*, 289–298.
20. Mason, J. S.; Morize, I.; Menard, P. R.; Cheney, D. L.; Hulme, C.; Labaudiniere, R. F. *J. Med. Chem.* **1999**, *42*, 3251–3264.
21a. Lamb, M. L.; Bradley, E. K.; Beaton, G.; Bondy, S. S.; Castellino, A. J.; Gibbons, P. A.; Suto, M. J.; Grootenhuis, P. D. J. *J. Mol. Graph. Model.* **2004**, *23*, 15–21.
21b. Savchuk, N.; Tkachenko, S. E.; Balakin, K. V. *Methods Principles Medicinal Chem.* **2005**, *23*, 287–314.
22. Lowrie, J. F.; Delisle, R. K.; Hobbs, D. W.; Diller, D. J. *Comb. Chem. High Throughput Screen.* **2004**, 7, 495–510.
23. Prien, O. *ChemBioChem* **2005**, *6*, 500–505.
24. Stewart, E. L.; Brown, P. J.; Bentley, J. A.; Wilson, T. M. *Abstracts of Papers*, 222nd ACS National Meeting, Chicago, IL, August 26–30, 2001.
25. Lang, S. A.; Kozyukov, A. V.; Balakin, K. V.; Skorenko, A. V.; Ivashchenko, A. A.; Savchuk, N. P. *J. Comput. Aided. Mol. Des.* **2003**, *16*, 803–807.
26. Schuffenhauer, A.; Zimmermann, J.; Stoop, R.; van der Vyver, J.-J.; Lecchini, S.; Jacoby, E. *J. Chem. Inf. Comput. Sci.* **2002**, *42*, 947–955.
27a. Aureus Pharmaceuticals, 174 Quai de Jemmapes, 75010 Paris, France; http://www.aureus-pharma.com (accessed April 2006).
27b. Jubilant Biosys Ltd, 8575 Window Latch Way, Columbia, MD 21045, USA; http://www.jubilantbiosys.com (accessed April 2006).
27c. Sertanty Inc., 1735 N. First St. #102, San Jose, CA 95112, USA; http://www.sertanty.com (accessed April 2006).
27d. Biowisdom Ltd, Babraham Hall, Babraham, Cambridge CB2 4AT, UK; http://www.biowisdom.com (accessed April 2006).
27e. GVK Biosciences Private Ltd., #210, 6–3–1192, Kundanbagh, Begumpet, Hyderabad – 500016, India; http://www.gvkbio.com (accessed April 2006).
27f. Chematica S*t*ARLIT*e* from Inpharmatica Ltd, 60 Charlotte St, London W1T 2NU, UK; http://www.inpharmatica.co.uk (accessed April 2006).
28. Gooding, O. *Curr. Opin. Chem. Biol.* **2004**, *8*, 297–304.
29. Debnath, A. K. Quantitative Structure–Activity Relationship (QSAR), A Versatile Tool in Drug Design. In *Combinatorial Library Design and Evaluation*; Ghose, A. K., Viswadhan, V. N., Eds.; Marcel Dekker: New York, NY, 2001, pp 73–155.
30. Oprea, T. I.; Zamora, I.; Ungell, A. *J. Combinat. Chem.* **2002**, *4*, 258–266.
31. Oprea, T. I.; Davis, A. M.; Teague, S. J.; Leeson, P. D. *J. Chem. Inf. Comput. Sci. J. Chem. Inf. Comput. Sci.* **2001**, *41*, 1308–1315.
32. Hann, M. M.; Oprea, T. I. *Curr. Opin. Chem. Biol.* **2004**, *8*, 255–263.
33. Hansch, C.; Leo, A. Calculation of Octanol-Water Partition Coefficients by Fragments. In *Exploring QSAR: Fundamentals and Applications in Chemistry and Biology*; Heller, S. R., Ed.; American Chemical Society: Washington, DC, 1995, pp 125–168.
34. Oprea, T. I.; Olah, M.; Mracec, M.; Rad, R.; Ostopovici, L.; Bora, A.; Hadaruga, N.; Bologa, C. G. Mapping bioactivity space for fragment-based lead discovery. Abstracts of Papers, 229th ACS National Meeting, San Diego, CA, USA, March 13–17, 2005.
35. Teague, S. J.; Davis, A. M.; Leeson, P. D.; Oprea, T. I. *Angew. Chem. Int. Edn.* **1999**, *38*, 3743–3748.
36. Kubinyi, H. *Comput. Med. Chem. Drug Disc.* **2004**, 539–570.
37. Topliss, J. G. *Perspect. Drug Disc. Des.* **1993**, *1*, 253–268.
38. Brannigan, L. H.; Grieshaber, M. V.; Schnur, D. M. *Chemtech* **1995**, *25*, 29–35.
39. Lipinkski, C. A. *Pharm. News* **2002**, *9*, 195–202.
40. Rose, S.; Stevens, A. *Curr. Opin. Chem. Biol.* **2003**, 7, 331–339.
41. Mitchell, T.; Showell, G. A. *Curr. Opin. Drug Disc. Dev.* **2001**, *4*, 314–318.
42. Gillet, V. J. *J. Comput. Aided Mol. Des.* **2002**, *16*, 371–380.
43. Diverse Solutions was developed by R. S. Pearlman and K. M. Smith, Laboratory for Molecular Graphics and Theoretical Modeling, College of Pharmacy, University of Texas at Austin, Austin TX 78712, USA; distributed by Tripos Associates, 1699 South Hanley Rd., St Louis, MO 63144, USA.
44. CombiGLIDE is available from Schroedinger, 120 West Forty-Fifth Street, 32nd Floor, Tower 45, New York, NY 10038, USA.
45. OptiveBenchware was developed by R. S. Pearlman, Laboratory for Molecular Graphics and Theoretical Modeling, College of Pharmacy, University of Texas at Austin, Austin, TX 78712, USA; and Optive Research, Inc.; distributed by Tripos Associates, 1699 South Hanley Rd., St Louis, MO 63144, USA.
46. Daylight Chemical Information Systems, Inc., 27401 Los Altos, Suite 360, Mission Viejo CA 92691, USA.
47. Tripos Associates, 1699 South Hanley Rd., St Louis, MO 63144, USA.
48. MDL Information Systems, San Leandro, CA, USA.
49. CambridgeSoft Corporation, 100 Cambridge Park Drive, Cambridge, MA 02140, USA.
50. Green, D. V. S.; Pickett, S. D. *Mini-Rev. Medicinal Chem.* **2004**, *4*, 1067–1076.
51. Gillet, V. J. *Comput. Med. Chem. Drug Disc.* **2004**, 617–639.
52. Matter, H. *J. Med. Chem.* **1997**, *40*, 1219–1229.
53. Zuccotto, F. *J. Chem. Inf. Comp. Sci.* **2003**, *43*, 1542–1552.
54a. Hall, L. H.; Kier, L. B. *Quant. Struct.–Act. Relat.* **1990**, *9*, 115–131.
54b. MolconnZ descriptors are available in most standard modeling software packages such as SYBYL (Tripos Associates), Cerius2 (Accelrys), and MOE (Chemical Computing Group).
55. Hall, L. H.; Mohney, B.; Kier, L. B. *Quant. Struct.–Act. Relat.* **1991**, *10*, 43–51.
56. Hassan, M.; Bielawski, J. P.; Hempel, J. C.; Waldman, M. *Mol. Div.* **1996**, *2*, 64–74.
57. Cerius2 is available from Accelrys, Inc., 9685 Scranton Road, San Diego, CA 92121–3752, USA.
58. Brown, R. D.; Hassan, M.; Waldman, M. *Methods Mol. Biol. (Totowa, NJ, United States)* **2004**, *275* (Chemoinformatics), 301–335.
59. Austel, V. *Methods Princ. Med. Chem.* **1995**, *2*, 49–62.
60. MODDE is available from Umetrics, Inc., Kinnelon, NJ, USA.
61. Linusson, A.; Gottfries, J.; Olsson, T.; Oernskov, E.; Folestad, S.; Norden, B.; Wold, S. *J. Med. Chem.* **2001**, *44*, 3424–3439.
62. Lee, A.; Breitenbucher, J. G. *Curr. Opin. Drug Disc. Dev.* **2003**, *6*, 494–508.
63. Skiles, J. W.; Gonella, N. C.; Jeng, A. Y. *Curr. Med. Chem.* **2001**, *8*, 425–474.
64. Pauls, H. W.; Ewing, W. R. *Curr. Top. Med. Chem.* **2001**, *1*, 83–100.

65. Lewis, R. A.; Pickett, S. D.; Clark, D. E. *Rev. Comp. Chem.* **2000**, *16*, 1–51.
66. Schnur, D.; Beno, B. R.; Good, A.; Tebben, A. *Methods Mol. Biol. (Totowa, NJ, United States)* **2004**, *275* (Chemoinformatics), 355–377.
67. Pearlman, R. S. *Network Sci.* **1996**, *2*.
68. Pearlman, R. S.; Smith, K. M. *Drugs Future* **1998**, *23*, 885–895.
69. Pearlman, R. S.; Smith, K. M. *Perspect. Drug Disc. Des.* **1997**, *9/10/11*, 339–353.
70. Pearlman, R. S.; Smith, K. M. *J. Chem. Inf. Comput. Sci.* **1999**, *39*, 28–35.
71. Stewart, E. L.; Brown, P. J.; Bentley, J. A.; Wilson, T. M. *Abstracts of Papers*, 222nd ACS National Meeting, Chicago, IL, August 26–30, 2001.
72. Wang, X.; Saunders, J. *Abstracts of Papers*, 222nd ACS National Meeting, Chicago, IL, August 26–30, 2001.
73. Mason, J. S.; Beno, B. R. *J. Mol. Graph. Model.* **2000**, *18*, 438–451.
74. Pirard, B.; Pickett, S. D. *J. Chem. Inf. Comput. Sci.* **2000**, *40*, 1431–1440.
75. Manallack, D. T.; Pitt, W. R.; Gancia, E.; Montana, J. G.; Livingstone, D. J.; Ford, M. G.; Whitley, D. C. *J. Chem. Inf. Comput. Sci.* **2002**, *42*, 1256–1262.
76. MDL Drug Data Report, MDL Information Systems, San Leandro CA, USA.
77. Bemis, G. W.; Murcko, M. A. *J. Med. Chem.* **1996**, *39*, 2887–2893.
78. Mason, J. S.; Morize, I.; Menard, P. R.; Cheney, D. L.; Hulme, C.; Labaudiniere, R. F. *J. Med. Chem.* **1999**, *42*, 3251–3264.
79. Classpharmer is available from Bioreason, Inc., 3900 Paseo del Sol, Santa Fe, NM 87507, USA; http://bioreason.com (accessed April 2006).
80. Schnur, D. M.; Hermsmeier, M. *Recent Approaches to Target Class Design*. 36th Middle Atlantic Regional Meeting of the American Chemical Society, 2003.
81. Horton, D. A.; Bourne, G. T.; Smythe, M. L. *Chem. Rev.* **2003**, *103*, 893–930.
82. Bleicher, K. H.; Green, L. G.; Martin, R. E.; Rogers-Evans, M. *Curr. Opin. Chem. Biol.* **2004**, *8*, 287–296.
83. Merlot, C.; Domine, D.; Cleva, C.; Church, D. J. *Drug Disc. Today* **2003**, *8*, 594–602.
84. Schnur, D. M.; Hermsmeier, M. A.; Tebben, A. J. *J. Med. Chem.* **2006**, *49*, 2000–2009.
85. Pozzan, A.; Feriani, F.; Tedesco, G.; Caplelli, A. M. In *3D Pharmacophoric Hashed Fingerprints. Rational Approaches to Drug Design*; Holtje, H. D., Sippl, W., Eds.; Prous Science: Barcelona, Spain, 2001, pp 224–228.
86. Schnur, D.; Beno, B. R.; Good, A.; Tebben, A. *Methods Mol. Biol. (Totowa, NJ, United States)* **2004**, *275* (Chemoinformatics), 355–377.
87. Cavallaro, C. L.; Schnur, D. M.; Tebben, A. J. *Chemoinf. Drug Disc.* **2004**, *23*, 175–198.
88. Ugi, I.; Steinbruckner, C. *Chem. Ber.* **1961**, *94*, 734–742.
89. Patchett, A. A.; Nargund, R. P. *Ann. Rep. Med. Chem.* **2000**, *35*, 289–298.
90. MDL Drug Data Report. MDL Information Systems: San Leandro, CA, USA.
91. Bradley, E. K.; Beroza, P.; Penzotti, J. E.; Grootenhuis, P. D. J.; Spellmeyer, D. C.; Miller, J. L. *J. Med. Chem.* **2000**, *43*, 2770–2774.
92. Zuckerman, R. N.; Martin, E. J.; Spellmeyer, D. C.; Stauber, G. B.; Shoemaker, K. R.; Kerr, J. M.; Figliozzi, G. M.; Goff, D. A.; Siani, M. A.; Simon, R. J. et al. *J. Med. Chem.* **1994**, *37*, 2678–2685.
93. Beno, B. R.; Mason, J. S. Combinatorial Library Design Using both Properties and 3D Pharmacophore Fingerprints. *Book of Abstracts*, 221st ACS National Meeting, San Diego, 2001.
94. Good, A. C.; Cho, S.-J.; Mason, J. S. *J. Comput. Aided Mol. Des.* **2004**, *18*, 523–537.
95. GRID, Molecular Discovery Ltd, London, UK.
96. Shoichet, B.; Bodian, D.; Kuntz, I. D. *J. Comput. Chem.* **1992**, *13*, 380–397.
97. Evensen, E.; Joseph-McCarthy, D.; Karplus, M. MCSS version 2.1, 1997. Harvard University: Cambridge.
98. Hendlich, M.; Rippman, F.; Barnickel, G. *J. Mol. Graph. Model.* **1997**, *15*, 359–363.
99. Eksterowicz, J. E.; Evensen, E.; Lemmen, C.; Brady, G. P.; Lanctot, J. K.; Bradley, E. K.; Saiah, E.; Robinson, L. A.; Grootenhuis, P. D. J.; Blaney, J. M. *J. Mol. Graph. Model.* **2002**, *20*, 469–477.
100. Brady, G. P.; Stouten, P. F. W. *J. Comput. Aided Mol. Des.* **2000**, *14*, 383–401.
101. DOCKIT, **2000**, Metaphorics, Santa Fe, NM.
102. Arnold, J. R.; Burdick, K. W.; Pegg, S. C-H.; Toba, S.; Lamb, M. L.; Kuntz, I. D. *J. Chem. Inf. Comput. Sci.* **2004**, *44*, 2190–2198.
103. Wolber, G.; Langer, T. *J. Chem. Inf. Model.* **2005**, *45*, 160–169.
104. Catalyst, Version 4.7, Accelrys Inc., 9685 Scranton Road, San Diego, CA 92121-3752, USA.
105. Grant, J. A.; Pickup, B. T. *J. Chem. Inf. Comput. Sci.* **1995**, *99*, 3503–3510.
106. Grant, J. A.; Gallardo, M. A.; Pickup, B. T. *J. Comput. Chem.* **1996**, *14*, 1653–1666.
107. ROCS (Rapid Overlay of Chemical Structures), version 2.0, Openeye Scientific Software: Santa Fe, New Mexico, USA, 2004.
108. Haigh, J. A.; Pickup, B. T. *J. Chem. Inf. Model.* **2005**, *45*, 673–684.
109. Putta, S.; Lemmen, C.; Beroza, P.; Greene, J. *J. Chem. Inf. Comput. Sci.* **2002**, *42*, 1230–1240.
110. Bradley, E. K.; Miller, J. L.; Saiah, E.; Grootenhuis, P. D. J. *J. Med. Chem.* **2003**, *46*, 4360–4364.
111. Srinivasan, J.; Castellino, A.; Bradley, E. K.; Eksterowicz, J. E.; Grootenhuis, P. D. J.; Putta, S.; Stanton, R. V. *J. Med. Chem.* **2002**, *45*, 2494–2500.
112. Available from Tripos Inc. www.tripos.com (accessed April 2006).
113. Leardi, R. *J. Chemometrics* **2001**, *15*, 559–569.
114. Sutter, J. M.; Jurs, P. C. *Data Handling Sci. Technol.* **1995**, *15*, 111–132.
115. Sheridan, R. P.; Kearsley, S. K. *J. Chem. Inf. Comput. Sci.* **1995**, *35*, 310–320.
116. Sheridan, R. P.; SanFeliciano, S. G.; Kearsley, S. K. *J. Mol. Graph. Model.* **2000**, *18*, 320–334.
117. Weber, L.; Wallbaum, S.; Broger, C.; Gubernator, K. *Agnew. Chem. Int. Ed. Engl.* **1995**, *34*, 2280–2282.
118. Illgen, K.; Enderle, T.; Broger, C.; Weber, L. *Chem. Biol.* **2000**, 7, 433–441.
119. Yokobayashi, Y.; Ikebukuro, K.; McNiven, S.; Karube, I. *J. Chem. Soc. Perkin Trans.* **1996**, *20*, 2435–2437.
120. Singh, J.; Ator, M. A.; Jaeger, E. P.; Allen, M. P.; Whipple, D. A.; Soloweij, J. E.; Chowdhary, S.; Treasurywala, A. M. *J. Am. Chem. Soc.* **1996**, *118*, 1669–1676.
121. Gillet, V. J.; Willett, P.; Bradshaw, J.; Green, D. V. S. *J. Chem. Comput. Sci.* **1999**, *39*, 169–177.
122. Gillet, V. J.; Khatib, W.; Willett, P.; Fleming, P. J.; Green, D. V. S. *J. Chem. Inf. Comput. Sci.* **2002**, *42*, 375–385.
123. Gillet, V. J.; Willett, P.; Fleming, P. J.; Green, D. V. S. *J. Mol. Graph. Model.* **2002**, *20*, 491–498.
124. Agrafiotis, D. K. *J. Comput. Aided Mol. Des.* **2002**, *16*, 335–356.
125. Good, A. C.; Lewis, R. A. *J. Med. Chem.* **1997**, *40*, 3926–3936.
126. Bohm, H.-J.; Stahl, M. *Curr. Opin. Chem. Bio.* **2000**, *4*, 283–286.

127. Beavers, M. P.; Chen, X. *J. Mol. Graph. Model.* **2002**, *20*, 463–468.
128. Kick, E. K.; Roe, D. C.; Skillman, A. G.; Liu, G.; Ewing, T.; Sun, Y.; Kuntz, I. D.; Ellman, J. A. *Chem. Biol.* **1997**, *4*, 297–307.
129. Haque, T. S.; Skillman, A. G.; Lee, C. E.; Habashita, H.; Gluzman, I. Y.; Ewing, T. J. A.; Goldberg, D. E.; Kuntz, I. D.; Ellman, J. A. *J. Med. Chem.* **1999**, *42*, 1428–1440.
130. Roe, D. C. *Mol. Divers. Drug Des.* **1999**, 141–173.
131. Rarey, M.; Lengauer, T. *Perspect. Drug Disc. Des.* **2000**, *20*, 63–81.
132. Makino, S.; Todd, J. A.; Kuntz, I. D. *J. Comput. Aided Drug. Des.* **1999**, *13*, 513–532.
133. Murray, C. W.; Clark, D. E.; Auton, T. R.; Firth, M. A.; Li, J.; Sykes, R. A.; Waszkowycz, B.; Westhead, D. R.; Young, S. C. *J. Comput. Aided Mol. Des.* **1997**, *11*, 193–207.
134. Sun, Y.; Ewing, T. J. A.; Skillman, A. G.; Kuntz, I. D. *J. Comput. Aided Mol. Des.* **1998**, *12*, 597–604.
135. Sprous, D. G.; Lowis, D. R.; Leonard, J. M.; Heritage, T.; Burkett, S. N.; Baker, D. S.; Clark, R. D. *J. Comb. Chem.* **2004**, *6*, 530–539.
136. Frye, L. L.; Murphy, R. B.; Reboul, T. M.; Shenkin, P. S.; Mainz, D. T.; Chambers, E. W.; McDonald, D. Q.; Friesner, R. A. Structure-Based Design of Focused Drug-Like Combinatorial Libraries. *Abstracts of Papers*, 229th ACS National Meeting, San Diego, CA, March 13–17, 2005.
137. Kuntz, I. D.; Blaney, J. M.; Oatley, S. J.; Langridge, R.; Ferrin, T. E. *J. Mol. Biol.* **1982**, *161*, 269–288.
138. Lamb, M. L.; Burdick, K. W.; Toba, S.; Young, M. M.; Skillman, A. G.; Zou, X.; Arnold, J. R.; Kuntz, I. D. *Prot. Struct. Funct. Genet.* **2001**, *42*, 296–318.
139. Rarey, M.; Kramer, B.; Lengauer, T.; Klebe, G. *J. Mol. Biol.* **1996**, *261*, 470–489.
140. Friesner, R. A.; Banks, J. L.; Murphy, R. B.; Halgren, T. A.; Klicic, J. J.; Mainz, D. T.; Repasky, M. P.; Knoll, E. H.; Shelley, M.; Perry, J. K. et al. *J. Med. Chem.* **2004**, *47*, 1739–1749.
141. Minranker, A.; Karplus, M. *Proteins* **1991**, *11*, 29–34.
142. McCarthy-Joseph, D.; Tsang, S. K.; Filman, D. J.; Hogle, J. M.; Karplus, M. *J. Am. Chem. Soc.* **2001**, *123*, 12758–12769.
143. Krier, M.; Araujo-Junior, J. X.; Schmitt, M.; Duranton, J.; Justiano-Basaran, H.; Lugnier, C.; Bourguignon, J.-J.; Rognan, D. *J. Med. Chem.* **2005**, *48*, 3816–3822.
144. Aronov, A. M.; Bemis, G. W. *Proteins* **2004**, *57*, 36–50.
145. Hilbert, M.; Bohm, G.; Jaenicke, R. *Proteins* **1993**, *17*, 138–151.
146. Fanelli, F.; De Benedetti, P. G. *Chem. Rev.* **2005**, *105*, 3297–3351.
147. Bissantz, C.; Bernard, P.; Hilbert, M.; Rognan, D. *Proteins Struct. Funct. Genet.* **2003**, *50*, 5–25.
148. Evers, A.; Klabunde, T. *J. Med. Chem.* **2005**, *48*, 1088–1097.
149. Rees, D. C.; Congreve, M.; Murray, C. W.; Carr, R. *Nat. Rev. Drug. Disc.* **2004**, *3*, 660–672.
150. Carr, R. A. E.; Congreve, M.; Murray, C. W.; Rees, D. C. *Drug Disc. Today* **2005**, *10*, 987–992.
151. Erlanson, D. A.; McDowell, R. S.; O'Brien, T. *J. Med. Chem.* **2004**, *47*, 3463–3482.
152. Hopkins, A. L.; Groom, C. R.; Alex, A. *Drug Disc. Today* **2004**, *9*, 430–431.
153. Shuker, S. B.; Hajduk, P. J.; Meadows, R. P.; Fesik, S. W. *Science* **1996**, *274*, 1531–1534.
154. Hajduk, P. J.; Gomtsyan, A.; Didomenico, S.; Cowart, M.; Bayburt, E. K.; Solomon, L.; Severin, J.; Smith, R.; Walter, K.; Holzman, T. F. et al. *J. Med. Chem.* **2000**, *43*, 4781–4786.
155. Szczepankiewicz, B. G.; Liu, G.; Hajduk, P. J.; Abad-Zapatero, C.; Pei, Z.; Xin, Z.; Lubben, T. H.; Trevillyan, J. M.; Stashko, M. A.; Ballaron, S. J. et al. *J. Am. Chem. Soc.* **2003**, *125*, 4087–4096.
156. Hajduk, P. J.; Sheppard, G.; Nettesheim, D. G.; Olejniczak, E. T.; Shuker, S. B.; Meadows, R. P.; Steinman, D. H.; Carrera, G. M.; Marcotte, P. A.; Severin, J. et al. *J. Am. Chem. Soc.* **1997**, *119*, 5818–5827.
157. Erlanson, D. A.; Braisted, A. C.; Raphael, D. R.; Stroud, R. M.; Gordon, E. M.; Wells, J. A. *Proc. Natl. Acad. Sci.* **2000**, *97*, 9367–9372.
158. Braisted, A. C.; Oslob, J. D.; Delano, W. L.; Hyde, J.; McDowell, R. S.; Waal, N.; Yu, C.; Arkin, M. R.; Raimundo, B. C. *J. Am. Chem. Soc.* **2003**, *125*, 3714–3715.
159. Hartshorn, M. J.; Murray, C. W.; Cleasby, A.; Frederickson, M.; Tickle, I. J.; Jhoti, H. *J. Med. Chem.* **2005**, *48*, 403–413.
160. Gill, A.; Cleasby, A.; Jhoti, H. *ChemBioChem* **2005**, *6*, 506–512.
161. Gill, A. L.; Frederickson, M.; Cleasby, A.; Woodhead, S. J.; Carr, M. G.; Woodhead, A. J.; Walker, M. T.; Congreve, M. S.; Devine, L. A.; Tisi, D. et al. *J. Med. Chem.* **2005**, *48*, 414–426.
162. Wyss, P. C.; Gerber, P.; Hartman, P. G.; Hubschwerlen, C.; Locher, H.; Marty, H.-P.; Stahl, M. *J. Med. Chem.* **2003**, *46*, 2304–2312.
163a. Holenz, J.; Mercè, R.; Díaz, J. L.; Guitart, X.; Codony, X.; Dordal, A.; Romero, G.; Torrens, A.; Mas, J.; Andaluz, B. *J. Med. Chem.* **2005**, *48*, 1781–1795.
163b. Shultz, M. D.; Ham, Y.-W.; Lee, S.-G.; Davis, D. A.; Brown, C.; Chmielewski, J. *J. Am. Chem. Soc.* **2004**, *126*, 9886–9887.
163c. Sagara, Y.; Sagara, T.; Mase, T.; Kimura, T.; Numazawa, T.; Fujikawa, T.; Noguchi, K.; Ohtake, N. *J. Med. Chem.* **2002**, *45*, 984–987.
163d. Wang, G. T.; Chen, Y.; Wang, S.; Gentles, R.; Sowin, T.; Kati, W.; Muchmore, S.; Giranda, V.; Stewart, K.; Sham, H. *J. Med. Chem.* **2001**, *44*, 1192–1201.
164a. Bohm, H. J. J. *Comput. Aided Mol. Des.* **1992**, *6*, 61–78.
164b. Bohm, H. J. J. *Comput. Aided Mol. Des.* **1992**, *6*, 593–606.
165. Lew, A.; Chamberlin, A. R. *Bioorg. Med. Chem. Lett.* **1999**, *9*, 3267–3272.
166a. Aiyar, J.; Withka, J. M.; Rizzi, J. P.; Singleton, D. H.; Andrews, G. C.; Lin, W.; Boyd, J.; Hanson, P.; Simon, M.; Dethlefs, B. et al. *Neuron* **1995**, *15*, 1169–1181.
166b. Aiyar, J.; Rizzi, J. P.; Gutman, G. A.; Chandy, K. G. *J. Biol. Chem.* **1996**, *271*, 31013–31016.
167. Im, I.; Webb, T. R.; Gong, Y.-D.; Kim, J.-I.; Kim, Y.-C. *J. Comb. Chem.* **2004**, *6*, 207–213.
168. Duart, M. J.; Antón-Fos, G. M.; Alemán, P. A.; Gay-Roig, J. B.; González-Rosende, M. E.; Gálvez, J.; García-Domenech, R. *J. Med. Chem.* **2005**, *48*, 1260–1264.
169a. Bleicher, K. H.; Wüthrich, Y.; De Boni, M.; Kolczewski, S.; Hoffmann, T.; Sleight, A. J. *Bioorg. Med. Chem. Lett.* **2002**, *12*, 2519–2522.
169b. Bleicher, K. H.; Wüthrich, Y.; Adam, G.; Hoffmann, T.; Sleight, A. J. *Bioorg. Med. Chem. Lett.* **2002**, *12*, 3073–3076.
170. Edvinsson, K. M.; Herslöf, M.; Holm, P.; Kann, N.; Keeling, D. J.; Mattsson, J. P.; Nordén, B.; Shcherbukhin, V. *Bioorg. Med. Chem. Lett.* **2000**, *10*, 503–507.
171. Naya, A.; Kobayashi, K.; Ishikawa, M.; Ohwaki, K.; Saeki, T.; Noguchi, K.; Ohtake, N. *Bioorg. Med. Chem. Lett.* **2001**, *11*, 1219–1223.
172. Molteni, V.; Penzotti, J.; Wilson, D. M.; Termin, A. P.; Mao, L.; Crane, C. M.; Hassman, F.; Wang, T.; Wong, H.; Miller, K. J. *J. Med. Chem.* **2004**, *47*, 2426–2429.

173. Wipf, P.; Cunningham, A.; Rice, R. L.; Lazo, J. S. *Bioorg. Med. Chem.* **1997**, *5*, 165–177.
174. Huang, P.; Loew, G. H.; Funamizu, H.; Mimura, M.; Ishiyama, N.; Hayashida, M.; Okuno, T.; Shimada, O.; Okuyama, A.; Ikegami, S. *J. Med. Chem.* **2001**, *44*, 4082–4091.
175. Huang, P.; Kim, S.; Loew, G. *J. Comput. Aided Mol. Des.* **1997**, *11*, 21–28.
176. Chen, Z.; Miller, W. S.; Shan, S.; Valenzano, K. J. *Bioorg. Med. Chem. Lett.* **2003**, *13*, 3247–3252.
177. Moitessier, N.; Dufour, S.; Chrétien, F.; Thiery, J. P.; Maigret, B.; Chapleur, Y. *Bioorg. Med. Chem.* **2001**, *9*, 511–523.
178. Khabnadideh, S.; Pez, D.; Musso, A.; Brun, R.; Pérez, L. M. R.; González-Pacanowska, D.; Gilbert, I. H. *Bioorg. Med. Chem.* **2005**, *13*, 2637–2649.
179. Poulain, R.; Horvath, D.; Bonnet, B.; Eckhoff, C.; Chapelain, B.; Bodinier, M.-C.; Déprez, B. *J. Med. Chem.* **2001**, *44*, 3378–3390.
180. Horvath, D. ComPharm-Automated Comparative Analysis of Pharmacophoric Patterns and Derived QSAR Approaches. Novel Tools in High-Throughput Drug Discovery. A Proof of Concept Study Applied to Farnesyl Transferase Inhibitor Design. In *QSPR/QSAR Studies by Molecular Descriptors*; Diudea, M., Ed.; Nova Science: New York, 2001, pp 395–439.
181. Schneider, G.; Nettekoven, M. *J. Comb. Chem.* **2003**, *5*, 233–237.
182. Schneider, G.; Neidhart, W.; Giller, T.; Schmid, G. *Angew. Chem. Int. Ed. Engl.* **1999**, *38*, 2894–2896.
183. Larsson, A.; Johansson, S. M. C.; Pinker, J. S.; Hultgren, S. J.; Almqvist, F.; Kihlberg, J.; Linusson, A. *J. Med. Chem.* **2005**, *48*, 935–945.
184. Rusinko, A., III; Young, S. S.; Drewry, D. H.; Gerritz, S. W. *Comb. Chem. High Throughput Scree.* **2002**, *5*, 125–133.
185. Honma, T.; Hayashi, K.; Aoyama, T.; Hashimoto, N.; Machida, T.; Fukasawa, K.; Iwama, T.; Ikeura, C.; Ikuta, M.; Suzuki-Takahashi, I. *J. Med. Chem.* **2001**, *44*, 4615–4627.
186. Honma, T.; Yoshizumi, T.; Hashimoto, N.; Hayashi, K.; Kawanishi, N.; Fukasawa, K.; Takaki, T.; Ikeura, C.; Ikuta, M.; Suzuki-Takahashi, I. *J. Med. Chem.* **2001**, *44*, 4628–4640.
187. Swann, S. L.; Bergh, J.; Farach-Carson, M. C.; Ocasio, C. A.; Koh, J. T. *J. Am. Chem. Soc.* **2002**, *124*, 13795–13805.
188. Smart, B. P.; Pan, Y. H.; Weeks, A. K.; Bollinger, J. G.; Bahnson, B. J.; Gelb, M. H. *Bioorg. Med. Chem.* **2004**, *12*, 1737–1749.
189. Kaila, N.; Somers, W. S.; Thomas, B. E.; Thakker, P.; Janz, K.; DeBernardo, S.; Tam, S.; Moore, W. J.; Yang, R.; Wrona, W. *J. Med. Chem.* **2005**, *48*, 4346–4357.
190. McMartin, C.; Bohacek, R. S. *J. Comput. Aided Mol. Des.* **1997**, *62*, 465–473.
191. Tedder, M. E.; Nie, Z.; Margosiak, S.; Chu, S.; Feher, V. A.; Almassy, R.; Appelt, K.; Yager, K. M. *Bioorg. Med. Chem. Lett.* **2004**, *12*, 3165–3168.
192. Nacro, K.; Bienfait, B.; Lee, J.; Han, K.-C.; Kang, J.-H.; Benzaria, S.; Lewin, N. E.; Bhattacharyya, D. K.; Blumberg, P. M. *J. Med. Chem.* **2000**, *43*, 921–944.
193. Wang, S.; Milne, G. W. A.; Nicklaus, M. C.; Marquez, V. E.; Lee, J.; Blumberg, P. M. *J. Med. Chem.* **1994**, *37*, 1326–1338.
194. Verras, A.; Kuntz, I. D.; Ortiz de Montellano, P. R. *J. Med. Chem.* **2004**, *47*, 3572–3579.
195. Liebeschuetz, J. W.; Jones, S. D.; Morgan, P. J.; Murray, C. W.; Rimmer, A. D.; Roscoe, J. M. E.; Waszkowycz, B.; Welsh, P. M.; Wylie, W. A.; Young, S. C. *J. Med. Chem.* **2002**, *45*, 1221–1232.
196a. South, M. S.; Case, B. L.; Wood, R. S.; Jones, D. E.; Hayes, M. J.; Girard, T. J.; Lachance, R. M.; Nicholson, N. S.; Clare, M.; Stevens, A. M. *Bioorg. Med. Chem. Lett.* **2003**, *13*, 2319–2325.
196b. Parlow, J. J.; Case, B. L.; Dice, T. A.; Fenton, R. L.; Hayes, M. J.; Jones, D. E.; Neumann, W. L.; Wood, R. S.; Lachance, R. M.; Girard, T. J. *J. Med. Chem.* **2003**, *46*, 4050–4062.
197. Choe, Y.; Brinen, L. S.; Price, M. S.; Engel, J. C.; Lange, M.; Grisostomi, C.; Weston, S. G.; Pallai, P. V.; Cheng, H.; Hardy, L. W. *Bioorg. Med. Chem.* **2005**, *14*, 2141–2156.
198. Ji, T.; Lee, M.; Pruitt, S. C.; Hangauer, D. G. *Bioorg. Med. Chem. Lett.* **2004**, *14*, 3875–3879.
199. Krier, M.; de Araújo-Júnior, J. X.; Schmitt, M.; Duranton, J.; Justiano-Basaran, H.; Lugnier, C.; Bourguignon, J.-J.; Rognan, D. *J. Med. Chem.* **2005**, *48*, 3816–3822.
200. Douguet, D.; Munier-Lehmann, H.; Labesse, G.; Pochet, S. *J. Med. Chem.* **2005**, *48*, 2457–2468.
201. Mitchell, D. D.; Pickens, J. C.; Korotkov, K.; Fan, E.; Hol, W. G. J. *Bioorg. Med. Chem.* **2004**, *12*, 907–920.
202. Linusson, A.; Gottfries, J.; Olsson, T.; Örnskov, E.; Folestad, S.; Nordén, B.; Wold, S. *J. Med. Chem.* **2001**, *44*, 3424–3429.
203. Svensson, A.; Larsson, A.; Emtenäs, H.; Hedenström, M.; Fex, T.; Hultgren, S. J.; Pinkner, J. S.; Almqvist, F.; Kihlberg, J. *ChemBioChem* **2001**, 915–918.
204. Emtenäs, H.; Åhlin, K.; Pinkner, J. S.; Hultgren, S. J.; Almqvist, F. *J. Comb. Chem.* **2002**, *4*, 630–639.

Biographies

Dora M Schnur is currently a Principal Scientist in computer-assisted drug design in the Pharmaceutical Research Institute of Bristol-Myers Squibb. She received a BS from Bucknell University and PhD from Temple University, where she worked with Dr David Dalton and with Dr Lou Allinger at the University of Georgia. After a 1-year postdoc with Dr Phillip Bowen at the Molecular Modeling Laboratory of the School of Pharmacy at the University of North Carolina, she spent 6½ years as a modeler in the Agricultural Division of Monsanto and ultimately was the team leader of that computational group. In 1996, she joined Pharmacopeia where she was responsible for combinatorial library design and other applied computational drug design. She joined Bristol-Meyers Squibb in 1999. Her research interests include library design, privileged substructure analyses, structure-based design, GPCR homology modeling, and QSAR.

Andrew J Tebben began his career as a medicinal chemist at Merck, making compounds for several ion channel and kinase targets. His interests eventually migrated from the bench to the computer, leading to a position in the Molecular Systems group at Merck doing computer-aided drug design. After 7 years at Merck, he moved to the CADD group at the Dupont Pharmaceutical Company, which was eventually acquired by Bristol-Myers Squibb. Andrew is currently an applications modeler in the CADD group at Bristol-Myers Squibb, focusing on library design, structure-based design, and GPCR modeling. He has a BS from the Ohio State University and an MS from Villanova University.

Cullen L Cavallaro began his studies in chemistry at Drew University in Madison, New Jersey. His chemistry education continued at Princeton University, where he received a PhD with the guidance of Prof Jeffrey Schwartz for research on the synthetic utility of organotitanium complexes. He then pursued research in combinatorial chemistry, solid-phase synthesis, automated synthesis, and parallel medicinal chemistry at Pharmacopeia and Coelacanth before joining Bristol-Myers Squibb in 2000. His current research is focused on high-throughput medicinal chemistry, parallel synthesis, and library design.

© 2007 Elsevier Ltd. All Rights Reserved
No part of this publication may be reproduced, stored in any retrieval system or transmitted in any form by any means electronic, electrostatic, magnetic tape, mechanical, photocopying, recording or otherwise, without permission in writing from the publishers

Comprehensive Medicinal Chemistry II
ISBN (set): 0-08-044513-6

ISBN (Volume 4) 0-08-044517-9; pp. 307–336

4.15 Library Design: Reactant and Product-Based Approaches

S D Pickett, GlaxoSmithKline, Stevenage, UK

© 2007 Elsevier Ltd. All Rights Reserved.

4.15.1 Introduction

Combinatorial chemistry has its origins in the solid-phase synthesis of peptides by Merrifield.[1] However, it was 20 years before Geysen[2] and Houghten[3] made the advances that enabled the rapid generation of large numbers of compounds. Lam *et al.*[4] used resin beads to produce millions of peptides that were evaluated against β-endorphin. This is generally recognized as the first truly combinatorial peptide synthesis. The potential of such an approach was realized with the extension to other chemistries and the rapid synthesis of large numbers of nonpeptidic compounds of the type more usually required within a drug discovery program. For example, Bunin and Ellman[5] were able to synthesize a library of benzodiazepines, a very important therapeutic class of molecules. Simon *et al.*[6] built upon the successes of peptide chemistry to design and synthesize peptoids, where the α-amino acid is replaced by an N-substituted glycine. DeWitt *et al.*[7] were able to synthesize hydantoins, dipepetides, and benzodiazepines.

The benzodiazepine library, as synthesized by DeWitt *et al.*[7] is illustrated in **Figure 1**. This scheme produced 40 benzodiazepine molecules related to valium, from reacting five resin-bound amino acids with eight 2-amino benzophenone imines. The compounds were assayed for inhibition of fluoronitrazepam and showed the expected profile. This example illustrates the power of combinatorial chemistry in making it possible to access relatively rapidly large numbers of compounds. The combinatorial nature of the approach means that doubling the size of the reagent lists to 10 amino acids and 16 2-amino benzophenone imines leads to a fourfold increase in the number of potential products to 160. The representation of the library as depicted by **1** is termed a Markush representation, familiar from

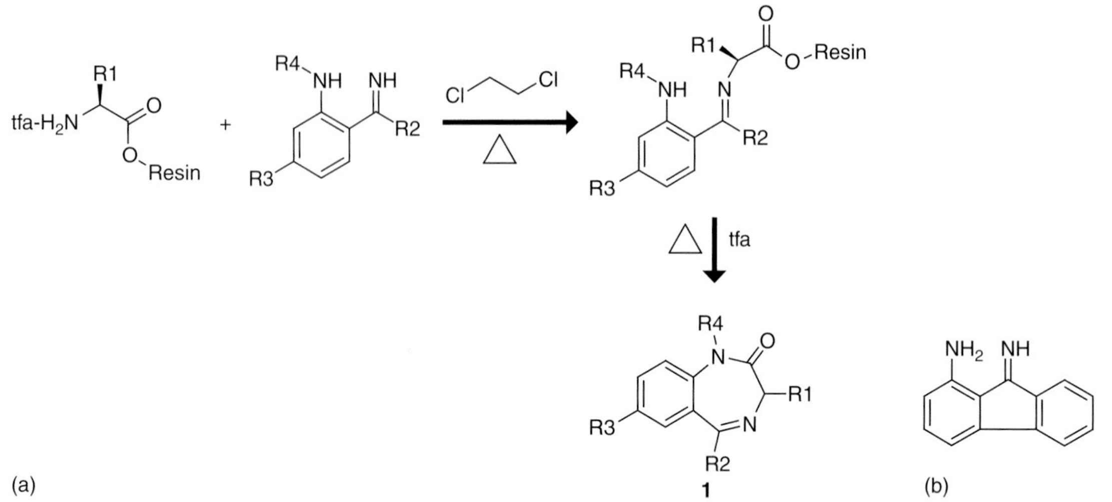

(a) (b)

Figure 1 (a) The benzodiazepine library synthesized by DeWitt *et al.*[7] Five amino acids were reacted with eight 2-amino benzophenone imines to give a library of 40 products. R^1 = H, Me, benzyl, 3-methylindole, isopropyl; R^2, R^3, R^4 = (Ph,H,H), (Ph,Cl,H), (4-MeOPh,H,H), (Ph,NO2,H), (Ph,Cl,Me), (cyclohexyl,H,H), (2-thienyl,H,H). (b) The eighth benzophenone imine used in the library.

the patent literature, and expanding this computationally to all plausible structures is termed enumeration. The theoretical list of products from a combinatorial reaction scheme is often termed the virtual library. The size of the virtual library can expand very rapidly to many millions of plausible structures with some well-known reaction schemes such as the Ugi reaction.[8]

An extensive review of published libraries and the trends observed in the size, scope, and application of combinatorial libraries in drug discovery is produced annually by Dolle.[9–16] Initially, combinatorial libraries were produced to enlarge corporate screening collections as high-throughput screening (HTS) methodologies became available that permitted the screening of large numbers of compounds very rapidly.[17] Many of the early approaches tended to generate mixtures of compounds and were screened as such. Whilst mixture screening is still being used to produce lead compounds,[18] pharmaceutical companies have tended to move to the synthesis of smaller numbers of well-characterized compounds[19] because of issues of mixture deconvolution and possible assay interference. It has also been recognized that early libraries tended to generate larger, more hydrophobic compounds than were generally the norm for druglike compounds.[20] Library design provides a means to select the most appropriate reagents so that the properties of the synthesized library are within acceptable limits and to provide appropriate direction for either a diverse or focused library as appropriate. Thus library design is an important facet of the combinatorial chemistry approach to lead generation and lead optimization.[21–29]

In this chapter we review the various aspects of library design. We begin with a brief review of the role of library design in drug discovery and where it can impact upon lead generation and lead optimization. We follow this with an overview of the whole library design process, taking the reader through the various stages from initial decision making on which chemistry to employ to the final design and subsequent iterations as building blocks fail in validation or are not available. This is followed by a detailed analysis of the various approaches that have been taken to designing combinatorial libraries and addressing the issues posed by the strength of the synthetic approach, that is, the large number of theoretically possible compounds that are available within a virtual library. The concepts of molecular diversity, chemical descriptors, similarity, protein and ligand-based virtual screening are covered in other chapters (*see* 4.08 Compound Selection Using Measures of Similarity and Dissimilarity; 4.12 Docking and Scoring; 4.14 Library Design: Ligand and Structure-Based Principles for Parallel and Combinatorial Libraries; 4.19 Virtual Screening; 4.24 Structure-Based Drug Design – The Use of Protein Structure in Drug Discovery); hence we shall review only the aspects pertinent for our discussion here. In Section 4.15.5 we review in detail some of the theoretical and practical considerations in designing libraries and discuss successful applications of library design in lead generation using both generic and focused libraries. We also discuss the role of library design in lead optimization where it is important to consider many aspects of the problem including absorption, distribution, metabolism, excretion, and toxicity (ADMET) and selectivity constraints alongside activity against the target.

4.15.2 Library Design in Drug Discovery

High-throughput screening and combinatorial chemistry are important components of the drug discovery process. However, contemporaneously with their introduction, pharmaceutical companies have seen a leveling off or even decline in productivity. Many plausible reasons for this have been proposed, mainly focusing on business factors such as the strengthening of the regulatory framework.[30–33] At the same time the industry has tended to move towards a more reductionist approach to drug discovery,[34] where HTS against specific targets has replaced the traditional physiology-based approach. One hypothesis for the productivity decline being that compounds discovered through more physiologically relevant cell-based or in vivo screening are one step ahead with regard to the requisite druglike properties of absorption and metabolic profile. Combinatorial chemistry too has taken its share of criticism. Undoubtedly, such reappraisals are not unexpected with new technologies, particularly when they lead to a large change in strategy for a multibillion dollar industry. Interestingly, similar criticisms have been directed at computational chemistry and structure-based drug design and the new high-throughput technologies were even heralded as a paradigm shift away from rational design to a more random approach.[35,36] Of course, on further reflection both diversity screening and focused approaches have their role in drug discovery.[37] It has been noted by several authors[38,39] that despite the apparent differences in philosophical approach between HTS and combinatorial chemistry on the one hand and rational drug design, with computational chemistry as one component, on the other, the two strategies in fact support each other. For example, clustering and data mining methods enable chemists to navigate HTS data to select the compounds to follow up and identify borderline hits.[40–44] Combinatorial chemistry benefits from computational methods for achieving appropriate property distributions and is very powerful when merged with structure-based drug design and other focused approaches.[39,45] On the other hand, computational methods can benefit from the ability of combinatorial chemistry to generate large numbers of compounds relatively quickly. Computational techniques are approximate and there is no model that is 100% accurate which makes application to individual compounds an issue. However, the selection of a large number of compounds for a combinatorial library has great merit even for approximate ADMET models.[46] Thus, the application of library design in drug discovery is wide-ranging and specific examples in a variety of application areas are given in Section 4.15.5.

High-throughput synthetic methodologies are used throughout the drug discovery cycle. At the earliest stages of the process, peptide[47] or peptidomimetic[48] libraries can be designed to deorphan receptors[49,50] and aid in target identification. Such libraries can provide tool compounds for target validation. Many of the early reviews on library design went hand in hand with discussions of molecular diversity,[22,26] reflecting the initial focus on the use of combinatorial chemistry to enhance corporate screening collections with large numbers of diverse compounds. Latterly, sophisticated mathematical models have been developed to help guide the generation of collection enhancement libaries.[51,52]

As companies evolved to a target class approach to lead generation,[53,54] so knowledge-based design methods have been developed to facilitate target class combinatorial library design.[55] G protein-coupled receptors (GPCRs) and kinases have been the main focus.[56–58] The former represent a significant proportion (around 50%) of current drug targets.[59] Kinases have emerged as an important class of enzymes for drug discovery with over 500 kinases in the human genome[60] and a wealth of three-dimensional protein structures from nuclear magnetic resonance (NMR) and x-ray crystallography to facilitate structure-based design. Following identification of a suitable hit structure combinatorial chemistry can be used to rapidly expand the hit to develop structure–activity relationships (SAR). In these hit expansion libraries, design can be used to ensure suitable diversity is explored whilst at the same time ensuring appropriate properties of the molecules and combining with structure-based approaches where possible.[45,61] In lead optimization similar approaches can be employed with a particular emphasis on optimizing the ADMET properties of the compounds.[62,63] Library design plays an important role in the emerging fields of fragment-based and NMR screening.[64,65] In the next section we describe the various approaches and algorithms that have been developed for library design.

4.15.3 Library Design Process

The virtual library represents all possible products from the combinatorial combination of appropriate reagents. Prior to a discussion of how best to optimize the choice of these reagents to generate the final library for synthesis, which is a (sometimes very small) subset of the virtual library, it is necessary to state assumptions that we are making at this point. These are discussed in further detail in Section 4.15.3.1 where we present the full library design process. Firstly, it is assumed that a decision has been made as to which chemistry is most appropriate for the problem at hand. This seemingly obvious question is rarely mentioned in reviews on library design. Secondly, reagent preselection is an important component of a successful library design. We shall assume that the chemist and/or designer have made a suitable choice of reagents from all available on the grounds of chemical tractability (the reagent is likely to react),

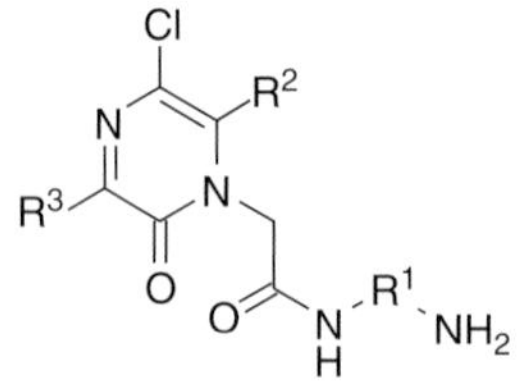

Cl
N
R^2
R^3
N
O
O
N
H
R^1
NH_2

Figure 2 Three-component library taken from Parlow *et al.*[66] The goal of library optimization is to select a subset of R^1, R^2, and R^3 so as to optimize the properties of the resulting combinatorial sublibrary.

an appropriate protection/deprotection strategy is in place to prevent undesirable side reactions, and the reagents are available at a cost that is not too prohibitive (as discussed later cost is one property that can be optimized in library design). In other words, the reagents selected by the design process are acceptable for further synthetic evaluation. Given these assumptions, library design can be cast as a constrained optimization problem with the goal of selecting a subset of reagents so as to optimize the properties of the resulting compounds.

Three general strategies have been adopted to solve this problem. In the first of these approaches, the product properties are ignored and the assumption is made that an appropriate design can be made by selecting each set of reagents independently. This is termed a reagent-based design. The alternative strategy, reagent-biased product design (RBPD), is to select the reagents based on the product properties of the resulting sublibrary in the context of the combinatorial constraint. That is, for a three-component library such as that shown in **Figure 2**[66] the goal is to select a subset of reagents, r^1, r^2, r^3 from the available set R^1, R^2, R^3 so as to find the best subset of products given by $r^1 \times r^2 \times r^3$. Best is here defined as a combination of some user-defined set of properties such as diversity, physical property distribution, docking score, absence of predicted ADMET liabilities as discussed further in Section 4.15.4. Of course, a third strategy is to ignore the combinatorial constraint altogether and simply choose the best set of products. This is known as 'cherry-picking' and would involve no special treatment over methods for compound selection and virtual screening described elsewhere (*see* 4.08 Compound Selection Using Measures of Similarity and Dissimilarity; 4.19 Virtual Screening; 4.20 Screening Library Selection and High-Throughput Screening Analysis/Triage). The designs that result are very inefficient from a combinatorial perspective as the products tend to be spread across the whole matrix of reagent space and cannot be synthesized efficiently. However, methods have been developed that attempt to benefit from the greater flexibility of a cherry-picked design whilst maintaining some of the benefits of the combinatorial synthesis.

4.15.3.1 Overview

In the following sections we explore in some detail the algorithms available for combinatorial library design. However, this is only one step in the library design process. First we look at the full library design process from selection of the initial chemistry through to the selection of the final set of compounds to be synthesized. A schematic of the library design work-flow is presented in **Figure 3**. A general discussion of each of these steps will be followed by a specific illustration from our own work. In this section we cover questions such as the choice of descriptors, reagent filtering and methods for evaluating the optimum size of a library.

The first and probably most important step is to decide upon which chemistry should be used in the design. Clearly, this choice will be influenced to a large degree by the purpose of the library. In lead optimization there will probably be a known active compound that can serve as the starting point. If a protein structure is available then docking or a de novo design algorithm could be used to suggest a possible template.[67] In the case of more general library designs, templates may be selected on the basis of perceived under-representation of a template in a screening collection. In this latter case there may be many chemistry options and an informed choice needs to be made between the templates. This question has been addressed by Pickett *et al.*[68] The procedure COMPLIB compares the pharmacophore coverage, as estimated from three-dimensional pharmacophore fingerprints, of a series of potential libraries. In the example cited 12 libraries were compared using a common set of reagents. A subset of the most informative libraries was selected for synthesis based upon pharmacophore coverage and diversity.

Once the chemistry has been decided upon, possible reagents must be identified. Generally these will come from searching databases such as the Available Chemicals Directory (ACD)[69] and in-house reagent repositories. For certain reagent classes, such as amines and carboxylic acids, there may be several thousand potential reagents. Several criteria may be used to reduce these numbers based both on chemistry and design decisions. Thus, it is important to take account of any possible problematic reagents as early as possible, particularly if docking or other computer-intensive

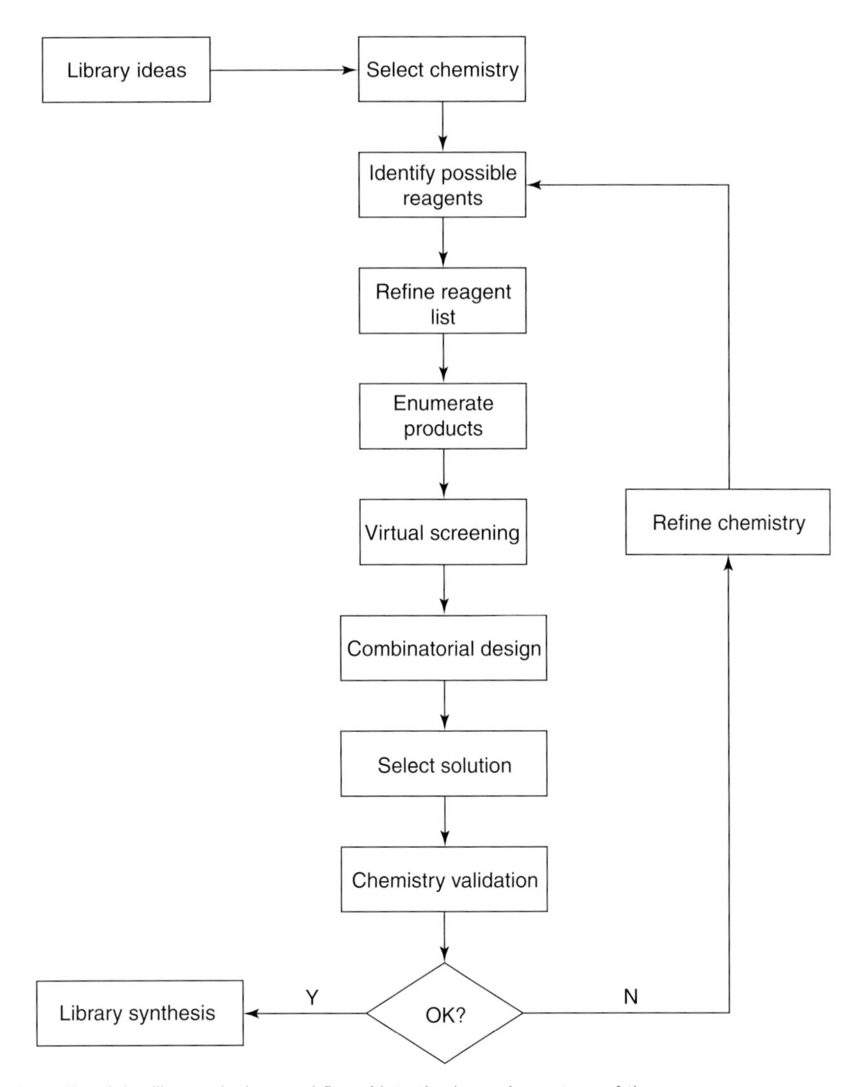

Figure 3 A schematic of the library design workflow. Note the interative nature of the process.

methods are to be used. The types of criteria to be considered include compatibility of the reagent in the reaction (for example certain classes of reagent may be less reactive than others), and protecting group strategies that may be accessible (so that, for example, amino acids can be included in the design). A number of substructure filters can also be used to significantly reduce the starting list of reagents.[70,71] Simple property filters on molecular weight and rotatable bond counts can be applied to remove reagents. For example, if the scaffold had a molecular weight in the region of 150, then any reagent in a multicomponent reaction with a molecular weight over about 350 is unlikely to lead to useful products in a design incorporating Lipinski type criteria. It is also the author's experience that it is very important for the chemist to visually inspect reagent lists prior to the start of what may be a lengthy design process. Cost and likely compound availability are also likely to be important. Certain, more esoteric reagents of the type often picked in diversity-based designs are frequently only available at high cost or with a lengthy delivery time. The exact nature of the library will determine the relative merits of such reagents. In addition any of the methods described in Section 4.15.3.2 may be used to further reduce the reagent list. The degree to which the reagent list needs to be reduced will depend upon the goal of the library and the size of the resulting virtual library.

The workflow described here assumes a standard design algorithm where appropriate properties are computed prior to the design algorithm, hence the requirement for enumeration. There are two general approaches to enumeration. Reagent clipping approaches involve the specification of the scaffold, clipping of all reagents, and subsequent combinatorial combination of the clipped reagents. This approach is straightforward provided that clipped reagents can be readily generated. Issues can arise, however, if some of the R-groups come from the same reagent which would have to be clipped twice. Clipping is also library-specific so reagents need to be processed for each library. Reaction-based enumeration involves the specification of the library using a series of reactions that govern the rules by which the reagents are combined. It is important to note that the reactions do not need to exactly match the chemistry to be employed in synthesis, for example protections and deprotections, use of resins, etc., and even the order of the reactions can be varied. The approach is more general as no modifications need be made to the reagents prior to enumeration and thus reagent lists can be reutilized from library to library. A reaction enumeration system has been described by Leach *et al.*[72] that uses the SMIRKS language[73] to define reactions in conjunction with the Daylight reaction toolkit for enumeration. Following enumeration, properties are calculated and additional virtual screening is carried out. This could include docking, pharmacophore searching, evaluating against a quantitative SAR (QSAR) model, and predicting ADMET liabilities as desired. How these properties are handled will depend upon the limitations of the design algorithm. For example, a simple mechanism to combine multiple properties is to flag compounds as passing or failing a property according to some cutoff and optimizing the number of compounds that pass all filters in the final design. Alternatively each property can be considered independently, but this leads to scaling issues and complex optimization functions as discussed in Section 4.15.3.3. The use of a multiobjective optimization methodology such as MoSELECT alleviates these issues as described in Section 4.15.3.6. Markush methods for storing combinatorial libraries can provide a very efficient mechanism for generating two-dimensional properties and fingerprints[74] and some of the methods described in Section 4.15.3.7 utilize the combinatorial nature of the virtual library to allow very efficient docking.

It is only at this stage that the product-based design is undertaken. An important point to note is the iterative nature of the process. The initial design may produce reagents that the chemist does not like – hence the need for an initial careful assessment of the reagents – or that are no longer available. As validation is carried out it may be that some reagents do not react as previously thought and so need to be eliminated from the design. In these cases it is not sensible to fully redesign the library – many of the reagents are already available and have been validated, and a full redesign could lead to a complete new set of reagents. Thus, the design software should provide the capability to preselect certain reagents so that the redesign is performed within a much smaller space. It may even be that the most pragmatic approach is a similarity search or equivalent to identify a replacement reagent within the virtual library. In this case the design software should rescore the new design to ensure that the design goals have not been severely compromised by the replacements.

As an example of this process we were presented with six possible libraries for prioritization and subsequent design. Potential reagent lists for each library were generated by searching of in-house and external databases using the program ADEPT.[72] These lists were filtered based on molecular property constraints and substructure filters appropriate to each library[70] and virtual libraries were enumerated. For each library, molecular properties (molecular weight, calculated log *P*, donor and acceptor counts) were computed. A sphere exclusion clustering algorithm[75] was used to cluster each library independently and so provide a basis for estimating the internal diversity of each library. In addition, the similarity of each product with the corporate screening collection was calculated to assess the novelty of the compounds with this external set and provide input for the screening collection model described in Section 4.15.5.1. The similarity profiles for each library, internally and with respect to the screening collection, are represented by the pie charts in **Figure 4a**. These plots show the proportion of compounds in the library with a defined number of nearest neighbors according to a similarity cutoff. One library, Lib4, was selected based on consideration of all these properties and was optimized using a multiobjective genetic algorithm implemented in the library design program MoSELECT.[76] This algorithm generates solutions according to a number of competing design objectives, with each solution being optimal according to the principle of Pareto optimality (*see* Section 4.15.3.6 for further details). Similarity profiles for several of the solutions are shown in **Figure 4b**. One of these solutions was subsequently selected for synthesis. An important point to note from **Figure 4b** is the great improvement in the profile of the designed library over the full virtual library. This can make for difficulties in interpreting plots such as those in **Figure 4a** for the complete virtual libraries and thus comparison of potential library proposals needs a great deal of careful consideration. This can be aided by a clear specification of the design goals and illustrates the need to compare libraries as a set rather than each individually – it is only in comparison with other ideas that the true value of the proposal can be assessed.

4.15.3.2 Reagent-Based Methods

Tripeptoid libraries,[6] mentioned briefly in Section 4.15.1, give access to potentially many millions of structures. One of the earliest published approaches to library design was developed in this context.[77] The approach taken was reagent-based; even today it would be impractical to enumerate all possible virtual products from such a library. A number of descriptors were calculated and combined using combinations of multidimensional scaling and principal components

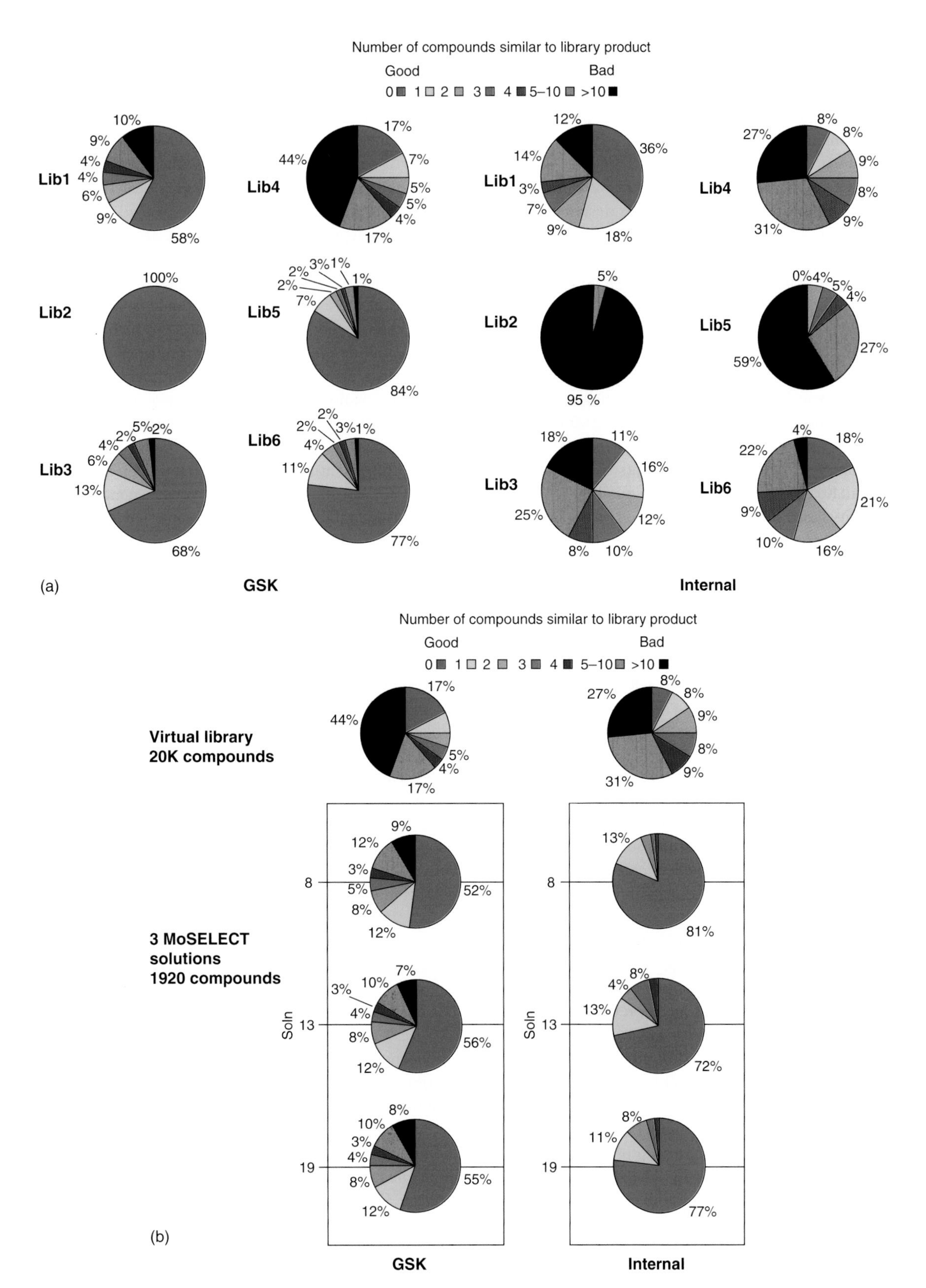

Figure 4 (a) Analysis of six virtual libraries according to their similarity with respect to the in-house screening collection and internally to the library. (b) Three solutions obtained with MoSELECT for Lib4. Note how the optimization has improved the diversity of the library both with respect to the corporate collection and internally.

analysis.[78] This transforms the substituent properties into a property space with specific coordinates allowing sophisticated experimental design approaches to be used that rely on an Euclidean distance measure. Chemical functionality, as described by the Daylight fingerprint,[73] was included by calculating the $N \times N$ dissimilarity (1 – Tanimoto coefficient) matrix between the reagents and converting to a small number of descriptors using multidimensional scaling. Lipophilicity was incorporated using a standard program such as CLOGP.[79,80] Shape and branching were accounted for by the use of topological indices[81–83] that can be calculated with programs such as MOLCONN-Z.[84] The 81 descriptors were reduced to latent variables with principal components analysis. For the amine and carboxylic acid reagent sets described in the work, the first five principal components explained over 85% of the variance in each case. A novel atom-layer descriptor was developed to describe the relationship of key pharmacophoric (or receptor recognition) features to the peptoid backbone. Five feature types were defined – acid, base, hydrogen bond donor, hydrogen bond acceptor, and aromatic – and the atomic radius of the atom was also taken into account. For each reagent a matrix is generated where each element represents the sum of atomic radii or the sum of each feature type at a particular topological distance from the reference point on the backbone. Two such tables can be compared to calculate the similarity between two reagents in this descriptor space. The similarity measure is represented by the ratio of the sum of minimum and maximum matrix elements and is calculated over all pairwise combinations of reagents. As for the fingerprint similarity measures, multidimensional scaling was applied to the resulting dissimilarity matrix. Finally, the various descriptor subsets were scaled to give equal weight to log P and the first principal component or multidimensional scaling component. The final set of 16–18 properties (depending on the number of principal components or multidimensional scaling dimensions) provides the frame of reference for design. Reagents can be selected by similarity to a side chain in a known active using a Euclidean distance measure. For diversity based selections an iterative D-optimal design procedure was used. D-optimal design is a method for selecting objects that are well spread out in a property space though it has been criticized in the context of chemistry-based applications for favoring selections from the edge of the space rather than sampling more uniformly through the space.[85] The advantage in this context is that a well-designed set of reagents should provide more information than one from a clustering exercise for example. In addition, it is possible to seed the D-optimal design algorithm with preselected reagents, derived from knowledge of the existing SARs, for example, and the speed of the algorithm allows for interactive analysis by the chemist to guide the selection.

The application of this methodology has led to the identification of several potent ligands.[86] However, it was also recognized[87] that many of the hits were outside the range of normal druglike molecules[20,62] in terms of flexibility, lipophilicity, or molecular weight, or were insoluble. Thus Martin and co-workers[87,88] extended the earlier work to allow for appropriate selection of the reagents to be more compliant with pharmaceutically relevant properties. Reagents are assigned to overlapping bins representing categorized properties such as rigid, polar, low molecular weight, expensive. The user selects the minimum or maximum number of reagents to be selected from each bin. The original optimization algorithm followed a 'greedy' approach, using D-optimal design[89] to select from each category iteratively, adding selected reagents to the starting set for the next optimization. A modification to the original Fedorov D-optimal algorithm[87] allowed for a parallel implementation where the entire property space is used throughout the optimization. An iterative single exchange algorithm starts from a random starting point, keeping the exchange that leads to the biggest increase in diversity score at each iteration. Several random starts can be used to ensure that the algorithm has not converged to a local minimum. As the libraries designed by such an approach are screened as mixtures, an interesting addition to this approach is the inclusion of an ambiguous molecular weight penalty. This is applied as an additional term to the diversity score to favor designs with no ambiguous molecular weights, allowing mass spectroscopy to identify a parent peak. A further refinement is a sensitivity analysis on the original user-defined profile, to monitor the tradeoff between diversity and each category.

The properties considered up to now in this approach have been two-dimensional in nature. Several groups have developed the use of three- and four-point pharmacophore descriptors to describe molecular diversity in an attempt to capture the three-dimensional nature of ligand–protein interactions.[90–93] Such descriptors have been employed in library design (*see* Section 4.15.4).[68,94–98] Martin and Hoeffel[97] have extended the work above to derive a pharmacophoric substituent descriptor termed OSPREYS (Oriented Substituent Pharmacophore PRopErtY Space). As the descriptors are calculated on the substituents rather than the enumerated products it is possible to generate a full similarity matrix and use multidimensional scaling to generate a Euclidean property space as for standard two-dimensional fingerprints mentioned above.

The reagents are clipped to the substituents and a dummy atom added to the attachment atom to represent the library template (for example the peptoid backbone). Conformational analysis is followed by pharmacophore perception. One edge of the pharmacophore is defined by the dummy atom and an orienting point, positioned along the bond connecting the substituent to the template. In this way one-, two-, and three-point pharmacophores are defined and stored in a fingerprint. All pairwise Tanimoto coefficients are calculated between the substituents and multidimensional scaling is used to generate a Euclidean space. These descriptors can be included with the two-dimensional property spaces described above. One problem with binary pharmacophore fingerprints is that discrete distance bin boundaries are used so that a small change in distance can lead to different bits being set. To account for this, an additional near neighbor fingerprint is constructed where each distance in turn sets adjacent bins. The similarity is a combination of the Tanimoto coefficient for the standard fingerprint, $T(\mathrm{fp})$ and the near neighbor fingerprint, $T(\mathrm{nnfp})$:

$$\text{Similarity} = 0.65 \times T(\mathrm{fp}) + 0.35 \times T(\mathrm{nnfp}) \qquad [1]$$

The generation of substituents pharmacophores then allows for a detailed conformational analysis with a more detailed energy function than is used for the generation of whole-molecule pharmacophores and the analyses can be stored and reused. The methodology can be applied to very large virtual libraries as the method scales by the number of substituents not products and the property space is Euclidean. In general, library designs based on whole-molecule pharmacophores have been limited to pharmacophore coverage calculations rather than similarity calculations (see the next section for a full discussion). In addition, considering one- to three-point OSPREYs for a three-component library to some extent contains information on up to nine-point whole-molecule pharmacophores. However, the methodology is limited to individual libraries and does not allow for comparison between libraries. In addition, some whole-molecule methods actually remove some of the smaller three-point pharmacophores that would be included in this analysis.[98] A fuller discussion of the relative merits of reagent based and reagent-biased product-based designs is given in Section 4.15.3.5.

The DIVSEL procedure of Pickett *et al.*[68] utilizes three-dimensional pharmacophore fingerprints as a descriptor for the reagents. The reagents are attached to the scaffold prior to calculation so that the conformational analysis takes account of the scaffold. In an iterative approach with a two-component library a representative subset of R^1 reagents were used in the analysis of R^2 groups so that the pharmacophore keys were the ensemble for R^2 with each R^1. An interactive dissimilarity based selection process was followed so that the chemist can decide on the suitability of the reagent whilst the algorithm is running. A user-defined similarity cutoff can remove similar compounds from consideration once a reagent is selected. To avoid flexible molecules dominating the selection the reagents are considered in an order defined by their complexity. The complexity is based on a functional group count and starts with the simplest reagents, adding more complex reagents as the algorithm proceeds. Dissimilarity was not the only factor considered during the analysis, the number of new pharmacophores being added by a reagent was also considered during the analysis thus combining the dissimilarity-based selection procedure with the partitioning aspects of the pharmacophore descriptors.

An alternative reagent based method specifically for targeted library design has been developed by Cramer and co-workers.[99–104] The starting point for this methodology is the concept of the topomer. The generation of a topomeric conformation for a substituent begins by generating a representative conformation using Concord[105,106] and aligning the scaffold attachment point with the origin. All other rotatable bonds are then adjusted sequentially according to a set of rules.[99] Steric or hydrogen bonding interaction fields are computed for each substitent as for the comparative molecular field analysis (CoMFA) approach, with the exception that the field values are reduced by a factor of 0.85 for each rotatable bond between the atom and the template attachment atom. ChemSpace[107] is a database system designed to hold substituents treated in this way. A number of two- (e.g., amide forming) and three-component (core and two reagents) libraries have been defined and suitable reagents abstracted from databases such as ACD.[69] These are stored in the ChemSpace system, which theoretically gives access to over 10^{13} enumerated products.[102]

To use this approach in library design, one or more query structures are fragmented according to the reaction schemes in the database. This could give rise to several different fragmentation patterns for one structure. Topomeric fields are calculated for all such fragments and used to search the substituent database. Top-ranking substituents can be retrieved and combined according to the reaction schemes to generate possible products for synthesis. This approach was validated prospectively, using four angiotensin II antagonists as query structures (**Figure 5a**).[101] A common reaction scheme could be used to synthesize the antagonists and this chemistry was used to populate a specific ChemSpace database (**Figure 5b**). The biphenyl acid (or tetrazole) moieties were selected from a set of readily available presynthesized templates, many of which would not be expected to be active. Topomer searching was used to select a set of nitrogen heterocycles. Of those selected a number were too expensive or unavailable, a common problem with library design as discussed in Section 4.15.3.1. Eventually, 425 compounds were synthesized and all seven which showed $>75\%$ binding were included in the 63 compounds selected by topomer shape similarity searching.

Leach *et al.*[70] have developed a procedure termed gridding and partitioning (GaP) for reagent selection. The main application of this approach is in reagent acquisition, ensuring that an appropriate range of starting materials is available

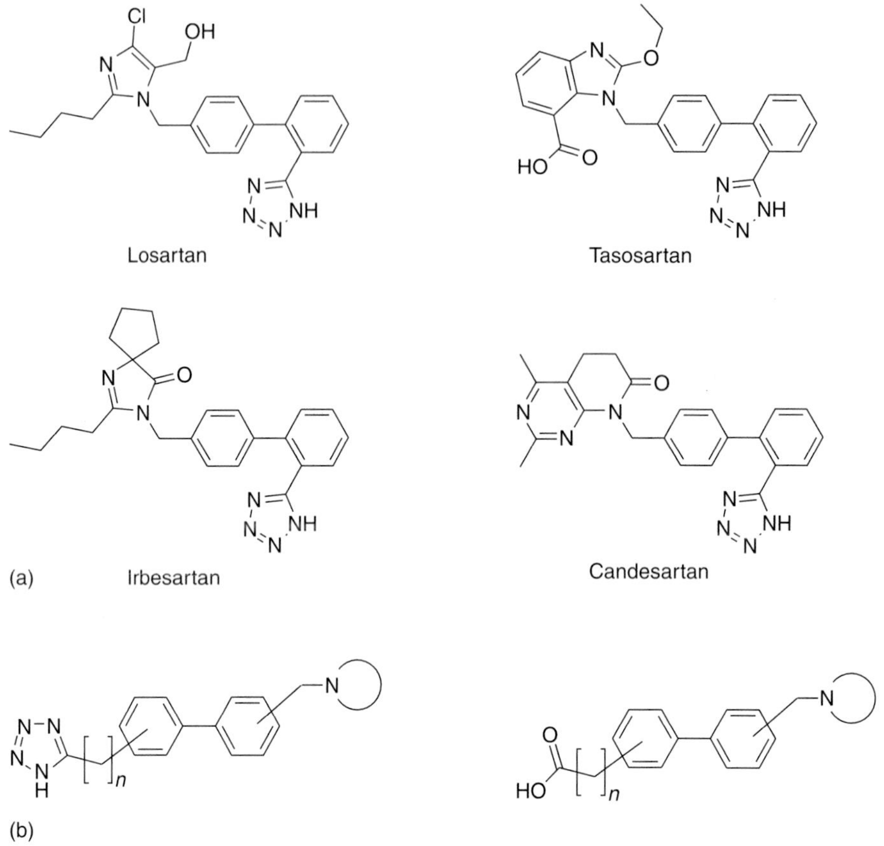

Figure 5 (a) Four angiotensin II antagonists used as query structures to search a ChemSpace database derived from the two virtual libraries shown in (b).

for a particular reaction scheme. However, the method also has applicability in reducing large reagent lists prior to a full product-based design. GaP is a pharmacophore-based method and the procedure is shown in **Figure 6**.

Reagents are aligned with the attachment atom at the origin and the adjacent nonhydrogen atom along the *x*-axis within a Cartesian coordinate space. In the case of groups such as secondary amines a dummy atom can be added where the appropriate template atom would be located (for example the carbonyl carbon for an amide forming reaction) and defined as the origin atom. A 1 Å grid is generated and a systematic conformational analysis is performed rooted at the origin. For each pharmacophoric atom type in the reagent, the pharmacophore cell occupied by the atom is recorded for every conformation. In addition, every conformation is freely rotated around the *x*-axis and occupied cells recorded. This is necessary as the same reagent can be involved in different reaction schemes, for example primary amines can combine with carboxylic acids to give amides or with aldehydes in reductive amination to leave the basic center. The location of all pairs of pharmacophore features is also recorded. Six pharmacophore features are recognized: hydrogen bond donor, hydrogen bond acceptor, combined hydrogen bond donor/acceptor groups (e.g., hydroxyl), aromatic ring, acid, and base. To make the method more specific for a particular combinatorial library, the template structure can be included and used as the reference point, with the attachment bond handled appropriately for its type (single, double, conjugated). If the structure of the protein target is available, the template may be modeled into the binding site and the reagent conformational analysis performed using the protein structure as a steric constraint.[108]

The selection process follows an iterative approach. A desired cell occupancy is defined and each reagent is scored against the current selection according to the number of underoccupied cells sampled. A weighting is applied based on the number of rotatable bonds otherwise the sequential selection process would tend to favor more flexible reagents. Additional properties such as molecular weight and calculated log *P* can be included into the reagent description using a binning procedure. Thus a hierarchy of scores is defined to score reagents: number of new pharmacophore cells filled, number of new profile bins, pharmacophore cell occupancy score, and profile occupancy score.

An alternative approach has been developed by Pastor and co-workers.[109] Anchor-GRIND is an extension of the grid independent descriptor (GRIND) method of Pastor *et al.*[110] In GRIND, a molecule is placed within a grid and

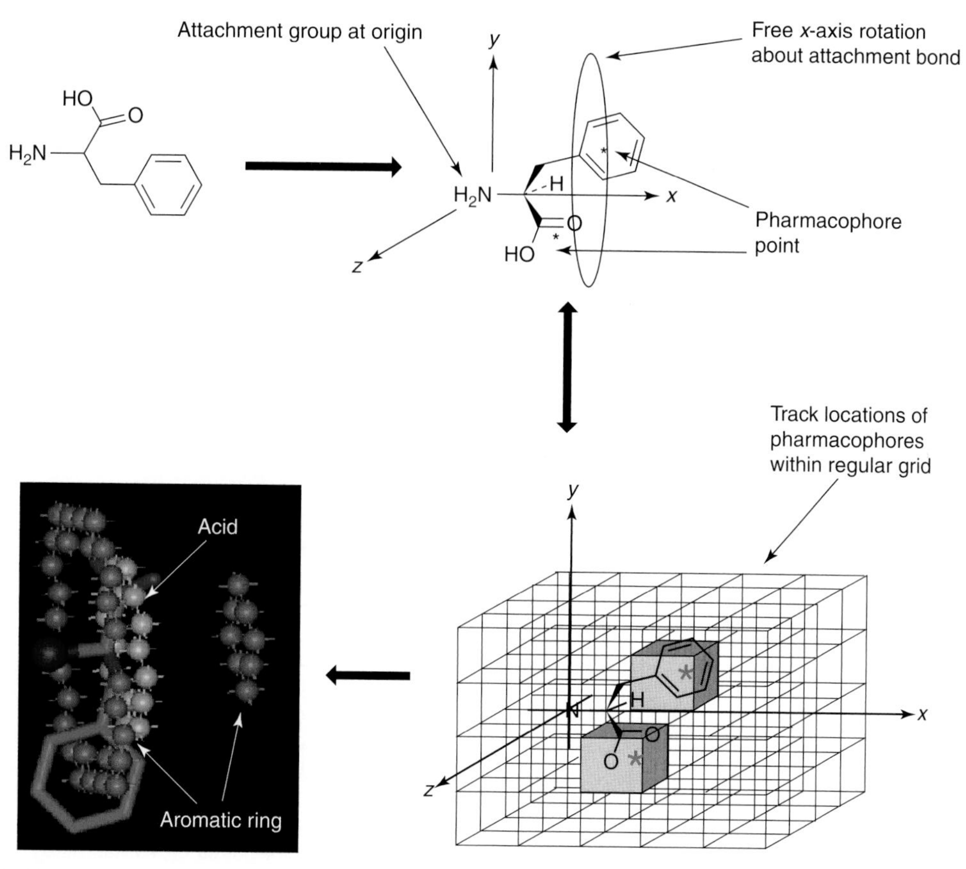

Figure 6 The GaP procedure illustrated with phenyl alanine as the reagent. The primary amine is the reaction center. (Reprinted from Leach, A. R.; Hann, M. M. *Drug Disc. Today, 5*, 326–336. Copyright (2000), with permission from Elsevier.)

molecular interaction fields (MIFs) are calculated with a series of probes at each point on the grid as in CoMFA[111] or GRID.[112] However, in a subsequent step a series of interaction terms are generated between pairs of MIFs using the program Almond,[113] developed specifically for this purpose. Local minima in each MIF are determined and a technique known as maximum auto cross correlation (MACC) is used to record the maximum interaction term between two minima within a set of predefined distance bins. Thus a vector is produced from the interaction fields containing maximum interaction terms as a function of distance. The MACCs are calculated between points of the same probe and between different probes. In Anchor-GRIND, a reference point is positioned on the scaffold so that an additional data block containing the anchor to probe MACC is calculated for each probe. Standard statistical methods such as partial least squares and principal components analysis can be applied to the resulting data matrix. The method has been validated from the standpoint of three-dimensional QSAR (3D-QSAR).

4.15.3.3 Reagent-Biased Product-Based Methods

Combinatorial library design based on product properties is a constrained optimization problem. The method will work in reagent space for selection but scores the solution based on the resulting combinatorial products. As a simple illustration, if the virtual library is profiled against a pharmacophore model, the algorithm would attempt to select reagents in such a way as to generate the combinatorial library subset with the largest number of products that match the pharmacophore. This will be achieved by algorithimically selecting a subset of the R-groups at each variable position and combining them combinatorially. This subset will then be scored by counting the number of products that match the pharmacophore. The procedure is iterated for a defined number of iterations or until some convergence criterion is reached. It should be noted that the search space being explored here is very large. For a single position there are

$$C(n,\ r) = n!/(r!(n-r)!) \qquad [2]$$

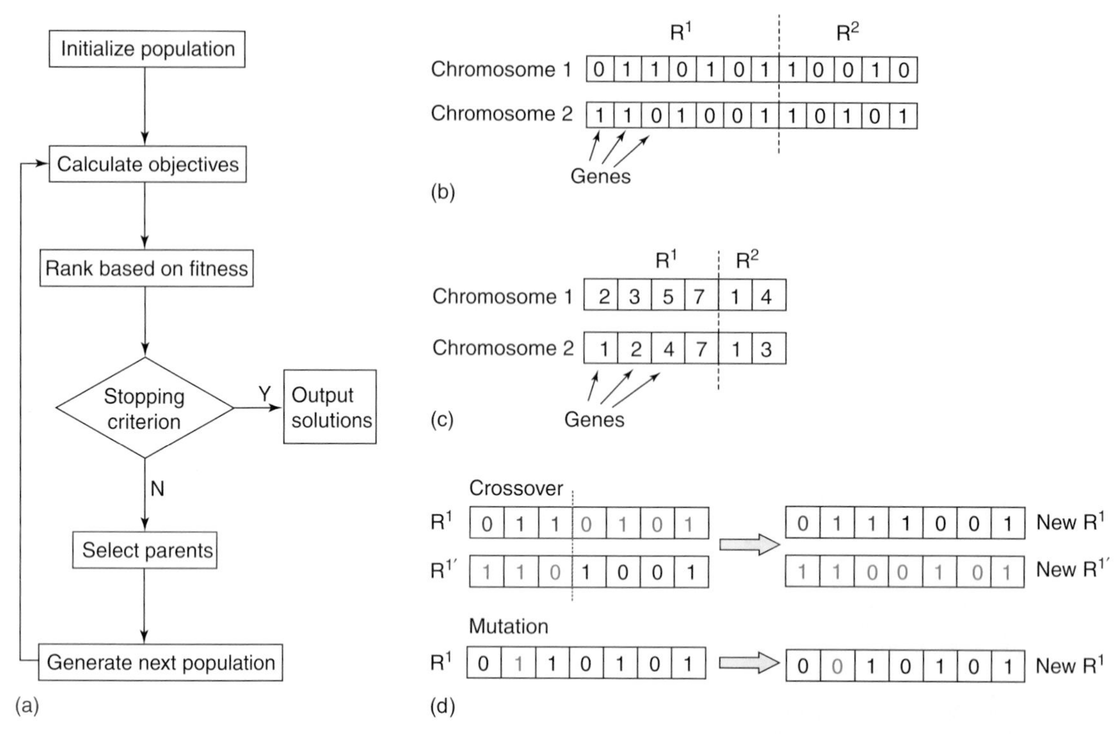

Figure 7 The basic workflow for a genetic algorithm. (a) An overview of the algorithm. (b) The relationship between chromosomes and genes. A chromosome is comprised of several partitions mapped to an R-group, indicated by the dashed line, and is composed of $NR^1 + NR^2$ genes where NR^1 is the number of available reagents at position R^1. Each gene indicates the absence (0) or presence (1) of a particular reagent in the sublibrary. Note that in this representation the library size need not be fixed. (c) In an alternative representation for a fixed library size (4×2) the chromosome has six genes, each integer relating to a specific reagent in the list for each R-group. (d) Genetic operators, crossover, and mutation. Each operator operates on a randomly selected gene in the chromosome.

ways of selecting r reagents from n possibilities. Thus there are 10 ways of selecting two reagents from 10 and over 2 million ways of selecting five from 50. More generally, for a library with p positions there are

$$n_{\text{poss}} = \prod_p (n_p!/r_p!(n_p - r_p)!) \qquad [3]$$

possible ways of selecting the subsets. For this reason it is generally not tractable to compute every possible subset and a variety of approaches have been applied to this optimization problem. The two most frequently studied methods are genetic algorithms (GAs) and simulated annealing/Monte Carlo searching. However, other methods have also been employed. Most methods are stochastic, that is they have a random component and do not guarantee to find the global minimum. In practice this is unlikely to be a problem, especially given all the assumptions implicit in the scoring functions used.

Genetic algorithms belong to the more general class of evolutionary algorithms that have become an important optimization methodology within computational chemistry.[114] GAs are loosely connected with the principles of Darwinian evolution as shown in the general process flow in **Figure 7** and much of the terminology is borrowed from this field.[115] Thus, there is the concept of a population that evolves under the constraint of some user-defined objective function. The objective function is used to compute a score or fitness for each individual within the population that is used, in turn, to assess the likelihood of that individual generating offspring at the next generation. The fitter the individual the more likely it is to breed. Each individual is represented by a chromosome (genotype) that encodes the particular problem to be optimized in a series of genes. At each generation chromosomes can mutate (one individual generates one offspring) or can breed with another individual via a crossover (two individuals, two offspring). Examples pertinent to library design are shown in **Figure 7**. Methods vary by the way that the problem is encoded.

Thus for a three-component library, each chromosome could be represented by three partitions each of size n_p genes. The presence or absence of a reagent in the final library would be marked by a 0 or 1 in the appropriate gene.

This binary representation has some potential drawbacks for combinatorial design as it is very difficult to control the number of on-bits (1s) in the chromosome. A lot of computational time can be wasted generating invalid chromosomes where a partition has fewer or more reagents selected than required. However, as we shall see later, this can also be used to our advantage. The more usual representation for fixed library sizes and configurations, say $8 \times 12 \times 12$, would be to use partitions of size, 8, 12, and 12 and use integer values to represent the reagents selected at each position (see **Figure 7**). In this case a mutation would swap a selected reagent for an unselected one. However, crossover could lead to one reagent being selected twice and so special schemes are used to avoid this. The probability that an individual is selected for breeding will be determined by its fitness. Several methods are available for selecting those individuals. In roulette wheel selection, the probability of selection is directly proportional to the individual's fitness. Each individual is assigned a proportion of the range 0–1 relative to its fitness and a random number generated in the same range will determine whether or not an individual is selected. In tournament selection, a number of individuals are selected and compete against each other, the winner being determined by its fitness. Several strategies are also available for replacing the population at each generation. The whole population can be replaced, perhaps with an elitist strategy to ensure that the best individuals survive from generation to generation. Alternatively, a fixed proportion of the least fit individuals can be replaced at each generation. Variations of all these strategies have been reported in library design though there has been no systematic exploration of the most appropriate strategy. The algorithm is run for a predefined number of generations or until some convergence criterion has been satisfied, for example if the fittest individual has not changed for a certain number of generations. The final population will represent a number of possible library designs ranked by their fitness. The approximations and uncertainties inherent in the fitness function itself mean that all individuals close to the fittest should probably be examined to select an appropriate library for synthesis. Being a stochastic process it may also be necessary to run the algorithm several times to ensure that solutions close to the global optimum are found.

Simulated annealing, as the name suggests, is based on the physical process of annealing where a system is heated to a high temperature and then allowed to cool slowly.[116,117] In methods such as molecular dynamics or Monte Carlo simulation with physically based potential energy functions,[118] the temperature is a property of the system and the method can be used to search for minima in potential functions, e.g., to find an optimum small-molecule conformation.[119] In the Monte Carlo approach, the system is randomly perturbed from one state to another (e.g., a torsion angle is changed) and the energy function recalculated in this new state. If the new state is at a lower energy than the original state then it is accepted. If the original state is of lower energy, the new state is accepted with a probability governed by the Boltzmann function:

$$\Pr(\text{new}) = \mathrm{e}^{-\Delta E/kT} \qquad [4]$$

where ΔE is the difference in energy between the two states, T is the temperature, and k the Boltzmann constant. At a high temperature the system has a higher probability of jumping over energy barriers. As the temperature is reduced slowly the system will anneal to a minimum in the potential function. Again due to the stochastic nature of the search the algorithm is run several times to find the minimum. In application to library design, the system will be represented by the appropriate number of reagent selections and perturbation will involve swapping one of the selected reagents for an unselected one (similar to a mutation in the GA) and re-evaluating the new library. The temperature of the system and Boltzmann constant become control variables that are set to ensure an appropriate number of accepted solutions early in the run.

The key feature of both the GA-based approach and simulated annealing is the need for a function to optimize and the variety of functions become clear in the examples given below. Sheridan and Kearsley developed one of the earliest applications of a GA in library design to optimize the selection of compounds from a large virtual library of peptoids.[120] The main aim of the work was in the design of targeted libraries, for example to select compounds similar to one or more probe molecules. Thus the fitness functions included descriptor similarity to the probes or the use of a trend vector. Of particular interest is the fact that the fitness function is evaluated on the fly. That is, a molecule is evaluated only if selected by the GA. This allows the exploration of very large virtual libraries where the precalculation of the similarity would be prohibitive. In their approach each chromosome represented an individual molecule from the virtual library, thus the final design is not combinatorial. However, it was suggested that a combinatorial design could be made by including reagents that occur frequently in high-scoring individuals. This hypothesis was validated in a subsequent publication[121] and was shown to be particularly valid for the two-dimensional descriptors. It was hypothesized that the three-dimensional scoring methods, SQ for molecule superposition to a target molecule,[122] and FLOG (flexible ligands oriented on grid) for molecular docking to a target protein,[123] produced a noisier fitness function where small changes in a molecule could lead to a large change in score. The molecule-based GA was also

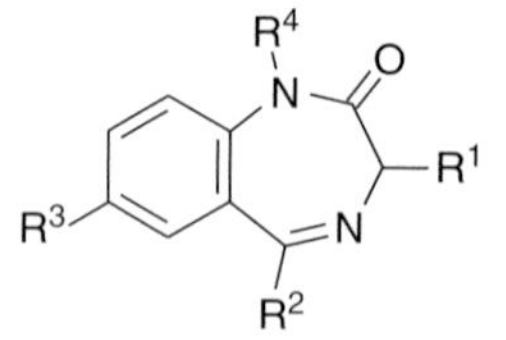

R^1 from Gly, Ala, Val, Phe, Trp, Asp, Asn, Glu, Gln, Thr, Lys
R^2 = Ph, 4-MeOPh, c-hexyl, 2-thienyl, 4-NO_2Ph, 4-NH_2Ph
R^3 = H, Cl, NO_2, OMe, NH_2
R^4 = H, Me, benzyl, *i*-Bu

Figure 8 The benzodiazepine library used to test a GA-based library design algorithm. The fitness function is the pharmacophore coverage of the selected library subset. The reagents selected in the final design are highlighted in red.

compared to a library-based GA where each chromosome represents a full combinatorial sublibrary. It was found that the two approaches gave similar quality solutions but the library-based GA took much longer to converge. This is related, at least in the part, to the different search spaces being explored by the two methods. The virtual library was composed of 5321 carboxylic acids, 1030 amino acids, and 2851 basic amines. This gives $\sim 1 \times 10^{10}$ individual molecules. Selecting a $5 \times 5 \times 5$ combinatorial subset from the virtual library gives $\sim 5 \times 10^{44}$ possible solutions according to eqn [3]. However, it was recognized that in a more general case the library-based approach could have advantages, such as in ensuring an appropriate property distribution across the library. An interesting implementation detail was the use of a neighbor mutation operator, where the mutation of a reagent occurs to one of 10 reagents most similar to it in a topological sense. Effectively this constrains the mutation to a more local jump, in a similar way to which a torsion angle change in a conformational search algorithm could be limited to a $\pm 30°$ range.

Pickett and co-workers[124] described a GA-based library design method which used coverage of three-dimensional pharmacophore space[90] as the prime fitness criterion. As the fitness function applies to the library as a whole each chromosome represented a library with a fixed library configuration. An example design, using a benzodiazepine library based on that first proposed by Dewitt,[7] is shown in **Figure 8** and serves to illustrate the concept of the optimization procedure. The virtual library covers 1320 products ($11 \times 6 \times 5 \times 4$) and a 96 compound subset ($4 \times 4 \times 3 \times 2$) was chosen. For R^1, R^2, and R^3 a range of hydrophobic and polar amino acids have been selected. For R^4, the selection of hydrogen introduces a new hydrogen bond donor atom and the isobutyl is chosen as a hydrophobic group.

As discussed in the Introduction, initial synthetic approaches to combinatorial chemistry resulted in mixtures of compounds. Screening is followed by deconvolution of the active mixtures to individual compounds that are then rescreened to find the active component. A variety of methods have been developed to identify the individual components within each mixture to enable the deconvolution.[125,126] These include the use of chemically reactive tags, the use of spectroscopically labeled beads and mass spectrometry techniques.[127] GALOPED[128] is a GA-based library design algorithm written specifically to optimize the diversity of the library and minimize the number of identical molecular weights within each pool. The chromosome representation used binary genes to mark the presence or absence of a reagent in the potential library. Upper and lower bounds are used to constrain the number of reagents selected at each position with any chromosomes created outside of these ranges being marked as invalid. In this way it is possible to design libraries of variable size and configuration. The fitness function includes a term to remove molecular weight redundancy by minimizing the number of molecular weight occurrences over a user-defined limit. The diversity is assessed at the substituent or product level based on two-dimensional clustering or a three-dimensional pharmacophore partitioning method.

Gillet and co-workers[129] published a GA-based library design method termed SELECT. In this method the chromosome represents an appropriately configured library and a variety of design objectives can be included via a weighted-sum fitness function:

$$\text{fitness} = a.\text{MW} + b.\text{ClogP} + c.\text{diversity} + \cdots \qquad [5]$$

Property scores, represented by MW and ClogP in eqn [5], are determined by a root mean square comparison of the property distribution of the library to that of a reference population such as the World Drug Index.[130] Internal diversity is computed from Daylight fingerprints[73] using a fast dissimilarity method based on the cosine coefficient[131] that computes the sum of pairwise dissimilarities over all molecules in the library. The same approach can be used to ensure that the selected compounds are maximally different from a reference collection. This diversity measure can favor outliers and result in some similar molecules being present in the selection.[132] Thus, an alternative measure, based on average nearest neighbor distance as calculated from the Tanimoto coefficient, was also implemented. However, this can be computationally expensive to compute and hence is used for smaller libraries. The coefficients *a*, *b*, *c*, etc. are selected empirically to ensure the appropriate balance of the objectives and of course also need to correct for differences in scale between the objectives. The GA implements a niching approach[133] as follows. The GA converges to a solution, which is taken as the center of a niche. The GA is rerun but solutions with a user-defined number of reagents in common with the solution at the center of the niche are removed. The number of niches and the niche radius are defined by the user. The use of niching helps to guard against convergence on suboptimal solutions and ensures that multiple runs explore different regions of the search space.

Agrafiotis has described in detail a simulated annealing approach to library design.[134] Chemical space is defined by a set of well-characterized topological descriptors.[81] Principal components analysis is used to generate a set of orthogonal variables that are further reduced to a two-dimensional set using an efficient, proprietary nonlinear mapping technique.[135–137] The algorithm can be used to maximize the similarity of the library to a set of leads by minimizing the average distance of each compound to its nearest lead. Diversity is estimated by maximizing the average nearest neighbor distance within the selection. To overcome the computational expense of this measure, mentioned above, the objective is calculated using an efficient *k–d* tree algorithm.[138] By including a reference set in this calculation, the same approach will lead to complementary designs where the selected compounds are optimized to fill the voids in a preexisting collection. Other simple functions can be used to optimize the predicted activity of the selection and to minimize the overlap of the selection with a reference set, for example when selecting multiple arrays from the same library in an iterative drug design approach. Compounds can be optimized to a particular property range by penalizing those that fall outside a user-defined range according to eqn [6]:

$$\text{penalty} = \max(p^{\min} - p_i, p_i - p^{\max}, 0) \qquad [6]$$

where $p^{\min}$ and $p^{\max}$ are, respectively, the minimum and maximum allowed values for property p, and p_i is the value of that property for compound i. This penalty is averaged over all compounds and properties. The Kolmogorov–Smirnoff statistic (KS) measures the maximum separation between two distributions and thus maximizing the function $1 - \text{KS}$ will lead to designs that most closely resemble a user-defined distribution,[139] for example the molecular weight profile of known drugs. These various objective functions are combined through a weighted sum fitness function similar to that in eqn [5]. An interesting implementation detail of the simulated annealing algorithm is the use of an adaptive algorithm whereby Boltzmann's constant, k in eqn [4], is varied throughout the run to control the acceptance probability.

In addition to the GA approach described above, researchers at Rhone-Poulenc also investigated simulated annealing approaches to library design.[98,124] The resulting program, HARpick, incorporated a number of objectives in the fitness function. The primary objective is similar to that employed in the GA, and optimizes pharmacophore coverage, the number of unique pharmacophores in the selected library. There are several interesting features of the pharmacophore generation that are specific to this implementation. With seven center types (hydrogen bond donor, hydrogen bond acceptor, joint donor and acceptor, e.g., hydroxyl, acid, base, aromatic, hydrophobe) and 17 distance bins there are $\sim$185 000 accessible pharmacophore triangles. To reduce storage space only the pharmacophores present in a particular molecule are stored. Larger molecules will cover a number of pharmacophores that are small relative to the largest pharmacophore in the molecule and hence a user-defined ratio for the minimum ratio between pharmacophore perimeter and the largest perimeter in the molecule was used to filter out these small pharmacophores. This is particularly important in coverage calculations which might otherwise favor larger, more promiscuous molecules. A rough estimate for molecular shape was achieved through the use of three properties – heavy atom count (ha), largest pharmacophore perimeter (pp), and largest pharmacophore area (pa). The property ranges were binned based on the virtual library and a score defined according to:

$$\text{PS} = \frac{\text{sd}_{\max} - \sum_{j=1}^{p} \sqrt{(\overline{\text{occ}} - \text{occ}_j)^2}}{\text{sd}_{\max}} \qquad [7]$$

UNIVERSITY COLLEGE Library CORK

In this partition function, $\text{sd}_{\max}$ is the maximum standard deviation when all molecules are in the same partition (bin), $\overline{\text{occ}}$ is the mean occupancy over all partitions and occ_j is the number of molecules in partition j. The average number of conformations sampled per molecule (the conformational search used a systematic procedure) provides a measure of flexibility, FS. In order to fill diversity voids, in this case under sampled pharmacophores in a reference set of compounds, a constraint score, CS was developed:

$$\mathrm{CS} = \sum_{i=1}^{a} O_i S_i \qquad [8]$$

where

$$S_i = [\max(0, \overline{\mathrm{cov}} - Oc_i)]^{v} \qquad [9]$$

O_i and Oc_i are the number of molecules that contain pharmacophore i in the library and reference set, respectively; $\overline{\mathrm{cov}}$ is the average pharmacophore coverage across all pharmacophore triplets in the reference set with the total occupancy of any pharmacophore bin being cut at a user-defined limit; v is a user-defined weight. To further reduce the influence of more flexible molecules in the pharmacophore calculations the total number of pharmacophores present in all selected molecules, TP, is included in the fitness function. User-defined property bounds on molecular weight, flexibility, etc. could mean that some molecules do not contribute to the score. Thus the total number of scoring molecules, TS, is also included. The overall scoring function is

$$\mathrm{score} = \frac{\mathrm{cov}^{w} \times \mathrm{CS} \times \mathrm{PS}_{\mathrm{pp}}^{x} \times \mathrm{PS}_{\mathrm{pa}}^{x} \times \mathrm{PS}_{\mathrm{ha}}^{x} \times \mathrm{TS}}{\mathrm{TP}^{y} \times \mathrm{FS}^{z} \times n} \qquad [10]$$

where w, x, y, z are user-defined weights and n is the number of selected molecules. This example serves to show the complexity of the weighted sum (or product in this case) fitness functions, with a number of user-definable weights and the need to appropriately normalize the individual score components. An equally complex function has been described by Brown *et al.*[140] that includes cost constraints in the design.

In summary, GAs and simulated annealing have found widespread use as optimization methods in library design, as shown by the number of academic, industrial, and commercial applications included in the examples above, and remain the choice of many when implementing a library design program.[141–143] Simulated annealing in particular is relatively straightforward to implement. A wide range of objective functions can be included for diverse or targeted libraries and to ensure that properties are within acceptable bounds (*see* Section 4.15.4 for further discussion). There have been few comparative studies of the relative merits of the two approaches. In one such study by Jamois *et al.*[144] the methods were found to be equivalent for the problems explored. However, as we see later, GAs are particularly applicable to novel ways of solving the multiobjective optimization problem of library design.

Library design methods have not been limited to simulated annealing or genetic algorithm optimization techniques, however. Agrafiotis and Lobanov[145] have described a very efficient greedy algorithm for combinatorial library design. The similarity, activity, and confinement objectives described above are used as these depend on individual molecule contributions. Thus the algorithm is particularly suited to focused libraries rather than libraries that depend upon properties of the molecule set as a whole. Results from the algorithm compare favorably to the more time-consuming stochastic approach in these cases. Taking a three-component library as an example, a subarray is generated that contains one R^1 group and the required number of R^2 and R^3 groups. Each R^1 reagent is evaluated in turn in the context of the selected R^2 and R^3 groups and the best n_1 are added to the selection, where n_p is the number of reagents required at position p. The process is repeated for each position in turn until no improvement in overall score for the selected subset is obtained.

Bravi *et al.*[146] have described an iterative algorithm that maximizes the number of molecules in the designed set that satisfy a set of user-defined constraints. The constraints could include any computational or other property, such as fit to a pharmacophore, cost, and molecular weight range. The virtual library is enumerated and the favorable molecules are marked. At each iteration the worst monomer is removed. This is the monomer that leaves the best-scoring library once removed. The score is a combination of effectiveness, the ratio of the number of favorable compounds in the selected subset to the total available, and efficiency, the ratio of favorable compounds in the subset to the number of compounds in the subset (assuming a combinatorial expansion). The algorithm is iterated until only favorable molecules are left in the combinatorial subset. The best library will be that with the best score given by the weighted sum of efficiency and effectiveness. Interestingly, the user does not need to enter any constraints on library dimensions as the algorithm guides the user to the most appropriate library configuration. As for the algorithm of Agrafiotis and Lobanov described above, this approach is best suited to focused libraries as it takes no account of the properties of the set as a whole. PLUMS performs better than an equivalent monomer frequency analysis approach[147] and gives comparable results to a GA.

The alternating algorithm of Young *et al.*[148] combines elements of both these approaches. Starting with a random selection of reagents, $n_1 \times n_2$, a reagent from R^1 is selected that most improves the design, for example a uniform coverage design,[149] and a reagent is removed that least changes the design. This procedure is repeated for each position in turn until no overall improvement is achieved. The authors also suggest using the algorithm to select compounds based upon a predictive QSAR model. A fast exchange algorithm, termed WEALD, has been described by Le Bailly de Tilleghem *et al.*[150] again focused on lead optimization. This is somewhat similar to the other iterative algorithms described here in that reagents are swapped in iteratively until no further improvement is found. However, each reagent has a probability selection according to a series of rules. At each iteration the average fitness for compounds containing each reagent in the current selection are computed. At any step the least fit reagent is removed, that with the lowest average fitness. A new reagent is selected from the available pool of reagents based on a set of standardized selection probabilities. Reagents yet to be chosen are given a score of 1. For reagents already explored the score is the ratio of the average fitness of all compounds already explored containing the reagent to the maximum fitness yet found. Fitness is a molecule property based on a weighted product of properties normalized to 1. Eight properties were computed based on predictions from QSAR models. Each sublibrary of size N has an associated loss function defined as

$$\text{loss} = \frac{1}{N}\sum_{j=1}^{N}(1 - \text{fitness}_j)^2 \qquad [11]$$

A reagent is exchanged if this leads to a lower value of loss, otherwise another reagent is tried. The algorithm iterates until no reagents can be exchanged. The algorithm was compared to the ultrafast algorithm of Agrafiotis,[145] described above, and a simulated annealing approach.[151] Whilst all algorithms converged on similar loss values, the WEALD algorithm converges the fastest as the selection probabilities take account of previous iterations.

Informative design[152] aims to select the subset of compounds that, when screened, will give the most information about the target. In the context of pharmacophores for example, the selected compounds should enable one to determine the key features required for activity based upon the screening results. The approach is based upon information theory[153,154] and attempts to optimize the Shannon entropy of the system. Entropy in this context represents the uncertainty in the system, that is, each compound conveys maximally different information. The approach as described[155] uses a greedy algorithm to optimize a function of the form

$$\text{score} = H - D \qquad [12]$$

where H is the Shannon entropy term

$$H = -\sum_{i=1}^{C}\frac{||c_i||}{F}\ln\frac{||c_i||}{F} \qquad [13]$$

For a fingerprint of F features, C is the number of classes in the selected subset, that is, the number of distinct feature combinations and c_i is the size of feature class i. D is a cost term that, like other methods described in this section, ensures that the selected compounds fit as closely as possible to property distributions such as those in Lipinski's 'rule of 5.'[20] Properties are binned and the distributions compared to a reference profile:

$$D = \sum_{j=1}^{p} w_p \sum_{k=1}^{b_j}(\rho d_{j,k} - \rho c_{j,k})^2 \qquad [14]$$

where $\rho d_{j,k}$ is the desired fraction for property j and bin k; $\rho c_{j,k}$ is the equivalent fraction for the current selection; w_p is the weight assigned to property p in the optimization. A greedy algorithm is used for the selection of the reagents. The compound with highest entropy is selected, which also adds its reagents to the selection. The next compound is selected that contains one of the previously selected reagents. This continues until the desired number of reagents has been selected. In a second phase, each reagent is dropped and a better reagent sought that improves the score. This is continued until no further improvement is found. As an additional objective, a minimum tanimoto similarity between selected molecules can be defined to add a diversity component to the design.

In this and the previous section we have looked at product-based and reagent-based methods for the selection of fully combinatorial libraries. A large number of functional forms have been used to describe diversity within the chemical space and to ensure appropriate property distributions. In addition, a variety of algorithms have been explored for the optimization itself. However, the combinatorial constraint can be limiting in that it may force the inclusion of undesirable compounds. Several groups have developed noncombinatorial or sparse array design methods (small combinatorial blocks) to address some of these limitations and these are discussed in the next section.

4.15.3.4 Noncombinatorial Library Design

The algorithms described in the previous section were written with the primary objective of supporting the synthesis of fully combinatorial arrays. This approach is the most efficient from a synthetic perspective and in the use of the various reagents. At the opposite extreme is a noncombinatorial or cherry-picked array. This has the disadvantage that reagent use is inefficient but the advantage that the chemist has total control over the properties of the synthesized molecules. Lying between these two extremes is a technique commonly called sparsing, where smaller combinatorial subblocks are synthesized in the generation of a larger library, for example certain combinations of reagents may be excluded as they generate high molecular weight, high log *P* compounds. In this section we look at algorithms designed specifically to deal with these situations. This approach can lend itself to a greater degree of statistical design than a fully combinatorial selection.

Everett *et al.*[156] have described a system termed LiCRA (Library Creation, Registration and Automation system) implemented at Pfizer, UK. The group favors a totally noncombinatorial approach to library synthesis. The rationale is that in being able to cherry-pick compounds the library will sample all possible reagents (though clearly not all combinations of those reagents) whereas a fully combinatorial approach will, by definition, only sample a small subset. As seen with the GA-designed benzodiazepine example (**Figure 8**), at some point the combinatorial selection will lead to choices, aspartic acid versus glutamic acid, phenylalanine versus tryptophan, which are perhaps somewhat arbitrary and may be hard to rationalize in the most rigorous sense. On the other hand, it could be argued that not exploring all combinations also involves arbitrary and equally difficult decisions on the products. The design algorithm, S1D (selection in one dimension), attempts to select products from the virtual library within a user-defined range of molecular weight, ClogP, and other (approximately) additive properties so that each reagent is used an equal number of times. Everett *et al.* described the application of the approach to the identification of multiple leads for a kinase target. An initial library of 5000 compounds was synthesized from a virtual library of 300 000. Three hits were identified from just two templates and a couple of iterations of further focused arrays explored the SAR around these hits. The virtual library was then expanded with further reagents and further noncombinatorial selections made to follow up on the activity. The primary purpose of S1D in this approach appears to have been to ensure that the properties of the selected molecules fall within appropriate bounds. All the compounds were synthesized as singles, analyzed, and purified.

Rose and co-workers[157,158] have described a methodology termed Predictive Array Design (PAD), for the statistical design of libraries. The method is not an optimization process that selects the reagents to use but, rather, a methodology to select combinations of selected reagents for the final compound set. The method is based upon the principle of Latin Squares and is applicable to libraries with three sites of variation. The aim is to ensure that a subset of compounds from the full combinatorial array is synthesized that contains all pairwise combinations of reagents. In the case where $n_1 = n_2 = n_3$ (say 5) then synthesize $n_1 \times n_2 = 25$ intermediates and react with each of the n_3 using a diagonal design as shown in **Figure 9**. Lay out a 5 × 5 table for R^1 and R^2, the five R^3 are matched with each element of row 1. The R^3 reagents are reordered by moving the first reagent to the end of the list and matching with the row 2 intermediates. This is repeated for each row to give a total of just 25 products rather than the 125 from the full array. The approach can be extended to cases where there are different numbers of reagents at each position by repeating the Latin Square unit.

	R^2				
R^1	1_1_1	1_2_2	1_3_3	1_4_4	1_5_5
	2_1_2	2_2_3	2_3_4	2_4_5	2_5_1
	3_1_3	3_2_4	3_3_5	3_4_1	3_5_2
	4_1_4	4_2_5	4_3_1	4_4_2	4_5_3
	5_1_5	5_2_1	5_3_2	5_4_3	5_5_4

Figure 9 The predictive array design for a three-component library. Each row represents a single reagent at R^1 and each column a single reagent at R^2 with the R^3 choices overlaid, labeled as 1_2_3.

One advantage of this approach is that the activity contribution associated with each reagent can be estimated from the screening data. This SAR information can then be used in further iterations of synthesis. The Latin Square assignment is nonunique and Lipkin *et al.*[157] describe the use of a simulated annealing algorithm to optimize the design to sample uniformly across physicochemical property ranges. For four-component libraries a Graeco-Latin Square design can be used.[159]

The OptDesign approach of Clark *et al.*[160] extends the OptiSim *k*-dissimilarity methodology[161] to combinatorial library design. OptiSim iteratively selects compounds from a larger set to balance the two objectives of representation and diversity. The user defines the value for *k*, the number of compounds to be sampled at each step, *r*, the minimum distance to a previously selected compound, and *M*, the total number of compounds to be selected. The algorithm scales as kM^2 and so is relatively fast. OptDesign alternates between reagent pools when selecting the *k* subsamples thus allowing the selection of combinatorial blocks. Thus, whilst it is possible to generate full combinatorial designs the algorithm is particularly suited to generating sparse arrays of combinatorial subblocks. Additional constraints can be added, for example cost, by grouping reagents and biasing the selection to more favorable groups.

It is often the case in lead optimization that fewer compounds are synthesized at each iteration and parallel synthesis techniques allow a greater control over the individual compounds to be made. A fully combinatorial solution may not always be the best approach, particularly when there are tight constraints on compound properties. An approach to generate near combinatorial designs has been reported by Pickett *et al.*[162] Consider the selection of 400 products from a larger virtual library. A 20×20 selection would be the most efficient. However, this may lead to compounds with undesirable properties in the final design. Hence a Monte Carlo search procedure was implemented that would select the 400 products closest to a full combinatorial solution. Each reagent pool is represented by a bit-string to show the presence/absence of each reagent in the library. The full combinatorial library is generated from the selected reagents and a score generated based on the number of acceptable products and the number of times each reagent is used in the acceptable products. The user can specify the minimum and maximum occurrence for a reagent and the ideal occurrence (20 in this example). Reagents can also be preselected to include, for example, particular fragments known to be required for activity. The algorithm was applied successfully to enhance the pharmacokinetic properties of a kinase library,[46] described in more detail in Section 4.15.4.3.

4.15.3.5 Reagent-Based versus Product-Based Design

It is clear from the previous discussions that there are a number of choices open to the designer in the strategies available for library design. These range from reagent-based methods to product-based methods, fully combinatorial to sparse arrays to cherry-picked solutions optimized to minimize reagent usage. A number of optimization strategies can be used on a wide variety of scoring functions and descriptor types. It is fair to say that there has been very little objective comparison of any these approaches in the literature. We referred earlier to a limited comparison of simulated annealing and GA approaches[144] and there have been some attempts to compare or validate molecular descriptors[163–165] but none of these methods is without its problems.[98,166] However, one area has been more systematically studied. That is the comparison between reagent-based and product-based design approaches. Though here again there are different schools of thought.

Gillet *et al.*[167,168] and Jamois *et al.*[169] have reported on the comparison of reagent-based and product-based designs with a number of descriptors. From the point of view of diversity, there is a clear bias in the results with the descriptors used. Perhaps not too surprisingly, with local fragment-based descriptors such as ISIS keys there is little to choose between a reagent-based design (selecting diverse reagents at each position) and a product-based design. For descriptors that encode more of the molecule, such as Daylight fingerprints that encode all paths up to a length seven and thus are more likely to cross R-group boundaries, there is a clear preference for product-based designs. Another important component of product-based design is the ability to match properties across the ensemble of products, and Gillet *et al.* demonstrated that reactant-based designs can lead to compounds with poor physicochemical property profiles (some steps to alleviate some of these concerns are described above where we discussed reagent-based design methods). Thus the true benefit of a product-based design is the ability to optimize several whole molecule properties simultaneously. Linusson *et al.*[170] compared a statistical molecular design in building blocks to product-based selections. Principal components analysis of reagent or product properties was used to generate the principal properties. This will create an orthogonal property space which accounts for the correlations between variables (see also the work of Martin *et al.* described in Section 4.15.3.2). Several design methods were used on this space: D-optimal design,[171,172] a space-filling design that maximized the minimum Euclidean distance between the nearest neighbors,[173] and a cluster-based design.[174,175] For the product selections a fuzzy clustering technique was used. This allows compounds to belong to more than one cluster by assigning the compound a probability of cluster membership.[176,177] The various design methods were applied to a small 400-member amide library of 20 amines and 20 carboxylic acids. It was shown that for the case of a cluster-based reagent selection of seven amines and seven acids, with a further selection to give 25 products, similar designs were obtained to a 25 compound selection based on the products. It is claimed that this challenges the assertions of Gillet *et al.*[167] that product-based designs are preferred. It is clear that removing some redundancy in the building block space will tend to lead to little information loss when a product-based selection is made on this reduced set of building blocks. Thus the approaches are not really comparable as the Gillet work focuses on combinatorial designs whilst that of Linusson *et al.* focuses on cherry-picking. The point is that to achieve a combinatorial design as close as possible to the optimal cherry-picked product selection it is necessary to use a reagent selection methodology that incorporates the product properties, i.e., reagent-biased product-based selection methods.

This does not mean that reagent selection methods do not have their place. As stated by Linusson *et al.*[170] a proper statistical design of the reagents provides a much better starting point for QSAR methods. This will be the case particularly for small, focused libraries in lead optimization with the assumption, of course, of a two-dimensional alignment of the substituents. That is, a common binding mode for compounds. This point is equally valid in QSAR where the alignment problem is often cited as a problem with three-dimensional methods such as CoMFA and yet the same assumptions are implicit in any substituent-based QSAR method, even if an explicit alignment step is not required. As noted by others,[97,170] reagent-based selection methods can allow for a much more detailed analysis than is usually possible with product-based approaches. The designer must weigh up the benefits of a more advanced representation of the reagent properties, using quantum mechanical calculations, more detailed conformational sampling, more time-consuming computation of docking scores, etc., against the inherent assumptions of additivity between substituents, the implicit scaffold alignment, the potential poorer performance if the reagent selections are then to be combined combinatorially and the inability to compare between libraries or with other compound collections. Modern computational hardware and design software allows libraries of over 1 million products to be processed in a reasonable timeframe and so it is the author's opinion that reagent-based methods are best applied in very focused instances in lead optimization where small focused sets of reagents are examined in detail, more detailed calculations can be justified as there is a clear goal, and it is relatively straightforward to control resulting product properties such as molecular weight and log P without explicit calculation. For more general lead generation and hit expansion combinatorial library design reagent-biased product-based approaches are to be preferred.

4.15.3.6 Multiobjective Optimization

All the methods considered up to this point share a common element. They optimize a fitness function which comprises a set of weighted components. The effect is to turn the multiobjective problem into a single-objective optimization. There are a number of issues with this approach. The weights need to be defined empirically and may need to be varied from problem to problem. Whilst normalization of the properties may help this may in itself be nontrivial for certain properties. The very fact that weights are used will restrict the search space to particular regions. This is exemplified in **Figure 10a** which shows the results from multiple runs of the GA optimization algorithm SELECT with varying weights for the diversity and molecular weight components to the score. As a stochastic algorithm it is usual that a different result is obtained for multiple runs with the same settings and it could be argued that the algorithm has not been run for a sufficient number of iterations. On the other hand, it is highly likely that in such a large search space a number of equivalent solutions will exist. However, the main point to observe is that each setting of the weights optimizes to a different region of the space. Regions 1 and 2 vary little in molecular weight score but considerably in terms of diversity, regions 2 and 3 differ little in diversity but considerably in molecular weight score. Thus, a naive setting of equivalent of weights for the diversity and molecular weight components would not alert the user to these effects. Increasing the number of terms in the scoring function will serve only to exacerbate the problem.

Gillet *et al.*[76,178] have addressed the problems of the weighted sum approach through the use of a multiobjective optimization concept known as Pareto optimality.[179] The resulting program, MoSELECT, treats each objective independently. The population is ranked as shown in **Figure 11**. A Pareto optimal solution is one in which no other member of the population is better in all objectives and is termed nondominated. The Pareto ranks provide the fitness function to the GA. At the end of the GA run the user is presented with a number of Pareto optimal solutions to explore as shown by the example in **Figure 10b**. Several points are to be noted from this example. Firstly, one run of MoSELECT generates solutions that cover the full range of possible solutions obtainable by varying the weights in the weighted sum function and running SELECT many times. Secondly, the solutions from MoSELECT are actually better in many cases. This is presumably a result of the weighted sum function constraining the search space.

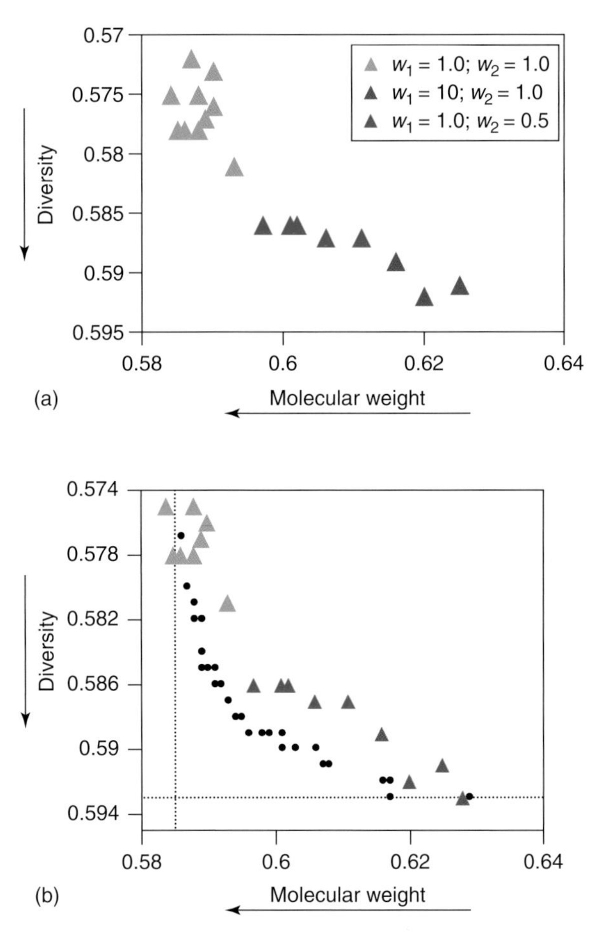

Figure 10 (a) Solutions obtained from multiple runs of the program SELECT with different weights for the diversity (w_1) and molecular weight (w_2) distribution components. The algorithm attempts to maximize diversity and minimize the molecular weight score as shown by the arrows. (Reprinted in part with permission from Gillet, V. J.; Khatib, W.; Willett, P.; Fleming, P. J.; Green, D. V. S. *J. Chem. Inf. Comput. Sci.* **2002**, *42*, 375–385. Copyright (2002) American Chemical Society.) (b) As for (a) with Pareto surface from a single run of MoSELECT shown as black circles. The multiobjective optimization covers all the solutions obtained from multiple runs of the weighted sum GA and leads to better solutions.

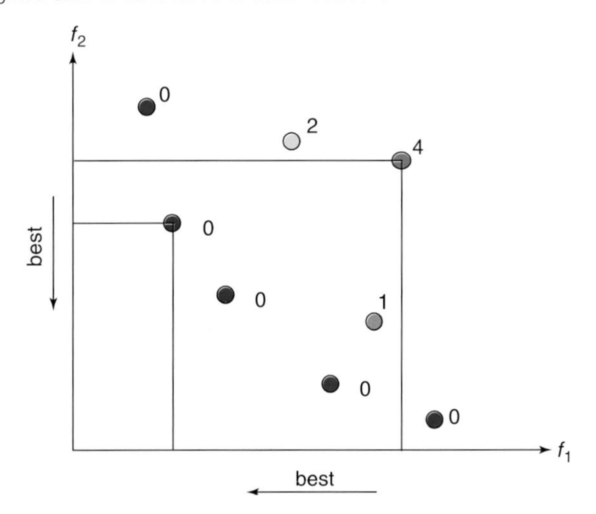

Figure 11 The principal of Pareto ranking. Solutions labeled 0 are nondominated as there are no solutions better in all of the objectives. Other solutions are ranked according to the number of solutions that dominate them. (Reprinted in part with permission from Gillet, V. J.; Khatib, W.; Willett, P.; Fleming, P. J.; Green, D. V. S. *J. Chem. Inf. Comput. Sci.* **2002**, *42*, 375–385. Copyright (2002) American Chemical Society.)

The true power of this methodology is demonstrated in a follow-up paper where the objectives are extended to allow for a variable library size and reagent configuration.[180] This would be an extremely difficult problem for a standard weighted-sum approach because of the difficulties of normalizing scores across the different library sizes. For example, a well-behaved diversity function will always increase as the library size increases.[181] A single solution would be too dependent upon the particular weights. For a multiobjective GA, however, this is a more natural problem. The user is presented with a range of solutions of varying size and configuration and can see the tradeoffs associated with the different solutions. The user is able to monitor the effect of varying library size, evaluate the relative improvement with respect to the additional compounds required and the behavior of other properties such as the molecular weight distribution. Thus the chemist is able to make an informed choice about the most appropriate library dimensions. There are several instances where use of this algorithm has allowed us to suggest different configurations to those initially proposed.

Thus the multiobjective approach offers several advantages over the weighted-sum approaches. Any reasonable number of objectives can be considered without needing to worry about defining appropriate weights upfront. The functional forms can be varied to suit the problem, combining compound counts with profile fits and cell coverage. Most importantly, the user interacts with the information in a manner that makes MoSELECT a decision-support tool as opposed to returning one or several solutions optimized against a somewhat arbitrary weighted-sum score.

4.15.3.7 Utilizing Protein Structure

The number of available protein structures continues to grow and initiatives such as the Structural Genomics Consortium[182] aim to provide protein structures in a number of therapeutically important areas. Docking methods can be used to dock the enumerated virtual library to provide a score for use in the design algorithms described above. This score could be derived from the docking score or some predefined filter on the score could be used in a binary sense to indicate compounds likely to bind to the target protein. Taylor *et al.*[183] have reviewed the major docking methods and we shall not discuss these in detail here. Rather, we shall focus on approaches developed specifically within the context of library design.[45,184] A number of methods have been developed that benefit from the combinatorial nature of the library, for example prealigning or docking the core scaffold and then placing the R-groups can significantly speed up the docking. Specific applications of these methods are given in Section 4.15.5.

An early successful application of combinatorial chemistry and structure-based design was reported by Kick *et al.*[185] A library of hydroxyethylamine inhibitors (**Figure 12**),[186] targeted against the aspartyl protease cathepsin D, was designed using CombiBUILD, a modification to the BUILDER program.[187,188] The design strategy involved an initial conformational search of the scaffold within the cathepsin D active site, based upon the binding mode of pepstatin.[189]

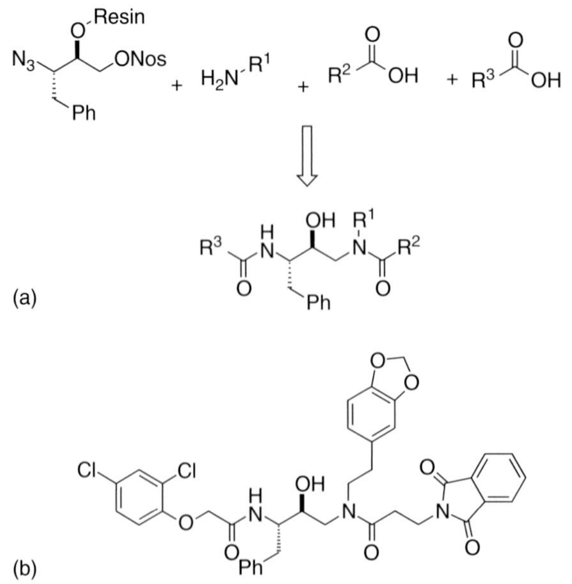

Figure 12 (a) Components and general structure of a hydroxyethylamine library designed as inhibitors of cathepsin D.[185,186] (b) The most active compound from a second-generation library with $IC_{50} = 14\,nM$ against cathepsin D.

This initial analysis allowed the identification of the stereochemistry indicated in **Figure 12** as the most favorable, which was confirmed by a pilot study. Representative scaffold conformations were selected as the basis for conformational sampling of the R^1 to R^3 components. Whilst each component was sampled independently in the site, a probabilistic clash grid was generated to remove for example R^1 conformers that have a high probability of clashing with an R^2 component. Further clustering of the products derived from the top-scoring components led to the final selection of 1000 combinatorial products. A diversity-based reagent selection was also made, clustering reagents using Daylight fingerprints. This served as a control experiment. The structure-based library showed a significant enrichment in active compounds at 1 μM (67 versus 26), 330 nM (23 versus 3), and 100 nM (7 versus 1). In a second iteration, a small focused library was synthesized from analysis of active components in the first library. This led to compounds with IC_{50} values below 20 nM. The most active compound, with an IC_{50} of 14 nM, is shown in **Figure 12b**.

The program DOCK[190–193] is one of the earliest and perhaps best-known docking programs and there have been several reports of the adaptation of the basic DOCK algorithm to combinatorial library design. In DOCK the receptor site is represented by a set of spheres that are used to match a subset of ligand atoms, thus orienting the ligand into the site. An interaction score between the ligand and the protein is then computed to rank the pose. CombiDOCK[194] is a modification to this DOCK algorithm whereby only the scaffold atoms are used in the initial matching phase. Fragments derived from the reagents are docked in relation to the scaffold and scored independently. The fragments at each scaffold position are sorted on the score and, in a final step, combinations of high-scoring fragments are made and checked for steric clashes. The algorithm was tested by designing a library from a 36 million virtual library of benzodiazepines (**Figure 1a**) with R^3 = OH, against dihydrofolate reductase (DHFR). The CombiDOCK approach was compared against a random reagent selection and a selection where each position on the scaffold is docked independently. Each selection method was evaluated based on the average score of top-ranking compounds with CombiDOCK selections clearly outperforming both of the other methods. Interestingly, the single-fragment approach performs worse than random selection. Optimizing at each position independently can lead to very poor solutions when the reagents are combined into the final products. In a second test the algorithm was used to rescore the hydroxyethylamine library design described above.[185] A reasonable enrichment factor of fourfold was found in selecting out the most active compounds from the set.

A similar philosophy is taken by PRO_SELECT which employs a more interactive approach.[195] The protein is described by a series of interaction sites and the core may be placed manually or in an automated fashion. Substituent scoring is followed by a user filtering of substituents prior to full enumeration. Importantly, the success of this approach was demonstrated by the successful design of a series of potent and selective Factor Xa inhibitors.[196] Several iterations of library design led to the discovery of a compound with 16 nM K_i (**Figure 13**). A crystal structure of this compound in a related serine protease, trypsin, confirmed the predicted binding mode (**Figure 13b**). The subsequent development of this series lead to the discovery of a related series with oral bioavailability. A candidate is currently in phase II clinical trials as an oral antithrombotic.[197]

Another well-known docking program is FlexX,[198–200] which uses an incremental construction algorithm to dock a ligand. The ligand is fragmented, fragments docked, and the ligand rebuilt from high-scoring fragments. It can be seen that this approach is readily adaptable to combinatorial library design. In the recursive combinatorial docking algorithm described by Rarey and Lengauer[201] the initial fragment is the core or scaffold. R-groups are then ordered in such a way that each R-group is always linked to one of the previous R-groups or the core. This data structure allows a very efficient tree-traversal algorithm to be used in docking each library molecule as the results from partially built molecules common to many molecules can be used. Whilst this method does not necessarily lead to the same results obtained with docking each molecule independently, very significant speed-ups of the order of 30-fold can be achieved in favorable circumstances.

An alternative strategy to the purely docking driven approach is to use the protein structure to generate descriptors for the design. Thus Mason and Cheney describe the use of GRID[112,202,203] to map the binding site of the related serine proteases thrombin, Factor Xa and trypsin. Three- and four-centre pharmacophore fingerprints were built from minima in the interaction fields. It was demonstrated that these active-site pharmacophore fingerprints can be used to identify selectivity features and commonality between the sites through comparison with ligand-derived pharmacophores and produce reasonable docking modes in Factor Xa.[204,205] Design in Receptor (DiR)[206] was developed specifically to use protein-derived interaction sites in pharmacophore-based docking. The application of this methodology enables the generation of site-derived ligand-based pharmacophore fingerprints for use in library design as described by Mason and Beno.[141] For example, the impact of other design criteria on the sampling of possible binding modes can be quantified.

Eksterowitz *et al.* extended the concept of informative library design, described earlier, to utilize site-derived pharmacophoric information.[95] The informative design is performed within the space of active site pharmacophores as

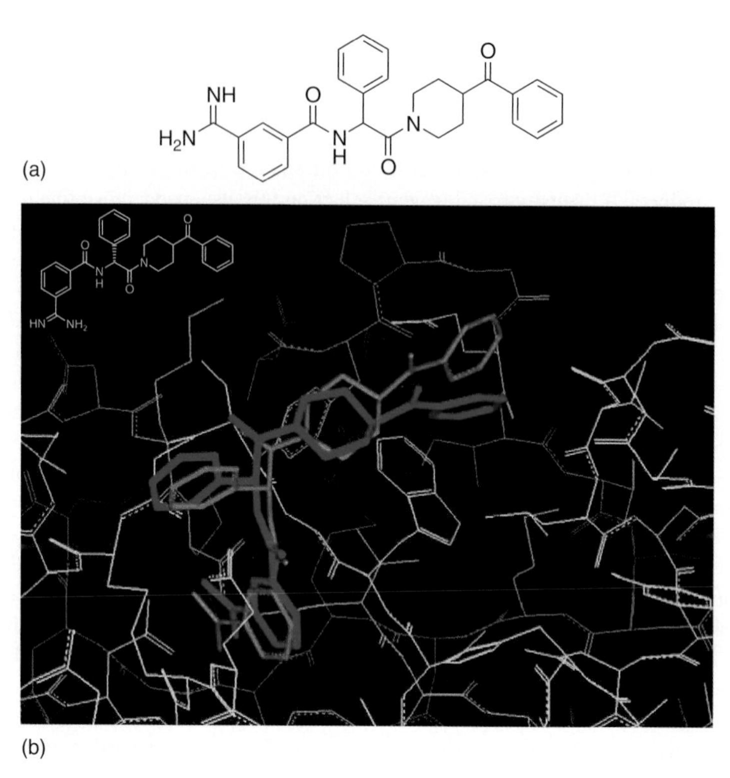

Figure 13 (a) 16 nM K_i inhibitor of Factor Xa designed using PRO–SELECT. (b) The crystal structure of the inhibitor in trypsin superimposed with the predicted binding mode in Factor Xa. (Reprinted in part with permission from Lieberschuetz, J. W.; Jones, S. D.; Morgan, P. J.; Murray, C. W.; Rimmer, A. D.; Roscoe, J. M. E.; Waszkowycz, B.; Welsh, P. M.; Wylie, W.; Young, S. C. *et al. J. Med. Chem.* **2002**, *45*, 1221–1232. Copyright (2002) American Chemical Society.)

illustrated in **Figure 14**. It is possible to account for receptor flexibility by utilizing multiple ligand bound protein structures in the generation of a union fingerprint.

SPROUT is a suite of programs for structure-based design that provides algorithms for identifying potential pharmacophore points, docking of small fragments to these points and subsequent connection of the fragments.[207–210] A successful application of this methodology has been reported in the design of Factor Xa inhibitors.[67] Incorporation of synthetic constraints into the ligand build up procedure has given rise to the program VLSPROUT, whereby structures are generated from a particular combinatorial chemistry scheme.[211]

4.15.4 Design Considerations

The synthesis of a successful combinatorial library requires the consideration of a number of different aspects, both chemical and computational. We mentioned some chemistry aspects in the previous section, compatibility of reagents with the reaction, protecting group strategies and so on, and will not consider these further here. In this section we discuss additional computational considerations when designing the library: choice of appropriate descriptors, methods to assess diversity or focus of the library, ADMET considerations.

4.15.4.1 Descriptors

A large number of descriptors have been developed to represent chemical structures, enabling similarity searching, the definition of a design space and the relationship of structure to biological activity. Structural keys such as the ISIS keys[212] encode the presence of specific molecular fragments into a bit-string, with presence or absence marked by a 1 or 0. Fingerprints such as those from Daylight[73] encode all paths within a molecule up to a user-defined maximum length. The paths are hashed into a bit-string so that each path sets several bits, though no two different paths will set exactly the same set of bits. Both methods were developed initially to improve the speed of structure retrieval from a database. Computing the key or fingerprint for a substructure provides an efficient filter on structures in the database

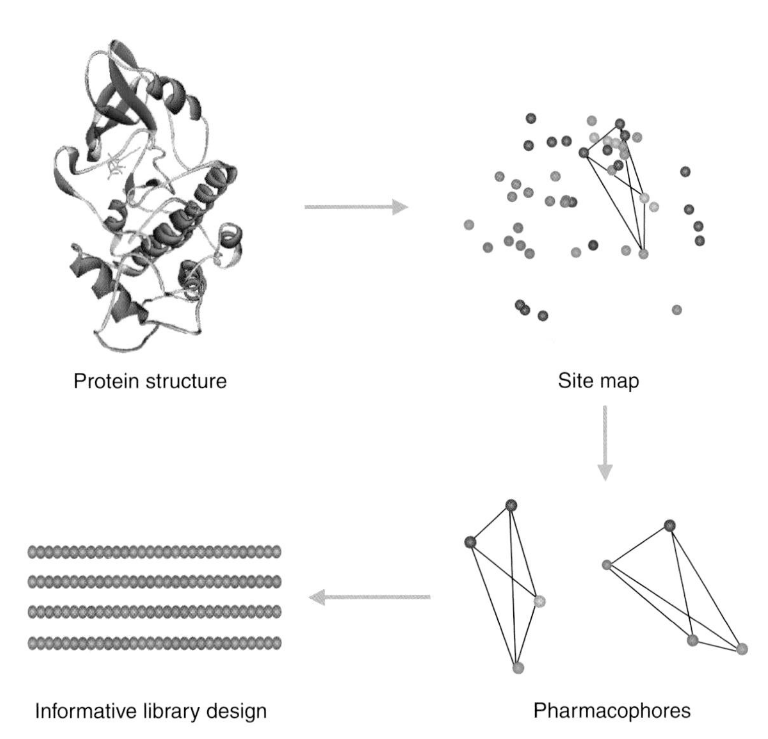

Figure 14 An illustration of informative library design using protein active site derived pharmacophore fingerprints. (Reprinted from Eksterowicz, J. E.; Evensen, E.; Lemmen, C.; Brady, G. P.; Lanctot, J. K.; Bradley, E. K.; Saiah, E.; Robinson, L. A.; Grootenhuis, P. D. J. *et al.*, *J. Mol. Graphics Model.* **2002**, *20*, 469–477. Copyright (2002), with permission from Elsevier.)

as any match must set at least the bits set by the substructure. Subsequently, alternative methods for encoding structural information into a fingerprint were introduced. Cahart *et al.*[213] introduced the concept of atom-pairs where the topological distances (number of bonds) between atoms of specified element type are encoded in a bit-string. This was extended to the topological torsion,[214] where elements on all paths of length four are encoded. Sheridan and coworkers developed atom pair descriptors that encode information about atom types (pharmacophoric type features such as donor or acceptor, charge) and the number of bonds between such features.[215,216] Binary fingerprints are an efficient way to store such information because it is straightforward to calculate similarities between molecules using any of a variety of similarity coefficients or distance measures.[217]

The CATS descriptors of Schneider *et al.*[218] store not only the occurrence but the count of each binding property pair in a correlation vector. With such real-valued descriptors alternative measures such as the Euclidean distance can be used to compute inter molecule distance. The CATS descriptors were shown to be useful in scaffold-hopping, identifying actives with a structural type distinct from that of the initial lead structure and have also been used as the basis for a de novo design program, TOPAS.[219]

Reduced graphs[44,220,221] provide a pharmacophore feature based representation of the molecular structure whilst maintaining the connectivity information of the molecular graph. In this approach the molecule is reduced to its key pharmacophoric features which become the nodes in the graph – hence the name. Representing the reduced graph features by rare earth or transition metal atoms that occur very infrequently in druglike molecules allows the reduced graph to be written in the SMILES notation and fingerprinted (**Figure 15**). However, these fingerprints are sparse and an alternative fingerprinting scheme has been described similar in concept to the atompair type fingerprints. Similarity can also be computed at the level of the graph itself using a dynamic programming approach to calculate the edit distance.[44]

An alternative to fingerprinting approaches is to compute a series of molecular descriptors such as molecular weight, hydrophobicity, and polar surface area. Such whole molecule descriptors give a broad perspective on protein–ligand recognition, bulk, hydrophobicity, hydrogen bonding ability, and so on, and are also related in a general sense to the question of druglikeness and general ADME properties such as in the Lipinski rule-of-five (see below). Other whole molecule descriptors with a closer relationship to the underlying chemical structure have been used particularly in the area of QSAR. These include a wide range of molecular connectivity indices, first proposed by Randić as a means of estimating physical properties of alkanes.[222] This formalism was quickly extended to other types of molecules[223,224] and since then a wide range of indices has been proposed, as reviewed by Hall and Kier[81,82] and Randić.[83] The indices

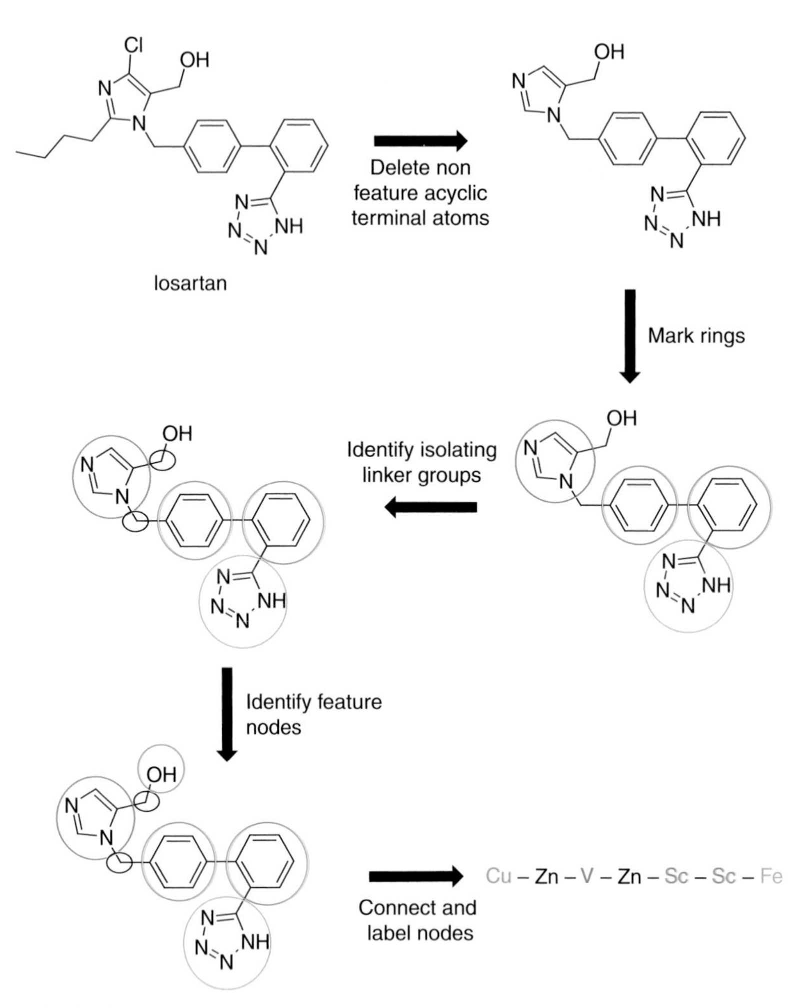

Figure 15 An example of reduced graph generation, starting from losartan. Pharmacophoric groups and rings become nodes in the graph with acyclic carbon chains becoming linker nodes. Fused rings are joined by a double bond. The use of heavy metal element names to represent the various features provides a convenient mechanism to display the reduced graph as a SMILES string and gives access to the full functionality of the SMILES language such as SMARTS searching. The feature nodes correspond to the following symbols: Cu, donor/acceptor feature; Zn, linker (any length); V, aromatic ring acceptor; Sc, aromatic ring; Fe, aromatic ring negatively ionizable.

are derived from a graph-theoretical representation of the structure where bonds are represented by the edges between nodes (atoms). They provide a direct representation of the topological structure of a molecule encoding information such as the degree of branching ($^1\chi$) and the adjacency of the branch points ($^3\chi$), flexibility, and shape. When using such descriptors it is generally necessary to first normalize and orthogonalize the property space using methods such as principal components analysis or nonlinear mapping methods.[225,226]

An alternative set of real-valued molecular descriptors are the BCUT (Burden CAS University of Texas) descriptors of Pearlman *et al.*[227,228] The BCUTs were derived from earlier work by Burden[229] to develop a unique signature for a molecule. Each molecule is described by a series of square matrices with atom labels defining the rows and columns. In a given matrix, the diagonal represents an atomic property such as charge, hydrogen bonding ability (donor/acceptor), or polarizability and the off-diagonal terms represent topological distance or other such property. Molecular descriptors are generated from the lowest and highest eigenvalues of these matrices, and describe the molecular surface distributions of positive or negative charge, hydrogen bond donors, hydrogen bond acceptors, high or low polarizability. A variety of descriptors can be calculated based on the nature of the diagonal and off-diagonal properties and the scaling between them. A number of authors have found BCUTs to be useful in a range of library design and QSAR applications.[94,141,230–234] The DiverseSolutions package[235] provides a suite of tools for the calculation and application of the BCUTs and other descriptors to library design and related design tasks. A particular application to the design of GPCR targeted libraries is described in more detail in Section 4.15.5.2.

The descriptors described above rely on the molecular graph only, that is, they contain no explicit information on the conformation or three-dimensional properties of a molecule. As for the two-dimensional fingerprints it is possible to generate three-dimensional pharmacophoric fingerprints where each bit marks the presence or absence of a particular triplet or quadruplet of pharmacophore points and the distances between them.[90,91,236] In general the distances are binned and up to seven pharmacophore features are represented: hydrogen bond donor, hydrogen bond acceptor, joint donor–acceptor feature, acid, base, hydrophobic, aromatic. The latter two tend to be represented by dummy atoms at the centroid of a phenyl ring for example rather the atoms that make up the ring itself. The fingerprints can contain many millions of bits and schemes have been developed to efficiently store and use the fingerprints.[68,98,237] They have found widespread application in library design, virtual screening, and related applications.[24,27,93,94,238] Other methods for incorporating three-dimensional and surface properties into design include the use of molecular fields as in GRIND/ALMOND described above[110] and VOLSURF[239–241] or the use of molecular shape either explicitly[242,243] or as descriptors.[244,245] In general, shape searching could be used in a manner analogous to other similarity methods. That is the virtual library could be ranked against one or more lead molecules and the library design algorithm could be used to maximize the number of high-scoring molecules in the final design. An alternative method has been described by Briem and Kuntz[246] whereby the program DOCK is used to dock molecules against a range of protein targets, thus generating a fingerprint that can be used in similarity searching. Attractive as these three-dimensional methods are they still pose a number of challenges when used for library design. The descriptors can be relatively slow to calculate as they generally require some form of conformational sampling. Whilst this is possible on large corporate collections as a series of one-off calculations, to perform routinely for large virtual libraries is still a major challenge. It is certainly the case for three-dimensional pharmacophores and shape-based descriptors that the conformation information is in someway merged into a single fingerprint or vector. This averaging out of the conformations is one reason why the three-dimensional descriptors do not always give the improvement over two-dimensional descriptors that might be expected.[247] However, they provide a complementary approach and a powerful methodology for locating under sampled areas of pharmacophore space when assessing potential libraries.

4.15.4.2 Diversity and Focus

There are three general strategies for analyzing and selecting compounds within a particular descriptor (chemistry) space: distance-based methods, clustering, and partitioning.[27,248] Distance-based methods utilize a similarity coefficient or distance metric such as Euclidean distance to measure the distance, similarity, or dissimilarity (generally taken as 1 − similarity) between two compounds. For a focused design compounds can be selected that are similar to one or more lead molecules. For a diverse design the compounds will be selected to be diverse with respect to each other and optionally a preexisting set of molecules such as a corporate screening collection. A number of diversity metrics have been developed for this particular problem and are applicable within the context of library design.[27,181,249] Waldman *et al.* have proposed a set of criteria that should be met by a diversity metric[181]:

1. Adding redundant molecules to a system does not change its diversity.
2. Adding nonredundant molecules to a system always increases its diversity.
3. Space-filling behavior of diversity space should be preferred.
4. Perfect (i.e., infinite) filling of a finite descriptor space should result in a finite value for the diversity function.
5. If the dissimilarity or distance of one molecule to all others is increased, the diversity of the system should increase. However, as this distance increases to infinity, the diversity should asymptotically approach a constant value.

The basic premise behind requirement 5 is that as the distance between two molecules increases the probability of the two sharing a specific biological activity decreases giving rise to a sampling or similarity radius.[247] However, the asymptotic behavior will not hold for any diversity function which uses a partition of molecular space into clusters, including all cell-based methods.

Clustering involves an additional step of grouping together similar compounds based on the similarity matrix. A variety of clustering algorithms have been proposed for chemistry applications, mostly adapted from the numerical taxonomy and information science literature.[250] In general, agglomerative schemes are used whereby each compound starts in a cluster of one (singleton) and pairs of clusters are selected sequentially based on optimizing a function of the similarities or distances between the compounds in the two clusters. Algorithms vary depending on the choice of function used. For example, in complete linkage the maximum distance between all pairs of compounds, one from each cluster, is used. By contrast, in single linkage the minimum distance between all pairs of compounds is used. It is clear that the choices made here affect the general shape of the clusters in the descriptor space, with complete linkage tending to give small compact clusters and single linkage long meandering clusters.

The use of clustering in library design would involve the selection of compounds in such a way as to cover a large number of clusters. Coclustering of the virtual library with a reference compound set would enable the selection of compounds from clusters under represented in the reference collection. However, most clustering algorithms have a computational complexity of at least $O(N^3)$ because of the need to calculate the $N \times N$ similarity matrix, where N is the number of compounds, and this has tended to limit the size of data sets that can be used even with efficient implementations.[251] Sphere exclusion clustering is a particularly efficient nonhierarchical algorithm that takes as input the nearest neighbor table.[75] The algorithm starts by selecting the compound with the largest number of nearest neighbors and holding out from selection all neighbors within a user-defined radius. The algorithm continues iteratively selecting compounds in this way until there are no more to select. With current computer resources it is possible to cocluster virtual libraries of a few million compounds with large corporate collections in a reasonable timeframe (1–2 days).

Partition-based methods[252] have the advantage that it is possible to identify directly where the gaps are in a particular compound set. The descriptor space is divided into a series of partitions or cells. Cell occupancy becomes the main driver for a diversity-based selection whilst for a focused selection compounds can be selected from cells occupied by a lead compound.[253] The methods generally require a low-dimensional descriptor space and it is necessary to decide where to make the partitions in each dimension. The DiverseSolutions approach of Pearlman *et al.*[227,228,235] uses a χ^2 metric to select a low-dimension (five or six BCUT descriptors) chemistry-space from the full set of BCUT descriptors. An extension of this approach allows for the generation of a receptor relevant subspace where a set of compounds active against a particular target are used to further guide the selection of the appropriate chemistry space.[254] An application of this approach to the design of GPCR-focused libraries is described below.

4.15.4.3 Developability

Increasingly, developability is being considered at the earliest stages of drug discovery. Analysis of marketed oral drugs led Lipinski to propose the rule-of-five which is obeyed by approximately 90% of marketed oral drugs.[20] Several other groups have analyzed the properties of marketed drugs to provide general guidelines.[62,255–258] The development of discriminant models to separate drugs from nondrugs provides a mechanism to rapidly evaluate large compound collections and virtual libraries for druglikeness.[259–262] The problem here is to define a set of nondrugs with workers generally considering general supplier data sets such as the ACD and comparing to drug databases such as the WDI.

More recently the debate has moved from druglikeness to lead-likeness as an important concept in lead generation.[263–265] Analysis of drugs and the leads that led to them suggests that leads should have a narrower range of properties, lower molecular weight and log *P*, than the final drug as these properties tend to increase during lead optimization as hydrophobic and bulkier groups such aromatic rings are incorporated to increase activity and modulate selectivity.[266–268] Such trends have been consistent over the past few decades.[258] Thus these studies provide guiding principles for the synthesis of libraries for lead generation.

Initial profiling of a lead molecule will likely identify potential developability concerns, cytochrome P450 metabolism, poor absorption, central nervous system (CNS) penetration (or not depending on the target) and so on. This has led to a plethora of in silico models based on a variety of modeling paradigms, QSAR analysis,[269,270] pharmacophore modeling,[271] and structure-based approaches.[272] One of the limitations to these models is the availability of large diverse data sets of compounds and the restrictions that this places on the prediction domain.[273,274] Recent reviews have discussed the state of the art in this area and highlight the role that such approaches can play when applied in an appropriate manner.[275,276]

Predictions from such models can be incorporated into library designs via an appropriate objective function. As an example, Pickett *et al.*[46,162] used relatively simple models for permeability based on polar surface area to improve the permeability of a kinase library. As mentioned earlier, a design algorithm was developed to generate a near-combinatorial solution as no combinatorial solution satisfied all the constraints. Evaluation of the library in a model for permeability showed a significant improvement in permeability when compared to a previous library based on the same scaffold.

4.15.5 Applications of Library Design in Lead Generation

Combinatorial chemistry is used routinely within the pharmaceutical industry for lead generation where large numbers of compounds are synthesized, largely based on diversity criteria, as a source of compounds for HTS. Combinatorial chemistry (or parallel synthesis) is then used to rapidly expand around hit molecules in the generation of SARs against both the target of interest and selectivity and developability targets. In lead optimization smaller focused sets of compounds will be made in conjunction with more traditional single compound synthesis to further explore and improve the compound properties.[25] The major strategy for lead generation when there is little or no information about the target has been HTS. High-throughput screening provides an efficient process to assess large numbers of compounds against a particular biological target. This has been seen as a main driver for the application of combinatorial chemistry to the generation of large diverse sets of compounds. Some of the current issues and challenges in this approach were discussed from a strategic perspective in Section 4.15.2. The major points of this debate revolve around how large the collection should be, what types of compounds should be included in the screen, and how many compounds of a specific type should be made. We discuss these questions in the following paragraphs, starting with summary of the latest thinking about how to design a collection. This is followed by examples of the application of systems-based approaches to library design, a system being defined biologically as a set of related protein target such as kinases or subgroups of 7-transmembrane (7TM) receptors. We finish this section with examples of specific target-based approaches to lead generation.

4.15.5.1 Building a General Screening Collection

The various methods for diversity-based selection discussed in Section 4.15.4.2 provide only part of the answer when it comes to designing a screening collection. Adding new compounds will always increase the diversity of a set of compounds and so they provide little information on which or how many compounds to make a priori. Nilakantan and colleagues at Wyeth-Ayerst have approached this problem by asking the question "How big should each potentially active series be so that there is a high likelihood (say 95%) that at least one member of the series is a hit in the relevant biological screen?"[51,277] In order to answer this question they analyzed in-house data to assess the hit rate within series, with a series defined using the concept of the ring-scaffold that describes a molecule by the rings and interconnections between them.[278,279] The knowledge of hit rates allows the generation of a probabilistic model to determine the number of compounds that should be sampled from any series (scaffold) such that there is a given probability that there will be at least one hit from the sample against any target for which the series as a whole will give a hit. This is given by:

$$s = \frac{\ln(1-P)}{\ln(1-p)} \qquad [15]$$

where s is the sample size, p is the probability that a compound in the series is a hit, and P is the probability that at least one of the compounds in s is a hit. This result is general and can be applied to any clustering algorithm. It was necessary to apply some heuristics in estimating the probabilities when using the ring-scaffolds, for example, setting a minimum ratio for atoms in the scaffold to total atoms to identify clusters such as that where benzene is the only ring. The estimate obtained for s is in the order of 70–100 compounds per ring-scaffold. Thus, the proposal is to build a screening collection from say 500 different ring-scaffolds each with $\sim$100 compounds. The Nilakantan work provides a rational approach to limit the number of compounds made or acquired around a particular scaffold. However, it says little about the relative merits of a scaffold with regard to an existing collection (other than presence or absence) and does not allow for the fusion with a typical pharmaceutical screening portfolio, which will generally include targets falling into particular target families. Some of these families will have a significant knowledge-base about the types of compounds likely to be active and/or are amenable to structure-based or other knowledge-based approaches. In looking to provide a rationale for the make-up of the screening collection at GlaxoSmithKline, Harper *et al.* were motivated by the problem of wishing to maximize the expected number of lead series from a screen.[52] In this approach we start with the premise that compounds may be clustered in some way. It is clear that just because cluster i contains a hit, not all compounds in the cluster will necessarily test active. There is therefore a probability, $\alpha_i < 1$, that any given compound, sampled at random from a cluster i containing actives, will itself turn out to be a hit. The other parameter in the model is the probability, π_i, that the cluster contains a lead molecule. Note that in this model, being a hit in a screen, and being a lead for the screen are very different concepts. The definition of lead must be realistic. At the minimum, a lead is a compound with a confirmed IC_{50}, which has been characterized by physical chemistry, is amenable to fairly rapid chemical expansion, and is patentable. Armed with these two parameters and the assumption that once a hit has been found in a cluster, the chemist will be able to find the lead by substructure searching or simple chemical modification, it is possible to derive an expression for the expected number of lead containing clusters[52]:

$$E = \sum_{i=1}^{p} \pi_i[1 - (1 - \alpha_i)^{N_i}] \qquad [16]$$

In building the collection it is desirable to maximize eqn [16]. Doing so gives a general solution to the number of compounds to be selected from any cluster as:

$$N_i = \begin{cases} \dfrac{\ln \lambda - \ln \pi_i - \ln[-\ln(1-\alpha_i)]}{\ln(1-\alpha_i)} & \text{whenever this is } \geq 0 \\ 0 & \text{otherwise} \end{cases} \quad [17]$$

as shown in Harper *et al.*[52] (λ is the Lagrange multiplier found via a line search.) It is this relationship between π_i and α_i that gives the model its power. The model provides a formalism for answering several highly relevant questions concerned with screening collection design.

Assuming a wide spectrum of biological targets will be screened, an average value for π can be assumed for all i. Similarly, taking a constant value for α gives the following equation as a diversity score when adding compounds to an existing collection[52] where N_i is the number of compounds in cluster i and p the total number of clusters in the combined collection after adding the new compounds:

$$D[(N_i)_{i=1}^{p}] = p - \sum_{i=1}^{p}(1-\alpha)^{N_i} \quad [18]$$

This is directly linked to an increase in the likelihood of finding a lead from any screening campaign and can be used as an objective function in combinatorial library design using the multiobjective approaches described in Section 4.15.3.6. This function also satisfies the first four of Waldman's criteria for a diversity metric.[181] As noted above, the fifth criterion cannot be satisfied by any partition or cluster-based method. The model can also be used to answer the question of how large a collection should be and provides a basis for applying property filters such as Lipinski's rules. In the latter case, the implication of a cluster of compounds failing property filters is that they are considered as less likely to give a lead. The extent to which this is considered to be the case will determine the relative value for π_i for these clusters in eqn [17]. Thus, the model would suggest the inclusion of a lower number of such compounds in the screening collection but not necessarily their complete absence. A similar argument can be made when considering targeted versus nontargeted designs, the model providing a bridge between the ideals of focused and diverse designs.

Note that these results are general and independent of any particular clustering methodology. However, to apply the model it is necessary to estimate values for π_i and α. Analysis of in-house screening results[52] suggests $\pi = 1 \times 10^{-5}$ as an appropriate value for a typical, filtered compound collection. An appropriate value for α can be determined from the work of Martin *et al.*[280] where 115 HTS results sets were analysed using sphere exclusion clustering on Daylight fingerprints. With a cluster radius of 0.85 there is a 30% probability of two compounds in the same cluster sharing the biological activity. Thus, if this method is used to cluster compounds $\alpha = 0.3$ is an appropriate value. The clustering produces compact, spherical clusters, one cluster does not equate to one chemical series, and the particular implementation cannot be compared directly to that of Nilakantan. (Nevertheless the two theoretical frameworks are in agreement provided they are parameterized appropriately for the chosen clustering methodology, and Harper *et al.'s* framework can be viewed as a generalization of the earlier work by Nilikantan *et al.*) It is evident that neighboring clusters will contribute to the SARs derived from screening to reduce the impact of screening false negatives. In effect, the model is being pessimistic in our ability to move from cluster to cluster.

In the absence of information distinguishing clusters from each other (taking common values of π_i and similarly of α_i for all clusters), the model suggests selecting one compound from each cluster (maximizing the diversity metric of eqn [18]). This may seem unintuitive, and it is clear from the model that selecting one compound from each cluster in no way guarantees finding everything in those clusters (in fact, there is only a 0.3 chance in a given lead-containing cluster of finding a hit). However, the lost opportunity represented by not sampling further from a given cluster is outweighed by the opportunity gained from sampling from other clusters, leading to an overall increase in the overall probability of finding a lead from the screening collection as a whole. In addressing this tradeoff, the model takes a holistic view that moves beyond the examination of individual clusters or scaffolds in isolation.

4.15.5.2 System-Biased Libraries

The screening collection models presented in the preceding paragraphs do not in themselves direct the chemist to a particular region of chemical space. In some senses this can be considered a strength of the models as a wide variety of knowledge-driven approaches can be brought to bear on this problem whilst the collection models provide a rational framework for deciding upon the numbers of compounds to be synthesized and screened. Two important groups of biological targets have attracted the most interest in the area of biologically targeted libraries or systems-biased design.

GPCRs are the biological targets for a number of marketed drugs[31] and comprise a significant proportion of the portfolio in most pharmaceutical companies.[281,282] Ser/threonine and tyrosine kinases represent another major focus of activity and the wealth of genomic and structural data available makes them ideally suited to a knowledge-driven approach.[53] There is considerable effort in these areas[56,283] and is not limited to pharmaceutical companies as many commercial compound suppliers market collections or libraries of compounds targeted towards particular groups of biological targets.[58] Some recent examples of the application of library design methods in this area are highlighted in the following paragraphs.

4.15.5.2.1 G protein-coupled receptor-biased libraries

Knowledge-based drug discovery for GPCRs has largely been driven by ligand-based approaches such as pharmacophore methods.[284–287] Homology modeling of the receptors is an important arm in the strategy though traditionally their use has been limited by the absence of high-quality crystal structures of these targets.[288] However, there are indications that the situation will improve as crystal structures begin to appear.[289–293] The concept of a privileged structure was first introduced by Evans *et al.* largely driven by experiences with GPCR targets.[294] It was recognized that some scaffold classes could lead, with appropriate modification, to active compounds across several receptors. This philosophy has been extended to a wide range of compound classes[295] and provides a chemically driven approach to the targeting of libraries towards GPCRs as reviewed by Guo and Hobbs.[296] An elegant example of this approach to GPCR-biased library design is described by Mason *et al.*[91] The MDL Drug Data Report (MDDR)[297] was used as a data source for information on GPCR ligands. Several privileged substructures were identified and used to represent the fourth point in a four-point pharmacophore fingerprint. This provides a frame of reference for the pharmacophores calculated from the relevant ligands and provide a design space for subsequent GPCR-focused designs. As a specific example, the Ugi reaction was chosen (**Figure 16**) with R^3 as a biphenyl tetrazole. Limited availability of aldehydes and isonitriles meant that the main selection criterion was applied for the acids. Pharmacophore keys were calculated for each product using the biphenyl tetrazole as the fourth point and were combined to create a union key for each acid in the initial list. A pharmacophore key was computed from 502 biphenyl tetrazole containing compounds in the MDDR and acids were selected iteratively based on the degree of coverage of the relevant product-based pharmacophore keys with the MDDR pharmacophore key. Subsequent analysis of the pharmacophores left underrepresented in the initial design led to modifications to the chemistry and reagent lists to incorporate protected acids and bases into the synthesis and so enable the coverage of previously missing pharmacophores.

An excellent example of the application of systems-based design to GPCRs is provided by Lavrador *et al.*[298] The design was focused on the GPCR superfamily activated by positively charged peptide ligands (abbreviated as GPCR-PA$^+$) that cover proteins in GPCR categories A and B for melanocortins, gonadotropin releasing hormone (GNRH), bradykinin, melanin concentrating hormone (MCH), calcitonin gene related peptide (CGRP), and vasoactive intestinal peptide (VIP). A set of 81 560 drugs taken from the literature and in-house programs at Neurocrine defined a drug space utilizing the BCUT metrics[227,228,254] and the DiverseSolutions package[235] (*see* Section 4.15.4.1). A five-dimensional chemistry space (hydrogen bond donor and acceptor, two metrics of polarizability, and charge) was calculated and each of the axes divided into 10 bins to give a total of 100 000 individual cells. A representative set of GPCR-PA$^+$ active compounds are shown projected in this space in **Figure 17** and were shown to occupy only a small region of the total drug space. A set of 19 templates were enumerated, covering over 9 million compounds, and evaluated against the GPCR-PA$^+$ space, selecting compounds that fell within the same or neighboring cells to the known active compounds. In total a set of 2025 compounds were synthesized combinatorially covering seven templates. Note that the combinatorial constraint will lead to some compounds being outside of the design space.

The confirmed hit rates (actives were screened again in duplicate on a second occasion) against three GPCRs are shown in **Table 1**. Results were compared to two other compound sets: a random set of screening plates (2024

Figure 16 The Ugi reaction, employed by Mason *et al.*[91] in the design of GPCR-focused libraries and Weber *et al.*[310] in the discovery of submicromolar thrombin inhibitors, utilizing a biological activity guided optimization.

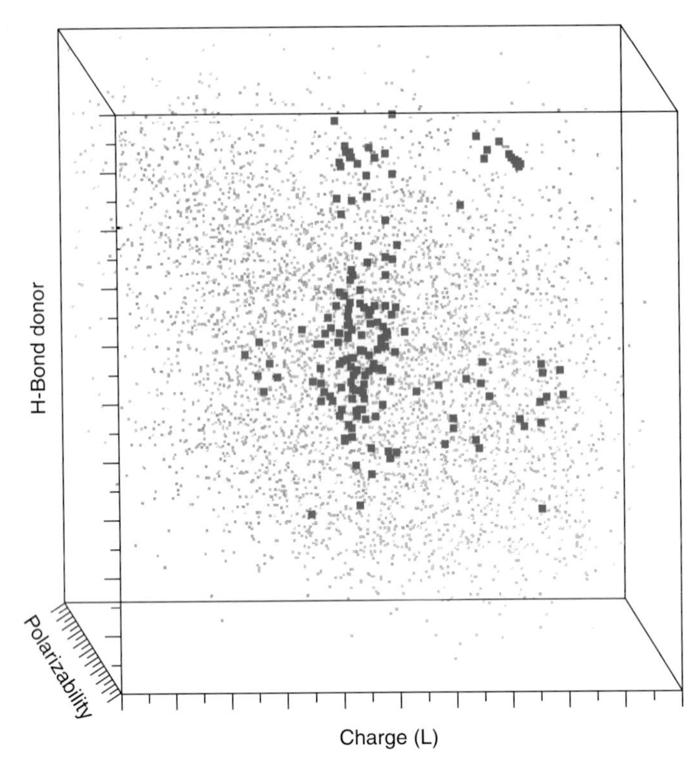

Figure 17 Three dimesions of a five-dimensional chemistry space calculated from 81 560 druglike compounds shown as grey dots. The blue circles represent 111 compounds active against GPCRs activated by positively charged peptides. (Reprinted in part with permission from Lavrador, K.; Murphy, B.; Saunders, J.; Struthers, S.; Wang, X.; Williams, J. *J. Med. Chem.* **2004**, *47*, 6864–6874. Copyright (2004) American Chemical Society.)

compounds) and a random set of compounds containing a positive charge (1401 compounds). In all cases the designed compound set gave significant enrichment over the random sets, thus providing validation for this design methodology. Interestingly, as shown in **Figure 18**, the active compounds against each target occupy discrete regions of the space and all but one of the hits occupy cells within the GPCR-PA$^+$ space or their neighbors.

Several groups have developed a chemogenomics approach to support GPCR targeted library design.[57,299,300] Jacoby *et al.*[299] utilized homology modeling in conjunction with mutagenesis information to study the binding of small molecule ligands of the monoamine GPCRs such as serotonin and propranolol. This allowed the identification of three binding pockets within the transmembrane region with distinct properties afforded by the amino acid residues at each site. This information permits the generation of combinatorial libraries that explores combinations of these binding regions in a systematic way. BioFocus have developed a methodology termed thematic analysis[57] to define the binding regions across a range of different GPCRs. Like the Novartis approach this involves sequence analysis, homology modeling, and use of mutagenesis information to locate binding sites for small-molecule ligands within the transmembrane regions. According to this analysis only 30–40 residue positions are critical to ligand-binding across a broad spectrum of drugs. These interaction regions define eight microenvironments with reasonably well-defined properties for each site dependent on the types of amino acids (the Theme) and small-molecule fragments that interact. These themes define a signature for each GPCR and can be used to group the different receptors. Libraries are designed to explore various combinations of these themes and have been reported to have hit rates between 1% and 13% across a range of GPCR classes.[57] In addition, the ability to analyze hits in the context of the themes provides a good starting point for subsequent lead optimization.

4.15.5.2.2 Kinase-biased libraries

Kinases represent a particularly attractive target class from the perspective of knowledge-based design because of the wealth of structural information,[301] the relative conservation of the ATP binding site, and the existence of over 500 kinases in the human genome[60,302] which provides ample opportunities for disease modifying pharmacology. However, they also present severe challenges: the conservation of the binding site can lead to issues of selectivity and the proteins themselves are highly flexible.

Table 1 Confirmed hit rates obtained screening three compound sets against three GPCR targets

Compound set[a]	*Compounds screened*	*Confirmed hit rate (%)*		
		hMC4	*rMCH1*	*GnRH*
Design	2025	3.00	6.10	0.44
Random	2024	0.05	0.30	0.10
Basic	1401	0.35	0.35	0.07

[a] The designed set used a five-dimensional chemistry space to design libraries based on the compound adjacency to known GPCR-PA$^+$ ligands. The random and basic sets were selected at random from a screening collection, with the latter set constrained to compounds containing a basic group.

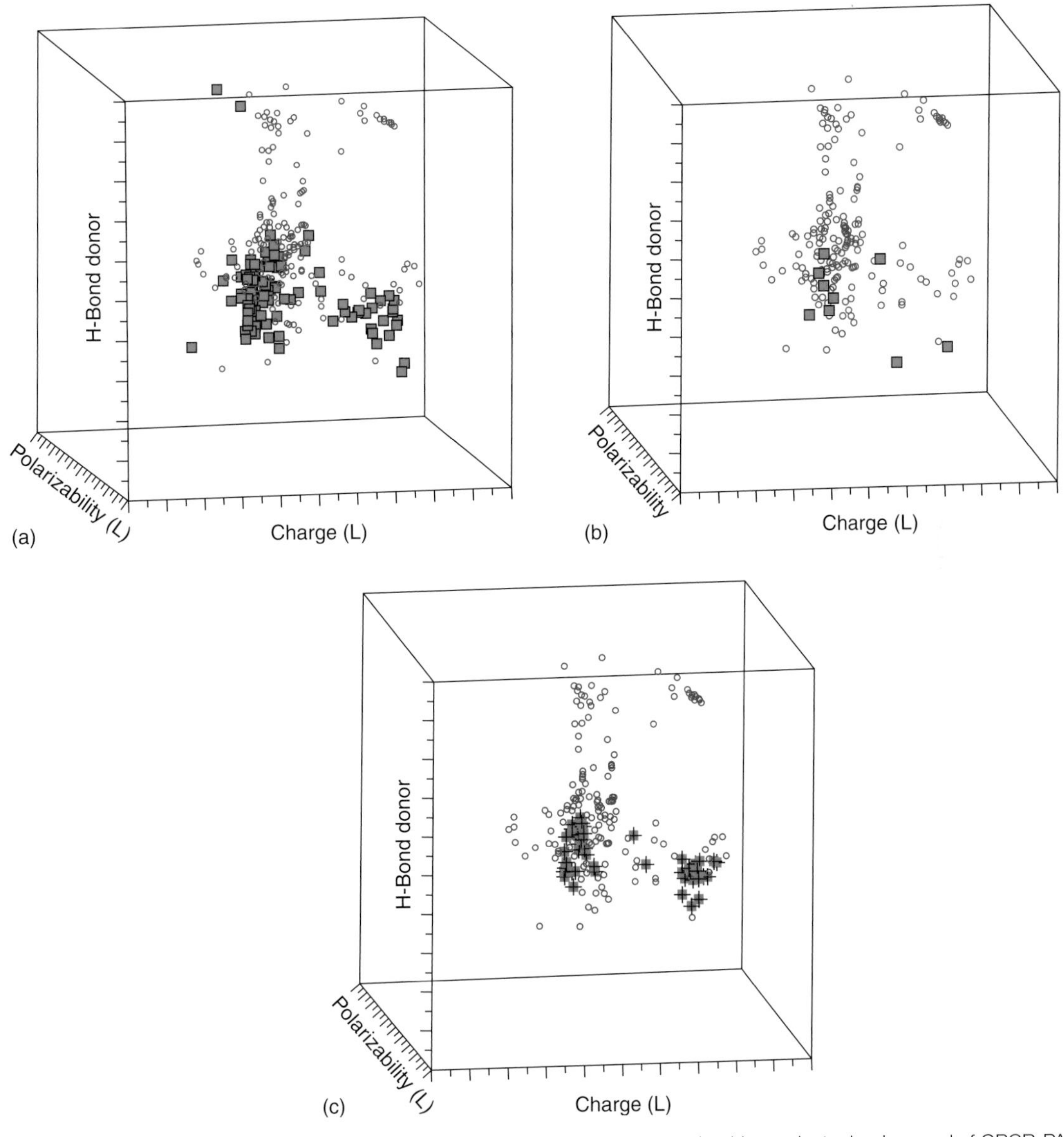

Figure 18 The location of designed compounds (red squares) that were screening hits against a background of GPCR-PA$^+$ ligands (open circles). The view is the same as in **Figure 17**. (a) MCH-R hits; (b) GnRH-R hits; (c) MC4-R hits. (Reprinted in part with permission from Lavrador, K.; Murphy, B.; Saunders, J.; Struthers, S.; Wang, X.; Williams, J. *J. Med. Chem.* **2004**, *47*, 6864–6874. Copyright (2004) American Chemical Society.)

We have already highlighted the importance of the choice of chemistry and hence scaffold or core to any successful library design. In this regard Stahura *et al.*[303] have described the prioritization of scaffolds for library design using a docking-based approach. Scaffolds were derived from known compounds and analysis of whole-molecule docking to identify scaffolds with shape complementarity to the ATP binding site. A related approach has been described by Lowrie *et al.*[56] Small virtual libraries are generated for each core and docked into the relevant proteins. The best cores will lead to consistently docked structures, analyzed using clustering techniques. Todorov *et al.*[304] have described a de novo design approach for combinatorial library design, taking the CDK family of proteins as an example. Putative ligands are designed using the validated de novo design program Skelgen.[305–307] Pharmacophore points define areas of interest in the binding site and structures are grown into the site using a build-up procedure, drawing fragments from a fragment library. A number of objectives are used to control the growing stage including fit to the pharmacophore points and the ScreenScore scoring function.[308] Protein flexibility was considered by using several crystal structures for the target of interest. Selected ligands are then docked across all targets and a specificity score is defined that allows the selection of ligands with a particular binding profile.

BioFocus have applied the strategy employed for GPCR-focused libraries to kinase-targeted libraries.[309] In this instance, however, the analysis is aided by the availability of detailed crystallographic data. Three-dimensional superposition of relevant proteins allows an analysis to be made of binding pockets and the relevant amino acids. This information is abstracted into a two-dimensional Roadmap as exemplified in **Figure 19**, which provides a tool for the analysis of the various subpockets and their properties as determined by the amino acids. Both backbone and sidechain interactions are presented. Library design involves the alignment of a scaffold into this view and the enumeration of products that sample different combinations of regions in the map with substituent properties appropriate for the pockets being explored.

An approach to aid in the design of selective kinase libraries was investigated by Pirard and Pickett.[233] A set of 770 compounds was compiled from literature and in-house sources based on their being active against at least one of five different kinases (JNK1, P38, CDK1, SYK, and EGFR) and a partial least squares discriminant analysis performed, using BCUT metrics as descriptors. The performance of this model was compared to clustering on Daylight fingerprints, using 52 novel EGFR compounds. The model correctly classified 48/52 compounds whilst clustering only grouped 8/52 compounds into EGFR clusters. In addition, three- and four-point pharmacophore fingerprints were calculated and used to investigate pharmacophores common within and between classes. The four-point pharmacophore fingerprints were shown to have a greater selectivity between classes and complement the partial least squares discriminate analysis model in that 75% of the misclassified compounds contain at least one four-point pharmacophore common to the appropriate class. Thus, the methodology provides a mechanism for prioritizing libraries targeted against specific kinases.

4.15.5.3 Activity-Guided Design

The methods and applications described up to this point have relied to a greater or lesser extent upon a computational model to guide the library design. An issue referred to above has been the vastness of chemical space; the Ugi reaction can give rise to potentially millions of products, there are 64 million possible hexapeptides from the 20 naturally occurring amino acids. Are there alternative strategies for the discovery of active compounds in this vast space? One such iterative approach to library design has been proposed and exemplified by several groups.[310–312] The idea is simple in principle: screen a subset of compounds from a library, measure the biological activity, input this information to an optimization algorithm, and generate the next set of compounds to synthesize and screen. This process is repeated until the desired activity level is reached or no improvement is seen. A GA has been the optimization method of choice. It relies upon a population of individuals, the standard genetic operators of crossover and mutation can be used at each iteration to suggest new molecules and the fitness function is highly configurable and does not rely on a continuous functional form. The power of this approach is exemplified by the work of Singh *et al.*[311] on the optimization of hexapeptides against stromelysin. The starting population of 60 compounds was biased by knowledge that proline is the favored amino acid at position two, with this constraint being removed for future generations. The GA was used to select subsequent generations based upon the screening data for the population. The reported data shows a significant improvement in activity for each of the five generations completed. Clearly the approach requires some initial knowledge of where to start the search. Weber *et al.*[310] reported on the optimization of Ugi products, **Figure 16**, against the serine protease thrombin. In this case the amines were selected from a subset of basic groups known to be likely P1 binders. The virtual library contained 160 000 products derived from 10 isocyanate, 40 aldehydes, 10 amines, and 40 carboxylic acids. Working with a population size of just 20, 16 generations were sufficient to generate compounds with submicromolar potency. The compound shown in **Figure 20**, with activity of 0.22 μM, was synthesized at generation 18. Thus, only 400 out of the possible 160 000 products were synthesized.

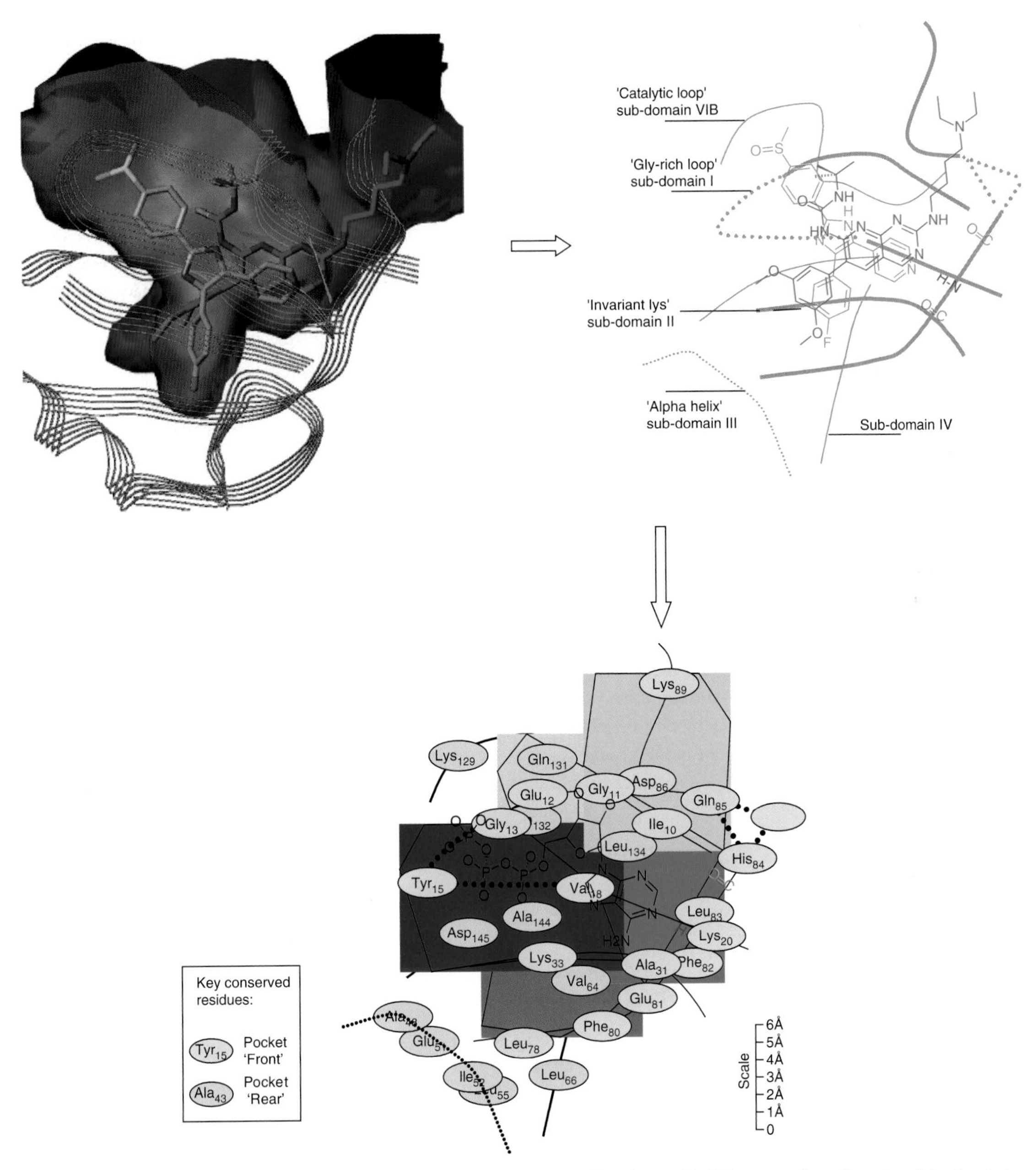

Figure 19 The strategy for generation of a kinase two-dimensional Roadmap. Proteins are aligned and an idealized view generated. Analysis of sequence alignments and interaction patterns with known inhibitors allows the generation of the roadmap. Residues circled in bold interact primarily through the side chain, other labeled residues will interact primarily through the backbone. (Reproduced with permission from Birault, V.; Harris, C. J.; Le, J.; Lipkin, M.; Nerella, R.; Stevens, A. *Curr. Med. Chem.* **2006**, *13*, 1735–1748. Copyright Bentham Science Publishers.)

Brauer *et al.*[313] have recently described a similar approach to discover novel inhibitors of glucose-6-phosphatase translocase (G6PT), a therapeutic target for the treatment of Type 2 diabetes. Several amide and sulfonamide inhibitors were discovered via HTS and provided the starting point for the virtual libraries. Seven generations of 352 compounds were synthesized and led to several compounds with activity below 10 μM. Interestingly, the information gleaned from these compounds were then used to direct a GA based library design to select compounds based on the two-dimensional similarity to the most potent compounds from the activity guided design, thus permitting exploration of SAR. In a final step, an imidazole template was used as a constrained analog of the amide and sulfonamide groups.

Figure 20 The most active compound found during an activity-guided GA optimization of a Ugi library, with an activity of 0.22 μM versus thrombin.[310]

These examples outline the potential of this iterative, data-driven approach to lead generation and optimization. It is thus intriguing that so few applications have appeared in the literature. Possible reasons for this could include the fact that a starting point is required to make the approach viable. In addition the chemistry must be very robust and ready to run with a very efficient screening process. The bottleneck in combinatorial chemistry is generally in the chemistry feasibility and validation steps rather than production so that making several thousand analogs in one batch may be perceived as a more efficient approach.

4.15.6 Conclusion

Combinatorial chemistry is an important component of medicinal chemistry in the generation of leads and the subsequent generation of drug candidates. In this chapter we have explored the role that computational library design plays in the process to ensure that the synthesized compounds satisfy appropriate goals for drug or lead-likeness, target focus and diversity. The multiobjective nature of the design task requires algorithms that allow the chemist to make rational decisions about which reagents to select and the final library to synthesize. Thus, design algorithms have themselves evolved over time to meet these needs from approaches that present a single solution that satisfies a weighted sum fitness function to multiobjective designs that return a family of equivalent solutions. This paradigm shift enables the medicinal chemist to employ their domain knowledge in selecting the library to make. Further, advances in computational design methodology allows for the targeting of libraries to specific proteins or families of related proteins as exemplified by the examples described here.

References

1. Merrifield, R. B. *J. Am. Chem. Soc.* **1963**, *85*, 2149–2154.
2. Geysen, H. M.; Meloen, R. H.; Barteling, S. J. *Proc. Natl. Acad. Sci. USA* **1984**, *81*, 3998–4002.
3. Houghten, R. A. *Proc. Natl. Acad. Sci. USA* **1985**, *82*, 5131–5135.
4. Lam, K. S.; Salmon, S. E.; Hersh, E. M.; Hruby, V. J.; Kazmierski, W. M.; Knap, R. J. *Nature* **1991**, *354*, 82–84.
5. Bunin, B. A.; Ellman, J. A. *J. Am. Chem. Soc* **1992**, *114*, 10997–10998.
6. Simon, R. J.; Kaina, R. S.; Zuckermann, R. N.; Huebner, V. D.; Jewell, D. A.; Banville, S.; Ng, S.; Wang, L.; Rosenberg, S.; Marlowe, C. K. et al. *Proc. Natl. Acad. Sci. USA* **1992**, *89*, 9367–9371.
7. DeWitt, S. H.; Kiely, J. S.; Stankovic, C. J.; Schroeder, M. C.; Cody, D. M. R.; Pavia, M. R. *Proc. Natl. Acad. Sci. USA* **1993**, *90*, 6909–6913.
8. Ugi, I.; Steinbruckner, C. *Chem. Ber.* **1961**, *94*, 734.
9. Dolle, R. E. *Mol. Divers.* **1997**, *3*, 199–233.
10. Dolle, R. E. *Mol. Divers.* **1998**, *4*, 233–256.
11. Dolle, R. E.; Nelson, K. H., Jr. *J. Comb. Chem.* **1999**, *1*, 235–282.
12. Dolle, R. E. *J. Comb. Chem.* **2000**, *2*, 383–433.
13. Dolle, R. E. *J. Comb. Chem.* **2001**, *3*, 477–517.
14. Dolle, R. E. *J. Comb. Chem.* **2002**, *4*, 369–418.
15. Dolle, R. E. *J. Comb. Chem.* **2003**, *5*, 693–753.
16. Dolle, R. E. *J. Comb. Chem.* **2004**, *6*, 623–679.
17. Wolcke, J.; Ullmann, D. *Drug Disc. Today* **2001**, *6*, 637–646.
18. Willoughby, C. A.; Hutchins, S. M.; Rosauer, K. G.; Dhar, M. J.; Chapman, K. T.; Chicchi, G. G.; Sadowski, S.; Weinberg, D. H.; Patel, S.; Malkowitz, L. et al. *Bioorg. Med. Chem. Lett.* **2002**, *12*, 93–96.
19. Lee, A.; Breitenbucher, J. G. *Curr. Opin. Drug. Disc. Dev.* **2003**, *6*, 494–508.
20. Lipinski, C. A.; Lombardo, F.; Dominy, B. W.; Feeney, P. J. *Adv. Drug Deliv. Rev.* **1997**, *23*, 3–25.
21. Agrafiotis, D. K.; Myslik, J. C.; Salemme, F. R. *Mol. Divers.* **1999**, *4*, 1–22.
22. Blaney, J. M.; Martin, E. J. *Curr. Opin. Chem. Biol.* **1997**, *1*, 54–59.
23. Bures, M. G.; Martin, Y. C. *Curr. Opin. Chem. Biol.* **1998**, *2*, 376–380.
24. Green, D. V. S.; Pickett, S. D. *Mini-Rev. Med. Chem.* **2004**, *4*, 1067–1076.
25. Laird, E. R.; Blake, J. F. *Curr. Opin. Drug. Disc. Dev.* **2004**, 7, 354–359.
26. Lewis, R. A.; Pickett, S. D.; Clark, D. E. Computer-Aided Molecular Diversity Analysis and Combinatorial Library Design. In *Reviews in Computational Chemistry*; Lipkowitz, K. B., Boyd, D. B., Eds.; John Wiley: New York, 2000; Vol. 16, pp 1–51.
27. Mason, J. S.; Pickett, S. D. Combinatorial Library Design, Molecular Similarity and Diversity Applications. In *Burger's Medicinal Chemistry and Drug Discovery*; Abraham, D. J., Ed.; John Wiley: New York, 2003; Vol. 1.

28. Matter, H.; Baringhaus, K.-H.; Naumann, T.; Klabunde, T.; Pirard, B. *Comb. Chem. HTS* **2001**, *4*, 453–475.
29. Young, S. S.; Ge, N. *Curr. Opin. Drug. Disc. Dev.* **2004**, *7*, 318–324.
30. Chanda, S. K.; Caldwell, J. S. *Drug Disc. Today* **2003**, *8*, 168–174.
31. Drews, J. *Science* **2000**, *287*, 1960–1964.
32. Drews, J. *Drug Disc. Today* **2003**, *8*, 411–420.
33. Handen, J. S. *Drug Disc. Today* **2002**, *7*, 83–85.
34. Sams-Dodd, F. *Drug Disc. Today* **2005**, *10*, 139–147.
35. Muller, K. *J. Mol. Struct. (Theochem.)* **1997**, *398–399*, 467–471.
36. Salemme, F. R.; Spurlino, J.; Bone, R. *Structure* **1997**, *5*, 319–324.
37. Valler, M. J.; Green, D. V. S. *Drug Disc. Today* **2000**, *5*, 286–293.
38. Schneider, G. *Curr. Med. Chem.* **2002**, *9*, 2095–2101.
39. Ghose, A. K.; Viswanadhan, V. N.; Wendoloski, J. *J. Recept. Signal Transduct. Res.* **2001**, *21*, 357–375.
40. Harper, G.; Bradshaw, J.; Gittins, J. C.; Green, D. V. S.; Leach, A. R. *J. Chem. Inf. Comput. Sci.* **2001**, *41*, 1295–1300.
41. Shen, J. *J. Chem. Inf. Comput. Sci.* **2003**, *43*, 1668–1672.
42. Diller, D. J.; Hobbs, D. W. *J. Med. Chem.* **2004**, *47*, 6373–6383.
43. Engels, M. F.; Wouters, L.; Verbeeck, R.; Vanhoof, G. *J. Biomol. Screen.* **2002**, *7*, 341–351.
44. Harper, G.; Bravi, G. S.; Pickett, S. D.; Hussain, J.; Green, D. V. S. *J. Chem. Inf. Comput. Sci.* **2004**, *44*, 2145–2156.
45. Kubinyi, H. *Curr. Opin. Drug Disc. Dev.* **1998**, *1*, 16–27.
46. McKenna, J. M.; Halley, F.; Souness, J.; McLay, I. M.; Pickett, S. D.; Page, K.; Ahmed, I. *J. Med. Chem.* **2002**, *45*, 2173–2184.
47. Cowell, S. M.; Gu, X.; Vagner, J.; Hruby, V. J. *Methods Enzymol.* **2003**, *369*, 288–297.
48. Senderowitz, H.; Rosenfeld, R. *J. Recept. Signal Transduct. Res.* **2001**, *21*, 489–506.
49. Wise, A.; Jupe, S. C.; Rees, S. *Annu. Rev. Pharmacol. Toxicol.* **2004**, *44*, 43–66.
50. Civelli, O. *Trends Pharmacol. Sci.* **2005**, *26*, 15–19.
51. Nilakantan, R.; Nunn, D. S. *Drug Disc. Today* **2003**, *8*, 668–672.
52. Harper, G.; Pickett, S. D.; Green, D. V. S. *Comb. Chem. HTS* **2004**, *7*, 63–70.
53. Frye, S. V. *Chem. Biol.* **1999**, *6*, R3–R7.
54. Jacoby, E.; Schuffenhauer, A.; Acklin, P. The Contribution of Molecular Informatics to Chemogenomics: Knowledge-Based Discovery of Biological Targets and Chemical Lead Compounds. In *Chemogenomics in Drug Discovery: A Medicinal Chemistry Perspective*; Kubinyi, H., Muller, G., Eds.; Methods and Principles in Medicinal Chemistry; Mannhold, R., Kubinyi, H., Folkers, G., Eds.; Wiley–VCH: Weinheim, Germany, 2004; Vol. 22, pp 139–166.
55. Schnur, D.; Beno, B. R.; Good, A. C.; Tebben, A. Approaches to Target Class Combinatorial Library Design. In *Chemoinformatics: Concepts, Methods, and Tools for Drug Discovery*; Bajorath, J., Ed.; Methods in Molecular Biology; Humana Press: Totowa, NJ, 2004; Vol. 275, pp 379–398.
56. Lowrie, J. F.; DeLisle, R. K.; Hobbs, D. W.; Diller, D. J. *Comb. Chem. HTS* **2004**, *7*, 495–510.
57. Crossley, R. *Curr. Top. Med. Chem.* **2004**, *4*, 581–589.
58. Prien, O. *ChemBioChem* **2005**, *6*, 500–505.
59. Klabunde, T.; Hessler, G. *ChemBioChem* **2002**, *3*, 929–944.
60. Manning, G.; Whyte, D. B.; Martinez, R.; Hunter, T.; Sundarsanam, S. *Science* **2002**, *298*, 1912–1934.
61. Beavers, M. P.; Chen, X. *J. Mol. Graph. Model.* **2002**, *20*, 463–468.
62. Clark, D. E.; Pickett, S. D. *Drug Disc. Today* **2000**, *5*, 49–58.
63. Davis, A. M.; Riley, R. J. *Curr. Opin. Chem. Biol.* **2004**, *8*, 378–386.
64. Lepre, C. A. Strategies for NMR Screening and Library Design. In *BioNMR in Drug Research*; Zerbe, O., Ed.; Methods and Principles in Medicinal Chemistry; Mannhold, R., Kubinyi, H., Folkers, G., Eds.; Wiley–VCH: Weinheim, Germany, 2003; Vol. 16, pp. 391–415.
65. Hartshorn, M. J.; Murray, C. W.; Cleasby, A.; Frederickson, M.; Tickle, I. J.; Jhoti, H. *J. Med. Chem.* **2005**, *48*, 403–413.
66. Parlow, J. J.; Case, B. L.; Dice, T. A.; Fenton, R. L.; Hayes, M. J.; Jones, D. E.; Neumann, W. L.; Wood, R. S.; Lachance, R. M.; Girard, T. J. et al. *J. Med. Chem.* **2003**, *46*, 4050–4062.
67. Han, Q.; Dominguez, C.; Stouten, P. F. W.; Park, J. M.; Duffy, D. E.; Galemmo, R. A.; Rossi, K. A.; Alexander, R. S.; Smallwood, A. M.; Wong, P. C. et al. *J. Med. Chem.* **2000**, *43*, 4389–4415.
68. Pickett, S. D.; Luttmann, C.; Guerin, V.; Laoui, A.; James, E. *J. Chem. Inf. Comput. Sci.* **1998**, *38*, 144–150.
69. *Available Chemicals Directory.* MDL Information Systems, Inc.: San Leandro, CA.
70. Leach, A. R.; Green, D. V. S.; Hann, M. M.; Judd, D. B.; Good, A. C. *J. Chem. Inf. Comput. Sci.* **2000**, *40*, 1262–1269.
71. Hann, M. M.; Hudson, B. D.; Lewell, X. Q.; Lifely, R.; Miller, L.; Ramsden, N. *J. Chem. Inf. Comput. Sci.* **1999**, *39*, 897–902.
72. Leach, A. R.; Bradshaw, J.; Green, D. V. S.; Hann, M. M.; Delany, J. J. *J. Chem. Inf. Comput. Sci.* **1999**, *39*, 1161–1172.
73. James, C. A.; Weininger, D.; Delaney, J. *Daylight Theory Manual, Daylight Chemical Information Systems.* http://www.daylight.com/dayhtml/doc/theory/theory.toc.html, **2005** (accessed April 2006).
74. Downs, G. M.; Barnard, J. M. *J. Chem. Inf. Comput. Sci.* **1997**, *37*, 59–61.
75. Taylor, R. *J. Chem. Inf. Comput. Sci.* **1995**, *35*, 59–67.
76. Gillet, V. J.; Khatib, W.; Willett, P.; Fleming, P. J.; Green, D. V. S. *J. Chem. Inf. Comput. Sci.* **2002**, *42*, 375–385.
77. Martin, E. J.; Blaney, J. M.; Siani, M. A.; Spellmeyer, D. C.; Wong, A. K.; Moos, W. H. *J. Med. Chem.* **1995**, *38*, 1431–1436.
78. Jackson, J. E. *A User's Guide to Principal Components*; John Wiley: New York, 1991.
79. Leo, A. J. *Chem. Rev.* **1993**, *93*, 1281–1306.
80. CLOGP. Daylight Chemical Information Systems Inc., 120 Vantis - Suite 550, Aliso Viejo, CA 92656.
81. Hall, L. H.; Kier, L. B. The Molecular Connectivity Chi Indexes and Kappa Shape Indexes in Structure-Property Modelling. In *Reviews in Computational Chemistry*; Lipkowitz, K. B., Boyd, D. B., Eds.; VCH: New York, 1991; Vol. 2, pp 367–421
82. Hall, L. H.; Kier, L. B. *J. Mol. Graph. Model.* **2001**, *20*, 4–18.
83. Randić, M. *J. Mol. Graph. Model.* **2001**, *20*, 19–35.
84. Hall Associates Consulting. MOLCONN-Z. eduSoft L.C. 2 Davis Street, Quincy, MA.
85. Agrafiotis, D. K. *J. Chem. Inf. Comput. Sci.* **1997**, *37*, 576–580.
86. Zuckermann, R. N.; Martin, E. J.; Spellmeyer, D. C.; Stauber, G. B.; Shoemaker, K. R.; Kerr, J. M.; Figliozzi, G. M.; Siani, M. A.; Simon, R. J.; Banville, S. C. *J. Med. Chem.* **1994**, *37*, 2678–2685.
87. Martin, E. J.; Wong, A. K. *J. Chem. Inf. Comput. Sci.* **2000**, *40*, 215–220.

88. Martin, E. J.; Critchlow, R. E. *J. Comb. Chem.* **1999**, *1*, 32–45.
89. Miller, A.; Nguyen, N.-K. *Appl. Stat.* **1994**, *43*, 669–678.
90. Pickett, S. D.; Mason, J. S.; McLay, I. M. *J. Chem. Inf. Comput. Sci.* **1996**, *36*, 1214–1223.
91. Mason, J. S.; Morize, I.; Menard, P. R.; Cheney, D. L.; Hulme, C.; Labaudinière, R. F. *J. Med. Chem.* **1999**, *42*, 3251–3264.
92. McGregor, M. J.; Muskal, S. M. *J. Chem. Inf. Comput. Sci.* **1999**, *39*, 569–574.
93. Pickett, S. D. The Biophore Concept. In *Protein–Ligand Interactions: From Molecular Recognition to Drug Design*; Bohm, H. J., Schneider, G., Eds.; Methods and Principles in Medicinal Chemistry; Mannhold, R., Kubinyi, H., Folkers, G., Eds.; John Wiley: New York, 2003; Vol. 19, pp 73–105.
94. Beno, B. R.; Mason, J. S. *Drug Disc. Today* **2001**, *6*, 251–258.
95. Eksterowicz, J. E.; Evensen, E.; Lemmen, C.; Brady, G. P.; Lanctot, J. K.; Bradley, E. K.; Saiah, E.; Robinson, L. A.; Grootenhuis, P. D. J. et al. *J. Mol. Graphics Model.* **2002**, *20*, 469–477.
96. McGregor, M. J.; Muskal, S. M. *J. Chem. Inf. Comput. Sci.* **2000**, *40*, 117–125.
97. Martin, E. J.; Hoeffel, T. J. *J. Mol. Graph. Model.* **2000**, *18*, 383–403.
98. Good, A. C.; Lewis, R. A. *J. Med. Chem.* **1997**, *40*, 3926–3936.
99. Cramer, R. D.; Clark, R. D.; Patterson, D. E.; Ferguson, A. M. *J. Med. Chem.* **1996**, *39*, 3060–3069.
100. Cramer, R. D.; Patterson, D. E.; Clark, R. D.; Soltanshahi, F.; Lawless, M. S. *J. Chem. Inf. Comput. Sci.* **1998**, *6*, 1010–1023.
101. Cramer, R. D.; Poss, M. A.; Hermsmeier, M. A.; Caulfield, T. J.; Kowala, M. C.; Valentine, M. T. *J. Med. Chem.* **1999**, *42*, 3919–3933.
102. Andrews, K. M.; Cramer, R. D. *J. Med. Chem.* **2000**, *43*, 1723–1740.
103. Jilek, R. J.; Cramer, R. D. *J. Chem. Inf. Comput. Sci.* **2004**, *44*, 1221–1227.
104. Cramer, R. D.; Jilek, R. J.; Guessregen, S.; Clark, S. J.; Wendt, B.; Clark, R. D. *J. Med. Chem.* **2004**, *47*, 6777–6791.
105. Pearlman, R. S. *Chem. Design Autom. News* **1987**, *2*, 1.
106. Rusinko, A., III; Skell, J. M.; Balducci, R.; McGarity, C. M.; Pearlman, R. S. CONCORD. Program available from Tripos, Inc., St. Louis, MO.
107. ChemSpace is a licensed Trademark of Tripos, Inc., St. Louis, MO.
108. Leach, A. R. *J. Mol. Graph. Model.* **1997**, *15*, 158–160.
109. Fontaine, F.; Pastor, M.; Zamora, I.; Sanz, F. *J. Med. Chem.* **2005**, *48*, 2687–2694.
110. Pastor, M.; Cruciani, G.; McLay, I. M.; Pickett, S. D.; Clementi, S. *J. Med. Chem.* **2000**, *43*, 3233–3243.
111. Cramer, R. D.; Patterson, D. E.; Bunce, J. D. *J. Am. Chem. Soc.* **1988**, *110*, 5959–5967.
112. Goodford, P. J. *J. Med. Chem.* **1985**, *28*, 849–857.
113. Cruciani, G.; Fontaine, F.; Pastor, M. ALMOND. Molecular Discovery Ltd, Perugia, Italy.
114. Clark, D. E. *Evolutionary Algorithms in Molecular Design*; Methods and Principles in Medicinal Chemistry; Mannhold, R., Kubinyi, H., Timmerman, H., Eds.; Wiley-VCH: Weinheim, Germany, 2000; Vol. 8.
115. Goldberg, D. E. *Genetic Algorithms in Search, Optimization, and Machine Learning*; Addison-Wesley: Reading, MA, 1989.
116. Kirkpatrick, S. *Science* **1983**, *220*, 671–690.
117. Van Laarhoven, P. J. M.; Aarts, E. H. L. *Simulated Annealing: Theory and Applications*; Reidel: Dordrecht, Netherlands, 1987.
118. Allen, M. P.; Tildesley, D. J. *Computer Simulation of Liquids*; Oxford University Press: New York, 1987.
119. Leach, A. R. A Survey of Methods for Searching the Conformational Space of Small and Medium-Sized Molecules. In *Reviews in Computational Chemistry*; Lipkowitz, K. B., Boyd, D. B., Eds.; VCH: New York, 1991; Vol. 2, pp 1–55.
120. Sheridan, R. P.; Kearsley, S. K. *J. Chem. Inf. Comput. Sci.* **1995**, *35*, 310–320.
121. Sheridan, R. P.; SanFeliciano, S. G.; Kearsley, S. K. *J. Mol. Graph. Model.* **2000**, *18*, 320–324.
122. Miller, M. D.; Sheridan, R. P.; Kearsley, S. K. *J. Med. Chem.* **1999**, *42*, 1505–1514.
123. Miller, M. D.; Kearsley, S. K.; Underwood, D. J.; Sheridan, R. P. *J. Comput. Aided Mol. Design* **1994**, *8*, 153–174.
124. Lewis, R. A.; Good, A. C.; Pickett, S. D. Quantification of Molecular Similarity and Its Application to Combinatorial Chemistry. In *Computer-Assisted Lead Finding and Optimization: Current Tools for Medicinal Chemistry*; van de Waterbeemd, H., Testa, B., Folkers, G., Eds.; Wiley-VCH: New York, 1997, pp 135–156.
125. Janda, K. D. *Proc. Natl. Acad. Sci. USA* **1994**, *91*, 10779–10785.
126. Choong, I. C. *Annu. Rep. Med. Chem.* **1996**, *31*, 309–318.
127. Loo, J. A. *Annu. Rep. Med. Chem.* **1996**, *31*, 319–325.
128. Brown, R. D.; Martin, Y. C. *J. Med. Chem.* **1997**, *40*, 2304–2313.
129. Gillet, V. J.; Willett, P.; Bradshaw, J.; Green, D. V. S. *J. Chem. Inf. Comput. Sci.* **1999**, *39*, 169–177.
130. *World Drug Index*. Thomson Scientific: London, 2006.
131. Holliday, J. D.; Ranade, S. S.; Willett, P. *Quant. Struct.-Act. Relat.* **1995**, *14*, 501–506.
132. Snarey, M.; Terrett, N. K.; Willett, P.; Wilton, D. J. *J. Mol. Graphics Model.* **1997**, *15*, 372–385.
133. Beasley, D.; Bull, D. R.; Martin, R. R. *Evol. Comput.* **1993**, *1*, 101–125.
134. Agrafiotis, D. K. *J. Comput. Aided Mol. Des.* **2002**, *16*, 335–356.
135. Agrafiotis, D. K. *Protein Sci.* **1997**, *6*, 287–293.
136. Agrafiotis, D. K.; Lobanov, V. S. *J. Chem. Inf. Comput. Sci.* **2000**, *40*, 1356–1362.
137. Rassokhin, D. N.; Lobanov, V. S.; Agrafiotis, D. K. *J. Comput. Chem.* **2000**, *22*, 373–386.
138. Agrafiotis, D. K.; Lobanov, V. S. *J. Chem. Inf. Comput. Sci.* **1999**, *39*, 51–58.
139. Rassokhin, D. N.; Agrafiotis, D. K. *J. Mol. Graph. Model.* **2000**, *18*, 370–384.
140. Brown, R. D.; Hassan, M.; Waldman, M. *J. Mol. Graph. Model.* **2000**, *18*, 427–437.
141. Mason, J. S.; Beno, B. R. *J. Mol. Graph. Model.* **2000**, *18*, 438–451.
142. Chen, G.; Zheng, S.; Luo, X.; Shen, J.; Zhu, W.; Liu, H.; Gui, C.; Zhang, J.; Zheng, M.; Puah, C. M. et al. *J. Comb. Chem.* **2005**, 7, 398–406.
143. Zheng, W.; Cho, S. J.; Waller, C. L.; Tropsha, A. *J. Chem. Inf. Comput. Sci.* **1999**, *39*, 738–746.
144. Jamois, E. A.; Lin, C. T.; Waldman, M. *J. Mol. Graph. Model.* **2003**, *22*, 141–149.
145. Agrafiotis, D. K.; Lobanov, V. S. *J. Chem. Inf. Comput. Sci.* **2000**, *40*, 1030–1038.
146. Bravi, G.; Green, D. V. S.; Hann, M. M.; Leach, A. R. *J. Chem. Inf. Comput. Sci.* **2000**, *40*, 1441–1448.
147. Zheng, W.; Cho, S. J.; Tropsha, A. *J. Chem. Inf. Comput. Sci.* **1998**, *38*, 251–258.
148. Young, S. S.; Wang, M.; Gu, F. *J. Chem. Inf. Comput. Sci.* **2003**, *43*, 1916–1921.
149. Lam, R. L.; Welch, W. J.; Young, S. S. *Technometrics* **2002**, *44*, 99–109.
150. Le Bailly de Tilleghen, C.; Beck, B.; Boulanger, B.; Govaerts, G. *J. Chem. Inf. Model.* **2005**, *45*, 758–767.

151. Zheng, W.; Hung, S. T.; Saunders, J. T.; Seibel, G. L. *Proc. Pacific Symp. BioComput.* **2000**, *5*, 588–599.
152. Teig, S. L. *J. Biomol. Screen.* **1998**, *3*, 85–88.
153. Shannon, C. E. *Bell System Tech. J.* **1948**, *27*, 379–423.
154. Shannon, C. E. *Bell System Tech. J.* **1948**, *27*, 623–656.
155. Miller, J. L.; Bradley, E. K.; Teig, S. L. *J. Chem. Inf. Comput. Sci.* **2003**, *43*, 47–54.
156. Everett, J.; Gardner, M.; Pullen, F.; Smith, G. F.; Snarey, M.; Terrett, N. *Drug Disc. Today* **2001**, *6*, 779–785.
157. Lipkin, M. J.; Rose, V. S.; Wood, J. *SAR QSAR Environ. Res.* **2002**, *13*, 425–432.
158. Wood, J.; Rose, V. S. Method of designing chemical substances 99/26901. PCT Int. Appl. (1999), WO1999026901.
159. Cochran, W. G.; Cox, G. M. *Experimental Designs*; John Wiley: New York, 1957.
160. Clark, R. D.; Kar, J.; Akella, L.; Soltanshahi, F. *J. Chem. Inf. Comput. Sci.* **2003**, *43*, 829–836.
161. Clark, R. D. *J. Chem. Inf. Comput. Sci.* **1997**, *37*, 1181–1188.
162. Pickett, S. D.; Clark, D. E.; McLay, I. M. *J. Chem. Inf. Comput. Sci.* **2000**, *40*, 263–272.
163. Brown, R. D.; Martin, Y. C. *SAR QSAR Environ. Res.* **1998**, *8*, 23–39.
164. Brown, R. D.; Martin, Y. C. *J. Chem. Inf. Comput. Sci.* **1997**, *37*, 1–9.
165. Patterson, D. E.; Cramer, R. D.; Ferguson, A. M.; Clark, R. D.; Weinberger, L. E. *J. Med. Chem.* **1996**, *39*, 3049–3059.
166. Good, A. C.; Cho, S. J.; Mason, J. S. *J. Comput. Aided Mol. Des.* **2004**, *18*, 523–527.
167. Gillet, V. J.; Willett, P.; Bradshaw, J. *J. Chem. Inf. Comput. Sci.* **1997**, *37*, 731–740.
168. Gillet, V. J.; Nicolotti, O. *Perspect. Drug Disc. Des.* **2000**, *20*, 265–287.
169. Jamois, E. A.; Hassan, M.; Waldman, M. *J. Chem. Inf. Comput. Sci.* **2000**, *40*, 63–70.
170. Linusson, A.; Gottfries, J.; Lindgren, F.; Wold, S. *J. Med. Chem.* **2000**, *43*, 1320–1328.
171. Mitchell, T. J. *Technometrics* **1974**, *2*, 203–210.
172. MODDE. Umetri AB, Box 7960, S-907 19 Umeå, Sweden.
173. Marengo, E.; Todeschini, R. *Chemom. Intell. Lab. Syst.* **1992**, *16*, 37–44.
174. Dunbar, J. B. J. *Perspect. Drug Disc. Des.* **1997**, *7/8*, 51–63.
175. Eriksson, L.; Johansson, E.; Muller, M.; Wold, S. *Quant. Struct.–Act. Relat.* **1997**, *16*, 383–390.
176. Bezdek, J. C. *Pattern Recognition with Fuzzy Objective Function Algorithms*; Plenum Press: New York, 1981.
177. Linusson, A.; Wold, S.; Norden, B. *Chemom. Intell. Lab. Syst.* **1998**, *44*, 213–227.
178. Gillet, V. J.; Willett, P.; Fleming, P. J.; Green, D. V. S. *J. Mol. Graph. Model.* **2002**, *20*, 491–498.
179. Fonseca, C. M.; Fleming, P. J. *Evol. Comput.* **1995**, *3*, 1116.
180. Wright, T.; Gillet, V. J.; Green, D. V. S.; Pickett, S. D. *J. Chem. Inf. Comput. Sci.* **2003**, *43*, 381–390.
181. Waldman, M.; Li, H.; Hassan, M. *J. Mol. Graph. Model.* **2000**, *18*, 412–426.
182. The Structural Genomics Consortium is a not-for-profit company that aims to determine the three dimensional structures of proteins of medical relevance, and place them in the public domain without restriction. http://www.sgc.utoronto.ca/ (accessed April 2006).
183. Taylor, R. D.; Jewsbury, P. J.; Essex, J. W. *J. Comput. Aided Mol. Design* **2002**, *16*, 151–166.
184. Beavers, M. P.; Chen, X. *J. Mol. Graph. Model.* **2002**, *20*, 463–468.
185. Kick, E. K.; Roe, D. C.; Skillman, A. G.; Liu, G.; Ewing, T. J. A.; Sun, Y.; Kuntz, I. D.; Ellman, J. A. *Chem. Biol.* **1997**, *4*, 297–307.
186. Kick, E. K.; Ellman, J. A. *J. Med. Chem.* **1995**, *38*, 1427–1430.
187. Lewis, R. A.; Roe, D. C.; Huang, C.; Ferrin, T. E.; Langridge, R.; Kuntz, I. D. *J. Mol. Graph. Model.* **1992**, *10*, 66–79.
188. Roe, D. C.; Kuntz, I. D. *J. Comput. Aided Mol. Des.* **1995**, *9*, 269–282.
189. Baldwin, E. T.; Bhat, T. N.; Gulnik, S.; Hosur, M. V.; Sowder, R. C.; Cachau, R. E.; Collins, J.; Silva, A. M.; Erickson, J. W. *Proc. Natl. Acad. Sci. USA* **1993**, *90*, 6796–6800.
190. Kuntz, I. D.; Blaney, J. M.; Oatley, S. J.; Langridge, R.; Ferrin, T. E. *J. Mol. Biol.* **1982**, *161*, 269–288.
191. Shoichet, B. K.; Bodian, D. L.; Kuntz, I. D. *J. Comput. Chem.* **1992**, *13*, 380–397.
192. Meng, E. C.; Shoichet, B. K.; Kuntz, I. D. *J. Comput. Chem.* **1992**, *13*, 505–524.
193. Ewing, T. J. A.; Kuntz, I. D. *J. Comput. Chem.* **1997**, *18*, 1175–1189.
194. Sun, Y.; Ewing, T. J. A.; Skillman, A. G.; Kuntz, I. D. *J. Comput. Aided Mol. Des.* **1998**, *12*, 597–604.
195. Murray, C. W.; Clark, D. E.; Auton, T. R.; Firth, M. A.; Li, J.; Sykes, R. A.; Waszkowycz, B.; Westhead, D. R.; Young, S. C. *J. Comput. Aided Mol. Des.* **1997**, *11*, 193–207.
196. Liebeschuetz, J. W.; Jones, S. D.; Morgan, P. J.; Murray, C. W.; Rimmer, A. D.; Roscoe, J. M. E.; Waszkowycz, B.; Welsh, P. M.; Wylie, W.; Young, S. C. et al. *J. Med. Chem.* **2002**, *45*, 1221–1232.
197. Liebeschuetz, J. W. pers. comm. 2005.
198. Rarey, M.; Kramer, B.; Lengauer, T.; Klebe, G. *J. Mol. Biol.* **1996**, *261*, 470–489.
199. Rarey, M.; Kramer, B.; Lengauer, T. *J. Comput. Aided Mol. Des.* **1997**, *11*, 369–384.
200. Kramer, B.; Rarey, M.; Lengauer, T. *Proteins: Struct., Funct., Genet.* **1999**, *37*, 228–241.
201. Rarey, M.; Lengauer, T. *Perspect. Drug Disc. Des.* **2000**, *20*, 63–81.
202. Boobbyer, D. N.; Goodford, P. J.; McWhinnie, P. M.; Wade, R. C. *J. Med. Chem.* **1989**, *32*, 1083–1094.
203. Wade, R. C.; Clark, K. J.; Goodford, P. J. *J. Med. Chem.* **1993**, *36*, 140–147.
204. Mason, J. S.; Cheney, D. L. *Proc. Pacific Symp. BioComput.* **1999**, *4*, 456–467.
205. Mason, J. S.; Cheney, D. L. *Proc. Pacific Symp. BioComput.* **2000**, *5*, 576–587.
206. Murray, C. M.; Cato, S. J. *J. Chem. Inf. Comput. Sci.* **1999**, *39*, 46–50.
207. Gillet, V.; Johnson, A. P.; Mata, P.; Sike, S.; Williams, P. *J. Comput. Aided Mol. Des.* **1993**, 7, 127–153.
208. Gillet, V. J.; Newell, W.; Mata, P.; Myatt, G.; Sike, S.; Zsoldos, Z.; Johnson, A. P. *J. Chem. Inf. Comput. Sci.* **1994**, *34*, 207–217.
209. Gillet, V. J.; Myatt, G.; Zsoldos, Z.; Johnson, A. P. *Perspect. Drug Disc. Des.* **1995**, *3*, 34–50.
210. Mata, P.; Gillet, V. J.; Johnson, A. P.; Lampreia, J.; Myatt, G. J.; Sike, S.; Stebbings, A. L. *J. Chem. Inf. Comput. Sci.* **1995**, *35*, 479–493.
211. Johnson, A. P., Boda, K., Marchland, J.-F., and Ting, A. De novo design of synthetically accessible ligands. *Book of Abstracts*, 219th National Meeting of the American Chemical Society, San Francisco, CA, Mar 26–30, 2000.
212. MDL Information Systems, Inc., San Leandro, CA.
213. Cahart, R. E.; Smith, D. H.; Venkataraghavan, R. *J. Chem. Inf. Comput. Sci.* **1985**, *25*, 64–73.
214. Nilakantan, R.; Bauman, N.; Dixon, J. S.; Venkataraghavan, R. *J. Chem. Inf. Comput. Sci.* **1987**, *27*, 82–85.
215. Kearsley, S. K.; Sallamack, S.; Fluder, E. M.; Andose, J. D.; Mosley, R. T.; Sheridan, R. P. *J. Chem. Inf. Comput. Sci.* **1996**, *36*, 118–127.

216. Sheridan, R. P.; Miller, M. D.; Underwood, D. J.; Kearsley, S. K. *J. Chem. Inf. Comput. Sci.* **1996**, *36*, 128–136.
217. Willett, P.; Barnard, J. M.; Downs, G. M. *J. Chem. Inf. Comput. Sci.* **1998**, *38*, 983–996.
218. Schneider, G.; Neidhart, W.; Giller, T.; Schmid, G. *Angew. Chem. Int. Ed. Engl.* **1999**, *38*, 2894–2896.
219. Schneider, G.; Clement-Chomienne, O.; Hilfiger, L.; Schneider, P.; Kirsch, S.; Bohm, H. J.; Neidhart, W. *Angew. Chem. Int. Ed. Engl.* **2000**, *39*, 4130–4133.
220. Barker, E. J.; Gardiner, E. J.; Gillet, V. J.; Kitts, P.; Morris, J. *J. Chem. Inf. Comput. Sci.* **2003**, *43*, 346–356.
221. Gillet, V. J.; Willett, P.; Bradshaw, J. *J. Chem. Inf. Comput. Sci.* **2003**, *43*, 338–345.
222. Randić, M. *J. Am. Chem. Soc.* **1975**, *97*, 6609–6615.
223. Kier, L. B.; Hall, L. H.; Murray, W. J.; Randić, M. *J. Pharm. Sci.* **1975**, *64*, 1971–1974.
224. Hall, L. H.; Kier, L. B.; Murray, W. J. *J. Pharm. Sci.* **1975**, *64*, 1974–1977.
225. Agrafiotis, D. K. *J. Chem. Inf. Comput. Sci.* **1997**, *37*, 841–851.
226. Hassan, M.; Bielawski, J. P.; Hempel, J. C.; Waldman, M. *Mol. Divers.* **1996**, *2*, 64–74.
227. Pearlman, R. S.; Smith, K. M. *Perspect. Drug Disc. Des.* **1998**, *9*, 339–353.
228. Pearlman, R. S.; Smith, K. M. *Drugs Future* **1998**, *23*, 885–895.
229. Burden, F. R. *Quant. Struct.-Act. Relat.* **1997**, *16*, 309–314.
230. Gao, H. *J. Chem. Inf. Comput. Sci.* **2001**, *41*, 402–407.
231. Stanton, D. T. *J. Chem. Inf. Comput. Sci.* **1999**, *39*, 11–20.
232. Schnur, D. *J. Chem. Inf. Comput. Sci.* **1999**, *39*, 36–45.
233. Pirard, B.; Pickett, S. D. *J. Chem. Inf. Comput. Sci.* **2000**, *40*, 1431–1440.
234. Schnur, D.; Venkatarangan, P. In *Combinatorial Library Design and Evaluation*; Ghose, A. K., Viswanadhan, V. N., Eds.; Marcel Dekker: New York, 2001, pp 473–501.
235. Pearlman, R. S.; Smith, K. M. *DiverseSolutions*. Distributed by Tripos Inc., St. Louis, MO.
236. Good, A. C.; Ewing, T. J. A.; Gschwend, D. A.; Kuntz, I. D. *J. Comput. Aided Mol. Des.* **1995**, *9*, 1–12.
237. Abrahamian, E.; Fox, P. C.; Nrum, L.; Christensen, I. T.; Thogersen, H.; Clark, R. D. *J. Chem. Inf. Comput. Sci.* **2003**, *43*, 458–468.
238. Good, A. C.; Mason, J. S.; Pickett, S. D. Pharmacophore Pattern Application in Virtual Screening, Library Design and QSAR. In *Virtual Screening for Bioactive Molecules*; Bohm, H. J., Schneider, G., Eds.; Wiley-VCH: New York, 2000, pp 131–159.
239. Cruciani, G.; Meniconi, M.; Carosati, E.; Zamora, I.; Mannhold, R. VOLSURF: A Tool for Drug ADME-Properties Prediction. In *Drug Bioavailability: Estimation of Solubility, Permeability and Absorption*; van de Waterbeemd, H., Lennernas, H., Artursson, P., Eds.; Methods and Principles in Medicinal Chemistry; Mannhold, R., Kubinyi, H., Folkers, G., Eds.; Wiley-VCH: Weinheim, Germany, 2003; Vol. 18, pp 391–415.
240. Cruciani, G.; Pastor, M.; Guba, W. *Eur. J. Pharm. Sci.* **2000**, *11*, S29–S39.
241. Zamora, I.; Oprea, T.; Cruciani, G.; Pastor, M.; Ungell, A. L. *J. Med. Chem.* **2003**, *46*, 25–33.
242. Grant, J. A.; Pickup, B. T. *J. Phys. Chem.* **1995**, *99*, 3505–3510.
243. Rush, T. S.; Grant, A. J.; Mosyak, L.; Nicholls, A. *J. Med. Chem.* **2005**, *48*, 1489–1495.
244. Hahn, M. *J. Chem. Inf. Comput. Sci.* **1997**, *37*, 80–86.
245. Haigh, J. A.; Pickup, B. T.; Grant, J. A.; Nicholls, A. *J. Chem. Inf. Model.* **2005**, *45*, 673–684.
246. Briem, H.; Kuntz, I. D. *J. Med. Chem.* **1996**, *39*, 3401–3408.
247. Matter, H. *J. Med. Chem.* **1997**, *40*, 1219–1229.
248. Computational Tools for the Analysis of Molecular Diversity; Willett, P., Ed.; *Perspectives in Drug Discovery and Design* 1997, *7/8*.
249. Agrafiotis, D. K.; Lobanov, V. S. *J. Chem. Inf. Comput. Sci.* **1999**, *39*, 51–58.
250. Barnard, J. M.; Downs, G. M. *J. Chem. Inf. Comput. Sci.* **1992**, *32*, 644–649.
251. Downs, G. M.; Barnard, J. M. *Rev. Comput. Chem.* **2002**, *18*, 1–40.
252. Mason, J. S.; Pickett, S. D. *Perspect. Drug Disc. Des.* **1997**, *7/8*, 85–114.
253. Lewis, R. A.; Mason, J. S.; McLay, I. M. *J. Chem. Inf. Comput. Sci.* **1997**, *37*, 599–614.
254. Pearlman, R. S.; Smith, K. M. *J. Chem. Inf. Comput. Sci.* **1999**, *39*, 28–35.
255. Walters, W. P.; Murcko, M. A. *Adv. Drug Deliv. Rev.* **2002**, *54*, 255–271.
256. Vieth, M.; Siegel, M. G.; Higgs, R. E.; Watson, I. A.; Robertson, D. H.; Savin, K. A.; Durst, G. L.; Hipskind, P. A. *J. Med. Chem.* **2004**, *47*, 224–232.
257. Vieth, M.; Lajiness, M. S.; Erickson, J. *Curr. Opin. Drug Disc. Dev.* **2004**, 7, 470–477.
258. Leeson, P. D.; Davis, A. M. *J. Med. Chem.* **2004**, *47*, 6338–6348.
259. Ajay, A.; Walters, W. P.; Murcko, M. A. *J. Med. Chem.* **1998**, *41*, 3314–3324.
260. Sadowski, J.; Kubinyi, H. *J. Med. Chem.* **1998**, *41*, 3325–3329.
261. Wagener, M.; van Geerestein, V. J. *J. Chem. Inf. Comput. Sci.* **2000**, *40*, 280–292.
262. Byvatov, E.; Fechner, U.; Sadowski, J.; Schneider, G. *J. Chem. Inf. Comput. Sci.* **2003**, *43*, 1882–1889.
263. Hann, M. M.; Leach, A. R.; Harper, G. *J. Chem. Inf. Comput. Sci.* **2001**, *41*, 856–864.
264. Proudfoot, J. R. *Bioorg. Med. Chem. Lett.* **2002**, *12*, 1647–1650.
265. Rishton, G. M. *Drug Disc. Today* **2003**, *8*, 86–96.
266. Oprea, T. I.; Davis, A. M.; Teague, S. J.; Leeson, P. D. *J. Chem. Inf. Comput. Sci.* **2001**, *41*, 1308–1315.
267. Teague, S. J.; Davis, A. M.; Leeson, P. D.; Oprea, T. I. *Angew. Chem. Int. Ed. Engl.* **1999**, *38*, 3742–3748.
268. Wenlock, M. C.; Austin, R. P.; Barton, P.; Davis, A. M.; Leeson, P. D. *J. Med. Chem.* **2003**, *46*, 1250–1256.
269. Abraham, M. H.; Ibrahim, A.; Zissimos, A. M.; Zhao, Y. H.; Comer, J.; Reynolds, D. P. *Drug Disc. Today* **2002**, 7, 1056–1063.
270. van de Waterbeemd, H.; Jones, B. C. *Prog. Med. Chem.* **2003**, *41*, 1–59.
271. de Groot, M. J.; Ekins, S. *Adv. Drug Deliv. Rev.* **2002**, *54*, 367–383.
272. Lewis, D. F. V. *J. Chem. Technol. Biotechnol.* **2001**, *76*, 237–244.
273. Stouch, T. R.; Kenyon, J. R.; Johnson, S. R.; Chen, X.-Q.; Doweyko, A.; Li, Y. *J. Comput. Aided Mol. Design* **2003**, *17*, 83–92.
274. Bruneau, P. *J. Chem. Inf. Comput. Sci.* **2001**, *41*, 1605–1616.
275. van de Waterbeemd, H.; Gifford, E. *Nat. Rev. Drug Disc.* **2003**, *2*, 192–204.
276. Davis, A. M.; Riley, R. J. *Curr. Opin. Chem. Biol.* **2004**, *8*, 378–386.
277. Nilakantan, R.; Immermann, F.; Haraki, K. *Comb. Chem. HTS* **2002**, *5*, 105–110.
278. Nilakantan, R.; Bauman, N.; Haraki, K. S. *J. Comput. Aided Mol. Des.* **1997**, *11*, 447–452.
279. Nilakantan, R.; Bauman, N.; Haraki, K. S.; Venkataraghavan, R. *J. Chem. Inf. Comput. Sci.* **1990**, *30*, 65–68.

280. Martin, Y. C.; Kofron, J. L.; Traphagen, L. M. *J. Med. Chem.* **2002**, *45*, 4350–4358.
281. Bailey, W. J.; Vanti, W. B.; George, S. R.; Blevins, R.; Swaminathan, S.; Bonini, J. A.; Smith, K. E.; Weinshank, R. L.; O'Dowd, R. F. *Expert Opin. Therapeut. Patents* **2001**, *11*, 1861–1887.
282. G-Protein-Coupled Receptors in Drug Discovery; Saunders, J., Ed.; *Bioorg. Med. Chem. Lett.* **2005**, 15.
283. Jimonet, P.; Jager, R. *Curr. Opin. Drug. Disc. Dev.* **2004**, 7, 334–341.
284. Astles, P. C.; Brown, T. J.; Handscombe, C. M.; Harper, M. F.; Harris, N. V.; Lewis, R. A.; Lockey, P. M.; McCarthy, C.; McLay, I. M.; Porter, B. *Eur. J. Med. Chem.* **1997**, *32*, 409–423.
285. Astles, P. C.; Brealey, C.; Brown, T. J.; Facchini, V.; Handscombe, C. M.; Harris, N. V.; McCarthy, C.; McLay, I. M.; Porter, B.; Roach, A. G. et al. *J. Med. Chem.* **1998**, *41*, 2732–2744.
286. Astles, P. C.; Brown, T. J.; Halley, F.; Handscombe, C. M.; Harris, N. V.; Majid, T. N.; McCarthy, C.; McLay, I. M.; Morley, A.; Porter, B. et al. *J. Med. Chem.* **2000**, *43*, 900–910.
287. Marriott, D. P.; Dougall, I. G.; Meghani, P.; Liu, Y. J.; Flower, D. R. *J. Med. Chem.* **1999**, *42*, 3210–3216.
288. Beeley, N. R. A.; Sage, C. *TARGETS* **2003**, *2*, 19–25.
289. Palczewski, K.; Kumasaka, T.; Hori, T.; Behnke, C. A.; Motoshima, H.; Fox, B. A.; Le Trong, I.; Teller, D. C.; Okada, T.; Stenkamp, R. E. et al. *Science* **2000**, *289*, 739–745.
290. Bosch, L.; Iarriccio, L.; Garriga, P. *Curr. Pharm. Des.* **2005**, *11*, 2243–2256.
291. Teller, D. C.; Okada, T.; Behnke, C. A.; Palczewski, K.; Stenkamp, R. E. *Biochemistry (Mosc).* **2001**, *40*, 7761–7772.
292. Evers, A.; Hessler, G.; Matter, H.; Klabunde, T. *J. Med. Chem.* **2005**, *48*, 5448–5465.
293. Evers, A.; Klabunde, T. *J. Med. Chem.* **2005**, *48*, 1088–1097.
294. Evans, B. E.; Rittle, K. E.; Bock, M. G.; DiPardo, R. M.; Freidinger, R. M.; Whitter, W. L.; Lundell, G. F.; Veber, D. F.; Anderson, P. S.; Chang, S. L. et al. *J. Med. Chem.* **1988**, *31*, 2235–2246.
295. Patchett, A. A.; Nargund, R. P. *Annu. Rep. Med. Chem.* **2000**, *35*, 289–298.
296. Guo, T.; Hobbs, D. W. *Assay Drug Dev. Technol.* **2003**, *1*, 579–592.
297. *MDL Drug Data Report*. MDL Information Systems, Inc., San Leandro, CA.
298. Lavrador, K.; Murphy, B.; Saunders, J.; Struthers, S.; Wang, X.; Williams, J. *J. Med. Chem.* **2004**, *47*, 6864–6874.
299. Jacoby, E.; Fauchère, J.-L.; Raimbaud, E.; Ollivier, S.; Michel, A.; Spedding, M. *Quant. Struct.-Act. Relat.* **1999**, *18*, 561–572.
300. Jacoby, E. *Quant. Struct.-Act. Relat.* **2001**, *20*, 115–123.
301. Williams, D. H.; Mitchell, T. *Curr. Opin. Pharmacol.* **2002**, *2*, 567–573.
302. The Kinome map at www.kinase.com/human/kinome (accessed May 2006).
303. Stahura, F. L.; Xue, L.; Godden, J. W.; Bajorath, J. *J. Mol. Graph. Model.* **1999**, *17*, 1–9.
304. Todorov, N. P.; Buenemann, C. L.; Alberts, I. L. *J. Chem. Inf. Model.* **2005**, *45*, 314–320.
305. Dean, P. M.; Lloyd, D. G.; Todorov, N. P. *Curr. Opin. Drug Disc. Dev.* **2004**, 7, 347–353.
306. Todorov, N. P.; Dean, P. M. *J. Comput. Aided Mol. Design* **1997**, *11*, 175–192.
307. Stahl, M.; Todorov, N. P.; James, T.; Mauser, H.; Boehm, H. J.; Dean, P. M. *J. Comput. Aided Mol. Design* **2002**, *16*, 459–478.
308. Stahl, M.; Rarey, M. *J. Med. Chem.* **2001**, *44*, 1035–1042.
309. Birault, V.; Harris, C. J.; Le, J.; Lipkin, M.; Nerella, R.; Stevens, A. *Curr. Med. Chem.* **2006**, *13*, 1735–1748.
310. Weber, L.; Wallbaum, S.; Broger, C.; Gubernator, K. *Angew. Chem., Int. Ed. Engl.* **1995**, *34*, 2280–2282.
311. Singh, J.; Ator, M. A.; Jaeger, E. P.; Allen, M. P.; Whipple, D. A.; Soloweij, J. E.; Chowdhary, S.; Treasurywala, A. M. *J. Am. Chem. Soc.* **1996**, *118*, 1669–1676.
312. Yokobayashi, Y.; Ikebukuro, I.; McNiven, K. I. *J. Chem. Soc. Perkin Trans.* **1996**, *1*, 2435–2439.
313. Brauer, S.; Almstetter, M.; Antuch, W.; Behnke, D.; Taube, R.; Furer, P.; Hess, S. *J. Comb. Chem.* **2005**, 7, 218–226.

Biography

Stephen D Pickett studied for a degree in Chemistry at Keble College, Oxford. He then took a PhD under Prof Sir John Meurig Thomas at the Royal Institution of Great Britain, studying the adsorption and diffusion of small molecules in zeolites using Monte Carlo and Molecular Dynamics methods. Postdoctoral studies followed at the Imperial Cancer Research Fund (now part Cancer Research UK) with Prof Michael Sternberg, looking at various aspects of protein folding including the derivation of a scale for side-chain entropy. From the ICRF he took up his first position in the pharmaceutical industry, working for Rhone-Poulenc Rorer at Dagenham, UK. Seven years at RPR included a two-year posting to the French research site at Vitry-sur-Seine, Paris. Eighteen months were spent at the former Roche research facility in Welwyn Garden City, UK before taking up his current position as a Group Leader in Cheminformatics at GlaxoSmithKline Stevenage. The group supports many aspects of HTS data analysis, library design, and data modelling globally. Dr Pickett is the author of over 30 peer-reviewed scientific articles and reviews and is a named contributor on six patent applications.

© 2007 Elsevier Ltd. All Rights Reserved
No part of this publication may be reproduced, stored in any retrieval system or transmitted in any form by any means electronic, electrostatic, magnetic tape, mechanical, photocopying, recording or otherwise, without permission in writing from the publishers

Comprehensive Medicinal Chemistry II
ISBN (set): 0-08-044513-6

ISBN (Volume 4) 0-08-044517-9; pp. 337–378

4.16 Quantum Mechanical Calculations in Medicinal Chemistry: Relevant Method or a Quantum Leap Too Far?

A A Alex, Pfizer Global Research and Development, Sandwich, UK

© 2007 Elsevier Ltd. All Rights Reserved.

4.16.1 Introduction

Drug discovery is a very complex and difficult scientific endeavor, which frequently results in extremely expensive failures. Virtually all the pharmaceutical companies have experienced the painful loss of new chemical entities (NCEs) in late stage preclinical or clinical development phases as well as the withdrawal of marketed products. Such late stage failures are very costly, both in terms of time and money.[1] A detailed analysis indicates that roughly one-third of the projects fail to enter the optimization phase, but of the ones that successfully pass this development stage, 95% never end up as drugs on the market due to late-stage failures. In addition, for those compounds that do make it to the market, the overall development costs are estimated to be in the region of $400–$800 million.[2] Therefore, extensive efforts and measures to minimize these high attrition rates are now taken at every stage of the discovery and development process throughout the pharmaceutical industry.[1] Some of those efforts are directed not at reducing attrition at the later stages of development but at the much earlier design stage in order to optimize the potency against a target receptor, introduce the appropriate selectivity or spectrum of activities, as well as to reduce the attrition risk by targeting the appropriate molecular properties even before the first compound in a drug discovery program is even synthesized. Is the prediction of attrition risks going to remain an aspiration or could it be a reality in the near future? Well, unfortunately one has to realize that it is not a reality yet, but due to the massive efforts in method and model development in the area of computational chemistry and drug design and the enormous improvements in computer hardware power it is rapidly becoming a more likely scenario as time progresses. The effective use of in silico methods in medicinal chemistry and drug design has enormous potential to reduce attrition at the very early stages of the discovery process and will have to play a role of increasing importance. This is particularly the case if one looks at the alternative option of continuing business as before, which is according to recent analysis of industry trends as outlined above, likely to end up as the failure of the whole pharmaceutical industry paradigm, as it stands today. Due to the increasing business pressures, drug discovery and therefore medicinal chemistry are transitioning to more industrialized, high-throughput activities. The important components of this industrialization are high throughput and combinatorial chemistry for compound libraries, high-throughput screening (HTS) technologies, integrated database and knowledge management, and in silico compound design. Computational chemistry is an integral part of this development, enabling fast design cycles with high success rates in terms of desired compound properties as well as relevant biological activity. Computational chemistry and in silico drug design have many facets, from quantitative structure–activity relationships (QSAR), virtual screening, protein structure modeling, and pharmacophore modeling, to structure-based drug design and the modeling of absorption and metabolism. Although quantum mechanical methods often play a part in these approaches, like for example for QSAR and structure-based drug design as well as the modeling of metabolism, they are rarely referred to as an independent approach in this context.

When looking for a fitting title for this chapter, there was a concern that if it contained the expression 'quantum mechanical,' many chemists and medicinal chemists might be tempted to immediately skip to the next chapter. The original title of just 'quantum mechanical calculations in medicinal chemistry' probably would have achieved exactly that result. In order to avoid this and to raise the interest level in this chapter, it was decided to extend the title to account for a common criticism that the author has frequently come across over the last 15 years of applying quantum mechanical calculations. The fact that you have got this far in reading the chapter could mean that this small change in the title has perhaps worked. In my experience, many medicinal chemists don't initially believe that quantum mechanical calculations can be useful and frequently dismiss them as a peripheral method or as outright irrelevant and effectively a theory that lacks any practical value or application, most certainly in areas related to medicinal chemistry. Just to be absolutely clear at this point, this review is primarily intended for the medicinal chemist, computational chemist, or other interested scientists who are directly involved in the advancement of medicine through discovery of pharmaceutically relevant small-molecule therapies. It is not the remit of this review to deal with the theoretical aspects of quantum mechanics or any approximation thereof in any detail, which has already been done extensively by others, and references will be included where appropriate. Therefore, the focus of this chapter will be entirely on the applications of quantum mechanical methods to answer questions in and related to medicinal chemistry and drug design. However, it is inevitable to at least give a broad outline of the types of methodologies that will be discussed throughout this chapter. This introduction to key methods in quantum chemistry from ab initio to density functional as well as semiempirical methods and other related methods is given in Section 4.16.3 below. This introduction will be kept to a minimum, and the reader will be referred to the relevant literature for more detailed reviews, where appropriate.

Despite the fact that quantum mechanical methods are the only approach amongst the computational chemistry techniques that is able to predict with very high accuracy molecular structures, properties and reaction mechanism without any prior knowledge of the system or any parameterization, they are not widely used and accepted amongst drug designers. Parameterization in this context means the building of computational models based on empirical knowledge or calculations. The classic example for this is the development of the so-called molecular force field methods, where various parameters are optimized to most closely simulate for example a known or high-level (quantum mechanically) calculated molecular structure. Any information that is generated from quantum mechanical calculations is derived purely from the calculated distribution of electrons in a molecule or molecular system like a protein–ligand complex. The high accuracy of the results comes of course at an increased expense in terms of computation time. However, this is no longer a bottleneck due to the enormous advances in computer hardware over the last 10 years in particular, in addition to software parallelization approaches, as well as very significant advances in computational algorithms that scale only with N^2 (with N being the number of electrons in a given system) or even almost linearly with the size of the model system under investigation, rather than with N^3 to even N^4 associated with earlier methods. Therefore, taking these advances into account, and projecting onto these the expected technological developments in the area of computing within the next 10 years, the potential for the application of quantum mechanical calculations to medicinal chemistry and drug design to answer questions immediately relevant to drug discovery seems literally endless.

Quantum mechanical calculations have been applied to solving scientific problems in chemistry for almost 80 years. However, it is only for the last 25 years that significant advances in computer power have allowed researchers to further develop and refine approaches to achieve the high levels of accuracy that we see in modern-day calculations. In spite of this, as mentioned above, quantum mechanical calculation are still generally considered by most chemists as a methodology that is far removed from being able to provide answers to real-life problems, particularly in life sciences and more specifically medicinal chemistry.[3] The greatest appeal of quantum mechanical methods is that they can, in principle, be used to calculate the entire range of properties that are necessary to understand the characteristics of a molecule which are responsible for all its properties and that allow its recognition by and activation of receptors. Quantum mechanical calculations are also able to extrapolate and predict properties of basically any compound without prior knowledge or parameterization, as mentioned above. This is in contrast to other computational methods like for example the classical force fields, which require extensive and complex parameterization and can only interpolate within the boundaries of that parameterization. However, quantum mechanics techniques are computationally very intensive and require significantly more specialist expertise for the correct and meaningful interpretation of the data they generate then do conventional molecular mechanics or force field methods.[4] This significant difference in required expertise combined with a lack of understanding of the potential of quantum mechanical calculations in the wider chemistry community and particular in medicinal chemistry may perhaps be responsible for the apparent underutilization of these very powerful methods. An important milestone for increasing the awareness of the achievements and the potential of quantum mechanical calculations was reached with the award of the Nobel Prize in Chemistry in 1998 to John A. Pople for the development of computational methods in quantum chemistry and to Walter Kohn for the development of density functional theory (DFT).[5] This was a great recognition for this relatively young field of research as a major contributor to the advancement of science, and particularly chemistry.

Although today the use of quantum mechanical calculations for applications in medicinal chemistry is more widespread as will be shown in this chapter, it is not as common as it could be if computers were faster and so-called compute-farms combining the power of hundreds of computers were more accessible. A widely held view is that quantum mechanical calculations are extremely valuable for small systems, which are not relevant for medicinal chemistry applications, but the additional accuracy compared to force field methods is not worth the massively increased computational cost for applications to larger molecular systems. This view of the perceived usefulness of quantum mechanical calculations has not changed significantly over the last 50 years, as highlighted by a comment by Albert Szent-Györgyi (Nobel Prize in 1937 for his work on the citric acid cycle), which was published in 1960[6]: "The distance between those abstruse quantum mechanical calculations and the patient bed may not be as great as believed." So how far have we come over these last five decades in bridging this perceived gap between medicine and quantum mechanics? As far back as the late 1970s and early 1980s scientists using computational tools have recognized and emphasized the opportunities and value that quantum mechanical calculations could bring to biological and medicinal science.[7] When researching the relevant scientific literature of the last 10–15 years, it becomes apparent that quantum mechanical calculations have been applied to a wide range of scientific questions related to medicinal chemistry and drug discovery in general, as highlighted in a very recent review[3]and in a number of chapters in a recently published book on quantum medicinal chemistry.[8–11] However, in many cases of pragmatic applications of quantum mechanical calculations they are not the dominant computational method that is applied to a particular problem, but are more of an essential contributor of atomic or molecular properties of molecules that are then being integrated with other methods like QSAR and structure-based drug design to address a medicinal chemistry question. When reviewing the literature, it also becomes apparent that the potential of quantum mechanical calculations has rarely been fully exploited, and that these methods often have much more to offer to medicinal chemistry than they are being given credit for. This chapter will give an overview of the area of quantum mechanical calculations in the context of computational medicinal chemistry, as outlined above. It will aim to review recent applications of these methods and also highlight the potential for further applications beyond what is currently common practice in terms of quantum mechanical calculations in medicinal chemistry and drug design.

4.16.2 Definition of Terms and Remit of this Chapter

There are many terms that are being used in the literature in connection with the application of computational chemistry to biological and medicinal problems and in order to define the remit of this chapter, it is necessary to clarify the meaning of terms that are going to be used throughout this review and also to emphasize the focus of this chapter. Such a definition of terms has been very clearly formulated in a recent review by Bard and Schaefer,[3] and we will for the purpose of this chapter largely adopt their definitions and views. They clearly distinguish between the terms 'computational quantum chemistry,' which refers to any method that uses computation to model a chemical system via the Schrödinger equation or a variation thereof, and 'computational medicinal chemistry,' which only refers to computational investigations dealing directly with the advancement of medicine. Although this differentiation may seem somewhat arbitrary, and there are clearly areas of overlap between the two definitions, it is a very useful distinction since the majority of applications of quantum mechanical methods in the literature fall within the first category, and a much smaller number into the second. This review will exclusively deal with the latter aspect of 'computational medicinal chemistry' and will aim to provide a clear focus on the direct applications of quantum mechanical calculations to problems relevant for medicinal chemistry and drug design. This also means that this chapter will not deal with applications of quantum mechanical methods for example to transition metal catalysis, mechanisms of organic reactions, predictions of material properties such as crystal structures, as well as applications to analytical chemistry and spectroscopy, like for example prediction of nuclear magnetic resonance (NMR) chemical shifts. Although it can be argued that these are applications that are essential for drug discovery, they are not directly related to medicinal chemistry and drug design in the context of this chapter and were therefore considered out of scope. It is clear that this differentiation between the uses of quantum mechanical methods and the focus on computational medicinal chemistry will exclude a large proportion of quantum mechanical calculations published over the years, but it will help focus this chapter and enable more detailed discussions of the key themes.

4.16.3 Theoretical Background for Quantum Mechanical Calculations

Although the purpose of this chapter is to review applications of quantum mechanical calculations relevant to medicinal chemistry and drug design, it is considered important to give at least a short introduction to provide context and also to point the interested reader toward more extensive reviews in the literature. These introductions are however kept to a minimum, focus on the main differences between methods and their known strengths and weaknesses in relation to medicinal chemistry applications, and will contain only one mathematical equation, probably much to the relief of most readers. For a short introduction to quantum mechanical calculations we refer to the review by Barden and Schaefer[3] for the background on quantum mechanics, ab initio calculations, density functional calculations, and semiempirical as well as quantum mechanical/molecular mechanical (QM/MM) methods.

The Schrödinger equation (eqn [1]), forms the basis of quantum mechanics and has the simple form for an eigenvalue problem:

$$H\psi = E\psi \qquad [1]$$

This famous equation cannot be solved explicitly for anything larger than hydrogen, even with modern computer power. Therefore, several approximations were introduced in order to enable the quantum mechanical treatment of molecules of more immediate interest to most chemists. These are for example the Born–Oppenheimer approximation, treating the nuclei of atoms as fixed, and the Hartree–Fock (HF) approximation where an effective potential replaces the true electron–electron potential description, effectively eliminating electron correlation. Another is the introduction of basis sets, designed to mimic the structure of orbitals, in place of actual electron integrals. The quality of these basis sets is essential for ab initio calculations, and as a general rule the larger they are and the more individual Gaussian functions they contain the better. Rather than discussing them here in detail we will refer to the landmark publication on the application of quantum mechanics by Hehre *et al.*[12] for further reading. Rather than always use the exact basis set nomenclature throughout this chapter, the level of any basis sets mentioned will be annotated in a more descriptive way, for example with large, medium, and small. The usefulness and accuracy of calculations using these approximations is mostly confirmed by comparison with experiment, particularly for molecular geometries and properties like for example dipole moments, and despite the use of these simplifications, the results are often surprisingly accurate. However, some approximations are more questionable and the magnitude of error introduced is often unclear. It is the aim of this chapter to show that despite using a number of approximations, quantum mechanical calculations are generally very accurate and useful in practically answering questions and describing molecular structures, properties, and interactions important for medicinal chemistry. It also has to be emphasized that for some of these questions, for example those relating to chemical reactivity, molecular properties like nucleophilicity, electrophilicity, charge distribution, spin–orbit coupling, dipole and higher multipole moments relating to polarizability, infrared, Raman and NMR chemical shifts, circular dichroism, and magnetic susceptibility,[13] quantum mechanical calculations are the only available option to the computational and medicinal chemist to obtain accurate predictions.

As pointed out above, the focus of this chapter is to review the application of quantum mechanical calculations in medicinal chemistry and drug design rather than to deal in detail with the theory of the methods. For the interested reader, there are short summaries for the most important approaches below (Sections 4.16.3.1–4.16.3.7). However, we will also provide at this point a very short summary. Those readers who would like to skip directly to the applications of quantum mechanical calculations should start with Section 4.16.4.

Quantum mechanical calculations, in contrast to the molecular mechanics approach, are directly derived from the physical principles that govern molecular structure, by solution of the Schrödinger equation (eqn. [1]) in an approximate way. The techniques can be divided into ab initio, DFT methods and semiempirical methods. While ab initio and density functional methods do not resort to parametrization to solve the Schrödinger equation, semiempirical methods contain parameters that avoid the computation of some time-consuming integrals required in ab initio and DFT calculations. Moreover, the semiempirical techniques take into account only the valence electrons. Although there are far fewer parameters in semiempirical methods, they are also less intuitive then those in molecular mechanics methods. All three methods (ab initio, DFT, and semiempirical) provide a wave function from which all electronic properties can be computed.[4]

4.16.3.1 Ab Initio Methods

This category of methods based on HF theory utilizing the self-consistent-field procedure (SCF), is the most widely used type of quantum mechanics calculation. It scales with about N^4, which means that when doubling the number of electrons in a calculation that it will take 16 times as long. This of course immediately sets a limit to the scope of this type of approach in terms of the size of molecule that can be calculated on a reasonable timescale in terms of medicinal chemistry and drug design, i.e., in a matter of a couple of days at best. Over the last 15 years, this size limit for molecules that are amenable to ab initio calculations has increased from about 10–15 heavy atoms at a moderate basis set (3-21G) to about 40–50 heavy atoms currently on even a high-end desktop computer. However, if larger timescales and more compute power is invested, very significant results can be achieved. Recently, a full geometry optimization on a 126-atom chain of 12 alanines has been performed at the HF 3-21G level.[14] Although it is unclear whether this level of theory provides an accurate enough description of the system in terms of for example hydrogen bonding geometries, it is certainly an important realization that calculations of this size are not only possible but also practical for addressing medicinal chemistry problems. It is also clear that for example the treatment of electron correlation, that would be provided by methods like perturbation methods like MP2 (Møller–Plesset level 2) is rather less practical, since they scale with N^5, which is also the case for the higher level correlation methods like coupled-cluster methods, scaling with N^7.[3] Although these higher level methods provide excellent accuracy and agreement with experiment in terms of geometries and relative energies, they are rather less useful for applications in computational chemistry and medicinal chemistry on pharmacologically relevant molecular systems, and will therefore not be discussed further in this chapter. However, there are methods that enable the treatment of electron correlation while remaining fast enough to be used for larger systems. They are the DFT methods and this approach will be discussed in the next section.

4.16.3.2 Density Functional Theory Methods

DFT is the latest addition to the field of quantum chemistry. It is probably an understatement to state that DFT has strongly influenced the evolution of quantum chemistry during the last 15 years, the term revolutionized is perhaps more appropriate.[15] DFT is based on the Hohenberg–Kohn paradigm,[16] which states that the electron density and electronic Hamiltonian have a functional relationship, which allows the computation of all ground-state molecular properties without a wave function. This means that it is possible to obtain the properties of a molecule after determination of only three coordinates, regardless of molecular size. However, we do not know exactly what the nature of this functional relationship is. The only approach is to build trial exchange-correlation functionals and assess their relevance and accuracy. In its current Kohn–Sham formulation DFT is a method still very much under development, and although it is a long way away from its promise, the modern-day Kohn–Sham DFT has still massive computational advantages over ab initio methods and can be applied just as easily via implementations in modern-day commercial software packages. Current implementations are for example the functionals B3LYP[17,18]and BP86[19,20] which have been shown to have significant advantages over ab initio approaches since their performance is roughly equivalent to the electron correlation MP2 method at the cost of only a HF/SCF level calculation.[3] Another way of utilizing this advantage is to use lower-level density functional equivalent in performance to HF/SCF approaches and trade off advantages in speed against increase in quality of for example size of basis sets to allow a more accurate description of molecular systems. With current compute power, in our experience full geometry optimizations of systems with around 30 heavy atoms are easily accessible with DFT methods like the functionals mentioned above (B3LYP), providing accurate and valuable information on molecular geometries and properties. In addition to the above-mentioned advantages in speed and performance of DFT over traditional ab initio methods, it is also differentiated by its ability to accurately describe the electronic properties of transition metals and their complexes.[21] This is important for calculations of for example some anticancer agents like *cis*-Platin, which will be described briefly later.[22] A more detailed introduction to DFT and its use in medicinal chemistry can be found in a recent review by Seminario,[23] Cavalli and co-workers,[8] and Raber and co-workers,[24] as well as Sebastiani and Röthlisberger.[25]

4.16.3.3 Semiempirical Methods

This category of methods has been developed in parallel with ab initio methods based on the realization that further simplifications were needed in order to be able to perform calculations on larger molecular systems and reactions. The main difference between semiempirical and ab initio as well as DFT is the additional use of parameters derived either empirically or from high-level ab initio calculations in place of the explicit calculation of some molecular integrals.[3] This has the obvious benefit of speeding up the calculations but comes at a significant cost in terms of accuracy of the results. Methods like AM1[26] and PM3[27] perform very well compared to ab initio and DFT methods for properties like atomic charges, electrostatic potentials, dipole moments and highest occupied molecular orbit/lowest unoccupied molecular orbit (HOMO/LUMO) energies. However, they have significant deficits in terms of accuracy of molecular structures, particularly hydrogen bond geometries (AM1) as well as the hybridization of for example nitrogen atoms in amide bonds and also heterocyclic aromatic ring systems.[28] Nevertheless, they are a very valuable addition to the tools available to computational chemists, particularly when dealing with larger molecular systems.

4.16.3.4 Molecular Mechanics or Force Field Methods

Molecular mechanics have virtually nothing in common with any of the methods described so far. They describe chemical bonds as a spring between two spheres and build a model of a molecular system based on classical mechanics and some empirical corrections.[3] Parameters and parameter sets, often called 'force fields,' are derived based on a training set to provide a best fit for specific bond types or molecular classes. Some examples for molecular force fields are the MM2/MM3/MM4[29] series, which is generalized for large organic systems, whereas others like AMBER[30] are specialized for certain classes of macromolecules like proteins. Others are the Tripos force field[31] and the Merck molecular force field (MMFF).[32] It has to be pointed out at this stage that some if not all force fields are developed by using quantum mechanical calculations to derive bond and torsion parameters. As an example, the parameterization of the MM3 force field makes extensive use of high-level ab initio calculations and the results of the force field calculations are compared with high-level ab initio calculations to assess the quality of the results.[33] Force fields are widely used in the combination with quantum mechanical methods in the so-called QM/MM approaches, which are described in the next section.

4.16.3.5 Combined Quantum Mechanics/Molecular Mechanics Methods

Molecular mechanics calculations can generate decent results close to ab initio accuracy for some conventional systems[34] for many thousands of atoms, provided nothing in the molecule needs accurate modeling, bond breaking, or formation or a description of polarization. Hybrid QM/MM methods have evolved largely driven by the need to explore reactions in biological systems and the characterization of transition states. For example, Warshel and Levitt's study of lysozyme initiated the field of QM/MM methods.[35] The unparalleled speed of molecular mechanics may be applied to the parts of the molecular system that have a negligible chemical impact, while some quantum mechanical theory of higher accuracy may be used for the difficult-to-model catalytic active site.[36] Such calculations are, in principle, capable of handling systems with several thousand atoms, which is why such studies hold a significant share of research in computational medicinal chemistry and drug design.[3] For a more in-depth introduction to QM/MM methods and their applications in medicinal chemistry please see the recent reviews by Peräkylä[37] and Perruccio *et al.*[9]

4.16.3.6 Molecular Dynamics (MD) Methods in Combination with Quantum Mechanical Calculations

First principles (or ab initio) MD, the direct combination of DFT with classical MD, was introduced in 1985 in a seminal paper by Car and Parrinello.[38] More recently, there have been some exciting new developments in this area, primarily around first principles (Car–Parrinello) MD (CP-MD) and its latest advancements into a mixed QM/MM scheme.[39–41] There have been several recent publications on applications of this relatively novel scheme to chemistry[42] and biology.[44] CP-MD offers the unique possibility of performing parameter-free MD simulations in which all the interactions are calculated on-the-fly within the framework of DFT or other alternative electronic structure methods. In this way, finite temperature and entropic effects are taken into account and simulations can be performed in realistic condensed-phase environments. Furthermore, this approach is also highly amenable to parallelization so that currently simulations of 100–1000 atoms can be performed. CP-MD offers promising perspectives for applications in medicinal chemistry, and a growing number of studies have emerged in recent years.[45] The CP-MD approach has also been combined recently with a mixed QM/MM scheme,[39–41] which enables the treatment of chemical reactions in biological systems with thousands of atoms.[40] Among the latest extensions of this method is also the calculation of NMR chemical shifts.[46] This relatively new field of quantum mechanical calculations has been reviewed recently by Sebastiani and Röthlisberger.[25]

4.16.3.7 Calculation of Solvent Effects Based on and in Combination with Quantum Mechanical Calculations

Understanding solvent effects is one of the most important aspects in the calculation of biological systems.[47] Solvation of ligands in a binding site is a very complex area, and extensive studies on x-ray structures have revealed some fascinating insights into the importance of for example hydrogen bonding patterns as well as conserved water molecules and their role[48] and influence on ligand binding interactions.[49–51] One of the strengths of quantum mechanical calculations is the ability to accurately describe solvent effects.[3] This can be done either by including solvent molecules explicitly in the calculation, or in an averaged fashion utilizing the polarizable continuum model[52] or the self-consistent reaction field (SCRF).[53,54] For semiempirical calculations, the so-called Solvent Model (SMx) series has been developed by Cramer and Truhlar.[55] This series of models offers a very efficient and fast alternative of calculating free energies of solvation based on empirical parameters, and they were implemented for the semiempirical AM1 and PM3 methods. Another approach similar to that of the SMx model series that is also used in conjunction with semiempirical and ab initio methods has been developed by Dixon *et al.*[56] which has a significant speed advantage over the SMx models. In addition, a solvent model based on the perfect, i.e., conductorlike, screening of the solute molecule and a quantitative calculation of the deviations from ideality appearing in real solvents has been developed.[57,58] It has been successfully used in combination with HF and DFT methods to rationalize and predict physicochemical properties of molecules.[59,60] The area of quantum mechanical continuum solvation models has been extensively reviewed by Tomasi *et al.* recently.[61]

4.16.4 Application of Quantum Mechanical Calculations to Medicinal Chemistry and Drug Design

One of the major challenges for computer-aided drug design is that it is not governed by the clear-cut rules of design in engineering, and hence, these methods do not produce a finished product by a fully prescribed procedure in the same sense that computer aided design (CAD) can produce other goods, like for example cars or aircraft. The limitations of the rational computer-aided drug design approach arise because of the complexity of the biological processes involved in drug action and metabolism at the molecular level and the level of approximation that must be used in describing molecular properties.[4] However, there is clear evidence in the literature that molecular modeling and computer-aided drug design methods and also data analysis and chemoinformatics approaches have become very important tools for drug discovery and that they have been successfully applied to medicinal chemistry,[62] particularly hit and lead generation as well as at the lead development stages.[63–65] Accepting that molecular modeling and chemoinformatics are useful techniques does however not sufficiently explain why one needs quantum mechanical methods. This has been done by Clark in a recent review, where he indicates that calculational techniques used to describe molecules should be able to describe the intermolecular interactions adequately.[66] He points out that this can only be achieved if the molecular electrostatics and the molecular polarizability are described well. The former is responsible for strong interactions and the latter is directly related to dispersion and other weak interactions. Therefore, following this argument, molecular interactions of any type can only be described adequately and accurately by using quantum mechanical calculations.

In the following sections we will focus on the application of quantum mechanical calculations to answer medicinal chemistry related questions in a drug design environment. These are divided into a total of four sections on: (1) the accurate calculation of molecular structure, (2) the calculation of quantum mechanical descriptors for prediction of molecular properties and QSAR,[67,68] (3) applications to chemical reactivity and the investigation of enzyme mechanisms, and (4) the calculation of interactions and binding energies of small molecules with proteins. This selection of topics is meant to reflect the main areas of interest to medicinal chemists working in the field of drug discovery. It is noted that although there are a great number of publications on the use of quantum mechanical calculations to medicinal chemistry, however, a large number of them are retrospective studies concerned with the validation of new technology rather then the prospective application to problem solving and design of NCEs. The following sections will therefore mainly focus on applications that either have a direct connection to the design of compounds or that are of immediate interest and applicable to medicinal chemistry.

4.16.4.1 Application of Quantum Mechanical Calculations to Generate Accurate Molecular Structures

Probably the most important and reliable application of quantum mechanical calculations is for the accurate description of the electronic and molecular structure of a given compound. One of the underlying fundamental principles of medicinal chemistry is that molecules with similar molecular properties have similar biological activities. The debate of what constitutes molecular similarity has been ongoing for probably more than two decades, and will probably continue for quite some time. One of the potential reasons is that similarity is usually defined by chemists in terms of the two-dimensional structures of a molecule or descriptors based on those two-dimensional parameters. Although we don't really understand fully how receptors recognize ligands, we can be almost certain that they do not identify them based on the chemical scaffold or Lewis structure, but rather based on their shape and surface properties and features. This is to some extent taken into account in the principle of bioisosterism,[69] which is essentially the expression of features of one functional group or fragment by another, perhaps even structurally unrelated one, that leads to similar activity at a given receptor. The other aspect that needs to be discussed in this context is the concept of a pharmacophore, which is defined as the minimum set of features in a molecule that are essential for activity at a certain receptor. This can be a very simple two-point pharmacophore, like for example a basic center at a certain distance from an aromatic ring, as found in the metabolizing enzyme cytochrome P450 2D6,[70] or it can be a rather complex, three-dimensional pharmacophore, like for example for the serine proteases, where several quite distinct pharmacophores have been identified. In theory, all that is needed to derive a meaningful pharmacophore from a single compound or a series of compounds is an accurate molecular geometry of a low energy, pharmacologically relevant structure of the ligand with the electrostatic features mapped onto the molecular surface. This information should be sufficient to determine whether a ligand could, at least in principle, bind to a given receptor. Predicting whether a molecule is active at a certain receptor is of course not that simple, but an accurate molecular structure and description of electrostatic features can provide invaluable information as to whether a compound is likely to bind to a receptor.

4.16.4.1.1 Accurate prediction of atom hybridization with quantum mechanical methods

One of the fundamental problems for molecular modeling is of course the generation of accurate molecular structures and conformations. Molecular mechanics methods achieve good structural accuracy for classical molecules, whereas their reliability for species with particular combinations of atoms may be questionable, particularly for molecules containing heteroatoms, which affect the geometry and conformation via the position of their lone-pairs. Force field programs, for example, often fail to calculate the geometry of particular nitrogen atoms. Force fields sometimes offer pure sp^3 or pure sp^2 hybridization for classification of the nitrogen in the NH_2 group. We can highlight this issue using aniline as an example. It is known that the nitrogen atom in aniline has intermediate geometry between pyramidal (ammonia) and trigonal planar (amide), because of a certain amount of sp^2 hybridization,[71] leading to an HNH angle in the amino group of around 113.6°, whereas the ideal sp^3 and sp^2 angles would be 109.5° and 120°, respectively. The fact that there seems to be a limited amount of conjugation of the nitrogen lone pair into the phenyl ring is not surprising, since full conjugation would effectively result in a delocalized 8-π-electron system, which would be an antiaromatic Huckel system and therefore energetically unfavorable. However, delocalized Lewis structures can easily be written for aniline, which indicates the mesomeric stabilization leading to a significant sp^2 character. This structural characteristic is only accurately reproduced by high-level ab initio and DFT methods, and perhaps surprisingly also by semiempirical methods like AM1 and PM3, which are known to overemphasize the sp^3 character of for example the amino group in amide bonds. To illustrate this, we have analyzed the structure of aniline at various levels of theory and compared it to the experimental structure of aniline derived from microwave spectroscopic studies.[72] The experimental and calculated data is summarized in **Table 1**.

The results show that in order to accurately represent the geometry of the NH_2 group as well as the overall polarity of aniline, represented by the dipole moment, a high level of theory including electron correlation and a large basis set

Table 1 Results from quantum chemical and molecular mechanical calculations for the structure and dipole moment of aniline

	HNH angle in degrees	*Dipole moment in Debye*
Experiment	113.6±2	1.53
Sybyl N sp^2	119.73	–
Sybyl N sp^3	109.30	–
MMFF	118.44	–
AM1	113.11	1.54
PM3	111.05	1.30
HF/3-21G	118.11	1.63
HF/6-31G*	110.63	1.54
HF/6-311+G**	111.31	1.44
SVWN/6-31G*	113.41	2.00
B3LYP/6-31G*	111.01	1.71
B3LYP/6-311+G**	112.03	1.59

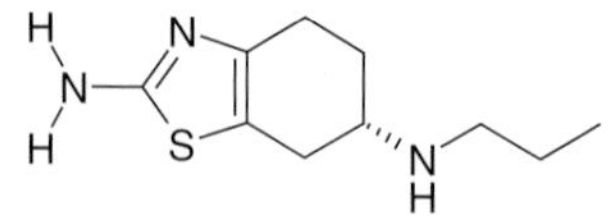

Figure 1 Structure of the D3 receptor agonist pramipexol.

is needed. Although the dipole moment at the HF level using the 6-31G* basis set is reproducing the dipole moment accurately, the geometry around the NH_2 group shows significant deviation from the experimental value. The semiempirical AM1 method shows excellent agreement with experiment on both, the geometry of the amino group as well as the dipole moment, which perhaps not surprising in terms of the dipole moment since fitting to experimental dipole moments is part of the parametrization of the method. However, caution is needed when applying semiempirical methods, since they are generally not very reliable for the prediction of amino group geometries and energetics, as exemplified in the significant underestimation of the *cis–trans* energy difference for amide bonds.[73] From our analysis, it becomes apparent that satisfactory agreement with experiment on both NH_2 geometry and dipole moment is achieved only at the B3LYP/6-311 + G** level. It is also clear from the data that force field calculations (for example the Tripos force field[31] and the Merck molecular force field, MMFF[32]) are inadequate for the accurate description of atoms with a more complex, noninteger hybridization state that significantly influences the overall molecular geometry, particulary angles and dihedral angles. Although the differences in hybridization at the NH_2 group between the different levels of theory could be considered relatively small and may seem insignificant, particularly for terminal amino groups like in aniline, in cases where such a group is located centrally in a molecule, the implications for the molecular geometry and a resulting three-dimensional pharmacophore derived from it can be substantial. Also, the resulting vectors for the hydrogen bond donor groups are different, which is very significant due to the known sensitivity of hydrogen bond donor group arrangements with respect to hydrogen bond acceptor groups, exemplified by the narrow range of hydrogen bond geometries at for example the N–H–O angle, which has a narrow observed range of between about 150–180° in crystal structures.[74] This consideration of hybridization was also found to be important in a recent study of the equilibrium geometry for a dopamine D3 receptor antagonist, pramipexol.[11] The structure of pramipexol is shown in **Figure 1**.

Therefore, in order to derive an accurate pharmacophore model with respect to the hydrogen bond donor vectors, the structure of pramipexol required geometry optimization. The NH_2 group in pramipexol is connected to an aminothiazole ring, which creates a similar aromatic environment to that of an NH_2 in aniline. Although crystal data for aminothiazoles is available, it is not conclusive and could not be used to solve the problem, because available structures tend to exhibit both planar and tetrahedral geometries.[75,76] The planar geometry of the NH_2 group is feasible due to a possible tautomeric exchange of hydrogen atoms between NH- and NH_2 group, whereas a more pyramidal geometry could be rationalized in which the sulfur potentially pushes electrons into the ring, which might cause greater sp^3 hybridization on the NH_2 group, because its lone pair is not attracted by the ring, as mentioned above in the context of the potential antiaromaticity of those systems. These assumptions were confirmed by quantum mechanical calculations. The 2-amino-1,3-thiazole molecule was geometry optimized starting from two different points (pyramidal NH_2 and planar NH_2) using the HF method and the 3-21G* and 6-31G** basis sets. Both basis sets yielded a slightly pyramidal geometry for the amino group as energetically favorable.[11] However, as we have shown above for aniline as a simple example, these basis sets (3-21G and 6-31G*) may not be sufficient in analyzing complex hybridization states of nitrogen atoms, and larger basis sets as well as electron correlation may be needed to achieve accurate geometries as well as physical properties for molecules of this type. The authors point out that solvation and the formation of hydrogen bonds can lead to a change in the energetically preferred tautomeric forms of heterocyclic compounds. In those circumstances a trigonal planar arrangement at the NH_2 group should be energetically more favorable.[11] Overall, a slightly pyramidal structure seems to be the most likely geometry for pramipexol. A further application of quantum mechanical calculations in the context of bioactive conformations is exemplified be the work of Schappach and Höltje on a series of 17α-hydroxylase-17,20-lyase inhibitors.[77] 17α-hydroxylase-17,20-lyase converts gestagens such as progesterone and pregnenolone to androgens by 17-α-hydroxylation, followed by the cleavage of the side chain. Because of its key role in the biosynthesis of androgens, inhibition of this enzyme results in a total blockade of androgen production, which makes the enzyme an interesting target in the treatment of prostate cancer. As pointed out above, a pharmacophore can be derived by superimposing energetically accessible or favorable conformations of some or ideally all of the ligands according to their consensus in structural features. The template structure for the superposition should fulfill at least two criteria. It should have high biological activity against the target, since this indicates that it has a significant number of nearly ideal interaction points with the receptor, and it should be reasonably rigid in order to limit the number of possible hypotheses. The highly potent compound MH3, a semirigid 17β-substituted aziridinyl steroid shown in **Figure 2** was chosen as template for a pharmacophore study.[77]

In this very rigid molecule only the bond between the steroid skeleton and the aziridine moiety can be considered as having the ability to rotate freely. An initial molecular mechanics conformational analysis resulted in two low-energy conformations, both of which had almost identical potential energy, as calculated with the Tripos force field.[31] The conformers differ in the position of the nitrogen lone pair. This is essential for enzyme inhibition, since the ligand interacts directly with the heme iron of the enzyme. Since the orientation of the lone pair is essential for a meaningful superposition, both conformations were studied in more detail at the HF level using the 3-21G basis set. Optimization of the two conformers highlighted a significant energy difference, and the lower energy conformation was chosen as a template structure.[77] Another structural feature to be investigated was the NH group in the aziridine ring, for which pyramidal inversion is normally observed. This NH is, however, part of a conformationally very restricted three-membered ring with a high inversion barrier, and the nitrogen inversion process of aziridine rings has been subject of quantum mechanical investigations.[78] Ab initio optimization of the geometry of the MH3 aziridine invertomer with the 3-21G** basis set revealed no significant energy difference between the configurations, therefore both structures had to be considered in the pharmacophore development.

In a further study, quantum mechanical calculations were applied to investigate the biologically relevant conformation for H_2 antagonists used for ulcer therapy.[11] The compounds occupy the histamine binding site of the histamine H_2 receptor, thus inhibiting histamine induced gastric acid secretion. Four main structural classes are used as drugs: (1) imidazole derivatives (cimetidine), (2) basically substituted furans (ranitidine), (3) guanidinothiazoles (famotidine), and (4) aminoalkylphenoxy derivatives (roxatidine). Their structures are shown in **Figure 3**. All these antagonists were supposed to bind at the same binding site in the receptor.[11] Because the three-dimensional structure of the H_2 receptor is not yet known, a pharmacophore model was derived to attempt to prove this hypothesis. Most of

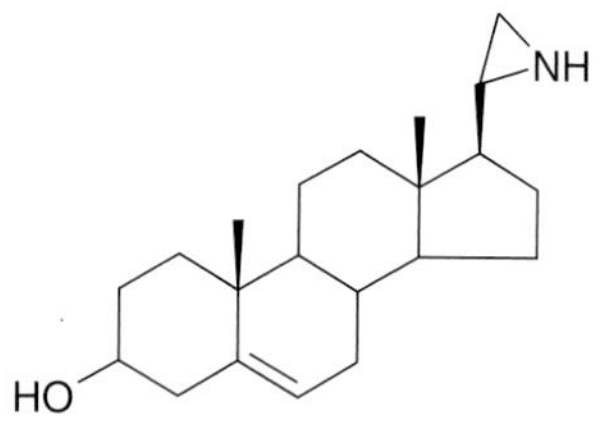

Figure 2 Structure of MH3, a highly potent, semirigid 17β-substituted aziridinyl steroid.

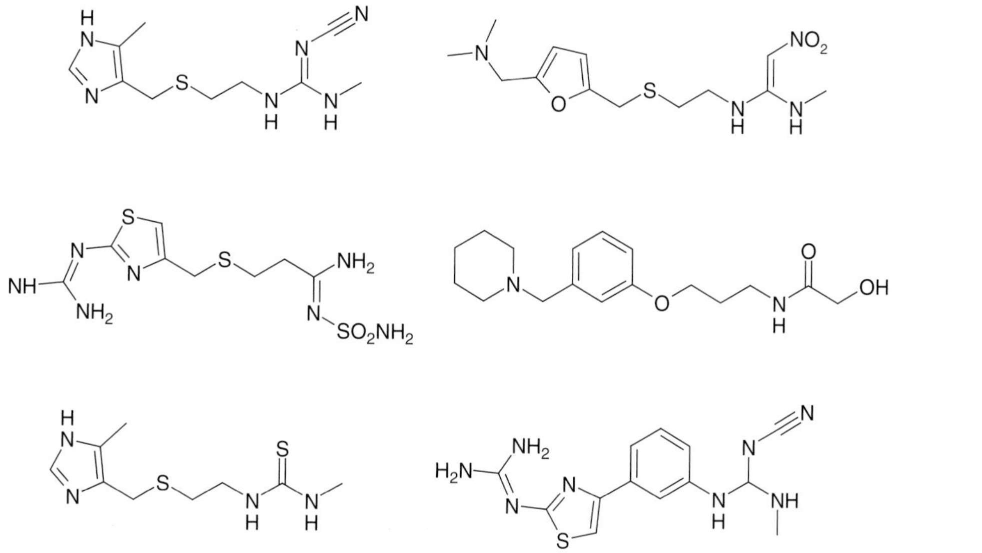

Figure 3 Structures of H_2 antagonists cimetidine (top left), ranitidine (top right), famotidine (middle left), roxatidine (middle right), metiamide (bottom left), and ICI27032 (bottom right).

the H_2 antagonists are conformationally flexible, and an evaluation of available crystal data and a systematic conformational search revealed mainly bent conformations, partly with intramolecular hydrogen bonds, as the most stable structures. This result is in accordance with an investigation of the H_2 antagonist metiamide. The authors used the HF method with three different basis sets (3-21G*, 6-31G*, and 6-31 + G**) to study the conformational properties of metiamide.[79] The calculations clearly indicate a preference for a folded conformation with a hydrogen bond between the imidazole ring and one of the NH groups, similar to the crystal structure. Calculations with one isolated molecule in vacuum often result in an overestimation of intramolecular contacts. Therefore, in gas phase calculations in general polar groups interact with each other in the absence of binding partners like protein residues or solvent molecules, often leading to folded structures. Although these folded structures are quite common in small-molecule x-ray structures, they are rarely observed in crystal structures of ligands bound to proteins, since they generally reduce the number of possible surface contacts between the ligand and the protein, thus reducing the binding contacts and therefore the achievable binding free energy. Therefore, folded conformations, especially of small molecules, can in general be disregarded in the context of protein interactions and pharmacophores, since they mostly constitute an artifact of the gas phase calculations. When those ligands are surrounded by solvent molecules and the geometry is again optimized, ligands tend to favor extended conformations, thus maximizing the polar interactions with their environment and particularly their dipole moment. Therefore, including solvent either explicitly or through continuum models even in optimization of small molecules can be essential to get meaningful geometries, especially with respect to simulating receptor-bound structures. In addition, the charge distribution can change significantly, influencing the dipole moment and polarizability, and therefore the electrostatic potential believed to be a property relevant for receptor recognition. One of the known competitive H_2 antagonists, ICI27032, cannot adopt a folded conformation due to a central aromatic ring. This highlights that the folded structures of for example metiamide are not biologically relevant. In fact, extended conformations for the H_2 antagonists cimetidine, ranitidine, famotidine, and roxatidine can be overlaid onto the extended conformation of ICI27032, and were found to exhibit similar electrostatic properties, indicating a common pharmacophore.

4.16.4.1.2 Quantitative structure–activity relationship using molecular orbitals from quantum mechanical calculations

A further application of quantum mechanical calculations focused on a SAR analysis of calcium channel-blocking 1,4-dihydropyridine (DHP) derivatives such as nifedipine, which are widely used in the therapy of cardiovascular disorders.[11] The structure of the receptor protein and its DHP binding site are not available and specific information about the binding interactions is not known. The main binding interactions are believed to be the hydrogen bond donor properties of the NH group and at least one further hydrogen bond accepted by the carbonyl groups of the ester side chains in addition to electrostatic attractions. It has been demonstrated by the authors that these binding elements alone cannot account for the high affinity of some compounds. The binding affinity of 23 nifedipine-like DHP has been determined experimentally and the pK_i values range over more than five log units although the single structural change was the varied substitution pattern of the 4-phenyl ring.[11] The DHF scaffold is shown in **Figure 4**. It has not yet been clarified whether the ring substituents interact directly with the binding site or affect the molecular characteristics of the DHP molecules. A recently used atomistic pseudoreceptor model for a series of DHP indicated a putative charge-transfer interaction was stabilizing the DHP binding site complex.[80] To prove this hypothesis, qualitative and quantitative analysis of the molecular orbitals of nine DHP derivatives was performed.[81] Charge transfer (or electron donor–acceptor) interactions are indicative of electronic charge transfer from the HOMO of a donor molecule ($HOMO_D$) to the LUMO of an accepting ($LUMO_A$) neighboring molecule. Small energy barriers between the $HOMO_D$ and the $LUMO_A$ increase the probability of charge transfer but with two further additions – the corresponding molecular orbitals must be able to overlap, and $HOMO_D$ and $LUMO_A$ must be energetically close. In a

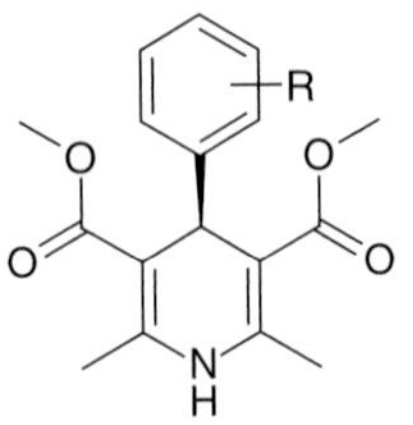

R
O O
O O
N
H

Figure 4 Scaffold of dihydropyridines: nifedipine carries an *ortho*-NO_2 substituent on the phenyl ring.

charge-transfer interaction for the stabilization of the DHP binding site complex the electron-accepting LUMO should be located on the 4-phenyl ring of the DHP, since highest binding affinities are found for derivatives with electron-withdrawing substituents at this position. Using the semiempirical AM1 method, the molecular structures of the DHP were optimized and the molecular orbitals were computed.[11] The reliability of the semiempirical AM1 method had been demonstrated by comparison of the results with high-level ab initio calculations. In nearly all cases the LUMO* (LUMO located mainly on the phenyl substituent, which could be the LUMO or LUMO + 1 and even LUMO + 2) of the DHP ligands was positioned at the 1,4-dihydropyridine heterocyle. To decide whether charge transfer interactions might play a major role in the receptor binding of DHP, the experimentally derived free binding energies (ΔG) were correlated with the calculated LUMO energies and a highly significant correlation was obtained with an r value of 0.91. The authors point out that due to the high similarity of the ligands investigated in this study, the obtained correlation should reflect the effect of potential 4-phenyl ring charge transfer interactions on the binding affinities of the DHP. To determine whether a reliable model had been found, the binding energy of a novel DHP was predicted. Calculations were performed to determine the LUMO* energy of the most active DHP isradipine, a compound which has a benzoxadiazole ring instead of the 4-phenyl ring. The LUMO* energy was indicative of a very strong charge transfer interaction and ranked the compound in accordance with the experimental data as the most potent.[81]

4.16.4.1.3 Calculation of relative energies of conformations of molecules

Another very important field for the application of quantum mechanical calculations is the accurate description of geometries and conformations of molecules, as described in a paper by Martin *et al.* on the design of novel human immunodeficiency virus-1 (HIV-1) proteinase inhibitors.[82] Ab initio HF and DFT calculations were performed on several conformers of a new HIV-1 inhibitor and the calculations were able to shed light on the relative energies of three conformers of interest that would be able to bind to the protein. The structure of the conformer found to be the most stable was in excellent agreement with NMR and x-ray structures, underpinning the validity of the quantum mechanical calculations for the accurate prediction of molecular structures. In a further study on the conformation of pharmacologically relevant compounds, Mora *et al.* have performed a thermodynamic conformational analysis and structural stability of the nicotinic analgesic ABT-594.[83] Current analgesics in use fall mainly in one of two classes, opiod analgesics and nonsteroidal anti-inflammatory drugs (NSAIDs).[84] These compounds are extensively used in the treatment of pain, but they exhibit unwanted side effects. Opioids, for example, produce physical addiction, respiratory depression, and constipation, among other undesirable effects. The nonspecificity of some NSAIDs for the COX-1 and COX-2 isoforms of the cyclooxygenase enzyme can lead to gastric irritation and renal dysfunction.[84] For these reasons the development of new approaches to the treatment of pain free from side effects is of great interest. This could potentially be achieved with nicotinic analgesics like for example ABT-594. Quantum mechanical calculations at a very high level including electron correlation (MP2) were used to investigate the conformational properties of ABT-594 in its neutral and protonated forms in vacuum and aqueous solution.[83] The structure of ABT-594 is shown in **Figure 5**. A conformational analysis was performed on the two torsional angles describing the orientation of the azetidinyl group and the azetidinylmethoxy moiety. To account for entropic effects, a thermostatistical study of conformational populations at physiological temperature was also carried out. For the neutral form of ABT-594, the conformation where the nitrogen of the azetidinyl group is far from the electron pairs of the oxygen and from the pyridinic nitrogen is the most stable structure. In the protonated form, that conformer which has the additional proton on the azetidinyl group oriented toward the electron lone pairs or the oxygen is the energetically most preferred structure. For the neutral form, the effect of solvent was found to increase the barriers of interconversion between conformers. On the contrary, in the protonated form the solvent reduces the interconversion barriers. Therefore, on energetic grounds, the active (protonated) form of ABT-594 is more flexible in solution. This result could influence the design of future compounds by providing a very detailed pharmacophore.

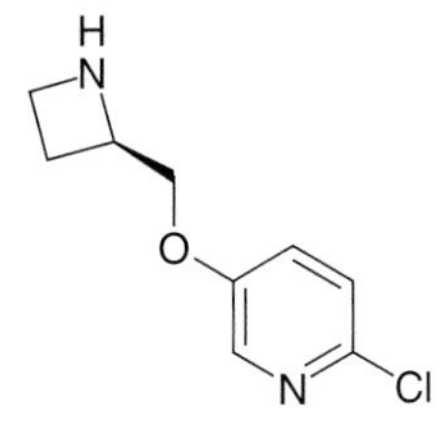

H
N
O
N
Cl

Figure 5 Structure of the nicotinic analgesic ABT-594.

Predicting the bioactive conformation of a given molecule, especially in the absence of structural information about the receptor, is an extremely difficult problem for molecular modeling in general due to the number of factors that contribute to ligand protein binding. In many cases, indirect methods like three-dimensional QSAR (3D-QSAR), which are always based on fundamental assumptions like for example that all molecules in a given data set bind to the same site in the receptor and can be at least partially overlaid, can nevertheless give valuable insights into aspects of ligand binding. There are a number of publications that have investigated the problem of bioactive conformations by comparing protein-bound and small-molecule crystal structures of ligands with their corresponding calculated geometries.[85–87] In many cases, the small-molecule crystal structure and the protein-bound structure are not closely related and can even have very different conformations. At the same time, due to the restrictions that are encountered when a ligand enters a binding site, the protein-bound conformation and the calculated low-energy conformations may also be very different, although they can be energetically very close. Therefore, deriving a pharmacophore is a very difficult and complex problem, and the computational approach to solving this problem and determining a meaningful pharmacophore from low-energy structures and ligand superposition can often generate a number of feasible solutions rather than just one most likely solution. This is highlighted in an example of a Erk2 complex of the inhibitor olomoucine (**Figure 6**) that we have investigated recently.

Several methods have been used to optimize the geometry of the ligand starting from the protein-bound conformation, like for example the Tripos force field, MMFF as well as the semiempirical AM1 and PM3 methods, and ab initio HF as well as a DFT approach (B3LYP). The receptor-bound conformation has been taken from the Brookhaven data bank (PDB code 4erk)[88] and is shown together with the MMFF, AM1, and HF/3-21G derived structures in **Figure 7**. All methods result in a geometry that exhibits a very different pharmacophore from that found in the protein, with the phenyl substituent moving out of the plane of the bicyclic ring system by about 4 Å from its position in the protein-bound conformation. If there was no information available about the shape of the ATP binding site in kinases, based on these calculations one would not necessarily assume that it would have a letterbox shape and derive a potentially irrelevant pharmacophore. These results perhaps exemplify the difficulties in describing any meaningful and pharmacologically relevant pharmacophore in the absence of binding information even for compounds with, like in this case, only a few rotational bonds. This does not mean that the results of the calculations are wrong, it is more likely that the receptor in this case induces a conformational change that results in a bound conformation that is not perhaps the global energy minimum. A similar result was obtained for a second kinase ligand, SB-218655 (PDB code 1bmk),[88] which effectively only has two rotatable bonds. In this case, one of the pharmacophoric points, the position of the cyclopropyl substituent, could not be determined accurately, since different calculations gave slightly different positions for this substituent in relation to the other pharmacophoric features. The calculated structures are compared to the protein bound conformation in **Figure 8**. Interestingly, almost all the computational methods used agree on the structure of this very rigid ligand, and there is very little difference between the force field structures and the high-level ab initio and DFT calculations. This indicates that in certain cases, even force field methods can give relatively accurate structures, but only the quantum mechanical calculations, starting at the lower end of the spectrum with the semiempirical methods, can provide information about the electronic structures and therefore the molecular properties. The resulting structures are shown in **Figure 9**.

In the example described above for olomoucine, it appears that even for a relatively small and rigid ligand it is difficult to derive pharmacologically relevant information in the form of a pharmacophore in the absence of ligand binding information from a protein–ligand x-ray structure or data on ideally a number of rigid ligands. Although highest level ab initio or DFT methods can provide very accurate descriptions of molecular geometries and their properties that can come very close to a true reflection of the structure and properties of a compound, they can provide little insight into whether a given conformer is relevant in the context of binding to a receptor. This knowledge can only be derived

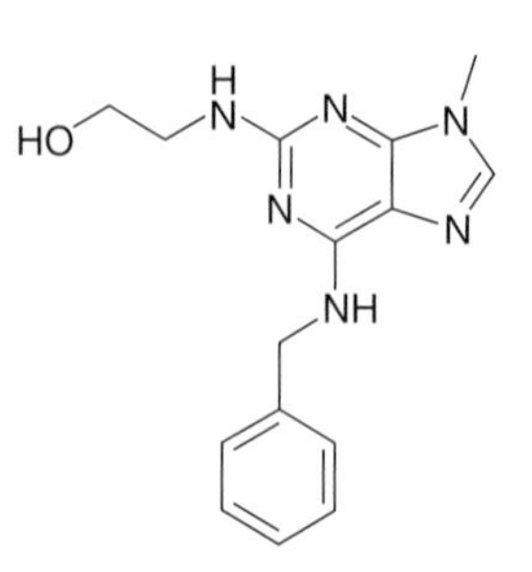

Figure 6 Erk2 inhibitor Olomoucine (PDB code 4erk).

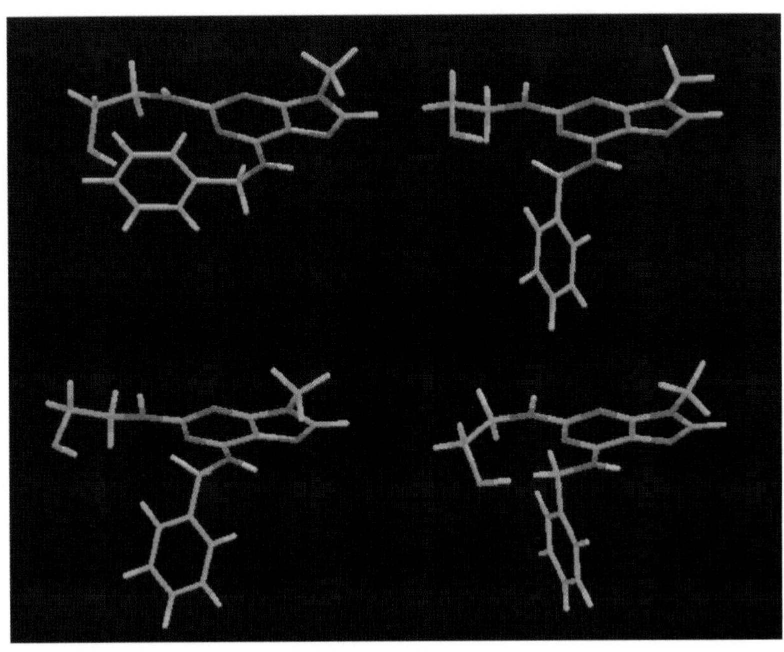

Figure 7 Comparison of the receptor bound conformation of olomoucine in Erk2 (top left) together with the MMFF (top right), AM1 (bottom left), and HF/3-21G (bottom right) derived structures.

Figure 8 p38 kinase inhibitor SB-218655 (PDB code 1bmk).

from activity data for a series of compounds in a QSAR or pharmacophore study, which leads to one of the more feasible binding mode hypotheses. This knowledge can then be applied to design molecules, synthesize them, and biologically screen them to provide feedback on the validity of the hypothesis. It is this iterative process of deriving a hypothesis based on available data, testing and updating it, and using the updated hypothesis for the next design effort, that can provide a way to rationally design pharmacologically relevant compounds in the absence of structural information about receptor binding.

4.16.4.2 Application of Quantum Mechanical Descriptors for the Prediction of Molecular Properties and Use in Quantitative Structure–Activity and Structure–Property Relationships

As mentioned in the previous section, it is a central and essential assumption of medicinal chemistry that structurally similar molecules have similar biological activity or physical properties. This concept has been validated through decades of experience within the medicinal chemistry discipline during countless drug discovery programs. Therefore, it became one of the key aspects in medicinal chemistry to understand and utilize QSARs and quantitative structure–property relationships (QSPRs) in compound design to enhance biological activity against the molecular target(s) or to improve properties related to absorption, distribution, metabolism, excretion, and toxicity (ADMET) related properties. Whereas traditionally, the driving force of drug discovery has been the synthesis of novel structures that show increased potency, current research strategies increasingly recognize the importance of pharmocokinetic information such as oral bioavailability and suitable half-life and safety margins over unacceptable side effects and

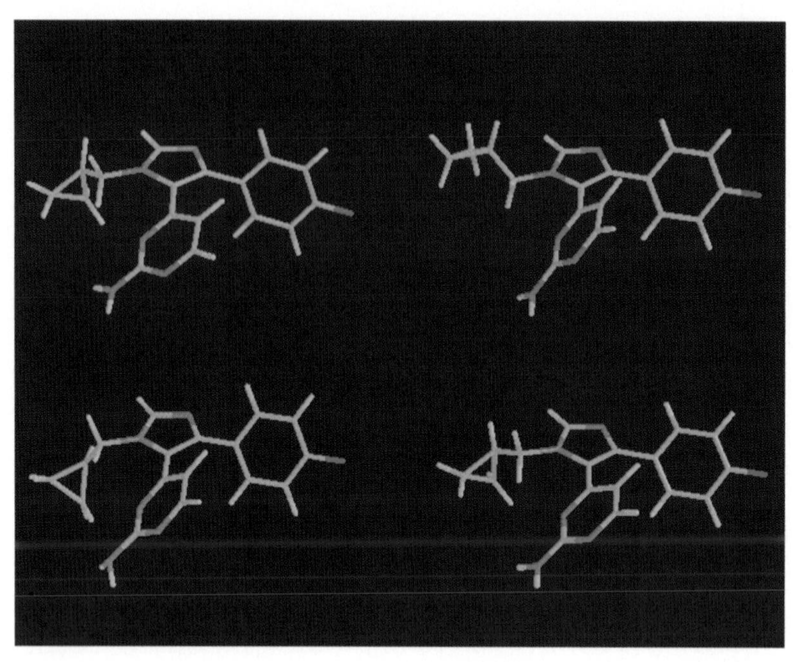

Figure 9 Comparison of the protein-bound conformation for SB-218655 (top left) with the calculated structures for MMFF (top right), AM1 (bottom left), and HF/3-21G (bottom right).

toxicity are often equally important as ligand specificity and selectivity.[89] These are key to some of the factors associated with the properties of drug candidates and their risk of attrition during drug development. Other factors that influence attrition are aqueous solubility, lipophilicity, and chemical stability or reactivity as well as photostability. There have been numerous attempts to derive models for the prediction of ADMET properties based on mainly empirical models utilizing classical QSAR descriptors like for example molecular weight, hydrogen bond acceptors and donors, lipophilicity, and so on. Some of these models are very successful in predicting properties for a given set of compounds, but they are unable to extrapolate to new chemical series that have not been used in the parameterization or training of the model. The main advantage of quantum mechanical applications that enables them to contribute in this area is of course the ability to generate new atomic or molecular descriptors, which are often orthogonal to existing classical descriptors, for the generation of QSAR models. Also, incremental values of properties like dipole moment or polarizability can be calculated for molecular fragments, which often form the basis of conventional QSAR models.

In principle, QSAR models built on quantum mechanical descriptors should be able to extrapolate to new molecules in contrast to conventional classical QSAR models, which are not valid for chemical structures that are not related to those in the training set used in the generation of the model. This is extremely powerful for example the prediction of physicochemical properties in the ADMET area, where it is very important to assess the potential ADMET liabilities of a given design prior to synthesis. This enables prioritization of a large number of designs and assists in the selection of the most promising structures for synthesis. The area of quantum mechanical descriptors in QSAR and QSPR studies has been extensively reviewed by Karelson *et al.*[90] as well as by Benigni.[91]

4.16.4.2.1 Quantum mechanics descriptors in quantitative structure–activity relationship for molecular properties and protein–ligand binding

Early applications of quantum mechanical calculations to derive descriptors that could be used to predict physicochemical properties include for example the prediction of the partition coefficient of molecules between octanol and water, log *P*, by Brinck *et al.*[92] Although they used small basis sets in their ab initio HF calculations, they achieve excellent correlations of log *P* with parameters like local polarity and the variability of the electrostatic potential over the molecular surface. Later work by Beck *et al.*[93] derived a model for log *P* based on experimental values for 1085 molecules. They utilized a neural network approach in combination with quantum mechanical descriptors and achieved excellent agreement with experimental values. The same group also used quantum mechanical descriptors to derive a QSAR model for the binding energies of small molecules in chymotrypsin[94] and achieved a good correlation with experimental values. Furthermore, Mu *et al.* have applied quantum mechanical descriptors to perform a QSPR analysis of the unified nonspecific solvent polarity scale.[95] This is of particular interest in areas like chemical reactivity and experimental design, where it is desirable to minimize the number of experiments required in order to obtain an optimum amount of information about for example the feasibility of a synthetic route or its optimum yield depending on the solvent used. They obtained a two-parameter correlation ($r^2 = 0.96$) for 48 diverse solvent molecules using two orthogonal descriptors, the dipolar density given by the total dipole moment of the molecule divided by the molecular volume, and the reciprocal of the calculated HOMO/LUMO energy gap. The correlation allows confident estimations of the solvent polarity S' from quantum mechanical calculations which can assist in the interpretation of nonspecific solvation effects and also enables the analysis of solvent conformation-dependent solvation influences.[95]

In a recent paper, Hemmateenejad *et al.* utilized quantum mechanical descriptors to derive a QSAR model for the biological activities of a set of 45 1,4-dihyropyridine-based calcium channel blockers.[96] They calculated a large number of descriptors, including HOMO and LUMO energies as well as various charge and electronegativity parameters, and derived models using genetic algorithms and principal components analysis as well as neural networks. The results indicate a significant improvement in the predictive value of their QSAR model over earlier work involving classical descriptors, again highlighting the additional information that quantum mechanical descriptors can provide.

The true value and uniqueness of quantum mechanical parameters has been highlighted again recently by the work of Clare on a structure–activity study of 89 phenylalkylamine hallucinogens.[97] This study showed that quantum mechanical descriptors like for example atomic charges and molecular orbital energies combined with hydrophobic and steric parameters were important for generating a predictive QSAR model. The model involves the energies of four π-like near-frontier orbitals and the orientation of their nodes, which illustrates the essential quantum mechanical nature of the interaction of a drug with its receptor, as the π-like orbitals involved are described as standing waves of probability of finding an electron in a given location in the field of the atomic nuclei. These parameters have no classical counterpart that would be able to describe this important contribution to receptor–ligand binding.

Focusing on steroid binding affinity and also antibacterial activity of nitrofuran derivatives, Smith and Popelier utilized QSAR studies based on optimized ab initio and semiempirical AM1 bond lengths.[98] They derived satisfactory QSAR models for the ligand–protein binding energy predictions of the steroid data set of 31 molecules for the semiempirical AM1 method with $r^2 = 0.708$ and $q^2 = 0.575$. In addition, QSAR models were derived for the antibacterial activity of a set of nine nitrofuran derivatives. They were again based on the calculated AM1 bond lengths and resulted in excellent correlations for the activities of these nitrofuran derivatives against two bacterial species, with r^2 of 0.92 and 0.96.

4.16.4.2.2 Quantum mechanical descriptors in quantitative structure–activity relationship for the prediction of pK_a

As mentioned above, semiempirical calculations have a distinct speed advantage over ab initio and DFT calculations and can therefore give very useful insights for larger molecules or data sets relatively quickly and at relatively low computational cost. As an example, we have performed a QSAR analysis of the acidic pK_a of 47 phenols and a variety of other heterocyclic, nonphenolic, organic acids using the semiempirical AM1 method. We included phenols and also aromatic heterocyclic structures like pyrazoles as well as hydantoins and barbituric acid derivatives in the data set. The best correlation was found between the experimental aqueous pK_a and the calculated AM1 gas phase proton affinity, followed closely by the AM1/SM2 calculated aqueous proton affinity. The model has very good statistics ($r = 0.95$, $r^2 = 0.90$, $r^2_{cv} = 0.89$) and the graph resulting from the regression analysis is shown in **Figure 10**. This result is somewhat surprising and counterintuitive in that the calculated proton affinity in water correlates less well with the measured aqueous pK_a than with the calculated gas phase proton affinity. The reason for this finding is not at all clear, and it could be speculated that it may potentially point toward inadequacies of the AM1/SM2 solvation model for this particular data set. Although the molecules used in this study were relatively small and could easily have been calculated using for example a DFT approach, for sets of larger, drug-size molecules containing 30–40 atoms this may not always be feasible. In these instances, the use of semiempirical approaches for the prediction of molecular properties like pK_a can provide very valuable information.

Quantum mechanical calculations by themselves can be a very useful tool, but as pointed out before, they are best applied in combination with other computational approaches. For example, quantum mechanical calculations have been successfully used in combination with free energy perturbation calculations for compound design. In an earlier study, Erion and Reddy have calculated the relative hydration free energy differences in combination with ab initio HF calculations for heteroaromatic compounds. This information was then used in the successful design of adenosine

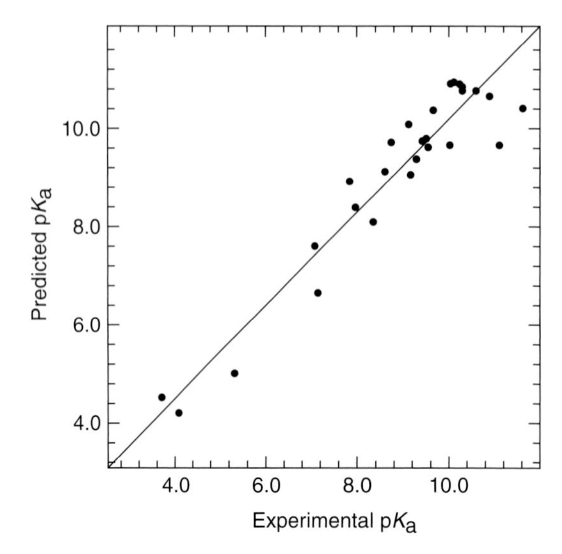

Figure 10 Graph representing the correlation of experimental and predicted pK_a for 47 organic noncarboxylic acids.

deaminase and cytidine deaminase inhibitors.[99] They achieved good agreement with experimental hydration equilibrium constants and found that the accurate calculation of relative hydration free energies were useful in their efforts toward the discovery of substrate analogs that act as potent adenosine and cytidine deaminase inhibitors.

4.16.4.2.3 Quantum mechanical calculations in quantitative structure–activity relationship for the prediction of toxicity

In a recent paper, He *et al.* utilized quantum mechanical descriptors like atomic charges, dipole moments as well as HOMO and LUMO energies derived at the semiempirical AM1 and PM3 levels in a QSAR study to predict the genotoxicity of polycyclic aromatic compounds.[100] The accurate prediction of genotoxicity based on quantum mechanical parameters is an application that would be very relevant for medicinal chemistry. Based on the descriptors mentioned above and also other additional classical descriptors, various models with high predictive power were developed, that were able to correctly classify between 80% and 90% of the 277 compounds used in the study. In a related paper, Mosier *et al.* predict the genotoxicity of 140 thiophene derivatives using a similar approach. They also generate highly predictive models for this class of compounds based on a variety of parameters, including those generated from semiempirical quantum mechanical calculations.[101] More recently, He and Jurs applied a similar approach to the prediction of fathead minnow acute aquatic toxicity for a data set of 322 compounds,[102] achieving predictive models with good accuracy. Very recently, Smieško and Benfenati applied high-level DFT calculations in the gas phase and solvent phases to derive thermodynamic and charge-related descriptors for the development of QSAR models for the prediction of aquatic toxicity of a series of 53 substituted phenols.[103] They found a good correlation between experimental and calculated phenol pK_a with a correlation coefficient of $r^2 = 0.93$, similarly to our work on the calculation of phenol pK_a mentioned above. Interestingly, it was the combination of the calculated gas phase deprotonation enthalpy as a descriptor for reactivity together with Clog*P* as a descriptor of the membrane penetration abilities of a compound that resulted in models for the prediction of the toxicity for two aquatic species. The results were in very good agreement with experiment ($r^2 = 0.83$ and 0.85). In some related work, Popelier *et al.* used QSAR based on quantum topological descriptors to investigate the relationships of the mutagenic activity of 23 triazenes and 24 halogenated hydroxyfuranones (mutagen-X) derivatives.[104] They used a newly developed method, called quantum topological molecular similarity (QTMS), which generates bond descriptors from high-level geometry-optimized ab initio wave functions derived at the HF and also DFT level. The QSAR models were then derived using a combined genetic algorithm (GA)/partial least squares method. The authors emphasize that QTMS improves the understanding of the mutagenic activity beyond that achieved with existing QSAR models using classical parameters, since it is able to select the bonds directly involved in the biological activity or chemical reactivity of the molecules in the data set. For both data sets used in this work, the triazenes and also the halogenated hydroxyfuranones (mutagen-X derivatives) the results indicate that the quality of the QSAR model improves with increasing level of theory of the quantum chemically derived parameters, going from the semiempirical AM1 toward HF and DFT level using extensive basis sets. The

QSAR models that were generated for both datasets have r^2 of 0.86 and 0.75, respectively. In addition to the statistical models, information about the actual mechanism of mutagenicity for both substance classes is generated by the QTMS method, highlighting the active and reactive center of the mutagens.[104] In a related study on the cell growth inhibitory properties of a set of 17 substituted (*E*)-1-phenylbut-1-en-3-ones, O'Brien and Popelier applied QTMS based on high-level DFT calculations, and derived a QSAR model with $r^2 = 0.91$.[105] In addition, the QTMS method was also applied to the prediction of the toxicity of polychlorinated dibenzo-*p*-dioxins (PCDDs).[106] It was again shown that the QTMS approach competes in terms of achieved model quality with traditional QSAR approaches using classical descriptors. However, the identification of the active center would in this case require further and more detailed work. Although the QTMS approach based on high-level HF and DFT calculations applied for toxicity prediction more recently, it was also used by Chaudry and Popelier for the estimation of pK_a of 40 carboxylic acids, 36 anilines, and 19 phenols.[107] Again, they derived high-quality QSAR models with r^2 of 0.92 for acids, 0.97 for anilines, and 0.95 for phenols. The authors point out that their work improved on previous attempts at modeling pK_a and highlighted the increasing use of quantum chemically derived descriptors over empirical parameters such as the Hammett descriptors and also descriptors obtained from semiempirical calculations. The QTMS method is described as being capable of highlighting important bonds that are responsible for the observed property and reactivity. The analysis is described as being unbiased, since no bonds are picked to build the QSAR model, the method allowed the data to provide the chemical insight as to which bond is the most relevant in relation to the biological activity. This could be very useful for the investigation of systems where the modes of action are not yet known. QTMS may also be applied to compounds that are not amenable to analysis via Hammett parameters, for example when the values are not obtainable for the fragments under observation. Even though electron correlation was necessary in some cases to produce good models at significantly increased computational cost, such as with the anilines and phenols, a QTMS analysis of larger and industrially relevant drug-size molecules is considered to be perfectly feasible from a computational point of view given modern hardware.[107]

4.16.4.2.4 Application of quantum mechanics methods to hydrogen bonding, polarizability, and solubility

In addition to the use of quantum mechanical descriptors like dipole moments, polarizability, or molecular orbital energies in QSAR studies, it is also possible to derive parameters for hydrogen bond acidity and basicity, as highlighted in a recent paper by Cacelli *et al.* using ab initio DFT methods and solvation models.[108] They used a data set of 55 compounds to determine the Abraham descriptors of hydrogen bond acidity and basicity and achieved a high degree of accuracy for a large number of classes of molecules with correlation coefficients for acidity and basicity parameters of 0.99 and 0.97, respectively. This is again a very useful application for medicinal chemistry, since it provides information on molecules for which the Abraham descriptors are not available.

One of the key questions that is relevant for many of these QSAR studies is around the significance of solvation. In some cases, like for example for the pK_a prediction of phenols, gas phase parameters are capable of describing a molecular property very accurately, at least within a series of relatively small molecules. However, it can be assumed that solvation is important when comparing molecules from different chemical classes and with very different properties, where a fortuitous cancellation of errors cannot be expected. The importance of the inclusion of an adequate treatment of solvation in the calculation and prediction of physicochemical properties like for example pK_a, acidity, basicity, and solubility was also highlighted by Klamt *et al.* in a paper on the prediction of aqueous solubility of drugs and pesticides using a combined DFT and COSMO approach.[109] They achieve excellent agreement with experiment for a set of 150 drugs and 107 pesticides. The authors point out that the real power and significant advantage of these models compared to conventional QSAR approaches using classical descriptors is that it is able to predict aqueous solubilities of almost arbitrary structural classes, therefore being able to extrapolate beyond the boundaries of classical QSAR.

4.16.4.2.5 New developments for quantum chemical descriptors

A new class of molecular descriptors for use in QSAR has been described recently by Ehresmann *et al.*, which are based on statistical descriptions of the local ionization potential, local electron affinity, and the local polarizability at the surface of a molecule. These descriptors were derived using the semiempirical AM1 and PM3 wavefunctions[110] and were used in addition to a set of 26 descriptors reported previously by the same group.[111] It was shown that the 13 new descriptors showed little correlation with those 26 derived earlier. This extended set of descriptors has been used in a neural net based approach to derive a boiling point model, which is relevant to medicinal chemistry in the context of predicting solubility. The models were derived for a data set of 5455 compounds with a boiling point range of around

150 to 700 K using between 10 and 18 quantum mechanical descriptors, achieving very good agreement with experimental values. The results for the validation set of 606 compounds were also in good agreement with experiment with a mean unsigned error of between 17 and 20 K, indicating the excellent predictive power of these models. The authors also noted that although these quantum mechanical descriptors cannot be used to predict biological activity because they do not address specific binding sites in molecules, the local properties can be used analogously to generate alignment-free models, enabling their use in traditional SAR studies.[110] A very similar approach using surface-integral QSPR models for the prediction of free energies of solvation based on local energy properties was recently described by Ehresmann *et al.* based on the semiempirical AM1 method.[112] A parameterized function of four local properties calculated at the isodensity surface, the molecular electrostatic potential (MEP), local ionization energy, electron affinity, and polarizability, was integrated over a triangulated surface area of the molecule. Data sets of 387 compounds with measured aqueous solubility, 88 compounds with measured chloroform solubility as well as 168 compounds with measured log $P_{\text{octanol/water}}$ were used in this study. The resulting models for aqueous and chloroform solubility as well as log $P_{\text{octanol/water}}$ show good agreement with experiment. The authors point out that there are two fundamental approximations underlying their approach. One is that the target properties can be treated using local contributions at the surface of the molecule and the other that gas phase semiempirical molecular orbital electron densities can be used to derive properties that depend on the presence of a polar medium. Their conclusion is that the results from the surface-integral model derived in their study suggests that both those approximations are relatively well justified. However, the authors also mention that their models are not as accurate as the most reliable approaches currently available, like the SMx models introduced by Cramer and Truhlar.[55]

4.16.4.2.6 Application to absorption, blood–brain barrier penetration, metabolism, bioavailability, and druglikeness

In addition to the prediction of physicochemical properties and toxicity, quantum mechanical descriptors have also been used to derive models for the prediction of compound metabolism. This is a very important area in drug design, since understanding, predicting, and eliminating the metabolic liabilities of a compound especially at the design stage would be a very powerful tool for the medicinal chemist. In a recent study, Sorich *et al.* have described their work on the rapid prediction of chemical metabolism by human UDP-glucuronosyltransferase isoforms using quantum mechanical descriptors based on 523 chemical structures.[113] A combined model using two-dimensional and quantum mechanical descriptors was derived that was capable of predicting substrates of UGT isoforms with approximately 84% success rate.

A further key area of interest for medicinal chemistry is the accurate prediction of the intestinal absorption as a major component of the bioavailability of compounds.[114] It is essential to design compounds that have the appropriate molecular properties that enable absorption, as highlighted by Lipinski *et al.* in his landmark paper in 1997 where he introduced the 'Rule of five.'[115] Although the problem can be addressed with HTS methods, the ideal solution would be a virtual ADME prediction tool that would require no sample. The properties linked to absorption are primarily numbers of hydrogen bond acceptors and donors or polar surface area, lipophilicity in the form of a calculated log P (alog P, clog P), as well as molecular weight. In addition to permeability and intestinal absorption, the absorption of drug molecules into the brain by crossing what is commonly referred to as the blood–brain barrier also constitutes a very important property for drug design. Absorption also requires molecules to have specific physicochemical properties, expressed for example by a certain proportion of polar surface area and a lower limit for lipophilicity in terms of log P and log D. There have been a number of publications in recent years on the generation of models for both, intestinal absorption[116–118] and the blood–brain barrier[119,120] using QSAR approaches and conventional, classical descriptors. Both, intestinal absorption and blood–brain barrier penetration are considered to be passive diffusion processes for most molecules. In contrast, there are also active transport mechanisms known across the same membranes that affect small molecules, and this currently constitutes an area of immense interest for the application of molecular modeling methods, like for example for P-glycoprotein.[121]

Quantum mechanical methods have been applied particularly to the prediction of the blood–brain barrier with relatively good success. In an early publication in 1996, Lombardo *et al.* reported the application of semiempirical AM1 calculations in combination with the SMx model[55] to the prediction of brain–blood partitioning of organic solutes via the calculation of their solvation free energies.[122] The ratio of blood–brain partitioning, $\log(C_{\text{brain}}/C_{\text{blood}})$ (log BB) of a series of 55 compounds, ranging from simple organic solutes to druglike histamine H_2 antagonists, was found to correlate with the free energy of solvation in water. A function was developed for these 55 compounds with a correlation coefficient of $r = 0.82$. The authors indicate that this type of approach is valuable due to its direct practical application in the design and discovery of new therapeutic molecules that are able to penetrate the blood–brain barrier. They also point out that these types of models can serve as a powerful tool for rank-ordering compounds prior to synthesis.[122]

More recently, Hutter reported a new model for blood–brain barrier pentration that also utilized semiempirical AM1 calculations.[123] He was able to describe the polar surface area of compounds with six descriptors derived from the MEP, which showed a strong correlation with log BB for a set of 90 compounds. Additional quantum chemically computed descriptors that also contributed to the final model were the ionization potential and the covalent hydrogen bond basicity, resulting in a model using 12 descriptors based on 90 compounds with $r^2 = 0.76$.

In addition to understanding and predicting absorption and bioavailability, it would be of interest to medicinal chemists to have computational methods that would be able to differentiate between druglike compounds and those that are not druglike. This could be very important approach for example to scan compound libraries prior to synthesis or before purchasing libraries from commercial vendors. This area of druglikeness filters has also been investigated extensively with some success. The most common approach to this problem has been to use some type of molecular descriptors linked with a pattern of recognition or interpolation technique, such as neural nets, to distinguish between a data set of drugs and one of nondrugs.[124–126] However, quantum mechanical calculations have only very recently been used to differentiate between drugs and nondrugs.[111] This work indicates that semiempirical quantum mechanical descriptors (AM1) in combination with a neural network approach can be used successfully to differentiate drugs from nondrugs in a set of more than 42 000 compounds.

4.16.4.2.7 Applications of molecular electrostatic potentials

In addition to properties derived from quantum mechanical calculations like for example atomic charges, HOMO and LUMO energies, etc., the MEP is a property that is also easily derived from quantum mechanical calculations and can be used to provide insights into structure–activity relationships and molecular properties, including binding interactions. It reflects the potential at a certain point on the electron density surface of a molecule when interacting with a positive charge at a certain distance from the atoms in the molecule. The MEP is a very widely applicable property that can be used to rationalize and predict molecular interactions, reactivity as well as molecular properties. The application of MEPs in medicinal chemistry has been reviewed recently by Murray and Politzer highlighting early applications from the late 1980s and early 1990s on the rationalization of carcinogenicity of halogenated olefins and their epoxides as well as toxicity of dibenzo-*p*-dioxins and analogs.[10] Also, MEPs have been used in a 3D-QSAR analysis on 37 benzodiazepines, and good-quality models ($r^2_{cv} = 0.76$) were derived even at the semiempirical level compared to higher level ab initio HF methods.[127] MEPs have also been applied to rationalize cation–π interactions like for example the binding of actylcholine, and have been found to be a useful guide for the prediction of these types of interactions.[128]

More recently, electrostatic potential maps have been used in the structure-based design of new p38 mitogen-activated protein kinase (MAPK) inhibitors in order to understand the hydrogen bond capabilities of various heterocycles and the long-range electrostatic effects of their dipoles moments and MEPs on the binding process in the protein.[129] The aim of this work was to replace an undesirable benzimidazolone group with other heterocycles with better overall properties but similar hydrogen bond acceptor functionality, which would potentially lead to improved physicochemical properties in the newly designed molecules overall. The MEPs were calculated using DFT and their hydrogen bond potential was compared to water as a benchmark. From the results two heterocycles, benzotriazole and triazolopyridine, were identified as displaying superior electrostatic characteristics for hydrogen bonding compared to water for binding to the ATP binding site of p38 MAPK. However, these two heterocycles were also found to have the highest calculated solvation energies, which indicates that they might have a higher desolvation penalty upon binding, therefore counteracting the advantage gained from the increased hydrogen bond capability. Nevertheless, as a result of these studies, significantly more potent, atypical kinase inhibitors were identified that have superior calculated physicochemical properties compared to the parent compound.

4.16.4.3 Applications of Quantum Mechanical Calculations to Reactivity and Enzyme Mechanisms

Understanding the chemical reactivity of proteins as well as small molecules and enzyme mechanisms are key aspects in medicinal chemistry for a variety of reasons. For example, a better understanding of chemical reactivity can shed light on the behavior of specific molecules in organic synthesis, enabling rationalization and optimization of reaction conditions, solvents, and therefore potentially also yields from those reactions. Although a potentially very useful application of quantum chemistry, this area was considered largely out of scope for this chapter for two reasons. One was the sheer volume of publications on quantum mechanical calculations of organic reactions, which would easily fill a separate chapter. The other reason is that this chapter focuses mainly on the application to medicinal chemistry and the design of compounds, not on the actual synthesis of molecules. However, we will include examples of chemical reactivity when it is considered relevant to molecules or classes of molecules that are of interest to current pharmaceutical therapies of human diseases.

The detailed understanding of enzyme mechanisms can be an essential component in the design of small molecule inhibitors. Therefore, studying enzyme mechanisms has attracted the attention of many researchers for decades. Early examples during the 1980s utilized ab initio calculations with small basis sets focusing on individual steps in the reactions, rather than entire reaction mechanisms due to the limited computer power available at the time.[130,131] In the late 1980s, the newly available semiempirical AM1 method was utilized to model for the first time entire reaction mechanisms using enzyme model systems explicitly incorporating some active site residues, notably for carbonic anhydrase[132] and carboxypeptidase A.[133] The status of computational studies of enzyme-catalyzed reactions has been reviewed extensively by Kollman *et al.*[134] and Noodleman *et al.*[135] more recently.

Since those early examples, there have been a large number of applications of QM/MM methods to the study of enzyme mechanisms that are relevant to problems in medicinal chemistry and the area has already been reviewed extensively.[9] Due to the sheer number of publications in the literature, we will only be able to review some selected examples in this chapter, but we will refer to further examples that will not be discussed in detail here toward the end of this section.

4.16.4.3.1 Applications to beta-lactamase, elastase, and hydrolysis reactions

In order to understand the reaction mechanism of beta-lactamases, Massova and Kollman have studied the effect of solvation on the barriers of reaction and the stability of transition states and reaction intermediates for the nucleophilic attack on the carbonyl group in beta-lactams.[136] As a background to the relevance of this work for medicinal chemistry, a short introduction on the beta-lactamases is needed at this point. The beta-lactamases are a class of enzymes that are of fundamental importance for the discovery of new antibiotics. The antimicrobial function of the beta-lactam antibiotics is determined by their ability to bind specifically and covalently to enzymes in bacteria that synthesize cell walls (penicillin-binding proteins, PBPs). These PBPs catalyze the cross-linking reaction of the peptidoglycans, an important structural component of the bacteria cell wall, rendering it into a three-dimensional rigid network. This rigid cell wall is what allows the bacteria to survive in a harsh environment. Deprived of these essential PBPs, bacteria cannot sustain the rigidity that is crucial to the stability of their cell walls, which means that they can be ruptured by osmotic forces, thereby killing the bacteria. In order to protect themselves against beta-lactam antibiotics, such as penicillins and cephalosporins, bacteria have developed beta-lactamases as a defense mechanism. These enzymes protect the bacteria from beta-lactam antibiotics by hydrolyzing their beta-lactam ring and converting them into inactive compounds. The products of beta-lactam hydrolysis by beta-lactamases have no noticeable affinity for PBPs, and therefore cannot interfere with the stability of bacterial cell walls. Thus, beta-lactamases effectively destroy the antimicrobial function of the antibiotic drug and provide the bacteria with very efficient resistance to it. One of the major problems with respect to resistance is the widespread use of beta-lactam antibiotics in hospitals, agriculture, and household products, which has caused the emergence of new multiple antibiotic resistant (MAR) strains of bacteria that carry beta-lactamase genes.[137,138] Nevertheless, beta-lactam pharmaceuticals still remain extremely useful for many infections.[139] One of the potential solutions to the MAR resistance problem is to develop new classes of beta-lactamase inhibitors, which can then be administered together with the antibiotics to protect them from hydrolytic inactivation by beta-lactamases.[140,141] Therefore, extensive knowledge about the mechanism of beta-lactamase catalysis is very important for understanding the function and interactions of these enzymes with known substrates and inhibitors. This kind of information would then assist in the development and evaluation of novel beta-lactams to try and overcome the resistance problem (for a review of antibacterial resistance to beta-lactam antibiotics see the review by Fisher *et al.*[142]) In order to shed light on the mechanism of beta-lactam hydrolysis quantum mechanical calculations were carried out at the ab initio HF level and solvent effects were included using the polarizable continuum dielectric model mentioned in Section 4.16.3.7. The mechanism for lactam hydrolysis is shown in **Figure 11**. The calculations were shown to reproduce the observed 30–500-fold increase in rates of hydrolysis of beta-lactams compared to the corresponding acyclic amides. The results also show that the length of the C=O bond in the lactam rings correlates well with the calculated activation barrier of the first step in the hydrolysis reaction, which is the attack of for example a hydroxide or methoxide ion on the carbonyl bond of the lactam ring. More importantly, the calculated reaction barrier or activation energy also correlates with the experimental second-order rate constants for hydroxide ion catalyzed hydrolysis, and therefore with beta-lactam hydrolytic stability. This information can be used as criteria for evaluation of the susceptibility of beta-lactams toward hydrolysis and, as a consequence, toward hydrolysis by beta-lactamases. This may then also relate to the lack of beta-lactam inhibitory activity with PBPs, causing resistance. The work also sheds light on the contributions of the various structural elements in the lactam ring and attached substituents to the reactivity of beta-lactams, information that can prove useful in the design of novel potent antimicrobials.[136]

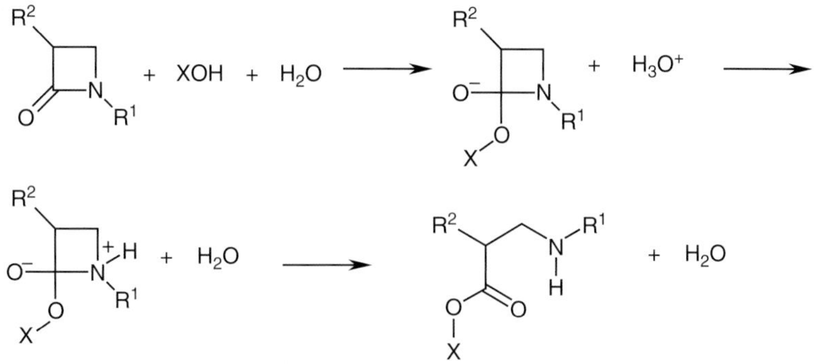

Figure 11 Simplified reaction scheme for the hydrolysis of beta-lactams.

Diaz and co-workers have also recently published a mechanistic study that typifies the modern paradigm of computational medicinal chemistry. They have investigated the problem of penicillin resistance from a slightly different angle. The purported mechanisms for benzyl penicillin acylation of class A TEM-1 beta-lactamase (PDB entry 1BTL) follows a number of pathways. To investigate these pathways, the relevant conformation of the reactive part of the enzyme that is involved in binding and reaction with beta-lactams, was optimized using semiempirical QM/MM (PM3/AMBER, with 66 atoms in the QM area). The Ser70 residue was considered essential to the proper catalytic activity, reacting with the carbonyl group in the penicillin. The target penicillin was optimized at the B3LYP/6-31 + G* level, and transition states for the reaction pathways were computed at MP2/6-31 + G* as well as DFT B3LYP levels. Short-range solvent effects were treated with explicit solvent molecules where practical, while the rest of the solvent effects were explicitly included in the QM/MM treatment, but approximated using SCRF in the ab initio treatment. The authors concluded from their results that the acylation of class A beta-lactamases by penicillin proceeds through a hydroxyl- and carboxylate-assisted mechanism.[143]

In a related study on the reactive mechanism of ester hydrolysis in solution rather than catalyzed by a specific enzyme, Chaudry and Popelier used quantum mechanical descriptors derived from HF and DFT calculations to predict experimental rate constants.[144] The rate of hydrolysis of esters can potentially be of importance since esters are relatively reactive groups that can be introduced into so-called prodrugs, compounds that are designed to release an active compound after an activation step, like for example hydrolysis, when administered. Chaudry and Popelier found that particularly the bond length in the carbonyl and C–O single bond in ester groups are the most significant descriptors for rationalizing the rate of hydrolysis of esters, yielding an excellent QSAR model with r^2 of 0.969 and q^2 of 0.948. Since the bond parameters were calculated with ab initio methods, these models are very likely to be predictive for molecules outside the training set, and are therefore useful in the design of new compounds and potential prodrugs.

An important question related to the work described above on the hydrolysis of lactams and esters is whether one can use quantum mechanical methods to predict the potency of reactive ligands when they are involved in covalent interactions with enzymes. This is of particular relevance for the design of enzyme inhibitors, for serine proteinases, and particularly elastase, where so-called mechanism-based inhibitors have been characterized. It is believed that some of these inhibitors, particularly activated ketones, form tetrahedral intermediates with the protein through reaction with a catalytic serine residue and therefore it was assumed that the electrophilicity of those ketone groups present in inhibitors could be influencing the potency of compounds.[145] The term 'mechanism-based inhibitor' describes in this context those compounds whose mode of action is based on or closely linked with the mechanism of the enzymatic substrate hydrolysis and which inactivate an enzyme by forming a covalent adduct with at least one of the catalytic residues within the protein active site. In addition to the irreversible suicide inactivators, reversible and transition-state analog inhibitors are also included in this definition. In this context, Edwards *et al.* have reported an important relationship of Hammett substituents, particularly σ-values (for an extensive review of Hammett substituent constants see the review by Hansch *et al.*[146]) with the biological activity of a series of activated ketone-containing inhibitors of human neutrophil elastase.[147] The type of elastase inhibitor described in this publication is shown in **Figure 12**.

The authors find that the potency of elastase inhibitors increases with increasing σ-values of the substituents attached to the carbonyl group of the inhibitor, which is believed to interact with the catalytic serine in elastase. However, when applying this relationship to the design of new compounds, it becomes immediately apparent that σ-values are only available for a relatively small number of for example hetercycles,[148] which then significantly limits the usefulness and scope of any property–activity relationship beyond a relatively small set of compounds. In order to

Figure 12 Series of elastase inhibitors used for the correlation with σ-values.

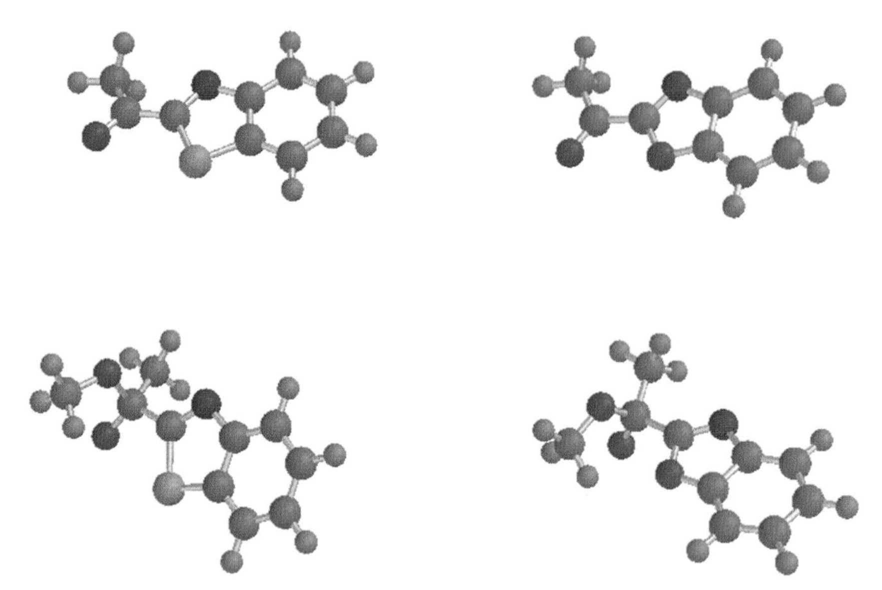

Figure 13 Benzthiazole and benzoxazole derivatives and their products after reaction with a methoxy-anion from AM1 calculations.

eliminate this restriction, we have attempted to find a mechanism that would allow the accurate prediction of σ-values through relatively inexpensive semiempirical quantum mechanical calculations. In addition to semiempirical calculations, DFT calculations could also be used in this context, whereas ab initio HF calculations are not suitable for anionic species, unless very extensive basis sets are used. For the purpose of the modeling, the mechanism by which the ligand interacts with the enzyme has been described as a nucleophilic attack of the catalytic serine residue on the carbonyl group of the ligand. The feasibility of this approach was initially tested using a methanolate ion (a model for the catalytic serine) as the nucleophile attacking a compound that constitutes an inhibitor with a central carbonyl group, substituted with a methyl on one side and on the other side with a benzoxazole or benzthiazole, respectively. We found that there was a significant difference in the calculated AM1 reaction enthalpy between the two model compounds leading to the product of nucleophilic attack. The reaction between the methanolate and the benzoxazole compound was significantly more exothermic (168.6 kJ mol^{-1}) than the reaction for methanolate with the benzthiazole compound (146.2 kJ mol^{-1}), indicating an increased reactivity of the carbonyl group when attached to a benzoxazole compared to a benzthiazole. This result also reflected their relative order of Hammett σ-values of 0.41 and 0.37, respectively. The reactants and products after nucleophilic attack of methanolate are shown in **Figure 13**.

Having established a potential link between the calculated reaction enthalpy and the Hammett σ-values, we then calculated the AM1 reaction enthalpies for 27 compounds for which experimentally derived σ-values were available in the literature.[148] All compounds contained heterocyclic groups attached to the carbonyl in the same way as described above. The resulting correlation is of excellent quality ($r = 0.96$, $r^2 = 0.92$, $r^2_{cv} = 0.90$), and the graph for the regression analysis is shown in **Figure 14**.

This result clearly indicates that it is possible to accurately predict σ-values from calculated semiempirical AM1 reaction energies for simple model systems, for example if experimental values are not available for a particular substituent. A further key advantage of this method is that the calculated σ-values fit in very well with the existing scale of Hammett σ-values, and can be used to augment that scale. In order to take this study one step further and

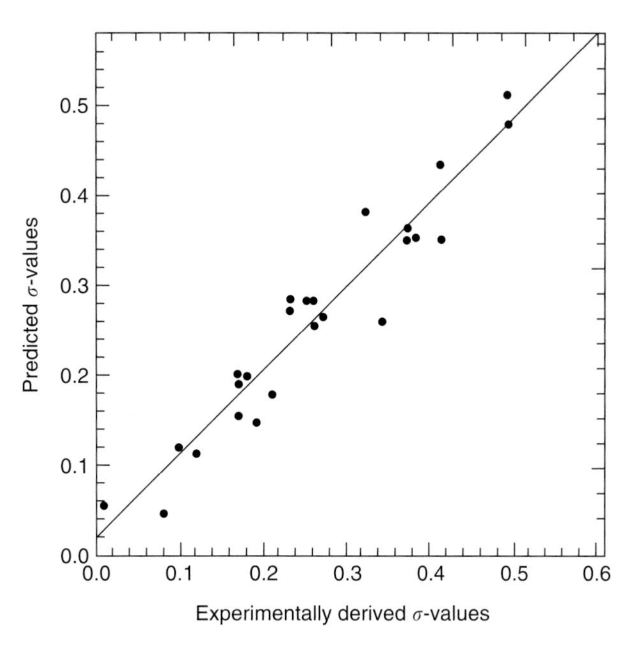

Figure 14 Graph representing the correlation of experimental and predicted σ-values based on AM1 calculated reaction enthalpies for 27 carbonyl compounds.

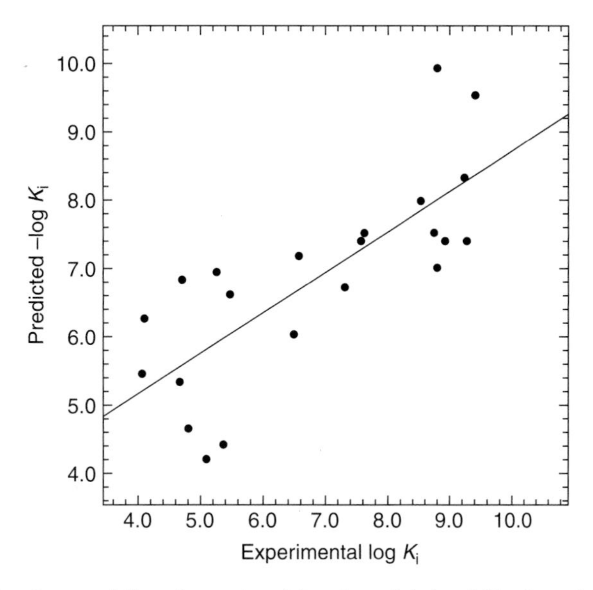

Figure 15 Graph representing the correlation of experimental and predicted activities based on calculated σ-values for 22 elastase inhibitors.

establish the validity of this method for real enzyme systems, we have applied this approach to a QSAR study on a series of 22 elastase inhibitors[147] with a wide variety of substituents attached to the reactive carbonyl group. A very reasonable correlation was found between experimental activity and calculated σ-value ($r = 0.77$, $r^2 = 0.59$, $r^2_{cv} = 0.51$). The resulting graph from the regression analysis is shown in **Figure 15**. This work clearly indicates that it is possible to use quantum mechanical calculations to predict binding affinities of certain classes of mechanism-based enzyme inhibitors based on their calculated reactivities, making quantum mechanical calculation a useful tool in the context of compound design. Furthermore, based on this work above, we were able to identify novel, mechanism-based serine proteinase inhibitors with increased potency for a closely related enzyme target.

4.16.4.3.2 Applications to Cdc-42-catalyzed GTP hydrolysis, uracil-DNA glycosylase, aldose reductase, glutathione S-transferases, neuraminidases, and platelet-derived growth factor receptor (PDGFR)

Quantum mechanical calculations have also been applied to enzyme reaction mechanisms that are linked to diseases. For example, Cavalli and co-workers have shown applied DFT methods to the phosphoryl transfer reaction in the Cdc-42-catalyzed GTP hydrolysis.[8] This type of reaction is relevant since it is believed to be involved in some pathology like cancer and Alzheimer's disease.[149,150] In biological systems this reaction is catalyzed by several different enzymes, e.g., hydrolases, kinases, and nucleoside monophosphate kinases. The calculations reveal that the key step in the reaction mechanism is the formation of a bond between the oxygen of the catalytic water and the γ-phosphorus atom of GTP leading to GDP and inorganic phosphate, rather than an initial deprotonation step of the catalytic water by the γ-phosphate. These findings are fully consistent with x-ray data and site-directed mutagenesis data.[8]

Dinner *et al.*[151] have recently studied uracil-DNA glycosylase (UDG), and have provided detailed insight into the catalytic mechanism of the enzyme. UDG excises uracil from DNA, where they are either a result of misincorporation of deoxyuridine or deamination of cytosine. In this work, the QM region was treated by using the semiempirical AM1 method. In contrast with the concerted associative mechanism proposed initially, the calculations showed that UDG catalyzes base excision by a stepwise dissociative mechanism. Interestingly, the corresponding solution reaction occurs via a concerted mechanism in contrast to the reaction in the enzyme. The calculated activation energies for the reaction of 62.3 kJ mol^{-1} in the enzyme and 144.3 kJ mol^{-1} in solution is in good agreement with the corresponding measured activation enthalpies of 50.6 kJ mol^{-1} and 134.3 kJ mol^{-1}, respectively.

Human aldose reductase (ALR2) is a monomeric, NADPH-dependent oxidoreductase catalyzing the reduction of a wide variety of carbonyl compounds to the corresponding alcohols. It is known that under hyperglycemic conditions it participates in the polyol pathway to reduce D-glucose to D-sorbitol. It has been found that when tissues contain a high level of glucose, sorbitol builds up and apparently damages the membranes lining body tissues. Because this process is thought to be one of the factors contributing to diabetic neuropathy, a nerve disorder caused by diabetes, much effort is under way to develop effective inhibitors.[152] Although crystal structures and kinetic data reveal some aspects of the reaction mechanism of ALR2, details of the catalytic step are still unclear. The mechanism has been shown to involve stereospecific hydride attack from NADPH on the carbonyl carbon of the substrate, and the protonation of the carbonyl oxygen from the nearby proton donor residue in the active site. Ternary complex crystal structures enabled identification of a specific anion-binding site in the active site formed by NADPH, His110 and Tyr48, suggesting that the reaction proceeds through a negatively charged species. Semiempirical calculations were performed by Varnai *et al.*[153] to elucidate whether the protonation of the carbonyl oxygen of the substrate precedes or follows hydride attack from NADPH on the carbonyl carbon of the substrate, and the identity of the proton donor to the carbonyl oxygen. The nature of the proton donor in ALR2 is of great importance in understanding the catalytic mechanism of the enzyme. Potential proton donor residues in the active site are Tyr48 (part of a conserved triad Tyr48-Lys77-Asp43) and His110. Several reaction pathways were investigated on the basis of two models in which either Tyr48 or protonated His110 act as proton donor. The catalytic step of the reduction of the substrate, D-glyceraldehyde, to glycerol was analyzed first in model calculations on fragments of the active site using the semiempirical AM1 and PM3 as well as HF/3-21G* methods to obtain information about the catalytic region utilizing a QM/MM approach. The simulated system was an 18 Å radius sphere containing 174 protein residues, the cofactor, the substrate, and 611 water molecules, including water molecules from the crystal structures plus addition of solvent water molecules by superimposition of a 20 Å radius sphere of pre-equilibrated water molecules. The quantum mechanical region consisted of the substrate, the nicotinamide ring of the cofactor, and the side chains of residues Tyr48, His110, Lys77, and Asp43. The rest of the system was treated by molecular mechanics force field. The results showed that the substrate binds to the enzyme by hydrogen bonding in an orientation that facilitates the stereospecific catalytic step in both models. The catalytic mechanism with Tyr48 as the proton donor proceeds through a negatively charged intermediate, in agreement with experimental results, but has a high activation energy. In this case hydride transfer occurs before protonation. The calculations indicate that the His110 in the protonated form is a better proton donor than Tyr48, the lowest energy pathway in this mechanism proceeds via a protonated intermediate, a conclusion that is apparently not in agreement with experimental data. With His110 as the proton donor, the lowest-energy pathway has the protonation step preceeding hydride transfer. According to these calculations, if His110 is present in its protonated form in the native complex it is the energetically favored proton donor compared with Tyr48 in the active site with neutral His110.

The family of glutathione *S*-transferases (GST) has an important place in the large array of biotransformation enzymes that metabolize and detoxify drugs and other xenobiotics. Biotransformation enzymes are of increasing toxicological and pharmacological interest, because they determine to a large extent how fast, and via which metabolic pathway, xenobiotics are metabolized. In general these enzymes have broad substrate specificities, are present as multiple classes of isoenzymes, often subject to polymorphism, and sometimes (e.g., the enzymes cytochrome P450 and GST) catalyze multiple types of reactions.[9] In a recent application,[154] the conjugation of glutathione to phenanthrene 9,10-oxide, catalyzed by GST M1-1 from rat, was studied by QM/MM-based umbrella sampling. Phenanthrene 9,10-oxide is the model substrate for epoxide ring opening reactions by GST. Some aspects of this reaction are of particular interest. First, as is often the case for biotransformation enzymes, the active site is highly solvent-accessible and proper inclusion of solvent effects is required for accurate modeling. In this example solvent was included explicitly, by use of a stochastic boundary approach. A second aspect of interest is the use of a genetic algorithm to calibrate the semiempirical AM1 treatment of the QM region. The QM region of the GST model contained the thiolate sulfur of glutathione, which is central in the conjugation reaction. Sulfur is a versatile, and therefore difficult, element in the context of semiempirical methods. In this specific case it was shown that parametrization of just the sulfur, keeping the standard AM1 parameters for the other elements present in the system, could significantly improve the accuracy of the results for the conjugation reaction studied. Free energy profiles for the conjugation of glutathione to the epoxide moiety of phenanthrene 9,10-oxide were obtained by sampling along an approximate reaction coordinate, which was represented by the difference between the $S_{thiolate}$–$C_{epoxide}$ and $C_{epoxide}$–$O_{epoxide}$ distances. The barriers in the calculated free energy profiles agree with the experimental rate constant for the overall reaction, which supported the QM/MM model of this reaction and which confirmed the epoxide ring-opening step to be the rate-limiting in the enzyme-catalyzed reaction. The model was analyzed to obtain detailed insight into several aspects of the reaction. First, an atomic structure for the transition state was obtained as an average structure of the dynamic trajectory restrained to the top of the energy profile. The transition state structures are indicative of interactions with key active-site residues, indicating important catalytic effects. This gave valuable information to supplement insight obtained from x-ray structures. Analysis of hydrogen bonding by solvent molecules along the reaction coordinate indicated a significant change in solvation of both, the thiolate moiety of glutathione and the epoxide oxygen of phenanthrene 9,10-oxide, indicating that solvent effects have a dramatic effect on the energetics of the reaction, in agreement with experimental results. Finally, approximate effects of mutations were established. One mutation (Asp8Asn) represents a difference between two isoenzymes (M1-1 and M2-2) with markedly different stereoselectivity with respect to the products formed. The modeled mutation seemed to have a differential effect on the barriers toward these products, which suggests that this mutation is an important determinant of different diastereoselectivity between the isoenzymes. This illustrates the potential of QM/MM modeling in enabling understanding of the phenotypical consequences of genetic variation.

Neuraminidases are enzymes present in viruses, bacteria, and parasites. They are implicated in serious diseases such as cholera, meningitis, and pneumonia. Neuraminidase from influenza virus aids the transmission of the virus between cells and maintains viral infectivity. In different strains of influenza several amino acids are conserved, especially in the active site, giving rise to hopes of finding a single inhibitor (and so a drug) for all the neuraminidase enzymes from influenza strains. A crucial question is whether a covalent bond is formed between the enzyme and the reaction intermediate. Thomas *et al.* investigated the reaction catalyzed by neuraminidase from influenza virus using QM/MM calculations.[155] The calculations found that there was no covalent intermediate formed in the viral neuraminidase reaction and the intermediate was more likely to be hydroxylated directly. Because there is only a small energy difference between the two options, formation of a covalent bond or direct hydroxylation, the authors proposed it might be possible to design novel inhibitors that covalently bound to the enzyme.

Guha and Jurs also developed a number of linear, nonlinear, and ensemble models utilizing quantum mechanical descriptors based on the semiempirical PM3 method for the prediction and interpretation of the biological activity of a set of 79 PDGFR inhibitors.[156] They derive models with excellent predictive capabilities, which differentiate clearly between active and inactive compounds.

4.16.4.3.3 Application of quantum mechanical calculations to the stereoisomerism of a nicotinic acetylcholine receptor ligand

Recently, Hammond *et al.* investigated the protonation-induced stereoisomerism in nicotine using molecular mechanics conformational studies and ab initio CP-MD calculations.[157] Nicotinic acetylcholine receptors (nAChRs) are ligand-gated ion channels (LGIC) that influence a variety of biological functions and are believed to play important roles in various neurological and mental disorders. Compounds that modulate these receptors (nicotine agonists and antagonists) are under investigation as potential therapeutics for pain, cognitive deficits, Alzheimer's and Parkinson's diseases, schizophrenia, ulcerative colitis, anxiety, depression, and Tourette's syndrome among others.[158] Ligands known to modulate a variety of neuroreceptors of importance often contain procationic centers of the type illustrated in nicotine. As is the case for many other druglike compounds containing an asymmetrical tertiary amine, protonation of

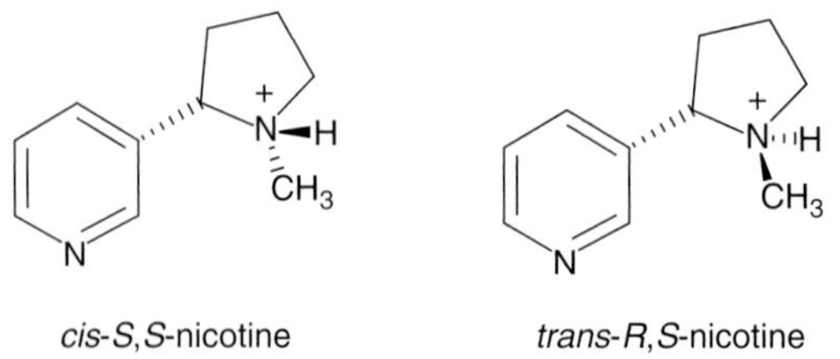

cis-S,S-nicotine *trans-R,S*-nicotine

Figure 16 Diastereomeric structures of protonated *S*-nicotine.

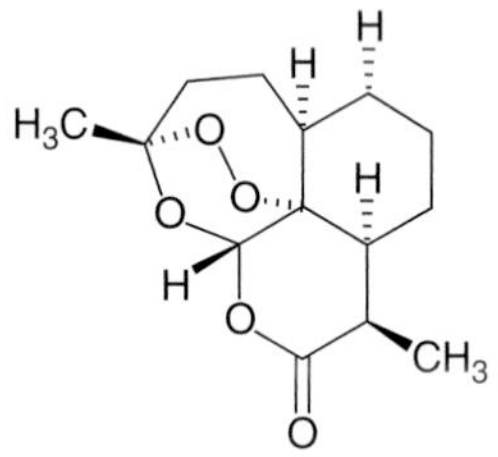

Figure 17 Structure of artemisinin.

nicotine results in chiral induction. For nicotine, this protonation-induced stereoisomerism leads to two different diastereomers, each presumably having different receptor or host binding properties. The structures of nicotine are shown in **Figure 16**. Whether one or both of these diastereomers contribute to the binding affinity of nicotine to nAChRs is not known. The crystal structure of protonated nicotine shows that the *trans* form exists in the solid phase,[159] whereas NMR experiments suggest that the *cis* form predominates in solution.[160] The results of the quantum mechanical calculations involving MD show that protonation-induced isomerism produce significant differences with regards to the relative geometric relationship of the cationic center and the pyridyl nitrogen atom. The calculations around a single conformer provide the potential energy surface and the curvature of the energy surface around that single point, while relative distributions from MD provide a general picture of the free energy surface, with the explicit inclusion of entropy. These differences have implications for the nAChRs via both, conformational preference and electronic effects, through ligand–receptor π-cation interactions that are likely to influence the behavior of each diastereomer in its interaction with the target protein. Also, the results clearly indicate that there is a significant difference in the frequency of occurrence of conformers between the *cis* and *trans* form of protonated nicotine, indicating entropic differences that are not captured in calculations of static geometries representing a single conformer. This could have significant implications for the binding affinities of both diastereomers of nicotine to nACHR.[157] However, there does not seem a clear indication as to the predominant diastereomer for protonated nicotine based on these calculations.

4.16.4.3.4 Application to the discovery of new antimalarial compounds

In the next paragraph, we will focus on work around artemisinin, a compound that was isolated in 1972 from an herb (quinghao, *Artemisia annua*) and that was shown to have significant antimalarial activity. This is very significant due to the fact that artemisinin is structurally very different from the standard family of antimalarial drugs, which are based on quinine and its synthetic analogues.[161] In a recent QSAR study of 179 artemisinin analogs Guha and Jurs[162] find that the HOMO–LUMO gap energy is an important descriptor for the accurate description of biological activity. The fact that electronic parameters seem important is perhaps not surprising since the antimalarial activity of artemisinin is linked to the reactivity of its peroxide linker group. The structure or artmisinin is shown in **Figure 17**. Although artemisinin showed high activity in early clinical studies against both drug-resistant and drug-sensitive malaria, it has a low solubility in water and nonoptimal pharmacokinetic properties, reducing its effectiveness as a drug. Therefore, many efforts have been made to modify the chemical structure of artemisinin and to create analogs that would have better potency and better pharmacokinetic properties compared to the parent molecule, while retaining its biologically crucial functionality.[163] Artemisinin was found to generate a carbon-centered free radical intermediate in the parasite on interaction with heme iron by opening the endoperoxide functionality of the molecule.[164,165] A recent study by

Rafiee *et al.* investigated the effect of substitution on the charge density changes and on the nuclear quadrupole coupling constants (NQCC) of ^{17}O in artemisinin and some of its derivatives.[166] Based on their results, they suggest new analogs of artemisinin substituted at the carbon atom on the seven-membered ring next to the peroxide bridge with absolute *S* configuration, that are expected to yield improved antimalarial activity.

4.16.4.3.5 Application to a binding model for α_1-adrenoceptors

Based on quantum mechanical calculations, we have recently developed a hypothesis for the binding of a new series of nonbasic α_1-adrenoceptor antagonists. The α_1-adrenoceptors are a member of the alpha family of receptors, which belongs to the family of G-protein coupled receptors (GPCRs). There are several functional subtypes of the α_1-adrenoceptor family known (A, B, D, and L). Members of this receptor family are relevant for the treatment of hypertension and also benign prostatic hyperplasia (BPH). Some of the known α_1-antagonists contain a 2,4-diaminoquinazoline core structure.[167–169] For this type of compound a binding hypothesis has been developed which was based on the interaction of the protonated form of the 2,4-diaminoquinazoline core ($pK_a \sim 7$), with an aspartic acid in the alpha adrenoceptor, as shown in **Figure 18**.[67]

The compounds are believed to protonate on N_1 of the diaminoquinazoline ring based on a small-molecule x-ray structure of an analog.[167] The basis for this hypothesis was the apparent similarity in the pharmacophores of the protonated 2,4-diaminoquinazoline and the endogenous ligand, norepinephrine, as shown in **Figure 19**.[167]

During an in-house drug discovery program that aimed at finding an α_{1L}-selective compound for the treatment of BPH, it was discovered that changing the 4-amino group to a hydroxyl group led to potent compounds with enhanced selectivity for the α_{1L}-receptor subtype. The two substituted quinazolines are shown in **Figure 20**.

However, due to the significantly lower pK_a of the 4-hydroxyl compound ($pK_a \sim 4$), protonation as in the 4-amino derivative is less likely and the receptor binding model requiring a hydrogen bond donor functionality interacting with an aspartic acid is not feasible. This brought into question the validity of the binding hypothesis and therefore the rational for the design strategy for α_1-antagonists. However, the hydroxyl derivative can adopt several pyridone tautomeric forms, one of which would have an NH group pointing in the direction of the aspartic acid in the receptor when overlaid onto the protonated form of the 2,4-diaminoquinazoline. The three tautomeric forms considered in the quantum mechanical calculations at the DFT B3LYP/6-31G* level using Spartan'02[170] are shown in **Figure 21** together with their electrostatic potentials.

The calculations indicate that for the gas phase, tautomer 1 is clearly the most stable, whereas in water tautomer 3 is the most stable by 16.6 and 18.4 kJ mol^{-1}. Therefore, the binding hypothesis for the interaction with an aspartic acid

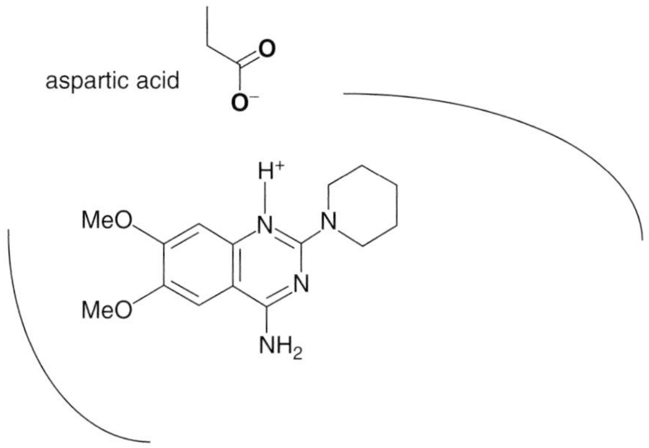

Figure 18 Schematic of the binding hypothesis for 2-4-diaminoquinazolines in α_1-adremoceptors.

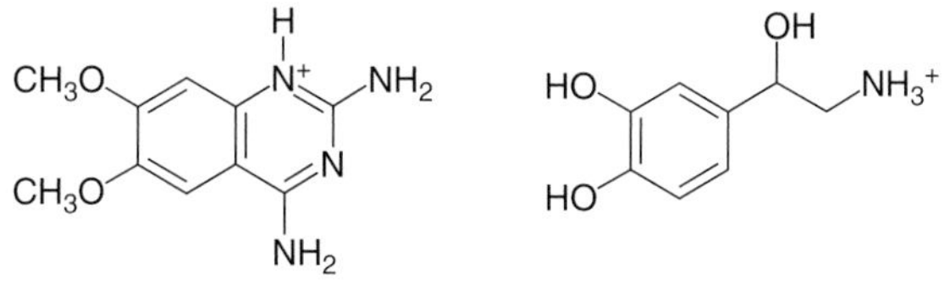

Figure 19 Structural similarity of the protonated forms of 2,4-diaminoquinazoline (left) and norepinephrine (right).

Figure 20 Protonated 2,4-diamino-6,7-dimethoxyquinazoline and 2-amino-4-hydroxyl-6,7-dimethoxyquinazoline.

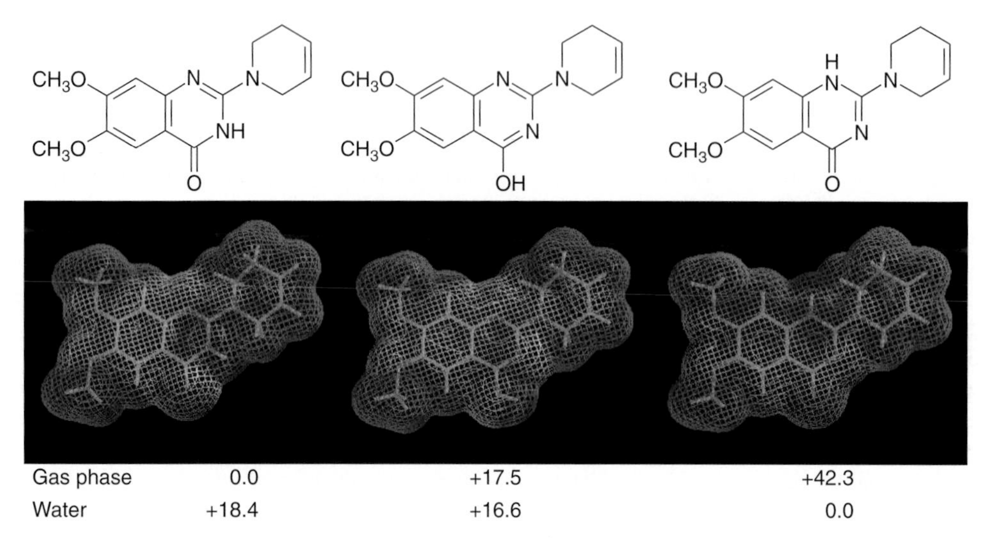

Figure 21 Three tautomeric forms and relative energies in kJ mol^{-1} of the 2-amino-4-hydroxyquinazoline model compound considered in the quantum chemical calculations (DFT B3LYP/6-31G*, solvent calculations using SM model in Spartan'02).

Figure 22 Binding hypothesis for tautomer 3 of the 4-hydroxyl-2aminoquinazoline.

in the receptor is still valid, although the 4-hydroxy derivative due to its low pK_a of around 4 is not significantly protonated at physiological pH, as shown in **Figure 22**.

The quantum mechanical calculations were able to verify the existing binding hypothesis and they have rationalized why nonbasic aminoquinazolines are able to bind to the α_1-receptor in a similar way to their protonated diaminoquinazoline analogs. In addition, the results pointed out new opportunities for the design of novel α_1-antagonists.

4.16.4.3.6 **Applications to protein kinases**

A protein family that is of considerable interest for the development of anticancer drugs are the cyclin-dependent kinases (CDKs). They are serine threonine kinases consisting of a structurally characteristic catalytic subunit complexed to an activating cyclin that each function as key regulators during progressive phases of the eukaryotic cell

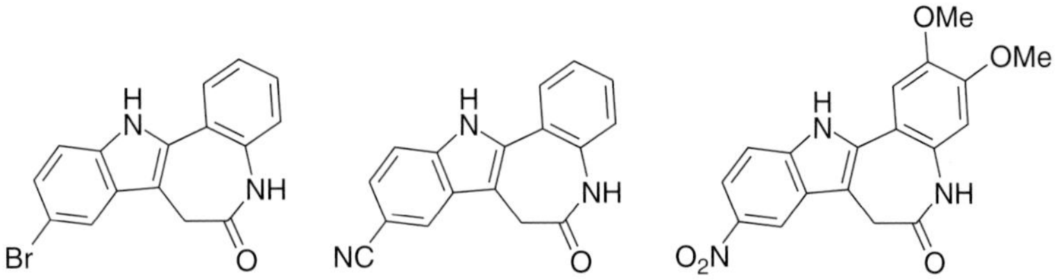

Figure 23 Structures of the CDK-1 inhibitor kenpaullone (left) and the designed inhibitors 9-cyanopaullone (middle) and 2,3-dimethoxy-9-nitropaullone.

division cycle.[171] They are attractive targets for the development of cancer therapeutics due to the observation that CDK regulators are frequently altered in malignancies. Full activation of CDKs typically results after complexation of the catalytic subunit followed by phosphorylation of a key residue and loss of inhibitory phosphorylations near the ATP binding pocket. Inhibition of CDK is of therapeutic interest for the development of anticancer drugs and some CDK inhibitors have entered clinical trials as anticancer agents, like for example the flavone flavopiridol.[172] In a recent study, Gussio *et al.*[173] have performed a 3D-QSAR study using descriptors derived from quantum mechanical calculations that led to the design of novel, active CDK inhibitors. The work focused on the optimization of the scaffold of the CDK inhibitor paullone[174] by combining quantum mechanical descriptions of a congeneric series of paullones to formulate a QSAR for CDK1-cyclin B inhibition data. The descriptors used in the QSAR study were HOMO and LUMO energies as well as dipole moment and electrostatic potentials. The QSAR model was used to predict the activities of paullone derivatives, like for example 9-cyanopaullone, which after synthesis proved to be one of the most active paullones known to date. The structures of 9-cyanopaullone, 9-nitropaullone and 2,3-dimethoxy-9-nitropaullone are shown in **Figure 23**. The authors note that the combination of classical descriptors with molecular mechanics geometries are sufficient to design overt steric and chemical complementarity of the ligands.[173] However, the electronic properties derived from quantum mechanical calculations helped direct synthetic efforts to produce ligands that promote better charge transfer and strengthen hydrogen bonding as facilitated by resonance stabilization. Compounds with low affinity for CDK-1 were poor charge acceptors and made less than ideal hydrogen bonding arrangements with the receptor. These considerations led to the prediction of structures such as 9-cyanopaullone would be considerably more potent than the parent compound, a finding that was supported by enzyme inhibition data. In addition, 2,3-dimethoxy-9-nitropaullone was also predicted to have good CDK-1 activity, which was also confirmed by the enzyme in vitro assay data. Also, 9-nitropaullone emerged as a new paullone derivative, which also had similar potency to known paullones as well as favorable antiproliferative activity profile in living cells.[174] Due to the large amount of current work in this area, it is not possible to discuss all relevant papers on enzyme mechanisms relevant in medicinal chemistry in detail, but they are acknowledged here. They include studies on neuraminidase,[175] human thrombin,[176] and HIV protease.[177]

4.16.4.3.7 **Applications to cytochrome P450 and drug metabolism**

In addition to predicting the interactions with a desired target molecule, quantum mechanical calculations can also be used to rationalize and predict interactions and properties related to undesirable effects, for example those related to absorption, distribution, metabolism, excretion, and toxicity (ADMET). There has been a considerable increase in interest in this area over the last 10 years mainly driven by the clear need to reduce attrition of compounds in drug discovery and development and to improve the number of NCEs reaching the market, improving the productivity in the pharmaceutical industry. There have been significant advances in screening techniques for properties related to attrition like for example inhibition of cytochrome P450s involved in metabolism, providing valuable data for the computational chemist to develop and calibrate computational models, improving their predictive capabilities further. Clearly, the onset of high-throughput screens for ADMET related endpoints has enabled the efficient screening of relatively large numbers of compounds with good turnaround times; however, this does not address the need for improved design methods that enable to 'screen out' undesired effects even before synthesis of a compound is considered. This is where the real value of computational models lies, assisting in the design of compounds that have the desired activity against the biological target of interest and lack the unwanted liabilities related to for example the lack of absorption and cytochrome P450 metabolism as well as inhibition that are potentially going to cause attrition. This is now widely recognized by many medicinal chemists.

Prediction of the metabolism of compounds by cytochrome P450 enzymes has been of interest to medicinal chemists for many years. Several models have been developed based on pharmacophore and structure-based approaches utilizing protein homology modeling as well as quantum mechanical calculations (*see* Chapter 4.31.2.2). De Groot *et al.* have generated models for cytochrome P450 2D6 and 2C9, utilizing all of those methodologies.[70,178] The quantum mechanical calculations applied in this work provided information on the likelihood of hydrogen abstraction and stability of the corresponding radical as well as stability of the hydroxylated potential metabolite resulting from oxidation by the P450 enzyme. Starting from the lowest energy conformation of each compound, determined using the semiempirical AM1 method, all likely radicals, hydroxylated products, and the radical cation were geometry optimized using the AM1 method. For each substrate, the chemically most likely products were compared with the experimentally observed metabolic products, where available. The predictions are predominantly providing qualitative information about the likelihood of a particular position in a molecule being attacked by a cytochrome P450. However, within related chemical groups, for example for oxidations of alkyl or aromatic groups as well as *N*-oxidations, the relative energies obtained from the calculations can be used for direct and quantitative comparison of radicals and products, since they are known to reproduce experimental trends very well, particularly for radical stabilities.

Quantum mechanical calculations have also been applied to understand the oxidative cycle of cytochrome P450s, and provide useful insights into the function and the reaction steps involved in the metabolism of compounds. However, this field of research will not be reviewed in detail and the reader is referred to some recent publications and reviews on the subject,[179–181] with a particular focus on DFT calculations.[182,183]

4.16.4.3.8 Applications to DNA interactions

A further application of quantum mechanical calculations related to medicinal chemistry is related to transition metal containing therapeutics, like for example the recent work by Burda *et al.* on *cis*-platin, *cis*-$[Pt(NH_3)_2Cl_2]$, a well-known anticancer agent.[22] They investigated the thermodynamic and kinetic aspects of hydration reactions of *cis*-platin, and were able to rationalize the hydrolytic activation step of exchange of chloro-ligand(s) with water important for activation of the drug in order to interact with the DNA. In a related study, Lau and Deubel investigated the loss of ammine from platinum(II) complexes using a DFT approach including solvent calculations.[184] They were able to shed light on substitution reactions at platinum complexes using a variety of ligands, and conclude that theory does not support for example a postulated key mechanism in the mode of action of *cis*-platin, the platination of sulfur ligands and subsequent metal release to DNA.[184]

More recently, quantum mechanical calculations were also applied to the rationalization of binding interactions of small molecules with DNA. For example, Xiao and Cushman have described the binding of potential new anticancer agents based on the inhibition of topoisomerase I (top-1) using high-level ab initio calculations.[185] The recent discovery of the cytotoxic agent camptothecin (CPT) has provided a lead for the development of new anticancer drugs and has also led to the validation of top-1 as a chemotherapeutic target.[186] Recently, the ternary complex of top-1 with DNA and a compound structurally related to CPT, topotecan (TPT) has been dertermined.[187] The structures of CPT and TPT are shown in **Figure 24**. This structure has been used as a basis to investigate a variety of possible binding modes for the structurally related compound CPT to DNA. They concluded that the preferred binding orientation determined from the calculations for CPT is consistent with the experimental x-ray structure of TPT in the ternary complex with DNA and top-1. Since the calculations were primarily based on π–π stacking interactions and are capable of predicting binding orientation and binding site selectivity, it was also concluded that hydrogen bonding of the ligand to the surrounding amino acid residues of the protein, or the base pairs, is of minor significance. The authors indicate that the calculations should be applicable to other polycyclic top-1 inhibitors to generate similar models for further structure-based drug design.[185]

In another recent study of small molecule interactions with DNA, Tuttle *et al.* investigated the rationale for biological activity of dynemicin A using a high-level QM/MM DFT approach.[188] The authors claimed that the insights gained in their study of the docking of dynemicin A will be used to design more powerful, nontoxic antitumor leads.

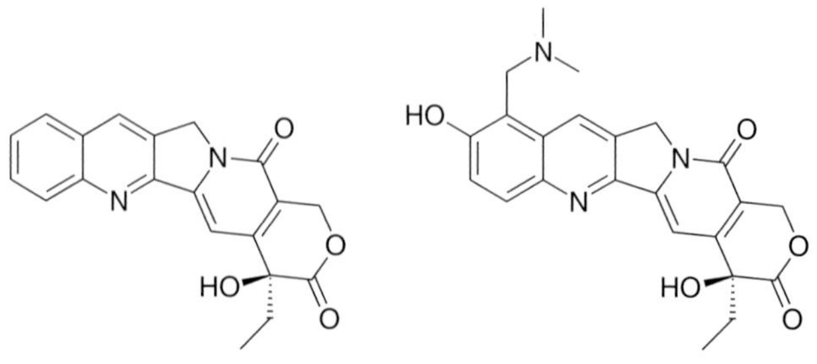

Figure 24 Structures of camptothecin (CPT, left) and topotecan (TPT, right).

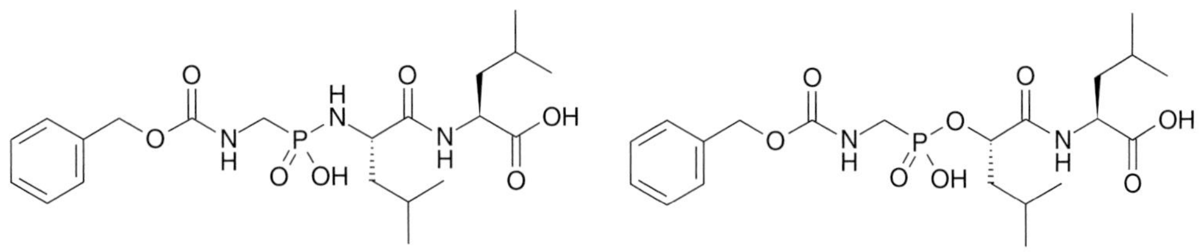

Figure 25 Structures of two inhibitors with virtually identical binding modes (PDB codes 5tmn and 6tmn) in thermolysin but with very different potencies due to an unfavorable electrostatic contact with the protein. The compound on the left has a K_i of 16.5 nM in thermolysin, whereas the compound on the right has a K_i of 13 000 nM. This corresponds to a difference in binding energy of 17.6 kJ mol^{-1}.

4.16.4.4 Calculation of Interactions and Binding Energies of Small Molecules with Proteins

A key requirement for the rational design of new molecular therapies is a clear understanding of the interactions between small molecules and the target receptor(s). This is also one of the most difficult and complex questions that medicinal chemistry is trying to answer as part of the process of successfully designing NCEs. Early work on the semiquantitative estimation of binding constants by Williams *et al.* has provided important insights into the various components of the interaction energy between a ligand and a protein, and has highlighted the enormous complexity that is needed for even the most basic understanding of these interactions.[189] Although very significant progress has been made over the last 25 years in this area of rationalizing and predicting protein–ligand binding energies, a lot of this progress was based on the early development of empirical scoring functions included in ligand docking software programs like for example LUDI,[190,191] FlexX,[192,193] and GOLD,[194] which are based entirely on the analysis of binding interactions like hydrogen bonds and hydrophobic contacts that are observed in protein–ligand x-ray structures (*see* Chapters 4.12.2 and 4.19.2). Although the pool of available protein–ligand x-ray structures is rapidly expanding, the improvements in empirical scoring functions have been rather incremental over the last decade, and are still largely only qualitatively understood.[195] This could be largely due to the fact that most empirical scoring functions are very good at recognizing favorable interactions, but are very poor at recognizing and adequately penalizing unfavorable interactions like for example electrostatic repulsion between two hydrogen bond acceptor or two hydrogen bond donor functions. This is easily rationalized since there are not many of these interactions observed, with good reason, and therefore empirical functions are not trained appropriately. In fact, to our knowledge there is only one such interaction present in the entire Brookhaven protein data bank. This interaction is a close contact between two oxygen atoms in the ligand and protein in a small molecule inhibitor bound to thermolysin (PDB code 6tmn).[196] The two inhibitors are shown in **Figure 25**.

Also, entropic factors can only be estimated through such parameters as number of rotatable bonds, and rather crude estimates of solvation energies. Also, protein flexibility has only very recently been incorporated into some of these scoring functions. At the other end of the spectrum, the computationally very expensive free-energy perturbation approaches have been used for the calculation of binding free energies for about 20 years,[197,198] very often achieving excellent agreement with experimental results.[199] These approaches take into account entropic contributions and solvation effects; however, other important effects like for example changes of charge distribution in both the ligand and the protein upon binding as well as polarization can not be taken into account due to the classical nature of the approach. This can only be achieved through quantum mechanical calculations, and there has been significant progress in this area, particularly with the development of advanced QM/MM approaches as summarized in Section 4.16.3.5. It has been recognized very early on that QM/MM approaches have enormous potential in this area, and early applications include studies on the binding of thermolysin inhibitors.[200] The area of the development and applications of QM/MM methods has been reviewed extensively by others.[9] More recently, there have been a very significant number of applications of quantum mechanical calculations in this area and this section will review some representative recent examples. The focus will be mainly on calculations of interactions and binding energies of small organic molecules with a receptor, however, we will also discuss applications in the area of predicting SAR through accurate calculations of structures and properties of small organic molecules as well as transition metal complexes where relevant in the treatment of diseases.

4.16.4.4.1 Applications to *p*-hydroxybenzoate hydroxylase, haloalkane dehalogenase, and thymidine kinase

In some relatively early work utilizing ab initio and semiempirical calculations to investigate the ligand binding by *p*-hydroxybenzoate hydroxylase, Peräkylä and Pakkanen found a good correlation between the binding energies for seven ligands based on a static active site model of the enzyme which consisted only of six amino acid residues surrounding the ligand. However, despite the relatively small enzyme model, they found a good correlation between experimental and calculated binding energies with a correlation coefficient of 0.90.[201] It has to be pointed out that the number of compounds used in this study is very small, the result is nevertheless considered significant.

A combination of QM/MM with an artificial neural network approach was used by Mlinsek *et al.* to predict binding of small molecule inhibitors to thrombin.[202] They implemented a method that enabled the calculation of the MEP at the van der Waals surfaces of atoms in the protein active site, and used these descriptors to derive a model for 18 thrombin inhibitors with a correlation coefficient of 0.96.

A recent paper on the binding analysis of 18 haloalkane dehalogenase substrates using quantum mechanical calculations uses part of the protein active site in the calculation of interaction energies.[203] Although the application is not directly relevant to medicinal chemistry, the method used in this work provides an interesting example of a combination of classical QSAR and quantum mechanically derived parameters directly related to ligand binding.

Quantum mechanical parameters have also been used in combination with neural networks for the prediction of protein–ligand binding energies based on the ligand structure and its respective biological activity.[204,205] The electrostatic potential surfaces of 21 known inhibitors were calculated and used to train a neural network. The resulting model was used to predict 18 unknown inhibitor molecules, and was able to achieve an accuracy to within fivefold of the measured activity, and the authors suggest that this method could be a useful approach in the design of new inhibitors.

In a case study on herpes simplex virus type 1 thymidine kinase (HSV1 TK) substrates and inhibitors, Cavalli and co-workers[8] have used DFT calculations to answer questions related to substrate diversity and catalysis as well as compound SAR that remained unanswered even after several x-ray structures of HSV1 TK with different prodrugs bound have become available. They were able to discriminate between substrates and inhibitors based on these calculations based on the calculated dipole moment of the bound ligand within the active site of HSV1 TK.

4.16.4.4.2 Application to thrombin

In order to predict binding modes of novel inhibitors of the serine proteinase thrombin, we have applied quantum mechanical HF and also semiempirical as well as QM/MM calculations. During an in-house drug discovery program, we found that thrombin inhibitors containing halogen-substituted phenyl rings at the P1 position exhibit excellent potency against thrombin. These compounds were lacking the basic group that would form a salt bridge interaction with the aspartic acid residue Asp219 at the bottom of the S1 pocket in the active site, which had been believed to be essential for inhibitory activity against thrombin.

There was no rationale at this time for the binding mode of these compounds and what type of interactions would be formed between the inhibitor and the enzyme active site. Similar types of inhibitors and their x-ray structures bound to thrombin were described in the literature[206,207] after we had concluded our quantum mechanical studies, with our conclusions largely being confirmed. The structures of these compounds are shown in **Figure 26**.[206,207] We postulated at the time that the dichlorophenyl ring would form an electrostatic interaction with the two hydrogens on the dichlorophenyl pointing toward Asp219. This interaction would of course not constitute a hydrogen bond, but rather a polarized electrostatic interaction. In order to show the effect of polarization and to underpin the binding mode hypothesis, we performed ab initio HF/6-31G* geometry optimizations on a number of substituted benzenes shown in **Figure 27** and determined the MEP-derived charges. The results clearly show that compared to unsubstituted benzene there is a significant polarization effect of the two chlorines on the ring, increasing the positive charge on the

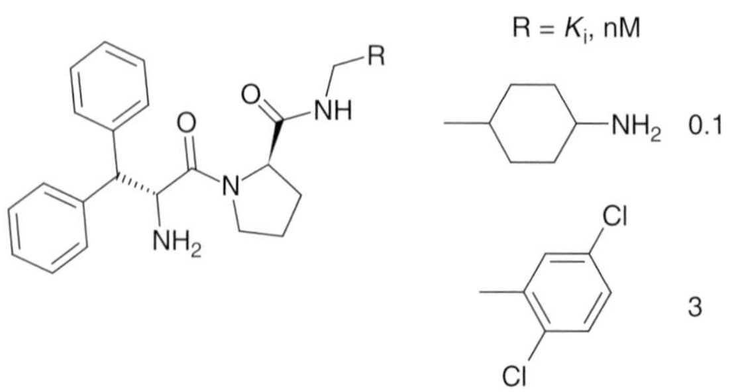

R = K_i, nM
O NH R O N NH_2
NH_2 0.1
Cl Cl 3

Figure 26 Comparison of activities and structure of thrombin inhibitor with and without a basic P1 substituent.

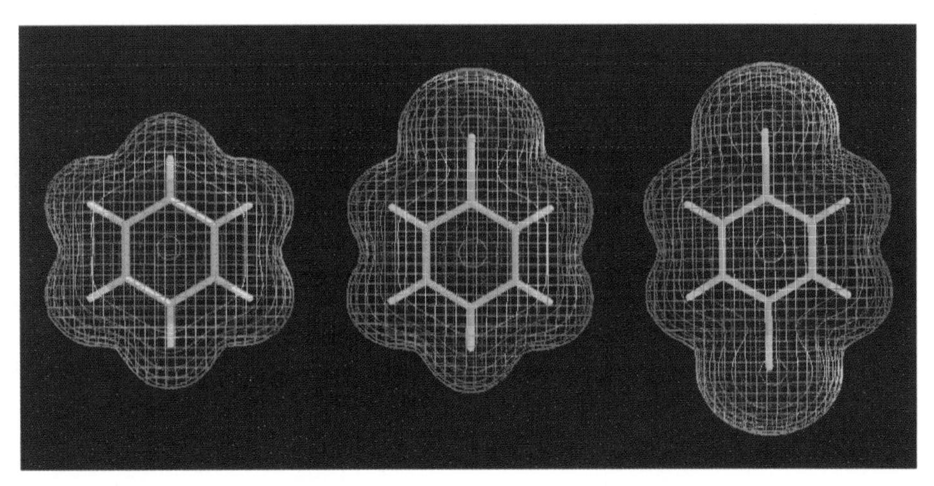

Figure 27 Benzene derivatives used to highlight the polarization effect of chlorine substitution on the neighboring ring hydrogen atoms. Electrostatic potentials are shown from -63 kJ mol^{-1} (red) to $+63$ kJ mol^{-1} (blue). The blue (positive) regions on the ring hydrogens significantly increase in size with increasing chlorine substitution.

phenyl hydrogen atoms from 0.12 to 0.14, therefore strengthening the proposed electrostatic interaction with the aspartic acid at the bottom of the S1 pocket in thrombin. Charges on the hydrogen atoms next to the chlorine substituents increase significantly, therefore the electrostatic interaction with the aspartic acid anion would also increase, making this a significantly more attractive interaction compared to an unsubstituted phenyl group. Therefore, we concluded that a direct interaction between the dichlorophenyl P1 substituent would have a favorable electrostatic component in addition to the good fit in the binding site, indicating a favorable van der Waals interaction, underlining the proposed binding mode. Utilizing this binding hypothesis, we were able to design other novel nonbasic P1 substituents with activity against thrombin. More recently and in order to understand the importance of polarization effects, we also carried out additional semiempirical AM1 QM/MM calculations using the semiempirical program VAMP[208] on a selection of 12 compounds containing basic and nonbasic P1 substituents within the series outlined in **Figure 26**. We achieved a good correlation between calculated and experimental binding energies ($r = 0.84$, $r^2_{cv} = 0.69$). In addition, the calculations showed a significant polarization effect of the S1 aspartic acid on the hydrogens of the phenyl ring of the 2,5-dichloro substituent. In fact the charges on the two hydrogen atoms increased from 0.161 and 0.153 in the gas phase to 0.201 and 0.197 when bound to the enzyme, further increasing the electrostatic binding interaction by about 25%. Therefore, it was concluded that in addition to the van der Waals binding interactions there are two significant electrostatic and polarization effects that contribute to the dichlorophenyl substituent forming favorable interactions with the enzyme. One is the increase in positive charge on the phenyl hydrogens upon chlorine substitution, the other is a significant increase in the positive charge on the same hydrogen atoms when binding to the negatively charged Asp219 when bound to thrombin.

4.16.4.4.3 New developments in the application of quantum mechanics to protein–ligand binding prediction

There have been a very significant number of QM/MM applications on the prediction of protein–ligand binding energies published recently, indicating the increasing applicability of the QM/MM methods to this very important scientific problem. This trend is quite clearly supported by the ever-increasing computer power and also the application of parallelization of software code.

An important advantage of quantum mechanical calculations, as pointed out previously, is the fact that atomic charges are calculated based on the wave function of a given molecule. This wave function is influenced by the environment as shown above and therefore the atomic charges change as a result. This can have very significant effects on for example the accuracy of calculating protein–ligand binding energies, as was recently found by Cho *et al.* in a QM/MM study for a set of 40 protein–ligand complexes.[209] They conclude that the use of accurate charges and environmental polarization effects lead to dramatic improvements in docking accuracy for a wide variety of protein–ligand complexes. In addition, they point out that rather than use quantum mechanical calculations, the same effect could be achieved by using polarizable force fields at much lower cost, enabling the treatment of much larger systems.

In a further study related to docking and the prediction of binding energies, Khandelwal *et al.* have used a combination of docking, QM/MM, and MD simulations for the estimation of binding affinities for 28 metalloprotein ligands (MMP-9).[210] Their work addresses particularly the inadequacies in the force field description of coordination bonds to zinc ions in proteins and metal interactions in general. Using the combined approach, they achieve an excellent correlation for 28 hydroxamate MMP-9 inhibitors ($r^2 = 0.90$). They apply a four-step process starting from initial complex structures derived using the docking program FlexX, followed by QM/MM geometry optimization, MD simulation, and further QM/MM geometry optimization, with every step providing a very significant improvement over the correlation from the previous step. In addition to the QM/MM interaction energy, the solvent-accessible surface was also found to contribute significantly, accounting for solvation, and was included in the correlation.

Recently, Gräter *et al.* have used QM/MM calculations in combination with the Poisson–Boltzmann/surface area model of continuum solvation (QM/MM-PB/SA). They have tested the feasibility of their method using a set of 47 benzamidine derivatives binding to trypsin as well as three ligands binding to FKBP-12.[211] In this QM/MM approach, only the ligand is treated quantum mechanically. Although the experimental range of absolute binding energies for all 47 benzamidine ligands is reproduced well with a root-mean-square error of 5.0 kJ mol^{-1}, the overall correlation coefficient is very low ($r = 0.20$), potentially due to the narrow range of binding energies. The correlation coefficient increases to 0.56 when the three FKBP-12 ligands are included in the analysis. A key point in this work is that conformational flexibility of the protein as well as a number of different docking modes for each ligand were considered, highlighting this approach as a promising new way for docking and scoring of flexible ligands, although further improvements in accuracy are needed.[211]

In addition to protein–ligand binding energies, the understanding of protonation energies and prediction of protonation states of ligands and proteins is of fundamental importance in the design of drugs. Very recently, Cummins and Gready have shown that a QM/MM plus MD simulation scheme using the perturbed quantum mechanical atom (PQA) method is capable of predicting relative protonation free energies in enzyme active sites.[212] Their results demonstrate the importance of including the effects of the enzymic environment in the calculations. In addition, their investigation of different choices of QM region indicate the importance of including the whole ligand in the QM region. One of the limitations of this method mentioned in the paper is the fact that it is based on the less accurate semiempirical quantum mechanical calculations. The authors note that the PQA/MM thermodynamic integration approach should in general provide more reliable estimates of relative protonation free energies.[212]

In addition to the QM/MM approach, a number of other methods have been used recently to calculate ligand–protein binding energies. One of these new methods, the ab initio fragment molecular orbital method (FMO), has been applied very recently to the calculation of relative binding energies of 10 estrogen receptor ligands to the alpha subtype of the human estrogen receptor ligand binding domain.[213] The receptor protein was modeled by 50 amino acid residues surrounding the bound ligands. The number of atoms in these model complexes is about 850, including hydrogen atoms. For four of the ligands, the entire estrogen receptor binding domain consisting of 241 residues and 4000 atoms was included in the calculations. Due to the size of the calculations, the method only uses the small STO-3G basis set at the HF level. The calculated binding energies correlate well with the experimental values ($r = 0.84$) for the 10 ligands studied. The authors claim that this is the first reported ab initio quantum mechanical study of an entire protein–ligand complex to date.[213] In a related study, Xiang *et al.* have recently published a study of the protein–ligand binding structure of a adipocyte lipid-binding protein complexed with propanoic acid.[214] They used the recently developed method (molecular fractionation with a conjugate caps method, MFCC) for the calculation of the protein–ligand interaction energy of a model system consisting of the entire adipocyte lipid-binding protein with 2057 atoms and the ligand, propionic acid, with 10 atoms. However, the calculations are carried out at the HF level using the relatively small 3-21G basis set. Nevertheless, given the size of the system, the achievement is quite remarkable, pointing the way to what will be possible in the near future in the field of protein–ligand binding energy calculations. One of the main advantages of the fragmentation algorithms like FMO and MFCC is their potential for parallelization of the program code, therefore significantly speeding up computations.

Raha and Merz have also recently published a quantum mechanics based scoring function for zinc ion mediated ligand binding that for the first time treats the electrostatic part of the protein–ligand interactions entirely quantum mechanically using the AM1 method. This work indicates that describing whole protein–ligand systems quantum mechanically is now computationally feasible. They have applied their method to 18 carbonic anhydrase and five carboxypeptidase A complexes and achieve a very good correlation with experimentally measured binding energies ($r^2 = 0.69$).[215] In a further publication, Raha and Merz reported more recently the development of the first-generation quantum mechanics based scoring function for predicting the binding affinity and binding mode for a broad range of protein–ligand complexes.[216] They used the semiempirical AM1 and PM3 approaches to estimate the binding energies for 165 protein ligand complexes, and achieved a reasonable correlation ($r^2 = 0.55$). However, models that were derived for one protein only showed much better accuracy, as for example for 33 complexes of HIV-1 protease. This new method was compared to results from existing empirical scoring functions from the literature, and the quantum mechanical methods appears to perform significantly better across a data set of 56 complexes used in the study.

4.16.5 Summary

The purpose of this chapter was to review the field of quantum mechanical calculations and to show their application, usefulness, and impact on research in medicinal chemistry and in particular drug design. There have been a large number of relevant applications of quantum mechanical methods over the last two decades, and the number is rising at an increasing rate. For example, the *Journal of Medicinal Chemistry* now regularly publishes papers that contain quantum mechanical calculations applied to medicinal chemistry problems, which was not the case even around 10 years ago. One of the obvious reasons for this is the easy access to relatively cheap computer power on everyday PC hardware for both academic institutions and the pharmaceutical industry. Also, the wider acceptance of computational simulations and models as a tool in drug design has increased significantly within the medicinal chemistry community over the last decade. The use of desktop software has now become an everyday routine activity for many medicinal chemists who understand the value additional information from such methods can bring. The barriers between medicinal chemistry and computational chemistry, and quantum mechanical methods in particular, are slowly being eroded away based on the successes that are achieved. Also, the quality of available computational tools, whether of commercial origin or from academic institutions, has generally improved very significantly over recent years, for example in the area of docking and virtual screening as well as ADMET predictions, giving medicinal chemists more confidence in the computational results. Interestingly, although quantum mechanical calculations never suffered from the same problems that empirical methods have with their systematic errors and their limitation to only be able to interpolate, quantum mechanics was generally seen as very much the same type of unreliable computational methodology that only tells the medicinal chemists what they already knew. Also, quantum chemists did perhaps not do a particularly great job at publicizing the distinct advantages of the methods they themselves invented and developed. This has possibly hampered a more general acceptance and therefore widespread application in medicinal chemistry and drug design up to today. However, with computational methods now being generally more widely used, quantum mechanical methods also seem to be increasingly utilized as a result, and their value is being recognized as a contributor to solving medicinal chemistry problems. Although still a 'black box' method for most chemists, quantum mechanical calculations are probably the only computational tool in existence that is able to extrapolate beyond known data, something that cannot be emphasized and publicized enough. They are, in principle, able to tell the medicinal chemists what they do not know yet, and sometimes are not willing to believe either. As pointed out above, the real power of quantum mechanics is the fact that no parametrization is necessary to gain insights into chemical problems. So if they are so powerful and very often give the right answer, why are they not used more? One reason is clearly that they are applicable only to a relatively small number of areas where they are able to provide accurate answers, for example structures of small molecules and their physicochemical properties like charge distributions, MEPs, and orbital structures. Also, there are only a relatively small number of computational chemists and medicinal chemists who are adquately trained in the use and more importantly the interpretation of quantum mechanical methods. However, as outlined in many examples throughout this chapter, quantum mechanical calculations, when utilized appropriately, can provide valuable information for drug design, and they are clearly able to broaden the arsenal of computational methods that can help advance the design of new small molecule pharmaceutical therapies.

4.16.6 Outlook

The future of quantum mechanical calculations looked bleak only 10 years ago,[217] when advances in methodology seemed to focus more on accuracy than increases in performance, and an application of these methods to systems of several thousand atoms seemed unrealistic, requiring a 600-fold speed-up in compute power in order to even consider conventional HF or DFT approaches. To put this in perspective, for example desktop compute power has increased by around 1000-fold in the last 25 years. However, recent advances in particular in the fields of QM/MM and around new methods that only scale with N^2 (with N being the number of atoms in the calculation), compared to the traditional HF methods that scale with N^4, have opened up new opportunities in particular for research in the area of enzyme catalytic mechanisms as well as binding energy predictions, which require the inclusion of at least part of the protein into the model systems to adequately represent the molecular environmnent. In addition, recent advances in massive parallel distributed computing have allowed unprecedented calculations on whole proteins using ab initio and DFT methods as well as combined QM/MM methods.

One important objective of this chapter was also to attempt to dissolve some of the quite understandable ignorance, but also the suspicion and resistance toward quantum calculations that seems to persist in the medicinal chemistry community, perhaps originating from the perception that quantum mechanics is a quirk of science that is somewhat irrelevant for solving life sciences problems. Let's just remind ourselves that quantum mechanics is currently the only theory that is able to explain the workings of for example the transistor, clearly a very real and everyday item. When one overcomes this initial barrier to understanding quantum mechanics, it presents itself to the interested researcher as a field of amazing and exciting new opportunities and enormous potential, ready to be exploited further for the advancement of medicinal chemistry and drug design and the discovery of new small molecule pharmaceutical therapies.

References

1. Bleicher, K. H.; Nettekoven, M.; Peters, J.-U.; Wyler, R. *Chimia* **2004**, *58*, 588–600.
2. Demasi, J. A.; Hansen, R. W.; Grabowski, H. G. *J. Health Econ.* **2003**, *22*, 151–185.
3. Barden, J. C.; Schaefer, H. F., III. In *Computational Medicinal Chemistry for Drug Discovery*; Bultnik, P., Ed.; Marcel Dekker: New York, 2004, pp 133–149.
4. Loew, G. H.; Villar, H. O.; Alkorta, I. *Pharmaceut. Res.* **1993**, *10*, 475–486.
5. Miller, J. *J. Chem. Educ.* **1999**, *76*, 12–14.
6. Chasse, G. A.; Rodriguez, A. M.; Mak, M. L.; Deretey, A.; Perczel, A.; Sosa, C. P.; Enriz, R. D.; Csizmadia, I. G. *J. Mol. Struct. (Theochem.)* **2001**, *537*, 319–361.
7. *Annals of the New York Academy of Sciences*; Weinstein, H., Green, J. P., Eds.; New York Academy of Sciences: New York, 1981; Vol. 367.
8. Cavalli, A.; Folkers, G.; Recanatini, M.; Scapozza, L. In *Methods and Principles in Medicinal Chemistry*; Carloni, P., Alber, F., Eds.; Wiley: Weinheim, Germany, 2003; Vol. 17, pp 41–71.
9. Perruccio, F.; Ridder, L.; Mulholland, A. J. In *Methods and Principles in Medicinal Chemistry*; Carloni, P., Alber, F., Eds.; Wiley: Weinheim, Germany, 2003; Vol. 17, pp 177–198.
10. Murray, J. S.; Politzer, P. In *Methods and Principles in Medicinal Chemistry*; Carloni, P., Alber, F., Eds.; Wiley: Weinheim, Germany, 2003; Vol. 17, pp 233–254.
11. Höltje, H.-D.; Höltje, M. In *Methods and Principles in Medicinal Chemistry*; Carloni, P., Alber, F., Eds.; Wiley: Weinheim, Germany, 2003; Vol. 17, pp 255–274.
12. Hehre, W.; Radom, L.; Schleyer, P. v. R.; Pople, J. A. *Ab Initio Molecular Orbital Theory*; Wiley: New York, 1986.
13. Yamaguchi, Y.; Osamura, Y.; Goddard, J. D.; Schaefer, H. F., III. *A New Dimension to Quantum Chemistry: Analytic Derivative Methods in Ab Initio Molecular Electronic Structure Theory*; Oxford University Press: New York, 1994.
14. Topol, I. A.; Burt, S. K.; Deretey, E.; Tang, T. H.; Perczel, A.; Rashin, A.; Csizmadia, I. G. *J. Am. Chem. Soc.* **2001**, *123*, 6054–6060.
15. Geerlings, P.; De Proft, F.; Langenaeker, W. *Chem. Rev.* **2003**, *103*, 1793–1873.
16. Hohenberg, P.; Kohn, W. *Phys. Rev. B* **1964**, *136*, 864.
17. Lee, C.; Yang, W.; Parr, R. G. *Phys. Rev. B* **1988**, *37*, 785–789.
18. Becke, A. D. *J. Chem. Phys.* **1993**, *98*, 5648–5652.
19. Becke, A. D. *Phys. Rev. A* **1988**, *38*, 3098–3100.
20. Perdew, J. P. *Phys. Rev. B* **1986**, *33*, 8822–8824.
21. Ziegler, T.; Autschbach, J. *Chem. Rev.* **2005**, *105*, 2695–2722.
22. Burda, J. V.; Zeizinger, M.; Leszczynski, J. *J. Comput. Chem.* **2005**, *26*, 907–914.
23. Seminario, J. M. In *Modern Density Functional Theory: A Tool for Chemistry: Theoretical and Computational Chemistry*; Seminario, J. M., Politzer, P., Eds.; Elsevier Science: New York, 1995; Vol. 2, pp 1–27.
24. Raber, J.; Liano, J.; Eriksson, L. A. In *Methods and Principles in Medicinal Chemistry*; Carloni, P., Alber, F., Eds.; Wiley: Weinheim, Germany, 2003; Vol. 17, pp 113–153.
25. Sebastiani, D.; Röthlisberger, U. In *Methods and Principles in Medicinal Chemistry*; Carloni, P., Alber, F., Eds.; Wiley: Weinheim, Germany, 2003; Vol. 17, pp 5–39.
26. Dewar, M. J. S.; Zoebisch, E. G.; Healy, E. F.; Stewart, J. J. P. *J. Am. Chem. Soc.* **1985**, *107*, 3902–3909.
27. Stewart, J. J. P. *J. Comput. Chem.* **1989**, *10*, 209-220, 221–264.
28. Shaffer, A. A.; Wierschke, S. G. *J. Comput. Chem.* **1993**, *14*, 75–88.
29. Allinger, N. L.; Chen, K. S.; Lii, J. H. *J. Comput. Chem.* **1996**, *17*, 642–668.
30. Cornell, W. D.; Cieplak, P.; Bayly, C. I.; Gould, I. R.; Merz, K. M.; Ferguson, D. M.; Spellmeyer, D. C.; Fox, T.; Caldwell, J. W.; Kollman, P. A. *J. Am. Chem. Soc.* **1995**, *117*, 5179–5197.
31. Sybyl 6.4, Tripos Associates, St. Louis, MO, USA.
32. Halgren, T. *J. Comput. Chem.* **1996**, *17*, 490–519.
33. Todebush, P. M.; Bowen, J. P. In *Free Energy Calculations in Rational Drug Design*; Reddy, M. R., Erion, M. D., Eds.; Plenum Publishers: New York, 2001, pp 37–59.
34. Rappé, A. K.; Casewit, C. J.; Colwell, K. S.; Goddard, W. A; Skiff, W. M. *J. Am. Chem. Soc.* **1992**, *114*, 10024–10035.
35. Warshel, A.; Levitt, M. *J. Mol. Biol.* **1976**, *103*, 227–249.
36. Åqvist, J.; Warshel, A. *Chem. Rev.* **1993**, *93*, 2523–2544.
37. Peräkylä, M. In *Methods and Principles in Medicinal Chemistry*; Carloni, P., Alber, F., Eds.; Wiley: Weinheim, Germany, 2003; Vol. 17, pp 157–176.
38. Car, R.; Parrinello, M. *Phys. Rev. Lett.* **1985**, *55*, 2471.
39. Laio, A.; Van de Vondele, J.; Rothlisberger, U. *J. Chem. Phys.* **2002**, *116*, 6941–6948.
40. Colombo, M. C.; Guidoni, L.; Laio, A.; Magistrato, A.; Maurer, P.; Piana, S.; Röhrig, U.; Spiegel, K.; Sulpizi, M.; Van de Vondele, J. et al. *Chimia* **2002**, *56*, 11–17.
41. Laio, A.; Van de Vondele, J.; Röthlisberger, U. *J. Phys. Chem. B* **2002**, *106*, 7300–7307.

42. Röthlisberger, U. *Comput. Chem. Rev. Curr. Trends* **2001**, *6*, 33–68.
44. Carloni, P.; Röthlisberger, U.; Parrinello, M. *Acc. Chem. Res.* **2002**, *35*, 455–464.
45. Carloni, P.; Alber, F. *Perspect. Drug Dis.* **1998**, *9/11*, 169–179.
46. Sebastiani, D.; Parrinello, M. *J. Phys. Chem. A* **2001**, *105*, 1951–1958.
47. van Gunsteren, W. F.; Luque, F. J.; Timms, D.; Torda, A. E. *Annu. Rev. Biophys. Biomol. Struct.* **1994**, *23*, 847–863.
48. Ladbury, J. E. *Chem. Biol.* **1996**, *3*, 973–980.
49. Poornima, C. S.; Dean, P. M. *J. Comput.-Aided Mol. Design* **1995**, *9*, 500–512.
50. Poornima, C. S.; Dean, P. M. *J. Comput.-Aided Mol. Design* **1995**, *9*, 513–520.
51. Poornima, C. S.; Dean, P. M. *J. Comput.-Aided Mol. Design* **1995**, *9*, 521–531.
52. Tomasi, J.; Persico, M. *Chem. Rev.* **1994**, *94*, 2027–2094.
53. Wong, M. W.; Frisch, M. J.; Wiberg, K. B. *J. Am. Chem. Soc.* **1991**, *113*, 4776–4782.
54. Christiansen, O.; Mikkelsen, K. V. *J. Chem. Phys.* **1999**, *110*, 1365–1375.
55. Cramer, C. J.; Truhlar, D. G. *J. Comput.-Aided Mol. Design* **1992**, *6*, 629–666.
56. Dixon, R. W.; Leonard, J. M.; Hehre, W. J. *Isr. J. Chem.* **1993**, *33*, 427–434.
57. Klamt, A.; Scheuermann, G. *J. Chem. Soc. Perkin Trans.* **1999**, *2*, 799–805.
58. Klamt, A. *J. Phys. Chem.* **1995**, *99*, 2224–2235.
59. Klamt, A.; Eckert, F.; Diedenhofen, M.; Beck, M. E. *J. Phys. Chem. A* **2003**, *107*, 9380–9386.
60. Klamt, A.; Eckert, F.; Hornig, M. *J. Comput.-Aided Mol. Design* **2001**, *15*, 355–365.
61. Tomasi, J.; Mennucci, B.; Cammi, R. *Chem. Rev.* **2005**, *105*, 2999–3093.
62. Ooms, F. *Curr. Med. Chem.* **2000**, 7, 141–158.
63. Bleicher, K. H.; Böhm, H.-J.; Müller, K.; Alanine, A. I. *Nat. Rev. Drug Dis.* **2003**, *2*, 369–378.
64. Klebe, G. *Drug Disc. Today: Technol.* **2004**, *1*, 225–230.
65. Goodnow, R. A., Jr.; Gillespie, P.; Bleicher, K. In *Chemoinformatics in Drug Discovery*; Oprea, T., Ed.; Wiley-VCH: Weinheim, Germany, 2005, pp 381–435.
66. Clark, T. In *Rational Approaches to Drug Design*; Höltje, H.-D., Sippl, W., Eds.; Prous Science: Barcelona, Spain, 2001, pp 29–40.
67. Hansch, C. *Drug Devel. Res.* **1981**, *1*, 267–309.
68. Hansch, C. *Acc. Chem. Res.* **1993**, *26*, 147–153.
69. Patani, G. A.; LaVoie, E. J. *Chem. Rev.* **1996**, *96*, 3147–3176.
70. De Groot, M. J.; Ackland, M. J.; Horne, V. A.; Alex, A. A.; Jones, B. C. *J. Med. Chem.* **1999**, *42*, 1515–1524.
71. Gillespie, R. G.; Morton, M. J. *Inorg. Chem.* **1970**, *9*, 616–618.
72. Lister, D. G.; Tyler, J. K. *J. Mol. Struct.* **1974**, *23*, 253–264.
73. Luque, F. J.; Orozco, M. *J. Org. Chem.* **1993**, *58*, 6379–6405.
74. Murray-Rust, P.; Glusker, J. P. *J. Am. Chem. Soc.* **1984**, *106*, 1018–1025.
75. Luger, P.; Griss, G.; Hurnaus, R.; Trummlitz, G. *Acta Crystallogr.* **1986**, *B42*, 478–490.
76. Caranoni, P. C.; Reboul, J. P. *Acta Crystallogr. Sect. B* **1982**, *38*, 1255–1259.
77. Schappach, A.; Höltje, H.-D. *Pharmazie* **2001**, *56*, 835–842.
78. Jennings, W. B.; Boyd, D. R. In *Cyclic Organonitrogen Stereodynamics*; Lambert, J. B., Takeuchi, Y., Eds.; Verlag Chemie: Weinheim, Germany, 1992, pp 105–158.
79. Martins, J. B. L.; Taft, C. A.; Perez, M. A.; Stamato, F. M. L.; Longo, E. *Int. J. Quant. Chem.* **1998**, *69*, 117–128.
80. Schleifer, K.-J. *J. Med. Chem.* **1999**, *42*, 2204–2211.
81. Schleifer, K.-J. *Pharmazie* **1999**, *54*, 804–807.
82. Martin, S. F.; Dorsey, G. O.; Gane, T.; Hillier, M. C.; Kessler, H.; Baur, M.; Mathä, B.; Erickson, J. W.; Bhat, T. N.; Munshi, S. et al. *J. Med. Chem.* **1998**, *41*, 1581–1597.
83. Mora, M.; Munoz-Caro, C.; Nino, A. *J. Comput.-Aided Mol. Design* **2003**, *17*, 713–724.
84. MacPherson, R. D. *Pharmacol. Therapeut.* **2000**, *88*, 163.
85. Vieth, M.; Hirst, J. D.; Brooks, C. L., III. *J. Comput.-Aided Mol. Design* **1998**, *12*, 563–572.
86. Nicklaus, M. C.; Wang, S.; Driscoll, J. S.; Milne, G. W. A. *Bioorg. Med. Chem.* **1995**, *3*, 411–428.
87. Ricketts, E. M.; Bradshaw, J.; Hann, M.; Hayes, F.; Tanna, N.; Ricketts, D. M. *J. Chem. Inf. Comput. Sci.* **1993**, *33*, 905–925.
88. Wang, Z.; Canagarajah, B. J.; Boehm, J. C.; Kassisa, S.; Cobb, M. H.; Young, P. R.; Abdel-Meguid, S.; Adams, J. L.; Goldsmith, E. J. *Structure* **1998**, *6*, 1117–1128.
89. Wenlock, M. C.; Austin, R. P.; Barton, P.; David, A. M.; Leeson, P. D. *J. Med. Chem.* **2003**, *46*, 1250–1256.
90. Karelson, M.; Lobanov, V. S.; Katritzky, A. R. *Chem. Rev.* **1996**, *96*, 1027.
91. Benigni, R. *Chem. Rev.* **2005**, *105*, 1767–1800.
92. Brinck, T.; Murray, J. S.; Politzer, P. *J. Org. Chem.* **1993**, *58*, 7070–7073.
93. Beck, B.; Breindl, A.; Clark, T. *J. Chem. Inf. Comput. Sci* **2000**, *40*, 1046–1051.
94. Beck, B.; Glen, R. C.; Clark, T. *J. Mol. Graphics* **1996**, *14*, 130–135.
95. Mu, L.; Drago, R. S.; Richardson, D. E. *J. Chem. Soc. Perkin Trans. 2* **1998**, 159-167.
96. Hemmateenejad, B.; Safarpour, M. A.; Miri, R.; Taghavi, F. *J. Comput. Chem.* **2004**, *25*, 1495–1503.
97. Clare, B. W. *J. Comput.-Aided Mol. Design* **2002**, *16*, 611–633.
98. Smith, P. J.; Popelier, P. L. A. *J. Comput.-Aided Mol. Design* **2004**, *18*, 135–143.
99. Erion, M. D.; Reddy, M. R. *J. Am. Chem. Soc.* **1998**, *120*, 3295.
100. He, L.; Jurs, P. C.; Custer, L. L.; Durham, S. K.; Pearl, G. M. *Chem. Res. Toxicol.* **2003**, *16*, 1567–1580.
101. Mosier, P. D.; Jurs, P. C.; Custer, L. L.; Durham, S. K.; Pearl, G. M. *Chem. Res. Toxicol.* **2003**, *16*, 721–732.
102. He, L.; Jurs, P. C. *J. Mol. Graphics Model.* **2005**, *23*, 503–523.
103. Smieško, M.; Benfenati, E. *J. Chem. Inf. Model.* **2005**, *45*, 379–385.
104. Popelier, P. L. A.; Smith, P. J.; Chaudry, U. A. *J. Comput.-Aided Mol. Design* **2004**, *18*, 709–718.
105. O'Brien, S. E.; Popelier, P. L. A. *J. Chem. Soc. Perkin Trans. 2* **2002**, 478-483.
106. Popelier, P. L. A.; Chaudry, U. A.; Smith, P. J. *J. Chem. Soc. Perkin Trans. 2* **2002**, 1231-1237.
107. Chaudry, U. A.; Popelier, P. L. A. *J. Org. Chem.* **2004**, *69*, 233–241.
108. Cacelli, I.; Campanile, S.; Giolitti, A.; Molin, D. *J. Chem. Inf. Model.* **2005**, *45*, 327–333.

109. Klamt, A.; Eckert, F.; Hornig, M.; Beck, M. E.; Bürger, T. *J. Comput. Chem.* **2002**, *23*, 275–281.
110. Ehresmann, B.; de Groot, M. J.; Alex, A.; Clark, T. *J. Chem. Inf. Comput. Sci.* **2004**, *44*, 658–668.
111. Brüstle, M.; Beck, B.; Schindler, T.; King, W.; Mitchell, T.; Clark, T. *J. Med. Chem.* **2002**, *45*, 3345–3355.
112. Ehresmann, B.; de Groot, M. J.; Clark, T. *J. Chem. Inf. Model.* **2005**, *45*, 1053–1060.
113. Sorich, M. J.; McKinnon, R. A.; Miners, J. O.; Winkler, D. A.; Smith, P. A. *J. Med. Chem.* **2004**, *47*, 5311–5317.
114. Martin, Y. C. *J. Med. Chem.* **2005**, *48*, 3164–3170.
115. Lipinski, C. A.; Lombardo, F.; Dominy, B. W.; Feeney, P. J. *Adv. Drug Deliv. Rev.* **1997**, *23*, 3–25.
116. Yoshida, F.; Topliss, J. G. *J. Med. Chem.* **2000**, *43*, 2575–2585.
117. Raevski, O. A.; Fetisov, V. I.; Trepalina, E. P.; McFarland, J. W.; Schaper, K.-J. *Quant. Struct. –Act. Relat.* **2000**, *19*, 366–374.
118. Platts, J. A.; Abraham, M. H.; Hersey, A.; Butina, D. *Pharm. Res.* **2000**, *17*, 1013–1018.
119. Crivori, P.; Cruciani, G.; Carrupt, P.-A.; Testa, B. *J. Med. Chem.* **2000**, *43*, 2204–2216.
120. Clark, D. E. *Drug Disc. Today* **2003**, *8*, 927–933.
121. Seelig, A. *Eur. J. Biochem.* **1998**, *251*, 252–261.
122. Lombardo, F.; Blake, J. F.; Curatolo, W. J. *J. Med. Chem.* **1996**, *39*, 4750–4755.
123. Hutter, M. C. *J. Comput.-Aided Mol. Design* **2003**, *17*, 415–433.
124. Sadowski, J.; Kubinyi, H. A. *J. Med. Chem.* **1998**, *41*, 3325–3329.
125. Ajay; Walters, W. P.; Murcko, M. A.; *J. Med. Chem.* **1998**, *41*, 3314-3324.
126. Wagener, M.; van Geersestein, V. J. *J. Chem. Inf. Comput. Sci.* **2000**, *40*, 280–292.
127. Kroemer, R. T.; Hecht, P.; Liedl, K. R. *J. Comput. Chem.* **1996**, *17*, 1296–1308.
128. Mecozzi, S; West, A. P.; Dougherty, D. A. *Proc. Natl. Acad. Sci. USA* **1996**, *93*, 10566–10571.
129. McClure, K. F.; Abramov, Y. A.; Laird, E. R.; Barberia, J. T.; Cai, W.; Carty, T. J.; Cortina, S. R.; Danley, D. E.; Dipesa, A. J.; Donahue, K. M. et al. *J. Med. Chem.* **2005**, *48*, 5728–5737.
130. Nakagawa, S.; Umeyama, H.; Kitaura, K.; Morokuma, K. *Chem. Pharm. Bull.* **1981**, *29*, 1.
131. Lambros, S. A.; Richards, G. W.; Marchington, A. F. *J. Mol. Struct. (Theochem.)* **1984**, *109*, 61–71.
132. Merz, K. M., Jr.; Hoffmann, R.; Dewar, M. J. S. *J. Am. Chem. Soc.* **1989**, *111*, 5636.
133. Alex, A.; Clark, T. *J. Comput. Chem.* **1992**, *13*, 704–717.
134. Kollman, P. A.; Kuhn, B.; Peräkylä, M. *J. Phys. Chem. B* **2002**, *106*, 1537–1542.
135. Noodleman, L.; Lovell, T.; Han, W.-G.; Li, J.; Himo, F. *Chem. Rev.* **2004**, *104*, 459–508.
136. Massova, I.; Kollman, P. A. *J. Phys. Chem. B* **1999**, *103*, 8628–8638.
137. Levy, S. B. *The Antibiotic Paradox: How Miracle Drugs Are Destroying the Miracle*; Plenum Press: New York, 1992.
138. Garrett, L. *The Coming Plague: Newly Emerging Diseases in a World out of Balance*; Farrar, Strauss and Giroux: New York, 1994.
139. Quinn, J. P. *Diagn. Microbiol. Infect. Dis.* **1998**, *31*, 389–395.
140. Livermore, D. M. *J. Antimicrob. Chemother. (Suppl. D)* **1998**, *41*, 25–41.
141. Rosen, H.; Hajdu, R.; Silver, L.; Kropp, H.; Dorso, K.; Kohler, J.; Sundelof, J. G.; Huber, J.; Hammond, G. G.; Jackson, J. J. et al. *Science* **1999**, *283*, 703–706.
142. Fisher, J. F.; Merouch, S. O.; Mobashery, S. *Chem. Rev.* **2005**, *105*, 395–424.
143. Díaz, N.; Sordo, T. L.; Merz, K. M., Jr.; Suárez, D. *J. Am. Chem. Soc.* **2003**, *125*, 672–684.
144. Chaudry, U. A.; Popelier, P. L. A. *J. Chem. Phys. A*, **2003**, *107*, 4578–4582.
145. Edwards, P. D.; Bernstein, P. R. *Med. Res. Rev.* **1994**, *14*, 137–194.
146. Hansch, C.; Leo, A.; Taft, W. *Chem. Rev.* **1991**, *91*, 165–195.
147. Edwards, P. D.; Wolanin, D. J.; Andisik, D. W.; Davis, M. W. *J. Med. Chem.* **1995**, *38*, 76–85.
148. Taylor, P. J.; Wait, A. R. *J. Chem. Soc. Perkin Trans.* **1986**, *2*, 1765–1770.
149. Cohen, P. *Curr. Opin. Chem. Biol.* **1999**, *3*, 459–465.
150. Patrick, G. N.; Zukerberg, L.; Nikolic, M.; De la Monte, S.; Dikkes, P.; Tsai, L. H. *Nature* **1999**, *402*, 615–622.
151. Dinner, A.; Blackburn, G. M.; Karplus, M. *Nature* **2001**, *413*, 752–755.
152. Mulholland, A. J.; Richards, W. G. *Proteins: Struct. Funct. Genet.* **1997**, *27*, 9–25.
153. Varnai, P.; Richards, W. G.; Lyne, P. D. *Proteins: Struct. Funct. Genet.* **1999**, *37*, 218–227.
154. Ridder, L.; Rietjens, I. M. C. M.; Verhoort, J.; Mulholland, A. J. *J. Am. Chem. Soc.* **2002**, *124*, 9926–9936.
155. Thomas, A.; Jourand, D.; Bret, C.; Amara, P.; Field, M. J. *J. Am. Chem. Soc.* **1999**, *121*, 9693–9702.
156. Guha, R.; Jurs, P. C. *J. Chem. Inf. Comput. Sci.* **2004**, *44*, 2179–2189.
157. Hammond, P. S.; Wu, Y.; Harris, R.; Minehardt, T. J.; Car, R.; Schmitt, D. *J. Comput.-Aided Mol. Design* **2005**, *19*, 1–15.
158. Bencherif, M.; Schmitt, J. D. *Curr. Drug Target CNS Neurol. Disord.* **2002**, *1*, 349.
159. Barlow, R. B.; Howard, J. A. K.; Johnson, O. *Acta Crystallogr., Sec. C* **1986**, *C42*, 853–856.
160. Chynoweth, K. R.; Ternai, B.; Simeral, L. S.; Maciel, G. E. *Mol. Pharmacol.* **1973**, *9*, 144–151.
161. Haynes, R. K.; Vonwiller, S. C. *Acc. Chem. Res.* **1997**, *30*, 73–79.
162. Guha, R.; Jurs, P. C. *J. Chem. Inf. Comput. Sci.* **2004**, *44*, 1440–1449.
163. Dong, Y.; Matile, H.; Chollet, J.; Kamisky, R.; Wood, J. K.; Vennerstrom, J. L. *J. Med. Chem.* **1999**, *42*, 1477–1480.
164. Meshik, S. R. *Med. Trop.* **1998**, *58*, 13–17.
165. Posner, G. H.; Wang, D.; Cumming, J. N.; Ho Oh, C.; French, A. N.; Bodley, A. L.; Shapiro, T. A. *J. Med. Chem.* **1995**, *38*, 2273–2275.
166. Rafiee, M. A.; Hadipour, N. L.; Naderi-manesh, H. *J. Chem. Inf. Model* **2005**, *45*, 366.
167. Campbell, S. F.; Davey, M. J.; Hardstone, D. J.; Lewis, B. N.; Palmer, M. J. *J. Med. Chem.* **1987**, *30*, 49–57.
168. Alabaster, V. A.; Campbell, S. F.; Danilewicz, J. C.; Greengrass, C. W.; Plews, R. M. *J. Med. Chem.* **1987**, *30*, 999–1003.
169. Campbell, S. F.; Plews, R. M. *J. Med. Chem.* **1987**, *30*, 1794–1798.
170. Spartan'02, Wavefunction Inc., Irvine, CA, USA.
171. Pavletich, N. P. *J. Mol. Biol.* **1999**, *287*, 821–828.
172. Senderowicz, A. M.; Headlee, D.; Stinson, S. F.; Lush, R. M.; Kalil, N.; Villalba, L.; Hill, K.; Steinberg, S. M.; Figg, W. D.; Tompkins, A. et al. *J. Clin. Oncol.* **1998**, *16*, 2986–2999.
173. Gussio, R.; Zahrevitz, D. W.; McGrath, C. F.; Pattabiraman, N.; Kellog, G. E.; Schultz, C.; Link, A.; Kunick, C.; Leost, M.; Meijer, L. et al. *Anti-Cancer Drug Design* **2000**, *15*, 53–66.

174. Schultz, C.; Link, A.; Leost, M.; Zaharevitz, D. W.; Gussio, R.; Sausville, E. A.; Meijer, L.; Kunick, C. *J. Med. Chem.* **1999**, *42*, 2909–2919.
175. Williams, I. H.; Barnes, J. A. *Biochem. Soc. Trans.* **1996**, *24*, 263–268.
176. Mlinsek, G.; Novic, M.; Hodoscek, M.; Solmajer, T. *J. Chem. Inf. Comput. Sci.* **2001**, *41*, 1286–1294.
177. Liu, H.; Müller-Plathe, F.; van Gunsteren, W. F. *J. Mol. Biol.* **1996**, *261*, 454–469.
178. De Groot, M. J.; Alex, A. A.; Jones, B. C. *J. Med. Chem.* **2002**, *45*, 1983–1993.
179. De Groot, M. J.; Havenith, R. W. A.; Vinkers, H. M.; Zwaans, R.; Vermeulen, N. P. E.; van Lenthe, J. H. *J. Comput.-Aided Mol. Design* **1998**, *12*, 183–193.
180. Rovira, C. In *Methods and Principles in Medicinal Chemistry*; Carloni, P., Alber, F., Eds.; Wiley: Weinheim, Germany, 2003; Vol. 17, pp 73–112.
181. Harris, D. L. *Curr. Opin. Drug. Disc. Devel.* **2004**, 7, 43–48.
182. Mennier, B.; de Visser, S. P.; Shaik, S. *Chem. Rev.* **2004**, *104*, 3947–3980.
183. Shaik, S.; Kumar, D.; de Visser, S. P.; Altum, A.; Thiel, W. *Chem. Rev.* **2005**, *105*, 2279–2328.
184. Lau, J. K.-C.; Deubel, D. V. *Chem. Eur. J.* **2005**, *11*, 2849–2855.
185. Xiao, X.; Cushman, M. *J. Am. Chem. Soc.* **2005**, *127*, 9960.
186. Thomas, C. J.; Rahier, N. J.; Hecht, S. M. *Bioorg. Med. Chem.* **2004**, *12*, 1585–1604.
187. Staker, B. L.; Hjerrild, K.; Feese, M. D.; Behnke, C. A.; Burgin, A. B., Jr.,; Stewart, L. *Proc. Natl. Acad. Sci. USA* **2002**, *99*, 15387–15392.
188. Tuttle, T.; Kraka, E.; Cremer, D. *J. Am. Chem. Soc.* **2005**, *127*, 9469–9484.
189. Williams, D. H.; Cox, J. P. L.; Doig, A. J.; Gardner, M.; Gerhard, U.; Kaye, P. T.; Lal, A. R.; Nicholls, I. A.; Salter, C. J.; Mitchell, R. C. *J. Am. Chem. Soc.* **1991**, *113*, 7020–7030.
190. Böhm, H.-J. *J. Comput.-Aided Mol. Design* **1992**, *6*, 61–78.
191. Böhm, H.-J. *J. Comput.-Aided Mol. Design* **1992**, *6*, 593–606.
192. Rarey, M.; Kramer, B.; Lengauer, T.; Klebe, G. *J. Mol. Biol.* **1996**, *261*, 470–489.
193. Rarey, M.; Wefing, S.; Lengauer, T. *J. Comput.-Aided Mol. Design* **1996**, *10*, 41–54.
194. Jones, G.; Willett, P.; Glen, R. C.; Leach, A. R.; Taylor, R. *J. Mol. Biol.* **1997**, *267*, 727–748.
195. Davis, A. M.; Teague, S. J.; Kleywegt, G. J. *Angew. Chem., Int. Ed. Engl.* **2003**, *42*, 2718–2736.
196. Morgan, B. P.; Scholtz, J. M.; Ballinger, M. D.; Zipkin, I. D.; Bartlett, P. A. *J. Am. Chem. Soc.* **1991**, *113*, 297–307.
197. Jorgensen, W. L. *Acc. Chem. Res.* **1989**, *22*, 184–189.
198. Kollman, P. *Chem. Rev.* **1993**, *93*, 2395–2417.
199. Kollman, P. A.; Merz, K. M., Jr. *Acc. Chem. Res.* **1990**, *23*, 246–252.
200. Alex, A.; Finn, P. *J. Mol. Struct. (Theochem.)* **1997**, *398-399*, 551–554.
201. Peräkylä, M.; Pakkanen, T. A. *Proteins: Struct. Funct. Genet.* **1995**, *21*, 22–29.
202. Mlinsek, G.; Novic, M.; Hodoscek, M.; Solmajer, T. *J. Chem. Inf. Comput. Sci.* **2001**, *41*, 1286–1294.
203. Kmuníeek, J.; Boháè, M.; Luengo, S.; Gago, F.; Wade, R. C.; Damborskỳ, J. *J. Comput.-Aided Mol. Design* **2003**, *17*, 299–311.
204. Braunheim, B. B.; Miles, R. W.; Schramm, V. L.; Schwartz, S. D. *Biochemistry* **1999**, *38*, 16076–16083.
205. Braunheim, B. B.; Bagdassarian, C. K.; Schramm, V. L.; Schwartz, S. D. *Int. J. Quant. Chem.* **2000**, *78*, 195–204.
206. Lumma, W. C., Jr.,; Witherup, K. M.; Tucker, T. J.; Brady, S. F.; Sisko, J. T.; Naylor-Olsen, A. M.; Lewis, S. D.; Lucas, B. J.; Vacca, J. P. *J. Med. Chem.* **1998**, *41*, 1011–1013.
207. Tucker, T. J.; Brady, S. F.; Lumma, W. C.; Lewis, S. D.; Gardell, S. J.; Naylor-Olsen, A. M.; Yan, Y.; Sisko, J. T.; Stauffer, K. J.; Lucas, B. J. et al. *J. Med. Chem.* **1998**, *41*, 3210–3219.
208. Clark, T.; Alex, A.; Beck, B.; Burkhardt, F.; Chandrasekhar, J.; Gedeck, P.; Horn, A. H. C.; Hutter, M.; Martin, B.; Rauhut, G. et al. VAMP 8.1, Erlangen, 2002.
209. Cho, A. E.; Guallar, V.; Berne, B. J.; Friesner, R. *J. Comput. Chem.* **2005**, *26*, 915–931.
210. Khandelwal, A.; Lukacova, V.; Comez, D.; Kroll, D. M.; Raha, S.; Balaz, S. *J. Med. Chem.* **2005**, *48*, 5437–5447.
211. Gräter, F.; Schwarzl, S. M.; Dejaegere, A.; Fischer, S.; Smith, J. C. *J. Phys. Chem. B* **2005**, *109*, 10474–10483.
212. Cummins, P. L.; Gready, J. E. *J. Comput. Chem.* **2005**, *26*, 561–568.
213. Fukuzawa, K.; Kitaura, K.; Uebayasi, M.; Nakata, K.; Kaminuma, T.; Nakano, T. *J. Comput. Chem.* **2005**, *26*, 1–10.
214. Xiang, Y.; Zhang, D. W.; Zhang, J. Z. H. *J. Comput. Chem.* **2004**, *25*, 1431–1437.
215. Raha, K.; Merz, K. M., Jr. *J. Am. Chem. Soc.* **2004**, *126*, 1020–1021.
216. Raha, K.; Merz, K. M., Jr. *J. Med. Chem.* **2005**, *48*, 4558–4575.
217. Head-Gordon, M. *J. Phys. Chem.* **1996**, *100*, 13213–13225.

Biography

Alexander A Alex is Director, Computational Chemistry at Pfizer Global Research & Development, Sandwich, UK. He joined Pfizer in 1993 after receiving a doctorate in chemistry from the University of Erlangen in Germany. Dr Alex has extensive pharmaceutical drug discovery experience, and his main scientific interests are in the areas of virtual screening and binding energy predictions, structure-based drug design, ADMET modeling, and predictions of chemical reactivity through molecular orbital calculations.

© 2007 Elsevier Ltd. All Rights Reserved
No part of this publication may be reproduced, stored in any retrieval system or transmitted in any form by any means electronic, electrostatic, magnetic tape, mechanical, photocopying, recording or otherwise, without permission in writing from the publishers

Comprehensive Medicinal Chemistry II
ISBN (set): 0-08-044513-6

ISBN (Volume 4) 0-08-044517-9; pp. 379–420

4.17 Chemogenomics in Drug Discovery – The Druggable Genome and Target Class Properties

A L Hopkins and G V Paolini, Pfizer Global Research and Development, Sandwich, UK

© 2007 Published by Elsevier Ltd.

4.17.1 Introduction

Over the past 100 years since Paul Ehrlich's first systematic search for drugs to discover arsphenamine (Salvarsan), medicinal chemistry has continuously sought more effectively means to navigate the vastness of chemical space in the search for new therapies. Arguably the greatest contributions to the changing practice of medicinal chemistry in recent decades have come from the influence of molecular biology and protein crystallography. Advances in molecular biology, culminating in whole genome sequencing, provide modern drug discoverers with the entire palette of proteins that are the past and future drug targets. From the genomics scale to the atomic scale, insights from protein crystallography enable drug designers to observe in atomic resolution the details of the interaction between ligands and drug targets. Modern medicinal chemistry is now capable of synthesizing knowledge from structure–activity relationships (SARs), large-scale screening campaigns, and insights from structure-based drug design to find the intersects between protein sequences and chemical structure. Chemogenomics attempts to integrate chemical space with biology on a genome scale. In the following chapter we outline how insights from chemogenomics can be directly applied in medicinal chemistry in the target discovery and lead discovery stages.

4.17.2 Pharmacological Target Space

One of the key questions for molecular approaches to medicinal chemistry is what are all the proteins which current leads and drugs act upon? The list of molecular targets for which small-molecule chemical matter has been discovered has been difficult to ascertain, because of the lack of integrated and accessible databases for pharmacological information. Overington *et al.*[1] from a comprehensive survey of the literature identifies 196 human protein drug targets that the current pharmacopoeia of US Food and Drug Administration (FDA) approved small-molecule drugs act on. In terms of the number of protein targets for which lead matter has been identified, Paolini *et al.* have attempted the large-scale integration of proprietary and published screening data to identify the number of unique molecular targets for which chemical tools, leads, or drugs have been discovered.[2] The Paolini *et al.* global survey of the data from Pfizer, Warner-Lambert, and Pharmacia integrated with a large body of medicinal chemistry SAR results published in the literature (*J. Med. Chem.* 1980–2003 and *Bioorg. Med. Chem. Lett.* 1990–2003)[3] identified for 1306 proteins from 55 organisms, with biologically active chemical matter. These include a nonredundant list of 836 genes in the human genome for which small-molecule chemical tools have been discovered, of which 727 human targets have at least one compound with binding affinity below 10 μM compliant with Lipinski's 'rule-of-five' criteria for oral drug absorption[4] and 529 human targets have at least one 'rule-of-five' compound below 100 nM (**Table 1**).

Table 1 Pharmacological target space[2]

Gene taxonomy	*Human targets at* $<10\,\mu M$	*Human targets at* $<1\,\mu M$	*Human targets at* $<10\,\mu M$ *Ro5*[a] $n>1$	*Human targets at* $<100\,\mu M$ *Ro5*[a] $n>1$
Protein kinases	105	99	98	83
Peptide GPCRs	63	59	59	42
Transferases	49	42	36	24
Aminergic GPCRs	35	35	35	35
GPCRs Class A others	44	44	40	32
Oxidoreductases	40	36	38	25
Metalloproteases	44	41	41	35
Hydrolases	36	29	30	21
Ion channels ligand gated	29	28	24	22
Nuclear hormone receptors	24	24	22	19
Serine proteases	30	30	28	21
Ion channels others	18	16	16	11
PDEs	19	19	19	18
Cysteine proteases	16	16	14	13
GPCRs Class C	10	10	10	6
Kinases others	12	9	11	5
GPCRs Class B	7	7	4	3
Aspartyl proteases	7	7	4	4
Others	139	119	108	63
Enzymes others	109	97	90	47
Total	**836**	**767**	**727**	**529**

[a] Compounds passing Lipinski's 'rule-of-five' criteria.[4]

4.17.3 Chemical Properties of Drugs and Leads

Essential to the design of a drug are the physicochemical characteristics of the lead compound. A balance of solubility and polar/hydrophobic properties is crucial for specific routes of absorption and membrane permeabilities and other biological barriers that a drug needs to penetrate to reach the desired site of action, in order to affect the biological equilibrium of a whole organism. The presence of such biological barriers limits the range of molecular properties, and thus the chemical space the medicinal chemists can design within.

Lipinski's analysis of the Derwent World Drug Index introduced the concept of physicochemical property limits, with respect to solubility and permeability of drugs. Lipinski *et al.* demonstrated that orally administered drugs are far more likely to reside in areas of chemical space defined by a limited range of molecular properties. Lipinski's[4] analysis showed that drugs with molecular weights of less than 500 Da, fewer than 5 hydrogen-bond donors (such as the combined OH and NH group count), fewer than 10 hydrogen-bond acceptors (such as the combined nitrogen and oxygen atom count), and lipophilicity less than calculated log P (ClogP) of 5 were far more likely to be orally absorbed. The multiples of five observed in the molecular properties of drugs led to the coining of the term Lipinski's 'rule-of-five' (Ro5). Several methods of predicting 'druglikeness' have been proposed, in which a defined range of molecular properties and physicochemical descriptors can discriminate between drugs and nondrugs for such characteristics as oral absorption, aqueous solubility, and permeability.[4–16]

Since our current data on the properties of drugs point to a range of molecular properties within which the likelihood of compound becoming an oral drug is increased, it is interesting to ask how do the range of properties of ligands binding to specific targets overlap with 'druglike' space. Paolini *et al.* have investigated the relationship between target class and the physicochemical properties of ligands by calculating a set of physicochemical descriptors for over a quarter of a million biologically active compounds, across over 1300 targets, where the protein sequences assigned to each of the pharmacological targets were classified into gene families. Distinct differences in the distribution of molecular properties between sets of compounds active against different gene families were observed (**Table 2**, **Figure 1**). For example, ligands for the nuclear hormone receptors are significantly most lipophilic, as measured by ClogP, mirroring the properties of steroids. In comparison the mean molecular weight (MW) of ligands binding to aminergic G protein-coupled receptors (GPCRs) is 378 Da (SD = 93 Da), close to the mean MW of approved drugs (383 Da, SD = 155 Da), while the mean MW of peptide GPCR ligands is greater at 514 Da but with a wider spread (SD = 202 Da), significantly over Lipinski's 'rule-of-five' limits of 500 Da.

Development of the ideas in druglikeness has lead to the proposal of the concept of 'degrees of druggability.' Degrees of druggability proposed druggability and druglikeness can be measured as a probabilistic continuum, where two protein targets may be both classified as druggable but may exhibit differences in the probabilities of success, due to the physicochemical properties of their respective ligands. One proposed measure of the degree of druggability is proposed as the distance of the centroid in reduced chemical space (e.g., MW, ClogP, number of hydrogen bond donors, and number of hydrogen bond acceptors) for all of the potent actives (i.e., binding affinities <100 nM) associated with each target, to that of the centroid of the probabilistic clustering of approved oral drugs. Over 65% of targets for oral drugs are within a distance of 0.4 from the centroid of oral drug space and 87% of oral drug targets are within a distance of 0.6. Within these degrees of druggability, approximately 200 human targets with potent leads, including the current drug targets, are within a distance of 0.4 from the oral drug centroid but have yet to produce approved drugs.

4.17.4 Trends in Molecular Properties

Over the past two decades the number of targets (including selectivity counterscreens) published in the medicinal chemistry literature has been growing steadily. In recent years screening data on nearly 900 proteins have been published, with around 500 molecular targets reported with potent chemical matter with binding affinities below 100 nM.[2] Chemical tools and leads for approximately 80 to 100 new molecular targets are first disclosed each year (**Figure 2**). No doubt this is a conservative estimate as many new compounds and targets are only disclosed in patents, which are not included in this initial analysis, which was based on published journal data.[2] In comparison, the rate of first disclosure for novel targets with new leads has doubled from an average of 30 new targets with leads being disclosed in the 1980s to an average of 60 new targets per year in the 1990s. Over the same time period there have also been some significant trends in the changing character of the industry's portfolio of targets and targets classes (**Figure 3**) such as rise of interest across the industry in protein kinases and the relative decline in proportion of aminergic GPCRs in the industry's target portfolio.

Interestingly over the past two decades there has been a steady rise in the mean and median molecular weight of reported medicinal chemistry compounds (**Figure 4**) by around 20% with the median MW of all reported medicinal chemistry compounds in the literature rising 68 Da from 354 to 422 Da, for the periods 1986–1990 to 1999–2003, respectively. Interestingly, this growth is also reflected across the board in the increase of the median MW of disclosed ligands for several gene families. Aminergic GPCRs compounds increased in molecular weight by 56 Da from 337 to 393 Da between the two 5-year periods. In contrast to the changes in the properties of medicinal chemistry compounds over the past two decades, Vieth *et al.*[17] have observed that the distribution of mean molecular properties of approved oral (small-molecule) drugs has changed little in the past 20 years, despite differences in the range of indications and targets. Interestingly, a steady decline in MW through each subsequent stage of clinical development and increase in the proportion of compounds that are 'rule-of-five' compliant has also been observed.[18,19] The relative difference in molecular properties between the gene families is also reflected in compounds in clinic development; however, even within a gene family, the median MW of compounds surviving subsequent clinical phases exhibits a slight decline (**Figure 5**).[2]

In order to reduce the MW of leads and clinical candidates, and improve their chances in clinical development, the metric of 'ligand efficiency'[20,21] is gaining popularity amongst medicinal chemists as a means to assess the potential of a low molecular weight but low-affinity lead to be optimized into a high-affinity clinical candidate. The binding energy of the ligand per atom,[22] or ligand efficiency (Δg)[20] of a compound can be calculated by converting the K_d into the free energy of binding (eqn [1]) at 300 K and dividing by the number of 'heavy' (i.e., non-hydrogen atoms) atoms (eqn [2]):

Table 2 Physicochemical properties of ligands by gene family

Gene taxonomy	*MW (Da) (Mean)*	*MW (Da) (SD)*	*MW (Da) (Median)*	*90% limit of MW (Da)*	*ClogP (Mean)*	*ClogP (SD)*	*ClogP (Median)*	*90% limit of ClogP*
Aminergic GPCRs	378	93	376	460	3.8	1.6	3.9	5.6
Ion channels ligand gated	359	91	362	430	3.0	1.8	3.2	4.7
Metalloproteases	428	103	429	530	3.0	1.9	3.1	4.8
Nuclear hormone receptors	398	96	396	495	5.1	1.7	5.0	7.3
Peptide GPCRs	514	202	477	752	4.3	2.3	4.6	6.5
Phosphodiesterases	400	65	397	465	3.7	1.4	3.7	5.2
Protein kinases	407	109	402	505	3.8	1.8	3.9	5.7
Serine proteases	467	145	463	572	2.7	2.1	2.7	4.8

	No. of hydrogen bond acceptors (Mean)	*No. of hydrogen bond acceptors (SD)*	*No. of hydrogen bond acceptors (Median)*	*90% limit of no. of hydrogen bond acceptors*	*No. of hydrogen bond donors (Mean)*	*No. of hydrogen bond donors (SD)*	*No. of hydrogen bond donors (Median)*	*90% limit of no. of hydrogen bond donors*
Aminergic GPCRs	4	2	4	6	1	1	1	2
Ion channels ligand gated	4	2	4	6	2	1	2	3
Metalloproteases	6	2	6	8	3	1	2	4
Nuclear hormone receptors	4	2	4	6	1	1	1	2
Peptide GPCRs	5	4	4	10	2	3	1	8
Phosphodiesterases	6	2	6	8	1	1	1	2
Protein kinases	5	2	5	7	2	1	2	4
Serine proteases	5	3	5	8	3	2	2	4

	No. of rotatable bonds (Mean)	*No. of rotatable bonds (SD)*	*No. of rotatable bonds (Median)*	*90% limit of no. of rotatable bonds*	*Ligand efficiency (kcal mol^{-1} per non-H atoms) (Mean)*	*Ligand efficiency (kcal mol^{-1} per non-H atoms) (SD)*	*Ligand efficiency (kcal mol^{-1} per non-H atoms) (Median)*
Aminergic GPCRs	6	3	6	8	0.4	8.0E-02	0.4
Ion channels ligand gated	5	3	4	7	0.4	0.1	0.4
Metalloproteases	8	4	8	13	0.4	0.2	0.3
Nuclear hormone receptors	6	3	6	10	0.3	6.E-02	0.3
Peptide GPCRs	9	7	8	17	0.2	7.E-02	0.2
Phosphodiesterases	6	3	6	9	0.3	3.E-02	0.3
Protein kinases	6	3	5	9	0.3	7.E-02	0.3
Serine proteases	8	5	7	12	0.3	9.E-02	0.3

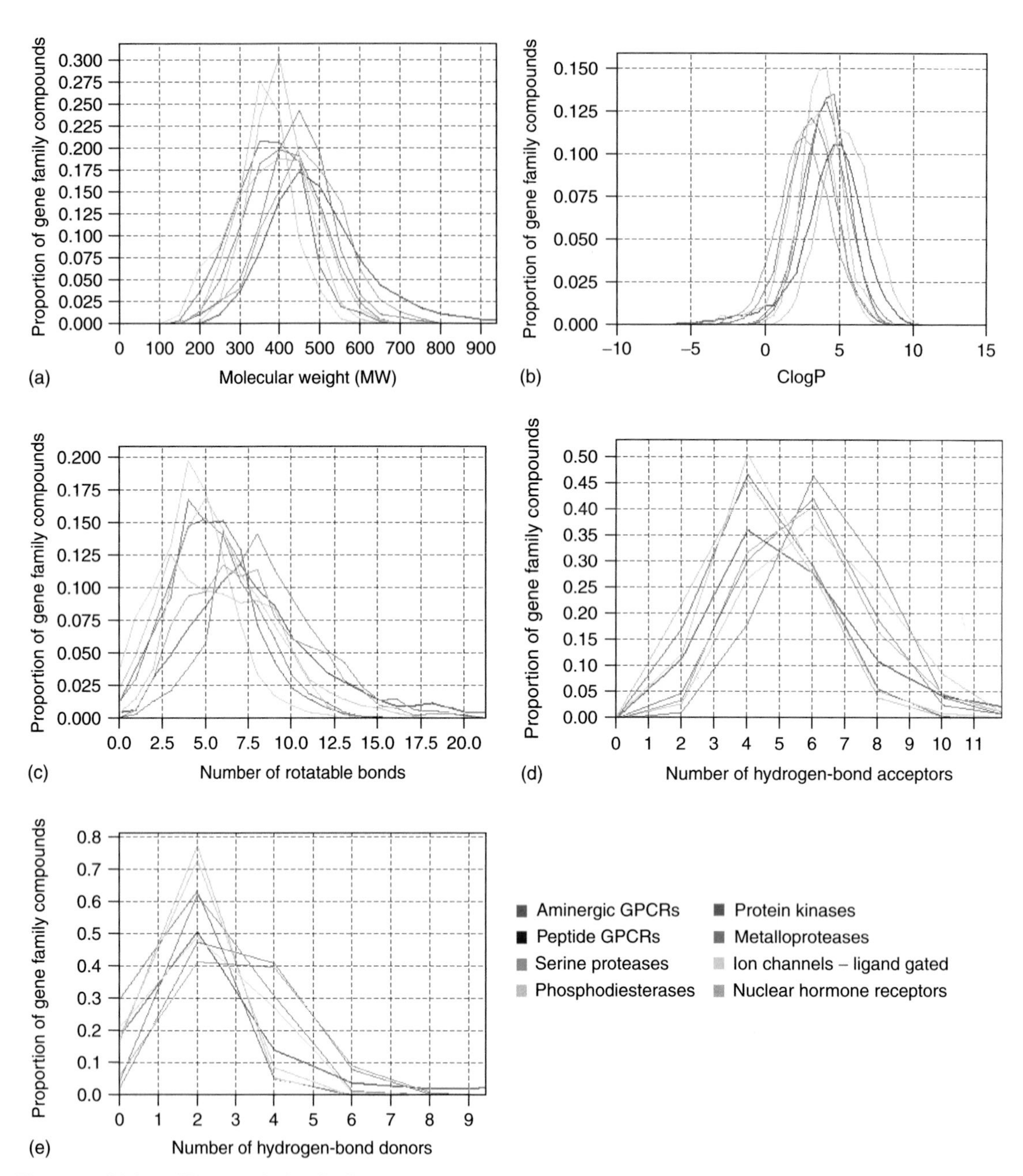

Figure 1 Distinct differences in the distribution of molecular properties between sets of compounds active against different gene families by (a) molecular weight, (b) ClogP, (c) number of rotatable bonds, (d) number of hydrogen bond acceptors, (e) number of hydrogen bond donors.

Free energy of ligand binding:

$$\Delta G = -\mathrm{RT}\ \ln K_{\mathrm{i}} = 1.4\ \log K_{\mathrm{i}} \qquad [1]$$

where R = gas constant = 1.986 cal mol^{-1} K^{-1}

Binding energy per atom (ligand efficiency):

$$\Delta g = \Delta G/\mathrm{N}_{\text{non-hydrogen atoms}} \qquad [2]$$

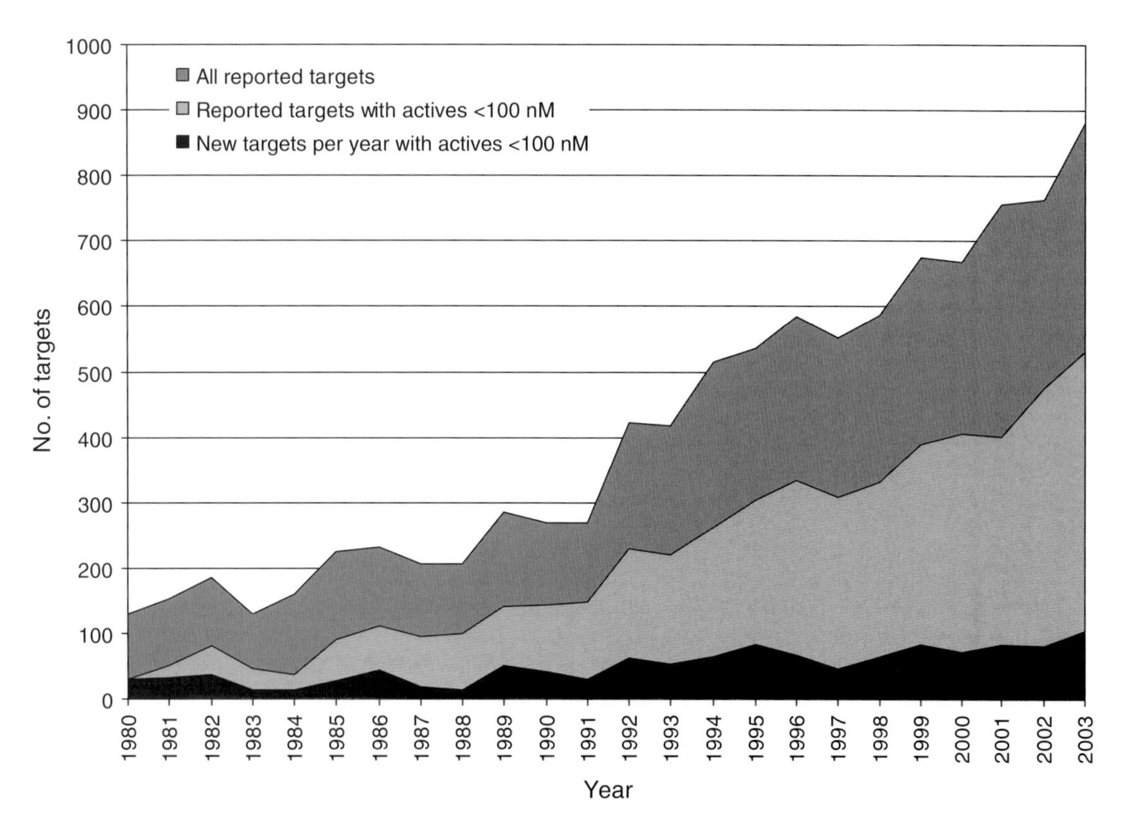

Figure 2 Number of protein targets with small-molecule leads reported in the medicinal chemistry literature per year.

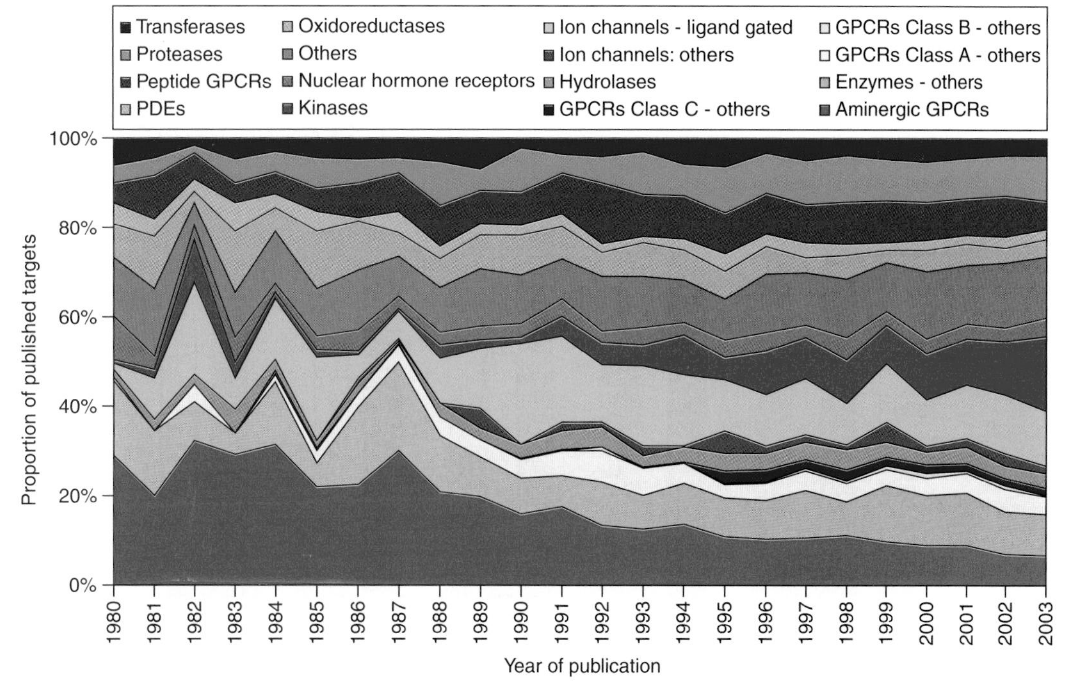

Figure 3 Changes in the pharmaceutical industry's portfolio of targets classes (as disclosed in medicinal chemistry literature per year).

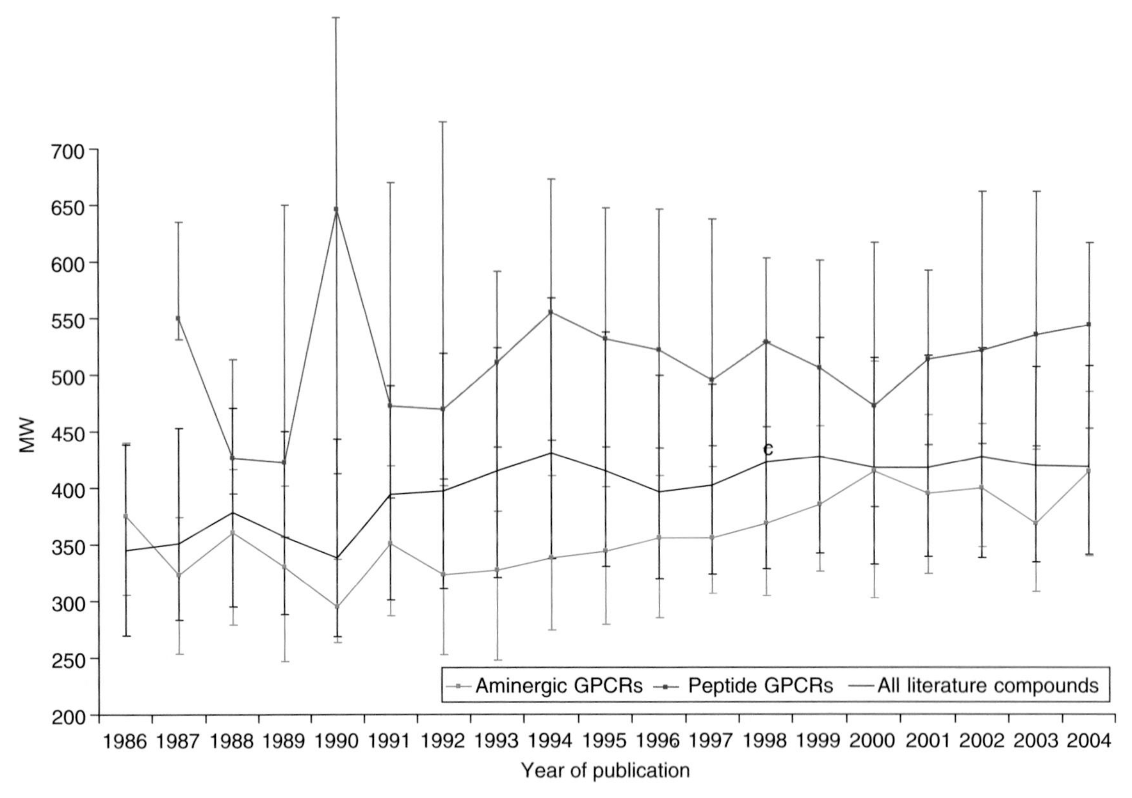

Figure 4 Steady rise in the median MW of reported medicinal chemistry compounds over time.[2]

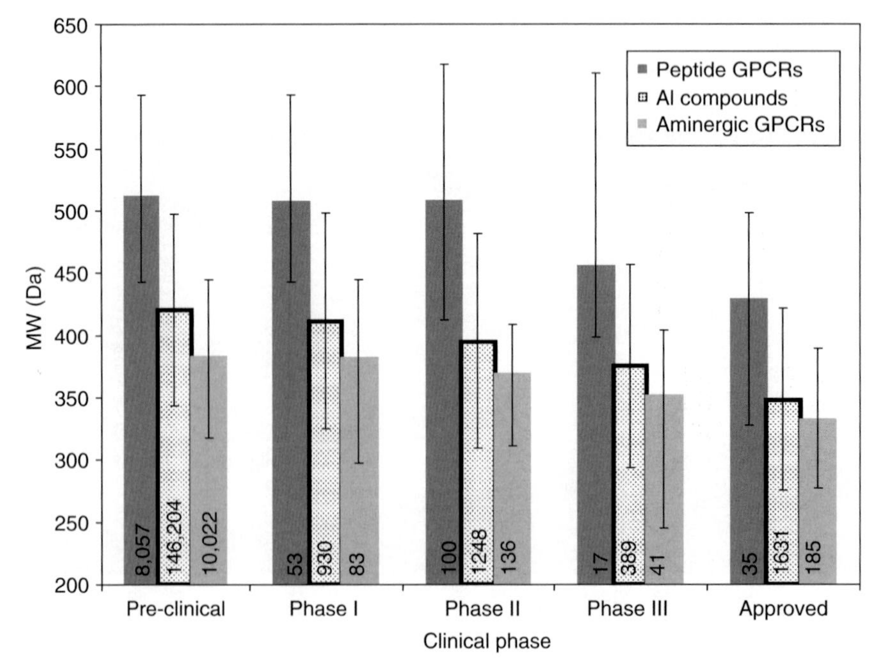

Figure 5 Decline in MW of drugs in development. Median MW between aminergic GPCRs, peptide GPCRs, and all compounds through subsequent stages of clinical development. The number of compounds for each class at each stage is labeled.[2]

4.17.5 Identifying the Druggable Genome

The knowledge of which proteins current medicinal chemistry has developed drugs and leads against can be used to infer the subset of the proteins expressed by the human genome that have a high probability of being potentially druggable, i.e., capable of binding druglike small molecules with high affinity. The first systematic estimate of the number of druggable proteins – the 'druggable genome' – following the publication of the draft human genome,[23,24] was based on a search for membership of an extensive list of druggable gene families.[25] Gene family-based analysis assumes that the sequence and functional similarities underlie a conservation of binding site architecture between protein family members. Thus the explicit assumption being that if one member of a gene family is modulated by a drug molecule, other members of the druggable protein domain family are likely to also be able to bind a compound with similar physicochemical properties. Thus analysis based on druggable protein families or domains is likely to overestimate the number of druggable targets. Following the construction of a drug target sequence database of 399 targets of approved and experimental drugs and leads, 376 sequences could be assigned to 130 drug-binding domains, as captured by their InterPro domain annotation. Of these, 125 InterPro domains have orthologs present in the human proteome. At the time of the initial draft of the human genome[23,24] 3051 genes were identified as belonging to the 125 druggable protein domains and thus predicted to encoded proteins that are inferred to bind a drug like molecules.

Further refinements of the Hopkins and Groom analysis has been published by Orth *et al.*[26] and Russ and Lampel[27] reflecting how the number of predicted protein expressing genes in the human genome has been modified since the initial draft. Orth *et al.*[26] estimate that there are 3080 genes belonging the druggable genome with over 2950 druggable gene sequences in public databases in 2004 based on an estimate of the InterPro domain assignments of druggable gene families.[25] Russ and Lampel[27] conducted and analysis of the 120 druggable protein domains using InterPro and Pfam[35] on the final assembly of the human genome.[28] Overall the Pfam protein domain annotation predicted fewer false positives than the InterPro classification used. When corrected for the overestimate of olfactory and taste GPCRs the authors identify, again, 3050 druggable genes from the previously defined set of druggable protein domains,[25] but with some significant changes within individual gene families. Using more stringent predictions for enzymes, proteases, and other subfamilies a conservative estimate of around approximately 2200 druggable genes are identified.[27]

In order to expand the homology analysis methodology for identifying which targets expressed from the human genome are likely to be druggable, it is necessary to expand our survey to identify all the known biological targets of drugs and lead compounds. Al-Lazikani and Overington (Inpharmatica, London) have conducted the most extensive analysis, to date, on identifying the druggable genome, based on the homology to chemically tractable drug targets.[29] Using the BLAST sequence alignment algorithm to search each of the sequences against the human genome, Al-Lazikani and Overington identified 945 distinct genes that show homology to 170 human proteins of approved small-molecular drugs,[3] at a cutoff of 30% sequence identity and E-value less than or equal to 10^{-5}. Expanding the BLAST analysis to include human proteins from the known small-molecule chemical leads, Al-Lazikani and Overington expanded the sequence set to include Inpharmatica's StARLITe database of medicinal chemistry journal data (i.e., *J. Med. Chem.* 1980–2004, *Bioorg. Med. Chem. Lett.* 1990–2004) containing 1155 protein targets known with at least one drug or lead compound with a binding affinity below 10 μM, 707 of which are human molecular targets. BLAST sequence analysis of this database of medicinal chemistry literature[3] identified 2921 protein sequences within the same sequence identify cutoffs, which are predicted to be druggable proteins expressed by the human genome.

4.17.6 Molecular Recognition Basis for Druggability

The hypothesis that the druggability of a protein can be assessed a priori derives from the biophysical basis of molecular recognition.[30–32] The binding energy (ΔG) of a ligand to a molecular target such as a protein, RNA, DNA, or carbohydrate is defined in eqn [1]. Van der Waals and entropy components predominately drive the binding energy by the burying of hydrophobic surfaces. A low-affinity 'hit' from a high-throughput screen of $K_i = 1\ \mu M$ affinity equates to $-8.4\ kcal\ mol^{-1}$. A high-affinity drug molecule binding with an affinity of $K_i = 10\ nM$ requires a binding energy (ΔG) of $-11\ kcal\ mol^{-1}$. Thus $1.36\ kcal\ mol^{-1}$ of binding energy is equivalent to a 10-fold increase in potency. The binding energy potential of a ligand is approximately proportional to the available surface area and its properties, assuming there are no strong covalent or ionic interactions between the ligand and the protein. Analysis of nearly 50 000 biologically active druglike molecules reveals a linear correlation between molecular surface area and molecular weight (**Figure 6**). The van der Waals attractions between atoms and the hydrophobic effect from the displacement of water contributes approximately $0.03\ kcal\ mol^{-1}\ Å^2$. Thus, assuming there are no strong ionic interactions between the protein and the ligand, a ligand with a 10 nM dissociation constant would be required to bury $370\ Å^2$ of hydrophobic surface area. The contribution of the hydrophobic surface to binding energy is demonstrated by the medicinal chemistry phenomenon of

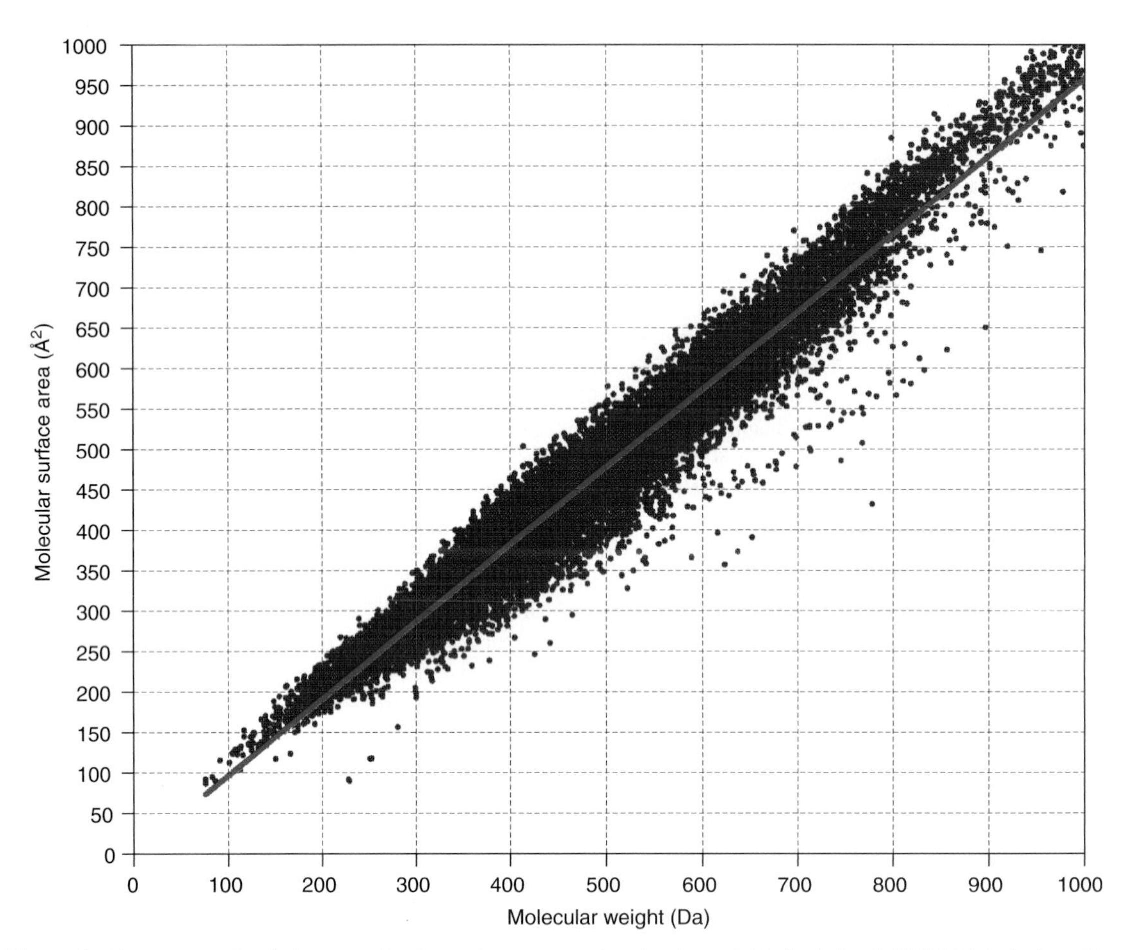

Figure 6 Linear correlation between molecular surface area and molecular weight. Analysis of 49 456 biologically active, druglike compounds with $IC_{50} \leq 100\,nM$. Molecular weight was calculated from the chemical structures represented as desalted, canonical SMILES strings. The calculated molecular surface area N, O, P, and S atoms. was estimated using the fast Ertl method[34] using a 2D approximation. All other atom types (excluding hydrogen atoms) were estimated using an overlapping spheres method.

the 'magic methyl,' where a single methyl group placed in the correct position, can increase ligand affinity by tenfold. The accessible hydrophobic surface area of a methyl group is approximately $46\,Å^2$ (if one assumes that all of the hydrophobic surface area is encapsulated by the protein binding site and thus forms full contact with the protein) with a hydrophobic effect of $0.03\,kcal\,mol^{-1}\,Å^2$ equal approximately to $1.36\,kcal\,mol^{-1}$ equivalent to the observed tenfold affinity increase: approximately the maximal affinity per non-hydrogen atom.[22] In addition to the predominantly hydrophobic contribution to the binding of many drugs, ionic interactions, such as those found in zinc proteases (such as angiotensin-converting enzyme (ACE) inhibitors) contribute to the binding energy. The attraction of complementary polar groups contributes up to up to $0.1\,kcal\,mol^{-1}\,Å^2$, with ionic salt bridge approximately three times greater, allowing low molecular weight compounds to bind strongly. Unlike hydrophobic interactions complementary polar interactions are dependent on the correct geometry.

4.17.7 Analysis of Protein Structures as Drug Targets

The physicochemical and energetic constraints of molecular recognition lead to the conclusion that a drug target needs a 'pocket,' whether the pocket is predefined or formed on binding by allosteric mechanisms. Druggable cavities on proteins that are complementary with the high-affinity binding of noncovalent, small-molecule, 'rule-of-five' compliant ligands (whose binding energy is predominantly driven by the entropic, hydrophobic, and van der Waals contributions) are predominately apolar cavities of $400–1000\,Å^3$, where over 65% of the pocket is buried or encapsulated, with an accessible hydrophobic surface area of at least 350 $Å^2$.[31] Encapsulated cavities maximize the ratio of the surface area to the volume and are thus capable of binding low molecular weight compounds with high affinities.

The hypothesis that the physicochemical properties of cavities on protein structures can be analyzed a priori to predict the druggability of a protein has been developed further into automatic algorithms to assess the protein structures in the Protein Data Bank (PDB) and the stream of novel structures determined by the structural genomics initiatives.[29,30–33] Empirical druggability predictions have been explored experimentally at Abbott using heteronuclear nuclear magnetic resonance (NMR) to identify and characterize the binding surfaces on protein by screening ~10 000 low molecular weight molecules (average MW 220, average ClogP 1.5).[31,32] In a small sample of 23 proteins, the screening results reveal that about 90% of the ligands binds to sites known to be small-molecule ligand binding sites. Only in three out of the 23 proteins were distinct uncompetitive new binding sites discovered. In the relatively small sample of proteins studied, Hajduk *et al.* noted a high correlation between experimental NMR hit rates and the ability to find high-affinity ligands. From the experimental screening hit rates, Hajduk *et al.* constructed a simple model that included physicochemical property descriptor such as cavity dimensions, surface complexity, and polar and apolar surface area that accurately predicts the experimental screening hit rates with an R^2 of 0.72, and an adjusted R^2 of 0.65.

A decision-tree approach to assessing the druggability of protein structure has been developed by Inpharmatica by Al-Lazikani and Overington.[29] A range of physicochemical properties of the identified binding sites and cavities were calculated from the protein structures including volume, depth, curvature, accessibility, hydrophobic surface area, and polar surface area. The algorithm was trained set against a test set of 400 protein complexes binding small molecule, 'rule-of-five' compliant ligands. From this analysis a decision-tree was derived to predict the druggability of a binding site or cavity from calculated physicochemical properties. The decision-tree predicts whether a cavity is druggable within the statistical confidence of the tree. A success rate of 91% when predicting druggability on the protein drug targets has been claimed for this approach.[29] The method requires either an experimentally derived structure or a high-quality homology model. Ideally, because of the inherent flexibility of many protein–ligand binding sites, a sample of multiple conformations is preferred. The decision-tree method was applied to the entire PDB (December 2004 release). Following a clean-up process, 27 409 files were suitable for analysis, further classified into 76 322 structural domains (using SCOP[33]) of which 28% (21 522) were found to have at least one site predicated to have some degree of druggability. From this analysis a nonredundant set of 427 human proteins were predicted to contains a druggable binding site, of which 281 had no prior known compounds or drugs developed against them. In a similar analysis Hudjuk *et al.* calculated the druggability of 1000 nonredundant human proteins derived from the PDB, of which 35% of entries contained at least one site predicted to be highly druggable; slightly higher but comparable with Al-Lazikani's prediction.

4.17.8 Conclusions

The palette of potential drug targets for modern medicinal chemistry can now be efficiently derived from searching entire genomes. Knowledge-based methods enable the mapping of chemical space to protein structure and protein sequences to predict druggable targets. The observed relationships between the physicochemical properties of ligands and the targets they bind to identifies not only potential druggable targets but also the degree of druggability – a means of assessing their probabilistic likelihood of success. Understanding the degrees of druggability between protein targets can aid the medicinal chemist in the selection of a portfolio of drug targets, in the design of the screening strategy, in identifying the likely region of chemical space target ligands may reside in and in the probability of success through clinical development relative to disease indication. Advances in molecular biology and structural biology have had great impact on the practice of modern medicinal chemistry by enabling a detailed understanding of the atomic basis of SARs on individual protein targets. The next wave of advances in modern medicinal chemistry is likely to benefit from the effective integration of pharmacological and chemogenomic knowledge gained from the past 100 years in a collective whole to aid the practice of target discovery and compound design.

References

1. Overington, J.; Al-Lazikani, B.; Hopkins, A. L. *Nat. Rev. Drug Disc.* **2006**, *5*, in press.
2. Paolini, G. V.; Shapland, R. H. B.; Van Hoorn, W. P.; Mason, J. S.; Hopkins, A. L. *Nat. Biotech.* **2006**, *24*, 805–815.
3. Overington, J.; Al-Lazikani, B. Inpharmatica, StARlite database; Inpharmatica Ltd: London, 2005.
4. Lipinski, C. A.; Lombardo, F.; Dominy, B. W.; Feeney, P. J. *Adv. Drug Deliv. Rev.* **1997**, *23*, 3–25.
5. Ajay, A.; Walters, W. P.; Murcko, M. A. *J. Med. Chem.* **1998**, *41*, 3314–3324.
6. Wang, J.; Ramnarayan, K. *J. Combin. Chem.* **1999**, *1*, 524–533.

7. Walters, W. P.; Ajay, A.; Murcko, M. A. *Curr. Opin. Chem. Biol.* **1999**, *3*, 384–387.
8. Lipinski, C. A. *J. Pharmacol. Toxicol. Methods* **2000**, *44*, 3–25.
9. Podlogar, B. L.; Muegge, I.; Brice, L. J. *Curr. Opin. Drug Disc. Dev.* **2001**, *4*, 102–109.
10. Muegge, I.; Heald, S. L.; Brittelli, D. *J. Med. Chem.* **2001**, *44*, 1841–1846.
11. Veber, D. F.; Johnson, S. R.; Cheng, H. Y.; Smith, B. R.; Ward, K. W.; Kopple, K. D. *J. Med. Chem.* **2002**, *45*, 2615–2623.
12. Proudfoot, J. R. *Bioorg. Med. Chem. Lett.* **2002**, *12*, 1647–1650.
13. Walters, W. P.; Murcko, M. A. *Adv. Drug Deliv. Rev.* **2002**, *54*, 255–271.
14. Egan, W. J.; Walters, W. P.; Murcko, M. A. *Curr. Opin. Drug Disc. Dev.* **2002**, *5*, 540–549.
15. Muegge, I. *Med. Res. Rev.* **2003**, *23*, 302–321.
16. Lajiness, M. S.; Vieth, M.; Erickson, J. *Curr. Opin. Drug Disc. Dev.* **2004**, 7, 470–477.
17. Vieth, M.; Siegel, M. G.; Higgs, R. E.; Watson, I. A.; Robertson, D. H.; Savin, K. A.; Durst, G. L.; Hipskind, P. A. *J. Med. Chem.* **2004**, *47*, 224–232.
18. Wenlock, M. C.; Austin, R. P.; Barton, P.; Davis, A. M.; Leeson, P. D. *J. Med. Chem.* **2003**, *46*, 1250–1256.
19. Blake, J. F. *Biotechniques* **2003**, *June*, 16-20.
20. Hopkins, A. L.; Groom, C. R.; Alex, A. *Drug Disc. Today* **2004**, *9*, 430–431.
21. Cele, A. Z.; Metz, J. T. *Drug Disc. Today* **2005**, *10*, 464–469.
22. Kuntz, I. D.; Chen, K.; Sharp, K. A.; Kollman, P. A. *Proc. Natl. Acad. Sci. USA* **1999**, *96*, 9997–10002.
23. International Human Genome Sequencing Consortium. *Nature* **2001**, *409*, 860–921.
24. Venter, J. C.; Adams, M. D.; Myers, E. W.; Li, P. W.; Mural, R. J.; Sutton, G. G.; Smith, H. O.; Yandell, M.; Evans, C. A.; Holt, R. A. et al. *Science* **2001**, *291*, 1304–1351.
25. Hopkins, A. L.; Groom, C. R. *Nat. Rev. Drug Disc.* **2002**, *1*, 727–730.
26. Orth, A. P.; Batalov, S.; Perrone, M.; Chanda, S. K. *Expert Opin. Ther. Targets* **2004**, *8*, 587–596.
27. Russ, A. P.; Lampel, S. *Drug Disc. Today* **2005**, *10*, 1607–1610.
28. Lander, E. S.; Linton, L. M.; Birren, B.; Nusbaum, C.; Zody, M. C.; Baldwin, J.; Devon, K.; Dewar, K.; Doyle, M.; FitzHugh, W. et al. *Nature* **2004**, *409*, 860–921.
29. Al-Lazikani, B.; Gaulton, A.; Paolini, G.; Lanfear, J.; Overington, J.; Hopkins, A. Chemical Biology. In *From Small Molecules to Systems Biology and Drug Design*; Schreiber, S. L., Kapoor, T., Wess, G., Eds.; Wiley: New York, 2007, pp 1–20.
30. Hopkins, A. L.; Groom, C. R. *Ernst Schering Res. Found. Workshop* **2003**, *42*, 11–17.
31. Hajduk, P. J.; Huth, J. R.; Tse, C. Predicting Protein druggability. *Drug Disc. Today* **2005**, *10*, 1675–1682.
32. Hajduk, P. J.; Huth, J. R.; Fesik, S. W. *J. Med. Chem.* **2005**, *48*, 2518–2525.
33. Murzin, A. G.; Brenner, S. E.; Hubbard, T.; Chothia, C. *J. Mol. Biol.* **1995**, *274*, 536–540.
34. Ertl, P.; Rohde, B.; Selzer, P. *J. Med. Chem.* **2000**, *43*, 3714–3717.
35. Finn, R. D.; Mistry, J.; Schuster-Bockler, B.; Griffiths-Jones, S.; Hollich, V.; Lassman, T.; Moxon, S.; Marshall, M.; Khanna, A.; Durbin, R. et al. *Nucleic Acids Res.* **2006**, *34*, D247–D251.

Biographies

Andrew L Hopkins is presently Associate Research Fellow and Head of Chemogenomics at the Sandwich site of Pfizer Global Research and Development. He joined Pfizer in 1998 at Sandwich, Kent, UK. Over the years, he has established various new functions for Pfizer including, Target Analysis in 1999, Indications Discovery in 2001 and most recently Knowledge Discovery in 2004. He won a British Steel scholarship to attend the University of Manchester from where he graduated with first class honours in 1993 with a BSc in Chemistry. Following a brief spell in the steel industry he won a Wellcome studentship to attend the University of Oxford, working with Prof David I Stuart FRS. He received his DPhil in Structural Biology from the University of Oxford in 1998. During his doctorate research Dr Hopkins designed a new class of anti-HIV agents which were developed to drug candidates by Glaxo-Wellcome. Following his interest in drug discovery he then joined Pfizer directly after graduating from Oxford. At Pfizer,

Dr Hopkins' research involves combining chemical and biological knowledge to identify new targets or other new opportunities for medicines. His work has involved the design and construction of major informatics systems, including literature-mining system and a large-scale chemogenomics knowledge-base. He is the author of over 6 patents and 25 scientific publications, two of which have been cited as Hot Paper by the Thomson ISI citation index. Dr Hopkins lives in Canterbury, Kent, UK.

Gaia Paolini is Senior Principal Scientist at the Sandwich site of Pfizer Global Research and Development. Gaia joined Pfizer in 2002 at Sandwich, Kent, UK. Gaia received her degree (laurea) in Physics at the University of Rome, with a research thesis on statistical mechanics computer simulation of condensed matter systems. In her career, Gaia has held a number of positions in industry and academia, combining roles of research scientist, software specialist, and business systems analyst. At Pfizer, Gaia designed and developed a LIMS system for structural biology, and, in her current role, led the design, development and mining of a large chemogenomics knowledge base. For her contribution she has won a Pfizer Achievement Award in 2006. Gaia is the author of 18 peer-reviewed scientific publications, spanning the fields of applied mathematics, materials science, classical density functional theory, and chemogenomics. Gaia currently lives in Canterbury, Kent, UK.

© 2007 Elsevier Ltd. All Rights Reserved
No part of this publication may be reproduced, stored in any retrieval system or transmitted in any form by any means electronic, electrostatic, magnetic tape, mechanical, photocopying, recording or otherwise, without permission in writing from the publishers

Comprehensive Medicinal Chemistry II
ISBN (set): 0-08-044513-6

ISBN (Volume 4) 0-08-044517-9; pp. 421–433

4.18 Lead Discovery and the Concepts of Complexity and Lead-Likeness in the Evolution of Drug Candidates

M M Hann and A R Leach, GlaxoSmithKline R&D, Stevenage, UK
J N Burrows, AstraZeneca R&D, Södertälje, Sweden
E Griffen, AstraZeneca R&D, Macclesfield, UK

© 2007 Elsevier Ltd. All Rights Reserved.

4.18.1 Introduction

The process by which new drugs have been discovered has at one level not changed over the millennia of human interest in disease-modifying substances. Thus, it has always involved gathering data to support the idea that a chemical entity (in varying degrees of purity) can modify or alleviate a disease state. In ancient times, extracts of plants and fungi were the principal sources of such substances. The last two hundred years or so have seen an increasing emphasis on an hypothesis-led approach to drug discovery, whereby an understanding of the disease has led directly to hypotheses of how to intervene to alleviate medical problems. But there has also been an important underlying opportunistic character to drug discovery. While serendipity and opportunism remains a vital 'tool' for drug discovery, the past 10 years have seen a conscious move toward the industrialization of the drug discovery process. There are many reasons for this, which include the exhaustion of the easy pickings, the mapping of the human genome, the development of high-throughput synthesis and screening processes, and the use of enabling robotics to name just a few.

Prior to the current period of industrialization, the pharmaceutical industry had sought and embraced what it believed would be other panaceas to help in the quest for new drugs. These included both trends in approaches (e.g., QSAR) and breakthroughs in technology (e.g., those enabling genomic sequencing and production of proteins as needed). To reflect this, it is instructive to consider how the subject of medicinal chemistry has evolved over the past 50 years. **Figure 1** illustrates the breadth of the subjects that medicinal chemistry now embraces. At each stage over these decades most of the components shown have been considered as transforming technologies that will radically enhance our ability to find new drugs.

Since the 1990s the trend has been increasingly toward enhancing the number of experiments that can be done in parallel with a view to dramatically increasing the throughput and hence the output of the drug discovery industry. This has in part been enabled by the availability of pure proteins, which has also allowed the reductionist approach to biology and drug discovery to be exploited. Such individual protein targets provide the mainstay of the assays that are currently

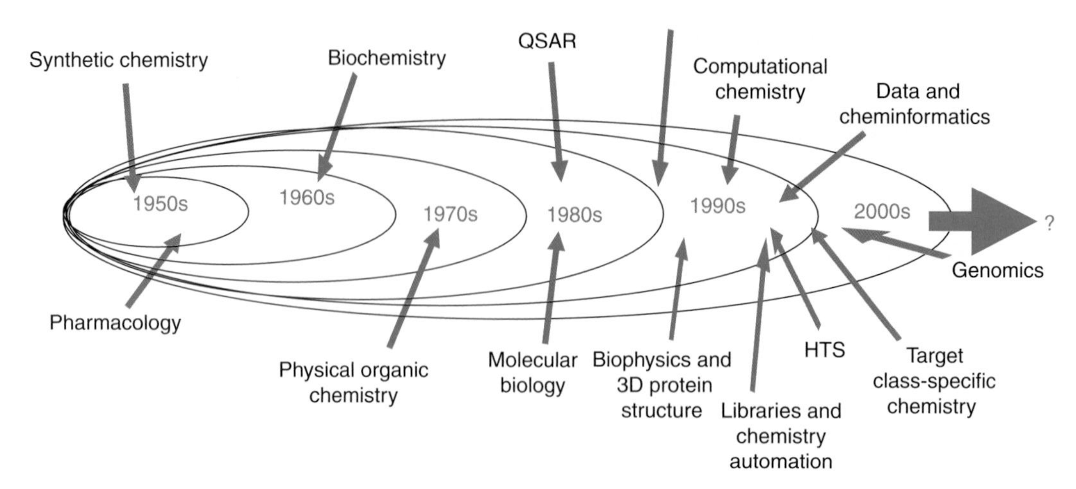

Figure 1 The evolution and incorporation of new technologies into the expanding subject of medicinal chemistry and drugs.

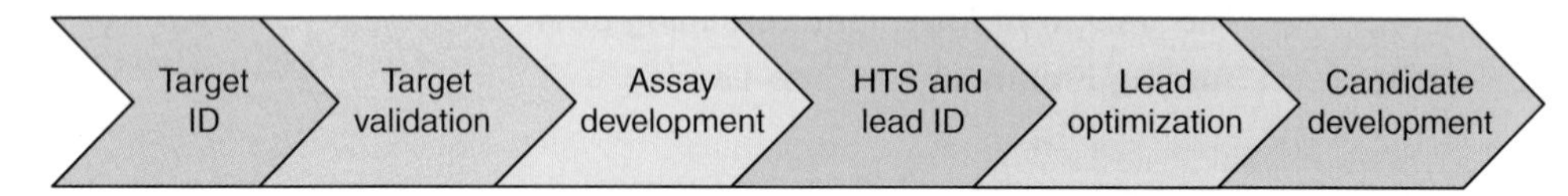

Figure 2 Process stages in a typical drug discovery organization.

used throughout the industry to find new starting points for drug discovery programs. These assays are usually conducted as cell-free assays to facilitate both implementation and interpretation. One of the many consequences of this is that while it has become easier to find compounds with potency against a given target, adsorption, distribution, metabolism, excretion, and toxicity (ADMET) issues have become more prominent when compared to the earlier paradigm of carrying out initial drug discovery in animals.

As the 1990s progressed it became increasingly apparent that the output of a chemist could be greatly enhanced by the methods of high-throughput chemistry. Whether as large libraries or as multiple smaller libraries the trend in chemistry for lead generation and optimization became focused on how many compounds could be made rather than what particular ones should be made. This was compounded by moving the filter of in vivo or even intracellular activity further down the screening cascade to exploit 'simple assay' screening capacities (and hence match compound synthesis rates). However, this approach began to change as a result of a 'bottleneck' in the discovery of new drugs. There are, of course, successful drug discovery stories using these paradigms (as exemplified in other chapters of this volume) but at the same time new realizations have emerged in our search for how best to discover important new drugs.

The entire process of the development of a new therapeutic creates many challenges; a drug has to possess an extraordinary profile for it to be an effective medicine and to satisfy today's high standards of safety and efficacy. To discover within one molecule all the properties needed to ensure target specificity and potency, bioavailability, appropriate duration of action, efficacy, and lack of toxicity is a very tough challenge and unfortunately more often than not one that ends in failure even after many years of work. In addition, higher standards of drug specificity and efficacy are continually being demanded as are drugs for new diseases that test our current understanding of science. Frustratingly, it is often only in phase II and III clinical trials that the lack of efficacy in man is discovered for what previously had seemed good candidate compounds based on potency in vitro or in animal models.

Most drug discovery organizations now consider the drug discovery process using a process map that looks broadly like the 'chevron diagram' shown in **Figure 2**. The organization of drug discovery into phases such as target identification and validation, lead generation, lead optimization, clinical candidate selection and evaluation, and finally late-stage clinical trials is often mirrored by organizational structures within a company. Such divisions are responsible for the different phases. At the interfaces between the different phases something tangible has to be handed on to the next part of the business. This tangible entity is the essence of research and is derived knowledge about a target or compounds that potentially or definitively can be shown to interact with it. As such there is a series of milestones (often associated with metrics) that are created along the chevron diagram for the purposes of helping to understand the capacity and effectiveness of different aspects of the various organizational components.

In an ideal scenario it is clear that starting as close as possible to the desired endpoint (i.e., a new effective drug) should be a good strategy for drug discovery. This would mean that early stages could be either shortened or completely eliminated. However, the sheer diversity and number of molecular species that are probably available to be assessed as potential drug molecules, coupled with the complexity of the dynamic and intersecting pathways of biology, suggests that a process of evolution by selection will invariably play some part in the way in which new drugs are discovered.

In earlier decades little attention was paid to the origin of the starting points for drug discovery. This was because research into new drug leads was less intense and refinement of preexisting leads tended to be the norm. Those new leads that were found often arose from research around endogenous small molecule ligands or serendipitous biological observations rather than systematic attempts to survey all potential targets with a large number of chemical entities. Also in this 'cottage industry' era there was no attempt to understand how to enhance the processes involved.[1] Pressure on the industry to increase its productivity has completely changed the situation, such that a much more industrialized paradigm now exists. As a result of this pressure, the need to feed the pipeline with good starting points has become a key practical and intellectual part of the drug discovery process. However important the transition from a 'cottage industry' to a fully industrialized approach for drug discovery becomes, we should still continue to review and challenge aspects of how things are done. Because technology enables something to be done should not necessarily be taken as a reason for doing it.

While the aim is to find drugs in as few steps as possible, it is clear that the challenge of achieving the right profile in a single compound is becoming increasingly difficult. If direct design remains impossible (as seems likely), it would seem that what has been learnt about Darwinian evolution in biology may have parallels in how we find molecules that satisfy the requirements of a complex profile. With this in mind, it is interesting to note how the terminology of drug discovery mirrors that of Darwinian evolution (**Table 1**). Those terms that relate to Darwinian evolution (such as neutral changes) and the consequential chance of survival (e.g., fitness) are particularly pertinent to understanding how the activities of the drug discovery/evolution process could be adjusted in order to provide more drugs in a more timely manner. A key factor in shortening the cycle time of evolving a new drug is the rate at which changes in structure can be made and effectively related to yield structure–activity relationships (SARs).

It is clear that the pressure for discovering new and better drugs and the survival of the pharmaceutical industry in a more demanding environment are the external pressures that drive the development of methodology and understanding in drug discovery. The idea that we take an evolutionary approach to finding the desired molecular

Table 1 A concordance of evolutionary biology and drug discovery terms[a]

Evolutionary biology	*Drug discovery*
Chromosome	Molecular structures that can be modified
Phenotype	Properties of the structure in a biological setting
Neutral changes and drift	Bio-isosterism or ability to replace parts of structure with other entities while maintaining biological activity that may later reveal new opportunities
Fitness	Quality of a lead or lead series to be optimized as a drug
Parallel evolution	Multiple lead series for same target
Mutations/crossover	Structural changes made by a medicinal chemist
Genetic pool	Size and diversity of screening collection
Rate of random genetic drift	Rate at which analogs can be made
Rate of phenotypic change	Rate of progress toward the properties desirable for a new drug
Speciation	Drugs that make it all the way
Extinction	Compounds that fail to make it

[a]The most challenging process to cross correlate in this analogy is that of the 'true' starting points in the processes being compared. While this chapter is about the choice of starting points for drug discovery, the corresponding comparator is the subject of many theologies, both ancient and modern, which are not appropriate for discussion here.

profile means that the wider the range of 'appropriate' starting points that we can consider and their quality (i.e., fitness) become key issues in enabling success. Of course, in the drug discovery process, we do not rely on a totally random approach to evolving new structures. Thus, we are able to use our medicinal chemistry knowledge and experience to propose those changes out of a huge number of possibilities that are likely to work, and then we use real experiments to find which of these hypotheses actually do move us toward the drug (i.e., phenotype) that we require. Increasingly, these real experiments allow us to explore more possibilities (e.g., through array synthesis and high-throughput screening (HTS)). The prefiltering through in silico selection of what is sensible to consider is really only another aspect of the evolutionary selection process.

This chapter addresses these points by considering what makes good leads and especially how different levels of complexity in their structure have been considered as key aspects of discovering leads and their fitness for progression.

4.18.2 The Origin of Drug-Likeness as a Concept

The modern technologies referred to above for the synthesis and screening of large numbers of compounds have provided some unique challenges and opportunities in drug discovery. In the 1990s new methods of drug discovery, such as high-throughput chemistry (HTC) for chemical libraries generation and HTS for biological testing, dramatically altered how much of chemical space could be made and assessed, and thus yielded some valuable improvements in the capacity of the drug discovery processes. However, it soon became clear that although these methods enabled higher output, they were also creating some new problems. One was that the purity and integrity of compounds made by early parallel synthesis methods (whether by solid phase or solution phase) was not as good as the earlier 'hand-sculpted' approaches and it took some time for analytical capacities to catch up with the chemists' aspirations for higher output. Furthermore, it was realized in the late 1990s that the new approaches were not yielding a higher output of good clinical candidate molecules, despite the huge investments made. In hindsight it is clear that the change to the complete reductionist approach of studying protein targets isolated from their cellular environment and the ability to make new compounds on an unprecedented scale had removed some of the checks and balances present in more traditional medicinal chemistry approaches. The realization that production technology was not the simple answer to productivity issues meant that renewed consideration was given to what should be made rather than what can be made.[2]

By the 1990s computational chemistry and informatics methods were being used for retrospective and proactive analyses of the data that had been accumulated over the less intense earlier decades of drug discovery. A seminal contribution was made by Lipinski and colleagues who examined a series of clinically tested drug molecules to try to determine whether they possessed any distinguishing properties when compared to those that had failed (or become extinct).[3,4] This led Lipinski to derive the 'rule of fives,' which constitutes a set of very simple rules designed to suggest whether or not a molecule is likely to have oral absorption problems due to poor solubility and/or poor permeability. The rule of fives (note that there are only four of them!) states that poor oral absorption and/or distribution are more likely when one or more of the following rules are 'broken':

- the molecular weight is greater than 500 Da;
- the ClogP is greater than 5 (ClogP is the calculated lipophilicity as calculated by the methods of Leo and Hansch)[5];
- there are more than 5 hydrogen bond donors (defined as the sum of OH and NH groups); and
- there are more than 10 hydrogen bond acceptors (defined as the number of N and O atoms).

The rule of fives is usually implemented by flagging compounds that exceed two or more of the above parameters. Lipinski and colleagues found that fewer than 10% of their data set of clinical drug candidates violated two or more of these rules.

The simplicity of Lipinski's rules was seized upon by many in the drug discovery business as a useful rule of thumb that could be simply calculated from the chemical structure. Following Lipinski's publication, several other groups also reported analyses of collections of nondrugs and drugs with the aim of identifying the most likely properties that distinguish 'druglike' molecules. The purpose of these studies was to provide computational models which could be used by medicinal chemists to help decide what not to make on the basis that compounds that were predicted to have poor intracellular or in vivo properties should not be synthesized. These analyses were usually done by selecting and comparing sets of drug and nondrug molecules available in the literature or from in-house compilations. For example, Veber proposed that the number of rotatable bonds (less than or equal to 10) and the polar surface area (lower than 140 $Å^2$) were two important properties to enable oral bioavailability in the rat.[6] The polar surface area (PSA) is defined as that part of the molecular surface that is composed of polar atoms (usually oxygen or nitrogen atoms) or from polar hydrogen atoms (i.e., those attached to nitrogen or oxygen atoms). PSA has become widely used in assessing compounds (or suggestions for compounds) for further progression.[7] In another study, Oprea found that most druglike compounds have between zero and two hydrogen bond donors, between two and nine hydrogen bond acceptors, between two and eight rotatable bonds, and between one and four rings.[8]

Other types of in silico filters have been similarly proposed and used.[9,10] Some of these filters are used to spot compounds that contain reactive or otherwise undesirable functionality, such as Michael acceptors, alkyl halides, or aldehydes. The power of all these methods is that they can be applied both to 'real' compounds (e.g., from those in a company's screening collection or being offered for purchase) and 'virtual' compounds (i.e., molecules that have not yet been made).

Other types of filter use more mathematical models, often still based on simple calculated properties, to score or classify molecules according to their degree of drug-likeness. Several types of mathematical model have been used and the book by Leach and Gillet gives more detailed discussion of these models.[11] Multiple linear regression, neural networks, and genetic algorithms have all been used to construct such models. Many studies have shown similar results and a general consensus has evolved which clearly provides good ground rules for the characteristics of molecules that are likely to be cellular penetrant. Attention is now increasingly being turned to models which help explain molecular behavior in relation to other ADMET issues. Models for blood–brain barrier penetration, volume of distribution, various metabolism phenomena, and transporters are all now widely available and used with varying degrees of success in helping to define a priori whether a compound is likely to have the potential to be a drug, i.e., is it druglike?

4.18.3 Lead-Likeness and Fragment Screening Concepts

As the various concepts of drug-likeness gained credence they were incorporated into many areas of drug discovery. However, it was also realized that the processes of medicinal chemistry need good starting points (i.e., leads) to allow exploration of chemical and associated bioactivity space in order to produce the final drugs. An examination of the properties that contribute to the assessment of leads as starting points for drug discovery programs was introduced in 1999 by Teague and colleagues at AstraZeneca.[12] They analyzed a series of compounds from the literature to identify the original leads that the medicinal chemists had started from to develop the final chosen drug entity. This provided pairs of 'leads' and the corresponding drugs. They then calculated a number of molecular properties for these pairs of molecules so as to ascertain whether the properties of leads might differ (if at all) from those of the final evolved drugs. They did find that many properties showed a statistically significant change in their value and this gave rise to the concept of lead-likeness as something distinct from drug-likeness. Several properties increased in the optimized drugs are more complex than their initial leads. For example, the molecular weight and log P increase in going from a lead to a drug, as do the numbers of hydrogen bond donors and acceptors. In a further publication the same group considerably expanded their literature survey and published the differences shown in **Table 2** between the median values of the various derived properties.[8]

Almost simultaneously with this AstraZeneca paper, Hann, Harper, and Leach from GlaxoSmithKline published their analysis of a much larger data set comprising a different collection of lead/drug pairs.[13] The data that Hann and colleagues studied was derived from the extensive compendium previously published in a book by Sneader[14] (see **Table 3**). Sneader's terminology is slightly different in that he refers to leads as drug prototypes but the terms are effectively synonymous with 'lead' and 'lead-likeness' now being the accepted terms in the literature.

Table 2 Observed increases in derived properties in going from leads to drugs for a set of compounds from the published literature

Property	*Increment*
Molecular weight	69 Da
Hydrogen bond acceptors	1
Rotatable bonds	2
Number of rings	1
ClogP	0.43
Hydrogen bond donors	0

Table 3 Average property value changes between leads and drugs based on the data set drawn from Sneader's book

Property	*Average value for leads*	*Average value for drugs*	*Increment*
Molecular weight	272.0 Da	314.0 Da	42.0
H-bond donors	0.8	0.8	0
H-bond acceptors	2.2	2.5	0.3
ClogP	1.9	2.4	0.5
Number of heavy atoms	19.0	22.0	3.0

	1	2	3	4	5	6	7	8
Position	1	2	3	4	5	6	7	8
Receptor features	+	+	−	−	+	+	−	+
Ligand features								
Correct match	−	−	+					
Correct match					−	−	+	
Incorrect match*		−	−	+				

* There are numerous incorrect matches but only two correct matches with these randomly chosen features.

Figure 3 In the simple Hann model an exact correspondence between ligand and receptor features has to take place for a successful interaction to be recorded. In this example, a ligand of complexity of 3 with points of interaction (− − +) is matched in various positions against a receptor whose complexity is 8 features in length and has the pattern (+ + − − + + − +).

Despite covering different data sets, both these studies showed that retrospective analysis of many drug discovery programs demonstrates that the initial hits have statistically different properties to those of the final drugs. A number of possible explanations for these observations have been suggested. Initial hits from screening are often less potent than the ultimate drug needs to be, and improving potency is often most easily achieved by adding further chemical functionality to make new interactions. This in turn increases the molecular weight together with properties such as the numbers of donors and acceptors. Thus, medicinal chemists tend to add mass to a compound in pursuit of potency. log P is another property that often increases during lead optimization. Two possible reasons for this exist. One reflects again the need for the addition of specific interactions, this time hydrophobic, with the target to give increased potency. The other is that more nonspecific hydrophobic interactions can yield increased potency due to the phenomenon of increasing the apparent concentration of a lipophilic drug in the lipophilic environment of a membrane-bound target. A further reason for the differences between leads and drugs, at least in the case of the Sneader data set, is that many of the leads included in his set were small hormones such as biogenic amines. These starting points are of such low complexity (i.e., very low molecular weight) that adding mass was inevitable in evolving the drug especially when trying to target different receptor subtypes at which the natural hormone is pan-active.

Complementary to the analysis of historical data in the manner described above, increasing consideration has been given to these issues from a more theoretical perspective. One analysis that has been widely cited as helping to understand the underlying issues is also in the paper of Hann and colleagues.[13] They presented an idealized model to explore how the probability of finding a hit varies with the complexity of the number of potential interactions between the ligand molecule and a receptor site. In Hann's model the ligand and its binding site are represented as simple bitstrings of interaction points. The numbers of interaction points in the ligand and the binding site are considered to be measures of their complexity. The bitstrings represent molecular properties of the ligand that might influence binding but are not directly related to the more specific interactions that computational chemists usually study. However, they do represent such concepts as shape, electrostatics, and lipophilicity. In the model, the bitstring of the ligand has to exactly match that of the binding site for the interaction to be 'allowed.' Thus, each 'positive' element in the ligand must match a 'negative' element in the binding site and vice versa. **Figure 3** illustrates a number of examples of successful and unsuccessful matches in the case of a ligand with three features and a receptor with eight. It is then possible to calculate the probability that a ligand of a given size will match a binding site also of a specified size. A different system is shown graphically in **Figure 4** for a binding site of size 12 and varying sizes of ligand. Thus, the probability that a ligand with complexity elements represented with lengths of 2, 3, 4... bits will match the binding

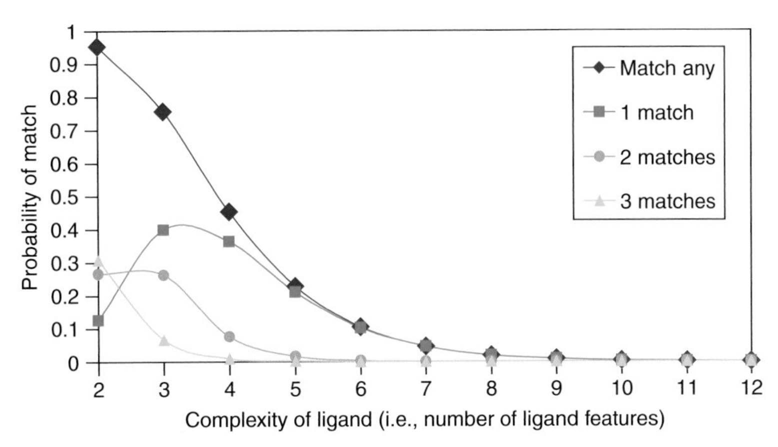

Figure 4 Plot shows the probability of finding 1, 2, or more matches (and total matches by summation) for varying ligand complexity using a receptor with 12 interaction sites in the Hann model. (Reprinted with permission from Hann, M. M.; Leach, A. R.; Harper, G. *J. Chem. Inf. Comp. Sci.* **2001**, *41*, 856–864. Copyright (2001) American Chemical Society.)

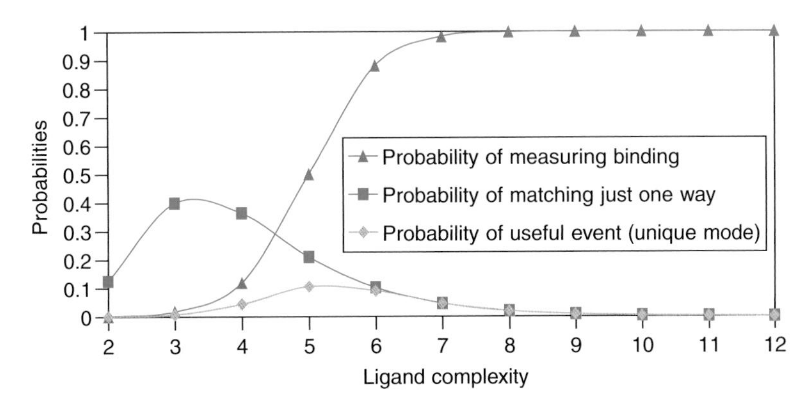

Figure 5 Shows the product of multiplying the probability of a single match (green line) by the probability of being able to detect experimentally the binding event (red line). The resulting probability of a 'useful event' (orange line) passes through a maximum due to the nature of the two contributing and underlying probabilities. (Reprinted with permission from Hann, M. M.; Leach, A. R.; Harper, G. *J. Chem. Inf. Comp. Sci.* **2001**, *41*, 856–864. Copyright (2001) American Chemical Society.)

site in one, two, three, four… ways can be calculated. Also shown is the total probability that a ligand can match in any way, which is simply the sum of these individual matches. As shown in the figure, the chance that a ligand can match at all shows a smooth and rapid decay to zero as the complexity increases. From the point of view of drug discovery, the probability that the ligand has a unique match is important, as this helps in the interpretation of SARs and the development of an unambiguous binding mode. In the example shown, this probability passes through a maximum at a ligand complexity of 3.

In the next part of the Hann model, the likelihood of being able to detect experimentally the binding of a ligand as the complexity of the interaction increases is considered. Thus, counting the number of interactions made (equivalent to the degree of complexity for any given ligand that matches the receptor) can be used as a crude indicator of the likely strength of the interaction. As the number of pairwise matches increases then the probability of measuring experimentally the interaction must also increase. This is represented in **Figure 5** as a hyperbolic curve. The choice of this curve shape reflects that if the number of interactions is below a certain number then it is not possible to measure the binding because it will be very weak. There is then a rapid increase in the probability, consistent with the notion that once the potency exceeds some threshold the interaction will be measurable and so the probability finally equals 1.

The final part of the model is to combine the two probabilities derived so far into the 'probability of a useful event.' This is defined as the product of the two previously defined probability distributions. This then reflects the true balance of the probability of the ligand and protein having matching interactions and also of being able to measure the interaction that results.

The two probabilities from the first two parts of the model can be seen to have competing distributions in that the probability of finding a match decreases as the complexity of the interactions increases while the probability of measuring the resulting interaction increases as the complexity of the recognition event increases. These distributions will clearly vary for different model criteria and for real systems. However, the combined effect is that the 'probability of a useful event' will always have a maximum somewhere in the middle with lower probabilities at increasing or decreasing complexity. At the high end of the range of possible interaction complexities the chance of getting a complete match is very small. However, if such a match did occur then it would be easily measured because so many individual correct interactions would contribute. At the low end of the complexity spectrum, the probability of a useful event is asymptotic to zero because even though there will be a high probability that the simple ligand can find a way to match something in the receptor there will not be sufficient to contribute to an observable binding in a real experimental assay. In the intermediate region there is the highest probability of a useful event being found. Here there is a maximum in the probability curve and hopefully the probability is high enough in both matching and being able to measure it.

Jacoby and co-workers have taken these ideas and used them to analyze the results of different types of screening campaigns at Novartis.[15] They compared HTS summary data with that derived from a method such as nuclear magnetic resonance (NMR), which is more appropriate for detecting the weaker interactions typical of low-complexity hits. While it is always difficult to make comparisons based on subjective terms such as 'screening hits' they found that the hit rate for micromolar hits from generic fragment libraries in HTS is in the 0.001–0.15% range while the hit rate for millimolar hits found by NMR screening of fragment libraries is in the 3% range. As the authors are careful to point out these hit rates do not necessarily translate into ultimate success but the data they present does seem to support the underlying premises of the Hann model.

As interest in these ideas developed the terminology has evolved and the terms fragment and fragment screening have become synonymous with the concept of screening less complex molecules to increase the probability of finding hits and how to evolve them into larger more druglike molecules. In essence, the term 'fragment' reflects the medicinal chemists' appreciation that the compounds being used are probably not of the size that will reflect the complete needs of an optimized drug but will provide a suitable starting point. The numerical ranges for descriptors that are now currently used to define fragment concepts for different screening paradigms are discussed later in this chapter.

Another theoretical aspect of the problem of molecular complexity and why the use of complex ligands or fragment approaches may provide additional ways of finding starting points for drug discovery programs concerns the effectiveness with which the extent of chemical space relevant to biological activity can be sampled. A number of groups have estimated how many potential druglike molecules exist. This refers to molecules that might be considered to generally fall within what is considered druglike space. This can be defined in various ways and as a result estimates[16,17] vary quite widely, but all agree that druglike chemical space is very large – many orders of magnitude greater than the number of compounds that have been made to date and indeed probably large enough to use up more material than is available on earth or probably in the known universe!

The challenge in drug discovery is to explore effectively the huge number of potential compounds in order to identify those that not only possess the necessary activity at the target(s) but will also have appropriate ADMET properties to enable development into an effective drug. Again, these concepts can best be explored with simplified model systems; **Figure 6** shows a representation of a target protein with two binding sites and two different approaches to finding a compound that fits both sites. Consider a simple five-component fragment library that could be evaluated using some biophysical or biochemical method for binding to the two binding sites. An exhaustive synthetic campaign would mean making the 5×5 ($=25$) fully combinatorial library and doing 10 assays. The alternative fragment approach would involve doing five assays (or possibly 10 if each binding site required a different experiment) of the fragments (or related simple derivatives). If in an idealized result, two of the five compounds were found to be weakly active they could then be linked synthetically to give a compound which would, of course, have been found as one of the fully combinatorial synthetic library if it had been made. However, the latter fully combinatorial and exhaustive screening approach clearly uses more synthetic and screening resources to find the best compound. It could also be the case that none of the 25 complex molecules bind because of the reasons explained above in the Hann model. However, one of the five fragments might bind, giving at least a starting point for trying to elaborate this compound with fragments other than those already tested.

When working with trivially small numbers of molecules the case for lower complexity may not be immediately obvious. But the combinatorial versus additive dilemma becomes more dramatic when we consider that a 1000-member fragment library with 20 additional linker chemistries explored would combinatorially give a 20 million-member library for a two-site target. However, this could be surveyed initially with 1000 assay points if the initial screening was done with just the building blocks.

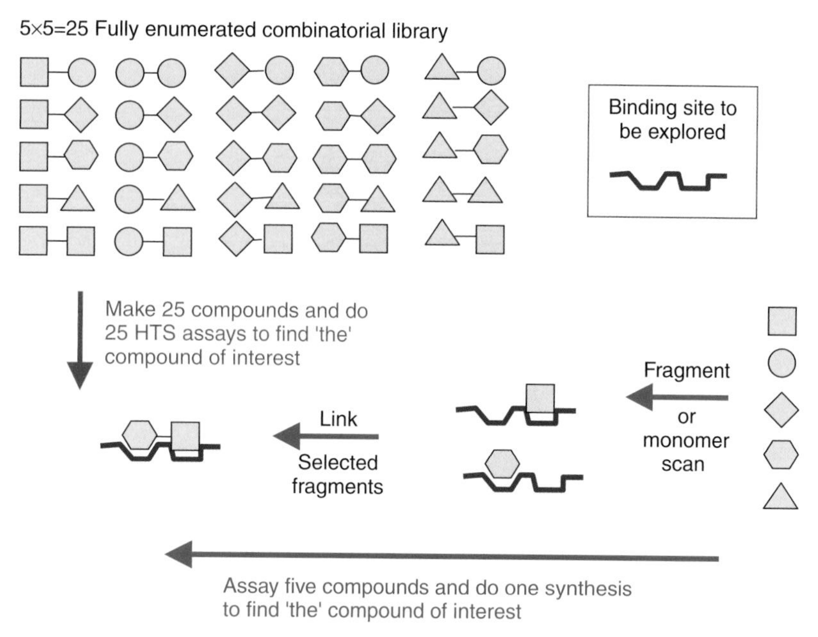

Figure 6 For a set of five fragments there are 25 different pairwise combinations. By contrast, if the fragments are screened individually then the desired combination can be obtained more directly.

Combinatorial chemistry is also of value in its potential for making thousands of molecules that can be re-used for further screening campaigns, and so the investments may possibly be recouped in the future. Therefore, the dilemma is deciding what to include in a screening collection in terms of whether to be proactive by exploring as many large molecules as possible or reactive with synthesis after finding smaller (i.e., less complex) compounds with weaker activity. The choice of what level of complexity to have in molecules in a screening collection for lead discovery does not have a universal solution and is probably best handled by taking a balanced approach. The essence of the fragment approach to lead discovery is that it provides a way to break the combinatorial explosion that is the Achilles' heal of chemical space exploration. At the same time it is important to recognize that fragments represent a reductionist approach in chemical space and there is no guarantee that a simple molecule will express the biological phenotype of a more complex molecule that structurally contains the simple fragment.

To explore further the scale of this problem it is useful to consider the amount of sampling that can be done for a given level of complexity. Provided biological properties do follow an additive (or quasi-additive) behavior then fragment screening approaches provides an effective and alternative way of sampling the enormity of chemical space. This can be illustrated as follows. **Figure 7** shows the number of compounds registered in the GlaxoSmithKline collection that contain a carboxylic acid functional group.[18] The distribution is plotted as a function of binned molecular weight. As can be seen the number of carboxylic acids in a particular molecular weight band initially increases rapidly following an approximately exponential curve. At about 150 Da this exponential behavior stops. Although the most populated bin is at a molecular weight of around 400 Da (maximum point in curve) the growth in numbers per bin relative to the previous one is maximal at about 150 Da. This can be seen by considering the lower curve, which is the differential change (up or down) from the number in the previous bin. Therefore, the GlaxoSmithKline acid set is significantly undersampling the virtual space of carboxylic acids in a way that gets progressively worse as the molecular weight increases. Thus, when operating at the lower molecular weight region (e.g., <350 Da, typical of many fragment sets) the set of available acids provides a more effective sampling than at a higher molecular weight (say 450 Da). This is schematically illustrated in **Figure 8**, which also includes the increase that would exponentially continue from the initial rate of increase if all carboxylic acids were available for consideration. Fink and co-workers showed in an exhaustive enumeration of possible structures that there are *c.* 145×10^6 compounds that could be considered with only 12 nonhydrogen atoms.[19] If the number of atoms is increased to 25 (still equivalent to a molecular weight of only *c.* 350 Da) then there are *c.* 10^{25} possible structures that can be considered.

One of the key consequences of using a fragment-based lead discovery approach is that the binding affinities of the molecules initially identified will often be much weaker than larger, more druglike molecules. This could lead to an

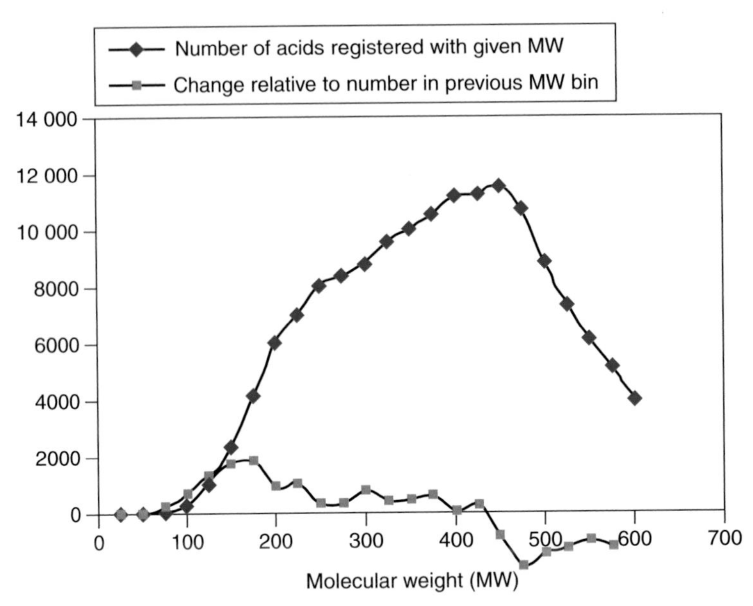

Figure 7 The number of carboxylic acids registered at GlaxoSmithKline with a given molecular weight (blue) together with the change in 25 Da increments (magenta) relative to previous bin.

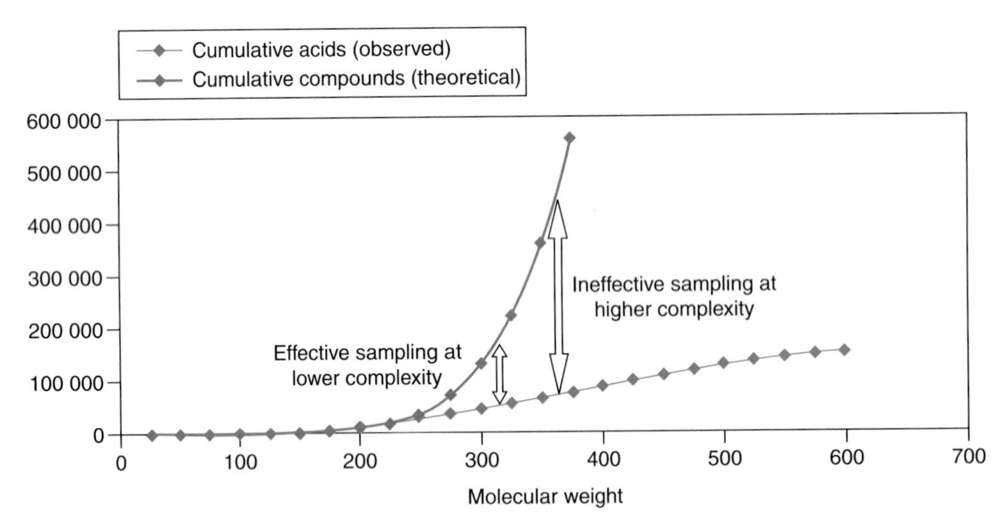

Figure 8 Graph to illustrate that the carboxylic acids available at GlaxoSmithKline reasonably represent the total number possible for low molecular weight; at higher molecular weight values the divergence of the two curves means that there is very poor representation.

interesting fragment of relatively weak potency being passed over in favor of more complex molecules that may have higher initial potency but which are ultimately less developable. Several groups have therefore proposed the use of binding affinities normalized by molecular weight so as to have ways to compare molecules that takes account of their size and helps ensure that useful but small starting points are not missed or ignored as potential leads.

The origins of this idea can be traced back to a publication from 1984.[20] Andrews proposed the concept of a ligand having a 'maximal affinity' based on an analysis of the functional groups in a ligand. This affinity is equal to the maximum free energy of interaction that a compound might be expected to express in an interaction with a biological macromolecule if all the available groups in the ligand are utilized. The analysis was based on a set of ligands with known binding affinities taken from the literature. The functional groups present in each ligand were identified and counted and a multiple linear regression analysis performed in order to determine the contributions of each functional

Table 4 Binding energy contributions for various functional groups as calculated by Andrews

Functional group	*Energy (kcal^{-1} per functional group)*
sp^2 C	0.7
sp^3 C	0.8
N^+	11.5
N	1.2
CO_2^-	8.2
OH	2.5
C=O	3.4
O, S	1.1
Halogens	1.3
PO_4^{2-}	10.0

group to the observed binding affinity, together with a fixed entropy term (14 kcal mol^{-1}) related to the freezing of translational and rotational degrees of freedom of the molecule and a variable entropy term (0.3 kcal per bond) summed over all rotatable bonds in the ligand. The values found are shown in **Table 4**. Although there is a large statistical variation in these values, Andrews proposed that summing the appropriate contributions for any novel ligand would give an estimate of the maximal binding energy that might be expected if all functional groups make their optimal contribution – in other words the binding efficiency of the compound.

The classification of functional groups in a ligand (i.e., whether a group is neutral or basic) and subsequent assignment of values can be seen to make large differences in the results. Thus, the difference between a protonated and a nonprotonated amine is very high (*c.* 10 kcal) meaning that the choice of whether an amine is protonated or not in the binding site can alter the estimated binding affinity by 8 log units of potency – essentially making the prediction very unreliable in its absolute sense. However, as an intellectual tool, the Andrews approach remains a useful way to help explore the efficiency of compounds in their binding.

More recently, Kuntz and colleagues analyzed a data set of 160 ligands.[21] Although they observed that the initial slope of a plot of free energy of binding versus number of heavy atoms in the ligand has a value of approximately 1.5 kcal mol^{-1}, it is clear that this initial slope is dominated by some very nondruglike entities that contain fewer than six atoms. When these compounds are removed and the data set is re-analyzed then a different conclusion can be drawn for those compounds that remain with less than 25 heavy atoms (i.e., less than *c.* 330 Da) (see **Figure 9**).

Now the initial high binding per atom drops quite quickly and almost asymptotically to around 0.3 kcal per heavy atom, and this fits better with results from a much larger set of data based on HTS by researchers at Pfizer. They coined the term 'ligand efficiency' for the experimental binding affinity per heavy atom and have proposed that it is a useful parameter to use when prioritizing the output from HTS or other screening strategies.[22] They suggest that a lower limit on the ligand efficiency can be estimated by assuming that the goal is to achieve a binding constant of 10 nM in a molecule with a molecular weight of 500 (to be consistent with Lipinski's rules). An analysis of the Pfizer screening collection revealed that the mean molecular mass for a heavy atom in their 'druglike' compounds is 13.3 and so a molecule with a molecular weight of 500 and a binding constant of 10 nM would have 38 heavy atoms and therefore a ligand efficiency of 0.29 kcal mol^{-1} per heavy atom. This is also consistent with the asymptotic value in the analysis of the more druglike molecules found in the Kuntz data set (**Figure 9**). The Pfizer proposal was that the hits with the highest ligand efficiencies are the best ones to consider for optimization, provided that all other factors such as synthetic accessibility are equal.

An extension of these ideas enables other properties to be taken into account. Thus, to achieve compounds with a not too high log *P* while still retaining potency, the difference between the log potency and the log *D* can be utilized. Burrows and colleagues at AstraZenecca have proposed that when this term is greater than 2 log units then it is likely that the compound will be a good lead compound.[23] Further analysis and comparison of a number of potential ligand efficiency metrics from a survey of drug hunting projects that delivered clinical candidates suggests two additional metrics with statistical validity: potency (pIC_{50})/nonhydrogen atom $>$ 0.2 and potency minus serum protein binding

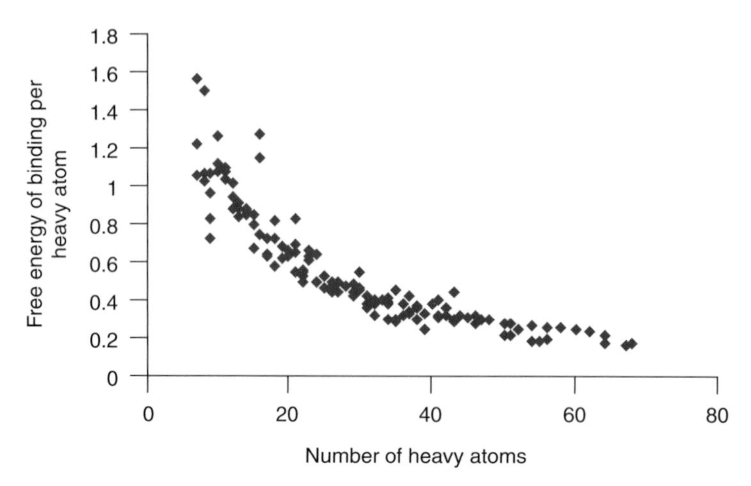

Figure 9 Plot of binding free energy per heavy atom versus number of heavy atoms for more restricted set of fragmentlike compounds.

affinity constant (log K_{app}) > 1. Simultaneous expansion of these ideas was provided by Abad-Zapatero and Metz who in addition to normalizing the binding affinity by molecular weight (i.e., ligand efficiency) also defined a surface-binding efficiency index in which the polar surface area was used as the normalizing factor.[24]

4.18.4 Finding Fragments by Screening

Advances in HTS technologies has allowed increasingly large compound collections to be effectively screened for the desired activity. Typically HTS is done with ligand concentrations around 10 μM and is regularly successful in hit generation; however, a number of screens do still fail to deliver attractive hit series for the biological targets concerned. Some of these targets are, a priori, not believed to be chemically intractable, i.e., the molecular properties necessary to bind to the target protein and elicit a biological response are compatible with those properties necessary for oral dosing.[25] This can be explained in various ways including: (1) assays or assay technology configured inappropriately; (2) problems with screening collections (solubility, stability)[26]; or (3) lack of appropriate compounds in the screening library, i.e., no potent actives in the library 'waiting' to be found. This is unfortunate given the time, effort, and expense in building, running, and analyzing such libraries and screens. One way to avoid these two latter issues is to ensure that screening collections are continually updated with new compounds that are derived from novel and 'lead-like' libraries of compounds (through synthesis or acquisition) and which cover the chemical space that is missing (as discussed above).[12] In the absence of any idea of where to start it is often difficult to design novel enriching libraries other than to ensure chemical diversity compared to that which is already available and that the compounds are chemically tractable for further analog constructions and have favorable drug metabolism and pharmacokinetic (DMPK) properties.

One alternative approach to this whole problem is to screen libraries of 'ultra-lead-like' fragments (which are much smaller than compounds typically screened in HTS) at much higher concentrations allowing for the detection of only weakly binding compounds but which are small and novel as starting points from which new areas of chemistry can be developed. Not only is this attractive due to the theoretical arguments outlined above but it can be extremely valuable in identifying truly novel cores, scaffolds, and warheads, which can then give competitive advantage in terms of patentability. Analogs or libraries prepared from these fragment hits can then significantly enhance the diversity of the original library and hopefully provide a way of dealing with related targets that may have been previously intractable.

There are essentially two ways in which fragment screening can be carried out. The first approach involves biochemical screening (often referred to as high-concentration screening (HCS)) and the second approach uses biophysical and direct structure-based screening (using for example NMR or x-ray methods) (*see* 4.31 New Applications for Structure-Based Drug Design; 3.41 Fragment-Based Approaches).[27–29]

4.18.4.1 High-Concentration Screening Using a Biochemical Assay

This involves the use of typical biochemical assays but running them in such a way that they are robust to the higher concentrations of ligands that are to be tested. This concentration is typically in the 1 mM range. The major advantages of this approach are that the assays are fast, quantitative in principle, and use widely available technologies for detection. In addition, only small amounts of protein are necessary and so assays involving G protein-coupled receptor (GPCR) or ion channel targets may be considered. However, there are also many potential problems with this approach. Not every assay is suitable for this approach; for instance, the concentrations of added ligand may interfer with the assay through undesirable mechanisms or may be toxic to a cell, if the assay is cell based. In addition, there can be problems with the identification of false positives as a result of compound aggregation at the concentration of ligands used,[30] interference with the assay endpoint (e.g., optical interference fluorescence, quenching, toxicity, etc.),[9,31] or disruption of the protein by unfolding or precipitation. Additionally, false negatives can occur due to lack of effective solubility of compounds.

4.18.4.2 Biophysical and Direct Structure Determination Screening

An alternative approach is to screen at high concentration using a more direct biophysical assay or structure determination.

4.18.4.2.1 Screening by crystallography

The strategy of directly obtaining x-ray crystallographic data on small fragments bound to proteins is well documented.[28,32] A clear advantage of this approach is that false positives are reduced because if a compound is seen by crystallography then an immediate assessment can be made of how to enhance binding by using modeling techniques. On the downside, these techniques can be very time and resource intensive in that a large (milligram) amount of a protein construct is needed that is compatible with crystallization to yield robust crystals and diffracts well and is compatible with ligand binding. Also, those fragment ligands being considered will need to be soluble in the crystallization medium. No affinity information is obtained from crystallography unlike biochemical assays or NMR (see below). False negatives can still occur in that there may be kinetic or crystallographic reasons why a compound does not get into the binding site in the crystallographic disposition of the protein.

NMR has been used for a number of years as a very useful tool for identifying weak binders at high concentration, particularly in the technique developed at Abbott termed SAR-by-NMR.[33] Advantages include being able to observe either the protein or the ligand, measuring binding at high ligand concentrations (up to 10 mM), and there is the additional possibility of obtaining at least some structural information and affinities. It is also very difficult for there to be false positives or negatives. However, NMR requires a large amount of protein, not all proteins are suitable (based on solubility and etc.), and the technique does not work well with membrane-bound proteins. In addition for 2D methods labeled protein is required, which results in further expense, time, and effort. It is much slower than a biochemical assay but can be faster than x-ray crystallography if appropriate protein is available.

In general, these biophysical and structure-based methods are more robust than biochemical screening though not always technically feasible. Alternative direct biophysical approaches include affinity detection by mass spectrometry[34] and surface plasmon resonance (Biacore).[27]

Another method that has been pioneered at Sunesis involves the introduction of tags (usually individual cysteine residues) into the binding site. These are then used to capture (by disulfide formation) probes that also contain a free thiol moiety. These probes are taken from a library of fragments that can be screened against the protein and binding is detected by a mass spectrometric procedure. The disposition of bound fragments is then found by protein crystallography. The Sunesis group have shown that fragments can often adopt novel insertion modes into the protein surface and that these fragments can then be grown (directed by structure-based design) to give larger molecules with more interactions. Eventually the disulfide tag is dispensed with so as to leave a noncovalent compound with specific and novel interactions. This method has most recently been exploited to explore the design of novel GPCR inhibitors.[35]

In practice, screening for fragments is often performed using a variety of the above approaches, e.g., biochemical screening followed by x-ray crystallography on the hits. The real value of the NMR and/or x-ray methods is that they give structural insights that aid in decisions about what to make next in the search for increased potency and specificity.

4.18.5 The Design of Fragment Screening Sets

Three key issues need to be considered when designing and implementing a fragment screening library: (1) how many molecules to include in the set; (2) which molecules to include in the set; and (3) which method (or methods) are going to form the basis of the detection of binding. These issues are broadly the same as those that need to be considered in the creation of HTS sets but the impact of trying to find compounds that are representative of larger molecules is an added challenge. Many of these issues are closely related and a compromise often has to be reached. All

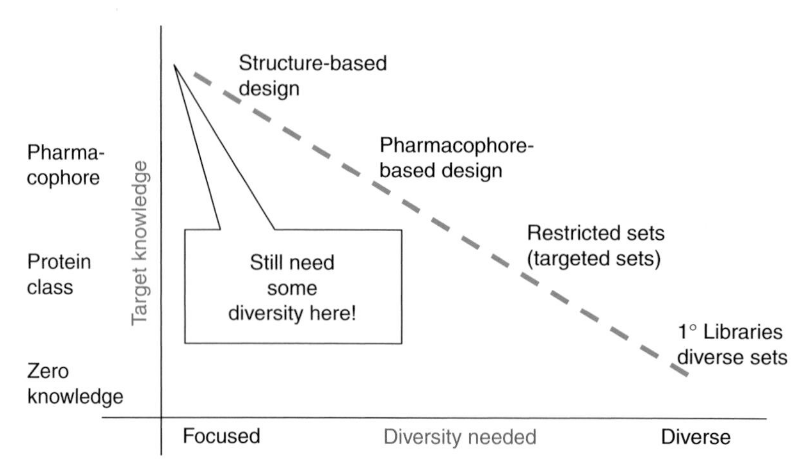

Figure 10 The knowledge plot demonstrates that the degree of diversity needed has an inverse relationship to the amount of knowledge available about that target.

assay techniques impose some kind of constraint on the number of the molecules that can be screened. The size of a library will be critically dependent on the expected testing concentration as this will influence the number of compounds that can be followed up. For instance, an HTS library (screened at 10 μM) might include 10^6 compounds, a high concentration library (screened at 100 μM) might include 10^4 compounds, and a small fragment library (to be screened by NMR or x-ray at 1 mM) might include 10^3 compounds. There are many approaches that can be taken to the design of the content of all screening sets but each one involves a number of pragmatic or subjective choices. Generally though, the objective is to cover appropriate chemical space with a testable number of compounds and to use existing knowledge to weight compound selection.[17]

A distinction is often made between the large, 'diverse' sets such as those used in HTS and the smaller, 'focused' sets that are selected with a specific target or group of related targets in mind. The methods that can be used to construct diverse and focused screening sets have been discussed and reviewed extensively in the literature and so will not be considered in detail here except where there are specific issues arising from the use of fragments (*see* 4.08 Compound Selection Using Measures of Similarity and Dissimilarity; 4.20 Screening Library Selection and High-Throughput Screening Analysis/Triage). One factor worth considering here is the balance between diversity and focus. The knowledge plot shown in **Figure 10** is a particularly useful way of representing this balance.[36] This relates the different levels of knowledge about targets to the level of diversity required in the screening set. This diagram suggests that the need for diversity is inversely proportional to the knowledge that is available about the biological target.

The key factors to be considered when constructing a fragment library include:

1. *Availability.* Screening compounds at 1 mM will require a 100 mM dimethyl sulfoxide (DMSO) concentration if a final DMSO concentration of 1% is acceptable. Depending on the volume required this can easily require 25 mg of compound.
2. *Availability of close analogs.* It is a great advantage to have available close analogs of any hits in the screening library to help confirm actives, to build up clusters, and define SARs. Alternatively, ready access to analogs by array synthesis is useful.
3. *Solubility.* Although related to lipophilicity, the other factors contributing to solubility are difficult to predict and model and therefore this property can only be safely obtained through measurement.
4. *Purity.* Given compound concentrations and the sensitivities of many assays, levels of contaminants or impurities need to be kept to a minimum to avoid identification of false positives particularly those impurities that could lead to irreversible inhibition. This is usually ascertained through liquid chromatography-mass spectrometry (LC/MS) and NMR on each sample in the library.
5. Molecular size/weight, lipophilicity, and other parameters need to be constrained (see below).[37]
6. *Absence of reactive functionality.* Compounds containing such moieties can be removed using computationally derived filters.[37]
7. *Opportunity for synthetic elaboration.* Clearly it is important that the opportunity for elaboration is not compromised by impossible synthetic handles or templates. However, there is an additional dilemma that must be considered. For example, a carboxylic acid may provide a handle for further elaboration but if a fragment hit possessing an acid is

identified it is likely that the acid provides a key interaction per se that would be destroyed by further chemistry (e.g., amide formation). Usually the inclusion of synthetic handles on library compounds is a decision based on an organizational rationale (i.e., hits with functional groups that can be readily derivatized are more likely to be followed up for operational reasons). The pragmatic solution to this is to include, for example, the oxime and its methylated analog as prototypes for further structural evolution.[39] The situation is obviously greatly alleviated if early structural insights can be gained that show the binding mode of a fragment and then allow directed evolution (including fragment joining) to exploit the synthetic potential of a compound.

8. *Reduced chemical complexity.* As has been stated a complex molecule is less likely to be able to use all its features for binding and the number of compounds required to cover chemical space increases with molecular size and complexity.[13]

Several computational techniques have been used in the selection of fragments to include in lead-like screening sets. A key parameter is the physicochemical properties of the ligands and a more restrictive set of parameters than typical Lipinski criteria is often used. Within the lead-likeness arena an analogous rule has also been adopted at Astex.[37] These workers analyzed the hits obtained by screening their own collections of fragments using x-ray crystallography at a variety of targets (e.g., kinases and proteases). From this analysis evolved the idea that a 'rule of threes' might be appropriate to help select good fragments to try. This rule requires that the molecular weight be less than 300, that the number of hydrogen bond donors and hydrogen bond acceptors should be less than or equal to 3, and that the calculated octanol/water partition coefficient (using ClogP) should be less than or equal to 3. Three or fewer rotatable bonds and a polar surface area of 60 $Å^2$ or less were also proposed as useful criteria.

As some of the screening techniques used in fragment-based discovery are limited in capacity (compared to HTS) it is usually necessary to refine further the initial set of compounds that meet such simple filters; some form of further selection is required. One useful way to do this is to identify fragments related to those that commonly occur in druglike molecules. Certain fragments (often referred to as privileged structures) are those that appear frequently in drug molecules. While some of the most common of these privileged structures have been defined 'by hand' a number of computational methods have also been developed to systematically identify appropriate fragments from collections of druglike molecules. The fragments may be sorted by frequency and after removal of trivial examples, such as simple alkyl groups, the highest scoring fragments are selected.

Bemis and Murcko defined a clear hierarchical approach whereby a molecule is simplified into its graph representation and then broken down into embedded rings, linker atoms, and side chains.[40,41] The ring systems and linkers together define 'frameworks' as illustrated in **Figure 11**. The most prevalent frameworks identified by applying this algorithm to the Comprehensive Medicinal Chemistry database are shown in **Figure 12**. It was found that just 32 frameworks accounted for approximately 50% of the 5120 drug molecules in the entire set. An alternative approach that is widely used is the retrosynthetic combinatorial analysis procedure (RECAP).[42] In RECAP the fragmentation is performed by successively cleaving bonds that can be easily formed in a reaction sequence, such as amides and ethers. These methods can be further used to help in identifying more appropriate synthetic fragments that may be directly used in further array chemistries. The advantage of this method is that it does retain some relationship to synthetic accessibility when fragments are rejoined in other ways.

Several groups have published their approaches to generate fragment screening collections. At GlaxoSmithKline the term reduced complexity screening (RCS) is used to cover the fragment screening activities. For this purpose a screening set was developed by taking a large set of available in-house and external compounds and applying a series of 2D substructure and property to identify potential candidates for inclusion in the set (heavy atoms <22, rotatable bonds <6, donors <3, acceptors <8, ClogP <2.2). The selection criteria also required there to be a synthetic handle present in order to facilitate the rapid synthesis of further analogs. Note the use of heavy atoms rather than molecular weight as a criterion as this avoids deselecting molecules that may contain, for example, bromine atoms, which may prove to be useful synthetic handles at a later stage. The GaP diversity measure based on 3D pharmacophore keys was then used to select a subset of compounds from the initial filtered selection.[18,43] This set has been used for screening against targets at a ligand concentration of 800 μM. Those targets that have not yielded sufficient tractable hits in the more standard HTS run at 10 μM and for which structural studies can be used to help progress the RCS hits are chosen. The approach has yielded several interesting hits that have contributed to progressing targets.

Scientists at AstraZeneca have used a broadly similar approach to select a set of 2000 compounds for what they term HCS. This set was designed to have a roughly equal proportion of acidic, basic, and neutral compounds (with a small number of zwitterionic molecules) and with a predefined physicochemical property distribution.[23]

Scientists at Astex have described the construction of screening sets for use in x-ray crystallographic fragment screening[43] (*see* 4.31 New Applications for Structure-Based Drug Design). Again, sets directed against a specific target

Figure 11 The definition of rings, linkers, side chains, and frameworks from the molecular graph as proposed by Bemis and Murcko.

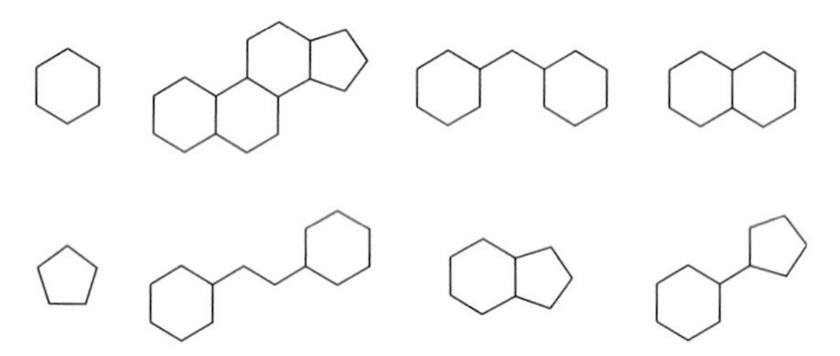

Figure 12 Top-scoring frameworks in drug molecules as identified by Bemis and Murcko.

or groups of related targets have been constructed together with a more general-purpose set. The starting point for the latter was a fragmentation analysis of drug molecules, which identified a small set of commonly found, simple organic ring systems. These ring systems were then combined with a set of desirable side chains. Three sources of side chains were used: those observed frequently in drug molecules, lipophilic/secondary side chains (intended to pick up hydrophobic interactions in a protein binding site), and a set of nitrogen substituents. Each of the relevant side chains was combined with the ring systems to give a virtual library (of size 4513); the structures in this virtual library were then compared against databases of compounds available from external sources giving a final set of 327 compounds.

At Vernalis, scientists have described four generations of a low molecular weight fragment library for use in NMR-based screening.[45] As in other examples, in silico property calculations were developed to automate the selection process. Both 'general purpose' sets together with those directed toward particular protein targets (kinases) were constructed. Three-dimensional-based descriptors were also used (analogous to those employed by the GlaxoSmithKline group) as a measure of diversity and complexity.

4.18.6 Turning Fragment Hits into Leads

If, in the retrosynthetic sense (as compared to the objective of using fragments to build new molecules), a molecule with high potency is split into two fragments the individual potencies of the derived fragments will be considerably less than half of the potent molecule. The corollary, and hence value to the fragment approach, is that if one can successfully link two weakly potent fragments then this may afford more than the sum of the parts. The reason why the expected affinity of the joined molecule should be larger than the sum of the affinities of the two individual fragments has been extensively discussed in the literature.[46,47] Any molecule loses significant rigid body translational and rotational entropy when it forms an intermolecular complex. This unfavorable entropic term is only weakly dependent on the molecular weight of the components. Therefore, two unfavorable terms have to be overcome when two fragments bind at the same time and this is replaced by just one unfavorable term for the combined molecule. However, this ignores the fact that combining the two separate binding entities may not be completely compatible with the binding configuration and conformation of the individual entities as first identified. However, the possibility of regaining as much as possible of this rotational and translational entropy when joining or extending fragments is one of the driving forces for using this approach.

As the bioactivity of initial fragment hits will be weak (sometimes very weak), a key requirement is to increase the potency. This is done by synthetic manipulations to the structure to give a sustainable lead series, which will be more akin to a series that might be found from traditional HTS. Sometimes this can be achieved by linking different fragments that have been found to bind to nonoverlapping sites and more commonly by growing the hits in an iterative manner driven by structural insights from crystallography or NMR analysis. Other properties besides potency will need to be balanced as the structure is evolved (e.g., selectivity against other targets and ADMET properties). As these nonefficacy parameters need to be balanced against the potency/complexity of the compound it is useful to have starting points that are as simple as possible. In this way the number of off-target pharmacophores can be limited because the amount of 'excess baggage' in a molecule's structure is also kept to a minimum.

A number of scenarios for getting potent druglike molecules from fragment lead-like hits have been identified[29] and these include fragment evolution, fragment linking, fragment self-assembly, and fragment optimization.

4.18.6.1 Fragment Evolution

Fragment evolution is most like standard lead optimization, requiring the addition of functionality that binds to additional parts of the target protein. The fact that the starting point is a small molecule means that there should be plenty of opportunity for this approach before violating the 'Lipinski rules.' Where structural information is available on the binding mode of the initial fragment hit, e.g., from x-ray crystallography[29] or NMR spectroscopy[48] then structure-based design approaches can give rapid direction and progress. When structural information is not available then the screening of appropriate analogs of the original hit would be performed in order to try and establish a traditional SAR.

Scientists at AstraZeneca have presented data on the profiles of fragment 'hits' versus a range of different target classes that have contributed 'hits' from fragment approaches (**Table 5**).[23] In each of these cases evolutionary growth of the initial fragment hits was used as the strategy. X-ray crystallography and NMR data were also used in some cases to enable this evolution.

In one example where no structural information was available, a set of 600 compounds was screened against a class A GPCR and around 10% of these were found to have reproducible activity at around 1 mM. One of these actives (pIC_{50} 3.2) had a near-neighbor search performed against it and in the next screening round a more complex and potent neighbor (pIC_{50} 5.4) was identified. In another example, the same set of 600 compounds was screened versus a class B GPCR and this led to 29 actives, of which nine repeated from solid retest. Near-neighbor similarity screening of a weak hit (pIC_{50} 3.2) then helped rapidly identify a more complex neighbor as a submicromolar antagonist ($pIC_{50} > 6$).

An example of the fragment evolution approach is in the design and synthesis of DNA gyrase inhibitors.[49] In this example a set of potential low-molecular-weight inhibitors (termed 'needles' at Roche) was computationally selected by docking and pharmacophore approaches to give a set of *c.* 3000 compounds for testing. This resulted in the identification of 150 initial hits that were subjected to a variety of techniques to identify those suitable for synthetic elaboration. Using x-ray structure analysis as the guiding hand the potency was further increased by four orders of magnitude as summarized in **Figure 13**.

Another example from AstraZeneca illustrates the power of the fragment approach coupled with computational chemistry design. In a program to develop novel inhibitors of the phosphatase PTP-1B, medicinal and computational chemistry design identified the sulfahydantoin motif (compound 1, **Figure 14**) as a potential phophotyrosine mimetic.[50] After synthesis it was shown that this fragment demonstrated a weak inhibitory effect in an NMR-based assay ($\sim$1–3 mM). When the x-ray crystal structure of this fragment bound to the target enzyme was solved (**Figure 15**) it suggested that further potency could be found by introducing the *o*-methoxy group as a conformational constraint. Further structure-inspired changes led to the biphenyl compound with activity of 3 μM.

Table 5 'Fragment' refers to initial weak fragment hits and 'hit' refers to compound profile of more typical series post-HTS appropriate for the start of hit-to-lead work

Target class	*Concentration of screening*	*Technology*	*Hit rate %*	*Initial fragment potency per heavy atom*	*Fragment potency–ClogD*	*Derived hit potency per heavy atom*	*Hit potency–ClogD*
Asparyl protease	300 μM	NMR		0.27	2.34	0.25	3.35
Serine protease	1 mM	NMR		0.4	4.28	0.23	8.85
Metalloprotease	1 mM	HCS	1	0.3	2.75	0.43	3.5
Enzyme	300 μM	HCS	26	0.41	2.98	0.39	7.65
Kinase	600 μM	NMR		0.42	3.05	0.41	3.2
Phosphatase	1 mM	NMR	33	0.18	2.5	0.25	4.62
ATPase	1 mM	NMR	6.3	0.27	0.77	0.35	0.1
Protein–protein interaction	50 μM	HCS	0.2	0.29	6.7	0.31	7.14
GPCR class A	1 mM	HCS	9	0.2	1.25	0.18	2.89
GPCR class B	1 mM	HCS	1.5	0.18	3.1	0.27	5.2

Ligand efficiencies for the fragments and hits are calculated and shown in units of pIC50/non-H atom. Additionally the potency relative to ClogD for initial fragment hit and derived lead is shown.

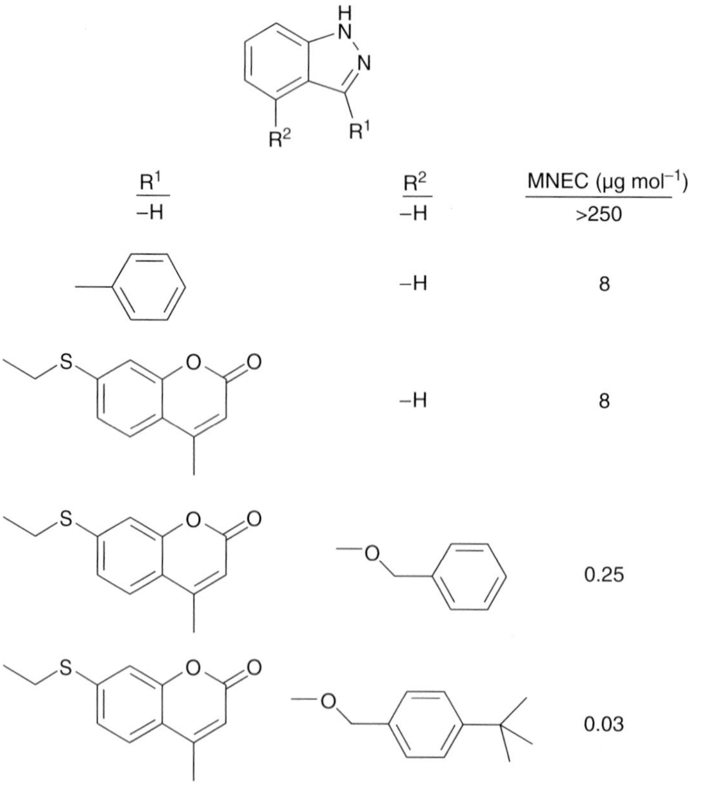

Figure 13 Evolution of indazole inhibitors of DNA gyrase by scientists at Roche using the 'needles approach.' MNEC, maximal noneffective concentration (a measure of activity).

Conformational lock compound 2 (150 μM)

Compound 1 (3 mM)

Compound 3 (3 μM)

Hydrophobic *m*-subst. (130 μM)

Figure 14 PTP-1B inhibitors evolved by structure-based design from an initial 3 mM fragment hit.

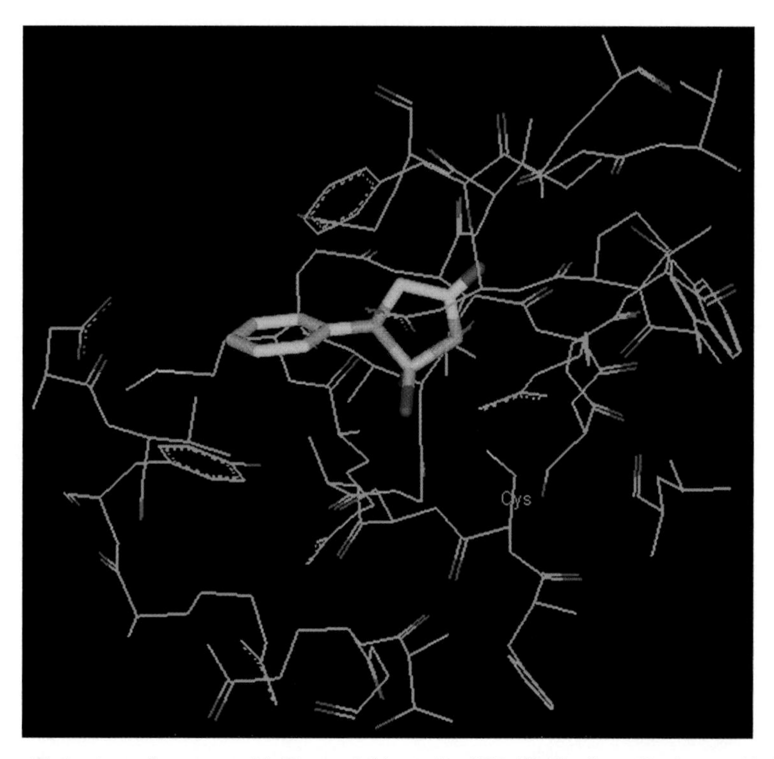

Figure 15 X-ray crystal structure of compound 1 (**Figure 14**) bound to PTP-1B. The ligand is shown with thick bonds and the active site cysteine is labeled.

4.18.6.2 Fragment Linking

Fragment linking is illustrated in **Figure 16**; this involves joining two fragments that have been identified to bind at adjacent sites. Even in those cases where it is possible to find fragments binding to more than one site, the linking step can also be difficult to achieve. Having access to structural information is vital to avoid having to do a large combinatorial and random search in order to find the optimal linking scheme.

An example of the fragment-based linking approach was the identification of a potent inhibitor of *cis–trans* isomerase FKBP (FK506 binding protein) using the SAR-by-NMR method developed at Abbott.[51] Compounds that bound weakly to FKBP included a trimethyoxyphenyl pipecolic acid derivative ($K_d = 2.0\,\mu M$). A second round of screening was carried out using the same library but in the presence of saturating amounts of the pipecolic acid fragment identified initially. This led to the identification of a benzanilide derivative that bound with an affinity of 0.8 mM. Screening of close analogs enabled the SAR to be expanded, and thus a structural model for the binding of these fragments was developed. Four compounds that linked the two sites were then synthesized and found to have nanomolar activities (see **Figure 17**).

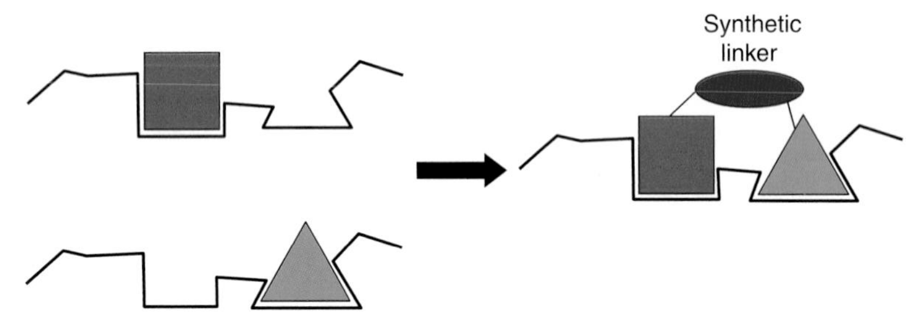

Figure 16 Fragment linking schema that is dependent on structural insight such as relative disposition so as to enable rapid progress.

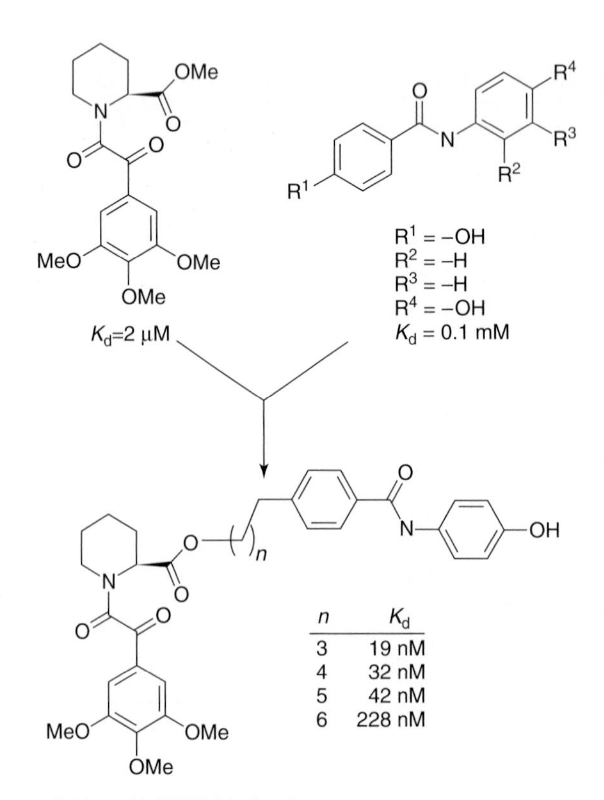

n	K_d
3	19 nM
4	32 nM
5	42 nM
6	228 nM

Figure 17 Example of fragment linking with FKBP binding fragments.

4.18.6.3 Fragment Self-Assembly

Fragment self-assembly involves the use of reactive fragments that link together to form an active inhibitor in the presence of the protein target. The essence of the approach is that the protein serves as a template and therefore selects those combinations of reagents that act as inhibitors. One of the first examples of this method was the reaction between four amines and three aldehydes to give imines (subsequently reduced to amines).[52] Although 12 possible amines could arise from this reaction when performed in the presence of carbonic anhydrase, the proportion of one specific amine was increased at the end of the reaction and this was presumed to correspond to the most active inhibitor (**Figure 18**).

4.18.6.4 Fragment Optimization

This involves the optimization or modification of only a part of the molecule, often to enhance properties other than the inherent potency of the original molecule or to deal with some problem. An example of this approach is the incorporation of alternative S1-binding fragments into a series of *trans*-lactam thrombin inhibitors.[13] The complexity of the synthesis of the *trans*-lactam system made it desirable to have a mechanism to prioritize potential S1 substituents in advance of committing chemistry resource (**Figure 19**).

A novel proflavin displacement assay was developed to identify candidate fragments that bound at S1. This was possible because proflavin had been shown by x-ray crystallography to bind into the S1 pocket of thrombin. This provided the basis for a simple absorbance-based assay to allow HCS of fragments which might bind just in S1. One fragment so discovered was 2-aminoimidazole whose binding mode in this region of the enzyme was then confirmed using x-ray crystallographic analysis.[53] It was subsequently incorporated into the translactam series of inhibitors.

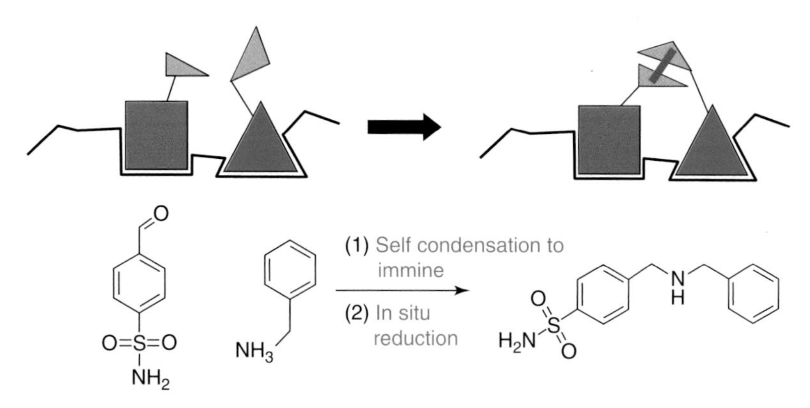

Figure 18 Fragment self-assembly enables components that bind at the same time in the active site to be linked together. The example shown is for inhibitors of carbonic anhydrase.

S1 binder detected by proflavin assay

Fully elaborated *trans*-lactam example

Figure 19 2-Amino imidazole was identified as a novel group that binds into the thrombin S1 binding group by using a proflavin displacement acid. It was then successfully introduced into the previously discovered *trans*-lactam template.

4.18.7 Conclusions

The discovery of new drugs remains a highly challenging and complex process. In this chapter, aspects of the process by which leads evolve into drugs have been discussed. Screening at higher concentrations and/or using direct biophysical approaches such as x-ray crystallography can be used to find smaller and less complex starting points. Theoretical analysis suggests that these may provide highly effective starting points for exploring novel chemical space that may not have been exemplified sufficiently, or at all, in more complex molecules. While the approach provides an alternative entry point into the drug discovery pathway it is not an universal panacea as it is highly dependent on good structural insights to move initial results forward. Like most of the techniques developed and deployed for drug discovery, an appropriate role for this approach will take time to establish, although early indications are that it can have impact in several different guises. But it is already clear from publications by scientists working in design and analysis of screening collections for use in general HTS screening and fragment approaches[15,44,45,54] that the various aspects discussed in this chapter are being increasingly considered and are contributing to the processes by which we look for good starting points and from which we continue to facilitate the evolution of the drugs of the future.[55]

References

1. Goodfellow, P. N. *Chem. Br.* **2002**, *38*, 3.
2. Horrobin, D. F. *J. R. Soc. Med.* **2000**, *93*, 341–345.
3. Lipinski, C. A.; Lombardo, F.; Dominy, B. W.; Feeney, P. J. *Adv. Drug Deliv. Rev.* **1997**, *23*, 3–25.
4. Lipinski, C. A. *J. Pharmacol. Toxicol. Methods* **2001**, *44*, 235–249.
5. Leo, A.; Hansch, C. *Perspect. Drug Disc. Des.* **1999**, *17*, 1–25.
6. Veber, D. F.; Johnson, S. R.; Cheng, H.-Y.; Smith, B. R.; Ward, K. W.; Kopple, K. D. *J. Med. Chem.* **2002**, *45*, 2615–2623.
7. Clark, D. E.; Pickett, S. D. *Drug Disc. Today* **2000**, *5*, 49–58.
8. Oprea, T. I.; Davis, A. M.; Teague, S. J.; Leeson, P. D. *J. Chem. Inf. Comp. Sci.* **2001**, *41*, 1308–1315.
9. Rishton, G. M. *Drug Disc. Today* **2003**, *8*, 86–96.
10. Walters, W. P.; Murcko, M. A. *Methods Principles Med. Chem.* **2000**, *10*, 15–32.
11. Leach, A. R.; Gillet, V. J. *An Introduction to Cheminformatics*; Kluwer: Dordrecht, 2003.
12. Teague, S. J.; Davis, A. M.; Leeson, P. D.; Oprea, T. *Angew. Chem. Int. Ed. Engl.* **1999**, *38*, 3743–3748.
13. Hann, M. M.; Leach, A. R.; Harper, G. *J. Chem. Inf. Comp. Sci.* **2001**, *41*, 856–864.
14. Sneader, W. *Drug Prototypes and Their Exploitation*; John Wiley and Sons Ltd.: New York, 1996.
15. Jacoby, E.; Schuffenhauer, A.; Popov, M.; Azzaoui, K.; Havill, B.; Schopfer, U.; Engeloch, C.; Stanek, J.; Acklin, P.; Rigollier, P. et al. *Curr. Topics Med. Chem.* **2005**, *5*, 397–411.
16. Oprea, T. I. *J. Comput.-Aided Mol. Des.* **2000**, *14*, 251–264.
17. Ertl, P. *J. Chem. Inf. Comp. Sci.* **2003**, *43*, 374–380.
18. Hann, M. M.; Leach, A. R.; Green, D. V. S. Computational Chemistry, Molecular Complexity and Screening Set Design. In *Chemoinformatics in Drug Discovery*; Oprea, T., Ed.; Wiley-VCH: New York, 2005.
19. Fink, T.; Bruggesser, H.; Reymond, J.-L. *Angew. Chem. Int. Ed.* **2005**, *44*, 1504–1508.
20. Andrews, P. R.; Craik, D. J.; Martin, J. L. *J. Med. Chem.* **1984**, *27*, 1648–1657.
21. Kuntz, I. D.; Chen, K.; Sharp, K. A.; Kollman, P. A. *Proc. Natl. Acad. Sci. USA* **1999**, *96*, 9997–10002.
22. Hopkins, A. L.; Groom, C. R.; Alex, A. *Drug Disc. Today* **2004**, *9*, 430–431.
23. Burrows, J. N. *High Concentration Screening: Integrated Lead Generation*. Oral Presentation at the Society for Medicines Research: Trends in Drug Discovery, 2004.
24. Abad-Zapatero, C.; Metz, J. T. *Drug Disc. Today* **2005**, *10*, 464–469.
25. Deprez-Poulain, R.; Deprez, B. *Curr. Top. Med. Chem.* **2004**, *4*, 569–580.
26. Lipinski, C. A. *Chemistry Quality and the Medicinal Chemistry–Biology Interface*. Oral Presentation at the 6th Winter Conference on Medicinal and Bioorganic Chemistry, 2004.
27. Erlanson, D. A.; McDowell, R. S.; O'Brien, T. *J. Med. Chem.* **2004**, *47*, 3463–3482.
28. Carr, R.; Jhoti, H. *Drug Disc. Today* **2002**, *7*, 522–527.
29. Rees, D. C.; Congreve, M.; Murray, C. W.; Carr, R. *Nat. Rev. Drug Disc.* **2004**, *3*, 660–672.
30. McGovern, S. L.; Helfand, B. T.; Feng, B.; Shoichet, B. K. *J. Med. Chem.* **2003**, *46*, 4265–4272.
31. Rishton, G. M. *Drug Disc. Today* **1997**, *2*, 382–384.
32. Lesuisse, D.; Lange, G.; Deprez, P.; Bénard, D.; Schoot, B.; Delettre, G.; Marquette, J.-P.; Broto, P.; Jean-Baptiste, V.; Bichet, P. et al. *J. Med. Chem.* **2002**, *45*, 2379–2387.
33. Hajduk, P. J.; Sheppard, G.; Nettesheim, D. G.; Olejniczak, E. T.; Shuker, S. B.; Meadows, R. P.; Steinman, D. H.; Carrera, G. M.; Marcotte, P. A.; Severin, J. et al. *J. Am. Chem. Soc.* **1997**, *119*, 5818–5827.
34. Moy, F. J.; Haraki, K.; Mobilio, D.; Walker, G.; Powers, R.; Tabei, K.; Tong, H.; Siegel, M. M. *Anal. Chem.* **2001**, *73*, 571–581.
35. Buck, E.; Wells, J. A. *Proc. Natl. Acad. Sci. USA* **2005**, *102*, 2719–2724.
36. Hann, M. M.; Green, R. *Curr. Opin. Chem. Biol.* **1999**, *3*, 379–383.
37. Congreve, M.; Carr, R.; Murray, C.; Jhoti, H. *Drug Disc. Today* **2003**, *8*, 876–877.
39. Maly, D. J.; Choong, I. C.; Ellman, J. A. *Proc. Natl. Acad. Sci. USA* **2000**, *97*, 2419–2424.
40. Bemis, G. W.; Murcko, M. A. *J. Med. Chem.* **1996**, *39*, 2887–2893.
41. Bemis, G. W.; Murcko, M. A. *J. Med. Chem.* **1999**, *42*, 5095–5099.
42. Lewell, X.-Q.; Judd, D. B.; Watson, S. P.; Hann, M. M. *J. Chem. Inf. Comp. Sci.* **1998**, *38*, 511–522.
43. Leach, A. R.; Green, D. V. S.; Hann, M. M.; Judd, D. B.; Good, A. C. *J. Chem. Inf. Comp. Sci.* **2000**, *40*, 1262–1269.
44. Hartshorn, M. J.; Murray, C. W.; Cleasby, A.; Frederickson, M.; Tickle, I. J.; Jhoti, H. *J. Med. Chem.* **2004**, *48*, 403–413.

45. Baurin, N.; Aboul-Ela, F.; Barril, X.; Davis, B.; Drysdale, M.; Dymock, B.; Finch, H.; Fromont, C.; Richardson, C.; Simmonite, H. et al. *J. Chem. Inf. Comput. Sci.* **2004**, *44*, 2157–2166.
46. Page, M. I.; Jencks, W. P. *Proc. Natl. Acad. Sci. USA* **1971**, *68*, 1678–1683.
47. Murray, C. W.; Verdonk, M. L. *J. Comput.-Aided Mol. Des.* **2002**, *16*, 741–753.
48. Schade, M.; Oschkinat, H. *Curr. Opin. Drug Disc. Dev.* **2005**, *8*, 365–373.
49. Boehm, H.-J.; Boehringer, M.; Bur, D.; Gmuended, H.; Huber, W.; Klaus, W.; Kostrewa, D.; Kuehne, H.; Luebbers, T.; Muenier-Keller, N. et al. *J. Med. Chem.* **2000**, *43*, 2664–2674.
50. Black, E.; Breed, J.; Breeze, A. L.; Embrey, K.; Garcia, R.; Gero, T. W.; Godfrey, L.; Kenny, P. W.; Morley, A. D.; Minshull, C. A. et al. *Bioorg. Med. Chem. Lett.* **2005**, *15*, 2503–2507.
51. Shuker, S. B.; Hajduk, P. J.; Meadows, R. P.; Fesik, S. W. *Science* **1996**, *274*, 1531–1534.
52. Huc, I.; Lehn, J.-M. *Proc. Natl. Acad. Sci. USA* **1997**, *94*, 2106–2110.
53. Conti, E.; Rivetti, C.; Wonacott, A.; Brick, P. *FEBS Lett.* **1998**, *425*, 229–233.
54. Davis, A. M.; Keeling, D. J.; Steele, J.; Tomkinson, N. P.; Tinker, A. C. *Curr. Top. Med. Chem.* **2005**, *5*, 421–439.
55. Fattori, D. *Drug Disc. Today* **2004**, *9*, 229–238.

Biographies

Mike M Hann received his PhD in 1981 from the City University, London (under the supervision of Professor Peter Sammes) for his studies on the synthesis of bioisosteric enkephalin peptides. He subsequently worked as a medicinal chemist at Wyeth and GD Searle in the UK before moving into the area of computational chemistry. He joined Glaxo in 1986 and is currently Director of Structural and Biophysical Sciences UK for GlaxoSmithKline. He has a long-standing interest in computational and biophysical approaches to lead generation and optimization to enable effective medicinal chemistry.

Jeremy N Burrows studied for his first and second degrees in chemistry/organic chemistry at Oxford University and did his DPhil with Dr Jeremy Robertson on natural product synthesis (1993–96). He joined ZENECA Pharmaceuticals in January 1997 as an Associate Team Leader in Medicinal Chemistry and was promoted first to Team Leader in 1999 and then to Associate Director in January 2003, leading the Lead Generation Chemistry section in Inflammation at Alderley Park. In October 2005 he moved to AZ, Södertälje on secondment. During his career, he has worked in infection, cardiovascular, inflammation, and neurology research areas. He has worked on several projects that have delivered candidate drugs but most of the research focus has been in the hit identification and hit-to-lead phases including projects from the following target families: ATPases, metalloproteinases, kinases, enzymes, GPCR antagonists, integrins, nuclear hormone receptors, and aspartyl proteases. Because of his keen interest in methods of 'hit identification,' he has become one of the champions of fragment-based and high concentration screening within AstraZeneca.

Ed Griffen obtained his BSc and PhD from Imperial College under the supervision of Professors Charles Rees and Chris Moody and went on to carry out postdoctoral research at the University of Waterloo, Canada with Prof Victor Snieckus. He joined Zeneca in 1994 and has worked in a variety of areas including analgesics, anti-infectives, and currently oncology; the majority of this being in the lead generation phase. Ed has been an advocate of new synthetic technologies, QSAR and library design which he has promoted at Alderley Park and other AstraZeneca sites. Promoted to Principal scientist in 2003, Ed has taught a number of medicinal chemistry courses within AstraZeneca, and gives a lecture course at the University of Manchester.

© 2007 Elsevier Ltd. All Rights Reserved
No part of this publication may be reproduced, stored in any retrieval system or transmitted in any form by any means electronic, electrostatic, magnetic tape, mechanical, photocopying, recording or otherwise, without permission in writing from the publishers

Comprehensive Medicinal Chemistry II
ISBN (set): 0-08-044513-6

ISBN (Volume 4) 0-08-044517-9; pp. 435–458

4.19 Virtual Screening

A Good, Bristol-Myers Squibb, Wallingford, CT, USA

© 2007 Elsevier Ltd. All Rights Reserved.

4.19.1 Ligand-Based Virtual Screening

Virtual screening (VS) based on constraints imposed by ligand template structures comprise the largest single family of VS techniques.[1–9] Such methods have been spawned from much research into computational techniques for ligand description and storage. Many of these methodological developments are detailed below.

4.19.1.1 Machine-Readable Descriptions of Chemical Structure

In order to undertake virtual screening, it is first necessary to convert chemical structure into an easy to interpret machine readable format. While a number of methods have been proposed for two-dimensional (2D) structure depiction of chemical entities,[10] connection tables are the most important representation to emerge. Perhaps the most widely applied of these is the structure data (SD) file. This file was designed to permit the movement of large numbers of molecules and their associated data between databases. Chemical structures are stored in a connection table which houses x and y atom coordinates based on bond lengths (z coordinates can be added when three-dimensional (3D) data is to be stored) together with associated atom type, chirality, and bond connection data.[11] A connection table is in essence a graph containing the complete and explicit description of molecular topology and forms an easily analyzable repository of 2D chemical data for VS. Graph theory forms the mathematical model at the core of topology description,[12–15] and many of the key concepts of molecular graph theory are highlighted in **Figure 1**. SD files and other extended connection table formats (e.g., mol2 files[16]) provide a perfectly usable means of structure data transport. Their inflexible format requirements and somewhat inefficient storage needs led to efforts to devise other methods for chemical structure interpretation. The most widely applied of these is the Simplified Molecular Input Line Entry System (SMILES).[17–18] SMILES is a line notation (a typographical method using printable characters) for entering and representing molecules. While a SMILES string contains the same information as an extended connection table, it is in essence a chemical language with a vocabulary (atom and bond symbols) and grammatical rules (e.g., for substitution pattern recognition). SMILES representations of structure can in turn be used as 'words' in the vocabulary of other languages designed for chemical storage. A typical SMILES will take 50% to 70% less space than an equivalent connection table, even in binary format (typically under two bytes per molecule). Other chemical languages include Sybyl Line Notation (SLN),[19] which was designed as an extension of SMILES capable of substructure search and Markush[20] structure specification. **Figure 2** shows an example of melatonin converted to a number of machine readable connectivity formats.

4.19.1.2 Three-Dimensional Structure Generation

A key step for VS techniques requiring 3D structural information is the ability to create multiple 3D models for large databases of potential ligand molecules. The geometrically invariant nature of topological structure data is such that 2D connection tables are sufficient for descriptor generation and subsequent screening to proceed. Many of the basic tenets of 2D structure storage extrapolate naturally to the 3D world (e.g., the addition of a z coordinate in the SD file). Nevertheless, the geometric complexity of 3D models is such that a number of issues must be carefully considered during their construction. Suitable techniques must be fast, since large numbers of compounds both real and virtual need to be converted on a regular basis. The tool needs to be robust enough to handle a wide variety of chemical and structural subtypes with a high rate of conversion (>95%). It must also be able to handle stereochemistry, both from the perspective of model generation and the explosion of potential stereo center options when chirality is not explicitly specified. Where high-quality scoring functions are to be applied, the ability to generate multiple ionization and tautomeric states also takes on significant importance. For conformational searching, the capacity to extensively sample conformational search space in a rapid time frame is critical.

The difficulties engendered by these requirements are a quantum leap relative to those of topological model generation and considerable research has been undertaken to illicit their solution. Such efforts have been dominated by the development of knowledge-based techniques which began in the early 1980s when Corey and Feiner undertook

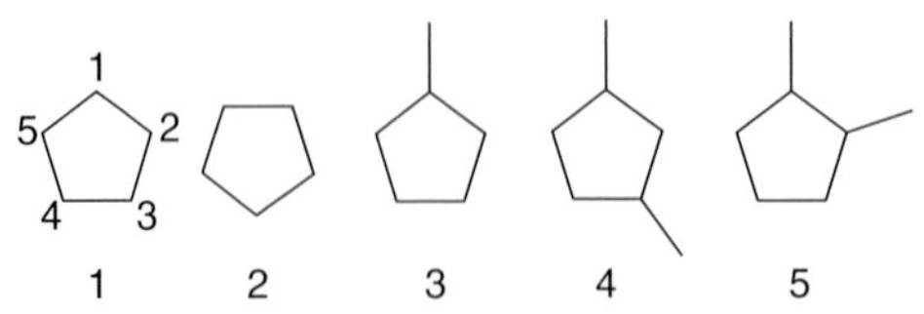

	1	2	3	4	5
1	0	1	0	0	1
2	1	0	1	0	0
3	0	1	0	1	0
4	0	0	1	0	1
5	1	0	0	1	0

Figure 1 Graph theory provides the mathematical translation of chemical connectivity, with atoms corresponding to graph vertices and bonds to edges. Sample graphs shown here illustrate some core graph theory definitions. Graph 1 is isomorphic to graph 2; graph 1 is a subgraph of graph 3; both graph 1 and graph 3 are subgraphs of graphs 4 and 5; however, only graph 3 is a maximum common subgraph; graph 4 can not include graph 5 and vice versa. The translation of graph 1 to its corresponding adjacency matrix is also shown (right). Subgraph matching is at the heart of substructure searching.

```
(1) -MTS- 04300517352D 0 0.00000 0.00000 0

 17 18 0 0 0 0 0 0 0 0 1 V2000
   8.2566  0.9979  0.0000 C 0 0 0 0 0 0 0 0 0 0 0 0
   7.8503  0.0804  0.0000 C 0 0 0 0 0 0 0 0 0 0 0 0
   8.4392 -0.7278  0.0000 O 0 0 0 0 0 0 0 0 0 0 0 0
   6.8610 -0.0245  0.0000 N 0 0 0 0 0 0 0 0 0 0 0 0
   6.4547 -0.9420  0.0000 C 0 0 0 0 0 0 0 0 0 0 0 0
   5.4654 -1.0468  0.0000 C 0 0 0 0 0 0 0 0 0 0 0 0
   4.8769 -0.2384  0.0000 C 0 0 0 0 0 0 0 0 0 0 0 0
   5.1869  0.7216  0.0000 C 0 0 0 0 0 0 0 0 0 0 0 0
   4.3769  1.3016  0.0000 N 0 0 0 0 0 0 0 0 0 0 0 0
   3.5669  0.7216  0.0000 C 0 0 0 0 0 0 0 0 0 0 0 0
   2.5877  0.9244  0.0000 C 0 0 0 0 0 0 0 0 0 0 0 0
   1.9195  0.1804  0.0000 C 0 0 0 0 0 0 0 0 0 0 0 0
   2.2326 -0.7729  0.0000 C 0 0 0 0 0 0 0 0 0 0 0 0
   3.2065 -0.9756  0.0000 C 0 0 0 0 0 0 0 0 0 0 0 0
   3.8769 -0.2384  0.0000 C 0 0 0 0 0 0 0 0 0 0 0 0
   1.5663 -1.5186  0.0000 O 0 0 0 0 0 0 0 0 0 0 0 0
   0.5873 -1.3143  0.0000 C 0 0 0 0 0 0 0 0 0 0 0 0
  1  2  1  0  0  0  0
  2  3  2  0  0  0  0
  2  4  1  0  0  0  0
  4  5  1  0  0  0  0
  5  6  1  0  0  0  0
  6  7  1  0  0  0  0
  7 15  1  0  0  0  0
  7  8  2  0  0  0  0
  8  9  1  0  0  0  0
  9 10  1  0  0  0  0
 10 15  1  0  0  0  0
 10 11  2  0  0  0  0
 11 12  1  0  0  0  0
 12 13  2  0  0  0  0
 13 14  1  0  0  0  0
 14 15  2  0  0  0  0
 13 16  1  0  0  0  0
 16 17  1  0  0  0  0
M END
$$$$

(2) CC(=O)NCCC2=CNc1ccc(cc12)OC

(3) CH3C(=O)NHCH2CH2C[15]:CH:NH:C[20]:CH:CH:C(:CH:C:@15:@20)OCH3
```

Figure 2 Melatonin structure converted into the three primary machine-readable formats for chemical structure: (1) SDF, (2) SMILES, (3) SLN.

efforts to predict the preferred conformations of six member rings as part of the LHASA program.[21,22] The SCA program of Hoflack and De Clerq[23] is perhaps the first program with direct relevance to the virtual screening tools in use today. It is still available through QCPE,[24] and finds application within the popular FLEXX docking program.[25,26] At around the same time the programs COBRA[27] and WIZARD[28] were created for systematic conformational analysis exploiting symbolic logic and artificial intelligence (AI) techniques. These systems break molecules down into substructures for which the AI system contains knowledge of conformational behavior. The substructures are then linked together to form a unit graph, the edges of which represent the type of junction found between two units (**Figure 3**).[29] Using conformational templates assigned to these or similar units from a library, combined with the unit graph junction data, suggestions for conformations are generated. The resulting conformational space is then searched using directed search strategies such as the A* algorithm,[30] which criticizes the symbolic suggestions using predefined or learned rules to rapidly trim search space.

Knowledge-based techniques such as those described above are capable of rapidly constructing high-quality models and dominate the field of 3D molecule generation. This quality is, however highly dependent on the number of suitable templates present in the constituent libraries and databases for a given molecule. An alternative constraint-based approach that has found wide application in the construction of 3D models is the technique of distance

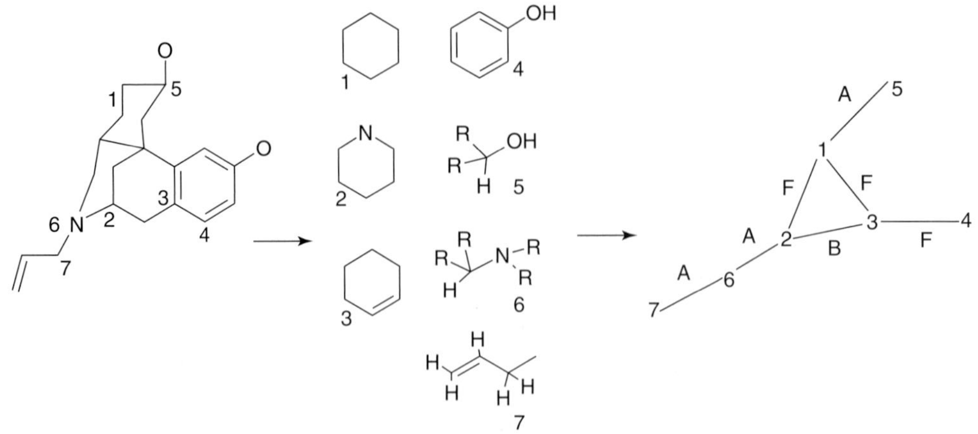

Figure 3 For a systematic conformational search the number of possible conformer is given by the equation $N_{\text{conf}} = \prod_{i=1}^{N} \prod_{j=1}^{ninc} \frac{360}{\theta_{i,j}}$ where N is the number of rotatable bonds and $\theta_{i,j}$ is the size of incremental rotational angle j assigned to bond i. A popular method to circumvent the resultant potential for combinatorial explosion with increasing conformational flexibility is through molecular fragmentation. In the example above *N*-allyl-3,6β-dihydroxy morphinan has been deconstructed into its constituent conformational units and associated unit graph representation. Graph function types: A, acyclic; F, fused; B, bridged. Variants of this fragmentation technique form the basis of many conformational search techniques. (Adapted from Leach, A. R.; Prout, K.; Dolata, D. P. *J. Comput.-Aided Mol. Des.* **1990**, *4*, 271–282.)

geometry.[31–33] The essential premise of distance geometry is that the set of accessible conformations can be aptly described through appropriate distance and chirality constraints. The first are simply lower and upper bounds on the interatomic distances; the second include the handedness of the asymmetric centers in the molecule, together with the planarity of any sp^2 hybridized centers. The resultant description is essentially the computer analog of a CPK model. The technique was made famous by the freely available program DGEOM,[34] which has been widely applied in the generation of molecular conformers. Distance geometry techniques have the advantage of being fast and easy to parameterize. The models produced can be crude, however, due to the ambiguity inherent in the distance matrix.

The techniques described above form the backbone for much of the technology applied today. With the subsequent development of easily convertible 2D connectivity files, high-throughput 3D structure generation software development was able to go mainstream. The resulting software tool families developed in response to this advance are described below.

4.19.1.2.1 Two-dimensional to three-dimensional (2D–3D) structure conversion

The CONCORD program[35,36] has historically been the most popular of the 3D conversion softwares. The program utilizes a knowledge base of rules combined with energy minimization to generate a single low-energy 3D conformation for each structure. Cyclic and acyclic portions are considered separately, with two rings regarded as one if they are attached by at least two common atoms (spiro systems are thus considered as distinct). Resulting substructures are fused to form the complete molecule. Optimum acyclic bond lengths, angles, and torsions are extracted from a table of published values. Similarly, bond lengths and torsion angles of single cyclic portions are built using precalculated topological rules. Ring systems are constructed through the assignment of gross conformations of each ring. Each constituent ring is then fused into the system in order, with a strain minimization function employed to create an acceptable geometry. Once constructed, the structure can be further optimized through a final energy minimization. Two supplemental programs used as preprocessors for CONCORD come in the form of STEREOPLEX[37] and PROTOPLEX.[38] STEREOPLEX is designed to enumerate the stereochemistry at undefined stereo centers. The system allows user prioritization of stereo centers (acyclic double bonds, ring fusion atoms, ring bridgehead atoms, etc.) allowing control over which and how many of the $2n$ possible stereomers will be generated. Sterically and topologically impossible stereo isomers are automatically eliminated and control is also permitted over the naming convention applied to the output. PROTOPLEX provides multiple protonation and tautomeric states for each input structure according to user-specified limits and priority rules. These tell the program which states to generate if the complete set would exceed the user's limit.

Another popular 3D conversion program developed by Gasteiger and Sadowski is CORINA.[39–41] Of particular note with this program is its extensive and flexible handling of ring systems. Rings are classified as small (eight atoms or fewer), rigid macrocycles (large rings containing bridged and/or fused rings), and flexible macrocycles. The system contains a particularly extensive table of small and medium ring conformations characterized and ordered by strain energy. A backtracking system is used to reconstruct fused and bridged systems from their constituent ring fragments. Additional flexibility is incorporated through the application of pseudo force field minimization to optimize systems that are highly strained or contain unusual heteroatom combinations. Rigid macrocycles are built up from crude superstructures that maintain molecule shape and symmetry. Flexible macrocycles are constructed using rapid conformational search procedures derived from linear notation of ring features that permits an energy ranking of potential conformers in 1D.[42] In both cases rings are further refined using pseudo force field minimization. The system also permits the creation of multiple models for systems containing rings up to nine atoms. This can be valuable for programs which only handle conformational searching of acyclic torsions (e.g., FLEXX,[26] for which a CORINA interface has been written). Examples of rings built by CORINA are shown in **Figure 4**. Acyclic torsions for the molecule are assigned from libraries constructed from statistical analysis of conformational preferences contained within the Cambridge Structural Database (CSD).[43,44] These are applied upon molecular reconstruction with additional conformational analysis and minimization executed to remove any remaining repulsive nonbonded interactions.

The CONVERTER program[45] applies the distance geometry variant DGII[46] to undertake 2D to 3D conversion. Based on the EMBED algorithm,[47] the system creates a set of distance matrix bounds via a process known as bound smoothing. A random assignment of interatomic distance values is made within said bounds obtained by bound smoothing, and 'best-fit' mapping of atomic coordinate are made to this guess. Since this best fit is computed by a projection method, this step is termed embedding. Coordinate deviations to the distance bounds are then minimized, with the option to apply an addition '4D' error function to overcome local mimima problems, for example with respect to the local chirality of molecules.

OMEGA[48–50] software is another program that exploits distance geometry techniques for the generation of 3D structures,[51] subsequently refining the resultant structures using the Merck Molecular Force Field (MMFF).[52] Its sister program QUAPAC enumerates 3D models to create potential tautomeric and ionization states using knowledge-based protocols. The resultant models can then be charged up using a variety of different charge models. The system also comes with the program PKATYPER, which provides a rudimentary approach to pK_a prediction amenable to the enumeration of all reasonable charge states of a wide variety of small-molecule chemistries based on analysis of atomic environment.

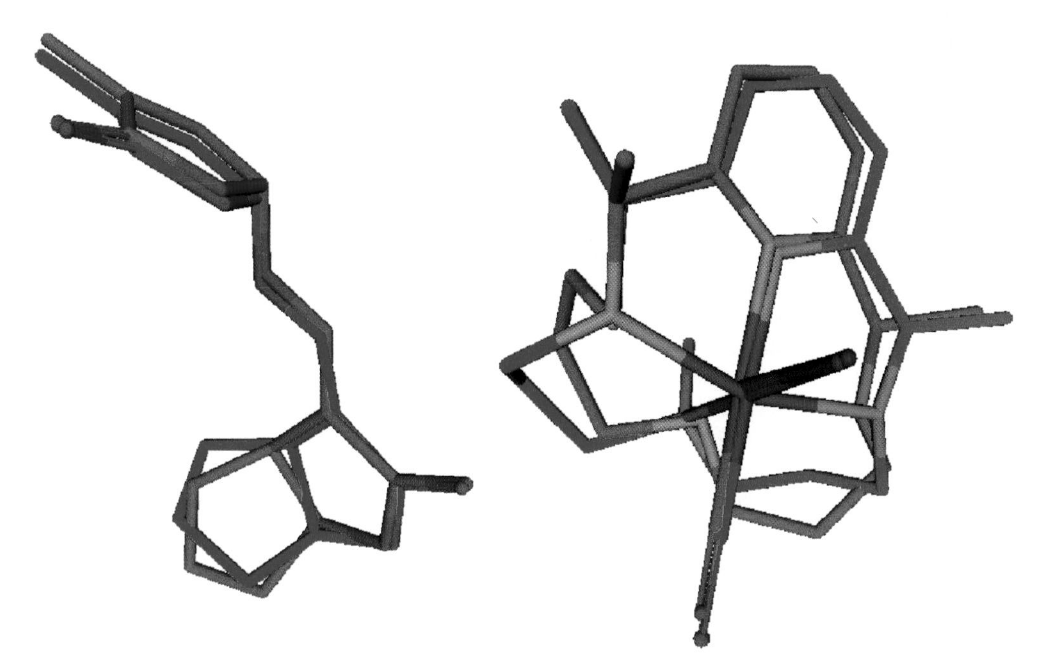

Figure 4 Examples of CORINA's[41] (magenta) ability to reproduce the experimentally determined 3D structures from the CSD (green) from small organics (cyclopentafuranone derivative, CSD RefCode: HIYCUN, RMSD = 0.41 Å) through to complex organometallic macrocycles (ruthenium complex, CSD RefCode: DIJKAI, RMS = 0.20 Å).

LIGPREP[53] processes ligands in stages using a number of utilities. The program first adds explicit hydrogens (APPLYHTREAT) and strips salts or other cocrystallites, which are often present in 2D structure databases (DESALTER) from the 2D connectivity data. The hydrogens are adjusted (NEUTRALIZER) to attain neutrality, then protonated forms for each input molecule are generated (IONIZER) consistent with a user specified pH range (typically 7 ± 2). Once ionization states have been determined, the TAUTOMERIZER utility generates one or more tautomeric forms for each input molecule. Tautomers are specified by SMARTS[18] patterns from a user-modifiable data file. Tautomeric probabilities are based either on experiment or Jaguar solvated QM calculations.[54] STEREOIZER is then run in conjunction with a series of user-definable options to generate the required stereochemical combinations. Once chiralities are defined ring conformations are generated using the commonly from a library of precalculated templates (RING_CONF). Templates for several hundred ring systems (one or more fused rings at least one of which is flexible) have been compiled, each consisting of one of more low-energy ring conformations pregenerated using MACROMODEL conformational searches.[55,56] When a ring system is recognized, ligand energies for the templated ring conformations are estimated using the inherent ring energy together with estimates of the axial versus equatorial energies of ring attachments. By default the combination of ring conformations giving the lowest overall ring energy is selected. Multiple ring conformations may also be requested, however. A constrained MACROMODEL[57] minimization (bmin) is then run on the resultant structures. Chiral properties introduced in the stereochemistry step are enforced using additional improper torsional potentials. The input structure is then minimized, and if distorted bond lengths or angles are found (a sign of tangled structures, e.g., intertwined rings) a short conformation search is automatically performed to untangle the structure.

4.19.1.2.2 Explicit conformer generation

A number of programs have been devised for the population of databases with 3D conformations. CONFIRM[58–60] uses fast mode conformation generation for the rapid creation of conformers. As with many of the structure generation algorithms discussed previously, the fast algorithm deals with cyclic and acyclic portions of a molecule separately. Ring conformations are drawn from a library of predefined structures, while a modified pseudo exhaustive search is applied for the open chain portions. This is achieved through a set of torsion rules that permits up to six predefined allowed torsion settings depending on the hybridization states of the two connected atoms. This is a common approach, variants of which have been applied in a variety of other conformational programs.[61,62] The fast algorithm includes a number of additional features, however, to improve conformational coverage. Rather than simply rejecting conformations for which internal clashes are detected, torsions causing the clashes are adjusted upto $\pm 60^\circ$ and the clashes are monitored using a simple distance function. If the contact is relieved the clash minimization is ended and the conformation accepted. This technique provides an efficient method to extend the conformational space explored by the torsion rules.

CONFORT[63] is designed to undertake speedy conformational analysis through the rapid identification of the local energy minima bounds of each rotatable bond. For each combination of these ranges, the bounded region hyperspace centroid is applied to create an initial 'raw conformation.' This is then relaxed through the application of analytic gradients to the internal coordinate subspace. The resultant population of 'relaxed conformations' can be optimized and filtered by energy and diversity to determine a set of diverse low energy conformers for the system.

RUBICON[64] combines distance geometry with knowledge-based constraints to generate 3D conformers. The distance geometry constraints are augmented by user-customizable SMARTS[18] based rules to overcome the limited chemical geometry information and hard coded chemical intelligence inherent in distance geometry algorithms. Sets of constraint rules can be derived automatically from a given set of training conformations using the sister program AUTORULES. This allows RUBICON rules to be altered to generate structures that have geometries similar to those found in training sets of interest, for example crystal structures, docked structures, or computed low-energy structures.

OMEGA applies a depth-first divide-and-conquer algorithm for torsion driving to generate conformations. Molecules are broken into small fragments and preferred torsion angles and ring conformations extracted from a series of libraries. The default ring library ringlib.txt contains 1808 common ring templates. These were generated through abstraction, conformation generation, and minimization of all unique ring system present in a set of 2.5 million molecules (other rings can be added to a customized variant of the library if required). These libraries are used to generate conformations for each fragment. The program then builds the final conformations by reconstructing whole molecules from the multiconformer fragments. The conformations accepted are filtered based on an energy window from the conformational global minimum are limited by their total energy (as measured by the Drieding force field[65]) Conformational diversity is ensured through the removal of conformers found to be too close in RMSD space. In addition it is possible to optimize each conformation using the MMFF force field. To prevent a gas-phase force field induced molecular collapse, an optional torsion constraint and coloumbic buffering term[66] can be used to prevent excessive torsion deviation on minimization.

4.19.1.2.3 On-the-fly conformation generation techniques

While this section has dealt primarily with explicit conformer generation, many programs used in virtual screening apply on-the-fly techniques to this problem. Such an approach saves on database storage space and provides user-controlled tuning over the level of conformational sampling at search time based on available CPU. It also permits conformational search strategies that directly complement the design of the screening algorithm. The price paid for such an approach is that in general the only the more rudimentary of techniques described above are applied due to time constraints. For example conformers rarely undergo full minimization subsequent to initial generation. In addition many of the more CPU-intensive and conformationally variant properties that can be associated with an explicit conformation (for example qauntum mechanically derived partial charges) cannot in general be assigned to conformations generated at search time. Nevetherless these are trade-offs that many software packages are willing to make.

A popular on-the-fly approach to conformation generation is torsion fitting designed expressly for single pharmacophore model screening. Rather than undertaking an explicit conformational, each candidate molecules undergoes pharmacophore constrained torsion optimization instead. Specific molecular torsions are tweaked to determine whether all the pharmacophore constraints can be met simultaneously, with an optional check for internal van der Waals (vdW) clashes. Variants of this approach have been implemented within CHEM-DBS3D,[61] UNITY,[67,68] ISIS/3D,[69,70] and CATALYST.[45] The technique permits extremely efficient sampling but can lead to highly strained conformers when the pharmacophore constraints are particularly onerous.

Many structure-based virtual screening techniques use the popular molecular fragmentation approach to conformational search as a time-saving device. This technique has also been applied in some ligand-based virtual screening approaches such as FLEXS,[71–73] but will be covered in more depth in Section 4.19.2.1.3.

4.19.1.2.4 Structure quality measures

Numerous studies have been undertaken in an attempt to gain perspective regarding the relative performance of many of the techniques described above. In the arena of 3D structure generation, a number of groups have compared conversion rates and structure quality.[74–78] CONVERTER and CORINA were found to have the most robust successful conversion rates (>98%) compared to CONCORD (~90%), while CONCORD exhibited the most rapid conversion rates (0.14 seconds per molecule). CORINA was the most successful at reproducing crystallographically determined conformations (28–46% versus 20–38% for CONCORD) and ring geometries (78–90% versus 71–89% for CONCORD). It should be noted that LIGPREP, RUBICON, and OMEGA were created subsequent to these studies and are thus not included in the comparison, though comparable RUBICON tests have been conducted elsewhere.[79] It should be noted that CSD structure RMSD values can deviate from their Protein Data Base (PDB)[80] bioactive structure equivalents by as much as conformations created by the structure generators already discussed above.[81] The relevance of being able to reproduce CSD structures in their entirety is thus open to debate.

Investigations have also been undertaken to compare the ability of conformer generators to reproduce ligand bioactive conformations.[82–84] A number of conclusions can be drawn from these analyses. Firstly, with the sampling levels typically applied using such searches (typically fewer than 100 conformers per molecule), molecules with more than eight rotatable bonds represent a significant challenge from the perspective of accurate reproduction of the bioactive conformation. Application of supposedly more sophisticated algorithms do not necessarily help. For example the fast algorithm of CONFIRM was found to perform a little better than its more CPU-intensive CONFIRM counterpart best. In general, however, comparison of relative performance is somewhat ambiguous. This is highlighted by **Figure 5**, where relative performance is rated based on average heavy atom RMS deviation to ligand bioactive conformation and average internal rank of the RMS deviations. As expected the worst performance is seen with the initial generated 3D model (CONCORD generated structure on the right) and the best with the minimized bioactive conformation (on the left). Were a system to prove the best performer in all cases its average rank would be 1. Interestingly, however, even the minimized x-ray structures average rank is greater than 4, highlighting the fact that accuracy is inherently limited by the underlying force field used by any of the conformer generators. As sampling increases the results often improve (CONFORT 500 is approaching minimized x-ray performance) but the average ranks for the different techniques at comparable sampling levels are relatively similar. Overall the techniques described above all perform well on many systems and each has undoubted utility in the context of structure generation. The validation experiments undertaken so far point toward some small advantages for certain techniques, though it is unclear how well these differences translate between databases. The procedures described often share many traits (knowledge-based templating with underlying force field refinements for example), so strengths and weaknesses also overlap to a significant extent. Knowledge-base quality and customizability is clearly a key factor in system performance, and it is those programs still undergoing active development in this context that will likely provide increasingly superior performance in the years to come.

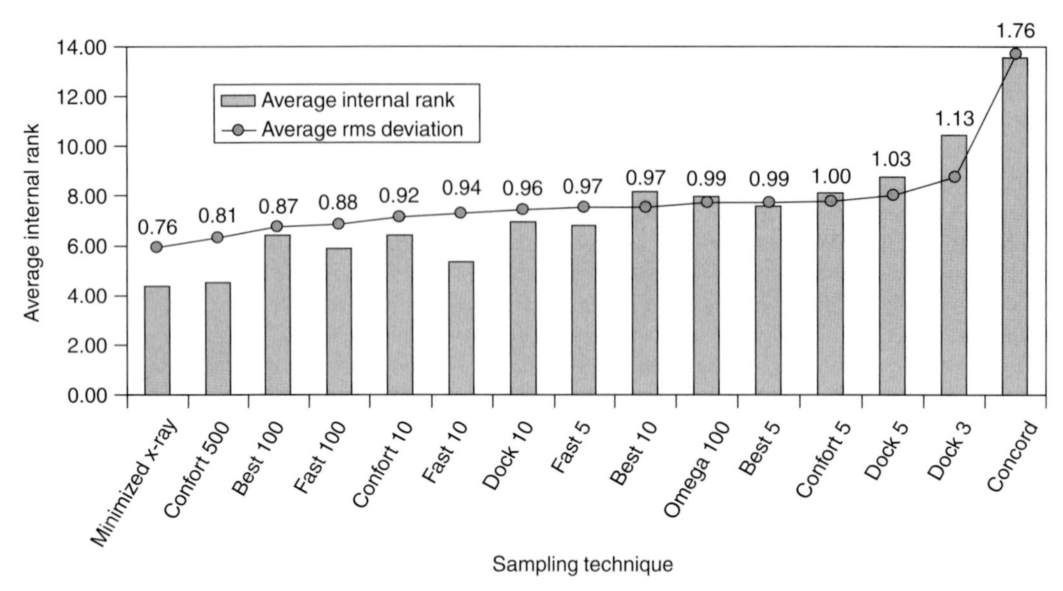

Figure 5 Relative performance of multiple sampling techniques, based on average heavy atom RMS deviation to ligand bioactive conformation and average internal rank of the RMS deviations. Average internal rank is determined using the relative rank of the closest conformer of each technique to the bioactive conformer as calculated by heavy atom RMSD. The technique with the closest conformer thus has rank 1, the next 2, etc., down to 15 (the number of conformational techniques/sampling level combinations compared. Ranks are averaged across all ligands studied to determine the average internal rank. (Adapted from Good, A. C.; Cheney, D. L. *J. Mol. Graph. Model.* **2003**, *221*, 23–30.)

4.19.1.3 Substructure Searching

With suitable structure databases available for processing, their exploitation through VS is the next natural step. Exploitation of these concepts began with the venerable discipline of substructure searching (SS), which has been applied successfully for many years in both 2D and 3D data.

4.19.1.3.1 Topological substructure searching

2D substructure analysis (or subgraph isomorph matching; see **Figure 1**) is still perhaps the most routinely used of all screening techniques with a history dating back to the 1950s with the development of the backtracking algorithm.[85,86] Backtracking, also known as atom-by-atom searching, operates by matching an initial node of the substructure query with a database structure before exploring the nodes off it. The algorithm moves from atom to atom until the last node is successfully matched, or until a mismatch is found, in which case it backtracks to the last correct choice point. Refinements such as early pruning of fruitless branches have been devised to speed the backtracking process.[87,88] Even with such changes the size of the search space rendered by its recursive matching algorithm can prove time-consuming to traverse. Nevertheless the technique is one of few able to guarantee to find the answer for any query–target combination. As such it still forms a common fallback for difficult searches.

In order to search large databases of structures a number of additional methods had to be devised to permit rapid data set prescreening before more exhaustive techniques such as backtracking could be employed on more manageable data set sizes. The most popular prescreens are predefined structural keys used to characterize structures in a database. As Lynch observed in his database analyses in the 1970s,[89,90] the creation of keys that describe the structural features of an average molecule are the most content-rich. Database fragment distributions are typically skewed so that a few fragments occur most of the time and a lot of fragments occur rarely. For screen set selection, the most efficient use of a limited number of bit positions is to choose descriptors that occur neither too frequently nor too infrequently. Frequent descriptors can be eliminated from the selected set, whereas infrequent descriptors can be generalized or grouped so that they become more frequent. An excellent example of this come in the form of the keys developed for the CAS ONLINE screen dictionary.[91] Twelve fragment types are included as screens, including augmented, hydrogen-augmented, and twin-augmented atoms, and atom, bond, and connectivity sequences. To reduce the number of screens to a reasonable number atom and bond types are generalized in some fragments. Generalized fragments also help when the screens are used to support similarity analyses. The fragments selected for use as screens were largely chosen on an

Table 1 Ligand-based virtual screening successes

Source	*Target*
Wang *et al.*[105]	HIV-1 protease
Hong *et al.*[106]	HIV-1 integrase
Marriott *et al.*[107]	Muscarinic M3 receptor
Singh *et al.*[108]	Very late antigen-4(VLA-4)
Clark *et al.*[109]	Melanin-concentrating hormone 1 receptor
Pirard *et al.*[110]	Voltage-dependent potassium channel kv1.5
Ekins *et al.*[111]	Human proton-coupled small peptide carrier 1
Steindl *et al.*[112]	Human rhinovirus coat protein

empirical basis through extensive manual analysis of fragments generated from a test data set. Keys are generally implemented as bit strings which are set for each molecule upon registration of its structure into the database. The percentage of the database that can typically be ignored after key screening can be as much as 99.5%,[88] rendering subsequent analysis by techniques such as atom-by-atom far more manageable. A variety of substructure search tools have been creating using variations of this approach.[92–94]

4.19.1.3.2 Three-dimensional substructure searching

In the arena of 3D ligand structure, substructure isomorph searching is dominated by the field of pharmacophore mapping.[95–100] A pharmacophore is commonly defined as a critical 3D geometric arrangement of molecular features or fragments forming a necessary but not sufficient condition for biological activity.[101] These features most typically represent functional groups (donors, acceptors, hydrophobes, etc.), but can also include larger 2D substructures, planes, vectors, exclusion volumes, and other features. 3D substructure searching was first put to use in a virtual screening context by van Drie *et al.* using ALADDIN.[102] Since then pharmacophore searching has become a main stay for lead discovery, with many applications written to support such screens.[45,61,68,70,103,104] **Table 1** provides a list of recent pharmacophore map VS successes.[105–112] A main prerequisite for substructure searching is the determination of the key substructure that will act as the search query. With topological substructures, a natural connection exists with the graphical chemical structure language used daily by chemists. Consequently key substructures can often be picked out from a visual analysis of structure–activity data. Nevertheless a number of algorithms have been developed to aid in this process of automated maximum common substructure elucidation.[113,114] Because of the geometric nature of pharmacophore maps and the intrinsic flexibility of many query ligands, visual perception of 3D substructures is a much trickier proposition. As such the use of software to aid in their elucidation becomes a critical component of most such screens. In the studies of Marriott *et al.* 3D conformer models of known muscarinic M3 antagonists were used in conjunction with the DISCO program[115,116] to determine the required pharmacophore model. DISCO takes as its input a series of low-energy conformations for each active molecule, and potential pharmacophore site points are generated automatically for each conformer. A useful feature of DISCO is its ability to define the locations of potential protein hydrogen bond donors and acceptors. This is of potential important since ligands are able to approach the same polar site point from different directions, something not easily accounted for when purely atomic superimposition is used. The program then utilizes clique detection[117] to determine matching site points distances between the conformers of a specified number of molecules within the study set, with the most rigid molecule used as reference. The resulting pharmacophore map is shown in **Figure 6**, together with a peptide-mimetic derived pharmacophore also applied successfully in virtual screening. The automated 3D substructure perception of pharmacophore model determination has received much attention, with many alternative techniques devised for their elucidation.[45,73,103,118–122]

4.19.1.4 Fingerprint and Dataprint-Based Descriptors

The need to determine specific subgraph queries for substructure searches adds significant complexity to the VS process. One way in which this has been mitigated is through the application of whole-molecule similarity comparisons which remove the need for specific feature selection. Substructural keys have formed an important source of

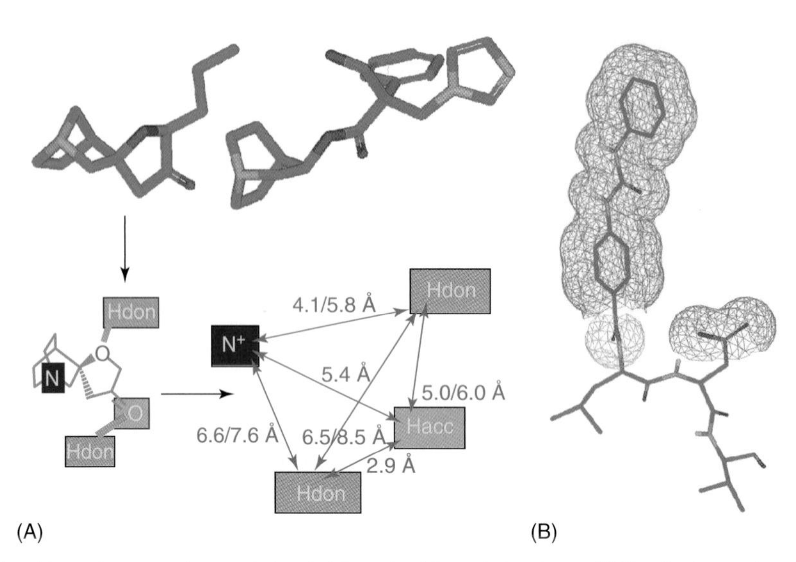

Figure 6 Example of two pharmacophores abstracted from lead ligand structures. (A) Two hydrogen bond donor vectors, a hydrogen bond acceptor, and a base, abstracted from known muscarinic M3 receptor antagonists. (Adapted from Marriott, D. P.; Dougall, I. G.; Meghani, P.; Liu, Y.-J.; Flower, D. R. *J. Med. Chem.* **1999**, *42*, 3210–3216.) (B) A superstructure fragment PUPA, acceptor and acid from the crystal structure of peptidic VLA-4 inhibitor PUPA-Leu-Asp-Val. (Adapted from Singh, J.; van Vlijmen, H.; Liao, Y.; Lee, W.-C.; Cornebise, M.; Harris, M.; Shu, I.-H.; Gill, A.; Cuervo, J. H.; Abraham, W. M.; Adams, S. P. *J. Med. Chem.* **2002**, *45*, 2988–2993.)

descriptors for such searches. While initially designated as techniques for speeding substructure searching, the potential of such keys as descriptors in their own right soon became apparent, and a plethora of descriptors that exploit the technique began to see the light of day. A number of these are described below.

The transformation of structural keys into representations that permit rapid comparison form one of the fundamental data handling tenets of virtual screening. Fingerprints form the most widely applied techniques for key storage. A fingerprint is a bit string indicating the presence or absence of the chosen descriptors. When descriptors are assigned based on stored screen numbers (bit positions, usually stored in a dictionary form to provide correspondence between the descriptor and bit string position), then the fingerprint is considered to be 'keyed.' Such a fingerprint permits a direct correspondence between an 'on' bit and a particular descriptor or group of descriptors. Keyed fingerprints allow descriptors selection and assignment considered relevant to particular activities or properties. As with substructure keys, careful descriptor selection is crucial, since if relevant descriptors are omitted from the dictionary the desired features are not represented in the fingerprint.[94,123–127] An alternative approach is to create a 'hashed' fingerprint, which generates all descriptors of a particular type, and then superimposes several descriptor bits onto identical bit string positions on the basis of a hash function. The possibility of descriptor collisions (different descriptors setting the same bit position) thus exists, and there is no direct correspondence between a particular bit position being 'on' and a particular descriptor or group of descriptors, rendering the emphasis or subsequent decomposition of desired features tricky at best. The major advantage of a hashed fingerprint is that it avoids the issue of feature selection, since all descriptors of the chosen type are represented in the bit string.[128,129]

An extrapolation of the keyed descriptor approach comes in the form of the data print. This family of descriptors uses array(s) of real numbers indicating the value of the chosen descriptors. The array can be anything from 1D (encoding a simple frequency measure for a single structure key[130]), to a multidimensional array encompassing diverse structural property data.[131] **Figure 7** provides a schematic highlighting an example of molecular decomposition into a variety of structural keys and hence onto fingerprints/data prints.

The most common application of such descriptors is in the context of straight molecular similarity comparisons between a single query molecule and compounds in the target database. This is undertaken using an equation to quantify similarity, typically known as a similarity index. By far the most widely applied of these indices is the Tanimoto coefficient:

$$\frac{C}{T + D - C} \qquad [1]$$

where C is the number of bits in common between two fingerprints, and T/D are the total number of nonzero bits in the target molecule and database molecules respectively.

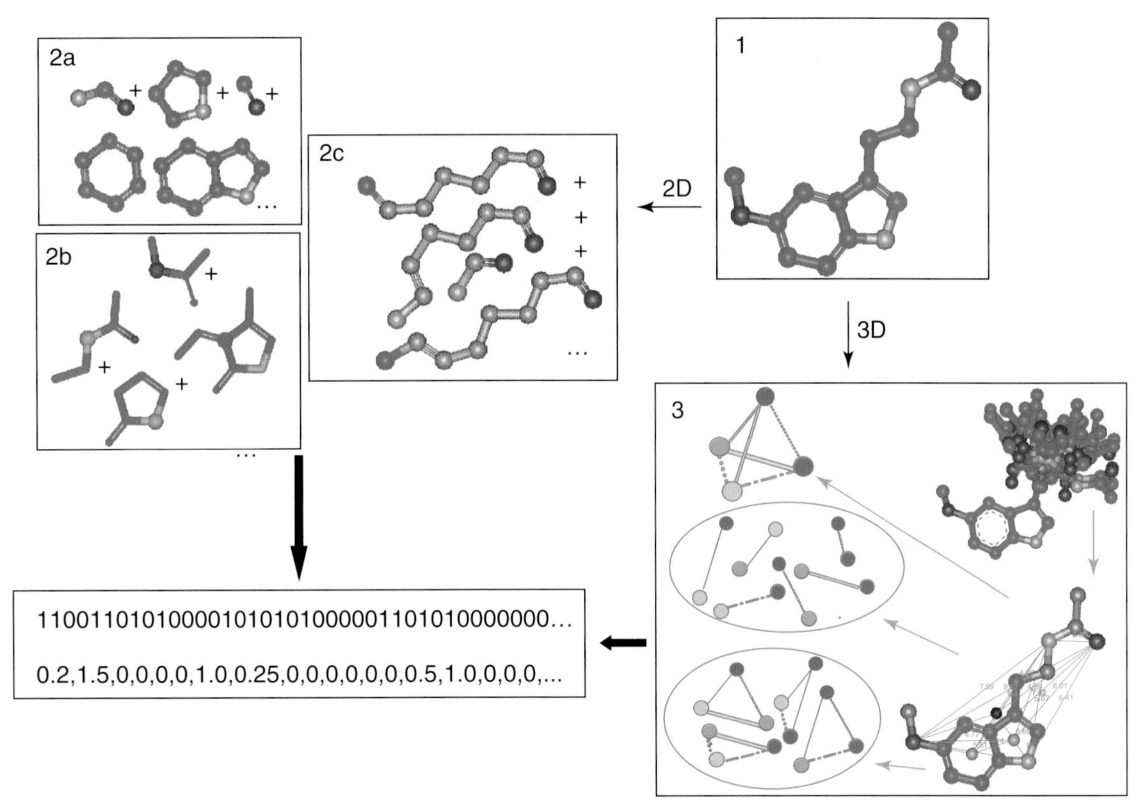

Figure 7 Schematic illustration of methods used in molecular fingerprint creation (1). (2a) Substructural dictionary fragmentation,[92,124] (2b) atom centered circular substructures,[128] (2c) topological atom pairs,[127] (3) 2/3/4 center 3D pharmacophores. [125,126,130]

In a comparison of similarity indices, this equation was found to be the most suitable for 2D database searches.[132] Variants of this index have also been applied to 3D fingerprint searches.[126] Unless otherwise mentioned descriptors mentioned below all apply some variant of this approach in their application.

4.19.1.4.1 Topological fingerprints/dataprints

A wide array of topological fingerprints have been developed for virtual screening. Perhaps the most widely applied of these are MDL's MACCS keys, which use a substructure dictionary approach of varying bit length complexity (166–960).[92] Variants of these keys have been widely applied, and though they have never been explicitly published (the 166-bit key is defined in the user manual for use in searching), information on their construction and optimization is available.[133,134] Barnard Chemical Information (BCI) fingerprints also apply a substructure dictionary approach to fingerprint generation.[124,135] With BCI's fingerprint generation software, bit mappings are defined through dictionaries of user-defined fragments. It is possible to create a dictionary of 'generalized' fragments as well as customized dictionaries, based on a statistical analysis of the fragments actually present in the structures of a particular data set. In this way the user has direct control on the optimization problem highlighted by Lynch.[89]

The most widely applied of the hashed fingerprints are those utilized within DAYLIGHT software.[128] These encode atom types, augmented atoms, and atom path sequences of two to seven atoms in length. The resulting descriptors are then typically hashed into a fingerprint comprising 1024 or 2048 bits. UNITY fingerprints use a similar approach,[94,134] though atom sequence paths also contain explicit hydrogens with length two to four, and ordinary atom sequence paths have lengths four to six allocated to different regions of a 988-bit string.

Scitegic software utilizes a circular substructure approach to create extended and functional connectivity fingerprints (ECFPs and FCFPs).[129,136] ECFP atoms are defined based on their DAYLIGHT invariant atom code (type, charge, hydrogen count, and connectivity), while FCFP atoms apply physicochemical definitions (donor, acceptor, aromatic, halogen, anion, cation). The procedure makes use of a modified version of the Morgan algorithm[137] to undertake an extended connectivity calculation. On every algorithm iteration, each atom acts as the center of a circular substructure cut from the molecule of given path length (in increments of two), with the code of the atom and its neighbors hashed together. The results are then either hashed into a 1024-bit fingerprint, or integrated into a fingerprint whose length is defined by the number of unique fragments features present in the database of interest.

Carhart *et al.*[127] created fingerprints based on atom pairs and their separation (based on the atom count of the shortest path separating them). Atoms are binned based on type, connectivity and π electron count. Using the same atom type definitions, Nilakantan *et al.*[138] devised the topological torsions descriptor, in which keys are defined by linear sequences of four connected atoms. Variants of these approaches have also been devised in which atoms environments are modified to incorporate physicochemical property descriptions (donor, acceptor, cation, anion, polar, hydrophobic, and other).[139] Further extensions have also recently been made to both the physicochemical descriptions and their application in three atom combinations to form SIMILOG keys.[140] In addition histogram data prints of topological atoms pairs descriptors have also been created for scaffold hopping screens.[141,142]

4.19.1.4.2 Geometric fingerprints/dataprints

The techniques described for topological fingerprint creation extrapolate easily to 3D descriptors. With this in mind, a variety of shape-based fingerprints/dataprints have been devised for database screening.[143–145] Bemis and Kuntz[143] designed a simplified version of a distance triplet, measuring the perimeter of each atom triplet in a molecule, and using the resultant distance to augment the appropriate bin of molecular shape histogram. Nilakantan *et al.*[144] also developed a shape-based search fingerprint based on atom triplet distances within a molecule. Atom triplet distances are defined for every combination of three atoms. The distances are sorted and scaled by length, integerized, and packed into a 2048-bit signature in order to save space. The signatures of the template molecules and database molecules are compared during the first stage of a database search, and the triplets of those molecules deemed similar enough are regenerated on the fly. Triplets matches are used as the molecular shape property to quantify similarity. Sheridan and co-workers[146] extrapolated their atom pair descriptor to geometric atom pair dataprint equivalent. The method works by using a series of precalculated (10–25) conformations for each molecule, with atoms defined using two pharmacophoric centers types. These are based either on core binding properties (donor, acceptor, acid, base, hydrophobic, polar, and other) or atomic environment (element type, number of neighbors, and π electron count). For each conformation, all combinations of atom pairs are analyzed, with the atom pair and distance combination corresponding to a particular bin in the fingerprint. Interatomic distances are partitioned into a series of 30 continuous bins from 1 to 75.3 Å. The contribution of each atom pair to the descriptor is divided across these bins according to the position of the distance relative to the bin centers (**Figure 8**).

The atom pair descriptors described above have been extended by a number of research groups to cover pharmacophore triplets and quartets.[126,147,148] With such descriptors, triangles (tetrahedra) are formed from all combinations of three (four) pharmacophoric points for all conformations of a given molecule. As before, each descriptor bin represents a particular combination of pharmacophore points (donor–acceptor–aromatic, donor–basic–aromatic, etc.) and distances. Unlike the atom pair measures described, such descriptors are generally applied in a

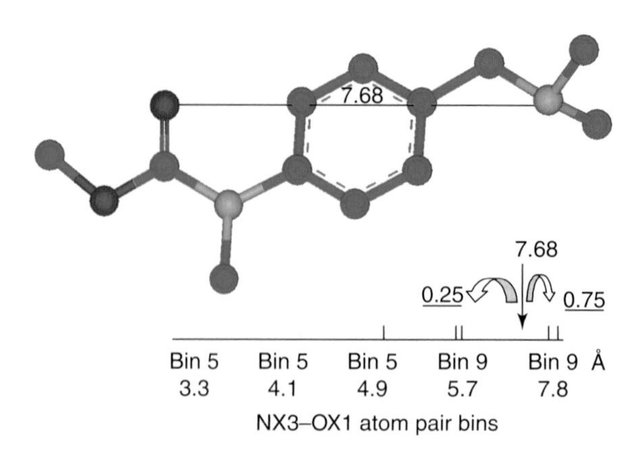

Figure 8 Bin assignment for geometric atom pair data print. Contribution of each atom pair (shown here underlined) is inversely dependent to the distance of the neighboring bins (here bins 8 and 9). The resultant descriptor is thus a histogram distribution of atom pairs with a sum total bin contribution of $n(n-1)/2$ per molecule. (Adapted from Sheridan, R. P.; Miller, M. D.; Underwood, D. J.; Kearsley, S. K. *J. Chem. Inf. Comp. Sci.* **1996**, *361*, 128–136.)

binary manner, with the resultant bit string equally weighting all pharmacophores from the whole conformational ensemble of the molecule. One exception to this is has been proposed by Good *et al.*,[149] who construct pharmacophore dataprints by normalizing pharmacophore contributions by their relative frequency of occurrence during conformation generation and include the facility to remove low information content pharmacophores (e.g., all examples of 4* lipophilic pharmacophores) during conformation generation.

McGregor and Muskal have proposed a variant of three-point pharmacophore fingerprint analysis for library design. A relatively standard pharmacophore fingerprint is generated for each molecule, consisting of seven pharmacophore types (donor, acceptor, cationic, anionic, aromatic, hydrophobic, and other) combined with six distance bins.[125] This provides a total of 10 549 accessible pharmacophores passing the triangle rule (the length of one side cannot exceed the length of the other two) and/or redundant by symmetry. Rather than screening using individual pharmacophore fingerprint queries, however, a partial least-squares (PLS)[150] quantitative structure–activity relationship (QSAR) analysis of all actives molecules is undertaken to weight pharmacophores for relevance, in essence focusing in on the key pharmacophore mappings. The system was found to perform well at locating active estrogen antagonists actives from a library of decoys.

Another variant of weighted pharmacophore descriptors has recently been proposed by Renner and Schneider with their program SQUID.[151] The program converts an alignment of active reference molecules into a 3D distribution of generalized pharmacophore interaction points. The interaction points are clustered to form potential pharmacophore points (PPPs) based on the feature density distribution of the respective features and the defined resolution of the resulting model. Each PPP represents a local maximum in the local feature density distribution. The standard deviation of the PPP that dictates it size is based on the median distance of the cluster's constituent atoms to the PPP. The resultant PPPs are then combined into a correlation vector for database screening. Other variants of this approach have also been proposed.[152]

4.19.1.4.3 Multiproperty fingerprints/data prints

The data prints we have considered so far have been one-dimensional in the properties they consider. A number of techniques have also been developed that partition property space to create a data print consisting of multiple physicochemical properties. Perhaps the best-known multiproperty descriptor is Lipinski's 'Rule of 5',[153] which applies four whole-molecule properties (molecular weight, log P, number of hydrogen bond donors, and number of hydrogen bond acceptors) as a screen for 'druglikeness.' More typically descriptors of this sort are created by selecting a small number of orthogonal descriptors (typically <10) from a large array of calculated properties. For example, the Diverse Property-Derived (DPD) code developed at Rhone-Poulenc Rorer for directed screening subset selections is based on six descriptors (number of hydrogen bond acceptors and donors, a flexibility index, the normalized sum of squared electrotopological indices, ClogP, and aromatic density) statistically chosen from an initial selection of 49 descriptors (including topological indices, physicochemical properties, and functional group counts).[154] Xue *et al.* have linked a genetic algorithm with principal component analysis to produce 'mini fingerprints' (short-length keyed fingerprints).[155] Sheridan *et al.* applied trend vector analysis to an array of properties correlated to biological activity.[156]

Perhaps the most widely method for multiproperty data print generation is the BCUT (Burden, CAS, and University of Texas) methodology developed by Pearlman and Smith.[157] These metrics form an extension to Burden's work[158] in which the heavy atom connection table of a molecule is represented as an $N \times N$ matrix, with atomic numbers along the diagonal and bonding information in the off-diagonal elements. BCUTs extend the approach to a multidimensional space by considering the highest and lowest eigenvalues of four classes of BCUT matrices, which contain atomic properties significant for ligand–receptor interactions on their diagonals: atomic charges, polarizabilities, hydrogen bond acceptor abilities, and hydrogen bond donor abilities. These properties can also be weighted by atomic surface areas. Bonding information on the off-diagonal elements is extended to include topological and interatomic distances functions. The combinations of possible diagonal, off-diagonal, and scaling factor choices lead to over 60 possible BCUT descriptors per chemical structure. The DiverseSolutions[159] software has been written to manipulate BCUTs, four to six of which can be automatically selected by a squared-based 'auto-choose' algorithm to best represent the structural diversity of a given compound collection.[160] The ensuing low-dimension chemistry space is then partitioned into cells for compound selection and library comparison. The resultant descriptors have been widely applied in virtual screening[161] and library design.[162]

4.19.1.4.4 Reduced graph representations of structure

Perhaps the simplest of ligand-based descriptors come in the form of reduced graph chemical structure representations, which are in essence 1D descriptors. Perhaps the simplest of these has been developed by Gillett *et al.*,[163,164]

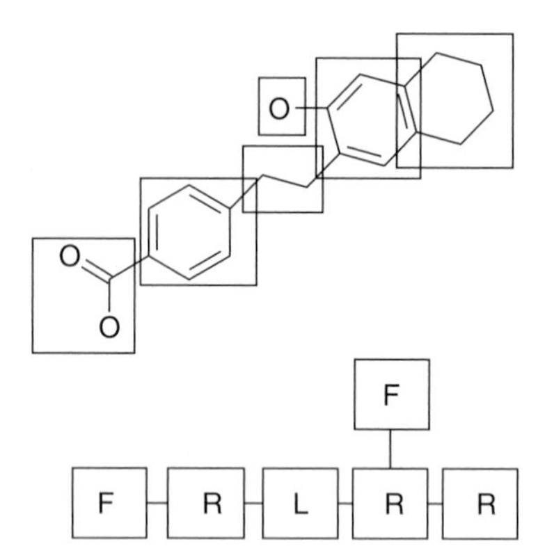

Figure 9 Example of a reduced graph. Descriptor is generated as follows: (i) Identify hydrogen bonding atoms and remove terminal non hydrogen bonding atoms. (ii) Determine smallest rings and reduce each to a ring node (R). (iii) Reduce isolated connecting carbons to linker nodes (L). (iv) Set the remaining non connected sections of the molecule to feature nodes (F). (v) Link nodes by edges to form an R/F reduced graph. (Adapted from Gillett, V. J.; Willett, P.; Bradshaw, J. *J. Chem. Inf. Comput. Sci.* **2003**, *43*, 338–345.)

who represent molecules using descriptors such as the R/F reduced graph (**Figure 9**) which consists of three simple node types: ring, linker, and feature, with structure decomposition undertaken using the DAYLIGHT toolkit.[165] Similarity comparisons were undertaken through the conversion of reduced graphs into Pseudo SMILES mapping each node type to a unique element and each edge to single bond. Fingerprints can then be generated as with normal molecules thus permitting standard molecular similarity comparison.

The most widely applied of reduced graph measures is the Feature Tree descriptor[166] as implemented in the FTREES software.[167] As in **Figure 9** molecules are described using a tree representation, rather than a more familiar bit string or vector. The features of the chemical building blocks (steric properties such as atomic volume and ring closures, plus a chemical property profile) of the molecule are stored at the nodes of the tree, while the edges between the nodes retain the topological relationship of these building blocks. The similarity between two Feature Trees is defined as the score for the best possible alignment or superposition of the two trees. FTREES provides a number of techniques for making such comparisons, for example via the use of a dynamic match search algorithm which also calculates alignments between descriptor pairs in a manner similar to algorithms designed for aligning protein sequences. Dixon and Merz[168] also created such a descriptor by mapping pairwise distances of a 2D/3D descriptor into 1D space while optimizing the preservation of distances between key features. Molecules are then compared using a sequence alignment method derived from amino acid comparison techniques.

4.19.1.5 Virtual Screening by Explicit Molecular Alignment

At the opposite end of descriptor space from the reduced graphs stand the techniques that screen databases based on explicit molecular alignment. This is perhaps the most accurate but also the most CPU intensive approach for ligand-based VS. The techniques applied are in many ways analogous to structure-based VS, but rather than placing the ligand into the binding site, the molecule is aligned onto a query compound known to bind to the target protein (for example a natural substrate or another inhibitor). Relative to the well-known lock-and-key principle, here the comparison is between two flexible keys.

In the early years of computer-aided molecular design (CAMD), continuous molecular properties were often approximated through the application of rectilinear grids or quantum chemical properties.[169–171] The CPU-hungry nature of such calculation typically limited their application to QSAR set analysis and superposition. In the early 1990s van Geerestein and Perry[172,173] extended such calculations to database searching using the gnomonic projection techniques presented by Dean and Chau[174] with their program SPERM. The technique of gnomonic projection extends the molecular properties of a structure onto the surface of a sphere. The sphere is approximated by a tessellated icosahedron, or an icosahedron and dodecahedron oriented so that the vertices of the dodecahedron lie on the vectors from the center of the sphere through the midpoints of the icosohedral faces. The resultant dramatic simplification of how to map two irregular surfaces to one another permits its application in virtual screening. At the same time Moon and Howe[175] developed a 3D database search system in which queries are built up from single or multiple overlapped active ligands. Atoms from the query are defined as required or optional: database molecules must have an atom match with required atoms in order to be retrieved as a hit, while optional atoms matches can further augment the shape match score. Since these initial efforts a number of approaches have been devised to allow virtual screening by molecular alignment. These are detailed below.

4.19.1.5.1 Gaussian function evaluations

While grid-based similarity evaluation techniques are common, their numerical foundations impart inherent drawbacks. The largest of these problems is that, to gain computation speed, the grids employed are normally coarse, with the consequence that resulting evaluations of spatial properties are somewhat rough. In particular, the similarity optimization through the modification of relative molecular position is coarse and crude. It is for example very difficult for a grid-based similarity optimization to superimpose a molecule on top of itself, since the program tends to converge prematurely at some discrete point.

The mathematical structure of similarity coefficients employing product-based numerators are such that they can be evaluated using analytical Gaussian functions. For example, molecular electrostatic potential (MEP) calculations employing the standard point charge approach, where the charges (q_i) assigned to each atom (i) create an electrostatic potential at point r for a molecule of n atoms according to the following equation:

$$\text{Pr} = \sum_{i=1}^{n} \frac{q_i}{(r - R_i)} \qquad [2]$$

where R_i is the nuclear coordinate position of atom i.

Substituting the inverse distance dependence term of eqn [2] with a Gaussian function approximation[176] and its subsequent insertion into the Carbo Similarity index[177] leads to eqn [3]:

$$R_{AB} = \frac{\sum_{i=1}^{n}\sum_{j=1}^{m} q_i q_j \int (G_1^i + G_2^i + \cdots G_k^i)(G_1^j + G_2^j + \cdots G_k^j)\,\mathrm{d}v}{\left(\sum_{i=1}^{n}\sum_{i=1}^{n} q_i q_i \left(\int (G_1^i + G_2^i + \cdots G_k^i)^2\,\mathrm{d}v\right)^{1/2}\right)\left(\sum_{j=1}^{m}\sum_{j=1}^{m} q_j q_j \left(\int (G_1^j + G_2^j + \cdots G_k^j)^2\,\mathrm{d}v\right)^{1/2}\right)} \qquad [3]$$

where $G_k^i = \gamma e^{-\alpha_k (r - R_i)}$.

The integral terms shown in eqn [3] expand into a series of two-center Gaussian overlap integrals. A two-center Gaussian overlap integral has a simple solution based on the exponent values and distances between atom centres shown in eqn [4]:[178]

$$\int e^{-\alpha_1 (r - R_i)^2} e^{-\alpha_2 (r - R_j)^2}\,\mathrm{d}v = \left(\frac{\pi}{\alpha_1 + \alpha_2}\right)^{3/2} \exp\left(\frac{\alpha_1 \alpha_2}{\alpha_1 + \alpha_2} |Ri - Rj|^2\right) \qquad [4]$$

The similarity calculation can thus be broken down into a succession of readily calculable exponent terms. As a result of this, it is possible to evaluate similarity rapidly and analytically. Willett and coworkers[179,180] exploited the speed of Gaussian-based evaluations with efficient methods for undertaking molecular alignment (genetic algorithms, MEP field graphs, and bit climbers) to allow the rapid comparison of MEPs for virtual screening. Using such an approach, it has been possible to compare whole databases of molecules against given lead structures, searching for similar MEP distributions. As with many 3D screening techniques, molecules found using these screens often show little structural resemblance to each other, unlike those located using traditional substructure search techniques. As such this form of screen is of particular interest when high scaffold novelty is key.

The use of Gaussian functions need not be limited to MEP calculations. In principle, any property which can be approximated to a set of Gaussians could be compared in this way. Molecular shape calculations have received a lot of attention to this end.[181,182] Good and Richards[181] proposed a more elementary approach to shape matching, with electron density simply approximated to the square of the STO-3G atomic orbital wave functions. Mestres *et al.* combined a simple single Gaussian volume measure with Gaussian MEPs to pharmacophore alignment studies.[183] Grant and Pickup[182] applied Gaussian approximations to the hard sphere model of molecular shape, further improving on the approach by exploiting the fact that the product of multiple Gaussians is a single Gaussian centered at a coalescence point. This allows molecular shapes based on Gaussians to be collapsed into increasingly simple elliptical representations or 'steric multipoles,' further increasing calculation speed. These techniques have been harnessed and made available in the ROCS program,[184] which was recently applied successfully in a VS context for the discovery of

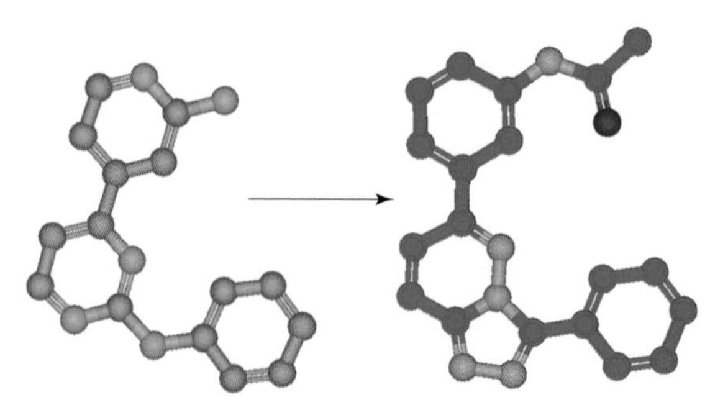

Figure 10 One of two template structures used in the ROCS[184] ZipA search (orange), together with an active molecule extracted from the subsequent hit list (green).

novel antibacterials[185] (**Figure 10**). Using the binding conformer abstracted from the crystal structure of an HTS inhibitor of bacterial protein ZipA, a lead-like (molecular weight 180–400) subset of the Wyeth corporate deck was screened for novel scaffolds. From this search 29 molecules were selected. Three molecules of interest were determined, all exhibiting significant scaffold novelty and selectivity properties distinct from the lead.

4.19.1.5.2 Feature-based alignment

A number of techniques have been developed that exploit pharmacophoric feature constraints in order to direct ligand superposition calculations. While such calculations have been used for sometime in pharmacophore development,[118,121,122] variants have also been proposed for explicit virtual screening calculations. The FLEXS[71–73] system uses an incremental construction procedure (detailed in Section 4.19.2.1.3) to build up a flexible test molecules onto a rigid reference structure. The program applies a procedure called RIGFIT to superimpose rigid fragments of the test molecule selected automatically using a heuristic algorithm onto the reference. Superpositions are optimized using a Gaussian function representation of physicochemical property space (as with the approach of Grant and Pickup above,[182] these Gaussian representations can also be coalesced for speed improvement). The flexible superpositioning part of FLEXS relies on the matching of hydrogen bonding partners in both molecules. Regions in space that describe the preferred position of protein counter atoms are required to intersect in order to form a valid matching. Both the hydrogen bonding geometries as well as directional hydrophobic interactions are modeled by point sets in space. A combinatorial optimization procedure enumerates triangles of such interaction points on the base fragment and searches for compatible triangles. Several alternative rank-ordered superpositions are given for each molecule pair according to a similarity score. Enrichment factors of between 10- and 55-fold have been reported using FLEXS.[186]

More recently, Pitman *et al.* and Krämer *et al.* reported on their program FLASH.[187,188] Adjacent pairs of fragments in two molecules being compared undergo conformational sampling and the resulting variety of binding features presented is stored in a lookup table. These property fields may range from simple, phenomenological fields to fields derived from quantum mechanical calculations. VS using a query compound then constitutes an on-the-fly reassembly of the fragmented compounds, using the wide-spread, graph-based procedure of clique detection[117] on the feature patterns in the lookup table.

Candidate Ligand Identification Program (CLIP)[189] is another program that applies alignment-based virtual screening, on this occasion through exploitation of the Bron–Kerbosch clique detection algorithm. Molecules are characterized by their constituent pharmacophore points, this representation being chosen on the basis of the very successful results that have been obtained in previous studies of the use of three-point and four-point pharmacophores for similarity searching. The similarity between two sets of pharmacophore points is calculated from a graph-theoretic fitting algorithm that identifies the maximum common substructure (MCS) in 3D. The size of the MCS (in terms of the number of matching pharmacophore points) is the principal component of a similarity measure that also incorporates the interpoint distance tolerances used during the MCS generation. The advantage of this approach is that, unlike in pharmacophore fingerprint descriptor comparisons, the geometric relationship between substructural features in maintained.

4.19.1.5.3 Shape-based alignment

A variety of methods have been developed to permit alignment based on various measures of molecular shape. Hahn[190] has developed a rapid grid-based method for shape evaluation. The technique uses rapid volume and principal axes

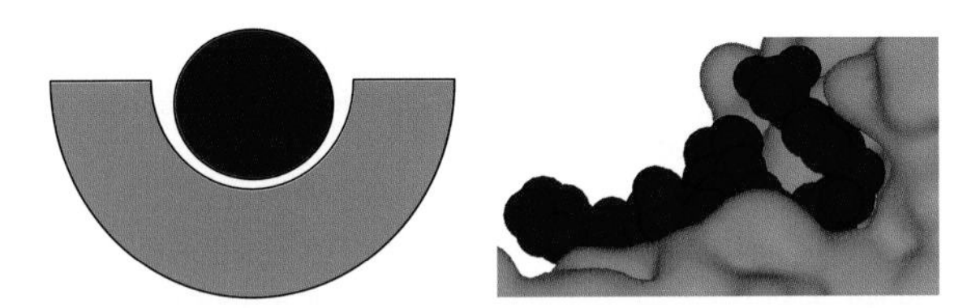

Figure 11 It is sometimes the case that significant portions of a ligand are exposed to solvent and not explicitly interacting with the binding site. This is shown schematically on the left with the inhibitor in (blue) partially matching to the protein (in green) binding site. Volumetric shape comparisons by their nature typically match the whole of the query molecule but it is only shape similarity in the orange zone that is critical to binding. An example of a solvent-exposed active site is shown on the right (1RGQ.pdb[80] – HCV NS3 protease).

indices to prescreen for molecules similar in shape to a given query. Molecules passing this screen are then aligned using their principal axes and symmetrically equivalent superpositions. Shape similarity is then evaluated via volumetric comparison. Optimization, flexible fitting, and electrostatic similarity comparison functions are also included as optional additional measures of fit quality. The use of such rapid prescreens and coarser grids allows this approach to be used for full-scale 3D database searching.

Putta *et al.*[191] rapidly detect shape matches using a rough alignment that is derived from second-order moments of conformer shape, followed by a binary comparison of steric occupancy on a grid. Most shape methods by their nature rely on molecules being similar in size in order for their shapes to be regarded as similar. It is often the case, however, that not all parts of a query molecules shape is critical to binding (**Figure 11**). To address this tricky problem the system has been extended[192] to allow the comparison of critical regions using a skeletal representation of the shape topologically unrelated to molecule bond connectivity. Initial possible alignments are determined by matching similar geometric triangles between terminal points (mapping shape edges) in the query molecule with terminal and skeleton points (more evenly spaced exhaustive distribution of shape points to permit partial matching) on the target molecule. This triangle-matching filter rapidly eliminates most geometrically impossible matches. Surviving matches are filtered further in successive stages. These stages involve direction, feature, and shape matching procedures.

4.19.1.6 Virtual and Iterative Screening Tailored to Large Data Sets

Molecular similarity calculations involving the majority of descriptors described above are applied in a classic single molecule query to template comparison. With the advent of HTS, however, scientists are often presented with a wealth of activity data available from molecules belonging to multiple structural classes. With these opportunities and issues in mind, numerous techniques have been proposed both to improve the discrimination of the molecular similarity calculations and augment HTS efficiency. These methodologies are being exploited in multiple ways to create a variety of strategies for high throughput iterative screening paradigms.[193,194–196] A number of core techniques underlying said strategies are described below.

4.19.1.6.1 Descriptor weighting paradigms

One of the most popular ways to exploit multiple active molecule data is through the merging and subsequent weighting of their respective descriptors. Techniques of this type have already been alluded to, for example in the work of McGregor and Muskal[125] (Section 4.19.1.4.2). A number of other weighting schemes have been proposed for the exploitation of multiple actives. Shemetulskis[197] defined a modal fingerprint designed to capture the common chemical features present in a training set. Any given bit in the modal fingerprint is turned on if it is present in more than a threshold percentage of the training set molecules, thus weighting the fingerprint toward features common to molecules exhibiting activity. The resulting fingerprints can then be used to molecular similarity calculations in a variety of ways, including a modified application of the Tanimoto coefficient (see eqn [1]). Singh *et al.*[198] proposed a weighted fingerprint approach that does not require the use of such a threshold value. Rather, the algorithm computes a weighted fingerprint from the set of actives, where the weight of the jth bit is the number of actives with that bit set. Database molecules are then ranked using a version of the Tanimoto coefficient adapted for continuous variables.

Another alternative for exploiting data from multiple actives and/or multiple descriptors is through the application of data fusion techniques. This name is given to a range of methods that combine inputs from different sources, with the expectation that more effective decisions can then made compared to the use of a single source.[199] When used in chemoinformatics applications (where it is often referred to as consensus scoring) the fusion is effected by combining the results of multiple screens using different descriptors.[200–203]

A common method in this technique family involves fusion of ranks.[200] Rank fusion assumes that a specific database molecule appears at rank position r_i when the ith similarity measure ($1 \leq i \geq n$) is used to match the reference structure against each of the database structures. Given n individual rankings of the database, the final score for that molecule is calculated using a fusion rule such as the maximum or the sum of the ranks, r_i. The technique is particularly useful when rankings are generated by disparate similarity measures (e.g., 2D/3D and physicochemical property-based descriptors). Such similarity measure combinations result in radically different types of similarity score that cannot readily be combined without the introduction of bias. The use of ranks, rather than the scores underlying those rank positions, provides a way of standardizing the data so that all of the similarity searches are comparable. Other weighting schemes include binary kernel discrimination (BKD) method.[204,205] This approach uses a training set comprising actives and inactives weighting data using a kernel with a smoothing parameter. The use of BKD requires calculating these scores while optimizing the smoothing parameter such that it gives the best results in terms of ranking the actives toward the top of the compound list. A variety of these procedures has been recently been tested, with data fusion providing the best balance of ease and accuracy.[206,207]

4.19.1.6.2 Classification techniques

Perhaps the most widely used technique for focused deck screening is unsupervised data clustering.[208] A nice example of this is presented by Böcker *et al.*[209,210] who apply hierarchical k-means clustering, an algorithm able to hierarchically cluster very large (>1 million data points) data sets in a high-dimensional descriptor space. Clusters can be graphically represented by a dendrogram, and activity data (e.g., IC_{50} values, class labels) can be assigned to the individual clusters. A retrospective study highlighted the ability of the method to find novel chemotype scaffolds via analysis of MDDR data sets described using Molecular Operating Environment (MOE) 2D descriptors and topological pharmacophore atom types.

Rusinko *et al.*[211] extended the statistical classification technique of recursive partitioning (RP)[212] to analyze large biological data sets. The technique works by weighting descriptors through the construction of a classification tree. At each branch of the tree the program selects descriptors whose presence or absence permits differentiation between inactive and active molecules. These descriptors and the resultant splitting of actives and inactives are described at each tree node, allowing for easy visual interpretation of the results analysis. For this implementation the recursive partitioning technique (SCAM[213]) was constructed to analyze a variety of binary fingerprint descriptors.

An example of the technique in action using atom triplet descriptors (somewhat analogous to SIMILOG keys) is shown in **Figure 12**. For this study 1650 monoamine oxidase (MAO) inhibitors were studied.[214] A variety of molecular descriptors were analyzed and VS calculations were run on the resulting classification tree to select further compounds for screening. In one such calculation, the WDI[215] (35 631 structures) was searched for MAO inhibitors using the RP tree classification rules derived from the training set, 227 structures were identified as having potential MAO activity. Of these, seven were found to be classified as such (a 3% 'hit' rate or 15-fold enrichment over random selection).

Nicolau *et al.* developed CLASSPHARMER, an alternative to clustering based on phylogenetic trees that provide an adaptive data driven organization and analysis of large screening data sets.[216–218] The procedure applies its analysis on MACCS-like keys, in a multistep process to determine the key substructure relationships to activity. Data sets (typically taken from HTS) are initially clustered using a self-organizing neural net.[219] The resulting clusters are then analyzed and a set of 'natural clusters' are selected.[220] From these clusters common substructures are identified. A series of expert rules are then applied to the resultant substructures to filter out those that do not add knowledge or are redundant. Once selection is complete, each remaining substructure is used as a query for all active molecules in the parent node. Molecules that hit form a new set placed within a new node. Each new node is appended to the tree through the construction to and from its parent. All the resulting nodes of the existing tree structure are then analyzed via breadth-first or diversity-first node selection of leaf nodes, and subsequent iteration through the above procedure. Process termination occurs once predefined termination criteria are met, for example the remaining nodes contain too few compounds or tree depth has reached a predefined level. Once the tree is complete automated statistical methods and expert rules driven hybrid analysis is undertaken to extract the classes of interest.

4.19.1.7 Insights into Descriptor Selection

With the myriad of descriptors available for virtual screening, users can be excused for any confusion they might have with regard to which is most suitable for their particular screening requirement. Unfortunately there is no single VS

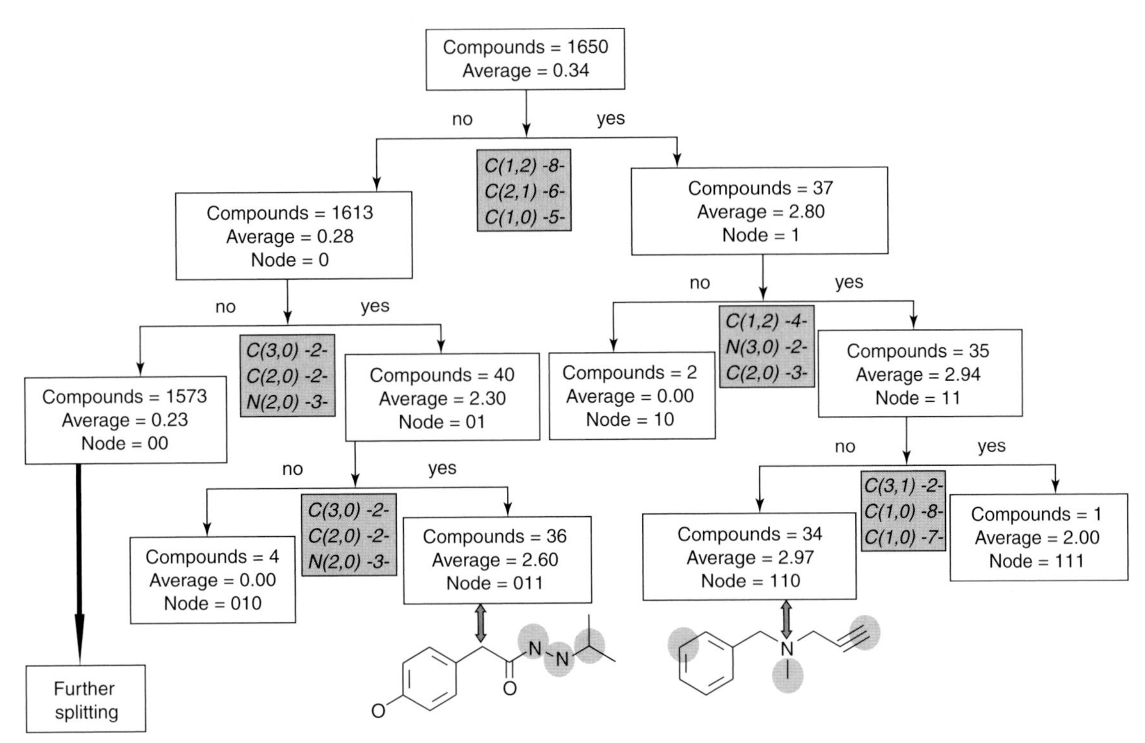

Figure 12 Results of SCAM monoamine oxidase (MAO) inhibitor calculation using atom triple descriptors. Node split atom triple information and selected associated statistics are included. Compounds, number of compounds in split; Average, average activity; Node, node number. Two active molecules associated with specific nodes are included in the display, together with the selected atom triples associated with the classification nodes. (Adapted from Rusinko, A., III; Farmen, M. W.; Lambert, C. G.; Brown, P. L.; Young, S. S. *J. Chem. Inf. Comput. Sci.* **1999**, *39*, 1017–1026.)

solution that continually delivers superior enrichment values.[221,222] While articles describing novel descriptors will often provide associated enrichment data to illustrate their utility, interpretation of said data can prove difficult. Good *et al.* highlight many of the pitfalls inherent in enrichment graph interpretation in an extensive analysis of VS enrichment techniques and descriptors.[223] A pertinent illustration of potential problems comes from an analysis of a CDK2 data set containing a high number of closely related analogs (**Figure 13**). When analyzed using a typical overall enrichment graph, DAYLIGHT fingerprints produce the best overall enrichment. If, however, chemotype enrichment is investigated, DAYLIGHTenrichment falls to the bottom of the list. This is a classic illustration of the importance of technique selection, since a descriptor well suited to substructure or analog searching is often poor when looked at in the context of scaffold hopping.

While there are no hard and fast rules to descriptor selections, a number of guidelines exist that can help. When searching for close analogues of a set of actives, descriptors with high substructure information content will tend to perform best (e.g., DAYLIGHT, UNITY, and ECFP fingerprints). When attempting to scaffold hop, however, descriptors that divorce binding properties from underlying molecular topology are typically more useful (e.g., pharmacophores, atom pairs, and FTREES). When templates are relatively rigid or the binding conformation is known, the highly focused techniques of explicit molecular alignment becomes particularly attractive (Section 4.19.1.5), since the ambiguity in potential binding conformations is attenuated. Where template structures are highly branched, 3D descriptors become preferable to 2D, since the typical through bond measure of distance inherent to 2D measures does not account for the compact nature of the molecule.

Perhaps the most relevant strategy of all is to use all the information at one's disposal. When an array of active structures is available, techniques capable of analyzing all the actives to create a filtered template descriptor can appreciably improve the signal to noise ratio. The resultant improvement in enrichment can prove equally significant. Such techniques include pharmacophore elucidation (Section 4.19.1.3.2), descriptor weighting (Section 4.19.1.6.1), and classification (Section 4.19.1.6.2).

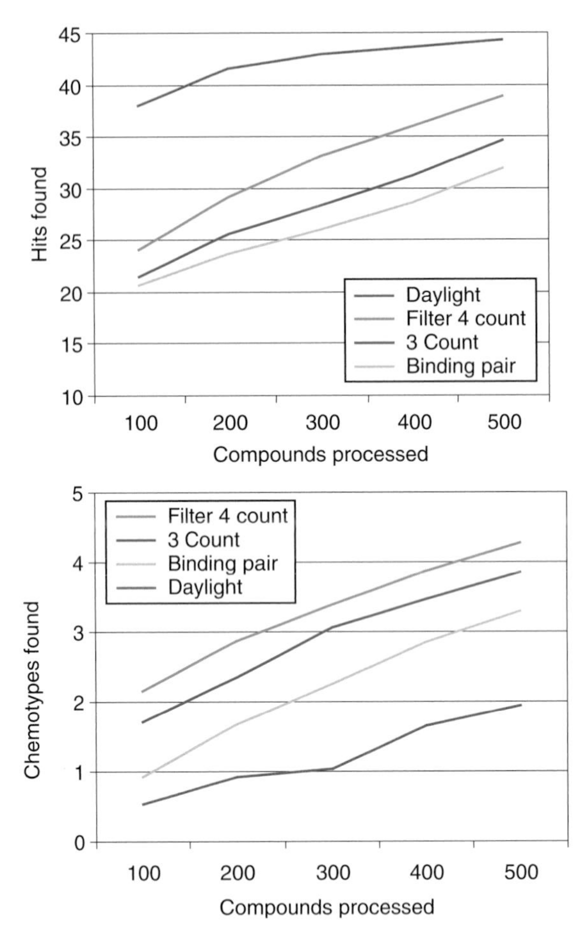

Figure 13 Enrichment graphs for CDK2 virtual screens using a number of different fingerprint descriptors. The data highlight what can happen with compound collections that contain a large number of closely related analogues. When only overall hit enrichment is considered (top graph) Daylight fingerprints perform best. When Adjustments are made to only include novel chemotypes in the enrichment score (bottom graph), Daylight fingerprint move to the bottom of the rankings. (Adapted from Good, A. C.; Hermsmeier, M. A.; Hindle, S. A. *J. Comput.-Aided Mol. Des.* **2004**, *18*, 529–536.)

4.19.2 Structure-Based Virtual Screening

So far we have focused on approaches that exploit ligand structural information to create queries for database searching. Perhaps the most dynamic branch of virtual screening, however, is structure-based virtual screening (SVS), which focuses on compound database screening within the confines of a target active site. An outstanding feature of such approaches is that searches are no longer confined to binding modes inferred from known ligands. Instead, whole receptor sites can be explored, allowing a large array of potential receptor–ligand interactions to be sampled. New chemotypes can thus be discovered that bind with the receptor in hitherto unforeseen ways. Such an advantage also poses a significant challenge for SVS, one of time. This is a much more significant problem for SVS relative to most ligand-based virtual screens, since the large number of potential binding modes creates a huge search space within which to work. Searches must be rapid (no more than 5–10 min per compound for even for the most computationally well endowed, significantly less for those less well equipped), to keep pace with the ever-increasing speed of biological screening methods. To meet this demand, computational docking and scoring techniques have evolved to be used in high-throughput virtual screening protocols for in silico lead identification. As in most disciplines, accuracy is compromised for speed. Sophisticated energy functions, for example those that apply first principle methods for protein–ligand affinity calculations[224] are too time-consuming for high-throughput predictions. Therefore, simplified scoring functions have been used to describe protein–ligand interactions. These are combined with a series of rapid binding mode docking techniques to create a variety of SVS tools. The opportunities and challenges of SVS are such that the topic been extensively reviewed.[225–240] Many of the docking protocols and the associated technology that pervades the discipline are described below.

4.19.2.1 Docking Methodology

Protein–ligand docking is in essence a geometric search problem. While the conformation of a given protein target active site is reasonably well known (ignoring the significant issue of conformational change that can occur on ligand binding), the protein-bound ligand conformation is usually not. Consequently the majority of docking techniques focus on the issue of ligand flexibility and binding mode exploration while keeping the protein rigid. Protein–ligand docking experiments require the availability a 3D structure of the target protein at atomic resolution. The most reliable sources of such structures are provided by the Protein Data Bank (PDB)[241] or from in-house efforts. Homology models[242] can also be used, though care must be taken that the resolution of the docking technique applied is in keeping with the perceived accuracy of the model, since small changes in structure can significantly alter the outcome of a computational docking experiment.[243,244] SVS searches typically break down into two primary phases, ligand docking followed by scoring of the resultant interactions. The main techniques applied in these protocols are detailed below.

4.19.2.1.1 Clique detection

The docking of small fragments or ensembles of explicitly generated ligand conformations lend themselves well to rigid docking techniques such as clique search techniques.[117,245] These techniques can be used to search for distance-compatible matches of protein and ligand features,[246] for example complementary hydrogen bonding interactions, distances, or volume segments. The most widely applied of clique detection-based docking techniques is the DOCK program.[247–249] DOCK applies distance-compatible match searches incorporating clique detection algorithms[250] for rigid-body docking (**Figure 14**). Site points that map the molecular features of the binding site are matched to the ligand atom centers. Initial orientations of the ligand in the receptor site are generated using distance-compatible matches of user definable size (usually 3–4). The final position of the ligand is then determined through optimization against the selected scoring function. In addition chemical properties can be assigned to the site points[251,252] in conjunction with critical clusters of points, allowing clique matching to be converted into multiple simultaneous pharmacophore searches within the constraints of the binding site which simultaneously increasing search speed. The FLEXX program[25,26] applies pose clustering to the docking problem using an algorithm developed to detect objects in 2D scenes with unknown camera location. The algorithm matches each feature triplet of the first object to feature triplets of the second object. Any resulting triangle matches can then be used to locate the first object with respect to the second through triangle superposition. Locations are stored and clustered, and large clusters are used to map locations with a high number of matching features. Matches are constrained by compatibility of interaction (e.g., a hydrogen bond donor can interact only with an acceptor) and the length of triangle edges.

Another pattern recognition method useful in docking was originally developed for recognizing partially occluded objects in camera scenes.[253,254] A hashing function is used to map the addresses of the data entry into a smaller address space. In the case of geometric hashing, distance features are used to create the hashing key. As a result, objects with certain geometric features (for example the sphere representation of DOCK) can be rapidly accessed through a geometric hashing table.

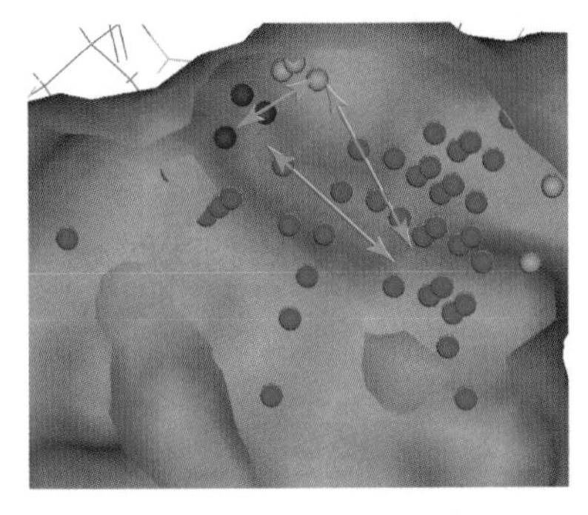
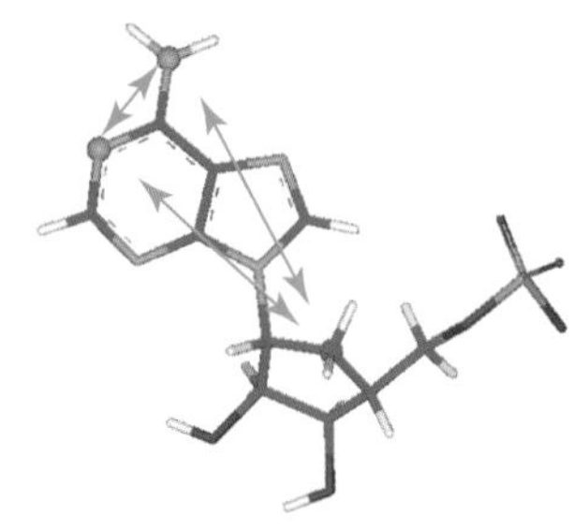

Figure 14 Schematic of clique detection as applied to the receptor site point/ligand atom paradigm used in DOCK. Distances and optionally atom chemistries need to correspond to produce a match. In addition atom clusters can be assigned to critical regions which must be matched during clique detection. It should be noted that, while this figure relates specifically to receptor–ligand site points, it nevertheless provides a reasonable schematic representation of clique detection in general.

4.19.2.1.2 Stochastic approaches

A large number of docking programs apply stochastic techniques to the problem of ligand sampling and docking. In contrast to the combinatorial approaches for docking discussed above, simulation methods start with a given configuration and move to configurations with more favorable energies. Monte Carlo simulation techniques have found wide use in docking programs in this regard. Monte Carlo methods refer, in a very general sense, to any simulation of an arbitrary system which uses a computer algorithm explicitly dependent on a series of (pseudo) random numbers. The name, which derives from the famous Monaco casino, emphasizes the importance of chance in the method. In a system with D degrees of freedom, for example, the thermal average of a quantity A associated with each microstate of the system in equilibrium at absolute temperature T is given by

$$\langle A \rangle = \frac{1}{Z} \int A(x) e^{-\frac{E(x)}{T}} \, dx \qquad [5]$$

where x is a point in D-dimensional space representing the state of the system, $E(x)$ is the energy of state x, and $Z = \int e^{-E(x)/T} \, dx$ the partition function (units set so that Boltzmann K_B is set to 1).

Monte Carlo simulations choose conformations in such a way that the selection is biased towards conformations that are significantly populated at equilibrium. This is typically achieved by weighting the probability of occurrence of a given conformation to its Boltzmann factor via application of the Metropolis criterion.[255] If the difference between the energy of the resulting conformation and the energy of the current conformation, ΔE, is negative (i.e., the energy of the resulting conformation is smaller than the energy of the current conformation), then the resulting conformation is accepted and it becomes the new conformation in the chain. If ΔE is positive, however, a (pseudo) random number R is generated with a value between 0 and 1. The resulting conformation is only accepted if $e^{-\Delta E/T} > R$. Whenever the conformation resulting from the attempted move is refused, then the current conformation becomes the new conformation of the chain.

Abagyan *et al.* have combined efficient internal coordinate representations of protein and ligand with a Monte Carlo optimization protocol in their program ICM.[256,257] The software uses a biased Monte Carlo algorithm to minimize an energy function in torsional space. A pseudo random set of ligand torsion angles is selected and then the local energy minimum about those angles is determined. The selection of torsion angles to manipulate during Monte Carlo steps is not completely random but rather biased toward maximizing search speed. Once the angles to be manipulated have been chosen, a new conformation is then adopted using the Metropolis criteria. The Monte Carlo steps are then repeated, in conjunction with methods to foster sampling of unexplored conformational space during this Monte Carlo process, a stack of low-energy conformations is created, and if the same conformation is visited a certain number of times, the simulation temperature T is doubled. Thus, the algorithm allows for the escape from local minima. The QXP program uses a similar approach that can also be applied to the superimposition of small molecules.[258,259] An alternative search technique called 'tabu search' is used in the program PRO-LEADS.[260] Starting from a random structure, new structures are created by random moves. A tabu list is maintained containing the best and most recently found binding configurations. New configurations generated resembling those of the tabu list are rejected unless they exhibit superior scores to those configurations in the tabu region. As a result the sampling performance is improved since previously sampled configurations are avoided. AUTODOCK[261,262] uses a variant of Monte Carlo approaches called simulated annealing[263] as a simulation technique to search potential ligand docking poses. Simulated annealing has also been used by Yue to optimize the distance constraints for rigid body docking.[264] The technique uses a decaying temperature function to gradually reduce the probability of higher energy docking mode acceptance in an effort to focus searching in more appropriate regions of docking space.

Genetic algorithms have been used as an alternative stochastic docking approach in programs such as AUTODOCK and GOLD.[265,266] In GOLD, two bit strings are used to represent a docking configuration. The first string contains the ligand conformations defining the torsion angle of each rotatable bond. The second string contains the hydrogen bond mapping between the relevant protein and ligand atoms. The fitness function takes into account the evaluation of hydrogen bonds, internal energy of the ligand and the protein–ligand vdW energy, together with any scoring functions associated with the docking run.

4.19.2.1.3 Ligand flexibility

Ligand flexibility is tackled as part of the docking problem using a variety of techniques. A common approach is to dock conformers generated from whole molecule systematic search. The Flexibases used in the FLOG docking program[267] are a prime example of this. Similar approaches are also applied with DOCK. The conformational ensembles stored for DOCK have been further extended by other groups to improve docking efficiency. DOCK can also apply whole molecule systematic searches at runtime through the application of allowed torsion range rules. Due to the combinatorial explosion the system defers to random sampling for more flexible molecules (typically five to eight rotatable bonds for default sampling levels). The technique still has significant utility, however, as whole molecule conformers permit easy pharmacophore constraint mapping, which can both enrich and speed up the docking portion of virtual screening protocols significantly.[268]

Lorber and Shoichet[269] introduced an ensemble storage approach to conformation database construction, with about 300 conformations per molecule with a predefined substructure superimposed. The DOCK algorithm is then applied to the rigid part and all conformations are tested and scored. This docking protocol results in a significant speed-up compared to single-molecule docking. Thomas *et al.*[270] introduced a variant of this approach in which conformational ensembles are superimposed based on commonalities in predetermined pharmacophore fingerprints. These are then docked into the active site searching chemical match constraints[251] that map to the same pharmacophores.

Fragment construction provides a popular alternative for ligand conformer generation designed to improve the amount of conformational sampling possible during docking; 'incremental construction' is the most commonly applied of the fragment construction techniques. Incremental construction algorithms are generally made up of three steps: (1) anchor fragment selection, for example through dissection at a ligand's flexible bonds, (2) placement of the anchors into the binding site, based on factors like anchor placement score or hydrogen bonding complementarity, (3) the incremental construction phase, where the ligand is rebuilt around the docked anchor with full torsion sampling of the reattached fragments. The GLIDE program[271–273] uses a variant of incremental construction, with heuristic screening parameters optimized via extensive testing on cocrystallized PDB complexes. Each ligand divided into a 'core' region and some number of attached 'rotamer groups' (**Figure 15**). The core is defined by what remains when each terminus of the ligand is severed at the 'last' rotatable bond (the directly attached atom of each rotamer group is also considered to be part of the core). Carbon and nitrogen end groups terminated with hydrogen ($-CH_3$, $-NH_2$, $-NH_{3+}$) are ignored as they contain little conformational variation content. Each core region is represented by a set of core conformations defined by the number of rotatable bonds, conformationally flexible five- and six-member rings, and asymmetric pyramidal trigonal nitrogen centers. Every rotamer state for each rotamer group is enumerated, and the core plus all possible rotamer–group conformations is docked as a single object in Glide. Because cores typically contain many rotamer–group combinations, the effective number of conformations docked can easily number in the thousands or tens of thousands for molecules with several rotatable bonds. The procedure also applies heuristic screens to rapidly remove conformations deemed unsuitable for receptor binding, for example conformations containing long-range internal hydrogen bonds. High-energy conformers are eliminated through torsion energy cutoff using a truncated version of the OPLS-AA (optimized potentials for liquid simulations all atom) molecular mechanics potential function. The result of such techniques is that the number of conformations that can be analyzed is in the thousands per molecule. The method was tested on 796 cocrystallized ligands taken from the PDB. 93% of ligands with 1–10 rotatable bonds, 77% of ligands having 11–15 rotatable bonds, and 48% of ligands having 16–20 rotatable bonds were found to have RMSDs <1.5 Å. This is an impressive result and might be thought of as a generally suitable method for all forms of conformational search. There are implications for certain virtual techniques and descriptors, however, as orders of magnitude more conformational noise has to be created to generate the correct conformation. This is tenable in the context of SVS calculations since active site search space is highly constrained. In contrast, the resultant effect on the signal-to-noise ratios for unconstrained and noisy descriptor spaces such as pharmacophore fingerprints is not inconsequential.

DOCK[247–249] is also able deconstruct molecules down to a core fragment which is positioned into the active site of a screening target based on steric fit. Flexible side chains are then grown a bond at a time constrained to predefined

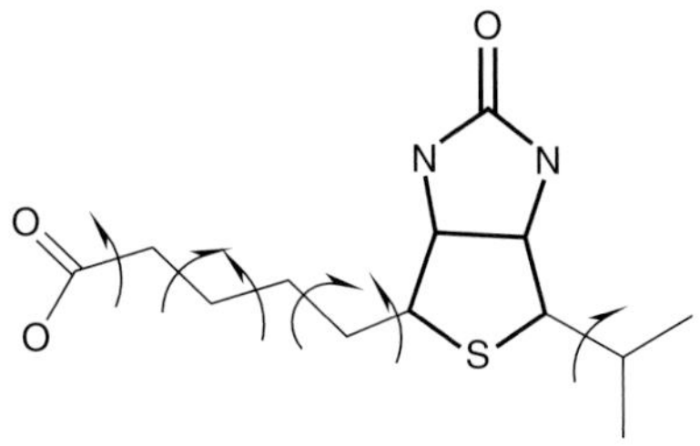

Figure 15 An example of incremental construction. The core region of the molecule is highlighted in bold, along with the attached 'rotamer groups.'

preferred torsion ranges, with unfavorable conformations pruned from the search space early on to reduce search space. FLEXX[25,26] applies a similar approach, except that initial core positioning is based on complementary hydrophobic and hydrogen bonding interactions, and torsion ranges are based on CSD[44] distributions rather than force field energies.

The EHITS program also uses a fragment-based approach to ligand flexibility.[274,275] Its approach is somewhat different in that each fragment is docked exhaustively throughout the site, with graph-matching algorithms subsequently being used to reconstruct the molecule.

4.19.2.1.4 Protein flexibility

Numerous examples have highlighted the ability of a protein to adjust its conformation upon ligand binding,[276,277] and consequently a number of lead docking techniques have been devised to explicitly handle protein flexibility.[278,279] Given the CPU limits imposed by VS calculations, however, such exhaustive approaches are typically not applicable to SVS, which generally treat the protein as rigid. While the incorporation of protein flexibility without the concomitant introduction of noise into the SVS continues to prove a difficult problem,[233] a number of techniques have been developed that attempt to address this issue in a computationally efficient manner. Knegtel *et al.*[280] handle protein flexibility by docking to ensembles of protein structures. Broughton has developed an elegant technique for incorporating protein flexibility implicitly into SVS calculations.[281] The method analyzes short target site molecular dynamics runs, using the results to weight SVS scoring towards regions showing less flexibility. In initial tests SVS runs undertaken using this methodology exhibited improved performance relative to using the rigid crystal structure alone.

A number of attempts have been made to include elements of explicit protein flexibility calculations into SVS searching.[265–284] FlexE is an extension module of FlexX designed to take protein flexibility into account.[283,285] The flexibility of the protein is represented by an ensemble of protein structures which is combined to create a so-called united protein description. The program then recombines elements from different ensemble structures to create new valid protein structures using graph search algorithms, thus extending the conformational space of the ensemble structures.

Cavasotto and Abagyan have created the program IFREDA to account for protein flexibility in virtual screening.[282] By docking flexible ligands to a flexible receptor, IFREDA generates a discrete set of receptor conformations, which are then used to perform subsequent flexible ligand–rigid receptor docking and scoring runs. This is followed by a merging and shrinking step, where the results of the multiple virtual screenings are condensed to improve the enrichment factor. The approach allows both side chain rearrangements and essential backbone movements to be taken into consideration.

More recently in an extensive analysis of crystal structures of CDK2 and HSP90, Barril and Morley[286] determined that enrichments could be improved using a consensus scoring approach to flexible docking. This is accomplished by weighting ligands by their ability to score well to a wide range of receptor poses.

4.19.2.2 Scoring Techniques

Once docked, the quality of a ligand's resulting interactions must be determined using some form of scoring function. Scoring functions attempt to provide measures that relate to the Gibb's binding free energy G°. This quantity is related to the experimentally determined association constant K_A (or its reciprocal dissociation or inhibition constants, K_D or K_i, respectively). G° is composed of an enthalpic (H°) and an entropic (TS°) portion. T refers to the absolute temperature:[287]

$$K_A = K_D^{-1} = K_i^{-1} = \Delta G^\circ = -RT\ln K_A = \Delta H^\circ - T\Delta S^\circ \qquad [6]$$

A variety of scoring functions have been created to evaluate enthalpy and/or entropy either directly or indirectly, and hence estimate the binding affinities between proteins and their putative ligands. A range of different classes of function have been devised, many of which have been extensively reviewed elsewhere.[288–295] A number of these functions are discussed below.

4.19.2.2.1 Force field scoring

Force field scoring (FFS) has historically been one of most widely applied scoring functions. FFS variants have been applied in multiple virtual screening packages including DOCK, AUTODOCK, and GOLD. FFS most often relies on the nonbonded electrostatic and vdW interaction energy terms of standard force fields:

$$E_{\text{ff}} = E_{\text{nonbonded}} = \sum_{i}^{\text{lig}} \sum_{j}^{\text{prot}} \left[\frac{A_{ij}}{R_{ij}^{12}} - \frac{B_{ij}}{R_{ij}^{6}} + 332 \frac{q_i q_j}{\varepsilon R_{ij}} \right] \qquad [7]$$

A_{ij} and B_{ij} are constants, R_{ij} is the distance between atoms, q_i and q_j are the atoms' charges, and ε is the effective dielectric constant.

For example, DOCK makes use of the intermolecular terms of the AMBER energy function[296] with the exception of an explicit hydrogen bonding term.[297] Soft vdW potentials are often used in simulations of whole-ligand docking approaches to decrease the sensitivity of the function in order to compensate for shortcoming in binding orientation brought about due to sampling limitations. The CHARMM energy function was recently explored by Vieth *et al.*[298] in the context of lead docking, who concluded that a soft core vdW potential is needed for the kinetic accessibility of the binding site. FLOG[267] uses a 6–9 (as opposed to the 6–12 powers used in eqn [6]) Lennard-Jones function for vdW interactions and local dielectric constants in a Coulomb representation of the electrostatic interactions and additional terms for hydrogen bond potentials and hydrophobic potential. GOLD[265,266] combines a soft intermolecular Lennard-Jones 4–8 potential with hydrogen bonding terms precalculated using model fragments.

A major shortcoming of the application of simple in vacuo force field functions is that they reflect purely enthalpic contributions to the free enthalpy of binding.[299] To overcome this, efforts have been made to incorporate solvation terms. The importance of solvation is such that it alone can on occasion produce a significant correlation with binding affinities.[300] The effect of water on electrostatic interactions can be determined by solving the linearized Poisson–Boltzmann equation.[301] The Poisson–Boltzmann approach is somewhat time consuming, however, so to permit the use of solvation terms in virtual screening Zou *et al.*[302] incorporated the generalized Born model (GB/SA)[303] into the DOCK program. According to the GB equation, the electrostatic interaction energy between two charges depends on both the intercharge distance and the effective solvation radii of the charges as a measure of their solvent exposure. This approach can also account for the hydrophobic effect in terms of the change in solvent-accessible surface area (SA) during binding. The force field scoring functions are typically modified through simple addition of the solvation term

$$E_{\text{ff}} = E_{\text{nonbonded}} + G_{\text{sol}} \qquad [8]$$

The solvation free energy (G_{sol}) of a molecule consists of three terms: a solvent–solvent cavity term (G_{cav}), a solute–solvent vdW term (G_{vdW}), and an electrostatic polarization term (G_{pol}):

$$G_{\text{sol}} = G_{\text{cav}} + G_{\text{vdW}} + G_{\text{pol}} \quad \text{where} \quad G_{\text{cav}} + G_{\text{vdW}} = \sum \sigma_i SA_i \qquad [9]$$

The nonelectrostatic terms, G_{cav} and G_{vdW}, are approximated by a linear dependence on the solvent–accessible surface area (SA). SA_i is the solvent-accessible surface area of atom i; σ_i is an empirical atomic solvation parameter. G_{pol} is the change in electrostatic energy when a molecule is transferred from vacuum to solvent.

While such calculations can be successfully applied to SVS calculations,[304] they still tend to be somewhat time consuming when applied over a large database. With this in mind, attempts have been made to incorporate terms such as solvation into scoring functions based on rapidly calculable empirical approximations.[305] A number of such functions are described below.

4.19.2.2.2 Regression-based scoring

Another popular method for quantifying ligand poses comes in the form of regression-based scoring functions. These are derived from fitting coefficients of 3D protein–ligand structure-derived terms of a binding energy equation (e.g., hydrogen bonding energy and lipophilic contact energy) to reproduce the experimental binding affinities of a training set of known protein–ligand complexes. To illustrate the design of regression-based scoring functions, we describe here the CHEMSCORE scoring function as developed by Eldridge *et al.*[306] and Murray *et al.*[307] and implemented in the program PRO-LEADS.[260] CHEMSCORE is one of the most widely applied empirical scoring functions available today and is written as:

$$\begin{aligned} \Delta G_{\text{binding}} = {} & \Delta G_o + \Delta G_{\text{hbond}} \sum_{il} g_1(\Delta r) g_2(\Delta\alpha) + \Delta G_{\text{metal}} \sum_{aM} f(r_{aM}) \\ & + \Delta G_{\text{lipophilic}} \sum_{aM} f(r_{iL}) + \Delta G_{\text{rot}} \Delta H_{\text{rot}} \end{aligned} \qquad [10]$$

Δr is the deviation from the ideal hydrogen bond length of 1.85 Å (HO/N) and α is the deviation from the ideal angle of 180°. $g_1(\Delta r) = 1$ if $\Delta r \leq 0.25$ Å, $1 - (\Delta r - 0.25)$ if 0.25 Å $< \Delta r \leq 0.65$ Å and 0 if $\Delta r > 0.65$ Å. $g_2\Delta\alpha = 1$ if $\Delta\alpha \leq 30°$, $1 - (\Delta\alpha - 30)/50$ if $30° < \Delta\alpha \leq 80°$ and 0 if $\Delta\alpha > 0.65°$. Hydrogen bonds are calculate between all ligand atom i and protein atoms I. No attempt is made to differentiate ionic from nonionic hydrogen bonds. The lipophilic and metal terms are calculated as simple contact terms, while frozen rotatable bonds are identified as those for which the atoms on both sides of the bond are in contact with the receptor according to:

$$H_{rot} = (1 - 1/N_{rot}) \sum_r \frac{(P_{nl}(r) + P'_{nl}(r))}{2} \qquad [11]$$

where N_{rot} is the number of frozen rotatable bonds and $P_{nl}(r)$ and $P'_{nl}(r)$ are the percentages of nonlipophilic heavy atoms on either side of the rotatable bond, respectively. CHEMSCORE has achieved a statistically significant correlation between prediction and experiment of protein–ligand binding affinities with a standard error of 2.1 kcal mol^{-1} for the set of 82 complexes on which it was trained. This function has subsequently found use in a number of other virtual screening tools, including GOLD, CSCORE,[308] and standard precision GLIDE.[272]

A number of other studies have been undertaken into the creation of empirical scoring functions. Wang *et al.*[309] derived a function using 170 complexes in their training set. The primary difference compared to approach of Böhm is the classification of hydrogen bonds as strong, moderate, and weak. including the occurrence of interstitial water molecules. Head *et al.*[310] created VALIDATE using AMBER electrostatic and steric interaction energies, octanol/water partition coefficients, polar and nonpolar contact surfaces, and an intramolecular flexibility term. The function was derived from 55 protein–ligand complexes using partial least-squares and neural net analyses, resulting in regression equations that are hard to interpret in physical terms. Jain developed a continuously differentiable function[311] using hydrophobic and polar complementarity terms modeled on Gaussian and sigmoidal functions. Only ligand-dependent contributions are used for handling entropic considerations. The analysis is based on a 34 protein–ligand complex training set.

Krammer *et al.*[312] employed a genetic algorithm to explore regression equations compromising multiple scoring term combinations across a training set of 118 protein–ligand complexes. The resulting LIGSCORE2 function includes a 6–9 vdW term, an attractive polar contact surface metric, and most interestingly a desolvation penalty incorporating buried polar surface area terms for both protein and ligand.

More recent efforts in enhancing scoring function performance have been undertaken by Murphy *et al.* in their development of Extra Precision GLIDE.[272,313] This scoring function incorporates a number of systematic enhancements to the CHEMSCORE function based on extensive protein–ligand complex analysis. A hydrophobic enclosure term extends the CHEMSCORE lipophilic atom–atom pair function by weighting for details of the local geometry for proximal lipophilic protein atoms. This term quantifies the observation that intermolecular water hydrogen bond perturbation in a hydrophobic cavity is maximized by hydrophobic contact encompassing multiple protein atoms. In essence the function encodes the fact that connected lipophilic ligand groups of at least three atoms, enclosed on two sides (at a 180° angle) by lipophilic protein atoms, contribute to binding free energy beyond that encoded in the atom–atom pair term. This is further extended with the special neutral–neutral hydrogen bond term. The central concept of such hydrogen bonds is to locate positions in the protein cavity where a water molecule forming a hydrogen bond to a polar group would have particular difficulty making its complement of additional hydrogen bonds. Forming a hydrogen bond with a protein atom imposes nontrivial geometrical constraints on the water molecule. The hydrophobic enclosure term analyses suggest that the environment will be significantly more challenging if the water molecule has hydrophobic protein atoms on two faces. Replacement of such water molecules by the ligand will be particularly favorable if the donor or acceptor atom of the ligand achieves its full complement of hydrogen bonds in such an environment. **Figure 16** highlights these terms in the context of specific ligand–protein interaction examples.

Other terms used to augment this scoring function include a charge–charge hydrogen bond term and a desolvation penalty term that incorporates a crude explicit water model. Analysis of the new function across 120 complexes with known binding affinities produced an average error of 1.8 kcal mol^{-1}, approximately half that produced by the CHEMSCORE implementation variant of standard precision GLIDE on the same complexes.

4.19.2.3 Knowledge-Based Scoring

The principal idea behind knowledge-based approaches is that analysis of a sufficiently large data sample of a given database can serve to derive rules and general principles for the molecules stored therein. The functions are based on the derivation of statistical preferences in the form of potentials for protein–ligand atom pair interactions. As such the

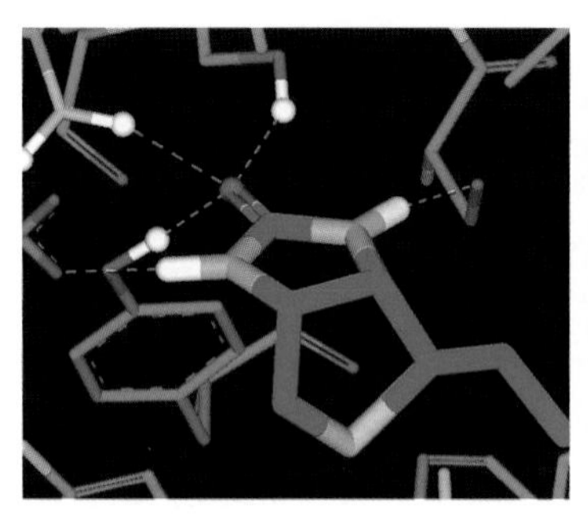
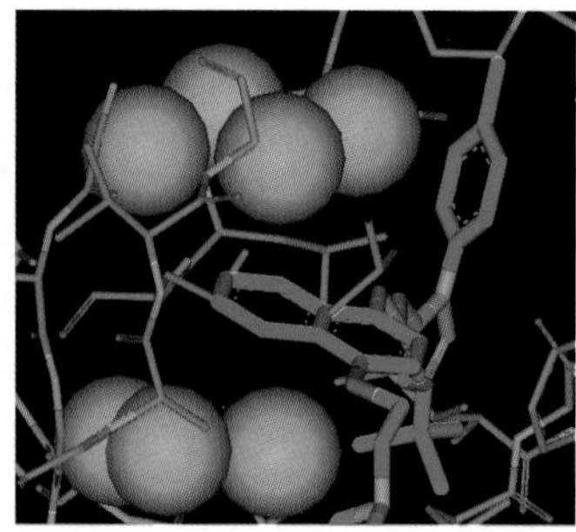

Figure 16 Example of hydrophobic enclosure for inhibitor of PDB entry 1 kv2. The naphthyl group receives a $-4.5\,\text{kcal mol}^{-1}$ hydrophobic packing reward. Biotin bound to streptaviden provides an excellent example of a special neutral–neutral hydrogen bond network (PDB entry 1 mk5). The triplet of hydrogen bond pairs to the ligand ring and the three hydrogen bonds to the ligand carbonyl oxygen contribute a $-6\,\text{kcal mol}^{-1}$ reward to this tightly bound complex ($G = -18.3\,\text{kcal mol}^{-1}$, XP binding $= -18.8\,\text{kcal mol}^{-1}$). (Adapted from Glide XP, developed and distributed by Scrödinger Inc.)

development of a knowledge-based scoring function at an atomic level is based upon observed frequency distributions of typical interactions in experimentally determined structures. Applying the inverse Boltzmann law,[314] interatomic interaction frequency distributions derived from protein crystal structures are converted into potentials of mean force. Only interactions close to the frequency maxima of knowledge base interactions are considered favorable. This is akin to potentials derived for protein folding and protein structure evaluation.[315] The approach is attractive since it offers the hope that the implicit treatment of all relevant contributions to binding will 'automatically' provide the correct balance between solvation and enthalpy contributions.

A number of different variants of this approach have been presented in the literature. Verkhivker *et al.*[316] created a focused scoring function combining desolvation terms and knowledge-based pair potentials from a data set of 30 HIV and SIV protease–inhibitor complexes. In addition protein side chain conformational immobilization terms were also incorporated using a method introduced by Pickett and Sternberg.[317] The scoring function is able to explain the differences in binding affinities of several HIV-1 protease–inhibitor complexes. Another HIV focused scoring function was created by Wallqvist *et al.*,[318] based on the calculated frequency distributions interacting atom pair buried surface patches for a set of 38 protein–ligand crystal structures. Atom-based statistical preferences are produced by normalizing with the product of buried surfaces of the corresponding individual atoms.

DeWitte and Shakhnovich[319] developed interatomic-interaction free energies (SMoG-Score) for ligands that bind to the surface of a protein or into binding pockets. Using a Metropolis-Monte Carlo-based[255] construction procedure, ligands are generated and energetically ranked in the binding pocket. The method has been applied to multiple systems including complexes of purine nucleoside phosphorylase and HIV-1 protease.

The program BLEEP has been developed by Mitchell *et al.*[320] using pair potential analysis derived from 820 protein–ligand atom-pair distributions based on the inverse Boltzmann approach. The analysis includes hydrogens positioned by the program HBPlus,[321] and the inclusion of water molecules as part of the protein has also been studied. BLEEP achieved a correlation coefficient of 0.74 for 90 diverse protein–ligand complexes with experimentally determined affinities.[322]

Muegge and Martin[323] produced potential of mean force (PMF) score using 697 crystallographically determined proteins using Helmholtz free interaction energies for 16 protein and 34 ligand atom types. Pair-distribution functions were produced using atom sampling distances up to 12 Å, with water contributions treated implicitly using a specific volume correction term.[324] For a test set of 77 protein–ligand complexes studied crystallographically, a deviation of 1.8 log units in reproducing the experimentally determined binding constants was found.

Gohlke *et al.* developed the scoring scheme DRUGSCORE.[325] The function was derived using distance-dependent pair and solvent-accessible surface dependent singlet potentials from 1376 protein–ligand complexes stored in ReliBase.[326] DRUGSCORE produced a deviation of 1.8 log units from experimentally determined inhibition constants for 55 diverse protein–ligand complexes.

Muryshev *et al.* created a hybrid function combining empirical and knowledge-based approaches.[327] Beginning with a knowledge-based potential based on seven atom types, the resulting functional form is approximated via fitting using multivariate regression to a smooth analytical function. Multivariate regression is again used to recalibrate the variable parameters of the resulting functions against the binding energy data of the training set (164 protein complexes from the PDB). The resulting scoring function was found to outperform GOLD, DOCK, and FLEXX for the 19 rigid complex test set studied. The EHITS program applies a similar methodology in its scoring function creation.[275] In addition the function can be customized by allowing the user to select input protein data sets for function generation. Zhang *et al.* have recently created the knowledge-based statistical energy function DFIRE using 19 atom types and a distance-scale finite ideal-gas reference (DFIRE) state.[328] When compared with 12 other scoring functions over 100 protein–ligand complex test set, DFIRE was found to be only moderately successful in ranking native or near-native conformations. Nevertheless it yielded the strongest binding affinity correlations and energy separations relative to docking decoys.

4.19.2.4 Consensus Scoring, Scoring Function Comparisons, and Other Tricks and Traps

The search for a single scoring function that can reliably score protein–ligand complexes continues to prove elusive. Consequently, as in ligand-based VS, researchers have tried to combine scoring functions to enhance the performance. Consensus scoring can outperform single scoring since statistically the mean value of repeated samplings tends to be closer to the true value[202] (assuming individual scoring functions exhibit relatively high performance and the scoring functions combined are not highly correlated[329]). The most generally applied consensus approach evaluates the ranking of binding modes measured with different scoring functions and favors those that rank consistently high in several. For example Gohlke *et al.*[325] generated a limited number of possible binding modes with FLEXX, which were then subjected the resulting putative complexes to ranking using DRUGSCORE to select the correct binding mode. Similarly Stahl[330] used FLEXX to generate protein–ligand conformations, with PMF applied to post scoring. Wang *et al.* analyzed 11 scoring functions over 100 ligand-protein complexes studying the ability of different functions combinations to identify the correct binding modes (RMSD$\leq$2.0 Å).[331] Any combinations of two or three of the six best-performing single functions pushed the ability to find the correct binding mode from 66–76% to >80%. A number of studies have been undertaken into the ability of consensus scoring schemes to improve binding affinity and virtual screening ranking.[201,332–335] Charifson *et al.* reported the first extensive consensus scoring study involving the docking programs DOCK and GAMBLER in combination with 13 scoring functions.[204] The study involved p38 MAP kinase, inosine monophosphate dehydrogenase, and HIV protease. The intersection of the top-scoring compounds of each scoring hit list led to a significant reduction in the list of false positives (inactive compounds that have high predicted scores). Weakly active compounds were found with a hit rate between 2% and 7%, significantly better than the scoring hit rates in many virtual screens. A comparison of the different scoring functions revealed that CHEMSCORE,[306,307] PLP,[337] and DOCK[297] energy score performed best as single scoring functions and also in consensus combination.

Consensus scoring experiments reported by Bissantz *et al.* found that docking consensus scoring performances varied widely among targets.[332] In contrast, Stahl and Rarey suggested that the combinations of FLEXX and PLP scores work best for consensus scoring for a variety of targets including COX-2, ER, p38 MAP kinase, gyrase, thrombin, gelatinase A, and neuraminidase.[333] Klon *et al.* developed a protocol to rescue poor docking results using a combination of rank-by-median consensus scoring and naive Bayesian categorization.[335] Consensus scoring was undertaken against protein tyrosine phosphatase 1B (PTP-1B) and protein kinase B/Akt (PKB) with FLEXX, with subsequent rescoring using the scoring functions in CSCORE[308] (DOCK, CHEMSCORE, GOLD, and PMF). Enrichment results are shown in **Table 2**. For the PKB test in particular significant improvement was found.

Xing *et al.* also used CSCORE in consensus scoring experiments involving factor Xa.[336] In these studies consensus scoring by pairwise intersection failed to enrich the hit rate yielded by single scoring terms. The authors highlighted

Table 2 Area under ROC curves corresponding to the enrichment after each step in the protocol of Klon *et al.*[335]

Method	*Target*	
	PTP-1B	*PKB*
FLEXX	0.89	0.62
FLEXX/naive Bayes	0.97	0.18
FLEXX/rank-by-median	0.80	0.80
FLEXX/rank-by-median/naive Bayes	0.96	0.82

the fact that reported successes of consensus scoring in hit rate enrichment could be due to overfitting, since comparisons were based on selected single scoring subsets and markedly reduced subsets of double or triple scoring.

The work of Xing provides a nice segue into the difficulties inherent to enrichment comparisons and scoring function design. There are many potential pitfalls the reader should be aware of when attempting to compare docking, scoring and enrichment performance. Often slightly different input files controls are applied from one comparison to another. For a program like DOCK which contains many input parameters, this can have a major effect on performance. Selection of target can have a major effect on the study conclusions, as can be seen through comparison of the conclusions of Bissantz *et al.*[332] and Stahl and Rarey.[333] Selection of training sets can have a similar effect scoring function design and performance. These issues have been highlighted by Smith *et al.* in their scoring function design efforts.[338] There is a tendency to consider crystal structure data as a form of absolute truth when undertaking computational design, scoring functions included. Davis *et al.* elegantly highlight the dangers of such a philosophy in their review on the utility and limitations of crystallographic data.[339] Efforts to improve available training sets through careful PDB complex selection, data collation and compound decoy choices have recently become the subject of significant effort.[340–342] These and other problems, including issues with the use of RMSD-based binding mode comparisons and enrichment studies decoy selection, have been nicely highlighted in a recent perspective on scoring function comparison.[343] Readers are encouraged to study these articles carefully before drawing there own conclusions.

One final point to note is that in this chapter as in many others on the subject, docking and scoring have been divided into separate sections. It should be noted, however, that given the limitations inherent in current scoring functions, pharmacophore constraints provide an implicit crossover point between these two key aspects of SVS calculations. Such constraints allow the user to define docking constraints that are in essence binary modifications to the scoring function. These constraints force the creation of key hydrogen bond motifs or the occupation of important hydrophobic pockets, forcing the scoring function to focus on biologically relevant regions of active site space. The potential utility of such constraints have been illustrated by Good *et al.* with reference to DOCK calculations,[252] and their importance is such that they have been incorporated into virtually all SVS packages available commercially. Indeed the two techniques have even been combined to create a pharmacophore-based scoring function in the program GEMDOCK.[344] Such approaches are widely used by pharmaceutical laboratories, for example within Hoffman La Roche.[233] These groups will often take the final hit lists one stage further, subjecting the top hits to a scoring function that is still hard to beat, the eye of the experienced computational chemist.

4.19.2.5 Structure-Based Virtual Screening Successes

Despite the challenges inherent to SVS techniques, they have been applied successfully to a wide array of targets.[345–355] **Table 3** lists a number of such successes, illustrating the diversity of targets and techniques applied. A nice example of SVS utility can be found in studies undertaken against type 2 diabetes target protein tyrosine phosphatase 1B (PTP1B).[354] In the HTS experiments, a 400 000 compound in-house library was screened against the target. For the SVS campaign, libraries of commercially available compounds were docked against the x-ray structure of PTP1B. From these compound collections 365 high-scoring docked compounds were tested, yielding a hit rate 1700-fold higher than that found by HTS. Furthermore, the hits found were also more druglike in their physical properties. The complementarity of the two techniques is highlighted by the fact that no overlap exists among the HTS and docking hit lists.

Another nice example of SVS comes from a screen of the Merck chemical collection against the tuberculosis target dihydrodipicolinate reductase.[349] With binding defined as an IC_{50} $<100\,\mu M$, HTS produced a hit rate of 0.2% compared to 6% hit rate for the SVS campaign.

Table 3 also highlights the potential utility of homology models in SVS campaigns. Recent studies have shown that when active site sequence identity surpasses 50%, SVS enrichments approach those seen in native crystal structures.[356] The work of Evers *et al.* on the G protein-coupled receptor (GPCR) NK1 and alpha1A receptors provides excellent examples of homology model utility and the application of an integrated approach to the use of computations design tools (**Figure 17**).[351–357] Homology model generation for GPCR structures such as the alpha1A receptor are particularly challenging, since sequence identity relative to the rhodopsin template structure is only 27%. Considerable deviations from the native structure may be obtained upon homology model creation with such low levels of sequence conservation.[358] A modified version of the MOBILE approach[358] was thus applied to generate an ensemble of crude homology models of the target protein. The models were subsequently refined and ranked for quality via constrained ligand docking analysis and minimization using amino acid mutation mapping. The resulting binding pocket characteristics were then condensed into two ligand-based pharmacophore models, which were used as a first filter for database screening using CATALYST.[45] Next, an alpha1A antagonist spiked data set was docked into the preferred homology model using GOLD[266] and poses ranked using multiple scoring functions. Database compounds matching

Table 3 Examples of successful SVS campaigns

Target	*Docking program used*	*Active site source*
Carbonic anhydrase II[345]	FLEXX	X-ray
B-Tubulin[347]	DOCK	X-ray
Retinoic acid receptor[346]	ICM	Homology model
Dihydrodipicolinate reductase[349]	FLOG	X-ray
CDK4[346]	LEGEND/SEEDS	Homology model
Farnesyl transferase[350]	EUDOC	X-ray
NK1 GPCR[351]	UNITY / FLEXX-PHARM	Homology model
Estrogen receptor antagonists[352]	PRO_LEADS	X-ray
Casein kinase II[353]	DOCK	Homology model
Protein tyrosine phosphotase 1B[354]	DOCK	X-ray
TRNA-guanine transglycosylase[355]	UNITY/FLEXX	X-ray

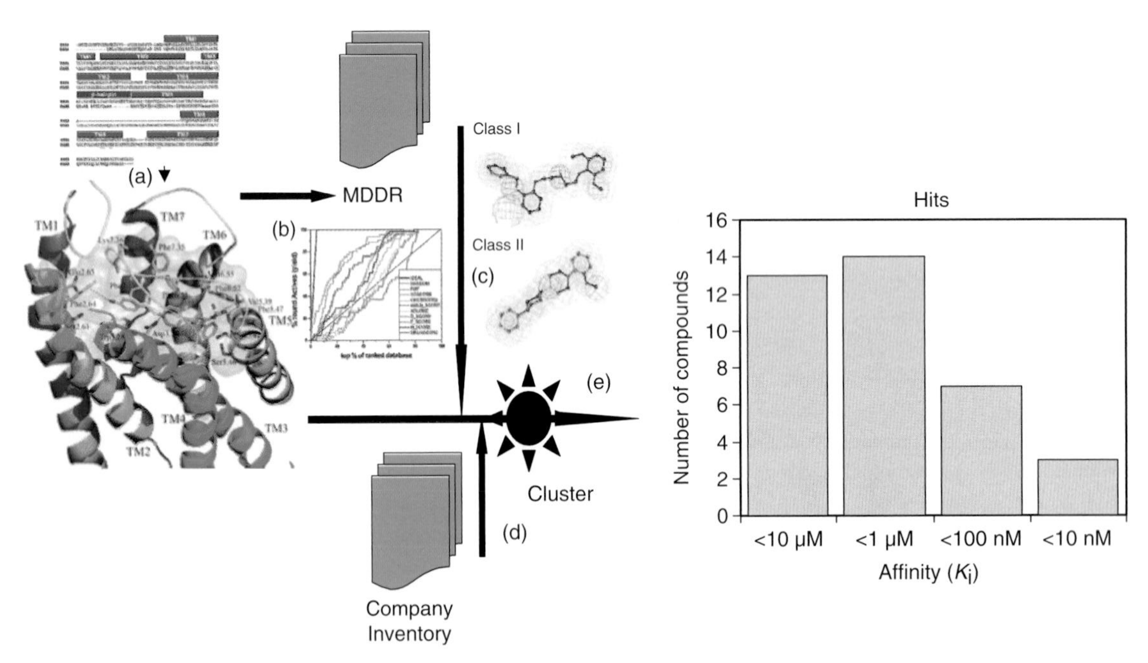

Figure 17 Excellent example of a successful integrated approach to virtual screening as applied to GPCR targets. (a) Construct homology model. (b) Use spiked data set from MDDR to analyze performance of various scoring functions. Select best for use in subsequent SVS calculations. (c) Derive pharmacophore models from existing SAR and homology model and apply as a first filter to database screen. (d) Dock molecules passing this filter and score using selected scoring function. (e) Cluster top hits and select diverse subset for screening. (Adapted from Evers, A.; Klabunde, T. *J. Med. Chem.* **2005**, *48*, 1088–1097.)

the pharmacophore models were subsequently docked to the homology model using GOLD and rescored using the scoring function showing the highest enrichments (PMF[323]). The top-scoring 300 compounds were clustered according to their UNITY[68] fingerprint similarity and a diverse set of 80 compounds submitted for experimental testing. Of these, 38 compounds showed percent inhibition values >50% at 10 μM, with 10 of these having K_is below 100 nM. While the corporate databases of companies with a long history of GPCR research will be biased toward such targets, a hit rate near 50% is nevertheless an impressive result by any standards.

4.19.3 Conclusions

Virtual screening techniques have become an increasingly important tool for lead discovery. Whether used in conjunction with HTS or stand alone, VS techniques provide a quick and economical method for the discovery of novel actives. Further, for smaller companies without access to routine HTS resource, such methods are crucial in the selection of molecules for internal screening. With the ever-improving accuracy of VS and SVS technology and the costs associated with unconstrained HTS campaigns, this importance will only increase with time. As a result an understanding of VS technology application will likely prove a useful asset to researchers engaged in the lead discovery process.

References

1. *Virtual Screening for Bioactive Molecules*; Böhm, H.-J., Schneider, G., Kubinyi, H., Mannhold, R., Timmerman, H., Eds.; Wiley-VCH: New York, 2000.
2. *Virtual Screening: An Alternative or Complement to High Throughput Screening*; Klebe, G., Ed.; Springer: New York, 2000.
3. Good, A. C.; Krystek, S.; Mason, J. S. *Drug Disc. Today* **2000**, *5*, 61–65.
4. Seifert, M. H. J.; Wolf, K.; Vitt, D. *Biosilico* **2003**, *1*, 143–149.
5. Oprea, T. I.; Matter, H. *Curr. Opin. Chem. Biol.* **2004**, *8*, 349–358.
6. Stahura, F.; Bajorath, *Comb. Chem. & H.T.S.* **2004**, 7, 259-269.
7. Lengauar, T.; Lemmen, C.; Rarey, M.; Zimmerman, M. *Drug Disc. Today* **2004**, *9*, 27–34.
8. Olah, M. M.; Bologa, C. G.; Oprea, T. I. *Curr. Drug Disc. Tech.* **2004**, *1*, 211–220.
9. *Virtual Screening in Drug Discovery*; Alvarez, J., Schoichet, B., Eds.; CRC Press: Boca Raton, FL, 2005.
10. Willett, P. *J. Chemometrics* **1987**, *1*, 139–146.
11. Dalby, A.; Nourse, J. G.; Hounshell, W. D.; Gushurst, A. K. I.; Grier, D. L.; Leland, B. A.; Laufer, J. *J. Chem. Inf. Comput. Sci.* **1992**, *32*, 244–255.
12. Cramer, R. D.; Redl, G.; Berkoff, C. E. *J. Med. Chem.* **1974**, *17*, 533–535.
13. Tarjan, R. E. Graph Algorithms in Chemical Computation. In *Algorithms for Chemical Computations*; American Chemical Society: Washington, DC, 1977, pp 1–20.
14. Barnard, J. M. *J. Chem. Inf. Comput. Sci.* **1993**, *33*, 532–538.
15. Milne, G. W. A. *J. Chem. Inf. Comput. Sci.* **1997**, *374*, 639–644.
16. See mol2 at http://www.tripos.com/ (accessed April 2006).
17. Weininger, D. *J. Chem. Inf. Comput. Sci.* **1988**, *28*, 31–37.
18. SMILES and SMARTS, developed and distributed by Daylight Chemical Information Systems: http://www.daylight.com/dayhtml/ (accessed April 2006).
19. Ash, S.; Cline, M. A.; Homer, W. R.; Hurst, T.; Smith, G. B. *J. Chem. Inf. Comput. Sci.* **1997**, *371*, 71–79.
20. Lynch, M. F.; Downs, G. M. Chemical Patent Database Systems. In *Chemical Structure Systems: Computational Techniques for Representation, Searching, and Processing of Structural Information*; Ash, J. E., Warr, W. A., Willett, P., Eds.; Ellis Horwood: Chichester, UK, 1991, pp 126–153.
21. Corey, E. J.; Feiner, N. E. *J. Org. Chem.* **1980**, *45*, 757–764.
22. Corey, E. J.; Feiner, N. F. *J. Org. Chem.* **1980**, *45*, 765–780.
23. Hoflack, J.; De Clercq, P. J. *Tetrahedron* **1988**, *4421*, 6667–6676.
24. SCA, QCPE Program QCMP079. Bloomington, IN, USA, Quantum Chemistry Program Exchange. Indiana University, 1989. http://qcpe.chem.indiana.edu/ (accessed April 2006).
25. Rarey, M.; Kramer, B.; Lengauer, T.; Klebe, G. *J. Mol. Biol.* **1996**, *261*, 470–489.
26. FlexX, developed and distributed by BioSolveIT gmbh: http://www.biosolveit.de (accessed April 2006).
27. Leach, A. R.; Prout, K.; Dolata, D. P. *J. Comput.-Aided Mol. Des.* **1990**, *4*, 271–282.
28. Dolata, D. P.; Leach, A. R.; Prout, K. *J. Comput.-Aided Mol. Des.* **1987**, *1*, 73–85.
29. Leach, A. R.; Dolata, D. P.; Prout, K. *J. Chem. Inf. Comput. Sci.* **1990**, *30*, 316–324.
30. Leach, A. R. *J. Chem. Inf. Comput. Sci.* **1994**, *34*, 661–670.
31. Havel, T. F.; Kuntz, I. D.; Crippen, G. M. *J. Theor. Biol.* **1983**, *1043*, 359–381.
32. Crippen, G. M.; Havel, T. F. Distance Geometry and Molecular Conformations. In *Chemometrics*; Bawden, D., Ed.; *Research Studies Series 15*; Research: New York, 1988, pp 101–122.
33. Crippen, G. M.; Havel, I. F. *J. Chem. Inf. Comput. Sci.* **1990**, *30*, 222–227.
34. DGEOM, QCPE Program No. 590. Bloomington, IN, USA, Quantum Chemistry Program Exchange, Indiana University, 1995. http://qcpe.chem.Indiana.edu/ (accessed April 2006).
35. CONCORD, STEREOPLEX, PROTOPLEX, and CONFORT are distributed by Tripos Inc. http://www.tripos.comsol (accessed April 2006).
36. Pearlman, R. S. *Chem. Des. Autom. News* **1987**, *2*, 5–6.
37. Pearlman, R. S.; Stewart, E. L.; Smith, K. M.; Balducci, R. *Book of Abstracts*, 211th American Chemical Society National Meeting, New Orleans, LA, March 24–28, 1996; CINF-065.
38. Pearlman, R. S.; Khashan, R.; Wong, D.; Balducci, R. *Book of Abstracts*, 224th American Chemical Society National Meeting, Boston, MA, Aug 18–22, 2002; COMP-232.
39. Sadowski, J.; Gasteiger, J. *J. Chem. Rev.* **1993**, 7, 2567–2581.
40. Sadowski, J. Three Dimensional Structure Generation: Automation. In *Encyclopedia of Computational Chemistry*; Schleyer, P. R., Allinger, N. L., Clark, T., Gasteiger, J., Kollman, P. A., Shaefer, H. F., III., Scheiner, P. R., Eds.; John Wiley: Chichester, UK, 1998, pp 2976–2988.
41. CORINA, developed and distributed by the University of Erlangen. http://www2.chemie.uni-erlangen.de/software/ (accessed April 2006).
42. Dale, J. *Acta Chem. Scand.* **1973**, *27*, 1115–1129.
43. Klebe, G.; Mietzner, T.; Weber, F. *J. Comput.-Aided Mol. Des.* **1999**, *13*, 35–49.
44. Cambridge Structural Database. http://www.ccdc.cam.ac.uk/products/csd/ (accessed April 2006).
45. Confirm and Catalyst, developed and distributed by Accelrys Inc. http://www.accelrys.com/catalyst/ (accessed April 2006).
46. DGII, distributed by Accelrys Inc. http://www.accelrys.com (accessed April 2006). See also http://www.scripps.edu/rc/softwaredocs/msi/insight2K/CONVERTER/02_Theory.html.#661284 (accessed April 2006).

47. Havel, T. F. *Biopolymers* **1990**, *2912*, 1565–1585.
48. Omega and QuACPAC, developed and distributed by Openeye Scientific Software. http://www.eyesopen.com/ (accessed April 2006).
49. Stahl, M. T.; Nicholls, A.; Anthony, R. A; Grant, A. J. *Book of Abstracts*, 217th American Chemical Society National Meeting, Anaheim, CA, March 21–25, 1999; COMP-026.
50. Stahl, M.; Skillman, G. *Abstracts of Papers*, 229th American Chemical Society National Meeting, San Diego, CA, March 13–17, 2005; COMP-141.
51. Spellmeyer, D. C.; Wong, A. K.; Bower, M. J.; Blaney, J. M. *J. Mol. Graph. Model.* **1997**, *15*, 18–36.
52. Halgren, T. A. *J. Comp. Chem.* **1999**, *20*, 730–748.
53. LIGPREP, developed and distributed by Schrodinger Inc. http://www.schrodinger.com/ (accessed April 2006).
54. JAGUAR quantum mechanical software, developed and distributed by Schrodinger Inc. http://www.schrodinger.com (accessed April 2006).
55. Chang, G.; Guida, W. C.; Still, W. C. *J. Am. Chem. Soc.* **1989**, *111*, 4379–4384.
56. Saunders, M.; Houk, K. N.; Wu, Y. D.; Still, C. W.; Lipton, M.; Chang, G.; Guida, W. C. *J. Am. Chem. Soc.* **1990**, *112*, 1419–1429.
57. MACROMODEL, developed and distributed by Schrodinger Inc. http://www.schrodinger.com/ (accessed April 2006).
58. Smellie, A.; Kahn, S. D.; Teig, S. L. *J. Chem. Inf. Comput. Sci.* **1995**, *35*, 285–294.
59. Smellie, A.; Kahn, S. D.; Teig, S. L. *J. Chem. Inf. Comput. Sci.* **1995**, *35*, 295–304.
60. Smellie, A.; Teig, S. L.; Towbin, P. *J. Comput. Chem.* **1995**, *16*, 171–187.
61. Murrall, N. W.; Davies, E. K. *J. Chem. Inf. Comput. Sci.*, **1990**, *30*, 312–316.
62. Ewing, T. J. A.; Kuntz, I. D. *J. Comput. Chem.* **1997**, *189*, 1175–1189.
63. Balducci, R.; Pearlman, R. S. *Book of Abstracts*, 217th American Chemical Society National Meeting, Anaheim, CA, March 21–25 1999; COMP-011.
64. RUBICON, developed and distributed by Daylight Chemical Information Systems. http://www.daylight.com/ (accessed April 2006).
65. Mayo, S. L.; Olafson, B. D.; Goddard, W. A., III. *J. Phys. Chem.* **1990**, *9426*, 8897–8909.
66. Sykes, M.; Pickup, B; Grant, J. A. The Sheffield Solvation Model, In preparation.
67. Hurst, T. *J. Chem. Inf. Comput. Sci.* **1994**, *34*, 190–196.
68. UNITY, developed and distributed by Tripos Inc. http://www.tripos.com/ (accessed April 2006).
69. Moock, T. E.; Henry, D. R.; Ozkabak, A. G.; Alamgir, M. *J. Chem. Inf. Comput. Sci.* **1994**, *34*, 184–189.
70. ISIS-3D, developed and distributed by MDL Information Systems Inc. http://www.mdl.com/ (accessed April 2006).
71. Lemmen, C.; Lengauer, T.; Klebe, G. *J. Med. Chem.* **1998**, *4123*, 4502–4520.
72. Lemmen, C.; Zimmermann, M.; Lengauar, T. *Perspect. Drug Disc. Des.* **2000**, *20*, 43–62.
73. FlexS, developed and distributed by BioSolveIT gmbh: http://www.biosolveit.de/ (accessed April 2006).
74. Sadowski, J.; Schwab, C. H.; Gasteiger, J. http://www2.chemie.uni-erlangen.de/ (accessed July 2006).
75. Sadowski, J.; Gasteiger, J.; Klebe, G. *J. Chem. Inf. Comput. Sci.* **1994**, *34*, 1000–1008.
76. Green, D. V. S. In *Des. Bioact. Mol.*; Martin, Y. C.; Willet, P., Eds.; American Chemical Society: Washington, DC, 1998, pp 47–71.
77. Van Geerestein, V. J.; Verwer, P. *Book of Abstracts*, 211th American Chemical Society National Meeting, New Orleans, LA, March 24–28, 1996, CINF038.
78. Sadowski, J.; Schwab, C. H.; Gasteiger, J. 3D Structure Generation and Conformational Searching. In *Computational Medicinal Chemistry for Drug Discovery*; Bultinck, P., de Winter, H., Langenaeker, W., Tollenaere, J. P., Eds.; Marcel Dekker: New York, 2004, pp 151–212.
79. Daylight Chemical Information System Inc. http://www.daylight.com/meetings/ (accessed April 2006).
80. Berman, H. M.; Westbrook, J.; Feng, Z.; Gilliland, G.; Bhat, T. N.; Weissig, H.; Shindyalov, I. N.; Bourne, P. E. *Nucl. Acids Res.* **2000**, *28*, 235–242.
81. Ricketts, E. M.; Bradshaw, J.; Hann, M. M.; Hayes, F.; Tanna, N.; Ricketts, D. M. *J. Chem. Inf. Comput. Sci.* **1993**, *33*, 905–925.
82. Boström, J. *J. Comput.-Aided Mol. Des.* **2001**, *1512*, 1137–1152.
83. Good, A. C.; Cheney, D. L. *J. Mol. Graph. Model.* **2003**, *221*, 23–30.
84. Perola, E.; Charifson, P. *S. J. Med. Chem.* **2004**, *47*, 2499–2510.
85. Ray, L. C.; Kirsch, R. A. *Science* **1957**, *126*, 814–819.
86. Downs, G. M.; Willett, P. Similarity Searching in Databases of Chemical Structures. In *Reviews in Computational Chemistry*; Lipkowitz, K. B., Boyd, D. B., Eds.; VCH: New York, 1996; Vol. 7, pp 1–66.
87. Ullmann, J. R. *J. Assoc. Comput. Mach.* **1976**, *23*, 31–42.
88. Downs, G. M.; Lynch, M. E.; Willett, P.; Manson, G. A.; Wilson, G. A. *Tetrahedron Comput. Methodol.* **1988**, *1*, 207–217.
89. Lynch, M. E. Screening Large Chemical Files. In *Chemical Information Systems*; Ash, J. E., Hyde, E., Eds.; John Wiley: Chichester, UK, 1974, pp 177–194.
90. Bartmann, A.; Maier, H.; Roth, B.; Walkowiak, D. *J. Chem. Inf. Comput. Sci.* **1990**, *33*, 539–541.
91. Graf, W.; Kaindl, H. K.; Kneiss, H.; Schmidt, B.; Warszawski, R. *J. Chem. Inf. Comput. Sci.* **1982**, *22*, 177–181.
92. Isis and Crossfire, developed and distributed by MDL Information Systems. http://www.mdl.com/ (accessed April 2006).
93. Merlin, developed and distributed by Daylight Chemical Information Systems Inc. http://www.daylight.com/products/ (accessed April 2006).
94. Unity, developed and distributed by Tripos Inc. http://www.tripos.com/ (accessed April 2006).
95. Good, A. C.; Mason, J. S. In *Reviews in Computational Chemistry*; Lipkowitz, K. B., Boyd, D. B., Eds.; VCH: New York, 1996; Vol. 7, pp 67–127.
96. Milne, G. W. A.; Nicklaus, M. C.; Wang, S. *SAR QSAR Environ. Res.* **1998**, *9*, 23–38.
97. Warr, W. A.; Willett, P. In *Des. Bioact. Mol.*; Martin, Y. C.; Willet, P., Eds.; American Chemical Society: Washington, DC, 1998, pp 73–95.
98. *Pharmacophore Perception, Development, and Use in Drug Design*; Guner, O. F., Ed.; International University Line: La Jolla, CA, 2000.
99. Good, A. C. Mason, J. S.; Pickett, S. D. Pharmacophore Pattern Application in Virtual Screening, Library Design and QSAR. In *Virtual Screening for Bioactive Molecules*; Böhm, H.-J., Schneider, G., Ed.; John Wiley: New York, 2000; Vol. 10, pp 131–154.
100. van Drie, J. H. *Curr. Pharm. Des.* **2003**, *920*, 1649–1664.
101. Marshall, G. R. Binding Site Modeling of Unknown Receptors. In *3D QSAR in Drug Design*; Kubinyi, H., Ed.; ESCOM: Leiden, The Netherlands, 1993, pp 80–116.
102. Van Drie, J. H.; Weiniger, D.; Martin, Y. C. *J. Comput.-Aided Mol. Des.* **1989**, *3*, 225–251.
103. Phase, developed and distributed by Schrodinger Inc. http://www.schrodinger.com/ (accessed April 2006).
104. MOE, Molecular Operating Environment. Distributor: Chemical Computing Group. http://www.chemcomp.com/ (accessed April 2006).
105. Wang, S.; Milne, G. W. A.; Yan, X.; Posey, I. J.; Nicklaus, M. C.; Graham, L.; Rice, W. G. *J. Med. Chem.* **1996**, *39*, 2047–2054.
106. Hong, H.; Neamati, N.; Wang, S.; Nicklaus, M. C.; Mazumder, A.; Zhao, H.; Burke, T. R., Jr.; Pommier, Y.; Milne, G. W. A. *J. Med. Chem.* **1997**, *40*, 930–936.

107. Marriott, D. P.; Dougall, I. G.; Meghani, P.; Liu, Y.-J.; Flower, D. R. *J. Med. Chem.* **1999**, *42*, 3210–3216.
108. Singh, J.; van Vlijmen, H.; Liao, Y.; Lee, W.-C.; Cornebise, M.; Harris, M.; Shu, I.-H.; Gill, A.; Cuervo, J. H.; Abraham, W. M.; Adams, S. P. *J. Med. Chem.* **2002**, *45*, 2988–2993.
109. Clark, D. E.; Higgs, C.; Wren, S. P.; Dyke, H. J.; Wong, M.; Norman, D.; Lockey, P. M.; Roach, A. G. *J. Med. Chem.* **2004**, *47*, 3962–3971.
110. Pirard, B.; Brendel, J.; Peukert, S. *J. Chem. Inf. Model.* **2005**, *452*, 477–485.
111. Ekins, S.; Johnston, J. S.; Bahadduri, P.; D'Souza, V. M.; Ray, A.; Chang, C.; Swaan, P. W. *Pharm. Res.* **2005**, *224*, 512–517.
112. Steindl, T. M.; Crump, C. E.; Hayden, F. G.; Langer, T. *J. Med. Chem.* **2005**, *48*, 6250–6260.
113. Sheridan, R. P.; Miller, M. D. *J. Chemut. Inf. Comput. Sci.* **1998**, *385*, 915–924.
114. Chen, L. Substructure and Maximum Common Substructure Searching. In *Computational Medicinal Chemistry for Drug Discovery*; Bultinck, P., de Winter, H., Langenaeker, W., Tollenaere, J. P., Eds.; Marcel Dekker: New York, 2004, pp 483–511.
115. Martin, Y. C.; Bures, M. G.; Danaher, E. A.; DeLazzer, J.; Lico, I.; Pavlik, P. A. *J. Comput.-Aided Mol. Des.* **1993**, 7, 83–102.
116. DISCO, distributed by Tripos Inc. http://www.tripos.com/ (accessed April 2006).
117. Brint, A. T.; Willett, P. J. *J. Chem. Inf. Comput. Sci.* **1987**, *27*, 152–158.
118. Jones, G.; Willett, P.; Glen, R. C. *J. Comput.-Aided Mol. Des.* **1995**, *96*, 532–549.
119. Lemmen, C.; Lengauer, T.; Klebe, G. *J. Med. Chem.* **1998**, *4123*, 4502–4520.
120. Miller, M. D.; Sheridan, R. P.; Kearsley, S. K. *J. Med. Chem.* **1999**, *429*, 1505–1514.
121. HypoGen, developed and distributed by Accelrys Inc. http://www.accelrys.com/ (accessed April 2006).
122. Gasp, distributer by Tripos Inc. http://www.tripos.com/ (accessed April 2006).
123. Durant, J. L.; Leland, B. A.; Henry, D. R.; Nourse, J. G. *J. Chem. Inf. Comput. Sci.* **2002**, *42*, 1273–1280.
124. BCI fingerprints, part of the BCI toolkit, developed and distributed by Barnard Chemical Information Ltd. http://www.bci.gb.com/ (accessed April 2006).
125. McGregor, M. J.; Muskal, S. M. *J. Chem. Inf. Comput. Sci.* **1999**, *39*, 569–574.
126. Mason, J. S.; Morize, I.; Menard, P. R.; Cheney, D. L.; Hulme, C.; Labaudiniere, R. F. *J. Med. Chem.* **1999**, *4217*, 3251–3264.
127. Carhart, R. E.; Smith, D. H.; Venkataraghavan, R. *J. Chem. Inf. Comput. Sci.* **1985**, *25*, 64–73.
128. Daylight Chemical Information System Inc. http://www.daylight.com/ (accessed April 2006).
129. ECFP and FCFP fingerprints, part of Pipeline Pilot, developed and distributed by Scitegic. http://www.scitegic.com/ (accessed April 2006).
130. Sheridan, R. P.; Miller, M. D.; Underwood, D. J.; Kearsley, S. K. *J. Chem. Inf. Comput. Sci.* **1996**, *36*, 128–136.
131. Pearlman, R. S.; Smith, K. M. *J. Chem. Inf. Comput. Sci.* **1999**, *39*, 28–35.
132. Willett, P.; Winterman, V.; Bawden, D. *J. Chem. Inf. Comput. Sci.* **1986**, *26*, 36–41.
133. Christie, A. D.; Leland, B. A.; Nourse, I. G. *J. Chem. Inf. Comput. Sci.* **1993**, *33*, 545–547.
134. Brown, R. D.; Martin, Y. C. *J. Chem. Inf. Comput. Sci.* **1996**, *36*, 572–584.
135. Barnard, J. M.; Downs, G. M. *J. Chem. Inf. Comput. Sci.* **1997**, *37*, 141–142.
136. 2003 User Group Meeting http://www.scitegic.com/news_event/ (accessed April 2006).
137. Willett, P. *J. Chem. Inf. Comput. Sci.* **1979**, *19*, 159–162.
138. Nilakantan, R.; Bauman, N.; Dixon, S. J.; Venkataraghavan, R. *J. Chem. Inf. Comput. Sci.* **1987**, *27*, 82–85.
139. Kearsley, S. K.; Sallamack, S.; Fluder, E. M.; Andose, J. D.; Mosley, R. T.; Sheridan, R. P. *J. Chem. Inf. Comput. Sci.* **1996**, *361*, 118–127.
140. Schuffenhauer, A.; Floersheim, P.; Acklin, P.; Jacoby, E. *J. Chem. Inf. Comput. Sci.* **2003**, *432*, 391–405.
141. Klein, K.; Lehmann, C. W.; Schmidt, H.-W.; Maier, W. F. *Angew. Chem.* **1998**, *110*, 3369–3372.
142. Schneider, G.; Neidhart, W.; Giller, T.; Schmid, G. *Angew. Chem. Int. Ed. Engl.* **1999**, *38*, 2894–2896.
143. Bemis, G. W.; Kuntz, I. D. *J. Comput.-Aided Mol. Des.* **1992**, *66*, 607–628.
144. Nilakantan, R.; Bauman, N.; Venkataraghavan, R. *J. Chem. Inf. Comput. Sci.* **1993**, *331*, 79–85.
145. Good, A. C.; Ewing, T. J. A.; Gschwend, D. A.; Kuntz, I. D. *J. Comput.-Aided Mol. Des.* **1995**, *9*, 1–10.
146. Sheridan, R. P.; Miller, M. D.; Underwood, D. J.; Kearsley, S. K. *J. Chem. Inf. Comput. Sci.* **1996**, *361*, 128–136.
147. Mason, J. S.; Cheney, D. L. *Pac. Symp. Biocomput.* **1999**, 456–467.
148. McGregor, M. J.; Muskal, S. M. *J. Chem. Inf. Comput. Sci.* **1999**, *39*, 569–574.
149. Good, A. C.; Cho, S.-J.; Mason, J. S. *J. Comput.-Aided Mol. Des.* **2004**, *18*, 523–527.
150. Wold, S.; Sjöström, M.; Eriksson, L. In *Encyclopedia of Computational Chemistry*; John Wiley: New York, 1998, pp 2006–2021.
151. Renner, S.; Schneider, G. *J. Med. Chem.* **2004**, *47*, 4653–4664.
152. Bradley, E. K.; Beroza, P.; Penzotti, J. E.; Grootenhuis, P. D. J.; Spellmeyer, D. C.; Miller, J. L. *J. Med. Chem.* **2000**, *4314*, 2770–2774.
153. Lipinski, C. A.; Lombardo, F.; Dominy, B. W.; Feeney, P. J. *Adv. Drug Deliv. Rev.* **2001**, *46*, 3–26.
154. Lewis, R. A.; Mason, J. S.; McLay, I. M. *J. Chem. Inf. Comput. Sci.* **1997**, *37*, 599–614.
155. Xue, L.; Stahura, F. L.; Godden, J. W.; Bajorath, J. *J. Chem. Inf. Comput. Sci.* **2001**, *41*, 394–401.
156. Sheridan, R. P.; Nachbar, R. B.; Bush, B. L. *J. Comput.-Aided Mol. Des.* **1994**, *83*, 323–340.
157. Pearlman, R. S.; Smith, K. M. *J. Chem. Inf. Comput. Sci.* **1999**, *391*, 28–35.
158. Burden, F. R. *J. Chem. Inf. Comput. Sci.* **1989**, *29*, 225–227.
159. *DiverseSolutions*, v4.0.5; University of Texas, Austin; distributed by Tripos, Inc. http://www.tripos.com/ (accessed April 2006).
160. Pearlman, R. S.; Smith, K. M. Novel Software Tools for Chemical Diversity. In *Perspective in Drug Discovery and Design*; Kubinyi, H., Folkers, G., Martin, Y. C., Eds.; Kluwer/Escom: Dordrecht, the Netherlands, 1998; Vols. 9/10/11, pp 339–351.
161. Lee, M. R.; Sage, C. R. *Abstracts of Papers*, 223rd American Chemical Society National Meeting, Orlando, FL, April 7–11, 2002, COMP-094.
162. Mason, J. S.; Beno, B. R. *J. Mol. Graph. Model.* **2000**, *18*, 438–451.
163. Gillett, V. J.; Willett, P.; Bradshaw, J. *J. Chem. Inf. Comput. Sci.* **2003**, *43*, 338–345.
164. Barker, E. J.; Gardiner, E. J.; Gillett, V. J.; Kitts, P.; Morris, J. *J. Chem. Inf. Comput. Sci.* **2003**, *43*, 346–356.
165. Daylight toolkit, developed and distributed by Daylight Chemical Information Systems: http://www.daylight.com/ (accessed April 2006).
166. Rarey, M.; Dixon, J. S. *J. Comput.-Aided Mol. Des.* **1998**, *12*, 471–482.
167. FTREES, developed and distributed by BioSolveIT. http://www.biosolveit.de/ (accessed April 2006).
168. Dixon, S. L.; Merz, K. M. *J. Med. Chem.* **2001**, *44*, 3795–3809.
169. Good, A. C. Application of 3D Molecular Similarity Index Calculations to QSAR studies. In *Molecular Similarity in Drug Design*; Dean, P. M., Ed.; Blackie: Glasgow, UK, 1995, pp 24–56.
170. Good, A. C.; Richards, W. G. *Perspect. Drug Disc. Des.* **1998**, *9*, 321–338.
171. Carbo, R.; Besalu, E. A. *Theomchem*, **1998**, *4511*, 11–23.

172. van Geerestein, V. J.; Perry, N. J.; Grootenhuis, P. D. J.; Haasnoot, C. A. G. *Tetrahedron Comput. Methodol.* **1992**, *3*, 595–613.
173. Perry, N. C.; van Geerestein, V. J. *J. Chem. Inf. Comput. Sci.* **1992**, *32*, 607–616.
174. Dean, P. M.; Chau, P.-L. *J. Mol. Graph.* **1987**, *5*, 152–158.
175. Moon, J. B.; Howe, W. J. *Tetrahedron Comput. Methodol.* **1990**, *3*, 697–711.
176. Good, A. C.; Hodgkin, E. E.; Richards, W. G. *J. Chem. Inf. Comput. Sci.* **1992**, *32*, 188–191.
177. Carbo, R.; Domingo, L. *Int. J. Quantum Chem.* **1987**, *32*, 517–545.
178. Szabo, A., Ostland, N. S. In *Modern Quantum Chemistry*; Macmillan: London, 1982, pp 410–412.
179. Wild, D. J.; Willett, P. *J. Chem. Inf. Comput. Sci.* **1996**, *36*, 159–167.
180. Thorner, D. A.; Wild, D. J.; Willett, P.; Wright, P. M. *Perspect. Drug Disc. Des.* **1998**, *9*, 301–320.
181. Good, A. C.; Richards, W. G. *J. Chem. Inf. Comput. Sci.* **1993**, *33*, 112–116.
182. Grant, J. A.; Gallardo, M. A.; Pickup, B. T. *J. Comput. Chem.* **1996**, *17*, 1653–1666.
183. Mestres, J.; Rohrer, D. C.; Maggiora, G. M. *J. Mol. Graph. Model.* **1997**, *15*, 114–121.
184. ROCS, developed and distributed by OpeneyeScientific Software Inc. http://www.eyesopen.com/ (accessed April 2006).
185. Rush, T. S., III; Grant, A. J.; Mosyak, L.; Nicholls, A. *J. Med. Chem.* **2005**, *485*, 1489–1495.
186. Lemmen, C.; Zimmermann, M.; Lengauer, T. *Perspect. Drug Disc. Des.* **2000**, *20*, 43–62.
187. Pitman, M. C.; Huber, W. K.; Horn, H.; Krämer, A.; Rice, J. E.; Swope, W. C. *J. Comput.-Aided Mol. Des.* **2001**, *15*, 587–612.
188. Kraemer, A.; Horn, H. W.; Rice, J. E. *J. Comput.-Aided Mol. Des.* **2003**, *17*, 13–18.
189. Rhodes, N.; Willett, P. *J. Chem. Inf. Comput. Sci.* **2003**, *43*, 443–448.
190. Hahn, M. *J. Chem. Inf. Comput. Sci.* **1997**, *37*, 80–86.
191. Putta, S.; Lemmen, C.; Beroza, P.; Greene, J. *J. Chem. Inf. Comput. Sci.* **2002**, *42*, 1230–1240.
192. Putta, S.; Eksterowicz, J.; Lemmen, C.; Stanton, R. *J. Chem. Inf. Comput. Sci.* **2003**, *435*, 1623–1635.
193. Engels, M. F. M.; Thielemans, T.; Verbinnen, D.; Tollenaere, J. P.; Verbeeck, R. *J. Chem. Inf. Comput. Sci.* **2000**, *40*, 241–245.
194. Bajorath, J. *Nat. Rev. Drug Disc.* **2002**, *1*, 882–894.
195. Karnachi, P. S.; Brown, F. K. *J. Biolmol. Screen.* **2004**, *9*, 678–686.
196. Shanmugasundaram, V.; Maggiora, G. M.; Lajiness, M. S. *J. Med. Chem.* **2005**, *481*, 240–248.
197. Smemetulskis, N. E.; Weiniger, D.; Blankley, C. J.; Yang, J. J.; Humblet, C. *J. Chem. Inf. Comput. Sci.* **1996**, *36*, 862–871.
198. Singh, S. B.; Sheridan, R. P.; Fluder, E. M.; Hull, R. D. *J. Med. Chem.* **2001**, *44*, 1564–1575.
199. Klein, L. A. *Sensor and Data Fusion Concepts and Applications,* 2nd ed.; International Society for Optical Engineering: Bellingham, WA, 1999.
200. Ginn, C. M. R.; Willett, P.; Bradshaw, J. *Perspect. Drug Disc. Des.* **2000**, *20*, 11–16.
201. Charifsen, P. S.; Corkery, J. J.; Murcko, M. A. *J. Med. Chem.* **1999**, *42*, 5100–5109.
202. Wang, R.; Wang, S. *J. Chem. Inf. Comput. Sci.* **2001**, *41*, 1422–1426.
203. Salim, N.; Holliday, J. D.; Willett, P. *J. Chem. Inf. Comput. Sci.* **2003**, *43*, 435–442.
204. Wilton, D. J.; Willett, P.; Lawson, K.; Mullier, G. *J. Chem. Inf. Comput. Sci.* **2003**, *43*, 469–474.
205. Harper, G.; Bradshaw, J.; Gittins, J. C.; Green, D. V. S.; Leach, A. *J. Chem. Inf. Comput. Sci.* **2001**, *41*, 1295–1300.
206. Hert, J.; Willett, P.; Wilton, D. J.; Acklin, P.; Azzaoui, K.; Jacoby, E.; Schuffenhauer, A. *J. Chem. Inf. Comput. Sci.* **2004**, *44*, 1177–1185.
207. Hert, J.; Willett, P.; Wilton, D. J.; Acklin, P.; Azzaoui, K.; Jacoby, E.; Schuffenhauer, A. *Org. Biomol. Chem.* **2004**, *2*, 3256–3266.
208. Downs, G. M.; Barnard, J. M. Clustering Methods and Their Uses in Computational Chemistry. In *Reviews in Computational Chemistry*; Lipkowitz, K. B., Boyd, D. B., Eds.; Wiley-VCH: Weinheim, Germany, 2002; Vol. 18, pp 1–40.
209. Böcker, A.; Derksen, S.; Schmidt, E.; Teckentrup, A.; Schneider, G. *J. Chem. Inf. Model.* **2005**, *45*, 807–815.
210. This Java-based software is freely available: http://www.modlab.de (accessed April 2006).
211. Rusinko, A., III.; Farmen, M. W.; Lambert, C. G.; Brown, P. L.; Young, S. S. *J. Chem. Inf. Comput. Sci.* **1999**, *39*, 1017–1026.
212. Quinlan, J. R. In *C4.5 Programs for Machine Learning*; Morgan Kaufmann: San Francisco, 1992.
213. Scam has subsequently been developed into a suite of products developed and distributed by Golden Helix Inc. http://www.goldenhelix.com/ (accessed April 2006).
214. Brown, R. D.; Martin, Y. C. *J. Chem. Inf. Comput. Sci.* **1996**, *36*, 572–584.
215. World Drug Index, Developed and distributed by Thomson Scientific: http://scientific.thomson.com/ (accessed April 2006).
216. Nicolaou, C. A.; Tamura, S. Y.; Kelley, B. P.; Bassett, S. I.; Nutt, R. F. *J. Chem. Inf. Comput. Sci.* **2002**, *42*, 1069–1079.
217. ClassPharmer, developed and distributed by Bioreason Inc. http://www.bioreason.com/ (accessed April 2006).
218. Classpharmer application presentation: http://cisrg.shef.ac.uk/shef2004/talks/CNicolaou.pdf (accessed April 2006).
219. Wagener, M.; Sadowski, J.; Gasteiger, J. *Angew. Chem. Int. Ed. Engl.* **1995**, *34*, 2674–2677.
220. Wild, D.; Blankley, C. J. *J. Chem. Inf. Comput. Sci.* **2000**, *40*, 155–162.
221. Pickett, S. D.; McLay, I. M.; Clark, D. E. *J. Chem. Inf. Comput. Sci.* **2000**, *40*, 263–269.
222. Sheridan, R. P.; Kearsley, S. K. *Drug Disc. Today* **2002**, 7, 903.
223. Good, A. C.; Hermsmeier, M. A.; Hindle, S. A. *J. Comput.-Aided Mol. Des.* **2004**, *18*, 529–536.
224. Kollman, P. *Chem. Rev.* **1993**, 7, 2395–2417.
225. Muegge, I.; Rarey, M. Small Molecule Docking and Scoring. In *Reviews in Computational Chemistry*; Boyd, D. B., Lipkowitz, K. B., Eds.; Wiley-VCH: New York, 2001; Vol. 17, pp 1–60.
226. Abagyan, R.; Totrov, M. *Curr. Opin. Chem. Biol.* **2001**, *5*, 372–382.
227. Langer, T.; Hoffmann, R. D. *Curr. Pharm. Des.* **2001**, *11*, 509–527.
228. Good, A. C. *Curr. Opin. Drug Disc. Dev.* **2001**, *4*, 301–307.
229. Waszkowycz, B.; Perkins, T. D. J; Sykes, R. A.; Li, J. *J. Res. Dev.* **2001**, *40*, www.research.ibm.com/journal/sj/402/ (accessed April 2006).
230. Shoichet, B. K.; McGovern, S. L.; Wei, B.; Irwin, J. J. *Curr. Opin. Chem. Biol.* **2002**, *6*, 439–446.
231. Lyne, P. D. *Drug Disc. Today* **2002**, 7, 1047–1055.
232. Taylor, R. D.; Jewsbury, P. J.; Essex, J. W. *J. Comput.-Aided Mol. Des.* **2002**, *16*, 151–166.
233. Tuma, R. *Drug. Disc. Dev.* http://www.dddmag.com/ (accessed July 2006).
234. Kitchen, D. B.; Deconrez, H.; Furr, J. R.; Bajorath, J. *Nat. Rev. Drug Disc.* **2004**, *3*, 935–949.
235. Alvarez, J. C. *Curr. Opin. Chem. Biol.* **2004**, *8*, 365–370.
236. Shoichet, B. K. *Nature* **2004**, *432*, 862–865.
237. Barril, X.; Hubbard, R. E.; Morley, S. D. *Mini Rev. Med. Chem.* **2004**, *13*, 779–791.
238. Jansen, J. M.; Martin, E. J. *Curr. Opin. Chem. Biol.* **2004**, *17*, 359–364.

239. Krovat, E. M.; Steindl, T.; Langer, T. *Curr. Comp. Aided Drug Des.* **2005**, *1*, 93–102.
240. Mohan, V.; Gibbs, A. C.; Cummings, M. D.; Jaeger, E. P.; DesJarlais, R. L. *Curr. Pharm. Des.* **2005**, *11*, 323–333.
241. Berman, H. M.; Westbrook, J.; Feng, Z.; Gilliland, G.; Baht, T. N.; Weissig, H.; Shindyalov, L.; Bourne, P. E. *Nucl. Acids Res.* **2000**, *28*, 235–242.
242. Sander, C.; Schneider, R. *Proteins* **1991**, *9*, 56–62.
243. Muegge, I. *Med. Chem. Res.* **1999**, *9*, 490–500.
244. Jones, G.; Willett, P.; Glen, R. C.; Leach, A. R. *J. Mol. Biol.* **1997**, *267*, 727–748.
245. Bron, C.; Kerbosch, I. *Commun. Assoc. Comput. Mach.* **1973**, *16*, 575–577.
246. Kuhl, F. S.; Crippen, G. M.; Friesen, D. K. *J. Comput. Chem.* **1984**, *3*, 524–534.
247. Ewing, T. J. A.; Kuntz, I. D. *J. Comput. Chem.* **1997**, *18*, 1175–1189.
248. DOCK, developed and distributed at UCSF: http://dock.compbio.ucsf.edu/ (accessed April 2006).
249. Kuntz, I. D.; Blaney, I. M.; Oatley, S. I.; Langridge, R. L. *J. Mol. Biol.* **1982**, *161*, 269–288.
250. Kuhl, F. S.; Crippen, G. M.; Friesen, D. K. *J. Comput. Chem.* **1984**, *5*, 24–34.
251. Shoichet, B. K.; Kuntz, I. D. *Prot. Eng.* **1993**, *6*, 723–732.
252. Good, A. C.; Cheney, D. L.; Sitkoff, D. F.; Tokarski, J. S.; Stouch, T. R.; Bassolino, D. A.; Krystek, S. R.; Li, Y.; Mason, J. S.; Perkins, T. D. J. *J. Mol. Graph. Model.* **2003**, *22*, 31–40.
253. Fischer, D.; Norel, R.; Wolfson, H.; Nussinov, R. *Proteins* **1993**, *16*, 278–292.
254. Fischer, D.; Lin, S. L.; Wolfson, H. L.; Nussinov, R. *J. Mol. Biol.* **1995**, *248*, 459–477.
255. Metropolis, N.; Rosembluth, A.; Rosembluth, M.; Teller, A. *J. Chem. Phys.* **1953**, *21*, 1087–1092.
256. Abagyan, R.; Totrov, M.; Kuznetsov, D. *J. Comput. Chem.* **1994**, *15*, 488–506.
257. ICM, developed and distributed by Molsoft Inc. http://www.molsoft.com/ (accessed April 2006).
258. McMartin, C.; Bohacek, R. S. *J. Comput.-Aided Mol. Des.* **1997**, *11*, 333–339.
259. QXP, part of the FLO package, developed and distributed by Thistlesoft: http://cmcma@ix.netcom.com/ (accessed April 2006).
260. Westhead, D. R.; Clark, D. L.; Murray, C. W. *J. Comput.-Aided Mol. Des.* **1997**, *11*, 209–228.
261. Goodsell, D. S.; Olson, A. L. *Proteins* **1990**, *8*, 195–202.
262. AUTODOCK, developed and distributed by Scripps University: http://www.scripps.edu/mb/ (accessed April 2006).
263. Kirkpatrik, S.; Gelatt, C. D. J.; Vecchi, M. P. *Science* **1983**, *220*, 671–680.
264. Yue, S. *Prot. Eng.* **1990**, *4*, 177–184.
265. Jones, G.; Willett, P.; Glen, R. C.; Leach, A. R. *J. Mol. Biol.* **1997**, *267*, 727–748.
266. Gold, developed and distributed by the CCDC. http://www.ccdc.cam.ac.uk/products/ (accessed April 2006).
267. Miller, M. D.; Kearsley, S. K.; Underwood, D. I.; Sheridan, R. P. *J. Comput.-Aided Mol. Des.* **1994**, *8*, 153–174.
268. Good, A. C.; Cheney, D. L.; Sitkoff, D. F.; Tokarski, J. S.; Stouch, T. R.; Bassolino, D. A.; Krystek, S. R.; Li, Y.; Mason, J. S.; Perkins, T. D. J. *J. Mol. Graph. Model.* **2003**, *22*, 31–40.
269. Lorber, D. M.; Shoichet, B. K. *Protein Sci.* **1998**, 7, 938–950.
270. Thomas, B. E.; Joseph-McCarthy, D.; Alvarez, J. C. Pharmacophore-Based Molecular Docking. In *Pharmacophore Perception, Development and Use in Drug Design*; Guner, O. F., Ed.; International University Line: La Jolla, CA, 2000, pp 351–367.
271. Friesner, R. A.; Banks, J. L.; Murphy, R. B.; Halgren, T. A.; Klicic, J. J.; Mainz, D. T.; Repasky, M. P.; Knoll, E. H.; Shelley, M.; Perry, J. K. et al. *J. Med. Chem.* **2004**, *477*, 1739–1749.
272. GLIDE, developed and distributed by Schrodinger Inc. http://www.schrodinger.com/ (accessed April 2006).
273. Halgren, T. A.; Murphy, R. B.; Friesner, R. A.; Beard, H. S.; Frye, L. L.; Pollard, W. T.; Banks, L. *J. Med. Chem.* **2004**, *47*, 1750–1759.
274. Zsoldos, Z.; Johnson, A. P.; Simon, A.; Szabo, I.; Szabo, Z. *Abstracts of Papers*, 224th American Chemical Society National Meeting, Boston, MA, Aug 18–22, 2002; CINF-005.
275. *EHITS* is developed and distributed by Simbiosys Inc. http://www.simsbiosys.ca/ (accessed April 2006).
276. Muegge, I. *Med. Chem. Res.* **1999**, *9*, 490–500.
277. Najmanovich, R.; Kuttner, I.; Sobolev, V. *Proteins* **2000**, *39*, 261–268.
278. Leach, A. R. *J. Mol. Biol.* **1994**, *235*, 345–356.
279. Wasserman, Z. R.; Hodge, C. N. *Proteins* **1996**, *24*, 227–237.
280. Knegtel, R. M. A.; Kuntz, I. D.; Oshiro, C. M. *J. Mol. Biol.* **1997**, *266*, 424–440.
281. Broughton, H. B. *J. Mol. Graph. Model* **2000**, *18*, 247–257.
282. Cavasotto, C.; Abagyan, A. *J. Mol. Biol.* **2004**, *337*, 209–225.
283. Claussen, H.; Buning, C.; Rarey, M.; Lengauer, T. *J. Mol. Biol.* **2001**, *308*, 377–395.
284. FlexE, developed and distributed by BioSolveIT. http://www.biosolveit.de/FlexE, and Tripos Inc. http://www.tripos.com/ (accessed April 2006).
285. Schnecke, V.; Swanson, C. A.; Getzoff, E. D.; Tainer, J. A.; Kuhn, L. A. *Proteins* **1998**, *33*, 74–87.
286. Barril, X.; Morley, S. D. *J. Med. Chem.* **2005**, *48*, 4432–4443.
287. *Drug-Receptor Thermodynamics: Introduction and Applications*; Raffa, R. B., Ed.; John Wiley: Chichester, UK, 2001.
288. Gohlke, H.; Klebe, G. *Angew. Chem. Int. Ed. Engl.* **2002**, *41*, 2644–2676.
289. Boehm, H.-J. Prediction of Non-Bonded Interactions in Drug Design. In *Methods and Principles in Medicinal Chemistry*; Mannhold, R.; Kubinyi, H.; Folkers, G., Eds.; Vol. 19, pp 3–20.
290. Brooijmans, N.; Kuntz, I. D. *Annu. Rev. Biophys. Biol. Struct.* **2003**, *32*, 335–373.
291. Muegge, I.; Enyedy, I. In *Computational Medicinal Chemistry for Drug Discovery*; Bultinck, P., de Winter, H., Langenaeker, W., Tollenaere, J. P., Eds.; Marcel Dekker: New York, 2004, pp 405–436.
292. Krovat, E. M.; Steindl, T.; Langer, T. *Curr. Comput. Aided Drug Des.* **2005**, *1*, 93–102.
293. Nielsen, D. L.; Aadal, P.; Hedstroem, M.; Norden, B. *Curr. Comput. Aided Drug Des.* **2005**, *1*, 275–306.
294. Muegge, I. *J. Med. Chem.* ACS.
295. Cole, J. C.; Nissink, J.; Willem, M.; Taylor, R. *Virtual Screen. Drug Disc.* **2005**, *14*, 379–415.
296. Weiner, S.; Kollman, P. A.; Nguyen, D. T.; Case, D. A. *J. Comput. Chem.* **1986**, 7, 230–239.
297. Meng, E. C.; Shoichet, B. K.; Kuntz, I. D. *J. Comput. Chem.* **1992**, *13*, 505–524.
298. Vieth, M.; Hirst, J. D.; Kolinski, A.; Brooks, C. L., III. *J. Comput. Chem.* **1998**, *14*, 1612–1622.
299. Joseph-McCarthy, D. *Pharm. Ther.* **1999**, *84*, 179–191.
300. Weber, P. C.; Pantoliano, M. W.; Simons, D. M.; Salemme, F. R. *J. Am. Chem. Soc.* **1994**, *116*, 2717–2727.

301. Honig, B.; Nicholls, A. *Science* **1995**, *268*, 1144–1149.
302. Zou, X.; Sun, Y.; Kuntz, I. D. *J. Am. Chem. Soc.* **1999**, *121*, 8033–8043.
303. Still, W. C.; Tempczyk, A.; Hawley, R. C.; Hendrickson, T. *J. Am. Chem. Soc.* **1990**, *112*, 6127–6129.
304. Kalyanaraman, C.; Bernacki, K.; Jacobson, M. P. *Biochemistry* **2005**, *44*, 2059–2071.
305. Kellogg, G. E.; Fornabaio, M.; Spyrakis, F.; Lodola, A.; Cozzini, P.; Mozzarelli, A.; Abraham, D. J. *J. Mol. Graph. Model.* **2004**, *22*, 479–486.
306. Eldridge, M. D.; Murray, C. W.; Auton, T. R.; Paolini, G. V.; Mee, R. P. *J. Comput.-Aided Mol. Des.* **1997**, *11*, 425–445.
307. Murray, C. W.; Auton, T. R.; Eldridge, M. D. *J. Comput.-Aided Mol. Des.* **1998**, *12*, 503–519.
308. Cscore, part of SYBYL, developed and distributed by Tripos Inc. http://www.tripos.com/ (accessed April 2006).
309. Wang, R.; Liu, L.; Lai, L.; Tang, Y. *J. Mol. Model.* **1998**, *4*, 379–394.
310. Head, R. D.; Smythe, M. L.; Oprea, T. I.; Waller, C. L.; Green, S. M.; Marshall, G. R. *J. Am. Chem. Soc.* **1996**, *118*, 3959–3969.
311. Jain, A. N. *J. Comput.-Aided Mol. Des.* **1996**, *10*, 427–440.
312. Krammer, A.; Kirchhoff, P. D.; Jiang, X.; Venkatachalam, C. M.; Waldman, M. *J. Mol. Graph. Model.* **2005**, *23*, 395–407.
313. Glide XP, developed and distributed by Scrodinger Inc. http://www.schrodinger.com/ (accessed April 2006).
314. Sippl, M. J. *Curr. Opin. Struct. Biol.* **1995**, *5*, 229–235.
315. Sippl, M. J. *J. Mol. Biol.* **1990**, *213*, 859–883.
316. Verkhivker, G.; Appelt, K.; Freer, S. T.; Villafranca, J. E. *Protein Eng.* **1995**, *8*, 677–691.
317. Pickett, S. D.; Sternberg, M. J. *J. Mol. Biol.* **1993**, *231*, 825–839.
318. Wallqvist, A.; Jernigan, R. L.; Covell, D. G. *Protein Sci.* **1995**, *4*, 1881–1903.
319. DeWitte, R. S.; Shakhnovich, E. I. *J. Am. Chem. Soc.* **1996**, *118*, 11733–11744.
320. Mitchell, J. B. O.; Laskowski, R. A.; Alex, A.; Thornton, J. M. *J. Comput. Chem.* **1999**, *20*, 1165–1176.
321. McDonald, I. K.; Thornton, J. M. *J. Mol. Biol.* **1994**, *238*, 777–793.
322. Mitchell, J. B. O.; Laskowski, R. A.; Alex, A.; Forster, M. J.; Thornton, J. M. *J. Comput. Chem.* **1999**, *20*, 1177–1185.
323. Muegge, I.; Martin, Y. C. *J. Med. Chem.* **1999**, *42*, 791–804.
324. Muegge, I. *J. Comput. Chem.* **2001**, *22*, 418–425.
325. Gohlke, H.; Hendlich, M.; Klebe, G. *J. Mol. Biol.* **2000**, *295*, 337–356.
326. Hendlich, M. *Acta Crystallogr. Sect. D* **1998**, *54*, 1178–1182.
327. Muryshev, A. E.; Tarasov, D. N.; Butygin, A. V.; Butygina, O. Y.; Aleksandrov, A. B.; Nikitin, S. M. *J. Comput.-Aided Mol. Des.* **2003**, *17*, 597–605.
328. Zhang, C.; Liu, S.; Zhu, Q.; Zhou, Y. *J. Med. Chem.* **2005**, *48*, 2325–2335.
329. Yang, J.-M.; Chen, Y.-F.; Shen, T.-W.; Kristal, B. S.; Hsu, D. F. *J. Chem. Info. Model.* **2005**, *45*, 1134–1146.
330. Stahl, M. *Perspect. Drug Disc. Des.* **2000**, *20*, 83–98.
331. Wang, R.; Lu, Y.; Wang, S. *J. Med. Chem.* **2003**, *46*, 2287–2303.
332. Bissantz, C.; Folkers, G.; Rognan, D. *J. Med. Chem.* **2000**, *43*, 4759–4767.
333. Stahl, M.; Rarey, M. *J. Med. Chem.* **2001**, *44*, 1035–1042.
334. Marsden, P. M.; Puvanendrampillai, D.; Mitchell, J. B. O.; Glen, R. C. *Org. Biomol. Chem.* **2004**, *2*, 3267–3273.
335. Klon, A. E.; Glick, M.; Davies, J. W. *J. Med. Chem.* **2004**, *47*, 4356–4359.
336. Xing, L.; Hodgkin, E.; Liu, Q.; Sedlock, D. *J. Comput.-Aided Mol. Des.* **2004**, *18*, 333–344.
337. Gehlhaar, D. K.; Verkhivker, G. M.; Rejto, P. A.; Sherman, C. J.; Fogel, L. J.; Freer, S. T. *Chem. Biol.* **1995**, *2*, 317–324.
338. Smith, R.; Hubbard, R. E.; Gschwend, D. A.; Leach, A. R.; Good, A. C. *J. Mol. Graph. Model.* **2003**, *22*, 41–53.
339. Davis, A. M.; Teague, S. J.; Kleywegt, G. J. *Angew. Chem. Int. Ed. Engl.* **2003**, *42*, 2718–2736.
340. Nissink, J.; Willem, M.; Murray, C.; Hartshorn, M.; Verdonk, M. L.; Cole, J. C.; Taylor, R. *Proteins: Struct. Funct. Genet.* **2002**, *49*, 457–471.
341. Hu, L.; Benson, M. L.; Smith, R. D.; Lerner, M. G.; Carlson, H. A. *Proteins: Struct. Funct. Bioinform.* **2005**, *60*, 333–340.
342. Graves, A. P.; Brenk, R.; Shoichet, B. K. *J. Med. Chem.* **2005**, *48*, 3714–3728.
343. Cole, J. C.; Murray, C. W.; Nissink, J. W. M.; Taylor, R. D.; Taylor, R. *Proteins: Struct. Funct. Bioinform.* **2005**, *60*, 325–332.
344. Yang, J. M.; Shen, T. W. *Proteins: Struct. Funct. Bioinform.* **2005**, *59*, 205–220.
345. Gruneberg, S.; Wendt, B.; Klebe, G. *Angew. Chem. Int. Ed. Engl.* **2001**, *40*, 389–393.
346. Honma, T. Structure-Based Design of Potent and Selective Cdk4 Inhibitors Protein Crystallography in Drug Discovery. In Babine, R. E., Abdel-Meguid, S. S., Eds.; John Wiley: New York, 2004; http://www3.interscience.wiley.com/cgi-bin/ (accessed April 2006).
347. Wu, J. H.; Batist, G. *Anti-Cancer Drug Des.* **2001**, *16*, 129–133.
348. Schapira, M.; Raaka, B. M.; Samuels, H. H. *BMC Struct. Biol.* **2001**, *1*, 1–13.
349. Kelly, T. M. *Biochim. Biophys. Acta* **2001**, *1545*, 67–77.
350. Perola, E.; Xu, K.; Kollmeyer, T. M. *J. Med. Chem.* **2000**, *43*, 401–408.
351. Klebe, G. *J. Med. Chem.* **2004**, *47*, 5381–5392.
352. PRO_LEADS Docking presentation. http://www.lib.uchicago.edu/cinf/ (accessed April 2006).
353. Vangrevelinghe, E.; Zimmermann, K.; Schoepfer, J.; Portmann, R.; Fabbro, D.; Furet, P. *J. Med. Chem.* **2003**, *46*, 2656–2662.
354. Doman, T. N.; McGovern, S. L.; Witherbee, B. J.; Kasten, T. P.; Kurumbail, R.; Stallings, W. C.; Connolly, D. T.; Shoichet, B. K. *J. Med. Chem.* **2002**, *45*, 2213–2221.
355. Brenk, R.; Naerum, L.; Gradler, U.; Gerber, H.-D.; Garcia, A. G.; Reuter, G. K.; Stubbs, M. T.; Klebe, G. *J. Med. Chem.* **2003**, *46*, 1133–1143.
356. Oshiro, C.; Bradley, E. K.; Eksterowicz, J.; Evensen, E.; Lamb, M. L.; Lanctot, J. K.; Putta, S.; Stanton, R.; Grootenhuis, P. D. J. *J. Med. Chem.* **2004**, *47*, 764–767.
357. Evers, A.; Klabunde, T. *J. Med. Chem.* **2005**, *48*, 1088–1097.
358. Evers, A.; Gohlke, H.; Klebe, G. *J. Mol. Biol.* **2003**, *334*, 327–345.

© 2007 Elsevier Ltd. All Rights Reserved
No part of this publication may be reproduced, stored in any retrieval system or transmitted in any form by any means electronic, electrostatic, magnetic tape, mechanical, photocopying, recording or otherwise, without permission in writing from the publishers

Comprehensive Medicinal Chemistry II
ISBN (set): 0-08-044513-6

ISBN (Volume 4) 0-08-044517-9; pp. 459–494

4.20 Screening Library Selection and High-Throughput Screening Analysis/Triage

J B Dunbar Jr, Pfizer Inc., Michigan Laboratories, Ann Arbor, MI, USA

© 2007 Published by Elsevier Ltd.

4.20.1 Introduction

Compound collections in the pharmaceutical industry have been growing by leaps and bounds over the past few years. The rapid growth in these collections has been fueled by a number of sources including the mergers of companies each with a large compound collection, aggressive compound acquisition programs, and the widespread use of combinatorial chemistry. These collections are now in the millions of compounds. Screening these compounds in drug discovery efforts is a major expense involving a lot of expensive factors including time, reagent cost, personnel, equipment, and storage and handling facilities. Even with miniaturization, screening millions of compounds can be time-consuming and potentially wasteful of compound resources. Some assays are more amenable to large-scale screening than others, which may have significant time or reagent cost constraints. Other methods are available as alternatives to single compound per well full file screening and may be more appropriate depending on the situation. One method would be to do multiple compounds per well using all of the compounds in the collection. Another of these alternatives would be to use a subset of the full file in either a single or multiple compound per well format.

Subsetting compounds is the task of selecting a smaller portion of compounds from a much larger list with the hope of providing increased efficiency in some form in a screening effort. Subsets can be found in many different forms for a number of different uses. They can be transient and very specific in nature, such as those constructed for use with a particular assay and then discarded when there is no further need. Or they can also be of more long-term service and general purpose in nature, such as those constructed as a general surrogate for a much larger set and screened in many different types of assays. This is best illustrated by a subset screened in lieu of a full compound file mass screen. These two concepts of subsets tend to define the boundaries of subset usage, although there are also many that are intermediate in nature, such as subsets of compounds known to be active in a given gene family. Taken to an extreme form, subsetting can be thought of as finding a series of compounds that are specific for a given biological target, forming the basis for finding a new pharmaceutical agent – a standard medicinal chemistry project. The techniques and approaches for constructing these compound subsets will have a lot in common, but also will have significant differences based on the task the subset is to be used in. These subsets are tools, and appropriate subsets are needed for particular tasks. By analogy, while one can drive a nail with a pipe wrench, a much better job can be done with a hammer. This chapter will discuss the ways in which subsets have been or could be generated focusing on what the subsets will be used for and how they are constructed to do that job.

Subsets, by definition, do not have all of the available chemical matter. One aspect of subset selection is finding methods to assess what is found and what is missed relative to a full file screen. This will vary with the assay, but providing some estimates of expected performance can also be valuable in deciding what projects to do with a full file screen and those that can be addressed only with a subset screen. This is directly analogous to how stocks perform on the stock market. While past performance of a stock is not a guarantee of future performance, it will help provide some context for managing expectations. Another important aspect of subset screening is how to follow up the hits. In a full file screen, one theoretically has all of the compound activity information, but in a subset screen, only a portion of the information is available. What can be done after the initial screen to find additional compound activity information based on the initial hits in an efficient manner will also be discussed.

To a first approximation, subsets will probably perform much like a full file screen. Some subset screens will result in a lot of good chemical matter, but other screens will not perform as well. This opens the door for the subset, and its constructors, to be both lauded and cursed within your organization. Therefore in situations such as this, where the performance will vary widely, it is imperative to appropriately manage expectations. The larger the organization, the more difficult and time-consuming those crucial communications will be. One cannot just create a subset and toss it over the wall to the rest of the organization. What it is, and when and how it is to be used must be explained in the context of the business rules that your corporate organization follows. A carefully crafted, well-documented, and permanently available roll-out of the product should be done.

Two recent papers one by Davis *et al.*[1] in 2005 and another by Harper *et al.*[2] in 2004 both cover some of the general strategies in the design of subsets for use in drug discovery operations. These two papers provide a brief overview of what their respective organizations considered, at that time, important in the process and some of the characteristics that were desired in the subsets for screening. The subject matter of these two publications and the current date illustrate the fact that there are still no hard and fast rules for constructing subsets and that the design and use of subsets along with iterative or sequential screening is still an area of active research.

4.20.2 Initial Considerations for Subset Composition and Use

There are any number of reasons to contemplate using a subset in lieu of a full file mass screen. The assay may not be amenable to the large scale of a full screen (up to millions of compounds). The assay might require costly and/or difficult-to-obtain reagents or biologicals. The assay may be of speculative nature regarding potential use in a therapeutic area and the team just wants to obtain some tools for further exploring the target in terms of confidence in rational or confidence in mechanism. Budgetary constraints are always a factor and a subset will most likely be cheaper to run than a full file mass screen. As such, it can provide alternatives to the therapeutic areas regarding what targets would warrant a full file screen and those which could be done with a subset, allowing more efficient use of screening dollars.

Figure 1 provides a simplified view of the process of selecting a subset and using a subset. This selection process starts with the larger, for example full corporate, compound set. This set is then filtered to remove compounds that are unwanted. There are lots of types of unwanted chemical matter and removing these early will improve the efficiency of the overall process. A lot of the ensuing discussion will be on the details of what and how compound collections are filtered. Once the set has been filtered, the subset is selected and the general characteristics are compared to those in the filtered set to be sure the subsets are generally representative of the original filtered collection. This is for selecting a subset that is supposed to be representative, if this is not the intended use, this will be dropped or modified. This will also be discussed in more detail later in the chapter. After the selection of the subset, the process is documented and the subset performance is monitored for use in the construction of the next generation. This is an important step, even if the particular subset does have a limited lifetime and use, there is a high probability of performing another subset selection in the future. Appropriate performance monitoring allows one to feed forward relevant information for use in future subsets.

As with most projects, there is a substantial amount of preparatory work that needs to be done early on in the subsetting workflow in order to have an appropriate compound set for a starting point that does not include unwanted chemical matter that complicates the analysis of screening results. A number of the tasks associated with generating the starting point set would not at first glance seem relevant, or only peripherally so. Why all the additional work? Just grab a random set of plates and go! The goal of screening a subset of compounds is exactly the same as a full file screen: to have an efficient process for obtaining suitable chemical matter to enable the team to progress in its mission. In a full file screen, one has the luxury of being able to say: We have screened it all and here is your answer. What is the deliverable in a subset screen? With a subset screen, by definition one has screened only a subset and there will always be the question of: Just what did we get relative to the full file? This is a particularly important point in that there is no way, a priori, to determine which compound in a series will be the most active for any given target. What is meant by a series is a collection of structurally very similar compounds that have the same biological action with differing potency which provides structure–activity relationships (SARs). Coming back to subset screening, a case can be made, that the compound found as a hit from the subset screening is probably not the most active compound available in the series. How can we ensure that the subset screen will not miss an entire series of compounds? While it may not be possible to find all series of compounds, what can be done to maximize our chances of finding as many series as the subset constraints will allow?

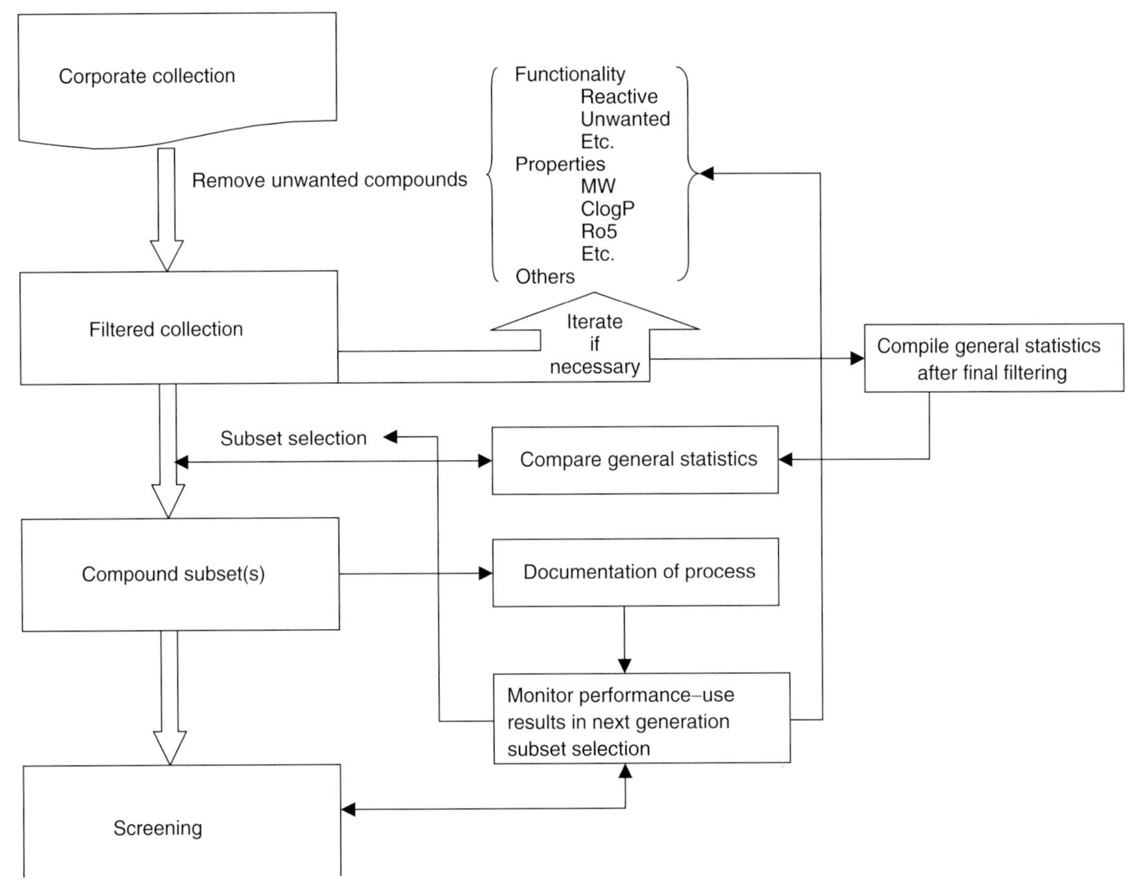

Figure 1 Compound subset selection.

The most efficient method of subsetting would be to subset those compounds in the collection that are directly available for testing. By selecting compounds that are readily available, one can shorten the turnaround time for the entire screening and hit follow-up process. Here the most likely source of chemical matter is the solvated store that is available from within your organization. These available compounds can usually be further subdivided into liquid-only samples, exemplified by compounds obtained from combinatorial synthetic methods and those compounds for which there is some dry powder available in addition to solvated sample. These compounds with dry powder available have usually been synthesized in traditional medicinal chemistry efforts. It may be of benefit to select compounds from both of these two subdivisions, but the selection methods may be different for each. With compounds synthesized in a combinatorial fashion the advantage is in the existing synthetic protocol and ease of repeat synthesis if needed. Pharmaceutical organizations, both large and small, have recognized this and as such a large portion of any individual organizations compound collection may consist of combinatorially synthesized compounds in liquid-only (solvated) sample.

But it may also be very important to have compounds for which there is dry powder sample available. For example, some secondary biological testing may require solid sample to be used to create the test samples in different solvent systems to get confirmatory results. Other secondary biological tests may require more sample than is available from the combinatorial methods. If the solvated sample stores contain dimethyl sulfoxide (DMSO), and this interferes with the primary screen, one could create another subset collection with the same chemical matter, but with another solvent system. By having dry powder available, you also provide the opportunity to go back and check with traditional analytical techniques the identity, quality, and purity of the sample if any questions arise from the screening assays regarding the identity of the compound. Fresh samples from dry powder can also be used when necessary to confirm the hits in the assay. Another related purity or concentration issue is the number of freeze–thaw cycles the solvated store has undergone. If one can obtain that information, it could be used in the selection process of the starting compound collection. By selecting on compounds that have undergone some minimal number of freeze–thaw cycles, one can attempt to improve the quality of the compounds in the set.

Combinatorially synthesized sets of compounds from a single synthetic protocol can be thought of as a single series, with a smaller sample size and much larger numbers of compounds than those typically synthesized via traditional medicinal chemistry. This concept of combinatorial synthetic compounds as a series can be used in the subsetting process to help ensure adequate coverage of synthetic space in the collection, by having compounds from each protocol included in the subset. The concept of a series can also be used to help assess compound coverage in a subset. While a subset will not have all the hits that could be in a full file screen, one can asses the quality of the subset coverage by the series hit rate instead of the compound hit rate. The question would be: Does one find all or a major number of the series available based on a subset screen relative to the full file screen? Compounds from traditional therapeutic area projects also have the concept of series, but here the definition is far less straightforward. In the more mature projects, this can be even more difficult as teams tend to subdivide series even further, defining even more distinctions between series with large common components and core structures.

Compounds do not necessarily have to be from an in-house collection. For some organizations, the purchase of compounds from an external company is also a viable source of chemical matter. There are a number of suppliers of compounds that have arisen in the last few years that market compounds for exactly this purpose. Some of the considerations that arise are the suppliers ability to replenish depleted material, plate the compounds to specifications provided by the purchaser, supply predesigned sets of compounds, and in some cases provide analytical and spectroscopic data on the compound collection. These can be attractive features for organizations with limited resources or an interest in developing chemical probes and tools. An example of a subset used specifically as a chemical tool would be the construction of a small molecular weight compound collection to use in spectroscopic studies of a biological target – fragment-based nuclear magnetic resonance (NMR) screening.[3–5] These attractive features need to be balanced by the organization's need for proprietary chemical matter. Some of the suppliers will work with the organization to help design and provide exclusive chemical matter, but as exclusivity increases, usually the cost increases also. Suppliers, depending on the nature of the collections they have for sale, can often supply dry powder.

Often chemical series revolve around a mixture of core structure and subtle pendant R-groups differences which are impossible to codify for automatic, machine-based perception. There are some methods in use to automatically define series, but most of the time, these series definitions will not correspond to those decided upon by the members of a project team, and it can be hard to arrive at consensus. One method is to employ Murcko fragment analysis to define common cores used in medicinal chemistry. These cores were defined based on an analysis of the MDDR[34] compound database, but the concepts could also be used to obtain any additional cores in the corporate compound database and then derive a definition of series based on the largest common core fragment. Pipeline Pilot[36] has a script that can be used for Murcko fragment analysis.

Another method to analyze for fragments is to use the Daylight[35]-supplied ringsmi program to search for rings in the corporate database, then use these separately, or in conjunction with the Murcko[6,7] fragments, to arrive at an automated series definition. Barnard Chemical Inc.[31] also has a means of deriving generalized fingerprints representing fragments that could be adapted to automatically derive a series definition. Software from Pannanugget,[39] using the concept of molecular equivalents, might also be used to define what they refer to as 'license plates' which can be utilized to arrive at a series definition process. Clustering[8] could also serve as a mechanism for defining a series in a automated and nonsubjective manner. If the number of the initial set of compounds is very large then one would be limited to those clustering methods, such as Jarvis–Patrick,[8,9] which are amenable to large collections on the order of several hundred thousand to several million compounds. The splitting into the two types of sets in a large corporate collection (liquid-only and dry powder available) allows one to use different methods to classify compounds into series, and can reduce the number of compounds that need to be analyzed by any single method. The methods mentioned above all rely on the use of some structural characteristics to provide the descriptors used in the series classification. But another radically different method would some form of retrosynthetic processing, to determine the series definition in a synthetic protocol method, much like one that is available for combinatorially synthesized compounds.

The large compound collections of pharmaceutical organizations contain compounds that are starting materials for synthesis, intermediates in a synthesis, finished compounds, and compounds from various acquisitions. These acquired compounds may be purchased from another supplier or acquired through a merger with another company. Even the finished compounds usually reflect the state of knowledge at the time of synthesis and as such have different historical prospectives embedded in their design. Compounds synthesized several years ago did not have the benefit of the absorption, distribution, metabolism and excretion (ADME) and safety knowledge we have today. In this regard, there has been a lot of effort and scholarship around what constitutes an ethical pharmaceutical agent, i.e., a drug. This includes the 'Rule of 5'[10,11] and other analyses of those characteristics present in known drugs. The initial 'Rule of 5' study has led other groups to perform analyses of the physical and/or calculated properties of collections of drugs, contributing to a host of publications on the distributions of the characteristics of successful pharmaceutical agents.

Examples of this work directly applied to the creation of subsets is the paper by Viswanadhan[12] and the review by Egan[13] in 2002. The studies of Vieth[14] and the review by Leeson[15] in 2004 are representative of the ongoing research of the physical properties of drugs. Vieth's work on physical properties associated with 1729 marketed drugs was even categorized by administration route (oral, injectable, topical, and absorbent).

There has been an increasing amount of literature discussing reactivity, metabolic toxicity, drug–drug interactions, solubility, cell permeability, stability, and tissue or organ selectivity that can be brought to bear in determining which structural fragments can cause problems. Compounds containing these fragments can then be removed from the screening stream by use of a series of filters. These filters are constructed based on the unwanted structural fragments. Starting in about the mid-1990s organizations began to routinely use automated structural filters to identify unwanted chemical matter and remove it; either physically from the screening stream so it is not present in the screening plates, or electronically after a screen has been run during the very early stages of the triage/analysis step. For example, Hann *et al.*[16] published a list of SMARTs[36] strings used for removing unwanted compounds from the high-throughput screening (HTS) stream. These are found in the supporting information pages associated with the publication. Modelers[17] have worked and continue to work extensively with senior medicinal chemists to define structural filters for use in screening, triage, and compound acquisition. Constructing the filters usually falls into three stages. The first stage is to write filters to remove the obvious reactive material. The second stage is then to write filters to remove the nonobvious, by its very nature, not an easy task due to the vastness of chemistry space. Therefore, for practical reasons, the filter construction process then focuses down to removing chemical matter you have seen and do not want. This process is highly dependent on the chemical content of the databases and the accessible information on that content. Structural filters can be considered to be living documents that need to modified as information on the compounds changes or increases (i.e., new ADME data), or new compounds are added to the collection. The third stage, is then, determining when to apply the various filters during the drug discovery cycle. What filters to apply when may change with time based on increasing information and with current cultural and risk–benefit elements in play at the organization. For subsetting purposes, this may also depend on the perceived use of the subset. For example, if this is to be a limited use subset that is representative of the full screening file, and the filtering of the full file was not stringent, a less stringent set of filters might be applied. One example of this would be a validation subset. On the other end of the spectrum, the subset could be the only screening opportunity for a project, and therefore should only provide hits that could be pursued and thus more stringently filtered. Work on filters, both property-based and structural, is a continuous process and a recent review by Davis[1] on successful lead generation lists some of AstraZeneca's chemistry filter definitions employed to remove unwanted chemical matter. Chemical matter with ADME or toxicological implicated groups needs to be highlighted in the triage process but not necessarily removed from the initial screen. If the groups are not involved in a covalent bond to the target, and if the unwanted group is not actively interfering with the assay results causing a false positive, these groups may be synthetically replaced with acceptable chemical functionality or even removed entirely, without adverse effect on potency. In the normal course of following up with additional synthesis in a standard medicinal chemistry project, the chemist would then know which groups to engineer out early on in the optimization process, potentially turning a compound unacceptable as a clinical candidate into one that is. The chemical structure filter then would serve the purpose of highlighting chemical functionality that needed to be replaced, augmenting the risk–benefit analysis of the screening hits.

Most organizations have sets of structural filters and/or structural alerts that are applied to compounds to determine suitability for various uses. In the early screening stages, one may or may not chose to be very conservative in the application of filtering techniques. Some, probably most, corporate cultures previously allowed for a very lenient set of initial screening filters for full file screening. As everything was being screened, one could afford to be inclusive. Once chemical matter was obtained, then one progressively applied more stringent criteria as the project approached the declaration of a compound as a lead clinical candidate. This may or may not be applicable to subset creation. Here one may very well want more stringent structural filtering to maximize the return on subset screening. With a limited number of compounds in the subset, the ability to maximize the chances of getting a productive hit and not waste a single well with unwanted material would be the primary consideration. As the size of the subset gets smaller and the number of uses for the subset increases, the need to maximize every well will probably also increase. Screening unwanted chemical matter takes valuable resources that could be devoted to useful compounds, and each one has to count as a potential shot on goal. But agreeing on what structural features to remove is a very difficult task and is very subject to personal preferences, experiences, and prejudices. The larger the organization, the greater and more complex the task of agreeing on these filters becomes. A paper by Lajiness *et al.*[18] provides informative reading on the perception of attractiveness of chemical matter for lead discovery use and the vast differences in opinion of that chemical matter. If your organization does not already have a set of globally-agreed-upon filters, this is a task that should be taken on very early and with a wide net cast for interested parties. This discussion should center on just what filtering should be applied to the subset to be constructed and the nature of the use of the subset must be determined beforehand. Getting consensus on these points can be a long and arduous task, but very well worth it to remove the possibility of tainting the subset before its initial use and to provide the organization the maximum chances of success with the current set of business rules.

Along with various structural features that can be filtered out, other characteristics can be used to limit compound access to the screening stream. Things such as ClogP,[38] molecular weight, number of rotatable bonds, number of ionizable groups, and other calculated or tabulated characteristics can be used to remove undesirable compounds. Suitable ranges for these properties would also need to be discussed and agreed upon prior to the creation of the subset. A compound could also be deemed undesirable for features other than those mentioned above. For example, a company might not want to work with peptides, sugars, or natural products for valid business reasons specific to that company. A decision needs to made regarding what to remove from consideration as a member of the initial set of compounds available for subsetting. If this subset is to be used over a number of years, or will be used by a number of different assays, and will involve major effort in the actual construction and distribution of the subset, then the issues involving the initial compound set and subset criteria are best made by all the affected parties arriving at consensus.

As part of the initial selection of the subset, some thought should be given as to how the screening information will be followed up. After the screen, will the hits from the screen be independently confirmed or will the screen be run in duplicate or triplicate and, based on those results, not go to a confirmation step? After the confirmation of the hits, will these hits be used as is, without additional mass screening efforts, or will these hits serve as seeds to be used in subsequent mass screening, with analysis, compound orders, and similarity testing cycles to look for additional hits? Thinking about these issues can help in the choices made as to the number of compounds to be included in the subset and also the process used to make the selection. For example, Abt *et al.*[21] have suggested that about 10% of compounds in a set of 70 000 would be enough to use as an initial subset for iterative screening, but how well this scales with increasing compound collection size is unknown at this point. That particular study only involved a set of about 70 000 compounds and the study should be repeated with the much larger screening collections available today. For example, 10% of 1000 000 compounds may be acceptable for a subset, but 10% of 10 000 000 might be far more than your organization could utilize as a subset.

Other pragmatic issues such as the capacity of current and future screening laboratorie can impact the total number of compounds selected. Are there other natural classes or sets in the compound collection that can be used other than the ones mentioned above (liquid-only or dry powder) that can be used? In thinking about how many compounds are in the total subset, given the existence of different classes, how many compounds per class should be selected in order to get the largest subset the market will bear? What weights or value does the organization give to the various classes of compounds in the corporate collection? Remembering back to the earlier points on combinatorial compounds and compounds with existing dry powder, how does your organization view the relative value of (1) rapid resupply with existing protocol-based, newly synthesized compounds versus (2) traditionally synthesized compounds available as dry powder? How easily can your organization determine if older compounds synthesized by traditional medicinal chemistry means are amenable to combinatorial synthesis? These types of value-based judgements need to be made by the organization and used in the selection of the compounds for the subset. It would be useful to check with the business units that handle the storage and distribution of compounds, along with the screening units to see if any increases in capacity or automation are forthcoming. Communication with these two groups may reveal potential downtime which could delay the construction and distribution of the subset, or indicate increased capacity, allowing for a larger subset.

Other issues to be examined early on in the subsetting process would include how much sample would is needed in the mother plate and how often the mother plates are subjected to a freeze–thaw cycle. What type of plates would be used for the sample collection and how many wells would be in each plate? Does there need to be one type of plates, multiple types of plates, or one type of plate with the capability to be converted via robot to other types? By type of plate, I refer to both the material used to construct the plates, and the number of wells contained within the plate. How would the plates be laid out? Would there be an advantage to place compounds in a specific order based on physical characteristics or calculated properties such as ClogP,[38] molecular weight order, or chemical characteristics such as acid or base? Would there be room on each daughter plate for control wells and other blank wells for use by the screening team? What concentration(s) would the samples have? If some sort of rudimentary dose response would be desired, this would be a reasonable place to think about how one would accomplish it. This impacts what concentrations would need to be used to construct the master plates as it is easier to dilute down from a more concentrated solution. Solubility may be an issue and could be addressed in a number of ways, ranging from a solubility predictor, to removing marginally soluble compounds, to various mixing schemes in the creation of the solvated sample.

Would there be a need for some form of quality assessment of the chemical matter going into the collection? This quality assessment might impact the makeup of the set of initial compounds for the subset, particularly in the area of sufficient sample for the tests deemed necessary. Is there enough material to test in the quality control (QC) studies and still have sufficient sample to use for the subset? Extensive QC studies can be resource, and time-intensive, consuming a lot of material and consequently impacting on the size of the subset and on the completion date.

Another major decision will be to determine if the total subset is to be comprised of smaller subsets based on some classification scheme. Whatever the case, detailed presentations (product roll-out) on the nature, expectations, and characteristics should be made to the interested parties. Will the roll-out of the final subset collection be phased or staged based on the construction of the smaller subsets? If yes, what would be the order of the smaller subsets, which one to do first, second, etc.? A divide and conquer strategy can be used to advantage, given appropriate time frames and expectations.

Would the plates be in a compressed format or in a singleton format? Compressed format plates would have multiple compounds in each well with a mechanism for deconvoluting the resulting hits to the individual compounds. If a compressed format is to be used how many compounds per well would there be? What mechanism would be used to minimize problems from potential intrawell interactions? With the increased prevalence of combinatorial compounds, compound collections tend to have series of compounds with very minor changes to the structure. Because of the existence of these compound series, issues associated with compounds of very similar structures being in the same well can arise. These compounds may both be active in the original well and also in the deconvolution wells and, as such, one has no additional information to say which one or both may be responsible for the activity. Techniques to spread out the similar compounds among wells to reduce this problem might need to be employed at the plating stage. Another problem that could arise is that compounds might be marginally soluble in the solvent system and multiple compounds in a single well might cause precipitation of one or more of the compounds in the mixture. Is more stringent solubility criteria needed for multiple-compound wells?

As noted in this section, there are many considerations that should be taken into account while formulating a subsetting strategy. These include determining how the subset will be used, how many compounds should be in the subset, and whether there are any natural classifications, e.g., a series definition, that can be use to help organize the data. Is the subset format to be single or multiple compounds per well? Measured or calculated properties of the compounds can be used to remove compounds that are not suitable or as an aid in constructing multiple-compound-per-well plates. Structural filters can be used to remove unwanted compounds. But the process of obtaining consensus within your organization on the appropriate property and structure-based filtering for the subset is arguably the most important.

4.20.3 Methods for Subset Selection

Once all the upfront work on the subsetting initiative has been accomplished, the next phase of the process is to actually do the subset selection. There are a number of ways to actually select compounds from a larger collection. A paper has been published by Hamprecht *et al.*[19] on selection algorithms. The authors have examined the general theoretical design methods and have cast a number of the methods into a single conceptual framework for sampling compounds. This paper provides a frame of reference for a number of the sampling methods, and integrates what at first glance would appear to be disparate methods for compound selection. This framework allows a very high level view of some of the methods and how these methods compare and contrast. The authors also comment on representative versus diverse selections and random versus rational selections, in all, a very interesting introduction to subset selection methods.

The reason for broad range of selection mechanisms is due to the clumpy, unevenly distributed nature of compound collections. If the compounds were all evenly related to one another and evenly represented chemical space, the selection process would merely involve a uniform selection of the number of desired compounds for the subset. Compound collections typically have clumps of related compounds; some of the clumps are large and some are small. Compound collections also have a lot of singletons, compounds not closely related to any other compound. Never do these collections evenly distribute in chemical space and cover only small amounts of the theoretically available chemical space. Couple nonuniform distributions with the myriad definitions of what could constitute chemical space, there have been a lot of ways devised to organize compounds and subsequently select within that organization. This portion of the chapter will discuss some of the methods used to define chemical space, organize compounds within that space, and select compounds.

Some of the earliest means of selecting compounds for subsetting were random selections. In the absence of any other information relating to the target, such as a mass screen, this is still a viable form of selecting compounds.

Random selection should on average not have a bias toward any given types of compounds and should sample from series within the larger collection in a consistent manner. Some methods may bias toward outlier compounds in the collection because they tend to make selections by maximizing the differences between compounds. This defeats the purpose of the representative sample selection. A mechanism probably should also be provided for including singletons in the clustering type of sampling methods. Singletons are organizational elements consisting of one compound, i.e., a cluster of one compound. If one used another method to select from the larger collection, in general, one would expect singletons. By lumping all singletons into a single collection, a random selection method can then be used to sample this singleton collection, helping to increase the diversity of the samples chosen. Random selections make updating, augmenting, or reselecting the subset a simple process. One randomly selects the necessary number of compounds for the update and moves on. Random selection can also be part of a more directed sampling method. Here the compounds would be organized in some fashion and random selection within the collections would take place. An example of this type of random selection can be found in a paper by Reynolds *et al.*,[20] where the algorithm (stochastic cluster analysis) selected compounds randomly if they fell outside a predetermined similarity or distance range. The authors found that this method was superior to straight random selection and was applicable to very large data sets where other more complicated and computationally demanding methods were unable to accomplish the selections in a reasonable amount of time. Also included in this paper is a simulated annealing method for selecting those compounds which optimized a diversity metric and is included in the comparisons of methods. Some thought should be given to what random sampling method to use. Examples would include random sampling both with and without replacement of duplicates, stratified random sampling with sample size based on the size of the bins the collection was partitioned into, and systematic sampling where one selects every *n*th compound. If using the stratified random sampling method, some thought needs to go into what characteristics would be used to generate the partitions, as these characteristics could be chemical in nature (molecular weight, ionic, basic, neutral, etc.), calculated properties (ClogP,[38] molar refractivity, solubility, etc.), structural based, or combinations thereof.

Intuitively, random selection would be expected to give fairly representative subsets when the percentage of compounds selected for the subset is relatively large compared to the number of total compounds in the initial set. For example, if one were selecting one in every 10 compounds for the subset, one would expect to obtain adequate coverage of the chemical space defined by the larger set. If on the other hand, one were selecting one in every 1000 compounds, one would worry about not having adequate coverage, and would want to have a more active role in ensuring that every compound selected for the subset would give the maximum information possible. Abt *et al.*[21] have performed a study that suggest that where a relatively small set of initial compounds are selected from the initial screening set, random or systematic methods worked equally well. Abt *et al.*[21] used initial starting points of 5000 and 10 000 compounds out the ~70 000 compounds used in the study. A study by Spencer[22] also provides evidence that a purely random selection of compounds for subsets is a viable alternative to systematic selection. However, most other literature provides evidence that rational selection mechanisms are more effective than a purely random selection. The term 'effective,' as in *x* is more effective than *y*, can mean many different things. In the course of this chapter, when a literature reference makes that type of statement, it will be highlighted, with the details of how and why it is considered more effective.

One method that allows for a more active role in attempts to ensure maximum information content within the limited set of compounds in a subset is to use some type of clustering to limit near neighbors in structure. One basic tenet in medicinal chemistry is that similar compounds tend to have similar biological action; if this were not the case then SARs would not exist. Clustering methods have long been used to generate subsets of compounds from larger collections.[8] These methods typically rely on generating fingerprints, i.e., bit-string representations, based on two-dimensional structure. Fingerprints are a method of defining chemical space. These fingerprints fall into three general categories. One category is based on the presence and number of predefined structural features in a given molecule, exemplified by the fingerprints used in the maximum auto cross correlation (MACCS) keys from MDL.[35] If a structural feature exists in the compound of interest but is not included in the fingerprint definition, that structural feature will be ignored by the clustering software because it will be undefined. A second type of fingerprint is generated based on a graph-based analysis of the atoms and connectivity and as such represents all of the structural features in the molecule. This second type of fingerprint would be exemplified by those fingerprints used in the Daylight[36] software package and will describe any molecule completely. The third type of fingerprint is a hybrid of the two previously mentioned fingerprints and uses a graph-based analysis of the atoms and connectivity to construct part of the fingerprint but also includes the presence and counts of various predefined structural features. This third type of fingerprint would be exemplified by those used in the UNITY database package from Tripos Associates.[31] Other fingerprints, such as those comprised of pharmacophore elements as exemplified by the MOE package from Chemical Computing Group[34] and the BCI fingerprints from Barnard Chemical,[32] are also commonly used and fall within the three general categories described above. These fingerprints are used to determine the similarity of compounds to each other and to group similar compounds. There are a number of mathematical ways to calculate the similarity between two compounds. Cosine[8,40] and Tanimoto[23,40] are just two ways of defining a similarity coefficient, but, to a first approximation, all the methods are equivalent: the larger the similarity coefficient are the more similar the two molecules are. SciTegic[37] also has two fingerprints that are in common use. A number of fingerprints, including most of the major ones listed above, have been systematically compared in a recent paper by Hert *et al.*[45] and SciTegic's ECFP and FCFP fingerprints were found to be marginally better than the rest of the fingerprints used in this particular study of similarity based virtual screening.

Along with the choice of descriptor(s) to use in the clustering process, there are two basic clustering methodologies that can be applied to subset selection: hierarchical or nonhierarchical clustering. In a hierarchical method, one generates the complete family tree, so to speak, for the compounds in the collection. At the base of the organizational tree is the single cluster with all compounds contained in it, and at its leaves is a collection of clusters of singletons, each compound being its own cluster. In the devise method, one starts at the base and works toward the leaves. While the agglomerative method one starts with the leaves and works toward the base. Having these relationships between compounds can be useful in determining what compounds to test next, after the initial screening results for the subset are known. There are however, a few drawbacks to using this type of clustering. One drawback of hierarchical clustering is that it is not, in general, amenable to extremely large collections, those on the order of millions of compounds. Research is being done in this area and the size of compound collections able to be clustered via hierarchical methods is increasing.[32] Another drawback is that, with hierarchical clustering, it can be difficult to find the natural clusters in the collection. Often very pragmatic elements, such as the maximum practical size of the subset, will drive the level of clustering chosen. As the number of compounds in the subset will not likely coincide with the number of clusters generated, the cluster level is chosen that is just beyond what the desired subset size is, and a means of paring it down to the required number is applied. Once again there is research in this area and there are methods that can be used to help define the natural breaks in the hierarchy. One could also use hierarchical clustering as a means to stratify the compound collection and then use a random method to select within the strata using a proportional sampling strategy based on the population of each strata. One means of increasing the speed of hierarchical clustering is to use the reciprocal nearest neighbor method, which is available in a number of software packages. Here the speed increase is obtained by using only a single nearest neighbor and other compromises. This will potentially impact the ordering of the clusters within the hierarchy, producing differences from a traditional complete hierarchical clustering method.

The other general clustering method is nonhierarchical clustering, which is also typically based on a fingerprint constructed from the two-dimensional structure of the compounds. One example of this method would be Jarvis–Patrick[9] clustering. Here one builds ordered lists of near neighbors based on similarity. If two compounds have a given number of the same near neighbors in common, then they are considered to be in the same cluster, another sort of guilt-by-association similarity-based clustering mechanism. This method of clustering also has it drawbacks. One is the possibility of the single linkage problem, where compound A looks like compound B, which looks like compound C, which looks like compound D but compound D looks nothing like compound A. These types of long, stringlike clusters often do not produce groups of compounds that are meaningful to chemists trying to organize them. Nonhierarchical clustering also has difficulty in that one needs, once again, to define how many clusters one wants or what occupancy one wants and the clustering method then clusters to arrive at those predefined characteristics, which may not coincide with the natural groupings. One crucial advantage associated with Jarvis–Patrick clustering is that it is amenable to very large numbers of compounds, on the order of tens to hundreds of millions of compounds. An example of the utility of Jarvis–Patrick clustering is given in a paper by Menard *et al.*,[24] who using cascaded clustering to select a screening subset.

Once one has generated the clusters, where in the cluster should you sample from, and what strategy should be employed? Should one select the centroid, or an outlier, or randomly within the cluster? Typically one finds the centroid is usually highly decorated (complex) and the more structurally representative member of the cluster, while the very simplest structure in the cluster (scaffoldlike) will be located somewhere on the edge. The more complex the molecule, the more chances it has to pick up an additional interaction with the target and give a hit. However this needs to be balanced by the fact that the additional features might also cause a bad interaction and just miss being a hit.

There is also a correlation of molecular weight and binding. As one adds molecular weight, particularly aromatic or hydrocarbons, there is an associated increase in binding. One can increase the chances of getting a hit by choosing the largest representative of a cluster. This is counterbalanced by an associated concept, that of ligand efficiency.[41] Here by selecting the smallest, least molecular weight, representative(s) from the cluster, any compound hitting in the assay would be the most efficient of the cluster and a better candidate for follow-up by a medicinal chemistry team. There also may be business rules that need to be followed regarding the choices made within a cluster, based on the concepts above, and additional ones particular to the individual project needs.

Once the subset has been selected and/or constructed using a clustering method, updating or modifying the subset becomes more problematic than for those subsets constructed via a random selection method. This problem may arise much quicker than expected. There will probably be some point in the definition of the original collection to be used in generating the subset, where one has to freeze the date for inclusion into the input set. In the time between the freeze date and the actual selection of the subset, there will be some attrition of sample and you may then have to select substitutes for the subset members for which there is now insufficient sample, or too many freeze–thaw cycles, or any number of other criteria that render compounds unavailable for construction of the actual screening plates. If one can then cast the new compounds into the original clusters, the original mechanisms used to select within clusters will work. This is a very likely scenario for the substitution problem, but at some point in time there will be sufficient change in the original collection such that a complete reclustering or reanalysis will need to take place. This could lead to a long-term maintenance problem for this type of selection and update process.

Another means of selecting a diverse set of compounds from a larger collection that avoids clustering is to use the maximally diverse selection method. This method typically uses structural fingerprints to measure the similarity of molecules and make selections. A compound is selected at random from the large initial collection. This molecule is compared to all the others in the collection and the compound that has the least similarity to that compound is then chosen. These two compounds are then compared to the large collection and the compound that has the least similarity to the first two is selected. This process goes on until you reach the specified cutoff in similarity or the desired number of compounds has been obtained. One problem with this process is that it tends to pick from only the outliers in the collection and is therefore not as representative as one would hope. The subset chosen by this method will get more representative as the number of compounds in the subset increases relative to the original, input collection. A very viable modification to this process is given by the optimizable *K*-dissimilarity selection (Optisim[25] method available from Tripos[31]). By varying the size of the subsamples chosen, one can adjust the balance between diversity and representation in the subset. By default, Optisim chooses more representative subsets that do not tend to contain only outliers and singletons. By increasing the size of the subsamples, the diversity of the subset chosen will increase, and conversely by decreasing the subsample size the more representative the subset will be, i.e., choosing more compounds from the larger clusters. Optisim is not limited to using only fingerprints and can use other descriptors in the selection process such as molecular weight and ClogP[38] with varying weights applied to the individual descriptors. Ashton *et al.*[26] have published a study comparing MaxMin, a dissimilarity-based selection and *K*-means cluster-based similarity. This study concluded that MaxMin was a better method and that property-based descriptors were somewhat better than fragment-based descriptors. Once again, in this study, the authors found that both approaches were better than a purely random selection.

Another method that uses as an integral part of the descriptor process, a range of characteristics, is Diverse Solutions (DVS)[27,28] from Tripos.[31] DVS[27,28] is primarily a cell-based, low-dimensional method with the capability to also work in high-dimensional space (fingerprint space). The basic unit associated with DVS is the BCUT. This metric is a combination of atom type, connectivity, atomic charge, polarizability, and hydrogen bonding ability attributed to ideas from Burden, Chemical Abstracts Service, and the authors at the University of Texas.[27,28] These values can be at the two- or three-dimensional level, and also could include molecular orbital and semiempirical derived information. This does lead to a large number of metrics. To cope, the software provides a mechanism for automatically determining the best descriptors and the best dimensionality for the input compound set. Generally, for any sort of compound collection on the order of 100 000 or more, the dimensionality will be six. This combination of dimensionality and metrics then defines a chemistry space that describes the compounds in a relatively uniform manner. The chemistry space can be further subdivided by either increasing or decreasing the number of cells per bin that comprise the space. By progressively increasing the number of cells in the space one obtains an increasingly finer distinction between the compounds. With large cells an entire cluster can be contained within one cell, and with more, smaller cells a single cluster may be represented by several cells. By having the compounds spread evenly over the chemistry space and by using evenly distributed cells, a fairly representative selection of a subset of compounds without bias toward outliers should be possible. There is also a distance-based method that can be employed in DVS low-dimensional space but it typically requires more time to run.

DVS contains tools for subset selection. Typically the low-dimensional, cell-based tools are used. By telling DVS how many compounds are to be selected, the software automatically increases the number of cells to get a close match with the desired number of compounds. However, the software sometimes undershoots in selecting because of the interaction of compound distributions and cells. In that case, one needs to increase the target value and then have some means of removing compounds from the selection to reach the exact target number. Because of the same interaction of distribution of compounds and cells, the software may also overshoot, so typically it almost never selects the exact number of compounds specified. Several methods can be employed to further refine the selections to arrive at the target number. One method would be to work down the list in order based on molecular weight, for example, until the desired number is chosen. Another method would be to select every *N*th compound until that target value is reached. Yet another is to use a random selection from the list until the desired number of compounds is reached. If the target number of compounds is not rigidly fixed, for example, there is no need to fit into a predefined number of plates with a predefined order, one can just take the selections as produced by DVS and move on. The issue of obtaining a fixed number of selections is not found in the distance-based method previously mentioned. With the distance-based method, the DVS software will return the exact number requested.

The prior discussion gave a general overview of DVS, but there are many options available in the software and different strategies that could be employed. The first decision is what compounds to use in defining the chemistry space(s). One method is to take the full compound collection, appropriately filter the collection to remove unwanted compounds, and then create a single, all-encompassing chemistry space. The initial collection can also be broken up in many ways to create chemistry spaces that can be more discriminating to select on finer details of chemical characteristics. One method mentioned earlier on is the natural split of the compounds from combinatorial sources and traditional medicinal chemistry. Another variant could be the need to subdivide a larger subset into smaller subsets based on specialized screening capacity needs for a few assays. For example, you already have a diverse subset, but that subset is still too large for an operational reason; the 50 000 compound subset needs to be reduced to a 20 000 compound set and this smaller set still needs to be evenly representative of the chemistry space. One could then create a new chemistry space for just the 50 000 subset designed to provide a relatively even distribution of compounds and use the new chemistry space to divide up the larger subset into a smaller 20 000 set that would be amenable to the specific screening needs. This would need the associated caveats to be strongly emphasized to appropriately manage the performance expectations from using a smaller subset of a larger subset.

One could envision even finer splits, such as, the individual combinatorial libraries could each have their own chemistry space allowing for fine distinctions in selection within libraries. These libraries would need to be very large, because it is not a good idea to do individual library chemistry spaces on a few hundred compounds. DVS needs a relatively large number of compounds, typically in the tens of thousands, to have enough information to create a chemistry space that has a high probability of accurately describing the compounds. If too few compounds are used as input for the chemistry space generation, the software will create one but warn the user with the appropriate caveats on its use. Still, given the size of corporate collections, there are a large number of possible chemistry spaces that could be generated and used for valid reasons. This rapidly becomes a balance between the time available for selection and the time necessary to do the validation studies based on the large number of possible strategies. Once the strategy has been selected and used to select compounds for subsetting, DVS has built into the software update mechanisms for adding in new compounds into a given chemistry space and subsequently finding any new cells occupied by those additional compounds. Augmenting and or updating a subset is very straightforward process within DVS. There is also provision for finding alternatives to compounds already selected to provide substitutes if necessary.

The DVS operations discussed above are in the low-dimensional chemistry space. DVS can also operate in high-dimensional (fingerprint) space. DVS can generate Daylight or MACCS fingerprints, convert ASCII fingerprints to binary and vice versa from sources such as Tripos[31] or BCI,[32] report bit statistics, and enrich fingerprints in a chemical space. Enriching fingerprints is the process of removing those bits that are information-poor and keeping those that are information-rich. DVS high-dimensional chemistry space work can involve using a distance-based method to select a diverse subset, find near neighbors, and assess those near neighbors statistically. DVS can also be used in high-dimensional space to assess clusters within a set. This functionality may be of use when means orthogonal to the cell-based space can help make selection decisions.

Pipeline Pilot[37] is a software package that can be used to create workflows by linking together existing components or protocols that can accomplish a wide variety of tasks. Pipeline Pilot protocols can be used to filter large numbers of compounds based on combinations of numeric and categorical data, including calculated physical properties, sample availability, or any other available data deemed important. One could also filter on the presence or absence of chemical fragments that could be deemed attractive or unwanted. Another task could be the actual selection of the subset compounds. Pipeline Pilot can work with its own internal fingerprints and use existing clustering techniques included as components within the system. One can also write one's own components thereby increasing the range of possibilities for the system. A random number generator is included and protocols could be written to do random selections from the full file of compounds or from a subset to reduce the number of compounds to a given target value. Another possibility is the ability to hook up to external software in an automated fashion, incorporating external programs as units within your workflow via web services using Simple Object Access Protocol (SOAP) services, a text-based data encoding protocol that is vendor-independent. The software works with large volumes of data, handles missing data, and provides a means of sharing within an organization exactly how a task was accomplished.

Any subset that is chosen will almost surely, have to have some estimates made of its predicted performance. Predicted performance can be used to determine which of the theoretical subsets chosen by different methods should be the one actually constructed and put into service. This is usually accomplished by a retrospective analysis of historical screening data, just as various methods are compared in the literature. Typically the subset would be screened and those hits from the initial screen would form seeds with which to find additional active compounds. Therefore, the retrospective analysis should follow the proposed use of the subset as closely as possible in order to provide the best estimates. Within those active compounds one would look for various series to become evident. A series is a collection of structurally similar compounds with a range of biological activity at the target indicating the presence of SARs. As part of the predicted performance, one should also include some estimate of the number of series found in addition to individual hit rates. Some methods for defining series were touched upon earlier in this chapter. If the compounds were selected at random, the follow-up to the initial testing will not have the benefit of prior organizational elements available from the other selection methods from which additional compounds to probe any follow-up processes could be selected. The similarity search on each active from a subset chosen via a purely random selection will have to be generated after the fact. This may also influence whether a purely random selection is appropriate. One must also be cognizant of the overlap of the subset and the historical dataset. If most of the compounds in the subset are also in the historical screen, this screen will then be representative of the performance of the subset in the screen. If there is very little overlap, then that particular screen probably will not reflect what the subset would have done. Another factor to consider is whether to use the raw hits from the historical screen or the confirmed data. Confirmed data will have more of what one would ideally want, those hits that are truly real, but confirmed data can suffer from being incomplete. Often, only a few selected compounds from the initial hits were actually run through the confirmation assay and it may not be easy to determine an accurate, unbiased prediction for hit rates. Another problem that may exist with confirmed data is that there may be more than one definition of what is considered confirmation data. One group may consider compounds confirmed if it repeats in the same assay, others may only consider it confirmed if it is also active in a different secondary assay, or yet another group might consider it confirmed only if it is active upon resynthesis and retest. More on the subject of retrospective analysis will be discussed in the next section dealing with iterative follow-up.

4.20.4 Subset Screening Analysis, Iterative Follow-Up, and Triage

At some point, a subset will have been selected and distributed for use. What will be the recommended use, how will the screen be triaged, and how will it be followed up? One way to approach these questions would be to define various general categories of screens and how they best fit into the subset screening scenarios. Once these have been defined, this information can serve as guidelines to prospective users of the subset to help manage expectations.

The first scenario would be for projects and screens that are in very early stages of development. For these, very little may be known about the mechanism of action, relevance to disease targets, and very little chemical matter is available to use in assessing the target and probing for additional information. In this type of situation, a risk–benefit analysis may recommend that just the subset, or a selected portion thereof, be screened for chemical tools to use in furthering the project. In subset screening, probably the worst scenario is that the screen does not uncover any chemical matter, that there are no hits. If nothing is found, then it might not be worth the effort or money to take this into a full file screen at this time, but rather go back to assay development to improve the assay. Or one might even consider dropping the target altogether or, at least, waiting and monitoring the literature for a development that might change the status quo for the target. Some combination of the above actions could be taken. If a subset screen did not provide any hits and if the target is known to be hit-poor, then the project may rather decide to run a full file screen if the target was worth the expense and the assay was well validated. Subset testing might reveal a problem in the assay, requiring refinement or retooling portions of the assay that did not surface in the initial assay development and validation work.

At the other end of the spectrum, there are well-validated assays for targets with a number of existing pharmaceutical agents on the market. Here subset screening could be used to take a serious look for any additional unknown chemical matter that could be used in this project. This would be a scaffold-hopping exercise (also known as island or series hopping) with a limited scope in a cost effective manner. One would be looking for relatively weak interaction with the target in the screen, and follow up with something like NMR or crystallography to confirm the hits at the appropriate site. The idea here to screen the subset, and if anything new is discovered, it could potentially open up completely new avenues in the treatment of the disease. If on the other hand, one does not find any new chemical matter, it might be cost-effective to stop at this time, shelve the assay, and wait for a new subset, before revisiting.

In the middle of the scenario spectrum, the decisions become less clear-cut. If few or no hits were found after screening the subset, one could stop testing and make the decision to move to another target. Or, one could also decide to then go and do a full file screen. Here the attractiveness of the target, the relevance/importance of the therapeutic area or disease to your organization, and the results of the subset screen could warrant looking at every compound in the file. A middle ground could be to do some sort of iterative screening approach. Here one seeks to test some number of additional compounds and use this information to feedback into the selection mechanism(s), to select additional compounds for testing. One iteration or loop would involve selection of compounds, testing, and feedback of the information into the triage process. Iteration zero would be the initial test of the subset and be followed by several additional iterations to find additional hits outside of those in the initial subset and also possibly find additional series. One caution would be in overselling this concept of series- or scaffold-hopping with subset screening and iterative follow-up, particularly for less-developed, less-mature targets. If the information available regarding the target and associated chemical matter is sparse and contains mixed binding modes, it is more difficult to model the target–ligand interaction and the predictive power is often low. With less mature targets and projects, the focus should probably be on finding five to ten viable chemical series to allow the team to effectively prosecute the target, gaining valuable knowledge about the target, particularly if it is not amenable to full file screening, in a cost-effective manner. In other words, the focus should be on finding the team sufficient and viable chemical matter, and not on finding all possible chemical matter.

Iterative follow-up is the process of finding compounds similar in structure or properties to active compounds and testing them to find additional actives in a looping fashion. These active compounds could also be used to construct a mathematical models, etc., to make predictions. When performing an iterative follow-up to a subset screen, choices in the detailed process of how to accomplish the iterations will need to be made. For instance, how many iterations should be done to maximize return on investment? These very pragmatic issues and questions should be addressed early on. A lot of these will involve the business practices of various lines within your organization, such as the groups responsible for maintaining and distributing compounds from the corporate collection, screening groups, and analysis groups. These groups will potentially define upper limits and guidelines that the individual teams will need to work within, but the teams may have considerable latitude depending on the outcomes of screening and the needs of the project. Some of the questions that should be asked early on are: How fast does the whole iterative screening process have to be accomplished, and how much time does one have to provide the team with chemical matter? What priority does an iterative screen have relative to the other screening activities concurrently in operation? How many compounds can one order for follow-up screening at any given time? Is there a per-order limit on compounds or a per-assay limit, or a site/group/department limit? How fast can you get the compounds? How fast can the screeners run the assay on those compounds? Where do the data go, what data are needed, and how fast can they get there? How much time does the analysis, with appropriate feedback, of the results take and how much time and resources have been allotted to the analysis? Factors involving a mixture of these types of questions are usually to be expected. As an example how fast an iterative screening process could be prosecuted would depend on whether you went with the percent effect from the original screen (fastest), to using confirmed data with a second set of percent effect taken on new samples using the same screen (medium), to working only with confirmed IC_{50}s and information from a secondary assay (slowest). This would also involve how many iterations and how fast can one analyze the iterations and turn around a new set of compounds for the next iteration.

Figure 2 illustrates the desired goal in an iterative follow-up of a subset screen; finding a large percentage of actives with the minimum of compounds screened. This is a generalized plot of the type often found in publications describing the enrichment of actives found by various rational means relative to the number of actives found via random screening. The rational means would be those methods looking for similar compounds either in structure or properties. Here the line labeled 'Random' shows the expected linear increase in the number of actives found relative to the number of compounds tested. There are two other lines or curves labeled 'Enrichment.' These lines show a much greater slope at the beginning illustrating one finding a higher percentage of actives per number of compounds tested. In some publications or presentations in the literature, one often sees a line showing a dramatic enrichment early on with a rapid turn over to another line, while in other presentations one sees a less dramatic enrichment looking more like the curved line. At some point with the rational methods the slope decreases, often dramatically, and one is now not experiencing enrichment in the number of actives and may even be performing worse than random. For efficient iterative screening, one would want to rationally select and test when there is significant enrichment and probably stop when the enrichment rate slows, where one has reached the level of diminishing returns.

A lot of the potential follow-up issues can be worked through by retrospective analysis. In this process, you take several assays that have already been run and treat them as a subset run with iterative follow-up. By varying a number of operational elements, the subsequent analysis will begin to provide information on the details of how to do the iterative

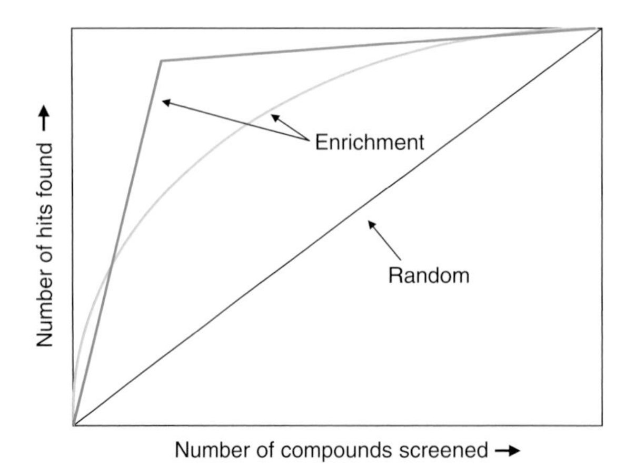

Figure 2 Enrichment due to rational selection versus random selection.

screening. Karnachi and Brown[29] have used a method of orthogonal similarity descriptors within a clustering paradigm to select additional compounds for testing based on similarity to hits found in the initial subset screening. This study uses a retrospective analysis, and the process has been in use at Johnson & Johnson for several years. Here three types of descriptors were used to reduce local minimum effects and to increase the potential for clustering hopping events to occur. The descriptors chosen were the 24 electrotopological properties from Cerius2,[33] MACCS keys, and Daylight fingerprints. These descriptors produced bit-string data that could be used for clustering, the common organizational element in the analysis. There are other choices of descriptors that could be used to increase the orthogonality of the descriptors and the associated chemical space descriptions. But this increase in orthogonality would most likely result in having to use combinations of several different organizational methods, for example clustering and cell-based methods. A study by Fechner and Schneider[30] evaluating various distance metrics also concludes that a fusion of metrics results in distinctly improved performance in finding actives.

Karnachi and Brown[29] used from three to five rounds of iteration with the goals of finding actives and inactives for use in lead identification, and also creating mathematical models to use in predicting actives for lead optimization. The clustering software used was Opticlus from BCI which has a method for determining optimum clustering levels. The clustering was done for each descriptor, optimized, and the intersection of the results taken forward into screening. This intersection method was also compared with the combined-descriptor method, where the results of each individual descriptor clustering were pooled, and the union of all the compounds selected by the various methods was the result. The union of the individual clustering methods results in much larger sampling of the compound space. After a round or two of clustering, the results were used to create mathematical models for use in predicting additional active compounds and for potential scaffold-hopping. There was a second feedback loop into the model building to allow for improvement of the model before the final predictions were made. The concepts of using intersections and unions of results from the organizational methods can be utilized with either a single one, such as clustering, or to combine the results from many different types of organizational elements, such as DVS and clustering.

This iterative process tested a subset of about 30% of the available compounds and resulted in the finding of about 80% of the actives. The authors do state that this is only one study and that this level of performance was not observed for all screens. The authors comment that as the number of hits in a screen decreases, the performance of the iterative screen also decreases. They also concluded that by using a near neighbor method, substitution for depleted members in the cluster is valid and did not require reclustering. One mechanism to reduce the possibility of picking depleted members is to select only those compounds for which a sample is known to be available. While availability could change rapidly, the probability that those selected will be available is dramatically greater. The downside is that the clustering needs to be done on a regular basis. If the compound collection is changing significantly with time, due to mergers or the turnover of combinatorially synthesized compounds, etc., it might be worthwhile to do the near neighbor search each time on the currently available compounds rather than try to replace depleted members from a dated near neighbor analysis.

Another example of a retrospective analysis of a theoretical iterative screen was presented by Yeap *et al.*[46] This screen was performed on a G protein-coupled receptor (GPCR), a popular gene family for pharmaceutical targets. Originally a full file screen of several hundred thousand compounds had been previously performed. This dataset of known screening results was used to predict the performance of a subset in this same assay. A subset was generated from the large, several hundred thousand compound collection. The large collection was initially very stringently filtered and then the remaining compounds were clustered such that a target value about 12 000 clusters was obtained. Compounds were chosen from the clusters based on priority of first is there an active compound in the cluster. If there was a compound known to be active in any screen, it was chosen preferentially. If there were no known actives, then the compound closest to the center of the cluster was chosen. This process provided a subset of 12 500 relatively diverse compounds. The subset and subsequent iterative follow up were then performed electronically, but as if it were a real screen. **Figure 3** summarizes the results found in the retrospective screen. In the initial screen (iteration 0) the hit rate was 1.6%. The iterative process employed for this study was to then choose via a single similarity method the top 1000 near neighbors (NN) and test those. Here the second set screened (iteration 1) provided another 204 hits for a hit rate of 20%. The graph shows the expected enrichment of hits based on the iterative approach. The iterations were carried out to iteration 10 and the results were that 40.4% of the known actives were found by screening approximately 5% of the compound collection. In choosing the number of near neighbors to select, both 500 and 1000 near neighbors seemed to work well in this particular analysis. While the time necessary to actually do ten iterations might not save time over a full file screen and will require good cherry-picking capability, testing only about 5% of the full file collection could result in considerable cost savings.

Triage of the subset screen will be somewhat different than that usually performed for a full file screen. Compounds going into a subset are very carefully chosen. In a full file screen, the compounds in the screening set do not tend to be stringently filtered before entering the screening stream, so a major task is to define the actives and remove the unwanted chemical matter electronically before proceeding. Next one would apply additional filters to highlight those remaining compounds containing risky elements or functionality that would need to be synthetically removed or before the compound would be an appropriate drug candidate. Once the filtering is done, the next task is to determine what compounds are considered to be active in the screen. In a full file screen, each series should be fully represented. This minimizes the risk of missing any given series and should, in theory, provide some indication of SAR if they exist or indicate if the series has a flat SAR. A flat SAR would indicate that the series might be difficult to optimize.

Continuing with the theme of a triage of full file or very large compound subset HTS, Gribbons *et al.*[42] have published a recent paper on 'Evaluating real-life high-throughput screening data.' Understanding the errors and random noise factors is important in the full file screens. An error rate of one in 1000 can add up to a large number when dealing with several million data points. The authors begin by examining what the Z' factor is, how the Z' factor can be determined, from control wells and also from whole plates, and some of the pitfalls in their use in determining the

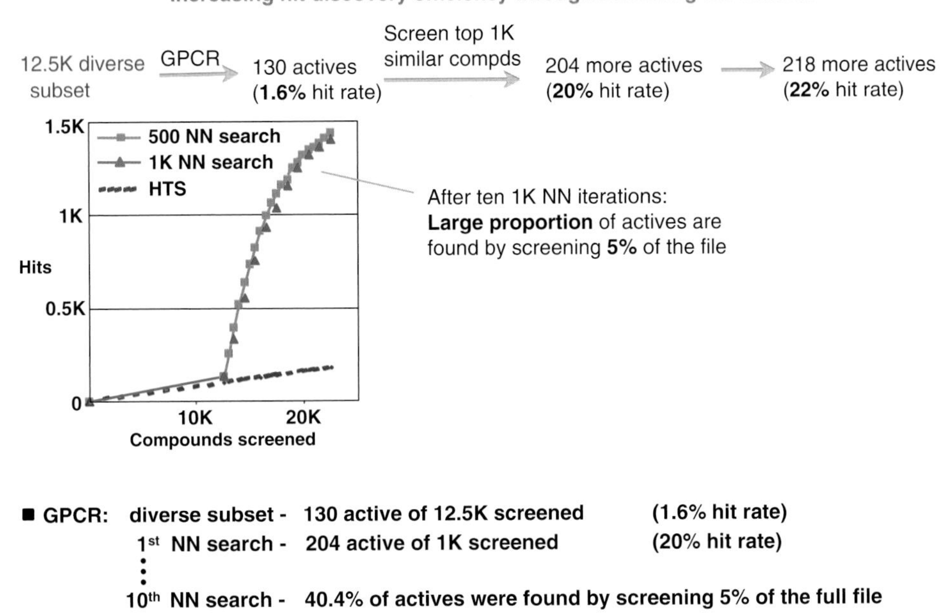

Figure 3 Iterative screening results using a structurally diverse subset.[46]

quality of a screen. The authors then move to discussion of systematic errors and examples of how to visually look for those. They then finish with a discussion of well-level QC and median polishing. Median polishing will remove the error associated with systematic differences within row and columnar data. Another paper on examining systematic errors, by Kevorkov *et al.*,[43] which provides another mechanism for identifying systematic errors and removing them by using surfaces based on data variation versus row and column data. All of this information figures into determining what is considered active in a full file screen.

The two papers mentioned in the previous paragraph and others use the definition of activity as those compounds at or greater than 3σ based, in part, on work by Brideau *et al.*[44] At the 3σ level, one would be at about the noise floor for the assay and should be reasonably generous definition in terms of a several million compound full file screen. But as indicated in the Brideau paper,[44] even with this cutoff, one runs the risk of not finding all of the actives. In a typical full file screen, one could then take all of the compounds at or greater than the 3σ level and consider them active and if the number of hits was very large there are a number of way to organize the data and test compounds representative of a group of similar compounds. However one does not have to apply the same cutoff in the definition of an active to all of the compounds. An example of applying different cutoffs for different structural classes is given by Yan *et al.*[47] Here the authors perform a statistical analysis of the SARs provided by testing a number of very similar compounds. The reported that this method did increase the confirmation rate and reduced the false negative rate. This type of situation would exist for a full file screen where you are testing everything but generally not for a subset, which is typically composed of entirely diverse compounds.

Figure 4 provides a sample workflow that could be used to triage and report out on the analysis of a full file compound screen. The active compounds would have been identified by some mechanism, possibly one of the ones discussed above. Those active compounds would then be subjected to a filtering step to remove unwanted chemical matter. Once the actives have been filtered, they would move on to a confirmation step. As all of the active compounds have tested active at least once in a full file screen, one would expect to have a fairly high confirmation rate. Those compounds confirming would then move on to a step where secondary screens would be warranted, compound integrity tests to be sure of the

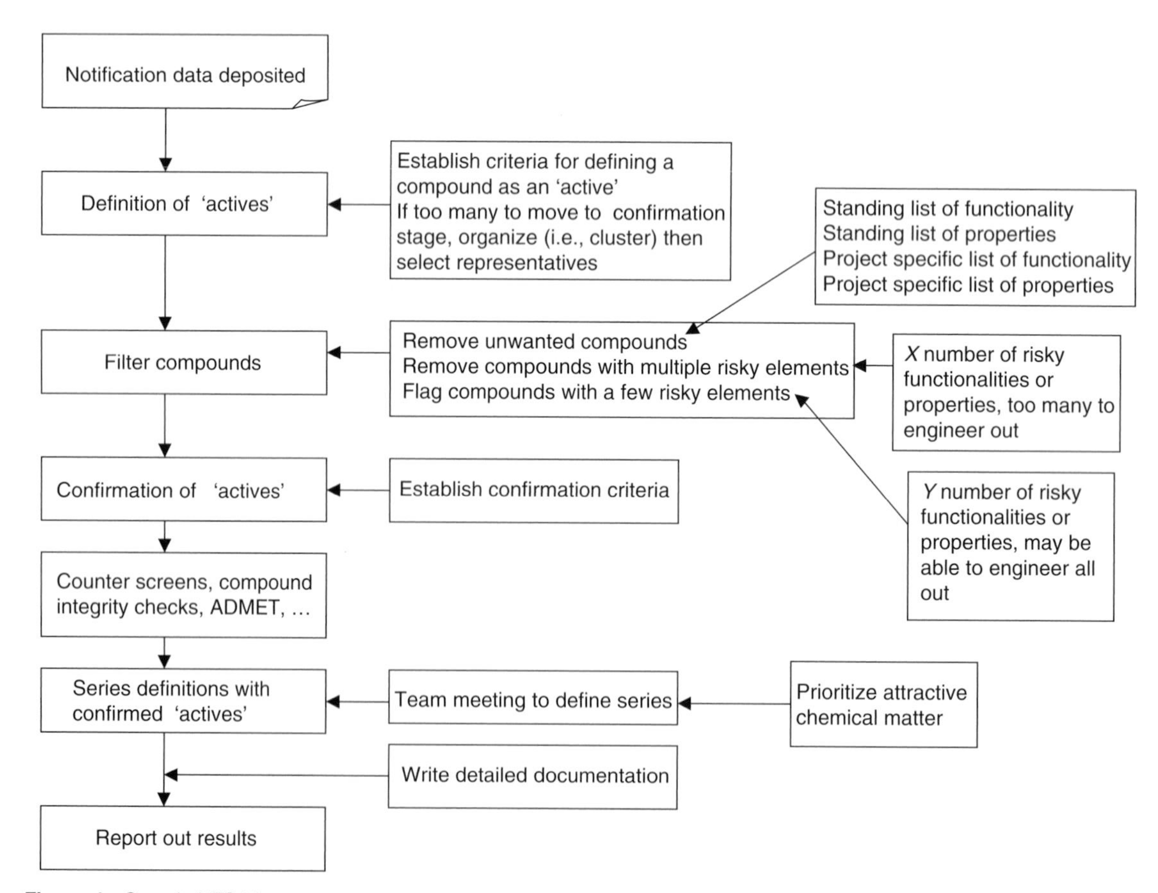

Figure 4 Sample HTS triage process.

chemical identity, and ADME would be determined. By now in the workflow, one should be able to better prioritize attractive chemical matter and begin the series definition process. After the readout of the results and assuming sufficient attractive chemical matter, traditional medicinal chemistry would kick in and the HTS would be considered complete. The full file HTS data could be reanalyzed for the project if necessary, but typically it is not revisited except in the context of a retrospective study or in relation to information that might impact another project. **Figure 4** provides a sample workflow that could be used to triage and report out on the analysis of a full file compound screen.

In terms of using the 3σ level as a definition of an active in the triage of a subset screen, this is not as generous as one would expect based on a superficial comparison with the full file screen. In a subset screen one would have to be very fortunate indeed to have as members of the subset the most active compounds for the screen being run. Most likely you will not have the very active compounds, but rather ones that give a hint of activity. An analogy would be in finding lost money; it is fairly common to find small change such as pennies or nickels, but very rare to find $20 bills. Here at the 3σ level definition of active, the compounds would give a hint of activity that similar compounds would also possess. By finding the compounds similar to those considered active and testing them, one would hope that you would find a few similar compounds with better activity and thereby have more potent compounds at the end of the analysis process and by the nature of having similar compounds would be able to define series of compounds. What one would also find is that there will be similar compounds that have less activity also, those considered to be truly inactive and what could happen is that most of the similar compounds tested will be in the inactive category. This will have an impact on the confirmation rate, a higher percentage of compounds being found as inactive, and create the false impression that the confirmation rate is not as good as that for a full file HTS screen. **Figure 5** is a stylized illustration of what could happen in this scenario. The initial hit is shown as an asterisk (*), which lies just above the 3σ cutoff. This cutoff is depicted by the labeled line and indicating its relationship to the curve showing the initial results of the screen. Those compounds with percent effect equal to or greater than 3σ are considered active in the screen. After a similarity search the close neighbors were found and tested. The results for those compounds are depicted by the ovals, which encompass the results of the subsequent testing in relation to the curve generated by the initial screening data. These ovals indicate a few compounds that are significantly more active than the original hit, several that are as active or possibly marginally more active than the original hit, and a large number of similar compounds that are inactive. At this point one should have additional active compounds some with more potency than those found in the original screen. One then has a situation in which the definition of what constitutes an active compound can be changing with each subsequent iteration, a rolling definition based on obtaining increasing numbers of actives and increasing potencies. This situation can have a major impact on how one defines and compares hit rates, confirmation rates, and other metrics along with what information is feed into subsequent analysis processes. These issues should be considered up front in the project and have a plan in place for this situation.

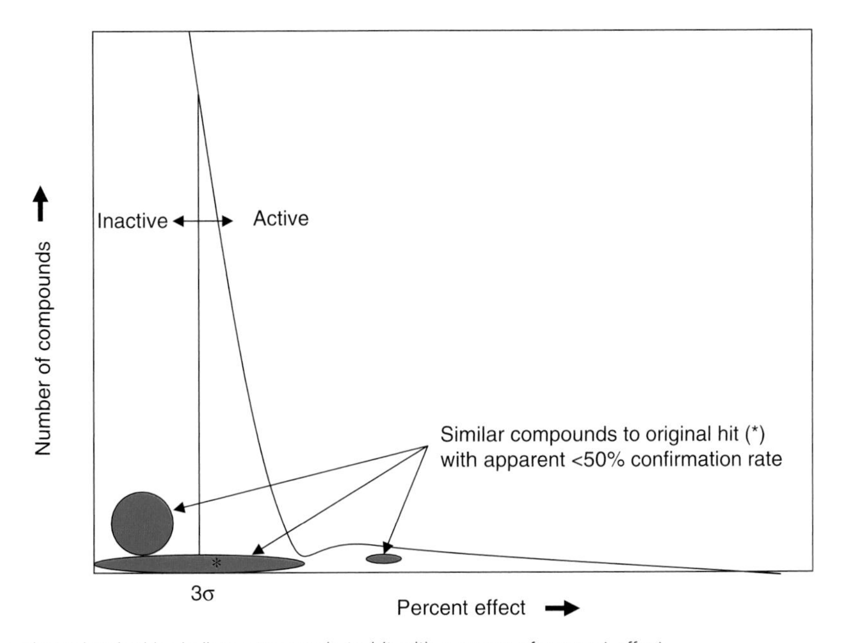

Figure 5 Three sigma level with similar compounds to hit with a range of percent effect.

In creating the subset, a lot of preparatory work should have gone into the filtering procedures and this work will not need to be repeated in the triage. All of the compounds in the subset are potential valid hits suitable for follow-up. The triage will most likely consist of determining what level of biological activity is necessary to define any given compound as being active in the assay. Once the determination of what is to be considered active has been done, and if the hit rate for the subset is non zero, then the search for similar compounds should produce additional active compounds. As discussed previously in the chapter, the hits found are most likely not the most active compound that is available. However, by examining the SAR of near neighbors in the follow-up, you can determine whether a weak hit indicates a truly viable series to take on into secondary testing or is just noise.

References

1. Davis, A.; Keeling, D.; Steele, J.; Tomkinson, N.; Tinker, A. *Curr. Top. Med. Chem.* **2005**, *5*, 421–439.
2. Harper, G.; Pickett, S.; Green, D. *Combin. Chem. HTS* **2004**, 7, 63–70.
3. Jacoby, E.; Davies, J.; Blommers, M. *Curr. Top. Med. Chem.* **2003**, *3*, 11–23.
4. Huth, J. R.; Sun, C. *Combin. Chem. HTS* **2002**, *5*, 631–643.
5. Liu, G; Huth, J.; Olejniczak, E.; Mendoza, R.; DeVries, P.; Reiley, E.; Leitza, S; Okasinski, G; Fesik, S.; von Galdern, T. *J. Med. Chem.* **2001**, *44*, 1202–1210.
6. Murcko, M.; Bemis, G. *J. Med. Chem.* **1996**, *39*, 2887–2893.
7. Murcko, M.; Bemis, G. *J. Med. Chem.* **1999**, *42*, 5095–5099.
8. Willett, P. *Similarity and Clustering in Chemical Information Systems*; Research Studies Press: Letchworth, Hertfordshire, England, 1987.
9. Jarvis, R.; Patrick, E. *IEEE Trans. Comp.* **1973**, *C22*, 1025–1034.
10. Lipinski, C.; Lombardo, F.; Dominy, B.; Feeney, P. *Adv. Drug Deliv. Rev.* **1997**, *23*, 3–25.
11. Lipinski, C. *J. Pharmacol. Toxicol. Methods* **2000**, *44*, 235–249.
12. Viswanadhan, V.; Balan, C.; Hulme, C.; Cheetham, J.; Sun, X. *Curr. Opin. Drug Disc. Dev.* **2002**, *5*, 400–406.
13. Egan, W.; Walters, W.; Murcko, M. *Curr. Opin. Drug Disc. Dev.* **2002**, *5*, 540–549.
14. Vieth, M; Siegel, M; Higgs, R.; Watson, I.; Robertson, D.; Savin, K; Durst, G.; Hipskind, P. *J. Med. Chem.* **2004**, *47*, 224–232.
15. Leeson, P.; Davis, A.; Steele, J. *Drug Disc. Today: Technol.* **2004**, *1*, 189–195.
16. Hann, M.; Hudson, B.; Lewell, X.; Lifely, R.; Miller, L.; Ramsden, N. *J. Chem. Inf. Comput. Sci.* **1999**, *39*, 897–902.
17. Dunbar, J. In *Pacific Symposium on Biocomputing 2000*; Altman, R., Dunker, A., Hunter, L., Lauderdale, K., Klein, T., Eds.; World Scientific: Singapore, 1999, pp 555–565.
18. Lajiness, M.; Maggiora, G.; Shanmugasundaram, V. *J. Med. Chem.* **2004**, *47*, 4891–4896.
19. Hamprecht, F.; Thiel, W.; van Gunsteren, W. *J. Chem. Inf. Comput. Sci.* **2002**, *42*, 414–428.
20. Reynolds, C.; Tropsha, A.; Pfahler, L.; Druker, R.; Chakravorty, S.; Ethiraj, G.; Zheng, W. *J. Chem. Inf. Comput. Sci.* **2001**, *41*, 1470–1477.
21. Abt, M.; Lim, Y.; Sacks, J.; Xie, M.; Young, S. *Technical Report Number 105*; National Institute of Statistical Sciences: Research Triangle Park, NC, 2000.
22. Spencer, R. *Biotechnol. Bioeng.* **1998**, *61*, 61–67.
23. Tanimoto, T. *IBM Internal Report*, **1958**.
24. Menard, P.; Lewis, R.; Mason, J. *J. Chem. Inf. Comput. Sci.* **1998**, *38*, 497–505.
25. Clark, R. *J. Chem. Inf. Comput. Sci.* **1997**, *37*, 1181–1188.
26. Ashton, M.; Barnard, J.; Casset, F.; Charlton, M.; Downs, G.; Gorse, D.; Holliday, J.; Lahana, J.; Willett, J. *Quant. Struct.–Act. Relat.* **2002**, *21*, 598–604.
27. Pearlman, R.; Smith, K. *Perspect. Drug Disc. Design* **1998**, *9/10/11*, 339–353.
28. Pearlman, R.; Smith, K. *J. Chem. Inf. Comput. Sci.* **1999**, *39*, 36–45.
29. Karnachi, P.; Brown, F. *J. Biomol. Screen.* **2004**, *9*, 678–686.
30. Fechner, U.; Schneider, G. *ChemBioChem* **2004**, *5*, 538–540.
31. Tripos, Inc. St. Louis, MO, USA.
32. Barnard Chemical Information (BCI) Ltd, Sheffield, UK.
33. Accelrys, Inc., San Diego, CA, USA.
34. Chemical Computing Group, Montreal, Quebec, Canada.
35. Elsevier MDL, San Leandro, CA, USA.
36. Daylight Chemical Information Systems, Mission Viejo, CA, USA.
37. Scitegic, San Diego, CA, USA.
38. Biobyte Corporation, Claremont, CA, USA.
39. Pannanugget Consulting L.L.C., Kalamazoo, MI, USA.
40. Ellis, D.; Furner-Hines, J.; Willett, P. In *Perspectives in Information Management*; Oppenheim, C., Citroen, C., Griffiths, J., Saur, K. G., Eds.; Munich: Germany, 1990; Vol. 3, pp 128–149.
41. Hopkins, L.; Groom, C.; Alex, A. *Drug Disc. Today* **2004**, *9*, 430–431.
42. Gribbons, P.; Lyons, R.; Laflin, P.; Bradley, J.; Chambers, C.; Williams, B.; Keighley, W.; Sewing, A. *J. Biomol. Screen.* **2005**, *10*, 99–107.
43. Kevorkov, D.; Makarenkov, V. *J. Biomol. Screen.* **2005**, *10*, 557–567.
44. Brideau, C.; Gunter, B.; Pikounis, W.; Pajni, N.; Liaw, A. *J. Biomol. Screen.* **2003**, *8*, 634–647.
45. Hert, J.; Willett, P.; Wilton, D.; Acklin, P.; Azzaaoui, K.; Jacoby, E.; Schuffenhauser, A. *Org. Biomol. Chem.* **2004**, *2*, 3256–3266.
46. Yeap, S.; Walley, R.; Mason, J. Presented at 230th ACS National Meeting of the American Chemical Society, Washington, DC, Aug 28–Sept 1, 2005.
47. Yan, F.; Asatryan, H.; Li, J.; Zhou, Y. *J. Chem. Inf. Model.* **2005**, *45*, 1784–1790.

Biography

James B Dunbar is a Research Fellow working in the Computer Assisted Drug Discovery (CADD) group, Chemistry Technologies Section of the Chemistry Department at the Ann Arbor Michigan Laboratories of Pfizer Global R&D. Dunbar works on site-based Therapeutic Area modeling support for project teams and on chemical informatics and database technology aspects associated with computer assisted drug discovery for both site-based and global projects. His project work has encompassed all areas of modeling support in all of the therapeutic areas.

Dunbar joined Parke Davis, Ann Arbor Laboratories, Warner Lambert Company in 1989 working in the CADD group and has continued on in that group through the merger with Pfizer. Prior to his work at Pfizer, Jim had two postdoctoral positions: one at Tripos Associates, a software company specializing in modeling software, and with Prof Garland Marshall at the Department of Pharmacology at Washington University in St Louis Medical School, continuing on as a staff scientist there for two more years. He received his BS in chemistry from West Virginia University in 1977, MA in chemistry from Washington University in St Louis in 1979, and his PhD in chemistry from Washington University in St Louis in 1983.

© 2007 Elsevier Ltd. All Rights Reserved
No part of this publication may be reproduced, stored in any retrieval system or transmitted in any form by any means electronic, electrostatic, magnetic tape, mechanical, photocopying, recording or otherwise, without permission in writing from the publishers

Comprehensive Medicinal Chemistry II
ISBN (set): 0-08-044513-6

ISBN (Volume 4) 0-08-044517-9; pp. 495–513

4.21 Pharmacophore Modeling: 2 – Applications

Y C Martin, Abbott Laboratories, Abbott Park, IL, USA

© 2007 Elsevier Ltd. All Rights Reserved.

4.21.1 Overview

This chapter will review the use of pharmacophores and three-dimensional (3D) searching. Although it is interesting to derive a pharmacophore hypothesis for the 3D requirements for a biological activity of interest, the real utility of such a hypothesis is to lead to the discovery of structurally novel or biologically selective compounds. This chapter will illustrate the use of pharmacophores to discover activity or selectivity in as yet untested molecules with a series of figures. Each figure contains representative structures that led to the pharmacophore, typically those 'most similar' to the new molecules. The figures also indicate the type of software used and, if a database was searched, which one was searched. If it is available, the biological potency of the structures shown will be provided. Surveying the figures will provide the reader with an impression of the power of pharmacophore hypothesis generation and 3D database searching.

Pharmacophores have been used to discover new molecules in two ways: The original approach is to use the pharmacophore hypothesis, usually derived from a diverse set of molecules, to guide the synthesis in an un-going series or to 'lead-hop' to a new series. Although using a pharmacophore hypothesis in lead optimization may not yield totally novel structures, it can make the discovery of potent molecules more efficient. The second approach is to use the pharmacophore as a 3D search query to identify in a database other molecules that contain the pharmacophore, a technique that first became available in 1989. Searching a 3D database is more likely to yield totally novel compounds, to allow one to 'lead hop,' but frequently these hits require optimization to attain acceptable potency. In addition, pharmacophores can be used to design out undesired features. For example, Klabunde and Evers describe the generation of pharmacophore models for biogenic amine binding GPCRs to support the design of compounds that avoid GPCR-mediated side effects.[1,2] Ekins has explored the use of pharmacophores to understand the structure–activity relationships of drug metabolism[3–5] and hERG.[6] Although such uses of pharmacophores are valuable, they are outside the scope of this review (see Chapter 4.01 tables for cross references to in silico ADMET approaches).

Figures 1, 2, and **3**[7–9] show examples from the 1980s of the successful use of a pharmacophore hypothesis to design novel compounds. In all three cases the design process was manual; this work was done before the advent of 3D searching methods and automated pharmacophore perception so the only software used was molecular graphics. Although the designed compounds are novel, they still bear a strong resemblance to the original leads. In contrast, **Figure 4** shows the compounds discovered by three different groups of researchers[10–12] who started with the same protein active site, but searched different databases using their own pharmacophore and different software. The diversity of structures illustrates the power of 3D pharmacophore searching.

4.21.2 Manual Design of Novel Compounds Using a Pharmacophore Hypothesis

Figures 5–11, arranged by date of publication, illustrate the use of a pharmacophore to design a new compound. In five of the six cases (as well as those in **Figures 1** and **2**), the designed compound is more potent than the starting

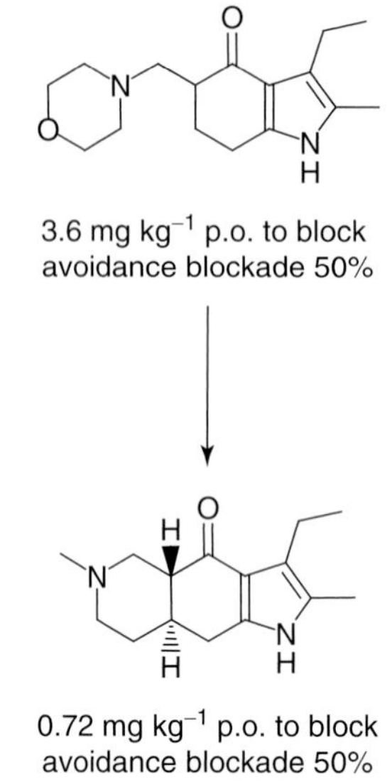

3.6 mg kg^{-1} p.o. to block avoidance blockade 50%

0.72 mg kg^{-1} p.o. to block avoidance blockade 50%

Figure 1 The use of a model of the dopamine receptor to design novel antipsychotics.[7] The hypothetical model, based on the x-ray structures of five diverse antipsychotics, included functional groups from the receptor with which key groups in the molecules interact. The pharmacophore contains a basic nitrogen and the center of an aromatic ring with a distance of 4.9–6.1 Å between them; an 84–104° angle between the former line and the normal to the aromatic ring; an 88–108° angle between the former line and the nitrogen lone-pair direction; and a −100° to 65° torsion angle between the lone pair direction and the normal to the aromatic ring.

pIC_{50} 6.9

pIC_{50} 7.7

Manual pharmacophore hypothesis

Manual design

pIC_{50} 9.2

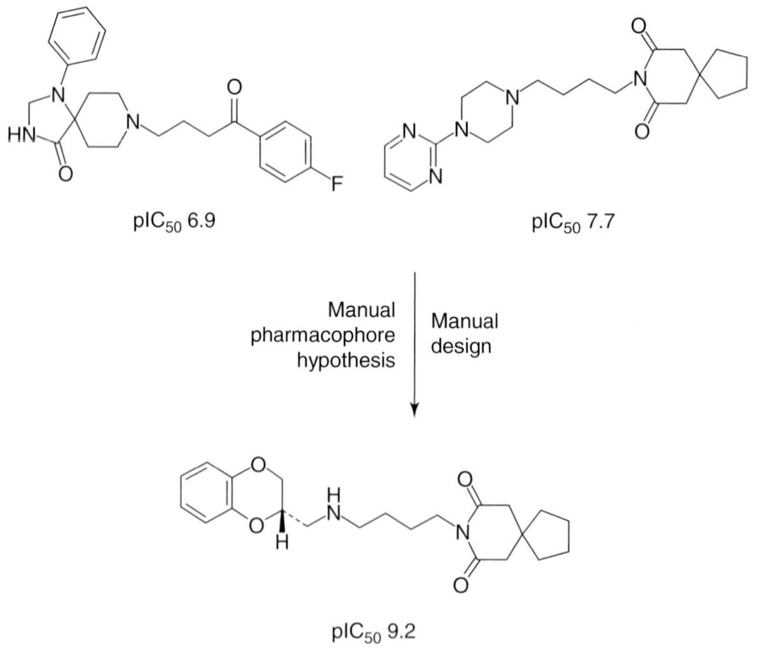

Figure 2 The use of a pharmacophore model to design $5HT_{1A}$ antagonists.[8] The pharmacophore contains a basic nitrogen and the center of an aromatic ring at a distance of 5.6 Å. In addition, the basic nitrogen is 1.6 Å above the plane of the aromatic ring.

compound. The design may involve merging structural features of two leads (**Figures 5**, **7**, **8**, and **9**[13–18]) or it might involve total redesign based on the pharmacophore (**Figures 6**, **10**, and **11**[19–22]). The biological endpoints include ion channel openers, GPCR ligands, and enzyme inhibitors.

IC_{50} 3.5 × 10^{-8}

Sybyl conformational search and manual superposition

Manual design

IC_{50} 2.4 × 10^{-7}

Figure 3 The use of a pharmacophore model to design novel aldose reductase inhibitors.[9] No exact pharmacophore is provided.

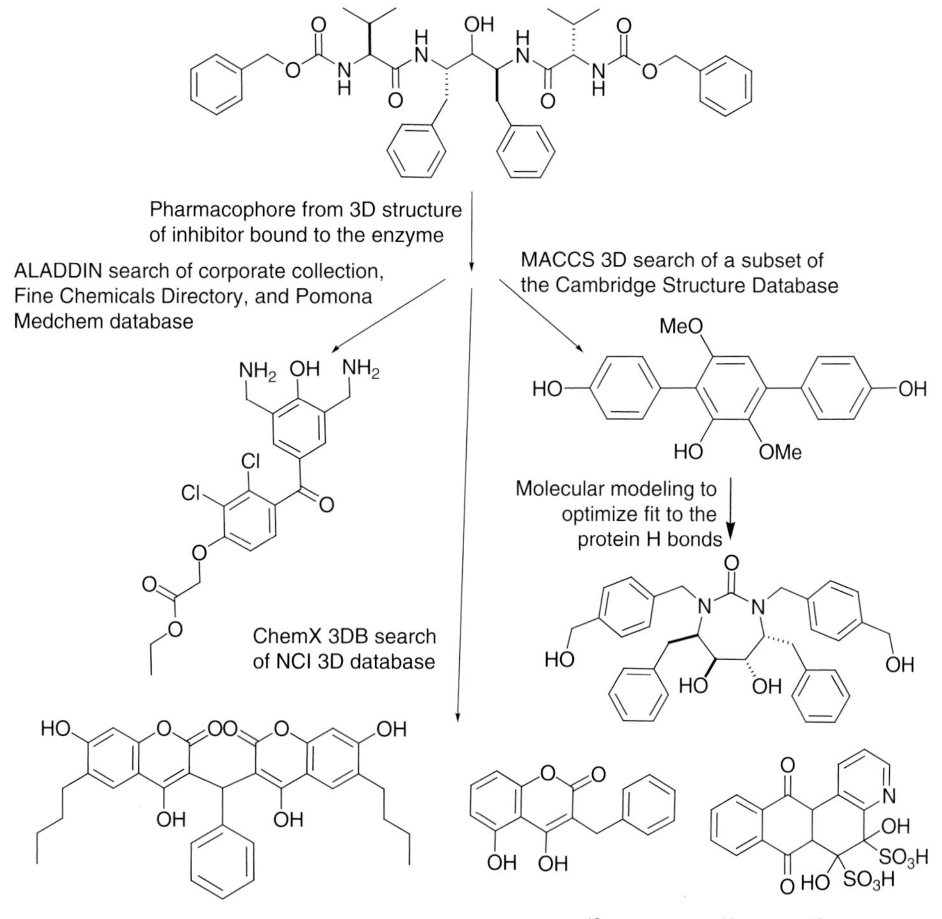

Figure 4 The compounds identified by 3D database searches (ALADDIN,[10] MACCS 3D,[11] ChemX[12]) based on the binding interactions of a peptidergic inhibitor of HIV protease. The pharmacophore for the Aladdin search contains (1) the oxygen of an aliphatic hydroxyl group, (2) a hydrogen bond donor, (3) a hydrogen bond acceptor, (4) an optional hydrogen bond donor, and (5) an optional hydrophobic group at the following distances: 1–2 and 1–4, 2.4–3.8 Å; 1–3, 5.6–7.0 Å; 2–3, 3.7–6.0 Å; 2–5, 4.9–6.6 Å. The pharmacophore for the MACCS-3D search contains two hydrophobic groups separated by 8.5–12.0 Å and a hydrogen bond donor/acceptor 3.5–6.5 Å from each hydrophobic group. The pharmacophore for the ChemX search contains (1) a hydrogen bond acceptor and (2,3) the oxygens of two hydroxyl groups at the following distances: 1–2, 4.4–6.4 Å; 1–3, 4.1–6.1 Å; and 2–3, 1.8–3.8 Å.

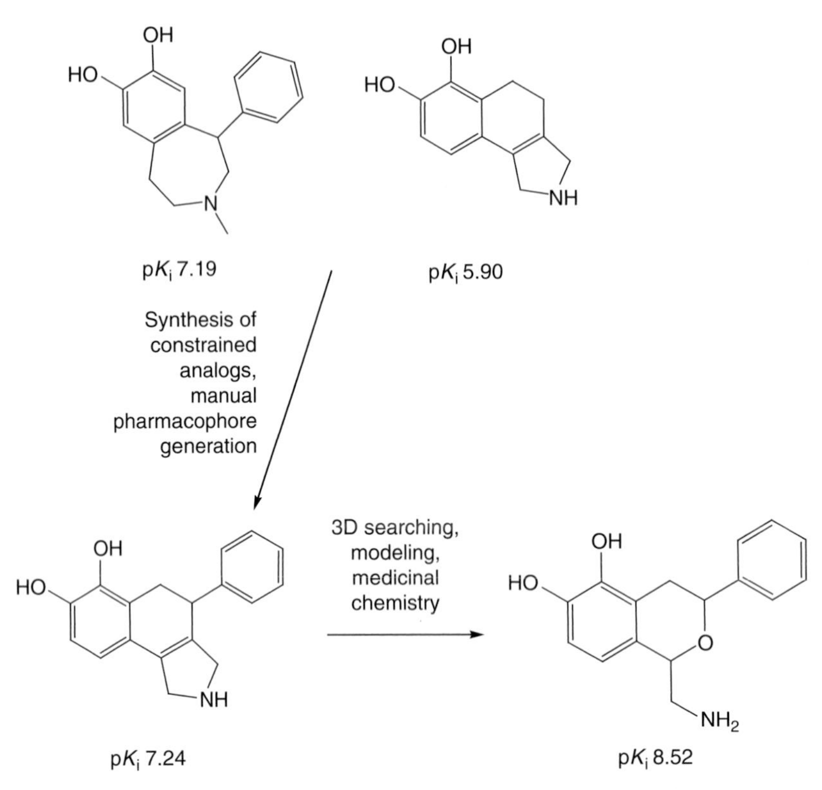

Figure 5 The use of a pharmacophore model to design potent D1 dopaminergic agonists.[13,15] The pharmacophore contains (1) a basic nitrogen, (2) the proton of a hydrogen bond donor, and (3) an aromatic group located in the molecular plane, 10.7 Å from 1 and 8.6 Å from 2 at the following distances: 1–2, 6.8–8.3 Å, and 3 > 4 Å from any atom in the molecule.

pEC_{50} 7.0 pEC_{50} 6.8

Manual pharmacophore hypothesis | Manual design

pEC_{50} 7.6

Figure 6 The use of a pharmacophore model to design potassium channel openers.[19] The pharmacophore model contains two lipophilic and two hydrogen bond acceptors. No constraints are given.

4.21.3 Discovery of Novel Compounds Using a Pharmacophore Hypothesis and Three-Dimensional Database Searching

3D database searching (*see* 4.06 Pharmacophore Modeling: 1 – Methods; 4.19 Virtual Screening) is a powerful method to discover novel compounds that match a pharmacophore hypothesis. In the figures, sample database

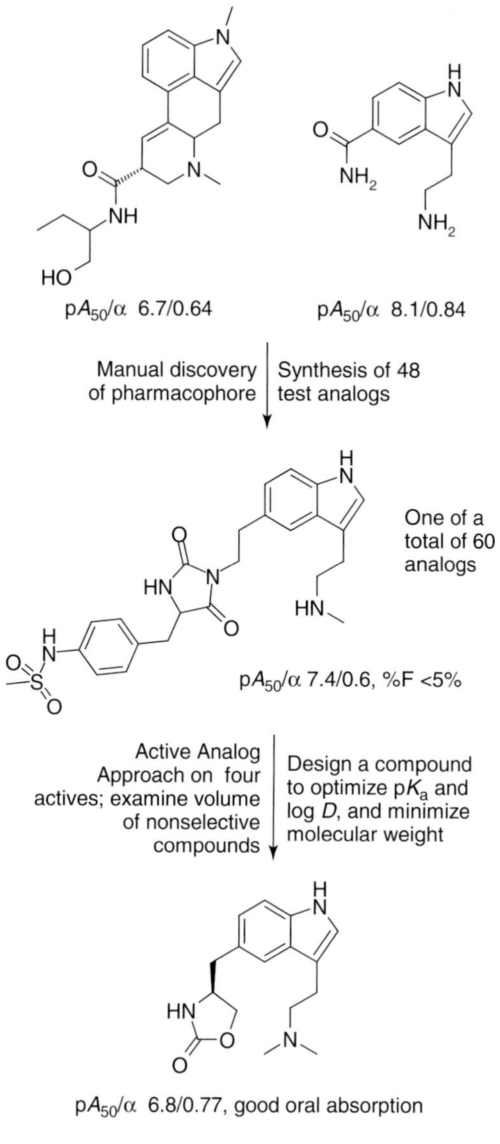

pA_{50}/α 6.7/0.64

pA_{50}/α 8.1/0.84

Manual discovery of pharmacophore | Synthesis of 48 test analogs

One of a total of 60 analogs

pA_{50}/α 7.4/0.6, %F <5%

Active Analog Approach on four actives; examine volume of nonselective compounds | Design a compound to optimize pK_a and log D, and minimize molecular weight

pA_{50}/α 6.8/0.77, good oral absorption

Figure 7 The use of a pharmacophore model and physical property considerations to design an orally active $5HT_{1D}$ agonist.[16] The pharmacophore model contains four points: (1) a protonated amine, (2) an aromatic region, (3) a hydrogen-bond acceptor, and (4) a hydrogen-bond donor or acceptor at the following distances: 1–2, 5.2 Å; 1–3, 4.2–4.7 Å; 1–4, 6.7 Å; 2–3, 5.1–7.1 Å; and 3–4, 4.8 Å with a hydrophobic group bracketing 3 and 4.

structures will be shown. Frequently, not all of the hits from a search have been examined experimentally: Compounds might not be tested because they lack some feature that wasn't included in the pharmacophore, because they aren't drug-like or don't pass the rule-of-five, or because other similar compounds will be tested. Because of such omissions, it may not be possible to evaluate how many diverse hits might have been found. Another important factor that affects the number and quality of the hits is the quality and diversity of the database searched.

Figures 12–17 include examples of searching 3D databases using pharmacophores derived from HIV-1 integrase inhibitors.[23–28] **Figures 12–15** represent the work from one group. Their investigations led to two distinct pharmacophore hypotheses and at least nine distinct series that are equal to or more potent than the original lead. The second research group later searched for HIV-1 integrase inhibitors and identified three additional series. These results illustrate the power of 3D database searching.

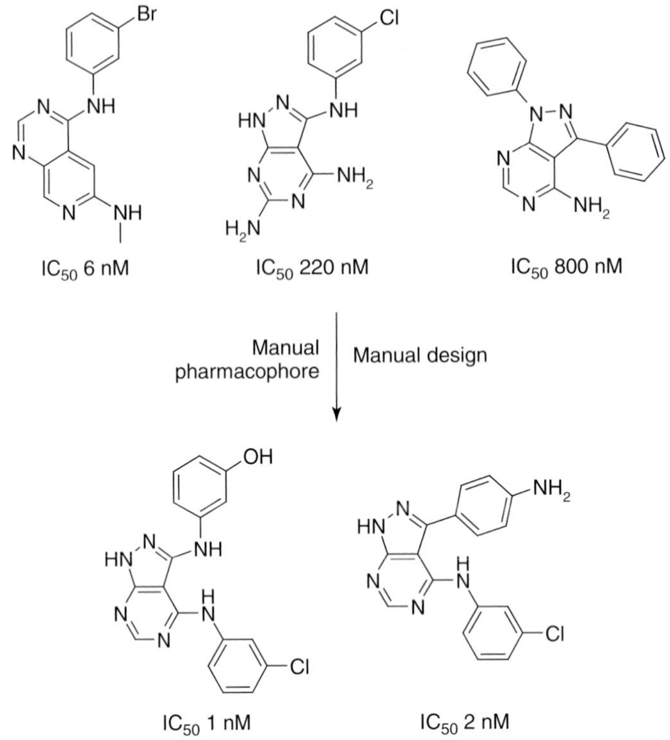

Figure 8 The use of a pharmacophore model to design EDG-4 tyrosine kinase inhibitors.[17] The pharmacophore model contains an NH2-C = N group with a 'large' pocket closer to the amine group and a sugar pocket opposite to it. No constraints are given.

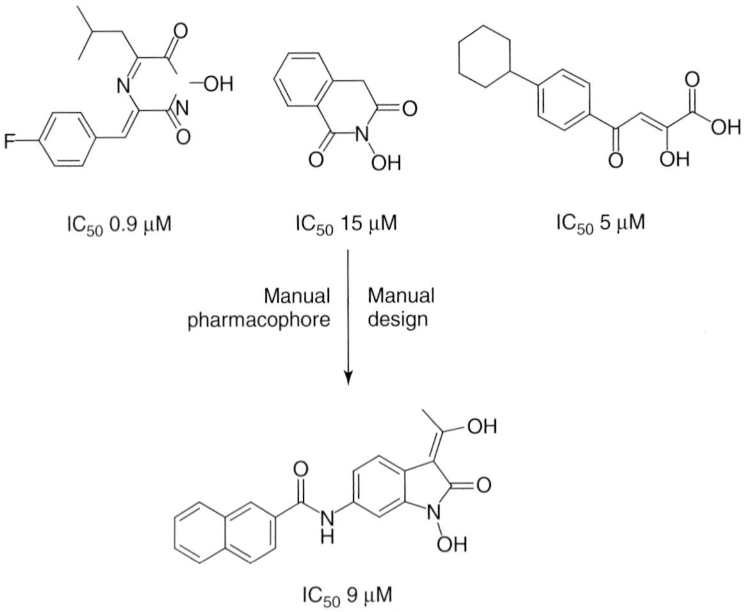

Figure 9 The use of a pharmacophore model to design influenza endonuclease inhibitors.[18] The pharmacophore contains three oxygen atoms, one or two of which may be ionized. Two distances between the oxygen atoms are 2.6–2.8 Å and the third distance is 4.5–5.5 Å.

pK_i 8.5

pK_i 8.2

pK_i 8.4

Catalyst HipHop

Manual design

pK_i <8.0

pK_i 9.0

Figure 10 The use of a pharmacophore model to design $5HT_7$ antagonists.[20] The pharmacophore contains (1) a hydrogen bond acceptor; (2) a site point for 1; (3) a hydrophobic group; (4) a basic center at the following distances: 1–2, 3 Å; 1–3, 4.7 Å; 1–4, 6.3 Å; 2–3, 7.8 Å; 2–4, 7.2 Å; 3–4, 7.3 Å. No tolerances are listed.

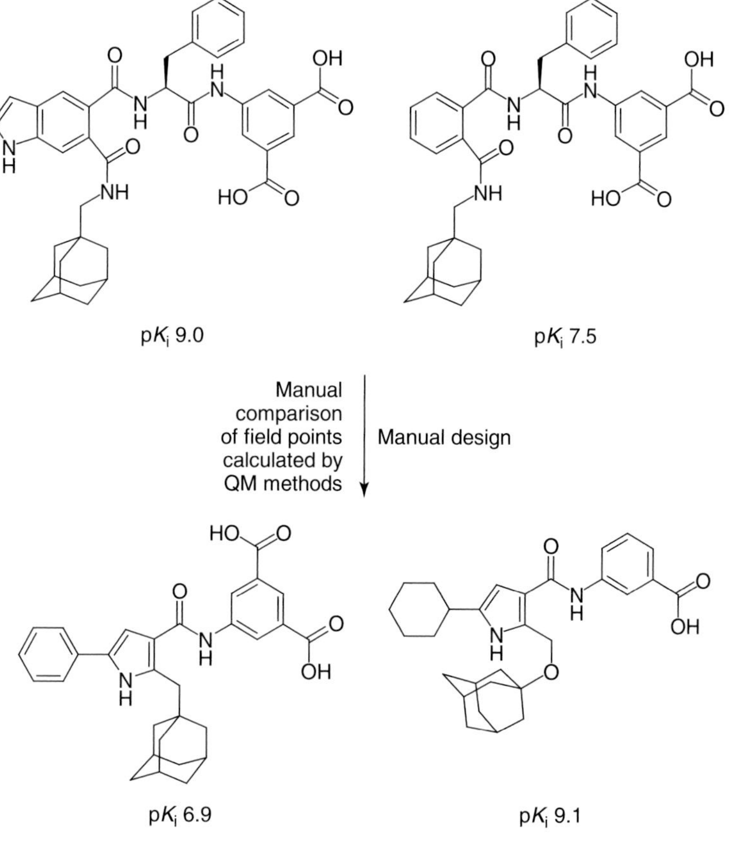

Figure 11 The use of a pharmacophore model to design cholecystokinin-2 (CCK2 antagonists).[21,22]

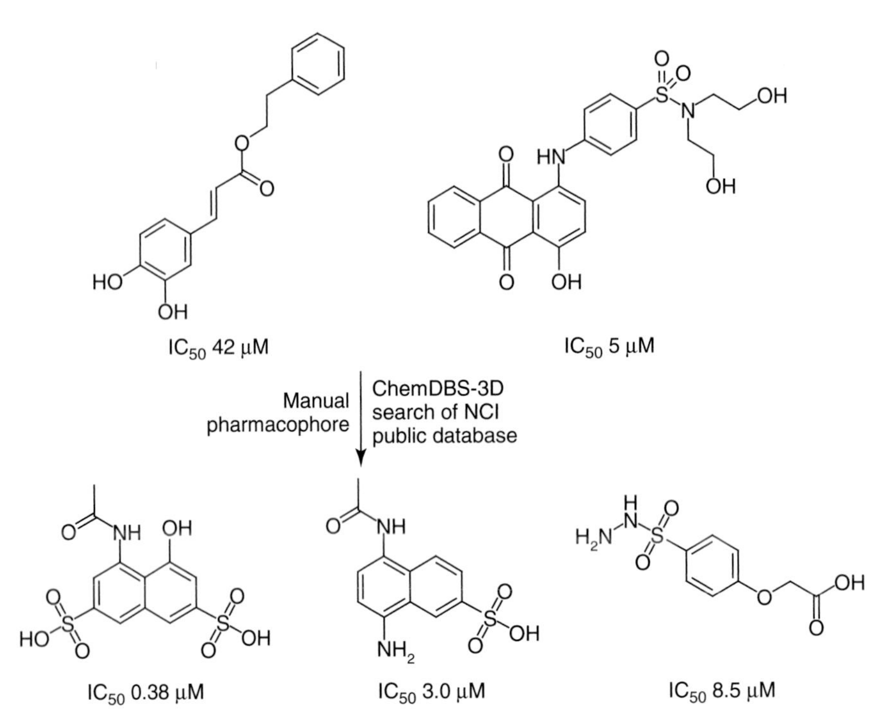

Figure 12 The use of a pharmacophore model and 3D searching to identify novel inhibitors of HIV integrase.[23] The pharmacophore contains two oxygen or nitrogen hydrogen bond acceptors separated by 2.55±0.3 Å, an oxygen hydrogen bond acceptor 9.1±0.1 Å from the first and 8.71±0.4 Å from the second, and an exclusion sphere of radius 0.8 Å positioned 1.711 Å from the third hydrogen bond donor.

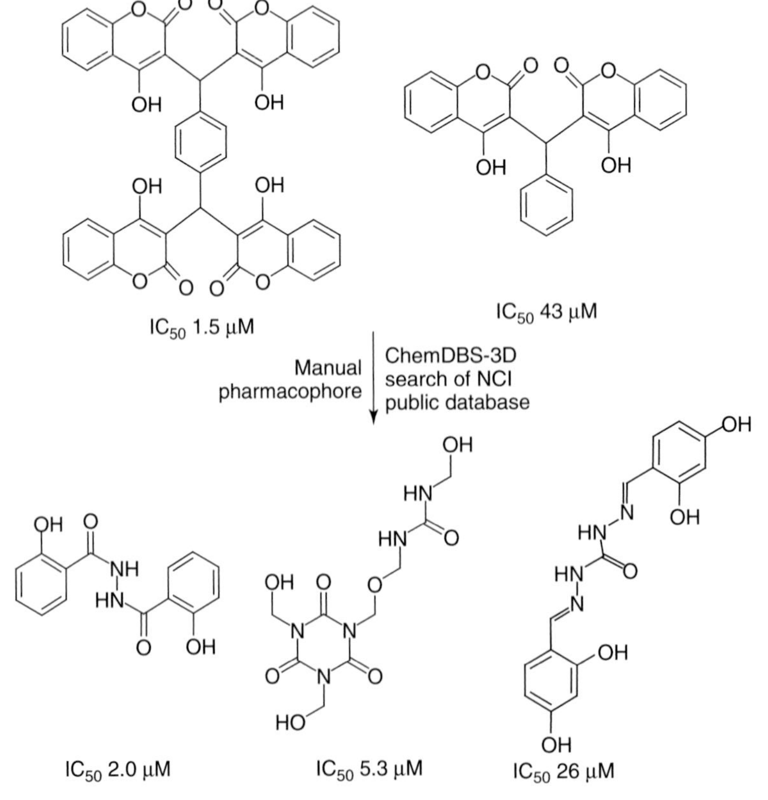

Figure 13 The use of a pharmacophore model and 3D searching to identify novel inhibitors of HIV integrase.[24] The pharmacophore contains three oxygen or nitrogen atoms separated by 2.87±0.7 Å, 4.76±0.4 Å, and 5.62±0.7 Å.

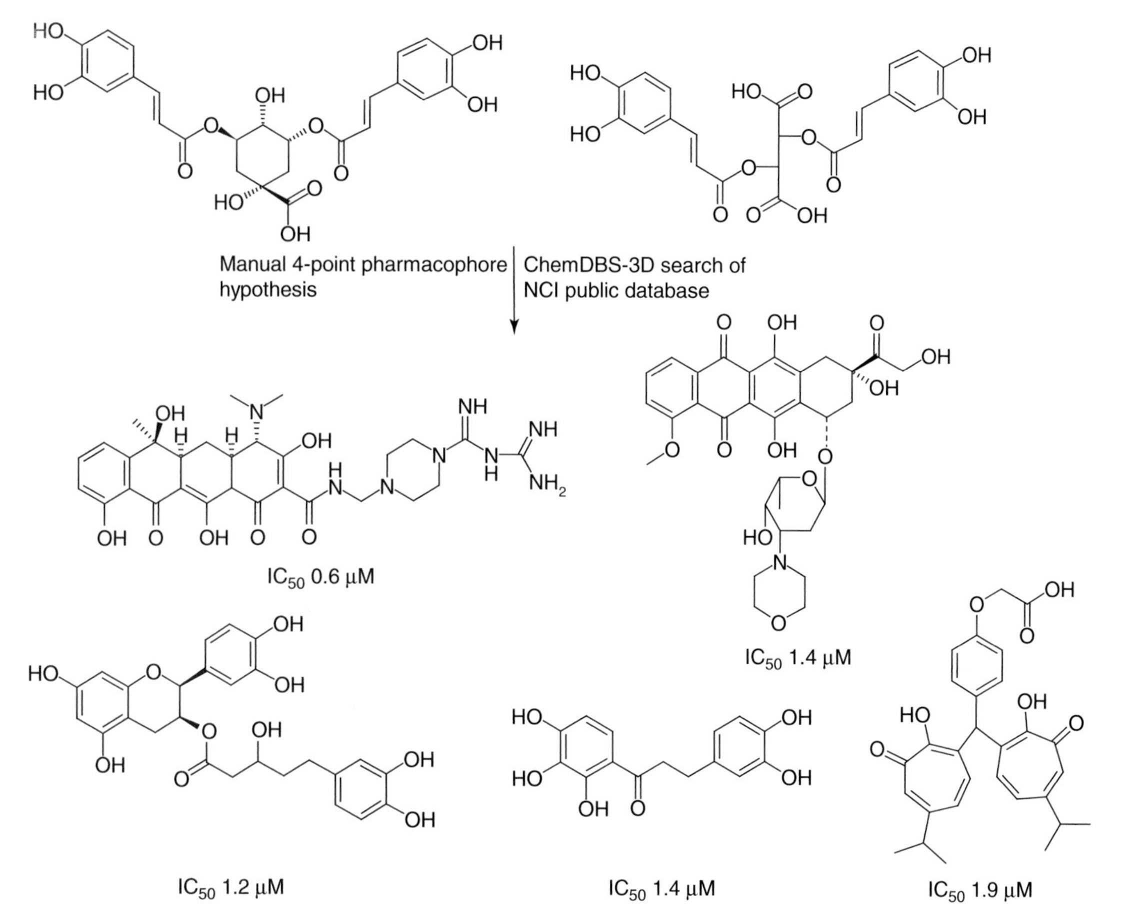

Figure 14 The use of a pharmacophore model and 3D searching to identify novel inhibitors of HIV integrase.[25] The pharmacophore contains six oxygen or basic nitrogen atoms at the following distances: 1–2, 11.5 Å; 1–3, 2.6 Å; 1–4, 10.5 Å; 2–3, 9.5 Å; 2–4, 2.6 Å; 3–4, 8.6 Å. A tolerance of 0.7 Å was used in every case.

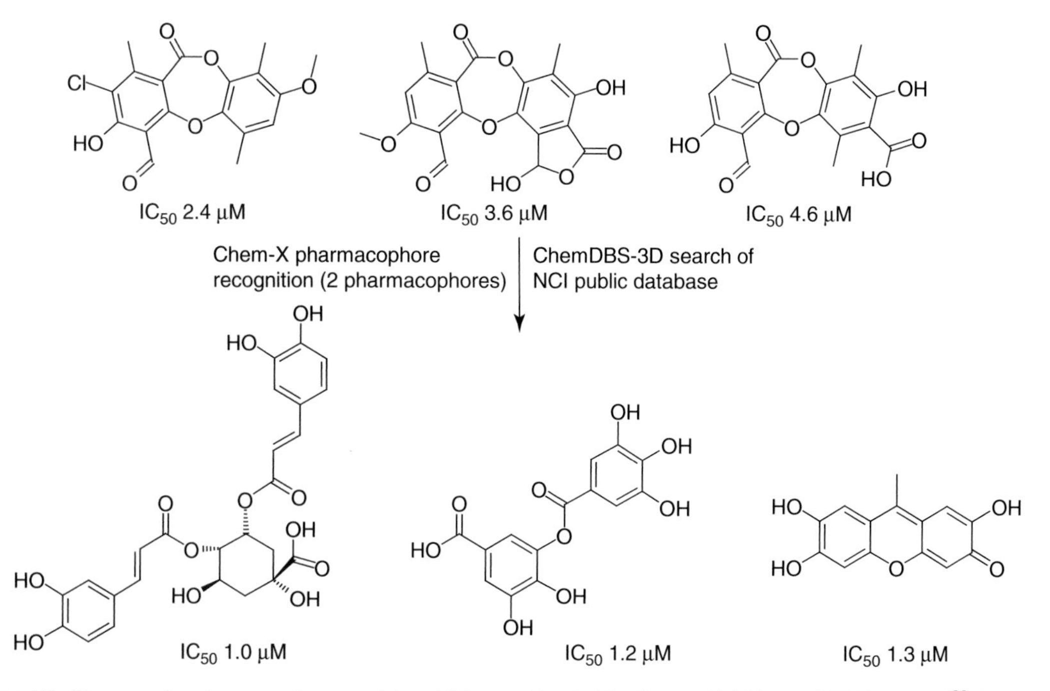

Figure 15 The use of a pharmacophore model and 3D searching to identify novel inhibitors of HIV integrase.[26] Two similar three-point pharmacophores are reported. Each contains two oxygen atoms and an oxygen or nitrogen atom. For the first model distances of 2.7 Å, 8.0 Å, and 8.7 Å were used; for the second, the corresponding distances are 3.0 Å, 7.4 Å, and 8.7 Å. Each distance has a tolerance of 0.7 Å.

Catalyst HipHop 4-point pharmacophore hypothesis | Catalyst search of vendor database

IC_{50} 0.9 μM

IC_{50} 1.9 μM

IC_{50} 2.8 μM

Figure 16 The use of a pharmacophore model and 3D searching to identify novel inhibitors of HIV integrase.[27] The pharmacophore contains (1) a hydrophobic aromatic region, (2,3) two hydrogen bond acceptor sites and (4) a hydrogen bond donor site at the following distances: 1–2, 6.2 Å; 1–3, 8.82 Å; and 1–4, 10.9 Å. All distances have a 1 Å, tolerance.

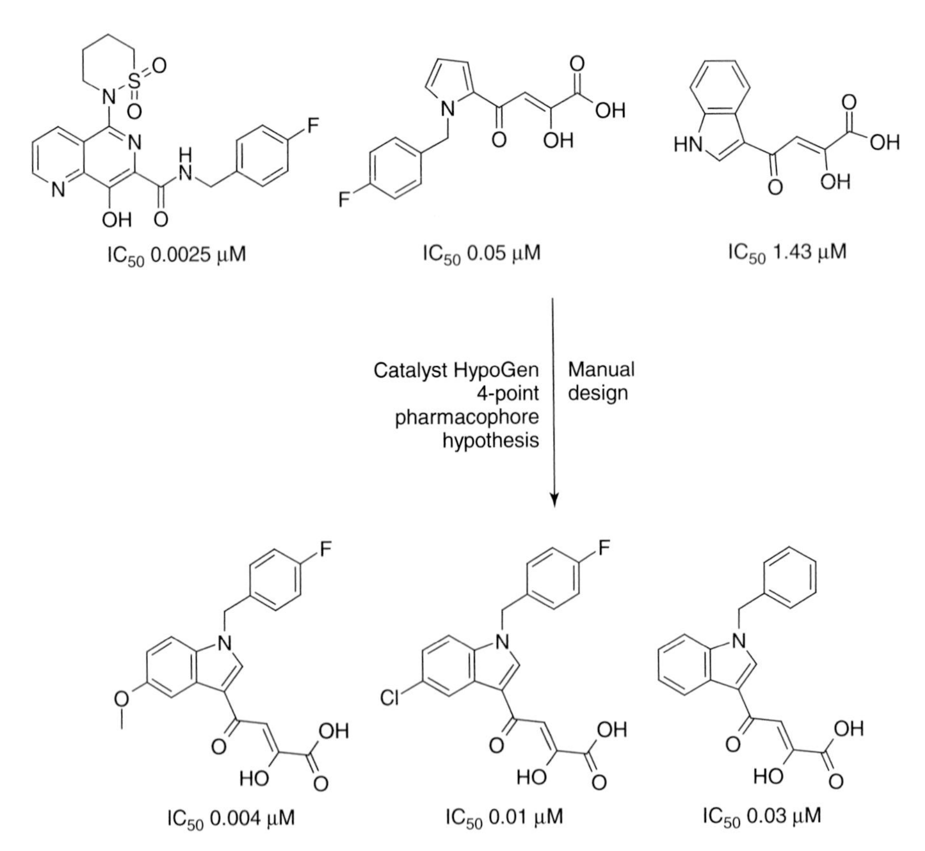

Figure 17 The use of a pharmacophore model and 3D searching to identify novel inhibitors of HIV integrase.[28] The pharmacophore model contains (1-3) three hydrogen bond acceptors and (4) a hydrophobic aromatic region at the following distances: 1–2, 2.5–3.0 Å; 1–3, 5.0–5.6 Å; 1–4, 3.5–4.2 Å; and 2–3, 2.5–3.2 Å.

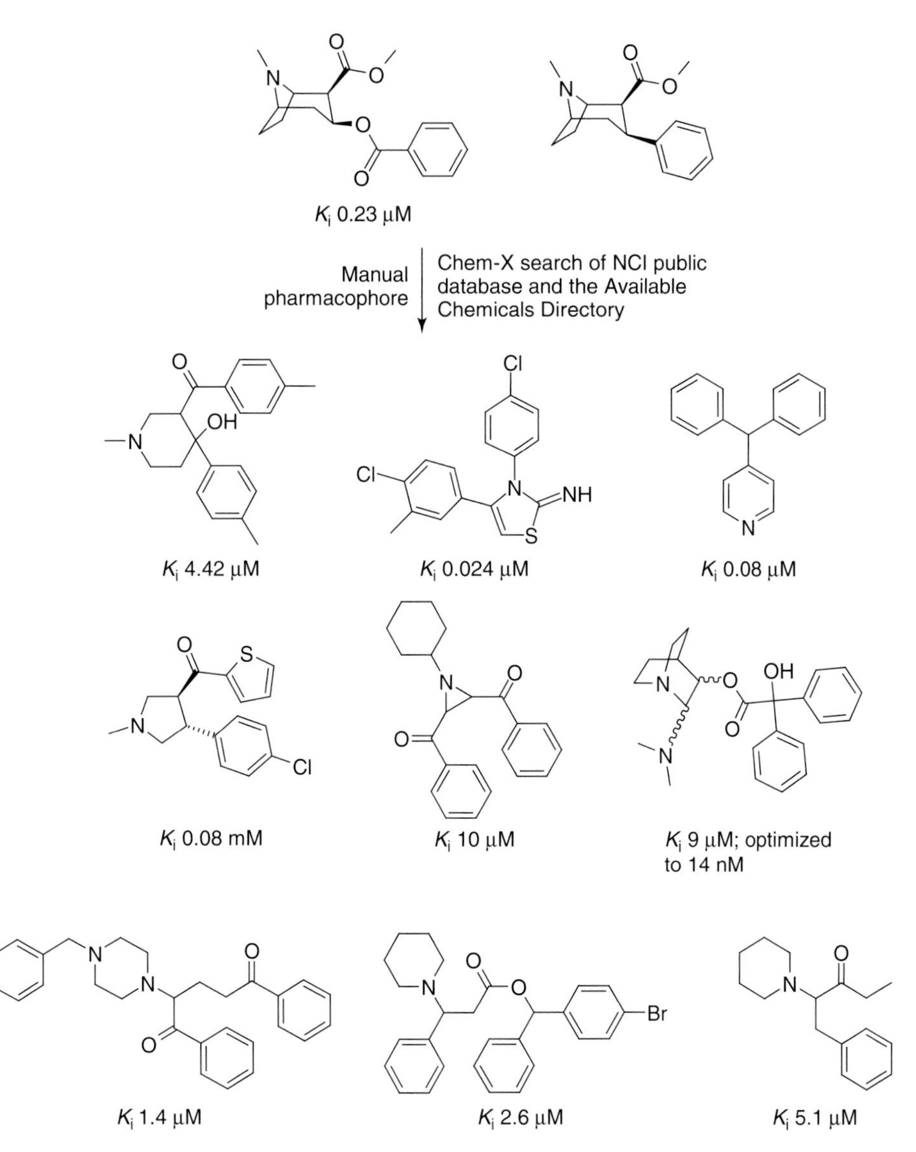

Figure 18 The iterative use of pharmacophore identification and 3D searching to identify novel dopamine transport inhibitors.[29–33] The first pharmacophore[29] contains (1) a basic or quaternary nitrogen, (2) a carbonyl oxygen, and (3) a centroid of an aromatic ring at the following distances: 1–2, 2.2–4.5 Å; 1–3, 5.0–7.0 Å; and 2–3, 3.4–6.1 Å. The second pharmacophore[30] contains (1) a basic or quaternary nitrogen and (2,3) two aromatic ring centroids at the following distances: 1–2, 5.1–6.1 Å; 1–3, 5.7–6.7 Å; and 2–3, 5.6–6.5 Å. The third pharmacophore[31] contains (1) an sp2 nitrogen and (2,3) two aromatic ring centroids at the following distances: 1–2, 3.4–4.4 Å; 1–3, 4.7–5.7 Å; and 2–3, 4.4–5.4 Å. A fourth pharmacophore[32] contains (1) an sp2 nitrogen and (2,3) two aromatic ring centroids at the following distances: 1–2, 5.1–6.7 Å; 1–3, 3.7–6.1 Å; and 2–3, 4.0–6.5 Å. The final pharmacophore[33] is the same as the first except that atom 1 can also be an aliphatic carbon.

In a similar vein, **Figure 18** shows the types of compounds identified by 3D searching for dopamine transport inhibitors as a potential treatment for cocaine abuse.[29–33] At least nine series of active compounds were identified; most series have members that are more potent than cocaine.

Figure 19 shows compounds that were identified by 3D searching based on a proposed pharmacophore for inhibition of protein kinase C.[34] Interestingly, a number of other groups also proposed somewhat different pharmacophores for this enzyme.[35–38] The subsequent determination of the structure of protein kinase C bound to a phorbol ester shows that none of the pharmacophore models was correct in every detail, not even the one that led to the discovery of novel active molecules.[39]

Figures 19–31 illustrate the variety of structures that result from a 3D search.[34,40–54]

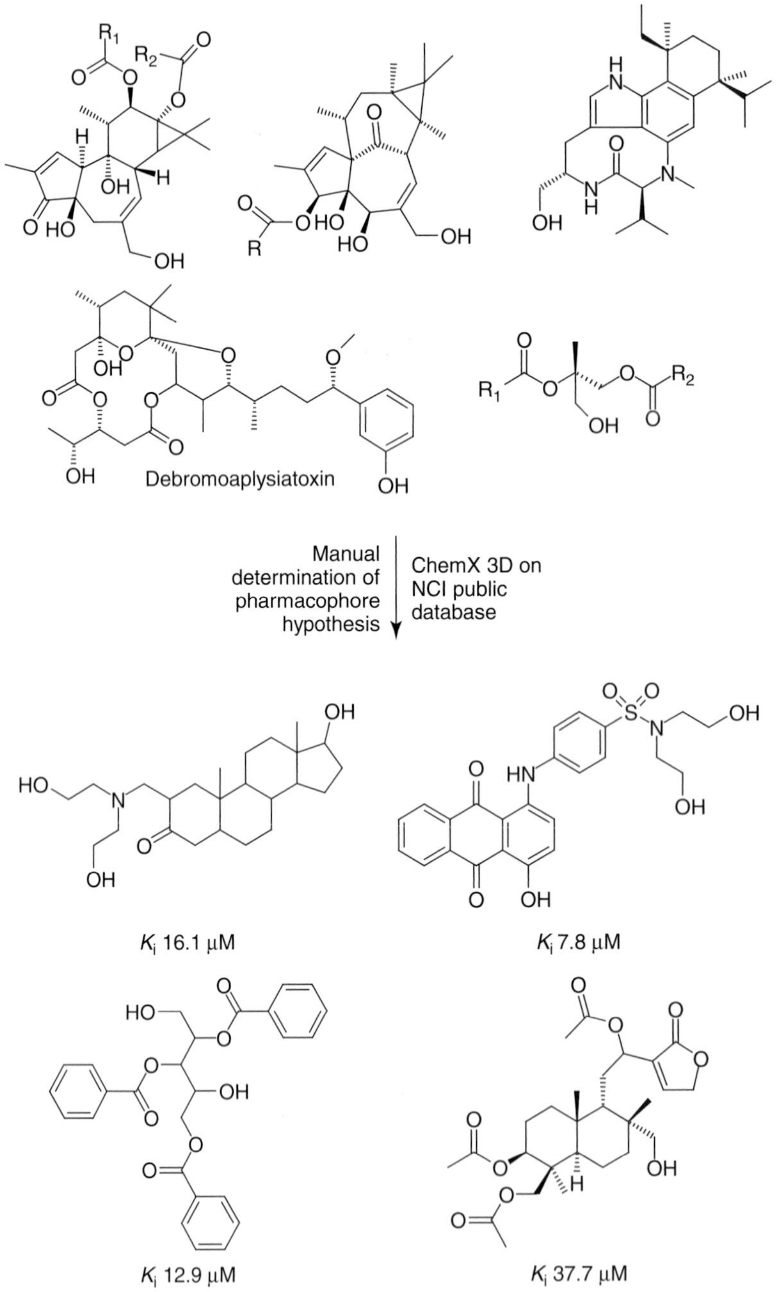

Debromoaplysiatoxin

Manual determination of pharmacophore hypothesis

ChemX 3D on NCI public database

K_i 16.1 μM

K_i 7.8 μM

K_i 12.9 μM

K_i 37.7 μM

Figure 19 The use of a pharmacophore model and 3D searching to identify novel inhibitors of protein kinase C.[34] The pharmacophore contains three points: (1) a carbonyl carbon atom, (2) an oxygen atom, and (3) a hydroxyl oxygen at the following distances: 1–2, 6.00±0.25 Å; 1–3, 5.70±0.6 Å; and 2–3, 6.40±0.6 Å.

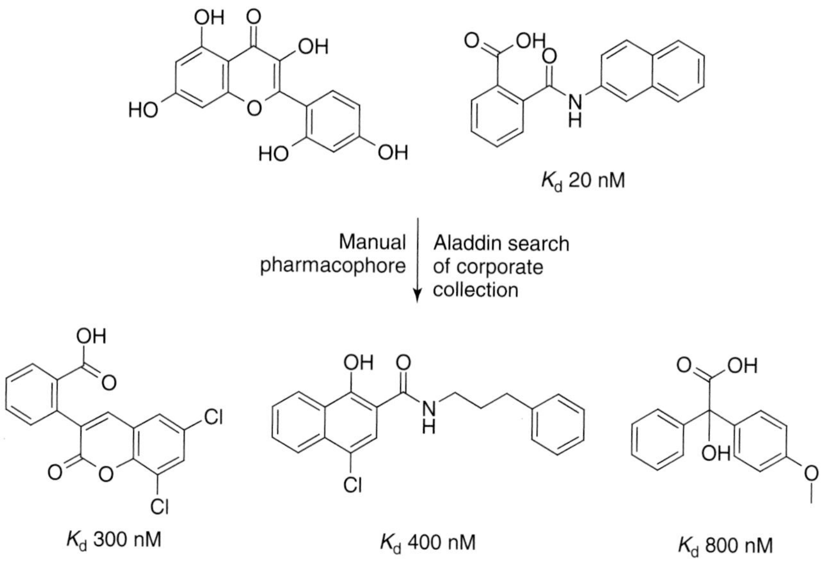

K_d 20 nM

Manual pharmacophore | Aladdin search of corporate collection

K_d 300 nM

K_d 400 nM

K_d 800 nM

Figure 20 The use of a pharmacophore model and 3D searching to identify novel auxin transport inhibitors.[40] Three related three-point pharmacophores contain (1) an acidic oxygen atom, (2) a tertiary carbon (pharmacophores 1 and 2) or an aromatic atom (pharmacophore 3), and (3) an aromatic atom (pharmacophores 1 and 2) or aromatic ring centroid (pharmacophore 3). The following distances apply to the first two models: 1–2, 2.7–3.7 Å; 2–3, 4.0–5.9 Å; 1–3, 4.6–6.8 Å. For the third model the following distance ranges apply: 1–2, 5.7–7.3 Å; 1–3, 4.9–7.0 Å; and 2–3, 4.9–7.0 Å.

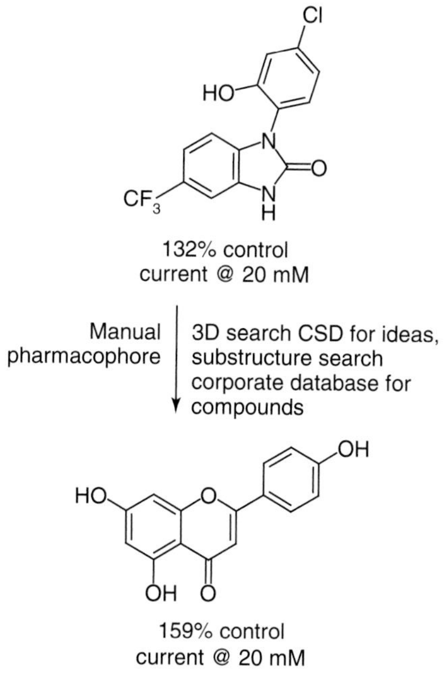

132% control current @ 20 mM

Manual pharmacophore | 3D search CSD for ideas, substructure search corporate database for compounds

159% control current @ 20 mM

Figure 21 The use of a pharmacophore model and 3D searching to identify novel openers of Ca^{2+}-dependent large-conductance potassium channels.[41] The pharmacophore contains four points; (1) a carbonyl oxygen, (2) an oxygen or nitrogen hydrogen bond donor, and (3,4) two aromatic ring centroids at the following distances: 1–2, 2.5–3.0 Å; 1–3, 3.0–5.0 Å; 1–4, 3.0–7.0 Å. Point (2) is directly bonded to the aromatic ring that generates centroid 3.

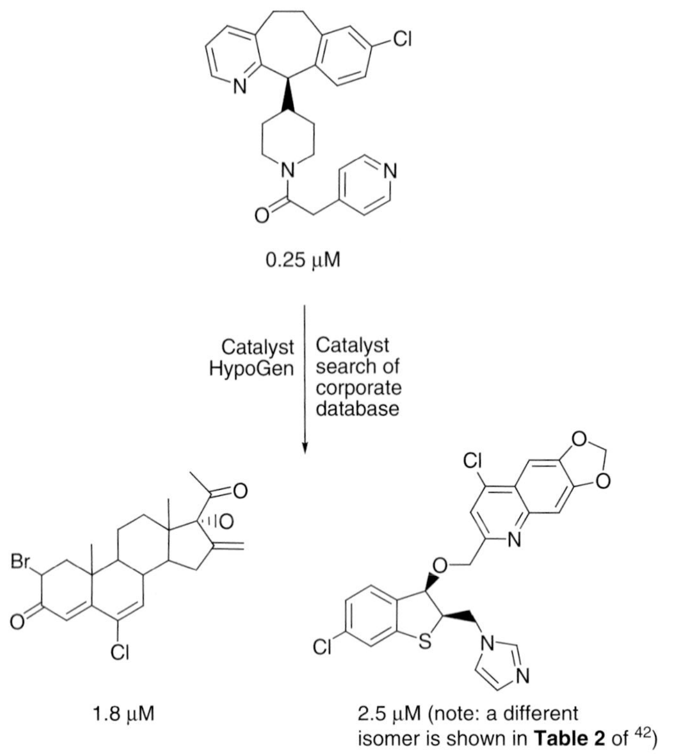

0.25 μM

Catalyst HypoGen | Catalyst search of corporate database

1.8 μM

2.5 μM (note: a different isomer is shown in **Table 2** of [42])

Figure 22 The use of a pharmacophore model and 3D searching to identify novel farnesyl protein transferase inhibitors.[42] The pharmacophore model contains four hydrophobic regions and one hydrogen bond acceptor site. No constraints are provided.

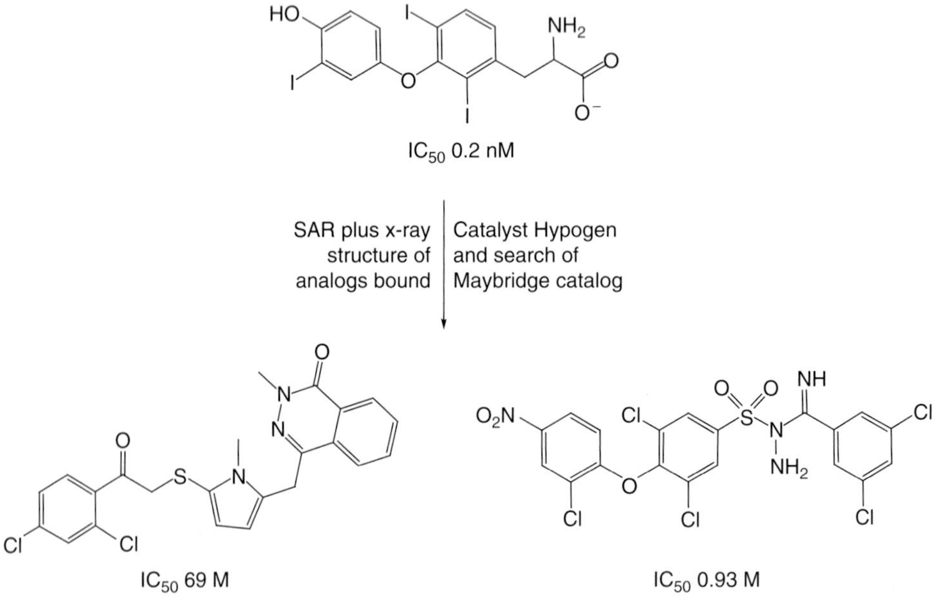

IC_{50} 0.2 nM

SAR plus x-ray structure of analogs bound | Catalyst Hypogen and search of Maybridge catalog

IC_{50} 69 M

IC_{50} 0.93 M

Figure 23 The use of a pharmacophore model augmented with excluded volumes from the protein structure to search a 3D database for novel inhibitors of thyroid hormone binding.[43] Ten different pharmacophore models, all based on the bound conformation of thyroxin, were used for database searching: all contain hydrophobic points at the positions of the 3-I and 5-I groups and the centroids of the inner and outer rings. They differ in whether or not the 3′-I group was included, whether the 4′-OH is considered a donor or acceptor, and whether the COOH is included.

DISCO pharmacopho research | Unity search of company plus vendor databases

pA_2 6.7 pA_2 5.5 pA_2 4.8

Figure 24 The use of a pharmacophore model and 3D searching to identify novel muscarinic M_3 receptor antagonists.[45] The pharmacophore contains (1) a tertiary nitrogen, (2) a hydrogen bond acceptor; and (3,4) two hydrogen bond donors. The first pharmacophore has the following distances: 1–2, 5.4 Å; 1–3, 5.8 Å; 1–4, 7.6 Å; 2–3, 5.0; 2–4, 2.9 Å; and 3–4, 6.5 Å. The second pharmacophore has the following distances: 1–2, 5.4 Å; 1–3, 4.1 Å; 1–4, 6.6 Å; 2–3, 6.0; 2–4, 2.9 Å; and 3–4, 8.5 Å. Tolerances of $\pm$0.3 Å apply to all distances.

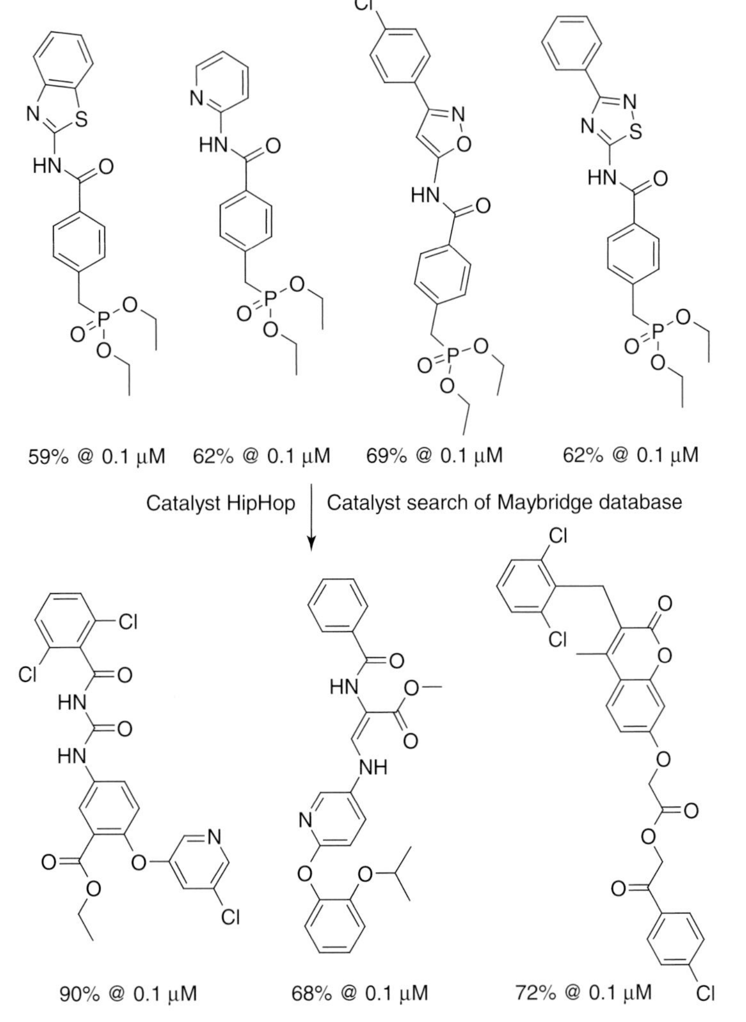

Figure 25 The use of a pharmacophore model and 3D searching to identify novel inhibitors of mesangial cell proliferation.[46] The pharmacophore contains two hydrophobic regions, two hydrophobic aromatic regions, and three hydrogen bond acceptors. No distances or tolerances are given.

20 Catalyst pharmacophore models

Catalyst search of Available Chemicals Directory, tested 32 compounds

>30% inhibition @ 10 μM

>30% inhibition @ 10 μM

Figure 26 The use of multiple pharmacophore models and 3D searching to identify novel EDG-3 antagonists.[48] Twenty catalyst models were generated manually. Each contains a positive ionizable feature, a negative ionizable feature, a hydrogen bond acceptor, and a hydrophobic group centered at a double bond. A tolerance of 1.5 Å was considered.

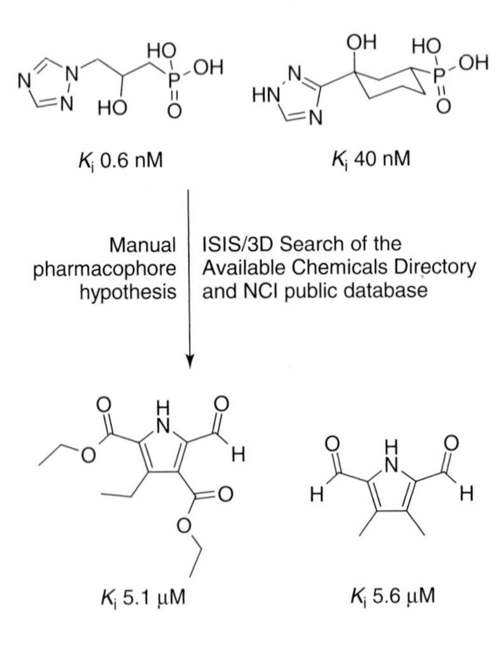

Figure 27 The use of a pharmacophore model and 3D searching to identify novel imidazole glycerol phosphate dehydratase inhibitors.[49] The pharmacophore contains (1) a hydrogen bond donor oxygen, nitrogen, or sulfur, (2) a carbon or nitrogen in a five-membered aromatic ring, (3) an aromatic nitrogen in a five-membered ring, (4) an oxygen singly bonded to carbon, phosphorous, or surfur, and (5) the corresponding carbon, phosphorous, or sulfur at the following distances: 1–2, 4.2–5.2 Å; 1–3, 2.6–3.6 Å; 1–4, 4.1–5.1 Å; 1–5, 3.5–4.5 Å; 2–4, 5.0–6.0 Å; 3–4, 5.6–6.6 Å.

K_i 1.9 nM

Catalyst HypoGen model

Catalyst 3D search of Maybridge and the Available Chemicals Directory

K_i 121 nM

Figure 28 The use of a pharmacophore model and 3D searching to identify novel ligands for the benzodiazepine binding site of the $GABA_A$ receptor.[50] The pharmacophore emphasizes the characteristics of the protein-binding site, which is proposed to include a hydrogen bond acceptor, an hydrogen bond donor, a bifunctional hydrogen bond donor/acceptor site, three lipophilic pockets and three regions of steric repulsion. No distances or tolerances are given.

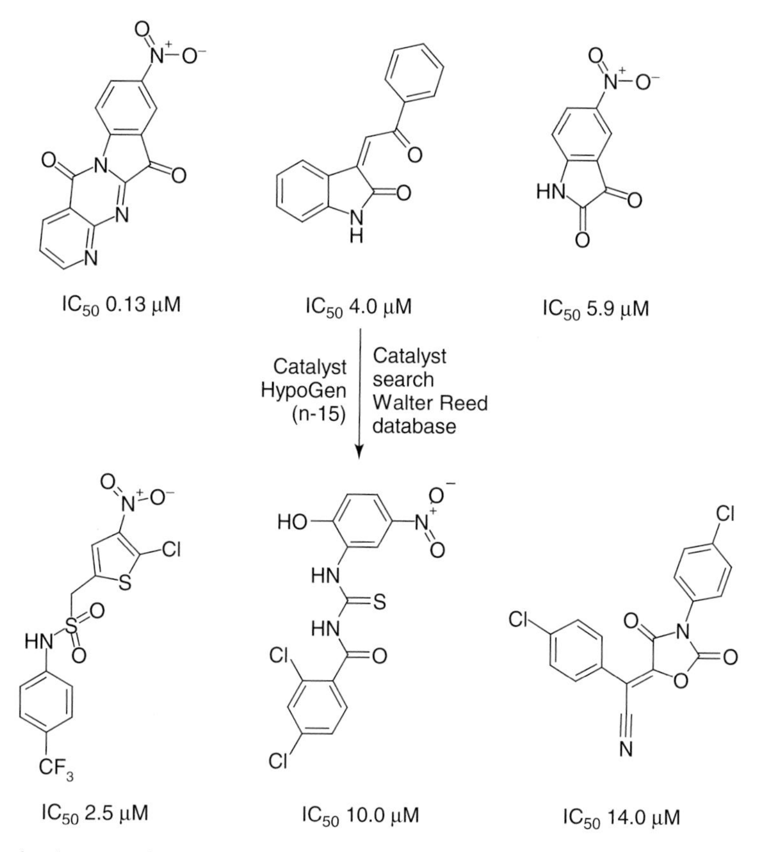

Figure 29 The use of a pharmacophore model and 3D searching to identify novel inhibitors of *Plasmodium falciparum* cyclin-dependent kinases.[51] The pharmacophore contains two hydrogen bond acceptors, one hydrophobic aromatic ring centroid, and a generic hydrophobic site. Only one distance was indicated, 3.4 Å between a hydrogen-bond acceptor and the aromatic ring centroid.

IC_{50} 0.2 μM

IC_{50} 2.8 μM

Conformational search followed by DISCO pharmacophore detection

Unity search of company collection

IC_{50} 5.8 μM

IC_{50} 0.5 μM

Figure 30 The use of a pharmacophore model and 3D searching to identify novel blockers of the Kv1.5 channel.[52] The pharmacophore contains three hydrophobic centers spaced 6.56 Å, 6.58 Å, and 12.62 Å apart.

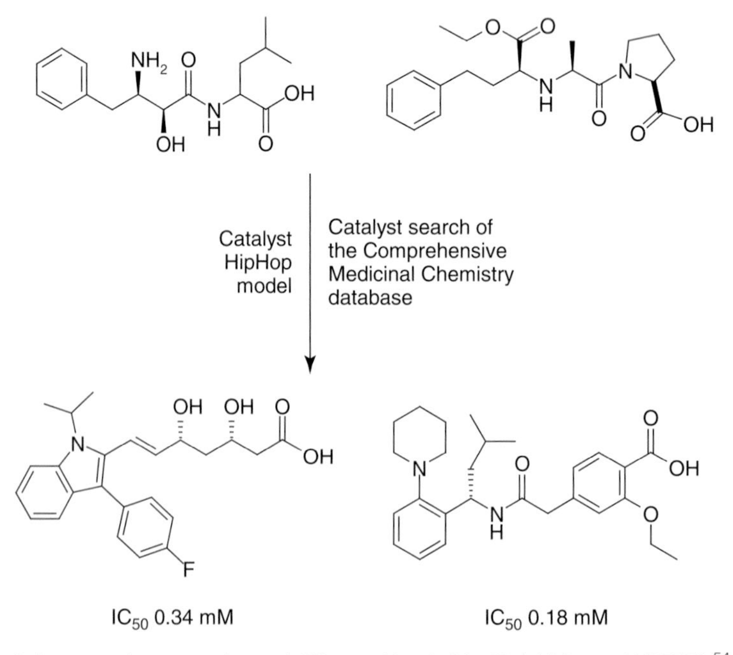

Figure 31 The use of pharmacophore mapping and 3D searching to identify inhibitors of hPEPT1.[54] The pharmacophore contains a negative ionizable group, a hydrogen bond acceptor, a hydrogen bond donor, and two hydrophobes at unspecified distances.

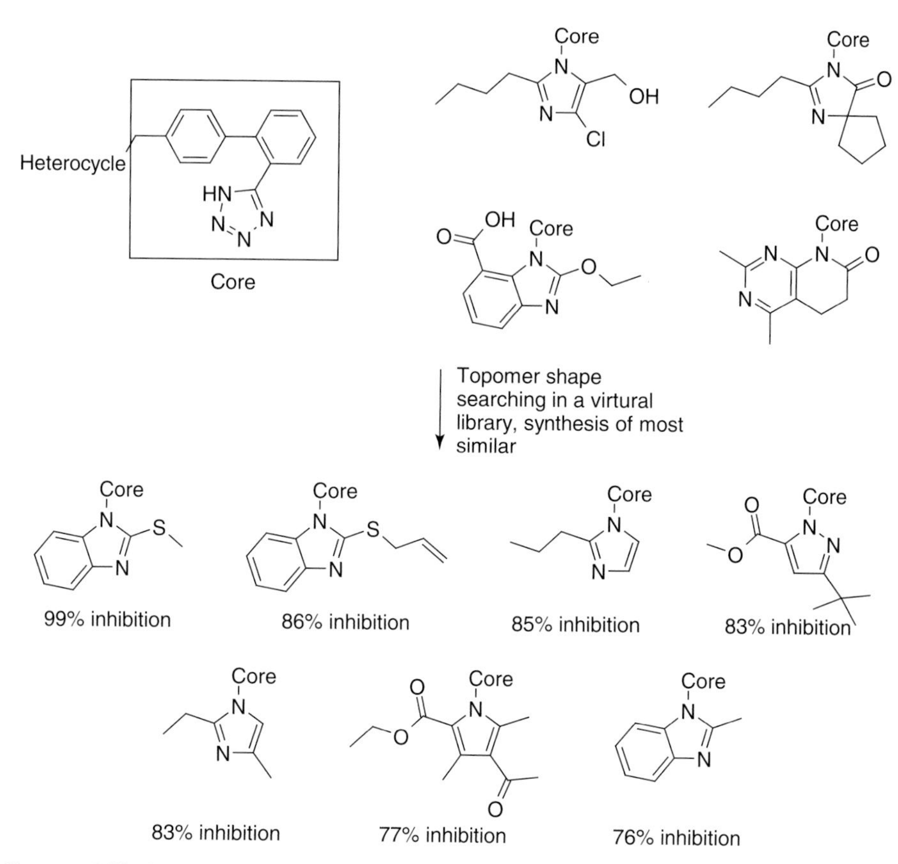

Figure 32 The use of 3D shape searching to identify novel angiotensin II antagonists.[55]

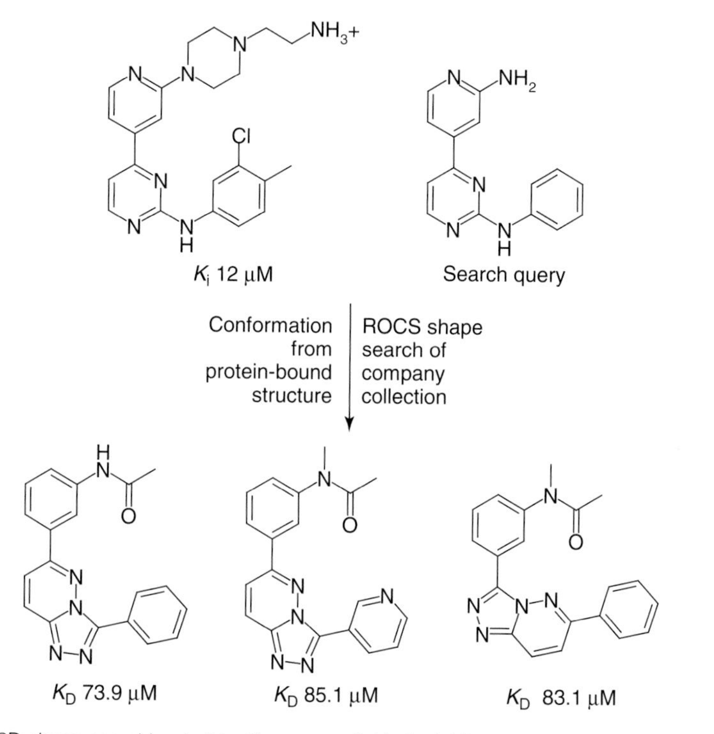

Figure 33 The use of 3D shape searching to identify new scaffolds for inhibitors of ZipA-FtsZ protein–protein interaction.[56]

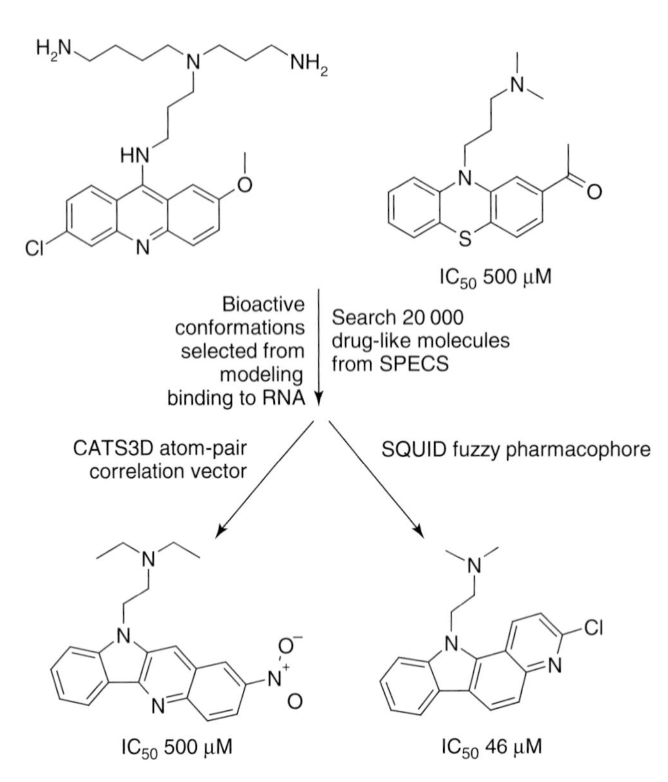

Figure 34 The use of two 3D searching methods to identify novel inhibitors of the Tat-TAR RNA interaction.[57]

Table 1 Further examples of novel compounds discovered from pharmacophores

Type of compound sought	*Method of pharmacophore identification*	*Method of compound design*	*Compound class identified*
Dopamine antagonists[65]	Manual	Manual	3*H*-benz[*e*]indol-8-amines
ET_A (endothelin A) selective endothelin antagonists[66]	Manual	3D searching, program and database not specified	Sulfonamide analogs of nonselective antagonist
EGFR protein tyrosine kinase inhibitors[67]	Manual	Manual	Isoflavones and 3-phenyl-4(1H)-quinolones
ALS inhibitors[68]	DISCO	ISIS/3D search of the Available Chemicals Directory	Sulfonamide analogs
5HT (5-hydroxytryptamine, serotonin) reuptake inhibitors[69]	Catalyst HypoGen	Manual	Analogs of known tricyclic $5HT_3$ antagonist
$5HT_4$ antagonists[70]	Catalyst HypoGen	Manual	Analogs of known tricyclic $5HT_3$ antagonist
α4β1 (VLA-4) antagonists[71]	Manual based on bound conformation of a related antigen	Catalyst search of a virtual library	Substituted benzamide replacement for a tripeptide portion
Chymase inhibitors[72]	Catalyst HypoGen	Catalyst search of the Available Chemicals Directory	Benzo[*b*]thiophen-3-sulfonamide
Binding to the nicotinic receptor[73]	Manual	Cambridge Structural Database for ideas	Quinoline derivatives
D3 dopaminergic binding[58]	Support vector machine	3D search of Specs IBS collection	Benzanilide derivatives
Modulators of mGlu R5[74]	Manual pharmacophore	3D search of a subset of SPECS using CATS3D atom-pair correlation vectors	Vinyl ketones and amides

Figures 32 and **33**[55,56] illustrate the use of two different approaches to shape-based screening. For the search shown in **Figure 32** 'standard' conformations are used, but for that shown in **Figure 33** the shape is based on the presumed bioactive conformation of the ligand. Such recently developed ligand shape-based methods hold promise for further development. Lastly, **Figure 34** involved innovative methods to search databases.[57] Other methods from the same group are listed in **Table 1**, including strategies to automatically generate the search query.[58] These newer methods complement traditional pharmacophore-based searching. The legends of the figures highlight the fact that many published pharmacophores and 3D search queries are incompletely described. As noted previously,[75] the resulting ambiguities make it difficult to benchmark newer methods with existing ones, which in turn hinders developments in the field. It is also difficult to benchmark the results of 3D searching if the contents of the database and details of its preparation are not clearly described.

4.21.4 Discussion

Given the success of pharmacophore-based methods to aid the discovery of novel series of bioactive molecules, why have these methods been slow to penetrate the day-to-day workings of medicinal chemists? One factor may be the challenge of the discovery of one or more pharmacophore hypotheses: Usually, a computational chemist must do this, with the resulting culture gap between the computational and experimental disciplines. The competition for 3D database screening (virtual screening) is high-throughput screening (HTS), to which experimentalists are more attuned. The most useful databases for 3D searching are unlikely to be those of corporations, but are more likely to be the NCI (National Cancer Institute) database,[59] vendor databases,[60–62] or virtual libraries[63,64] because they contain compounds that would not be part of HTS.

References

1. Klabunde, T.; Evers, A. *ChemBioChem* **2005**, *6*, 876–889.
2. Ekins, S. *Drug Disc. Today* **2004**, *9*, 276–285.
3. Ekins, S.; Boulanger, B.; Swaan, P. W.; Hupcey, M. A. Z. *J. Comput.-Aided Mol. Des.* **2002**, *16*, 381–401.
4. Zhang, E. Y.; Phelps, M. A.; Cheng, C.; Ekins, S.; Swaan, P. W. *Adv. Drug Deliv. Rev.* **2002**, *54*, 329–354.
5. Ekins, S. *Biochem. Soc. Trans.* **2003**, *31*, 611–614.
6. Ekins, S.; Crumb, W. J.; Sarazan, R. D.; Wikel, J. H.; Wrighton, S. A. *J. Pharmacol. Exp. Ther.* **2002**, *301*, 427–434.
7. Olson, G. L.; Cheung, H.-C.; Morgan, K. D.; Blount, J. F.; Todaro, L.; Berger, L.; Davidson, A. B.; Boff, E. *J. Med. Chem.* **1981**, *24*, 1026–1034.
8. Hibert, M. F.; Gittos, M. W.; Middlemiss, D. N.; Mir, A. K.; Fozard, J. R. *J. Med. Chem.* **1988**, *31*, 1087–1093.
9. Butera, J.; Bagli, J.; Doubleday, W.; Humber, L.; Treasurywala, A.; Loughney, D.; Sestanj, K.; Millen, J.; Sredy, J. *J. Med. Chem.* **1989**, *32*, 757–765.
10. Bures, M. G.; Hutchins, C. W.; Maus, M.; Kohlbrenner, W.; Kadam, S.; Erickson, J. *Tetrahedron Comput. Methodol.* **1990**, *3*, 673–680.
11. Lam, P. Y.; Jadhav, P. K.; Eyermann, C. J.; Hodge, C. N.; Ru, Y.; Bacheler, L. T.; Meek, J. L.; Otto, M. J.; Rayner, M. M.; Wong, Y. N. et al. *Science* **1994**, *263*, 380–384.
12. Wang, S. M.; Milne, G. W. A.; Yan, X. J.; Posey, I. J.; Nicklaus, M. C.; Graham, L.; Rice, W. G. *J. Med. Chem.* **1996**, *39*, 2047–2054.
13. Schoenleber, R. W.; Kebabian, J. W.; Martin, Y. C.; DeNinno, M. P.; Perner, R. J.; Stout, D. M.; Hsiao, C.-N. W.; DiDomenico, S., Jr.; DeBernardis, J. F.; Basha, F. Z. et al. *Dopamine Agonists*; US Patent 4,963,568; Abbott Laboratories: IL, 1990.
14. DeNinno, M. R.; Schoenleber, R.; MacKenzie, R.; Britton, D. R.; Asin, K. E.; Briggs, C.; Trugman, J. M.; Ackerman, M.; Artman, L.; Bedzarz, L. et al. *Eur. J. Pharmacol.* **1991**, *199*, 209–219.
15. Martin, Y. C.; Kebabian, J. W.; MacKenzie, R.; Schoenleber, R. Molecular Modeling-Based Design of Novel, Selective, Potent D1 Dopamine Agonists. In *QSAR: Rational Approaches on the Design of Bioactive Compounds*; Silipo, C., Vittoria, A., Eds.; Elsevier: Amsterdam, 1991, pp 469–482.
16. Glen, R.; Martin, G.; Hill, A.; Hyde, R.; Woollard, P.; Salmon, J.; Buckingham, J.; Robertson, A. *J. Med. Chem.* **1995**, *38*, 3566–3580.
17. Traxler, P.; Bold, G.; Frei, J.; Lang, M.; Lydon, N.; Mett, H.; Buchdunger, E.; Meyer, T.; Mueller, M.; Furet, P. *J. Med. Chem.* **1997**, *40*, 3601–3616.
18. Parkes, K. E. B.; Ermert, P.; Fassler, J.; Ives, J.; Martin, J. A.; Merrett, J. H.; Obrecht, D.; Williams, G.; Klumpp, K. *J. Med. Chem.* **2003**, *46*, 1153–1164.
19. Koga, H.; Ohta, M.; Sato, H.; Ishizawa, T.; Nabata, H. *Bioorg. Med. Chem. Lett.* **1993**, *3*, 625–631.
20. Lepailleur, A.; Bureau, R.; Lemaître, S.; Dauphin, F.; Lancelot, J. C.; Contesse, V.; Lenglet, S.; Delarue, C.; Vaudry, H.; Rault, S. *J. Chem. Inf. Comput. Sci.* **2004**, *44*, 1148–1152.
21. Low, C. M. R.; Buck, I. M.; Cooke, T.; Cushnir, J. R.; Kalindjian, S. B.; Kotecha, A.; Pether, M. J.; Shankley, N. P.; Vinter, J. G.; Wright, L. *J. Med. Chem.* **2005**, *48*, 6790–6802.
22. Buck, I. M.; Black, J. W.; Cooke, T.; Dunstone, D. J.; Gaffen, J. D.; Griffin, E. P.; Harper, E. A.; Hull, R. A. D.; Kalindjian, S. B. et al. *J. Med. Chem.* **2005**, *48*, 6803–6812.
23. Nicklaus, M. C.; Neamati, N.; Hong, H. X.; Mazumder, A.; Sunder, S.; Chen, J.; Milne, G.; Pommier, W. A. Y. *J. Med. Chem.* **1997**, *40*, 920–929.
24. Hong, H. X.; Neamati, N.; Wang, S. M.; Nicklaus, M. C.; Mazumder, A.; Zhao, H.; Burke, T. R.; Pommier, Y.; Milne, G. W. A. *J. Med. Chem.* **1997**, *40*, 930–936.
25. Neamati, N.; Hong, H. X.; Mazumder, A.; Wang, S. M.; Sunder, S.; Nicklaus, M. C.; Milne, G.; Proksa, B.; Pommier, Y. *J. Med. Chem.* **1997**, *40*, 942–951.
26. Neamati, N.; Hong, H. X.; Sunder, S.; Milne, G. W. A.; Pommier, Y. *Mol. Pharmacol.* **1997**, *52*, 1041–1055.

27. Barreca, M. L.; Ferro, S.; Rao, A.; De Luca, L.; Zappalà, M.; Monforte, A. M.; Debyser, Z.; Witvrouw, M.; Chimirri, A. *J. Med. Chem.* **2005**, *48*, 7084–7088.
28. Barreca, M. L.; Rao, A.; De Luca, L.; Zappalà, M.; Gurnari, C.; Monforte, P.; De Clercq, E.; Van Maele, B.; Debyser, Z.; Witvrouw, M. et al. *J. Chem. Inf. Comput. Sci.* **2004**, *44*, 1450–1455.
29. Wang, S. M.; Sakamuri, S.; Enyedy, I. J.; Kozikowski, A. P.; Deschaux, O.; Bandyopadhyay, B. C.; Tella, S. R.; Zaman, W. A.; Johnson, K. M. *J. Med. Chem.* **2000**, *43*, 351–360.
30. Enyedy, I. J.; Wang, J. S.; Zaman, W. A.; Johnson, K. M.; Wang, S. M. *Bioorg. Med. Chem. Lett.* **2002**, *12*, 1775–1778.
31. Enyedy, I. J.; Sakamuri, S.; Zaman, W. A.; Johnson, K. M.; Wang, S. M. *Bioorg. Med. Chem. Lett.* **2003**, *13*, 513–517.
32. Enyedy, I. J.; Zaman, W. A.; Sakamuri, S.; Kozikowski, A. P.; Johnson, K. M.; Wang, S. *Bioorg. Med. Chem. Lett.* **2001**, *11*, 1113–1118.
33. Sakamuri, S.; Enyedy, I. J.; Zaman, W. A.; Tella, S. R.; Kozikowski, A. P.; Flippen, A.; Judith, L.; Farkas, T.; Johnson, K. M.; Wang, S. *Bioorg. Med. Chem.* **2003**, *11*, 1123–1136.
34. Wang, S. M.; Zaharevitz, D. W.; Sharma, R.; Marquez, V. E.; Lewin, N. E.; Du, L.; Blumberg, P. M.; Milne, G. W. A. *J. Med. Chem.* **1994**, *37*, 4479–4489.
35. Jeffrey, A. M.; Liskamp, R. M. J. *Proc. Natl. Acad. Sci. USA* **1986**, *83*, 241–245.
36. Wender, P. A.; Koehler, K. F.; Sharkey, N. A.; Dell'Aquila, M. L.; Blumberg, P. M. *Proc. Natl. Acad. Sci. USA* **1986**, *83*, 4214–4218.
37. Itai, A.; Kato, Y.; Tomoika, N.; Iitaka, Y.; Endo, Y.; Hasegawa, M.; Shudo, K.; Fujiki, H.; Sakai, S.-I. *Proc. Natl. Acad. Sci. USA* **1988**, *85*, 3688–3692.
38. Nakamura, H.; Kishi, Y.; Pajares, M. A.; Rando, R. *Proc. Natl. Acad. Sci. USA* **1989**, *86*, 9672–9676.
39. Zhang, G.; Kazanietz, M. G.; Blumberg, P. M.; Hurley, J. H. *Cell* **1995**, *81*, 917–924.
40. Bures, M. G.; Black-Schaefer, C.; Gardner, G. *J. Comput.-Aided Mol. Des.* **1991**, *5*, 323–334.
41. Li, Y.; Starrett, J. E.; Meanwell, N. A.; Johnson, G.; Harte, W. E.; Dworetzky, S. I.; Boissard, C. G.; Gribkoff, V. K. *Bioorg. Med. Chem. Lett.* **1997**, *7*, 759–762.
42. Kaminski, J. J.; Rane, D. F.; Snow, M. E.; Weber, L.; Rothofsky, M. L.; Anderson, S. D.; Lin, S. L. *J. Med. Chem.* **1997**, *40*, 4103–4112.
43. Greenidge, P. A.; Carlsson, B.; Bladh, L. G.; Gillner, M. *J. Med. Chem.* **1998**, *41*, 2503–2512.
44. Liu, J.; Li, Z. M.; Yan, H.; Wang, L. X.; Chen, J. P. *Bioorg. Med. Chem. Lett.* **1999**, *9*, 1927–1932.
45. Marriott, D. P.; Dougall, I. G.; Meghani, P.; Liu, Y. J.; Flower, D. R. *J. Med. Chem.* **1999**, *42*, 3210–3216.
46. Kurogi, Y.; Miyata, K.; Okamura, T.; Hashimoto, K.; Tsutsumi, K.; Nasu, M.; Moriyasu, M. *J. Med. Chem.* **2001**, *44*, 2304–2307.
47. Singh, J.; van Vlijmen, H.; Liao, Y. S.; Lee, W. C.; Cornebise, M.; Harris, M.; Shu, I. H.; Gill, A.; Cuervo, J. H.; Abraham, W. M. et al. *J. Med. Chem.* **2002**, *45*, 2988–2993.
48. Koide, Y.; Hasegawa, T.; Takahashi, A.; Endo, A.; Mochizuki, N.; Nakagawa, M.; Nishida, A. *J. Med. Chem.* **2002**, *45*, 4629–4638.
49. Schweitzer, B. A.; Loida, P. J.; CaJacob, C. A.; Chott, R. C.; Collantes, E. M.; Hegde, S. G.; Mosier, P. D.; Profeta, S. *Bioorg. Med. Chem. Lett.* **2002**, *12*, 1743–1746.
50. Kahnberg, P.; Howard, M. H.; Liljefors, T.; Nielsen, M.; Nielsen, E. Ø.; Sterner, O.; Pettersson, I. *J. Mol. Grap. Model.* **2004**, *23*, 253–261.
51. Bhattacharjee, A. K.; Geyer, J. A.; Woodard, C. L.; Kathcart, A. K.; Nichols, D. A.; Prigge, S. T.; Li, Z.; Mott, B. T.; Waters, N. C. *J. Med. Chem.* **2004**, *47*, 5418–5426.
52. Peukert, S.; Brendel, J.; Pirard, B.; Strubing, C.; Kleemann, H. W.; Bohme, T.; Hemmerle, H. *Bioorg. Med. Chem. Lett.* **2004**, *14*, 2823–2827.
53. Guandalini, L.; Martini, E.; Dei, S.; Manetti, D.; Scapecchi, S.; Teodori, E.; Romanelli, M. N.; Varani, K.; Greco, G.; Spadola, L. et al. *Bioorg. Med. Chem.* **2005**, *13*, 799–807.
54. Ekins, S.; Johnston, J. S.; Bahadduri, P.; D'Souza, V. M.; Ray, A.; Chang, C.; Swaan, P. W. *Pharm. Res.* **2005**, *22*, 512–517.
55. Cramer, R. D.; Poss, M. A.; Hermsmeier, M. A.; Caulfield, T. J.; Kowala, M. C.; Valentine, M. T. *J. Med. Chem.* **1999**, *42*, 3919–3933.
56. Rush, T. S., III.; Grant, J. A.; Mosyak, L.; Nicholls, A. *J. Med. Chem.* **2005**, *48*, 1489–1495.
57. Renner, S.; Ludwig, V.; Boden, O.; Scheffer, U.; Gobel, M.; Schneider, G. *ChemBioChem* **2005**, *6*, 1119–1125.
58. Byvatov, E.; Sasse, B. C.; Stark, H.; Schneider, G. *ChemBioChem* **2005**, *6*, 997–999.
59. Milne, G. W. A.; Nicklaus, M. C.; Driscoll, J. S.; Wang, S. *J. Chem. Inf. Comput. Sci.* **1994**, *34*, 1219–1224.
60. Mdl Acd: Available Chemicals Directory, Elsevier MDL: San Leandro, CA.
61. Iresearch Library. *ChemNavigator*; Iresearch Library: San Diego, CA.
62. Irwin, J. J.; Shoichet, B. K. *J. Chem. Inf. Model.* **2005**, *45*, 177–182.
63. Cramer, R. D.; Patterson, D. E.; Clark, R. D.; Soltanshahi, F.; Lawless, M. S. *J. Chem. Inf. Comput. Sci.* **1998**, *38*, 1010–1023.
64. Andrews, K. M.; Cramer, R. D. *J. Med. Chem.* **2000**, *43*, 1723–1740.
65. Asselin, A. A.; Humber, L. G.; Voith, K.; Metcalf, G. *J. Med. Chem.* **1986**, *29*, 648–654.
66. Chan, M. F.; Okun, I.; Stavros, F. L.; Hwang, E.; Wolff, M. E.; Balaji, V. N. *Biochem. Biophys. Res. Commun.* **1994**, *201*, 228–234.
67. Traxler, P.; Green, J.; Mett, H.; Sequin, U.; Furet, P. *J. Med. Chem.* **1999**, *42*, 1018–1026.
68. Liu, J.; Li, Z. M.; Yan, H.; Wang, L. X.; Chen, J. P. *Bioorg. Med. Chem. Lett.* **1999**, *9*, 1927–1932.
69. Bureau, R.; Daveu, C.; Lancelot, J.-C.; Rault, S. *J. Chem. Inf. Comput. Sci.* **2002**, *42*, 429–436.
70. Bureau, R.; Daveu, C.; Lemaître, S.; Dauphin, F.; Landelle, H.; Lancelot, J.-C.; Rault, S. *J. Chem. Inf. Comput. Sci.* **2002**, *42*, 962–967.
71. Singh, J.; van Vlijmen, H.; Liao, Y. S.; Lee, W. C.; Cornebise, M.; Harris, M.; Shu, I. H.; Gill, A.; Cuervo, J. H.; Abraham, W. M.; Adams, S. P. *J. Med. Chem.* **2002**, *45*, 2988–2993.
72. Koide, Y.; Tatsui, A.; Hasegawa, T.; Murakami, A.; Satoh, S.; Yamada, H.; Kazayama, S.; Takahashi, A. *Bioorg. Med. Chem. Lett.* **2003**, *13*, 25–29.
73. Guandalini, L.; Martini, E.; Dei, S.; Manetti, D.; Scapecchi, S.; Teodori, E.; Romanelli, M. N.; Varani, K.; Greco, G.; Spadola, L.; Novellino, E. *Bioorg. Med. Chem.* **2005**, *13*, 799–807.
74. Renner, S.; Noeske, T.; Parsons, C. G.; Schneider, P.; Weil, T.; Schneider, G. *ChemBioChem* **2005**, *6*, 620–625.
75. Kristam, R.; Gillet, V. J.; Lewis, R. A.; Thorner, D. *J. Chem. Inf. Model.* **2005**, *45*, 461–476.

© 2007 Elsevier Ltd. All Rights Reserved
No part of this publication may be reproduced, stored in any retrieval system or transmitted in any form by any means electronic, electrostatic, magnetic tape, mechanical, photocopying, recording or otherwise, without permission in writing from the publishers

Comprehensive Medicinal Chemistry II
ISBN (set): 0-08-044513-6

ISBN (Volume 4) 0-08-044517-9; pp. 515–536

4.22 Topological Quantitative Structure–Activity Relationship Applications: Structure Information Representation in Drug Discovery

L H Hall, Eastern Nazarene College Quincy, MA, USA
L B Kier, Virginia Commonwealth University, Richmond, VA, USA
L M Hall, Hall Associates Consulting, Quincy, MA, USA

© 2007 Elsevier Ltd. All Rights Reserved.

4.22.1 Introduction

Modern drug design has evolved into a multifaceted approach linking (1) molecules, (2) measured properties and biological activities, and (3) receptors, active sites, or binding sites. The interplay of these ingredients is drug action or effect that has an ultimate clinical significance.

This complex system[1] presents a challenge to derive information about a molecule that is pertinent to information about its properties and biological activity, so that eaningful relationships can be developed for the list of molecules in a data set. In this chapter we focus on the molecules in the relationship and the associated information called structure. The molecule is the only member of the biological trio that can be overtly manipulated by the scientist. The receptors and the active sites are handed down to us through evolutionary processes.

In drug design we exploit the structure of the molecule in two broad categories of application. In one case we study the influence of structure on properties and biological activities to developing a relationship. In the other application, the molecular structure is used for the evaluation of similarity by screening a large library of structures for the purpose of identifying targeted molecules. Both applications require a structure description that is uniform, reproducible, and clearly interpretable. These applications will be described in the following sections followed by the description of a structure description paradigm called structure information representation.

4.22.1.1 Relating Structure to Activity

One approach to the development of new drug candidates is the development of a relationship between the structure of molecules and their pharmaceutical properties. Historically, a variety of methods have been utilized to encode and incorporate molecular structure information for the purpose of developing quantitative structure–activity relationships (QSARs). One of the significant methods for encoding structure information which we have developed is known as structure information representation (SIR).[2,3] In the historical QSAR tradition, this concept of molecular structure contributed in two significant ways. First, SIR encodes electronic and steric attributes of molecular structure in a unified way, in contrast to most other methods. Second, structure information is encoded directly with structure descriptors rather than indirectly through the use of calculated physicochemical properties selected on the basis of mechanistic assumptions.

To provide a useful context for the presentation of SIR, we will consider the kinds of information necessary in drug development. The identification of molecular structure and the set of experimental values from biological testing are the twin features of information that are exploited. These provide the entry point for QSAR.

4.22.1.2 The Structure–Activity Relationship

An examination of the typical data developed in a drug development project reveals certain aspects that are significant for this presentation. In earlier publications we have described the challenges in developing a model relating structure and activity/property as the QSAR problem.[2,3] Typical data may be presented as a set of structures along with an associated set of measured property values as illustrated in **Table 1**. Early in the design process the set may be small in number; later a larger set is developed. In addition, more diverse data sets may be used for broad-based models. **Table 1**, binding affinities of dopamine D_3 ligands, may be used to illustrate information for which a relationship is desired for data with a common core skeleton.[4]

A useful way to gain perspective on a data set is to rank order the list on the property values, placing the most potent at the top of the list. By most potent we mean smallest IC_{50}, largest binding constant, smallest inhibition constant. The list of structure and property values, illustrated in **Table 1**, reveals more than the mere magnitude and range of activity values. The position of a compound in the rank-ordered list reveals the level of activity for the corresponding structure. A significant aspect of rank-ordered QSAR data is that the highest values (most potent, most tightly bound, etc.) in the list indicate compounds that have structure features that enable them to produce the phenomenon more effectively. By

Table 1 Binding affinity (log(*K*)) for selected dopamine D_3 ligands

Molecular structure	*log(K) Binding affinity*	*Molecular structure*	*log(K) Binding affinity*
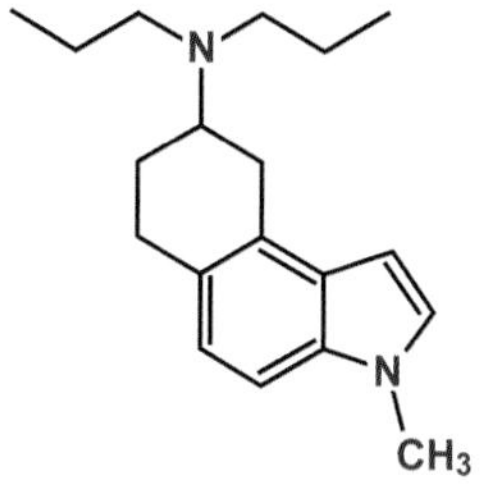	3.01	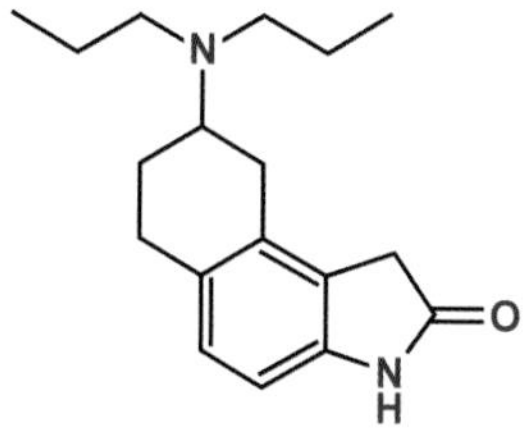	1.36
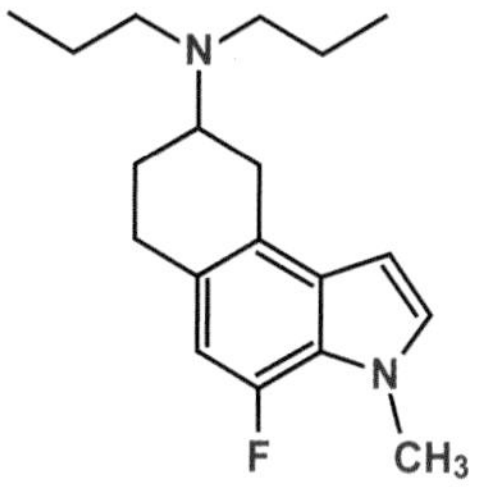	1.72	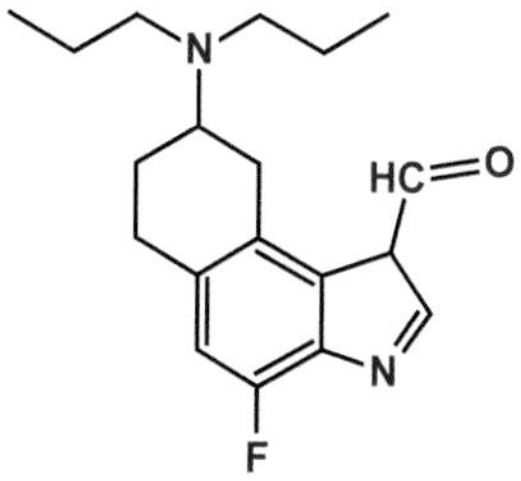	0.20
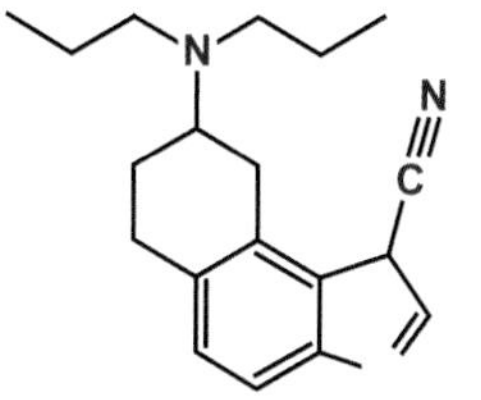	1.63		

contrast, those entries at the bottom (least potent, least tightly bound, etc.), possess different structure features and are least effective for the property. For data such as binding or inhibition data, strong effects arise from molecules that have higher complementarity to the receptor or active site. Structures with the lowest values have less complementarity to the receptor. Thus, these activity values lead the investigator to the structures that have features that are significantly related to the phenomenon and, hence, important for design.

Variation in structure in the data set is shown in **Table 1** as the second column. This particular data set is built around a common core skeleton. Structure variation arises from both type and position of substituent on an otherwise constant skeleton of the common core. Examination of the structures and activities reveals that placement of a methyl or a fluoro group alters the activity. Even these seemingly small changes lead large differences in activity. A QSAR model must account for this relationship.

When we consider another data set with a very diverse set of structures, a different picture emerges.[2,3] Examination of the structures reveals several features that vary, including number of rings, chemical types of rings, polarity, hydrogen bond donors/acceptors, bioisosteres, chemotype, branching, size, chemical elements, bond orders, and so forth.

In drug development, modeling addresses the question of which structure feature variations actually parallel the property variation. Many structure feature variations can be listed but how can the important ones be selected? What form of relationship is most appropriate? These questions define the modeling task.

The observations on structure–activity variation described above have been combined into an empirical generalization: variation in the activity parallels the variation in structure. Furthermore, this generalization leads to the relationship in QSAR (or computer-aided drug design (CADD)): property is a function of structure. Even when mechanistic assumptions are made, no way is known for direct computation of the biological activity.[2,3] One way to describe the task of QSAR is the establishment of methods that facilitate the search for the significant relationship between structure and activity. These methods are statistical in nature.

In some structure–property relations, the influence of structure is largely topological in nature. For example, the relation of $\log P$ with alkane structure shows a variation with counts of carbon groups such as methyl, methylene, etc. For alkanes a simple group additive scheme may suffice to predict $\log P$ values. For other properties, such as pK_a, relation to structure appears to be primarily electronic in origin. A substituent-dependent descriptor scheme may be adequate for pK_a prediction of substituted acetic acids. In general, however, as indicated by physical and biological data for pharmacologically active compounds, both electronic and topological aspects of structure vary through a list of molecular structures and both relate to property variation. For a structure–information system to be useful, combinations of electronic and topological information must be properly encoded.

The SIR approach capitalizes upon well-known qualitative physical-organic intuition based on experience with molecular structure. It also develops the information in a quantitative fashion useful for modeling.[2,3,5–10] In the approach based on SIR, QSAR models encode structure attributes derived from the SIR formalisms, validated with external test sets. They reveal molecular structure features significant to the property under investigation needed for prediction of new candidate structures.

The structure of a molecule must be quantitated in order to relate it to the behavior of a drug molecule or ligand. Structure is a many-faceted and hierarchical feature of a molecule that Testa and Kier have described in detail.[1] In this chapter we will consider what structure information is needed to create predictive models for molecules in relation to the activity of a biological endpoint.

4.22.1.3 Structure Information in the Library Screening for Similarity

A second major application of SIR is the use of the information to screen virtual libraries and search compound databases for structures that are similar to a lead compound or that meet certain structural criteria. The structure descriptors in the SIR system encode the kind of information that is well suited for screening structure libraries in an efficient manner.[11–15]

Drug design often begins with the identification of a particular macromolecule (usually a protein) implicated in a disease or pathology. Further research reveals an active site or receptor on the macromolecule, that responds to a molecule (ligand) functioning as a transmitter, substrate, or inhibitor. Further studies identify this ligand as a lead compound, a potential archetype for the design of a clinically useful drug. The next step in this process is to identify the salient features of the ligand that produce the chemical events of the observed activity. These features are called the pharmacophore, a pattern of atoms, groups, or fragments positioned in precise locations within the ligand. The pharmacophore may be considered to be a template for the reproduction or, alternatively, the search for other, similar molecules with comparable responses at the receptor.

The next step in the process is to create or find additional compounds to expand and improve upon the lead compound in the search for a new drug. The synthesis of additional compounds is an iterative process of synthesis, and testing, followed by more synthesis. An alternative approach to the enrichment of the pool of candidate molecules for evaluation is to make use of the many large libraries or databases that have come on the scene in recent years. These computer-stored files contain hundreds, thousands, millions of molecules in the form of codes that can be selected using molecular structure descriptors. Using the pharmacophore or other significant structure features as a working model of desirable molecular features, the descriptors that characterize the pharmacophore or significant features are created and used to search for similar molecules in the database.

To accomplish this task, a set of molecular descriptors must quantify atom, group, and fragments in an unambiguous way that relates to intermolecular interactions. The structure information resident in the E-state descriptors is appropriate for screening and searching tasks. Intermolecular interactions are governed by the electron accessibility at atoms and in groups of atoms. As is described in the next section, electron accessibility is the information encoded in E-state indices. The application of E-state to both QSAR and library screening and searching will be presented later in the chapter.

4.22.1.4 Molecular Structure as an Information Network

In the analysis of intermolecular interactions, the conventional approach has been to dissect the encounter of molecules into two separate phenomena. In this classic reductionist approach the electronic structure and the topology of molecules have been treated as two distinct entities. In this approach each attribute is quantified separately and entered independently into the modeling process which is aimed at acquiring a provisional model of structure-engagement in the encounter. Success by this approach over the years is undenied. Our position, however, is that we achieve greater insight if we accept the reality that molecular structure is the representation of a complex system made up of atoms and functional groups with internal patterns describable by information from both topological and electronic content in an integrated manner.

One useful way to view a molecule is as an information network. In this molecular network, atoms are considered as the nodes and bonds as the connections. As a node, each atom possesses three characteristics. First, the atom has a state; that is, it is a chemical element with electrons and protons and a set of bonds, usually characterized by the valence state. Second, the atom has a set of relationships to other atoms in the network directly through the bonds connecting neighboring atoms and indirectly, to atoms beyond. The valence state represents the electronic information associated with the set of directly bonded atoms or, in the case of aromaticity, atoms beyond those directly bonded but connected through π electron delocalization. Finally, each atom influences and is influenced by its connected (bonded) neighbors and, to varying extents, more remote atoms. SIR seeks to encode this information in several ways that facilitate QSAR development and interpretation.

These influences are summarized in the electron distribution which arises from the effect of electronegativity of the sigma-bonded neighbors directly through the bonds and indirectly from remote neighbors as well. The effect of electronegativity is a complex emergent property of the valence state of the atoms and their connectedness in the network, that is, the electronic structure arising from the bonded relationships in the molecule. From this starting point, we can define two structure information systems which are known collectively as SIRs.[2,3] These include the electrotopological state (E-state) and molecular connectivity (chi) indices.

In the interaction between molecules in noncovalent encounters, leading to physicochemical and biological properties, parts of molecules directly encounter each other. The results of these encounters and interactions are understood in terms of attractions and repulsions between the molecules. These forces include dispersion, dipolar interaction, hydrogen bond, electrostatic, and others. These are mitigated by interference from bulky parts of molecules termed steric effects. Both the electronic and steric aspects of these interactions actually arise from the electron distribution. In our approach both the electronic and steric aspects are unified in the representation as they are in the actual electronic structure.

4.22.2 A General Concept of the E-State

Both the electronic and steric (topological) aspects of molecular structure arise from the electron distribution across the molecule. In SIR these two aspects of structure are represented in a unified manner. The E-state formalism is developed on this foundation.

The electronic structure of a molecule is a critical factor in its interaction with another molecule. Based on the principle of logical depth,[16] it is clear that a molecule cannot engage any other entity but another molecule. For example, it is pointless to speak of a molecule engaging a whole cell. In the actual interaction event, a molecule engages another molecule that is part of the cell surface. This level of interaction is what can be quantified. By extension of this principle, only that part of a molecule, atom, functional group, or fragment that is engaging a comparable level feature on another molecule is suitable for characterization and investigation. This principle of logical depth provides the motivation for us to attempt to characterize molecular features so that we can model the possibility, extent, and specificities of a molecule engaging another molecule. An approach that fulfills these expectations must define features of a molecule, encode both electronic and topological attributes, and also encode the influence by these features on neighboring positions. This is the philosophy behind the creation of the electrotopological state (E-state) system over a decade ago.

4.22.2.1 Encoding Electron Counts in a Molecule

In order to characterize the important features of a molecule that are involved in intermolecular encounters, we begin by considering each atom or hydride group ($-CH_3$, –NH–, –OH, etc.) in the molecule as a distinct entity. Two characteristics of each fragment are encoded into a single value. This unification brings together the electron richness, derived from the electronegativity, and the topology, derived from a quantification of the adjacency of the fragment. Each of these attributes has been addressed and quantified by us in past work.[5,17] The sigma orbital electronegativity of an atom or hydride fragment has been shown by Kier and Hall to be describable with an expression based on electron counts. These electron counts are encoded into two atom indices, the delta (δ, δ^v) values. These delta values are defined as counts of electrons assigned to sigma orbitals and also to all valence electrons, not including the number of bonds to hydrogen atoms in the case of a hydride group, as follows[5,17]:

$$\delta = \sigma - h = \text{count of skeletal neighbors in the molecular skeleton}$$

$$\delta^v = \sigma + \pi + n - h = \text{count of valence electrons for an atom}$$

Table 2 Selected δ values of organic groups

Group	δ	δ^v	$\delta^v - \delta^a$
$-CH_3$	1	1	0
$-CH_2-$	2	2	0
$>CH-$	3	3	0
$>C<$	4	4	0
$=CH_2$	1	2	1
$=CH-$	2	3	1
$=C<$	3	4	1
$\equiv CH$	1	3	2
$\equiv C-$	2	4	2
$-NH_2$	1	3	2
$-NH-$	2	4	2
$-N<$	3	5	2
$-OH$	1	5	4
$-O-$	2	6	4
$=O$	1	6	5
$-F$	1	7	6
$-SH$	1	5	4
$-S-$	2	6	4
$=S$	1	6	5
$-Cl$	1	7	6
$-Br$	1	7	6

[a] Equal to the count of π and lone pair electrons on the atom.

In these definitions, the following conventions are used:

σ is the number of electrons assigned to sigma orbitals on an atom
π is the number of electrons assigned to pi orbitals
n is the number of electrons assigned to lone pair orbitals
h is the number of hydrogen atoms bonded to an atom

Each count refers to an atom or hydride group in the molecular skeleton, such as $>N-$, $=N-$, $-O-$, $=O$, $-Cl$, $-CH_3$, $-NH-$, etc.[5,17] (See **Table 2** for a list of atoms and groups common in organic chemistry along with their delta values.)

We have shown that these two counts, simple and valence electrons, are related to very important properties of atoms, the valence state electronegativity and volume.[5,17] These electron count descriptors form the basis for the development of both the electrotopological state and molecular connectivity. The starting point for the E-state development is the analysis of valence state electronegativity based on electron counts. Also listed in **Table 2** is the difference between the simple and valence delta values. This difference is shown to have a significant meaning for valence state electronegativity because it is equal to the count of pi and lone pair electrons.

4.22.2.2 Valence State Electronegativity

Valence state electronegativity (VSE) is defined as the ability of an atom, in its valence state, to attract electron density to itself from neighboring atoms through the sigma bonding network of the molecule. An experimental measure of VSE

Table 3 Delta values and experimental valence state electronegativity values

Group	*N*	δ	δ^v	$\delta^v - \delta$	X_{MJ} *(eV)*[a]
C(sp^3)	2	1	1	0	7.98
C(sp^2)	2	2	3	1	8.79
C(sp)	2	2	4	2	10.39
N(sp^3)	2	3	5	2	11.54
N(sp^2)	2	2	5	3	12.87
N(sp)	2	1	5	4	15.68
O(sp^3)	2	2	6	4	15.25
O(sp^2)	2	1	6	5	17.07
F(sp^3)	2	1	7	6	17.63
Si(sp^3)	3	1	1	0	7.30
Si(sp^2)	3	2	3	1	7.90
P(sp^3)	3	3	5	2	8.90
S(sp^3)	3	2	6	4	10.14
S(sp^2)	3	1	6	5	10.88
Cl(sp^3)	3	1	7	6	11.84
>As–	4	3	5	2	8.26
–Se–	4	2	6	4	9.08
–Br	4	1	7	6	9.90
–I	5	1	7	6	9.02

[a] $r^2 = 0.965$, $s = 0.55$, $n = 19$. See text for discussion.

has been developed by Hans Jaffé and co-workers,[18] based on the Mulliken definition of electronegativity.[19] In the Mulliken definition, electronegativity is based on the average of the ionization potential (I_p) and electron affinity (E_A) for an atom. Hinze and Jaffé extended the Mulliken definition to valence states by using I_p and E_A data taken from electronic spectroscopic data for atomic orbitals; they developed numerical values for valence states. This approach permits the evaluation of X_{MJ} values for various hybrid valence states.

The Mulliken–Jaffé valence state electronegativity values (X_{MJ}) characterize an atom in its several valence states unlike the less sensitive Pauling values which are the same for all valence states. For example, the Pauling value for carbon is the same for carbons atoms in ethane, ethylene, and acetylene.[20] However, the X_{MJ} values are different for these three hydrocarbons, corresponding to the increasing acidity of the hydrogen atoms in these three molecules. The larger valence state electronegativity for the acetylenic carbon correlates with the greater acidity of the hydrogen atom in acetylene and the greater polarity of the C–H bond. Similar information is resident in the X_{MJ} values for all atoms in their valence state (see **Table 3**).

In the development of the E-state formalism, we incorporated valence state electronic information because it characterizes the organic molecule effectively. For this reason, we adopted the Mulliken–Jaffé valence state electronegativity values as a reference for atomic properties that are important to electron density. The first step in this development is the finding of a relationship between X_{MJ} and molecular structure information. The linkage can be understood from the point of view of the source of electronegativity in a molecule. This effect arises from the effective nuclear charge on the atom, that is, the attraction of electrons arising from unshielded protons in the nucleus of the atom.[5,17]

As a reference point, we can consider that a carbon atom in an sp^3 valence state has four valence electrons in four sigma orbitals. Each valence electron in its sigma orbital, as part of the sigma bond, effectively shields one nuclear proton. The effective nuclear charge for this sp^3 carbon is taken to be zero and the corresponding VSE is also taken to be zero.

For an sp^2 carbon atom, one valence electron is in a pi orbital and three are in sigma orbitals. In this case, the pi electron, whose electron density is primarily outside of the sigma bonding orbital, does not effectively screen the core charge. The effective nuclear charge may be taken as one. The resulting electronegativity of the sp^2 carbon is higher than for the sp^3 carbon. Following the same line of argument, the effective nuclear charge for an acetylenic carbon (sp) is much higher than that of the sp^2 carbon, based on two unshielded protons. More generally, for carbon valence state electronegativity, the following ranking is observed in properties and described by valence state electronegativity: $\equiv$C–H$>$ $=CH_2$$>$–$CH_3$.

This approach to valence state electronegativity can be generalized. The ineffective shielding arises from the fact that pi and lone pair electron density has low or zero probability along the line of the bond axis. As a result, pi and lone pair electrons do not screen nuclear protons as effectively as electrons in sigma orbitals that are directed along the bond axis.[5,17]

The count of the pi and lone pair electrons on a bonded atom can be justified as a model of the electronegativity of its bound atom on both experimental and theoretical grounds. The work of Slater has shown that the pi and lone pair electrons, being further from the core than the sigma electrons, result in less shielding hence a greater influence of the core on the sigma bonding electrons.[23] This is the essence of electronegativity.

The count of pi and lone pair electrons may be readily related to the effective nuclear charge and, as a result, to the valence state electronegativity:

$$X_{MJ} \rightarrow \delta^v - \delta = \sigma + \pi + n - h - (\sigma - h) = \pi + n$$

This expression, for second-row elements, contains the count of the extrajacent electrons on a sigma-bonded atom, that is, the count of pi (π) and lone pair (n) electrons. For elements beyond fluorine, the additional screening of the additional core electrons must be taken into account. To include valence states for elements beyond fluorine, the principal quantum number of the valence electrons (N) is included, as follows:

$$X_{MJ} = 7.44(\delta^v - \delta)/N^2 + 6.57 \quad r^2 = 0.965, \ s = 0.55, \ n = 19 \qquad [1]$$

This expression was used in a model of VSE for 19 valence states, leading to a high correlation with X_{MJ} with a standard error of regression that approaches the known experimental errors, primarily in the determination of electron affinity values.[5,17] **Table 3** shows values for VSE for 19 atom valence states with the corresponding values for the delta values.

4.22.2.3 Intrinsic State

In the E-state formalism we encode the potential for noncovalent intermolecular interaction.[5] The E-state formalism is based on three hypotheses, set out in Sections 4.22.2.3.1–4.22.2.3.3.

4.22.2.3.1 First hypothesis

Our first hypothesis states that the interaction potential for an individual atom depends upon the electron accessibility at the atom. Electron accessibility here is defined as the accumulation/depletion of electron density mitigated by the steric (topological) accessibility of that electron density for intermolecular interaction. This potential unifies both the electron density and the topological environment. The accumulation/depletion of electron density arises from valence state electronegativity differences among the atoms in the molecule. The potential for accumulation/depletion of electron density is represented by the valence state electronegativity. The topology of an atom (or hydride group) is characterized by the count of its sigma bonds to atoms other than hydrogen (δ). The accessibility of the atom or hydride group is characterized by the topological information resident in the δ value.

4.22.2.3.2 Second hypothesis

The second hypothesis in the E-state formalism is that each atom in its valence state is characterized by an intrinsic state which encodes its electron accessibility in the isolated valence state and simultaneously encodes the relation between the electronic state and the topological state. The electron accessibility may be interpreted in one of two ways that leads to the same final equation. The actual topological accessibility is the reciprocal of the simple delta value, δ (count of skeletal neighbors). The encoding of the intrinsic state may then be considered as the product of the valence state electronegativity ($\pi + n$) and the topological accessibility ($1/\delta$). Alternatively, the intrinsic state may be considered as the ratio of the VSE to the steric crowding at that atom, δ. In either way, the provisional relation for intrinsic state, I, is as follows:

$$I \rightarrow (\pi + n)/\delta$$

The expression $(\pi + n)/\delta$ encodes the relative electron accessibility of the particular molecular fragment in isolation from bonded atoms. This expression results in zero values for all C(sp^3) atoms $(\pi + n = 0)$. The equation may be modified with a constant value of 1 in the numerator, giving $(\pi + n + 1)/\delta$, to provide a different value for $-CH_3$, $-CH_2-$, $>CH-$, and $>C<$ groups with their varying delta values. Finally, to scale all values greater than 1.0, the expression is modified to give an intrinsic state value of the noncovalent intermolecular interaction potential, I:

$$I = [(\delta^v + 1)/\delta] + 1 \qquad [2]$$

Table 4 presents the intrinsic state values for typical organic atom types. An examination of the table indicates that the intrinsic state values encode both the valence state electronegativity and the steric or topological environment of each atom. For example, the ranking of intrinsic state I values for some common groups is as follows:

$$\begin{array}{ccccc} \equiv CH(4.000) & > & =CH-(2.000) & > & -CH<(1.333) \\ -NH_2(4.000) & > & -NH(2.500) & > & -N<(2.000) \\ \equiv N(6.000) & > & =NH-(5.000) & > & -NH_2(4.000) \\ =O(7.000) & > & -OH(6.000) & > & -O-(3.500) \end{array}$$

In these examples, the intrinsic state values mirror known organic chemistry (see **Figure 3a** below for the intrinsic state values for *N*-methylpropanamide).

This expression, eqn [2], holds for second-row atoms but must be extended for higher rows of the periodic table. In considering atoms in higher rows, we note that electronegativity depends upon the total electron count, Z, as well as the valence electron count, Z^v. Equation [2] may be modified to deal with the attributes of higher-level atoms and groups. Oxygen and sulfur have the same δ values (by the definition above) but they have different valence state electronegativity values. The difference is even more pronounced when the halogen atoms are compared. We have accounted for these differences by modifying δ^v in the equation, since this quantity carries the information about the valence electrons.

Table 4 Intrinsic state values for organic atom types

Atom type	*I value*	*Atom type*	*I value*	*Atom type*	*I value*
$-CH_3$	2.000	$-NH_2$	4.000	–OH	6.000
$-CH_2-$	1.500	–NH–	2.500	–O–	3.500
>CH–	1.333	>N–	2.000	⩦O⩦	3.500
>C<	1.250	=NH	5.000	=O	7.000
$=CH_2$	3.000	=N–	3.000		
=CH–	2.000	⩦NH⩦	2.500	–F	8.000
	2.000	⩦N⩦	3.000	–Cl	4.111
=C<	1.667	≡N	6.000	–Br	2.750
⩦C(–)⩦	1.667			–I	2.120
≡CH	4.000	–SH	3.222		
≡C–	2.500	–S–	1.833		
		⩦S⩦	1.833		
>P–	1.074	=S	3.667		
>P=	0.806	>S=	0.917		
		=S(–)(–)=	0.917		

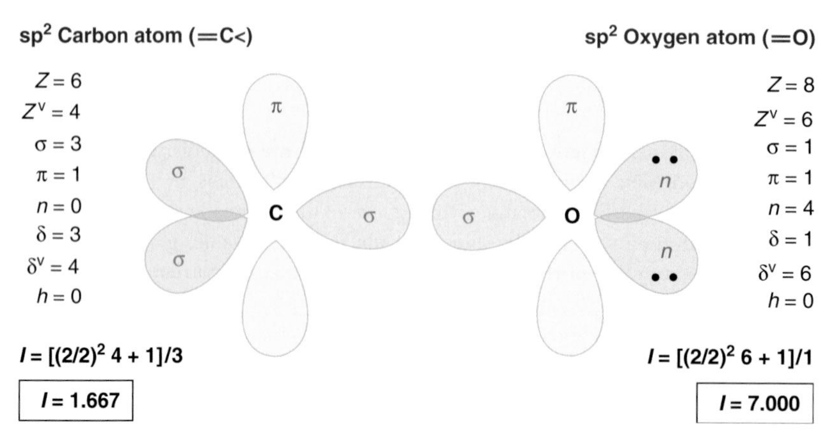

Figure 1 Intrinsic state for sp^2 carbon and sp^2 oxygen atoms. The sp^2 carbon atom has low electronegativity and low topological accessibility (three skeletal neighbors). The sp^2 carbon atom is assigned a smaller I value than sp^2 oxygen atom which has a high electronegativity and high steric accessibility (a terminal group with one skeletal neighbor). Definitions of symbols: σ, π, n are the count of electrons assigned to sigma, pi, and lone pair orbitals; δ is the simple connectivity delta value (count of skeletal neighbors); δ^v is the valence connectivity value; h is then number of hydrogen atoms in the group.

4.22.2.3.2.1 Higher row quantum states

The principal quantum number, N, reflects the influence of the quantum level on the properties of valence electrons. Therefore, N is used to modify the valence delta value, leading to the difference in intrinsic states among different atoms. In order to retain the values derived from this equation for second-row atoms, a term including N should be used as a coefficient of the δ^v value. The value of $(2/N)\delta^v$ equals δ^v when $N=2$. Based on the inverse square relationship for radial distribution in atoms, we encode the influence of N on the diminished electronegativity of higher level atoms as $(2/N)^2$ to modify the δ^v value. The equation for the intrinsic state, I_i, of any atom or group, I, becomes[5]:

$$I_i = [(2/\mathrm{N_i})^2\delta_i^v + 1]/\delta_i \qquad [3]$$

This modification results in an expression for the intrinsic state values for atoms in covalent organic compounds, including those in higher quantum levels shown in **Table 4**. A schematic representation of the information ingredients is found in **Figure 1** for the $C(sp^2)$ and $O(sp^2)$ valence states. Included in this figure are all the electron counts that are part of the intrinsic state formula. Also given is the actual calculation of the intrinsic state I values for these two valence states.

4.22.2.3.2.2 E-state formalism

The intrinsic state, I, of an atom or hydride group represents the structure unperturbed by the other atoms in the molecule. The final expression for electron accessibility, the E-state, must include the perturbations arising from all the other atoms. Each atom in a molecule is in a field of all the atoms in the whole molecule. The field is an information field; the information derives from the influences of all the other atoms, encoded by the intrinsic state (unified electronic and steric effects) values of other atoms. For the E-state value of each atom, this influence is encoded in a series of perturbations in units of the intrinsic state of each atom.

4.22.2.3.3 Third hypothesis

The third hypothesis of the E-state formalism states that this perturbation, ΔI_{ij}, is given by the difference between a reference atom intrinsic state and the intrinsic states of each other atoms.[5] The influence is mitigated by the distance, r, between the reference atom and each other atom in the molecule. These perturbations are summed and applied to the reference atom intrinsic state, as follows:

$$S_i = I_i + \sum_j (I_i - I_j)/r_{ij}^2 = I_i + \sum_j \Delta I_{ij} \qquad [4]$$

In this relation, r_{ij} is the count of atoms in the shortest path of atoms connecting the reference atom, i, to the perturbing atom, j. The sum of the intrinsic state and its perturbations from all other atoms in the molecule is the electrotopological state (E-state), S_i for atom i in the molecule under study.

Figure 2 depicts the E-state formalism for the >C=O bond formation. The intrinsic state I value for =O is much larger than that of >C=; hence, the resulting E-state value for =O is much larger than for >C=. The actual computation of the E-state values is shown in **Figure 2**, based on formaldehyde, $H_2C=O$.

An examination of the formalism as well as inspection of computed E-state values clearly indicates that this index encodes information about both the valence electron structure of an atom and its neighbors in the network making up the molecule. The E-state values are rich in information about the electron accessibilities of the atoms (hydride groups) in the molecule.

Variations of E-state values with structure changes are illustrated in **Figure 3** for three isoconnective molecules. In **Figure 3a**, the intrinsic state values are given for *N*-methylpropanamide. The carbonyl oxygen has the largest value because of its high electronegativity and terminal topological status. The two methyl groups have the same intrinsic

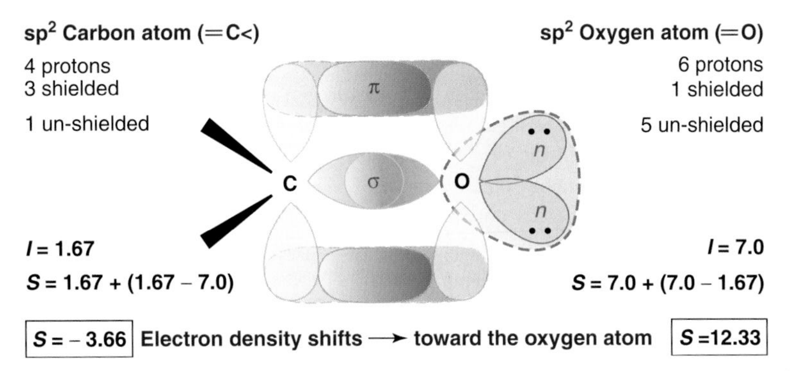

Figure 2 Illustration of the effects leading to the computed E-states value of an sp^2 carbon atom and an sp^2 oxygen atom. The oxygen atom experiences the effect of the one nuclear proton unshielded by the sigma electrons in the carbon sp^2 atom. The carbon atom experiences the effect of the five nuclear protons unshielded by the sigma electrons in the oxygen sp^2 atom. An electronic potential exists between the carbon and oxygen resulting in the electron density shifting toward the oxygen. The steric and electronic characters are encoded for an atom in a unified descriptor based on electron counts as defined by the delta values. The computation is for the formaldehyde molecule.

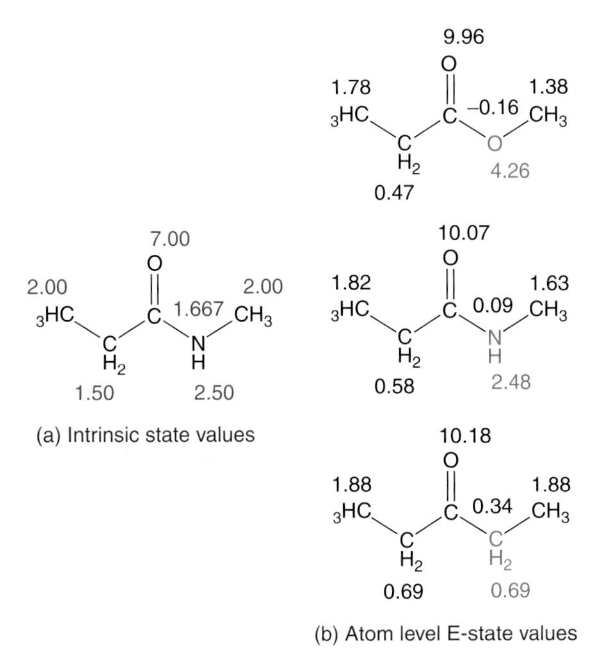

Figure 3 (a) Structure of *N*-methylpropanamide with intrinsic states given for each atom. (b) Computed E-state values for each atom group in three structures to illustrate E-state variation with atom electronegativity variation: methylpropanoate, *N*-methylpropanamide, and 3-pentanone. Red color indicates atom variation with corresponding E-state variation. See text for discussion.

state values, as do methyl groups in all structures. The –NH– group has an intermediate value with its intermediate electronegativity and topological status.

Figure 3b shows the final computed E-state S_i values that result from intrinsic state perturbations. In *N*-methylpropanamide the carbonyl oxygen E-state value is the largest because its I value is the highest in the structure, making all its perturbations positive. The two methyl groups have different E-state values reflecting the expected relationship known in chemistry. In the top structure in **Figure 3b**, methylpropanoate, the two methyl groups are even more different, reflecting the greater electronegativity of the ester oxygen (compared to –NH–). In the bottom structure of **Figure 3b**, the two methyl groups have equal E-state values, reflecting their symmetry in the structure.

In the E-state formalism, the E-state value is computed for each atom and hydride group. That is, for the –OH group, the E-state values implicitly includes the hydrogen atom. Before unfolding the information-rich nature of the E-state, we consider the possibility and potential utility of a separate E-state formalism for hydrogen atoms.

4.22.2.4 Specific Consideration of Hydrogen

Since the E-state values are derived for a molecule with a hydrogen-implied representation (with hydrogen atoms included in the 'atom') the numerical values of groups subsume the contributions from the separate atoms into a single index. This may be a tolerable situation in the case of the alkanes where the carbon and hydrogen atoms are very close in electronegativity. There is no appreciable polarity in the C–H bonds and so the group concept is a legitimate consolidation of the two atoms. This is clearly not the case with polar groups such as the –OH and =NH and others where the difference in electronegativity between the two atoms is very high. The hydrogen atom in this case is much more interactive than in the case of an alkane.

In general, hydrogen atoms in applications of the E-state require an independent consideration of their electrotopological state. This independence reveals itself in situations where a hydrogen bond may form or where an intermolecular interaction involving separate atoms in a group is to be considered in a structure–activity model. In these situations the strength of these interactions is dependent upon the electronic and topological nature of the individual atoms in the group. However, each atom plays a different role and so the standard E-state group treatment described above falls short of a useful E-state description of the hydrogen atoms. Attention to this matter has produced a useful extension of the E-state method.

In an approach to quantifying hydrogen atoms, Hall and Vaughn have focused on the electronegativity of each atom or hydride group.[6] All hydrogen atoms encountered in organic molecules occupy a terminal position, that is, a topology described with as $\delta = 1$. Topology is not a crucial factor in determining the E-state value for a hydrogen atom. Whether the hydrogen atom is part of hydride group in which the X–H bond is polar or nonpolar, relative electronegativity plays the dominant role in determining the character of that bond. For this reason, the intrinsic state value for the hydrogen atom arises from the relative electronegativity of the attached atom, X, together with perturbations of all other atoms. The Kier–Hall relative electronegativity is used to describe the electronegativity of each atom or hydride group in the molecule. For hydrogen atoms, this value is taken to be -0.200, based on experimental values.

4.22.2.5 Parameterization of Onium Groups

As presented thus far, the E-state formalism applies only to neutral organic molecules. Quaternary nitrogen atoms in the sp^3 hybrid state (e.g., tetramethylammonium) or sp^2 state (*N*-methylpyridinium) do appear in molecules of biological interest. Consideration of this group must be included in some E-state studies in order to make the method as broadly applicable as possible. An extension of the E-state formalism is needed which is consonant with the overall E-state formalism.[5] Our intention is to provide intrinsic state values for onium groups which permit an approximation of the intermolecular and intramolecular characteristics of the onium groups.

A useful extension should encode the observations that onium groups are quite electronegative, have varying topological status in a molecule, and are highly interactive across space. The approach to this issue begins with the observation that the ammonium group, $-NH_3^+$, is more electronegative than the amine group, $-NH_2$.[5,7] This effect arises from the effective core charge being higher in the onium group, taking into account the overall positive charge on the group. In this approach, representation of the core of the onium group must be defined to account for this difference. This definition must take into account the presence of one more hydrogen atom on the onium nitrogen relative to the corresponding amine group.

In this approach the $-NH_3^+$ onium group is regarded as a composite group or pseudoatom made up of five valence electrons from the nitrogen atom and one electron each from two hydrogen atoms.[5] That is, the core of the pseudoatom is made up of the nitrogen atom core plus the cores of the two hydrogen atoms sharing sigma bond electrons with the

Table 5 Intrinsic states of onium pseudoatoms

Onium group	*Onium pseudoatom*	Z^v	δ^v	δ	I
RNH_3^+	$[NH_3]$	7	6	1	7.00
$R_2NH_2^+$	$[NH_2]$	6	5	2	3.00
R_3NH^+	$[NH]$	5	4	3	1.67
R_4N^+ or $=N^+<$	$[N]$	5	4	4	1.50

nitrogen. The third hydrogen atom is formally considered to be outside of this pseudoatom model, sharing the (original) nitrogen atom lone-pair electrons to form its sigma bond. Based on the general equation for the intrinsic state, the intrinsic state value for the primary pseudoatom $[NH_3]$ is found to be 7.000, as shown in **Table 5**. For the secondary onium pseudoatom, $[NH_2]$ for $R_2NH_2^+$, there are six valence electrons, leading to $\delta^v = 5$, along with two sigma bonds, leading to $\delta = 2$. The intrinsic state of the pseudoatom $[NH_2]$, based on the equation for I, is found to be $I = 3.000$ as shown in **Table 5**. For the tertiary onium group, R_3NH^+, there are five valence electrons and three sigma bonds resulting in $\delta^v = 4$ and $\delta = 3$. The intrinsic state calculated is therefore $I = 1.67$, for this group designated as [NH].

In the final case, the quaternary nitrogen atom, R_4N^+, the pseudoatom is depicted by [N]. Here there are five valence electrons resulting in a value $\delta^v = 5$ and four sigma bonds leading to $\delta = 4$. The intrinsic value calculated is $I = 1.50$ for this pseudoatom, [N] (see **Table 5**). This approach succeeds, at least in a relative sense, in developing parameters for these onium groups so that they reflect the electronegativity within a molecule and possibly their interactive potential with other molecules.

4.22.3 Characterization of the E-State and its Chemical Significance

The E-state formalism produces calculated values for each atom in the molecule. The E-state has several characteristics of importance to drug research. As a computed value, the E-states have special significance among quantities available for modeling. Furthermore, the range of values for a given atom type is important in assessing the domain of applicability of a model. Finally the E-state has been related to others properties such as nuclear magnetic resonance (NMR) chemical shift and free valence. These topics are treated separately in the following sections.

4.22.3.1 Special Character of E-State

The E-state descriptors possess a unique feature that is not found in other, nonempirical indices such as molecular orbital, molecular dynamics, and Monte Carlo values or in empirical indices estimating molecular volume, surface area, or conformation. That feature is the exactness of the E-state value, not depending upon various approximated parameters. The indices are numerically exact because they are derived from cardinal numbers representing an exact count of electrons, number of neighbors, and distances measured in atom counts. If we assume that the assignment of these counts is correct, then the calculated index from them is exact; it is without error or probabilistic variation. The E-state formalism does not rest upon any assumptions of geometry, distance, energy, etc. The same numerical value is obtained regardless of other situations in the overall QSAR problem.

The E-state computed for an atom is based purely on electron and neighbor counts in the molecule. Because of this purity of formalism, the E-states for a molecule may be inverted from values found in a model to a structure. Such an inversion is difficult for property values such as log P or molar refraction because of the high number of possible structures that correspond to a property value.

We can say that the E-state indices encode a very high quality of structure information within the network model of a molecule. The quality of the structure information in the SIR approach has supported the development of QSAR models for data that involves pharmacophoric or other specific interactions, such as dopamine binding,[21] HIV-1 protease inhibitors,[24] or fish toxicity.[25] Further estimates of the structure as an extension in space may yield information not obtainable from the E-state analysis of the network model of a molecule. But this comes about with a loss of some accuracy because of the necessity to make approximations. This characteristic of the E-state formalism is a very useful feature because the apparent success or failure of a study using these indices cannot be ascribed to a poor choice of a parameter in the E-state calculation.

4.22.3.2 The E-State Spectrum Significance

Information associated with E-state values lies in numerical ranges shown in **Table 6**. For each E-state descriptor, the range of values arises from the interactions within the structure, according to the E-state formalism given by eqn [4]. The range of computed values represents the range of electron accessibility for that atom type. Some ranges in this data set are relatively small, such as the acetylenic carbon, ≡CH. Others such as the quaternary carbon, >C<, are very large because of the wide range of substituent atoms and groups possible in organic structures. The table also indicates the increasing set of values for atoms with increasing intrinsic state values, such as the series for C, N, O, and F. This information further reinforces the E-state concept that electron accessibility is related to both the

Table 6 Table of atom-type E-state descriptors with their range of values found in a data set of 21 000 compounds,[14,15] along with an indication of the type of intermolecular interaction usually considered to be involved. This array is called the E-state spectrum. The atom symbol is placed at the average value corresponding to the scale given on the y-axis

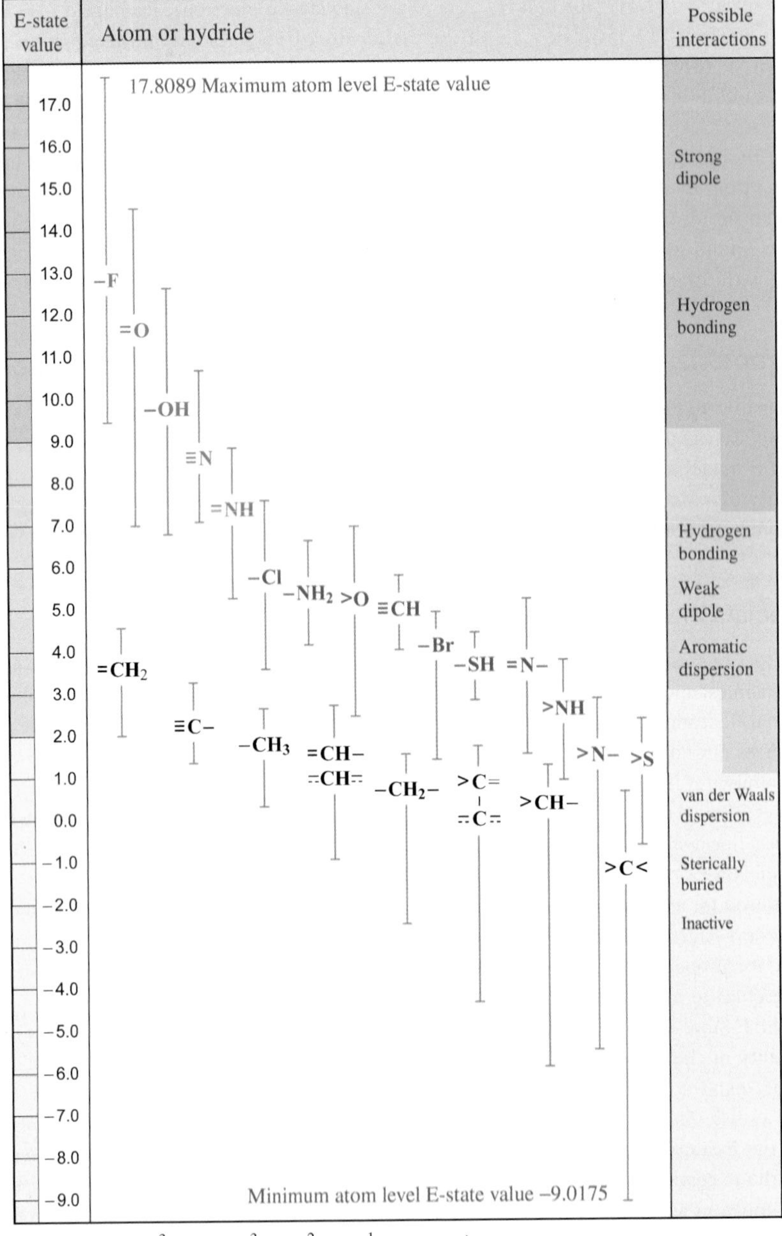

Bond symbols: sp^3, −; two sp^3, >; sp^2, =; sp^1, ≡; aromatic, ⩦.

electronic aspect (from electronegativity) and steric aspects of structure, the average value corresponding to the scale given on the y-axis.

We observe that stronger forces correspond to higher average E-state values.[5] Weak hydrogen bonding and dipolar forces occur for atoms and groups at intermediate levels of E-state in which these fragments are less available topologically and have a more modest electronegativity.

A set of fragments lying below this category are groups associated with van der Waals, dispersion, and hydrophobic bonding. The lowest category includes fragments that have low electronegativity and are topologically buried. The case of a quaternary carbon is that of a noninteractive atom. This analysis demonstrates that the intrinsic state values and the derived E-state descriptors encode information about the potential of a fragment to engage in some type of intermolecular interaction. This description is the basis for the term electron accessibility for E-state descriptors.

4.22.3.3 Nuclear Magnetic Resonance Relation to E-State

In confirmation of the E-state formalism, we have examined experimental data that might provide direct indication of its validity. A potentially good test of the validity of the unification of attributes in the E-state can be found in a test of the correlation between the E-state index and the NMR chemical shifts. This property reflects the environment of an atom in a molecule due to electronic and steric influences. Two studies have been reported in which the E-state descriptors (atom-level E-state value of oxygen atoms) in a series of ethers and also in a series of carbonyl compounds were calculated.[8,9] These indices were compared to the ^{17}O chemical shifts revealing good correlations between the E-state values and the chemical shifts.

The atom-level E-state has been shown to correlate to experimental measures of atom electronegativity. The results of two comparisons between atom-level E-state values and ^{17}O NMR shift, $\Delta[^{17}\mathrm{O}]$, are given in eqns [5] and [6] as examples.

For 10 ether compounds:

$$\Delta[^{17}\mathrm{O}] = 92.56\ S(-\mathrm{O}-) - 441.65 \quad r^2 = 0.99,\ s = 4.3,\ n = 10 \qquad [5]$$

For the carbonyl in 10 aldehydes and ketones:

$$\Delta[^{17}\mathrm{O}] = -27.77\ S(=\mathrm{O}) + 834.48 \quad r^2 = 0.97,\ s = 3.67,\ n = 10 \qquad [6]$$

$S(-\mathrm{O}-)$ and $S(=\mathrm{O})$ are the atom-level E-state values for ether and carbonyl oxygen atoms, respectively. These results compare very favorably with models based on quantum mechanical partial charges.[8,9]

4.22.3.4 The E-State and Free Valence

The E-state index may be compared with the concept of free valence introduced by Coulson as a molecular orbital description of potential reactivity at an atom.[24,26] Free valence is defined as:

$$F = N_{\max} - N_r \qquad [7]$$

in which N_r is the sum of bond orders joining atom r. The constant $N_{\max}$ is the maximum possible value of N which is usually taken to be 4.73, based the Hückel molecular orbital calculation of N_r as the central atom in trimethylenemethane.

This number encodes the relative amount of left-over or residual bonding capability available for intermolecular interactions. Kier and Hall have reported on a comparison of the E-state and free valence.[27] Using free valence values from a compilation of Hückel molecular orbital calculations by Coulson and Streitwieser,[28] a comparison with the computed E-state values reveals that there is a significant relationship between these descriptors. This leads us to believe that the E-state indices carry significant information that may be interpreted as the potential for intermolecular interaction.

4.22.3.5 The E-State and How it Encodes Structure Information

The E-state system of molecular description encodes structure information in a different form than commonly utilized fragment description systems use. Fragment systems encode information as counts for a given structure feature. A model based on such a description system includes a coefficient to the value for the feature count such that 'X' count of a fragment will result in 'Y' variation in activity. For example 'X' count of aniline groups will be assigned 'Y' contribution.

Variation in ‘Y’ is taken to be a function of ‘X’ count of the fragment. In contrast, the E-state structure description system encodes information in the form of electron accessibility at a given atom, thus a model based on the E-state descriptors will include a coefficient to the build-up of electrons at an atom in a given hybridization state and topological environment that will result in ‘Y’ variation in activity. For example, ‘X’ electron accessibility of the valence electrons on aniline nitrogen atoms will be assigned ‘Y’ contribution to activity. For the E-state, variation in ‘Y’ is taken to be a function of ‘X’ electron density and topological environment at a given atom type. Because an atom interacts with its environment through its valence electrons, it is reasonable to assert that a quantification of the build-up or depletion of valence electron density modified by topological constraints at an atom may provide a more information-rich parameter than the simple integer feature count of that atom type. A more elementary visualization would be to assume that in some structural configurations, a count of ‘1 aniline’ would be inaccurate because some valence electron density has been depleted from the nitrogen by a nearby withdrawing group causing a change in its potential to participate in intermolecular interactions. Other cases may involve electron buildup as in 2-methyl-aniline or local steric interference such as in the case of 2-*t*-butyl-aniline. The E-state formalism provides a powerful mechanism for encoding the potential of a molecule to participate in noncovalent intermolecular interactions.

4.22.4 Extensions of the E-State Formalism

The E-state formalism was developed for individual atoms and is well suited for common core skeleton problems. For diverse data sets, the formalism needs to be extended. The formalism of atom typing, bond typing, and groups has been adopted for this purpose.

4.22.4.1 The Atom-Type E-State Index

For data sets with no common core skeleton, the individual atom-level E-state indices cannot be used because individual atom values in one molecule do not correspond to any one particular atom in other molecules. However, what is common to a diverse list of structures are the atom types and functional groups in the molecules. These atom types and functional groups are expected to engage in similar interactions across a list of structures. For this reason, E-state values have been assigned to each atom type in organic molecules. Hall and Kier have introduced this extension of the E-state, called the atom-type indices, similar to group additive schemes, in which an index appears for each atom type in the molecule.[5,29–34] The atom-type E-state description provides the basis for application to a wider range of problems in which the E-state formalism is used without the need for superposition.

In this scheme, atom types include the familiar groups encountered in organic structural chemistry, such as $-CH_3$, –NH–, =O, –F, –S–, etc. Each atom in the molecule is identified by its valence state including the number of attached hydrogen atoms. Classification is based on four hydride group characteristics: (1) atom (element) identification; (2) valence state, including aromaticity indication; (3) number of bonded hydrogen atoms; and (4) in a few cases, the identity of other bonded atoms.[29]

The atom-type E-state value for type X, $S^T(X)$, is defined as the sum of all the atom-level E-state values, $S_i(X)$, for that atom type in the structure:

$$S^T(X) = \sum_i S_i(X) \quad [8]$$

The atom-type E-state scheme makes possible an important combination of structure information for QSAR analysis as well as for similarity and diversity analyses. The atom-type E-state indices combine three very important aspects of structure information:

1. The electron accessibility associated with each atom type.
2. An indication of the presence or absence of a given atom type.
3. The count of the number of atoms, i, of a given atom type.

Atom-type E-state descriptors contain atom count information but encode much more information than mere counts of atoms as indicated above.

Table 7 lists the bond symbols along with atom type symbols used. For example, in the symbol for methylene group atom-type E-state, SssCH2, S stands for the sum of E-state values for all the $-CH_2-$ groups in the molecule, ss stands for the two single bonds in that group, and CH_2 represents the formula for the hydride group. In this manner, it is possible to distinguish between $-CH_2-$ and $=CH_2$ (SdCH$_2$.) Also, SaaCH stands for sum of E-state indices for the CH

Table 7 Symbols used for atom type structure descriptors

A. Bond symbols		
Symbol	*Bond type*	
s	single (sp^3)	
d	double (sp^2)	
t	triple (sp^1)	
a	aromatic (sp^2)	
B. Atom-type E-state		
Atom type	*Atom-type E-state symbols*	*Definition Sum of atom-level E-state values for*
$-CH_3$	SsCH3 $S^T(-CH_3)$	all methyl carbon atoms
–OH	SsOH $S^T(-OH)$	all OH oxygen atoms
–NH–	SssNH $S^T(-NH-)$	all secondary amine nitrogen atoms
	SaaCH $S^T(=CH=)$	all unsubstituted aromatic carbon atoms
C. Hydrogen atom-type E-state		
Atom type	*Atom-type HE-state symbols*	*Definition Sum of atom-level E-state indices*
–OH	SHsOH $SH^T(-OH)$	all OH group hydrogen atoms
$-NH_2$	SHssNH2 $SH^T(-NH_2)$	all on primary amine hydrogen atoms
C–H	SHother SH^T(other)	all C–H hydrogen atoms, except acetylenic

in an aromatic ring: CH with two aromatic bonds; SsOH stands for the sum of E-state indices for –OH groups in the molecule.[29–31] A second set of symbols has also been employed. For example, the atom-type E-state symbol for –OH is $S^T(-OH)$ and for the carbonyl oxygen, $S^T(=O)$.[5,27–29] (A glossary (**Tables 12** and **13**) of E-state descriptor symbols is given at the end of the chapter.) The atom-typing concept includes the hydrogen E-state index as well. For example, for the hydrogen in an –OH group, SHsOH [$SH^T(-OH)$] and the secondary amine, SHssNH [($SH^T(-NH-)$] (see **Table 7** for a summary of the two symbols systems that have been used in the literature).

The atom level E-state quantifies the build up or depletion of valence electrons modified by the steric accessibility as the electron accessibility at the atom. The atom type E-state qualifies the kinds of interactions that the specified electron accessibility may participate in. This is of fundamental importance because a build-up of electrons on the oxygen of an acid group will have a different qualitative and quantitative impact than the same electron accessibility on the nitrogen of a secondary amide. Electron accessibility at an aromatic carbon will have an entirely different impact. The use of atom typing provides a computational tool to allow the modeling algorithm to assign the appropriate impact for each of these functional configurations based on the availability of valence electrons to participate in the given interaction.

The ability of E-state indices to encode information focused on a small number of atoms is a powerful advantage in their application. The fact that these descriptors encode important electronic and topological information endows them with the ability to relate to biological data for database characterization. The addition of atom typing extends the usefulness of the E-state indices so that large, diverse data sets may be examined. Recent studies attest to the potential of atom type E-states to the E-state methodology for the modeling of absorption, distribution, metabolism, excretion, and toxicity (ADME-Tox) properties (*see* 5.24 Molecular Fields to Assess Recognition Forces and Property Spaces, for a discussion of these applications).

4.22.4.2 The Group Type E-State Index

Group types allow for combining the contributions of related atom types into a broad-scope parameter, allowing a structure feature to appear in a model in more than one form to reflect multiple modes of interaction. If the atom-type E-state for primary amines appears in a model, it is plausible that secondary amines may also be important to the property. It may be that secondary amines are underrepresented in the data set causing the secondary amine descriptor to be overlooked by the feature selection algorithm. The creation of an amine group type will allow the combined primary and secondary amine information (SsNH2 + SssNH) to be evaluated to see if its use improves validation statistics. In many cases, an approximation of the contribution of a feature is an improvement over ignoring its presence. This technique is especially important when modeling small data sets where the total number of input parameters that can be used is limited by statistical concerns.

Group types can be created for any set of features that can be assumed to undergo similar interactions relative to the property or activity being described. Since hydrogen bonding is an important aspect of intermolecular interactions, E-state descriptors have been developed for both donor and acceptors (see **Table 12** and **13** at the end of this chapter for examples). SHHBd is the sum of the hydrogen E-state values for all hydrogen bond donor (HBd) groups. SHBa is the sum of E-state values for all hydrogen bond acceptor (HBa) groups in a molecule. For internal hydrogen bonding, a composite index was developed as the product of the hydrogen E-state value for the donor group and the E-state value for the acceptor. SHBint4 is the internal hydrozen bond descriptor for the potential hydrogen bond when the donor group, XH, is separated from the acceptor group by four skeletal bonds. The X–H bond and hydrogen bond itself complete the six-membered hydrogen bond ring. SHBint3 encodes similar information for the potential five-membered hydrogen bond ring. In both cases only the strongest internal hydrogen bond information is encoded in the index, based on the product of the donor hydrogen E-state and the acceptor E-state. The index SHBint2 has also proven to be a useful descriptor although it clearly does not stand for any kind of internal hydrogen bond. It encodes information about polar group content, especially those in close proximity in the structure. A group type hydrogen E-state index was also developed for all carbon-hydrogen fragments, using the symbol SHother, as the sum of hydrogen E-state values for all C–H groups. Since such fragments are considered to be nonpolar, this descriptor encodes the structure attributes of hydride groups that are nonpolar. Such a descriptor may represent parts of the structure that may participate in hydrophobic-like interactions.[31–33]

E-state values may be used collectively to characterize groups other than hydride groups. For example, group type E-state indices and hydrogen E-state descriptors have been developed for acids, including phenols as well as carboxylic, phosphoric, and sulfuric acids. Nitrogen groups, including amides, ureas, and thioureas, and sulfonamides have also been encoded.[33–37] The use of group types is also important in enhancing global predictability. Acid groups other than carboxyl are often represented at a minimal level in drug data. An E-state acid group type combining the sulfuric, phosphoric, and phosphorus acid atom types has been used to combine limited information on these groups.

4.22.4.3 Electrotopological State Index for Bonds and Bond Types

The characteristics of an –OH group are similar from molecule to molecule. The character and behavior of an –OH group depends on its immediate surroundings. Molecular structure representation may be improved by specifying the bond of which the –OH group is a part. For example, primary, secondary, and tertiary alcohols require separate identification and characterization for good models. The –OH in the carboxyl group plays a very different role than in an alcohol. An E-state value may be computed for each bond type in the molecule.

4.22.4.3.1 Bond E-state formalism

A bond can be classified in terms of the types of the two atoms which define the bond. The symbol used for these designations is based on the graph edge: e_{ij}. In the graph of 2-pentanol, there are several bond types (see **Figure 4**). Classification may be based on an $n \times n$ matrix of the atom types. Any combination of atom types which does not

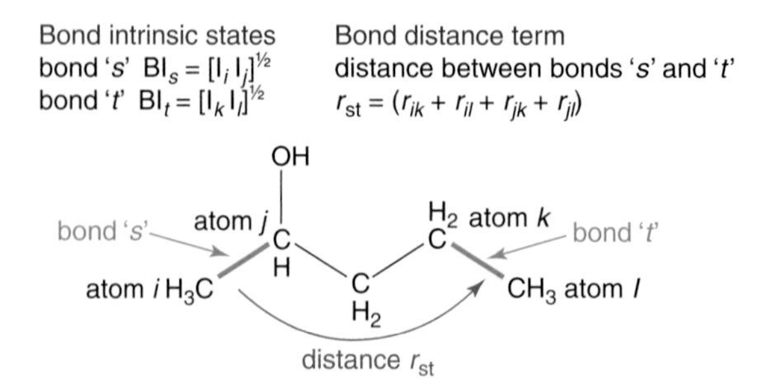

Figure 4 Illustration of the terms and symbols involved in the definition and computation of the bond E-state values for 2-pentanol. The bond intrinsic state (BI) is based on the intrinsic state of the atoms in the bond. The separation distance (r, in the perturbation term) is based on the distances between atoms in the two bonds. See text for discussion.

correspond to an actual bond is eliminated. For example, the $-CH_3$ and $=O$ entry does give rise to an entry in the matrix of bond types. At present Molconn-Z computes E-state values for 52 atom types commonly found in organic molecules along with 450 bond types.

The E-state formalism for bonds is parallel to that for atoms.[5] The bond E-state is the sum of its intrinsic state plus perturbations by all other bonds in the molecule. In this section we develop relations for the bond intrinsic state, $BI(e_{ij})$, and the bond perturbation, $\Delta BI(ij)$.

We propose for the intrinsic state value of a bond, BI, that each edge, e_{ij}, be represented as the geometric mean of the atom intrinsic state, I, for the two atoms in the bond:

$$BI(ij) = [(I_i\ I_j)]^{1/2} \qquad [9]$$

Based on this relation, intrinsic state values can be computed for each bond type in the molecule. In the same manner as for atom-level E-state, we propose the perturbation on bond ij by bond kl be of the difference form:

$$\Delta BI(ij) = BI(ij) - BI(kl) \qquad [10]$$

Perturbation is considered to increase with the difference between intrinsic states of the two bonds.

As in the atom-level E-state formalism, we consider the actual bond E-state value, SB, to be the perturbed intrinsic state, that is, the intrinsic state plus the sum of the perturbations by all other bonds in the molecule. The perturbation decreases with increasing distance between the two bonds. The distance is counted as the average atom–atom distance for the four atoms in the two bonds. The bond E-state value (SB) for a given bond, s, is determined as follows:

$$SB_s = BI(ij) + \sum_s \Delta BI(ij)_s / r^2 \qquad [11]$$

The graph distance term r, indicated in **Figure 4**, can be obtained as the average of the atom–atom distance terms:

$$r_{ij} = (r_{ik} + r_{il} + r_{jk} + r_{jl})/4 \qquad [12]$$

4.22.4.3.2 Classification into bond types

A systematic symbolism has been developed to express each of the 450 organic bond types. Each symbol is of the following form, $eiAT_1AT_2$, in which the symbol AT_i is a designation of atom type and ei is the designation for bond order. For example, the bond order is indicated as e1, e2, e3, and ea for single, double, triple, and aromatic bond orders.

Each atom type consists of the element symbol and an indication of the number and type of skeletal bonds involving that atom type. Simple and familiar organic bond types can be symbolized simply. For example, for the CH_3–CH_2– bond, the symbol is e1C1C2. Thus, the symbol C1 refers to the $-CH_3$ group with one skeletal bond and C2 refers to the $-CH_2-$ group with two skeletal bonds. For the tertiary alcohol bond, >CH–OH, the symbol is e1C3O1; a single bond (e1) between the tertiary carbon (C3) and the –OH (O1). For the carbonyl group, the bond symbol is e2C3O1s; e2 for the double bond. In this case, the terminal symbol s indicates that one of the atoms also has a single bond not used in this bond type. In this case that single bond must be on the carbon atom since the keto oxygen only has one double bond (see **Table 8** for illustration of the E-state bond-type symbols).

4.22.4.4 Molecular Polarity Index

The relative polarity of a molecule may be represented by the values of the intrinsic states of the atoms. The same approach can yield a numerical value for a fragment or group within the molecule. One approach is to derive a numerical value that is relative to a very low and a very high polarity value. Such an approach was used in the derivation of the kappa shape indices. The polarity of an atom or atom hydride is certainly related to its intrinsic state, I, if we define polarity as the electron richness and the topological availability of this electronic complement. Therefore the sum of I values for a molecule may be a good descriptor of this attribute.

Polarity is best defined as a relative value; thus pentane is intuitively less polar than diethyl ether even though they are of close molecular weight. To capture this relative relationship, we have defined the polarity of an atom (or hydride group) as the relationship of its intrinsic state value to the intrinsic state value of the most and least polar atom with the same topological structure. Accordingly, the polarity-weighted I value for the –NH– group is somewhere between the intrinsic state value I values for –O– and $-CH_2-$, that have the same topological structure. To arrive at a numerical value of the polarity of a molecule, i, we relate the sum of its I_i values to the sum of I_{max} values for a hypothetical

Table 8 Illustration of E-state bond-type symbols

Depiction	*Index name*	*Illustration*
$-CH_2-CH_3$	eIC2C1	CH_3–CH_2 ← sum of bond E-state values
$-CH_2-NH_2$	eIC2N1	NH_2–CH_2 ← sum of bond E-state values
$-CH_2-NH$	eIC2N2	NH–CH_2 ← sum of bond E-state values
>CH–OH	eIC3O1	OH–CH ← sum of bond E-state values
>CH=NH	e2C3N1	NH=C ← sum of bond E-state values
–CH=O	e2C2O1	O=CH ← sum of bond E-state values
$-CH=CH_2$	e2C2C1	CH_2=CH ← sum of bond E-state values
$>CH=CH_2$	e2C3C1	CH_2=C ← sum of bond E-state values
–C≡N	e3C3N1	N≡C ← sum of bond E-state values
···CH···CH···	eaC2C2	CH···CH ← sum of bond E-state values
···CH–NH—	eIC3N2a	NH– ← sum of bond E-state values

molecule with the highest polarity I_{min} values at each position, and the molecule with the lowest I values at each position[5]:

$$\text{Sum } I_{max} \geq \text{Sum } I_i \geq \text{Sum } I_{min} \quad [13]$$

The polarity, Q_v, expressed in this relationship can be transformed into a descriptor by the technique used in deriving the kappa shape indices:

$$Q_v = (\text{Sum } I_{max})\ (\text{Sum } I_{min})/(\text{Sum } I_i)^2 \quad [14]$$

The Q_v values have appeared in several QSAR models where a role is played by the hydrophobic character of the molecules.[5]

4.22.5 Molecular Connectivity

Biological properties arise from the interaction of electron densities distributed across molecular skeletons. For a comprehensive system of SIR, both the electron accessibility at each atom and the whole skeleton description must be

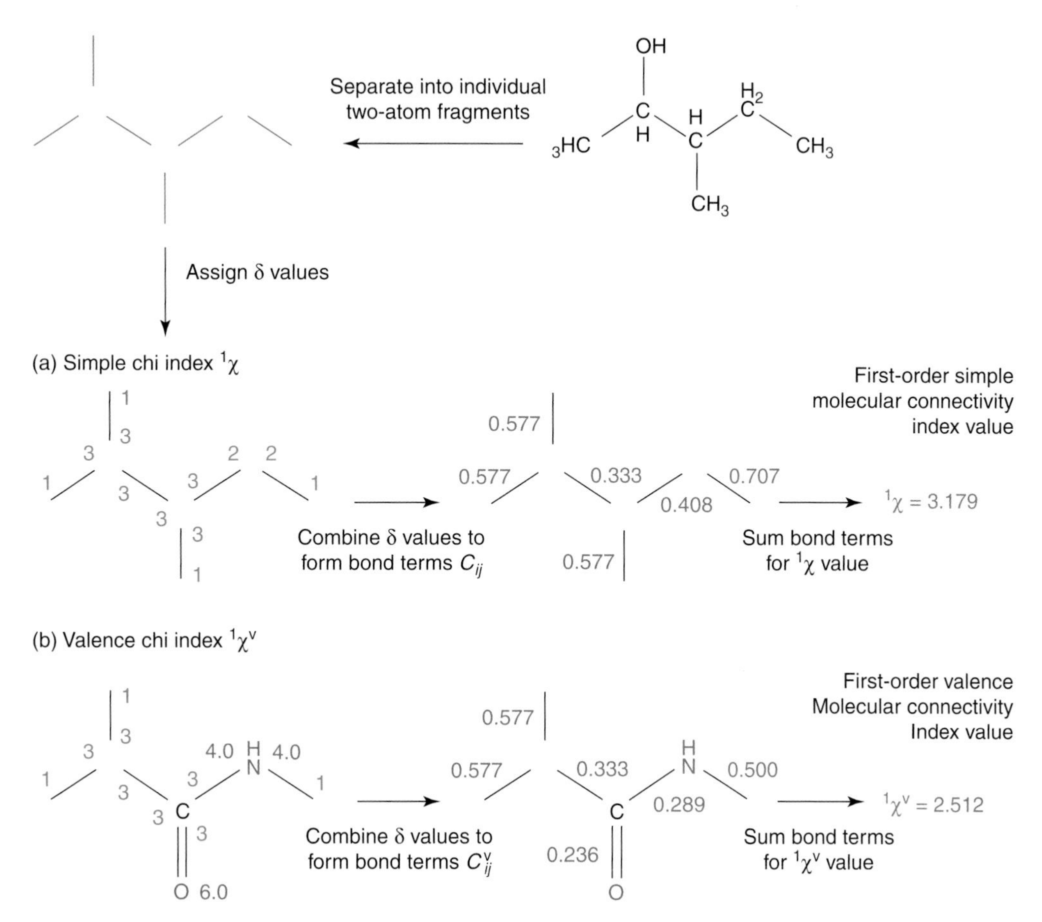

Figure 5 Illustration of the process for computation of the first-order molecular connectivity chi index, including breaking the skeleton into bonds, assignment of delta values, computation of bond values, and summation to final index value. (a) $^1\chi$ value for 2,3-dimethylpentane, based on simple delta values; (b) $^1\chi^v$ value for *N*-methyl-2-methylpropanamide, based on valence delta values.

encoded. The electrotopological state description of molecular structure was designed to encode information at the atom level. As a companion of this system, it is necessary to encode whole-molecule attributes in order to fully exploit relationships between structure and properties or to seek similarity in library searches. The molecular connectivity indices fulfill this need for whole molecule description as part of the SIR system.

Molecular connectivity indices are derived from the molecule represented as a sigma-bonded network. The carbon skeleton or principal atom skeleton is used, leaving the hydrogen atoms as implied parts of the network. Using both the simple and valence delta values described in Section 4.22.2.1, the atoms are labeled according to their relationships to other atoms in the molecule. A topological profile of the molecule can then be developed by dissecting the molecule into sets of fragments consisting of one-bond, two-bond, or fragments with more bonds.

Figure 5 illustrates the dissection of the skeleton into two-atom bonds in 2,3-dimethylpentane as shown in the top line of the figure along with the subsequent computation of values for each bond and then sum of the bond values. This collection of skeletal bonds is the basis for representing the structure information resident in the molecular skeleton. The number of skeletal bonds in the fragment is called the order of the fragment and of the corresponding chi index. For the first-order chi index $^1\chi$, the number of fragments is equal to the number of skeletal bonds in the structure.

The molecular topology may be quantified on several levels depending on the choice of fragment size selected. To quantify the topology using these sets of fragments, an algorithm proposed by Randić is adopted.[38] For the first-order index, each skeletal bond is characterized as the reciprocal square root of the product of the two atom delta values, δ_i and δ_j. The skeletal fragment term, C_{ij}, is computed as follows[38–40]:

$$c_{ij} = (\delta_i\ \delta_j)^{-0.5} \qquad [15]$$

For higher orders, m, of fragmentation, each fragment, k (containing $m+1$ atoms $i, j, k, \ldots$), is quantified with a single number, ${}^{m}c_k$ according to the equation:

$$ {}^{m}c_k = (\delta_i\, \delta_j \ldots)_k^{-0.5} \qquad [16] $$

To encode the topological information conveyed by this order of fragments, the skeletal bond terms ${}^{m}c_k$ are summed over the entire molecule to give a connectivity chi index of order m, ${}^{m}\chi$:

$$ {}^{m}\chi = \sum_{k}^{m} c_k $$

(see **Figure 5a** for an example of this index calculation for 2,3-dimethylpentane). For inclusion of all valence electrons for each atom, the valence delta values, δ^{v}, are used. **Figure 5b** illustrates the algorithm for ${}^{1}\chi^{v}$ for *N*-methyl-2-methylpropanamide. For second-row atoms, the valence chi first-order index, ${}^{1}\chi^{v}$, is smaller than the corresponding simple chi index value.

The nature of the structure information encoded in the molecular connectivity indices is skeletal variation or ramification. For example, the simple first-order molecular connectivity index ${}^{1}\chi$ provides a measure of the degree of branching in the molecular skeleton. For heptane isomers, the ${}^{1}\chi$ index decreases in value as the degree of branching increases: *n*-heptane > 2-methylhexane > 2,3-dimethylpentane > 3,3-dimethylpentane > 2,2,3-trimethylbutane.[38–40] When this index appears in a model, the chemist can use this structure information in the molecular modification process for design of candidate drug molecules. If the ${}^{1}\chi$ index makes a positive contribution to predicted property values, structures with decreased degree of branching will tend to have larger predicted property values. This quantity is a first-order simple index for the network, encoding the sum of first-neighbor relationships throughout the skeleton of the molecule. This index represents the topology of the molecule as derived from the undifferentiated connections (skeletal bonds) in that molecule, that is, the set of bonded relationships.

Other higher-order chi descriptors encode various aspects of skeletal variation. For example, the third-order path term, ${}^{3}\chi_{P}$, encodes the adjacency of branch points in the skeleton. This specific aspect of structure information will be exploited in an example application in the next section.

This first-order index is the prototype index in this SIR system that has been greatly expanded to describe multiple neighbors and atom states representing various types of structure information in a molecule.[38–41] In addition to the structure information described above, it has been shown that the first-order index encodes information about a molecule as it impacts on nearby molecules in intermolecular interactions. More detail on the nature of molecular connectivity indices is available.[42]

4.22.6 Applications

The formalisms for E-state and molecular connectivity descriptors provide the basis for a wide range of applications. The essence of SIR is the encoding of molecular structure information so that the SIR descriptors can readily be used in biological and chemical applications. These applications are taken from the published literature (a selected list of publications is available on the Internet[43]). The applications presented below are divided into two areas: screening of databases using similarity methods (Section 4.22.6.1) and development of QSAR models (Section 4.22.6.2).

4.22.6.1 Use of Structure Information Representation for Library/Database Screening

The ability of E-state and molecular connectivity descriptors to facilitate library screening may be illustrated in several ways. The first is the use of the SIR descriptors to create a space in which structures are arrayed in chemically meaningful patterns. Both the atom-type E-state descriptors and the molecular connectivity indices have been used in this way. These SIR structure spaces provide the basis for similarity searching of a library in which the user can tailor-make the similarity screen by selecting appropriate atom types and skeletal characteristics.

4.22.6.1.1 Atom-type E-state descriptors as a basis for molecular structure space

To illustrate the ability of the E-state descriptors to represent structure information, consider the dichlorobiphenyl isomers as shown in **Figure 6**.[11–15,30,31] The 12 isomers appear in the plot according the calculated values of the two atom-type E-state descriptors, SsCl [sum of E-state values for chlorine atoms, S^{T}(–Cl)] and SaaCH [sum of E-state

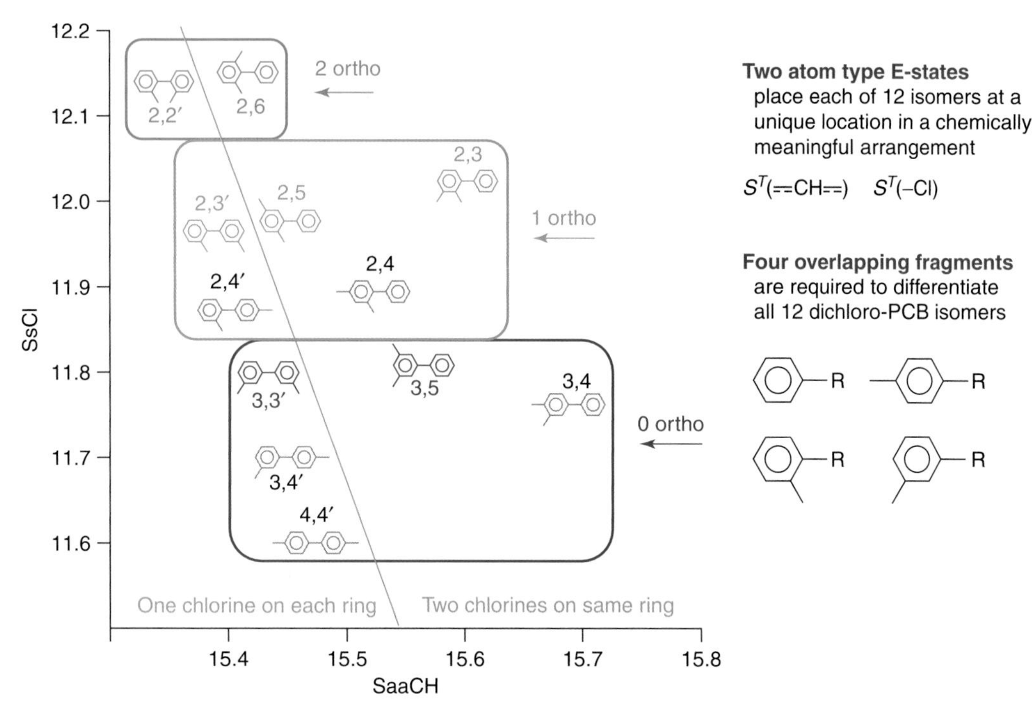

Figure 6 Dichlorobiphenyl (PCB) isomers arrayed in the space described by two atom-type E-state descriptors, chlorine [SsCl or S^T(–Cl)] and aromatic CH [SaaCH]. Colored structures may be considered as a training set. Each isomer with at least one *para*-substituted chlorine is shown in gray and may be considered as part of a prediction set. If (overlapping) fragments were used to organize this data set, four fragments would be necessary.

values for aromatic CH atoms, $S^T()$].[31] Consider seven of the isomers, shown in color, as a training set and the five isomers shown in gray as a prediction set.

The arrangement in E-state space clearly provides chemically meaningful relationships among the structures. As indicated in **Figure 6**, the two dichlorobiphenyl isomers with both chlorines as *ortho*-substituents (2,2′ and 2,6 in red) are placed at the top of the plot. In addition, the three isomers (in magenta) with only one *ortho* substituent (2,3′, 2,3, 2,5) occupy a lower position, with somewhat higher values for SaaCH. Also, the two isomers (in blue) with no *ortho*-substituents occupy the lowest position in the plot with even higher values for SaaCH. In addition to information about *ortho*-, *meta*-, and *para*- substitution patterns, the positions on the two rings are also encoded. The diagonal line is the plot indicates that all those isomers with one chlorine atom on each ring lie to the left of the line; those with both chlorines on the same ring lie to the right.

The significance of this information is clear. The structures near any given structure are chemically related so that an inquiry will lead to the appropriate location in the structure space. The impression gained from this information clearly indicates that arrangement in atom-type E-state space is a chemically meaningful pattern. Both structure similarity screening and diversity selection are facilitated by the properties of the E-state described here. It should be noted here that four overlapping fragments would be required to represent these 12 PCB isomers whereas only two E-state descriptors are required. This fact indicates further the information richness of the E-state indices.

Five of the structure drawings in the plot in **Figure 6** are given in gray. When the seven structures in color are considered as a training set, can the five structures, 2,4′, 4,4′, 3,4, 3,4′, 4,4′ (as a prediction set, shown in gray), be properly placed for purposes of prediction of a property? Placement of the 2,4′ and 2,4 isomers is a type of interpolation, since it has neighbors among the training set structures. Its position is determined by computation of the E-state equations and is properly placed as indicated in **Figure 6**. By contrast, the 3,4, 3,4′, and 4,4′ isomers represents a type of extrapolation. Clearly, their structure lies outside of the training set of the seven isomers shown in color. However, the range of the E-state algorithm extends beyond the perimeter of the training set structures. Therefore, the positions of the 3,4, 3,4′, and 4,4′ isomers are properly given because of the nature of the E-state formalism.

Because of this feature of the E-state, in contrast to the use of fragment or group counts, no missing-fragment problem arises nor is there need to estimate a value for a group absent from the training set. This feature of possible extrapolation has recently been demonstrated in the prediction of the human serum protein binding of eight compounds that are outside of the parameter space of the training set of the QSAR model developed for protein binding (based on 800 compounds).[44] This aspect of the E-state formalism is a valuable asset for library screening and for property prediction.

Other E-state studies present structure space creation for two data sets.[14,15] In one study, a set of substituted methylacetate esters was arrayed as a two-dimensional space. In another study polysubstituted benzenes are investigated. In both cases, results to similar the PCB investigation are obtained.

4.22.6.1.2 Molecular connectivity indices as a basis for molecular structure space

The simple molecular connectivity indices provide structure information that represents the types of molecular skeletons and variations within a set of skeletons. The collection of path and chain indices has been used to cluster molecular skeletons in chemically meaningful ways.[32–37] These indices encode structure information including degree of branching, adjacency of branching, ring size, and nature of fused ring systems.

This structure information can be demonstrated by considering the arrangement of molecular structures in a connectivity structure space. Consider the octane isomers, including acyclic, mono-, and bicyclic molecules displayed in a space consisting of only two molecular connectivity indices, ${}^{1}\chi$ and ${}^{1}\chi_{P}$ (**Figure 7**). This plot may be considered as a projection from a higher-dimension space based on several simple chi indices. In this plot three structure attributes are clearly revealed by the pattern of structures: degree of branching, adjacency of branching, and number of branch points. Variation of these attributes is indicated in the figure.

To emphasize the utility of this chemically meaningful structure information, seven of the structures in **Figure 7** are shown in red. The set of structures shown in black may be considered as a training set (with no cyclobutanes); the structures in red may be considered as a prediction set, consisting entirely of cyclobutanes. These seven structures,

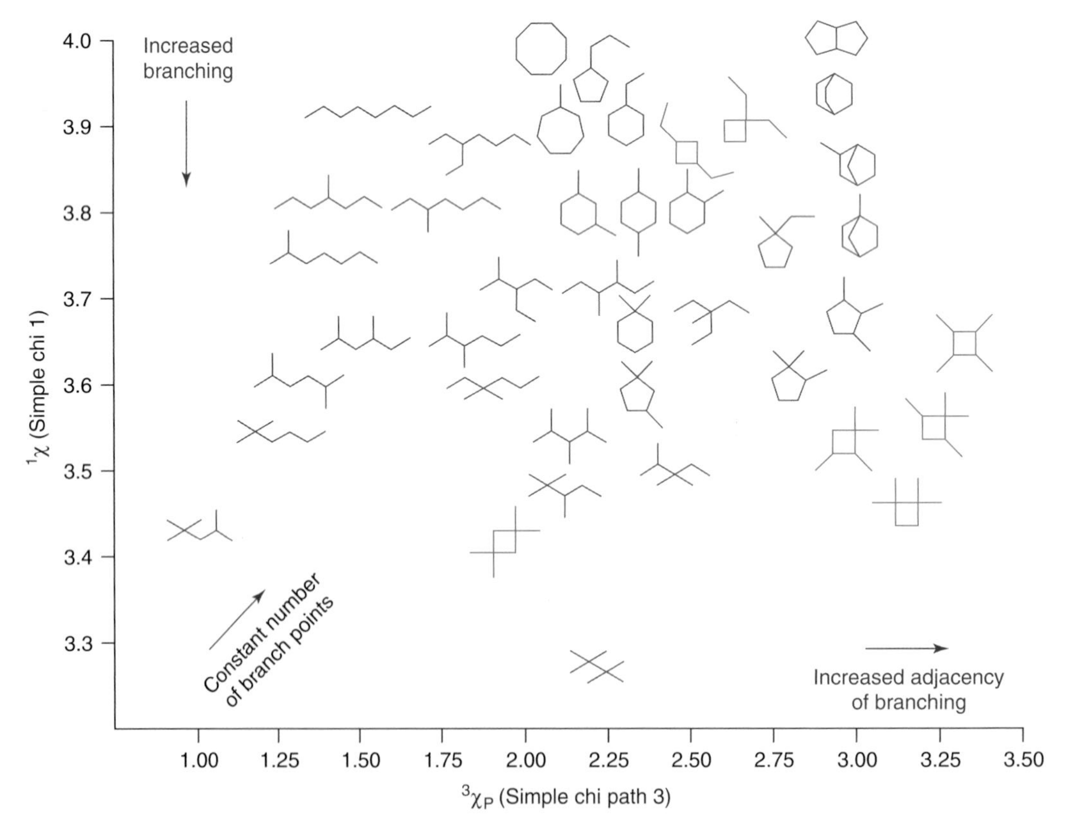

Figure 7 Several structurally diverse isomers of octane are displayed in molecular connectivity space using two simple chi descriptors: the simple ${}^{1}\chi$ index and the simple ${}^{3}\chi_{P}$ index. The use of two connectivity descriptors places each isomer in a unique position in a chemically meaningful pattern. Degree of branching decreases along the *y*-axis and adjacency of branching increases along the *x*-axis. The number of branch points is constant along *xy* diagonal directions. The gray structures may be considered a training set and the red structures as a prediction test set.

diethylcyclobutanes and tetramethylcyclobutanes, are shown to be placed in chemically meaningful positions in the plot, based entirely on the results obtained from the molecular connectivity formalism. These seven substituted cyclobutanes are arrayed in the pattern as the other octanes. This example further supports the idea that the SIR descriptors provide a useful basis for representing and using chemical structure information.

The whole set of simple chi indices represents structure information that organizes molecular structures into chemically meaningful patterns. Based on this information, one can navigate through this structure space. The immediate neighborhood of any structure in this space consists of similar structures. Library screening in this space takes advantage of this structure information and provides a basis for similarity screening. This example further illustrates the manner in which SIR descriptors provide a chemically meaningful pattern that facilitates similarity searching and diversity selection.

4.22.6.1.3 The quantitation of molecular fragments, groups, and pharmacophores

The E-state formalism can be adapted to produce values characteristic of various kinds of molecular fragments and groups of fragments. These characteristic values may be used in screening libraries for structures with the desired fragments.

4.22.6.1.3.1 Molecular fragment characterization

We define a fragment in a molecule as a substructure consisting of two atoms or hydride groups together with the path separating those two atoms by the smallest number of atoms comprising the path. The atoms or groups are described by their atom type which includes their valence states.[5,30] The fragment, NH–C–C–C=O, illustrates this idea as follows. We specify this fragment using the terminal atoms as codes in addition to a numerical value for the count of atoms in the path. For this fragment the specification is 5[(–NH–)/(=O)]. The ingredients or variables of this fragment specification are: (a) an sp^3 nitrogen atom in a secondary amine group, –NH–; (b) an sp^2 oxygen atom in a carbonyl group, =O; (c) a count of five atoms in the path separating (and including) the terminal atoms inthe chemical fragment.

The derivation of a numerical code specifying this fragment is taken from the electrotopological state algorithm.[9,30] From this we calculate the characteristic value as the perturbation term in the E-state algorithm (ΔI_{ij}; see eqn [4]). This term is interpreted as an electronegativity-based perturbation of the intrinsic state values of the two terminal atoms in the fragment. In the case of the fragment shown above, we use the calculation of the E-state perturbation values of the terminal atoms of the fragment, –NH– and =O. We calculate the specific perturbation of one terminal atom in this fragment upon the other terminal atom. Thus, the calculation of the influence of –NH– on =O, for F is as follows:

$$F = [I(=\mathrm{O}) - I(-\mathrm{NH}-)]/r^2 \quad [17]$$

This expression, given here by the symbol F (adopted for this particular application as the numerical value encoding the fragment), is identical to the E-state perturbation term, ΔI_{ij}. Using the intrinsic state values, $I(\text{–NH–}) = 2.5$, $I(\mathrm{O}) = 7.0$, and distance, $r = 5$, we calculate for the perturbation of –NH– on =O the following value: $F = +0.1800$. It is apparent that the perturbation of the =O atom on the –NH– group is -0.1800. The positive value is adopted in all cases as the characteristic F_{ij} value for any fragment. This calculation produces a code for any fragment of five atoms that includes the terminal atoms (hydride groups) –NH– and =O. The number 0.1800 is thus a characteristic number encoding the fragment five(–NH–/=O). As a result, the F_{ij} index value, 0.1800, can be the basis of a search and molecules identified, whenever this fragment is the reference.

4.22.6.1.3.2 Group characterization

A functional group can be identified in any molecule simply by searching for the appropriate F_{ij} value or set of F_{ij} values which corresponds to it.[5,30] Consider as an example the carboxyl group: –C(=O)OH. (Atom 1 in this group is the carbonyl carbon; atom 2 is the carbonyl oxygen; atom 3 is the –OH group.) The three fragments in the group may be coded using the convention for fragments described above, as follows:

$$F_{12} = [I(=\mathrm{O}) - I(=\mathrm{C}<)]/(2)^2 = (7.00 - 1.67)/4 = 1.333 \quad [18]$$

$$F_{13} = [I(-\mathrm{OH}) - I(=\mathrm{C}<)]/(2)^2 = (6.00 - 1.67)/4 = 1.083 \quad [19]$$

$$F_{23} = [I(=\mathrm{O}) - I(-\mathrm{OH})]/(3)^2 = (7.00 - 6.00)/9 = 0.111 \quad [20]$$

The atom identity numbers are used to determine whether the three two-atom fragments are connected in the manner of a carboxyl group. Atom 1 in the structure must be the same atom in both F_{12} and F_{13}. These three F_{ij} values (representing the ΔI_{ij} values for the three bonds in a carboxyl group) form a constellation of three codes which characterize the group. All three fragments must be present with the appropriate set of atoms numbers to indicate the presence of the group. A structure library may be searched for this group by using these three values.

For the ester group, similar in structure to the carboxyl, a different set of values is found. For two of the three codes, distinct values are obtained, and the ester is separately characterized as follows:

$$F_{12} = [I(=\mathrm{O}) - I(=\mathrm{C}<)]/(2)^2 = (7.00 - 1.67)/4 = 1.333 \quad [21]$$

$$F_{13} = [I(-\mathrm{O}-) - I(=\mathrm{C}<)]/(2)^2 = (3.50 - 1.67)/4 = 0.458 \quad [22]$$

$$F_{23} = [I(=\mathrm{O}) - I(-\mathrm{O}-)]/(3)^2 = (7.00 - 3.50)/9 = 0.388 \quad [23]$$

These three F_{ij} values may be the basis for library screening for ester groups, providing that atom 1 in for the F_{12} term is the same atom as atom 1 in the F_{13} term, etc.

These structure group codes can be the basis for rapid library screening and can be used in conjunction with screening codes for other fragments and/or pharmacophore patterns.

4.22.6.1.3.3 Pharmacophore pattern characterization

One of the most active areas of database management is in the search for molecules with structural features mimicking a pattern in a reference molecule considered to be a pharmacophore. The method described here is very useful for this objective.[5,30] As an example, consider the pharmacophore proposed for histamine:

$_1NH_2$

$_2HN$ N_3

The three nitrogen atoms, each in a different hybrid state form a pattern, represented by three bonded fragments:

$$\mathrm{NH_2{-}C{-}C{-}C{-}C{-}NH}$$
$$\mathrm{>N{-}C{-}NH}$$
$$\mathrm{NH_2{-}C{-}C{-}C{-}N<}$$

The coding for this pharmacophore embraces the three fragments, as encoded by the three F_{ij} values[9]:

$$F_{12} = [I(-\mathrm{NH_2}) - I(-\mathrm{NH}-)]/r^2 = (4.00 - 2.50)/36 = 0.04167 \quad [24]$$

$$F_{23} = [I(>\mathrm{N}-) - I(-\mathrm{NH}-)]/r^2 = (3.00 - 2.50)/9 = 0.05556 \quad [25]$$

$$F_{13} = [I(-\mathrm{NH_2}) - I(>\mathrm{N}-)]/r^2 = (4.00 - 3.00)/25 = 0.04000 \quad [26]$$

Note that the ordering of the terms in each fragment is designed to give positive values for the F_{ij} index. This is an arbitrary choice since there are both negative and positive values for each fragment value in all molecules. The search for the histamine pharmacophore is now possible using this constellation of three F_{ij} values in a database search. The atom identity numbers are used to determine whether the three two-atom fragments are connected in the manner of the pharmacophore.

4.22.6.1.4 Atom-type E-state descriptors in a library similarity search

The utility of the atom-type E-state descriptors can be illustrated with a similarity search of a compound library.[11–15,29–31] For this purpose we used a library of 21 000 structures from the Pomona MedChem database.[20,30] A reference structure was selected and its similarity to structures in the library was determined, based on the atom-type E-state values for the reference and library structures. The E-state values were first converted to z-scores.

Similarity was determined with two metrics: generalized Euclidean distance, d (Minkowski distance) and the Tanimoto coefficient (cosine) formula, T, as follows:

$$d_{\text{Euclidean}} = \left[\sum_j (Z_{j,\text{ref}} - Z_j)^2\right]^{1/2} \qquad [27]$$

$$T = \text{cosine} = \sum_j (Z_{j,\text{ref}} \cdot Z_j)^2 \Big/ \left[\sum_j (Z_{j,\text{ref}})^2\right]^{1/2} \left[\sum_j (Z_j)^2\right]^{1/2} \qquad [28]$$

The z-score $z_{j,\text{ref}}$ refers to the jth E-state descriptor for the reference structure; z_j refers to the jth E-state descriptor for one of the candidate structures in the library. Each sum is performed over the specified number of dimensions, that is, the number of atom-type E-state descriptors used in the similarity search. For the Albuterol search, E-state descriptors were used for all eight atom types in the Albuterol structure: [$S^T(\text{–CH}_3)$, $S^T(\text{–CH}_2\text{–})$, $S^T(\text{–CH<})$, $S^T(\text{>CH<})$, $S^T()$, $S^T(\overset{|}{\text{=C=}})$, $S^T(\text{–NH–})$, $S^T(\text{–OH})$].

Similarity library screening results have been published for the following drugs: Albuterol, Chlorzoxazone, Iproniazid, Lormetazepam, Mefloquine, Prednisone, and Sulfisoxazole. In all cases, the structures found in the library as highly similar to the reference drug are considered to be actually similar in both structure and biological activity.

4.22.6.1.4.1 Library screening for structures similar to albuterol

Rapid screening of structure databases have been carried out among the 21 000 organic structures for several drugs.[11–15,29–32] For example, the results found for the first five structures found in the screening for structures similar to Albuterol are listed in **Table 9**. The first five, the most similar structures, are all known bronchodilators like Albuterol.[14] The rank ordering in similarity is not necessarily expected to reflect the rank ordering in activity since the structure similarity is not based on a model of activity. Subsequent to library screening, rank ordering could be done with a QSAR model.

4.22.6.1.4.2 Library screening for structures similar to mefloquine

A feature of the atom-type E-state library screening method is the ability to tailor-make the search screen. The user can select the atom types considered significant for the screening process.

The selected atom types can focus on a limited portion of the reference structure or the whole set of atom types found in the reference structure. For example, in the published screening for Mefloquine, all ten atom types were used. The three most similar structures, based on Euclidean distance, are shown in **Figure 8a**. The biological activity of this antimalarial drug Mefloquine is known not to depend upon the presence of a ring containing the secondary nitrogen.[45] A second screening was performed with the atom type for the carbon atoms in the ring, methylene [$S^T(\text{–CH}_2\text{–})$ or SssCH2], left out. In this screening, a different ordering of structures is found as shown in **Figure 8b**. In this second case, the structure found most similar in structure is also known to be most similar biologically.[45] This example indicates the utility of this approach to library screening.

4.22.6.1.4.3 Library screening for structures with molecular connectivity indices

Galvez and colleagues have demonstrated in several publications that the SIR descriptors can be used effectively in discriminant analysis.[22,46,47] In addition, they have shown that screening a structure library found compounds that proved to be active.

A database with compounds in seven different pharmacological classes of activity was used for development of discriminant models of each activity class. The classes included analgesic, antiviral, bronchodilator, antifungal, hypolipidemic, hypoglycemic, and beta-blocking activity. Galvez developed separate discriminant models for each class by using molecular connectivity indices. Based on each model, activity was predicted for a list of structures as both a prediction and a subsequent experimental test. In some cases, compounds were also tested experimentally. Compounds predicted to be active were generally known from the literature to be active or tested in the laboratory.

For example, for antiviral activity, a library of over 12 000 commercial compounds was screened. Seventeen compounds predicted to be active were selected to be tested. In an in vitro assay (herpes simplex-1virus on cellular cultures) 12 compounds were found to significantly active, with activity similar to that of phoscarnet. These compounds included nitrofurantoine, 1-chloro-2,4-dinitrobenzene, 5-methylcytidine, 1,2,3-triazol-4,5-dicarboxylic acid, cordicepine, nebularine, and inosine. Galvez concluded that some of these compounds might be considered as lead candidates for new drugs. Similar results were reported for the antibacterials and analgesia.

Table 9 Similarity screening results for Albuterol, based on eight atom-type E-state descriptors for all eight atom types in the target structure, Albuterol

Structure[a]	*Name*	*Euclidean distance*[b]	*Tanimoto (cosine)*
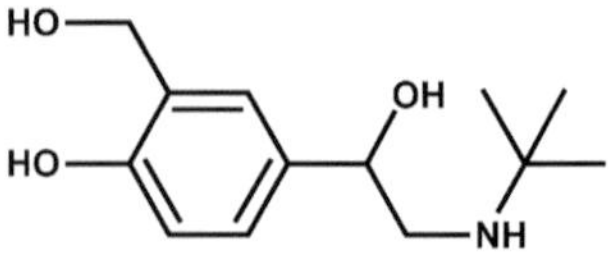	Albuterol	0.00	0.000
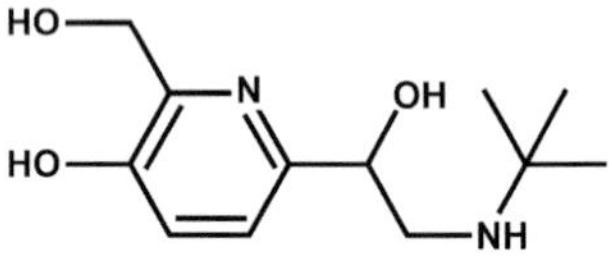	Pirbuterol	0.52	0.988
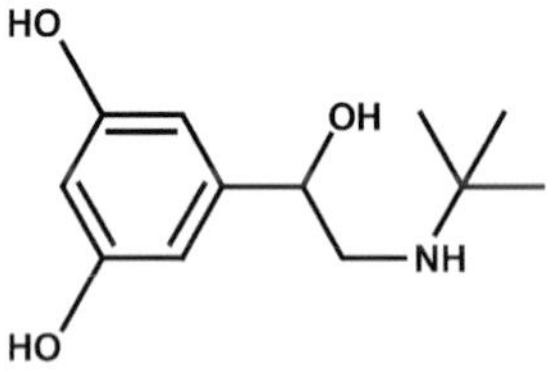	Terbutaline	0.53	0.987
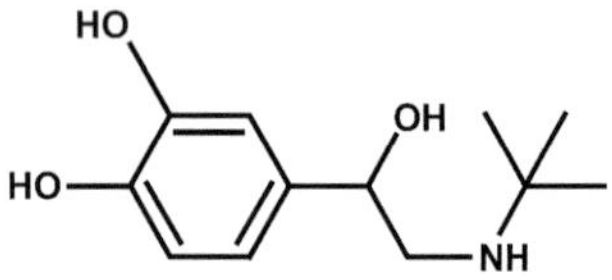	Colterol	066	0.980
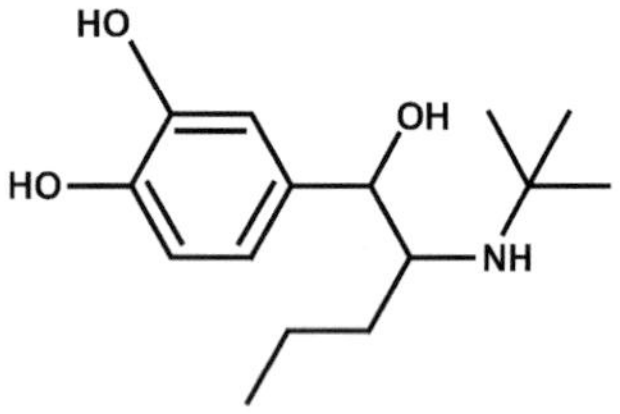	Isoetharine	0.67	0.980
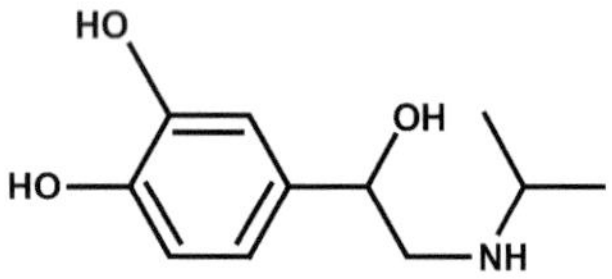	Metaproternol	0.67	0.980
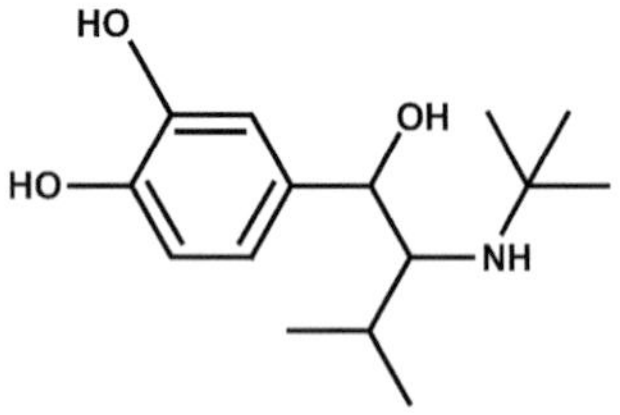		1.02	0.967

[a] Data set consisted of 21 000 organic compounds, including drug and druglike substances.
[b] Distance based on arbitrary units arising from the *z*-cores of the atom-type descriptor values.

4.22.6.2 Use of Structure Information Representation in Quantitative Structure–Activity Relationship Model Development and Interpretation

Many studies have been reported in which E-state descriptors were found useful in the development of a QSAR model. Several of these studies are briefly summarized in this section.

Series (a)	Distance	Series (b)	Distance
	0.00		0.00
	0.48		0.13
	0.53		0.14
	0.61		0.48

Figure 8 (a) Library screening results for Mefloquine. The similarity screen used all 10 atom-type E-state types found in Mefloquine. (b) Screening results for Mefloquine when only nine atom type E-state descriptors were used as the basis for the similarity screening; the atom type for methylene, SssCH2 [S^T(–CH_2–)], was not included.

4.22.6.2.1 Use of E-state in a topological superposition model for hERG inhibitors

Recently David Diller and colleagues have reported the use of the E-state descriptors in a method that utilizes topological superposition for a series of hERG inhibitors.[48] Over 300 literature sources were examined to provide the data set which included 190 compounds with measured hERG IC_{50} values, all based on patch clamp data, together with 317 known inactive compounds. Diller estimated that the experimental error in the data might permit the development of a model with $r^2 \sim 0.7$. Several methods for model development were employed, including pattern recognition, traditional QSAR, and pharmacophore methods. None of these methods were found successful for model development or for prediction.

A successful model was developed, however, by using E-state values in a topological superposition technique developed by Dixon.[49] Based on the simplicity, ease of use, and demonstrated ability to explain structure–activity data, Dixon and co-workers developed a methodology for creating a one-dimensional representation of structure–activity information.[49] The method is essentially done by multidimensional scaling.[50,51] Diller found in a scan of MDDR that at least 90% of the topological distance information is retained in 95% of the molecules in the database.[50,51]

For the hERG data set, each molecule is assigned a one-dimensional representation and each atom is assigned its E-state value. **Figure 9** presents a representation of the one-dimensional representations for a set of ten structures, all centered on a nitrogen atom shown in blue.[48] Specific values were included for aliphatic carbon, aromatic carbon, tertiary nitrogen, amide nitrogen, and ether oxygen. The one-dimensional topological mapping provides an association between the E-state value and the topological distance in the structure. This aspect of the method is illustrated for cisapride in **Table 10**. The contribution to activity for each atom is calculated as the product of its atom descriptor(s) and the value of its corresponding function at the one-dimensional coordinate of that atom.

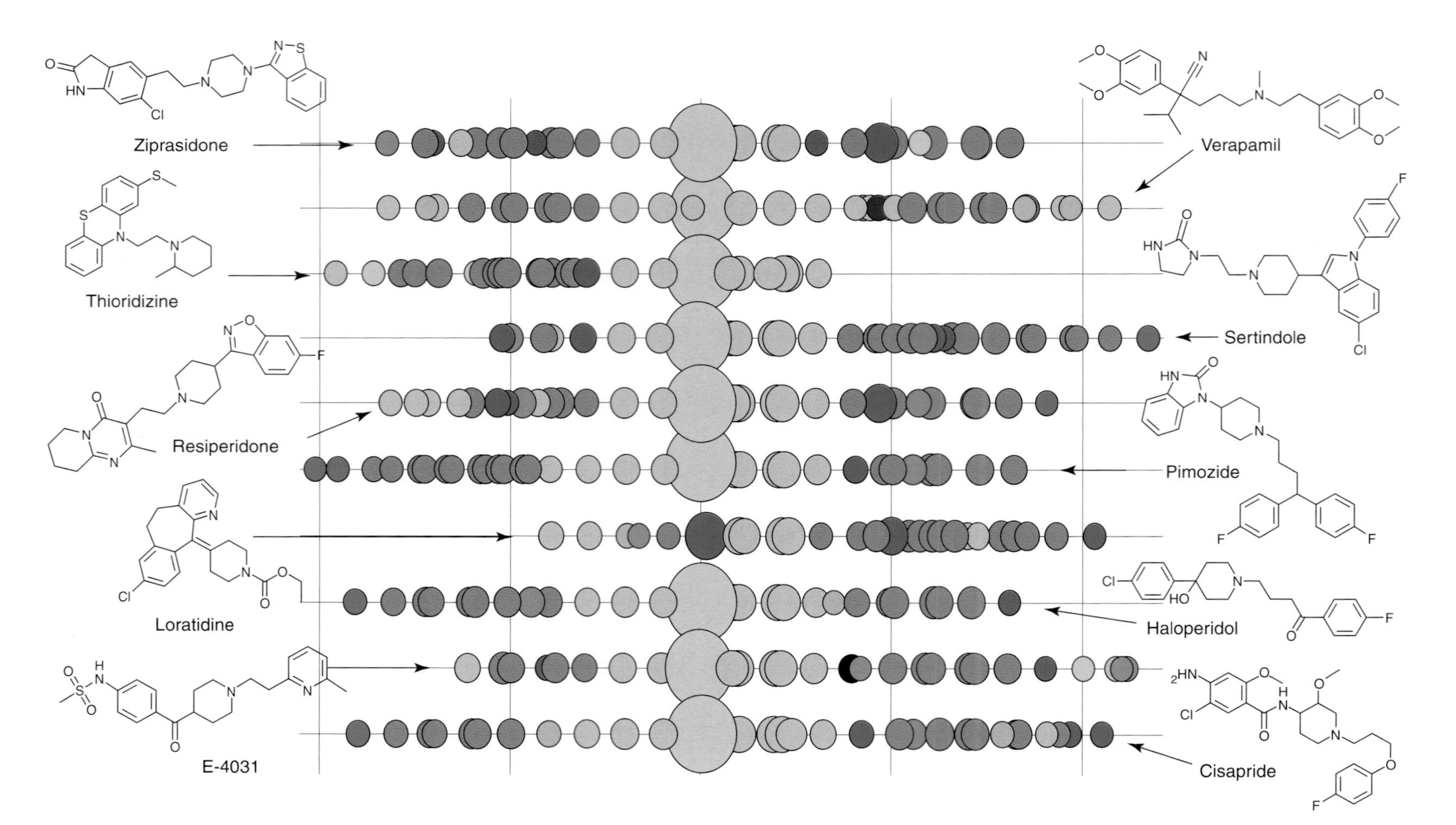

Figure 9 A set of ten hERG blockers shown in their reduced one-dimensional representations, all aligned on the 'central' tertiary amine that is selected to be coincident for all structures.

Diller developed a fast small-molecule similarity method with multiple alignment profiles, based on the one-dimensional topological representation of each molecule in the data set.[50,51] The model for the training set produced the following statistics: $r^2 = 0.70$, rms error $= 0.90$, $n = 190$. Diller reported results for three test sets which are considered useful for the modeling process. Finally, another result of this method is that each atom in a structure can be assigned a contribution to the overall activity of the molecule. In Terfenadine, for example, a mild hERG blocker, the model highlights the tertiary amine, the aromatic rings and also two OH groups as shown in **Figure 10**. This visualization technique permitted the development of Terefenedine analogues with decreased hERG activity.[51]

Table 10 Illustration for cisapride, displaying its atom E-state values along with the one-dimensional topological distance X found for each atom in the reduced representation

Cisapride	*Atom number*	*Type*	*X*	*Size*	*Aliphatic carbon E-state*	*Aromatic carbon E-state*	*Nitrogen acceptor E-state*	*Nitrogen donor E-state*	*Oxygen acceptor E-state*
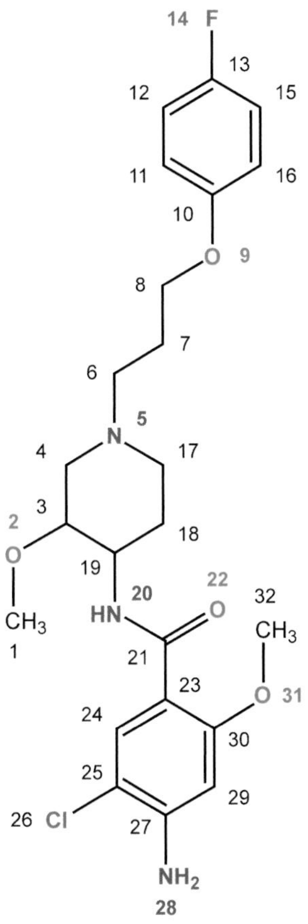	14	F	10.7	1	0	0	0	0	0
	13	C	9.7	1	0	−0.28	0	0	0
	15	C	8.8	1	0	1.38	0	0	0
	12	C	8.7	1	0	1.38	0	0	0
	16	C	7.7	1	0	1.62	0	0	0
	11	C	7.6	1	0	1.62	0	0	0
	10	C	6.7	1	0	0.65	0	0	0
	9	O	5.7	1	0	0	0	0	5.85
	8	C	4.7	1	0.54	0	0	0	0
	7	C	3.7	1	0.83	0	0	0	0
	6	C	2.7	1	0.84	0	0	0	0
	5	N	1.7	1	0	0	7.72	0	0
	17	C	0.9	1	0.82	0	0	0	0
	4	C	0.7	1	0.69	0	0	0	0
	18	C	−0.2	1	−0.74	0	0	0	0
	3	C	−0.4	1	−0.15	0	0	0	0
	2	O	−0.6	1	0	0	0	0	5.86
	1	C	−0.7	1	1.64	0	0	0	0
	19	C	−1.5	1	−0.14	0	0	0	0
	20	N	−2.5	1	0	0	0	3.05	0
	21	C	−3.5	1	−0.28	0	0	0	0
	22	O	−3.9	1	0	0	0	0	12.87
	23	C	−4.6	1	0	0.34	0	0	0
	24	C	−5.3	1	0	1.52	0	0	0
	30	C	−5.5	1	0	0.37	0	0	0
	31	O	−6.2	1	0	0	0	0	5.29
	25	C	−6.6	1	0	0.30	0	0	0
	29	C	−6.6	1	0	1.54	0	0	0
	32	C	−7.4	1	1.48	0	0	0	0
	27	C	−7.7	1	0	0.35	0	0	0
	26	Cl	−8.0	1	0	0	0	0	0
	28	N	−8.8	1	0	0	0	3.80	0

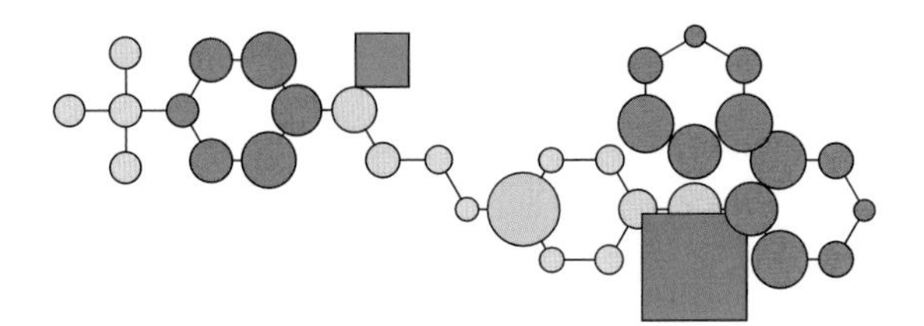

Figure 10 Illustration of the relative contribution of each atom to the hERG blocking activity computed by the model for Terfenedine. Atoms represented by circles indicate a positive contribution to activity; squares indicate a negative contribution to activity.

Overall this example further illustrates the power of the E-sate representation of molecular structure and its ability to contribute creatively to the drug candidate development process. This method has been partially automated.[50,51]

4.22.6.2.2 Summary of E-state quantitative structure–activity relationship (QSAR) studies

Applications of the E-state and other SIR descriptors for development of QSAR models are presented here in summary form. Information for several studies is given in **Table 11** including nature of activity, and a reference. This summary list indicates the wide range of properties and activities for which SIR descriptors provide the basis for successful QSAR models. A larger list can be obtained on the Internet.[36]

Table 11 Selected list of QSAR studies with references

Activity/property	*Compound class*	*Author*	*Ref.*
Antiproliferative tumor activity	Murine colon 38	Roy	52
Antileukemic potency	Carboquinones	Gough	53
Receptor binding	Salicylamides	Kier	54
Herbicide toxicity	Amides	Gough	55
Monamineoxidase inhibition	Aryloxyhydizides	Mohney	56
Analgesic potency	Diverse data set	Galvez	22
Dopamine transporter binding	Tropane derivatives	Maw	57
HIV-1 protease inhibition	Peptidomimetics	Maw	21
Toxicity to fish	Phenols	Vaughn	6
Boiling point	298 organic compounds	Hall	58
Boiling point (372 organics)	Alcohols, chloroalkanes, alkanes	Story	59
Antithyroid agents	Diverse data set	Abou-Shaaban	60
Antimicrobial activity	Heterogeneous organics	Garcia-Domenech	46
Dermal penetration	Poly-aromatic hydrocarbons	Gute	61
Anti-adrenergic activity	*N,N*-Dimethyl-2-bromo-2-phenylethylamines	Roy	62
HIV-1 integrase inhibition	Flavone derivatives	Buolamwini	63
HIV-1 reverse transcriptase inhibiton	2-Amino-6-arylsulfonyl benzonitriles	Roy	64
Receptor binding	Corticosteroids	Maw	65
Antimicrobial activity	*Toxoplasma gondii*	Gozalbes	47
Human factor Xa inhibitor	N2-aroylanthranilamides	Roy	66
Estrogen receptor modulation	Tetrahydroquinoline derivatives	Mukerjee	67
Human MT_1 and MT_2 binding	Melatonins	Sengupta	68
Anticonvulsant activity	Functionalized amino acids	Tropsha	69
Binding to cyclodextrin	Barbiturates	Kier	70
Carcinogenicity	Drugs	Contrera	33
Protein binding	Beta-lactams	Hall	34
Human oral absorption	Drugs	Votano	35
Ames mutagenicity	Drugs	Votano	36
Aqueous solubility	Organic compounds	Votano	37
Fish toxicity	Organic compounds	Rose	25
Albumin binding affinity	Drugs	Hall	71
Receptor binding	Corticosteroids	Kellogg	72

Table 12 Definition and illustrations for atom type and group type E-state descriptors, taken from a model for human serum protein binding[44]

	Index	*Descriptor definition*	*Descriptor illustration*
Atom type E-state	SssCH2	Sum of atom level E-state values of all methylene carbons	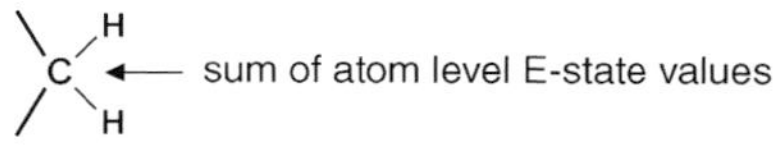 sum of atom level E-state values
	SsNH2	Sum of atom level E-state values of all primary amine nitrogen atoms	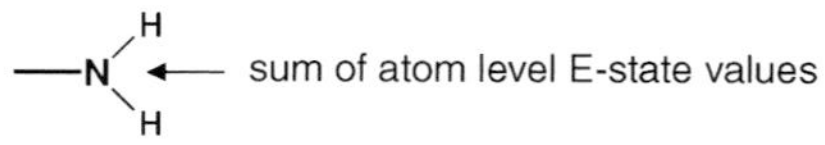 sum of atom level E-state values
	SsssN	Sum of atom level E-state values of all tertiary amine nitrogen atoms	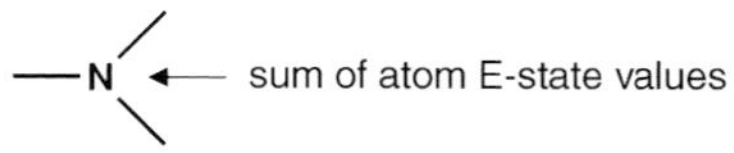 sum of atom E-state values
	SssO	Sum of atom level E-state values of all ether oxygen atoms	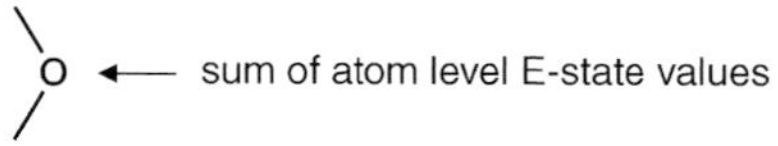 sum of atom level E-state values
	SssS	Sum of atom level E-state values of all thioether sulfur atoms	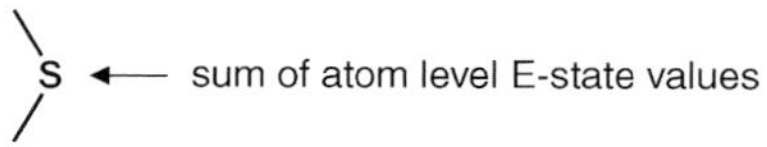 sum of atom level E-state values
Single atom E-state	SCarOH1	Largest E-state value in the molecule of all carboxyl oxygens (acidic oxygen)	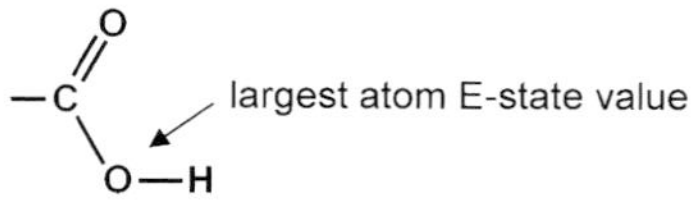 largest atom E-state value
	SHCarOH1	Largest HE-state value in the molecule of all hydrogen atoms bonded to carboxyl oxygen (acidic oxygen)	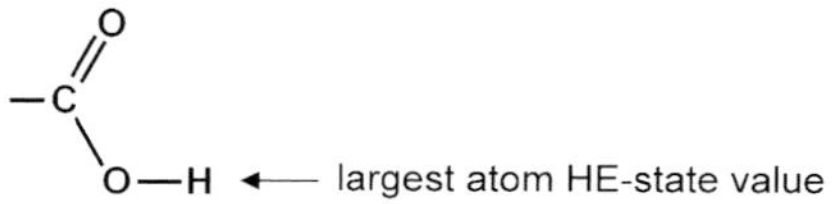 largest atom HE-state value
	SPheOH1	Largest E-state value in the molecule of all phenolic oxygens	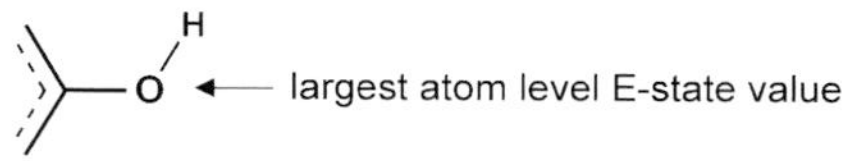 largest atom level E-state value
	SHBint2	Largest product of E-state and HE-state from all acceptor (A) and donor (D) pairs separated by 2 skeletal bonds	 product of E-state values
	SHBint4	Largest product of E-state and HE-state from all acceptor (A) and donor (D) paris separated by 4 skeletal bonds	 product of E-state values

continued

Table 12 Continued

	Index	*Descriptor definition*	*Descriptor illustration*
	Ssp3NH	Sum of E-state values of all secondary amines bonded to two sp^3 carbons	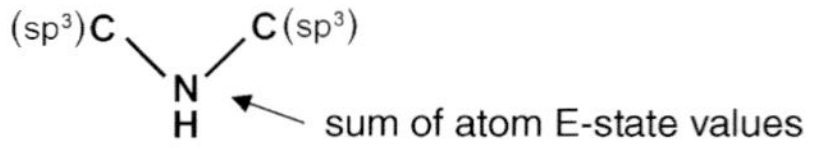 $(sp^3)C$–NH–$C(sp^3)$ ← sum of atom E-state values
	Ssp3N	Sum of E-state values of all tertiary amines bonded to three sp^3 carbons	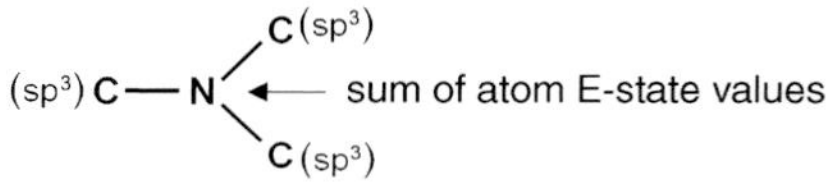 $(sp^3)C$–N($C(sp^3)$)($C(sp^3)$) ← sum of atom E-state values
Group type E-state	SCarom	Sum of E-states values for all substituted and unsubsituted aromatic carbon atoms	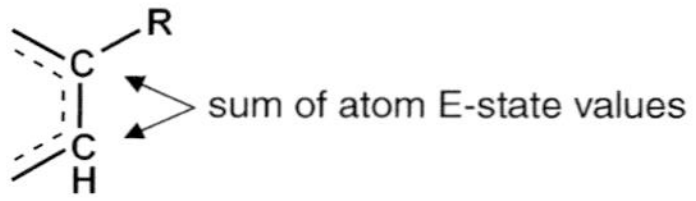 C–R, C–H ← sum of atom E-state values
	SHCsats	Sum of E-state values of hydrogen atoms on sp^3 carbons that are only bonded to other sp^3 carbons	 $(sp^3)C$, $(sp^3)C$, C, H, H ← sum of HE-state values
	SallNp	Sum of E-state values for all quaternary, pyridinium and aminium nitrogens	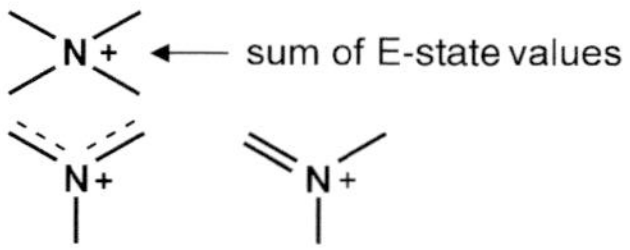 N+ ← sum of E-state values; N+; N+

Table 13 Definitions and illustrations of bond type and bond group E-state descriptors, taken from a model for human serum protein binding[44]

	Index	*Descriptor definition*	*Descriptor illustration*
	eaC2C2a	Sum of the bond E-state values for carbon aromatic bonds where both carbon atoms are unsubstituted	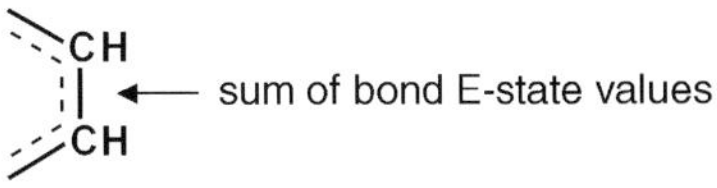 CH ← sum of bond E-state values CH
	eaC2C3s	Sum of the bond E-state values for carbon to carbon aromatic bonds between substituted and unsubstituted carbon	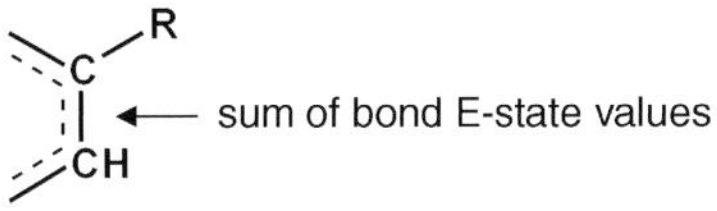 R C ← sum of bond E-state values CH
	eaC3C3s	Sum of the bond E-state values for carbon to carbon aromatic bonds between two substituted carbons	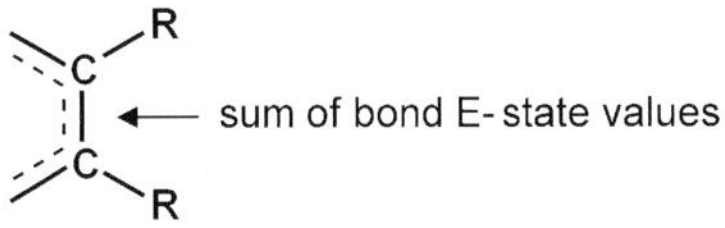 R C ← sum of bond E- state values C R
	e1C2N2	Sum of bond E-states for single bonds between secondary amine nitrogen and methylene carbon	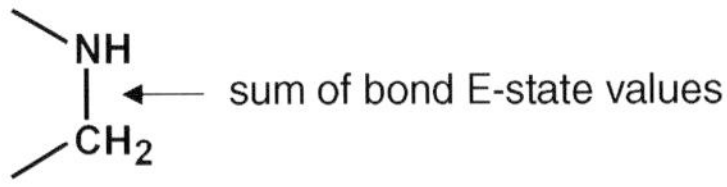 NH ← sum of bond E-state values CH_2
Bond-type E-state	e1C3N3	Sum of bond E-states between tertiary amines and >CH– carbon atoms	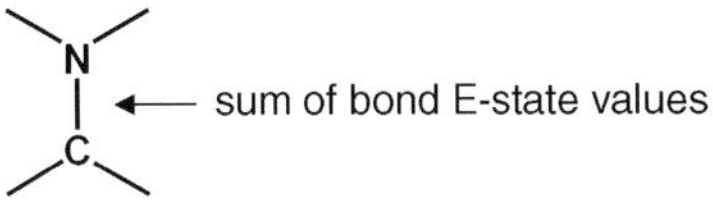 N ← sum of bond E-state values C
	eaC2N2	Sum of the bond E-state values for carbon to nitrogen aromatic bonds where the carbon is unsubstituted	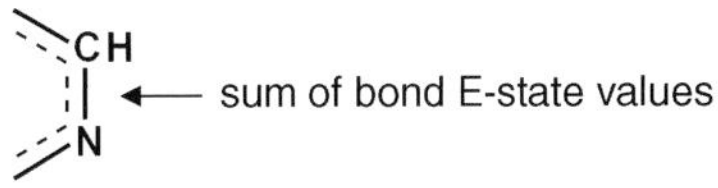 CH ← sum of bond E-state values N
	e2C3O1s	Sum of the bond E-state values for double bonds between oxygen and >C– carbon atoms	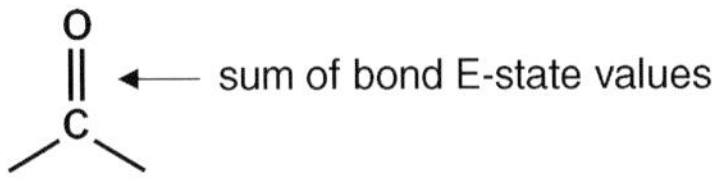 O ← sum of bond E-state values C
	e1C3Fa	Sum of the bond E-state values for fluorine to aromatic carbon bonds	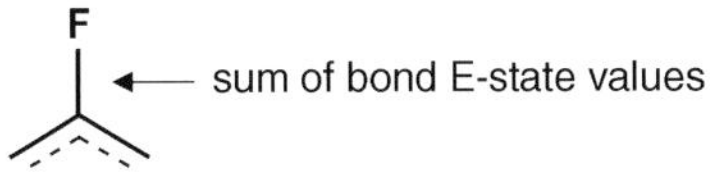 F ← sum of bond E-state values
	e1C3Cla	Sum of the bond E-state values for chlorine to aromatic carbon bonds	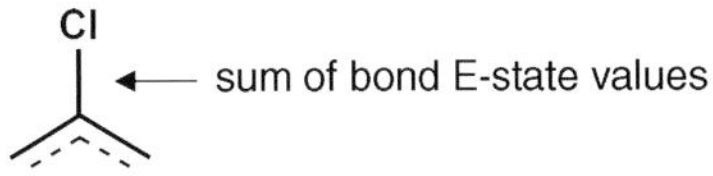 Cl ← sum of bond E-state values
Bond Group E-state	g1CalF	Sum of the bond E-state values for single bonds between fluorine and any aliphatic carbon	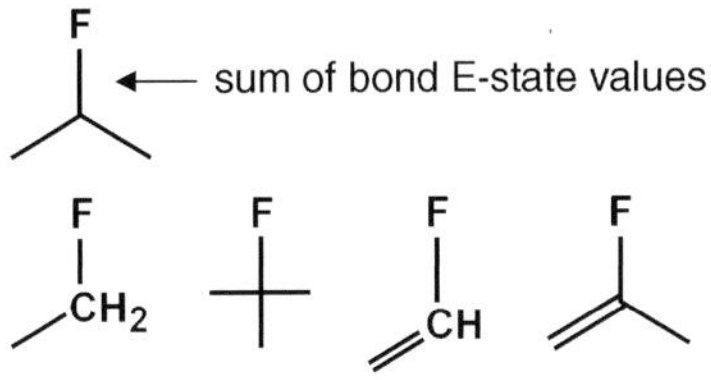 F ← sum of bond E-state values F F F F CH_2 CH

Chapter 5.24 is devoted to modeling of ADME/Tox properties; reports of several studies are given there along with appropriate references.

The **Tables 12** and **13** include a selected list of SIR descriptors that appear in a human plasma protein binding model on a large set of drugs. The tables includes a corresponding definitions and illustration along with the descriptor name. A more comprehensive list can be found on the Internet.[43]

References

1. Testa, B.; Kier, L. B. *Med. Chem. Rev.* **1991**, *11*, 35–48.
2. Hall, L. H. *Chem. Biodivers.* **2004**, *1*, 183–201.
3. Hall, L. H.; Hall, L. M. *SAR QSAR Environ. Res.* **2005**, *16*, 13–41.
4. Stjernlof, P.; Ennis, M. D.; Hansson, L. O.; Hoffman, R. L.; Ghazal, N. B.; Sundell, S.; Smith, M. W.; Svensson, K.; Carlsson, A.; Wikstrom, H. *J. Med. Chem.* **1995**, *38*, 2202–2216.
5. Kier, L. B.; Hall, L. H. *Molecular Structure Description: The Electrotopological State*; Academic Press: San Diego, CA, 1999.
6. Hall, L. H.; Vaughn, T. A. *Med. Chem. Res.* **1997**, 7, 407–416.
7. Kier, L. B.; Hall, L. H. *Adv. Drug Res.* **1993**, *22*, 1–3.
8. Kier, L. B.; Hall, L. H. *Pharm. Res.* **1990**, *72*, 801–807.
9. Hall, L. H. *Quant. Struct.–Act. Relat.* **1991**, *10*, 43–51.
10. van der Waterbeemd, H.; Carrupt, P.-A.; Testa, B.; Kier, L. B. In *Trends in QSAR and Molecular Modeling*; Wermuth, C. G., Ed.; Escom: Leiden, The Netherlands, 1993, pp 69–75.
11. Hall, L. H.; Kier, L. B.; Brown, B. B. *J. Chem. Inf. Comput. Sci.* **1995**, *35*, 1074–1080.
12. Kier, L. B.; Hall, L. H. *Il Farmaco* **1999**, *54*, 346–353.
13. Hall, L. H.; Kier, L. B. *J. Chem. Inf. Comput. Sci.* **2000**, *40*, 784–791.
14. Kier, L. B.; Hall, L. H. *SAR QSAR Environ. Res.* **2001**, *12*, 55–74.
15. Kier, L. B.; Hall, L. H. Database Organization and Similarity Searching with E-state Indices. In *The Fundamentals of Molecular Similarity*; Carbo-Dorca, R., Ed.; Kluwer: Amsterdam, 2001, pp 33–49.
16. Bennett, C. H. Logical Depth and Physical Complexity. In *The Universal Turing Machine: A Half-Century Survey*; Herken, R., Ed.; Oxford University Press: Oxford, UK, 1988, pp 227–257.
17. Kier, L. B.; Hall, L. H. *J. Pharm. Sci.* **1981**, *70*, 583–589.
18. Hinze, J.; Jaffe, H. *J. Am. Chem. Soc.* **1962**, *84*, 540–546.
19. Mulliken, R. S. *J. Chem. Phys.* **1934**, *2*, 782–793.
20. Pauling, L. *The Nature of the Chemical Bond*; Cornell University Press: Ithaca, NY, 1960.
21. Maw, H. H.; Hall, L. H. *J. Chem. Inf. Comput. Sci.* **2000**, *40*, 1270–1275.
22. Galvez, J.; Garcia-Domenech, R.; de Julian-Ortiz, T. V.; Soler, R. *J. Chem. Inf. Comput. Sci.* **1995**, *35*, 272–284.
23. Slater, J. *Phys. Rev.* **1930**, *36*, 57–67.
24. Maw, H. H.; Hall, L. H. *J. Chem. Inf. Comput. Sci.* **2001**, *42*, 290–298.
25. Rose, K.; Hall, L. H. *SAR QSAR Environ. Res.* **2003**, *14*, 113–129.
26. Coulson, C. A. *Valence*, 2nd ed.; Oxford University Press: New York, 1961.
27. Kier, L. B.; Hall, L. H. *J. Chem. Inf. Comput. Sci.* **1997**, *37*, 548–552.
28. Streitweiser, A. *Molecular Orbital Theory for Organic Chemists*; John Wiley: New York, **1961**, pp 329–341.
29. Hall, L. H.; Kier, L. B. *J. Chem. Inf. Comput. Sci.* **1995**, *35*, 1039–1045.
30. Kier, L. B.; Hall, L. H. The Electrotopological State: Structure Modeling for QSAR and Database Analysis. In *Topological Indices and Related Descriptors in QSAR and QSPR*; Devillers, J., Balaban, A. T., Eds.; Gordon and Breach: Reading, UK, 1999, pp 491–562.
31. Kier, L. B.; Hall, L. H. *Commun. Math. Comput. Chem.* **2001**, *44*, 215–235.
32. Hall, L. H.; Kier, L.B. The Molecular Connectivity Chi Indexes and Kappa Shape Indexes in Structure–Property Relations. In *Review of Computational Chemistry*; Boyd, D.; Lipkowitz, K. Eds.; VCH Publishers Inc., 1991; Chapter 9, pp 367–422.
33. Contrera, J. F.; Hall, L. H.; Kier, L. B.; MacLaughlin, P. *Curr. Drug Disc. Technol.* **2005**, *2*, 55–67.
34. Hall, L. M.; Kier, L. B.; Hall, L. H. *J. Comput.-Aided Mol. Des.* **2003**, *17*, 103–118.
35. Votano, J. R.; Parham, M. E.; Hall, L. H.; Kier, L. B. *J. Mol. Divers.* **2004**, *8*, 385–397.
36. Votano, J. R.; Parham, M. E.; Hall, L. H.; Kier, L. B; Orloff, S.; Tropsha, A.; Xie, Q.; Tong, W. *Mutagenesis* **2004**, *19*, 365–378.
37. Votano, J. R.; Parham, M. E.; Hall, L. H.; Kier, L. B.; Hall, L. M. *Chem. Biodivers.* **2004**, *1*, 1829–1841.
38. Kier, L. B.; Hall, L. H.; Randić, M. *J. Pharm. Sci.* **1975**, *64*, 1971–1974.
39. Kier, L. B.; Hall, L. H. *Molecular Connectivity in Chemistry and Drug Research*; Academic Press: New York, 1976.
40. Kier, L. B.; Hall, L. H. *Molecular Connectivity in Structure–Activity Analysis*; John Wiley: New York, 1986.
41. Kier, L. B.; Hall, L. H. Molecular Connectivity Chi Indices for Database Analysis and Structure–Property Modeling. In *Topological Indices and Related Descriptors in QSAR and QSPR*; Devillers, J., Balaban, A. T., Eds.; Gordon and Breach: Reading, UK, 1999, pp 307–360.
42. Kier, L. B.; Hall, L. H. *J. Chem. Inf. Comput. Sci.* **2000**, *40*, 792–795.
43. Molconn. http://www.molconn.com/ (accessed April 2006).
44. Hall, L. H.; Hall, L. M.; Kier, L. B.; Parham, M. E. Votano, J. R. Proceedings Solvay Conference on ADME/Tox Modeling, May 2005 in press.
45. R. Munson, private communication.
46. Garcia-Domenech, R.; de Julian-Ortiz, J. V. *J. Chem. Inf. Comput. Sci.* **1998**, *38*, 445–449.
47. Gozolbes, R.; Galvez, J.; Garcia-Domenech, R.; Derouin, F. *SAR QSAR Environ. Res.* **1999**, *10*, 47–60.
48. Diller, D. Development and Application of Methods Based on a 1-Dimensional Molecular Representation. *Book of Abstracts*, 230th American Chemical Society Meeting, Washington, DC, 2005.
49. Dixon, S. L.; Merz, K. M. *J. Med. Chem.* **2001**, *44*, 3795–3809.
50. Delisle, R. K.; Lowrie, J. F.; Hobbs, D. W.; Diller, D. J. *Curr. Comput.-Aided Drug Des.* **2005**, *1*, 65–72.
51. Wang, N.; Delisle, R. K.; Diller, D. *J. Med. Chem.* **2005**, *48*, 6980–6990.
52. Roy, K.; Pal, D. K.; Sengupta, C. *Drug Des. Disc.* **2001**, *17*, 207–218.
53. Gough, J.; Hall, L. H. *J. Chem. Inf. Comput. Sci.* **1999**, *39*, 356–361.
54. Kier, L. B.; Hall, L. H. *Med. Chem. Res.* **1992**, *2*, 497–502.
55. Gough, J.; Hall, L. H. *Environ. Tox. Chem.* **1999**, *18*, 1069–1075.
56. Hall, L. H.; Mohney, B. K.; Kier, L. B. *Quant. Struct.–Act. Relat.* **1993**, *12*, 44–48.
57. Maw, H. H.; Hall, L. H. *J. Chem. Inf. Comput. Sci.* **2000**, *40*, 1270–1275.
58. Hall, L. H.; Story, C. T. *J. Chem. Inf. Comput. Sci.* **1996**, *36*, 1004–1014.
59. Hall, L. H.; Story, C. T. *SAR QSAR Environ. Res.* **1997**, *6*, 139–161.
60. Abou-Shaaban, R. R.; al-Khamees, H. A.; Abou-Auda, H. S.; Simonelli, A. P. *Pharm. Res.* **1996**, *13*, 129–136.

61. Gute, B. D.; Grunwald, G. D.; Basak, S. C. *SAR QSAR Environ. Res.* **1999**, *10*, 1–16.
62. Roy, K.; De, A. U.; Sengupta, C. *Indian J. Chem.* **1999**, *38B*, 942–949.
63. Buolamwini, J.; Raghavan, K.; Fesen, M.; Pommier, Y.; Kohn, K.; Weinstein, J. *Pharm. Res.* **1996**, *13*, 1892–1895.
64. Roy, K.; Leonard, J. T. *Bioorg. Med. Chem. Lett.* **2004**, *14*, 745–754.
65. Maw, H. H.; Hall, L. H. *J. Chem. Inf. Comput. Sci.* **2001**, *41*, 1248–1254.
66. Roy, K; Sengupta, C. *Drug. Des. Disc.* **2002**, *18*, 33–43.
67. Mukherjee, S.; Saha, A; Roy, K. *Bioorg. Med. Chem. Lett.* **2005**, *15*, 957–961.
68. Sengupta, C.; Leonard, J. T.; Roy, K. *Bioorg. Med. Chem. Lett.* **2004**, *14*, 3435–3439.
69. Shen, M.; LeTiran, A.; Xiao, Y.; Kohn, H.; Tropsha, A. *J. Med. Chem.* **2002**, *45*, 2811–2823.
70. Kier, L. B.; Hall, L. H. *Pharm. Res.* **1990**, *7*, 801–807.
71. Hall, L. M.; Hall, L. H.; Kier, L. B. *J. Chem. Inf. Comput. Sci.* **2003**, *43*, 2120–2128.
72. Kellogg, G. E.; Kier, L. B.; Gaillard, P.; Hall, L. H. *J. Comput.-Aided Mol. Des.* **1996**, *10*, 513–520.

Biographies

Lowell H Hall is a physical chemist who pioneered methods of molecular structure representation for the past 30 years. He received the PhD degree in physical chemistry from The Johns Hopkins University and pursued postdoctoral work at The National Bureau of Standards and Oak Ridge National Laboratory in single crystal x-ray crystallography. For 38 years Hall has been Professor of Chemistry at Eastern Nazarene College, a liberal arts college in Quincy, MA. Hall began his research in QSAR during a sabbatical leave with Lemont B Kier at Massachusetts College of Pharmacy in 1974. Together Hall and Kier have written four books, eight book chapters, and more than 100 papers. Their most recent book, *Molecular Structure Description: The Electrotopological State* (Academic Press, 1999) presents their E-state approach to molecular structure representation. Hall and Kier are the codevelopers of the Structure Information Representation method as well as topological QSAR.

Hall has been a consultant to pharmaceutical companies, a regular participant in the Gordon Conference on QSAR/CADD, and is currently a consultant with MDL Information Systems. Hall is the creator of the Molconn software and also operates Hall Associates Consulting. He is a partner in ChemSilico LLC, a producer of model predictors for properties of interest in the pharmaceutical area.

Lemont B Kier received a BS in pharmacy from Ohio State University and a PhD in medicinal chemistry from The University of Minnesota. He has been a pioneer in the development of several theoretical methods used for rational drug design, based on molecular structure. He introduced the use of molecular orbital theory to compute the preferred conformation of molecules. New insight into the pharmacophores of several neurotransmitters and drug molecules were predicted by Kier using these calculations. In another application, Kier pioneered in the use of molecular orbit calculated energies to predict intermolecular interactions, now called molecular docking.

In the mid-1970s, Prof Kier and Prof Hall developed nonempirical molecular descriptors named molecular connectivity indices. Information derived from these indices is a nonempirical estimate of the valence state electronegativity, called the Kier–Hall electronegativity, closely correlated with the Mulliken–Jaffé electronegativity values. In the mid-1980s Kier introduced the kappa indices, encoding information about several aspects of molecular shape and flexibility. Another approach to structure description was developed by Kier and Hall, in the early 1990s. This set of descriptors is called the electrotopological state (E-state) and is based upon the Kier–Hall electronegativity and the topological characteristics of an atom or group in a molecule.

L Mark Hall received a BS in biology with minor degrees in chemistry and writing from Eastern Nazarene College where he did undergraduate research developing artificial neural network QSAR models on pesticide toxicity and human plasma protein binding. He has worked as a consultant to vendors that develop computational chemistry software tools. This consulting has included the development of product user guides, software specifications, and quality assurance reports. Hall has also worked in Product Development for ChemSilico, helping to develop commercial ANN predictors for human intestinal absorption, human plasma protein binding, and blood–brain barrier partition. He has published QSAR studies for human intestinal absorption, human plasma protein binding, albumin binding affinity, and aqueous solubility, as well as Ames genotoxicity.

Hall has worked for Hall Associates Consulting on the continuing development of the Molconn molecular descriptor software. These projects include the development of a new class of descriptors designed for use in modeling the acid ionization constant of drugs, the development of a system for recognizing and correcting aromaticity encoding errors in computer-based molecular structure files, and development of new approaches to error analysis of QSAR models using novel applications of hierarchical clustering. Recently, Hall was elected to serve as the Communications Director for the Boston area Group for Informatics and Modeling.

© 2007 Elsevier Ltd. All Rights Reserved
No part of this publication may be reproduced, stored in any retrieval system or transmitted in any form by any means electronic, electrostatic, magnetic tape, mechanical, photocopying, recording or otherwise, without permission in writing from the publishers

Comprehensive Medicinal Chemistry II
ISBN (set): 0-08-044513-6

ISBN (Volume 4) 0-08-044517-9; pp. 537–574

4.23 Three-Dimensional Quantitative Structure–Activity Relationship: The State of the Art

A M Doweyko, Bristol-Myers Squibb, Princeton, NJ, USA

© 2007 Elsevier Ltd. All Rights Reserved.

4.23.1 Introduction

4.23.1.1 Development of Quantitative Structure–Activity Relationship (QSAR)

The relationship between a molecule and its properties has long been recognized to involve elements of its structure. In 1868, Crum-Brown and Fraser suggested that the physiological action of a substance was a function of its chemical composition and constitution, in other words, its structure.[1] It was 95 years after the publication of the first structure–activity relationship (SAR), describing the effect of molecular weight on the narcotic properties of alcohols in 1869,[2] that quantitative relationships between structure and activity were formulated. These entailed electronic, hydrophobic, and steric properties of phenyl substituents and were developed as a predictive mathematical tool by Hansch and Fujita in 1964,[3] heralding the arrival of QSARs. That same year the methodology was complemented by Free and Wilson with approaches identifying important positional features in a molecular scaffold.[4] Pseudo-3D steric features (parameters used in the STERIMOL program) were introduced in 1976 by Verloop[5,6] as an aid in describing simple alkyl geometries. Although the structural descriptors were limited in scope, this 2D-QSAR methodology was effectively applied to the quantitation and prediction of both molecular activities and properties, culminating in the successful development of a number of commercial drugs and pesticides.[7–9]

4.23.1.2 Birth of Three-Dimensional Quantitative Structure–Activity Relationship (3D-QSAR)

The first truly 3D approach applied to understanding SAR followed the introduction of a 3D description of the essential molecular elements required for activity proposed by Kier in 1971, which introduced the concept of a pharmacophore.[10] Interaction analyses conducted by Holtje and Kier were aimed at mathematically modeling potential interactions. Such pharmacophoric elements provided a means to quantitate the contribution of individual 3D structural features to activity.[11,12] This work eventually led to the active analog approach of Marshall in 1979, wherein the SAR for a series of conformationally flexible molecules was used to derive a common pharmacophore.[13,14]

Alignment-independent methods also began to evolve during this period of time. In 1975 Kier and co-workers reported that it was possible to correlate molecular properties and activities to molecular connectivity.[15,16] Very significant linear correlations were found between the connectivity index and molecular polarizability, cavity surface areas calculated for water solubility of alcohols and hydrocarbons, biological potencies of nonspecific local anesthetics, and octanol/water partition coefficients ($\log P$). These approaches were extended to include QSAR (*see* 4.22 Topological Quantitative Structure–Activity Relationship Applications: Structure Information Representation in Drug Discovery).[17–19]

The 3D paradigm was further advanced with the work of Hopfinger in 1981 with the introduction of molecular shape analyses (MSA), wherein metrics which include common overlap steric volume and potential energy fields between pairs of superimposed molecules were successfully correlated to the activity of the series under study.[20–23] The MSA using common volumes also provided some insight regarding the receptor-binding site shape and size. The closely related approach, minimum steric difference or minimum topological difference (MTD), developed at about the same time used a 'hypermolecule' concept for molecular alignment which correlated vertices (atoms) in the hypermolecule (a superposed set of molecules having common vertices) to activity differences in the series.[24–27] The use of the 3D structure, conformational energy, and important atom-based physicochemical properties to model the hypothetical binding site cavity was introduced by Ghose *et al.*, in 1985 in the form of a modeling paradigm called REMOTEDISC.[28–30]

The analysis of potential interactions between a ligand molecule and a receptor-binding site was formalized with the introduction of GRID by Goodford[31] in 1985, wherein the interaction of a probe atom/group with a protein of known structure is computed at regularly spaced sample positions throughout and around the macromolecule, yielding an array of energy values (*see* 4.11.3 Characterization of Protein-Binding Sites and Ligands Using Molecular Interaction Fields). In the initial formulation of GRID these probes included water, the methyl group, amine nitrogen, carboxy oxygen, and hydroxyl. Subsequent developments led to the calculation of additional probe energetics which included hydrogen-bonding[32] and hydropathic interactions.[33,34] The use of molecular overlays and interaction fields served as the groundwork for the development of DYLOMMS in 1983–87 by Cramer, Wise and Bunce,[35–37] which was further expanded to comparative molecular interaction field analysis (CoMFA) in 1988,[38] a grid-based interaction field approach utilizing partial least squares (PLS)[39,40] to correlate steric and electrostatic interactions surrounding an overlaid set of molecules with their activities. In the same year, Doweyko introduced hypothetical active site lattice (HASL), a grid-based method that captured the location of molecular features (atom types) and correlated them to activity using a simple linear combination of terms representing molecular occupancy.[41] The intervening decades witnessed the further evolution of grid-based methods dependent upon molecular alignment, as well as the introduction of alignment-independent approaches. The general subject of 3D-QSAR has been reviewed extensively.[42–49] The details of recent developments in both alignment-dependent and -independent 3D-QSAR methods serve as the subject matter of this review, with emphasis placed on those methods that are 'defined by an unexpected combination of imagination and insight.'[50]

4.23.2 Current Methods

4.23.2.1 Alignment-Dependent Methods

4.23.2.1.1 Comparative molecular field analysis

Since its advent in 1988, CoMFA has become the gold-standard 3D-QSAR methodology for the past two decades. Although details have been reviewed elsewhere,[51–56] it is worthwhile highlighting the basic strategy and assumptions inherent in the method (*see* 4.07 Predictive Quantitative Structure–Activity Relationship Modeling). A set of molecules are aligned (using any of a variety of methods described later in this chapter), a uniform 3D grid surrounding the aligned molecules is used to develop probe interaction energies at each grid point (typically using a Csp^3 steric probe and using a $H+$ electrostatic probe), and the changes in interaction energies are correlated with changes in activity using PLS. One of the strengths of CoMFA, as currently embodied in SYBYL,[57] is the capability of visualizing these interaction/activity correlations by means of color-coded contours reflecting the strongest association between changes in a probe interaction and activity. One such example is shown in **Figure 1a**.

Several variable selection procedures have been developed to focus on the descriptor space most relevant to the CoMFA model. One such example is the q^2-guided region selection (q^2-GRS)[58,59] method which ameliorates some of the method's major deficiencies, namely, problems related to overall orientation, lattice placement, and step size. A genetic algorithm-based region selection (GARGS) approach was devised to identify the most significant regions within a CoMFA model.[60] The generating optimal linear PLS estimation (GOLPE)[61–63] approach entails a preliminary variable selection routine by means of D-optimal designs in which the effect of each variable on model predictivity is iteratively evaluated.

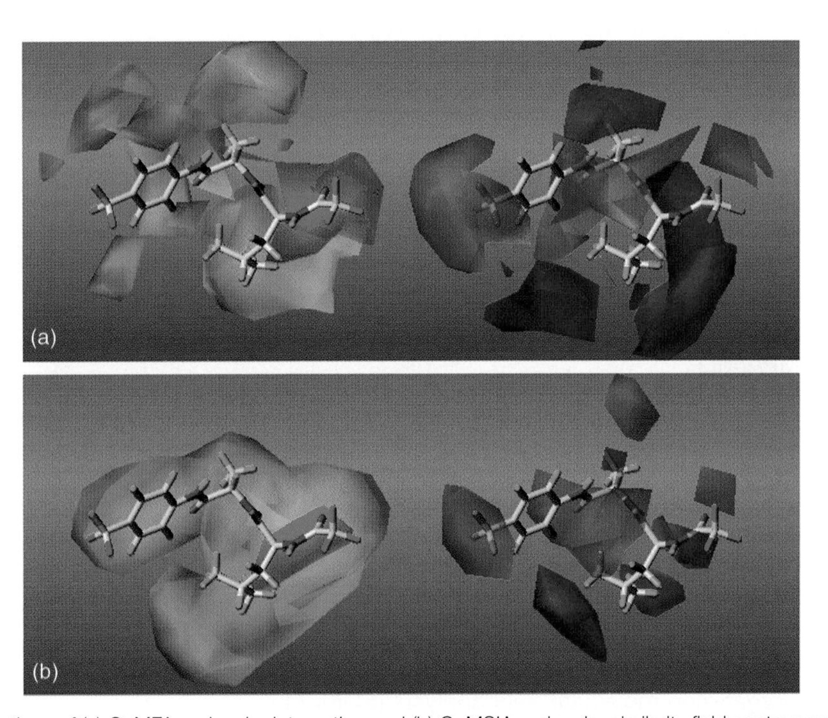

Figure 1 Illustrations of (a) CoMFA molecular interaction and (b) CoMSIA molecular similarity field contour surfaces depicting steric (yellow/green) and electrostatic (red/blue) changes correlated to activity for a series of 32 elastase inhibitors ($pK_i = 2.54–7.6$) overlaid on one of the active congeners in the series.[150] Both models were derived from MCP-based overlays of each molecule and yielded good LOO cross-validation coefficients (CoMFA $r^2_{cv} = 0.638$, 3 PLS components; CoMSIA $r^2_{cv} = 0.708$, 2 PLS components). The analyses point to a consistent positive effect on activity due to steric bulk (green) at the *N*-trifluoracetyl group and due to an electron-rich effect at the *p*-trifluoro phenyl group, underscoring the similarity of results when using similar probes.

Besides standard grid-based steric and electrostatic descriptors, probes designed to obtain estimates of the desolvation free energy at or near a molecular surface have been explored with varying success. Examples include the use of boundary element method (BEM) within a continuum dielectric model[64] and an approach entailing a numerical solution to the Poisson–Boltzmann equation incorporated into the program DELPHI.[65–67] Other types of probes utilized in the generation of molecular interaction fields (MIFs) are discussed below.

4.23.2.1.2 Comparative molecular similarity indices analysis (CoMSIA)

Molecular similarity indices can serve as a set of field descriptors in a novel application of 3D-QSAR referred to as CoMSIA.[68,69] The indices are generated using a probe atom at regularly spaced grid intersections which surround and penetrate an aligned set of molecules, wherein a metric is computed as a Gaussian distribution of similarity between the probe atom and its surroundings at each grid point. Probe atoms with differing physical chemical properties provide different sets of indices reflecting those properties in the training set of molecules. The subsequent application of a PLS methodology identical to that used in CoMFA can provide a correlation between the field values and observed activities. The CoMSIA method may have some advantages over CoMFA in several aspects: (1) in the use of the Gaussian distribution of similarity indices, avoiding the abrupt changes in grid-based probe–atom interactions that can occur when using Lennard-Jones and Coulomb potential fields; (2) in the choice of similarity probe, not limited to either steric or electrostatic potential fields; and (3) in that CoMSIA can attempt to capture solvent entropic terms by means of a hydrophobic probe. Features deemed significant in the analysis can be contoured and inspected visually in a manner similar to that for CoMFA, as exemplified in **Figure 1b**. Other philosophically similar approaches that use some type of MIF analysis to generate a set of descriptors have been reported.

4.23.2.1.3 Hydropathic intermolecular field analysis (HIFA)

HIFA[33,34] is a structure-based 3D-QSAR method that employs empirically derived hydropathic fields generated by hydrophobic interactions (HINT).[33,34,70] Whereas CoMFA measures the interaction between ligands and a probe atom, HIFA captures the interaction between the ligand and the receptor. HINT calculates empirical atom-based hydropathic parameters that are believed to encode all significant intermolecular and intramolecular noncovalent interactions implicated in drug binding or protein folding. Coulombic, hydrogen-bonding, dispersion, as well as hydrophobic effects may be extrapolated from the hydrophobic atom constant. In addition, since hydrophobicity is defined in terms of solubilities, the effects of solvent are also encoded within the HIFA.

4.23.2.1.4 Voronoi field analysis (VFA)

VFA[71–73] utilizes a nonregularly spaced grid based on steric and electrostatic potential indices calculated or estimated at lattice points, but the indices are manipulated to be assigned to superposed molecular regions defined by the Voronoi polyhedral division.

4.23.2.1.5 COMPA/MLP

Molecular lipophilicity potential (MLP)[74–77] represents a method of calculating lipophilicity at the surfaces of molecules based on atomistic log *P*.[78] COMPA relies on the use of MLP in a grid-based paradigm.[77]

4.23.2.1.6 Molecular quantum similarity measures (MQSM)

MQSM[79–82] is a method that relies on vectors of the electron density function as descriptors and utilizes a similarity measure based on that metric to describe the system. Although not exactly a 3D-QSAR method, this approach relies on the accurate ab initio determination of molecular structure and generates a unique set of similarity metrics derived from comparisons made between electron density functions for each molecule.

4.23.2.1.7 Self-organizing molecular field analysis (SOMFA)

SOMFA[83,84] is a method with similarities to the HASL (discussed below) and the Free-Wilson 3D-QSAR approaches. Utilizing a self-centered activity, i.e., dividing the molecule set into actives (+) and inactives (−), and a grid probe process that penetrates the overlaid molecules, the resulting steric and electrostatic potentials are mapped onto the grid points and are correlated to activity using linear regression. There are a number of other molecular field-based similarity approaches that have been used to generate 3D-QSAR models, including the molecular field similarity (MFS) method.[85–87]

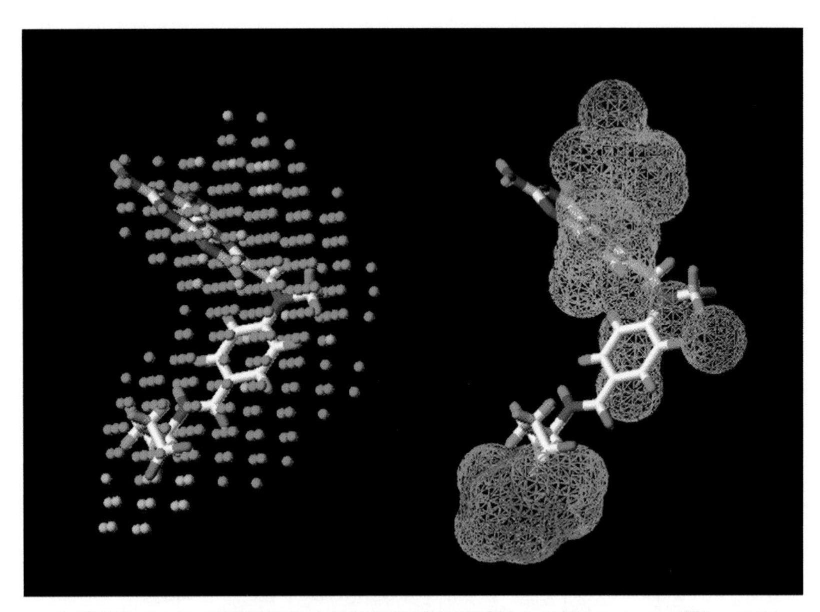

Figure 2 Using the HASL model generated from a series of 23 superposed DHFR inhibitors,[205] the grid points corresponding to electron-neutral (steric) effects are illustrated as overlaid on the protein-bound structure of methotrexate (mtx). The alignment paradigm entailed the separate docking and minimization of each member of the series into the DHFR substrate-binding site. HASL grid spacing was 1.5 Å. The yellow and cyan points (left) correspond to molecular features that affect activity negatively and positively, respectively. The negative steric effect is shown right, wherein the strongest effects are shown in red, weakest, in blue.

4.23.2.1.8 Hypothetical active site lattice

The HASL methodology developed in 1986–88 is based on correlating the presence or absence of features within molecules to their activity, and thus represents a kind of inverse grid-based methodology compared to more common MIF approaches.[41,88–93] Although HASL can utilize any number of atom type descriptors, three are typically sufficient (electron-rich, electron-poor, and neutral atoms, roughly representing H-bond acceptors, donors, and lipophylic atom types, respectively). Correlation between structure and activity is achieved through an iterative solution to a set of simultaneous equations, each representing a single molecule's occupied grid points (dimensioned as x, y, z, atom type, and partial activity) whose coefficients (partial activities) are to be determined. Once the model is generated, significant features are readily apparent by their grid point coefficients and these features can be contoured for inspection within Sybyl using the HASL SPL module[94] or illustrated as a grid of significant atom types. An example of a HASL model is shown in **Figure 2**.

4.23.2.1.9 Pseudoreceptors

The simultaneous use of a receptor-fitting and receptor-mapping paradigm is at the heart of the pseudoreceptor approach embodied in programs like YAK[95–98] and PrGem.[99] After defining a pharmacophore model, a pseudoreceptor is constructed as an explicit molecular-binding pocket for the bioactive conformation of a series of ligands with high affinity for a particular receptor subtype. YAK constructs a peptidic pseudoreceptor around any single small-molecule or molecular ensemble of interest by locating 'nucleation' sites defined as ligand-derived vectors. Three types of site are identified: (1) hydrogen extension vectors (HEV: ideal H-bond acceptor sites); (2) lone pair vectors (LPV: H-bond donor sites); and (3) hydrophobicity vectors (HPV: hydrophobic interaction sites). An example of a series of potential interaction sites surrounding a set of superposed dihydrofolate reductase (DHFR) inhibitors is shown in **Figure 3**. An important assumption here is that all the molecules under consideration are assumed to be equally buried within the receptor. A correlation-coupled minimization of the pseudoreceptor residue interactions with ligands is conducted in an iterative manner in order to maximize a linear regression-derived correlation between calculated binding energies and observed binding affinities for the training set. Subsequent validation is conducted with a test set following a similar protocol using the calculated binding energies to generate predicted binding affinities. A conceptually analogous approach known as receptor surface models (RSM)[100,101] generates a surface loosely enclosing the common volume of the most potent ligands. Points on the surface are described by the complement of the average partial charge,

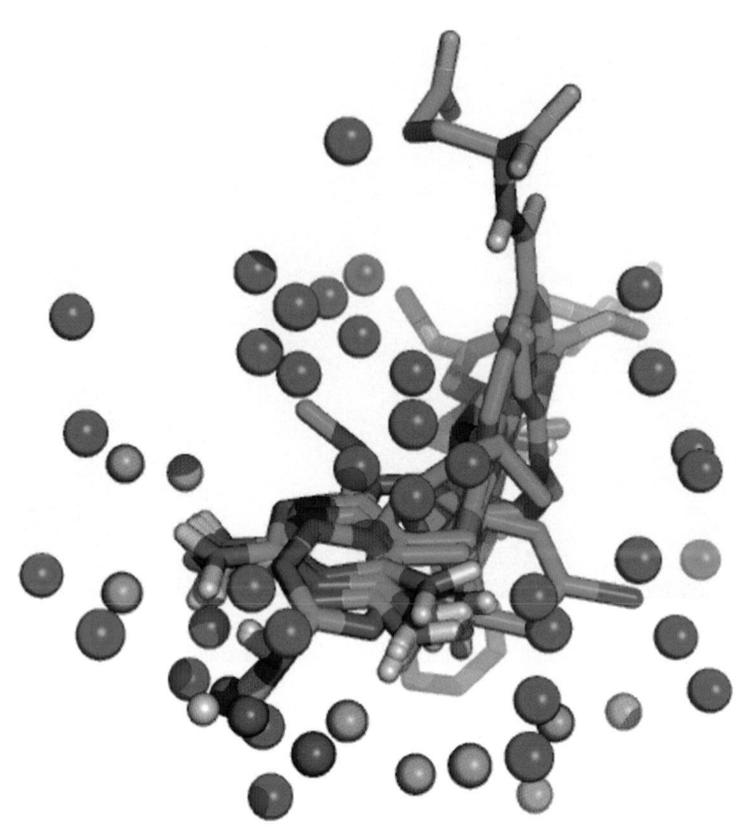

Figure 3 A schematic representation of a pseudoreceptor composed of atoms arranged about 22 overlaid DHFR inhibitors (inhibitor structures from [205]). Atom types are hydrophobic (carbon, green), H-bond acceptors (oxygen, red), and H-bond donors (nitrogen, blue). Computed interactions between atoms and ligands are used to generate a predictive binding model for the superposed set of DHFR inhibitors.

electrostatic potential or hydrogen-bonding ability, or by some hydrophobicity measure. Structures can be energy-minimized within the receptor surface model to arrive at conformations that are consistent with the model, and interaction energies can be estimated. Genetic algorithms (GAs) can be used to automatically detect those molecules producing the most predictive model. Another receptor-building approach which searches for a match between a subset of ligand alignments and possible receptor shape and property (using probe spheres) is embodied in biological substrate search (BIS).[102] The method is self-superposing and based on the assumption that the formation of a receptor–ligand complex depends on the probability of contact between receptor and ligand atoms. Contact probabilities are derived from atom charge, radius, and accessible area.

4.23.2.1.10 Genetically evolved receptor models (GERM)/comparative receptor surface analysis (CoRSA)

The GERM approach was developed by Walters in 1994.[103,104] The method relies on the correlation of biological potency of molecules with the potential energy of the interaction of the ligands with probes placed on the union surface of the superposed molecules. The location and nature of the probes are evolved using a GA which is designed to optimize the selection of probes using the correlation of biological activity of each molecule with its calculated interaction energy as a scoring function. The net result after many generations is a population of 3D-QSAR models. A related methodology known as CoRSA[105] uses the common steric and electrostatic features of the most active members of a series of compounds to generate a virtual receptor model, represented as points on a surface complementary to the van der Waals or Wyvill steric surface of the aligned compounds. 3D energetics descriptors are calculated from the receptor surface model–ligand interaction and these 3D descriptors are used in genetic PLS data analysis to generate the 3D-QSAR model. A further evolution of GERM, known as pseudo atomic receptor model (PARM), was developed; this uses a combination of GA and cross-validation techniques to produce an atomic-level pseudoreceptor model based on a set of known SARs.[106,107]

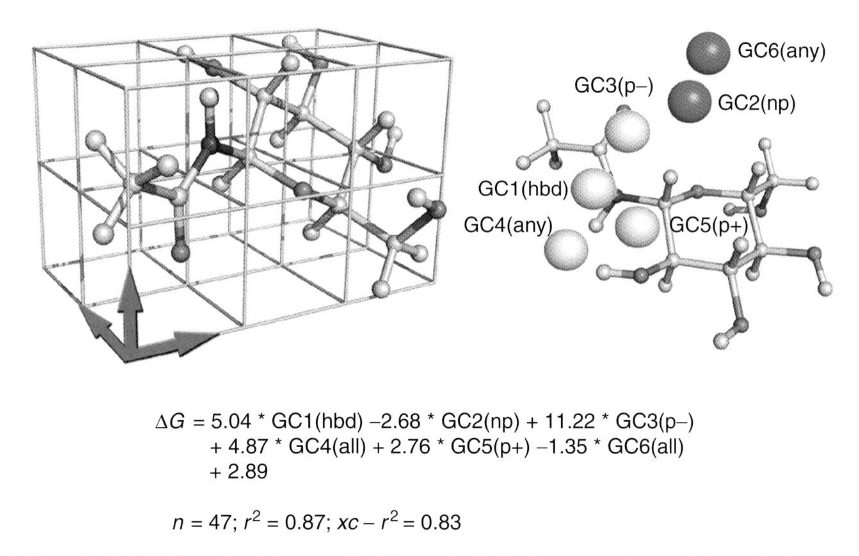

$$\Delta G = 5.04 * GC1(hbd) - 2.68 * GC2(np) + 11.22 * GC3(p-) + 4.87 * GC4(all) + 2.76 * GC5(p+) - 1.35 * GC6(all) + 2.89$$

$$n = 47;\ r^2 = 0.87;\ xc - r^2 = 0.83$$

Figure 4 Illustration of 4D-QSAR methodology applied to a set of 47 glucose analog inhibitors of glycogen phosphorylase B.[221] Grid cell (GC) occupancies for various types of atoms are determined for the superposed series using dynamics and are correlated to observed activities (as ΔG). In this example, six grid cells are identified (np, nonpolar; p+, polar positive; p–, polar negative; hbd, hydrogen bond donor) as significantly correlated with activity, yielding an equation which indicates that occupancy at GC2 by nonpolar atoms and at CG6 by any atoms results in a loss of activity. (Courtesy of Anthony J. Hopfinger.)

4.23.2.1.11 Four-dimensional-quantitative structure–activity relationship (4D-QSAR)

Aligned molecules are usually treated as a static set having conformations essentially frozen in an arbitrary manner. A significant departure from this paradigm is represented by the 4D-QSAR approach.[108–110] In this method each molecule in the aligned training set is represented by a large number of possible conformations in such a way that the occupancy of submolecular features (e.g., nonpolar atoms, hydrogen bond donors, etc.) becomes a metric focused on the distribution of such features within arbitrarily defined cells (derived from a regular-spaced 3D grid). These grid cell occupancy descriptors (GCODs) are then used as independent variables in constructing 3D-QSAR models after data reduction.[108,111,112] One possible shortcoming of this procedure is that it can produce a number of seemingly equivalent 3D-QSAR models depending on the number of alignment paradigms used (typically, pharmacophoric overlays) which require further statistical filtering and comparison with other 3D-QSAR methods to narrow the models down to a useful few. An illustration of the method is provided in **Figure 4**.

4.23.2.2 Methods of Alignment

The traditional technique used for the alignment of molecules with each other has been to superpose similar molecular features (e.g., atom types) upon one another in some systematic manner. Since molecules can be aligned using a flexible conformer strategy wherein any combination of rigid or flexible conformers is used, an enormous number of alignment paradigms can be envisioned. In an attempt to narrow the possibilities, often a single molecule is chosen as a rigid template upon which the remaining molecules are overlaid. Alternative approaches include the generation of a number of feasible conformers using a short run of molecular dynamics (MD) or a Monte Carlo scheme designed to generate a random set of accessible conformers. The conformer generating system, conformational analyzer with molecular dynamics and sampling (CAMDAS),[113] performs MD calculations for a target molecule and samples conformers from the trajectory of the MD. The program then evaluates the similarities between each of the sampled conformers in terms of the root-mean-square deviations (rmsd) of the atom positions, clusters similar conformers, and provides a list of the clustered conformers. From a practical standpoint, the best choice of a 'template' molecule is one that has good activity and is as conformationally constrained as possible. These requirements are often not completely met, resulting in arbitrary choices of 'template' and conformation. Nonetheless, the ensuing 3D-QSAR model can be quite informative and useful as long as the assumptions used in the alignment paradigm are kept in mind. In cases where the binding site is known (e.g., by x-ray structural information) or inferred (e.g., by modeling), it may be possible to use the binding-site model as a constraint in developing a reasonable set of aligned ligand molecules. But even here there is a potential for distortion, namely, that weakly binding molecules may not adopt the same overall binding motif observed in the x-ray structure. A number of alternative methods for the alignment of molecules have been reported in the literature which may be of interest within the context of developing predictive 3D-QSAR models.

4.23.2.2.1 Atoms

Besides a time-consuming manual and arbitrary alignment of molecules focusing on the superpositioning of similar atoms, there have been a number of automated procedures described. The SEAL[114] program was developed to optimize the alignment of two 3D structures using their atomic partial charges and steric volumes as factors. In addition, this method performs the superposition many times using randomly generated starting configurations and keeps only the best unique results based on the value of the alignment function. Recent modifications to SEAL permit the use of a variety of physicochemical properties as similarity descriptors. A similar paradigm is found in the implementation of Flo/QXP,[115] wherein an inverse force field is used to coax similar atoms of separate, flexible molecules together. The program, SUPERPOSE,[116] utilizes four types of physicochemical properties ((1) hydrophobicity; (2) presence of a hydrogen-bonding donor; (3) presence of a hydrogen-bonding acceptor; and (4) presence of a hydrogen-bonding donor/acceptor) in the form of a pseudomolecule consisting of functional atoms instead of a real molecule. The program, FLEXS,[69,117] represents an automatic method for structurally superposing pairs of ligands, approximating their putative binding-site geometry. In this approach, one of the ligands is treated as flexible, while the other one is kept as a rigid reference structure. FLEXS is an incremental construction procedure based on molecular fragments. SLATE,[118,119] a program for the superposition of flexible molecules, uses simulated annealing to minimize the difference between the distance matrixes calculated from the hydrogen-bonding and aromatic ring properties of two ligands. A molecular stack is generated using multiple pairwise matches which are used by the program DOH to predict the relative positions of receptor atoms that could form hydrogen bonds to two or more ligands in the data set. An approach to atom-based alignment which avoids the bias inherent in a pairwise process is embodied in the generalized procrustes analysis (GPA),[120,121] wherein a consensus alignment of all molecules is obtained without the need for an arbitrarily selected template molecule.

4.23.2.2.2 Fields

Molecules can be aligned using their computed MIFs.[53,92,122,123] This has been found to be a useful alternative to arbitrary atom-on-atom procedures, which can be time-consuming. Molecules are thought to recognize each other through their surfaces and the properties of these surfaces, suggesting that alignments of the molecules such that they present consistent surface properties in 3D space would set the stage for a more realistic QSAR analysis. Several examples of such field-fitting methods include: (1) the field-fit option implemented within SYBYL,[57] which uses grid-based steric and electrostatic MIFs; (2) the program SUPER,[124] which optimizes the 3D overlap of molecules based on the electrostatic potential mapped on their van der Waals surfaces; (3) methods utilizing integration over a number of analytical Gaussian functions to approximate electrostatic potentials, which avoid singularities at atomic positions common to a Lennard-Jones or Coulombic potential calculations[125]; and (4) related approaches involving solvent-accessible electrostatically mapped surfaces[126] and pattern-matching of surfaces engendered by rotational sampling.[127]

4.23.2.2.3 Pharmacophores

In addition to using an atom-by-atom process or calculated field effects for the alignment of molecules, a hypothetical pharmacophore can serve as an effective common target template. Namely, each molecule can be conformationally directed to assume the shape necessary for its submolecular features to coincide with either a known pharmacophore or one that is generated during conformational analysis. One of the first programs to search automatically for conformers consistent with a potential pharmacophore was DISCO (DIStance COmparisons),[128,129] developed by Martin *et al.*, which is designed to discover: (1) how many pharmacophores explain the data by using selected conformations and superposition rules; (2) the trade-off between a low rmsd for superposition and including more points in the model; and (3) the trade-off between having a low rmsd and including higher-energy conformations in the model. As is a common limitation in such approaches, data sets containing molecules with a high degree of conformational flexibility may be problematic from a combinatorial point of view. Other approaches more suited to comparing multiple conformers have been successfully used in database searching, but are not directly applicable in generating pharmacophore overlays needed for 3D-QSAR analysis.[130] One of the more widely used pharmacophore-based fitting programs is CATALYST,[131–133] which represents an approach that focuses on modeling the ligand–receptor interactions from the point of view of the receptor, using information derived only from ligands, which are described as collections of chemical functions arranged in 3D space. Conformational flexibility is modeled by creating multiple conformers, judiciously prepared to emphasize representative coverage over a specified energy range. A CATALYST 3D-QSAR functionality creates a number of pharmacophore hypotheses from which the activities of separately superposed molecules is correlated by comparing the fit between molecular functionalities and the pharmacophores.[134] Since a pharmacophore can be deduced from a set of molecules, such a procedure may exploit a 3D-QSAR analysis of a set of aligned molecules, identifying significant and predictive features from the model.[135] Programs such as APEX-3D[136–138] are designed to infer a pharmacophoric pattern from a set of molecules based on either 2D-topological or 3D-topographical patterns and inductive logic. HASL has also been successfully used to devolve a 3D-QSAR model to a small set of pharmacophoric features.[135]

4.23.2.2.4 Multiple conformers

The generation of a predictive 3D-QSAR model which is alignment-based is not necessarily limited to the consideration of a single conformer per molecule. As it happens, there are often cases where ligand molecules may bind to a receptor in multiple ways,[139–144] or the exact binding mode is unknown for ligands having a fair degree of conformational flexibility. Generally, these limitations can represent a considerable source of confusion in generating a meaningful 3D-QSAR model, typically resulting in poor models based on a single binding mode assumption. In spite of these seemingly gloomy prospects, a few methods have been developed which have shown promise in dealing with multiple conformers (multiple binding modes). For example, an iterative procedure relying on 3D-QSAR (CoMFA) models developed by Nicklaus *et al.*[145] used only those compounds that possessed a symmetrical substituent pattern on a phenyl ring to select the active conformers of the less symmetrical compounds in the set Allowing multiple conformers for each compound in the data set yielded higher r^2_{cv} (cross-validated correlation coefficient). Multiple binding modes directly incorporated into a 3D-QSAR model have also been shown to lead to 3D-QSAR models having higher r^2_{cv} when compared with those derived from single binding modes.[145,146] Although not strictly a 3D-QSAR approach, the dynamic QSAR method of Mekenyan *et al.*[147–149] requires a least-squares fit to be applied on multiple predictor data sets, and the method can include 3D-based descriptor sets. The correlation between the experimental and calculated values is determined by three terms: (1) the experimental error if multiple observations are taken into account; (2) within-group deviations if multiple predictors are taken into account; and (3) lack of fit between experimental and calculated means. Doweyko[150] was able to demonstrate that subsets of statistically likely binding modes or preferred ligand-binding modes could be identified based on intermediate 3D-QSAR analyses of subset models which point to those conformers handled well (as determined by subset r^2_{cv} values) in a procedure called the multiple conformer protocol (MCP). A method which statistically determines the potential distribution of likely binding modes for each molecule has also been described in a ligand-based CoMFA procedure by Lukacova and Balaz.[151] A further extension of this general technique using three-way PLS/CoMFA and multiway PLS applied to both conformer selection and alignment rules was demonstrated by Hasegawa *et al.*[152,153] Multiple ligand conformers are used to discover a reasonable alignment paradigm in the program, COMPASS,[154,155] developed by Jain *et al.* The program is designed to automate the selection of an optimal alignment, to use descriptors that are less sensitive to steric misalignment, and to analyze SAR with a neural net. The descriptors are based on steric, hydrogen-bond donor, and hydrogen-bond acceptor distances between the van der Waals envelope surrounding all aligned compounds and a set of sampling points outside the envelope.

4.23.2.3 Alignment-Independent Methods

The approaches to 3D-QSAR modeling described thus far have completely relied on some form of an alignment rule in order to allow for a systematic evaluation of atom positions and overlays which can lead to the identification of significant features and their spatial arrangement (i.e., the pharmacophore). The alignment rule is key to any grid-based methods which rely on the computation of MIFs. In addition, the alignment rule facilitates the construction of putative receptor topology and properties surrounding the set of molecules under study. However, if there is no information available to guide an alignment, this so-called rule could easily lead to an arbitrary 3D relationship disconnected from actual binding modes, thus negatively affecting the potential for the construction of a predictive model. In addition, conformationally promiscuous molecules would naturally tend to exacerbate the chances for a failed model. These concerns and an underlying interest in moving beyond the need for molecular alignment have led to the development of several novel approaches that are considered as alignment-independent. Such methods are somewhat misclassified herein. Since they are characteristically based on metrics developed from 3D properties and avoid any requirement for molecular overlays, they may more accurately reflect something akin to 2D+ approaches. These methods are also particularly amenable to the treatment of large numbers of compounds and have found utility in modeling both structure–activity and structure–property relationships.

4.23.2.3.1 Autocorrelation/correlograms

One of the first forays into an alignment-free 3D analysis is represented by the autocorrelation methodology of Broto *et al.*[156–159] The representation of each molecule is based on a single conformer and on the correlation of properties between pairs of atoms in that molecule expressed as a distance. The original versions of autocorrelation vectors were examples of 2D-QSAR descriptors wherein distances were computed based on numbers of bonds between atoms. Distances between atoms (or properties of atoms) as they exist in each 3D molecular structure brought the technique a step closer to a bona fide 3D-QSAR approach, wherein correlations between such distances were made to observed activities or properties. Both points on a CoMFA-like lattice and points on the molecular surface have been used for these calculations.[160,161] GRid-INdependent Descriptors (GRIND), introduced by Pastor *et al.*,[162] represent a further extension of the autocorrelation paradigm using a set of MIFs computed by the program GRID, or by other means. The procedure for computing the descriptors involves a first step, in which the fields are simplified, and a second step, in which the results are encoded into alignment-independent variables using a particular type of autocorrelation transform. The molecular descriptors so obtained can be used to develop graphical diagrams called correlograms and can be used in different chemometric analyses, such as principal component analyses or PLS. The program ALMOND provides for the facile correlation of the GRIND autocorrelograms with activity, and their subsequent analysis and interpretation.[162–165] An example correlogram is shown in **Figure 5**.

4.23.2.3.2 Weighted holistic invariant molecular (WHIM) descriptors

The complexity of chemical information embodied in a 3D structure is taken into account in the WHIM[166,167] descriptors. These are 3D molecular indices that contain a variety of information, including weighted terms for size, shape, symmetry, atom distribution, atomic mass, atomic volume, and electronegativity, generally calculated from the Cartesian coordinates of the energy-minimized molecule. The autoscaled descriptor matrices for a set of molecules are correlated to activity using PLS. Grid-WHIM[168] descriptors are calculated from the coordinates of the grid points of a MIF, each point being weighted by the field value. This procedure has been applied using fields related to nonbonding, electrostatic, and H-bonding interaction energies, evaluated by classical potentials. MS-WHIM was developed to consider the contribution of molecular surface in specific ligand–receptor interactions.[169–172]

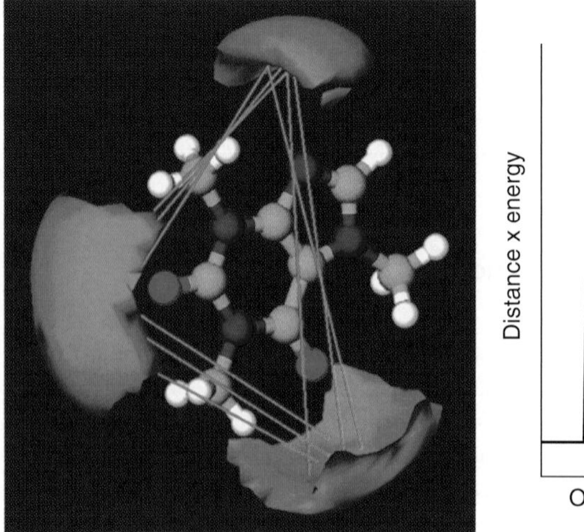

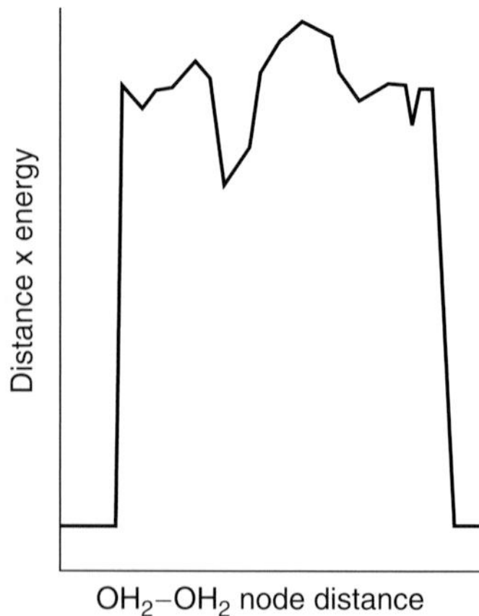

Figure 5 An example of an autocorrelogram (right) created from a GRID analysis of the caffeine molecule using a water (OH_2) probe. The regions reflecting the strongest interactions between water and caffeine are shown (left), along with several lines depicting distances between such interactions. The autocorrelogram is created by plotting the product of distance and energy of interaction against the distance between a pair of OH_2 interaction nodes.

4.23.2.3.3 Comparative molecular moment analysis (CoMMA)

CoMMA[173,174] represents a unique alignment-independent 3D-QSAR approach which utilizes moments of the molecular mass (shape) and charge distributions up to and including second order in the development of descriptors such as the principal moments of inertia, the magnitude of the dipole moment, and the principal quadrupolar moment. In order to maintain an alignment-independent system, these descriptors are obtained after translation to the center of mass as well as the center of dipole for each molecule. Correlation of activity with these molecular moment descriptors is performed using PLS.

4.23.2.3.4 Eigen-value analysis (EVA)/comparative spectral analysis (CoSA)

The use of computed spectral data as a source for descriptors in the generation of 3D-QSAR models represents yet another unique approach to alignment-independent correlations. Turner *et al.*[175–177] introduced the EVA descriptor which is derived from calculated infrared or Raman spectra. The normal coordinate EVA and eigenvectors (corresponding to vibrational frequencies and atomic displacements) are calculated using standard quantum mechanical or molecular mechanics methods. The projection of the eigenvalues on to a bounded frequency scale ultimately yields a Gaussian-smoothed set of peaks in a spectrum representing the EVA descriptor (**Figure 6**). The CoSA[178] approach entails the calculation of ^{1}H- or ^{13}C-nuclear magnetic resonance chemical shifts using quantum mechanical approaches as a viable source of descriptors.

4.23.2.3.5 Hologram quantitative structure–activity relationship (HQSAR)

A 3D-QSAR approach based on fingerprints, i.e., submolecular connectivity, is embodied in HQSAR.[179,180] Utilizing an extended fingerprint known as a molecular hologram, HQSAR encodes more information than the traditional 2D fingerprint by including branched and cyclic fragments (as well as stereochemistry), all possible fragments within a molecule (including overlapping fragments), and the number of times each unique fragment occurs. This extra set of descriptors enables HQSAR implicitly to encode 3D structural information.

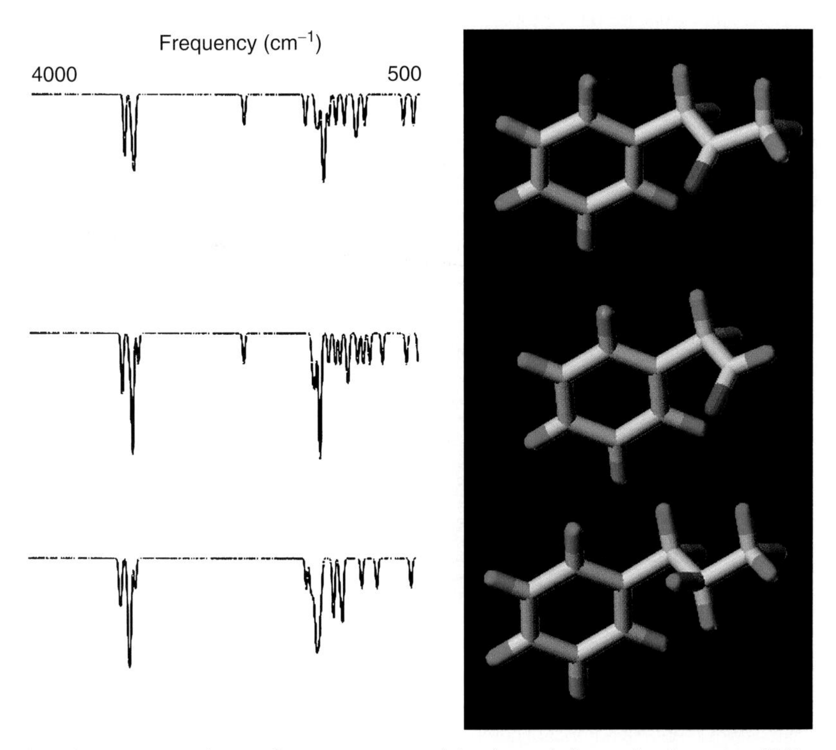

Figure 6 EVA infrared spectra are shown above as generated for three similar molecules using AM1, peak resolution of 10 cm^{-1} ranging from 200 to 4000 cm^{-1} and a sampling interval (δ) of 5 cm^{-1}. Comparisons between the intensities and positions of peaks characteristic for each molecule provide the means to generate useful correlations to molecular structure or property.

4.23.3 Mathematics

Early work in QSAR, which focused on a small set of molecules described by a handful of physicochemical descriptors, relied primarily on the use of multiple linear regression (MLR) analyses to generate meaningful correlation equations. This technique relies on a linear combination of descriptor terms obtained through a least-squares fit of the data (molecular descriptors to observed activities). Although utilized with great effect, MLR is vulnerable to descriptors which are correlated to one another, making it incapable of deciding which correlated sets may be more significant to the model. 3D-QSAR models often present a large number of descriptors for a small number of molecules, so that besides the strong likelihood of co-linearity among descriptors, there is a strong tendency to overfit the data, which can result in the failure of linear regression techniques to provide predictive models. In such cases, approaches are used which can handle both the correlated descriptor issue and the overdescribed nature of the problem.

4.23.3.1 Principal Component Analysis (PCA)/Partial Least Squares

Correlative methods designed to deal effectively with issues of descriptor co-linearity and overdescription are based on the statistical manipulation (covariance matrices) of either the descriptor data PCA or both descriptor and observed data (PLS). Comprehensive tutorials for PCA[181] and PLS[40] have been published. The use of PLS in the context of 3D-QSAR is well documented, serving as the correlation engine in many 3D-QSAR approaches.[34,39,51,61,63,171,174,182–185]

4.23.3.2 Neural Nets

As an alternative to the fitting of data to an equation and reporting the coefficients derived therefrom, a neural net or artificial neural network (ANN)[186–188] is designed to process input information and generate a hidden model of the relationships. The network algorithm consists of a set of input units, layers of 'neurons,' and an output, wherein each neuron performs calculations on its input signal to produce an output signal (**Figure 7**). The choice of arithmetic operations is arbitrary, but is usually sigmoidal in nature. In the case of QSAR applications,[189,190] the outputs are compared to observations and errors are back-propagated into the neural net to adjust weights of various inputs to the neurons. In this case the neurons often represent the various descriptors utilized in the QSAR analysis. The neural net is a learning machine that eventually, given sufficient cycles of back-propagation, settles upon a set of weighted neural input/output signals which provide a correlation between actual (input) and predicted (output) observations. One advantage of neural nets is that they are naturally capable of modeling nonlinear systems. Disadvantages include a tendency to overfit the data, and a significant level of difficulty in ascertaining which descriptors are most significant in the resulting model – a consequence of the interconnected character of a multilayered neural network.[191–193]

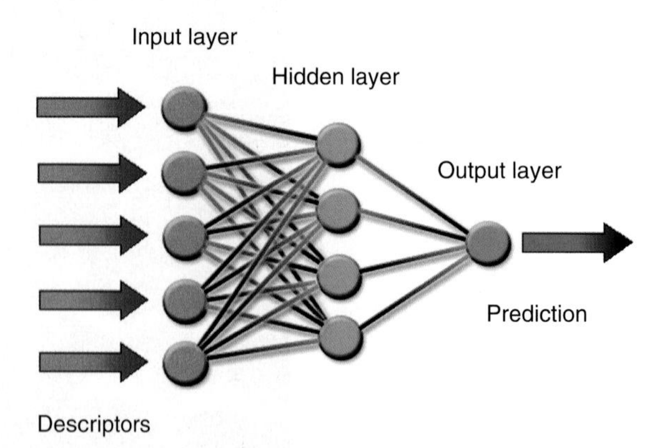

Figure 7 A schematic diagram of an artificial neural network (ANN) with one hidden layer (magenta) and a single output layer (green). QSAR applications would entail a number of numerical descriptors for each molecule, which correspond to the 'signals' sent to each node in the input layer (red). Weighted 'signals' are then sent to all the hidden layer nodes and a simple summed 'signal' reaches the output layer (green), essentially representing the neural net's estimate of activity. The error in prediction is then corrected by means of a nonlinear adjustment of the weights connecting all nodes. The process is repeated for each molecule, and cycled through all the molecules until a reasonably low prediction error is achieved.

4.23.3.3 Genetic Algorithms

As indicated previously, the large descriptor sets typical of a 3D-QSAR application present a significant problem in variable (descriptor) selection and model building which can easily lead to a time-consuming process yielding ambiguous models. GAs have been demonstrated to provide a fast and useful way of identifying the most important variables, particularly as applied to 3D-QSAR modeling.[194–198] They work by encoding the variable choices under investigation into 'genes' which are then subjected to various permutations of length, mutation, recombination, and cross-over, with every generation of genes scored by a user-defined fitness function (e.g., model predictivity). The results can be further skewed to artificially increase the population of genes which score well. Although not expected to identify the absolute best set of variables, GAs often lead to informative sets of models which clearly work to identify most noteworthy variables.

4.23.3.4 Support Vector Machines (SVM)

A class of algorithms called SVM can simultaneously provide a remarkably powerful way of classifying points in multidimensional descriptor space and linear regression.[199–202] The basic principle involves finding a separating 'plane' that divides descriptor space in such a way as to induce a maximal margin, i.e., the plane has a maximal distance to the closest points (object vectors) on both sides. In this way, the optimal plane represents a linear combination of object vectors closest to the plane. These object vectors are called support vectors.

4.23.4 Caveats

4.23.4.1 Assumptions

There are a number of assumptions inherent to the development of most 3D-QSAR models which are important to remember as they will limit both the construction of the models and subsequent interpretations[43] (some of these assumptions may not apply depending on the descriptor set and specific 3D-QSAR paradigm): (1) a single bioactive conformation (or one binding mode) is responsible for the activity observed; (2) the observed activity comes from the molecule and not from a degradation product or metabolite; (3) all the molecules bind to the same site; (4) the observed activity is largely explained by enthalpic processes, e.g., H-bonding, electrostatics, van der Waals interactions; (5) entropic processes and desolvation energetics are similar for all compounds under study; (6) observed activity measurements are conducted at equilibrium; (7) other extrinsic effects are not considered, e.g., diffusion or transport; and (8) the resulting QSAR model may represent one of potentially many solutions to the descriptor–activity correlation problem. The list of assumptions is long and important to keep in mind when trying to interpret subsequent results.

4.23.4.2 Validation

A major issue confronting the use of any 3D-QSAR method is validation, i.e., addressing the question of the reliability or predictivity of the resulting model. One indirect approach to determine the generality of a model relies on consensus modeling, wherein several different 3D-QSAR methods are used on the same data set.[110,203,204] A comparison of the significant features in each model provides a qualitative measure of confidence in such features, and may offer the opportunity to explore them in subsequent analyses. Often 'leave-some-out' methods are cited as useful tools in assessing predictivity, as they involve the creation of a 3D-QSAR model minus the 'left-out' molecule(s). For example, a leave-one-out (LOO) procedure which creates N models (in each case with $N-1$ molecules) can be used to predict the activity of each 'left-out' molecule, yielding an r^2_{cv} or q^2 (cross-validated correlation coefficient) whose value is often assumed to reflect the predictivity of the model. Although this may be the case, a significant q^2 value (for example, >0.3) may just as easily reflect redundancy in the data set, i.e., the presence of a large number of similar molecules. Conversely, a poor q^2 value (for example, <0.3) may not necessarily reflect a poor model, but indicate a lack of redundancy in the data set, i.e., the presence of a diverse set of molecules. In fact, the reliance on a q^2 value as an indicator of model predictivity is, at best, imperfect, and at worst, misleading. Recent surveys have shown that in about half of the 3D-QSAR models reported in the past 10 years the q^2 value has no relationship to model predictivity as assessed by performance on external test sets.[205] Since the best assessment of model predictivity is to gauge its performance on an external test set, model building should be viewed as an iterative process, where the test set is eventually incorporated into the model, with new predictions guiding the synthesis of a new test set and expectations that the model be further improved.

4.23.4.3 Prediction/Interpretability

Determining the predictivity of a 3D-QSAR model is not a simple process. As noted previously, cross-validation is likely to provide a measure of self-consistency but cannot speak to the model's true utility. Predicting the activities of an external test set of compounds is generally regarded as the best approach, although even here some care should be taken. The significant descriptors for the test set should fall within the bounds covered by the training set. Otherwise, any predictions made are based on extrapolation, and therefore subject to the vagaries associated with unexplored terrain. A prudent modeling system should provide the user with predictions along with an estimate of a molecule's extent of inclusion or similarity to the descriptor space correlated by the model's training set.

Many of the correlative methods described in this chapter are very capable of fitting the descriptor space to observed activities for the training set. This circumstance makes it profoundly easy to overfit the data by including sufficient descriptors, obtaining impressive r^2 (correlation coefficient) values, which may have the undesired effect of eroding the model's actual predictivity by the utilization of globally unimportant descriptors. Typical approaches leading to a prudent addition of descriptors include cross-validation (LOO) to limit the inclusion of new descriptors to the minimum required for a self-consistent model. An additional approach is to examine the error in the activity values. There is a limiting effect of experimental error on the theoretically achievable r^2. This is based on the standard error inherent in experimental data and assumes a normalized distribution of that error. For example, a perfect model predicting a data set consisting of 17 observations of activity spanning 3 orders of magnitude having a standard error of 0.7 log units would exhibit an r^2 (predictions) of only 0.49–0.71.[205] A separate study found that an observational error of about twofold (typical of many biological assays) applied to a data set of 19 compounds is equivalent to a standard error of 0.2–0.3 log units, which then limits the perfect QSAR model to an r^2(predictions) of 0.77–0.88.[206] The statistical bounds of the data can therefore provide significant guidance in the prudent construction of a QSAR model embodying a minimum set of the most relevant set of descriptors.

There are several implied goals in the construction of a 3D-QSAR model: to explain the SAR in the training set of molecules, pointing out the significant molecular features leading to changes in observed activities, and to predict accurately the activities of novel compounds outside the training set and, better yet, have the model suggest which compounds should be made next. Of immediate interest is whether the descriptors chosen for the analysis provide an interpretable model, i.e., can the user understand the physicochemical relationship? When the objective of the model is to search structural databases, this issue is less important than when the objective is to guide the synthesis of the next congener. Interpretable descriptors are key in the lifecycle of a 3D-QSAR model, as they lead to the synthesis of novel compounds designed to test the model. The new data can be fed back into the model to enhance its predictivity or generate new hypotheses. In addition, interpretable descriptors may point to critical features external to the molecule, i.e., the possible functionalities and their actual positions in the receptor. When a receptor structure is available (e.g., x-ray structure of a bound complex), this type of validation is most gratifying, since it serves both to validate the model and to point out which of the receptor features represent potentially important interactions.

One of the challenges of 3D-QSAR modeling is de novo design or inverse QSAR, i.e., the computer-aided use of the QSAR model to derive new structures of potential interest. In earlier times, 2D-QSAR models typically produced equations containing a few descriptors, or physicochemical parameters, which facilitated the translation of these correlations to simple changes in the original molecular scaffold. With the advent of 3D-QSAR and its attendant profusion of descriptors, the derived models not only typically point to a significantly expanded set of simple structural changes to the original chemotype, but can now include major changes to the scaffold, including entirely new scaffolds. Pharmacophores,[118,133,207] receptor structures,[208–210] and 3D-QSAR models[211–214] have been successfully used as platforms for the de novo design of new structures. These latter methods invoke computer-assisted molecular construction algorithms which utilize clever ways of positioning functionalities into the locations identified by the 3D-QSAR model as important while constraining the search to synthetically feasible structures.

4.23.5 Guidelines

The construction of a meaningful QSAR model depends heavily on several critical components of the model: (1) the nature of the observations; (2) the selection of compounds; and (3) the selection of molecular descriptors.[185] In spite of the attention given to these fundamental components, it will remain likely that the emergent 3D-QSAR model will retain a component-derived bias that will limit its usefulness. For example, compounds selected to demonstrate the possible effect of changes to a specific portion of their scaffold will result in a QSAR model which is insensitive to changes elsewhere on the scaffold. It will never be possible to include the number and type of compounds into a 3D-QSAR model necessary to cover every possible structural question raised in the future. This is a basic fact of statistical life. Nonetheless, such models can serve a useful function and the discussion of the components listed below will serve to detail their critical involvement in model construction.

4.23.5.1 The Observations

The best activity observations belong to a homogeneous class, i.e., all observations reflect the same biological or chemical mechanism of action. For example, in the case of enzyme inhibitors, all compounds should exhibit the same type of inhibition, preferably competitive. This constraint provides some level of assurance that the compounds in the study bind in the same way to the same site. In contrast, cell-based inhibition data, which are often used, represent an example of an inhomogeneous class, wherein the extent of observed inhibition is clearly subject to other factors (e.g., cell wall permeation, age of cells, number of cells) besides binding to the target enzyme or receptor protein. The use of in vivo data may represent the worst-case scenario of inhomogeneity, as such data introduce issues of transport, metabolism, and multiple mechanisms of action, to name just a few sources of correlative noise. The use of inhomogeneous data does not necessarily preclude the construction of a useful QSAR model; however, its limitations and the assumptions used in its construction should be kept in mind.

Activity values utilized in a QSAR analysis are typically of a form wherein larger values represent better activity. A convenient transformation of activity given as molar K_i, IC_{50}, and ED_{50} values is to log units in the form of the reciprocal, e.g., $\log(1/K_i) = -\log(K_i) = pK_i$. This is likely an inherited transform from the early days of QSAR (where activity was correlated much more directly to linear free-energy terms). Activity values provided in a percentage control form, e.g., 36% inhibition at 10 mM, are less desired and usually not amenable to extensive QSAR modeling because such values are highly variable and their conversion to an IC_{50}-equivalent form is fraught with potential error since the transform assumes a constant shape for the dose–response curve. Activity values of the form 'greater than' or 'less than' represent major issues in a QSAR application, as almost all methods cannot use this type of data. As conversion to an arbitrary low- or high-activity value can skew the resulting 3D-QSAR model, such compounds may best serve model validation by being part of the test set.

4.23.5.2 The Compounds

At one time, the rule of thumb in a 2D-QSAR analysis was to use about five compounds for every descriptor in the final correlation equation. This was loosely arrived at by a statistical analysis of typical Hansch-like correlations seeking a sufficient representation for any given physicochemical parameter. Given that many 2D-QSAR analyses entailed only several descriptors, the compound requirement was easily met, although even then, some consideration was needed to insure a relatively well-spread set of compounds for each parameter. The selection of number and type of compounds is likely to be the single most difficult and critical step in a 3D-QSAR analysis. There is a good reason for this. In principle, the compounds selected should be as different as possible from each other in terms of the descriptors to be used in the construction of the QSAR model. This is so that the correlative analysis makes use of the greatest extents available for each significant descriptor. Since it is not normally known what descriptors will be important to the final model, it is essentially impossible to select the proper compounds! However, the selection of a reasonable set could be carried out in an iterative manner, i.e., after the initial QSAR model is built, a new set of compounds can be selected based on the descriptors suggested by the initial model. Even here, it will be likely that the initial compound set was limited in scope and biased the model, and iteration in this context may lead to perpetuating the bias.

Since the 3D-QSAR descriptors are often correlated to each other to varying extents, a proper set of compounds could represent a balanced and well-spread set in latent variable space (i.e., along the axes of the descriptor principal components), making the selection even more challenging. The attractive feature of this design is that fewer compounds are necessary, since latent variable space is generally composed of fewer axes.

Assuming a large number of compounds to choose from, an additional and rather important consideration to keep in mind is the range of activity values associated with the choice of a data set. It is generally acknowledged that a narrow range, e.g., spanning three orders of magnitude or fewer, is apt to lead to spurious and artifactual correlations. In addition, it may be worthwhile choosing compounds that represent a fairly even distribution of activities as an indirect way of assuring some level of diversity in the selection. Given no clue regarding which descriptors will be important to the initial modeling analysis, compounds are often randomly chosen based on this criterion alone. The size of typical training sets varies in the literature, averaging 48 and ranging from 15 to 191 in number (based on a random sampling of 61 reported 3D-QSAR studies between 1994 and 2003).[205] In practice, compounds are rarely selected for an analysis, but rather, emerge as a consequence of an active synthetic program. In this case, every compound is often assimilated into the model-building process, and the predictivity of the model is assessed by synthesis and testing. New compound data serve as the input to an iteratively refined model. In spite of the almost unavoidable data set-related bias expected to exist in a 3D-QSAR model (at least initially), in a discovery setting, the model serves well as a hypothesis-generating engine.

If there are a sufficiently large number of compound data available, then that number should be split into a training set and test set using similar criteria, basically insuring that the test set is randomly chosen and has a similar range of activities as the training set. The ratio of the two normally runs 1/3 or 1/4 test/training. The 3D-QSAR model is built using only training set data, and should never be influenced by any information from the test set. This is an error that often occurs in practice in an effort to 'check' that the model is predictive by running the test set through, and then making adjustments to the model so that the test set is better predicted!

4.23.5.3 The Descriptors

As is evident from the number and variety of 3D-QSAR methods available, there are many ways of describing the relationship between structure and activity. If the objective of the modeling exercise is to help direct the synthesis of future analogs, then interpretability of descriptors should be paramount. Grid-based, alignment-dependent methods represent those most amenable to interpretation, since MIFs or atom types can be visualized surrounding or embedded in a template structure, giving rise to idea structures through either de novo computer-assisted means or direct observation. If the modeling objective is to search a structural database for potentially active analogs, then any of the alignment-independent methods would work especially well, as they are tailor-made for efficient searching. Such methods are also preferred in the development of quantitative structure–property relationships, since properties like solubility, permeability, and log P are not expected to be dependent on an alignment.

The plethora of descriptors is both a wonder and a bane. It is essentially possible to create viable 3D-QSAR models for the same data set using any number of approaches. This means that a binding model can be explained in terms of MIFs using a wide variety of probes, similarity measures developed from molecular connectivities of various kinds, position-dependent atom types of any variety, calculated interactions with artificial receptors made of surfaces or atoms, correlograms, molecular dipole moments, computed spectrograms, and combinations of molecular fragments, to name a few. The reason for such apparent ambiguity lies in the correlated nature of 3D-descriptors. Descriptor properties captured by one method, e.g., using atom types, can easily overlap with descriptor properties from another method, e.g., MIFs. In the end, it is far less important to determine which set of descriptors works best than to understand that models are models, and that it is their potential use and interpretation that set them apart.

4.23.6 Conclusions

3D-QSAR represents a fledgling first step in our attempt at understanding the obscure relationship between the actual structure of molecules and their behavior in biological systems. Recent efforts in the field have added multiple dimensions to the paradigm, offering to capture variations in conformation/orientation, induced-fit in receptors, and solvation/desolvation effects. Surprisingly, only vague consideration has been given to the entropic impact produced when ligand meets receptor, particularly as it applies to conformational freedom and, even more so, to the hydrophobic effect, i.e., the change in solvent entropy, which has been demonstrated to play an enormous but hidden role in the final determination of binding affinity for intermolecular interactions in solution.[33,34,215–220] It is expected that, with the help of a continued increase in computational power, future 3D-QSAR methods will embrace these effects, and others, perhaps not as yet considered, in clever and unexpected ways, producing models that are explanatory, predictive and above all, a joy to use.

References

1. Crum-Brown, A.; Fraser, T. R. *Trans. R. Soc. Edinb.* **1868**, *25*, 151.
2. Richardson, B. J. *Med. Times Gazette* **1869**, *2*, 703.
3. Hansch, C.; Fujita, T. *J. Am. Chem. Soc.* **1964**, *86*, 1616–1626.
4. Free, S. M., Jr.; Wilson, J. W. *J. Med. Chem.* **1964**, *7*, 395–399.
5. Verloop, A.; Hoogenstraaten, W.; Tipker, J. *Med. Chem.* **1976**, *11*, 165–207.
6. Verloop, A. *Proc. Int. Congr. Pestic. Chem.* **1983**, *1*, 339–344.
7. Boyd, D. B. *Review of Computational Chemistry*; Lipkowitz, B.; Boyd, D. B., Eds.; VCH: New York, 1990, pp 355–371.

8. Fujita, T. The Role of QSAR in Drug Design. In *Drug Design: Fact or Fantasy?* Jolles, G., Wooldridge, K. R. H., Eds.; Academic Press: London, 1984, pp 19–33.
9. Hansch, C.; Leo, A.; Khoekman, D. *Exploring QSAR: Hydrophobic, Electronic and Steric Constants*; ACS: Washington, DC, 1995.
10. Kier, L. B. *Molecular Orbital Theory in Drug Research*; Academic Press: New York, 1971.
11. Holtje, H. D.; Kier, L. B. *J. Pharm. Sci.* **1974**, *63*, 1722–1725.
12. Holtje, H. D.; Kier, L. B. *J. Med. Chem.* **1974**, *17*, 814–819.
13. Marshall, G. R. *Med. Chem., Proc. Int. Symp.* **1979**, 225–235.14
14. Marshall, G. R.; Barry, C. D.; Bosshard, H. E.; Dammkoehler, R. A.; Dunn, D. A. *ACS Symp. Ser.* **1979**, *112*, 205–226.
15. Murray, W. J.; Hall, L. H.; Kier, L. B. *J. Pharm. Sci.* **1975**, *64*, 1978–1981.
16. Kier, L. B.; Hall, L. H.; Murray, W. J.; Randic, M. *J. Pharm. Sci.* **1975**, *64*, 1971–1974.
17. Kier, L. B.; Murray, W. J.; Hall, L. H. *J. Med. Chem.* **1975**, *18*, 1272–1274.
18. Murray, W. J.; Kier, L. B.; Hall, L. H. *J. Med. Chem.* **1976**, *19*, 573–578.
19. Kier, L. B.; Hall, L. H. *Medicinal Chemistry, Molecular Connectivity in Chemistry and Drug Research*; Academic Press: New York, 1976; Vol. 14.
20. Hopfinger, A. J. *J. Med. Chem.* **1983**, *26*, 990–996.
21. Hopfinger, A. J. *Arch. Biochem. Biophysics* **1981**, *206*, 153–163.
22. Hopfinger, A. J.; Burke, B. J. *Prog. Clin. Biol. Res.* **1989**, *291*, 151–159.
23. Tokarski, J. S.; Hopfinger, A. J. *J. Med. Chem.* **1994**, *37*, 3639–3654.
24. Montsenigos, A.; Ciubotariu, D.; Chiriac, A.; Simon, Z. *Rev. Roumaine Chimie* **1989**, *34*, 2101–2106.
25. Simon, Z.; Dragomir, N.; Plauchitiu, M. G.; Holban, S.; Glatt, H.; Kerek, F. *Eur. J. Med. Chem.* **1980**, *15*, 521–527.
26. Simon, Z. MTD and Hyperstructure Approaches. In *3D QSAR Drug Design*; Kubinyi, H., Ed.; ESCOM: The Netherlands, 1993, pp 307–319.
27. Simon, Z.; Chiriac, A.; Holban, S.; Ciubotam, D.; Mihalas, G. I. *Chemometrics Series, Minimum Steric Difference. The MTD Method for QSAR Studies*; Wiley: New York, 1984; Vol. 7.
28. Ghose, A. K.; Crippen, G. M. *Mol. Pharmacol.* **1990**, *37*, 725–734.
29. Ghose, A. K.; Crippen, G. M. *Quant. Struct.–Act. Relat. J. Med. Chem.* **1985**, *28*, 333–346.
30. Ghose, A. K.; Crippen, G. M.; Revankar, G. R.; McKernan, P. A.; Smee, D. F.; Robins, R. K. *J. Med. Chem.* **1989**, *32*, 746–756.
31. Goodford, P. J. *J. Med. Chem.* **1985**, *28*, 849–857.
32. Boobbyer, D. N.; Goodford, P. J.; McWhinnie, P. M.; Wade, R. C. *J. Med. Chem.* **1989**, *32*, 1083–1094.
33. Semus, S. F. *Med. Chem. Res.* **1999**, *9*, 535–547.
34. Semus, S. F., Kellogg, G. E., Application of HINT Interaction Scores and Hydropathic Intermolecular Field Analysis (HIFA) to the Prediction of Ligand Binding Affinity. *Abstracts of Papers, 228th ACS National Meeting*, Philadelphia, PA, United States, August 22–26, 2004; American Chemical Society: Washington, DC, 2004, COMP-102.
35. Cramer, R. D., III; Bunce, J. D. *Pharmacochem. Libr.* **1987**, *10*, 3–12.
36. Wise, M.; Cramer, R.D.; Smith, D.; Exman, I. Progress in Three-Dimensional Drug Design: The Use of Real Time Colour Graphics and Computer Postulation of Bioactive Molecules in DYLOMMS. In *Quantitative Approaches to Drug Design*; Proceedings of the 4th European Symposium on Chemical Structure–Biological Activity: Quantitative Approaches; Dearden, J., Ed.; Elsevier: Amsterdam, 1983, pp 145–146.
37. Wise, M. Evolution of QSAR Methodology and the Role of Newer Computational Techniques. In *QSAR and Strategies in the Design of Bioactive Compounds*; Proceedings of the 5th European Symposium on QSAR; VCH: Weinheim, 1985, pp 19–29.
38. Cramer, R. D., III; Patterson, D. E.; Bunce, J. D. *J. ACS* **1988**, *110*, 5959–5967.
39. Skagerberg, B.; Dunn, W. J., III; Hellberg, S.; Wold, S. The PLS Data Analytic Method in QSAR. In *Fed. Rep. Ger. QSAR Strategies in the Design of Bioactive Compounds*; Seydel, J. K., Ed.; 1985, 305–310.
40. Geladi, P.; Kowalski, B. R. *Anal. Chim. Acta* **1986**, *185*, 1–17.
41. Doweyko, A. M. *J. Med. Chem.* **1988**, *31*, 1396–1406.
42. Green, S. M.; Marshall, G. R. *Trends Pharm. Sci.* **1995**, *16*, 285–291.
43. Oprea, T. I.; Waller, C. L. *Rev. Comput. Chem.* **1997**, 127–182.
44. Greco, G.; Novellino, E.; Martin, Y. C. *Rev. Comput. Chem.* **1997**, 183–240.
45. Lewis, R. A. *Chem. Model.* **2002**, *2*, 271–292.
46. Akamatsu, M. *Curr. Top. Med. Chem.* **2002**, *2*, 1381–1394.
47. Kellogg, G. E.; Semus, S. F. 3D QSAR in Modern Drug Design. In *Modern Methods of Drug Discovery (Experientia Supplementum Vol. 93)*; Hillisch, A., Hilgenfeld, R., Eds.; Verlag: Birkhauser, Switzerland, 2003, pp 223–241.
48. Cruciani, G.; Carosati, E.; Clementi, S. Three-Dimensional Quantitative Structure–Property Relationships. In *The Practice of Medicinal Chemistry*, 2nd ed.; Wermuth, C. G., Ed.; Elsevier: London, 2003, pp 405–416.
49. Oprea, T. I. 3D-QSAR Modeling in Drug Design. In *Computational Medicinal Chemistry for Drug Discovery*; Bultinck, P., Ed.; Marcel Dekker: New York, 2004, pp 571–616.
50. Greener, M. QSAR: Predictions Beyond the Fourth Dimension. In *Drug Discovery and Development*; Reed Business Information, div of Reed Elsevier, Inc., 2005.
51. Cramer, R. D., III; Wold, S. B. Comparative molecular field analysis (CoMFA); U.S. Patent 5025388A, June, 18, 1991, 22pp.
52. Kim, K. H. *Perspect. Drug Disc. Design* **1998**, *12–14*, 317–338.
53. Clark, M. R. C.; Jones, D.; Patterson, D.; Simeroth, P. *Tetrahedron Comp. Methods* **1990**, *3*, 47.
54. Sun, H.; Xie, Q.; Zhou, J.; Xu, Z. *Huaxue Jinzhan* **1996**, *8*, 79–85.
55. Kim, K. H.; Greco, G.; Novellino, E. *Perspect. Drug Disc. Design* **1998**, *12/13/14*, 257–315.
56. Norinder, U. *Perspect. Drug Disc. Design* **1998**, *12/13/14*, 25–39.
57. SYBYL Molecular Modeling Software; Tripos, Inc.: St. Louis, MO 63144, US.
58. Cho, S. J.; Tropsha, A. *J. Med. Chem.* **1995**, *38*, 1060–1066.
59. Tropsha, A.; Cho, S. J. *Perspect. Drug Disc. Design* **1998**, *12/13/14*, 57–69.
60. Kimura, T.; Hasegawa, K.; Funatsu, K. *J. Chem. Inf. Comp. Sci.* **1998**, *38*, 276–282.
61. Baroni, M. G. C.; Cruciani, G.; Riganelli, D.; Valigi, R.; Clementi, S. *Quant. Struct.–Act. Relat.* **1993**, *12*, 9.
62. Cruciani, G.; Watson, K. A. *J. Med. Chem.* **1994**, *37*, 2589–2601.
63. Nilsson, J.; Wikstrom, H.; Smilde, A.; Glase, S.; Pugsley, T.; Cruciani, G.; Pastor, M.; Clementi, S. *J. Med. Chem.* **1997**, *40*, 833–840.
64. Sulea, T.; Purisima, E. O. *Quant. Struct.–Act. Relat.* **1999**, *18*, 154–158.
65. Nicholls, A.; Honig, B. *J. Comput. Chem.* **1991**, *12*, 435–445.

66. Waller, C. L.; Marshall, G. R. *J. Med. Chem.* **1993**, *36*, 2390–2403.
67. DELPHI; Accelrys, Inc.: San Diego, CA.
68. Klebe, G.; Abraham, U.; Mietzner, T. *J. Med. Chem.* **1994**, *37*, 4130–4146.
69. Bringmann, G.; Rummey, C. *J. Chem. Inf. Comp. Sci.* **2003**, *43*, 304–316.
70. Kellogg, G. E.; Semus, S. F.; Abraham, D. J. *J. Comput.-Aided Mol. Design* **1991**, *5*, 545–552.
71. Aurenhammer, F. *ACM Comput. Surv.* **1991**, *23*, 345–405.
72. Srivastava, S. W. W. R.; Bradley, M. P.; Crippen, G. M. Three-Dimensional Receptor Modeling Using Distance Geometry and Voronoi Polyhedra. In *3D QSAR in Drug Design: Theory, Methods and Applications*; Kubinyi, H., Ed.; ESCOM: Leiden, 1993, pp 409–430.
73. Chuman, H.; Karasawa, M.; Fujita, T. *Quant. Struct.–Act. Relat.* **1998**, *17*, 313–326.
74. Audry, E.; Dallet, P.; Langlois, M. H.; Colleter, J. C.; Dubost, J. P. *Prog. Clin. Biol. Res.* **1989**, *291*, 63–66.
75. Gaillard, P.; Carrupt, P. A.; Testa, B.; Boudon, A. *J. Comput.-Aided Mol. Design* **1994**, *8*, 83–96.
76. Carrupt, P. A.; Billois, F.; Weber, P.; Testa, B.; Meyer, C.; Perez, S. The Molecular Lipophilicity Potential (MLP): A New Tool for LogP Calculation and Docking, and in Comparative Molecular Field Analysis (CoMFA). In *Lipophilicity in Drug Action and Toxicology*; Pliska, V., Testa, B., van de Waterbeemd, H., Eds.; VCH: Weinheim, 1995, pp 195–215.
77. Floersheim, P.; Nozulak, J.; Weber, H. P. Experience with Comparative Molecular Field Analysis. In *Trends in QSAR and Molecular Modelling 92*; Proceedings of the European Symposium in Structure–Activity Relationships: QSAR and Molecular Modelling; Wermuth, C.-G., Ed.; ESCOM: Leiden, The Netherlands, 1993, pp 227–232.
78. Ghose, A. K.; Crippen, G. M. *J. Comput. Chem.* **1986**, *7*, 333–346.
79. Amat, L.; Robert, D.; Besalu, E.; Carbo-Dorca, R. *J. Chem. Inf. Comp. Sci.* **1998**, *38*, 624–631.
80. Girones, X.; Amat, L.; Carbo-Dorca, R. *SAR QSAR Environ. Res.* **1999**, *10*, 545–556.
81. Carbo, R.; Besalu, E.; Amat, L.; Fradera, X. *J. Math. Chem.* **1995**, *18*, 237–246.
82. Fradera, X.; Amat, L.; Besalu, E.; Carbo-Dorca, R. *Quant. Struct.–Act. Relat.* **1997**, *16*, 25–32.
83. Robinson, D. D.; Winn, P. J.; Richards, W. G. Self-Organizing Molecular Field Analysis. *Book of Abstracts, 216th ACS National Meeting, Boston, August 23–27*; American Chemical Society: Washington, DC, 1998, COMP-082.
84. Winn, P. J.; Robinson, D. D.; Richards, W. G. Self-Organizing Molecular Field Analysis (SOMFA): A Tool for Structure–Activity Studies. In *Fundamental of Molecular Similarity*; Carbo-Dorca, R., Girones, X., Mezey, P. G., Eds.; Kluwer: New York, 2001, pp 321–332.
85. Horwell, D. C.; Howson, W.; Higginbottom, M.; Naylor, D.; Ratcliffe, G. S.; Williams, S. *J. Med. Chem.* **1995**, *38*, 4454–4462.
86. Klebe, G.; Abraham, U. *J. Med. Chem.* **1993**, *36*, 70–80.
87. Mestres, J.; Rohrer, D. C.; Maggiora, G. M. *J. Mol. Graphics Model.* **1997**, *15*, 114–121, 103–116.
88. Doweyko, A. M. *ACS Symposium Ser.* **1989**, *413*, 82–104.
89. Nowakowski, M.; Tishler, M.; Doweyko, A. M. *Phosphorus Sulfur Silicon Relat. Elements* **1989**, *45*, 183–188.
90. Doweyko, A. M. *J. Math. Chem.* **1991**, *7*, 273–285.
91. Doweyko, A. M.; Mattes, W. B. *Biochemistry* **1992**, *31*, 9388–9392.
92. Kaminski, J. J.; Doweyko, A. M. *J. Med. Chem.* **1997**, *40*, 427–436.
93. Woolfrey, J. R.; Avery, M. A.; Doweyko, A. M. *J. Comput.-Aided Mol. Design* **1998**, *12*, 165–181.
94. Doweyko, A. M.; Kellogg, G. E. HASL SPL Module Available for Use within Sybyl; Tripos, St. Louis, MO; HASL version 4.00s. www.edusoft-lc.com (accessed Aug 2006).
95. Vedani, A.; Zbinden, P.; Snyder, J. P. *J. Receptor Res.* **1993**, *13*, 163–177.
96. Schmetzer, S.; Greenidge, P.; Kovar, K. A.; Schulze-Alexandru, M.; Folkers, G. *J. Comput.-Aided Mol. Design* **1997**, *11*, 278–292.
97. Sippl, W.; Stark, H.; Hoeltje, H. D. *Pharmazie* **1998**, *53*, 433–437.
98. Gurrath, M.; Muller, G.; Holtje, H.-D. *Perspect. Drug Disc. Design* **1998**, *12/13/14*, 135–157.
99. Zbinden, P.; Dobler, M.; Folkers, G.; Vedani, A. *Quant. Struct.–Act. Relat.* **1998**, *17*, 122–130.
100. Hahn, M. *J. Med. Chem.* **1995**, *38*, 2080–2090.
101. Hahn, M.; Rogers, D. Receptor surface models. *Perspect. Drug Disc. Design* **1998**, *12/13/14*, 117–133.
102. Potemkin, V. A.; Bartashevich, E. V.; Grishina, M. A.; Guccione, S. An Alternative Method for 3D-QSAR and the Alignment of Molecular Structures: BiS (Biological Substrate Search). In *Rational Approaches to Drug Design*; Proceedings of the European Symposium on Quantitative Structure–Activity Relationships, 13th, Duesseldorf, Germany; Höltje, H.-D., Sippl, W., Eds.; Prous Science: Philadelphia, 2001, pp 349–353.
103. Walters, D. E. *Perspect. Drug Disc. Design* **1998**, *12/13/14*, 159–166.
104. Walters, D. E.; Hinds, R. M. *J. Med. Chem.* **1994**, *37*, 2527–2536.
105. Hirashima, A.; Eiraku, T.; Kuwano, E.; Eto, M. *Combinat. Chem. High Throughput Screen.* **2004**, *7*, 83–91.
106. Santagati, M.; Doweyko, A.; Santagati, A.; Modica, M.; Guccione, S.; Chen, H.; Barretta, G. U.; Balzano, F. *5-HT1A Receptors Mapping by Conformational Analysis (2D NOESY/MM) and Molecular Modeling and Prediction of Bioactivity*; Proceedings of the European Symposium on Quant. Struct.–Act. Relat.: Molecular Modeling and Prediction of Bioactivity, 12th, Copenhagen, Denmark; Kluwer: New York, 2000, pp 183–194.
107. Chen, H. M.; Zhou, J. J.; Ren, T. R.; Xie, G. R. *Chin. Chem. Lett.* **1997**, *8*, 975–978.
108. Albuquerque, M. G.; Hopfinger, A. J.; Barreiro, E. J.; de Alencastro, R. B. *J. Chem. Inf. Comp. Sci.* **1998**, *38*, 925–938.
109. Klein, C. D.; Hopfinger, A. J. *Pharm. Res.* **1998**, *15*, 303–311.
110. Ravi, M.; Hopfinger, A. J.; Hormann, R. E.; Dinan, L. *J. Chem. Inf. Comput. Sci.* **2001**, *41*, 1587–1604.
111. Hopfinger, A. J.; Tokarski, J. S.; Jin, B.; Albuquerque, M.; Madhav, P. J.; Duraiswami, C. *J. Am. Chem. Soc.* **1997**, *119*, 10509–10524.
112. Hopfinger, A. J.; Tokarski, J. 3D-QSAR Analysis. In *Practical Application of Computer-Aided Drug Design*; Charifson, P. S., Ed.; Dekker: New York, 1997, p 105.
113. Tsujishita, H.; Hirono, S. *J. Comput.-Aided Mol. Design* **1997**, *11*, 305–315.
114. Kearsley, S. K.; Smith, G. M. *Tetrahedron Comput. Methodol.* **1990**, *3*, 615–633.
115. McMartin, C.; Bohacek, R. S. *J. Comput.-Aided Mol. Design* **1997**, *11*, 333–344.
116. Iwase, K.; Hirono, S. *J. Comput.-Aided Mol. Design* **1999**, *13*, 499–512.
117. Lemmen, C.; Lengauer, T.; Klebe, G. *J. Med. Chem.* **1998**, *41*, 4502–4520.
118. De Esch, I. J.; Mills, J. E.; Perkins, T. D.; Romeo, G.; Hoffmann, M.; Wieland, K.; Leurs, R.; Menge, W. M. P. B.; Nederkoorn, P. H. J.; Dean, P. M. et al. *J. Med. Chem.* **2001**, *44*, 1666–1674.
119. Mills, J. E. J.; De Esch, I. J. P.; Perkins, T. D. J.; Dean, P. M. *J. Comput.-Aided Mol. Design* **2001**, *15*, 81–96.
120. Kroonenberg, P. M.; Dunn, W. J.; Commandeur, J. J. F. *J. Chem. Inf. Comput. Sci.* **2003**, *43*, 2025–2032.
121. Commandeur, J. J. F.; Kroonenberg, P. M.; Dunn, W. J., III. *J. Chemometrics* **2004**, *18*, 37–42.

122. Kearsley, S. K.; Smith, G. M. *Tetrahedron Comput. Methodol.* **1990**, *3*, 615–633.
123. Waller, C. L.; Marshall, G. R. *J. Med. Chem.* **1993**, *36*, 2390–2403.
124. Hermann, R. B.; Herron, D. K. *J. Comput.-Aided Mol. Design* **1991**, *5*, 511–524.
125. Good, A. C.; Hodgkin, E. E.; Richards, W. G. *J. Chem. Inf. Comput. Sci.* **1992**, *32*, 188–191.
126. Namasivayam, S.; Dean, P. M. *J. Mol. Graph.* **1986**, *4*, 46–50.
127. Dean, P. M.; Chau, P. L. *J. Mol. Graph.* **1987**, *5*, 152–164.
128. Martin, Y. C. *IUL Biotechnol. Ser.* **2000**, *2*, 49–68.
129. Martin, Y. C.; Bures, M. G.; Danaher, E. A.; DeLazzer, J.; Lico, I.; Pavlik, P. A. *J. Comput.-Aided Mol. Design* **1993**, *7*, 83–102.
130. Mason, J. S. *Pharmacochem. Library* **1993**, *20*, 147–156.
131. Halova, J.; Zak, P.; Strouf, O.; Uchida, N.; Yuzuri, T.; Sakakibara, K.; Hirota, M. *Organic Reactivity (Tartu)* **1997**, *31*, 31–43.
132. Kristam, R.; Gillet, V. J.; Lewis, R. A.; Thorner, D. *J. Chem. Inf. Model.* **2005**, *45*, 461–476.
133. Sprague, P. W. *Perspect. Drug Disc. Design* **1995**, *3*, 1–20.
134. CATALYST *3D-QSAR Program*; Accelrys, Inc.: San Diego, CA.
135. Doweyko, A. M. *J. Med. Chem.* **1994**, *37*, 1769–1778.
136. Golender, V.; Vesterman, B.; Vorpagel, E. APEX-3D expert system for drug design. *Network Science [Electronic Publication]* 1996, *2*. www.netsci.org (accessed Aug 2006).
137. Golender, V. E.; Vorpagel, E. R. *3D QSAR Drug Des.* **1993**, 137–149.
138. Hongmei, S.; Xie, Q.; Xie, G.; Jiaju, Z.; Zhihong, X.; Zhengmin, L.; Guofeng, J.; Lingxiu, W. *Wuli Huaxue Xuebao* **1995**, *11*, 773–776.
139. Chow, M. M.; Meyer, E. F., Jr.; Bide, W.; Kam, C. M.; Radhakrishnan, R.; Vijayalakshimi, J.; Powers, J. C. *J. Am. Chem. Soc.* **1990**, *112*, 7783–7789.
140. Bolin, J. T.; Matthews, D. J.; Hamlin, D. A.; Kraut, J. *J. Biol. Chem.* **1982**, *257*, 13650–13662.
141. Kester, W. R.; Matthews, B. W. *Biochemistry* **1977**, *16*, 2506–2516.
142. Case, D. A.; Karplus, M. *J. Mol. Biol.* **1979**, *132*, 343–368.
143. Shoichet, B. K.; Stroud, R. M.; Santi, D. V.; Kuntz, I. D.; Perry, K. M. *Science* **1993**, *259*, 1445–1449.
144. Mattos, C.; Ringe, D. Multiple Binding Modes. In *3D QSAR in Drug Design: Theory, Methods and Applications*; Kubinyi, H., Ed.; ESCOM: Leiden, 1993, pp 226–254.
145. Nicklaus, M. C.; Milne, G. W.; Burke, T. R., Jr. *J. Comput.-Aided Mol. Design* **1992**, *6*, 487–504.
146. Guccione, S.; Doweyko, A. M.; Chen, H.; Barretta, G. U.; Balzano, F. *J. Comput.-Aided Mol. Design* **2000**, *14*, 647–657.
147. Dimitrov, S. D.; Mekenyan, O. G. *Chemometrics Intelligent Lab. Systems* **1997**, *39*, 1–9.
148. Mekenyan, O. *Curr. Pharm. Design* **2002**, *8*, 1605–1621.
149. Mekenyan, O. G.; Ivanov, J. M.; Veith, G. D.; Bradbury, S. P. *Quant. Struct.–Act. Relat.* **1994**, *13*, 302–307.
150. Doweyko, A. M., Multiple Conformer Protocol: A New Method for the Identification of Preferred Ligand Binding Motifs Using Cross-validated 3D-QSAR models. In *Rational Approaches to Drug Design*; Proceedings of the European Symposium on Quantitiative Structure–Activty Relationships, 13th, Duesseldorf, Germany; Höltje, H.-D., Sippl, W., Eds.; Prous Science: Philadelphia, 2001, pp 307–315
151. Lukacova, V.; Balaz, S. *Incorporation of Multiple Binding Modes into 3D-QSAR Methods. Rational Approaches to Drug Design*; Proceedings of the European Symposium on Quantitiative Structure–Activty Relationships, 13th, Duesseldorf, Germany, 2001, pp354–358.
152. Hasegawa, K.; Arakawa, M.; Funatsu, K. *Chemometrics Intell. Lab. Systems* **2000**, *50*, 253–261.
153. Hasegawa, K.; Arakawa, M.; Funatsu, K. *Curr. Comput.-Aided Drug Design* **2005**, *1*, 129–145.
154. Jain, A. N.; Harris, N. L.; Park, J. Y. *J. Med. Chem.* **1995**, *38*, 1295–1308.
155. Jain, A. N.; Koile, K.; Chapman, D. *J. Med. Chem.* **1994**, *37*, 2315–2327.
156. Broto, P.; Moreau, G.; Vandycke, C. *QSAR Des. Bioact. Compd.* **1984**, 393–401.
157. Broto, P.; Moreau, G.; Vandycke, C. *Eur. J. Med. Chem.* **1984**, *19*, 61–65.
158. Moreau, G.; Broto, P. *Nouv. J. Chimie* **1980**, *4*, 757–764.
159. Moreau, G.; Broto, P. The autocorrelation of a topological structure: a new molecular descriptor. *Nouv. J. Chimie* **1980**, *4*, 359–360.
160. Clementi, S.; Cruciana, G.; Riganelli, D.; Valigi, R.; Constantino, G.; Baroni, M.; Wold, S. *Pharm. Pharmacol. Lett.* **1993**, *3*, 5–8.
161. Wagener, M.; Sadowski, J.; Gasteiger, J. *J. ACS* **1995**, *117*, 7769–7775.
162. Pastor, M.; Cruciani, G.; McLay, I.; Pickett, S.; Clementi, S. *J. Med. Chem.* **2000**, *43*, 3233–3243.
163. Afzelius, L.; Zamora, I.; Masimirembwa, C. M.; Karlén, A.; Andersson, T. B.; Mecucci, S.; Baroni, M.; Cruciani, G. *J. Med. Chem.* **2004**, *47*, 907–914.
164. Gratteri, P.; Cruciani, G.; Scapecchi, S.; Romanelli, M. N. GRID Independent Descriptors (GRIND) in the Rational Design of Muscarinic Antagonists. *Rational Approaches to Drug Design*; Proceedings of the European Symposium on Quantitiative Structure–Activity Relationships, 13th, Duesseldorf, Germany, 2001, 241–243.
165. *ALMOND*; Molecular Discovery: Ponte San Giovanni, PG, Italy.
166. Todeschini, R.; Lasagni, M.; Marengo, E. *J. Chemometrics* **1994**, *8*, 263–272.
167. Todeschini, R.; Gramatica, P. *SAR QSAR Environ. Res.* **1997**, *7*, 89–115.
168. Todeschini, R.; Moro, G.; Boggia, R.; Bonati, L.; Cosentino, U.; Lasagni, M.; Pitea, D. *Chemometrics Intell. Lab. Systems* **1997**, *36*, 65–73.
169. Bravi, G.; Gancia, E.; Mascagni, P.; Pegna, M.; Todeschini, R.; Zaliani, A. *J. Comput.-Aided Mol. Design* **1997**, *11*, 79–92.
170. Gancia, G. B.; Mascagni, P.; Zaliani, A. *J. Comput.-Aided Mol. Design* **2000**, *14*, 293–306.
171. Bravi, G.; Wikel, J. H. *Quant. Struct.–Act. Relat.* **2000**, *19*, 29–38.
172. Bravi, G.; Wikel, J. H. *Quant. Struct.–Act. Relat.* **2000**, *19*, 39–49.
173. Silverman, B. D.; Platt, D. E. *J. Med. Chem.* **1996**, *39*, 2129–2140.
174. Silverman, B. D.; Platt, D. E.; Pitman, M.; Rigoutsos, I. *Perspect. Drug Disc. Design* **1998**, *12/13/14*, 183–196.
175. Turner, D. B.; Willett, P.; Ferguson, A. M.; Heritage, T. *J. Comput.-Aided Mol. Design* **1997**, *11*, 409–422.
176. Heritage, T. W.; Ferguson, A. M.; Turner, D. B.; Willett, P. *Perspect. Drug Disc. Design* **1998**, *9/10/11*, 381–398.
177. Turner, D. B.; Ferguson, A. M.; Heritage, T. W. *J. Comput.-Aided Mol. Design* **1999**, *13*, 271–296.
178. Asikainen, A.; Ruuskanen, J.; Tuppurainen, K. *J. Chem. Inf. Comput. Sci.* **2003**, *43*, 1974–1981.
179. Hurst, T. HQSAR – A Highly Predictive QSAR Technique Based on Molecular Holograms. *Book of Abstracts, 213th ACS National Meeting, San Francisco, April 13–17*; American Chemical Society: Washington, DC, 1997, CINF-019.
180. Tong, W.; Lowis, D. R.; Perkins, R.; Chen, Y.; Welsh, W. J.; Goddette, D. W.; Heritage, T. W.; Sheehan, D. M. *J. Chem. Inf. Comput. Sci.* **1998**, *38*, 669–677.

181. Wold, S.; Esbensen, K.; Geladi, P. *Chemometrics Intell. Lab. Systems* **1987**, *2*, 37–52.
182. Kubinyi, H.; Hamprecht, F. A.; Mietzner, T. *J. Med. Chem.* **1998**, *41*, 2553–2564.
183. Filipponi, E.; Cruciani, G.; Tabarrini, O.; Cecchetti, V.; Fravolini, A. *J. Comput.-Aided Mol. Design* **2001**, *15*, 203–217.
184. Cianchetta, G.; Singleton, R. W.; Zhang, M.; Wildgoose, M.; Giesing, D.; Fravolini, A.; Cruciani, G.; Vaz, R. J. *J. Med. Chem.* **2005**, *48*, 2927–2935.
185. Wold, S.; Johansson, E.; Cocchi, M. PLS – Partial Least Squares Projections to Latent Structures. In *3D-QSAR in Drug Design: Theory, Methods and Applications*; Kubinyi, H., Ed.; ESCOM: Leiden, 1993, pp 523–550.
186. Bhagat, P. *Chem. Eng. Prog.* **1990**, *86*, 55–60.
187. Wythoff, B. J. *Chemometrics Intell. Lab. Systems* **1993**, *18*, 115–155.
188. Zupan, J.; Gasteiger, J. *Anal. Chim. Acta* **1991**, *248*, 1–30.
189. Polanski, J. *J. Chem. Inf. Comput. Sci.* **1997**, *37*, 553–561.
190. Livingstone, D. J.; Manallack, D. T. *QSAR Comb. Sci.* **2003**, *22*, 510–518.
191. Zhang, X.; Zhou, J. *Jisuanji Yu Yingyong Huaxue* **1995**, *12*, 186–191.
192. Tetko, I. V.; Livingstone, D. J.; Luik, A. I. *J. Chem. Inf. Comput. Sci.* **1995**, *35*, 826–833.
193. Agrafiotis, D. K.; Cedeno, W.; Lobanow, V. S. *J. Chem. Inf. Comput. Sci.* **2002**, *42*, 903–911.
194. Kubinyi, H. *Quant. Struct.–Act. Relat.* **1994**, *13*, 285–294.
195. Clark, D. E.; Westhead, D. R. *J. Comput.-Aided Mol. Design* **1996**, *10*, 337–358.
196. So, S.-S.; Karplus, M. *J. Med. Chem.* **1996**, *39*, 1521–1530.
197. Gillet, V. J. *Struct. Bond. (Berl. Germany)* **2004**, *110*, 133–152.
198. Holland, J. H. *Sci. Am.* **1992**, *267*, 66.
199. Boser, B. E.; Guyon, I. M.; Vapnik, V. N. A Training Algorithm for Optimal Margin Classifiers. In *5th Annual ACM Workshop on COLT*; Haussler, D., Ed.; ACM Press: Pittsburgh, PA, 1992, pp 144–152.
200. Platt, J. Fast Training of Support Vector Machines Using Squential Minimum Optimisation. In *Advances in Kernel Methods – Support Vector Learning*; Scholkopf, B., Burges, C. J. C., Smola, A. J., Eds.; MIT Press: Cambridge, MA, 1998, pp 185–208.
201. Czerminski, R.; Yasri, A.; Hartsough, D. *Quant. Struct.–Act. Relat.* **2001**, *20*, 227–240.
202. Apostolakis, J.; Hofmann, D.; Lengauer, T. Using Simple Learning Machines to Derive a New Potential for Molecular Modeling. In *Rational Approaches to Drug Design*; Höltje, H.-D., Sippl, W., Eds.; Prous Science: Philadelphia, 2001, pp 125–134.
203. Datar, P. A.; Coutinho, E. C.; Srivastava, S. *Lett. Drug Design Disc.* **2004**, *1*, 115–120.
204. Votano, J. R.; Parham, M.; Hall, L. H.; Kier, L. B.; Oloff, S.; Tropsha, A.; Xie, Q.; Tong, W. *Mutagenesis* **2004**, *19*, 365–377.
205. Doweyko, A. M. *J. Comput. Aided Mol. Design* **2004**, *18*, 587–596.
206. Doweyko, A. M.; Bell, A. R.; Minatelli, J. A.; Relyea, D. I. *J. Med. Chem.* **1983**, *26*, 475–478.
207. Martin, Y. C. *Pharmacochem. Library* **1993**, *20*, 129–137.
208. Bohacek, R. S.; McMartin, C. *ACS Symposium Ser* **1995**, *589*, 82–97.
209. DeWitte, R. S.; Shakhnovich, E. I. *J. ACS* **1996**, *118*, 11733–11744.
210. Vinkers, H. M.; de Jonge Marc, R.; Daeyaert, F. F. D.; Heeres, J.; Koymans, L. M. H.; van Lenthe, J. H.; Lewi, P. J.; Timmerman, H.; Aken, K. V.; Janssen, P. A. *J. Med. Chem.* **2003**, *46*, 2765–2773.
211. Kubinyi, H. *Pharmazie unserer Zeit* **1994**, *23*, 281–290.
212. Oprea, T. I.; Ho, C. M. W.; Marshall, G. R. *ACS Symposium Ser.* **1995**, *589*, 64–81.
213. Erickson, J. A.; De Novo Design Using a 3-D QSAR Derived Receptor. *Book of Abstracts, 212th ACS National Meeting*, Orlando, FL, August 25–29; American Chemical Society: Washington, DC, 1996, COMP-165.
214. Pearlman, R. S.; Balducci, R.; Smith, K. M.; Brusniak, M. Y. Dirty Comfa, Dirty Docking, and a Clean Method for Inverse-QSAR-based Drug Design. *Book of Abstracts, 217th ACS National Meeting*, Anaheim, Calif., March 21–25; American Chemical Society: Washington, DC, 1999, COMP-048.
215. Doweyko, A. M.; Johnson, S. R. Q. A Novel Method to Simulate the Solvation of Hydrophobic Surfaces by Water and Its Application to the Estimation of Protein-Ligand Binding Constants. *Abstracts of Papers, 229th ACS National Meeting*, San Diego, CA, United States, March 13–17, 2005; American Chemical Society: Washington, DC, 2005, COMP-218.
216. Steinberg, I. Z.; Scheraga, H. A. *J. Biol. Chem.* **1963**, *238*, 172–181.
217. Tunon, I.; Silla, E.; Pascual-Ahuir, J. L. *Protein Eng.* **1992**, *5*, 715–716.
218. Soda, K. *Adv. Biophysics* **1993**, *29*, 1–54.
219. Lazaridis, T. *J. Phys. Chem. B* **2000**, *104*, 4964–4979.
220. Scheraga, H. A. *J. Phys. Chem.* **1961**, *65*, 1071–1072.
221. Venkatarangan, P.; Hopfinger, A. J. *J. Chem. Inf. Comput. Sci.* **1999**, *39*, 1141–1150.

Biography

Arthur M Doweyko After earning a PhD in Bioorganic Chemistry from Rutgers University in1975, Arthur was a Research Scholar/Postdoctoral Fellow in Oncology working with Charles Heidelberger at the USC-LAC Comprehensive Cancer Research Center investigating mechanism-based enzyme inhibitors as potential cancer chemotherapeutic agents. He joined Uniroyal Chemical in 1978, where he supported the discovery of novel crop protection agents as both a synthetic and a xenobiotic metabolism chemist. In 1989 he joined Ciba-Geigy, where he continued working in registration chemistry in support of agrochemical research. After a short stint with the Schering-Plough veterinary drug group, he joined the Bristol-Myers Squibb CADD group in 1997. During the course of his research career, Arthur has developed an expertise in synthesis, bioorganic chemistry, the metabolism of xenobiotics, and computational aspects of drug discovery. He was one of the first to develop a simple 3D-QSAR methodology, the hypothetical active site lattice (HASL), at a time when such methods were without precedent. He has authored 36 papers and 10 patents. His extracurricular interests include competitive soccer, table tennis, painting, and music.

© 2007 Elsevier Ltd. All Rights Reserved
No part of this publication may be reproduced, stored in any retrieval system or transmitted in any form by any means electronic, electrostatic, magnetic tape, mechanical, photocopying, recording or otherwise, without permission in writing from the publishers

Comprehensive Medicinal Chemistry II
ISBN (set): 0-08-044513-6

ISBN (Volume 4) 0-08-044517-9; pp. 575–595

4.24 Structure-Based Drug Design – The Use of Protein Structure in Drug Discovery

G Lange, Bayer CropScience, Frankfurt, Germany

© 2007 Elsevier Ltd. All Rights Reserved.

4.24.1 Introduction

Over the past 10–15 years, several drugs have been developed with the help of structure-based drug design. Established successes, such as the drugs designed against human immunodeficiency virus-1 (HIV-1) protease and neuraminidase (NA), have demonstrated that structure-based design can lead to successful marketed products. Historically, the approach has suffered from low throughput since the determination of a three-dimensional (3D) protein structure required 1–10 years depending on the difficulties encountered. Thus, compared to the progress in a chemical synthesis program, the determination of protein structures with inhibitors complexed to their binding sites was slow and only few structures were obtained, even in structure-based drug design projects. Within the last 10 years, several breakthroughs have helped to overcome some of the bottlenecks in the structure determination process.[1,2] New technologies have resulted in a significant speed-up of the structure determination process and in good cases it is now possible to determine the structures of several inhibitor complexes within a day. Thus the speed of crystal structure elucidation is now within the same timeframe as chemical synthesis, thereby making a complete cycle of a structure-based design project possible within a short time period.

Protein production is still one of the major bottlenecks in the structure determination process, even though the molecular biology tools that became available in the 1980s and 1990s have significantly shortened the time required to obtain pure protein. Methods currently exist for parallel expression and purification of large numbers of gene products which allow researchers to investigate multiple gene constructs, protein homologs, and variants for specific protein targets.[3,4] In addition, methods are now available that allow selenomethionine to be easily incorporated into overexpressed protein, thereby yielding protein which can be used in multiple-wavelength anomalous dispersion (MAD) experiments.[5] Although several crystallization robots were developed in the 1980s, the systems were not always reliable and crystallization remained one of the rate-limiting steps in protein structure elucidation. Since the end of the 1990s, however, efforts have been made to develop more robust robots to process crystallization trials in a more automated fashion. For example, standalone workstations are now available for specific tasks such as protein drop dispensing.[6] Even fully integrated systems have been developed with throughput capacities of 2500–140 000 experiments per day, making large-scale, automated crystallization trials possible. Automated crystallization trials incorporating smaller, higher-density drop-plating configurations (96-well, 384-well, and 1536-well formats) enable a more condensed experimentation scale, minimizing plate storage space requirements and maximizing the number of experiments that can be pursued in a given time. Crystallization trials require periodic examination of the individual experiments to monitor the crystallization process. Manual inspection is tedious, error-prone, and impractical for the number of plates generated robotically. As a consequence, systems for robotic plate storage and automated image acquisition and analysis were designed and are now available.

Another major advance has been in the structure determination process. The use of high-brilliance beam-lines at synchrotrons and flash-cooling[7] allow x-ray data to be collected on very small crystals, which had not been possible earlier. Thus, valuable time can be gained because the time needed for crystallization optimization has become shorter. Beam-line robotics including automatic sample changes[8] have resulted in a significant speed-up of data collection, making it possible to collect the required x-ray data for a protein complex within minutes. Due to the high quality of the x-ray data obtained, in many cases it is possible to use MAD for determination of the phases.[9] The resulting high-quality x-ray data also allow the regular use of programs that either support the protein chain tracing or even do the initial chain tracing automatically.[10–12] In addition, refinement programs are now available that in many cases replace the time-consuming manual refinement by an automatic procedure.[13]

Drugs have been designed using protein 3D structural information for targets such as proteases, kinases, and an expanding number of other biological macromolecules. In particular, drugs that were found using structure-based design for HIV protease and NA are used in the clinic. Other protein targets include renin,[14] factor Xa,[15,16] α-thrombin,[17] factor VIIa/tissue factor (TF),[18] urokinase,[19] cathepsin B,[20] cathepsin L and S,[21] β-secretase,[22] Abelson tyrosine kinase,[23] Cdk4,[24] epidermal growth factor receptor (EGFR) tyrosine kinase,[25] Grb2-SH2,[26] aldose reductase,[27] phospholipase,[28] interleukin-1β converting enzyme,[29] thymidylate synthase,[30] acetylcholinesterase,[31] matrix metalloproteinase,[32] monoamine oxidase,[33] DNA gyrase,[34] peptide deformylases,[35] farnesyltransferase,[36] and Chk1.[37] Many of these structures are deposited in the Protein Data Bank.[38] In the following chapter, four target proteins have been chosen in order to illustrate the process of rational drug design. The examples were chosen to cover different procedures or different starting points, such as ab initio design or starting with a screening lead, and different objectives, such as obtaining a high-affinity inhibitor, a bioavailable drug, or drugs against resistant proteins. For other applications of SBDD see also Chapters 4.02, 4.04, 4.16, 4.25–4.29, and 4.31.

4.24.2 Design of Human Immunodeficiency Virus Protease Inhibitors against Acquired Immune Deficiency Syndrome (AIDS)

AIDS was first reported in the US in 1981. As a result of the alarming spread of HIV, the etiologic agent of AIDS,[39,40] research was initiated during the 1980s in order to comprehend and control this disease. A few years later, rapid advances in molecular, viral, and cell biology had already led to the identification of targets for potential intervention with synthetic drugs. HIV is a member of the Lentiviridae subfamily of retroviruses that contain three major genes (*gag*, *pol*, *env*).[41–43] Products of the *gag* gene include structural proteins of the virus nucleocapsid. The *pol* gene encodes three enzymes: a protease, a reverse transcriptase, and an endonuclease. The *env* gene encodes the membrane proteins of the mature virus. The *gag* and *pol* gene products are expressed as polyproteins that need to be processed. In 1988, it was reported that an aspartic protease was required for HIV replication.[44] Inhibition of this protease prevents the maturation and replication of the virus in cell culture since the polyproteins are no longer cleaved into sequences that can fold into the active HIV proteins. Structure-based drug design has led to the fast discovery and approval of drugs inhibiting the HIV protease. Already in 1993 antiviral effects of protease inhibitors seen in human clinical trial were reported. Today, there are seven approved drugs and several others in advanced clinical trials. These include ritonavir (Abbott), saquinavir (Hoffman-La Roche), nelfinavir (Agouron), indinavir (Merck), amprenavir (Vertex), atazanavir (BMS), and lopinavir (Abbott). Additional compounds such as tipranavir[45] are in clinical development. Since their introduction into the clinic in late 1995, HIV protease inhibitors have proven to provide an advantageous extension to the existing antiretroviral medication, which at the time consisted solely of nucleoside analog reverse transcriptase inhibitors. Their clinical use has been marked by a profound decrease in mortality rate associated with HIV infection. Today, HIV protease inhibitors are considered as essential components of antiretroviral therapy.[46,47] Despite this remarkable success, the emergence of HIV

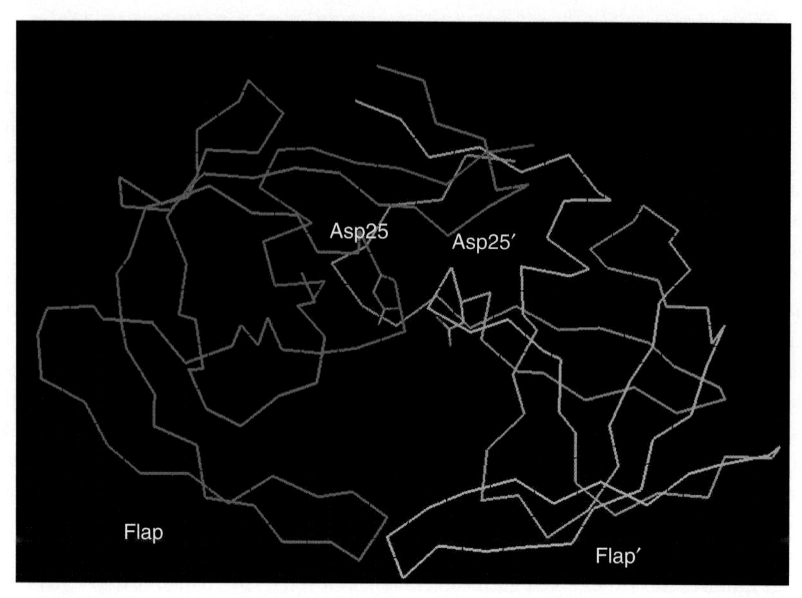

Figure 1 Cα-trace of the apo structure of HIV protease (3HVP). Monomer 1 is shown in green and monomer 2 in yellow. In addition, the catalytic aspartates, i.e., residues 25 and 25′, are displayed in order to mark the active site of the HIV protease.

mutants that are resistant to available drugs remains a critical factor in the clinical failure of antiviral therapy. For all approved HIV protease inhibitors, the emergence of HIV protease mutants that confer multidrug resistance has been reported.

4.24.2.1 First Crystal Structure of Human Immunodeficiency Virus Protease

On the basis of the conserved Asp-Thr/Ser-Gly sequence at the putative protease active site, retroviral proteases had been predicted quite early to belong to the family of pepsin-like aspartic proteases. It had been suggested that the aspartic proteases evolved from a smaller, dimeric ancestral protein.[48] Thus, an early model of HIV protease postulated that HIV protease was a symmetric dimer with a similar fold to pepsin.[49] Indeed, the crystal structure[49–52] revealed that the HIV protease consisted of a C2-symmetric homodimer (**Figure 1**). Even though the details in the interface are different to pepsin-like proteases, the overall topology is similar. The topology of the monomer can be described as a β-sheet barrel and includes an antiparallel β-sheet consisting of four strands. These sheets form the bottom of the dimer, on top of which the active site loops are situated. Two 'flap' arms, one from each monomer, project over the active site in the HIV protease, whereas in the pepsin-like proteases only one prominent flap is found. There is a large cleft situated perpendicular to the twofold symmetry axis and between the active site and the flaps: this flap was recognized as the substrate binding pocket. Similarly to the conserved water molecule in pepsin, there is electron density for a water molecule which is located between the catalytic aspartates. The active site residues, i.e., the sequence Asp25-Ser26-Gly27, are typical for an aspartic protease. The residue following the triplet is always an alanine in retroviral proteases, while in all pepsin-like proteases this residue is either serine or threonine. The crystal structure made evident that inhibitors from pepsin-like protease such as renin would be good lead structures for the HIV proteases. However, the differences between HIV protease and the aspartic proteases, such as the hydroxyl function next to the active site residues, need to be accounted for. In the following description the amino acid numbering is arranged such that monomer 1 counts from 1 to 99 and monomer 2 from 1′to 99′. HIV-1 and HIV-2 protease will be collectively called HIV protease.

4.24.2.2 Crystal Structure of Human Immunodeficiency Virus Protease with a Substrate-Based Inhibitor

The close similarity between the 3D structures of dimeric retroviral proteases and the monomeric pepsins has led to the assumption that both have a common mode of interaction with the peptidic substrate. Initial modeling based on x-ray complexes of aspartic protease with transition-state isosteres had suggested that the peptidic HIV substrate probably binds pseudosymmetrically in the active site cleft using topologically equivalent H-bond functions and specificity pockets on each side of the scissile bond.[53] This was confirmed in the first crystal structure of HIV protease complexed with a substrate-based inhibitor which had the sequence *N*-acetyl-Thr-Ile-Nle-P[CH_2-NH]-Nle-Gln-Arg-amide.[54] Here, the scissile bond has been replaced by a reduced analog. Interestingly, there is a substantial change in the HIV protease upon binding of the substrate-like peptide (**Figure 2a**). In particular, the residues forming the flaps are now well ordered and contribute significantly to the binding of the inhibitor. Despite the symmetric nature of the uncomplexed enzyme, the asymmetric inhibitor is bound in a single conformation and makes extensive interactions to both subunits. The details of the interactions are summarized in **Figure 2b**. Six direct H-bonds between the protease backbone belonging to amino acids Asp29/Ap29′, Gly27/Gly27′, and Gly48/48′, and backbone H-bond functions from the substrate-like peptide were observed in the 3D structure. The backbone amides of Ile50 and 50′ bind via a bridging water molecule to the inhibitor carbonyls at position P2 and P1′ (**Figure 2c**). In addition, there is a substantial burial of hydrophobic inhibitor atoms at positions P2, P1, P1′, and P2′ into hydrophobic protease pockets, giving rise to a stabilizing hydrophobic effect.

Tight-binding inhibitors of HIV protease were discovered fairly quickly because the principles for inhibiting this class of enzyme were known from earlier studies on inhibitors of renin and other aspartic proteinases.[55,56] Several strategies for the development of HIV protease inhibitors have been employed. Many of these utilize substrate-based inhibitor design

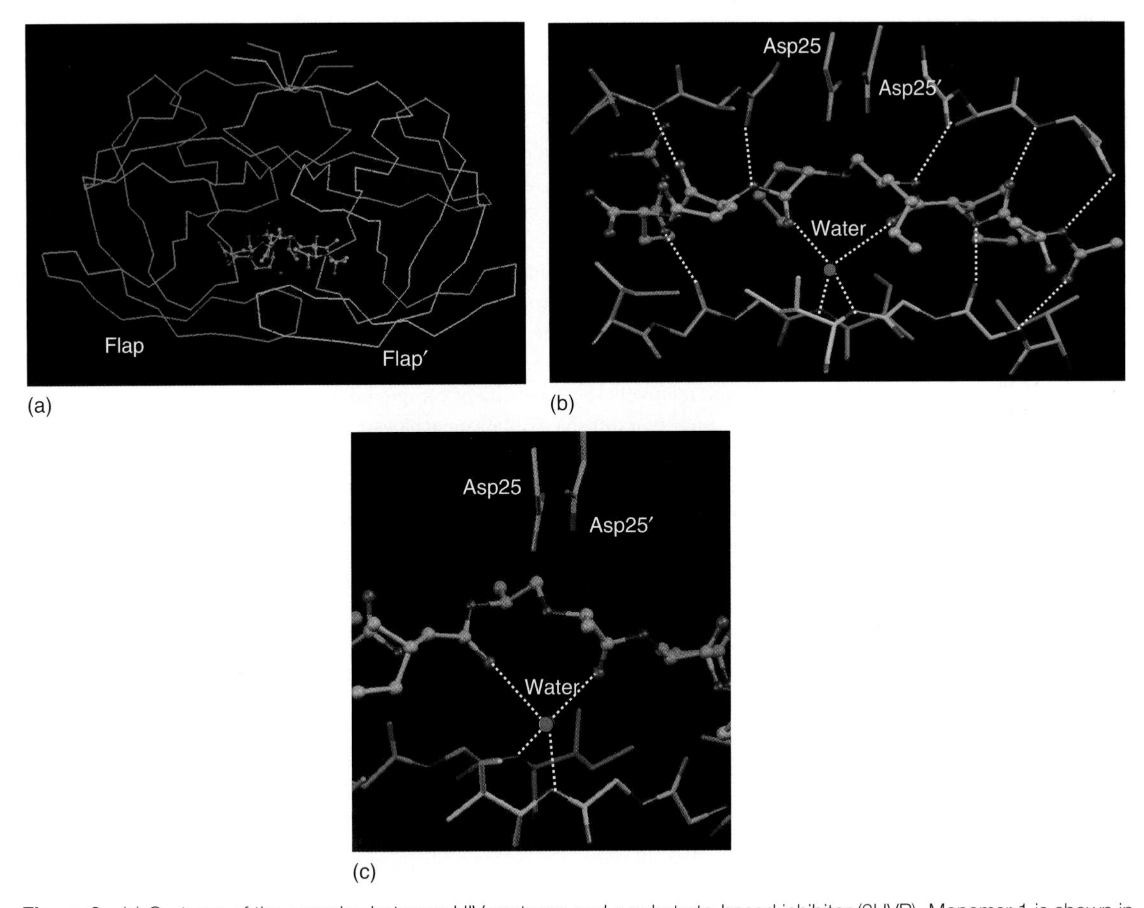

Figure 2 (a) Cα-trace of the complex between HIV protease and a substrate-based inhibitor (3HVP). Monomer 1 is shown in green and monomer 2 in yellow. The inhibitor with the sequence *N*-acetyl-Thr-Ile-Nle-P[CH_2-NH]-Nle-Gln-Arg-amide is represented in blue. (b) Interaction between HIV protease and a substrate-based inhibitor with the sequence *N*-acetyl-Thr-Ile-Nle-P[CH_2-NH]-Nle-Gln-Arg-amide (3HVP). Monomer 1 is shown in green and monomer 2 in yellow. The inhibitor is represented in blue. The H-bonds between the HIV protease and the inhibitor, including those involving the conserved water molecules, are indicated. (c) Close-up of the interaction of the conserved water molecule with the tip of the flaps of the HIV protease and the substrate-based inhibitor with the sequence *N*-acetyl-Thr-Ile-Nle-P[CH_2-NH]-Nle-Gln-Arg-amide (3HVP). Monomer 1 is shown in green and monomer 2 in yellow. The inhibitor is represented in blue.

in which the scissile bond is replaced with a noncleavable isostere, e.g., a reduced amide,[57] hydroxyethylene,[58] or hydroxyethylamine group.[59] Transition-state mimetics such as statine analogs, phosphinates, and difluoro ketone derivatives have also been incorporated into HIV protease inhibitors.[60,61] The molecular interactions of many of these types of inhibitors have been identified using x-ray crystallographic and molecular modeling techniques. These studies have allowed the optimization of lead compounds through an iterative cycle of protein crystallography, biological assay, and medicinal chemistry. Some selected studies are presented in more detail below.

4.24.2.3 Structure-Based Design Leading to Ritonavir and Lopinavir: Exploiting the Symmetry of the Human Immunodeficiency Virus Protease

The x-ray crystal structure of HIV protease shows the enzyme to consist of a C2-symmetric homodimer. Initially, strategies that exploit this element of symmetry in the design and synthesis of HIV protease inhibitors were pursued and these resulted in the discovery of C2-symmetric inhibitors that are potent and specific for HIV protease.[62] Based upon modeling studies of a reduced peptide inhibitor, it was proposed that the symmetric core element of compound 1 (**Figure 3a**) should make reasonable interactions with the catalytic aspartates. In addition, the structure of the S1 and S1′ pockets in the HIV protease should be indistinguishable in the ligand-free protease and thus two identical substituents were positioned at positions P1 and P1′. Phenyl substituents were used since they are frequently observed at these positions in naturally occurring substrates. The prototype compound was a weakly binding inhibitor but supported the hypothesis that a C2-symmetric compound could lead to a potent HIV protease inhibitor. Careful examination of the substrate binding site indicated that only the P1 and P2 substituents would likely be buried in subsites of the enzyme, while the P3 substituents were likely to be exposed. Modeling suggested the addition of a carbobenzyloxy-valine attached to both sides of compound 1. The resulting compound 2 turned out to be a competitive inhibitor with a K_i of 4.5 nM. The C2 symmetry of the inhibitor is also seen in the crystal structure of HIV protease with bound compound 2 (**Figure 3b**). The conserved water molecule bridging the inhibitor carbonyls to the two backbone amides of the flap was clearly visible. This water lies within about 0.2 Å of the molecular twofold axis. The central hydroxyl group made the expected good interactions with the catalytic aspartates. However, the bioavailability of this compound was insufficient due to its poor aqueous solubility.

Thus, the goal of discovering inhibitors with good oral pharmacokinetic properties had not been realized and a study was undertaken to analyze the relationship between the pharmacokinetic profiles of HIV protease inhibitors and a variety of physicochemical features.[62,63] This analysis, combined with the information from the 3D structures, led to the pseusdosymmetric compound 3 (**Figure 3a**), which seemed to represent an optimal compromise between oral bioavailability and potency. Unfortunately, the relatively short half-life of compound 3 limited its ability to maintain the required blood plasma concentration. Subsequently, extensive studies on analogs were carried out, which gave valuable insights into the relationship of chemical structure to antiviral activity, aqueous solubility, and hepatic metabolism. These studies resulted in the design of compound 4.[64] Several factors seem to contribute to the superior properties of compound 4 but the enhancement of antiviral potency over compound 3 predominantly results from better inhibition of HIV protease. Earlier modeling studies had proposed an additional favorable hydrophobic contact between the isopropyl substituent on the P3 thiazolyl group of compound 4 and the side chains of Pro81 and Val82. This was confirmed by the crystal structure of compound 4 bound to HIV protease (**Figure 3c**) and by the emergence of Val82 mutants upon in vitro passage of HIV in the presence of increasing concentrations of compound 4. Compound 4 (ritonavir) was the first HIV protease inhibitor from the Abbott research program to advance to late-stage clinical trials and was licensed for antiretroviral therapy in 1996.

The requirements defined as important for their second-generation inhibitor included: (1) improvement of potency against wild-type HIV in vitro; (2) inclusion of structural features that would offset the effects of specific resistance

Figure 3 (a) Schematic drawing of compounds 1–5. (b) Interaction between HIV protease and compound 2 (9HVP). Monomer 1 is shown in green and monomer 2 in yellow. Compound 2 is seen in blue. The H-bonds between the HIV protease and the inhibitor, including those involving the conserved water molecules, are indicated. (c) Interaction between HIV protease and compound 54 (1HXW). Monomer 1 is shown in green and monomer 2 in yellow. Compound 4 is represented in blue. The H-bonds between the HIV protease and the inhibitor, including those involving the conserved water molecules, are indicated. (d) Interaction between HIV protease and compound 5 (lopinavir) (1MUI). Monomer 1 is shown in green and monomer 2 in yellow. Compound 5 is represented in blue. The H-bonds between the HIV protease and the inhibitor, including those involving the conserved water molecules, are indicated. (e) Superposition of compound 4 (blue) and compound 5 (green) using the coordinates of HIV protease with bound compound 5. Monomer 1 is shown in green and monomer 2 in yellow. The position of Pro81 and Val82 is indicated.

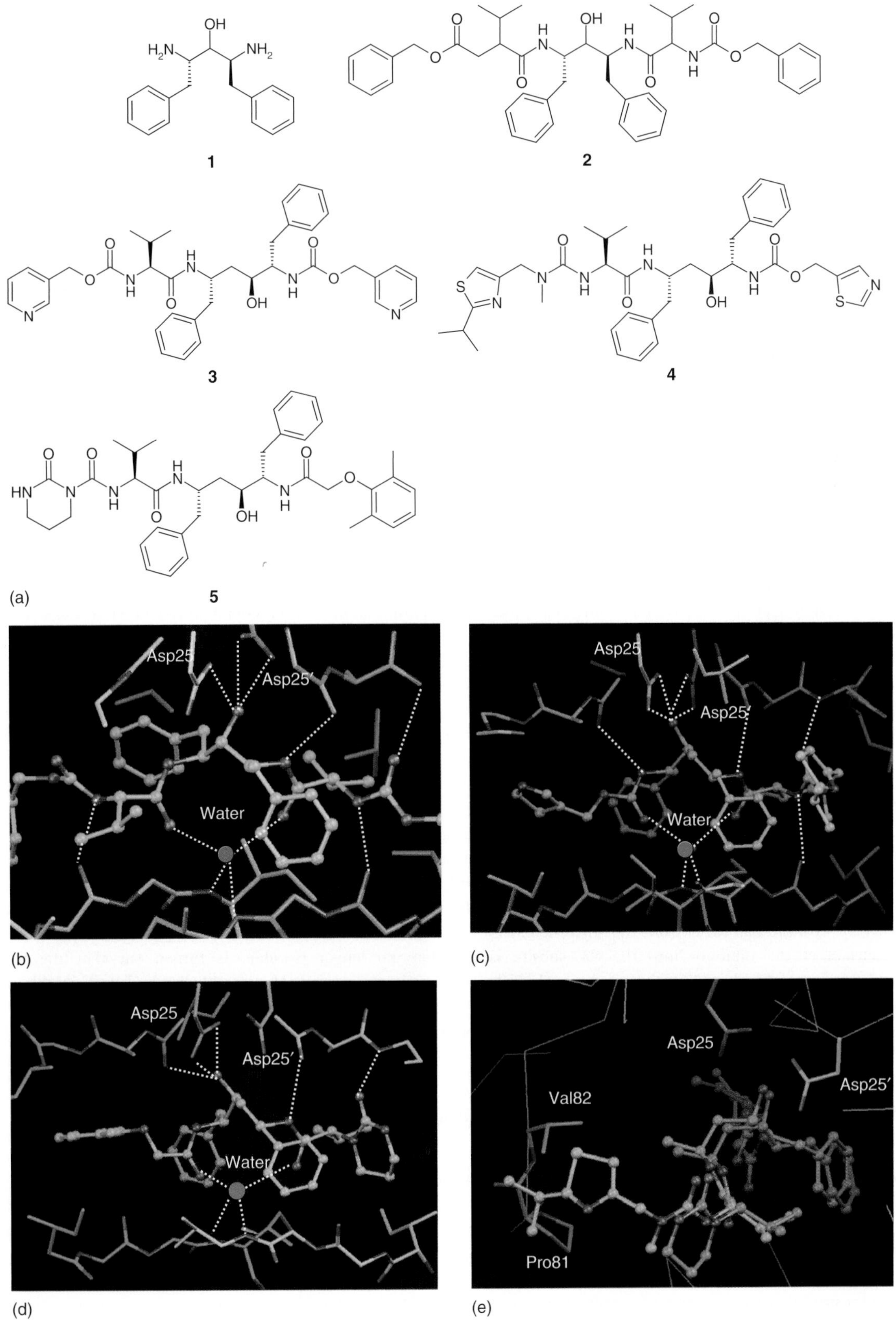

1

2

3

4

(a) 5

(b)

Asp25
Asp25′
Water

(c)

Asp25
Asp25′
Water

(d)

Asp25
Asp25′
Water

(e)

Asp25
Asp25′
Val82
Pro81

mutations such as mutations involving Val82; and (3) pharmacokinetic properties able to sustain plasma levels well in excess of the in vitro EC_{50}. An intense rational drug design program led to the discovery of compound 5 in **Figure 3a.**[65] Compound 5 is an extremely potent inhibitor (K_i 1.3 pM) of HIV protease which occupies the active site much as had been anticipated by modeling studies (**Figure 3d**).[66] The inhibitor has hydrophobic substituents that fill the four hydrophobic pockets closest to the catalytic aspartates in the binding site. Two clear binding modes of the inhibitor were seen in the crystal structure, which represents a twofold disorder with both orientations similarly occupied. This twofold occupancy has been observed in many inhibitor complexes of HIV protease. The superposition of compounds 4 and 5 showed that, in contrast to compound 4, compound 5 made no close contact to Val82 (**Figure 3e**), since the isopropylthiazolylmethyl urea unit in compound 4 had been replaced by a significantly shorter cyclic urea unit in order to generate an inhibitor which inhibits both the wild-type and Val82 mutants. This lack of interaction with Val82 agrees well with the retention of wild-type inhibitory potency of compound 5 toward single-site mutated HIV protease in biochemical assays. The cyclic urea unit of compound 5 provides a novel H-bonding arrangement with Asp29 and ensures bioavailability. In addition to the structural aspects described here, other pharmacological properties, such as serum protein binding, half-life, and metabolism, were simultaneously optimized in an extensive medicinal chemistry effort. On the basis of demonstrated efficacy in clinical studies, compound 5 (lopinavir) was licensed for antiretroviral therapy in 2000.

4.24.2.4 Structure-Based Drug Design Leading to Sasquinavir: Exploitation of Earlier Experience with Renin Inhibitors

Incorporation of a hydroxyethylamine transition-state mimetic led quite early on to very potent inhibitors of HIV protease. One such inhibitor is compound 6 (**Figure 4a**) from the laboratories of Hoffman-La Roche.[67] Interestingly, in this series of inhibitors, the stereochemistry of the carbon atom bearing the hydroxyl function appears to be dependent both upon the length of the inhibitor and upon the nature of individual residues. This observation is in agreement with earlier findings in the field of renin inhibitors. Molecular modeling studies suggested that the binding mode adopted by inhibitors containing the DIQ group would require an (*R*)-hydroxyl group and would not allow the presence of further residues at the C-terminus. This was confirmed experimentally for compound 6 and other analogs with additional residues. Compounds which have additional residues inhibit HIV protease only weakly, irrespective of the stereochemistry of the hydroxyl group. The proposed binding mode of compound 6 to HIV protease has been confirmed by an x-ray structure. Two clear binding modes of the inhibitor were seen, and this represents a twofold disorder, with both orientations similarly occupied (**Figure 4b**). The x-ray crystal structure shows that the inhibitor binds in an extended conformation, forming the characteristic set of H-bonds with the enzyme (**Figure 4c**). The backbone in the peptidic compound 6 has the same conformation as seen in other HIV protease inhibitors and the Asn and Phe side chains make similar contacts with the enzyme to those previously observed. Except for the tips of the flaps, the twofold symmetry of the enzyme has been largely preserved, resulting in essentially equivalent S1 and S1′ subsites. The (*R*)-hydroxyl group is located between the catalytic aspartic acids. The [(4aS,8aS)-decahydroisoquinolin-3(S)-yl]carbonyloxy (DIQ) moiety occupies almost the entire S1′ subsite and makes good hydrophobic contacts with both the flap regions and the core of the enzyme. The *tert*-butyl amide group fits tightly into the S2′ subsite with the methyl groups completely buried. The carbonyl of the DIQ group can still bind to the water molecule connecting the inhibitor with the flap regions but the adjacent nitrogen atom occupies a location distinct from that observed earlier. Extension of the inhibitor into the S3′ subsite is therefore no longer possible. It turned out that the S,S stereochemistry of the DIQ ring is optimal for good inhibitory activity. Hydroxyethylamine-based inhibitors incorporating the (*R*)-hydroxyl group thus represent a new class of small and highly potent HIV protease inhibitor. In addition to its good affinity to the HIV protease, compound 6 turned out to be bioavailable and active in human clinical studies. Compound 6 (sasquinavir) was approved for use in the clinic in 1995.

4.24.2.5 Structure-Based Drug Design Leading to Indinavir: Creating a Bioavailable Drug

Starting from a renin inhibitor-based lead, researchers at Merck reported the discovery of a highly potent and selective inhibitor containing a hydroxyethylene isostere in 1991 – compound 7 in **Figure 5a.**[59,68] Although very potent, the optimized molecules of this series lacked aqueous solubility and an acceptable pharmacokinetic profile. The researchers decided to try and improve the physical properties of this peptidomimetic lead structure by including structural elements from an orally available HIV protease inhibitor, compound 6, published by Hoffmann-La Roche.[67] It was hypothesized that incorporation of a basic amine into the scaffold of compound 7 might improve the bioavailability.[69] Modeling studies suggested that replacement of the *tert*-butyl carbamate and Phe moieties of compound 7 with the

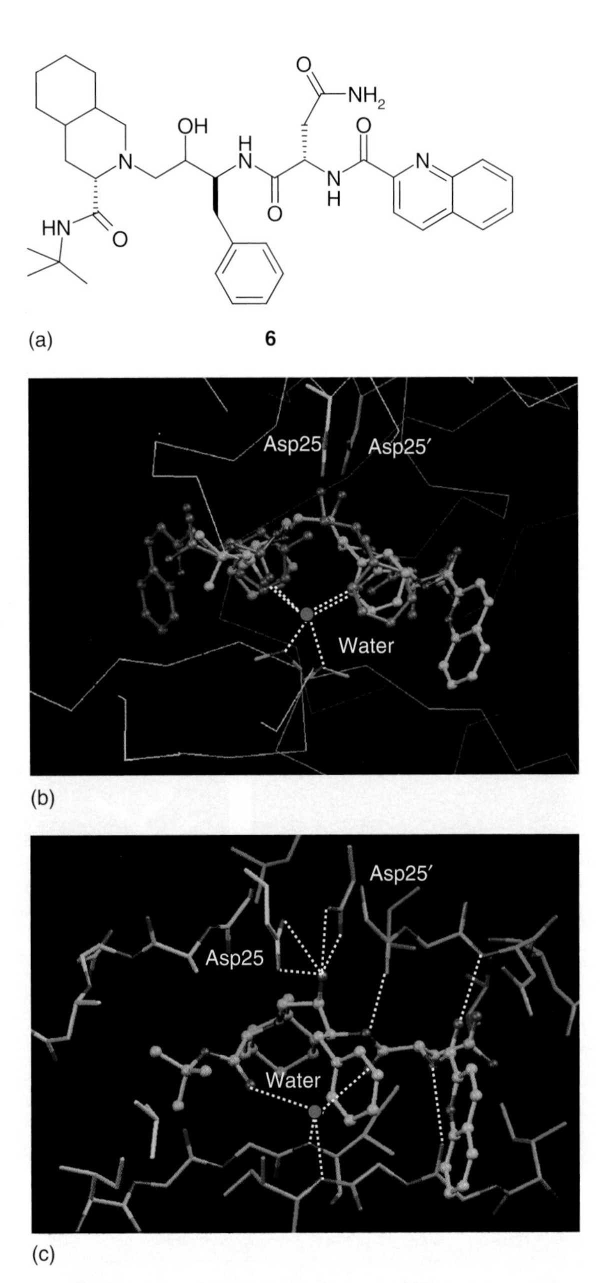

Figure 4 (a) Schematic drawing of compound 6. (b) Interaction between HIV protease and compound 6 (sasquinavir) (1HXB). The Cα-trace of monomer 1 is shown in green and that of monomer 2 in yellow. Clearly, there are two symmetry-related inhibitors located in the active site. The inhibitor molecules are shown in blue and magenta. The catalytic aspartates and the interactions of the conserved water molecules are indicated for easier orientation. (c) Interaction between HIV protease and compound 6 (sasquinavir) (1HXB). Monomer 1 is shown in green and monomer 2 in yellow. Compound 6 is represented in blue. The H-bonds between the HIV protease and the inhibitor, including those involving the conserved water molecules, are indicated.

corresponding substituents of compound 6 might generate a novel class of bioavailable transition-state isosteres, as exemplified by compound 8 (**Figure 5b** and **c**). Incorporation of the decahydroisoquinoline *tert*-butylamide group should provide two advantages. The amine would provide much-needed aqueous solubility and the incorporation of the amine into a ring should limit the conformational freedom of the inhibitor, thereby decreasing the entropy change upon binding to HIV protease. Indeed, compound 8 had a high intrinsic potency (IC_{50} 7.6 nM). However, the inhibition of the spread of viral infection in MT4 human T-lymphoid cells was relatively weak (CIC_{95} = 400 nM) and thus other amine-containing analogs such as piperazine analogs were investigated. In addition to achieving improved aqueous solubility,

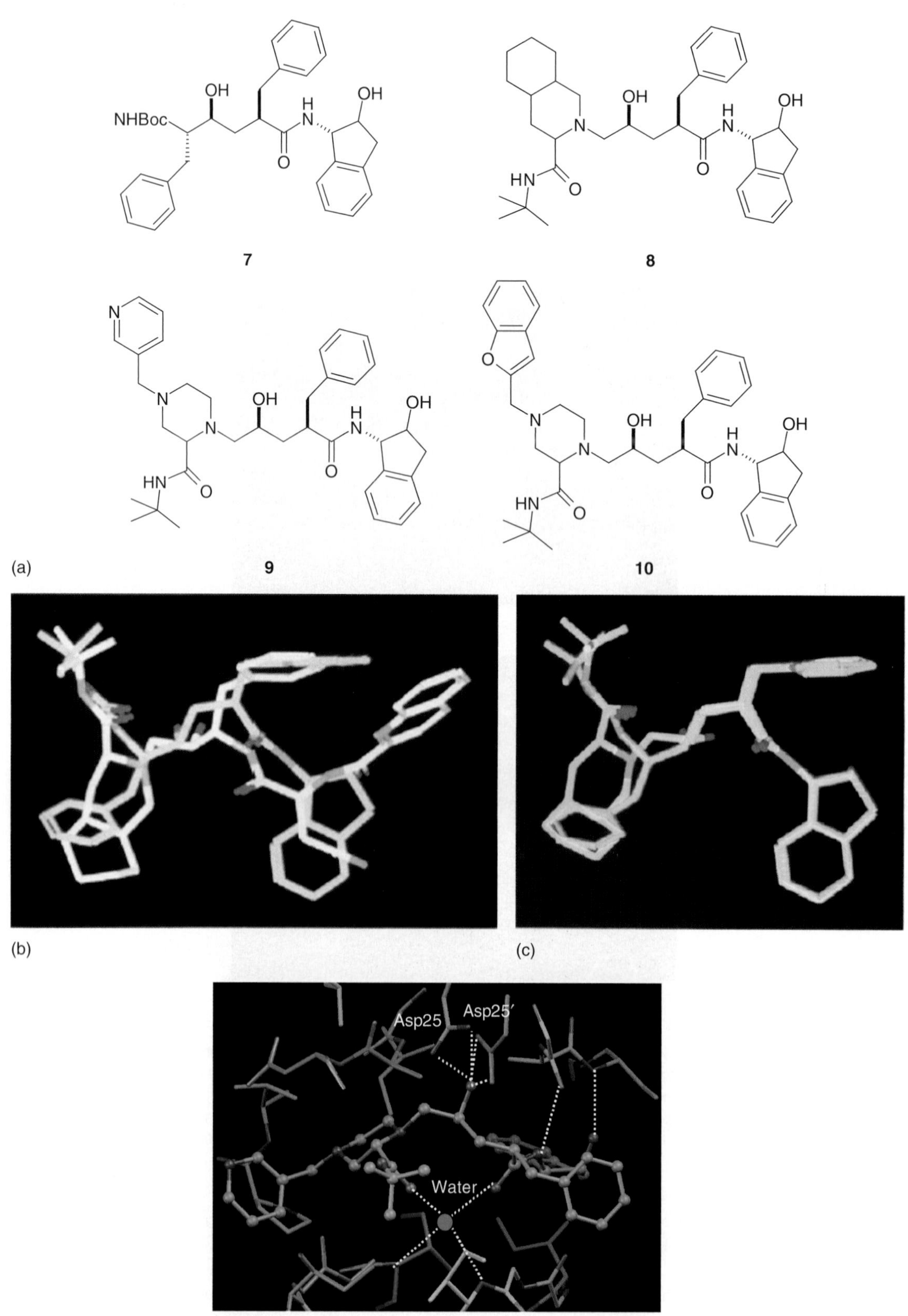

NHBoc
OH
H
N
OH
O
7
N
OH
H
N
OH
HN
O
O
8
N
N
OH
H
N
OH
N
HN
O
O
9
O
N
OH
H
N
OH
N
HN
O
O
10
(a)
(b)
(c)
Asp25
Asp25′
Water
(d)

the nitrogen in the 4-position of the piperazine ring could be easily functionalized. This allowed for the introduction and optimization of a P3 substituent which was used to balance both the hydrophobic and hydrophilic requirements of the target molecules. One of the first compounds prepared in the piperazine series, compound 9, possessed a benzyloxycarbonyl moiety attached to the N4 position of the piperazine ring. To understand better the resulting significant increase in potency of compound 9, the crystal structure of a HIV protease as a complex with compound 9 was determined. The inhibitor is bound in a single orientation in the crystal lattice. Consistent with prior modeling, the substituents at the P2–P2′ positions bind tightly into the S2–S2′ pockets of the HIV protease (**Figure 5d**). Related by a pseudo-twofold rotation, the *t*-butyl and indanyl groups of compound 9 are bound in the S2 and S2′ pockets of the enzyme while the pyridyl methyl piperidine and benzyl rings are situated in the S3/S1 and S1′ pockets. The transition-state isoster hydroxyethylene is H-bonded to the side chains of Asp25 and Asp25′. Located between the tips of the flap loops and two backbone amides is the conserved water molecule with its oxygen tetrahedrally coordinated. Compound 9 competitively inhibits HIV-1 protease and HIV-2 protease, with K_i values of 0.52 and 3.3 nM, respectively. This compound prevented the spread of viral infection in the genetically diverse SIVmaC251-infected cells at concentrations of less than 100 nM. In 1996 compound 9 (indinavir) was approved for clinical use.

Further improvements in this transition-state isostere series of HIV protease inhibitors were reported in 2000.[70] Human clinical studies had established that indinavir must be dosed every 8 h at a dose of 800 mg and that the pharmacokinetic half-life of indinavir was quite short. The metabolites found in both the urine and feces included the oxidative *N*-dealkylated 3-pyridylmethyl ligand, the quaternary pyridine *N*-glucuronide and the pyridine *N*-oxide. One strategy to try and improve the pharmacokinetic properties of indinavir would be to block or avoid these metabolic pathways. Therefore, replacement of the metabolically labile 3-pyridylmethyl with either a large lipophilic heterocycle or a modified pyridine ring was tried in order to increase the in vivo half-life of the inhibitor. A series of 3-pyridylmethyl replacement analogs possessing a wide range of physical properties were designed using structure-based drug design and synthesized.[70] The most interesting compound was a benzofuran derivate, compound 10 (**Figure 5a**). This compound is potent, selective, and competitively inhibits HIV-1 protease with a K_i value of 0.049 nM. It stops the spread of the HIV-IIIb-infected MT4 lymphoid cells at 25.0–50.0 nM, even in the presence of R1 acid glycoprotein, human serum albumin, normal human serum, or fetal bovine serum. In particular, compound 10 has a longer half-life in several animal models (rats, dogs, and monkeys) than indinavir and has advanced to human clinical trials.

4.24.2.6 Structure-Based Design Leading to a Novel Nonpeptidic Scaffold

In general, peptidomimetics retain a substantial amount of peptidic character and as a result have a high likelihood of exhibiting inadequate oral bioavailability. These difficulties have limited their usefulness as therapeutic agents. Instead of optimizing the bioavailability of peptidomimetics, Lam *et al.* have taken a different route.[71]

Using 3D chemical databases in combination with information derived from the 3D structure of HIV protease, these researchers intended to identify a scaffold that should serve as starting point for the design of nonpeptidic inhibitors. In their 3D search, scaffolds were preferred which could replace the conserved water molecule and which contained rigid cyclic structures. In addition, structure–activity relationships (SARs) established for C2 symmetric inhibitors suggested that a central diol gave rise to higher potency compared to corresponding mono-ol transition-state analogs and thus the desired scaffold should contain a diol. Based upon the structure of a C2 symmetric diol docked into the HIV protease, a 3D pharmacophore model for the search in the database was derived. The database search yielded a hit which was subsequently modified in order to incorporate the desired diol function. In the next step, the carbonyl moiety of the aliphatic seven-membered ring was replaced by urea in order to strengthen those H-bond functions that were supposed to replace the conserved water molecule. Subsequently, molecules with this scaffold were synthesized and indeed turned out to be potent nonpeptidic HIV inhibitors (compounds 11–13 in **Figure 6a**). The crystal structure of compound 12 bound to HIV protease confirms that the water had been replaced by the central carbonyl function of the cyclic urea ring. In addition, the H-bonds between the inhibitor and the flap tips have an excellent geometry,

Figure 5 (a) Schematic drawing of compounds 7–10. (b) Superposition of the energy-minimized structure of compound 7 (magenta) and the x-ray crystal structure of 6 (green) in the conformation found in the crystal structure. (c) Superposition of the energy-minimized structures of compounds 8 (yellow) and 7 (magenta). (d) Interaction between HIV protease and compound 9 (indinavir) (1HSG). Monomer 1 is shown in green and monomer 2 in yellow. Compound 9 is represented in blue. The H-bonds between the HIV protease and the inhibitor, including those involving the conserved water molecules, are indicated. (b, c: Reprinted with permission from Dorsey, B. D.; Levin, R. B.; McDaniel, S. L.; Vacca, J. P.; Guare, J. P.; Darke, P. L.; Zugay, J. A.; Emini, E. A.; Schleif, W. A.; Quintero, J. C. *et al. J. Med. Chem.* **1994**, *37*, 3443–3451. Copyright (1994) American Chemical Society.)

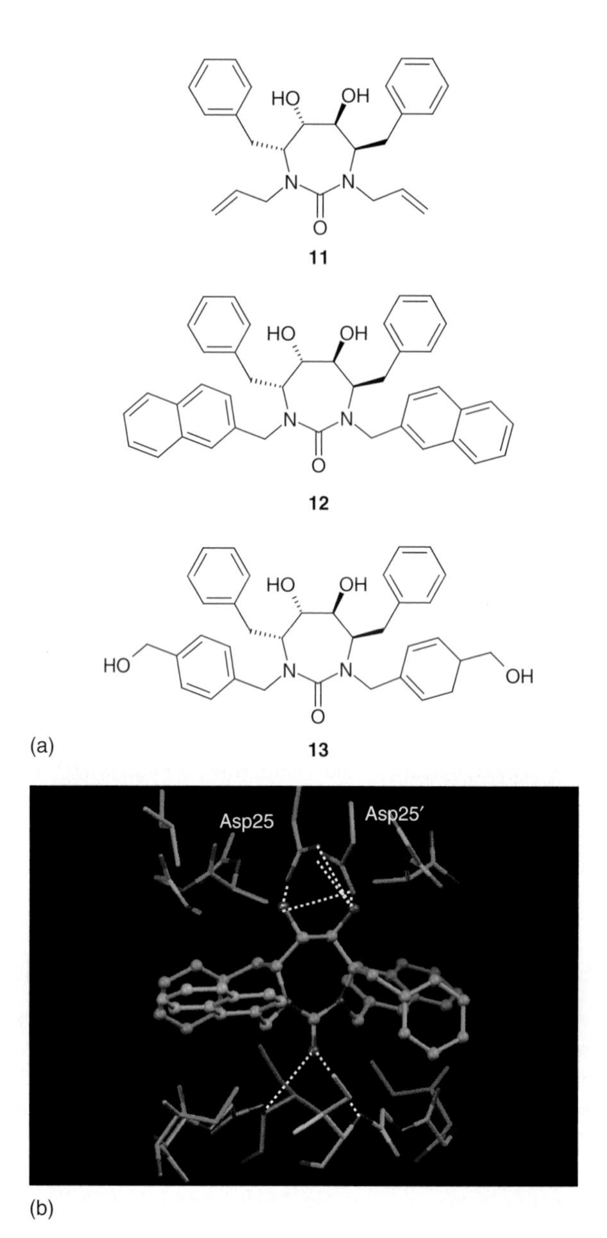

Figure 6 (a) Schematic drawing of compounds 11–13. (b) Interaction between HIV protease compound 12 (1HVR). Monomer 1 is shown in green and monomer 2 in yellow. Compound 12 is represented in blue. The H-bonds between the HIV protease and the inhibitor, including those involving the conserved water molecules, are indicated.

suggesting that they contribute to the overall binding affinity (**Figure 6b**). Based on the crystal structure, different derivates were proposed and synthesized. Again, the authors found a crucial balance between the hydrophobic nature of the inhibitor and its bioavailability. Compound 13 combines high potency against HIV protease and bioavailability and was submitted for clinical trials.

4.24.2.7 Structure-Based Design Leading to a Substrate-Based Inhibitor with a Central Diol as Transition-State Analog

The possession of a twofold axis passing through the active site has led to the design of C2-symmetrical inhibitors in the form of substrate-based transition-state analogs. Besides those reported by Abbott, other transition-state isosteres containing a vicinal diol as central unit have been designed and have shown substantial affinity to HIV protease.[72–75] One of the most active compounds of this class of inhibitors is compound 14, shown in **Figure 7a** (HOE/BAY 793.

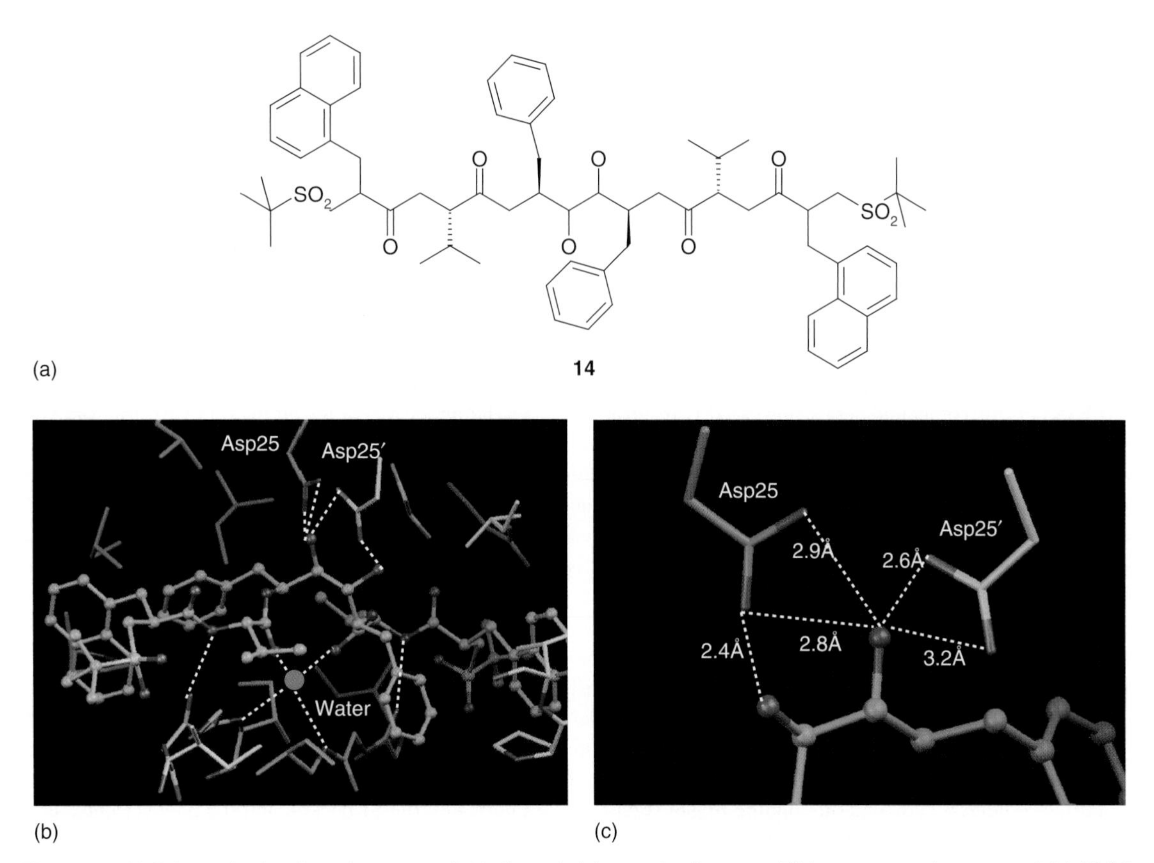

Figure 7 (a) Schematic drawing of compound 14. (b and c) Interaction between HIV protease and compound 14 (1VIK). Monomer 1 is shown in green and monomer 2 in yellow. Compound 14 is represented in blue. The H-bonds between the HIV protease and the inhibitor, including those involving the conserved water molecules, are indicated.

Compound 14 is a C2 symmetric peptidic inhibitor which has an IC_{50} of 0.3 nM and an EC_{50} of 3 nM in cell culture experiments against HIV. In order to determine the binding mode of the vicinal diol to the catalytic aspartates, the crystal structure was solved in two space groups.[76] Earlier reports had either observed a symmetric binding mode to the catalytic aspartates[74] or an asymmetric binding mode.[72,73] Although the protease is symmetric, compound 14 was observed in both crystal forms to bind in only one direction. Compound 14 is bound in a β-strand conformation forming a β-sheet comprised from the flaps on one side of the inhibitor and the residues located in the active site of the protease on the other side. H-bonds were formed between NAsp29′, OGly48′, OGly27′, OGly49, NAsp29, and the backbone of the peptidic inhibitor (**Figure 7b**). Additional H-bonds were seen between the sulfonyl groups and the protease, though the interaction of the sulfonyl with the HIV protease does not seem to be entirely symmetrical. There is a substantial hydrophobic effect due to the burial of the hydrophobic substituents P1, P2, P3, and P1′, P2′, and P3′ into the hydrophobic pockets provided by the protein. The vicinal hydroxyl group was found to interact in an asymmetric binding mode with one hydroxyl group within H-bonding distance of both catalytic aspartates, while the second hydroxyl group is close to only one aspartate (**Figure 7c**). For the orthorhombic crystal form, the root mean square deviation between corresponding Cα atoms in the two protease monomers was 0.67 Å. Most differences were located in flexible surface loops which have different crystal contacts. Within the flap regions, several differences were observed, indicating that the binding of the C2 symmetrical inhibitor has broken the symmetry of the protease itself. An analysis of the differences in the hexagonal crystal form showed an even greater difference between corresponding residues of the two monomers. Most surprisingly, the flaps displayed a significant degree of asymmetrical behavior, with one flap having significantly moved away from the core of the protease. In a general analysis of the mobility of the flap regions, it was found that in most structures deposited in the Brookhaven database,[38] the flexibility of the flaps is limited by local crystal contacts. However, in the structure of compound 14 determined using the hexagonal crystal form, no significant crystal contacts to the flap regions were present and as a result the flexibility of the flaps has increased significantly. This suggests that the mobility and conformational flexibility of the flap residues are important in the functioning of HIV protease and must be considered in the design of drugs against HIV protease.

HIV protease in itself is pseudo C2 symmetric. However, upon binding the nonsymmetric peptide substrate an asymmetry is introduced into the system, indicating that the actual protease–substrate complex was not designed to be symmetric in evolution. The structures of the C2 symmetric diols have shown that asymmetric binding might be predominant not only for the substrate but also for inhibitor complexes.

4.24.2.8 Structure-Based Drug Design Starting from High-Throughput Screening (HTS) Hit

An alternative to the strategy of starting from a substrate-based inhibitor is the use of lead compounds detected in enzyme-based HTS programs. Using this approach, the group at Glaxo had identified a novel series of penicillin-derived C2-symmetric inhibitors of HIV protease.[77,78] A crystal structure of one of these dimeric C2-symmetric inhibitors, compound 15 (**Figure 8a**), complexed to HIV protease, showed a symmetrical binding mode (**Figure 8b**). The SARs of this compound series suggested that optimization of the linker region for interaction with the catalytic aspartates may require the removal of the second penicillin-like unit.[78,79] Thus, shortened inhibitors were synthesized which contain only one penicillin-like moiety such as compound 16. Subsequently, the crystal structure of compound 16 was determined.[80] There appears to be no direct interaction between compound 16 and the side chains of Asp25 and Asp25′. However, an indirect interaction via a water molecule appeared likely since a strong positive density peak between the catalytic aspartates was observed. This water is positioned asymmetrically between the aspartates and is closer to Asp25′ (2.5 Å), but is more than 3 Å away from Asp25. This water molecule is reminiscent of the water molecule found at a similar position in crystal structures of uncomplexed fungal aspartyl proteases[81] and in the structure of uncomplexed HIV protease.[52] Furthermore, compound 16 forms only one direct H-bond to the protein. In order to exploit the potentially favorable interaction with Asp25/Asp25′, analogs bearing a hydroxyl group were proposed based on molecular modeling studies. In addition, an extra lipophilic group supposed to occupy the S1′ pocket was added.[82] One such proposed compound is compound 17. Its improved potency indicated that the desired interactions had been achieved. To identify precisely the interactions of compound 17 with HIV protease, the crystal structure was determined. The thiazolidine ring of compound 17 points into the S1 pocket, and the phenylacetate side chain occupies the S2 pocket (**Figure 8c**). In addition, strong electron density for the hydroxyl group of compound 17 indicates that a good H-bonding with the Asp25/Asp125′ had been achieved. The hydroxyl oxygen atom of compound 17 is located equidistant, about 2.8–3.2 Å, between the oxygen atoms of the two aspartic acid carboxylates. Furthermore, the complex with compound 17 shows, as had been predicted by modeling, the binding of a benzyl group into the S1′ pocket. Also present is an electron density peak corresponding to the conserved water molecule, which is somewhat surprising as there is only one H-bond acceptor from the inhibitor molecule in this region. In an attempt to improve inhibitor binding further, molecular modeling was used to propose modifications that would improve the interaction with the water molecule by providing an additional H-bond acceptor.[82] In addition, it was suggested that positioning a suitable substituent pointing into the S2′ pocket should provide extra favorable interactions. On the basis of this strategy, analogs of compounds 17 were synthesized that contained an additional amide and an extra lipophilic group. This approach led to the synthesis of compound 18, which contains a benzimidazole substituent as well as an additional amide. This inhibitor shows improved potency against HIV protease ($IC_{50} = 3.8$ nM) suggesting an effective binding mode for this series of compounds.

4.24.2.9 Structure-Based Drug Design Leading to Amprenavir: Nonpeptidic Scaffold with a Novel Binding Mode to the Catalytic Aspartates

The design goal of the researcher at Vertex included: (1) low molecular weight; (2) high potency; (3) low cellular toxicity; and (4) aqueous solubility without obligate charges.[83] In addition, the Vertex group favored compounds that require minimal reorganization upon binding to HIV protease. In order to enhance inhibitory potency the binding of the inhibitor to the catalytic aspartates of the enzyme and to the conserved water molecule that mediates inhibitor interactions with the flap was reevaluated. All goals were met by compound 19 (**Figure 9a**), which was the lead compound in a novel class of *N,N*-disubstituted (hydroxyethyl) amino sulfonamides. Compound 19 has a molecular weight of 506 Da and has a higher aqueous solubility than other members of its class. It inhibits the HIV-1 and HIV-2 proteases competitively with K_i, values of 0.60 and 19 nM, respectively. The structure of HIV protease in complex with compound 19 shows clear electron density for all atoms of the inhibitor. Compound 19 binds in a single extended conformation and occupies the S2 to S2′ binding pockets of the enzyme (**Figure 9b**). Extensive interactions are

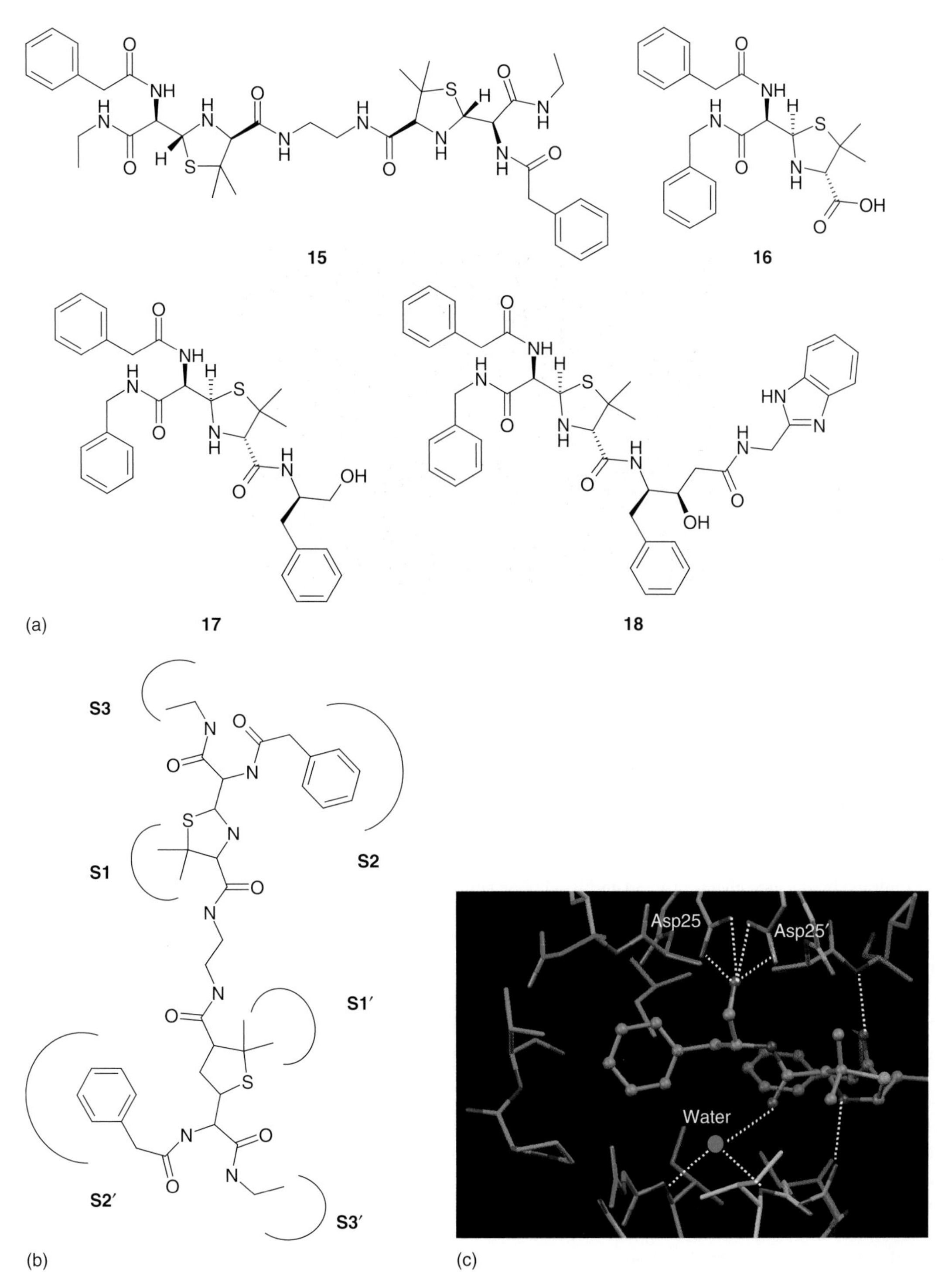

Figure 8 (a) Schematic drawing of compounds 15–18. (b) Schematic drawing of the interaction between compound 15 and HIV protease. The binding pockets S3–S3′ are indicated. (c) Interaction between HIV protease and compound 17 (1HTF). Monomer 1 is shown in green and monomer 2 in yellow. Compound 17 is represented in blue. The H-bonds between the HIV protease and the inhibitor, including those involving the conserved water molecules, are indicated.

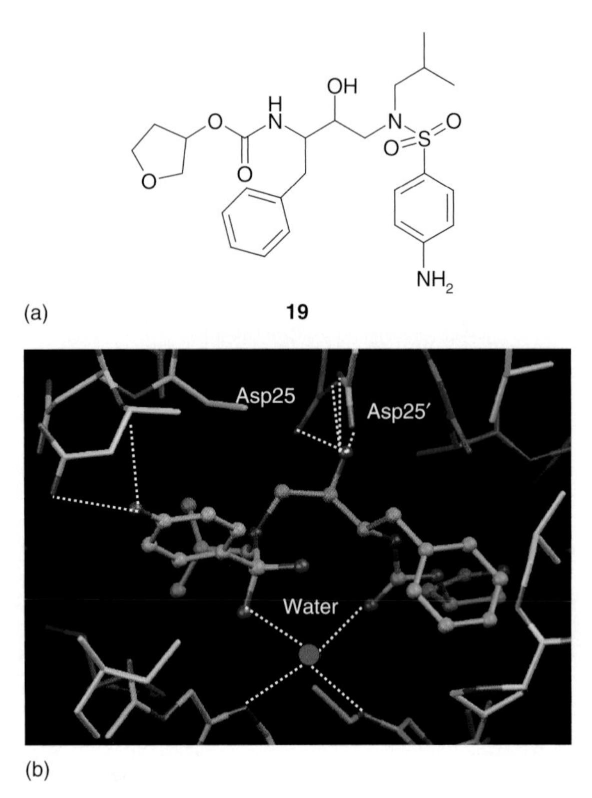

Figure 9 (a) Schematic drawing of compound 19. (b) Interaction between HIV protease and compound 19 (amprenavir) (1HPV). Monomer 1 is shown in green and monomer 2 in yellow. Compound 19 is represented in blue. The H-bonds between the HIV protease and the inhibitor, including those involving the conserved water molecules, are indicated.

evident, with a total of 400 $Å^2$ in solvent-accessible surface area excluded on complex formation. Nearly 60% (235 $Å^2$) of the buried surface area is due to apolar atoms and thus contributes to the hydrophobic binding effect. Strong hydrogen bonding to the well-defined flap water molecule is observed for the carbonyl oxygen of the inhibitor and one of the two sulfonyl oxygens. The second sulfonyl oxygen is positioned asymmetrically and is partially buried in a hydrophobic pocket formed by the side chains of Ile50 and Ile84. The central hydroxyl group of compound 19 forms H-bonds to the side chains of the catalytic Asp25 and Asp25′. The higher aqueous solubility is facilitated by the 4-amino-substituent on the arylsulfonamide. This was designed to form H-bonds to the side chains of Asp30. Compound 19 is orally available and has an IC_{50} of 40 nM in CEM cells infected by HIV-IIIB virus, as assayed by extracellular p24 levels. Compound 19 (amprenavir) was approved for use in the clinic in 1999.

4.24.2.10 Structure-Based Drug Design Leading to Nelfinavir

In the course of the structure-based work at Agouron, a nonpeptidic inhibitor of HIV protease containing a novel 2-methyl-3-hydroxybenzamide moiety was discovered.[84–86] Although compound 20 was highly potent, it was not seen as a viable drug candidate due to its suboptimal antiviral activity (ED_{50} 970 nM) and poor aqueous solubility. However, its relatively low molecular weight and nonpeptidic structure prompted further research using a combination of iterative structure-based design and an analysis of oral pharmacokinetics and antiviral activity. Based on the crystal structure of compound 20 in a complex with HIV protease, together with other structural information as a guide, the Agouron group aimed to improve the in vitro activity and the aqueous solubility of 20.[87] Superposition of the crystal structures of HIV protease complexed to 20 and to compound 6 suggested that the known tertiary amine-containing dipeptide isostere could be combined with a 2-methyl-3-hydroxybenzoic group leading to compound 21 in **Figure 10a**. This resulted in a slightly weaker enzyme inhibitor but markedly improved antiviral agent when compared to compound 20. Additional optimization work focused on the P1 phenylalanine moiety. In earlier crystal structures it had been observed that the *S*-aryl substituents effectively spanned the S1 to S3 sub-sites of HIV protease, providing compounds with substantially improved enzyme-inhibitory activity relative to their phenylalanine

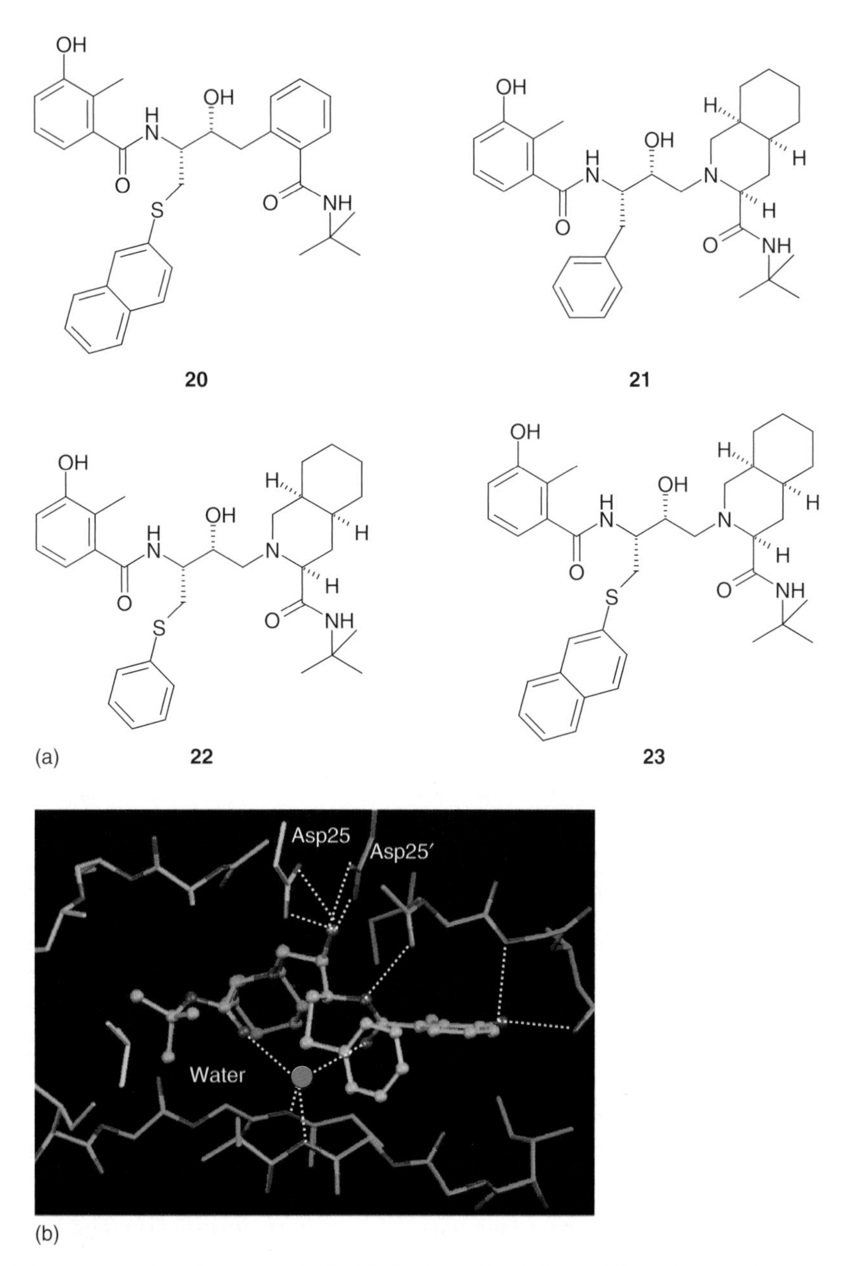

Figure 10 (a) Schematic drawing of compounds 20–23. (b) Interaction between HIV protease and compound 22 (nelfinavir) (1OHR). Monomer 1 is shown in green and monomer 2 in yellow. Compound 22 is represented in blue. The H-bonds between the HIV protease and the inhibitor, including those involving the conserved water molecules, are indicated.

counterparts.[88] The *S*-phenyl analog (compound 22) turned out to be a potent inhibitor of HIV protease (K_i 2 nM). Interestingly, it binds approximately 10-fold better than the phenyl analog (compound 21) and significantly better than the 2-*S*-naphthyl analog (compound 23), which has the same substituent as present in compound 20. In an effort to understand this unanticipated trend in binding affinities, the crystal structure of compound 22 complexed with HIV protease was determined. As predicted from the structure of 20, compound 22 binds to the enzyme in an extended conformation (**Figure 10b**). The *tert*-butylcarboxamide moiety occupies the S2 subsite of HIV protease and the lipophilic dodecahydroisoquinoline ring system fits into the hydrophobic S1 pocket. The central hydroxyl group binds to the catalytic aspartates of the enzyme. The *S*-phenyl group resides in the S1 site and partially extends into the S3 region. The 2-methyl-3-hydroxybenzamide moiety of the inhibitor occupies the S2 pocket, with the *o*-methyl substituents burying Val32 and Ile84. The *m*-phenol function H-bonds to the side chain of Asp30. As seen in many structures, a conserved water molecule is mediating the interactions between the inhibitor and the two backbone amide carbonyls of Ile50. Compound 22 is a potent, nonpeptidic inhibitor of the HIV protease with a desirable combination of potent in vitro and antiviral activity. In vivo studies indicate that compound 22 is orally well absorbed and possesses excellent pharmacokinetic properties in humans. Compound 22 (nelfinavir) was approved for the treatment of AIDS in 1997.

4.24.3 Design of Neuraminidase Inhibitors against Influenza

The threat of a catastrophic outbreak of viral infection by the influenza virus is ever present, as has been recently experienced with avian flu, which it is feared could be transmitted to humans.[89] Vaccines are only partially effective since the influenza virus has a very rapid rate of antigenic variation. Quite regularly, the virus undergoes a major antigenic transformation, resulting in a pandemic strain, such as the strain which killed more than 20 million people in 1918–1919. As a consequence, a pandemic can strike human and other populations, such as seal, horse, or fowl, at any time. Efficient prevention of a pandemic requires a simply administered drug that is effective against all types of influenza, has minimal side effects, and does not induce fast viral resistance. Up until 1998 only two compounds, amantidine and rimantidine, were used against influenza in the clinic. These compounds act by blocking the ion channel function of the virus protein M2.[90] However, due to their side effects and the rapid emergence of resistant influenza strains, they were of only limited use and an alternative target, the influenza NA, has been explored. As a result of this research, there are now two drugs inhibiting NA approved for use in the clinic: zanamivir from GSK and oseltamivir from Hoffmann-La Roche and Gilead Sciences. A crucial breakthrough in this process was the determination of the crystal structure of influenza NA and its exploitation in structure-based drug design programs.

There are three types of influenza virus as classified by their serological cross-reactivity with viral matrix proteins and soluble nucleoprotein (A, B, and C). Only types A and B are known to cause severe diseases to humans. Type B is only found in humans, while type A occurs naturally in birds and mammals such as pigs and horses. Influenza, an orthomyxovirus, is a 100-nm lipid-enveloped virus. On the surface of the influenza virus there are two glycoproteins, hemagglutinin (HA) and NA, which appear as spikes protruding out of the viral envelope. There are between 50 and 100 NA spikes per virus.[91] Electron microscopic images of the NA spikes reveal a mushroom-shaped molecule made up of a box-like head of about $80 \times 80 \times 40$ Å. It has a narrow centrally attached stalk (15 Å wide and 100 Å long) which terminates in a hydrophobic knob anchored in the viral envelope.[92] The spikes can be released by detergents and digested by pronase to release the NA 'heads,' which retain full antigenic and enzyme activity.[93] NA was found to be a tetramer of molecular weight 240 kDa, which reduces to 200 kDa when treated with pronase.[94] NA cleaves terminal sialic acid residues (compound 24) from glycoconjugates,[95] promoting the release of newly formed virus particles from infected cells. Studies with a NA-deficient influenza virus have shown that the mutant virus is still infective but the budding virus particles form aggregates or remain bound to the infected cell surface,[96] showing that NA is an attractive target against influenza.

4.24.3.1 Drug Design before Availability of the Crystal Structure of Neuraminidase

Random screening done before the crystal structure of NA became available did not result in any potent inhibitors of NA. Instead, a classical mechanistic route was used in enzyme inhibitor design (**Figure 11a**). The design of a transition-state analog resulted in a promising lead compound, 2-deoxy-2,3-dehydro-*N*-acetylneuraminic acid or Neu5Ac2en (compound 25), with an ~100-fold greater binding affinity than the product of the reaction, sialic acid or Neu5Ac.[97]

Several analogs of compound 25 were synthesized. However, it was not possible to increase the potency significantly compared to compound 25. The most potent analog, a halogenated derivative, had only a micromolar K_i, and none of these analogs showed antiviral activity in animals.[98]

4.24.3.2 First Crystal Structure of Neuraminidase

The first 3D molecular structure of NA heads was determined in 1983.[99,100] Since then, crystal structures from different subtypes such as A/Tokyo/3/67 and A/RI/5+/57, avian N9 subtypes and influenza type B/Beijing/1/87 have been reported.[101–103] They all have an identical protein fold. Sixty out of 390 residues are conserved including 16 structurally important cysteine residues. The protein fold consists of a symmetrical arrangement of six four-stranded antiparallel β-sheets arranged like the blades of a propeller (**Figure 11b**). The propeller axis is approximately parallel to but tilted away from the circular fourfold axis of the tetramer. This tilt angle varies between the known subtypes. Since 1983, more than 60 structures of NA with and without ligands have been deposited in the Protein Data Bank.[38]

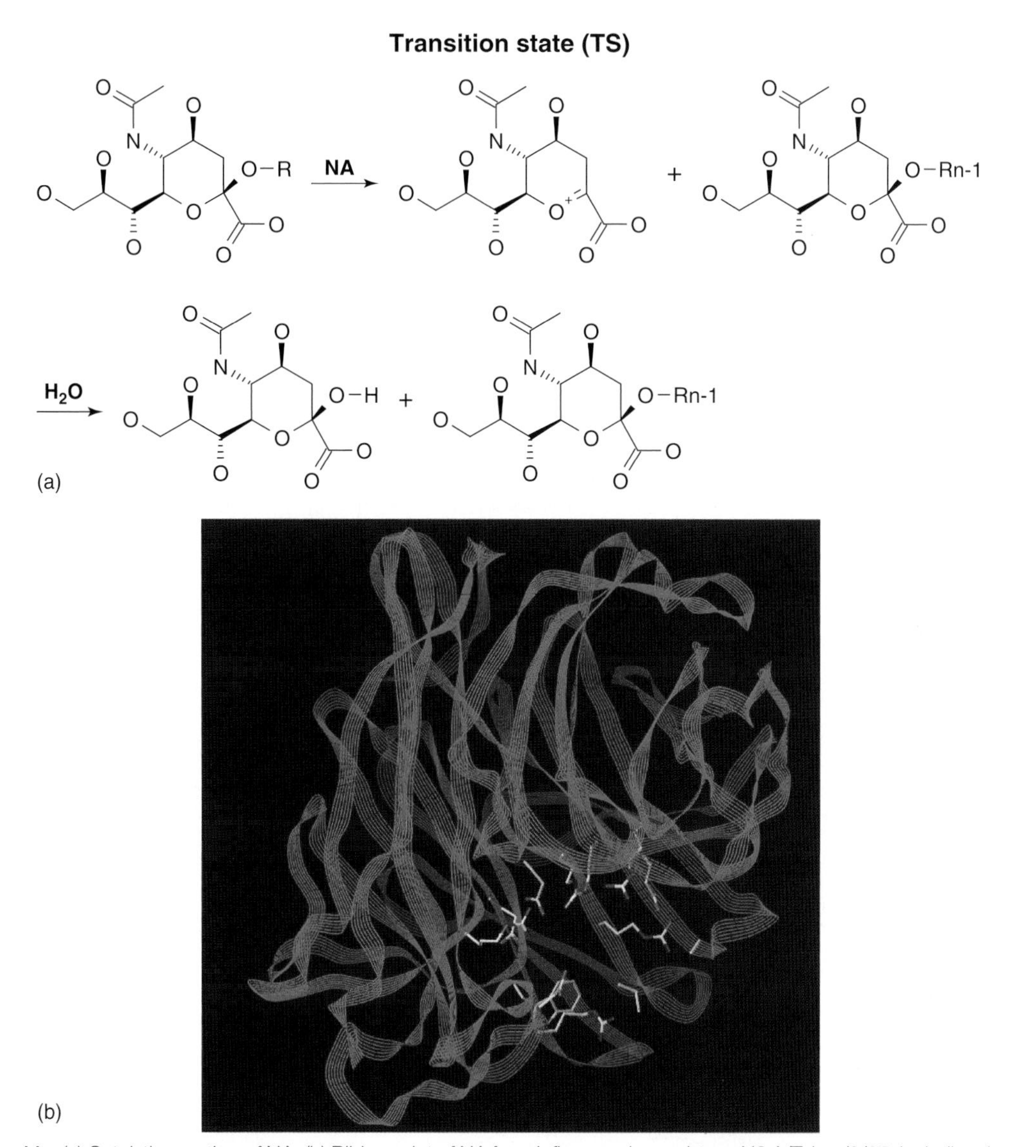

Figure 11 (a) Catalytic reaction of NA. (b) Ribbon plot of NA from influenza virus subtype N2 A/Tokyo/3/67, including the active site residues (1NN2).

Although the identity of amino acid sequences among NA from both type A and type B virus strains has been found to be less than 30%, the enzyme activity of NA is the same among the different strains. A closer analysis of the active site reveals that the amino acids which line and surround the walls of the binding pocket are highly conserved among all influenza strains, increasing the likelihood that active site analogs inhibit all different strains. In addition, the high identity suggests that these amino acids have an important function, making it less likely that resistant viral strains would quickly emerge.

4.24.3.3 Crystal Structure of Sialic Acid Complexed to Neuraminidase

The structure of sialic acid (**Figure 12a**) bound to the active site has been determined for various strains.[103,104] It shows that the α-anomer is bound to the active site in a distorted half-chair conformation (**Figure 12b**). The glycosidic oxygen interacts with the side chain of Asp151. The carboxylate function interacts with the three side chains of Arg118, Arg292, and Arg371 and has an equatorial conformation with respect to the sugar ring. The 4-hydroxyl group is oriented toward Glu119. The NH group of the 5-*N*-acetyl side chain interacts with the active site cavity via a bound water molecule. The oxygen of the 5-*N*-acetyl side chain is H-bonded to Arg152, while the methyl group is located in a hydrophobic pocket

Figure 12 (a) Schematic representation of sialic acid (compound 24). (b) Interaction of NA with sialic acid (1NSC). The side chains of NA are colored in green. Sialic acid is shown in magenta.

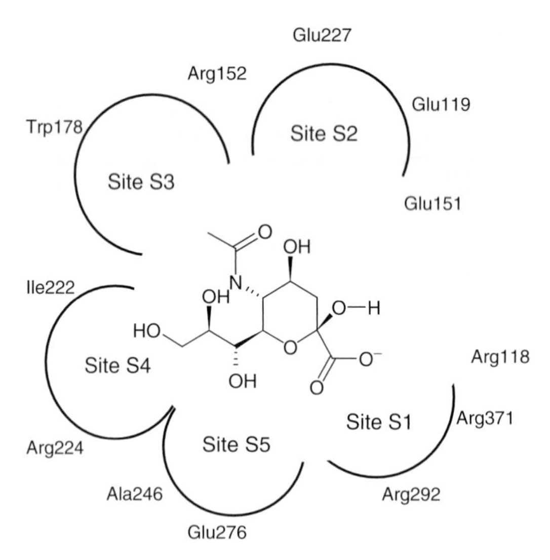

Figure 13 Schematic representation of the interaction of NA with sialic acid. The binding sites S1–S5 are indicated.

near Ile222 and Trp178. The last two hydroxyl groups of the 6-glycerol moiety form H-bonds to the side chain of Glu276. There are no significant movements of side chains between the apo structure and the sialic acid complex, suggesting that the active site is fairly rigid. In addition, a comparison of type A and type B NA with sialic acid complexed to the active site[104] shows that there are no significant differences between active site orientations, except for some minor displacements of Arg224 and Glu276, where the major interactions with the 6-glycerol group of sialic acid occur.

To facilitate the discussion of the binding modes of the inhibitors described below, the active site is divided schematically into five regions, termed subsites S1–S5. These subsites are shown diagrammatically in **Figure 13** and are numbered in counterclockwise fashion using the crystal structure of sialic acid bound to the active site as standard. Site 1 (S1) consists of three residues, Arg118, Arg292, and Arg371, and provides a positively charged electrostatic and hydrogen-bonding environment for anionic substituents from the inhibitor, such as carboxylates. Site 2 (S2) is a negatively charged region of the active site and includes Glu119 and Glu227. Site 3 (S3) is a small hydrophobic region formed by the side chains of Trp178 and Ile222. Site 4 (S4) is not occupied by any portion of sialic acid and is primarily a hydrophobic region formed by the side chains of Ile222, Ala246, and the hydrophobic face of Arg224. Site 5 (S5) is a region of mixed polarity and is comprised of the side chain of Glu276 and the methyl of Ala246. Glu276 can exist in an alternative conformation with its carboxylate ion-paired with Arg224. In this case, Glu276 forms together with Ala246 a hydrophobic pocket within S5. In the following, the amino acids are numbered according to strain B/Lee/40 in order to compare different structural studies more easily.

4.24.3.4 Crystal Structure of Neuraminidase with a Transition-State-Like Analog

The x-ray crystal structures of the transition-state-like analog compound 25 (**Figure 14a**) complexed to NA from different strains such as type A/N2,[105] type B,[106] and N9[107] have been reported. In all structures, compound 25 binds in the active site of NA with the carboxylate oxygen atoms placed in the same location as the carboxylate of sialic acid (**Figure 14b**). In addition, the 5-*N*-acetyl, 4-hydroxy, and 6-glycerol are positioned isosterically in the active site when compared to the positioning of the sialic acid. Not surprisingly, the affinity of compound 25 is similar for all types of sialidases, including mammalian ones, confirming again that inhibitors binding to the conserved region of the active site will be active against the different influenza strains.

4.24.3.5 Structure-Based Design Leading to Zanamivir: Exploring the Binding Pocket of the Transition State-Like Analog

A close analysis of the binding pocket was performed using the GRID program in order to detect energetically favorable binding sites for different functional groups (*see* 4.11.3 Characterization of Protein-Binding Sites and Ligands Using Molecular Interaction Fields).[108] This analysis revealed a conserved negatively charged pocket in the complex of NA

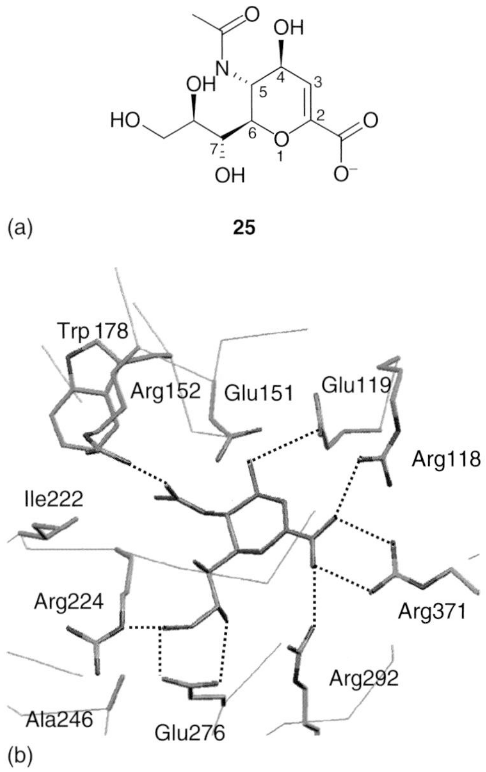

Figure 14 (a) Schematic representation of the transition-state-like analog Neu5Ac2en (compound 25). (b) Interaction of NA with the transition-state analog compound 25 (1F8B). The side chains of NA are colored in green. Compound 25 is shown in magenta.

with compound 25 that could be filled by a basic substituent at the 4-position of compound 25. This substituent would interact with Glu119 and Glu227. As a result of this finding, 4-amino-Neu5Ac2en (compound 26 in **Figure 15a**) and 4-guanidino-Neu5Ac2en (compound 27 in **Figure 15a**) were synthesized and indeed turned out to be very potent inhibitors against influenza of types A and B. While the K_i of compound 25 was only 4 μM, compound 26 has a K_i of 10^{-8} and compound 27 a K_i of 10^{-10}. This nicely shows how the exchange of a single side chain can turn a weakly binding into a very potent inhibitor. Subsequently, the crystal structures of compounds 26 and 27 complexed to NA were determined. The binding mode was very similar to that found in the x-ray crystal structure of compound 25 complexed with NA. As predicted, the new 4-amino moiety of compound 26 forms a salt-bridge to Glu119 (**Figure 15b**). The 4-guanidino function in compound 27, on the other hand, forms charged H-bonds to the backbone of Trp178 and the side chain of Glu227 (**Figure 15c**).

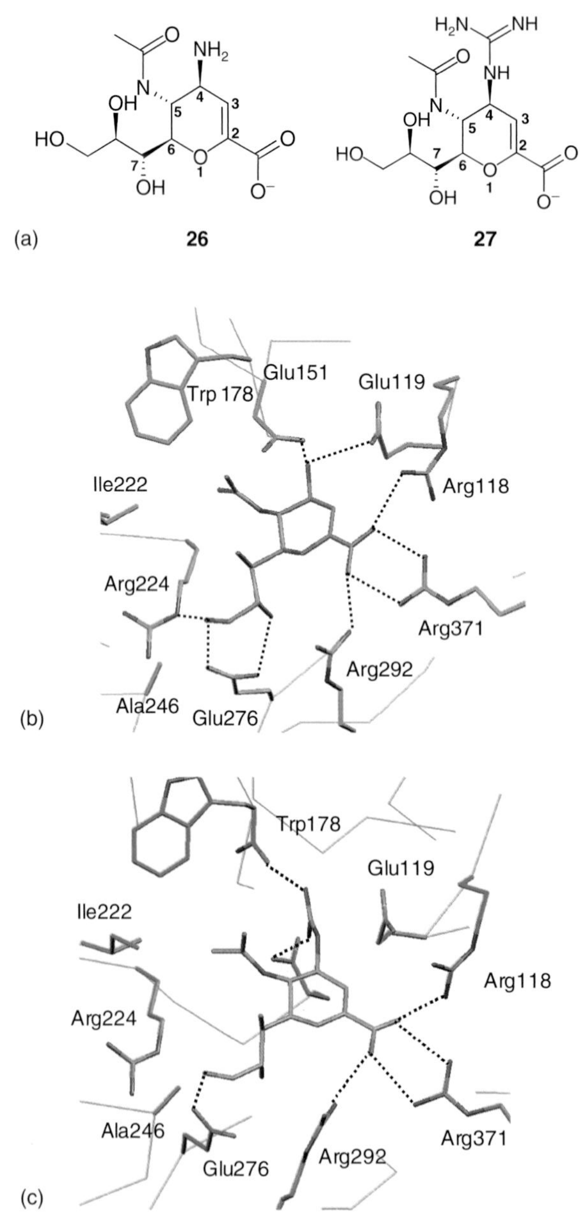

Figure 15 (a) Schematic representation of compounds 26 and 27. (b) Interaction of NA with compound 26 (1F8C). The side chains of NA are colored in green. Compound 26 is shown in magenta. (c) Interaction of NA with compound 27 (1A4G). The side chains of NA are colored in green. Compound 27 is shown in magenta.

Compound 27 has been evaluated in human clinical trials and has shown efficacy in phase II studies in both prophylaxis and treatment of influenza virus infections. It was approved in 1998 by the US Food and Drug Administration and is available in the clinic (zanamivir or Relenza, Glaxo-Wellcome). However, poor oral bioavailability and rapid excretion precluded compound 27 as a potential oral agent against influenza infection and the compound has to be administered by either intranasal or inhaled routes. Further research has focused on the design of compounds possessing good oral bioavailability.

4.24.3.6 Structure-Based Drug Design Leading to Oseltamivir: Identification of a New Apolar Pocket

The first move toward drugs with improved bioavailability involved the design of a new class of compounds with a carbocyclic scaffold in place of the dihydropyran ring of the Neu5Ac2en system. It had been expected that the carbocyclic ring would be chemically more stable than the dihydropyran ring and easier to modify for further optimization of the antiviral and pharmacological properties of the compound class. Considering the flat oxonium cation in the transition state, the cyclohexene scaffold was selected as a replacement for the oxonium ring so as to keep conformational changes to a minimum.[109,110]

Crystallographic studies of sialic acid and its analogs bound to NA indicate that the C7 hydroxyl of the glycerol side chain does not interact with any amino acids of NA and can therefore be eliminated. Thus, the CHOH group at the C7 position of the glycerol side chain in the Neu5Ac system was replaced with an oxygen atom. It was also noted that some carbons of the glycerol chain make hydrophobic contacts with the protein and it was hoped that the optimization of this hydrophobic interaction would lead to new NA inhibitors with increased lipophilicity while maintaining potent NA-inhibitory activity. This consideration is especially important for designing orally bioavailable drugs since balancing lipophilicity and water solubility is thought to be as critical as the size of the molecule when considering absorption from the intestinal tract. For optimization of the hydrophobic effect, the size and shape of the substituents play an important role in the binding affinity. Therefore, a systematic variation adding various aliphatic side chains was carried out. Some of these compounds, such as the three pentyl analog, compound 28 in **Figure 16a**, turned out to be

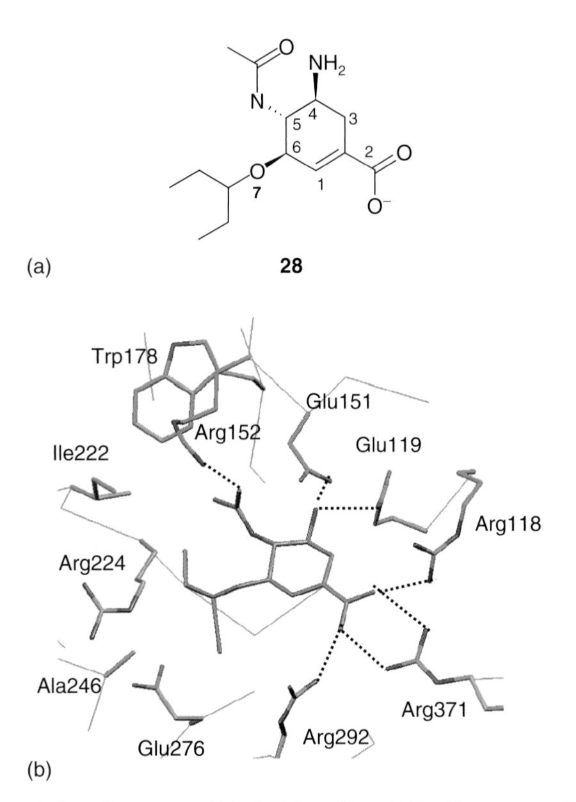

Figure 16 (a) Schematic representation of compound 28. (b) Interaction of NA with compound 28 (2QWK). The side chains of NA are colored in green. Compound 28 is shown in magenta.

very potent NA inhibitors. Subsequently, the structure of compound 28 complexed to NA was determined. As shown in **Figure 16b**, the interaction between NA and the carboxylate at the 2-position (S1) and the amide at 3-position (S3) were not very different from those found in the complexes between NA and compound 25 or other sialic acid-based inhibitors. However, the 3-pentyloxy side chain at the 4-position was positioned against a large hydrophobic surface created by the hydrocarbon chains of Glu276, Ala246, Arg224, and Ile222 (S4 and S5 pocket). Compared to other complexes with sialic acid-based inhibitors, the side chain of Glu276 is forced to change its conformation in order to accommodate the large 3-pentyl group. The hydrophobic effect caused by the burial of the aliphatic moieties in the NA active site explains well the SARs observed for the different alkyl chains. There is a remarkable correlation between the shape of the aliphatic side chains and NA-inhibitory activity resulting from the different size of the hydrophobic effect due to the burial of these aliphatic side chains.

Compounds that exhibited potent NA-inhibitory activity were further evaluated in cell culture by a plaque reduction assay using an influenza A (H1N1) strain. The ethyl ester of compound 28 exhibited good oral bioavailability in several animals and demonstrated oral efficacy in the mouse and ferret influenza model. On the basis of its potency, its in vitro and in vivo activity and very favorable pharmacological properties, the ethylester of compound 28 (oseltamivir) had been selected as a clinical candidate for the oral treatment and prophylaxis of influenza infection[109] and was approved for use in the clinic in 1999.

4.24.3.7 Structure-Based Drug Design Leading to an Orally Available Drug: Exploring a New Scaffold

A compound with a cyclopentane scaffold (compound 29 in **Figure 17a**) with NA-inhibitory potency had been described earlier by the researchers BioCryst Pharmaceuticals.[111] Even though the central ring quite different from that in compound 25, a similar affinity to NA was observed for compounds 25 and 29. The superposition of the NA complexes with bound compounds 25 and 29, respectively, showed that the cyclopentane ring is significantly displaced from the pyranose ring of compound 25. However, the interacting groups superimpose quite nicely, demonstrating that it is not the absolute position of the central ring but rather the relative positions of the interacting groups that is of importance (**Figure 17b**).

By analogy to the derivates of compound 25, the hydroxyl group was replaced with a guanidine group, leading to compound 30. Further cycles of structure-based drug design in which hydrophobic substituents were added resulted in compounds with several asymmetric carbon atoms. In order to identify the most active isomer, NA crystals were soaked in a solution containing all isomers and subsequently the x-ray structure of the soaked crystal was determined. The observed difference in electron density in the active site of NA displayed unambiguously the stereochemistry of the active isomer. The structure of the bound isomer in the active site revealed that the carboxylic acid and 1-acetylaminopentyl group are *trans* to each other while the guanidino and carboxylic acid groups are *cis* to each other. In the next step, the possibility of attaching a hydrophobic substituent was explored, leading to compound 31. X-ray crystallographic studies of compound 31 bound to influenza NA showed that the same isomer is bound in both influenza A and B NA active sites. Interestingly, the *n*-butyl side chain of compound 31 adopts two different binding modes in the two structures (**Figure 17c**). The *n*-butyl side chain in the influenza B NA active site is positioned against a hydrophobic surface formed by Ala246, Ile222, and Arg224 (S4). However, in the active site of strain A, the *n*-butyl side chain occupies a hydrophobic pocket formed by the reorientation of the side chain of Glu276 (S5). Compound 32 was designed to take advantage of both hydrophobic pockets in the active site. Again, this compound was synthesized as a racemic mixture and the crystal structure was used to identify the active isomer. The crystal structure confirmed the predicted binding mode (**Figure 17d**). Compound 32 turned out to be a highly potent compound which retained inhibitor activity against a zanamivir-resistant Glu119Gly variant of influenza A NA.

The oral efficacy of compound 32 was shown in a mouse influenza model. Infected mice were treated with compound 32 4 h prior to virus exposure. Seven out of 9 mice treated with compound 32 survived, while only 1 out of 9 mice in the control group survived. Compound 32 (BCX-1812) was efficacious at a dose as low as 0.1 $mg\,kg^{-1}\,day^{-1}$ b.i.d.[111]

4.24.3.8 Structure-Based Drug Design Leading to Novel Inhibitors with a Cyclopentane Scaffold: Combining Combinatorial Chemistry and Structure-Based Drug Design

The first studies of complex structures of NA had identified the unique and complementary binding site for carboxylate-based inhibitors in which the carboxylate was H-bonded to the arginines in the S1 pocket. Therefore, a subset of

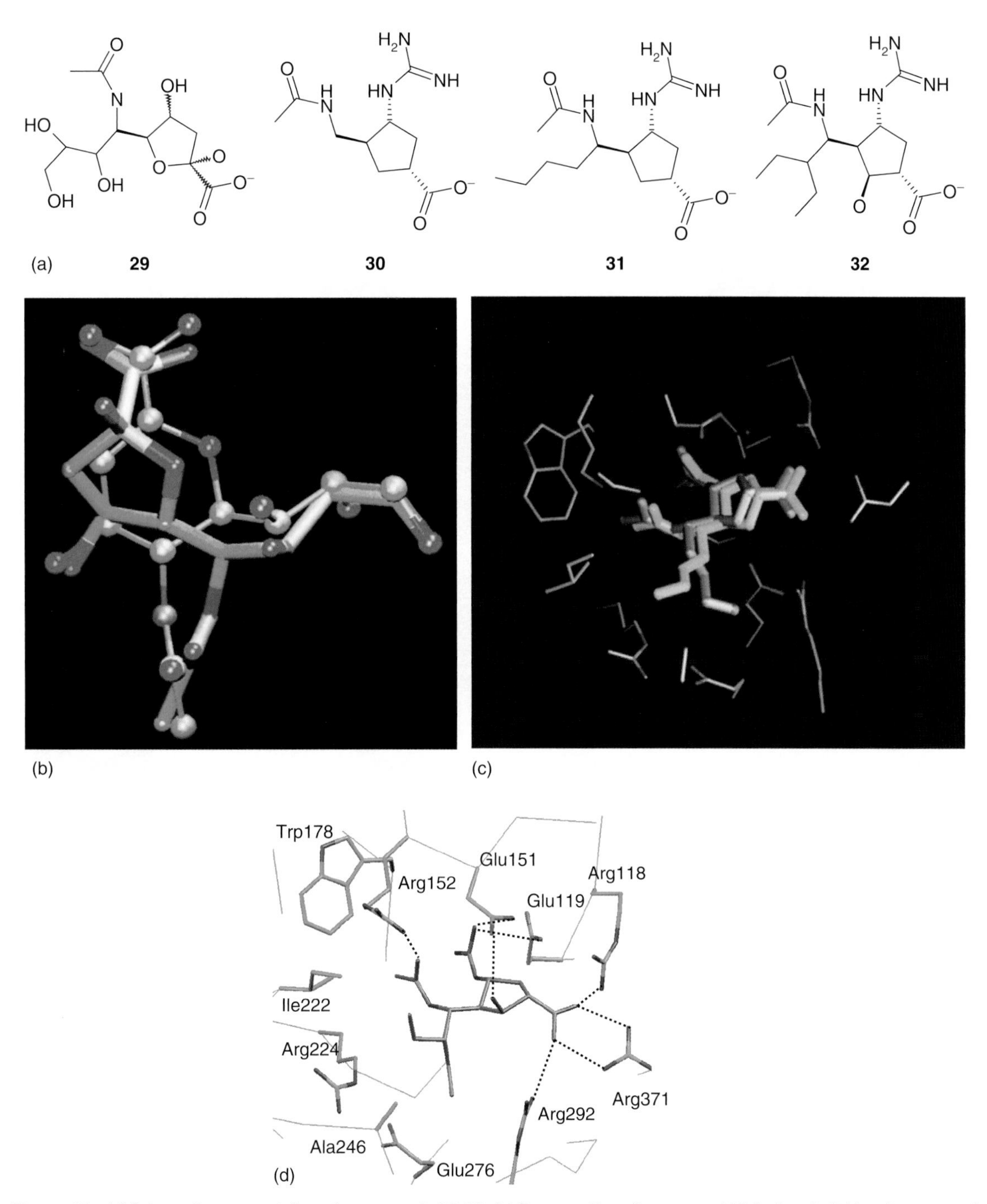

Figure 17 (a) Schematic representation of compounds 29–32. (b) Superposition of compound 25 (ball-and-stick) and compound 29 (liquorice) in the conformations of the molecules when bound to influenza NA. (c) Superposition of compound 31 bound to influenza A (green) and influenza B (yellow) NA. The protein side chains are from influenza A NA N9 active site and are shown in blue. (d) Interaction of NA with compound 32 (1L7F). The side chains of NA are colored in green. Compound 32 is shown in magenta. (b, c: Reprinted with permission from Babu, Y. S.; Chand, P.; Bantia, S.; Kotian, P.; Dehghani, A.; El-Kattan, Y.; Lin, T.-H.; Hutchison, T. L.; Elliott, A. J.; Parker, C. D. *et al. J. Med. Chem.* **2000**, *43*, 3482–3486. Copyright (2000) American Chemical Society.)

300 organic carboxylates from the Abbott compound collection was screened against NA.[112] The most potent lead compound (compound 33 in **Figure 18a**) had a K_i of 58 μM against NA and contained a five-membered ring scaffold. The first reported efforts to improve the potency of compound 33 using a combination of molecular modeling and combinatorial chemistry had led to only a 50-fold improvement of the inhibitor potency with a best K_i of 1.1 μM.

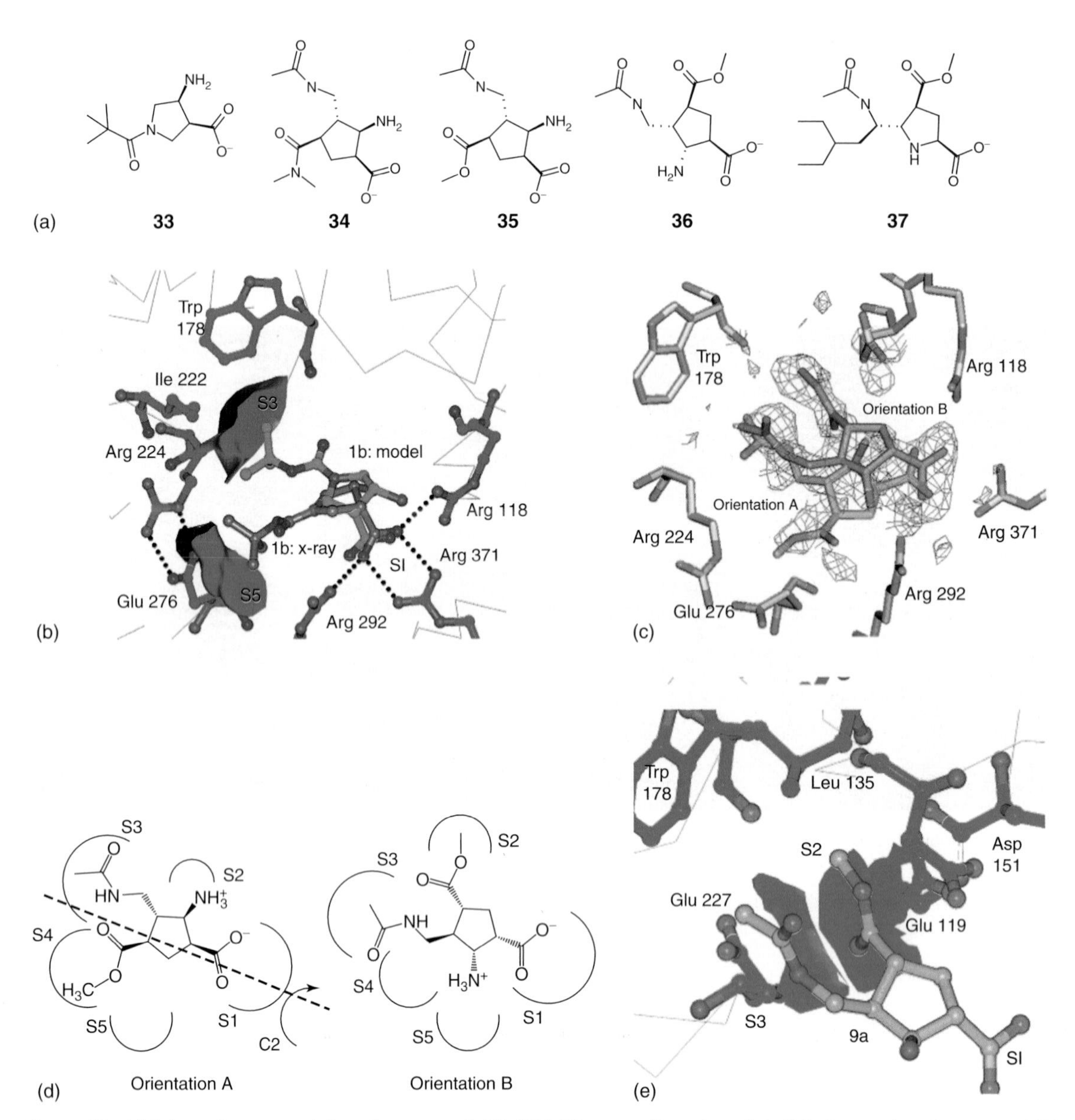

Figure 18 (a) Schematic representation of compounds 33–37. (b) Superposition of predicted binding modes of compound 33 (blue) and its binding mode observed in x-ray structure (orange). (c) Two different binding modes are observed in case of compound 35. The observed electron density for compound 35 with orientation A (pink carbon atoms) and orientation B (orange) fit to the electron density. (d) Diagram of binding orientations A and B for compound 35. The orientations are approximately related by a C2 axis (shown in orientation A as a dashed line). (e) Interaction of NA with compound 36. The side chains of NA are colored in green. Compound 36 is shown in magenta. (b, c, d, e: Reprinted with permission from Stoll, V.; Stewart, K. D.; Maring, C. J.; Muchmore, S.; Giranda, V.; Gu, Y.-G.; Wang, G.; Chen, Y.; Sun, M.; Zhao, C. *et al. Biochemistry* **2003**, *42*, 718–727. Copyright (2003) American Chemical Society.)

Additional improvement was not possible and the SAR of the compound class was inconsistent.[112] As a consequence, more crystal structures of analogs were determined. During the work of the Abbott group, approximately 3–10 high-resolution (1.9–2.8 Å) compound enzyme crystal structures were determined per week, typically with 12–24 h turnaround time. Ultimately > 120 individual crystal structures were obtained in the NA research program at Abbott.[113]

In order to optimize the lead, models of the screening lead bound to the NA active site were prepared by manual docking and subsequent energy minimization of both enantiomers into a rigid active site. The carboxylate was supposed to fit into S1, while the *t*-butyl group was oriented within the hydrophobic portion of S3 so as to overlap with the hydrophobic region of the acetyl group of compound 25. In the models of both enantiomers there were suboptimal contacts. On the basis of this initial modeling, the proposed *t*-butyl interaction within S3 was judged to be suboptimal and a combinatorial library of urea analogs with alternative alkyl groups was evaluated. This produced optimized N-Et, N-iPr analogs with 55-fold increased K_i potency.

In the meantime, the crystal structure of compound 33 complexed to NA was determined. In contrast to the prediction, the crystal structure showed that, while the carboxylate bound as anticipated, the *t*-butyl group did not bind within S3 as modeled above, but instead was located in the S5 pocket (**Figure 18b**). Interestingly, the computer docking experiment using a rigid active site could not have anticipated this binding mode because of a conformational change seen for Glu276. It was observed that S3 was unoccupied in the crystal structure of the complex. In order to increase the potency, a new series of tetra-substituted pyrrolidines was designed that would add the acetamide fragment located at the 5-position in Neu5Ac2en which points into the S3 subsite on to the scaffold. Unlike the six-membered ring inhibitors, an extra methylene is required to span the required distance between the acetamide moiety. Adding this functional group to the best compounds of the first series did not lead to a breakthrough although the crystal structure confirmed that the compounds bound as had been anticipated. An example is compound 34 which has a K_i of 260 μM. However, in striking contrast to the methyl amides above, a compound with a methyl ester at this position (compound 35) was as potent as the most potent amide even though it had no hydrophobic substituents presumably filling the S5 pocket. Interestingly, the crystal structure of compound 35 showed that it binds in two quite different binding modes (**Figure 18c** and **d**), suggesting that the methyl ester makes in one orientation equally good interactions in the S2 pocket, thereby leaving the S5 pocket empty. To understand better the two different binding modes of compound 35, a diastereomeric analog was prepared (compound 36). With a K_i of 0.7 μM, this compound exhibited a 30-fold gain in potency relative to compound 35. The crystal structure of this compound confirmed that it binds in the anticipated binding mode, i.e., only in orientation B (**Figure 18e**). Thus, compound 36 was the first compound which had unambiguously a methylester in the S2 pocket. Subsequent optimization of this compound by adding an alkyl moiety such that it points into the hydrophobic S3 and the neighboring S5 pockets yielded inhibitors with K_i values up to 0.0008 μM (compound 37).

In summary, combinatorial and structure-based medicinal chemistry strategies were used together to advance a lead compound with K_i 58 μM to give an analog with K_i of 0.8 nM against influenza NA (A/Tokyo/67), representing a $>$70 000-fold enhancement in potency. Protein crystal structures revealed that inconsistent SAR data resulted from different binding orientations of the five-membered ring inhibitor cores from one series to another. Binding modes for a series of compounds showed up to 180° variation in orientation of the five-membered ring within the active site. Potent analogs were only achieved with chemical series that were observed to bind in the same orientation and yielded consistent SAR. This explains why earlier attempts to achieve high-affinity inhibitors starting from weak screening leads were not successful.

4.24.3.9 Structure-Based Drug Design Investigating Differences in the Active Site: Exploring Resistant Strains

It has been observed in structure-based drug design projects that a single amino acid difference within the binding pocket can have a dramatic impact on the potency of inhibitors. These single-site differences can occur either naturally or alternatively can be induced by the extensive use of a particular inhibitor. Therefore, the NA inhibitors described above have been designed such that they interact only with the conserved active site residues. Although the active site of the influenza virus has been conserved in all known field strains, the possibility of induced drug resistance needs to be addressed. NA inhibitors, if used extensively as antiinfluenza drugs, will apply selection pressure on the active site residues of the virus for the first time in its evolutionary history. Experience with influenza and other viruses, in particular HIV, have shown that drug-resistant mutants arise very rapidly, resulting in the effectiveness of antiviral drugs being short-lived.[114] In order to locate possible resistant mutations ahead of them appearing in a large virus population, drug-resistant mutants have been created by in vitro screening of the virus in the presence of NA inhibitors.

A zanamivir-resistant NA mutation has been isolated that stems from a single-site residue mutation. In this strain Glu119 is mutated to Gly119. The crystal structure of this mutant and of its complex with compound 27 has been determined.[115] The location of compound 27 in the complex with the mutant enzyme is isosteric compared to the drug/wild-type complex. The only difference is in the interaction of compound 27 with residue 119. The structures suggest that the decrease in inhibitor binding arises from the loss of stabilizing interaction of the 4-guanidino group of compound 27 with the resistant NA and alterations in the solvent structure of the active site. In contrast, compound 28 has an amine moiety at the 4-position and does not form the close association with Glu119 seen with the guanidinium group of compound 27. Therefore, it is unaffected by the Glu119Gly mutation.

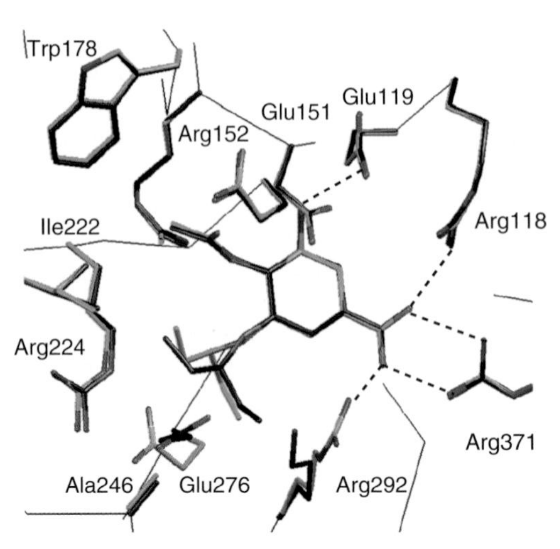

Figure 19 Superposition of compound 28 bound to wild-type NA (green) (2QWK) and to the R292 K mutant (black) (2QWH).

Another resistant strain which has been reported has a mutation at position 292. Here, the wild-type arginine has mutated to a lysine. This Arg292Lys mutation affects the binding of the bulky hydrophobic side chain in compound 28 by several orders of magnitude. In contrast, its effect on the binding of compound 27 is only small. A comparison of the crystal structures of compound 28 bound to the wild-type and the mutant protein,[116] respectively, shows that the binding of compound 28 requires reorientation of Glu276 to form a salt link to Arg224 in order to accommodate the hydrophobic substituent (**Figure 19**). In the case of the Arg292Lys variant, the salt link between Lys292 and Glu276 must be broken in order to allow the reorientation of Glu276. However, the energy penalty for achieving this conformation seems to be too great. As a consequence, the hydrophobic binding pocket for the pentyl ether of compound 28 is not created, resulting in 10 000-fold resistance to compound 28 in an enzyme assay.

Many more reports describing the design of NA inhibitors using structure-based drug design have been published.[117–119] For instance, attempts were made to replace the pyranose ring of Neu5Ac2en with a flat benzene ring[120] and have resulted in nanomolar binding inhibitors.[121] A sialic acid-derived phosphonate analog has been reported that inhibits different strains of influenza virus NA with different efficiencies.[122] Efforts are still ongoing (see, for instance, [123]) and may result in new compound classes for use in the clinic. In addition, investigations have been initiated to find inhibitors of NA from different organisms such as *Vibrio cholerae*[124] and inhibitors of microbial sialidases (see, for review, [125]).

4.24.4 Design of Inhibitors Binding to SH2 Domains

Protein tyrosine kinases and phosphatases have been implicated in mediating a variety of different intracellular activities such as cell proliferation, migration, and differentiation. Many such proteins are able to bind their cognate protein ligands selectively and initiate a cascade of signaling events through key modular domains that control protein–protein interactions.[126] One such domain, the SH2 domain, has been determined to play a crucial role in many signaling pathways by recognizing the phospho-tyrosine (pTyr) sequences of cognate proteins.[127–129] The fact that short pTyr-containing peptides are sufficient to compete with larger protein ligands for SH2 domain binding has prompted researchers in both academia and industry to develop inhibitors targeting the SH2 domains of clinically relevant proteins such as Grb2, STAT3, Lck, Zap70, PI3 K, and Src.

Src is the prototype of a nonreceptor protein tyrosine kinase. Its activation in human tumors such as breast cancer has been reported.[130] In addition, Src-deficient mice exhibit osteoporosis or hypertrophy of bones, indicating disturbed osteoclast function.[131,132] Thus, inhibiting Src function in vivo may inhibit tumor growth and be useful for treatment of osteoporosis. Nonreceptor tyrosine kinases generally require SH2 domains for efficient substrate phosphorylation.[129] Therefore SH2 inhibition is likely to suppress phosphorylation and downstream signaling. Src kinase is normally held in an inactive conformation via an intramolecular interaction between its SH2 domain and a tyrosine-phosphorylated C-terminal residue, pTyr527. This is shown in **Figure 20a**, which shows the 3D structure of src kinase as a ribbon plot, including the interaction of the phosphorylated tyrosine pTyr527 with the SH2 domain.[133] The disruption of this interaction leads to an activation of the kinase. Tyr527 mutants of Src, such as

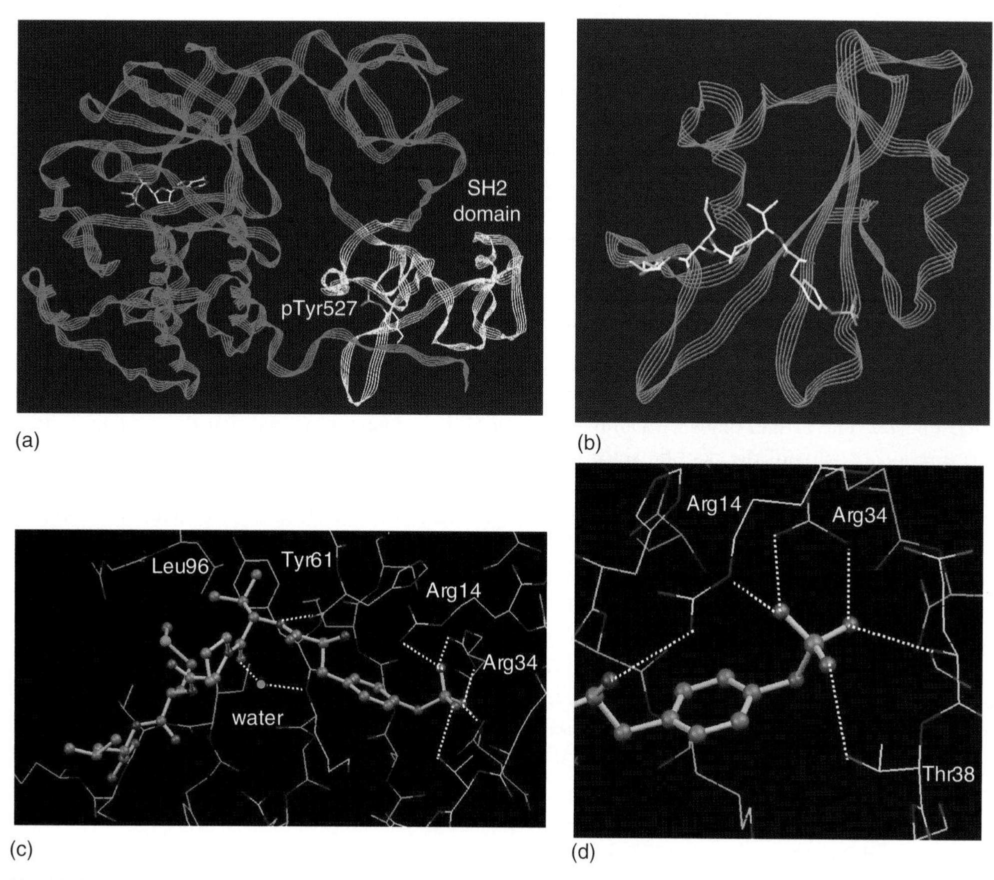

Figure 20 (a) Ribbon plot of src (1FMK) including the interaction between pTyr527 at the C-terminal and the SH2 domain. The SH2 domain is indicated in white. (b) Ribbon plot of the SH2 domain of src (1SHA). The low-affinity peptide is shown in white. (c) Interaction of peptide pTyrValProMetLeu with the SH2 domain of src (1SHA). (d) Close-up of the interaction of pTyr with phospho-tyrosine pocket of the SH2 domain of src (1SHA).

Tyr527Phe, are known to transform an inactivated Src kinase into the active state. Thus, small molecules designed to inhibit the SH2-mediated protein–protein interactions have promise as pharmaceutical agents to block the signaling pathway of src.

4.24.4.1 First Crystal Structure of SH2 Domain

SH2 domains are small protein modules of approximately 100 amino acids. The amino acid numbering throughout this chapter will be done based upon the structures in Lange *et al.*[134] The first crystal structure of the SH2 domain of pp60src was reported in 1992.[135] The fold of the SH2 domain belongs to the $\alpha + \beta$ class and forms a compact flattened hemisphere. The core structural elements comprise a large central hydrophobic antiparallel β-sheet and two flanking α-helices (**Figure 20b**). The first SH2 domain structure was obtained by crystallization in the presence of a low-affinity phosphopeptide. This crystal structure showed that there are two major interaction sites (**Figure 20c**). A positively charged phospho-tyrosine binding site (pTyr-pocket) is formed by three strands of the β-sheet, a loop connecting two strands of the β-sheet (residues 34–42), and an arginine from a flanking α-helix (Arg14). The phosphate makes H-bonds to the side chains of Arg14, Arg34, Ser36, Thr38, and the backbone amide of Glu37 (**Figure 20d**). Remarkably, this pTyr pocket is highly conserved in all SH2 sequences, reflecting the fact that all SH2 domains bind pTyr. The second important binding site within the SH2 domain is a so-called specificity-determining pocket in which side chains of peptide residues to the C-terminal end of pTyr are located. In case of the SH2 domain of src, the specificity pocket is the pocket into which the pTyr + 3 substituent points. This pocket is formed by amino acids Tyr61, Ile73, Thr74, Tyr89, and Leu96 and thus has a mainly hydrophobic character. An analysis of SH2 sequences indicates that the amino acids pointing into the pTyr + 3 pocket differ significantly between individual SH2 domains.

This explains the selectivity by which a given SH2 domain is able to recognize its cognate protein peptide, even though the common recognition site, i.e., the pTyr pocket, is so highly conserved. In addition to the interactions described above, there are H-bonds between the SH2 domain and the backbone of the bound peptide. The peptide P+1 backbone amide forms a H-bond with the backbone carbonyl of His60. The backbone carbonyl at the peptide P+1 position forms H-bonds to well-ordered water molecules, which in turn bind to the protein backbone amide of Lys62. Since 1992, crystal structures have been described from a variety of different SH2 domains, including the SH2 domains of syp,[136] p56 lck kinase,[137] and Grb2 adaptor protein,[138,139] and all have shown a similar binding of their substrates to the respective SH2 domain.

4.24.4.2 Crystal Structure of the SH2 Domain with a Cognate High-Affinity Peptide

In 1993 a crystal structure of the src SH2 domain bound with a high-affinity peptide has been reported.[140] The 11-residue phosphopeptide contains the motif pTyrGluGluIle and thus includes the cognate sequence recognized by the *scr* SH2 domain. The 11-mer binds with an IC_{50} of 800 nM. It had been proposed that the binding mainly involves the cognate sequence parts of the peptide since the affinity of the much smaller Ac-pTyrGluGluIle-NH_2 peptide is with an IC_{50} of 4300 M,[141] very similar to that of the crystallized 11-mer peptide. The crystal structure confirms that, indeed, only these four residues make extensive contacts with the protein (**Figure 21**). As found in the first structure with the low-affinity peptide,[135] the phosphate group forms part of an intricate H-bond network with the pTyr pocket of the protein. The glutamate side chains at positions P+1 and P+2 point toward the solvent and seem not to contribute directly to the binding of the peptide. The ileucine at the peptide position P+3 points into the hydrophobic pocket formed by residues Tyr61, Ile73, Thr74, Tyr89, and Leu96. The H-bond network between the SH2 domain and its cognate peptide is similar to that found for the low-affinity peptide.[135] The amide at the peptide position P+1 forms an H-bond with the backbone carbonyl of His60. The backbone carbonyl at the peptide P+1 position and the backbone amide at position P+3 form H-bonds to well-ordered water molecules, which in turn bind to the protein backbone residues Lys62N and Ile73O, respectively.

Since the first crystal structure, a considerable number of structures of SH2 domains with peptidic and nonpeptidic inhibitors have been reported. With few exceptions,[142,143] the peptidic and nonpeptidic ligands bind in an extended conformation resembling a two-pronged plug. A flat protein surface is located between the two major interaction sites, i.e., the hydrophilic pTyr pocket and the hydrophobic pTyr+3 pocket. Hence, most of the corresponding inhibitors can be described as scaffolds which interact with the flat protein surface and possess two substituents pointing into the pTyr and the pTyr+3 pockets. In addition, the H-bond network, involving the H-bond between the amide of the pTyr+1 residue and the peptide carbonyl of His60 and the water molecules bridging the backbone atoms of the peptide inhibitor and SH2, is highly conserved.

4.24.4.3 Structure-Based Design Leading to Nonpeptidic Inhibitors

Initially, few inhibitors had been described which combined a nonpeptidic character with an affinity comparable to the natural substrate pTyrGluGluIle. Among the first nonpeptidic inhibitors were those reported by Lunney *et al.*[144]

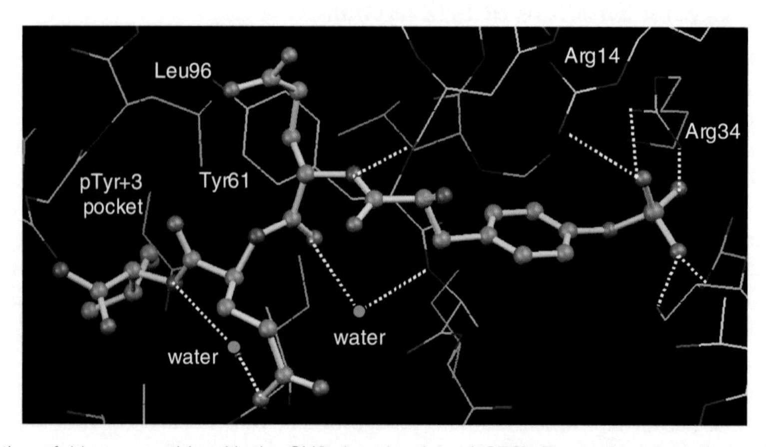

Figure 21 Interaction of 11-mer peptide with the SH2 domain of src (1SPS). For easier orientation, only the pTyrGluGluIle sequence is shown.

Their initial strategy was to satisfy the pTyr and P + 3 pockets and bridge these binding moieties using a nonpeptidic template. In addition, they intended to replace the conserved water molecules, since water molecules mediating the binding of inhibitors to the protein were considered for entropic reasons to be less favorable and they had been successfully replaced in projects such as the HIV-1 protease.[71] Thus, a fairly rigid scaffold was sought which could bind directly to Lys62N and Ile73O. A search in the Cambridge Crystallographic Database[145] suggested the benzoxazinone bicycle as a potential candidate for the linking template. However, some modifications were needed in order to make the synthesis more practicable and, as a result, compound 38 was synthesized (**Figure 22a**). The IC_{50} value of 9800 nM seemed to indicate that the design strategy was successful. Replacement of the phenyl moiety pointing into the pTyr + 3 pocket with either a dimethylphenyl group (compound 39) or a cyclohexyl ring (compound 40) resulted in somewhat increased affinities. The crystal structure of SH2 with bound compound 40 showed that the water molecule which binds in the peptide structure to Lys62N had indeed been replaced by an H-bond acceptor from the inhibitor scaffold. However, very much to their surprise, the phosphorylated tyrosine binds in a novel conformation and as a consequence the dimethylphenyl moiety points deeper into the pTyr + 3 pocket than modeling had predicted.

The SH2 binding site consists of two binding pockets which are separated by a flat and solvent-accessible surface. When this region was carefully visualized using a probe colored to display both hydrophobic and H-bonding sites, a hydrophobic region created by the phenyl ring of Tyr61 became apparent. The stacking of aromatic rings, either parallel or perpendicular, is a well-known phenomenon observed in complexes between ligands and proteins. Since it is considered to contribute significantly to the affinity, compounds were designed which would position an aromatic ring next to Tyr61.[146] An analysis of the accessible surface next to the central benzamide ring of compound 40 complexed to SH2 showed that the phenyl ring did not extend beyond Tyr61 and thus did not fully cover the hydrophobic surface.[146] Therefore, a seven-membered ring was added to the phenyl ring of compound 40 so that the hydrophobic surface created by Tyr61 would be better complemented, while at the same time maintaining the other intermolecular contacts. This approach led to compounds 41 and 42, with IC_{50} values of 200 and 300 nM, respectively. The subsequently obtained crystal structure of compound 41 confirmed that the seven-membered ring indeed adopts a concave conformation which complements the shape of the hydrophobic surface created by Tyr61. The benzamide carbonyl forms a hydrogen bond with the backbone amide of Lys62 displacing one of the two water molecules observed in the phosphopeptide complex. The second water molecule is not displaced and is H-bonded to the benzamide amide moiety and the backbone carbonyl of Ile73. The cyclohexyl group, although predicted to extend more deeply into the hydrophobic pTyr + 3 pocket, was not fully extended (**Figure 22b**).

4.24.4.4 Structure-Based Design Resulting in High-Affinity Inhibitors: Exploration of Different Scaffold Binding Modes

Structure-based drug design is usually based on a limited number of experimentally obtained protein–ligand complexes. The binding affinity of related compounds is subsequently predicted using molecular modeling tools. However, sometimes there are difficulties in correlating the predicted binding affinities with their corresponding experimental values. In some cases, for which subsequently an experimental 3D structure was obtained, it turned out that these discrepancies were clearly due to different binding modes of the inhibitors.[147,148] An example is the case of nonpeptidic inhibitors binding to the SH2 domain of src.[134] A total of 11 x-ray structures of SH2 complexes with inhibitors containing two closely related seven-membered ring scaffolds (**Figure 23a**) were investigated. Scaffold II differed from scaffold I by the insertion of an acetamidomethylene moiety. A superposition of these structures shows that the binding mode differs significantly between individual compounds, resulting in a considerably different localization of the inhibitor within the binding region. Closer inspection revealed that there are three main binding modes that may be represented by compounds 43–45 (**Figure 23b**). The interaction of compound 43 (IC_{50} 9 nM) with the SH2 domain of src is shown in **Figure 23c**. The phosphate group forms salt bridges to the arginine side chains of Arg14 and Arg34 and H-bonds to the amide nitrogen of Glu37 and the side chain of Thr38. Additional H-bonds are formed between an amide nitrogen of compound 43 and the carbonyl oxygen of His60, and between the carbonyl oxygen of the acetamido-moiety of compound 43 and the side chain of Arg14. Further stabilization is derived from two water molecules which link the acetamido moiety to His60N and the lactam carbonyl group to Lys62N via hydrogen bonds. Mayor hydrophobic interactions exist between the biphenyl moiety and protein residues 94–96 and 73–74. A second, minor hydrophobic interaction is found between the caprolactam scaffold itself and Tyr61. The superposition of compound 43 with the pTyrGluGluIle peptide[140] shows that both compounds display the same binding mode. The side chain of the first glutamate of the cognate peptide is replaced in compound 43 by a lactam ring which

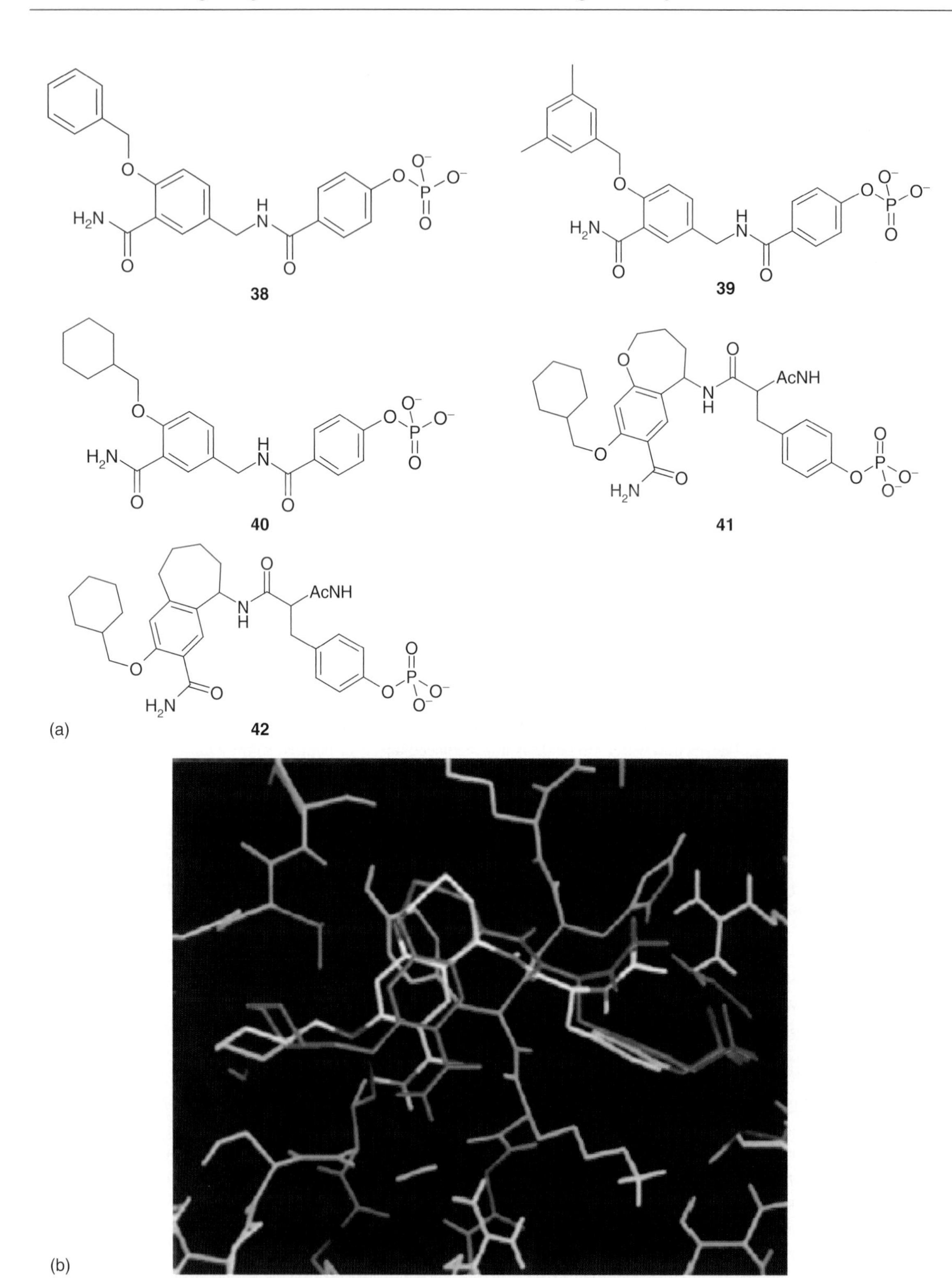

Figure 22 (a) Schematic representation of compounds 38–42. (b) Superposition of the model of compound 41 (white) and the crystal structure of the 41 (green). The water-mediated hydrogen bonds seen in the crystal structure between the carboxamide of 4 and the protein were not predicted, which explains the differences between model and crystal structure in the pY+3 pocket. (Reprinted with permission from Shakespeare, W. C.; Bohacek, R. S.; Azimioara, M. D.; Macek, K. J.; Luke, G. P.; Dalgarno, D. C.; Hatada, M. H.; Lu, X.; Violette, S. M.; Bartlett, C. *et al. J. Med. Chem.* **2000**, *43*, 3815–3819. Copyright (2000) American Chemical Society.)

interacts favorably with Tyr61. Compared to the isoleucine side chain in the pTyrGluGluIle peptide, the biphenyl moiety in compound 43 fits better in the hydrophobic pocket, resulting in a total of 15 carbon–carbon interactions of less than 4 Å, compared to six interactions for the corresponding isoleucine side chain in the pTyrGluGluIle peptide. The interaction of the phosphate group in compound 44 (IC_{50} 290 nM) with the side chains of Arg14 and Arg34 is different compared to compound 43 (**Figure 23d**). The salt-bridges are not formed by two terminal oxygens as in compound 43 but by a terminal oxygen and the ester oxygen of the phosphate ester. This results in an alternative conformation of the phospho-tyrosine ring which is accompanied by a novel interaction of the acetamido moiety to the protein. The acetamido moiety of compound 44 interacts with the His60N and an amide forms an H-bond to His60O. The interaction of the lactam carbonyl oxygen with Lys62N is bridged by a water molecule. The lactam ring, including the phenyl ring, is shifted compared to compound 43 in a way such that the terminal benzyl ring is comfortably positioned in the hydrophobic pocket. The conformation of the phenylphosphate moiety found in compound 44 corresponds to the conformation found in the complex structure with phenylphosphate[149] and to that described for compound 40 above.[144] The interaction of compound 45 within the pTyr + 3 pocket and the pTyr binding site is very similar to that found in compound 43, even though compound 45 (IC_{50} 2700 nM) is shortened by two carbon atoms compared to compound 43 (**Figure 23e**). In order to compensate for this difference the inhibitor adopts an alternative binding mode of the scaffold. There are no water molecules bridging the interaction between compound 45 and the protein. Instead, H-bonds are formed between the amide nitrogen of the inhibitor 45 and His60O and the lactam carbonyl and Lys62N. As a result, the terminal ring is comfortably positioned in the hydrophobic pocket, resulting in a total of 11 carbon–carbon interactions of less than 4 Å between the inhibitor and the protein.

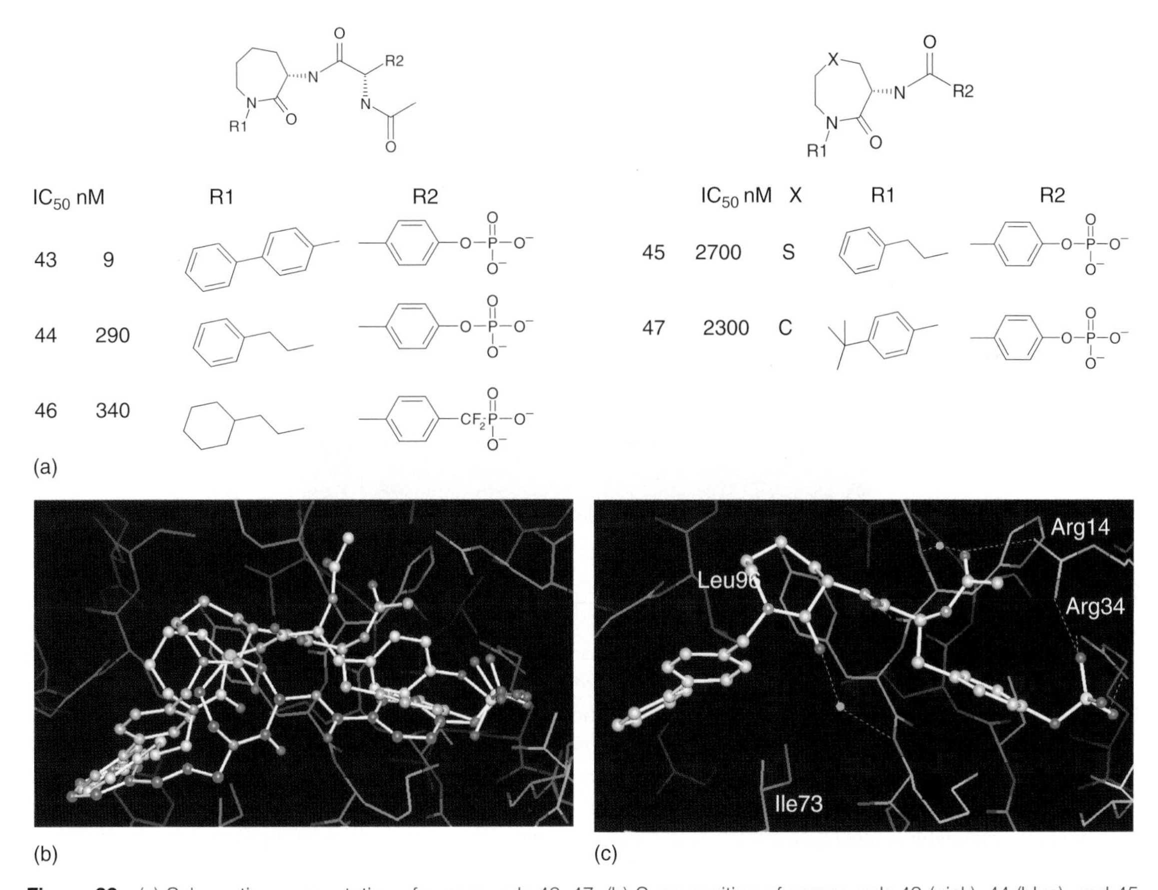

Figure 23 (a) Schematic representation of compounds 43–47. (b) Superposition of compounds 43 (pink), 44 (blue), and 45 (green) bound to the SH2 domain of src. (c) Interaction of compound 43 with the SH2 domain of src. (d) Interaction of compound 44 with the SH2 domain of src. (e) Interaction of compound 45 with the SH2 domain of src. (f) Superposition of the electron density of compounds 44 (green) and 45 (blue). (g) Superposition of compounds 43 (pink) and 44 (blue) bound to the SH2 domain of src. (h) Superposition of compounds 45 (green) and 47 (yellow) bound to the SH2 domain of src. (b, c, d, e, f, g, h: Reprinted with permission from Lange, G.; Lesuisse, D.; Deprez, P.; Schoot, B.; Loenze, P.; Bénard, D.; Marquette, J.-P.; Broto, P.; Sarubbi, E.; Mandine, E. *J. Med. Chem.* **2002**, *45*, 2915–2922. Copyright (2002) American Chemical Society.)

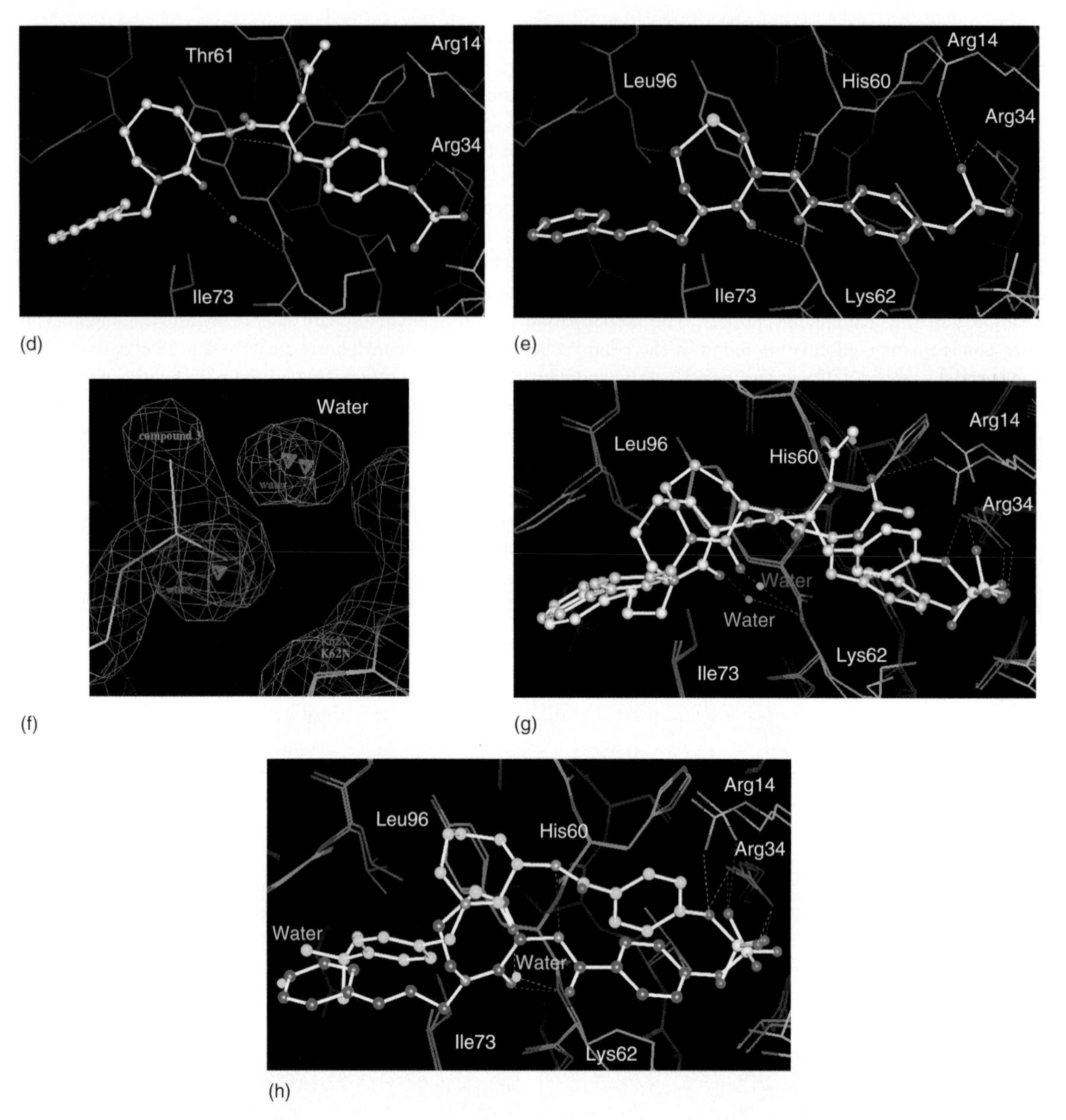

Figure 23 Continued.

An analysis of the superimposed structures suggests that the binding modes of the scaffolds can be characterized based on the H-bond pattern formed between the individual scaffold and the protein. SH2 has three main chain H-bond donors/acceptors which need to be satisfied by the scaffold: His60N, His60O, and Lys62N. The individual binding modes differ in the details of the H-bond pattern formed between the inhibitor scaffold and the SH2 domain of src. While the scaffold in compound 45 interacts in all cases directly with the protein backbone, the interaction of the scaffold of compound 44 with the protein backbone includes a bridging water molecule and that of the scaffold of compound 43 includes two bridging water molecules. The superposition of the crystal structures shows that, for those acceptors interacting directly with the protein, the inhibitor H-bond acceptors are located in two well-defined clusters. These clusters are quite limited in their size due to a combination of a rigid protein surface and the angular and distance restraints of H-bonds. When a water molecule mediates the inhibitor–protein interaction, the water molecule is positioned exactly at the position in which otherwise the H-bond partner of the inhibitor scaffold would be located (**Figure 23f**). However, the positions of those inhibitor H-bond acceptors which interact with the SH2 domain via a bridging water molecule are much more diffuse, indicating that the positional requirement for these second-layer H-bond acceptors is not as harsh as for those interacting directly with the protein. As a consequence, if the scaffold binds in a mode including bridging water molecules, it has significantly more degrees of freedom to position its hydrophobic substituent optimally within the rigid pTyr + 3 affinity pocket.

As mentioned above, there are two conserved interactions of high importance: the binding of a negatively charged group in the phosphate recognition site and the hydrophobic effect due to the burial of a hydrophobic moiety in the pTyr + 3 affinity pocket. Some inhibitors, such as compounds 43–45, reach fairly deeply into the pocket, while other inhibitors, such as compounds 46 and 47, are not able to penetrate very deeply into the pocket. The superposition of all complex structures shows that the pTyr + 3 pocket is fairly rigid, suggesting that the shape and size of the substituent are essential features for high-affinity inhibitor binding and that only limited deviations from the ideal shape are tolerated by the protein. Interestingly, all inhibitors seem to aim at filling the pTyr + 3 pocket as much as possible and achieve this by different interactions of the inhibitor scaffold with the protein. However, there are some restrictions as to which binding modes can be assumed by an individual inhibitor. These restrictions become visible if pairs of crystal structures are compared. Compounds 43 and 44 differ only in their substituent at the pTyr + 3 position. While compound 43 has a biphenyl at this position, in compound 44 there is a much shorter phenyl propyl moiety present. A comparison of the two x-ray structures shows that the hydrophobic substituents penetrate similarly deeply into the hydrophobic pocket, with the phenethyl group taking the place of the terminal phenyl ring of compound 43 (**Figure 23g**). Compound 44 achieves this by adopting a different conformation of the phenylphosphate group. The main difference between compounds 45 and 47 is also in the substituent pointing into the pTyr + 3 pocket. As described above, compound 45 adopts a short-cut binding mode which buries the aromatic ring deeply into the pTyr + 3 pocket. However, the superposition of the two x-ray structures (**Figure 23h**) shows that compound 47 has a different binding mode, namely that of compound 43. A closer analysis of the complex structure with compound 47 revealed that the tertiary butyl group is too bulky to slide deeper into the pocket, which puts the scaffold in a position favoring the compound 43 binding mode. These results suggest that the inhibitors bind in a mode which maximizes the hydrophobic interaction in the pTyr + 3 pocket. This is achieved by either including water molecules and/or a change in the conformation of the substituent pointing into the phospho-tyrosine pocket. As found in other cases,[150] the water molecules form an integral part of the protein–ligand interface and increase the flexibility of the interface between a rigid inhibitor part and a rigid protein surface.

4.24.4.5 Structure-Based Design Aiming at Replacing pTyr with a pTyr Mimetic

The first SH2 inhibitors were phosphopeptides or peptidomimetics, which incorporate pTyr itself or mimics closely related to pTyr. However, inhibitors with pTyr moieties suffer from poor cell penetration and are often rapidly degraded by phosphatases. Unfortunately, attempts to design metabolically stable pTyr mimics have generally resulted in compounds with weaker affinity for the target SH2 domain. Therefore, different approaches were tried in order to replace the pTyr. These used either structure-based drug design[151,152] or an approach whereby pTyr-mimicking fragments were screened using different experimental techniques.

An example leading to an osteoclast-selective, nonpeptide inhibitor with in vivo antiresorptive activity has been presented by Shakespeare *et al.*[151] Based on the observation that the crystal structure of Src-SH2 contained a well-defined citrate molecule in the pTyr pocket, different pTyr replacements were designed. This was done by analyzing the crystal structure carefully and flexibly docking the candidates into the SH2 structure. Subsequently, the most promising mimetic was attached to their potent bicyclic benzamide scaffold described above (compound 40) such that the mimetic would replace the pTyr moiety. The crystal structure of the resulting compound 48 (**Figure 24a**) confirmed that the bisphosphonate contributes to an H-bond network similar to that found for citrate (**Figure 24b** and **c**). The interaction of compound 48 with the SH2 domain is shown in **Figure 24d**. Compound 48 exhibits increased Src SH2 binding affinity (IC_{50} 300 nM). Furthermore, compound 48 inhibits rabbit osteoclast-mediated resorption of dentine in a cellular assay and exhibits bone-targeting properties based on a hydroxyapatite adsorption assay. While direct experimental evidence is not available, a plausible explanation involves accumulation of these inhibitors in bone matrix followed by uptake into osteoclasts by phagocytosis.[153] Interestingly, bisphosphonates themselves are known to inhibit osteoclast function and bone metastasis.[154,155] The osteoclast-inhibitory abilities of compound 48, however, depend on its binding potency to the src SH2 domain.

4.24.4.6 Structure-Based Design Using a Fragment Approach for Replacing Phosphotyrosine

Traditionally, structure-based drug design plays a major role in the optimization of hits found by screening of compound libraries, either company-owned or derived from other sources. More recently it has emerged that, particularly for

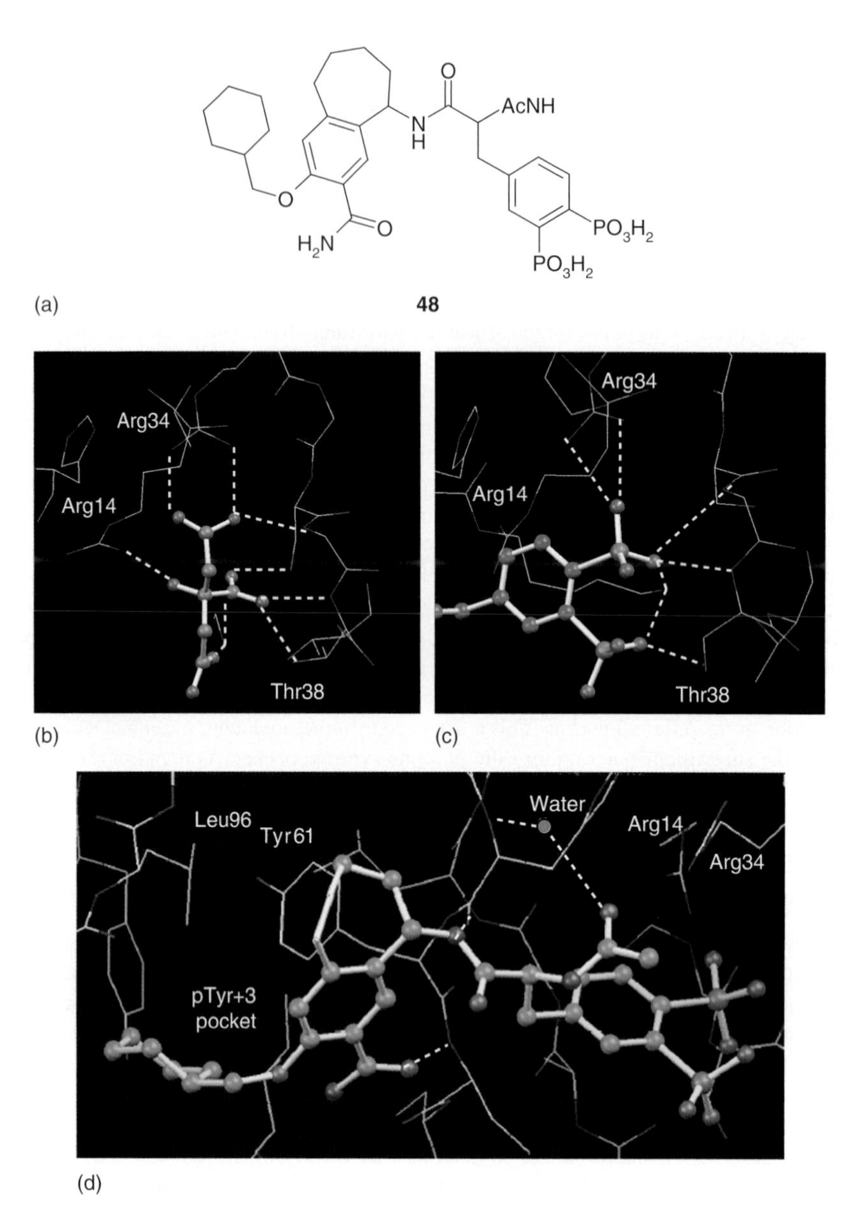

Figure 24 (a) Schematic representation of compound 48. (b and c) Interaction of citrate (b) and compound 48 (c) with the SH2 domain of src. (d) Interaction of compound 48 with the SH2 domain of src (1FBZ).

targets involving protein–protein interactions, there is a high probability of not finding a promising hit from HTS[156] and alternative approaches for lead generation are required. With no hits available, computer programs have been used to generate leads for structure-based drug design (*see* 4.12 Docking and Scoring; 4.19 Virtual Screening). These approaches range from programs such as DOCK[157] and FlexX,[158] which dock full-length inhibitors into putative binding sites, to programs such as GRID[159] which scan a putative binding site using probes representing H-bond donors/acceptors or hydrophobic elements in order to identify H-bond partner positions in the binding pocket and evaluate the size of hydrophobic pockets. Somewhat in between are programs such as LUDI[160] which selects fragments from a virtual database and places them into the binding pocket. Connecting these fragments ideally results in a full-length inhibitor. Alternatively, fragments can be identified using experimental techniques such as nuclear magnetic resonance (NMR) measurement of the target protein mixed with inhibitor fragments.[161,162] Based on the difference in the NMR spectra of free and complexed protein, inhibitor fragments are identified which bind to pockets in the target protein (SAR by NMR). In addition, fragment screening using protein crystallography has been used for lead generation (*see* 4.31 New Applications for Structure-Based Drug Design).[163,164] Inhibitor fragments are soaked into the binding pocket of the crystallized protein and, in case of specific binding, the corresponding x-ray structures are determined. Starting from the high-resolution structure a full-length inhibitor can be obtained by attaching an inhibitor scaffold to the fragment using structure-based drug design. Compared to NMR, protein crystallography has less high-throughput potential since collecting a complete data set takes several hours at a home source. This lack has been compensated for by either using a cocktail of compounds[163] or by filtering the fragment library using a Biacore assay.[164] Also fragments found by other techniques such as mass spectrometry,[165] NMR experiments,[166] and, in particular, computer prediction[167] can be used as input for crystallographic screening.

Fragment libraries can also be screened with the intention of replacing fragments in already-existing full-length inhibitors. These replacements can be selected in order to modify a variety of features, such as binding affinity, cell penetration, solubility, or selectivity. Examples include the combination of a fragment approach with traditional structure-based drug design, which has resulted within little more than a year in nonpeptidic inhibitors of the SH2 domain of pp60src with IC_{50} values in the low nanomolar range.[149,164] The fragment library used in this case included 150 compounds typically consisting of 6–30 atoms. The compounds were either commercially available or were part of small libraries prepared by parallel synthesis. In order to reduce the number of compounds to a feasible amount for the crystallographic experiments, a Biacore assay was used as a filter. Fragments that displayed comparable or better binding affinity than phenyl phosphate were kept for soaking. Soaking of src SH2 crystals was tried for about 200 compounds, including fragments and full-length inhibitors. Since soaking is a fast but not always reliable method, phenyl phosphate was used as an internal standard. A total of 200 x-ray data sets were collected and analyzed, resulting in a total of 45 structures with bound inhibitor fragments or full-length inhibitors. They included 14 structures with fragments found using the Biacore assay and two structures with fragments predicted by LUDI. These two fragments showed no biochemical activity within the range measured. The citrate fragment was found bound to the protein after a citrate buffer had been used during purification and displayed no measurable affinity. The successfully soaked fragments, including their IC_{50} values, are listed in **Figure 25a**. The omit electron density of several fragments is shown in **Figure 25b**.

The general experience was that soluble fragments with a substantial number of putative H-bond partners had a fairly good chance of binding in a specific binding mode. Except for phenyl phosphate, which was clearly bound at two sites, all other fragments were only bound in the phospho-tyrosine pocket and are therefore phospho-tyrosine mimics (**Figure 25c**). The fragments identified differed significantly in their chemical nature and included oxalate, malonate, phenylmalonate, sulfates, and various compounds from an aldehyde library. The superposition of the fragment structures showed that the protein in all structures superimposed excellently with root mean square deviations of typically 0.2 Å for most main chain atoms. However, closer inspection revealed that there were significant differences in the pTyr pocket. The structure of the phospho-tyrosine binding loop is stabilized by H-bonds between the backbone amides of Ser36 and Ala42 and backbone carbonyls of Ala42 and Thr39, respectively. In addition, the side chain of Ser36 interacts via its two free electron pairs with the amide backbones of Thr38 and Thr39, while the proton of Ser36OG acts in all fragment structures as an H-bond donor to the fragment. It seems that a rotation of the serine side chain induces a conformational change of the loop 36–42. For instance, comparison of the oxalate and malonate structures showed that both fragments interact in an identical manner. In both structures, one carboxylate moiety interacts with the side chains of Arg14 and Arg34 and the main chain nitrogen of Glu37, while the second carboxylate moiety forms H-bonds to the side chains of Ser36, Thr38, and the main chain nitrogen of Thr38. The superposition of all H-bond acceptors from the fragments within the citrate SH2 structure showed that these acceptors form well-defined clusters when interacting with rigid protein H-bond donors such as Arg34 (**Figure 25d**). However, the cluster formed by the H-bond acceptors to Ser36OG was quite diffuse, indicating that the exact location of the H-bond partner of Ser36OG within the pTyr pocket is less crucial. A closer inspection of the fragment binding revealed that in all cases complex H-bond networks are formed between the fragments and SH2. All networks are identical in the rigid part of the pocket and include H-bonds/salt-bridges between the fragments and the side chains of Arg14, Arg34, Ser36, and the backbone amide of Glu37. The observed conformation of the phospho-tyrosine binding loop does not depend on the size or the affinity of the particular ligand. However, it seemed that a loop conformation is assumed which allows the system to complete the H-bond network.

In parallel with the identification of fragments pointing into the phosphotyrosine pocket, a nonpeptidic inhibitor containing a phospho-tyrosine unit was designed and optimized.[168,169] The best inhibitor with a phospho-tyrosine had an activity of 9 nM (compound 43). Subsequently, replacements for pTyr were selected based on the x-ray structures of the fragments and the corresponding inhibitors were synthesized (compounds 49–56). In particular, different phenyl malonate fragments incorporating the structural information obtained from the malonate and citrate complexes were synthesized. Replacement of pTyr by 2-carboxy phenyl malonate resulted in an inhibitor with an IC_{50} value of 3 nM (compound 49). The structures of the complexes with compound 43 and compound 49, including their common inhibitor atoms, superimpose extremely well. As expected, there are some changes within the phospho-tyrosine pocket

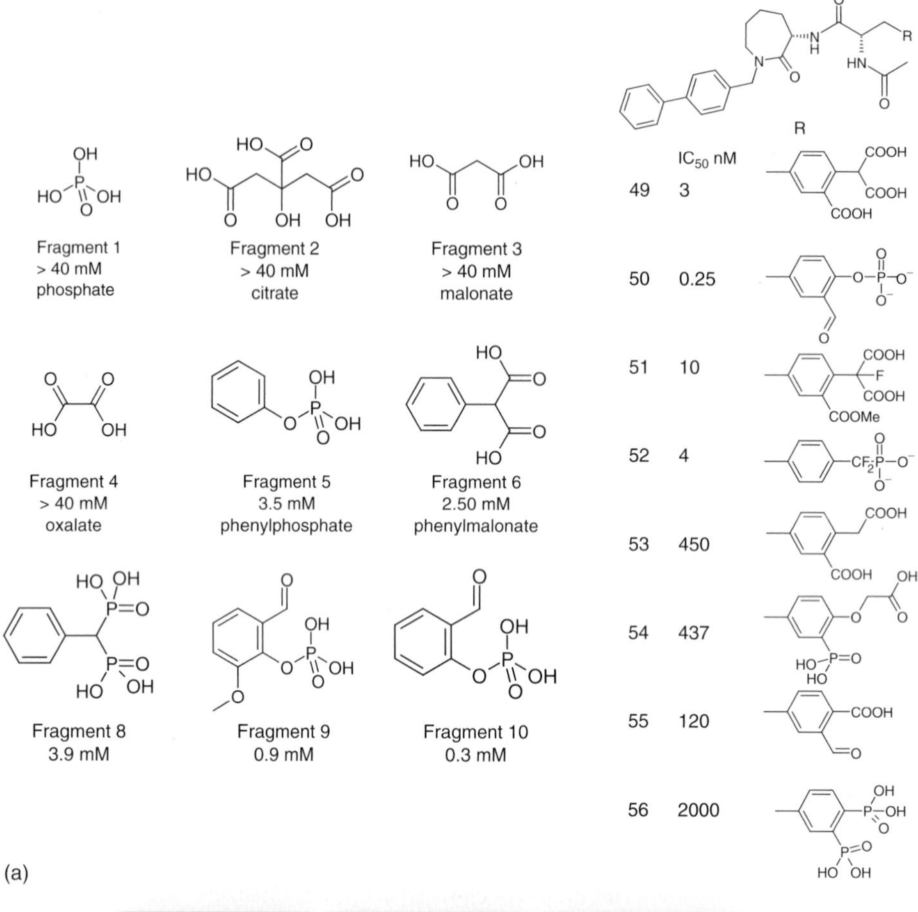
Fragment 1
> 40 mM
phosphate
Fragment 2
> 40 mM
citrate
Fragment 3
> 40 mM
malonate
Fragment 4
> 40 mM
oxalate
Fragment 5
3.5 mM
phenylphosphate
Fragment 6
2.50 mM
phenylmalonate
Fragment 8
3.9 mM
Fragment 9
0.9 mM
Fragment 10
0.3 mM
IC50 nM
R
49 3
50 0.25
51 10
52 4
53 450
54 437
55 120
56 2000
(a)

Fragment 1
Fragment 2
Fragment 3
Fragment 4
Fragment 5
Fragment 6
Fragment 8
Fragment 9
Fragment 10
(b)

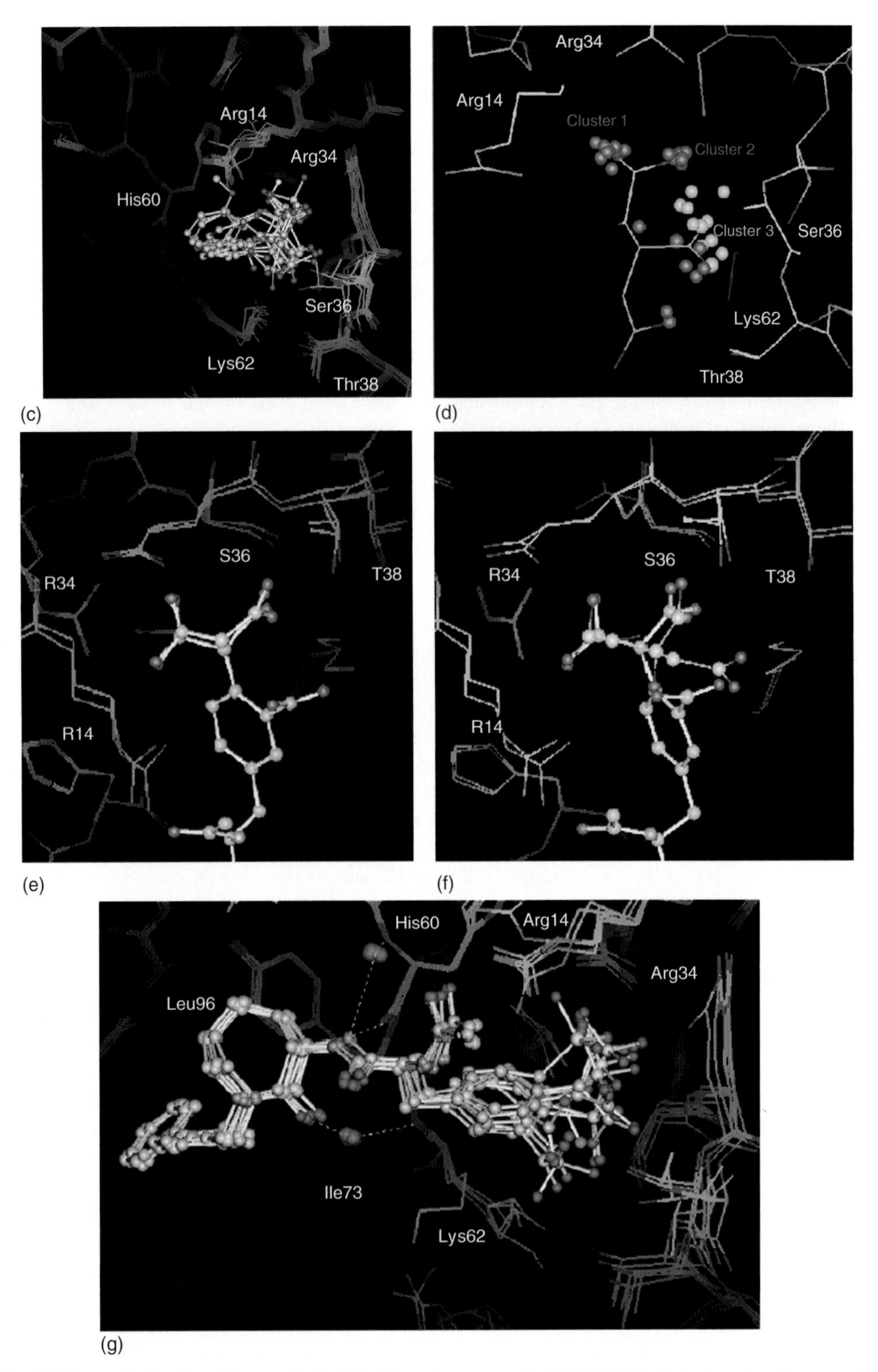

Figure 25 (a) Schematic drawing of fragments 1–6, 8–10, and compounds 49–56. (b) Omit electron density maps of fragments bound to the SH2 domain of src contoured at 3σ. (c) Superposition of x-ray structures with fragments bound to the SH2 domain of src. (d) The superposition of the fragment of H-bond acceptors shows that they form well-defined clusters when interacting with rigid protein parts (red) and less well-defined clusters when interacting with flexible protein parts (pink). For easier orientation only, the full structure of the citrate complex is displayed. (e and f) Superposition of the x-ray structures of compound 39 (pink) and (e) malonate and (f) citrate bound to the SH2 domain of src. (g) Superposition of x-ray structures of 18 fragments and 10 full-length inhibitors based on inhibitor. (b, c, d, e, f: Reprinted with permission from Lange, G.; Lesuisse, D.; Deprez, P.; Schoot, B.; Loenze, P.; Benard, D.; Marquette, J.-P.; Broto, P.; Sarubbi, E.; Mandine, E. *J. Med. Chem.* **2003**, *46*, 5184–5195. Copyright (2003), American Chemical Society; g: from Lange, G.; Lesuisse, D.; Deprez, P.; Schoot, B.; Loenze, P.; Bénard, D.; Marquette, J.-P.; Broto, P.; Sarubbi, E.; Mandine, E. *J. Med. Chem.* **2002**, *45*, 2915–2922. Copyright (2002) American Chemical Society.)

due to the different induced fit of the pTyr and 2-carboxy phenyl malonate fragment. Also the superpositions of the structure with compound 49 and the malonate, citrate, and phenyl malonate structures, respectively, showed that all bind identically to the pTyr pocket, confirming that the fragments were bound in a mode relevant for drug design (**Figure 25e** and **f**).

The affinity of compounds 43 and 49–56 differed by several orders in magnitude, i.e., between micromolar and picomolar, even though the superposition of all fragment and inhibitor structures showed that the common atoms interact identically within the SH2 domain (**Figure 25g**). Thus, the differences in affinity could only be correlated to the different interaction of the inhibitors within the pTyr pocket. The positions of all inhibitor H-bond acceptors projected into the protein structure of compound 49 showed that most of the H-bond acceptors of the full-length inhibitors are located within the clusters defined by the position of the fragment H-bond acceptors (**Figure 25d**). In particular, the H-bond acceptors of high-affinity inhibitors were located well within these clusters and formed complete H-bond networks with the phospho-tyrosine pocket. In contrast, in complexes with poor inhibitors such as compounds 53 and 56, the central H-bond to the side chain of Ser36 was not made. In other complex structures with low-affinity inhibitors, the electron density of the pTyr replacement was disordered even though the electron density of the common inhibitor atoms was well defined. In these cases, the inhibitor H-bond acceptors were not located within the clusters defined by the fragments and were thus unable to complete the H-bond network. An example is compound 56 with an IC_{50} of 2000 nM. The superposition of a multitude of x-ray structures with different inhibitor fragments reveals at which positions within the binding site specific functional groups such as H-bond donors/acceptors or hydrophobic groups are required. If a binding site is rigid, the positions of functional groups should form a well-defined cluster. However, for flexible ligand-binding sites, there is additional information encoded in the size and shape of the clusters for a given functional group. The topology of the individual clusters and their relative position will reflect the flexibility of the protein and show to what extent the protein can stretch itself in order to accommodate a ligand into the binding site. The x-ray structures indicated that neither ligand affinity nor the exact positioning of functional groups within the pharmacophore are an essential requirement for fragment binding and thus good selection criteria for polar fragments to the SH2 domain. However, it seems important that the fragments complete the H-bond network within the pTyr pocket. The importance of the H-bond network is highlighted by the significantly lower affinity of full-length inhibitors which form incomplete H-bond networks within the phospho-tyrosine pocket. This is also true for the phospho-tyrosine mimics described in the literature, such as carboxymethylphenylalanine,[170] which lack an H-bond acceptor to Ser36OG. Benzylmalonates do not seem to be able to position their H-bond acceptors within the required limits.[171] In addition, it was shown that bisphosphonates such as compound 48 maintained their activity, compared to their pTyr analogs, if the correct scaffold is attached[151] and are thus able to complete the H-bond network in the pTyr pocket.

4.24.5 Design of PTP-1B Inhibitors Against Diabetes

Diabetes is increasingly becoming an epidemic not only in industrialized countries but also in developing nations. If the current trend continues, 300 million people worldwide may become diabetic by the year 2025, according to a prediction by the World Health Organization (WHO). Although the causes of diabetes are not clearly understood and appear to be multifactorial, insulin resistance seems to be an underlying factor in the progression of the disease. Attempts to reverse this physiological condition would thus go a long way toward ameliorating the disease. Recent therapies, including the medication with thiazolidinediones that target peroxisome proliferator-activated receptor-gamma and act as insulin-sensitizers, have been found to be quite efficacious against diabetes.[172] Another target that is receiving strong attention as a candidate for the development of drugs against diabetes is the protein tyrosine phosphatase PTP-1B.[173]

Protein tyrosine phosphatases (PTPases) form a large, structurally diverse family of enzymes expressed in all eukaryotes. Recent reports indicate that there are about 60 PTPase genes encoded within the human genome, including transmembrane, receptor-like, and nonreceptor-like enzymes. Six PTPases have been found to have mutations that contribute to inherited human diseases, such as familial stroke, hypercholesteremia, coronary artery disease, Alzheimer's disease, and diabetes.[174] Each PTPase is composed of at least one conserved domain characterized by a unique 11-residue sequence motif ((I/V)HCXAGXXR(S/T)G) containing a cysteine and an arginine known to be essential for catalytic activity. Their catalytic mechanism involves a nucleophilic attack by the conserved cysteine on a phospho-tyrosine substrate, resulting in a covalent phosphocysteine intermediate that is subsequently hydrolyzed by an activated water molecule. PTP-1B was the first PTPase to be isolated in homogeneous form.[175,176] Since then, a number of biological and enzyme kinetic studies on PTP-1B have suggested that PTP-1B is an important negative regulator of the insulin-signaling pathway. In particular, PTP-1B seems to regulate negatively insulin signaling by dephosphorylating the phospho-tyrosine residues of the tissue insulin receptor kinase. Recent studies on PTP-1B knockout mice[177,178] provided significant support for the view that PTP-1B is a key regulator of insulin signaling. PTP-1B–/– mice showed increased insulin sensitivity and obesity resistance. Thus, PTP-1B is an attractive target for the treatment of type 2 diabetes and obesity, and selective PTP-1B inhibitors could be of significant therapeutic utility.[179]

4.24.5.1 First Crystal Structure of PTP-1B

The first crystal structure of the catalytic domain of PTP-1B was published in 1994.[180] A 10-stranded mixed β-sheet that adopts a highly twisted conformation forms the core of the protein (**Figure 26**). Sodium tungstate was used to form a heavy-metal derivative in the structure determination since it was found to bind tightly to PTP-1B. It presumably acts as a phosphate mimic, and thus the structure of tungstate complexed to PTP-1B provides the first details of the interactions between PTP-1B and its active site ligands. The catalytic site is located at the base of a shallow cleft. The phosphate recognition site is created from a loop that is located at the N-terminus of an α-helix. This site consists of the 11-residue sequence motif which is typical for PTPases and contains the catalytically active cysteine and arginine. There are H-bonds between the tungstate and the backbone amides of Ser216, Ala217, Gly218, Ile219, Gly220, Arg221 and the side chains of the conserved Arg221, suggesting that these H-bond functions are recognizing the phosphate in the phosphorylated substrate peptide. The position of the conserved cysteine within the phosphate binding site is consistent with its role as a nucleophile in the catalytic reaction.

4.24.5.2 Crystal Structure of PTP-1B with a Peptide

A close analysis of the interaction of phospho-tyrosine peptides with PTP-1B provided the structure of a catalytically inactive Cys215Ser mutant PTP-1B in complex with a peptide substrate.[181] The peptide AspGlu(pTyr)Leu corresponds to an autophosphorylation site on the epidermal growth factor receptor. Free pTyr is a relatively poor substrate compared with pTyr-containing peptides, suggesting that, though the primary determinant of specificity is a pTyr side chain, the flanking amino acids contribute to the binding affinity of the peptide. The crystal structure shows that apolar amino acids, including Ala217, Ile219, Tyr46, Val49, and Phe181, form the hydrophobic pocket for the aromatic ring in pTyr. In addition, the phosphate of pTyr makes H-bonds with the main chain amides of Ser216-Arg221 and the side chain of Arg221 (**Figure 27a**). Peptide binding is accompanied by a conformational change of a surface loop, the so-called WD loop, that completes the pTyr recognition site by positioning Phe180 and Asp181 into the active site (**Figure 27b**). In addition its has been observed that a putative H-bond between the phenolic oxygen of pTyr and Asp181 requires a protonated Asp181, suggesting that Asp181 may function as general base by protonating the phenolic oxygen and thereby facilitating bond cleavage. A primary determinant of the peptide conformation at the pTyr site is

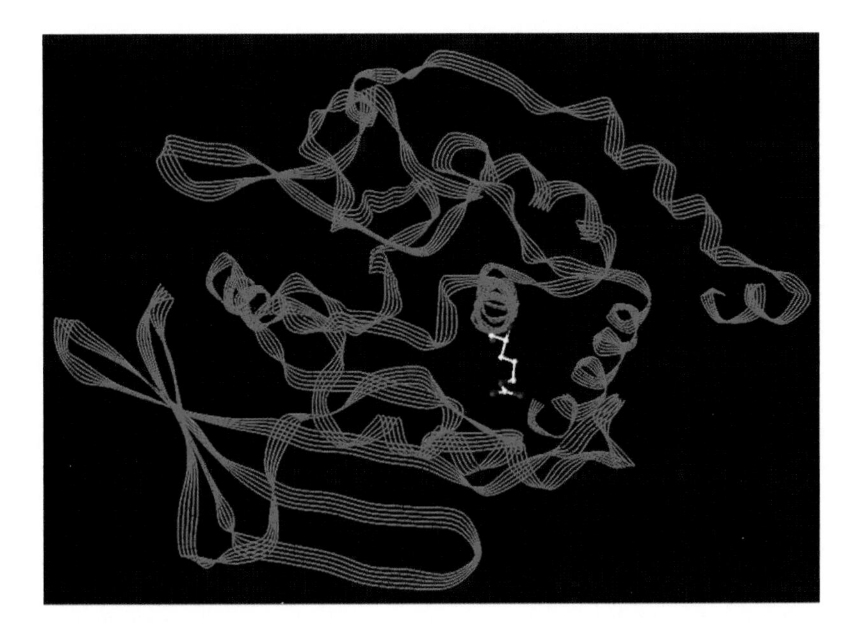

Figure 26 Ribbon plot of PTP-1B (2HNP). The active site (Arg221) is indicated.

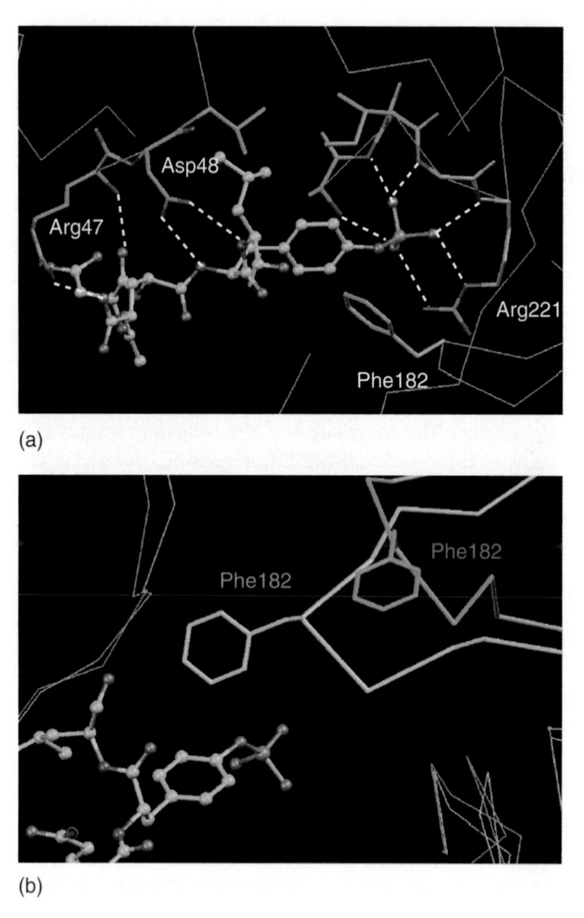

Figure 27 (a) Interaction of the AspGlu(pTyr)Leu peptide with PTP-1B(1PTT). PTP-1B is shown in green. The peptide is displayed in blue. (b) Superposition of the apo structure of PTPT-1B (green) (2HNP) and its structure with bound AspGlu(pTyr)Leu peptide (blue) (1PTT). The WD loop, including Phe182, assumes in the peptide structure a conformation in which the Phe182 is in close contact with the tyrosine phosphate.

the formation of two H-bonds between the main chain amides of the pTyr and the P + 1 substrate residue and the side chain of Asp48 in PTP-1B. These H-bonds seem to be important because they stabilize a helical main-chain conformation at the pTyr site of the substrate. The general rolé of Asp48 in peptide recognition is also supported by the fact that there is an aspartic acid at this position in many PTPases. Additional H-bonds between the side chain of Arg47 of PTP-1B and acidic side chains of the AspGlu(pTyr)Leu-peptide are consistent with kinetic data, suggesting a preference for peptidic substrates with acidic amino acids N-terminal to pTyr.

4.24.5.3 Crystal Structure of PTP-1B with a Cognate Peptide

In contrast to the EGFR peptide which contains a single pTyr residue, the activation segment of the insulin receptor is phosphorylated at three sites. In order to define the molecular specificity of the insulin activation segment, Salmeen *et al.*[182] investigated three different peptides derived from the insulin receptor kinase using an approach integrating crystallographic, kinetic, and peptide-binding studies. Their data suggest that the GluThrAsp(pTyr1162) (pTyr1163)Arg peptide is recognized by PTP-1B. The crystal structure with this peptide shows that pTyr1162 forms similar H-bonds with PTP-1B to those which had been seen in the structure with the monophosphorylated peptide. The second pTyr residue, pTyr1163, points into a shallow groove on the protein surface (**Figure 28a**). Its interactions with PTP-1B seem to be dominated by salt-bridges between the phosphate and the side chains of Agr24 and 254. The specificity of this site for a pTyr residue C-terminal to the substrate pTyr, rather than shorter pSer and pThr residues, is primarily due to the length of pTyr which has the correct size to position its phosphate so as to reach Arg24 and Arg254. Similarly to the structure of the monophosphorylated peptide, two H-bonds between the side chain of Asp48 and the peptide amides of

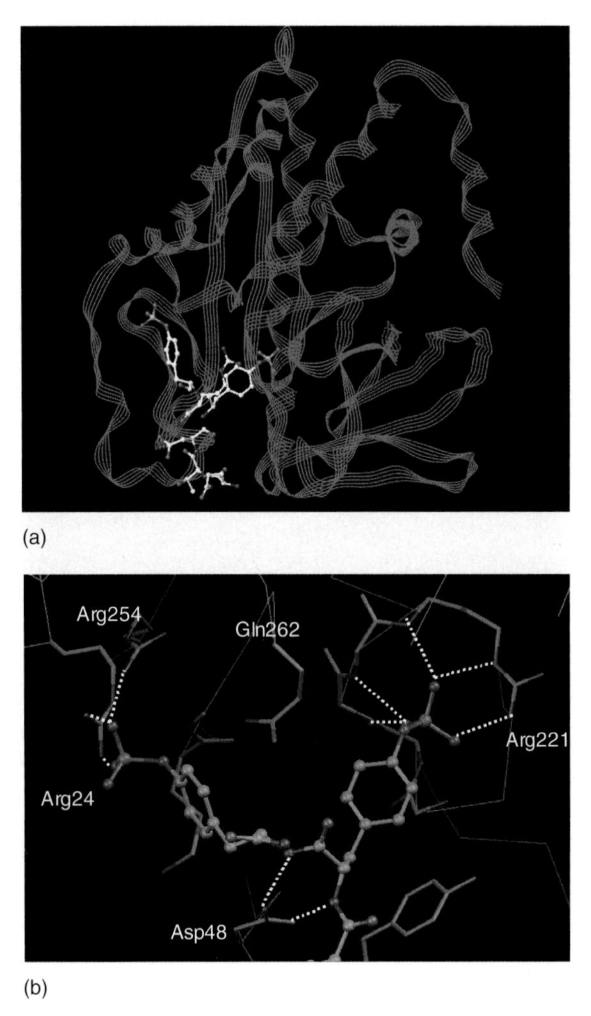

Figure 28 (a) Ribbon plot of PTP-1B with GluThrAsp(pTyr1162)(pTyr1163)Arg peptide (1GIF). The peptide is shown in white. (b) Interaction of the GluThrAsp(pTyr1162)(pTyr1163)Arg peptide with PTP-1B (1GIF). PTP-1B is shown in green. The peptide is displayed in blue.

pTyr1162 and pTyr1163 are seen to orient the cognate peptide and to determine its conformation. Again, there is an H-bond between Arg47 of PTP-1B and the side chain of Asp1161, giving further evidence that PTP-1B prefers peptides with acidic residues N-terminal to the substrate pTyr residue. Additional specific interactions involve residues C-terminal to the substrate. The interactions are summarized in **Figure 28b**.

4.24.5.4 Identification of a Second Aryl Phosphate-Binding Site by Soaking Studies

Bis-(*para*-phosphophenyl) methane and pTyr (compounds 57 and 58 in **Figure 29a**) have been found to be a low-molecular weight and nonpeptidic inhibitor.[183] Subsequently, the crystal structure of compound 58 complexed to a catalytically inactive mutant (C215S) of PTP-1B was determined.[184] Surprisingly, difference Fourier maps showed that compound 58 binds in two different binding modes. There was electron density for compound 58 in the pTyr-binding site located in the active site and additional density corresponding to a second molecule of compound 58 in a pocket approximately 4 Å apart. Similarly, two identical binding modes were observed in PTP-1B crystals which were grown in a saturating concentration of pTyr (**Figure 29b**). The pTyr found in the active site has a binding mode identical to that observed in the structures with peptides.[181] Electron density and refinement statistics indicated that the tyrosine phosphate located in the second aryl phosphate-binding site has a lower occupancy and larger temperature factors than that located in the active-site tyrosine phosphate pocket, suggesting that it is less tightly bound and more disordered. The most important residues of the second aryl phosphate binding pocket seem to be Gln262 and two arginines, Arg24

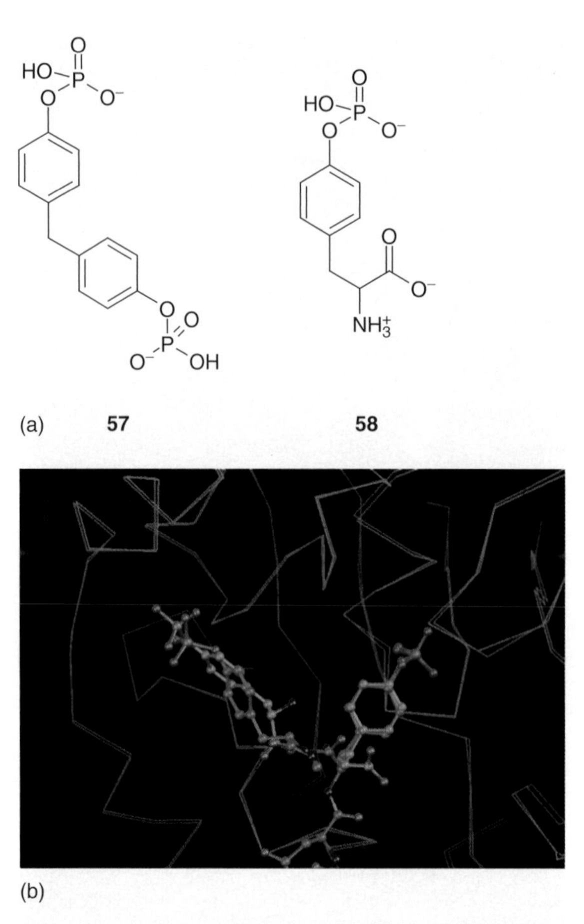

Figure 29 (a) Schematic representation of compounds 57 and 58. (b) Two binding modes of tyrosine phosphate (blue) superimposed with the GluThrAsp(pTyr1162)(pTyr1163)Arg peptide (green). The Cα-trace of PTP-1B complexed to pTyr is shown in blue (1PTY), while that of the peptide is shown in green (1GIF).

and Arg254, which form H-bonds to the phosphate. The hydrophobic effect may also play a role for the binding of the second aryl phosphate, because 75% of its apolar surface area is buried upon binding to PTP-1B.

Puius *et al.*[184] proposed that the identification of these alternate binding modes suggests a new strategy for the design of inhibitors with enhanced affinity and specificity. Prior studies of PTPase inhibitors have generally ignored the issue of specificity because the majority of the invariant amino acid residues conserved in PTPases are located in the enzyme active site.[180,185] However, incorporating fragments identified for the second aryl phosphate-binding site should give rise to nonhydrolyzable aryl phosphate analogs that occupy simultaneously both sites A and B and thus should have higher affinity and specificity. A sequence analysis showed that Arg254 and Gln262 are conserved amongst many PTPases.[185] Amino acid sequence alignment also suggests that an Arg residue equivalent to Arg24 of PTP1B may also exist in T-cell phosphatase (TCPTP), tyrosine-protein phosphatase nonreceptor type 12 (PTP-PEST), and the *Yersinia* PTPase, but not in PTPα, PTPε, leukocyte antigen-related protein tyrosine-phosphatase (LAR), or PTPγ phosphatases. Thus, it appears that the residues that form the second site in PTP-1B are less conserved among PTPases than residues that form the active site and it might be possible to achieve selectivity by utilizing the interactions involved in this secondary aryl phosphate-binding site.

4.24.5.5 Structure-Based Design Leading to Novel Bioavailable and Nonpeptidic Inhibitors Based on a Screening Hit

Significant progress was made toward developing high-affinity PTP inhibitors. However, most compounds had features that made them unsuitable as starting points for optimization to orally active drugs. As an example, peroxovanadium compounds had contributed significantly to the understanding of insulin signaling, but appeared to be too toxic.[186,187] Bisphosphonates, such as alendronate, had been shown to inhibit PTPases, but their inherent affinity for bone prevented their general use in other target tissues.[188] Furthermore, many inhibitors were time-dependent and seemed to act through covalent modification of the catalytic cysteine in PTPases.[189,190] Thus, most inhibitors reported prior to the year 2000 were not suited for clinical use due to their lack of oral bioavailability and metabolic instability.[191] The researchers at Novo embarked on identifying a general, reversible, competitive PTP inhibitor that could be used as a common scaffold for lead optimization for specific PTPases. It was thought important that it should: (1) be a general inhibitor mimicking the binding of pTyr; (2) be a competitive, reversible active-site inhibitor; and (3) have a molecular weight below 300 Da to leave room for further optimization for potency and selectivity.

HTS of a diverse compound library using PTP-1B and a synthetic 33P-phosphorylated peptide as substrate was initiated. Using this approach, 2-(oxalylamino)-benzoic acids, such as compound 59 (**Figure 30a**), were identified.[192,193] Kinetic studies showed that compound 59 acts as a classical, time-independent, active-site-directed, reversible competitive inhibitor that does not covalently modify PTP-1B. As a first attempt to improve the potency of compound 59, structure-based design suggested analogs that included either naphtlyl or indole moieties. In agreement with the predictions, both the naphthyl-(compound 60) and the indole-(compounds 61 and 62) derivatives showed increased affinity for PTP-1B. These compounds were tested against a diverse set of catalytic domains representing six different PTP families. They inhibit most PTPases, with LAR as a notable exception. Significant differences in the inhibitor profiles against the different PTPases were observed for these compounds. As an example, compound 61 shows a 30-fold increase in potency against *src* homology domain 2 (SH2)-containing tyrosine phosphatase-1 (SHP-1), but only a twofold increase against PTP-1B compared to compound 59. These differences are particularly noteworthy since all compounds most likely address the active site-binding pocket only, due to their small size.

The crystal structures show that the overall ligand conformation of compounds 59–64 is almost planar, with both carboxylic acid groups possessing small twists out of the plane. As seen in **Figure 30b**, the oxalylamino part of compound 62 superimposes quite nicely with the pTyr residues in the phosphorylated peptides. The carboxylate group is H-bonded to the side chain of Arg221 and to the main-chain amides of Arg221 and Ser216 (**Figure 30c**). The carbonyl function forms an H-bond with the main-chain amide of Gly220. As reported earlier, the binding of inhibitors such as pTyr, tyrosinephosphorylated peptide, or vanadate induces a dramatic conformational change in the loop containing Trp179Pro180Asp181(WPD) loop which brings the conserved Asp181 into a position where it can participate in substrate binding and serve as a general acid in substrate hydrolysis. A similar movement of the WPD loop is observed when compounds 59–64 bind to PTP-1B. The closure of the WPD loop brings the conserved Asp181 into an apparently unfavorable position, i.e., only 2.9 Å from the *o*-carboxyl moiety of the compounds, which is unfavorable if Asp181 is deprotonated. However, biochemical studies have shown that Asp181 functions as a general acid and thus must be protonated in a significant proportion of the molecules.[194] Hence, Asp181 can form H-bonds with the *o*-carboxyl moiety of the compounds. In comparison to other x-ray crystallographic structures of PTP-1B, Lys120 has moved approximately 1 Å to be within about 2.8 Å distance from the *o*-carboxylate moiety of compound 61. The pK_a values for the compounds have been determined to be between 3.8 and 4.8, respectively. Thus, the *o*-carboxylate moiety of compound 61 is likely to be fully deprotonated at neutral pH. This will allow the formation of a salt-bridge between Lys120 and compound 61. Clearly the *o*-carboxylate of the compounds provides additional interactions within the active site, compared to the binding of pTyr, thus giving rise to high-affinity inhibitors. Although these inhibitors are charged, good oral bioavailability has been observed in rats for some compounds using prodrug concepts.

A key issue is to obtain selectivity for PTP-1B over other PTPases. A detailed comparison of all published PTPases was made, with the aim of identifying unique combinations of amino acid residues that could be used in a structure-based approach to design selective inhibitors.[195] The analysis was based on both primary-sequence alignments of PTP domains and low-resolution homology modeling. In particular, residues in the vicinity of compounds 59–62 were noted which are different between individual phophatases. Several potential selectivity-determining areas were identified, including a region defined by amino acids 47, 48, 258, and 259. Asp48 is believed to play an important role in positioning substrates relative to the active site[182] and seems a particularly attractive position for the generation of selective PTP-1B ligands, since this residue is an aspartic acid in PTP-1B and in many other PTPases an asparagine. This hypothesis was tested using compound 63 which, according to modeling, should place an H-bond donor next to the Asp48.[193] The crystal structure shows that the secondary amine forms an H-bond to Asp48 in PTP1B (**Figure 30d**). In other PTPases with an asparagine in the equivalent position, the basic nitrogen would cause repulsion and hence lead to selectivity against these enzymes. In addition, this compound showed remarkable

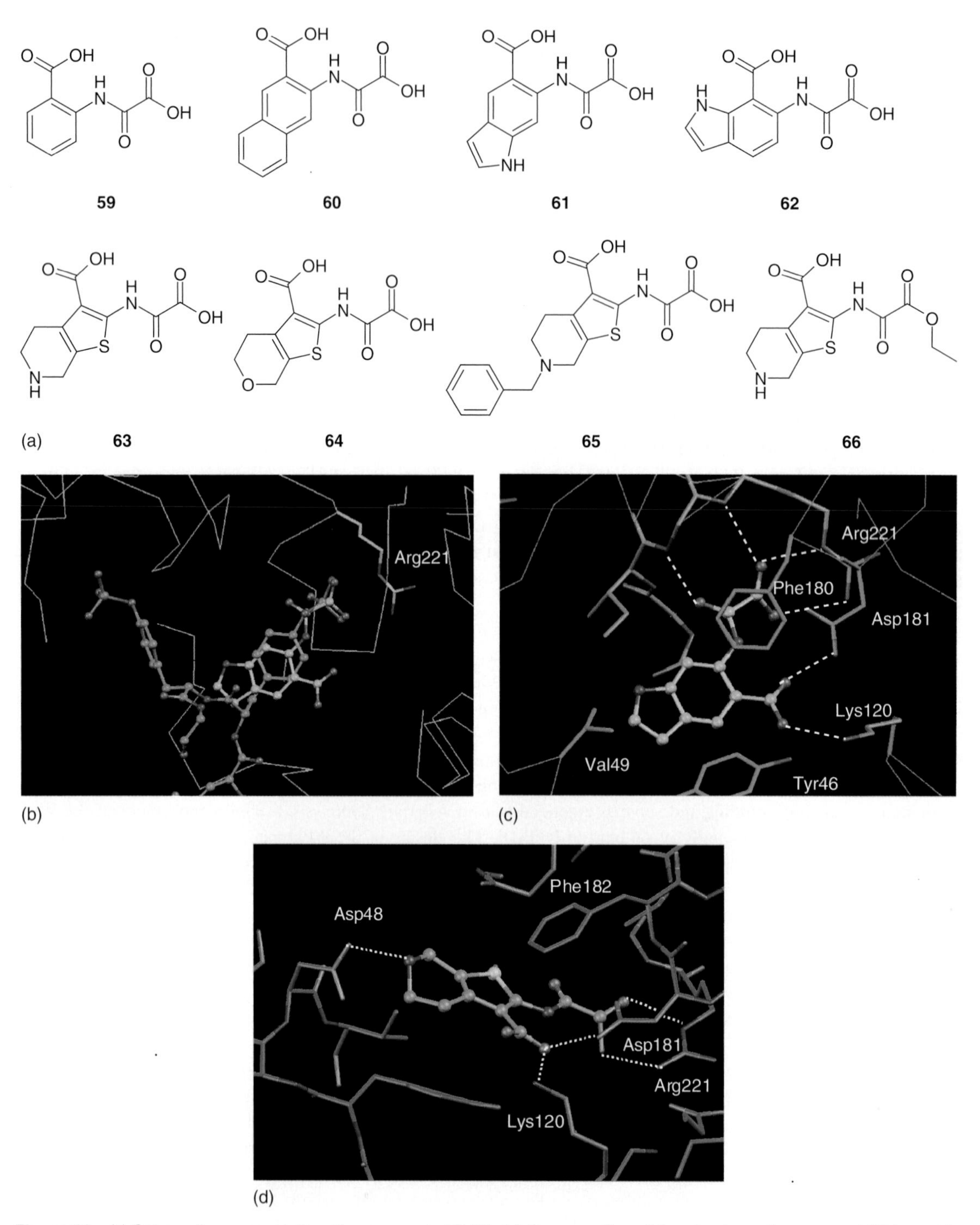

Figure 30 (a) Schematic representation of compounds 59–66. (b) Superposition of the structure of compound 62 bound to PTP-1B (blue) with the structure of the GluThrAsp(pTyr1162)(pTyr1163)Arg peptide (green) (1GIF). (c) Interaction of compound 62 with PTP-1B (1C83). PTP-1B is shown in green. Compound 62 is displayed in blue. (d) Interaction of compound 63 with PTP-1B (IC87). PTP-1B is shown in green. Compound 63 is displayed in blue.

selectivity for PTP-1B versus all other PTPases tested. Further cycles of structure-based drug design[196] led to compounds such as compounds 64 and 65. These compounds have reasonable bioavailability. In addition, prodrug analogs such as compound 66 have given rise to an enhancement of 2-deoxy-glucose accumulation in C2C12 cells and this confirms the validity of PTP-1B as a target in diabetes.

4.24.5.6 Structure-Based Drug Design Using Fragment Screening and X-ray Structure-Based Assembly

Liu *et al.*[197] used an NMR-based fragment screening technique in order to identify fragments binding into the pTyr pocket. The general idea of using low-molecular-weight fragments as starting points had already led to the discovery of oxamic acid-based PTP-1B inhibitors at Abbott, such as compound 67, shown in **Figure 31a**.[198] While potent, these highly charged inhibitors were not cell-permeable and the lack of cellular permeability and activity needed to be circumvented through an ester prodrug approach. In order to make cell permeability more likely, the researcher at Abbott screened only monocarboxylic acid or noncarboxylic acid-based fragments for potential active-site ligands. Since *N*-phenyloxamic acid appears to be the most potent nonphosphorus-containing pTyr mimetic, a series of heterocycle carboxylic acids as potential *N*-phenyloxamic acid mimetics with reduced pK_a was designed (compounds 68–71). These

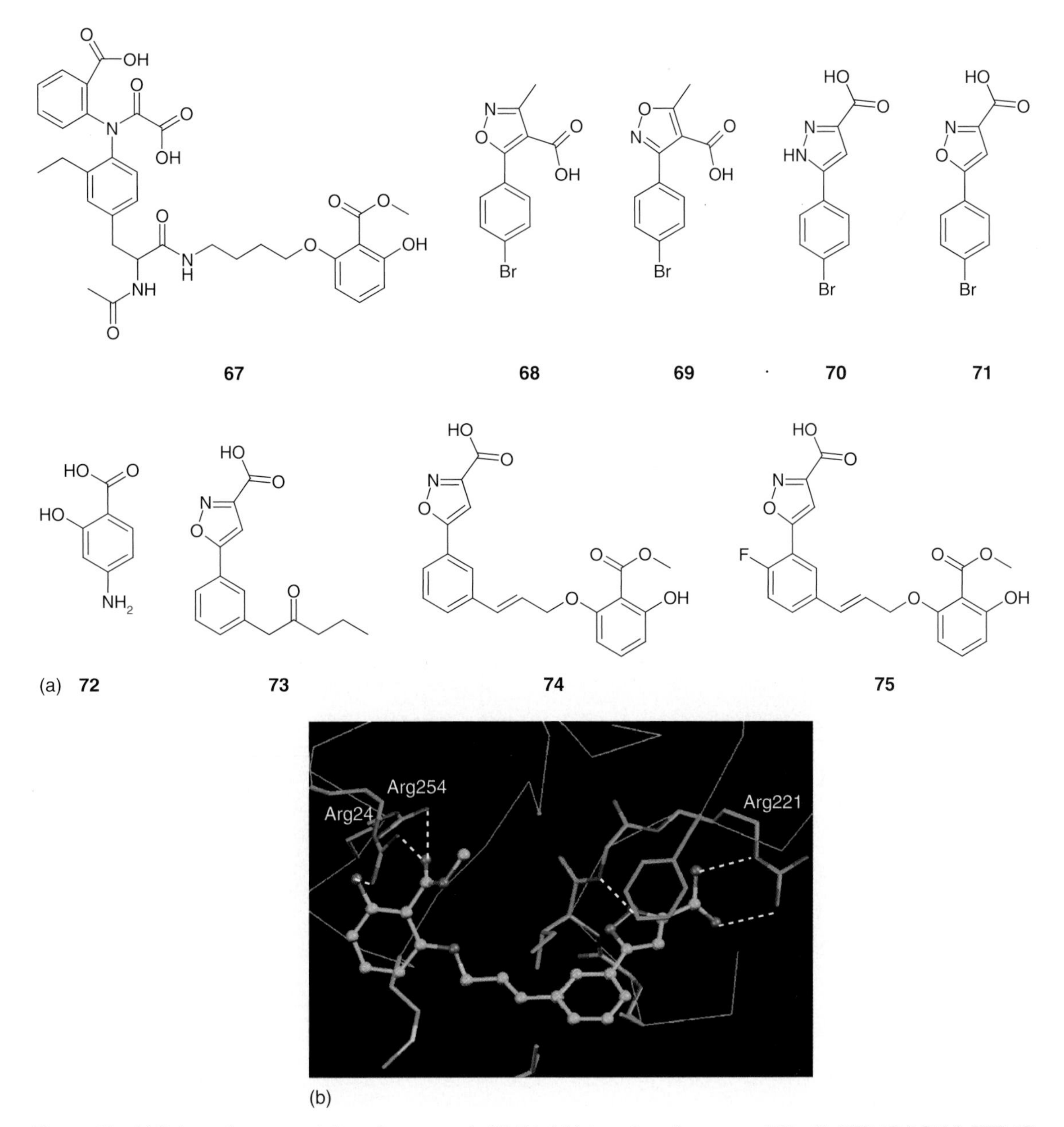

Figure 31 (a) Schematic representation of compounds 67–74. (b) Interaction of compound 74 with PTP-1B (1Q1M). PTP-1B is shown in green. Compound 75 is displayed in blue.

compounds were measured using NMR-based screening for a possible interaction with the catalytic site. Thus, compound 71 was identified as a weak binder to the pTyr pocket, with a dissociation constant K_d of 800 μM.

Subsequently, structure-based design was used to link compound 71 to fragments which had been identified earlier for the second aryl phosphate-binding site, such as compound 72.[198] To link the fragments more effectively, the binding mode of compound 73 was determined since it was not possible to obtain the structure of compound 71. This x-ray structure suggested that a four-atom linker off the meta-position of the phenyl ring should be optimal for linking the fragments. Improvement of the initially synthesized molecules by conformational modification of the linker led to the identification of compounds 74 and 75, which have a K_i against PTP-1B of 5.7 μM and K_i of 6.9 μM, respectively. The x-ray crystal structure of compound 74 complexed to PTP-1B confirms that the isoxazole carboxylic acid binds to the active site of PTP-1B (**Figure 31b**). The isoxazole and phenyl rings are located in the hydrophobic pocket normally occupied by the phenyl ring of pTyr. The H-bond network between compound 75 and PTP1B is essentially the same as that between compound 67 and PTP1B. The salicylate carboxylate group adapts an out-of-plane conformation and is within H-bonding distance of Arg254. The hydroxyl moiety forms H-bonds with Arg24 and Arg254. The aromatic portion of the salicylate lies on top of the hydrophobic side chain of Met258, giving rise to a significant hydrophobic effect. Both compounds 74 and 75 demonstrated a greater than 30-fold selectivity over TCPTP, the most homologous PTPase, and showed no inhibition of LAR, CD45, cdc25, and SHP-2 at the highest concentration tested. In addition, the researchers at Abbott reported good cellular activity in COS-7 cells.

4.24.5.7 Structure Elucidation of TCPTP: Toward Selective PTP-1B Inhibitors

Several groups had reported structurally diverse PTP-1B inhibitors that showed a high degree of selectivity for PTP-1B over several PTPases (including PTPR, LAR, CD45, and VHR (VH1-related dual-specific protein phosphatase)) but quite often not over the TCPTP. Despite its name, TCPTP is an ubiquitous enzyme. The TCPTP cDNA encodes a 45-kDa protein that displays 65% sequence identity overall and 72% identity within the conserved catalytic domain with PTP-1B.[199] TCPTP-deficient mutant mice exhibit specific defects in bone marrow stromal cells, B-cell lymphopoiesis, and erythropoiesis, as well as impaired T- and B-cell functions.[200] These studies suggest that TCPTP plays a significant role in both hematopoiesis and immune function, and indicate that a therapeutically useful PTP-1B inhibitor should display a significant selectivity over TCPTP. The structures of PTP-1B in complex with pTyr[181] and with a peptide derived from the insulin receptor[182] revealed a PTP-1B-specific second aryl-phosphate-binding site adjacent to the catalytic site. Many authors have provided convincing evidence that at least some differences in substrate recognition between the two enzymes are related to differences in this pocket. Also, Iversen *et al.*[201] reported that their active-site inhibitors inhibited PTP-1B and TCPTP with almost identical potency. In order to determine if

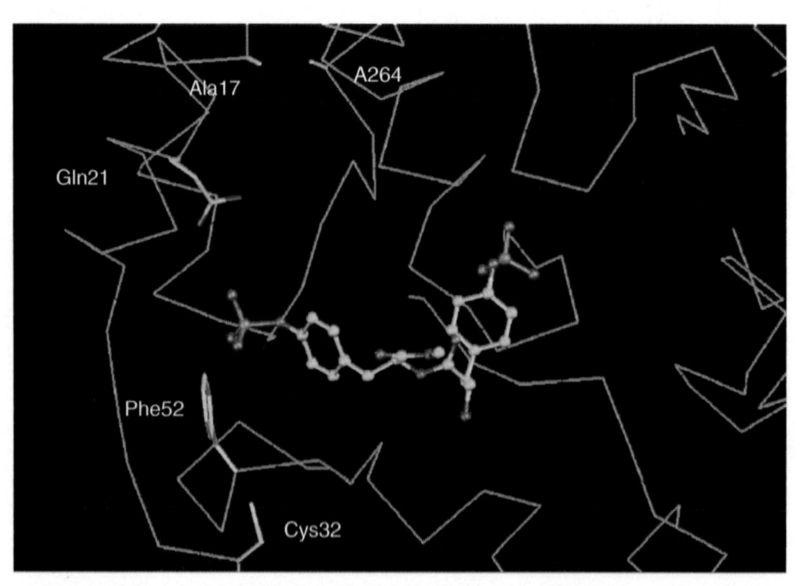

Figure 32 Location of amino acids within the PTP-1B binding pocket which have a different sequence in TCPTP using the structure of pTyr complexed to PTP-1B. PTP-1B is shown in green. The pTyr residues are displayed in blue.

there are differences between PTP-1B and TCPTP which can be exploited and where they are located within the active site, the researchers at Novo determined the crystal structure of the TCPTP.[201]

The crystal structure of TCPTP shows that the pTyr and secondary aryl phosphate-binding sites are nearly identical to those in PTP-1B. Nevertheless, small differences were found around the secondary aryl-binding sites. PTP-1B inhibitors that bind there may, therefore, be expected to achieve some degree of selectivity between the two enzymes. Two areas of potential interest were identified. The first region, termed the 258/259 gateway, consists of His34, Glu41, and Tyr54 in TCPTP and Cys32, Lys39, and Phe52 in PTP-1B. The second cluster is formed by Gln19, Leu23, and Pro262 in TCPTP and Ala17, Gln21, and Ala 264 in PTP-1B (**Figure 32**). Even though these differences may seem small, it should be noted that subtle differences in the adenosine triphosphate (ATP)-binding sites of kinases have previously been used successfully to develop selective inhibitors.[202] Also, a recent structural comparison of the closely related insulin-like growth factor 1 receptor kinase and the insulin receptor kinase (80% sequence identity) led to the identification of similar minor differences that the authors hypothesized could be used to develop selective ATP-competitive inhibitors of each kinase.[203]

4.24.6 Summary

These examples show that structure-based drug design can successfully contribute to the discovery process at different stages. It can be used at a very early stage at which no leads are available. Protein structures have been used to generate leads either by taking advantage of the similarity of the target protein to a protein with known inhibitors or by ab initio design. An example for the first case represents the modification of renin inhibitors such that they bind to HIV protease.[59,62,67] An example for ab initio design leading to HIV protease inhibitors was published by the group at DupontMerck for the HIV protease.[71] Alternatively, screening fragments using protein crystallography has resulted in leads which were successfully combined to yield high-affinity inhibitors. Examples include the fragment approaches reported for the src SH2 domain[149] and PTP-1B.[197] The most common use of protein structures represents the optimization of lead structures. These lead structures can be either HTS hits or naturally occurring ligands such as substrates or transition-state analogs. HTS hits which were optimized using structure-based drug design include the penicillin-derived inhibitors of the HIV protease[80] and the 2(-oxalylamino)-benzoic acid inhibitors of PTP-1B.[192,193] Neu5Ac2en, a transition-state analog, was optimized using structure-based drug design, leading to zanamivir.[108] In addition, structure-based drug design has been shown to support the drug discovery at a very late stage. Improving the physical chemical properties based on the protein structures has led in many cases to bioavailable inhibitors. Examples here are the optimization of HIV inhibitors[64,70] and NA inhibitors.[109] Protein structures have also been shown to be a useful tool for understanding the selectivity within protein families, such as the phosphatases[201] and the target-based resistance observed for selected inhibitors.[66,115,116] For instance, researchers at Merck[66] have replaced an isopropylthiazolylmethyl urea by a much shorter cyclic urea, thereby achieving that the modified inhibitor does not interact directly with a target-based mutation site of the HIV protease. As a result, this shorter compound also inhibits HIV protease, for which point mutants at this site have been reported to be resistant.

References

1. Abola, E.; Kuhn, P.; Earnest, T.; Stevens, R. C. *Nat. Struct. Biol.* **2000**, *7*, 973–977.
2. Blundell, T. L.; Jhoti, H.; Abell, C. *Nat. Rev. Drug Disc.* **2002**, *1*, 45–54.
3. Stevens, R. C. *Curr. Opin. Struct. Biol.* **2000**, *10*, 558–563.
4. Gilbert, M.; Albala, J. S. *Curr. Opin. Chem. Biol.* **2002**, *6*, 102–105.
5. Hendrickson, W. A.; Horton, J. R.; LeMaster, D. M. *EMBO J.* **1990**, *9*, 1665–1672.
6. Santarsiero, B. D.; Yegian, D. T.; Lee, C. C.; Spraggon, G.; Gu, J.; Scheibe, D.; Uber, D. C.; Cornell, E. W.; Nordmeyer, R. A.; Kolbe, W. F. et al. *J. Appl. Crystallogr.* **2002**, *35*, 278–281.
7. Garman, E. *Acta Crystallogr. D* **1999**, *55*, 1641–1653.
8. Muchmore, S. W.; Olson, J.; Jones, R.; Pan, J.; Blum, M.; Greer, J.; Merrick, S. M.; Magdalinos, P.; Nienaber, V. L. *Struct. Fold Des.* **2000**, *8*, R243–R246.
9. Guss, J. M.; Merritt, E. A.; Phizackerley, R. P.; Hedman, B.; Murata, M.; Hodgson, K.; Freeman, H. C. *Science* **1988**, *241*, 806–811.
10. Adams, P. D.; Grosse-Kunstleve, R. W. *Curr. Opin. Struct. Biol.* **2000**, *10*, 564–568.
11. Oldfield, T. J. *Acta Crystallogr. D* **2002**, *D58*, 487–493.
12. Oldfield, T. J. *Acta Crystallogr. D* **2003**, *D59*, 483–491.
13. Perrakis, A.; Morris, R.; Lamzin, V. S. *Nat. Struct. Biol.* **1999**, *6*, 458–463.
14. Rahuel, J.; Rasetti, V.; Maibaum, J.; Rueger, H.; Goschke, R.; Cohen, N. C.; Stutz, S.; Cumin, F.; Fuhrer, W.; Wood, J. M. et al. *Chem. Biol.* **2000**, *7*, 493–504.
15. Maignan, S.; Mikol, V. *Curr. Top Med. Chem.* **2001**, *1*, 161–174.
16. Liebeschuetz, J. W.; Jones, S. D.; Morgan, P. J.; Murray, C. W.; Rimmer, A. D.; Roscoe, J. M.; Waszkowycz, B.; Welsh, P. M.; Wylie, W. A.; Young, S. C. et al. *J. Med. Chem.* **2002**, *45*, 1221–1232.

17. Chirgadze, N. Y.; Sall, D. J.; Briggs, S. L.; Clawson, D. K.; Zhang, M.; Smith, G. F.; Schevitz, R. W. *Protein Sci.* **2000**, *9*, 29–36.
18. Young, W. B.; Kolesnikov, A.; Rai, R.; Sprengeler, P. A.; Leahy, E. M.; Shrader, W. D.; Sangalang, J.; Burgess-Henry, J.; Spencer, J.; Elrod, K. et al. *Bioorg. Med. Chem. Lett.* **2001**, *11*, 2253–2256.
19. Hajduk, P. J.; Boyd, S.; Nettesheim, D.; Nienaber, V.; Severin, J.; Smith, R.; Davidson, D.; Rockway, T.; Fesik, S. W. *J. Med. Chem.* **2000**, *43*, 3862–3866.
20. Greenspan, P. D.; Clark, K. L.; Tommasi, R. A.; Cowen, S. D.; McQuire, L. W.; Farley, D. L.; van Duzer, J. H.; Goldberg, R. L.; Zhou, H.; Du, Z. et al. *J. Med. Chem.* **2001**, *44*, 4524–4534.
21. Katunuma, N.; Murata, E.; Kakegawa, H.; Matsui, A.; Tsuzuki, H.; Tsuge, H.; Turk, D.; Turk, V.; Fukushima, M.; Tada, Y. et al. *FEBS Lett.* **1999**, *458*, 6–10.
22. Ghosh, A. K.; Bilcer, G.; Harwood, C.; Kawahama, R.; Shin, D.; Hussain, K. A.; Hong, L.; Loy, J. A.; Nguyen, C.; Koelsch, G. et al. *J. Med. Chem.* **2001**, *44*, 2865–2868.
23. Schindler, T.; Bornmann, W.; Pellicena, P.; Miller, W. T.; Clarkson, B.; Kuriyan, J. *Science* **2000**, *289*, 1938–1942.
24. Honma, T.; Hayashi, K.; Aoyama, T.; Hashimoto, N.; Machida, T.; Fukasawa, K.; Iwama, T.; Ikeura, C.; Ikuta, M.; Suzuki-Takahashi, I. et al. *J. Med. Chem.* **2001**, *44*, 4615–4627.
25. Ghosh, S.; Liu, X.-P.; Zheng, Y.; Uckun, F. M. *Curr. Cancer Drug Targets* **2001**, *1*, 129–140.
26. Fretz, H.; Furet, P.; Garcia-Echeverria, C.; Schoepfer, J.; Rahuel, J. *Curr. Pharm. Des.* **2000**, *6*, 1777–1796.
27. Iwata, Y.; Naito, S.; Itai, A.; Miyamoto, S. *Drug Des. Disc.* **2001**, *17*, 349–359.
28. Mihelich, E. D.; Schevitz, R. W. *Biochim. Biophys. Acta* **1999**, *1441*, 223–228.
29. Shahripour, A. B.; Plummer, M. S.; Lunney, E. A.; Albrecht, H. P.; Hays, S. J.; Kostlan, C. R.; Sawyer, T. K.; Walker, N. P.; Brady, K. D.; Allen, H. J. et al. *Bioorg. Med. Chem.* **2002**, *10*, 31–40.
30. Webber, S. E.; Bleckman, T. M.; Attard, J.; Deal, J. G.; Kathardekar, V.; Welsh, K. M.; Webber, S.; Janson, C. A.; Matthews, D. A.; Smith, W. W. *J. Med. Chem.* **1993**, *36*, 733–746.
31. Doucet-Personeni, C.; Bentley, P. D.; Fletcher, R. J.; Kinkaid, A.; Kryger, G.; Pirard, B.; Taylor, A.; Taylor, R.; Taylor, J.; Viner, R. et al. *J. Med. Chem.* **2001**, *44*, 3203–3215.
32. Schroder, J.; Henke, A.; Wenzel, H.; Brandstetter, H.; Stammler, H. G.; Stammler, A.; Pfeiffer, W. D.; Tschesche, H. *J. Med. Chem.* **2001**, *44*, 3231–3243.
33. Gritsch, S.; Guccione, S.; Hoffmann, R.; Cambria, A.; Raciti, G.; Langer, T. *J. Enzyme Inhib.* **2001**, *16*, 199–215.
34. Boehm, H. J.; Boehringer, M.; Bur, D.; Gmuender, H.; Huber, W.; Klaus, W.; Kostrewa, D.; Kuehne, H.; Luebbers, T.; Meunier-Keller, N. et al. *J. Med. Chem.* **2000**, *43*, 2664–2674.
35. Guilloteau, J.-P.; Mathieu, M.; Giglione, C.; Blanc, V.; Dupuy, A.; Chevrier, M.; Gil, P.; Famechon, A.; Meinnel, T.; Mikol, V. *J. Mol. Biol.* **2002**, *320*, 951–962.
36. Hucke, O.; Gelb, M. H.; Verlinde, C. L. M. J.; Buckner, F. S. *J. Med. Chem.* **2005**, *48*, 5415–5418.
37. Foloppe, N.; Fisher, L. M.; Howes, R.; Kierstan, P.; Potter, A.; Robertson, A. G. S.; Surgenor, A. E. *J. Med. Chem.* **2005**, *48*, 4332–4345.
38. Berman, H. M.; Westbrook, J.; Feng, Z.; Gilliland, G.; Bhat, T. N.; Weissig, H.; Shindyalov, I. N.; Bourne, P. E. *Nucleic Acids Res.* **2000**, *28*, 235–242.
39. Gallo, R. C.; Montagnier, L. *Sci. Am.* **1988**, *259*, 41–48.
40. Ratner, L.; Haseltine, W.; Patarca, R.; Livak, K. J.; Starcich, B.; Josephs, S. F.; Doran, E. R.; Rafalski, J. A.; Whitehorn, E. A.; Baumeister, K. et al. *Nature* **1985**, *313*, 277–284.
41. Dickson, C.; Eisenman, R.; Fan, H.; Hunter, E.; Teich, N. Protein Biosynthesis and Assembly. In *RNA Tumor Viruses, Molecular Biology of Tumor Viruses*; Weiss, R., Teich, N., Varmus, H., Coffin, J., Eds.; Cold Spring Harbor: New York, 1984, pp 513–648.
42. Henderson, L. E.; Sowder, R.; Copeland, T. D.; Smythers, G.; Oroszlan, S. *J. Virol.* **1984**, *52*, 492–500.
43. Jacks, T.; Power, M. D.; Masiarz, F. R.; Luciw, P. A.; Barr, P. J.; Varmus, H. E. *Nature* **1988**, *331*, 280–283.
44. Kohl, N. E.; Emini, E. A.; Schleif, W. A.; Davis, L. J.; Heimbach, J. C.; Dixon, R. A. F.; Scolnick, E. M.; Sigal, I. *Proc. Natl. Acad. Sci. USA* **1988**, *85*, 4686–4690.
45. Turner, S. R.; Strohbach, J. W.; Tommasi, R. A.; Aristoff, P. A.; Johnson, P. D.; Skulnick, H. I.; Dolak, L. A.; Seest, E. P.; Tomich, P. K.; Bohanon, M. J. et al. *J. Med. Chem.* **1998**, *41*, 3467–3476.
46. Bartlett, J. A.; DeMasi, R.; Quinn, J.; Moxham, C.; Rousseau, F. *AIDS* **2001**, *15*, 1369–1377.
47. Gulick, R. M.; Mellors, J. W.; Havlir, D.; Eron, J. J.; Meibohm, A.; Condra, J. H.; Valentine, F. T.; McMahon, D.; Gonzalez, C.; Jonas, L. et al. *Ann. Intern. Med.* **2000**, *133*, 35–39.
48. Tang, J.; James, M. N. G.; Hsu, I. N.; Jenkins, J. A.; Blundell, T. L. *Nature* **1978**, *271*, 618–621.
49. Lapatto, R.; Blundell, T.; Hemmings, A.; Overington, J.; Wilderspin, A.; Wood, S.; Merson, J. R.; Whittle, P. J.; Danley, D. E.; Geoghegan, K. F. *Nature* **1989**, *342*, 299–302.
50. Navia, M. A.; Fitzgerald, P. M.; McKeever, B. M.; Leu, C. T.; Heimbach, J. C.; Herber, W. K.; Sigal, I. S.; Darke, P. L.; Springer, J. P. *Nature* **1989**, *337*, 615–620.
51. Miller, M.; Jaskolski, M.; Rao, J. K.; Leis, J.; Wlodawer, A. *Nature* **1989**, *337*, 576–579.
52. Wlodawer, A.; Miller, M.; Jaskolski, M.; Sathyanarayana, B. K.; Baldwin, E.; Weber, I. T.; Selk, L. M.; Clawson, L.; Schneider, J.; Kent, S. B. *Science* **1989**, *245*, 616–621.
53. Weber, I. T.; Miller, M.; Jaskolski, M.; Leis, J.; Skalka, A. M.; Wlodawer, A. *Science* **1989**, *243*, 928–931.
54. Miller, M.; Schneider, J.; Sathyanarayana, B. K.; Toth, M. V.; Marshall, G. R.; Clawson, L.; Selk, L.; Kent, S. B.; Wlodawer, A. *Science* **1989**, *246*, 1149–1152.
55. Rich, D. H. Proteinase Inhibitors. In *Research Monographs in Cell and Tissue Physiology*; Barrett, A. J., Salvesen, G., Eds.; Elsevier Science: Amsterdam, 1986, pp 179–217.
56. Rich, D. H. Peptidase Inhibitors. In *Comprehensive Medicinal Chemistry*; Sammes, P. G., Ed.; Pergamon Press: Oxford, 1990; Vol. 2, pp 391–441 and references therein.
57. Billich, S.; Knoop, M. T.; Hansen, J.; Strop, P.; Sedlacek, J.; Mertz, R.; Moelling, K. *J. Biol. Chem.* **1988**, *263*, 17905–17908.
58. Rich, D. H.; Green, J.; Toth, M. V.; Marshall, G. R.; Kent, St. H. *J. Med. Chem.* **1990**, *33*, 1285–1288.
59. Vacca, J. P.; Guare, J. P.; deSolms, S. J.; Sanders, W. M.; Giuliani, E. A.; Young, S. D.; Darke, P. L.; Zugay, J.; Sigal, I. S.; Schleif, W. et al. *J. Med. Chem.* **1991**, *34*, 1225–1228.
60. Huff, J. R. *J. Med. Chem.* **1991**, *34*, 2305–2314.
61. Martin, J. A. *Antiviral Res.* **1992**, *17*, 265–278.

62. Erickson, J.; Neidhart, D. J.; VanDrie, J.; Kempf, D. J.; Wang, X. C.; Norbeck, D. W.; Plattner, J. J.; Rittenhouse, J. W.; Turon, M.; Wideburg, N. et al. *Science* **1990**, *249*, 527.
63. Kempf, D. J.; Norbeck, D. W.; Codacovi, L.; Wang, X. C.; Kohlbrenner, W. E.; Wideburg, N. E.; Paul, D. A.; Knigge, M. F.; Vasavanonda, S.; Craig-Kennard, A. *J. Med. Chem.* **1990**, *33*, 2687–2689.
64. Kempf, D. J.; Marsh, K. C.; Denissen, J. F.; McDonald, E.; Vasavanonda, S.; Flentge, C. A.; Green, B. E.; Fino, L.; Park, C. H.; Kong, X.-P. et al. *Proc. Natl. Acad. Sci. USA* **1995**, *92*, 2484–2488.
65. Sham, H. L.; Kempf, D. J.; Molla, A.; Marsh, K. C.; Kumar, G. N.; Chen, C.-M.; Kati, W.; Stewart, K.; Lal, R.; Hsu, A. et al. *Antimicrob. Agents Chemother.* **1998**, *42*, 3218.
66. Stoll, V.; Qin, W.; Stewart, K. D.; Jakob, C.; Park, C.; Walter, K.; Simmer, R. L.; Helfrich, R.; Bussiere, D.; Kao, J. et al. *Bioorg. Med. Chem.* **2002**, *10*, 2803–2806.
67. Krohn, A.; Redshaw, S.; Ritchie, J. C.; Graves, B. J.; Hatada, M. H. *J. Med. Chem.* **1991**, *34*, 3340–3342.
68. Lyle, T. A.; Wiscount, C. M.; Guare, J. P.; Thompson, W. J.; Anderson, P. S.; Darke, P. L.; Zugay, J. A.; Emini, E. A.; Schleif, W. A.; Quintero, J. C. et al. *J. Med. Chem.* **1991**, *34*, 1228–1230.
69. Dorsey, B. D.; Levin, R. B.; McDaniel, S. L.; Vacca, J. P.; Guare, J. P.; Darke, P. L.; Zugay, J. A.; Emini, E. A.; Schleif, W. A.; Quintero, J. C. et al. *J. Med. Chem.* **1994**, *37*, 3443–3451.
70. Dorsey, B. D.; McDonough, C.; McDaniel, S. L.; Levin, R. B.; Newton, C. L.; Hoffman, J. M.; Darke, P. L.; Zugay-Murphy, J. A.; Emini, E. A.; Schleif, W. A. et al. *J. Med. Chem.* **2000**, *43*, 3386–3399.
71. Lam, P. Y. S.; Jadhav, P. K.; Eyermann, C. J.; Hodge, C. N.; Ru, Y.; Bacheler, L. T.; Meek, O. M. J.; Rayner, M. M. *Science* **1994**, *263*, 380–384.
72. Tanki, N.; Rao, J. K. M.; Foundling, S. I.; Howe, W. J.; Moon, J. B.; Hui, J. O.; Tomasselli, A. G.; Heinrikson, R. L.; Thaisrivongs, S.; Wlodawer, A. *Protein Sci.* **1992**, *1*, 1061–1072.
73. Dreyer, G. B.; Boehm, J. C.; Chenera, B.; DesJarlais, R. L.; Hassell, A. M.; Meek, T. D.; Tomaszek, T. A., Jr.; Lewis, M. *Biochemistry* **1993**, *32*, 937–947.
74. Hosur, M. V.; Bhat, T. N.; Kempf, D. J.; Baldwin, E. T.; Liu, B.; Gulnik, S.; Wideburg, N. E.; Norbeck, D. W.; Appelt, K.; Erickson, J. W. *J. Am. Chem. Soc.* **1994**, *116*, 847–855.
75. Budt, K.-H.; Peyman, A.; Hansen, J.; Knolle, J.; Meichsner, C.; Paessens, A.; Ruppert, D.; Stowasser, B. *Bioorganic Med. Chem.* **1995**, *3*, 559–571.
76. Lange-Savage, G.; Berchtold, H.; Liesum, A.; Budt, K.-H.; Peyman, A.; Knolle, J.; Sedlacek, J.; Fabry, M.; Hilgenfeld, R. *Eur. J. Biochem.* **1997**, *248*, 313–322.
77. Humber, D. C.; Cummack, N.; Coates, N. J. A. V.; Cobley, K. N.; Orr, D. C.; Storer, R.; Weingarten, G. G.; Weir, M. P. *J. Med. Chem.* **1992**, *35*, 3080–3081.
78. Humber, D. C.; Bamford, M. J.; Bethell, R. C.; Cammack, N.; Cobley, K.; Evans, D. N.; Gray, N. M.; Hann, M. M.; Orr, D. C.; Saunders, J. et al. *J. Med. Chem.* **1993**, *36*, 3120–3128.
79. Holmes, D. S.; Clemens, I. R.; Cobley, K. N.; Humber, D. C.; Kitchin, J.; Orr, D. C.; Patel, B.; Paternoster, I. L.; Storer, R. *Bioorg. Med. Chem. Lett.* **1993**, *3*, 503–508.
80. Jhoti, H.; Singh, O. M. P.; Weir, M. P.; Cooke, R.; Murray-Rust, P.; Wonacott, A. *Biochemistry* **1994**, *33*, 8417–8427.
81. James, M. N. G.; Sielecki, A. R. *J. Mol. Biol.* **1983**, *163*, 299–361.
82. Holmes, D. S.; Bethell, R. C.; Cammack, N.; Clemens, I. R.; Kitchin, J.; McMeekin, P.; Mo, C. L.; Orr, D. C.; Patel, B.; Paternoster, I. L. et al. *J. Med. Chem.* **1993**, *36*, 3129–3136.
83. Kim, E. E.; Baker, C. T.; Dwyer, M. D.; Murcko, M. A.; Rao, B. G.; Tung, R. D.; Navia, M. A. *J. Am. Chem. Soc.* **1995**, *117*, 1181–1182.
84. Kaldor, S. W.; Hammond, M.; Dressman, B. A.; Fritz, J. E.; Crowell, T. A.; Hermann, R. A. *Bioorg. Med. Chem. Lett.* **1994**, *4*, 1385–1390.
85. Dressman, B. A.; Reich, S.; Pino, M.; Nyugen, D.; Appelt, K.; Musick, L.; Wu, B. *Bioorg. Med. Chem. Lett.* **1995**, *5*, 727–732.
86. Reich, S. H.; Melnick, M.; Davies, J.; Appelt, K.; Lewis, K.; Fuhry, M. A.; Pino, M.; Trippe, A. J.; Nguyen, D.; Dawson, B. W. et al. *Proc. Natl. Acad. Sci. USA* **1995**, *92*, 3298–3302.
87. Kaldor, S. W.; Kalish, V. J.; Davies, J. F.; Shetty, B. V.; Fritz, J. E.; Appelt, K.; Burgess, J. A.; Campanale, K. M.; Chirgadze, N. Y.; Clawson, D. K. et al. *J. Med. Chem.* **1997**, *40*, 3979–3985.
88. Kaldor, S. W.; Appelt, K.; Fritz, J. E.; Hammond, M.; Crowell, T. A.; Baxter, A. J.; Hatch, S. D.; Wiskerchen, M.; Muesing, M. A. *Bioorg. Med. Chem. Lett.* **1995**, *5*, 715–720.
89. Butler, D. *Nature* **2005**, *436*, 614–615.
90. Hay, A. J.; Thompson, C. A.; Geraghty, A.; Hayhurst, S.; Grambas, S.; Bennett, M. S. The Role of the M2 Protein in Influenza Virus Infection. In *Options for the Control of Influenza Virus II*; Hannoun, C., Kendal, A. P., Eds.; Excerpta Medica: Amsterdam, 1993, pp 281–288.
91. Bucher, D. J.; Palese, P. The Biologically Active Proteins of Influenza Virus: Neuraminidase. In *Influenza Virus and Influenza*; Kilbourne, E. D., Ed.; Academic Press: New York, 1975, pp 83–123.
92. Laver, W. G.; Valentine, R. C. *Virology* **1969**, *38*, 105–119.
93. Drzenick, R.; Frank, H.; Rott, R. *Virology* **1968**, *36*, 703–707.
94. Blok, J.; Air, G. M.; Laver, W. G.; Ward, C. W.; Lilley, G. G.; Woods, E. F.; Roxburgh, C. M.; Inglis, A. S. *Virology* **1982**, *119*, 109–121.
95. Klenk, E.; Faillard, H.; Lempfrid, H. Z. *Physiol. Chem.* **1955**, *301*, 235–246.
96. Palese, P.; Compans, R. W. *J. Gen. Virol.* **1976**, *33*, 159–163.
97. Meindl, P.; Tuppy, H. *Hoppe-Seyler's Z. Physiol. Chem.* **1969**, *350*, 1088–1092.
98. Meindl, P.; Bodo, G.; Palese, P.; Schulman, J.; Tuppy, H. *Virology* **1974**, *58*, 457–463.
99. Colman, P. M.; Varghese, J. N.; Laver, W. G. *Nature* **1983**, *303*, 41–44.
100. Varghese, J. N.; Laver, W. G.; Colman, P. M. *Nature* **1983**, *303*, 35–40.
101. Baker, A. T.; Varghese, J. N.; Laver, W. G.; Air, G. M.; Colman, P. M. *Proteins* **1987**, *1*, 111–117.
102. Tulip, W. R.; Varghese, J. N.; Baker, A. T.; Van Donkelaar, A.; Laver, W. G.; Webster, R. G.; Colman, P. M. *J. Mol. Biol.* **1991**, *221*, 487–497.
103. Burmeister, W. P.; Ruigrok, R. W. H.; Cusack, S. *EMBO J.* **1992**, *11*, 49–56.
104. Varghese, J. N.; Epa, V. C.; Colman, P. M. *Protein Sci.* **1995**, *4*, 1081–1087.
105. Varghese, J. N.; McKimm-Breschkin, J. L.; Caldwell, J. B.; Kortt, A. A.; Colman, P. M. *Proteins: Struct. Funct. Genet.* **1992**, *14*, 327–332.
106. Burmeister, W. P.; Henrissat, B.; Bosso, C.; Cussack, S.; Ruigrok, R. W. H. *Structure* **1993**, *1*, 19–26.
107. Bossart-Whitaker, P.; Carson, M.; Babu, Y. S.; Smith, C. D.; Laver, W. G.; Air, G. M. *J. Mol. Biol.* **1993**, *232*, 1069–1083.
108. von Itzstein, M.; Wu, W. Y.; Kok, G. B.; Pegg, M. S.; Dyason, J. C.; Jin, B.; Phan, T. V.; Smythe, M. L.; White, H. F.; Oliver, S. W. *Nature* **1993**, *363*, 418–423.
109. Kim, C. U.; Lew, W.; Williams, M. A.; Zhang, L.; Liu, H.; Swaminathan, S.; Bischofberger, N.; Chen, M. S.; Tai, C. Y.; Mendel, D. B. et al. *J. Am. Chem. Soc.* **1997**, *119*, 681–690.

110. Kim, C. U.; Lew, W.; Williams, M. A.; Wu, H.; Zhang, L.; Chen, X.; Escarpe, P. A.; Mendel, D. B.; Laver, W. G.; Stevens, R. C. *J. Med. Chem.* **1998**, *41*, 2451–2460.
111. Babu, Y. S.; Chand, P.; Bantia, S.; Kotian, P.; Dehghani, A.; El-Kattan, Y.; Lin, T.-H.; Hutchison, T. L.; Elliott, A. J.; Parker, C. D. et al. *J. Med. Chem.* **2000**, *43*, 3482–3486.
112. Wang, G. T.; Chen, Y.; Wang, S.; Gentles, R.; Sowin, T.; Kati, W.; Muchmore, S.; Giranda, V.; Stewart, K.; Sham, H. et al. *J. Med. Chem.* **2001**, *44*, 1192–1201.
113. Stoll, V.; Stewart, K. D.; Maring, C. J.; Muchmore, S.; Giranda, V.; Gu, Y.-G.; Wang, G.; Chen, Y.; Sun, M.; Zhao, C. et al. *Biochemistry* **2003**, *42*, 718–727.
114. Kimberlin, D. W.; Crumpacker, C. S.; Straus, S. E.; Biron, K. K.; Drew, W. L.; Hayden, F. G.; McKinlay, M.; Richman, D. D.; Whitley, R. J. *Antiviral Res.* **1995**, *26*, 423–438.
115. Blick, T. J.; Tiong, T.; Sahasrabudhe, A.; Varghese, J. N.; Colman, P. M.; Hart, G. J.; Bethell, R. C.; McKimm-Breschkin, J. L. *Virology* **1995**, *214*, 475–484.
116. Varghese, J. N.; Smith, P. W.; Sollis, S. L.; Blick, T. J.; Sahasrabudhe, A.; McKimm-Breschkin, J. L.; Colman, P. M. *Structure* **1998**, *6*, 735–746.
117. Finley, J. B.; Atigadda, V. R.; Duarte, F.; Zhao, J. J.; Brouillette, W. J.; Air, G. M.; Luo, M. *J. Mol. Biol.* **1999**, *293*, 1107–1119.
118. Howes, P. D.; Cleasby, A.; Evans, D. N.; Feilden, H.; Smith, P. W.; Sollis, S. L.; Taylor, N.; Wonacott, A. J. *Eur. J. Med. Chem.* **1999**, *34*, 225–234.
119. Chand, P.; Kotian, P. L.; Dehghani, A.; El-Kattan, Y.; Lin, T.-H.; Hutchison, T. L.; Babu, Y. S.; Bantia, S.; Elliott, A. J.; Montgomery, J. A. *J. Med. Chem.* **2001**, *44*, 4379–4392.
120. Chand, P.; Babu, Y. S.; Bantia, S.; Chu, N.; Cole, L. B.; Kotian, P. L.; Laver, W. G.; Montgomery, J. A. P.; Ved, P.; Petty, S. L. et al. *J. Med. Chem.* **1997**, *40*, 4030–4052.
121. Atigadda, V. R.; Brouillette, W. J.; Duarte, F.; Ali, S. M.; Babu, Y. S.; Bantia, S.; Chand, P.; Chu, N.; Montgomery, J. A.; Walsh, D. A. et al. *J. Med. Chem.* **1999**, *42*, 2332–2343.
122. White, C. L.; Janakiraman, M. N.; Laver, W. G.; Philippon, C.; Vasella, A.; Air, G. M.; Luo, M. *J. Mol. Biol.* **1995**, *245*, 623–634.
123. Maring, C. J.; Stoll, V. S.; Zhao, C.; Sun, M.; Krueger, A. C.; Stewart, K. D.; Madigan, D. L.; Kati, W. M.; Xu, Y.; Carrick, R. J. et al. *J. Med. Chem.* **2005**, *48*, 3980–3990.
124. Moustafa, I.; Connaris, H.; Taylor, M.; Zaitsev, V.; Wilson, J. C.; Kiefel, M. J.; von Itzstein, M.; Taylor, G. *J. Biol. Chem.* **2004**, *279*, 40819–40826.
125. Streicher, H. *Curr. Med. Chem. Anti-Infect. Agents* **2004**, *3*, 149–161.
126. Schultz, J.; Milpetz, F.; Bork, P.; Ponting, C. P. *Proc. Natl. Acad. Sci. USA* **1998**, *95*, 5857–5864.
127. Schaffhausen, B. *Biochim. Biophys. Acta* **1995**, *1242*, 61–75.
128. Brown, M. T.; Cooper, J. A. *Biochim. Biosphys. Acta* **1996**, *1287*, 121–149.
129. Machida, K.; Mayer, B. *J. Biochim. Biophys. Acta* **2005**, *1747*, 1–25.
130. Ottenhoff-Kalff, A. E.; Rijksen, G.; Van Beurden, E. A. C. M.; Hennipman, A.; Michels, A.; Staal, G. E. *J. Cancer Res.* **1992**, *52*, 4773–4778.
131. Soriano, P.; Montgomery, C.; Geske, R.; Bradley, A. *Cell* **1991**, *64*, 693–702.
132. Boyce, B. F.; Yoneda, T.; Lowe, C.; Soriano, P.; Mundy, G. *J. Clin. Invest.* **1992**, *90*, 1622–1627.
133. Xu, W.; Doshi, A.; Lei, M.; Eck, M. J.; Harrison, S. C. *Mol. Cell* **1999**, *3*, 629–638.
134. Lange, G.; Lesuisse, D.; Deprez, P.; Schoot, B.; Loenze, P.; Bénard, D.; Marquette, J.-P.; Broto, P.; Sarubbi, E.; Mandine, E. *J. Med. Chem.* **2002**, *45*, 2915–2922.
135. Waksman, G.; Kominos, D.; Robertson, S. C.; Pant, N.; Baltimore, D.; Birge, R. B.; Cowburn, D.; Hanafusa, H.; Mayer, B. J.; Overduin, M. *Nature* **1992**, *358*, 646–653.
136. Lee, Ch.-H.; Kominos, D.; Jacques, S.; Margolis, B.; Schlessinger, J.; Shoelson, S. E.; Juriyan, J. *Structure* **1994**, *2*, 423–438.
137. Mikol, V.; Baumann, G.; Keller, T. H.; Manning, U.; Zurini, M. G. M. *J. Mol. Biol.* **1995**, *246*, 344–355.
138. Maignan, S.; Guilloteau, J. P.; Fromage, N.; Arnoux, B.; Becquart, J.; Ducruix, A. *Science* **1995**, *268*, 291–293.
139. Gao, Y.; Wu, L.; Luo, J.; Guo, R.; Jang, D.; Zhang, Z.-Y.; Burke, T. *Bioorg. Med. Chem. Lett.* **2000**, *10*, 923–927.
140. Waksman, G.; Shoelson, S. E.; Pant, N.; Cowburn, D.; Kuriyan, J. *Cell* **1993**, *72*, 779–790.
141. Gilmer, T.; Rodriguez, M.; Jordan, S.; Crosby, R.; Alligood, K.; Green, M.; Kimery, M.; Wagner, C.; Kinder, D. *J. Biol. Chem.* **1994**, *269*, 31711–31719.
142. Rahuel, J.; Gay, B.; Erdmann, D.; Strauss, A.; Garcia-Echeverria, C.; Furet, P.; Caravatti, G.; Fretz, H.; Schoepfer, J.; Gruetter, M. G. *Nat. Struct. Biol.* **1996**, *3*, 586–589.
143. Rahuel, J.; Garcia-Echeverria, C.; Furet, P.; Strauss, A.; Caravatti, G.; Fretz, H.; Schoepfer, J.; Gay, B. *J. Mol. Biol.* **1998**, *279*, 1013–1022.
144. Lunney, E. A.; Para, K. S.; Rubin, J. R.; Humblet, C.; Fergus, J. H.; Marks, J. S.; Sawyer, T. K. *J. Am. Chem. Soc.* **1997**, *119*, 12471–12476.
145. Allen, F. H. *Acta Crystallogr.* **2002**, *B58*, 380–388.
146. Shakespeare, W. C.; Bohacek, R. S.; Azimioara, M. D.; Macek, K. J.; Luke, G. P.; Dalgarno, D. C.; Hatada, M. H.; Lu, X.; Violette, S. M.; Bartlett, C. et al. *J. Med. Chem.* **2000**, *43*, 3815–3819.
147. Mathews, I. I.; Padmanabhan, K. P.; Ganesh, V.; Tulinsky, A.; Ishii, M.; Chen, J.; Turck, C. W.; Coughlin, S. R.; Fenton, J. W., II. *Biochemistry* **1994**, *33*, 3266–3279.
148. Strickland, C. L. *Acta Crystallogr.* **1998**, *D54*, 1207–1215.
149. Lange, G.; Lesuisse, D.; Deprez, P.; Schoot, B.; Loenze, P.; Benard, D.; Marquette, J.-P.; Broto, P.; Sarubbi, E.; Mandine, E. *J. Med. Chem.* **2003**, *46*, 5184–5195.
150. Minke, W. E.; Diller, D. J.; Hol, W. G.; Verlinde, C. L. *J. Med. Chem.* **1999**, *42*, 1778–1788.
151. Shakespeare, W. C.; Bohacek, R. S.; Azimioara, M. D.; Macek, K. J.; Luke, G. P.; Dalgarno, D. C.; Hatada, M. H.; Lu, X.; Violette, S. M.; Bartlett, C. et al. *Proc. Natl. Acad. Sci. USA* **2000**, *97*, 9373–9378.
152. Kawahata, N.; Yang, M. Y.; Luke, G. P.; Shakespeare, W. C.; Sundaramoorthi, R.; Wang, Y.; Johnson, D.; Merry, T.; Violette, S.; Guan, W. et al. *Bioorg. Med. Chem. Lett.* **2001**, *11*, 2319–2323.
153. Shakespeare, W. C. *Curr. Opin. Chem. Biol.* **2001**, *5*, 409–415.
154. Sohara, Y.; Shimada, H.; Scadeng, M.; Pollack, H.; Yamada, S.; Ye, W.; Reynolds, C. P.; DeClerck, Y. A. *Cancer Res.* **2003**, *63*, 3026–3031.
155. Yoneda, T.; Hashimoto, N.; Hiraga, T. *Calcif. Tissue Int.* **2003**, *73*, 315–318.
156. Spencer, R. W. *Biotechnol. Bioeng.* **1998**, *61*, 61–67.
157. Kuntz, I. D.; Blaney, J. M.; Oatley, S. J.; Langridge, R.; Ferrin, T. E. *J. Mol. Biol.* **1982**, *161*, 269–288.

158. Rarey, M.; Kramer, B.; Lengauer, T.; Klebe, G. *J. Mol. Biol.* **1996**, *261*, 470–489.
159. Goodford, P. J. A. *J. Med. Chem.* **1985**, *28*, 849–857.
160. Boehm, H. J. *J. Comput.-Aided Mol. Des.* **1992**, *6*, 61–78.
161. Shuker, S. B.; Hajduk, P. J.; Meadows, R. P.; Fesik, S. W. *Science* **1996**, *274*, 1531–1534.
162. Hajduk, P. J.; Zhou, M.-M.; Fesik, S. W. *Bioorg. Med. Chem. Lett.* **1999**, *9*, 2403–2406.
163. Nienaber, V. L.; Richardson, P. L.; Klighofer, V.; Bouska, J. J.; Giranda, V. L.; Greer, J. *Nat. Biotechnol.* **2000**, *18*, 1105–1107.
164. Lesuisse, D.; Lange, G.; Deprez, P.; Schoot, B.; Benard, D.; Delettre, G.; Marquette, J.-P.; Broto, P.; Jean-Baptiste, V.; Bichet, P. et al. *J. Med. Chem.* **2002**, *45*, 2379–2387.
165. Kelly, M. A.; Liang, H.; Sytwu, I.-I.; Vlattas, I.; Lyons, N. L.; Bowen, B. R.; Wennogle, L. P. *Biochemistry* **1996**, *35*, 11747–11755.
166. Hajduk, P. J.; Meadows, R. P.; Fesik, S. W. *J. Med. Chem.* **2000**, *43*, 3443–3447.
167. Hartshorn, M. J.; Murray, C. W.; Cleasby, A.; Frederickson, M.; Tickle, I. J.; Jhoti, H. *J. Med. Chem.* **2005**, *48*, 403–413.
168. Lesuisse, D.; Deprez, P.; Albert, E.; Duc, T. T.; Sortais, B.; Gofflo, D.; Jean-Baptiste, V.; Marquette, J.-P.; Schoot, B.; Sarubbi, E. et al. *Bioorg. Med. Chem. Lett.* **2001**, *11*, 2127–2131.
169. Deprez, P.; Baholet, I.; Burlet, S.; Lange, G.; Schoot, B.; Vermond, A.; Mandine, E.; Lesuisse, D. *Bioorg. Med. Chem. Lett.* **2002**, *12*, 1291–1294.
170. Tong, L.; Warren, T. C.; Lukas, S.; Schembri-King, J.; Betageri, R.; Proudfoot, J.; Jakes, S. *J. Biol. Chem.* **1998**, *273*, 20238–20242.
171. Charifson, P. S.; Shewchuk, L. M.; Rocque, W.; Hummel, C. W.; Jordan, S. R.; Mohr, C.; Pacofsky, G. J.; Peel, M. R.; Rodriguez, M.; Sternbach, D. D. et al. *Biochemistry* **1997**, *36*, 6283–6293.
172. Murphy, G. J.; Holder, J. C. *Trends Pharmacol. Sci.* **2000**, *21*, 469–474.
173. Evans, J. L.; Jallal, B. *Exp. Opin. Investig. Drugs* **1999**, *8*, 139–160.
174. Uhlik, M. T.; Temple, B.; Bencharit, S.; Kimple, A. J.; Siderovski, D. P.; Johnson, G. L. *J. Mol. Biol.* **2004**, *345*, 1–20.
175. Tonks, N. K.; Diltz, C. D.; Fisher, E. H. *J. Biol. Chem.* **1988**, *263*, 6722–6730.
176. Tonks, N. K.; Diltz, C. D.; Fisher, E. H. *J. Biol. Chem.* **1988**, *263*, 6731–6737.
177. Elchebly, M.; Payette, P.; Michaliszyn, E.; Cromlish, W.; Collins, S.; Loy, A. L.; Normandin, D.; Cheng, A.; Himms-Hagen, J.; Chan, C.-C. et al. *Science* **1999**, *283*, 1544–1548.
178. Klaman, L. D.; Boss, O.; Peroni, O. D.; Kim, J. K.; Martino, J. L.; Zabolotny, J. M.; Moghal, N.; Lubkin, M.; Kim, Y. B.; Sharpe, A. H. et al. *Mol. Cell. Biol.* **2000**, *20*, 5479–5489.
179. Ukkola, O.; Santaniemi, M. *J. Intern. Med.* **2002**, *251*, 467–475.
180. Barford, D.; Flint, A. J.; Tonks, N. K. *Science* **1994**, *263*, 1397–1404.
181. Jia, Z.; Barford, D.; Flint, A. J.; Tonks, N. K. *Science* **1995**, *268*, 1754–1758.
182. Salmeen, A.; Andersen, J. N.; Meyers, M. P.; Tonks, N. K.; Barford, D. *Mol. Cell* **2000**, *6*, 1401–1412.
183. Montserat, J.; Chen, L.; Lawrence, D. S.; Zhang, Z.-Y. *J. Biol. Chem.* **1996**, *271*, 7868–7872.
184. Puius, Y. A.; Zhao, Y.; Sullivan, M.; Lawrence, D. S.; Almo, S. C.; Zhang, Z.-Y. *Proc. Natl. Acad. Sci. USA* **1997**, *94*, 13420–13425.
185. Zhang, Z.-Y.; Maclean, D.; McNamara, D. J.; Dobrusin, E. M.; Sawyer, T. K.; Dixon, J. E. *Biochemistry* **1994**, *33*, 2285–2290.
186. Drake, P. G.; Posner, B. I. *Mol. Cell. Biochem.* **1998**, *182*, 79–89.
187. Posner, B. I.; Faure, R.; Burgess, J. W.; Bevan, A. P.; Lachance, D.; Zhang-Sun, G. Y.; Fantus, I. G.; Ng, J. B.; Hall, D. A.; Lum, B. S. et al. *J. Biol. Chem.* **1994**, *269*, 4596–4604.
188. Murakami, H.; Takahashi, N.; Tanaka, S.; Nakamura, I.; Udagawa, N.; Nakajo, S.; Nakaya, K.; Abe, M.; Yuda, Y.; Konno, F. et al. *Bone* **1997**, *20*, 399–404.
189. Schmidt, A.; Rutledge, S. J.; Endo, N.; Opas, E. E.; Tanaka, H.; Wesolowski, G.; Leu, C. T.; Huang, Z.; Ramachandaran, C.; Rodan, S. B. et al. *Proc. Natl. Acad. Sci. USA* **1996**, *93*, 3068–3073.
190. Skorey, K.; Ly, H. D.; Kelly, J.; Hammond, M.; Ramachandran, C.; Huang, Z.; Gresser, M. J.; Wang, Q. J. *Biol. Chem.* **1997**, *272*, 22472–22480.
191. Burke, T. R., Jr.; Zhang, Z.-Y. *Biopolymers (Pept. Sci.)* **1998**, *47*, 225–241.
192. Andersen, H. S.; Iversen, L. F.; Jeppesen, C. B.; Branner, Sv.; Norris, K.; Rasmussen, H. B.; Moller, K. B.; Moller, N. P. H. *J. Biol. Chem.* **2000**, *275*, 7101–7108.
193. Iversen, L. F.; Andersen, H. S.; Branner, S.; Mortensen, S. B.; Peters, G. H.; Norris, K.; Olsen, O. H.; Jeppesen, C. B.; Lundt, B. F.; Ripka, W. et al. *J. Biol. Chem.* **2000**, *275*, 10300–10307.
194. Lohse, D. L.; Denu, J. M.; Santoro, N.; Dixon, J. E. *Biochemistry* **1997**, *36*, 4568–4575.
195. Andersen, J. N.; Mortensen, O. H.; Peters, G. H.; Drake, P. G.; Iversen, L. F.; Olsen, O. H.; Jansen, P. G.; Andersen, H. S.; Tonks, N. K.; Møller, N. P. H. *Mol. Cell. Biol.* **2001**, *21*, 7117–7136.
196. Andersen, H. S.; Olsen, O. H.; Iversen, L. F.; Sorensen, A. L. P.; Mortensen, S. B.; Christensen, M. S.; Branner, S.; Hansen, T. K.; Lau, J. F.; Jeppesen, L. et al. *J. Med. Chem.* **2002**, *45*, 4443–4459.
197. Liu, G.; Xin, Z.; Pei, Z.; Hajduk, P. J.; Abad-Zapatero, C.; Hutchins, C. W.; Zhao, H.; Lubben, T. H.; Ballaron, S. J.; Haasch, D. L. et al. *J. Med. Chem.* **2003**, *46*, 4232–4235.
198. Liu, G.; Xin, Z.; Liang, H.; Abad-Zapatero, C.; Hajduk, P. J.; Janowick, D. A.; Szczepankiewicz, B. G.; Pei, Z.; Hutchins, C. W.; Ballaron, S. J. et al. *J. Med. Chem.* **2003**, *46*, 3437–3440.
199. Flint, A. J.; Tiganis, T.; Barford, D.; Tonks, N. K. *Proc. Natl. Acad. Sci. USA* **1997**, *94*, 1680–1685.
200. Tiganis, T.; Bennett, A. M.; Ravichandran, K. S.; Tonks, N. K. *Mol. Cell. Biol.* **1998**, *18*, 1622–1634.
201. Iversen, L. F.; Moller, K. B.; Pedersen, A. K.; Peters, G. H.; Petersen, A. S.; Andersen, H. S.; Branner, S.; Mortensen, S. B.; Moller, N. P. H. *J. Biol. Chem.* **2002**, *277*, 19982–19990.
202. Dalgarno, D. C.; Metcalfe, D. D.; Shakespeare, W. C.; Sawyer, T. K. *Curr. Opin. Drug Disc. Dev.* **2000**, *3*, 549–564.
203. Favelyukis, S.; Till, J. H.; Hubbard, S. R.; Miller, W. T. *Nat. Struct. Biol.* **2001**, *8*, 1058–1063.

Biography

Gudrun Lange studied chemistry from 1980–86 at the University of Freiburg (Germany). In her master thesis she looked at the molecular structure of liquids and calculated the intermolecular atom-pair correlation functions of $Si(CH_3)_4$, $Ge(CH_3)_4$, and $Sn(CH_3)_4$ using x-ray data obtained form wide angle scattering experiments. Her doctoral work was performed between 1986 and 1989 at the Max-Planck Institute for Structural Biology in Hamburg with E Mandelkow. She analyzed the self-association of a protein by combining time-resolved small x-ray scattering and a kinetic modeling of this nonlinear process. From 1989 to 1990 she undertook a postdoc at the European Molecular Biology Laboratory (EMBL) with K Wilson where she looked at the impact of pH on the active site of a protease using protein crystallographic techniques. In 1990, she moved to York University where she determined the crystal structure of CD4. From 1994 to 2000 she worked as structural biologist at Aventis Pharma (formerly Hoechst) determining crystal structures of protein–ligand complexes, modeling and predicting protein structures, and analyzing protein structures and their relationships. In 2000, she moved to the Scientific Computing department at Bayer CropScience (formerly Aventis CropScience). The main emphasis of her research is structure-based drug design, protein–ligand interactions, and relationships within protein families.

© 2007 Elsevier Ltd. All Rights Reserved
No part of this publication may be reproduced, stored in any retrieval system or transmitted in any form by any means electronic, electrostatic, magnetic tape, mechanical, photocopying, recording or otherwise, without permission in writing from the publishers

Comprehensive Medicinal Chemistry II
ISBN (set): 0-08-044513-6

ISBN (Volume 4) 0-08-044517-9; pp. 597–650

4.25 Applications of Molecular Dynamics Simulations in Drug Design

C Oostenbrink and M M H van Lipzig, Vrije Universiteit, Amsterdam, The Netherlands
W F van Gunsteren, Eidgenössische Technische Hochschule, Zürich, Switzerland

© 2007 Elsevier Ltd. All Rights Reserved.

4.25.1 Biomolecular Simulation

Computer simulations on models of physical systems have been carried out for more than 50 years now. The first Monte Carlo simulations on liquids represented by spheres and hard disks were reported in 1953.[1] The classical equations of motion for such systems were first solved in a molecular dynamics simulation in 1957,[2] after which it still took several years to simulate a Lennard-Jones fluid in 1964[3] and liquid water in 1971.[4] Only 6 years later was the first molecular dynamics simulation of a protein published opening the way to true biomolecular simulation.[5] After this landmark, the field has seen tremendous improvements, both in terms of methods, algorithms, and parameters that describe the physics behind the models[6–9] and in terms of system sizes, system complexities, and simulation timescales.[10,11]

These developments have turned molecular dynamics simulations into a valuable tool, complementary to experimental investigations, to probe into the structure, dynamics, and activity of large biologically relevant molecules and molecular complexes. After careful validation of the methods and parameters that are involved, it offers the possibility to explore such biomolecular systems at a time and space resolution that is often inaccessible to experiment. It has been proven invaluable in the structure determination of proteins and nucleic acids from both x-ray and nuclear magnetic resonance (NMR) experiments and is regularly used to explore the flexibility and dynamical behavior of such molecules. The thermodynamic information that can be obtained from computer simulations allows for an analysis and understanding of it in terms of molecular processes and for a prediction of molecular properties. More and more often, molecular dynamics techniques are being used by medicinal chemists and become integrated in the multidisciplinary research of molecular medicine. Molecular flexibility and the rigorous inclusion of entropic terms in the computational estimate of binding affinity are more and more being recognized as essential parts of computational modeling.

Free energy can be seen as the driving force of virtually all molecular processes. Already at an early stage, simulations were used to calculate free energies and free enthalpies of molecular systems.[12–15] The underlying thermodynamic and statistical–mechanical theory had been developed years earlier[16,17] and many methods to calculate these thermodynamic quantities have been suggested over the years.[18–28] This chapter briefly reviews the numerical integration of the equations of motion and the statistical mechanics behind the simulations. It then continues to discuss several practical methods to calculate free energies from the simulations, focusing mainly on methods that have a theoretical foundation in statistical mechanics. To give a specific example, we will discuss various aspects of drug design where molecular dynamics simulations can make significant contributions.

4.25.2 Equations of Motion

Consider a system that contains N particles that are treated explicitly. Often, these particles are atoms, but also groups of atoms can be treated as a single particle, such as for instance an aliphatic CH, CH_2, or CH_3 group. In classical simulation, such a system is fully defined by the positions, $\mathbf{r}$, and the conjugate momenta, $\mathbf{p}$, of the individual particles, where $\mathbf{r}$ and $\mathbf{p}$ represent $3N$ dimensional vectors. We will use $\mathbf{r}_i$ and $\mathbf{p}_i$ for the three-dimensional vectors describing the position and momentum of particle i. In the absence of constraints and velocity dependent forces, the Hamiltonian of the system can be written as

$$H(\mathbf{r},\mathbf{p}) = K(\mathbf{p}) + V(\mathbf{r}) \qquad [1]$$

where $K(\mathbf{p})$ is the kinetic energy which can be calculated from

$$K(\mathbf{p}) = \sum_i \frac{\mathbf{p}_i^2}{2m_i} = \sum_i \tfrac{1}{2} m_i \mathbf{v}_i^2 \qquad [2]$$

Here, we have used the definition of the momentum $\mathbf{p}_i \equiv m_i \mathbf{v}_i$, with m_i the mass of particle i and the velocity $\mathbf{v}_i$ the time derivative of the position of the particle, $\mathbf{v}_i = \mathrm{d}\mathbf{r}_i / \mathrm{d}t$.

$V(\mathbf{r})$ in eqn [1] is the potential energy, describing the interactions between the particles in the system and possibly external influences on the system. It is a function of the particle positions $\mathbf{r}$. The functional form and parameters describing $V(\mathbf{r})$ is called a force field. There are several well-known force fields for biomolecular simulation described in the literature, such as AMBER,[29–31] CHARMM,[32–34] CHARMm,[35] ECEPP/3,[36] ENCAD,[37,38] GROMOS,[39–41] and OPLS.[42,43] Assuming that the potential energy has been properly defined and parameterized, one can calculate the force $\mathbf{f}_i$ on particle i as the negative derivative of the potential energy with respect to its position,

$$\mathbf{f}_i = -\frac{\partial}{\partial \mathbf{r}_i} V(\mathbf{r}) \qquad [3]$$

Using this force we can write down the equation of motion according to Newton,[44] which states that the second time derivative, or acceleration, of the position of a classical particle is equal to the force exerted on it divided by its mass,

$$\frac{\mathrm{d}^2}{\mathrm{d}t^2}\mathbf{r}_i = \frac{1}{m_i}\mathbf{f}_i \qquad [4]$$

One can now integrate the equations of motion for all particles simultaneously and follow the movements of the particles in time. One, but not the only, way of doing so is through the leapfrog algorithm,[45] where the positions and velocities are propagated numerically at shifted times using a time step Δt,

$$\mathbf{r}_i(t + \Delta t) = \mathbf{r}_i(t) + \mathbf{v}_i(t + \tfrac{1}{2}\Delta t) \cdot \Delta t \qquad [5]$$

$$\mathbf{v}_i(t + \tfrac{1}{2}\Delta t) = \mathbf{v}_i(t - \tfrac{1}{2}\Delta t) + m_i^{-1}\mathbf{f}_i(t) \cdot \Delta t \qquad [6]$$

The application of eqns [5] and [6] will lead to a trajectory of all particle positions in the system. Alternative integration schemes such as the Verlet[46] or Beeman[47] algorithms can be shown to produce the same positional trajectory.[48] The time step should be chosen sufficiently small to correctly integrate the fastest motion in the system. In biomolecular simulation, the fastest motions would typically be the bond vibrations, requiring a time step of 0.5 fs. For this reason, bonds are often not treated as flexible degrees of freedom, but the bond lengths are constrained to a given value.[49] Bond length constraints do not greatly affect the overall dynamics of the system and allow for an increase of the time step to typically 2 fs.[50,51] Whether the particle motion represents the physically relevant dynamics of the system depends directly on the quality of the force field.

4.25.3 Statistical Mechanics

According to the ergodic hypothesis[52] one can simulate a single molecule with its surroundings for a period of time and get time-averaged molecular properties that approach the experimentally measurable ensemble averages. This means that from a simulation of a system in time, we can get conformations that correspond to a thermodynamic ensemble or state point. In a simulation that is performed according to eqns [5] and [6] the total energy of the system, H or E, will be conserved as well as the number of particles, N, and the volume, V. This corresponds to the microcanonical (N,V,E) ensemble. Other relevant ensembles are the canonical ensemble (N, V, T), in which the temperature, T, rather than the energy is kept constant and the isothermal–isobaric ensemble (N, p, T), where the pressure p is constant additionally. Simulations corresponding to these ensembles can also be carried out by adding the proper algorithms to the simulation protocol.[7,9,53]

The probability to find a specific configuration of the system defined through the positions and momenta of the (indistinguishable) particles is defined by the phase-space probability $P(\mathbf{r},\mathbf{p})$. In the canonical ensemble one can write the phase-space probability as

$$P_{NVT}(\mathbf{r},\mathbf{p}) = \frac{e^{-H(\mathbf{r},\mathbf{p})/k_B T}}{\int\int e^{-H(\mathbf{r},\mathbf{p})/k_B T}\mathbf{dp}\ \mathbf{dr}} = \frac{e^{-H(\mathbf{r},\mathbf{p})/k_B T}}{h^{3N}N!Z(N,V,T)} \qquad [7]$$

where k_B is the Boltzmann constant, h is Planck's constant, and $Z(N,V,T)$ is the canonical partition function, defined as

$$Z(N,V,T) = \frac{1}{h^{3N}N!}\int\int e^{-H(\mathbf{r},\mathbf{p})/k_B T}\mathbf{dp}\ \mathbf{dr} \qquad [8]$$

In the isothermal–isobaric ensemble the partition function is written as

$$Z(N,p,T) = \frac{1}{Vh^{3N}N!}\int\int\int e^{-(H(\mathbf{r},\mathbf{p})+pV)/k_B T}\mathbf{dp}\ \mathbf{dr}\ \mathrm{d}V \qquad [9]$$

and the phase-space probability also depends on the volume of the system,

$$P_{NpT}(\mathbf{r},\mathbf{p},V) = \frac{e^{-(H(\mathbf{r},\mathbf{p})+pV)/k_B T}}{Vh^{3N}N!Z(N,p,T)} \qquad [10]$$

For any experimentally measurable property Q, that may depend on $\mathbf{r}$ and $\mathbf{p}$, the measured value will be equal to the expectation value of the property over the whole phase-space. For the canonical ensemble we write

$$\langle Q\rangle_{NVT} = \int\int Q(\mathbf{r},\mathbf{p})\cdot P(\mathbf{r},\mathbf{p})\mathbf{dp}\ \mathbf{dr} = \frac{\int\int Q(\mathbf{r},\mathbf{p})\cdot e^{-H(\mathbf{r},\mathbf{p})/k_B T}\mathbf{dp}\ \mathbf{dr}}{\int\int e^{-H(\mathbf{r},\mathbf{p})/k_B T}\mathbf{dp}\ \mathbf{dr}} \qquad [11]$$

where angular brackets indicate an ensemble average. The ergodic hypothesis can now be described mathematically by stating that such an ensemble average should in the case of sufficient sampling be equal to the long-time average of the time dependent quantity Q.

$$\langle Q\rangle_{NVT} = \lim_{\tau\to\infty}\frac{1}{\tau}\int_0^{\tau} Q(\mathbf{r}(t),\mathbf{p}(t))\mathrm{d}t \qquad [12]$$

A major limitation in molecular dynamics simulations is often the fact that for many biologically interesting properties the timescales reached in computer simulations are still not long enough to obtain convergence to this limit.

4.25.4 Calculation of Free Energy

A key equation from statistical mechanics links the Helmholtz free energy, A, of a system to the canonical partition function,

$$A(N,V,T) = -k_B T \ln Z(N,V,T) \qquad [13]$$

Similarly the Gibbs free enthalpy, G, can be calculated from the isothermal–isobaric partition function,

$$G(N,p,T) = A(N,V,T) + pV = -k_B T \ln Z(N,p,T) \qquad [14]$$

From these equations we can see that all one needs to do in order to calculate the free energy of the system is to calculate the integral over all phase space in eqns [8] or [9]. Or in terms of a simulation, one needs to sample until the system has visited all of the conformational or configurational space. If we now recall that **r** and **p** in these integrals represent $3N$-dimensional vectors, it becomes clear that we can not hope to calculate the absolute free energy of a system containing more than a handful of particles.[54]

4.25.4.1 Overlapping Ensembles

Fortunately, in order to understand the behaviour of a system one does not need to know the absolute free energy, but rather the change in free energy corresponding to some process or state change. For example in the case of solvating a small molecule in water, we are interested in the free energy of the solvated system relative to the molecule in the gas phase. In the case of a ligand binding to a protein we compare the free energy in the bound state to that of the ligand in solution and in the case of DNA base pairing we compare the free energy in the paired state to that of a situation in which both bases are unpaired. For the relative free energy between two states A and B of the system we can write,

$$\begin{aligned}
\Delta A_{\mathrm{BA}} &= A_{\mathrm{B}} - A_{\mathrm{A}} = -k_{\mathrm{B}}T\ln\frac{Z_{\mathrm{B}}(N,V,T)}{Z_{\mathrm{A}}(N,V,T)} \\
&= -k_{\mathrm{B}}T\ln\frac{\int\int \mathrm{e}^{-H_{\mathrm{B}}(\mathbf{r},\mathbf{p})/k_{\mathrm{B}}T}\mathbf{dp}\,\mathbf{dr}}{\int\int \mathrm{e}^{-H_{\mathrm{A}}(\mathbf{r},\mathbf{p})/k_{\mathrm{B}}T}\mathbf{dp}\,\mathbf{dr}} \\
&= -k_{\mathrm{B}}T\ln\frac{\int\int \mathrm{e}^{-(H_{\mathrm{B}}(\mathbf{r},\mathbf{p})-H_{\mathrm{A}}(\mathbf{r},\mathbf{p}))/k_{\mathrm{B}}T}\mathrm{e}^{-H_{\mathrm{A}}(\mathbf{r},\mathbf{p})/k_{\mathrm{B}}T}\mathbf{dp}\,\mathbf{dr}}{\int\int \mathrm{e}^{-H_{\mathrm{A}}(\mathbf{r},\mathbf{p})/k_{\mathrm{B}}T}\mathbf{dp}\,\mathbf{dr}} \\
&= -k_{\mathrm{B}}T\ln\left\langle \mathrm{e}^{-(H_{\mathrm{B}}-H_{\mathrm{A}})/k_{\mathrm{B}}T}\right\rangle_{\mathrm{A}}
\end{aligned} \qquad [15]$$

where Z_{A} and Z_{B} indicate the canonical partition function of the system while it is in the corresponding state. We have used the short notation H_{A} for $H_{\mathrm{A}}(\mathbf{r},\mathbf{p})$ and likewise for H_{B}. It is clear that the free energy of state B relative to that of state A can be calculated from an ensemble average of the Boltzmann factor, $\mathrm{e}^{-(H_{\mathrm{B}}-H_{\mathrm{A}})/k_{\mathrm{B}}T}$ at state A. Equation [15] is a formulation of the well-known perturbation formula, due to Zwanzig.[17] Making use of ergodicity, the ensemble average in this equation can be obtained from the time average of the exponential from a simulation at state A. The sampling problem that was hinted at in the previous section, is reduced for the calculation of a relative free energy. One no longer needs to sample all of conformational space, but only those parts of conformational space that are relevant to and energetically different for both states A and B. If the conformational space relevant to state B is very different to that of state A, it is not likely that a simulation at state A will yield an accurate estimate of the ensemble average. Only if the ensembles of states A and B show considerable overlap, can one hope to reach convergence in the ensemble average within reasonable time.

Graphically, this is depicted in the top panel of **Figure 1**. From a simulation of state A, that samples configurations in the left-hand side of a one-dimensional conformational space, one will not expect to accurately reproduce the ensemble average for the free energy difference with state B, which corresponds mainly to the right hand side of the conformational space. For a state B′ (dashed curve) the ensemble average in eqn [15] is more likely to reach convergence.

4.25.4.2 Thermodynamic Cycles

As a first attempt to calculate free energy differences one can try to formulate the problem in terms of free energies between states which are as close as possible in conformational space. This can for instance be done by recognizing the fact that a computer simulation is not limited to physical processes, but can also be used to calculate free energy differences for nonphysical processes. Knowing that the free energy is a state function and does not depend on the pathway connecting the two states of interest, one can try to reformulate the problem using a thermodynamic cycle.[55] Consider for example a case where one is interested in the free energy of binding of a compound to a receptor. For pharmaceutical purposes, one is often mostly interested in the free energy of binding relative to another compound. In **Figure 2** the binding processes for compounds **1** and **2** to a receptor are depicted. The relative free energy of binding can be calculated as

$$\Delta\Delta A_{\mathrm{bind}}(\mathit{2},\mathit{1}) = \Delta A_{\mathrm{bind}}(\mathit{2}) - \Delta A_{\mathrm{bind}}(\mathit{1}) \qquad [16]$$

One can imagine that the ensembles corresponding to the end-states of the binding processes are extremely different since the compound is in the free state physically not even near the protein. Simulations that cover the spontaneous binding of a compound to a protein are still very much beyond currently accessible timescales. From the fact that the free energy is a state function, however, one can also calculate the same relative free energy of binding from the difference between two mutation free energies,

$$\Delta\Delta A_{\mathrm{bind}}(\mathit{2},\mathit{1}) = \Delta A_{\mathit{21}}(\mathrm{bound}) - \Delta A_{\mathit{21}}(\mathrm{free}) \qquad [17]$$

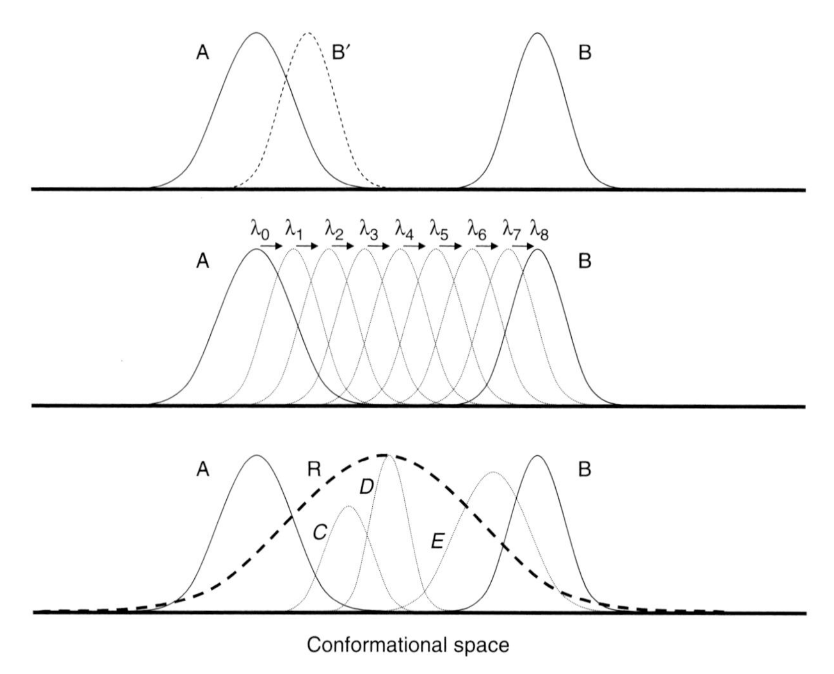

Figure 1 Pictorial representation of the distribution in conformational space for ensembles of different states. The free energy difference between states A and B cannot be computed directly from eqn [15] in case of finite sampling (top panel). Multistep approach (middle panel). One-step from an unphysical reference state (lower panel). See text for details.

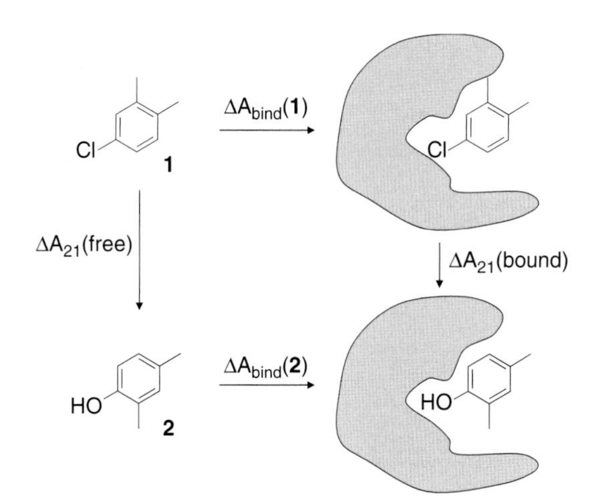

Figure 2 Thermodynamic cycle for the calculation of the relative free energy of binding of compounds **1** and **2** to a common receptor.

Even though the mutation free energies, as indicated by the vertical arrows in **Figure 2** do not correspond to any direct physical process, the end-states of the respective 'processes' can be expected to be much closer in conformational space. By reformulating the problem, the same relative free energy can be calculated by comparing states A and B in the top panel of **Figure 1** that are closer in conformational space.

Unfortunately, a reformulation of the problem does often still not bring the ensembles between which the free energy difference needs to be calculated close enough to expect the ensemble average in eqn [15] to converge within reasonable time. The conformational space relevant to compound **1** may still be very different from the conformational space relevant to compound **2**. In the following section different approaches will be discussed to calculate the free energy difference between two states A and B that do not overlap in their conformational spaces.

4.25.4.3 Free Energy Perturbation

The middle panel of **Figure 1** suggests one possibility to obtain the free energy difference between states A and B, commonly known as the free energy perturbation method (FEP).[14,19] The free energy difference that will not converge can be divided into smaller steps for which the free energy change can be calculated individually. This requires the definition of a system that is in an intermediate state between A and B, which can be done using the coupling parameter approach in which the Hamiltonian is written as a function of a coupling parameter, λ, e.g.,

$$H(\mathbf{r},\mathbf{p},\lambda) = (1-\lambda)\cdot H_{\mathrm{A}}(\mathbf{r},\mathbf{p}) + \lambda\cdot H_{\mathrm{B}}(\mathbf{r},\mathbf{p}) \qquad [18]$$

Using this definition we ensure that at $\lambda = 0$, $H(\mathbf{r}, \mathbf{p}, 0) = H_{\mathrm{A}}(\mathbf{r}, \mathbf{p})$ and at $\lambda = 1$, $H(\mathbf{r}, \mathbf{p}, 1) = H_{\mathrm{B}}(\mathbf{r}, \mathbf{p})$. At intermediate λ-values the Hamiltonian is a linear combination of the two. A coupling parameter between different states of a system was already used by Kirkwood in 1935[16] and goes back to the work of De Donder 8 years earlier.[56] Different functional forms and pathways interpolating the two end-state Hamiltonians can be used as well.[57–59] As depicted in the middle panel of **Figure 1**, the free energy between states A and B can now be calculated as the sum of $N_\lambda - 1$ free energy differences between adjoining λ-points

$$\Delta A_{\mathrm{BA}} = \sum_{i=0}^{N_\lambda-2} A_{\lambda_{i+1}} - A_{\lambda_i} = \sum_{i=0}^{N_\lambda-2} -k_{\mathrm{B}}T\ln\left\langle \mathrm{e}^{-(H(\mathbf{r},\mathbf{p},\lambda_{i+1})-H(\mathbf{r},\mathbf{p},\lambda_i))/k_{\mathrm{B}}T}\right\rangle_{\lambda_i} \qquad [19]$$

This requires a simulation to be carried out at every (except for one) of the N_λ λ-values, in order to calculate the ensemble average at that λ-point.

4.25.4.4 Thermodynamic Integration

A slightly different approach is taken in the thermodynamic integration method (TI).[16] It also makes use of the coupling parameter approach, as in eqn [18]. The free energy difference can be calculated as the integral from $\lambda = 0$ (state A) to $\lambda = 1$ (state B) of the derivative of the free energy A with respect to λ,

$$\Delta A_{\mathrm{BA}} = \int_0^1 \frac{\mathrm{d}A(\lambda)}{\mathrm{d}\lambda}\mathrm{d}\lambda \qquad [20]$$

Direct application of eqn [13] yields for the derivative

$$\begin{aligned}\frac{\mathrm{d}A(\lambda)}{\mathrm{d}\lambda} &= \frac{\mathrm{d}}{\mathrm{d}\lambda}\left(-k_{\mathrm{B}}T\ln\frac{1}{h^{3N}N!}\int\int \mathrm{e}^{-H(\mathbf{r},\mathbf{p},\lambda)/k_{\mathrm{B}}T}\mathrm{d}\mathbf{p}\,\mathrm{d}\mathbf{r}\right)\\ &= -k_{\mathrm{B}}T\frac{\frac{\mathrm{d}}{\mathrm{d}\lambda}\int\int \mathrm{e}^{-H(\mathbf{r},\mathbf{p},\lambda)/k_{\mathrm{B}}T}\mathrm{d}\mathbf{p}\,\mathrm{d}\mathbf{r}}{\int\int \mathrm{e}^{-H(\mathbf{r},\mathbf{p},\lambda)/k_{\mathrm{B}}T}\mathrm{d}\mathbf{p}\,\mathrm{d}\mathbf{r}}\\ &= \frac{\int\int \frac{\partial H(\mathbf{r},\mathbf{p},\lambda)}{\partial\lambda}\mathrm{e}^{-H(\mathbf{r},\mathbf{p},\lambda)/k_{\mathrm{B}}T}\mathrm{d}\mathbf{p}\,\mathrm{d}\mathbf{r}}{\int\int \mathrm{e}^{-H(\mathbf{r},\mathbf{p},\lambda)/k_{\mathrm{B}}T}\mathrm{d}\mathbf{p}\,\mathrm{d}\mathbf{r}}\\ &= \left\langle\frac{\partial H(\lambda)}{\partial\lambda}\right\rangle_\lambda\end{aligned} \qquad [21]$$

being exactly the ensemble average of the derivative of the Hamiltonian with respect to λ from an ensemble generated at that λ-value. The free energy difference can be calculated from a numerical integration of this ensemble average obtained from N_λ simulations at different λ-values. Like for the free energy perturbation approach, several simulations at intermediate values are required. The approach of sampling different areas of conformational space at different λ-values, is depicted in **Figure 3** in the upper left panel. In this plot, state A in **Figure 1** corresponds to the lower part of the one-dimensional conformational space, while state B corresponds to the upper part of the vertical axis.

4.25.4.5 Slow Growth

The slow growth approach can be seen as a modification of the FEP or TI approaches. Historically, TI calculations were originally carried out in a slow growth manner. Instead of simulating at discrete λ-points, the value of λ is increased at every time step such that $\lambda = 0$ (state A) at the start of the simulation and $\lambda = 1$ (state B) at the end of the simulation. This is depicted by the bold line in the upper right-hand corner of **Figure 3**. It can easily be shown that if eqn [18] is

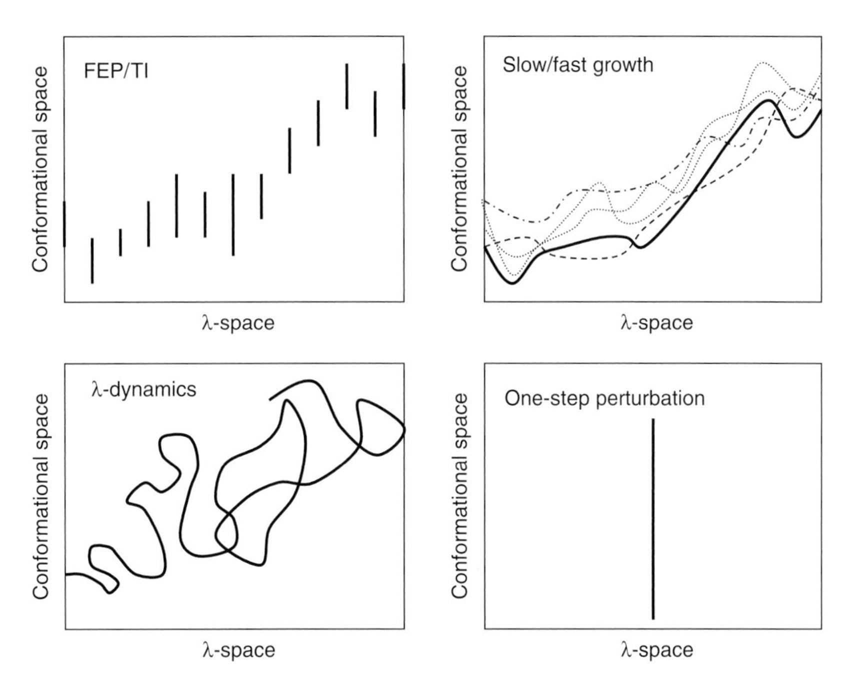

Figure 3 Pictorial representation of motion in conformational and λ-space for different approaches to free energy calculations. See text for details.

used to interpolate the Hamiltonian between states A and B, applying the perturbation formula [19] or the thermodynamic integration formula [20] will yield exactly the same estimate of the free energy difference. For different λ-dependencies of the Hamiltonian, the different approaches might result in different numerical results.

The major drawback of the slow growth approach is that the ensemble averages at every λ-value in eqns [19] or [21] are estimated from a single configuration. Obviously, this will be a very poor estimate if one considers the ergodic long time limit of the average in eqn [12]. In other words, because the λ-value is continuously changing, the system is never allowed to reach equilibrium. The motion of the particles will never be able to adapt to the new, slightly changed Hamiltonian, because the latter is constantly changing. Only if the transition from state A to state B is performed very slowly, using very long simulations, will the resulting free energy estimate converge toward the actual value, even though the conceptual error of approximating an ensemble average by a single value will remain even under such conditions. A slow growth estimate of the free energy difference is fundamentally a nonequilibrium estimate.[30] For this reason, the pure slow growth approach is no longer applied on a large scale. However, it has found a new application in the so-called fast growth approach, which will be discussed in Section 4.25.4.7.

4.25.4.6 λ-Dynamics

Reminiscent of a slow growth approach is a method that is depicted in the lower left-hand panel of **Figure 3**. In the λ-dynamics approach,[60] the value of λ is also continuously changing, but is treated as another degree of freedom in an extended Hamiltonian. The system is free to dynamically optimize the value of λ during the course of a simulation, possibly overcoming barriers that would occur in a slow growth simulation. The free energy is obtained by calculating the potential of mean force as a function of λ. To make sure that the system gradually moves from $\lambda = 0$ to $\lambda = 1$, and that all values of λ are sampled sufficiently, λ is often restrained to stepwise increasing values, through an umbrella sampling technique.[61] As in the FEP and TI approaches, several simulations at different λ-values will thus be required to obtain the free energy difference. The λ-dynamics method is particularly used to have one or several coupling parameters scale between different ligands bound to a common receptor. From the average value of the λ-values, the preference of the system for one of the ligands can be determined. This raises the question of what is the physical meaning of a system that prefers a ligand that consists for 80% of compound **1** and 13% of compound **2** and 7% of compound **3**. A similar approach is the chemical Monte Carlo approach, where a single ligand is bound to the protein at all times, but the ligand changes character based on a random, Monte Carlo method.[62–64]

4.25.4.7 Fast Growth

Recently, the slow growth method to obtain free energy differences has received renewed attention thanks to the work of Jarzynski.[65,66] It was shown that from a canonical collection of N_A nonequilibrium free energy estimates ΔA^i, such as those obtained from slow growth simulations, one can obtain an equilibrium estimate for the free energy from

$$\Delta A_{\mathrm{BA}} = -k_{\mathrm{B}}T \ln \frac{1}{N_{\mathrm{A}}} \sum_{i=1}^{N_{\mathrm{A}}} \mathrm{e}^{-\Delta A^i_{\mathrm{BA}}/k_{\mathrm{B}}T} \qquad [22]$$

The upper right-hand corner of **Figure 3** demonstrates this approach based on slow growth simulations. Many different slow growth simulations from a canonical ensemble of starting configurations lead to many different trajectories. The required sampling of conformational space, represented by the vertical lines in the upper left-hand corner of **Figure 3** is not obtained sequentially at discrete λ-points, but is acquired as more slow growth trajectories are sampled. This procedure has become known as the fast growth method.[67,68] Even though the Jarzynski equation is often applied to slow growth estimates to obtain the true free energy difference, it is not restricted to it. In principle it can be applied to any canonical ensemble of (nonequilibrium) free energy estimates.

4.25.4.8 One-Step Perturbation

The methods to calculate the free energy difference between states A and B in **Figure 1** that have been discussed so far were all aimed at gradually changing the system A into system B, be it in a limited number of discrete steps, or in a continuous, but nonequilibrium manner. The one-step perturbation method takes a different approach. It is based on the free energy perturbation formula [15], but tackles the problem of nonoverlapping ensembles by generating a single, broad ensemble, as displayed in the lower panel of **Figure 1**. The underlying idea is that we use a single, not necessarily physical, reference state R of which the ensemble shows overlap with the ensembles of both states A and B. For each of these states the free energy difference from the reference state can be calculated directly from a single simulation of the reference state using eqn [15], or

$$\Delta A_{\mathrm{AR}} = A_{\mathrm{A}} - A_{\mathrm{R}} = -k_{\mathrm{B}}T \ln \left\langle \mathrm{e}^{-(H_{\mathrm{A}}-H_{\mathrm{R}})/k_{\mathrm{B}}T} \right\rangle_{\mathrm{R}} \qquad [23]$$

giving rise to the name one-step perturbation.[69] The free energy difference between states A and B is then simply

$$\Delta A_{\mathrm{BA}} = \Delta A_{\mathrm{BR}} - \Delta A_{\mathrm{AR}} \qquad [24]$$

One might argue that we have now only shifted the problem from obtaining overlap between states A and B to generating a reference state that covers both of these states simultaneously. A solution to this problem can be found by recognizing that the formulae leading to ΔA_{BA} in eqn [24] do not require the reference state R to actually correspond to a physically meaningful state. In the example of calculating the relative free energies of two species, it is not required that the reference state corresponds to a real molecule itself. In fact, a considerable broadening of the ensemble can be obtained by making some of the atoms in the reference state 'soft.'[70,71] That is, let the interaction of these atoms level off to a finite value at the origin, rather than to an infinitely high repulsion. See for instance **Figure 3** in Oostenbrink and van Gunsteren[72] in which it is shown that such a reference state samples both configurations in which an atom is present as well as configurations that closely resemble situations where the atom is not there at all. The ensemble that is obtained by simulating an nonphysical reference state shows overlap with both the ensemble that corresponds to a simulation of the real state A and the ensemble of the real state B. In fact, the method is not restrictive to states A and B. As indicated in the lower panel of **Figure 1**, the reference state can also show overlap between additional states of interest, C, D, and E. From the single simulation of the reference state, free energies between all these states can be obtained. In addition, structural information about the real states can be obtained by a close inspection of those configurations from the simulation of R that contribute most to the ensemble average in eqn [23]. Obviously these configurations will belong to the most relevant configurations for that particular real state.

Equation [23] can be seen as an application of the Jarzynski eqn [22] with an infinitely fast sampling of the individual free energy estimates, $\Delta A^i = H_{\mathrm{A}} - H_{\mathrm{R}}$, taken at every configuration of the simulation of state R.

4.25.4.9 Other Methods

In this section several methods to obtain free energies will be discussed which have a more empirical background. These commonly calculate the free energy difference between states A and B from simulations at these states, either by estimating the absolute free energies for both end-states or by computing specific energetic differences between the states.[73,74] The advantage of such approaches is that one only needs to simulate systems with physical relevance, as opposed to the previously discussed methods that all involve simulations of nonphysical states of the system. However, compared to the one-step approach in which the number (few) of simulations is independent of the number of real states, here this number of simulations is at least equal to the number of real states.

The first method that will be discussed is the linear interaction method (LIE) due to Åqvist *et al.*[75,76] It has found widespread application to calculations of ligand–protein complexation free energies[77] and of free energies of solvation for small compounds.[78] The free energy difference between the states A and B is written as

$$\Delta A_{\mathrm{BA}} = \alpha\, \Delta\langle V^{vdw}\rangle_{\mathrm{BA}} + \beta\, \Delta\langle V^{el}\rangle_{\mathrm{BA}} + \gamma \qquad [25]$$

where $\Delta\langle V^{vdw}\rangle_{\mathrm{BA}}$ is the difference of the ensemble average of the van der Waals energy in states B and A, and $\Delta\langle V^{el}\rangle_{\mathrm{BA}}$ is the difference of the ensemble average of the electrostatic energy in states B and A. Often only the contribution to these energies that are expected to change most are taken into account. In the case of a ligand binding to a protein, where state A represents the ligand in water and state B the ligand bound to the protein, this would typically include only the interaction energy of the ligand with its surroundings. α, β, and γ are empirical parameters that are obtained by fitting calculated free energy differences to experimental data for a set of compounds. In some cases γ depends on the solvent accessible surface area (SASA) of the molecule of interest. This means that a learning set of experimental data to obtain these empirical parameters is required to calculate free energies for similar systems. From linear response theory, a theoretical value for β of 1/2 can be derived, which seems to hold well for ionic compounds, but is less accurate for dipolar and potential hydrogen bonding compounds.[79]

Another popular method is aimed at estimating the absolute free energy for the end-states of a simulation (excluding the kinetic energy). In the MM-PBSA method, the free energy is estimated from four contributions, using[80]

$$\langle A\rangle = \langle V^{\mathrm{MM}}\rangle + \langle V^{\mathrm{PB}}\rangle + \langle V^{\mathrm{SA}}\rangle - TS^{\mathrm{MM}} \qquad [26]$$

V^{MM} is the molecular mechanics potential energy as in eqn [1], V^{PB} is the electrostatic solvation (free) energy of the solute, as calculated from a numerical solution of the Poisson–Boltzmann equation, V^{SA} is a nonpolar contribution to the solvation (free) energy, estimated from the SASA of the solute, T is the temperature, and S^{MM} is the configurational solute entropy, which can be estimated from a quasiharmonic analysis or a normal-mode analysis of the molecular dynamics simulation. The accuracy of the MM-PBSA method is generally lower than conventional free energy perturbation calculations due to the numerous approximations and assumptions that go into the calculation of especially the last three terms. However, the computational gain is significant and impressive correlations with experimental free energies are often reported.

Empirical free energy scoring functions are popular for computer-aided molecular design purposes.[81] In general, the free energy of ligand binding is estimated from an equation involving several terms and empirical parameters. For example, such terms can depend on the presence of hydrogen bonds or hydrophobic contacts between the protein and the ligand, or can be related to the overall partition coefficient ($\log P$) of the ligand, the solvent-accessible surface of the ligand or size of the protein–ligand interface. The parameters are usually obtained from a regression analysis on a number of protein–ligand complexes with known binding affinity. Scoring functions are frequently used in automated ligand design programs,[82,83] or by programs that dock a given ligand into the binding site of a protein.[84–86] Typically, only a single or a few rigid conformations of the protein (or even the ligand) are considered, which severely limits the reliability and the theoretical support of these approaches. On the other hand, examples are known where the conformational changes from molecular dynamics simulations seem to have a limited influence on the estimated free energy values.[87]

4.25.5 Example: The Estrogen Receptor

This section describes examples of usage of the methods that were introduced above. Molecular dynamics simulations have proven useful for drug design purposes on many different targets.[88–103] Here, we will focus on the ligand-binding domain of the estrogen receptor. The estrogen receptor is a member of the nuclear hormone receptors and plays a key role in the growth, development, and maintenance of a diverse range of tissues. It consists of an N-terminal DNA-binding domain, a ligand-binding domain, and a C-terminal activation domain. Upon binding of an agonist to the ligand-binding domain a conformational change in the estrogen receptor takes place, allowing it to homodimerize and subsequently to translocate to the nucleus. Here, the DNA-binding domain directly interacts with response elements on the DNA, thereby activating or repressing transcription.[104] Apart from the physiological effects, the estrogen receptor is involved in a range of diseases such as breast cancer, osteoporosis, endometrial cancer, and prostate hypertrophy.[105]

Naturally occurring estrogens are steroid hormones, such as the endogenous ligand 17β-estradiol. Apart from this endogenous ligand, the estrogen receptor is known to show affinity for a wide range of structurally diverse compounds, including synthetic estrogens, phytoestrogens, pesticides, polychlorinated biphenyls (PCBs), and polycyclic aromatic hydrocarbons (PAHs).[106–110] PAHs are products of incomplete combustion of fossil fuels, wood, and other organic matter and as such are ubiquitous in the environment. Oxidative biotransformation by cytochrome P450 introduces hydroxyl groups, which results in structures that can mimic estradiol.[111–113] Everyday foodstuffs are another source of estrogenlike compounds. Some foods are known to contain a wide range of phytoestrogens, such as the isoflavonoids genistein and daidzain that show estrogenic activity in vitro and in vivo. The most important human sources of these isoflavonoids are soybeans and soybean products.[114] Although the production and use of polychlorinated biphenyls was banned in the late 1970s, they have still been observed in all kinds of tissues and species.[115–117] Again, one of the metabolic pathways of the PCBs involves the hydroxylation of (one of the) aromatic rings,[118] after which the affinity for the estrogen receptor is altered.[119]

Several crystal structures of the estrogen receptor ligand-binding domain in complex with various agonists and antagonists are available from which it becomes clear that the ligand occupies a deeply buried, largely hydrophobic cavity within the ligand-binding domain. In most structures there is no clear path or channel connecting this cavity with the exterior of the protein, in contrast to typical protein–ligand complexes. In addition, the structure of an antagonist bound to the estrogen receptor ligand-binding domain shows that a bulky side chain moiety of the antagonist serves to prevent the formation of the active tertiary structure of helices 11 and 12, known to be crucial for activity.[120,121]

The overall topology of the estrogen receptor ligand-binding domain, the wide structural diversity of known agonists, its physiological relevance, and the availability of crystal structures of the target make the estrogen receptor an ideal candidate to apply and develop computational methods for the prediction of binding affinities, to be used for drug design purposes. The following sections will describe some of our experiences of molecular dynamics simulations and free energy calculations of the estrogen receptor ligand-binding domain, the prediction and structural interpretation of binding affinities, and the design of a new ligand binding to the estrogen receptor.[72,101,122,123]

4.25.5.1 Simulation of Known Ligands

A detailed understanding of the structure and dynamics of protein–ligand complexes and the interactions between proteins and ligands has proven to be very helpful in designing new potent drugs. Producing crystallographic structures of many different ligands is still very time-consuming and only provides structural information about the complex. In order to capture and understand the flexibility of protein and ligand, molecular dynamics simulations are especially well suited. Such simulations have been performed on the estrogen receptor ligand-binding domain, complexed with the synthetic estrogen diethylstilbestrol (DES) using the crystal structure 3ERD[124] from the RCSB Protein Data Bank.[125] This crystal structure contains residues Ser305 to Leu549, while the side chains of 18 residues were not resolved and modeled according to standard side chain configurations from the GROMOS96 simulation package.[39,126] The estrogen receptor ligand-binding domain–DES complex was solvated in a periodic rectangular box of 520 nm^3, filled with 15824 explicit SPC water molecules.[127] After proper equilibration the system was simulated for 1 ns, at a constant temperature of 298 K and a constant pressure of 1 atm.

Before exploring the protein–ligand dynamics and the interactions between them, the simulation needs to be validated in terms of energetic and structural stability. The first is done by ensuring that the potential and kinetic energy have converged to reasonably constant values, while the latter is often expressed in terms of the atom-positional root-mean-square deviation from the initial (crystal) structure.[122] The presence and stability of secondary structure elements over the course of the simulation can be calculated as well, where it should be noted that these are not primary experimental data. The crystal structure from which the secondary structure elements are inferred is merely a model that represents the primary experimental data (electron density or rather structure factor amplitudes) best. One could also compare to primary structural data, such as NMR derived upper bounds to proton–proton distances (or rather NOE intensities).

A well-established estrogenic pharmacophore model consists of two hydroxyl groups, at a distance of 1.0 to 1.2 nm. At least one of the hydroxyl groups is bound to an aromatic ring and in between the two hydroxyl moieties, the estrogen is mainly hydrophobic.[111] From the crystal structures these pharmacophoric features could be explained by a hydrogen bonding network consisting of Glu353, Arg394, a water molecule, and the aromatic hydroxyl group of the ligand on the one (proximal) side and a hydrogen bond between His524 and the other hydroxyl group on the other (distal) side (**Figure 4**). From the simulations, the occurrence of protein–ligand hydrogen bonds was monitored. This analysis revealed that the hydrogen bonding interactions are highly dynamic; hydrogen bonds are formed and broken

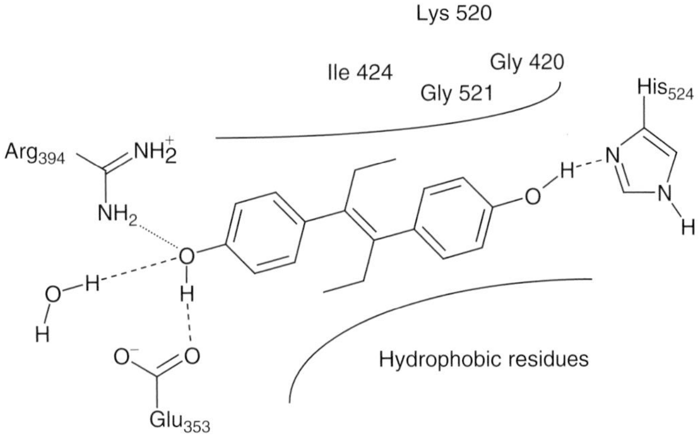

Figure 4 Schematic representation of ligand diethylstilbestrol (DES) bound to the estrogen receptor active site.

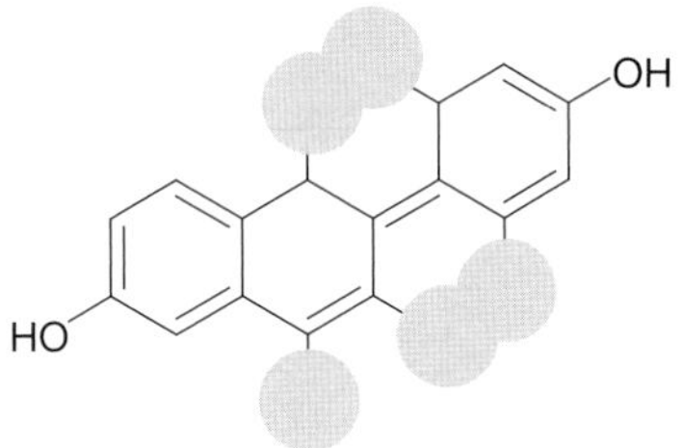

Figure 5 Reference compound (REF) used in the free energy calculations. Soft atoms are represented as gray circles.

continuously over time. While none of the hydrogen bonds at the proximal side of the molecule was present for more than 60% of the time, this hydroxyl group is always involved in at least one hydrogen bond, and forms on average 1.5 hydrogen bonds. The distal hydroxyl group forms a hydrogen bond to His524 for 97% of the simulation time.

During protein solvation in the computational box, two noncrystallographic water molecules were added near one of the ethyl groups of the DES in a hydrophobic cavity. These waters leave the binding site past His524 and toward the charged moiety of Lys520. No other water molecules enter the binding site, and the crystallographic water molecules found in the hydrogen bonded network remain trapped next to the ligand over the entire simulation. Once the added water molecules have left the binding site (after 300 ps), the ethyl group of DES shows much more flexibility in the cavity as was observed from the fluctuations of the dihedral angle. This is relevant information that can be used in the design of alternative ligands.

4.25.5.2 Calculation of Binding Free Energies

Thermodynamic integration was used to calculate the free energy of ligand binding between DES, estradiol, and genistein. These compounds are quite different to one another and relative binding affinities were obtained that agree well with experimental data. However, for drug design purposes TI is often too inefficient to screen multiple compounds on their binding affinity. In order to estimate the potential toxic effects of modified hydroxylated polyaromatic hydrocarbons through the estrogen receptor, the one-step perturbation method was applied. As a reference state, the compound (indicated with REF) in **Figure 5** was used, in which the indicated atoms were described by the so-called soft-core potential.[70] For these atoms, the repulsive singularity at the origin was absent, allowing surrounding atoms to show considerable overlap with these atoms from time to time. In this way, the broad ensemble in the lower panel of **Figure 1** was obtained, that overlaps with the ensembles that would be generated for the compounds listed in **Figure 6**.

By applying eqn [23] over a 1-ns trajectory of this reference state in the unbound state (solvated in water) and when bound to the protein (solvated in water), the relative free energies of binding that are indicated in **Figure 6** were obtained. From a comparison of computed free energies of binding relative to the reference state to the experimental values, the free energy of binding for the (unphysical) reference state can be estimated as about $-57\ \mathrm{kJ\ mol^{-1}}$. It can be seen from **Figure 7** that the calculated relative free energies of binding compare reasonably well to the

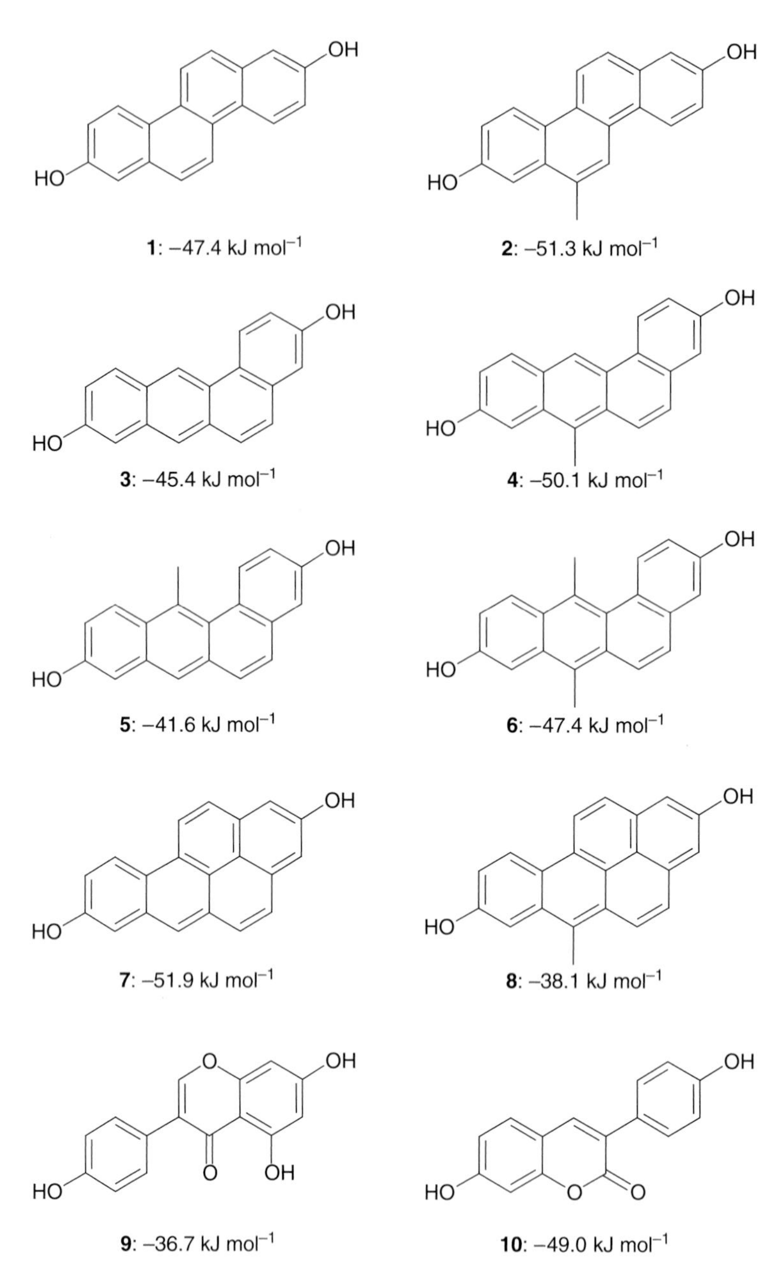

Figure 6 Compounds for which the relative free energy of binding was calculated. Indicated are free energies of binding using an estimated binding affinity of the reference compound of $-57\,\mathrm{kJ\,mol^{-1}}$.

experimental values, with an absolute mean error of $1.7\,\mathrm{kJ\,mol^{-1}}$. The maximum deviation occurs for compound **4** and is still only $3.1\,\mathrm{kJ\,mol^{-1}}$.

It is important to note the efficiency gain of the one-step perturbation method as compared to the other methods described in Section 4.25.4. Here, nine relative free energies of binding were calculated from two 1 ns simulations. Alternative methods, such as TI or FEP, would require at least eight sets of simulations, each consisting of typically 10–20 simulations at different intermediate λ-values of several hundreds of picoseconds.

4.25.5.3 Structural Interpretation

Apart from the efficiency of the free energy calculation obtained at the expense of a somewhat reduced accuracy, an important asset of the one-step perturbation method is that it allows for an interpretation of the results at a molecular

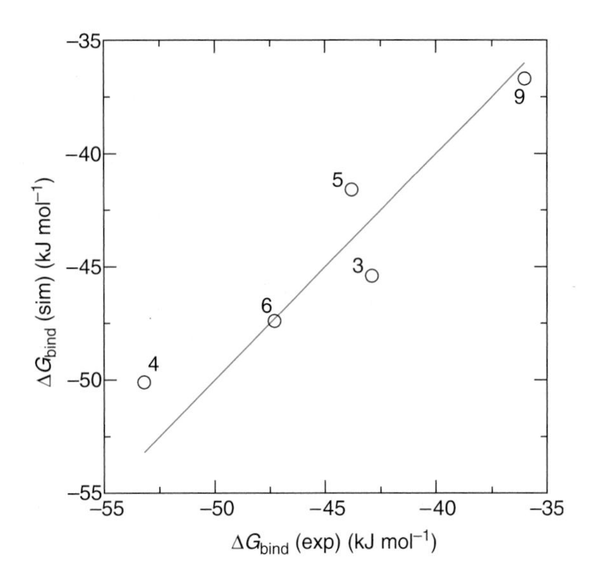

Figure 7 Graphical comparison of experimental and calculated results for the binding free energies of the indicated ligands.

basis. From the trajectory of the reference state, one can extract those configurations of the protein–ligand complex that contribute most to the ensemble average in eqn [23]. The configurations that show the largest Boltzmann probability can then be compared for different real ligands and a structural explanation for different free energy values can be obtained.

Consider for example the polyaromatic hydrocarbons and calculated free energies of binding in **Figure 6**. It is interesting to note that the addition of a methyl to structures **1**, **3**, and **5** to obtain structures **2**, **4**, and **6** is calculated to improve the binding free energy by about 4–6 kJ mol^{-1}, while introducing a methyl group in the same location in structure **7**, yielding **8**, reduces the affinity by 14 kJ mol^{-1}. It can be expected that three-dimensional quantitative structure–activity relationship (3D-QSAR) methods that do not take the protein structure explicitly into account will have a hard time predicting such nonadditive behaviour. By overlaying the protein–ligand conformation that contributes most to the free energy of binding for compounds **4** and **8** (**Figure 8**), a structural explanation can be given. In compound **8** (in green), all soft atoms in the reference compound are real atoms, forcing an unfavorable interaction of the methyl group with a cysteine (in orange) and leading to an increased binding free energy. For compound **4** (in red), however, two soft atoms in the reference state are not there in the real compound. In this case, the most favorable conformation is one in which the soft atoms show considerable overlap with an Arg side chain (in blue), allowing the methyl group to shift away from the cysteine and fit snugly into a niche in the binding site, reducing the binding free energy as compared to compound **3**.

For rational drug design purposes it is not only important to predict accurately relative binding affinities, but also to understand why one compound interacts more favorably with a receptor or solvent than another compound. In the example described here, we have come to a detailed understanding of the subtle differences between different compounds, and of the details that could be used in the development of novel ligands interacting with the receptor.

4.25.5.4 Drug Design

So far, the example that we have described has been mainly involved with the interpretation of the structure and dynamics of ligand–protein complexes and the reproduction of experimental binding free energies. The model of the reference compound in **Figure 5** has also been used to rationally design a new compound binding to the ligand-binding domain of the estrogen receptor. Experimentally, accurate screening of estrogenic activity in, e.g., environmental samples can be performed using the phytoestrogen coumestrol.[128] This fluorescent compound has the remarkable property that its fluorescence is increased when bound to the estrogen receptor, allowing for a competitive screening of other compounds driving coumestrol away from the binding site, which can be observed directly as a reduction of the fluorescence. For practical purposes, there was a demand for additional fluorescent compounds with a relatively high affinity to the estrogen receptor, but preferably fluorescing at slightly different wavelengths.

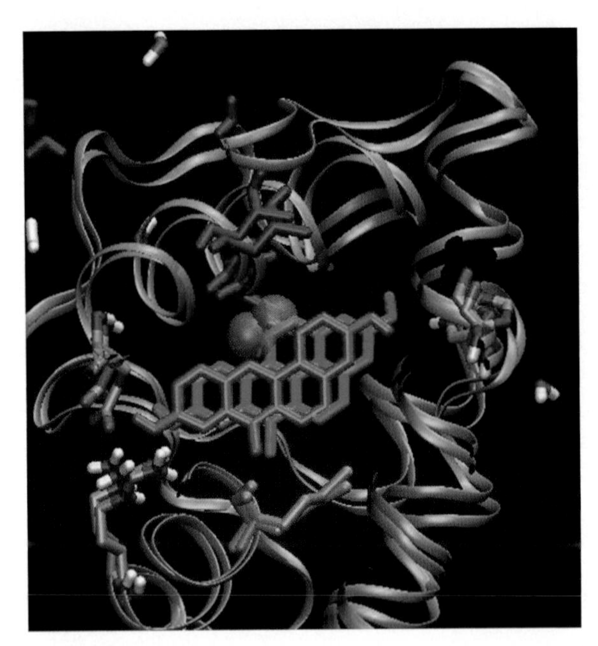

Figure 8 The most contributing configuration of the reference compound for the case when the real compound is compound **4** (in red) and for the case when the real compound is compound **8** (in green) are drawn. In dark blue Leu347 and Thr348 are indicated that show overlap with the soft atoms in the red structure (transparent spheres), allowing it to shift its methyl group away from Met389 (in orange). Protein backbone in cyan. Interacting residues Glu353, Arg394, and His524 in stick representation (compare **Figure 4**).

Based on the fact that coumarines are known to be highly fluorescent and using the knowledge that was obtained in the examples described above, compound **10** in **Figure 6** was proposed. Using the estimated free energy of binding of the reference compound of $-57\,\mathrm{kJ\,mol^{-1}}$, the binding free energy for compound **10** was predicted to be $-49.0\,\mathrm{kJ\,mol^{-1}}$. The compound was then synthesized[129] and subsequently tested to yield an experimental binding affinity of $-44\,\mathrm{kJ\,mol^{-1}}$, only $5\,\mathrm{kJ\,mol^{-1}}$ from the predicted value. Subsequent experimental analysis revealed that the fluorescence of this compound is also increased upon binding to the protein.

Even though the compound that was designed in this example was not designed as a potential drug, but rather for analytical purposes, it is clear that accurate free energies of binding can be predicted from molecular dynamics simulations. It is also interesting to note that the predicting strength of the method described here does not depend on training a model to known data. Even without the estimated free energy of binding of the reference compound, compound **10** would have been predicted to bind comparably to compound **6**. Moreover, because the protein environment is taken into account explicitly, the method can be used to accurately predict binding affinities of compounds with a more hydrophobic character (compounds **1–8**) as well as for compounds that have a more hydrophilic scaffold (compounds **9**, **10**).

Overall, these examples show that the one-step perturbation method is an accurate and efficient means to predict the free energies of binding for a series of compounds. The protein environment is taken into account explicitly allowing for a detailed structural explanation of calculated differences between compounds. Other than the parameters that go into a (relatively standard) force field, no empirical parameters are required and a statistically sound prediction of the binding affinity is obtained.

4.25.6 Conclusions

This chapter has briefly described methods that can be used to perform molecular dynamics simulations of biomolecular systems. For drug design purposes, this will mainly be receptor–ligand complexes. In addition, we have described methods to calculate the free energies of binding on a sound statistical mechanical basis. From the examples in Section 4.25.5, it should have become clear how these methods can be applied in rational drug design.

First, simulations of protein–ligand complexes give insight in the structure and dynamics of the ligand when bound to the protein. A detailed analysis of the protein–ligand interactions reveals the relative importance of different interaction sites and allows us to predict the required functional groups in the ligand. From the dynamics of the complex and the average shape of the binding site, possible sites of modification to known ligands can be suggested.

Second, we have shown that molecular dynamics simulations can be used to calculate (relative) binding affinities of a series of compounds with reasonable accuracy. The methods described in Sections 4.25.4.3–4.25.4.8 have in common that no experimental data are required to predict the binding affinities, other than the parameters from a common biomolecular force field. With an average absolute error smaller than k_BT ($\sim 2.5\,\mathrm{kJ\,mol^{-1}}$) the example that was described shows that such a force field can describe the protein–ligand interactions quite accurately.

Third, the big advantage of molecular dynamics simulation and in particular the one-step perturbation approach is that an explanation of the structural basis of the calculated free energies can be obtained at an atomic level. Structural details can be obtained that explain why one compound binds more tightly to the protein than another.

Fourth, the one-step perturbation methodology allows the calculation of tens of thousands[130] to hundreds of millions[131] of relative free energies for not too different ligands from only a handful of simulations.

Finally, we have shown how the knowledge of the protein and ligand structure and dynamics, an understanding of the structural reasons behind ligand binding, and the machinery to calculate binding free energies can be used to rationally propose novel compounds that will bind to the same target and to predict the binding affinity of such compounds.

With current availability of computational power, molecular dynamics simulations of protein–ligand complexes have become feasible. The methods described offer insight in structure and dynamics at a resolution often not accessible experimentally and the advantages of taking the protein environment of ligands into account will be evident from the results presented in this chapter.

References

1. Metropolis, N.; Rosenbluth, A. W.; Rosenbluth, M. N.; Teller, A. H.; Teller, E. *J. Chem. Phys.* **1953**, *21*, 1087–1092.
2. Alder, B. J.; Wainwright, T. E. *J. Chem. Phys.* **1957**, *27*, 1208–1209.
3. Rahman, A. *Phys. Rev. A* **1964**, *136*, 405–411.
4. Rahman, A.; Stillinger, F. H. *J. Chem. Phys.* **1971**, *55*, 3336–3359.
5. McCammon, J. A.; Gelin, B. R.; Karplus, M. *Nature* **1977**, *267*, 585–590.
6. Allen, M. P.; Tildesley, D. J. *Computer Simulation of Liquids*; Clarendon Press: Oxford, UK, 1987.
7. van Gunsteren, W. F.; Berendsen, H. J. C. *Angew. Chem., Int. Ed. Engl.* **1990**, *29*, 992–1023.
8. Frenkel, D.; Smit, B. *Understanding Molecular Simulation*; Academic Press: San Diego, CA, 2002.
9. Tuckerman, M. E.; Martyna, G. J. *J. Phys. Chem. B* **2000**, *104*, 159–178.
10. Hansson, T.; Oostenbrink, C.; van Gunsteren, W. F. *Curr. Opin. Struct. Biol.* **2002**, *12*, 190–196.
11. Norberg, J.; Nilsson, L. *Q. Rev. Biophys.* **2003**, *36*, 257–306.
12. Sarkisov, G. N.; Dashevsky, V. G.; Malenkov, G. G. *Mol. Phys.* **1974**, *27*, 1249–1269.
13. Mruzik, M. R.; Abraham, F. F.; Schreiber, D. E.; Pound, G. M. *J. Chem. Phys.* **1976**, *64*, 481–491.
14. Mezei, M.; Swaminathan, S.; Beveridge, D. L. *J. Am. Chem. Soc.* **1978**, *100*, 3255–3256.
15. Postma, J. P. M.; Berendsen, H. J. C.; Haak, J. R. *Faraday Symp. Chem. Soc.* **1982**, *17*, 55–67.
16. Kirkwood, J. G. *J. Chem. Phys.* **1935**, *3*, 300–313.
17. Zwanzig, R. W. *J. Chem. Phys.* **1954**, *22*, 1420–1426.
18. Mezei, M.; Beveridge, D. L. *Ann. NY Acad. Sci.* **1986**, *482*, 1–23.
19. Beveridge, D. L.; DiCapua, F. M. *Annu. Rev. Biophys. Biophys. Chem.* **1989**, *18*, 431–492.
20. van Gunsteren, W. F. Methods for Calculation of Free Energies and Binding Constants: Successes and Problems. In *Computer Simulation of Biomolecular Systems, Theoretical and Experimental Applications*; van Gunsteren, W. F., Weiner, P. K., Eds.; Escom Science Publishers: Leiden, the Netherlands, 1989, pp 27–59.
21. Straatsma, T. P.; McCammon, J. A. *Annu. Rev. Phys. Chem.* **1992**, *43*, 407–435.
22. Kollman, P. *Chem. Rev.* **1993**, *93*, 2395–2417.
23. van Gunsteren, W. F.; Beutler, T. C.; Fraternali, F.; King, P. M.; Mark, A. E.; Smith, P. E. Free Energy via Molecular Simulation: A Primer. In *Computer Simulation of Biomolecular Systems, Theoretical and Experimental Applications*; van Gunsteren, W. F., Weiner, P. K., Eds.; Escom Science Publishers: Leiden, the Netherlands, 1993, pp 315–348.
24. Lamb, M.; Jorgensen, W. L. *Curr. Opin. Chem. Biol.* **1997**, *1*, 449–457.
25. Mark, A. E. Free Energy Perturbation (FEP). In *Encyclopedia of Computational Chemistry*; John Wiley: New York, 1998, pp 1070–1083.
26. van Gunsteren, W. F.; Daura, X.; Mark, A. E. *Helv. Chim. Acta* **2002**, *85*, 3113–3129.
27. Warshel, A. *Acc. Chem. Res.* **2002**, *35*, 385–395.
28. Brandsdal, B. O.; Österberg, F.; Almlöf, M.; Feierberg, I.; Luzhkov, V. B.; Åqvist, J. *Adv. Protein. Chem.* **2003**, *66*, 123–158.
29. Weiner, P. K.; Kollman, P. A. *J. Comput. Chem.* **1981**, *2*, 287–303.
30. Pearlman, D. A.; Kollman, P. A. *J. Chem. Phys.* **1989**, *91*, 7831–7839.
31. Cornell, W. D.; Cieplak, P.; Bayly, C. I.; Gould, I. R.; Mertz, K. M.; Ferguson, D. M.; Spellmeyer, D. C.; Fox, T.; Caldwell, J. W.; Kollman, P. A. *J. Am. Chem. Soc.* **1995**, *117*, 5179–5197.
32. Brooks, B. R.; Bruccoleri, R. E.; Olafson, B. D.; States, D. J.; Swaminathan, S.; Karplus, M. *J. Comput. Chem.* **1983**, *4*, 187–217.
33. MacKerell, A. D., Jr.; Wiórkiewicz-Kuczera, J.; Karplus, M. *J. Am. Chem. Soc.* **1995**, *117*, 11946–11975.
34. MacKerell, A. D., Jr.; Bashford, D.; Bellot, M.; Dunbrack, R. L., Jr.; Evanseck, J. D.; Field, M. J.; Fischer, S.; Gao, J.; Guo, H.; Ha, S. et al. *J. Phys. Chem. B.* **1998**, *102*, 3586–3616.
35. Momany, F. A.; Rone, R. *J. Comput. Chem.* **1992**, *13*, 888–900.
36. Nemethy, G.; Gibson, K. D.; Palmer, K. A.; Yoon, C. N.; Paterlini, G.; Zagari, A.; Rumsey, S.; Scheraga, H. A. *J. Phys. Chem.* **1992**, *96*, 6472–6484.

37. Levitt, M. *J. Mol. Biol.* **1983**, *168*, 595–620.
38. Levitt, M.; Hirshberg, M.; Sharon, R.; Daggett, V. *Comput. Phys. Commun.* **1995**, *91*, 215–231.
39. van Gunsteren, W. F.; Billeter, S. R.; Eising, A. A.; Hünenberger, P. H.; Krüger, P.; Mark, A. E.; Scott, W. R. P.; Tironi, I. G. *Biomolecular Simulation: The GROMOS96 Manual and User Guide*; Vdf Hochschulverlag AG an der ETH Zürich: Zürich, Switzerland, 1996.
40. Schuler, L. D.; Daura, X.; van Gunsteren, W. F. *J. Comput. Chem.* **2001**, *22*, 1205–1218.
41. Oostenbrink, C.; Villa, A.; Mark, A. E.; van Gunsteren, W. F. *J. Comput. Chem.* **2004**, *25*, 1656–1676.
42. Jorgensen, W. L.; Tirado-Rives, J. *J. Am. Chem. Soc.* **1988**, *110*, 1657–1666.
43. Jorgensen, W. L.; Maxwell, D. S.; Tirado-Rives, J. *J. Am. Chem. Soc.* **1996**, *118*, 11225–11236.
44. Goldstein, H. *Classical Mechanics*; Addison-Wesley: Reading, MA, 1980.
45. Hockney, R. W. *Meth. Comput. Phys.* **1970**, *9*, 136–211.
46. Verlet, L. *Phys. Rev.* **1967**, *159*, 98–103.
47. Beeman, D. *J. Comput. Phys.* **1976**, *20*, 130–139.
48. Berendsen, H. J. C.; van Gunsteren, W. F. Practical Algorithms for Dynamic Simulations. In *Molecular-Dynamics Simulation of Statistical-Mechanical Systems*; Ciccotti, G., Hoover, W. G., Eds.; North-Holland: Amsterdam, the Netherlands, 1986, pp 43–65.
49. Ryckaert, J.-P.; Ciccotti, G.; Berendsen, H. J. C. *J. Comput. Phys.* **1977**, *23*, 327–341.
50. van Gunsteren, W. F.; Berendsen, H. J. C. *Mol. Phys.* **1977**, *34*, 1311–1327.
51. van Gunsteren, W. F.; Karplus, M. *Nature* **1981**, *293*, 677–678.
52. Ehrenfest, P.; Ehrenfest, T. *Enzyklopädie der Mathematischen Wissenschaften IV*; Teubner: Leipzig, 1912; Vol. 2, pp 3–90.
53. Hünenberger, P. H. *Adv. Polym. Sci.* **2004**, *173*, 105–149.
54. Owicki, J. C.; Scheraga, H. A. *J. Am. Chem. Soc.* **1977**, *99*, 7403–7412.
55. Tembe, B. L.; McCammon, J. A. *Comput. Chem.* **1984**, *8*, 281–283.
56. De Donder, T. *L'Affinité*; Gauther-Villars: Paris, 1927.
57. Mark, A. E.; van Gunsteren, W. F.; Berendsen, H. J. C. *J. Chem. Phys.* **1991**, *94*, 3808–3816.
58. Resat, H.; Mezei, M. *J. Chem. Phys.* **1993**, *99*, 6052–6061.
59. Pearlman, D. A. *J. Phys. Chem.* **1994**, *98*, 1487–1493.
60. Kong, X.; Brooks, C. L., III. *J. Chem. Phys.* **1996**, *105*, 2414–2423.
61. Torrie, G. M.; Valleau, J. P. *Chem. Phys. Lett.* **1974**, *28*, 578–581.
62. Bennett, C. H. *J. Comput. Phys.* **1976**, *22*, 245–268.
63. Tidor, B. *J. Phys. Chem.* **1993**, *97*, 1069–1073.
64. Pitera, J. W.; Kollman, P. *J. Am. Chem. Soc.* **1998**, *120*, 7557–7567.
65. Jarzynski, C. *Phys. Rev. Lett.* **1997**, *78*, 2690–2693.
66. Jarzynski, C. *Phys. Rev. E* **1997**, *56*, 5018–5035.
67. Hendrix, D. A.; Jarzynski, C. *J. Chem. Phys.* **2001**, *114*, 5974–5981.
68. Hummer, G. *J. Chem. Phys.* **2001**, *114*, 7330–7337.
69. Liu, H. Y.; Mark, A. E.; van Gunsteren, W. F. *J. Phys. Chem.* **1996**, *100*, 9485–9494.
70. Beutler, T. C.; Mark, A. E.; van Schaik, R. C.; Gerber, P. R.; van Gunsteren, W. F. *Chem. Phys. Lett.* **1994**, *222*, 529–539.
71. Zacharias, M.; Straatsma, T. P.; McCammon, J. A. *J. Chem. Phys.* **1994**, *100*, 9025–9031.
72. Oostenbrink, C.; van Gunsteren, W. F. *Proteins* **2004**, *54*, 237–246.
73. Hummer, G.; Szabo, A. *J. Chem. Phys.* **1996**, *105*, 2004–2010.
74. Bürgi, R.; Läng, F.; van Gunsteren, W. F. *Mol. Sim.* **2001**, *27*, 215–236.
75. Åqvist, J.; Medina, C.; Samuelsson, J. E. *Protein Eng.* **1994**, *7*, 385–391.
76. Åqvist, J.; Luzhkov, V. B.; Brandsdal, B. O. *Acc. Chem. Res.* **2002**, *35*, 358–365.
77. Hansson, T.; Marelius, J.; Åqvist, J. *J. Comput.-Aided Mol. Des.* **1998**, *12*, 27–35.
78. Duffy, D. E.; Jorgensen, W. L. *J. Am. Chem. Soc.* **2000**, *122*, 2878–2888.
79. Åqvist, J.; Hansson, T. *J. Phys. Chem.* **1996**, *100*, 9512–9521.
80. Kollman, P. A.; Massova, I.; Reyes, C.; Kuhn, B.; Huo, S. H.; Chong, L.; Lee, M.; Lee, T. S.; Duan, Y.; Wang, W. et al. *Acc. Chem. Res.* **2000**, *33*, 889–897.
81. Oprea, T. I.; Marshall, G. R. *Perspect. Drug Disc.* **1998**, *9*, 35–61.
82. Böhm, H. J. *J. Comput.-Aided Mol. Des.* **1992**, *6*, 61–78.
83. Lewis, R. A.; Leach, A. R. *J. Comput.-Aided Mol. Des.* **1994**, *8*, 467–475.
84. Kuntz, I. D.; Meng, E. C.; Shoichet, B. K. *Acc. Chem. Res.* **1994**, *27*, 117–123.
85. Morris, G. M.; Goodsell, D. S.; Halliday, R. S.; Huey, R.; Hart, W. E.; Belew, R. K.; Olson, A. J. *J. Comput. Chem.* **1998**, *19*, 1639–1662.
86. Sotriffer, C. A.; Flader, W.; Winger, R. H.; Rode, B. M.; Liedl, K. R.; Varge, J. M. *Methods* **2000**, *20*, 280–291.
87. Marelius, J.; Ljungberg, K. B.; Åqvist, J. *Eur. J. Pharm. Sci.* **2001**, *14*, 87–95.
88. Rami Reddy, M.; Viswanadhan, V. N.; Weinstein, J. N. *Proc. Natl. Acad. Sci. USA* **1991**, *88*, 10287–10291.
89. Rao, B. G.; Tilton, R. F.; Singh, U. C. *J. Am. Chem. Soc.* **1992**, *114*, 4447–4452.
90. Hulten, J.; Bonham, N. M.; Nillroth, U.; Hansson, T.; Zuccarello, G.; Bouzide, A.; Aqvist, J.; Classon, B.; Danielson, U. H.; Karlen, A. et al. *J. Med. Chem.* **1997**, *40*, 885–897.
91. Marelius, J.; Hansson, T.; Åqvist, J. *Int. J. Quantum Chem.* **1998**, *69*, 77–88.
92. McCarrick, M. A.; Kollman, P. A. *J. Comput.-Aided Mol. Design* **1999**, *13*, 109–121.
93. Eriksson, M. A. L.; Pitera, J.; Kollman, P. A. *J. Med. Chem.* **1999**, *42*, 868–881.
94. Lee, T. S.; Kollman, P. A. *J. Am. Chem. Soc.* **2000**, *122*, 4385–4393.
95. Okimoto, N.; Tsukui, T.; Kitayama, K.; Hata, M.; Hoshino, T.; Tsuda, M. *J. Am. Chem. Soc.* **2000**, *122*, 5613–5622.
96. Daura, X.; Haaksma, E.; van Gunsteren, W. F. *J. Comput.-Aided Mol. Des.* **2000**, *14*, 507–529.
97. Pak, Y.; Enyedy, I. J.; Varady, J.; Kung, J. W.; Lorenzo, P. S.; Blumberg, P. M.; Wang, S. *J. Med. Chem.* **2001**, *44*, 1690–1701.
98. Graffner-Nordberg, M.; Kolmodin, K.; Åqvist, J.; Queener, S. F.; Hallberg, A. *J. Med. Chem.* **2001**, *44*, 2391–2402.
99. Rami Reddy, M.; Erion, M. D. *J. Am. Chem. Soc.* **2001**, *123*, 6246–6252.
100. Brandsdal, B. O.; Aqvist, J.; Smalas, A. O. *Protein Sci.* **2001**, *10*, 1584–1595.
101. van Lipzig, M. M. H.; ter Laak, A. M.; Jongejan, A.; Vermeulen, N. P. E.; Wamelink, M.; Geerke, D.; Meerman, J. H. N. *J. Med. Chem.* **2004**, *47*, 1018–1030.

102. Park, H.; Lee, S. *J. Comput.-Aided Mol. Des.* **2005**, *19*, 17–31.
103. Kuhn, B.; Gerber, P.; Schulz-Gasch, T.; Stahl, M. *J. Med. Chem.* **2005**, *48*, 4040–4048.
104. Parker, M. G.; Arbuckle, N.; Dauvois, S.; Danielian, P.; White, R. *Ann. NY Acad. Sci.* **1993**, *684*, 119–126.
105. Crisp, T. M.; Clegg, E. D.; Cooper, R. L.; Wood, W. P.; Anderson, D. G.; Baetcke, K. P.; Hoffman, J. L.; Morrow, M. S.; Rodier, D. J.; Schaeffer, J. E. et al. *Environ. Health Perspect.* **1998**, *106*, 11–56.
106. Soto, A. M.; Sonnenschein, C.; Chung, K. L.; Fernandez, M. F.; Olea, N.; Serrano, F. O. *Environ. Health Perspect.* **1995**, *103*, 113–122.
107. Waller, C. L.; Oprea, T. I.; Chae, K.; Park, H. K.; Korach, K. S.; Laws, S. C.; Wiese, T. E.; Kelce, W. R.; Earl Gray, L. *Chem. Res. Toxicol.* **1996**, *9*, 1240–1248.
108. Breinholt, V.; Larsen, J. C. *Chem. Res. Toxicol.* **1998**, *11*, 622–629.
109. Garner, C. E.; Jefferson, W. N.; Burka, L. T.; Matthews, H. B.; Newbold, R. R. *Toxicol. Appl. Pharmacol.* **1999**, *154*, 188–197.
110. Blair, R. M.; Fang, H.; Branham, W. S.; Hass, B. S.; Dial, S. L.; Moland, C. L.; Tong, W. D.; Shi, L. M.; Perkins, R.; Sheehan, D. M. *Toxicol. Sci.* **2000**, *54*, 138–153.
111. Anstead, G. M.; Carlson, K. E.; Katzenellenbogen, J. A. *Steroids* **1997**, *62*, 268–303.
112. Charles, G. D.; Bartels, M. J.; Zacharewski, T. R.; Gollapudi, B. B.; Freshour, N. L.; Carney, E. W. *Toxicol. Sci.* **2000**, *55*, 320–326.
113. van Lipzig, M. M. H.; Vermeulen, N. P. E.; Gusinu, R.; Legler, J.; Frank, H.; Seidel, A.; Meerman, J. H. N. *Environ. Toxicol. Pharmacol.* **2005**, *19*, 41–55.
114. Adlercreutz, H.; Mazur, W. *Ann. Med.* **1997**, *29*, 95–120.
115. Bergman, A.; Klassonwehler, E.; Kuroki, H. *Environ. Health Perspect.* **1994**, *102*, 464–469.
116. Jansson, B.; Andersson, R.; Asplund, L.; Litzen, K.; Nylund, K.; Sellstrom, U.; Uvemo, U. B.; Wahlberg, C.; Wideqvist, U.; Odsjo, T. et al. *Environ. Toxicol. Chem.* **1993**, *12*, 1163–1174.
117. Dewailly, E.; Nantel, A.; Weber, J. P.; Meyer, F. *Bull. Environ. Contam. Toxicol.* **1989**, *43*, 641–646.
118. Sipes, I. G.; Schnellmann, R. G. Biotransformation of PCBS: Metabolic Pathways and Mechanisms. In *Polychlorinated Biphenyls (PCBs): Mammalian and Environmental Toxicology*; Safe, S., Ed.; Springer-Verlag: Heidelberg, Germany, 1987, pp 97–110.
119. Matthews, J.; Zacharewski, T. *Toxicol. Sci.* **2000**, *53*, 326–339.
120. Brzozowski, A. M.; Pike, A. C. W.; Dauter, Z.; Hubbard, R. E.; Bonn, T.; Engstrom, O.; Ohman, L.; Greene, G. L.; Gustafsson, J. A.; Carlquist, M. *Nature* **1997**, *389*, 753–758.
121. Danielan, P. S.; White, R.; Lees, J. A.; Parker, M. G. *EMBO J.* **1992**, *11*, 1025–1033.
122. Oostenbrink, B. C.; Pitera, J. W.; Van Lipzig, M. M. H.; Meerman, J. H. N.; van Gunsteren, W. F. *J. Med. Chem.* **2000**, *43*, 4594–4605.
123. Oostenbrink, C.; van Gunsteren, W. F. *Proc. Natl. Acad. Sci. USA* **2005**, *102*, 6750–6754.
124. Shiau, A. K.; Barstad, D.; Loria, P. M.; Cheng, L.; Kushner, P. J.; Agard, D. A.; Greene, G. L. *Cell* **1998**, *95*, 927–937.
125. Berman, H. M.; Westbrook, J.; Feng, Z.; Gilliland, G.; Bhat, T. N.; Weissig, H.; Shindyalov, I. N.; Bourne, P. E. *Nucleic Acids Res.* **2000**, *28*, 235–242.
126. Scott, W. R. P.; Hünenberger, P. H.; Tironi, I. G.; Mark, A. E.; Billeter, S. R.; Fennen, J.; Torda, A. E.; Huber, T.; Krüger, P.; van Gunsteren, W. F. *J. Phys. Chem. A* **1999**, *103*, 3596–3607.
127. Berendsen, H. J. C.; Postma, J. P. M.; van Gunsteren, W. F.; Hermans, J. Interaction Models for Water in Relation to Protein Hydration. In *Intermolecular Forces*; Pullman, B., Ed.; Reidel: Dordrecht, the Netherlands, 1981, pp 331–342.
128. Oosterkamp, A. J.; Villaverde Herraiz, M. T.; Irth, H.; Tjaden, U. R.; van der Greef, J. *Anal. Chem.* **1996**, *68*, 1201–1206.
129. Buu-Hoï, N. P.; Ekert, B.; Royer, R. *J. Org. Chem.* **1954**, *19*, 1548–1552.
130. Oostenbrink, C.; van Gunsteren, W. F. *Chem. Eur. J.* **2005**, *11*, 4340–4348.
131. Yu, H. B.; Amann, M.; Hansson, T.; Köhler, J.; Wich, G.; van Gunsteren, W. F. *Carbohydr. Res.* **2004**, *339*, 1697–1709.

Biographies

Chris Oostenbrink was born in 1977 in Amsterdam (The Netherlands). In 2000, he obtained MSc degrees in Chemistry and in Medicinal Chemistry at the Free University of Amsterdam. In 2004, he defended his PhD on free energy calculations in biomolecular simulation at the ETH in Zurich. From 2004 on he holds a position as assistant professor in the field of Computational Medicinal Chemistry and Toxicology at the Free University in Amsterdam.

Marola M H van Lipzig was born in Delft (The Netherlands). In 1998, she received her MSc degree in environmental studies from the Wageningen Agricultural University. Until 2003 she worked on her PhD thesis on environmental estrogens at the Free University in Amsterdam, followed by a postdoc in Computational Molecular Toxicology in 2004 and a postdoc in Medicinal Chemistry in 2005 at the same university.

Wilfred F van Gunsteren was born in 1947 in Wassenaar (The Netherlands). In 1968, he gained a BSc in physics at the Free University of Amsterdam; in 1976 he was awarded a 'Meester' in Law, and in 1976 a PhD in nuclear physics. After postdoc research at the University of Groningen and at Harvard University he was, 1980–87, senior lecturer and, until August 1990, professor for physical chemistry at the University of Groningen. In 1990, he became professor of computer chemistry at the ETH Zurich. He is holder of a gold medal for research of the Royal Netherlands Chemical Society. His main interests center on the physical fundamentals of the structure and function of biomolecules.

© 2007 Elsevier Ltd. All Rights Reserved
No part of this publication may be reproduced, stored in any retrieval system or transmitted in any form by any means electronic, electrostatic, magnetic tape, mechanical, photocopying, recording or otherwise, without permission in writing from the publishers

Comprehensive Medicinal Chemistry II
ISBN (set): 0-08-044513-6

ISBN (Volume 4) 0-08-044517-9; pp. 651–668

4.26 Seven Transmembrane G Protein-Coupled Receptors: Insights for Drug Design from Structure and Modeling

N Barton, F E Blaney, S Garland, B Tehan, and I Wall, GlaxoSmithKline Pharmaceuticals plc, Harlow, UK

© 2007 Elsevier Ltd. All Rights Reserved.

4.26.1 Introduction

Integral membrane-bound proteins form a huge class of receptors and transport effectors of great importance to the body, and hence to the pharmaceutical industry. They include the ion channels and transporter proteins as well as the subject of this chapter, the G protein-coupled receptors (GPCRs). All contain a number of membrane-spanning regions as well as extracellular and cytoplasmic domains. There is a wide variation in the number and size of these regions but within each family, the topography is well defined and maintained (**Figure 1**). While this topography varies considerably within the channel and transporter subfamilies, the GPCRs all share a common fold with a total of seven transmembrane (7TM) segments, three extracellular and three cytoplasmic loop regions, an extracellular N-terminal domain, and a cytoplasmic C-terminal region (**Figure 1a**). Although some 'all beta' membrane-bound proteins, such as the bacterial outer membrane porins, are known,[1] most membrane-spanning segments are α-helical. This is because an amphipathic α-helix is the most stable structural element that can span a lipophilic membrane, presenting a hydrophobic face to the lipids while at the same time having polar residues on the opposite face which can form some functional assembly when packed together in a helix bundle. These helices generally contain a minimum of 20 amino acid residues, are tilted with respect to each other, and pack in an antiparallel manner so as to maintain the best alignment of the large helix dipoles. The earliest evidence for the existence of a seven-helical bundle topography in GPCRs came from protein chemistry studies on the opsins and this led to the conclusion that there was a structural similarity between GPCRs and the protein bacteriorhodopsin.[2] Although the latter is not G protein-coupled, it shares a similar functional role with the opsins, and its structure, solved by cryoelectron diffraction, does contain a seven-helical bundle.[3] This formed the basis on which much of the earliest modeling work of GPCRs was done.

4.26.2 Classification of G Protein-Coupled Receptors

The first GPCR sequences appeared in 1979.[4] Since then, the number of known receptors has grown astoundingly and now exceeds 1000 (not including the odorant receptors, which are also part of this family). There are a vast number of stimuli triggering these receptors. They range from light photons in the case of the opsins, through all the substances recognized by the odorant and taste receptors, to small-molecule neurotransmitters, amino acids, nucleosides, prostaglandin and other lipid molecules, a bewildering array of peptide hormones, and even larger proteins such as the chemokines and glycoprotein hormones (**Figure 2**). Most of these stimuli come from outside the cell membrane, although in one small subfamily, the protease activated receptors such as thrombin, the signaling molecule is actually part of the N-terminal domain which is proteolytically cleaved off prior to activation. The binding of the signal molecule triggers a conformational change which is transmitted to the cytoplasmic domains, where the heterotrimeric guanidine nucleotide binding protein complex is bound. In the resting state this is complexed with GDP but upon activation it is replaced by GTP and the protein complex detaches from the receptor. The α-subunit breaks off from the βγ assembly and goes on to activate one of a variety of secondary message systems within the cell.

GPCRs represent the single largest mechanism by which signals are passed into cells. They thus affect a vast number of biological processes and as such represent the largest class of targets for drug intervention. Indeed it is estimated that between 30% and 60% of known drug molecules act on GPCR targets.[5]

As the number of known sequences rose, it was observed that despite a general lack of sequence identity, each domain contained one or more highly conserved residues. Thus, in the majority of the earliest sequences: (1) TM1 contained an asparagine toward the cytoplasmic side; (2) TM2 had an aspartate at the same level in the helix; (3) TM3 contained a cysteine at the extracellular end of the helix and a D(E)RY(FHC) motif at the cytoplasmic side; (4) TM4 had a highly conserved tryptophan in the middle of the helix; (5) TMs 5, 6, and 7 each had a conserved proline residue. In addition TM6 had an invariant phenylalanine six residues below the proline and TM7 had an asparagine immediately before its proline.

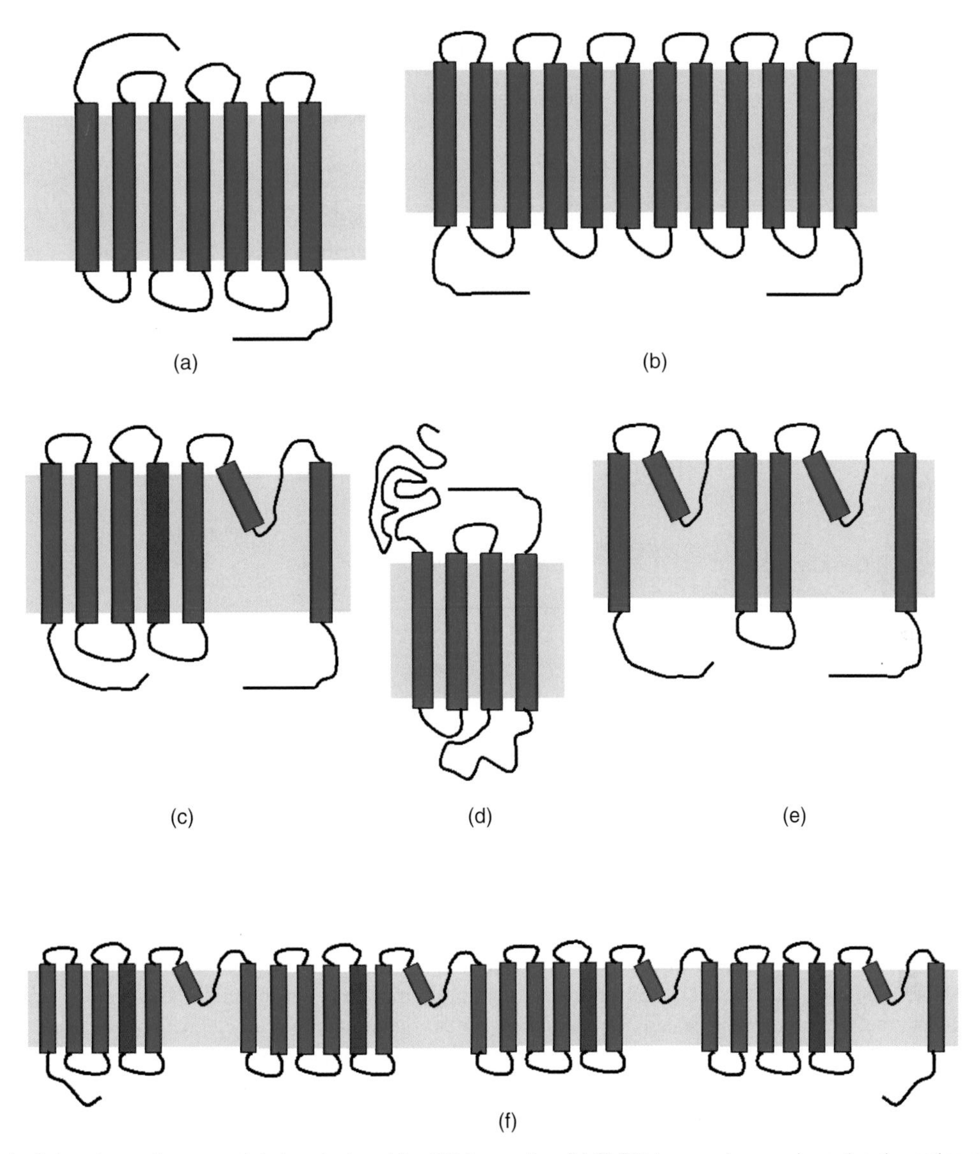

Figure 1 Integral membrane protein topologies: (a) a 7TM receptor; (b) SLC6 transporters such as the dopamine or GABA proteins; (c) a voltage-gated potassium channel which functions as a tetramer; (d) an extracellular ligand-gated ion channel such as the nicotinic acetylcholine channel. Which acts as a pentamer; (e) a twin-pore potassium channel such as TREK (a dimer); (f) a monomeric voltage-gated channel with four repeat domains, the sodium and calcium channels adopt this topology.

The existence of all these motifs in the sequences made alignment of the helical regions straightforward and was a valuable aid in the modeling studies described below. Other conserved features could also be identified. For example, there was a highly conserved cysteine in the second extracellular loop, which forms a disulfide bond with the TM3 cysteine.

The presence of all these motifs in the opsins led to the family being named the 'rhodopsin family' (also known as Family A or Family 1 receptors). However several other receptor sequences were identified which had few or none of the above features. One group of peptide hormone receptors which had a high degree of sequence conservation within its members, was characterized by a larger N-terminal domain of 120–150 amino acids. Among these were six cysteines which were shown to form three disulfide bonds. Members of this family include the secretin, glucagon, calcitonin, and parathyroid receptors. The family became known as the 'secretin family' or Family B GPCRs.

In yet another subgroup known as the 'Family C' receptors, the agonist ligand binds in the very large (900–1200 residues) N-terminal domain. The structure of this domain was predicted,[6] and subsequently proven by crystallography,[7] to be similar to the bacterial periplasmic binding proteins. Upon binding of the agonist, the bilobal structure

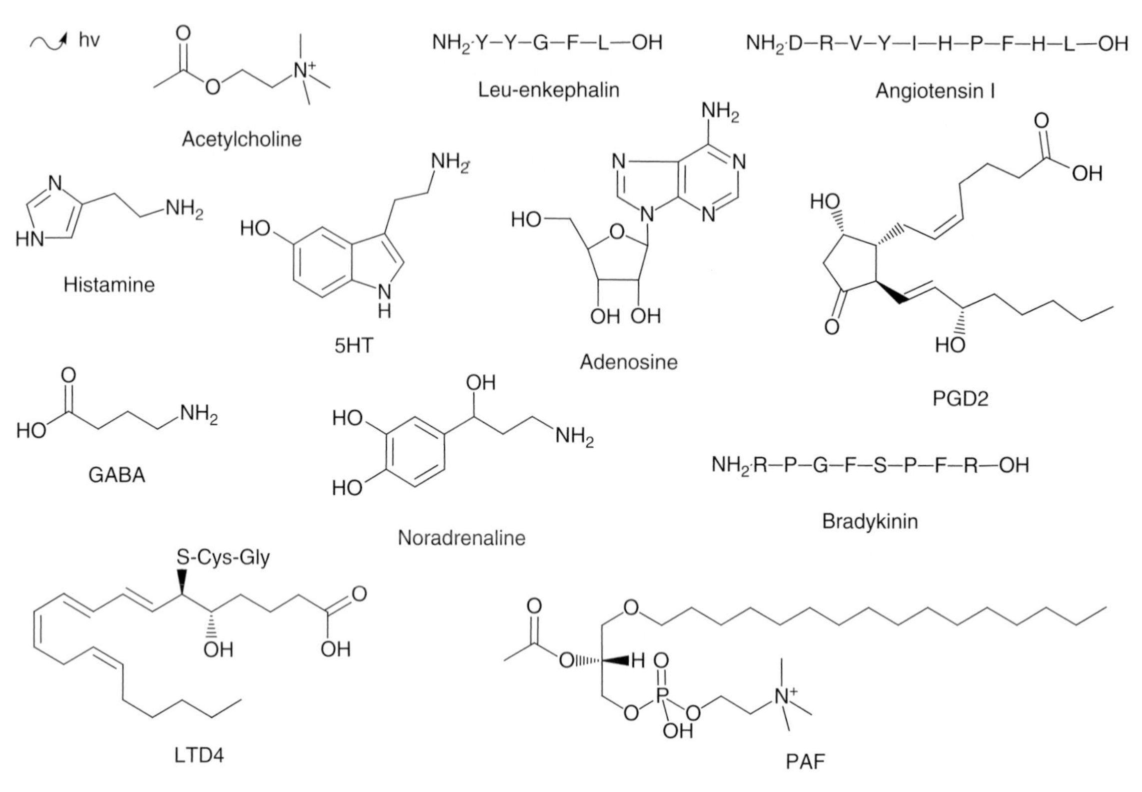

Figure 2 Some typical diverse ligands for 7TM receptors.

of the N-terminal domain switches from an 'open' to a 'closed' state, presumably then causing some allosteric conformational change in the TM domain. Relatively few members of this family are known, among them being the metabotropic glutamate, the γ-amino-butyric acid B (GABA-B), and the calcium sensing receptors.

Other families have also been identified. The yeast pheromone and the slime mold cyclic AMP families are not found in higher species. The 'Frizzled' receptor family[8] may well prove to be an important therapeutic target in the future, although the Wnt proteins, which are the natural ligands for these receptors, are likely to be too large for normal small molecule intervention. Other more recent subfamilies include the vomeronasal and taste receptors.[9] Obviously a fuller description of all the sequences and properties of the GPCR superfamily is beyond the scope of this chapter. The reader is however referred to an excellent database, GPCRDB,[30] where full details of sequence, classification, and original literature references can be found.

4.26.3 Computational Modeling of G Protein-Coupled Receptors

4.26.3.1 Early Modeling Attempts

By 1991, several groups had started to construct molecular models of 7TM receptors.[2,11] In most of the published examples it was assumed that there was some perceived homology, or at least structural similarity, with bacteriorhodopsin (BR), which, as noted earlier, was in fact a 7TM helical membrane protein for which a three-dimensional structure was known.

There was no real sequence identity between BR and the GPCRs but it is well known that proteins with no sequence identity can often adopt the same fold. The assumption that BR could be used as a template for homology modeling of GPCRs therefore seemed reasonable at that time. Findlay *et al.* were among the first to describe a model of bovine rhodopsin.[2] However the largest impact to medicinal chemistry came perhaps from Hibert *et al.*[12,13] Their models of the dopamine and adrenergic receptors suggested that the catecholamine agonist ligands bound their primary amino groups to an aspartate on TM3 while the catechol hydroxyls hydrogen-bonded to serines on TM5. This binding mode was in agreement with early site-directed mutagenesis (SDM) work in the β2 adrenergic receptor.[14] The postulation of an aromatic cage of residues from TM5, TM6, and TM7 which surrounded the aromatic ring of the ligand and stabilized the positive charge on the its amino group is still a feature of many current models today.

In a second approach used in early GPCR models, the lack of homology between BR and GPCRs was clearly recognized and was given as an argument for using some de novo helix packing methodology. Hydropathy plots, using the scales of Kyte–Dolittle,[15] Goldman–Engelman-Steitz,[16] or Hopp–Woods[17] were used to predict the helical regions. Use could also be made of conservation because it was often found that the more important functional residues occurred in the TM helices where the ligands presumably bound. This same conservation, as observed in multiple sequence alignments, could be used to calculate a conservation 'moment' that would define the inner face of the helix.[18] Using a measure of hydrophobicity from one of the scales mentioned above, a hydrophobic moment could also be calculated which should be an indicator of the outer face of the helix. These moment calculations were based on the recognition that regular helices were periodic and could therefore be analyzed by Fourier transforms. In our group we combined both measurements to generate helical wheels containing a conservation moment and a hydrophobic arc which was a measure of the degree of exposure of each helix to the lipid bilayer. The seven helical wheels could then be manipulated on a graphics screen to give a two-dimensional (2D) packing which was readily converted to a three-dimensional (3D) model.[19,20]

This method of course could not correctly predict the interhelical tilts which were vitally important for any model. With the use of de novo folding approaches, helix packing was often carried out manually, usually followed by molecular mechanics/dynamics simulations for refinement. The modeling of loop regions in early models was generally ignored because ligand binding was assumed to occur in the TM bundle. The problem of predicting interhelical tilts however could not be ignored and this came to the fore in 1993 when Schertler *et al.* published a 9 Å electron diffraction map of rhodopsin.[21] This showed clear differences in the helix packing of rhodopsin (a real GPCR) and BR, and led to a spate of renewed efforts in GPCR model construction.

Up to this time most of the models were broadly similar although some researchers did suggest more unusual topographies. Lybrand proposed an adrenergic receptor model in which the generally accepted sequential anticlockwise arrangement of the helices (as viewed from the extracellular side) was reversed.[22] In one of their earlier models, the group of Weinstein *et al.* argued for a nonsequential arrangement of the helices compared with BR.[23] Both these of course are now known to be incorrect although the arguments in favor of each were quite strong at the time.

Many of the early models were of aminergic neurotransmitter receptors such as dopamine and norepinephrine. This was undoubtedly because this was and still remains, an area of particular importance to the pharmaceutical industry. A lot of the early SDM studies in GPCRs had been carried out on these receptors and there was also a wealth of ligand structure–activity relationship (SAR) available for them. Despite this there were still debates about the exact binding sites of key ligands. For example Hutchins *et al.* argued against the SDM findings of the TM5 serines and suggested that dopamine was actually binding between the two aspartates on TM2 and TM3.[24] In their molecular dynamics simulations of antagonist, Dahl *et al.* also used the TM2 aspartate as the primary binding site.[25]

These early models were generally accepted as an important step within the scientific community, both in academia and in particular in the pharmaceutical area. They were often used successfully to explain known SAR from ligand binding or SDM data, or alternatively, to suggest potential binding modes or mutagenesis experiments. There was, however, a great deal of uncertainty at the atomic level, which led to disagreements between research groups about even coarse details of the models. In retrospect they probably raised even more questions than they answered and many people therefore rejected them as too crude to be of use in real structure-based drug design.

4.26.3.2 The Problems Associated with De Novo Integral Membrane Protein Structure Prediction

It is worth looking in closer detail at the problems involved in the de novo structure prediction of GPCRs. These can be summarized as:

- definition of the TM domains, i.e., the length of the helices;
- definition of the inward-facing and lipid exposed faces, i.e., the vertical orientation of the helices;
- definition of the interhelical tilt angles.

Somewhat separate problems are:

- modeling the cytoplasmic and extracellular loops;
- modeling the N- and C-termini;
- defining the residues involved in the ligand-binding site.

hydropathic analysis remains the best way of estimating the extent of the TM helix. Crystal structure determinations[26] have shown, however, that there is considerable variation in the length of these helices (e.g., TM3 of bovine rhodopsin is 36 residues long whereas TM4 contains only 24 residues) and that they can protude from the bilayer to quite a large extent.

Donnolly[27] suggested that the change in direction of the conservation moment might be useful as a guide to where TM helices cross the bilayer. Secondary structure prediction algorithms can also be useful and one method in particular, TMPred, was developed especially with TM helices in mind.[28] A special Hidden Markov model has also been developed for GPCRs.[29]

The vertical orientation of the helices can still be best estimated by Fourier-based moment calculations. They are very dependent on the degree to which the individual helices are buried and as a result are only accurate to within $\approx \pm 60^\circ$.

As has already been alluded to, the other main problem in modeling integral membrane proteins is the prediction of the interhelical tilt angles. Atom-level simulations using molecular mechanics and/or dynamics are not really suitable for these types of predictions as they do not reflect the large-scale movements necessary to change from a starting conformation. Observations by one group on the spread of interhelical tilt angles found in globular proteins led to the suggestion that there was a likelihood of similarity of helix packing between GPCRs and the four-helix bundle proteins.[30] In an early attempt by our group to solve the helix packing problem, we tried to optimize the bundle by maximization of the hydrophobic energy. The reasoning was that the helices would pack in such a way as to maximize their interactions with the lipid bilayer. The Eisenberg method of calculating the hydrophobic energy was used[31]; in this, the energy is the sum over all atoms of the product of each atom's atomic solvation parameter and its accessible surface area. It was found, however, that the helices tended to drift apart and remain essentially parallel, so the method was abandoned.

As an extension to this procedure the hydrophobic potential energy could be calculated and plotted on a surface projection of each helix. This potential energy at a point on the surface, P_i, is defined as:

$$P_i = \varepsilon_i + \sum_{j \neq i}^{\text{num\_atoms}} \frac{\varepsilon_j(1+\cos\theta)/2.0}{r_{ij}} \qquad [1]$$

This is analogous to the classical electrostatic potential definition, i.e., the sum of all atomic contributions divided by that atom's distance from the point in question. The hydrophobic energy, ε_i, associated with each atom is again defined by the Eisenberg term, i.e., the atom's product of its atomic solvation parameter and its accessible surface area. As hydrophobic energy is a surface-based term, in a helix partially buried within a bundle, atoms on the same face of the helix will have a much greater effect than those closer in distance but lying on the opposite side. The angle θ is the angle between the two atoms perpendicular to the main longitudinal axis of the helix; thus, this angular term varies from 1 when θ is 0° to 0 when it is 180°, i.e., on the opposite face. In **Figure 3** the hydrophobic potential is plotted onto the surface of TM3 of the $5HT_{2C}$ receptor. Color coding shows that the extracellular side is very hydrophobic (blue) compared to the cytoplasmic end (yellow–red) suggesting that this helix is highly tilted with the top exposed and the bottom deeply buried in the bundle. Again manipulation of these surfaces on a graphics workstation allows real-time examination of potential packing modes.

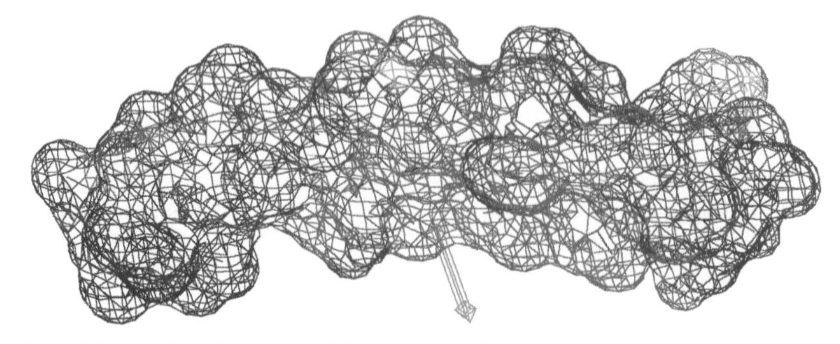

Figure 3 The hydrophobic potential of TM3 of the $5HT_{2C}$ receptor. The conservation moment is attached to the surface. This shows the very hydrophobic extracellular end in blue with the more buried hydrophilic end in green/red. (Adapted from *Membrane Protein Models*, edited by J. B. C. Findlay, Bios Scientific Publishers Ltd, Oxford, UK.)

A completely de novo method for helix bundle packing was developed by Adcock.[32] Use was made of a specially developed genetic algorithm with the incorporation of energy terms, rotamer libraries, and distance constraints. Distance constraints have also been used by a number of other groups to refine their models by distance geometry algorithms[33,34] or by inclusion of these restraints in constrained dynamics simulations.[35] The constraints themselves usually come from a large number of biophysical and biochemical experimental data. While a full description of these experiments is beyond the scope of this chapter, it is worth looking at some of the techniques involved in a little more detail, as much of our knowledge of GPCR structure has been derived from them.

4.26.4 Experimental Techniques Used in G Protein-Coupled Receptor Structure Determination

4.26.4.1 Diffraction Methods

In the absence of an x-ray crystal structure of a GPCR, the major source of direct structural information has undoubtedly been cryoelectron diffraction. This technique can in theory produce high-resolution data, but the extreme high energy of the beams means that the 'crystals' are quickly destroyed during data collection. A further difficulty is that the 'crystals' are generally 2D, which means that many thousands of data sets from different directions need to be collected, and the data then converted into a 3D image using special computer algorithms.[36] Thus, electron diffraction structures generally take much longer to generate than conventional x-ray equivalents.

The earliest structure on which most GPCR models were based was bacteriorhodopsin. However, a breakthrough occurred in 1993 when Schertler published his 9Å cryoelectron diffraction map of rhodopsin, which was a true G protein-coupled receptor.[21] While this was well beyond atomic resolution, it did show that the packing of the helices in rhodopsin was quite different from that of bacteriorhodopsin. Hibert briefly argued that the observed differences in the two cryo-electron microscopy maps could be explained by a different tilt of the whole TM bundle,[37] but most groups just started rebuilding their models based on the limited information from Schertler's findings. In addition to the low resolution, the main problem was that those data were for a single 2D slice in the membrane.

In 1997 this problem was largely overcome when Schertler published details of a 6Å electron diffraction map of frog rhodopsin.[38] These data were collected as 4Å slices through the plane of the membrane and, while still not at atomic resolution, they gave direct information on the interhelical tilts of the TM bundle.

Baldwin, working in the same laboratory, described a C-alpha model for GPCRs based on this density data,[39] and Vriend, using his WhatIF program, quickly followed on this by making available a large number of receptor models on the 7TM website.[30]

In our group, we combined the diffraction data with a wealth of other experimental results described in the following sections, to generate numerous models which differed from Baldwin's rhodopsin template. These were used extensively in structure drug design work, some of which will be described later.[40,41]

The importance of Schertler's cryoelectron diffraction work can never be underestimated, but it was largely superseded when, finally, the first x-ray structure of a GPCR, that of bovine rhodopsin, was published in 2000.[26] It was now possible to use direct homology methods to construct 7TM models. Confidence in these models was increased because rhodopsin itself contains the conserved residues in each TM helix which play key functional or structural roles in Family A receptors. Alignment of other GPCR sequences is therefore straightforward in the TM bundle except for the occasional receptor where one or more of these conserved motifs are missing.

4.26.4.2 Nuclear Magnetic Resonance (NMR) Studies

Conventional solution state NMR experiments have been widely used in protein structure determination. Much of this work has involved the use of 2D techniques such as NOESY or COSY which can give interatomic distances between pairs of nuclei. These can then be used as constraints for molecular dynamics (MD) or distance geometry (DG) calculations as described earlier. With 7TM receptors, solution NMR has been used for the structure determination of individual domains such as the N-terminal domain of the corticotropin–releasing factor (CRF) receptor.[42] It can also be used for conformational analysis of loop fragments,[43] although to be useful, it has to be assumed that the independent loop folds in the same manner as when it is incorporated into the whole protein.

In a combination of ^{13}C NMR studies with quantum charge density calculations, Han and Smith concluded that the proton of the Schiff base formed between retinal and the TM7 lysine (Lys296) of rhodopsin, did not directly interact with the TM3 glutamate as originally thought. Instead, this proton formed a hydrogen bonding network through a lone

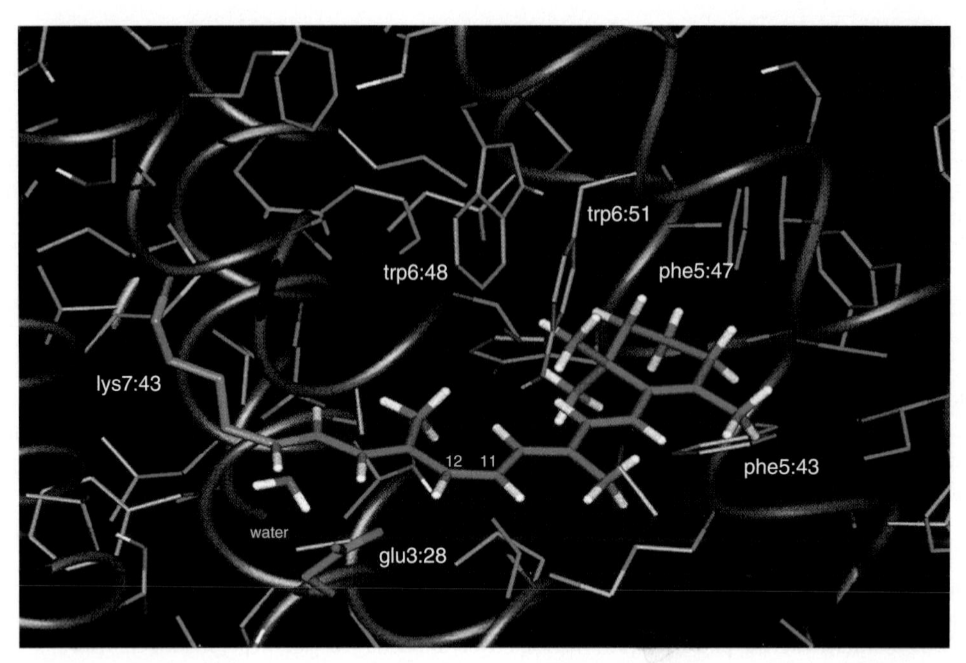

Figure 4 The fourth-generation model of bovine rhodopsin with bound retinal. This shows the lysine in an extended conformation with the protonated Schiff base forming a hydrogen bond through a water molecule to glutamate 3:28. The residue numbering system is that of Ballesteros and Weinstein[45] where the first number refers to the TM helix and the second number to the position in that helix with the most conserved residue being numbered as 50.

water molecule to one oxygen of the glutamate, with the second oxygen being within 3 Å of C12 of retinal.[44] We were able to use this information to refine the relative orientations of TM3 and TM7 in our model of rhodopsin based on the 6 Å diffraction data described above (**Figure 4**).[45]

Nuclear Overhauser effect (NOE) measurements have also been used to look at the interaction of the complement glycoprotein C5a with its receptor. Peptide fragments of the receptor were studied and their interactions with the C5a protein confirmed by chemical shift and line width changes. The results showed that residues 1–34 in the N-terminal domain of the receptor were important for protein binding.[46]

More recently, solid-state NMR has played a role, particularly in the determination of ligand-bound conformations and environments. Use can be made of the special anisotropic properties of certain nuclei such as ^{2}H, ^{13}C, and ^{15}N. The orientational constraints derived from solid-state NMR gives accurate information about the bound conformation, but the technique has the not inconsiderable drawback that extensive isotopic labeling of ligands and/or protein is necessary. Despite this, notable successes have been described. Thus, the bound conformation of retinal in rhodopsin in both the light (meta I) and ground (dark) states, was solved using deuterium solid-state NMR.[47] A short but useful review of NMR studies in ligand-membrane receptor interactions has been published by Watts.[48]

4.26.4.3 Site-Directed Mutagenesis and Ligand Structure–Activity Relationship

In the absence of direct experimental data, the single most important technique in the structural elucidation of GPCRs is undoubtedly SDM. SDM, first developed by Nobel laureate Michael Smith in 1978,[49] involves the systematic alteration of one amino acid in a protein to another, in order to better understand the function of that residue through measurement of the change in activity of various ligands. SDM has been used extensively with GPCRs since the earliest sequence information became available. A notable example was the finding by Strader *et al.* that several serine residues in TM5 of the adrenergic receptors were responsible for the binding of the catechol hydroxyls of norepinephrine.[14]

The process of rational drug design with GPCR models typically involves a cyclical process by which ligands are docked into a receptor model and initial hypotheses are developed for the binding mode. SDM experiments are performed on key residues involved in these hypotheses and the hypotheses are confirmed or altered in light of the results. Some examples of this process are discussed later. A fuller discussion of the many thousands of SDM experiments carried out with GPCRs is beyond the scope of this chapter, but the reader is referred to the excellent web-based databases GRAP[50] and tinyGRAP[51] which contain full details and references to many of these. Again, although direct structural data are largely unavailable, the affinities and efficacy of many thousands of ligands have been measured and details of these data, especially in association with SDM results, have given valuable further information on key residues involved in ligand–receptor interactions. In addition to single point mutations, similar experiments have been carried out which involve multiple substitutions or even entire domains of receptors. Chimeric constructs with domains from two related receptor subtypes are often useful in pinpointing areas of importance for receptor selectivity.[52]

As an alternative to specific mutations, many groups have carried out a systematic search of all residues in areas of interest in a sequence. The most common approach is the so-called 'alanine scanning' technique. Each residue in turn is mutated to alanine, a residue which is not likely to give rise to misfolding, unlike glycine, but which lacks much of a side chain with which to interact with the ligand. The effects of the alanine mutations are then measued in an appropriate assay. The differences between the mutant and native receptor can be attributed to the side chain of the native receptor.

In an early review Schwartz described some of these systematic approaches as they were applied to GPCRs.[53] A more recent alternative to alanine scanning was described by Holst *et al.*, using the NK1 receptor as a target.[54] In this 'steric hindrance mutagenesis,' it was found that substitution of alanine at position 115 by valine led to an increase in affinity of a diverse set of 13 ligands. Finally the process of mutation, often to cysteine residues, followed by chemical modification through a variety of biophysical probes, can give valuable additional information on receptor structure. Some of these experiments are described in rather more detail below.

4.26.4.4 Substituted Cysteine Accessibility Method

One of the most commonly used of the aforementioned techniques is the substituted cysteine accessibility method (SCAM).[55] Cysteine, being a small residue, is not likely to perturb the structure and has the added advantage that it has no secondary structural preference. However its main property of interest is that it has a labile sulfhydryl group which can undergo a variety of chemical reactions, especially in an aqueous environment. Thus it is possible to map out those residues which are exposed to water in the channels or binding cavities of membrane-bound proteins, by mutating each residue in turn to cysteine and then establishing whether this cysteine can react with the appropriate reagent. Residues which are buried in the protein or which are exposed to the lipid bilayer are incapable of reaction.

One class of commonly used reagents is based on substituted maleimides (**Figure 5**). These can undergo irreversible nucleophilic Michael addition by a cysteine sulfur. If, for example, the maleimide substituent is a biotin analog, the cysteine reaction can then be quantified by immunoprecipitation followed by a Western blot using streptavidin linked to horseradish peroxidase.[56] The most commonly used reagents, however, are derivatives of

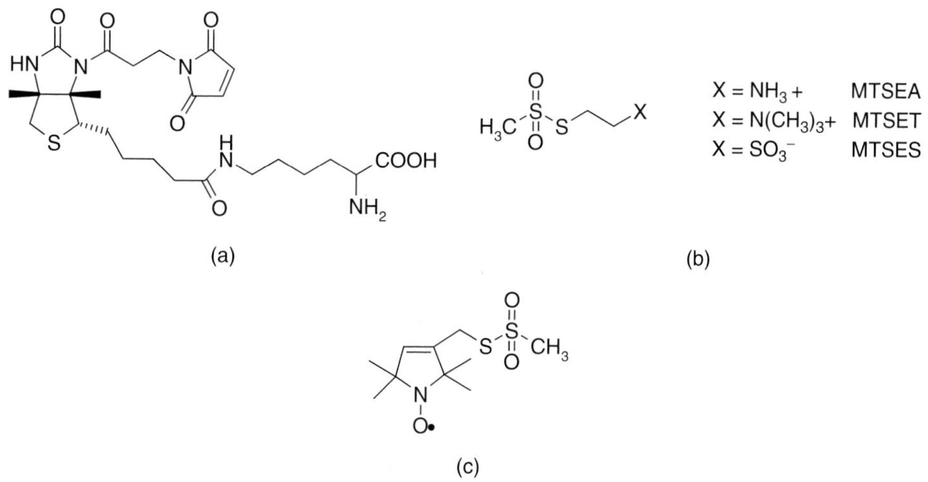

X = NH_3+ MTSEA
X = $N(CH_3)_3$+ MTSET
X = SO_3^- MTSES

Figure 5 Various thiolating reagents used as biophysical probes. (a) Biocytin maleimide; (b) various hydrophilic derivatives of methylthiosulphonate (MTS), either the positive ethylammonium (MTSEA) and ethyltrimethylammonium (MTSET) analogs or the negative ethylsulphonate (MTSES); (c) a spin-labeled nitroxide reagent.

methanethiosulfonate (MTS). These react with cysteines to form a cross-linked disulfide bond. They are charged (either positive or negative) and can vary in size. The widely used MTS-ethylammonium analog (MTSEA) is the smallest of these, but the size is considerably increased with other ammonium derivatives. The ethyltrimethylammonium derivative (MTSET) is approximately the size of dopamine. Replacement of the ammonium group with a sulfonic acid gives the negatively charged reagent MTSES (**Figure 5**).

To interpret the results of SCAM mutations properly, it must be assumed that only those cysteines in an aqueous environment can undergo reaction and that these are most likely therefore to be found in the channel or binding site. Thus incorporation of the cross-linked reagent should result in a large change in ligand or substrate affinity. Conversely, prior incubation with a potent inhibitor before introduction of the reagent should slow down the rate of incorporation. A major problem that often arises is the presence of more than one accessible cysteine which makes interpretation very difficult. It is often necessary therefore, to mutate native wild-type cysteines, prior to SCAM experiments.

In an elegant series of papers, Javitch described the complete SCAM analysis of the TM bundle in the dopamine D2 receptor.[57–65] The helical periodicity was very evident, although one interesting finding was that this broke down at the extracellular end of TM5.[58] Our model of the D2 receptor, based on the rhodopsin crystal structure, shows that the sequence length of ECL2 between the disulfide bond and TM5 is quite short, therefore necessitating some unfolding of this helix.

Another unexpected result from Javitch's work was that MTSEA reaction with the native Cys118 (TM3) resulted in a massive loss of binding of the antipsychotic ligand sulpiride, but had practically no effect on the binding of haloperidol, another widely used antipsychotic antagonist.[57] This would clearly suggest that the two antagonists bind in different pockets.

In our construction of a dopamine D2 receptor model, based on the 6 Å density slice data described earlier, we were able to make use of all of Javitch's SCAM results to improve the orientation of the helices. **Figure 6b** shows the resultant packing of TM3 with the exposure of various residues corresponding perfectly with the SCAM experiments (**Figure 6a**).

In addition to its use in binding site identification, SCAM can also be used to detect changes in the local environment of a residue. This has been important, for example, in the determination of conformational changes in switching from an antagonist to an agonist state of a receptor. Thus Javitch *et al.* showed that in a constitutively active β_2-adrenergic receptor, the largely conserved cysteine in TM6 is accessible to MTSEA, whereas it is not in the wild-type receptor.[66]

Use can also be made of the electrostatic properties of the cross-coupled reagents to probe structure. It is known for example that the positive headgroup of MTSEA favors negatively charged (i.e., acidic) environments. Thus by systematic alanine replacement of the native cysteines in the NK2 receptor, we were able to show that only MTSEA coupling to Cys167 (TM4) affected binding of the potent antagonist, [^{3}H]SR48968. Subsequent sequential mutation of the aspartate and glutamate residues in the extracellular part of the receptor showed that mutation of Asp5 in the N-terminal domain abolished the effects of MTSEA on [^{3}H]SR48968 binding. This suggested that the ammonium headgroup was forming a salt bridge with this aspartate, and thus gave us additional information on the placement of the N-terminal domain.[67]

4.26.4.5 Spin Labeling, Fluorescence, and Photoaffinity Mutations

Cysteine mutations have also been extensively used to introduce more specific types of probes into GPCRs. One common type of reagent involves a nitroxide species which, when cross-linked to a cysteine, provides a spin label in the protein. Analysis is then performed by means of electron paramagnetic resonance spectroscopy. The depth of a residue within the lipid bilayer can be measured using collision gradient calculations.[68] The interaction of the spin label with paramagnetic reagents can be used to determine the environment of the residue. Typically reagents of opposite polarity such as nickel diethylamine *N,N′*-diacetic acid (NiEDDA) (polar) and molecular oxygen (nonpolar) are used to detect asymmetric solvation.[69] Changes in movements of a protein can also be time-resolved. In a series of papers from Khorana and Hubbell, this was used to look at the activation process of rhodopsin.[70–74] Numerous movements in both the TM helices and loops were described. Probably one of the most useful aspects of site-directed spin labeling is in the calculation of interresidue distances. These usually involve the introduction of two specific nitroxide labels or alternatively one nitroxide and a bound metal ion. Very accurate distances can be found with these experiments.[75,76] For further information the reader is referred to some excellent reviews by the Hubbell group.[69,77]

As an alternative to spin labeling, Gether *et al.* used the cysteine cross-linking reagent, *N,N*-dimethyl-*N*-(iodoacetyl)-*N*-(7-nitrobenz-2-oxa-1,3-diazol-4-yl)ethylenediamine (IANBD) to introduce a fluorescent label into the β_2-adrenergic

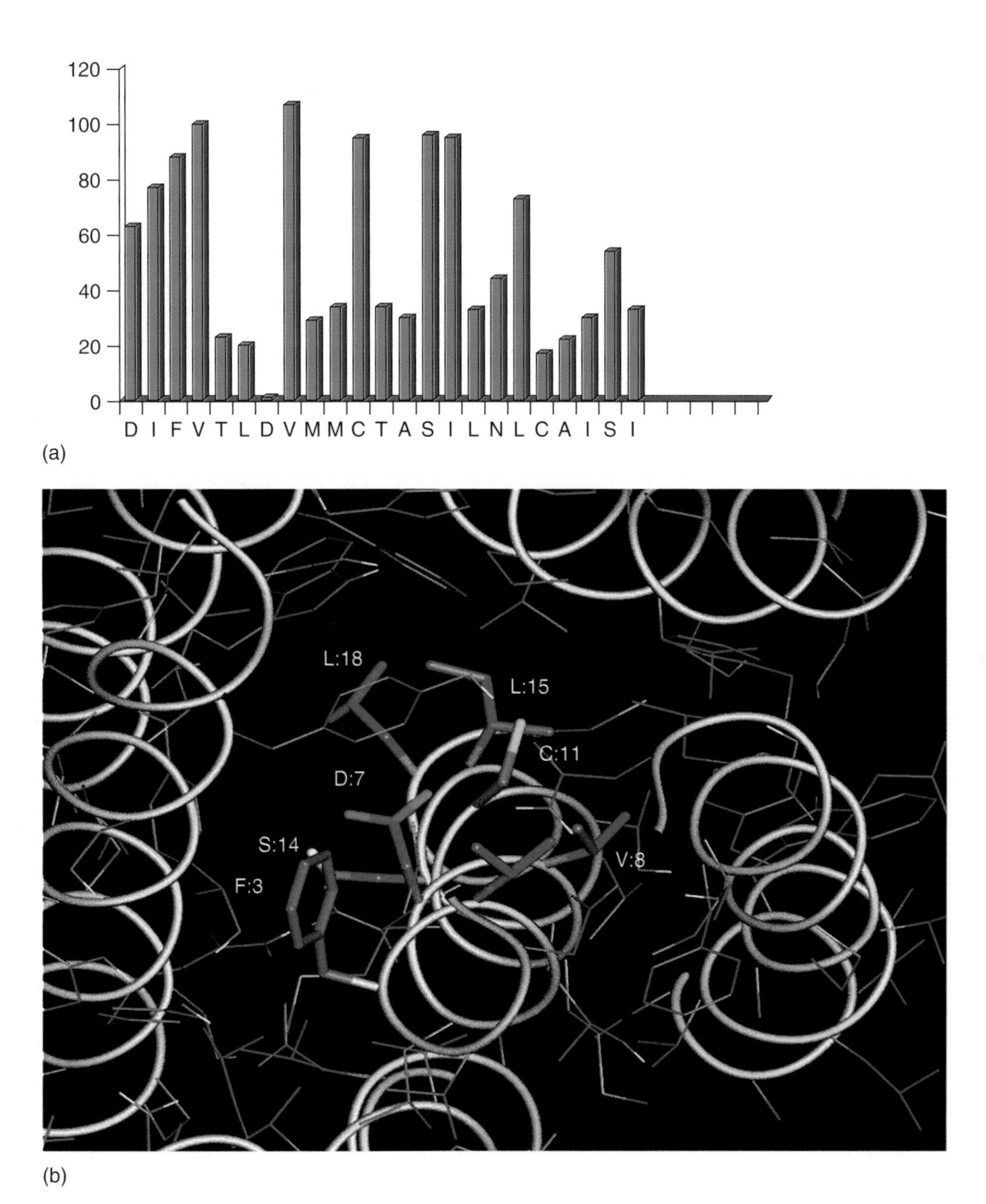

Figure 6 (a) Percentage effect of MTSEA cysteine scanning on antagonist binding in TM3 of the dopamine D2 receptor; each residue in the helix is mutated in turn to a cysteine. (b) The exposed residues in the fourth-generation model of D2 showing an excellent agreement with the SCAM results.

receptor.[78] The fluoresence is highly sensitive to the polarity of the local environment, thus enabling information to be derived on the binding of several agonist and antagonist molecules.

Coupling to cysteine (and lysine) residues has been a common method of incorporating biophysical probes. However, a more recent development which has proven to be extremely valuable is the use of nonsense stop codon suppression techniques to introduce unnatural amino acids into a protein sequence. Typically, at the desired position of the coding sequence, a stop codon such as UAG is introduced. In the following translation step, a supressor tRNA, misacylated with the desired unnatural amino acid, is introduced. This contains the anticodon CUA, which recognizes the UAG stop codon. In the field of GPCRs, André Chollet has been a pioneer in this field and his excellent review is highly recommended to the reader.[79] In this, he describes the use of various unnatural fluorescent and spin-labeled amino acid mutations in the neurokinin receptors. In an extension to this work, fluorescent receptor mutants were studied using fluorescence resonance energy transfer (FRET) to gain information on ligand protein distances within the NK2 receptor.[80] Photoaffinity labeling has not been used as widely as might be expected in the field of GPCRs.

This is undoubtedly due to the experimental difficulty, not only with the chemical synthesis of the labeled peptides or proteins, but also with the analysis of the results. The latter requires extensive protein purification, usually followed by enzymatic digest and extensive chromatographic–mass spectral analysis, to identify the sites of photolabel incorporation. One area, however, where it has met with notable success is in the study of the interactions of Family B receptors with their peptide ligands. In their recent review, Pham and Sexton discuss this in some detail, with the ultimate view of generating a template for rational drug design.[81] Another specific example worth mentioning is that of Dong *et al.* who introduced a photolabile *p*-benzoylphenylalanine residue at positions 16 and 26 of human calcitonin. The labeled peptides were potent agonists of the calcitonin receptor and upon excitation, were found to interact with Phe137 and Thr30 of the receptor respectively.[82]

4.26.4.6 Engineered Metal Binding Sites

One of the most elegant techniques to be used in the probing of interresidue distances is that of engineered metal-binding sites. In a series of experiments, Elling *et al.* mutated a number of residues in the neurokinin and opioid receptors to histidines. In some cases, where the histidines were in a specific spatial orientation, the mutated receptors were then able to bind zinc atoms, often with quite high affinity.[83,84] Since these early papers appeared in 1995–96, a large number of other examples using this technique have been published. Elling *et al.* were able to use the chelating properties of histidine and cysteine to generate a mutant β_2-adrenergic receptor which was agonized by zinc. The agonist state mutant receptor had a histidine in place of the binding aspartate in TM3 and a cysteine in TM7 replacing an asparagine.[85] A similar mutation in the NK1 receptor, however, showed only partial agonism to zinc.[86] In an interesting twist to this story, Schwartz's group have recently incorporated zinc into a chelating bis-cryptand structure. This metal complex, AMD3100, is a very potent ligand for the CXCR4 chemokine receptor.[87] The accurate interresidue distances established from these experiments have been of vital importance in the relative placements of pairs of helices in modeling. They were used for example in refining our model of the NK2 receptor, based on the 6 Å density slice data described earlier. **Figure 7** shows our model of the mutant NK2 receptor with the zinc atom bound by histidines in TM5 and TM6 in an almost identical coordination geometry to that found in carbonic anhydrase.

4.26.5 Model Generations

From the above description of modeling and experimental techniques it can be recognized that GPCR models have evolved through a number of 'generations.' The earliest modeling efforts were based either on de novo technologies or on an alignment of sequence and/or structure with bacteriorhodopsin. The 'third generation' of models emerged in 1993 following the publication of the 9 Å diffraction map of rhodopsin by Schertler. It was these models that were first used in some rational drug design and which led to hypotheses which could be tested by SDM experiments. The results from other biophysical data were also starting to emerge and these could be used to refine the models. However, the big step in model improvement came again from Schertler in 1997, when he published the 6 Å electron diffraction data for frog rhodopsin. These 'fourth-generation' models were routinely used by several groups in rational drug design, where the ligand SAR could be tested by SDM work. The biophysical results from the techniques described above could also be incorporated into the models in a meaningful manner. There were some notable successes in structure-based design with these models, some of which will be discussed below. The obvious conclusion to this nomenclature was the publication in 2000 of the crystal structure of bovine rhodopsin, a real GPCR, and this has formed the basis for most of the modeling work over the last 5 years. The conserved nature of key functional/structural residues in the TM bundle of Family A GPCRs has given a great deal of confidence in the reliability of these models. However, the binding of many ligands occurs within domains such as the NTD and extracellular loops where no structural assignment is readily possible.

4.26.6 Loop Modeling of G Protein-Coupled Receptors

Early-generation models of GPCRs ignored the extracellular loops (ECLs), despite their known involvement in ligand binding, especially in peptide receptors. As early as 1993, however, some authors were including the loop regions, with the obvious caveat that they were more speculative.[88–91] Chini *et al.* concluded from modeling and mutagenesis that Tyr115 in ECL1 was responsible for agonist selectivity at the vasopressin V1a receptor,[89] although recent models would place this residue at the top of TM2. Most early loop modeling made use of loop library searches often with molecular dynamics refinement. These libraries have been derived from analysis of the Brookhaven Protein Data Bank (PDB) and

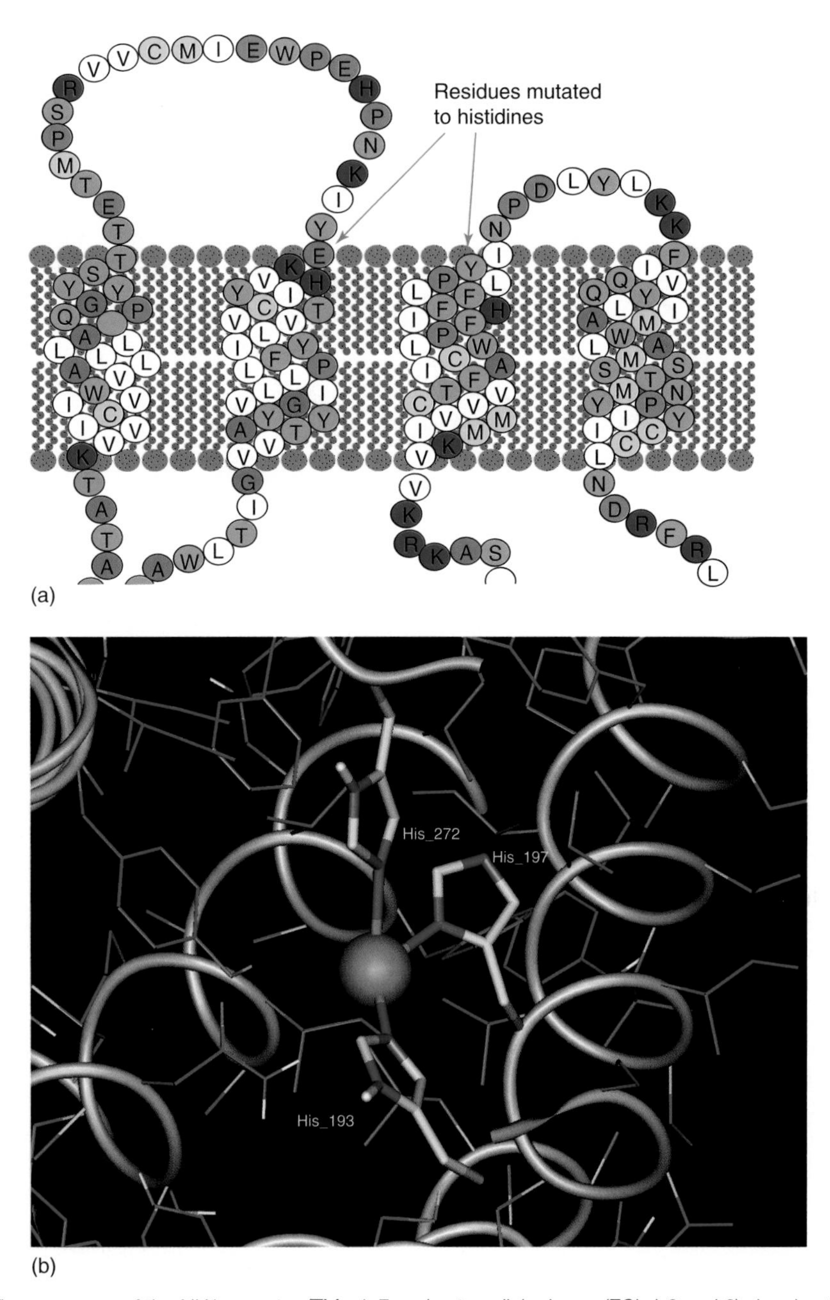

Figure 7 (a) The sequence of the NK1 receptor (TMs 4–7 and extracellular loops (ECLs) 2 and 3) showing the residues that were mutated to histidines; (b) the fourth-generation model of the double-mutant NK1 receptor showing the zinc-binding site.

are typically less than 10 residues in length. However, the loops of GPCRs are often much longer than this and have complicating features such as the conserved disulfide in ECL2.

In recent years within our group a semiautomated procedure for the generation of extracellular loops has been developed. Initial conformational sampling is carried out with the distance geometry program DGEOM95.[92] Intra- and interhelical distance constraints are automatically generated for the five helical residues at each N- and C-terminus of the loop. Further constraints come from the disulfide bond information in ECL2, the proximity of previously generated adjacent loops, and consensus secondary structure prediction information calculated for that loop region. For example, if a typical stretch of loop was predicted to exist as a β-strand using several different prediction algorithms, then one could set up consensus constraints for the strand backbone distances automatically (see **Figure 8**). In a typical loop calculation, 500–1000 random conformations are generated. Each of these is then minimized by molecular mechanics

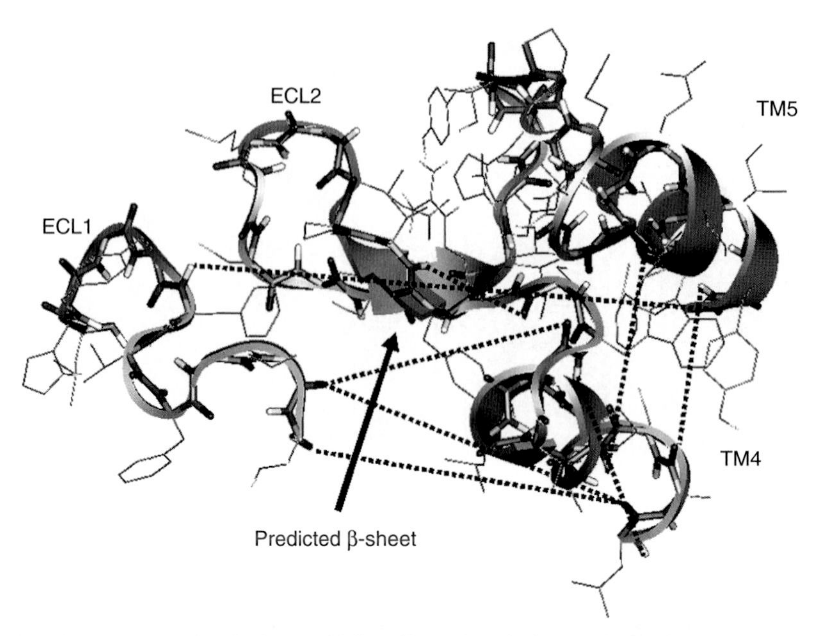

Figure 8 Generating distance constraints for loop modeling. (Reproduced with permission from *Ligand Design for G Protein-Coupled Receptors*, edited by D. Rognon, Wiley-VCH Verlag.)

and the conformations are analyzed by goodness of backbone phi-psi angles using the criteria described by Wilmot *et al.*[93] They are clustered into backbone dihedral 'families' and further ranked by energy. The best conformations are chosen for subsequent molecular dynamics analysis using the locally enhanced sampling (LES) method of Elber,[94,95] as implemented in CHARMm.[96,97]

In modeling the RF-amide receptor, GPR10, with loops generated using the above method, it was predicted that two adjacent glutamate residues in ECL2 formed crucial interactions with the peptide ligand. This was subsequently confirmed by SDM.[98]

Attempts have also been described in the literature to model other domains of GPCRs. The N-terminal domain of the luteinizing hormone receptor was modeled based on homology with some serine protease inhibitors.[99] The N-terminal domains of Families B and C receptors have also been built by homology although direct structural data are now available for both these. More will be said about these later. Lesser attention has been given to the intracellular loops and to the C-terminal domain.[100] This is justified by the fact that ligands do not bind in this region and hence there is less chance for pharmaceutical intervention. There is of course a great deal of interest in the conformational changes occurring following activation, and how this is relayed to the G-protein heterotrimer. The structure of the latter complex is known[101,102] and a large amount of SDM data are available for these domains.[51] No doubt modeling of this region will increase in the future.

4.26.7 Automated Modeling

The sheer number of receptors, their orthologs, and their mutants makes the construction of individual receptor homology models a daunting task, and efforts have been made to give some degree of automation to the process. General automated homology packages such as Modeler can and have been used but there is such a wealth of specialized knowledge associated with GPCRs which makes the development of bespoke programs a worthwhile task. In 1994 Blaney described a semiautomated program, GPCR_Builder, which made use of a variety of preconstructed templates, and user-selected approximate TM regions from hydropathy plots, to generate CHARMm scripts from which models could be readily built.[19] As already mentioned, Vriend made use of the homology features of his WhatIf program to automatically generate models of many Family A GPCRs based on the C-alpha template of Baldwin.[39] Bissantz *et al.* used alignments of the TM domains with a set of nine template structures of the TM bundle and two 'in-house' rotamer libraries, to construct a total of 277 human receptor models in an automated procedure.[103] Their models, however, do not contain any loop regions.

The PREDICT method developed by Becker and co-workers[104–106] produces models of GPCRs without using any homology modeling. The method, which is described in detail elsewhere,[106] carries out a detailed analysis of a very large number of possible receptor conformations. To start, the relative orientations of the TM helices are determined in two-dimensions using a hydrophobicity analysis. The two-dimensions structures are then converted to a reduced representation 3D model. These models are used to systematically refine the packing by looking at the relative helical positions in both the plane of the membrane and perpendicular to it, i.e., the helical rotation and the helical tilt, in conjunction with Monte Carlo optimization of the side chains. All resultant models are scored with an energy function and those with the lowest energy are clustered using a principal component analysis (PCA). Representatives from each cluster are then analyzed and those consistent with experimental observations are carried forward to a second phase of refinement. In the second phase, a large number of new models are generated that are close in conformational space to the best models from the first phase, and the above procedure is repeated for each of these. The final models selected by cluster analysis, energy score, size of binding pockets, and agreement with experimental data are expanded to an all atom representation. In the next stage, the all atom models are subjected to a molecular dynamics simulation, to refine side chain orientation and introduce helical kinks. Finally, a known ligand of the relevant target is docked in a mode consistent with experimental information about key interactions. Then, the ligand-protein interactions are set to zero and gradually increased during the course of an molecular dynamics simulation until the interactions reach their full potential. The idea behind this is to let the ligand freely explore different binding orientations during the early stages of the simulation. The removal of the ligand from this final structure gives the model with a well-defined binding site.

In our group, we have developed a fully automated program to generate homology models of Families A and B receptors in their antagonist and agonist (Family A only) states. The central core of this program is a knowledge-based database containing the prealigned TM sequence domains for all known human and rodent orthologs, their full length sequences and the ECL2 disulfide position (if present). The rhodopsin crystal structure is used as a template for the TM bundle of Family A antagonist models while agonist state and Family B models are constructed from templates developed 'in-house' from separate modeling studies. A library of polyalanine extracellular loops has been built from locally enhanced sampling molecular dynamics (LES-MD) simulations, with ECL2 being split into regions before and after the disulfide cysteine. Special algorithms have been added for exceptional domains such as the very short ECL1 loop of chemokines or the short stretch of ECL2 found between the cysteine and the top of TM5 in many aminergic neurotransmitter receptors. An 'in-house' rotamer library is used to set up the initial side chain conformations. As with its predecessor program, GPCR_Builder, the output is a starting structure and a CHARMm script which the user can modify for any further combination of molecular mechanics and/or dynamics calculations. As the final purpose of these models is for drug design, the scripts are ideally set up for careful ligand docking studies with little user intervention required.

4.26.8 Agonist State Modeling

The x-ray crystal structure of rhodopsin represents an inactive form of a GPCR and models based on it are therefore considered to be more appropriate for the study of antagonists. There is of course a great deal of interest in the design of agonist ligands and, therefore, in recent years attention has turned to the modeling of the agonist state of GPCRs. All the biophysical methods described earlier have been used to produce evidence for conformational changes occurring upon activation. They in turn have generated a large set of distance or other geometrical constraints which can be incorporated into distance geometry calculations or constrained molecular dynamics simulations. In this way a number of groups have generated agonist models of GPCRs. The details of these will not be discussed here. Instead the reader is referred to the excellent review by Bissantz[107] and to the paper of Gouldson *et al.*,[35] which describes a comprehensive list of the original experimental references and the constraints that were generated from them. The use of agonist state models in ligand design will be discussed in further detail, below.

4.26.9 Family B Modeling

The Family B (secretin-like) receptors comprise a group of pharmacologically important peptide hormone receptors which are characterized by a large N-terminal domain with three conserved disulfide bonds. The peptide hormones themselves are generally 30–40 residues in length and bind mainly to the N-terminal domain of their receptors. NMR studies have shown (at least in the solvents used) that the hormones largely form an amphipathic helix toward the C-terminal end with a more disordered N-terminal region. Despite their importance, relatively little modeling work has been carried out on this family. Tams *et al.* described a 2D model of the family based on simple helical wheels with considerations of residue size and hydrophilicity.[108,109] Donnelly did construct a 3D model based on his Fourier conservation moment program, PERSCAN.[110] The most detailed description of a Family B model was, however, given by Frimurer and Bywater, who combined a number of theoretical techniques with available SDM and ligand-binding data to construct a model of the GLP1 receptor.[111]

It has often been stated that there is no perceived homology between Family A and Family B receptors. However, we have noted a number of conserved residues in the latter family which do have equivalents in Family A. Thus, the conserved TM3-ECL2 disulfide bond is found in both families. The functionally important DRY motif in TM3 of Family A can be replaced by an EXXY with the arginine being provided by an invariant residue in TM2 (this TM2_R-TM3_EXXY motif was also noted by Frimurer and Bywater in their GLP1 model). In-house SDM studies on this arginine have shown that it is essential for activation. TM4, TM6, and TM7 also have a conserved tryptophan, proline, and asparagine respectively, in equivalent positions to Family A.

Based on these findings, and on a presumed geometric alignment with rhodopsin, various models have been constructed within our group. They have suggested a number of key interresidue interactions which may play functional roles in this family. SDM experiments have therefore been performed on these residues in the glucagon and calcitonin receptors which are in broad agreement with the proposed interactions (unpublished results).

It was noted earlier that binding of the agonist peptides occurs largely in the N-terminal domain. A considerable amount of modeling effort has therefore gone into studying this domain. Until fairly recently however this was hindered by a lack of knowledge of the disulfide bonding pattern.[112,113] In 2003 Taylor *et al.* used a combination of secondary structure prediction and distance geometry to generate a model of the GLP1 N-terminal domain.[114] However, this was quickly followed by an NMR structure of the CRF N-terminal domain.[42] This is already proving to be a useful template for homology modeling of other Family B receptors of interest.

4.26.10 Family C Modeling

The Family C receptors are an important class of GPCRs, characterized in particular by an especially large N-terminal domain. The pharmacologically important members of this family, i.e., the metabotropic glutamate (mGluR), GABA-B, and calcium sensing (CaSR) receptors, all bind their ligands in this domain. As early as 1993, O'Hara *et al.* had noted a similarity in structure between the N-terminal domain of GABA-B and the bacterial periplasmic binding proteins (PBPs) LBP and LIVBP. Their model of the GABA N-terminal domain had the typical bilobal structure in its 'open' form.[6] When the agonist ligand was bound, it was reasoned that the two lobes closed over rather like a Venus fly-trap. A second model, this time of a mGluR N-terminal domain, was published by Costantino and Pellicciari.[115] In this, they suggested that Ser165 and Thr188 were involved in ligand binding by homology with the PBPs. Tyr236 was also implicated and these residues have subsequently been shown to be important. Their suggestion, however, that Arg358 was the salt bridge partner for the ω-carboxylate of glutamate is now known not to be true. A large amount of SDM work, largely due to the group of Pin has been used to support the various models of GABA-B.[116,117] This group were the first to publish a 'closed' form model of the N-terminal domain using an unpublished coordinate dataset of LIVBP from O'Hara's group.[117] An alternative 'closed' form model was also published by Hibert using their 'open' form which was manually refolded around the arabinose binding protein structure.[118] More recently, the whole field of Family C N-terminal domain modeling has been aided by the publication of crystal structures of the mGluR1 subtype N-terminal domain in both 'open' and 'closed' states. Suprisingly however, because of the low sequence identity between the GABA-B and mGluR N-terminal domains, we have found that the PBPs remain better templates on which to base our models of GABA-B.

Unlike Family B, the TM domain really has no sequence identity with Family A. However, interest in this domain has arisen from the discovery of a number of molecules which do bind in the TM domain (from SDM studies) but which act as allosteric modulators rather than direct agonists or antagonists.[119–121] Several papers have appeared describing the modeling of Family C TM domains. The group of Malherbe *et al.* at Roche have constructed models of the mGluR5[120] and mGluR1[121] receptors. Miedlich *et al.* generated a similar model of the calcium sensing receptor.[122] Both these groups have been guided by some very useful SDM studies.

A wealth of experimental data has shown that many receptors, particularly the Family C members, act as homo- or heterodimers (or even oligomers). A few other GPCRs require additional protein partners for functional action. For example, the Family B receptor CGRP requires the presence of the associated RAMP1 protein.[123] A fuller discussion of these topics is beyond the scope of this chapter. In this context, however, it is worth noting that the requirement for dimerization has had no impact on the drug design process. A few groups have attempted to synthesize large dimeric drugs with long chain linkers, and have found some evidence of improved biological activity.[124] Likewise, it may be possible to disrupt a protein–protein interaction, such as a dimer, a CRLR–RAMP1 pair (CGRP) or even a GPCR–Gα protein assembly, but this is a far more difficult task than designing a small-molecule ligand.

4.26.11 Inward-Facing Residues and Microdomains

No recent review of GPCR modeling would be complete without a mention of 'inward-facing residues.' The concept of this is extremely simple and certainly not new. In GPCRs, (nonpeptide small) ligands generally bind on the inside of the TM bundle and toward the extracellular half. This has been well established from SDM and other biophysical studies. Parts of the ligand will therefore interact with a small subset of residues (typically two to four in number), known as a microdomain. A good example of this is the well-known aromatic cage (Trp6.48, Phe6.51, and Phe5.47) surrounding the catechol ring of the monoamine neurotransmitters.[12] As these residues will define the ligand specificity and selectivity, it makes some sense to redefine the receptor sequence in terms of these inward-facing residues or microdomains alone. Two broad approaches have been used to define these residues. Crossley *et al.* in the Biofocus company have used the wealth of information from SDM to define their residues which they call 'themes.' A thorough examination of this mutagenesis literature revealed that a maximum of 50 residues were ever involved in ligand binding with only 30–40 of these considered as critical for most drug binding.[125] As will be discussed later, they have used this 'thematic analysis' extensively in the design of targeted GPCR libraries. We and others have chosen to define the microdomains on the basis of homology models and the results of ligand docking instead. These definitions of the microdomains will, of course, be similar as the same SDM results were used to a large degree to refine the models.

Using our homology models we have identified the positions of all inward-facing residues in a typical TM bundle. Rather than just using residue type, we associated four residue properties with each position. These were hydrophobicity, principal ellipsoid volume (shape), and maximum and minimum electrostatic potential. In the user interface it is possible to interactively exclude any individual position and/or associated property. It is recognized that ECL regions and the N-terminal domain may also be involved in ligand binding so to circumvent the problem of widely variable length, bulk properties such as total charge, size, and so on were also calculated for these regions and could be interactively included in the data matrix. A number of novel phylogenetic trees were calculated from these inward-facing residues and were used as an aid in identifying potential ligands (or not!) for orphan receptors. For example, a number of orphan receptors were classified as purinergic from a full sequence phylogenetic analysis. However, when using a phylogeny from the inward-facing residues in the top (extracellular) third of the TM bundle, it was found that the orphans formed a separate group. They were subsequently found not to bind the standard purinergic ligands.[126]

Another use of this method is in the identification of orthologous pairs that do not cluster together. This may be, for example, an indication of receptors which have different functions in human and rodents, and will therefore be important in the development of animal models for a particular disease. In a similar approach Kratochwil *et al.* defined a 'ligand pocket vector' (LPV) consisting of properties associated with the potential set of inward-facing binding residues. A phylogenetic analysis of the trace amine family based on these LPVs showed that the human and mouse receptors clustered together but were separate from the rat orthologs. This is not evident from a standard full-length sequence phylogeny.[127] The rat pharmacology is indeed different from that of the human and mouse. In an extension of their methodology, they have used these LPVs to define pharmacophores for high-throughput database searching. It is evident that further applications of inward-facing residues will emerge in the future.

4.26.12 Structure-Based Drug Design

Structure-based drug design (SBDD) as elaborated on elsewhere (*see* 4.24 Structure-Based Drug Design – The Use of Protein Structure in Drug Discovery) is the use of 3D knowledge of protein–drug complex to guide the drug discovery process. The first example of SBDD was published in 1976 with the design of a number of aromatic dialdehydes based upon the 3D structure of the hemoglobin 2,3-diphosphoglycerate(2,3-DPG) complex.[128] Unfortunately these dialdehydes could not be engineered into suitable drugs for human therapy. However, shortly after this Squibb, now Bristol-Myers Squibb, succeeded in rationally designing an angiotensin-converting enzyme (ACE) inhibitor, captopril, from a complex of the closely related zinc proteinase carboxypeptidase A with benzylsuccinate.[129]

Not long after this first successful example of SBDD, a significant amount of work was already going into the processes deemed necessary for successful structure-based drug design using GPCR homology models. In the late 1980s and early 1990s[2,12,130] the first 3D GPCR homology models were constructed, and although not initially used in rational drug design, this was certainly one of the intentions for their use. As eloquently stated in one of these papers by Hibert *et al.*[12]: 'For molecular biologists and medicinal chemists, the models constitute an important working hypothesis useful to orient site-directed mutagenesis studies and rationalize drug design.' Thus began the era of SBDD for GPCRs.

In contrast to typical SBDD problems, there are additional considerations when performing SBDD with GPCRs. The first is that no complete 3D structures of GPCRs exist, with the exception of bovine rhodopsin.[26] This is a problem that has been overcome to some extent by combining homology modeling, loop building, and other techniques. However, it should be noted that most of the successes utilizing rhodopsin have, not surprisingly, been with antagonists and antagonist-state based models. This is slowly changing with the determination of more detail regarding metarhodopsin[131] and the activated state of other GPCRs.[35,107] A second, and very important, consideration is the lack of direct observation of the binding site of the drug in the receptor. Confidence in the binding mode is a prerequisite for de novo ligand binding or docking studies. As was discussed earlier, a number of methods exist for binding site elucidation, with SDM being the most significant. By performing such work, it is possible to work out which residues are critical for ligand binding. Ideally, the SBDD process of GPCR ligands should, as with typical SBDD, be an iterative model building process with constant feedback from chemistry SAR and biological work (in this case screening mutant constructs as well as the native receptor).

There are a number of ways to dock compounds into a GPCR with probably the most popular being the high-throughput docking (HTD), discussed further below. Although with the wealth of structural information of GPCRs increasing there is a greater amount of low-throughput manual docking taking place. This method has a number of benefits in that chiral centers, tautomers, ionization sites, and many other important aspects of a compound can be given consideration. In addition to this active feedback from SAR and SDM experiments can be taken into account to refine models to a greater degree.

Our group first attempted SBDD on GPCRs in the early 1990s. In these experiments, a number of homology models of the $5HT_{2C}$ receptor were built with the subtype selective compound, SB-200646, bound. From this docking, synthetic modification was undertaken and a sub-nanomolar $5HT_{2C}$ antagonist, SB-221284, resulted.[132] As a result of our successes we undertook an SDM project in an attempt to provide experimental proof of the proposed binding mode. This experiment involved mutating a number of serine residues to cysteine to remove what was thought to be an important hydrogen bond. Interestingly, an increase in binding affinity of SB-221284 was seen which contradicted the receptor/ligand docking model. Thus, after the 1997 publication by Schertler,[38] which gave some indication as to the interhelical tilts of frog rhodopsin, the model building was revisited and the dockings modified. The new model and docking indicated that the serines of interest were not accessible to the ligand. From this new docking model, further compounds were proposed for synthesis and an even more potent and selective compound, SB-243213, was discovered.[41] This demonstrated the importance of SDM work, not only in aiding the docking placement and subsequent SBDD, but also in the validation of receptor models. We also utilized our models based on frog rhodopsin in the stepwise modulation of N-3 and N-2 receptor affinity and selectivity.[40] Here the NK3 selective ligand, talnetant, is docked into the NK2 and NK3 receptor models, chemical synthesis is directed by the modeling and both a mixed NK3/NK2 antagonist and an NK2 selective ligand were discovered. More recently we have been utilizing the crystal structure of rhodopsin in our homology models for the characterization of the CCR2 antagonist binding site[133] and subsequent design of compounds.

More recently, the López-Rodriguez group have also been using the rhodopsin crystal structure for SBDD. In an interesting series of publications, they describe the design, synthesis and pharmacological evaluation of $5HT_{1A}$ ligands. Their initial work on arylpiperazines pre-dated publication of the rhodopsin structure and no attempt was made to rationalise the SAR in the context of a homology model.[134–138] However, shortly after the rhodopsin structure was published, they constructed models, utilized existing ligand SAR, and synthesized further compounds to explore the 3D structure of these models.[139]

The specifically designed test ligands had different distances between pharmacophoric features and were combined with previous SDM work on related biogenic amine receptors, to aid in the refinement of different homology models. The different homology models were generated from a molecular dynamics simulation used to dock the test ligands and to show which was the more likely model. The most consistent homology model showed a marked difference from the rhodopsin template and the authors proposed that the polar residue at position 3.37 may be forming an intrahelical hydrogen bond with the backbone and bending the helix. From this initial SBDD work, additional compounds were proposed and synthesized, culminating in a number of publications, one of these showing the generation of a $5HT_{1A}$ agonist[140] and another using arylpiperazine $5HT_{1A}$ analogs[141] and SAR in conjunction with recent crystallographic work on metarhodopsin[131] to predict a differing binding mode for the series of agonist ligands. The authors also propose the key to agonism for this series is a hydrogen bond between one oxygen atom of the diketopiperazine group and Trp6.48, enabling Trp6.48 to adopt a *trans* conformation as is experimentally observed in metarhodopsin. This group have also written a number of other papers[142–149] utilizing SDM and SAR to derive docking modes within receptor models to aid in rational design. In some of these papers, comparisons have also been made between homology models and other methods such as pharmacophores derived using CATALYST,[144] classical quantitative SAR (QSAR),[149] artificial neural networks (ANNs),[149] and comparative molecular field analysis (CoMFA).[145]

A succinct example of SBDD is presented in the publication of Löber *et al.*[150] Here, previous work on the molecular mechanisms of agonist receptor interactions[60,151,152] in combination with SDM studies[64,153] aided in the docking of the lead D4 partial agonist L345,740. Using this as basis, further ligands were rationally designed to interact with specific areas of the receptor and exclude specific hydrogen bonds that could lead to receptor activation and agonist behavior, resulting in a potent subtype-selective D4 antagonist.

In a publication by Moro *et al.*,[154] receptor-based 3D QSAR (3D-QSAR) is included in their iterative process of SBDD. They use a rhodopsin-based homology model of the adenosine A_3 receptor with SDM to identify the binding site to generate a receptor-based pharmacophore as the basis for 3D-QSAR models. From this, compounds are designed, synthesized, and characterized. The data are fed back into the model building to refine it in an iterative manner. A number of nanomolar adenosine A_3 compounds are described, all generated from differing scaffolds.

Derian *et al.* have described the use of a wide range of different techniques in the design of novel nonpeptide antagonists of the protease-activated receptor PAR1.[155] This involved identifying amino acids in the peptide necessary for activation of the receptor and gaining further information on the peptide conformational flexibility by NMR and circular dichroism (CD) studies. Essential hydrogen bonding of the backbone was identified via alanine and proline scanning. The knowledge gained from this work was then used in conjunction with known SAR for PAR1 agonist peptides and SDM work to improve the homology model of PAR1. A receptor-based pharmacophore was derived, to which different molecular templates were docked. Libraries were designed around these templates and a novel, selective, PAR1 antagonist identified.

Whilst there are a number of papers using low-throughput rational design of GPCR ligands, there are relatively few examples of SBDD utilizing GPCRs for automated molecular assembly. One example has been presented by Ali *et al.*[156] showing the de novo design of NK2 antagonists using SPROUT.[157,158] Here, a rhodopsin-based receptor model is built, automated docking followed by site specific SAR used to pinpoint the binding site, and then the de novo molecular design program SPROUT used to design novel antagonists. A number of the proposed compounds are synthesized and a compound with low micromolar affinity is obtained. The designed compounds are also compared back to the docked series used to identify the binding site.

The successes of low-throughput SBDD are becoming more and more common due to the wealth of information on the differing classes of receptors increasing, and the care that is taken in considering all aspects of the compounds being docked. However, rigorous initial evaluation of the receptors 'structural features and the compounds' SAR, combined with continual feedback, is still essential for any SBDD project to succeed.

4.26.13 Virtual Screening Using G Protein-Coupled Receptor Models

A major use of experimental protein structural information in virtual screening is through the application of HTD. The methodologies for carrying out such experiments have been extensively reviewed and evaluated elsewhere.[159–165] Although HTD has resulted in the successful discovery of small-molecule hits and leads in a number of cases,[166–168] the technique has a number of approximations. It is widely accepted that whilst the scoring methods associated with HTD can select subsets of compounds enriched with active molecules, a good correlation between predicted score and experimental activity can rarely be achieved.[169] When the approximations inherent with homology models and the approximations associated with HTD are considered, many people believe that attempts to combine the two are futile. Conversely, others believe that the huge importance of GPCRs as drug targets, and the potential impact of being able to apply structure-based virtual screening against them, warrants a large investment of time in developing such methodologies.

As has been discussed previously the only x-ray structure of a GPCR is that of rhodopsin in the inactive state.[26] Therefore, models of GPCRs in the inactive state are likely to be more reliable than those of active states. Hence, the use of HTD to search for antagonists is likely to have a better chance of success than searching for agonists. We will discuss attempts to identify antagonists first.

The simplest way of carrying out such an experiment is to build a rhodopsin-based homology model of the receptor of interest, define a binding site, and use standard docking tools for the docking and scoring. Bissantz *et al.*[170] recently investigated their ability to preferentially select known antagonists from a database of inactive molecules using the dopamine D3, muscarinic M1, and vasopressin V1a receptors. It is noteworthy that the receptors chosen have experimental evidence available to guide the binding site definitions. This information is used to manually dock a representative ligand into the binding site of each of the receptors prior to minimization, producing a model for the HTD experiments. The authors state that efforts to automatically dock ligands into a model built with no ligand present were not successful. This is easy to understand as a model built in the absence of a ligand will naturally result in the binding site collapsing to some extent. Most HTD methods do not allow protein flexibility and therefore cannot regenerate an appropriate binding pocket. Docking and scoring was carried out with a variety of algorithms using either the single scoring functions, or consensus scoring methods. Enrichments of up to 40-fold were achieved, although more typically around 5- to 10-fold. Consequently, the authors conclude that virtual screening can be carried out successfully against GPCRs. It is also noted that attempts to identify agonists with an analogous procedure were not successful (see below).

Some workers have looked at other methods, in addition to homology modeling, in an effort to generate the most appropriate receptor conformation for ligand docking. Varady *et al.*[171] have carried out virtual screening experiments against the dopamine D3 receptor using a combination of computational methods including HTD. A rigorous assessment of receptor conformations was carried out by subjecting their initial homology model to extensive molecular dynamics simulation (2 ns) in an explicit lipid and water environment. Representative snapshots were then taken throughout the trajectory and clustered, revealing four major clusters of conformations. Potential ligands were screened in a three-stage process: (1) consistency with a pharmacophoric constraint; (2) HTD using Cerius 2; and (3) novelty assessment. From an initial database of about 250 000 compounds, 6727 compounds progressed to the HTD stage. These compounds were also laced with 20 known D3 ligands (mainly antagonists, and some partial agonists). Analysis of the ability to preferentially identify these 20 ligands showed that docking into a representative conformation from any of the four clusters resulted in some enrichment. The conformation that occurred most commonly, which also agreed best with experimental data, gave the best enrichments. Requiring a single molecule to be identified as a hit against at least two of the conformations resulted in a further improvement in enrichment. Finally, after having been assessed for novelty with respect to known D3 antagonists, 20 compounds that scored in the top one-third were purchased and screened. Of these 20, four have K_is $< 0.1\,\mu$M, four are in the range 0.1–1 μM, three have activity between 1 and 10 μM, and the remaining nine compounds were not active at the concentrations tested. By comparison eight compounds from the middle one-third of the docking score range were purchased and tested. In this case two showed activity in the range 1–10 μM and the other six showed no activity at the concentration screened. This indicates that the docking and scoring methodology is prioritizing active compounds for this receptor.

The MOBILE (MOdeling Binding sites Including Ligand information Explicitly) method[172,173] also attempts to identify the most appropriate receptor conformation for a docking study by considering a number of possibilities. In this approach, a large number of different homology models are constructed (100 in reported studies), and a key antagonist is then docked (possibly in multiple binding orientations) to each model, taking into consideration experimental information about the binding mode. The interactions between the ligand and each active site amino acid are then scored, the best scoring amino acid geometry from all models taken, and these combined to give a consensus model. This consensus model is then minimized to give the final model for the HTD experiments. Application of the methodology to NK1 receptor gave some indication that the method may be able to preferentially select novel antagonists. Seven compounds were purchased and screened and one showed activity at 0.25 μM. When the methodology was applied to the α-1a adrenergic receptor[174] the results were much more conclusive, from 80 compounds screened 37 had K_is $<10\,\mu$M, $24<1\,\mu$M, $10<0.1\,\mu$M, and $3<0.01\,\mu$M. The discrepancy in these results for the two receptors may reflect the relative difficulty of applying these methods to peptide receptors (e.g., NK1) compared to aminergic receptors (e.g., α1a).

Evers recently built on this work by comparing the performance of a number of virtual screening tools in identifying known antagonists of the α1a adrenergic receptor, $5HT_{2A}$, dopamine D2, and muscarinic M_1 receptors. The structure-based approach used the MOBILE procedure described above achieving enrichments in the region of 5- to 10-fold depending on the receptor and the point at which the enrichment is calculated. Interestingly, the results also show that in a number of cases, ligand-based methods outperform the docking-based methods in terms of enrichment factors. This is an important message when considering the correct approach to take to a given virtual screening problem. However, as is highlighted by the authors, the consideration of these headline enrichment factors is not the full story. The receptors studied here have been of interest to the pharmaceutical industry for a long time and consequently there is a lot of information about their ligands. In cases where there is a lot less information about known ligands, the performance of the ligand-based approaches would be expected to be a lot poorer; indeed in the extreme case of an orphan receptor nothing is known about the ligands. Additionally, structure-based methods can also provide key insights into the receptor–ligand interactions, which cannot be offered by ligand-based methods. Such information can be very useful in understanding how to progress any initial hits. Finally, ligand-based methods work on the assumption that all ligands studied bind to the receptor in the same way, an assumption which isn't implied in structure-based methods. Hence structure-based methods allow the identification of ligand classes binding to the receptor in novel ways.

The PREDICT method, described earlier, is a detailed de novo model building procedure. One of its key uses is to build models for HTD studies. HTD experiments are carried out using the DOCK software,[175] repeating the docking of each ligand 10 times with different random number seeds to encourage extensive coverage of possible docking solutions. Virtual screening experiments have been carried out against a number of targets. These experiments took two forms. Analysis of the ability to identify known actives spiked into a set of presumed inactives, and true virtual screening where collections of available compounds are screened and high scoring examples purchased and biologically tested. The first type of experiment has been performed on $5HT_{1A}$, $5HT_4$, dopamine D2, NK1, CCR3, and NPY-Y1, giving typical enrichments in the range 20- to 50-fold, and in extreme cases as good as 350-fold (enrichment calculated at the point where 50% of actives have been identified). Compounds identified by PREDICT have been purchased for the $5HT_{1A}$, $5HT_4$, D2, NK1, and CCR3 receptors. Typical hit rates show approximately 20% of screened compounds are active. In some cases, these compounds have K_is of <10 nM. Hence success has been shown to varying degrees with aminergic, peptidergic, and chemokine receptors. The PREDICT procedure is quite complex and relatively time-consuming to run (3 days on a 60 CPU LINUX farm for the initial model building and 2 weeks on an SGI Octane for the refinement), but the authors state they have seen superior performance with these models compared to standard homology models.

Virtual screening for GPCR agonists adds yet another layer of complexity to an already complex problem. Most importantly, a viable model of the activated states of the receptor must be obtained before the virtual screening can be carried out. The issues surrounding this problem have been discussed previously.

Bissantz *et al.*[170] compare two different approaches to the problem. Firstly they simply dock an agonist into an inactive state model of the receptor of interest and allow the minimization procedure to drive the receptor into the active state. Alternatively, they use a more complex procedure where multiple ligands are superimposed to overlap key pharmacophoric points, and then the receptor is minimized in the presence of all the overlaid agonists. These procedures are carried out on the D3, β2-adrenergic and δ-opioid receptors. The authors find that enrichment is not often achieved using the first method and that HTD tools are often not able to reproduce the key interactions identified by experimental studies. However, using the second method, good enrichments are found in most cases, up to 30% in the best examples, and key interactions that could not be modeled using the first procedure could be reproduced.

Gouldson *et al.*[35] built models of active and inactive forms of rhodopsin and the β2-adrenergic receptor using molecular dynamics simulations with constraints taken from a large number of experimental studies including site-directed cross-linking, engineered zinc binding, site-directed spin labeling, and cysteine accessibility methods. The resulting models show some substantial differences between the active and inactive conformations. The authors then assess whether the active state model can preferentially identify known agonists and the inactive model can successfully identify antagonists. This assessment is carried out by docking a set of compounds consisting of 14 β-adrenergic agonists, 50 β-adrenergic antagonists, 23 other adrenergic ligands, and 80 other nonpeptide GPCR ligands. These ligands are docked into the receptor models using the LigandFit program.[176] Enrichments for the identification of antagonists from inactive models are around 1.5- to 2-fold depending at which point the enrichment is calculated. Those for identification of agonists from active models are around 5- to 10-fold. The authors explain that this difference in enrichments may be attributed partly to the fact that agonists tend to bind to both the active and inactive structure, whereas antagonists tend to bind only to the inactive structure. It should also be noted that due to the make up of the ligand set, the theoretical maximum enrichment for agonists is higher that for antagonists. These results are particularly impressive because the known actives are identified from a set of known GPCR ligands that have similar structure.

The ability to predict ligand binding affinities from HTD experiments remains an area of development and great difficulty for the computational chemistry community. When the difficulties of such a methodology are combined with the uncertainties of receptor modeling the area becomes one that is on the boundary of what is achievable with state-of-the-art modeling technologies. This is reflected by the relatively small number of publications in this area. That having been said, this section has summarized some clear examples of where this technology has been applied successfully. Common themes between those studies that have reported success using these methods are the use of a ligand in the model building process, and extensive reference back to experimental data to validate the model and binding modes.

In our opinion, such studies should be undertaken with caution, but can be successful, particularly when the protein conformation and ligand-binding site have been well characterized by experimental methods. The potential impact of such experiments on the drug discovery process means this will continue to be an active area of research, and as understanding of GPCR structure and conformation improves the number of successes looks set to increase.

4.26.14 Library and Focused Screening Set Design for G Protein-Coupled Receptors

Whilst computational drug design for other classes of targets can use a variety of structure based and HTD approaches, the limited structural information for GPCRs necessitates more inventive techniques for library design (*see* 4.14 Library Design: Ligand and Structure-Based Principles for Parallel and Combinatorial Libraries; 4.15 Library Design: Reactant and Product-Based Approaches). There is a subtle distinction between a bespoke designed library and a focused subset of a much larger virtual library. Rather than the more elegant notion of a bespoke library design in which every compound is designed toward a specific target, the more common approach is to design or obtain a comprehensive data set, such as a large virtual array designed around a single chemotype or chemistry, or perhaps a commercially available catalog or corporate compound collection, from which a focused subset is selected. This selection or focusing approach is almost always multifaceted and iterative, and general approaches to library design or focused screening set selection will be covered elsewhere (*see* 4.20 Screening Library Selection and High-Throughput Screening Analysis/Triage; 4.14 Library Design: Ligand and Structure-Based Principles for Parallel and Combinatorial Libraries; 4.15 Library Design: Reactant and Product-Based Approaches). Here, we will attempt to review some of the different approaches which have been used in the design of focused screening sets for 7TM receptors starting from very broad property-based approaches to more highly refined pharmacophores for individual receptors. The different sections suggest the level of knowledge about the molecules required to apply a particular design rationale; in the case of one-dimensional library design it is only necessary to know the properties of the compounds being investigated and nothing about the structure, for 2D designs, 2D structural information is required and for 3D designs, the techniques use full 3D representations of the compounds under consideration.

4.26.14.1 One-Dimensional Methods

4.26.14.1.1 Property-based approach

An explosion of information available toward the end of the twentieth century, coinciding with advances in information handling, has made the calculation and analysis of physicochemical properties an area of huge interest for drug discovery. Balakin and co-workers suggested that an analysis of properties including molecular weight, Clog P at pH 7.4, number of hydrogen bond donors and acceptors, number of rotatable bonds, and an aqueous solubility measure could be used to discriminate GPCR and non-GPCR actives.[177] This analysis suggested that GPCR-like compounds were significantly less flexible, less polar, and more hydrophobic for any given molecular weight. A subsequent comparison with an in-house fragment-based similarity method for the identification of actives in a test showed an overall improved enrichment for the property-based methods, although this was not universally more effective.[178] The authors have coupled together property and similarity-based methods to optimise the selection of GPCR-like compounds from virtual libraries for commercial sale.[179]

4.26.14.2 Two-Dimensional Methods

The next level of complexity in the design of GPCR compound libraries uses 2D information from the molecular structures of the compounds in the library. Whilst inherently more complex by the nature of molecular descriptions, 2D methods still avoid the issues and challenges presented by the calculation, storage, and analysis of large 3D conformational databases. This reduction in complexity also generates inherent improvements in the speed with which comparisons can be made but still reflects a significant reduction in the degree of complexity of the problem. Many methods have been developed to try and distil the key features of molecular structures into simplified concepts or representations as outlined in some of the examples below.

4.26.14.2.1 Privileged structure approach

Simple inspection of the structures of small molecules acting at GPCRs has resulted in the identification of common structural motifs, referred to as privileged substructures which can be found in ligands active at a selection of different receptors. The benzodiazepine system was originally identified as a privileged structure by Evans *et al.*[180] and lead to the definition of "privileged structures [which] are capable of providing useful ligands for more than one receptor and that judicious modification of such structures could be a viable alternative in the search for new receptor agonists and antagonists."

A logical explanation for the existence of these promiscuous fragments would be that they are interacting with conserved, or at least similar pockets on the targeted receptors. An investigation by Bondensgaard and co-workers aimed to address this hypothesis and the authors have identified conserved binding regions for three privileged fragments acting at different Family A receptor types.[181] The privileged structures hypothesis has been applied to a variety of GPCR library designs, reviewed extensively by Guo *et al.*[182] A solid-phase split pool library was developed by a team from Merck based around an indole template prominent in both natural products and marketed GPCR drugs, which showed activity at a variety of different GPCR targets.[183] Bleicher and co-workers have also demonstrated a parallel synthetic approach in which a 3,5-di-trifluoromethyl fragment associated with the N1 receptor was positioned in different positions around two related spiropiperidine scaffolds to give six starting templates, each incorporating two different privileged fragments.[184,185] This example of combining multiple features of interest, lead directly to compounds with nanomolar affinity for the NK1 receptor.

An obvious consequence of the incorporation of privileged fragments into 7TM library designs is that the resulting compounds may not be selective between targets. Screening of the pools of compounds from the indole library of Willoughby *et al.* demonstrated that several pools had similar activity profiles when screened at a panel of 7TM receptors; however, there were some pools which had selective profiles showing activities specific to particular monomers. Bleicher and co-workers' spiropiperidine NK1 array demonstrates that potency and selectivity can be enhanced through the use of multiple privileged fragments. Mason *et al.* have developed a library design methodology in which a privileged fragment is incorporated as a required feature in a pharmacophoric fingerprint and shows great improvement over more general three-feature methodologies.[186] This suggests the possibility of developing highly selective library profiles, built around the most promiscuous of privileged fragments.

A computational tool has also been developed to help identify privileged fragments, known as the recursive combinatorial analysis procedure (RECAP) algorithm.[187] Klabunde and co-workers have used this approach to fragment 20000 GPCR compounds at 11 combinatorially tractable positions in the molecule such as amides, sulfonamides, and esters, analyzing the occurrence of the resulting fragments.[188] A 4-phenyl piperazine fragment was the most commonly occurring fragment found in 32 different compounds acting at 13 separate GPCR targets. The original RECAP paper also outlines how these fragments can be used to design building blocks for use in combinatorial libraries, and stored for use in structure-based designs.

A classical fragment-based approach to a specifically targeted GPCR library design is exemplified by Ashton *et al.* in the identification of potent oxytocin antagonists from an HTS hit.[189] In this paradigm, an oxime ether proline analog, identified from an HTS campaign, forms the central feature of an array designed in several different layers. Firstly, a small array of analogs is designed focused on the original hit, using other privileged fragments for oxytocin. Secondly, a less-focused array, using only the core of the original hit compound, is enumerated using more general GPCR privileged fragments. Thirdly, a more diverse set of compounds designed around the central core are included to extend the library according to druglike properties and explore greater diversity. In this example, the 3500-member array generated enough active compounds to suggest SAR to drive lead optimization resulting in bioavailable compounds for clinical trials.

4.26.14.2.2 **Two-dimensional similarity/knowledge-based approaches**

Empirical methods for comparing small molecules based on their 2D frameworks are both numerous and well established (*see* 4.08 Compound Selection Using Measures of Similarity and Dissimilarity; 4.19 Virtual Screening). These are most commonly used as fingerprints encoding the fragments present in a molecule and used to retrieve nearest neighbors from a database on the basis of a Tanimoto or other such similarity coefficient of the reference and test fingerprints. The application and performance of these techniques have been reviewed in depth elsewhere, including their applicability to different target classes such as GPCRs.[190] A variation on this concept is the use of reduced graphs or topological pharmacophores[191,192] which describe the arrangement and connectivity of pharmacophoric features in a molecule. A further comparison of these fragment (e.g., Daylight) and feature (e.g., Reduced Graph) based fingerprints by Willet *et al.* suggested that the fragment-based methods were generally more successful at enriching datasets than the feature-based topological methods, although they did observe significant variation within the two classes with different implementations of the technologies.[193]

4.26.14.2.3 **Pharmacophoric fingerprint techniques**

The design and application of pharmacophoric fingerprint-based technologies has been an area of considerable interest for some time now. This kind of technology is one of many approaches which generates ensemble 3D structural information for small molecules, in the absence of any real information of the true bioactive conformation of the molecules of interest. An example of library design using an explicit 3D pharmacophore is discussed below (*see* Section 4.26.14.3) but here we focus on the application of pharmacophore fingerprint technologies, of which there are many (*see* 4.08 Compound Selection Using Measures of Similarity and Dissimilarity; 4.19 Virtual Screening).

In these approaches, 3D conformations are generated for the molecules and then for each conformation, the presence or absence of a three-point pharmacophore, with a particular geometry between the features, can be encoded in a bit string or fingerprint. Fingerprints can then be compared using a Tanimoto similarity or other score and by generating a complete set of fingerprints for known actives a receptor specific drug relevant space can be developed.

In the approach adopted by Bradley *et al.*,[194] α-adrenergic antagonists were successfully obtained from a virtual library. Four-point pharmacophores (four features, six inter-feature distances) are calculated for a set of active and inactive compounds and bit strings are derived detailing the presence or absence of each feature/geometry. Importantly, both the presence and the absence of the features are compared in an information content score, which is used to filter a virtual library. In a simpler approach described by Pickett *et al.*,[195] three- or four-feature pharmacophore keys (similar to the bit strings described above) are compared by counting the number of bits set in common, or the Tanimoto similarity between a probe and a test molecule in a virtual library.

The most recent demonstration of the successful application of this approach to drug discovery for GPCRs has been published by the group at AstraZeneca.[196] In this application, support vector machine technologies (SVMs) are used to learn the difference between active and inactive molecules from a training set of fingerprints detailing the binned distances between three-point SMARTS[197]-defined pharmacophoric features. 3D conformations for the molecules were generated using Corina[198] which were enumerated into a conformation model with OMEGA.[199] In their most extreme validation experiment, the authors use an MDDR-derived dataset[200] to try and identify adrenergic-α_2 antagonists: molecules from the most abundant chemotype, responsible for 40% of the actives, were removed from the training set and put back into the test set. The model was recalculated and the test set reranked using the SVM pharmacophoric fingerprint model and over 90% of the known active molecules were ranked in the top 10% of the test set. This represents an excellent example of the technology to select novel chemotypes from a set of active training data, with a knowledge-based approach, a technique referred to here and elsewhere as 'lead-hopping.'

Many other examples of this type of technology have been developed using different methods of conformer generation, pharmacophore definition, and fingerprint derivation and encoding which include 3DMill,[201] Family-based pharmacophore descriptors for design of an opioid family library,[202] and Gridding and Partitioning (GaP) for selection of diverse ligands and monomers.[203]

4.26.14.2.4 Topomers

Another reduced representation of small molecules, which has been applied to the design of commercial libraries for GPCRs, is the Topomer technology from Tripos. The technology is described by Jilek *et al.* whereby an invariant three-dimensional topological description of a molecular fragment is derived using deterministic rules from a 2D structure.[204] The suggestion from these authors is that molecules can be represented realistically as in three-dimensions, but the use of the deterministic approach means that it does not have the full complexity. The only reported application of this technology is an example of 'lead-hopping' by Cramer *et al.* in which the topomer description is able to identify potent CCR5 antagonists with different chemistries but similar topomer shapes.[205] This technology has already been applied by Tripos to the design of serotonin (5HT) libraries derived from published active compounds with plans to extend the methodology to chemokines, other Family A targets, and reportedly Family B receptors.[206]

4.26.14.2.5 Burden, CAS, and University of Texas (BCUT) topological molecular descriptors

The so-called 'BCUT' (Burden, CAS, and University of Texas) descriptors are a novel way of describing the features of small molecules, incorporating information about the connectivity and separation of the constituent atoms and also physicochemical information from a 2D representation of the molecule. The geometries and properties of CONCORD[207]-generated 3D conformations of a molecule are encoded into a set of multidimensional, square matrices, the eigenvalues of which can be used as molecular descriptors.[208]

One of the main applications of this technology has been to identify 'holes' in diversity space (*see* 4.14 Library Design: Ligand and Structure-Based Principles for Parallel and Combinatorial Libraries; 4.15 Library Design: Reactant and Product-Based Approaches). Virtual libraries are generated in an attempt to fill these voids and then selected compounds are synthesized and added to screening collections to make them more comprehensive. However, an analysis of a 2D space projection of the BCUT descriptors for the compounds from the MDDR, identified that ACE inhibitors were tightly clustered in a 'receptor-relevant subspace.'[209] This lead to identification of additional receptor-relevant subspaces makes it possible to identify regions of a proposed virtual library which overlap with the compounds which are known to be active at the target(s) of interest.[210]

In an alternative application of BCUT descriptors Manallack[211] and co-workers explored the application of neural networks to identify BCUT descriptors which could be used to build a model which correctly identified over 80% of

GPCR or kinase molecules from a test set. In this case, the model was used to select compounds from third-party suppliers for testing but this model could equally well be applied to virtual libraries in the design of focused arrays for GPCRs or even individual targets.

4.26.14.3 Three-Dimensional Methods

Various library design techniques which incorporate 3D information from known actives are commonly used to target design toward GPCRs in general, or specific receptors and subfamilies using a variety of different techniques. These approaches are generally somewhat more sophisticated than the approaches outlined above, requiring proprietary tools and often substantial knowledge bases of potent compounds. Here, the techniques are not described in detail, but the particular application of a variety of different library design techniques demonstrates some of the possibilities and problems for GPCR library design.

4.26.14.3.1 Pharmacophores

Classically, a standard pharmacophore approach would derive a relatively explicit multifeature pharmacophore based on a set of highly potent, selective, and preferably conformationally restrained compounds for a given receptor[212,213] resulting in a conformationally explicit model. Where possible, additional information would be incorporated from a receptor model or crystal structure to refine the hypothesis (e.g., Koide *et al.*[214]).

An interesting approach is described by Nordling *et al.*, in which chemoinformatics is used to build a new pharmacophore for $5HT_7$ library design, but using a homology model for the related $5HT_{2A}$ receptor, which has a well-described pharmacophore.[215]

4.26.14.3.2 De novo techniques

General de novo design of GPCR ligands is discussed earlier in this chapter, but it is useful to discuss the approach taken by DeNovo Pharmaceuticals[216] in the design of GPCR-targeted libraries available from Peakdale Molecular.[217] This strategy makes use of so-called 'extended pharmacophores' coupled with de novo design technologies applied to novel PeakDale chemotypes.[218] This design uses a maximal similarity overlay of conformations of a set of diverse molecules with known activity for a single receptor of interest. Pharmacophoric features shared by the molecules in the overlay are identified and a molecular surface is derived resulting in a model of the receptor binding site. The de novo design tool SkelGen[219–221] is then used to identify novel chemotypes to fit the features of the receptor binding site models. In the final stage, virtual libraries constructed around the novel chemotypes are screened against the binding site models before finally being selected for synthesis.

4.26.14.3.3 Peptidomimetic design

One of the challenges in targeting GPCRs is hinted at by the broad array of endogenous ligands of the receptors, ranging from the smallest, a photon which activates the rhodopsin receptor, through to large protein agonists such as chemokines and glycoprotein hormones (**Figure 2**). Whilst it is clear that the true bioactive conformation of a small molecule can be difficult to identify and isolate, that problem becomes all the more apparent when trying to design compounds, and libraries to mimic polypeptide molecules such as the agonists of oxytocin, somatostatin, and chemokine receptors, even when the structures of the endogenous agonist peptide is known. An excellent review by Hruby *et al.* details how the subtleties of small changes to GPCR peptide agonists can have significant functional effects on behavior, demonstrating the complexity of this challenge.[222] However, Jones and co-workers describe a variety of success stories in the identification of synthetically and sterically constrained peptide mimics[223] as well as some so-called Type III[224] small molecules more relevant for library design. At GlaxoSmithKline, we have also demonstrated success in the design of small-molecule peptidomimetic antagonists from known endogenous agonist structures[225] and developing focused libraries for the identification of SAR and lead molecules.[226]

4.26.14.3.4 Beta-turn peptidomimetics

The β-turn conformation represents an energetically favourable arrangement for a tetrapeptide segment and is a convenient size for potential mimicry by a druglike small molecule. Furthermore, the fact that the peptide segment is a turn and reverses the overall direction of the chain means that it is often found on the surface of larger proteins and hence implicated in molecular recognition. Even for relatively small peptides, a β-turn conformation is often thought to be present in the bioactive conformation.[227] Hence, this is an attractive target for GPCR ligand design.

Cho and co-workers have demonstrated that diversity screening can return hits which yield interesting core moieties with substitution patterns reminiscent of a β-turn conformation.[228] Optimization was then performed using knowledge of the endogenous peptide. A more comprehensive overview of successes in the design of β-turn mimetic small molecules is provided by Kee *et al.*[229]

A more specific approach to such designs was suggested by Garland and Dean.[230] They performed an analysis of the protein structure database to identify the distances between the four α carbon atoms involved in the β-turn units. This provided a set of constraints which could be used to identify small molecule templates for decoration with fragments resembling the amino acid side chains. This approach was extended to incorporate the vectors between the Cα–Cβ bonds of a β-turn to further refine the templates.[231] This technique has successfully been employed by Chianelli *et al.* in the identification of somatostatin-2 receptor agonists.[232]

4.26.14.3.5 Receptor-based approaches

Methods that endeavor to couple together the wealth of information on small molecules and the amino acid sequences of the receptors at which they are active are nowadays routinely referred to as chemogenomics approaches. Described as an extension to systems-based discovery research,[233] chemogenomics approaches endeavor to identify all possible ligands at all target families.[234] Building on bioinformatics efforts to translate and interpret target information contained in the human genome, the activity data curated through cheminformatics for known targets can be used to infer likely activity of compounds at appropriate homologs of the original receptor.

4.26.14.3.6 Microdomains or 'themes'

Crossley and co-workers at BioFocus have developed a successful receptor sequence based approach to library design which enables specific design toward binding subpockets or microdomains[235] (*see* Section 4.26.14.2.1 above). A consensus binding domain for the 7TM receptor has been defined, supported by mutagenesis data for the family which identified only 50 residue positions involved in ligand binding. Consideration of the relative equivalence of many amino acid side chains such as Ser/Thr or Ile/Leu/Val further reduces the complexity of the problem. The consensus binding site is divided into eight microenvironments with different 'flavors' as defined by their constituent amino acid types. 'The set of amino acids at the specific sequence positions that define a particular microenvironment with a certain flavour is called a Theme.' The presence or absence of 14 such themes in the consensus binding site is then transcribed into a fingerprint which has been used to reclassify GPCR space.[125] Compound libraries can then be designed and scored on the basis of their ability to interact with the themes present in a particular receptor or receptor subfamily. The authors report hit rates of between 1% and 13% for the desired receptors using this approach.

4.26.14.3.7 Cross-family screening approaches

Rather than using an analysis of sequence and structure data, Cerep's BioPrint database contains experimental assay data for a diverse set of marketed and failed drugs at a broad panel of GPCRs and other molecular targets. At publication, some 2225 compounds had been tested in a panel of well-validated assays, including molecular targets, ADME tests, and cell-safety based assays.[236] This formidable database extends the familiar paradigm of structure–activity relationships into a new dimension where trends in not only activities across the family of GPCR targets, but also safety related properties, can be tracked and parameterized. The associated BioPrint software facilitates the construction of complex SAR models which can be used to predict activities for structural neighbors. In this way, virtual libraries can be constructed and optimized relative to comprehensive experimentally derived models as part of the library design process.

Researchers at Pfizer have developed a so-called Biological Spectra analysis,[237] in which 1567 compounds (a subset of the BioPrint database) have been screened in 92 ligand binding assays (42 of which were GPCR assays) to develop fingerprints of the activity profile. A validation exercise using an additional four compounds with incomplete assay data sets provided confidence in the ability of the model to predict the biological spectra of compounds with a high degree of structural similarity to others in the dataset. The direct application of this technology to GPCR library design is not discussed at length, although the knowledge gained from this exercise is clearly of high value in the design of focused arrays. Scientists at GlaxoSmithKline have also explored methods for interrogating SAR data from screening compounds at a panel of GPCRs.[238]

4.26.14.4 Summary of G Protein-Coupled Receptor Library Design Methods

In this section, we have attempted to provide a comprehensive review of methods which are currently being applied to the design of GPCR directed compound libraries. It is impossible to suggest which are the best and worst, as any library design strategy should be developed in accordance with the amount of information which can reasonably be used to meet the requirements of the library. By grouping the techniques according to the amount of structural and target related information available, it is easy to see how comprehensive information about a receptor subfamily could be used to focus design around an individual chemotype to enrich hit rates at receptors of interest. Similarly, we have also hopefully suggested that even the most general GPCR library can be designed using property and privileged fragment information to help improve on standard Lipinski-style diversity libraries and obtain more hits at targets of interest.

4.26.15 Ligand-Based Design

The core concepts and techniques of ligand-based design, covered extensively elsewhere (see tables in Chapter 4.01) are often also integral parts of structure-based design. Indeed the terms 'ligand-based' and 'structure-based' are only indications as to the amount of information of a receptor-ligand complex that is utilized. In GPCR drug design there are numerous examples of ligand-based design, from pure conformational analysis being used in the discovery of $5HT_6$ ligands[239] to more elaborate 3D similarity methods. In our overview of structure-based design a number of these ligand-based methods have been used in conjunction with SBDD giving notable successes. It is our opinion that these methods and techniques will continue to be merged further.

References

1. Jap, B. K.; Walian, P. J. *Physiol. Rev.* **1996**, *76*, 1073–1088.
2. Findlay, J.; Eliopoulos, E. *Trends Pharmacol. Sci.* **1990**, *11*, 492–499.
3. Henderson, R.; Baldwin, J. M.; Ceska, T. A.; Zemlin, F.; Beckmann, E.; Downing, K. H. *J. Mol. Biol.* **1990**, *213*, 899–929.
4. Schechter, I.; Burstein, Y.; Zemell, R.; Ziv, E.; Kantor, F.; Papermaster, D. S. *Proc. Natl. Acad. Sci. USA* **1979**, *76*, 2654–2658.
5. Wise, A.; Gearing, K.; Rees, S. *Drug Disc. Today* **2002**, *7*, 235–246.
6. O'Hara, P. J.; Sheppard, P. O.; Thogersen, H.; Venezia, D.; Haldeman, B. A.; McGrane, V.; Houamed, K. M.; Thomsen, C.; Gilbert, T. L.; Mulvihill, E. R. *Neuron* **1993**, *11*, 41–52.
7. Kunishima, N.; Shimada, Y.; Tsuji, Y.; Sato, T.; Yamamoto, M.; Kumasaka, T.; Nakanishi, S.; Jingami, H.; Morikawa, K. *Nature* **2000**, *407*, 971–977.
8. Foord, S. M.; Jupe, S.; Holbrook, J. *Biochem. Soc. Trans.* **2002**, *30*, 473–479.
9. Kristiansen, K. *Pharmacol. Ther.* **2004**, *103*, 21–80.
10. Horn, F.; Weare, J.; Beukers, M. W.; Hörsch, S.; Bairoch, A.; Chen, W.; Edvardsen, Ø; Campagne, F.; Vriend, G. *Nucleic Acids Res.* **1998**, *26*, 277–281.
11. Mirzadegan, T.; Humblet, C.; Ripka, W. C.; Colmenares, L. U.; Liu, R. S. *Photochem. Photobiol.* **1992**, *56*, 883–893.
12. Hibert, M. F.; Trumpp-Kallmeyer, S.; Bruinvels, A.; Hoflack, J. *Mol. Pharmacol.* **1991**, *40*, 8–15.
13. Trumpp-Kallmeyer, S.; Hoflack, J.; Bruinvels, A.; Hibert, M. *J. Med. Chem.* **1992**, *35*, 3448–3462.
14. Strader, C. D.; Candelore, M. R.; Hill, W. S.; Sigal, I. S.; Dixon, R. A. *J. Biol. Chem.* **1989**, *264*, 13572–13578.
15. Kyte, J.; Doolittle, R. F. *J. Mol. Biol.* **1982**, *157*, 105–132.
16. Engelman, D. M.; Steitz, T. A.; Goldman, A. *Annu. Rev. Biophys. Biophys. Chem.* **1986**, *15*, 321–353.
17. Hopp, T. P.; Woods, K. R. *Mol. Immunol.* **1983**, *20*, 483–489.
18. Donnelly, D.; Overington, J. P.; Blundell, T. L. *Protein Eng.* **1994**, *7*, 645–653.
19. Blaney, F. E.; Tennant, M. *Membrane Protein Models*, Leeds, UK, Proceedings of a Conference, Mar/Apr 1994, 161–176, 110.
20. Blaney, F. E. *Genomics: Commercial Opportunities from a Scientific Revolution*, Paper presented at the Society of Chemical Industry (SCI) Conference, Cambridge, UK, June 30–July 2, 1997, 77–104.
21. Schertler, G. F.; Villa, C.; Henderson, R. *Nature* **1993**, *362*, 770–772.
22. Kontoyianni, M.; Lybrand, T. P. *Med. Chem. Res.* **1993**, *3*, 407–418.
23. Pardo, L.; Ballesteros, J. A.; Osman, R.; Weinstein, H. *Proc. Natl. Acad. Sci. USA* **1992**, *89*, 4009–4012.
24. Hutchins, C. *Endocrine J.* **1994**, *2*, 7–23.
25. Dahl, S. G.; Edvardsen, O.; Sylte, I. *Proc. Natl. Acad. Sci. USA* **1991**, *88*, 8111–8115.
26. Palczewski, K.; Kumasaka, T.; Hori, T.; Behnke, C. A.; Motoshima, H.; Fox, B. A.; Le Trong, I.; Teller, D. C.; Okada, T.; Stenkamp, R. E. et al. *Science* **2000**, *289*, 739–745.
27. Donnelly, D.; Cogdell, R. J. *Protein Eng.* **1993**, *6*, 629–635.
28. Hofmann, K.; Stoffel, W. *Biol. Chem. Hoppe-Seyler* **1993**, *347*, 166.
29. Krogh, A.; Larsson, B.; von Heijne, G.; Sonnhammer, E. L. *J. Mol. Biol.* **2001**, *305*, 567–580.
30. GPCRDB. http://www.gpcr.org/7tm/ (accessed April 2006).
31. Eisenberg, D.; Wilcox, W.; McLachlan, A. D. *J. Cell Biochem.* **1986**, *31*, 11–17.
32. Adcock, S. Ph.D. Dissertation, University of Oxford, UK, 2001
33. Herzyk, P.; Hubbard, R. E. *Biophys. J.* **1995**, *69*, 2419–2442.
34. Pogozheva, I. D.; Lomize, A. L.; Mosberg, H. I. *Biophys. J.* **1997**, *72*, 1963–1985.
35. Gouldson, P. R.; Kidley, N. J.; Bywater, R. P.; Psaroudakis, G.; Brooks, H. D.; Diaz, C.; Shire, D.; Reynolds, C. A. *Proteins* **2004**, *56*, 67–84.
36. van Heel, M.; Gowen, B.; Matadeen, R.; Orlova, E. V.; Finn, R.; Pape, T.; Cohen, D.; Stark, H.; Schmidt, R.; Schatz, M. et al. *Q. Rev. Biophys.* **2000**, *33*, 307–369.
37. Hibert, M. F.; Trumpp-Kallmeyer, S.; Hoflack, J.; Bruinvels, A. *Trends Pharmacol. Sci.* **1993**, *14*, 7–12.
38. Unger, V. M.; Hargrave, P. A.; Baldwin, J. M.; Schertler, G. F. *Nature* **1997**, *389*, 203–206.
39. Baldwin, J. M. *EMBO J.* **1993**, *12*, 1693–1703.
40. Blaney, F. E.; Raveglia, L. F.; Artico, M.; Cavagnera, S.; Dartois, C.; Farina, C.; Grugni, M.; Gagliardi, S.; Luttmann, M. A.; Martinelli, M. *J. Med. Chem.* **2001**, *44*, 1675–1689.

41. Bromidge, S. M.; Dabbs, S.; Davies, D. T.; Davies, S.; Duckworth, D. M.; Forbes, I. T.; Gaster, L. M.; Ham, P.; Jones, G. E.; King, F. D. *J. Med. Chem.* **2000**, *43*, 1123–1134.
42. Grace, C. R.; Perrin, M. H.; DiGruccio, M. R.; Miller, C. L.; Rivier, J. E.; Vale, W. W.; Riek, R. *Proc. Natl. Acad. Sci. USA* **2004**, *101*, 12836–12841.
43. Chung, D. A.; Zuiderweg, E. R.; Fowler, C. B.; Soyer, O. S.; Mosberg, H. I.; Neubig, R. R. *Biochemistry* **2002**, *41*, 3596–3604.
44. Han, M.; Smith, S. O. *Biochemistry* **1995**, *34*, 1425–1432.
45. Ballesteros, J. A.; Weinstein, H. *Methods Neurosci.* **1995**, *25*, 366–428.
46. Chen, Z.; Zhang, X.; Gonnella, N. C.; Pellas, T. C.; Boyar, W. C.; Ni, F. *J. Biol. Chem.* **1998**, *273*, 10411–10419.
47. Grobner, G.; Choi, G.; Burnett, I. J.; Glaubitz, C.; Verdegem, P. J.; Lugtenburg, J.; Watts, A. *FEBS Lett.* **1998**, *422*, 201–204.
48. Watts, A. *Curr. Opin. Biotechnol.* **1999**, *10*, 48–53.
49. Hutchison, C. A., III.; Phillips, S.; Edgell, M. H.; Gillam, S.; Jahnke, P.; Smith, M. *J. Biol. Chem.* **1978**, *253*, 6551–6560.
50. Kristiansen, K.; Dahl, S. G.; Edvardsen, O. *Proteins* **1996**, *26*, 81–94.
51. Edvardsen, O.; Reiersen, A. L.; Beukers, M. W.; Kristiansen, K. *Nucleic Acids Res.* **2002**, *30*, 361–363.
52. Gearing, K. L.; Barnes, A.; Barnett, J.; Brown, A.; Cousens, D.; Dowell, S.; Green, A.; Patel, K.; Thomas, P.; Volpe, F. et al. F. *Protein Eng.* **2003**, *16*, 365–372.
53. Schwartz, T. W. *Curr. Opin. Biotechnol.* **1994**, *5*, 434–444.
54. Holst, B.; Zoffmann, S.; Elling, C. E.; Hjorth, S. A.; Schwartz, T. W. *Mol. Pharmacol.* **1998**, *53*, 166–175.
55. Javitch, J. A. *Methods Enzymol.* **1998**, *296*, 331–346.
56. Seal, R. P.; Leighton, B. H.; Amara, S. G. *Methods Enzymol.* **1998**, *296*, 318–331.
57. Javitch, J. A.; Li, X. C.; Kaback, J.; Karlin, A. *Proc. Natl. Acad. Sci. USA* **1994**, *91*, 10355–10359.
58. Javitch, J. A.; Fu, D. Y.; Chen, J. Y. *Biochemistry* **1995**, *34*, 16433–16439.
59. Javitch, J. A.; Fu, D. Y.; Chen, J. Y.; Karlin, A. *Neuron* **1995**, *14*, 825–831.
60. Javitch, J. A.; Ballesteros, J. A.; Weinstein, H.; Chen, J. Y. *Biochemistry* **1998**, *37*, 998–1006.
61. Javitch, J. A.; Ballesteros, J. A.; Chen, J.; Chiappa, V.; Simpson, M. M. *Biochemistry* **1999**, *38*, 7961–7968.
62. Javitch, J. A.; Shi, L.; Simpson, M. M.; Chen, J. Y.; Chiappa, V.; Visiers, I.; Weinstein, H.; Ballesteros, J. A. *Biochemistry* **2000**, *39*, 12190–12199.
63. Shi, L.; Simpson, M. M.; Ballesteros, J. A.; Javitch, J. A. *Biochemistry* **2001**, *40*, 12339–12348.
64. Simpson, M. M.; Ballesteros, J. A.; Chiappa, V.; Chen, J.; Suehiro, M.; Hartman, D. S.; Godel, T.; Snyder, L. A.; Sakmar, T. P.; Javitch, J. A. *Mol. Pharmacol.* **1999**, *56*, 1116–1126.
65. Fu, D. Y.; Ballesteros, J. A.; Weinstein, H.; Chen, J. Y.; Javitch, J. A. *Biochemistry* **1996**, *35*, 11278–11285.
66. Javitch, J. A.; Fu, D.; Liapakis, G.; Chen, J. *J. Biol. Chem.* **1997**, *272*, 18546–18549.
67. Bhogal, N.; Blaney, F. E.; Ingley, P. M.; Rees, J.; Findlay, J. B. *Biochemistry* **2004**, *43*, 3027–3038.
68. Altenbach, C.; Greenhalgh, D. A.; Khorana, H. G.; Hubbell, W. L. *Proc. Natl. Acad. Sci. USA* **1994**, *91*, 1667–1671.
69. Hubbell, W. L.; Gross, A.; Langen, R.; Lietzow, M. A. *Curr. Opin. Struct. Biol.* **1998**, *8*, 649–656.
70. Altenbach, C.; Yang, K.; Farrens, D. L.; Farahbakhsh, Z. T.; Khorana, H. G.; Hubbell, W. L. *Biochemistry* **1996**, *35*, 12470–12478.
71. Altenbach, C.; Klein-Seetharaman, J.; Hwa, J.; Khorana, H. G.; Hubbell, W. L. *Biochemistry* **1999**, *38*, 7945–7949.
72. Altenbach, C.; Cai, K.; Khorana, H. G.; Hubbell, W. L. *Biochemistry* **1999**, *38*, 7931–7937.
73. Langen, R.; Cai, K.; Altenbach, C.; Khorana, H. G.; Hubbell, W. L. *Biochemistry* **1999**, *38*, 7918–7924.
74. Yang, K.; Farrens, D. L.; Altenbach, C.; Farahbakhsh, Z. T.; Hubbell, W. L.; Khorana, H. G. *Biochemistry* **1996**, *35*, 14040–14046.
75. Steinhoff, H. J.; Radzwill, N.; Thevis, W.; Lenz, V.; Brandenburg, D.; Antson, A.; Dodson, G.; Wollmer, A. *Biophys. J.* **1997**, *73*, 3287–3298.
76. Hustedt, E. J.; Smirnov, A. I.; Laub, C. F.; Cobb, C. E.; Beth, A. H. *Biophys. J.* **1997**, *72*, 1861–1877.
77. Hubbell, W. L.; Altenbach, C.; Hubbell, C. M.; Khorana, H. G. *Adv. Protein Chem.* **2003**, *63*, 243–290.
78. Gether, U.; Lin, S.; Kobilka, B. K. *J. Biol. Chem.* **1995**, *270*, 28268–28275.
79. Chollet, A.; Turcatti, G. *J. Comput.-Aided Mol. Des.* **1999**, *13*, 209–219.
80. Turcatti, G.; Nemeth, K.; Edgerton, M. D.; Meseth, U.; Talabot, F.; Peitsch, M.; Knowles, J.; Vogel, H.; Chollet, A. *J. Biol. Chem.* **1996**, *271*, 19991–19998.
81. Pham, V. I.; Sexton, P. M. *J. Pept. Sci.* **2004**, *10*, 179–203.
82. Dong, M.; Pinon, D. I.; Cox, R. F.; Miller, L. J. *J. Biol. Chem.* **2004**, *279*, 31177–31182.
83. Elling, C. E.; Nielsen, S. M.; Schwartz, T. W. *Nature* **1995**, *374*, 74–77.
84. Thirstrup, K.; Elling, C. E.; Hjorth, S. A.; Schwartz, T. W. *J. Biol. Chem.* **1996**, *271*, 7875–7878.
85. Elling, C. E.; Thirstrup, K.; Holst, B.; Schwartz, T. W. *Proc. Natl. Acad. Sci. USA* **1999**, *96*, 12322–12327.
86. Holst, B.; Elling, C. E.; Schwartz, T. W. *Mol. Pharmacol.* **2000**, *58*, 263–270.
87. Rosenkilde, M. M.; Gerlach, L. O.; Jakobsen, J. S.; Skerlj, R. T.; Bridger, G. J.; Schwartz, T. W. *J. Biol. Chem.* **2004**, *279*, 3033–3041.
88. Brann, M. R.; Klimkowski, V. J.; Ellis, J. *Life Sci.* **1993**, *52*, 405–412.
89. Chini, B.; Mouillac, B.; Ala, Y.; Balestre, M. N.; Trumpp-Kallmeyer, S.; Hoflack, J.; Elands, J.; Hibert, M.; Manning, M.; Jard, S. *EMBO J.* **1995**, *14*, 2176–2182.
90. Underwood, D. J.; Strader, C. D.; Rivero, R.; Patchett, A. A.; Greenlee, W.; Prendergast, K. *Chem. Biol.* **1994**, *1*, 211–221.
91. Yamano, Y.; Ohyama, K.; Kikyo, M.; Sano, T.; Nakagomi, Y.; Inoue, Y.; Nakamura, N.; Morishima, I.; Guo, D. F.; Hamakubo, T. *J. Biol. Chem.* **1995**, *270*, 14024–14030.
92. Blaney, J. M., Crippen, G. M., Dearing, A., Dixon, J. S., and Spellmeyer, D. C. DGEOM95. 1995.
93. Wilmot, C. M.; Thornton, J. M. *Protein Eng.* **1990**, *3*, 479–493.
94. Elber, R.; Karplus, M. *J. Am. Chem. Soc.* **1990**, *112*, 9161–9175.
95. Roitberg, A.; Elber, R. *J. Chem. Phys.* **1991**, *95*, 9277–9287.
96. CHARMm. (25.2), 1999. Distributed by Accelrys Inc. http://www.accelrys.com/ (accessed April 2006).
97. Brooks, B. R.; Bruccoleri, R. E.; Olafson, B. D.; States, D. J.; Swaminathan, S.; Karplus, M. *J. Comp. Chem.* **1983**, *4*, 187–217.
98. Blaney, F. E.; Langmead, C. J.; Bridges, A.; Evans, N.; Herdon, H. J.; Jones, D. N. C.; Ratcliffe, S. J.; Szekeres, P. G. Submitted for publication.
99. Grewal, N.; Talwar, G. P.; Salunke, D. M. *Protein Eng.* **1994**, 7, 205–211.
100. Mahmoudian, M. *J. Mol. Graph.* **1994**, *12*, 22–834.
101. Wall, M. A.; Coleman, D. E.; Lee, E.; Iniguez-Lluhi, J. A.; Posner, B. A.; Gilman, A. G.; Sprang, S. R. *Cell* **1995**, *83*, 1047–1058.
102. Lambright, D. G.; Sondek, J.; Bohm, A.; Skiba, N. P.; Hamm, H. E.; Sigler, P. B. *Nature* **1996**, *379*, 311–319.
103. Bissantz, C.; Logean, A.; Rognan, D. *J. Chem. Inf. Comput. Sci.* **2004**, *44*, 1162–1176.

104. Becker, O. M.; Shacham, S.; Marantz, Y.; Noiman, S. *Curr. Opin. Drug Disc. Dev.* **2003**, *6*, 353–361.
105. Becker, O. M.; Marantz, Y.; Shacham, S.; Inbal, B.; Heifetz, A.; Kalid, O.; Bar-Haim, S.; Warshaviak, D.; Fichman, M.; Noiman, S. *Proc. Natl. Acad. Sci. USA* **2004**, *101*, 11304–11309.
106. Shacham, S.; Marantz, Y.; Bar-Haim, S.; Kalid, O.; Warshaviak, D.; Avisar, N.; Inbal, B.; Heifetz, A.; Fichman, M.; Topf, M. et al. *Proteins* **2004**, *57*, 51–86.
107. Bissantz, C. *J. Recept. Signal. Transduct. Res.* **2003**, *23*, 123–153.
108. Tams, J. W.; Knudsen, S. M.; Fahrenkrug, J. *Ann. NY Acad. Sci.* **1998**, *865*, 375–377.
109. Tams, J. W.; Knudsen, S. M.; Fahrenkrug, J. *Receptors Channels* **1998**, *5*, 79–90.
110. Donnelly, D. *FEBS Lett.* **1997**, *409*, 431–436.
111. Frimurer, T. M.; Bywater, R. P. *Proteins* **1999**, *35*, 375–386.
112. Grauschopf, U.; Lilie, H.; Honold, K.; Wozny, M.; Reusch, D.; Esswein, A.; Schafer, W.; Rucknagel, K. P.; Rudolph, R. *Biochemistry* **2000**, *39*, 8878–8887.
113. Qi, L. J.; Leung, A. T.; Xiong, Y.; Marx, K. A.; Abou-Samra, A. B. *Biochemistry* **1997**, *36*, 12442–12448.
114. Taylor, W. R.; Munro, R. E.; Petersen, K.; Bywater, R. P. *Comput. Biol. Chem.* **2003**, *27*, 103–114.
115. Costantino, G.; Pellicciari, R. *J. Med. Chem.* **1996**, *39*, 3998–4006.
116. Galvez, T.; Parmentier, M. L.; Joly, C.; Malitschek, B.; Kaupmann, K.; Kuhn, R.; Bittiger, H.; Froestl, W.; Bettler, B.; Pin, J. P. *J. Biol. Chem.* **1999**, *274*, 13362–13369.
117. Galvez, T.; Prezeau, L.; Milioti, G.; Franek, M.; Joly, C.; Froestl, W.; Bettler, B.; Bertrand, H. O.; Blahos, J.; Pin, J. P. *J. Biol. Chem.* **2000**, *275*, 41166–41174.
118. Bernard, P.; Guedin, D.; Hibert, M. *J. Med. Chem.* **2001**, *44*, 27–35.
119. O'brien, J. A.; Lemaire, W.; Chen, T. B.; Chang, R. S.; Jacobson, M. A.; Ha, S. N.; Lindsley, C. W.; Schaffhauser, H. J.; Sur, C.; Pettibone, D. J. et al. *Mol. Pharmacol.* **2003**, *64*, 731–740.
120. Malherbe, P.; Kratochwil, N.; Zenner, M. T.; Piussi, J.; Diener, C.; Kratzeisen, C.; Fischer, C.; Porter, R. H. *Mol. Pharmacol.* **2003**, *64*, 823–832.
121. Malherbe, P.; Kratochwil, N.; Knoflach, F.; Zenner, M. T.; Kew, J. N.; Kratzeisen, C.; Maerki, H. P.; Adam, G.; Mutel, V. *J. Biol. Chem.* **2003**, *278*, 8340–8347.
122. Miedlich, S. U.; Gama, L.; Seuwen, K.; Wolf, R. M.; Breitwieser, G. E. *J. Biol. Chem.* **2004**, *279*, 7254–7263.
123. Poyner, D. R.; Sexton, P. M.; Marshall, I.; Smith, D. M.; Quirion, R.; Born, W.; Muff, R.; Fischer, J. A.; Foord, S. M. *Pharmacol. Rev.* **2002**, *54*, 233–246.
124. Portoghese, P. S. *J. Med. Chem.* **2001**, *44*, 2259–2269.
125. Crossley, R.; Slater, M. J.; de Zoysa, P. *Drugs Future* **2002**, *27*, 29–96.
126. Holbrook, J.; Abramo, M.; Blaney, F. E.; Foord, S. M. HUGO Human Genome Meeting, Cancún, Mexico, Apr 27–30, 2003.
127. Kratochwil, N. A.; Malherbe, P.; Lindemann, L.; Ebeling, M.; Hoener, M. C.; Muhlemann, A.; Porter, R. H.; Stahl, M.; Gerber, P. R. *J. Chem. Inf. Model.* **2005**, *45*, 1324–1336.
128. Beddell, C. R.; Goodford, P. J.; Norrington, F. E.; Wilkinson, S.; Wootton, R. *Br. J. Pharmacol.* **1976**, *57*, 201–209.
129. Kubinyi, H. *Curr. Opin. Drug Disc. Dev.* **1998**, *1*, 4–15.
130. Venter, J. C.; Fraser, C. M.; Kerlavage, A. R.; Buck, M. A. *Biochem. Pharmacol.* **1989**, *38*, 1197–1208.
131. Ruprecht, J. J.; Mielke, T.; Vogel, R.; Villa, C.; Schertler, G. F. *EMBO J.* **2004**, *23*, 3609–3620.
132. Forbes, I. T.; Dabbs, S.; Duckworth, D. M.; Ham, P.; Jones, G. E.; King, F. D.; Saunders, D. V.; Blaney, F. E.; Naylor, C. B.; Baxter, G. S. et al. *J. Med. Chem.* **1996**, *39*, 4966–4977.
133. Berkhout, T. A.; Blaney, F. E.; Bridges, A. M.; Cooper, D. G.; Forbes, I. T.; Gribble, A. D.; Groot, P. H.; Hardy, A.; Ife, R. J.; Kaur, R. et al. *J. Med. Chem.* **2003**, *46*, 4070–4086.
134. Lopez-Rodriguez, M. L.; Morcillo, M. J.; Fernandez, E.; Rosado, M. L.; Pardo, L.; Schaper, K. J. *J. Med. Chem.* **2001**, *44*, 198–207.
135. Lopez-Rodriguez, M. L.; Morcillo, M. J.; Rovat, T. K.; Fernandez, E.; Vicente, B.; Sanz, A. M.; Hernandez, M.; Orensanz, L. *J. Med. Chem.* **1999**, *42*, 36–49.
136. Lopez-Rodriguez, M. L.; Morcillo, M. J.; Fernandez, E.; Porras, E.; Murcia, M.; Sanz, A. M.; Orensanz, L. *J. Med. Chem.* **1997**, *40*, 2653–2656.
137. Lopez-Rodriguez, M. L.; Rosado, M. L.; Benhamu, B.; Morcillo, M. J.; Fernandez, E.; Schaper, K. J. *J. Med. Chem.* **1997**, *40*, 1648–1656.
138. Lopez-Rodriguez, M. L.; Rosado, M. L.; Benhamu, B.; Morcillo, M. J.; Sanz, A. M.; Orensanz, L.; Beneitez, M. E.; Fuentes, J. A.; Manzanares, J. *J. Med. Chem.* **1996**, *39*, 4439–4450.
139. Lopez-Rodriguez, M. L.; Vicente, B.; Deupi, X.; Barrondo, S.; Olivella, M.; Morcillo, M. J.; Behamu, B.; Ballesteros, J. A.; Salles, J.; Pardo, L. *Mol. Pharmacol.* **2002**, *62*, 15–21.
140. Lopez-Rodriguez, M. L.; Morcillo, M. J.; Fernandez, E.; Benhamu, B.; Tejada, I.; Ayala, D.; Viso, A.; Olivella, M.; Pardo, L.; Delgado, M. et al. *Bioorg. Med. Chem. Lett.* **2003**, *13*, 1429–1432.
141. Lopez-Rodriguez, M. L.; Morcillo, M. J.; Fernandez, E.; Benhamu, B.; Tejada, I.; Ayala, D.; Viso, A.; Campillo, M. et al. *J. Med. Chem.* **2005**, *48*, 2548–2558.
142. Ortega-Gutierrez, S.; Lopez-Rodriguez, M. L. *Mini. Rev. Med. Chem.* **2005**, *5*, 651–658.
143. Lopez-Rodriguez, M. L.; Benhamu, B.; de la, Fuente, T.; Sanz, A.; Pardo, L.; Campillo, M. *J. Med. Chem.* **2005**, 48, 4216–4219.
144. Lopez-Rodriguez, M. L.; Porras, E.; Morcillo, M. J.; Benhamu, B.; Soto, L. J.; Lavandera, J. L.; Ramos, J. A.; Olivella, M.; Campillo, M.; Pardo, L. *J. Med. Chem.* **2003**, *46*, 5638–5650.
145. Lopez-Rodriguez, M. L.; Murcia, M.; Benhamu, B.; Viso, A.; Campillo, M.; Pardo, L. *J. Med. Chem.* **2002**, *45*, 4806–4815.
146. Lopez-Rodriguez, M. L.; Benhamu, B.; Morcillo, M. J.; Murcia, M.; Viso, A.; Campillo, M.; Pardo, L. *Curr. Top. Med. Chem.* **2002**, *2*, 625–641.
147. Lopez-Rodriguez, M. L.; Murcia, M.; Benhamu, B.; Olivella, M.; Campillo, M.; Pardo, L. *J. Comput.-Aided Mol. Des.* **2001**, *15*, 1025–1033.
148. Lopez-Rodriguez, M. L.; Murcia, M.; Benhamu, B.; Viso, A.; Campillo, M.; Pardo, L. *Bioorg. Med. Chem. Lett.* **2001**, *11*, 2807–2811.
149. Lopez-Rodriguez, M. L.; Morcillo, M. J.; Fernandez, E.; Rosado, M. L.; Pardo, L.; Schaper, K. *J. Med. Chem.* **2001**, *44*, 198–207.
150. Lober, S.; Hubner, H.; Utz, W.; Gmeiner, P. *J. Med. Chem.* **2001**, *44*, 2691–2694.
151. Sansom, M. S.; Weinstein, H. *Trends Pharmacol. Sci.* **2000**, *21*, 445–451.
152. Wiens, B. L.; Nelson, C. S.; Neve, K. A. *Mol. Pharmacol.* **1998**, *54*, 435–444.
153. Schetz, J. A.; Benjamin, P. S.; Sibley, D. R. *Mol. Pharmacol.* **2000**, *57*, 144–152.
154. Moro, S.; Spalluto, G.; Jacobson, K. A. *Trends Pharmacol. Sci.* **2005**, *26*, 44–51.
155. Derian, C. K.; Maryanoff, B. E.; Andrade-Gordon, P.; Zhang, H. C. *Drug Dev. Res.* **2003**, *59*, 355–366.
156. Ali, M. A.; Bhogal, N.; Findlay, J. B. C.; Fishwick, C. W. G. *J. Med. Chem.* **2005**, *48*, 5655–5658.

157. Gillet, V.; Johnson, A. P.; Mata, P.; Sike, S.; Williams, P. *J. Comput.-Aided Mol. Des.* **1993**, 7, 127–153.
158. Gillet, V. J.; Newell, W.; Mata, P.; Myatt, G.; Sike, S.; Zsoldos, Z.; Johnson, A. P. *J. Chem. Inf. Comput. Sci.* **1994**, *34*, 207–217.
159. Bissantz, C.; Folkers, G.; Rognan, D. *J. Med. Chem.* **2000**, *43*, 4759–4767.
160. Stahl, M.; Rarey, M. *J. Med. Chem.* **2001**, *44*, 1035–1042.
161. Wang, R.; Lu, Y.; Fang, X.; Wang, S. *J. Chem. Inf. Comput. Sci.* **2004**, *44*, 2114–2125.
162. Wang, R.; Lu, Y.; Wang, S. *J. Med. Chem.* **2003**, *46*, 2287–2303.
163. Ferrara, P.; Gohlke, H.; Price, D. J.; Klebe, G.; Brooks, C. L., III. *J. Med. Chem.* **2004**, *47*, 3032–3047.
164. Perola, E.; Walters, W. P.; Charifson, P. S. *Proteins* **2004**, *56*, 235–249.
165. Kellenberger, E.; Rodrigo, J.; Muller, P.; Rognan, D. *Proteins* **2004**, *57*, 225–242.
166. Boehm, H. J.; Boehringer, M.; Bur, D.; Gmuender, H.; Huber, W.; Klaus, W.; Kostrewa, D.; Kuehne, H.; Luebbers, T.; Meunier-Keller, N. et al. *J. Med. Chem.* **2000**, *43*, 2664–2674.
167. Schapira, M.; Raaka, B. M.; Das, S.; Fan, L.; Totrov, M.; Zhou, Z.; Wilson, S. R.; Abagyan, R.; Samuels, H. H. *Proc. Natl. Acad. Sci. USA* **2003**, *100*, 7354–7359.
168. Pickett, S. D.; Sherborne, B. S.; Wilkinson, T.; Bennett, J.; Borkakoti, N.; Broadhurst, M.; Hurst, D.; Kilford, I.; McKinnell, M.; Jones, P. S. *Bioorg. Med. Chem. Lett.* **2003**, *13*, 1691–1694.
169. Warren, G. L.; Andrews, C. W.; Capelli, A.-M.; Clark, B.; Lalonde, J.; Lambert, M. H.; Lindvall, M.; Nevins, N.; Semus, S. F.; Senger, S.; *J. Med. Chem.* **2006**.
170. Bissantz, C.; Bernard, P.; Hibert, M.; Rognan, D. *Proteins* **2003**, *50*, 5–25.
171. Varady, J.; Wu, X.; Fang, X.; Min, J.; Hu, Z.; Levant, B.; Wang, S. *J. Med. Chem.* **2003**, *46*, 4377–4392.
172. Evers, A.; Klebe, G. *J. Med. Chem.* **2004**, *47*, 5381–5392.
173. Evers, A.; Klebe, G. *Angew. Chem. Int. Ed. Engl.* **2004**, *43*, 248–251.
174. Evers, A.; Klabunde, T. *J. Med. Chem.* **2005**, *48*, 1088–1097.
175. Ewing, T. J.; Makino, S.; Skillman, A. G.; Kuntz, I. D. *J. Comput.-Aided Mol. Des.* **2001**, *15*, 411–428.
176. Venkatachalam, C. M.; Jiang, X.; Oldfield, T.; Waldman, M. *J. Mol. Graph. Model.* **2003**, *21*, 289–307.
177. Balakin, K. V.; Tkachenko, S. E.; Lang, S. A.; Okun, I.; Ivashchenko, A. A.; Savchuk, N. P. *J. Chem. Inf. Comput. Sci.* **2002**, *42*, 1332–1342.
178. Balakin, K. V.; Lang, S. A.; Skorenko, A. V.; Tkachenko, S. E.; Ivashchenko, A. A.; Savchuk, N. P. *J. Chem. Inf. Comput. Sci.* **2003**, *43*, 1553–1562.
179. ChemDiv Website. http://www.chemdiv.com/ (accessed April 2006).
180. Evans, B. E.; Rittle, K. E.; Bock, M. G.; DiPardo, R. M.; Freidinger, R. M.; Whitter, W. L.; Lundell, G. F.; Veber, D. F.; Anderson, P. S.; Chang, R. S. *J. Med. Chem.* **1988**, *31*, 2235–2246.
181. Bondensgaard, K.; Ankersen, M.; Thogersen, H.; Hansen, B. S.; Wulff, B. S.; Bywater, R. P. *J. Med. Chem.* **2004**, *47*, 888–899.
182. Guo, T.; Hobbs, D. W. *Assay. Drug Dev. Technol.* **2003**, *1*, 579–592.
183. Willoughby, C. A.; Hutchins, S. M.; Rosauer, K. G.; Dhar, M. J.; Chapman, K. T.; Chicchi, G. G.; Sadowski, S.; Weinberg, D. H.; Patel, S.; Malkowitz, L. et al. *Bioorg. Med. Chem. Lett.* **2002**, *12*, 93–96.
184. Bleicher, K. H.; Wuthrich, Y.; Adam, G.; Hoffmann, T.; Sleight, A. J. *Bioorg. Med. Chem. Lett.* **2002**, *12*, 3073–3076.
185. Bleicher, K. H.; Wuthrich, Y.; De Boni, M.; Kolczewski, S.; Hoffmann, T.; Sleight, A. J. *Bioorg. Med. Chem. Lett.* **2002**, *12*, 2519–2522.
186. Mason, J. S.; Morize, I.; Menard, P. R.; Cheney, D. L.; Hulme, C.; Labaudiniere, R. F. *J. Med. Chem.* **1999**, *42*, 3251–3264.
187. Lewell, X. Q.; Judd, D. B.; Watson, S. P.; Hann, M. M. *J. Chem. Inf. Comput. Sci.* **1998**, *38*, 511–522.
188. Klabunde, T.; Hessler, G. *ChemBioChem* **2002**, *3*, 928–944.
189. Ashton, M.; Charlton, M. H.; Schwarz, M. K.; Thomas, R. J.; Whittaker, M. *Combin. Chem. High-Throughput Screen.* **2004**, 7, 441–452.
190. Hert, J.; Willett, P.; Wilton, D. J.; Acklin, P.; Azzaoui, K.; Jacoby, E.; Schuffenhauer, A. *J. Chem. Inf. Comput. Sci.* **2004**, *44*, 1177–1185.
191. Gillet, V. J.; Willett, P.; Bradshaw, J. *J. Chem. Inf. Comput. Sci.* **2003**, *43*, 338–345.
192. Barker, E. J.; Gardiner, E. J.; Gillet, V. J.; Kitts, P.; Morris, J. *J. Chem. Inf. Comput. Sci.* **2003**, *43*, 346–356.
193. Hert, J.; Willett, P.; Wilton, D. J.; Acklin, P.; Azzaoui, K.; Jacoby, E.; Schuffenhauer, A. *Org. Biomol. Chem.* **2004**, *2*, 3256–3266.
194. Bradley, E. K.; Beroza, P.; Penzotti, J. E.; Grootenhuis, P. D.; Spellmeyer, D. C.; Miller, J. L. *J. Med. Chem.* **2000**, *43*, 2770–2774.
195. Pickett, S. D.; McLay, I. M.; Clark, D. E. *J. Chem. Inf. Comput. Sci.* **2000**, *40*, 263–272.
196. Saeh, J. C.; Lyne, P. D.; Takasaki, B. K.; Cosgrove, D. A. *J. Chem. Inf. Model.* **2005**, *45*, 1122–1133.
197. SMARTS: Daylight Chemical Information Systems. http://www.daylight.com/ (accessed April 2006).
198. Sadowski, J.; Gasteiger, J.; Klebe, G. *J. Chem. Info. Comput. Sci.* **1994**, *34*, 1000–1008.
199. *Omega*; OpenEye Science Software. http://www.eyesopen.com/ (accessed April 2006).
200. *MDL Drug Data Report*, version 2002.2. MDL ISIS/HOST software. (2002.2). 2002. MDL Information Systems, Inc. MDL Drug Data Report. www.mdli.com (accessed Aug 2006).
201. Pozzan, A. M.; Capelli, A.-M.; Feriani, A.; Tedesco, G. 3D Pharmacophoric Hashed Fingerprints, Porous Science. *Euro QSAR 2000 Rational Approached to Drug Design*, 2001.
202. Lamb, M. L.; Bradley, E. K.; Beaton, G.; Bondy, S. S.; Castellino, A. J.; Gibbons, P. A.; Suto, M. J.; Grootenhuis, P. D. *J. Mol. Graph. Model.* **2004**, *23*, 15–21.
203. Leach, A. R.; Green, D. V.; Hann, M. M.; Judd, D. B.; Good, A. C. *J. Chem. Inf. Comput. Sci.* **2000**, *40*, 1262–1269.
204. Jilek, R. J.; Cramer, R. D. *J. Chem. Inf. Comput. Sci.* **2004**, *44*, 1221–1227.
205. Cramer, R. D.; Jilek, R. J.; Guessregen, S.; Clark, S. J.; Wendt, B.; Clark, R. D. *J. Med. Chem.* **2004**, *47*, 6777–6791.
206. Tripos Lead Discovery GPCR Brochure. http://www.tripos.com/ (accessed April 2006).
207. CONCORD; developed by R.S. Pearlman, A. Rusinko, J.M. Skell, and R. Balducci at the University of Texas, Austin, TX and distributed by Tripos, Inc., http://www.tripos.com/ (accessed April 2006).
208. Pearlman, R. S.; Smith, K. M. *Perspect. Drug Disc. Des.* **1998**, *9*, 339–353.
209. Pearlman, R. S.; Smith, K. M. *J. Chem. Inf. Comput. Sci.* **1999**, *39*, 28–35.
210. Wang, X. C.; Saunders, J. *Abstracts of Papers*, 222nd American Chemical Society National Meeting, Chicago, IL, Aug 26-30, 2001; MEDI-012.
211. Manallack, D. T.; Pitt, W. R.; Gancia, E.; Montana, J. G.; Livingstone, D. J.; Ford, M. G.; Whitley, D. C. *J. Chem. Inf. Comput. Sci.* **2002**, *42*, 1256–1262.
212. Webb, T. R.; Melman, N.; Lvovskiy, D.; Ji, X. D.; Jacobson, K. A. *Bioorg. Med. Chem. Lett.* **2000**, *10*, 31–34.
213. Mottola, D. M.; Laiter, S.; Watts, V. J.; Tropsha, A.; Wyrick, S. D.; Nichols, D. E.; Mailman, R. B. *J. Med. Chem.* **1996**, *39*, 285–296.
214. Koide, Y.; Hasegawa, T.; Takahashi, A.; Endo, A.; Mochizuki, N.; Nakagawa, M.; Nishida, A. *J. Med. Chem.* **2002**, *45*, 4629–4638.
215. Nordling, E.; Homan, E. *J. Chem. Inf. Comput. Sci.* **2004**, *44*, 2207–2215.

216. De Novo Pharmaceuticals Website. http://www.denovopharma.com/ (accessed April 2006).
217. Peakdale Molecular Website. http://www.peakdale.co.uk/ (accessed April 2006).
218. Källblad, P.; Lloyd, D. G.; Manallack, D. T.; Willems, H. M. G. Ligand-Based De Novo Design of GPCR Libraries. *18th Symposium on Medicinal Chemistry 2004*, Copenhagen, Denmark, Aug 15, 2004.
219. Stahl, M.; Todorov, N. P.; James, T.; Mauser, H.; Boehm, H. J.; Dean, P. M. *J. Comput.-Aided Mol. Des.* **2002**, *16*, 459–478.
220. Todorov, N. P.; Dean, P. M. *J. Comput.-Aided Mol. Des.* **1997**, *11*, 175–192.
221. Todorov, N. P.; Dean, P. M. *J. Comput.-Aided Mol. Des.* **1998**, *12*, 335–349.
222. Hruby, V. J. *Acc. Chem. Res.* **2001**, *34*, 389–397.
223. Jones, R. M.; Boatman, P. D.; Semple, G.; Shin, Y. J.; Tamura, S. Y. *Curr. Opin. Pharmacol.* **2003**, *3*, 530–543.
224. Ripka, A. S.; Rich, D. H. *Curr. Opin. Chem. Biol.* **1998**, *2*, 441–452.
225. Borthwick, A. D.; Davies, D. E.; Exall, A. M.; Livermore, D. G.; Sollis, S. L.; Nerozzi, F.; Allen, M. J.; Perren, M.; Shabbir, S. S.; Woollard, P. M. et al. *J. Med. Chem.* **2005**, *48*, 6956–6969.
226. Wyatt, P. G.; Allen, M. J.; Borthwick, A. D.; Davies, D. E.; Exall, A. M.; Hatley, R. J.; Irving, W. R.; Livermore, D. G.; Miller, N. D. et al. *Bioorg. Med. Chem. Lett.* **2005**, *15*, 2579–2582.
227. Tyndall, J. D.; Pfeiffer, B.; Abbenante, G.; Fairlie, D. P. *Chem. Rev.* **2005**, *105*, 793–826.
228. Cho, N.; Harada, M.; Imaeda, T.; Imada, T.; Matsumoto, H.; Hayase, Y.; Sasaki, S.; Furuya, S.; Suzuki, N.; Okubo, S. et al. *J. Med. Chem.* **1998**, *41*, 4190–4195.
229. Kee, K. S.; Jois, S. D. *Curr. Pharm. Des.* **2003**, *9*, 1209–1224.
230. Garland, S. L.; Dean, P. M. *J. Comput.-Aided Mol. Des.* **1999**, *13*, 469–483.
231. Garland, S. L.; Dean, P. M. *J. Comput.-Aided Mol. Des.* **1999**, *13*, 485–498.
232. Chianelli, D.; Kim, Y. C.; Lvovskiy, D.; Webb, T. R. *Bioorg. Med. Chem.* **2003**, *11*, 5059–5068.
233. Frye, S. V. *Chem. Biol.* **1999**, *6*, R3–R7.
234. Jacoby, E.; Schuffenhauer, A.; Floersheim, P. *Drug News Perspect.* **2003**, *16*, 93–102.
235. Crossley, R. *Curr. Top. Med. Chem.* **2004**, *4*, 581–588.
236. Krejsa, C. M.; Horvath, D.; Rogalski, S. L.; Penzotti, J. E.; Mao, B.; Barbosa, F.; Migeon, J. C. *Curr. Opin. Drug Disc. Dev.* **2003**, *6*, 470–480.
237. Fliri, A. F.; Loging, W. T.; Thadeio, P. F.; Volkmann, R. A. *Proc. Natl. Acad. Sci. USA* **2005**, *102*, 261–266.
238. Jones-Hertzog, D. K.; Mukhopadhyay, P.; Keefer, C. E.; Young, S. S. *J. Pharmacol. Toxicol. Methods* **1999**, *42*, 207–215.
239. Hirst, W. D.; Abrahamsen, B.; Blaney, F. E.; Calver, A. R.; Aloj, L.; Price, G. W.; Medhurst, A. D. *Mol. Pharmacol.* **2003**, *64*, 1295–1308.

Biographies

Nicholas Barton, born in Sheffield, studied at University of York where he obtained a BSc in 1998 from the Department of Chemistry at the University of York, which included a year's study with Dr Ulli Enlgert in the AKS of the RWTH in Aachen, Germany. He went on to study for a DPhil in the York Structural Biology Laboratory under Dr Leo Caves, investigating the conformational flexibility and dynamics of the Calmodulin system which was completed in 2002. He then took a position as a Computational Chemist at Millennium Pharmaceuticals in Cambridge before joining the 7TM Systems Modelling Group at GlaxoSmithKline in 2003, where he supports drug discovery projects across the company portfolio.

Frank E Blaney is a research manager in the Computational and Structural Sciences division of Discovery Research in GlaxoSmithKline. He obtained his PhD from Queen's University, Belfast, in 1974, under the direction of Prof Tony McKervey. He then moved to USA where he spent the next three years at the University of Illinois, Urbana, as a NATO and subsequently NSF postdoctoral fellow, working in the field of natural product synthesis. He joined Beecham Pharmaceuticals upon his return to the UK in 1978 where he quickly developed an interest in QSAR and molecular modeling. In the early 1980s, he formed the computational chemistry group at Beecham and this has grown considerably in size following several mergers. Since 1979, he has worked not only in the development of novel modeling techniques, but also in a wide number of therapeutic areas, with particular emphasis in CNS and metabolic diseases. In particular, however, in the last 15 years, he has specialized in the computational modeling of membrane-bound proteins, especially 7TM receptors, ion channels, and transporter proteins, as well as mechanistic studies in cytochrome P450 enzymes and nuclear hormone receptors.

Steve Garland, born in London, studied at Oxford University, where he obtained a BA in chemistry in 1993 with his Part II under the direction of Prof Graham Richards and then at University of Cambridge for his PhD which was completed in 1996 under the direction of Dr Philip Dean. After spending the next 18 months as a Rhone-Poulenc Rorer Research Fellow in the Pharmacology Department at Cambridge, he joined the Molecular Modelling group at SmithKline Beecham in Harlow. Upon the merger to form GlaxoSmithKline in 2000, he was appointed to his current position as leader of the 7TM Systems Modelling Group. He is also a member of the 7TM Target Class Committee core group where he represents the Computational, Analytical and Structural Sciences department and has a number of other matrix management roles, particularly program leadership. His scientific interests include low tractability 7TM receptors, particularly those from Family B and C, chemical diversity and library design.

Benjamin Tehan, born in Seymour, Australia, studied at Swinburne University, where he obtained a Bachelor of Applied Science in 1998. He then completed a PhD in computational/medicinal chemistry at the Victorian College of Pharmacy, Monash University, in 2003 under the direction of Dr Edward Lloyd and Dr Margaret Wong. This included a 3-month sabbatical at Celltech Pharmaceuticals UK under the guidance of Dr David Manallack. He completed a 1-year Postdoctoral Research Fellowship with Victorian Partnership for Advanced Computing (VPAC) and Monash University before working for VPAC and the Victorian Infectious Disease Research Laboratories (VIDRL). He then accepted his current position in the 7TM group at GlaxoSmithKline. His scientific interests include membrane-bound receptors, homology modeling, structure-based design, ligand-based design, database searching, and toxicity targets related to ADME prediction.

Ian Wall, born in Surrey, studied at Southampton University, where he obtained BSc's in chemistry and maths and a PhD in the development of free energy methods under the direction of Dr Jonathan Essex. He joined Roche in Welwyn Garden City in 2000 where he worked on viral proteins with particular interest in docking and pharmacophore methods. In 2001, he took up his current position in the 7TM Systems Modelling Group at GlaxoSmithKline. His scientific interests include modeling 7TM receptors, docking methods, pharmacophore analysis, free energy methods, and the application of these techniques to drug discovery.

UNIVERSITY COLLEGE Library CORK

© 2007 Elsevier Ltd. All Rights Reserved
No part of this publication may be reproduced, stored in any retrieval system or transmitted in any form by any means electronic, electrostatic, magnetic tape, mechanical, photocopying, recording or otherwise, without permission in writing from the publishers

Comprehensive Medicinal Chemistry II
ISBN (set): 0-08-044513-6

ISBN (Volume 4) 0-08-044517-9; pp. 669–701

4.27 Ion Channels: Insights for Drug Design from Structure and Modeling

S Haider and M S P Sansom, University of Oxford, Oxford, UK

© 2007 Elsevier Ltd. All Rights Reserved.

4.27.1 Introduction: Ion Channels

Biological membranes are formed from a lipid bilayer that presents a hydrophobic barrier to the movement of ions and other solutes. Ion channel proteins have evolved to provide a route for the passage of ions across the cell membrane. These integral membrane proteins span the bilayer, and have an aqueous pore running through them that acts as a pathway for ions to traverse the membrane. Channels are passive in that ions move down their electrochemical gradients. Channels are generally selective, that is, only certain ions can pass through a given channel. Ion channels are present in a wide range of organisms, including viruses, bacteria, plants, and animals. In animals they play a key role in excitable cells, such as neurons and muscle.[1]

Ion channels are generally able to switch between an open (ion-permeable) and a closed (ion-impermeable) state. The change in the conformation of the protein that results in the opening and closing of the pore is referred to as gating. Channel gating thus controls the permeability to ions, and may be regulated by a variety of cellular events, such as changes in extracellular or intracellular ligand concentrations, or changes in transmembrane (TM) voltage. Once open, the passage of ions through the channel depends upon the electrochemical gradients of ions across the membrane. Ion flow across the membrane is of the order of 10^7–10^8 ions per second per channel. This is close to diffusion limited, and is

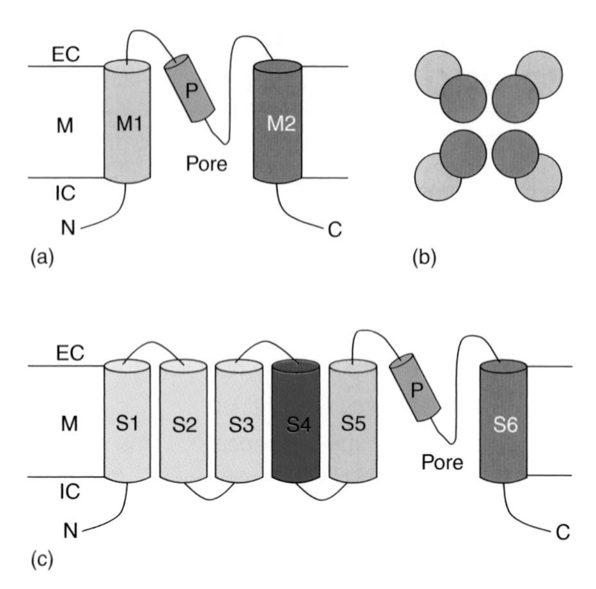

Figure 1 Schematic illustration of the K^+ channel topologies. (a) 2TM helix topology of the pore-forming domain (M1-P-M2) as seen in KcsA, MthK, and Kir channels. The filter signature sequence lies between P and M2. (b) Four subunit domains associate together to form a pore running down the center of the tetramer. (c) The 6TM helix of Kv channels. Helices S1–S4 form the voltage sensor, and S5-P-S6 are equivalent to M1-P-M2.

approximately four orders of magnitude faster than can be achieved by the active transport mechanisms. This indicates that the movement of ions through the open pore is not tightly coupled to large changes in protein conformation.

The pore of an ion channel is selective, only permitting the passage of certain ions. Indeed, under physiological conditions, many channels only permit a single ionic species to flow, and so selectivity is the basis for the classification of channels.[1,2] Ion selectivity depends on the structure of the pore and on the energetics of interactions between the ion and the protein and water microenvironment as it traverses the channel.[3,4] Channel families may be further subdivided on the basis of their gating properties and/or biological functions.

In this review we will focus on K^+ channels. The K^+ channels share a common architecture within their TM pore domains (**Figure 1**), as revealed by topological, spectroscopic,[5–9] and crystallographic (see below) studies. They also may contain a range of extra-membraneous domains (and subunits) that are concerned with channel gating and regulation. In most K^+ channels the pore is formed by four subunits associated in a symmetrical fashion around the central axis (**Figure 1b**). In the simplest K^+ channels each subunit consists of two TM α helices (M1 and M2), in between which is a pore loop containing a short α-helical segment (the P helix) and an extended segment of polypeptide chain, which forms the selectivity filter. The filter contains the signature sequence (TVGYG) that is characteristic of K^+ ions. A more complex TM domain is seen within, for example, the voltage-gated (Kv) channels, which contain four additional TM α helices (S1 to S4), and which form the voltage-sensing domain. Kv channels may also contain β subunits in the cytoplasm, which contribute to the regulation of the channels. Moreover, β subunits have been implicated in coupling the conduction of ions to intracellular signaling.[10] A further variation on the theme of K^+ channel architectures is seen in the TWIK and related 2P domain K^+ channels,[11] in which a channel is formed by two subunits, with each subunit containing two copies of the M1-P-M2 domain.[12]

4.27.1.1 K^+ Channels as Targets for Drug Design

It is estimated that approximately 25% of all genes encode membrane proteins,[13] corresponding to perhaps 10 000 proteins in the human genome. Membrane proteins account for about two-thirds of known protein drug targets.[14] Of these known targets, ion channels contribute over 5%. Furthermore, ion channels form one of the major classes of potential targets for new drugs.[15] This is perhaps not surprising, as ion channels play a central role in a number of key functions such as electrical excitability, regulation of cell volume, regulation of cytoplasmic ion concentration and pH, and hormone and neurotransmitter secretion. Furthermore, incorrect function of specific ion channels may lead to a wide range of diseases (sometime referred to as channelopathies).[16,17] K^+ channels in particular are encoded by about 75 genes in humans.[18]

Given the importance of K^+ channels as potential drug targets, a knowledge of their structures should enable us to understand structure–function relationships in this important class of channel protein, and thus open new avenues for structure-based drug design. In the following sections we review experimental studies on K^+ channel structure, before describing how computational studies have been used to help relate static x-ray structures to their dynamics and biological function.

4.27.2 Structures of Ion Channels

The number of high-resolution structures that have been determined for membrane proteins remains small (≈ 100) largely because of problems of over-expression and crystallization.[19] The two main approaches to formation of ordered crystals suitable for x-ray diffraction are based on the formation of protein–detergent complexes or upon co-crystallizing channel proteins with an antibody fragment that forms a scaffold protein. Despite these difficulties, there has been considerable progress in structural studies of ion channels in recent years. However, over-expression of mammalian membrane proteins is currently more difficult than for bacterial membrane proteins, and so (with one exception – see below) bacterial homologs of human ion channels have been used to study the structure of ion channels. In **Table 1** we summarize the structures of ion channels that have been determined. In the following sections we will review how these structures combined with computational studies have provided detailed information about ion selectivity, conduction, and gating processes in K^+ channels.

4.27.2.1 KcsA: A pH-Dependent K^+ Channel

The 3.6 Å resolution x-ray crystallographic structure of KcsA,[20] a K^+ channel from *Streptomyces lividans*, was a landmark in ion channel studies as it furnished the first structure of an ion channel protein, providing the basis for understanding the molecular basis of potassium conduction and selectivity. KcsA is highly selective for K^+ ions,[21] and is activated by low pH.[22] Although not activated by changes in transbilayer voltage, it shares many open-channel properties (e.g., ion selectivity and block) with Kv channels. The structure of KcsA was subsequently redetermined at 2.0 Å resolution (via co-crystallization with antibody fragments), to provide detailed information on ion–protein and ion–water interactions within the channel.[23] The KcsA structure (**Figure 2a**) corresponds to the simpler K^+ channel architecture described above. Each of the four identical subunits contains two TM α helices connected by an approximately 30 amino acid pore region that contains a turret, pore helix, and selectivity filter. The subunits are packed within the tetramer such that the outer (M1) helix faces the surrounding lipid bilayer while the inner (M2) helix lines the central pore. The inner helices are tilted with respect to the membrane normal by 25°, and packed in the form of an inverted truncated cone, with the wider end of the cone at the extracellular side of the membrane. The extracellular end of the M2 helix bundles thus accommodates the structure formed by tetramer of pore loops formed by the P helices and the selectivity filter. This latter region contains the signature sequence that determines the selectivity of K^+ channels and their more distant relatives.[24]

Table 1 X-ray structures of K^+ channels

Channel	*Function*	*Resolution (Å)*	*Species*	*PDB code*
KcsA	pH-gated K^+ channel	2	*Streptomyces lividans*	1K4C
MthK	Ca^{2+}-gated K^+ channel	3.3	*Methanobacterium thermoautotrophicum*	1LNQ
KvAP	Voltage-gated K^+ channel	3.2	*Aeropyrum pernix*	1ORQ
KvAP (S1–S4)	Voltage sensor domain of KvAP	1.9	*Aeropyrum pernix*	1ORS
Kv1.2	Mammalian Shaker K^+ channel	2.9	*Rattus norvegicus*	2A79
KirBac1.1	Inward rectifier K^+ channel	3.7	*Burkholderia pseudomallei*	1P7B
KirBac3.1	Inward rectifier K^+ channel	2.8	*Magnetospirillum magnetotacticum*	1XL6
Kir2.1 IC domain	Intracellular domain of Kir2.1 (Irk1)	2.4	*Mus musculus*	1U4F
Kir3.1 IC domain	Intracellular domain of Kir3.1 (GIRK1)	1.8	*Mus musculus*	1N9P

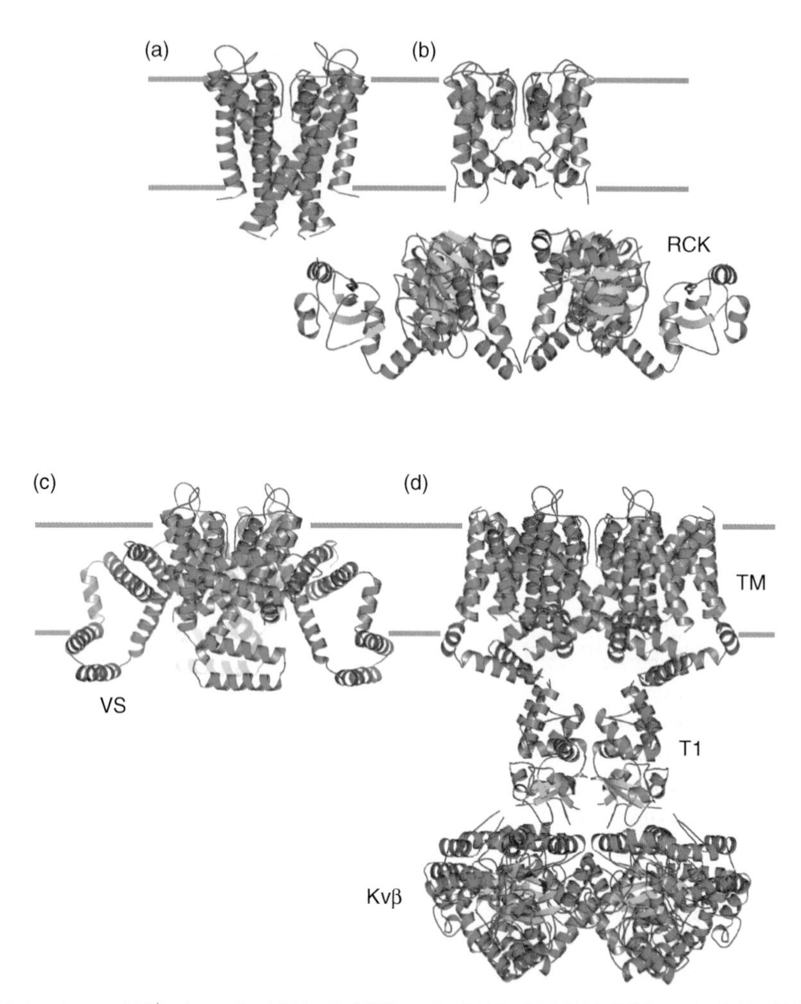

Figure 2 Crystal structures of K^+ channels: (a) KcsA, PDB code 1K4C; (b) MthK, PDB code 1LNQ (RCK, Ca^{2+}-binding domain); (c) KvAP, PDB code 1ORQ (VS, voltage sensor domain, i.e., S1 to S4); and (d) Kv1.2, PDB code 1QDB (T1, tetramerization domain; Kvβ, β subunit). The horizontal gray lines represent the approximate location of the membrane, and in each case the extracellular end of the channel protein is uppermost.

The four M2 helices pack closely with each other near the intracellular face of the membrane, providing a constriction that closes the channel. The P helices are located in between the M2 helices at the extracellular mouth. This arrangement of the P helices provides several intersubunit contacts, and may also play a key role in the electrostatic properties of the channel.[25] The overall length of the pore in KcsA is approximately 45 Å, with considerable variation in radius along the length of the pore. Thus, in KcsA (which is crystallized in a closed state), the intracellular entrance to the pore is very narrow (radius ≈ 1 Å) and hydrophobic. This opens into a wide cavity (radius ≈ 5 Å) near the middle of the membrane. The selectivity filter separates the cavity from the extracellular environment, and is narrow (radius ≈ 1 Å), such that the K^+ ion (hydrated radius ≈ 4.3 Å) would have to shed its solvation shell in order to enter it.

Within the filter there are five potential ion-binding sites. Under physiological conditions, either two or three K^+ ions are present within the filter, in a single file with water molecules in between them. Each K^+ ion forms interactions with eight oxygen atoms, either from the backbone carbonyl oxygen atoms of the filter (which protrude into the pore), or with four carbonyl oxygen atoms and four threonine side chain oxygen atoms (at the intracellular end of the filter) or with four carbonyl oxygen atoms and water oxygen atoms (at the extracellular end of the filter). The square antiprismatic ion coordination of K^+ ion with carbonyl oxygen atoms compensates energetically for the dehydration process, thus stabilizing a K^+ ion to an approximately equal extent at each of the five binding sites along the length of the filter. On the basis of the x-ray structure it is suggested that, under physiological conditions, two K^+ ions separated by a water molecule move in a synchronized manner along the selectivity filter until a third ion enters from one side of the filter, and thus displaces another ion from the opposite side.[26]

It was initially suggested that a relative rigid selectivity filter provided an exact fit for K^+ ions as opposed to Na^+ ions.[20] However, higher-resolution crystal structural studies revealed that the selectivity filter is flexible, and may adopt one of two different conformations, depending upon the concentration of K^+ ions.[23,26] In the presence of a low K^+ ion concentration, the filter adopts a 'collapsed' conformation, in which not all of the carbonyl oxygen atoms are directed toward the center of the pore. The conformation adopted by the selectivity filter in the presence of a high concentration of K^+ ions is thought to be the functionally active form, with all of the carbonyl oxygen atoms pointing inward, thus providing optimal interaction sites for K^+ ions. On this basis it has been suggested that K^+ channels may achieve high conduction rates by utilizing some of the binding energy of the ion to the selectivity filter, to change the conformation from a collapsed state to an active state. In addition to electrostatic repulsions between the closely spaced ions within the filter, this utilization of binding energy to distort the filter has been suggested to weaken interactions of the protein with the ions, so that they may pass rapidly along the filter.[27] Computational approaches offer a way in which to test such hypotheses derived from structural data.

The KcsA structure contains a water-filled cavity halfway across the membrane. The cavity is approximately 5 Å in radius, and accommodates a hydrated K^+ ion at its center.[23] Simple electrostatic calculations show that when an ion is moved through a membrane, the energy barrier reaches a maximum at the center of the membrane.[28,29] Thus, the functional significance of the central water-filled cavity in the KcsA structure is to overcome the electrostatic destabilization resulting from this low dielectric bilayer by surrounding an ion with water. The helix dipole effect generated by the pore helices whose C-termini are directed toward the central cavity may also help in lowering the electrostatic barrier,[25] thus enabling cations to cross the membrane.

As noted above, the crystal structure of KcsA corresponds to a closed state of the channel. The hydrophobic residues in the M1 helices come close together at the intracellular mouth. This narrow hydrophobic region has been implicated as a putative intracellular gate of the channel. Although a crystal structure of an open state of KcsA has not been determined, site-directed spin labeling studies suggest that upon channel activation by lowering of pH, the M2 helices move away from one another, thus opening up the intracellular mouth of the channel.[6,8] The energetic consequences of this are discussed below.

4.27.2.2 MthK: A Ca^{2+}-Activated K^+ Channel

The crystal structure of a calcium-activated K^+ channel (MthK) from *Methanobacterium thermoautotrophicum*[30] provides a snapshot of a K^+ channel in an open state, albeit at medium resolution (**Figure 2b**). As with KcsA, there are two TM helices per subunit, with four such subunits coming together to form the tetrameric channel. In addition to the TM domain, the C-terminus of the MthK sequence contains an extensive regulator of K^+ conductance (RCK) domain that binds Ca^{2+} ions. The MthK tetramer is also associated with four additional RCK domains. Thus, the overall structural arrangement of the MthK channel is such that eight RCK domains are connected to the TM pore of the channel. RCK domains have been found in both eukaryotes and prokaryotes.[31] These domains bind ligands that in turn control channel gating, and thus in MthK the RCK domains are referred to as a gating ring. In MthK, Ca^{2+} ions are bound to conserved residues that contribute to the ligand-binding cleft, which lies between the main lobes of two adjacent RCK domains. It is thought that Ca^{2+} binding to the cleft triggers a series of conformational changes that lead to the opening of the channel. Since the protein has been crystallized in the presence of 200 mM Ca^{2+}, it is thought that the channel structure corresponds to a Ca^{2+}-activated state.

The overall architecture of the MthK TM domain closely resembles that of KcsA. For example, comparing MthK and KcsA structures reveals that the structure adopted by the selectivity filter is the same in both proteins. However, the structures differ in the conformation and orientation of their M2 helices. As described above, in KcsA the M2 helices cross at the intracellular mouth of the pore, to form a narrow hydrophobic region that corresponds to a closed gate. In contrast, in MthK the M2 helices form an open gate.[30,32] This is achieved by a kink in the center of the M2 helices, which allows them to splay outward such that their C-termini are more distant from one another than in KcsA. A highly conserved glycine residue (G99 in KcsA, G83 in MthK) is found in K^+ channels, located in the middle of the M2 helix. Due to the greater conformational flexibility associated with a glycine residue, this conserved glycine has been implicated as a hinge point within the M2 helix. Thus, significant differences in the open and closed conformations are observed at the intracellular mouth of the channel pore. In the open-conformation MthK structure, the M2 helices are bent at the glycine hinge, resulting in a wide mouth of the channel pore, with a radius of about 6 Å. In the closed-conformation KcsA structure, the M2 helices come together at the intracellular mouth of the channel pore, to form a crossed helix bundle with a radius of approximately 1.2 Å.

Unfortunately, the polypeptide chain linking the C-termini of the M2 helix to the adjacent RCK domain is unresolved in the x-ray structure of MthK. This means that from a structural perspective the mechanism of coupling of Ca^{2+} binding to the RCK domains to the conformational change that results in channel opening is unclear. However, KcsA and MthK provide structural templates for closed versus open conformations of K^+ channels, and indicate the importance of a molecular hinge in the pore-lining M2 helix in the gating mechanism.

4.27.2.3 KvAP: A Bacterial Voltage-Dependent K^+ Channel

Voltage-dependent (Kv) channels form a large group of K^+ channels that are activated by membrane depolarization. Each subunit of a Kv channel contains six TM helices per subunit in addition to possible N- and/or C-terminal domains (**Figure 1b**). Four subunits come together to form the canonical tetrameric K^+ channel assembly. The pore is formed by a tetramer of an S5-P-S6 domain that is homologous to the M1-P-M2 domain in the simpler K^+ channels. The S1–S4 helices form the voltage-sensing domain in Kv channels. In particular, the S4 helix contains a cluster of positively charged (Arg and Lys) side chains that play a major role in voltage sensing. As a result of the conformational flexibility of the voltage sensor domain, crystallographic analyses of this family of ion channels have proved to be especially difficult.

The crystal structure of a (bacterial) voltage-dependent K^+ channel from *Aeropyrum pernix* (KvAP) was solved to 3.5 Å resolution (**Figure 2c**).[33] The channel was crystallized using antibody fragments as a structural scaffold. The structure of the TM pore domain, and especially of the filter, was similar to that of the KcsA and MthK structures. The inner (S6) helices of KvAP resemble those of MthK, in that they are bent at the conserved glycine hinge point, midway along the helix. Thus, the KvAP channel appears to be in a functionally open conformation.

The voltage sensor domain, formed by the S1–S4 helices, adopts an unusual conformation in the KvAP crystal structure. The S1 and S2 helices lie in the center of the bilayer, oriented almost parallel to the plane of the membrane. The S4 (and S3b) helix appear to be close to the intracellular surface of the presumed bilayer. This resulted in the proposal of a 'paddle' model, in which the S4 helix was suggested to undergo a substantial reorientation upon depolarization of the membrane.[34] However, if the KvAP structure has the voltage sensor in a closed-state conformation (i.e., before moving in response to a change in voltage), then it is puzzling that the pore domain is in an open conformation.

A number of observations suggest that the conformation of KvAP in the crystal structure may not represent a physiologically important state of the channel. First, the voltage-sensing domain (S1–S4) TM topology in the KvAP structure is inconsistent with a range of physiological and biophysical data on Kv channels.[35,36] Secondly, the KvAP structure does not correlate well with electron microscopy images of Kv channels[37,38] or with site-directed spin label data.[9] That the voltage sensor domain is flexible is evidenced by the crystal structure of the isolated KvAP voltage sensor domain (at 1.9 Å resolution),[33] which has a more 'canonical' structure with all four helices (S1–S4) in a presumed TM orientation. Together, this suggests that the conformation of the sensor domain in the intact KvAP structure may represent either a distorted or a minor conformation of the channel, perhaps corresponding to an intermediate state in the gating mechanism. Thus, the KvAP structure illustrates some of the complexities that may arise in crystallographic studies of channel proteins that are inherently flexible and which may adopt multiple conformations. In particular, it may prove difficult to directly relate the observed crystal structure with an inferred conformational state derived from biophysical studies of channel gating mechanisms.

4.27.2.4 Kv1.2: A Mammalian Voltage-Dependent K^+ Channel

A more recent structure of a mammalian Kv channel (Kv.1.2[39]) helps to answer some of these questions. This structure is of the intact Kv1.2 channel in complex with the cytoplasmic Kvβ subunit (**Figure 2d**). The channel is in an open conformation, and the TM pore-forming domain (i.e., S5-P-S6) resembles that of the KvAP structure. In mammalian Kv channels the S6 helix contains a conserved Pro-Val-Pro motif, in addition to the glycine residue discussed above. The Pro-Val-Pro sequence seems to provide an alternative S6 hinge that enables the channel to switch between a closed and open conformation.[40–43]

Significantly, the voltage sensor (S1–S4) domain of Kv1.2 adopts a conformation and orientation more consistent with biophysical and mutational data than does that of the KvAP channel. Thus, the voltage sensor conformation in the intact Kv1.2 channel is similar in fold to that of the isolated KvAP channel, i.e., a bundle of four approximately antiparallel α helices. Furthermore, the Kv1.2 voltage sensor, although rather loosely packed against the pore-forming domain, is such that the S4 helix is in a TM orientation. However, the S4 helix is relatively exposed to the surrounding membrane, raising the question of how the positively charged side chains may interact with lipid molecules (see below).

The structure of Kv1.2 also contains the intracellular T1 domain, which controls tetramerization of Kv channels. The structure of an isolated T1 domain tetramer had been determined previously, both in isolation[44] and in complex with a Kvβ tetramer.[45] Between the T1 domain and the TM pore is a relatively empty space that may permit the unhindered movement of the S6 helices during gating, and also provide a low-resistance permeation pathway to ions between the pore and the cytoplasm. The T1 domain in turn interacts with the Kvβ subunit tetramer. The intracellular Kvβ subunit contains the cofactor $NADP^+$. The exact function of the β subunit is still unknown. It is thought that the subunit might couple the redox state of the cell to regulation of membrane electrical activity.

4.27.2.5 KirBac1.1: A Bacterial Inward Rectifier K^+ Channel

The inward rectifier (Kir) family of K^+ channels is so called because they preferentially conduct K^+ ions in the inward direction. Kir channels play an important role in cellular physiology by maintaining the resting membrane potential near the K^+ equilibrium potential.[46] They are involved in a variety of functions, including the modulation of electrical activity in the cardiac and neuronal cells, secretion of insulin from the β cell in the pancreas, and transport across the epithelial membrane. The crystal structure of a bacterial homolog of Kir channels (KirBac1.1) has been solved to 3.7 Å resolution,[47] and a further bacterial homolog (KirBac3.1; Protein Data Bank (PDB) code 1XL4) has been solved at 2.6 Å (**Figure 3**). There is relatively little functional characterization of either protein, which injects a note of caution into the interpretation of structural features. However, KirBac1.1 has been shown to induce K^+ ion permeation of liposomal membranes, and this permeation is inhibited by divalent metal ions,[48] consistent with known Kir channel properties. Both the KirBac structures are of complete proteins, containing both TM and cytosolic intracellular domains. KirBac is a homotetramer, with each subunit containing two TM α helices and a large intracellular domain made up of mostly β strands. The TM pore domain is similar to that in KcsA, and is in an apparently closed conformation. However, in KirBac a short (11 residue) amphipathic α helix precedes the M1 helix. (A similar helix has been inferred to be present in KcsA on the basis of spin label data.[7]) In KirBac this N-terminal extension is referred to as a 'slide' helix, and runs approximately parallel to the cytoplasmic membrane–water interface.

The structure of KirBac1.1 appears to have been solved in a closed state. The M2 helices cross at the intracellular mouth to form a tight hydrophobic constriction, with phenylalanine side chains (F146) occluding the pore. The pore helices in the closed state are somewhat differently oriented to those in KcsA, and do not point directly toward the central cavity. The functional significance of this difference remains unclear. Gating of KirBac1.1 has been hypothesized to involve coupling of motions of the TM helices (M1 and M2) via the amphipathic slide helix to a presumed conformational change in the intracellular domains. Interestingly, the overall structure of KirBac1.1 exhibits a dimer-of-dimers symmetry, with the TM domain displaying fourfold symmetry, while the cytoplasmic intracellular domains interact via a twofold symmetry axis.

4.27.2.6 The Kir3.1 (mGIRK) Intracellular Domain

In addition to the structures of the bacterial homologs, the crystal structure of the intracellular domain of a mammalian inwardly rectifying K^+ channel (Kir3.1) has been determined at 1.8 Å.[49] The structure is a homotetramer exhibiting

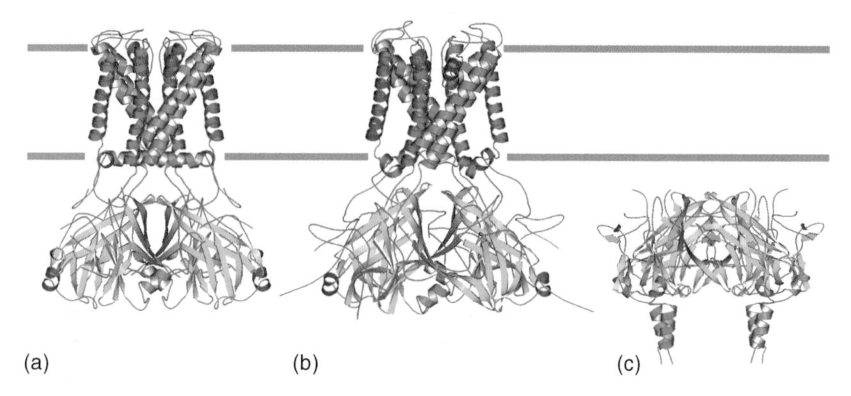

Figure 3 Crystal structures of Kir channels: (a) KirBac1.1, PDB code 1P7B; (b) KirBac3.1, PDB code 1XL6; (c) Kir3.1 intracellular domain tetramer, PDB code 1NPQ.

a conventional K^+ channel fourfold symmetry. The intracellular domain is made up mainly from β strands plus a C-terminal α helix, which points 'down' away from the membrane (**Figure 3c**). At the center is a pore that varies between 3.5 and 7.5 Å in radius, and may serves to conduct a hydrated K^+ ion from the membrane to the cytoplasm. This pore is lined by negatively charged glutamate and aspartate residues, which may help to concentrate K^+ ions, thus aiding channel conductance. Kir3.1 is gated by binding of G-protein to the channel.[50] The C-terminal α helices are thought to mediate protein–protein interactions with the G-protein subunits.

Recently, two further crystal structures of intracellular domains of mammalian inwardly rectifying K^+ channels (mouse Kir2.1 and rat Kir3.1[51]) have been determined. These structures (which share the same fold as the earlier Kir3.1 IC domain structure) reveal the presence of loops between the intracellular and the TM domains that form a girdle around the central pore axis. These loops are extremely flexible, and can adopt different conformations. It is suggested that these loops may play modulatory roles in gating and inward rectification of Kir channels.

In summary, recent advances in structural biology have provided a number of K^+ channel structures that can be used as templates for modeling studies of human homologues of these channels, and as the basis of computational studies of the relationship between ion channel structure and function.

4.27.3 Computational Methodologies for Studying K^+ Channels

Computational approaches provide a valuable complement and extension to structural biology in order to study structure–function relationships in ion channels and related membrane proteins.[3] These methodologies can be broadly categorized into two groups: (1) molecular modeling and (2) molecular simulations. Most of the structures that have been determined for ion channels are of bacterial proteins. In order to study structure–function relationships and structure-based drug design, we require structures of their human homologs. One way in which this may be achieved is via molecular modeling techniques. Simulation techniques can be used to predict the dynamic behavior of channel proteins in a membrane environment at approximately physiological temperatures, starting with the static structures obtained either by x-ray diffraction or homology modeling.

In the following sections we will briefly describe some of the computational methods that have been used to study ion channels. Details are provided in other chapters in this volume and in standard texts on molecular modeling,[52] in recent reviews of biomolecular simulations[53] and of protein modeling,[54] and in reviews of modeling[55] and simulations[56] as applied to ion channels.

4.27.3.1 Homology Modeling

Homology or comparative modeling is used to generate structural models of proteins, based on sequence alignment to a protein whose structure has been determined experimentally.[57,58] There are two practical requirements to carry out homology modeling: (1) a suitable template structure; and (2) an accurate amino acid sequence alignment between the template and the target (unknown) structure that is to be modeled. The conformational stability of homology models of channels may be assessed by subsequent molecular dynamics simulations.[59,60]

4.27.3.2 Molecular Dynamics Simulation

Current molecular dynamics simulations can provide a description of the dynamics of a membrane protein on a time-scale of the order of 10–100 ns.[61] This time-scale is comparable to that for a single ion to pass through a channel. Thus, such simulations can describe the conformational dynamics of ion channels on a time-scale relevant to aspects of their physiological function. However, to reach time-scales comparable to those of gating (1 μs and above), less direct simulation approaches must be used. Simulations of channel must be performed with the protein embedded in either a lipid bilayer or a simple membrane mimetic (e.g., an octane slab).[62] This enables the modeling of the membrane environment of the channel, but increases the computational cost of the simulation.

Molecular dynamics simulations can be performed using a number of simulation codes (e.g., GROMACS,[63,64] AMBER,[65] CHARMM,[66] or NAMD).[67] The different codes and forcefields all have their proponents: as yet, no systematic study of the same ion channel system simulated using different codes and forcefields has been performed. The major limitations of such studies are the limited time-scale that can be addressed[68] and the absence of any model of electronic polarizability[69] in the forcefields employed. However, despite these limitations, they have been extensively applied to K^+ channels,[70,71] and have yielded a number of important insights into their structure and function.

4.27.3.3 Brownian Dynamics

Brownian dynamics simulations enable one to address a longer time-scale ($\approx 1\,\mu s$) when simulating, for example, permeation of ions through channels. This is achieved by treating the diffusion of ions through a static protein pore, with the surrounding solvent (water) treated as a continuum. A number of researchers have used Brownian dynamics to address ion permeation through K^+ channels, both directly[72,73] and via a hierarchical simulation approach.[74] One limitation of the direct approach is the problem of how to incorporate filter flexibility (see below) in such simulations.

4.27.3.4 Coarse Grain Methodologies

The limitation of molecular dynamics simulations in terms of time-scales for describing protein motions has led to the development of a number of more coarse-grained methodologies for describing protein motions.[75] Of these, two classes have found applications to ion channels, especially in terms of trying to understand the conformational changes underlying channel gating. Normal mode analysis (NMA)[76] is based on a molecular mechanics forcefield, and attempts to identify the low-frequency motions that are expected to dominate slow conformational changes. The Gaussian network model (GNM) assumes that a protein is elastic in nature, and is constructed from nodes (defined by the Cα atoms of the constituent amino acids) connected to one another with uniform springs with a defined force constant and located within a defined interaction cut-off distance.[77–80] An improved version of the GNM is the anisotropic network model (ANM), which allows for spatial anisotropy in the network.[79] NMA, GNM, and ANM have all been used to explore ion channels and membrane proteins (see below).

4.27.3.5 Automated Ligand Docking

Automated ligand docking may be used to predict the interactions of small molecules with macromolecular targets.[81,82] A number of codes are available (*see* 4.12 Docking and Scoring; 4.19 Virtual Screening; and [83] for a recent review). The major methodological issues are how to include flexibility of the ligand, and, more challengingly, of the receptor (i.e., channel protein).[84] One approach may be to combine docking with molecular dynamics simulations, but such an approach needs to be explored in more detail.

4.27.4 Insights from Molecular Modeling and Simulations

4.27.4.1 KcsA: Permeation, Gating, and Lipid Interactions

Atomistic molecular dynamics simulations of KcsA have been used for detailed analysis of a variety of channel functions, including ion permeation and block, channel gating, and interactions of the channel with its lipid environment. Although KcsA is not of interest as a drug target, these studies provide a paradigm for computational studies of models of mammalian K^+ channels.

There have been many simulation studies of ion permeation mechanisms of KcsA, with different levels of computational complexity (see [70,73,85–88] for some recent reviews). Early simulation studies, based on the low-resolution structure of KcsA, provided an important 'reality check' on ion channel simulations. Several such studies[62,86,89,90] suggested the existence of a K^+-binding site just extracellular to the filter, which was not identified in the low resolution x-ray structure, but was revealed in the higher-resolution structure as determined subsequent to the simulations.[23] A key result emerging from molecular dynamics simulation studies of ion permeation is the importance of concerted motion of a single file of alternating K^+ ions and water molecules through the filter of KcsA, a result of general importance for all K^+ channels.

There have been a number of studies using molecular dynamics simulations to explore the energetics of ion movement through the selectivity filter.[88,90,91] These have emphasized the importance of the movement of an ion–water–ion single file through the filter. Although attempts have been made to explore the energetic basis of ion selectivity for K^+ over Na^+, these are made difficult by the limitations of the forcefields.[92] Density functional theory calculations[69] suggest that electronic polarizability may play an important role, thus limiting the accuracy of molecular mechanics forcefields.

One consistent feature that has emerged from a number of simulations, both of KcsA[86,90,93–96] and of other K^+ channels[97] and models,[98] is the importance of filter flexibility.[99] During permeation, a small degree of flexibility is required to enable movement of ions between adjacent sites in the filter. However, larger-scale motions (i.e., peptide carbonyl group flipping) is also seen, which appears to be related to the substantial changes in filter conformation seen in x-ray diffraction studies in response to changes in K^+ ion concentration.[23,27]

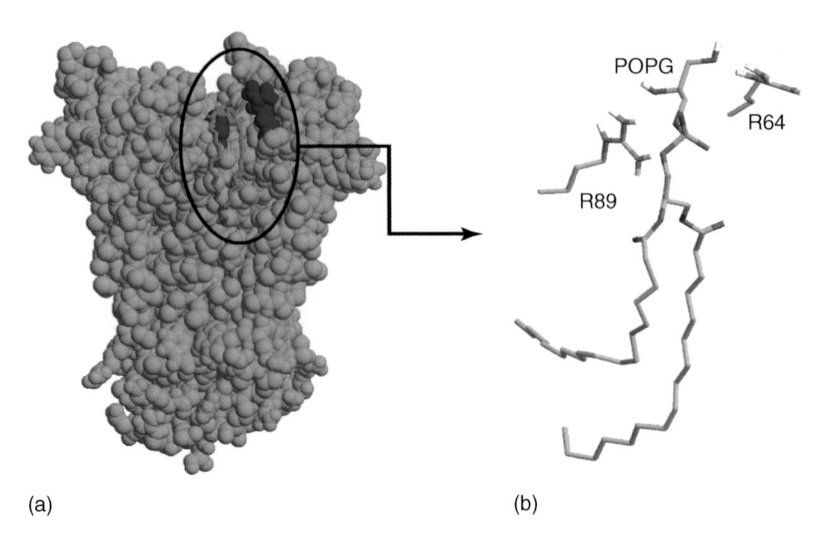

Figure 4 KcsA interactions with anionic lipids as revealed by molecular dynamics simulations. (a) KcsA, from a simulation in a mixed POPE–POPG bilayer. The channel molecule is in purple, with the two basic side chains that form the lipid binding site in blue, and a bound lipid molecule (POPG) in green. (b) Close-up view of the interactions of the bound POPG molecule with the two basic side chains (R89 from one subunit and R64 from the adjacent subunit) that form the lipid binding site.

The time-scale of molecular dynamics simulations is too short to allow direct simulation of channel gating. However, simulations of KcsA provided some clues as to channel gating mechanisms, revealing transient openings of the cytoplasmic mouth of the pore related to M2 helix flexibility.[100] This correlated with spin label studies that implicated M2 helix motions in KcsA channel opening.[8,101] Both normal mode analysis[102] and nonequilibrium molecular dynamics simulations[103] supported this model. Following the determination of the MthK structure, it was possible to model an open state conformation of KcsA using the MthK structure as a template.[32] Molecular dynamics simulations comparing the behavior of KcsA in open and closed states indicated that the conserved glycine residue in the M2 helix could indeed act as a molecular hinge.[104] As a result of the hinge bending of the M2 helices, the intracellular barrier to ion permeation formed by hydrophobic rings of valine residues is removed. This is supported by theoretical studies[105–108] that indicate the sensitivity of whether a gate is closed or open to small changes in gate radius and polarity. When open, the central cavity within the channel continuous with the intracellular environment. This enables intracellular blockers of the channel to access the central cavity and the intracellular end of the filter.

Molecular dynamics simulations have also been used to explore specific interactions of anionic phospholipids with an intersubunit binding site on the surface KcsA. Crystallographic data had revealed a diacyl glycerol fragment at this site,[109] and a number of biochemical studies have indicated the importance of interactions with lipids to KcsA function.[22,109–112] An extended (>25 ns) molecular dynamics simulation of KcsA in a lipid bilayer containing both neutral (POPE) and acidic (POPG) lipid molecules[113] revealed specific and preferential interactions of POPG with KcsA (**Figure 4**). This study indicates that (long) molecular dynamics simulations can help to identify and characterize sites for specific lipid interactions on an ion channel surface.

4.27.4.2 Modeling Studies of Inwardly Rectifying K^+ Channels

As described above, there are now two complete structures of bacterial homologs of mammalian Kir channels, in addition to structures of isolated mammalian Kir intracellular domains. Given this relative wealth of structural data plus their physiological importance, a number of computational studies have focused on Kir channels.

Early studies on the TM pore domain of Kir6.2 used a homology model based on KcsA.[59,114] Molecular dynamics simulations were used to explore the conformational stability of such models, and the sensitivity of results to the simulation protocol was explored. Modeling and simulation was also used[98] to explore changes in filter conformation in response to mutations within the P loop of Kir6.2 that changed its open channel properties.[115] These simulations reinforced the conclusions obtained from KcsA simulations concerning the importance of local changes in filter conformation in function, especially with respect to the possible role of the filter in 'fast gating' of the channel. Homology modeling based on KcsA was also used to explore other members of the Kir family (e.g., Kir2.1).[116]

The determination of the structure of the bacterial homologue KirBac1.1 enabled further molecular dynamics studies of the role of filter flexibility.[97] Analysis of simulations in the absence of K^+ ions revealed that the filter in KirBac1.1 underwent conformational changes similar to those as observed in the low $[K^+]$ structure of KcsA. This strengthened the suggestion that changes in filter conformation might provide a second ('filter' or 'fast') gate for Kir channels.

Extensive molecular dynamics simulations of the TM domain of KirBac1.1 and KirBac3.1 have also been used to study the role of the M2 (inner) helices in Kir channel gating.[117] Principal component analysis has revealed that bending of M2 occurs at a conserved glycine residue (G134 in KirBac1.1 and G120 in Kirbac3.1) that has been implicated as a molecular hinge based on the comparisons between the structures of KcsA and MthK. The analysis further suggested that the helix bundle exhibited a dimer-of-dimers motion in which the opposite pair of helices moves together. Thus, the intrinsic flexibility of the M2 helices in the TM domain supports a model of Kir gating similar to that suggested for KcsA based on comparison of the KcsA and MthK structures (see above).

The structures of the intact KirBac1.1 channel and of the mammalian Kir3.1 intracellular domain have been used to construct a model of Kir6.2 (**Figure 5**) and related mammalian Kir channels.[118] Thus, the intracellular domain of Kir6.2 was homology modeled based on the crystal structure of the corresponding domain from Kir3.1 while the TM domain was based on KirBac1.1. The intact model was constructed in two segments, because while the overall sequence identity between Kir6.2 and KirBack1.1 is quite low (27%), their sequence identity rises to 36% over the TM domains, whereas the sequence identity between the intracellular domains of Kir6.2 and Kir3.1 is 48%. The three different segments of the model (N-terminal domain, C-terminal domain, and TM domain) were modeled separately, and then reassembled. The spatial orientation of the intracellular domain with respect to the TM region was determined from the location of conserved residues in the intracellular domain of KirBac1.1.

The structural stability of the Kir6.2 model has been confirmed using multiple 10 ns molecular dynamics simulations in a lipid bilayer (**Figure 6**), which have yielded a degree of conformational drift from the initial model comparable to that seen in simulations, starting from an x-ray structure of a membrane protein (Cα RMSD ≈ 3 Å). Essential dynamics of the protein reveal a dimer-of-dimers motion of the IC domain coupled to the bending of the inner helices within the TM region via the horizontal movement of the slide helix (Haider, Ashcroft, and Sansom, unpublished data). The Kir6.2 model is in good agreement with mutational data,[118] and so has been used as the basis of ligand docking related analysis.

Ligand-docking studies carried out on Kir6.2 models (both of the isolated intracellular domain and of the intact channel protein) have been successful in identifying the ATP-binding site.[119] The model is consistent with a large amount of functional data, and has been tested by mutagenesis.[118–120] Ligand binding occurs at the interface between two subunits. The ATP molecule is positioned in the binding site such that there is a separation of charges. The phosphate backbone interacts with two positively charged side chains (R201 and K185) from the C-terminal domain from one subunit and another positively charged residue from the N-terminal domain (R50) from another subunit (**Figure 7**). ATP sensitivity to channel inhibition was reduced when the residues lining the binding site were mutated.

Kir channels are also modulated by lipidic ligands such as phosphatidylinositol 4,5-bisphosphate (PIP_2).[121–123] Based on the success of simulations of KcsA–lipid interactions, and of modeling Kir6.2–ATP interactions (see above), simulations have been used to explore the nature of the interactions of Kir6.2 with PIP_2 (Haider, Ashcroft, and Sansom, unpublished data) (**Figure 8**). As a first step, simulations of PIP_2 molecules in a phosphatidylcholine bilayer were used

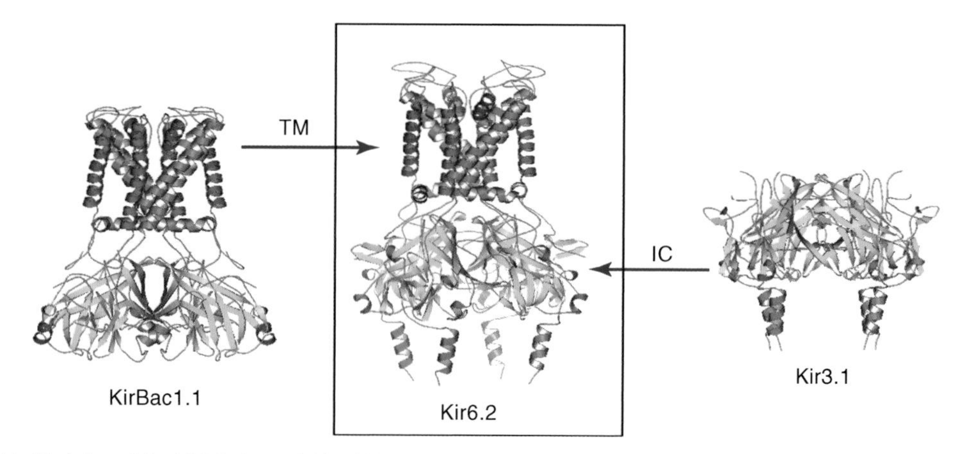

Figure 5 Modeling of the Kir6.2 channel. The KirBac1.1 structure is used as a template for the TM domain, and the structure of the intracellular domain of Kir3.1 as a template for the intracellular domain.

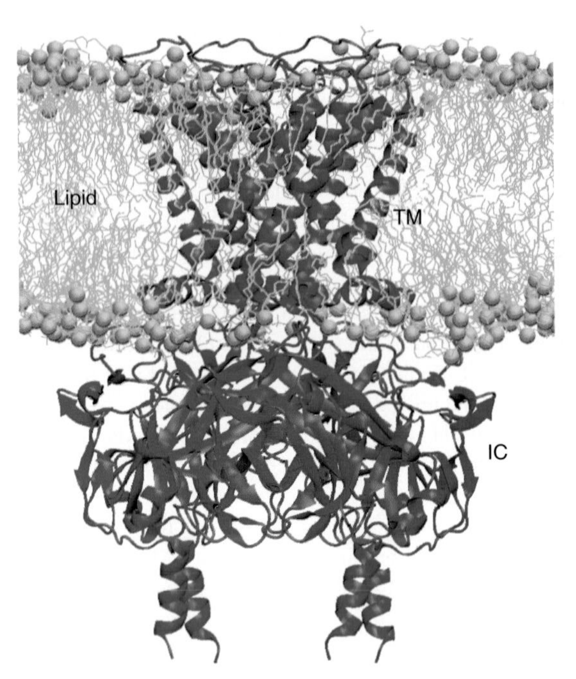

Figure 6 Model of a K^+ channel (Kir6.2, in blue) embedded in a lipid bilayer (POPC, in green, with headgroup phosphorus atoms in yellow). This system (including several thousand water molecules – not shown) is the basis of molecular dynamics simulations of the channel.

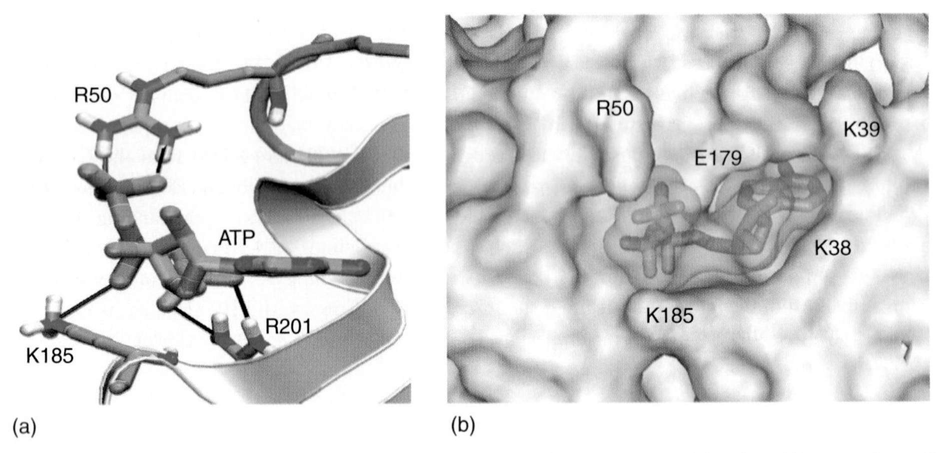

Figure 7 The ATP binding site on Kir6.2 as predicted by homology modeling and ligand docking. (a) Interaction with positively charged residues (R50, K185, and R201). (b) Surface representation of binding site, showing the close fit of ATP to the surface pocket on the Kir6.2 molecule.

to explore possible conformations of the inositol head group that can be adopted by PIP_2. In parallel, continuum electrostatics were employed to identify the regions on Kir6.2 that were likely to interact with the negatively charged headgroup of PIP_2. PIP_2 molecules from the bilayer simulation were then be used in docking studies with Kir6.2 in order to generate a model of PIP_2 bound with Kir6.2 in a similar manner done for ATP. This model is being used in simulations to explore in detail the interactions between the ligand and the protein complex and how these modulate the dynamic behavior of the Kir6.2 channel molecule (Haider and Sansom, unpublished data).

Molecular dynamics simulations have also been used to investigate the intrinsic flexibility and dynamics of the intracellular domain tetramers of Kir3.1 (a crystal structure) and Kir6.2 (a homology model).[124] Principal component analysis of the simulations revealed a pattern of motion conserved between the two structures that informs our thinking concerning gating mechanisms of the parent channel. In particular, a loss of fourfold symmetry was observed in the movement of both intracellular domains. Instead, the motions of the tetramer seemed to correspond to those of a

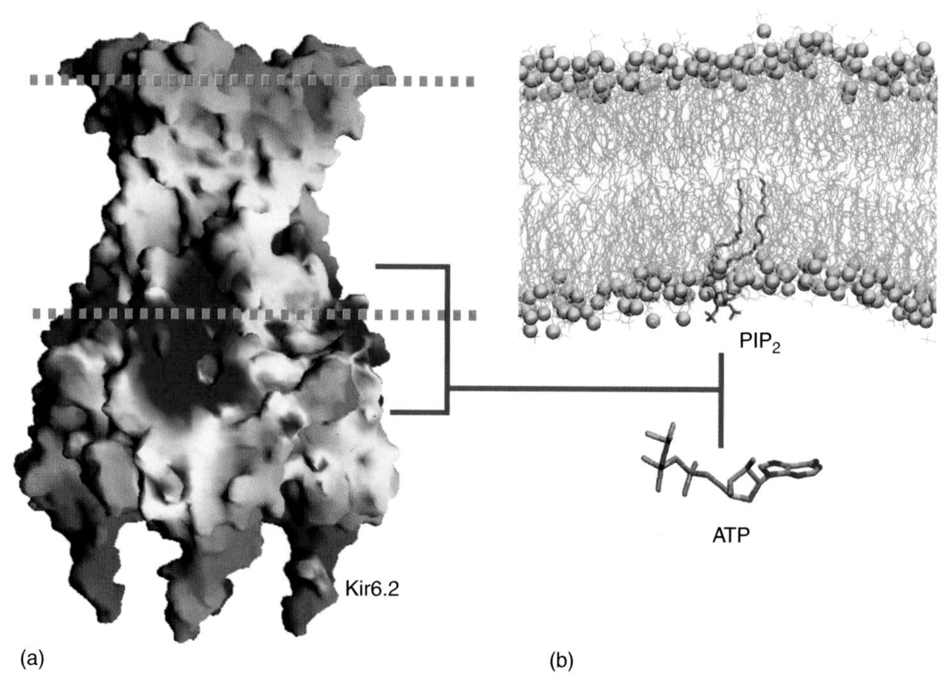

Figure 8 (a) Surface charge representation of Kir6.2. Blue represent positive charge and red negative charge. (b) Two ligands that modulate the activity of Kir6.2 are membrane-bound PIP_2 and intracellular ATP. The residues implicated in binding of these ligands are located in the positively charged surface region of Kir6.2.

dimer-of-dimers (**Figure 9**). This presents an interesting correlation with the dimer-of-dimers motion observed in the KirBac TM domain simulations (see above), and the dimer-of-dimers symmetry in the intracellular domain region of the KirBac1.1 crystal structure.

Of course, the time-scale of molecular dynamics simulations of the intracellular domain is several orders of magnitude shorter than that of channel gating. In order to explore the significance of the simulated motions for longer time-scales, an ANM calculation was performed on the intracellular domains. Significantly, the ANM results revealed a dimer-of-dimers motion similar to that observed in molecular dynamics simulations. This suggests the short-time-scale molecular dynamics results are indicative of the intrinsic larger-scale flexibility of the intracellular domain tetramer.

By combining the results of molecular dynamics simulations on the intrinsic flexibility of (1) the TM domain of KirBac and (2) the intracellular domain tetramers of Kir3.1 and of Kir6.2, one can propose a model for the gating mechanism of Kir channels (**Figure 10**).[124] The assumption underlying this is that molecular dynamics simulations reveal the intrinsic short-time-scale flexibility of the constituent domains of Kir channels. In the intact channel the coupling of the intracellular (regulatory) domain motions to the TM (channel) domain motions slows the time-scale, and links the conformational change to binding of ligands (e.g., ATP) to the intracellular domain. In the open state of the channel, the M2 helices are kinked (e.g., as in MthK). These interact with a fourfold-symmetric intracellular domain tetramer, and are thus stabilized in an open conformation. Asymmetric (i.e., dimer-of-dimers) movements of the intracellular domain lead to the closure of the channel via the adoption of a dimer-of-dimers-like packing by the M2 helices. The helices are no longer kinked upon closure. Such a model provides a template about which to design further experiments to explore changes in symmetry associated with Kir channel gating.

4.27.4.3 Modeling Studies of Voltage-Dependent (Kv) Channels

Voltage-dependent K^+ channels present a more complex challenge to modeling. As previously mentioned, the presence of a Pro-Val-Pro motif in the pore-lining S6 helix provides a hinge motif additional to the conserved glycine residue. Interestingly, this provides a valuable test of the predictive capability of modeling and simulation studies of channels (**Figure 11**). A number of simulation studies of both isolated S6 helices[125–128] and of model pore domains containing S6 helix bundles[129] had indicated that the Pro-Val-Pro motif could generate a molecular hinge. In combination with mutational[130,131] and chemical modification[40–42] studies of Kv channels, this had led to the

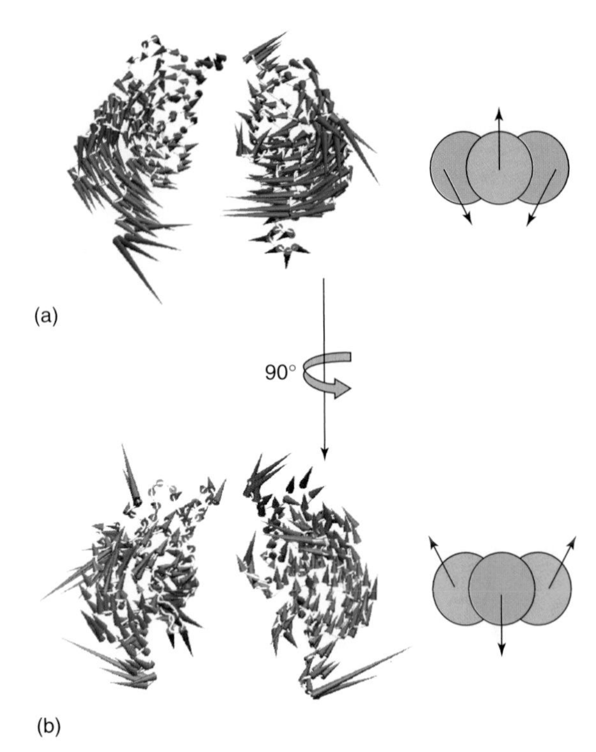

Figure 9 The first eigenvector from principal components analysis of a molecular dynamics simulation of the intracellular domain of Kir6.2. The arrows attached to each Cα atom indicate the direction of the eigenvector and the magnitude of the corresponding eigenvalue. (a) Subunits A and D (blue) and (b) subunits B and C (red). The two views are related by a 90° rotation about the vertical axis. The schematic diagrams illustrate the dimer-of-dimers motion of the tetramer.

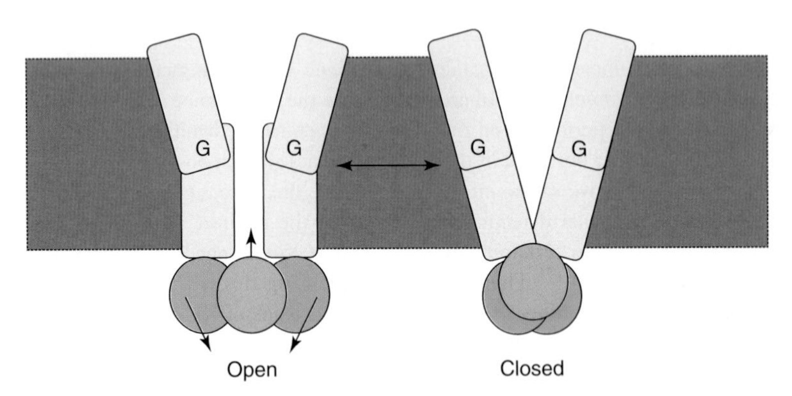

Figure 10 Proposed gating mechanism for Kir channels. The pore-lining M2 helices of two opposite subunits are shown. In the open state, the helices are kinked at the hinge-forming glycine residues. Asymmetric movements of the intracellular domains (circles), as indicated by the arrows, lead to the closure of the channel due to loss of kinking of the M2 helices.

suggestion that the Pro-Val-Pro hinge might play a key role in Kv channel gating.[132] The determination of the x-ray structure of a Kv1.2 channel (see **Figure 2d**) has demonstrated that the S6 helix is indeed kinked in the region of the Pro-Val-Pro motif, and has strengthened the suggestion that this plays a role in the gating mechanism of the channel.[43] In particular, the crystal structure of an open state of the Kv1.2 channel shows that the pore-lining S6 helices are splayed open by the Pro-Val-Pro motif.

The functional importance of S6 helix bending becomes evident in calculations of the energetic barrier heights to ion permeation at the hydrophobic gates comparing closed and open models of Shaker Kv channels, based on the structures of the KcsA and KvAP pore domains, respectively (**Figure 12**).[133] In the closed conformation, the S6 helices are not kinked at the Pro-Val-Pro motif, and the pore radius at the gate is approximately 1.4 Å, extending over about 20 Å. Estimating the energetic barrier using Born energy calculations reveal a barrier height of approximately 100 kT.

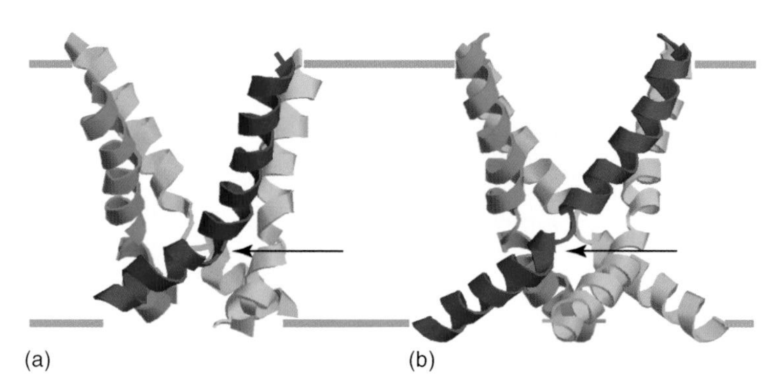

Figure 11 Comparison of the inner (S6) helix bundle in Kv channels from (a) molecular dynamics simulations of the (S5-P-S6)$_4$ domain in a membrane mimetic and (b) the x-ray structure of Kv1.2. In each case the Pro-Val-Pro hinge is indicated by an arrow.

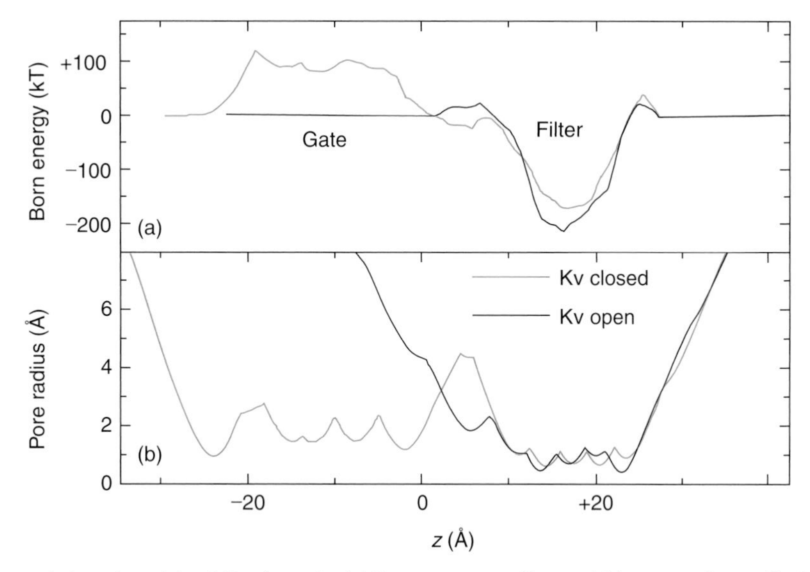

Figure 12 Open and closed models of Kv channels: (a) Born energy profiles and (b) pore radius profile for TM domains of open-state (black) and closed-state (red) Kv models.

It has been shown that Born energy calculations overestimate energies by twofold.[108] Thus, correcting this yields an estimated barrier height of approximately 40 kT for the channel in the closed conformation. When the channel is in the open conformation, there is no barrier present. Thus, S6 helical bending motions about the Pro-Val-Pro hinge are more than sufficient to switch the Kv pore from a closed to an open conformation.

Molecular dynamics simulation studies have also been used to explore the intrinsic flexibility of the voltage sensor (VS) domain from KvAP.[134] The simulations were of the isolated VS in a detergent micelle environment (**Figure 13**), corresponding to the conditions under which the VS was isolated and purified. Extended (30–40 ns) molecular dynamics simulations of the VS in dodecyl maltoside micelles at two different temperatures (300 K and 368 K) revealed an intrinsic flexibility/conformational instability of the S3a region (i.e., the C-terminus of the S3 helix). Also, the S4 helix underwent hinge bending and swiveling about its center. The conformational instability of the S3a region facilitates the motion of the N-terminal segment of S4 (i.e., S4a). Overall, these simulations support a model of an S3b–S4a voltage-sensing 'paddle' that can move relative to the rest of the VS domain. Such movement may form the underlying basis of voltage sensing by Kv channels.

4.27.4.4 K^+ Channel–Blocker Interactions

In addition to computational studies of channel function per se, there have been a number of simulation and modeling studies of ion channel block (**Figure 14**). Channel blockers inhibit K^+ channel function by either physically occluding

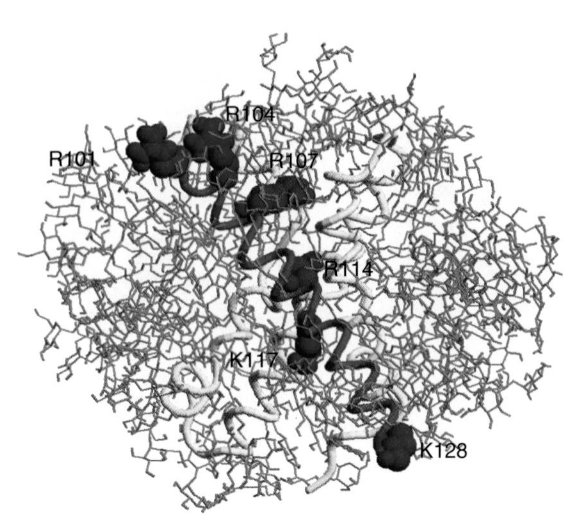

Figure 13 The isolated voltage sensor domain (gray coil) from KvAP (PDB code 1ORS). A snapshot from a simulation of the VS domain in a detergent (decyl maltoside, in green) micelle. The S4 helix and its positively charged side chains are shown in blue.

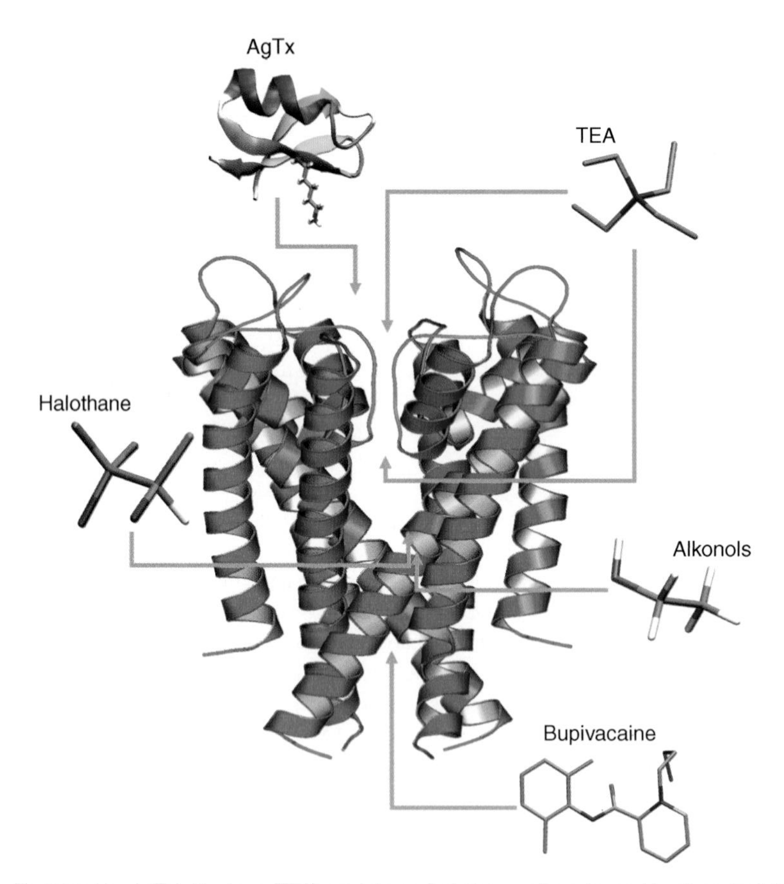

Figure 14 Toxins (illustrated by AgTx), blockers (TEA), and drugs (halothane, alkanols, and bupivacaine) that interact with K^+ channels, illustrated with KcsA. The arrows indicate the approximate site of interaction (note that bupivacaine binds to the open conformation of the channel).

the pore and/or competing with K^+ ions for binding sites. The crystal structures of the TM domains of K^+ channels reveal that the binding sites for channel blockers on intracellular and extracellular sides are rather different. Blockers from the intracellular side are expected to bind within the central water-filled cavity (as revealed by, for example, the x-ray structure of KcsA complexed with tetrabutyl ammonium[135]), and require the channel to open in order for the blocker to reach its binding site. In contrast, on the external side, the blocker would interact with residues and structural features near the entrance of the pore, as seen in, for example, the x-ray structure of KcsA complexed with tetraethyl arsonium[135]).

Mutational studies have indicated that tetraethyl ammonium (TEA) interacts with aromatic residues present on the extracellular side of Kv channels.[136,137] TEA also interacts with the equivalent tyrosine residues positioned in KcsA.[138] Two simulation studies have been reported on the interactions of KcsA with extracellular TEA,[139] using automated docking along with molecular dynamics simulation to study KcsA–TEA complexes. TEA was suggested to bind to the external side only when an ion or a water molecule did not occupy the first binding site. In this case, TEA was tightly positioned at the entrance of the channel, with the central nitrogen atom lying on the pore axis and all four ethyl groups in contact with the channel. Simulations of TEA binding within the internal cavity showed that TEA binds effectively when one K^+ ion is present in the selectivity filter at either position 1 or 2. The calculated binding energies for both internal and external modes were similar.[140] Molecular dynamics simulations and free energy calculations suggest that TEA binds to four tyrosine residues, one from each subunit. This correlates well with the x-ray structure of KcsA complexed with tetraethyl arsonium (PDB code 2BOC), lending support to computational approaches to identify channel–blocker interactions.

4.27.4.5 K^+ Channel–Toxin Interactions

A number of small peptide toxins are potent inhibitors of K^+ channel function.[141,142] They bind to the pore entrance at the extracellular mouth of the channel, blocking the passage of K^+ ions, and thereby inhibiting the ion flux. These toxins have high binding affinities for their specific channels, and are effective at nanomolar concentrations. As a result of these properties, these toxins serve as powerful investigative tools in studying K^+ channels, including applications in pharmacological studies, where they have been used to differentiate different channel and subtypes. Several nuclear magnetic resonance and crystal structures of toxins have been determined,[142,143] but a structure of a K^+ channel–toxin complex remains elusive. In the absence of such a complex, computational studies have been used to generate molecular models of channel–toxin complexes.

Comparative molecular dynamics simulations of three different toxins (Tc1, AgTx, and ChTx) demonstrated that the conserved lysine residue that is thought to bind close to the extracellular mouth of the pore has low backbone flexibility in solution.[144] Principal component analysis of simulation results indicated that the toxins shared not only the same structural fold but also a common pattern of conformational dynamics. This has since been extended to >20 toxins from the scorpion toxin family (Bemporad and Sansom, unpublished data). Thus, with a shared fold and dynamics, we may expect that toxins from this family have a conserved mode of action on their target K^+ channels.

Models of an agitoxin (AgTx)–K^+ channel complex have been studied.[145] Several models of the complexes were generated, and were selected between on the basis of distance restraints between specific toxin–channel residue pairs, as suggested by mutagenesis analysis.[146] Unrestrained molecular dynamics simulations were performed on the best structure, and the difference in binding free energy between the wild type and mutants were calculated. The free energies thus obtained were in good agreement with the results from the mutagenesis studies.

Brownian dynamics simulations have also been used to study channel–toxin interactions.[147] By combining extensive Brownian dynamics simulations with structural refinement using molecular mechanics, a three-dimensional model of a channel–toxin complex consistent with experimental data was obtained, in which the conserved lysine residue blocked the pore (see above). Similar studies using Brownian dynamics have identified interactions between maurotoxin and a voltage-gated K^+ channel,[148] and between P05 toxin and calcium-activated K^+ channels.[149]

More recently, complexes between a Ca^{2+}-activated K^+ channel (KCa) and a number of toxins (including leiurotoxin 1, maurotoxin, tamapin, apamine, and P05) have been modeled.[150] Models of the closed conformation of the pore domain of the KCa were homology modeled on KcsA. The toxins were docked to the channel, and the model complexes refined employing molecular mechanics and dynamics. The docking energies calculated showed a reasonable correlation with the experimental dissociation constants.

Overall, the correlations between the results of computational studies of channel–toxin interactions and available experimental data suggest that molecular modeling and simulation approaches can indeed provide insights into the modes of action of channel blockers, especially if some experimental data are available to filter possible models.

4.27.4.6 K^+ Channel–Drug Interactions

To date there are relatively few published computational studies of channel–drug interactions. However, in combination with studies of channel–blocker and channel–toxin interactions (see above), these studies provide an indication of how a combined structural and computational approach may be used to investigate channel–drug interactions (**Figure 14**).

A number of modeling studies of Kv channels have used homology models to study channel–ligand complexes.[151–153] For example, models of mammalian Kv4.2 channels (based on KcsA) have been used to investigate interactions of halothane with the channel.[154] The results suggested that halothane binds to the hydrophobic regions within the channel, where it may block the binding site for the inactivation domain. This correlated with experimental studies[155] that suggested that the same region was involved in the binding of 1-alkanols to dShaw2 K^+ channels.

Homology modeling has also been used to build an open conformation of the mammalian Kv1.5 channel, which has been used in docking and molecular dynamicssimulations to study block by the local anaesthetic bupivacaine.[156] Bupivacaine binds preferentially to the open state of the channel from its intracellular side. Electrophysiological studies have shown that when bupivacaine is bound, the channel is unable to close. It is suggested that binding of the anesthetic is near the highly conserved Pro-Val-Pro hinge region in the S6 of the TM domain. Such a binding mode places the quaternary nitrogen in the ligand at the center of the cavity. Electrostatic interactions between charges on the ligand and on the pore helices help to stabilize bupivacaine at this position.

K^+ channels encoded by the human ether-a-go-go (hERG) genes are of general relevance to drug design. A number of ligands (e.g., anticancer agents, antihypertensives, antipsychotics, and antihistamines) have all shown to have high binding affinities for the hERG K^+ channel, causing impairment of cardiac repolarization and leading to cardiac arrhythmia.[157] As a result of this, screening against the hERG channel is of major importance in toxicity testing of possible drugs. A computational study of hERG channels[158] used homology models of partially open and fully open conformations. This model was used study a set of ligands previously used to define a hERG pharmacophore.[159] A total of 31 ligands were used in docking studies, of which 21 ligands preferred the open conformation, and the remaining 11 ligands the partially open state. The binding modes were in agreement with the experimental mutational data and the established pharmacophore models. This computational approach offers a new approach to screening compounds for potential interactions with K^+ channels. A combined modeling and simulation study[160] of the binding of a series of hERG blockers reproduced the relative binding affinities of sertindole and its analogs. The binding mode suggested was comparable to that identified in a parallel homology modeling and pharmacophore analysis study of hERG interactions with sertindole analogs and other inhibitors.[161] Both studies indicated the importance of the Phe656 side chain in the central cavity with the bound ligands. Although further studies, both computational and experimental, will be needed, the promising convergence of these studies indicates the future importance of computational screening for interactions of compounds with hERG.

4.27.5 Conclusions and Future Directions

We have seen how molecular modeling and simulations can be used to extend our understanding of the relationship between structure and function in K^+ channels, starting from multiple x-ray structures of members of this family of channels. A major trend in the future is likely to be the determination of more structures of K^+ channels, trapped in different conformational states. By combining such structures with more extensive simulations, especially those combining coarse-grained and atomistic approaches, it should be possible to arrive at a better understanding of the structures and dynamics of multiple channel conformations. This, in turn, will greatly improve models of drug docking to channels.

References

1. Hille, B. *Ionic Channels of Excitable Membranes*; Sinauer Associates Inc.: Sunderland, MA, 2001.
2. Alexander, S. P.; Mathie, A.; Peters, J. A. *Br. J. Pharmacol.* **2005**, *144*, S73–S94.
3. Tieleman, D. P.; Biggin, P. C.; Smith, G. R.; Sansom, M. S. P. *Q. Rev. Biophys.* **2001**, *34*, 473–561.
4. Roux, B.; Allen, T.; Berneche, S.; Im, W. *Q. Rev. Biophys.* **2004**, *37*, 15–103.
5. Perozo, E.; Cortes, D. M.; Cuello, L. G. *Nat. Struct. Biol.* **1998**, *5*, 459–469.
6. Perozo, E.; Cortes, D. M.; Cuello, L. G. *Science* **1999**, *285*, 73–78.
7. Cortes, D. M.; Cuello, L. G.; Perozo, E. *J. Gen. Physiol.* **2001**, *117*, 165–180.
8. Liu, Y.; Sompornpisut, P.; Perozo, E. *Nat. Struct. Biol.* **2001**, *8*, 883–887.
9. Cuello, L. G.; Cortes, D. M.; Perozo, E. *Science* **2004**, *306*, 491–495.
10. Yellen, G. *Nature* **2002**, *419*, 35–42.
11. Biggin, P. C.; Roosild, T.; Choe, S. *Curr. Opin. Struct. Biol.* **2000**, *10*, 456–461.
12. Lesage, F.; Reyes, R.; Fink, M.; Duprat, F.; Guillemare, E.; Lazdunski, M. *EMBO J.* **1996**, *15*, 6400–6407.
13. Wallin, E.; von Heijne, G. *Protein Sci.* **1998**, *7*, 1029–1038.
14. Hopkins, A. L.; Groom, C. R. *Nat. Rev. Drug Disc.* **2002**, *1*, 727–730.

15. Terstappen, G. C.; Reggiani, A. *Trends Pharmacol. Sci.* **2001**, *22*, 23–26.
16. Ashcroft, F. M. *Ion Channels and Disease*; Academic Press: San Diego, CA, 2000.
17. Hubner, C. A.; Jentsch, T. J. *Hum. Mol. Genet.* **2002**, *11*, 2435–2445.
18. Venter, J. C.; Adams, M. D.; Myers, E. W.; Li, P. W.; Mural, R. J.; Sutton, G. G.; Smith, H. O.; Yandell, M.; Evans, C. A.; Holt, R. A. et al. *Science* **2001**, *291*, 1304–1351.
19. White, S. H. *Protein Sci.* **2004**, *13*, 1948–1949.
20. Doyle, D. A.; Cabral, J. M.; Pfuetzner, R. A.; Kuo, A.; Gulbis, J. M.; Cohen, S. L.; Cahit, B. T.; MacKinnon, R. *Science* **1998**, *280*, 69–77.
21. LeMasurier, M.; Heginbotham, L.; Miller, C. *J. Gen. Physiol.* **2001**, *118*, 303–313.
22. Heginbotham, L.; Kolmakova-Partensky, L.; Miller, C. *J. Gen. Physiol.* **1998**, *111*, 741–749.
23. Zhou, Y.; Morais-Cabral, J. H.; Kaufman, A.; MacKinnon, R. *Nature* **2001**, *414*, 43–48.
24. Heginbotham, L.; Abramson, T.; MacKinnon, R. *Science* **1992**, *258*, 1152–1155.
25. Roux, B.; MacKinnon, R. *Science* **1999**, *285*, 100–102.
26. Morais-Cabral, J. H.; Zhou, Y.; MacKinnon, R. *Nature* **2001**, *414*, 37–42.
27. Zhou, Y.; MacKinnon, R. *J. Mol. Biol.* **2003**, *333*, 965–975.
28. Parsegian, A. *Nature* **1969**, *221*, 844–846.
29. Parsegian, V. A. *Ann. N Y Acad. Sci.* **1975**, *264*, 161–171.
30. Jiang, Y.; Lee, A.; Chen, J.; Cadene, M.; Chait, B. T.; MacKinnon, R. *Nature* **2002**, *417*, 515–522.
31. Jiang, Y. X.; Pico, A.; Cadene, M.; Chait, B. T.; MacKinnon, R. *Neuron* **2001**, *29*, 593–601.
32. Jiang, Y.; Lee, A.; Chen, J.; Cadene, M.; Chait, B. T.; MacKinnon, R. *Nature* **2002**, *417*, 523–526.
33. Jiang, Y.; Lee, A.; Chen, J.; Ruta, V.; Cadene, M.; Chait, B. T.; Mackinnon, R. *Nature* **2003**, *423*, 33–41.
34. Jiang, Y.; Ruta, V.; Chen, J.; Lee, A. G.; Mackinnon, R. *Nature* **2003**, *423*, 42–48.
35. Swartz, K. J. *Nat. Rev. Neurosci.* **2004**, *5*, 905–916.
36. Elliott, D. J. S.; Neale, E. J.; Aziz, Q.; Dunham, J. P.; Munsey, T. S.; Hunter, M.; Sivaprasadarao, A. *EMBO J.* **2004**, *23*, 4717–4726.
37. Sokolova, O.; Kolmakova-Partensky, L.; Grigorieff, N. *Structure* **2001**, *9*, 215–220.
38. Jiang, Q. X.; Wang, D. N.; MacKinnon, R. *Nature* **2004**, *430*, 806–810.
39. Long, S. B.; Campbell, E. B.; MacKinnon, R. *Science* **2005**, *309*, 897–902.
40. Camino, D. D.; Holmgren, M.; Liu, Y.; Yellen, G. *Nature* **2000**, *403*, 321–325.
41. Camino, D. D.; Yellen, G. *Neuron* **2001**, *32*, 649–656.
42. Webster, S. M.; del Camino, D.; Dekker, J. P.; Yellen, G. *Nature* **2004**, *428*, 864–868.
43. Long, S. B.; Campbell, E. B.; MacKinnon, R. *Science* **2005**, *309*, 903–908.
44. Kreusch, A.; Pfaffinger, P. J.; Stevens, C. F.; Choe, S. *Nature* **1998**, *392*, 945–948.
45. Gulbis, J. M.; Zhou, M.; Mann, S.; MacKinnon, R. *Science* **2000**, *289*, 123–127.
46. Reimann, F.; Ashcroft, F. M. *Curr. Opin. Cell Biol.* **1999**, *11*, 503–508.
47. Kuo, A.; Gulbis, J. M.; Antcliff, J. F.; Rahman, T.; Lowe, E. D.; Zimmer, J.; Cuthbertson, J.; Ashcroft, F. M.; Ezaki, T.; Doyle, D. A. *Science* **2003**, *300*, 1922–1926.
48. Enkvetchakul, D.; Bhattacharyya, J.; Jeliazkova, I.; Groesbeck, D. K.; Cukras, C. A.; Nichols, C. G. *J. Biol. Chem.* **2004**, *279*, 47076–47080.
49. Nishida, M.; MacKinnon, R. *Cell* **2002**, *111*, 957–965.
50. Reuveny, E.; Slesinger, P. A.; Inglese, J.; Morales, J. M.; Iniguez-Lluhi, J. A.; Lefkowitz, R. J.; Bourne, H. R.; Jan, Y. N.; Jan, L. Y. *Nature* **1994**, *370*, 143–146.
51. Pegan, S.; Arrabit, C.; Zhou, W.; Kwiatkowski, W.; Collins, A.; Slesinger, P. A.; Choe, S. *Nat. Neurosci.* **2005**, *8*, 279–287.
52. Leach, A. R. *Molecular Modelling. Principles and Applications*; Longman: Harlow, UK, 2001.
53. Karplus, M. J.; McCammon, J. A. *Nat. Struct. Biol.* **2002**, *9*, 646–652.
54. Marti-Renom, M. A.; Stuart, A.; Fiser, A.; Sanchez, R.; Melo, F.; Sali, A. *Annu. Rev. Biophys. Biomol. Struct.* **2000**, *29*, 291–325.
55. Giorgetti, A.; Carloni, P. *Curr. Opin. Chem. Biol.* **2003**, *7*, 150–156.
56. Krishnamurthy, V.; Chung, S. H.; Dumont, G. *IEEE Trans. Nanobiosci.* **2005**, *4*, 1–2.
57. Sali, A.; Blundell, T. L. *J. Mol. Biol.* **1993**, *234*, 779–815.
58. Sanchez, R.; Sali, A. *Methods Mol. Biol.* **2000**, *143*, 97–129.
59. Capener, C. E.; Shrivastava, I. H.; Ranatunga, K. M.; Forrest, L. R.; Smith, G. R.; Sansom, M. S. P. *Biophys. J.* **2000**, *78*, 2929–2942.
60. Law, R. J.; Capener, C.; Baaden, M.; Bond, P. J.; Campbell, J.; Patargias, G.; Arinaminpathy, Y.; Sansom, M. S. P. *J. Mol. Graph. Model.* **2005**, *24*, 157–165.
61. Ash, W. L.; Zlomislic, M. R.; Oloo, E. O.; Tieleman, D. P. *Biochim. Biophys. Acta* **2004**, *1666*, 158–189.
62. Guidoni, L.; Torre, V.; Carloni, P. *Biochemistry* **1999**, *38*, 8599–8604.
63. Berendsen, H. J. C.; van der Spoel, D.; van Drunen, R. *Comp. Phys. Commun.* **1995**, *95*, 43–56.
64. Lindahl, E.; Hess, B.; van der Spoel, D. *J. Mol. Model.* **2001**, *7*, 306–317.
65. Pearlman, D. A.; Case, D. A.; Caldwell, J. W.; Ross, W. S.; Cheatham, T. E.; Debolt, S.; Ferguson, D.; Seibel, G.; Kollman, P. *Comp. Phys. Commun.* **1995**, *91*, 1–41.
66. Brooks, B. R.; Bruccoleri, R. E.; Olafson, B. D.; States, D. J.; Swaminathan, S.; Karplus, M. *J. Comp. Chem.* **1983**, *4*, 187–217.
67. Kalé, L.; Skeel, R.; Bhandarkar, M.; Brunner, R.; Gursoy, A.; Krawetz, N.; Phillips, J.; Shinozaki, A.; Varadarajan, K.; Schulten, K. *J. Comp. Phys.* **1999**, *151*, 283–312.
68. Faraldo-Gómez, J. D.; Forrest, L. R.; Baaden, M.; Bond, P. J.; Domene, C.; Patargias, G.; Cuthbertson, J.; Sansom, M. S. P. *Proteins Struct. Funct. Bioinform.* **2004**, *57*, 783–791.
69. Guidoni, L.; Carloni, P. *Biochim. Biophys. Acta* **2002**, *1563*, 1–6.
70. Roux, B.; Bernèche, S.; Im, W. *Biochemistry* **2000**, *39*, 13295–13306.
71. Roux, B.; Schulten, K. *Structure* **2004**, *12*, 1343–1351.
72. Chung, S. H.; Allen, T. W.; Kuyucak, S. *Biophys. J.* **2002**, *82*, 628–645.
73. Chung, S. H.; Kuyucak, S. *Eur. Biophys. J.* **2002**, *31*, 283–293.
74. Mashl, R. J.; Tang, Y. Z.; Schnitzer, J.; Jakobsson, E. *Biophys. J.* **2001**, *81*, 2473–2483.
75. Tozzini, V. *Curr. Opin. Struct. Biol.* **2005**, *15*, 144–150.
76. Krebs, W. G.; Alexandrov, V.; Wilson, C. A.; Echols, N.; Yu, H.; Gerstein, M. *Proteins Struct. Funct. Genet.* **2002**, *48*, 682–695.
77. Bahar, I.; Atilgan, A. R.; Erman, B. *Fold. Des.* **1997**, *2*, 173–181.

78. Keskin, O.; Jernigan, R. L.; Bahar, I. *Biophys. J.* **2000**, *78*, 2093–2106.
79. Atilgan, A. R.; Durell, S. R.; Jernigan, R. L.; Demirel, M. C.; Keskin, O.; Bahar, I. *Biophys. J.* **2001**, *80*, 505–515.
80. Yang, L. W.; Liu, X.; Jursa, C. J.; Holliman, M.; Rader, A. J.; Karimi, H. A.; Bahar, I. *Bioinformatics* **2005**, *21*, 2978–2987.
81. Goodsell, D. S.; Morris, G. M.; Olson, A. J. *J. Mol. Recog.* **1996**, *9*, 1–5.
82. Morris, G. M.; Goodsell, D. S.; Halliday, R. S.; Huey, R.; Hart, W. E.; Belew, R. K.; Olson, A. J. *J. Comp. Chem.* **1998**, *19*, 1639–1662.
83. Campbell, S. J.; Gold, N. D.; Jackson, R. M.; Westhead, D. R. *Curr. Opin. Struct. Biol.* **2003**, *13*, 389–395.
84. Taylor, R. D.; Jewsbury, P. J.; Essex, J. W. *J. Comp. Chem.* **2003**, *24*, 1637–1656.
85. Sansom, M. S. P.; Shrivastava, I. H.; Ranatunga, K. M.; Smith, G. R. *Trends Biochem. Sci.* **2000**, *25*, 368–374.
86. Sansom, M. S. P.; Shrivastava, I. H.; Bright, J. N.; Tate, J.; Capener, C. E.; Biggin, P. C. *Biochim. Biophys. Acta* **2002**, *1565*, 294–307.
87. Domene, C.; Bond, P.; Sansom, M. S. P. *Adv. Prot. Chem.* **2003**, *66*, 159–193.
88. Roux, B. *Annu. Rev. Biophys. Biomol. Struct.* **2005**, *34*, 153–171.
89. Guidoni, L.; Torre, V.; Carloni, P. *FEBS Lett.* **2000**, *477*, 37–42.
90. Bernèche, S.; Roux, B. *Nature* **2001**, *414*, 73–77.
91. Åqvist, J.; Luzhkov, V. *Nature* **2000**, *404*, 881–884.
92. Roux, B.; Bernèche, S. *Biophys. J.* **2002**, *82*, 1681–1684.
93. Shrivastava, I. H.; Sansom, M. S. P. *Biophys. J.* **2000**, *78*, 557–570.
94. Shrivastava, I. H.; Tieleman, D. P.; Biggin, P. C.; Sansom, M. S. P. *Biophys. J.* **2002**, *83*, 633–645.
95. Noskov, S. Y.; Bernèche, S.; Roux, B. *Nature* **2004**, *431*, 830–834.
96. Domene, C.; Sansom, M. S. P. *Biophys. J.* **2003**, *85*, 2787–2800.
97. Domene, C.; Grottesi, A.; Sansom, M. S. P. *Biophys. J.* **2004**, *87*, 256–267.
98. Capener, C. E.; Proks, P.; Ashcroft, F. M.; Sansom, M. S. P. *Biophys. J.* **2003**, *84*, 2345–2356.
99. Allen, T. W.; Andersen, O. S.; Roux, B. *J. Gen. Physiol.* 2004, *124*, 679–690.
100. Shrivastava, I. H.; Sansom, M. S. P. *Eur. Biophys. J.* **2002**, *31*, 207–216.
101. Perozo, E.; Liu, Y. S.; Smopornpisut, P.; Cortes, D. M.; Cuello, L. G. *J. Gen. Physiol.* **2000**, *116*, 5a.
102. Shen, Y. F.; Kong, Y. F.; Ma, J. P. *Proc. Natl. Acad. Sci. USA* **2002**, *99*, 1949–1953.
103. Biggin, P. C.; Shrivastava, I. H.; Smith, G. R.; Sansom, M. S. P. *Biophys. J.* **2001**, *80*, 514.
104. Holyoake, J.; Domene, C.; Bright, J. N.; Sansom, M. S. P. *Eur. Biophys. J.* **2003**, *33*, 238–246.
105. Beckstein, O.; Sansom, M. S. P. *Proc. Natl. Acad. Sci. USA* **2003**, *100*, 7063–7068.
106. Beckstein, O.; Biggin, P. C.; Bond, P. J.; Bright, J. N.; Domene, C.; Grottesi, A.; Holyoake, J.; Sansom, M. S. P. *FEBS Lett.* **2003**, *555*, 85–90.
107. Beckstein, O.; Sansom, M. S. P. *Phys. Biol.* **2004**, *1*, 42–52.
108. Beckstein, O.; Tai, K.; Sansom, M. S. P. *J. Am. Chem. Soc.* **2004**, *126*, 14694–14695.
109. Valiyaveetil, F. I.; Zhou, Y.; MacKinnon, R. *Biochemistry* **2002**, *41*, 10771–10777.
110. Alvis, S. J.; Williamson, I. M.; East, J. M.; Lee, A. G. *Biophys. J.* **2003**, *85*, 3828–3838.
111. Demmers, J. A. A.; van Dalen, A.; de Kruijiff, B.; Heck, A. J. R.; Killian, J. A. *FEBS Lett.* **2003**, *541*, 28–32.
112. van Dalen, A.; van der Laan, M.; Driessen, A. J. M.; Killian, A.; de Kruijff, B. *FEBS Lett.* **2002**, *511*, 51–58.
113. Deol, S. S.; Domene, C.; Bond, P. J.; Sansom, M. S. P. *Biophys. J.* **2006**, *90*, 822–830.
114. Capener, C. E.; Sansom, M. S. P. *J. Phys. Chem. B* **2002**, *106*, 4543–4551.
115. Proks, P.; Capener, C. E.; Jones, P.; Ashcroft, F. *J. Gen. Physiol.* **2001**, *118*, 341–353.
116. Thompson, G. A.; Leyland, M. L.; Ashmole, I.; Sutcliffe, M. J.; Stanfield, P. R. *J. Physiol.* **2000**, *526*, 231–240.
117. Grottesi, A.; Domene, C.; Sansom, M. S. P. *Biochemistry* **2005**, *44*, 14586–14594.
118. Antcliffe, J. F.; Haider, S.; Proks, P.; Sansom, M. S. P.; Ashcroft, F. M. *EMBO J.* **2005**, *24*, 229–239.
119. Trapp, S.; Haider, S.; Sansom, M. S. P.; Ashcroft, F. M.; Jones, P. *EMBO J.* **2003**, *22*, 2903–2912.
120. Haider, S.; Antcliff, J. F.; Proks, P.; Sansom, M. S. P.; Ashcroft, F. M. *J. Mol. Cell. Cardiol.* **2005**, *38*, 927–936.
121. Lopes, C. M. B.; Zhang, H. L.; Rohacs, T.; Jin, T. H.; Yang, J.; Logothetis, D. E. *Neuron* **2002**, *34*, 933–944.
122. Schulze, D.; Krauter, T.; Fritzenschaft, H.; Soom, M.; Baukrowitz, T. *J. Biol. Chem.* **2003**, *278*, 10500–10505.
123. Du, X. O.; Zhang, H. L.; Lopes, C.; Mirshahi, T.; Rohacs, T.; Logothetis, D. E. *J. Biol. Chem.* **2004**, *279*, 37271–37281.
124. Haider, S.; Grottesi, A.; Ashcroft, F. M.; Sansom, M. S. P. *Biophys. J.* **2005**, *88*, 3310–3320.
125. Kerr, I. D.; Son, H. S.; Sankararamakrishnan, R.; Sansom, M. S. P. *Biopolymers* **1996**, *39*, 503–515.
126. Shrivastava, I. H.; Capener, C.; Forrest, L. R.; Sansom, M. S. P. *Biophys. J.* **2000**, *78*, 79–92.
127. Tieleman, D. P.; Shrivastava, I. H.; Ulmschneider, M. B.; Sansom, M. S. P. *Proteins Struct. Funct. Genet.* **2001**, *44*, 63–72.
128. Bright, J. N.; Shrivastava, I. H.; Cordes, F. S.; Sansom, M. S. P. *Biopolymers* **2002**, *64*, 303–313.
129. Bright, J. N.; Sansom, M. S. P. *IEE Proc. Nanobiotechnol.* **2004**, *151*, 17–27.
130. Hackos, D. H.; Chang, T. H.; Swartz, K. J. *J. Gen. Physiol.* **2002**, *119*, 521–531.
131. Labro, A. J.; Raes, A. L.; Bellens, I.; Ottschytsch, N.; Snyders, D. J. *J. Biol. Chem.* **2003**, *278*, 50724–50731.
132. Sansom, M. S. P.; Weinstein, H. *Trends Pharm. Sci.* **2000**, *21*, 445–451.
133. Grottesi, A.; Domene, C.; Haider, S.; Sansom, M. S. P. *IEEE Trans. Nanobiosci.* **2004**, *4*, 112–120.
134. Sands, Z.; Grottesi, A.; Sansom, M. S. P. *Biophys. J.*, in press.
135. Lenaeus, M. J.; Vamvouka, M.; Focia, P. J.; Gross, A. *Nat. Struct. Mol. Biol.* **2005**, *12*, 454–459.
136. MacKinnon, R.; Yellen, G. *Science* **1990**, *250*, 276–279.
137. Heginbotham, L.; MacKinnon, R. *Neuron* **1992**, *8*, 483–491.
138. Heginbotham, L.; LeMasurier, M.; Kolmakova-Partensky, L.; Miller, C. *J. Gen. Physiol.* **1999**, *114*, 551–559.
139. Luzhkov, V. B.; Åqvist, J. *FEBS Lett.* **2001**, *495*, 191–196.
140. Crouzy, S.; Berneche, S.; Roux, B. *J. Gen. Physiol.* **2001**, *118*, 207–217.
141. Miller, C. *Neuron* **1995**, *15*, 5–10.
142. Possani, L. D.; Merino, E.; Corona, M.; Bolivar, F.; Becerril, B. *Biochimie* **2000**, *82*, 861–868.
143. de la Vega, R. C. R.; Possani, L. D. *Toxicon* **2004**, *43*, 865–875.
144. Grottesi, A.; Sansom, M. S. P. *FEBS Lett.* **2003**, *535*, 29–33.
145. Eriksson, M. A. L.; Roux, B. *Biophys. J.* **2002**, *83*, 2595–2609.
146. Gross, A.; MacKinnon, R. *Neuron* **1996**, *16*, 399–406.

147. Cui, M.; Shen, J. H.; Briggs, J. M.; Luo, X. M.; Tan, X. J.; Jiang, H. L.; Chen, K. X.; Ji, R. Y. *Biophys. J.* **2001**, *80*, 1659–1669.
148. Fu, W.; Cui, M.; Briggs, J. M.; Huang, X.; Xiong, B.; Zhang, Y.; Luo, X.; Shen, J.; Ji, R.; Jiang, H. et al. *Biophys. J.* **2002**, *83*, 2370–2385.
149. Cui, M.; Shen, J.; Briggs, J. M.; Fu, W.; Wu, J.; Zhang, Y.; Luo, X.; Chi, Z.; Ji, R.; Jiang, H. et al. *J. Mol. Biol.* **2002**, *318*, 417–428.
150. Andreotti, N.; di Luccio, E.; Sampieri, F.; De Waard, M.; Sabatier, J. M. *Peptides* **2005**, *26*, 1095–1108.
151. Lew, A.; Chamberlin, A. R. *Bioorg. Med. Chem. Lett.* **1999**, *9*, 3267–3272.
152. Legros, C.; Pollmann, V.; Knaus, H. G.; Farrell, A. M.; Darbon, H.; Bougis, P. E.; Martin-Eauclaire, M. F.; Pongs, O. *J. Biol. Chem.* **2000**, *275*, 16918–16924.
153. Hanner, M.; Green, B.; Gao, Y. D.; Schmalhofer, W. A.; Matyskiela, M.; Durand, D. J.; Felix, J. P.; Linde, A. R.; Bordallo, C.; Kaczorowski, G. J. et al. *Biochemistry* **2001**, *40*, 11687–11697.
154. Haider, S.; Westhead, D. R.; Davies, L.; Hopkins, P. M.; Boyett, M. R.; Harrison, S. M. *Biophys. J.* **2000**, *78*, 221A.
155. Harris, T.; Shahidullah, M.; Ellingson, J. S.; Covarrubias, M. *J. Biol. Chem.* **2000**, *275*, 4928–4936.
156. Luzhkov, V. B.; Nilsson, J.; Arhem, P.; Aqvist, J. *Biochim. Biophys. Acta* **2003**, *1652*, 35–51.
157. Fermini, B.; Fossa, A. A. *Nat. Rev. Drug Discov.* **2003**, *2*, 439–447.
158. Rajamani, R.; Tounge, B. A.; Li, J.; Reynolds, C. H. *Bioorg. Med. Chem. Lett.* **2005**, *15*, 1737–1741.
159. Cavalli, A.; Poluzzi, E.; De Ponti, F.; Recanatini, M. *J. Med. Chem.* **2002**, *45*, 3844–3853.
160. Osterberg, F.; Aqvist, J. *FEBS Lett.* **2005**, *579*, 2939–2944.
161. Pearlstein, R. A.; Vaz, R. J.; Kang, J.; Chen, X. L.; Preobrazhenskaya, M.; Shchekotikhin, A. E.; Korolev, A. M.; Lysenkova, L. N.; Miroshnikova, O. V.; Hendrix, J. et al. *Bioorg. Med. Chem. Lett.* **2003**, *13*, 1829–1835.

Biographies

Shozeb Haider was born in Aligarh, India and studied at the Aligarh University, India, where he obtained his BSc and MSc in Biochemistry in 1998. He then went on to do MRes in Bioinformatics from the University of Leeds and his PhD in Biophysics from The Institute of Cancer Research, London, in 2002 under the supervision of Prof S Neidle. He spent the next few years as a Postdoctoral Research Fellow (2003–2005) in the laboratory of Prof M S P Sansom at the University of Oxford, working on mammallian ion channels. He took his present position as Research Fellow in Biological and Pharmaceutical Chemistry at The London School of Pharmacy in January 2006. His scientific interests include molecular modeling and simulations of Inward Rectifying Ion Channels and rational drug design based on G-quadruplex DNA.

Mark S P Sansom is a Professor in the Department of Biochemistry at the University of Oxford, where he is director of the Structural Bioinformatics and Computational Biochemistry Unit (website http://sbcb.bioch.ox.ac.uk). He received his DPhil in Molecular Biophysics from Oxford in 1983, and worked for 7 years in the University of Nottingham, before returning to Oxford. His research is concerned with computational studies of membrane proteins with special interests in ion channels, bacterial outer membrane proteins, and the development of methods for simulation and modeling of complex membrane systems.

© 2007 Elsevier Ltd. All Rights Reserved
No part of this publication may be reproduced, stored in any retrieval system or transmitted in any form by any means electronic, electrostatic, magnetic tape, mechanical, photocopying, recording or otherwise, without permission in writing from the publishers

Comprehensive Medicinal Chemistry II
ISBN (set): 0-08-044513-6

ISBN (Volume 4) 0-08-044517-9; pp. 703–724

4.28 Nuclear Hormone Receptors: Insights for Drug Design from Structure and Modeling

J-P Renaud, AliX, Illkirch, France
D Moras and J-M Wurtz, Institut de Génétique et de Biologie Moléculaire et Cellulaire, Illkirch, France

© 2007 Elsevier Ltd. All Rights Reserved.

4.28.1 Introduction

4.28.1.1 The Nuclear Receptor Superfamily

Nuclear receptors (NRs) form the largest family of transcription regulators in metazoans. In mammals, they control most aspects of physiology from development and cellular life to homeostasis. NRs bind to specific sequences on DNA called 'response elements' through a DNA-binding domain (DBD) and are activated by small lipophilic ligands through a ligand-binding domain (LBD). The ligand-dependent modulation of their transcriptional activity makes them attractive drug targets. Indeed, in 2002, among the top 100 world best-selling drugs, eight targeted an NR.[1] There are 48 NRs in humans and 49 in mice, sharing a common organization built around the highly conserved DBD and the LBD, which is well conserved structurally but moderately in sequence, allowing for cognate ligand specificity. Additional regions comprise the highly divergent N- and C-terminal domains responsible for cell and tissue specificities, and a hinge domain between the DBD and the LBD, the flexibility of which is important functionally. Most NRs function as homodimers or as heterodimers with the promiscuous partner retinoid X receptor (RXR), according to a sequence signature.[2]

4.28.1.2 Transcriptional Regulation through Cofactor Recruitment

NRs contain two transactivation functions, a ligand-independent one in the N-terminal domain named activation function 1 (AF-1), often involving phosphorylation site(s), and a ligand-dependent one in the LBD, named AF-2. The LBD is thus the most prominent domain in the regulation of transcription, since it is responsible for ligand recognition and cofactor recruitment. Cofactors include corepressors, coactivators, and the mediator complex. The interplay between them, finely tuned by the ligand, results in the transcriptional status of the receptor (**Figure 1**).

4.28.1.3 The Ligand-Binding Domain Architecture

4.28.1.3.1 The ligand-binding domain fold

The NR LBD fold, first revealed by the structure of apo-RXRα LBD,[3] is an 'antiparallel alpha-helical sandwich' composed of 12 helices and a β-hairpin (sometimes including additional β-strands) distributed over three layers (**Figure 2**).

The first layer contains H1, H2, and H3; the second one H4, H5, the β-hairpin, H8, and H9; and the third one H6, H7, H10, and H11. The position of the terminal helix H12, which contains the AF-2 sequence motif, is controlled by the ligand. The structure of RARγ LBD in complex with all-*trans* retinoic acid revealed the conformation of the active state and suggested a 'mousetrap' mechanism for ligand binding.[4] The superposition of both structures allowed a global alignment of NR LBD sequences which led to the discovery of an NR-specific signature motif and the proposal of a canonical LBD fold.[5] These results have been confirmed by all subsequent structure determinations of NR LBDs (**Table 1**).

The sequence alignment also revealed several hypervariable regions containing the residues lining the ligand-binding pocket (LBP). The specificity of ligand recognition arises from the nature of these residues for each receptor, as well as from local shifts and distortions of the secondary structures containing these residues, namely H3, H5, the β-hairpin, H6, H7, H11, and H12. The functional unit of NRs is mostly a dimer. Homo- and heterodimers share a common spatial organization with a nearly identical interface involving helices H7, H9, and H10 (**Figure 2b**).

4.28.1.3.2 The different ligation states

Structural as well as biochemical and biophysical studies have shown that NR LBDs undergo a conformational change upon agonist binding, making them transcriptionally active. Indeed, in the absence of ligand, the apo-receptor is either inactive (estrogen receptor (ER), RXR) or acts as a strong transcriptional repressor (retinoic acid receptor (RAR), thyroid hormone receptor (TR)), in the latter case through the recruitment of a corepressor complex. The binding of an agonist ligand triggers the remodeling of the LBD surface, involving large movements of helix 12, which allows a switch from an inactive or a corepressor-binding surface to a coactivator-binding surface.[6] The coactivator-binding surface, also called the AF-2 surface, is a hydrophobic groove composed of H3, H4 (static part), and the relocated H12 (dynamic part). The coactivator peptide, called 'NR box,' contains an LXXLL motif that was shown to be essential for coactivator binding.[7] Several crystal structures revealed the helical structure of the LXXLL motif; the three leucines were accommodated in the hydrophobic groove, making close van der Waals contacts with residues contributing to the AF-2 surface.[8–10]

4.28.1.4 Ligand Dependence of Nuclear Receptors

4.28.1.4.1 Constitutive receptors

The transcriptional activity of constitutive receptors is independent of any added ligand. This group includes receptors such as Nurr1 and NGFI-B, whose LBP is filled with hydrophobic side chains,[11,12] and for which the mechanism of cofactor recruitment is unknown at present. Rev-erb4α and Rev-erb4α-related receptor (RVR) probably belong to the same category as their LBP has been predicted to be filled with hydrophobic side chains.[13] Rev-erb and RVR are constitutive repressors acting through the corepressor NCoR, and their ligand independence would be in good

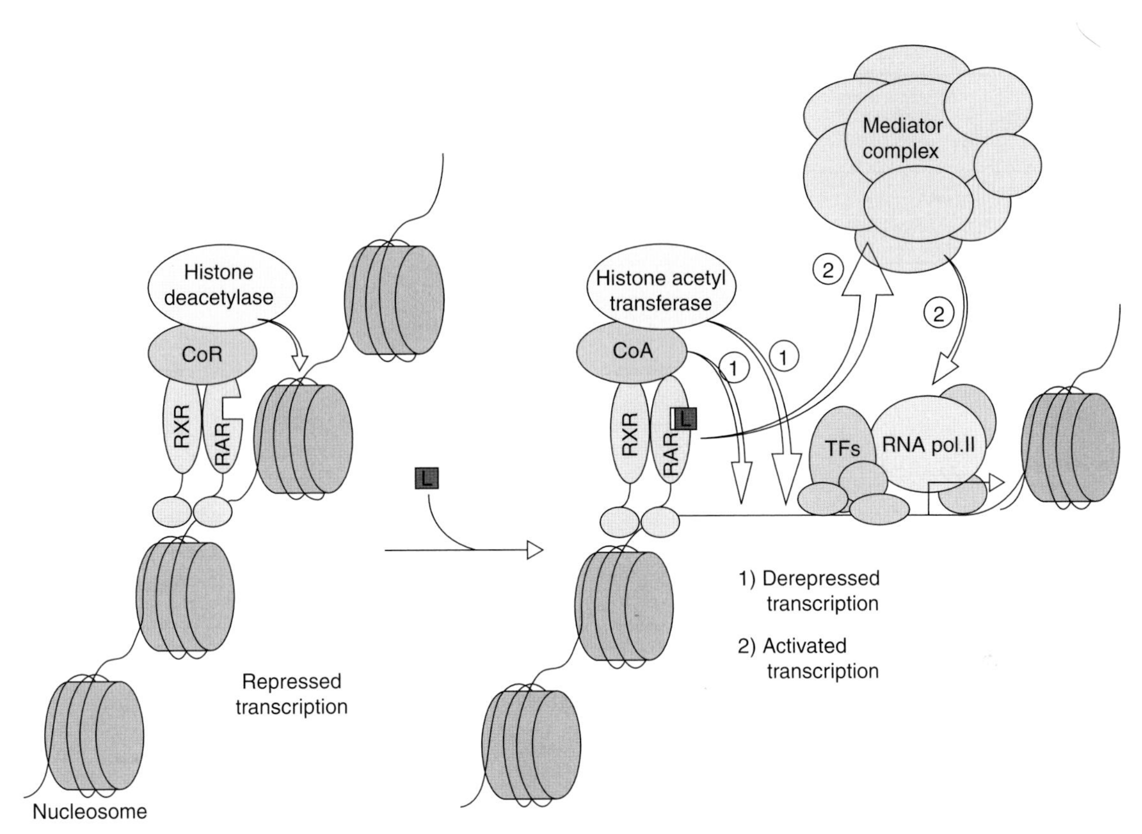

Figure 1 Ligand-dependent nuclear receptor activation. In the absence of ligand, nuclear receptors such as RAR or TR, bound to their response element on DNA as heterodimers with RXR, recruit corepressor complexes that exhibit a histone desacetylase activity which contributes to maintain chromatin in a compact, transcriptionally silent state. Transcriptional activation comprises two steps. First, binding of an agonist ligand (L) changes the LBD surface, precluding corepressor binding but promoting coactivator recruitment. The coactivator complex exhibits a histone acetyl transferase activity contributing to a local chromatin unpacking (derepressed transcription). Second, the coactivator complex is dissociated and the agonist-bound nuclear receptor now recruits a mediator complex which helps to assemble and/or stabilize the pre-initiation complex (activated transcription).

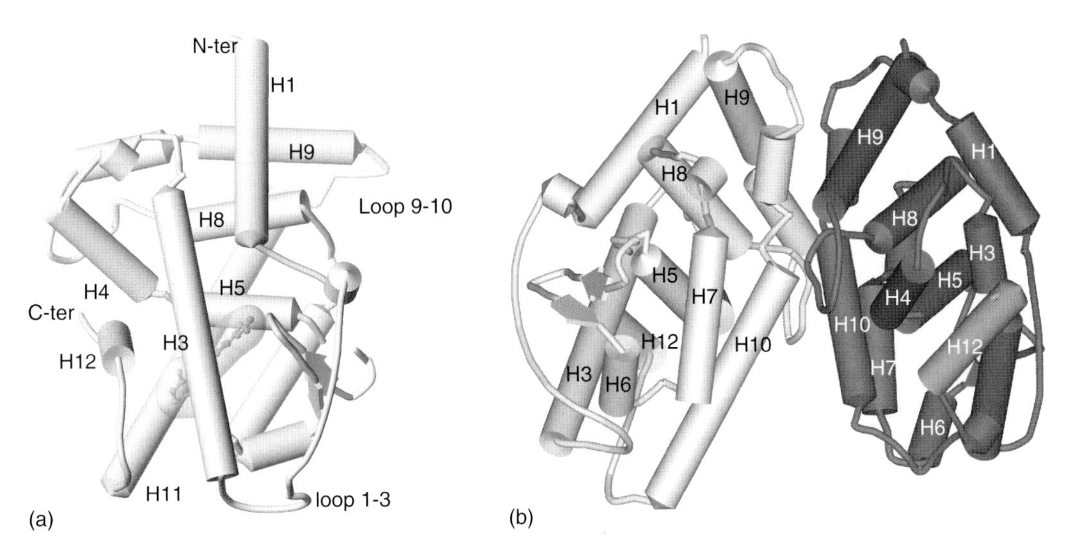

Figure 2 (a) Nuclear receptor LBDs exhibit an alpha-helical sandwich fold. The structure shown is the complex of all-*trans* retinoic acid with human RARγ LBD (Protein Data Bank code 2LBD). (b) An example of LBD heterodimer, formed between (mouse RXRα LBD/oleic acid) and (human RARα LBD/BMS614) (Protein Data Bank code 1DKF).

Table 1 Natural ligands for human nuclear receptors

Systematic nomenclature	*Common name*	*Known natural ligands*	*Other natural ligands*	*LBD structure available*	*Seminal/main references*
NR1A1	TRα	Thyroid hormones		+	Wagner *et al.*, 1995[143]
NR1A2	TRβ	Thyroid hormones		+	Darimont *et al.*, 1998[9]
NR1B1	RARα	All-*trans* retinoic acid		+	Bourguet *et al.*, 2000[144]
NR1B2	RARβ	All-*trans* retinoic acid		+	Germain *et al.*, 2004[33]
NR1B3	RARγ	All-*trans* retinoic acid		+	Renaud, 1995[4]
NR1C1	PPARα	Fatty acids	Eicosanoids (e.g., leukotriene B_4)	+	Cronet *et al.*, 2001[66]
NR1C2	PPARβ	Fatty acids	Eicosanoids (e.g., prostacyclin)	+	Xu *et al.*, 1999[65]
NR1C3	PPARγ	Fatty acids	Eicosanoids (e.g., 15-deoxy-Δ12,14-prostaglandin J_2)	+	Nolte *et al.*, 1998[10]
NR1D1	Rev-erbAα	–		–	–
NR1D2	Rev-erbAβ	–		–	–
NR1F1	RORα	Cholesterol		+	Kallen *et al.*, 2002[145]
NR1F2	RORβ	All-*trans* retinoic acid ?		+	Stehlin *et al.*, 2001,[17] Stehlin-Gaon *et al.*, 2003[146]
NR1F3	RORγ	All-*trans* retinoic acid ?		–	–
NR1H2	LXRβ	Oxysterols		+	Williams *et al.*, 2003,[76] Färnegardh *et al.*, 2003[21]
NR1H3	LXRα	Oxysterols		+	Svensson *et al.*, 2003[78]
NR1H4	FXR	Bile acids	Farnesoids	+	Downes *et al.*, 2003,[147] Mi *et al.*, 2003[148]
NR1I1	VDR	1,24 dihydroxyvitamin D_3	Lithocholic acid	+	Rochel *et al.*, 2000[38]
NR1I2	PXR	5β-pregnane-3, 20-dione	Pregnenolone and its metabolites, lithocholic acid, bile acid precursors	+	Watkins *et al.*, 2001[149]
NR1I3	CAR	Androstanol		+	Shan *et al.*, 2004,[150] Xu *et al.*, 2004[23]
NR2A1	HNF4α	Endogenous fatty acids		+	Dhe-Paganon *et al.*, 2002[14]
NR2A2	HNF4γ	Endogenous fatty acids		+	Wisely *et al.*, 2002[15]
NR2B1	RXRα	9-*cis* retinoic acid ?	Docosahexaenoic acid ?	+	Bourguet *et al.*, 1995,[3] Egea *et al.*, 2000[19]

Table 1 Continued

Systematic nomenclature	*Common name*	*Known natural ligands*	*Other natural ligands*	*LBD structure available*	*Seminal/main references*
NR2B2	RXRβ	9-*cis* retinoic acid ?		+	Love *et al.*, 2002[151]
NR2B3	RXRγ	9-*cis* retinoic acid ?		–	–
NR2C1	TR2	–		–	–
NR2C2	TR4	–		–	–
NR2E1	TLX	–		–	–
NR2E3	PNR	–		–	–
NR2F1	COUP-TFα	–		–	–
NR2F2	COUP-TFβ	–		–	–
NR2F6	COUP-TFγ	–		–	–
NR3A1	ERα	Estradiol		+	Brzozowski *et al.*, 1997[42]
NR3A2	ERβ	Estradiol		+	Pike *et al.*, 1999[45]
NR3B1	ERRα	–		+	Kallen *et al.*, 2004[54]
NR3B2	ERRβ	–		–	–
NR3B3	ERRγ	–		+	Greschik *et al.*, 2002[52]
NR3C1	GR	Cortisol		+	Bledsoe *et al.*, 2002[152]
NR3C2	MR	Aldosterone		+	Fagart *et al.*, 2005,[153] Li *et al.*, 2005,[154] Bledsoe *et al.*, 2005[155]
NR3C3	PR	Progesterone		+	Williams and Sigler, 1998[156]
NR3C4	AR	Dihydrotestosterone		+	Matias *et al.*, 2000[157]
NR4A1	NGFI-B	–		+	Flaig *et al.*, 2005[12]
NR4A2	Nurr1	–		+	Wang *et al.*, 2003[11]
NR4A3	NOR1	–		–	–
NR5A1	SF1	Phospholipids ?		+	Krylova *et al.*, 2005,[158] Li *et al.*, 2005[159]
NR5A2	LRH-1	Phospholipids ?		+	Sablin *et al.*, 2003,[160] Krylova *et al.*, 2005[158]
NR6A1	GCNF	–		–	–
NR0B1	DAX-1	–		–	–
NR0B2	SHP	–		–	–

? Not proven as a physiological ligand.

agreement with their lack of H12, since the role of a ligand is to trigger a conformational change of H12. HNF4 remains a controversial case. The crystal structures of HNF4α[14] and γ[15] LBDs have revealed the presence of endogenous fatty acids in the LBP. These fatty acids apparently act as LBD-stabilizing cofactors that cannot be displaced without denaturing the fold. However, other studies suggest that HNF4 is ligand-dependent.[16]

4.28.1.4.2 Ligand-dependent receptors

Ligand-dependent receptors belong to two categories: those activated by specific, well-defined ligands and those activated by numerous, loosely-defined ligands.

4.28.1.4.2.1 Nuclear receptors with well-defined ligands

This group comprises RARs, TRs, the vitamin D receptor (VDR), and the steroid receptors (SRs). Their ligands all exhibit a very high affinity (most of the time subnanomolar) and tightly regulated cellular levels. Accordingly, these cognate endogenous ligands are very potent biologically active molecules. At the transcriptional level, these receptors act as 0/1 switches according to the presence or absence of ligand. To achieve this stringent transcriptional control, their LBP is constrained and well suited to the size and chemistry of the cognate ligands.

4.28.1.4.2.2 Nuclear receptors with loosely defined ligands

This group includes peroxisome proliferator activated receptors (PPARs), liver X receptors (LXRs), the farnesoid X receptor (FXR), the liver receptor homolog-1 (LRH-1), the steroidogenic factor 1 (SF1), the constitutive androstane receptor (CAR), and the pregnane X receptor (PXR). Receptors of this group bind families of less-specific, lower-affinity ligands that can be either endogenous molecules and/or nutrients (fatty acid and their derivatives for PPARs, oxysterols for LXRs, bile acids for FXR) and xenobiotics such as drugs (CAR and PXR). Their ability to respond to global levels of related ligands suggests they function as 'sensors.' They trigger a transcriptional response proportional to the concentrations of ligands present in the cell, more like a cursor than a switch. For instance, when fatty acid levels are high, PPARα and γ induce genes involved in fatty acid catabolism and storage, respectively. Receptors of this group possess large, flexible LBPs allowing the promiscuous binding of chemically diverse ligands.

4.28.1.5 Two Functional Classifications

4.28.1.5.1 Homodimers versus heterodimers

A multiple alignment of NR LBD sequences revealed two sequence motifs that partition the superfamily into two classes related to their oligomeric behavior (**Figure 3**).[2] Class I encompasses all homodimer-forming NRs. Members of this class exhibit a specific conserved path that connects the cofactor-binding surface to the dimer interface. An invariant tryptophan in the LBP complements the class I signature motif. Two conserved salt bridges constitute the signature of class II, containing all NRs known to function as heterodimers with RXR or ultraspiracle (USP). These salt bridges connect loops 8–9 to loop 9–10. RXR, the ubiquitous partner of all class II NRs, exhibits the class I-specific motifs. Site-directed mutagenesis confirms the functional importance of these residues for the transcriptional activity and/or the dimerization process.

4.28.1.5.2 Activity versus ligand dependence

NRs ligands can be classified according to their transcriptional effect (**Figure 4**): (1) total agonists trigger a full transcriptional response; (2) weak agonists elicit a partial transcriptional response due to a weak affinity for the receptor; (3) partial agonists do not stabilize the active conformation of H12, leading to an equilibrium between the active and inactive (nonproductive) conformations; (4) antagonists that prevent the binding of coactivators; (5) antagonists that recruit corepressors; (6) antagonists that prevent the binding of corepressors in the case of constitutive transcriptional repressors; (7) inverse agonists which deactivate the receptor in the case of constitutive transcriptional activators. Note that, in some cases, fortuitous ligands can bind without affecting the transcriptional activity of the receptor.[15,17,18]

4.28.2 The Ligand-Binding Mechanisms

Ligand binding occurs through an induced-fit mechanism that impacts the size and shape of the LBP and the conformation of the ligand.

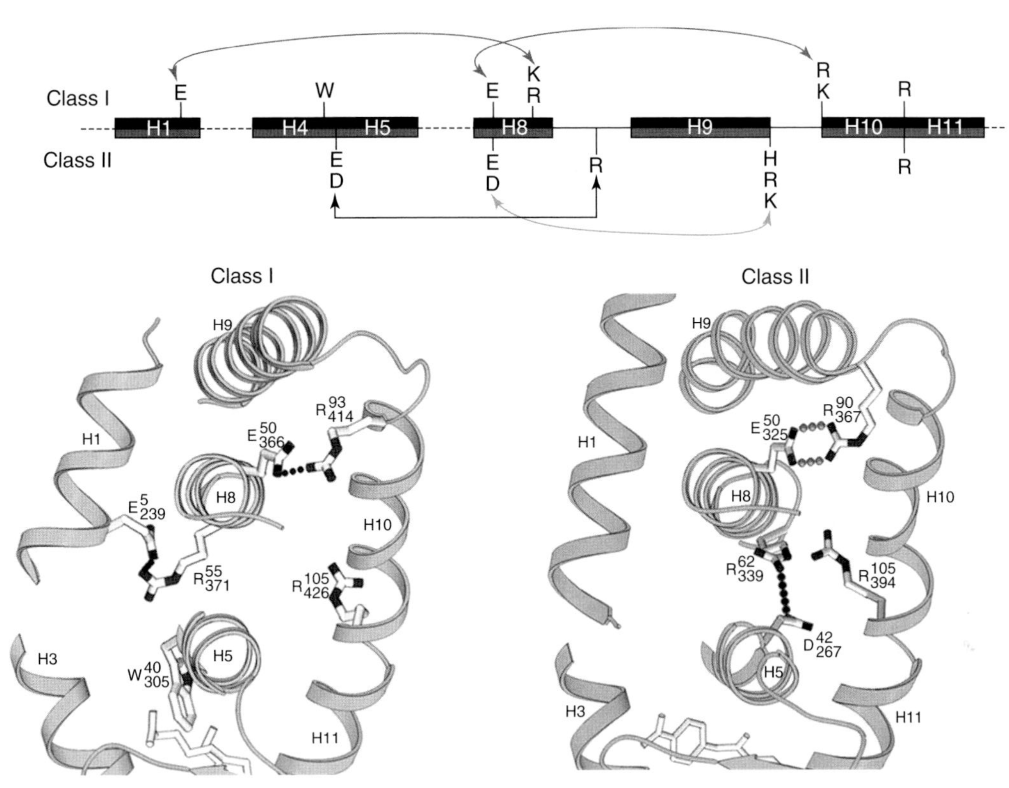

Figure 3 Top: secondary structure diagram showing nuclear receptor class-specific signature motifs. Arrows indicate the salt bridges found in the 3D structures. Bottom left: close-up view of the class I-specific residues of hRXRα, including the two salt bridges forming the class I communication pathway. Bottom right: close-up view of the two class II-specific salt bridges of hRARα. The residue numbers refer to the full-length receptors (subscripts) or an arbitrary scale (superscripts).

4.28.2.1 Protein Plasticity

The necessity of a large conformational change upon ligand binding is obvious when considering the structure of any holo structure. In most, if not all, complexes the ligand is buried inside the LBD and shielded from the solvent. The existence of a conformational change between an 'open' apo form and a compact 'closed' holo form has been highlighted by the direct comparison between the x-ray structures of apo-RXRα LBD[3] and RXRα LBD bound to 9-*cis* retinoic acid (9C-RA)[19] showing the same receptor in two extreme conformational states. Similar compact holo conformations have been observed in several heterodimers, the only variable module being H12, which has been observed in the canonical agonist position or near/in the coactivator binding cleft in the so-called antagonist location. The crystal structure of RARγ LBD bound to its natural ligand all-*trans* retinoic acid (ATRA) suggested the concept of a 'mousetrap' binding mechanism, where the hydrophobic ligand would enter the LBP and anchor itself through its carboxylate moiety, while triggering the bending of H3 and the flipping of the Ω-loop connecting H1 to H3.[4] This mechanism may hold for most receptors, though the precise conformational changes may differ. The available structural data stress the crucial role of H3 in tightly embracing the ligand.

An interesting feature of some NRs is their property to adapt their LBPs to different ligands while maintaining their functional structure, i.e., they are transcriptionally active. Several cases illustrate this situation. The most extreme one is that of the ecdysone receptor (EcR) where two totally different, only partially overlapping LBPs could be observed with two different agonists.[20] Other significant variations of the LBP shape in order to fit different ligands have been shown with LXR[21] and VDR.[22]

4.28.2.2 Ligand Adaptability

A large number of natural or synthetic ligands, i.e., the steroid family, are rigid or marginally flexible. Selectivity is thus mainly due to the anchoring capability of the receptor LBPs. On the opposite side of the flexibility range, molecules like fatty acids can easily adapt to different hydrophobic LBPs (i.e. RXRα,[3,23] the retinoic acid receptor-related orphan

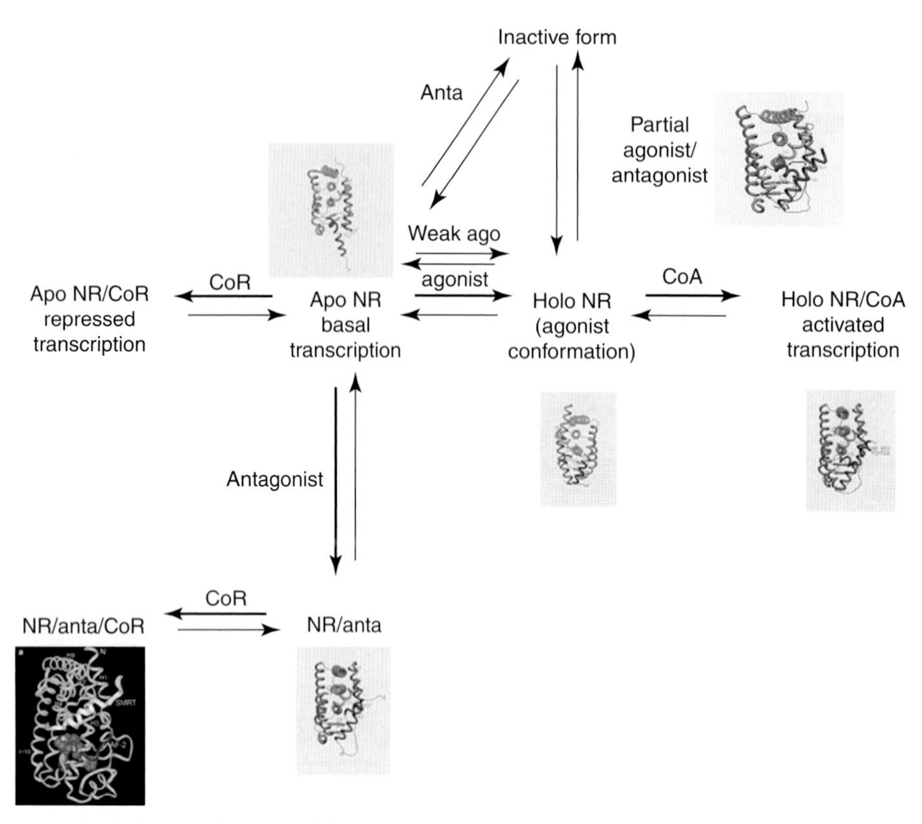

Figure 4 Nuclear receptor transcriptional activity versus ligand dependence.

receptor β (RORβ)[17], and the hepatocyte nuclear factor-4α or γ (HNF4α,γ)[14,15]). Between numerous ligands are molecules that can adapt, but with a limited number of possible conformations. Among these ligands, one can mention retinoids, vitamin D analogs, and most ligands of receptors linked to metabolic pathways (PPARs, FXR, PXR, LXR ...). For this last category of ligand, the entropic cost of ligand adaptability is part of its efficiency, but the benefit of a better fit can be enormous.

Retinoids offer a good illustration of the mutual induced-fit mechanism: whereas RXR exclusively binds 9C-RA, RAR binds both ATRA and 9C-RA stereoisomers. The comparison of the crystal structures of ATRA and 9C-RA bound to RARγ[4,24–26] addresses the question of recognition of different ligands by the same receptor. The intrinsic flexibility of retinoic acid isomers allows them to adapt a structurally unique LBP. In RXR, 9C-RA exhibits a pronounced bend, whereas in RAR its shape is closer to that observed for ATRA (**Figure 5a, b**). In contrast, the shape of the RXR LBP selects the 9-*cis* isomer but excludes the all-*trans* isomer due to its flexure limit.[27]

4.28.3 Structure-Based Drug Design for Nuclear Receptors

Natural ligands of NRs are chemically very diverse, including steroids, retinoids, amino acid derivatives (thyroid hormones), fatty acids and fatty acid derivatives (prostaglandins, leukotrienes), bile acids, and phospholipids (**Table 1**). Even heme was shown very recently to be the ligand of *Drosophila* E75.[28] Initially ligands were not known for NRs that were either cloned by homology or found in the human genome, hence these newly identified receptors were called orphans. For many of them, ligands were subsequently found, most of the time by high-throughput screening. Orphan NRs for which ligands could be identified are called 'adopted,' for instance the LXRs (oxysterols) and FXR (bile acids).

Although NRs represent only ~2% of all current drug targets (**Figure 3** in [29]), eight among the 100 top prescription drugs targeted NRs and their worldwide sales amounted to over US$9 billion in 2002.[1] Examples of currently marketed drugs targeting various NRs are given in **Table 2**.

Intense efforts are being undertaken by both academic and industrial groups on classical receptors (validated targets) and adopted receptors (promising targets). For classical receptors, the aim is to design drugs with reduced side effects. For adopted receptors, the first goal is to validate them as therapeutic targets (for instance, LXRs for

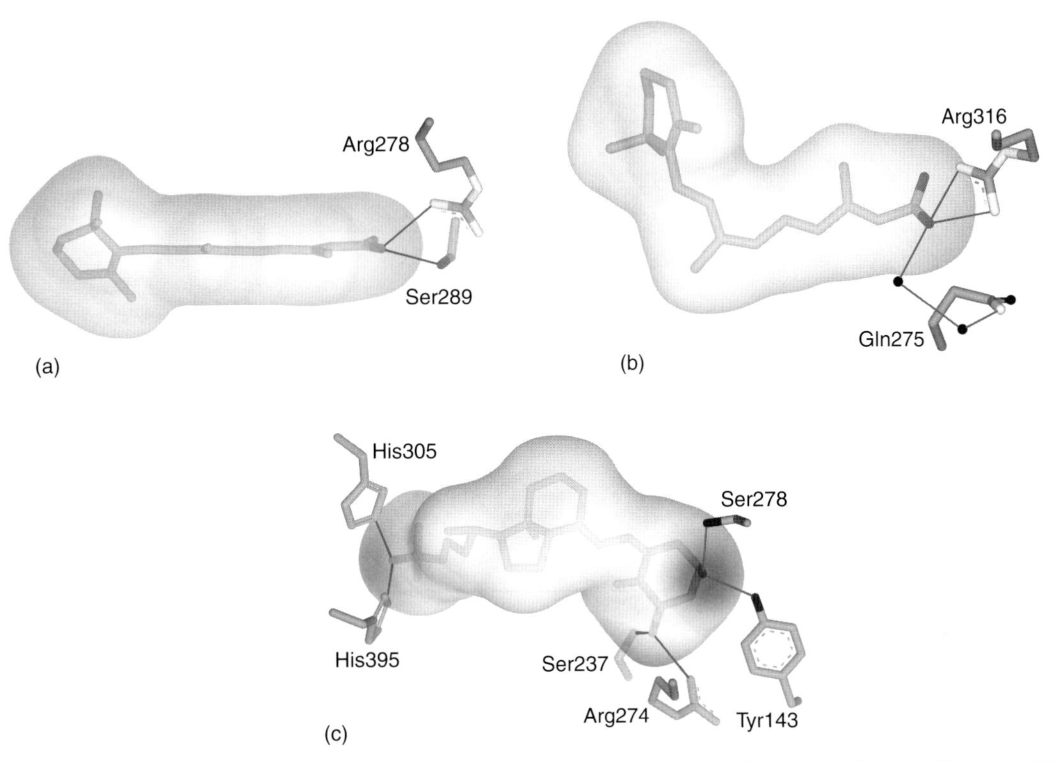

Figure 5 Nuclear receptor ligand-binding pocket (LBP). (a) LBP in the complex of all-*trans* retinoic acid with human RARγ (2LBD). (b) LBP in the complex of 9-*cis* retinoic acid with human RXRα (1FBY). (c) LBP in the complex of 1,25 $(OH)_2$ vitamin D_3 with human VDR (1DB1). The key residues contacting the ligands are shown for each ligand. The molecular surface is shown around the ligand. Thin lines represent hydrogen bonds between polar groups.

cardiovascular diseases, inflammation, diabetes, and neurodegenerative diseases, FXR for bile acid- and lipid-related diseases, and RORα for metabolic diseases).

Unwanted side effects of drugs targeting NRs arise from the receptor pleiotropic effects. Quite often an optimized drug will be a selective NR modulator, which is a partial agonist/antagonist that either recruits selectively a given cofactor or recruits other cofactors too weakly to activate other pathways. The term was coined first in the case of estrogen receptor (selective estrogen receptor modulators or SERMs).[30]

4.28.3.1 Retinoic Acid Receptors and Retinoid X Receptors

Retinoids are used in the treatment of various skin diseases, including psoriasis and acne, and in the treatment or chemoprevention of cancer, such as acute promyelocytic leukemia and skin, cervical, and breast cancer.[31] RARβ gene is frequently deleted or its expression is epigenetically silenced during cancer progression and RARβ re-expression can restore retinoic acid-mediated growth control, suggesting that the anticancer action of retinoids is mediated by RARβ. RARβ has been viewed as a tumor suppressor.

Numerous synthetic agonist and antagonist retinoids differentiate RXR from RAR or are selective for the RARα, β, or γ isotypes. A sequence alignment of RARα, β, and γ shows that all but three residues in the LBP are conserved. These divergent residues are the most important discriminatory elements for synthetic retinoid selectivity (**Figure 6**).[4,32]

Existing isotype-selective ligands dissociate mostly RARα from RARβ/γ by exploiting the presence of RARαS232, which can establish hydrogen bonds with suitable ligands; this was predicted for ligands harboring an amide group, such as Am80 and Am580.[24] The crystal structures of a RARβ LBD complexed to β-selective ligands[33] clarify the selectivity potential of the I263$\rightarrow$M272 replacement in RARγ. The main consequence of methionine is to allow specific H bonds as, for example, in the case of BMS270394 (see below). The side-chain orientation of I263 opens a cavity that is closed in RARγ. Ligands with a bulky substituent require such a cavity to accommodate acquire β-selectivity.

The different activity of RARγ-specific enantiomers has been described for two retinoids: BMS270394 and the inactive BMS270395.[25,34,35] Synthesis and characterization of the individual enantiomers showed that the biological

Table 2 Examples of marketed nuclear receptor-targeting drugs

Target receptor	*Class of drug*	*Example*	*Therapeutic indications*
TR	Thyroid hormone and analogs	Synthroid	Hypothyroidism, thyroid cancer, goiter
RAR	Retinoids	Vesanoid	Acute promyelocytic leukemia
RAR	Retinoids	Tazorac, Accutane	Psoriasis, acne
PPARα	Fibrates	Lipanthyl, gemfibrozil	Dyslipidemia
PPARγ	Thiazolidinediones	Actos, Avandia	Type 2 diabetes
VDR	1α,25 dihydroxyvitamin D_3 and analogs	Rocaltrol	Vitamin D-resistant rickets, osteoporosis, renal osteodystrophy
VDR	1α,25 dihydroxyvitamin D_3 and analogs	Dovonex	Psoriasis
RXR	Rexinoids	Targretin	Cutaneous T-cell lymphoma
ER	Estrogens	Premarin	Osteoporosis, menopausal symptoms
ER	Selective estrogen receptor modulators	Nolvadex	Breast cancer
ER + PR	Estrogens + progestins	Lunelle	Contraception
PR	Progestins	Provera	Menstrual cycle disorders
GR	Glucocorticoids	Advair, Pulmicort	Asthma
MR	Aldosterone antagonists	Aldactone, Inspra	Hypertension, heart failure
AR	Androgens	AndroGel	Hypogonadism
AR	Antiandrogens	Casodex, Flutamide	Prostate cancer

activity resides in the *R*-enantiomer (BMS270394), whereas the *S*-enantiomer (BMS270395) shows no measurable binding affinity and transactivation. Without an experimental structure, one would have probably predicted that the *S*-enantiomer does not bind to RARγ.

The low affinity can be explained by the unfavorable contacts between the ligand and residues of the pocket and by the energetically unfavorable conformation adopted in the complex. The energy contribution of the hydrogen bond between the hydroxyl moiety of the ligands and the Met-272 sulfur atom allows partial compensation of the energetically unfavorable conformation of BMS270395. It leads to the unexpected result that the position of the oxygen atom of the hydroxyl group is almost conserved in both enantiomer complexes. Moreover, this hydrogen bond is the molecular basis for the RARγ selectivity of BMS270394, enhanced by the presence of the fluorine atom.

In all cases, antagonists are characterized by the presence of bulky substitutions that interfere with the holo conformation of H12.

4.28.3.2 Vitamin D Receptor

VDR and its cognate ligand 1α,25-dihydroxyvitamin D_3 (1α,25$(OH)_2D_3$), the active form of vitamin D_3, are important regulators of calcium homeostasis and bone metabolism.[36] In addition, 1α,25$(OH)_2D_3$ exerts antiproliferative and proapoptotic effects on transformed mammary cells such as MCF7 cells.[37] These antitumoral features show the potential of VDR ligands in adjuvant therapy of breast cancer.

Active vitamin D exhibits a chair B conformation with the 19-methylene 'up' and the 1α-OH and 3β-OH groups in an equatorial and axial orientation, respectively.[38] A chair A conformation of the A-ring would disrupt the hydrogen bonds formed by the hydroxyl group and the protein. The conjugated triene system connecting the A-ring to the C- and D-rings accommodates an almost *trans* conformation, with the C6–C7 bond exhibiting a torsion angle of $-149°$ that deviates significantly from the planar geometry which results in the curved shape of the ligand bound to the receptor.

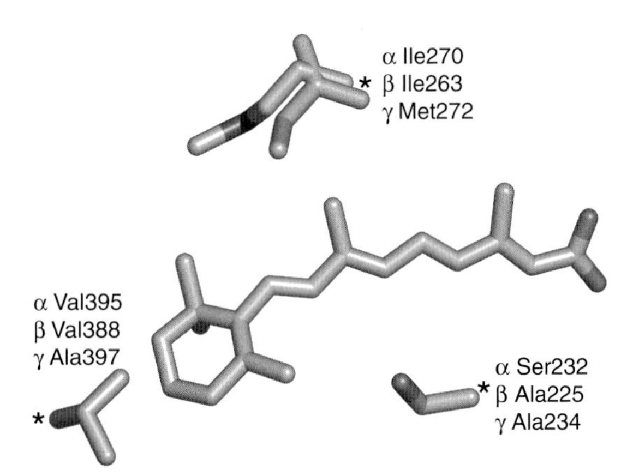

Figure 6 RARα, β, γ isotype selectivity. Ligand-binding pocket residues differing between the RARα, β, and γ isotypes illustrated with the (hRARγ LBD/all-*trans* retinoic acid) complex (2LBD). The residues are drawn in the stick representation with carbon atoms colored in light gray, the sulfur atom in dark gray, and the oxygen atoms in medium gray. The residues at the three positions are given for the three isotypes.

The deviation explains the lack of biological activity of the analogs with a *trans* or a *cis* conformation of the C6–C7 bond.[39] The ligand-binding cavity of the hVDRΔ is large (697 Å^3), with the ligand occupying only 56% of this volume. A channel of water molecules near position 2 of the A-ring makes an additional space that can accommodate ligands with a methyl group at position 2. The fourfold increase in binding affinity of the 2α-methyl analog is in agreement with this observation (Hourai *et al.*, unpublished results). Additional space around the aliphatic chain is also observed.

Hypercalcemia is an undesirable side effect of pharmacological doses of 1α,25(OH)$_2$D$_3$. Several promising vitamin D$_3$ analogs with potent antiproliferative but reduced side effects have been developed.[37] Among these, the 20-epi compounds have attracted considerable attention. These compounds activate VDR and exert antiproliferative effects at concentrations at least 100-fold lower than the natural ligand. Limited proteolytic digestion studies suggested that the binding of a 20-epi compound to VDR induces a distinct LBD conformation which selectively modulates the interaction with cofactors.[40] Surprisingly, comparison of the crystal structures of the VDR LBD bound to 1α,25(OH)$_2$D$_3$ or the 20-epi analogs MC1288 or KH1060 revealed identical LBD conformations.[38,41] As already observed for RARs and RXRs, the ligands adapt to the LBP of VDR. Superimposition of 1α,25(OH)$_2$D$_3$, MC1288, and KH1060 as positioned in the VDR LBP showed that the ligands are tighly bound around the A-, seco B-, C-, and D-ring, but are less restrained around the aliphatic chains. As a consequence, the aliphatic chains follow different paths to adapt to the VDR pocket, resulting in different numbers of protein–ligand contacts and more or less favorable ligand conformations. It appears that these differences result in higher complex stability in the case of the 20-epi compounds, which may explain the longer half-life of VDR–20-epi compound complexes and account for their increased transcriptional potency.[41]

4.28.3.3 Estrogen Receptors

The structure of ERα LBD in complex with estradiol (E$_2$) showed the LBD homodimer in the agonist conformation and revealed a closed LBP.[42] E$_2$ recognition is mediated through a combination of a few specific hydrogen bonds to the two hydroxyl groups and numerous van der Waals contacts, especially with the A and D rings. E$_2$ occupies only 54% of the LBP volume, with empty spaces around the B and C rings that can accommodate various hydrophobic groups, allowing the binding of many E$_2$ analogs, including nonsteroidal molecules (**Figure 7a–c**). Indeed, the requirement for an aromatic ring in high-affinity ligands can be explained by the pincerlike arrangement around the A ring. Binding of the antagonist raloxifene induces a large conformational change; its bulky side chain extension prevents H12 from adopting the agonist conformation.[42]

A deeper understanding of the mechanism of antagonism for this ligand was provided by the comparison of the structure of the ERα LBD in complex with the agonist diethylstilbestrol (DES) and an LXXLL motif-containing peptide from the coactivator GRIP1 (glucocorticoid receptor interacting protein-1) with that of the ERα LBD in complex with the antagonist 4-hydroxytamoxifen (4OHT).[8] It appears that the role of the agonist ligand is to assemble the AF-2 surface composed of H3 and H4 (the AF-2 static part) and H12 (the AF-2 dynamic part) which is repositioned upon ligand binding, allowing the coactivator peptide to bind to the hydrophobic cleft thus

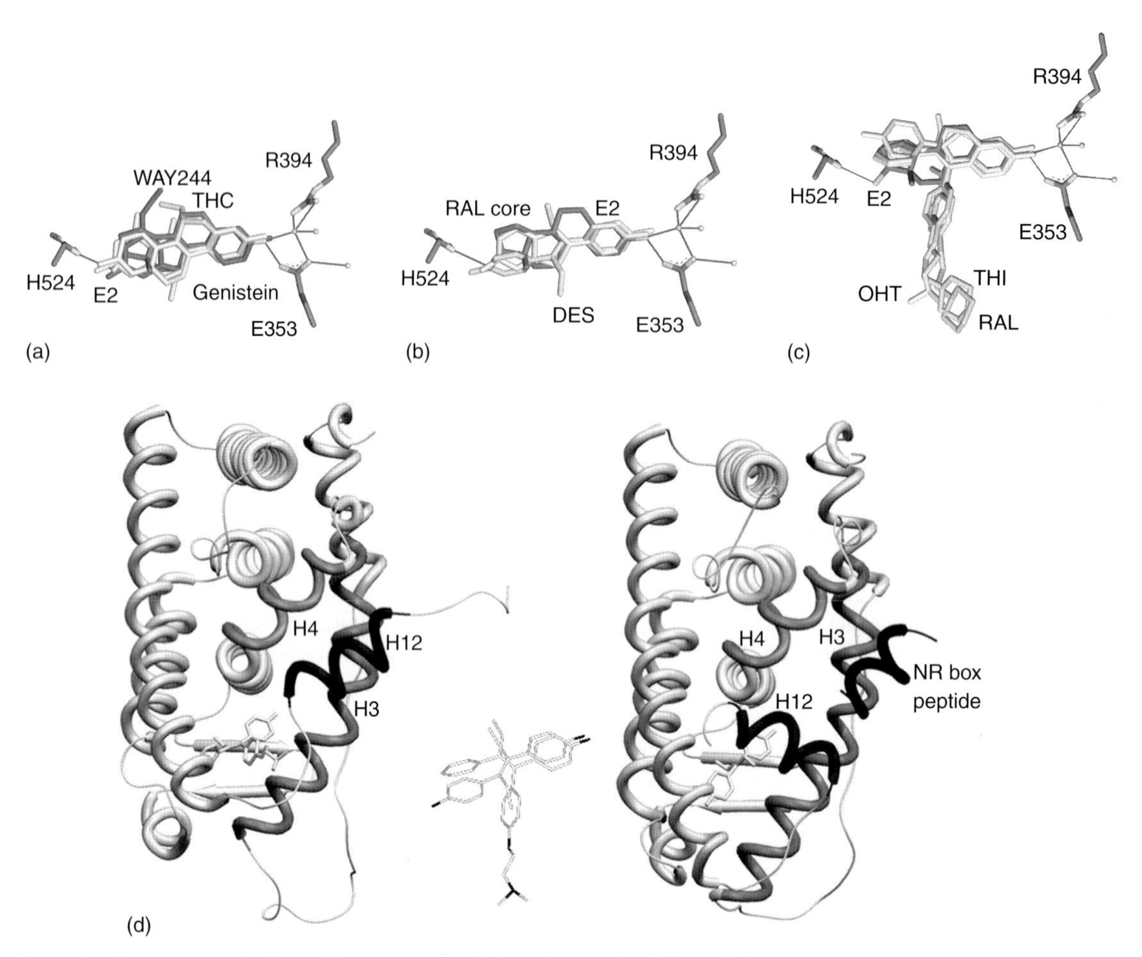

Figure 7 Estrogen agonist (a and b) or antagonist (c) ligands bound to ERα or ERβ. The orientation of the ligands is obtained by superimposing the protein Cα traces. Estradiol (E_2) is used as a reference in all three pictures. (a) Estradiol, genistein, THC ((R,R)-5,11-*cis*-diethyl-5,6,11,12-tetrahydrochrysene-2,8-diol) and WAY-244 are superimposed ((hERα LBD/E_2): 1ERE; (hERβ LBD/genistein): 1QKM; (hERα LBD/THC): 12LI; (hERα LBD/WAY-244): 1X7E). (b) Estradiol, raloxifene core, and diethylstilbestrol (DES) are superimposed ((hERα LBD/DES/GRIP1 peptide): 3ERD; (hERα LBD/RAL core): 1GWQ). (c) Estradiol, 4-hydroxytamoxifen (OHT), raloxifene (RAL), and tetrahydroisochiolin (THI) are superimposed ((hERα LBD/4OHT): 3ERT; (rERβ LBD/RAL): 1QKN; (hERα LBD/THI): 1UOM). (d) Comparison of the (hERα LBD/4OHT) complex (left; 3ERT) and the (hERα LBD/DES/GRIP1 peptide) complex (right; 3ERD).

generated (**Figure 7d**, right). Binding of the antagonist 4OHT (similar but not identical to the raloxifene mode of binding) prevents H12 from adopting the agonist conformation and properly completing the hydrophobic, coactivator-binding AF-2 surface. Instead, H12 relocates in the uncomplete cleft, occupying the position of the coactivator peptide in the agonist-bound complex (**Figure 7d**, left). This alternative position of H12 is permitted by the flexible character of the H11-loop 11–12 region (indeed, H11 is unwound by one turn in the 4OHT complex) and by the LXXLL-like character of the ERα H12 sequence. In this perturbed conformation, H12 hinders the binding of a coactivator, explaining the mechanism of this class of antagonists. The mixed partial agonist–partial antagonist character of antiestrogens such as 4OHT comes from the fact that they inhibit only the AF-2 function, explaining the observed promoter and tissue specificities.[43] On the other hand, pure antiestrogen such as ICI 164,384 inhibits transactivation through both the AF-1 and AF-2 functions. The crystal structure of the (rERβ LBD–ICI 164,384) complex[44] revealed the mode of action of a pure antiestrogen: the protruding of the bulky side-chain substituent out of the LBP sterically prevents the binding of H12 either in the agonist conformation or in the 'AF-2 antagonist' conformation. H12 is completely dissociated from the LBD core and is probably disordered since it is not visible in the electronic density.

Genistein represents another class of partial agonists/antagonists. These are smaller ligands that do not act through steric hindrance of H12 agonist position but via a suppression of some interactions which lock H12 in the active conformation. In the crystal structure of the (hERβ LBD–genistein) complex,[45] H12 is found in an antagonist conformation. However, the genistein-bound hERβ LBD should be able to sample the agonist conformation from time to time, since genistein is a partial ERβ agonist. Indeed, H12 is found in agonist conformations when the genistein-bound hERβ LBD is cocrystallized with a coactivator peptide which brings the additional stabilization needed to lock H12 in the transcriptionally active conformation.[46] At the extreme, 'passive antagonists' completely shift the equilibrium toward the inactive conformation and thus act as pure antagonists, such as 5,11-*cis*-diethyl-5,6,11,12-tetrahydrochrysene-2,8-diol for hERβ.[47]

Since the second estrogen receptor, ERβ, was cloned,[48] intense efforts have been dedicated to the design of ligands with high α or β specificity. Several groups have used the comparison of the ERα and ERβLBD crystal structures to understand the molecular basis of isotype discrimination.[46,47,49,50] There are two obvious substitutions in the LBP – ERα Leu384/ERβ Met336 and ERα Met421/ERβ Ile373 – but the observed differences cannot be explained on this basis alone. The contribution of regions farther from the ligand was thus investigated, pointing out the role of long-range interactions in determining the ligand selectivity of the two isotypes.[51]

4.28.3.4 Estrogen Receptor-Related Receptors

The first crystal structure was obtained for ERRγ, in which the LBD exhibited a stable agonist conformation in the absence of ligand, in good agreement with the constitutive activity reported by several groups.[52] The structure revealed a residual LBP smaller than that of the closely related ER but still significant ($220\,\text{Å}^3$), leaving open the possibility of the existence of naturally occurring ligands. But what could then be the role of such ligands, since the receptor is already transcriptionally active in the absence of any ligand? One can think of two possibilities: superagonists that further stabilize the active conformation of the LBD by creating or strengthening a network of interactions with H12 or antagonists that interfere with H12 positioning through steric hindrance. In line with this last hypothesis, some synthetic ligands such as DES and 4OHT were reported as ERR deactivators (inverse agonists). Docking of these ligands in the apo structure led to the proposal of a mechanism for ligand deactivation of ERRγ through a rotation of F435 required to accommodate the ligands.[52] This rotation provokes a clash with H12 that has to move from the transcriptionally active conformation. The crystal structures of ERRγ LBD in complex with DES and 4OHT confirmed the proposed mechanism.[53] This phenylalanine is a leucine in ERα. A single mutation F435L suppresses the steric clash and restores the agonist character of DES in ERRγ.[53]

The crystal structure of the ERRα LBD in complex with a peptide from the PPAR gamma coactivator 1 alpha (PGC-1α) was recently reported, also bringing evidence of ligand-independent transcriptional activation by ERRα, since the LBD is in agonist conformation in the absence of ligand and the LBP is almost filled up with side chains.[54] A recent study permitted the identification of a selective inverse agonist for ERRα, XCT790 (**Figure 8**).[55] ERRα is a potential new target for the treatment of type 2 diabetes, obesity, and other metabolic diseases. The NR coactivator PGC-1α has been identified as a key regulator of energy metabolism in the liver and in skeletal muscle through the control of genes involved in oxidative phosphorylation. PGC-1α stimulates ERRα expression and acts synergistically with ERRα to upregulate many of these genes. Thus ERRα ligands could restore oxidative phosphorylation, which is impaired in type 2 diabetes. Indeed, XCT790 was shown to regulate PGC-1α signaling, validating ERRα as a promising new target for the treatment of metabolic disorders.[56]

4.28.3.5 Peroxisome Proliferator-Activated Receptors

PPARs were named after the initial characterization of PPARα as an NR activated by peroxisome proliferators.[57] There are three PPAR isotypes in vertebrates: α, β (also called δ), and γ, each displaying a distinct expression pattern. PPARα is abundantly expressed in the liver, PPARγ in adipose tissue, intestine, and spleen; PPARβ expression is more ubiquitous. All three PPARs bind with varying affinities and are activated by naturally occurring fatty acid and fatty acid

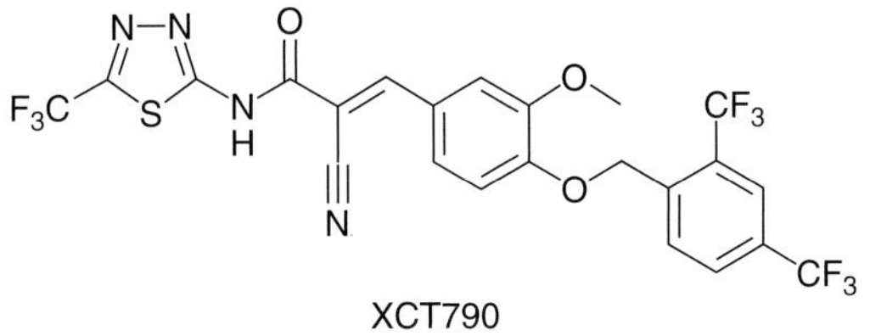

XCT790

Figure 8 ERRα inverse agonist XCT790.

derivatives, and were thus proposed to function as lipid sensors.[58–60] PPARs bind PPREs in the promoter region of target genes as heterodimers with RXR and regulate the expression of numerous genes involved in lipid and glucose metabolism: PPARα is involved in fatty acid catabolism and inflammation; PPARγ is a master regulator of adipocyte differentiation (and thus fat storage), and is also involved in insulin sensitization (and thus glucose storage) and in inflammation. Although the role of PPARβ is less well known, its importance in lipid metabolism has been increasingly recognized and PPARβ has recently emerged as a potential new target for the treatment of atherosclerosis and cardiovascular disease.[61–63]

PPARs are also activated by synthetic ligands, some of them being already marketed drugs, such as fibrates, targeting PPARα as hypolipidemic drugs in the preventive treatment of coronary artery disease, and thiazolidinediones, targeting PPARγ as hypoglycemic drugs in the treatment of type 2 diabetes. However, due to the pleiotropic effects of PPARs, these drugs present unwanted side effects, such as weight gain and edema in the case of PPARγ agonists. Designing selective modulators that could differentially recruit coactivators in order to minimize side effects constitutes a real challenge and an important therapeutic goal. Dual PPARα/γ agonists and PPAR panagonists are currently being developed to improve the management of type 2 diabetes and dyslipidemia, and for the treatment of the metabolic syndrome.[64]

The first crystal structures of PPARγ LBD, in the apo form and in complex with rosiglitazone,[10] revealed one of the largest LBPs among NRs ($\sim 1300\,\text{Å}^3$), displaying a T shape, the end contact points being H12, H5 (upper distal cavity), and the cleft between H3 and the β-hairpin (lower distal cavity). The large LBP may explain the wide variety of PPARγ ligands. Several crystal structures of complexes have been published. In contrast, PPARα and β LBDs have proven to be more difficult to crystallize and the available structural information is scarce. The crystal structures of the PPARβ LBD in the apo form and in complex with a polyunsaturated fatty acid and with a synthetic ligand were published in 1999,[65] while that of the PPARα LBD was published in 2001.[66] Comparison of agonist-bound PPAR LBD structures suggested a general model for ligand activation in the PPAR family[66]: many PPAR agonists share features such as an acidic head group interacting directly with the AF2 helix (for instance the TZD head in thiazolidinediones[10] or a carboxylate group[66,67]), a central hydrophobic part forming van der Waals contacts, and a flexible tail extending toward the lower or upper distal cavity. The role of the acidic head group is crucial for locking H12 in agonist conformation. Indeed, in the (PPARγ LBD–GW0072) complex structure, the weak partial agonist GW0072 occupies the upper and lower distal cavities and makes no direct interaction with H12, resulting in a suboptimal conformation of H12.[68] Based on the same principle, the work by Xu *et al.*[69] constitutes a nice example of structure-based design of a PPARα antagonist involving H12 displacement from the agonist conformation by substituting an ethylamide group for the carboxylate group of the agonist ligand GW409544 to disrupt the hydrogen bond with Tyr464 in H12. The resulting compound, GW6471, is a potent PPARα antagonist that promotes the binding of a corepressor peptide which prevents H12 from assuming the agonist conformation (**Figure 9**).

However, Östberg *et al.*[70] recently described a new class of PPARγ agonists, 5-substituted 2-benzoylaminobenzoic acids. As shown by x-ray crystallography, these compounds utilize a novel mode of binding, without direct interaction with H12, though some of them are potent PPARγ agonists in a cell-based reporter gene assay.

4.28.3.6 Liver X Receptors

LXRα and LXRβ function as heterodimers with RXR and regulate genes involved in cholesterol and fatty acid metabolism in a ligand-dependent fashion upon binding to LXREs in the promoter region of target genes. LXRα was identified from a liver cDNA library using oligonucleotides as probes, hence the name liver X receptor.[71] Indeed, LXRα is highly expressed in the liver. By contrast, LXRβ expression is much more ubiquitous. LXRs were discovered as orphan receptors but soon oxysterols were identified as natural ligands.[72–74] Since an LXRE was found in the rat *cyp7a1* gene encoding cholesterol 7α-hydroxylase, the rate-limiting enzyme in the oxidation of cholesterol to bile acids, it was tempting to speculate that LXRs play a key role in the metabolism of cholesterol.[74] This was confirmed by knockout experiments: in LXRα null mice, *cyp7a1* expression and subsequently bile acid synthesis are no longer increased in response to a high-cholesterol diet.[75]

The structure of the LXRβ LBD bound to the synthetic agonist T0901317[76,77] permitted the identification of a new mechanism for receptor activation (a cation-π interaction holding H12 in active conformation) as well as the determination of the binding mode of the natural ligand 24(*S*),25-epoxycholesterol.[76] Interestingly, T0901317 is able to adopt two different conformations in the large LXRβ LBP. As in the case of PPARs and PXR, it seems that a large pocket allows LXRs to sense multiple ligands of variable affinities rather than a specific high-affinity ligand.

Another structural study of the LXRβ LBD in complex with two very different ligands has revealed an additional feature – the LBP is flexible and can accommodate differently sized ligands.[21] This plasticity of the pocket, which is of

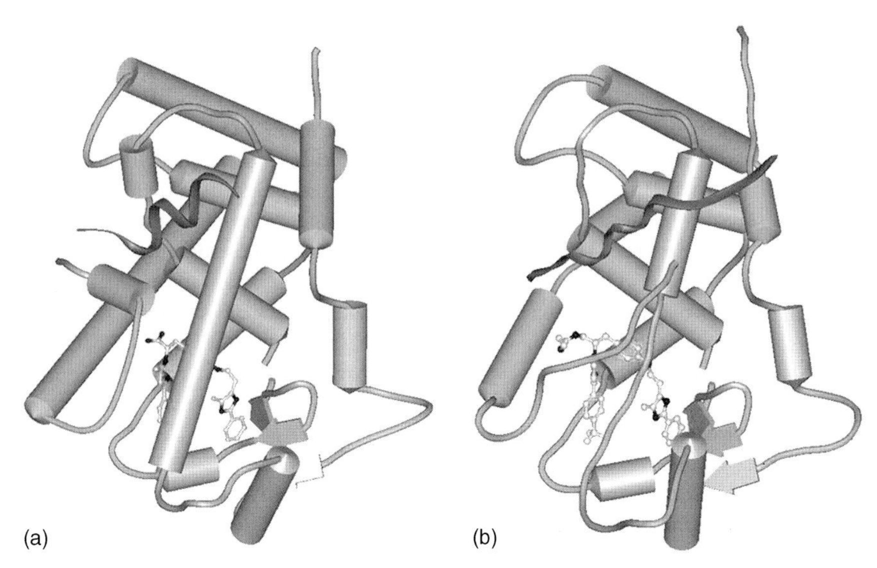

Figure 9 (a) In the (hPPARα LBD–GW409544–SRC1 peptide) complex (1K7L), H12 is the agonist conformation and the coactivator peptide is bound in the coactivator groove. (b) In the (hPPARα LBD–GW6471–SMRT peptide) complex (1KQQ), the bulkier ligand prevents H12 from assuming the agonist conformation. Instead, H12 is distorted and loosely packed against H3. The SMRT peptide adopts a three-turn α-helical conformation and binds to the modified hydrophobic groove.

fundamental relevance for structure-based drug design, has also been observed in the case of the insect receptor EcR, where structural comparison of the EcR LBD in complex with steroidal and nonsteroidal ligands revealed radically different, only partially overlapping LBPs.[20] Clearly, homology modeling would fail to predict correctly these variations in the volume and shape of the LBP, and experimental structure determination is still essential.

A single crystal structure of LXRα LBD is available at present, that of a heterodimer with RXRβ LBD in a fully agonist conformation.[78] The availability of experimental structures for both α and β isotypes has opened the way to structure-based design of isotype-selective LXR ligands, though this approach could prove to be quite challenging since all residues lining the LBP in the first shell are conserved between the two isotypes.

4.28.4 Computer-Assisted Drug Design (CADD) Applied to Nuclear Receptors: Potential and Limits

4.28.4.1 Homology Modeling

NR model structures are still constructed, for example to understand mutations observed in patients or to study NRs from model organisms.

The human ER first model was based on a human RXRα template in order to predict the complex of the estradiol bound to hERα.[79] Similarly, Wurtz *et al.* generated a model structure of hERα, docked estradiol in different orientations, and scored the orientations according to mutants affecting the binding properties.[80] Later on, with the publication of more NR experimental complexes, Egner *et al.* compared different hERα complexes with agonists and antagonists to identify structural conserved and flexible regions. This information was then exploited to fit a number of ligands in the LBP.[81]

More recently, Costache *et al.* modeled the LBDs of zebrafish estrogen receptors α, β_1, and β_2, suggesting that zebrafish would constitute a good model system to study human ER-binding drugs.[82]

Other steroid members have been studied, e.g., the androgen receptor (AR), to understand partial androgen insensitivity due to point mutations in this receptor. The mutation N758T was analyzed with the help of a three-dimensional (3D) model generated using hRXRα as template.[83] Poujol *et al.* identified key residues for binding agonist or antagonist based on an analysis combining a 3D model constructed after the progesterone receptor and selected mutants in the AR LBP.[84] The mechanism of antagonism in the human mineralocorticoid receptor (MR) has been studied by Fagart *et al.* who showed that ligands smaller than the natural ligand (e.g., progesterone versus adolsterone) can act as antagonists.[85] The 3D model of the MR and mutants of the loop between helix 11 and helix 12, the activation helix, revealed the crucial role of this loop in stabilizing the active conformation of the human MR.[86] Another

MR study analyzed the close structural similarity between the mineralocorticoid and glucocorticoid receptors and their binding of de-oxycorticosterone.[87] Using an ERα-based model of the glucocorticoid receptor (GR), Lewis *et al.* performed a quantitative structure–activity relationship (QSAR) study of a series of cytochrome P450 3A4 (CYP3A4) inducers, where induction is mediated by GR.[88] They did similar work on PPARα.

More recently, Honer *et al.* investigated GR antagonism by cyproterone acetate and RU486 and suggested that cyproterone acetate may inhibit GR transactivation by a molecular mechanism described as passive antagonism.[89] Jacobs *et al.* published homology models of hERβ and the constitutive androstane receptor based on the hERα structure, and of hPPARα based on the hPPARγ structure. They used these models as tools to understand the physiologic response to xenobiotics by examining the ligand-binding interactions in the LBP.[90] Lewis *et al.* already modeled the PPARα of human, mouse, and rat and exploited them in a QSAR analysis.[91] More recently, Pinelli *et al.* in a molecular modeling study, investigated new chiral fibrates with PPARα and PPARγ[92] LXRβ, involved in the metabolism of cholesterol, has also been extensively studied. Urban *et al.* showed the key role of F268 in ligand binding.[93] Wang *et al.* analyzed the molecular determinants of LXRα agonism and identified key residues involved in binding steroidal and nonsteroidal ligands.[94]

VDR is another member of the NR superfamily which has been the focus of numerous studies. Norman *et al.* generated a model structure based on the atomic coordinates of the thyroid hormone receptor. They analyzed the orientation of the natural ligand and related ligands in the LBP.[95] Adachi *et al.* studied the binding of lithocholic acid (LCA), a secondary bile acid, to the VDR and proposed a docking model for the LCA–VDR complex.[96] Within the orphan receptor subgroup, Renaud *et al.* performed a structure–function analysis using homology models of Rev-erb and RVR. The models revealed a ligand-binding cavity occupied by side chains and the authors suggested that these receptors are ligand-independent, in good agreement with their constitutive repressing activity and the absence of H12.[13] DAX1 (dosage-sensitive sex-reversal adrenal hypoplasia congenita on the X chromosome, gene 1) and SHP (small heterodimeric partner) are two other intriguing and close orphan receptors, exhibiting for the first one a very different DBD and for the second no DBD. Mutations in DAX1 in humans result in impaired biological function of the receptor. A 3D model of DAX1 has been proposed and exploited in a structure–function analysis to understand the role of the mutated residues.[97,98] For small heterodimer partner, Macchiarulo *et al.* investigated the repression mechanism of this receptor. With protein–protein docking experiments they identified the surface region of the protein likely to interact with the corepressor EID-1 (E1A-like inhibitor of differentiation 1).[99]

4.28.4.2 Molecular Dynamics (MD)

MD calculations sample the receptor–ligand complex conformational space over time and generate a so-called MD trajectory. Often a homology model of the target receptor is built with the ligand of interest docked in the LBP. One of the most studied receptor is ER. Abou-Zeid *et al.* used an MD simulation to study the differential recognition of resveratrol isomers by the hERα. Different MD simulations revealed the key features responsible for the stereo-selective ligand binding.[100] A different application of the MD technique has been used by Sivanesan *et al.* who selected 51 energetically favorable ERα LBD structures, extracted from a 3 ns MD simulation, to describe the flexibility of the LBP. These were then used in an in silico screening to identify ERα putative ligands.[101]

Nam *et al.* applied MD and molecular mechanics generalized Born surface area (MMGBSA) to simulate the different biological activity of DES on two closely related receptors, ERα and ERRγ. DES acts as an agonist ligand on ERα and as an antagonist on ERRγ.[102] Furthermore, van Lipzig *et al.* designed a model based on MD techniques and linear interaction methods to predict the binding affinity of xenoestrogen ER. Such a model could be helpful in identifying environmental estrogens that could represent a risk factor as endocrine disrupters.[103] Other members of the steroid subgroup have been the subject of detailed studies, e.g., von Langen *et al.* analyzed the binding of five steroids to hGR. They measured the in vitro binding affinity of aldosterone, cortisol, estradiol, progesterone, and testosterone in competition with the ligand dexamethasone, and tried to predict the theoretical relative binding affinities of these ligands. They performed separate 4-ns MD simulations of these complexes and calculated the binding affinities with two different approaches (molecular mechanics Poisson–Boltzmann surface area (MMPBSA), FlexX).[104]

For the androgen receptor, Wu *et al.* studied a pathogenic mutation in humans (R774C). This mutation results in a complete androgen-insensitivity syndrome. They investigated this mutation by MD simulations and showed that the mutation caused local structural distortions that are most likely responsible for the thermal instability of the mutated receptor.[105] For another subgroup, the PPARs, Yue *et al.* studied the role of a conserved phenylalanine residue in PPARγ and α (Phe282 and Phe273; respectively). They analyzed the wild-type receptor versus the Phe→Ala mutated receptor and showed that ligand binding is significantly decreased in the mutant receptor. In order to understand this difference, they launched 5-ns MD simulations that showed the mutation did not influence the flexibility of the receptor and concluded that solvent effects are the major source of the decrease.[106] More recently, the same team studied the ligand-binding regulation of LXR/RXR and LXR/PPAR heterodimerization. They characterized the kinetic features of these dimer complexes by surface plasmon resonance, and correlated these data with MD simulations of the complexes.[107]

The MD technique has also been applied by different groups to gain some insight on the entry/exit of the ligand from the LBP. Blondel *et al.* applied a multiple-copy MD investigate the binding/escape mechanism of all-*trans* retinoic acid with respect to the LBD of the NR RARγ.[108] Kosztin *et al.* realized a similar study on the same complex where they used steered MD simulations to explore the binding/unbinding process of the hormone from the LBP.[109] More recently, on the same theme, Martinez *et al.* used locally enhanced sampling by MD simulations to study the thyroid hormone receptor LBD in complex with its natural ligand (3,5,3′-triiodo-L-thyronine). Their aim was to predict molecular motions of the TR LBD and determine events that would permit ligand escape.[110]

4.28.4.3 Drug Design in the Field of Nuclear Receptors

The field of structure-based drug design is a rapidly growing area. The major challenge in today's drug design is the prediction of new biologically active compounds on the basis of known data. Over the last decade, an enormous amount of data concerning NRs has been published covering genomic, proteomic, and structural information. These data, especially the structural data, have provided the means to understand the receptor/ligand structure–function relationships and helped to identify new targets for structure-based drug/lead discovery. Numerous theoretical approaches have been applied in the field of NRs, with some success. The methods that have been developed fall into two main categories – the indirect ligand-based and the direct receptor-based approach. The first set comprises the numerous QSAR approaches and pharmacophores by virtual screening, whereas the second set of methods encompasses ligand-docking algorithms and ligand screening using a virtual screening with the help of the known LBP. Many studies combine more than one technique to predict ligand binding.[111]

4.28.4.3.1 Ligand-based approaches: quantitative structure–activity relationship and pharmacophores

Traditional QSAR approaches are statistical methods that rely solely on the ligand features. The goal of these studies is to identify structural characteristics necessary for high binding affinity to the cognate receptor. These approaches can be used to capture the effects of structural variations on the activity of a compound. However, applying a QSAR model for identifying compounds that are structurally divergent, for example with a different scaffold, is generally unreliable. Novel methods have been developed, such as 3D QSAR method, especially the comparative molecular field analysis (CoMFA), and have been widely applied to NR targets. In order to derive a correct model, all ligands need to be superimposed. The success of a molecular field analysis is therefore completely determined by the quality of the superimposition. The combined model generates a rationale of the binding site model, and allows for the quantification of the interactions between ligand molecules and protein, by simulating electrostatic forces, hydrogen bonds, or van der Waals interactions. In contrast to the real biological receptor, where the binding site is defined by a 3D arrangement of amino acids, most 3D QSAR models typically represent this binding site by mapping physicochemical properties on to a surface or a grid surrounding the ligand molecules, superimposed in 3D space.

The ER isotypes have been the focus of many studies. Waller[112] compared different QSAR approaches using ER as its target. Earlier, Sippl predicted the binding affinity of novel estrogens based on 3D QSAR.[113] Katzenellenbogen *et al.* developed subtype-selective ligands based on their understanding of structure–activity relationships in these two ERs.[114] More recently, Wolohan *et al.* developed a similar approach to predict isotype specificity.[115] Numerous studies have also been published where ERs are used as a test case to validate the methodological approach.[116,117]

The AR has also been studied extensively by Bohl *et al.* to identify structural features necessary for high binding affinity[118] and by Lill *et al.* both to screen and monitor xenobiotics expected to modulate endocrine effects.[119] A recent relationship study has also been published for the VDR, whose aim was to capture the relevant molecular 3D information regarding the molecular size, shape, symmetry, and atom distribution with respect to some invariant reference frame.[120] Khanna *et al.* developed three CoMFA models to capture the α, γ, or dual α/γ activity exhibited by some compounds.[121] A multidimensional QSAR/COMFA approach has been applied to ecdysteroid agonist in order to address the question of relative binding orientation of two ligand families and used as a virtual screen to identify new chemotypes.[122]

4.28.4.3.2 Pharmacophore models

A primary goal of 3D similarity searching is to find compounds with similar bioactivity to a reference ligand but with different chemotypes. A pharmacophore represents the 3D arrangement of chemical features that are shared by molecules exhibiting activity at a protein receptor. In the absence of structural information, the 3D arrangement of atoms or functional groups that participate in key ligand–receptor interactions, or *pharmacophore*, can be deduced from the alignment of (diverse) compounds that bind to the same receptor. The alignment or pharmacophore can be obtained manually or with an automated approach. The pharmacophore can be used to search a database of compounds or to design 'focused' combinatorial libraries. Pharmacophores are used routinely in 3D database searching for identifying potential lead compounds.

The first pharmacophore for an NR has been designed by Spencer *et al.* for the analysis of the nuclear oxysterol receptor LXRα.[123] They developed a ligand-sensing assay (LiSA) that measures ligand-dependent recruitment of a peptide from the SR coactivator 1 (SRC1) to LXRα by fluorescence resonance energy transfer. The resulting structure–activity relationships were used to develop a pharmacophore model of LXRα to design new ligands for this orphan receptor. Soon after, Ekins *et al.* proposed a pharmacophore for human pregnane X receptor ligands. This receptor is involved in the transcriptional regulation of multiple cytochromes P450 and multidrug resistance-associated protein (MDR1), which encodes for the drug transporter P-glycoprotein. Their pharmacophore captures the hydrophobic nature of the LBD recently deduced from the x-ray crystal structure. The model could be used in drug design as a high-throughput filter for identifying compounds that may bind to PXR.[124] More recently, Rathi *et al.* proposed a pharmacophore model to identify specifically the essential structural and electronic features important for the PPARγ agonistic activity and in particular in the study of the *N*-(2-benzoyl phenyl)-L-tyrosines as PPARγ agonists. More methodological developments have been applied to ER modulators, where the authors developed a pharmacophore-based evolutionary approach for virtual screening. They combine an evolutionary approach with a new pharmacophore-based scoring function with the aim of developing new SERMs with fewer side effects, such as benign and malignant uterine lesions. Other developments include either estrogen[125] or retinoic acid[126] receptor ligand sets as one of the test cases.

4.28.4.3.3 Ligand design and docking

Numerous studies have been published that combine structure-based optimization in combination with docking approaches. The aim of a docking algorithm is to sample conformations of small molecules in the ligand-binding site and with the help of a scoring function to evaluate which of these orientations fits best in the protein-binding site.[127] Structure–function relationships to design novel ligands have been published for the VDR some time ago[128] and revisited more recently in the light of the experimental complex structures.[129] Adachi *et al.* recently proposed a model for LCA, a bile acid, bound to the VDR that is supported by mutagenesis data.[96,130] But the largest subgroup studied is the steroid subgroup, including ER. Different studies designed isotype-selective ligands based on the protein–structure information.[46,131–133] In another study the authors designed an estrogen antagonist showing an unusual structure.[134] Cavasotto *et al.* also studied the transcriptional antagonism in RXR.[135] Khanna *et al.* designed a dual model for PPARα and PPARγ that combines features from both members to design and dock ligands.[121] The EcR and ultraspiracle receptor, two insect NRs, have also been the focus of two studies.[136] More recently, Folkertsma *et al.* did a family-based structure analysis of the NR LBD, identifying for every NR member key residues in the LBP.[137] Sheu *et al.* applied computational solvent-mapping methods to PPARγ in order to identify the most favorable binding positions. The solvent-mapping techniques move molecular probes (small functional groups) around the protein surface to find favorable positions. Their conformations are then clustered and ranked according to some score. That analysis revealed 10 binding 'hot spots' located in the coactivator-binding groove, the dimerization interface, in the LBP and around the ligand entrance site.[138]

4.28.4.3.4 Virtual screening

In essence, virtual screening exploits the 3D structure of the LBD to screen a large number of ligands in a way similar to the previously described docking algorithms. Other approaches routinely used to screen a large number of ligands are ligand-based approaches that use pharmacophore or QSAR models[111] (see above). Bissantz *et al.* performed an evaluation of different docking/scoring combinations based on ERα antagonists as a test case.[139] Paul *et al.* developed an approach that permits the recovery of the true target of specific ligands by virtual screening of the Protein Data Bank and applied it as a test case to 4OHT targets.[140] Abagyan *et al.* applied a virtual screening approach to identify new antagonist ligands of the human RARα.[141] Thyroid hormone receptor antagonists by high-throughput docking have also been identified by Schapira *et al.* They screened a library of over 250 000 compounds to select 100 TR antagonists.[142] More recently, Sivanesan *et al.* selected 51 structures of the LBP of the human ERα that were extracted from an MD run. They then screened 3500 endocrine-disrupting ligands against these targets.[101]

4.28.5 Conclusion

NRs form a family of important therapeutic targets that has now reached maturity for rational drug design approaches. The contribution of structural studies has been crucial for understanding the mechanisms of ligand-dependent transcriptional activity and paves the way for the structure-based design of innovative and more selective drugs. The wealth of experimental structures constitutes a gold mine for medicinal chemistry, and the availability of high-throughput techniques for functional screening and structure determination allows us to tackle more efficiently the remaining bottlenecks associated with toxicity and tissue specificity.

References

1. *Med. Ad News,* Internet communication May 2004; www.pharmalive.com/special_reports/index.cfm?start = 41#nav (accessed April 2006).
2. Brelivet, Y.; Kammerer, S.; Rochel, N.; Poch, O.; Moras, D. *EMBO Rep.* **2004**, *5*, 423–429.
3. Bourguet, W.; Ruff, M.; Chambon, P.; Gronemeyer, H.; Moras, D. *Nature* **1995**, *375*, 377–382.
4. Renaud, J. P.; Rochel, N.; Ruff, M.; Vivat, V.; Chambon, P.; Gronemeyer, H.; Moras, D. *Nature* **1995**, *378*, 681–689.
5. Wurtz, J. M.; Bourguet, W.; Renaud, J. P.; Vivat, V.; Chambon, P.; Moras, D.; Gronemeyer, H. *Nat. Struct. Biol.* **1996**, *3*, 206.
6. Glass, C. K.; Rosenfeld, M. G. *Genes Dev.* **2000**, *14*, 121–141.
7. Heery, D. M.; Kalkhoven, E.; Hoare, S.; Parker, M. G. *Nature* **1997**, *387*, 733–736.
8. Shiau, A. K.; Barstad, D.; Loria, P. M.; Cheng, L.; Kushner, P. J.; Agard, D. A.; Greene, G. L. *Cell* **1998**, *95*, 927–937.
9. Darimont, B. D.; Wagner, R. L.; Apriletti, J. W.; Stallcup, M. R.; Kushner, P. J.; Baxter, J. D.; Fletterick, R. J.; Yamamoto, K. R. *Genes Dev.* **1998**, *12*, 3343–3356.
10. Nolte, R. T.; Wisely, G. B.; Westin, S.; Cobb, J. E.; Lambert, M. H.; Kurokawa, R.; Rosenfeld, M. G.; Willson, T. M.; Glass, C. K.; Milburn, M. V. *Nature* **1998**, *395*, 137–143.
11. Wang, Z.; Benoit, G.; Liu, J.; Prasad, S.; Aarnisalo, P.; Liu, X.; Xu, H.; Walker, N. P.; Perlmann, T. *Nature* **2003**, *423*, 555–560.
12. Flaig, R.; Greschik, H.; Peluso-Iltis, C.; Moras, D. *J. Biol. Chem.* **2005**, *280*, 19250–19258.
13. Renaud, J. P.; Harris, J. M.; Downes, M.; Burke, L. J.; Muscat, G. E. *Mol. Endocrinol.* **2000**, *14*, 700–717.
14. Dhe-Paganon, S.; Duda, K.; Iwamoto, M.; Chi, Y. I.; Shoelson, S. E. *J. Biol. Chem.* **2002**, *277*, 37973–37976.
15. Wisely, G. B.; Miller, A. B.; Davis, R. G.; Thornquest, A. D., Jr.,; Johnson, R.; Spitzer, T.; Sefler, A.; Shearer, B.; Moore, J. T.; Miller, A. B. et al. *Structure (Camb.)* **2002**, *10*, 1225–1234.
16. Bogan, A. A.; las-Yang, Q.; Ruse, M. D., Jr.; Maeda, Y.; Jiang, G.; Nepomuceno, L.; Scanlan, T. S.; Cohen, F. E.; Sladek, F. M. *J. Mol. Biol.* **2000**, *302*, 831–851.
17. Stehlin, C.; Wurtz, J. M.; Steinmetz, A.; Greiner, E.; Schule, R.; Moras, D.; Renaud, J. P. *EMBO J.* **2001**, *20*, 5822–5831.
18. Potier, N.; Billas, I. M.; Steinmetz, A.; Schaeffer, C.; Van, D. A.; Moras, D.; Renaud, J. P. *Protein Sci.* **2003**, *12*, 725–733.
19. Egea, P. F.; Klaholz, B. P.; Moras, D. *FEBS Lett.* **2000**, *476*, 62–67.
20. Billas, I. M.; Iwema, T.; Garnier, J. M.; Mitschler, A.; Rochel, N.; Moras, D. *Nature* **2003**, *426*, 91–96.
21. Farnegardh, M.; Bonn, T.; Sun, S.; Ljunggren, J.; Ahola, H.; Wilhelmsson, A.; Gustafsson, J. A.; Carlquist, M. *J. Biol. Chem.* **2003**, *278*, 38821–38828.
22. Ciesielski, F.; Rochel, N.; Mitschler, A.; Kouzmenko, A.; Moras, D. *J. Steroid Biochem. Mol. Biol.* **2004**, *89–90*, 55–59.
23. Xu, R. X.; Lambert, M. H.; Wisely, B. B.; Warren, E. N.; Weinert, E. E.; Waitt, G. M.; Williams, J. D.; Collins, J. L.; Moore, L. B.; Willson, T. M. et al. *Mol. Cell* **2004**, *16*, 919–928.
24. Klaholz, B. P.; Renaud, J. P.; Mitschler, A.; Zusi, C.; Chambon, P.; Gronemeyer, H.; Moras, D. *Nat. Struct. Biol.* **1998**, *5*, 199–202.
25. Klaholz, B. P.; Mitschler, A.; Moras, D. *J. Mol. Biol.* **2000**, *302*, 155–170.
26. Klaholz, B. P.; Mitschler, A.; Belema, M.; Zusi, C.; Moras, D. *Proc. Natl. Acad. Sci. USA* **2000**, *97*, 6322–6327.
27. Egea, P. F.; Mitschler, A.; Rochel, N.; Ruff, M.; Chambon, P.; Moras, D. *EMBO J.* **2000**, *19*, 2592–2601.
28. Reinking, J.; Lam, M. M.; Pardee, K.; Sampson, H. M.; Liu, S.; Yang, P.; Williams, S.; White, W.; Lajoie, G.; Edwards, A. et al. *Cell* **2005**, *122*, 195–207.
29. Drews, J. *Science* **2000**, *287*, 1960–1964.
30. Lewis, J. S.; Jordan, V. C. *Mutat. Res.* **2005**, *591*, 247–263.
31. Sun, S. Y.; Lotan, R. *Crit. Rev. Oncol. Hematol.* **2002**, *41*, 41–55.
32. Wurtz, J. M.; Bourguet, W.; Renaud, J. P.; Vivat, V.; Chambon, P.; Moras, D.; Gronemeyer, H. *Nat. Struct. Biol.* **1996**, *3*, 87–94.
33. Germain, P.; Kammerer, S.; Perez, E.; Peluso-Iltis, C.; Tortolani, D.; Zusi, F. C.; Starrett, J.; Lapointe, P.; Daris, J. P.; Marinier, A. et al. *EMBO Rep.* **2004**, *5*, 877–882.
34. Chen, S.; Ostrowski, J.; Whiting, G.; Roalsvig, T.; Hammer, L.; Currier, S. J.; Honeyman, J.; Kwasniewski, B.; Yu, K. L.; Sterzycki, R. et al. *J. Invest. Dermatol.* **1995**, *104*, 779–783.
35. Reczek, P. R.; Ostrowski, J.; Yu, K. L.; Chen, S.; Hammer, L.; Roalsvig, T.; Starrett, J. E., Jr.; Driscoll, J. P.; Whiting, G.; Spinazze, P. G. et al. *Skin Pharmacol.* **1995**, *8*, 292–299.
36. van Leeuwen, J. P.; van den Bemd, G. J.; van, D. M.; Buurman, C. J.; Pols, H. A. *Steroids* **2001**, *66*, 375–380.
37. Colston, K. W.; Hansen, C. M. *Endocrinol. Rel. Cancer* **2002**, *9*, 45–59.
38. Rochel, N.; Wurtz, J. M.; Mitschler, A.; Klaholz, B.; Moras, D. *Mol. Cell* **2000**, *5*, 173–179.
39. Norman, A. W.; Okamura, W. H.; Hammond, M. W.; Bishop, J. E.; Dormanen, M. C.; Bouillon, R.; van Baelen, H.; Ridall, A. L.; Daane, E.; Khoury, R. et al. *Mol. Endocrinol.* **1997**, *11*, 1518–1531.
40. Maenpaa, P. H.; Vaisanen, S.; Jaaskelainen, T.; Ryhanen, S.; Rouvinen, J.; Duchier, C.; Mahonen, A. *Steroids* **2001**, *66*, 223–225.
41. Tocchini-Valentini, G.; Rochel, N.; Wurtz, J. M.; Mitschler, A.; Moras, D. *Proc. Natl. Acad. Sci. USA* **2001**, *98*, 5491–5496.
42. Brzozowski, A. M.; Pike, A. C.; Dauter, Z.; Hubbard, R. E.; Bonn, T.; Engstrom, O.; Ohman, L.; Greene, G. L.; Gustafsson, J. A.; Carlquist, M. *Nature* **1997**, *389*, 753–758.
43. Berry, M.; Metzger, D.; Chambon, P. *EMBO J.* **1990**, *9*, 2811–2818.
44. Pike, A. C.; Brzozowski, A. M.; Walton, J.; Hubbard, R. E.; Thorsell, A. G.; Li, Y. L.; Gustafsson, J. A.; Carlquist, M. *Structure (Camb.)* **2001**, *9*, 145–153.

45. Pike, A. C.; Brzozowski, A. M.; Hubbard, R. E.; Bonn, T.; Thorsell, A. G.; Engstrom, O.; Ljunggren, J.; Gustafsson, J. A.; Carlquist, M. *EMBO J.* **1999**, *18*, 4608–4618.
46. Manas, E. S.; Xu, Z. B.; Unwalla, R. J.; Somers, W. S. *Structure (Camb.)* **2004**, *12*, 2197–2207.
47. Shiau, A. K.; Barstad, D.; Radek, J. T.; Meyers, M. J.; Nettles, K. W.; Katzenellenbogen, B. S.; Katzenellenbogen, J. A.; Agard, D. A.; Greene, G. L. *Nat. Struct. Biol.* **2002**, *9*, 359–364.
48. Kuiper, G. G.; Enmark, E.; Pelto-Huikko, M.; Nilsson, S.; Gustafsson, J. A. *Proc. Natl. Acad. Sci. USA* **1996**, *93*, 5925–5930.
49. Sun, J.; Baudry, J.; Katzenellenbogen, J. A.; Katzenellenbogen, B. S. *Mol. Endocrinol.* **2003**, *17*, 247–258.
50. Manas, E. S.; Unwalla, R. J.; Xu, Z. B.; Malamas, M. S.; Miller, C. P.; Harris, H. A.; Hsiao, C.; Akopian, T.; Hum, W. T.; Malakian, K. et al. *J. Am. Chem. Soc.* **2004**, *126*, 15106–15119.
51. Nettles, K. W.; Sun, J.; Radek, J. T.; Sheng, S.; Rodriguez, A. L.; Katzenellenbogen, J. A.; Katzenellenbogen, B. S.; Greene, G. L. *Mol. Cell* **2004**, *13*, 317–327.
52. Greschik, H.; Wurtz, J. M.; Sanglier, S.; Bourguet, W.; Van, D. A.; Moras, D.; Renaud, J. P. *Mol. Cell* **2002**, *9*, 303–313.
53. Greschik, H.; Flaig, R.; Renaud, J. P.; Moras, D. *J. Biol. Chem.* **2004**, *279*, 33639–33646.
54. Kallen, J.; Schlaeppi, J. M.; Bitsch, F.; Filipuzzi, I.; Schilb, A.; Riou, V.; Graham, A.; Strauss, A.; Geiser, M.; Fournier, B. *J. Biol. Chem.* **2004**, *279*, 49330–49337.
55. Busch, B. B.; Stevens, W. C., Jr.; Martin, R.; Ordentlich, P.; Zhou, S.; Sapp, D. W.; Horlick, R. A.; Mohan, R. *J. Med. Chem.* **2004**, *47*, 5593–5596.
56. Willy, P. J.; Murray, I. R.; Qian, J.; Busch, B. B.; Stevens, W. C., Jr.,; Martin, R.; Mohan, R.; Zhou, S.; Ordentlich, P.; Wei, P. et al. *Proc. Natl. Acad. Sci. USA* **2004**, *101*, 8912–8917.
57. Issemann, I.; Green, S. *Nature* **1990**, *347*, 645–650.
58. Forman, B. M.; Chen, J.; Evans, R. M. *Proc. Natl. Acad. Sci. USA* **1997**, *94*, 4312–4317.
59. Kliewer, S. A.; Sundseth, S. S.; Jones, S. A. et al. *Proc. Natl. Acad. Sci. USA* **1997**, *94*, 4318–4323.
60. Krey, G.; Braissant, O.; L'Horset, F.; Kalkhoven, E.; Perroud, M.; Parker, M. G.; Wahli, W. *Mol. Endocrinol.* **1997**, *11*, 779–791.
61. Oliver, W. R., Jr.,; Shenk, J. L.; Snaith, M. R.; Russell, C. S.; Plunket, K. D.; Bodkin, N. L.; Lewis, M. C.; Winegar, D. A.; Sznaidman, M. L.; Lambert, M. H. et al. *Proc. Natl. Acad. Sci. USA* **2001**, *98*, 5306–5311.
62. Desvergne, B.; Michalik, L.; Wahli, W. *Mol. Endocrinol.* **2004**, *18*, 1321–1332.
63. Muscat, G. E.; Dressel, U. *Curr. Opin. Investig. Drugs* **2005**, *6*, 887–894.
64. Staels, B.; Fruchart, J. C. *Diabetes* **2005**, *54*, 2460–2470.
65. Xu, H. E.; Lambert, M. H.; Montana, V. G.; Parks, D. J.; Blanchard, S. G.; Brown, P. J.; Sternbach, D. D.; Lehmann, J. M.; Wisely, G. B.; Willson, T. M. et al. *Mol. Cell* **1999**, *3*, 397–403.
66. Cronet, P.; Petersen, J. F.; Folmer, R.; Blomberg, N.; Sjoblom, K.; Karlsson, U.; Lindstedt, E. L.; Bamberg, K. *Structure (Camb.)* **2001**, *9*, 699–706.
67. Gampe, R. T., Jr.; Montana, V. G.; Lambert, M. H.; Wisely, G. B.; Milburn, M. V.; Xu, H. E. *Genes Dev.* **2000**, *14*, 2229–2241.
68. Oberfield, J. L.; Collins, J. L.; Holmes, C. P.; Goreham, D. M.; Cooper, J. P.; Cobb, J. E.; Lenhard, J. M.; Hull-Ryde, E. A.; Mohr, C. P.; Blanchard, S. G. et al. *Proc. Natl. Acad. Sci. USA* **1999**, *96*, 6102–6106.
69. Xu, H. E.; Stanley, T. B.; Montana, V. G.; Lambert, M. H.; Shearer, B. G.; Cobb, J. E.; McKee, D. D.; Galardi, C. M.; Plunket, K. D.; Nolte, R. T. et al. *Nature* **2002**, *415*, 813–817.
70. Ostberg, T.; Svensson, S.; Selen, G.; Uppenberg, J.; Thor, M.; Sundbom, M.; Sydow-Backman, M.; Gustavsson, A. L.; Jendeberg, L. *J. Biol. Chem.* **2004**, *279*, 41124–41130.
71. Willy, P. J.; Umesono, K.; Ong, E. S.; Evans, R. M.; Heyman, R. A.; Mangelsdorf, D. J. *Genes Dev.* **1995**, *9*, 1033–1045.
72. Janowski, B. A.; Willy, P. J.; Devi, T. R.; Falck, J. R.; Mangelsdorf, D. J. *Nature* **1996**, *383*, 728–731.
73. Forman, B. M.; Ruan, B.; Chen, J.; Schroepfer, G. J., Jr.; Evans, R. M. *Proc. Natl. Acad. Sci. USA* **1997**, *94*, 10588–10593.
74. Lehmann, J. M.; Kliewer, S. A.; Moore, L. B.; Smith-Oliver, T. A.; Oliver, B. B.; Su, J. L.; Sundseth, S. S.; Winegar, D. A.; Blanchard, D. E.; Spencer, T. A. et al. *J. Biol. Chem.* **1997**, *272*, 3137–3140.
75. Peet, D. J.; Turley, S. D.; Ma, W.; Janowski, B. A.; Lobaccaro, J. M.; Hammer, R. E.; Mangelsdorf, D. J. *Cell* **1998**, *93*, 693–704.
76. Williams, S.; Bledsoe, R. K.; Collins, J. L.; Boggs, S.; Lambert, M. H.; Miller, A. B.; Moore, J.; McKee, D. D.; Moore, L.; Nichols, J. et al. *J. Biol. Chem.* **2003**, *278*, 27138–27143.
77. Hoerer, S.; Schmid, A.; Heckel, A.; Budzinski, R. M.; Nar, H. *J. Mol. Biol.* **2003**, *334*, 853–861.
78. Svensson, S.; Ostberg, T.; Jacobsson, M.; Norstrom, C.; Stefansson, K.; Hallen, D.; Johansson, I. C.; Zachrisson, K.; Ogg, D.; Jendeberg, L. *EMBO J.* **2003**, *22*, 4625–4633.
79. Maalouf, G. J.; Xu, W.; Smith, T. F.; Mohr, S. C. *J. Biomol. Struct. Dyn.* **1998**, *15*, 841–851.
80. Wurtz, J. M.; Egner, U.; Heinrich, N.; Moras, D.; Mueller-Fahrnow, A. *J. Med. Chem.* **1998**, *41*, 1803–1814.
81. Egner, U.; Heinrich, N.; Ruff, M.; Gangloff, M.; Mueller-Fahrnow, A.; Wurtz, J. M. *Med. Res. Rev.* **2001**, *21*, 523–539.
82. Costache, A. D.; Pullela, P. K.; Kasha, P.; Tomasiewicz, H.; Sem, D. S. *Mol. Endocrinol.* **2005**, *19*, 2979–2990.
83. Yong, E. L.; Tut, T. G.; Ghadessy, F. J.; Prins, G.; Ratnam, S. S. *Mol. Cell. Endocrinol.* **1998**, *137*, 41–50.
84. Poujol, N.; Wurtz, J. M.; Tahiri, B.; Lumbroso, S.; Nicolas, J. C.; Moras, D.; Sultan, C. *J. Biol. Chem.* **2000**, *275*, 24022–24031.
85. Fagart, J.; Wurtz, J. M.; Souque, A.; Hellal-Levy, C.; Moras, D.; Rafestin-Oblin, M. E. *EMBO J.* **1998**, *17*, 3317–3325.
86. Hellal-Levy, C.; Fagart, J.; Souque, A.; Wurtz, J. M.; Moras, D.; Rafestin-Oblin, M. E. *Mol. Endocrinol.* **2000**, *14*, 1210–1221.
87. Dey, R.; Roychowdhury, P. *J. Biomol. Struct. Dyn.* **2002**, *20*, 21–29.
88. Lewis, D. F.; Ogg, M. S.; Goldfarb, P. S.; Gibson, G. G. *J. Steroid Biochem. Mol. Biol.* **2002**, *82*, 195–199.
89. Honer, C.; Nam, K.; Fink, C.; Marshall, P.; Ksander, G.; Chatelain, R. E.; Cornell, W.; Steele, R.; Schweitzer, R.; Schumacher, C. *Mol. Pharmacol.* **2003**, *63*, 1012–1020.
90. Jacobs, M. N.; Dickins, M.; Lewis, D. F. *J. Steroid Biochem. Mol. Biol.* **2003**, *84*, 117–132.
91. Lewis, D. F.; Jacobs, M. N.; Dickins, M.; Lake, B. G. *Toxicol. In Vitro* **2002**, *16*, 275–280.
92. Pinelli, A.; Godio, C.; Laghezza, A.; Mitro, N.; Fracchiolla, G.; Tortorella, V.; Lavecchia, A.; Novellino, E.; Fruchart, J. C.; Staels, B. et al. *J. Med. Chem.* **2005**, *48*, 5509–5519.
93. Urban, F., Jr..; Cavazos, G.; Dunbar, J.; Tan, B.; Escher, P.; Tafuri, S.; Wang, M. *FEBS Lett.* **2000**, *484*, 159–163.
94. Wang, M.; Thomas, J.; Burris, T. P.; Schkeryantz, J.; Michael, L. F. *J. Mol. Graph. Model.* **2003**, *22*, 173–181.
95. Norman, A. W.; Adams, D.; Collins, E. D.; Okamura, W. H.; Fletterick, R. J. *J. Cell Biochem.* **1999**, *74*, 323–333.
96. Adachi, R.; Shulman, A. I.; Yamamoto, K.; Shimomura, I.; Yamada, S.; Mangelsdorf, D. J.; Makishima, M. *Mol. Endocrinol.* **2004**, *18*, 43–52.
97. Lalli, E.; Bardoni, B.; Zazopoulos, E.; Wurtz, J. M.; Strom, T. M.; Moras, D.; Sassone-Corsi, P. *Mol. Endocrinol.* **1997**, *11*, 1950–1960.

98. Lehmann, S. G.; Wurtz, J. M.; Renaud, J. P.; Sassone-Corsi, P.; Lalli, E. *Hum. Mol. Genet.* **2003**, *12*, 1063–1072.
99. Macchiarulo, A.; Rizzo, G.; Costantino, G.; Fiorucci, S.; Pellicciari, R. *J. Mol. Graph. Model.* **2005**, *24*, 362–372.
100. Abou-Zeid, L. A.; El-Mowafy, A. M. *Chirality* **2004**, *16*, 190–195.
101. Sivanesan, D.; Rajnarayanan, R. V.; Doherty, J.; Pattabiraman, N. *J. Comput.-Aided Mol. Des.* **2005**, *19*, 213–228.
102. Nam, K.; Marshall, P.; Wolf, R. M.; Cornell, W. *Biopolymers* **2003**, *68*, 130–138.
103. van Lipzig, M. M.; ter Laak, A. M.; Jongejan, A.; Vermeulen, N. P.; Wamelink, M.; Geerke, D.; Meerman, J. H. *J. Med. Chem.* **2004**, *47*, 1018–1030.
104. von Langen, J.; Fritzemeier, K. H.; Diekmann, S.; Hillisch, A. *ChemBioChem* **2005**, *6*, 1110–1118.
105. Wu, J. H.; Gottlieb, B.; Batist, G.; Sulea, T.; Purisima, E. O.; Beitel, L. K.; Trifiro, M. *Hum. Mutat.* **2003**, *22*, 465–475.
106. Yue, L.; Ye, F.; Xu, X.; Shen, J.; Chen, K.; Shen, X.; Jiang, H. *Biochimie* **2005**, *87*, 539–550.
107. Yue, L.; Ye, F.; Gui, C.; Luo, H.; Cai, J.; Shen, J.; Chen, K.; Shen, X.; Jiang, H. *Protein Sci.* **2005**, *14*, 812–822.
108. Blondel, A.; Renaud, J. P.; Fischer, S.; Moras, D.; Karplus, M. *J. Mol. Biol.* **1999**, *291*, 101–115.
109. Kosztin, D.; Izrailev, S.; Schulten, K. *Biophys. J.* **1999**, *76*, 188–197.
110. Martinez, L.; Sonoda, M. T.; Webb, P.; Baxter, J. D.; Skaf, M. S.; Polikarpov, I. *Biophys. J.* **2005**, *89*, 2011–2023.
111. Smith, P. A.; Sorich, M. J.; McKinnon, R. A.; Miners, J. O. *J. Med. Chem.* **2003**, *46*, 1617–1626.
112. Waller, C. L. *J. Chem. Inf. Comput. Sci.* **2004**, *44*, 758–765.
113. Sippl, W. *Bioorg. Med. Chem.* **2002**, *10*, 3741–3755.
114. Katzenellenbogen, B. S.; Sun, J.; Harrington, W. R.; Kraichely, D. M.; Ganessunker, D.; Katzenellenbogen, J. A. *Ann. NY Acad. Sci.* **2001**, *949*, 6–15.
115. Wolohan, P.; Reichert, D. E. *J. Comput.-Aided Mol. Des.* **2003**, *17*, 313–328.
116. Lill, M. A.; Vedani, A.; Dobler, M. *J. Med. Chem.* **2004**, *47*, 6174–6186.
117. Sippl, W. *J. Comput.-Aided Mol. Des.* **2002**, *16*, 825–830.
118. Bohl, C. E.; Chang, C.; Mohler, M. L.; Chen, J.; Miller, D. D.; Swaan, P. W.; Dalton, J. T. *J. Med. Chem.* **2004**, *47*, 3765–3776.
119. Lill, M. A.; Winiger, F.; Vedani, A.; Ernst, B. *J. Med. Chem.* **2005**, *48*, 5666–5674.
120. Gonzalez, M. P.; Suarez, P. L.; Fall, Y.; Gomez, G. *Bioorg. Med. Chem. Lett.* **2005**, *15*, 5165–5169.
121. Khanna, S.; Sobhia, M. E.; Bharatam, P. V. *J. Med. Chem.* **2005**, *48*, 3015–3025.
122. Hormann, R. E.; Dinan, L.; Whiting, P. *J. Comput.-Aided Mol. Des.* **2003**, *17*, 135–153.
123. Spencer, T. A.; Li, D.; Russel, J. S.; Collins, J. L.; Bledsoe, R. K.; Consler, T. G.; Moore, L. B.; Galardi, C. M.; McKee, D. D.; Moore, J. T. et al. *J. Med. Chem.* **2001**, *44*, 886–897.
124. Ekins, S.; Mirny, L.; Schuetz, E. G. *Pharm. Res.* **2002**, *19*, 1788–1800.
125. Sutherland, J. J.; O'Brien, L. A.; Weaver, D. F. *J. Med. Chem.* **2004**, *47*, 3777–3787.
126. Jenkins, J. L.; Glick, M.; Davies, J. W. *J. Med. Chem.* **2004**, *47*, 6144–6159.
127. Warren, G. L.; Andrews, C. W.; Capelli, A. M.; Clarke, B.; LaLonde, J.; Lambert, M. H.; Lindvall, M.; Nevins, N.; Semus, S. F.; Senger, S. et al. *J. Med. Chem.*, Internet communication August 13, 2005; http://dx.doi.org/10.1021/jm050362n (accessed April 2006).
128. Bouillon, R.; Okamura, W. H.; Norman, A. W. *Endocrinol. Rev.* **1995**, *16*, 200–257.
129. Yamada, S.; Shimizu, M.; Yamamoto, K. *Endocrinol. Dev.* **2003**, *6*, 50–68.
130. Adachi, R.; Honma, Y.; Masuno, H.; Kawana, K.; Shimomura, I.; Yamada, S.; Makishima, M. *J. Lipid Res.* **2005**, *46*, 46–57.
131. Hillisch, A.; Peters, O.; Kosemund, D.; Muller, G.; Walter, A.; Schneider, B.; Reddersen, G.; Elger, W.; Fritzemeier, K. H. *Mol. Endocrinol.* **2004**, *18*, 1599–1609.
132. Yang, J. M.; Shen, T. W. *Proteins* **2005**, *59*, 205–220.
133. Tan, Q.; Blizzard, T. A.; Morgan, J. D.; Birzin, E. T.; Chan, W.; Yang, Y. T.; Pai, L. Y.; Hayes, E. C.; DaSilva, C. A.; Warrier, S. et al. *Bioorg. Med. Chem. Lett.* **2005**, *15*, 1675–1681.
134. Zhou, H. B.; Comninos, J. S.; Stossi, F.; Katzenellenbogen, B. S.; Katzenellenbogen, J. A. *J. Med. Chem.* **2005**, *48*, 7261–7274.
135. Cavasotto, C. N.; Liu, G.; James, S. Y.; Hobbs, P. D.; Peterson, V. J.; Bhattacharya, A. A.; Kolluri, S. K.; Zhang, X. K.; Leid, M.; Abagyan, R. et al. *J. Med. Chem.* **2004**, *47*, 4360–4372.
136. Sasorith, S.; Billas, I. M.; Iwema, T.; Moras, D.; Wurtz, J. M. *J. Insect Sci.* **2002**, *2*, 25.
137. Folkertsma, S.; van Noort, P. I.; Brandt, R. F.; Bettler, E.; Vriend, G.; de Vlieg, J. *Curr. Med. Chem.* **2005**, *12*, 1001–1016.
138. Sheu, S. H.; Kaya, T.; Waxman, D. J.; Vajda, S. *Biochemistry* **2005**, *44*, 1193–1209.
139. Bissantz, C.; Folkers, G.; Rognan, D. *J. Med. Chem.* **2000**, *43*, 4759–4767.
140. Paul, N.; Kellenberger, E.; Bret, G.; Muller, P.; Rognan, D. *Proteins* **2004**, *54*, 671–680.
141. Abagyan, R.; Totrov, M. *Curr. Opin. Chem. Biol.* **2001**, *5*, 375–382.
142. Schapira, M.; Raaka, B. M.; Das, S.; Fan, L.; Totrov, M.; Zhou, Z.; Wilson, S. R.; Abagyan, R.; Samuels, H. H. *Proc. Natl. Acad. Sci. USA* **2003**, *100*, 7354–7359.
143. Wagner, R. L.; Apriletti, J. W.; McGrath, M. E.; West, B. L.; Baxter, J. D.; Fletterick, R. J. *Nature* **1995**, *378*, 690–697.
144. Bourguet, W.; Vivat, V.; Wurtz, J. M.; Chambon, P.; Gronemeyer, H.; Moras, D. *Mol. Cell* **2000**, *5*, 289–298.
145. Kallen, J. A.; Schlaeppi, J. M.; Bitsch, F.; Geisse, S.; Geiser, M.; Delhon, I.; Fournier, B. *Structure (Camb.)* **2002**, *10*, 1697–1707.
146. Stehlin-Gaon, C.; Willmann, D.; Zeyer, D.; Sanglier, S.; Van, D. A.; Renaud, J. P.; Moras, D.; Schule, R. *Nat. Struct. Biol.* **2003**, *10*, 820–825.
147. Downes, M.; Verdecia, M. A.; Roecker, A. J.; Hughes, R.; Hogenesch, J. B.; Kast-Woelbern, H. R.; Bowman, M. E.; Ferrer, J. L.; Anisfeld, A. M.; Edwards, P. A. et al. *Mol. Cell* **2003**, *11*, 1079–1092.
148. Mi, L. Z.; Devarakonda, S.; Harp, J. M.; Han, Q.; Pellicciari, R.; Willson, T. M.; Khorasanizadeh, S.; Rastinejad, F. *Mol. Cell* **2003**, *11*, 1093–1100.
149. Watkins, R. E.; Wisely, G. B.; Moore, L. B.; Collins, J. L.; Lambert, M. H.; Williams, S. P.; Willson, T. M.; Kliewer, S. A.; Redinbo, M. R. *Science* **2001**, *292*, 2329–2333.
150. Shan, L.; Vincent, J.; Brunzelle, J. S.; Dussault, I.; Lin, M.; Ianculescu, I.; Sherman, M. A.; Forman, B. M.; Fernandez, E. J. *Mol. Cell* **2004**, *16*, 907–917.
151. Love, J. D.; Gooch, J. T.; Benko, S.; Li, C.; Nagy, L.; Chatterjee, V. K.; Evans, R. M.; Schwabe, J. W. *J. Biol. Chem.* **2002**, *277*, 11385–11391.
152. Bledsoe, R. K.; Montana, V. G.; Stanley, T. B.; Delves, C. J.; Apolito, C. J.; McKee, D. D.; Consler, T. G.; Parks, D. J.; Stewart, E. L.; Willson, T. M. et al. *Cell* **2002**, *110*, 93–105.
153. Fagart, J.; Huyet, J.; Pinon, G. M.; Rochel, M.; Mayer, C.; Rafestin-Oblin, M. E. *Nat. Struct. Mol. Biol.* **2005**, *12*, 554–555.
154. Li, Y.; Suino, K.; Daugherty, J.; Xu, H. E. *Mol. Cell* **2005**, *19*, 367–380.
155. Bledsoe, R. K.; Madauss, K. P.; Holt, J. A.; Apolito, C. J.; Lambert, M. H.; Pearce, K. H.; Stanley, T. B.; Stewart, E. L.; Trump, R. P.; Willson, T. M. et al. *J. Biol. Chem.* **2005**, *280*, 31283–31293.

156. Williams, S. P.; Sigler, P. B. *Nature* **1998**, *393*, 392–396.
157. Matias, P. M.; Donner, P.; Coelho, R.; Thomaz, M.; Peixoto, C.; Macedo, S.; Otto, N.; Joschko, S.; Scholz, P.; Wegg, A. et al. *J. Biol. Chem.* **2000**, *275*, 26164–26171.
158. Krylova, I. N.; Sablin, E. P.; Moore, J.; Xu, R. X.; Waitt, G. M.; MacKay, J. A.; Juzumiene, D.; Bynum, J. M.; Madauss, K.; Montana, V. et al. *Cell* **2005**, *120*, 343–355.
159. Li, Y.; Choi, M.; Cavey, G.; Daugherty, J.; Suino, K.; Kovach, A.; Bingham, N. C.; Kliewer, S. A.; Xu, H. E. *Mol. Cell* **2005**, *17*, 491–502.
160. Sablin, E. P.; Krylova, I. N.; Fletterick, R. J.; Ingraham, H. A. *Mol. Cell* **2003**, *11*, 1575–1585.

Biographies

Jean-Paul Renaud, born 1960, is co-founder and CEO/CSO of AliX S.A. (Illkirch, France), a structure-based drug design company created in 2002 as a spin-off from the Department of Structural Biology and Genomics of IGBMC headed by Dino Moras. His field of expertise is the structural biology of nuclear receptors. He was previously:

- CNRS research director (2000–2002) and CNRS research scientist (1993–2000) at IGBMC where he initiated and developed structural studies on orphan nuclear receptors after studying the mechanism of the ligand-dependent activation of transcription by nuclear receptors at the structural level; he solved in 1995 one of the two first structures of a liganded nuclear receptor ligand-binding domain, that of RAR gamma in complex with all-*trans* retinoic acid;
- CNRS research scientist at the Laboratoire de Chimie et Biochimie Pharmacologiques et Toxicologiques of University Paris 5 (1988–1993) where he was working on the expression in yeast of individual forms of hepatic cytochrome P450s as a tool for the prediction of the early steps of drug metabolism.

Jean-Paul Renaud was a postdoc at the MRC Laboratory of Molecular Biology (Cambridge, UK) under the supervision of Kiyoshi Nagai (1987–1988). He obtained his PhD in organic chemistry from University Paris 6 (1986) under the supervision of Daniel Mansuy and holds an engineering diploma from the Ecole Nationale Supérieure de Chimie de Paris (ENSCP, 1982).

Dino Moras, Research Director at the CNRS, is the head of the Structural Biology Laboratory at the IGBMC. His research has been mainly related to the expression of genetic information, with two major topics: translation of the genetic code and regulation of transcription by nuclear receptors.

His major contributions in the field of translation are: (1) the partition of aaRS in two classes, based on structural and functional correlations (1990); and (2) the first structure determination of a class II complex and the elucidation of the mechanism of aminoacylation in the aspartic acid system (1991). More recently, he elucidated an important problem related to the fidelity of the translation of the genetic code: the solution of the Pauling paradox in the case of threonine via a two-step mechanism involving a zinc atom to discriminate valine plus an editing reaction to correct for tRNA mischarging by serine (1999/2002).

In 1995, he reported the first crystal structures of the ligand-binding domains of two NRs of retinoids (RXR and RAR) in their apo and liganded forms respectively. The study of the structure–function relationships continued with several other NRs, notably the NR of vitamin D and orphan receptors like ROR and ERR. In 2004, using a structure-based sequence analysis, the discovery of two mutually exclusive signature motifs led to the partition of the NRs superfamily into two classes (RXR and functional homodimers on one side, functional heterodimers on the other).

Jean-Marie Wurtz received his MSc in computer science from the University Louis Pasteur (Strasbourg), and in 1988 received his PhD in theoretical chemistry and informatics from the same university. He has been working for several years in the computer-aided drug design field as project leader in software development with Biosym Inc., San Diego, USA, and Biostructure SA, Strasbourg, France. In 1994, he joined the Laboratoire de Biologie et de Génomique Structurale at the IGBMC (Strasbourg, France) to study the structure/function relationship of a large family of transcription factors: nuclear receptors. Since 2001, he is professor of bioinformatics at the University Louis Pasteur (Strasbourg).

© 2007 Elsevier Ltd. All Rights Reserved
No part of this publication may be reproduced, stored in any retrieval system or transmitted in any form by any means electronic, electrostatic, magnetic tape, mechanical, photocopying, recording or otherwise, without permission in writing from the publishers

Comprehensive Medicinal Chemistry II
ISBN (set): 0-08-044513-6
ISBN (Volume 4) 0-08-044517-9; pp. 725–747

4.29 Enzymes: Insights for Drug Design from Structure

M M Flocco, D G Brown, and A Pannifer, Pfizer Global Research and Development, Sandwich, UK

© 2007 Published by Elsevier Ltd.

4.29.1 Introduction

This chapter discusses and provides a few examples of how knowledge of the three-dimensional structure of enzymes and their substrate, cofactor, and/or inhibitor complexes contributes to the understanding of enzyme function, mechanism, specificity, and inhibition. Structure-based drug design (SBDD) is discussed elsewhere (*see* 4.04 Structure-Based Drug Design – A Historical Perspective and the Future; 4.24 Structure-Based Drug Design – The Use of Protein Structure in Drug Discovery).

At least 16% of the human genome comprises enzymes (just over 4000 enzymes out of around 25 000 genes). However, many genes are still of unknown function, so the actual number of enzymes could turn out to be higher. Enzymes are involved in the chemical reactions that sustain life in all organisms, from viruses to humans. Thus, enzymes are an important class of drug targets. Around 47% of all launched small-molecule drugs target enzymes.[1]

A wealth of structural knowledge on enzymes has accumulated since the first three-dimensional structure of a protein, myoglobin, was determined about 50 years ago. There are nearly 15 000 crystal structures of enzymes in the Protein Data Bank (PDB),[2] roughly 50% of all crystal structures deposited. This number includes approximately 2400 crystal structures of human enzymes, with about 600 crystal structures representing unique human enzymes (sequence identity <90%), while the remaining crystal structures represent enzyme–ligand complexes. The ever-increasing use of structural information in drug discovery is illustrated in **Table 1**, where it is shown that a significant proportion of world drug sales are accounted for by drugs for which the structures of their targets are now available, with about 75% of these being enzymes.

The impact of structural information on the elucidation of reaction mechanisms as well as understanding inhibition and regulation for enzyme families such as proteases, dehydrogenases, and protein kinases has been extensively discussed in numerous publications. In this chapter, we focus on three examples: protein tyrosine phosphatases (PTPs), pyruvate dehydrogenase kinase (PDHK), and phosphodiesterases (PDEs). The PTP case exemplifies the use of structural information to: provide a detailed description of the reaction mechanism; identify the determinants of specificity toward phosphotyrosine (pTyr) substrate over phosphothreonine (pThr) and phosphoserine (pSer); and reveal a previously unknown oxidation state of the catalytic cysteine, the sulfenyl–amide intermediate, which is relevant in PTP redox regulation (see Chapter 4.24.5 for the design of PTP-1B inhibitors). In the case of PDHK, crystallographic studies have identified a number of sites for drug interaction, namely the adenosine triphosphate

Table 1 Launched drugs with known protein structure[a]

Drug[b]	*Compound*	*Protein target*	*PDB*
Lipitor	**Atorvastatin**	**HMG CoA reductase**	**1HWK**
Lescol	Fluvistatin	HMG CoA reductase	1HWI
Zocor	**Simvastatin**	**HMG CoA reductase**	**1HW9**
Pravachol/Mevalotin	**Pravastatin**	**HMG CoA reductase**	
Crestor	Rosuvastatin	HMG CoA reductase	1HWL
Avandia	**Rosiglitazone**	**PPARγ**	**2PRG**
Tricor	Fenofibrate	PPARα	1I7G
Celebrex	**Celecoxib**	**COX-2**	**1CX2**
Vioxx	**Rofecoxib**	**COX-2**	
Bextra	**Valdecoxib**	**COX-2**	
Mobic	Meloxicam	COX-2	
Voltaren	Diclofenac	COX-2	1PXX
Neurofen	Ibuprofen	COX-2	
Actonel	**Risedronate**	**Farnesyl pyrophosphate FPP synthase**	**1YV5**
Fosamax	**Alendronic acid**	**Farnesyl pyrophosphate FPP synthase (also broad-spectrum MMP inhibitor)**	
Zometa	**Zoledronic acid**	**MMP (broad-spectrum)**	
Viagra	**Sildenafil**	**PDE5**	**1TBF**
Levitra	Vardenafil	PDE5	1XP0
Cialis	Tadalafil	PDE5	1XOZ
Prograf	**FK-506 (macrolide)**	**FKBP-12**	**2FKE**
Lo/Ovral	Norgestrel	Estrogen receptor	
Triphasil	Norgestrel	Estrogen receptor	
Nolvadex	Tamoxifen	Estrogen receptor	3ERT
Evista	**Raloxifene**	**Estrogen receptor**	**1ERR/1QKN**
Depo-provera	Medroxyprogesterone	Estrogen receptor	
Casodex	**Bicalutamide**	**Androgen receptor**	**1E3G**
Prempro	**Mefipristone**	**Progesterone receptor glucocorticoid receptor**	
Advair/seretide	Fluticasone	Glucocorticoid receptor	1QKT
Plumicort	Budesonide	Glucocorticoid receptor	
Inspra	Eplerenone	Mineralocorticoid receptor	
Gleevec	**Imatinib**	**Abl kinase/C kit/syk tyrosine kinase**	**1XBB**
Iressa	Gefitnib	EGFR tyrosine kinase	
Tarceva	Erlotinib	EGFR tyrosine kinase	1M17
Capoten	Captopril	ACE	1UZF
Prinivil	Lisinopril	ACE	1O86
Enalaprilat	Enalapril	ACE	1UZE

Table 1 Continued

Drug[b]	*Compound*	*Protein target*	*PDB*
Accupril	Quinapril	ACE	
Monopril	Fosinopril	ACE	
Delix/Tritace	**Ramipril**	**ACE**	
Lotrel	**Benazepril**	**ACE**	
Zovirax	Aciclovir	Thymidine kinase (HSV)	2KI5
Valtrex	**Valacyclovir**	**Nonspecific viral polymerase and NNRTi**	
Zerit	Stavudine	HIV reverse transcriptase	
Sustiva	Efavirenz	HIV reverse transcriptase (NNRTi)	1FKO
Viramune	Nevirapine	HIV reverse transcriptase (NNRTi)	1LW0
Rescriptor	Delavirdine	HIV reverse transcriptase (NNRTi)	1KLM
Viracept	Nelfinavir	HIV protease	
Agenerase	Amprenavir	HIV protease	
Kaletra	Lopinavir	HIV protease	
Crixivan	Indinavir	HIV protease	1HSH
Invirase	Sequinavir	HIV protease	1HXB
Norvir	Ritonavir	HIV protease	1HXW
Sporanox	Itraconazole	Lanosterol 14α-demethylase	
Diflucan	Fluconazole	Lanosterol 14α-demethylase	1EA1
Trusopt	Dorzolamide	Carbonic anhydrase	1CIL
Azopt	Brinzolamide	Carbonic anhydrase	3ZNC
Aricept	**Donepezil**	**Acetylcholine esterase**	**1EVE**
Exanta	Melagatran	Thrombin (structure versus trypsin)	1K1P
Relenza	Zanamivir	Influenza neuraminidase	1A4G
Tamiflu	Oseltamivir	Influenza neuraminidase	
Augmentin	Clavulanate	Beta-lactamase	1BLC
Cellcept	**Mycophenolate mofetil**	**Inosine monophosphate dehydrogenase**	**1JCN**
Zithromax	**Azithromycin**	**50S ribosome**	**1M1K**
Biaxin/Claricid	**Clarithromycin**	50S ribosome	1J5A
Ciprobay	**Ciproflaxacin**	**Bacterial DNA gyrase**	**1AJ6/1AB4**
Lequavin	**Levofloxacin**	**Bacterial DNA gyrase**	

ACE, angiotensin-converting enzyme; CoA, coenzyme A; COX, cyclooxygenase; EGFR, epidermal growth factor receptor; FKBP, FK-binding protein; HIV, human immunodeficiency virus; HMG, hydroxymethylglutaryl; HSV, herpes simplex virus; NNRTi, non-nucleoside reverse transcriptase inhibitor; PDE, phosphodiesterase; PPAR, peroxisome proliferator-activated receptor.

Reproduced from Hubbard, R. E., Ed. Structure-Based Drug Discovery: An Overview, with permission of the Royal Society of Chemistry.

[a] For a comprehensive table of antibiotic/ribosome structures see Table 4.1 from [70].

[b] Drugs whose sales were or the top 200 in 2004 are shown in bold type.

(ATP) binding site, the E2 lipoyl-binding domain site, and two additional allosteric sites that may be the sites of regulation by pyruvate and coenzyme A (CoA).

The PDE section of this chapter exemplifies the use of x-ray data to infer the catalytic mechanism and mode of inhibition of existing drugs. The work on PDEs also illustrates the employment of x-ray structures of a number of complexes of diverse ligands to understand subfamily selectivity, which can then be used to improve drug specificity.

Three-dimensional structure elucidation of proteins can be accomplished by x-ray crystallography or nuclear magnetic resonance (NMR) methods. However, x-ray crystallography is the most frequently used approach. Approximately, 85% of all structures currently deposited in the PDB have been determined by x-ray crystallography (Section 4.29.5 presents basic concepts that are useful to bear in mind when working with crystal structures. Structural Biology and x-ray crystallography are extensively discussed in Volume 3).

4.29.2 Protein Tyrosine Phosphatases

The first PTP was cloned in 1982, and since then the importance of the PTPs as key regulatory enzymes in metabolism and cell replication has been established.[3,4] The structural characterization initiated by the first PTP crystal structure just over a decade ago,[5] together with kinetic work,[6] has provided a powerful insight into the catalytic mechanism underlying these regulatory functions.

PTPs catalyze the hydrolysis of phosphotyrosine via a two-step mechanism (**Figure 1**): (1) a nucleophilic attack on the phosphate moiety by the thiolate side chain of a conserved catalytic cysteine and (2) hydrolysis of the resulting thiophosphate intermediate to regenerate the enzyme.[7]

The first PTP crystal structure[5] identified the catalytic cysteine (Cys215 in PTP-1B) at the base of a deep cleft in the enzyme surface (**Figure 2**). The sides of the cleft are formed by Phe182 and Tyr46, and the rim by the residue 48–52 loop. This structure provided an immediate explanation for the specificity of PTPs toward pTyr over pSer and pThr – the shorter side chains cannot reach the cysteine at the base of the cleft. The structure also showed the Cys215 sulfur atom within hydrogen bonding distance of a series of main chain amide hydrogen atoms from the 214–221 loop.

$$\text{Cys–S}^- + \text{(}^-\text{O)(O)(O)P–O–C}_6\text{H}_4\text{–} \xrightarrow{H^+} \text{Cys–S–PO}_3 + \text{tyrosine}$$

$$\text{Cys–S–PO}_3 + H_2O \xrightarrow{H^+} \text{Cys–S}^- + \text{phosphate}$$

Figure 1 Hydrolysis of phosphotyrosine by PTP via a two-step mechanism.

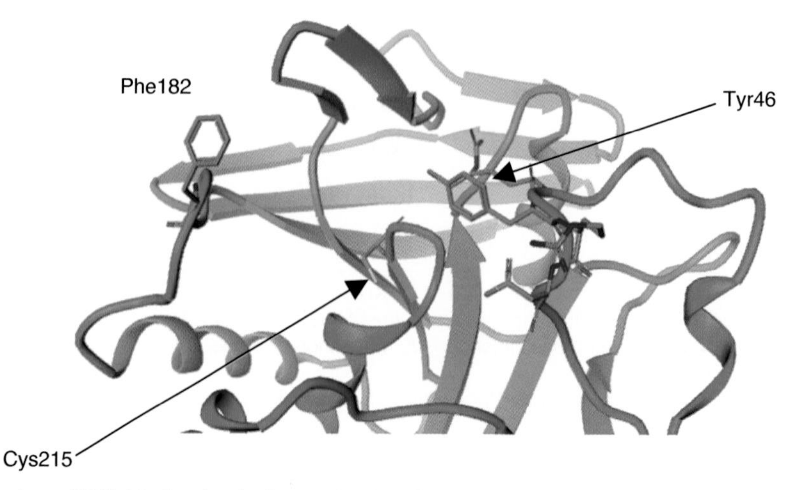

Figure 2 The structure of PTP-1B showing the key nucleophile Cys215 and the aromatic residues Phe182 and Tyr46, which form the walls of the substrate binding site. Phe182 moves approximately 8 Å upon substrate binding.

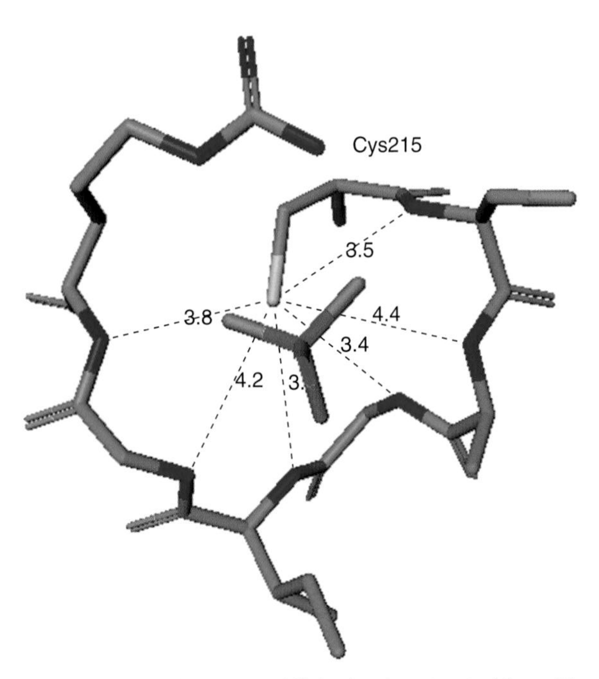

Figure 3 Cys215 at the focus of six amide hydrogens, stabilizing its deprotonated form. These amide hydrogens, together with Arg221, also provide the binding site for the phosphate moiety of the substrate.

These δ-positive hydrogens are thought to stabilize the deprotonated form of the Cys side chain, increasing its nucleophilicity (**Figure 3**). In turn, the negative charge on the cysteine is required to maintain the structural integrity of the enzyme.[8] Mutation of Cys215 to Ser does not simply inactivate the enzyme by replacement with a poorer, neutral nucleophile but results in a major loop shift that completely disrupts the catalytic site. It is likely that the longer hydrogen bonds from the Ser hydroxyl to the amide hydrogens and absence of a negatively charged atom destabilize the tight 214–221 loop. Another function of the catalytic cysteine in regulating the activity of the enzyme was demonstrated by the structural studies of Salmeen *et al.*[9] and van Montfort *et al.*[10] The sulfur atom can undergo reversible oxidation through a number of states, revealed in the crystal structure, inactivating the enzyme. The oxidation of Cys215 is correlated with PTP-1B activity in the cell under oxidative stress.

The original PTP structure and later structures of the Michaelis complex[11] also provided information regarding the identity of the general acid required for the reaction. Site-directed mutagenesis had indicated that Glu115 was a candidate but the crystal structure showed that this was not possible. Glu115 is too far from the scissile bond to play this role. Instead, Glu115 forms a bidentate salt bridge to Arg221, maintaining it in the correct conformation to bind the phosphate during the catalytic cycle.

Comparison of the unliganded PTP structure with the structure of the Michaelis complex[11] also showed the large conformational changes that take place over the catalytic cycle. A loop that forms one side of the active site (the WPD loop) undergoes a movement of about 8 Å upon substrate binding driven by stacking interactions between Phe182 and the phenyl group of the substrate and the formation of a hydrogen bond between Asp181 and the phenolic oxygen atom of the pTyr (**Figure 4**).

This movement of Asp181 toward the scissile bond is crucial to the reaction as it is both the general acid that protonates the tyrosine leaving group and the general base that deprotonates the hydrolytic water molecule in the second step of the reaction as demonstrated in the closely related enzyme PTP1.[12] Indeed, knowledge of the structure of the catalytic site has allowed mutants to be created, which uncouple the two steps of the reaction and allow the intermediate phosphorylated form of the enzyme to be investigated.[7,12] This intermediate form is normally a transient state in which the catalytic cysteine is covalently phosphorylated prior to hydrolysis. Mutations at two sites (Ser222 and Glu262) have been shown to trap this covalent intermediate. Lohse *et al.*[12] mutated Ser222 to Ala and showed by NMR the accumulation of a ^{31}P signal consistent with the formation of the intermediate. Analysis of the structure shows Ser222 hydrogen bonding to the sulfur of Cys215. It is likely that the δ-positive hydrogen of the Ser222 hydroxyl stabilizes the build-up of negative charge on the sulfur atom of Cys215 as the intermediate is hydrolyzed. The intermediate has also been trapped in a crystal structure. Kinetic and structural work indicated that Gln262 was important in intermediate hydrolysis. Mutation of this residue to Ala, crystallization of the mutant, and soaking in the

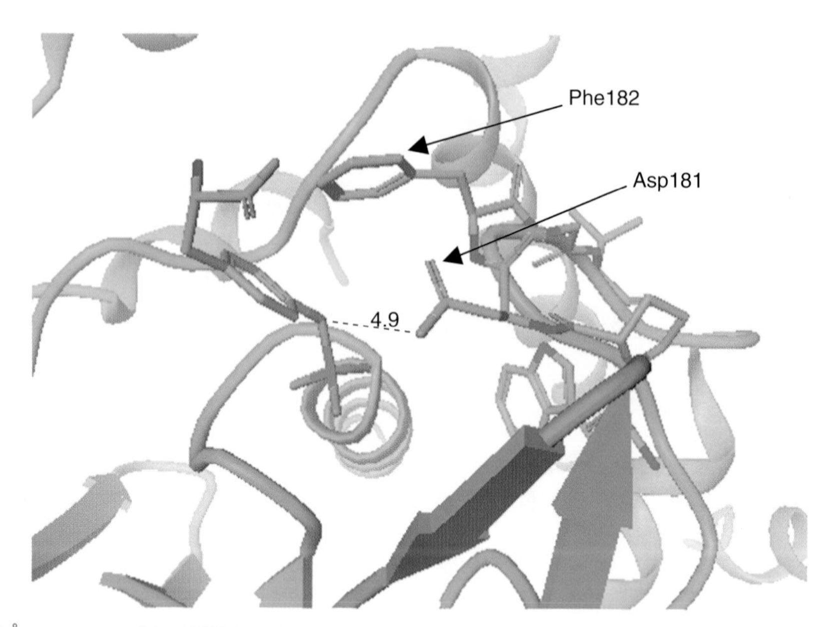

Figure 4 An 8 Å movement of the WPD loop buries Phe182 in a herringbone-type interaction with the aromatic ring of pTyr and brings Asp181 close to the scissile bond.

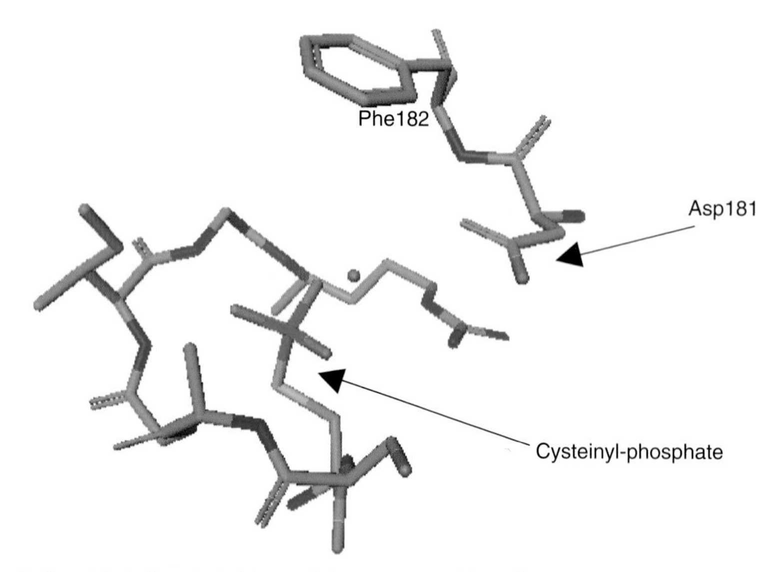

Figure 5 The covalent cysteinyl-phosphate intermediate was trapped by a Q262A mutation, slowing the rate of hydrolysis of this short-lived state. None of the water molecules visible in the electron density map adjacent the phosphoryl moiety is well positioned for an in-line nucleophilic attack, possibly resulting in the slow hydrolysis.

artificial substrate pNPP allowed the intermediate to be trapped. The structure suggested that Gln262 was required for aligning the nucleophilic water correctly for intermediate hydrolysis (**Figure 5**).

In addition to providing structural insights into the nature of the covalent intermediate, it has also been possible to probe the structure of the transition state. The vanadate ion is a potent inhibitor of tyrosine phosphatases, with a low micromolar K_i, and is much more potent than other oxyanion inhibitors such as molybdate and tungstate. The reason for this increased potency is that vanadate has some of the properties of a transition state analogue,[13,14] able to adopt a trigonal bipyramidal geometry and make a partial covalent bond to the cysteine sulfur while both molybdate and tungstate are product analogues and maintain their tetragonal geometry. The vanadate-bound crystal structure has been solved and shows clear evidence of a covalent bond to the sulfur. It also shows Asp181 forming a strong charge–charge interaction with the apical oxygen of the vanadate ion. This oxygen atom would occupy the position of the incoming hydrolytic water molecule in the second step of the reaction providing further evidence for its role in deprotonating the water.

These crystal structures of PTPs, together with the structure of the PTP-1B/tungstate complex,[5] provide a series of snapshots at key points in the catalytic cycle from unliganded protein structure, through Michaelis complex, transition state, intermediate, and product complex. This body of structural information combined with kinetic analysis and biophysical studies, has allowed for a detailed description of the protein phosphatase catalytic mechanism (see also Chapter 4.24.5 for the design of PTP-1B inhibitors).

4.29.3 Pyruvate Dehydrogenase Kinase

The formation of acetyl-CoA and reduced nicotinamide adenine dinucleotide (NADH) by oxidative decarboxylation of pyruvate is catalyzed by pyruvate dehydrogenase (PDH), a component of the pyruvate dehydrogenase multienzyme complex (PDC). PDH (E1) is regulated by product inhibition and reversible phosphorylation/dephosphorylation by four isoforms of PDHK and PDH phosphatase, respectively (**Figure 6**).[15,16] Phosphorylation of PDH at three serine residues downregulates PDC activity and is mediated by noncovalent binding of PDHK to the acetylated lipoyl domain of the dihydrolipoyl acetyltransferase (E_2L_2) component of the PDC.[15–20] The PDC also contains a third component, dihydrolipoyl dehydrogenase (E3).

Insufficient PDC activity leads to accumulation of pyruvate, which is converted into lactate and H^+. This lactic acidosis causes pain and weakness in ischemic skeletal muscles, which are typical symptoms for intermittent claudication. Inhibition of PDHK would increase PDC activity and, consequently: (1) reduce the amount of lactic acid formation and (2) reduce the demand of oxygen for ATP production by shifting metabolism from fatty acid oxidation to glycolysis as primary acetyl-CoA source. Therefore, PDHK inhibition provides a route for therapeutic intervention in diabetes and cardiovascular disorders.[21–28]

PDHK phosphorylates serines on a protein substrate, however PDHK is not a standard serine/threonine kinase. Initial sequence analysis identified a similarity with the GHKL superfamily.[29–37] The family is structurally identified as two domain proteins with an ATP binding domain and a four-helix bundle domain. Although the proteins have similar overall topology, they have a variety of functions and highly diverse ATP-binding sites.

A few PDHK inhibitors had been previously disclosed when Pfizer performed high-throughput screening (HTS) based on a protein phosphorylation assay to identify potential new agents against this pharmaceutical target. A number of structurally diverse ligands were found to have a significant inhibitory effect on PDH phosphorylation. While some of the inhibitors were found to be ATP-competitive, a number were shown to be noncompetitive, and a structural analysis to understand the modes of inhibition of these different inhibitors was undertaken.

X-ray crystal structures of PDHK-2 with a number of these inhibitors have recently been published, which together with the structure of a PDHK-3 complex with the lipoyl domain have allowed for a greater understanding of the allosteric regulation of PDHK.[38–41] Although no structure has yet been obtained with either the protein substrate or a peptide analogue, the substrate-binding site is proposed to be an extended groove formed between the catalytic and regulatory domains, which is consistent with proximity to the ATP site and dichloroaceate (DCA)/pyruvate regulatory site.

Crystal structures of complexes of PDHK-2 with the inhibitors identified in HTS were solved using both a soaking and cocrystallization approach to ensure that any protein rearrangement could be rationalized based on inhibitor effect alone. For some of the inhibitors, the structures were also solved in the presence of ATP in order to determine the influence of the inhibitors on both the substrate and cofactor binding site.

Unexpectedly, these crystal structures uncovered three different sites in the regulatory domain of the enzyme, where a small ligand can induce a significant effect on PDHK phosphorylation of E1. Close analysis of each of the

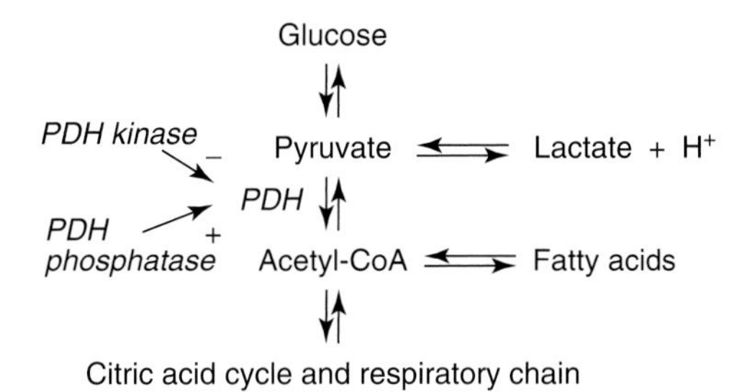

Figure 6 Regulation of PDH via reversible phosphorylation/dephosphorylation by PDH kinase and PDH phosphatase.

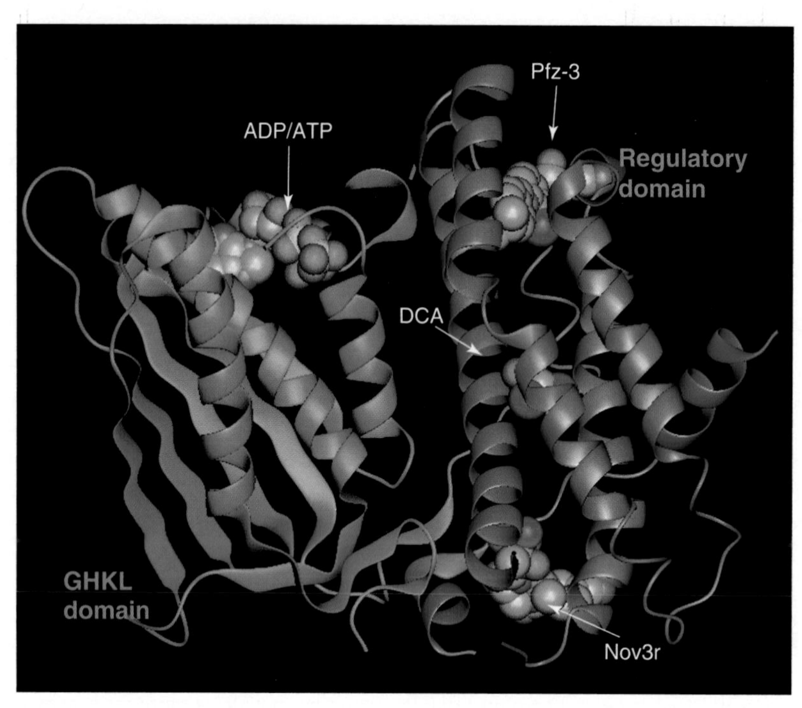

Figure 7 Ribbon diagram of the x-ray structure of PDHK-2 illustrating the domain structure and the ligand binding sites derived from x-ray crystal structures of HTS hits (PDB codes 2BTZ, 2BU2, 2BU5, 2BU6, 2BU7, and 2BU8[41]).

complexes reveals subtle changes in protein conformation, which have a marked effect on enzyme activity (**Figure 7**). Although there is no obvious structural effect on the ATP binding site upon binding of small molecules at the Pfz3 site, the observed inhibitory and kinetic results may be due to an effect on PDHK dimerization. The Pfz3 site appears to be conserved among the PDHK's, which also raises the possibility that it may be the site of binding for another natural physiological effector such as CoA.

Small-molecule binding at the Nov3r site was postulated to inhibit E2 elevated ATP-dependent phosphorylation of E1 by blocking the association between PDHK and the acetylated lipoyl domain (L2) of E2. This hypothesis was originally based on the similarity of the inhibitors to the acetylated lipoyl moiety of L2. Recent studies of Kato *et al.*[40] have provided direct evidence of the interaction of the closely related PDHK-3 with the lipoyl domain of E2 at this site.

Compound binding to the DCA site may have a direct effect on substrate recognition but may also involve disruption of PDHK dimer formation through C-terminal cross-arms as observed in a number of x-ray structures.[42]

The four different classes of inhibition that have been revealed by x-ray studies are thus described as: (a) direct ATP competition by compounds that bind at the ATP site; (b) allosteric regulation by compounds that bind at the Pfz3 site of the regulatory domain and may regulate PDC activity by effecting PDHK dimerisation; (c) direct competition with binding of the L2 domain by compounds that bind at the Nov3r site of the regulatory domain; and (d) direct regulation of substrate binding by compounds that bind at the DCA site.

4.29.4 Phosphodiesterases Inhibitors

A wide variety of biological processes, including cardiac muscle contraction, regulation of blood flow, neural transmission, glandular secretion, cell differentiation and gene expression, are affected by steady-state levels of the cyclic nucleotide biological second messengers cyclic adenosine monophosphate (cAMP) and cyclic guanosine monophosphate (cGMP). Intracellular receptors for these molecules include cyclic nucleotide-dependent protein kinases,[43] cyclic nucleotide-gated channels, and class I PDEs.[44] PDEs are a large family of proteins, which were first reported by Sutherland and co-workers.[45–47] The family of cyclic nucleotide phosphodiesterases catalyzes the hydrolysis of 3′, 5′-cyclic nucleotides to the corresponding 5′ monophosphates. Current literature shows that there are 11 related, but biochemically distinct, human PDE gene groups and that many of these groups include more than one gene subtype, giving a total of 21 genes. PDE catalytic domains recognize and hydrolyse cAMP and cGMP as substrates. PDE4, PDE7, and PDE8 are highly specific for cAMP[48,49] whereas PDE5, PDE6, and PDE9 are cGMP specific. PDE1, PDE2, PDE3, PDE10, and PDE11 have mixed cAMP and cGMP specificity.[50–52]

The PDE gene family has for many years been attracting widespread attention as therapeutic intervention points in a number of disease states. Most recently, three-dimensional structures of the catalytic domains of PDE4, PDE5, PDE3, PDE9, PDE10, and PDE7 in complex with inhibitors have been published, allowing us for the first time the opportunity to understand the interactions of a wide range of chemotypes with PDEs at a molecular level. In this section we focus on the structural aspects of PDE inhibition and selectivity; a broader discussion on PDE inhibitors, chemical classes is presented in Chapter 2.25.

The crystal structures now available (and their respective PDB code) are PDE1B (1TAZ); PDE3B (1SOJ, 1S02); PDE4B, PDE4D (1MKD, 1PTW, 1OYN, 1Q9M, 1FOJ, 1ROR, 1RO6, 1RO9, 1RKO, 1TB5, 1TB7, 1TBB, 1XLX, 1XLZ, 1XMU, 1XM4, 1XM6, 1XMY, 1XOM, 1XON, 1XOQ, 1XOR, 1XOS, 1XOT, 1Y2B, 1Y2C, 1Y2D, 1Y2E, 1Y2H, 1Y2J, 1Y2K, 1ZKN, 2FM0, 2FM5); PDE5A (1RKP, 1UDU, 1UDT, 1UHO, 1T9R, 1T9S, 1TBF, 2CHM, and PDE5 structural Patent[53,54]); PDE7A (1ZKL); PDE9A (1TBM); PDE10 (Pfizer Structural Patent[55]). Given the similar overall nature of the catalytic domains of the whole gene family, the structures of these representative examples allow us to build models of inhibitors of other gene family members and to predict the interactions at a molecular level, thus enabling the design of PDE and subtype specific inhibitors.

All catalytic domain structures of PDEs solved to date unsurprisingly illustrate the conservation of the general fold of the domain. The catalytic domain is all alpha-helical (**Figure 8**) and can be divided into three subdomains with helices H1 to H7 forming the N-terminal subdomain, helices H8 to H11 forming the small subdomain, and a C-terminal subdomain comprising helices H12 to H16. The small subdomain contained a partially disordered region in the first crystal structures solved at Pfizer (patent 'wild-type PDE5' structure),[53] while this region appeared to be in an extended nonphysiologically relevant conformation in the structure solved by Sung.[56] In each of the catalytic domain structures solved so far, the active site region contains two metal-binding sites where the two divalent metal ions are coordinated with the conserved sequence (and structural) motif $[H(X)_3H(X)_{69}E]$. In the structures of PDE4B, PDE4D, PDE5, PDE7A, and PDE10, these ions have been modeled as one Zn^{2+} and one Mg^{2+}; however, in the recent publication of the PDE3 catalytic domain structure the ions have been modeled as two Mg^{2+}.[57]

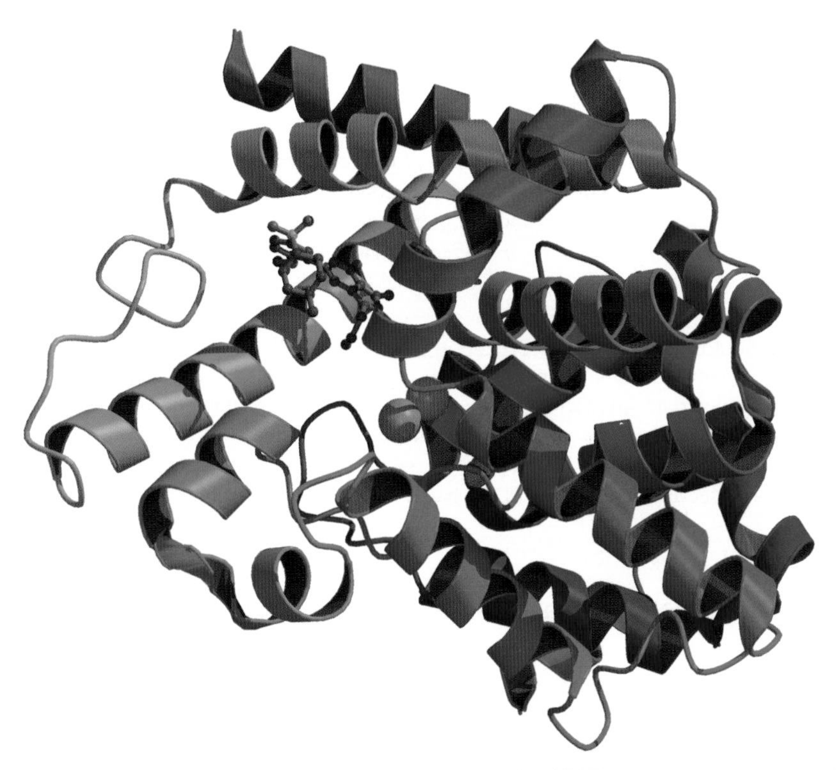

Figure 8 Three-dimensional x-ray crystal structure of the catalytic domain of PDE5* in complex with sildenafil[53] showing the overall fold of this domain and its all helical nature.

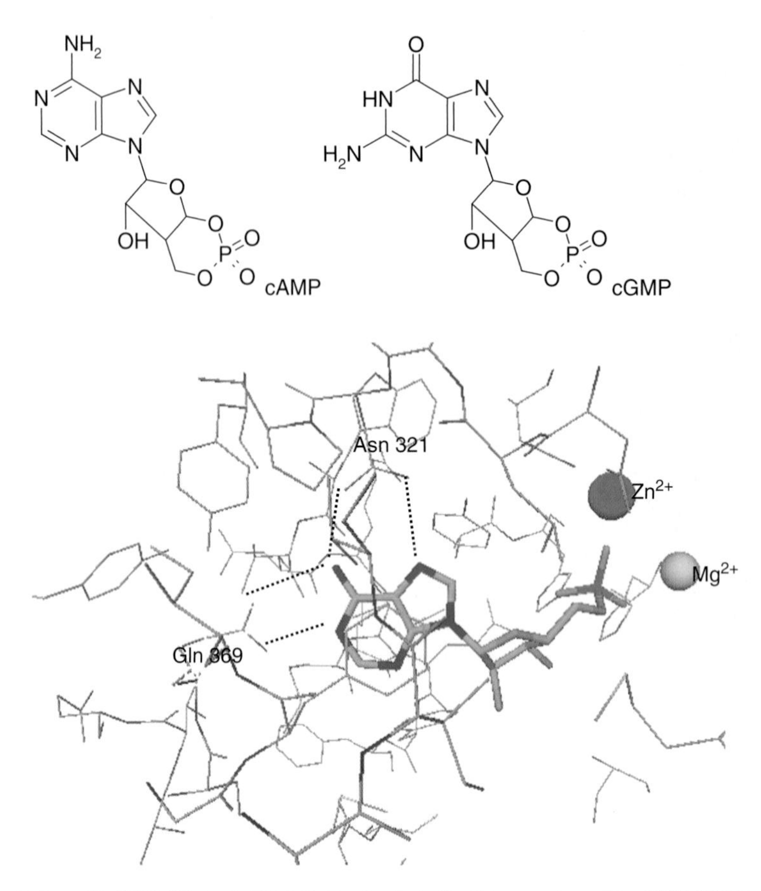

Figure 9 PDE4 complex with AMP. Although there is no structure for the complex with cAMP, 1PTW contains the product AMP which is presumed to form similar interactions with PDE4. Note the two hydrogen bonds between N3 and N4 of AMP and Gln369 and in addition the two hydrogen bonds from Asn321 to N2 and N3 of AMP. Residue Asn321 is absolutely conserved in all cAMP-specific PDE isoforms and is likely to be a key determinant for cAMP recognition.[48,59]

The active site lies mainly within the third subdomain of the protein and is bounded by helices H15 and H14, the C-terminus of H13, and the C-terminus of H11, along with the loop region between H11 and H12. The majority of the interactions between the substrates/inhibitors and the protein are hydrophobic in nature, with only one or two direct hydrogen bonds observed. The key hydrogen bond interactions are with the absolutely conserved glutamine (Gln817 in PDE5, Gln988 in PDE3, and Gln339 in PDE4) with additional H-bonds formed with Asn321 in the complex of PDE4 with AMP (**Figure 9**).[58]

Each of the structures of wild-type PDE5 has inherent disorder (**Figure 10**). This disorder is likely to be due to the absence of the regulatory domain in the protein construct used in the crystallographic studies. On publication of the PDE4 structure, Pfizer adopted an approach of replacing the sequence corresponding with the small subdomain with that of PDE4 in order to produce a more robust crystallographic system for SBDD. The resulting chimeric construct (dubbed PDE5* and later reported in Zhang *et al.*[59]) retained catalytic activity and PDE5-like inhibitor binding profiles. The subdomain structure added from PDE4 has since been shown to be conserved among other PDE gene family members.

4.29.4.1 Substrate Recognition and Selectivity among Phosphodiesterase Family Members

The overall shape and size of the active site pocket in all PDEs are similar. The pocket is deep and wide and predominantly lined with hydrophobic residues that, if not identical, are conserved in terms of their properties across the whole family. Although no three-dimensional structures are available for the substrates cAMP and cGMP, models for cGMP-bound PDE5 and cAMP-bound PDE4, generated based on the crystal structures of PDE5 in complex with AMP, sildenafil and vardenafil (see below), indicate that the major hydrophobic interactions with the substrate above and below the bicyclic ring are provided by the conserved aromatic hydrophobic residues, Phe820 and Phe786 (PDE5

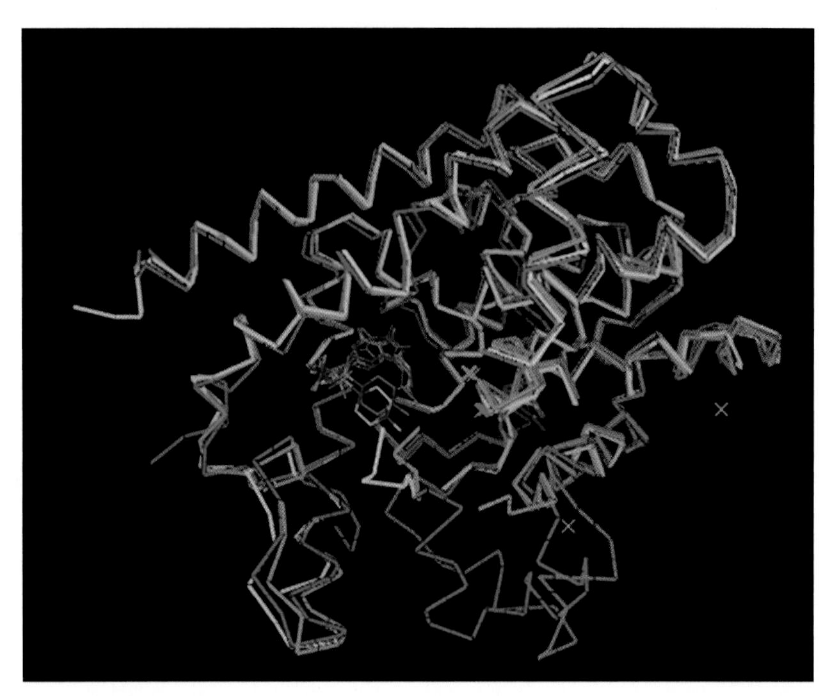

Figure 10 X-ray structures of human PDE5A. White with sildenafil (Pfizer patent) chain break 665–685. Khaki with IBMX (1RKP) chain break 792–807. Purple with sildenafil (1UDT) chain break 664–675. Green chimeric PDE5 (Pfizer patent) residues 657–682 of PDE5 replaced with equivalent PDE4 residues forming ordered subdomain replacing disordered region)

numbering). It is also clear from these models and from the crystal structures of published PDE complexes and structures with other inhibitors (proprietary data) that the hydrogen-bonding network to the absolutely conserved Gln817 (PDE5 numbering) is key to specific substrate recognition. The conserved active site glutamine (Gln817 in PDE5) forms two hydrogen bonds with the nucleotide base of the substrate. In this arrangement, the cyclic phosphate moiety can interact directly with the two metal ions and would be positioned for nucleophilic attack by the hydoxyl ion located between the two metal ions, in accordance with the proposed catalytic mechanism.[50] The presence of the hydroxyl ion has been observed in all high-resolution structures solved so far.

In PDE5, the position of Gln817 is stabilized by Gln775 and Trp853, which allows the recognition of cGMP by two hydrogen bonds (**Figure 10**). Similarly in PDE4 the equivalent glutamine Gln369 is rotated by 180° about the C_γ–C_δ bond to allow two H bonds to cAMP as exemplified in the structure of PDE4 in complex with AMP[60] (**Figure 9**) and the structures of PDE5 and PDE4 complexed with IBMX (1RKP and 1RKO, respectively).[61] In PDE4, Gln369 is stabilized by Tyr329 and Tyr406.

The stabilizing residues for PDE5's Gln817 are not absolutely conserved in all cGMP specific PDEs. However, Trp853 (PDE5 numbering) is conserved in all PDEs apart from PDE4. In PDE9, the residue structurally equivalent to PDE5's Gln775 is an alanine, which does not interact with the conserved glutamine (Gln817 in PDE5); however, this residue (Gln768) interacts directly with NH_2 of cGMP, accounting for the substrate specificity for PDE9. In the structure of PDE10 and models of PDE2, Gln775 (PDE5 numbering) is substituted with Thr, and the position equivalent to Ala783 is a Tyr, which would allow this residue to further stabilize Gln817. In PDE3, however, the conserved Trp forms a hydrogen bond with an adjacent His and does not stabilize the conserved Gln, which shows conformational flexibility in this enzyme allowing it to recognize and bind both cAMP and cGMP as substrates. Similar interactions may also occur in other PDEs with mixed substrate specificity.[59,62]

It is noteworthy that Asn321 (PDE4 numbering) is conserved in all PDEs that exclusively hydrolyse cAMP, as is Trp332, which packs against Asn321, and Pro322, which is contiguous to Asn321. This pattern of residue conservation may imply that the cAMP purine ring may interact with this residues.

In summary, the catalytic mechanism is conserved among the PDE family and the substrate recognition is predominantly tailored by the relative orientation of the conserved glutamine (Gln817 in PDE5) and the stabilization of this orientation by surrounding residues in the substrate pocket.

4.29.4.2 Phosphodiesterase-Binding Site Promiscuity

IBMX is a nonspecific PDE inhibitor. The available crystal structures of its complex with PDE3, PDE4, PDE5, PDE7, and PDE9 constitute a valuable tool when investigating the conserved interactions as well as determinants of inhibitor binding and PDE selectivity (**Figure 11**). IBMX exhibits different binding modes for different PDEs, and the same is likely to be true for other nonspecific PDE inhibitors.

As discussed earlier, the hydrophobic nature of the pocket designed for recognition of the bicyclic core of the substrates dominates the energetics of interaction and it is therefore no surprise that a diverse set of inhibitors with a planar moiety have been shown to bind PDEs.

It is important to note that none of the structures solved to date of complexes of PDE inhibitors with their target enzyme has demonstrated direct interaction with the metal ions. In all cases, the metals are surrounded by a highly ordered network of water molecules, which is also highly conserved between structures.[62] This observation is of particular note when considering how the compounds related to sildenafil bind in the active site. There would appear to be space in the binding site to include a methyl at the N1 position of the sildenafil molecule. The x-ray structure of UK-371800, a compound closely related to sildenafil with a methyl substitution at the N1 position, revealed that the N1 substituent induced a 'flipped' binding mode relative to sildenafil (**Figure 12**).[63] The major interactions are still hydrophobic interactions with the highly conserved Phe820 and Phe786. These two inhibitors also make two specific hydrogen bonds, one to Gln817 and another to Gln775.

The N1 substituted inhibitors are actually more selective for PDE5 over PDE6 than either sildenafil or vardenafil, which exhibit similar hydrophobic interactions with PDE5. This is thought to be due to the interactions with Leu804, which has a different side chain conformation in the structures of complexes with more PDE5 selective compounds including tadalafil (**Figure 13**). The conformational change of Leu804 leading to interaction with the inhibitor could be key to the greater selectivity of cialis for PDE5 as the residue equivalent to Leu804 is a methionine in PDE6 (although Met816, located on the other side of the pocket entrance, is a leucine in PDE6). Additionally, Ile813 in the alkoxy pocket is a leucine in PDE6, which could be important in achieving additional selectivity. The phosphodiesterase examples presented in this section highlight the use of multiple chemotypes and three-dimensional structures of gene family members to reveal structural features that give rise to compound selectivity at an atomic level, and how this information can be utilized in drug design.

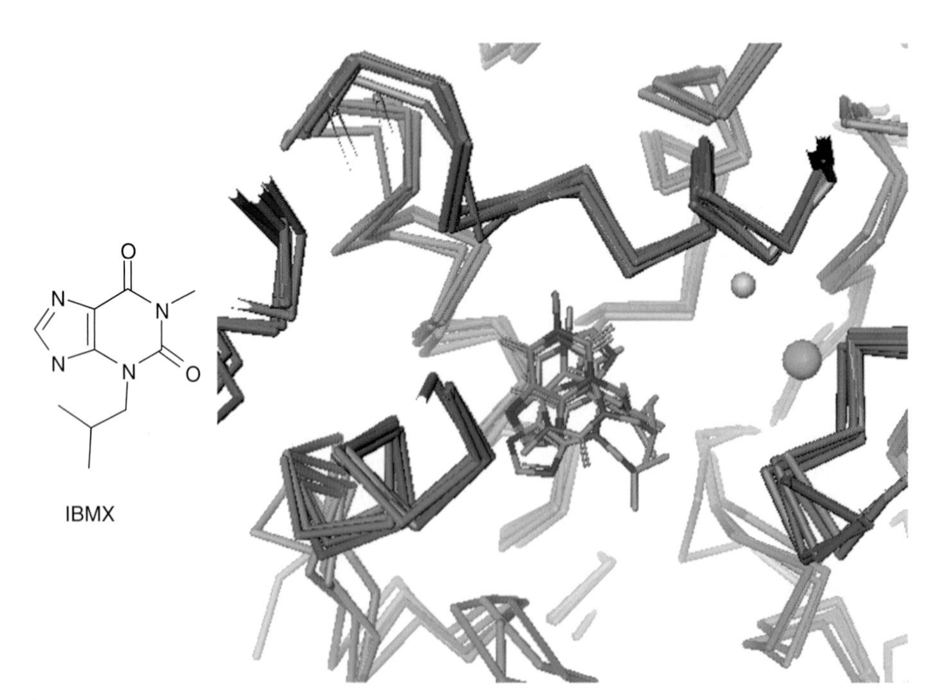

Figure 11 Overlay of the crystal structures of IBMX with PDE5A (khaki), PDE3B (purple), PDE4D (green), PDE7A (light blue), and PDE9A (brown). IBMX is recognised by two hydrogen bonds through Gln817 PDE5, Gln998 in PDE3 and Gln453 in PDE9 to N4 and O4 of IBMX. Only a single direct H bond to N4 of IBMX are made by Gln369 in PDE4 and Gln413 in PDE7.

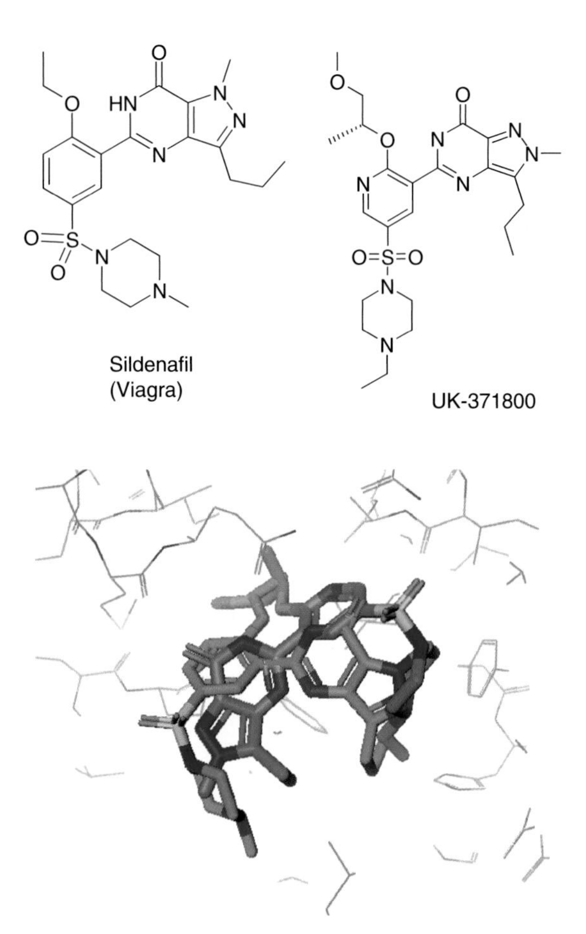

Figure 12 Overlay of the x-ray structures of the complexes of PDE5 with UK-371800 (purple) and sildenafil (green) highlighting the 'flipped' binding mode of N1 substituted compounds.

4.29.5 The Structure Determination Process

4.29.5.1 X-ray Crystallography

The basis of x-ray crystallography is the interaction of matter (proteins, nucleic acids, organic compounds, etc.) in the crystalline state with x-rays. Since the wavelengths of x-rays (in the nanometer range) are of the same order of magnitude as the interatomic distances within the crystal, the interaction results in x-ray diffraction. The diffraction pattern, which can be recorded by x-ray-sensitive detectors, is uniquely determined by the atomic arrangement in the crystal and, thus, contains information about the three-dimensional structure of the molecules making up the crystal. The use of computer software to analyze the x-ray diffraction data allows the crystallographer to derive an electron density map of the protein (or any other molecule) in the crystal. From the electron density map, the position of all the atoms (excluding hydrogen) can be deduced and a first three-dimensional model of the protein built. The process of generating and recording x-ray diffraction data from a crystal is referred to as data collection, while the phrase 'to solve a structure' means to obtain the three-dimensional structure, as a set of cartesian coordinates (x, y, z), from the diffraction data. The first molecular model may contain inaccuracies due to experimental errors and approximations inherent to the method. The iterative process of improving this model to make it as close a representation of the 'true' protein as possible is called refinement; it is carried out using computer programs and interactive computer graphics.

4.29.5.2 From Clone to Crystal

The starting point of the structure determination process is the generation of suitable protein reagent by recombinant DNA technology (occasionally, by isolation of the protein from the natural source). A construct is designed, cloned into

Tadalafil (Cialis)

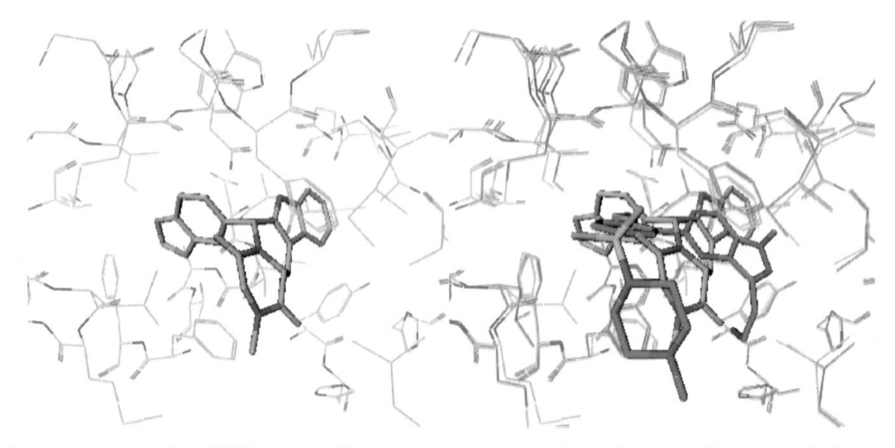

Figure 13 Stucture of tadalafil in PDE5 (grey, 1UDU) and comparison to sildenafil (green, Pfizer patent). The conformation of Gln817 changes to make a single H bond to NH of tadalafil. The benzodioxane moiety of tadalafil occupies the same pocket as the alkoxy substituent of sildenafil and vardenafil. This conformation of Leu804 changes to narrow the entrance to this PDE5 specific "alkoxy pocket" and form additional hydrophobic interactions with the inhibitor.

a suitable vector, and expressed in an appropriate expression system. The expressed protein is purified to a high degree of purity and homogeneity and subject to crystallization screening.

A crystal of the protein of interest is the basic requirement for protein structure determination by crystallography. A crystal is a regular (three-dimensional and periodic) arrangement or packing of 'identical' molecules (or groups of molecules). Because of its regular and periodic nature, the crystal displays symmetry. The smallest repeating motif from which the complete crystal can be derived by use of symmetry operations (including translations) is called the asymmetric unit. The motif can be a molecule or a group of molecules. If the motif is made of multiple copies of the molecule or protein complex, there will be local symmetry operators that can be applied to relate these molecules within the motif. This is known as noncrystallographic symmetry.[64,65] The unit cell is the basic building block, which repeated in space along three (noncolinear) directions forms the complete crystal. The unit cell may contain one or more asymmetric units.

For structure determination, a crystal is suitable if it is single and has a sufficiently large volume to produce measurable diffracted intensities. Crystallization is the process of slowly transferring the protein molecules present in a solution to a crystal. Usually, this is accomplished by the addition of organic substances and/or salts in order to bring the protein solution to supersaturation. Many variables such as quality of the protein preparation, chemical composition of the solution, temperature, and many other factors (some of them unknown) affect crystallization. In practice, given its multiparametric nature and complexity, crystallization is largely a trial-and-error process.

The path from clone to crystal may be lengthy and may involve work with a multiplicity of protein variants in order to achieve success. In general, this part of the process can be accomplished in time intervals varying from a few days to months. Once crystals are available, to obtain a full diffraction data set can take from minutes to hours, depending on the x-ray source, the quality of the crystal, and the symmetry of the crystal.

4.29.5.3 From X-ray Data to the Atomic Model of the Protein

The diffraction pattern recorded during a crystallographic experiment measures a set of x-ray intensities corresponding to the x-rays scattered by the crystal. These intensities are proportional to the square of the amplitude of the scattered x-rays (which are electromagnetic waves), but the phase of the scattered waves cannot be derived from the measured intensities. Since the electron density in the crystal is related to both the amplitudes and the phases of the scattered waves, the information obtainable from the measured intensities is incomplete to elucidate the electron density of the molecules in the crystal. This fact is known as the 'phase problem' in crystallography. Methods that have been developed to circumvent the phase problem are covered in many crystallography books[66,67] and will not be described here (see also Chapter 3.21).

The electron density, computed with the amplitudes derived from the intensities measured in the crystallographic experiment and the phases obtained by one of the standard methods of phase determination, is interpreted by the crystallographer. Using appropriate computer programs and interactive computer graphics as well as his/her chemical knowledge, the crystallographer builds an atomic model of the protein that fits the observed electron density. The initial model of the protein is refined in an iterative fashion using computer programs. The process of refinement consists of a variable number of cycles in which the structure factors are calculated from the model, the calculated structure factors (F_c) are compared to the observed structure factor amplitudes (F_o, obtained from the crystallographic experiment), and the model is modified to find a closer agreement between calculated and observed structure factors.

The electron density is represented as a three-dimensional object, called the electron density map, on the screen of the computer graphics. Crystallographers use a variety of linear combinations of the observed and calculated structure factors to compute electron density maps that are used to refine the model. The most frequently used maps are referred to as F_o–F_c and $2F_o$–F_c. The F_o–F_c map is computed as the difference between structure factor amplitudes calculated from the model and those observed in the experiment, using calculated phases. The difference map shows those atoms that have not been accounted for by the model as positive electron density, while those atoms not present in the molecule but wrongly placed in the model by the crystallographer are seen as negative density. The F_o–F_c map can show small changes in structure such as the inclusion or removal of a water molecule. The $2F_o$–F_c map is computed as the difference between twice the observed structure factor amplitudes and the calculated amplitudes ($|F_o| + |F_o| - |F_c|$), using calculated phases. This map shows both the electron density of the model and the differences between the actual structure and the model (as in the difference map).

The improvement of the atomic model during crystallographic refinement is monitored by the reliability index R, also referred to as crystallographic R-factor, which measures the agreement between observed (experimental) and calculated (derived from the model) structure factor amplitudes. The R-factor is usually expressed as per cent. Values between 15% and 25% indicate a well-refined structure. If, instead of the model, a random distribution of atoms were used to calculate F_c then the R-factor would be approximately 59%.[68] The 'free R-factor' is a reliability index introduced by Brünger in 1992,[69] based on the statistical method of cross-validation. Its value should not be more than 10% higher than the crystallographic R-factor.

4.29.5.4 Useful Crystallographic Terminology When Utilizing Crystal Structures

Resolution is the minimal distance between two objects at which they can be observed as separate entities. The higher the resolution the smaller the distance between the objects (2 Å resolution is higher than 3 Å). In protein crystal structure, the higher the resolution, the better the information content of the structure (interatomic distance of C–C bond = 1.54 Å). For SBDD, it is desirable to obtain crystal structures at resolutions better than 2.5 Å. Crystal structures determined at 3.0–3.5 Å resolution contain valuable information about the main features of the protein and positioning of the ligand but, in general, do not provide information detailed enough to unambiguously describe all protein–ligand interactions.

Crystal structures, which are usually presented as ribbon diagrams or ball and stick models on a computer graphics screen or in pictures, are stored as files containing the cartesian coordinates (x, y, z) for the position of each atom in the structure. In addition to the positional coordinates, two other parameters are included for each atom to describe the structure. These are the occupancy and the temperature factor.

Atoms in a crystal are not static, they vibrate around their equilibrium positions; this is called thermal vibration. The temperature factor or B-factor is a parameter that models thermal vibration. The lower the value of an atom's B-factor, the better defined the electron density for that atom (well-defined atoms have B $< 30\,Å^2$), while high values of B-factor imply poorly defined electron density and low confidence in atomic position. Regions of the protein where the atoms have high B-factors are often flexible regions such as loop regions. In extreme cases, where the electron density is not significantly above the noise level, atoms are omitted from the structure. These regions are referred to as 'disordered.'

In high-resolution structures, sometimes multiple conformations can be observed for amino acid side chains; these are modeled in the protein structure by utilization of partial occupancy for the specified atoms (e.g., two equally observed conformations would be modeled with two sets of coordinates, one for each conformation, each atom with an occupancy of 0.5).

References

1. Hopkins, A. L.; Groom, C. R. *Nat. Rev. Drug Disc.* **2002**, *1*, 727–730.
2. Berman, H. M. et al. *Nucleic Acids Res.* **2000**, *28*, 235–242.
3. Tonks, N. K. *FEBS Lett.* **2003**, *546*, 140–148.
4. Zhang, Z. Y.; Zhou, B.; Xie, L. *Pharmacol. Therapeut.* **2002**, *93*, 307–317.
5. Barford, D.; Flint, A. J.; Tonks, N. K. *Science* **1994**, *263*, 1397–1404.
6. Zhang, Z. Y. *Prog. Nucleic Acid Res. Mol. Biol.* **2003**, *73*, 171–220.
7. Pannifer, A. D. et al. *J. Biol. Chem.* **1998**, *273*, 10454–10462.
8. Scapin, G. et al. *Prot. Sci.* **2001**, *10*, 1596–1605.
9. Salmeen, A. et al. *Nature* **2003**, *423*, 769–773.
10. van Montfort, R. L. M. et al. *Nature* **2003**, *423*, 773–777.
11. Jia, Z. et al. *Science* **1995**, *268*, 1754–1758.
12. Lohse, D. L. et al. *Biochemistry* **1997**, *36*, 4568–4575.
13. Davies, D. R.; Hol, W. G. *FEBS Lett.* **2004**, *577*, 315–321.
14. Zhang, M. et al. *Biochemistry* **1997**, *36*, 15–23.
15. Patel, M. S.; Roche, T. E. *FASEB J.* **1990**, *4*, 3224–3233.
16. Roche, T. E. et al. *Prog. Nucleic Acid Res. Mol. Biol.* **2001**, *70*, 33–75.
17. Yeaman, S. J. et al. *Biochemistry* **1978**, *17*, 2364–2370.
18. Korotchkina, L. G.; Patel, M. S. *J. Biol. Chem.* **1995**, *270*, 14297–14304.
19. Kolobova, E. et al. *Biochem. J.* **2001**, *358*, 69–77.
20. Patel, M. S.; Korotchkina, L. G. *Exp. Mol. Med.* **2001**, *33*, 191–197.
21. Espinal, J. et al. *Drug Dev. Res.* **1995**, *35*, 130–136.
22. Aicher, T. D. et al. *Bioorg. Med. Chem. Lett.* **1999**, *9*, 2223–2228.
23. Aicher, T. D. et al. *J. Med. Chem.* **1999**, *42*, 2741–2746.
24. Aicher, T. D. et al. *J. Med. Chem.* **2000**, *43*, 236–249.
25. Bebernitz, G. R. et al. *J. Med. Chem.* **2000**, *43*, 2248–2257.
26. Mann, W. R. et al. *Biochim. Biophys. Acta* **2000**, *1480*, 283–292.
27. Morrell, J. A. et al. *Biochem. Soc. Trans.* **2003**, *31*, 1168–1170.
28. Mayers, R. M. et al. *Biochem. Soc. Trans.* **2003**, *31*, 1171–1173.
29. Bower-Kinley, M.; Popov, K. M. *Biochem. J.* **1999**, *344*, 47–53.
30. Wynn, R. M. et al. *J. Biol. Chem.* **2000**, *275*, 30512–30519.
31. Dutta, R.; Inouye, M. *Trends Biochem. Sci.* **2000**, *25*, 24–28.
32. Wigley, D. B. et al. *Nature* **1991**, *351*, 624–629.
33. Brino, L. et al. *J. Biol. Chem.* **2000**, *275*, 9468–9475.
34. Obermann, W. M. et al. *J. Cell Biol.* **1998**, *143*, 901–910.
35. Tanaka, T. et al. *Nature* **1998**, *396*, 88–92.
36. Bilwes, A. M. et al. *Cell* **1999**, *96*, 131–141.
37. Ban, C.; Junop, M.; Yang, W. *Cell* **1999**, *97*, 85–97.
38. Steussy, C. N. et al. *J. Biol. Chem.* **2001**, *276*, 37443–37450.
39. Machius, M. et al. *Proc. Natl. Acad. Sci. USA* **2001**, *98*, 11218–11223.
40. Kato, M. et al. *EMBO J.* **2005**, *24*, 1763–1774.
41. Knoechel, T. R. et al. *Biochemistry* **2006**, *45*, 402–415.
42. Bao, H. et al. *Biochemistry* **2004**, *43*, 13432–13441.
43. Lohmann, S. M. et al. *Trends Biochem. Sci.* **1997**, *22*, 307–312.
44. Charbonneau, H. Structure-Function Relationships among Cyclic Nucleotide Phosphodiesterases. In *Cyclic Nucleotide Phosphodiesterases: Structure, Regulation and Drug Action*; Beavo, J. A., Houslay, M. D., Eds.; John Wiley Inc.: New York, 1990, pp 267–296.
45. Rall, T. W.; Sutherland, E. W. *J. Biol. Chem.* **1958**, *232*, 1065–1076.
46. Sutherland, E. W.; Rall, T. W. *J. Biol. Chem.* **1958**, *232*, 1077–1091.
47. Butcher, R. W.; Sutherland, E. W. *J. Biol. Chem.* **1962**, *237*, 9021–9258.
48. Houslay, M. D.; Milligan, G. *Trends Biochem. Sci.* **1997**, *22*, 217–224.
49. Beavo, J. A. *Physiological Rev.* **1995**, *75*, 725–748.
50. Kenan, Y. et al. *J. Biol. Chem.* **2000**, *275*, 12331–12338.
51. Ho, Y.-S. J. et al. *EMBO J.* **2000**, *19*, 5288–5299.
52. Fawcett, L. et al. *Proc. Natl. Acad. Sci. USA* **2000**, *97*, 3702–3707.
53. Brown, D. G. et al. The Crystal Structure of Phosphodiesterase 5 and Use Thereof. WO 2003038080.
54. Brown, D. G. et al. Crystal Structures of Phosphodiesterase 5 and Use Thereof. WO 2004097010, 2004.
55. Pandit, J. Crystal Structure of 3′,5′-Cyclic Nucleotide Phosphodiesterase (PDE10A) and Uses Thereof. Pfizer, Inc. US2005/0202550, 2005.
56. Sung, B.-J. et al. *Nature* **2003**, *425*, 98–102.
57. Scapin, G. et al. *Biochemistry*, **2004**.
58. Xu, R. X. et al. *J. Mol. Biol.* **2004**, *337*, 355–365.
59. Zhang, K. Y. J. et al. *Mol. Cell.* **2004**, *15*, 279–286.
60. Huai, Q.; Colicelli, J.; Ke, H. *Biochemistry* **2003**, *42*, 13220–13226.

61. Huai, Q. et al. *J. Biol. Chem.* **2004**, *279*, 13095–13101.
62. Manallack, D. T. et al. *J. Med. Chem.* **2005**, *48*, 3449–3462.
63. Allerton, C. M. N. et al. *J. Med. Chem.* **2006**, *49*, 3581.
64. Kleywegt, G. J. *Acta Crystallogr. Sect. D* **1996**, *52*, 842–857.
65. Rossmann, M. G. *Acta Crystallogr. Sect. D.* **2001**, *57*, 1360–1366.
66. Drenth, J. *Principles of Protein X-Ray Crystallography*; Springer-Verlag: New York, 2002.
67. Glusker, J. P.; Lewis, M.; Rossi, M. *Crystal Structure Analysis for Chemists and Biologists*; VCH Publishers: New York, 1994.
68. Wilson, A. J. C. *Acta Crystallogr.* **1950**, *3*, 397–398.
69. Brünger, A. T. *Nature* **1992**, *355*, 472–474.
70. Hansen, J. L. Antibiotics and the Ribosome. In *Protein Crystallography in Drug Discovery*; Babine, R. E., Abdel-Megid, S. S., Eds.; Wiley-VCH: Weinheim, 2004, pp 99–125.

Biographies

Maria M Flocco is Director and Head of Structural Biology and Biophysics at Pfizer Global Research and Development, Sandwich Laboratories, UK. Before joining Pfizer, she was Head of Structural Chemistry at Pharmacia Corp., Nerviano, Italy, and a Senior Scientist in Structural Chemistry at Pharmacia & Upjohn, Stockholm, Sweden. Previous to her career in the pharmaceutical industry, Dr Flocco was a Lecturer at the Karolinska Institute, Stockholm, and held a research position at the Uppsala Biomedical Center, Sweden. Dr Flocco received a PhD in physical chemistry from the City University of New York, NY, and postdoctoral training in crystallography at the Fox Chase Cancer Center, Philadelphia, and the Biomedical Center of the University of Uppsala, Sweden. Dr Flocco's main scientific interests are in the areas of protein structure and its relation to drug design, and biophysical methods and their applications to new paradigms in lead discovery.

David G Brown is head of the Crystallography Team at Pfizer Global Research and Development, Sandwich Laboratories, UK. Dr Brown graduated in biophysics from the University of Leeds. He received a PhD from the University of London (Institute of Cancer Research) in crystallography. From 1991 to 1992, Dr Brown held the position of Postdoctoral Research Fellow at the Institute of Cancer Research, moving on to the Kings College University, London, in 1992. During his time in these positions he studied the molecular interactions of anticancer agents targeting both DNA and proteins. In 1995, Dr Brown joined Pfizer Global Research and Development. He is a practicing x-ray crystallographer solving three-dimensional structures of proteins of therapeutic interest, with inhibitors to enable SBDD. He has been involved in SBDD on numerous projects of diverse therapeutic interest.

Andrew Pannifer is a Principal Scientist at Pfizer Global Research and Development, Sandwich Laboratories, UK. Dr Pannifer graduated from the University of Oxford in biochemistry and received a DPhil in crystallography from the University of Oxford (Laboratory of Molecular Biophysics) where he was supervised by Dr David Barford. His thesis included the structures of several PTP-1B/ligand complexes to gain insights into the catalytic mechanism of tyrosine phosphatases. In 1997, Dr Pannifer moved to The University of Leicester to take up a postdoctoral position with Prof Bob Liddington before moving to AstraZeneca and its spin-off company Syngenta, where he worked on pharmaceutical and agrochemical SBDD projects. In 2004, Dr Pannifer joined Pfizer Global Research and Development. He is a practicing x-ray crystallographer solving three-dimensional structures of proteins of therapeutic interest, with inhibitors, to enable SBDD. He has been involved in SBDD on numerous projects of diverse therapeutic interest.

© 2007 Elsevier Ltd. All Rights Reserved
No part of this publication may be reproduced, stored in any retrieval system or transmitted in any form by any means electronic, electrostatic, magnetic tape, mechanical, photocopying, recording or otherwise, without permission in writing from the publishers

Comprehensive Medicinal Chemistry II
ISBN (set): 0-08-044513-6

ISBN (Volume 4) 0-08-044517-9; pp. 749–766

4.30 Multiobjective/Multicriteria Optimization and Decision Support in Drug Discovery

M Afshar and A Lanoue, Ariana Pharmaceuticals, Paris, France
J Sallantin, CNRS, Montpellier, France

© 2007 Elsevier Ltd. All Rights Reserved.

4.30.1 Introduction

Decision making in real world problems often involves consideration of multiple objectives and uncertain outcomes, which has already generated a vast and growing literature (see for example [1]). While we give a brief overview of the field, our paper is not intended to cover all aspects of multiple objective optimization/decision making and instead focuses on the few published applications in cheminformatics, where it is beginning to be exploited, and discusses future directions.

A drug must act effectively but without any serious side effects. These two requirements are often in conflict and at the more complicated functional level, a design solution will be acceptable only if it meets a number of different objectives at the same time. Thus, the design of effective and safe new drugs involves optimizing multiple objectives (also called multiple criteria) at the same time, a much harder problem than optimizing a series of individual ones. These objectives are often dependent and to make matters worse, they can be conflicting or competing. It is a common situation that the individual optima corresponding to the distinct objective functions are very different; therefore, one cannot optimize the competing objectives simultaneously and must make tradeoffs among these objectives. Further difficulties arise when particular states of the objectives conflict. Simple examples are the trade-off of risk versus profit in investment, potency of a molecule versus its pharmacokinetic profile, or in treating osteoporosis, the benefit of using HRT against the risk of breast cancer.

4.30.2 Scalar Objective

In an attempt to simplify multiple objective problems, one can consider that although the various characteristics or objectives that need be optimized are distinct, it is possible to combine these objectives into one, for example, by defining a score that is the weighted sum of all objectives. Although this strategy is severely biased by the structure of the scoring function, i.e., the weights and the operations between the objectives (for example a sum), it opens up the possibility of using classical methods such as genetic algorithm, random search, simulated annealing, etc. to solve multiple objective problems.

Quantitative structure–activity relationship (QSAR) equations are standard examples in cheminformatics where an overall fitness score is developed as a weighted sum of numerous descriptors. In docking, the score includes ligand internal energy, interaction energy, and entropic considerations in the form of a weighted sum of terms. Typically, the score is developed empirically by analyzing a set of examples and deriving a weighted sum. The weights are 'fitted' to the learning set and may not necessarily be relevant or precise for other complexes. A precautionary approach is to use the individual contributions from distinct terms for ranking, e.g., one seeks 'good' van der Waals and electrostatic contributions but one is not quite sure of how these two should be scaled to become additive.[2] However, an effective method for working with multiple objectives is desirable.

4.30.3 Multiobjective

The principle of multiobjective optimization was first formalized by Vilfredo Pareto, a late nineteenth century French born Italian economist. In his *Manual of Political Economy*, first published in 1906, he introduced (and further formalized in the revised 1909 version) the concept of the *Pareto optimum*, a standard of judgment in which the optimum allocation of the resources of a society is not attained as long as it is possible to make at least one individual better off while keeping others as well off.

This concept has been subsequently formalized for reaching the best compromises: the Pareto-optimal set or 'Pareto front' is reached when a set of solutions have been obtained where no objective can be further improved without at the same time worsening another one.

This paradigm has led to variety of applications in different industries, including manufacturing, R&D management, product development, finance, transportation, power systems management, and capacity planning.[3]

4.30.4 Structure of Multiobjective Problem Solving

Once the objective functions have been defined, quantitative multiple objective problems require a search method in order to identify the Pareto-set and ideally a 'decision maker' to select solutions from the Pareto-set.

Using stochastic methods, the generation of scenarios can be performed by discretizing the continuous probability distributions or by Monte Carlo type simulation techniques.[3] Multiobjective evolutionary algorithms (MOEAs) represent perhaps the most popular alternative in multiple objective optimization. MOEA methods use genetic algorithm approaches to generate all the Pareto-front solutions.[4,5] The selection of individuals is based on their Pareto optimality. There are several ways of performing this Pareto ranking, leading to several well-known approaches under the category MOEA. These include multiobjective genetic algorithm (MOGA).[6,7] Multiple objectives are treated independently, and the fitness ranking of a genetic algorithm is replaced by Pareto ranking, which is based on the concept of dominance. A nondominated solution is one where an improvement in one objective results in deterioration in one or more of the other objectives when compared with the other solutions in the population. This approach differs from conventional genetic algorithms with respect to the selection operator emphasizing the nondomination of solutions. Nondomination is tested at each generation in the selection phase, thus defining an approximation to the Pareto optimal set.

Other popular algorithms are niched Pareto genetic algorithm (NPGA),[8] Pareto archived evolutionary strategy (PAES),[9] strength Pareto archived evolutionary algorithm (SPEA),[10] elitist nondominated sorting genetic algorithm (NSGA-II),[11] and region-based selection (PESA-II).[12]

An important aspect of MOGA is how diversity in the population is maintained and encouraged. In PESA-II if the external population becomes too large, then some nondominated solutions are discarded. The method discards solutions that are close to others in the multiple objective fitness space in order to maintain diversity. Similarly, the selection of internal population that undergoes the mutations, etc. are done in a way where unique or isolated solutions are selected more often than those that populate densely crowded areas of fitness space.[12]

4.30.5 Applications in Cheminformatics

Applications in complex chemical engineering or analytical instrumentation optimization problems have become increasingly popular (see, for example, polymerization optimization using NSGA-II with three objectives,[13] optimization in time-of-flight mass spectrometry using PESA-II with three objectives[14]). Applications of MOEA in cheminformatics are growing in numbers.

4.30.5.1 Maximum Common Substructure

The first application of the MOEA in cheminformatics has been the use of MOGA for flexible superposition of three-dimensional structures, using two objectives: 'number of atoms in substructure' and 'fit of the matching atoms'.[15] These two objectives compete: as the size of the substructures increase, the geometric fit would tend to get worse. Harik[16] improves the use of MOGA by restricted tournament selection (RTS). Rather than producing one best overlay, for each possible size of the common substructure, an optimal geometric fit is produced.

4.30.5.2 Combinatorial Library Design

Combinatorial library design, where libraries need to be optimal over a number of competing properties, such as size and diversity, has attracted several multiple objective studies. Agrafiotis[17] uses simulated annealing as search method, while Gillet *et al.*[18] and Wright *et al.*[19] both use a MOGA framework.

Wright *et al.*[19] investigate the effect of optimizing library size simultaneously with other library characteristics such as diversity and drug-like physicochemical properties. Their system, MoSELECT.II, has been applied to two different virtual libraries: a two-component aminothiazole library consisting of 12 850 products generated from 74 R-bromo-ketones coupled with 170 thioureas, and a four-component benzodiazepine library consisting of 256 036 products. The method uses a niching technique that uses fitness sharing to improve diversity of the solutions thus avoiding MOGA's tendency to genetic drift where they converge toward a single solution. Fitness sharing and its advantages relating to MOGA are discussed by Horn *et al.*[8] MoSELECT.II's implementation of fitness sharing ensures that nondominated individuals are evenly distributed on the Pareto surface.

As argued by Fonseca and Fleming,[20] many and highly competing objectives can lead to a nondominated set that is too large for a finite population to sample effectively. Furthermore, only a very small portion of the nondominated set is of practical relevance, which further substantiates the need to supply preference information or constraints to the genetic algorithm. Wright *et al.* use constraints to ensure the validity of the pareto-set of solutions. The constraints include: plate constraints, specified as the minimum percentage of plate coverage that is allowed; size constraints, specified as upper and lower bounds on the final number of products in a library; and combinatorial efficiency constraints, specified as upper and lower limits on the number of reactants allowed for each library component. The method ensures that the search will favor nondominated feasible solutions over nondominated unfeasible solutions.

Nevertheless, introduction of constraints can emphasize the nonconvexity of the search space (feasible solutions could be reached through a path including unfeasible solutions) and degrade the performance of the search engine.

Wright *et al.* present aminothiazole libraries designed when simultaneously and effectively optimizing over six objectives: library size, diversity, and profiles of molecular weight; rotatable bonds; hydrogen bond donors and acceptors.

4.30.5.3 De Novo Design

Another exciting application of MOGA is reported by Brown *et al.*[21] The authors present a graph-based genetic algorithm for the evolution of molecules 'toward' a number of objective molecules. Their methodology combines sophisticated graph-based mutation operations with Pareto ranking. The Tanimoto similarity of each candidate molecule with a number of objective molecules is calculated and is Pareto ranked. Examples are given when using two objective molecules that are either similar or diverse.

4.30.5.4 Improving Quantitative Structure–Activity Relationships and Pharmacophore-Based Techniques

MOGA framework can also be used for the design of predictive tools, i.e., as a means of choosing more desirable QSAR or pharmacophore models.

4.30.5.4.1 Selecting quantitative structure–activity relationship models

QSAR attempts to relate a numerical description of molecular structure/properties to known biological activity. Large numbers of readily computable descriptors are available that in combination with sophisticated techniques improve the initial linear regression analysis methods used in deriving QSAR equations. In general, QSAR equations relate one objective (such as activity) with a number of descriptors. They could be improved in several ways: first by trying to relate multiple objectives rather than a single objective to a set of descriptors; second, within the framework of mono-objective QSAR equations, deriving models that are accurate, reliable, and easily interpretable, i.e., improving their quality. Accuracy, reliability, and interpretability can be considered three simultaneous objectives that would need to be balanced when choosing a QSAR equation.

Nicolotti *et al.*[22] report on the application of Pareto ranking to deriving a family of QSAR models that are simultaneously optimized on accuracy, complexity, and interpretability. This is 'hybrid' thinking between multiple objective strategies and traditional QSAR (i.e., identifying linear models that represent a balance between variance and number of descriptors). Since QSAR equations are constructed by the combination of a number of weighted terms (descriptors), Nicolotti *et al.* introduce an additional objective, which is Desirability. Desirability is a scalar derived from the desirability of each individual descriptor, as set by the user: desirability of each descriptor can be 3(excellent), 2(fair), or 1(poor). Desirability rewards the presence of more desirable descriptors in any given QSAR equation. On several test cases, they show that their method, MoQSAR, selects a number of QSAR models that are at least as good as the models derived using standard statistical approaches, but importantly, at the same time, yields models that allow a medicinal chemist to trade statistical robustness for chemical interpretability.

4.30.5.4.2 Selecting pharmacophores

Pharmacophores are a set of methods related to QSAR: they produce three-dimensional arrangements of functional group that are required for activity. Pharmacophores are usually used to derive relation between the structure and the activity of a set of molecules. Similarly to the QSAR situation, pharmacophores are traditionally used to predict only one activity (i.e., one objective). However, pharmacophores can have multiple 'quality' descriptors such as conformational energy (mean internal van der Waals energy), volume score (measures overlap between the molecules), and feature score (correspondence between particular properties of the molecules summed over all pharmacophore points). Cottrell *et al.*[23] use MOGA to generate pharmacophore hypothesis that optimize these three 'quality' objectives. The method presented can sometimes lead to a large number of pareto-optimal pharmacophoric models that may be difficult to analyze However, this is a better alternative to the standard approach that reduces the number of models based on nonoptimal criteria.

4.30.6 Limitations of Numerical Multiobjective Algorithms

All applications that we have discussed so far focus on quantitative machine learning methods. They use numerical methods for both calculating the objective functions and searching for the pareto-front. All have to avoid local optima problems, ensure the effective selection of the pareto-front set while maintaining a good diversity of the solutions (good spread). Obviously, the calculations need to be done in a timely manner.

Several situations can generate problems that can limit the applicability of these methods and in what follows we shall briefly mention a few. First, the need to have a large number of fitness evaluations requires a rapid simulation method. In situations where one lacks the possibility of performing fast simulations (i.e., experimental context), the number of evaluations of the fitness scores need to be dramatically reduced, hence impacting the viability of the convergence of the above discussed methods. On the contrary, problems with large numbers, such as library design problems where the method is asked to select out 250 000 compounds, a subset that optimizes a few objectives, are well adapted. However, even in the latter, if the number of objectives grows, then again nonconvexity bias will interfere with the search and adversely impact the quality of the pareto-set. Overall, in problems where it is not easy to simulate a particular potential solution in a timely fashion, MOEAs are less suitable.

Another limitation is that most methods are highly vulnerable to nonlinear interdependencies between objectives, i.e., when the ranking of a given objective changes depending upon the values of other objectives. A further issue is linked to Intransitivity where A B and B C does not lead to A C.

Search strategies can be compromised when confronted with nonconvex solution fronts, i.e., a solution 'between' two valid solutions could be invalid. Furthermore, scale invariance is not always true, i.e., even for a continuous property such as molecular weight, its use and therefore significance is distinct for different ranges (for example, 200–600 Da range corresponds to small molecules, a molecular weight greater than 2000 Da does not). This is to say that some relations are sensitive to scale. More generally, qualities can be converted into quantities (binning) but the reverse is not always true. This leads us to the necessity of defining domains of validity for all parameters, in both the search and the objective spaces. In turn, the notion of domain is linked to boundaries and hence allows characterization of paradoxical combinations or conflicts. Here conflicts are real mutual exclusions rather than a competition between several continuous parameters.

Finally, as discussed earlier, once the pareto-set has been generated, choosing between the pareto-front solutions can still be nontrivial. Are all pareto-front solutions equivalent? Could they be further 'ranked' based on the stability of each solution in response to small changes? Foncesca and Fleming[20] argue the need for preference articulation in cases where many and highly competing objectives lead to a nondominated set too large for a finite population to sample effectively. In several examples they show that only a very small portion of the nondominated set is of practical relevance, which further substantiates the need to supply preference information to the genetic algorithm. This can be performed in the form of additional constraints, but then leading to nonconvexity and affecting the search.

4.30.7 New Directions

Another related and highly active area in the field of computer science and artificial intelligence is constraint satisfaction programming (CSP).[24,25] They are applied to real life multiple objective optimization problems that can be as difficult and dynamic as air traffic control. A constraint-based problem is defined by a set of unknowns, the variables with domains, subject to constraints that must be satisfied by these variables. In comparison with our previous discussion on evolutionary multiple objective methods, the CSP paradigm does not distinguish between the objective functions and the parameters. They are all considered as variables that need to satisfy a set of constraints. Variables have valid domains. A domain may be a continuous range, an integer range, or enumerated set of values (that is, a range with holes). The solution is given by the set of values for each variable such that all the constraints are satisfied.

4.30.7.1 Constraint Propagation

Constraint propagation is one of the techniques central to the success of constraint programming. There are a large number of constraint propagation methods.[26] A simple algorithm to search for a solution is to reduce each domain sequentially, until it either leads to a solution or fails to find one. Every time a domain is restricted, the constraints acting on the domain are checked to see if a value belonging to another domain has become forbidden. The process is repeated recursively reducing domains. Upon failure (a domain becomes empty), 'backtracking' is used. The stack of decisions is considered bottom to top until the system proves there are no solutions or finds one. Maintaining global consistency during the search ensures a solution is found, or fails early. Many more sophisticated search methods have been developed.[27] CSP has been applied to various synthetic chemistry problems as well substructure searching.[28]

4.30.7.2 Rule-Based Machine Learning

CSPs and rule-based reasoning can be effectively combined. Machine learning methods that generate logical rules (such as 'if properties A, B, and not C are present then properties F and G are present too') can be used as constraints and the CSP paradigm used to propagate them. There are a large number of rule-based machine learning methods, amongst which the inductive logic paradigm[29] (which generates rules of the type 'if A and B and C then D') and Gallois lattice-based methods.[30] A few direct drug design applications of rule-based methods have been described.[31,32] This contrasts with the large activity in rule-based toxicity related drug design issues.[33,34]

Combined with CSPs, logical rules are used as a set of constraints that define the boundaries of convexity of the search space.

Over the last 2 years, our group at Ariana Pharma has been focusing on the implementation of novel rule-based methods combined with CSPs and their applications to multiobjective drug design issues. To the best of our knowledge, this is the first attempt in this direction using these technologies. The program KEM (knowledge extraction and management) has been applied to a number of problems such as multiobjective lead optimization and decision where one is interested in models that cover activity as well as other terms such as selectivity, absorption, P450 metabolism, protein binding, half-life, etc. The main advantages of this novel approach are its ability to dynamically handle a large number of objectives, its ability to work with a small number of examples that could be incomplete (i.e., with missing parameters), the exhaustivity of the constraint set and the solution set, as well as the ability of the user to examine the constraints and manipulate them in order to dynamically modify the knowledge base, while maintaining consistency. In the logic-based CSP environment, decisions can be explained and the analysis of contradictions allows the user to further understand the data, leading to an interactive exploration of the solution space. The origin of contradictions can be precisely identified. The identification of contradictions is put at the heart of the scientific discovery process. This is a rupture compared to existing approaches in this field. The detailed methodology will be described elsewhere.

4.30.8 Conclusions

Given the inherently complex and multiobjective nature of drug discovery problems, computational multiobjective methods are set to attract a growing interest and deliver high impact in drug design. The first applications in combinatorial library design, where up to six objectives are optimized (e.g., library size, diversity, and several profile features), show great promise when a large number of valid molecules can be constructed and can be searched effectively. The multiple objective paradigm can also be used to significantly improve the selection of scalar functions such as QSAR and pharmacophore-based models. Novel rule-based methods show promise when handling a large number of objectives in noncontinuous search spaces and situations such as lead optimization where large number of simulations is not possible. In a new paradigm, it is the identification and analysis of contradictions that could directly lead to scientific discovery.

References

1. Ehrgott, M.; Gandibleux, X., Eds. *Multiple Criteria Optimization: State of the Art Annotated Bibliographic Surveys*; International Series in Operations Research & Management Science; Kluwer Academic Publishers: USA, 2002.
2. Morley, S. D.; Afshar, M. *J Comput. Aided Mol. Des.* **2004**, *18*, 189–208.
3. Cheng, L.; Subrahmanian, E.; Westerberg, A. W. *Ind. Eng. Chem. Res.* **2005**, *44*, 2405–2415.
4. Deb, K. *Multiobjective Optimization Using Evolutionary Algorithms*; Wiley: Chichester, UK, 2001.
5. Coello Coello, C. A.; Carlos, A.; Van Veldhuizen, D. A.; Lamont, G. B. *Evolutionary Algorithms for Solving Multi-Objective Problems*; Kluwer Academic Publishers: New York, 2002.
6. Foncesca, C. A.; Fleming, P. J. In *Proceedings of the Fifth International Conference on Genetic Algorithms*, San Mateo, CA, 1993; Forrest, S., Ed.; Morgan Kauffman Publishers: San Francisco, CA, 1993, pp 416–423.
7. Fonseca, C. M.; Fleming, P. J. *IEEE Transactions on Systems, Man and Cybernetics* **1998**, *28*, 26–37.
8. Horn, J.; Nafpliotis, N.; Goldberg, D. E. In *Proceedings of the First IEEE Conference on Evolutionary Computation*; IEEE World Congress on Computational Intelligence, Vol. 1; IEEE Service Center: Piscataway, NJ, 1994; pp 82–87.
9. Knowles, J. D.; Corne, D. W. *Evol. Comput.* **2000**, *8*, 149–172.
10. Zitzler, E.; Thiele, L. *IEEE Trans. Evol. Comput.* **1999**, *3*, 257–271.
11. Deb, K.; Agrawal, S.; Pratab, A.; Meyarivan, T. *KanGAL Report 200001*; Indian Institute of Technology: Kanpur, India, 2000.
12. Corne, D. W.; Jerram N. R.; Knowles, J. D.; Oates, M. J. In *Proceedings of the Genetic and Evolutionary Computation Conference (GECCO'2001)*; Morgan Kaufmann Publishers: San Francisco, CA, 2001, pp 283–290.
13. Mitra, K; Majumdar, S; Raha, S *Ind. Eng. Chem. Res.* **2004**, *43*, 6055–6063.
14. O'Hagan, S.; Dunn, W. B.; Brown, M.; Knowles, J. D.; Kell, D. B. *Anal. Chem.* **2005**, 77, 290–303.
15. Handschuh, S.; Wagener, M.; Gasteiger, J. *J. Chem. Inf. Comput. Sci.* **1998**, *38*, 220–232.
16. Harik, G. R. In *Finding Multimodal Solutions Using Restricted Tournament Selection*, Proceedings of the 6th International Conference on Genetic Algorithms; Eshelman, L. J., Ed.; Morgan Kaufmann: San Francisco, CA, 1995, pp 24–31.
17. Agrafiotis, D. K. *IBM J. Res. Dev.* **2001**, *45*, 545–566.
18. Gillet, V. J.; Khatib, W.; Willett, P.; Fleming, P. J.; Green, D. V. S. *J. Chem. Inf. Comput. Sci.* **2002**, *42*, 375.
19. Wright, T.; Gillet, V. J.; Green, D. V. S.; Pickett, S. D. *J. Chem. Inf. Comput. Sci.* **2003**, *43*, 381.
20. Fonseca, C. M.; Fleming, P. J. Multiobjective Optimization and Multiple Constraint Handling with Evolutionary Algorithms. In *Practical Approaches to Multi-Objective Optimization*; Branke, J., Deb, K., Miettinen, K., Steuer, R. E., Eds.; Dagstuhl Seminar Proceedings Series; Denmark, 2005, p 237.
21. Brown, N.; McKay, B.; Gilardoni, F.; Gasteiger, J. *J. Chem. Inf. Comput. Sci.* **2004**, *44*, 1079.
22. Nicolotti, O.; Gillet, V. J.; Fleming, P. J.; Green, D. V. S. *J. Med. Chem.* **2002**, *45*, 5069–5080.
23. Cottrell, S. J.; Gillet, V. J.; Taylor, R.; Wilton, D. J. *J. Comput. Aided Mol. Des.* **2004**, *18*, 665–682.
24. Tsang, E. P. K. *Foundations of Constraint Satisfaction*; Academic Press: London, 1993.
25. Achlioptas, D.; Naor, A.; Peres, Y. *Nature* **2005**, *435*, 759–764.
26. Dechter, R. *Constraint Processing*; Morgan Kaufmann Publishers: San Francisco, CA, 2003.
27. Bessiere, J. C. C.; Régin, R. H. C.; Yap, Y. Zang *Artif. Intell.* **2005**, *165*, 165–185.
28. Barnad, J. M. *J. Chem. Inf. Comput. Sci.* **1993**, *33*, 532–538.
29. Srinivasan, A.; King, R. D. *Data Min. Knowl. Disc.* **1999**, *3*, 37–57.
30. Liquière, M.; Sallantin, J. In ICML'98; Morgan Kaufmann Publishers: Madison, Wisconsin, 1998, pp 305–317.
31. King, R. D.; Srinivasan, A.; Sternberg, M. J. E. *New Gen. Comput.* **1995**, *13*, 411–433.
32. Marchand-Geneste, N; Watson, K. A; Alsberg, B. K.; King, R. D. *J. Med. Chem.* **2002**, *45*, 399–409 (erratum in: *J. Med. Chem.* **2003**, **46**, 653).
33. Duquesne, M.; Sallantin, J. *C. R Seances Acad. Sci. III* **1981**, *292*, 495–498.
34. Richard, M.; Benigni, R. *SAR QSAR Environ. Res.* **2002**, *13*, 11–19.

Biographies

Mohammad Afshar is the CEO and co-founder of Ariana Pharmaceuticals, Pasteur Institute, Paris, France, a drug discovery company focusing on accelerating discovery and development of novel, small molecule drug therapies using its novel multiobjective decision support technologies. Prior to joining Ariana, he was one of the founding members of the senior management group at RiboTargets, Cambridge, UK, where, as the Director of IT and Head of Drug Design, he set up and managed RiboTargets' structure-based discovery platform RiboDock–rDock. The patented technology developed within his team (RiboDock–rDock) allowed the identification and validation of novel therapeutic molecules, becoming the central focus of RiboTargets. He left RiboTargets to launch Ariana Pharmaceuticals at the end of 2002, prior to the company's successful merger with British Biotech and Vernalis. Before joining RiboTargets he was a reserach fellow at the Department of Chemistry of the University of York. He obtained a Medical Degree (DCEM), MPhil in computer science, a PhD in structural biochemistry, and a 'Habilitation doctorate' from the University of Montpellier, France. He is currently a member of the scientific committee of the French Cystic Fibrosis Association (VLM).

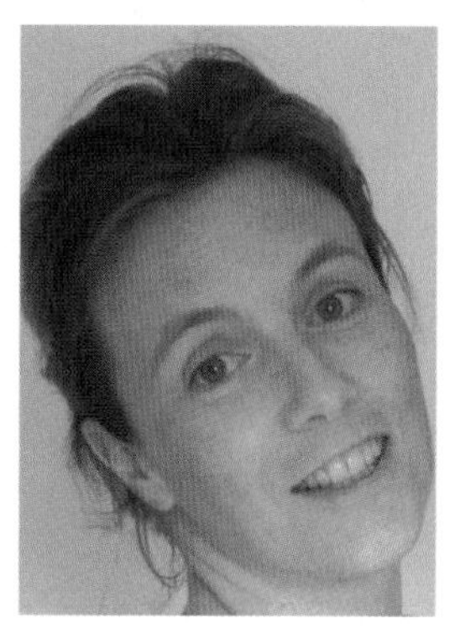

Astrid Lanoue is senior scientist at Ariana Pharmaceuticals. Prior to joining Ariana, she was a research fellow at the Laboratory of Molecular Biology of the MRC at Cambridge, UK, from 1998 to 2004, working with M Neuberger and A McKenzie. She obtained a PhD in immunology in 1998, under the supervision of Prof H von Boehmer and from the University of Paris V (Necker, INSERM U 373), France.

Jean Sallantin, PhD, an expert in artificial intelligence, is a Director of Research at the LIRMM (Laboratoire d'Informatique, de Robotique et de Micro-électronique de Montpellier) of the CNRS in Montpellier, France. One of the pioneers of bioinformatics in France, he established the bioinformatics laboratory at the Curie Institute, Paris (1983–85), and was the director of the GSDIARL research consortium (1985–89) involving Sanofi, the CNRS Pharmacology Institute, the CNRS centre for macromolecular biochemistry and the Montpellier University Medical Centre. From 1993 to 1996, he was appointed the Ministry of Research correspondent on artificial intelligence and cognitive sciences. He currently heads the 'Rationality and machine learning' team that develops and studies the applications of machine-learning techniques in a scientific discovery environment. The methods and algorithms developed in his group have led to a number of patents and collaborations with large companies such as Areva, BNP Paribas, and Fidal KPMG.

© 2007 Elsevier Ltd. All Rights Reserved
No part of this publication may be reproduced, stored in any retrieval system or transmitted in any form by any means electronic, electrostatic, magnetic tape, mechanical, photocopying, recording or otherwise, without permission in writing from the publishers

Comprehensive Medicinal Chemistry II
ISBN (set): 0-08-044513-6

ISBN (Volume 4) 0-08-044517-9; pp. 767–774

4.31 New Applications for Structure-Based Drug Design

C W Murray and M J Hartshorn, Astex Therapeutics, Cambridge, UK

© 2007 Elsevier Ltd. All Rights Reserved.

4.31.1 Introduction

In the last two decades, the pharmaceutical industry has invested heavily in research and development but has failed to generate increases in productivity.[1,2] The current situation in drug discovery research could be summarized as 'target-rich, lead-poor,' where a significant amount of attrition occurs in the early phases of drug discovery when researchers are focused on the identification and optimization of lead compounds. Understandably, there has been substantial interest in technologies aimed at reducing this attrition. In the 1980s, structure-based drug design (SBDD) and computational chemistry were touted as methods that would revolutionize drug discovery, but whilst these methods have firmly established themselves as important and useful disciplines in drug discovery, they have not revolutionized the process. Instead, the 1990s saw substantial investment in high-throughput screening (HTS) and combinatorial chemistry, with HTS becoming the dominant approach to early drug discovery in major pharmaceutical companies. Despite the predominant position of HTS, there remains a need for alternative approaches.[3,4] To this end, there has been a resurgence of interest in SBDD and a recognition that access to structure can improve the chances of delivering high-quality lead compounds, regardless of whether the original hits came from high-throughput methods or from other sources.

It should also be noted that SBDD is now a much more effective paradigm for drug discovery compared to applications 20 years ago, for a number of reasons.[5] Firstly, it is now a significantly more mature technology and there have been important improvements in equipment and associated software. Secondly, a much wider number of drug targets and drug target classes have been crystallized so that it can no longer be considered a niche technology. Thirdly, structural biologists, modelers, and medicinal chemists have a much greater body of experience on which to base their approaches in SBDD projects. The most obvious illustration of these facts lies with the number of structures deposited in the Protein Data Bank (PDB), which, at the end of 2005, stood at over 34 000, compared with fewer than 600 at the end of 1990.[6]

SBDD has traditionally been used to establish the binding mode of lead molecules so as to allow the design of molecules with improved affinity for the target.[5,7,8] The binding mode of a lead compound has also been used to guide the design of compounds with improved physicochemical properties; for example, by suggesting where solubilizing groups might best be placed on the lead compound so as to avoid significant losses in affinity.[9] Another established application of SBDD is virtual screening where a database of compounds is docked against a protein target of known structure to identify inhibitors.[10–15] The first successful application of virtual screening was published in 1993.[16] Other chapters in *Comprehensive Medicinal Chemistry* have reviewed the established methods and applications of SBDD; this chapter is concerned with new application areas for SBDD within drug discovery.

The first part of this chapter will be concerned with the expansion of SBDD into the area of absorption, distribution, metabolism, and excretion (ADME). Here the focus will be on applications of structural biology to determine the binding mode of lead compounds in proteins associated with an ADME-related property. Such structures are, in principle, useful to design out the ADME liabilities of a lead compound whilst maintaining the activity against the target enzyme. The ADME-related proteins considered in detail will be HSA and the CYPs. HSA is one of the main proteins implicated in plasma protein binding (PPB), and extensive binding can often be undesirable because it limits the amount of unbound drug available to inhibit the target protein.[17] The CYPs are responsible for the metabolism of over 90% of clinically used drugs[18] and the modulation of the binding of compound to, or the metabolism of compound by, P450 enzymes is often one of the goals in lead optimization projects.

The second part of this chapter will be on the use of x-ray crystallography much earlier in the drug discovery process. The approach is to use crystallography to screen small libraries of up to a few thousand fragments (i.e., molecules with molecular weight between 100 and 250 Da) against the target protein. Attractive fragment hits are characterized by having weak affinity but high values for the ratio of the free energy of binding to the number of heavy atoms (i.e., high ligand efficiency[19,20]). It is usually possible to optimize fragment hits into attractive lead compounds using SBDD. Overall this represents a new and exciting application of SBDD but also presents a number of experimental and informatics challenges. The informatics challenges in particular will be discussed in this chapter and some examples of the successful application of this approach will be given. It is not the intention to review noncrystallographic approaches to fragment screening here as this is covered in Chapter 3.41 and is also the subject of a number of recent reviews.[19,21,22]

The chapter will end with a summary of the themes discussed and will give a perspective on the expanding applications of SBDD.

4.31.2 Applications of Structure-Based Drug Design to Absorption, Distribution, Metabolism, and Excretion Properties

4.31.2.1 Human Serum Albumin

4.31.2.1.1 Introduction to human serum albumin

HSA is an abundant transport protein found in plasma and is responsible for binding and transporting a number of relatively insoluble endogenous compounds such as fatty acids, bilirubin, and bile acids.[17] Binding to HSA is also a

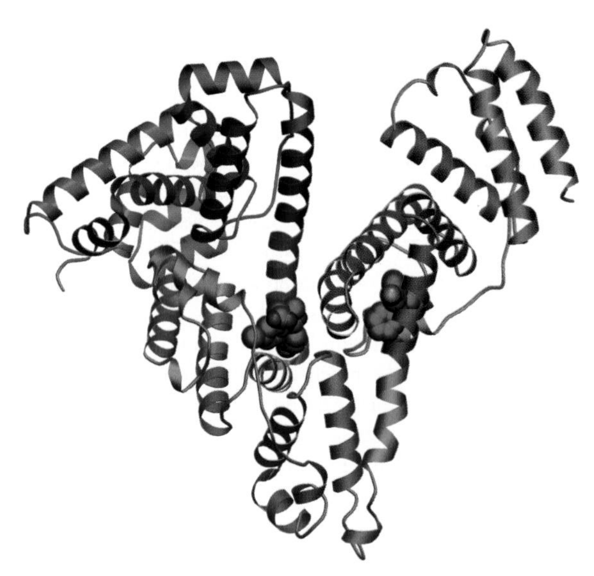

Figure 1 A schematic of the structure of HSA derived from the PDB file, 2BXM. HSA is a heart-shaped molecule composed of three homologous domains which are colored in magenta (domain 1, residues 5–190), cyan (domain 2, residues 191–383), and orange (domain 3, residues 384–end). Indometacin is shown in a space-filling representation bound to drug site I in domain 2 on the center right-hand side of the figure. Also shown is the superimposed position of diazepam (taken from the PDB file, 2BXF) which binds in drug site II in domain 3 on the center left-hand side.

major contributor to the PPB of drug molecules. PPB limits the amount of free drug available in plasma, restricting both the concentration of drug available to inhibit the targeted protein and the concentration of drug available to elimination processes. In the former case, it is often highly desirable to reduce the extent of PPB so as to increase the amount of free drug available. This is particularly important for targets where high concentrations of free drug are required for efficacy (e.g., antibiotic targets[23]).

The percentage of compound bound to plasma is often used to explain poor performance of tight binding inhibitors in cell assays or in vivo. In series where there is a PPB liability, it has often been observed that there is a positive correlation between binding and increasing lipophilicity,[24–26] so a common strategy for medicinal chemists is to increase the polarity of molecules. However, despite this correlation, it would be wrong to assume that binding to HSA is entirely nonspecific.

4.31.2.1.2 Structure of human serum albumin

HSA is a 585-residue protein monomer containing three homologous helical domains each split into A and B subdomains and arranged to form a heart-shaped molecule (**Figure 1**).[27,28] It contains two primary drug-binding sites.[29] Site I (or the warfarin site) is on domain 2 and prefers to bind large heterocyclic and negatively charged compounds, whereas site II (or the indole-benzodiazepine site) is located on domain 3 and is the preferred site for small aromatic carboxylic acids.[24] **Figure 1** shows these sites occupied by representative ligands.

There are now over 40 crystal structures of HSA deposited in the PDB. Structural work has also been performed using nuclear magnetic resonance (NMR) on domain 3 of HSA,[30,31] allowing the elucidation of the binding mode of drug molecules to site II of the protein.

4.31.2.1.3 Quantitative structure–activity relationship (QSAR) approaches to predicting plasma protein binding

A number of computational approaches have been used to try to predict PPB.[23,32–36] Most work has focused on trying to fit affinities for HSA, as measured by a variety of different techniques (Kratochwil *et al.*[17] discuss different approaches to using and obtaining binding data). It should be understood that low values for binding to HSA do not always imply low PPB because the compound may bind to another plasma protein, such as α_1-acid glycoprotein.[37] Similarly, models derived from binding data to domain 3 of HSA[33] will not necessarily be predictive of binding to other domains on HSA.

There is an overall tendency for PPB to increase with increasing lipophilicity, especially within a series that already shows a protein-binding liability. However, for diverse collections of molecules, the evidence seems to suggest a moderately weak correlation and that this correlation is very data set-dependent.[17] For example, Kratochwil *et al.*

showed a poor correlation (Pearson's correlation factor (r) = 0.43) between Clog P and affinity to HSA for a diverse set of 138 compounds,[23] whilst a somewhat better correlation (r = 0.75) was obtained by Hajduk *et al.* in a study of 889 diverse compounds binding to domain 3 of HSA.[33] Researchers have therefore had to use additional descriptors to obtain good models for binding of diverse compounds to HSA. Colmenarejo *et al.* obtained an $r^2 = 0.83$ and $q^2 = 0.79$ for a diverse set of 95 drugs using 12 electronic, topological structural and thermodynamic descriptors (with five outliers removed from the data set).[32] Using the same data set, Hall *et al.* obtained $r^2 = 0.77$ and $q^2 = 0.69$ with a six-variable model composed of various topological and structural descriptors.[34] Using property and topological descriptors, Xue *et al.* also obtained good correlations with this data set.[36] In a study on binding to domain 3 of HSA, Hajduk *et al.* fitted a much larger set of 889 compounds using a group contribution model based on 74 chemical fragments, and obtained $r^2 = 0.94$ and $q^2 = 0.90$.[33] The paper is particularly interesting because it gives tables of common chemical fragments that are correlated with increases or decreases in binding to HSA, and these tables can be directly used by medicinal chemists to suggest designs. However, it should be noted that the original data set comprised 1826 compounds and about half of the compounds were removed from this analysis either because their measured affinities were ambiguous or because they contained functional groups that are not represented in the 74 chemical fragments. Kratochwil *et al.* have measured binding of 151 diverse compounds to HSA and have applied partial least squares using pharmacophoric fingerprints as descriptors. This yielded a model with $r^2 = 0.72$ and $q^2 = 0.48$.[23]

As these reports indicate, there have been many attempts to obtain global models for binding to HSA and these will have utility in general assessments of the PPB liabilities of compounds. However, for a particular compound series with a known PPB liability, it appears advisable to construct a local model using interpretable descriptors in order to drive the medicinal chemistry forward. The reader is referred to the review by Colmenarejo for a summary of local models that have appeared in the literature.[24]

4.31.2.1.4 Structure-based approaches to predicting plasma protein binding

The advent of increasing amounts of structural information on binding to HSA means that it is now practical to use structure-based methods as an alternative to classical QSAR. For example, the many crystal structures recently deposited in the PDB[38] provide an excellent test set for the validation and improvement of docking methods. Accurate predicted binding modes can be used directly to suggest changes to lead compounds that are likely to disrupt binding to HSA, whilst knowledge of the SAR against the target of interest can focus the designs toward compounds that are more likely to retain target activity. Perhaps even more attractive is the possibility of experimentally determining the binding mode of a lead compound in HSA. Such a structure-based approach is particularly intellectually appealing when the structure of the lead compound can also be determined in its target protein.

Mao *et al.* have rationally designed diflunisal derivatives with the aim of maintaining cyclooxygenase-2 (COX-2) activity whilst reducing binding affinity for domain 3 of HSA.[30] The NMR structure of diflunisal with domain 3 of HSA indicated that the salicylic acid group formed hydrogen bonds with side chains of Arg410 and Tyr411 of HSA whilst the difluorophenyl ring bound in a tight hydrophobic pocket (**Figure 2**). SAR on simpler analogs indicated that the

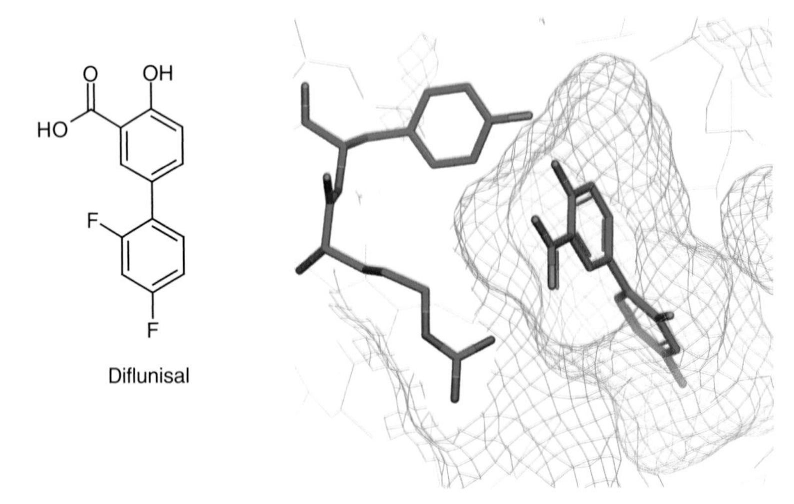

Figure 2 The chemical structure of diflunisal and its binding mode in HSA. The figure shows a clipped molecular surface for the protein and illustrates that the difluorophenyl group is in a deep hydrophobic pocket. Also shown in thick sticks are the two protein residues (Arg410 and Tyr411) whose side chains form good hydrogen bonds with the salicylic acid group of diflunisal.

binding to domain 3 of HSA was indeed sensitive to removal of either of these features but, critically, modeling of diflunisal in COX-2 (based on the crystal structure of a related analog) predicted that such changes were unlikely to retain affinity for COX-2. Instead, the modeling suggested that changes at the 5-position of the difluorophenyl ring might be tolerated in COX-2, and that they would be detrimental to HSA binding. A small number of analogs were synthesized on this basis and met the design goals with mixed success. Firstly, the designed analogs all had lower affinities for domain 3 of HSA but, frustratingly, some of these analogs still bound to full-length HSA. Secondly, the analogs were all at least an order of magnitude less potent than diflunisal against COX-2. It is not clear whether this is a good or bad result because there are many situations where sacrificing some potency for improved ADME properties would be very useful, especially if there is a clear design strategy for regaining affinity, but this work outlines the basic principles associated with using structural information to improve the PPB properties of lead compounds.

The same group at Abbott have also recently outlined another example[39] in which an inhibitor of the Bcl-2 family of proteins was discovered. SAR by NMR[40] was used to find fragments that bound to the antiapoptotic protein, Bcl-X_L. The relatively small lipophilic acid (**1**) was identified as a 300-μM binder and its binding mode was determined. Small lipophilic acids of this type conform to the known binding motif for site II on HSA which is located on domain 3.[24] A significant amount of structure-based optimization on the hit fragment led to compound (**2**), shown in **Figure 3**; this compound has 36-nM affinity for Bcl-X_L but its affinity is attenuated by >280-fold in the presence of 1% human serum, owing to tight binding to domain 3 of HSA. The crystal structure of Bcl-X_L with **2** was determined and is shown in **Figure 4a**. **Figure 3** shows the chemical structure of a close analog (**3**), for which the complex with domain 3 of HSA has been determined by NMR. In the Bcl-X_L complex, the terminal *S*-phenyl group of the inhibitor is folded back on itself and stacks under the central nitrophenyl ring, leaving the linker region between these two ring systems exposed to solvent. In contrast the terminal *S*-aryl group in the HSA complex (**Figure 4b**) is in a more extended conformation and parts of the linker region are in van der Waals contact with the protein. Oltersdorf *et al.*[39] reasoned that appending a basic dimethylaminoethyl group on to the linker region would perturb the HSA binding without affecting the binding to Bcl-X_L. Similarly, examination of the binding of the biphenyl group in HSA (**Figure 4c**) reveals that it is in a deep hydrophobic pocket, whereas in Bcl-X_L (**Figure 4a**) this group is more solvent-exposed. This led to the idea of replacing the terminal phenyl group with substituted piperidines, which, along with subsequent optimization, eventually led to the identification of lead candidate ABT-737 (compound **4** in **Figure 3**). In the presence of 10% human serum, this molecule retains nanomolar affinity for Bcl-X_L and also shows promising in vivo antitumor activity.

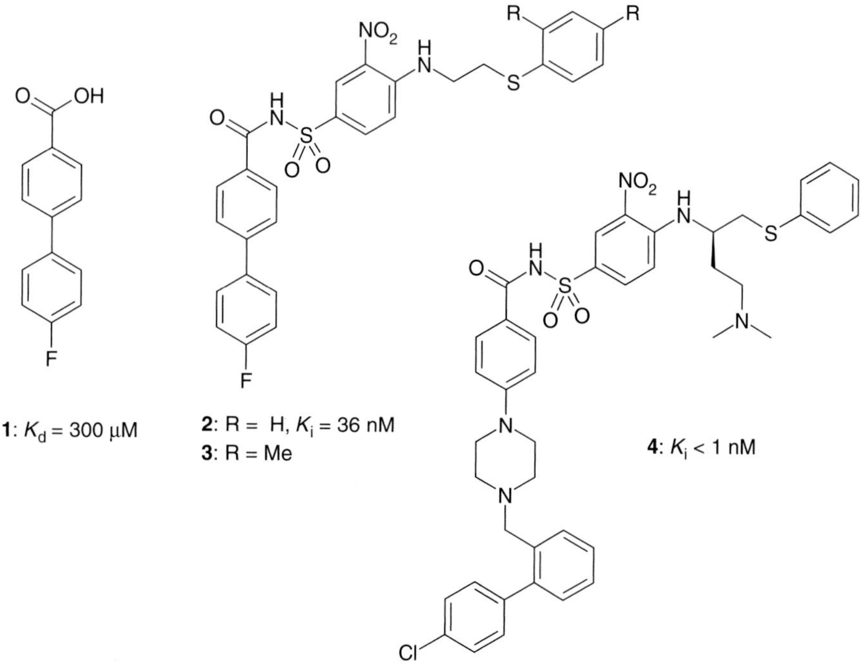

1: K_d = 300 μM

2: R = H, K_i = 36 nM
3: R = Me

4: K_i < 1 nM

Figure 3 The chemical structures of four compounds relevant to the design of Bcl-X_L inhibitors with reduced PPB liabilities.

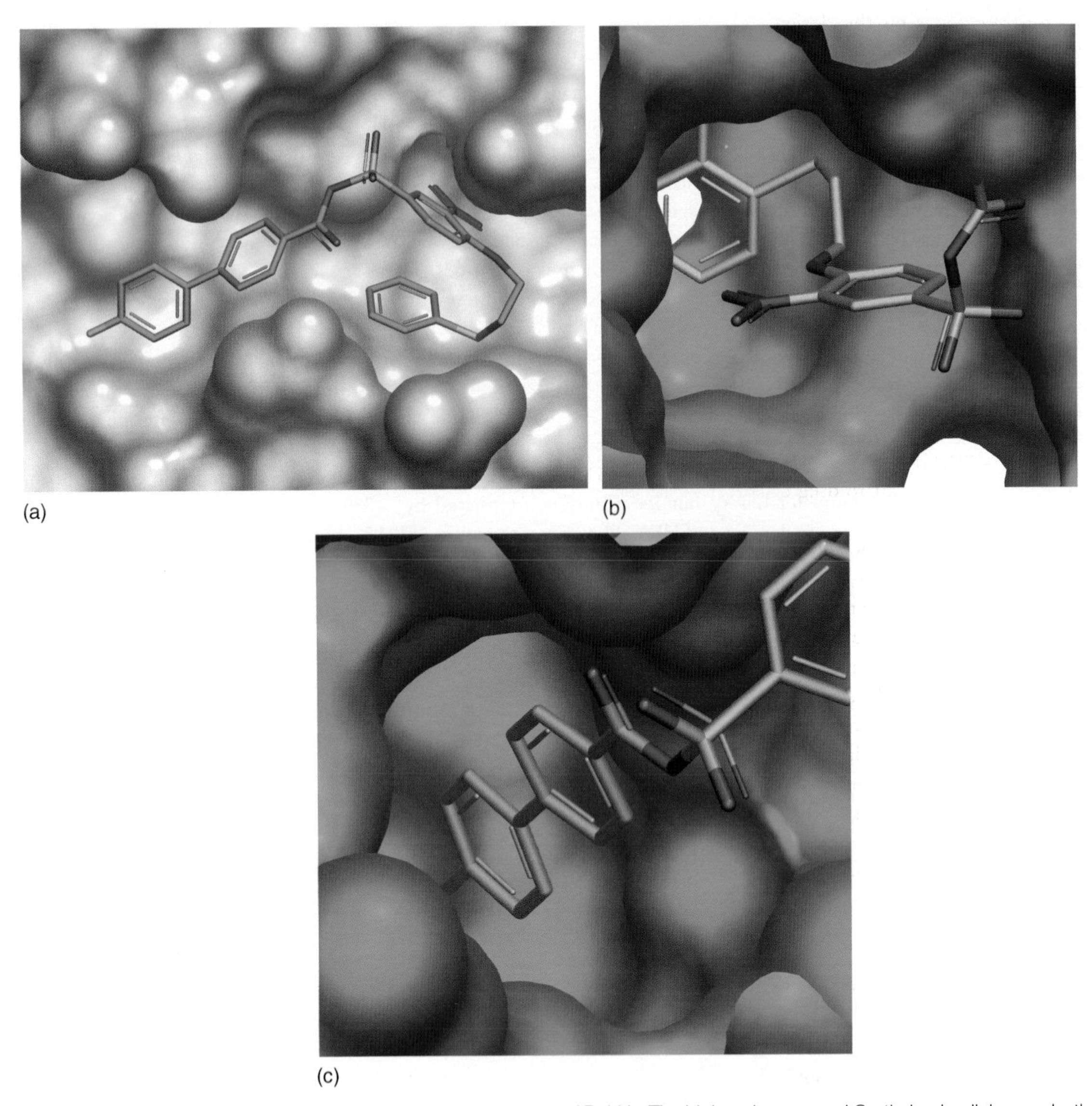

Figure 4 (a) Complex of compound **2** in surface representation of Bcl-X_L. The biphenyl group and *S*-ethylamino linker are both surface-exposed, implying that substitution at these positions may retain affinity for Bcl-X_L. (b) Complex of compound **3** in surface representation of HSA. The surface has been clipped to show how the *S*-ethylamino linker region is buried in a pocket. Substitution at the linker region would be predicted to have a detrimental affect on HSA affinity. (c) Different view of the same complex where the surface has been clipped to show how the 4-fluorobiphenyl group is buried in a deep lipophilic pocket. Substitution of the terminal aryl group with bulkier or more polar groups would be expected to affect binding affinity to HSA adversely.

The above example shows how structure-based design can be used to remove protein-binding liabilities. It is interesting, although perhaps futile, to compare this structure-based method with what an informed medicinal chemistry approach might have achieved. Faced with the structure of the lead compound (**2**) in the target protein (Bcl-X_L), a medicinal chemist might look to append groups to affect the PPB properties. Knowledge of the literature on PPB would point toward the introduction of polar groups to modify the $\log P$ (or the $\log D$) of the molecule. More detailed knowledge of the literature might, for example, point toward the group contribution method of Hajduk *et al.*,[33] in which secondary and tertiary amines are highlighted as the best chemical fragments for the reduction of binding to domain 3 of HSA. In this way, the medicinal chemist (or modeler) might arrive at designs similar to the ones described above. However, we would argue that such hindsight can be misleading. SBDD is rarely used to propose designs that medicinal chemists could not have thought of in the absence of structure; it is more concerned with allowing medicinal chemists to focus on the designs that are supported by structural hypotheses and to deprioritize the very much larger set of design ideas that are supported by weaker evidence. In this case, it is clear that the structure-based approach to HSA binding has enabled Oltersdorf *et al.*[39] to arrive at new sets of compounds with improved properties and has facilitated the design of a potential candidate molecule. It remains to be seen whether the increasing structural awareness of HSA will allow this kind of application to be more generally applied across the pharmaceutical industry.

4.31.2.2 Cytochromes P450

4.31.2.2.1 Introduction to cytochromes P450

The CYPs are a superfamily of heme-containing monooxygenases involved in a variety of functions, including the metabolism of exogenous and endogenous substances in the human body.[41] Although there are several drug targets that are CYP enzymes (e.g., aromatase), their most general interest to drug design lies in their fundamental role in the metabolism of many drug molecules. Of the 57 known human CYPs,[42] seven isoforms are responsible for the metabolism of over 90% of the drugs in clinical use,[18] in the approximate order of importance CYP3A4, CYP2D6, CYP2C9, CYP2C19, CYP1A2, CYP2E1, and CYP2C18. Understanding and modulating the binding to these key enzymes are important to drug discovery for a number of reasons:

1. Reducing the clearance of compounds is often a goal of lead optimization projects and, if the clearance is mediated by a CYP (as it often is), then reduction of P450 binding becomes a focus of optimization.
2. When compounds are potent inhibitors or substrates of a particular CYP, they can interfere with the pharmacokinetics of existing co-administered clinically used compounds which are metabolized by that particular isoform, leading to undesirable drug–drug interactions.
3. Metabolism by CYPs sometimes yields toxic compounds.
4. Polymorphisms of particular isoforms (e.g., CYP2D6 and CYP2C9) exist which give rise to differential metabolism of drugs amongst individuals.[43] Metabolism via such isoforms can therefore be undesirable, particularly for compounds that have a narrow therapeutic window (e.g., the anticoagulant warfarin).
5. For some drugs a short half-life, or a half-life no longer than 24 h, may be a necessary or desirable property so it is important that drugs are effectively metabolized (e.g., by CYPs).
6. It is therefore standard practice to monitor the P450 binding of compounds synthesized in drug discovery projects and compounds are chosen for progression with regard to their activities toward relevant CYPs.

It is known that different CYP isoforms exhibit different substrate preferences[18] and knowledge of these can be used to drive medicinal chemistry. One frequently employed strategy is to introduce a fluorine atom into aromatic rings; this has the dual effect of making the aromatic less electron-rich and directly blocking the site of metabolism.[44] Similarly, it is known that P450 binding is broadly correlated with lipophilicity[45] so strategies for reducing lipophilicity are often successful. More sophisticated strategies are discussed in detail below.

4.31.2.2.2 Structure of cytochromes P450

Structural work on CYPs has been reviewed before.[46–51] In this subsection the focus will be on mammalian CYPs and in particular on the structures of the enzymes implicated in the metabolism of drug molecules.

For many years, determining the structure of key CYP isoforms associated with metabolism proved elusive, primarily because the proteins are membrane-associated and difficult to handle. As a result, a large amount of computational work was based on comparative modeling using soluble bacterial CYPs as templates. These templates have weak sequence homology (around 20%) with the key human isoforms (**Table 1**) and it is not generally possible to obtain structures accurate enough for SBDD by this mechanism. The situation changed upon the determination of the first mammalian P450 isoform CYP2C5 in 2000.[52,53] This provided an improved basis for comparative modeling of CYP2C9 and CYP2C19 because of the higher sequence identity[54] and a somewhat better starting point for the modeling of other key isoforms in the CYP2 family such as CYP2D6.[55] Comparative modeling methods for P450 structures have been reviewed recently.[56]

In 2003, the crystal structure of the first human CYP was determined, CYP2C9 at Astex.[57] **Figure 5** shows a protein schematic of the CYP2C9 structure which has the expected overall fold characteristic of CYPs. It is a two-domain protein consisting of a small, predominantly beta-strand N-terminal domain (on the left-hand side of **Figure 5**) and a larger helical C-terminal domain which contains the heme and the active site. The CYP2C9 complex with *S*-warfarin positions the drug molecule distal to the heme and cannot be used to explain directly why site-directed mutagenesis of

Table 1 Sequence identities amongst a number of key isoforms as measured by number of identical residues divided by the length of the shortest sequence

	BM3	*2C5*	*2B4*	*2C9*	*3A4*	*2C8*	*2A6*	*2D6*	*2C19*	*1A2*
Source	Bacterial	Rabbit	Rabbit	Human	Human	Human	Human	Human	Human	Human
Structure	Yes	Yes	Yes	Yes	Yes	Yes	Yes	Yes	No	No
Metabolizer	No	No	No	Yes	Yes	No	No	Yes	Yes	Yes
BM3	100.0	18.7	17.2	17.8	24.7	17.6	18.7	17.4	18.5	20.3
2C5		100.0	50.7	76.6	22.8	73.3	50.5	39.8	76.6	27.9
2B4			100.0	49.6	23.6	52.7	51.7	41.1	50.4	27.9
2C9				100.0	23.3	78.0	49.0	39.0	91.4	27.3
3A4					100.0	24.1	21.9	20.3	23.7	19.3
2C8						100.0	49.0	39.8	78.2	27.8
2A6							100.0	34.6	51.0	28.3
2D6								100.0	39.4	28.8
2C19									100.0	26.5
1A2										100.0

Also included are the species or species type from which the P450 are derived, whether the isoforms are major metabolizers of drug molecules, and whether the crystal structures have been solved. Also note that the human isoforms, 2A6 and 2C8, are involved in some human metabolism.

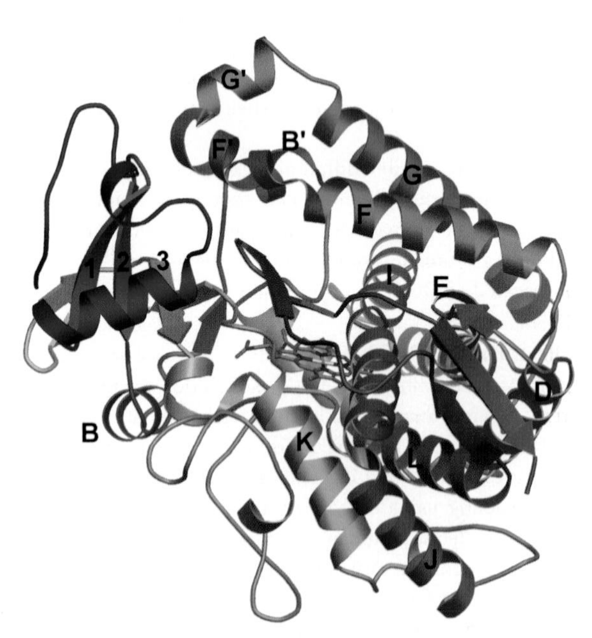

Figure 5 A schematic of the crystal structure of the human CYP2C9 colored from blue at the N-terminus to red at the C-terminus. The heme group is shown as a thick stick model. The helices have been labeled using the standard convention for CYPs.

Arg108 affects warfarin metabolism. A subsequent CYP2C9 complex with flurbiprofen exhibits a different conformation of the protein in which flurbiprofen interacts with Arg108 and is in a position suitable for metabolism by the enzyme.[58] The two results together indicate the complexity of substrate recognition by CYPs and underline the value of obtaining more structural data, especially on the compounds of specific interest.

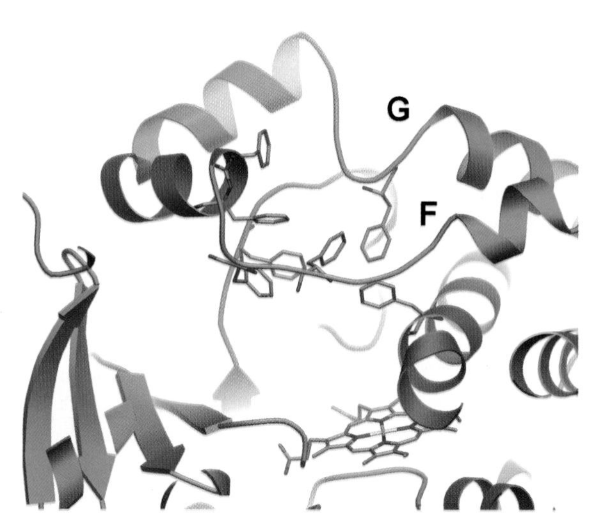

Figure 6 A view of the active site and Phe cluster region in CYP3A4. The heme group and the phenylalanine residues in the Phe cluster are represented in thick stick whilst the rest of CYP3A4 is represented as a schematic. The F and G helices are marked. Note how the formation of the Phe cluster has caused a conformational change in the F helix above the cluster compared to the corresponding region in the structure of CYP2C9 (**Figure 5**).

Recently the structure of CYP3A4 has been determined.[59,60] The result is of great significance because CYP3A4 is responsible for the metabolism of more drug molecules than the other major human drug-metabolizing isoforms combined, and is able to recognize the widest range of chemically diverse molecules.[61] Additionally, CYP3A4 is distant in sequence from the other drug metabolizing enzymes (**Table 1**) so comparative modeling of CYP3A4 did not have appropriate templates of known structure. Again, the structure and associated complexes revealed some interesting features which underline how much there is still to learn about these enzymes. Firstly, it is known that CYP3A4 can metabolize large substrates and yet the active site was clearly not big enough to accommodate such ligands without substantial conformational adjustment of the enzyme. Secondly, progesterone was observed to bind at a peripheral site away from the heme and it was suggested that this site may have functional relevance. Thirdly, the structure contains a cluster of seven phenylalanine residues (**Figure 6**) which might play a part in the conformational plasticity of the active site in response to larger substrates.

Very recently, the unliganded structure of the important drug-metabolizing enzyme, CYP2D6, has been determined.[164] The authors report a structure that agrees with much of the site-directed mutagenesis data on CYP2D6 and is broadly in agreement with previous homology models. Further crystallographic studies will be useful in clarifying the recognition of substrates and inhibitors.

The human P450, CYP2C8, is responsible for the metabolism of a number of drug molecules and its structure has been determined recently.[62] The molecule crystallizes as a dimer with two palmitic acid molecules bound at the dimer interface, close to the region where progesterone binds in CYP3A4. The authors suggest that this fatty acid-binding site may contribute to drug–drug interactions in P450 metabolism. The structure for human CYP2A6 has also become available;[63] this P450 is important for the metabolism of nicotine.

In the unliganded structure of rabbit CYP2B4 there is a substantially different conformation in which there is clear access to the active-site region for large ligands.[64,65] This conformation appears to have been imposed by the entry into the active site of a neighboring molecule of CYP2B4 in the crystal structure so may represent an artefact of the crystallization. A more appealing explanation is that this conformation provides a model for how active sites are accessed by substrates in mammalian CYPs and also a model for how very large substrates might be accommodated in the relatively small active sites so far observed for key human metabolizing enzymes.

It is clear that great strides are being made in the structural biology of mammalian P450 enzymes and that the field is moving fast. This looks set to continue with structure determination of additional isoforms. However, so far, the structural data have indicated that there is still a lot to be learnt about the molecular recognition in P450 isoforms associated with drug metabolism.

4.31.2.2.3 Quantitative structure–activity relationship approaches to predicting P450 inhibitors or substrates

The importance of P450 metabolism and the paucity of structural information have meant that an enormous amount of effort has been expended on ligand-based models for predicting P450 binding and sites of metabolism. The emphasis of this review is on new applications of structure-based design so a thorough description of ligand-based approaches is beyond its scope. Here a brief overview of the types of methods used and references to recent examples are given and, for more detail, readers are referred to more comprehensive, recent reviews.[43,66–68]

Quantum chemistry calculations are often used to predict the sites of metabolism of drug molecules.[68–70] Such approaches rely on estimating the substrate's electronic susceptibility to attack by the metabolizing species. In isolation, the methods cannot account for the molecular recognition which accompanies P450 turnover; for example, diclofenac is metabolized by CYP2C9 at a different position compared to CYP3A4 and this is usually ascribed to specific protein–ligand interactions in the CYP2C9 active site.[71] Recent quantum mechanical/molecular mechanical (QM/MM) work attempts to model the rate-determining step of the oxidation within the environment of specific CYPs,[72–74] although such approaches are not currently suitable for everyday use in drug design.

Classical pharmacophore methods have also been extensively used and are produced by extracting key features common to substrates or inhibitors.[18,43,66–68,72–75] QSAR or related classification methods have used topological descriptors to fit experimental binding data for different P450 isoforms.[43,68,76–81] Three-dimensional (3D) QSAR methods have also been extensively studied.[43,68,82–86]

This subsection and the next subsection draw a distinction between different ligand-based approaches and methods that consider the 3D structure of the active site. In some cases this distinction is difficult to make. De Groot *et al.* pioneered methods that combined several strategies for predicting P450 binding and metabolism.[54,87] For example, a model for metabolism by CYP2C9 was constructed using a combination of pharmacophore modeling, comparative modeling based on the 2C5 structure, and molecular orbital calculations.[54] The combined model worked well on a training set and test set of substrates and was also consistent with site-directed mutagenesis data. Subsequently, when the CYP2C9 crystal structures became available, it was confirmed[43] that the model was in good qualitative agreement with the experimentally determined complex of flurbiprofen in CYP2C9.[58]

Another example of a combined approach has been described by Cruciani and co-workers[88–90] and forms the basis of the MetaSite program. The method starts with analysis of the active site of a P450 isoform of interest (either a crystal structure or a comparative model). A key quantity in the analysis is the distance between the site in the protein where metabolism occurs and the preferred positions for a particular functional group (or probe) in the enzyme (as determined by the program GRID.[91] A variety of probe types are sampled and observed distances are binned and stored as a fingerprint describing the molecular recognition associated with metabolism in that P450 isoform. An analogous fingerprint can be constructed for each potential site of metabolism on the substrate,[89] and by comparing substrate fingerprints to the protein fingerprint, each metabolism site on the substrate can be scored for consistency with molecular recognition by the isoform of interest. This score is combined with data from quantum mechanical calculations so that potential sites of metabolism can be ranked according to their reactivity as well as their complementarity to the active site. The method was tested at four pharmaceutical companies and shown to predict correctly 85% of the observed sites of metabolism for over 200 proprietary compounds metabolized by CYP2C9, CYP2D6, or CYP3A4. The beauty of the MetaSite program is that protein (and substrate) flexibility can be taken into account during the construction of the fingerprints. On the other hand, it does not offer the atomistic detail of molecular recognition that one has with a docking method which in some circumstances will be a disadvantage.

4.31.2.2.4 Structure-based approaches to predicting P450 inhibitors or substrates

There is an increasing amount of structural information becoming available on the key CYPs associated with human metabolism so the possibility of using structure-based design to modulate the metabolism properties of key compounds is becoming a reality. A fully integrated example of structure-based metabolic design would start with the determination of a particular CYP complexed with a lead molecule. The lead molecule would possess a liability associated with that particular CYP which medicinal chemists had found difficult to overcome through the use of standard approaches such as blocking sites of metabolism and reducing lipophilicity. The CYP complex would then be used to suggest successful designs that would affect the binding to that P450 without affecting the activity of the lead molecule against its target. Such an example of structure-based metabolic design has not yet been described in the literature, primarily because the first structure of a human metabolizing CYP only became available two years ago.

Despite the lack of fully integrated examples of structure-based metabolic design, there has still been a lot of work exploiting CYP crystal structures. This work has mainly used docking to predict the P450 binding of substrates and inhibitors or used virtual screening to identify molecules that bind to a specific CYP.

Much docking work has been performed on comparative models of human P450 structures.[43,68] Here, a few recent examples are briefly described based on comparative models derived from mammalian P450 templates. Lewis has pioneered the use of manual docking approaches against the different isoforms and has also used these docking modes to fit an empirical scoring function.[92–98] This scoring function can be used to predict the binding affinity of new molecules after determination of their binding mode by the skilled modeler. Kemp *et al.* have prepared a homology model of CYP2D6 based on a CYP2C5 template and validated it against charged and uncharged substrates of CYP2D6.[99,100] Docking using the ChemScore function in GOLD[101–103] was used to predict the binding mode and affinity of a number of CYP2D6 binders. This included a set of 33 compounds chosen from the National Cancer Institute database for which the authors experimentally determined the CYP2D6 affinity. Comparison with the computational affinities obtained via the automated docking procedure yielded an $r^2 = 0.61$ and $q^2 = 0.59$. This is surprisingly good, especially given that the experimental values do not show an obvious molecular-weight dependency. However it should be cautioned that this test set was very small and no comparison has been made with simpler 2D approaches to assess whether the relatively complicated docking approach offered a substantial advantage. In another comparative modeling application to CYP2D6, Vermeulen and co-workers have examined a series of amphetamine analogs to understand the SAR and binding modes for the wild-type CYP2D6 and an active-site mutant.[55,104] A variety of automated docking methods were used and followed up with molecular dynamics simulations on the neutral forms of the amphetamines. These simulations rationalized the differences in the metabolic fate of these compounds between the wild-type and a mutant enzyme.

Docking methods have also been extensively tested on CYPs where the structures of the proteins are known and do not need to be modeled.[105–111] Most of this work has been performed on bacterial CYPs, reflecting the number of these structures in the public domain and the length of time over which the structures have been available. The most extensive validation study has been carried out by Kirton *et al.* using a test set of 45 crystal structures of ligand complexes with heme-containing proteins, of which over a third were CYPs.[105] The performance of GOLD using two scoring functions[102,103] was assessed (**Table 2**) and shown to be inferior when compared to GOLD's performance against a diverse set of proteins. The metal-acceptor term was reparameterized using weighted probability distributions for metal-acceptor distances from small-molecule or protein crystal structures. The lipophilic term in one of the scoring functions was also amended to treat aromatic nitrogen atoms as lipophilic. These changes led to an improvement in the overall success rate against heme-containing proteins.[105]

Subsequent unpublished work has applied the heme-targeted function to docking against CYP2C9. The 2D structures and names of the molecules considered are displayed in **Figure 7**. **Figure 8a** shows how the docking successfully reproduces the experimental binding modes for the flurbiprofen molecule.[58] **Figure 8b** shows the docked binding mode for diclofenac in the flurbiprofen CYP2C9 crystal structure in which the acid group forms a salt-bridge with the side chain of Arg108, and in which the known site of metabolism is positioned close to the heme iron. It is interesting that this docked binding mode is within 0.8 Å heavy-atom root mean square deviation (RMS) to that seen with diclofenac in the crystal structure with CYP2C5, which does not contain an equivalent salt-bridge to the acid on the ligand.[112] **Figure 8c** shows how the docking successfully reproduces the experimental binding mode of warfarin.[57] Interestingly, it is not possible to obtain convincing binding modes for diclofenac when docked against the warfarin crystal structure, suggesting the warfarin-binding site is not involved in the metabolism of diclofenac. On the other hand, **Figure 8d** indicates that warfarin can be docked reasonably well in an orientation suitable for metabolism against the flurbiprofen crystal structure. The dockings suggest that diclofenac binds and is metabolized in a one-step mechanism, whereas the warfarin initially binds to the distal site before being metabolized in a proximal site in a two-step mechanism. This hypothesis is supported by metabolism data with diclofenac and warfarin against mutants of CYP2C9 (J. Cosme, personal communication).

The program DOCK has been applied to the identification and synthesis of substrates and inhibitors of the P450cam enzyme and its L244A mutant.[108–110] P450cam is a bacterial P450 which is not itself of interest as a drug or ADME-related target but has the advantage of being well characterized structurally. It therefore represents a useful

Table 2 Percentage of best scoring docked structures within 2.0 Å heavy-atom RMS of the experimental binding mode for different choices of scoring function against different test sets of experimentally determined complexes.[105] Using the standard scoring functions, the docking program GOLD performs worse against heme-containing proteins compared to a diverse set of 139 protein complexes with drug-like ligands. The heme-targeted scoring functions offer an improvement in the success rates

Scoring	*Standard scoring*	*Standard scoring*	*Heme targeted*
Test set	139 diverse proteins	45 heme proteins	45 heme proteins
GoldScore success rate	79%	57%	65%
ChemScore success rate	79%	64%	74%

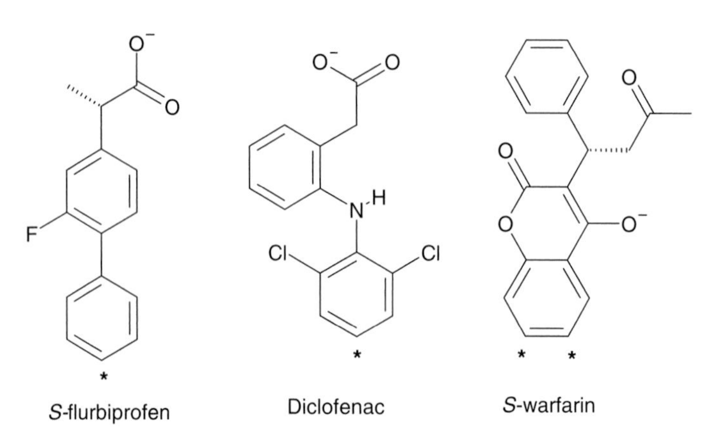

Figure 7 The chemical structures and names of the molecules docked into CYP2C9. Molecules are shown with the charge states that were adopted in the docking runs and the sites of metabolism are indicated by an asterix. Note that *S*-warfarin was docked in the anionic, open form.

surrogate in understanding how SBDD might be used to help modify ADME parameters. The wild-type protein has a very compact active site which cannot accommodate structures much bigger than the natural substrate camphor and the active site of the mutant protein is expected to be slightly larger (**Figure 9**). Virtual screening of 20 000 available compounds against the wild-type protein identified 11 compounds as potential substrates, seven of which were confirmed to be substrates.[110] In later work, the same authors constructed a virtual library of over 3000 1-substituted, 5-substituted, and 1,5-disubstituted imidazoles.[108] These molecules were docked against wild-type P450cam and the results indicated a preference for mono-substituted imidazoles. Four of these compounds were synthesized and all were shown to be low-micromolar inhibitors of the wild-type enzyme (IC_{50}s ranged between 0.8 and– 16 μM), with some selectivity over the mutant enzyme. Virtual screening against the L244A mutant enzyme showed a preference for disubstituted compounds and the best disubstituted compounds docked poorly against the wild-type protein. Five examples were synthesized and shown to be inhibitors of the mutant enzyme (IC_{50}s ranged between 6 and 130 μM) and additionally these five compounds showed excellent selectivity over the wild-type protein.

4.31.2.3 Perspectives on Future of Structure-Based Drug Design in Absorption, Distribution, Metabolism, and Excretion Applications

Structural information on HSA has been used to design out a PPB liability on a lead series aimed at Bcl-X_L inhibition. This has demonstrated the principles of using SBDD for ADME-related properties. It is likely that HSA structures will be used more in drug discovery in the future.

The use of CYP structures for addressing the P450 liabilities of compounds is less advanced, primarily because the structural information has only recently become available. It is to be expected that more CYP complexes and different isoforms will be crystallized in the near future. Such progress will have a profound effect on the way optimization of P450 binding will be performed. The previous section illustrated how, even now, a large amount of useful docking and design work is being based on the existing structural information and it will be very interesting to see how this field develops over the next few years.

The use of SBDD for ADME-related properties therefore represents a new and exciting application of SBDD which can be expected to grow in the next few years. In the second part of this chapter we discuss another emerging application for x-ray crystallography in which the method is used for screening fragments.

4.31.3 Structure-Based Fragment Screening

4.31.3.1 Introduction to Structure-Based Fragment Screening

For the past decade the predominant method for finding lead compounds against a particular target has been HTS. The method is resource-intensive and requires extensive infrastructure to facilitate the screening of hundreds of thousands of diverse compounds against targets of interest. Despite the successes of HTS, it does not always deliver good starting

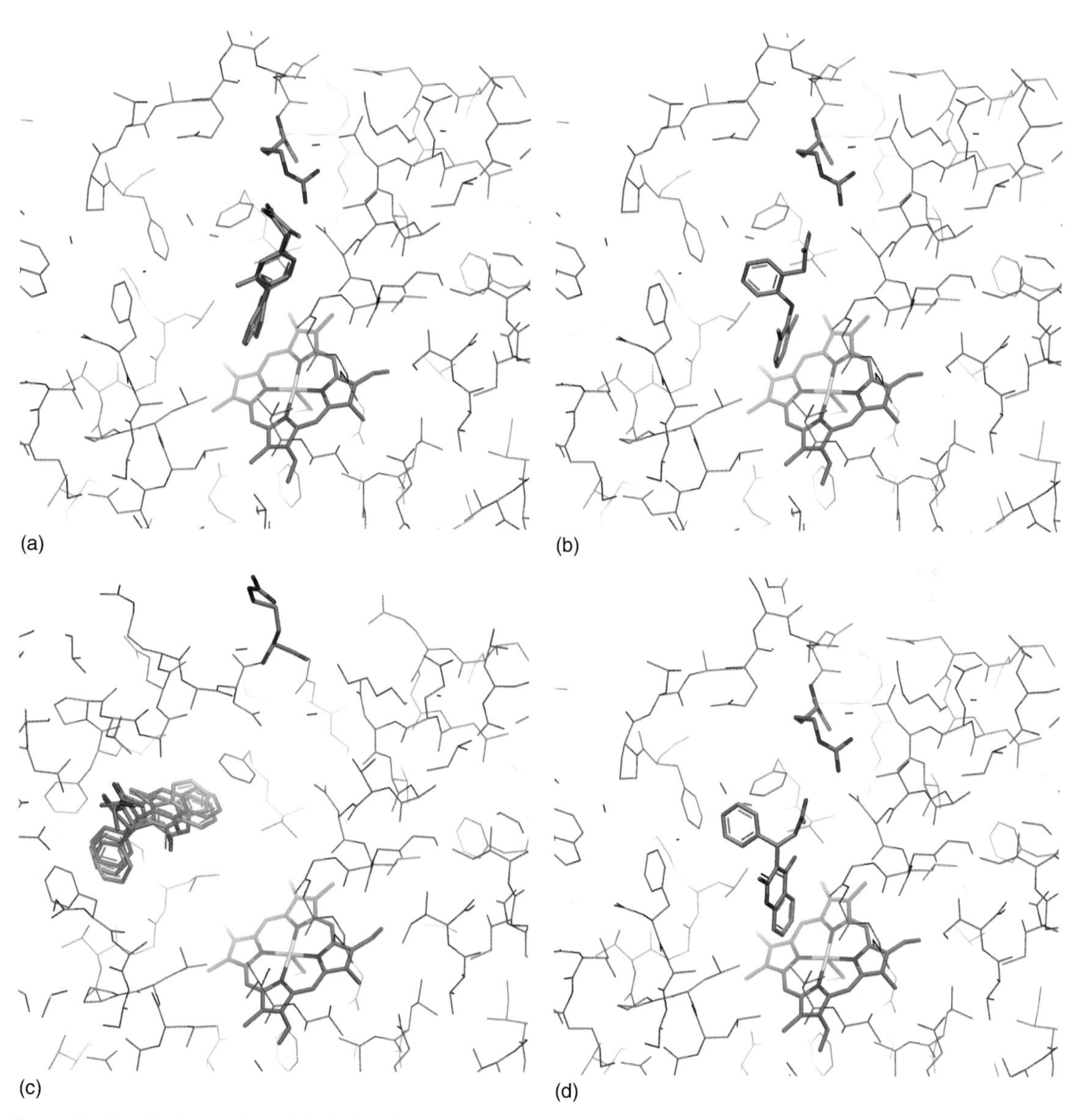

Figure 8 The binding modes of docked molecules against either the protein structure derived from the flurbipofen complex (1R9O) or the protein structure derived from the *S*-warfarin complex (1OG5). All protein structures are shown in green and are in the same frame of reference with the heme group and the side chain of Arg108 in a thick stick representation. (a) The best-scoring docked binding mode for flurbiprofen (shown in orange) against 1R9O compared with the experimental binding mode (shown in cyan). The molecule is successfully docked with a heavy atom RMS of 0.72 Å. (b) The docked binding mode for diclofenac (shown in orange) against 1R9O. (c) The best-scoring (shown in orange) and second-best-scoring (shown in magenta) docked binding modes of *S*-warfarin against 1OG5 compared with the experimental binding mode (shown in cyan). The molecule is successfully docked with heavy-atom RMSs of 2.19 Å and 0.93 Å, respectively. (d) The best-scoring docked binding mode for *S*-warfarin (shown in orange) against 1R9O.

points for a drug discovery program.[3,4] Sometimes this deficiency is obvious from inspection of the HTS hits but at other times the problems are only apparent after a significant amount of chemistry and biology has been invested in a particular hit. One key issue is that the hits often display flat SAR and their affinity is spread evenly through the molecule, making them inherently difficult to optimize. Another related problem is that the HTS hits are often too large relative to their potency (i.e., they have low ligand efficiency[19,20]) and subsequent optimization by medicinal chemists tends to increase the molecular weight and lipophilicity of the series.[113–115] The resulting candidates often fall outside the physical property ranges that are desirable for oral compounds (e.g., as specified in Lipinski rules[116]).

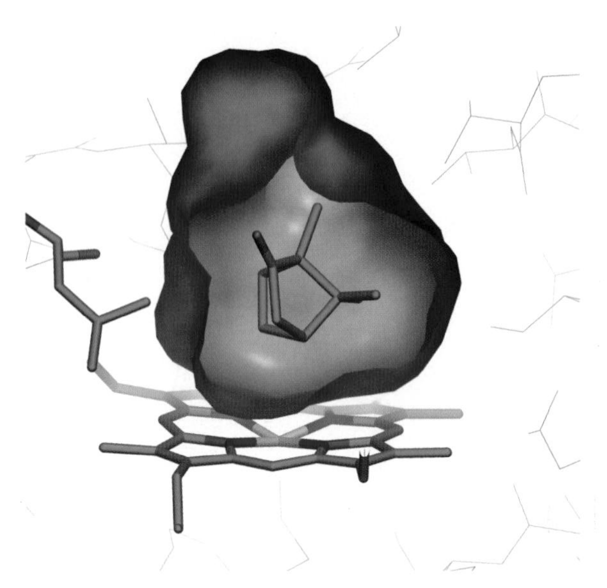

Figure 9 The crystallographic binding mode of camphor in P450cam taken from the PDB file, 1DZ4. Camphor (shown as a thick-stick model) fits tightly into the enzyme and is completely enclosed by the protein. The protein surface has been cropped to allow the ligand to be visible. Also shown in thick stick is the side chain of L244, most of which will be absent in the L244A mutant. This protein mutant is assumed to have the same structure but with a slightly larger active site.

It might be expected that larger or more diverse HTS libraries could significantly improve performance by allowing the identification of more potent drug-like compounds. However, a simple yet elegant statistical model developed by Hann *et al.* shows that, for a compound to achieve sufficient potency to be therapeutically useful, it must be able to make a number of high-quality interactions with the protein-binding site.[113] The model goes on to demonstrate that, as the compounds become complex enough to make these interactions, the probability of finding this arrangement by chance becomes vanishingly small.

In recent years, a new approach that addresses the points raised above has been developed for lead discovery. This approach is known as fragment screening and involves the use of libraries (or sets) of molecules of much lower molecular weight than those found in HTS. One of the principal advantages of fragment screening can be seen from Hann's simple model.[113] This shows that the chances of a fragment having the correct arrangement of features to bind to a particular binding site are much higher because of the smaller number of interactions that need to be made. So, if one had a screening technique that could detect these weaker binding events, it should be easier to find good-quality fragment hits than HTS hits. This also means that fragment libraries can be much smaller than HTS libraries and should still be able to yield hits against the majority of targets.

Most bioassays are not suitable for detecting fragment-binding events, as the K_is are often around the millimolar mark and the false-positive rates are relatively high. For this reason there has been a focus on biophysical techniques for the detection of fragments and, in particular, x-ray crystallography has been used both as a screening method and as a way of characterizing fragment hits obtained by other techniques. The direct experimental observation of a fragment's binding mode excludes the possibility of false positives and provides a viable, efficient strategy for optimizing the potency of millimolar fragments. The whole area of fragment-based drug discovery has been reviewed in detail elsewhere;[19,21,22,117] the focus of this review will be on the application of x-ray crystallography to fragment screening.

4.31.3.2 Fragment Screening

There are various methods that can be used to detect the binding of a compound to a protein. Most of these methods provide no direct information on the binding mode, or even on the binding site, of the compounds. In a structure-based design approach that uses information on the 3D binding modes of compounds throughout the project, such methods are mainly used as a prescreening filter, and we feel that an indepth description of these techniques is outside the scope of this chapter. Therefore, we describe crystallographic approaches to fragment-based screening in the next section, and cover some of the methods that have been used for prefiltering in less detail after that.

4.31.3.2.1 Fragment screening with x-ray crystallography

The original experiments to identify hits via crystallographic screening were described by Verlinde *et al.*,[118] who used the approach to try to find new inhibitors for the treatment of parasitic diseases.

More recently, Nienaber *et al.* have described an application of crystallographic screening against the cancer target, urokinase.[119] The design goal was to find fragments that did not contain a highly basic amidine or guanidine moiety because leads containing these groups often exhibit unfavorable absorption properties. A library of weakly basic compounds was screened, allowing the identification of the fragment hit shown in **Figure 10a**. This fragment was optimized using structure-based design, resulting in an orally bioavailable, submicromolar lead (**Figure 10a**). The same authors have subsequently applied the crystallographic screening approach to the discovery of dihydroneopterin derivatives.[120]

Lesuisse *et al.* described a crystallographic fragment screening approach for the development of inhibitors of the SH2 domain of ^{pp60}Src.[121] The work followed on from a previous study in which they developed mimetics of

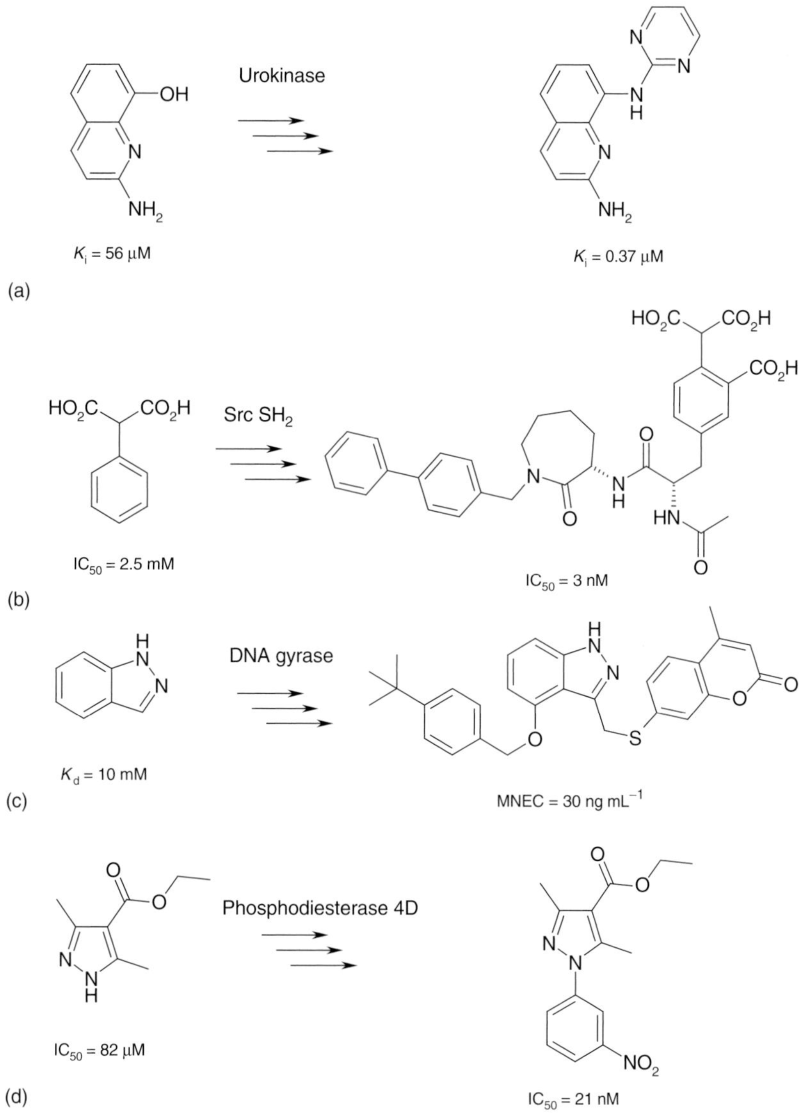

(a) Urokinase: K_i = 56 μM → K_i = 0.37 μM

(b) Src SH_2: IC_{50} = 2.5 mM → IC_{50} = 3 nM

(c) DNA gyrase: K_d = 10 mM → MNEC = 30 ng mL^{-1}

(d) Phosphodiesterase 4D: IC_{50} = 82 μM → IC_{50} = 21 nM

Figure 10 Examples discussed in the text of fragments (on the left-hand side) that have been used as starting points in the design of leads (on the right-hand side): (a) and (b) are examples of crystallographic screening (Figures 15 and 17 show other examples); (c) is an example of virtual fragment screening; and (d) is an example of bioassay fragment screening. The potencies of the molecules and the associated targets are also given. MNEC, maximum noneffective concentration in DNA gyrase inhibition.

the phosphorylated region of the activated receptor (pYEEI). These compounds were potent inhibitors but the phosphate group is an unattractive feature for drug candidates as it is rapidly hydrolyzed by phosphatases, and the high charge gives poor cell permeation.

The work commenced with the observation that phenyl phosphate shows measurable binding affinity to the SH2 domain. A number of related compounds with different substitution patterns around the phenyl ring were soaked into crystals, and those that could displace a bound citrate molecule were used to discover preferred scaffold geometries. Virtual screening using LUDI[122] was used to suggest replacements for the phosphate group and it was found that both malonic and oxalic acids could displace the bound citrate, despite neither of these compounds having measurable affinity. Combining these fragment-binding modes, the initial inhibitors were developed to remove the phosphate liability. The resulting inhibitors showed much improved stability in rat and human plasma. Additionally much more is known about the preparation of prodrugs of carboxylic acids than phosphates and so the final compounds are much more suitable as drug candidates.

In a continuation of this work, Lange *et al.* describe efforts to improve their nonpeptidic ^{pp60}Src SH2 inhibitors[123] (**Figure 10b**). They prepared a library of phospho-tyrosine mimetic fragments and some compounds suggested by the program LUDI. These were assayed using BiaCore surface plasmon resonance (SPR) and those that were more active than phenyl phosphate were soaked into crystals of the SH2 domain. Some of the fragments are able to induce a conformational change in the phosphate-binding region.

Later in the chapter, further examples of cystrallographic screening will be presented from our own work at Astex.

4.31.3.2.2 Fragment screening with noncrystallographic methods

A number of additional methods have been used to identify fragments that bind (or may bind) to a protein target. These are described briefly here, with more attention directed toward more recent activity in this area.

4.31.3.2.2.1 Protein-based nuclear magnetic resonance (structure–activity relationship-by-nuclear magnetic resonance)

In target-based NMR screening, chemical shift changes in the protein spectrum, caused by the binding of a ligand, are measured, hence determining not only whether a compound binds to the protein, but also where it binds. The most widely known form of target-based NMR screening is the SAR-by-NMR approach, which was pioneered by Fesik and co-workers,[40,117,124,125] although other techniques have been reported.[126] The main disadvantage of target-based NMR screening is that it requires large amounts (100–1000 mg) of isotope-labeled protein and sequence-specific resonance assignments, and is restricted to comparatively small protein targets, typically those <25 kDa. Ligand-based NMR screening, which detects changes in the ligand spectrum upon binding to the protein, is free of many of these disadvantages.

4.31.3.2.2.2 Mass spectrometry

Mass spectrometry (MS) occupies a similar position to NMR. Some applications detect only ligand-binding and provide no structural information on the interaction.[127–129] However, MS can also be used to detect and characterize changes in the H/D exchange rates of amide protons of a protein on ligand-binding, from which the location of the ligand-binding site may be inferred.[130,131] Another MS-based screening approach has been developed by Wells and co-workers, in which a library of drug fragments containing a disulfide bond is reacted with a cysteine residue in the target.[132–135] The covalent complexes are then analyzed using MS. Although very elegant, the technique is quite specialized; the protein active site needs to contain (or be engineered to contain) a cysteine residue that is more reactive than cysteines elsewhere in the protein; and the libraries screened are limited to compounds containing a disulfide bond.

4.31.3.2.2.3 Virtual screening

One of the earliest structure-guided fragment-screening applications is described by Boehm *et al.*, who were seeking novel inhibitors of DNA gyrase B.[136] Gyrase is a well-established antibacterial target, yet all of the known antiinfectives which target this enzyme have significant limitations. HTS had failed to yield new lead chemistries, so these authors set about using detailed knowledge of the active site to provide new start points.

Their approach commenced by identifying a simple pharmacophore that consisted of a hydrogen bond donor to Asp73 in the gyrase adenosine triphosphate (ATP)-binding site, a hydrogen bond to a conserved water molecule, and a lipophilic group. Virtual screening using LUDI and Catalyst generated a small number of fragments (needles) that could match this pharmacophore. These needles and other compounds that were identified by database searching were subjected to a cascade of experimental assays.

Those compounds that were active in the assays were investigated by spin–spin J-coupled heteronuclear $^{1}H/^{15}N$ correlation NMR spectroscopy. The compounds that were shown to interact with residues in the ATP-binding site of gyrase were co-crystallized with a loop deletion mutant. The crystal structures revealed seven distinct chemical classes that bound specifically in the active site. Using these chemical classes as start points, several iterations of structure-guided optimization led to the development of an indazole series that was 10 times more potent than novobiocin (**Figure 10c**).

4.31.3.2.2.4 Ligand-based nuclear magnetic resonance

The SHAPES strategy reported by Fejzo and colleagues represents an early example of ligand-based NMR screening,[137] but many other techniques have been reported.[138] Ligand-based NMR screening does not require isotope-labeled protein, nor isotope-labeled ligands, and the technique is sensitive, which makes it particularly suitable for measuring weak binding events like fragment binding. It does not, however, provide any direct structural information on the binding site.

Van Dongen and colleagues reported a structure-based fragment screening application to the obesity target human adipocyte fatty acid-binding protein (FABP4).[139] In their approach, a library of 531 low-molecular-weight (<350 Da) compounds was screened against FABP4 using an NMR ^{1}H 1D $T_{1\rho}$-relaxation filter experiment (a ligand-based NMR screening method, particularly suitable for relatively large proteins like FABP4). This resulted in 52 hits, which were further classified as strong binders (38) and weak binders (14) using a follow-up NMR experiment. The potencies of the strong binders were then measured using a fluorescence polarization assay. For two of these compounds, x-ray structures of the complex with FABP4 could be determined, and one compound with a 590 μM EC_{50} against FABP4 was selected for hit optimization. Initially, four commercially available related compounds were purchased and an additional 12 analogs were synthesized. This process resulted in a lead compound with a 10 μM EC_{50} against FABP4 and 26-fold selectivity over FABP3.

4.31.3.2.2.5 Bioassay

Card *et al.* describe a scaffold-based approach to fragment screening.[140] This approach involves the screening of a large library of chemical scaffolds that would be missed by conventional screening because their potency is too low. The library is screened by a low-affinity scintillation proximity assay against a set of proteins from the target family. The most potent binders that are observed to bind to several of the members of the target panel are then co-crystallized against the targets of interest and the characteristic features of the binding mode are identified. The scaffold-binding mode is then validated by confirming that it is tolerant to small substitutions. To do this, analogs of the observed scaffold that have a few additional substituents are synthesized and co-crystallized with the protein. The scaffold is considered validated if the conserved portion of the new molecules makes the same interactions as the scaffold itself.

The final step in the process is to optimize the validated scaffolds into potent and selective inhibitors using the co-crystal structure of the scaffold as a guide. This is done by making virtual libraries around the scaffold and scoring them using scaffold-anchored docking and energies from simulation studies. Using this method the authors claim to be able to add the maximum potency at every stage, whilst carefully controlling the molecular weight that is added to the lead compound.

The authors describe the development of a series of potent and selective inhibitors for phosphodiesterases. The process starts with an initial carboxy-pyrazole scaffold hit and proceeds through a small number of iterations to a 4000-fold more potent lead compound (**Figure 10d**).

The authors note that the approach is similar to other fragment-based screening methods, but believe that the scaffold validation stage (requiring the initial scaffold to bind to several members of the family, and that it is tolerant to small substituions) adds to the robustness of the technique. The fact that the scaffold-binding mode is consistent in all structures makes the SAR simpler to understand. On the other hand, the use of a medium-throughput assay is likely to mean that some weak but valuable fragment hits could be lost at the screening stage.

4.31.3.3 Libraries

A key component of fragment-based discovery is the set of compounds that are screened against the target. The basis of the approach is to keep the screening compounds (fragments) small and simple so that the chances of them binding is maximized. Most literature regarding fragment screening addresses the problem of library generation to some degree.

Two main techniques for library design are described in the literature. The first of these can be thought of as targeted (or biased) libraries and the second as general-purpose libraries.

4.31.3.3.1 Targeted libraries

Targeted libraries are developed from some knowledge of natural substrates, known inhibitors, or the structure of the protein-binding site. Nienaber and colleagues describe a targeted library of weakly basic fragments[119] that were used to discover binders to the S1 pocket of urokinase. It had proved difficult to find inhibitors lacking a highly basic group and so a targeted library of weakly basic compounds was prepared to follow up on a hit from a larger-scale crystal screening. This library yielded a number of useful start points for lead discovery.

Additionally, the binding site of the protein can be used to identify fragments that may bind. Boehm *et al.*[122,136] describe the selection of needles using an in silico approach with the programs LUDI and Catalyst. In this fashion the number of available compounds was restricted to those that could be expected to form the desired interactions with the protein-binding site.

Other in silico approaches can be adopted for virtual screening for fragment binders. In our work we use the protein/ligand docking program GOLD[102,103,141,142] to screen libraries of available compounds against the protein of interest. This has proved to be a useful source of targeted compounds but there are some difficulties in docking fragments that must be addressed.

The biggest problem is the fact that empirical scoring functions tend to reward larger molecules with higher binding scores because they make more interactions with the protein.[143,144] To address this it is useful to normalize the binding score with respect to the size of the docked ligand. This has analogy with the concept of ligand efficiency[20] for the analysis of physical binding data, and allows molecules of varying sizes to be compared more easily. Alternatively, the docking energies of fragments should only be compared with those of other fragments.

An additional problem is that fragments will frequently dock in regions away from the main area of interest in the binding site. This is because the scoring functions are not sensitive enough to recognize the dominant interaction features. One method to address this is to use simple pharmacophores[144] to guide the docking of virtual fragment libraries to the appropriate region of the binding site (for instance, the hinge region of a kinase).

4.31.3.3.2 General-purpose libraries

In contrast to targeted libraries, general-purpose libraries are intended to be used against a variety of protein-active sites. For example, Fejzo *et al.* describe the SHAPES methodology which uses a small library of compounds that contain only common drug scaffolds and side chains.[137] This library was screened using a ligand-based NMR procedure, and the binders were used to select compounds from larger libraries for enzymatic assaying.

Baurin *et al.* describe their work putting together a set of screening libraries for NMR-based fragment screening (SeeDs).[145] While this screening technique has some differences in application to x-ray screening, the general principles they adopted are of interest. The SeeDs library consists of a number of sublibraries, two of which are for general-purpose screening. The basic principle for selecting compounds for the general-purpose libraries was that of filtering compounds available from various suppliers. The set of compounds was reduced by limiting molecular weight to the range 110–250 Da (350 Da for compounds containing a sulfonamide). Compounds were then removed that had long unbranched carbon chains or that contained reactive functionality. Finally, inspection by medicinal chemists led to the removal of compounds that were likely to be insoluble or that were unsuitable for chemical evolution.

The second generation of their general-purpose library[145] formalized the lists of unwanted functionalities, and also encompassed a set of desirable side chains that were known to be well tolerated in drug molecules. The compounds were clustered on the basis of three-point topological pharmacophores, and solubilities were estimated using an in silico method.[145,146]

The scaffold-based drug design approach[140] developed by Card and colleagues uses a somewhat larger starting library than most other fragment-screening methods. This is possible as the screening starts with a high-throughput, low-affinity assay to remove the majority of fragments that do not bind. The library consists of about 20 000 scaffolds that have been selected on the basis of chemical diversity and that have molecular weight in the range 125–350 Da. The details of how a scaffold is defined are not presented, but it is worth noting that their library contains compounds that are somewhat bigger than most groups' definition of a fragment.

The next section describes the work at Astex concerned with developing a general-purpose fragment library. Most of the other library generation processes can be thought of as based on filtering, i.e., removing sets of compounds with unwanted functionality or physicochemical properties. The main problem with this is that the final set of compounds can be somewhat arbitrary and can require clustering and filtering to produce a well-defined set of compounds.

4.31.3.3.2.1 Drug fragment set

Our approach can be thought of as building up desirable compounds from observed scaffolds and side chains and using them if they are available.[147] In this way we have explicit control over the make-up of the cores and functional groups of the compounds that are screened.

4.31.3.3.2.2 Version 1.0

This set was designed to provide a diverse range of compounds that contain ring systems and functionalities often found in known drug molecules. The first stage was to identify a small set of simple organic ring systems that occur in drug molecules so that they are likely to have reduced toxicity liabilities and are amenable to development by medicinal chemistry practices. Several analyses have demonstrated that only a small number of simple organic ring systems (sometimes known as scaffolds or frameworks in other work) occur in many drug molecules.[137,148,149] Low-molecular-weight ring systems were selected and are shown in **Figure 11**. In addition to the drug rings a set of simple carbocyclic and heterocyclic ring systems were chosen as the basis for additional fragments. The ring systems chosen are shown in **Figure 11**.

The virtual library, from which the Drug Fragment Set is chosen, was generated by combining the ring systems shown above with a set of desirable side chains. The side chains used in this process are divided into the three categories shown on the right-hand side of **Figure 11**. The virtual library was generated by substituting each of the relevant side chains on to each of the ring systems. Each carbon atom was substituted by the side chains from the functional groups and by those from the secondary substituents. The nitrogen atoms were only substituted from the group of *N*-substituents. Each ring system was only substituted at one position at a time, with the exception of benzene and imidazole.

The virtual library was generated as SMILES strings, which were searched for in a database of available compounds. A total of 4513 compounds were generated by the virtual enumeration stage, of which 401 were available from commercial suppliers. Manual inspection to remove known toxophores and unavailability of some compounds resulted in a final set of 327 compounds (the Drug Fragment Set).

4.31.3.3.2.3 Version 2.0 and the rule of three

The first version of the Drug Fragment Set was a useful start point for our fragment-screening approach. After screening several targets, we carried out a retrospective analysis of the hit rates and the kinds of compounds that were present in, and absent from, our initial screening set. In light of this analysis a second generation of the screening set was developed.

We have observed that the most useful fragments have physical properties that lie in a defined range. These are shown below and we term these properties the rule of three,[150] by analogy with Lipinski's criteria for the useful range of physicochemical properties for drug-like compounds.[116]

- MW $\leqslant$300Da
- number of hydrogen bond donors $\leqslant$3
- number of hydrogen bond acceptors $\leqslant$3
- Clog P $\leqslant$3.0

These rules are thought to encompass many of the requirements for fragment binding that have been outlined above, namely, that the compounds should be small, simple, and reasonably soluble. In addition to the criteria outlined above, we have also found that it is useful to apply a complexity filter that forces them to be made up of a small number of preferred side chains.

The basic process is similar to that described for version 1.0 above. A revised set of scaffolds and side chains was identified from known drugs and advanced leads. These scaffolds and side chains are combined using a set of substitution patterns to generate a virtual library of approximately 5000 compounds. These were then filtered using the rule of three to give a library of 3000 fragment-like compounds. Compounds were selected from this fragment-like virtual library if they were commercially available or easily synthesized by functional group interconversion from an available analog.

This process resulted in a final set of approximately 1000 compounds. In order to maximize our coverage of fragment interaction space, these compounds were clustered using topological fingerprints[151] as the metric. Topological fingerprints capture the relative proximity of different kinds of functional groups (donor, acceptor, lipophile, etc.) within a molecule. Once the fingerprints have been generated for a whole library, the Tanimoto score can be used to generate a measure of similarity for each pair of molecules. This similarity matrix was clustered using Ward's algorithm,

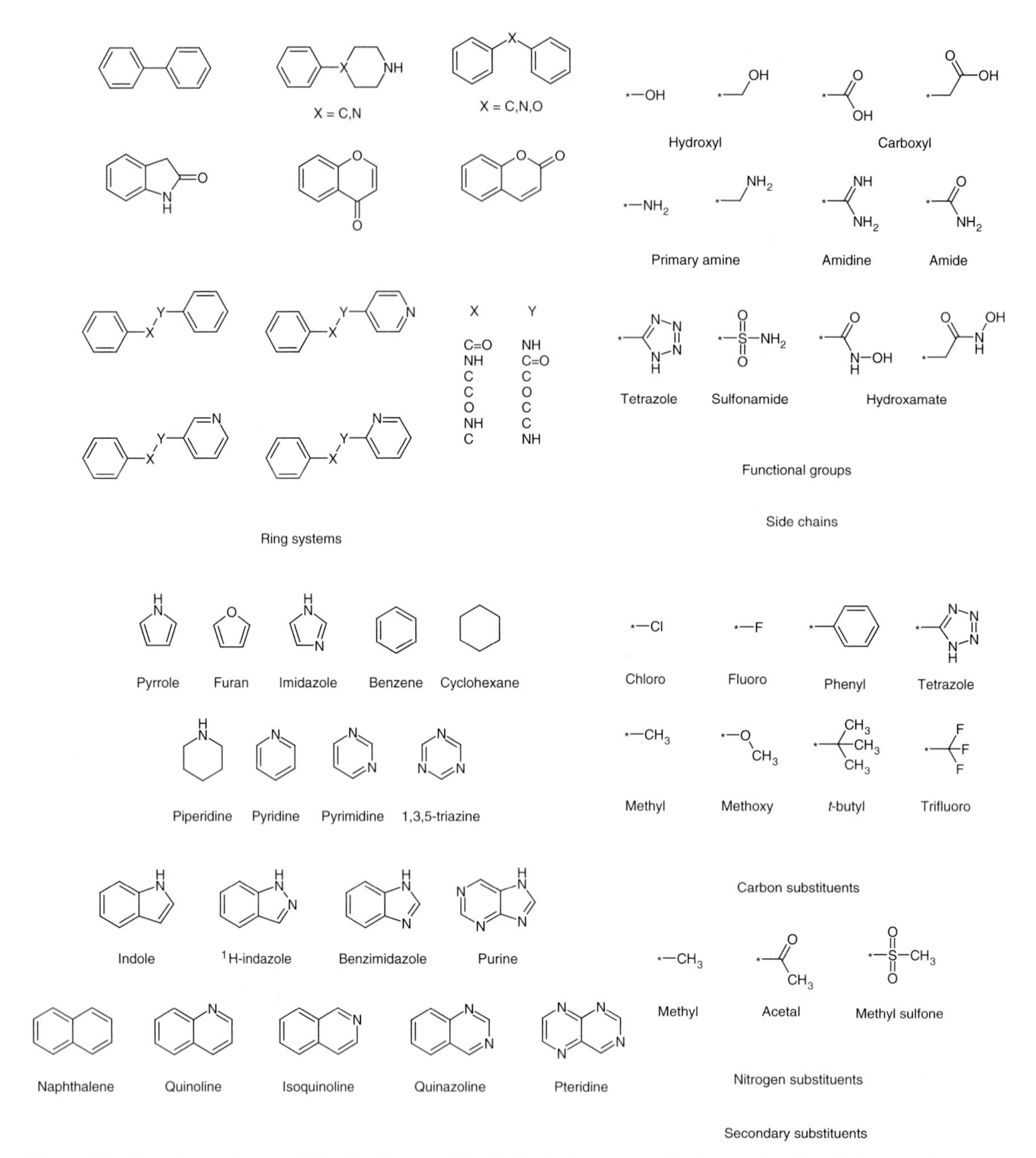

Figure 11 Drug ring systems and functional group side chains that were used in the construction of the general-purpose screening library (Drug Fragment Set).

and a system was developed to allow interactive browsing and cherry-picking of compounds. Fragments from the previous version of the Drug Fragment Set were included in the clustering stage and in this fashion it was possible to observe areas in which our fragment library was under- or overrepresented. Rather than use an automated procedure for selecting the new compounds, they were cherry-picked by medicinal chemists.

This process yielded approximately 400 compounds with the following average physicochemical properties: molecular weight 176, Clog P 1.51, *n* donors 1.44 and *n* acceptors 2.19. Once the library was assembled, all of the compounds were checked for being at least 90% pure, and that they met minimum solubility requirements in dimethyl sulfoxide (DMSO).

While we have not rigorously assessed differences between the two fragment libraries, it is noted that hit rates with the second set are much higher than with the first set. This may be due to a number of factors, such as increased diversity in the range of fragments or improved solubility.

4.31.3.4 Cocktailing

Typically, fragment-screening approaches use biophysical detection techniques with long detection times. Both NMR and x-ray crystallography can take several hours to collect the data for a single fragment screen. This should be compared with HTS where many hundreds of data points can be collected in a short period of time. The time constraints that arise from the use of crystallography or NMR mean that much use is made of cocktailing to increase the rate at which fragment screens can be carried out. Cocktailing is the process by which several screening compounds are mixed together and exposed to the protein system simultaneously, but some care has to be taken to make sure that it is as straightforward as possible to determine which compound in a cocktail has bound.

Nienaber's group at Abbott describe how they used cocktailing[119] to improve the throughput of their crystallographic fragment screening. They simply stated that they group their fragments together into shape diverse groups. As there is very little detail in the literature about the methods that can be used to make it easy to interpret cocktail binding, we will cover some methods that we have developed for x-ray screening.

4.31.3.4.1 Crystallographic shape similarity

The first step in our approach to this problem is the derivation of a method for chemical similarity. The chemical functionalities were categorized by counting the number of hydrogen bond donors and acceptors, and ionizable groups. The size and shape of the molecule were assessed by counting the total number of nonhydrogen atoms, counting the number of five- and six-membered rings, and distinguishing between substitution patterns on di-substituted aromatic rings. The dissimilarity of two molecules is calculated by finding the distance between the feature vectors of the two molecules.

4.31.3.4.2 Cocktailing procedure

A computational procedure was devised to minimize the chance of more than one compound in the same cocktail binding to the protein. This also allowed increased shape diversity in each cocktail that would aid the automated interpretation of ligand electron density by AutoSolve[152] (described in Section 4.31.3.5.1). The number of unique ways of partitioning N compounds into cocktails each containing c compounds is given by:

$$\frac{N!}{n!(c!)^n}$$

where $n = N/c$ is the number of cocktails. This number can increase rapidly, for example, there are 10^{860} unique ways of partitioning 400 compounds into 50 groups of 8. Therefore, any algorithm can sample only a small fraction of the possible partitioning space and needs to be extremely efficient to provide good solutions to the problem.

The partitioning into cocktails proceeds from a matrix that describes the dissimilarity of each pair of molecules to be partitioned. The compounds are initially assigned to random cocktails. The cocktail score is calculated by summing the dissimilarity measures, $d(i,j)$, of all pairs of compounds in a particular cocktail, c. The overall score is calculated by adding the scores of all n cocktails.

$$S = \sum_{c=1,n} \sum_{i,j \in c} d(i,j)$$

This function is maximized using a procedure that swaps pairs of compounds in different cocktails. Swaps are accepted if the score improves or does not change – this second step was found to be very useful in preventing premature termination. The search was terminated after 10 000 000 iterations, or after 100 compound swaps that did not yield an improvement in the overall score. Before use, the algorithm was tested on the largest partitioning problem that could be solved exhaustively (i.e., 24 compounds into three cocktails of eight) and was found to produce solutions rapidly at or very close to the global minimum.

4.31.3.5 Informatics Requirements for Fragment Screening

Fragment screening with x-ray crystallography naturally generates many crystallographic data sets. Once the crystallization conditions for the protein are determined and the soaking experiments have been performed, the processing of the x-ray data and analysis of the results can become the rate-determining step.

Most groups do not describe their efforts to streamline this section of the process, but suggest that they use data analysis techniques that are built around conventional crystallographic software.[153,154] We have found it extremely beneficial to develop specialized software to automate the process of structure determination and have developed a sophisticated internal database for communicating the resulting structures to crystallographers, molecular modelers, and medicinal chemists. These software developments will be briefly described in the next sections.

4.31.3.5.1 AutoSolve

To streamline crystallographic interpretation we found it extremely useful to develop an automated procedure for the interpretation of electron density that is due to a ligand binding to a protein.[152] The automated procedure has the additional benefit of removing the somewhat subjective nature of interpreting which compound from a cocktail has bound.

The procedure automates essentially all stages of the data processing, interpretation, and refinement associated with solving the structure of a protein–ligand complex. Once the refinement is complete the results are placed into our company database and immediately become available for inspection by crystallographers and for use by the lead discovery project teams.

As well as refining and interpreting the electron density due to a ligand binding to a protein, AutoSolve ensures that the ligand makes sensible hydrogen-bonding contacts with the protein. In order to do this it will consider all simple protonation and tautomeric states of the compound.

4.31.3.5.2 Making use of structures

The approaches for lead discovery described here require the generation of a large number of protein–ligand complexes. When these structures have been solved they must be made available to molecular modelers and medicinal chemists to make design decisions for the project.

Despite the enormous advances in computing power and graphics capabilities, it is still relatively difficult to examine a protein–ligand complex. In 2000, the process would have required the use of a molecular graphics program running on an expensive Silicon Graphics workstation. This is problematic for a number of reasons. Firstly, the graphics workstation probably used the UNIX operating system, which is unfamiliar to many medicinal chemists. Secondly, the conceptually simple task of loading a protein–ligand complex and centering on the binding site required several steps in most molecular graphics software. This combination of problems meant that it was very difficult for medicinal chemists to get access to the binding mode of their lead series without the help of crystallographers or modellers skilled in the use of such software.

In order to give greater access to the protein structures and to raise the awareness of medicinal chemistry groups we decided to invest in software development to make it easy to search for particular complexes and display them on a conventional desktop PC.

4.31.3.5.2.1 Pyramid database

As described above, the vast majority of protein–ligand complexes are automatically interpreted using our proprietary software system AutoSolve. Once the structure is solved, the resulting coordinate files for the protein and the ligand are placed into a standardized directory structure, and these locations are recorded in the company database. The protein structure is cross-referenced back to the original purification and crystallization experiments that led to the crystals that were used. The ligand is cross-referenced back to the company compound registry.

Using simple web interfaces, users can query the Pyramid database to look at particular structures and associated electron densities. Users can ask questions like the following:

- Show me the structure for a particular compound
- Show me all structures for this protein target
- Search for complexes that contain this chemical feature

4.31.3.5.2.2 AstexViewer – structure visualization

To simplify the viewing of protein–ligand complexes we developed a molecular graphics program that could be used to view structures and electron density maps on a standard PC. The resulting software is called AstexViewer and is implemented as a Java applet that can run in web-browsers.[155] AstexViewer has been specifically designed to load structures from the company database across the network. It will also automatically display the protein-binding site and a bound ligand, without the need for the user to navigate to this part of the protein structure. In this way the binding modes of ligands can be shown to nonexpert users with minimal skills on the part of the user.

AstexViewer provides a very simple user interface. The structure can be rotated by pressing the mouse button and dragging the picture about. Distances can be measured by selecting pairs of atoms. Additionally, electron density maps can be shown and the contour levels easily changed. Reference structures, for example, other lead compounds or structures from the public domain PDB,[6] can also be overlaid on top of the structure of interest.

AstexViewer supports a number of display styles for the protein and the ligand. These include ball and sticks, cylinders, and spheres for the ligand and schematics for the protein structure. An illustrative example is shown in **Figure 12**. The graphics are of sufficient quality that publication-quality images can be prepared directly from the web interfaces (indeed, all of the molecular graphics images in this chapter are generated with AstexViewer2.0). This drastically simplifies the time-consuming task of preparing figures for talks and publications. AstexViewer has been made available to the community free of charge[155] and has been incorporated into systems within a number of companies.

4.31.3.5.2.3 Project pages

In order to maximize the impact of structural information in drug development projects we have also developed interfaces for bringing together all of the information that relates to a particular project. These are known as project pages and provide a simple-to-use, yet extensive overview of the structural and assay results for a project.

Project pages consist of a set of superimposed protein–ligand complexes along with auxiliary information such as bioassay results (**Figure 12**). The pages are typically curated by a project modeler but the superpositions and the layout of the page are constructed automatically. The project pages consist of a simple hierarchical tree of folders that lets structures be grouped according to characteristics of interest. For example, there might be a folder per chemical

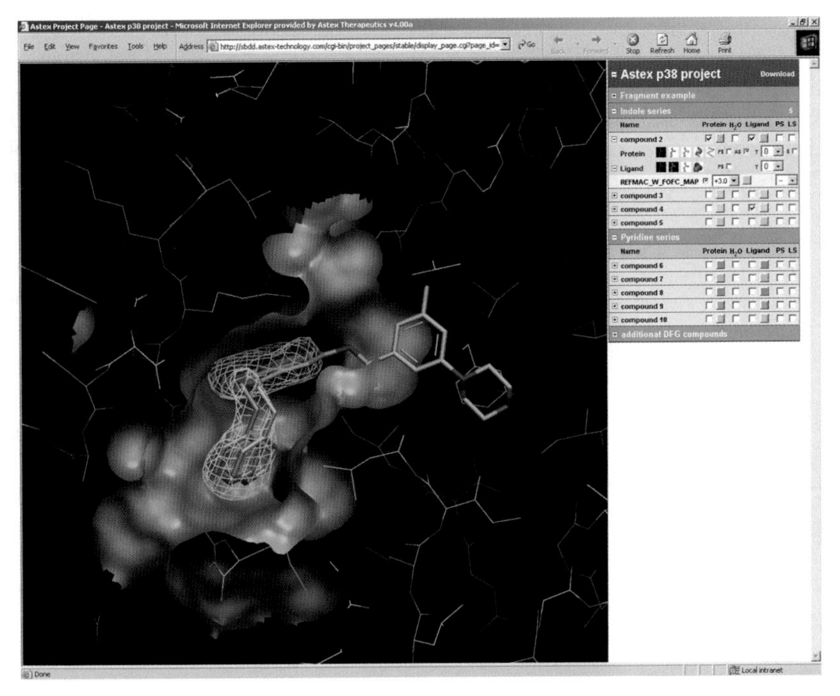

Figure 12 Overlay page showing some of the structures from a p38 project. The structures are arranged hierarchically into folders on the right-hand side of the interface. Each complex has an entry in the folder, and the protein, ligand, and water atoms can be displayed independently. The display style for each ligand or protein structure can be changed by expanding the entry for the complex to reveal more controls. Molecular surfaces can be generated for the ligand, the active site, or the whole protein by selecting the appropriate control. The protein structure can be shown as a schematic or a Cα trace. The colors of the molecules and surfaces can be changed independently, but each complex is given different colors by default. The pages have been carefully designed such that the most frequently used controls are always accessible: more advanced controls, such as the display of electron density maps, are accessed by expanding the protein or ligand entry in each complex. The page is currently displaying a fragment hit (in orange) with its electron density (in yellow), and a more advanced compound (in gray). The molecular surface of the protein-binding site is shown in orange.

series in the project, along with a folder of reference structures from the public domain. The complexes can be derived from real crystal structures, docking studies, homology models, or any other source of structural information that is useful for understanding the SARs of the project.

When a structure is added to a project page, it is possible that it will be in a different frame of reference to the structures already included in the page. The system provides a number of methods for superimposing the new structure on the old ones. For structures that are very similar to those that are in the project page, they can be fitted by identifying the sequence relationship between the new structures and the existing ones. Sometimes it is useful to add more distantly related structures to a project page (for instance, to understand issues of selectivity/cross-reactivity), and in this case simple sequence relationships may not be sufficient to align the structures correctly. For these cases we have developed a structural superposition algorithm (I. Tickle, personal communication) that can identify structures with a common shape but dissimilar sequence alignments. The combination of these two techniques means that essentially all structures can be imported and aligned with the rest of the structures in the project page.

The project pages are an extremely simple concept but have proved to be an invaluable aid in the discussion of lead development for chemical series. The pages have simple controls that allow users to change the color and display styles of the different structures. Additionally, electron density maps can be displayed and positions of hydrogen-bonding regions can be shown with Superstar[156] maps.

4.31.3.6 Examples of Fragment Screening

Earlier in this chapter we described the outcome of fragment screening carried out by various groups. Here we describe the results from applying our Pyramid fragment screening[147,157] using examples from p38 MAP kinase,[158] CDK-2,[159] thrombin,[160] RNase A,[161] and PTP1B.[162] The chemical structures of the fragment binders we will discuss are shown in **Figure 13**.

At the outset of development of fragment screening there were a number of questions that needed to be answered before it was clear that this approach would become a useful method for drug discovery.

One of the principal issues was whether fragment screening was capable of picking up the full range of interactions that are important in the binding of drug molecules to targets. There was initially some belief that only fragments that made extremely strong interactions with the protein (i.e., charge/charge) would ever bind well enough to be observed in fragment-screening experiments. Our first set of screening experiments showed that this was not the case and that fragments could be observed making essentially all types of interactions that are useful for ligand-binding in larger

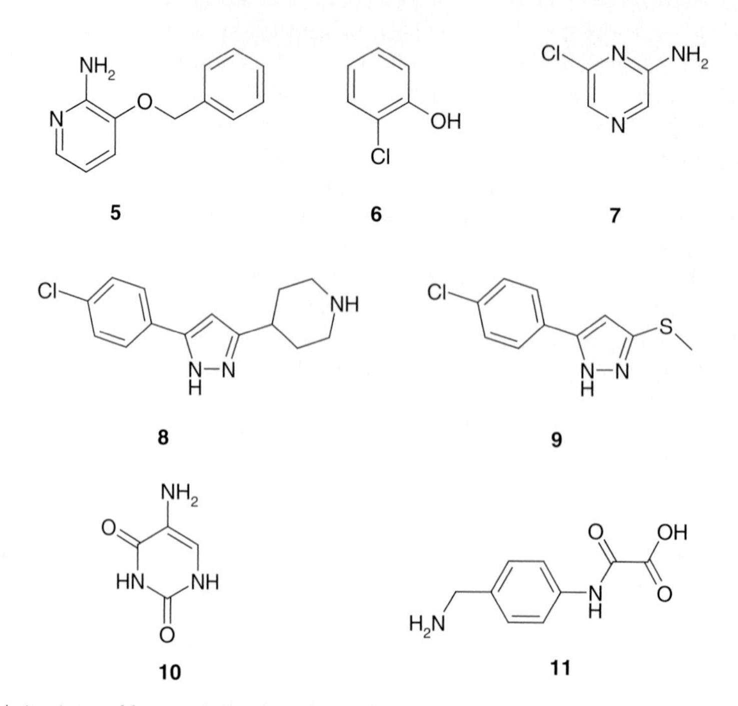

Figure 13 Chemical structures of fragments that have been observed to bind experimentally against a variety of targets. The binding modes and targets are discussed in more detail in the caption to **Figure 14**.

molecules. **Figure 14** shows the binding modes of a number of fragment hits from our screening approach. It can be seen that fragments bind to make neutral hydrogen bonds (**Figure 14a**: p38 and **14c**: CDK-2), lipophilic interactions (**Figure 14b**: p38 and **14d**: thrombin), charge/charge interactions (**Figure 14f**: ptp1b), and combinations of these (**Figure 14a**: p38). This gave us confidence that fragment screening was a source of useful start points for targeting interactions in protein active sites.

These examples also show that fragment screening is widely applicable against a variety of protein systems, including kinases, proteases, and phosphatases. Other groups have identified fragment binders for a number of other protein families, including gyrase,[136] phosphodiesterase,[140] and other proteases.[119] These observations suggest that fragment screening is a generally applicable technique for protein targets for which the crystal structure is available.

A related issue to do with fragment interactions is if fragments can induce conformational changes in the proteins to which they bind. This is sometimes important for the mode of action of drug molecules. **Figure 14** demonstrates that fragments are able to induce conformational changes in proteins. The first example shows that the fragment causes a single side chain in p38 to be moved (**Figure 14b**), but the second example shows that a whole loop in PTP-1B can be repositioned by fragment binding (**Figure 14f**). This is an extremely valuable observation as it suggests that even weakly binding inhibitors can stabilize different conformations in protein structures.

4.31.3.7 Hits to Leads

Previously we presented a description of several fragment-screening approaches that have produced hits against a number of targets. In this section we will round up our discussion of fragment screening with a brief description of two of our applications in which fragment hits have been evolved to potent lead compounds.

4.31.3.7.1 Application 1: p38 kinase

Gill *et al.* describe fragment screening against p38α MAP kinase,[157] which is implicated in the progression of many inflammatory diseases, such as rheumatoid arthritis, psoriasis, and others. Crystallographic screening against p38 yielded two promising hits (**12** and **13**), shown in **Figure 15**. Both fragments were found to make a hydrogen bond with the key backbone nitrogen of Met109.

Starting from fragment **12**, SBDD was used to optimize contacts in the lipophilic pocket where the phenyl group binds. For example, the naphthyl derivative was discovered to be about 30 times more potent than the original phenyl compound. The next breakthrough occurred after synthesis and structure determination of a literature compound,[163] which was found to induce a significant rearrangement of the residues Asp168-Phe169-Gly170 (DFG loop). Structural overlays comparing the binding mode of **12** with this literature compound suggested that further modifications around the phenyl ring of **12** would allow access to the polar channel formed by the movement of the DFG loop. Additionally, the movement of Phe169 revealed a hydrophobic pocket which could be exploited for specificity relative to other kinases that cannot adopt the DFG out conformation. This resulted in compound **14**, which was 20 000 times more potent than the starting fragment.

Fragment **13** was also an interesting starting point for design because the binding mode of the indole moiety was previously unknown in p38 kinase, despite the huge amount of work done in this area. Using structural information obtained from the development of **14**, it was possible to obtain potent and selective inhibitors such as **15** in a small number of design cycles. **Figure 16** shows the binding mode of the starting fragment **13** and the final compound **15**. The key interactions observed with the starting fragment are maintained in the final compound despite the substantial conformational shift of the protein.

These p38 examples illustrate how crystallographic screening followed by structure-based optimization can be used to generate potent lead compounds efficiently.

4.31.3.7.2 Application 2: thrombin

Howard *et al.*[165] have described a fragment-screening approach for thrombin, which is an anticoagulant target. A primary goal was to avoid strongly basic moieties such as benzamidines in the S1 pocket of the enzyme, since it was known that these moieties were detrimental to oral bioavailability. Virtual screening was used to define a set of 80 compounds that were screened using the x-ray structure. The uncharged S1 binder, **16**, was identified and shown to bind in the S1 pocket of thrombin with an IC_{50} of 330 μM (**Figures 17 and 18a**). Unexpectedly, another fragment, **17**, was found to bind in the S2–S4 region of the enzyme with an IC_{50} of 100 μM and forms two hydrogen bonds with Gly216 (**Figure 18b**).

It is known that if weak binding fragments can be joined together in an optimal manner, there can be a superadditivity of binding affinity, resulting in highly potent molecules. SBDD was used to suggest compounds based on linking of the fragments. This resulted in compound **18**, which has an IC_{50} of 3.7 nM, contains no peptide bonds,

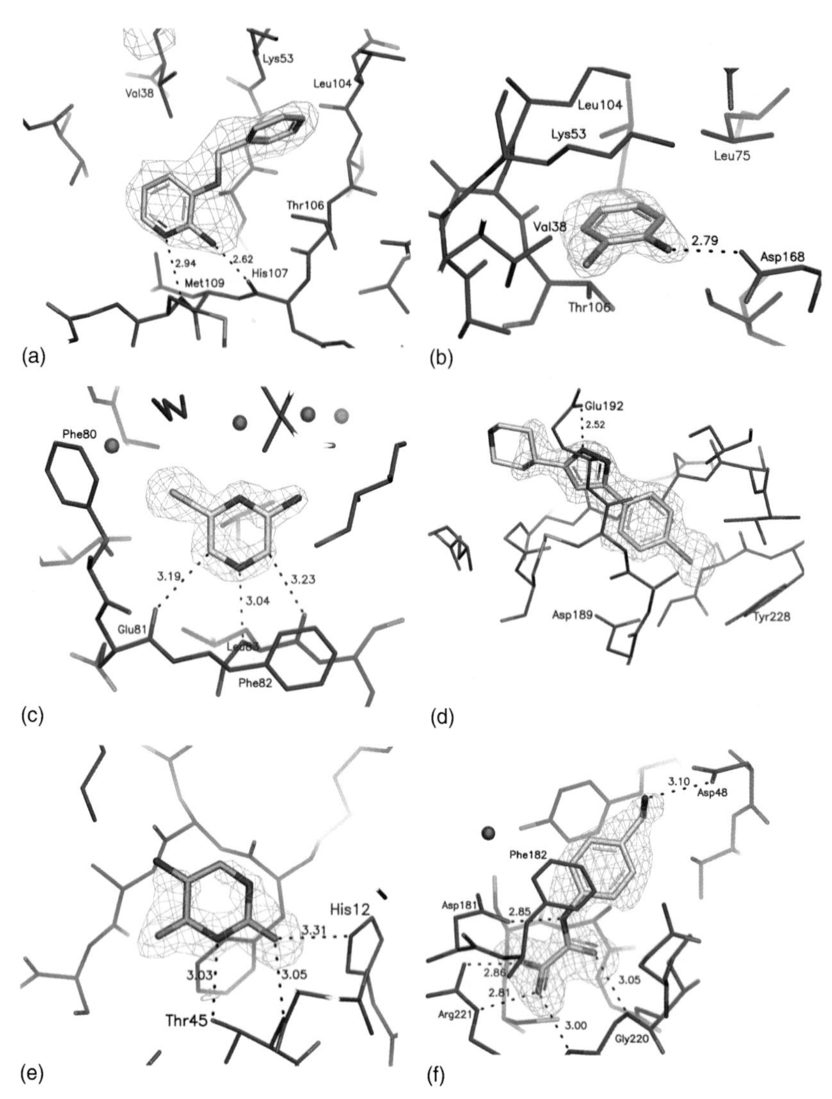

Figure 14 Fragments from **Figure 13** shown bound to different protein targets. All electron density is shown as F_o–F_c maps contoured at 3σ except for (E), which is contoured at 1.7σ. The electron density maps have been clipped to the region around the ligands for clarity. (a) Compound **5** (thick stick) is shown bound to p38 MAP kinase (thin stick). The fragment binds to the hinge region, forming hydrogen bonds with the N–H of Met109 and the C = O of His107. The phenyl ring is bound in a lipophilic pocket formed by the side chains of Lys53, Leu75 (omitted for clarity), Leu104, and Thr106. (b) Compound **6** (thick stick) is shown bound to p38 MAP kinase (thin stick). The chlorophenyl ring is bound in the lipophilic pocket formed by the side chains of Lys53, Leu75, Leu104, and Thr106. The phenol makes a hydrogen bond to the side chain of Asp168, which has moved over 2 Å to form this interaction. (c) Compound **7** (thick stick) is shown bound to CDK-2 (thin stick). The heterocycle binds at the ATP-binding cleft, forming a good uncharged hydrogen bond with the N–H of Leu83. There are also strong C–H aromatic contacts with the carbonyls of Leu83 and Glu81. The ligand makes lipophilic contact with Ala31 (omitted for clarity), Phe80, Phe82, and Leu134 (unlabeled, directly behind ligand). (d) Compound **8** (thick stick) is shown bound to thrombin (thin stick). The key interactions are the chlorophenyl group buried deep in the hydrophobic portion of the S1 pocket above Tyr228. There is also a hydrogen bond between the N–H of the pyrazole and the flexible side chain of Glu192. The piperidine forms no significant contacts with the enzyme and is disordered, although its presence probably helps solubilize the compound. (e) Compound **10** (thick stick) is shown bound to RNase A (thin stick). The heterocycle forms good uncharged hydrogen bonds with the backbone and side chain of Thr45 and with the side chain of His12. (f) Compound **11** (thick stick) is shown bound to PTP1B (thin stick). The carboxylate forms a salt-bridge with the side chain of Arg221 and also with Arg221's backbone N–H. The ligand carbonyl forms a hydrogen bond with the backbone of Gly220, whilst the ligand N–H forms a hydrogen bond with the side chain of Asp181. The amine on the ligand also forms a salt-bridge with Asp48. The ligand induces a movement of the 'WPD loop' of the enzyme, allowing the formation of additional interactions (only the final conformation of the WPD loop is shown).

12: IC_{50} = 1300 μM

14: IC_{50} = 0.065 μM

13: IC_{50} = 35 μM

15: IC_{50} = 0.63 μM

Figure 15 The chemical structures for p38 compounds, discussed in the text. The two fragments on the left were used as starting points in the design of the compounds on the right.

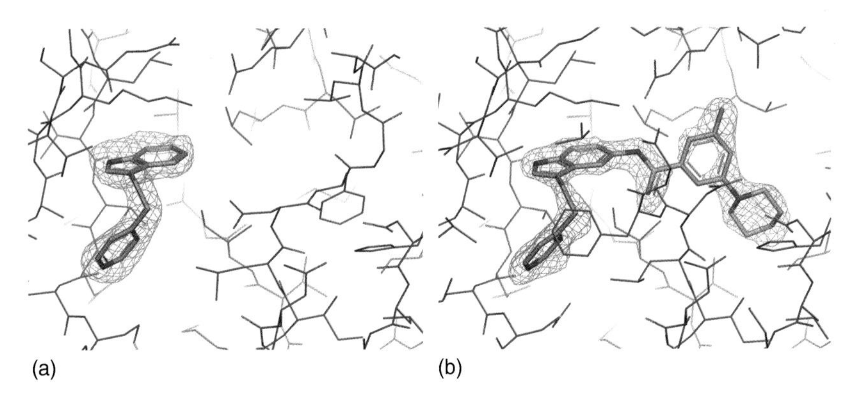

Figure 16 Binding modes of p38 structures. All ligands are shown with gray carbon atoms. Electron density maps are contoured at 3σ and clipped to the ligand for clarity. (a) The binding mode of **13** complexed with p38. The pyridine interacts with Met109 NH and the indole binds in the lipophilic specifity pocket. (b) The binding mode of **15** complexed with p38. The pyridine and indole still occupy the same positions but the additional parts of the molecule have moved the 'DFG' loop into the out conformation.

and retains the uncharged group in the S1 pocket (**Figure 18c**). It may therefore represent an interesting starting point for the design of oral anticoagulants.

Despite the introduction of additional rotatable bonds, the free energy of binding (as estimated by the logarithm of the IC_{50}) for the linked compound is approximately the sum of the free energies of binding of the fragments. The binding mode of compound **18** is grossly similar to that of the starting fragments but the two original hydrogen bonds with Gly216 have been lost in the joined molecule, suggesting that the linking process has not been achieved in an ideal fashion. In conclusion, the work indicates the potential of the fragment-linking approach to generate inhibitors for proteases but also illustrates the problems associated with linking fragments whilst maintaining key interactions. Our general experience has been that fragment growing can be a more successful and efficient strategy for generating lead molecules from starting fragments.

16: IC_{50} = 330 μM **17**: IC_{50} = 100 μM **18**: IC_{50} = 3.7 nM

Figure 17 The chemical structures for the thrombin molecules discussed in the text. The two fragments on the left were used to design the linked compound shown on the right. (Note that the linked compound contains an anisole group instead of the naphthyl group that is present in the original fragment.)

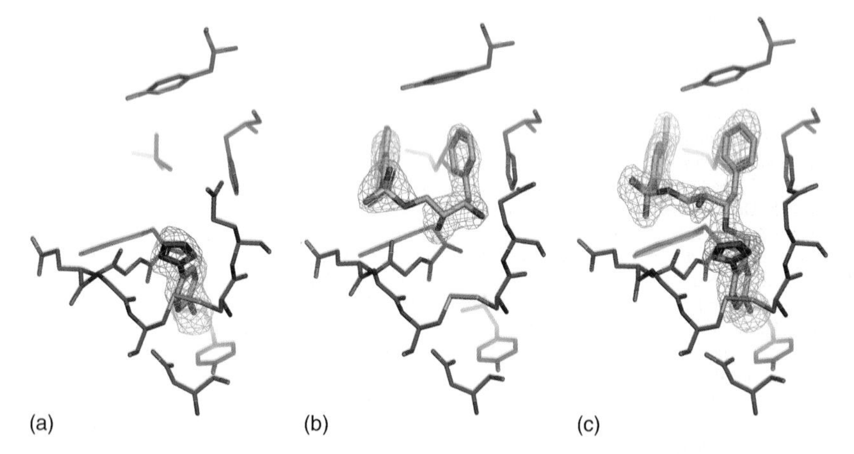

Figure 18 The binding modes in thrombin of the compounds shown in **Figure 17**. The ligands are shown with carbons in gray with the protein carbons in green. (a) The uncharged fragment **16** is shown buried deeply in the S1 pocket. (b) The larger fragment **17** binds in the S2/S4 region of the protein, forming a β-sheet with the enzyme. It does not impinge on the S1 region, where fragment **16** binds. (c) The joined molecule **18** shows grossly the same binding mode spanning the S1, S2, and S4 regions of the protein.

4.31.4 Perspective on New Applications to Drug Design

In this chapter, two new areas of applications for SBDD have been reviewed. The first was in the area of ADME, where it was demonstrated that the structures of HSA have been successfully used to solve liabilities associated with PPB. It is expected that applications to PPB will now be used from time to time in drug discovery. More exciting, but less advanced, is the possibility of using SBDD to address ADME problems associated with CYP binding, because CYP binding is a problem that is routinely addressed in drug discovery projects. It will be interesting to see how things develop over the next few years because there has certainly been massive progress in the last 5 years.

The second new application area for SBDD considered in this chapter has been the use of crystallography to screen and/or detect fragments. We believe that there is enough evidence that this is an efficient way to do drug discovery, and we anticipate that the next few years will see a number of clinical candidates emerging from this approach. Taken as a whole, the evidence presented in this review demonstrates that SBDD continues to expand its utility to drug discovery and that researchers in SBDD can look forward to an exciting future.

References

1. Campbell, S. F. *Clin. Sci.* **2000**, *99*, 255–260.
2. Kola, I.; Landis, J. *Nat. Rev. Drug Disc.* **2004**, *3*, 711–715.
3. Mestres, J.; Veeneman, G. H. *J. Med. Chem.* **2003**, *46*, 3441–3444.

4. Gribbon, P.; Andreas, S. *Drug Disc. Today* **2005**, *10*, 17–22.
5. Congreve, M.; Murray, C. W.; Blundell, T. L. *Drug Disc. Today* **2005**, *10*, 895–907.
6. Berman, H. M.; Westbrook, J.; Feng, Z.; Gilliland, G.; Bhat, T. N.; Weissig, H.; Shindyalov, I. N.; Bourne, P. E. *Nucleic Acids Res.* **2000**, *28*, 235–242.
7. Babine, R. E.; Bender, S. L. *Chem. Rev.* **1997**, *97*, 1359–1472.
8. Bohacek, R. S.; Mcmartin, C.; Guida, W. C. *Med. Res. Rev.* **1996**, *16*, 3–50.
9. Edwards, P. D.; Andisik, D. W.; Bryant, C. A.; Ewing, B.; Gomes, B.; Lewis, J. J.; Rakiewicz, D.; Steelman, G.; Strimpler, A.; Trainor, D. A. et al. *J. Med. Chem.* **1997**, *40*, 1876–1885.
10. Abagyan, R.; Totrov, M. *Curr. Opin. Struct. Biol.* **2001**, *5*, 375–382.
11. Bajorath, J. *Nat. Rev. Drug Disc.* **2002**, *1*, 882–894.
12. Jain, A. N. *Curr. Opin. Drug Disc. Dev.* **2004**, *7*, 396–403.
13. Kitchen, D. B.; Decornez, H.; Furr, J. R.; Bajorath, J. *Nat. Rev. Drug Disc.* **2004**, *3*, 935–949.
14. Lyne, P. D. *Drug Disc. Today* **2002**, *7*, 1047–1055.
15. Shoichet, B. K. *Nature* **2004**, *432*, 862–865.
16. Shoichet, B. K.; Stroud, R. M.; Santi, D. V.; Kuntz, I. D.; Perry, K. M. *Science* **1993**, *259*, 1445–1450.
17. Kratochwil, N. A.; Huber, W.; Muller, F.; Kansy, M.; Gerber, P. R. *Curr. Opin. Drug Disc. Dev.* **2004**, *7*, 507–512.
18. Smith, D. A.; Jones, B. C. *Biochem. Pharmacol.* **1992**, *44*, 2089–2098.
19. Carr, R. A. E.; Congreve, M.; Murray, C. W.; Rees, D. C. *Drug Disc. Today* **2005**, *10*, 987–992.
20. Hopkins, A. L.; Groom, C. R.; Alex, A. *Drug Disc. Today* **2004**, *9*, 430–431.
21. Erlanson, D. A.; McDowell, R. S.; O'Brien, T. *J. Med. Chem.* **2004**, *47*, 3463–3482.
22. Rees, D. C.; Congreve, M.; Murray, C. W.; Carr, R. *Nat. Rev. Drug Disc.* **2004**, *3*, 660–672.
23. Kratochwil, N. A.; Huber, W.; Muller, F.; Kansy, M.; Gerber, P. R. *Biochem. Pharmacol.* **2002**, *64*, 1355–1374.
24. Colmenarejo, G. *Med. Res. Rev.* **2003**, *23*, 275–301.
25. Koizumi, K.; Ikeda, C.; Ito, M.; Suzuki, J.; Kinoshita, T.; Yasukawa, K.; Hanai, T. *Biomed. Chromatogr.* **1998**, *12*, 203–210.
26. Deschamps-Labat, L.; Pehourcq, F.; Jagou, M.; Bannwarth, B. *J. Pharm. Biomed. Anal.* **1997**, *16*, 223–229.
27. He, X. M.; Carter, D. C. *Nature* **1992**, *358*, 209–215.
28. Carter, D. C.; Ho, J. X. *Adv. Protein Chem.* **1994**, *45*, 153–203.
29. Sudlow, G.; Birkett, D. J.; Wade, D. N. *Mol. Pharmacol.* **1976**, *12*, 1052–1061.
30. Mao, H.; Hajduk, P. J.; Craig, R.; Bell, R.; Borre, T.; Fesik, S. W. *J. Am. Chem. Soc.* **2001**, *123*, 10429–10435.
31. Mao, H.; Gunasekera, A. H.; Fesik, S. W. *Protein Expr. Purif.* **2000**, *20*, 492–499.
32. Colmenarejo, G.; Alvarez-Pedraglio, A.; Lavandera, J. L. *J. Med. Chem.* **2001**, *44*, 4370–4378.
33. Hajduk, P. J.; Mendoza, R.; Petros, A. M.; Huth, J. R.; Bures, M.; Fesik, S. W.; Martin, Y. C. *J. Comput. Aided Mol. Des.* **2003**, *17*, 93–102.
34. Hall, L. M.; Hall, L. H.; Kier, L. B. *J. Chem. Inf. Comput. Sci.* **2003**, *43*, 2120–2128.
35. Valko, K.; Nunhuck, S.; Bevan, C.; Abraham, M. H.; Reynolds, D. P. *J. Pharm. Sci.* **2003**, *92*, 2236–2248.
36. Xue, C. X.; Zhang, R. S.; Liu, H. X.; Yao, X. J.; Liu, M. C.; Hu, Z. D.; Fan, B. T. *J. Chem. Inf. Comput. Sci.* **2004**, *44*, 1693–1700.
37. Fournier, T.; Medjoubi, N.; Porquet, D. *Biochim. Biophys. Acta* **2000**, *1482*, 157–171.
38. Ghuman, J.; Zunszain, P. A.; Petitpas, I.; Bhattacharya, A. A.; Otagiri, M.; Curry, S. *J. Mol. Biol.* **2005**, *353*, 38–52.
39. Oltersdorf, T.; Elmore, S. W.; Shoemaker, A. R.; Armstrong, R. C.; Augeri, D. J.; Belli, B. A.; Bruncko, M.; Deckwerth, T. L.; Dinges, J.; Hajduk, P. J. et al. *Nature* **2005**, *435*, 677–681.
40. Shuker, S. B.; Hajduk, P. J.; Meadows, R. P.; Fesik, S. W. *Science* **1996**, *274*, 1531–1534.
41. Schlichting, I.; Berendzen, J.; Chu, K.; Stock, A. M.; Maves, S. A.; Benson, D. E.; Sweet, R. M.; Ringe, D.; Petsko, G. A.; Sligar, S. G. *Science* **2000**, *287*, 1615–1622.
42. Lewis, D. F. *Pharmacogenomics* **2004**, *5*, 305–318.
43. de Groot, M. J.; Kirton, S. B.; Sutcliffe, M. J. *Curr. Top. Med. Chem.* **2004**, *4*, 1803–1824.
44. Van de Waterbeemd, H.; Smith, D. A.; Beaumont, K.; Walker, D. K. *J. Med. Chem.* **2001**, *44*, 1313–1333.
45. Lewis, D. F.; Jacobs, M. N.; Dickins, M. *Drug Disc. Today* **2004**, *9*, 530–537.
46. Johnson, E. F.; Stout, C. D. *Biochem. Biophys. Res. Commun.* **2005**, *338*, 331–336.
47. Johnson, E. F. *Drug Metab. Dispos.* **2003**, *31*, 1532–1540.
48. Poulos, T. L. *Biochem. Biophys. Res. Commun.* **2005**, *338*, 337–345.
49. Li, H.; Poulos, T. L. *Curr. Top. Med. Chem.* **2004**, *4*, 1789–1802.
50. Poulos, T. L. *Curr. Opin. Struct. Biol.* **1995**, *5*, 767–774.
51. Hasemann, C. A.; Kurumbail, R. G.; Boddupalli, S. S.; Peterson, J. A.; Deisenhofer, J. *Structure* **1995**, *3*, 41–62.
52. Williams, P. A.; Cosme, J.; Sridhar, V.; Johnson, E. F.; McRee, D. E. *J. Inorg. Biochem.* **2000**, *81*, 183–190.
53. Williams, P. A.; Cosme, J.; Sridhar, V.; Johnson, E. F.; McRee, D. E. *Mol. Cell* **2000**, *5*, 121–131.
54. de Groot, M. J.; Alex, A. A.; Jones, B. C. *J. Med. Chem.* **2002**, *45*, 1983–1993.
55. Venhorst, J.; ter Laak, A. M.; Commandeur, J. N.; Funae, Y.; Hiroi, T.; Vermeulen, N. P. *J. Med. Chem.* **2003**, *46*, 74–86.
56. Kirton, S. B.; Baxter, C. A.; Sutcliffe, M. J. *Adv. Drug Deliv. Rev.* **2002**, *54*, 385–406.
57. Williams, P. A.; Cosme, J.; Ward, A.; Angove, H. C.; Matak, V. D.; Jhoti, H. *Nature* **2003**, *424*, 464–468.
58. Wester, M. R.; Yano, J. K.; Schoch, G. A.; Yang, C.; Griffin, K. J.; Stout, C. D.; Johnson, E. F. *J. Biol. Chem.* **2004**, *279*, 35630–35637.
59. Yano, J. K.; Wester, M. R.; Schoch, G. A.; Griffin, K. J.; Stout, C. D.; Johnson, E. F. *J. Biol. Chem.* **2004**, *279*, 38091–38094.
60. Williams, P. A.; Cosme, J.; Vinkovic, D. M.; Ward, A.; Angove, H. C.; Day, P. J.; Vonrhein, C.; Tickle, I. J.; Jhoti, H. *Science* **2004**, *305*, 683–686.
61. Guengerich, F. P. *Annu. Rev. Pharmacol. Toxicol.* **1999**, *39*, 1–17.
62. Schoch, G. A.; Yano, J. K.; Wester, M. R.; Griffin, K. J.; Stout, C. D.; Johnson, E. F. *J. Biol. Chem.* **2004**, *279*, 9497–9503.
63. Yano, J. K.; Hsu, M. H.; Griffin, K. J.; Stout, C. D.; Johnson, E. F. *Nat. Struct. Mol. Biol.* **2005**, *12*, 822–823.
64. Scott, E. E.; White, M. A.; He, Y. A.; Johnson, E. F.; Stout, C. D.; Halpert, J. R. *J. Biol. Chem.* **2004**, *279*, 27294–27301.
65. Scott, E. E.; He, Y. A.; Wester, M. R.; White, M. A.; Chin, C. C.; Halpert, J. R.; Johnson, E. F.; Stout, C. D. *Proc. Natl. Acad. Sci. USA* **2003**, *100*, 13196–13201.
66. de Groot, M. J.; Ekins, S. *Adv. Drug Deliv. Rev.* **2002**, *54*, 367–383.
67. Ekins, S.; de Groot, M. J.; Jones, J. P. *Drug Metab. Dispos.* **2001**, *29*, 936–944.
68. de Graaf, C.; Vermeulen, N. P.; Feenstra, K. A. *J. Med. Chem.* **2005**, *48*, 2725–2755.

69. Jones, J. P.; Mysinger, M.; Korzekwa, K. R. *Drug Metab. Dispos.* **2002**, *30*, 7–12.
70. Singh, S. B.; Shen, L. Q.; Walker, M. J.; Sheridan, R. P. *J. Med. Chem.* **2003**, *46*, 1330–1336.
71. Mancy, A.; Antignac, M.; Minoletti, C.; Dijols, S.; Mouries, V.; Duong, N. T.; Battioni, P.; Dansette, P. M.; Mansuy, D. *Biochemistry* **1999**, *38*, 14264–14270.
72. Schoneboom, J. C.; Cohen, S.; Lin, H.; Shaik, S.; Thiel, W. *J. Am. Chem. Soc.* **2004**, *126*, 4017–4034.
73. Bathelt, C. M.; Zurek, J.; Mulholland, A. J.; Harvey, J. N. *J. Am. Chem. Soc.* **2005**, *127*, 12900–12908.
74. Park, J. Y.; Harris, D. *J. Med. Chem.* **2003**, *46*, 1645–1660.
75. van De, W. H.; Smith, D. A.; Beaumont, K.; Walker, D. K. *J. Med. Chem.* **2001**, *44*, 1313–1333.
76. Chohan, K. K.; Paine, S. W.; Mistry, J.; Barton, P.; Davis, A. M. *J. Med. Chem.* **2005**, *48*, 5154–5161.
77. Susnow, R. G.; Dixon, S. L. *J. Chem. Inf. Comput. Sci.* **2003**, *43*, 1308–1315.
78. Yap, C. W.; Chen, Y. Z. *J. Chem. Inf. Model.* **2005**, *45*, 982–992.
79. Borodina, Y.; Sadym, A.; Filimonov, D.; Blinova, V.; Dmitriev, A.; Poroikov, V. *J. Chem. Inf. Comput. Sci.* **2003**, *43*, 1636–1646.
80. Molnar, L.; Keseru, G. M. *Bioorg. Med. Chem. Lett.* **2002**, *12*, 419–421.
81. Korolev, D.; Balakin, K. V.; Nikolsky, Y.; Kirillov, E.; Ivanenkov, Y. A.; Savchuk, N. P.; Ivashchenko, A. A.; Nikolskaya, T. *J. Med. Chem.* **2003**, *46*, 3631–3643.
82. Ridderstrom, M.; Zamora, I.; Fjellstrom, O.; Andersson, T. B. *J. Med. Chem.* **2001**, *44*, 4072–4081.
83. Afzelius, L.; Zamora, I.; Ridderstrom, M.; Andersson, T. B.; Karlen, A.; Masimirembwa, C. M. *Mol. Pharmacol.* **2001**, *59*, 909–919.
84. Korhonen, L. E.; Rahnasto, M.; Mahonen, N. J.; Wittekindt, C.; Poso, A.; Juvonen, R. O.; Raunio, H. *J. Med. Chem.* **2005**, *48*, 3808–3815.
85. Locuson, C. W.; Suzuki, H.; Rettie, A. E.; Jones, J. P. *J. Med. Chem.* **2004**, *47*, 6768–6776.
86. Locuson, C. W.; Rock, D. A.; Jones, J. P. *Biochemistry* **2004**, *43*, 6948–6958.
87. de Groot, M. J.; Ackland, M. J.; Horne, V. A.; Alex, A. A.; Jones, B. C. *J. Med. Chem.* **1999**, *42*, 1515–1524.
88. Afzelius, L.; Zamora, I.; Masimirembwa, C. M.; Karlen, A.; Andersson, T. B.; Mecucci, S.; Baroni, M.; Cruciani, G. *J. Med. Chem.* **2004**, *47*, 907–914.
89. Zamora, I.; Afzelius, L.; Cruciani, G. *J. Med. Chem.* **2003**, *46*, 2313–2324.
90. Cruciani, G.; Carosati, E.; De Boeck, B.; Ethirajulu, K.; Mackie, C.; Howe, T.; Vianello, R. *J. Med. Chem.* **2005**, *48*, 6970–6979.
91. Goodford, P. J. *J. Med. Chem.* **1985**, *28*, 849–857.
92. Lewis, D. F.; Dickins, M.; Lake, B. G.; Goldfarb, P. S. *Drug Metab. Drug Interact.* **2003**, *19*, 257–285.
93. Lewis, D. F. *Curr. Drug Metab.* **2003**, *4*, 331–340.
94. Lewis, D. F.; Lake, B. G.; Dickins, M.; Goldfarb, P. S. *Drug Metab. Drug Interact.* **2002**, *19*, 115–135.
95. Lewis, D. F.; Dickins, M. *Drug Disc. Today* **2002**, 7, 918–925.
96. Lewis, D. F. *Arch. Biochem. Biophys.* **2003**, *409*, 32–44.
97. Lewis, D. F. *J. Inorg. Biochem.* **2002**, *91*, 502–514.
98. Lewis, D. F. *Biochem. Pharmacol.* **2000**, *60*, 293–306.
99. Kemp, C. A.; Flanagan, J. U.; van Eldik, A. J.; Marechal, J. D.; Wolf, C. R.; Roberts, G. C.; Paine, M. J.; Sutcliffe, M. J. *J. Med. Chem.* **2004**, *47*, 5340–5346.
100. Kirton, S. B.; Kemp, C. A.; Tomkinson, N. P.; St Gallay, S.; Sutcliffe, M. J. *Proteins* **2002**, *49*, 216–231.
101. Eldridge, M. D.; Murray, C. W.; Auton, T. R.; Paolini, G. V.; Mee, R. P. *J. Comput. Aided Mol. Des.* **1997**, *11*, 425–445.
102. Jones, G.; Willett, P.; Glen, R. C.; Leach, A. R.; Taylor, R. *J. Mol. Biol.* **1997**, *267*, 727–748.
103. Verdonk, M. L.; Cole, J. C.; Hartshorn, M.; Murray, C. W.; Taylor, R. D. *Proteins* **2003**, *52*, 609–623.
104. Keizers, P. H.; de Graaf, C.; de Kanter, F. J.; Oostenbrink, C.; Feenstra, K. A.; Commandeur, J. N.; Vermeulen, N. P. *J. Med. Chem.* **2005**, *48*, 6117–6127.
105. Kirton, S. B.; Murray, C. W.; Verdonk, M. L.; Taylor, R. D. *Proteins* **2005**, *58*, 836–844.
106. de Graaf, C.; Pospisil, P.; Pos, W.; Folkers, G.; Vermeulen, N. P. *J. Med. Chem.* **2005**, *48*, 2308–2318.
107. Keseru, G. M. *J. Comput. Aided Mol. Des.* **2001**, *15*, 649–657.
108. Verras, A.; Kuntz, I. D.; Ortiz De Montellano, P. R. *J. Med. Chem.* **2004**, *47*, 3572–3579.
109. Zhang, Z.; Sibbesen, O.; Johnson, R. A.; Ortiz De Montellano, P. R. *Bioorg. Med. Chem.* **1998**, *6*, 1501–1508.
110. De Voss, J. J.; Sibbesen, O.; Zhang, Z.; Ortiz De Montellano, P. R. *J. Am. Chem. Soc.* **1997**, *119*, 5489–5498.
111. Rosenfeld, R. J.; Goodsell, D. S.; Musah, R. A.; Morris, G. M.; Goodin, D. B.; Olson, A. J. *J. Comput. Aided Mol. Des* **2003**, *17*, 525–536.
112. Wester, M. R.; Johnson, E. F.; Marques-Soares, C.; Dijols, S.; Dansette, P. M.; Mansuy, D.; Stout, C. D. *Biochemistry* **2003**, *42*, 9335–9345.
113. Hann, M. M.; Leach, A. R.; Harper, G. *J. Chem. Inf. Comput. Sci.* **2001**, *41*, 856–864.
114. Oprea, T. I.; Davis, A. M.; Teague, S. J.; Leeson, P. D. *J. Chem. Inf. Comput. Sci.* **2001**, *41*, 1308–1315.
115. Teague, S. J.; Davis, A. M.; Leeson, P. D.; Oprea, T. *Angew. Chem. Int. Ed. Engl.* **1999**, *38*, 3743–3748.
116. Lipinski, C. A.; Lombardo, F.; Dominy, B. W.; Feeney, P. J. *Adv. Drug Deliv. Rev.* **2001**, *46*, 3–26.
117. Hajduk, P. J.; Meadows, R. P.; Fesik, S. W. *Q. Rev. Biophys.* **1999**, *32*, 211–240.
118. Verlinde, C. L. M. J.; Rudenko, G.; Hol, W. G. *J. Comput. Aided Mol. Des.* **1992**, *6*, 131–147.
119. Nienaber, V. L.; Richardson, P. L.; Klighofer, V.; Bouska, J. J.; Giranda, V. L.; Greer, J. *Nat. Biotechnol.* **2000**, *18*, 1105–1108.
120. Sanders, W. J.; Nienaber, V. L.; Lerner, C. G.; McCall, J. O.; Merrick, S. M.; Swanson, S. J.; Harlan, J. E.; Stoll, V. S.; Stamper, G. F.; Betz, S. F. et al. *J. Med. Chem.* **2004**, *47*, 1709–1718.
121. Lesuisse, D.; Lange, G.; Deprez, P.; Benard, D.; Schoot, B.; Delettre, G.; Marquette, J. P.; Broto, P.; Jean-Baptiste, V.; Bichet, P. et al. *J. Med. Chem.* **2002**, *45*, 2379–2387.
122. Bohm, H. J. *J. Comput. Aid. Mol. Des.* **1992**, *6*, 61–78.
123. Lange, G.; Lesuisse, D.; Deprez, P.; Schoot, B.; Loenze, P.; Benard, D.; Marquette, J. P.; Broto, P.; Sarubbi, E.; Mandine, E. *J. Med. Chem.* **2003**, *46*, 5184–5195.
124. Hajduk, P. J.; Meadows, R. P.; Fesik, S. W. *Science* **1997**, *278*, 497–499.
125. Hajduk, P. J.; Sheppard, G.; Nettesheim, D. G.; Olejniczak, E. T.; Shuker, S. B.; Meadows, R. P.; Steinman, D. H.; Carrera, G. M., Jr.; Marcotte, P. A.; Severin, J. et al. *J. Am. Chem. Soc.* **2004**, *119*, 5818–5827.
126. Meyer, B.; Peters, T. *Angew. Chem. Int. Ed. Engl.* **2003**, *42*, 864–890.
127. Moy, F. J.; Haraki, K.; Mobilio, D.; Walker, G.; Powers, R.; Tabei, K.; Tong, H.; Siegel, M. M. *Anal. Chem.* **2001**, *73*, 571–581.
128. Tsang, S. K.; Cheh, J.; Isaacs, L.; Joseph-McCarthy, D.; Choi, S. K.; Pevear, D. C.; Whitesides, G. M.; Hogle, J. M. *Chem. Biol.* **2001**, *8*, 33–45.

129. Swayze, E. E.; Jefferson, E. A.; Sannes-Lowery, K. A.; Blyn, L. B.; Risen, L. M.; Arakawa, S.; Osgood, S. A.; Hofstadler, S. A.; Griffey, R. H. *J. Med. Chem.* **2002**, *45*, 3816–3819.
130. Woods, V. L., Jr.; Hamuro, Y. *J. Cell. Biochem.* **2001**, *84*, 89–98.
131. Zhu, M. M.; Rempel, D. L.; Gross, M. L. *J. Am. Soc. Mass Spectrom.* **2004**, *15*, 388–397.
132. Braisted, A. C.; Oslob, J. D.; DeLano, W. L.; Hyde, J.; McDowell, R. S.; Waal, N.; Yu, C.; Arkin, M. R.; Raimundo, B. C. *J. Am. Chem. Soc.* **2003**, *125*, 3714–3715.
133. Erlanson, D. A.; Braisted, A. C.; Raphael, D. R.; Randal, M.; Stroud, R. M.; Gordon, E. M.; Wells, J. A. *Proc. Natl. Acad. Sci. USA* **2000**, *97*, 9367–9372.
134. Erlanson, D. A.; Lam, J. W.; Wiesmann, C.; Luong, T. N.; Simmons, R. L.; DeLano, W. L.; Choong, I. C.; Burdett, M. T.; Flanagan, W. M.; Lee, D. et al. *Nat. Biotechnol.* **2003**, *21*, 308–314.
135. Erlanson, D. A.; Wells, J. A.; Braisted, A. C. *Annu. Rev. Biophys. Biomol. Struct.* **2004**, *33*, 199–223.
136. Boehm, H. J.; Boehringer, M.; Bur, D.; Gmuender, H.; Huber, W.; Klaus, W.; Kostrewa, D.; Kuehne, H.; Luebbers, T.; Meunier-Keller, N. et al. *J. Med. Chem.* **2000**, *43*, 2664–2674.
137. Fejzo, J.; Lepre, C. A.; Peng, J. W.; Bemis, G. W.; Ajay; Murcko, M. A.; Moore, J. M. *Chem. Biol.* **1999**, *6*, 755–769.
138. Meyer, B.; Peters, T. *Angew. Chem. Int. Ed. Engl.* **2003**, *42*, 864–890.
139. van Dongen, M. J.; Uppenberg, J.; Svensson, S.; Lundback, T.; Akerud, T.; Wikstrom, M.; Schultz, J. *J. Am. Chem. Soc.* **2002**, *124*, 11874–11880.
140. Card, G. L.; Blasdel, L.; England, B. P.; Zhang, C.; Suzuki, Y.; Gillette, S.; Fong, D.; Ibrahim, P. N.; Artis, D. R.; Bollag, G. et al. *Nat. Biotechnol.* **2005**, *23*, 201–207.
141. Jones, G.; Willett, P.; Glen, R. C.; Leach, A. R.; Taylor, R. Further Development of a Genetic Algorithm for Ligand Docking and its Application to Screening Combinatorial Libraries. In *Rational Drug Design, Novel Methodology and Practical Applications*; Parrill, A. L., Reddy, M. R., Eds.; American Chemical Society: Las Vegas, USA, 1999, pp 271–291.
142. Verdonk, M. L.; Berdini, V.; Hartshorn, M.; Mooij, W. T.; Murray, C. W.; Taylor, R. D.; Watson, P. *J. Chem. Inf. Comput. Sci.* **2004**, *44*, 793–806.
143. Pan, Y.; Huang, N.; Cho, S.; MacKerrel, A. D. *J. Chem. Inf. Comput. Sci.* **2003**, *43*, 267–272.
144. Verdonk, M. L.; Berdini, V.; Hartshorn, M. J.; Mooij, W. T. M.; Murray, C. W.; Taylor, R. D.; Watson, P. *J. Chem. Inf. Comput. Sci.* **2004**, *44*, 793–806.
145. Baurin, N.; Aboul-Ela, F.; Barril, X.; Davis, B.; Drysdale, M.; Dymock, B.; Finch, H.; Fromont, C.; Richardson, C.; Simmonite, H. et al. *J. Chem. Inf. Comput. Sci.* **2004**, *44*, 2157–2166.
146. Baurin, N.; Baker, R.; Richardson, C.; Chen, I.; Foloppe, N.; Potter, A.; Jordan, A.; Roughley, S.; Parratt, M.; Greaney, P. et al. *J. Chem. Inf. Comput. Sci.* **2004**, *44*, 643–651.
147. Hartshorn, M. J.; Murray, C. W.; Cleasby, A.; Frederickson, M.; Tickle, I. J.; Jhoti, H. *J. Med. Chem.* **2005**, *48*, 403–413.
148. Bemis, G. W.; Murcko, M. A. *J. Med. Chem.* **1996**, *39*, 2887–2893.
149. Bemis, G. W.; Murcko, M. A. *J. Med. Chem.* **1999**, *42*, 5095–5099.
150. Congreve, M.; Carr, R.; Murray, C.; Jhoti, H. *Drug Disc. Today* **2003**, *8*, 876–877.
151. Schuffenhauer, A.; Floersheim, P.; Acklin, P.; Jacoby, E. *J. Chem. Inf. Comput. Sci.* **2003**, *43*, 391–405.
152. Blundell, T. L.; Abell, C.; Cleasby, A.; Hartshorn, M. J.; Tickle, I. J.; Parasini, E.; Jhoti, H. High-Throughput X-Ray Crystallography for Drug Discovery. In *Drug Design: Special Publication*; Flower, D. R., Ed.; Royal Society of Chemistry: Cambridge, UK, 2002, pp 53–59.
153. Brunger, A. T.; Adams, P. D.; Clore, G. M.; DeLano, W. L.; Gros, P.; Grosse-Kunstleve, R. W.; Jiang, J. S.; Kuszewski, J.; Nilges, M.; Pannu, N. S. et al. *Acta Crystallogr. D* **1998**, *54*, 905–921.
154. The collaborative computational project number 4. *Acta Crystallogr. D* **1994**, *50*, 760–763.
155. Hartshorn, M. J. *J. Comput. Aided Mol. Des.* **2002**, *16*, 871–881.
156. Verdonk, M. L.; Cole, J. C.; Taylor, R. *J. Mol. Biol.* **1999**, *289*, 1093–1108.
157. Gill, A. L.; Frederickson, M.; Cleasby, A.; Woodhead, S. J.; Carr, M. G.; Woodhead, A. J.; Walker, M. T.; Congreve, M. S.; Devine, L. A.; Tisi, D. et al. *J. Med. Chem* **2004**, *48*, 414–426.
158. Wang, Z.; Harkins, P. C.; Ulevitch, R. J.; Han, J.; Cobb, M. H.; Goldsmith, E. J. *Proc. Natl. Acad. Sci. USA* **1997**, *94*, 2327–2332.
159. Schulze-Gahmen, U.; De Bondt, H. L.; Kim, S. H. *J. Med. Chem.* **1996**, *39*, 4540–4546.
160. Jhoti, H.; Cleasby, A.; Reid, S.; Thomas, P. J.; Weir, M.; Wonacott, A. *Biochemistry* **1999**, *38*, 7969–7977.
161. Leonidas, D. D.; Shapiro, R.; Irons, L. I.; Russo, N.; Acharya, K. R. *Biochemistry* **1997**, *36*, 5578–5588.
162. Andersen, H. S.; Iversen, L. F.; Jeppesen, C. B.; Branner, S.; Norris, K.; Rasmussen, H. B.; Moller, K. B.; Moller, N. P. *J. Biol. Chem.* **2000**, *275*, 7101–7108.
163. Wang, Z.; Canagarajah, B. J.; Boehm, J. C.; Kassisa, S.; Cobb, M. H.; Young, P. R.; Abdel-Meguid, S.; Adams, J. L.; Goldsmith, E. J. *Structure* **1998**, *6*, 1117–1128.
164. Rowland, P.; Blaney, F. E.; Smyth, M. G.; Jones, J. J.; Leydon, V. R.; Oxbrow, A. K.; Lewis, C. J.; Tennant, M. G.; Modi, S.; Eggleston, D. S. *J. Biol. Chem.* **2006**, *281*, 7614–7622.
165. Howard, N.; Abell, C.; Blakemore, W.; Chessari, G.; Congreve, M.; Howard, S.; Jhoti, H.; Murray, C. W.; Seavers, L. C.; van Montfort, R. L. *J. Med. Chem.* **2006**, *49*, 1346–1355.

Biographies

Chris W Murray studied for his undergraduate degree and doctorate at the University of Cambridge, UK. After his postdoctorate at Indiana University, he joined the biotech company, Protherics, where he helped develop docking methods and applied structure-based drug design to serine proteases such as factor Xa. Since 2000 he has been Director of Computational Chemistry and Informatics at Astex Therapeutics, where he has worked on establishing fragment-based approaches to drug discovery. His current research interests include the exploitation of cytochrome P450 structures, the design of enzyme inhibitors, and the development of technologies aimed at improving structure-based drug design.

Mike J Hartshorn joined Astex Therapeutics in 1999, where he has developed systems for fragment-based screening and software for the interpretation of protein–ligand complexes determined by x-ray crystallography. In addition he has overseen the development of much of the company's computational infrastructure and database systems. Before joining Astex, Dr Hartshorn was at MDL Information Systems, where he worked on their next generation of desktop chemistry software. He obtained his DPhil in 1992 in the group of Prof Rod Hubbard at the University of York Structural Biology Group in the UK.

© 2007 Elsevier Ltd. All Rights Reserved
No part of this publication may be reproduced, stored in any retrieval system or transmitted in any form by any means electronic, electrostatic, magnetic tape, mechanical, photocopying, recording or otherwise, without permission in writing from the publishers

Comprehensive Medicinal Chemistry II
ISBN (set): 0-08-044513-6

ISBN (Volume 4) 0-08-044517-9; pp. 775–806

4.32 Biological Fingerprints

R V Stanton and Q Cao, Pfizer Research Technology Center, Cambridge, MA, USA

© 2007 Elsevier Ltd. All Rights Reserved.

4.32.1 Chemical and Biological Similarity

The concepts of: (1) biological activity space and (2) chemical diversity space are fundamental to the modern process of drug discovery. But what do these terms really mean and how can computational data analysis techniques be used in this context? Biological activity space can be usefully defined as the region of chemical space (defined below) that produces a specific biological response, be that enzyme inhibition, the disruption of protein–protein binding, or interference of a cell-signaling pathway. Biological activity space for a given response can encompass a very specific/restricted portion of chemical space such as in the case of biotin and streptavadin, where compounds differing only slightly from streptavadin can lose many orders of magnitude in activity.[1] In contrast, transporter proteins such as multidrug-resistant transporter 1 (MDR1)/P-glycoprotein are known for the diversity of chemotypes they recognize.[2]

The definition of chemical space can be slightly trickier, but might be defined as a conceptual representation of the relative structural differences between compounds. The extent/size of chemical space is only restricted by the synthetic feasibility of compounds. When quantified, distances in chemical space are typically taken to be some type of Euclidean or Tanimoto distance in a set of calculable chemical descriptors. These descriptors might include one-dimensional (1D) properties (such as ClogP, number of rotatable bonds, or the presence of a structural motif), two-dimensional (2D) connectivity information, or even three-dimensional (3D) pharmacophores and shapes. The area of chemical descriptors has already been reviewed in this volume (*see* 4.08 Compound Selection Using Measures of Similarity and Dissimilarity), so it will not be discussed further here. However, it is important to note that the ability to reduce a chemical structure to a list of descriptors, or fingerprint, is fundamental to many of the cheminformatic and computational chemistry methods discussed in this book. The success of these methods is highly dependent on the accuracy and completeness of such descriptors/fingerprints in describing those factors that influence biological activity.

A number of studies have attempted to quantify the information captured by various fingerprint types as well as the additional data presumably included with increasing fingerprint complexity in 1D, 2D, and 3D descriptors.[3–5] Unfortunately, chemical fingerprints are often not able to resolve differences in biological activity. This is to say, compounds very close in chemistry space may have very different activity profiles, while compounds far apart may have similar biological activities. This is in spite of the 'neighborhood behavior' principle that is a fundamental tenet of medicinal chemistry. Horvath and Jeandenans define this principle as follows: "a subset of the structurally nearest neighbors of an active compound is likely to include more actives than found in any other, randomly chosen, set of the same size."[4,6] This behavior was also seen by Martin *et al.* but at a much lower level than might be expected.[7] They found that, with a typical definition of neighborhood ($\geq$0.85 similar in Tanimoto distance using Daylight fingerprints[8]), there is only a 30% chance that a neighboring compound will be active in the same assay. Many additional examples exist of compounds that differ by only very small changes in structure yet have dissimilar activities.[9,10]

This lack of correlation between chemical fingerprints and activity is one reason researchers have begun turning to the use of biological descriptors. Biological fingerprints are typically the result of a screening panel of enzymatic, cell-based, phenotypic, or absorption, distribution, metabolism, and excretion (ADME) assays (at times in the presence of known chemical modulators). The resulting profile (dubbed by various research groups affinity fingerprints, biological fingerprints, biospectra, or as a component of multidimensional screening) can then be used in much the same way as chemical fingerprints for similarity searching, model generation, etc. Although expensive to obtain, biological fingerprints contain a wealth of information on how a compound interacts with real biological systems instead of calculated chemical descriptors. However, the same caution must be used in interpretation for either chemical or biological descriptors, as they might be highly correlated or even noisy.

In the remainder of this chapter, we discuss many of the ways various research groups have recently used biological fingerprints. These will include: (1) similarity searching and clustering, done in a descriptor space completely abstracted from chemical structure; (2) activity modeling using only a small panel of assays; (3) clustering of proteins using descriptors different from the traditional sequence-based analysis; and (4) chemical genomics analysis, including additional dimensions of data focusing on functional and pathway effects. The study of biological fingerprints is in its nascent stage as it has only been since the mid-1990s that advances in molecular biology have made it possible to create the necessary data sets in a cost-effective manner.

4.32.2 Use Cases

4.32.2.1 Activity Modeling

4.32.2.1.1 General discussion of method

One of the early uses of biological fingerprints introduced by Kauvar *et al.* was modeling the activity of a new protein target by comparison to assay results for a reference panel of proteins.[11,12] In this method, termed TRAP (target-related affinity profiling) by Telik,[13] activity measured against a small (~18), carefully selected panel of proteins is used as the descriptors, creating affinity fingerprints. Affinity fingerprints can be considered a subset of biological fingerprints in which only specific protein–ligand interaction assays are used as descriptors, in contrast to the use of cell-based or functional assays discussed elsewhere in this chapter. The panel is chosen for orthogonality of the assay results, minimizing interassay correlation and thereby maximizing the information content provided by each assay. The authors note that they have profiled ~500 proteins for possible inclusion in the panel but have found that moving beyond 20 there is little additional information to be gained.[12,13] An additional feature of affinity fingerprints as implemented by the Telik group is that they are created through screening with high ligand concentrations. The authors postulate that, at these concentrations, the binding interactions can be driven by those elements common to gene families causing the results of binding to a single protein to be more generalizable.

Having developed the reference panel of assays, a diverse training set of compounds is selected as an initial probe of the new target. While a training set consisting of 40–70 compounds seems tiny in comparison to modern high-throughput screening (HTS) standards where 1536-well plates are becoming common, the small size of the training set is determined by what is necessary in the restricted descriptor space. A larger assay panel might require a larger training set of compounds to span the potential interactions fully. The screening data for the training set with the target protein are then used to create a 'computational surrogate'[12] or activity model for the target. The simplest possible model for the activity of the target protein in terms of the activity of the reference panel proteins is linear (eqn [1]), for which the authors demonstrate impressive correlation over several orders of magnitude in activity for their example studies (discussed more fully below).

$$\mathrm{pIC}_{50} = C_o + \sum_{k=1}^{N} C_k \mathrm{pIC}_{50}[k] \qquad [1]$$

The authors report that even a two-component model for aldehyde dehydrogenase (trained from the activities of 120 compounds for snake venom phosphodiesterase and *S*-transferase rat 8–8) can show a correlation coefficient of 0.86.[13]

4.32.2.1.2 Example

The group at Telik have published several examples of models created using affinity fingerprints. In a 2004 study of cyclooxygenase-1 (COX-1) inhibitors,[14] results from a panel of 12 assays with 19 known nonsteroidal anti-inflammatory drugs were used to create an initial linear model for COX-1 activity. Sixteen compounds selected using this initial model proved to be inactive. After updating the model substantially by including the information from the inactive compounds, 46 additional compounds were selected for testing. Five of the 46 compounds were active ($<100\,\mu mol\,L^{-1}$), with three more potent than ibuprofen ($76\,\mu mol\,L^{-1}$ under the same assay conditions).[14] This study is interesting as it shows the importance of the inclusion of inactive data for these models, something that has been demonstrated previously with chemical fingerprints.[15,16]

4.32.2.1.3 Significance/discussion

The reported effectiveness of models created using affinity fingerprints poses a number of interesting questions. One question raised indirectly by Dixon and Villar is whether the technique will be effective for highly specific binding sites.[10] It seems unlikely that the proteins in the screening panel would encode enough information in such a case. This would be analogous to using a chemical fingerprint to do a substructure search (SSS) in a corporate library using a unique chemotype for which the library has no representatives. If the affinity fingerprint comes back with no signal (nothing binds to the query protein), it will be impossible to make a model. Given this, one application that might be made of such a panel is the evaluation of target druggability. A carefully selected group of compounds with a known broad-affinity fingerprint profile might help quickly estimate the potential hit rate of an HTS. A second question is how effective this strategy will be at generating attractive leads amenable to parallel chemistry. While comparatively expensive, traditional HTS of a corporate collection has the advantage of being complete (i.e., every compound is tested, including parallel amenable compounds). It would seem that the two techniques are complementary, each having advantages depending on the type of screen being run. Models created from affinity fingerprints might be particularly appropriate in low-throughput expensive screens, or for targets with known high hit rates.

4.32.2.2 Computational 'Biological' Fingerprints

Given the utility of biological fingerprints, could they be generated computationally? Initially this may seem somewhat of an oxymoron in that one of the principal benefits of biological fingerprints is that they are based on the experimentally measured responses of real biological systems. However, several authors have explored techniques that attempt to mimic biological responses and exploit the benefits of affinity fingerprints. Briem and Kuntz[17] explored this concept by creating 'computational affinity fingerprints' using DOCK[18]-generated scores. A diverse panel of eight proteins with known structure (except for one homology model of a G protein-coupled receptor) was selected as the descriptor panel, along with reference sets of compounds from the MDL Drug Data Report (MDDR).[19] The sets of MDDR compounds were chosen so as to create groups of known inhibitors to specific targets. After docking, the scores against the panel were used to create virtual-affinity fingerprints, within which similarity and nearest-neighbor calculations could then be conducted. The results found virtual fingerprints that are able to cluster compounds active against a specific target and were seen to be complementary to the results obtained with Daylight 2D canonical fingerprints.[8] While the chemical fingerprints were able to demonstrate a better enrichment in every case when compared to the docking-based fingerprint, DOCK was able to identify several compounds that were structurally dissimilar but shared biological activity. Briem and co-workers have expanded these studies to include both flexible docking (FLEXSIM-X)[20] and fragment similarity (FLEXSIM-R).[21]

Krejsa *et al.*[22] take a combined theoretical–experimental approach to calculating affinity fingerprints with their generalized neighborhood behavior (GNB) models. These models use structural or pharmacophore similarity to define a set of nearest neighbors to a query molecule from their database of profiled molecules (BioPrint). The affinity values from the neighbors for each assay in the profile are then used to estimate an affinity fingerprint for the test compound. Given the work of Martin *et al.* around the relation of chemical and biological similarity,[7] it would seem that the selection of nearest neighbors (i.e., that they be very near neighbors) would be critical to the success of this approach. The authors, though, are able to show a reasonable relation between the calculated and experimental fingerprints for several examples.

4.32.2.3 Searching for Selectivity or Differentiation

With the sequence of the human genome complete and the finite expanse of the proteome now better understood, pharmaceutical companies are increasingly focusing on the selectivity of their drug candidates. Selectivity can define the success of a 'fast-follower' compound or a lack of selectivity can lead to the withdrawal from the market of a long-established drug. An example of off-target activity is shown by Fliri *et al.* in their comparison of the antifungal agents clotrimazole and tioconazole[23] (a similar analysis is shown in **Figure 1**). For the 181 assays shown, these drugs

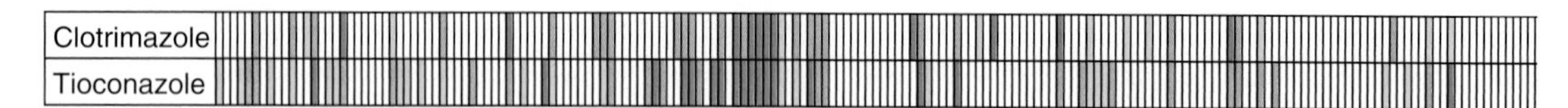

Figure 1 Biological fingerprints for clotrimazole and tioconazole over a set of 180 assays (shown in columns). The color scale runs from white (inactive) to red (very active).

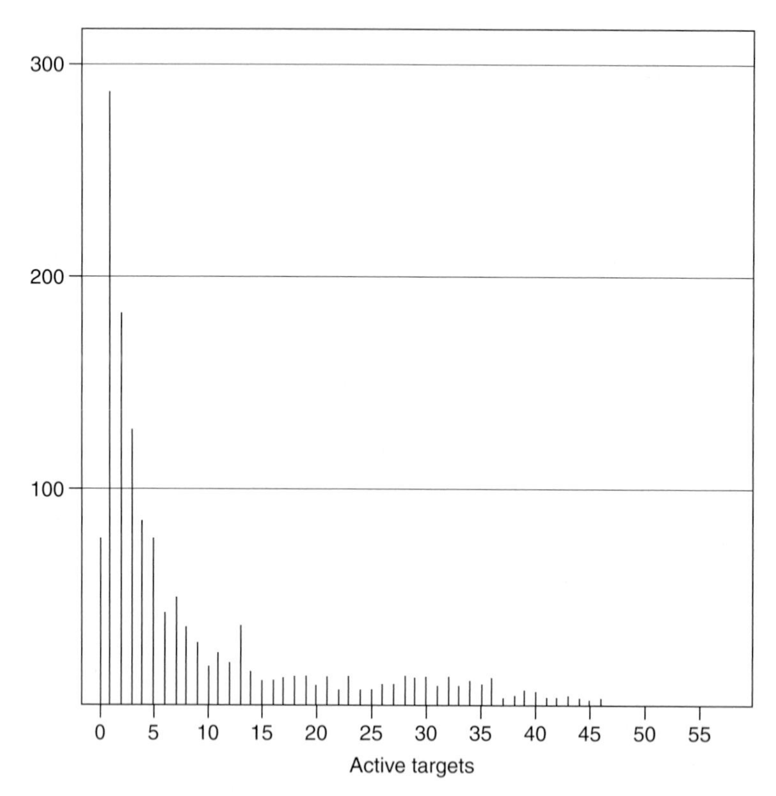

Figure 2 Histogram of compound target activity distribution. The *y*-axis is the number of active compounds, and the *x*-axis is the number of targets. A compound is considered active against a target when primary assays had an average of 50% or higher activity at the selected concentration (mostly 10 μmol L^{-1}, with some exceptions ranging from 1 to 200 μmol L^{-1}).

demonstrate potent activity against a number of calcium channels and cytochromes. Interestingly, the putative target of these drugs, YP51, is not contained in the panel.

As a demonstration that off-target activity is not restricted to antifungal agents, **Figure 2** shows statistics for 1338 marketed drugs that have been profiled in Cerep's BioPrint database.[24] These compounds have been screened against 120–130 targets. Using a definition of activity of ≥50% inhibition or activation at the tested concentration (typically 10 μmol L^{-1}), less than one-third (365) of the compounds are specific (active against 0 or 1 target), while one-third of the compounds have activity against 2–5 targets, and 37% against 6 or more targets. The least selective compound has shown activity against as many as 46 targets in the panel.

An example of the use of biological fingerprints for predicting selectivity is the Greenbaum *et al.*[25] study of amino acid positional scanning libraries linked to an irreversible epoxide electrophile scaffold. Mixtures of these compounds were profiled against a set of 19 papain family cysteine proteases. Each mixture held one amino acid position fixed while allowing the others to vary. An initial screening of mixtures for 19 natural amino acids identified the P2 pocket as the site of greatest potential specificity. The authors go on to interrogate the P2 pocket further using mixtures of 41 non-natural amino acids chosen for their hydrophobicity as well as changes in the linker stereochemistry. A clustering of the resulting fingerprints shows that it is possible to achieve selectivity across the family of papain enzymes with clusters of compounds showing behavior as: (1) universal inhibitors; (2) poor inhibitors; and (3) specific inhibitors.[25] Studies of this type could help identify those enzyme pockets that can be targeted through medicinal chemistry to help achieve selectivity and provide insight into what chemical fragments and stereochemistries may define selective chemical moieties.

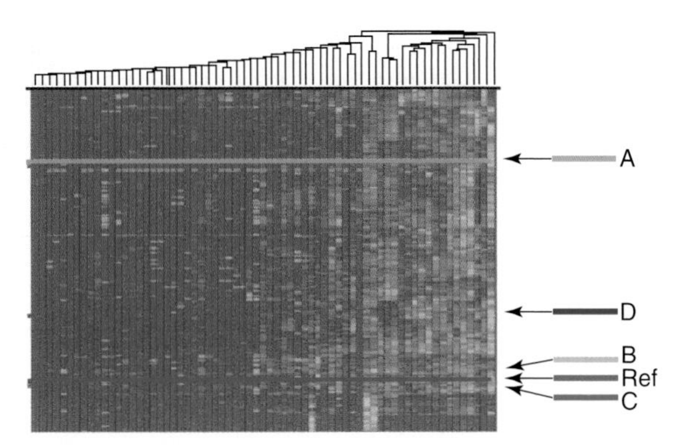

Figure 3 Clustering of potential backup compounds based on biological fingerprints along with the clinical reference (Ref). The rows in this figure are compounds while the columns are assays. Primary inhibition data are color-coded from green (inactive) to red (very active).

An example of using biological fingerprints to differentiate different hit/lead series comes from a recent project at Pfizer.[26] The goal in lead differentiation is to focus chemistry efforts as early as possible on the best series, to avoid the situation where potential liabilities are discovered at a late stage, when the only practical option is to choose the 'best' candidate from what may not be the 'best' series. The differentiation can be in terms of identifying series that are either more selective and/or have a different polypharmacological profile to compounds already in development or marketed (in-house or competitor). This provides an approach to trying to orthogonalize attrition risks for multiple drug candidates (or versus competitor compounds), as even if the potential in vivo effects of particular in vitro pharmacological binding activities (single or patterns/profiles) are not known, the likelihood of multiple candidates attriting for the same reason is reduced, and differentiation from other drugs may be possible in terms of side/adverse effects.

In this project, in which some polypharmacology was to be expected, the problem was not identifying a binding site that allowed selective interactions but rather differentiating a set of backup compounds for further clinical study. In this case four relatively similar compound series (base–linker–aromatic), with nearly equivalent potency, were seen to have distinct biological fingerprint profiles. These profiles are shown in **Figure 3**, clustered with other known drugs from the BioPrint database,[24] and with a reference competitor compound that was in clinical development. Given that the other factors were equal, in an effort to minimize risk in the backup structure, the compound with the most distinct affinity profile (from that of the one already in clinical development), that had the additional benefit of being much cleaner, was selected for further study. In this way, a very different off-target profile could be tested and compared to the primary candidate, helping to establish the true effects of target inhibition. The analysis also provided the project team with early information on a key selectivity screen to set up, and of potential cytochrome P450 (CYP) inhibition issues.

If these same five compounds (four active series + clinical reference compound) are clustered with the BioPrint drug set using a common 2D structure-based method such as Daylight fingerprints, then a very different clustering is obtained. The much cleaner compound with the differentiated biological profile clusters with a compound that has a biological profile very similar to the clinical reference compound, and thus could have been missed for follow-up studies. In fact, a singleton from this clustering also has a similar biological profile to the clinical reference compound, and the compound that clusters close, but differentiated, to the clinical reference compound in biological space is clustered in structural space to that reference compound. Thus it is unlikely that the two more promising series would have been the primary choices for follow-up.

Use of BioPrint profiling in lead generation projects has led to many examples where small changes to substituents on otherwise quite similar 'chemotypes' has led to quite different biological profiles. While the main desired biological activity is generally retained (a basic principle of medicinal chemistry that similar compounds tend to have similar activity), the off-target activity difference can be quite dramatic, changing from quite clean to promiscuous, for example through an alkyl to alkoxy change to a substituent on a distinctive base–linker–heteroaromatic-substituent system.[27]

4.32.2.4 Structural Clustering

The concept of neighborhood behavior and molecular similarity is central to many of the most commonly used medicinal and computational chemistry techniques in drug discovery.[28] Substructure searching, analoging, quantitative structure–activity relationships (QSAR), and virtual screening are just a few of the techniques rooted in molecular similarity. However, as mentioned in the introduction to this chapter, depending on how they are measured, distances in chemical and biological space can be quite different. As one demonstration of this we show in **Figure 4** the comparison of similarity, for Daylight structural fingerprints[8] and biological fingerprints created from a panel of 154 assays from the BioPrint database (measured by Tanimoto distance). A pairwise comparison of 347 compounds (drugs with MW = 200–600) gives 60031 points, plotted in **Figure 4a**. The overall scatter of the data demonstrates the poor correlation ($R^2 = 0.13$) of the two measures. Interestingly, this is even true in the range where we would expect neighborhood behavior to be strong, in the region where the structural similarity is >0.85. In this region Martin *et al.* noted only a 30% chance of activity against the same target,[7] and this is supported by the correlation of our data in this region, only giving an R^2 of 0.05. Moreover, in the converse comparison, where we look in the region of high biological fingerprint similarity (Tanimoto >0.7), we find that the correlation is still only $R^2 = 0.04$.

The question of whether chemical similarity infers biological similarity is further muddied by the many possible definitions of chemical similarity. Horvath and Jeandenans[4,29] investigate which chemical fingerprints best correlate with biological similarity. The authors studied descriptors (2D topological, shape, three- and four-point pharmacophores, and fuzzy bipolar pharmacophores) as well as similarity metrics. They found that 3D descriptors have improved performance in replicating the biological similarities calculated from a panel of 42 targets. However, while three- and four-point pharmacophores worked well for structures of similar chemotypes, once the scaffolds of compounds were

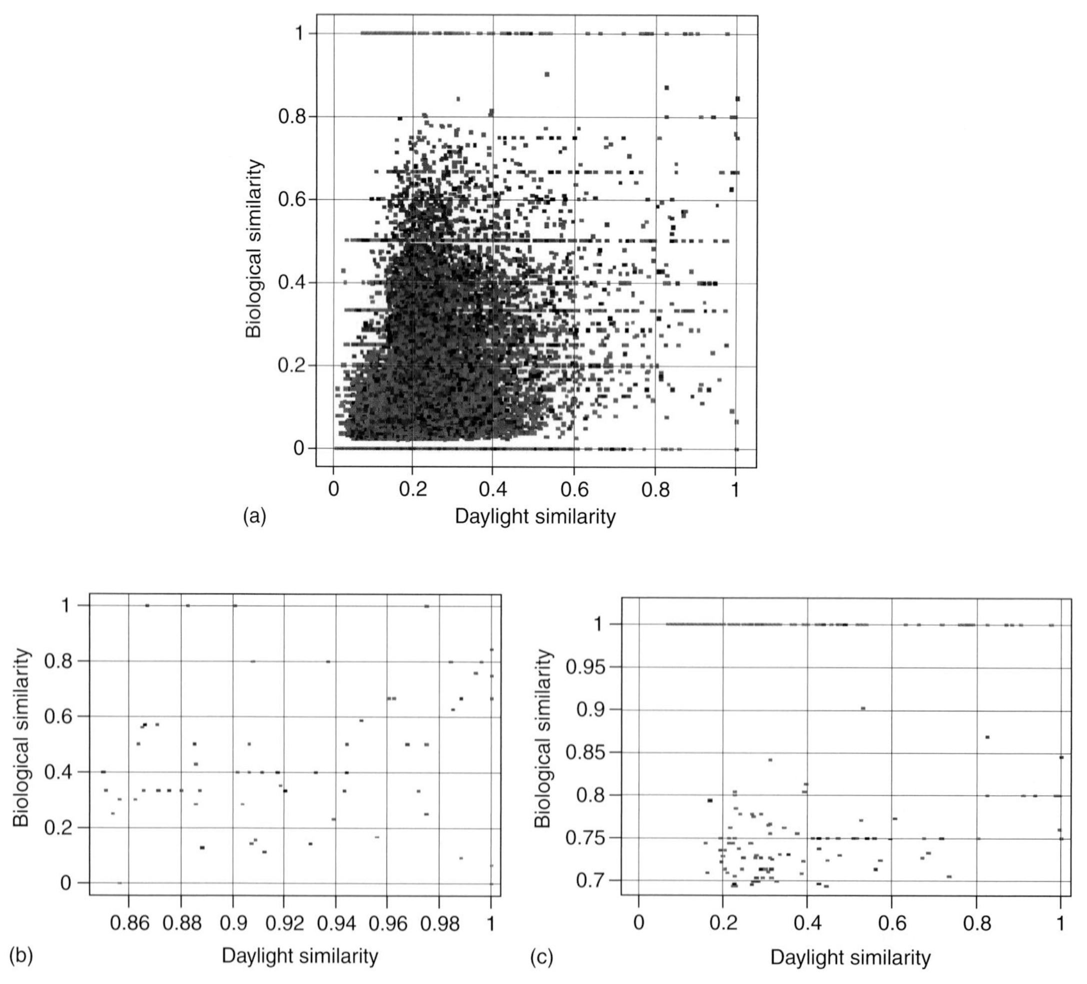

Figure 4 Compound biological activity similarity versus chemical structure similarity. (a) Pairwise Tanimoto distance of 347 drugs (MW = 200–600) that has been screened in 154 assays. The biological activity fingerprint was constructed so that the compound is considered active for an assay if it showed IC_{50} (EC_{50}) $\leq 100\,\mu M$. Similarity = 0–1 (1 = identical). (b) Enlarged view of (a) in the region where Daylight fingerprint similarity is >0.85. (c) Enlarged view of (a) in the region where biological activity similarity is >0.7.

significantly different, the performance of these descriptors decreased. In comparison, the 'fuzzy pharmacophores'[30] studied were able to correlate much more closely with the biological similarity. (The Cerep 'fuzzy pharmacophore' descriptors are counts of the number of feature pairs of types a and b separated by a binned distance. Feature types include hydrophobes, aromatic groups, hydrogen bond donors/acceptors, and positive/negative charges.) This study should be contrasted with that of Brown and Martin, who found several 2D descriptors were better able to separate active and inactive compounds.[31]

As one of the first groups to collect an extensive data set, the National Cancer Institute (NCI) has published a number of studies using biological fingerprints to search for compounds that demonstrate similar activity. Their assay panel was functional inhibition of growth in a set of 60 tumor cell lines.[32] The results of these studies showed the ability of biological descriptors to abstract from chemotype, returning structurally diverse compounds. Using the COMPARE[33] algorithm, similarity searches for new agents were run against existing data sets of profiled compounds. In one example[32] redoxal (**1**) was used as a query in the COMPARE algorithm, returning **2**, **3** and brequinar (**4**). **3** and **4** are known to be potent inhibitors of pyrimidine biosynthesis and further studies of **1** and **2** confirmed that they were also inhibitors of this pathway. This result is particularly intriguing as it demonstrates a specific biological activity (inhibition of pyrimidine biosynthesis) that correlates with similarity in a biological fingerprint space composed solely of functional assays.

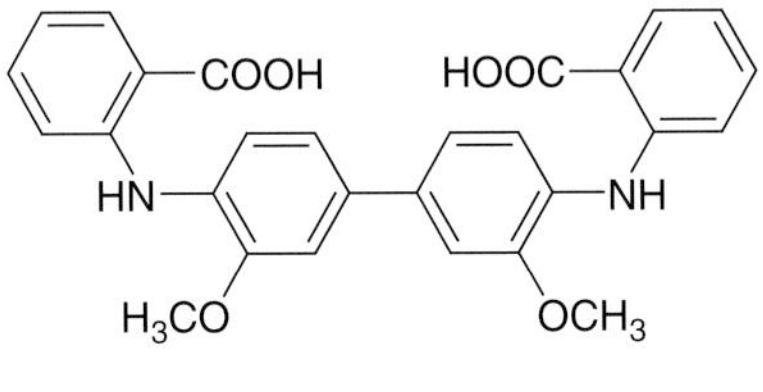

COOH HOOC HN NH H_3CO OCH_3

1

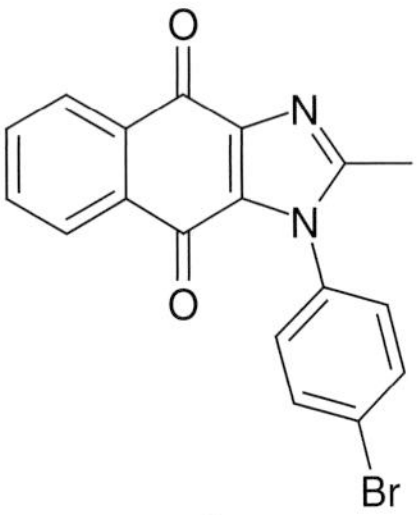

O N N O Br

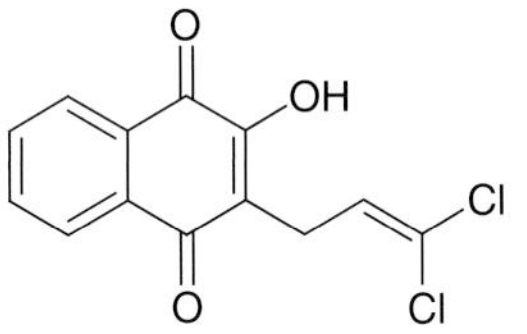

2

O OH Cl O Cl

3

COOH F N F

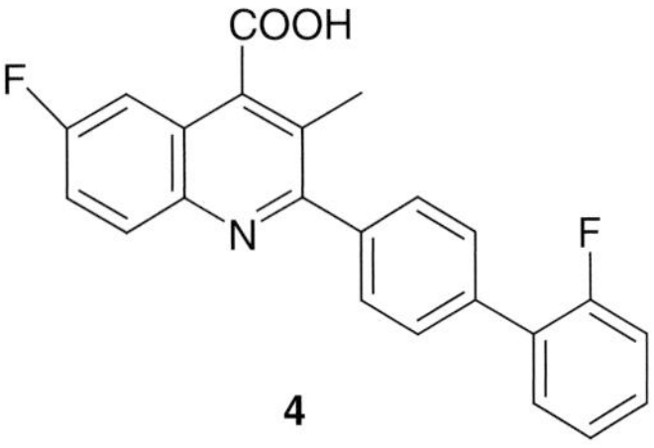

4

In contrast to this example from the NCI that finds structurally dissimilar molecules with similar biological fingerprints, Fliri *et al.* demonstrated a strong correlation between biological fingerprints and structural similarity.[23] Using a BioPrint data set of 1597 compounds and 92 assays, they reported the ability to form 73 tight clusters of 317 molecules. Within these clusters they saw both a strong structural and functional correlation. This demonstrates why neighborhood behavior can be used as a key component in drug discovery: compounds of close structural similarity often have a strong biological similarity, although it has been demonstrated in this chapter that this is not always the case. **Figure 5** shows an analysis similar to one from a separate study by these authors, clustering of a set of potent dopamine inhibitors.[34] While all of the compounds are similarly active against the dopamine isozymes, the structural chemotypes are clustered by their off-target activity. One rationale for the different results seen in the NCI and Fliri *et al.* studies is the vastly different assay panels used in each study. The panel used by Fliri *et al.* primarily consists of measures of specific target interactions, whereas the NCI uses functional tumor suppression assays.

4.32.2.5 Clustering of Sequences

Pharmaceutical companies are increasingly leveraging the efficiency of scale by pursuing targets within gene families as compared to single targets. The knowledge and infrastructure (assays, intellectual property space, medicinal chemistry experience, and directed screening libraries) generated make the pursuit of two targets in the same protein family significantly easier than pursuing targets in different families. Example of work in kinase[35–38] and G protein-coupled receptor[39,40] families are abundant in the literature. Unfortunately, selectivity concerns also increase as closely related

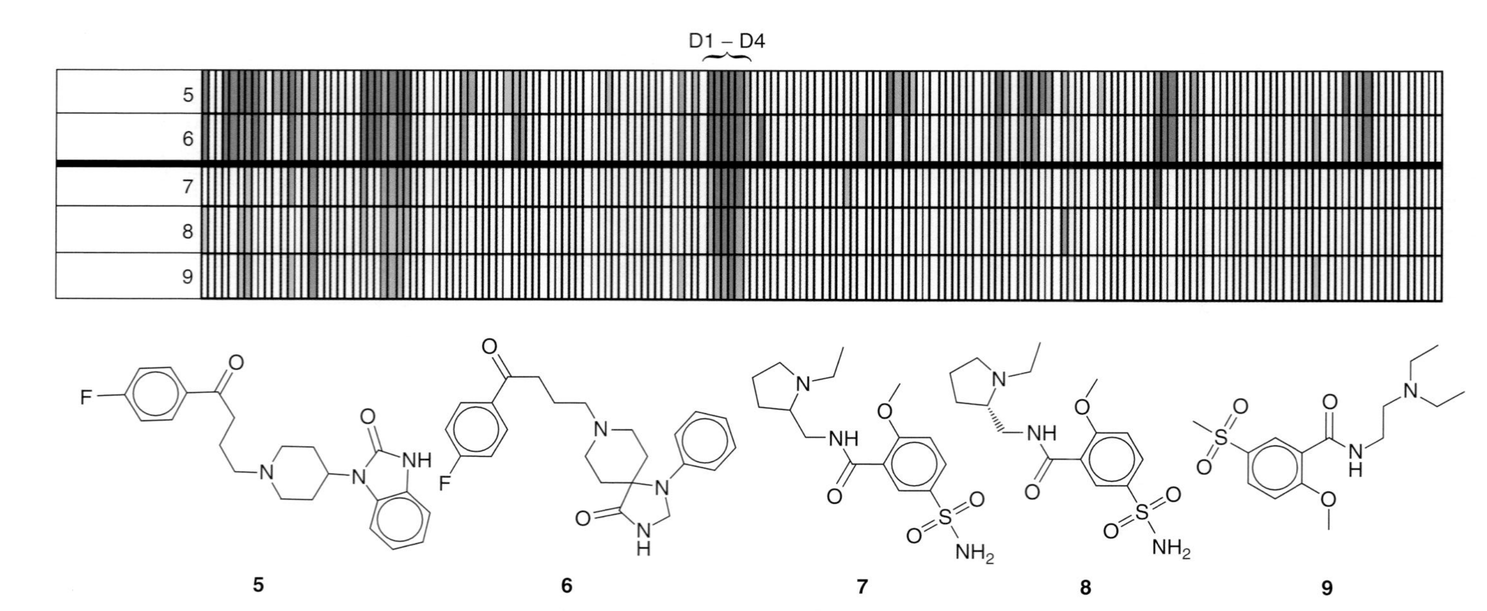

Figure 5 A partial clustering of dopamine active compounds from the BioPrint data set is shown using the data from 180 assays. Activity is color-coded in a range from white (inactive) to red (very active). The columns associated with dopamine isozymes active (D_1–D_4) are indicated. Examples from two of the clusters formed are shown (**5** and **6** and **7**–**9**), demonstrating how the off-target activity drives the clustering of these chemotypes.

targets are pursued. A typical question faced by a therapeutic project team is: which proteins represent a selectivity concern and are of such a risk as to warrant the expense of developing an assay? This question is commonly approached through sequence-based clustering of the proteome. Proteins can be clustered based on amino acid sequence (domain or active site), with near neighbors being a particular risk for cross-reactivity. An alternative to sequence clustering is the use of affinity data. In contrast to the biological fingerprints described thus far, the descriptors used here are the assay data for multiple compounds against a single target. Using these data (simply the column information instead of the rows in a matrix composed of compounds by assays, as shown in **Figure 3**), a new target clustering can be created which is based upon the measured interaction characteristics of a protein. Greenbaum *et al.* demonstrated this in a comparative clustering of 12 cysteine proteases using sequence-based and biological fingerprints.[25] The resulting clusters show several significantly different groupings where sequence-similar proteins cluster far apart due to differences in ligand affinities. This result provides a note of caution against the choice of selectivity panels solely based on sequence-based clusters.

The protein kinases offer a gene family for which selectivity and the implications of off-target activity are being closely scrutinized. With the ~518 members of this protein family identified with the sequencing of the genome, their sequence-based clustering has been extensively studied.[41] However, only recently have affinity fingerprinting methodologies been used to derive compound interaction-based clustering maps for segments of the kinome, although it had been suggested previously.[42] Vieth *et al.* demonstrate that, in a clustering of 43 kinases using a sparse data set from 22 inhibitors, the clustering differs significantly from sequence-based clustering.[43] In contrast to the sparse data set approach, Fabian *et al.* have generated a complete matrix of 20 known kinase inhibitors (marketed drugs, compounds in clinical trial, or tool compounds) against 119 kinases.[44] While the relatively small data set of compounds does not allow for an accurate clustering of the kinase panel (several kinases are only hit by staurosporine), it does provide interesting insight into the selectivity profiles of current kinase drugs and clinical candidates. It also emphasizes the potential utility of biological fingerprints to help differentiate compound profiles and mitigate risk among candidate compounds.

4.32.2.6 Chemogenomics

The 'omics' revolution has not bypassed the world of biological fingerprints. The data sets of biological fingerprints are ideally suited for expansion and use as one component of more complex modeling (such as systems biology). As an example, chemical genomics (chemogenomics) is a technique in which small molecules (chemical genetic modifiers) take the place of mutations in traditional genetics to perturb the functions of cellular pathways.[45–47] The advantage of using small molecules is that inhibition/disruption of a pathway can be controlled both temporally and spatially. For profiling, chemogenomic assay panels usually consist of functional readouts that monitor a specific cell function or pathway (for example, cell progression) and might contain deletion strains to probe a specific pathway further. The basic concept of chemogenomics has now been used in a number of studies, through analysis of differing multidimensional data sets. Dimensions studied in various combinations include compounds, compound mixtures, phenotypic assays, and enzyme assays.

In one study, Haggarty *et al.* were able to increase the dimensionality of traditional biological fingerprints by studying mixtures of two compounds assayed against a series of 10 phenotypic cell growth assays. The 10 assays consisted of one wild type, with each of the other nine missing a different component of the spindle assembly cell polarity network.[48] The resulting data set contains 5760 measurements in a $24 \times 24 \times 10$ matrix (compound $\times$ compound $\times$ deletion strain). In a manner analogous to the clustering of affinity fingerprint data, the authors used graph theory to analyze the resultant data sets for each deletion strain. They were able to identify two functionally related but redundant genes/proteins as being close in the space defined by the networks of small molecules. The multidimensionality of their data also allowed them to cluster the small molecules based on their inhibitions of cell growth in the various strains. While different clusterings were observed for each strain, a few molecules remained tightly clustered in all of the strains. The molecules seen to cluster together were structurally diverse (including a macrocycle, a steroid, and simpler druglike compounds). This result suggests once again the difficulty of correlating distances in chemical and biological space. This point is emphasized in a study by Kim *et al.* examining the effects of stereochemistry and rigidification in a set of otherwise very similar molecules.[49] The panel used four phenotypic assays chosen to assess broadly the impact of small molecules on the cell. The data set was further increased through the inclusion of 17 known 'small-molecule modulators of biological response'[49] (compounds known to be active in various cell processes). This resulted in a much more complex but information-rich data set, composed of (four phenotypic assays) $\times$ (17 known small-molecule modulators + 1 control) $\times$ (244 compounds). On analysis, the data showed that conformational restriction reduced the number of assays hit by a compound (increased specificity). Results for fixing the stereochemistry were less conclusive but also correlated with overall specificity.[49]

As mentioned previously, the NCI has one of the largest multidimensional data sets. For more than a decade the NCI has been profiling compounds against a panel of 60 cancer cells lines, with tens of thousands of compounds currently screened. The NCI has further enhanced this data set by profiling the cell lines for protein expression and cell characteristics that might correlate with compound activity. This multidimensional data set has been used to look for correlations among targets and compounds using the Discovery program. Discovery is a data visualization and clustering program which allows the correlation coefficient to be displayed in a type of heat map for various analyses of the data. One example presented by Weinstein *et al.* examines the correlation of compounds with targets (i.e., cell characteristics and expression profiles) based upon activity in the panel of 60 cancer growth screens.[50] The analysis shows close correlation of both compounds with a known mechanism of action, and targets which are indices for similar pathways. As an example, of 35 known MDR1 substrates in the compound set, 18 were tightly clustered, even though they had very diverse structures and reported mechanisms of action.

Data sets like that available at the NCI are rare, even in large pharmaceutical companies, leaving the creation of multidimensional data sets as one of the major hurdles to work in this area. Another group that has accumulated such a data set is Cerep, with its BioPrint collection.[24] This data set is composed of more than 2000 known marketed and withdrawn drugs. Associated with the biological fingerprints (>200 ADME, binding, enzyme, and safety assays) are adverse drug relations, which can be used to correlate the structural and in vitro data to in vivo effects. One study using this data set shows the ability to identify the correlation between muscarinic (M_1) and tachychardia, dopamine (D_1) activity and tremor, as well as histamine (H_1) activity and somnolence.[22] While these relations were previously known, the ability to correlate such associations quantitatively within a robust data set is the first step toward mining this type of multidimensional data.

4.32.3 Conclusion: Information-Rich Modeling

Advances in the areas of molecular biology and informatics have made the field of information intensive drug discovery'[50] possible. Biological fingerprinting is one key component of these information-rich approaches to systems modeling. Broad profiling of chemical entities against panels of carefully selected assays has provided novel data sets that have not yet been fully exploited.

The future of biological fingerprinting is in the further expansion of the data sets into associations with genomic and proteomic data. The chemogenomic examples discussed show the power of this type of analysis. Understanding the complexity and redundancy of biological systems and pathways has pushed researchers to increase the complexity and scope of their models. However, biological fingerprinting is a component of information-intensive drug discovery that is available today, and will only see expanded use as analysis tools and techniques evolve further.

References

1. Dixon, R. W.; Radmer, R. J.; Kuhn, B.; Kollman, P. A.; Yang, J.; Raposo, C.; Wilcox, C. S.; Klumb, L. A.; Stayton, P. S.; Benhke, C. et al. *J. Org. Chem.* **2002**, *67*, 1827–1837.
2. Chiba, P.; Ecker, G. F. *Exp. Opin. Ther. Patents* **2004**, *14*, 499–508.
3. Brown, R. D.; Martin, Y. C. *J. Chem. Inf. Comput. Sci.* **1997**, *37*, 1–9.
4. Horvath, D.; Jeandenans, C. *J. Chem. Inf. Comput. Sci.* **2003**, *43*, 680–690.
5. Oprea, T. I. *J. Braz. Chem. Soc.* **2002**, *13*, 811–815.
6. Petterson, D. E.; Cramer, R. D.; Ferguson, A. M.; Clark, R. D.; Weinberger, L. E. *J. Med. Chem.* **1996**, *39*, 3049–3059.
7. Martin, Y. C.; Kofron, J. L.; Traphagen, L. M. *J. Med. Chem.* **2002**, *45*, 4350–4358.
8. Daylight Fingerprints. Daylight Chemical Information Systems, Inc. www.daylight.com: Irvine, CA (accessed April 2006).
9. Kubinyi, H. *Perspect. Drug Disc. Design* **1998**, *9*, 225–252.
10. Dixon, S. L.; Villar, H. O. *J. Chem. Inf. Comput. Sci.* **1998**, *38*, 1192–1203.
11. Kauvar, L. M.; Villar, H. O.; Sportsman, J. R.; Higgins, D. L. *J. Chromatogr. B* **1998**, *715*, 93–102.
12. Kauvar, L. M.; Higgins, D. L.; Villar, H. O.; Sportsman, J. R.; Engqvist-Goldstein, A.; Bukar, R.; Bauer, K. E. *Chem. Biol.* **1995**, *2*, 107–118.
13. Beroza, P.; Villar, H. O.; Wick, M. M.; Martin, G. R. *Drug Disc. Today* **2002**, 7, 807–814.
14. Hsu, N.; Cai, D.; Damodaran, K.; Gomez, R. F.; Keck, J. G.; Laborde, E.; Lum, R. T.; Macke, T. J.; Martin, G.; Schow, S. R. et al. *J. Med. Chem.* **2004**, *47*, 4875–4880.
15. Miller, J. M.; Bradley, E. K.; Teig, S. *J. Chem. Inf. Comput. Sci.* **2003**, *43*, 47–54.
16. Eksterowicz, J. E.; Evensen, E.; Lemmen, C.; Brady, G. P.; Lanctot, J. K.; Bradley, E. K.; Saiah, E.; Robinson, L. A.; Grootenhuis, P. D. J.; Blaney, J. M. *J. Mol. Graph Mod.* **2002**, *20*, 469–477.
17. Briem, H.; Kuntz, I. D. *J. Med. Chem.* **1996**, *39*, 3401–3408.
18. Kuntz, I. D.; Blaney, J. M.; Oatley, S. J.; Landgridge, R.; Ferrin, T. E. *J. Mol. Biol.* **1982**, *161*, 269–288.
19. MDL Drug Data Report (MDDR). www.mdl.com (accessed April 2006).
20. Lessel, U. F.; Briem, H. *J. Chem. Inf. Comput. Sci.* **2000**, *40*, 246–253.
21. Weber, A.; Teckentrup, A.; Briem, H. *J. Comput.-Aided Mol. Des.* **2002**, *16*, 903–916.

22. Krejsa, C. M.; Horvath, D.; Rogalski, S. L.; Penzotti, J. E.; Mao, B.; Barbosa, F.; Migeon, J. C. *Curr. Opin. Drug Disc. Dev.* **2003**, *6*, 470–480.
23. Fliri, A. F.; Loging, W. T.; Thadelo, P. F.; Volkmann, R. A. *Proc. Natl. Acad. Sci.* **2005**, *102*, 261–266.
24. BioPrint. Cerep SA, Inc.: Paris, France.
25. Greenbaum, D. C.; Arnold, W. D.; Lu, F.; Hayrapetian, L.; Baruch, A.; Krumrine, J.; Toba, S.; Chehade, K.; Bromme, D.; Kuntz, I. D. *Chem. Biol.* **2002**, *9*, 1085–1094.
26. Mason, J. S.; Mills, J. E.; Barker, C.; Loesel, J.; Yeap, K.; Snarey, M. *Higher-Throughput Approaches to Property and Biological Profiling, Including the Use of 3-D Pharmacophore Fingerprints and Applications to Virtual Screening and Target Class-Focused Library Design*; Abstracts of Papers, 225th ACS National Meeting, New Orleans, LA, United States, March 23–27, 2003, COMP-343.
27. Mason, J. S. Understanding Leads and Chemotypes from a Biological Viewpoint: Chemogenomic, Biological Profiling and Datamining Approaches. In *Book of Abstracts of First European Conference on Chemistry for Life Sciences: Understanding the Chemical Mechanisms of Life*. October 4–8, 2005. Rimini, Italy.
28. Bender, A.; Glen, R. C. *Org. Biomol. Chem.* **2004**, *2*, 3204–3218.
29. Horvath, D.; Jeandenans, C. *J. Chem. Inf. Comput. Sci.* **2003**, *43*, 691–698.
30. Horvath, D. Throughput Conformational Sampling and Fuzzy Similarity Metrics: A Novel Approach to Similarity Searching and Focused Combinatorial Library Design and its Role in the Drug Discovery Laboratory. In *Principles, Software Tools and Applications*; Ghose, A., Viswandhan, V., Eds.; Marcel Dekker: New York, 2001, pp 429–472.
31. Brown, R. D.; Martin, Y. C. *J. Chem. Inf. Comput. Sci.* **1996**, *36*, 572–584.
32. Cleavland, E. S.; Monks, A. P.; Vaigro-Wolff, A.; Zaharevitz, D. W.; Paull, K.; Ardalan, K.; Cooney, D. A.; Ford, H. *Biochem. Pharmacol.* **1995**, *49*, 947–954.
33. Paull, K.; Shoemaker, R. H.; Hodes, L.; Monks, A. P.; Scudiero, D. A.; Rubinstein, L. V.; Plowman, J.; Boyd, M. R. *J. Natl. Cancer Inst.* **1989**, *81*, 1088–1092.
34. Fliri, A. F.; Loging, W. T.; Thadeio, P. F.; Volkmann, R. A. *J. Med. Chem.* *48*, 6918–6925.
35. Lamb, M. L. Targeting the Kinome with Computational Chemistry. In *Annual Reports in Computational Chemistry; Spellmeyer, D. C.*, Ed.; Elsevier: Amsterdam, 2005, pp 185–202.
36. Ahsen, O. V.; Bomer, U. *Chem. Biochem.* **2005**, *6*, 481–490.
37. Wesche, H.; Xiao, S.-H.; Young, S. W. *Comb. Chem. HTS* **2005**, *8*, 181–195.
38. Dumas, J. *Exp. Opin. Ther. Patents* **2001**, *11*, 405–429.
39. Attwood, T. K.; Croning, M. D. R.; Gaulton, A. *Prot. Eng.* **2001**, *14*, 7–12.
40. Lamb, M. L.; Bradley, E. K.; Beaton, G.; Bondy, S. S.; Castellino, A. J.; Gibbons, P. A.; Suto, M. J.; Grootenhuis, P. D. J. *J. Mol. Graph Mod.* **2004**, *23*, 15–21.
41. Manning, G.; Whyte, D. B.; Martinez, R.; Hunter, T.; Sudarsanam, S. *Science* **2002**, *298*, 1912–1934.
42. Frye, S. V. *Chem. Biol.* **1999**, *6*, R3–R7.
43. Vieth, M.; Higgs, R. E.; Robertson, D. H.; Shapiro, M.; Gragg, E. A.; Hemmerle, H. *Biochim. Biophys. Acta* **2004**, *1697*, 243–257.
44. Fabian, M. A.; Biggs, W. H.; Treiber, D. K.; Atteridge, C. E.; Azimioara, M. D.; Benedetti, M. G.; Carter, T. A.; Ciceri, P.; Eden, P. T.; Floyd, M. et al. *Nat. Biotechnol.* **2005**, *23*, 329–336.
45. Shcreiber, S. L. *C&E News* **2003**, *81*, 51–61.
46. Strausberg, R. L.; Schreiber, S. L. *Science* **2003**, *300*, 294–295.
47. Haggarty, S. J.; Koeller, K. M.; Wong, J. C.; Butcher, R. A.; Schreiber, S. L. *Chem. Biol.* **2003**, *10*, 383–396.
48. Haggarty, S. J.; Clemons, P. A.; Schreiber, S. L. *J. Am. Chem. Soc.* **2003**, *125*, 10543–10545.
49. Kim, Y.-K.; Arai, M. A.; Arai, T.; Lamenzo, J. O.; Dean, E. F.; Patterson, N.; Clemons, P. A.; Schreiber, S. *J. Am. Chem. Soc.* **2004**, *126*, 14740–14745.
50. Weinstein, J. N.; Myers, T. G.; O'Connor, P. M.; Friend, S. H.; Fornace, A. J.; Kohn, K. W.; Fojo, T.; Bates, S. E.; Rubinstein, L. V.; Anderson, N. L. *Science* **1997**, *275*, 343–349.

Biographies

Robert V Stanton received his BS at the College of William and Mary in 1991 and went on to complete a PhD at Penn State University (1995), and a Post-Doctoral Fellowship at the University of California at San Francisco. For the next 5 years he worked at several biotechnology companies in the San Francisco area before moving to Pfizer's Research Technology Center in 2002.

Qing Cao received her BS degree from Nankai University (China) in 1993, and MS degrees in biotechnology and computer sciences from the Center for Biotechnology at Northwestern University in 1996 and Loyola University in 2000, respectively. She went on to work for 2 years as a bioinformatics software engineer at the Department of Basic Sciences at Northwestern University before moving to Pfizer's Research Technology Center in 2000.

© 2007 Elsevier Ltd. All Rights Reserved
No part of this publication may be reproduced, stored in any retrieval system or transmitted in any form by any means electronic, electrostatic, magnetic tape, mechanical, photocopying, recording or otherwise, without permission in writing from the publishers

Comprehensive Medicinal Chemistry II
ISBN (set): 0-08-044513-6

ISBN (Volume 4) 0-08-044517-9; pp. 807–818

INDEX FOR VOLUME 4

Notes

Abbreviations

CADD – computer-assisted drug discovery
GPCRs – G protein-coupled receptors
QSAR – quantitative structure–activity relationship

Cross-reference terms in italics are general cross-references, or refer to subentry terms within the main entry (the main entry is not repeated to save space). Readers are also advised to refer to the end of each article for additional cross-references – not all of these cross-references have been included in the index cross-references.

The index is arranged in set-out style with a maximum of three levels of heading. Major discussion of a subject is indicated by bold page numbers. Page numbers suffixed by T and F refer to Tables and Figures respectively. *vs.* indicates a comparison.

Names of scientists included in subentries refer to their development role, unless otherwise specified.

Radioactive isotopes are listed under the chemical symbol e.g. ^{131}I

This index is in letter-by-letter order, whereby hyphens and spaces within index headings are ignored in the alphabetization. Prefixes and terms in parentheses are excluded from the initial alphabetization.

Any method, model or other subject, associated with the name of the developer (e.g. name's model) does NOT imply that Elsevier, nor the indexers, have assumed the right to name models/methods after the authors of the papers in which they are described. This is merely a succinct phrase to refer to a model/method developed/described by the relevant author, so that the subentry could be alphabetized under the most pertinent name.

A

B

C

D

E

F

G

I

M

N

O

P

Q

R

T

U

V

W

X

Y

Z